**Fourth
Edition**

THE
STUDY
OF
BIOLOGY

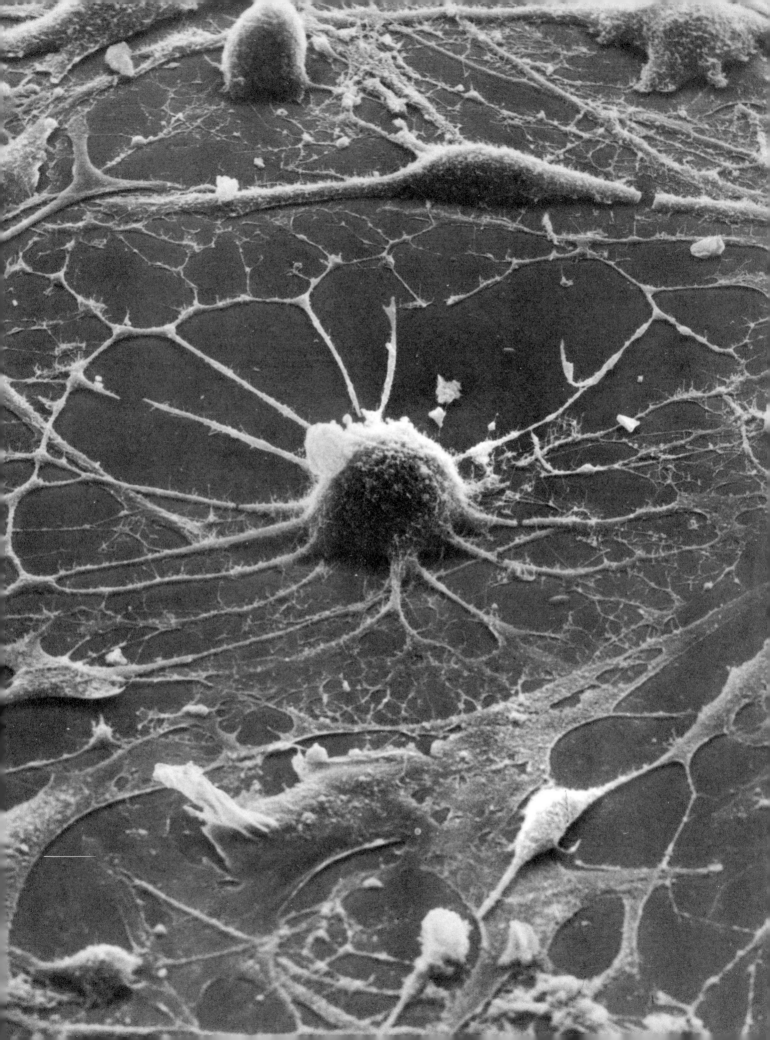

Fourth Edition

THE STUDY OF BIOLOGY

Jeffrey J. W. Baker
Wesleyan University

Garland E. Allen
Washington University

ADDISON-WESLEY PUBLISHING COMPANY

Reading, Massachusetts • Menlo Park, California
London • Amsterdam • Don Mills, Ontario • Sydney

Credits

Sponsoring Editor: James Funston

Production Editors: Margaret Pinette and Jerrold Moore

Designer: Catherine Dorin

Illustrator: Oxford Illustrators, Ltd.

Art Coordination: Robert Gallison

Cover Design: Richard Hannus

Cover Photograph: Talbot Lovering

This book is in the
Addison-Wesley Series in Life Science

Also by Baker and Allen:

A Course in Biology, Third Edition (1979)

Hypothesis, Prediction, and Implication in Biology (1968)

Matter, Energy, and Life: An Introduction to Chemical Concepts,
Fourth Edition (1981)

The Process of Biology, Primary Sources (1970)

The Study of Biology foreign-language editions

Spanish: *Biología e Investigación Científica*
             First Edition

Portuguese: *Estudo da Biología*
             Second Edition

The cell shown opposite the title page was infected with a tumor virus called polyoma. The virus caused the cell to undergo changes in both morphology and growth properties (a process called transformation) so that now it is more like a cancer cell than a normal cell. This particular cell is enormous relative to the surrounding cells. Such cells occur fairly frequently in this type of cell culture. The cell shown has rounded up in the early stages of cell division (mitosis). Originally the cell was spread out and covered the entire area. Then, as mitosis began, the cell cytoplasm withdrew, forming the rounded central ball and leaving behind the many fine "arms" seen in the picture. After dividing, the cell will respread as the cytoplasm flows back out into the "arms," again assuming a large, flat form. The other cells pictured are also polyoma-transformed 3T3 cells. The original line was derived from mouse embryos by Howard Green and George Todero. (Photo by Dr. Paul B. Bell, Jr., and Dr. Jean-Paul Revel, California Institute of Technology.)

Library of Congress Cataloging in Publication Data

Baker, Jeffrey J. W.
     The study of biology.

     Includes bibliographies and index.
     1. Biology.   I. Allen, Garland E.   II. Title.
QH308.2.B33 1982   574   81-17550
ISBN 0-201-10180-7   AACR2

ISBN 0-201-10180-7
ABCDEFGHIJ-DO-898765432

Preface

To us it seems hardly possible that fifteen years have elapsed since the first edition of *The Study of Biology*. Its philosophy, considered almost radical at the time, was that it is the *process*, not merely the content or "facts" of biology, that should form the basic thrust of introductory courses. By process of science we mean *how we know what we know;* how experiments are designed, data analyzed, and conclusions drawn—in short the logic and method of science. This same philosophy continued to guide the preparation of all subsequent editions, including the fourth. The widespread acceptance of this approach, literally throughout the world, has indeed been gratifying. Although the size, format, and content of the fourth edition were enlarged significantly, a major emphasis on the process of science was rigorously maintained.

In preparing the fourth edition we made three basic types of alterations. The first and most obvious is the updating and correcting of older ideas and the addition of new information that has emerged since publication of the third edition (1977). Such updating includes new material on cell metabolism and photosynthesis, the process of initiation and termination of protein synthesis, and the discovery that genes in eukaryotic cells frequently exist in isolated segments, or what are called "interrupted sequences." These alterations enable the student to stay abreast of the most recent discoveries and hypotheses of modern biology.

A second area of change involved expanding the treatment of certain areas of biology that were dealt with only briefly in previous editions. These include:

Cell Biology. The addition of a new chapter on the structure and function of membranes emphasizes the important role they are now thought to serve in the general metabolic life of the cell.

Plant and Animal Physiology. The much-expanded treatment of both neurophysiology and endocrinology reflects the increasing importance of these fields over the past decade. The original two chapters on animal physiology in the third edition were increased to six.

Genetics and Development. We expanded considerably the treatment of genetic control mechanisms, protein synthesis, the genetics of cancer, recombinant DNA, and genetic engineering.

Evolution and Ecology. This material was increased from two to five chapters and reorganized into a more complete and integrated unit. We included expanded treatment of taxonomy (including the cladist–pheneticist controversy and numerical taxonomy), adaptation, community structure, biomes, and population and island biogeography. In addition, we treated more fully the process of natural selection, especially that of speciation, than in previous editions.

A third and final area of revision included the filling in of major gaps, that is, subject areas that were omitted completely or treated only in passing in the third edition:

The Diversity of Life. Three completely new chapters on this subject were added. They present a survey of the living world from the Monera and Protista through plants, fungi, and animals—approached from an evolutionary point of view. Because many introductory courses now include some discussion of diversity, we felt that addition of this material would be far more useful than the general overviews of plant and animal phylogeny provided in the previous editions. To make this material more realistic and dynamic, we included numerous and sometimes unusual photographs of organisms in their natural surroundings (rather than line drawings or photographs of preserved specimens).

A completely new section on the exciting field of immunology, a topic that was completely omitted in previous editions, has been added to Chapter 20 (Developmental Biology). Since so many significant advances have been made in this field at both the cellular and molecular levels during the past decade, we felt strongly that a new, in-depth discussion was required. We included the material on immunology in the chapter on developmental biology because the cellular response to antigen has provided biologists with a model for gene regulation, and thus differentiation, at the DNA level.

As in the third edition, we included numerous "Supplements" to introduce historical views of a particular discovery: "How Do We Know" passages describing the way a particular idea has been established, and presenting controversial scientific and social issues. In these latter supplements, as before, we often cast the discussion in the form of a debate, presenting different perspectives on issues such as sociobiology, race and I.Q., *XYY* chromosomes and crime, and the human population issue. Many users of the third edition commented enthusiastically on this use of the supplementary essays, and urged that we continue and expand this treatment of such topics. Some even suggested new topics of debate for additional supplements. We appreciate very much these encouraging thoughts and suggestions.

As introductory biology courses have become more comprehensive in the past five years, increasing length is a problem with which authors of all college biology textbooks have had to grapple. *The Study of Biology* is no exception, and we have aimed from the start to keep the length of the fourth edition within bounds. Thus, someone's pet subject may have been given shorter shrift than he or she would have liked, or omitted altogether. For any particular trespasses of this sort, we offer our apologies. We tried to steer a sensible course between a reasonably wide scope of coverage (meeting most course demands) and enough in-depth coverage to avoid superficiality. We believe that it is important for students to gain an understanding of not only *what* we know about modern biology, but also *how* we know it.

We wish to stress again that the spirit and intent of the first three editions has been scrupulously maintained. The opening chapter, dealing with the nature and logic of science (hypothesis formulation and testing, experimental design, and analysis of data) sets the stage for the rest of the book. Throughout, emphasis is placed on experimental design and detail so that the student can examine the evidence on which current theories are based; we believe that one of the most important opportunities an introductory science course offers is that of *learning how to think independently and critically.* To aid the student in acquiring critical attitudes and practice in data analysis, many of the exercises at the end of the chapters and in the instructor's guide are inquiry-based and problem-solving in nature. We have found that these exercises challenge the student to interpret scientific data, draw conclusions, and choose between alternative hypotheses.

In preparing this edition we have had considerable help from a large number of individuals. Among those

who read various parts of the manuscript and offered invaluable suggestions regarding content, emphasis, and scientific accuracy were: Penelope Haunchey Bauer, Colorado State University; Robert Shaw Egan, Texas A&M University; Stanley Gartler, University of Washington; Richard Goldberg, Temple University; Richard Goldsby, University of Maryland; Judith Goodnough, University of Massachusetts; Thomas Gray, University of Kentucky; John R. Gregg, Duke University; Brian Hazlett, University of Michigan; Alice Holtz, Columbia University Medical School; Miles F. Johnson, Virginia Commonwealth University; K. W. Joy, Carleton University; Norman Kerr, University of Minnesota; Attila Klein, Brandeis University; Lewis Kleinsmith, University of Michigan; Joseph H. McCulloch, Normandale Community College; Richard Mintel, University of Virginia; Elizabeth Nicotri, University of Washington; David O. Norris, University of Colorado; Gary Ogden, Moorpark College; R. Douglas Powers, Boston College; Deborah Rabinowitz, University of Michigan; Roger R. Ragonese, Nassau Community College; James L. Seago, Jr., State University of New York; Joanne White, University of North Carolina; and David S. Woodruff, University of California, San Diego. In addition, Dr. Otto Solbrig, Harvard University, provided an especially valuable contribution with regard to all of the material on evolution and ecology. In all cases, of course, the final responsibility for what appears in print rests with us.

Many other people assisted us at various stages in the production of the manuscript. Diane Pitochelli assisted in developmental editing. Larry Bennett provided invaluable help in organizing the records of illustrations; Jack Diani and Regina Birchem provided several particularly fine photographs; Eve Fine contributed to several of the supplements, and took major responsibility for obtaining permissions for many photographs. Andy Futterman stimulated many discussions about the best way to approach the teaching of process; and Randy Bird made numerous valuable suggestions on material in many of the supplements. Very special thanks go to Nina Plurad for help with routine problems in preparing the manuscript, for her strong support on basic issues of teaching, and for maintaining an interest in the larger view of biology. Deborah Ann Baker undertook the tedious task of preparing the index, and Linda Lazar reworked and added to the glossary. To all of these individuals we owe a considerable debt of gratitude.

The staff at Addison-Wesley assisted in countless ways in the often confusing and seemingly thankless task of organizing and producing a book of this length. We owe particular thanks to James Funston, our editor for the past two editions. He advised, stimulated, cajoled, and in many ways guided us through the revision process with a keen eye to the value of preserving the book's basic philosophy while also incorporating new material to meet the ever-changing demands of modern biology courses. Considerable help also was provided by Kathy Gregg, Margaret Pinette, Jerrold Moore, Diann Korta, and Bob Gallison who helped in editing and coordinating the entire production process.

Middletown, Connecticut   J.J.W.B.
St. Louis, Missouri   G.E.A.
January, 1982

Contents in Brief

Contents

PART VI
THE ORIGIN
AND DIVERSITY
OF LIFE   797

CHAPTER 26
THE ORIGIN OF LIFE   799

CHAPTER 27
THE DIVERSITY OF LIFE I:
THE MONERA AND
PROTISTA   812

CHAPTER 28
THE DIVERSITY OF LIFE II:
FUNGI AND PLANTS   841

CHAPTER 29
THE DIVERSITY OF LIFE III:
ANIMALS   876

Fourth Edition

THE STUDY OF BIOLOGY

PART I
BIOLOGY AND THE
SCIENTIFIC ENTERPRISE

"The philosopher's task [is] well compared to that of a mariner who must rebuild his ship on the open sea. [We are not] stuck with the conceptual schemes that we grew up on. We can change them bit by bit, plank by plank, though meanwhile there is nothing to carry us along but the evolving conceptual scheme itself."

—*W. V. O. Quine*

All fields of human activity have a philosophical and methodological base. That no single philosophy or method distinguishes the experimental sciences from all other fields does not mean no philosophy or methodology of science exists. Modern science rests on distinct philosophical biases or presuppositions, and there *are* acceptable and unacceptable methods and procedures.

Gaining some insight into the philosophy and methods of modern science will serve two purposes. It will promote a more thorough understanding of the major concepts of biology by showing *how* they were derived. And it will help to clarify some basic principles of problem-solving applicable beyond the realm of biology in particular or even science in general.

Chapter 1
The Nature and
Logic of Science

1.1
INTRODUCTION

One way to approach the study of biology would be to present it as a group of accumulated facts and general principles that biologists today believe to be true. You would study how evolution works, how living cells get energy, how nerves conduct an impulse, and so forth. Such a traditional approach to science has a major drawback: it quickly makes what you have learned obsolete. Today's scientific truths are almost as likely as not to be tomorrow's errors.

In science, keeping abreast of changing ideas requires understanding the *process* by which they developed. In short, it means learning how to think. Scientific thinking includes such techniques as framing questions, reasoning logically, and testing ideas against experience. Because this book is about biology, many of these specific techniques—such as designing and analyzing experiments—will be illustrated with biological examples. Yet none of the techniques discussed here is restricted to biology, nor to the natural sciences. The tools of clear thinking are applicable in all aspects of human thought, from nuclear physics and medical research to fixing a car or preparing a dinner. In an age when human beings are confronted with perplexing social and political problems of profound importance, there is reason for learning how to use one's own brain to solve problems.

Biology, the study of living things, is a particularly appropriate area in which to learn the problem-solving methods. We are all living human beings who experience curiosity about ourselves and our biological processes. At the same time, many of today's most immediate problems have an explicitly biological base. Pollution, depletion of our natural resources, drugs, bacterial and chemical warfare, genetic engineering, overpopulation, birth control, abortion, and ideas about the inheritance of intelligence are only a few of many examples. The principles of biology and the methods of deriving those principles have an important bearing on understanding and dealing with these current problems.

1.2
BIOLOGY AS A SCIENCE

Science, to paraphrase the words of biologist Albert Szent-Gyorgyi, is seeing what everyone else has seen and thinking what no one else has thought. Certainly this epithet would apply to history and sociology as well as to science. The most important difference is that in the natural sciences we can often distinguish between several possible answers to a problem by experimentation. Because of practical and ethical restraints, scientific experimentation is often not possible in the social sciences and humanities.

Suppose, for example, a biologist wanted to know if light influences the development of plant seedlings. Of course the biologist could observe seedlings growing in the wild; if he or she found a group of seedlings developing in the dark under a log, it might be possible to compare them with normal seedlings growing in the field. But there would be many additional variables. The seedlings growing under the log might be a different species from those growing in the field. Those in the field might be competing strongly with other plants around them while those under the log were not. The amount of water available to each might be different. Without more information or more carefully controlled conditions, what is "normal" development might be difficult to determine.

All human learning involves problem-solving. There are fewer differences in problem-solving methods between the "sciences" and the social sciences or humanities than is generally believed.

A more reliable answer can be obtained by conducting an experiment. The biologist can plant two groups of seedlings, say 500 in each group, in the laboratory, exposing both to the same soil, temperature, humidity, and water. The only difference between the two is the amount of light. Under these conditions, any difference in the development of the seedlings can be attributed to the difference in available light. If any difference is demonstrated, it can be safely concluded that, in the laboratory at least, light influences seedling development.

Consider the problems faced by a sociologist who wants to determine what factors influence crime rate. Like the biologist, the sociologist can observe existing situations and compare similarities and differences. But the sociologist cannot usually take the next logical step and set up a controlled laboratory experiment. If the sociologist suspects that housing facilities, unemployment, or inflation are responsible for crime, it is not possible to vary these factors experimentally in different human populations for the purpose of observing their effects on the crime rate.

Under these circumstances sociologists have two alternatives. They can abandon the problem because an adequate method for getting a scientific experiment going is not available. Or they can, for example, survey two cities that have many social conditions in common but differ in one respect—unemployment levels, for instance. It may then be possible to compare those two cities for differences or similarities in crime rate. If the crime rate is lower in the city with low unemployment, then relating crime to unemployment is justified. The crucial point, however, is that social scientists can only compare differences and similarities between two existing situations. They cannot advance their conclusions with as much certainty as natural scientists, because they cannot usually carry out controlled experiments.

If it is agreed that the differences traditionally emphasized between the natural sciences on the one hand and the social sciences and humanities on the other are not as great as they are often made out to be, what is the so-called scientific method we hear so much about? The scientific method consists simply of making accurate observations and of logical thinking—a process, as has been pointed out, that can be applied in all realms of thought. What are the elements of scientific observation and logical thinking?

1.3
OBSERVATION AND SCIENTIFIC "FACTS"

The foundation of all modern science is observation. Our entire view of the natural world depends on the accurate recording of sensory data and the organization of these data into general concepts.

Fig. 1.1
(*a*) Photograph showing chromosomes from a stained spleen cell culture of a four-month-old human fetus. Note how the chromosomes clump, making accurate counts very difficult. (*b*) Camera lucida drawing of the same group of chromosomes. Such drawings help to elucidate detail more clearly, since they represent a composite of observations at different depths of focus—something no single photograph could provide. (Courtesy T. C. Hsu, from *Journal of Heredity* 43 (1952), p. 168. Reprinted with permission.)

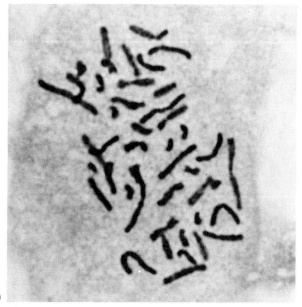

(a)

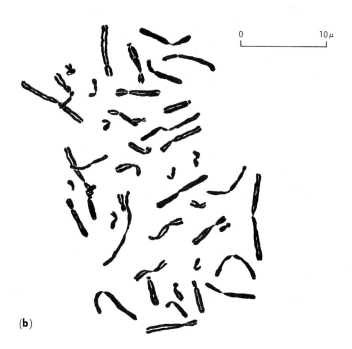

(b)

At the turn of the century, it became an intriguing problem to determine the number of chromosomes present in the body cells of human beings. To observe chromosomes it is necessary to fix and stain cell preparations for viewing under the microscope. Chromosomes so treated appear as dark, oblong objects surrounded by other partially stained material in the nucleus. A quick glance at Fig. 1.1(a) will suggest some of the problems observers encountered in making accurate chromosome counts. For one thing, the chromosomes appear clumped together in such a way that it is not always easy to tell where one ends and the next begins. For another, male and female cells seemed to early observers to have a different number of chromosomes, the female having one more than the male. (It is now known that the female has two *X* chromosomes while males have one *X* and one *Y*.) Still a third problem is that chromosomes are curled and twisted. When one part of the chromosome is in focus, another part may not be. Thus it is easy to count as two chromosomes what is really a single chromosome appearing at different focal planes of the microscope.

In the early decades of this century, a new method of observing chromosomes was introduced. The **camera lucida** is a device that projects a microscope slide onto a flat surface and allows the observer to trace the pattern of the projected image. A camera lucida drawing of the same preparation shown in Fig. 1.1(a) is given in Fig. 1.1(b). It was much easier to count chromosomes in camera lucida drawings than actual chromosome preparations, but the drawings were still not free of ambiguity. The three camera lucida tracings shown in Fig. 1.2 are drawings of the same cell preparation by three different observers, who were experienced in examining chromosome preparations. Yet they still produced three different representations, which frequently can show up as different chromosome counts.

The earliest counts, by German cytologist von Winniwarter in 1907, reported 47 chromosomes as the total in human beings: 23 pairs plus an extra "accessory" chromosome, now called the *X* chromosome. It was not completely clear at the time that human females have two *X* chromosomes and males only one. In 1921, however, an American named T. S. Painter developed new techniques for preparing and observing chromosomes. With these methods he reported a count of 48 chromosomes, or 24 pairs. Painter observed that the *X* chromosome had a partner in females, another *X* chromosome, and a different-looking partner in males, the *Y* chromosome. Between 1932 and 1952, Painter's count of 48 chromosomes was confirmed by at least five other observers. This count gained reliability—so much so, in fact, that when another observer in the 1940s claimed to have spotted only 47 chromosomes, confirming von Winniwarter's original figure, the count was dismissed as inaccurate. By the early 1950s it was generally accepted dogma in biology that the chromosome number for the human species was 48. All textbooks agreed.

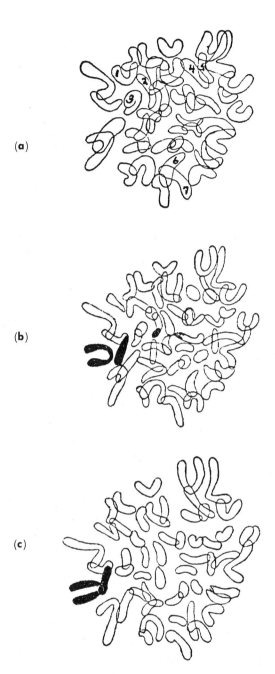

(a)

(b)

(c)

Fig. 1.2
Three different camera lucida tracings of the same chromosome preparation; (*a*) the preparation drawn by Evans, (*b*) the same preparation drawn by von Winniwarter, and (*c*) by Oguma. Note the difficulty in determining whether some parts of chromosomes are attached to, or separate from, other parts. To see how this affects actual counting, observe the chromosome labeled 2 in (*a*). Evans saw number 2 as a long, crescent-shaped chromosome, while both von Winniwarter and Oguma saw it as two shorter, separate chromosomes. Evans saw chromosome number 3 as a single chromosome, whereas von Winniwarter and Oguma saw it as two separate ones. Such problems greatly confused the issue of human chromosome counts. (From King and Beams, *Anatomical Record* 65 (1936), p. 169. Reprinted with permission.)

SUPPLEMENT 1.1
FIXING AND STAINING

Preparation of tissue for microscopic observation involves three processes: fixing, dehydrating, and staining. Fixing involves placing the tissues in a solution of such agents as formalin or heavy metal salts. Such solutions precipitate proteins and thus render cell structures more stable. Dehydration involves placing the tissue in a graded series of alcohol solutions of increasing concentration. This process removes water from the tissue and helps preserve the specimen. The fixed tissue may then be freeze-dried or embedded in some form of wax or paraffin to give it rigidity. The tissue is then sliced by an instrument called a microtome into very thin sections. The thin slices are stained with dyes. Some dyes stain certain parts of the cell while others stain other parts, a property known as differential specificity. Two stains, aceto-carmine and orcein, specifically stain chromosomal material. Since these stains do not color other parts of the cell, they make the stained chromosomes stand out against their unstained background.

The "certainty" of Painter's count had two negative effects. First, it stifled further investigation. People simply stopped counting chromosome preparations for the purpose of determining the total number for the species. A second negative effect was that it prejudiced the few chromosome counts that *were* made. Nearly everyone who counted human chromosomes found 48: what people expected to find, they found.

Then several new techniques were introduced. One was phase-contrast microscopy (see p. 92), which made it possible to observe *living* cells, in which the chromosomes do not clump together so much. Another involved new methods of fixing and staining cells, allowing the chromosomes to spread out within the nucleus. Still a third innovation was the preparation of **karyotypes**. In this procedure, a chromosome preparation is photographed, and the individual chromosomes are cut out of the photographic print. The chromosome images are then arranged on a sheet in a systematic fashion, making it possible to account for each individual chromosome and match it with its partner (see Fig. 1.3).

In 1955, two researchers in Sweden were using karyotypes to study the chromosomes of cancer cells in mammals, including humans.* In their chromosome counts, they consistently came up with a total of 46 chromosomes for males and females. In addition, they found that other medical researchers had also found only 46 chromosomes but had let the matter go. For example, Dr. Eva Hansen-Melander in cooperation with Yugre Melander and Stig Kullander had been studying human liver tissue the previous spring and had also counted 46 chromosomes. They had discontinued their work, feeling that their "failure" to observe the expected 48 chromosomes reflected poorly on them as observers. The two Swedish workers finally challenged the long-established notion that human beings have 48 chromosomes. By 1960, after many confirmations of the number 46, the biological community was ready to admit that the older count was wrong.

What can we conclude about the process of observation from this case study?

1. Observations depend upon sensory data. If the material being observed has some ambiguity, such as clumped chromosomes, the observations will reflect that ambiguity in one way or another.

2. Observations contain some subjective input. Determining whether a particular stained mass represents one or two chromosomes often requires the exercise of prejudgment.

3. People often find what they expect to find. The expectation that human cells contain 48 chromosomes caused many workers actually to "see" 48 chromosomes!

4. Even when something different from the expected is observed, the force of accepted dogma may cause investigators to disbelieve their own observations.

5. New techniques often make possible the refinement of observations. New techniques can make observations more accurate and thus add new sensory data for the solution of a problem.

The case of the human chromosome number also illustrates the important role of anomalies in science. An **anomaly** is an unexpected observation; either it does not agree with other people's observations, or it is not expected on the basis of an accepted conceptual scheme. Given general acceptance of the idea that human cells have 48 chromosomes, the observation that particular cells had only 46 was considered an anomaly. Many times investigators ignore anomalies, thinking that their

---

*Abnormality in chromosome structure, if not total number, is associated with certain types of cancer, such as chronic granulocytic leukemia (cancer of the blood cells known as granulocytes). For a more detailed discussion of the relationship between chromosome abnormalities and cancer, see Section 18.6.

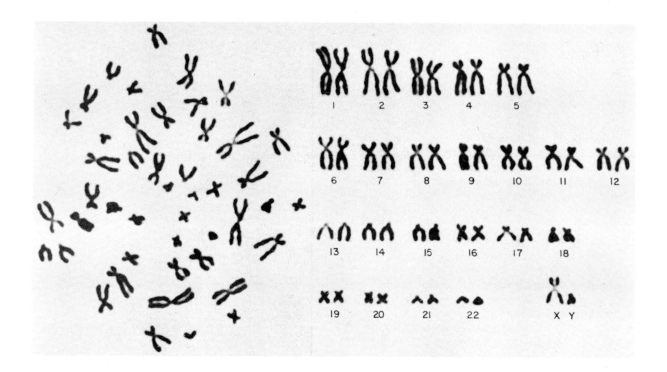

Fig. 1.3
On the left is a photomicrograph of a full complement of human chromosomes, showing their random positions after fixing and staining. On the right is a karyotype of the same cell, showing the chromosomes lined up in order of decreasing size. Each chromosome pair is given a particular number. The first 22 pairs (44 chromosomes) are the autosomes; the twenty-third pair (the *XX* or *XY* pair) represent the sex chromosomes. In the above photograph, the sex chromosomes are *X* and *Y*, indicating the cell was from a male. (Photo from A. M. Winchester, *Human Genetics.* Columbus, Ohio: Charles Merrill, 1974.)

own observations must be incorrect. Yet it is also true that, in the history of science, many new ideas have ultimately been developed because some workers refused to ignore an anomaly.

Most important, the human chromosome case illustrates that even a "scientific" statement based upon the observations of many different people may still be subject to correction or revision. As we will see in the next section, the same holds true for statements based upon hypothetical or theoretical constructs.

1.4
THE "SCIENTIFIC METHOD"

Even among scientists there is wide disagreement about what is meant by "the scientific method." Some science textbooks list a series of six or seven steps involved in

the scientific method. Such a formal description is quite unrealistic. No research scientist follows any such formalized ritual in performing experiments. Other writers, however, have gone to the opposite extreme in their description of the scientific method. One states that "science is simply doing one's damndest with one's mind with no holds barred." This view certainly conveys that the means used by scientists in solving problems are not necessarily unique to science. As a definition of the scientific method, however, the statement is not very useful. Followed to its logical conclusion, it indicates that artists, philosophers, mechanics, mathematicians, plumbers, and all other persons who work diligently and creatively to solve problems are also scientists. This renders the terms "science" and "scientist" meaningless.

Induction

Science proceeds by formulating and testing hypotheses. **Hypotheses** are simply tentative explanations put forth to account for observed phenomena. Formulating testable hypotheses draws heavily upon the scientist's creativity and imagination. Attempts to pin down these qualities often seem doomed to failure. How does one account for the genius of a scientist like Albert Einstein, whose most brilliant ideas often appeared to be nothing more than an intuitive leap of immense proportions? Perhaps one cannot—or should not—even attempt to do so. Not all scientists are Einsteins, however. For most, the means of hypothesis formulation can be described, at least in general terms. One general pattern of

Inductive logic proceeds from the specific to the general; deductive logic proceeds from the general to the specific.

thought often recognized in the creation of hypotheses is known as induction or **inductive logic.**

Consider a person who tastes a green apple and finds it sour. Tasting a second, third, and fourth green apple yields the same result. From this experience, the person concludes that all green apples are sour. One physical entity (greenness) has been correlated with another (sourness). The conclusion, "all green apples are sour," is an example of a generalization formed by induction: it is an inductive generalization.

The inductive generalization "all green apples are sour" becomes a hypothetical statement when expressed as an "if" statement, the "if" denoting the tentative nature of the hypothesis. But is the hypothesis "true"? We can test it by tasting yet another green apple to see if it is sour; the hypothesis predicts that it will be. This tasting of another apple after forming the hypothesis "all green apples are sour" by induction represents a test of the hypothesis; that is, it is an experiment.

Hypotheses that arise from inductive generalizations simply summarize a set of specific data and allow that summary to be tested. The second and, to science, far more important type of hypothesis is an explanatory hypothesis. Explanatory hypotheses do more than merely generalize from a set of similar observations; they attempt to get at the *causes* for the phenomena in question, the *reasons* behind certain occurrences or processes.

An explanatory hypothesis goes beyond merely stating that "all green apples are sour" to propose reasons for their sourness. For example, one explanatory hypothesis might propose that the green apples are sour because they contain high concentrations of a particular acid. Such a hypothesis is readily testable by comparing the amount of that acid found in the sour green apples with the amount found in sweet apples. Since it tests the explanatory hypothesis, such a chemical analysis of the apples is, by definition, an experiment. *The primary purpose of scientific experiments is to test hypotheses.* Any hypothesis selected by a scientist must meet a very important requirement: *it must be testable.*

Deduction

How do experiments test hypotheses? The answer is quite simple. *Experiments test hypotheses by testing the correctness of the predictions that can be derived from them.* This process involves the use of deduction, or **deductive logic.**

Deductive logic (often called *if . . . then* reasoning) is the heart and soul of mathematics. It becomes most evident in plane geometry: "*If* two points of a line lie in a plane, *then* the line lies in the same plane." Deduction plays no lesser role in other fields of mathematics: "If $a < b$ and $x \le y$, then $a + x < b + y$" (the addition law), and "If $x < y$ and $a > 0$ then $ax < ay$" (the multiplication law).

In science (and therefore in biology) deduction is just as vital as it is in mathematics, but there are important differences between the way deduction is used in mathematics and the way it is used in experimental science. Mathematicians generally deal with symbols. They are not concerned with such physical entities as living organisms. Furthermore, the mathematician can manipulate symbols at will. He or she can create situations in proofs that make certain that only one hypothesis is being tested, only one question being asked. Not so the biologist. The organism being studied cannot be so easily manipulated. The biologist can never be absolutely certain that his or her experiment has eliminated all the variables that might influence the results. A major problem in biological research, then, becomes one of experimental design (to which considerable attention will be devoted later in this chapter).

The "if" portion of deductive logic represents the hypothesis; the "then" portion, the predictions that *must* follow from acceptance of the hypothesis. The stress placed upon the word "must" emphasizes that once a hypothesis is formulated, the predictions follow automatically.

Testing Hypotheses

Scientific experiments are designed to test hypotheses by determining the correctness of the prediction that

Table 1.1
Comparison of Inductive and Deductive Logic

Inductive	Deductive
Begins with observations; leads to hypothesis	Begins with hypothesis; leads to predictions
Proceeds from specific to general	Proceeds from general to specific
A method of discovery	A method of verification

A TRUTH TABLE

HYPOTHESIS	CONCLUSION OR PREDICTION
TRUE	TRUE
FALSE	TRUE or FALSE

Fig. 1.4
The "truth table" shows the relation between a hypothesis and its predictions. Note that a true prediction may be derived from a false hypothesis as well as from a true one. Therefore true predictions do not constitute proof of the truth of a hypothesis.

follows once a hypothesis is advanced. This implies that there is some sort of consistent relationship between hypotheses and the predictions they automatically generate. Indeed there is. This relationship is shown in the "truth table" (see Fig. 1.4).

An example from scientific research will illustrate how the truth table applies. It has been shown many times that exposure of certain strains of mice to X-ray beams of 600 roentgens or more (a roentgen is a unit of measure of the amount of energy delivered in X-ray beams) causes death within two weeks or less. The death might be due to either secondary or primary effects of the radiation. A primary effect would be damage to tissues by the radiation itself, rendering them nonfunctional. A secondary effect might be that death was caused by bacterial infection resulting from a migration of bacteria through the intestinal epithelium (lining), which histological (tissue) examination showed had been severely damaged by the X-rays. In order to test this latter hypothesis, antibiotics of various types were administered to the irradiated mice in several different ways to see whether this had any effect on the time of death. No such effect was shown; the mice still died in the same length of time as the control animals, which had been irradiated under the same conditions but given no antibiotics. It was tentatively concluded, therefore, that death in the period tested (from one to five days after exposure) was not due to bacterial infection.

Note the deductive *if . . . then* reasoning here. The experimental logic can be simply stated as follows:

Hypothesis: *If* the deaths of irradiated mice within one to five days after exposure are due to bacterial infection . . .

Prediction: *then* the administration of antibiotics should lower the death rate of mice that receive them.

The experimental results showed the prediction to be false. The mice still died in the same length of time after exposure to the X-rays. Thus we know, barring experimental errors, that the hypothesis explaining the deaths as due to bacterial infection is also false and must be either discarded or modified.*

Suppose that the administration of antibiotics *had* caused a lengthening of life. Would this have shown that our hypothesis must be the correct explanation? Absolutely not, although this result would have lent support to the *probability* of its being correct.

Examine again the truth table in Fig. 1.4. Note that the word "conclusion" as used by a mathematician is interchangeable with the word "prediction" as used by a biologist, for predictions that can be made from a hypothesis are simply the conclusions that one must draw from accepting it. In the case of the irradiated mice, it must be predicted that the mice will live longer or recover if the bacterial-infection-as-cause-of-death hypothesis is correct. This did not occur. The mice died as before; the prediction was false.

From the truth table, we see that this automatically means that our hypothesis is false, for *a true hypothesis can never give rise to a false prediction.* In other words, predictions derived from a true hypothesis should never lead to contradictions.

The truth table also shows that we can never *prove* a hypothesis true. While a true hypothesis always gives rise to true predictions, so also may a false hypothesis. The importance of this last fact cannot be overemphasized. It shows that *science can only deal with its "truths" in terms of probabilities and never in terms of certainties.*

In the past, many false hypotheses have been held by scientists and lay people alike, simply because accurate predictions could be made from these hypothe-

---

*Note that it could not be said that death from radiation in *animals* is due to other causes than bacterial infection, for the world "animals" includes many more forms than just mice. Nor could it be stated that death from radiation *in mice* is due to some other cause than bacterial infection, for not all strains of mice were tested. When the research paper is written for publication, the biologist will carefully word the interpretations, limiting them to the precise strains of mice tested and to the time period of death (one to five days after exposure) with which she or he worked.

Truth and validity have no necessary connection. Validity is a concept in logic, referring to conclusions derived by deductive syllogisms. Truth is that which agrees with people's experience with the world around them.

A true hypothesis never gives rise to a false conclusion (prediction). However, a false hypothesis can give rise to either a true or a false conclusion (prediction). Hence a false conclusion (prediction) leads to the certain rejection of the hypothesis from which it was derived, but a true conclusion (prediction) does not prove the hypothesis true beyond doubt.

ses despite their falsity. Acceptance of the belief that the sun orbits the earth leads one to predict that the sun will rise on one horizon, cross the sky, and set on the other horizon—and so it does. The fact that this prediction turns out to be correct does not, of course, mean that the sun *does* orbit the earth. In order to demonstrate that this hypothesis is false, other tests must be devised.

One such test is to predict the future relative positions of the sun, earth, and other planets, given that the sun does orbit the earth. Such predictions are invariably shown to be false, thus forcing rejection of the hypothesis. On the other hand, accepting the hypothesis that the earth, along with the other planets, orbits the sun leads to very accurate predictions about the relative positions of the sun, earth, and other planets at any particular time.

Although the truth table shows that a true hypothesis never gives rise to a false conclusion (prediction), only in mathematics does obtaining just one false conclusion spell certain death for the hypothesis. Biologists rarely deal with cases in which *every* prediction made by a hypothesis turns out to be correct. The question then becomes one of *how many* or *what proportion* of a given number of predictions must be verified in order to make the hypothesis a useful one. For this reason, experimental data are often subjected to **statistical analyses,** in which mathematics is employed to determine whether deviations from the pattern predicted by the hypothesis are significant.

1.5
EXPLANATIONS IN SCIENCE

Like all areas of human problem-solving, the natural sciences seek answers in the form of explanations.

A few years ago biologist Ernst Mayr of Harvard University observed that a warbler living all summer in a tree next to his house in Maine began its southern migration on August 25. This single observation raised a question in his mind. *Why* did the warbler begin its migration on that date?

What does the word "why" mean in this question? Two types of answers can be given to two versions of this "why" question:

1. A teleological answer: *for what purpose* did the warbler begin to migrate?

2. Causal answers: *what caused* the warbler to start migrating?

While it may be appropriate to ask teleological questions about purposeful human behavior, it is futile to ask such questions about natural phenomena. To ask "For what purpose did the warbler begin to migrate?" implies that the warbler made the same kind of conscious, goal-oriented choice on August 25 that a person might make in deciding to go downtown.

In modern science, *causal* answers are sought, since they attempt to explain natural phenomena in terms of the events that may have given rise to them. For example, Dr. Mayr's original question could be rephrased as a second, more precise question: *"What caused* the warbler to begin its migration on August 25?" It is possible to answer this question without making the ungrounded assumption that the warbler made a rational choice about when to migrate and in what direction.

The physicist Albert Einstein once said: "The formulation of a problem is often more essential than its solution." We need further clarification before we try to answer Dr. Mayr's rephrased question scientifically. The question is phrased in the singular: "What caused the warbler to begin its migration on August 25?" In most scientific situations, such questions should be posed in another form: "What causes *warblers* to begin

Facts are very different from hypotheses. Facts are observations that by general agreement have been established as true. Hypotheses are ideas that go beyond the observations and attempt to explain why certain phenomena occur. Two or more people may often develop different hypotheses from the same set of observations.

their migration *around* August 25?" Seldom in science do questions about individual, particular events find a satisfactory answer. No two situations are absolutely identical. No two warblers are the same. While warblers may in fact respond to the same general types of stimuli—heat, cold, day length, and the like—in the same general ways, they will not *as individuals* respond in exactly the same way nor at exactly the same time. Warbler A might begin its migration on August 25; warbler B might already have departed on August 23; and warbler C might not depart until August 26. In general, scientists are less interested in explaining particular events than in the principles underlying collections of events.

In studying any collection of events in nature, whether dealing with organisms, cells, atoms, molecules, or whatever, it is necessary to deal with *statistically significant* numbers. Statistical significance means that a large enough collection of events or objects is being observed to eliminate the distorting effects of individual differences. Observing five birds might be enough to make a generalization about the onset of bird migratory behavior. In general, however, larger samples are necessary, and the greater the number of cases studied, the more likely the hypothetical explanation derived from studying them will be correct.

Causal Explanations

To return now to Dr. Mayr's original question, let us assume that he was indeed interested in what triggered the migratory response of warblers in general, as exemplified by the particular warbler observed in the tree outside his window. In response to the general question of what causes warblers to begin migrating on August 25, Dr. Mayr points out that at least four different kinds of causal explanations are possible.

An Ecological Explanation. Warblers must move south in response to a decline in food supply. As insect eaters, they would starve to death if they remained in Maine where the insects begin dying out around the end of August. This explanation causally relates warbler migration patterns to general environmental changes, especially those concerned with food supply.

A Historical Explanation. Warblers begin moving south because through the course of evolution they have acquired a hereditary constitution that programs them to respond to certain environmental changes associated with the end of summer. Other birds, such as screech owls, do not have this same genetic constitution and hence do not respond to end-of-summer environmental

stimuli by migrating. This explanation relates warbler migration to genetic causes resulting from evolutionary processes.

An Internal Explanation. Warblers begin migrating because of their particular physiological state on August 25. The physiological mechanism triggers certain parts of the nervous system, resulting in migratory flight response. This internal physiological explanation is closely associated with the fourth type of explanation.

An External Explanation. One or more specific environmental factors activate an intrinsic physiological mechanism in warblers. This explanation puts the emphasis on the external factors that do the triggering, whereas the third explanation deals with the nature of the trigger itself.

These explanations are not mutually exclusive, and the most complete explanation would consist of all of them. They fall into two categories. The last two involve the interaction of environmental factors with the organism's immediate, internal makeup. They can be called the *immediate* or *proximate causes.* The first two explanations, on the other hand, involve interaction of the long-term history of the warblers with their environment, both physical and biological. These can be called *ultimate causes.* They act continuously in the present but have arisen by historical development.

Cause and Effect

Several other features are worth noting about the general character of explanations as they apply to any field, but especially to biology. For one, explanations tend to be most successful when they relate one kind of particular event to more general processes or events that are already understood. Thus the particular event of warbler migration is "explained" by showing how it relates to ecological, hereditary, physiological, and evolutionary principles. For another, the most successful causal explanations are those that can be tested *empirically:* the particular causal relationship pointed out in the explanation is subject to some kind of experimental test. It is possible, for example, to breed animals and learn something about their genetic constitution; physiological mechanisms can be studied in the laboratory; and determining whether those mechanisms are triggered by one or another external factor is also subject to experimental analysis. This testability criterion for an acceptable scientific explanation is particularly important. As suggested earlier, a causal explanation stating that warblers fly south because of an inherent desire to do so would not be very acceptable, since it

---

There are two types of hypotheses in science: generalizing hypotheses (generalizations) and explanatory hypotheses (explanations).

Experiment and observation are means of testing the truth of predictions derived from hypotheses.

is stated so vaguely that it cannot be tested in any way.

Modern science, then, is built upon belief in cause and effect: for every observed effect, there is some cause or causes. Bird migration is not the result of any single factor (temperature, moisture, or day length) alone, but a complicated interaction of many of these factors. Such a multicausal viewpoint is characteristic of most sciences today.

It is obvious that causal explanations have distinct limitations. Some of these limitations arise from the methods of observing and gathering information. Others are limitations in our technological ability to test certain types of explanations. The explanation that warbler migration is the result of a delicate change in hormones might be a reasonable hypothesis, but if the technique for measuring small changes in hormones were not available, it could not be tested. This hypothesis would not be unacceptable, but it might have distinct limitations at the time it was proposed.

Many cause-effect relationships may be more apparent than real. This usually results from our having no way to sort out the various interactions involved. Suppose, for example, that a cold air mass from Canada arrived in Maine on August 25, just when day length was appropriate to trigger the warbler's migratory response. It might appear that the primary cause for onset of migration was the drop in temperature. In this case, since it is known the actual cause is day length, the drop in temperature would be an example of a *spurious cause*. Such spurious relationships are common; many events occur simultaneously and thus appear to be related.

In many science textbooks and other sources, the term theory is used in place of, or along with, the term hypothesis. In this book **theory** refers to a hypothesis that has survived being tested a great many times in a number of different ways. Thus, if the hypothesis proposing that day length is the primary cause of bird migration is repeatedly tested and supported, it may eventually gain the status of a theory. This distinction between hypothesis and theory is quite arbitrary and is not necessarily how the terms are distinguished by other authors. It will serve, however, as a useful working definition for our purposes.

1.6
TESTING HYPOTHESES THROUGH OBSERVATION AND EXPERIMENT

There are an almost endless number of hypotheses possible about any phenomenon. To distinguish those scientific hypotheses that are true (correspond to the way things happen in the real world) from those that are not true (do not correspond to the real world), it is necessary to test them. In science, testing hypotheses means finding out whether the predictions that must follow from accepting the hypotheses are correct or incorrect, or, in terms of the truth table, True or False (see Fig. 1.4).

There are two general methods of testing hypotheses: seeking additional, ready-made observations or generating those observations by conducting an experiment.

1.7
THE MYSTERY OF HOMING IN SALMON

The silver salmon, *Oncorhyncus kisutch*, hatches in the freshwater streams of the Pacific Northwest. The young fishes swim downstream to the Pacific Ocean, where they may spend five years attaining full size and sexual maturity. Then, in response to some undetermined stimulus, they return upstream to freshwater, often jumping incredible heights up waterfalls to spawn, or lay their eggs (see Fig. 1.5).

Generalizing

By tagging salmon as young smolts (two- to three-year-old salmon that are ready to go from their home streams into the sea to mature), biologists were able to determine that *the fish return to their home streams to spawn*. Notice that this statement is *not* an explanation and that it is based on a finite number of specific observations—a sample from the populations studied. Thus it is an inductive generalization.

A deeper question now arises: *How* do the salmon find their way to the stream of their birth? In other words, what *causes* the salmon to return precisely to their home streams, as opposed to any other freshwater stream?

Testing Possible Explanations

Among the explanatory hypotheses that might be proposed are:

1. Salmon find their way back by the sense of sight (by recognizing certain objects they saw when they passed downstream on their way to the sea).

2. Salmon find their way back by odor (recognizing certain characteristic substances found in their home streams).

Several other hypotheses are possible, of course.* Working with just these two, the next step is to find a method of determining whether either of these hypotheses is correct or whether either is incorrect.

If the first hypothesis is correct, then salmon with shields placed over their eyes should be unable to find their way home. This reasoning can be expressed more formally as follows: If *Oncorhyncus kisutch* salmon use visual stimuli alone to find their way to their home streams to spawn, then blindfolded salmon of this species should not be able to find their way home. In fact, blindfolding experiments reveal that the fish find their way home without sight just as well as they did before.

If we assume that no factors (or variables) that might influence the results have been overlooked, have the experimental results *disproved* our hypothesis? Yes. Suppose, on the other hand, that the blindfolded fish did *not* find their way to their home streams. Would these results *prove* the visual-stimulus hypothesis? No. The experimental results can only be said to *support* the hypothesis. While supporting the visual hypothesis, such results would not rule out the role of smell.

Now let us test the second hypothesis. If salmon find their way to their home stream by following its distinctive odor upstream, then blocking the olfactory sacs (with which the fish detect odors) should prevent the salmon from finding their home stream.

This experiment was performed by biologist A. D. Hasler and his associates of the University of Wisconsin. They studied salmon from two streams near Seattle, Washington: Issaquah Creek and its East Fork, which join together just before emptying into Lake Sammanish (see Fig. 1.6). A total of 302 salmon were captured for

Fig. 1.5
Salmon leaping up a waterfall on their way to spawn. Despite the difficulties in swimming up the Columbia River with its many waterfalls and rapids, large percentages of salmon not only reach the quiet waters upstream, but also find the very stream in which they hatched several years earlier. (Devaney Stock Photos.)

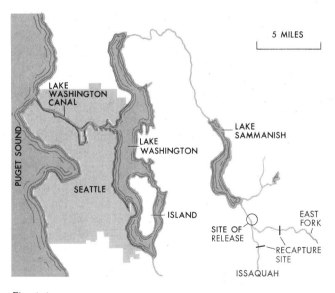

Fig. 1.6
Map showing the area near Seattle, Washington, where A. D. Hasler did his field studies on the factors influencing homing behavior in the silver salmon. (After A. D. Hasler, *Underwater Guideposts.* Madison: University of Wisconsin Press, 1966, p. 39. Copyright © 1966 by the University of Wisconsin Press and reprinted with their permission.)

---

*It may also be true that both hypotheses are correct and that salmon use their senses of sight *and* smell. However, it is more practical to start by investigating one possibility at a time.

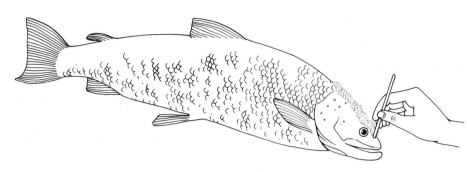

Fig. 1.7
Inserting a cotton wad (visible above the eye of the fish) permeated with petroleum jelly or benzocaine ointment into the olfactory pit of a tagged, anesthetized silver salmon (*Oncorhyncus kisutch*). (After A. D. Hasler, *Underwater Guideposts.* Madison: University of Wisconsin Press, p. 39. Copyright © 1966 by the University of Wisconsin Press and reprinted with permission.)

the experiment. Approximately equal numbers came from each river. The fish were then divided into four groups: two **control groups** (one from the Issaquah and one from the East Fork), and two experimental groups (one from the Issaquah and one from the East Fork). The control groups were tagged on the back, indicating which stream they were from. The **experimental groups** were also tagged. In addition, however, their olfactory sense was disrupted by the insertion of cotton plugs coated with petroleum jelly or benzocaine ointment (an anesthetic) into their olfactory pits (see Fig. 1.7), in which are located the sensory nerve endings responsible for their sense of smell. The fish were then released three-quarters of a mile downstream from where the Issaquah and the East Fork join. The fish were then recaptured at traps set about one mile above the junction of the two streams.

Tabulating the Results

As Table 1.2 shows, not all the released fish were recaptured upstream. Some swam downstream after release, while others swam upstream but missed the traps. Only 50 percent of the controls and 45 percent of the experimental fish were recaptured. More important, however, were the results of the distribution of fish according to the stream of their origin. As Table 1.3 indicates, there was a significant difference between the control and experimental groups in terms of finding their way home. Among the controls, 100 percent of the Issaquah and 71 percent of the East Fork salmon were able to find their

way back to their home streams; for the experimental group, only 77 percent of the Issaquah and 16 percent of the East Fork salmon were able to find their stream of origin. The hypothesis maintaining that the olfactory sense is responsible for salmon being able to identify their home streams as they swim upstream to spawn is supported.

Notice, however, that *some* fish from each experimental group did find their way back to their home streams. This might seem to contradict our initial hypothesis. Here the laws of probability come into play. Statistical analysis shows that a certain number of the fish could be expected to return to their home streams purely by chance. In this case, the number of experimental salmon that found their way to the home stream was *not* significantly greater than the number that could be expected to get there by chance, so the odor hypothesis is still supported.

Table 1.3
Distribution of Recaptured Silver Salmon, Comparing Control and Experimental Groups

Capture site	Recapture site			
	Issaquah		East Fork	
	Actual number	Percent	Actual number	Percent
Controls				
Issaquah (46 fish)	46	100	0	0
East Fork (27 fish)	8	29	19	71
Experimentals				
Issaquah (51 fish)	39	77	12	23
East Fork (19 fish)	16	84	3	16

Data from A. D. Hasler, *Underwater Guideposts.* Madison: University of Wisconsin Press, 1966, p. 42.

Table 1.2
Total Salmon Tagged Compared to Number Recaptured from Experimental and Control Groups

	Total tagged	Recaptured		Not recaptured	
		Actual number	Percent	Actual number	Percent
Control	149	73	49.6	76	50.4
Experimental	153	70	45.0	83	55.0

Further Questions

The salmon experiment demonstrates one characteristic of any problem-solving activity: the answer to one question often raises many more. Hasler's conclusion immediately raises another question. What *is* it that salmon detect by their sense of smell? What factor or factors in the stream prove to be such an infallible guide? Again, a number of hypotheses come to mind:

1. Salmon smell dissolved minerals in the water.

2. Salmon smell dissolved organic material.

3. Salmon smell other salmon of their own ancestral population.

Again, using deduction, each of these hypotheses could lead to a specific prediction that could, in turn, be tested. And that might well raise even more questions. Studies with other closely related migratory organisms may also help. It has been shown that eels, which also migrate from freshwater to saltwater and return (they spawn in the saltwater) are enormously sensitive to dissolved minerals and organic material in water. A single eel may detect the presence of a substance dissolved in water and diluted to the extent that only two or three molecules of the substance were present per liter. This observation might lead us to favor hypotheses 1 and 2 for the salmon.

Fortunately, some experimental data are available to help distinguish between the above hypotheses. Dr. D. J. Solomon of the Ministry of Agriculture, Fisheries, and Food in Great Britain studied salmon migration around three rivers, the Usk, Wye, and Severn, which empty into the Bristol Channel (on the west coast of England) as shown in Fig. 1.8. The adult salmon breed and the young mature in the upstream portions of each of these rivers. Several other rivers also empty into the Bristol Channel, but they do not contain salmon populations.

Solomon was curious about what attracted adult salmon into the Bristol Channel and thence into the three rivers and, conversely, what kept them from invading other rivers not inhabited by salmon.

Solomon hypothesized that if something as general as dissolved minerals and organic matter were the clue, then one would expect to find salmon in the Bristol Channel selectively swimming into the estuaries (point of entrance of freshwater into saltwater) of *all* the rivers emptying into the Channel. Since ocean water contains far more dissolved minerals and organic matter *of the same general type* than is found in freshwater rivers, fish would have to swim part way up a river to be able to determine whether it was "home" or not. In a survey of 23,000 tagged smolts, only *six* were found swimming into the rivers without salmon. It appeared that the salmon were able to detect their home rivers long before

they entered the mouths of those rivers. Some other factor or factors must be at work.

Solomon now formulated another hypothesis. Suppose that the main factor attracting salmon to a river is the presence of other salmon. Most animal species are known to produce distinctive substances called **pheromones** by which members of the species recognize one another. Solomon hypothesized that the factor attracting salmon to one salmon-containing river over another is the pheromone produced by that river's particular population of young salmon. Solomon needed a way to test this hypothesis. Designing an experiment in nature with such a huge population would be difficult—there would be innumerable practical obstacles, not the least of which would be expense.

As it happened, Solomon discovered data from a 20-year-old fish breeding experiment that helped test his hypothesis. In the 1950s, the British government had financed an attempt to establish a salmon breeding ground in a tributary of the Parrett River, which also empties into the Bristol Channel. Ultimately the attempt met with failure; the population did not establish itself in the Parrett. But records showed that in the three years immediately after the first breeding population was

Fig. 1.8
Map of the west coast of England showing the Bristol Channel, with the rivers Usk, Wye, and Severn emptying into it. These three rivers normally contain salmon. Other rivers also empty into the Bristol Channel but normally contain no salmon. (From *Nature* 244 [July 27, 1973], p. 231. Crown copyright. Reproduced by permission of the Controller of Her Brittanic Majesty's Stationery Office.)

introduced, the number of adult salmon found swimming up the Parrett increased severalfold. Most of these fish had not come from the newly established fishery population, but were from the Usk, Wye, and Severn populations. They had suddenly found the Parrett interesting!

Using deduction, in this case retroactively, Solomon made a prediction based upon his pheromone hypothesis, a prediction already verified by the Parrett River study done 20 years before. The formalized logic of Solomon's work looks something like this: If salmon recognize pheromones produced by other salmon, then introducing young salmon into a river without salmon ought to increase the chances of adult salmon swimming into that river. As noted, the Parrett River data supported the hypothesis. As the truth table shows, however, a true prediction does not necessarily mean a true hypothesis. Still other experiments would have to be carried out to further substantiate Solomon's hypothesis. Nonetheless the experiment *does* lend support to the hypothesis by verifying the prediction.

Testing the Odor Hypothesis Further

In 1938 it was shown that salmon transplanted at the presmolt stage from their native river to a second river returned as adults to the second river. In a contrasting set of experiments, salmon kept in a hatchery until well past the smolt stage and then released into different streams failed to return as adults to their transplant stream; they returned randomly to a number of different streams. These results suggested that a *critical period* must exist in a young salmon's life during which it learns to recognize its "home" stream.

Further experiments conducted between 1955 and 1971 by a number of investigators found that young salmon are especially receptive to learning the chemical "odors" of their home stream just at the time they enter the smolt stage. In fact, during this critical period, exposure of a young salmon to the chemical odors for as short a time as four hours is long enough for it to remember this odor for life. Latching onto a sensory impression early in life is sometimes called "imprinting" (see Chapter 30). Thus, it could be hypothesized that at a critical period in life, young salmon imprint on specific chemical "odors" in their native streams; they later use this imprinted memory to guide them back to their native stream as adults.

In the mid-1970s, along with Allen Scholz and Ross Harrel, Hasler decided to study the imprinting process directly. They exposed a group of young smolts to a nontoxic organic substance called morpholine (C_4H_9NO). They then took two groups of salmon that had been raised in a hatchery until the smolt stage; one group (experimental group) they exposed to low dosages of morpholine (1×10^{-5} mg per liter) for a month, while the other (control group) received no mor-

pholine. Both groups remained in the hatchery for ten more months, after which the fish were brought to the laboratory and tested for olfactory response to morpholine. Both control and experimental groups were placed in tanks of morpholine and their brain wave responses recorded. If the fish sensed morpholine, there would be an electrical signal recorded; and if they did not, there would be no signal. As Hasler and his co-workers predicted, the salmon that had been exposed to morpholine as young smolts showed a significantly higher amount of brain wave activity than those that had not been exposed to morpholine. In terms of neurophysiological response, salmon exposed as smolts to morpholine were shown to be highly sensitized to its presence as adults.

Hasler and his co-workers then proceeded to ask another question: Will fish imprinted on morpholine swim toward streams or areas of water into which the substance has been placed? They took the experimental and control groups of salmon described above and released them into Lake Michigan, about equidistant between the outlets of two creeks. Into one outlet, Owl Creek, they placed a concentration of morpholine about equal to that maintained in the hatching tanks; into the other stream, they placed nothing. The fish were all marked, and eighteen months later the investigators looked to see what fish were found in each stream. In Owl Creek a total of 246 tagged fish were recaptured (out of a starting tagged population of 16,000, about the expected percentage of recapture in such experiments). Out of the 246, however, 218 were fish that had been exposed to morpholine, while only eighteen were from the control group. In the other creek most of the fish were from the control group. It seemed clear that the cue that had attracted such a disproportionate number of the experimental group to Owl Creek was the presence of morpholine in the water.

However, could something else in Owl Creek have served as the cue to the experimental fish, masking the effects of morpholine? Representatives of the control and experimental groups were released near the two streams, but no morpholine was added to either stream. If the experimental fish had found their way into Owl Creek by some other cues, they ought to do so in this second experiment as well as in the first one. When Hasler and his co-workers carried out this experiment, they found about equal numbers of both control and experimental fish in each creek. The hypothesis that salmon imprint onto a specific chemical substance early in life seemed to be borne out.

In another interesting experiment, Hasler and his associates attempted to track the movement of adult salmon through the water in response to the presence of morpholine. They implanted ultrasonic transmitters into forty salmon, twenty of which had been imprinted on morpholine, and twenty of which had not. After releasing the fish into Lake Michigan the researchers set up an "odor barrier" by pouring a line of morpholine from a boat in the water along the shore. Using ultrasonic tracking devices (like radio receivers) the investigators found that the morpholine-imprinted fish would stop swimming when they encountered the odor barrier and mill around for up to four hours within the odor barrier until the scent had been washed away by water currents. Non-morpholine-imprinted salmon swam right through the barrier without stopping. From all these results Hasler felt that he could draw three major conclusions:

1. Fish use the olfactory sense to detect the presence of familiar substances in water.

2. Fish imprint onto the familiar chemical odors of their home streams when in the early smolt stage.

3. The specific chemical odors on which young salmon imprint guide them as adults to return to their home streams.

Hasler's and Solomon's work with the salmon illustrates several important features of framing and testing hypotheses in science (or any other subject, for that matter). Note that both investigators framed their questions in ways that were testable. They did not put their question in a general form such as, "How do salmon find their way home?" Rather, they broke such a large question down into several smaller, more directly answerable questions, for example, "Do salmon use the olfactory sense to find their way home?" Or, "Will salmon imprint on an artificial substance if they are exposed to it early in life?" Both of these questions suggest immediate observations or experiments that might provide an answer.

A second point about scientific research illustrated by this case is the importance of approaching a given problem from a number of different directions. Hasler and his team were not content simply to do one or two experiments testing the olfactory hypothesis. They not only showed that salmon use the olfactory sense to find their way home as adults (that is, in the first set of experiments, by blocking the olfactory lobes), they also showed that: (1) salmon imprint on specific substances early in life; (2) salmon show a distinct neurological reaction to substances on which they have been imprinted; and (3) in natural situations salmon, imprinted on a specific artificial substance, select streams containing this substance when they go to spawn. All these lines of evidence converged to support the major, initial hypothesis: namely, that young salmon learn to smell the chemical peculiarities of their home streams, and use these smells as a signpost to guide them back to the same stream as adults.

A final point about scientific methodology illustrated by the salmon experiments comes from Solomon's work. It is not always possible in science to perform an experiment to test a hypothesis. Yet, it frequently happens that an observation or a group of observations will serve the same purpose. Solomon's *observation* that adult salmon appeared in the Parrett River only after young fish had been introduced there supported an aspect of the olfactory hypothesis—in this case the idea that part of the olfactory stimuli to adult salmon is chemical products (pheromones or other substances) produced by other salmon. Solomon did not do an *experiment* to come to this conclusion; in a way, an experiment had already been done for him. He simply took advantage of some results that bore on the question he was asking.

1.8
ANALYSIS AND INTERPRETATION OF DATA

Quantitative data are the cornerstone of modern science, and thus of biology. Yet it is obvious that a large collection of quantitative data is of little or no value if it is not arranged in such a way as to demonstrate important relationships. The individual data for each tagged fish in Dr. Hasler's experiments yield little information if they are presented randomly. In collecting the data in the field, of course, Hasler and his associates had to tabulate it as it came in—that is, as they captured a fish they would record where it was from, whether it belonged to an experimental or control group, and where it had been found. Their field notebooks might have contained information arranged as shown in Table 1.4.

Information collected from observations or experiments may be of two kinds: quantitative or qualitative. **Quantitative** data are the result of measurement and can be expressed in some definite and precise form, usually in numbers. **Qualitative** data do not lend themselves to precise numerical expression.

Table 1.4
Hypothetical Table of Raw Data as It Might Have Been Collected for Salmon Migration by A. D. Hasler and His Associates

Fish number (in order of recapture)	Control group		Experimental group	
	Where released	Where recaptured	Where released	Where recaptured
1	Issaquah	Issaquah		
2	East Fork	Issaquah		
3			Issaquah	East Fork
4	East Fork	East Fork		
5			East Fork	East Fork
6			East Fork	Issaquah
7	Issaquah	Issaquah		
8			Issaquah	East Fork
etc.				

In the above discussion of the salmon experiments, both quantitative and qualitative data were collected. Dr. Hasler's experiment involved quantitative data. He counted the number of each fish population recaptured in each river to express the amount of recapture in terms of percentages of those released. Had he chosen to express the data qualitatively, he could have reported simply that "more" salmon in the control groups than in the experimental groups returned to their home streams. In general, scientists prefer quantitative to qualitative data. For one thing, qualitative data are vague and communicate less information than quantitative data. "More" could refer to anything from a slight increase to a manyfold increase. For another, qualitative data cannot be checked so readily by other experimenters who might wish to repeat the work. Repeatability of results is an important aspect in the verification of scientific hypotheses, and quantitative data are more useful for repeating experimental work. And third, quantitative data give the reader the experimental results directly, without anyone's subjective interpretation entering between. Quantitative data allow the reader to form an independent judgment.

Organization of Data

As presented thus far, the data do not allow us to determine readily whether the "smell" hypothesis has been confirmed or rejected. The first step in analyzing the data would be to arrange it in some order, as Hasler did in Table 1.3. By arranging the data so that he could compare experimental and control groups, Hasler was able to determine that a difference did exist between them. It may seem like a very simple point, but it is one of the most important aspects of data analysis in science: *data must be organized in order to yield useful information.* In this case, the most logical form of organization is to draw a comparison between the two groups. The comparison makes it possible to answer the original question. That salmon with their olfactory sacs plugged found their way back home less frequently than those with a functional sense of smell becomes apparent only when the data have been organized in the manner shown in Table 1.3.

What, then, is the function of Table 1.2? This table records the difference between the number of fish released in each group and the number recaptured. Note that for both groups the percent of fish recaptured (or conversely, not recaptured) is approximately the same (49.6 percent of controls, 45 percent of experimentals). This is important information. If there were a great difference in recapture rate between experimentals and controls, comparison of the data shown in Table 1.3 would not be very useful. There are two reasons for this.

1. Suppose that the experimental animals had found their way home with considerably less frequency than the controls. We would have to conclude that the plug-

ging off of their olfactory pits must have affected the general ability of the fish to navigate, or perhaps even to swim. A similar recapture rate indicates that the two groups must have had about equal ability to swim and navigate.

2. Table 1.2 also tells us whether or not both experimental and control groups were subject to the same sampling error. For example, suppose that only 10 percent of the total experimental group had been recaptured, compared to, let us say, 50 percent of the controls. That is a big difference. Especially since one of the figures, 10 percent, is very small, the number of fish known to have arrived at either their home or another stream is subject to severe sampling error, which enters in when we consider only a very small number of cases out of a very large total population. If the sample is very small, the individual items (organisms) measured might be atypical: unrepresentative of the whole. The larger the sample taken, the less significant the sampling error is likely to be. In Hasler's experiment, sampling error would not be much of a worry if the difference were between 80 percent and 60 percent recapture. It would also be less of a problem if the number of fish in both groups were very large (let us say thousands). However, since Hasler was dealing with a total of only about 150 fish in each group (control and experimental), a big difference in recapture rate between the groups could easily lead to sampling error in the smaller group. Knowing recapture rates were about the same for both groups eliminates this problem. Comparatively speaking, the two groups have about the same degree of sampling error.

Remember that *some* of the experimental fish did, in fact, end up in their nonhome stream. And only in the case of the Issaquah control group did 100 percent of the fish manage to get back home. Among the East Fork controls, only 71 percent returned to the East Fork (the other 29 percent ended up in the Issaquah). This tells us right away that the generalization that "salmon return to their home streams" is not true all the time. It is only true in a statistically significant number of cases. This is true of most situations in the real world. Although from a logical point of view a single exception to a generalization is enough to make the generalization untrue, in real life this is not the case. The generalization that salmon find their way home by sense of smell is still considered true—even though it is contradicted by a certain percentage of the data.

Many factors—a predator, water current, degree of fatigue, the effectiveness of its own sense of smell (not all salmon smell with equal accuracy), the behavior of other fish nearby—all could contribute to the choice a salmon makes when it comes to the juncture of the Issaquah and the East Fork. Thus, although it might be generally true that salmon find their way home by the sense of smell, the statement does not say that no other factors can influence an individual salmon's choice of direction. The generalization that salmon find their

SUPPLEMENT 1.2
IT'S ALL IN WHAT YOU WANT TO SHOW

Statistics can be very useful in showing trends and relationships. How data are presented, numerically or graphically, affects the kind of information they convey. What we can call well-designed presentations allow a reader to obtain the most important information from a set of data quickly and accurately. Subjective judgment is usually involved in deciding how to present data, and the same data can be made to suggest very different ideas.

Consider the question of changes in annual income. This is something that most people are interested in understanding and predicting. The graph shown as Diagram A is very simple in design: on the vertical axis is plotted amount of money taken in as national income (in billions of dollars); on the horizontal axis are the twelve months of a particular year. During one twelve-month period, national income increased from 20 to 22 billion (per capita). This demonstrates economic growth and suggests that on the average people are getting wealthier. Nevertheless, this may not be true. The increase may be largely among a very few wealthy people, with very little increase going to average- or low-income families. Diagram A shows an increase, but it certainly does not appear to be a very dramatic one.

The same data graphed in another way, on a different vertical scale, would show a very different picture (see Diagram B). Choice of the vertical scale in this case would depend upon what the person drawing the graph wanted to show. A government or business spokesperson might prefer Diagram B because it is good public relations. A labor spokesperson might prefer Diagram A, which suggests that income is not rising so fast after all. For these data to be useful in answering a question about whether income is getting better or worse, more information is needed. We would need to know how income was distributed, whether the increase in annual income was greater than, equal to, or less than the inflation rate, whether unemployment went up, down, or stayed the same during the time period, and so on. The important point here is that the way in which data are presented (1) determines a great deal about what conclusions are conveyed and (2) is thus dictated by what the person presenting the data wants to show.

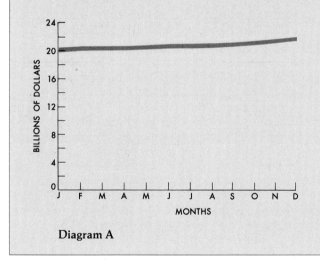

Diagram A

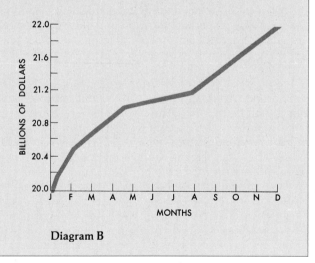

Diagram B

way home by smell is said to be true in a large number of cases. But it should not be understood—nor should any generalization—to mean that the results must be predictable 100 percent of the time.

Statistical Significance

How is it possible to distinguish between exceptions that actually contradict a hypothesis and exceptions that occur more or less by chance? There is no absolutely certain way to make this decision. Biologists and other scientists have statistical methods for estimating whether or not the difference between expected and actual results is significant. This test is called, not surprisingly, the *significance test*. It starts with the assumption that two situations (two populations, two drugs, an experimental and control group) are identical and proceeds to try and disprove the assumption. If the difference between two groups or situations is large, it is said to be a *significant* difference; if small, it is said to be *insignificant*.

Let us consider an example. Table 1.2 shows that of all the fish released, about 50 percent of the controls and 45 percent of the experimentals were recaptured. Is this difference significant? Under some circumstances it might be. To Hasler, knowing the conditions under which fish were tagged, released, and recaptured, such a difference appeared to be *insignificant.* He was confident that the rate of recapture between the experimental and control groups was virtually the same. The significance test verified this conclusion, but that does not mean that Hasler was absolutely right. Sometimes small but unexpected differences from expectation can lead to new discoveries. The highly subjective element in science lies partly in knowing when to ignore and when to pay attention to such differences. No statistical calculation can provide that insight.

The same test of significance could be applied to the data in Table 1.3. If the smell hypothesis were correct, for example, we might expect that with very large samples, approximately 50 percent of the Issaquah group would end up in the Issaquah and 50 percent in the East Fork, and similarly for the East Fork group. With no sense of smell to help them, the fish should choose streams more or less at random. The data show that among the experimental groups, more Issaquah fish end up in the Issaquah than in the East Fork (77 percent as compared to 23 percent). There is about a 25 percent error between the actual and the expected results. Among the East Fork experimental subgroup, the situation is very different. Here 84 percent end up in their nonhome stream, the Issaquah, whereas only 16 percent end up in their home stream, the East Fork.

Are these differences from expectation (50:50 in both cases) significant or not? Hasler did not think so. The situation is most clear-cut in relation to the East Fork experimental group. Of the fish, 84 percent end up in their nonhome stream, whereas by chance we would expect about 50 percent. Comparing the East Fork experimentals with their counterparts in the controls we see that in the latter only 29 percent ended up in the wrong stream. No statistical test is really needed to say conclusively that the difference between 29 percent and 84 percent is almost certainly a significant one.

With regard to the Issaquah experimental and control groups, the situation has an interesting twist to it. It is probably unique that 100 percent of the Issaquah controls were recaptured *where they were predicted to be.* Among the experimentals only 77 percent were recaptured in the home stream—considerably above the 50 percent we might expect on the basis of chance. After all, only 71 percent of the East Fork controls were recaptured in their home stream.

Strictly speaking, the Issaquah experimental group with no sense of smell actually did better than the East Fork controls with a functional sense of smell! Obviously, Hasler's feeling that the data still support the smell hypothesis comes to some extent from other consider-

ations than hard data. This example emphasizes the many arbitrary and subjective elements that enter into analyzing data and into evaluation of hypotheses based on these data. Is is an old cliché, but statistics can be used to make almost any point one wishes to make—on almost any side of an argument! This does not mean that statistics are not useful. Quite the contrary—they are one of the cornerstones of all scientific methodology. But it is important to understand the limitations the use of statistics imposes on our interpretations.

1.9
THE PHILOSOPHICAL SIDE OF BIOLOGY

For over a century a debate has raged on and off within the biological community between the viewpoints of "vitalism" and "mechanism." This debate has been part of a much larger philosophical dispute that developed during the later nineteenth century, the dispute between philosophical *idealism* and *materialism.* Since certain very different biological assumptions come from each of these philosophical views, it will be important to distinguish between them.

Vitalism

The controversy between vitalists and mechanists has always focused on the question of what causes living organisms to differ from nonliving material. Even the most complex crystal structure is less complex than molecules in living systems. The abilities to move, reproduce, take in and assimilate material from the environment, grow, and have consciousness all seem qualitatively different from anything observed in the nonliving world. Vitalists explain this unique feature by postulating, in one form or another, the existence of a "vital force." This force is assumed to be wholly different from other known physical forces and ultimately to be unknowable—not subject to physical and chemical analysis. This vital force departs from a cell or organism at death. Vitalists do not necessarily deny that chemical analyses of living organisms are valuable. But they feel that "life" involves something more than the principles of physics and chemistry.

An example of vitalist thinking appears in the memoirs of Assistant Surgeon Edward Curtis of the Washington, D.C. Army Medical Museum. After President Abraham Lincoln died on April 15, 1865, an autopsy was performed in the Northeast Corner guest room of the White House. In the words of Dr. Curtis,

Silently, in one corner of the room, I prepared the brain for weighing. As I looked at the mass of soft gray and white substance that I was carefully washing, it was impossible to realize that it was that

"In nature we never see anything isolated but everything in connection with someone else. . . ."

—*Johann Wolfgang von Goethe (1749–1832)*

mere clay upon whose workings, but the day before, rested the hopes of the nation. I felt more profoundly impressed than ever with the mystery of that unknown something which may be named "vital spark" as well as anything else, whose absence or presence makes all the immeasurable difference between an inert mass of matter owing obedience to no laws but those governing the physical and chemical forces of the universe and, on the other hand, a living brain by whose silent, subtle machinery a world may be ruled.

This example may be somewhat more dramatic than most, but it illustrates clearly the vitalist's belief in some superphysical factor differentiating living from nonliving matter.

Mechanism

Essentially, mechanists view organisms as very complicated machines. We can find out how an organism works by taking it apart, studying the components one at a time, and then puting it back together again, so to speak. Basic to the mechanist view is the idea that organisms function according to fundamental laws of physics and chemistry.

The conflict between mechanism and vitalism is in many ways an artificial problem for both are limited ways of viewing living systems. Vitalism is limited because it presupposes the existence of a mystical, unknowable force animating all living beings, a force that lies by definition beyond physical and chemical investigation. Since one cannot ask questions about something ultimately unknowable, such a viewpoint limits research. Thus vitalism is an example of **metaphysics**—a mode of thinking that transcends present, material reality to seek nonmaterial causes. Modern biology avoids metaphysical statements as untestable. Biologists may not know exactly what processes make an organism so unique and different from nonliving matter, but as scientists they assume this can ultimately be determined.

Mechanism has its limitations, too. Biologists, especially, have always sensed that living organisms are so complex that their *organization* is the key to their unique functions. Biologists have attempted to discover that organization by breaking the complex whole down into component parts and studying each part separately. Physiologists in the midnineteenth century, for

example, often studied just one organ at a time. They would remove the liver or stomach from an anesthetized animal and bathe it in a solution containing various substances. The effect of these substances on the functioning of the organ could then be determined and measured.

A whole school of thought emerged to claim that such mechanistic experiments revealed all there is to know about life processes. It was assumed that an organ functioned inside the intact animal exactly as it did in the laboratory. Other physiologists disagreed. They knew that blood pressure, state of the central nervous system, diet, and a host of other factors could greatly alter the organ's functioning inside the animal. Studies of organs in isolation did not necessarily give wrong information, but the results were *incomplete*. A limitation of mechanism as a way of viewing living systems is that it often oversimplifies to the extent of error.

The Modern View

Must we choose between two limited perspectives, vitalism or mechanism? Obviously not. Modern biologists accept the basic tenets of mechanism without accepting the analogy of the machine. They view organisms as very different from machines but do not postulate any vital or metaphysical force to account for this difference.

What are the characteristics of this modern view of living systems?

1. No part of any biological system exists in isolation. For example, the liver is affected by the sugar content of the blood passing through it; the blood is affected (its sugar is removed) as it passes through the liver. Thus while both blood and liver can be described partially in isolation, neither can be described fully except in their interactions with each other and all the other organs.

2. Unlike machines, living systems are constantly in a process of change: growth, development, and degradation. This change results from the constant interaction of internal and external forces. For example, a seed sprouts, grows, and matures into a plant, which then flowers, produces fruit, and eventually dies.

3. The internal processes of any living system undergo change as a result of the interaction of opposing forces

(see Fig. 1.9). All living organisms, for example, carry out both anabolic (build-up) and catabolic (breakdown) chemical reactions. Thus the growth and development of a seed represents a change in which the overall effect of anabolic reactions is greater than that of catabolic reactions. Maturity occurs when the two are balanced, and aging and death result from the dominance of catabolic over anabolic processes. Far from being accidental, this developmental process is programed into the genetic and physiological makeup of the organism. At all stages the overall process can best be studied by investigating the interaction of anabolic and catabolic process, rather than just anabolic or catabolic processes alone.

4.   The accumulation of many small quantitative changes can eventually lead to a large-scale, qualitative change. For example, the heating of water from 90° to 91°C represents a quantitative, but not a qualitative, change in temperature, since the water has not been fundamentally changed (it is only one degree hotter). However, when the temperature goes from 99° to 100°C, the water begins to boil, turning into steam. This represents a qualitative change; water and steam have very different physical properties. The accumulation of many quantitative changes has led to an overall qualitative change.

This principle is as true of the biological as of the physical world. If a nerve attached to a muscle is stimulated with a low-voltage electric shock, the muscle will not respond. If we keep increasing the voltage, the nerve eventually reacts and causes the muscle to twitch. The point at which the reaction takes place is called the **threshold**, the point at which quantitative changes give rise to qualitative change. With the gradual build-up of voltage (a quantitative change), a qualitative change is achieved in the nerve, and it transmits a message.

5.   In living systems, the whole is greater than the sum of its parts. Realizing that all the parts of a system—living or nonliving—interact, and that one cannot accurately consider any one part in isolation, it might appear that studying any complex system is hopeless. How can *all* the parts of a system be studied at once, in all their interactions? There are two answers to this dilemma. First, it *is* still possible to study individual components of a system in isolation if the components are also studied as part of the whole. Second, the computer and the mathematical theory of systems analysis represent important new techniques for studying complex interacting systems.

These views are characteristic of the way increasing numbers of biologists are coming to see living systems. It is important to be familiar with them, since an investigator's philosophical viewpoint often affects the kinds

Fig. 1.9
The life of a plant is shown from seed through death, illustrating the two opposing processes of anabolism (build-up of substances) and catabolism (breakdown of substances). Both go on all the time in all organisms. In the first part of its life, the plant grows and produces seeds because anabolic processes occur in greater number than catabolic processes. In the last part of its life, the plant withers and dies because catabolic processes become the more frequent. This interacting set of processes produces developmental patterns, rather than merely random changes.

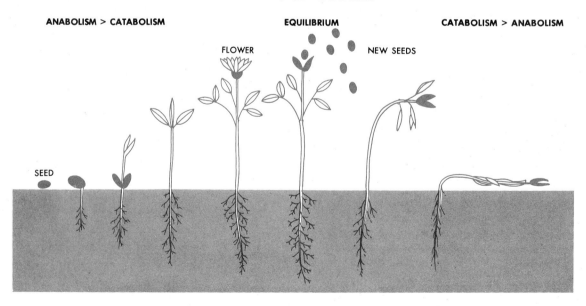

Fig. 1.10
The consequences of logical thinking do not always lead to "correct" or even logical conclusions. It is a well-known fact of observation that snakes can and sometimes do swallow other snakes, even larger ones than themselves. If it is assumed that the one that is swallowed disappears from sight, the situation depicted is the logical conclusion! (Courtesy of Johnny Hart and Publishers Newspaper Syndicate.)

of questions asked and the kinds of answers considered acceptable.*

1.10
THE LIMITATIONS OF SCIENCE

Science is one of our most productive ways of exploring and trying to understand our environment. Yet, despite its many contributions to human intellectual growth, health, and general welfare, science does have serious limitations. Oddly, one of these stems from one of its greatest strengths. By dealing only with sets of phenomena that can be experienced directly or indirectly through the senses, science is necessarily excluded from other sets of phenomena.

The philosopher George Boas has stated that ". . . what science wants is a rational universe, by which I mean a universe in which the reason has supremacy over both our perceptions and our emotions." Yet what science wants and what science gets are often two different things. For one, nothing in the foregoing statement should be read to imply that scientists as individuals are any more rational, objective, or unemotional than anyone else. Nor is science itself necessarily independent in its thought and action from the society that surrounds and supports it; often the directions in which scientific research swings may be dependent upon the availability of funds allocated by governmental agencies to support that research.

Further, the solutions to purely scientific problems worked out by a scientist may be used by society in a manner that is repugnant to him or her. For example, Yale biologist Dr. Arthur Galston's research on the chemical basis of the shedding of leaves by plants was later used for defoliation of forests in Vietnam.

Despite the logical basis of science, it would be a mistake to give the impression that scientists are never wrong. Astronomer Johannes Kepler once wrote, "How

many detours I had to make, along how many walls I had to grope in the darkness of my ignorance until I found the door which lets in the light of truth." It is doubtful, in fact, that there has ever been a scientist who has not made mistakes. James Bryant Conant said,

One could write a large volume on the erroneous experimental findings in physics, chemistry, and biochemistry which have found their way into print in the last one hundred years; and another whole volume would be required to record the abortive ideas, self-contradictory theories, and generalizations recorded in the same period.

Lord Rutherford stated that humans would never tap the energy in the atomic nucleus. The first atomic bomb exploded a few years after his death. The famous nineteenth-century physiologist Johannes Müller asserted that the speed of the nerve impulse would never be measured. Six years later, Hermann von Helmholtz measured it in a frog nerve only a few inches long.

Nor is it safe to think that scientists always reason correctly. As a matter of fact, some scientists are notorious for "going off the deep end," particularly when writing in areas outside their own specialty. Furthermore, if scientists can be wrong, so can science. Ask a physicist about ether, a chemist about phlogiston, or a biologist about Lamarckism. Science has had incorrect theories in the past, it has them now, and it will continue to have them in the future. The strength of science does not lie in any infallibility. Nor does it lie in its logical basis, for the conclusion of a perfectly logical argument can be utter nonsense (see Fig. 1.10). Rather, it lies in the self-criticizing nature of science—the constant search for "truth" by eliminating experimentally established error.

The scientist A. J. Lotka stated in 1925 that "Science does not explain anything . . . science is less pretentious. All that falls within its mission is to observe phenomena and to describe them and the relations between them." Any absolute truth is beyond the reach of science. *In science, "truth" is a well-supported hypothesis.* Should the hypothesis fall, a new "truth" takes its place.

---

*The five points outlined above are recognizable as basic tenets of "dialectical materialism." This philosophical school is contrasted with the philosophy of mechanistic materialism, or "mechanism," to which it has been opposed throughout the nineteenth and twentieth centuries.

SUPPLEMENT 1.3
BIAS IN SCIENCE

While nonscientists are expected to be strongly influenced by opinion, subjective feelings, and biases, scientists are supposed to be relatively free from these human failings. Such a view, however, like most stereotypes, is far from accurate. Science is practiced by people, and therefore is subject to prejudice and bias, like all human activity.

There are two kinds of bias which appear in scientific work. One is intentional bias—that is, deliberate and conscious manipulation or alteration of data to support a favorite or preconceived idea. Conscious or intentional bias is, in reality, simple dishonesty. The second form of bias is called unconscious bias, that is, bias of which the individual is not fully aware. Unconscious bias is the tendency among researchers to slant their data or their conclusions toward their own ends. Both types of bias are quite prevalent in scientific work. We will give only a few examples to illustrate the role such biases have played and continue to play in the development of scientific ideas.

Conscious or intentional bias covers a wide range of problems, from omission of unrepresentative data to faking measurements and juggling decimal points. In the 1920s a controversy raged among biologists about the inheritance of acquired characteristics, the idea that traits acquired by an organism in its own lifetime (for example, loss of a digit or building up of big muscles) could be passed on to that organism's offspring. Although most biologists rejected the idea, some workers persisted in trying to prove that acquired characteristics could be inherited. One was Austrian biologist Paul Kammerer, who studied *Alytes obstetricans*, the "midwife toad," in the years immediately after World War I.

Most toads and frogs mate in the water. In order to grasp onto the female, the males develop what have been called "nuptial pads" on their palms and fingers during the mating season. The midwife toad, however, mates on land; the female's skin is rough and dry, and the males do not develop nuptial pads. Kammerer was able to induce midwife toads to mate in water. After only a few generations he claimed that the males showed nuptial pads and transmitted this trait to their male offspring. Kammerer's results were criticized from many quarters. When the American biologist G. K. Noble finally examined Kammerer's one specimen (there was only one!), it turned out that the so-called nuptial pad was a blotch on the skin induced by the injection of India ink under the surface. Some evidence suggests that Kammerer himself did not fake the data but that the injections were done by one of his assistants. Even so, had Kammerer not been such a zealous advocate for the idea of inheritance of acquired characters, he would have been more likely to have verified his own data more carefully. Kammerer's case was an embarassment to the myth of science as pure, objective, and beyond suspicion.

Cases of unconscious bias are even more prominent. One comes from the work of American craniologist Samuel S. G. Morton in the 1840s and 1850s. Morton was convinced that the white and the nonwhite races had separate origins, and that the whites were superior. Among nonwhite groups he studied the American blacks and the Indians of South and Central America. Morton collected a large number of measurements of cranial capacity—the volume of the skull—supposedly a measure of brain size and, thus, innate capacity. His method of measuring cranial volume was to fill skull samples with lead shot and then measure the volume of the shot. It is extremely difficult to determine exactly when a skull is filled, since shot is composed of little metal balls (of the size called BB) and one can always add a few more without obviously overfilling the cavity. Morton knew whether he was measuring a white or a nonwhite skull. Thus it is quite possible that unconsciously he may have tended to add a little more shot to Caucasian skulls, and a little less shot to others, in order to get his expected result. Not surprisingly, Morton's data showed that whites had significantly larger skull capacities than nonwhites.

Another source of error in Morton's work came from his statistical analysis. Morton included in the native samples a large number of Peruvian Inca skulls, the smallest of all Indian groups; and he excluded from the Caucasian samples all but a very few Hindu skulls, because, by his own admission, they represented the smallest Caucasian types. Morton measured each Indian group separately; he must have been aware of the fact that Inca skulls were smaller than the others, and thus, if present in greater numbers, would lower the average of the Indian cranial volume. Along with eliminating the Hindu skulls, this arrangement would, of course, magnify the differences between Indian and Caucasians. (Recent corrections for sample size, using Morton's own data, show that the mean differences between Indian and Caucasian skulls are negligible.) It appears that Morton knew what he wanted to prove and chose his data accordingly. How conscious he was of his own transgressions is hard to tell. Since he published all his raw data (actual measurements), thus making it easy to see his errors, it seems unlikely that he was consciously aware of biasing his conclusions.

Morton's published raw data allowed other scientists to check his conclusions against their own figures and values. It appears, however, that many scientists do not keep such good records. In an interesting experiment, a sociologist wrote to thirty-seven authors asking for the raw data on which they had based recent journal articles. He found that of the thirty-two who replied, twenty-one reported their data to be either misplaced, lost, or inadvertently destroyed. He was able to analyze only seven sets of data supplied by nine of the original thirty-seven authors. And, of these, three had errors of such magnitude as to invalidate the authors' conclusions! So much for objective, rigorous science.

Perhaps the most extensive case of fraudulent results in recent years has come to light in the wake of publications about racial differences in I.Q. By analysis of I.Q. scores of many pairs of identical twins reared apart, California psychologist Arthur Jensen concluded in 1969 (and reaffirmed in a new book in 1979) that I.Q. is about 80 percent heritable in the human population. Noting that blacks as a group consistently received lower average scores on I.Q. tests than whites, Jensen concluded that blacks must be genetically inferior in intelligence to whites. Needless to say this conclusion generated considerable controversy.

One outcome of the controversy was the investigation of the original data on I.Q. in the separately reared twins. Those data were not collected by Jensen, but by Sir Cyril Burt, an eminent and respected British psychologist. Investigations into Burt's data revealed several interesting facts. First, Burt published his data correlating I.Q. scores between identical twins (that is, comparisons of the similarity of I.Q. scores—the closer the two scores were for any twin pair, the higher the correlation was said to be) over a forty-year span. Although he added many new pairs of twins to the total sample over this period of time, the correlation values remained exactly the same—to the third decimal point—a rather incredible coincidence. Second, it turned out that many of the twin pairs had never been given I.Q. tests at all; their I.Q.'s were *estimated* from personal interviews. Third, it later came to light that much of the I.Q. data had not been obtained by even so much as an interview, but had been totally fabricated; even one of Burt's published co-authors turned out to have been invented, apparently for the purpose of authenticating Burt's procedures by showing that other researchers were involved. Burt's raw data cannot be found (some of it, at least, could have been lost in the bombing of London during the war), lending further credence to the suspicion that it may never have existed at all. Why Burt, who was a man of prestige and position, would have resorted to such fabrications is difficult to imagine. His biographer offers a psychological explanation: that Burt was so concerned with the "correctness" of his position that he may have simply deemed it unnecessary to collect actual data on the point.

Burt's data have been largely discredited, and as a result the studies based on it. But the problem illustrated by this incident has not departed. Science is no more value-free than anything else human beings do. To Morton, for example, the superiority of the Caucasian over the nonwhite was taken for granted—his writings make that extremely clear. Ranking human groups as superior or inferior, however, is a value judgment, not a fact. Both collection and interpretation of data are bound to reflect those values, consciously or unconsciously. The point is not so much to try and make science value-free or to eliminate all subjectivity. That would be impossible. Nor is the point to excuse subjectivity and bias as "natural" and therefore acceptable. Clear thinking about any subject involves differentiating between one's own views, feelings, or biases and the world outside. It is important to recognize that bias is always present in science and that someone who puts on a white coat does not shed his or her subjectivity. Moreover, biases can, at times, be creative—they can lead investigators to persist on a problem despite continual discouragement. But by recognizing our own biases explicitly, we are in a better position to use them creatively when necessary and reduce, if not eliminate, their worst effects when they are blocking our vision.

Despite its limitations, science is a remarkably successful way of accumulating knowledge. In applied research, scientists may use the methods of science for the purpose of developing products to improve human comfort and welfare. In pure or basic research, the scientist searches for knowledge—knowledge for its own sake—regardless of whether or not discoveries will benefit humankind. It is important to note that the results of basic research have contributed as much as or more than those of applied research; indeed, the former often leads to the latter. In science, applications seem to flow naturally from understandings, and the field itself seems to be inherently productive.

Summary

1. Two types of logical thinking, induction and deduction, can be identified in science.

a) Inductive logic is a creative process in which general statements or generalizations are derived from a specific number of discrete observations. Observing that a few green apples are sour, one might arrive inductively at the generalization "All green apples are sour."

b) Deductive logic, or *if . . . then* reasoning, is the process of inferring from a general hypothetical statement (hypothesis) some specific consequence or prediction that must follow. *If* all green apples are sour, *then* the next green apple tasted will be sour. If the prediction is not verified, the hypothesis is false. If the prediction is verified, the hypothesis is supported but not proved.

2. Valid statements are those that follow logically from hypotheses. True statements are those that conform to people's experience. A statement may be valid yet still untrue.

3. There are two types of hypotheses. "All green apples are sour" is a generalizing hypothesis. "Green apples are sour because they have a higher concentration of a certain acid than do sweet apples" is an explanatory hypothesis. There is constant interaction between observation and hypothesis in science. Typically, observations lead to the formulation of hypotheses that, in turn, suggest further observations.

4. There are two ways to test hypotheses: observations and experiments. A hypothesis suggests new observations that might be made or experiments that might be carried out. According to the truth table (Fig. 1.4) an observation or experiment that contradicts a prediction based on a hypothesis shows the original hypothesis to be invalid. An observation or experiment that verifies a prediction based on a hypothesis supports the original hypothesis, although that hypothesis may still turn out to be invalid.

5. To be conclusive, experiments must
    a) test one specific prediction of the hypothesis at a time.
    b) use a control as a standard for comparison with the experimental group.
    c) collect quantitative, rather than qualitative, data whenever possible. Quantitative data can be checked more readily by other investigators and be more thoroughly and systematically analyzed.

6. Raw data from field or laboratory studies must first be organized into a table or graph. Then they can be analyzed in a number of different ways (tested for significance, for example).

7. All data in science represent samples of a whole. Failure to get a large or representative sample produces sampling error.

8. Statistics are a very useful and important tool in all phases of science. Yet statistics can be manipulated to prove almost any point if they are used incorrectly, irresponsibly, or without proper precaution.

9. Mechanism and vitalism are two very different philosophical positions from which to view living organisms. Vitalists claim that some undetectable vital force makes the difference between living and nonliving things. Mechanists claim that nothing unknowable, no special force, resides in living things. The modern view is that though living systems do not contain any special force or vital property, they are not simply organized collections of chemicals. Living organisms are seen as entities containing many separate but interacting parts. This interaction modifies the parts continually, so that all living systems change through time.

Exercises

1. Distinguish between basic (pure) and applied research. Is it important for both kinds of research to be supported? If so, why?

2. Devise a hypothesis to explain each of the following observations. Then outline an experiment to test your hypothesis.

    a) There are more automobile accidents at dusk than at any other time.

    b) When glass tumblers are washed in hot soapsuds and then immediately transferred face downwards onto a cool, flat surface, bubbles at first appear on the outside of the rim, expanding outwards. In a few seconds they reverse, go under the rim, and expand inside the glass tumbler.

    c) It has been noticed that one species of mud-dauber wasp will build its nest from highly radioactive mud, although the radiation received by the developing young may be enough to kill them. Under the same environmental conditions, another species of mud-dauber wasp avoids this mud and selects nonradioactive mud to build its nests.

    d) In mice of strain A, cancer develops in every animal living over 18 months. Mice of strain B do not develop cancer. If the young of each strain are transferred to mothers of the other strain immediately after birth, cancer does not develop in the switched strain-A animals, but it does develop in the switched strain-B animals living over 18 months.

3. P. F. and M. S. Klopfer of Duke University have studied maternal behavior in goats. The following facts have been established: A mother goat (doe) will reject her young (kid) if deprived of it immediately after birth, but given it back an hour later. If allowed contact with her own kid for five minutes after birth and then separated from it, the doe immediately accepts the kid and its littermates (if any) when returned an hour later. If allowed contact with her kid immediately after birth and then deprived of it, the doe shows obvious signs of distress. However, if this early contact is denied, the doe acts as if she had never mated or given birth to young.

    a) Propose a hypothesis to account for these observations, along with an experiment to test your hypothesis. Do not read the remainder of the exercise until you have finished this part.

Now consider the following additional facts about maternal behavior in goats established by the Klopfers: Under the conditions described above, the doe will not accept a kid of the correct age that is not her own (an alien kid) if it, instead of her own kid, is returned to her. Denied her own kid immediately after birth, but allowed five minutes with an alien kid, the doe will not only accept the alien kid but also her own when returned. However, only the alien kid with which she is allowed contact will be accepted; all other alien kids are rejected.

    b) How do these facts affect your hypothesis? (Do not change your hypothesis as given in the previous question, even if it did not fare too well; only consistency with the facts given and experimental design are of primary importance.)

c) If necessary, propose a new hypothesis to account for all the Klopfers' data as given above and suggest an experiment to test this hypothesis. If your original hypothesis stands, O.K. If it needs modifying, do so.

4. What is the relation between a hypothesis and an experiment?

5. A biologist finds that removal of organ A, an endocrine gland, from an adult mammal causes organs B and C to cease functioning. Organ B is also an endocrine gland. The three possible explanations for this occurrence are diagrammed below. (For $A \rightarrow B$, read "A is necessary to B," and so on.)

$$A \nearrow^{B}_{\searrow C} \qquad A \rightarrow B \rightarrow C \qquad A \rightarrow C \rightarrow B$$

Design an experiment or experiments to test these possibilities and distinguish between them.

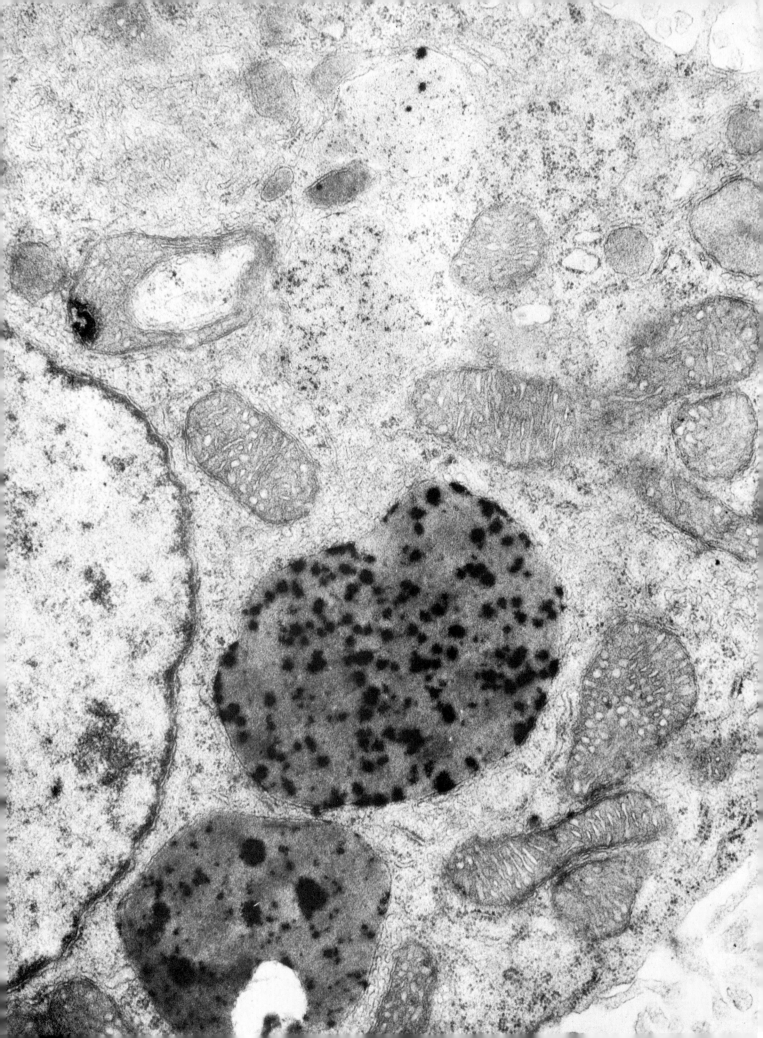

PART II
THE STRUCTURE
AND FUNCTION OF
CELLS

Advances in cytology, the study of cells, have been directly tied to improvements in the tools and techniques used in examining cells. Several different subcellular structures, or organelles, can be distinguished in this electron micrograph ($\times$ 38,000) of a kidney tubule cell of a thirteen-day-old chick embryo.

(Migrograph courtesy of Dr. E. Sue Lumb, Department of Biology, Vassar College.)

There is a constant flow of energy through all living systems. Green plants and certain kinds of bacteria can use radiant energy from the sun to produce energy-rich molecules by the process of photosynthesis. Thus a store of chemical energy accumulates, which then serves as the energy source for most other chemical reactions.

Animals are not capable of photosynthesis. They must acquire their energy in the form of the energy-rich carbohydrates originally produced by plants. Just as animals are dependent on plants for all their food, plants are dependent on the sun. Without the sun, the flow of energy through the living world would soon come to a halt.

When fuel molecules are broken down inside the cells of animals or plants, energy is released for doing biochemical work. Biochemical work is necessary for breaking down molecules and building up other types of molecules, such as proteins and fats; it is necessary to pump certain molecules in and out of the cell and to prevent the passage of others; it is necessary for all the processes involved in growth, replication, and general physiological maintenance. Cells are restless, ever-active units, through which matter and energy constantly flow. Living organisms can be viewed as specifically structured entities constantly engaged in capturing energy from the sun, directly or indirectly, and transforming it into useful chemical work.

This unit begins with an investigation of matter and energy: the kinds of atoms and molecules found in living systems and the nature of biochemical reactions and energy flow. Chapter 2 is an introduction to some basic principles of chemistry in living systems. Chapters 3 and 4 outline the structure of cells, the architecture of these basic "factory units" in the living world. Chapters 5 and 6 discuss the two most general sets of biochemical processes involved in the living world: respiration and photosynthesis.

Chapter 2
Matter, Energy, and Chemical Change

2.1
INTRODUCTION

"Getting and expending energy: that is the basic function of life." This statement captures one of the most important aspects of all living systems. Regardless of what else they do, all organisms must first and foremost provide an energy base for their activities. An energy base is nothing more than a set of chemical and physical processes by which the organism obtains energy from raw materials in its environment. All activity associated with life involves the use of energy—quite considerable amounts of it over a period of time. To keep alive—to replace parts, build new substances, obtain food, eliminate wastes, and reproduce—all organisms must extract energy from raw materials, such as sugars, starches, and other foods, and convert that energy into useful work. Without energy there would be no such thing as life.

Living organisms are constantly involved in *energy flow.* They take in energy from outside themselves and convert it into some useful form. A portion of that useful energy is then released to carry out some life process, after which it leaves the organism once again and is dispersed into the environment. The nature of energy intake is different for different kinds of organisms. Photosynthetic organisms, such as green plants and some kinds of bacteria, are able to use solar energy directly to produce useful energy. Nonphotosynthetic organisms, including animals, certain bacteria, and plants such as yeasts, must get their energy in the form of energy-containing substances in the environment (sugars, proteins, and the like). In either case, the organism is part of a flow system in which a certain portion of the energy taken in is extracted and released for the organism's own life processes.

The energy flow that characterizes life involves the interaction of many kinds of substances in a variety of complex chemical reactions. The study of these reactions is sometimes referred to as **biochemistry,** though the boundaries between biochemistry, general chemistry, and biology are becoming less distinct. By definition, biochemistry is concerned with the chemistry of life; its province is all the chemical reactions that occur in living systems. In recent years it has become clear that many of these chemical reactions are intimately connected to biological *structures*—for instance, many reactions occur on the surface of membranes, and others are localized in compartments within cells. The more we learn about biochemical processes, the more difficult it becomes to study them in isolation from living organisms.

Since all biological phenomena depend upon the energy-getting and energy-using processes in organisms, an understanding of these processes is essential for investigating living systems. It is first necessary to establish some general principles of how chemical reactions occur: the atoms, molecules, or ions involved and the energy exchanges taking place. The present chapter will focus on the interaction of matter and energy. The general chemical principles developed will then be applied to the study of biochemical processes in the basic unit of life, the cell (Chapters 4 and 5).

2.2
MATTER AND ENERGY

Matter can be defined as anything that has mass and occupies space. The fundamental unit of matter is the *atom,* which in turn consists of particles known as *protons, neutrons,* and *electrons. Energy,* on the other hand, must be defined *indirectly* in terms of the movement of matter, that is, in terms of work.

It is often useful to speak of energy as existing in one of two states, kinetic and potential. **Potential energy** is stored or inactive, is capable of performing **work,** and is frequently referred to as *free energy* (symbolized as G).* A stick of dynamite represents a great deal of potential energy; in quantitative terms it can be said to have a high free-energy value. In a biological context, sugar or fat also represents potential energy, though the free-energy value of either of these sub-

---

*The letters F and G are both used to represent free energy. F is an older term taken from free energy; G is a more recently used symbol from the name of J. Willard Gibbs (1790–1861), one of the founders of the field of thermodynamics.

All forms of energy are interrelated and interconvertible. The process of interconversion is called the transformation of energy. It always occurs with some loss of usable energy—energy transformations are never 100 percent efficient.

stances would be considerably less than that of an equivalent quantity of dynamite.

Kinetic energy is energy in action. It is energy in the process of having an effect on matter, and thus in the process of doing work. A boulder perched on the top of a hill has potential energy. If the boulder is given a slight push and begins to roll down the hill, the potential energy is released as kinetic energy. By the time the boulder has reached the bottom of the hill, all its potential energy (in relation to the hill) has been released. For the boulder to get back to the top of the hill, its potential energy must be restored. This can occur only if an input of energy comes from some outside agent (such as someone pushing the boulder back up the hill).

There are five forms of energy, each of which exist in either the potential or kinetic state. These forms are chemical energy, electrical energy, mechanical energy, radiant energy, and atomic energy. The last of these has little direct relationship to the normal functioning of the individual organism, but the others are all directly involved in living systems.

Chemical energy is involved in putting atoms together into compounds, or in breaking down compounds to form individual atoms. Compounds such as gasoline, for example, represent considerable amounts of potential chemical energy. The burning of gasoline inside an engine releases the potential energy of the fuel molecule.

Electrical energy is produced by the flow of electrons along a conductor. A charged battery contains potential electrical energy in the sense that it has the power to produce a flow of electrons. When the battery is activated, allowing the electrons to flow along a conductor, kinetic electrical energy is released.

Mechanical energy is directly involved in moving matter. The rolling of a boulder down a hill and the movement of a piston in a motor are examples of kinetic mechanical energy.

Radiant energy travels in waves. Well-known examples of radiant energy are light and heat. However, the category of radiant energy also includes radio waves, infrared and ultraviolet light, X-rays, gamma rays, and cosmic rays.

Atomic energy originates within atoms. It may be released spontaneously by radioactive atoms (for example, radium or uranium) when atomic nuclei are split apart (nuclear fission) or when protons and neutrons are joined together to form a new nucleus (nuclear fusion).

During atomic fission or fusion a certain quantity of mass is lost, that is, converted into energy. Thus, in proportion to the number of atoms involved, nuclear reactions release (or consume) far more energy than normal chemical reactions. Like other forms of energy, nuclear energy can be stored as well as released.

All these forms of energy are interrelated and interconvertible. The conversion of one form of energy to another goes on continually. It is the basis on which all living organisms maintain themselves. For example, kinetic radiant energy from the sun is converted into potential chemical energy in green plant cells by the food-making process, **photosynthesis.** When an animal eats the plant, it transforms the potential chemical energy of the plant substance into further kinds of chemical energy (by building its own kind of molecule) or into mechanical energy (for movement). Energy transformation is the basis of all life.

If the total amount of radiant energy transmitted to a green plant in a given period of time is measured, it will be found that only a small percentage of the total available energy is captured as potential chemical energy. The same is true of any step in the transformation from one form of energy to another. Careful recordings show that a leaf exposed to the sunlight has a higher temperature than one in the dark. Much of the solar energy striking the leaf is transformed into heat and lost to the environment. This illustrates the fact that *the transformation of energy is never 100 percent efficient.* Of course, none of the energy in such transformations is actually lost in the sense of being unaccounted for. Long ago physicists formulated the very important law of conservation of energy or, as it came to be called, the **first law of thermodynamics.** This law states that, during ordinary chemical or physical processes, energy is neither created nor destroyed; it is only changed in form.*

The first law of thermodynamics provides a useful framework for studying energy transformations in living systems. Some of the energy an organism derives from its food is recaptured in a useful form as potential chemical energy, while the rest is lost. In one sense, the success of an organism in its struggle for existence depends on the effectiveness (or efficiency) with which it

---

*This generalization does *not* apply to atomic fission or fusion reactions, where a very small amount of mass is converted into energy.

The first law of thermodynamics (law of conservation of energy) states that in any chemical or physical change, the total amount of energy involved remains the same. Energy is neither created nor lost, but only changed in form.

can use the energy available to it, converting as much as possible into potential chemical energy. The more energy lost, the more food the organism must consume in order to accomplish the same amount of work.

2.3
ATOMIC STRUCTURE

As stated earlier, atoms are composed of three primary building blocks: protons, neutrons, and electrons. The only exception is hydrogen-1 (protium), the lightest element, which has no neutrons.

Early in this century it was hypothesized that atoms were composed of a small, dense, central portion—the nucleus—surrounded by various numbers of other particles—the electrons. In 1913, Niels Bohr (1885–1962) suggested that the atom resembled a tiny solar system, with the nucleus representing the sun and the electrons the orbiting planets. Later, the nucleus was shown to contain two types of particles: protons, which are positively charged, and neutrons, which carry no charge. The nucleus carries a positive charge because of the protons. The electrons circling the nucleus have a negative charge, exactly offsetting the net positive charge of the protons. The mass difference between the electrons and the protons and neutron of the nucleus is immense,

however; one electron has approximately 1/1840 the mass of a proton or neutron (which are almost equal in mass). Figure 2.1(a) shows this general conception of the atom in diagrammatic form.

The modern view of the atom views the electrons, moving at extremely high velocities, as being located *most of the time* in a volume of space known as an **orbital**. The stress on the words "most of the time" simply emphasizes that, while particular electrons could theoretically be found anywhere from just outside the nucleus to an infinite distance away from it, an orbital merely represents the region where they will *most probably* be found (see Fig. 3.1b). In some cases orbitals may be spherical (*s* orbitals); in others, they may be dumbbell-shaped (*p* orbitals). Orbitals with still other shapes, located farther away from the nucleus, are also known.

The electron is the part of the atom most directly involved in chemical reactions. Atoms have equal numbers of electrons and protons. Under certain conditions, however, an atom can gain or lose electrons, acquiring a negative or positive charge and becoming an ion. When atoms interact with one another by giving up, taking on, or even sharing electrons, a chemical reaction occurs. The exchanges and interactions of electrons among atoms form the basis of chemical reactions and thus of all life processes.

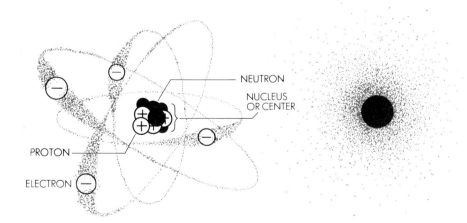

NEUTRON

NUCLEUS OR CENTER

PROTON

ELECTRON

(a)   (b)

Fig. 2.1
(*a*) Diagrammatic sketch of the atom representing its parts. Negatively charged electrons circle the nucleus of protons and neutrons. This diagram represents a working model of the atom. It is not, in any real sense, a "picture" of the atom. (*b*) The modern conception of the atom, showing the dense, centrally located nucleus and the outer haze of electrons.

Electrons and Energy

The electron cloud around the nucleus of an atom is composed of electrons in different *energy levels* or *shells.* Energy levels are roughly analogous to the planetary orbits suggested by Niels Bohr. The term "energy level" is an expressive one, since it leads to the idea of electrons as possessing certain amounts of potential energy. Electrons can be pictured as moving in spe-

cific energy levels outside the atomic nucleus (see Fig. 2.2). The electrons neither absorb nor radiate energy so long as they remain in these energy levels. However, should one or more electrons fall from the energy level they occupy to a lower one, they will *radiate* a precise amount of energy. If energy is *absorbed* by the atom, one or more electrons may jump from a lower energy level to a higher one.

The amount of energy an electron possesses depends on the energy level it occupies in an atom. The energy levels are analogous to successively higher steps cut into a cliff. The electrons in the energy levels are analogous to rocks of equal size distributed among the various steps, from the ground up. It takes a certain amount of energy to get individual rocks to each higher level. Thus the position each rock occupies represents a certain amount of potential energy.

An atom can have a large number of energy levels. Indeed, it is possible to recognize *seven* energy levels in which the electrons can be found. These seven energy levels are known as the *k, l, m, n, o, p,* and *q* levels. Each energy level has a certain maximum number of electrons that it can hold. If we assign a number to each of the energy levels, as follows:

Energy level	k	l	m	n	o	p	q
Number (n)	1	2	3	4	5	6	7

the rule for the maximum number of electrons in any outer energy level (other than k) is $M = 8$. For k or any inner energy level, the number is given by $M = 2n^2$.

Increasing the energy of an atom does not just increase its vibrational motion. It also serves to move electrons from lower to higher energy levels. It takes a definite amount of energy to move an electron from one level to another. This process is often referred to as "exciting" an atom. Exciting an atom involves capturing a certain amount of energy for each upward jump of an electron. The term *electron transition* or *quantum shift* is used to describe the movement of electrons from lower to higher energy levels and vice versa.

In an atom, however, electrons can jump to a large number of energy levels, depending on the amount of energy supplied. A small but definite amount of energy will cause an electron to jump only to the next higher level. The right amount of additional energy, however, may cause the electron to jump farther, perhaps so far from the nucleus that it escapes completely from the atom to which it originally belonged. In the case of the atom, the loss of an electron gives the atom a charge of +1, since it now has one more proton than electrons. The electrically charged atom is now an ion. Loss of two electrons would give the atom a charge of +2, and so on. Ions can also be formed by the *gain* of electrons. Such ions would be negatively charged.

Raising electrons to a higher energy level produces some gaps below; some of the lower energy levels are

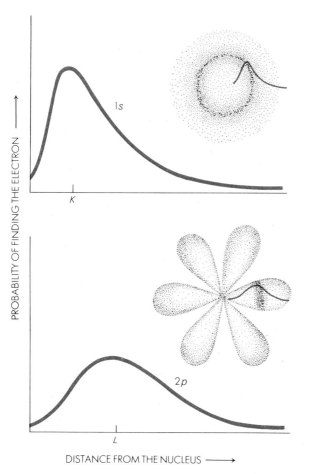

PROBABILITY OF FINDING THE ELECTRON ———→

1s

K

2p

L

DISTANCE FROM THE NUCLEUS ———→

Fig. 2.2
The graphs above represent the probability of finding an electron in the first (top) and second (bottom) energy levels. While the electrons could theoretically be found at any position, from just outside the nucleus to an infinite distance from it, the most *likely* distances are the ones we label the *k, l, m, n, o, p,* and *q* orbitals. The orbital of an electron, then, is simply the region of space where an electron would be *expected* to be found *most of the time.* A representation of the entire atom is shown to the right of each graph; in the atom with 2p orbitals, three are shown to show their three-dimensional orientation in space.

Atoms store potential energy when electrons are raised to higher energy levels; they release energy when electrons fall to lower energy levels.

then incomplete. These are usually filled by other electrons, which fall down from higher energy levels. Just as an electron absorbs energy to jump to a higher energy level, so it releases the same amount of energy in falling back to a lower position.

The movement of electrons to lower energy levels may release energy as X-rays, visible light, or other radiation. The wavelength of the radiation emitted by an electron in making a downward transition depends on the distance it falls and the type of atom in which the transition occurs (see Fig. 2.3). To pass from the *m* to the *l* level in atoms of one element involves emitting a certain amount of energy. Going from the *l* to the *k* level in the same atom produces a different amount. Going from the *m* to the *k* level produces still a third amount, and so on. Transitions of the same sort in atoms of other elements would produce a somewhat different series of energy emissions. When light from excited atoms is passed through a spectroscope (a device that bends light through a prism), it is possible to identify many specific electron transitions by the characteristic bright lines they produce. Every time an electron falls from a higher energy level to a lower one, a specific wavelength of energy is emitted. This shows up as a line in the observed spectrum. Such a spectrum analysis is based on the principle that for each distance of fall (say from the *m* to the *l* level), a specific line will appear. The number of electrons making any given downward transition in a given period of time (and under specified conditions) determines the brightness of the line. Furthermore, the number and kind of transitions that occur depend on the electron configuration of the atoms or molecules involved. Hence, study of such spectra gives a good clue to the electron structure of atoms, molecules, or ions.

When an electron makes a transition, it absorbs or releases distinct amounts of energy. The quantum theory, first proposed by the German physicist Max Planck in 1900, holds that energy can be described as coming in discrete packets, or **quanta**. The light emitted by an incandescent lamp, for example, consists of millions of discrete quanta or *photons* that, because they have some physical properties, can interact with matter. Although the reality of energy "packets" is debated among physicists, the quantum model is useful in understanding how energy can cause or result from electron transitions. If an atom absorbs one or more quanta of radiant energy, electrons at certain energy levels jump to a higher level. The energy is thus temporarily captured by the atom. It may then be released in ways that allow electrons in other atoms to make transitions also. Thus energy can be transferred among groups of atoms by the movement of electrons from one energy level in one atom to a different energy level in another.

Atoms are often combined with other atoms. Such a combination of two or more atoms is called a **molecule**. When atoms unite to form molecules, the energy levels of the individual atoms interact to form molecular energy levels. Electron transitions are possible in the energy levels of molecules, just as they are among energy levels of individual atoms. When a molecule of chlorophyll, for example, absorbs a photon of light energy from the sun, electrons are raised to higher energy levels. The electrons are temporarily lost to the molecule as a whole. But when the electrons make downward transitions, they release the same amount of energy they absorbed. This energy is captured by the plant cell and used to power certain chemical reactions during photosynthesis.

The atoms making up a molecule are held together by **chemical bonds**. A chemical bond is not a physical structure, but simply an energy relationship between atoms that holds them together in a molecule. The chemical bond that unites the atoms can therefore be measured in units of energy. The units of energy are fre-

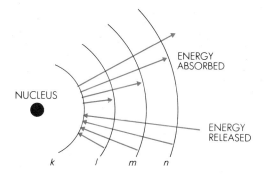

Fig. 2.3
Electron jumps occur when an atom is supplied with energy. Outward jumps may occur from any energy level to any other, depending on how much energy the electron absorbs. In the diagram, outward jumps are shown only from the *k* level. When electrons fall inward they give off precise amounts of energy, which appears as X-rays, visible light, and other types of radiant energy.

quently expressed as kilocalories (kcal)* of potential energy per mole of compound.† Thus kilocalories can be used to compare the potential energy that the chemical bonds of various chemical compounds contain.

To understand how chemical bonds are formed, consider the interaction of atoms A and B to form molecule AB. First, A and B must come close enough together for their electron clouds to overlap; rearrangements of electrons between the two atoms is the basis for the formation of chemical bonds. As A and B approach each other, however, they exhibit mutual repulsion as the result of their negatively charged electron clouds (like charges repel, unlike attract). We have to do work (put in energy) to push the atoms still closer together. It is not until the atoms are forced so close together that an electron from one is attracted to the positively charged nucleus of the other that the rearrangements, which make the two atoms have a net attraction for each other, can take place.

All interactions between atoms to make or break chemical bonds involve the exchange of energy. To break any chemical bond, a certain input of energy (called dissociation energy) is required to overcome the stable state and move the atoms far enough apart so that their mutual repulsion again takes over and sends them on their opposite ways. This energy varies from one type of atom to another. For example, it takes a large amount of energy (104.2 kcal/mole) to break hydrogen-to-hydrogen bonds, but a smaller amount (50.9 kcal/mole) to break sulfur-to-sulfur bonds.

The overlapping of electron clouds causes a rearrangement of electrons in the outermost energy level of each atom. This rearrangement involves one of two possibilities: (1) one atom will tend to *give up* one or more of its electrons to the other; or (2) each atom will tend to *share* one or more electrons with the other. In either case, the total electric charge of one atom may be either less positive or more positive than that of the other.

The interaction between outer electrons is the result of a process in which each atom approaches a stable outer electron configuration. This process requires closer examination. It is central to an understanding of current thinking about the formation of a chemical bond.

With the exception of the k or innermost energy level, a stable outer electron configuration is achieved with eight electrons. The atoms of any element with eight outer electrons are stable. Neon, argon, krypton, xenon, and radon are all examples of such elements.‡ Such atoms do not generally react with other atoms. Most atoms, however, have fewer than eight electrons in their outer energy level. These atoms tend to reach the stable configuration by giving up, taking on, or sharing electrons. The driving force behind any chemical reaction originates in the tendency of atoms to attain a stable outer energy level.

2.4
CHEMICAL BONDING

Ionic Bonds

There are several types of chemical bonds found in chemical compounds.

One of these, found most frequently in inorganic compounds, is the *ionic* or *electrovalent* bond. In the formation of this type of bond, one atom gives up its outermost electrons to one or more other atoms. When this occurs, the outermost energy level of each atom becomes more stable.

The formation of lithium chloride from the elements lithium and chlorine provides an example of ionic bonding (see Fig. 2.4). The electron configuration for chlorine, counted from the nucleus outward, is 2, 8, and 7. Atoms of lithium have a single electron in the l level. Lithium can reach stability by giving up its one l electron to chlorine. This gives the chlorine atom a total of eight electrons in its m level—which represents stability for this element. Both atoms now have a stable electron configuration in their outer energy levels.

When lithium gives up its electron, the atom has one less negative charge. Hence it is positively charged (+1). Chlorine, by accepting an electron, now has one more negative charge than positive charges and thus is negatively charged (−1). Since opposite charges attract, the positively charged lithium atom and the negatively charged chlorine atom attract each other. This attraction holds the two atoms together. In this way a molecule of lithium chloride, LiCl, is formed.

There is no 100 percent ionic bond. Though one atom tends to give its electrons to another, this handing over is not complete. The donated electron may still occasionally orbit the nucleus of the donor atom.

---

*A **calorie** is the amount of heat energy required to raise the temperature of 1 g of water (at 15°C) one degree Celsius. A **kilocalorie** would thus be the amount of heat required to raise the temperature of 1000 g of water the same amount. In biology, the kilocalorie is often written as Calorie, with the first letter capitalized to distinguish it from the smaller physical calorie used by physical scientists.

†A **mole** is the amount of substance defined as containing a fixed number of particles (called Avogadro's number). Avogadro's number is given as 6.024×10^{23}. Thus, a mole of sodium chloride contains $1 \times 6.024 \times 10^{23}$ sodium ions and $1 \times 6.024 \times 10^{23}$ chloride ions. Two moles of sodium chloride would contain $2 \times 6.024 \times 10^{23}$ sodium ions and $2 \times 6.024 \times 10^{23}$ chloride ions, and so on. Moles of two different substances will usually weigh different amounts or occupy different volumes of space. But each mole contains the same number of particles (Avogadro's number).

---

‡Until a few years ago, it was thought that none of these elements formed any compounds. For this reason they were known as inert gases. Their lack of chemical activity is due to the fact that they have a stable electron configuration in their energy levels. It is now known that certain compounds, such as xenon tetrafluoride, can be formed under special conditions.

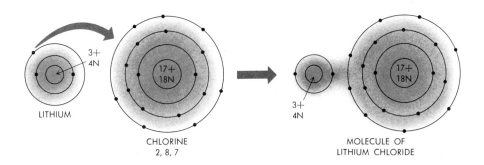

Fig. 2.4
The formation of a molecule of
lithium chloride by ionic bonds.

In the formation of ionic bonds, which atoms will give up electrons and which will receive them? In general, atoms with fewer than four electrons in the outer energy level tend to lose electrons. Those with more than four tend to gain electrons. Atoms such as sodium, potassium, hydrogen, calcium, and iron possess three or fewer outer electrons. Atoms such as oxygen, chlorine, sulfur, and iodine need one or two electrons to complete their outer energy levels. Thus these atoms tend to take on electrons.

Covalent Bonds

What about an atom such as carbon, which has four electrons in its outer energy level? Does such an atom tend to give up or take on electrons when combining with other atoms?

Carbon combines with atoms of many other elements by forming *covalent* chemical bonds, a sort of compromise between the giving up and the taking on of one or more pairs of electrons between atoms. In such bonding, atoms combine by undergoing a rearrangement of electrons in their outer energy levels. Neither atom loses its electrons completely. Instead the electrons are shared and may orbit the nucleus of any atom in the molecule.

The formation of a molecule of the gas methane illustrates the principle of covalent bonding. Under suitable conditions carbon reacts with hydrogen to form molecules of methane. Four atoms of hydrogen react with each atom of carbon to produce a symmetrical molecule, CH_4:

$$\begin{array}{c} H \\ | \\ H-C-H \\ | \\ H \end{array}$$

Each line between the carbon atom and a hydrogen atom represents a single pair of shared electrons. The pair consists of one electron from the carbon atom and one from the hydrogen atom. This may be shown more clearly by writing the molecular formula in the following manner:

$$\begin{array}{c} H \\ ox \\ H\,{}^{o}_{x}C{}^{o}_{x}\,H \\ ox \\ H \end{array}$$

The open dots represent the outer electrons originally in the l energy level of carbon. The crosses represent the electrons originally in the k level of each hydrogen atom.

Consider why this type of bonding takes place. The carbon atom has four electrons in its outer energy level. To attain stability, carbon needs eight electrons. Each hydrogen atom has one electron in its k level. Hydrogen can reach stability by either losing or gaining one electron. In the formation of the covalent bond between carbon and hydrogen, the electrons in the outer energy levels of each atom orbit the nuclei of both hydrogen and carbon. As a result, each of the four hydrogen atoms has its own electron plus one electron from the carbon to orbit its nucleus. In turn, the carbon atom has not only its own four electrons but also one from each of the hydrogen atoms to orbit its nucleus. This completes the requirements for stability in the outer energy levels of both atoms. The sharing of these outer electrons produces the covalent chemical bond.

Weak Bonds

As their name suggests, **weak bonds** are chemical bonds that require very little energy to break. In this regard both ionic and covalent bonds are strong bonds, though

All chemical bonds represent stored chemical energy. Some bonds contain more energy than others. To release this energy bonds must be broken, and this in turn requires the investment of some energy.

ionic bonds in solution (as is the case in virtually all parts of living systems) are much weaker than covalent bonds and are usually considered weak bonds in biochemical studies. There are also, however, other types of weak bonds found in living systems. Of these, three deserve special attention here.

Hydrogen bonds are often formed when hydrogen is covalently bonded to a larger atom (usually oxygen or nitrogen). The larger atom tends to draw the shared electrons toward it and away from the smaller hydrogen atom. This results in the hydrogen portion of the molecule developing a slight positive charge which, in turn, may result in a weak electrostatic attraction between the hydrogen atom and another molecule with a region negatively charged. It is this electrostatic force that is the hydrogen bond.

Hydrogen bonding is perhaps most easily demonstrated by water molecules, where the hydrogen atoms of each molecule form hydrogen bonds with the oxygen atoms of the adjacent molecules (see Fig. 2.5). As will be seen shortly, hydrogen bonds play an important role in maintaining the molecular shape of other considerably larger molecules found in living systems, most notably proteins.

Hydrophobic interactions occur when nonpolar, insoluble molecules are surrounded by water. This type of weak bond is actually the result of repulsion by the solvent rather than the sort of attractions between atoms or molecules generally called bonds; this is the reason for the term "interactions." Like hydrogen bonds, hydrophobic interactions play an important role in retaining the molecular structure of proteins and other important molecules.

Van der Waal's interactions (or forces) between molecules result from slight shiftings in the distribution of electrons due to molecular orbital interactions. Even electrically neutral molecules or portions of them show such weak forces. Van der Waal's interactions join with other weak bonds in helping to maintain the structural configurations of large molecules such as proteins.

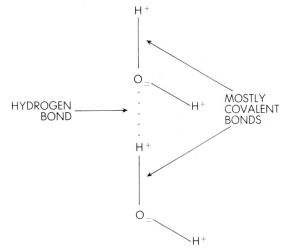

Fig. 2.5
Hydrogen bonding between two water molecules. The hydrogen bond results from the tendency of the larger oxygen atom to draw electrons away from the smaller hydrogen atoms, thereby making the area around the oxygen atom negatively charged. The nearest hydrogen atom in the next water molecule, since it is slightly positively charged, has a slight electrostatic attraction for the oxygen atom. It is this slight attraction that results in the hydrogen bond. (See also Fig. 2.10 and Diagram A, Supplement 2.1.)

Table 2.1 compares the strength of the various types of weak bonds or interactions.

Bond Angles

When two or more atoms combine to form a molecule, a predictable geometric relationship is established between the atoms involved. The molecule takes on a definite shape. When three atoms combine, an angle is formed; for example, $104.5°$ between the two hydrogen atoms and one oxygen atom of a water molecule (Fig. 2.6). Such angles are called *bond angles*. By certain physical techniques, bond angles can be accurately measured. The determination of bond angles allows the scientist to establish the relative positions of atoms in a molecule.

The way in which one molecule reacts with another depends in part on the shape of each molecule. The geometric configuration of water molecules (with a molecular weight of only 18) is crucial to the properties of water that make life possible. Similarly, the chemical characteristics of some proteins (with molecular weights in the millions) may be completely changed by altering the physical shape of one small part of the molecule.

For the most part, molecules are represented in diagrams as if they were flat, two-dimensional structures. In reality, they are three-dimensional, having depth in addition to length and breadth. For example, the

Table 2.1

Approximate Strengths of Some Bonds or Interactions Common in Living Systems, Based on the Amount of Energy (ΔG) Necessary to Break Them

Type of bond or interaction	Range of strength, in kilocalories per mole (approximate)
Covalent	50–110
Ionic (aqueous solution)	5
Hydrogen	4–5
Hydrophobic	1–3
van der Waal's	1–2

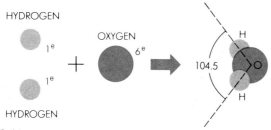

HYDROGEN

1^e

1^e

HYDROGEN

OXYGEN

6^e

104.5

O

Fig. 2.6
The formation of a water molecule from two hydrogen atoms and one oxygen atom. The outer orbital of oxygen contains only six electrons; oxygen therefore needs two more electrons to achieve the stability requirement of eight electrons in the outer orbital. Since each hydrogen atom has only one electron to share, two are required to satisfy the stability requirements of oxygen. When united, the atoms assume a configuration leading to the establishment of distinct bond angles (see Section 2.4).

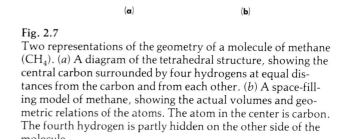

Fig. 2.7
Two representations of the geometry of a molecule of methane (CH_4). (a) A diagram of the tetrahedral structure, showing the central carbon surrounded by four hydrogens at equal distances from the carbon and from each other. (b) A space-filling model of methane, showing the actual volumes and geometric relations of the atoms. The atom in the center is carbon. The fourth hydrogen is partly hidden on the other side of the molecule.

organic compound methane, shown in a face-on view on p. 39, actually forms a solid, four-sided pyramid known as a tetrahedron (Fig. 2.7a). Rather than 90°, the bond angles between the four hydrogen atoms are actually 109.5°. It is therefore desirable to show the three-dimensional structure with a space-filling model (Fig. 2.7b). Here the space occupied by the atoms, as well as their orientation within the molecule, is taken into consideration.

The concepts of molecular configuration and bond angles play a role in explaining chemical reactions between molecules. These concepts will be important in later consideration of the larger molecules found in living organisms.

2.5
THE POLARITY OF MOLECULES

As has been suggested in the previous section, a frequent consequence of the geometric shapes of molecules is a distinct separation of electric charge. This means

that one portion of a molecule is positive or negative in relation to another portion of the same molecule. When such an uneven distribution of charge occurs, the molecule is said to be **polar**. The molecule has a positive and a negative end, separated from each other like the poles of a bar magnet. The charge results because the nuclei of individual atoms in the molecule attract more electrons to orbit them than do the nuclei of other atoms in the same molecule. Thus the area surrounding these electron-attracting nuclei becomes negatively charged, while the area around the electron-deprived nuclei becomes positively charged.

The water molecule illustrates this point. Although the water molecule, as a whole, is electrically neutral, it does have a positive and a negative end (Fig. 2.8). The geometric configuration of the molecule places both hydrogen atoms at one end. The nucleus of the oxygen atom attracts electrons more than the nuclei of the hydrogen atoms. This results in two positively charged regions on one end of the molecule and a single negatively charged region on the other. The result is a mole-

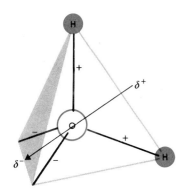

Fig. 2.8
The geometry of a water molecule contributes to its polarity. Because of the strong force of attraction by the oxygen atom for the four outer electrons (which are unpaired), the centers of negative charge lie on the opposite side of the oxygen from the hydrogen atoms (that is, to lower left). There is thus a separation of a positive and negative charge such that one side of the molecule (the O—H side) is designated δ^+, while the other side is designated δ^-. The arrow pointing from δ^+ to δ^- indicates the direction of the charge distribution, which passes along a gradient from most positive to most negative. The Greek letter δ (delta) indicates "change in." (Modified from Fig. 6.2 of A. G. Loewy and P. Siekevitz, *Cell Structure and Function*, 2d ed. New York: Holt, Rinehart and Winston, 1969, p. 91.)

cule with a positive and a negative end, or two poles; the molecule is therefore polar.

Molecular polarity is significant to the biological sciences in two ways. First, polar molecules tend to become oriented in precise spatial patterns with respect to other molecules (either polar or nonpolar). Because of this, polar molecules are important in certain structural elements of organisms, for example, cell membranes (see Fig. 2.9).

Second, polarity is important in understanding both the geometry and the chemical characteristics of large molecules. Polarity thus tends to bring small molecules, or specific regions of large molecules, into definite geometric relation with each other. In this way, the chemical bonding or interaction between individual molecules or between specific regions on the same molecule is brought about more easily (see Fig. 2.10).

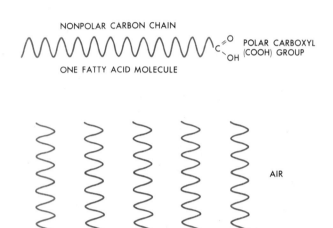

Fig. 2.9
Polarity of fatty acid molecules determines their orientation at a water–air interface. The carboxyl group is a polar region of the fatty acid molecule. When carboxyl groups touch water, one hydrogen dissociates, leaving the carboxyl group as a whole negatively charged. It thus dissolves well in the polar water molecules. The carbon chain is nonpolar and does not dissociate in water; hence the carbon chains do not dissolve, and project out into the air. This is why all fats and oils float.

Fig. 2.10
(a) Hydrogen bonding between five water molecules, showing the precise patterns of bond formation (represented by the dotted line). The five molecules of water form a tetrahedronal lattice. (b) A more complex organization of many water molecules, held in place by hydrogen bonds, forms a crystal of ice. The bonding patterns are identical to the patterns shown in (a) for just five molecules of water. Note the large spaces formed by the latticework. These spaces make the density of ice lower than that of water. (Modified from Linus Pauling, *General Chemistry*, 2d ed. San Francisco: W. H. Freeman, 1953.)

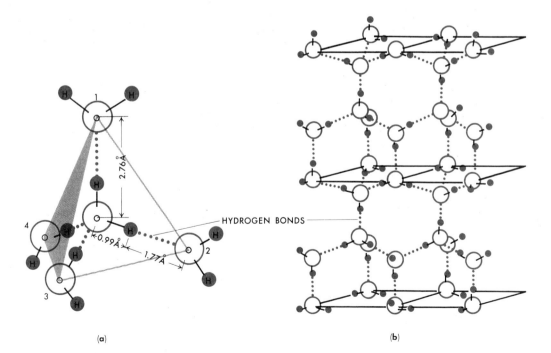

(a)   (b)

SUPPLEMENT 2.1
WATER AND LIFE

The living cell is approximately 90 percent water. Life as we know it occurs in a water medium and could not exist otherwise. The chemical and physical properties of water make it a unique medium for living organisms. In all cases this uniqueness stems from the polar nature of water molecules.

1. *Water is a good solvent.* Water serves as an excellent solvent for a wide range of substances. Any molecule that contains atoms or groups of atoms that are polar (have unpaired electrons or exposed protons) will attract, and be attracted by, polar water molecules. Hence, water molecules can snuggle around or between many other types of atoms and/or molecules. This snuggling dissociates like atoms or molecules from one another, just as it can dissociate an atom or group of atoms (as ions) from other groups of atoms. The solvent property of water makes it a universal agent for transporting substances to organisms. Accordingly, water is the liquid medium in which all single-celled organisms and the individual cells of multicelled organisms exist.

2. *Water molecules are highly cohesive.* Because of their polarity, water molecules tend to stick together, they even come to lie in some general patterned relationship to each other, held in position by the hydrogen bonds formed between positive and negative ends of the molecules (see Diagram A). The cohesiveness of water has several effects. One is that water tends to have a high *surface tension.* This is be-

cause the topmost layer of molecules is pulled from beneath, but not from above. Consequently it packs more densely to the layer beneath it than do other layers of molecules under the surface. This densely packed top layer creates the surface tension. Because of this tension, many insects can walk on water, but not on other liquids such as alcohol or ammonium. Thus the surface layer of water bodies serves as a habitat for many organisms. Cohesiveness also allows a column of water to be raised over greater vertical distances than noncohesive liquids. A noncohesive liquid column (even a thin column) tends to break apart as it is moved upward. Water columns maintain their integrity as they are moved up the conducting tubes of trees, partly by the force of cohesion.

3. *Water has a high heat capacity and heat of vaporization.* Heat capacity is a measure of the amount of heat required to raise one cubic centimeter (1 cm³) of a substance one degree Celsius (1°C). The higher the heat capacity of a substance, the more heat it will absorb to show a temperature rise. With its high heat capacity, water can serve to insulate the organism (or cell) from abrupt changes in temperature. Heat of vaporization is defined as the quantity of heat required, per unit mass, for the vaporization of a substance measured at a given temperature (most substances vaporize slowly at any temperature; generally, the higher the temperature, the more rapid the vaporization process). Water

Diagram A

ELECTROSTATIC
ATTRACTIONS (HYDROGEN BONDS)

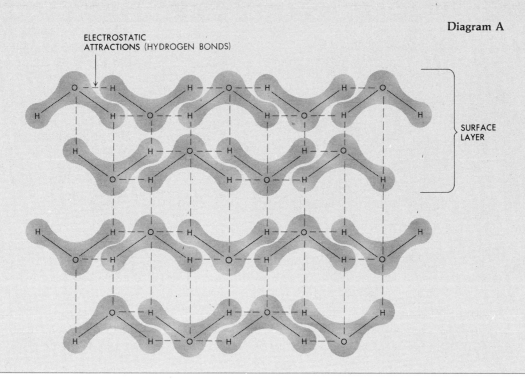

SURFACE
LAYER

requires much more heat to vaporize than most other liquids. This means that water can absorb much more heat, in proportion to the number of molecules passing into the vapor state, than most other liquids. If water had a low heat of vaporization, organisms would lose considerably more water than they do. The high heat capacity and heat of vaporization characteristic of water result from the cohesiveness of the molecules. Because of their electrostatic attraction, it takes more energy to separate the molecules of water than for nonpolar, noncohesive molecules.

4. *Ice is less dense than water.* Because of the particular structure of hydrogen bonds between water molecules, the latticework of the ice crystal has large openings in the center (Fig. 2.10). As these openings are empty, the ice crystal has less density than the water from which it is formed. This is why ice floats. The advantage of floating ice is considerable to living things. If ice sank, most of the world's rivers, lakes, and oceans would be permanently iced in, with only a small surface of melted water in warm weather. With its high heat capacity, the surface water would provide a good insulator, preventing the ice at the bottom of lakes or oceans from ever absorbing enough heat to melt. Such a situation would profoundly alter the pattern of life on earth. Interestingly, other liquids possessing some properties similar to water, such as ammonia, form ice that is denser than the liquid. Water is unique in having a solid state less dense than its liquid state.

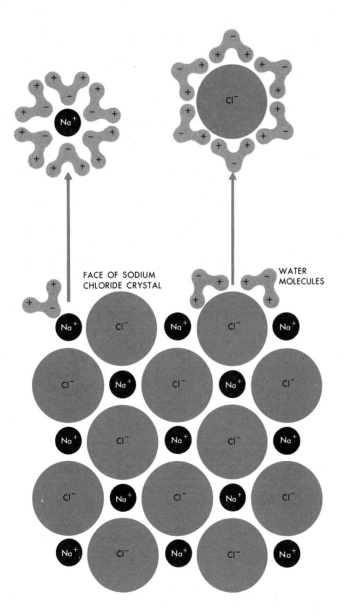

FACE OF SODIUM CHLORIDE CRYSTAL

WATER MOLECULES

2.6
IONS AND RADICALS

At room temperature, the compound hydrogen chloride (HCl) is a gas. If molecules of this gas are dissolved in water, the hydrogen separates from the chlorine. The separation, or ionization, occurs in such a way that the hydrogen atom does not take back the electron it loaned to chlorine in forming the bond. Thus the hydrogen atom, now a hydrogen ion or simply one proton, bears a charge of $+1$. The chlorine retains the extra electron it received from the hydrogen atom. Since it has one more electron than protons, the chlorine bears a negative charge of -1 and is thus a chloride ion. The hydrogen ion is written as H^+, and the chloride ion as Cl^-.

Recall that ions are formed whenever an atom loses or gains electrons. The dissolving of sodium chloride in water results in a separation, or *dissociation*, of the sodium and chloride ions (see Fig. 2.11). Since crystalline sodium chloride is a mostly ionic compound, its component atoms are already in the ionic state. Substances such as hydrogen chloride and acids, on the other hand, undergo ionization when entering into solution. As soon as the water is removed, the oppositely charged particles recombine to form the same molecules or *ion pairs.* The molecule of lithium chloride shown in Fig. 2.4 is an example of an ion pair.

Fig. 2.11
A representation of sodium chloride dissolving in water. The negative ends of the polar water molecules are attracted to the positive sodium ions and pull them off the crystal lattice. The positive ends of the water molecules are attracted to the negative chloride ions and pull them off. As indicated, each sodium and chloride ion is probably surrounded by at least six water molecules.

The process of ionization can be represented by an ionization equation. In chemical language, an equation indicates what goes into and what comes out of a certain reaction. For example, the ionization equation for HCl indicates that one molecule of this compound yields, upon ionization, a positively charged hydrogen ion (H^+) and a negatively charged chloride ion (Cl^-):*

$$HCl \rightarrow H^+ + Cl^-.$$

Similarly, the dissociation of sodium chloride (table salt) gives

$$NaCl \rightarrow Na^+ + Cl^-.$$

When calcium chloride ionizes, it produces two chloride ions for every ion of calcium:

$$CaCl_2 \rightarrow Ca^{++} + 2Cl^-.$$

This equation tells us several things. First, it shows that a molecule of calcium chloride consists of one atom of calcium and two atoms of chlorine. The subscript after the symbol for any atom indicates the number of atoms of that element in the molecule. Second, in writing the ionization equation for this compound, the two atoms of chlorine must be accounted for by showing that there are two chloride ions in the solution. To indicate that these two ions do occur, a 2 is placed in front of the Cl^-. The equation is now balanced. Each of the atoms on the left side of the equation is accounted for on the right.

When some compounds ionize, one of the products is a *complex ion.* Complex ions are associations of two or more atoms that bear an overall positive or negative charge. They are thus collections of two or more atoms that act as a single ion. For example, the ionization of sulfuric acid yields two hydrogen ions (protons) and a sulfate ion:

$$H_2SO_4 \rightarrow 2H^+ + SO_4^{2-}.$$
$$\text{(sulfate ion)}$$

The sulfate ion is composed of one atom of sulfur and four atoms of oxygen. As indicated, these five atoms have an overall electric charge of -2.

Molecules of calcium nitrate contain two nitrate ions bonded to one calcium atom. They are written as

$$Ca(NO_3)_2.$$

Aluminum sulfate, which contains three sulfate ions bonded to two aluminum atoms, is written as

$$Al_2(SO_4)_3.$$

The parentheses enclose the complex ion itself. The subscript number after the parentheses indicates the number of groups (such as sulfate, SO_4, of which there are three) contained in the molecule.

Not all compounds that ionize in water do so with equal readiness. In all the ionization equations listed above, nearly 100 percent of the molecules dissociate to release the appropriate ions. However, water molecules ionize only very slightly, so that the reaction

$$H_2O \rightarrow H^+ + OH^-$$
$$\text{(hydroxyl ion)}$$

occurs in approximately one out of every 554 million molecules.

Carbonic acid ionizes more than water, but still only about one percent of the molecules dissociate:

$$H_2CO_3 \rightarrow H^+ + HCO_3^-.$$
$$\text{(bicarbonate ion)}$$

Solubility, the degree to which a molecule will enter into solution, depends on several characteristics of the molecule, such as size, the presence or absense of charged or uncharged regions of the molecule, and the types of chemical bonds holding the atoms in the molecule together. Most molecules bound by ionic bonds are soluble. Molecules demonstrating polarity also tend to be soluble.

Atoms held together by ionic bonds can separate more easily because one atom has tended to give up electrons while another has tended to accept them. In this way, the outer energy level of each atom has been satisfied. When such molecules dissociate, no further exchange of electrons is required. Dissociation in this case merely involves overcoming the electrostatic attraction between positive and negative particles. The action of water molecules accomplishes this dissociation (see Fig. 2.11).

In covalent bonds, the outer energy level of each atom is saturated only as long as the shared electrons revolve about both nuclei. For this to be possible, the atoms must remain close together. It is very difficult to separate one from another if the atoms are forced to assume unstable outer electron configurations. For this reason water molecules generally cannot force covalently bonded atoms apart. Such molecules fail to show ionization in water (although a few covalent compounds do ionize, that is, are soluble).

2.7
ACIDS, BASES, AND THE pH SCALE

The concept of acids and bases is important in a discussion of chemical reactions in living systems—so important that a special scale has been devised to conveniently indicate the acidic or basic character of solutions.

---

*Hydrogen ions (H^+) do not exist in solution. As soon as a hydrogen ion (that is, a proton) is removed from a molecule, or dissociates from another ion as in the dissociation of HCl, it is picked up by some other molecule. In dissociation reactions such as that for HCl, the other molecule is inevitably water. This forms a hydronium ion: $H^+ + H_2O \rightarrow H_3O$ (hydronium). The complete reaction for the dissociation of HCl would thus be:

$$HCl + H_2O \rightarrow H_3O + Cl^-.$$

For practical purposes we will usually write hydrogen ions as H^+, but keep in mind that this is convenient notation, not chemical reality.

Acids and Bases

An **acid** is defined as any substance that can *donate* a proton or hydrogen ion. This property is represented by the generalized equation:

$$HA \rightarrow H^+ + A^- \qquad (2.1)$$

where HA stands for any acid, H^+ the proton that the acid can release, and A^- the negative ion to which the proton is bound in the acid molecule. Especially common as proton donors in living systems are carboxyl groups, written $-COOH$. Found in many organic molecules, $-COOH$ groups dissociate to yield a proton and a negatively charged carboxyl ion (COO^-).

$$-C{\overset{O}{\underset{OH}{\diagup\hspace{-0.5em}\diagdown}}} \rightarrow -C{\overset{O}{\underset{O^-}{\diagup\hspace{-0.5em}\diagdown}}} + H^+ \qquad (2.2)$$

A **base** is any substance that *accepts* protons. Bases are thus the opposite of acids in their chemical properties. Their dissociation reactions can be written in general as:

$$B^- + H^+ \rightarrow BH. \qquad (2.3)$$

It can be surmised from this equation that whatever is designated as B^- in reality is equal to what we called A^- in Eq. 2.1, since the product of combining a proton with B produces a molecule that can release the proton again. Many substances known as hydroxides (because they have a $-OH$ group attached to them) are bases because they dissociate in water into hydroxyl ions, which are very powerful proton acceptors:

$$BOH \rightarrow B^+ + OH^- \qquad (2.4)$$

or

$$OH^- + H^+ \rightarrow HOH \text{ (that is, } H_2O). \qquad (2.5)$$

Water acts as both an acid and a base, and consequently mediates most acid–base reactions. In fact, when an acid gives up a proton, it always gives it up to water first. When an acid is added to water, the acid dissociates and gives its protons to water molecules, creating hydronium ions (H_3O^+). If there is a stronger base around than water (one that has a stronger affinity for protons than water does), the hydronium ion will pass the extra protons to the stronger base. If no stronger base is around, the number of hydronium ions increases in the solution.

Because acids and bases are defined as opposites, it is apparent that when put together in the same solution they would tend to counteract or neutralize each other's effects. One of the neutralization products is a salt, and the other is water. Consider the neutralization of hydrochloric acid (HCl) by sodium hydroxide (NaOH):

$$HCl + NaOH \rightarrow NaCl + H_2O. \qquad (2.6)$$

The products in this case are a salt, sodium chloride, and water.

The pH Scale

So far we have discussed acids and bases in general, qualitative terms. However, scientists have devised a scale to measure how acidic or basic a given solution is. Called the **pH scale,** it is based upon the concentration of hydrogen ions (or protons) in a liter of solution.*

The pH scale runs from 0 to 14. The lower numbers refer to acid solutions. The higher numbers refer to basic solutions. The midpoint in the scale is 7, the pH of water. At this point the concentration of hydrogen ions equals the concentration of hydroxide ions. Any solution with a pH of less than 7 has more hydrogen ions than hydroxide ions in solution. Conversely, any solution with a pH of more than 7 has fewer hydrogen ions than hydroxide ions in solution.

The pH scale is based on actual calculations of the number of hydrogen ions in a solution. The concentration of hydrogen ions in solution is expressed as moles of hydrogen ion per liter.

Why was the number 7, the pH of water, chosen as the midpoint on the pH scale? Careful measurements show that a liter of pure water contains one ten-millionth of a mole of hydrogen ions, or $1/10,000,000 \times 6.024 \times 10^{23}$ particles of H^+. It is awkward to write all of the zeros involved in such numbers as $1/10,000,000$. It is much easier to use an exponential system and write this number as 10^{-7}. The pH scale is actually a logarithmic one, however, and, indeed, pH is defined in terms of logarithms:

$$pH = log_{10}(H^+).$$

Thus, in the case of water, in which the $(H^+) = 10^{-7}$, the conversion to a positive number occurs as follows:

$$pH = -log_{10}(10^{-7});$$
$$pH = -(-7);$$
$$pH = 7.$$

Thus, water has a pH of 7. Similarly, hydrogen ion concentrations of $1/10,000$ (10^{-4}) mole per liter and $1/100,000,000$ (10^{-8}) mole per liter are written as pH 4 and pH 8 respectively. Since the number is actually a negative exponent, *the smaller the exponential value, the greater the concentration of hydrogen ions.* A diagram of the pH scale, showing both actual concentrations and corresponding pH values, is given in Fig. 2.12.

---

*You should not become confused by the interchanging of the terms "proton" and "hydrogen ion." Recall that a hydrogen atom consists only of a nucleus of one proton with one electron orbiting around it. The removal of that electron, or ionization, leaves only the proton, which therefore represents the hydrogen ion. While most modern texts define acids and bases in terms of their being proton donors or acceptors, the measurement of the strength of particular acids and bases on the pH scale is usually given in terms of hydrogen ion (that is, proton) concentration. Thus we shall use the terms interchangeably here. In actuality, of course, the pH scale measures the concentration of hydronium ions.

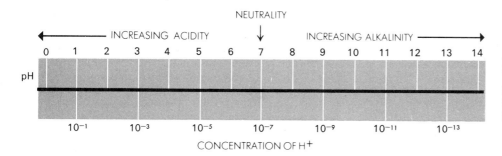

Fig. 2.12
The pH scale. The pH value is at the top. At the bottom is the actual concentration of hydrogen ions expressed in moles per liter.

The pH scale gives values as high as one-tenth (10^{-1}) of a mole of hydrogen ions per liter and as low as one hundred-trillionth (10^{-14}) mole per liter. The pH of most acids and bases falls somewhere within this span.

Since the pH scale is a logarithmic one, numerical values on the scale are not like an arithmetical progression, where the value of two is twice that of one, or the value of three is three times that of one. Rather, numbers on the pH scale are based on powers of ten. Logarithmic scales are quite useful when large changes in quantity must be measured. On the pH scale, a pH of 2 indicates ten times fewer hydrogen ions than a pH of 1, rather than half as many. A pH of 3 indicates ten times fewer hydrogen ions than a pH of 2, and a hundred times fewer hydrogen ions than a pH of 1. A pH of 4 indicates ten times fewer hydrogen ions than a pH of 3, a hundred times fewer than a pH of 2, and a thousand times fewer than a pH of 1. Thus the number of hydrogen ions released by compounds of successively higher pH decreases by a factor of ten with each step up the scale.

A 0.1M solution of hydrochloric acid would have a pH of 1, and a 0.01M solution a pH of 2. The 0.1M means one-tenth of a mole of hydrogen ions (hydronium ions) per liter, while 0.01M means one-hundredth of a mole per liter. Thus, 0.01 represents a tenfold *decrease* in hydrogen ion concentration over 0.1, and is represented by the difference between the expressions 10^{-1} and 10^{-2}. Similarly, an HCl solution of 0.000001M/liter (10^{-6}) would represent a millionfold dilution of a solution at 0.1M/liter concentration. Each step up the pH scale thus represents a tenfold decrease in the hydrogen ion concentration, and each step down a tenfold increase.

The following statements summarize how pH scale values relate to the actual strength or weakness of acidic and basic solutions:

1. The pH scale ranges from 0 to 14. Solutions having a pH of less than 7 are acidic; those with a pH greater than 7 are basic.

2. The midpoint of a scale, pH 7, represents neutrality. Here the number of hydrogen ions equals the number of hydroxide ions. The pH of water is 7.

3. The pH scale is a logarithmic one. A change of one unit in pH corresponds to a tenfold change in hydrogen ion concentration.

4. The pH scale is a standardized means of expressing the acidity or alkalinity of any solution. It provides a frame of reference by which the concentration of hydrogen ions in various solutions can be judged.

It is quite possible to have a pH outside of the normally encountered range of 1 to 14. These extremes rarely occur in biochemical work, however. They are certainly not found in living organisms. As a matter of fact, the large majority of pH measurements within living organisms lie well between 6 and 8. It is true that the stomach, with large quantities of hydrochloric acid, has a pH of 1 or 2, and that certain bacteria require a very acid medium in which to live. But these are exceptions to the rule. The majority of plants and animals are restricted to an internal environment that varies only slightly to one side or the other of neutrality, or pH 7.

2.8
OXIDATION AND REDUCTION: REDOX REACTIONS

As we have seen, some atoms tend to give up electrons to others, while other atoms tend to gain electrons from others. The process of giving up electrons is called **oxidation,** and the atom that does so is said to be *oxidized.* The process of gaining electrons is called **reduction;** the atom that gains them is said to be *reduced.* It must be stressed, however, that the degree of oxidation (electron loss) or reduction (electron gain) is dependent upon the situation. For example, a carbon atom in the compound CH_3OH has lost more electrons (that is, is more oxidized) than the carbon atom in CH_4.

The process of oxidation does not necessarily involve the element oxygen. The name "oxidation" was originally derived from the class of reactions involving the combination of various elements (mostly metals) with oxygen. Now, however, the term oxidation is used more broadly to refer to any tendency to give up electrons in a chemical reaction, whether or not oxygen is involved.

Oxidation and reduction are useful terms when employed to describe what happens when two atoms, such as sodium and chlorine, combine to form a compound—in this instance, sodium chloride. Sodium atoms undergo a change from a neutral to an electrically charged condition (from 0 to +1) by losing an electron. Chlorine goes from a neutral to a negatively charged condition (0 to −1) by gaining an electron. Thus the sodium is oxidized and the chlorine reduced. The formation of ionic chemical bonds often (though not always) involves an oxidation-reduction or **redox** reaction.

By contributing electrons that reduce chlorine, sodium acts as a *reducing agent*. By accepting electrons from sodium, chlorine acts as an *oxidizing agent*. An oxidizing agent, then, tends to accept electrons, while a reducing agent tends to give up electrons. Once again, however, the degree to which an atom or compound acts as a reducing or oxidizing agent is relative. Relative to potassium, for example, hydrogen is an oxidizing agent, while relative to fluorine it is a reducing agent.

When most biological molecules are oxidized, electrons are removed in combination with protons, rather than alone. In other words, a whole hydrogen atom, in the form of one proton and one electron, is removed during biological oxidation. Thus biological oxidation is frequently associated with hydrogen removal, or "hydrogen transfer," as it is sometimes called. This fact should not obscure the important point that the process is still one of oxidation—the loss of electrons.

2.9
THE COLLISION THEORY AND ACTIVATION ENERGY

No one knows exactly what causes atoms or molecules to interact during a chemical reaction. As a model hypothesis, however, the collision theory has been of much value in formalizing our ideas about how reactions occur. This hypothesis offers an explanation of how atoms and molecules actually interact and leads to accurate predictions. It also helps explain how factors such as temperature, concentration of reactants, and catalysts (substances that speed up a reaction but are not themselves used up during the reaction) affect reaction rates.

The collision theory is derived from the idea that all atoms, molecules, and ions in any system are in constant motion. Particles that are to interact chemically must first come into contact so that electron exchanges or rearrangements are possible. The collision of any two particles is considered to be a completely random event. If two negatively charged particles approach each other, each will mutually repel the other, and a direct collision is not likely. The same will be true of two positive particles. If a positive and a negative particle approach each other, however, a collision is more likely. Furthermore, this collision may be successful in the sense that it produces an interaction and hence a chemical change.

Not every collision between oppositely charged particles will produce a chemical interaction. Several other factors are involved. First, the average velocity of the particles determines what percentage of collisions will be successful for any given kind of reactants. More rapidly traveling particles will be more likely to yield successful collisions.

Second, particles of each element or compound have their own minimum energy requirements for successful interaction. Imagine a system in which molecules of A and B interact to produce C and D. For any collision between A and B to produce a reaction, each molecule must have a certain minimum kinetic energy. This energy is usually referred to in terms of particle velocity. Greater kinetic energy of a particle means greater velocity. Greater velocity means greater probability that a collision will be successful. If the average kinetic energy of a system is increased, the number of successful collisions will also generally be increased.

Third, molecular geometry plays a role in determining whether or not a collision is successful. If a molecule collides with an atom or another molecule in such a way that the reactive portion of the molecule is not exposed to the other particle, no reaction will occur. This is true in spite of the fact that the particles may have possessed the proper amount of kinetic energy. For this reason, molecular geometry, though a factor in any chemical reaction, is particularly important in reactions between very large molecules. Here the relative position of two colliding molecules is crucial to successful interaction. Living systems have developed means of holding large molecules in specific positions which aid in exposing the reactive portion of the molecule. This is one of the main functions of organic catalysts, or **enzymes** (Section 3.6).

The minimum kinetic energy required by any system of particles for successful chemical reaction is known as the **activation energy,** which is a characteristic of any reacting chemical system. If the average energy of the particles is below this minimum, the reaction will proceed slowly or not at all. If the average is above the minimum, the reaction will proceed more rapidly.

As an example, the velocities of the molecules in a gas will vary considerably with some showing relatively low velocity, some very high, while others are somewhere in between. Depending upon the level of activation energy of a particular reaction, perhaps only those molecules possessing the highest velocity will be able to

---

Oxidation is the loss of electrons; reduction is the gain of electrons.

Chemical reactions can occur only when two or more interacting atoms collide with enough "energy of activation." This energy is necessary to overcome the natural tendency of the negatively charged electron clouds of the atoms to repel one another.

participate. In Fig. 2.13 we can see that a hypothetical reaction requiring an activation energy greater than that impacting a molecular velocity of 2000 meters per second could not occur at a temperature of 270° Kelvin (K), but could for some molecules at the higher temperatures of 1273°K and 2273°K, since some (though considerably less than half) of the molecules possess the required velocity. The graph in Fig. 2.13 emphasizes the fact that *chemical interaction between atoms or molecules can be discussed only in terms of probability.* The rate of a chemical reaction is influenced by factors that increase or decrease the probability that collisions between particles will be successful.

2.10
FREE-ENERGY EXCHANGE AND
CHEMICAL REACTIONS

All chemical reactions involve an exchange of free energy. On the basis of these exchanges, chemical reactions can be divided into two classes. Reactions that absorb more energy than they release are called **endergonic** reactions. Those that release more free energy than they absorb are called **exergonic** reactions.

Endergonic and exergonic reactions can be compared in terms of the energy hill analogy. Endergonic reactions occur in an uphill direction. Exergonic reactions occur in a downhill direction. This means that, like rolling a stone uphill, endergonic reactions require an input of free energy. And, like a stone rolling downhill, exergonic reactions release free energy (see Fig. 2.14).

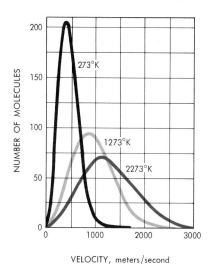

Fig. 2.13

The distribution of molecular velocities (Maxwell-Boltzmann distribution) in nitrogen at given temperatures. Only those molecules possessing sufficient velocity have enough activation energy to enter into chemical combination. The peak of each curve represents the most probable velocity of the molecules.

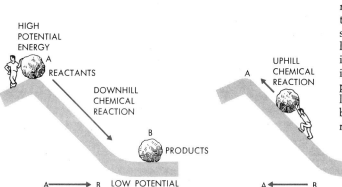

Fig. 2.14

In this analogy, a chemical reaction that releases free energy is compared to a stone rolling down a hill. When the stone has reached the bottom, it has less potential energy than when at the top. To go in the reverse direction, back up the hill, the stone will have to absorb the same amount of energy that it released while rolling down. Under natural conditions, absorbing this amount of energy is quite unlikely. Only very rarely, if at all, would a stone ever get back to the top unless it was pushed. The same is true of chemical reactions that release large amounts of energy. They are considered to be irreversible. Unless energy is supplied from the outside, the reverse reaction will not occur.

Change in Free Energy as ΔG

As pointed out earlier, the energy available in any particular chemical system for doing useful work is known as free energy (G). If a net change in free energy occurs during a chemical reaction, the system has either more or less free energy after the reaction than before. An exergonic reaction always involves the loss of free energy. We can say that such a system shows negative free-energy change, $-\Delta G$, where Δ means "change." An endergonic reaction takes in free energy. Thus it shows an increase in free energy, $+\Delta G$. It is possible, therefore, to show whether a given reaction involves an overall increase or decrease in free energy simply by putting the symbol $-$ or $+$ after the equation.

Like the energy in chemical bonds, free energy exchange in reactions is measured in kilocalories per mole of reactant. For example, in the reaction between hydrogen and oxygen to produce water, we find*

$$H_2 + \tfrac{1}{2}O_2 \rightarrow H_2O, \qquad \Delta G = -56.56 \text{ kcal/mol}. \qquad (2.7)$$

The reaction has a negative ΔG and hence gives off energy. The more negative the value for ΔG, the more energy the reaction releases. Consider the reaction of the sugar glucose and oxygen, which releases in several steps the energy for many life processes. These steps can be summarized in the following equation:

$$C_6H_{12}O_6 + 6O_2 + 6H_2O \rightarrow 6CO_2\uparrow + 12H_2O,$$
$$\Delta G = -690 \text{ kcal/mol}. \qquad (2.8)$$

This overall reaction releases a great deal more energy than the reaction shown in Eq. 2.7.

In a similar manner, the numerical value for reactions with a positive ΔG indicates how much energy the reaction requires. In Eq. 2.9 below, iodine reacts with hydrogen to form the compound hydrogen iodide:

$$\tfrac{1}{2}I_2 + \tfrac{1}{2}H_2 \rightarrow HI, \qquad \Delta G = +0.315 \text{ kcal/mol}. \qquad (2.9)$$

This reaction requires a small amount of energy, as shown by the low positive value for ΔG. On the other hand, the process of photosynthesis, in which green plants produce carbohydrates from carbon dioxide and water, requires a large intake of energy. This energy is supplied by light. The overall process can be written as

$$6CO_2 + 12H_2O \rightarrow C_6H_{12}O_6 + 6H_2O + 6O_2\uparrow,$$
$$\Delta G = +690 \text{ kcal/mol}. \qquad (2.10)$$

Knowing the value of ΔG makes it possible to compare the amounts of energy that various reactions absorb or release.

All exergonic reactions show an overall loss of free energy. Many of these same reactions, however, require an energy input to get them started. If left to themselves, many reactions will never show any measurable chemical activity. However, if the right amount of energy is supplied, the reaction begins. It then goes to completion without the addition of more energy from the outside.

How can this be explained? In such chemical systems, the reactants have relatively high activation energies. The addition of energy gets a larger percentage of particles in the system up to the required kinetic energy. In absorbing this energy, the particle becomes activated. When particles are in an activated state, a successful reaction is much more probable.

A specific example will clarify this point. Formic acid, HCOOH, is the pain-causing substance in wasp and bee stings. Under certain conditions, formic acid decomposes into carbon monoxide (CO) and water, with a slightly positive ΔG. For this reaction to occur, however, a formic acid molecule must first absorb enough energy to become activated (see Fig. 2.14). In being activated, the molecule undergoes a rearrangement of one hydrogen atom. The molecular structure is changed, and along with it the stability of the whole molecule. Thus it splits into two parts, carbon monoxide and water:

$$HCOOH \rightarrow CO + H_2O. \qquad (2.11)$$

Entropy and Free Energy†

We have seen that energy changes are of great importance as a driving force in many chemical reactions. We have also seen that, in general, chemical reactions tend to proceed from higher to lower energy states. In other words, a chemical reaction can be visualized as proceeding down the energy hill, eventually reaching a point where the slope of the hill levels off.

However, some chemical reactions seem to roll *up* the energy hill in going to completion. It is apparent that such reactions are driven by something other than the tendency to proceed from higher to lower energy states. These and other considerations have led to the development of the *second law of thermodynamics*. It is useful to consider biological processes in light of this generalization.

The second law of thermodynamics relates changes in the free energy of a chemical system with changes in *organization* or *orderliness* of the parts of that system. The concept of entropy is used to represent the amount of *disorganization* that any system shows, or the degree to which it is *disorderly*.

Consider a room filled with molecules of one kind of gas. We can view the distribution of the gas molecules from a standpoint of probability. It is highly improbable that all of the molecules would be located in one corner of the room. Instead, it is much more proba-

---

*In equations where energy equivalents are given, the numbers before each molecule refer to numbers of moles. The one-half O_2 thus means one-half mole of oxygen.

†Material in this section is adapted from Baker and Allen, *Matter, Energy, and Life: An Introduction to Chemical Concepts*, 4th ed., © 1981. Reading Mass.: Addison-Wesley, pp. 83–87. Reprinted with permission.

ble that these molecules will be equally distributed, and that there will be no more molecules in one area of the room than in another.

If all the molecules of gas are located in one corner of the room, the degree of organization in this system is rather high. Recall that the molecules of any substance are constantly in motion. Left to themselves, they tend to spread out, distributing themselves uniformly over the entire space. Thus, it is highly improbable that gas molecules will accumulate in the corner of a container by their own random motions. It would be necessary to make an organized effort, that is, expend free energy, to force all the molecules into one area. If the attempt were made, it would bring about an increase in the organization of the system.

If the system is so organized and then left to itself, the molecules begin scattering about the rest of the container. The system begins to become more disorganized almost immediately. It is characteristic of physical systems that, left to themselves, they tend to become more and more disorganized.

Everyone is familiar with this principle from day-to-day experience. It is well known that if a house is not cleaned and kept in repair, it tends to become more messy. We might say that it becomes more disorderly or disorganized. A constant expenditure of energy is necessary to keep things from becoming disorganized. A human being is a very complex organization of specific parts. Yet, that organization can be maintained only as long as a certain amount of energy is expended to overcome the tendency to assume a more random state. The energy for maintaining this order comes from the food we eat.

We are now in a position to understand what entropy means. Since it is a numerical measure of disorder, entropy can be defined in mathematical terms and given the symbol S. This definition states that the less probable a given distribution of molecules or atoms in a system, the less entropy that system contains. The greater the disorganization, the greater the value for S; the more organization (that is, the more order), the less the value for S. In other words, the more probable a given distribution, the greater the entropy in that system.

It follows, therefore, that physical or chemical systems can do work as they proceed from a state of low entropy to a state of high entropy. For example, as the molecules of gas in a room move from one corner into the rest of the room, the pressure which their diffusion exerts can be used to move matter. Consider the box shown in Fig. 2.15. The only way that a molecule of gas can get from one compartment to another is to go through the opening shown in the center partition. This opening is blocked by a paddle-wheel device. As molecules hit the blades of the paddle wheel, they exert enough pressure to turn the wheel.

In Fig. 2.15(a), compartment 1 contains more molecules than compartment 2. As a result, there will be greater movement of molecules from compartment 1 into compartment 2. This will turn the paddle wheel to the right. This, in turn, will run the generator, doing work. In time, however, the molecules will become

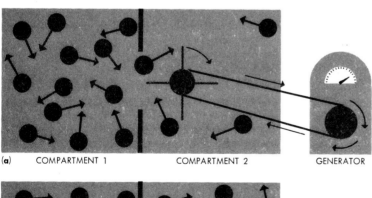

(a)   COMPARTMENT 1   COMPARTMENT 2   GENERATOR

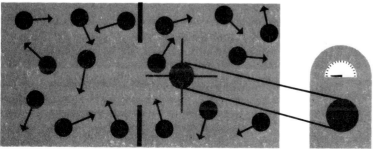

(b)

Fig. 2.15
The closed box (a) has two compartments containing a gas, and represents one complete physical system. There are more molecules in compartment 1 than in compartment 2. This represents an organized state, since gases tend to diffuse evenly throughout a container until a completely random distribution is achieved. Thus there is an *increase* in entropy. While this happens, the system performs work by turning the wheel connected to an electric generator. (b) When the difference in energy levels between two parts of the system disappears, the system no longer performs work.

evenly distributed on each side of the partition, as shown in Fig. 2.15(b). Hence, there is no greater movement in one direction than in another. The system is at equilibrium and cannot yield any more work. The driving force behind this reaction is the tendency of the system to proceed from a more organized to a less organized state. In other words, *the driving force is toward an increase in entropy.*

Chemical or physical reactions proceed only from states of high organization to states of low organization, unless energy is supplied. Or, stated in more precise terms, *reactions proceed from lower to higher entropy states.* The relationship between entropy values and degree of organization is thus an *inverse* one. The higher the degree of organization, the lower the entropy, and vice versa.

The concept of entropy is also related to that of free energy. In terms of chemical reactions, this relationship can best be shown by a simple equation:

$$\Delta G_0 = \Delta H - T \Delta S, \tag{2.12}$$

where ΔG stands for the change in free energy, ΔH for the change in heat content (the heat content of a system is often referred to as *enthalpy*), T for the absolute temperature, and ΔS for the change in entropy. In other words, this equation informs us that the change in free energy of a system is equal to the change in heat content minus the product of the absolute temperature and the change in entropy.

From Eq. 2.12, it can be seen that free-energy changes in any chemical system are inversely related to changes in entropy. A system with high entropy has little free energy and thus little organization. Conversely, a system with low entropy has a greater degree of free energy and a greater degree of organization. Therefore, a system with a high degree of organization is capable of performing more work than a system with a low degree of organization. Recall that free energy is generally equated with the ability to do work. A highly organized system has the capability of showing a greater change in free energy than one that is less organized.

Why do heat and temperature (the measure of heat) appear in this equation? Recall that heat involves the random motion of molecules. In a very real sense, heat is chaotic energy. It is obvious that heat will have a very great effect on the order or disorder of a system, that is, on its entropy. The greater the amount of heat given off, the greater the free-energy change. A large free-energy change indicates that a greater amount of potential energy is released. Therefore, there would be less free energy in the system after reaction than before. Correspondingly, as we just saw, the entropy of the system after reaction would be greater than before.

Consideration of chemical or biological processes in light of the second law of thermodynamics emphasizes two general points. First, for ordinary chemical reactions to proceed, there must be some driving force, whether it is an energy change from a higher to a lower state, or one from lesser to greater entropy. This driving force may be defined as the difference in energy levels between two parts of a system. The difference in energy level between glucose and its end products, carbon dioxide and water, is sufficient to keep the biochemical reactions of a cell going.

Second, all processes in the universe run toward an increase in entropy. This means that the universe as a whole is moving toward greater disorganization. Like a giant clock, wound up sometime in the past, the universe is gradually running down. The heat which is given off from the sun and all other stars passes off into space and is lost. The energy is not destroyed. It is simply reduced to a nonusable form. As entropy of the universe increases, free energy decreases. The total amount of usable energy in the universe thus decreases with passage of time.

In the past, various writers have claimed that living organisms defy the second law of thermodynamics. Since living systems grow, reproduce, and metabolize, it has been said that they actually show an increase in free energy and a decrease in entropy. This is true, of course, if the living organism is considered as an isolated system. During periods of growth, for example, an

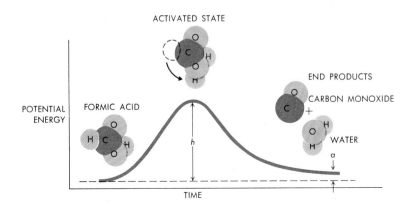

Fig. 2.16
Changes in potential energy during the decomposition of formic acid. The original molecule, to the left, is in a relatively low energy state. By collision with another molecule of high kinetic energy, this molecule becomes activated. The potential energy of such an activated molecule is greater. During activation, a molecular arrangement occurs and the molecule splits. The end products, carbon monoxide and water, are at a higher energy state (*a* on the graph) than the original molecule. Distance *h* represents the height of the energy barrier.

organism builds complex molecules, increases the number of its cells, and shows specialization of certain tissues. The organism, indeed, becomes more highly organized. But no organism is independent of outside energy sources. The energy for all life on earth ultimately comes from the sun. Thus the sun, green plants, and animals must all be considered as parts of a single system if we are to make any meaningful statements about the thermodynamics of life. Living organisms show increases in free energy only because other parts of the universe show a decrease in free energy.

The balance sheet of the earth-sun system shows that the free energy of the system as a whole is decreasing at an enormous rate. Only a very small fraction of this energy is actually captured by living organisms and used to maintain their high levels of organization. Thus, despite activities of living organisms, the second law of thermodynamics still applies to all processes, living or nonliving, in the known physical world.

Graphing Energy Exchanges

Energy exchanges in chemical systems are often given on a graph that shows the changes in potential energy during the course of reaction. These changes are then compared with the time it takes the reaction to go to completion. Such a graph for the formic acid reaction is shown in Fig. 2.16.

Analysis of this graph reveals some important things about this chemical reaction. The graph describes the changes in energy for one molecule as that molecule undergoes the decomposition reaction shown in Eq. 2.11. Before reacting, an individual molecule is in a relatively low energy state. By absorbing energy, this molecule passes to a higher potential-energy level. It is now in an activated state. The appropriate rearrangement occurs, and the product molecules are formed. Note that the product molecules are at a slightly higher energy state than the original molecule of formic acid. This indicates that the overall reaction absorbed a small amount of energy. The reaction is endergonic.

The distance h on the graph indicates the energy of activation for this chemical system. The height of the graph line can thus be considered an **energy barrier**: a "hill" over which the molecule has to climb before it can roll down the other side to completion. After absorbing the required activation energy the reaction proceeds spontaneously, just as a stone rolls down a hill once it is pushed over a rise at the top.

Under standard conditions* hydrogen and oxygen exist together in a single container without the least indication of reacting to produce water. The molecules simply do not have the necessary energy of activation.

Standard conditions means standard temperature (25° Celsius) and one atmosphere of pressure (760 millimeters of mercury), as opposed to the standard conditions necessary for free-energy change.

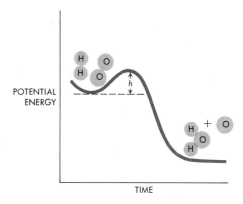

Fig. 2.17
A spontaneous exergonic reaction, the formation of water from H_2 and O_2. The end products are at a lower energy state than the two starting reactants, indicating that the reaction releases energy. This reaction has a small energy barrier, as indicated by the size of h.

However, if a small electric spark is introduced into the chamber, an explosive reaction takes place. Thus hydrogen and oxygen react quickly to form water if given the necessary push to get them started (see Fig. 2.17). The fact that a spark is all that is needed to provide this push shows that the energy barrier for this reaction is not very high. Passing the spark through hydrogen and oxygen provides enough energy to put some molecules of each in the activated state. The activated molecules spontaneously react to form molecules of water. From Eq. 2.7 we see that this reaction releases energy. The energy released from one reaction is enough to get several other molecules of each element over the energy barrier. The spark provides an initial push. The rest of the reaction occurs by a chain-reaction effect, just as a house of cards collapses when one card is disturbed. The products of such exergonic reactions are always at lower potential-energy states than the reactants.

Energy Exchange and Chemical Bonds

How does the energy involved in chemical reactions relate to the formation and breaking of chemical bonds? In exergonic reactions, the end products are at a lower energy state than the reactants; in endergonic reactions, the end products are at a higher energy state. Some exergonic reactions result in the formation of chemical bonds, while others result in the breaking of bonds. There is no necessary correspondence between exergonic reactions and the breaking of bonds or between endergonic reactions and the building of bonds. However, quite frequently those reactions that result in the building or synthesis of a large molecule from smaller components are, indeed, endergonic, while those that result in the breaking down of larger molecules into smaller parts are exergonic.

As we saw earlier, the fact that activated atoms always release a certain amount of energy in the formation of chemical bonds does not mean that all chemical reactions in which bonds are formed show an overall release of energy. The net energy exchange of *any* reaction is a result of the energy required to activate the atoms (move them to the top of the rise), combined with the energy released in attaining stability. If more energy is required to move the reacting atoms to the top of the rise than is released as the atoms combine to form a molecule, then the reaction as a whole will be *endergonic*. If less energy is required, then the reaction as a whole will be *exergonic*. Thus, every particular set of possible reactants will have its own energy characteristics. It is important to remember, however, that net energy exchange must always account for both the energy of activation necessary to drive the two or more atoms close

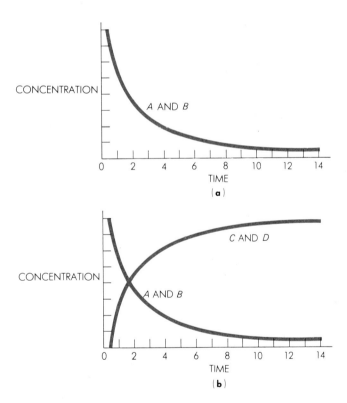

Fig. 2.18
Changes in concentration of reactants A and B and products C and D during the course of a chemical reaction. This relationship between change of concentration and time indicates the rate of reaction. Graph (a) shows only change in concentration of reactants. Graph (b) also shows the change in concentration of products. Change in concentration of either reactants or products is most rapid during the first few minutes of reaction.

enough together to interact and the energy released as the atoms interact in the formation of chemical bonds.

2.11
RATES OF REACTION

The rate of any chemical reaction is defined as *the amount of reaction in a given period of time.* The amount of reaction is generally measured in terms of the change in concentration of reactants or products. The basic relationship between amount of reaction and time can be expressed as a word equation:

$$\text{rate of reaction} = \frac{\text{change in concentration}}{\text{change in time}}. \qquad (2.13)$$

The concept of rate in chemical reactions is vital to an understanding of chemical equilibrium. In addition, knowledge about reaction rates allows a clearer understanding of the mechanisms by which a particular chemical process occurs. Under given conditions, the rate of a reaction is a predictable characteristic of chemical systems. Such systems can thus be described in terms of reaction rates as well as direction or energy exchange.

A specific example will help in understanding the concept of rate as it applies to chemical reactions. Consider a reaction in which molecules of A and B combine to yield the product C and D:

$$A + B \rightarrow C + D. \qquad (2.14)$$

If we begin with molecules of A and B only, the rate of reaction at the outset is very high. Rate in this case can be measured in terms of how rapidly the reactant molecules A and B disappear in specific units of time. If we were to stop the reaction every thirty seconds and determine the amount of A and B present, these data plotted on a graph would give a line like that shown in Fig. 2.18(a).

We can see that the rate at which A and B disappear from solution changes with time. For instance, the concentration of A and B decreases most rapidly in the first two minutes and begins to level off about the sixth minute. By the ninth minute nearly all of A and B have been used up in the reaction. At this point, the rate at which A and B combine to yield C and D is almost zero.

If we also plot the rate of appearance of products C and D in this reaction, the curve we obtain is just the opposite of that for the disappearance of A and B, as shown in Figure 2.18(b). This is not surprising, since the rate at which C and D appear depends directly on the rate at which A and B interact.

The molecular explanation for the change in rate of chemical reactions goes back to the collision theory. The rate at which a chemical reaction progresses toward completion depends on the number of effective collisions between reacting molecules or atoms. This number is determined for any given reaction by the concentration of reactants. More molecules or atoms of reactants mean a greater number of effective collisions.

During the course of any chemical reaction in a closed system, the concentration of reactants decreases as the chance of collision between a molecule of A and a molecule of B decreases. At the same time, the concentration of product molecules C and D is increasing. This means that collisions between C and D molecules will become more frequent.

Many factors influence the rate of chemical reactions: temperature, concentration of reactants, pH, and the presence or absence of catalysts. Each of these factors influences the rate of chemical reactions by increasing or decreasing the number of effective collisions. Temperature affects reaction rates by changing the average velocity of particles in a chemical system. It also affects reaction rates by changing the fraction of molecules possessing minimum energy for reaction. Higher temperatures mean faster velocities and a greater fraction of molecules possessing the minimum energy; lower temperatures mean slower velocities and a smaller fraction of molecules possessing minimum energy. This works as follows: change in velocity means a change in the possibility of collision; the faster the velocities of the reacting particles, the greater the frequency of collision. In general, for every rise of $10°C$, the rate of a chemical reaction is doubled; conversely, for a fall of $10°C$, the rate is cut in half.

The concentration of reactants has a similar effect. A greater concentration of one or both reactants means a greater chance that a collision will occur. A smaller concentration of reactants means less chance of collision, hence a slower rate of reaction. **Catalysts** affect the rate of reaction by increasing the effectiveness of any given collision once it has occurred or by increasing the chances of the reaction occurring by providing a surface on which the reactants can meet. Catalysts are molecules (or sometimes atoms) that facilitate a particular reaction without themselves being permanently changed in the reaction. Catalysts do not cause a reaction to occur that would not occur on its own. They do, however, speed up the rate of reactions that would occur anyway, and reduce the amount of activation energy necessary to begin them.

2.12
REVERSIBLE AND IRREVERSIBLE REACTIONS

It follows from the collision theory that chemical reactions are **reversible**: the reaction can go in either direction. In any chemical system, then, two reactions are usually taking place:

$$A + B \rightarrow AB \tag{2.15}$$

and the reverse,

$$AB \rightarrow A + B. \tag{2.16}$$

The forward and reverse equations can be combined into one, with the reversibility indicated by double arrows:

$$A + B \rightleftharpoons AB. \tag{2.17}$$

In principle, all chemical reactions are reversible. There is no reaction known that, under suitable conditions, cannot proceed (however slowly) in the reverse direction. Yet under ordinary conditions some reactions are far less reversible than others. In these cases, the reaction from right to left occurs so slowly that its rate is barely detectable. Such reactions are said to be **irreversible.**

What conditions tend to produce irreversibility? Two factors are important. First, there is the consideration of energy. Some reactions release a great deal of energy going in one direction. Such reactions will tend to go in the reverse direction only if the same amount of energy can be absorbed. An irreversible reaction can be compared to rolling a stone down a hill. The downward path releases potential energy. To get back up the hill, the stone requires the input of the same amount of energy. Obviously, it is likely that far more stones will roll down a hill than will be pushed back up.

Second, chemical reactions are irreversible if one of the products leaves the site of reaction. This may occur if the product escapes as a gas,

$$A + B \rightarrow C + D\uparrow, \tag{2.18}$$

or if one product is a **precipitate** (an insoluble substance that settles out of solution),

$$A + B \rightarrow C + D\downarrow. \tag{2.19}$$

In each case the reverse reaction is inhibited by the removal of one of the product substances (D). Since removal of their products occurs often, many biochemical reactions can be considered irreversible.

2.13
CHEMICAL EQUILIBRIUM

Within a certain period of time, reactions reach a state of equilibrium. When this condition is reached, the *proportion* of reactants in relation to products remains the same. Notice that this does *not* mean that the *amounts* of reactant and of product are necessarily equal.

An example will illustrate this point. Consider the reaction in which molecules A and B yield products C and D:

$$A + B \rightleftharpoons C + D. \tag{2.20}$$

If the reaction begins with only molecules of A and B present, it will occur at first only to the right. The

Chemical equilibrium is established when the forward rate of any chemical reaction is equal to the reverse rate. The equilibrium point of any reaction is measured as the ratio of product to reactant after equilibrium is established.

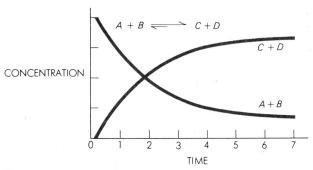

Fig. 2.19
Graph showing the change in concentration of reactants and products in a reversible reaction. The point of chemical equilibrium is reached where the two curves level off (at about the fifth minute). At this point, the rate of the forward reaction is equal to the rate of the reverse reaction, the defining criterion of chemical equilibrium.

graph in Fig. 2.19 plots change in concentration of reactants and products. It is apparent that the change in concentration ceases after about the fifth minute. Beyond this point, there is no net change. When these curves level off, the concentration of C and D is greater than the concentration of A and B. The reaction is thus directed or shifted to the right, as the relative lengths of the arrows indicate.

As long as molecules of reactant and product are still present in a system, chemical reactions never cease. In the reaction above, C and D accumulate because of the relatively high activation energy required to convert them back into A and B. Eventually the number of collisions between molecules of C and D will be higher than the number of collisions between molecules of A and B. As a result, the rate of the reverse reaction will increase, in spite of its higher energy of activation. At the same time, the rate of the forward reaction will decrease because of the decreasing concentrations of A and B.

Eventually a point will be reached at which the forward rate equals the reverse rate. When this condition is reached, we say that a state of *chemical equilibrium* exists. Note that the concentration of reactants does *not*

have to equal the concentration of products in order for equilibrium to be established. In the above case, the concentrations of C and D are much greater than the concentration of either A or B at the equilibrium point. The important point is that at equilibrium *the rates of forward and reverse reactions are the same*. This means that per unit of time as many molecules of A and B are being converted into C and D as molecules of C and D are being converted into A and B.

The condition that exists at chemical equilibrium is referred to as **dynamic equilibrium.** While the overall characteristics of a given system at dynamic equilibrium remain relatively constant, its individual parts are in a state of continual change.

Left to itself, the direction any reaction takes is toward a condition of equilibrium. The point where chemical equilibrium lies is characteristic for any given chemical reaction.

What happens when a chemical reaction at equilibrium is disturbed by the removal or addition of substances on either side? Suppose that the reaction

$$A + B \rightleftharpoons AB \qquad (2.21)$$

exists in perfect equilibrium. If we add a quantity of either substance A or substance B, or both simultaneously, we will push the reaction to the right so that more molecules of AB will be formed. We can accomplish the same effect by another method. Without adding more of either A or B, we can shift the reaction to the right by removing some AB. If we add molecules of AB to the system in equilibrium, the reaction is shifted to the left. The point at which equilibrium will eventually be reached in such reactions has not been changed by adding or removing substances. Only the direction or rate of the reaction has been changed momentarily. Reversible reactions eventually return to dynamic equilibrium.

The concept of equilibrium, then, can be approached in terms of the net transformation of $A \rightarrow B$ or $B \rightarrow A$, depending on concentrations. Catalysts are able to catalyze both forward and backward transformations, hastening the approach to equilibrium.

Summary

1. Living organisms are highly organized entities, and as such they represent thermodynamically improbable states. They are able to maintain their improbable condition only by an enormous and continual expenditure of energy.

2. Atoms are composed of electrons, protons, and neutrons. The protons and neutrons, located in the nucleus, determine the element's atomic number (protons) and weight (protons plus neutrons) and thus its basic physical proper-

ties. Electrons, located in a cloud outside the nucleus, determine the atom's chemical properties—the ways it interacts with other atoms. Since the number of electrons equals the number of protons for any atom, the atom is electrically neutral. Loss or gain of an electron creates a positive or negative (respectively) charge for the atom as a whole.

3. Atoms can store energy by upward transitions of electrons; they release energy by downward transitions. Electron transitions occur in discrete steps.

4. Chemical bonds represent electron redistribution among the atoms involved. Bonds contain energy. Two or more atoms joined together by such bonds are called molecules.

5. Among the kinds of chemical bonds important to living systems are covalent bonds, in which electrons are mostly shared, ionic bonds, in which the electrons tend to be more or less donated by one atom to another, and the various forms of weak bonds or interactions. Among the latter are found ionic bonds in aqueous solution, hydrogen bonds, and hydrophobic and van der Waals interactions.

6. The electronic structure of atoms determines the bond angles they form with other atoms in a molecule. Bond angles determine molecular geometry, and molecular geometry determines chemical reactivity and specificity.

7. Polar molecules have differential charge distribution, so that some regions are more positively and others more

negatively charged. Several important properties of water are due to molecular polarity.

8. The partial gain of electrons by an atom or molecule is called reduction; partial loss of electrons is called oxidation. The loss or gain of one or more electrons produces an overall net charge (positive or negative, respectively) on the atom or molecule. Such charged atoms or molecules are called ions.

9. Endergonic chemical reactions are those that absorb more energy than they liberate; exergonic reactions are those that give off more energy. Endergonic and exergonic refer to the net overall energy balance in a chemical reaction. Endergonic reactions need an overall net input of energy to keep them going; exergonic reactions usually need only an initial input, since the energy they release as they proceed is in excess of that needed to keep the reaction going.

10. All chemical reactions are reversible. Reversibility means that the reaction proceeds in two directions: from reactants to products, and back from products to reactants. For practical purposes, some reactions may be considered irreversible, since their reverse rates are infinitesimally small compared to their forward rates.

11. Chemical reactions reach equilibrium when the rate of forward reaction is equal to the rate of reverse reaction.

Exercises

1. Describe the difference between an endergonic and an exergonic reaction. How does this feature relate to the equilibrium of the reaction (that is, is there any relation between whether the equilibrium is shifted to the right or left, and whether the reaction is endergonic or exergonic)? How does it relate to how spontaneous the reaction is (spontaneous reactions require very little to get them started)? How does this relate to whether the reaction involves synthesis (the building of chemical bonds) or breakdown (the breaking of chemical bonds)?

2. What is a chemical bond? Why is the term "bond" misleading?

3. Discuss the three graphs in Fig. 2.20 in terms of the chemical events they represent. What does the change in concentration of reactants and products indicate? Relate these graphs to the concept of chemical equilibrium.

4. How do upward transitions represent a form of potential energy? How is such potential energy released as kinetic energy?

5. What is the difference between an acid and a base? What is important about the concept of acid and base in relation to living systems?

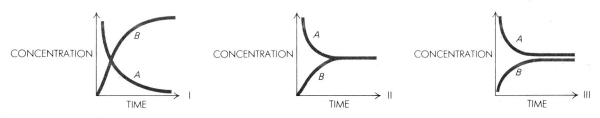

Figure 2.20

Chapter 3
Chemistry and
Life

INTRODUCTION

The principles of matter and energy discussed in the previous chapter were illustrated mostly with examples from *inorganic chemistry*, which deals with reactions, energy exchanges, and the like between atoms and molecules that generally do not contain the element carbon. The field of inorganic chemistry thus covers chemical interactions among a very large number of different elements including the metals; the common gases such as nitrogen, oxygen, and hydrogen; the generally nonreactive or so-called noble gases such as neon and freon; and the rare earths, to name only a few.

The study of chemical reactions involving the element carbon is known as *organic chemistry*. Although it might seem that, because it deals only with reactions containing carbon, organic chemistry would be a far more restricted field than inorganic chemistry, such is not the case. Carbon is capable of forming more varieties of compounds with more different elements than almost any other atom in the periodic table.

Organic chemistry is a very diverse field. It includes the study of such substances as plastics, polyesters, and petroleum and its derivatives, all of which are produced outside of living systems (though some, such as petroleum, are derived directly from the remains of living organisms). It also includes the study of all the organic compounds, and their reactions, that are found within living systems.

That branch of organic chemistry specifically concerned with the chemical reactions within living organisms is called *biochemistry*. Both biochemistry and organic chemistry as a whole involve the principles of chemical reactions, bond formation, and energy exchange described in Chapter 2. There is nothing fundamentally different in principle about the chemical reactions within living organisms and those in nonliving systems. In general, however, the chemical reactions within living organisms are more complex and involve many more individual catalyzed steps and specialized energy sources than those that occur in the inorganic or nonliving organic world. This greater complexity is important, for it imparts to living systems their unique abilities to organize and regulate chemical reactions in a way that produces the characteristics of life. But greater complexity does not imply the operation of fundamentally different physical and chemical laws. All modern developments in biochemistry argue against any vitalistic claim that living systems defy the laws of physics and chemistry.

3.2
CLASSES OF MOLECULES IN LIVING SYSTEMS

Living organisms are composed mostly of only a few chemical elements: carbon, hydrogen, oxygen, nitrogen, phosphorus, and sulfur.* These elements, in turn, are used to construct small building blocks, such as amino acids, simple sugars (monosaccharides), nitrogenous bases, and the like, which in turn are assembled into macromolecules such as proteins, polysaccharides, and nucleic acids. It is in the latter, macromolecular (*macro* means large) sizes that most of the compounds differ from each other from organism to organism; the smaller units are the same from one species to another.

The following sections will deal with those types of compounds currently considered among the most important in living systems. As varied as all these types may seem, it is important to recognize, however, that *all* can be chemically synthesized by some species organism from a relatively small number of basic building blocks. From monosaccharides, a wide variety of polysaccharides can be synthesized; add to the pool of monosaccharides a few other simple building blocks (for example, amino acids, fatty acids, nitrogenous bases, and the like) and the synthesis of proteins, lipids, and nucleic acids becomes possible. This build-up process of *biosynthesis*, the assembling of macromolecules from simpler parts, and the reverse process of biological degradation are constantly occurring in living systems.

---

*In 1936 it was calculated that all of the chemicals in the human body, if collected and sold at market value, would be worth about 98 cents. According to the American Council of Life Insurance, by 1980 inflation had increased this figure to $5.60.

3.3
CARBOHYDRATES

The term **carbohydrate** denotes a general category of two major types of substance, sugars and starches. Sugars serve as the primary fuel molecules for all living cells. Starches, composed of many simple sugar units linked together, serve as reserve fuels. Carbohydrates generally contain hydrogen and oxygen in the ratio of 2:1. A generalized carbohydrate formula can thus be written as $(CH_2O)_n$, where the subscript n means that molecules of carbohydrates are multiples of this basic unit. The formula for glucose, one of the most common carbohydrates in living systems, is $C_6H_{12}O_6$ ($n = 6$). Another common carbohydrate is ribose, whose formula is $C_5H_{10}O_5$ ($n = 5$). A structured formula, showing the arrangement of atoms in a molecule of glucose, a simple sugar, is as follows:

GLUCOSE
(CHAIN FORM)

GLUCOSE
(RING FORM)

(a)　　(b)

Glucose, ribose, and several other five- or six-carbon sugars are the basic building blocks of all types of larger carbohydrate molecules. Glucose is accordingly called a simple sugar, or **monosaccharide**. Two simple sugars joined together (such as glucose and glucose, or glucose and fructose) are called double sugars, or **disaccharides**. Large carbohydrate molecules, such as starch or cellulose, are composed of many simple glucose units joined end-to-end, forming very large, complex carbohydrates called **polysaccharides**. Large starch molecules can have molecular weights up to 500,000 or more. The term **macromolecule** is sometimes used to refer to any type of large molecule.

The following example illustrates how glucose units can be joined together to produce a larger molecule, in this case, the disaccharide maltose. The equation shows that during this process one molecule of water is eliminated:

$$C_6H_{12}O_6 + C_6H_{12}O_6 \rightarrow C_{12}H_{22}O_{11} + H_2O.$$
$$\text{(maltose)}$$

This process of joining, with the elimination of a water molecule, is called **dehydration synthesis** and is shown with the following molecular formulas:

GLUCOSE　　H_2O　　GLUCOSE

MALTOSE

The reverse process, whereby maltose is broken down by the chemical *addition* of water between two units, is called **hydrolysis**. Both dehydration synthesis and hydrolysis occur in the formation and degradation of all major macromolecules in the living cell. The chemical bond that joins two glucose units (or glucose and a maltose) is known as an **α-glycosidic bond**. Some disaccharides (such as lactose, to the right) have a **β-glycosidic bond**. This bond is formed in the same way as the α-bond, but it has a different orientation in space.

Carbohydrates are a major quick-energy source for living systems. The most immediately available forms are sugars; many carbohydrates are stored as polysaccharides such as starches. Some polysaccharides also serve as structural elements, as in the case of cellulose plant cell walls.

Whereas the α-glycosidic bond projects below the plane of the two rings, the β-bond projects above it:

α-LINKAGE

α-MALTOSE

β-LINKAGE

β-D-GALACTOSE   β-D-GLUCOSE

β-LACTOSE

Disaccharides bound together by α-glycosidic bonds thus have a different overall three-dimensional shape from those bound together by β-bonds; accordingly, the two types of molecules behave differently in cellular metabolism.

As pointed out earlier, when several monosaccharides are joined together, the result is a polysaccharide. In plants, for example, the polysaccharide cellulose is used as a major structural element in cell walls, while starch is an insoluble polysaccharide used for storage. In animals, the polysaccharide counterpart to starch is glycogen, used for glucose storage in liver and muscle tissue. Polysaccharides are also found in the adhesive coats on cell membranes, as components of fluids in bone joints, and in many other structures.

3.4
LIPIDS

The lipids are a group of organic chemical compounds which includes the fats (in their liquid state, commonly called "oils") and sterols.* Fats are an important component of living tissue and a major foodstuff.

Fats are an organism's most concentrated source of biologically usable energy. Most provide approximately twice as many calories per gram as do carbohydrates. This is because, while both carbohydrates and fats contain carbon, hydrogen, and oxygen, fats contain considerably more hydrogen per molecule than do carbohydrates; that is, they are more highly reduced. These hydrogen atoms supply electrons for the energy-releasing chemical processes in living systems; recall from pp. 48–49 that the removal of electrons—oxidation—is the basic process in energy release. Relative to fats, therefore, carbohydrates are partially oxidized already, and thus release less energy when they undergo the oxidation process.

In terms of long-term energy *storage*, fats have an even greater advantage, due to their being far less polar than either proteins or carbohydrates. Fats thus have less water associated with them. For example, the most common form of carbohydrate energy storage in animals, glycogen, yields less than one-sixth the energy per gram than does an equal amount of fat.

Each fat molecule consists of two building blocks, an *alcohol* (usually glycerol) and one or more representatives of a group of compounds known as *fatty acids.* During digestion, as with sugars, fats are broken down into two parts by hydrolysis.

The glycerol portion of a fat molecule has the following structural formula:

The arrows point to the alcohol groups, to which the fatty acids attach during dehydration synthesis.

While there are many kinds of fatty acids, all share in common the possession of a carboxyl group (—COOH). This, of course, gives them their acid characteristics. To the carboxyl group are attached varying numbers of carbon and hydrogen atoms. A

---

*The terms "lipid" and "fat" are often used interchangeably, but their meanings are not identical. Fats are a specifically defined class of molecules, composed mostly of fatty acids combined with glycerol. Lipid is a more general category of substances having no structural features in common but made up of molecules soluble in nonpolar solvents. Thus all fats are lipids, but not all lipids are fats.

two-dimensional drawing of a molecule of one fatty acid, stearic acid, $C_{17}H_{35}COOH$, is shown below:

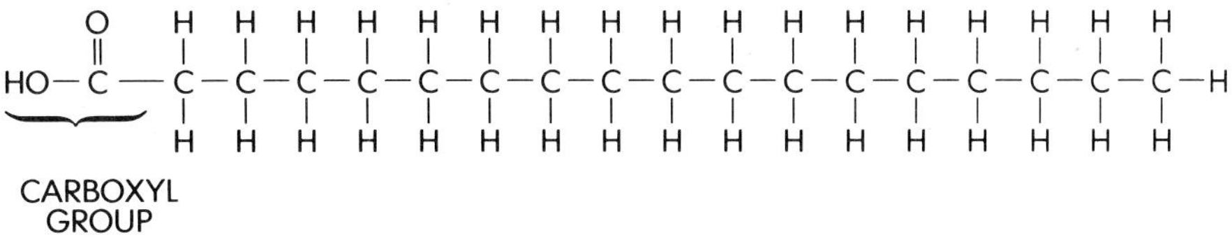

CARBOXYL
GROUP

SUPPLEMENT 3.1
STRUCTURAL REPRESENTATIONS OF MOLECULES: GLUCOSE

There are several different ways of writing the structural formula for glucose molecules in the ring form.

The conventional ring structure, as shown in Section 3.3, is drawn with certain bond lines heavier than others. These lines are used to give some idea of perspective. As the line widens, it suggests that that bond line is facing toward the viewer; conversely, the narrowing of the line suggests that the bond points away from the viewer.

Even this attempt to show something of the three-dimensional structure of glucose is not completely accurate. When in the ring form, glucose actually exists in a bent plane known as the "chair configuration" (from its similarity to a reclining chair). The chair configuration for one type of glucose molecule (β-D-glucose) is as illustrated in Diagram A.

The numbers beside each C indicate the conventional numbering pattern for the carbon atoms in the glucose ring. This diagram is thought to be a more accurate representation of the actual three-dimensional structure of glucose. For simplicity, however, glucose molecules will be shown in the more schematic ring form throughout this book.

A common shorthand procedure employed in writing the formulas of all organic compounds can also be illustrated with glucose. It is customary to leave out the symbols for carbon (C) in the structural formulas, indicating the presence of a carbon atom by an angle where two or more bond lines join together. Thus β-D-glucose, shown in Diagram A, would be written in shorthand as shown in Diagram B (for the chair and the ring diagrams).

Again, the numbers provide the traditional method of designating each particular carbon atom in the ring. In all structural diagrams used throughout this book, the carbon atoms making up a ring or chain of any sort will be left unlabeled. Only those carbon atoms that are not part of a ring or chain (as in carbon number 6) will be indicated with the letter C.

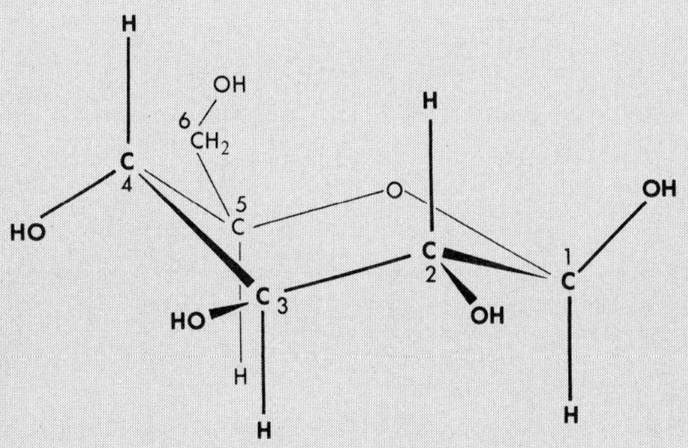

Diagram A

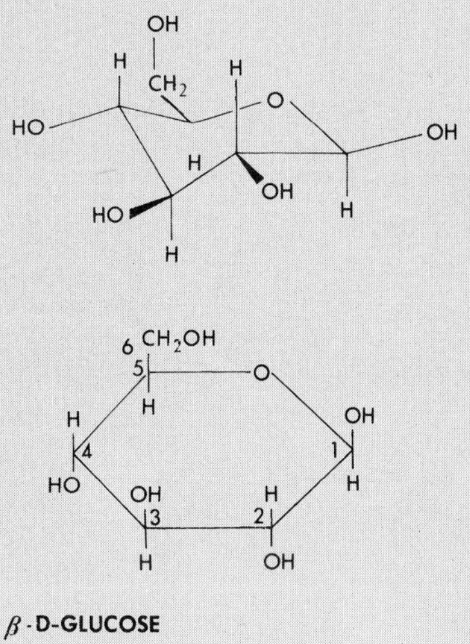

β-D-GLUCOSE

Diagram B

Carbohydrates composed of more than two subunits are called polysaccharides (many sugars joined together). Cellulose is generally a long, straight-chain molecule of glucose units joined together by β-glycosidic linkages. Starch, on the other hand, is a highly branched molecule of glucose joined together by α-glycosidic linkages. Where branches occur, they take off from a number-6 carbon on the straight chain.

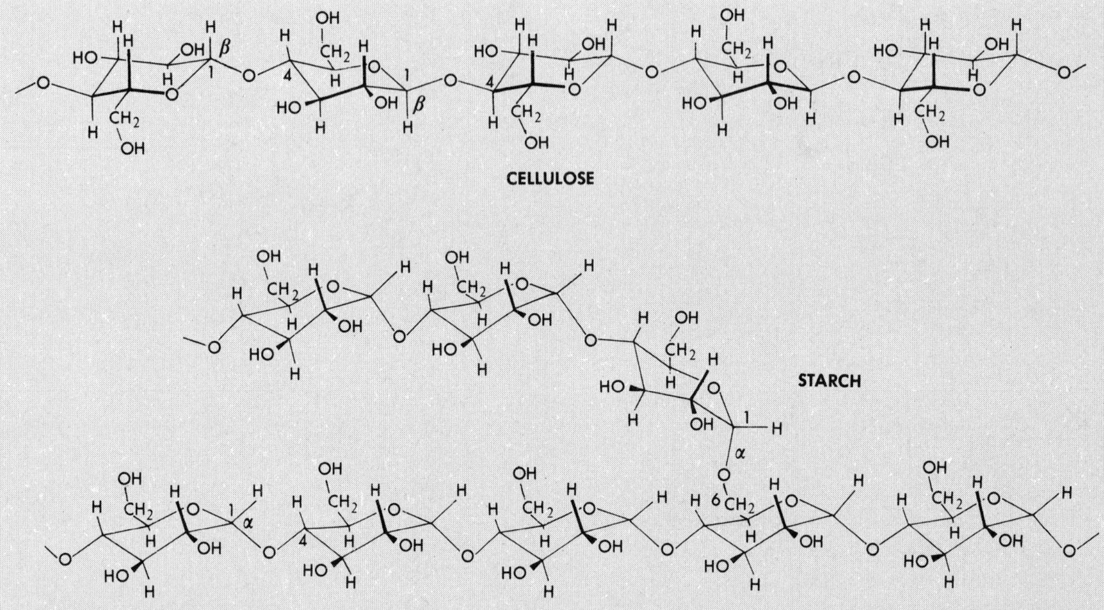

Diagram C

(From A. G. Loewy and P. Siekevitz, *Cell Structure and Function*, 2d ed. New York: Holt, Rinehart and Winston, 1969, p. 82).

Since every carbon atom in the carbon-chain portion of the molecule has each of its four bonds attached to another atom, the molecule is said to be *saturated;* there are no more bonds available for other atoms to be attached. Many fatty acids, however, have one or more places on their carbon chains where double bonds occur between carbon atoms. It would be possible, therefore, for more hydrogen atoms to be attached, and therefore such fatty acid molecules are said to be *unsaturated.* Some fatty acids are singly unsaturated, some doubly, while some have so many spaces open for additional hydrogen atoms they are commonly called polyunsaturates. The role of saturated and unsaturated fats in causing or preventing the build-up of fatty deposits in the arteries (atherosclerosis) is still being investigated, as is the role of another lipid, cholesterol, to be discussed shortly.

Stearic acid can be converted into oleic acid, $C_{17}H_{33}COOH$, by simply removing two hydrogen atoms:

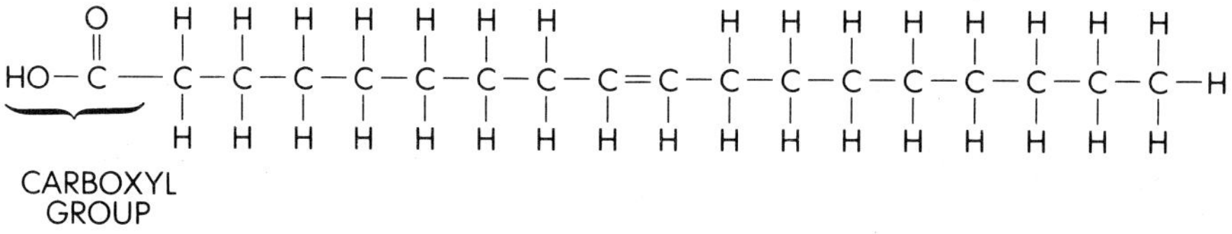

As can be seen, oleic acid is a singly unsaturated fatty acid.

Since fats are digested into glycerol and fatty acids by hydrolysis, the reverse process, dehydration synthesis, synthesizes fats from these two substances. Since the glycerol molecule has three alcohol groups available, there is room for three fatty acids to be attached. In the representation of this reaction below, the R represents the long carbon chain of the fatty acids:

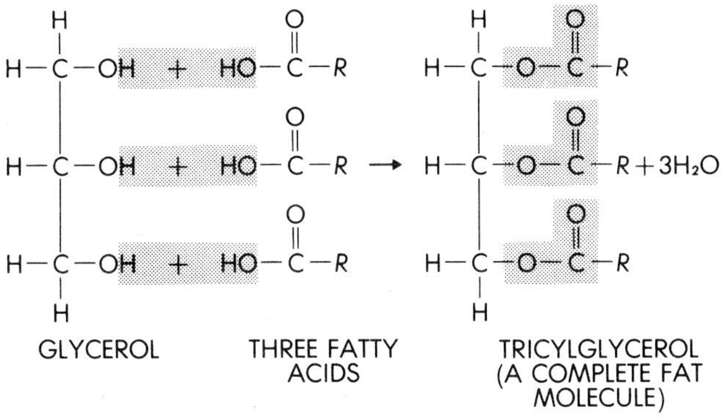

$$
\begin{array}{ccc}
\text{GLYCEROL} & \text{THREE FATTY ACIDS} & \text{TRICYLGLYCEROL (A COMPLETE FAT MOLECULE)}
\end{array}
$$

The resulting covalent bond between an alcohol and an acid is called an **ester bond.** The ester bond is broken by digestion, the reaction being catalyzed by the enzyme pancreatic lipase.

Lipids of particular importance to living systems because of their role in cell membranes are called phospholipids. The molecule below represents a phospholipid. Note that it can be formed from the fat molecule shown above by merely replacing one of the three fatty acids with a phosphate group that is often bound to a nitrogen-containing group:

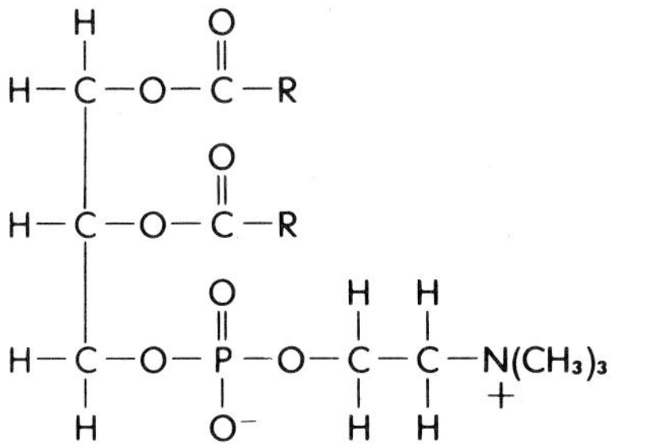

The phosphate group, by virtue of its tendency to release a hydrogen ion (proton), becomes negatively charged. The nitrogenous end, on the other hand, tends to attract the hydrogen ion, and thus becomes positively charged. As was pointed out in Section 2.5, this results in this end of the molecule exhibiting strong polarity and hence being soluble in water. The other end of the molecule, represented by the R, is still composed of the two remaining long carbon chains (that is,

$-CH_2-CH_2-CH_2-CH_2-CH_2-$, and so on), which are nonpolar and insoluble in water. It is this quality of solubility at one end and insolubility at the other, as well as their size, that appears to make phospholipid molecules well suited for the role they play in the structure and function of cell membranes (see Section 4.6).

One other important group of molecules often classified as lipids are the *steroids.* They are so classified, however, because their solubility characteristics are similar; structurally, their molecules are quite distinctively different, with no alcohol or fatty acid components. Steroids consist of four interlocking hexagonal (six-sided) or pentagonal (five-sided) rings of carbon atoms. They are distinguished from each other by the several side groups that may be bonded to these rings. A molecule of perhaps one of the best known steroids, cholesterol, is shown below:

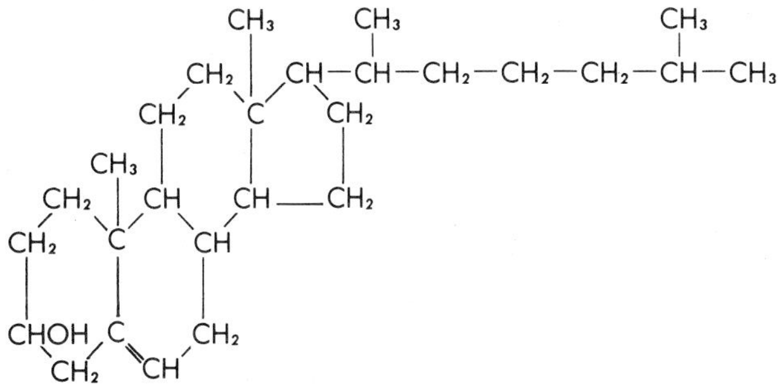

Steroids play a wide variety of roles in living organisms. Like phospholipids, they are found as structural elements of cells, especially the membranes. Some other steroids or steroid derivatives are hormones (for example, the sex hormone testosterone) or vitamins (such as the "sunshine" or "milk" vitamin, Vitamin D).

3.5 PROTEINS

Proteins play as varied a role as any substances in living systems. Protein molecules are found as structural elements in the contractile fibers of muscle, supporting tissue, hair, nails, skin, and many parts of living cells. As hemoglobin, protein transports oxygen to the body tissues; as antibodies, proteins help ward off infections. Certain other proteins (including ribosomal proteins and repressors) help regulate gene activity. Finally, as enzymes proteins catalyze the biochemical reactions that collectively comprise most of the activities we associate with life.

The fundamental building block of protein is the amino acid. Each amino acid possesses a nitrogen-containing amino group ($-NH_2$) and a carboxyl group ($-COOH$). The presence of the carboxyl group gives them the acidic properties from which their name is derived.

A two-dimensional representation of a generalized amino acid is shown below:

Note that both the amino group and the carboxyl group are attached to the same carbon atom. This carbon atom, called the alpha carbon, always serves as the attachment point for a characteristic group of atoms, symbolized by the R. *It is in the number and arrangement of atoms comprising the R group that one amino acid differs from another.*

From about twenty different amino acids (see Appendix 3 for structural formulas), all the proteins known to exist in plants and animals are constructed. (Very recently a twenty-first one was identified.*) Amino acids are to proteins as letters of the alphabet are to words. A group of amino acids can be joined together in a specific order to produce a given protein, just as a group of letters can be arranged to form a specific word. For this reason, amino acids are often referred to as the "alphabet" of proteins.

Amino acids are linked end-to-end to form long chains. *The variety found among proteins is the result of the types of amino acids composing each and the order or sequence in which these types are arranged.*

The comparison of amino acids to letters of the alphabet is helpful in representing how a small change in a protein molecule can completely change its chemical properties. Changing one letter in a word may cause that word to become meaningless. For example, "skunk" conveys one idea, while "skank" means nothing at all. Likewise, a change or substitution of one amino acid for another may make an entire protein molecule meaningless to the cell. The molecule is no longer able to carry out its function. For example, a certain portion of the human population possesses a condition known as sickle-cell anemia. Hemoglobin molecules (protein) in afflicted persons differ from normal hemoglobin in only one amino acid out of about 300. Such hemoglobin molecules will not combine as readily with oxygen. Persons with this type of hemoglobin generally have a shortened life span. Thus a change in one amino acid in one type of protein can have far-reaching effects on the entire organism.

Sometimes the removal or addition of one or two letters in a word may change the meaning without making the word senseless. For example, "live" can become "liver" or "olive." Similarly, one protein may be changed into another by the removal or addition of one or a few amino acids.

In one way, however, the comparison of proteins to words falls shorts. This is in the matter of length. Words are generally composed of relatively few letters. Even the name of the New Zealand village Taumatawhakatangihangakoauauotamateapokaiwhenuakitanatahu only approaches the complexity and length of a small to average protein. Proteins are macromolecules, often consisting of several hundred to over a thousand amino acids. Their molecular weights range from 6000 to 2,800,000. When one compares the formula for glucose, $C_6H_{12}O_6$, with that of the protein β-lactoglobulin, $C_{1864}H_{3012}O_{576}N_{468}S_{21}$, the size difference becomes perhaps more vivid.

The great size of proteins gives them added versatility in cell chemistry. They can take on a variety of shapes and sizes, each of which may serve very specialized functions. For this reason, biochemists speak of *chemical specificity* as being characteristic of many proteins. It is through the use of such chemically specific proteins as enzymes that living organisms are able to carry out their many different reactions so efficiently.

When amino acids unite to form proteins, the amino end of one amino acid molecule forms a chemical bond with the carboxyl end of the other:

Note that this process involves the loss of a water molecule between the two amino acids being joined, and is thus a dehydration synthesis. The result is the formation of a connecting link between the two. The linkage process continues until all the amino acids necessary to form the protein are joined together in the characteristic order for whatever particular protein is being synthesized.

The linkage between each two amino acids is called a **peptide bond.** A peptide bond involves partial bonding (indicated by the solid and dotted lines) between a carbon atom and nitrogen atom on one side, and the same carbon atom and oxygen on the other. This bonding so constrains the atoms in the vicinity that a flat plane (represented by the colored area) results:

---

*Called citric amino acid.

The flatness of this plane is the same regardless of what two amino acids are joined together in the peptide bond. As we shall see later, this has considerable importance for our understanding of protein synthesis and breakdown.

The uniting of several amino acids by peptide bonds results in the formation of *polypeptide chains.* In large proteins, several hundred amino acids may be involved in a polypeptide chain, while in such small molecules as the neuropeptides, believed to be involved in some as yet undetermined way with the transmission of nerve signals, the polypeptide chain may be only two or three amino acids long. For any particular protein, however, the number and sequence of amino acids is the same. In some proteins, the polypeptide chains line up end-to-end to form one long molecule. In others, the chains line up side-to-side, with various sorts of interconnecting chemical bonds between them.

The molecular structure of a peptide consisting of four amino acids is shown below:

Note that the polypeptide chain bears numerous positive and negative charges along its backbone. At the left and right ends of the chain (called the N-terminal and C-terminal, respectively) the charges are a result of the interaction of those groups (the amino and carboxyl, respectively) with water. The NH_2 terminal tends to pick up a hydrogen ion from hydronium (H_3O^+) to produce an overall net positive charge on the amino group. The carboxyl terminal tends to lose a hydrogen ion to water, creating H_3O^+ and an overall net negative charge at the C-terminal. The peptide bond region also shows charged groups. Because of the nature of the double bond between the carbon and oxygen atoms and the carbon and nitrogen atoms, the oxygen gets more of the shared electron pairs, while the hydrogen gets less. Consequently the $C\!=\!O$ group is negative, and the $C\!=\!N\!=\!H$ group is positive.

While the sugars and fats of one group of organisms are like those of another, the same is not true of proteins. Indeed, individual organisms often have proteins significantly different from those of even a closely related individual of the same species. The heart of this uniqueness lies in the structure of protein molecules. Since protein molecules vary widely in complexity from simple polypeptide chains to highly complicated interconnections between these chains, it has been found convenient to view them as consisting of up to four levels of structure.

The *primary structure* of a protein molecule is the precise linear sequence in which the various kinds of amino acids characteristic of the protein are found. (A discussion of how the primary structure of a protein is determined appears in Supplement 3.2.) The conformation of the folding of the linear chain of amino acids in space and how that conformation is attained determine the other levels of protein structure. (Without proper conformation, a protein molecule cannot carry out its function.) The *secondary structure* of a protein, for example, involves a conformation that results from the formation of hydrogen bonds between atoms involved in the peptide bonds (see Fig. 3.1). *Tertiary structure* involves folding within a single polypeptide chain caused by chemical bonds forming between amino acid side chains; these bonds may be of any type (for example, hydrogen, ionic, covalent, and the like). (See Diagram C, Supplement 3.2.) *Quaternary structure* may also involve various sorts of bonding between amino acid side chains, but here the amino acids involved are located in *different* polypeptide chains. The several polypeptide chains in a protein molecule may be held together in a particular conformation in this way. An alpha helix is produced by a spiral twisting of the amino acid chain. To visualize the geometry of a helix, think of a ribbon as a straight-chain polypeptide. A helical structure can be formed by twisting the ribbon several times around a pencil. The spiral that remains after the pencil is removed has the general shape of an alpha helix. The alpha helix is held in position by the formation

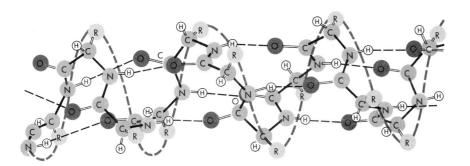

Fig. 3.1
A diagram of a peptide backbone wound into an alpha-helix configuration. The heavy dashed curve traces the helical structure. The broken lines going from the C—O to the N—H groups represent the hydrogen bonds that hold the chain in the alpha-helix form. (After R. B. Corey and Linus Pauling, *Proc. Intern. Wood Textile Research Conf.* B. 249, 1955.)

SUPPLEMENT 3.2
DETERMINATION OF PROTEIN STRUCTURE

One of the most difficult and challenging problems in twentieth-century biology has been the determination of protein structure. In the early years of the century, it was not clear that proteins even had definite or precise structures. At least one prominent school of biochemists claimed that proteins were simply indeterminate aggregates of smaller molecules, themselves of indefinite composition. Gradually, however, it became clear that proteins *were* substances with definite and regular composition. Although several workers in the nineteenth and early twentieth centuries had isolated specific proteins and determined something about the elements composing them, the problem confronting protein chemists was how to determine the exact chemical composition. Because of the growing interest among biochemists in that class of catalytic proteins known as enzymes, much of the work on the anatomy of protein molecules was carried out on specific enzymes. Also prominent in this work were studies on the hormone insulin, one of the smallest proteins of biological significance.

An early problem confronting protein anatomists was obtaining pure samples of a given protein with which to work. In 1926 James B. Sumner succeeded in crystallizing the enzyme urease, which breaks down urea, from the jack bean. In 1930 John Northrop crystallized the enzyme pepsin from gastric juice; pepsin catalyzes the breakdown of proteins. These workers established techniques for crystallizing other proteins, and their work made possible the obtaining of pure samples on which chemical analyses could be carried out.

The Chemical Determination of Amino Acid Sequence

Once a protein is crystallized, it can be studied by breaking down its primary structure a bit at a time and analyzing the products. Breakdown can be accomplished by several means. (1) In acid hydrolysis, the protein is subjected to strong acids that cleave the peptide bonds. Since acids act indiscriminately, affecting all peptide bonds, the result is a mixture of free amino acids. (2) In enzymatic hydrolysis, specific enzymes are used to break the long polypeptides down into fragments. The enzymes used break only certain peptide bonds; trypsin, for example, hydrolyzes the bond formed at the C-terminal of lysine or arginine units. Thus, a sample of protein treated with trypsin will contain fragments terminated at one end by arginine or lysine. Another sample of the same protein, treated with the enzyme carboxypeptidase (which cleaves off only the final C-terminal amino acid, especially when that amino acid contains an aromatic ring

structure), yields a different set of fragments. By using all these methods, a single protein may be broken down into a variety of smaller units. The problem then lies in determining by analysis something about the molecular structure of these fragments.

This problem was attacked by the British protein chemist Frederick Sanger in the period during and following World War II. Sanger and his co-workers chose to study the hormone insulin because its relatively small size (molecular weight 5733; most proteins are at least double that) made the task appear at least feasible. The first problem was separating out the fragments containing several amino acids from a set of protein fragments produced by acid or enzyme hydrolysis. Classical methods of chemical separation (distilling and precipitating) were extremely laborious and often altered the chemical composition of the fragments. Sanger and his collaborators used the newly perfected technique known as *chromatography* to effect the separation of fragments easily. Chromatography is based on the principle that different molecules will adhere to a solid surface (such as paper or a column of starch grains) with different degrees of intensity, based on differences in molecular structure. Thus if a solution of several different kinds of molecules is allowed to seep up through the paper or down through a packed column, each molecular type will pass from the solution to adhere to the solid material at a different rate. If the flow of liquid is kept constant, each type of molecule will congregate more or less at the same point on the solid surface. The molecules collecting in one area will all be alike and can be extracted separately from the others and chemically analyzed.

Another separation technique, developed since Sanger and his group did their work, is **electrophoresis**, which is based on the principle that different molecules (especially individual amino acids or peptide fragments) bear different net electric charges at a specified pH. A mixture of protein fragments can be placed at one end of flat gel plate and an electric current (DC, or direct current) applied to the plate so that one end of the gel is positively charged, the other negatively. Various fragments of the protein mixture begin to migrate in the electric field created at the gel surface. A protein with an overall net positive charge will migrate toward the negative pole; one with an overall net negative charge will migrate toward the positive pole. Each fragment will migrate only so far, depending on its overall net charge. For example, a fragment with a net positive charge of +2 will migrate farther toward the positive pole than one with a net charge of +1, and so on. Electrophoresis has proved to be an extremely valuable technique for the study of protein structure.

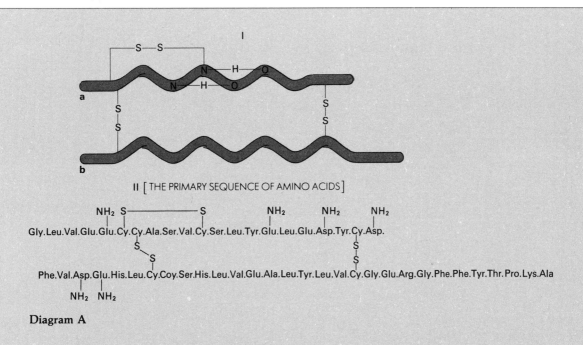

II [THE PRIMARY SEQUENCE OF AMINO ACIDS]

Gly.Leu.Val.Glu.Glu.Cy.Cy.Ala.Ser.Val.Cy.Ser.Leu.Tyr.Glu.Leu.Glu.Asp.Tyr.Cy.Asp.

Phe.Val.Asp.Glu.His.Leu.Cy.Coy.Ser.His.Leu.Val.Glu.Ala.Leu.Tyr.Leu.Val.Cy.Gly.Glu.Arg.Gly.Phe.Phe.Tyr.Thr.Pro.Lys.Ala

Diagram A

Once the fragments of a protein are separated and their chemical nature determined, the next step is to try to determine how the fragments fit together in the intact protein molecule. Sanger has likened the problem to trying to infer from pieces in a junk yard what an intact automobile looks like if one has never seen an automobile. As Sanger's analogy points out, in a junk yard one might find a piece of axle joined to a wheel and a piece of axle joined to steering column. One could thus conclude by deduction that steering column, axle, and wheel were all joined together in the intact car. Similarly, persistent analyses of all the various kinds of fragments obtained by hydrolyzing insulin with different methods begin to construct a picture of the exact amino acid sequence. The complete sequence was eventually determined, showing insulin to consist of two chains joined together by disulfide bonds, as shown in Diagram A.

The work of Sanger and his group was tedious; it could never have been accomplished with the same precision without chromatography. The history of science is full of examples of theoretical advances being made possible by the development of new techniques.

The Physical Study of Molecular Three-Dimensional Structure

The work of Sanger's group established for the first time the primary structure (essentially the amino acid sequence) of a natural protein. This work did not, however, reveal the overall three-dimensional structure of insulin: it did not indicate the solid geometry of the molecule. At about the same time that Sanger was beginning work on the chemical analysis of insulin, another group of scientists in Britain, led by an Austrian émigré Max Perutz and later by John Kendrew at Cambridge, started to study protein crystals by the technique known as *X-ray diffraction.* Perutz and Kendrew chose to work on hemoglobin and myoglobin, two oxygen-carrying proteins. They used crystals because the molecules in a crystal are, by definition, lined up in exactly the same planes, in an orderly array. A beam of X-rays is passed through the crystal, the X-rays being diffracted by the molecules making up the crystal. Since the molecules in the crystal are lined up precisely, all deflections from the same part of each molecule will hit the same point on a photographic plate placed on the opposite side of the crystal. The pattern is thus multiplied, but in an orderly way; the crystalline structure makes it possible to study the additive effects of many molecules as if they were one large molecule. Moreover, the angle of deflection is determined by the shape of the portion of the molecule an X-ray beam strikes. It is rather like casting a shadow (though the X-ray diffraction pattern does *not* bear the same simple correspondence to its protein that a shadow does to its object).

From the X-ray diffraction results, an electron density map is constructed; this map measures where certain densities of electrons would have to be in the protein molecule to produce a certain kind of deflection of the X-ray beam. An electron density map for a portion of a myoglobin molecule

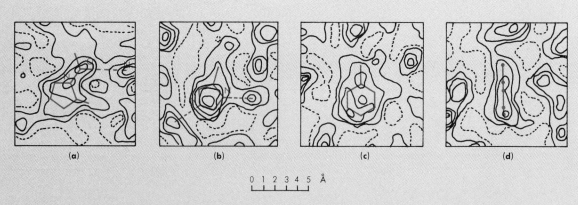

Diagram B
(Courtesy of John Kendrew, *Nature* 190, 663, 1961.)

is shown in Diagram B. The four diagrams show four different regions: (a) a proline side-chain; (b) a histidine side-chain; (c) and (d) phenylalanine side-chains. X-rays have very short wavelengths; consequently they can penetrate into small spaces between atoms that visible light, which has far longer wavelengths, would never be able to get through. An electron density map built up from X-ray diffraction data is able to distinguish between the electron clouds of adjacent but still separate atoms in a molecule.

By the detailed study and analysis of electron density maps, the three-dimensional outline of a polypeptide chain can be discerned. In this way the precise position of all atoms and groups of atoms making up a complete protein can be determined. A schematic diagram of the myoglobin molecule's three-dimensional structure is shown in Diagram C. Note that some parts of the molecule are in an alpha-helix configuration, while others are more or less straight or bent chains. The active site is indicated by the disc and bridge region in the center of the diagram. This portion of the molecule binds oxygen for use by muscle cells (myoglobin is the muscle cell's counterpart of hemoglobin in the blood).

The analysis of X-ray diffraction data to produce electron density maps requires considerable mathematical calculation. When Perutz and Kendrew first began their work, the calculations had to be done with desk calculators. This was laborious and made the work extremely slow. With the development of computers after World War II, the process was greatly speeded up. By the early 1960s, it was possible to construct the first three-dimensional model of a protein. More recently, techniques have been developed using computer graphics, in which the signals generated by diffracted electrons are converted into three-dimensional molecular models and projected onto a screen. This system even allows for color coding of the various parts of the molecule, as well as rotation, tilting, and other spatial movements, so the molecule can be "seen" from various angles. It is hoped that the present resolution limitation of 6 angstrom units will be

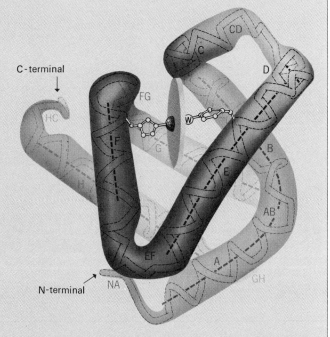

Diagram C
(Reproduced with permission, from *The Structure and Action of Proteins* by Richard Dickerson and Irving Geis, W. A. Benjamin, Menlo Park, California © 1969 Richard Dickerson and Irving Geis.)

improved to around 3.5. Thus far, results of studies carried out on a few proteins by a team of scientists under the direction of biophysicist Nguyen Huu Xuong of the University of California, San Diego, and using a much improved system designed by Xuong, have nicely verified conclusions drawn from the earlier, less precise techniques. Xuong's technique holds particular promise for further refining our knowledge concerning enzyme-catalyzed reactions, especially the precise nature of the enzyme-substrate complex (see Supplement 3.3).

of hydrogen bonds between amino acids on one part of the chain with those on another part of the same chain. The hydrogen bonds are formed between the carboxyl group of one amino acid and an amino group nearby.

Among the proteins exhibiting a helical structure is the one most prevalent in body tissues, *collagen.* Collagen forms the major fibrous component of the skin, tendons, cartilage, ligaments, and teeth. In addition to this role in mature tissues, collagen plays a vital role in developing tissue, and the precise nature of this role is the subject of much research in developmental biology.

Collagen's distinctive feature is that it forms insoluble fibers that have a high tensile strength, yet it has a structural base that can be modified to meet the special needs of a wide variety of tissues. In its primary structure, collagen has an unusually high proportion of the amino acids, glycine (about one-third) and proline, as well as containing two amino acids, hydroxyproline and hydroxylysine, present in very few other proteins. There appears to be a remarkable uniformity in the sequence of amino acids, with nearly every third one (called a *residue*) being glycine. The long sequence of amino acids is found in three polypeptide chains, two identical and one only slightly different. Study of the secondary structure of these chains reveals them to be helical and wound around each other. Projecting out from each of the three helices are the *R* groups.

The *keratins* are fibrous proteins that compose such specialized structural elements of the skin as nails and hooves. In these forms, they may vary from flexible to brittle. Keratins also form the major portion of hair which, of course, is often soft and flexible. The secondary structure of such keratin is simpler than that of collagen in that it is composed of two helices not wound around each other, as in collagen. Yet keratin molecules are also found in a beta structure, in which the polypeptide bonds are located side-by-side in a sort of pleated sheet, with the chains being held together by many hydrogen bonds. While this structure imparts considerable lateral strength, the polypeptide chains are extended nearly to the breaking point and little more stretching is possible (as anyone who has tried to stretch silk, an important keratin, will testify). The scales, claws, and beaks of reptiles and birds are other examples of material containing keratins.

An important factor in maintaining tertiary structure in some proteins is the presence of the amino acid cysteine. Cysteine is the only amino acid containing sulfur, and two cysteine molecules can form a covalent, **disulfide bond** between the two sulfur atoms. The result is a new molecule, cystine:

The cystine molecule(s) and the resulting disulfide bond(s) are formed after the primary sequence structure of the protein is laid down, that is, after the protein has been synthesized.

Often a single protein molecule is composed of several semi-independent polypeptide chains, each with its own particular configuration; it is the specific manner in which these subunits fit together to form the complete molecule that is referred to as its quaternary structure. Quaternary structure is usually maintained by the combined forces of several different weak bonds and interactions. Perhaps the best known protein exhibiting quaternary structure is the oxygen-transporting blood protein hemoglobin. As well as being composed of more than one polypeptide chain, hemoglobin, myoglobin, and many other proteins contain nonamino organic or inorganic components called prosthetic groups. The prosthetic group of such conjugated proteins as hemoglobin and myoglobin is a heme group, a ring-shaped formation of molecules that, in the case of myoglobin and hemoglobin, contains one iron atom in the center. Other prosthetic groups may be organic (for example, carbohydrates or lipids) or inorganic metallic ions such as zinc (Zn^{+2}) or magnesium (Mg^{+2}). The prosthetic group of conjugated proteins plays an important role in the function of the protein. For example, the presence of their heme groups accounts for the ability of hemoglobin and myoglobin to transport oxygen.

Proteins play the most varied roles of all macromolecules in living systems. Normally they serve as enzymes for all biochemical reactions; they are important structural components in cell membranes, skeletal structures, hair, nails, muscle, and connective tissue such as tendons and ligaments; under extreme conditions they can be used as an energy source. Proteins are also found as hormones, receptors, and antibodies, as well as molecules involved in genetic regulation.

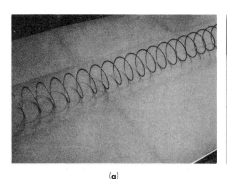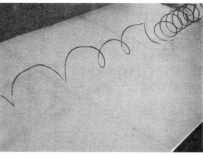

(a)   (b)

Fig. 3.2
(a) A coiled spring toy known as a Slinky. This represents the alpha-helix structure of certain proteins. When the Slinky is pulled out of shape, as shown in photograph (b), it cannot recoil into the original helical structure. It has been irreversibly altered. This is analogous to denaturing a protein by such physical means as heating or by such chemical means as placing the protein in a concentrated solution of urea.

Denaturation

As might be expected, factors that disrupt the weak bonds and interactions responsible for maintaining the secondary, tertiary, and quaternary structure of a protein may change that structure and thereby disrupt biological function. When such changes occur, the proteins are said to be denatured (see Fig. 3.2). Among the factors that can cause denaturation are extremes of temperature and pH. In some cases, denaturation is irreversible; the egg white protein albumin, after being hard boiled, does not revert to its original form upon cooling—you cannot unboil an egg! On the other hand, some proteins that have been denatured can, under favorable conditions, regain their normal structure and function. The entropy of such molecules is actually lower in the folded than unfolded state, and thus the process tends to take place of its own accord. In fact, an input of energy into the system is necessary to *prevent* this folding. The energy holding the tertiary and/or quaternary structure is directly attributable to the bonding and interaction forces already discussed. These forces, in turn, are the result of the kinds and sequences of the amino acids making up the primary structure of the protein. It is in this primary structure, therefore, that the secret to the other aspects of protein structure must lie.

3.6
ENZYMES

It is now necessary to pay special attention to one of the most important functions of proteins in living systems—that of acting as organic catalytic agents called enzymes. Without these protein catalysts, life as we know it could not exist.

As mentioned in Section 2.10, catalysts are molecules that speed up a chemical reaction without themselves being changed chemically. After participation in one reaction, a catalyst can go on to participate in a second, a third, and so on. As catalysts, enzymes participate in almost all the chemical reactions that keep or-

ganisms alive. Since catalysts are not destroyed during chemical reactions, a few enzyme molecules go a long way. This means that the organism does not have to expend a great deal of energy in order to constantly resynthesize enzymes at a rate proportional to the rate of the reactions they catalyze. The substance upon which an enzyme works is called its **substrate.**

Effect of Temperature

The rate of enzyme-catalyzed reactions is greatly affected by temperature. If the initial velocity of a specific enzyme-catalyzed reaction is measured at a number of different temperatures (with enzyme and substrate concentrations held constant), a curve like that shown in Fig. 3.3 is obtained. Note that at low temperatures the rate of reaction is quite slow and that the rate increases with increasing temperature, up to about 36°C. Beyond 36°C, however, the rate begins to slow down again even though the temperature is raised. From our knowledge of chemical reactions, we can explain the first half of this curve: the higher the temperature, the more rapidly the reacting molecules (in this case, substrate and en-

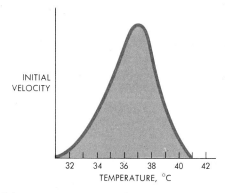

Fig. 3.3
The effect of temperature on the activity of one enzyme.

Enzymes can catalyze the reaction for a large number of substrate molecules in a very short period of time—in some cases up to 10,000,000 molecules of substrate per enzyme molecule per second.

zyme molecules) move about, and the greater the fraction of molecules that possess minimum energy of activation.

Should not this principle apply to temperatures above 36°C as well? In non-enzyme-catalyzed reactions, initial velocity does indeed increase as the temperature is raised beyond 36°C. Why, then, are enzyme-catalyzed reactions so sensitive? At temperatures greater than 40°C, the enzyme molecule denatures and, as will be seen shortly, the geometric configuration of the molecule is essential to its catalytic activity.

The initial velocity/temperature curve shown in Fig. 3.3 varies from one type of enzyme to another. Every specific enzyme has its so-called *optimum temperature*, at which the enzyme achieves its maximum rate (initial velocity). For the particular reaction shown in Fig. 3.3, 38°C is the optimum temperature; for another enzyme-catalyzed reaction, 25°C or 40°C might be the optimum.

The term optimum temperature is somewhat misleading and should be viewed not only in a chemical but also in a biological context. Seldom in nature do biochemical reactions operate at their maximum rate. While a particular enzyme may work at a maximum rate at 38°C, we cannot assume that this is the temperature at which the enzyme usually functions. The enzyme might normally exist inside a mammal that maintains a fairly constant internal temperature of 37°C, or it might exist in a bacterium that lives in the soil, where the temperature varies considerably. In general the optimum temperature of an enzyme-catalyzed reaction, as measured in a chemical sense, is often close to the average temperature at which the enzyme functions in nature. But we must not assume that this is always true. And, more important, we must not assume that whatever temperature (or other condition) may be optimal when the system is measured in a test tube is the "optimum" at which the system operates in terms of the survival of organisms. The fastest rate is not always the best from the standpoint of an organism's operating efficiency.

Poisons

Enzymes can be poisoned. Certain compounds, such as bichloride of mercury or hydrogen cyanide, are deadly poisons to all living organisms. They exert their effect by inactivating one or many enzymes. Hydrogen cyanide blocks one of the enzymes involved in the chemistry of respiration. The way in which this is believed to occur will be discussed shortly.

Specificity

Enzymes are specific in their action. This is one of the most distinctive characteristics of enzymes. They will often catalyze only one particular reaction. For example, the enzyme sucrase will catalyze only the breakdown of sucrose to glucose and fructose. It will not catalyze the reactions that involve lactose or maltose; lactase and maltase, respectively, must be used as the catalytic agent for these two sugars.* With other enzymes, specificity of action is not quite so obvious. Trypsin, for example, is a *proteolytic* (protein-splitting) enzyme that catalyzes proteolysis involving many different proteins.

Why should some enzymes work on only one compound while others will work on several? Enzymes act upon a specific chemical linkage group. In the case of trypsin, only those peptide linkages of protein molecules formed with the carboxyl group of the amino acids lysine or arginine are acted on by trypsin:

Since peptide linkages involving the carboxyl group of lysine and arginine are found in many proteins, it is not surprising that trypsin will act on a broad range of protein molecules.

*The *-ase* ending indicates that the compound is an enzyme. Other enzymes, such as trypsin, end with *-in*. This signifies that they, like all enzymes, are proteins. Enzymes ending in *-in* were discovered and named before an international ruling was made in favor of the *-ase* ending. A few changes have been made. For example, the mouth enzyme ptyalin is now called salivary amylase. Enzymes are also named after the compounds they attack. Thus peptides are attacked by peptidases; peroxides by peroxidases; lipids by lipases; ester linkages by esterases; hydrogen atoms are removed by dehydrogenases; and so on.

3.7
THEORIES OF ENZYME ACTIVITY

We now come to a crucial question in the study of enzymes: By what mechanism do they operate?

The most important hypothetical model proposes that enzymes have certain surface configurations produced by the three-dimensional folding of their polypeptide chains. On this surface, there is an area to which the substrate molecule is fitted. This area is called the **active site** of the enzyme. It is thought that when the substrate molecule becomes attached to the enzyme at this site, the internal energy state of the substrate molecule is changed, thereby bringing about the reaction (Fig. 3.4).

As long ago as the 1890s biochemist Emil Fischer was proposing that the high specificity of certain enzymes in relation to their substrates was analogous to the complementary relationship of a lock and its key. The results of X-ray diffraction studies of the enzyme lysozyme lend support to Fischer's hypothesis. The structure of the complex that results when this enzyme reacts with a molecule very similar in molecular configuration to its normal substrate shows that the substitute substrate molecule fits very snugly into a groove or cleft in the lysozyme molecule's surface, where the

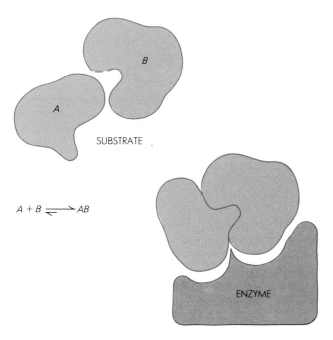

$$A + B \rightleftharpoons AB$$

Fig. 3.4
It is believed that enzymes may serve as orientation surfaces on which substrate molecules are able to bring their reactive groups close together, thereby increasing the efficiency of their reactivity.

enzyme's catalytic groups are held in just the right position to break the substrate's bonds.

There is some evidence however that the lock-and-key model of enzyme action is not completely satisfactory. It does not always yield accurate predictions with all enzymes. For example, the enzyme specific for the amino acid isoleucine must often "choose" between its proper substrate (isoleucine) and valine. The only difference between these two molecules is one methylene ($-CH_2-$) group:

ISOLEUCINE   VALINE

The chemist and Nobel laureate Linus Pauling estimated that if an enzyme discriminated between these two highly similar molecules purely on their ability to form a complex with it (in the manner described by the lock-and-key model), then from these physicochemical grounds alone the enzyme should make a mistake (bind to valine instead of isoleucine) about one time in twenty. However, it has been shown experimentally that mistakes are made at a frequency of less than one in 3000.

This discrepancy can be accommodated by hypothesizing that the correct substrate molecules, besides themselves being changed, also induce changes in groups located in the region of the enzyme's active site. This **induced-fit hypothesis** suggests that enzymatic specificity is only partly due to complementarity of structure between enzyme and substrate molecules, and that the ability of substrates to induce the structural changes in the enzyme necessary for catalysis must also be taken into account (see Fig. 3.5).

Recent X-ray diffraction work on the proteolytic enzyme carboxypeptidase A provides strong evidence in favor of the induced-fit hypothesis. It has been shown that the binding of this enzyme to its substrate causes movement of the side-chain amino acid 248 (tyrosine, known to be involved in the actual catalysis) some eight angstrom units toward the substrate so that its hydroxyl ($-OH$) group is near the substrate peptide bond to be split. Furthermore, the positively charged group of the arginine (145) moves two angstrom units toward the substrate carboxyl group, where it binds it. It has further been shown that an inhibition of carboxypeptidase A binds to the enzyme as does its normal substrate, but does not bring about a change in the tyrosine 248 side-chain. Further evidence concerning the induced-fit hypothesis and the exact mechanism by which another enzyme catalyzes the breakdown of its substrate is described in detail in Supplement 3.3.

These models for enzyme action help us interpret the characteristics of enzyme action described in the previous section. We know that most proteins are heat sensitive. High temperatures denature the proteins by breaking hydrogen bonds or other types of bonds that hold the molecule in its specific three-dimensional shape. Thus, when an enzyme system is exposed to increasing temperatures, more and more of the enzyme molecules become denatured. Although more molecules are colliding with each other, the number of effective collisions between enzyme and substrate declines.

A similar situation exists for the effects of different pH ranges on the action of enzymes. Changing pH affects hydrogen bonds of proteins; at high or low pH values many hydrogen bonds are broken, and thus the enzyme molecules' three-dimensional structure and charge distribution are changed. The effect of both pH and heat on enzymes emphasizes the importance of a specific molecular shape to the biochemical activity of enzymes.

Inhibition of Enzyme Activity

Especially interesting in light of the lock-and-key analogy is the effect of **inhibitors** on enzyme activity. For example, one step in the breakdown of sugars involves the conversion of a four-carbon molecule, succinic acid, to another four-carbon molecule, fumaric acid, with the removal of two electrons and two protons (see Fig.

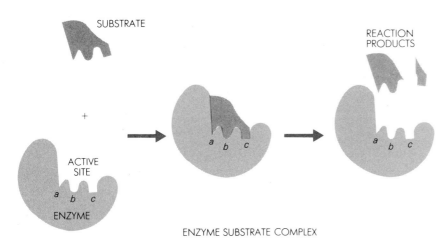

(a)

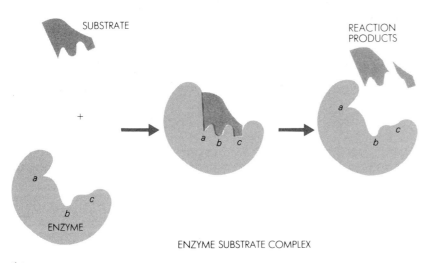

(b)

Fig. 3.5
A comparison of the lock-and-key model (*a*) of the interaction of substrates and enzymes with the induced-fit model (*b*). Note that the lock-and-key model envisons the active site as being complementary in shape to that of the substrate molecule. The induced-fit model envisions that the enzyme molecule changes shape during the binding interaction and becomes complementary in shape to the substrate molecule only *after* the substrate is bound.

SUPPLEMENT 3.3
HOW AN ENZYME WORKS

The enzyme lysozyme is secreted by many tissues in the human body, and it is part of the body's defense mechanism against bacterial infection. Back in 1922 Alexander Fleming, one of the discoverers of penicillin, noted that mucus containing lysozyme dissolved bacteria—that was the secret of how the enzyme helped ward off bacterial infection. Later work established that lysozyme is specific for the polysaccharide chains making up the bacteria's protective coating. Lysozyme catalyzes the reaction hydrolyzing this carbohydrate, dissolving the coat and leaving the naked bacterial cell exposed to other enzymes and destructive agents in the body. The polysaccharide is the substrate on which lysozyme acts.

The polysaccharide component of the cell wall in many bacteria consists of alternating rings of two types of sugars:

N-acetylglucosamine (NAG for short), and N-acetylmuramic acid (NAM), shown in Diagram A. Three of the rings A, C, and E are NAG residues, while the other three B, D, and F are NAM. Only six residues are shown, although the entire polysaccharide chain consists of many hundreds of such alternating units. Out of every six residues, starting with NAM as shown above, lysozyme cleaves the molecule between rings D and E (dashed line). Thus a single long polysaccharide molecule is broken down into a number of six-unit pieces; when this happens all over the bacterial surface, the bacterial polysaccharide coat falls apart.

Investigators using X-ray crystallography and other techniques have worked out the three-dimensional structure of lysozyme. It consists of a single long polypeptide chain folded back on itself in a number of places. Approximately

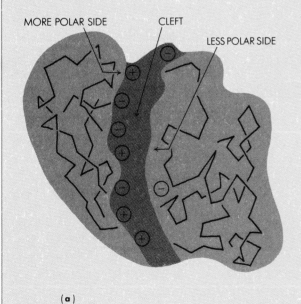

Diagram A

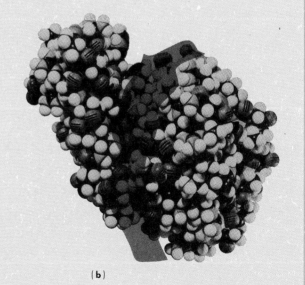

(a)   (b)

Diagram B
(a) Schematic view of a lysozyme molecule, (b) space-filling molecular model of lysozyme. Compare with the schematic view at left. The color denotes the cleft. (Courtesy Dr. J. A. Rupley, The University of Arizona.)

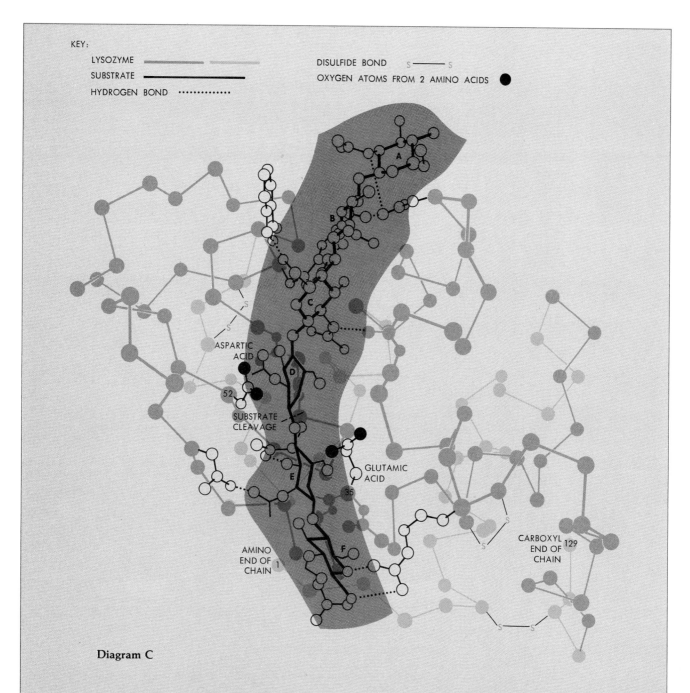

KEY:

LYSOZYME

SUBSTRATE

HYDROGEN BOND

DISULFIDE BOND   S———S

OXYGEN ATOMS FROM 2 AMINO ACIDS ●

A

B

C

ASPARTIC
ACID

D

52

SUBSTRATE
CLEAVAGE

E

GLUTAMIC
ACID

35

AMINO
END OF
CHAIN   1

F

CARBOXYL
END OF   129
CHAIN

Diagram C

75 percent of the chain exists in the alpha-helix configuration. The chain is held in its folded position by relatively weak interactions, such as electrostatic attractions, and by stronger covalent forces, such as disulfide bonds. The folding occurs in such a way that side groups of nonpolar amino acids (those having no net electric charge) point toward the center of the molecule, while side groups of polar amino acids (those possessing net positive or negative charge) point toward the outside. This makes lysozyme particularly soluble, since its polar groups are easily surrounded by water. The folding creates a cleft along one side of the mole-cule, a surface indentation that makes up the active site. Because of the sequence of amino acids and their three-dimensional arrangement produced by the folding, one side of the cleft is outlined by a series of polar amino acids, while the other side has only a few polar amino acids, as shown schematically in Diagram B.

Six rings of the polysaccharide (starting with an NAG ring, like A in Diagram A) fit into this cleft (active site) so that A is near the top and F at the very bottom. It is hypothesized that these six rings are held in the active site by hydrogen bonds, as shown in Diagram C by the broken line.

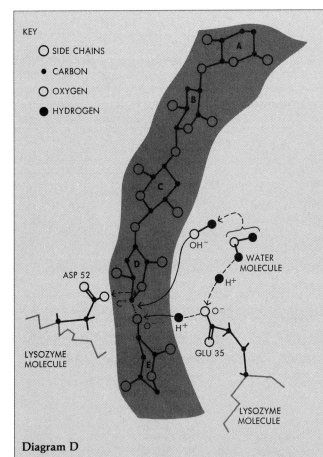

KEY
○ SIDE CHAINS
• CARBON
○ OXYGEN
● HYDROGEN

OH⁻

WATER MOLECULE

H⁺

ASP 52

C⁺

H⁺

O⁻

LYSOZYME MOLECULE

GLU 35

LYSOZYME MOLECULE

Diagram D

stretching between enzyme and substrate in two places. Due to the interaction of several nonpolar side groups on the enzyme and the B-ring of the substrate, the whole left side of the cleft closes a little, bringing the oxygen atoms (black

spheres) of amino acids #53 (aspartic acid) and #35 (glutamic acid) much closer to ring D. As a result, ring D of the polysaccharide chain is distorted from the normal chair configuration. This distortion is crucial since it apparently helps break the glycosidic bond joining rings D and E on the substrate.

The details of exactly how the glycosidic bond in the polysaccharide is broken have been worked out by David Phillips, Peter Dunhill, Louise Johnson, and others at Oxford University. The process is summarized schematically in Diagram D. The basic feature of the process derives from the distortion of both the enzyme and substrate once the latter has fit into the active site. Distortion of the protein brings amino acid #35 (glutamic acid) into proximity to the glycosidic bond joining rings D and E. Since glutamic acid is an "acid," it has an extra hydrogen ion to donate to anything that will accept it. When the amino acid is brought close to the glycosidic bond, it can transfer the H^+ to the oxygen of the glycosidic bond. The bond is broken (other changes occur to complete the breakage), and the two segments of the polysaccharide fall apart. The glutamic acid regains its hydrogen from hydronium (H_3O^+) ions in the water. Water from the medium is therefore necessary for this reaction to occur. Since a molecule of water from the medium is actually consumed, the splitting of the glycosidic bond is an example of hydrolysis.

When the two parts of the substrate are split, the cleft on the enzyme opens up again and the substrate components fall away. The cleft (active site) is now free to receive another substrate molecule and catalyze another hydrolysis. This example shows that enzyme and substrate do not fit together like a rigid lock-and-key, but actually change their shapes when interacting in such a way as to facilitate a specific chemical change.

3.6a). This reaction is catalyzed by the enzyme succinic dehydrogenase, which is highly specific for its substrate, succinic acid. If the reaction is carried out in a test tube with just enzyme and substrate, a particular initial velocity can be observed. If, along with succinic acid, malonic acid is added to the test tube, the rate of formation of fumarate is greatly reduced. Malonic acid is a three-carbon molecule whose overall molecular shape is very similar to that of succinic acid (see Fig. 3.6b). Apparently the malonic acid molecule can fit into the active site of succinic dehydrogenase and "fool" the enzyme. However, because malonic acid is slightly different from succinic, the enzyme cannot catalyze its conversion into fumaric. Hence, whenever a malonic acid molecule gets into the active site of succinic dehydrogenase, no reaction occurs. Thus, because its molecular conformation is similar to that of the normal substrate, succinic acid, malonic acid *inhibits* the enzyme

system. Malonic acid molecules can fall out of the active site, however. If many succinic acid molecules are around, it is possible for them to enter the site once it is free. Malonic acid is thus a **competitive inhibitor**: it *competes* with the normal substrate for the enzyme's active site, yet does not permanently deactivate the enzyme.

Unlike competitive inhibitors, in which the competition for the enzyme's active site is at least a "fair" one (theoretically each type of molecule has a chance to "win"), certain types of molecules can inactivate an enzyme's active site, even though they do not act on the site directly, by disrupting the enzyme's proper configuration and thereby disturbing the active site indirectly. Such molecules are called **noncompetitive inhibitors** (see Fig. 3.7). Two examples of the results of noncompetitive inhibition are mercury and organophosphorus poisoning. An important special case of

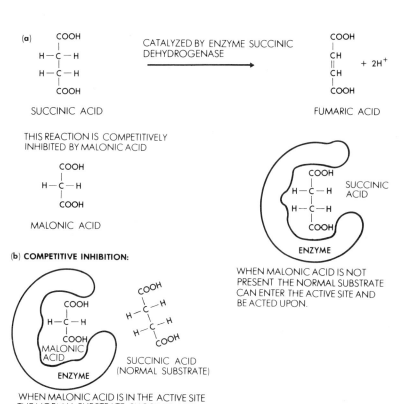

(a)

SUCCINIC ACID → CATALYZED BY ENZYME SUCCINIC DEHYDROGENASE → FUMARIC ACID + 2H⁺

THIS REACTION IS COMPETITIVELY INHIBITED BY MALONIC ACID

MALONIC ACID

(b) COMPETITIVE INHIBITION:

WHEN MALONIC ACID IS IN THE ACTIVE SITE THE NORMAL SUBSTRATE, SUCCINIC ACID, CANNOT GET IN. THE ENZYME DOES NOT AFFECT MALONIC ACID (DOES NOT BREAK IT DOWN)

SUCCINIC ACID (NORMAL SUBSTRATE)

ENZYME

WHEN MALONIC ACID IS NOT PRESENT THE NORMAL SUBSTRATE CAN ENTER THE ACTIVE SITE AND BE ACTED UPON.

Fig. 3.6
Competitive inhibition in the reaction converting succinic acid to fumaric acid. The reaction shown in (a) is catalyzed by the enzyme succinic dehydrogenase. The similarity in molecular structure between succinic acid and malonic acid enables the latter to occupy the active site of succinic dehydrogenase (b), thereby temporarily denying the succinic acid molecules access to it.

noncompetitive inhibition, allosteric inhibition, will be dealt with shortly.

Small Molecule Requirements for Enzyme Function

Besides requiring environmental conditions such as proper temperature and pH, many enzymes also need the presence of certain other substances before they will work. For example, salivary amylase will work on amylose (a sugar) only if chloride ions (Cl^-) are present. Magnesium ions (Mg^{++}) are needed for many of the enzymes involved in the breakdown of glucose.

Some enzymes require other organic substances in the medium in order to function properly. In a few cases, enzymes actually consist of two molecular parts. One of these is a protein, called an apoenzyme. The

ACTIVE SITE
SUBSTRATE MOLECULE
COMPETITIVE INHIBITOR

COMPETITIVE INHIBITION

NONCOMPETITIVE INHIBITOR

NONCOMPETITIVE INHIBITION

Fig. 3.7
A comparison of competitive and noncompetitive inhibition in enzyme-substrate interactions. Note that in competitive inhibition the inhibitor prevents the normal substrate molecule from binding to the active site by occupying it itself. Thus a competitive inhibitor reduces the rate of catalysis by reducing the proportion of enzyme molecules that can bind to the proper substrate. Noncompetitive inhibition, on the other hand, involves the inhibitor molecule binding to the enzyme at a different site than that occupied by the regular substrate molecule. Noncompetitive inhibitors act by reducing the number of enzyme molecules that can bind substrate, possibly by changing the enzyme molecule's shape and making it less efficient. (More complex patterns of inhibition are known that combine aspects of both types of inhibition).

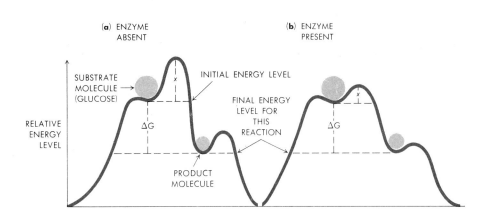

Fig. 3.8
The effect of an enzyme on the activation energy requirements of a molecule undergoing chemical degradation. The amount of activation energy required is represented by the distance x. The net gain in free energy is symbolized by ΔG (it is assumed that the molecule falls to the lowest energy level). Note that the uncatalyzed reaction (a) has a far higher activation energy barrier than the catalyzed reaction (b) on the right. The new molecule must overcome another energy barrier, requiring more activation energy, before it can be broken down and release more energy. In a living organism another enzyme, specific for this second reaction, would be needed.

other molecular part is a smaller, nonprotein molecule called a **coenzyme.** Its name signifies that it works with the main apoenzyme molecule as a co-worker in bringing about a reaction.

In an apoenzyme–coenzyme, the two molecular parts are chemically bonded to each other. In other cases, the coenzyme is combined only briefly with the enzyme. In either case, the coenzyme must be present for any catalytic activity to take place.

Chemical analysis of the smaller coenzymes has shown that they often contain a vitamin as part of the molecule. This finding has led to the idea that vitamins serve as coenzymes. This would explain why the absence of certain vitamins causes such remarkable physical effects on the organism. The enzyme that works with the vitamin-based coenzyme cannot work by itself. Therefore when the coenzyme is missing, an entire series of important physiological reactions may be blocked. The hypothesis that vitamins serve as coenzymes also explains why only a small supply of vitamins is sufficient to fulfill the requirements for good health. Like enzymes, coenzyme molecules need be replaced only from time to time, at a relatively slow rate.

Small molecules such as coenzymes could affect enzyme activity by interacting with specific active sites (not necessarily those for the substrate) on the enzyme molecule. It is thought that such an interaction could cause the three-dimensional shape of the enzyme to shift slightly, opening up the substrate active sites for easier access.

The Energetics of Enzyme Catalysis

Since most biochemical reactions are to one degree or another reversible, it is not surprising that enzymes can catalyze reactions in either direction. In a completely reversible reaction, the enzyme can catalyze the reverse reaction as readily as the forward. If the equilibrium is shifted to the right, the enzyme can catalyze the forward direction more easily than the backward, and so forth. This observation suggests that enzymes do not make reactions occur that would not occur on their own, but only increase the rate at which the reactions take place. Enzymes speed up reactions by changing the energy requirements for getting the reaction started. They do not affect the net energy changes (the ΔG) of any reaction; that would be thermodynamically impossible.

How, then, do enzymes change these energy requirements? The energy hill analogy, discussed earlier in the book, will help to elucidate this point. The glucose molecule at the top of the energy hill shown in Fig. 3.8 represents a certain amount of potential chemical energy. However, glucose may be stored indefinitely without ever releasing its chemical energy. In terms of the energy hill, there is nothing to lift it over the hump at the top and start it rolling down.

This is where enzymes fit into the picture. Their presence lowers the amount of activation energy (x) needed to start the reaction. Enzymes accomplish this by putting the substrate molecule into a constrained position where the *probability* that certain electron exchanges will take place is greatly increased. Enzymes bring particular chemical bonds in a substrate molecule into proximity with specific reactive groups of the enzyme. This process converts a relatively unlikely event into a likely one.

3.8
ENZYMES AND THE CONTROL OF BIOCHEMICAL REACTIONS

So far we have discussed proteins in general, and enzymes in particular, as if they were composed of single polypeptide chains folded into globular and other tertiary configurations. Many proteins, especially enzymes, also appear to have a quaternary structure; that is, they are composed of several polypeptide chains called *subunits* or *monomers* bound together by weak interactions such as hydrogen bonds and electrostatic forces. Evidence suggests that the subunits will join together spontaneously, with a negative free-energy change ($-\Delta G$). Changing the pH, ionic concentration, or temperature of a solution of such proteins will dissociate the subunits. The polypeptide chains making up each subunit are often of at least two types. For example, each of the two subunits of the enzyme malate dehydrogenase consists of a different type of polypeptide chain. The existence of two or more subunits, each composed of a different type of polypeptide chain, provides an important clue as to how enzymes can actually regulate the rate of biochemical reactions.

Regulatory Enzymes and Feedback Pathways

The presence of two or more subunits makes it possible to regulate an enzyme's activity. Regulation plays an extremely important part in the maintenance of life. Since external conditions and the level of activity of any organism are bound to vary from time to time, it is important for the organism's chemical processes to be adjustable. Adjustment involves regulation of the rates of chemical reactions within the organism. If a given reaction occurred all the time at a high rate, there would be a considerable waste of energy. If the rate could never be speeded up, the organism would be at a disadvantage when increased activity was required. The ability to speed up and slow down rates of individual chemical reactions spells greater efficiency of operation for the organism as a whole. Let us consider how enzymes are crucial to controlling rates of reactions in living systems.

A classic example of an enzymatic control system is illustrated by the biochemical reactions involved in the synthesis of cytosine triphosphate (CTP), an essential building block of nucleic acids, from carbamyl phosphate and aspartic acid. This reaction is diagrammed in Fig. 3.9. The initial step involves carbamyl phosphate in a reaction catalyzed by the enzyme aspartate transcarbamylase (ATCase). Through additional reactions, carbamyl aspartate is further transformed into cytosine triphosphate. Such a reaction series is called a *biochemical pathway* because the sequence of individual reactions leads from one place to another along a well-defined route.

How are such biochemical pathways regulated? Apparently the concentration of the cellular end product has a profound effect on the rate at which new CTP molecules are formed. CTP acts as some sort of inhibitor to ATCase, the enzyme catalyzing the initial step in the reaction series. An enzyme, such as ATCase, capable of being regulated by its end product is called a *regulatory enzyme*. Where the regulatory enzyme is inhibited by one of its own end products, the process is known as **negative feedback inhibition** (the effect of the end product is a negative one on the enzyme).

In the biochemical pathway in our example, as the amount of cytosine triphosphate builds up, increasing numbers of ATCase molecules are inhibited, so that the pathway as a whole is slowed down. Conversely, as the amount of aspartic acid and carbamyl phosphate builds up, ATCase begins to catalyze more reaction. The rate at which the enzyme acts is thus greatly affected by the amount of either substrate or end product available. However, the control process is far more finely tuned than could be accounted for by the equilibrium effects of increasing concentration of end product or reactants. Something else must be involved. In the middle and late 1950s the following information was available regarding regulatory enzymes and feedback pathways:

1. Inhibition by the end product is reversible; as more substrate is made available, the negative effects of the end product can be overcome.

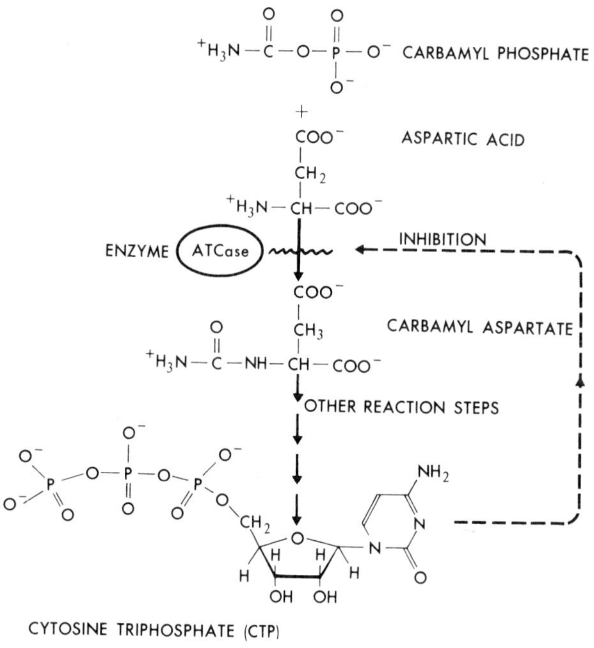

Fig. 3.9

The biochemical pathway for cytosine triphosphate synthesis from carbamyl phosphate and aspartate transcarbamylase. The end product, CTP, acts as an inhibitor for the enzyme catalyzing the first step in the pathway, asparate transcarbamylase. Note that the molecular geometries of end product and original substrates are quite different, implying that CTP does not inhibit ATCase competitively. By comparison, see Fig. 3.10.

$$
\underset{\text{1,3-DIPHOSPHOGLYCERATE}}{
\begin{array}{c}
\overset{\displaystyle O}{\overset{\|}{C}}-OPO_3^{2-} \\
| \\
H-C-OH \\
| \\
H_2C-OPO_3^{2-}
\end{array}}
\quad \rightleftharpoons \quad
\underset{\text{2,3-DIPHOSPHOGLYCERATE}}{
\begin{array}{c}
\overset{\displaystyle O}{\overset{\|}{C}}-O^- \\
| \\
H-C-OPO_3^{2-} \\
| \\
H_2C-OPO_3^{2-}
\end{array}}
$$

Fig. 3.10
The similarity in molecular structure between 1,3-diphosphoglycerate and the compound into which it can be converted, 2,3-diphosphoglycerate, allows the latter compound to act as a competitive inhibitor for the reaction. Thus the more 2,3-diphosphoglycerate produced, the more competitive inhibitor molecules there are and the more the reaction rate is slowed. Such a phenomenon is an example of negative feedback inhibition.

2. No structural relationship could be found between substrate and end-product molecules. They often appeared to be very different in size and molecular structure.

The first observation suggests that the end-product molecule might be a competitive inhibitor for the enzyme. If that were true, however, from what we know about competitive inhibition the end-product molecule would be expected to be very similar to the original substrate. The second observation contradicts that prediction, so it seems clear that simple competitive inhibition is not involved in this case. In some other reactions, it appears to be (see Fig. 3.10).

Allostery

Some regulatory enzymes regulate reaction rates by shuttling back and forth between the active and inactive states within a time span of only a few minutes. Yet even this is not fast enough for many reaction systems. Some regulatory enzymes, called **allosteric enzymes**, have been shown to possess the ability to make such changes in only a few seconds.

In the mid-1960s, a team of French investigators (Jacques Monod, Jean-Pierre Changeaux, Jeffries Wyman, and Francois Jacob) proposed a model to explain how such allosteric regulatory enzymes work. The term "allosteric" means "another space." The French scientists suggested that, besides their active sites, allosteric enzymes possess another region to which specific *modulators* or *effectors*, either positive or negative in their effect on the enzyme's activity, may become bound. This binding is reversible. Some allosteric enzymes have only positive modulators, some only negative, and still others possess both. Those possessing only positive modulators become converted from their normal active state to a still more active form; with negative modulators, the activity is slowed by a reverse

conversion and, with both, the conversion (and thus the activity) may move in either direction (see Fig. 3.11).

To explain the precise mechanism of allosteric inhibition, which, as mentioned on pp. 76–77, is a special case of noncompetitive inhibition, Monod and his co-workers proposed that allosteric enzymes are composed of two or more independent parts or subunits that, in the intact molecule, exist in two shapes or conformations. In one of these conformations, called the *R* state, the enzyme is active; in the other, called the *T* state, it is less so, if at all. When, for example, a negative modulator becomes bound to an allosteric enzyme, it is thought to change the molecular configuration of its active site, thereby inhibiting the enzyme's normal catalytic activity (see Fig. 3.12).

As would be expected, allosteric enzyme activity does not conform to the curve demonstrated by line (b) in the graph in Fig. 3.11, since the presence of activating (positive) or inhibiting (negative) modulators will have an effect on the degree of activity of such enzymes. Indeed, it was the observation of these differences that led to the development of the concept of allostery. As a way of accounting for the observed reactions catalyzed by this group of regulatory enzymes, as well as suggesting a

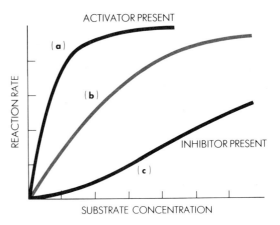

Fig. 3.11
A comparison of enzyme activity. If the velocity of an enzyme-catalyzed reaction is plotted against substrate concentration, a hyperbolic curve results (*b*), enabling a constant to be determined (called the Michaelis-Menten constant, Km). However, some enzymes do not follow this pattern, and because of this, the concept of allostery was developed. In line (*a*) the increased activity is accounted for by assuming that the proportion of enzyme molecules in the active or *R* form increases progressively as more substrate is added. At the peak of the line, all the active sites are fully saturated and all of the enzyme molecules are in the active form. In curve (*c*) the effects of an allosteric inhibitor are shown; in this case, the majority of the enzyme molecules are held in the inactive or *T* state.

Organisms gain economy and flexibility by constructing large proteins out of several identical parts. Less genetic information is needed to code for the amino acid sequences, and a protein composed of such parts is more flexible in the geometric configurations it can assume than one consisting of a single long chain.

possible way that living systems can so quickly adjust their vital reaction rates in a manner that meets their needs, allostery provides a most useful hypothesis.

Several advantages derive from the fact that proteins in general, and regulatory enzymes in particular, consist of individual parts instead of a single long polypeptide chain. One is that conformational changes are more easily brought about among separate parts than along a single, tightly wound chain. Geometric flexibility is introduced through interacting parts. A second advantage is that much less information needs to be coded in the genetic program if a large molecule is composed of several sets of identical parts. With its two α and two β chains, the hemoglobin molecule requires only half the genetic information that would be necessary if the molecule were a single polypeptide equal to its four parts in size. A third advantage is that the parts can be produced at several sites in the cell and joined together elsewhere. This streamlines the process of protein construction. Since the parts come together spontaneously, the cell gains further economy in not having to develop elaborate biochemical pathways for joining pieces of polypeptide together. All in all, the nature of protein structure adds considerable efficiency to the organism's overall biochemical processes.

3.9
NUCLEIC ACIDS

Though nucleic acids were first discovered by the Swiss chemist Friedrich Miescher in 1869, approximately a century passed before their structure and significance were determined. One type, **deoxyribonucleic acid,** whose initials **DNA** have become virtually a household word, composes the genetic material—the genes of all organisms (with the exception of a few viruses). The other, **ribonucleic acid,** or **RNA,** includes among its members those molecules that carry encoded information from the genes to the rest of the cell, information that determines both cell structure and function.

Like proteins, nucleic acids are composed of distinct building blocks, with the counterpart to the amino acids of proteins termed **nucleotides.** Nucleotides, in turn, are composed of three distinct molecular parts, a pentose or five-carbon sugar, a phosphate group, and a nitrogenous (nitrogen-containing) base. The sugar serves as the central or focal point of the nucleotide, with the nitrogenous base and phosphate group bonded to it. Three of the five nitrogenous bases, *cytosine, thymine,* and *uracil,* belong to a class of compounds called **pyrimidines,** and are single-ring molecules. The

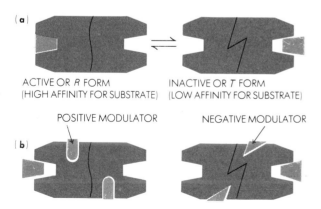

ACTIVE OR *R* FORM
(HIGH AFFINITY FOR SUBSTRATE)

INACTIVE OR *T* FORM
(LOW AFFINITY FOR SUBSTRATE)

POSITIVE MODULATOR

NEGATIVE MODULATOR

Fig. 3.12
(*a*) schematic representation of allostery and the role of positive and negative modulators. Each allosteric enzyme is composed of two parts and these parts, joined together, can exist in two different configurations. One of these configurations is the active or *R* form; the other, the inactive or *T* form. The two forms, under the influence of positive or negative modulators, may shift from one form to another; an allosteric inhibitor shifts the form from the active to the inactive state, while a positive modulator shifts the form from the inactive to the active state. (*b*) Some models propose that positive and negative modulators work by locking the enzyme into the active or inactive state, respectively. (Another model proposes that the enzyme exists in either one state or the other, that is, active or inactive, and the modulator molecules merely convert them to the other form by becoming bound to them.)

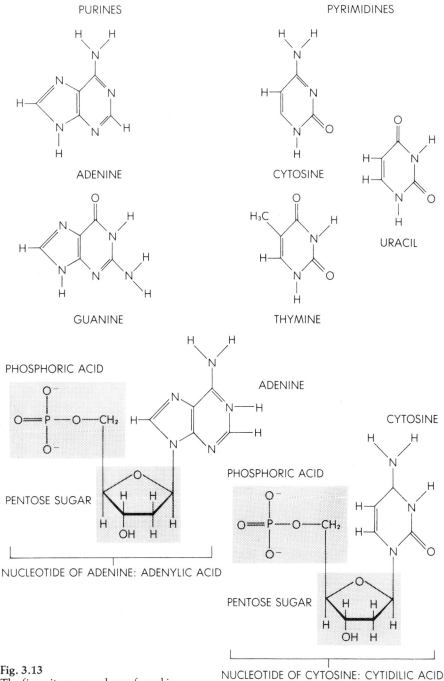

Fig. 3.13
The five nitrogenous bases found in nucleic acids and two of the five nucleotides they form when united with a phosphate group and pentose sugar. The joining together of many different nucleotides makes the chains of nucleic acid. Nucleotides are designated as the acid forms of the base from which they are formed: for example, adenylic acid from adenine, guanidilic acid from guanidine, and uridilic acid from uridine.

other two, *adenine* and *guanine*, are **purines,** and are double-ring molecules (see Fig. 3.13).

Just as the number and sequence of amino acids distinguish a protein, so does the number and sequence of nucleotides distinguish the nucleic acids. Indeed, the amino acids in a protein owe their precise sequence, in turn, to the precise sequence of bases in DNA and certain types of RNA.

As was suggested, RNA is found in several forms, most notably **messenger RNA (*m*RNA), transfer RNA (*t*RNA)**, and **ribosomal RNA (*r*RNA)**. The structure and function of each of these will be discussed in detail in Chapter 19. It is sufficient at this point simply to state that these various types of RNA carry instructions from the genes (DNA) to the sites of cellular protein synthesis (the ribosomes), bring amino acids to be used in this synthesis to the ribosomes, and are structural components of the ribosomes themselves.

Besides differences in their function, DNA and RNA differ from each other structurally in several important respects. For one, RNA contains uracil as the pyrimidine other than cytosine, whereas DNA contains thymine. For another, as the names suggest, the pentose sugar in RNA is ribose, while that in DNA is deoxyribose. Finally, RNA is generally single-stranded, while DNA is often (though by no means always) double-stranded. The double strands, composed of alternating sugar and phosphate groups, are joined by hydrogen bonds between the paired nitrogenous bases, two between adenine and thymine and two between cytosine and guanine. The ladder-like structure formed is twisted to form a double helix (see Figure 19.7).

The unique double-stranded structure of DNA gives it its equally unique function. Discussion of the significance of this structure, as well as how it was determined, will be postponed until Chapter 19.

3.10
ENERGY IN PACKETS: ATP

All biochemical reactions require energy. Even reactions that are exergonic and release energy generally require at least a small amount of energy to get them going, just as a match or spark is needed to start a fire. One of the most important energy-supplying molecules in living systems (to which reference will often be made in the pages to come) is **adenosine triphosphate**, or **ATP**. ATP is so widely used in so many different biochemical reactions it is often referred to as the cell's chief "energy currency."

ATP's molecular structure consists of three parts already discussed: the nitrogenous base adenine, the five-carbon sugar ribose, and, as the "triphosphate" of ATP's name suggests, three phosphate groups:

Adenosine triphosphate
(ATP)

Two of the three negatively charged phosphate groups possess bonds that, when broken, release a considerable amount of free energy (approximately seven kilocalories per mole) in comparison to other phosphate ester bonds. For this reason, these bonds are often referred to as "high-energy bonds" and symbolized by a wavy line in the molecular formula. It should be stressed, however, that this does *not* mean that the energy released is all locked within the bond itself. Rather, much of it is due to changes in the electron orbitals of the molecule that result from the breaking of the bonds.

ATP is converted into **adenosine diphosphate (ADP)** by the hydrolysis of one high-energy bond, and this reaction is a common means of providing energy for the cell. The ADP can then be converted back into ATP by the input of a sufficient quantity of energy. ADP may occasionally be further broken down into **adenosine monophosphate (AMP)** with the release of another seven kilocalories of energy.

ATP is by no means the only source of immediate energy in the cell. For example, another compound, *guanosine triphosphate* (GTP) is often used in certain reactions by some forms, being converted to guanosine diphosphate (GDP) and guanosine monophosphate (GMP) in the process. But ATP is by far the most immediate source of energy found in living systems. In Chapters 5 and 6 more will be said about the nature of exergonic and endergonic reactions involving ATP.

Summary

1. The major classes of large molecules (macromolecules) of great importance to living systems include carbohydrates (sugars and starches), lipids, proteins, and nucleic acids. Carbohydrates serve primarily as fuel molecules (energy sources) but also as structural elements; lipids serve as reserve fuel and as phospolipids also take part in membrane structure; proteins serve as structural components (hair, nails, skeletal systems, muscle, and connective tissue) and as enzymes. Nucleic acids serve as the genetic material in organisms, both carrying the genetic code and transcribing it.

2. Carbohydrates, lipids, proteins, and nucleic acids are all constructed by piecing together various special com-

ponent parts. A great variety of each type of molecule is achieved by the variable arrangement of essentially a few basic elements.

  a) The basic structural unit of carbohydrates is the sugar molecule (three-carbon, four-carbon, five-carbon, or six-carbon sugars), joined together by glycosidic bonds.

  b) The basic structural units of lipids are fatty acids and glycerol, joined together by ester bonds.

  c) The basic structural units of proteins are the amino acids (of which there are approximately twenty types), joined together by peptide bonds.

  d) Glycosidic, ester, and peptide bonds of carbohydrates, lipids, and proteins, respectively, are formed by dehydration synthesis (removal of a molecule of water between two units); the bonds are broken by hydrolysis (addition of water between two units).

3. Enzymes are organic catalysts. They are highly specific for their reactants, or substrates. The enzyme molecule consists of one or more polypeptide chains folded into more or less globular shapes. The substrate fits into a particularly structured region on one part of the enzyme molecule, known as the active site. Fitting together of enzyme and substrate causes a shift in three-dimensional structure of the protein (known as a conformational or configurational

shift), which brings certain active groups of the enzyme into contact with specific regions of the substrate. This weakens certain chemical bonds and promotes a reaction.

4. The rate at which enzymes catalyze reactions among their substrates is often subject to allosteric control. Allosteric enzymes have two types of active sites, the catalytic site for substrate, the regulatory site for end-product and other regulatory molecules. It is hypothesized that when the end-product molecule fits into its active site, a conformational shift occurs that causes distortion of the substrate active site.

5. Nucleic acids are composed of phosphate, nitrogenous bases, and a sugar, either ribose (RNA) or deoxyribose (DNA). DNA serves as the genetic material in most organisms, while the various types of RNA (messenger, transfer, and ribosomal) serve as transcribers of the genetic message in protein synthesis.

6. Adenosine triphosphate (ATP) is composed of the nitrogenous base adenine, ribose, and three phosphate groups. Two of the three phosphate groups are joined by high-energy bonds, with the ATP being converted to adenosine diphosphate (ADP) and then to adenosine monophosphate (AMP) as the energy is released. ATP is by far the most common form of energy currency used in living cells.

Exercises

1. The reaction $A + B \rightarrow AB$ takes place slowly at $20°C$ unless either compound x or y is present. Compound x is a metallic catalyst and y is an enzyme; both compounds catalyze the reaction. Ten milliliters of a solution of A and B are placed in each of four test tubes, to which varying amounts of x or y are added, as shown below:

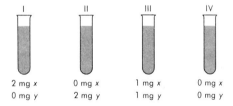

I	II	III	IV
2 mg x	0 mg x	1 mg x	0 mg x
0 mg y	2 mg y	1 mg y	0 mg y

  a) Predict which tubes would show the greatest and which the least rate of reaction at $20°C$. Explain your reasons.

  b) If A and B are heat-stable at $100°C$, in which tube(s) would the reaction rate be greatest at this temperature? Least? Why?

  c) Increasing the temperature from $20°C$ to $30°C$ will probably double the reaction rate in which tube(s)? How do you know?

  d) If the reaction is allowed to reach equilibrium, in which tube(s) will the amount of AB be greatest? Least? What are your reasons?

  e) The contents of tubes I, II, III, and IV are poured into separate dialyzing sacks made from cellophane. The sacks are placed in separate containers of distilled water. The reaction slows down in sacks I and III but continues at the previous rate in sacks II and IV. If A and B are added to the distilled water from outside of sacks II and IV (each tried separately), the reaction proceeds at a very slow rate. When A and B are added to distilled water from outside sacks I and III, the reaction speeds up. Explain the results (assume that A, B, and AB are all large molecules).

2. What are the major differences between carbohydrates, fats, and proteins in terms of (a) molecular weight, (b) molecular geometry (size and shape), and (c) general biological uses (in cell metabolism)?

3. Explain the lock-and-key model of enzyme function. How is this model useful? In what ways could it be misleading?

4. Proteins are said to display optimal ranges of activity within any environment, for example, within a range of pH values, temperatures, and the like. In terms of the graph in Fig. 3.14 on p. 85, explain how changes in pH could affect the chemical reactivity of protein molecules. Include in your answer a discussion of the following points:

a) Where on a protein molecule (and at what level of organization of the molecule—primary, secondary, or tertiary structure) changes in hydrogen ion (H^+) concentration in the medium have their effect.

b) How changes in the protein brought about by change in H^+ concentration could affect chemical reactivity of the protein.

c) Why the curve is relatively steep on either side of the optimal peak.

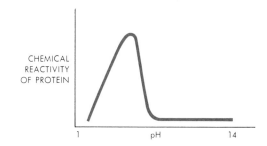

Fig. 3.14

Chapter 4
Cell Structure and Function I

4.1
INTRODUCTION

Near the end of the nineteenth century the German physiologist Max Verworn wrote:

It is to the cell that the study of every bodily function sooner or later drives us. In the muscle cell lies the problem of the heartbeat and that of muscular contraction; in the gland cell reside the causes of secretion; in the lining cells of the digestive tract lies the problem of the absorption of food; and the secrets of the mind are hidden in the ganglion [nerve] cell.

Though written more than seventy-five years ago, Verworn's words have a distinctly modern ring. In present-day terms, the problems of heredity, embryonic development, immunity, and cancer have all been studied on the level of the **cell** and its components. What are cells? What different kinds of cells are found throughout the animal and plant kingdoms? What are the structural components of cells? And finally, in what ways do cells interrelate to form higher levels of organization—the tissues—of multicellular animals and plants? These questions are the focus of the next three chapters.

The "typical" cell often pictured in biology textbooks simply does not exist. Even among closely related cell types, there is wide variation. While, for example,

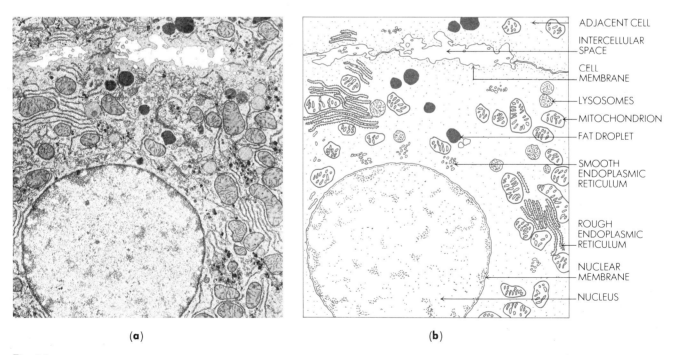

(a)   (b)

Fig. 4.1
Electron micrograph (a) and diagram (b) of a representative animal cell. Like Fig. 5.5(b), this cell is from a rat liver. The dark globules are fat droplets; the smaller dark granules are glycogen, a polysaccharide known as "animal starch," which is synthesized and stored in liver cells. (Micrograph courtesy Albert L. Jones, M.D.)

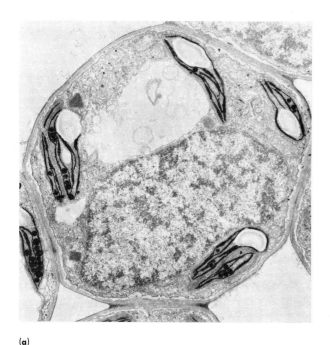

(a)

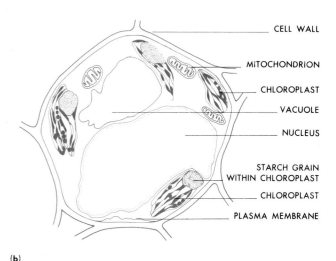

CELL WALL

MITOCHONDRION

CHLOROPLAST

VACUOLE

NUCLEUS

STARCH GRAIN
WITHIN CHLOROPLAST

CHLOROPLAST

PLASMA MEMBRANE

(b)

the cells of plants and animals show many similarities (see Figs. 4.1 and 4.2), they also exhibit significant differences; and even among closely related cell types, there are considerable differences. This chapter will review a brief history of our knowledge of cells and the techniques used to study them. It will then consider cells in terms of their membranes and the role they play in cell function.

4.2
THE CELL CONCEPT

The "cell concept" (sometimes called "the cell theory," though it is *not* a theory) can be summarized in the following four propositions:

1. *Cells are the structural units of virtually all organisms.* From simple one-celled organisms like an *Amoeba* or bacterium to multicelled organisms like a human being or tree, all organisms are composed of cells. Cells are the building blocks in the architecture of life. As we have seen, however, cells themselves are composed of several types of molecular building blocks which are, in turn, composed of several kinds of atoms.

2. *Cells are functional (and dysfunctional) units of virtually all organisms.* All the chemical reactions necessary to the maintenance and reproduction of living systems take place within cells. The chemical processes (metabolism) that provide the energy for contraction of a muscle cell, for example, take place within the muscle cell itself. At the same time, failures of functions, or sicknesses, are also cellular in nature. Certain types of

Fig. 4.2
Electron micrograph (a) and drawing (b) of a representative plant cell. Note the presence of a thick cell wall, in addition to the plasma membrane, at the boundary of the cell. Note also the large vacuole, typical of most kinds of plant cells. (Photo courtesy H. J. Arnott, University of Texas at Arlington.)

cancer, for example, appear to be the result of a failure of certain cells to regulate the reproductive process.

3. Since the first cells evolved, cells have arisen only from preexisting cells. Cells do not spontaneously generate. A multicellular organism grows by duplication of its individual cells. By special cell divisions, some organisms produce gametes, or specialized sex cells such as eggs or sperm, capable of generating a whole new organism. The idea of biogenesis—that all living cells originate only from preexisting living cells, or that life originates from life—has been fundamental to the modern cell concept since around 1860.

4. Cells contain hereditary material (nucleic acid) through which specific characteristics are passed on from parent cell to daughter cell. The hereditary material contains a "code" that ensures continuity of species from one cell generation to the next.

The modern cell concept is not the work of any one person. Cells were first observed and described by some of the early seventeenth-century microscopists. Robert Hooke's *Micrographia* (1665) contains some of the first

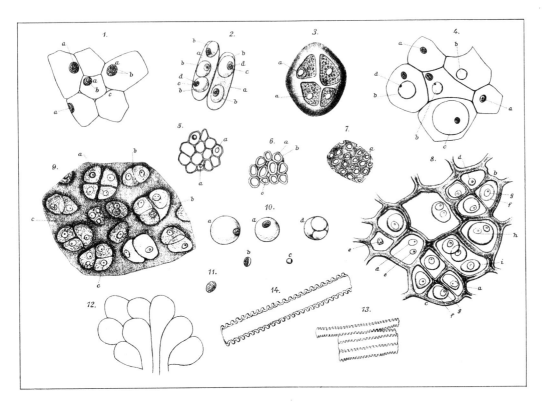

Fig. 4.3
Drawings from Schwann's *Microscopical Investigations* (1839). This work shows that by the middle of the nineteenth century a great deal was known about the structure of different kinds of cells.

clear drawings of plant cells, made from observations of thin sections of cork.* Hooke coined the term "cell" to refer to the boxlike structures he saw. The fact that virtually all living organisms are composed of cells was not recognized, however, until the nineteenth century. The most important generalized statement (our first proposition) about the cellular nature of living organisms was made by two German biologists, Matthias Schleiden and Theodor Schwann, in 1838 and 1839. Schleiden, a botanist, and Schwann, a zoologist, studied many types of tissue in their respective fields. Both came to the conclusion that the cell was the basic structural unit of all living things. Some of Schwann's diagrams of various animal cells can be seen in Fig. 4.3. The second and third propositions were added by the German pathologist and statesman Rudolf Virchow (1821–1902). In his work *Cellular Pathology* (1858), Virchow spoke of the cell as the basic metabolic as well as structural unit. In this same work he emphasized the continuity in living or-

ganisms with the statement: *"omnis cellula e cellula"* —"all cells come from [preexisting] cells."

4.3
THE SIZES OF THINGS

Since cells themselves are generally quite small (see Supplement 4.1), the structures they contain, the **organelles,** are smaller still. Thus units of measurement smaller than those generally used in the metric scale (for example, meters, centimeters) must be used to convey their size. Yet, since these organelles are composed of many of the macromolecules discussed in the previous chapters, it is more convenient to measure them with larger units than the metric unit used to measure atoms and molecules, the angstrom (symbolized Å). The units in most common use today are either the *micrometer* (formerly called the micron), symbolized μm, and the *nanometer,* symbolized nm. Their size in relation to the meter and angstrom are shown below:

1 meter (m) = 10^2 centimeter (cm) = 10^3 millimeter (mm) = 10^6 micrometer (μm) = 10^9 nanometer (nm) = 10^{10} angstrom (Å).

---

*"Cork" refers to a portion of the bark, or outer covering layer, of any woody plant. The cork used for thermos bottles is part of the bark of the cork oak tree, found mostly in Spain.

Recall that the superscript number indicates the number of zeros to follow the first number; for example, one meter is 100 centimeters (or a centimeter is 1/100 of a meter) but 1,000,000,000 nanometers (or a nanometer is 1/1,000,000,000 of a meter). For consistency, whenever possible, the dimensions of the various cell parts to be discussed in the following pages will be given in nanometers.

4.4
TOOLS AND TECHNIQUES FOR STUDYING CELLS

Advances in our knowledge of cellular anatomy are tied directly to improvements in the techniques and instruments used to examine cells. The nineteenth and early twentieth centuries saw the introduction and refinement of certain techniques, such as staining and microtoming.

SUPPLEMENT 4.1
UPPER LIMIT ON CELL SIZE: A MATTER OF GEOMETRY

Cells vary widely in size and shape. The difference in size between the smallest and largest is comparable to the difference in size between a mouse and a whale. The smallest cells, known as pleuropneumonia-like organisms (PPLO), or mycoplasmas, range in size from 100 to 280 nm. They are responsible for a highly infectious pneumonia in cattle and also for some forms of human pneumonia. The largest free-living cells in terms of volume include protozoans such as the giant amoeba *Chaos chaos*, which reaches a length of 5 cm. In terms of length, in multicelled organisms some nerve cells run the length of an entire arm or leg: for an elephant such a cell may be more than 8 ft long. Diagram A offers a range of cell sizes drawn to scale. In life, the large amoeba is easily

visible to the naked eye; the human cheek cell is barely visible; the others are invisible without a microscope.

How large can cells get? The answer to this question comes partly from solid geometry. All cells depend for their continued existence on the passage of materials (such as nutrients and wastes) in and out through their boundaries, the cell membrane. From solid geometry we know that as a sphere becomes greater in size, its volume increases as the *cube* of the radius. (The formula for determining the volume of a sphere is $V = 4/3\pi r^3$.) The surface area, on the other hand, increases only as the *square* of the radius. (The formula for determining the surface area of a sphere is $S = 4\pi r^2$.) This simple principle applies to the other three-dimensional

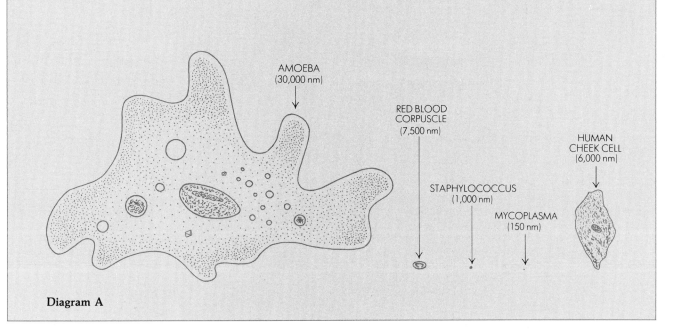

AMOEBA
(30,000 nm)

RED BLOOD
CORPUSCLE
(7,500 nm)

STAPHYLOCOCCUS
(1,000 nm)

MYCOPLASMA
(150 nm)

HUMAN
CHEEK CELL
(6,000 nm)

Diagram A

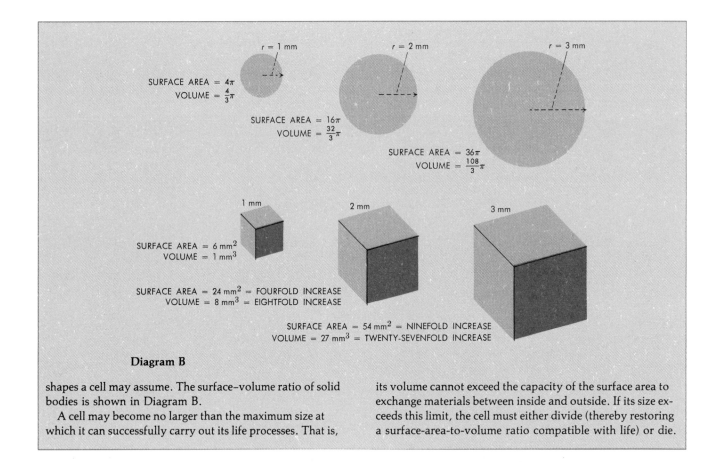

$r = 1$ mm

SURFACE AREA $= 4\pi$
VOLUME $= \frac{4}{3}\pi$

$r = 2$ mm

SURFACE AREA $= 16\pi$
VOLUME $= \frac{32}{3}\pi$

$r = 3$ mm

SURFACE AREA $= 36\pi$
VOLUME $= \frac{108}{3}\pi$

1 mm

SURFACE AREA $= 6$ mm^2
VOLUME $= 1$ mm^3

SURFACE AREA $= 24$ mm^2 = FOURFOLD INCREASE
VOLUME $= 8$ mm^3 = EIGHTFOLD INCREASE

2 mm

3 mm

SURFACE AREA $= 54$ mm^2 = NINEFOLD INCREASE
VOLUME $= 27$ mm^3 = TWENTY-SEVENFOLD INCREASE

Diagram B

shapes a cell may assume. The surface–volume ratio of solid bodies is shown in Diagram B.

A cell may become no larger than the maximum size at which it can successfully carry out its life processes. That is, its volume cannot exceed the capacity of the surface area to exchange materials between inside and outside. If its size exceeds this limit, the cell must either divide (thereby restoring a surface-area-to-volume ratio compatible with life) or die.

THE PRINCIPLE OF SECTIONING WITH A MICROTOME

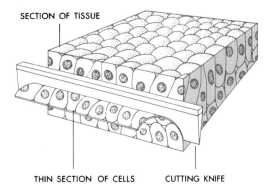

SECTION OF TISSUE

THIN SECTION OF CELLS   CUTTING KNIFE

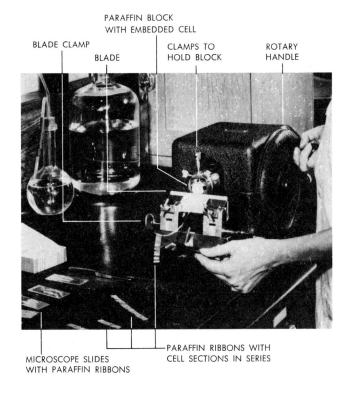

PARAFFIN BLOCK WITH EMBEDDED CELL

BLADE CLAMP

BLADE

CLAMPS TO HOLD BLOCK

ROTARY HANDLE

MICROSCOPE SLIDES WITH PARAFFIN RIBBONS

PARAFFIN RIBBONS WITH CELL SECTIONS IN SERIES

Fig. 4.4
A rotary microtome in use. The cells are permeated with paraffin to hold them rigid when cut by the blade. When the wheel is turned, the cells are pushed against the blade and a section is cut. The next turn of the wheel moves the cells forward a predetermined number of microns, and another section is cut. The entire section series is then stained and studied under the microscope. From such studies, it is possible to recreate the cell in three-dimensional perspective. (Photo courtesy Douglas C. Anderson, Department of Fisheries and Forestry, Ontario, Canada.)

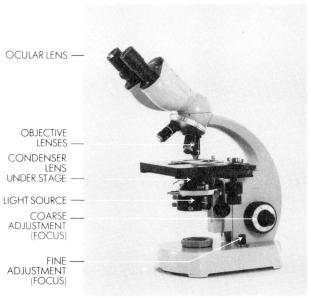

OCULAR LENS

OBJECTIVE LENSES

CONDENSER LENS UNDER STAGE

LIGHT SOURCE

COARSE ADJUSTMENT (FOCUS)

FINE ADJUSTMENT (FOCUS)

(a) **LIGHT MICROSCOPE**

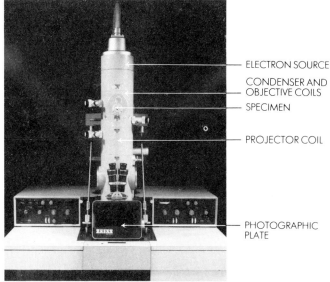

ELECTRON SOURCE

CONDENSER AND OBJECTIVE COILS

SPECIMEN

PROJECTOR COIL

PHOTOGRAPHIC PLATE

(b) **ELECTRON MICROSCOPE**

In *staining,* parts of cells and tissues are colored with dyes, causing them to stand out in sharp contrast to other structures left unstained or stained different colors. In sectioning with a *microtome,* cells and tissues are cut into thin sections so that they may be examined under a microscope (see Fig. 4.4).

The Light Microscope

The primary instrument for the study of cell structure is the light microscope (Fig. 4.5, a and c). From the seventeenth century to the 1920s, detailed knowledge of cellular anatomy increased as the resolving power of the light microscope was increased. *Resolving power* is a microscope's capacity for separating to the eye two points that are very close together.* Resolving power is inversely proportional to the wavelength of radiant energy used (as well as such factors as the opening size of the microscope lens). This means that as wavelength *decreases* (using light more toward the violet end of the spectrum), resolving power *increases.* The best modern light microscopes have resolving powers of 170 nm or 1700 Å, using violet light; two dots less than 170 nm apart will appear as a single dot when viewed under violet light. If light of longer wavelengths is used, for example white light (a mixture of many wavelengths, most of which are longer than violet), the resolving

*The naked human eye has a resolving power of approximately 0.1 mm. Thus two lines placed closer together than 0.1 mm will appear as a single line to the naked eye, no matter how close to them the observer may get. Since most cells are smaller than 0.1 mm across, the need for a microscope is obvious.

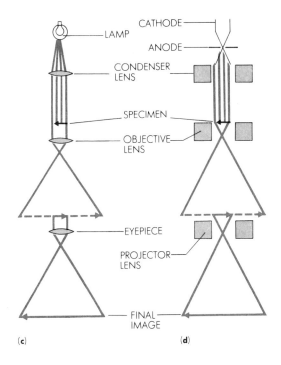

LAMP

CATHODE

ANODE

CONDENSER LENS

SPECIMEN

OBJECTIVE LENS

EYEPIECE

PROJECTOR LENS

FINAL IMAGE

(c)   (d)

Fig. 4.5
Comparison of light (*a* and *c*) with electron (*b* and *d*) microscopes. The principles on which they work are very similar. The light microscope uses radiation within the visible spectrum, whereas the electron microscope uses beams of electrons at much shorter wavelengths. For purposes of comparing the optics of these microscopes, the light microscope is inverted from its normal position in (*c*). (Photos courtesy Carl Zeiss, Inc., New York.)

(a) BRIGHT-FIELD

(b) DARK-FIELD

(c) PHASE-CONTRAST

power is limited to 250 nm. By using very short wavelengths of radiation (such as ultraviolet "light"—invisible to the human eye but visible to photographic plates) and quartz lenses, light microscopes can obtain a resolving power of 100 nm but no more.

Other techniques have been employed to increase the amount of detail a light microscope can disclose other than simply size of object. One method is known as **dark-field microscopy.** In this process, light rays are bent by the lenses in such a way that only the light passing through the specimen passes upward to the viewer's eye; the background remains dark. This contrast makes the object being viewed, and especially its boundaries, more distinct, as shown in Fig. 4.6(b).

Another technique is known as **phase-contrast microscopy** (Fig. 4.6). The human eye responds best to wavelengths of light that pass easily through water; much of the cell's content is composed of water. Hence, the majority of cell components appear transparent to the human eye. This difficulty can be overcome by taking advantage of the fact that in *all* light microscopes, light waves undergo slight changes in *phase:* a part of the beam of light passing through the cell (or any object being studied) is momentarily slowed down and thus rendered out of phase with the rest of the beam. In the normal light microscope this retardation is so slight that it cannot be detected by the human eye. In the phase-contrast microscope, the retardation is amplified by a set of mirrors so that it becomes visible. Hence, light passing through the cell reaches the eye at different times. As the light passes through a thicker portion of cell those waves are more slowed down. The result is that details become visible because each structure slows the light down (makes it out of phase with other parts of the light beam) in a very specific way.

The Electron Microscope

In 1924, the physicist Louis de Broglie theorized that electrons had a wave nature. The length of that wave was found to be only 0.05 Å. Therefore, it was reasoned, if a beam of electrons were substituted for a beam of light, the shorter wavelength of the electrons should cause a corresponding increase in resolving power. This prediction proved to be correct. In 1934, an electron microscope was used to obtain pictures with

Fig. 4.6
Three microscope techniques are shown here in viewing the same organism, *Dynobryon sertularia.* Bright-field microscopy directs light up through the specimen toward the viewer. Dark-field microscopy sends light in at an angle from the side; thus only light bent upward by the specimen is seen in the field. (Photos courtesy Dr. Earl Hanson, Wesleyan University.)

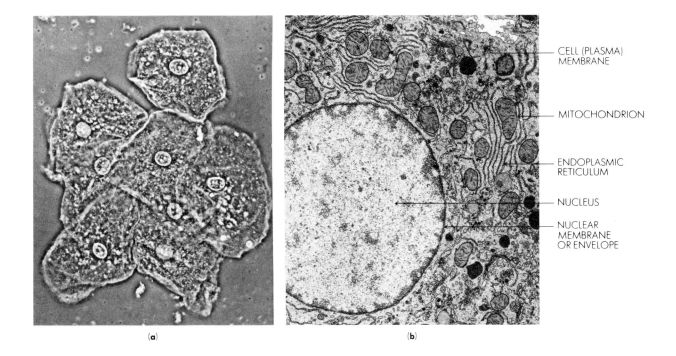

(a)   (b)

CELL (PLASMA)
MEMBRANE

MITOCHONDRION

ENDOPLASMIC
RETICULUM

NUCLEUS

NUCLEAR
MEMBRANE
OR ENVELOPE

greater resolution than those that could be obtained through the light microscope. The electron microscope uses a beam of electrons rather than a beam of light. Magnetic coils serve to focus the beam, much as the objective lenses focus light beams in the light microscope (see Fig. 4.5c and d). The electron microscope has revealed more new knowledge of cell structure in 30 years than was discovered with the light microscope in the previous 300.

As mentioned above, the limit of resolving power of the light microscope is about 170 nm to 250 nm (depending on wavelength used); that for the electron microscope is about 0.5 nm (5 Å). Furthermore, while the light microscope can magnify an object only at a maximum of about 2000 times, an electron microscope can magnify up to 1,000,000 times! It is obvious why the electron microscope gives an enormous advantage to modern biologists.

Figure 4.7 shows a comparison between a light microscope picture and an electron microscope picture, called a *photomicrograph,* of animal cells. The difference in the detail between the two is striking. The greater detail revealed by the electron microscope is a result of both its greater magnifying power and its greater resolving power.

From a distance, a forest-covered mountain appears a solid green or bluish color. Close to the mountain, however, one can see that this apparently solid green blanket is highly diverse, composed of individual trees differing widely in size and kind. The development of the electron microscope enabled the biologist to view

Fig. 4.7
View of two mammalian cells under light (*a*) and electron (*b*) microscopy. (*a*) A human check cell viewed under phase-contrast light microscopy (× 400). Note that the only part clearly visible inside the cell is the nucleus. The cytoplasm appears granular and can be seen to contain numerous particles, but few details are visible. (*b*) Electron micrograph (× 17,000) of a rat liver cell. Note the large, dominant nucleus, the many mitochondria, the cell and nuclear membranes, and the endoplasmic reticulum. (Micrograph courtesy Albert L. Jones, M.D.)

the cell up close. Like the blanket of trees, the cytoplasm showed itself not to be a uniform substance, but a highly structured region, with intricate design and detail.

The Scanning Electron Microscope

Recently a new instrument called the *scanning electron microscope* has been used to study the fine structure of cells and organisms. (The type of electron microscope already discussed is technically called a transmission electron microscope.) In the scanning electron microscope, a beam of electrons scans and bounces off the surface of an opaque specimen, rather than passing through an extremely thin section. This new instrument greatly facilitates the study of whole specimens or surface structures, such as those found in pollen grains, red blood cells, kidney tubule cells, bone and teeth surfaces, and cells in culture (grown outside the organism in a special medium in the laboratory). Scanning electron micrographs appear in Fig. 4.8.

Finally, a technique known as *freeze-etching*, in combination with electron microscopy, has provided still more detail concerning cell and cell organelle structure. The specimen is first frozen and then fractured. Scanning electron microscopy is then used to make a micrograph. If the fracture occurs along certain planes, a great deal of structural detail can be learned (see Fig. 4.9).

The Acoustic Microscope

Recently, a research team at Stanford University, headed by physicist Dr. Calvin F. Quate, has developed an acoustic microscope that uses sound waves instead of light waves to make an image. The acoustic microscope depends upon the density, elasticity, and viscosity of the object being viewed (properties actually far more vital to the nature of living matter than its optical refractive index upon which the light microscope relies). In terms of resolution, current models compete favorably with light microscopes. Within a short time, there is hope that they will be able to challenge electron microscopes, but without the latter's disadvantage of having to work in a vacuum and thus viewing material that is no longer alive.

Other Techniques

Besides the use of various forms and techniques of microscopy, there are several other techniques used to study cells. One of these, centrifugation, is described in Supplement 5.1.

Particularly significant is labeling with *isotopes*, which are different forms of the same element. For example, the element carbon normally contains six protons and six neutrons in its nucleus and six electrons in its orbitals. Since it is the number of protons and electrons that determines the type of atom, the addition of one more neutron changes carbon 12 into carbon 13; two more, into carbon 14. The carbon is still carbon in its chemical behavior, since it still has the four electrons in its outer orbital that give it its tendency to form covalent bonds. Its atomic *mass*, however, has increased from 12 to 13 or 14. This being the case, by using a tech-

Fig. 4.8
The scanning electron microscope allows high-magnification three-dimensional viewing of objects such as cells, in this case, pollen grains. Shown below are pollen grains of two different species of plants: the small herb *Cosmos bipinnatus* (left), × 6500; and the lily *Lilium longiflorum* (right), × 2000. (Photos courtesy J. Heslop-Harrison, University of Wisconsin.)

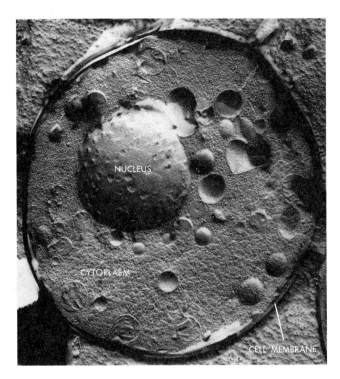

Fig. 4.9
Cell ultrastructure as revealed by freeze-etching. The specimen is fractured after freezing, and a micrograph is made using scanning electron microscopy. Some of the fractures demonstrate a considerable amount of detail concerning cell organelles and their parts. In this specimen, pores in the membrane of the cell nucleus show up clearly. (Micrograph courtesy of Samuel F. Conti, University of Massachusetts, Amherst.)

nique called *mass spectrometry*, lighter isotopes can be separated from heavier ones. Imagine rolling steel ball bearings down an incline so that they must pass a bar magnet. Clearly a lighter ball bearing will be deflected more than a heavy one, and thus will end up in a different position at the end of the incline. A similar technique, using a charged field instead of a magnet, is involved in mass spectrometry. Mass spectrometry played a major role in solving a major problem in plant cell physiology—that of the source of oxygen released during photosynthesis.

Some isotopes, such as carbon 14, are radioactive. This means that if a carbon-containing compound is "labeled" by synthesizing it using C_{14} instead of C_{12}, the compound can be traced through the cell or organism by a Geiger counter or some other radiation detecting instrument, depending upon the type of radiation being released (for example, alpha, beta, gamma, and so on). Still other compounds can be labeled with fluorescent dyes that give off light of various colors when exposed to, say, ultraviolet light. The pigments in a "black light" poster provide a familiar example of how fluorescence

can mark the presence of fluorescent chemical compounds. The radioactively labeled compounds can be detected in a *scintillation counter* that counts tiny flashes of light energy released when the radiation emitted hits a screen containing photosensitive atoms.

Finally, certain chemical compounds may cause specific blockage of certain biochemical processes. Thus if one hypothesizes that a certain cell organelle is involved in one such process, say protein synthesis, the hypothesis can be tested by exposing the organelle to a compound known to prevent this synthesis, seeing if the process is stopped. The functions of several cell organelles have been determined by such techniques.

4.5
BIOLOGICAL MEMBRANES:
THE IMPORTANCE OF COMPLEXITY

Biologists divide cells into two general types, **prokaryotic** and **eukaryotic** (they are often spelled with the letter *c* replacing the letter *k*), and organisms possessing one or the other type are called **prokaryotes** and **eukaryotes** respectively. Attention will be paid to the significant differences between these cell types later. For now you should simply keep in mind that, unless otherwise noted, most of the anatomical and functional descriptions that follow refer to eukaryotic cells exclusively.

Early views of the cell pictured it as composed of a formless, mysterious, undefinable stuff called protoplasm. Later, this cell protoplasm was seen to be divided into two parts, that within the nucleus (**nucleoplasm**) and that outside it (**cytoplasm**). One membrane separated the two parts, while the cytoplasm was held together by yet another membrane. Later, some cell parts (organelles) were found to be membrane-like in quality (such as the endoplasmic reticulum or Golgi complex to be discussed in Chapter 5), while membranes played a major role in the structure of others. Despite their widespread occurrence, however, they were still seen as "just membranes." The real action, it was assumed, lay beyond rather than within them.

In the past few years, this assumption has been shown to be quite false. Indeed, as the following few examples make clear, the cellular membranes appear to be the sites of many important happenings.

1. Far from being passive, sieve-like structures, cell membranes play an active selective role in determining what does and does not get into or out of cells.

2. The membranes of such cell organelles as mitochondria and chloroplasts (to be discussed in Chapter 5), as well as bacterial cell membranes, play a crucial role in the synthesis of the important cell energy compound adenosine triphosphate (ATP). (The precise manner in which this synthesis is thought to occur will be discussed in detail in Chapter 6.) All of the molecular

components and enzymes in both the anabolic (build-up) processes involved in photosynthesis and the catabolic (breakdown) processes involved in respiration are integral parts of the chloroplast and mitochondrial membranes.

3.   In the embryological development of organisms, cells must migrate from one region to another and recognize what kinds of cells are their proper "neighbors," and what kinds are not. This recognition ability appears to lie within a cell membrane coat or *glycocalyx*. Composed of short chains of sugars called oligosaccharides (see Fig. 4.13), this carbohydrate coat is covalently bonded to protein and lipid molecules of the plasma membrane itself, forming complexes called glycoproteins or glycolipids, respectively. Thus embryonic cells of, say, kidney will cease movement, growth, and reproduction when in proper contact with each other (*contact inhibition*). They will keep moving, however, if brought into contact with other types of cells, such as those from the liver. It has been suggested that the characteristic inability of some types of cancer cells to stop dividing may be due to some abnormality of the glycocalyx. Further, the tendency of certain cancer cells to *metastasize,* or form secondary or tertiary tumors by spreading from their primary formation site, is also at least partly determined by the nature of their cell membranes. Thus, once again, a major biological phenomenon, that of proper or improper development, is seen as being intimately related to membranes or membrane-associated structures.

4.   Closely related to the preceding discussion of the glycocalyx is the fact that membranes are now known to play a key role in the immune response. The glycocalyx oligosaccharides provide the recognition sites for proper and improper matching of the A, B, AB, and O blood types in human beings, as well as making foreign cells or viruses (antigenic agents) recognizable by an organism's antibodies. The presence of cholesterol in the plasma membrane tends to make fluid-like membranes more solid but solid-like membranes more fluid, as well as laterally spacing out lipids and other membrane components. This factor becomes important to the immune response in that it may make the foreign agent more recognizable and accessible to incoming antibodies.

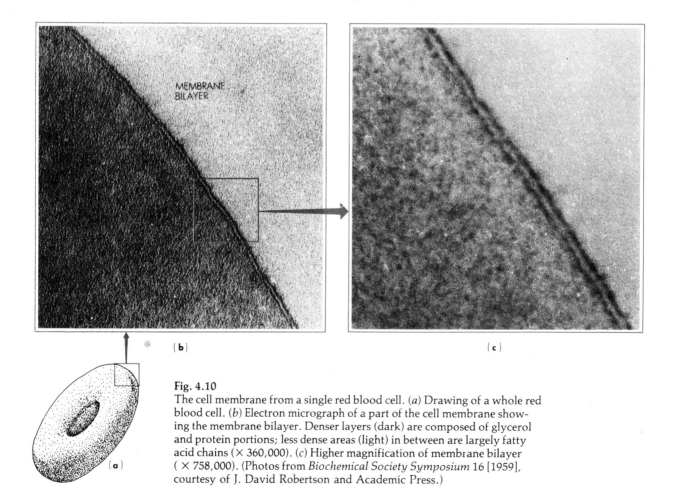

MEMBRANE BILAYER

(b)

(c)

(a)

Fig. 4.10
The cell membrane from a single red blood cell. (*a*) Drawing of a whole red blood cell. (*b*) Electron micrograph of a part of the cell membrane showing the membrane bilayer. Denser layers (dark) are composed of glycerol and protein portions; less dense areas (light) in between are largely fatty acid chains ($\times$ 360,000). (*c*) Higher magnification of membrane bilayer ($\times$ 758,000). (Photos from *Biochemical Society Symposium* 16 [1959], courtesy of J. David Robertson and Academic Press.)

In summary, for these and other reasons, biological membranes and the various molecular complexes embedded in them must now be viewed as playing a central role in both cell structure and function. For example, when the first edition of this textbook came out fifteen year ago, only a few paragraphs were devoted to the cell membrane itself. Now it will be necessary to spend more time discussing this structure than discussing any other cell part—stark testimony to the rapidity with which our knowledge has grown.

4.6
THE CELL MEMBRANE

Structure

Materials pass in and out of the cell through the **cell membrane** or **plasma membrane** (Fig. 4.10). Chemical studies of cells show their isolated plasma membranes to be composed mainly of phospholipids and proteins. As pointed out in Section 3.4, phospholipids consist, at one end, of a long hydrocarbon chain with nonpolar groups, and at the other end of a shorter polar group (see Fig. 4.11). Molecules with polar properties tend to dissolve in water. Thus phospholipid molecules in contact with water tend to line up with their polar ends in the water and their lipid ends away from it.

Based upon their knowledge of these chemical and physical properties of phospholipids, H. Davson and J. F. Danielli hypothesized that the polar glycerol phosphate ends of the phospholipid molecules were oriented

Fig. 4.11
Three prominent phospholipids found in many types of animal and plant membranes. All three molecules consist of two hydrocarbon chains (fatty acids) bound to a glycerol. In place of a third fatty acid chain, phospholipids have a phosphate compound bound to carbon 1 of the glycerol.

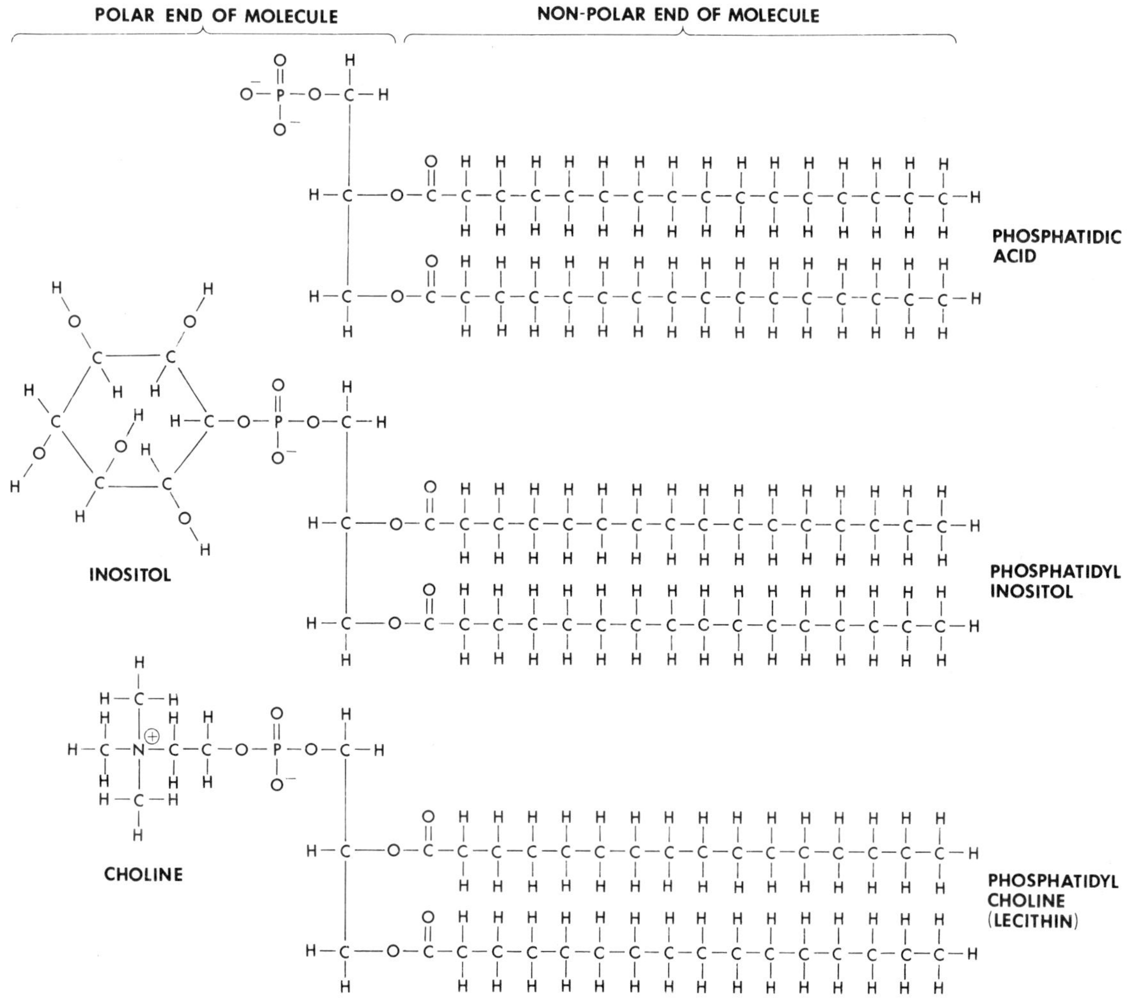

POLAR END OF MOLECULE   NON-POLAR END OF MOLECULE

PHOSPHATIDIC ACID

INOSITOL   PHOSPHATIDYL INOSITOL

CHOLINE   PHOSPHATIDYL CHOLINE (LECITHIN)

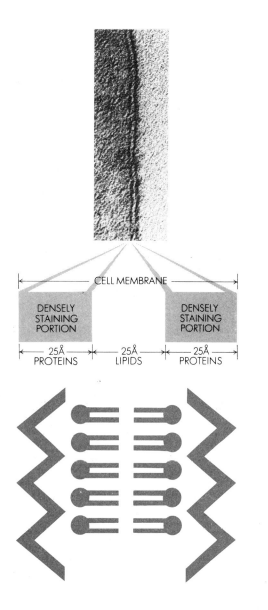

Fig. 4.12
The basic structure of cell membranes as proposed by the Davson-Danielli model. The upper illustration shows a high-power electron micrograph (× 24,000) of a membrane (belonging to a cell in the intestine), showing its structure. The middle and bottom illustrations show schematically the relationship between the membrane structure (as it appears in the electron microscope) and its molecular composition. The densely staining region of each membrane consists of the polar head of phospholipids and surrounding protein. The less densely staining region in the middle consists of the hydrocarbon chains of the fatty acids. The Davson-Danielli model has been refined and expanded in recent years, though much of its basic idea remains intact. (Micrograph courtesy Professor E. De Robertis; reproduced from E. De Robertis, *Cell Biology*, 4th ed. Philadelphia: W. B. Saunders, 1970; by permission of Holt, Rinehart and Winston.)

toward the outer membrane surface, with nonpolar fatty acid chains oriented inwards. According to the Davson-Danielli model, the membrane was a kind of bilayer (two-layer), whose structure derived in large part from the chemical and physical nature of its phospholipid components. Later, electron micrographs seemed to support this model, revealing the membrane to rather resemble a railroad track, with the "tracks" or darker lines being areas of greater density and the lighter space between them an area of lesser density (see Fig. 4.10). Among the forces tending to keep this structure intact would be those associated with the hydrophobic nature of the hydrocarbon chains; with both the inside and outside of the cell being largely water, these chains would orient themselves as far as possible away from it. The polar ends of the phospholipid molecules, on the other hand, are hydrophilic. Layers of protein, both inside and out, further serve to keep this structure intact. Knowing the dimensions of the molecules Davson and Danielli believed were involved, as well as their hypothesized orientation, enabled the scientists to estimate the thickness of the membrane (see Fig. 4.12).

Davson and Danielli built their model mostly upon the known properties of plasma membranes, with special reference to the types of substances that move freely through them (for example, lipid versus nonlipid, or charged versus noncharged), as well as such physical features as its elasticity (which suggested protein was involved). In the 1950s, electron microscope studies indicated that the cell membrane was composed of two relatively thin, dense layers separated by a wider, lighter area. These observations were precisely consistent with predictions stemming from the Davson-Danielli model. Further, the thickness was quite close to the model's predictions. These studies led to formulation of the concept of the so-called **unit membrane**, which viewed both extra- and intracellular membranes as having this same essential structure. Later studies, using X-ray diffraction, indicated that the molecules in the lighter, "meat" portion of the membrane "sandwich" were oriented with their long axes perpendicular to the protein "bread" layers, just as the Davson-Danielli model postulated. Further, artificial membranes, made on the basis of the model, closely resembled real plasma membranes in both structure and function.

Revision of the Davson-Danielli Model

Later research cast doubt on the universality of the unit membrane concept of Davson and Danielli. It was noted that the membranes of differing cells vary widely in thickness and the amounts and kinds of lipids and other substances present, such as cholesterol. Further evi-

The cell membrane, a dynamic association of lipid (phospholipid and glycolipid) and protein, is the site of many biochemical reactions in the cell.

dence that the Davson-Danielli model needed further revision came from still other experimental observations:

1. Electron microscope examinations of membranes after treatment with the enzyme phospholipase, which catalyzes the hydrolysis of phospholipids, were carried out by A. Otholenghi and M. Bowman at Ohio State University. These examinations showed that the membrane is not hydrolyzed within all over, as the Davson-Danielli model would predict, but only in patches. Presumably these patches are areas where exposed phospholipid molecules are concentrated; where not hydrolyzed, some other molecules must be located.

2. M. Glazer and his colleagues, working at the University of California at La Jolla, showed that when red blood cell membranes are heated, the areas assumed to be those hydrolyzed by phospholipase-catalyzed reactions are left intact, but areas sensitive to temperatures of the range that affects proteins are changed.

3. Detailed electron micrographs of cell membranes treated in special ways that separate the two layers sandwiching the inner portion (for example, freeze-etching) reveal many sphere-shaped particles in the region

where the Davson-Danielli model predicts that only lipid hydrocarbon chains should be found.

4. Accurate measurements of the sizes of pores in cell membranes show that some molecules that are small enough to move easily through these pores are somehow prevented from doing so. It is difficult to account for such selectivity on the basis of the Davson-Danielli model.

It sometimes happens in the history of science that a once useful model is entirely discarded as new evidence comes along. In other cases, the old model is simply modified to account for the new evidence. The latter is the case here. The most recent model was put forth in the 1970s by S. J. Singer of the University of California in San Diego and C. L. Nicholson of the Salk Institute. Called the *fluid-mosaic* model, the double or bilayer of phospholipids is retained, oriented as in the Davson-Danielli model (see Fig. 4.13). However, consistent with the first two observations in the list above, the protein layers are not uniform, but rather form a pattern or mosaic on both the outer and inner surfaces of the membrane.

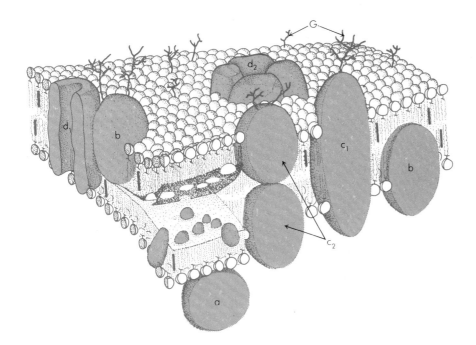

Fig. 4.13
The fluid-mosaic model of membrane structure showing protein molecules "floating" in the lipid bilayer. The heavier shorter objects within the bilayer represent molecules of cholesterol. (*a*) Extrinsic or surface proteins, which do not penetrate the bilayer. (*b*) and (*c*) Intrinsic or immersed proteins, which penetrate the bilayer either partially as a single molecule (*b*), completely (C_1), or as a two-part structure (C_2). Membrane pores, lined with proteins, are visualized as being either individual protein molecules (d_1) or tube-like passages lined with several protein molecules (d_2). The selectivity of the pores with regard to certain ions is accounted for by the differing properties of the R groups, which project into the pore cavity. In some portions of the membrane, glycocalyces (G) are shown (see p. 96).

As the other portion of the name "fluid mosaic" implies, since the membrane fatty acids are unsaturated and have a lower melting point than do saturated ones, the membrane is also viewed as being highly fluid. Within this fluid structure the protein molecules "float," the precise nature of flotation depending upon the protein's molecular weight, charge distribution, and the like. The use of fluorescent **antibodies** shows that these membrane proteins move about within the membrane, in some cases quite freely. Some, called *external* or *extrinsic proteins*, lie at the membrane surfaces and do not penetrate the lipid bilayer. Others, called *internal* or *intrinsic proteins*, may penetrate the lipid bilayer from either side, lie entirely within it, or even extend completely through the entire membrane. Whether a protein is completely extrinsic, intrinsic, or both, depends upon its properties. As one would expect, proteins with polar, hydrophilic chains tend to be extrinsic, while those possessing nonpolar hydrophobic regions tend to become oriented so that those portions of their molecules remain embedded within the lipid bilayer.

It has also been found that the degree of penetration of the lipid bilayer by any individual intrinsic protein varies with chemical changes at the membrane level. For example, the light-sensitive pigment rhodopsin is an intrinsic protein attached to cytoplasmic membranes in visual cells of the eye (in the retina, at the back inner wall of the eye). When retinal cells called rods are in the dark, rhodopsin is only one-third immersed in the lipid bilayer. When the cell is illuminated by a photon of light, the rhodopsin molecules become about one-half submerged (see Fig. 4.14). The membrane is thus intimately involved, structurally and functionally, with many biochemical reactions in the living cell.

The fourth observation is accounted for by assuming that the membrane pores are in some cases passages through individual protein molecules or, in others, tube-like passages lined with protein. (The previous unit membrane concept would envision them as lipid-lined). Pore selectivity can thus be accounted for by the properties of the R groups (for example, charged or noncharged or the like), which would project into the pore cavity.

As might be expected in such a fluid membrane, both vertical and lateral motion of molecules are possible in relation to the membrane's axis. Since only weak forces bind the lipid molecules to each other, almost unlimited lateral motion is possible, especially in the membranes of those forms such as bacteria, which contain no cholesterol to hamper movement. Since they have more specific roles to play than do lipids, the proteins are less mobile though, unless they are locked in a complex with other proteins or lipids (in which case both may be immobilized), even *they* seem to enjoy considerable freedom. As indicated, in general, the location of a protein in a membrane and its degree of mobility depend upon the role it plays. If, for example, a protein plays a role in moving particular substances from one cell to another, it must remain in the area where this work is done.

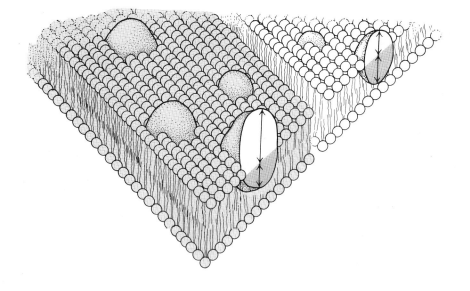

Fig. 4.14
Changes in degree of penetration of intrinsic proteins in the lipid bilayer of membranes are illustrated with the molecule rhodopsin (spherical unit). Rhodopsin is a visual pigment, responsible for receiving photons of light in animal retinas and converting it to electrical energy within retinal cells (called rods). In the dark, rhodopsin penetrates the cytoplasmic membrane (to which it is attached) about one-third of the diameter of the molecule. When exposed to light (right), the molecule sinks deeper into the lipid bilayer. The chemical function of the molecule is intimately related to its structural position in the lipid bilayer of the membrane. (Diagram from S. J. Singer and Nicholson, "The fluid mosaic model of the structure of cell membranes," *Science*, 175 (Feb. 1972), pp. 720–731. Copyright 1972 by the American Association for the Advancement of Science.)

SUPPLEMENT 4.2
CELL MEMBRANE ASSEMBLY: TWO HYPOTHESES

(Note: You may wish to postpone consideration of this supplement until you have completed the study of cellular anatomy and organelle function presented in this chapter and have read Chapter 19 dealing with the mechanism of protein synthesis. It is included here for convenient reference to the discussion of membranes presented in Sections 4.5 and 4.6.)

The development, improvement, and refinement of experimental techniques have increased our knowledge of cell membrane structure considerably. From the early view of the membrane as a simple, relatively static object to the intricate structure envisioned by the fluid-mosaic model is a giant step indeed.

Complex structures, however, usually require complex hypotheses to account for their existence. Granting, for example, that membranes are composed of a lipid bilayer with varying patterns or mosaics of protein often extruding through either surface, we must then ask: How was that structure assembled by the cell? As we have seen, the proteins found in the membrane may be polar and hydrophilic or nonpolar and hydrophobic. Yet the region in the cell associated with protein synthesis is distinctly watery (aqueous). How, then, can hydrophobic proteins be synthesized there

without aggregating or folding improperly? How do membrane proteins choose the right membrane in which to become embedded? In both bacteria and mitochondria, for example, they pass through one membrane on their way to the one in which they will reside. If they could not distinguish one bilayer from another, it would be expected that the proteins would stop at the first one they encounter.

Another problem is that the polar, hydrophilic ends of membrane proteins tend to be extrinsic, that is, to expose their polar ends to the surrounding aqueous medium. But in the case of a hydrophilic protein exposed to the cell exterior, how did the protein, first synthesized in the cell interior, manage to get through the inhospitable hydrocarbon center of the lipid bilayer? And, once there, what factors specify the protein's position in the membrane, in terms of both its transverse and lateral orientation and position?

Two hypotheses have been proposed to guide experimentation that might lead to some insight into the precise mechanism of membrane assembly. One, called the "signal hypothesis," was actually first put forth to suggest how cells could secrete proteins they had synthesized. Since such proteins must pass through the cell membrane, the mechanism by which they do so has obvious significance for problems

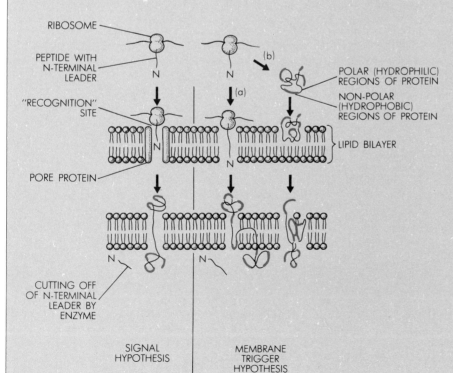

Diagram A

Schematic comparison of the signal and membrane trigger hypotheses. Note that the signal hypothesis involves the recognition and close association of the ribosome with a transport pore protein during protein synthesis. The membrane trigger hypothesis, on the other hand, does not require recognition by a transport pore protein and allows the incorporation of the new protein into the membrane either during (a) or after (b) protein synthesis. (After Wickner, W. Reproduced, with permission, from the *Annual Review of Biochemistry*, 48, 1979, p. 35. Copyright 1979 by Annual Reviews, Inc.)

dealing with those proteins that stop part of the way through to become part of the membrane. The signal hypothesis proposes that the key to both the transport of proteins across the membrane (secretion) and membrane protein assembly lies in the means by which the ribosome catalyzes the protein's synthesis.

Several observations led to the signal hypothesis. One was that cells that secrete proteins for use elsewhere outside the cell have more rough endoplasmic reticulum (that is, with ribosomes attached) than those who do not carry out such secretion. Biochemical observations indicate that those groups of ribosomes (polysomes) found free in the cytoplasm (not associated with the endoplasmic reticulum), synthesize mostly soluble (polar, hydrophilic) proteins. Thus polysomes isolated in association with the endoplasmic reticulum synthesize proteins that are destined for export, while the free, "soluble" polysomes do not. Yet on both secreted and membrane protein molecules, there is found the widespread occurrence of hydrophobic N-terminal "leader" (that is, coming off the ribosome first) peptides, consisting of approximately fifteen to thirty amino acid residues. Most significant, this N-terminal is removed by a specific protease enzyme during or shortly after the peptide's passage through the membrane.

The signal hypothesis proposes that this N-terminal region binds to a receptor in the bilayer and is recognized by a transport system built into the membrane. The ribosome catalyzing the peptide's synthesis binds to this transport system, and the force of the polypeptide chain elongation drives the peptide through this specific transport system pore. On emerging on the other side of the bilayer membrane, the leader peptide is removed by the protease enzyme. In the case of secreted proteins, the peptide is driven all the way through, while the integral membrane proteins are simply released from the peptide transport system at an earlier stage, while still within the membrane.

A second hypothesis, the "membrane-triggered folding hypothesis," lays less emphasis on the means of assembly and stresses the role of the N-terminal leader's reaction with the aqueous environment in allowing the growing peptide chain to fold into its normal functional configuration. In other words, according to this hypothesis, the thermodynamics of protein-folding alone can account for membrane assembly, and it is not necessary to complicate the explanation by involving the mechanism of protein synthesis and precise site recognition on the membrane by the ribosome. Instead, the recognition of the proper membrane by the protein is less specific and involves more than the N-terminal leader peptide portion of the new protein. Indeed, the membrane-recognizing element may be either protein, lipid, or even a combination of particular physical properties of the membrane at the reaction site. The interaction of the protein with the lipid components of the membrane results in the proper positioning of its hydrophobic and hydrophilic ends in relation to the bilayer's fatty acid chains, and this interaction may occur before the peptide's synthesis is complete. When the N-terminal leader is removed, the process becomes irreversible; the completed protein is now part of the fluid mosaic structure of the membrane.

A central distinction of the membrane trigger hypothesis in comparison to the signal hypothesis is the absence in the former of a peptide transport system and the proposed role

Table A
A Comparison of the Signal and Membrane Trigger Hypotheses. Each Difference Suggests to Cell Biologists Critical Experiments to Test Which of the Two Hypotheses Is the Most Valid.

Stage of synthesis	Signal hypothesis	Membrane trigger hypothesis
Site of initiation	soluble polysomes	soluble polysomes
Role of the leader peptide	recognition by the protein transport channel	to change the folding pathway
Association of the new protein with membrane	time: only when leader peptide is complete place: protein transport channel	time: during or after protein synthesis place: receptor protein or lipid portions of the bilayer
Specific ribosome associations	with the protein transport channel	none
Catalysis for assembly	a specific pore	the effect of the leader peptide on conformation
Driving force for assembly	polypeptide chain elongation	protein-protein and protein-lipid associations: self-assembly
Removal of leader peptide	during polypeptide extrusion	during or after polypeptide assembly into bilayer
Segregation of proteins into different cellular membranes	not clearly addressed	not clearly addressed
Final orientation	C-terminus in, N-terminus out	specified by the primary sequence of amino acids

of the N-terminal in helping the new or forming protein to assume its proper configuration in reaction to its exposure to the lipid bilayer. In these respects, and in the fact that it does not require any specific ribosome-membrane interactions, the membrane trigger hypothesis is the simpler.

Table A shows other differences between the signal hypothesis and the membrane trigger hypothesis. Both hypotheses illustrate nicely the fact that their mere formulation automatically suggests critical experiments that might be designed to distinguish between them. There is a great deal of experimental evidence that is consistent with the signal hypothesis. For example, if a specific protein transport pore recognizes an N-terminal signal peptide, then proteins that lie completely across the bilayer should do so with their N-terminal on the outside (since that is the part that is hypothesized to lead the molecule through the bilayer) and their

C-terminals on the inside, in the cytoplasm. This has, indeed, been shown to be the case for a large number of membrane proteins. Yet a protein has been identified in the human red blood cell membrane with precisely the opposite orientation. On another of its predictions, that proteins can insert into or cross membranes only while being synthesized, the signal hypothesis fares even worse; there are several examples that have been found in which this is clearly *not* true. These may, of course, be exceptional cases, but the fact remains that the signal hypothesis needs more conclusive evidence to become firmly established. One way would be to purify and rebuild the membrane components necessary for protein insertion and processing and then showing that a specific protein does, indeed, recognize N-terminal leader peptides, bond to the ribosomal subunit, and conduct peptides across the lipid bilayer.

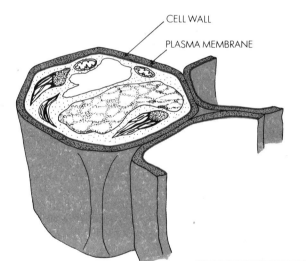

CELL WALL

PLASMA MEMBRANE

The Cell Wall

Plant cells, unlike animal cells, generally possess a cell wall, against which the plasma membrane presses internally (see Fig. 4.15). The cell wall is porous, allowing water and dissolved substances to pass through it (see Fig. 4.16).

Cell walls are the product of the cells they surround and are generally composed of layers. Adjoining cells both produce a **middle lamella**, composed of a group of polysaccharides called **pectins**. (The softening of ripen-

MIDDLE LAMELLA
PRIMARY CELL WALL
SECONDARY CELL WALL
INTERCELLULAR SPACE

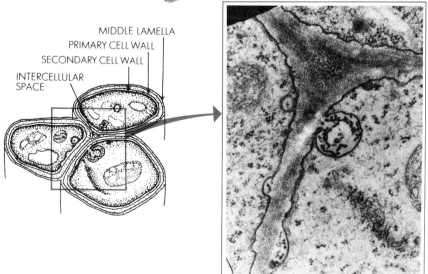

Fig. 4.15
The plant cell wall. Despite its rigidity, the wall is porous and allows many substances to pass through it. The electron micrograph shows the cell wall between three cells of the sugar pine, *Pinus lambertiana*. As the arrow indicates, material is often found filling the spaces caused by such cell wall junctures. The vacuole-like bodies touching the wall on either side between two cells are called lomasomes. They are commonly associated with strands of ribosomes (linear polysomes), and Golgi apparati are frequently found in their vicinity. It has been suggested that they may play a role in the construction and maintenance of the cell wall. (Micrograph courtesy of Dr. Graeme P. Berlyn, Yale University.)

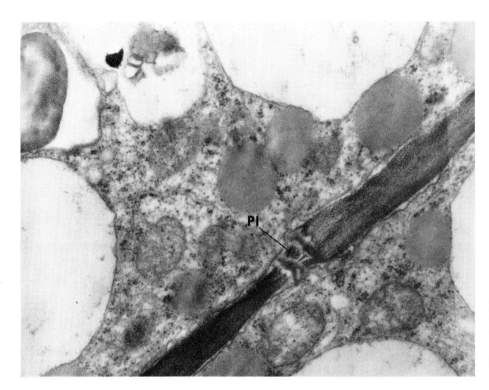

Fig. 4.16
Plasmodesmata (Pl) are openings in the cell walls between neighboring cells, allowing cytoplasmic bridges to form between them. (Micrograph courtesy of Dr. Graeme P. Berlyn, Yale University.)

ing fruit is due to the hydrolysis of pectins catalyzed by the enzyme pectinase.) In addition to the middle lamellae they build in conjunction with their neighboring cells, plant cells build a semiflexible primary **cell wall** on its own side of the middle lamellae out of the polysaccharide **cellulose**. In cells of the harder plant tissues, such as wood, a secondary cell wall is found, with the cellulose reinforced by a substance called **lignin**.

Since the cell wall material is both nonliving and rigid, with the individual plant cells joined to each other by their common lamellae, movement of the sort associated with animals is impossible for plants.

4.7
DIFFUSION, OSMOSIS, AND TRANSPORT ACROSS MEMBRANES

With the current model of the cell membrane firmly in mind, attention will now be directed to the various ways in which certain substances pass through it into the cell and the reasons why certain other substances do not.

If a spoonful of sugar is dropped into a cup of hot tea, without being stirred, *eventually* the dissolved sugar molecules disperse throughout the tea until it is no sweeter at the bottom than at the top or anywhere else. This movement of the sugar molecules is an example of *diffusion*; when first put in the tea, the sugar molecules are highly concentrated in the bottom of the cup where the sugar is dissolving; when the diffusion process

reaches an equilibrium, the concentration of sugar molecules is about the same throughout the tea.*

It is often said that all diffusing substances, whether atoms, ions, molecules, or whatever, *tend to move from an area of greater concentration to an area of lesser concentration.* However, there is another, more fruitful way of looking at it, a way first introduced in another context in Chapter 2, Section 10. We can change the italicized statement above to read *systems move from a state of high organization or orderliness to a state of maximum disorganization or disorderliness.* In other words, they move from a state of low entropy to one of high entropy. Recall from Chapter 2 that the relationship between entropy and free energy is an inverse one; the greater the entropy, the less free energy, all other things (such as enthalpy) being equal. This inverse relationship enables a restatement, in the most meaningful terms possible, of our original generalization: *the overall or net movement is from regions of greater to regions of less free energy.*

It may appear that each italicized statement is simply saying essentially the same thing using different words. However, in science generalizations that have

---

*Needless to say, the same phenomenon occurs in gases; pungent odors soon fill an enclosed room. However, since our eventual interest is in diffusion as it relates to cell membranes, and cell membranes deal with substances in solution, a liquid medium such as tea provides a better example.

the fewest exceptions are preferred. Consider the first italicized statement. While it seems that substances do usually move from a region of greater concentration to a region of less concentration, they may not *always* do so. Suppose, for example, the region with a lesser concentration of diffusing molecules possesses more heat, and thus more free energy. This greater temperature results in a higher velocity of the diffusing molecules, that is, more pressure. Thus, in this case, the diffusion occurs from the region of lesser to greater concentration. While this result contradicts the first italicized statement, it is entirely consistent with the last one.

Osmosis

With the preceding discussion firmly in mind, we will now direct our attention to a special case of diffusion of greatest biological interest—diffusion through a membrane. Consider the U-tubes below:

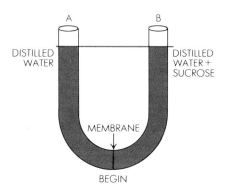

BEGIN

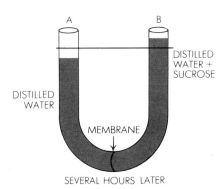

SEVERAL HOURS LATER

The two arms, A and B, are separated at the bottom by a membrane. Into arm A, distilled water is placed; into arm B, distilled water in which is dissolved 10 grams of the disaccharide sucrose (table sugar). Both arms are filled up to the colored line mark. After waiting several hours, tests of samples from each arm give positive results for sucrose in arm B but negative results in arm A.

Evidently, the membrane blocks the diffusion of sugar molecules across it into arm A; it is *impermeable* to sucrose. (Had the test been positive, the membrane would be said to be *permeable* to sucrose.) If we had dissolved some sodium chloride (table salt) in the sugar-containing water as well, and the tests for the presence of sodium and chloride ions been positive in arm A, the membrane would be referred to as *differentially permeable* or *semipermeable*.

Note, however, that despite the negative test results for sucrose, there *has* been a change in the system; the level of water has lowered in arm A and risen above the mark in arm B, thereby diluting the sugar solution. Clearly water *was* able to pass through the membrane, though the sucrose was not. The membrane thus allowed the passage of solvent (in this case, water) through it, but not solute (in this case, sucrose). The movement of a solvent across such semipermeable membranes is called **osmosis.** (The movement of *solute* across a semipermeable membrane will be discussed shortly.) Note also that the membrane now has a slight bulge in it pointing toward side A due to the greater **hydrostatic pressure** on its side B surface, exerted by the taller column of water. Indeed, theoretically, without this pressure osmosis would continue until there was no water on side A, since no matter how diluted side B became it would always have less free energy than side A.

What forces are involved in moving the water from A to B? Recall now the discussion in the previous section about the tendency of systems to move from a state of orderliness or organization (that is, possessing high free energy and low entropy) to a state of disorder or disorganization (that is, possessing low free energy and maximum entropy). In plain water, the molecules are in a state of high organization (see Fig. 2.10). The introduction of sugar into solution disrupts that organization, making side B more disorderly. Thus there is more free energy in side A than in side B, and the water flows across the membranes down this free-energy gradient.

Suppose now, instead of a U-tube, we have an expandable membrane system that is closed, like a sphere or box. Inside is a solvent (water) with several solutes (such as sugars, salts, proteins, and the like) at least some of which are unable to pass through the membrane. It follows from the discussion in the preceding paragraph that any time water with less solute particles comes in contact with the membrane, it will move into the cell by osmosis, since it has more free energy or, if you will, it is less disorderly. Since the membrane is a closed system, it will swell until the pressure is great enough to overcome the free-energy gradient difference between the outside and inside. Such pressure is called *osmotic pressure.* As you might expect, the amount of osmotic pressure is directly related to the *osmotic concentration*, that is, the number of osmotically active particles per unit volume of solvent (how concentrated

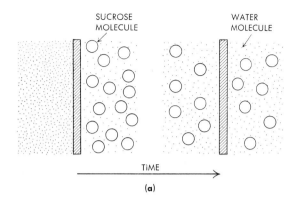

SUCROSE MOLECULE   WATER MOLECULE

TIME

(a)

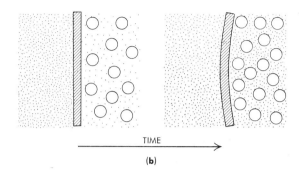

TIME

(b)

the solution is). The osmotic concentration on either side of the membrane determines both the direction and the rate of osmosis; under constant temperature and pressure, the greater the osmotic concentration, the greater the rate of osmosis (see Fig. 4.17).

Note that in moving from the U-tube to the closed membrane system, we are now dealing with an exact analog of the cell. It is now easy to understand why the cells of a plant's roots, with many substances dissolved in the *cytosol,* or liquid portion of the cytoplasm, may absorb rainwater from the soil. The entry of this water maintains an osmotic pressure within each cell; in plants this helps to maintain *turgor,* the pressure exerted by the cell contents against the cell membrane or cell wall to keep the stem and leaves rigid. Without water, the plant loses turgor, and wilts. If watered with a saturated solution of salt the plant also wilts, for now the free energy content of the water inside the cell is *greater* than that outside, and thus **exosmosis** or "reverse osmosis" occurs.

Biologists often find it easier to use the concept of *osmotic potential* than osmotic pressure, assigning distilled water an osmotic potential of zero. If a solute is introduced into this water on one side of a semipermeable membrane but not the other, the solute-containing water's osmotic potential is less than zero and the distilled water on the other side moves across the membrane by osmosis. Thus osmotic potential has an *inverse* relationship to osmotic concentration; the greater the osmotic concentration, the less the osmotic potential and vice versa. Indeed, since the solvent with which they deal in living systems is always water, physiologists usually refer to "water potential" rather than osmotic potential. Since osmotic potential is related to solute concentration alone, while water potential varies according to temperature and pressure, the latter is obviously a much more realistic concept when dealing with living organisms.

At first glance, it is tempting to view osmosis as simply a special case of diffusion by which water (or a solvent) passes through submicroscopic pores in semipermeable membranes. Certainly it is true that the

Fig. 4.17
(a) Diffusion of sucrose and water molecules through a membrane permeable to both. Eventually, the rates of each kind of molecule passing through the membrane from either side become equal. (b) A membrane impermeable to sucrose but permeable to water. The water molecules pass from left to right, since their free energy is greater at left. The resulting osmotic pressure or potential exerts a force on the membrane.

direction and equilibrium of osmosis can be accurately predicted by a hypothesis that interprets osmosis in this manner. Yet careful measurements using isotopically labeled water (H_2O^{18}) reveal a contradiction: water traverses a porous semipermeable membrane faster than the rate predicted, even if the theoretical maximum rate of diffusion is assumed to prevail. Current hypotheses concerning the nature of osmotic flow contain the concept of "bulk flow"; a steep free-energy gradient is envisioned at one end of the pore, resulting in such rapid diffusion that the water density decreases. The result is a pressure gradient in the pore bringing about the faster bulk flow.

It is now easy to see how a living cell can control the amount of water that enters it by osmosis. By simply concentrating solutes inside of it, the cell creates a situation in which the water outside now has more free energy and thus moves in; such a system is said to be *hypotonic* (see Fig. 4.18). If the cell has less solutes within it than the surrounding medium exosmosis occurs; such a system is said to be *hypertonic.* The fact that nonliving cells form balanced or *isotonic* systems suggests that cells must expend energy to maintain hypo- or hypertonic conditions. The evidence for believing this and the means by which cells are thought to be able to do so will be discussed next.

Transport across Membranes

Thus far everything that has been said about the movement of solvents and solutes across membranes relates to nonliving systems as well as to those that are living.

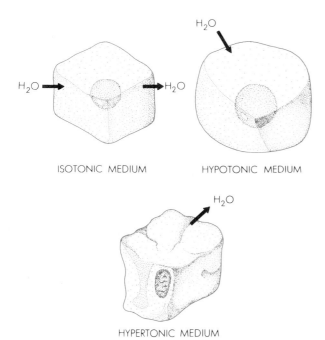

Fig. 4.18
A cell can exist in three different states in regard to the liquid medium that surrounds it. In an isotonic medium, the free energy situation on either side of the cell membrane is balanced and no more water moves in than moves out. In a hypotonic medium, a free energy imbalance exists; the water outside has more free energy outside than inside and thus enters the cell, causing it to swell. In a hypertonic medium, since the greatest free energy exists inside the cell, the reverse occurs and the resulting movement of water from the cell leads to its dehydration and shrinking.

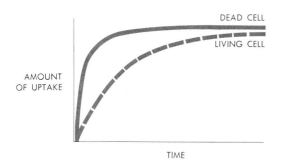

Fig. 4.19
This graph shows the uptake of ions introduced into a liquid medium surrounding dead and living cells. Note that the living cells are able to offer some resistance to diffusion of ions into the cell. A graph of the respiratory rates of such cells shows that they must expend energy to accomplish this.

There are some important membrane-transport mechanisms, however, that are strictly limited to living systems. Attention will now be directed to these transport mechanisms.

In general, proteins are not passed through the plasma membrane of cells. It is logical to wonder why this should be so. Since the plasma membrane is perforated by pores so small that they are measured in angstrom units, we might guess that only those molecules small enough to pass through the pores in the plasma membrane will pass into the cell. This hypothesis is supported by the fact that small molecules, such as those of water, generally pass freely in and out of cells, while larger molecules, such as proteins, generally do not.

Other observations, however, contradict this hypothesis. Ions of sodium are kept out of resting nerve cells, yet such ions are far smaller than the smallest pore in the plasma membrane. Conversely, under certain conditions large molecules of nucleic acid will pass into some cells. Thus the pore hypothesis may account for the movements of some substances in and out of cells, but it cannot account for others. The cells of many different organisms can concentrate molecules or ions in their interiors to a far higher degree than the surrounding medium can. Many of these molecules and ions are of such small dimensions that they could easily pass out of the cell, and since diffusion occurs from a region of greater to a region of lesser free energy, it seems they would do so. But they do not. Why *don't* they?

The answer is suggested by additional experiments. It has been shown that the respiratory rate of cells that are concentrating ions or molecules against the normal concentration gradient is far higher than the respiratory rate of similar cells that are not concentrating the same molecules or ions. It seems evident, then, that these cells can maintain a high concentration of dissolved substances within themselves only by expending energy. A similar group of cells, intact but nonliving, is unable to maintain a high concentration of salts. Not being alive, these cells are unable to expend the energy necessary to hold the ions and molecules within their cytoplasm. The laws of diffusion have the upper hand, and the free energy of the substances inside and outside these nonliving cells soon equalizes (see Fig. 4.19).

As reported by one investigator, the pores in the plasma membranes of red blood cells have a diameter of 7 Å. Yet glucose molecules 8 Å across are able to pass into the cells. Here again, experiments show that energy is being expended by the red blood cells; they must perform work in order to get the glucose inside.

Dialysis. **Dialysis** is the diffusion of a dissolved substance through a differentially (semi-) permeable membrane. Dialysis takes place as long as the membrane is permeable to the molecule or ion in question. If the membrane is impermeable to a molecule, the molecule cannot diffuse across it. Study of living cells has revealed that membranes play a crucial role not only in selecting what molecules will be allowed to pass, but in actually aiding the physical movement of certain substances into or out of the cell by the processes of **facilitated diffusion** and **active transport.** These forms of transport provide a good example of interrelationship between structure and function in biological membranes.

Facilitated diffusion.   A number of membrane-bound transport enzymes have been identified that aid in moving specific molecules across cell membranes, particularly the plasma membrane. For example, glucose enters some animal cells by the aid of specific enzymes at the cell surface. No energy is expended, so the process is one of facilitated diffusion. Yet sugars similar in size, structure, and properties to glucose are not transported in the same way. This indicates the highly specific nature of facilitated diffusion. The specificity is found to lie in the transport enzymes. Active transport also involves highly specific membrane proteins. In both cases the membrane proteins, called permeases, may act as carriers, facilitating the movement of molecules from one side of the membrane to another.

Active transport.   Active transport may follow the same general pattern. In fact, it may even be that the same carrier proteins are involved in both facilitated diffusion and active transport. It is known, for example, that such sugars as glucose and galactose can be moved by both facilitated diffusion and active transport. Hence, it is likely that their passage may involve the same carrier molecule. However, since active transport can move molecules across membranes against a con-

centration gradient, the process has several differences from facilitated diffusion, although the binding and translocation steps may be very similar. If the concentration of S is greater inside the cell than outside, the release step is different. With a high concentration of S inside, the new S molecule would have a tendency to stay bound to the carrier. Energy is used in the form of the energy-supplying molecule adenosine triphosphate (ATP) to remove S from the carrier protein. This frees the carrier, which can then pick up another S molecule at the outer surface of the membrane. In this way, the cell does work to increase the concentration of substances inside it to a degree considerably greater than the outside concentration.

Facilitated diffusion and active transport have many features in common. They also have some important differences. Their common features include: (1) movement of molecules and ions across membranes at rates faster than would occur by simple dialysis alone, and (2) a high degree of specificity; both are selective in terms of which molecules they transport. Their differences lies mainly in the fact that active transport requires the expenditure of energy by the cell, whereas facilitated transport does not. Active transport can actually move substances across a membrane *against* a concentration gradient (that is, from an area of lower to one of higher concentration), whereas facilitated diffusion cannot. Two models showing this distinction between facilitated diffusion and active transport appear in Fig. 4.20.

A hypothetical model has been developed to explain both facilitated and active transport in terms of membrane-bound carrier proteins. Figure 4.21 shows generalized schemes for facilitated diffusion and active

FACILITATED DIFFUSION

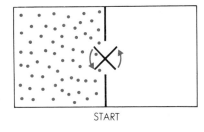

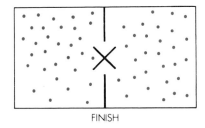

START   FINISH

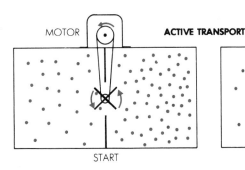

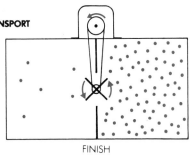

MOTOR   **ACTIVE TRANSPORT**

START   FINISH

Fig. 4.20
General model showing the chief differences between facilitated diffusion and active transport. Active transport involves the expenditure of energy (as in running the motorized paddle wheel in the diagram) to move substances against a concentration gradient. Facilitated diffusion does not involve expenditure of energy on the cell's part.

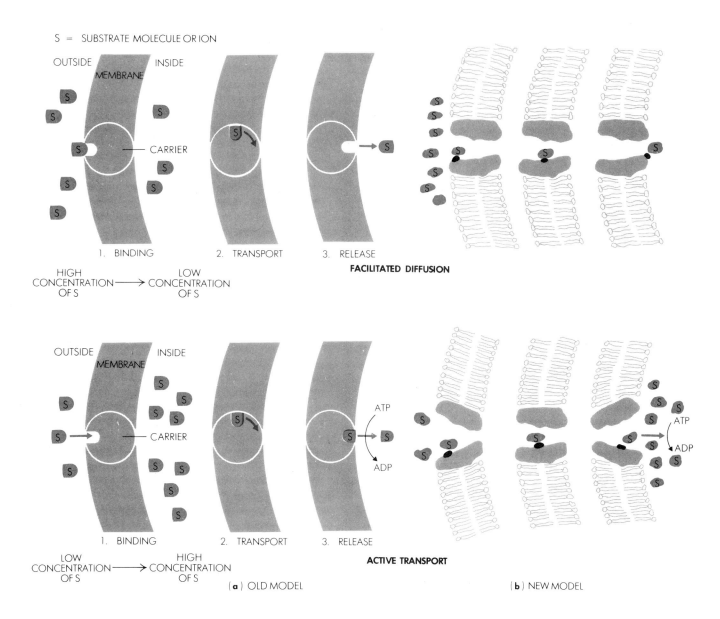

Fig. 4.21

The third edition of this textbook (1977) showed the model presented in (a) to account for facilitated diffusion and active transport across the cell membrane. The model was based upon the idea that the movement of a substance across a membrane was accompanied either by the movement of the transport or carrier protein as well, or by the rotation of the carrier protein as depicted in (a), so that its binding site was turned from one side of the membrane to the other. More recent evidence supports the concept of fixed pores formed by the association of transmembrane proteins (see also Fig. 4.13) to form channels entirely lined with protein (b). In the case of facilitated diffusion, the transport protein (permease) merely accelerates the movement of a substance by making it easier for it to penetrate the membrane. In the case of active transport, energy is expended in moving the substance against a free energy gradient, either by a shift in a conformational state that moves the active site and the associated substrate molecule to the other side of the membrane, as depicted above, or by shifting the substrate molecule from one binding site to another one located on the releasing side. The two processes of facilitated diffusion and active transport can be compared to simple diffusion, in which the molecules simply pass across a membrane permeable to them without either the involvement of permeases or the expenditure of energy.

Table 4.1
Ionic Concentrations in the Red Blood Cell and the Surrounding Blood Plasma

	Ions, concentration in milliequivalents per liter	
	Red blood cell	Blood plasma
K$^+$	150	5
Na$^+$	26	144
Cl$^-$	74	111
Ca^{++}	70.1	3.2

transport. Consider facilitated diffusion first. A molecule to be transported (S) must first come into contact with the membrane—specifically with a carrier protein component of the membrane. Carrier proteins are of the "intrinsic" type; that is, they are partially immersed in the lipid layer. When the specific diffusible S molecule contacts a carrier protein, the latter binds to it much like any enzyme binds to its substrate. The binding process causes a conformational shift in the protein, such that the protein may sink deeper into the lipid bilayer, or perhaps even flip completely over, undergoing a 180° change in position. The result is that the S molecule being aided in its passage is now on the inside of the membrane rather than the outside. The carrier protein simply ushers the S molecule through the lipid region of the molecule. Once on the other side of the membrane, the S molecule is released in a manner possibly similar to the release of a substrate from its enzyme after a chemical reaction.

Table 4.1 shows the enormous differences in concentration that can be achieved through active transport. The table illustrates several points. Active transport is selective; potassium (K$^+$) and calcium (Ca^{++}) are selectively transported into the cell against very high free-energy gradients, whereas sodium (Na$^+$) and chloride (Cl$^-$) are not. But, equally important, note that sodium and chloride are actively transported *out* of the cell. Active transport works in both directions, bringing substances into the cell and moving substances out of the cell—often against a free-energy gradient.

Active transport is sometimes referred to as a cell pumping system (for example, the "galactose pump" or the "sodium pump"). The term "pump" should not be taken too literally. It does signify, however, the fact that work is being done and that something is being moved in an "uphill" direction.

The molecular details of both facilitated diffusion and active transport are very poorly understood. That carrier proteins in the membrane exist is clear. Exactly how they work involves a great deal of speculation and model-building. Competitive inhibition, for example, has been shown to occur if solutes whose molecular structure is similar to the regular solute are introduced. Such a result would not occur if only simple diffusion was involved; presumably the introduced solute molecules compete for binding sites on the carrier molecules.

The actual existence of such carrier proteins—*permeases*—has been shown by the isolation of several types and description of their binding characteristics. However, hypotheses concerning the precise means by which the carrier proteins move the solutes for which they are specific from one surface of the membrane to the other remain highly speculative. Using the models, however, it is possible to make certain predictions: for example, that applying cyanide to cells (which stops certain energy-generating reactions in cells, and hence

Fig. 4.22
Phagocytosis in the giant amoeba, *Chaos chaos.* In more common terms, the amoeba is said to engulf the prey, a smaller protozoan called *Paramecium.* The amoeba extends parts of its cell body (a pseudopod) around the prey, enclosing it. The cell membrane then fuses to form a space, the vacuole, with the *Paramecium* inside. Digestion takes place by the passage of enzyme molecules from the prey's cytoplasm into the vacuole. At bottom right, the *Paramecium* is actually inside the vacuole, although the perspective makes it appear to be in the cytoplasm.

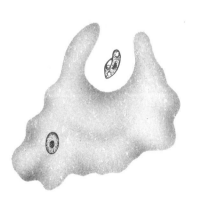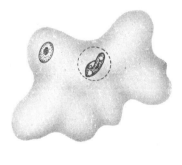

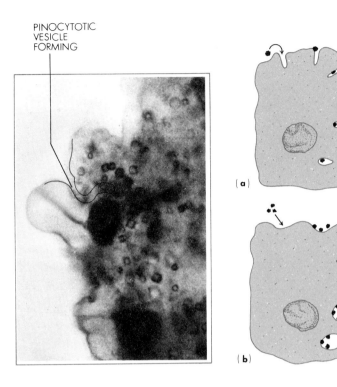

PINOCYTOTIC
VESICLE
FORMING

(a)

(b)

Fig. 4.23
Electron micrograph ($\times$ 315,000) of pinocytosis, showing the infolding of the membrane. The thin colored line
denotes the formation of the pinocytotic vesicle. The
cell shown here is very thin and narrow and located in
the smallest blood vessels, the capillaries. The drawings
to the right indicate two ways in which tiny particles
may enter the cell by pinocytosis, (a) by flowing into a
channel like the one in the electron micrograph and becoming pinched off, or (b) by simply becoming enclosed
in vesicles at the surface and being moved into the cell
interior. (Micrograph courtesy George E. Palade.)

Fig. 4.24
Reverse phagocytosis is depicted here as this pancreatic
cell secretes protein. Protein is synthesized in the cell
cytoplasm (around membranes containing the small
organelles called ribosomes). The protein (zymogen, a
precursor of a digestive enzyme) is concentrated within
the sacs in the cytoplasm near the bottom portion of the
cell. Small sacs are pinched off each large sac; these
small sacs fuse together forming zymogen-containing
sacs that are eventually extruded into the duct at the
upper end of the cell. (After Lowell E. and Mabel R.
Hokin, "The chemistry of cell membranes," *Sci. Am.*,
October 1965, 78–86.)

interferes with production of ATP) would soon bring a
stop to all active transport but would not affect facilitated diffusion. The prediction is borne out; soon after
application of cyanide to red blood cells, the ionic concentrations inside and outside come to equilibrium. Yet
much research still needs to be done in attempting to
isolate and characterize specific carrier proteins and to
understand the actual process by which they translocate
molecules across the membrane. The field of membrane
research is indeed an open and exciting area for future
investigation.

Bulk Transport.   Many cells are capable of moving
large masses, such as droplets of fluid, particles of food,
globules of protein, etc. in a process called **bulk transport.** If the substance is moved into the cell it is called
endocytosis; if out, *exocytosis.* The substance being
moved across the membrane is first surrounded in a
membrane vesicle; for example, the one-celled amoeba
uses its extremely flexible shape to surround its prey,
gradually enclosing it in a space with walls that fuse
together and shrink until the captured organism is
enclosed within a vesicle that is now *inside* the cell. Enzymes then catalyze digestion of the organism for absorption into the cytoplasm. White blood cells (leukocytes) perform a similar function when they engulf
infectious bacteria. This endocytotic process is often
called **phagocytosis,** meaning "cell eating" as opposed to
pinocytosis, meaning "cell drinking," when liquids are
involved (see Figs. 4.22 and 4.23). A representation of
"reverse phagocytosis" is shown in Fig. 4.24.

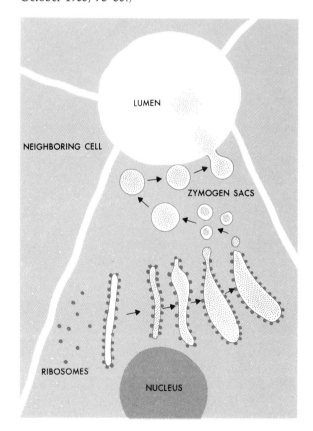

LUMEN

NEIGHBORING CELL

ZYMOGEN SACS

RIBOSOMES

NUCLEUS

SUPPLEMENT 4.3
MODELS IN SCIENCE

Whereas models are not necessarily a part of every scientific explanation, they have become such an integral part of many explanations, especially in contemporary biology, that the term has become commonplace.

Models are like analogies in that they attempt to represent the unfamiliar in terms of the familiar. Models usually involve some sort of physical representation of a complex process or structure whose details cannot be directly observed. An electron micrograph of a cell is not a model—it is a direct visual representation of the cell structure. The Davson-Danielli concept of membrane structure *is* a model because it attempts to represent the physical structure of something we cannot observe directly. Similarly, the Bohr concept of the atom as a miniature solar system is a model.

Most models attempt to relate their physical components to known functions. The Davson-Danielli membrane hypothesis postulates a physical structure that was in agreement with known functional properties of membranes at the time it was published (1937). Partly because they attempt to represent complex phenomena and structures with physical components, most models are admitted oversimplifications. By simplifying, and by finding physical components to represent what would otherwise be abstract notions, models can bring considerable coherence and easy comprehension into scientific theories.

The Davson-Danielli model of membrane structure is a part of what we can call an overall theory of membrane structure and function. The theory would encompass many more elements than the Davson-Danielli model. For instance, it would include concepts of ion diffusion (physical chemistry) and electrical properties of cells. Any broad theory is likely to encompass at least several models within its overall framework.

One of the chief functions of models is to aid in making predictions. The Davson-Danielli model predicts that dissolving away the lipid portion of a membrane should totally destroy the membrane's structure and function. This prediction can be (and was) put to the test. Because the results were negative, the model could be rejected, at least as it stood. The value of models is that they can be tested.

Like theories and hypotheses, models are not permanent. They are constantly being stated, modified, and sometimes rejected. Again, the Davson-Danielli model is a good example. Because it failed to predict certain phenomena accurately, the model had to be changed. This change did not reject all elements of the original model, but it did significantly alter the model. The new model (it does not really have a single name) accounts far more precisely than the old model for observational and biochemical data about membrane structure and function.

A persistent danger plagues the use of models in science. Sometimes workers or students tend to take the models too literally. The Bohr model of the atom is a good case in point. Planets in the solar system travel in specific orbits. It is possible to predict where a planet will be at almost any given moment for a long time in the future. Physicists have found that such predictions are not possible with electrons in atoms. The idea of electrons traveling in fixed orbits like planets became a hindrance to understanding how atoms are really put together. The quantum theory, which replaced Bohr's planetary model, is conspicuous in the absence of a mechanical model for atomic structure. Physicists now claim it is impossible to develop any model of atomic structure in physical terms. That is, they hold that no model (as we have been using the term) is possible, and now view atomic structure in mathematical and probabilistic terms.

Summary

1. The cell "theory" maintains that (a) cells are the structural units of virtually all organisms; (b) cells are also the functional units of all organisms and thus the seat of both health and disease; (c) cells arise only from preexisting cells and are not spontaneously generated from inorganic material; (d) cells contain hereditary material (nucleic acid) through which specific characteristics are passed on to the next generation of cells.

2. All cells, whether single-celled organisms or cells in multicelled organisms, face certain common problems: they must carry out metabolism (gain and use energy), reproduce, transport substances in and out of the cell, quite frequently show some form of motion (intracellular as well as overall cell motion), exhibit growth and development, and maintain themselves in the face of constantly changing external conditions.

3. Tools for studying cell structure include the microtome (for preparing thin tissue slices), the light microscope, and the electron microscope. Light microscopes can magnify only to a maximum of about 2000 times; electron microscopes can magnify up to 1,000,000 times.

4. The cell membrane is a lipid bilayer in which a variety of proteins float. It functions partly to determine what sub-

stances go in and out of the cell, and partly as a surface structure on which many reactions are localized. Membranes in living cells are physiologically very active, rather than being the passive barrier structures described in the past.

5.   Cells transport materials (water, ions, and small and large molecules) in and out by a variety of different processes. These include:

   a) Diffusion. A physical process in which materials tend to move from an area of greater free energy to one of lesser free energy. This can occur passively across cell membranes.

   b) Osmosis. A physical process in which water moves across a semipermeable membrane. The cell membrane appears to be relatively passive in osmotic flow.

   c) Facilitated diffusion. The movement of materials across the cell membrane from areas of greater to areas of less free energy, but at a rate faster than could occur by diffusion alone. Facilitated diffusion does not require ex-

penditure of cellular energy. It is particularly involved in the transport of sugars into the cell.

   d) Active transport. The movement of materials across the cell membrane from areas of less to areas of greater free energy. This process involves the expenditure of energy by the cell. Both facilitated diffusion and active transport involve membrane-bound carrier proteins as agents for translocating molecules across the membrane.

   e) Bulk transport. The process by which large, macroscopic portions of liquids or solids are moved into or out of cells. The membrane is intimately involved in bulk transport. Bulk transport involves at least two distinct kinds of processes: (i) Phagocytosis: the taking in (engulfing) of solid material by formation of a membrane enclosure around it, which becomes incorporated into the cytoplasm as a vacuole. (ii) Pinocytosis: the taking in of bulk quantities of water (or liquid). Both phagocytosis and pinocytosis appear to occur in the elimination of materials from the cell as well.

Exercises

1.   Describe the modern view of the plasma membrane of a living cell. In what way is the structure of the membrane closely correlated with its function?

2.   Distinguish between osmosis and diffusion.

3.   A particular membrane is said to be permeable to water and sodium chloride, yet impermeable to glucose molecules. Explain how this might be possible.

4.   Figure 4.25 is a diagram of an osmometer similar to those used to demonstrate osmosis. At the beginning of

a demonstration there is a 50 percent glucose solution within the membrane, which is impermeable to glucose. Draw a graph representing the rate of osmosis in this system, indicated by the rate at which the water climbs up the tube.

5.   A snail can be killed by throwing salt on it. You get very thirsty (indicating that your body cells are low on water) after eating a salty meal. In growing living cells in test tubes, physiologists are very careful to use a medium known as Ringer's solution, in which the salt concentration is exactly that found in body and cellular fluids. Explain all three of these facts in terms of osmosis and dialysis.

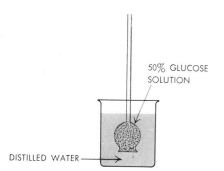

50% GLUCOSE
SOLUTION

DISTILLED WATER

Figure 4.25

Chapter 5
Cell Structure
and
Function II

5.1
INTRODUCTION

This chapter will be concerned primarily with the structure and function of the major cell organelles known to date. We will then turn our attention to a comparison of prokaryotic and eukaryotic cells and the kinds of organisms that possess them. Finally, we will consider the ways in which cells unite to form functional units called tissues and the types of these tissues found in plants and animals.

5.2
THE CELL NUCLEUS

Discovery of the cell nucleus was first reported in 1831 by Robert Brown.* Its early discovery was undoubtedly due to its prominence in many cells, where it stands out as slightly darker than the surrounding cytoplasm. Staining the cell with certain stains makes this distinction even greater and, indeed, due to its affinity for stains, especially in certain regions of its nucleoplasm, the substance in these darkly staining regions came to be called "chromatin," from the Greek word for "color." During cell division the chromatin becomes clearly visible as being organized into rod- or rope-shaped structures called **chromosomes** (that is, colored bodies), which are composed of the genetic material DNA, and associated proteins (see Fig. 5.1).

In eukaryotic cells, the nucleus possesses a limiting membrane, the **nuclear membrane** or **envelope** that separates its contents from the surrounding cytoplasm. This membrane bears the same relationship to the nucle-

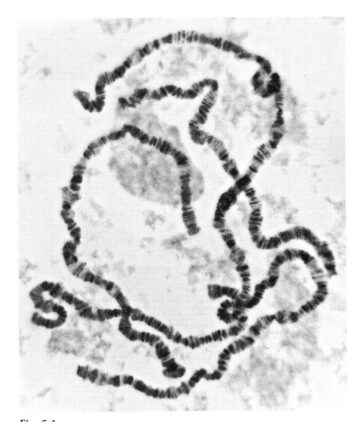

Fig. 5.1
Photo of chromosomes from the salivary gland cells of the fruit fly. The dark bands represent areas that preferentially take up stain. Chromosomes are composed mostly of protein and deoxyribonucleic acid (DNA). (Courtesy Turtox, Chicago.)

us as the plasma membrane does to the cytoplasm; it serves largely as a regulatory device through which materials pass in and out. Unlike the plasma membrane, however, the nuclear membrane is doubled. Electron micrographs reveal that the nuclear membrane has pores through which certain substances may pass from the nucleus into the cytoplasm and vice versa (see Fig.

---

*Brown's name is better known in physics; he discovered that dust particles in a suspension have a distinctive movement, thought to be the result of their being bombarded by the molecules surrounding them. This motion is still referred to as "Brownian motion." Robert Hooke, the discoverer and namer of cells, is also well known in physics for his work on coil springs, which led to the establishment of "Hooke's Law."

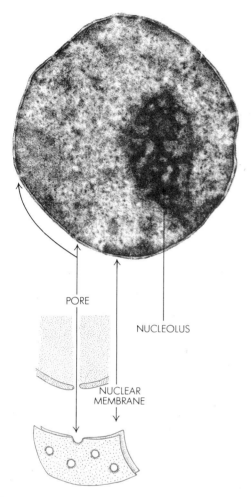

Fig. 5.2
The cell nucleus, showing the prominent nucleolus and the nuclear membrane containing pores (× 12,000). (Courtesy Dr. George E. Palade.)

5.2). There is now direct electron micrograph evidence that at least some of these pores connect with the endoplasmic reticulum, a cell organelle to be discussed shortly.

Some nuclei contain one or several rounded or oval bodies called *nucleoli* (singular, *nucleolus*). The fact that they are most prominent in cells active in protein synthesis suggests that nucleoli play an active role in the process, and experiments using labeled isotopes support this hypothesis. Indeed, since they, too, are composed of DNA and protein, and since they develop in association with certain regions of the chromosomes, they are now considered by many to be simply specialized parts of them. Nucleolar DNA codes for the synthesis of ribosomal RNA which, in combination with protein, is transported to the cytoplasm to become part of the ribo-

somes, or organelles, to be discussed shortly in greater detail.

Since most of the cell's genetic material (DNA) lies within the nucleus, it is not surprising that cells experimentally deprived of their nuclei or not possessing them normally (such as mammalian red blood cells) are unable to reproduce (red blood cells are produced from other nucleated cells in the bone marrow). The precise manner in which the genes within the chromosomes convey their instructions to the cytoplasm will be taken up in Chapter 19. It is easy to see, however, why its possession of most of the cell's genes conveys an image of the nucleus as a control center of cell structure and function.

5.3
THE ENDOPLASMIC RETICULUM

While it had long been suspected that the cytoplasm might have some sort of "skeletal" structure of its own, it was not until the mid-1940s that the phase contrast microscope enabled the identification of part of that skeleton—the **endoplasmic reticulum** (often abbreviated ER). The term "reticulum" refers to a series of interconnected spaces, while "endoplasmic" refers to its location within the cytoplasm. The ER membranes vary widely in appearance from one cell type to another, at times being merely membrane-enclosed, fluid-filled spherical or tubular spaces called *cisternae*; at other times being layers of membranes separated by fluid but with interconnections so that substances may pass from one layer to another. At times the ER is studded with darkly staining, rounded structures called ribosomes, the cell organelle to be discussed next. In this case the ER is called "rough" ER; with no ribosomes it is "smooth" ER, and in this form the ER plays an important role in neutralizing (detoxifying) poisons and such drugs as amphetamines, morphine, and the like. Like the cell membrane, the ER membranes are composed of phospholipid and protein (see Fig. 5.3).

In the third edition of this textbook (1977), the statement was made that the ER membranes "appear to connect with the nuclear membrane, but agreement on this point is not universal." The words "not universal" are probably still correct, but recent electron micrographs seem to show such connections via the nuclear membrane pores so clearly there seems little room left for doubt. Indeed, some investigators have suggested that the nuclear membrane be considered as only a specialized part of the endoplasmic reticulum. Since such a connection would allow for the efficient transfer of nuclear and ER-manufactured substances into the cytoplasm and beyond, the obvious convenience of the arrangement seemingly lends further support to this argument.

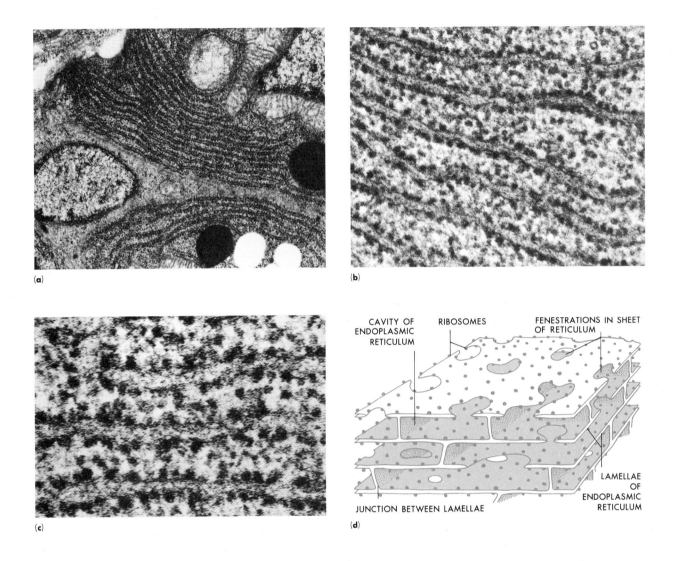

CAVITY OF
ENDOPLASMIC
RETICULUM

RIBOSOMES

FENESTRATIONS IN SHEET
OF RETICULUM

LAMELLAE
OF
ENDOPLASMIC
RETICULUM

JUNCTION BETWEEN LAMELLAE

Fig. 5.3
Electron micrographs of sections through chick embryo pancreas, showing endoplasmic reticulum at three different magnifications—(a) × 15,400, (b) × 75,600, (c) × 115,500. At the higher powers, the double nature of the lamellae and their associated ribosomes is clearly visible. Compare with the diagram (d). (Photos courtesy Elizabeth Johnson, Wesleyan University. Diagram after S. Hurry, *The Microstructure of Cells.* Boston: Houghton Mifflin, 1964, p. 10.)

It seems evident, however, that the endoplasmic reticulum provides cells with more than merely conducting channels through the cytoplasm. Simply by holding the ribosomes in a definite orientation, rough ER undoubtedly contributes to the efficiency of protein syn-

thesis, the major function of ribosomes. However, there is strong evidence that the ER membranes themselves contain, as an integral part of their structure, protein-enzyme complexes for the synthesis of several cell products. Smooth ER, for example, appears to have such built-in enzymes for the production of lipids, for smooth ER is abundant in cells heavily involved in lipid synthesis (particularly steroids), and it has not proven possible to separate this chemical ability from the structure of the smooth ER membranes themselves.

5.4
RIBOSOMES

The most numerous by far of the cell organelles, ribosomes may be found scattered throughout the cytoplasm or, as seen in Fig. 5.3, attached to the endoplasmic reticulum (thereby making it "rough"). Whether or not the ribosomes are located on the endoplasmic reticu-

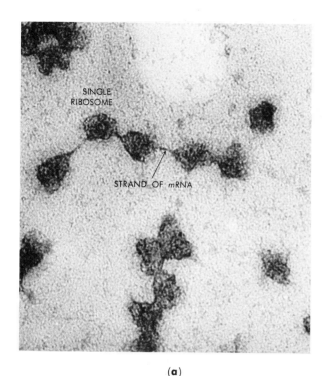

(a)

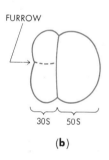

(b)

Fig. 5.4
Ribosomes are cell organelles responsible for protein synthesis. (a) Several ribosomes with what is thought to be a molecule of *mRNA* strung between them. Messenger RNA contains genetic information (from DNA) specifying amino acid sequence. Ribosomes are organelles that "read" the genetic information. In the ribosomes, specific amino acids are joined into polypeptides. Groups of ribosomes joined together by a strand of *mRNA* are called polysomes. (b) Diagram of a bacterial ribosome showing the two-part structure (30S and 50S). (Electron micrograph courtesy Alexander Rich, J. R. Warner, and C. E. Hall, from A. B. Novikoff and E. Holtzman, *Cells and Organelles.* New York: Holt, Rinehart and Winston, 1970, p. 69.)

lum appears to be related to the use of the completed product; if it is to be used by the cell itself, the ribosomes are free in the cytoplasm; if it is for export and use elsewhere, as in the case of hormones, digestive enzymes, and the like, the ribosomes are attached to the

endoplasmic reticulum. This, presumably, allows the latter's channels to be used to convey the finished product via other organelles to the plasma membrane for export.

Ribosomes are so named because they contain high concentrations of RNA. By "tagging" with radioactive atoms the smaller molecules (amino acids) of which proteins are composed, it can be shown that the amino acids go first to the ribosomes and later show up in protein molecules. This observation provides the basis for believing that proteins are synthesized either in or on the ribosomes. Since virtually all living cells synthesize proteins, ribosomes are one of the few subcellular particles that must be present in all living things. The hypothesis that ribosomes are the site of protein synthesis would predict their presence in larger numbers in cells that are very active in protein synthesis than in cells that are less active. This proves to be the case; a typical bacterium, growing rapidly (and thus synthesizing large quantities of protein), may contain some 15,000 ribosomes.

Electron microscope examination of bacterial ribosomes shows that they have a diameter of about 18 nm, whereas those from mammalian cells are a bit larger. Unfortunately, the electron microscope does not provide the detail necessary to study the *precise* structure of ribosomes. What is now known about ribosome structure has been established by indirect experimental evidence. It has been shown that the bacterial ribosome consists of two parts. These parts are designated 50S and 30S respectively, because of their sedimentation (settling out) characteristics in a centrifuge (see Fig. 5.4b).* Neither part can synthesize protein by itself, but it remains a mystery why this two-part structure is essential to ribosomal functioning. The parts themselves differ in structure, with the 50S bacterial ribosome subunit consisting of 23S ribosomal RNA and about thirty-five different proteins, and the furrowed 30S subunit consisting of 16S ribosomal RNA and about twenty different proteins. The proteins appear to be present as globular spheres with a diameter of 32 to 40 Å. Ribosomes are often found associated in groups known as **polyribosomes** or **polysomes**.

5.5
THE GOLGI COMPLEX

The **Golgi complex,** also often referred to as the **Golgi apparatus** or **Golgi body,** consists of a cluster of flattened, nearly parallel, smooth-surfaced vesicles located

---

*The S stands for Svedberg units. A Svedberg unit is a sedimentation constant—a measure of the rate of sedimentation of, for example, a ribosomal part in a centrifuge.

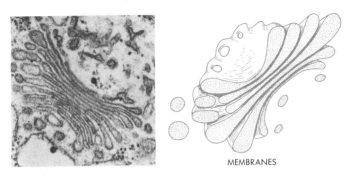

MEMBRANES

Fig. 5.5
Golgi complex, showing the stack of folded membranes called collectively the dictyosome (× 43,000). (Photo courtesy Dr. George E. Palade.)

within the cytoplasm (see Fig. 5.5). The vesicles, often referred to collectively as the **dictyosome**, contain numerous smaller infoldings or pockets, and at times rather resemble smooth endoplasmic reticulum. Indeed, it appears that the Golgi complex may have at least temporary connections with the endoplasmic reticulum membranes.

While for a long time the function of the Golgi complex was unknown, its prominence in cells known to secrete various products led to the belief that it played a role the secretion process. It was also noted that the Golgi complex's shape changed when such cells were highly active. The role of the Golgi complex in the storage and packaging of cell products for eventual export from the cell now seems assured. At times, the Golgi complex may modify the cell products it contains; proteins produced by the endoplasmic reticulum's ribosomes, for example, are believed to move to the Golgi complex via the smooth endoplasmic reticulum. The Golgi complex then adds to some of these proteins the polysaccharides it has synthesized from monosaccha-

rides, thereby producing glycoproteins. In a similar manner, it modifies lipids into glycolipids. Such products are wrapped into various types of vesicles (for example, in the case of digestive enzymes, lysosomes, to be discussed shortly) for movement to the cell membrane and bulk transport to the exterior. At times it appears that the membranes of these vesicles, after discharging their contents, have become modified to resemble the plasma membrane, for they often remain incorporated into it, thereby providing the cell with additional surface area.

5.6
LYSOSOMES

In the 1950s, the electron microscope revealed other structures in the cytoplasm that rather resembled mitochondria, but whose membranes were single rather than double. These are **lysosomes** (Fig. 5.6). Lysosomes are saclike structures probably produced by the Golgi complex and found widely in animal cells and fungi. They

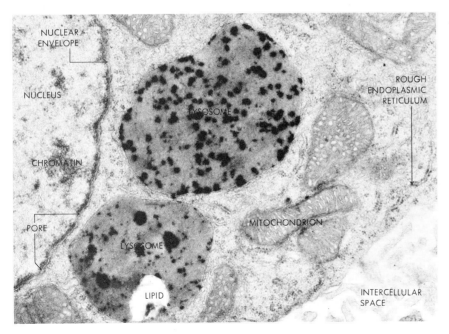

Fig. 5.6
Electron micrograph of a lysosome in a portion of a functional kidney tubule cell of a thirteen-day-old chick embryo (× 38,000). The dark spots in the lysosome are the sites of lead deposits that resulted from chemical tests for the activity of one of the enzymes present in the lysosome. Also labeled are the nucleus, nuclear envelope and pore, endoplasmic reticulum, and intercellular space into which cell processes extend. (Courtesy Dr. E. Sue Lumb, Department of Biology, Vassar College.)

contain hydrolytic enzymes that catalyze the breakdown of molecules such as fats, proteins, and nucleic acids into smaller molecules. These smaller molecules can then be used as raw materials for synthesis or as energy sources. It may be that the lysosomes serve to isolate these digestive enzymes from the cell cytoplasm, thereby keeping the cell from digesting itself. This hypothesis is supported by the fact that when the lysosome membrane is ruptured, the cell undergoes chemical breakdown, or **lysis.** There is experimental evidence linking the lysosomes to the muscle atrophy occurring after surgical denervation or disease-caused nerve paralysis. It has further been noticed that at death the mitochondria and lysosomes are broken down. It may be that the resulting release of enzymes contributes to the early, irreversible changes that occur after death.

In plants, lysosomal enzymes are associated with the vacuoles which, in dying plant cells, frequently contain partially digested parts of mitochondria and other cell organelles. On the other end of the life cycle, it is likely that lysosomes play a role in the destruction of that embryonic tissue that is either a vestigial product of evolution or that played a role in one stage of development but is no longer needed (for example, the vegetarian-adapted mouth parts of a tadpole, which as an adult frog is carnivorous).

It is significant to note that lysosomal enzymes are among the most stable components of living matter. Some can remain stable at temperatures that would denature almost any other large protein. We can hypothesize why this might be an important characteristic of lysosomal enzymes. Since these enzymes live in an environment of highly concentrated compartmentalized digestive enzymes, they must have various structural means (perhaps many covalent bonds holding the polypeptide chain of the enzyme in its tertiary form) to keep from being digested by each other. Lysosomal enzymes may therefore be adapted to exist in hostile environments.

5.7
PEROXISOMES AND GLYOXYSOMES

Cells contain various sorts of tiny vesicles similar to lysosomes, and it is often difficult to distinguish clearly between them. Often, such vesicles are simply termed "microbodies" or "microsomes."* However, recent cell fractionation studies have identified organelles called **peroxisomes** and **glyoxysomes.** Similar to lysosomes in general appearance, they are quite different in function, serving as containers for oxidative rather than hydrolytic enzymes. In green plants, peroxisomes are involved in puzzling phenomenon, *photorespiration*, to be discussed in Chapter 7. Their enzymes are also in-

volved in oxidation of amino acids and, as their name suggests, of hydrogen peroxide. Glyoxysomes are vesicles containing enzymes that catalyze the conversion of fat to carbohydrate.

5.8
MITOCHONDRIA

Mitochondria are highly complex cell organelles. They were first noted under the light microscope in the late nineteenth century; but it was not until the development of the electron microscope that their structure could be determined with any precision. Electron micrographs reveal mitochondria to be double-membraned structures, the outer one smooth and the inner one with many folds, called **cristae** (see Fig. 5.7). The membranes

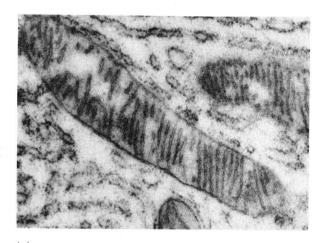

(a)

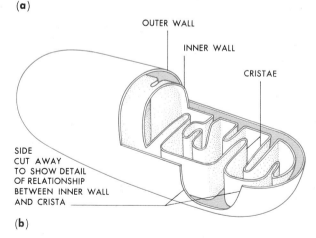

OUTER WALL

INNER WALL

CRISTAE

SIDE CUT AWAY TO SHOW DETAIL OF RELATIONSHIP BETWEEN INNER WALL AND CRISTA

(b)

Fig. 5.7
(a) Mitochondrion from a pancreas cell, shown in longitudinal section ($\times$ 20,000). Mitochondria are covered by an outer membrane and lined by an inner membrane that is infolded into numerous partitions, called cristae. Schematic drawing (b) illustrates the overall structure of the mitochondrion in three-dimensional perspective. (Photo courtesy Dr. George E. Palade.)

---

*"Microsomes" is a term first applied to ribosomes upon their discovery in 1938. Their name was changed as their composition and function became known.

The basic function of many organelles is to enclose in a limited space a set of chemical reactions that are interdependent.

are separated by a fluid of relatively low viscosity. The inner membrane's cristae possess stalked knobs that are exposed, with the cristae themselves, to a dense, viscous matrix that fills the inner cavity of the mitochondrion. An integral part of the cristae are the **respiratory assemblies,** highly organized collections of molecules such as respiratory enzymes and cytochromes, to which considerable attention will be devoted in the next chapter. Like the plasma membrane of the cell, mitochondrial membranes are composed of phospholipid and protein.

The precise functions of mitochondria remained unknown until the mid-1930s, when differential centrifugation (see Supplement 5.1) revealed high respiratory activity located in the fraction containing the mitochondria; it is in the mitochondria, then, that energy is released from certain breakdown products of ingested food and stored in a form easily usable by the cell. More will be said about how this is done in the next chapter.

As might be expected from their function, mitochondria vary widely in number from cell to cell, de-

SUPPLEMENT 5.1
THE FUNCTION OF CELL PARTS: HOW DO WE KNOW?

The electron microscope is extremely useful in studying cell structures. But how do we know what functions the various components—membranes, organelles, and the like—serve? Here the electron microscope is less useful. Cell biologists use the centrifuge as one tool for studying the function of cell components. Cells of a particular type (such as kidney or liver cells) are put into a blender so that the cell membranes are broken down. The cell contents are thus spilled out and can be suspended in liquid. This suspension is poured into a tube and spun in a high-speed ultracentrifuge. A centrifuge uses the power of centrifugal force to force particles in suspension to settle out, or sediment, at the bottom of the centrifuge tube. A centrifuge can effectively increase by many thousands the force of gravity that causes particles to settle out of a suspension if allowed to sit undisturbed for a period of

time (see Diagram A). Most particles the size of cell organelles would never settle out by simple gravitational force.

Increasing the speed of centrifuge rotation increases the centrifugal force. The denser the particle, the less centrifugal force is necessary to cause it to settle out of suspension. Thus a suspension of whole cell parts might first be spun at low speeds, forcing to the bottom of the tube only the densest particles. The fluid remaining at the top, the *supernatant,* is poured off, and the precipitate is resuspended. The resuspended precipitate can either be centrifuged again, to refine the separation (there will be mixed in with the dense particles some less dense particles that started out at the bottom of the tube), or studied biochemically. The first supernatant is then centrifuged at a higher speed, and the precipitate obtained from this procedure is resuspended and subjected to bio-

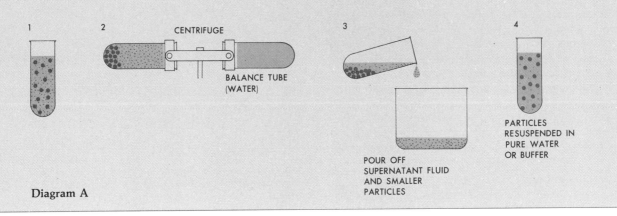

1

2   CENTRIFUGE

BALANCE TUBE (WATER)

3

POUR OFF SUPERNATANT FLUID AND SMALLER PARTICLES

4

PARTICLES RESUSPENDED IN PURE WATER OR BUFFER

Diagram A

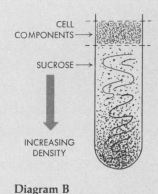

CELL
COMPONENTS →

SUCROSE →

INCREASING
DENSITY

Diagram B

chemical tests. This process can be repeated a number of times until the final precipitates (containing the least dense particles) are obtained. In this way the cell can be taken apart and each fraction separated from other fractions to be studied separately.

For further refinement of this technique, the cell components can be suspended in a specially prepared liquid called a *sucrose density gradient*, simply a sucrose solution that shows increasing density from the top of the tube to the bottom. The cell components are then added to the top of this tube as shown in Diagram B. In a liquid of variable density, particles will tend to settle out at various densities of the medium that more closely resemble their own density. (Sucrose is commonly used to form the density gradient because it is readily available and is not usually metabolized directly by most cell components.) After cell components are spun in a sucrose density gradient, particles of different density be-

come separated into relatively homogeneous layers. The density gradient thus allows the precise separation of numerous components in a single spin, as shown in Diagram C. The layers can then be drawn off one at a time by a special separatory process.

Once a relatively pure fraction is obtained, it can be subjected to biochemical tests to determine its function. The mitochondrial fraction, for example, has been found to metabolize pyruvate or acetate to produce carbon dioxide. This suggests that the function of mitochondria is to carry out cell respiration. The microsomal fraction (which consists of ribosomes attached to membrane fragments of the endoplasmic reticulum) is found to incorporate radioactively labeled amino acids into polypeptide chains. This suggests that the function of ribosomes is protein synthesis. Using these and other techniques, we can determine the role of each cell component.

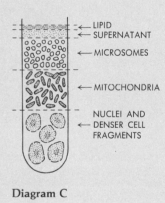

← LIPID
← SUPERNATANT
← MICROSOMES
← MITOCHONDRIA
NUCLEI AND
← DENSER CELL
FRAGMENTS

Diagram C

pending upon the degree of the cell's activity. Equally varied in number are the cristae; the cells of lining tissue (epithelial) have mitochondria with relatively few cristae infoldings, while the mitochondria in insect flight muscle cells have many.

In the mid-1960s, by tagging molecules of choline (a substance associated with mitochondrial lipids), D. L. Luck of the Rockefeller University demonstrated that mitochondria are, like cells, capable of growth and division, and are therefore the descendants of preexisting mitochondria. This observation suggested that mitochondria must possess their own DNA, and this has been shown to be the case; mitochondrial DNA appears to code for the synthesis of ribosomal RNA, transfer RNA, and some mitochondrial proteins. The reproductive independence of mitochondria has led to speculations concerning the origins of mitochondria during evolution. A discussion of some of these speculations can be found in Supplement 26.1.

5.9
PLASTIDS

Plastids are small structures found only in plant cells.* There are two kinds, those that are colorless, the **leukoplasts,** and those that are colored, the **chromoplasts.** The color varies according to the kind of pigment contained. By far the most significant chromoplasts in terms of all living things are the *chloroplasts.* Indeed, some plant biologists prefer to recognize the chloroplast as a third type of plastid, though most still consider that their green color makes them clearly a type of chromoplast.

---

*Certain marine animals, called slugs, are known to contain functional chloroplasts within the cells of their body tissues. However, the chloroplasts are obtained by feeding on algae; not all the chloroplasts are digested, and some end up inside the slug's body cells.

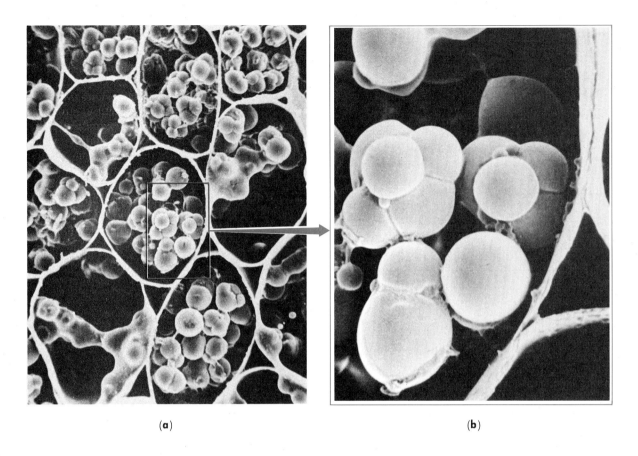

(a)   (b)

Fig. 5.8
(a) Photomicrograph (× 4300) of starch grains in root cells of the buttercup (*Ranunculus repens*). Leukocytes in many plants are primarily storage places for carbohydrates, particularly starches. Here the cells are filled with starch grains, which can be digested and used as an energy source for plant growth. Photograph (b) is an enlargement (× 1100) of the grains in photograph (a). (Photos courtesy Barbara Bole and Elizabeth Parsons, University of Bristol, Long Ashton Research Station.)

Leukoplasts are believed to serve as centers for the storage of certain foodstuffs such as starch (see Fig. 5.8), and they are often named after the type of food they store (for starch, amyloplasts). Chromoplasts, because of the various pigments they contain, give color to the flowers and fruits of many plants. The green color of many leaves and stems is, of course, due to the green pigment **chlorophyll** of the chloroplasts.

The close association of the chloroplast and its green pigment chlorophyll is one of the most significant in nature. The energy from sunlight (solar energy) is first captured by chlorophyll and, eventually, becomes trapped in the chemical bonds of organic molecules, pri-marily sugar. The precise manner in which this **photosynthesis** is done, to the degree that it is understood today, will be discussed in Chapter 7.

Function and structure are intimately related in nature, but nowhere is this more evident than in the structure of the chloroplast. Almost its entire anatomy can be viewed as a vast complex of interrelated membranes, to which chlorophyll and the other molecules involved in photosynthesis appear intimately bound. In this sense, then, chloroplasts are very similar to mitochondria, though of course the latter are involved with primarily catabolic or breakdown processes, rather than anabolic or building-up processes.

The similarity between chloroplasts and mitochondria does not stop here, however. Like mitochondria, chloroplasts are bounded by a double membrane and have a dense, semiliquid fluid, the **stroma,** that is quite different in composition from that of the cell cytoplasm surrounding the chloroplast.

As already indicated, besides the outer double membrane and the stroma it encloses, the rest of the chloroplast's anatomy is either membranous or membrane-related. These membranes or **lamellae** run in parallel series from one end of the organelle to the other (see Fig. 5.10). At times the membranes run through the stroma with little or no seeming connection between them; at others, they become united to form stacks of flattened sacs or discs called **thylakoids.** The molecules

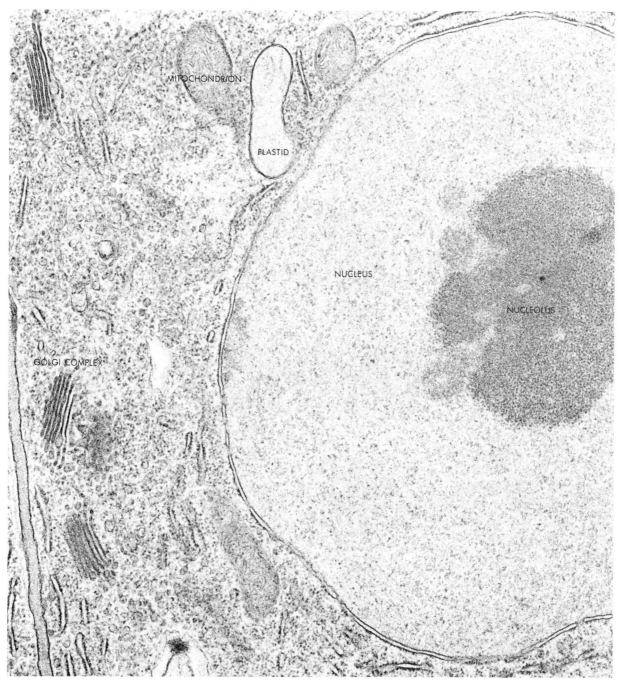

Fig. 5.9
Electron micrograph of a cell from the root tip of a plant, showing a number of cell organelles, including a prominent plastid (P) (× 28,000). (Photo courtesy Dr. M. C. Ledbetter, reproduced from T. P. O'Brien and M. McCully, *Plant Structure and Development*, New York: Macmillan, 1969, p. 7.)

of chlorophyll are embedded within the thylakoid membranes. This means that the actual green color of the chloroplasts is not uniform throughout their structure; where the lamellae run through the stroma between the thylakoids, they are colorless. Thus under high-powered light microscopes, the green color appears speckled or granular, and for this reason each stack of thylakoids is called a **granum** (plural, **grana**).

In a very real sense, the thylakoid discs can be compared to the panels of solar houses. Both are oriented to

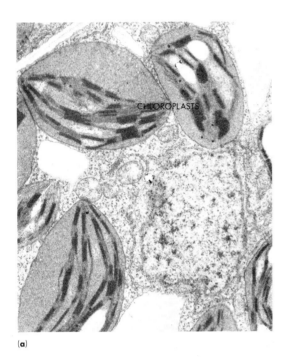

(a)

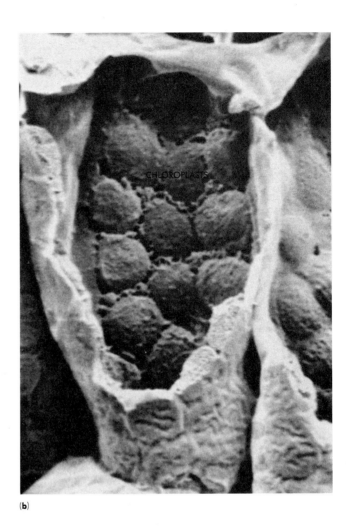

(b)

Fig. 5.10

Chloroplasts. (*a*) Electron microscope view of chloroplasts in cells of the bean plant, *Phaesolus* (× 1700). The chloroplasts are the football-shaped bodies within the cell boundaries. (Photo courtesy W. M. Laetsch). (*b*) Scanning electron microscope view of chloroplasts from the leaf of a French bean plant (× 2000). Note how tightly packed the chloroplasts are in these cells, compared to those shown in (*a*). (Photo courtesy Barbara Bole and Elizabeth Parsons, University of Bristol, Long Ashton Research Station). (*c*) Electron micrograph of a thin section of a single chloroplast from a leaf of *Froelichia* (× 45,550). Note how the internal membranes, the lamellae, run the length of the chloroplast. Periodically the lamellae are stacked together, producing a denser array of membranes known as grana. The space between lamellae is fluid-filled and is called the stroma. (Photo courtesy W. M. Laetsch). (*d*) Enlarged electron micrograph of a number of grana, showing the stacked lamellae from a single chloroplast of *Phyllorpadin* (× 61,750). (Photo courtesy W. M. Laetsch).

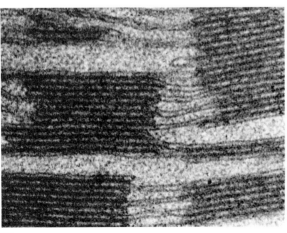

(d)

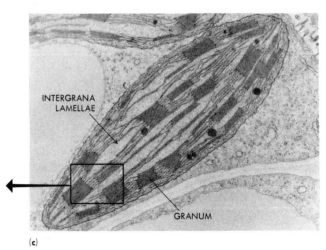

(c)

capture sunlight with maximum efficiency, though the thylakoids, through the slow movements of the plant's leaves, are able to do so far more efficiently. Further, each chlorophyll molecule, by virtue of its being held in position by the thylakoid membrane with which it forms a complex, has its most receptive portions held in position for the capture of the incoming quanta or photons of light energy, again resulting in a remarkable increase in efficiency. Finally, while solar panels only rather clumsily capture solar energy and convert it to heat, the photosynthetic apparatus of the chloroplast "locks up" solar energy into a myriad of biologically useful compounds.

To the striking similarities of chloroplasts and mitochondria already noted, which have led to many speculations about possible evolutionary relationships between these organelles, must be added one more: like mitochondria, chloroplasts contain their own DNA. This DNA appears to code for certain of the photosynthetic enzymes and for chloroplast ribosomal RNA. In their development, chloroplasts, like all plastids, are derived from small, unpigmented proplasts. Under the influence of light, some of these (including, under special conditions, leukoplasts) develop chlorophyll and become chloroplasts.

5.10 VACUOLES

Many kinds of nonliving inclusions are found within the cytoplasm. Plant cells, for example, contain bubblelike structures called **vacuoles** (see Fig. 5.11b). These often serve as reservoirs, holding sap or waste products. A **vacuolar membrane** separates the contents of the vacuoles from the surrounding cytoplasm. The vacuolar membrane has a structure similar to the outer plasma membrane of the cell, and it regulates the passage of materials into and out of the vacuole and the cytoplasm.

Some unicellular organisms called Protozoa (such as *Paramecium*) possess *contractile vacuoles*, which, when full, contract and force fluids from the cell. In other protozoans, small vacuoles or vesicles are formed when food particles are taken in by endocytosis.

The cell sap of a plant vacuole usually consists of water with several substances dissolved in it. Since the sap is generally hypertonic in relation to the surround-

Fig. 5.11
(*a*) A region of the cytoplasm of a rat hepatocyte (an epithelial cell of the liver) showing several large vacuoles. Four vacuoles are shown, within one of which a mitochondrion (M) is being digested; within another, a portion of endoplasmic reticulum (ER) appears; and within a third, peroxisomes (P), containing enzymes for the production and decomposition of hydrogen peroxide, are being broken down. Arrows indicate the membranes outlining the vacuoles. (*b*) Vacuoles in plant cells. (Electron micrograph (*a*) courtesy Dr. Luis Biempica, Albert Einstein College of Medicine (× 40,000); (*b*) courtesy Dr. William A. Jensen (× 6000).)

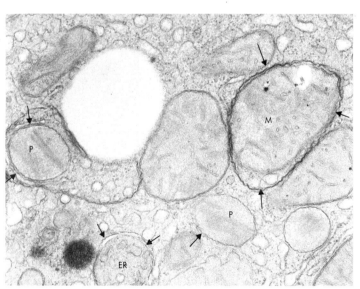

(a)

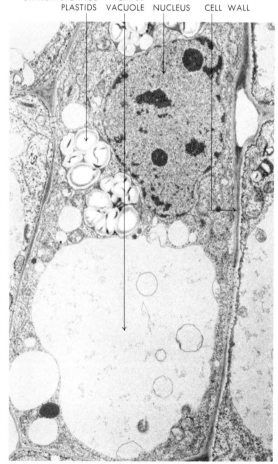

STARCH STORAGE PLASTIDS   LARGE VACUOLE   NUCLEUS   CELL WALL

(b)

ing medium, the vacuole tends to take in water by osmosis, swelling as it does so. This swelling causes the vacuolar membrane (usually called a *tonoplast*) to exert a pressure on the cytoplasm, which in turn presses against the cell membrane and thus the cell wall. As mentioned earlier, this pressure, *turgor*, plays a major role in keeping the nonwoody stems of plants firm; the wilting of a plant needing water is due to a loss of turgidity.

5.11
MICROTUBULES

In the early 1960s, reports of the presence of long, slender, hollow cylinders in cells were made by several investigators. Subsequent investigations confirmed that these structures, called **microtubules,** were, indeed, a distinct group of cell organelles (see Fig. 5.12).

Microtubules vary in thickness from 15 to 34 nm. The central core appears to contain an electron-dense

material, while the walls are constructed with globular protein parts called *tubulins.* A clear zone surrounding the wall is often observed around microtubules, and seems to separate them from the surrounding cytoplasm. Microtubules are generally divided into two groups, depending upon whether or not they are easily disrupted (by such agents as temperature, pressure, and the like) into *labile* and *stable* microtubules. The latter serve as the structural components of those whiplike projections that make certain cells highly motile (such as cilia and flagella) while the unstable, labile microtubules are more randomly distributed throughout the cytoplasm or compose the **spindle fibers,** temporary structures that aid the movement of the chromosomes during cell division. Evidence for this latter role comes from noticing the intimate association of microtubules with the chromosomes during the time of their movement to opposite poles of dividing cells. It has also been shown that certain chemicals, such as colchicine, and forces that disrupt labile microtubules will also result in no

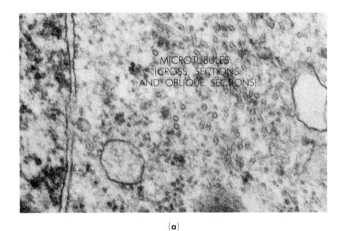

(a)

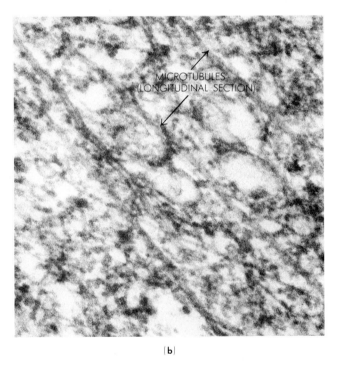

(b)

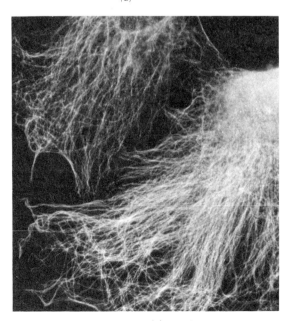

(c)

Fig. 5.12
Electron micrographs of microtubules. (*a*) Microtubules in cross section from sperm-producing cells of the scorpion (× 39,000). (*b*). Microtubules in longitudinal section from human kidney fibroblast cell (fibroblasts are cells that produce collagen and other structural proteins, expecially for connective tissue) grown in culture (× 47,000). (*c*) Photomicrograph of microtubules in cytoplasm of chick embryo cells treated with a fluorescent antibody to the protein tubulin, showing the extensive nature of the microtubule network. (Electron micrographs courtesy Dr. David M. Phillips; photomicrograph courtesy M. Osborn and K. Weber, reprinted from *Jour. Exp. Cell Research, 106,* 339 (1977).)

chromosome movement. The precise mechanism of this movement, however, is still uncertain.

The fact that colchicine will also cause cells to lose their shape indicates that microtubules also act as a supportive cytoskeleton. Colchicine also interferes with a wide variety of intracellular transport processes, suggesting that microtubules play a role in these, too. Finally, the observation that colchicine affects the distribution of immune response receptors on the membranes of lymphocytes (cells intimately involved in fighting infections), suggests that microtubules may play a role in holding these molecular receptors in place.

5.12 MICROTUBULES, BASAL BODIES, CILIA, AND FLAGELLA

As noted in the previous section, certain kinds of cells possess whip-like projections, ranging in number from one (for example, the human sperm) to many thousand (for example, *Paramecium*). These projections serve either to move the cell itself, as in the examples just provided, or, working in unison with other projections, enable attached cells to move things along across their surfaces (such as cells lining the respiratory ducts that move dust and other particles to the throat). The differences in size and number between, say, the tails of

sperm and the cilia of paramecia seemed so great they were given different names, **flagella** and **cilia.**

It now seems, however, that this distinction is not so clear. Electron micrographs reveal that the underlying structures of flagella and cilia are virtually identical, whether they are large or small, plant or animal. Each extends from a **basal body**, located within the cell body and composed of nine triplets of microtubules arranged in a ring. Flagella and cilia also have nine microtuble-composed structures in a ring, but in doublets rather than triplets (see Fig. 5.13). In the middle of this ring in

Fig. 5.13
Flagella, showing the microtubular composition. (*a*) The basal bodies (two centriole-like structures) that go to form a flagellum in the one-celled protozoan *Naegleria*. This species has a double flagellum, as shown here. Note the similarity to centriole structure (× 28,000). (*b*) Cross section of *Naegleria* flagellum, showing characteristic 9 + 2 arrangement of microtubules (× 120,000). (*c*) Cross section of sperm flagellum of the rat (× 255,000, showing the 9 + 2 structure. (*a* and *b* courtesy A. D. Dingle, "Control of flagellum number in *Naegleria*," *J. Cell Sci.* 7 (1970); (*c*) courtesy Dr. David M. Phillips.)

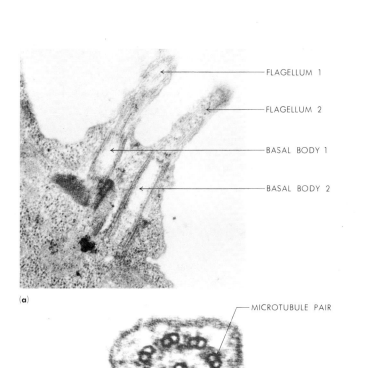

FLAGELLUM 1
FLAGELLUM 2
BASAL BODY 1
BASAL BODY 2

(a)

MICROTUBULE PAIR

(b)

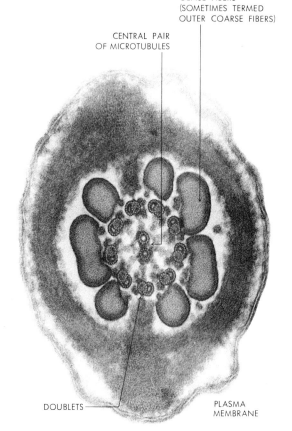

DENSE FIBERS (SOMETIMES TERMED OUTER COARSE FIBERS)

CENTRAL PAIR OF MICROTUBULES

DOUBLETS

PLASMA MEMBRANE

(c)

cilia and flagella, but not in the basal body, are two separate microtubules. The members of this pair are not identical; in *Paramecium* and another ciliated form, *Tetrahymena,* cross sections in certain regions reveal one member of the pair to possess a 20 nm spur. This spurred tubule also seems to originate lower within the cilium than its nonspurred partner. The basal body provides a stalk from which the cilia or flagella originate and at times even shows root-like structures extending into a cell cytoplasm.

In 1979, Drs. Charlotte K. Omoto and Ching Kung reported that in *Paramecium* the pair of central tubules rotates anticlockwise during ciliary beat, that the orientation of the central pair is correlated with beat direction, and that the amount of rotation is 360° with each beat. They suggest that the ciliary beat results from the adjacent tubules sliding over each other in the peripheral tubule doublets. Since for coordinated movement of the cilia it is necessary that a force generated on one side not be canceled by a force generated on the other (much as it is necessary that contraction of the muscle that raises the arm not be offset by contraction of the muscle that lowers it), Omoto and Kung propose that the central pair act much as the distributor of an automobile regulates the sending of electric charges to the spark plugs to ensure a uniform, balanced action. This hypothesis predicts that a correlation should exist between the rhythm of the ciliary beat and the period of central pair rotation, a prediction that has been verified in at least one form. Further, in forms with genetic damage that results in their missing one or both of the central tubules (or even accessory components of them) the cilia are paralyzed. There are connections via "radial spokes" between the peripheral doublets and projections from the central pair, and genetic mutants missing these also show ciliary paralysis.

The distinctiveness of the unique 9 + 2 structure of both cilia and flagella (as seen in Fig. 5.13) suggests that they are, in reality, different versions of the same thing. In this light, it is also significant to note that both possess the ability of movement independent of the cell; when detached and exposed to the energy-rich compound adenosine triphosphate (ATP) they are capable of movement.

It should be noted that while such prokaryotes as bacteria possess flagella, their structure is a hollow-cored, triple helix composed of a globular protein, *flagellin.* Bacterial flagella are thus so different from eukaryotic flagella and cilia that they may soon be referred to by another name.

5.13
MICROTUBULES AND CENTRIOLES

Though similar to basal bodies, being composed of nine triplets of microtubules arranged in a circle, **centrioles** are short, darkly stained structures located in close proximity to the nucleus. They occur in pairs, with the longitudinal axis of each member of the pair always per-

Fig. 5.14
Electron micrographs of centrioles. (*a*) Longitudinal section of the paired barrel components of a single centriole from cells of chick duodenum. (Courtesy S. P. Sorokin.) (*b*) Higher magnification of centriole in longitudinal section (× 120,000). (Courtesy J. André.) (*c*) Cross section of one of the barrel-shaped components, showing the filamentous tubules that run the length of the centriole. Tubules are arranged as nine groups of three units each (× 180,000). (Courtesy J. André.)

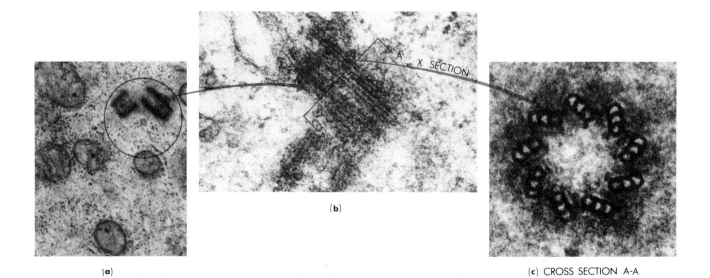

(a)

(b)

(c) CROSS SECTION A-A

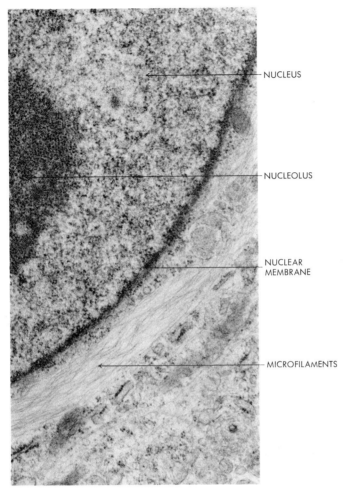

NUCLEUS

NUCLEOLUS

NUCLEAR
MEMBRANE

MICROFILAMENTS

Fig. 5.15
Electron micrograph of microfilaments from mouse fibroblast (cells that form collagen and other structural proteins) grown in culture (× 23,000). The microfilaments are shown in longitudinal section (lengthwise) and appear as long strands, almost hairlike. They seem to function in certain kinds of cell movements, such as the contraction of heart muscle cells. (Photo courtesy Dr. David M. Phillips.)

Fig. 5.16
Microfilaments in mouse cells, made visible by special treatments. The microfilaments are composed of two proteins, one as yet undetermined with a molecular weight of approximately 58,000 and the other actin, major component of muscle fibers. Note how the mircofilaments appear to cradle the nucleus into the position it occupies within the cell. (Reprinted from M. Osborn and K. Weber, *Exp. Cell Res. 106*, 339 [1977])

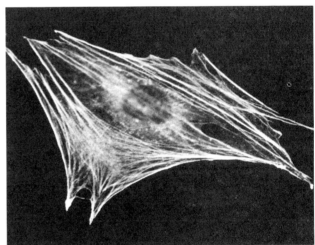

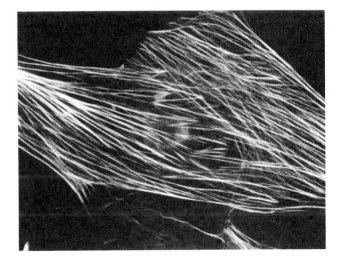

pendicular to that of the other (see Fig. 5.14a and b). Since centrioles become especially apparent at the formation of the spindle fibers during cell division, it would seem that, along with the labile microtubules, they may play a role in this formation. The centriole does not seem to be essential for spindle formation and cell division, however, as many cells not possessing them (most notably those of higher plant cells) manage to form spindles and divide without them.

5.14
MICROFILAMENTS

Microfilaments comprise a second class of fibrous structures found in the cytoplasm of eukaryotic cells (see Figs. 5.15 and 5.16). Microfilaments appear to be associated with such movements as cytoplasmic streaming, as they are found in the regions of both plants and animals where such movements occur. They are also associated with changes in cell shape during development or with the formation of the vesicles associated with eventual endo- or exocytosis. Precisely how they act to cause this motion, and indeed how they are produced by the cell, at present remains uncertain. The physical and chemical properties of the substance composing some microfilaments are similar to those of the thin skeletal muscle protein **actin**, and may indeed be identical with it. For example, treatment with the compound cytochalasin B, known to change actin's contractile properties, also inactivates microfilaments.

Finally, treatment of cells with certain antibodies to provide fluorescent markers, and then dissolve away all but the actin-like filaments, leaves a cytoskeleton very much resembling the original cell's shape (see Fig. 5.16). This suggests that microfilaments, along with microtubules and other intermediate-sized filaments, also serve as cytoskeletal components. Observations using such techniques also suggest that microfilaments may help hold the cell nucleus in position with the cytoplasm.

5.15
PROKARYOTIC AND EUKARYOTIC CELLS

As mentioned in Section 5 of Chapter 4, based primarily upon distinctive differences in cell anatomy, biologists have divided cells into two types, prokaryotic and eukaryotic, with the organisms possessing the former type of cell called prokaryotes; the latter, eukaryotes. Virtually all of the preceding discussion of cellular anatomy, organelles, and the like has been concerned with the cells of eukaryotes. The prokaryotes include bacteria, blue-green algae, and the smallest known cells, the mycoplasmas. Eukaryotes include all multicellular organisms such as plants, animals, and the fungi, as well as those unicellular and multicellular organisms called **protista**.

Prokaryotic cells generally lack many of the structures found in eukaryotic cells: for example, mitochondria, endoplasmic reticulum, chloroplasts, and Golgi complex. Since there is no nuclear membrane, a prokaryotic cell has no distinct nucleus, and its genetic material is contained in a single, ring-shaped DNA molecule usually found in a central region in the cell (see Fig. 5.17). While prokaryotic cells do possess a cell membrane, there are few if any cytoplasmic membranes. Some of the blue-green algae and photosynthetic bacteria do contain chlorophyll associated with membrane-like lamellae, but these are not found within chloroplasts. A membrane called the *mesosome* attached to the cell membrane is found in most prokaryotic cells, but its precise function remains uncertain. The similarities and differences among eukaryotic and prokaryotic cells are summarized in Table 5.1.

Because of their seemingly simpler structure, it was widely accepted for a long time that prokaryotic cells represented a more primitive stage of evolution than

Table 5.1
Comparison of Prokaryotic and Eukaryotic Cells

Feature	Eukaryotic		Prokaryotic bacteria
	Animal	Plant	
Size	10–60 nm	50–100 nm	1–5 nm
Cell Wall	no	yes	yes
Nucleus (with membrane)	yes	yes	no
Centriole	yes	yes (in lower plants only)	no
Genetic structure			
Chromosome number	>1	>1	1
Division by mitosis	yes	yes	no
Nuclear DNA bound to histones	yes	yes	no
DNA in cytoplasmic organelles	yes	yes	no
Cytoplasmic structure			
Nature of cytoplasmic ribosomes	80 S	80 S	70 S
Nature of organelle ribosomes	70 S	70 S	none
Mitochondria	yes	yes	none
Chloroplasts	no	yes	none
Golgi apparatus	yes	yes	none
Pinocytosis	yes	no	none
Phagocytosis	yes	no	none
Endoplasmic reticulum	yes	yes	none

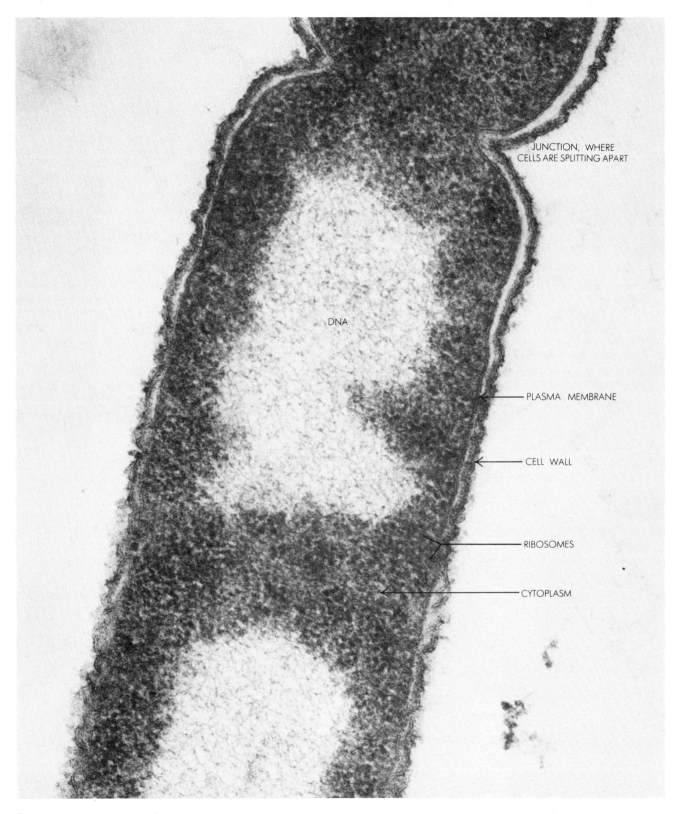

Fig. 5.17
Electron micrograph of a dividing bacterium, *Bacillus subtilis*,
a prokaryote, showing localization of DNA and absence of a
distinct nuclear membrane. The plasma membrane and cell
wall are also visible (× 120,000).

Two great developments have occurred in the evolution of cells: the prokaryotes, without a distinct nucleus, and the eukaryotes, with a membrane-bounded nucleus. Cells of most plants and animals are eukaryotic; cells of bacteria, mycoplasmas, and blue-green algae are prokaryotic.

eukaryotic cells. Indeed, the very name prokaryote means "before a nucleus" (eukaryote means "a true nucleus"), implying that prokaryotes have not yet evolved to the state of eukaryotes. However, certain very profound differences in the way they synthesize messenger RNA, and the structural variation in their DNA leading to these differences in *m*RNA synthesis, now suggest that the two cell forms may have evolved independently.

The only intracytoplasmic organelles occurring in both prokaryotic and eukaryotic cells are the ribosomes, though those of prokaryotic cells are smaller and differ in structure. As pointed out earlier, while tradi-

Fig. 5.18
Comparison between a prokaryotic and eukaryotic cell. Although the prokaryotic cell possesses a high degree of organization, it is not as complex and does not show as high a degree of specialization as the eukaryotic cell. (After E. O. Wilson et al., *Life on Earth*, Sunderland, Mass.: Sinauer, 1973.)

tionally some bacteria have been considered to possess flagella, their lack of microtubules suggest that the same term, flagella, is being applied to structures that internally are quite dissimilar.

Perhaps the most distinctive feature of the prokaryotic cell is its cell wall, composed of polysaccharide chains bonded covalently to shorter chains of amino acids. The entire cell wall is often regarded as a single huge molecule or molecular complex called *murein*. Fig. 5.18 shows most of the differences between prokaryotic and eukaryotic cells.

5.16
VIRUSES

In addition to prokaryotes and eukaryotes, there is another group of organisms that are not cellular. **Viruses** are composed of only two types of substances: protein and nucleic acid. The nucleic acid carries a program of information for producing more viruses; the protein forms a coat around the nucleic acid "core," which is either DNA or RNA. Figure 5.19 shows one

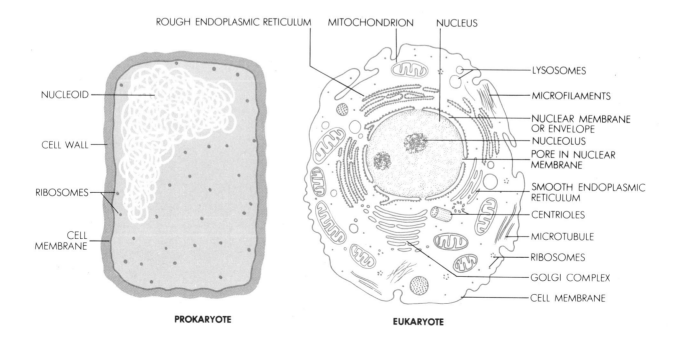

PROKARYOTE   EUKARYOTE

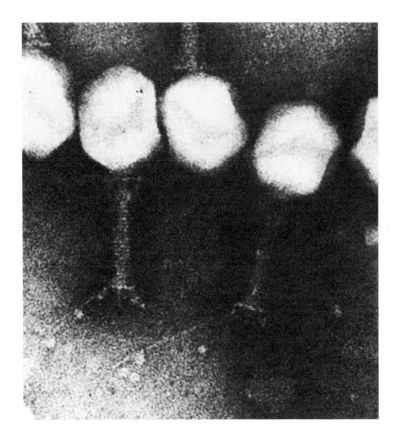

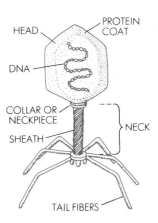

HEAD
PROTEIN COAT
DNA
COLLAR OR NECKPIECE
SHEATH
NECK
TAIL FIBERS

Fig. 5.19
Electron micrograph of a group of T$_2$ bacteriophages. Each consists of a head, neck, and tail region. The head outer coat and the neck and tail consist of protein. Inside the head is DNA carrying the virus's genetic information. Phages without the thickened neck-piece are "triggered," meaning they have released their DNA. DNA is shot out through the neck and tail into a bacterium to which the phage has attached itself. (Photo courtesy Dr. Sidney Brenner and Dr. R. W. Horne.)

type of virus, the so-called bacteriophage, which attacks bacterial cells. Viruses vary in size from less than 20 nm up to sizes larger than the smaller bacteria. Indeed, it was their size that first brought them to the attention of nineteenth-century researchers, when it was found that the infectious agents that caused smallpox could not be seen through the light microscope and could pass through filters that trapped all known bacteria. Viruses also vary widely in shape, from the unusually shaped bacteriophage to the rounded virus that causes influenza.

Viruses function by invading living cells. Inside the host cell, the virus reproduces itself by using the cell's chemical machinery. Often (but not always and depend-

ing upon the type of virus and pattern of infection) this destroys the host cell.

Viruses cannot reproduce outside of living cells, no matter what raw materials are available. Furthermore, some viruses can be crystallized, placed in a bottle, and stored for many years. When the crystals are resuspended in solution, they are capable of infecting cells once more, but while they are crystals, they do not carry out any of the chemical reactions associated with living cells. In view of these characteristics, many biologists have found it impossible to classify viruses as either living or nonliving. The terms simply have no meaning when applied to them. If the ability to reproduce constitutes living, then viruses are alive. On the other hand, if the ability to carry out respiration and synthesize one's own proteins is considered living, then viruses are not alive. Functioning only with the aid of intact living cells, they are like a set of blueprints with no factory of their own where their plans can be put into action. Yet, once inside a living cell, viruses can direct the synthesis of proteins, reproduce virus parts, and assemble those parts into new, intact viruses.

Viruses are a major cause of disease in both animals and plants. A mottled, deteriorating condition affecting tobacco plants is caused by the tobacco mosaic virus, TMV. Smallpox and polio are caused by viruses, as are chicken pox, mumps, measles, yellow fever, rabies,

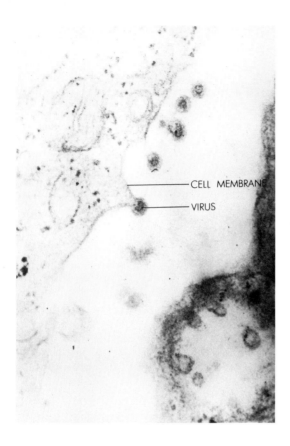

CELL MEMBRANE

VIRUS

Fig. 5.20
Shown above are rubella viruses emerging from an infected cell, from which they may go on to infect other cells. The rubella virus, which causes German measles, may be dangerous if it infects a woman in the first few weeks of pregnancy, when critical stages of embryological development occur. The results may be fatal to the embryo or cause it to have cataracts, deafness, and heart problems. (Micrograph courtesy of the Center for Disease Control, Atlanta, Georgia 30333).

viral pneumonia, poliomyelitis, infectious hepatitis, fever blisters, several types of venereal diseases, encephalitis, and many varieties of the common "cold" (see Fig. 5.20). Viruses are particularly difficult to combat because common antibiotics do not usually affect them. The most effective medical approach to viral diseases is prevention of the disease through immunization (for example, vaccination). Exposure of the human (or animal) body to a small dosage of virus allows the body's defenses against that specific virus to build up without the organism's succumbing to the disease. Then, when the virus is encountered in the environment, it can be eliminated from the body early because the body recognizes it as foreign. Once the virus has invaded the body's cells in an infection, it is much more difficult to eliminate.

While immune responses do occur (if you have had the mumps or measles once, you are generally immune for life), they usually are too slow to effect a cure for the disease if it occurs. However, the cells' production of the protein **interferon** in response to virus invasion is thought to play a role in recovery. One hypothesis proposes that a cell's infection by the virus induces interferon production which, though unable to save that cell, is transported to other as yet uninfected cells, conferring upon them an immunity, possibly by inducing the synthesis of another protein that blocks the replication of viral nucleic acid. Infection also stops cell multiplication, a fact that makes it of great interest in cancer research. Just how it does so is not certain. However, a team of researchers at Rockefeller University, headed by Dr. Lawrence Pfeffer, have evidence that interferon may act by "aging" cells, since young cells exposed to the antiviral protein act similarly to those that have gone through fifty generations *in vitro;* they concentrate their energy on growing larger rather than upon reproducing.

The genetic material of the influenza virus (RNA) appears to be particularly susceptible to mutation. Since these genetic changes alter its protein coat, the immunity acquired by previous infection is lost because the antibodies formed to protect against the old form no longer recognize it in its new disguise. Thus flu vaccines often fail, though they may at times lessen the severity of the infection. Influenza epidemics are often named after the place of origin (Asian flu, Hong Kong flu, and the like).

The general reproductive cycle for a virus is diagrammed in Fig. 5.21. The diagram shows a bacteriophage invading a rod-shaped bacterial cell. This entire cycle can be completed in less than half an hour under favorable conditions. The virus nucleic acid is injected into the host cell to be, leaving the protein coat attached to the outside of the cell wall. Inside the host cell, virus DNA commandeers the bacterial ribosomes, amino acid, and carbohydrate pools, and begins to direct the production of virus protein. Viral nucleic acid is also replicated prodigiously. When components have been completed, they self-assemble in the host cell's cytoplasm. The cell is ruptured and the new virus particles move to infect other cells nearby.

Some viruses do not destroy the host cell immediately after invading it. Certain viruses enter a host cell and simply deposit their nucleic acid in the cell but do not replicate new viruses. The viral nucleic acid either exists independently of the host cell's nucleic acid or can actually be incorporated into it. In either case, viral nucleic acid replicates at exactly the same rate as the host cell's. At some indefinite time in the future (for viruses in human cells this may be from a few months to twenty or more years), the viral nucleic acid does begin to take over the cell and produce new viruses. Thus there can be a long delay between infection and manifes-

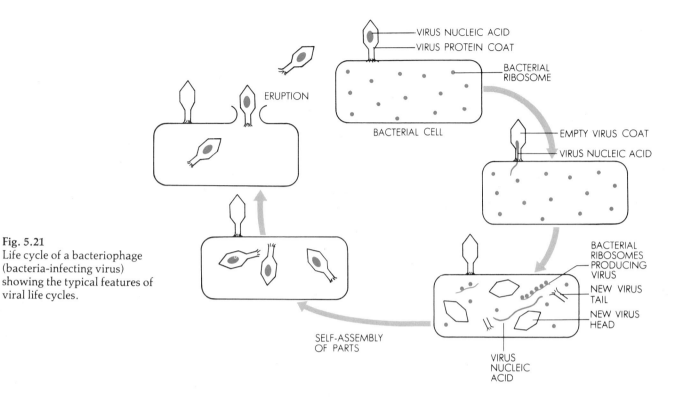

Fig. 5.21
Life cycle of a bacteriophage (bacteria-infecting virus) showing the typical features of viral life cycles.

tation of viral disease. It is thought, for example, that some forms of cancer may be caused by viral nucleic acid that gets into a person's cells early in life but is not activated until middle age or later.

5.17
CELLS AND TISSUES

While the cell may be the unit of structure and function, in organisms composed of more than one cell (multicellular organisms) cells generally work together as a unit, or **tissue**. In turn, several tissues are often joined together to form an **organ** (such as the stomach) and organs, in turn, united to form an **organ system** (such as the digestive system). Many multicellular organisms, including humans, are composed of several organ systems. We shall here provide a brief introduction to the various types of tissues found in multicellular plants and animals, leaving a more detailed discussion for when we deal with plant and animal physiology in Chapters 8 through 15.

5.18
PLANT TISSUES

Undoubtedly, one of the largest forward steps in the evolution of multicellular plants was the joining of individual cells into groups. Such a union allows cell differentiation and specialization. Some cells can serve for

protection, some for transport of food and waste materials, some for photosynthesis, some for reproduction, and so on.

During development, the cells of the plant divide, enlarge, and differentiate into many different kinds and become organized into complexes or associations, the tissues. The formation of various kinds of cells and tissues is closely correlated with the following four main functions that the land-dwelling plant must perform:

1. Production of new cells and tissues necessary for the continued life of the plant.

2. Restriction of water loss from the plant body, especially that portion above the ground, and protection from damage and disease.

3. Strengthening of the plant body to withstand such environmental forces as wind and gravity.

4. The distribution of water, minerals, and organic materials throughout the plant body.

These four functions are not necessarily performed exclusively by each of the four types of tissues found in plants. Tissues whose main function falls under (4), for example, also play an important role in (3). With this in mind, we shall now proceed to discuss the four plant tissue types in the same order that their main functions are listed above.

Higher plants are composed of four types of tissues: meristematic (embryonic), epidermal (protective), fundamental (parenchyma, collenchyma, and sclerenchyma), and vascular (conducting).

Fig. 5.22
Meristems, localized regions of new cell formation in the plant body. (Photos: shoot apical meristem courtesy of Triarch, Inc., Ripon, Wisc.; lateral meristem and root apical meristem courtesy of Turtox, Chicago. Drawing of root apical meristem from John W. Kimball, *Biology*, 2d ed. Reading, Mass.: Addison-Wesley, 1968.)

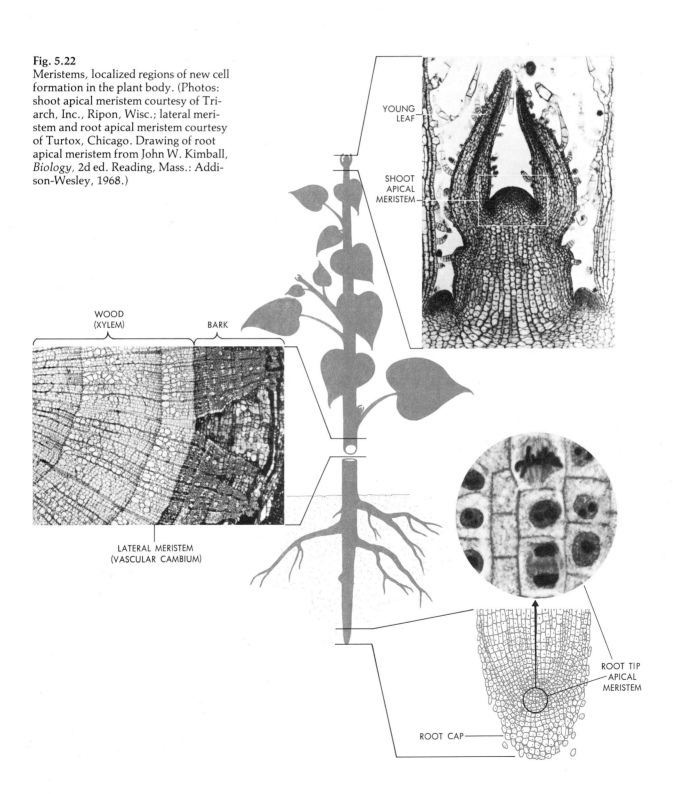

YOUNG LEAF

SHOOT APICAL MERISTEM

WOOD (XYLEM)

BARK

LATERAL MERISTEM (VASCULAR CAMBIUM)

ROOT TIP APICAL MERISTEM

ROOT CAP

Meristematic or Embryonic Tissues

From the time the zygote divides, the plant produces new cells, tissues, and organs throughout its entire life. In the embryo, the production of new cells occurs anywhere. As the embryo grows, however, the production of new cells and tissues is gradually limited to certain regions of the body, the **meristems** (Fig. 5.22). The remainder of the plant body becomes specialized for functions other than the formation of new cells. Thus, unlike an animal, which usually completes construction of all of its tissues and organs during an early period of embryonic development, a plant body is partially embryonic and partially adult throughout its entire life. Similarly, in a plant additional organs such as leaves are formed throughout the life span of the individual, while in the animal body the number of organs is fixed at a certain definite number early in the development of the embryo.

The meristems located at the apices of root and shoot (**apical meristems**) are often cone-shaped. In the root, the apical meristem is covered by a conical sheath, the **root cap,** which protects the embryonic cells as the root forces its way into the soil. In the shoot, the apical meristem is located at the very tip of the stem and receives some protection by the developing leaves. Just below the growing cone of cells, in the apical meristem of the shoot, the young leaves and branches are formed. The plant body produced by the apical meristems is called the *primary body,* composed of fully differentiated and mature primary tissues. In the **herbaceous** or *nonwoody* plants, most of the body is primary.

In addition to the primary body, many other species of plants have an increase in the thickness of the shoot and root due to the production of additional cells and tissues by a lateral meristem, the *vascular cambium.* Thus woody plants (trees and shrubs) have secondary tissues comprising a secondary body that is superimposed throughout most of the plant over the primary body produced by activity of the apical meristem.

Since they are constantly dividing, the metabolic rate of meristem cells is generally far higher than that of cells composing other plant tissues. Consequently, meristem cells must remain small enough to keep a high ratio of surface area per cell unit volume. They also have a smaller amount of endoplasmic reticulum, their plastids and mitochondria are less elaborate, and their cell walls are usually much thinner than those of fully differentiated living cells of the plant body.

Protective or Epidermal Tissues

As the names imply, protective tissues are generally located on the outer surfaces of the plant body for its protection. The entire body of young plants and fully mature herbs is covered with a protective tissue, the **epidermis** (Fig. 5.23). The epidermis, usually a single layer

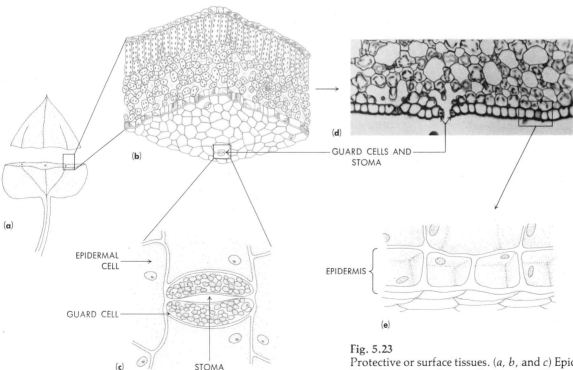

(b)

(a)

(d)

GUARD CELLS AND STOMA

EPIDERMAL CELL

GUARD CELL

STOMA

(c)

EPIDERMIS

(e)

Fig. 5.23
Protective or surface tissues. (*a, b,* and *c*) Epidermis of a leaf. (*d, e*) Epidermis of stem of buttercup. (Photo courtesy Triarch, Inc., Ripon, Wisc. Drawings *a* and *e* from John W. Kimball, *Biology,* 2nd ed. Reading, Mass.: Addison-Wesley, 1968.)

of living cells, forms a continuous covering over all portions of the primary body (the leaves and reproductive structures as well as the stems and roots). The epidermal cells of the aerial portion of the plant body usually produce a waxy substance, **cutin,** that impregnates the outer walls and forms a distinct layer, the cuticle, over the outer surface of the cells. In many species the waxy cuticle forms a light gray "bloom" that can be easily wiped off such fruits as unwashed plums and grapes and the leaves of the red cabbage. The epidermis helps to prevent the aerial parts of the plant body from losing excessive quantities of water.

The epidermis of aerial portions of the plant body possesses very distinctive, usually bean-shaped pairs of cells with a small opening between them. Since the size of the opening can vary through changes in the shape of this pair of cells, they are known as **guard cells** (Fig. 5.24c and d). The opening between the guard cells is called a **stoma** (plural: **stomata**). Stomata are adaptations that facilitate the exchange of gases between the interior of the plant body and the atmosphere.

Many species of vascular plants possess appendages extending from the epidermis of the stems and leaves. These **trichomes** (Fig. 5.24), which may consist of one cell or many cells, appear in the form of hairs or scales. In some species, the hairs may be many-branched and sufficiently dense to form a woolly covering over the epidermis. The hairs of many other species are glandular, producing and secreting droplets of oily materials. This phenomenon is especially noticeable in petunia and tobacco plants. The hairs of the nettle

(*Urtica*) have a spherical tip. If lightly touched with one's hand, this tip breaks off and penetrates the skin. The fluid contents of the hair escape into the wound. The fluid is poisonous and produces a stinging sensation and inflammation around the puncture. The epidermis of seeds of the cotton plant produces hairs (the cotton fibers) that may grow as long as ½ to 2½ inches. The length of these hairs, plus the high cellulose content in their thick walls, renders cotton very useful in making cloth.

The epidermal tissue of young roots has long slender appendages, the root hairs, at some distance from the tip where the root cells have reached their maximum length. Each root hair, a tubular outgrowth of a single epidermal cell, greatly increases the surface area of the root for absorption of water and minerals in solution. For example, one rye plant was calculated to have some 14 billion living root hairs with a total area of 4300 square feet. The same plant had 13,800,000 roots with a surface area of 2500 square feet. A single rye plant had a total of 6800 square feet of surface contact between its roots and root hairs and the soil. In fact, it had 130 times more surface in contact with the earth than it had exposed to the atmosphere by its aboveground parts. Root hairs commonly live only a few days. They are rapidly replaced, however, by new ones forming just back of the growing root apex.

In those species of plants in which the stems and roots increase in thickness by the formation of new cells and tissues by the cambium, the epidermis becomes stretched and finally ruptures. Before the epidermis

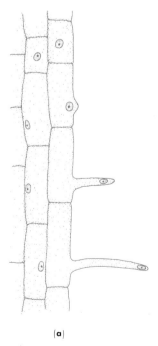

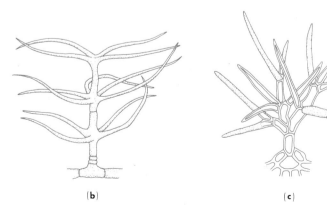

Fig. 5.24
The protective epidermal tissues of many plants possess appendages called trichomes that are merely specialized extensions of the cell. Most common are root hairs (*a*), which extend from the epidermal cells of root tips in certain regions and serve to increase the surface area for the absorption of water and minerals. Other, more intricate types of trichomes are represented by (*b*) a branched trichome and sycamore and (*c*) a dendroid hair from lavendar. (*b* and *c* after K. Esau, *Plant Anatomy.* New York: Wiley, 1953.)

(a)   (b)   (c)

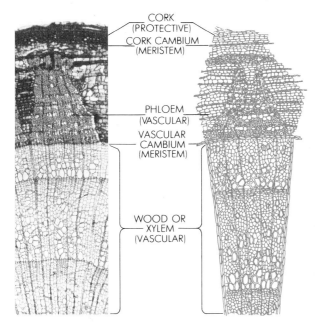

CORK (PROTECTIVE)
CORK CAMBIUM (MERISTEM)
PHLOEM (VASCULAR)
VASCULAR CAMBIUM (MERISTEM)
WOOD OR XYLEM (VASCULAR)

Fig. 5.25
Protective tissues as seen in a cross section of a basswood stem. As is the case with many trees, the outer epidermis is ruptured and shed as the tree grows. The cork cells are constantly being replaced from underneath by the cork cambium, a lateral meristem, as are the xylem and phloem (which are vascular, or conducting tissues) by the vascular cambium. (Photo courtesy Turtox, Chicago. Drawing from John W. Kimball, *Biology*, 2nd ed. Reading, Mass.: Addison-Wesley, 1968.)

is destroyed, however, a secondary protective layer tissue, the **periderm**, is developed through the activity of a special lateral meristem, the **cork cambium** (Fig. 5.25). When fully functional, the cells in the outer layer of the periderm (known as cork cells) are dead and usually filled with air (this characteristic gives cork its excellent insulating qualities and makes it an ideal choice for thermos bottle stoppers). Before they die, cork cells produce wax and a fatty substance, *suberin*, which accumulate as distinct alternating layers over the primary cellulose wall. Since the wax and suberin layers are practically impervious to water, the cork cells are well adapted to protect the plant body from excessive water loss.

Fundamental Tissues

Fundamental tissues generally compose the greatest mass of the plant body. The soft portions of the leaves, flowers, and fruits; the cortex regions of the stem and root; and the pith region of the stem are all composed of fundamental tissues. The chief function of fundamental tissues is the production and storage of food. In their

form and structure, the cells of the fundamental tissues are highly diverse. A brief examination of some of this diversity will be helpful in understanding the morphology and physiology of the plant body.

Parenchyma cells occur in continuous masses throughout the plant body. They are especially abundant in the region between the epidermis and the vascular tissues (see Fig. 5.26a, b, c), in the pith of the stem, in the photosynthetic tissue (*mesophyll*) of leaves, in the flesh of juicy fruits, and in the food storage region of the seed (usually the *endosperm*). Parenchyma cells are also present in the conducting tissues of both the primary and secondary body, that is, in both herbs and woody plants. Parenchyma cells are relatively unspecialized; even when fully mature, these cells are more similar in form and structure to the cells of the meristem than any other cells inside the plant body. In fact, parenchyma cells retain the ability to divide, a characteristic that enables the plant body to heal wounds and regenerate organs. Parenchyma cells usually have relatively thin primary walls; thick walls are present only in certain tissues (such as secondary xylem). A large vacuole, which may contain large quantities of water and other materials, is usually present. In fact, the water-filled vacuoles of parenchyma cells give considerable support and shape to the plant body.

Closely similar to parenchyma in both structure and function is *collenchyma* tissue (Fig. 5.26d, e, f). Like parenchyma cells, collenchyma cells are functional while living (some cells must die before they can function). Collenchyma cells have adaptations enabling them to help support the plant, not only its growing portions but also the fully mature primary body. These adaptive features include an elongate shape, irregularly thickened walls, and close packing. All these characteristics make collenchyma a very strong tissue, capable of considerable stretching yet still flexible and plastic. Such characteristics are especially valuable in the growing stem or leaf. Collenchyma tissue is usually located on the periphery of stems and leaves, often just beneath the epidermis.

Collenchyma tissue and water-filled parenchyma cells enable the young growing plant to maintain its characteristic shape and rigidity. In the mature herb, however, much of the strength and support of the body is provided by *sclerenchyma* tissue (Fig. 5.26g, h, i). Sclerenchyma cells have features enabling the plant body to withstand considerable bending and stretching without much damage to the softer cells of the parenchyma tissue. During its development, each sclerenchyma cell forms a very thick secondary wall as an additional deposit over the primary cell wall. The great strength of sclerenchyma cells comes from a framework of cellulose molecules (often reinforced with lignin) that composes the secondary wall.

Some cells of sclerenchyma, called *fibers*, may be conspicuously elongated from 7 cm (in the flax plant,

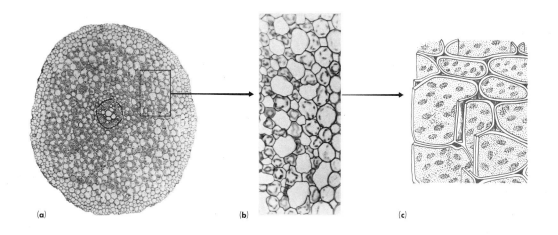

(a)   (b)   (c)

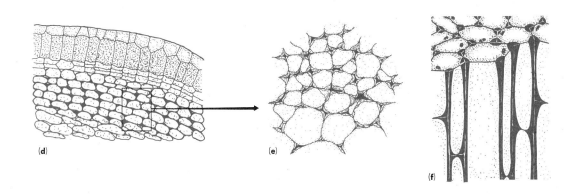

(d)   (e)   (f)

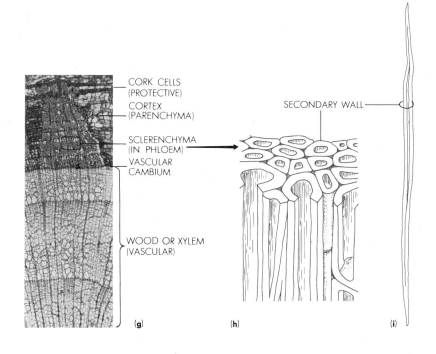

(g)   (h)   (i)

CORK CELLS
(PROTECTIVE)

CORTEX
(PARENCHYMA)

SCLERENCHYMA
(IN PHLOEM)

VASCULAR
CAMBIUM

WOOD OR XYLEM
(VASCULAR)

SECONDARY WALL

Fig. 5.26
Fundamental tissues. (*a*) Parenchyma tissues seen in a photograph of a cross section of a buttercup root. An enlargement of the rectangular area is shown in (*b*), while (*c*) gives an artist's re-creation of the cell structure. (*d*) Drawing of collenchyma tissue in a stem. An enlarged cross section of the rectangular area (*d*) is shown in (*e*) and a longitudinal view in (*f*). (*g*) Photograph of a cross section of a basswood stem, showing the location of sclerenchyma tissue among the food-conducting tissue (phloem). (*h*) Drawing shows the thickness of the cell walls. An individual schlerenchyma cell is shown in (*i*). (Photos: (*a*) and (*b*) courtesy Triarch, Inc., Ripon, Wisc.; (*g*) courtesy Turtox, Chicago. Drawings: (*c*), (*h*), and (*i*) from John W. Kimball, *Biology*, 2d ed. Reading, Mass.: Addison-Wesley, 1968; (*d*), (*e*), and (*f*) after K. Esau, *Anatomy of Seed Plants*. New York: Wiley, 1960.)

used in making linen cloth) up to 25 cm long (in the ramie plant, formerly used in making Chinese linen). Fiber cells strengthen plant parts no longer growing in length. They often occur as separate strands, bundles, or cylinders in various parts of the plant.

A second kind of sclerenchyma cell is the *sclereid*, a cell highly diverse in shape and size. Sclereid cells may occur anywhere in the plant body, either singly or in groups. The gritty texture of pear fruits is due to small clusters of sclereids known as "stone cells." The hardness of nut shells and seed coats is also partly due to sclereids.

Vascular or Conducting Tissues

The possession of special tissues for transporting water, food, and other materials throughout the plant body is a characteristic feature of most land-dwelling plants. The cells comprising these conducting tissues are usually tube-like and elongated in the direction in which the conduction occurs. Two kinds of conducting tissues are present in the body of most land plants: **xylem** and **phloem.** Each tissue consists of several kinds of cells.

Xylem. Xylem is the most abundant conducting tissue in the body of the vascular plant. It forms a continuous system running from near the root tips upward through the stem and out into the leaves. Xylem contains several kinds of cells, some of which are alive at functional maturity while others must die before they can do their main task. The most characteristic cells in the xylem are **tracheids** and **vessel elements;** both conduct water.* During their development, both types of cells produce secondary walls: these add extra strength

---

*Tracheids and vessel elements are both found in angiosperms, while in gymnosperms (conifers such as pines and firs) tracheids are the only conducting elements present.

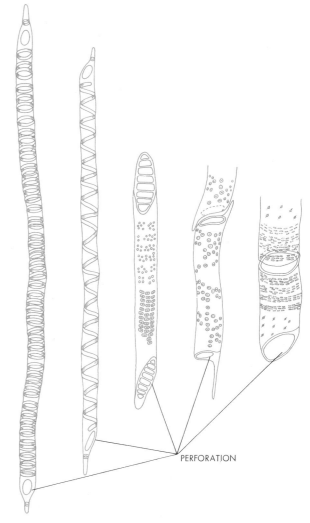

PERFORATION

Fig. 5.27
Diagrammatic sketches of vessel elements from xylem tissue. These cells display a diversity of secondary wall types similar to tracheids (see Fig. 5.28). Note the presence of one or more perforations in the ends of each vessel element.

and rigidity to the xylem tissue as well as to the entire plant body. These wall thickenings also prevent the water-conducting cells from collapsing during periods when much water is being lost from the plant by transpiration. When first formed by apical meristems or vascular cambium, both tracheids and vessel elements are living. When they are fully developed, however, their cytoplasm and nuclei disappear, leaving only the thickened cell wall. The result is either a long hollow structure (tracheid), admirably suited for the movement of water (Fig. 5.27), or the xylem vessel (a series of

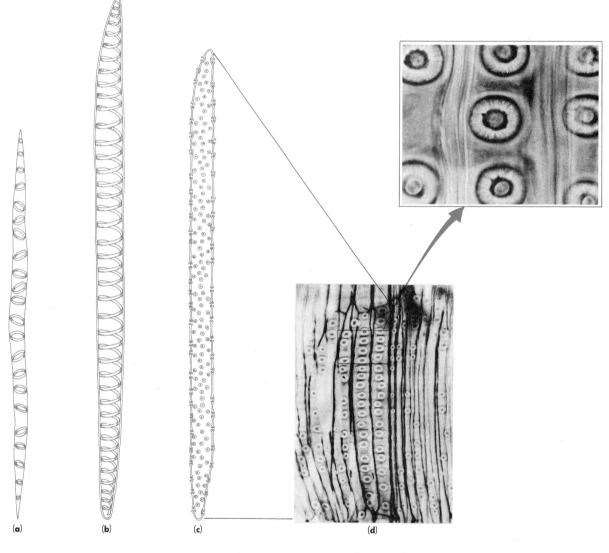

(a)   (b)   (c)   (d)

Fig. 5.28
Sketches of tracheids from xylem tissue. Note the variety of patterns: (a) rings, (b) continuous helix, (c) pitted. In the pitted tracheid, the secondary wall is a continuous layer, except for the pits. The micrograph in (d) shows a longitudinal section through some pitted tracheids, and (e) shows an enlarged view. (Micrograph (d) courtesy Dr. Fred H. Taylor, University of Vermont; micrograph (e) courtesy Dr. Graeme P. Berlyn, Yale University School of Forestry and Environmental Studies, New Haven, Connecticut.)

vessel elements joined end-to-end), which is an even more efficient structure for water movement. During the development of a vessel, the transverse walls separating adjacent cells of the series break down and are absorbed. The cytoplasm and nucleus also disappear.

The result is a long tubular structure that greatly accelerates the rate of water conduction (Fig. 5.28).

During the development of individual tracheids and vessel elements, the secondary wall thickening may be deposited over almost all of the primary wall except for scattered, isolated areas. These areas form small cavities or *pits* in the secondary wall. The pits of adjacent cells usually meet each other (Fig. 5.29). Separating the two adjacent pits is a *pit membrane*, composed of the primary wall of each cell and the cementing material that binds the cells together. Since the pit membrane is quite permeable, water can move easily from one cell to the next. Around each pit there is usually a circular border overhanging the pit cavity. Thus, bordered pit-pairs of adjacent tracheids and vessel elements constitute an adaptation for increasing the area of the thin, water-permeable parts of the primary walls without enlarging the cavities or pits in the secondary walls. Such an adaptation provides for rapid movement of water

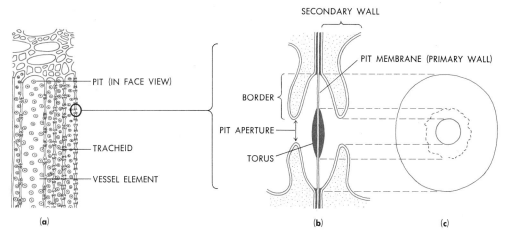

SECONDARY WALL

PIT MEMBRANE (PRIMARY WALL)

PIT (IN FACE VIEW)

BORDER

PIT APERTURE

TORUS

TRACHEID

VESSEL ELEMENT

(a)   (b)   (c)

Fig. 5.29
Diagrams of pairs of adjacent pits, each with a circular border composed of the overarching secondary wall. Bordered pits are present in both tracheids and vessel elements. (*a* from John W. Kimball, *Biology*, 2d ed. Reading, Mass.: Addison-Wesley, 1968; *b* and *c* adapted from K. Esau, *Plant Anatomy.* New York: Wiley, 1953.)

without greatly reducing the strength and rigidity of the tracheids and vessel elements.

While the pit membranes of the majority of vascular plants are rather thin, those of conifers have a conspicuous central thickening, or *torus*, which may function as a valve between cells. When the pressure in one cell becomes considerably greater than in the adjacent cell, the torus is forced against the border of the pit in the adjacent cell, blocking the pit opening and impeding the movement of water from one tracheid to another.

In addition to tracheids and vessel elements, xylem tissue also contains parenchyma cells. These living cells may be either scattered among the dead tracheids and vessel elements or aggregated into *rays*. Each ray is a radial system of cells in the xylem. They function in the lateral movement of materials across the xylem and in the storage of carbohydrates.

Finally, xylem tissue may contain elongated, very thick-walled *fiber cells* (see Figs. 5.30 and 5.31). Fibers add strength to xylem tissue, especially in species where the xylem contains many wide, relatively thin-walled

Fig. 5.30
Scanning electron micrograph ($\times$ 360) of the wood (xylem) of a maple tree, showing several vessels in both longitudinal and cross section. Note the pits lining the vessel membranes. The smaller cells surrounding the vessels are fiber cells, the main supporting and strengthening elements. The cells are generally long, with tapered ends, and in heavy woods tend to have very thick walls. The lateral rays are composed of parenchyma tissue. (Micrograph courtesy R. G. Kessel and C. Y. Shih, from *Scanning Electron Microscopy in Biology,* Springer-Verlag, New York, 1974).

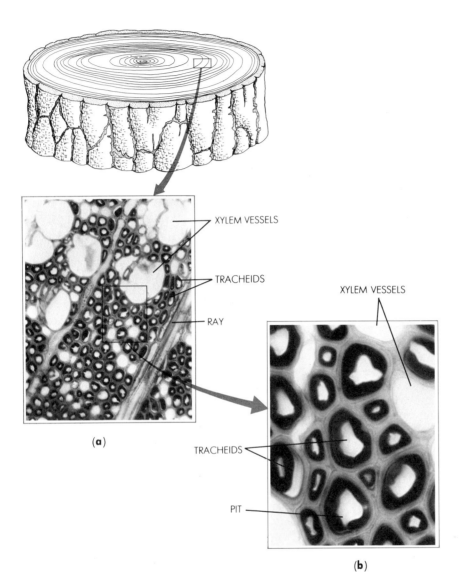

(a)

XYLEM VESSELS

TRACHEIDS

RAY

XYLEM VESSELS

TRACHEIDS

PIT

(b)

Fig. 5.31
(a) Cross section of xylem vessels and tracheids (× 350), also showing lateral rays containing what are believed to be starch grains. (b) Enlarged view (× 850) of a portion of (a). Note that some of the sections occur in the region of a pit, and thus show openings. (Micrographs courtesy of Dr. Graeme P. Berlyn, Yale University School of Forestry and Environmental Studies, New Haven, Connecticut.)

Fig. 5.32
Phloem tissue. Left, sieve-tube elements (in longitudinal section). Right, same in cross section of basswood stem. Note that the phloem tissue in basswood also contains thick-walled supporting fibers. (Both drawings from John W. Kimball, *Biology*, 2d ed. Reading, Mass.: Addison-Wesley, 1968.)

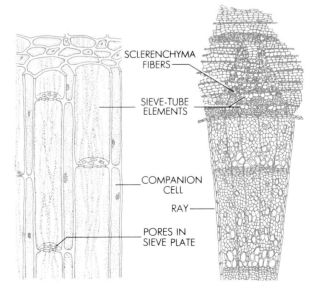

SCLERENCHYMA FIBERS

SIEVE-TUBE ELEMENTS

COMPANION CELL

RAY

PORES IN SIEVE PLATE

APICAL MERISTEM

EPIDERMAL TISSUE

VASCULAR OR CONDUCTING TISSUE

STARCH GRAINS

FUNDAMENTAL TISSUE

Fig. 5.33
Scanning electron micrograph of a potato tuber bud showing epidermal tissue, vascular or conducting tissue, apical meristem and fundamental tissue, with some of the cells containing starch grains. The peak at the left of the apical meristem region was in the process of developing into a leaf. (Micrograph courtesy R. G. Kessell and C. Y. Shih, from *Scanning Electron Microscopy in Biology*, Springer-Verlag, New York, 1974.)

vessels. In some species, fiber cells develop cross-walls after the secondary wall is deposited along their longitudinal walls. Fiber cells may also retain their cytoplasm and nucleus, and function in the storage of starch and other reserve foods.

Phloem. In addition to xylem, the conducting tissues include phloem. Like xylem, the phloem forms a continuous system, running from near the tips of the roots upward through the stem to the leaves. Also like xylem, phloem contains several kinds of cells; both fibers and parenchyma are present. The main conducting cells of the phloem are the **sieve cells** and **sieve-tube elements;** both transport food materials.* As implied by their name, these cells have characteristic clusters of small pores on their end walls that, in surface view, resemble a sieve (Fig. 5.32). Through these pores, strands of cytoplasm extend from one cell to the next.

Sieve cells and sieve-tube elements have one unique characteristic; during development, the nucleus becomes disorganized and disappears as a distinct organelle. Yet even without a nucleus, the cell cytoplasm continues to live and the cell performs its part in the conduction of food materials.

How can these cells continue to live and function without a nucleus? One hypothesis proposes that the cytoplasm of the sieve cells and sieve-tube elements may be controlled by nuclei-produced substances from nearby specialized parenchyma cells (**companion cells**). No satisfactory experimental test for this hypothesis has yet been performed, however.

Figure 5.33 shows a plant portion containing all four kinds of plant tissue.

*Sieve-tube elements are characteristic of angiosperms, while sieve cells are found in gymnosperms and lower forms.

5.19
ANIMAL TISSUES

Despite the fact that higher animals seem considerably more complex than plants, the tissues that compose them can be divided into only a few basic groups.

Epithelial Tissues

Epithelial tissues compose the surface and lining of the animal body. They form a continuous layer or sheet over the entire body surface and most of the body's inner cavities. On the external surface, epithelial tissues protect underlying cells from injury, bacteria, harmful chemicals, and drying. On the internal surface, epithelial tissues absorb water and food and give off waste products. In the digestive tract, glandular cells of the epithelial tissues secrete mucus to keep the passages damp and lubricated. Some places in which epithelial tissue always occurs are in the lining of the digestive tract, the air passages to the lungs, the lining of the kidney tubules, and the skin. In addition, several important glands in the animal body develop embryologically from epithelial tissues.

Epithelial tissues can be divided into three major groups, according to the shape of the cells composing them (Fig. 5.34a). **Squamous epithelium** is composed of flat cells, shaped rather like flagstones on a terrace. Squamous epithelium is found in the lining of the mouth and esophagus. In complex animals, squamous epithelium is found in stratified layers, with several of these flat cells piled one on top of the other. **Cuboidal epithelium,** as the name suggests, is made up of cube-shaped cells; it can be found lining the kidney tubules. **Columnar epithelium** also has a descriptive name. Its

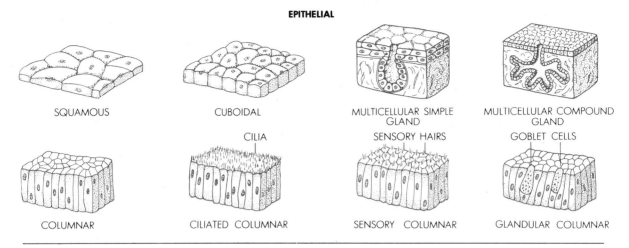

EPITHELIAL

SQUAMOUS

CUBOIDAL

MULTICELLULAR SIMPLE
GLAND

MULTICELLULAR COMPOUND
GLAND

CILIA

SENSORY HAIRS

GOBLET CELLS

COLUMNAR

CILIATED COLUMNAR

SENSORY COLUMNAR

GLANDULAR COLUMNAR

CONNECTIVE

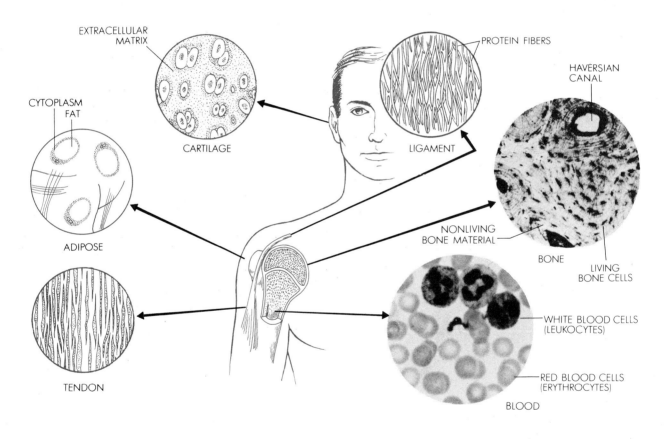

EXTRACELLULAR
MATRIX

PROTEIN FIBERS

HAVERSIAN
CANAL

CYTOPLASM
FAT

CARTILAGE

LIGAMENT

ADIPOSE

NONLIVING
BONE MATERIAL

BONE

LIVING
BONE CELLS

TENDON

WHITE BLOOD CELLS
(LEUKOCYTES)

RED BLOOD CELLS
(ERYTHROCYTES)

BLOOD

Fig. 5.34
Animal tissues. The organs of a complex animal are composed of at least three, and usually more, of these tissues. The stomach, for example, contains glandular epithelium, smooth muscle, blood, and nerves. (Photos: Bone cells courtesy Triarch, Inc., Ripon, Wisconsin; others courtesy Turtox, Chicago.)

cells resemble pillars or columns, with the nucleus usually located near the bottom of the cell. Columnar epithelium is widespread throughout the body, forming most of the lining of the digestive and respiratory tracts.

These various types of epithelial tissue are specialized for particular functions. Columnar cells, for example, often have small hairlike projections (*cilia*), which beat rhythmically, moving materials such as mucus in one direction. In the respiratory system,

MUSCULAR

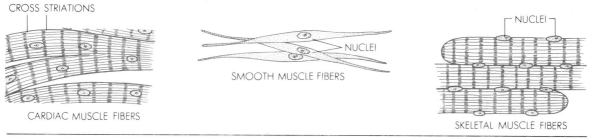

CROSS STRIATIONS

CARDIAC MUSCLE FIBERS

NUCLEI

SMOOTH MUSCLE FIBERS

NUCLEI

SKELETAL MUSCLE FIBERS

NERVOUS

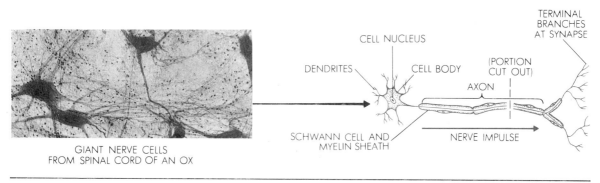

GIANT NERVE CELLS
FROM SPINAL CORD OF AN OX

CELL NUCLEUS

DENDRITES

CELL BODY

TERMINAL BRANCHES AT SYNAPSE

(PORTION CUT OUT)

AXON

SCHWANN CELL AND MYELIN SHEATH

NERVE IMPULSE

REPRODUCTIVE

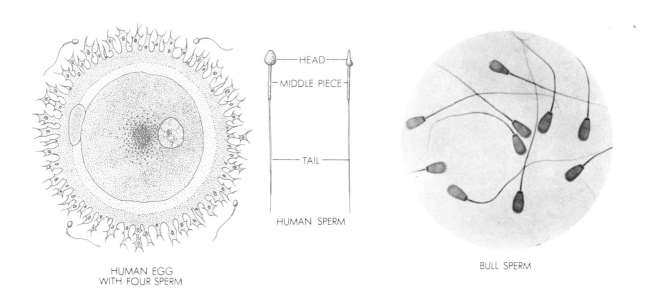

HUMAN EGG
WITH FOUR SPERM

HEAD

MIDDLE PIECE

TAIL

HUMAN SPERM

BULL SPERM

dust and other particles of foreign materials are removed by a ciliated columnar epithelium, which, by coordinating a wave-like beat of cilia from cell to cell, move the materials to the exterior.

Also scattered throughout the columnar epithelial cells of the respiratory passages are **glandular epithelial cells,** specialized to secrete substances such as mucus. In the lining of the ear canal, glandular columnar epithelial cells secrete wax. The milk with which female mammals

feed their young is produced by glandular epithelial cells. Within the skin, glandular epithelial cells secrete perspiration.

Finally, epithelial cells may be specialized into a *sensory epithelium* to receive stimuli. For example, olfactory epithelium, which lines the passages of the nose, is partly responsible for the sense of smell. Other sensory epithelial cells are scattered throughout the skin, where they may be sensitive to such stimuli as heat, cold, or pressure. In the sense that all incoming stimuli must pass through an animal's epithelial tissues, such tissues can truly be called the gateway to the body.

Connective Tissues

Perhaps the easiest tissues to identify, connective tissues (Fig. 5.34b) are composed of cells embedded in a non-living matrix, one which the cells themselves may often secrete. The nature of this matrix, rather than that of the cells themselves, determines the functions of the particular connective tissue. **Bone, cartilage, tendons,** and **ligaments** are all familiar examples of connective tissues that support and hold together the other cells of the animal body.

The versatility of connective tissues is due primarily to the protein collagen (see p. 69), acting as a structural element, and is achieved by collagen's ability to interconnect, through fine fibrils, with other types of molecules, thereby achieving great tensile strength, much as different metals combine in an alloy to achieve a strength neither metal has by itself. Diseases associated with a deficiency of collagen are arthritis and rheumatism, while cirrhosis of the liver and hardening of the arteries (arteriosclerosis) are the result of collagen overproduction.

Cartilage and bone are perhaps the best-known connective tissues. Living bone tissue is composed of the familiar material we call bone permeated by many cells and blood vessels. The true bone cells, or *osteocytes*, are actually imprisoned individually within the hard bone matrix. The central body of each osteocyte occupies a space (lacuna) from which it radiates out many fine cytoplasmic processes through tiny tunnels or *canaliculi*. Osteocytes are believed to regulate the flow of mineral ions, especially calcium, between the bony material and the bloodstream. Each bone is covered with a thin membrane, the **periosteum.**

During embryological development, bone is laid down first as fibrils of the protein collagen that, as development proceeds, becomes increasingly impregnated with hydroxyapatite, a calcium-containing substance that becomes cemented together with water, mucopolysaccharides, and several other organic and inorganic ions. In adult bone, the bone matrix is generally arranged in layers, or lamellae. While to the naked eye bone appears to be spongy on the inside but solid on the outside, besides the canaliculi the microscope reveals bone to be thoroughly permeated with other fine canals

(Haversian canals) containing blood vessels and bone-forming cells (*osteoblasts*) and bone-destroying cells (*osteoclasts*) responsible for the building up and breaking down of bone during maturation. Thus bones gradually change their shape in response to the stresses and strains of normal growth, as well as blood and tissue calcium availability.

Considering their light weight and slenderness, the strength of bones is remarkable; they are superior to such material as copper or concrete in their resistance to stress without breaking.

Cartilage, which is more elastic than bone, can be considered bone with lesser amounts of calcium deposits in the matrix. The skeleton of the outer regions of the nose and ears is composed of cartilage. In some animals, such as sharks, virtually the entire skeleton is cartilaginous. In vertebrates, the skeleton of the developing embryo is laid down in two different ways. Some bones are formed directly from special cells in the dermal (skin) layer. These bones lack any cartilage precursors and generally occur in the outer regions of the body such as the skull. Other bones form from a preexisting cartilaginous matrix. These bones are generally found in the deeper parts of the body, such as the bones of the arms, legs, and vertebral column.

In tendons and ligaments the matrix is a thick network of collagen and reticular fibers secreted by the connective tissue cells they surround. These stringlike tissues are found throughout the body. They hold bone to bone, muscle to bone, and skin to muscle, and bind many other structures together. The toughness of leather, formerly connective tissue of an animal, testifies how well the tissues are constructed to serve their purpose.

Blood may also be considered a connective tissue since it, too, is composed of living cells surrounded by a nonliving matrix, the *blood plasma.** Blood contains two main types of cells, red blood cells (erythrocytes), which carry oxygen, and a variety of white blood cells (leukocytes), which aid in fighting infection. Due to its liquid nature, blood cannot perform any supportive function, but in the sense that it serves as a transport medium between the different organs of the body, it is perhaps the most connective tissue of all.

---

*Some physiologists, in deference to its complexity, consider blood to be an organ.

Animal tissues consist of five major types: epithelial, connective, muscular, nervous, and reproductive.

Muscular Tissues

In terms of their structural and functional characteristics, muscle tissues can be divided into three types. Most of the body musculature is composed of **skeletal muscles,** which move the body appendages. The cells of skeletal muscles are long and fibrous, with many nuclei located close to the surface of the cell. Under the microscope, skeletal muscle cells show stripes, or *striations,* running across the width of the cell. Indeed, skeletal muscles are often called **striated muscles.** Skeletal muscle cells are capable of quite rapid contraction and can be controlled at will by the conscious portions of the brain. For this reason, the skeletal muscles are often referred to as *voluntary* muscles, as opposed to the types of muscle cells to be discussed next that operate mostly involuntarily. They cannot remain contracted for long, however, since they are rather easily fatigued.

Cardiac muscle is the muscle of the heart. Like those of skeletal muscle, the cells of cardiac muscle are long and striated. There are a few differences, however.

The nuclei of cardiac muscles are located in the central region of the cell. Furthermore, the long cells often branch and fuse with each other, so that there are interconnections between cells along their length. This structure helps to account for the rhythmic contractions characteristic of heart muscle. As might be expected, cardiac muscle does not fatigue easily. The period of rest between each contraction seems to be all that it needs.

Smooth muscle is not striated, and thus appears smooth under the microscope. Like those of the skeletal and cardiac muscles, smooth muscle cells are elongated, but they differ in having pointed ends and only one nucleus per cell. Smooth muscle cells are found within the walls of the stomach, intestines, esophagus, and blood vessels, as well as other places. The smooth muscle cells of vertebrates generally contract more slowly than skeletal muscles, but they can apparently remain contracted for considerable lengths of time without becoming fatigued. In some insects, smooth muscles operate the wings. Such muscles, of course, are capable of very rapid contraction.

SUPPLEMENT 5.2
BIOLOGICAL SELF-ASSEMBLY

Microtubules and microfilaments can be taken apart and rebuilt by simple chemical treatments. For example, microtubules can be broken down either by physical disruption of the cell or by addition of high concentrations of calcium ions. Under these conditions the microtubule breaks down into subunits of a protein called, appropriately, tubulin. These parts are all the same size and molecular weight for any given type of organism. If isolated tubulin parts are placed in a calcium-free medium and supplied with a phosphate compound such as adenosine triphosphate (ATP) or guanosine triphosphate (GTP), the parts reaggregate and form a section of intact microtuble. The structure can assemble itself if provided with the appropriate chemical agents.

Microfilaments can be disaggregated by a substance known as cytochalasin B, which does not appear to alter any other cell components. If cytochalasin B is subsequently removed, the microfilaments re-form. Aggregation and disaggregation are reversible. More important, the fact that reaggregation occurs spontaneously indicates that the parts carry within them the "instructions" for their own assembly. It is enough to produce the parts with a specific amino acid sequence, and the larger structure (microtubule) will form spontaneously.

Knowledge of disaggregation and reaggregation of tubulin parts makes it possible to learn something of the role microtubule structures may play in cell life. When cytochalasin B

is applied to dividing cells, the furrow that normally develops to cut the two parts of the cell in half never appears. We conclude that the furrow is formed by the action of microfilaments. Application of chemicals known specifically to disrupt the structure of microfilaments makes it possible to pinpoint the areas of cell activity in which these structures are involved.

Similar disruption experiments have been carried out with microtubules. Low temperature or the chemical colchicine disrupts microtubules in much the same manner in which cytochalasin B disrupts (disaggregates) microfilaments. When colchicine is applied to dividing cells, the duplicated chromosomes do not separate from each other. Formation of pseudopods in organisms such as *Amoeba* and the familiar streaming of cytoplasm within plant cells are also interrupted by colchicine. Removal of colchicine usually restores these functions. It is thus possible to conclude that there may be some distinct relationship between microtubular function and these various motions observed within cells. It is important to keep in mind that such evidence does not suggest an unequivocal cause-effect relationship. It might be that colchicine is disrupting a process that affects both microtubule structure and various cell processes. More and different kinds of evidence would be necessary before we could conclude that microtubules cause chromosome movements, pseudopod formation, or cytoplasmic streaming.

Figure 5.34(c) shows the structural characteristics of skeletal, cardiac, and smooth muscle.

Nervous Tissue

Nervous tissue is made up of several types of cells. **Neurons** are nerve cells specialized to conduct sensory or motor (movement-causing) signals or nerve impulses. Figure 5.35(d) shows a neuron and its parts. While neurons vary widely in size (and often in appearance), all have certain structural characteristics in common. The central cell body contains a nucleus surrounded by cytoplasm. Two types of projections, **axons** and **dendrites,** can be distinguished.

Some axons are covered with a *myelin sheath* laid down by specialized *Schwann cells.* The significance of this arrangement will be discussed when we deal with the nature of the nerve impulse in Chapter 13. Changes in axons and the myelin sheath that surrounds them are responsible for the symptoms seen in a large range of diseases, including multiple sclerosis, leprosy, and alcoholism. Axons generally conduct nerve impulses away from the cell body, and dendrites generally conduct them toward the cell body. Impulses travel from one cell to the next by passing across a small gap, the **synapse,** to the dendrite of the next neuron. The projections extending from neurons may vary considerably in length and complexity. A nerve cell in the spinal cord may send an axon down an arm or leg for three feet or more.

Most nerve fibers are microscopic in width. The sciatic nerve going down into the human leg is as thick as a pencil, but this is because it is composed of hundreds of nerve projections. Such a nerve cord is rather like a telephone cable in which many separate wires carry impulses and are insulated from each other along their length.

In addition to neurons, other nerve cell types include cells often found enveloping individual neurons called **neuroglia** or, more generally, **glia.** It has been hypothesized that the glia cells may provide nutrients for neurons or play a role in their embryological growth and orientation, but to date their precise function remains undetermined. **Neurosecretory cells,** as their name implies, secrete substances that may have an effect elsewhere in the body. We shall have more to say about such cells in Chapter 13.

Reproductive Tissue*

As its name implies, reproductive tissue is composed of those types of cells specialized to produce the next generation. In "higher" animals, the cells directly involved in reproduction (**gametes**) are specialized into the female egg cell, or **ovum,** and the male spermatozoa, or **sperm.** Egg cells are generally spherical in shape and contain yolk to feed the developing offspring from the moment of fertilization until it is able to obtain food in some other way. Sperm cells are far smaller than eggs and can swim by whipping their cytoplasmic tails. The shape of the sperm cell may vary widely from one species to another. Figure 5.34(e) shows details of these two types of reproductive tissue.

Summary

1. Biologists generally recognize two major types of cells: eukaryotic (possessing a nucleus surrounded by a nuclear membrane) and prokaryotic (lacking a nucleus, with no nuclear membrane). The cells of all protozoa and most types of algae and the higher plants and animals are eukaryotic. The cells of bacteria, blue-green algae, and mycoplasmas are prokaryotic. Viruses are not cellular, being composed only of nucleic acid surrounded by a protective coat of protein. Viruses are able to reproduce themselves inside living cells. Outside living cells, they behave very much like the crystals of any large molecule. Hence they cannot be grouped with either eukaryotes or prokaryotes.

2. Eukaryotic cells contain a number of structures, or organelles, each of which has its own particular function(s):

a) *Nucleus.* The nucleus is the center of hereditary processes in the cell, and contains nucleic acid as deoxyribonucleic acid (DNA) and ribonucleic acid (RNA). DNA is passed on from generation to generation and transmits the cell's hereditary information. RNA is involved in various aspects of translating the information contained in DNA into protein structure. Prokaryotes, which have no distinct nucleus, often have their genetic material localized within a central region of the cell.

b) *Endoplasmic reticulum.* The ER is a series of membrane-lined channels running through the cytoplasm of most cells. The endoplasmic reticulum facilitates transport within the cell; its membranes also serve as sites for many chemical reactions. There are two types of endoplasmic reticulum: rough endoplasmic reticulum, in which ribosomes are bound to the membranes; and smooth endoplasmic reticulum, in which there are no ribosomes attached to the membranes.

c) *Ribosomes.* Ribosomes are the site of protein synthesis in all cell types, prokaryotic as well as eukaryotic. In eukaryotic cells ribosomes are usually attached to endoplasmic reticulum; in prokaryotic cells they are freely dispersed in the cell cytoplasm.

d) *Golgi complex.* The Golgi complex is a cluster of flattened, parallel sacs found within the cytoplasm of many

---

*It is possible to question the use of the word "tissue" in connection with the reproductive cells, since they do not function as a unit.

cell types. It appears to be involved in the transport, packaging, and excretion of materials, particularly lipids.

e) *Lysosomes.* Lysosomes are spherical or rod-shaped organelles surrounded by a membrane boundary. Lysosomes contain a variety of enzymes, mostly those that act on macromolecules.

f) *Peroxisomes and glyoxysomes.* These are tiny vesicles containing oxidative enzymes. Peroxisome enzymes are involved in photorespiration and the oxidation of amino acids and hydrogen peroxide. Glyoxysome enzymes are involved in the conversion of fat to carbohydrate.

g) *Mitochondria.* Mitichondria are located throughout the cytoplasm of all eukaryotic (but not prokaryotic) cells. They are usually found in areas of the cell showing the greatest metabolic activity. Mitochondria have an outer membrane surrounding an inner membrane folded inward to form partitions called cristae that increase the inner surface area. Enzymes involved in the respiratory process are bound to the inner membrane. Mitochondria are the centers of capture of usable energy from oxidation of foodstuffs. Within the mitochondria, energy-rich molecules of adenosine triphosphate (ATP) are generated from energy released when glucose and other substances are oxidized.

h) *Plastids.* Found only in plant cells, plastids give color to plant leaves and fruits and often serve as storage centers. The most important plastids are chloroplasts, which are bounded by an external membrane and contain numerous internal membranes called lamellae. Lamellae are stacked at certain intervals into compact units called grana. Chlorophyll molecules are bound to lamellae. Chlorophyll absorbs light and converts it to a useful form of energy (ATP) for synthesizing carbohydrates. The process of energy capture and carbohydrate synthesis is known as photosynthesis; chloroplasts are the site of photosynthesis in all green plants.

i) *Vacuoles.* Vacuoles are bubble-like inclusions in cells that often serve as storage reservoirs. In animal cells, vacuoles are generally smaller and referred to as vesicles.

j) *Microtubules.* Microtubules are tubules composed of protein subunits, running throughout the cytoplasm in many cell types. They appear to function in relation to various cell movements, including cytoplasmic streaming in certain types of plant cells and especially the motion of cilia and flagella.

k) *Centrioles.* Centrioles are structures consisting of two sets of microtubular units resembling barrels, placed at right angles to each other. Centrioles are found in most cells except those of many kinds of higher plants. They appear to have a role in the process of cell division.

l) *Microfilaments.* These are bundles of protein threads found in certain cells. They appear to be related in some way to certain types of cell movements, such as the contraction of heart muscle.

3.   All eukaryotic cells undergo the processes of mitosis and meiosis. Mitosis involves replication of the existing set of chromosomes and subsequent division of these chromosomes equally between the two daughter cells. Mitosis preserves the chromosome number from one cell generation to the next. Meiosis is a sequence of two cell divisions, in the first of which the chromosome number is reduced by half. The process occurs in such a way that each daughter cell gets one complete set of chromosomes. Meiosis is particularly involved in the sexual phase of reproduction cycles in organisms, and prevents the repeated doubling of the chromosome number that would otherwise result from sexual reproduction.

4.   Prokaryotes reproduce by a process known as fission, which is asexual. Fission is analogous to mitosis, since it involves a replication of the nucleic acid and division of the two copies between daughter cells. Prokaryotes also engage in sexual reproduction, during which they exchange genetic material and subsequently undergo a division similar to meiosis.

5.   Viruses are dependent upon living cells to reproduce. They are composed of protein and nucleic acid, and constitute a major cause of disease.

6.   In both plants and animals, cells are grouped into function units called tissues. Plants have four types of tissues:

a) *Embryonic or meristematic tissue.* This tissue is found at root and stem tips and forming the cambium layer around the inside of plant stems and roots. Meristematic tissue is the growing region of the plant, where cell division is constantly taking place.

b) *Protective or epidermal tissue.* These cells form the outer protective layers of plants.

c) *Fundamental tissue.* The greatest part of the mass of plants, with rather generalized functions, fundamental tissue includes three types of cells. (1) Parenchyma cells carry out such functions as storage, photosynthesis, and secretion of special oils and other substances. (2) Collenchyma cells are long and thin with thick cell walls, and usually provide support for the young, growing plant. (3) Sclerenchyma cells provide support for the mature plant. There are two types of sclerenchyma cells, fibers and sclereids.

d) *Vascular or conducting tissue.* Vascular tissue is specialized for transporting water and dissolved substances. There are two basic types of vascular tissue: xylem (consisting of two cell types, vessels and tracheids), and phloem (consisting of sieve tubes). Xylem conducts water and ions, mostly upward; phloem conducts water and carbohydrates, mostly downward. The mature xylem cells are thick-walled and nonliving; they contribute to the rigidity of the plant body. Phloem cells are thin-walled and alive.

Animals have five kinds of tissue:

a) *Epithelial tissue.* Epithelial tissue is generally found lining passages such as the digestive tract, respiratory passages, and blood vessels. It also serves as a covering for various body parts and is of various shapes.

b) *Connective tissue.* Connective tissue is composed of living cells surrounded by a nonliving matrix, and includes bone, cartilage, tendons, ligaments, and blood.

c) *Muscular tissue.* Muscular tissue includes the skeletal muscles that produce voluntary movement, involuntary cardiac (heart) muscle, and smooth or visceral muscle.

d) *Nervous tissue.* Nervous tissue includes the neurons for the conduction of nerve impulses, registration of sensory impulses, and conduction of motor impulses. Other nerve cell types include the neuroglia, or glia cells, and neurosecretory cells.

e) *Reproductive tissue.* Reproductive tissue includes those cells involved in reproduction and the cells of the organs that give rise to them. Cells directly involved in the reproductive process in complex multicellular forms are the gametes, the female egg and the male sperm.

Exercises

1. Discuss briefly the functions of the following parts of a cell: (a) plasma membrane, (b) cell wall, (c) ribosomes, (d) endoplasmic reticulum, (e) mitochondria, (f) centriole, (g) plastids, (h) lysosomes.

2. What is significant about the fact that most cells contain almost all the same structures?

3. If electron micrographs show that mitochondria are grouped around a particular structure or region of the cell, what might you infer is happening in this region?

4. Table 5.2 shows the results of a series of differential centrifugations of a cell homogenate. At each step the pellet (or, in the case of the last step, the final supernatant) is resuspended and studied for its biochemical and structural properties (see last column on the right).

For the pellet fraction whose properties are described in the last column of Table 5.2, list which organelle or organelles (or parts of organelles) might be expected to appear in that fraction. Explain briefly what evidence leads you to identify particular organelles with particular fractions. In

Table 5.2

Sample	Centrifugation speed ($\times$ force of gravity)	Time centrifuged	Physiological and/or chemical characteristics of pellet fraction resulting from centrifugation
1. Cell homogenate	900 g	10 min	Contains much DNA and RNA, and *large* lipid-containing fragments.
2. Supernatant from no. 1	15,000 g	20 min	Is able to oxidize glucose and other carbohydrate intermediates. Consumes oxygen and produces CO_2. Contains particles that, if ruptured, are capable of digesting away most other cell compounds.
3. Supernatant from no. 2	25,000 g	30 min to 1 hr	Lipid and narrow tubular-shaped membranes; sometimes has ability to produce full-scale proteins from amino acids.
4. Supernatant from no. 3	50,000 g	30 min to 1 hr	Some lipid and tubular-shaped membranes; seldom shows ability to produce proteins from amino acids.
5. Supernatant from no. 4	100,000 g	2 hr	*Small* lipid fragments.
6. Supernatant from no. 5	[No centrifugation. Supernatant is studied directly.]		Contains particles with RNA and protein. Able to produce proteins from amino acids under proper conditions.

general, you should make use of two kinds of data supplied in the table: (1) the relative densities of the organelles as indicated by the centrifugation speeds at which they settle out, and (2) the biochemical, physiological, and structural information given in the last column on the right. It is possible for an organelle, or fragments of large organelles, to appear in more than one fraction. Taking Fraction 1 as an example, your answer might run like this:

1. *Probably nuclei (containing nucleoli), and large fragments of plasma or nuclear membranes. Nuclei are very large and are among the densest bodies in the cell; they contain DNA (in chromosomes) and RNA in nucleoli. Cell membranes contain much lipid, and large fragments have a relatively high density.*

Chapter 6
Cell Metabolism I:
Energy Release and Metabolic Pathways

INTRODUCTION

Cells can be compared to chemical machines. They consume fuels (sugars) and release waste products (carbon dioxide and water). The energy they extract from their fuel is used to perform **work**, which we will here define very generally as the energy required to change a system from one state to another. All cells perform general types of work, such as building up or synthesizing molecules, or actively transporting various substances across cell membranes. In addition, some cells also do specialized work, such as physical movement (as in the contraction of a muscle cell) or electrochemical work (such as the conduction of an impulse by a nerve cell).

Like machines, cells release energy from fuels in an orderly way to accomplish specific tasks. In both cases, the release of energy is accomplished by releasing some of the chemical energy of fuels in the process known as **oxidation.** But here the comparison between the cell and the machine begins to fail. When looked at more closely, the details of energy extraction and use show the cell to be a far more complex and delicate unit than any existing machine. In the long run, the differences between cell and machine are far more important than their similarities in adding to our understanding of the nature of life.

6.2
METABOLISM

The totality of the processes by which cells obtain and use energy to do work and maintain themselves is known as **metabolism.** All metabolic activity is ultimately chemical; it involves the interaction of atoms and molecules, although living organisms are also involved with interconversions of other forms of energy, such as radiant (light), thermal (heat), and electrical energy. The metabolism of cells is carried out by many series of interconnected chemical reactions known as *metabolic pathways.* Any particular metabolic pathway usually consists of many individual reactions whose total effect is the synthesis or degradation of molecules and the storage, or release, and use of energy.

The study of cellular metabolism is often perceived by students as the most difficult portion of a biology course. Yet the underlying principles are ones for which virtually everyone has at least an intuitive feel, if not, perhaps, a precise understanding. This chapter will be concerned with the way in which cells extract energy for their own use from the primary "fuel" molecules of glucose.* Cells extract this energy via many intermediate reactions, with the resulting production of many intermediate compounds. (The number and names of these intermediate compounds often make the study of cellular metabolism seem difficult.) This means that the energy is released in small amounts at each intermediate step, and thus more of it can be captured for useful cell work.

The energy is generally not used directly, however. Instead it is transferred and stored in the high-energy

---

*It should be emphasized that glucose is only one example of a number of compounds that can be starting points for these metabolic pathways. Humans and other mammals, for example, often use fats and proteins as significant fuels, and organisms such as bacteria can use an enormous diversity of organic compounds.

Metabolism is the sum total of biochemical reactions by which cells obtain and use energy to do work and maintain themselves. The particular sequence of biochemical reactions leading from one set of reactants through a series of intermediates to one or more final products is called a metabolic pathway.

If the glucose molecule were completely oxidized all at once (as in burning), its free energy would be released in an uncontrolled way and hence could not be captured by the cell for the generation of ATP.

bonds of the compounds first mentioned in Chapter 4 that serve as the energy currency of the cell, most especially adenosine triphosphate, or ATP. You might wish, therefore, to review the discussion of ATP in Section 3.9. Some of the energy within each glucose molecule is repackaged into small packets within each molecule of ATP, and can then be used for such cell work as the synthesis of certain molecules (see Fig. 6.1).

The formation of ATP occurs by the **phosphorylation**—the addition of a phosphate group—to adenosine diphosphate, or ADP. In the discussions of cellular metabolism to follow, two types of phosphorylation occur. In one, **substrate-level phosphorylation**, the phosphate group is donated by a molecule with a higher free energy than ADP. In the other, **oxidative phosphorylation**, no substrate donor molecule is involved—the phosphate ion (HPO_4^{-2}, often abbreviated P_i, with the i standing for "inorganic"), available in the cell, is added directly onto the ADP.

Reactions in which the breakdown of all or a portion of one molecule results in the synthesis of another are examples of *coupled reactions*. For example, the re-

action uniting two simple sugars (monosaccharides) is brought about by the formation of a chemical bond (glycosidic) between them. Such a reaction can be represented in an overall manner as follows:

GLUCOSE   FRUCTOSE

Specific enzymes

GLYCOSIDIC BOND $+ H_2O$

SUCROSE

This reaction is thermodynamically improbable as written; the ΔG is greater than zero ($\Delta G > 0$; actually

Fig. 6.1
The ATP cycle. Molecules of ATP, the "small-change energy currency" of the cell, give up one or two phosphate groups, releasing the energy locked in their high-energy bonds to perform cell work. Some of the types of work for which energy is needed are shown to the right. The energy-poor compounds ADP and AMP are built back up into ATP by addition of one or two phosphate groups respectively. The energy for this build-up is derived from the breakdown of glucose.

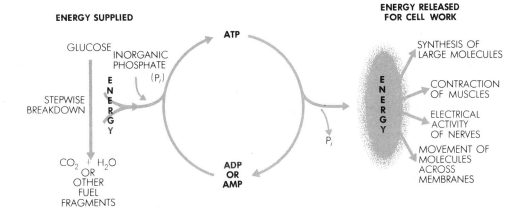

5.5 kcal/mol). The reaction is made probable, however, by coupling the process to the ATP→ADP reaction via two separate reactions, in both of which the ΔG is less than zero ($\Delta G < 0$).

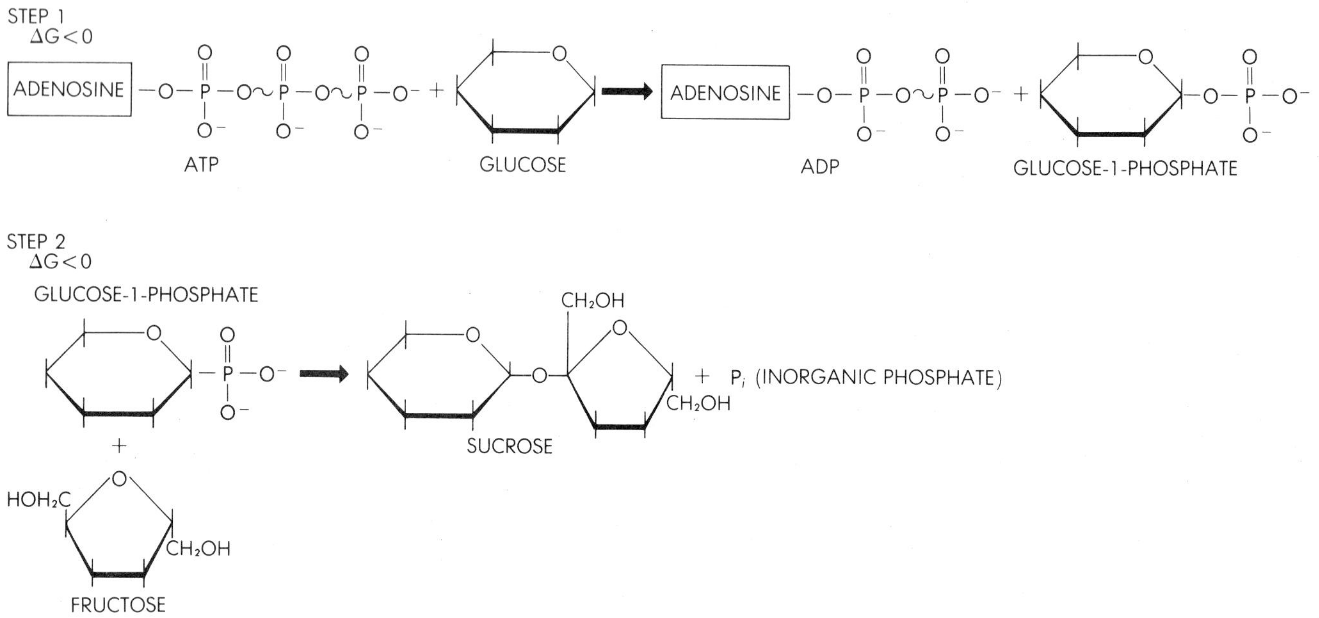

STEP 1
$\Delta G < 0$

STEP 2
$\Delta G < 0$

The changing of the thermodynamics situation from improbable to probable marks the significance of coupling. In the first reaction a molecule of ATP interacts with a molecule of glucose; during this reaction one phosphate group is transferred from the ATP to form glucose-1-phosphate, leaving ADP. Glucose-1-phosphate is now an energy-rich intermediate, containing some of the energy originally held in the high-energy phosphate bond of ATP. In the second reaction, hydrolysis of the phosphate bonds of glucose-1-phosphate yields about the same amount of energy as required to form the glycosidic bond. Thus the second reaction proceeds spontaneously as long as a supply of glucose-1-phosphate and fructose is present. In the transfer of the terminal phosphate of ATP to glucose, electrons are rearranged around carbon 1 of the glucose; the result is the formation of a relatively high-energy phosphate bond. The rearrangement of electrons forming this new chemical bond represents potential chemical energy. This potential energy is realized as kinetic chemical energy when the glucose-1-phosphate interacts with fructose to form a new chemical arrangement (the glycosidic bond).

Coupled reactions, then, involve the creation of a high-energy intermediate by rearrangement of atoms within at least one reactant. This first step in the process is exergonic. The activated, high-energy intermediate now contains the energy that drives the second, or endergonic, step. When exergonic and endergonic reactions are coupled in this way, relatively little of the total potential energy is lost.

Finally, you may wish to review Section 2.8 dealing with oxidation-reduction reactions, for they form the basis for the exergonic reactions of cellular metabolism (**catabolism**). The removal of electrons from atoms or compounds—oxidation—is always coupled with a gain of electrons—reduction—by other atoms or molecules, which gain the electrons lost by the oxidized atoms or molecules. (In the case of cellular metabolism, molecules rather than individual atoms are usually oxidized and reduced.) Indeed, as pointed out in Section 2.7, this coupling of oxidation with reduction is often emphasized by calling such interaction "redox" reactions. Since the electrons generally move from the higher energy levels they occupied in the molecules that become oxidized to lower energy levels in the newly reduced molecules, the process is accompanied by a release of energy from the oxidized molecule (see Fig. 6.2). Indeed, the entire catabolic portion of cellular metabolism can be viewed as the oxidation of complex molecules (such as glucose), with the production of carbon dioxide and water and the release of energy. For example,

$$C_6H_{12}O_6 + 6O_2 \rightarrow 6CO_2 + 6H_2O + \text{energy}$$

($\Delta G = 686$ kcal/mol).

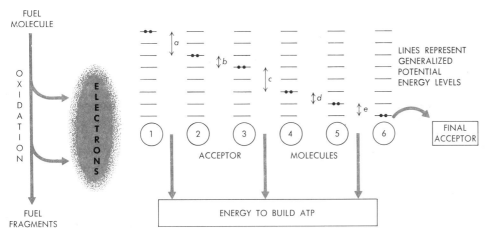

Fig. 6.2
Generalized diagram showing the principle involved in biological oxidation and the capturing of energy as ATP. The circular structures labeled 1 through 6 represent specific electron transport molecules. The electron drops in potential energy level as it passes from one transport molecule to another. These drops occur within the electron cloud of each transport molecule and are indicated by the distances *a*, *b*, *c*, *d*, and *e*. The drop in energy level with each transfer results in the release of small amounts of energy. This energy is captured in the formation of ATP.

In actuality, however, as the preceding discussion of intermediate products, coupled reactions, and the like suggests, the oxidation of glucose is not a one-step process. Instead, the overall process of cellular catabolism is divided into several stages, each of which consists of a number of individual metabolic reactions. Two of these stages, **glycolysis** and **respiration**, will be discussed next. It is important to recognize that, in any metabolic process, several optional biochemical pathways may be available, depending on such things as the species of organism or the internal or external environmental conditions.

6.3
GLYCOLYSIS

Glycolysis (*glyco*, sugar; *lysis*, breakdown) is that portion of cellular metabolism involved in the conversion of glucose molecules into a compound called **pyruvic acid.** The process of glycolysis occurs in the cell cytoplasm, or cytosol. For each glucose molecule ($C_6H_{12}O_6$), two molecules of the three-carbon compound pyruvic acid ($C_3H_4O_3$) are produced. This transformation of glucose to pyruvic acid occurs in nine reactions, each catalyzed by a specific enzyme.

For glycolysis to begin, various carbohydrate stores in the cell must be prepared for entry into the glycolytic pathway. The starting point involves a "priming" or **activation** process, and may involve either single, free molecules of glucose or glucose molecules bound together as starch (in plants) or **glycogen** (in animals) (*glyco*, sugar; *gen*, to give rise to). As its name suggests, glycogen can be easily converted into glucose and, in actuality, it serves as a sort of molecule reservoir of glucose.

Since, during the formation of glycogen from glucose by the joining together of many glucose molecules, some energy from ATP was locked into the compound, more ATP is not necessary to activate glycogen. The activation of glucose, however, *does* require an initial input of energy via ATP. Thus while the overall result of glycolysis is to capture some energy in a form (ATP) useful for cellular work, the initial steps of the activation process actually use up some ATP. There is nothing unusual in this, of course; we use the energy locked into a match to release the far larger amount of energy stored in the logs of a fireplace. Glycolysis is initiated by the transference of the terminal phosphate molecule of ATP to glucose, which thereby becomes glucose-6-phosphate (the *6* merely indicates to which of glucose's six carbon atoms the phosphate group becomes attached). The ATP, of course, is converted into ADP.

(two molecules)

GLUCOSE → HEXOKINASE → GLUCOSE-6-PHOSPHATE

Activation, sometimes called "mobilization," involves spending two ATPs per glucose molecule in order to start the sugar down the oxidative pathway.

You may recall that the formula for fructose, like that of glucose, is $C_6H_{12}O_6$. Yet its properties are quite different from those of glucose (fructose is much sweeter to the taste, for example). In the next step of glycolysis, the glucose-6-phosphate is converted into fructose-6-phosphate by the rearrangement of atoms from the six-sided glucose ring into the five-sided fructose ring:

GLUCOSE-6-PHOSPHATE   PHOSPHO-GLUCOSE ISOMERASE   FRUCTOSE-6-PHOSPHATE

This physical conversion of one molecular form into another is immediately followed by the expenditure of yet another ATP molecule, which gives up its terminal phosphate group to fructose-6-phosphate, thereby converting it to fructose-1,6-diphosphate:

FRUCTOSE-6-PHOSPHATE   ATP   ADP   PHOSPHO-FRUCTOKINASE   FRUCTOSE-1,6-DIPHOSPHATE

This step marks the end of the activation phase. Fructose-1,6-diphosphate is an energized molecule with more potential energy than the original glucose. Since

fructose-1,6-diphosphate represents a shove *up* the energy hill, activation is an endergonic process. As will be seen later, this step, involving an allosteric enzyme (see p. 80) is also a key one in controlling the rate of ATP production. A summary of the activation phase is shown in Fig. 6.3.

What happens next is both interesting and significant. The fructose-1,6-diphosphate is split into two three-carbon compounds, **glyceraldehyde 3-phosphate** or **phosphoglyceraldehyde (PGAL)** and **dihydroxyacetone phosphate:**

ALDOLASE (SPLITS FRUCTOSE-1,6-DIPHOSPHATE)   GLYCERALDEHYDE-3-PHOSPHATE OR PHOSPHOGLYCERALDEHYDE (PGAL)   TRIOSE PHOSPHATE ISOMERASE   DIHYDROXYACETONE PHOSPHATE

This process is interesting in that the two three-carbon sugars exist in chemical equilibrium with each other; thus the dihydroxyacetone phosphate can be converted directly into PGAL, thereby flowing down the same series of reactions followed by PGAL rather than requiring its own separate pathway. Efficiency is thus increased by reducing the number of specific enzymes required. The process is significant in that PGAL is precisely the same compound used by green plants to synthesize the glucose whose degradation we are now following.

Note that, in terms of energy, the entire process to this point is a losing proposition: two ATPs have been invested, and no dividends have been paid. The next step, however, changes all that. In this double reaction

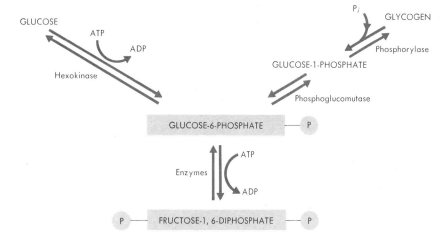

Fig. 6.3
The first phase of carbohydrate catalysis, called activation or mobilization. During this phase, an activated six-carbon molecule, fructose-1,6-diphosphate, is produced from either free glucose or glucose bound up as glycogen. Two ATP molecules are used for energy to convert each glucose molecule into fructose-1,6-diphosphate. The cell thus uses some stored energy of ATP to activate its carbohydrates for further oxidation, a sort of initial investment from which a considerable profit will eventually result.

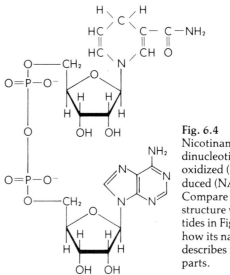

Fig. 6.4
Nicotinamide adenine dinucleotide (NAD) in its oxidized (NAD⁺) and reduced (NADH) forms. Compare its molecular structure with the nucleotides in Fig. 3.13 and note how its name accurately describes its molecular parts.

the PGAL becomes oxidized by a highly important co-enzyme, **nicotinamide adenine dinucleotide (NAD)**, which thereby becomes reduced (NADH) (see Fig. 6.4).* The energy that would ordinarily be released by such an oxidation, however, is captured by the substrate-level phosphorylation of PGAL:

$$2\times \begin{array}{c}\text{CH}_2\text{O}-\text{P}-\text{O}^-\\\text{CHOH}\\\text{C}\\\text{H}\quad\text{O}\end{array}\quad\xrightarrow[\text{DEHYDROGENASE}]{\text{TRIOSE-}\atop\text{3-PHOSPHATE}}\quad 2\times\begin{array}{c}\text{CH}_2\text{O}-\text{P}-\text{O}^-\\\text{CHOH}\\\text{C}\\\text{O}\sim\text{P}\end{array}$$

GLYCERALDEHYDE-3-PHOSPHATE (PGAL) P_i NAD   NADH

1,3-DIPHOSPHOGLYCERATE

Since the potential free energy that could be released by this high-energy-bond hydrolysis is even greater than that of ATP, the next reaction can transfer the phosphate groups to ADP, regaining the two ATPs invested in the initial phases of glycolysis:

$$2\times\begin{array}{c}\text{CH}_2\text{O}-\text{P}-\text{O}^-\\\text{CHOH}\\\text{C}\\\text{P}\sim\text{O}\end{array}\quad\xrightarrow[\text{KINASE}]{\text{PHOSPHO-}\atop\text{GLYCERATE}}\quad 2\times\begin{array}{c}\text{CH}_2\text{O}-\text{P}-\text{O}^-\\\text{CHOH}\\\text{C}\\\text{O}\end{array}$$

ADP ATP

1,3-DIPHOSPHOGLYCERATE   3-PHOSPHO-GLYCERATE (3-PHOSPHO-GLYCERIC ACID)†

---

*For the reader who wishes to delve deeper into the biochemistry of cell metabolism, it is important to be familiar with the various conventions for symbolizing reduced or oxidized compounds. Using NAD as an example, some writers simply symbolize the oxidized and reduced forms as NAD_{ox} and NAD_{re}, respectively. More commonly, oxidized and reduced NAD are written as NAD^+ (the positive charge sign indicating the loss of negatively charged electron, which results in the molecule being positively charged) and NADH, respectively. Or the reduced form may be written NADH or, more accurately, as NADH + H. This way of symbolizing reduced NAD emphasizes that what each NAD molecule picks up as a carrier is not two complete hydrogen atoms (each of which, you recall, is composed of one proton and one electron), but rather two electrons and one proton. The extra proton (the H^+) is released into the liquid medium, where it enters the general pool of protons resulting from the dissociation of water. When the second electron finally reacts with oxygen (as it does at the end of the metabolic process to be described here to form water) a proton is withdrawn from the general proton pool (there are actually two electrons per oxygen atom). Thus the reduction of NAD can be represented as:

$2[H] + NAD^+ \rightarrow NADH + \{H^+\} \rightarrow$ (lost in general proton pool)

or

$2[e^-] + NAD^+ + H^+ \rightarrow NADH$

and the oxidation of NAD as:

$\{2H^+\} + 2NADH + O_2 \rightarrow 2NAD^+ + 2H_2O$.
(from general proton pool)

In this textbook, we shall follow the standard notation of symbolizing oxidized NAD as NAD^+ and reduced NAD as NADH.

---

†The acids directly or indirectly involved with such metabolic pathways are often written with an -ate ending. For example, 3-phosphoglyceric acid is written as 3-phosphoglycerate; pyruvic acid as pyruvate; citric acid as citrate; ketoglutaric acid as ketoglutarate; fumaric acid as fumarate; malic acid as malate; oxaloacetic acid as oxaloacetate; lactic acid as lactate; and so on. The usual cell pH is such that the acids are found in the ionized state, signified by the -ate ending.

Nonetheless, a net profit is necessary to make it all worthwhile. In the following series of reactions, the phosphate groups remaining are added to two more ADP molecules to make two ATPs.

Thus the initial investment of two ATPs has yielded four, with a net profit of two. Beyond that, from each glucose molecule, two molecules of pyruvate (pyruvic acid) are formed. While not as rich in potential free energy as glucose, pyruvic acid, as will be seen shortly, still serves as an important source of energy. Finally, two molecules of reduced NAD (NADH) also result from the glycolytic process. This is important because of the vital role of NAD as an electron transport compound in cellular metabolism, sometimes taking on electrons and becoming reduced and, at others, giving up electrons and becoming oxidized.

Note the comparison that can be made here between the role of NADH and ATP:

NADH is an intermediate for coupled oxidation-reduction (redox) reactions, while ATP is an intermediate for coupled energy transfers.

To summarize glycolysis: after the activation phase and on its way to the production of pyruvate (pyruvic acid), the fructose-1,6-diphosphate is first broken down into two three-carbon sugars (glyceraldehyde-3-phosphate or phosphoglyceraldehyde [PGAL] and dihy-

droxyacetone phosphate). These two sugars, in solution, exist in chemical equilibrium with one another. Consequently both will eventually pass through the entire glycolytic pathway leading to the production of pyruvic acid (see Fig. 6.5).

During glycolysis two very important types of chemical events take place, each relating to the extraction of energy from the three-carbon phosphate sugar:

1. Electrons are removed from the substrate by an enzyme in conjunction with the electron acceptor molecule, NAD. This process occurs simultaneously with the addition of another phosphate group to the three-carbon phosphate sugar, creating a three-carbon diphosphate sugar. This removal of electrons is by definition an oxidation process similar to those associated with cellular respiration. A consequence of oxidation is that one of the phosphate groups on the three-carbon sugar is converted from a low-energy to a high-energy phosphate bond. This storage of energy in high-energy phosphate bonds marks the real importance of such substrate-level phosphorylations, which occur in several steps.

2. Catalyzed by specific enzymes, ADP molecules from the medium now interact with the high-energy phosphate sugar produced in the first step to generate ATP, the high-energy phosphate group being transferred from the sugar to the ADP. The process of generating ATP by direct interaction of ADP with a high-energy phosphate sugar is an example of the substrate-level phosphorylation mentioned earlier.

From one starting glucose molecule, the end products of glycolysis are two molecules of pyruvic acid, two molecules of reduced NAD (NADH) and four molecules of ATP. Since a total of two ATPs were used to start the process, the net energy gain for the cell is two ATPs. In other words, the cell has spent two ATPs and gotten back four.

We should emphasize that the net gain of two ATPs does not mean that cells defy the laws of thermody-

Table 6.1
The Consumption and Production of ATP in Glycolysis

Reaction	ATP change per glucose
Glucose $\longrightarrow$ glucose-6-phosphate	−1
Fructose-6-phosphate $\longrightarrow$ fructose-1,6-diphosphate	−1
(2) 1,3-Diphosphoglycerate* $\longrightarrow$ 2,3-phosphoglycerate	+2
(2) Phosphoenolpyruvate* $\longrightarrow$ 2 pyruvate (pyruvic acid)	+2
Net	+2

*The number (2) indicates the fact that two molecules are involved and thus each single conversion generates one ATP.

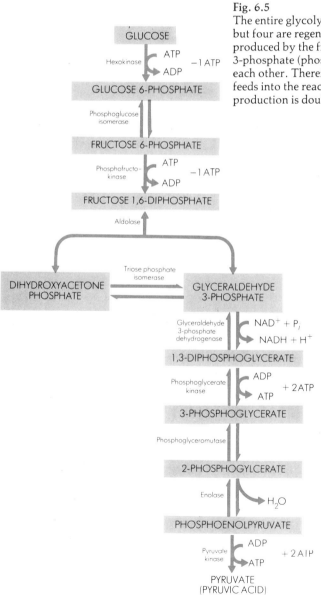

Fig. 6.5
The entire glycolytic pathway. Note that in the first phase two ATPs are used up, but four are regenerated in the second phase, for a profit of two. The two molecules produced by the first phase, dihydroxyacetone phosphate and glyceraldehyde-3-phosphate (phosphoglyceraldehyde or PGAL) exist in dynamic equilibrium with each other. Therefore, as the PGAL is used up, more dihydroxyacetone phosphate feeds into the reaction series via its conversion to more PGAL, and thus the ATP production is doubled (see also Table 6.1).

namics. Cells do not "create" energy from nothing through the process of respiration. The energy represented in the two additional ATPs comes solely from the potential energy of the glucose molecule itself. Through the enzyme-catalyzed, highly specific reactions of glycolysis, the cell is able to rearrange electrons within the glucose molecule so that electrons already at higher potential energy levels in the molecule are lowered. Pyruvic acid, however, the end point of glycolysis, does *not* represent the lowest potential energy stage to which glucose can be brought. Further processing results in the extraction of still more energy for useful work.

6.4
AEROBIC AND ANAEROBIC METABOLISM

In terms of the further processing of the pyruvate (pyruvic acid) referred to in the last section, there are two major alternative pathways that can be taken by the pyruvic acid and NADH produced during glycolysis. One of those pathways' processes involves the presence of oxygen and leads to what is called **aerobic** metabolism; those organisms using this pathway are called **aerobes**. We shall discuss aerobic metabolism shortly. The other pathway, leading to **anaerobic** metabolism, does not require the presence of oxygen; and indeed glycolysis itself is anaerobic, although not due to necessity. Organisms adapted to operate without oxygen are called **anaerobes**. Note that this does *not* mean oxidation doesn't occur in anaerobic metabolism, but only that some substance other than oxygen serves as the oxidizing agent. We shall have more to say later concerning anaerobic metabolism in terms of its efficiency and the organisms with which it is associated.

Several natural environments, such as the intestinal tract of an animal or wet soil, provide an environment in which the amount of oxygen available is very small or nonexistent. Anaerobes flourish in such environments. There are two types of anaerobic cells. *Obligate anaerobes* lack an electron transport system and cannot use molecular oxygen at all. Certain kinds of bacteria are obligate anaerobes; not only does their growth require oxygen, but in some instances they cannot survive if exposed to it. The other kind, *facultative anaerobes*, can exist anaerobically for long periods of time, even indefinitely, if oxygen is not present. Given a sufficient supply of oxygen, however, facultative organisms possess the molecular mechanism to carry out aerobic metabolism. Yeast provides an example of a facultative anaerobe: if oxygen is absent, yeast cells follow one metabolic pathway; if oxygen is present, they follow another. In this section we shall examine in some detail the various metabolic pathways followed by aerobes and anaerobes.

It should be stressed that in the initial stages of glucose metabolism, all organisms, whether aerobes or anaerobes, use glycolysis. Both aerobes and anaerobes share the glycolytic pathway, even to the extent of having most of the same enzymes involved. It is in the fate of the pyruvic acid molecule that anaerobes differ markedly from aerobes. Two of the more common mechanisms of anaerobic metabolism will now be described.

In the absence of oxygen some types of cells using anaerobic metabolism reduce the pyruvic acid (pyruvate) and oxidize reduced NAD to form lactic acid (lactate).

PYRUVATE   +   REDUCED NAD $\longrightarrow$ LACTATE
($C_3H_4O_3$)   (NADH)   ($C_3H_6O_3$)

                                                  + OXIDIZED NAD
                                                     (NAD^+)

Because lactic acid is the end product, this metabolic pathway is often called **lactic acid fermentation.** As will be seen shortly, even muscle cells, though generally aerobic, are forced to use this anaerobic pathway during the temporary absence of oxygen. A more familiar example of lactic acid fermentation, however, is the souring of milk by anaerobic bacteria; the accumulation of lactic acid causes the sour taste.

In other types of cells, notably yeast, a carboxyl group is removed from the pyruvic acid molecule (decarboxylation) to form an intermediate compound known as acetaldehyde. The two-carbon acetaldehyde picks up electrons (and accompanying protons) from

NADH to produce ethyl alcohol, as shown in the following equation:

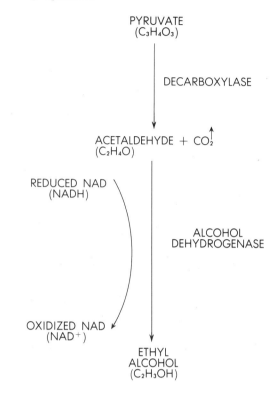

Because ethyl alcohol is the end product of this anaerobic pathway, the process is often referred to as **alcoholic fermentation.** Anyone who has left apple juice in a jug in a warm, dark place is aware of alcoholic fermentation, as the increase in alcohol turns the cider "harder." After a certain point, however, the concentration of alcohol poisons the anaerobic organism (yeast) and the conversion of alcohol to acetic acid (vinegar) by bacteria sours the cider (if, in the case of a sealed container, the gas pressure exerted by the carbon dioxide released in the decarboxylase-catalyzed reaction has not exploded it). Note that yeast is here acting anaerobically; as a facultative anaerobe it is carrying out the fermentation process because oxygen is not present. In the presence of oxygen, it can switch to aerobic metabolism.

Anaerobic organisms get their ATP by substrate-level phosphorylation during the conversion of glucose to pyruvic acid. The reactions from pyruvic acid onward are simply means of disposing of the end products of the earlier reactions (that is, pyruvic acid and NADH). Human beings have taken advantage of the anaerobic process of alcoholic beverages; the wine, beer, and distilling industries all depend on this fermentation process.

It should be stressed that the various types of fermentation are merely different solutions evolved by organisms to the problem posed by the need to reoxidize

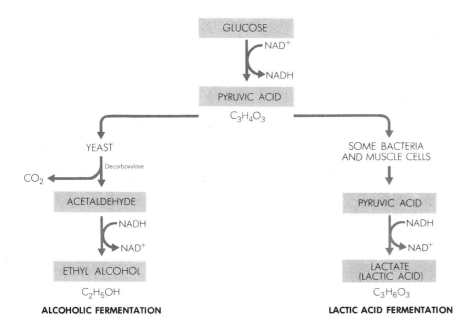

Fig. 6.6
Diagrammatic representation of two possible pathways of anaerobic respiration. The electrons made available through the production of reduced NADH in the conversion of glucose to pyruvate (pyruvic acid) are used to generate either ethyl alcohol (ethanol), as in yeast, or lactate (lactic acid), as in bacteria. Since the electrons from reduced NADH are used, NAD^+ is regenerated for use in the reaction in glycolysis that converts glyceraldehyde-3-phosphate to 1,3-diphosphoglycerate (see Fig. 6.5).

the recycled NADH so that glycolysis can continue. Ethyl alcohol (ethanol), lactic acid, and the other compounds produced by anaerobic bacteria are all the results of bringing about the same process—using pyruvate as the terminal electron acceptor to reoxidize NADH via slightly different means (see Fig. 6.6).

Figure 6.5 on p. 161 shows the steps of glycolysis in detail. The high free-energy content of each molecule of glucose is made even higher by the processes leading to the production of the two molecules of PGAL. After that, the process is strictly downhill, with the free energy released being used immediately both to produce the two NADH molecules and to become stored in the high-energy phosphate bonds of four ATP molecules. The two molecules of pyruvic acid remain for further processing. It is this "further processing" to which we will now turn our attention.

6.5
RESPIRATION

While the term "respiration" is often used synonymously with "breathing," the terms actually may refer to different processes. Breathing is the muscular activity of an organism that draws oxygen-containing air in and out of the body; respiration is the portion of cellular catabolism and ATP production that is dependent on the presence of that oxygen as a final electron acceptor. As indicated at the end of the preceding section, respiration in aerobic organisms begins where glycolysis left off; with the production of pyruvate (pyruvic acid).

You might wish to take a break from your reading at this point and try a simple observation; open and

close your hand over and over about 100 times, as fast as you can. After a while you should experience a feeling of fatigue in your lower arm.

Why does this occur? The muscles that pull the tendons that cause the fingers to move are located in the lower arm. As an aerobic organism, your blood supply to these muscles usually brings enough oxygen to act as a final electron acceptor and allow respiration to occur. However, with intense activity, the supply is not sufficient. In response to the "emergency" demand, NADH substitutes for oxygen, the muscle cells switch temporarily to the form of anaerobic metabolism described earlier, and the oxidation of NAD reduces the pyruvate (pyruvic acid) to lactate (lactic acid). The accumulation of lactic acid in the muscles causes the fatigue you experienced. When the intense muscular activity ceases, the reaction proceeds in the reverse direction; the lactic acid is reconverted into pyruvic acid, the NAD is reduced, and, with a sufficient supply of oxygen again being supplied by the blood, the pyruvic acid may now be further oxidized by cellular respiration.

6.6
THE CITRIC ACID CYCLE

The *complete* oxidation of pyruvate (pyruvic acid), with the consequent extraction of considerably more energy, occurs during the final stage of respiration, the **citric acid cycle.** This series of reactions is also known as the **Krebs cycle** (for the English biochemist Hans Krebs, who first worked out the chemical details of the pathway) or the **tricarboxylic acid cycle** (since a number of small molecules bearing three carboxyl groups are in-

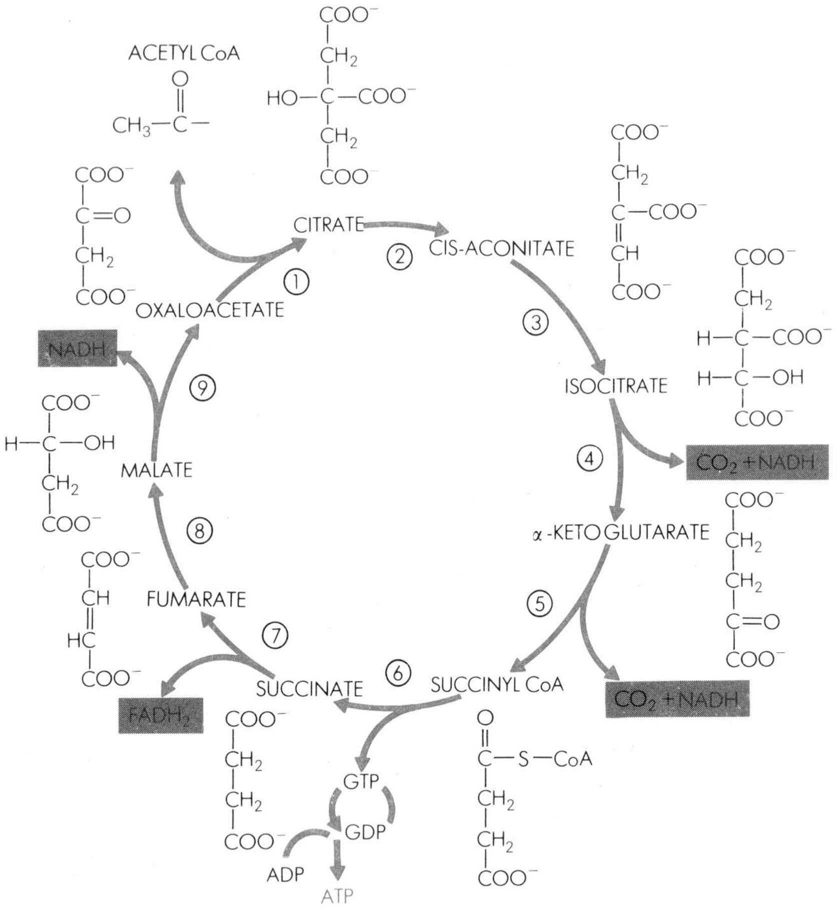

Fig. 6.7
The Krebs citric acid cycle. As the simplified insert at lower left shows, the cycle begins with the combination of two-carbon molecules (acetyl CoA molecules resulting from the conversion of the pyruvate produced by glycolysis) with four-carbon (oxaloacetate, or oxaloacetic acid molecules) to form six-carbon molecules (citrate or citric acid). During the cycle, the number of carbon atoms is reduced in each compound involved in the metabolic pathway from six, to five, to four, with the last step regenerating the four-carbon compound (oxaloacetate) to begin the cycle again. The two carbon atoms lost are given off in the oxidation of two molecules of carbon to produce carbon dioxide. In the diagram of the complete cycle, note that several steps release a pair of hydrogen atoms. The protons of these hydrogen atoms enter the liquid medium and become lost in the general pool of protons resulting from the breakdown of water into hydrogen ions (protons, H^+) and hydroxyl ions (OH^-). The electrons of these hydrogen atoms are used to reduce three molecules of NAD^+ to three molecules of NADH, and one molecule of FAD to $FADH_2$.

Note that in the entire cycle only one molecule of ATP results from substrate-level phosphorylation (via the production of another energy-rich compound, guanosine triphosphate, GTP, from guanosine diphosphate, GDP). This, however, represents the output from only one turn of the cycle. Since the complete oxidation of each glucose molecule involves *two* turns of the cycle, this actually means two molecules of ATP are produced in the citric acid cycle from each glucose molecule oxidized. However, it is in the production of the reduced NADH and $FADH_2$ that the biggest energy gain is realized. Since each turn of the cycle yields three molecules of NADH and one molecule of $FADH_2$, the two turns that result from the complete oxidation of each glucose molecule produce six NADH and two $FADH_2$ molecules. Since two more NADH molecules result from glycolysis and two more from the conversion of pyruvate (pyruvic acid) to acetyl CoA, the total number of reduced molecules is twelve: ten molecules of NADH and two of $FADH_2$. The hydrogen electrons these reduced compounds now possess represent the richest source of energy for ATP production.

volved). These reactions occur within the mitochondria into which the pyruvic acid produced by glycolysis enters.

Some of the major stages in the citric acid cycle are shown in Fig. 6.7. Specific enzymes are involved in the conversion of the three-carbon pyruvic acid into a two-carbon acetic acid derivative. During this conversion, one carbon dioxide molecule is removed and two electrons are picked up by NAD, forming a complex that combines immediately with coenzyme A, which is always present in the cell. This reaction forms acetyl coenzyme A (called acetyl CoA), which is able to enter directly the pathways involved in the citric acid cycle. The two-carbon acetyl fragment of acetyl CoA combines with the four-carbon oxaloacetic acid, also abundant to the cells, to yield the six-carbon citric acid.

Note that in this reaction the two-carbon acetate molecule is separated from coenzyme A, which is then returned to the medium and can be used for combination with another acetate. As a result of this process, two NADH molecules are generated, one from each pyruvic acid molecule entering the cycle.

The six-carbon citric acid molecule undergoes a series of oxidations and decarboxylations (removal of

SUPPLEMENT 6.1
HOW THE STEPS IN COMPLEX BIOCHEMICAL PATHWAYS ARE INVESTIGATED

The series of reactions involved in glycolysis and the citric acid cycle are numerous and intricate. How do we know the intermediate steps involved and the sequence in which they occur? Several examples will illustrate the way in which biochemists have learned about the respiratory reactions.

In 1935 the Hungarian biologist Albert Szent-Gyorgyi studied respiration in isolated pigeon breast muscle. He found that the rate of oxygen uptake, quite rapid at first, fell off slowly with time. Pigeon breast muscle respires very rapidly, producing little lactic acid. Its byproducts are CO_2 and H_2O. Szent-Gyorgyi noted something else of importance: the fall in rate of respiration paralleled the rate at which succinic acid disappeared from the muscle. He found, however, that the respiratory rate could be restored to its normal level by the addition of small amounts of succinic or fumaric acid. Szent-Gyorgyi concluded that these substances must somehow be involved in the oxidation of carbohydrate. Later studies showed that if succinic, fumaric, or oxaloacetic acid were added to muscle preparation, a large amount of citric acid would be produced. This gave such workers as Hans Krebs the idea that something like the following must be taking place:

SUCCINIC ACID
FUMARIC ACID   } → CITRIC ACID
OXALOACETIC ACID

However, the exact sequence of reactions was not at all clear.

The task of determining the exact reaction series was complicated by the fact that several of these substances are chemically interconvertible. For example, succinic acid can be converted into fumaric acid, and vice versa. This reversible reaction is catalyzed by the enzyme succinic dehydrogenase. When succinic acid is added to isolated muscle, fumaric acid and citric acid accumulate. When fumaric acid is added, succinic acid and citric acid accumulate. There is, therefore, no conclusive way to distinguish between the two alternative hypotheses for the sequence of reactions:

Hypothesis I: Succinic acid → fumaric acid → citric acid.

Hypothesis II: Fumaric acid → succinic acid → citric acid.

One experimental technique does make it possible to decide between these alternatives, however. The enzyme succinic dehydrogenase acts only in the chemical reactions by which succinic acid is converted to fumaric acid, and vice versa.

We know from an earlier chapter that a substance called malonic acid selectively inhibits succinic dehydrogenase. This means that when malonic acid is present in the system, only molecules of succinic dehydrogenase will cease to function. On the basis of the two hypotheses given above, it is possible to make two predictions.

Hypothesis I: If the correct sequence of reactions is succinic acid → fumaric acid → citric acid, then addition of malonic acid along with succinic acid should inhibit the production of citric acid.

Hypothesis II: If the correct sequence of reactions is fumaric acid → succinic acid → citric acid, then addition of malonic acid along with succinic acid should not inhibit the production of citric acid.

Performing the experiment shows that citric acid does not accumulate in the muscle tissue. These results thus support the predictions of hypothesis I and contradict those of hypothesis II.

The foregoing examples illustrate two major types of techniques involved in working out biochemical reaction series. The first technique involves the addition of large amounts of a suspected intermediate to a biochemical system, measurement of its disappearance rate, and identification of the new substance that will then begin to build up. This was the technique applied by Szent-Gyorgyi. The second technique involves the addition of some substance known to inhibit a certain enzyme. The alteration of the rate at which certain products accumulate will indicate at which point in the reaction series the particular enzyme acts. For example, consider this series:

                                          BLOCKED WITH
                                          METABOLIC POISON

A $\xrightarrow{\text{enzyme 1}}$ B $\xrightarrow{\text{enzyme 2}}$ C $\xrightarrow{\text{enzyme 3}}$ D $\xrightarrow{\text{enzyme 4}}$ E (END PRODUCT).

If enzyme 3 is put out of commission by a known inhibitor, the intermediate products B and C will accumulate while D disappears (it is converted to E). This indicates that C precedes D in the reaction series. It also indicates that enzyme 3 catalyzes the conversion of C to D. Several substances, such as cyanide and carbon monoxide, are known to block the action of certain cytochromes. By employing these poisons, it has been possible to work out the sequence of reactions in electron transport.

There are, of course, many other techniques frequently used in biochemical analysis. One such technique involves radioactively labeled molecules. These are fed to a biochemi-

cal system and the reaction series is stopped at various places. The distribution of radioactive atoms in various intermediate products gives an idea of the pathway taken by the series of reactions. An example of a labeling experiment will be discussed later with regard to the pathways of photosynthesis.

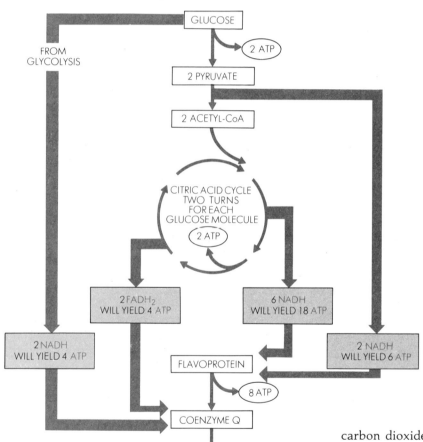

Fig. 6.8
The relationship among glycolysis, the formation of acetyl CoA, the citric acid cycle, the electron transport or carrier assembly, and ATP production. Note that eight of the ten reduced NAD (NADH) molecules (two from the conversion of pyruvic acid to acetyl CoA and six from the citric acid cycle) enter the electron carrier chain at flavoprotein, and thus benefit from the production of ATP resulting from the conversion of flavoprotein to coenzyme Q. The two NADH produced in glycolysis, however, as well as the two reduced FAD ($FADH_2$) molecules produced in the citric acid cycle enter the cycle at coenzyme Q, and thus produce only the ATP that can be generated from this point on. In summary, each NADH molecule entering the carrier chain at flavoprotein will yield three ATP molecules, while those entering at coenzyme Q, as well as each molecule of $FADH_2$, yield only two molecules of ATP. The final electron acceptor is oxygen.

carbon dioxide molecules) as shown in Fig. 6.7. After the removal of each carbon dioxide or each pair of electrons (equivalent to hydrogen atoms), internal rearrangements of the molecule are necessary to preserve its stability. Note that during the progress of the citric acid cycle, the six-carbon molecule is converted successively into a five-carbon molecule (α-ketoglutarate) and a series of four-carbon molecules (succinate, fumarate, malate, and oxaloacetate; see footnote, p. 159). Thus each turn of the cycle begins with a four-carbon compound combining with a two-carbon compound to yield a six-carbon intermediate, which is immediately broken back down to the original four-carbon compound by a series of reaction steps. Because one molecule of oxaloacetate (oxaloacetic acid) is regenerated from each molecule of oxaloacetate that origi-

Table 6.2
ATP Yield from the Complete Oxidation of Glucose*

Reaction sequence	
Glycolysis: glucose to pyruvate (in the cytosol)	
Phosphorylation of glucose	−1
Phosphorylation of fructose-6-phosphate	−1
Dephosphorylation of two molecules of 1,3-DPG	+2
Dephosphorylation of two molecules of phosphoenolpyruvate	+2
Two NADH are formed in the oxidation of two molecules of glyceraldehyde-3-phosphate	
Conversion of pyruvate to acetyl CoA (inside mitochondria)	
Two NADH are formed	
Citric acid cycle (inside mitochondria)	
Formation of two molecules of guanosine triphosphate from two molecules of succinyl CoA	+2
Six NADH are formed in the oxidation of two molecules of isocitrate α-ketoglutarate, and malate	
Two FADH$_2$ are formed in the oxidation of two molecules of succinate	
Oxidative phosphorylation (inside mitochondria)	
Two NADH formed in glycolysis; each yields two ATP (not three ATP each, because of the cost of the shuttle)	+4
Two NADH formed in the oxidative decarboxylation of pyruvate; each yields three ATP	+6
Two FADH$_2$ formed in the citric acid cycle; each yields two ATP	+4
Six NADH formed in the citric acid cycle; each yields three ATP	+18
Net yield per glucose	+36

*The oxidation of glucose yields 686 kcal, while the free energy stored in thirty-six ATP is 263 kcal. Thus the thermodynamic efficiency of ATP formation from glucose is 263/686, or 38 percent, under standard conditions. The actual degree of efficiency varies widely, of course, according to physiological conditions. Note that some of the processes occur outside the mitochondria and some within.

nally combined with acetyl CoA, this series of reactions is truly a cycle.

Two sources of potential chemical energy are produced during the citric acid cycle. A minor source is the high-energy phosphate produced directly by substrate-level phosphorylation, such as that found in glycolysis. The major energy component is in the production of reduced coenzymes; one of these coenzymes, NAD, has already been mentioned. The other is *flavin adenine dinucleotide*, or FAD. Through the process of oxidative phosphorylation, both NAD and FAD can generate a considerable amount of ATP. Each reduced NAD molecule (NADH) yields three ATPs by passing its two electrons along the complete electron transport chain.* In the one case where electrons are picked up from succinate by FAD, the step from NAD to flavoprotein (FP) is bypassed. Therefore each pair of electrons passing from reduced FAD (FADH$_2$) down the chain generates only two ATPs instead of three.

---

*An exception is the case of the two molecules of NADH produced from each molecule of glucose during glycolysis, which occurs outside of the mitochondria in the cytoplasm. This NADH cannot pass across the mitochondrial membrane (as can pyruvic acid), and hence does not enter the electron transport system directly. Such NADH converts the three-carbon sugar dihydroxyacetone phosphate (which, you recall, was produced during glycolysis with PGAL) into another compound, glycerol phosphate, which *can* pass into the mitochondria. Glycerol phosphate can be oxidized by one of the electron acceptor molecules (usually cytochrome *c*), but at a place that comes *after* the first ATP-generating site. Thus for each molecule of NADH generated in the cytoplasm during glycolysis, only two ATPs are produced.

Many different substances are involved in the citric acid cycle, and it is necessary in accounting for each of them to remember that each glucose molecule that enters the breakdown process generates two pyruvic acid molecules. Thus the complete oxidation of a single glucose results in two turns of the citric acid cycle. To obtain an accurate tally of the input and output of the many substances involved in glucose oxidation, it is necessary to remember that every reaction after the breakdown of fructose-1,6-diphosphate occurs twice for each original starting glucose.

Taking this into consideration, such tallies can account for most of the electrons, protons, atoms, and molecules involved, balancing input with output. An exception is water. Since for every mole of glucose respired, twelve pairs of electrons (plus twelve pairs of protons, H$^+$) reach molecular oxygen, yielding twelve molecules of water (H$_2$O), where do the extra six pairs come from? In this case, the citric acid cycle results in a deficit of six water molecules, a deficit that can be made up from the liquid medium surrounding the reaction sites within the cell.

The relationship between glycolysis, the citric acid cycle, and the generation of ATP is shown diagrammatically in Fig. 6.8 on p. 166. ATP is thus generated in the breakdown of glucose both by substrate-level phosphorylation and oxidative phosphorylation. Through both of these processes, but particularly through oxidative phosphorylation, an enormous amount of potential energy of glucose is harnessed as potential energy in high-energy phosphate bonds. Table 6.2 shows the

ATP is generated during glycolysis and the citric acid cycle by two means: substrate-level phosphorylation and oxidative phosphorylation during electron transport.

precise quantity of ATP generated by each process, and the particular steps where energy is harnessed. As the data show, for each starting glucose molecule, six ATPs are generated by substrate phosphorylation and thirty-two ATPs are generated by electron transport. Two ATPs were required to get the process going, so that from the complete oxidation of a single glucose molecule, a net total of thirty-six ATPs is produced. Or, put another way, from a mole of glucose thirty-six moles of ATP become available to the cell for work. Considering that only two ATPs were spent to get the process started, this represents a substantial return in potential energy.*

It should be mentioned that, besides the aforementioned case of water, various intermediate compounds and other molecules are constantly being exchanged between the pathway and the surrounding medium. It is very important to realize that cells contain pools of molecules of various sorts that are constantly available for interactions with one or another intermediate during a particular metabolic reaction. For example, coenzyme A is plentiful in the cell and is available to react with acetic acid to produce acetyl CoA. Coenzyme A is also used in other metabolic reactions, but it is always returned to the medium.

Like whole organisms, cells are "open systems." This means that they must continually exchange materials with the external environment. Cells cannot produce ATP without oxidizing some fuel molecule, and that fuel must come from somewhere. For animals, it

comes from the food they eat: plants and other animals. Like animals, plants also oxidize carbohydrates via glycolysis and the citric acid cycle. Unlike animals, however, plants can produce their own carbohydrates internally by photosynthesis. The energy to drive that uphill reaction is obtained externally from sunlight. Though more self-sufficient than animals, plants are still open systems in the sense that they require an external energy source. The ultimate source of all energy in living systems is external, whether in the form of specific fuel molecules that can be oxidized to regenerate ATP, or in the form of sunlight for the synthesis of carbohydrates. Without this continual input of energy from the environment, life would come quickly to a halt.

6.7
THE ROLE OF THE CYTOCHROMES

What molecules act as acceptors in the oxidative phosphorylation system, and what properties allow them to serve this special purpose? Of particular importance in the electron transport system are the **cytochromes**. In Fig. 6.2 the cytochromes are represented by the last four molecules in the electron transport system (numbers 3, 4, 5, and 6), of which the first two acceptors in the transport system are the coenzymes (NAD) and flavoprotein (FP). The cytochromes, sometimes referred to as "respiratory enzymes," are iron-containing molecules with certain molecular features in common. Each is composed of a basic ring structure (called a porphyrin ring) in the center of which is an iron atom bonded to four nitrogens. The porphyrin ring structure is also found in the hemoglobin and myoglobin molecules. In cytochromes, the porphyrin ring is bonded at several points to a polypeptide chain. The protein segment of the molecule stabilizes the active site in the region of the iron atom.

Reversible changes involved in cytochrome activity result from successive oxidations and reductions. When any tissue is active and using the energy from the breakdown of sugar to produce ATP, electrons are being transported at a rapid rate along the electron assembly. Each transport molecule (for example, one of the cytochromes) picks up an electron and thus becomes reduced. It passes the electron on to the next cytochrome molecule in the transport system and thus becomes oxidized. The ability of cytochromes successively to undergo oxidation and reduction in a reversible manner is a function of the electron configuration about the central iron atom (see Fig. 6.9). Iron is one of several types of atoms that can exist in one of several valence states: it

---

*The number of ATPs generated by complete oxidation of a single glucose molecule has been a matter of dispute for some years in textbooks and among biochemists. The total number produced has been claimed by some to be thirty-eight and by others to be forty. The reason for such discrepancy lies in the existence of a so-called "shuttle" system by which electrons removed from glucose in the cytoplasm during glycolysis are brought into the mitochondria and thus through the electron transport system. As mentioned in the footnote on p. 167, the mitochondrial membrane is impermeable to NADH produced in the cytoplasm during glycolysis. Electrons from cytoplasmically generated NADH are thus "shuttled" across the membrane by passing to a three-carbon sugar, which in its reduced form can pass into the mitochondrion. The reduced three-carbon sugar (known as glycerol phosphate) passes electrons to the electron transport system. Electrons shuttled into the electron transport system via glycerol phosphate enter at coenzyme Q; thus each pair of electrons shuttled in generates only two, rather than three, ATPs. Prior to knowledge of this shuttle system, it was assumed that cytoplasmically generated NADH produced three ATPs per molecule; hence the disparity in ATP count. The situation is further complicated by the discovery that malate may also act as a shuttle molecule and enter the electron transport chain before the site where ATP synthesis first occurs, thus allowing for the production of three rather than two ATP molecules. Most texts to date, however, assume that the glycerol phosphate is the one that is operative, and thus the lower ATP count is generally given.

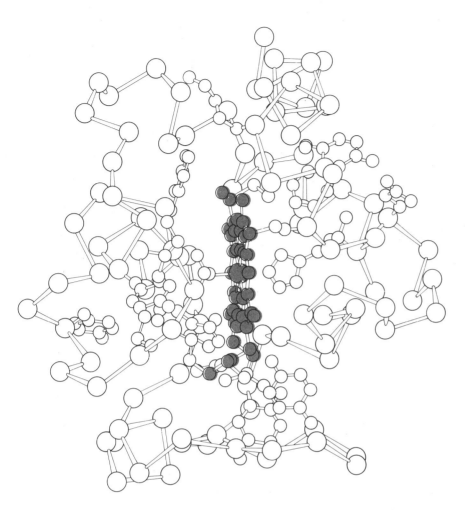

Fig. 6.9
A molecular model of reduced cyto-chrome *c* from tuna. The presence of the pigmented iron-containing heme group gives the cytochromes their name (*cyto,* cell; *chrom,* color). (Based on T. Ta-kano, O. B. Kallai, R. Swanson, and R. E. Dickerson. *J. Biol. Chem.* 248 [1973]: 5244)

SUPPLEMENT 6.2
CYTOCHROMES: HOW DO WE KNOW ABOUT THEM?

The existence of the cytochromes and the nature of their cel-lular function were first noted in 1925 by the British bio-chemist David Keilin. Keilin started out to study the chemi-cal composition of different kinds of living tissues by a pro-cess known as microspectroscopy. In this process, a piece of living tissue is placed on a microscope slide and light is passed through it. The ocular lens of the microscope is re-placed by a prism, however, so that light emerging from the tube is diffracted and a spectrum produced. Such spectra show a broad, continuous background ranging in color from red on one end to violet on the other. This represents the spectrum of light emitted by the light source. Onto this con-tinuous background are superimposed a number of dark ab-sorption lines, representing various types of molecules in the tissue being observed. Each type of molecule absorbs light at different specific wavelengths. Thus by examining the ab-sorption spectrum it is possible to identify some of the major components of living tissue.

Keilin was interested in studying a particular group of molecules (known at the time as "myohaematin") found in all types of living cells. In particular, Keilin was studying the characteristics of these molecules in bee thoracic muscles (which control wing movements). He placed the entire or-ganism on a microscope slide and allowed light to pass

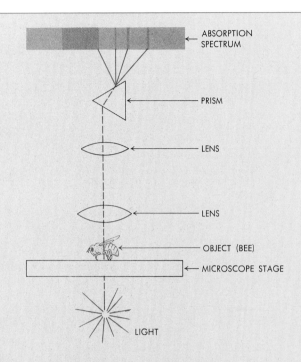

ABSORPTION SPECTRUM

PRISM

LENS

LENS

OBJECT (BEE)

MICROSCOPE STAGE

LIGHT

Diagram A

through the wing muscle at the joint between the wing and thorax (see Diagram A).

He noted that when the insect was quiet no particular absorption bands could be observed. When the insect moved its wings, however, several distinct absorption bands appeared on the spectrum, as shown in Diagram B. The continuous spectrum forming the background is produced by the light source and ranges from red on the left to violet on the right. The dark lines represent specific wavelengths of light that molecules in the tissue have absorbed. Each type of molecule absorbs at specific wavelengths. If the molecule is absent from the tissue, its corresponding lines on the spectrum are also absent.

Keilin's studies on myohaematin in yeast cells showed yet another interesting result. When examined with the microspectrograph, yeast growing normally in a suspension showed the major absorption bands. When oxygen was bubbled through the yeast culture, the absorption bands disappeared. When nitrogen was bubbled through the culture,

the absorption bands became intense. These observations suggested that myohaematin underwent reversible oxidation and reduction and that the absorption bands were characteristic of the reduced state and their absence characteristic of the oxidized state.

One more interesting piece of information was added to Keilin's observations. He found that the appearance and disappearance of absorption bands was markedly affected by the presence of such agents as carbon monoxide and cyanide. As was pointed out in Chapter 4, these same agents strongly affect the respiratory pigment hemoglobin. Keilin knew that hemoglobin was an iron-containing compound that could both pick up and discharge oxygen molecules, thereby serving as the oxygen-transporting molecule of the blood. When carbon monoxide or cyanide is added to a hemoglobin solution, however, the hemoglobin molecules are permanently reduced. While the mechanism of this hemoglobin-oxygen interaction is not a redox reaction, the effect of the pigment reduction of hemoglobin is to render it incapable of transporting the oxygen. Similarly, when carbon monoxide or cyanide is added to a solution of myohaematin, the dark absorption bands occur and persist, even if oxygen is added later. Oxygen is unable to reverse the effects of cyanide or carbon monoxide on myohaematin, just as it fails to reverse the effect of these molecules on hemoglobin.

Keilin drew the following specific conclusions from his observations:

1. The oxidation-reduction reactions of myohaematin are normally reversible: reduced myohaematin appears when oxygen is absent; it disappears when oxygen is present.

2. With specific poisons such as cyanide or carbon monoxide, molecules of myohaematin behave in some ways quite similar to those of hemoglobin. Thus myohaematin could be thought of as an iron-containing molecule with many of the structural properties of hemoglobin.

3. The existence of a number of different dark bands on the absorption spectrum of reduced myohaematin suggested that several different kinds of this molecule were present in the cell.

Fifteen years later, Keilin and one of his co-workers, Edward F. Hartree, showed that the dark lines of the absorption spectrum were produced by several different molecules

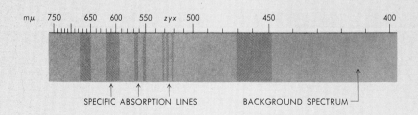

mμ   750   650   600   550   zyx   500   450   400

Diagram B

SPECIFIC ABSORPTION LINES   BACKGROUND SPECTRUM

rather than the single substance myohaematin. They renamed those molecules cytochromes. Each set of bands turned out to represent one of several types of cytochromes, called, in order of their sequence in the electron transport process, cytochrome b, cytochrome c_1, cytochrome c, cytochrome a, and cytochrome a_3 (or cytochrome oxidase).

Note the kind of reasoning involved in Keilin's and Hartree's work. Since it was difficult to isolate pure cytochrome, and since no techniques were then available for determining its exact molecular structure anyway, they were forced to reason deductively by analogy. *If* hemoglobin shows an absorption spectrum characteristic for reversible oxidation and reduction of iron, and *if* myohaematin shows similar absorption bands, *then* it is possible that myohaematin is an iron-containing compound that undergoes reversible oxidation and reduction. Keilin's initial studies did not provide direct evidence for this chemical property of myohaematin. Only more detailed studies on purified samples of myohaematin (cytochrome) itself would provide *direct* evidence. Lacking that, reasoning by analogy provided a start.

can give up either two or three electrons and exist as either the ferrous ion (Fe^{++}) or the ferric ion (Fe^{+++}). The ferric ion can pick up an electron and thus become reduced to the ferrous ion. The ferrous ion, in turn, can give up the electron and thus be oxidized back to the ferric ion. The ability of cytochrome molecules to pick up and release electrons easily allows them to function efficiently in the electron transport process.

The complete electron transport assembly is shown diagrammatically in Fig. 6.10. Electrons are removed

Fig. 6.10
The hydrogen electrons used to reduce NAD (and FAD) in glycolysis and the citric acid cycle are not passed directly to the final electron acceptor, oxygen, but instead from one electron carrier molecule to another, so that the energy is released gradually and packaged into the high-energy bonds of ATP by oxidative phosphorylation. Because some of the electron carrier compounds incorporate iron-containing pigments into their molecular structure, they are called cytochromes. The union of cytochromes b, c_1 and cytochromes a, a_3 indicates that it has not been possible to separate them in the laboratory.

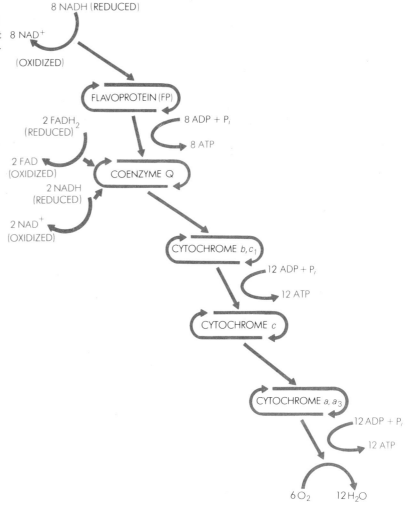

The electrons that pass through the electron transport system come from removal of electrons from glucose (in other words, oxidation of glucose). During their subsequent transport through the cytochrome system, electrons pass successively from higher to lower energy levels.

two at a time from glucose and picked up first by oxidized NAD or NAD^+, which thereby becomes reduced NAD (NADH). NADH passes its electrons to oxidized FP, which in turn becomes reduced FP (FPH_2). FPH_2 then passes electrons on to the cytochromes, first to b, c_1, which passes them on to c, a, and a_3.

Note that ATP is generated at only three points during electron transport. Why? As we have seen, energy is released as electrons pass from one carrier to the next as a result of the steps in the downward transition of electrons from higher to lower energy levels. To drive the uphill reaction involved in ATP formation, 7.3 kcal per mole are required. Only three specific points in the electron transport process generate more than 7.3 kcal per mole of electrons transported. These points occur in the transfer from NAD to a flavin containing protein, from cytochrome b, c_1 to cytochrome c, and from cytochrome a to a_3 to molecular oxygen. Only at these points, then, is there sufficient energy to generate ATP.

What is the ultimate fate of the electrons on cytochrome a_3? Obviously, if the electrons cannot be passed somewhere, the electron transport system will soon grind to a halt, since all the transport molecules will be in the reduced form, or "filled up." There must, therefore, be a final acceptor. *This is the function of oxygen in cellular respiration.* As shown in Fig. 6.10, oxygen picks up electrons from reduced cytochrome a_3 and thereby becomes negatively charged (O^{-2}). This negatively charged oxygen ion then picks up protons from the medium to produce water. Here is one of the chief differences between the use of gasoline as a fuel in an engine and the use of sugars as a fuel in a living organism. Both processes generally consume oxygen and release carbon dioxide. When gasoline burns, however, oxygen reacts directly with the molecules of the fuel. In cellular oxidation, oxygen does *not* unite directly with molecules of sugar. Rather, electrons are removed in successive stages from the fuel molecules and ultimately transferred to oxygen.

Although both cellular oxidation and the oxidation of fuel in an engine may produce the same end products (CO_2 and H_2O), the chemical mechanisms by which these processes occur are very different. This difference is vitally important in understanding the energy-har-

nessing process involved in the rebuilding of ATP. Release of energy in small steps, such as occurs in the cell, allows more energy to be captured in a usable form. The one-step oxidation of fuel through burning produces a great deal of heat. The many-step oxidation of sugars in the cell produces a minimum of heat. For organisms, as for the machine, heat represents wasted energy. Thus the steps in the energy-releasing and storing processes that occur in cells, although more intricate, are ultimately more efficient.

The electron transport system has a structural as well as a functional basis in cells. Found in virtually all types of cells, it is usually associated with membranes—in bacteria as parts of the cell membrane and in cells of higher organisms as part of the mitochondrial and chloroplast membranes as well. As was pointed out in Section 5.8, recent evidence has shown that the inside walls or cristae of mitochondria contain many small globular units believed to represent groups of respiratory enzymes called respiratory assemblies. Each respiratory assembly is thought to contain one set of cytochrome molecules, arranged in a specific order on a membrane. It is also probable (though not definitely established) that at least FP is also bound to the membrane with the cytochromes. The specific structural arrangement of the cytochrome molecules makes possible the effective passing of electrons from one acceptor to the other. Like workers on an assembly line, the electron transport molecules are arranged in a physical order corresponding to their function in transporting electrons. The generation of ATP by electron transport, then, occurs on the inner membranous walls of the mitochondria.

6.8
THE CHEMIOSMOTIC THEORY

We have seen that ATP synthesis involves either substrate or oxidative phosphorylation (see Section 6.2). In substrate phosphorylation, the energy for the addition of the phosphate group to ADP to produce ATP comes from the substrate molecule itself. In oxidative phosphorylation, however, no such substrate molecules are involved; free phosphate groups are added directly to

SUPPLEMENT 6.3
ELECTRON FLOW: A ONE-WAY STREET

How are electrons passed in a specific order along the electron transport system? Why, for example, does cytochrome c pass electrons only to cytochrome a, and not back to cytochrome b,c_1? Furthermore, why does cytochrome a_3 pass electrons to oxygen rather than back to cytochrome a? The answer to these question is obtained by measuring the so-called oxidation-reduction potentials in each molecule of the respiratory assembly. Essentially, oxidation-reduction potentials measure the affinity of a particular atom or molecule for electrons. The more positive the attraction for electrons in a given atom or molecule, the greater potential that

atom or molecule possesses. When oxidation-reduction potentials for the respiratory assembly molecules are measured, the data indicate that each successive molecule in the electron transport system has a greater affinity for electrons than the molecules preceding it (that is, has a stronger oxidation-reduction potential). Thus, as soon as any one electron transport molecule picks up electrons, they will be pulled away from it by the adjacent molecule, which possesses a more powerful oxidation-reduction potential. In this way the flow of electrons through the respiratory assembly is always maintained in one direction.

ADP. Such a reaction requires a large input of energy; that is, it is highly endergonic.

Where does this energy come from? It has for years been assumed that the oxidative phosphorylation of ADP was similar to the substrate phosphorylation of glycolysis, that it is an enzyme-catalyzed reaction directly coupled to the oxidation of electron carriers at appropriate points in the electron transport chain. The main problem with this hypothesis, however, is that it predicts the existence of some coupling factor or agent (to be specific, some high energy compound containing phosphate and capable of transfering this phosphate to ADP). Such factors were readily identified for the substrate phosphorylation reactions of glycolysis, but despite more than three decades of intensive research, all attempts to date to do the same for oxidative phosphorylation have been unsuccessful.

In the 1960s British biochemist Peter Mitchell proposed a new "chemiosmotic" hypothesis, for which he was eventually awarded a Nobel prize. In order to understand the premises on which Mitchell's hypothesis is based, a brief review of some principles may be helpful.

First recall from our discussion dealing with the movement of substances across the cell membrane (passive transport, see p. 104) that movement will occur from regions of high to low free energy. All other factors being equal, the side of the membrane with the highest concentration of particles (molecules, atoms, ions, and the like) will possess the highest free energy.

Suppose now that to the free-energy gradient across this membrane is added the force of electrical charge—that is, that the particles are negatively or posi-

tively charged. Since there are more particles on one side of the membrane than the other, an imbalance of electric charge also exists and this, too, will contribute to the tendency of the particles to move across the membrane. Since both electrical and chemical forces are involved here, such a phenomenon is referred to as an *electrochemical gradient*. The Mitchell hypothesis proposes that just such a gradient occurs across the mitochondrial membranes during respiration. The charged particles involved are hydrogen ions or protons (symbolized H^+). Since both electrical and chemical processes are involved in the Mitchell hypothesis, it is sometimes referred to as the "battery" theory.

Of course, for an electrochemical gradient of protons to exist, it must first be established. Just as it takes a source of energy (the sun) to put water behind a dam so that the potential energy that results can be used to generate electricity so, too, must there be a source of energy to move protons so there are more on one side of the mitochondrial membrane than on the other. This is the role Mitchell's hypothesis proposes that the electron transport chain performs; it moves the protons through the inner mitochondrial membrane from the inner, or M, compartment to the outer, or O, compartment (see Fig. 6.11). At special sites on the membrane, ATP synthesis–catalyzing molecules (ATPase) are found (see Fig. 6.12). Here, the protons pass back down the electrochemical gradient from the O to the M compartment. The energy released as they do so is coupled to the oxidative phosphorylation of ATP.

There are several reasons for considering the Mitchell hypothesis an attractive one. Foremost, perhaps, is the lack of success in finding the intermediate

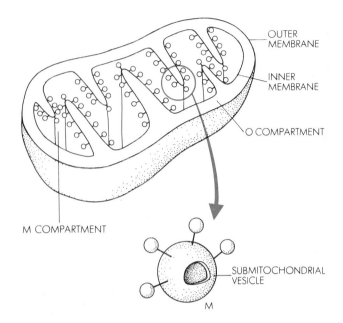

Fig. 6.11
When treated with detergents or exposed to high frequency sound waves, mitochondria can be broken into fragments for study of their individual parts. Among the fragments formed are spherical submitochondrial vesicles. Note that by virtue of the region of the mitochondrion from which they come, such vesicles have the inner, or M, compartment bearing the spherical particles on the outside and the outer, or O, compartment on the inside. This means that the innermost regions of the mitochondria are now in an exposed position where they can be studied more easily. Work with such vesicles has shown that one prediction of Mitchell's chemiosmotic theory—that the mitochondrial membrane can establish a proton gradient by actively transporting protons (H^+) from the M to the O side against a concentration gradient—is verified; under the appropriate conditions, such mitochondrial vesicles accumulate protons in their interiors.

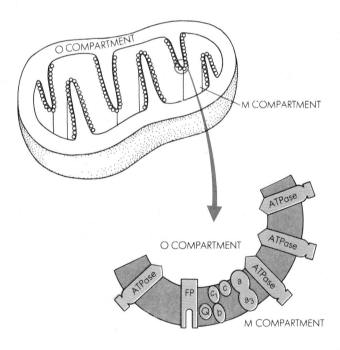

Fig. 6.12
A model to account for the establishment of the proton gradient proposed by the chemiosmotic theory and, when the potential energy of the gradient becomes kinetic, the locking of this energy into the high-energy bonds of ATP. When reduced NAD (NADH) transfers two electrons and two protons to the electron carrier flavoprotein (FP) on the M side of the mitochondrial membrane, FPH_2 is formed. The FPH_2 transfers the two electrons to coenzyme Q, but releases the protons (H^+) into the O compartment side of the membrane. FPH_2 is able to carry out this proton transfer directly because it penetrates the membrane surface of both sides. While coenzyme Q does not do so, the gaining of the two electrons from FP_2 leads it to attract two more protons from the M compartment to which it is exposed. These protons also make their way across the membrane and are released at a different site. Because of the higher concentration of protons in the O compartment than the M compartment that results, a free-energy gradient is established, caused by the expenditure of energy during electron transport from NADH to FP to Q. At specific points on the membrane the protons flow back to the M compartment, and their energy is captured by an enzyme (ATPase) that catalyzes the reaction converting ADP to ATP. The situation is analogous to pumping water behind a dam and then releasing the resulting potential energy by having the water flow through turbines to produce stored electrical energy.

coupling factor required by the older hypothesis; the Mitchell hypothesis predicts that no such substance exists. Beyond that, other predictions of the Mitchell hypothesis have been verified. It has been shown that mitochondrial membranes are, indeed, capable of establishing proton gradients by moving protons in the M to O direction. If the inner mitochondrial membranes are broken, ATP synthesis is halted, despite the fact that the chemical reactions of respiration continue; the Mitchell hypothesis predicts this would be the case. The Mitchell hypothesis depends on protons being able to pass through the mitochondrial membrane in the O to M direction *only* at the ATPase sites; the fact that detergents (which make the membrane more permeable and thus allow protons to pass through at several points rather than at the proper "turnstiles" only) also prevent ATP synthesis, though the respiratory reactions of the Krebs cycle go on, provides further support for the Mitchell hypothesis.

But *can* a proton gradient power the synthesis of ATP? Using the membranes of chloroplasts, structures playing a vital role in photosynthesis, this too has been verified. If the pH (recall pH is a measure of hydrogen ion or proton concentration) of the interior of chloroplasts is changed by soaking them in an acidic medium in the absence of light, the result is a large increase in protons (pH 4). A sudden change in the pH of the soaking medium to alkaline (pH 8.5) establishes a strong electrochemical gradient; the protons move rapidly out of the chloroplasts across their membranes, *and ATP synthesis occurs.*

To date, therefore, the supporting evidence for the Mitchell hypothesis seems strong. The sailing is not entirely clear, however. Dr. David Griffiths, at the University of Warwick, has evidence that appears to contradict the Mitchell hypothesis. Griffiths did his research working with the purple bacterium *Halobacterium halobium.* In this organism a specialized region, the purple membrane, is capable of absorbing light energy that can be used to make ATP. These absorbing regions, however, do not themselves contain the machinery for ATP synthesis; another region does. Since its ability to absorb light energy makes *H. halobium* a photosynthetic organism, the Mitchell hypothesis would propose that the energy of light absorption in the purple membrane (rather than energy from the electron transport chain) sets up a proton gradient in this other membrane region and that *this* gradient powers ATP synthesis. But Griffiths finds that he can obtain ATP synthesis by mixing fragments of purple membrane and fragments of the ATP-synthesizing regions of mitochondria and exposing the mixture to light. He reasons that, since a continuous membrane would not be formed under these conditions, the light absorption by the purple membrane and ATP synthesis by the mitochondrial machinery cannot be due to a potential or gradient across a membrane. Griffiths instead suggests that a fatty acid, lipoic acid, is the link. He notes that purple membranes contain high levels of lipoate, which is chemically reduced when exposed to light, and that the rate of this reduction is consistent with the rate of ATP synthesis observed in his purple-membrane-plus-mitochondrial-fragments experiments. Further, the addition of lipoic acid stimulates ATP synthesis. Thus the Griffiths hypothesis is somewhat of a return to the older idea of the coupling between respiration and ATP synthesis occurring via a chemical intermediate energy store rather than via a "battery" or chemiosmotic mechanism; the Griffiths hypothesis further suggests that this intermediate may be lipoic acid.

It is doubtful that Griffiths' findings to date will overthrow the Mitchell chemiosmotic hypothesis. Purple bacteria are rather unusual organisms, and their means of ATP synthesis may simply be different from that in other forms. Also, there may well be subtleties not yet detected in Griffiths' experimental methods that allow for alternative explanations for his results, which appear to fly in the face of the more overwhelming evidence in favor of the Mitchell hypothesis.

A third hypothesis has been advanced by Dr. David Green of the University of Wisconsin and Dr. Paul Boyer of the University of California. They propose that the energy released by electron transport causes temporary and reversible changes in the conformation of large molecules within the inner mitochondrial complex. When these molecules return to their lower-energy conformation, the energy released powers the phosphorylation of ADP. The primary evidence supporting this hypothesis is provided by electron micrographs indicating changes in mitochondria when undergoing active transport.

6.9
SOME CHARACTERISTICS OF METABOLIC PATHWAYS

Glycolysis and the citric acid cycle are but two of several kinds of metabolic pathways found in living systems. Since, in the chapters that follow, we shall be considering other examples, it is worthwhile noting certain features that all metabolic pathways have in common.

1.   The tendency of pathways to proceed in steps. Both glycolysis and the citric acid cycle involve a number of individual steps by which the original glucose molecule is broken down into smaller fragments and the potential energy extracted. As we have seen, these steps ensure that maximum energy can be extracted in a usable form from the fuel molecule. In order to extract energy in the most controlled (and therefore efficient) manner, the molecule is dismembered a few atoms at a time. Since life itself depends on the efficient use both of the substances present in a cell and of the available energy, the step-by-step breakdown or buildup of molecules is essential.

2.   The specificity of individual steps in metabolic pathways. Each step in the metabolic pathways of glycolysis and the citric acid cycle is catalyzed by specific enzymes that ensure that only certain chemical changes occur. For example, NAD can pick up electrons only

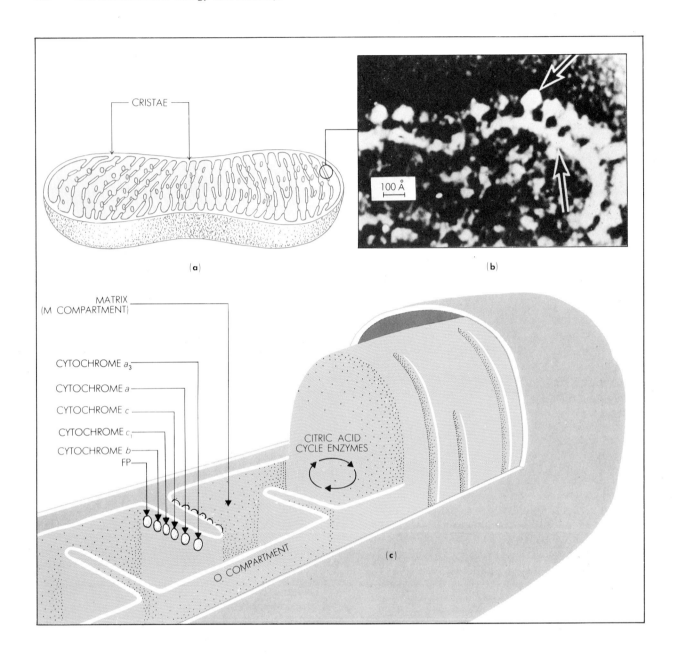

Fig. 6.13
Successively enlarged views of the internal structure of mitochondria, showing the localization of pathways. Note the intimate relationship between the mitochondrial membrane and
certain enzymatic reactions step-by-step, oxidative phosphorylation and electron transport in particular (see Fig. 6.14).
(Photo (b) courtesy Institute for Enzyme Research, University
of Wisconsin; diagram (c) after *Biology Today.* New York:
CRM/Random House, 1972, Interleaf 6.1.)

from certain intermediate products in the citric acid
cycle: it can interact with malate to form oxaloacetate
and NADH, but it cannot interact with fumarate. The
conversion of malate to oxaloacetate (with the production of NADH) is catalyzed by a specific enzyme (NAD-
malic dehydrogenase). If this enzyme is not present, the
reaction will not occur and the entire cycle can be
stopped.

Similarly, malate must be formed from fumarate in
order for the subsequent reactions of the citric acid cycle
to proceed. This is accomplished through the action of
an enzyme (fumarase) that catalyzes only this particular
reaction. Every step throughout the citric acid cycle and
glycolysis depends on such specific reactions. It is obvi-

ous that if fumarate, for example, could be converted into a very large number of intermediates other than malate, the specific production of ATP would be greatly retarded. All metabolic pathways depend on a similar high level of specificity of each component step to yield specific products.

Another aspect of specificity is apparent in glucose oxidation. Given the same starting reactants, the differences between two metabolic pathways are the result of the differences in enzymes present. For example, aerobic organisms can oxidize pyruvic acid to CO_2 and H_2O, with the production of a considerable amount of ATP, because they possess enzymes for the citric acid cycle and electron transport molecules, the cytochromes. Certain anaerobes lack all these enzymes and consequently cannot carry out the citric acid pathways. Similarly, anaerobes differ in the pathways by which they act upon pyruvic acid. As was seen in Section 6.4, yeast (a facultative anaerobe) has an enzyme (decarboxylase) that converts pyruvate to acetaldehyde. Acetaldehyde is then converted by another specific enzyme into ethyl alcohol. Some anaerobic bacteria, on the other hand, lack decarboxylase and convert pyruvic acid directly into lactic acid. Again, the specific fate of any molecule in a metabolic pathway depends on the specific enzymes present.

3. **The localization of pathways.** In the cells of most organisms, pathways are localized to some extent in various parts of the cell. The example of complete glucose oxidation illustrates this principle clearly. The glycolytic pathway occurs in the cytosol of cells and is catalyzed by soluble enzymes not bound to any particular cell organelles. The enzymes for the citric acid cycle, however, are not free in the cytoplasm, but are found in the liquid matrix inside mitochondria. These enzymes are apparently not bound to the mitochondrial membranes, but are too large to pass through the walls of the mitochondria into the cytoplasm. The electron transport molecules (the cytochromes, FP to some extent,

and possibly NAD) are generally bound to the cristae (Figs. 6.13, 6.14, and 6.15). Pyruvic acid, generated in the cytoplasm by glycolysis, can pass through the mitochondrial membranes and come into contact with the enzymes of the citric acid cycle. As soon as electrons are removed from intermediates in the citric acid cycle, they are carried by NAD (or FAD) to the respiratory assembly where electron transport can take place. The by-products of the citric acid cycle (CO_2, water, ATP, and heat) can pass out of the mitochondria into the cytoplasm for elimination or involvement in other metabolic pathways.

We see here, then, three levels of localization. Glycolysis is the least localized of the pathways, occurring throughout the cytoplasm. The citric acid cycle is much more localized, being confined to the liquid medium within the rather small mitochondria. Electron transport is the most highly confined of all the pathways, taking place within a membrane-bound respiratory assembly less than 50 Å in diameter. The advantages of this localization are obvious. When all the reactants and intermediates as well as the enzymes of a particular pathway are confined in a small space, the rate of reaction is markedly increased and efficiency is improved. The electron transport system shows one of the most highly developed localizations in all of biochemistry. Not only are the molecules tightly bound together in physical proximity, but they even seem to be arranged in an order appropriate to their functioning in the electron transport process. A diagram of one respiratory assembly, attached to the mitochondrial wall, is shown in Fig. 6.14. With this highly ordered spatial arrangement, electrons can be transmitted rapidly from one carrier to the next in line. In the reactions for glycolysis, the citric acid cycle, and electron transport, the relation between structure and function on a subcellular level is clearly evident. A summary of the localization involved in glucose breakdown is given in Fig. 6.15.

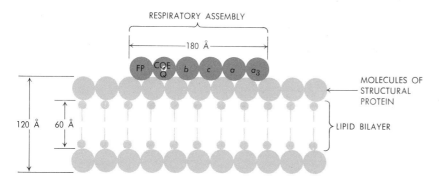

Fig. 6.14
A highly diagrammatic representation of the arrangement of respiratory assembly molecules along the mitochondrial membranes. The electron transport molecules, or cytochromes, are believed to be located side-by-side in the precise order in which they receive and pass along electrons to the next acceptor (see also Fig. 6.13). (Adapted from A. L. Lehninger, *Bioenergetics*. Menlo Park, California: W. A. Benjamin, 1965, p. 113.)

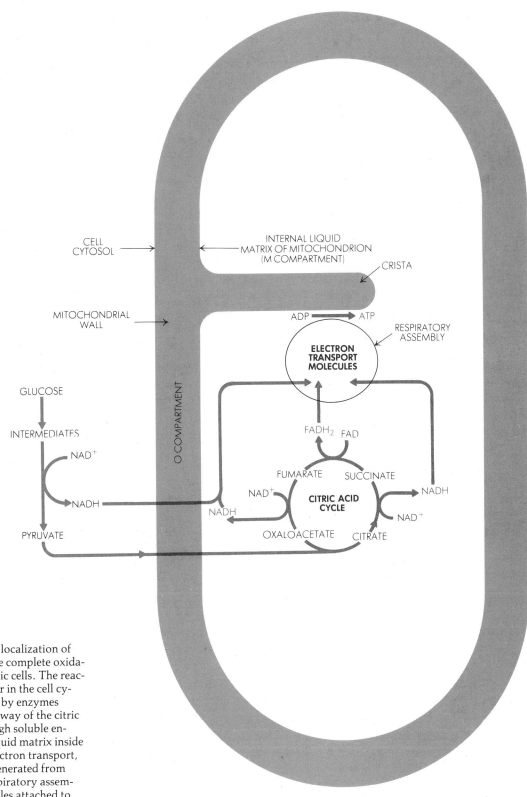

Fig. 6.15
Schematic summary of localization of various pathways in the complete oxidation of glucose in aerobic cells. The reactions of glycolysis occur in the cell cytorol and are catalyzed by enzymes located there. The pathway of the citric acid cycle occurs through soluble enzymes located in the liquid matrix inside the mitochondrion. Electron transport, during which ATP is generated from ADP, occurs in the respiratory assemblies, groups of molecules attached to the inside wall of the mitochondria (see also Fig. 6.13).

SUPPLEMENT 6.4
LOCALIZATION WITHIN MITOCHONDRIA:
THE BASIS FOR OUR KNOWLEDGE

Figures 6.11 and 6.12 illustrate the rather precise knowledge obtained about localization of respiratory pathways within mitochondria. Three separate pathways, for oxidizing glucose in the citric acid cycle, electron transport, and oxidative phosphorylation, have been localized in three separate parts of the mitochondrion.

How is it known that such precise positioning in the geometric space of the mitochondrion actually exists? As in all such biochemical questions, direct observation is not possible. True, the electron microscope has revealed some information about the fine structure of the inner regions of the mitochondrion. But the electron microscope cannot "see" molecules, nor can it ever reveal direct information about function. It is necessary to correlate structural observations with functional observations through experimental methods.

One such experimental approach is the disruption and reconstruction procedure illustrated in Diagram A.

If intact mitochondria, separated from other cell components, are subjected to osmotic treatment so that they take in water and swell, the inner membranes can be seen to contain numerous tiny, sphere-shaped particles, each on a short stalk. The osmotic treatment appears to cause the particles (called F_1 particles) to stick up out of the inner membrane surface where they are normally embedded. Further treatment of the mitochondria to sonic disruption (very high-frequency sound) breaks the inner membrane down into small vesicles. These vesicles are membrane bilayers from which the F_1 particles protrude.

Biochemical tests of the vesicles and attached F_1 particles indicate that the intact units have the following properties:

1. They show electron transport activity (that is, they have the electron transport system, the ETS, composed of the cytochromes and NAD^+).

2. They are capable of carrying out oxidative phosphorylation (they can effect the formation of ATP from ADP and inorganic phosphate), designated as OP (oxidative phosphorylation).

3. They show ATPase activity (they contain the enzyme ATPase, adenosine triphosphatase, which couples electron transport to oxidative phosphorylation). In other words, ATPase activity is necessary for the energy derived from electron transport to be joined to the formation of ATP from ADP.

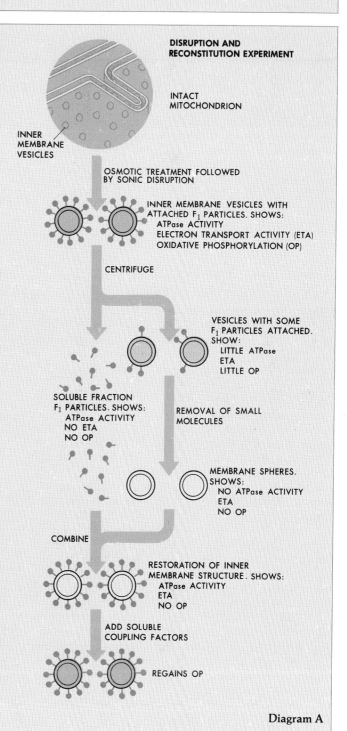

DISRUPTION AND RECONSTITUTION EXPERIMENT

INTACT MITOCHONDRION

INNER MEMBRANE VESICLES

OSMOTIC TREATMENT FOLLOWED BY SONIC DISRUPTION

INNER MEMBRANE VESICLES WITH ATTACHED F_1 PARTICLES. SHOWS:
ATPase ACTIVITY
ELECTRON TRANSPORT ACTIVITY (ETA)
OXIDATIVE PHOSPHORYLATION (OP)

CENTRIFUGE

VESICLES WITH SOME F_1 PARTICLES ATTACHED. SHOW:
LITTLE ATPase
ETA
LITTLE OP

SOLUBLE FRACTION F_1 PARTICLES. SHOWS:
ATPase ACTIVITY
NO ETA
NO OP

REMOVAL OF SMALL MOLECULES

MEMBRANE SPHERES. SHOWS:
NO ATPase ACTIVITY
ETA
NO OP

COMBINE

RESTORATION OF INNER MEMBRANE STRUCTURE. SHOWS:
ATPase ACTIVITY
ETA
NO OP

ADD SOLUBLE COUPLING FACTORS

REGAINS OP

Diagram A

If the vesicles and attached F_1 particles are further sonicated (subjected to high-frequency sound) and centrifuged, the F_1 particles can be removed and separated from the membranes. The separation of F_1 particles from the membranes by sonication takes time, so that after only a short exposure not all the F_1 particles are separated. Membrane vesicles with only a few F_1 particles are capable of carrying on electron transport (have ETS), can perform a small amount of oxidative phosphorylation (OP), and show the presence of a small amount of ATPase. Continued sonication removes all the particles. Membrane vesicles without F_1 particles have the ETS, but they show no ATPase activity nor any OP. The F_1 particles show a great deal of ATPase present, but no ETS and no OP. These data suggest that the membranes (which represent the inner membranes of the mitochondrion) contain the electron transport system, while ATPase (the so-called coupling factor) and other enzymes of oxidative phosphorylation are located on the F_1 particles.

The above hypothesis can be confirmed by reconstituting the intact vesicle-F_1 units (putting them back together). If the membrane vesicles and free F_1 particles are put together in a test tube and the ionic concentration adjusted in a very precise way, the F_1 particles recombine with the membrane to produce the intact respiratory unit. This unit now has full restoration of its three functions: it has ATPase, contains the ETS, and is capable of full OP activity.

Such reconstitution experiments are important steps in confirming the results of disruption experiments, for while it is important evidence to show that removal of a part *eliminates* the function of that part, it is still more convincing to show that replacing the part *restores* the lost function. Such experiments have shown that different parts of the mitochondrion are the sites for different parts of the biochemical pathways involved in glucose oxidation and generation of ATP.

6.10
THE ENERGETICS OF METABOLIC PATHWAYS

In looking at any metabolic pathway, it is important to distinguish between the energetics of the pathway as a whole and the energetics of the individual steps in that pathway. For example, some of the steps in the oxidation of glucose are endergonic, as is the conversion of glucose to glucose-6-phosphate or the conversion of glucose-6-phosphate to fructose-1,6-diphosphate, both of which require the input of ATP. Yet the overall oxidation of glucose is strongly exergonic. In looking at the energetics of glucose oxidation, we are concerned with (1) the total free-energy change and (2) the efficiency of the entire process—that is, the amount of energy recaptured in a usable form compared to the total free-energy change. In the case of glucose oxidation, the amount of work accomplished is equivalent to the number of high-energy phosphate bonds built for a given amount of glucose oxidized. As seen earlier, the total free energy change in the completed oxidation of a mole of glucose is 686 kcal. However, this figure is based upon the presupposition that all the reactants and products are in steady state at a one-molar concentration. In truth, intracellular conditions may be vastly different, varying widely as concentrations of the various substances involved increase and decrease. The figure of 686 kcal, then, actually represents the total *possible* potential energy that could be obtained from the process if 100% efficiency were possible.

Table 6.2 reveals that a total of 38 moles of ATP are generated per mole of glucose oxidized. At a value of 7 kcal per mole of ATP, roughly 266 out of 686 kcal are regained by the cell. Recall, however, that two moles of ATP are required to mobilize a mole of glucose for the oxidative pathways. The cell, therefore, shows a net gain of 36 moles of ATP, or 252 kcal per mole of glucose oxidized. Since 252 is approximately 37 percent of 686, glycolysis and the citric acid cycle combined extract more than one-third of the energy released during glucose oxidation. This is a remarkable degree of efficiency when compared to the best internal combustion engines, which usually achieve considerably less than 25 percent efficiency.

If the efficiency of glucose oxidation in aerobic and anaerobic organisms is compared, however, an astounding difference is observed. Anaerobic organisms,

Table 6.3
Efficiency of Anaerobic vs. Aerobic Respiration

Process	Anaerobic respiration (glycolysis)	Aerobic respiration (glycolysis plus citric acid cycle)
Total free-energy change during reaction	56 kcal	686 kcal
ATP synthesized (net gain)	2	36
Total free energy stored as high-energy phosphate bonds	14 kcal	252 kcal
Efficiency of recapturing usable energy from total energy released	25%	38%
Fraction of total available free energy recaptured as ATP	2%	38%

carrying out only glycolysis, generate a total of four moles of ATP per mole of glucose. Since two ATPs were required to start the process, the actual net gain is two ATPs. This is equivalent to 14 kcal per mole out of the 252 kcal that are released if the glucose molecule is broken down to its lowest energy states, carbon dioxide and water (as it is by aerobes). Thus anaerobic organisms are able to extract less than 2 percent of the energy theoretically available to them in the glucose molecule. This figure does not tell the whole story, however. Whereas anaerobic organisms do not break glucose down nearly as far as aerobic organisms do, the total energy released in the *partial* breakdown of glucose to alcohol or lactic acid is 56 kcal per mole. At a net gain of two moles of ATP per mole of glucose oxidized, anaerobic cells *can* recapture 25 percent of the energy they release. Thus, for the reactions they *do* carry out, anaero-

bic organisms are only slightly less efficient than aerobic. These calculations are recorded in Table 6.3.

Every metabolic pathway has a relative degree of efficiency which can be calculated in a similar fashion. It is particularly important to understand that the relatively high efficiency of cellular metabolic pathways is brought about by (a) their step-by-step nature, (b) the fact that they are frequently localized in specific regions of the cell, and (c) the role of enzymes in reducing the amount of activation energy necessary to get the reactions going.

From an energy point of view, therefore, it is clear that anaerobic cells gain far less ATP from the oxidation of glucose than do aerobic cells. Fig. 6.16 summarizes the total free-energy change (ΔG) in oxidation of glucose and compares anaerobic and aerobic processes. The total ΔG for the oxidation of a mole of glucose mol-

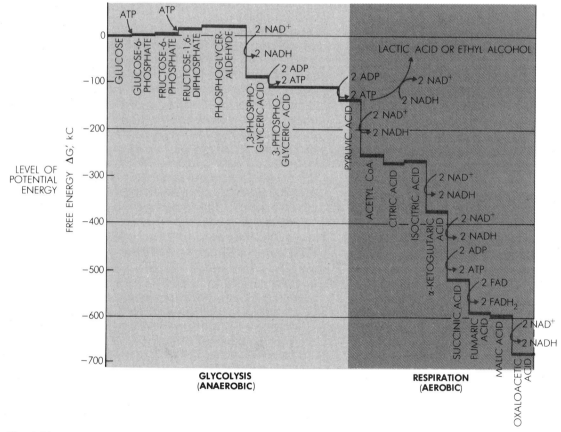

Fig. 6.16

Changes in potential free energy, ΔG, as glucose is successively oxidized through glycolysis and the Krebs citric acid cycle. Note that at the end of glycolysis only a small amount of the total energy of a glucose molecule has been released. The release of energy in successive steps occurs through electron rearrangements, a few at a time; at each step a few electrons are lowered from a higher to a lower energy level within the molecular orbitals composing the glucose and intermediate substances. The electrons picked up by acceptors such as NAD$^+$ and FAD are eventually passed to lower energy levels through the electron transport system associated with the cytochromes. (After E. O. Wilson et al., *Life on Earth.* Sunderland, Mass.: Sinauer Associates, 1973, p. 161.)

Aerobic organisms gain far more of the free energy available in glucose oxidation than do anaerobic organisms.

ecules to a mole of carbon dioxide and water is 686 kcal. The ΔG for oxidation of a mole of glucose to a mole of pyruvic acid is 56 kcal. Since anaerobic cells must reduce pyruvic acid to lactic acid or ethyl alcohol, the net ΔG for glycolysis is less than 50 kcal per mole. Thus anaerobic cells must use a great deal more sugar to get the same amount of energy as aerobic cells: about twelve to fifteen times as much! Because of their relative inefficiency, anaerobic cells face a constant "energy crisis" that aerobic cells do not face.

6.11
THE CONTROL OF METABOLIC PATHWAYS

Controls affect both the overall direction and the rate at which metabolic reactions take place. Such controls are usually exerted on specific steps of the pathway, and they can occur in a number of ways. One is by the **law of mass action** or **Le Châtelier's principle**, which states that the rate and direction of a reaction (or series of reactions) are affected by the relative concentrations of reactants and products. In a reversible reaction, the greater the concentration of reactants and the less the concentration of products, the more the reaction is shifted to the right. Conversely, the greater the concentration of products and the less the concentration of reactants, the more the reaction is shifted to the left. Concentration of either reactants or products is determined by several factors. One is the availability of reactants, in turn determined by what can get into and leave the cell. A second is the fate of the end products—whether they are removed from the site of the reaction, packaged up and stored in the cell, broken down, or simply transported out of the cell. In aerobic respiration, for example, H_2O diffuses out of the cell, thus "pulling" the reaction toward more sugar breakdown. An alternative fate is that the end products build up and accumulate in the cell. All these factors affect both the direction and the rate at which a pathway operates by exerting "pushes" and "pulls" on both reactants and products. However, a number of metabolic steps are essentially irreversible, proceeding with a large release of energy, and in these cases the law of mass action is not useful.

A second means by which metabolic pathways are controlled is through the energy requirement—the degree of reversibility—of the individual steps of the reaction series. For example, most of the individual re-

actions in the citric acid cycle are reversible, having relatively low free-energy changes in either direction. However, at least two steps—the conversion of oxaloacetate and acetyl CoA to citrate, and the conversion of α-ketoglutarate to succinate—are shifted heavily to the right (that is, in the forward direction). This means that the energy requirement for the forward reaction is considerably less than for the reverse. Thus, although it is theoretically possible for succinate to go back and reform α-ketoglutarate or for citrate to go back and reform acetyl CoA and oxaloacetate, neither of these reactions is very likely to occur. From the standpoint of equilibrium and energy requirements, both reactions are virtually irreversible. Thus the citric acid cycle is kept moving in a single direction, even though most of the steps are individually reversible. A few essentially irreversible steps in a metabolic pathway act in a manner analogous to a series of one-way turnstiles in a subway station: they prevent the whole reaction series (the passage of people) from backing up by stopping the reverse flow at specific points. The existence of irreversible steps does not change the *rate* of the overall reaction, which is controlled by other factors, including the concentration of reactants and products. It does, however, keep the reaction series moving in one direction.

Another example of how directionality is controlled occurs in the electron transport system. Each successive transport molecule has a greater affinity for electrons than the molecule preceding it in the series. There is thus little chance for the electrons to move in a reverse direction and avoid being passed in the sequence necessary to generate ATP.

The rate at which metabolic reactions take place can be controlled by a third and highly subtle device: controlling the activity of enzymes operating on specific steps in the pathway. This control can be exerted in two ways: (1) by controlling the rate at which existing molecules are synthesized, and (2) by controlling the rate at which existing molecules catalyze specific reactions with their substrate. The first method is essentially a genetic mechanism (that is, controlling the rate at which enzymes are synthesized by the genes) and will be dealt with in more detail in a later chapter. The second process operates by allosteric enzyme controls as described in Chapter 3. This interaction can serve either to increase (activate) or to decrease (inhibit) the rate at which the enzyme acts on the substrate.

A single example will illustrate both activation and inhibition of enzymes by products of the pathway. The

The rate at which the respiratory pathways function is controlled largely by feedback signals to allosteric enzymes located at key points along the pathway.

conversion of citric acid to α-ketoglutaric acid involves several steps, one of which is catalyzed by the enzyme isocitrate dehydrogenase, as follows:

CITRATE ⇌ → ⇌ →

isocitrate
dehydrogenase
⇌ → α-KETOGLUTARATE
+ REDUCED NAD
(NADH)
+ CO_2.

As part of the citric acid cycle, this reaction leads to the production of usable energy (NADH or, ultimately, ATP). Chemical investigation has shown that both these molecules are capable of inhibiting the action of the enzyme isocitrate dehydrogenase. In other words, in the presence of these two molecules the enzyme greatly slows down the conversion of citrate to α-ketoglutarate. As the cell builds up a greater concentration of energy-rich compounds, the pathway for energy production is inhibited. Conversely, it has been found that molecules of ADP or AMP stimulate isocitrate dehydrogenase to convert citrate into α-ketoglutarate. Thus, as the energy supply of the cell begins to run low, the inhibition of enzymes in the energy-producing pathway is removed.

Other control points can be found in the pathways of glucose oxidation (Fig. 6.17). Note that control occurs both by inhibition and activation of enzymes. With this number of sensitive "control spots," subject to either activation or inhibition by different reaction products, it is apparent that an extremely sensitive mechanism exists by which cells can constantly monitor and regulate their metabolic pathways. Such subtle control is highly advantageous to cells, which must always work to conserve as much energy as possible. A control system on metabolic processes allows the cell (1) to speed up necessary and vital reactions and thus not get caught short on crucial substances such as ATP, and (2) to slow down the production of unnecessary substances and thus not waste energy and valuable raw materials. Since the external conditions to which cells are exposed are constantly changing (especially for one-celled organisms), survival depends on being able to adjust internal biochemical reactions to meet new needs. The capacity to vary enzyme reaction rates brings about extremely effective controls within cells.

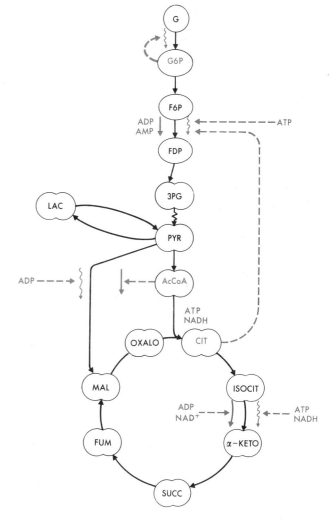

Fig. 6.17
The many control points in glucose oxidation. Each control point indicated here functions through enzyme inhibition or activation. Allosteric effector molecules are shown in color; heavy colored arrows indicate enzyme activation at that particular point in the pathway; wavy arrows indicate enzyme inhibition by the effector molecule. Dashed arrows point out the site of a control mechanism.

6.12
THE INTERRELATIONSHIP BETWEEN
METABOLIC PATHWAYS

Thus far, glycolysis and the citric acid cycle have been considered primarily as pathways for the breakdown of glucose and the production of energy. However, materials other than glucose, such as fats or proteins, may be used as respiratory substrates and these, as would be expected, feed into the overall pathway in different ways. Nearly all the intermediates in these two pathways are starting reactants, intermediates, or end products in other metabolic pathways within the cell. For example, citric acid cycle intermediates form connecting points with pathways for the synthesis (and breakdown) of all the major types of molecules found in the cell: fatty acids, amino acids, steroids, carbohydrates, and nucleotides (Fig. 6.18).

Consider, for example, the very important intermediate acetyl coenzyme A, the specific form in which the oxidation products of glucose enter the citric acid cycle. Acetyl CoA represents an intermediate in the breakdown of carbohydrates. In addition, the condensation of several acetyl CoA molecules (each containing two carbons) can produce the hydrocarbon chains, which are major structural components of fatty acid molecules. Thus, when fatty acids are broken down to be used as fuel, they can enter the citric acid cycle as acetyl CoA. Acetyl CoA can also be converted, by a series of reactions, into one of several amino acids. Furthermore, acetyl CoA is the starting point for the synthesis of steroids, molecules that form the basis of several important hormones.

The conversion of acetyl CoA to each of these other types of molecules occurs by reversible pathways, a very important factor in the regulation of cell metabolism. If a greater amount of carbohydrate comes into the cell than is needed to build ATP, much of the carbohydrate can be shunted off through acetyl CoA to form fatty acids. Fats are an important storage form of fuel molecules and can be held in reserve until needed. The existence of this pathway is responsible for the buildup of fat in human beings on a carbohydrate-rich diet.

Many intermediates in the citric acid cycle, such as α-ketoglutarate and oxaloacetate, can be converted into other types of molecules, principally amino acids. Pyruvate, too, can enter a pathway for the formation of several types of amino acids, and glucose can be converted into a variety of polysaccharides including, most prominently, glycogen. The existence of these many interconvertible steps between the citric acid cycle and other metabolic pathways allows for maximum efficiency in the use of molecules in the cells. If every pathway started with its own reactants, proceeded through its

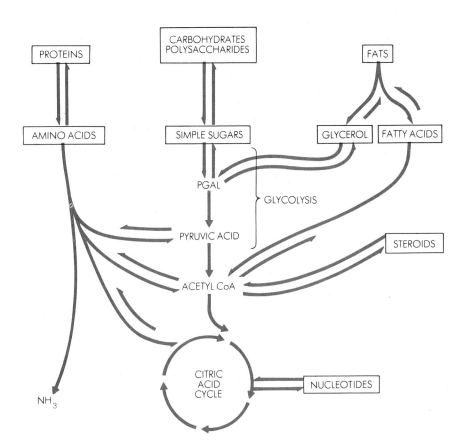

Fig. 6.18
The citric acid cycle acts as sort of a "metabolic hub" for the synthesis and degradation of many other substances, whose degradation products enter the cycle via several different entry points. The entry points for proteins and fats, as well as carbohydrates, are indicated above. Note that most of the reactions are reversible.

own particular set of intermediates, and produced its own individual set of products, far more types of molecules would be needed in a cell than are actually found there. That the end product of one pathway can serve as an intermediate or a reactant for another pathway conserves both the number and the different types of molecules necessary and reduces the amount of work the cell must do. Energy is usually at a premium. Thus the interconvertibility of pathways not only allows the exercise of more effective control but also increases the efficiency of metabolic activity.

Cells are amazing biochemical systems. With their high degree of structural organization and the close interrelationships between internal cellular structure and biochemical function, cells operate at a level of ac-

tivity and efficiency that the most elaborate machine human beings have devised cannot even approach. This does *not* mean that cells exhibit mystical or nonphysical properties defying explanation. It *does* mean that cells are far more intricate and subtle units of structure and function than people have sometimes thought.

In this chapter we have discussed the interrelationships between structure and function in the processes by which all types of cells, from simplest to most complex, gain usable energy by running the downhill or exergonic reactions of glucose oxidation. In the next chapter we will explore the other side of that energy hill: how cells of certain types of organisms, green plants, use the energy of sunlight to synthesize carbohydrates in an endergonic process, photosynthesis.

Summary

1. Metabolism is the sum total of chemical reactions involved in the degradation and synthesis of substances in the cell. Metabolism involves the reactions by which glucose is oxidized to liberate energy, as well as the reactions by which carbohydrates, fats, proteins, nucleic acids, and other important molecules are synthesized or degraded.

2. Adenosine triphosphate (ATP) is the energy currency of the cell. By hydrolysis of a phosphate group (yielding ADP + P_i), 7 kcal of energy are released. This is slightly more energy than is needed to synthesize a peptide, glycosidic, or ester bond. The reaction by which ATP is hydrolyzed to ADP is coupled to the reaction in which the energy from ATP is used. In this way, ATP provides virtually all the energy for cell reactions.

3. In animal cells, ATP can be generated in two main ways: (1) Through substrate-level phosphorylation, the process by which a phosphate group is added to an intermediate in a reaction series and then converted into a high-energy phosphate bond by electron rearrangements within the substrate. This high-energy phosphate group is then transferred to ADP, producing ATP. In cells of higher plants, a third method of ATP production is available: the use of light energy to join P_i onto ADP to produce ATP. This process is called photophosphorylation (see Chapter 7). (2) Through electron transport, in which three ATPs are normally generated for every pair of electrons passing down the acceptor chain, a process known as oxidative phosphorylation.

4. The oxidative phosphorylation system consists of two acceptors, nicotinamide adenine dinucleotide (NAD) and flavine adenine dinucleotide (FAD), and the cytochromes (cytochromes b, c_1, c, a, and a_3). The sequence of electron flow through this system is: NAD→FAD→Cyt b, c_1→Coenzyme Q→Cyt c→Cyt a→Cyt a_3→O_2. The final acceptor is oxygen, producing water. The flow of electrons goes in one direction, because each acceptor along the chain has a higher affinity for electrons than the molecule preceding it.

The electron transport system is bound to the inner membranes of mitochondria and is thus a highly localized system.

The oxidative phosphorylation system is coupled, through the enzyme adenosine triphosphatase (ATPase), to the generation of ATP from ADP and P_i. Energy is released during electron transport because the electrons pass successively from higher to lower energy levels in moving from one acceptor molecule to another. These downward transitions in energy levels release quantities of energy sufficient to produce the high-energy phosphate bonds of ATP.

5. Several hypotheses have been advanced concerning the precise mechanism of ATP synthesis during electron transport oxidative phosphorylation (and photosynthesis). The prevalent one to date, Mitchell's chemiosmotic theory, proposes that the energy produced by electron transport is used to establish an electrochemical gradient across the mitochondrial membrane. The flow of protons down this gradient at specific ATPase-containing sites on this membrane provides the energy for ATP synthesis.

6. There are three phases in the oxidation of glucose to yield energy for ATP production: (1) activation, during which glucose or its storage form glycogen (animals) or starch (plants) is activated by addition of two phosphate groups; (2) glycolysis, during which the activated sugar (called fructose-1,6-diphosphate) is broken down into two three-carbon sugars, which through oxidation (the removal of a pair of electrons) and internal rearrangement are converted into pyruvic acid. By substrate phosphorylation, four ATPs are generated, two from each three-carbon sugar; (3) citric acid cycle, during which pyruvic acid is further oxidized through a series of intermediate products that include citric acid; during the citric acid cycle, twenty-eight ATPs are produced by electron transport, and two by substrate phosphorylation. The total ATP produced by all processes of electron transport and substrate phosphorylation is thirty-eight. Since the cell spends two ATPs to "mobilize"

the glucose, the net gain is thirty-six ATPs per glucose molecule oxidized.

7.   Anaerobic respiration is a process of glucose breakdown involving only glycolysis. Anaerobic respiration occurs without the process of oxidative phosphorylation; hence it does not require oxygen. The electrons removed from the substrate during glycolysis are passed back to pyruvic acid to produce a waste product such as ethyl alcohol, butyric acid, or lactic acid.

8.   Aerobic respiration is the complete oxidation of glucose, beginning with mobilization and ending with the citric acid cycle. The final products of aerobic respiration are carbon dioxide and water, along with the ATP generated by substrate-level and oxidative phosphorylation.

Aerobic respiration produces far more ATP than anaerobic respiration. The aerobic process releases 686 kcal of energy per mole of glucose oxidized, while anaerobic respiration releases only 56 kcal, or about 8 percent of the total free energy of glucose. The efficiency of each process is measured as the total energy recaptured by the cell (as ATP) from the energy released. Aerobic organisms recapture 252 kcal per mole of glucose (a net gain of 6 ATP × 7 kcal), at about 38 percent efficiency (252/686 = .38). Anaerobic organisms recapture only 14 kcal per mole of glucose oxidized to pyruvic acid (a net gain of 2 ATPs × 7 kcal), for an efficiency of about 25 percent (14/56 = .25).

9.   The complex pathways involved in glucose oxidation are localized within specific parts of the cell. Glycolysis occurs in the cytoplasm and is mediated by enzymes not attached to membranes. The citric acid cycle occurs within the mitochondria, in the liquid matrix between the internal partitions, or cristae. These enzymes, too, appear to be unattached to membranes. Oxidative phosphorylation is carried out by the electron transport system, whose molecular components (the cytochromes) are bound to the inner walls of the mitochondrion. In addition, the enzyme adenosine triphosphatase (ATPase), which couples electron transport to oxidative phosphorylation, is bound with the cytochromes to the cristae.

10.   Control of the rate of metabolic pathways, such as glucose oxidation, comes about through three processes: (1) the concentration of reactants and products; (2) the degree of reversibility of individual steps; and (3) the feedback control exerted by allosteric enzymes. The latter is by far the most subtle and important of the control processes. Details of the mechanism of allosteric control are found in Chapter 3.

11.   The pathways for glucose oxidation are interconnected to the pathways for synthesis and breakdown of most of the other molecular components of the cell. Thus fats (including steroids) are synthesized from acetyl CoA and are broken back down to acetyl CoA; certain amino acids and nucleotides are synthesized from and broken down to α-ketoglutarate.

Exercises

1.   Explain why the equation

$$C_6H_{12}O_6 + 6O_2 + 6H_2O \rightarrow 6CO_2 + 12H_2O$$

is a more accurate representation of the process of respiration than the equation

$$C_6H_{12}O_6 + 6O_2 \rightarrow 6CO_2 + 6H_2O.$$

What is the role of water in the process of aerobic respiration?

2.   Give several reasons why energy-releasing reactions in living cells occur in a number of steps.

3.   Figure 6.19 relates the amount of carbon dioxide produced by yeast fermentation to time. Answer the following questions on the basis of the information contained in this graph.

   a) Why does the addition of inorganic phosphate at about the 75-min mark increase the amount of CO_2 produced? What does this show about the role of phosphate in the process of respiration?

   b) Compare the rates of CO_2 production for the first 25 min with those for the second 25 min (rate of respiration is determined by dividing the total amount of reaction, in this case measured as the number of ml of CO_2 produced, by duration of the reaction, 25 min for each period). Are the rates the same or different? Is this expected in terms

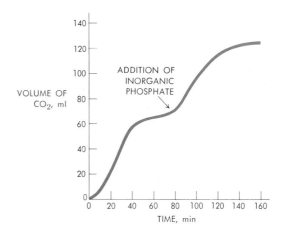

of the role we think phosphate plays in the process of respiration? Why or why not?

4.   How is it that the anaerobic respiration of glucose yields different end products in yeast and animal tissues? What must be the principal differences in the two types of cells?

5.   Which substance would you take for quick energy, glucose or sucrose? Why? Why is chocolate candy said to be a source of quick energy?

Chapter 7
Cell Metabolism II:
The Uphill Energy
Pathway—Photosynthesis

7.1
INTRODUCTION

The last chapter was concerned mostly with that portion of cellular metabolism dealing with the release of energy from glucose. Since the overall process involved for the most part the breakdown of glucose and other intermediate compounds, and thus moved from a state of higher to lower free energy, it was essentially catabolic. In this chapter, we shall be concerned with the other side of the coin; the processes to be described here are generally synthesizing or build-up reactions—that is, they are essentially **anabolic**.

We have seen that each glucose molecule represents a potential energy source, just as does water behind a dam. We know how the water got behind the dam: solar energy evaporated it from the sea and lakes, depositing some of it in upland or mountain regions where, as it ran back to the sea, it could be trapped behind the dam and released in a controlled way in order to generate electricity. But what is the source of the energy trapped in glucose? The answer turns out to be essentially the same as in the case of the water behind the dam—the sun. This chapter will be concerned primarily with our current state of knowledge concerning precisely how the synthesis of energy-rich compounds using sunlight —**photosynthesis**—occurs.

Photosynthesis and respiration represent two sides of the energy hill that characterizes the metabolic processes within the living world (see Fig. 7.1). Photosynthesis is an uphill, or endergonic, reaction. During photosynthesis, CO_2 is, in effect, "pushed up" the energy hill to a higher free-energy level in the form of carbohydrate. The carbohydrate stores free energy in its chemical bonds, which can be released by the oxidative processes of respiration. Respiration is thus a downhill, or exergonic, reaction. During respiration the free energy from glucose is released via gradual breakdown of the molecule through the removal of electrons (oxidation). The end products, CO_2 and H_2O, are the same as the starting reactants of photosynthesis. Since photosynthesis drives the uphill reactions that respiration pushes downhill, respiration could not occur without it. Photosynthesis winds up the world's biological motor, and respiration unwinds it in a controlled way. The energy source from the outside, nonbiological world that makes all this possible is, as noted, sunlight.

Fig. 7.1
Photosynthesis and respiration are opposite sides of the "energy hill" in the living world. Photosynthesis is an uphill, endergonic series of reactions, while respiration is a downhill, exergonic series. Both are essential for the continuance of life.

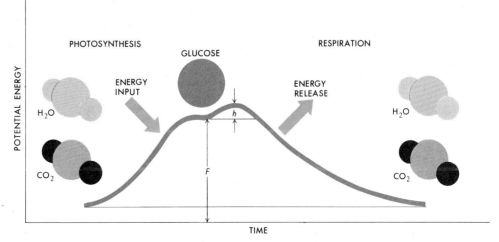

Respiration and photosynthesis represent opposite sides of the "energy hill," with respiration a "downhill" or catabolic process and photosynthesis an "uphill" or anabolic process.

7.2
PHOTOSYNTHESIS: AN OVERVIEW

It was pointed out in the previous chapter that the citric acid cycle results in a deficit of six water molecules that has to be made up for by extraction from the surrounding medium. An overall equation usually given for respiration is:

$$C_6H_{12}O_6 + 6O_2 \rightarrow 6CO_2 + 6H_2O.$$

Since photosynthesis involves using the waste products of respiration, carbon dioxide and water, as raw materials in the synthesis of glucose, it can be viewed as the reverse:

$$6CO_2 + 6H_2O \rightarrow C_6H_{12}O_6 + 6O_2.$$

There is a problem, however. It has been known for many years that all of the oxygen released into the atmosphere by photosynthesizing plants comes from the water molecules, and none from the carbon dioxide molecules. Yet a counting of the oxygen atoms available from the water molecules on the left side of the equation above reveals that only six are available, whereas twelve are released on the right-hand side of the equation. The overall equation for photosynthesis must be rewritten, therefore, as follows:

$$6CO_2 + 12H_2O \rightarrow C_6H_{12}O_6 + 6O_2 + 6H_2O.$$

Thus six molecules of water are produced along with oxygen as byproducts of the photosynthetic process, thereby indirectly balancing the deficit of six water molecules resulting from the citric acid cycle.

The modern view of photosynthesis is built upon investigations of plant physiology extending back over 300 years (see Supplement 7.1). But whereas in previous centuries plant physiologists were limited to varying experimentally the physical surroundings of whole plants or relatively large portions of them, the advent of molecular biology in the 1930s (and especially its development during the past twenty-five years) has greatly extended our knowledge.

For one thing, it is now known that photosynthesis is not all one process, but actually consists of two separate but interdependent phases, the **light reactions** and the light-independent or **dark reactions**. It should be stressed that the light reactions are so named because it is in this phase of photosynthesis that light energy is converted into chemical energy and that the name dark reactions does *not* mean that the processes involved in this phase occur only in the dark; the reactions are simply independent of light. While we shall discuss the light and dark reactions of photosynthesis in greater detail shortly, it will be helpful simply to remember at this point that the light reactions yield energy (in the form of ATP and reducing agents) which is then used to power the chemical reduction of carbon dioxide to carbohydrate during the dark reactions.

7.3
THE ORIGIN AND WAVE NATURE OF LIGHT

As has already been stressed, the direct or indirect source of light used in photosynthesis is starlight—in particular, that from the star of our solar system we call the sun. Of course, only a tiny fraction of the total light energy sent out by the sun actually strikes the earth. The rest goes elsewhere in the universe and is lost to us.

Light is a form of radiant energy. Physicists have proposed two theories to explain the behavior of light in certain situations. The first is the **wave theory of light**, which pictures light as traveling in waves, similar to the waves on the surface of a body of water. The color of the light varies according to the **wavelength**. Figure 7.2 shows how wavelengths are measured. *The longer the*

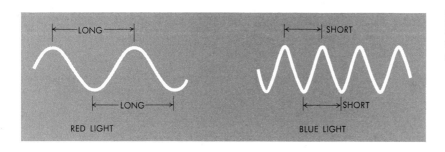

Fig. 7.2
Light energy travels in waves, the longer wavelengths conveying less energy than the shorter ones.

SUPPLEMENT 7.1
A HISTORY OF RESEARCH ON PHOTOSYNTHESIS

The earliest investigations of plant physiology were primarily directed toward the questions, how does a plant grow and where does it get the materials to build more plant matter? It seemed easy to understand how animals grew, since they could be observed devouring food which, it was assumed, they used to build more animal matter. With rare exceptions, however, plants do not feed in this manner. Since plants could be seen clinging to and penetrating the earth with their roots, it seemed logical to hypothesize that whatever they were using was somehow associated with the soil.

One of the first experiments designed to answer these questions was performed by Jan Baptista van Helmont (1577–1644). By carefully weighing a young willow tree, then putting it in a large container containing 200 pounds of dried earth and watering it with rainwater, van Helmont determined that after five years the tree had gained over 164 pounds. Noting that the earth had lost only "about two ounces" he concluded that the plant material had been produced from the water.

In order to understand Van Helmont's conclusion, it must be considered that he lived in a period when it was believed that dissimilar substances could be changed or "transmuted" into one another (for example, iron into gold or water into plant matter). Further, little or nothing was known of the nature of gases or, as they were then called, "aeroform substances." Thus it was natural for van Helmont to ignore any role that the air surrounding his willow tree might have played in his results. Finally, oxygen was unknown; the prevailing concept was the phlogiston theory, which was as important to the seventeenth-century scientist as the atomic

theory is to chemistry today. The phlogiston theory was advanced to explain many phenomena, especially burning. It was thought that the flames leaping upward and away from a burning object represented something escaping from the object. This unknown something was called "phlogiston." Today it is known that, far from giving anything up, a burning substance *unites* with oxygen. If all the products of the burning process are measured (including the gaseous compounds released), the weight is greater after burning than before. The additional weight can be accounted for by the amount of oxygen used in the burning process. The seventeenth-century scientist was also aware that a burning substance gained rather than lost weight, and one might think that this fact would raise serious questions about the phlogiston theory. To those early scientists, however, this posed no problem; phlogiston was simply assigned a negative weight! In other words, due to the loss of the negatively weighted phlogiston, a substance would weigh more after burning than before.

The phlogiston theory actually accounted for a considerable number of observable phenomena. It is a prime example of a false hypothesis that gives rise to many true conclusions. For example, the early scientists noted that a candle burned under a sealed bell jar eventually goes out. This, of course, is due to the fact that there is no more oxygen available in the air within the jar. To adherents of the phlogiston theory, however, the air was simply referred to as being "phlogisticated." Such phlogisticated air was said to be "fixed" and no longer capable of supporting burning. The air present under the bell jar before the candle burned was referred to as de-

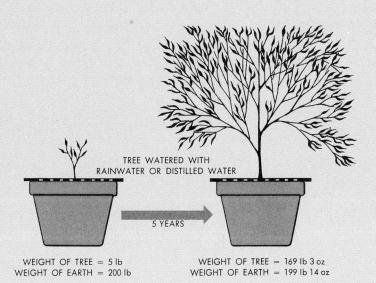

Diagram A
Van Helmont's experiment. He concluded that the gain in weight shown by the plant was due entirely to the water he had given it over the five-year period.

TREE WATERED WITH
RAINWATER OR DISTILLED WATER

5 YEARS

WEIGHT OF TREE = 5 lb
WEIGHT OF EARTH = 200 lb

WEIGHT OF TREE = 169 lb 3 oz
WEIGHT OF EARTH = 199 lb 14 oz

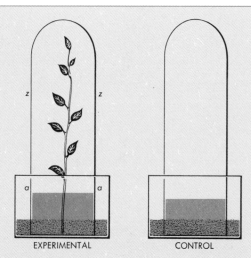

Diagram B
Stephen Hales's experiment. He concluded that plants remove something from the air. He also showed that the plant changed the composition of the air, although he was not sure what this change was.

phlogisticated air. The phlogiston theory thus adequately explained the phenomenon of burning without running into any contradictions that might disprove it. For this particular event, the burning of a candle, the phlogiston theory is just as good as the modern theory involving oxygen.

A belief in transmutation, a lack of any concept of the nature and significance of gases, and an acceptance of the phlogiston theory mark the mental perspective from which early plant physiologists approached their investigations of photosynthesis.

In the 1690s, van Helmont's conclusions were challenged by John Woodward of Cambridge University in England. Noting in his experiments that even very clean spring water had "terrestrial particles" in it and that the vast majority of water taken up by plants was "exhaled" by them into the atmosphere, he concluded that water only provided "a vehicle for the terrestrial matter which forms vegetables and does not itself make any addition to them."

The development of the microscope in the late seventeenth century led to the discovery that plants have many openings on the surfaces of their leaves. One early microscopist, Nehemiah Grew, suggested that these openings existed "either for the better avoleotion [evaporation] of superfluous sap, *or the admission of air*" (italics added for emphasis). Here, then, was the first recognition that there might be a significant interaction between a plant and the air surrounding it. Support for this idea was supplied by the British clergyman Stephen Hales (1677–1761). His experiments with mint shoots indicated to him that plants remove something from the air and that they changed its composition, though he was not sure what this change was. Hales's fellow countryman Joseph Priestley (1733–1804) followed with an important observation. Noting that a burning candle placed under an airtight bell jar will soon go out and that a mouse placed under a bell jar will eventually suffocate, he found that mint sprigs could restore the air so that, once again, the candle would burn and a mouse could live in it. He concluded that plants "reverse the effects of breathing," though he remained uncertain as to how. Today, of course, we recognize that plants do so by giving off oxygen, and taking up carbon dioxide.

Priestley's works were read by a Dutch physician, Jan Ingenhousz (1730–1799). He confirmed Priestley's observation that plants change the air in a manner rendering it "more fit for human life." But Ingenhousz made two other critical

1. CANDLE FLOATING ON CORK BURNS   2. CANDLE GOES OUT   3. GREEN PLANT PUT UNDER JAR   4. AFTER A FEW DAYS CANDLE BURNS AGAIN   1. LIVES   2. DIES

Diagram C
Priestley's experiments. From his results, he concluded that plants "reverse the effects of breathing."

observations: first, that light was necessary in order for them to do so and, second, that only the green portions of the plants were able to do so. The nongreen portions of plants, he noted, act just like burning candles or animals. Today we recognize that all living organisms carry on respiration, with plants being no exception. Illuminated green plants, however, contribute more oxygen to the air than they remove by their respiratory processes.

By the late eighteenth century, the element oxygen was isolated and the phlogiston theory was slowly abandoned. By this time, the steps of the photosynthetic process were suspected to be something like the following:

Note that no distinction was made between plant matter such as protein or fat and what we know today to be the actual final product of photosynthesis. Later, the process was written in a simplified form as:

Carbon dioxide + water→Glucose + Oxygen↑

or

$CO_2 + H_2O \rightarrow C_6H_{12}O_6 + O_2\uparrow$ (unbalanced).

Since only oxygen is released as a waste product in this reaction, the solution to the problem of where the "plant matter" (actually, carbohydrate) originates lies in discovering what parts of the water and carbon dioxide molecules go into making new plant materials. There are three possibilities:

1. The carbon of the carbon dioxide (CO_2) unites with water (H_2O) to form carbohydrate $(CH_2O)_6$, the oxygen of the CO_2 being released into the atmosphere.

2. The hydrogen of the H_2O unites with the CO_2, the oxygen of the H_2O being released into the atmosphere.

3. Both of these reactions occur, the oxygen being released from both the CO_2 and the H_2O.

These three alternatives actually serve to point up the main question: Where does the oxygen released by the plant come from: the water, the carbon dioxide, or both? The answer to this question indirectly tells us where the "plant matter" comes from, for whatever is left over after the oxygen is removed must go into the carbohydrate (glucose).

In the eighteenth century, two Frenchmen, M. Berthollet (1748–1822) and Jean Senebier (1742–1809) performed experiments which led to precisely opposite conclusions, Berthollet's results supporting possibility 2, and Senebier's re-

sults supporting 1.* A conclusive answer to which of the three possibilities was correct, however, was a long time in coming. The first truly significant evidence came from the work of the Dutch microbiologist C. B. Van Niel in the 1930s. Van Niel had studied photosynthesis in purple sulfur bacteria. Like the chlorophyll-containing cells of green plants, these bacteria use light energy to synthesize carbohydrate materials. However, purple sulfur bacteria use hydrogen sulfide (H_2S) instead of water. This fact suggests how the source of the oxygen given off by green plants during photosynthesis might be determined; if the oxygen released by plants during photosynthesis comes from the carbon dioxide molecule, then the purple sulfur bacteria will release oxygen as a result of their photosynthetic activity. On the other hand, if the oxygen released by plants during photosynthesis comes from a water molecule, then purple sulfur bacteria should release sulfur as a result of their photosynthetic activity.

From his observations, Van Niel already knew the answer: Photosynthesizing purple sulfur bacteria release sulfur, not oxygen, as a waste product. The process can be summarized as follows:

$CO_2 + 2H_2S \rightarrow (CH_2O)_n + H_2O + 2S.$

Van Niel reasoned that light decomposes the hydrogen sulfide into hydrogen and sulfur. The hydrogen atoms are then used to reduce the carbon dioxide to carbohydrate. It was an easy step to suggest that the same process occurs in green plants, except that water, rather than hydrogen sulfide, is decomposed by light. The hydrogen atoms so released can then be used to reduce carbon dioxide, while the oxygen is given off. Van Niel hypothesized, therefore, that the oxygen given off by green plants during photosynthesis comes from the water molecules, and not those of carbon dioxide.

There is one premise here, of course. It is assumed that, other than the raw materials involved, there is no difference between the photosynthetic process carried out by the purple sulfur bacteria and that performed by green plants. Purple sulfur bacteria are quite primitive organisms; most green plants are not. Therefore because of differences in evolutionary status, it might seem unlikely that the food-making processes in both forms are similar.

On the other hand, the conversion of light energy to chemical energy is truly a remarkable feat for a living system to perform, and so it is even less likely (though by no means impossible) that more than one way to accomplish such a conversion would have evolved. This consideration lends strength to Van Niel's extrapolation from the photosynthetic processes of the purple sulfur bacteria to the same processes

---

*While working on the problem, Senebier made another significant observation. He noted that shredded bits of leaves, when immersed in water and exposed to light, released oxygen just as well as whole leaves. This result indicated that the photosynthetic process is not performed by the leaf as an organ. Today it is known that photosynthesis is carried on within the chloroplasts found in green plant cells.

in green plants. The premise is still present, however, with all its accompanying doubts, and prevents Van Niel's observations of the purple sulfur bacteria from being conclusive.

It often happens that the crucial experiment that would resolve the issue between two conflicting hypotheses cannot be performed until the proper experimental instrument or technique is developed. Such was the case here. The introduction of tracer experiments by George Hevesy (1885–1966) in 1923 opened many new experimental pathways in biology. Hevesy used radioactive isotopes of lead to trace the pathways through which materials moved from place to place in plants.

In 1941, a team of scientists including Samuel Ruben and Martin Kamen at the University of California performed the crucial experiment that determined the source of oxygen released in photosynthesis. They exposed the green alga *Chlorella* to water which had been labeled with O^{18}. This nonradioactive isotope can be detected by a technique known as mass spectrometry. The experiment showed that the O^{18} turned up in the gas released during photosynthesis, while there was no O^{18} in the carbohydrate produced. A second experiment seemed obvious. The carbon dioxide was labeled (CO_2^{18}) and the water was not. In this case, Van Niel would predict that the O^{18} would appear in the carbohydrate rather than in the oxygen released during photosynthesis. The experimental data verified this prediction also. It now seemed clear that the oxygen released during photosynthesis came from the water, not the carbon dioxide.

A modern follow-up of the story indicates the difficulty of getting conclusive evidence in any field of science. The isotopic tracer studies by Ruben, Kamen, and others in 1941 contained one source of error: if cells are supplied with O^{18} as either carbon dioxide (CO_2^{18}) or water (H_2O^{18}), it is possible for there to be an exchange of oxygen atoms between the two molecules. That is, some of the O^{18} supplied in carbon dioxide can be exchanged with the normal oxygen (O^{16}) of the water, and vice versa. Thus, even though isotopic oxygen is administered to plant cells as either CO_2^{18} or H_2O^{18} (though never as both simultaneously), there is no way to guarantee that some exchange does not occur. As a result it is impossible to ensure that isotopic oxygen is being fed to the cells from only one source.

Only in 1975 did a way around this difficulty emerge. Biochemists Alan Stemler in Illinois and Richard Radmer in Maryland knew that the bicarbonate ion (HCO_3^-) plays a critical role in the initial stages of photosynthesis. So critical is this role, in fact, that when the ion is removed more than 90 percent of the oxygen evolution by the cells is stopped. Bicarbonate is a source of CO_2, since the ion interacts with water to produce CO_2. Stemler and Radmer decided to use isolated, broken chloroplasts rather than whole cells for their studies for two reasons: (1) Breaking up the chloroplasts allowed them to remove the enzyme carbonic anhydrase, which catalyzes the exchange of oxygen between CO_2 and H_2O. (2) It also allowed them to remove all bicarbonate ions already present in the chloroplasts. They then supplied the chloroplast fragments with isotopically labeled bicarbonate (HCO_3^{18-}). Oxygen evolution began almost immediately— but the oxygen was not labeled! Even though bicarbonate was the only source of CO_2, it was not the source of the evolved oxygen. The conclusion from this experiment seems inescapable: the source of oxygen evolved during photosynthesis must be water. Van Niel's hypothesis was thus verified in a conclusive way. The oxygen released in photosynthesis by green plants comes from the splitting of water molecules. The hydrogen electrons from the water are picked up by NADP and used to reduce carbon dioxide to carbohydrate.

---

wavelength, the less energy conveyed. The shorter the wavelength, the more energy conveyed.

While the wave theory of light is useful in describing the spectrum, a second theory is necessary for a complete understanding of the role of light in photosynthesis. For, while light may travel as a wave, it interacts with matter as a series of discrete particles or packets of energy called **quanta** or photons. The incoming photons, when absorbed, transfer their energy to the electrons of the absorbing molecules (such as chlorophyll), raising them to higher energy levels.

Much of the small portion of the energy radiated by the sun that interacts with the earth's atmosphere never reaches the surface. The highly dangerous short wave radiations are absorbed by oxygen and ozone (O_3), while carbon dioxide and water vapor screen out much of the long wave radiation. The sunlight that reaches us and green plants, then, is considerably modified from that which is encountered in space travel outside the earth's protective atmosphere.

Human beings and other animals use light energy primarily for vision. The various colors we detect (some animals are partially or totally colorblind) are the result of the various wavelengths the photosensitive portion of our eyes (the retina) absorbs. At one end of this "visible spectrum" of light is red light, which has long waves; at the other is violet, with a short wavelength. Between these extend the other colors we can see—orange, yellow, green, and blue—while beyond the red and violet on either end of the visible spectrum extend the longer and shorter wavelengths we cannot see.

Plants, of course, do not use light for vision; they use it as a source of energy for photosynthesis. They use only portions of the spectrum; certain wavelengths simply pass through (that is, are **transmitted**) while others are **reflected**. For most leaves, our eyes detect the reflected color as green. This wavelength is a significant one to us for our color vision; it is an insignificant one to green plants for photosynthesis.

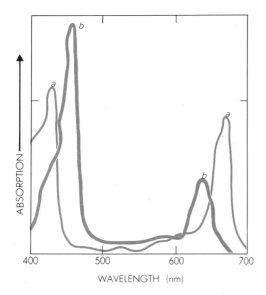

Fig. 7.3
Absorption spectra of chlorophyll *a* and *b*.

The significant wavelengths in photosynthesis, then, are not the ones transmitted or reflected, but those that are **absorbed**. If the most widespread forms of the green pigment chlorophyll (*chlorophylls a and b*) are dissolved in alcohol, and a beam of light passed through the solution, the resulting **absorption spectrum** distinguishes which wavelengths are absorbed (see Fig. 7.3). These, then, are the ones that power photosynthesis.

Chlorophylls *a* and *b*, however, are not the only type of pigment molecules in the plant cell. Besides chlorophyll, there are the pigments whose presence is often masked by the chlorophyll until its production ceases in the fall and the other colors of autumn foliage emerge, such as the orange and yellow *carotenoids* in the plastids and the red *anthocyanins* in the cell sap. Thus when an **action spectrum** of photosynthesis is determined that shows which wavelengths are most effective in performing the actual work of photosynthesis, more of these wavelengths are involved than the absorption spectra of chlorophylls *a* and *b* alone would predict. This suggests that the aforementioned accessory pigments, especially the carotenoids, also absorb some light energy and then transfer this energy to chlorophyll.

7.4
CHLOROPHYLL—THE PRIMARY ABSORBER

As just suggested, the primary key to the absorption of light energy for use in photosynthesis lies within the chlorophyll molecule, especially chlorophyll *a*. If chlo-

rophyll is isolated in a test tube and exposed to light, it re-emits this energy in a visible fluorescence. Clearly then, in the isolated state chlorophyll cannot convert the absorbed energy into a useful, chemical form. When exposed to light within intact chloroplasts, however, the light is *not* re-emitted. Instead some of its energy is captured and used to synthesize energy-rich chemical compounds in the series of reactions known collectively as the light reactions.

How is this energy capture effected? The chloroplast contains several photosynthetic units, each containing 200 to 300 pigment molecules, such as chlorophyll *a*, chlorophyll *b*, and carotenoid (see Fig. 7.4),

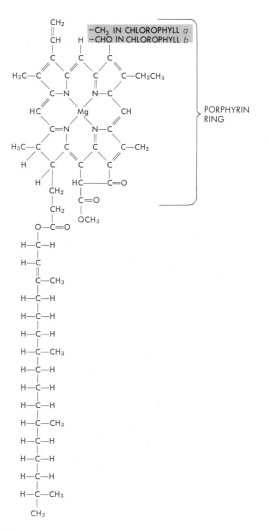

Fig. 7.4
Structural formula of chlorophyll *a*. Note the central magnesium (Mg) atom. Chlorophyll *b* has the same basic structure as chlorophyll *a*, but differs in having a —CHO group instead of the —CH₃ group at the location indicated by the colored circle.

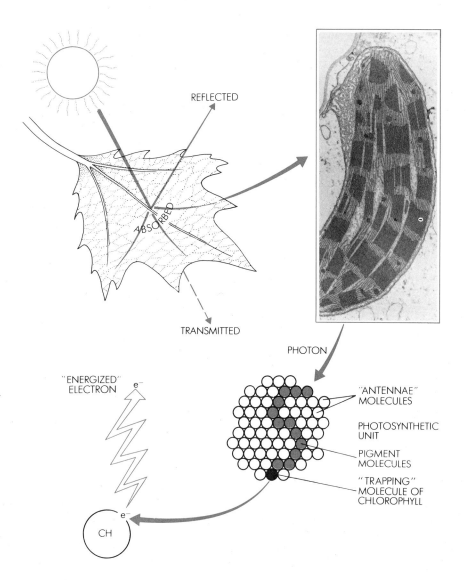

Fig. 7.5
Photons of light absorbed by "antenna" molecules of photosynthetic units within the chloroplasts are eventually captured by "trapping" molecules of chlorophyll *a*. The result is the raising of electrons to higher energy levels, from which they pass through a series of acceptor molecules where their kinetic energy is converted into potential energy stored within chemical compounds.

that act as "antennae" to the incoming radiation (see Fig. 7.5). Within each photosynthetic unit a specialized form of chlorophyll *a* seems to act as a "trap." It is hypothesized that an electron is raised to a higher energy level by the energy introduced into the photosynthetic unit by an absorbed photon, and is passed randomly from one antenna pigment molecule to another until it interacts with this "trapping" molecule. Once trapped the electron does not fall back to its normal energy level, as is the case with isolated chlorophyll. Instead, as in the respiratory chain discussed in Chapter 6, it passes to a series of acceptor molecules where its kinetic energy is converted into the form of potential chemical energy most usable to living cells.

The trapping molecules referred to above are specialized forms of chlorophyll *a*. There are actually two forms of these energy-trapping molecules, which differ only in the wavelengths of light for which they

show maximum absorption (700 nm for one; 680 nm for the other). Each of these molecules represents an entry point into two significantly different biochemical pathways involved in the light reactions of photosynthesis: **photosystem I** and **photosystem II**. The entry point to photosystem I consists of a complex of one trapping molecule of chlorophyll *a* that absorbs, at 700 nm or less (P_{700}), approximately 200 molecules of regular chlorophyll *a* and about 30 molecules of carotenoid. The entry point to photosystem II consists of a complex of one trapping molecule of chlorophyll *a* that absorbs, at 680 nm or less (P_{680}), about 200 molecules of regular chlorophyll *a* and about 200 molecules of chlorophyll *b*. Notice that in both photosystems I and II the energy-trapping molecules P_{700} and P_{680} represent significantly less than one percent of the total molecules in each complex. Their importance, however, in the role they perform in capturing the photon-transmitted energy for

SUPPLEMENT 7.2
CHLOROPHYLL AS THE ENERGY ABSORBER IN PHOTOSYNTHESIS

What evidence justifies the claim that chlorophyll is the energy absorber powering the photosynthetic process? How do we know that the selective absorption of certain wavelengths of light by chlorophyll has any direct connection to photosynthesis? A simple experiment performed as long ago as 1881 by the German plant physiologist T. Engelmann (1843–1909) provides an answer.

Engelmann had been interested in the ability of certain aerobic bacteria to move toward areas of high oxygen concentration within a liquid medium. He had also carried out some studies on photosynthesis in algae. From this work, Engelmann knew that bacteria will concentrate themselves in a region of the medium where oxygen is plentiful—in fact, in direct proportion to the amount of oxygen. The more oxygen present, the more bacteria will assemble there. He thus hit upon an ingenious way to measure the amount of photosynthesis green plant cells can carry on in different wavelengths of light.

Engelmann took a filament of the green alga *Cladophora* and placed it on a microscope slide. He then passed the light source of the microscope through a small prism, so that the white light was spread out into a spectrum. He focused the prism so that it fell along the length of the *Cladophora* filament. Thus, different portions of the filament were selectively exposed to different wavelengths of light. Knowing that photosynthesis releases oxygen as a byproduct, Engelmann reasoned that if different wavelengths of light were absorbed differently by the photosynthetic apparatus within the cells, then different parts of the filament should produce different amounts of oxygen. Oxygen production should be in direct proportion to the amount of the absorbed light of a given wavelength. By introducing bacteria onto the slide, Engelmann provided himself with an observable measure of oxygen production.

After exposing the *Cladophora* filament to the spectrum for some period of time in the presence of bacteria, Engelmann obtained the results shown in part a of Diagram A.

Comparing these results with the absorption spectrum of the pigments found in the *Cladophora* filament, Engelmann noted their exact correspondence (Diagram A, part b). In other words, where the bacteria accumulated in the greatest concentration were found the regions of the spectrum where the pigments were known to absorb most. Conversely, where the bacteria were least concentrated were found regions of the spectrum where the pigments absorbed least. This exact correspondence provided striking evidence to suggest that the plant pigments, chiefly the chlorophylls, are directly involved in absorbing the energy to power photosynthesis.

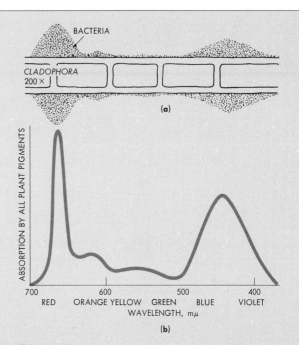

Diagram A

More recent research has confirmed and extended Engelmann's work. Diagram B shows the spectra for three different photosynthetic pigments, the chlorophylls (*a* and *b*), the carotenoids, and phycocyanin. The carotenoids and chlorophylls are present in the cells of all higher plants and the green algae. Phycocyanin is present in some groups of the algae, especially the blue-greens. Note that for the higher plants, the presence of both chlorophyll and carotenoids means that the cells can absorb light energy over a considerable range of the spectrum. If just one of the pigments were involved, the wavelengths that could be effectively absorbed would be greatly restricted. Note that with the three pigments shown in Diagram B, almost all of the visible spectrum is covered by an absorbing pigment. It is possible that

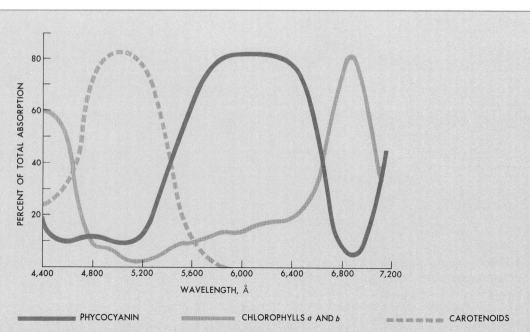

PHYCOCYANIN   CHLOROPHYLLS *a* AND *b*   CAROTENOIDS

Diagram B
Absorption spectra for several important plant pigments found in the green
alga *Chroococcus*. Note that the three major groups of pigments have absorption peaks at three different areas of the visible spectrum. Together, these pigments can absorb light energy from a considerable range of wavelengths.
Chlorophyll *a* and *b* and the caratenoids are found in varying proportions in
green plants, but phycocyanin is not common in the higher forms. (After *Biology Today*, New York: CRM/Random House, 1972, p. 96.)

such a distribution of pigments is especially suited to photosynthesis in the oceans, where there is a mixture of many
kinds of algae: green, blue-green, red, brown, and others.

With their specialized distribution of pigments, these algae
can collectively make maximum use of the solar energy striking the ocean surface.

useful work far outweighs their numerical minority
status.

In green plants both photosystems I and II absorb
light simultaneously and are connected to both enzymatically catalyzed reactions and electron acceptor molecules. Also, both photosystems normally act *in sequence*,
with photosystem II leading into photosystem I (see Fig.
7.6). Finally, in both photosystems the phosphorylation
of compounds is involved, most especially ATP.

However, the ultimate source of energy for the photosynthetic phosphorylation process is neither substrate-level
nor oxidative phosphorylation, as was the case with glycolysis and respiration. In photosynthesis, the ultimate
energy input for phosphorylation is supplied by light and
is referred to as **photophosphorylation**.

Nonetheless, despite the aforementioned characteristics that photosystems I and II have in common, there
are significant differences between them. For this

*Chlorophyll is the primary energy-trapping molecule of photosynthesis in
green plants.*

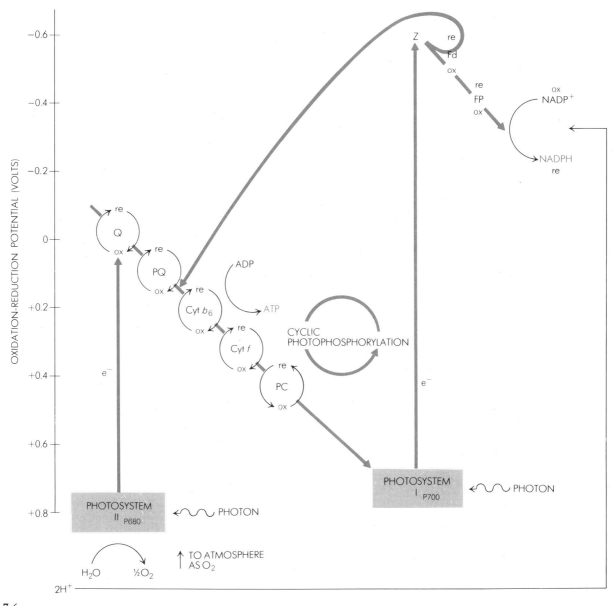

Fig. 7.6
The interrelationship of photosystems I and II. The numbers assigned to the two photosystems refer to the order in which they were worked out and, probably, evolved. Note that water is the primary source for the electrons that flow through the photosystems, and that these enter via P_{680} into photosystem II. The protons (H^+) that result from the loss of electrons become associated with the reduced NADP (NADPH) produced in photosystem I (you may recall that NADPH is often written as NADPH + H^+; see footnote, p. 159). Each water molecule split by the incoming photons yields one oxygen atom ($\frac{1}{2}O_2$), which unites with an oxygen atom released from another water molecule to form the molecular oxygen (O_2) released by the plant into the atmosphere. The two electrons released from each water molecule into photosystem II flow from acceptor Q through a sequence of acceptor or carrier molecules to photosystem I, and a small amount of ATP is produced along the way. These electrons are then raised again to higher energy levels by photons absorbed by the P_{700} of photosystem I. Acceptor Z, an unidentified ferredoxin-reducing substance (often called *FRS*), passes them on to ferredoxin, thereby reducing it. At this point, two pathways are open to the electrons. Some may pass back into the electron carrier sequence of photosystem II, thereby producing more ATP, before returning to reduce ferredoxin again via photosystem I and Z. Therefore this pathway, leading to the phosphorylation of ADP and powered by light, is a true cycle, and thus the term cyclic photophosphorylation. However, the electrons that reduce ferredoxin may also pass to another acceptor, in a pathway that leads to the production of the powerful reducing agent (see Fig. 7.8), NADPH, with which the protons from the water molecules split in photosystem II become associated. This NADPH and the ATP produced in photosystems I and II are used to power the so-called dark reactions of photosynthesis. Note from the oxidation-reduction scale that ATP is produced as the electrons flow through the carriers from a potential of only 0 to 0.4 V. How this is accomplished is not yet known, though it seems likely that the mechanisms of photophosphorylation and oxidative phosphorylation are similar. Note: If you compare this figure with other text sources you will find minor variations in symbols for the carrier molecules and in the number and names. This indicates there is much not yet completely understood about these pathways; thus there is as yet no uniform, agreed-upon digrammatic format. The overall principles *are* generally accepted, however, and it is these that are important to recall here.

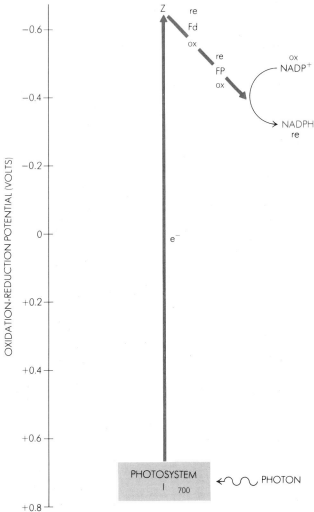

Fig. 7.7
Photosystem I. The electrons raised to higher energy levels in photosystem I lead, via several acceptor compounds, to the formation of reduced NADP (NADPH). For the relationship of photosystem I to the entire process of the light reactions, see Fig. 7.6.

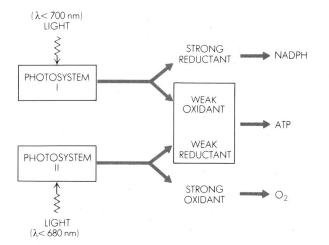

reason, it will be helpful to consider them separately. The discussion of photosystem I, which follows, will refer to Fig. 7.7.

7.5
THE LIGHT REACTIONS: PHOTOSYSTEM I

During the light reactions of photosystem I, incoming photons of light are absorbed by the various antenna pigment molecules. The "excited" electrons are passed randomly from one antenna molecule to another until trapped by P_{700}. From here, the electrons are passed to an acceptor molecule (yet to be clearly identified) labeled Z (in some schemes X, FX, or FRS). Z passes the electrons on to an iron-containing acceptor molecule, ferredoxin (Fd), thereby reducing it. From here, they may pass via an intermediate compound, flavoprotein (FP), to the highly important compound **nicotinamide adenine dinucleotide phosphate (NADP)**. As its name suggests, NADP's molecular structure is very similar to that of NAD, discussed in Chapter 6. However, recall that reduced NAD (NADH) is used mainly in oxidative phosphorylation, with the energy released being used to convert ADP to ATP. Reduced NADP (symbolized NADPH), on the other hand, is able to provide its powerful reducing energy (see Fig. 7.8) directly to those biosynthetic processes of the dark reactions that require large amounts of energy, albeit still via other intermediate compounds.

Another alternative pathway the electrons may follow in Photosystem I will be discussed after consideration of Photosystem II.

7.6
THE LIGHT REACTIONS: PHOTOSYSTEM II

As pointed out earlier, photosystem II is similar to photosystem I; both involve light-energy-absorbing molecules, a sequence of electron-accepting molecules, and the production of energy-rich molecules. In photosystem II, however, the light-energy-trapping molecule is P_{680} (see Fig. 7.9). The energized electrons are passed to a primary acceptor, Q. Acceptor Q represents the

Fig. 7.8
Photosystem I produces a strong reductant that leads to the formation of reduced NADP (NADPH). Photosystem II produces a strong oxidant that leads to the formation of molecular oxygen, O_2. The interaction of the accompanying weak oxidant and weak reductant produced by each photosystem leads to the formation of ATP. Note that both photosystems can be driven by wavelengths of light shorter than 680 nm, but only one of them, photosystem I, by light wavelengths of 700 nm.

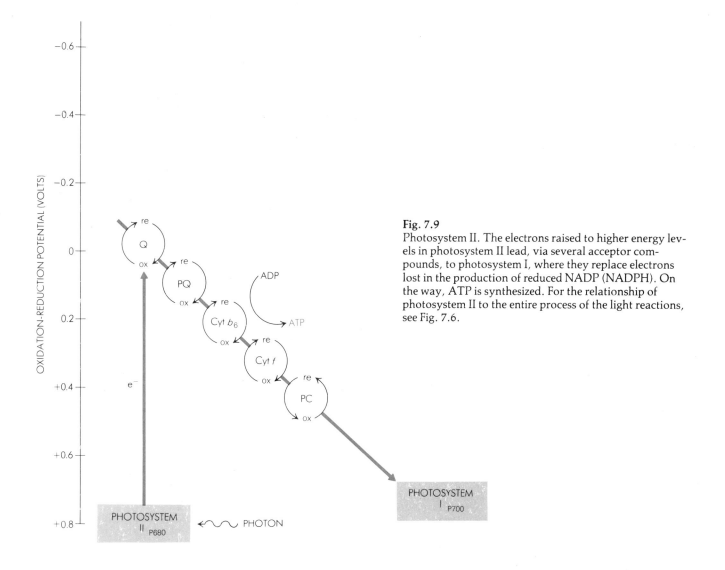

Fig. 7.9
Photosystem II. The electrons raised to higher energy levels in photosystem II lead, via several acceptor compounds, to photosystem I, where they replace electrons lost in the production of reduced NADP (NADPH). On the way, ATP is synthesized. For the relationship of photosystem II to the entire process of the light reactions, see Fig. 7.6.

first of a chain of four electron acceptors which, by accepting the electrons passed to them, become reduced. The energy released is used to synthesize some ATP from ADP (rather than NADPH as is the case with photosystem I). At the end of the electron acceptor series of photosystem II, the electrons feed into the events of photosystem I via its light-energy-absorbing molecule, P_{700}. Thus photosystem II is a noncyclic process; the energized electrons from its P_{680} pass via acceptor Q through a series of intermediate acceptors into P_{700} of photosystem I. This allows photosystem II to replace the electrons lost by the P_{700} of photosystem I when its electrons were energized by light and passed via acceptor Z through the sequence leading to the production of reduced NADP (NADPH). Note that this particular pathway of photosystem I is *also* a noncyclic one.

If the electrons lost in the production of NADPH in photosystem I are replaced by those contributed by photosystem II, this logically leads to the important question: where does photosystem II get the electrons to replace the ones it donates to photosystem I?

The answer for green plants is a simple one: water. In the light reactions, in a manner not yet entirely understood, P_{680} is able to attract electrons away from water and provide the replacement electrons in a process that some feel may involve an as-yet-unidentified intermediate compound and the synthesis of additional ATP. The process can be represented in two steps. First, the water molecule is dissociated into hydrogen ions (protons) and hydroxyl ions (OH^-):

$$H_2O \rightleftarrows H^+ + OH^-$$
(proton)   (hydroxyl ion)

The two products of this dissociation have differing fates. The hydroxyl ions appear to be the source of the electrons that enter photosystem II:

$$4OH^- \rightarrow 2H_2O + O_2\uparrow + 4e \text{ enter photosystem II via } P_{680}.$$

Thus the electrons released by the hydroxyl ions replace those lost by P_{680} in photosystem II. The protons (H^+), resulting from the dissociation of the water molecule, become associated with the NADPH produced in photosystem I. The single oxygen atoms released from each water molecule unite to form the oxygen molecules that green plants release into the atmosphere.

7.7
THE PHOTOSYSTEM SEQUENCE AND CYCLIC PHOTOPHOSPHORYLATION

As has been stressed, photosystems II and I act in sequence, with the energized electrons produced in photosystem II leading via intermediate acceptor compounds to photosystem I (see Fig. 7.6). Only by having two photoactivated processes in series and operating at different energy levels can sufficient energy be provided to remove an electron from the water molecule and deliver it with enough energy to reduce NADP to NADPH as well as produce some ATP (see Fig. 7.8).

Photosystem I, however, has another alternative pathway the electrons may follow instead of following the pathway leading to the production of NADPH. They may return via ferredoxin to the acceptor molecule cytochrome b_6 of photosystem II. Note that this means the electrons pass again through that region of the acceptor chain that results in the production of ATP. Thus additional ATP is generated for chloroplast and cellular requirements. Since this alternative pathway obviously involves a cyclical movement of electrons, with no replacement necessary, it is known as *cyclic photophosphorylation* (see Fig. 7.10).

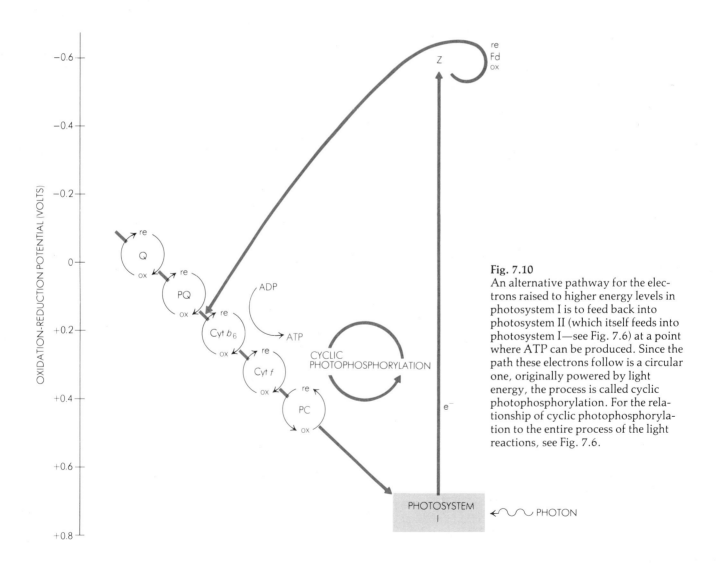

Fig. 7.10
An alternative pathway for the electrons raised to higher energy levels in photosystem I is to feed back into photosystem II (which itself feeds into photosystem I—see Fig. 7.6) at a point where ATP can be produced. Since the path these electrons follow is a circular one, originally powered by light energy, the process is called cyclic photophosphorylation. For the relationship of cyclic photophosphorylation to the entire process of the light reactions, see Fig. 7.6.

SUPPLEMENT 7.3
THE LIGHT REACTIONS: HOW DO WE KNOW?

Several lines of evidence have led over the years to our current understanding of the events in the light reactions.

One of the earliest experiments was that of Robin Hill in England in the mid-1930s. Hill found that isolated chloroplasts extracted from spinach leaves could, when illuminated, cause certain dyes like methylene blue to change color. Methylene blue is an oxidation-reduction indicator; it is blue when oxidized and colorless when reduced. Moreover, the chloroplasts evolved oxygen while reducing the dye. If the dye was not present in the medium, no oxygen was evolved (see Diagram A). From Hill's experiment came several important conclusions:

1. Chloroplasts contain all the necessary elements for splitting water and releasing oxygen. The rest of the plant cell is not necessary for this phase of photosynthesis.

2. When illuminated, chloroplasts have the ability to chemically reduce available electron acceptors. Without these acceptors, no oxygen is evolved.

3. There must be naturally occurring reducing agents in the chloroplast that act analogously to methylene blue.

In the 1950s a team of researchers headed by Daniel Arnon, at the University of California at Berkeley, began more detailed studies of Hill's isolated chloroplast system.

Arnon found that isolated chloroplasts could carry out complete photosynthesis (the evolution of oxygen and the reduction of CO_2 to carbohydrate) if supplied with the following substances: ADP, inorganic phosphate, (P_i), reduced NADP (NADPH), and light. This demonstrated for the first time that chloroplasts are the site of the complete photosynthetic process within cells.

Using this knowledge, Arnon then tried to dissect the various reactions involved by omitting one component at a time and observing the effects. For example, in his first experiment, Arnon supplied the chloroplasts all components except CO_2 (see Diagram B). Without CO_2, chloroplasts could carry out all aspects of photosynthesis except the production of carbohydrate. On the basis of these results, Arnon and others proposed that the photosynthetic process can be divided into two main sets of reactions: the light reactions, involving the splitting of water and the generation of ATP and reduced NADP, and the dark reactions, reduction of CO_2 by products of the light reactions.

Arnon then went on to reason that if photosynthesis was divided into a light and light-independent phase, then some additional predictions could be made. For example, if both CO_2 and reduced NADP were withheld, then the chloroplasts ought to be able to generate only ATP. The results are shown in Diagram C. The prediction was verified. Chloroplasts supplied with only ADP, P_i, and light could generate

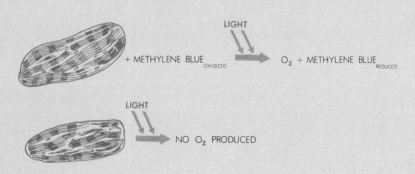

HILL'S EXPERIMENT (1937)

LIGHT

+ METHYLENE BLUE OXIDIZED → O_2 + METHYLENE BLUE REDUCED

LIGHT

→ NO O_2 PRODUCED

Diagram A

ARNON'S FIRST EXPERIMENT: NO CO_2 SUPPLIED

ADP
P_i
NADP+
LIGHT

LIGHT

→ ATP + NADPH + O_2

Diagram B

ARNON'S SECOND EXPERIMENT: NO CO_2 OR NADP SUPPLIED

ADP
P_i +
LIGHT

LIGHT

→ ATP

Diagram C

ARNON'S THIRD EXPERIMENT: NO LIGHT

CO_2 + ATP + NADH +   →   SUGARS + ADP

Diagram D

ATP but could produce neither oxygen nor reduced NADP. This experiment suggested to Arnon that the interaction of light with chloroplasts alone was enough to generate ATP. In terms of electrons, this meant there might be some cyclic flow, since no terminal acceptor such as oxidized NADP ($NADP^+$) was available. The generation of ATP by illuminated chloroplasts was thus seen to be independent of the generation of reduced NADP.

Now Arnon made another prediction—one with far-reaching implications. If the function of the light reactions was to produce ATP and reduced NADP, then chloroplasts kept in the dark ought to be able to produce carbohydrates if supplied with CO_2, ATP, and reduced NADP. He tried the experiment illustrated in Diagram D. The prediction was again borne out: chloroplasts kept from the light, but supplied with products of the light reaction plus the raw material CO_2, were still able to produce carbohydrates. This experiment seemed to confirm that the light reactions generate ATP and reduced NADP, and that these products are used to reduce CO_2 to carbohydrate during the dark reactions.

It will be seen, then, that the light reactions actually involve a flow of electrons from water via P_{680} through photosystem II and photosystem I. From this entry point, the electron flow is toward the production of the high-energy compound NADPH (with an occasional diversional movement through the cyclic photophosphorylation pathway leading to the production of a small amount of ATP). The production of NADPH marks both the significant product of the light reactions and the entry into the so-called dark reactions of photosynthesis.

7.8
THE "DARK" REACTIONS

Like the light reactions, the dark reactions comprise multistep biochemical pathways. And, like the pathways involved in glucose oxidation and photosystem I, part of the dark reactions pathway is cyclical. The term dark is somewhat misleading. Rather than implying that these reactions can only occur in the dark, the true meaning is that the reaction pathways can occur whether or not light is present; the dark reactions are simply *light independent.*

The dark reactions of photosynthesis are concerned with the reduction or "fixing" of environmental carbon dioxide (CO_2) to form energy-rich compounds. In most plants, the reactions consist of what is commonly called the **Calvin cycle**, for the investigator who was awarded a Nobel prize for working out many of the details. The Calvin cycle is also known as **C_3 photosynthesis**, because in the experiments done to work out the cycle using isotope labeling (C_{14}), the initial products of CO_2 fixation appear as three-carbon compounds.

As suggested in the last section, the pathway of CO_2 reduction is powered by the reduced NADP (NADPH) and ATP produced in the light reactions, the NADPH providing both energy and hydrogens (that is, protons and electrons), and the ATP energy only. But how do the electrons of the hydrogen donated by NADPH eventually lead to the addition of electrons onto CO_2, that is, to CO_2 reduction? The events of the Calvin cycle, during which the reduction of CO_2 takes place, are outlined in Fig. 7.11. A molecule of CO_2,

Light energy is absorbed into the biological system when electrons on chlorophyll are raised to higher energy levels and subsequently passed through an acceptor system. Downward transitions during electron transport release energy that is captured as high-energy phosphate bonds of ATP.

In the Calvin cycle, for every six molecules of CO$_2$ that combine with six molecules of RuDP, twelve molecules of PGAL are produced. PGAL is the primary product of the dark reactions. Two PGAL molecules are funneled off: the other ten are re-formed into six molecules of RuDP so that the cycle can begin again.

taken in from the atmosphere through openings in the leaves (stomata), is combined chemically with an already-existing five-carbon compound, *ribulose-1,5-diphosphate*, or *RuDP*, in a reaction catalyzed by the enzyme RuDP carboxylase. The product of this reaction, formed via a hypothetical intermediate compound, is two molecules of the three-carbon compound 3-phosphoglycerate (*phosphoglyceric acid*), or *PGA*. The PGA is then reduced via an intermediate, 1,3-diphosphoglycerate, to glyceraldehyde-3-phosphate, more commonly called *phosphoglyceraldehyde,* or PGAL. Recall that PGAL was also involved in glycolysis (see p. 158). This conversion of PGA to PGAL involves an input of energy from both ATP and NADPH.

Thus phosphoglyceraldehyde, the simplest carbohydrate, is a more energy-rich compound than phosphoglyceric acid. Some of the phosphoglyceraldehyde is converted into the three-carbon sugar dihydroxyacetone phosphate, which combines with some of the remaining phosphoglyceraldehyde to make the six-carbon compound fructose-6-diphosphate. Approximately one-third of the fructose-6-diphosphate is converted into glucose—the end product of the whole synthetic process. The remainder of the fructose-6-diphosphate

and the phosphoglyceraldehyde is involved in a complex set of cyclic reactions that eventually generate more ribulose diphosphate, which can then combine with another molecule of CO$_2$. The pathway is thus a true cycle, because it regenerates itself even while producing a net product that is subsequently funneled off.

7.9
AN ALTERNATIVE PATHWAY: THE PEP OR C$_4$ CYCLE

Whereas C$_3$ plants cannot remove *all* of the CO$_2$ from a sealed system, some plants, originally native to tropical rather than temperate zones, *can* do so. Two examples are sugar cane and corn. Early in this century it was discovered that such plants had certain anatomical features in common not shared by temperate zone plants. From one of these anatomical features—the wreath-like arrangement of the photosynthetic cells around the veins of the leaf—came the name **Kranz anatomy**, from the German word for "wreath" (see Fig. 7.13c). In the mid-1960s, an alternative pathway for photosynthesis was discovered that seemed to be associated with Kranz anatomy plants (though later it was discovered

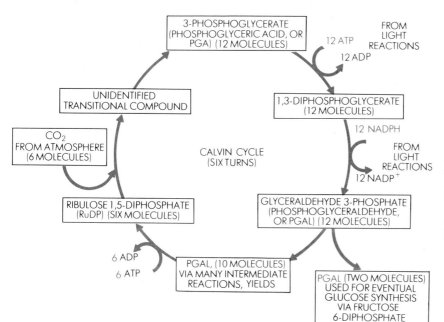

Fig. 7.11
The Calvin cycle. The numbers of each molecule needed for six turns of the cycle (the number required to produce one molecule of glucose) are indicated. At each turn, one molecule of carbon dioxide combines with one molecule of the five-carbon compound RuDP, to produce two three-carbon molecules of PGA. Of the twelve molecules of PGAL formed in six turns of the cycle, ten are converted to RuDP to begin the cycle again, while two are diverted for glucose synthesis. The ATP and reduced NADP (NADPH) that provide the energy to power the Calvin cycle are produced during the light reactions.

SUPPLEMENT 7.4
THE DARK REACTIONS: HOW DO WE KNOW?

The dark reactions of photosynthesis have been described as a cycle involving a number of intermediates that pass through a specific and predictable series of transformations. How has the sequence of events making up the Calvin cycle been determined? Much of the work was pioneered and carried out by Melvin Calvin and his colleagues at the University of California at Berkeley.

Calvin carried out his studies on the one-celled alga *Chlorella*, which is easy to culture and photosynthesizes at a great rate. *Chlorella* are grown in a large tank attached to a flow-coil system as shown in Diagram A. CO_2 is bubbled into the tank, which is constantly exposed to light. Liquid from the tank, containing algal cells, is made to flow through the coils at a fixed rate. At a certain point along the coils, radioactive CO_2, containing the C^{14} isotope of carbon, is injected into the flowing liquid. The cells absorb this CO_2 during the remainder of the flow through the coils. At the bottom, the cells are plunged into a beaker of boiling methyl alcohol (methanol) which kills them instantaneously. The cells can then be homogenized and their contents separated by paper chromatography. The compounds containing C^{14}-labeled carbon dioxide can then be identified by placing the dried chromatographic paper in the dark next to an X-ray film. The film is then developed, giving a typical "autoradiograph" such as that shown in Diagram B. The identity of each spot is determined by comparison with a standard or known chromatogram for each particular type of compound involved. The autoradiograph below, which involved an exposure of *Chlorella* cells to radioactive CO_2 for about 1 min, identifies the major intermediates involved in the dark reactions. It does not, however, reveal anything about the *sequence* in which these intermediates were formed.

Calvin and his associates realized that if the rate of flow of the *Chlorella*-containing medium through the coils was varied, they could increase or decrease the amount of time that cells were exposed to radioactive CO_2. When the rate of flow was increased so that the cells were exposed to CO_2 for only 30 sec, for example, the number of intermediates appearing on the autoradiographs decreased. When exposure was lowered to only 5 sec, still fewer spots appeared (see Diagram C). By varying the flow rate, and by using metabolic inhibitors that block certain steps in the process, Calvin and others were able to deduce the stages of CO_2 reduction. Using such techniques over the course of many years, a number of investigators have been able to piece together the complex series of reactions called the Calvin, or C_3, cycle.

Note from examining the autoradiographs in Diagrams B and C that many of the substances into which radioactive C^{14} finds its way are familiar to us in other contexts than photosynthesis. For example, there are a number of amino acids (aspartic or glutamic) and a number of intermediates in the oxidative pathways of the citric acid cycle (malic, citric, and pyruvic acids). Since molecules entering a cell's biochemical pathways during photosynthesis soon turn up in the building blocks of proteins, carbohydrates, or lipids, it seems clear that nearly all biochemical pathways are interrelated.

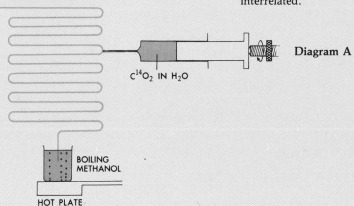

NORMAL AIR CONTAINING CO_2

$C^{14}O_2$ IN H_2O

Diagram A

BOILING METHANOL

HOT PLATE

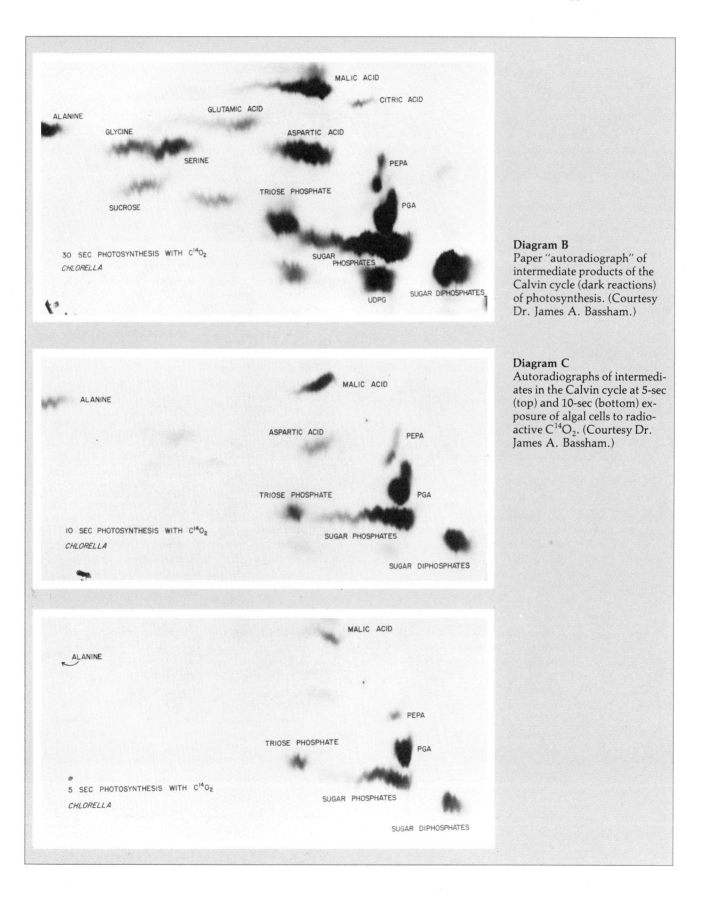

Diagram B
Paper "autoradiograph" of intermediate products of the Calvin cycle (dark reactions) of photosynthesis. (Courtesy Dr. James A. Bassham.)

Diagram C
Autoradiographs of intermediates in the Calvin cycle at 5-sec (top) and 10-sec (bottom) exposure of algal cells to radioactive $C^{14}O_2$. (Courtesy Dr. James A. Bassham.)

SUPPLEMENT 7.5
PHOTOSYNTHESIS AND THE CARBON CYCLE

As we have seen in this chapter, carbon dioxide (CO_2) is one of the two main raw materials of photosynthesis, the other one, of course, being water. Plants obtain this CO_2 from the atmosphere and their own respiration.

Compared to the relatively large amounts of such gases as nitrogen and oxygen, the percentage of CO_2 in the atmosphere is quite small, about .03 percent by volume. Yet this small amount is critically important. CO_2 absorbs the portion of the sun's radiant energy that comes at infrared wavelengths, thereby trapping heat. Should the amount of CO_2 in the atmosphere increase significantly, the result would be a warming of the earth's climate. Since 1850, such an increase in atmospheric CO_2 has occurred. During this period, the amount of carbon dioxide in the atmosphere has increased from approximately 290 parts per million (ppm) to 330 ppm.

The main hypothesis put forth to account for this increase is that it results from the industrial revolution of the nineteenth century. The burning of fossil fuels (first coal and, later, petroleum products such as oil and gasoline), was seen as the major reason for this increase in atmospheric carbon dioxide or, as it is often put, decrease in "fixed" carbon.

What does it mean to "fix" carbon? It means to secure it within organic molecules—carbohydrates, fats, proteins, and the like—as opposed to having it free as part of a gas such as CO_2. Since green plants are the major group of living organisms that can fix the carbon of carbon dioxide into such compounds, thereby passing it on in a fixed form to other, nonphotosynthesizing organisms, they obviously represent a major factor affecting atmospheric carbon dioxide concentrations. Recently, their importance in doing so has become still more evident.

It was stated earlier that since 1850 the amount of CO_2 in the atmosphere has increased from 290 to 330 ppm, and that the increase has been hypothesized to be due to the industrial revolution. There is little doubt that the industrial revolution was, indeed, a major factor. But is it still as significant today? The industrial revolution hypothesis predicts a leveling off of the increase of atmospheric CO_2 over the past few years (as the amount of fossil-fuel-burning industries has not increased appreciably) yet perhaps one-fourth of the total atmospheric CO_2 increase—about ten ppm—has occurred in the past ten years. If the current trend continues, the amount of CO_2 in the atmosphere at the end of another forty years could be twice what it is now.

It was mentioned earlier that green plants are a major source of carbon fixation. Within the roots, stems, trunks, bark, leaves, flowers, and so forth of plants, vast amounts of carbon remains locked—or fixed—into position. If it remains so, no free carbon as carbon dioxide is released into the atmosphere, other than the small amounts released by the plants and animals as a result of their respiration.

But in increasing amounts, this carbon is *not* remaining fixed. Instead, it is being released into the atmosphere as CO_2.

Why is this occurring? There is increasing evidence that the answer is deforestation. The exponentially increasing human population calls for more and more land to be put to use growing food crops. Thus, in alarming amounts, forest trees are being cut down and replaced with food plants that have nowhere near trees' ability to fix carbon. If the trees are burned for fuel, as they often are, their fixed carbon is immediately released. If the dead trunks, branches, and leaves are simply left on the ground, precisely the same thing occurs through the natural forms of oxidation we call decay, though the fixation processes of bacteria and other decay organisms make the loss significantly lower. In either case, however, a vast amount of carbon fixed by these plants ceases with their death, and the carbon they fixed while alive enters the atmosphere as CO_2 (and some other gases).

It was thought until relatively recently that the largest amount of net primary production occurred in the ocean by the algae that inhabit it. (Net primary production is the net amount of fixed carbon or organic matter left from photosynthesis after the needs of the plant for respiration are deducted). It now appears evident that this is *not* the case. Rather, the forests of the world have the highest total net primary production. Further, the tropical forests of the world are the most important in this regard; precisely the ones, unfortunately, that human population pressures are causing to be cut down to make more land available for food production.

For world-wide food production, deforestation is a short-sighted policy. The current rate of tropical deforestation—about one percent annually—could lead to a dramatic increase in atmospheric CO_2 content. In turn, because of CO_2's infrared absorption ability, this deforestation could significantly raise the earth's temperature. At first glance this may seem attractive—we generally associate the growing of food crops, to say nothing of pleasant weather, with warmer temperatures. However, the result would be to move the desert zones of the world toward the poles, *reducing* areas suitable for agriculture. Thus less, not more, food production would result. Considering projections of a doubling of the human population in the next three to four decades, the prospect, if any of these projections are correct, is not an encouraging one.

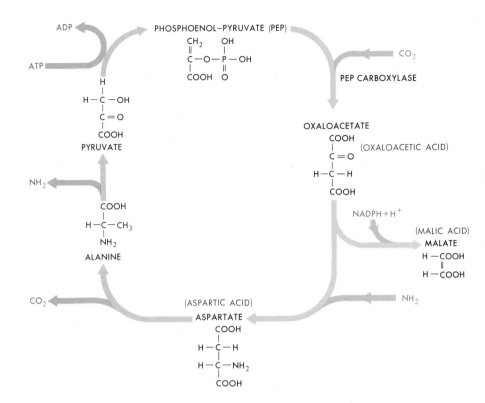

Fig. 7.12
The phosphoenol-pyruvate (PEP) cycle (also called the C$_4$ cycle, since four-carbon intermediates are involved). The PEP cycle is an alternative means of trapping CO$_2$ for reduction in the dark-reaction Calvin cycle. The CO$_2$ is captured by PEP to form oxaloacetate. Oxaloacetate is converted into either aspartate or malate, either of which can, by losing the CO$_2$, deliver it to the Calvin cycle.

that this association was not an *entirely* consistent one). This **Hatch-Slack pathway**, named after two of the primary investigators, is also known as the **C$_4$ photosynthesis** from the fact that it has four-carbon compounds, such as oxaloacetate, produced as the first products of CO fixation, rather than the three-carbon compounds that characterize the Calvin, or C$_3$, cycle. Alternatively, the C$_4$ cycle is often called the *PEP cycle*, after the compound phosphoenol pyruvate which, when combined with carbon dioxide, gives rise to oxaloacetate (see Fig. 7.12).

The C$_4$ cycle is particularly well adapted to the functioning of a desert plant and works in conjunction with the Calvin cycle. Desert plants, such as some members of the goosefoot family (*Atriplex*) of the southwestern United States, face two difficult and contradictory needs during the daylight hours. On the one hand, desert plants need to restrict water loss by keeping their stomata partially closed, since most water loss occurs by evaporation through these stomatal openings. When the stomata are partially closed, however, very little CO$_2$ can enter the leaf from the outside; hence, the rate of photosynthesis is greatly retarded. Under such conditions most plants, if they could survive at all, would have so low a concentration of CO$_2$ in their leaf cells that photosynthesis would proceed at a snail's pace. Yet *Atriplex*, with its stomata partially closed, photosynthe-

sizes at a maximum rate in the hottest part of the day, due to the remarkable characteristics of the PEP cycle.

What CO$_2$ is available is initially picked up not by RuDP (as in C$_3$ plants), but by phosphoenol pyruvate, as shown in Fig. 7.12. This reaction is catalyzed by the enzyme PEP carboxylase. Here is the first place we find an interesting feature of the C$_4$ system. PEP is far more reactive with CO$_2$ than is RuDP. In addition, whereas RuDP carboxylase is somewhat inhibited by high oxygen concentration, PEP carboxylase is not. This means that the PEP system is more effective at picking up CO$_2$ than the RuDP system when the proportion of carbon dioxide is low and that of oxygen is high. Together, these two features of the C$_4$ system mean that in low CO$_2$ concentrations, such as would prevail when the stomata are closed, PEP will take up CO$_2$ whereas RuDP will not.

However, none of the products of the C$_4$ cycle are useful in photosynthesis. Thus, the C$_4$ cycle itself does not lead to photosynthetic fixation of CO$_2$. But the C$_4$ cycle *does* facilitate the incorporation of CO$_2$ into the Calvin cycle. Figure 7.13(a) and (c) show the general structure of the leaves of two types of plants: (a) *Atriplex patula*, which grows in less dry environments and has only the C$_3$ (Calvin) cycle for CO$_2$ fixation; (c) the leaf of *Atriplex rosea*, a close relative, which grows in very hot places and utilizes both the C$_3$ and C$_4$ systems.

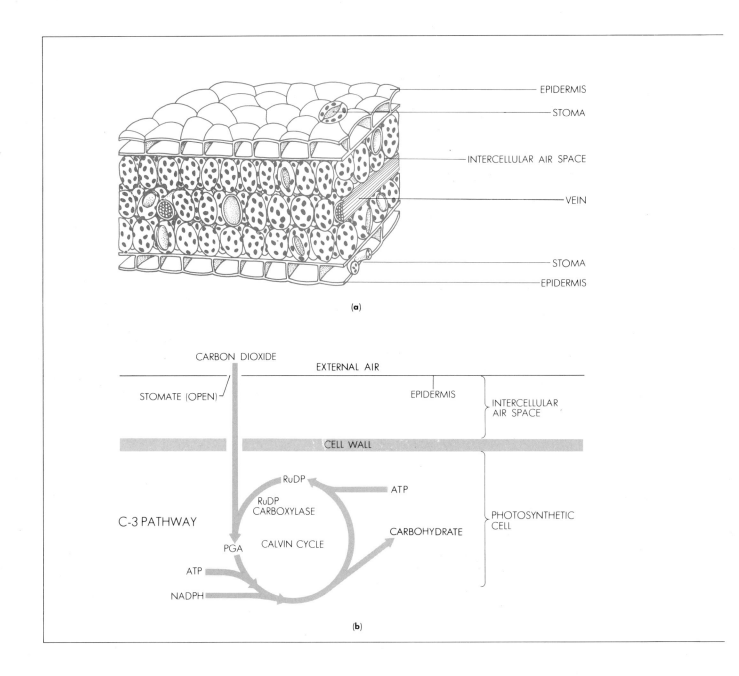

Fig. 7.13

(a and b) Diagram of leaf section (above) and photosynthetic pathway localization (below) of the desert plant *Atriplex patula*, a C_3 plant. This species lacks the enzymes for the PEP cycle; it also lacks cellular differentiation into mesophyll and bundle sheath layers of the leaf. (c and d) Diagram of leaf section (above) and photosynthetic pathway localization (below) in *Atriplex rosea*, a C_4 plant. *A. rosea* has both the C_4 (PEP) and the C_3 (Calvin) cycles. These are localized in different layers of cells. The PEP system is found in the mesophyll cells, near the outer surface of leaf, and near the stomata; the Calvin system is localized in the bundle sheath layer, deeper within the substance of the leaf, and surrounding the vascular tissues. The PEP system allows *A. rosea* to pick up and utilize CO_2 molecules even when the latter are in low concentration within the cellular fluids of the mesophyll layer. Thus, even on extremely hot days in the desert, *A. rosea* can carry on a high rate of photosynthesis; it simply uses every available CO_2 molecule, something its close relative *A. patula* (which lacks the PEP system) cannot do. (Diagrams after Olle Björkman and Joseph Berry, "High efficiency photosynthesis," *Sci. Am. 229*, no. 4, October 1973, p. 80.)

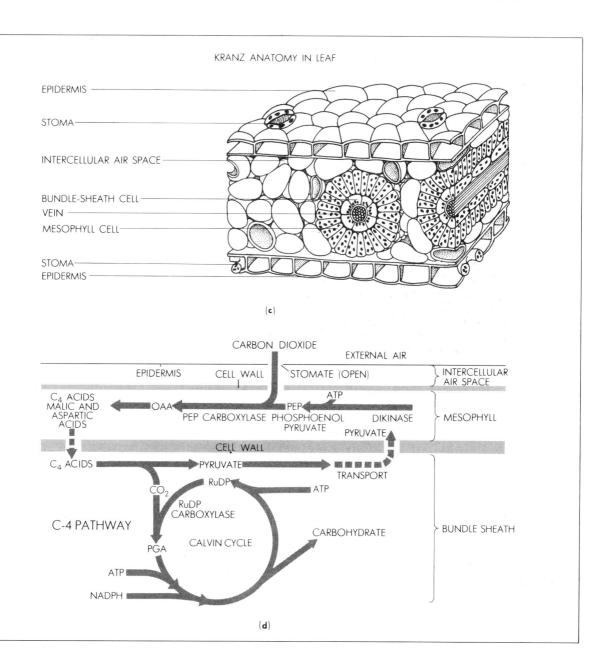

KRANZ ANATOMY IN LEAF

EPIDERMIS

STOMA

INTERCELLULAR AIR SPACE

BUNDLE-SHEATH CELL
VEIN
MESOPHYLL CELL

STOMA
EPIDERMIS

(c)

CARBON DIOXIDE

EXTERNAL AIR

EPIDERMIS   CELL WALL   STOMATE (OPEN)   INTERCELLULAR AIR SPACE

C_4 ACIDS
MALIC AND
ASPARTIC
ACIDS
← OAA ← PEP CARBOXYLASE   PHOSPHOENOL   DIKINASE
PYRUVATE
ATP
PEP
PYRUVATE   MESOPHYLL

CELL WALL

C_4 ACIDS → PYRUVATE → TRANSPORT

CO_2
RuDP
RuDP
CARBOXYLASE
ATP

C-4 PATHWAY

CALVIN CYCLE

PGA

ATP

NADPH

CARBOHYDRATE

BUNDLE SHEATH

(d)

Here emerges the second interesting feature of the C_4 system. The enzymes for the cycle are localized in the cells of the mesophyll, a layer of cells just beneath the epidermis of the leaves of most plants (see Fig. 7.13a and c). The mesophyll cells are close to air spaces, the openings to which are regulated by the stomata. Some investigators believe this arrangement means that two intermediates in the C_4 cycle, malate and aspartate, are easily transportable from the mesophyll into cells of a deeper layer in the leaf, the bundle-sheath cells, which contain enzymes for the C_3 Calvin cycle, and there is strong observational evidence to indicate that such transport does, indeed, take place.

In the bundle-sheath cells, malate and aspartate are converted into pyruvate by the removal of the CO_2 that had just been added. The pyruvate is then transported back into the mesophyll cells, where it is converted into PEP, thus rounding out the cycle. Meanwhile, the CO_2 delivered into the bundle-sheath cells enters the Calvin cycle by the usual path of combining with RuDP. The C_4 system picks up CO_2 at low concentrations and delivers it to where it can be used in the dark reactions.

It should be pointed out that cells of all green plants contain enzymes for the C_3 Calvin cycle. C_4 plants designated C_3 types engage in the C_4 cycle *in addition to* the C_3 Calvin cycle. The plants in which the C_4 system is found are precisely those species that thrive in dry, hot, and sunny regions. The C_4 system is a photosynthetic adaptation to hot, dry conditions under which the stomata of the plants are mostly closed. Accompanying the biochemical and physiological differences between C_3 and C_4 types of plants are anatomical differences. C_3 plants do not have the cell layers of the leaf differentiated into mesophyll and bundle sheath as do C_4 plants, but rather tend to have cells more or less resembling the mesophyll types. Biochemical and anatomical characteristics are clearly interrelated here.

7.10
PHOTORESPIRATION

It would seem obvious that no biochemical pathway would evolve that could "short circuit" another important biochemical pathway and prevent it from carrying out its function. Yet it seems that a nonspecificity of the Calvin or C_3 carboxylation (CO_2-adding) mechanism *does* allow such a phenomenon to occur. Rather than undergoing carboxylation and continuing in the cycle to produce PGAL for the synthesis of glucose, the high-energy intermediate RuDP may alternatively undergo addition of O_2 rather than CO_2. This oxidation produces glycolic acid or glycolate, which is then broken down to produce CO_2. This process, called **photorespiration**, is clearly counterproductive, since it leads to a net loss of carbon as CO_2, rather than an incorporation of carbon into the "backbone" of energy-rich compounds.

It should be emphasized that photorespiration is not the same as the ordinary respiration carried out by all living organisms, both plant and animal. Such respiration occurs in the mitochondria. Since photorespiration is not stopped by inhibitors that block mitochondrial respiration, it seems clear that it is a quite different biochemical process. It also does not yield any high-energy compound such as ATP.

Why does photorespiration occur? It has been hypothesized that CO_2 and O_2 compete for the active site of the enzyme that catalyzes either the carboxylation or oxidation of RuDP. Which one wins the competition depends upon the environmental conditions. If the temperature is high, for example, oxidation is favored. Thus if the leaves of C_3 plants get too hot, the production of photosynthesis slows or ceases. As would be predicted in cases where competitive inhibition is hypothesized to occur, concentration of the competing molecules is a factor; if CO_2 concentration is high and O_2 concentration low, carboxylation occurs and carbohydrate synthesis occurs. If the reverse conditions prevail—low CO_2 and high oxygen concentration—oxidation predominates and the photosynthetic process is thwarted. Indeed, a historically significant observation in the discovery of photorespiration was that C_3 plants

cannot remove all of the CO_2 from a sealed system and in fact *release* CO_2 if placed in a CO_2-free atmosphere.

What function could such a seemingly counterproductive process serve? Or, since in some C_3 plants the loss of photosynthetic production may be as great as 50 percent, how could such a "losing" system evolve? It is interesting to note that the C_4 pathway appears to allow Kranz anatomy plants to avoid photorespiration. Since so much of the world's food production is due to the photosynthetic activity of C_3 plants, the possibility of producing C_3 plants with the photorespiration avoidance mechanism of C_4 plants is a tempting idea, if perhaps still a rather remote one.

CAM Plants

CAM plants (the letters standing for *crassulacean acid metabolism*) are adapted to desert environments with extremes of heat during the day and coolness at night and very little water. The best known CAM plants, perhaps, are those coming under the general name of "cactus." The main functional characteristic of CAM plants is that they obtain their carbon dioxide by opening their stomata at night, when the loss of water is not so great. The CO_2 is stored as organic acids. During the day, the stomata close and the CO_2 is removed from the acids and undergoes C_3 photosynthesis. While CAM plants do not have high rates of photosynthesis, they are highly efficient in their use of water, as would be expected considering the arid environments with which they are generally associated.

7.11
LOCALIZATION OF THE
PHOTOSYNTHETIC COMPONENTS

Both the photosynthetic pigments and enzymes involved in the light reactions and the enzymes involved in the dark reactions are found within the plant cell organelles known as chloroplasts (see Fig. 5.10). They are found, however, both in different locations and circumstances. To understand these differences, it will first be necessary to describe the ultrastructure of chloroplasts as revealed by electron micrographs and other techniques for studying cell structure described in Chapter 4. It will be helpful to refer to Fig. 7.14 as the parts are described.

As shown in Chapter 4, chloroplasts when cut in sections are revealed to be an orderly assembly of membranes, called lamellae. Surrounding these lamellae is a medium in which they "float," the stroma. Closer inspection of the lamellae reveals that they are not uniform; at certain regions they appear to be flattened or folded into disc-shaped structures piled on top of each other like a stack of pancakes. Each of these discs is called a thylakoid; each pile of thylakoids, a granum (plural, *grana*). Recent evidence makes it appear likely that the thylakoids are interconnected via channels within the grana in a highly complicated manner. In turn, the grana themselves are interconnected via the intergrana or stromal lamellae.

Electron micrographs of freeze-etched thylakoid membranes reveal their inner surfaces to be roughened with spherical granules arranged in a rather regular pattern (see Fig. 7.14). Many investigators think that it is on these granules that the light reactions with their associated photosystems I and II occur, and that it is here that the electron transport molecules are bound. Electrons removed from P_{700} or P_{684} can thus be picked up readily by the various acceptor molecules because of the spatial relationships provided by their being bound to the thylakoid membranes. The end products of the light reactions, NADP and ATP, then pass into the stroma, where the enzymes for the dark reactions are located, to be used to produce carbohydrate. Thus the enzymes for the dark reactions, the Calvin cycle, are not membrane-bound, but are instead freely suspended in the liquid of the stroma. Because the whole chloroplast is surrounded by a membrane, these enzymes are kept localized within a very small space in the cell. They are available therefore to complete the job of photosynthesis started by the light reactions. The products of the light reactions—ATP and NADPH—can be used directly in the reduction of carbon dioxide in the free-floating stroma of the chloroplast.

The pattern of localization in the chloroplast is strikingly similar to that found in mitochondria, where the enzymes for electron transport and phosphorylation are bound to the cristae, while the enzymes for the citric acid cycle are free-floating within the liquid matrix of the mitochondria. Strong evidence suggests that the processes of photosynthetic phosphorylation and oxidative phosphorylation are very similar and may well have evolved from a common biochemical pathway. Such thinking is consistent with a hypothesis proposing that chloroplasts and mitochondria have evolved from a common ancestral organelle.

It should be noted that photosynthesis is not the only metabolic activity of chloroplasts, and that many other enzymes are present in the stroma. In particular, the primary reactions resulting in the synthesis of amino acids from inorganic nitrogen occur within chloroplasts, as do those resulting in the synthesis of more complex molecules such as proteins and lipids (see Section 7.14). These processes are driven by the same high-energy compounds produced in photosynthesis that are used to drive the reduction or fixation of CO_2 and carbohydrate synthesis.

7.12
CHEMIOSMOSIS AND PHOTOSYNTHESIS

In Chapter 6, both oxidative and substrate-level phosphorylation were discussed; here in Chapter 7 we have been concerned with photosynthetic phosphorylation (photophosphorylation), both cyclic and noncyclic. No matter what the type of phosphorylation, however, the results are the same: the phosphorylation of ADP to yield ATP. The only difference is the source of energy, glucose molecules instead of quanta of light (photons). It is not surprising, therefore, that the three hypotheses put forth to account for the phosphorylation that occurs in mitochondria are also appropriate for photophosphorylation. Indeed, as was pointed out in Chapter 4, significant experimental evidence supporting one of these—Mitchell's chemiosmotic theory—comes from experiments involving chloroplast membranes.

There is one significant difference, however. In mitochondria, the protons (H^+) move from the inner (M) to outer (O) region during electron transport; in chloroplasts, the movement is in the opposite direction, from outer to inner. When chloroplasts are exposed to light under experimental conditions that cause cyclic photophosphorylation, protons move from the surrounding medium (thereby making it less acidic) across the chloroplast membranes. When the light is turned off, the electrochemical balance is slowly restored, indicating that the protons are leaking back into the medium. Using very short light pulses (20 nanoseconds, ns) it is possible to get only a single electron pair "turnover" in the electron transport system. By such methods, the German biophysicist H. Witt and his colleagues have been able to establish very precise quantitative relationships between electron flow, proton movements across the chloroplast membranes, and ATP formation. Such observations provide still further support for the chemiosmotic hypothesis.

So, too, does the photosynthetic activity of the purple sulfur bacteria first mentioned in the work of C. B. Van Niel in Supplement 7.1. Using a pigmented molecule called bacteriorhodopsin (a combination of a protein with retinal, the visual pigment of the vertebrate eye) instead of chlorophyll, these bacteria are able to use light energy to form a proton gradient that is then used to synthesize ATP. It appears that no matter what energy-trapping molecule is used, the chemiosmotic mechanism becomes involved in the actual process of ATP synthesis.

7.13
LIMITING FACTORS OF PHOTOSYNTHESIS

While the role of light and the availability of the raw materials water and carbon dioxide have necessarily been stressed in this chapter, it should be obvious that there are many other factors that play a role in determining whether or not photosynthesis occurs at maximum efficiency if, indeed, it happens at all. Clearly, even with the optimum amounts of light and carbon dioxide concentration available, extreme cold or heat will prevent any photosynthesis from occurring. So, too, will a soil not properly enriched with minerals needed by the plant. All of these factors (light,

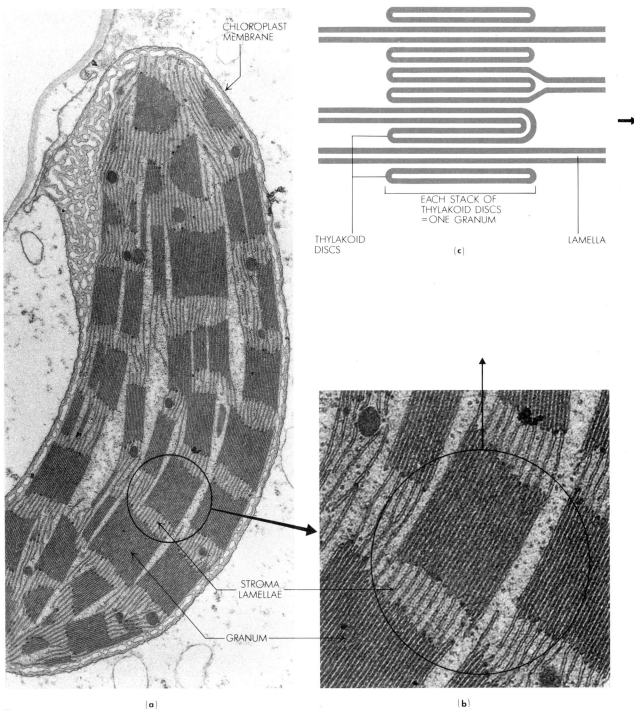

CHLOROPLAST
MEMBRANE

EACH STACK OF
THYLAKOID DISCS
= ONE GRANUM

THYLAKOID
DISCS

LAMELLA

(c)

STROMA
LAMELLAE

GRANUM

(a)   (b)

Fig. 7.14
Detailed structure of chloroplasts. (*a*) Single chloroplast from a leaf of *Froelichia*, showing chloroplast membrane and the internal membrane system composed of lamellae running lengthwise through the liquid matrix, the stroma. Periodically the lamellae are folded into a series of flat discs called thylakoids, which are stacked on top of one another to form grana. (Courtesy L. K. Shumway.) (*b*) Detail of granum showing stacking of thylakoid discs. (Courtesy L. K. Shumway.) (*c*) Schematic drawing of interrelationship between lamellae and the thylakoid discs. (After A. L. Lehninger, *Bioenergetics*, 2d ed. Menlo Park, Calif.: W. A. Benjamin, 1973, p. 120.) (*d*) Three-dimensional model of interrelationship between lamellae running through stroma and folded lamellae of the thylakoid discs. (Photo courtesy J. E. Heslop-Harrison, from "Structural features of the chloroplast," *Science Progress*, 54, 1966, 519–541). (*e*) Electron micrograph of several layers through a granum, showing granules, thought to contain the various pig-

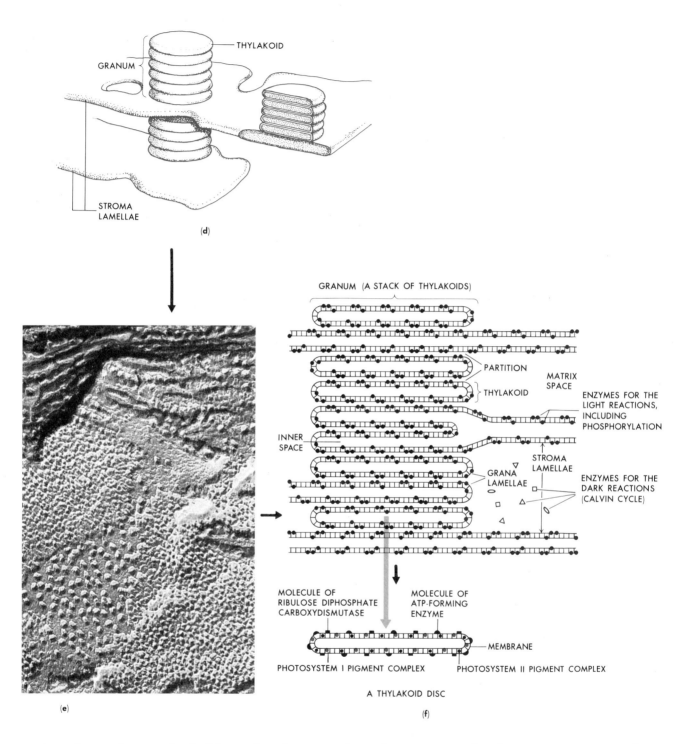

(d)

GRANUM (A STACK OF THYLAKOIDS)

PARTITION

THYLAKOID

MATRIX SPACE

ENZYMES FOR THE LIGHT REACTIONS, INCLUDING PHOSPHORYLATION

INNER SPACE

GRANA LAMELLAE

STROMA LAMELLAE

ENZYMES FOR THE DARK REACTIONS (CALVIN CYCLE)

MOLECULE OF RIBULOSE DIPHOSPHATE CARBOXYDISMUTASE

MOLECULE OF ATP-FORMING ENZYME

MEMBRANE

PHOTOSYSTEM I PIGMENT COMPLEX

PHOTOSYSTEM II PIGMENT COMPLEX

A THYLAKOID DISC

(e)

(f)

ment complexes and enzymes for the light reactions. Larger granules are the quantosomes, thought to contain pigment complexes. Smaller granules may be various enzyme systems for photophosphorylation and electron transport during the light reactions. (Courtesy Dr. R. B. Park.) (f) A hypothetical scheme for how the localization of various enzymes and pigment complexes could be distributed on the lamellae and within the grana. The enzymes and pigment molecules for the light reactions are bound to the membranes running through the chloroplasts, especially those concentrated in the grana (the thylakoid discs). The enzymes for the dark reactions (the Calvin cycle) are free-floating in the stroma, between the lamellae. (Diagram (d) after A. Trebst, "Energy conservation in photosynthesis electron transport of chloroplasts," *Annual Review of Plant Physiology* 25, 1974, p. 426. Diagram in (e) from A. L. Lehninger, *Bioenergetics*.)

temperature, carbon dioxide or nutrient concentration, and more) represent the *limiting factors* of photosynthesis; lacking any one of these, photosynthesis is slowed or stopped. Any farmer is familiar with the principle of limiting factors. Seeking maximum production from their crop plants, farmers are well aware that, no matter how many hours are spent in weeding, balancing fertilizer amounts, and the like, all may be for nought if there is a drought or too much rain; irrigation, for example, merely represents an attempt to bring the former of these natural limiting factors under human control.

The limiting factors to which any species of plant may be subjected vary widely from one environment to another. Such variations, over long periods of time, result in the adaptation of plants to whatever limiting factor or factors may be most prevalent in its environment. The result is to maximize the plant species' photosynthetic production under these conditions. In the desert, for example, two obvious limiting factors faced by green plants are wide temperature variations and dryness. Anyone who has ever seen a cactus will recall the thickness of the plant's green portions; the resulting low surface area per unit volume undoubtedly is important in cutting down the loss of precious water via evaporation. Thus cactuses are well adapted to the ecology of the region in which they live.

In terms of photosynthesis, so too are other plants. In some cases, this adaptation can be seen in an individual plant; an oak tree, for example, may have small, thick, highly lobed leaves at the top, where sunlight is abundant, and progressively larger, thinner leaves on the lower branches where the leaves tend to be shaded by those above. In other cases, the anatomical variation for efficient light capture is spread among several species; in a tropical rain forest, for example, each plant species has a characteristic leaf size, shape, coloration, or thickness that best adapts it to the region of the forest in which it lives.

7.14
CHEMOSYNTHESIS

As we have seen with the C_3 and C_4 systems, even within photosynthesis there is considerable variation on a theme. An even more radical departure is shown by the purple sulfur bacteria, as discussed in Supplement 7.1. These organisms use hydrogen sulfide (H_2S) rather than water (H_2O) as a source of electrons, thereby releasing sulfur rather than oxygen as a byproduct. Since light is still the source of energy, however, this activity of the purple sulfur bacteria is still a photosynthetic one. Indeed, even green plants, if experimentally prevented from using water as an electron source yet provided with other strong electron-donating compounds, continue noncyclic photophosphorylation and, like the purple sulfur bacteria, do not release oxygen as a byproduct.

There are some bacteria, however, that are capable of synthesizing organic compounds with high potential chemical energy *without* using light as their energy source. Such bacteria, found widely in the soil, are able to oxidize inorganic substances such as ammonia (NH_4^+), nitrate (NO_2^-), or even sulfur. The small amount of energy released is used to synthesize carbohydrate. Such **chemosynthetic** organisms are but a drop in the bucket in terms of a total carbohydrate synthesis comparison with green plants. However, because of the role such bacteria play in producing fertile soils for photosynthesizing green plants, their importance cannot be overestimated.

7.15
OTHER FORMS OF SYNTHESIS

Photosynthesis is such a unique and important process that it often tends to obscure the fact that green plants carry out many other biosynthetic reactions as well. Indeed, the "uphill" or anabolic reactions that provide such compounds as lipids, proteins, nucleic acids, and the like, upon which most of our food needs and those of other animals rely, are also carried out largely by plants. For example, the synthesis of certain amino acids absolutely essential for our health is carried out by many plants, as is the synthesis of lipids. In the case of lipids, the fatty acid portion of the molecule is synthesized from acetyl CoA (acetyl coenzyme A; see Fig. 7.18) by a process involving both ATP and NADPH, while the glycerol portions are synthesized from triose phosphate produced either by photosynthesis or by the catabolic breakdown of glucose.

The metabolic pathways of the synthesis of several types of lipid-containing substances are shown diagrammatically in Fig. 7.15. Such reactions, in plants as in animals, are chemical-energy-requiring processes rather than chemical-energy-producing, as is the overall case wth photosynthesis.

7.16
THE EFFICIENCY OF PHOTOSYNTHESIS

The amount of solar energy that strikes the surface of the earth every day is the equivalent of 100,000,000 atomic bombs of the size dropped on Hiroshima. Some of this energy evaporates water in the seas, water which may later fall back on the land and form sources of potential energy as it runs downward through streams and rivers on its way back to the sea.

Of much greater significance, however, is the solar energy captured by green plants. The fall of excited electrons in chlorophyll is far more important than all the waterfalls on earth. The trapping of chemical energy resulting from carbon dioxide fixation constitutes virtually all of the fuel for living matter. Without this constant conversion of light energy into a chemical form us-

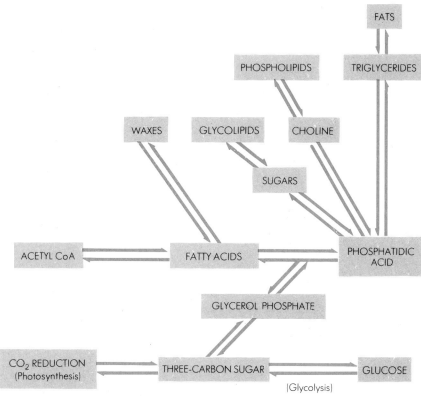

Fig. 7.15
Some of the pathways from the biosynthesis of lipids and their byproducts. Many intermediates are not shown. Note that lipids can feed into the Krebs citric acid cycle via acetyl CoA (compare with Fig. 6.18).

able by all kinds of organisms, virtually all life on earth would perish. For, directly or indirectly, most non-photosynthesizing organisms are dependent upon the photosynthesizers for the fuels that keep them going. Thus Joseph Priestley's original eighteenth-century concept of the need for plant life to "render air fit for breathing again" has been enlarged to include many other needs as well.

Since the photosynthetic process is of utmost importance to all forms of life, including humans, the efficiency of that process must also be a matter of great importance. There are several ways of looking at the subject. It can be noted, for example, that of the sun's energy reaching the earth (most is wasted by being radiated elsewhere in space as pointed out earlier), only one part in 2000 is actually captured by photsynthesizing plants. This very low efficiency would be far lower were it not for the fact that about 70 percent of the earth's surface is covered with water. Though we often tend to think of the familiar green land plants whenever we consider photosynthesis, a significant amount of photosynthesis is carried on by microscopic plants in the oceans, rivers, and lakes of the world. Indeed, *as far as humans are concerned*, many land plants are actually very inefficient food producers. A great deal of the energy they capture and store is used to build their own complex

bodies and a smaller proportion is harvested as a final useful product. A single-celled alga, on the other hand, is almost entirely food, and when it is eaten most of its body and stored energy enter the food chain. Compare this with the inefficiency of the apple tree, in which only the fruit is eaten. It must be stressed again, however, that the plant is inefficient *only* so far as humans are concerned. For the apple tree, fruit must provide an efficient means of seed dispersal or the structure would not have evolved. Though we have often acted as if it were otherwise, no plant or animal has evolved for the purpose of supplying *us* with food, efficiently or inefficiently.

The second law of thermodynamics states that whenever energy is transformed from one form to another, some of that energy is "lost"—cannot be made available for useful work. This problem is directly connected to that of photosynthesis and the use of photosynthetic products by all forms of life. Quite obviously (and fortunately) photosynthesis is an immensely profitable venture for the plant. The energy investment costs nothing; the dividends are so great that they can support those forms of life that do not contribute to them (that is, do not photosynthesize). The Biblical quotation "All flesh is grass" metaphorically, at least, is quite true.

SUPPLEMENT 7.6
PHOTOSYNTHESIS AND THE ENERGY CRISIS

As we have seen, the process of photosynthesis as carried out in plants takes carbon dioxide from the atmosphere and water from the surroundings, converting these into carbohydrates and releasing oxygen into the atmosphere as a waste product. There are now serious efforts to mimic a portion of the photosynthetic process—the portion that splits the water molecule into hydrogen and oxygen—using sunlight as the energy source. Such an achievement has the potential to exploit a clean, renewable fuel, hydrogen. Since solar energy is free, water abundant, and the fuel product, hydrogen, storable and nonpolluting, it is easy to see the attractiveness of the idea. This is especially true when it is considered that the process is a renewable one—when the hydrogen is consumed, water is regenerated. Finally, the process occurs at low temperatures (around 25° Celsius) and involves no poisonous intermediate compounds.

To duplicate a photobiological process that produces hydrogen, the idea is to do away with the dark or light-independent reactions and tap photosystems I and II. In the 1970s researchers showed that spinach chloroplasts and bacterial extracts containing a hydrogen-releasing enzyme (hydrogenase) would, when exposed to light, evolve hydrogen if an electron carrier such as ferredoxin were added:

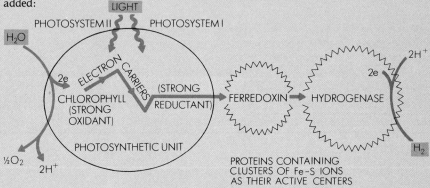

Since the hydrogenase involved deteriorates on exposure to oxygen, the reaction is usually carried out in an atmosphere of nitrogen, with "scavenger" molecules (glucose oxidase plus glucose) added to remove the oxygen evolved from the water molecules. This sensitivity to oxygen is one of several problems standing in the way of making significant progress in the area of hydrogen-producing solar energy conversion systems. With the severity of the energy crisis and the promise of such a system, however, research in this area will certainly continue.

Fortunately, once light is absorbed by chlorophyll molecules of the green plant, the synthetic process becomes remarkably efficient. More than half of the absorbed light energy ends up locked within glucose molecules as potential chemical energy. Any further chemical transformations of this energy are bound to draw on this stored chemical energy, sacrificing more of it. Feeding plant material to animals to fatten them for human consumption is, in terms of energy expenditure, an extremely wasteful process. Only the wealthiest nations can afford it. It is no accident that the poor nations of the world are primarily vegetarian, relying for their nourishment mostly on such crops as rice, corn, or other plants. Such nations cannot afford the considerable energy "leakage" that must inevitably accompany the transformation of plant matter into animal matter before it is eaten.

Summary

1. Photosynthesis is the process by which energy-rich compounds such as glucose are synthesized from low-energy precursors such as CO_2 and H_2O. Photosynthesis is an endergonic reaction driven by the energy of sunlight. Photosynthesis and respiration are opposite sides of the energy hill in the living world. Photosynthesis generates the high-energy compound glucose from CO_2 and H_2O; respiration breaks glucose down into CO_2 and H_2O. Photosynthesis is overall a reduction process; respiration is overall an oxidation process. Photosynthesis is absolutely essential for respiration to continue. Without photosynthesis there would be no glucose to oxidize, and the atmosphere would soon become exhausted of its oxygen. In turn, respiration supplies some of the CO_2 used by photosynthetic organisms as a raw material for photosynthesis. These two processes, on which all life depends, exist in a balanced state on earth. Preservation of that balance is essential for the continuance of life on earth.

2. Photosynthesis is characteristic only of green plants and those containing certain other types of pigments, not necessarily green. The pigments function as means of capturing the sun's light energy in a form the cell can use.

3. Photosynthesis takes place in two interconnected but separate pathways, the light reactions and light-independent or dark reactions. The light reactions is the name given to the pathway by which quanta or photons of light energy are used to raise electrons from the two forms of chlorophyll a, P_{700} and P_{680}, to higher energy levels. The products of the light reactions, ATP and reduced NADP (NADPH), are the important energy (and hydrogen) sources for driving the dark reactions.

4. Two photosystems, I and II, are involved in the light reaction phase of most plant photosynthesis: photosystem I consists of regular chlorophyll a and a special form of chlorophyll a, P_{700}; photosystem II consists of regular chlorophyll a, P_{680}, and chlorophyll b. The two pigment systems are able to use sunlight to carry out photophosphorylation, the use of light energy to produce ATP. Steps in the light reactions are as follows:

   a) Quanta of light strike molecules of regular chlorophyll a and b, which act as antennae, funneling their excitation energy into removing electrons from P_{700} and P_{680}. In photosystem I, two quanta of light raise two electrons from P_{700} to an excited state (to higher energy levels), from which they are picked up by an acceptor molecule, Z. Acceptor Z transfers the electrons to the iron-containing compound ferredoxin. From here they can follow one of two pathways. (1) Cyclic photophosphorylation: the electrons are passed through a group of acceptors (including cytochrome b, cytochrome f, and plastocyanin) back to P_{700}. During this process, at ATP is generated. (2) Noncyclic photophosphorylation: the electrons are passed from ferredoxin to NADP, thereby reducing it to NADPH.

   b) A pair of excited electrons from pigment P_{680} are picked up by the acceptor Z_1, from which they can follow only one course: they are passed through a series of acceptors including plastoquinone, cytochrome b_{599}, cytochrome f, and plastocyanin. These electrons are eventually passed to chlorophyll a. Every pair of electrons passed from chlorophyll b back to chlorophyll a generates four ATPs.

   c) Removal of electrons from pigments P_{700} and P_{680} leaves a "hole" in the chlorophyll molecules. The presence of this hole means the pigments have a tendency to pick up electrons to replace the lost electrons. Chlorophyll a gets its electrons back either from the cyclic process or by noncyclic flow from chlorophyll b. Thus, while chlorophyll a does not pass its electrons to chlorophyll b (to photosystem II), chlorophyll b does pass its electrons to chlorophyll a. Chlorophyll b replenishes these electrons by splitting H_2O, which releases molecular oxygen.

   d) The net products of the light reactions are ATP and NADPH. These energy-rich substances are used to drive the dark reactions. In addition to supplying energy, NADPH supplies hydrogens for the reduction of CO_2 in the dark reactions.

5. The dark reactions consist of the Calvin cycle, in which a prominent role is played by three-carbon sugars. Therefore the cycle is often called the C_3 system. Events of the dark reactions are as follows:

   a) A molecule of CO_2 is picked up by a molecule of the five-carbon ribulose diphosphate (RuDP), making an unstable six-carbon intermediate. It breaks down into two three-carbon sugars and phosphoglyceric acid (PGA).

   b) PGA is reduced by hydrogens from NADPH and by energy from ATP to phosphoglyceraldehyde (PGAL). PGAL undergoes a series of transformations (including pathways through four-, six-, and seven-carbon sugar intermediates).

c) From these transformations, one three-carbon sugar emerges for every three molecules of RuDP that combined with CO_2 in step *a*. This three-carbon sugar is actually PGAL. Two PGALs can combine to form fructose-1,6-diphosphate, which is easily convertible into glucose. PGAL is the basic photosynthetic product. The dark reactions compose a cycle because their intermediate compounds are regenerated—they show neither an overall increase nor an overall decrease in quantity within the chloroplast—during the process.

6.   An alternative very efficient pathway for carbon fixation occurs in the cells of certain green plants, especially those found in hot, arid climates. The C_4 pathway utilizes a number of four-carbon acids as intermediates. The function of this pathway is to pick up CO_2 at low concentrations in the outer layers of the leaf (the mesophyll) and transport it as a four-carbon intermediate (such as malic acid) to an inner layer of the leaf (the bundle-sheath layer). In the inner layer, the four-carbon intermediate breaks down, releasing the CO_2 it had picked up in the outer layer. This process delivers the CO_2 to the cells where it can be used in the Calvin cycle. Mesophyll has a high concentration of the enzymes for the C_4 pathway and a low concentration of enzymes for the C_3 pathway. Bundle-sheath cells are just the opposite. The existence of these two systems makes it possible for those desert plants that have the enzymes for the C_4 system to photosynthesize at a high rate even with the stomata partially closed, as would happen at midday when water loss through evaporation would be at its highest.

7.   Under some conditions, the enzyme responsible for adding carbon dioxide to RuDP may unavoidably react with oxygen instead of carbon dioxide, leading to the formation

of glycolic acid (glycolate). This process, known as photorespiration, occurs in Calvin or C_3 plants and is wasteful, since a net loss rather than net fixation of CO_2 results.

8.   Crassulacean acid metabolism (CAM) plants, such as cactuses, obtain their carbon dioxide at night when the air is cool and water loss is less, and chemically bind the carbon dioxide into organic acid molecules. During the day, the bound CO_2 is released and undergoes reduction in C_3 photosynthesis, and the stomata remain closed.

9.   Photosynthesis is localized within the chloroplasts of plant cells. Molecules of chlorophyll and the enzyme systems associated with photophosphorylation are bound to the lamellar membranes within the chloroplast. The enzymes associated with the Calvin cycle are free-floating in the intrachloroplast medium, the stroma. In terms of both general structure and functional organization, chloroplasts and mitochondria have much in common. Both are membrane-bounded organelles with an internal membrane system. Both are concerned with harnessing energy in a useful form for the cell. Both have the enzymes associated with electron transport and phosphorylation bound to internal membranes, while the enzymes involved in oxidation and reduction of intermediate compounds are free-floating in the spaces between the inner membranes. These similarities have led to some speculation that chloroplasts and mitochondria are evolved from a common ancestral cell organelle.

Chemiosmosis is thought to occur in chloroplasts as well as mitochondria, except that in the former the movement of protons is from outer to inner instead of the reverse. As in mitochondria, the flow of protons across the membranes is thought to be the power source for the production of ATP.

Exercises

1.   In what way was the origin of the oxygen evolved during photosynthesis eventually determined? Explain the technique involved.

2.   What does the fact that most leaves are green tell you about the wavelengths of the light used in photosynthesis?

3.   Observe the action spectra of two organisms (Fig. 7.16). Evaluate each of the following statements in relation to this graph.

a) The red alga and green plant, both exposed to roughly the same light source, would absorb the same wavelengths of light with the same degree of effectiveness.

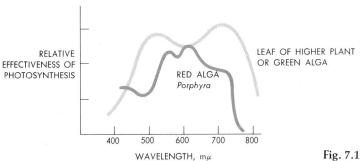

RELATIVE EFFECTIVENESS OF PHOTOSYNTHESIS

RED ALGA *Porphyra*

LEAF OF HIGHER PLANT OR GREEN ALGA

WAVELENGTH, mμ

Fig. 7.16

b) The single wavelength of light that would produce about equal photosynthesis in the red alga and the green plant is: (i) 450; (ii) 500; (iii) 510; (iv) 585; (v) 600; (vi) 650; (vii) 700.

c) If a green alga and a red alga were inhabiting the same region of the ocean, in general they (would, would not) compete for the same wavelengths of light energy.

d) The red alga is more effective in absorbing the total range of solar energy than the green plant.

4. The process of respiration is said to be the opposite of the process of photosynthesis. In what sense is this statement true? In what ways is it not true?

5. In the past, many textbooks have given the overall equation for photosynthesis as: $6CO_2 + 6H_2O \rightarrow C_6H_{12}O_6 + 6O_2\uparrow$. What is incorrect about this formulation of the equation? How should this overall equation be written?

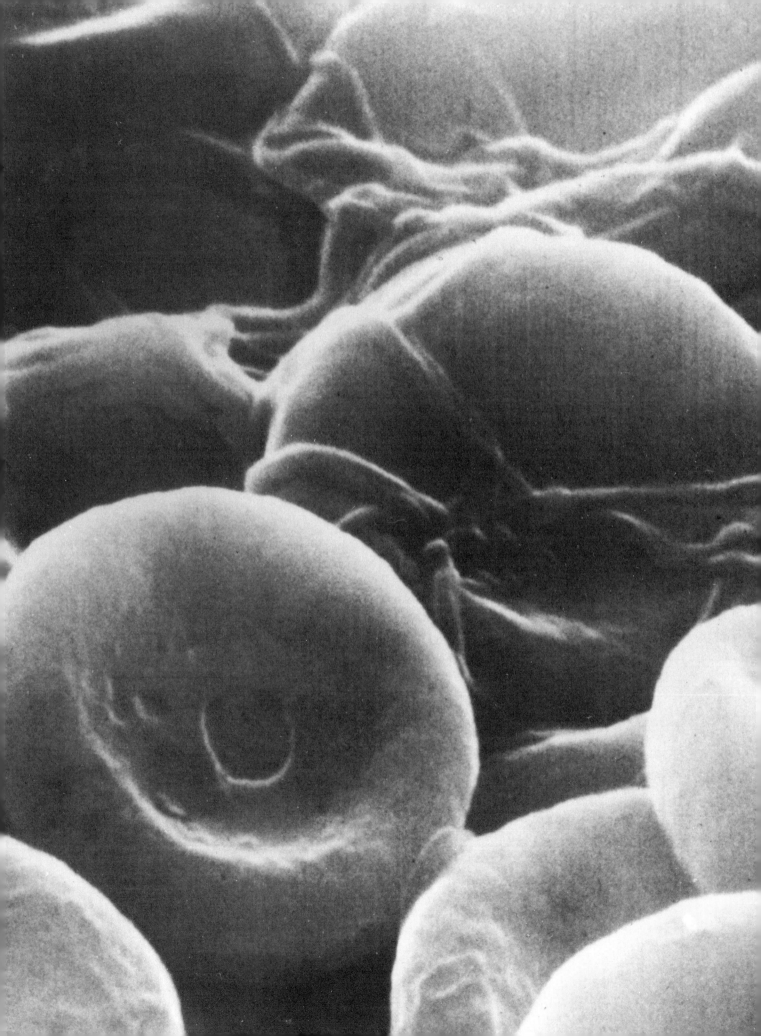

PART III
THE STRUCTURE AND
FUNCTION OF
WHOLE ORGANISMS

Under the scanning electron microscope, a blood clot appears as a network of long threads in which red blood cells are trapped. This fibrous network is composed of a protein known as fibrin. Clotting takes place whether or not the red blood cells are present; it is a property of the fluid plasma alone. The blood cells simply increase the effectiveness of the clot by becoming trapped in it and plugging up spaces between the fibers.

(Courtesy L. W. MacDonald, Lawrence Berkeley Laboratory, University of California, Berkeley.)

The properties of whole organisms are more than a sum of the properties of the individual cells composing them. Whole organisms are made up of organized, interacting components. Biologists distinguish between the structure (**anatomy**) of components and their function (**physiology**). This distinction is useful in analyzing and discussing the properties of living systems, and we can speak of structure and function at the cellular, tissue, organ, and system levels of organization. But really to understand either, we must visualize the interrelationships of structure and function. We must recognize that structure and function are two sides of a single coin, of a single reality.

Much discussion of physiology is in fact a discussion of the interactions between the various structural components making up the organism. To understand the life of the whole organism it is necessary to understand the interactions between cells, tissues, organs, and organ systems. It is clear that these interactions take place in specific, not random, ways. To say that interactions take place in specific ways is simply to say that all physiological processes are regulated. Regulation of these processes is an important aspect of life's organization and an important area of inquiry for modern biology.

Chapter 8
Plant Structure and Function I: The Movement of Materials Through Plants

8.1
INTRODUCTION

To primitive humans, the appearance of leaves and flowers in the spring must have signified an awakening, a return to life after the hardships of winter. Plants, dormant for so many months, seemed suddenly to become active. Even now this rapid growth and development of plants in the spring is highly symbolic, providing inspiration for countless works of music, poetry, and art.

To botanists—biologists interested specifically in **botany**, the study of plants—the activity of plant life in the spring has further significance. It puts them face to face with a number of important problems, and raises many questions. What forces are involved in the movement of materials through a plant? How are new leaves produced? What triggers the unfolding of leaves and flowers in the spring? What controls the growth and development of plants? Why do roots grow downward and stems upward? How do stimuli outside the plant (such as light) affect its growth and development?

Such questions demand an understanding of how a plant functions as a complete organism. So far, only the internal organization of individual cells and the types of tissues they form have been discussed. Although study of isolated cell organelles is important for understanding their special characteristics, we must view the organism as a whole—as a continuity of cells, tissues, and organs working together to carry out an integrated series of activities. Only when we explore the individual parts of an organism in relation to the whole can we begin to understand the intricate mechanisms by which living things maintain themselves.

Much of the dynamics of plant structure and function can be understood simply by viewing the various processes as ways in which the green plant survives as an organism. As we saw in Chapter 7, the process of photosynthesis enables it to use solar energy to build energy-rich molecules. Accordingly, there is great selective value in the development by a green plant of light-accessible surface area. Some of the ways in which the many problems of autotrophism are solved will be considered in this and the following chapter.

One of the most important problems a multicellular organism, whether plant or animal, must face is the movement of materials. All multicellular animals have some form of circulatory system by which materials taken in from the outside are transported to the cells in the body, and by which waste materials from the metabolism of those cells are removed. Most animals have an organ that pumps material through the circulatory system. The multicellular plant has no such pumping organ, yet it must solve the same problems—problems often made far more difficult by the huge size of many vascular plants (plants with conducting vessels). Some redwood trees, for example, are hundreds of feet tall, requiring that water and mineral materials taken in through the roots be transported up the entire length of the stem to the leaves.

The movement of materials from one part of a plant to another is called **translocation.** Basically, translocation involves all parts of the plant, since all parts contain the conducting or vascular tissues through which materials move. However, the principal organ systems that carry out translocation are the roots, stems, and leaves. After examining the organization and tissues of vascular plants, we will consider each of these organ systems and their interrelationships during translocation.

8.2
GENERAL ORGANIZATION OF A VASCULAR PLANT

A plant begins its life as a one-celled zygote, a cell produced by the union of an egg cell with a sperm cell (fertilization). The zygote grows and develops into the embryo and finally into a fully mature plant. Even in the very young embryo, there is already (in the vascular plants) a polarized differentiation into an axis composed of a **root** and a **shoot** with **leaves** attached to a stem. This basic organization of root and shoot is maintained

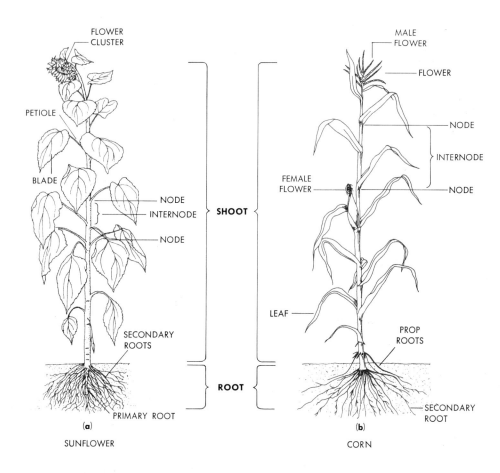

FLOWER CLUSTER
PETIOLE
BLADE
NODE
INTERNODE
NODE
SHOOT
SECONDARY ROOTS
ROOT
PRIMARY ROOT
(a)
SUNFLOWER

MALE FLOWER
FLOWER
NODE
INTERNODE
FEMALE FLOWER
NODE
LEAF
PROP ROOTS
SECONDARY ROOT
(b)
CORN

Fig. 8.1
The basic root-shoot organization of an adult vascular plant. (*a*) Sunflower, a dicotyledonous flowering plant. (*b*) Corn, a monocotyledonous flowering plant. (After Gilbert M. Smith et al., *A Textbook of General Botany*, 3d ed. New York: Macmillan, 1935.)

(a)

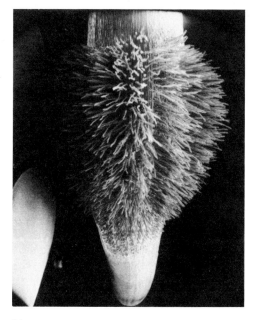

(b)

Fig. 8.2
Conspicuous differences in shoots and roots. (*a*) Leaves (and fruits) on stem of the basswood (*Tilia*) tree. (*b*) Root hairs on primary root of germinating radish. (Micrograph courtesy R. G. Kessel and C. Y. Shih, from *Scanning Electron Microscopy in Biology*. New York: Springer-Verlag, 1974.)

throughout the life of the plant (Fig. 8.1). Each of these parts performs special functions. Changes in the function of one part necessarily affect the other. For example, the amount of carbohydrate synthesis in the leaves of the shoot is dependent upon the amount of water uptake by the root, as well as the conduction of water upward by the stem of the shoot. Almost all land plants have this basic division of labor. In some species, however, a portion of the shoot may have become modified during evolution for specific functions (as in the modifications of stems into many kinds of thorns and tendrils).

In function and general organization, roots and shoots are conspicuously different systems of the plant body. Roots are naked axes, bearing no superficial appendages other than small, one-celled projections, the *root hairs* (Fig. 8.2). Shoots display conspicuous lateral appendages, the *leaves* (Fig. 8.2). Roots anchor the plant body firmly in the soil and absorb water and various minerals. In many species, the root system serves as a storage structure for carbohydrates and other molecules synthesized by the plant. The main functions of shoots include photosynthesis (especially in the leaves), transport of carbohydrates and other materials between the leaf-bearing portion and the roots, movement of water and dissolved minerals from root to leaves, storage of various molecules used in the normal functioning of cells in the stem itself, and the production of reproductive structures (such as flowers). Both externally and internally, the form and structure of roots and shoots are well adapted for the performance of these different tasks.

While roots and shoots differ significantly in structure and function, they are still similar in several important ways. Both have a common origin in the embryo. The tissues of each merge with the other almost imperceptibly; this arrangement obviously makes possible the performance of such joint functions as the transport of materials longitudinally from one part of the plant body to another. Both root and shoot appear to have evolved from a common structure; support for a hypothesis proposing such a common evolutionary origin comes from fossils of certain ancient plants that show no differentiation of the body into roots and leafy shoots.

8.3
THE ROOT SYSTEM

In patterns of growth, two types of root system are generally recognized.* Most trees and shrubs and many herbs have a root system that is basically a **taproot** and

*Some botanists recognize a third type, similar to the fibrous type, called diffuse.

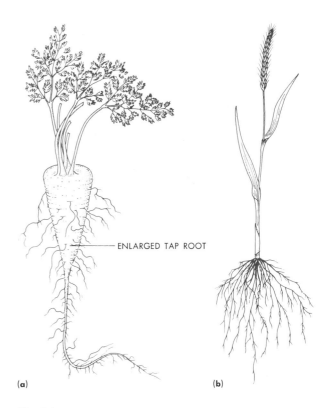

ENLARGED TAP ROOT

(a)   (b)

Fig. 8.3
Root systems of flowering plants. (*a*) Taproot of carrot. (*b*) Fibrous root system of foxtail grass.

its branches. Taproot systems are characterized by a large, central primary root that grows straight downward. From this taproot, lateral or branch roots develop in sequence, the youngest near the apical meristem and the oldest near the base. Taproots are especially effective in anchoring the plant in the earth, as anyone who has tried to pull up a dandelion from a lawn will readily attest. Since they can penetrate far into the soil, taproots can also reach deeper, more permanent sources of water. The taproots of many species may be specially adapted for the storage of the products of photosynthesis. The familiar edible portion of a carrot plant, for example, is a taproot packed with stored food (Fig. 8.3).

Grass plants, their relatives, and most ferns possess a **fibrous** root system, composed of numerous *adventitious roots* spreading from the stem. In plants (such as corn), these adventitious roots replace the primary root, which dies. While fibrous root systems do not anchor plants in the ground as well as taproot systems, they are better adapted for absorbing water near the surface.

Internally, fibrous roots differ little from taproots. For convenience, the taproot can be considered as a typical example of root structure. A longitudinal section

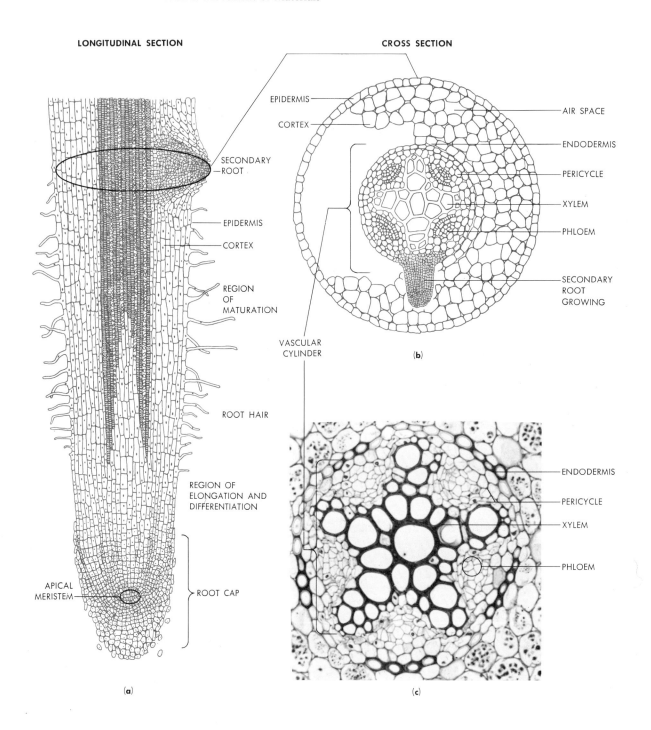

LONGITUDINAL SECTION

CROSS SECTION

EPIDERMIS

CORTEX

SECONDARY
ROOT

EPIDERMIS

CORTEX

REGION
OF
MATURATION

ROOT HAIR

REGION OF
ELONGATION AND
DIFFERENTIATION

APICAL
MERISTEM

ROOT CAP

VASCULAR
CYLINDER

AIR SPACE

ENDODERMIS

PERICYCLE

XYLEM

PHLOEM

SECONDARY
ROOT
GROWING

(b)

ENDODERMIS

PERICYCLE

XYLEM

PHLOEM

(a)

(c)

Fig. 8.4
Tissue organization of a root. (*a*) Longitudinal section (greatly
shortened in order to show the various stages in development). (*b*) Cross section showing location of various tissues.
(*c*) Vascular cylinder of root of buttercup. Compare with (*b*)
for identity of tissues. (Photo courtesy of Carolina Biological
Supply Company. Drawings adapted from John W. Kimball,
Biology, 2d ed. Reading, Mass.: Addison-Wesley, 1968.)

of the tip of a root is shown in Fig. 8.4(a). The end of the
root is covered by a root cap. As the root grows, the
cells of the root cap are worn away. They are continually replaced, however, by the apical meristem.

Behind the root cap is the apical meristem, a growing or embryonic region that provides new cells for the
tissues of the root as well as for the root cap itself. Behind the apical meristem is a region of elongation,
where cells increase in length. During elongation, the

cells also begin to differentiate—to become specialized. Farther up the root is the region of maturation, where the cells reach their full size and become completely specialized. In this region, root hairs project from many of the epidermal cells.

Seen in cross section, the cells and tissues of the root in the maturation region present a definite pattern (see Fig. 8.4b and c). Three principal regions are discernible: **epidermis, cortex,** and vascular cylinder or **stele.** The cortex tissue consists mainly of parenchyma cells characteristically separated from each other by spaces of various sizes. These intercellular spaces, which develop during the early growth of the root, appear to be adaptations for aeration of the root cells. The cortex helps move water and dissolved minerals across the root from the epidermis to the xylem, and it stores various molecules transported downward from the shoot. The innermost layer of cells in the cortex usually differentiates as an **endodermis,** composed of cells whose walls become thickened during their development. There is some evidence that the endodermis functions in the movement of water and dissolved minerals from the cortex to the xylem.

The vascular cylinder begins with one or more layers of parenchyma cells, the **pericycle,** immediately inside the endodermis. Cells of the pericycle tissue retain

the ability to undergo cell division, producing the secondary roots. On the inner surface of the pericycle are the vascular tissues, which in many species are arranged in a star-shaped pattern. Comprising the core of this star are the cells of the xylem. Between the points of the star are small groups of phloem cells. They are usually separated from the xylem by a layer of cambium which, in those species whose roots undergo increase in thickness, produces both secondary xylem and phloem.

8.4
MOVEMENT OF WATER AND MINERALS INTO THE PLANT

By the end of the nineteenth century (see Supplement 7.1), experiments had clearly demonstrated that green plants require three classes of materials from their environment: oxygen and carbon dioxide; water; and minerals. To be used by the plant, however, these materials have to be transferred from the environment into the plant body—they must be absorbed. In this section, the absorption of water and minerals will be considered; the obtaining of carbon dioxide and oxygen will be discussed in Section 8.10.

Absorption of Water

It is easy to hypothesize how absorption of water could take place. Examination of young roots with a microscope reveals numerous tiny, fingerlike projections (root hairs). The presence of root hairs greatly increases the surface area of the root, making it ideally suited for the absorption of materials (Fig. 8.5a). In the soil surrounding the root hairs, water is usually present in high concentration. Inside the cells, water is generally in much lower concentration. According to the simple laws of osmosis (in this case movement of solvent, water, across a differentially permeable membrane from a region of higher to a region of lower free energy), we would predict that water molecules would pass into

Fig. 8.5
Diagrammatic sketches of root structure and the pathways of water and mineral nutrient movement into the root. (a) In this longitudinal section, note the increased amount of surface provided by the root hair. (b) In this root cross section, water and nutrients are shown moving along pathway A through cell walls and intercellular spaces into the xylem. In pathway B, this movement takes place across cell membranes and living cells. (Drawings: from John W. Kimball, *Biology,* 2d ed. Reading, Mass.: Addison-Wesley, 1968; right, adapted from Peter M. Ray, *The Living Plant.* New York: Holt, Rinehart and Winston, 1963.)

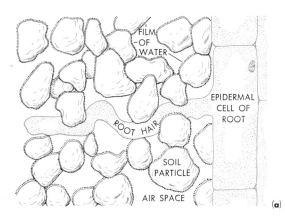

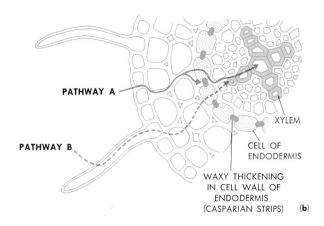

Water and ions pass into the root by osmosis, diffusion, and active transport.

the root cells. Tracer experiments using tritiated water (H_2^3O) verify this prediction.

It can be hypothesized that osmosis is the *only* process responsible for the absorption of water by the roots. An experiment can be designed to test this hypothesis. If narrow strips of epidermis from the middle of the root hair portion of young roots are placed in different concentrations of sodium chloride dissolved in distilled water (for example, 2, 2.5, 3, 3.5, 4, and 4.5 percent), microscopic examination shows that the contents of the root hair cells are contracted inward from the cell walls. In the most dilute solution, however, the root hairs are slightly swollen. In solutions of medium concentration (the 3 percent or 3.5 percent solution), the cell contents are only slightly withdrawn from the cell walls, while solutions of 2.5 percent concentration produce no visible effect. These observations are consistent with the hypothesis that only the physical process of osmosis is involved in the absorption of water into the root hair cells; evidently water does not pass into the cells immersed in the 2.5 percent solution because there is no concentration gradient in either direction. In other words, there is no visible change of the cell contents in the 2.5 percent solution, because the concentration of water inside the cell is exactly equal to the concentration of water outside in the soil. The contraction of root cell contents inward, away from the cell walls indicates that water has been removed by osmosis from these cells since, in solutions of higher sodium chloride concentration, the concentration of water is higher inside the cells than outside. The results are consistent with the hypothesis and indeed would be predicted by it.

Attempts to establish a hypothetical generalization from these experimental results to include *all* multicellular plant species, however, run into difficulties. For example, under situations in which the "osmosis alone" hypothesis would predict water absorption, it may actually be retarded in radish root hairs. This observation makes it reasonable to suspect that some form of active transport may be involved; it would seem that energy is expended directly or indirectly in the absorption of water—or, in this case, in keeping it out.

There is also increasing evidence supporting a hypothesis that considerable quantities of water may enter the root and move across the epidermis and cortex by diffusing along the cell walls without ever actually moving by osmosis across the cell membranes. To enter the vascular cylinder, however, the water must move across the membranes of the endodermal cell (in young roots)

and cambium cells (in older roots). A band-like layer of waxy thickening (Casparian strips) in the radial and transverse walls of the endodermal cells functions as a barrier to the further movement of water by any other means (Fig. 8.5b).

It was long assumed that absorption of water and the minerals it contains in solution took place only in the root hair region of young roots. Recently, however, several experiments have demonstrated that considerable absorption of water occurs in the roots mature enough to possess a corky bark. Since as much as 95 percent of the root surface of a shrub or a tree consists of bark-covered roots, it is possible that the major portion of water and mineral absorption may not occur through the root hairs, but perhaps enters through cracks and small areas of loosely organized, thin-walled cells.

Absorption of Minerals

In addition to their function as water-absorbing organs, roots are involved in absorption of the mineral elements necessary for plant growth. These elements enter the plant as ions, that is, as atoms or groups of atoms bearing an electric charge. For example, phosphorus is absorbed as phosphate ions (PO_4^{-3}), nitrogen as nitrate ions (NO_3^-) or ammonium (NH_4^+), and potassium as potassium ions (K^+). Some of these ions are dissolved in the soil water, while others are chemically bound to soil particles.

Since the roots of the plant are in a soil environment containing various ions, it is reasonable to hypothesize that ions might move into the root by diffusion. There is, indeed, some evidence suggesting that ions in the soil solution may move into roots as far as the endodermis by diffusing along the water-saturated cell walls (see Fig. 8.5). Some ions may also enter the roots simply by being swept along with the water that is being absorbed. These processes, however, cannot account for all the ion absorption that occurs. Ions continue to enter root cells even when their concentration there is many times greater than in the root's soil environment. In one 24-hr water culture experiment, for example, roots of barley plants absorbed all of the potassium and 75 percent of the nitrate originally present in the culture solution. Obviously, the absorption of these ions continued even in the face of an increasingly steep free-energy gradient.

SUPPLEMENT 8.1
MINERALS AND PLANT METABOLISM

The role minerals play in plant metabolism was not recognized until the latter part of the nineteenth century. The basic technique used to determine what minerals plants need, and for what purpose, is called *water culture* or *hydroponics*. In this process, plants are grown in distilled water to which are added various substances, including the salts (nitrates, chlorides, or sulfates) of metals such as potassium, sodium, calcium, or magnesium (see Diagram A). The physiological effects of the presence or absence of these minerals can then be observed. Hydroponics allows the investigator to control precisely and quantitatively the amounts of each substance to which a plant is exposed.

Using hydroponics in the 1860s, the German botanist Julius Sachs demonstrated that certain minerals were essential to plant growth and maintenance. Growing bean, corn, and buckwheat seeds in distilled water and others in solutions containing various combinations of salts, Sachs found that those growing in the presence of potassium nitrate, sodium chloride, calcium sulfate, magnesium sulfate, and calcium phosphate were most healthy and vigorous in appearance for the first three or four days. Those seedlings lacking all these minerals either failed to develop at all, or succumbed very quickly. Those lacking one or more showed various signs of deficiency: they formed no new roots, became yellow, and failed to grow. Further, when plants were grown in all the nutrients, the newly appearing leaves were completely white; they lacked chlorophyll. When Sachs examined the cells of these white leaves with the microscope, he saw no chloroplasts.

It was clear that Sachs's water solution of various chemicals still lacked something. Perhaps, Sachs hypothesized,

the missing nutrient element might be iron. To test this new hypothesis, he prepared a dilute solution of iron chloride and painted the surface of one of the white leaves with a small brush. In only a day or two, Sachs observed that the white leaf had begun to turn green.

To obtain additional support for his hypothesis that the absence of iron was causing the leaves to be lacking in chlorophyll, Sachs added a few drops of iron chloride solution to the water in the jars containing the plants with colorless leaves. Within two days, the white leaves of the seedlings had begun to turn green. After several more days, they became the normal green color. These experiments offered strong support for the hypothesis that the absence of iron results in the leaves' not producing chlorophyll. They also suggest that the iron may be involved in the synthesis of chlorophyll. In additional water culture experiments, Sachs found that only extremely small amounts of iron are necessary; in fact, in quantities larger than a few milligrams, the iron becomes a poison.

Using the water culture technique, Sachs and various other investigators of the late nineteenth century were able to establish that green plants require the elements phosphorus, potassium, nitrogen, sulfur, calcium, iron, and magnesium. With this knowledge about plant nutrition, botanists had conclusive evidence that the green plant obtains most of its nutrient elements from the soil. The problem was not completely solved, however. Soil in which plants grow contains numerous elements. When the chemical composition of the plant body was analyzed, several elements other than those mentioned in the preceding paragraph were found to be present (although usually in very small amounts). Were these elements actually used by plants in their metabolism?

Before this question could be answered conclusively, the techniques of water culture had to be more highly refined. In early studies using this technique, investigators used chemicals that they thought were pure, that is, free of chemicals other than the ones desired. During the 1920s, it was discovered that many so-called pure chemicals actually contain at least traces of other elements. In addition, it was discovered that minute amounts of certain elements may dissolve out of the glass of the culture containers. Even the distilled water was found to contain trace quantities of certain elements (presumably contamination from the apparatus used to distill the water). Once these sources of contamination were recognized, the investigators were able to refine the water culture. Within a few years, conclusive evidence was uncovered supporting the hypothesis that green plants require very small amounts of such elements as boron, manganese, copper, zinc, and molybdenum for normal growth. More recently, it has been found that some plants also require chlorine, sodium, cobalt, and vanadium.

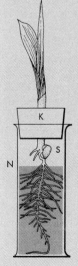

Diagram A
Water culture jar used by the plant physiologist Julius Sachs. A corn seedling is shown suspended by a cork (K). Note the seed (S), the level of the nutrient solution (N), and the roots immersed within the nutrient solution. (From Julius Sachs, *Lectures on the Physiology of Plants.* Oxford: Clarendon Press, 1887.)

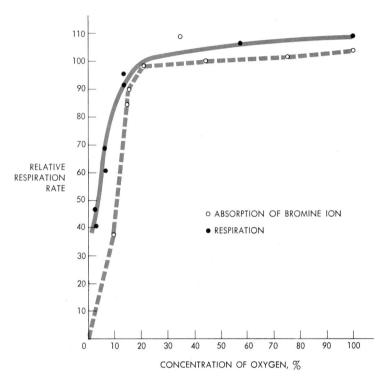

RELATIVE RESPIRATION RATE

○ ABSORPTION OF BROMINE ION
● RESPIRATION

CONCENTRATION OF OXYGEN, %

Fig. 8.6
The relationship between oxygen supply, respiration rate, and bromine ion absorption in thin discs of potato tuber (a greatly thickened stem with many root-like characteristics). (Data of F. C. Steward, from B. S. Meyer and D. B. Anderson, *Plant Physiology.* Boston: Willard Grant, 1939, p. 409.)

It seems evident, therefore, that there must be other forces at work in ion absorption. A hypothesis proposing that active transport may be involved is a logical next step: the barley plant's root cells may be performing work in order to move ions against the concentration gradient. Support for this active transport hypothesis comes from the observation that ions pass into the root cells against a free-energy gradient only so long as the root cells remain alive. Soon after death of the root cells, the ions diffuse out until an equilibrium is established between the tissue fluids and the external medium. Further evidence that root cells are expending energy to absorb ions comes from experiments in which the rate of ion absorption is greatly reduced or even stopped when the roots are deprived of oxygen (oxygen, of course, is necessary in respiration). Still other experiments demonstrate that when oxygen is bubbled through a culture solution in which roots are growing, the rate of ion absorption and accumulation increases. With an increase in the concentration of oxygen bubbled through the solution, the rate of respiration and the rate of absorption are increased (Fig. 8.6).

The Role of Leaves

While the roots are the principal water- and mineral-absorbing organs of the plant, some absorption of these materials is accomplished by the leaves. In fact, the only source of water and minerals for such plants as Spanish moss* (*Tillandsia usneoides*) and many species of orchids (which grow upon tree branches) is rain, water vapor, and airborne dust particles. Such plants are known as **epiphytes.** Epiphytes should not be confused with parasites; epiphytes make their own food and have no organic connection with the tissues of the tree on which they grow.

At times, water and minerals may also be absorbed through the stems and leaves of plants that normally grow with their roots in the soil. In one experiment, two leafy stems of a single raspberry plant were deprived of water until they wilted. After exposure of one leafy stem to a continuous spray of water, both stems recovered. It seems evident that the raspberry plant must have absorbed water through the leaves and stem of the sprayed stem and then transported the water into the unsprayed stem. Another investigator was able to keep plants of several different species alive and growing for several weeks with only one or two of the leaves immersed in water (the roots and the remainder of the leaves were kept exposed to the atmosphere). There is also evidence that leaves can absorb water from dew and fog.

In recent years it has become increasingly popular to apply minerals to shrubs and ornamental trees by spraying their leaves. For example, apples and pears are

---

*Spanish moss is a common name. It is misleading in that the plant is a member of the family *Bromeliaceae* of the Anthophyta (flowering plants) and not the division (or phylum) Bryophyta, which contains the true mosses.

sprayed with special preparations containing iron when the plants show symptoms of iron deficiency.

8.5
ORGANIZATION OF THE SHOOT SYSTEM

One of the most characteristic features of the shoot system is its differentiation into a central axis (the stem) to which a series of lateral appendages (the leaves) are attached. The leaves are arranged on the stem in patterns more or less characteristic of each species. The point of attachment of each leaf to the stem is called the **node**; the portion of the stem between the two nodes is the **internode** (Fig. 8.2). Depending upon the species, the internodes may be long and distinct or very short and indistinct. In plants bearing their leaves in a rosette, like the dandelion, internodes can be distinguished only with difficulty. The internodes are also very short in the spur shoots of fruit trees and the needle-bearing shoots of pines. In trees, the division of the stem into the nodes and internodes is eventually obscured by the growth in thickness, due to the activity of the vascular cambium. Thus, except for young twigs, external evidence for the spatial relationship between the stem and the leaves can no longer be seen in a tree.

New leaves and tissues of the shoot arise in the apex of the stem, a region called the shoot apical meristem. This meristem is usually enclosed by older leaves; the whole comprises a bud. If the older leaves are carefully removed, the very tip of the shoot appears as a minute dome or mound; this is the apical meristem proper, the ultimate source of the cells of the shoot. Embryonic leaves in various stages of development can be seen arising beneath this tip. The older leaves overarch the younger ones, thus protecting them (and the apical meristem proper) from *desiccation* (drying out).

Internally, the shoot apex presents an appearance considerably different from that of the root apex. The most obvious difference is the lack of a cap of tissue comparable to a root cap. Thus, in the shoot apex, the apical meristem itself terminates the shoot. In addition, while the root apical meristem adds new cells both outwardly to the root cap and inwardly to the root proper, the shoot apical meristem contributes new cells only to future leaf and stem tissues. Just as in the young root, after the new cells are formed by the meristem they gradually elongate and differentiate into the various kinds of cells characteristic of the mature tissues of the plant.

As in the root, the cells and tissues of the stem present a definite pattern when viewed in cross section. In some species (including certain ferns and a few aquatic flowering plants), three main regions can be distinguished: **epidermis, cortex,** and **vascular cylinder.** In such plants the vascular cylinder forms a solid rod in the center of the stem. In most plant species, however, the center of the stem axis is occupied by the pith, composed largely of parenchyma tissue.

Beneath the epidermis of the stem is the cortex which, as in the root, consists largely of parenchyma cells. Collenchyma and fibers are also present in some species. In ferns, the innermost layer of the cortex is usually differentiated into an endodermis. In most flowering plants, however, the endodermis does not develop.

The vascular or conducting system occupies the region of the stem inside the cortex. In most seed-producing plants, the stem vascular system has the form of either a continuous cylinder surrounding a central pith, or a split cylinder enclosing a pith. In the case of the split cylinder, the xylem and phloem tissues are in segments or strands called vascular bundles. If the cortex and the pith tissues are removed from the stem of such a plant, the vascular system may be seen as a network of individual vascular bundles that occasionally meet and fuse and then branch again (Fig. 8.7). At each node, one or more vascular bundles enter each leaf and each branch bud.

Within the stem, the vascular bundles of many species are situated so that the phloem tissue lies outside the xylem. In some plants (such as the potato, watermelon, milkweed, and sunflower), phloem may be present on both sides of the xylem. By looking at the arrangement of vascular bundles in the stem, we can identify two distinct patterns among the flowering plants (Fig. 8.7c and d). If the vascular bundles are arranged in a circular pattern around the central pith, like spokes radiating from the hub of a wheel, the plant is in a taxonomic category known as the **dicotyledonous** plants ("dicots"). The vascular bundles are replaced by cylinders early in the life of woody plants. If the vascular bundles are dispersed throughout the stem, the plant belongs to a group called the **monocotyledonous** plants ("monocots").

Clearly there are significant differences in the kind and arrangement of tissues between the stem and the root. In the root, the xylem is usually in the form of a star-shaped or fluted column occupying the central region. In the stem, however, the xylem is either a hollow or split cylinder with the central portion of the stem consisting of pith tissues. Despite these obvious differences in arrangement, the vascular systems of the root and stem meet and become adapted to one another in a transition region, generally located between the root and the cotyledons. The structure of this transition region varies from one group of plants to another and is often highly complex. In the flax plant, for example, the vascular tissue pattern in the lower part of the transition region is similar to that of the root. Some distance higher up, a pith develops in the center of the vascular cylinder. Just above this point, the vascular system becomes arranged into vascular bundles, some of which enter the cotyledons while others continue into the stem.

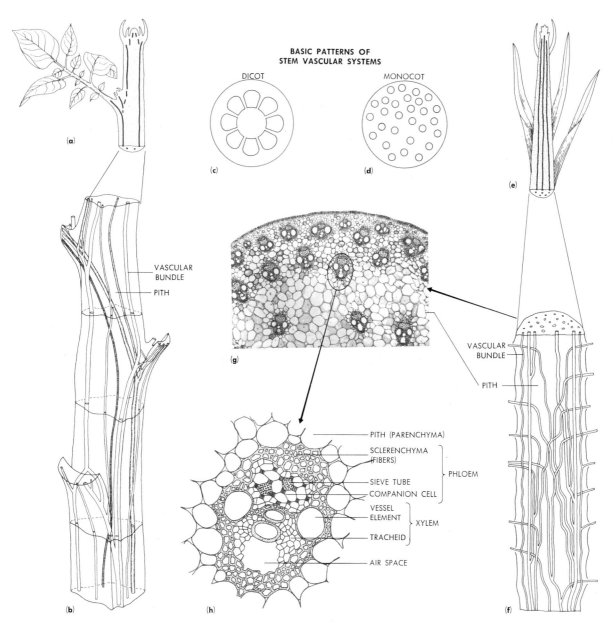

Fig. 8.7
Patterns of vascular tissues in a flowering plant. (*a–c*) Diagrams of the primary vascular system of the stem of potato. (*d–h*) Vasculature in stem of corn. (*h*) Cross section of a vascular bundle showing arrangement of tissues. Note that the phloem lies outside the xylem (with reference to the epidermis). (Photo courtesy Turtox, Chicago. Drawings: (*b*) adapted from E. F. Artschwager, *Journal of Agricultural Research* 14 (1918), pp. 221–252; (*f*) adapted from M. Kumazawa, *Phytomorphology* 11 (1961), pp. 128–139; (*h*) from John W. Kimball, *Biology,* 2d ed. Reading, Mass.: Addison-Wesley, 1968.)

8.6
TRANSLOCATION THROUGH THE SHOOT SYSTEM
The fibrous root system of a single plant may be divided into two halves and each half placed into a different beaker of water as shown in Fig. 8.8. One beaker con-tains only distilled water, while the other contains radioactive phosphorus (P^{32}) in solution. P^{32} is an iso-tope of nonradioactive phosphorus (atomic weight 30.975). Radioactive phosphorus emits low-energy radi-ation in the form of beta (β) particles. By using a Geiger

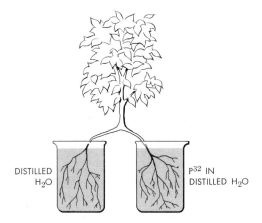

Fig. 8.8
Experiment to demonstrate the translocation of dissolved substances through a whole plant. Radioactive phosphorus (P³²), taken in by the portion of the root system to the right, is traced through the plant by means of a Geiger counter.

DISTILLED
H₂O

P³² IN
DISTILLED H₂O

counter, which is sensitive to this and other types of radiation, it is possible for the investigators to track the passage of P³² through the plant. Results of such an experiment show that after a while (the time depends upon the size and type of plant, as well as certain environmental conditions) the stem and then the leaves become radioactive. We can actually track the "wave" of radioactivity as it moves through the plant.

More interesting, however, is the fact that after a short while, radioactivity begins to appear in the roots exposed only to plain water. This implies that P³² from one-half of the root stem has entered the main stem of the plant and been carried down again to the roots in the other half. The fact that some radioactivity begins to appear in the left-hand portion of the root system *before* it can be detected in the leaves implies that substances do not have to move all the way to the top of a plant in order to be transported back to the roots.

Until the 1920s, plant physiologists believed that xylem tissue was concerned primarily with the upward movement of materials and phloem tissue with downward movement. Recent investigations show that there is more to the story than this—xylem and phloem are not simple, closed conducting tubes like the plumbing system in a building. We can get a more accurate picture of the way in which transport occurs in the living plant by considering two experiments.

One set of experiments is shown diagrammatically in Fig. 8.9. In the first experiment (Fig. 8.9a), the outer regions of the plant stem were cut away (a process known as girdling), removing both the cork and the phloem tissues. In the second experiment (Fig. 8.9b), the

xylem was removed but the phloem was left intact. The following data were recorded:

1. Organic substances collected principally at *M* and *P*, but some were found at *N* and *Q*, especially when *N* and *Q* occurred in a region where there were leaves both above and below the girdle.

2. Inorganic substances collected primarily at *S* and *V*, but some accumulated at *M*, *P*, *N*, and *Q*, especially when *N* and *Q* were located in a region with leaves both above and below the cut.

What generalizations can we draw from these data? First, it is obvious that inorganic materials must travel principally through the xylem. It is also clear that the principal direction of movement through the xylem is upward. Second, we can see that phloem carries both organic and inorganic substances, and that the chief direction of transport in the phloem is downward. However, as the data show, phloem transports materials upward as well. Substances such as glucose move from a lower leaf to a higher leaf or flower almost exclusively through the phloem, but the transport of glucose from leaves to the roots, where it is stored, also takes place through the phloem. Recent autoradiographic studies provide evidence that separate vascular bundles are involved in this two-directional movement.

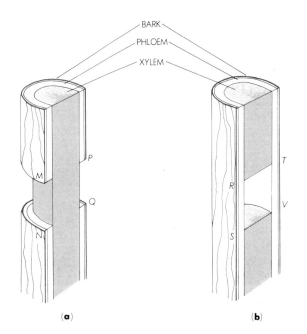

Fig. 8.9
Experiment to determine direction of flow of inorganic and organic subtances through xylem and phloem.

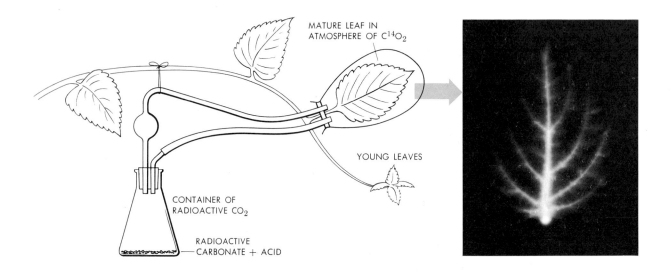

MATURE LEAF IN ATMOSPHERE OF $C^{14}O_2$

YOUNG LEAVES

CONTAINER OF RADIOACTIVE CO_2

RADIOACTIVE CARBONATE + ACID

Fig. 8.10
Radioactive carbon dioxide ($C^{14}O_2$) is fed to a plant leaf. The plant incorporates the radioactive carbon into the carbohydrates it produces by photosynthesis. The path of these "labeled" carbohydrates can then be followed throughout the plant either by a Geiger counter or by the autoradiograph technique (insert). (Photo courtesy Brookhaven National Laboratory.)

A second set of experiments, in which radioactive carbon (C^{14})* is used as a tracer, has helped to show the pathway carbohydrates and other organic substances may take throughout the plant. C^{14} may be given to a plant in the form of carbon dioxide ($C^{14}O_2$). The experimental setup for "feeding" radioactive carbon dioxide to leaves is shown in Fig. 8.10.

A leaf is exposed to $C^{14}O_2$ for a short period of time. This allows molecules of radioactive carbon dioxide to enter the plant and be incorporated into specific compounds. In 8 or 10 hr, other parts of the plant are sampled to determine where radioactive sugars may be found. The Geiger counter could again be used, but another technique is particularly useful. Leaves can be picked off the stem and placed between plates of photographic film for a period of several days. The radiation emitted by the C^{14} in each sample reacts with the photographic emulsion to produce a developed spot on the film. If radioactive carbon atoms are spread throughout an entire leaf, the leaf shape will appear as a picture on the film. This picture is an **autoradiograph** (Fig. 8.10, inset). The lighter the picture on the autoradiograph, the greater the number of radioactive carbon atoms in the leaf. This technique has the advantage of giving an ac-

curate record of exactly how the radioactive materials are distributed throughout the sample. For example, it can be shown that certain substances remain concentrated in the vascular system of the leaf (veins), whereas other substances become more concentrated in the nonvascular regions.

If we expose a young, growing leaf of a plant to radioactive carbon dioxide as described above, we find that all the radioactivity remains in that leaf. Autoradiographs of other leaves on the same stem show no image at all. No carbohydrate is translocated by a growing leaf because all the food produced is used up by the growth of the leaf itself. If we expose a mature leaf to $C^{14}O_2$, however, we find that some of the carbohydrate is indeed translocated. Autoradiographs of other leaves on the stem show that most of the carbohydrate translocated by the uppermost older leaves is passed to young, developing leaves and the stem apex. The mature leaf that was exposed to $C^{14}O_2$ still has a great deal of radioactivity, but it also passes a good deal of its newly manufactured carbohydrate along to those leaves that are still growing. This is of definite benefit to the young leaves, since they cannot produce enough food to supply their own needs completely until they are more mature. The lowermost leaves of a plant export carbohydrate primarily to the root. Intermediate leaves export to the roots and the stem apex.

It is evident from this research that the movement of materials throughout a plant's parts is controlled quite efficiently. In the early spring, carbohydrates pass upward from the stem and root to the developing flower and leaf buds. Later, when the leaves are mature, they manufacture carbohydrate that is transported to the developing fruit. In addition, some carbohydrate is moved down to the stem and root for immediate use and for the next season's reserve supply. The ability of a plant to regulate the direction in which substances are transported implies a high degree of organization.

*Carbon normally has an atomic weight of 12.

What forces are involved in both upward and downward translocation? What determines the rate and direction of flow? What factors account for the movement of dissolved substances such as ions or glucose independently of water flow? Attempts to answer some of these questions will be the subject of the next four sections.

8.7
UPWARD MOVEMENT THROUGH THE XYLEM

Plant physiologists have long wondered what force moves water upwards in plants. Many explanations have been proposed; none are completely satisfactory. To be acceptable, any hypothesis of water movement through the xylem must account for the movement of water upwards to the tops of such tall trees as the giant redwood (*Sequoia sempervirens*) or the Douglas fir (*Pseudotsuga taxifolia*). At normal atmospheric pressure, a vacuum pump attached to the top of a glass tube inserted into a bucket of water cannot lift a column of water higher than 34 ft. Yet the California redwoods reach a height ten times this value. It seems evident that there must be several factors involved in the upward movement of water and dissolved minerals through the xylem.

Perhaps water might be pumped upward in the plant by the living cells surrounding the xylem. This hypothesis was tested in some spectacular experiments carried out in the late 1880s by the German botanist Eduard Strasburger. Strasburger used a very long vine of *Wisteria* that climbed to his rooftop. He placed a 40-ft section of the stem of this vine (between its roots and leaves) in a large kettle and killed the cells with boiling water. Strasburger then replaced the section of vine in its normal location, extending from the ground to the rooftop 40 ft above. Since the leaves did not wilt, it seemed clear that water was continuing to rise through the long stem segment, even though the cells had been killed. Evidently, mechanisms independent of life must operate in water movement.

Capillarity

It was discovered about 1850 that water molecules have pronounced **cohesive** and **adhesive** properties (see Supplement 2.1). It is now hypothesized that water molecules cohere or stick to each other because the hydrogen atoms of one water molecule form a hydrogen bond

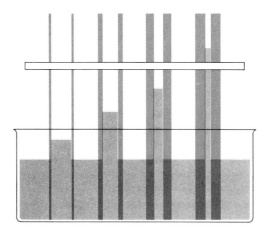

Fig. 8.11
Demonstration of capillarity. If the diameter of a tube is small enough, a column of water of considerable height may be maintained within it due to the adhesion and cohesion of water molecules.

with the oxygen atom of the adjacent molecule. Water molecules not only stick together, but also adhere to many other substances, such as the sides of a glass container or the sides of a xylem tracheid.* As the diameter of the container is decreased, this effect becomes more pronounced. If the diameter in the tube is reduced to one millimeter or less, a column of water can be maintained in the tube with no outside force required (Fig. 8.11). This effect, known as **capillarity**, is due to the combined forces of adhesion and cohesion. It is reasonable to hypothesize, therefore, that capillarity is involved in the upward movement of water and minerals in the xylem.

According to the capillary hypothesis, the tracheids and vessels of the xylem tissue (or capillary channels within the walls of the larger-bore xylem elements) would function as capillary tubes. Experiments have demonstrated, however, that water can be maintained by capillarity at a maximum height of only about 3 ft in the smaller xylem elements. Further, neither cohesion nor adhesion provides a force for moving water up-

---

*To observe this effect, hold a glass of water up so that your eye is on a level with the surface of the water. On either side of the glass, note how the surface turns upward. This is because molecules of water adhere to the glass.

Plants transport water and ions upward through the xylem; carbohydrates (mostly sugars from photosynthesis) are transported downward through the phloem. Lateral conduction of water, ions, and sugars also takes place between xylem and phloem.

Upward flow through the xylem occurs by transpiration pull (evaporation of water from leaves) and root pressure (osmotic pressure of incoming water from roots).

ward. These intermolecular attractions simply allow a column of water to withstand many pounds of pull without breaking. Like a fishing line, the column of water in a plant may take many pounds of pressure; however, it is the fisher and the fish who provide the pull. The capillary hypothesis is thus inadequate to account for the upward movement of water in plants.

Root Pressure

In the early spring, before the leaves have yet appeared, a watery fluid can frequently be seen running out of cracks or cut stems of grape vines. If the stem of a well-watered tomato plant is cut off, water will exude, or "bleed" from the stump. If a piece of glass tubing is attached to the stump, the water may rise inside the tube to a height of several feet (Fig. 8.12). The force producing this rise of water, called **root pressure,** causes water to be pushed up from the roots.*

Is root pressure an important factor in the upward movement of water? In 1938, Philip White of the Jackson Memorial Laboratory in Bar Harbor, Maine, showed that if single root tips are removed from the tomato plant and grown in sterile culture, each will produce root pressures sufficient to raise water above 200 ft high. This finding was startling to White and to the botanists of that time. Since tomato plants grow only a few feet in height, why should their roots possess the capacity to generate such high pressures? As a result of his experiments, White recognized root pressure as an "unappreciated force" in the ascent of water in plants. A hypothesis that root pressure is a major force in the upward translocation of water through the xylem became extremely attractive.

---

*The way in which the roots of a plant exude liquid under pressure is not yet fully understood. Involved in producing root pressure is the continuing movement of water from the soil, through the cells of the root, and into the xylem tissue. The water already in the cells of the xylem is thereby forced upward. The presence of living cells also seems to be necessary: if the roots are killed, root pressure stops. In fact, even if the cells remain living but are deprived of oxygen, root pressure ceases. Since oxygen is necessary for the respiratory production of ATP, this suggests that energy must be expended in the absorption of the water placed under pressure in the xylem.

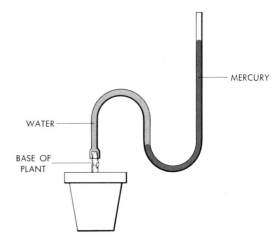

Fig. 8.12
A diagram illustrating root pressure due to the osmotic intake of water by the roots. In this apparatus, the height of the column of mercury indicates the amount of root pressure.

Yet in some species, especially conifers, little if any root pressure can be demonstrated. Further, when the humidity of the air is low, much water is evaporated from the leaves, and the water is moving upward rapidly in the xylem. Under such conditions, the water inside the xylem is rarely under pressure; when a xylem vessel is punctured with a tiny needle, the water usually does not flow outward. The root pressure hypothesis would predict otherwise. This hypothesis appears inadequate to account for the upward movement of water in many plants, especially tall trees.

Transpiration Pull

In one experiment, the tissues outside the xylem of a bean plant were removed and a vessel element was punctured with a tiny needle. As the investigator watched with a microscope, the water inside the vessel snapped apart when the needle entered the vessel. Also, a short hissing sound was heard frequently as air was drawn into the injured water-conducting cell. This experiment suggests that the water may actually be *pulled*

Transpiration pull is enormously effective in moving water through plants. One acre of lawn transpires over 27,000 gallons of water per week.

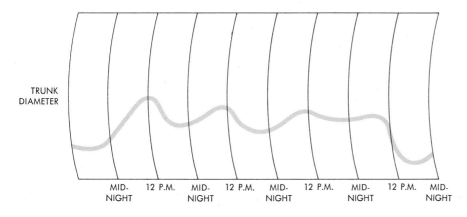

TRUNK DIAMETER

MID-NIGHT | 12 P.M. | MID-NIGHT | 12 P.M. | MID-NIGHT | 12 P.M. | MID-NIGHT | 12 P.M. | MID-NIGHT

Fig. 8.13
Curves produced by a dendrograph attached to a tree trunk. Note that the tree trunk decreases in diameter in the late afternoon and early evening and increases during the morning hours to a maximum diameter just before midday.

upward through the xylem, since like a stretched rubber band, such a column of water would be under considerable tension.

Perhaps the most dramatic support for this tension hypothesis, that the water in the xylem is being pulled upward, comes from D. T. MacDougal of the Carnegie Institute of Washington. MacDougal developed an ingenious experiment to test the hypothesis. He reasoned that if water in tree trunks is under tension while being raised, then the inward pull of water on the xylem vessels should decrease slightly the diameter of the tree trunk. MacDougal knew that more water is transported upward by a tree in the daytime than at night. In the daytime, the sun increases the rate of both photosynthesis and **transpiration** (loss of water from the leaves by evaporation). Thus, more water passes out of the tree during the sunlight hours.

To test his prediction, MacDougal designed a very sensitive instrument called a dendrograph. This apparatus is able to record very small changes in the diameter of a tree trunk over an extended period of time. A graph of the measurements shows that the diameter of the trunk *does* decrease in the day and increase at night (Fig. 8.13). The prediction is borne out and further validation given to the tension hypothesis.

What forces could provide the powerful upward pull of the water in the xylem, and where are those forces generated? In 1935, the German botanist Bruno Huber inserted small electric wires into tree trunks and heated the water inside the xylem elements. Then he measured the time it took for the warmed water to pass a thermocouple (a temperature-measuring device) placed a few inches higher on the tree. Huber found that, in the morning, water begins to move in the upper parts of the tree sooner than it does in the lower portion of the trunk (Fig. 8.14). This observation strongly suggested that the forces pulling the water upward are acting in the top of the tree, perhaps in the leaves.

Fig. 8.14
Velocity of water movement in a tree. Note that the water in the upper parts of the tree starts to move earlier in the morning than the water in the trunk.

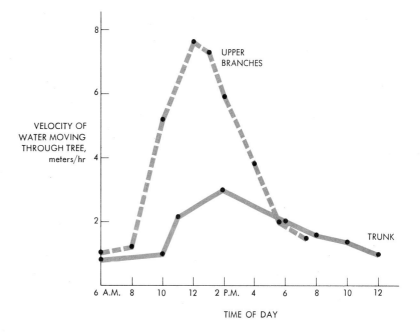

VELOCITY OF WATER MOVING THROUGH TREE, meters/hr

UPPER BRANCHES

TRUNK

6 A.M. 8 10 12 2 P.M. 4 6 8 10 12

TIME OF DAY

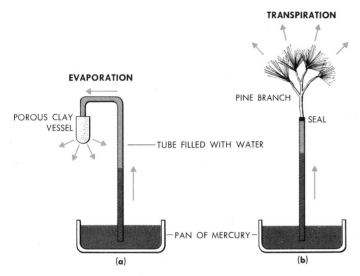

Fig. 8.15
The effect of transpiration on the upward movement of water in a plant is illustrated by these two experimental setups. In (a), a glass tube is filled with water, then inserted into a pan of mercury. The other end of the tube is covered with a porous clay vessel that allows water vapor, but not liquid, to pass. Evaporation through this vessel results in a column of mercury being lifted up the glass tube. In (b), a pine branch replaces the porous vessel, and evaporation of water from the pine "needles" (leaves) moves the column of mercury upward.

Might there be some relationship between the transpiration of water from the leaves (as well as its use in photosynthesis and other metabolic processes within the leaf cells) and the forces pulling water upward through the xylem? That such a correlation exists was demonstrated experimentally in the early 1890s by the botanists Josef Boehm of Austria, E. Askenasy of Germany, and H. H. Dixon of Ireland. Their experiments are diagrammed in Fig. 8.15. When water is evaporated from the top of a closed glass tube connected to a container of mercury, a column of this heavy metal can be lifted more than 100 cm in the tube. A branch from a pine tree sealed into the top of a similar system is able to accomplish the same task. This mechanical model of the xylem transport system of a tree shows that the forces pulling water upward are set in motion by transpiration of water from the leaves—**transpiration pull.** In addition, the experiment demonstrates that water is not being pushed upward by the roots (as the root pressure hypothesis maintains), since no roots are present.

Central to the hypothesis that the water is being pulled upward in the tree is the existence of great cohesive forces between the individual molecules of water, as well as adhesive forces between the water and the walls of the xylem cells. Due to the strong attraction between water molecules, the water in a closed tube, such as a xylem vessel or tracheid, behaves much like a wire and can be literally pulled upward. Measurements have

shown that a column of water in a thin, airtight tube can tolerate a pull of 300 pounds per square inch without breaking. This force would be enough to lift a column of plant sap to the top of the tallest trees known.

In the mechanical model of the xylem water transport system shown in Fig. 8.15, the water in the tube must be previously boiled to remove all dissolved air. The formation of an air bubble in the water may break the column and the water will not be lifted. In a living tree, also, there is a possibility that air bubbles will form and break the columns of water inside the tracheids and vessels, especially during periods of strong winds and/or rapid transpiration of water from the leaves. It seems clear, however, that if a water column is broken by the formation of a gas bubble, the resulting break remains confined to that column. Thus the presence of air in one part of the xylem does not interfere with water movement through other parts. Further, since an individual plant usually has far more conductive capacity in its xylem than necessary for its survival, even a substantial amount of breakage of its water columns produces no serious injury.

It is tempting to think of transpiration pull as suction, but this is really incorrect. Suction would involve removal of air from a tube above the water, allowing atmospheric pressure to force the water up the tube. As we know, a suction system can only lift a column of liquid about 34 ft—the limit of force for atmospheric pressure. Evaporation of water from the top of a 100-ft tree, however, can serve to lift a column of water all the way to the top. Transpiration pull depends upon evaporation, not suction, and is thus not dependent upon atmospheric pressure.

In trees growing in cold climates, however, the water in the xylem cells freezes during the winter. Since air is practically insoluble in ice, freezing causes bubbles to form in all the tracheids and vessels of the trees. How do these trees survive? Most deciduous trees (those that lose their leaves for the winter) retain very little water in their branches and trunks during cold weather, thus minimizing the rupture of vascular tissues. In the spring the sap rises in the trees—that is, the vascular tissues begin conducting again and a large volume of water begins to flow through the xylem. The sap of sugar maple trees is obtained during this period (later winter, early spring) by tapping into the xylem and phloem tissue. This sap is especially rich in proteins and sugars that give maple syrup its characteristic flavor. In nondeciduous trees such as conifers (pines, spruces, firs, and the like), however, conduction must occur continuously throughout the winter. Some breakage of vascular tissues obviously occurs because of freezing. However, there is some evidence that such trees produce substances that may act as antifreeze agents in the sap. Furthermore, the continuous (though slowed) movement of water through the vessels helps to hinder the freezing process. One hypothesis holds that in conifers the air inside the bubbles in the xylem cells goes into solution when the ice thaws and warm weather returns. Trees such as ashes, oaks, and elms, with vessels of large diameter, replace the ruptured water-conducting system of the previous season by growing a new ring of xylem in the early spring before the leaves appear. Upward movement of the water during the summer takes place almost entirely in this new growth ring.

The concept that the water is pulled upward through the xylem does not, of course, answer a very important question concerning the ascent of sap in tall trees. How did the water get up to the tops of these trees in the first place? The water cohesion hypothesis simply attempts to explain how the water transport system is maintained within the living plant. It says nothing about the relationship between the growth of a tree and the mechanism of water movement.

One viewpoint holds that the water in the top of a tall tree first got up there by growing there! In the spring, a new layer of potential xylem cells is produced by the vascular cambium. This new xylem is located just outside the layer formed the previous year. The cells of the cambium and the immature xylem cells (being still alive) absorb water laterally from the tracheids and vessels of the older xylem. Indeed, according to this hypothesis, the new cells absorb so much water that their contents actually develop a temporary positive pressure. Supporting this contention is the observation that when these cells are punctured with a fine needle, liquid may exude from them. Once the leaves have expanded sufficiently for water to transpire from them, however, the water in the new xylem cells loses its positive pressure and a tension develops in the column of xylem liquid. According to this hypothesis, therefore, the process that initially establishes the water columns in the new xylem each year seems to be the water-absorbing activity of the living immature xylem cells. Once

the water columns are formed and placed under tension generated by transpiration from the leaves and the metabolic use of water by the leaf cells, this initial role of the living cells in the water transport system ceases. Upward movement of the water columns is then maintained by the purely physical forces of transpiration pull and water cohesion.

8.8
MOVEMENT THROUGH THE PHLOEM: SOME OBSERVATIONS

We know far less about the mechanism of movement through the phloem. One of the reasons is the immense difficulty of studying it directly. Unlike nonliving xylem tissue, the very act of trying to insert a micropipette into living phloem to sample its contents affects normal function; the phloem elements shut down operation, closing off the perforated sieve plates at either end. Thus movement throughout the entire sieve tube (which of course is made up of the individual elements joined end-to-end) is also stopped.

During the 1950s this difficulty was eliminated by use of a new technique. Many organisms are known to feed on the contents of the phloem of trees and shrubs. One such organism is the aphid *Longistigma caryae* (Fig. 8.16a). This insect inserts its long feeding structure,

Fig. 8.16
(a) An aphid feeding on phloem of the basswood tree. The droplet of sugar solution shown here is being excreted by the insect and serves as food for ants. (b) Stylet of the aphid can be seen inserted into a single cell of the phloem. (From Martin H. Zimmermann, *Science* 133, 1961, p. 78. Copyright 1961 by the American Association for the Advancement of Science.)

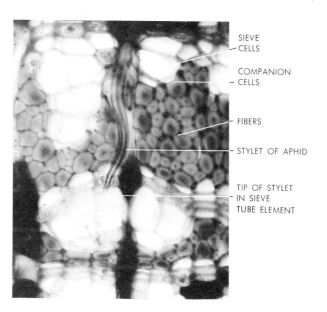

SIEVE CELLS

COMPANION CELLS

FIBERS

STYLET OF APHID

TIP OF STYLET IN SIEVE TUBE ELEMENT

(a)

(b)

Fig. 8.17
Evidence that the contents of the phloem are under pressure. Sugar solution continues to exude from the phloem through the aphid's stylet for many hours after the aphid has been cut away. (From Martin H. Zimmermann, *Science* 133, 1961, pp. 73–79. Copyright 1961 by the American Association for the Advancement of Science.)

known as a stylet, through the bark of a basswood tree and into an individual phloem element (Fig. 8.16b). The stylet does not seem to alter the normal functioning of the element; the aphid gets its fill from the rich contents of the phloem and then withdraws. The plant physiologist Martin H. Zimmermann of Harvard University found that if he anesthetized a feeding aphid, he could decapitate the insect and leave the stylet still inserted in the phloem. Material continues to come out of the stylet for many hours after it is severed from the body of the insect, indicating that the phloem contents are under some pressure (Fig. 8.17).

What materials are transported by the phloem? Analysis of phloem contents collected from aphid stylets has revealed that 10 to 25 percent of the fluid consists of sugar (largely sucrose). In some trees, including ash, elm, and basswood, more complex sugars such as stachyose and raffinose are present. Lesser amounts of amino acids and inorganic ions, especially phosphorus, are also present in phloem sap. There is also evidence that hormones are transported in the phloem.

In stem phloem, the movement of sugars is with a concentration gradient, that is, from a region of higher to a region of lower concentration. In white ash, for example, Zimmermann found a drop of some 20 percent phloem sugar concentration in fluid taken from aphid stylets between positions 9 m high on the trunk and 1 m above the soil. That this gradient in sugar concentration is due to leaf activity is strongly suggested by the fact that in the autumn, after the leaves have fallen, the gradient disappears. When Zimmermann removed the leaves during the growing season the gradient also disappeared. In contrast, however, the movement of sugar from the mesophyll cells into the phloem of the leaf veins occurs *against* a concentration gradient.

The fact that a concentration gradient exists within the stem phloem tempts the simple hypothesis that it moves by diffusion alone. This hypothesis is immediately contradicted, however; the movement of sugar molecules in phloem occurs many thousands of times faster than a simple diffusion hypothesis would predict. This suggests strongly that some sort of energy-expending process is involved in sugar transport in phloem. As might be expected, then, unlike the xylem, whose cells are dead at functional maturity, living elements are necessary for the phloem transport system to operate. In one set of experiments, the phloem elements of bean plants were killed in a local region of the stem by heating them (with steam, boiling water, or hot wax). While plants thus treated lived and photosynthesized for as long as fourteen days, no movement of sugars took place across the damaged but intact portion of the stem. These results can be contrasted with the results Strasburger obtained with *Wisteria* vines (Section 8.7).

Measurements of translocation using radioactive materials or dyes show that the maximum rate of phloem transport is about 50 to 100 cm per hour. This is a good deal slower than xylem, which conducts at a maximum value of 60 m (6000 cm or 200 ft) per hour. Further, different materials are moved at different rates in the phloem. In an autoradiographic experiment, plant physiologists O. Biddulph and R. Cory, at Washington State University, sprayed a solution of P^{32} in tritiated water (H_2^3O) on the lower surface of bean plant leaves and fed $C^{14}O_2$ to the upper surface. They discovered that the C^{14} was moved from the leaf down the stem at the rate of 107 cm per hour, while the P^{32} and H_2^3O were moved only 86.5 cm per hour.

Through the use of $C^{14}O_2$, Biddulph and Cory further found that when the flow of materials from a leaf reaches the phloem of the stem, it divides into an upward-moving and a downward-moving stream. When one leaf is located above another, the downward flow from the upper leaf passes the upward-moving stream from the leaf below. They also obtained evidence that this two-directional transport takes place in separate phloem bundles in the stem. In these same autoradiographic experiments, Biddulph and Cory also discovered that the upper leaves of a bean plant export materials primarily to the stem apex, the lower leaves to the root, and the intermediate leaves in both directions. Other experimenters have observed this same pattern of transport in cotton, soybean, and tomato plants, as well as various grasses.

SUPPLEMENT 8.2
THE TRANSPORT MECHANISM: SOME SCHOOLS OF THOUGHT

If materials move within the phloem as observations suggest they do, what is the mechanism behind this movement? In 1930, the German botanist Ernst Münch proposed that phloem translocation occurs because of the pressure of mass flow. According to this hypothesis, the force involved in moving materials through the phloem is the result of differences in hydrostatic pressure between leaf and root. Münch described a model system composed of two sacs, each constructed of membranes permeable only to water, connected by a glass tube. This system is shown in part a of Diagram A. One sac is filled with plain water, the other with a concentrated sugar solution. The model system is then placed in water. The water in the pan tends to flow by osmosis into the

sac containing the sugar solution, forcing a column of water-sugar solution through the glass tube into the other sac.

In part b of Diagram A, this principle is applied to an entire plant. The leaves, filled with carbohydrate produced by photosynthesis, are analogous to the membrane filled with concentrated sugar solution. The phloem elements can be compared to the glass tube, and the roots to the membrane filled with plain water. Because they are filled with carbohydrate, the leaves absorb water from the xylem. This, in turn, exerts a pressure, causing water to move into the phloem. Since the xylem always contains more water than the phloem, movement will always be in the direction of xylem to phloem.

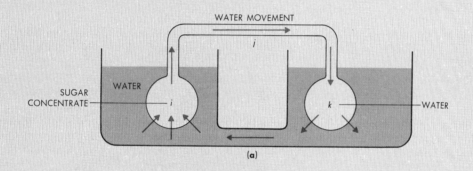

Diagram A
The Münch hypothesis to explain the movement of materials through the phloem. (*a*) Diagram of Münch's model. An increase in the sugar content at *i* would result in an increase in the osmotic pressure in that portion of the system. The solution would then flow through *i* toward *k*. With the increased osmotic pressure at *i*, water would enter at *i* and be lost at *k*. (*b*) The system at work in the whole plant. Here *i* is represented by the leaves, *k* by the roots, and *j* by the phloem tissue. (*c*) Details of the phloem, indicating differences in osmotic concentration in a group of cells.

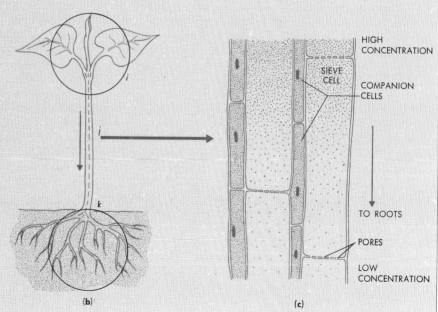

The major evidence in support of the pressure flow hypothesis comes from observations that sap containing a high concentration of sugar (10 to 30 percent) exudes from the phloem when an incision is made into the bark of various species of trees. Additional support is provided by the observation that when an aphid stylet is cut off while still inserted in the phloem, sap flows out under pressure for up to four days (see Fig. 8.17). Instead of sucking, the insect relies on the pressure in the sieve elements to force the phloem sap through the stylet into its body.

However, the pressure flow hypothesis calls for a concentration gradient of sugar down the stems of trees and within individual sieve cells, as shown in part c of Diagram A. Such gradients are indeed present in many species of trees during the growing season when phloem translocation is active. However, if there is a pressure flow of materials in the phloem, there should be a gradient of decreasing turgor pressure from the upper to the lower portions of this tissue. Recent experiments, however, have failed to demonstrate the existence of a gradient in turgor pressure in the phloem of tulip poplar and red maple at a time when the production and translocation of sugars would be expected to be high, during late July and early August.

Thus, the pressure flow hypothesis is not entirely satisfactory in accounting for the movement of materials in the phloem. Nor does it explain the mechanism by which sugars are moved from the mesophyll cells into the phloem (which must occur against a concentration gradient). Some experiments suggest that energy is required for this movement, and that an active transport mechanism may be operating to transfer sugars from the photosynthetic tissue to the conducting tissue.

Perhaps the most compelling objection to the pressure flow hypothesis, however, questions the capacity of the sieve tubes to pass the phloem sap along at sufficient rates. The phloem sap must pass from one sieve element to the next through extremely tiny openings in the sieve plates. From their minute size, it would appear that these openings would seriously hinder the flow of material from one sieve element to the next. Further, microscopic observations suggest that the sieve plate is covered with cytoplasm and the sieve pores filled with cytoplasmic filaments. Such obstacles would offer still more resistance to any mass flow in sieve tubes. This resistance, in turn, would require a pressure much greater than would be likely to exist in the sieve tubes of most plants.

Another objection to the pressure flow hypothesis is that of *bidirectional movement*—the demonstrated fact that substances move in opposite directions within the phloem *often at the same time.* This fact alone does not contradict the pressure flow hypothesis; it would do so if this bidirectional movement could be shown to occur *within a single sieve tube.* Such a movement would not be predicted to occur in a single sieve tube, since the driving force postulated by the pressure flow hypothesis is a gradient of hydrostatic pressure.

Experiments designed to demonstrate the absence or presence of bidirectional flow within a single sieve tube are, technically, very difficult to perform. Some experiments appear to demonstrate such a flow, whereas others do not.

Cytoplasmic Streaming

Recently John Field, at Nuneaton Technical College, Warwickshire, England, has offered the cytoplasmic streaming hypothesis as a direct challenge to Münch's mass flow concept. In delicately prepared thin sections of young wheat stems, Field claims to have observed rapid cytoplasmic streaming within the sieve elements themselves. He noted that large structures such as cell organelles and plastids were carried around with the streaming cytoplasm. He also noted that the streaming was two-directional: the cytoplasm streamed up and down within the sieve element. A two-directional streaming would make it extremely difficult for material to move in a single direction by mass flow. Certainly two-directional cytoplasmic flow within each individual element would not easily account for net one-directional flow through those elements. Field has also observed cytoplasm flowing, not only within the sieve elements themselves, but also through the sieve plates at the ends of the elements. This cytoplasmic flow carries with it plastids and various large cell inclusions. Field proposes that macromolecules could also be transported through the phloem in the same manner. Yet Field's observations do not establish the cytoplasmic streaming hypothesis very rigorously. It is not really clear how streaming could move the sugar-containing water downward at the rate observed in most plants. Field's observations actually do more to contradict predictions based on the mass flow hypothesis than they do to establish the cytoplasmic streaming hypothesis on a firm footing. (Any weakening of the Münch hypothesis would, of course, strengthen alternative hypotheses, including that of cytoplasmic streaming.)

Field's work is based on observations about which there is not yet full agreement among botanists. Not everyone has been able to observe cytoplasmic streaming to the extent Field claims to have done. Here again is a case where the difference between rival hypotheses is grounded, at least to some extent, in uncertainty about the observational data. It often happens that researchers may agree on the data yet still disagree on the best hypothesis to explain them. If researchers disagree on the data, it is almost certain they will disagree on the most acceptable hypothesis.

Cytoplasmic Pumping

The two hypotheses discussed thus far are by no means the only ones put forth to account for translocation through the phloem. A variant of the cytoplasmic streaming hypothesis has been advanced by the British plant physiologist R. Thaine and his associates, based upon the aforementioned strands of cytoplasm passing through the cavity of the sieve elements and through the sieve plate pores over the entire length of a sieve tube. They envision that a wavelike contraction and relaxation of protein filaments encircling the strands causes a sort of peristaltic wave, moving the contents over long distances much as food is moved along in the intestines of animals. While such a wavelike, pumping movement has not been observed directly, this hypothesis *is* consistent with reports of discontinuous, high velocity spurts of sugars near the front of translocation in the phloem.

This hypothesis of cytoplasmic pumping in transcellular strands is attractive in that it would account for the existence of bidirectional translocation, even within a single sieve tube. Needless to say, Thaine is one who believes reports of cytoplasmic particles moving in opposite directions in separate cytoplasmic strands within one sieve tube. Critics of the hypothesis maintain that the transcellular strands seen by Thaine are merely the result of a refraction of light during microscopic examination; they point out that such strands do not appear in electron micrographs of sieve elements (though it is admitted that they would be subject to destruction by the procedures of fixation, dehydration, and embedding that are a part of the preparation for EM work). Others maintain that the transcellular strands observed by Thaine and others are formed merely as a result of injury to the fragile sieve elements when prepared for light microscope examination, and do not exist in undamaged, living, and functioning sieve elements.

Electroosmosis

Another British plant physiologist, D. G. Spanner, has suggested that an electroosmotic mechanism may account for the translocation of sugar in sieve tubes. Electroosmosis refers to the flow of hydrated ions through the cell membrane; the term "hydrated" means that each ion passing across the membrane brings a "shell" of water molecules with it. Spanner hypothesizes that the shells of water surrounding each ion would have sugar molecules dissolved in them. Spanner's hypothesis is a complex one, involving the necessity of filamentous cytoplasmic protein, with spaces large enough to allow sugar molecules to pass through and an extended surface area for the pumping of potassium ions by the expenditure of ATP. The electroosmotic hypothesis also requires that each sieve plate be polarized in the same direction throughout the sieve tube, to act as an electroosmotic pumping station to boost the mass flow of the sugar solution along the way. Because it relies on the postulated existence of so many contingencies for which there is as yet little experimental evidence, the electroosmotic hypothesis has not received wide support.

Activated Diffusion

Almost fifty years ago, the facts that the movement of sugar in the phloem is in the same direction as it would diffuse (from greater to lesser concentration) and that the rate of translocation is proportional to the amount of concentration gradient and distances led T. G. Mason and E. J. Mandell, British plant physiologists working in Trinidad, to form a hypothesis involving a mechanism of activated diffusion. Since their own experimental results showed that the movement of sugars along the stems of cotton plants was 40,000 times faster than the diffusion of sugar molecules along a column of water in a glass tube, they recognized that some sort of activation mechanism had to be involved. They did not then have the experimental techniques to attempt to find out how this mechanism was connected to cellular metabolism.

In the 1960s, the Australian plant physiologist M. J. Canny and his associates used radioactively labeled sugars to show that the profiles of advancing radioactivity best fit the predictions of Mason and Maskell's hypothesis. In 1973, Canny postulated that the sieve element contains two phases, a stationary phase of sugar solution and a moving cytoplasmic phase that streams in both directions through the sap in long strands extending throughout the sieve element, much like the transcellular strands described by Thaine (but much smaller) in support of his cytoplasmic pumping hypothesis, described earlier. According to Canny, the pores of the sieve plate are completely filled with these strands, whose surfaces are partially permeable to sugars. Thus there can be an exchange of sugar molecules between the strands and the surrounding sap.

Canny envisions each sieve element having several such strands, each conveying sugar, *but some in one direction and some in another.* Strands in which the cytoplasm is moving downward will carry sugar in that direction and slowly release it to the sap in lower sieve elements. Those strands in which the movement of cytoplasm is upward will carry a lower sugar concentration than the surrounding sap, and thus will slowly take in sugar from it. The effect of such a diffusion would be to increase greatly the sugar concentration gradient between the upward- and downward-moving strands in each sieve element; in effect, the movement stirs the sugar solution in each one. Thus the sugar will diffuse longitudinally along the length of a sieve tube, from higher to lower concentration, but of course at a rate greatly accelerated over that of unstirred diffusion.

The main problem with the accelerated diffusion hypothesis is that, rather like Thaine's cytoplasmic pumping hypothesis, it relies on the existence of unusual subcellular anatomical features that have yet to be verified. For this reason, it remains at present a hypothesis that is not widely accepted.

All of the preceding hypotheses rest on sound physical principles, and none can be dismissed. It is possible that some partial truth is to be found in all of them, and that the final answer will incorporate aspects of each. Unlike the essentially passive transport of xylem translocation, phloem translocation involves active transport. Active transport, in turn, relies on a connection to the energy-releasing mechanisms of metabolism. How this connection is made and where it is applied remain unresolved at this time.

Though sieve elements were discovered in 1837, it was not until the 1960s that conclusive direct evidence of their function was obtained. In one of these experiments, performed by Sam Aronoff and R. S. Gage, the leaves of cucumber were fed tritium-labeled water. After 30 minutes, the petioles were frozen and dried and cross sections cut and placed in contact with a photographic film. The radioactivity was found to be concentrated in the phloem tissue, principally in a few sieve elements. None was present in the xylem. This last observation is extremely important, for a significant number of botanists, realizing the large quantities of material moved through a tree and the relatively small amount of sieve tubes available, found it difficult to imagine that so much material could be moved by so few tubes so rapidly, especially since phloem sieve tubes, unlike xylem, are not entirely open passages. Thus they believed that the xylem *had* to be involved. Yet it now seems clear that, with the exception in some species of some organic compounds, most of the transport of organic compounds occurs upward only in the phloem.*

8.9
LEAVES AND TRANSLOCATION: LEAF STRUCTURE

As described in the preceding two sections, the leaf system of a plant has an important role to play in the translocation process. Through its photosynthetic and transpiration activity, the leaf system appears to provide one of the several forces involved in both upward and downward movement of materials in the plant. We will first examine the structure of leaves and then, in the following section, the processes by which leaves affect translocation.

In its form and anatomy, the leaf portion of the shoot is the most variable organ of the plant. Due to this variability, several kinds of leaves have been distinguished, including foliage leaves, scale leaves, bracts, and cotyledons. **Foliage leaves** are the main photosynthetic structures. **Scale leaves** occur as bud scales and on underground stems; they protect tender growing parts and store reserve molecules. **Bracts** are usually associated with the flowers, where they too appear to play a protective role. Sometimes the bracts are colored (as in poinsettia, *Euphorbia pulcherrima*) and act much the same as petals. **Cotyledons** are the first leaves to be formed by the young plant, and the presence of one or

---

*It used to be thought that nitrogen, in the form of inorganic nitrates dissolved in soil water, was carried to the leaves for the synthesis of more complex nitrogen-containing compounds (such as nucleic acids or proteins) by the xylem. This does appear to be the case with most forms. However, it now appears that, in many species, the nitrogen is built into organic compounds such as amino acids in the roots, and then transported upwards in the phloem.

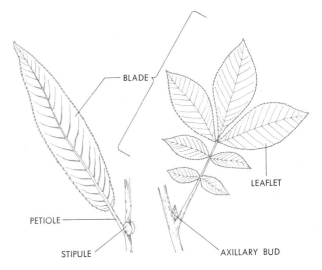

Fig. 8.18
Leaf forms. Left, a simple leaf of willow. Right, a compound leaf (with seven leaflets) of hickory. One or more lateral buds occur in the leaf axil, the angle formed by the leaf base and the internode above.

two of them gives the monocotyledonous and dicotyledonous plants their respective names.

A foliage leaf usually consists of three parts: blade, petiole, and stipule (Fig. 8.18). In the course of evolution, the blade has become greatly flattened, an adaptation that has greatly increased the efficiency of the surface areas in the capture of light energy and exchange of gases. The blade may be divided into *leaflets* (compound leaf) or it may be undivided (simple leaf), and its margin may be even or variously notched. In some species, the blade is directly attached to the stem and no petiole is present. In most plants, however, the blade is attached to the stem by a stalk (petiole), which serves to transport water and dissolved minerals from the stem to the blade, where they are used in photosynthesis. The photosynthetic products are sent back through the petiole into the stem for translocation to other parts of the plant. In some plants (such as corn), the base of the leaf is expanded into a sheath that surrounds the stem. This sheath often serves to protect the young terminal region of the shoot as well as the buds in the axil of the leaves. The leaves of many species of flowering plants (especially the dicots) have outgrowths of the leaf base called **stipules,** which function mostly to protect the young leaf before it unfolds.

Leaves grow from the **leaf primordia,** contained within the terminal bud of a plant and just beneath the dome of apical meristem cells. In the spring, rapid growth brings about the unfolding of mature organs. Once they reach their mature size, the leaves of most plants do not grow any longer or wider; they remain at their mature size until they die. Depending upon the species, leaves may live for only a few weeks (as in some desert annual herbs), a few months (as in most deciduous trees that drop their leaves in the fall), or for three or four years (as in the case of the needles of some evergreen plants such as pines).

Internally, foliage leaves are constructed of three main tissue systems: epidermis, mesophyll, and vascular tissue (Fig. 8.19). The outer cells on both surfaces form the protective layers as epidermis. The cells of both epidermal layers secrete a waxy material, **cutin.** Inside the epidermis is the **mesophyll,** a tissue composed of parenchyma cells rich in chloroplasts. In many species, especially the dicots, the mesophyll is differentiated into palisade parenchyma and spongy parenchyma. In cross section, the cells are seen to be elongated at right angles to the epidermis and arranged like a row of stakes (thus the name "palisade"). If a section is cut through the leaf just beneath the epidermis, the cells of the palisade parenchyma are seen to be more or less separated from each other by spaces, which facilitate aeration of these photosynthetic cells. Occupying the bottom half of the mesophyll is the spongy parenchyma, so called because of the conspicuous system of intercellular spaces that spreads throughout this tissue. Since these spaces connect with the numerous stomata present in the lower epidermis, the movement of water vapor, carbon dioxide, and oxygen is facilitated. Thus the leaf possesses an internal aerating system of impressive proportions. For example, one study of leaves from a *Catalpa* tree showed the internal surface to be some twelve times greater in extent than the external surface. The leaf, therefore, is well adapted for highly efficient photosynthesis.

As we have already seen, one to several branches of the vascular system of the stem enter the leaf at its attachment to the node. Within the leaf blade of dicots, the vascular bundles (veins) branch, fuse, and become smaller, forming a network; such a pattern of venation is said to be *reticulate,* or *netted.* In many monocots, such as corn, the venation is *parallel* throughout most of the entire leaf blade. Very small veins interconnect these parallel main veins. Surrounding the veins is a layer of tightly packed parenchyma cells, the bundle sheath, that extends to the ends of the veins, completely enclosing each one. The cells of the bundle sheath increase the contact between the cells of the xylem and phloem and the cells of the mesophyll. There is evidence that the bundle sheath takes part in the movement of materials between the veins and the mesophyll.

Much of the support for a leaf with a flat blade is provided by the system of veins that ramify throughout it. Additional support for the leaf blade is provided by the turgor pressure of the water inside the vacuoles of the mesophyll cells. Due to the compact arrangement of its cells and the strength of the cuticle layer, the epidermis also provides considerable support for the leaf. Collenchyma cells and sclerenchyma fibers may occur beneath the epidermis, close to the larger veins and along the blade margin; this arrangement also provides support for the leaf blade.

As pointed out in Supplement 7.1, in the seventeenth century Neremiah Grew observed that leaves contain small pores located mostly on their undersurface. These pores, called **stomata** (see Fig. 8.19), are sur-

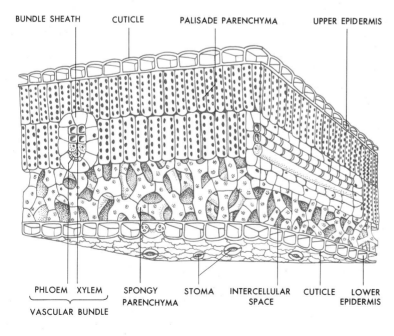

BUNDLE SHEATH   CUTICLE   PALISADE PARENCHYMA   UPPER EPIDERMIS

PHLOEM   XYLEM   SPONGY   STOMA   INTERCELLULAR   CUTICLE   LOWER
                  PARENCHYMA   SPACE   EPIDERMIS
VASCULAR BUNDLE

Fig. 8.19
Internal organization of a leaf as seen in three-dimensional section. Stomata are generally limited to or most numerous on the lower surface. Note that the vascular bundles are surrounded by protective tissue of the vascular bundle sheath.

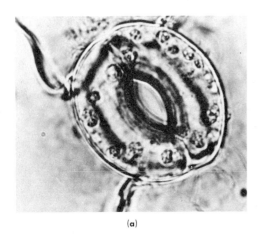

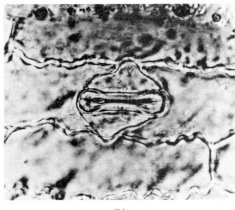

Fig. 8.20
(*a*) Stoma and guard cells of a bean leaf, open. (*b*) Stoma and guard cells of a corn leaf, closed. Both are highly magnified. (Photos provided by the United States Department of Agriculture.)

(a)   (b)

rounded by two specialized cells known as **guard cells.** Through the action of the guard cells, the stomata open and close (Fig. 8.20). When the stomata are open, gases and water vapor can be exchanged with the environment. When the stomata are closed, very little exchange is possible. Stomata tend to be closed when the atmosphere is dry, preventing excessive water loss from the plant.

8.10
LEAVES AND TRANSLOCATION: GAS EXCHANGE AND TRANSPIRATION

Leaves continuously absorb and give off gases. In the early seventeenth century, Stephen Hales found that in a 24-hr period, a sunflower plant lost seventeen times more water vapor per unit volume than a human being. This loss was largely through the leaves. More recently, another investigator has calculated that a 10,000-ft^2 lawn (an average house lot is about this size) loses 600 gallons of water a week. In the early twentieth century, several investigators found that leaves can absorb carbon dioxide at rates even faster than a solution of sodium hydroxide, which itself has a high affinity for CO_2 molecules. The special arrangement of cells within the leaf makes possible such unusual rates of gas and water exchange. Let us first examine a few characteristics of the leaf's gas exchange.

It should be emphasized that water loss from the leaves of a plant is an inevitable result of gas exchange. Thus, although we will have to discuss each process separately for the sake of simplicity, the two are inevitably interrelated.

Gas Exchange

The principal sites of gas exchange in leaves are the stomata. Investigators have shown that the number and spacing of stomata on the lower surface of leaves greatly facilitate their efficiency in gas exchange. The original experiments to determine this efficiency were performed in 1900 by two English botanists, H. T. Brown and F.

Escombe. Using the leaf of the tree *Catalpa bignonoides,* which has stomata only on the lower surface, these investigators found that under conditions favorable for photosynthesis, 1 cm^2 of leaf surface absorbed 0.07 cm^3 of carbon dioxide per hour. During the same time period, the absorption of carbon dioxide by 1 cm^2 of sodium hydroxide was found to be from 0.12 to 0.15 cm^3—just twice that of the plant. However, since the stomata comprise only about 1 percent of the area of the leaf surface (the remaining 99 percent being covered with cuticle), the absorption through the stomata was calculated to be at a rate fifty times higher than the rate of uptake by the sodium hydroxide solution.

To explain this seeming paradox, Brown and Escombe investigated the rate of diffusion in a model system composed of thin plates with small openings of known diameters. Using solutions of sodium hydroxide contained in test tubes covered with these perforated plates, they found that the smaller the opening, the more rapid the rate of diffusion of carbon dioxide through this opening per unit area.

In other experiments with this model system, Brown and Escombe discovered that diffusion was most rapid when the distances between the openings were at least eight to ten times their diameter. When the holes are closer together, the rate of diffusion is less. To explain this they hypothesized that, in still air, diffusing molecules tend to spread out like a fan as they leave small openings (Fig. 8.21a). If the holes are spaced at least eight to ten diameters from each other, the diffusing molecules of one hole do not interfere with those of the adjacent one (Fig. 8.21c).

Thus, by using a physical model, these investigators were able to calculate that stomata possess adequate capacity for carrying gases from and to the interior of the leaf. The open stomata of a sunflower leaf, for example, are able to permit the passage of three to six times as much water vapor as anyone had ever measured. Therefore, while stomata constitute only 1 percent to 3 percent of the epidermal surface, they possess the capacity to carry gases to a degree far

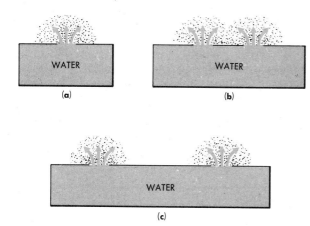

Fig. 8.21
Diffusion of water vapor from a water surface into the atmosphere. (*a*) Water molecules tend to form a dome-shaped diffusion shell as they leave a small opening. (*b*) If the openings are too close together, their diffusion shells overlap and the rate of diffusion is decreased. (*c*) A maximum rate of diffusion occurs when the openings are well separated from one another.

greater than their size suggests. The spacing of the stomata, and their ability to regulate the size of the opening outside the plant, make this efficiency possible.

Water Loss

The intracellular air spaces in the mesophyll contain air that is always saturated with water vapor, so as gas is lost from the leaves, it carries water vapor with it. As already noted, the process of water loss from leaves is called transpiration, and this evaporation provides a pull—transpiration pull—that lifts the column of water upward from the stem and root. But transpiration can be harmful to plants if it occurs to too great an extent. When plants are transpiring rapidly, as in the middle of the day when the temperature is highest and the sunlight is strongest (the period of high photosynthetic activity), plants may droop and become wilted. Wilting is simply the result of too much transpiration. This occurs when the plant is caught between two contradictory tendencies: to keep the stomata open for the exchange of gases necessary to photosynthesis, and to keep the stomata closed to prevent excessive water loss. Usually no permanent harm comes from wilting, since the proper water balance can be restored to the leaf cells during

the cooler period of the evening. Wilting is caused simply by the loss of turgidity in the leaf cells, producing a lack of firmness in the organ as a whole.

Regulation of Stomatal Size

Guard cells.   How do the guard cells regulate the rate of water loss from leaves—a process of crucial importance to the plant? During the opening and closing of the stomata, it can be seen that the two epidermal cells surrounding each opening or stoma have thicker walls on the side toward the stoma, than on the side away from it. Furthermore, when the stoma is open, each of the guard cells is shaped differently than when the stoma is closed (Fig. 8.22).

Fig. 8.22
Diagrams showing how guard cells function in regulating size of the stoma. Note that the inner walls next to the stoma are thicker than the outer walls (*a*). As the guard cells absorb water from the surrounding epidermal cells, they swell unevenly and the stoma is opened more widely (*b*). While the stoma is open (*c*), water vapor can diffuse out of the leaf, and exchange of oxygen and carbon dioxide between the leaf and the atmosphere can occur.

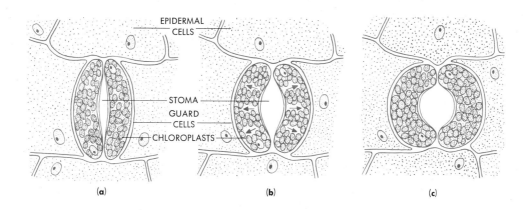

On the basis of these observations, the German botanist Hugo von Mohl hypothesized that these changes in guard cell shape are due to changes in the turgidity of the guard cells. In 1856 von Mohl designed an experiment to test this hypothesis. He isolated stomata of *Amaryllis* from the adjacent cells by making razor cuts parallel with the leaf surface. The small pieces of tissue thus freed were floated in plain water and in water in which sugar was dissolved. In plain water the stomata opened; in sugar solutions they closed.

It seemed likely that some sort of turgor mechanism was operating here. Still to be explained, however, was how changes in turgidity of guard cells could open and close the stomata. During the 1880s, the German botanist S. Schwendener concluded from anatomical and experimental studies that the thin wall of the guard cell (away from the stoma) elongates as the turgor increases. Since the thickened wall bordering the stoma cannot stretch as readily as the thinner outer wall, it is forced into a more semicircular shape by the expansion of the thinner wall (see Fig. 8.22b and c).

Schwendener's hypothesis poses other problems, however. If the opening and closing of the stomata are due to changes in the turgidity of the guard cells, then what controls changes in the turgor difference between the guard cells and the other epidermal cells lying nearby? Von Mohl hypothesized that, since in the light the chloroplasts in the guard cells produce sugar and starch, the concentration of water in these cells would be reduced. Water would then tend to pass by osmosis from adjacent epidermal cells (which do not contain chloroplasts) into the guard cells. This water movement would increase the turgidity of the guard cells, causing them to swell and open the stoma. At night, when photosynthesis is not occurring, the concentration of sugar in the guard cells would decrease as the food products were moved elsewhere in the plant. The concentration of water in them would increase, and water would move out of the guard cells, causing them to become less turgid and closing the stoma.

Of course, von Mohl's hypothesis rests squarely on the assumption that photosynthesis does indeed occur in the guard cells. This assumption was not experimentally challenged until nearly a century after von Mohl first proposed his hypothesis. In the early 1950s, autoradiography experiments showed that guard cells of onion and barley plants *do* take up greater amounts of $C^{14}O_2$ in light than in darkness. However, in order to be an effective factor in the production of osmotically active material, the rate of guard cell photosynthesis must be high enough to account for the rates of osmotic pressure change actually observed during stomatal opening. In onion and barley, however, the maximum rate of photosynthesis in the guard cells is only about 1/50 of the rate necessary to account for the observed osmotic pressure changes. It is obvious that von Mohl's hypothesis alone is not sufficient to account for the opening and closing of the stomata.

In 1908, the botanist F. E. Lloyd put forth a hypothesis proposing that the stomata may open in light because starch in the guard cells is enzymatically changed into sugar, thus increasing the osmotic pressure and raising the turgor. Lloyd's hypothesis was supported by his findings that, in the desert plant *Verbena ciliata*, the starch content of the guard cells was very low in the morning when the stomata opened. In the afternoon the starch content began to increase, culminating in closure of the stomata at night when the starch content of the guard cells was high.

In the 1920s, Lloyd's original hypothesis was extended by the botanist J. D. Sayre, who postulated that changes in the degree of acidity-alkalinity (pH) of the guard cell contents were involved. In experiments with *Rumex patientia*, Sayre showed that the opening of the stomata was accompanied by a decrease in guard cell starch, an increase in sugar, an increase in osmotic pressure, and a change in the pH toward alkalinity. During the night, the reverse occurred. In 1948, Sayre's hypothesis received strong support with the discovery that the guard cells of tobacco contain large amounts of phosphorylase, an enzyme that catalyzes the conversion of starch into glucose-1-phosphate. This reaction is pH-dependent: as the acidity is reduced, the activity of the enzyme is increased. The reduction in acidity would result from the photosynthetic removal of CO_2 which, in water, forms carbonic acid.

The Lloyd-Sayre hypothesis of stomatal movement was almost universally accepted for many years. In 1959, however, doubts began to arise. It was pointed out that the enzymatic conversion of starch to glucose-1-phosphate would produce no change in the osmotic pressure of the guard cells, because for each glucose-1-phosphate molecule formed, one inorganic phosphate ion would be taken out of solution. It was suggested that, in order to raise the osmotic pressure of the guard cells, a further reaction involving glucose-1-phosphate would be necessary, and that it would require energy for the re-formation of starch if the stomata were to close.

Recent evidence, however, suggests that energy *is* necessary for stomatal opening, but apparently not for stomatal closing. During the 1960s, the biochemistry of the guard cell mechanism was very intensively investigated, especially in the laboratory of Israel Zelitch at the Connecticut Agricultural Experiment Station in New Haven. Zelitch and his co-workers have shown evidence of a relationship between the stomatal mechanism and the metabolism of glycolic acid, an early byproduct of photosynthesis. Using chemicals that inhibit the enzymatic oxidation of glycolic acid, they have found that the stomata of tobacco plants no longer open in the light as usual. These results are interpreted as indicating that the oxidation of glycolic acid in the guard cells is necessary for stomata to open rapidly in the light. Glycolic acid may be involved as a necessary intermediate in the synthesis of carbohydrate by guard cells, which would

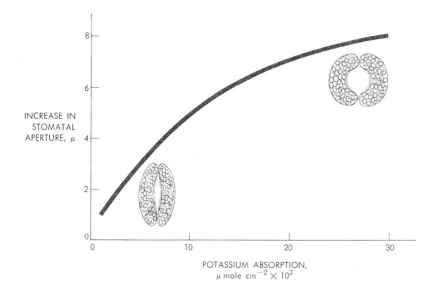

Fig. 8.23
Relationship between the absorption of potassium by guard cells and the opening of the stoma during a 3-hr period. (Adapted from R. A. Fischer, *Science* 170, 1968, p. 785, with permission.)

produce an increased osmotic pressure. Or glycolic acid might be involved in some kind of energy-requiring mechanism that increases the turgidity of guard cells (thus opening the stomata) through the use of ATP produced by the noncyclic reactions of photosynthesis in the guard cell chloroplasts.

The role of potassium.   That the opening of stomata may be related to the uptake of potassium was suggested in the mid-1940s; this idea has since gained increasing support. In 1968, R. A. Fischer floated very small strips from the lower epidermis of the broad bean plant (*Vicia faba*) on potassium chloride solutions of low concentrations. The stomata opened in the light when the air lacked carbon dioxide. The wider the

stomata became during the opening movement, the greater the amount of potassium that entered the guard cells (Fig. 8.23). From these experiments, Fischer hypothesized that in the light, plus air free of CO_2, the basic mechanism of stomatal opening is stimulation of potassium absorption by the guard cells. Additional support for this hypothesis comes from observations of relatively high concentrations of potassium in the guard cells. Fischer suggested that the starch-to-sugar conversion postulated by the classical theory of stomatal movement is of only secondary importance, though perhaps closely linked to potassium absorption.

Quite recent studies have yielded direct evidence that potassium ions (K^+) do get into the guard cells as swelling occurs. Figure 8.24 shows the results of studies

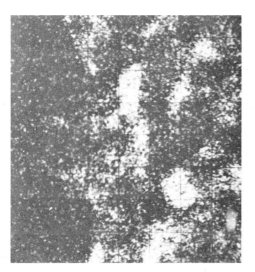

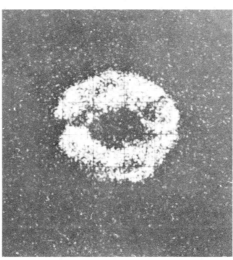

(a)   (b)

Fig. 8.24
X-ray images showing localization of K^+ in guard cells of closed (*a*) and open (*b*) stoma of the plant *Vicia faba*. The concentration and distribution of K^+ are indicated by the white spots. (Reprinted with permission from G. D. Humble and K. Raschke, "Stomatal opening quantitatively related to potassium transport," *Plant Physiology* 48, 1971.)

The size of stomatal openings is regulated by the interaction of two factors: photosynthetic rate and the humidity of the air surrounding the plant.

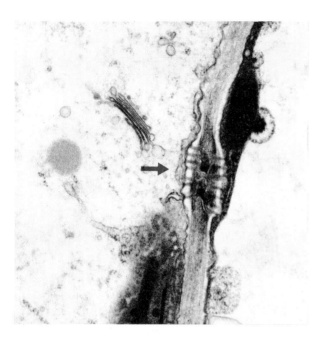

Fig. 8.25
The arrow points to plasmodesmata connecting a guard cell with neighboring epidermal cells in the leaf of the horsebean. (Courtesy Dr. James E. Pallas, Jr.)

with the electron probe microanalyzer. This technique involves using a high-energy electron beam at a frequency that will cause emission from a particular ion in solution—in this case K^+. The electron probe microanalyzer can be used for any element, of course. The sample to be investigated is scanned by the electron beam and then the material is exposed to X-ray film. Emissions from the ion in question (and hence the location of the ion) can then be detected on the developed film. As the photographs show, in a closed stoma there is no K^+ localized within the guard cells. But in an opened stoma, K^+ is clearly concentrated in the guard cells.

The Role of Plasmodesmata. Anatomical evidence has recently provided a further clue to how guard cells are able to control the stomatal opening so precisely. James E. Pallas of the U.S. Department of Agriculture Piedmont Conservation Research Center at Watkinsville, Georgia, has photographed cytoplasmic bridges (**plasmodesmata**) connecting guard cells with neighboring

epidermal cells (see Fig. 8.25). Plasmodesmata are known to form intercellular communication bridges between most kinds of cells, but until recently botanists thought guard cells were exceptions. Pallas's observation may allow us to understand the regulation of guard cell activity in a way more integrated with other cells in the leaf. Plasmodesmata may be a means by which guard cells get information from other nearby cells that influence their metabolism.

In summary, it is obvious that efforts to discover the mechanism of stomatal movement have met with only partial success. Researchers have learned a lot about the effects of light, carbon dioxide concentration, temperature, water content of the leaf, metabolic inhibitors, and other factors on the movements of the stomata. Still, the way the turgor of guard cells changes is not really known. Certainly both CO_2 concentration and the conversion of starch into sugar are involved. There is also good evidence that the mechanism of stomatal opening is different from that of stomatal closing, and that one is not simply a reversal of the other. For example, energy from ATP is probably used in the opening but not in the closing. Finally, the meaning of plasmodesmata connecting the guard and neighboring epidermal cells is still not clear.

This abbreviated view of the experimental work on a single problem—stomatal opening and closing—indicates the tortuous path often involved in investigating complex biological phenomena.

8.11
MOVEMENT OF LIQUID AND METABOLIC
WASTE OUT OF THE PLANT

If young corn seedlings are watered heavily and then covered with a glass jar, a watery liquid soon begins to collect as droplets at the tips of the leaves. The droplets enlarge and eventually fall off or run down the surface of the leaves. Similar droplets of water are frequently seen in the early mornings at the tips of grass leaves and along the margins of clover and other plants growing in the lawn, garden, or woods. (Some of the water droplets on the leaves at the same time are dew, which is simply water that has condensed from the moisture of the atmosphere.) This exudation of liquid water from the leaves of plants is known as **guttation** (from the Latin word *gutta*, meaning droplet). Guttation has been observed in plants of 333 genera belonging to 115 families of flowering plants (no examples of guttation are known in conifers).

The quantity of water lost by guttation from an individual plant varies greatly. A corn seedling may exude only a few drops of water. Elephant's-ear plants may lose 100 cm^3 during a single night from a young leaf tip. Chemical analysis of the exuded liquid shows it contains in solution such inorganic substances as nitrates and calcium, and such organic substances as enzymes and amino acids.

The liquid exuded during guttation escapes from the plant through structures called **hydathodes.** Hydathodes are composed of an enlarged, stoma-like opening leading into a large chamber. Near each hydathode are tracheids and xylem vessels at the ends of the leaf veins. It has been hypothesized that liquid exudes through hydathodes due to root pressure. The liquid is assumed to be forced from the vein endings, through intercellular spaces, and out of the leaf through the hydathodes.

What role does guttation play in the life of the plant? One hypothesis holds that it may be one way in which the turgor of leaf cells can be regulated, preventing it from becoming too high when transpiration is not occurring (as at night). There is no experimental evidence to support this hypothesis, however.

Liquids of various chemical compositions are exuded from various other parts of plants, including nectar from flowers and resin from glandular epidermal hairs.

Unlike animals, plants do not possess well-defined organs for removing potentially injurious metabolic byproducts from their bodies. They have, however, evolved other ways in which toxic chemicals are made nontoxic or transported to places in the plant where the toxicity will do little harm. One metabolic byproduct of respiration in all living cells of the plant, carbon dioxide, probably moves in solution into the xylem and through it to the leaves. Here the carbon dioxide may either be used again in photosynthesis or diffuse into the atmosphere via the stomata. Some wastes are simply excreted into cell vacuoles and stored there during the active life of that cell. For example, oxalic acid is immobilized as crystals of calcium oxalate inside the vacuole. These crystals may be highly distinctive in form.

Metabolic byproducts may also be excreted through the cell membrane into the nonliving cell wall. Here some of these byproducts may be enzymatically converted into lignin, one of the molecular components of the cell wall. The primary function of lignin may be simply that of storing some of the waste products of cell metabolism in a useful form. Evidence for this hypothesis has come from experiments showing that some of the precursors of lignin, such as phenolic glycosides, may be formed as byproducts of the metabolic pathways used by cells to produce the cellulose and other polysaccharide components of the cell wall. Lignin production by vascular plants, therefore, may have evolved as an adaptation enabling land plants to solve some of the problems of metabolic waste disposal. Since the presence of lignin in the cell walls gave added strength to the upright plant body, plants with the genetic ability to produce the enzymes for lignin synthesis were better adapted for land life.

Some toxic materials may be eliminated from the plant body by the leaves falling from the trees during the autumn and by the periodic shedding of bark. One metabolic byproduct, tannin, has been found in minimal amounts in the living cells of the cambium but in increased quantities in the parenchyma of the differentiating phloem and xylem. Periodically these tannin-containing cells are shed from the plant as part of the bark. Tannin is also produced in the leaves and eliminated from the plant during the autumnal leaf fall.

Many byproducts of plant metabolism are transported away from the cambium and living parenchyma of the xylem in the direction of the inner sapwood, where they are stored. Indeed, the formation of the heartwood in trees may be thought of as the result of such an excretory process. As the waste materials accumulate in the living parenchyma (in both the vertical and the horizontal rays) a toxic level is finally reached, the cells die, and the first cylinder of heartwood is produced. With time, the inward-moving waste products produce another layer of heartwood. In this manner, the sapwood-heartwood boundary moves outward as the diameter of the tree increases (see Fig. 9.3).

Summary

1. The vascular plants—those containing true vascular, or conducting, tissues—are organized into two main portions: root and shoot. The latter consists of stem, leaves, and flowers. Roots serve to anchor the plant and to absorb water and various ions. Stems serve to raise the leaves where they can obtain sunlight and air for photosynthesis.

2. The land-dwelling plant must perform four major functions in order to survive:

    a) Prevent excessive water loss.

    b) Absorb and distribute water.

    c) Withstand the force of gravity.

    d) Produce new cells and tissues, including the production of gametes for reproducing the species.

3. The roots of most vascular plants consist of a central cylinder, the stele, containing xylem and phloem and surrounded by a large layer of parenchyma tissue. Root hairs project from the epidermal surface and serve to greatly increase the surface area for absorption.

4. Water and minerals move into the root by osmosis, diffusion, and active transport.

5. The stem of vascular plants contains a large amount of xylem and phloem. The phloem generally lies external to the

xylem, the two vascular tissues being separated by a layer of meristematic tissue (the cambium) responsible for producing new xylem and phloem cells during the growing season.

6.   The movement of water and dissolved substances throughout the plant is called translocation. Materials absorbed by the roots travel in a variety of directions through the plant: upward, laterally, and downward.

7.   The vast majority of the water and dissolved ions taken in by the roots is transported upward by the xylem. The bulk of the carbohydrates produced by the leaves in photosynthesis is transported downward by the phloem.

8.   A number of factors interact to allow vascular plants to raise a column of water upward: cohesion, capillarity, root pressure, and transpiration pull.

9.   The forces responsible for downward movement through the phloem are not well understood. To some ex-

tent mass flow and cytoplasmic streaming in the sieve tubes may be involved. Biologists are not yet certain of the roles each force may play.

10.   Leaves consist of a blade, petiole, and stipule. Leaves develop from leaf primordia; generally vascular plants produce leaves only once a growing season.

11.   Leaves generally consist of three layers: an epidermal layer (on both sides of the leaf), a palisade layer, and a spongy mesophyll layer. Most photosynthesis takes place in the palisade and spongy mesophyll layers.

12.   Leaves function for the exchange of gases. Gases and water vapor move in and out of the leaves through the stomata. The rate of gas exchange is very efficient, facilitated partly by the spacing of the stomata across the leaf surface. Stomatal opening size depends upon the interaction of two factors: photosynthetic rate and humidity of the atmosphere surrounding the plant.

Exercises

1.   How does the internal structure of the root compare with that of the stem? How do the differences show the specialized functions each organ performs?

2.   In what important structural feature does a vessel element differ from a tracheid? How is this difference related to the function of transport in the plant?

3.   Compare and contrast the shoot and root in regard to their external structures.

4.   For each of the following conditions, state whether the guard cells of a leaf would be expected to be: (1) fully open,

(2) partially open, (3) completely closed.

    a) Atmosphere very moist, much water in soil, sunlight plentiful.
    b) Atmosphere dry, soil quite moist, sunlight plentiful.
    c) Atmosphere dry, not much soil water, cloudy.
    d) Atmosphere moist, plentiful supply of soil water, cloudy.

5.   List in correct order the cells and tissues through which a molecule of water travels from its point of entry into the root until it is used in photosynthesis.

Chapter 9
Plant Structure and Function II: Growth and Integration

9.1 INTRODUCTION

The essential functions of a plant are (1) the capture of life-sustaining energy by the leaves, and (2) the propagation of the species by the reproductive organs. The organs must be supported by a structure, the stem, that provides for their advantageous orientation in space and for the transport of materials involved in the metabolism of their various cells. How does such a complex organism grow and develop? How does it reproduce and respond to environmental factors such as gravity, light, and the changing seasons? Most important of all, how are these functions integrated and controlled in the whole organism?

In the present chapter we focus our attention on the problems of plant growth and response to environmental stimuli. These topics provide exciting and useful areas for investigating how the plant system is organized to function as a whole. A plant is not a static, inert structure. It is constantly interacting with its environment, adjusting itself to changing conditions, and responding in highly controlled and integrated ways.

9.2 GERMINATION

In the Bible, the word "seed" actually refers to the sperm cells that fertilize the egg (although the process of fertilization, as well as sperm cells themselves, was not as yet understood). In plants, however, the seed is a structure that results after fertilization has occurred. A plant seed usually consists of a protective coat, stored food, and the plant embryo. The seed may lie inactive, or **dormant,** for long periods of time before being stimulated by certain environmental factors (such as water and temperature) and beginning its initial growth or **germination.** Germination refers to all the steps from the time a seed takes in water (a process called *imbibition*) until the seedling is self-sustaining.

Anyone who has ever planted a garden knows that there are certain ways one must plant each type of seed to ensure proper germination. Some, for example, must not be planted too deep. This generally means that the amount of stored food in such seeds is only enough to get the seedling to a surface only a few centimeters away; if they are planted any deeper, the seeds "run out of gas" before the shoots can reach the light and begin to photosynthesize the energy-rich compounds necessary for further growth.

In wheat seedlings, germination involves the synthesis of growth hormones called gibberellins, to be discussed shortly. These hormones diffuse through the stored food or **endosperm** to its outermost layer, the **aleurone.** The aleurone, in response to the hormones, begins to secrete an enzyme (α-amylase) that converts the starch of the endosperm into simple sugars (monosaccharides). These sugars, in turn, become available to the embryo for the energy necessary to get its shoot to the surface and send down roots for additional nutrients.

The breakdown of sugar and other compounds occurs by hydrolysis. The water molecules needed for this hydrolysis are supplied by the water imbibed by the seeds. This water is also used for the translocation of reserve materials to the root and shoot. As the wet weight of the seeds increases, the dry weight—the portion of the germination seedling left after water is extracted—decreases. In soybeans, this decrease is about 25 percent in the seedling and 70 percent in the cotyledons, those "seed leaves" that store the food in the dicotyledonous plant. The decrease in dry weight in the whole seedling is due mainly to the loss of carbon atoms via carbon dioxide during respiration. Besides carbohydrates, fats, too, are hydrolyzed. The fats are converted to the dissaccharide sucrose, resulting in a slight temporary increase in soluble carbohydrates in whole seedlings and cotyledons. Reserve proteins in germinating seeds are hydrolyzed to amino acids. These amino acids are transferred to the tissues and used mainly for growth.

As one would expect, the DNA content of whole seedlings, roots, and shoots increases rapidly because of active cell growth, except in the cotyledons, where cell

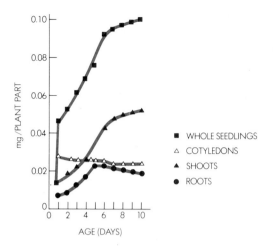

Fig. 9.1
Changes in DNA content in entire germinating seedlings and their constituent parts. Since both the shoots and roots are experiencing rapid cell division, there is a rapid increase in DNA production. This is not the case in the cotyledons, where cell division does not normally occur during germination.

division does not normally occur during germination (see Fig. 9.1). The RNA content of seedlings increases for a few days and then declines rapidly to show a net decrease. This is because RNA is involved in the synthesis of various types of cell constituents. Thus in a rapidly growing shoot, with many new cells and much biosynthesis, the percentage of RNA would be expected to be high. As the percentage of mature tissue to actively growing young cells increases, the proportional amount of RNA declines.

9.3
INCREASE IN SIZE OF THE PLANT BODY

"Great oaks from tiny acorns grow." This old saying is a capsule summary of a great deal of botany. A seedling may in time grow into an enormous tree with a trunk hundreds of feet in height and many feet in diameter. The longer a plant lives and the larger it grows, the greater must be its ability to withstand the increasing weight of its leaves and branches and the stress imposed by high winds. Further, in the young plant there is only a small amount of food-storage, water-conducting, and supporting tissue. As the seedling grows into a tree, some provision must be made for increasing the amount of conducting and supporting tissues. What pattern of growth is characteristic of such plants?

Plants grow upward and downward by adding new cells at their tips. This is true of roots as well as stems. Cell division, leading to increased height of the stem or depth of the root, occurs only at the tips of stems or

roots. For a few millimeters behind the apical meristem, in an elongation region, cells increase in length. Once the elongation process is completed, that portion of the root or stem does not change its distance from the surface of the ground. By and large, then, plants grow like a brick wall—by adding new building blocks at the top (or bottom, in the case of roots). As in the brick wall, the position of building blocks lower down is not altered by the increases at the top. A region of a tree trunk that was one foot above the ground in the sapling will still be one foot from the ground in the mature tree.

Growth in width, or thickness, of the plant is also an additive process. But it is a bit more complex than the type of growth associated with increase in length. Growth in width is accomplished largely by the cambium, a region of lateral (as opposed to apical) meristem. Most of the tissues of the mature plant are derived directly from the cambium rather than from the apical meristem.

The vascular cambium has a dual origin in the stem (Fig. 9.2). A portion arises in each embryonic vascular bundle, between the xylem and the phloem (see Fig. 9.3). The remainder originates from parenchyma cells lying between the vascular bundles. In the root, some of the cambium originates from cells located on the inner side of the phloem, in regions between the points of a core of primary xylem. Additional cambium is formed later by cells of the pericycle. In each organ the two areas of cambium become joined, forming a complete cylinder of meristematic cells throughout the length of the stem and root (except the very young twigs and root tips). The cambium begins to produce new cells about the time that the primary growth (resulting from cells formed by the apical meristem) is completed, usually during the first few weeks of the growing season. On the side next to the primary xylem the new cells differentiate into secondary xylem, while those on the inner side of the primary phloem become secondary phloem.

The cells of the secondary xylem usually have heavily lignified walls, a feature that makes them hard and nearly incompressible. Those of the secondary phloem, however, have thin walls and are delicate. The secondary xylem accumulates year after year. The secondary phloem, however, is progressively crushed and destroyed as the new secondary xylem presses out-

Fig. 9.2
Diagrams showing the development of the stem (left) and root (right) of a dicotyledonous plant. Note that the tissue patterns of both organs become closely similar after secondary growth has occurred. (Left: modified from R. M. Holman and W. W. Robbins, *Elements of Botany.* New York: Wiley, 1940. Right: from K. Esau, *Plant Anatomy,* 2d ed. New York: Wiley, 1965.)

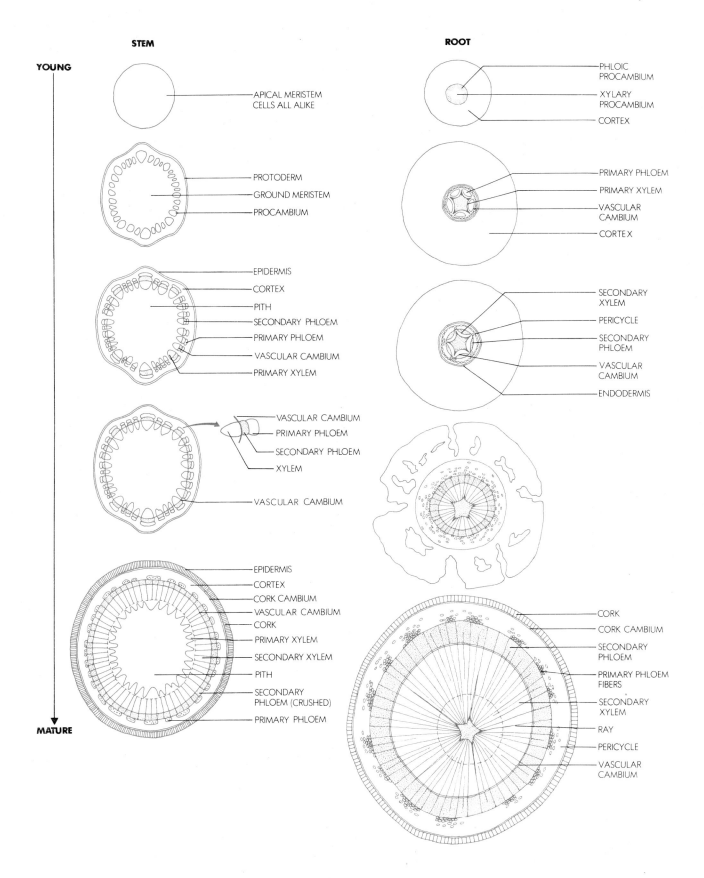

STEM

ROOT

YOUNG

APICAL MERISTEM
CELLS ALL ALIKE

PHLOIC
PROCAMBIUM

XYLARY
PROCAMBIUM

CORTEX

PROTODERM

GROUND MERISTEM

PROCAMBIUM

PRIMARY PHLOEM

PRIMARY XYLEM

VASCULAR
CAMBIUM

CORTEX

EPIDERMIS

CORTEX

PITH

SECONDARY PHLOEM

PRIMARY PHLOEM

VASCULAR CAMBIUM

PRIMARY XYLEM

SECONDARY
XYLEM

PERICYCLE

SECONDARY
PHLOEM

VASCULAR
CAMBIUM

ENDODERMIS

VASCULAR CAMBIUM

PRIMARY PHLOEM

SECONDARY PHLOEM

XYLEM

VASCULAR CAMBIUM

EPIDERMIS

CORTEX

CORK CAMBIUM

VASCULAR CAMBIUM

CORK

PRIMARY XYLEM

SECONDARY XYLEM

PITH

SECONDARY
PHLOEM (CRUSHED)

PRIMARY PHLOEM

CORK

CORK CAMBIUM

SECONDARY
PHLOEM

PRIMARY PHLOEM
FIBERS

SECONDARY
XYLEM

RAY

PERICYCLE

VASCULAR
CAMBIUM

MATURE

OUTER BARK (CORK)

INNER BARK (CORK CAMBIUM AND PHLOEM)

CAMBIUM CELL LAYER

SAPWOOD

XYLEM

HEARTWOOD

Fig. 9.3
The anatomy of a tree trunk. (Courtesy St. Regis Paper Company, New York.)

ward. The tree does not become devoid of secondary phloem, however, since the cambium is constantly producing new phloem during each growing season. As the secondary xylem accumulates, the cylinder of cambium increases in circumference through radial division of some of its cells. Thus the cambium can be said to move outward as the secondary xylem increases in amount.

In plants growing in temperate regions, the cambium ceases producing new cells toward the end of the growing season and enters a dormant state that usually lasts until the following spring. The xylem produced by the cambium during one growth period constitutes a layer which, in transverse section, appears as a ring. Since generally only one such layer is produced each year each is termed an **annual** or growth ring. One annual ring of growth can be distinguished from another because the cells produced by the cambium in the spring are larger in diameter than those formed during the

The annual rings of a tree provide a clue about the climates of past years: large rings indicate more rainfall than small rings.

Plants grow in length by the addition and expansion of cells at the tips, or apical meristems. They grow in width by adding new cells on either side of the cambium, the lateral meristem.

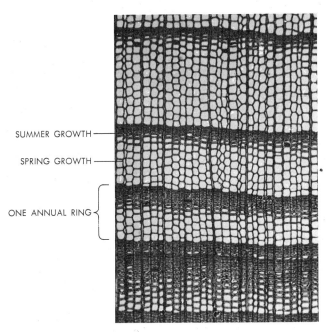

SUMMER GROWTH

SPRING GROWTH

ONE ANNUAL RING

Fig. 9.4
Cross section of stem of bald cypress (*Taxodium distichum*), a gymnosperm, showing four years' growth. Note the abrupt line of demarcation between the end of one season's growth and the beginning of the next. (Courtesy Turtox, Chicago.)

summer (see Fig. 9.4). The difference in growth rate is due in part to greater rainfall in the spring. Since the total width of any given ring depends on the climatic conditions of that particular growing season, climatologists can use borings from very old trees to determine the climate of the earth over the past 2000 or more years.

Although the secondary xylem produced each year appears as a ring in transverse sections, it is actually an open-ended cone that encloses and extends above the cone formed during the previous year. We can view the tree trunk as a series of successively longer cones of xylem stacked over each other with the big end toward the base of the tree. The cone formed last is the outermost and the longest. The secondary xylem in the roots may be visualized in the same manner, with the larger end of each xylem cone near the base of the trunk (Fig. 9.5).

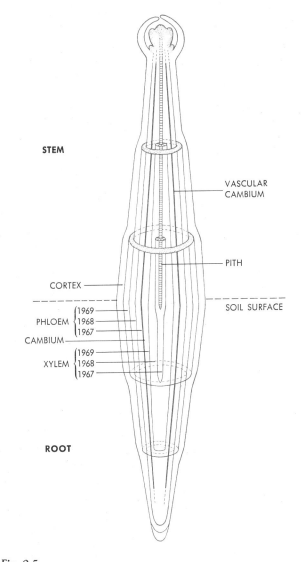

STEM

VASCULAR CAMBIUM

PITH

CORTEX

SOIL SURFACE

PHLOEM { 1969 / 1968 / 1967

CAMBIUM

XYLEM { 1969 / 1968 / 1967

ROOT

Fig. 9.5
Diagram showing relative positions of secondary xylem and phloem in a three-year-old plant.

9.4
PLANT HORMONES

For many centuries observers have noticed that plants on a windowsill grow toward the light. Intrigued by such observations, in 1880 Charles Darwin and his son Francis set out to discover why growing plants always bend toward a light source. Using seedlings of canary grass (*Phalaris canariensis*), they set up several experimental groups (see Fig. 9.6). One group of seedlings was allowed to germinate normally; these served as controls. The tip of each **coleoptile** (a rapidly growing sheath that envelops the young leaves) in the other group was covered with a cap of very thin tinfoil. The plants were placed near a window so they would receive light from only one direction. After 8 hr of daylight, the Darwins observed that the uncapped plants had curved strongly toward the light. The plants covered with tinfoil, however, grew straight up; they failed to show a response to the outside stimulus.

Such growth responses to external stimuli such as light or gravity are called **tropisms** (from the Greek word *tropos*, "to turn"). Response to light is called **phototropism;** to gravity, **geotropism;** to water, **hydrotropism;** to touch, **thigmotropism,** and so on.

The Darwins then removed the caps from ten of the straight seedlings. After an 8-hr exposure to the light, nine of these became greatly curved toward the light, while one curved only moderately. Coleoptiles whose tips were cut off remained upright. From these experiments it seemed clear that the tip of the plant was somehow influential in determining the reaction to light. The Darwins hypothesized that "some influence is transmitted from the upper to the lower part, causing the latter to bend." However, they did not know what the precise nature of the influence was or how it acted to cause its effects.

Demonstrating the Existence of Auxin

Thirty years later, Danish botanist P. Boysen-Jensen cut off the tips of oat coleoptiles, placed a block of gelatin on the cut surface of the stumps, replaced the tips, and illuminated the coleoptiles with a light from one side (Fig. 9.7a). The portion of the coleoptiles below the gelatin curved toward the light just like the intact coleoptiles. It seemed clear, therefore, that the influence hypothesized by the Darwins had moved downward from the tip through the gelatin to the portion of the stem below. In another experiment, Boysen-Jensen inserted a thin piece of mica, which is impermeable, into a trans-

Fig. 9.6
Experiments of Charles and Francis Darwin on phototropism of canary grass seedlings. (*a*) The coleoptile bends toward the light. (*b*) The coleoptile does not bend when its tip is covered with tinfoil.

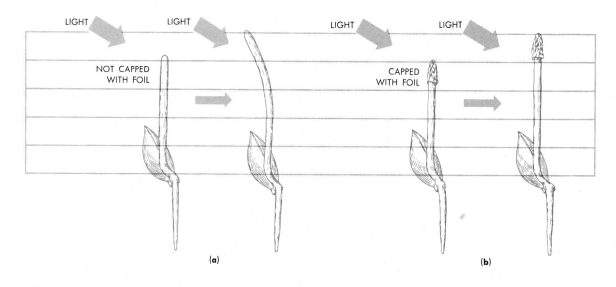

Hormones are "chemical messengers" produced in one part of the organism and carried to all other parts. Hormones generally have target tissues that they affect in specific ways. Auxins and gibberellins are two kinds of plant hormones that have meristematic tissue as their targets.

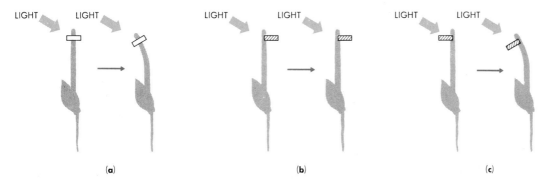

Fig. 9.7
Boysen-Jensen's experiments. (*a*) When the tip of the oat coleoptile is removed and a block of gelatin is inserted between it and the stump, the coleoptile curves toward the light. (*b*) If a mica plate is inserted into a transverse cut on the dark side, the coleoptile does not curve. (*c*) When the mica is placed on the illuminated side, the coleoptile curves toward the light.

verse cut halfway through the coleoptile. When the mica was on the dark side of the cut surface (away from the light), the coleoptile did not bend. When the mica was placed on the lighted side, however, normal curvature occurred. Boysen-Jensen concluded that the influence is most likely a diffusible chemical that moves down the dark side, accelerating growth curvature toward the light. Support for Boysen-Jensen's views was provided in 1918 by the Hungarian botanist A. Paál, who showed that the influence could move through gelatin, but not through mica, platinum foil, or cocoa butter.

In another experiment, Paál cut off the coleoptile tips, replaced them on one side of the stumps, and discovered that the coleoptiles curved away from the side with the tips *in the absence of light* (Fig. 9.8). It seemed clear from these experiments that the influence was indeed a material substance, probably some chemical or mixture of substances. Paál suggested that the growth-influencing material is produced in the coleoptile tip and normally moves downward along all sides. If the movement of the substance is disturbed on one side, as by light or a mica barrier, the growth of the cells on that side is decreased and the coleoptile grows in a curve.

Conclusive evidence that the influence that moves downward from the coleoptile tip is a chemical was finally provided in 1928 by Frits Went, a young Dutch botanist who later continued his work in the United States. From Paál's experiments, Went knew that the growth-promoting material can pass through a thin layer of gelatin between the coleoptile tip and stump. He reasoned that if the influence from the tip, first hypothesized by the Darwins, is chemical, then (1) the material should collect in the gelatin when excised coleoptile tips are placed upon it, and (2) later contact of this tip-exposed gelatin should result in growth of the coleoptile. To test the accuracy of these predictions, Went cut off a number of coleoptile tips of the oat plant and placed them on small blocks of gelatin (Fig. 9.9). After an hour

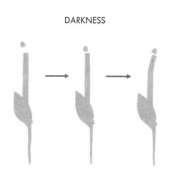

Fig. 9.8
Paál's experiment. When a coleoptile tip is cut off and replaced off center on the stump, the coleoptile curves away from the side with the tip. Such curvature occurs even in darkness.

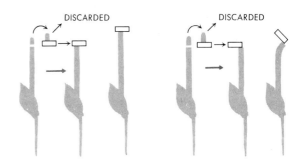

Fig. 9.9
Experiments of Frits Went to show that a growth-promoting substance produced in the coleoptile tip is responsible for the bending activity in plants. It was later found that this substance is one of the auxins.

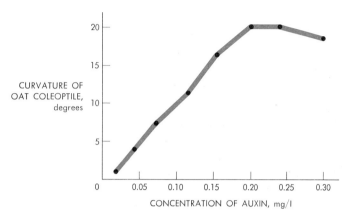

Fig. 9.10
Evidence of auxin action in oat coleoptiles. To obtain these data, auxin was unilaterally applied in gelatin blocks to cut stumps of coleoptiles (as in Fig. 9.9). The coleoptiles responded by curving away from the treated side. The degree of curvature was directly proportional to the concentration of auxin contained within the gelatin blocks. (After Frits Went and K. V. Thimann, *Phytohormones*. New York: Macmillan, 1937.)

or two, the tips were removed and the gelatin blocks placed on the cut surfaces of coleoptile tips. The stumps resumed growth shortly thereafter, but the stumps with plain gelatin blocks (the controls) did not grow very much. The predictions were found to be correct and Boysen-Jensen's hypothesis that the "influence" is a diffusible chemical was verified. Since the substance acts to increase growth it was named **auxin** (from the Greek work *auxein*, "to grow").

Went was also able to devise a simple test to determine the growth substance quantitatively. He placed the tips of numerous coleoptiles on a block of gelatin in order to increase the amount of growth substance collected by the block. After waiting a while to permit the material to diffuse into a gelatin, he cut the block into small cubes. Each cube was placed on only one side of the cut stump of other coleoptiles. The cells beneath the block began to grow more rapidly than those on the other side, and soon the decapitated coleoptile started to curve. Went was able to show that the amount of curvature was directly proportional to the concentration of the growth-promoting substance in the gelatin blocks (Fig. 9.10). This biological assay technique, which soon became known as the Avena Test (after the genus name

for the oat plant), has subsequently been used extensively to measure the quantities of growth-promoting substances present in plants.

It had taken nearly fifty years of painstaking work by numerous investigators to demonstrate conclusively the existence of the influence the Darwins postulated. But another problem remained: What is the chemical nature of this growth-promoting substance?

The Nature of Auxin

Went's work attracted the attention of investigators who were curious about the chemical nature of auxin. It was soon discovered that plant organs, including coleoptiles, are relatively poor sources of the material. The substance was found to be abundant, however, in malt extract (made from germinating grains), in cultures of the bread mold *Rhizopus*, and, surprisingly, in human urine. In 1934, after a long series of complex experiments, Dutch chemists K. Kögl and A. Haagen-Smit succeeded in isolating enough of the growth-promoting substance from human urine to permit a determination of its chemical composition. It turned out to be **indoleacetic acid (IAA)**, a substance long known to chemists but never connected with promoting growth in plants (see Fig. 9.11).

Is IAA the substance that enhances the growth of oat coleoptiles? If IAA is indeed the growth-promoting chemical, then it should be possible to isolate it from oats and other plants. Nevertheless, attempts to isolate IAA from plant tissues were unsuccessful until the 1940s, when minute quantities were extracted first from corn grains and later from young shoots of pineapple and from oat coleoptiles. The delay in confirmation was almost certainly due to the need of developing techniques and instruments able to detect very small amounts of substances in plant tissues; one investigator has calculated that the proportion of IAA present in 1000 g of plant tissue is something like the weight of a needle in 22 tons of hay. Perhaps the main reason for such a small amount of IAA in a plant is that this chemical is constantly being destroyed by an enzyme system (indoleacetic acid oxidase).

The auxins, of which indoleacetic acid (IAA) is apparently the most important, are synthesized in the youngest, growing parts of plants, such as the shoot apical meristem. One of their principal properties is the ability to stimulate the elongation and enlargement of cells (Fig. 9.12a). Auxins also stimulate cell division, including the initiation of roots on stem cuttings (Fig.

Up to a certain concentration, auxins stimulate plant growth. Beyond a certain concentration (around 0.20 mg/l), auxins may actually inhibit growth.

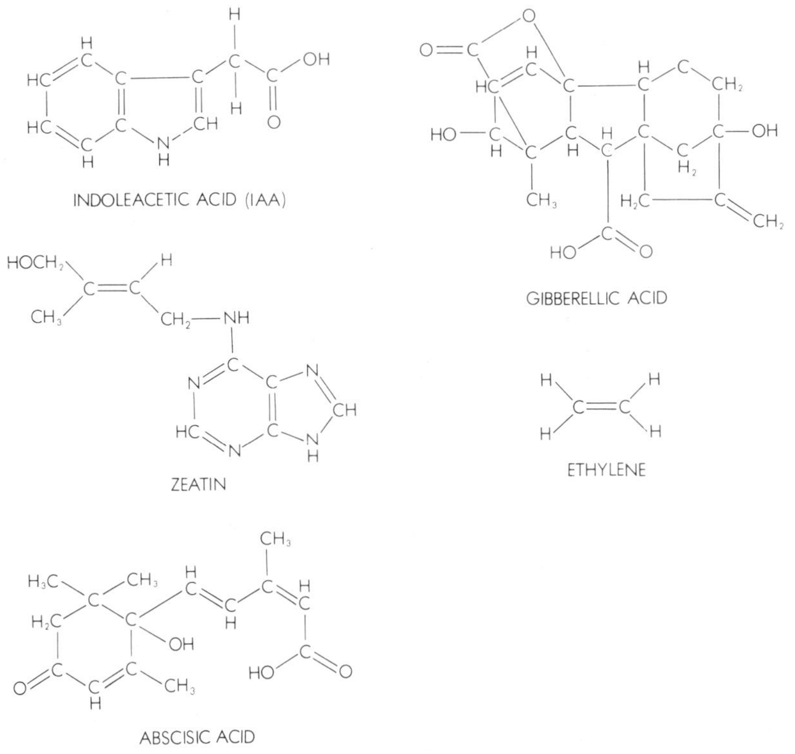

INDOLEACETIC ACID (IAA)

GIBBERELLIC ACID

ZEATIN

ETHYLENE

ABSCISIC ACID

Fig. 9.11
Structural formulas of five important plant growth-regulating substances. Note the presence of one or more rings in four of the molecules. Indoleacetic acid, abscisic acid, and the cytokinin zeatin all possess one six-atom ring, while gibberellic acid (GA_3) has two such rings. These rings give the molecule as a whole a definite shape and molecular architecture. The fifth molecule, ethylene, is a simple hydrocarbon with no rings.

9.13). This property is the basis of much of the plant propagation for commercial purposes carried out by floriculturists and horticulturists. Auxins are also involved in fruit development. As the seeds develop (following pollination and fertilization), auxin is formed first in the endosperm and later in the embryo, and it stimulates the enlargement of the cells of the ovary wall and other flower parts involved in producing the fruit. Tomato growers often deliberately initiate fruit formation by applying auxin to the flower. Auxins also have inhibitory properties and are used to prevent preharvest drop of apples. The effects of auxins are multiple and complex; probably they always act jointly with one or more of the other plant growth substances.

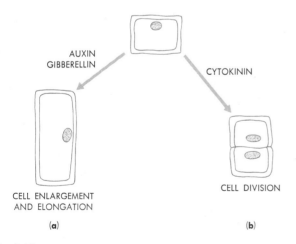

Fig. 9.12
Principal effects of growth substances on plant cells. (*a*) Auxins and gibberellins promote cell enlargement and elongation. (*b*) Cytokinins induce cells to divide.

Fig. 9.13
Promotion of root formation on cuttings of American holly by a synthetic auxin (indole-3-butyric acid). (Courtesy Paul C. Marth and the U.S.D.A., Beltsville, Md.)

Hormones are present in very small amounts in an organism. One hundred sunflower buds contain only 0.001 microgram (one billionth of a gram) of gibberellin.

Gibberellins

About the same time that Frits Went was providing convincing evidence that the growth of oat seedlings is promoted by auxin, the Japanese botanist E. Kurosawa became interested in a disease of rice plants. Some of the seedlings in the fields were greatly elongated and looked almost as if they had been stretched out, for the distance between the nodes was greatly exaggerated. In 1926 it was discovered that these elongated seedlings were infected by a fungus, *Gibberella fujikuroi.* To discover the relationship between the presence of the fungus and the abnormal elongation of the rice plant stems, Kurosawa performed an interesting experiment. He isolated the fungus from the infected rice plants and grew it in a culture medium. After adding drops of the cell-free medium surrounding the growing fungus to healthy rice seedlings, Kurosawa observed that their stems, too, soon became greatly elongated. In 1935 the active agent was crystallized by the Japanese biochemist T. Yabuta, who decided to name it **gibberellin** after the generic name of the fungus.

It was not until the early 1950s, however, that botanists in the Western world took notice of the Japanese research on gibberellin. Finally, in 1955, the exact chemical structure of gibberellin was determined. Subsequent research has shown that there are at least twenty-three gibberellins of closely similar structure. The best known of these, *gibberellic acid* (GA$_3$), is now commercially produced from fungus cultures by fermentation (see Fig. 9.11)

Like auxins, gibberellins are present in very small amounts within the plant. For example, 100 buds of sunflower seedlings contain only 0.001 microgram. Immature seeds and the fruits of some plants contain much larger quantities of gibberellins, however. They are synthesized in the embryo, shoot apical meristem, and young leaves. When applied externally to plants, gibberellins can produce striking and conspicuous effects. For example, when only a few drops of a very dilute solution of gibberellic acid are applied to the apex of a genetically dwarf bush bean plant, the plant will grow as tall as a pole bean within a short time, changing from the bush to the vine habit (Fig. 9.14). The main effect of gibberellins is to stimulate cell elongation and enlargement (Fig. 9.12).

Gibberellins are also involved in promoting the development of flowers in various plants. Even without pollination, the fruits of apples, figs, and grapes will develop following spray treatments of gibberellins (although they will contain no seeds). Gibberellins have also been found to promote seed germination. In many species of plants, the growth of the roots is inhibited by gibberellins.

The Cytokinins

A basic facet of plant growth is cell enlargement, a process regulated by both auxin and gibberellin growth substances. It seems logical for the plant to possess a system of control over cell division, for without the formation of new cells, continued growth of the plant body will be severely limited. The existence of such a control system for cell division was hypothesized in 1892 by the plant physiologist J. Wiesner. About twenty years later, the German botanist G. Haberlandt found that a preparation of crushed phloem cells applied to a wound in a potato tuber provoked the parenchyma cells of the potato to undergo division. In the early 1940s, the American plant physiologist J. van Overbeek and his associates discovered that the growth in culture of embryos of Jimson weed (*Datura stramonium*) was greatly enhanced by the addition of coconut milk (the liquid endosperm of the coconut seed). They hypothesized that the milk contained a growth substance unlike any then known. Attempts to isolate the material were abandoned after a year of work, due to the lack of suitable techniques for separating the active ingredient from the other constituents of the coconut milk.

In the late 1940s, F. C. Steward and his co-workers at Cornell University greatly increased the growth of carrot root tissue by adding coconut milk to the culture medium. During the early 1950s, Folke Skoog and his colleagues at the University of Wisconsin were using coconut milk to grow pieces of tobacco tissue in culture flasks and became interested in identifying the growth-promoting substance. Since coconut milk contains large amounts of sugar and other constituents (which increase the difficulties of chemically separating the growth factor), these workers explored other possible sources. An extract of yeast was found to stimulate their tobacco tissue cultures in much the same way as coconut milk. Absorption spectra and other properties of the yeast extract suggested that the active fraction might be one of a group of chemical substances known as purines.

Aware that nucleic acids contain purines, one of Skoog's co-workers, Carlos O. Miller, looked over the laboratory supply of chemicals, found one labeled "herring sperm DNA," and discovered that this material would also stimulate tobacco cells to divide and grow. He was astonished, later, to find that freshly prepared

(b)

(a)

Fig. 9.14
The effects of gibberellin on growth in two
different species of plants. The plants on
the left in each photograph are untreated.
Note that besides the obvious dramatic increase in size, there is also a tendency to
change to the vine form of growth. (Photo
courtesy S. H. Wittwer, Michigan State
University.)

DNA would not promote cell division. Yet when new
DNA was placed in an autoclave (a laboratory "pressure cooker" that creates high temperatures and pressures; macromolecules are broken down by being
placed in an autoclave) at 15 lb of pressure for 30 min, it
became extremely active. It was logical to conclude,
therefore, that the stimulant for cell division must be a
breakdown product of DNA. Finally, in 1955, these
workers managed to isolate a nucleic acid component, a
derivative of the purine adenine, that turned out to be a

long-sought promoter of cell division (Fig. 9.12b). They
named the new growth substance *kinetin*. Later, several
other adenine derivatives similar to kinetin were synthesized, and these substances are now known collectively
as **cytokinins.**

Due to their extremely small concentrations in the
plant body (for example, there are only 50 to 100 parts
per billion in the phloem sap of the grape plant), it was
not until 1964 that a natural cytokinin was isolated from
plant material. The chemical structure of this substance

SUPPLEMENT 9.1
AUTUMN LEAVES

In northern and southern temperate climates, autumn is a time when many plants shed their leaves in preparation for winter. However, leaf fall is not restricted to the autumn. In the tropics most plants shed their leaves continually, throughout the year, and even in north temperate zones the leaves of many evergreens—the conifers, principally—are being shed constantly.

What causes leaves to fall? It has been known for a long time that a thin layer of cells, called the **abscission layer,** can be found at the point where the leaf stem (petiole) joins the main plant stem (see Diagram A).

Similar abscission layers are also found where flower and fruit stalks join the main plant stem. Under certain conditions, the abscission layer cells die. The layer then becomes weak, and eventually the leaf falls off. The cells of the abscission layer thus become a target site for some sorts of changes within the plant or the individual leaf leading to leaf fall.

As long ago as the early 1930s it was found that cutting off or otherwise damaging the blade of a leaf, while leaving the petiole still intact, led to weakening of the abscission layer and consequent fall of the petiole. It was also found that applying auxin to the tip of the petiole after the blade had been severed would prevent abscission. Since the leaf normally produces small quantities of auxin as long as it is alive, botanists hypothesized that the presence of auxin retards leaf fall. Then in 1955, Dr. Daphne Osborne at Cambridge University in England found that a substance called *senescence factor* (SF) could be isolated from the petioles of dying leaves. When applied to excised abscission layer regions of bean leaves grown in culture, SF caused rapid abscission. Thus it seemed reasonable to conclude that abscission is initiated by diffusion of SF from dying or damaged leaves, and that it is prevented in young and undamaged leaves by the presence of auxin. Still, nothing about the mechanism of abscission could be deduced from these findings.

In 1965 two compounds called abscisin I and abscisin II were isolated from the cotton plant. Precise chemical determination showed that abscisin II was identical to a hormone already known, dormin. Dormin is produced by deciduous trees in response to short day length (as in the autumn), and it initiates many physiological responses preparatory to dormancy. (The name abscisic acid, or ABA, was given to what was formerly called dormin and abscisin II). While both would initiate leaf abscission, ABA was less active in producing abscission than SF.

Another clue came from previous studies of the effects of the hydrocarbon ethylene, C_2H_4, on plants. Ethylene is produced by various parts of the plant and controls seed germination, ripening of fruit, and a variety of other growth activities. Dr. Osborne then noted from the literature that

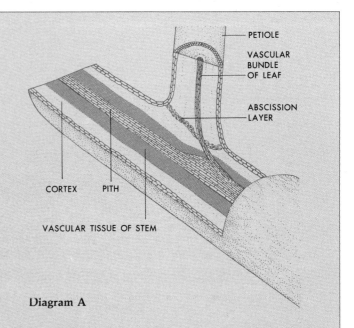

Diagram A

ethylene was an even more powerful promoter of abscission for flowers, fruits, or leaves than SF or ABA. Since ethylene was known to be produced by plants, Dr. Osborne hypothesized that somehow ethylene, SF, and ABA were all involved in the promotion of abscission.

What evidence did she use in developing this idea? Also in England, Dr. Osborne and two colleagues, Dr. Michael Jackson and Dr. Barry Milborrow, obtained abscission-promoting material from bean and other plant leaves by diffusion into dishes of water or by boiling in ethyl alcohol. They then subjected this material to electrophoresis and found several distinct bands. One contained ABA and promoted abscission to some extent. Another appeared to contain SF and promoted abscission at a very rapid rate. Material isolated from this band also stimulated ethylene formation in isolated abscission zone material. It appeared that three distinct substances were interrelated in producing abscission in leaves, flowers, and fruits: ABA, SF, and ethylene. How did they function?

Osborne, Jackson, and Milborrow knew that ABA depressed protein synthesis. They hypothesized the following chain of events in leaf fall: ABA depresses protein synthesis in leaf and petiole cells in response to the onset of cold weather or shorter days or simply by accumulating slowly in leaves as a result of age. Depression of protein synthesis would cause a decrease in the supply of crucial enzymes— those responsible for maintaining the structure and function of membranes, for example. Breakdown of membranes could lead to decompartmentalization of the cell, causing the breakdown of plastids and membrane-bounded sacs containing various substances, such as SF. Release of SF from a compartment into the general cell cytoplasm could promote the production of ethylene specifically in cells of the abscission layer. Ethylene, in turn, could trigger the activity of enzymes that produce rapid deterioration of the cells in the abscission zone. Although Osborne, Jackson, and Milborrow emphasize that this scheme is still largely hypothetical, it rests upon a firm foundation of observations about the existence of three different, interrelated substances in the abscission process: SF, ABA, and ethylene.

The Osborne-Jackson-Milborrow hypothesis is one example of how a scientific hypothesis is formulated. In an attempt to understand a complex process, such as abscission, the first stage of observation often focuses on one or two factors, such as SF. In trying to discover how the substance has its effects, Osborne found that other substances were involved: abscisins I and II, dormin, ABA, and finally ethylene. A considerable amount of chemical work was necessary to determine (1) that dormin and abscisin II were really the same thing, renamed as ABA, and (2) that while both SF and ABA caused abscission, their degrees of biological activity, in terms of causing abscission, were different. It was clear that whatever their exact functions, ABA and SF were different substances affecting the abscission process in different ways.

This case also illustrates the principle that developing any particular hypothesis usually involves drawing together the work of many other people. Osborne, Jackson, and Milborrow knew from the work of other colleagues that (1) ethylene had a powerful effect on many physiological processes in plants, and (2) ABA depressed protein synthesis. These observations were necessary to develop the full-scale hypothesis outlined above. Finally, this case illustrates the value to science of formulating hypotheses even if no evidence is yet available for showing that the hypothesis is valid. While the scheme that Osborne, Jackson, and Milborrow proposed for the chain of events in leaf fall is still largely hypothetical, it *is* experimentally testable.

(named *zeatin* since it was found in young seeds of corn, *Zea mays*) is shown in Fig. 9.11. Three other cytokinins have also been found in plants.

The cytokinins produce a wide variety of effects on the growth and development of plants. In addition to promoting cell division, cytokinins interact with auxin to induce the development of roots and shoots from tissue cultures of tobacco. If the concentration of indoleacetic acid in the culture medium is kept constant, then at low cytokinin levels roots develop. If the concentration of cytokinin is relatively high, however, shoots develop from the tissue. Cytokinins are also involved in promoting seed germination and fruit development, enhancing flowering, and preventing the senescence of plant organs. In fact, cytokinins may well play a role in controlling nearly every aspect of plant growth. They also interact with the auxins and gibberellins in influencing growth rate and differentiation of plants. It seems clear, therefore, that plant growth and development depend on the balance between cytokinins, auxins, and gibberellins instead of the presence or absence of any one of these growth-regulating substances.

Abscisic Acid

While the cytokinins, gibberellins, and auxins promote plant growth and development, it seems reasonable to hypothesize that the plant must also possess some means of inhibiting its growth and developmental processes. After all, a stem or any other plant organ can enlarge to only a certain extent before it becomes disadvantageous to the plant's welfare. Then, too, it is of survival value for a plant to be able to slow down or stop its growth or reproductive functions (as during the winter months in northerly regions of the earth, or during the dry season in the tropics).

During the 1960s several investigators offered evidence to support the hypothesis that plants contain ways of limiting their growth. In 1964 F. T. Addicott and his associates at the University of California at Davis extracted a substance from young fruits of cotton that they believed was responsible for the premature fall of the fruits from the plant; they named the substance "abscisin II." The next year, the Englishman P. F. Wareing and his co-workers isolated a substance from the leaves of birch, English sycamore, and other trees. Since it appeared to prepare the buds for winter dormancy, they named it "dormin." The chemical structure of both dormin and abscisin II was determined to be identical in 1965 by workers in both Addicott's and Wareing's laboratories (see Fig. 9.11). In 1967 both groups of researchers agreed to give the name **abscisic acid** to the substance.

In the relatively short time since the chemical nature of abscisic acid was revealed, the number and variety of responses to this hormone have been shown to be very large and complex. As demonstrated by the early experiments of Addicott and his group, applications of abscisic acid to young cotton fruits accelerated their premature fall (abscission) from the plant. In addition, Wareing's laboratory showed that leaves would drop from trees prematurely after application of abscisic acid. Other investigators have discovered that abscisic acid inhibits or retards the growth of such plant parts as embryos, coleoptiles, roots, and tissue cultures. For example, J. van Overbeek and his associates have found that the growth of the small, free-floating aquatic plant duckweed (*Lemna*) is materially reduced by the addition of a concentration of only one part per billion of abscisic acid to the culture medium. These investigators were able to keep duckweed plants in a permanent non-growing (dormant) state with a culture solution containing one part per million of abscisic acid. Abscisic acid also inhibits seed germination of many species and may be involved in the inhibition of flowering.

In 1910 it was discovered that the stems of pea seedlings exposed to the hydrocarbon **ethylene** (see Fig. 9.11), in a concentration as low as one part per million of air, underwent a decrease in length, swelled, and curved horizontally. About a quarter of a century later, several investigators demonstrated that ethylene is synthesized in plant tissues; for example, ripening apples were found to contain large amounts of ethylene, while immature fruits contained very little. This observation suggests that ethylene may be involved in the ripening process. The exact manner of this involvement is not yet known, however.

Ethylene has also been found to inhibit root growth and the formation of lateral roots. It breaks dormancy in bulbs and promotes their development. Pineapples can be induced to flower with ethylene. Flowering of the cocklebur plant, however, is inhibited by this chemical. Ethylene also acts in regulating the growth of pea seedlings subjected to physical stress. In one series of experiments, the normal elongation of pea seedlings was restricted by a mechanical barrier. Within 6 hr after the seedlings touched the obstruction, they were markedly curved and their production of ethylene had risen dramatically.

How do growth-regulating substances work in plants? This question has proved one of the most for-midable problems in the study of plant growth and development. A discussion of this important point must be postponed, however, until the central biological fields of genetics and development are discussed more fully in Chapters 16 through 20.

9.5 TROPISMS

As we saw in Section 9.5, tropisms are movements produced by growth of the plant in response to external stimuli, such as light and gravity.

Recall that in 1880 Charles Darwin and his son Francis studied the curvature of grass seedlings toward the light and concluded that something moving downward from the plant apex caused the seedling to bend toward the light. During the first quarter of the twentieth century, additional insight into the nature of the phototropic response was provided by several investigators, including Boysen-Jensen, Paál, and Went. These investigations led the Russian botanist N. Cholodny to propose a general theory of plant tropisms in the late 1920s. Cholodny hypothesized that curvatures of plant organs in response to external stimuli are due to the migration of a growth-promoting substance from one side of the stimulated plant to the other.

The Migration of Auxin

Almost simultaneously, Went proposed the same idea in the Netherlands. In addition, Went provided experimental verification. He placed coleoptile tips on razor blades so that the base of each tip was separated into two parts. The bases were then placed on top of agar blocks (a gelatin-like substance made from red algae), one block for each half of the separated coleoptile. Some of the tips were exposed to light from both sides, while

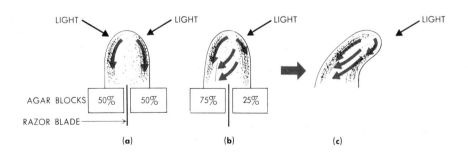

Fig. 9.15
Experiments to demonstrate that light causes an uneven distribution of auxin in coleoptiles. (*a*) The control group, illuminated from both sides. (*b*) The experimental group, illuminated from only one side. (*c*) Growth response of an intact coleoptile, due to auxin redistribution. The agar blocks show the relative amounts of auxin that diffuse into them from the stem tip.

Tropisms are plant growth responses to environmental stimuli such as light, gravity, chemicals, mechanical pressure, and the like. Tropisms are mediated by plant hormones, the most common being the auxins.

others were exposed to light coming from one direction. (All the previous preparation was done in a dark room with only red light, to which coleoptiles are relatively insensitive.) After exposure, the coleoptiles were placed in the dark for about 2 hr. When the agar blocks were examined, Went found that the auxin content was equal in both agar blocks in the control group, which received light from both sides. In the experimental group, however, the blocks under the bases that had been facing the light contained only 25 percent of the auxin (Fig. 9.15).

These results indicated that light stimulates a redistribution of auxin. Soon after Went's experiments, the Dutch botanist H. E. Dolk provided similar evidence for the lateral distribution of auxin in response to gravity. Dolk separated coleoptile tips into two parts and placed them in a horizontal position in contact with two agar blocks. After a short time in this position, the blocks on the lower side contained almost twice as much auxin as the upper ones. Thus, when a coleoptile is placed in a horizontal position, auxin accumulates on its lower side. This produces increased growth on that side, causing the stem to bend upward (Fig. 9.16). The results of Dolk's experiment were soon confirmed by several investigators using other organs of the plant. Recently K. V. Thimann and his colleagues have shown that when purified C^{14}-labeled indoleacetic acid is properly applied in small amounts to coleoptile tips, there is indeed a lateral movement of radioactive auxin during both geotropic and phototropic stimulation. It seems likely, therefore, that the Cholodny-Went theory of plant tropisms is essentially correct.

What triggers the lateral redistribution of auxin? In phototropism, it is natural to hypothesize that light is involved in some manner. As long ago as 1909, investigators of tropisms produced action spectra showing that blue light was most effective in inducing curvature of coleoptiles and that red light had little, if any, effect (hence the use of red light in dark rooms during the preparation of tropism experiments). If light is to be effective, it must be absorbed. This means that there must

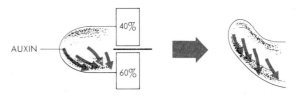

Fig. 9.16
Gravity also seems to affect auxin redistribution in stem tips. When a coleoptile is placed in a horizontal position, auxin accumulates on the lower part of the stem. It thus produces increased growth in this part, causing the stem to bend upward. It is difficult to explain the downward movement of roots entirely on the basis of this hypothesis, however.

be present in the plant some kind of light-receptor molecule. In 1933 it was hypothesized that this receptor might be a carotenoid pigment (carotenoids were known to exhibit action spectra similar to that which induced coleoptile curvatures). Support for this hypothesis was provided a few years later by the discovery that oat coleoptiles contain carotenoids, especially in the cells near the apex. However, the extreme apex, which had been shown some years earlier to be the region most sensitive to light, seemed to *lack* carotenoids.

In 1959 it was hypothesized that another substance, riboflavin (vitamin B_2), might be the phototropic light-receptor molecule. The plant physiologist A. W. Galston showed that in oat coleoptiles the action spectrum of riboflavin resembles that associated with phototropism (Fig. 9.17). He found that riboflavin is present in the coleoptile apex of a mutant strain of oats in which carotenoids are nearly absent. In addition, when Galston collected auxin from oat coleoptiles, he discovered that after riboflavin was added, the auxin was oxidized in the presence of light.

More recently, the action spectra for both riboflavin and carotenoid pigments have been shown to be

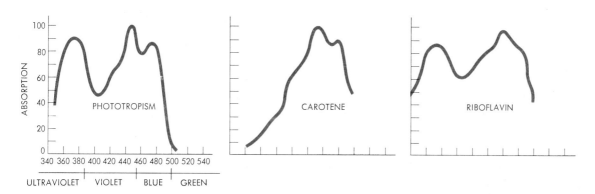

Fig. 9.17
Comparison of the action spectra for phototropism (left), carotene (center), and riboflavin (right).

Fig. 9.18
Geotropism in plants is the response of stem and root tips to gravity. Stems show negative geotropism; roots show positive geotropism.

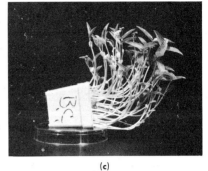

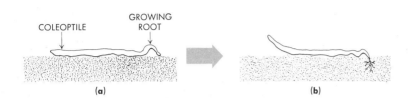

inexact fits for that of phototropism. The riboflavin spectrum more closely matches the near ultraviolet peak, but not those peaks in the visible light range. The spectrum for carotenoids fits the visible light peaks, but does not have a peak in the ultraviolet light (Fig. 9.17). It could be, of course, that *neither* of these molecules is involved in light reception during phototropism. While plausible, this conclusion is not very helpful. It implies that a third light-receptor molecule exists undetected. Though not impossible, this is quite unlikely. Finally, it has been suggested that perhaps both molecules may be functioning in the phototropic reaction, with carotenoid absorbing the visible light and riboflavin the ultraviolet light.

Whatever the situation, it seems clear that further study must be done before the question of light absorption in phototropism can be resolved. Then, once the nature of the light-absorbing system is determined, the mechanism whereby the light-activated chemical affects the lateral redistribution of auxin must be investigated.

Auxin's "Contradictory" Effects

The Cholodny-Went hypothesis holds that there is a lateral redistribution of auxin involved in the geotropic curvature of coleoptiles; when a coleoptile is placed horizontally, auxin accumulates on the lower side. This side then grows rapidly, causing the organ to curve upward. Can this hypothesis explain the geotropic response of roots? Since the early nineteenth century, it has been known that primary roots exhibit positive geotropism; they curve downward in response to gravity (Fig. 9.18). In the early 1930s it was found that when corn roots are placed horizontally for a short period of time, the lower half contains more auxin than the upper (just as in the horizontal coleoptiles). Yet such a horizontal root curves downward, while the coleoptile curves up. It seems clear, therefore, that while auxin promotes the growth of the shoot, it *inhibits* the growth of the root.

Why do auxins have an inhibitory effect on the growth of root cells and the reverse effect on stem cells? Recent experiments by Stanley P. Burg at the University

of Miami may shed some light on this paradoxical situation. Burg placed excised root tips of several species of flowering plants in solutions containing various concentrations of IAA plus an energy source, sucrose. The rate of production by the roots of one metabolic product, ethylene, was determined at intervals. The roots in the control group (without IAA) produced small amounts of ethylene. In those roots treated with a low concentration of IAA, the rate of ethylene formation doubled within 15 to 30 min and continued to increase rapidly thereafter. Burg also discovered that the greater the concentration of IAA, the greater the production of ethylene by the root cells. In addition, as the concentration of ethylene increases, the rate of growth in length and weight of the roots is gradually inhibited. This finding suggested that perhaps auxin inhibits root growth by inducing the synthesis of a growth inhibitor, ethylene. Burg also presented evidence that ethylene participates in the geotropic response of roots but not in that of stems.

This finding presents still another problem: Why does ethylene exhibit such different effects in the cells of the two organs? And there is still another problem: In what way is the auxin in a horizontally placed root redistributed? Obviously gravity must be involved. But how does the plant perceive gravitational forces? Around 1900, knowing that gravity affects mass, investigators postulated that perhaps the cells contain small particles that move under the influence of a gravitational field (one such type of particle could be starch grains).

A test of this hypothesis would reveal what response plants whose starch grains had been removed would make to gravity. In a recent experiment performed by K. V. Thimann, wheat coleoptiles were incubated for 34 hr at 30°C in a solution of two growth substances (gibberellin and cytokinin), both of which were known to promote growth by stimulating the enzymatic conversion of starch to sugar. After this treatment, the cells of the coleoptiles were found to be free of starch grains. When placed horizontally, these starch-free coleoptiles responded geotropically in much the same way as the controls, which contained starch grains. The rate of curvature of the destarched coleop-

The overall shape and size of a plant is controlled by the apical meristem, which regulates the production of hormones and their distribution to the lateral buds.

tiles was somewhat slower, however, than that of the coleoptiles with starch grains. Nevertheless it was concluded that starch grains are not necessary for wheat coleoptiles to undergo geotropic response. It has been suggested that the destarched coleoptiles responded more slowly because the postulated role of starch grains as gravity receptors was taken over by some other kind of particle inside the cell.

This last postulation of another gravity-perceptive particle is a good example of an attempt to save the hypothesis. It can be questioned whether science actually benefits from postulating the existence of such cell particles when the only evidence for them is negative. Much time can be lost in a search for particles of this kind. On the other hand, they *do* occasionally turn up.

9.6
THE CONTROL OF PLANT FORM

Many plants, especially trees, possess very distinctive body forms. For example, the Norway spruce (*Picea abies*) has a pointed, spire-like form with a single main trunk, while the sugar maple (*Acer saccharum*) is much more branched, and its central stem is eventually lost among the uppermost branches. The spruce tree is said to have an *excurrent* pattern of branching, whereas that of the maple is called *decurrent* or *deliquescent* (Fig. 9.19).

Apical Dominance?

It has long been known that when the growing apex of a young herb, shrub, or tree is removed, many of the buds in the axils of leaves lower down on the stem begin to grow, eventually taking the place of the main stem (Fig. 9.20). This observation is the basis of the art of pruning, which enables the horticulturist and gardener to modify plant form in various ways—witness the square-sided boxwood hedges so often seen in formal gardens. It can be hypothesized, therefore, that the presence of the terminal bud inhibits the growth and development of the lateral buds on the stem below it.

During the early 1930s K. V. Thimann and Folke Skoog became interested in this problem. They thought that perhaps the newly discovered auxins were involved in the domination of the apical bud over the lateral ones. They reasoned that if the terminal bud inhibits the growth of the lateral buds by producing a growth-inhibiting auxin, then decapitating the apex and replacing it with an auxin-containing agar block should result in inhibition of the lateral buds.

In experiments to determine if auxin is produced by the terminal bud, Thimann and Skoog cut off the tips of young healthy plants of broad bean (*Vicia faba*) and placed them upon agar blocks for 4 hr. Using the Avena coleoptile curvature bioassay, they discovered that the terminal buds produced considerable quantities of

Fig. 9.19
Patterns of branching in trees. Left, excurrent branching, as in Norway spruce. Right, decurrent or deliquescent branching, as in sugar maple.

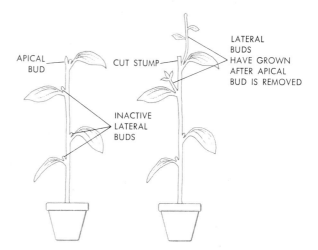

Fig. 9.20
After removal of the apical bud, some of the inactive axillary buds grow out.

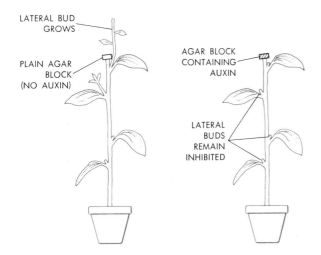

Fig. 9.21
Interpretation of experiments showing that the apical bud inhibits the lateral buds by producing auxin.

growth substances. The results supported their first prediction as a correct one.

In another series of experiments, Thimann and Skoog removed the terminal buds from broad bean plants and applied agar blocks with high concentrations of growth substance. The lateral buds on decapitated plants receiving auxin were found to be inhibited as completely as those on the intact control plants. Thus their second prediction was verified; by means of auxin, the apical bud exerts dominance over the lateral buds (Fig. 9.21).

Thimann and Skoog's findings on apical dominance in the herbaceous broad bean plant suggested that the characteristic growth forms of trees might also be due to a mechanism involving the terminal bud and auxin inhibition. It seemed logical that if the decapitation of bean plants leads to development of the lateral buds and bushy growth, then trees with the decurrent branching pattern must not possess apical dominance, while in those trees with an excurrent pattern of branching, apical dominance must be very strong.

Or Apical Control?

This rather simple interpretation of form in woody plants on the basis of weak or strong apical dominance was challenged in 1967 by the plant physiologist Claud L. Brown and his associates at the University of Georgia. Extensive observations of many species of trees showed that patterns of lateral bud inhibition are actually the reverse of what one would predict from the long-accepted version of the apical dominance theory. In trees with the decurrent branching habit, for example, the pecan (*Carya illinoensis*) and the oaks (*Quercus*), nearly all the lateral buds on the current

year's growth are completely inhibited, a pattern that suggests the action of strong apical dominance. Not until the second season of growth, after a period of winter dormancy, is the inhibition of the lateral buds broken and the much-branched stems characteristic of the decurrent growth form produced. In trees with the excurrent growth form such as spruce and some flowering plants, including sweetgum (*Liquidambar styraciflua*) and tulip poplar (*Liriodendron tulipifera*), the lateral buds on the current year's stem grow to varying degrees. Evidently apical dominance in these plants is rather weak.

Brown and his colleagues suggest that use of the term "apical dominance" should be restricted to the inhibition of buds on individual branches rather than extended to describe overall tree form. They propose that the term "apical control" might be more meaningful in describing the condition manifested by the excurrent or decurrent patterns of growth. Excurrent growth could be explained by hypothesizing that the terminal leader (the central or main stem) maintains apical control throughout the life of the tree by only partially suppressing the growth of the buds beneath it. Only in this way is it possible for truly spire-like forms to originate and be maintained throughout the life of the tree.

To produce the decurrent tree forms, it was hypothesized that apical control by the terminal leader would be lost very early in life. The large, well-developed, uppermost lateral buds on the current year's growth would be strongly inhibited the first season. By the second season, however, these large buds would be able to compete successfully with the terminal bud for food, water, and growth substances. In fact, this competition may be so severe that the terminal bud is killed by the midsummer of the second growing season, producing the distinctive much-branched growth form seen in maples. After a season or two, this now-dominant lateral bud will be replaced by one of its uppermost inhibited buds, and this cycle continues throughout the life of the tree.

How might the bud-inhibition mechanism operate? How is it possible for different lateral buds on the same plant to differ in their sensitivity to dominance? What causes some of the lateral buds to develop into branches while others remain inhibited? An early attempt to answer such questions was Thimann and Skoog's hypothesis that apical dominance operates through the inhibition of lateral buds by auxin transported from the stem apex. As we just saw, however, this hypothesis has been questioned by several investigators during the three decades since it was proposed.

In 1957 the English botanists F. G. Gregory and J. A. Veale attempted to evaluate the relative roles of auxin and nutrition using flax (*Linum*) instead of the pea plants utilized by Thimann and Skoog. In Gregory and Veale's experiments, all the buds tended to remain active as long as the supply of nitrogen was high, with apical dominance only poorly expressed. When the level of nitrogen was low, the lateral buds were inhibited. In addition, Gregory and Veale observed that the vascular tissue extending from the stem into the dormant buds was incomplete. Perhaps, they hypothesized, auxin from the apical bud inhibited the differentiation of the vascular tissue into buds, interfered with the movements of nutrients into them, and thus prevented their growth. Auxin, therefore, was visualized as having only an indirect role in lateral bud inhibition.

In the early 1950s, Skoog and his co-workers found that the formation of buds in tobacco pith tissue cultures is controlled by both the auxin and the cytokinin in the media. This discovery led Thimann and graduate student Margaret Wickson to study the role of cytokinin in the release of buds from dominance. Using sections of pea plant stems floating on a sugar solution, these investigators discovered that cytokinin could counteract the inhibition of lateral buds by auxin. They hypothesized that apical dominance in the intact plant depends on

interaction between auxin transported polarly from the growing terminal bud and cytokinin synthesized (most likely) in the inhibited buds themselves. Observations of other species, however, revealed that the inhibition was sometimes only partial; some lateral buds begin to grow while others on the same stem remain completely inhibited.

In 1964 Thimann and another colleague, Helen R. Sorokin, examined the relationship between the growth of lateral buds and their anatomy, especially the development of the xylem and phloem tissue connection with the stem. In pea stem sections floated on sugar-auxin solutions, the connection between the vascular system of the stem and that of the bud was not completed. When cytokinin was added to the culture solution, differentiation of the xylem and phloem into the bud continued to completion, and the bud began to grow. These experiments showed that Gregory and Veale were on the right track some years earlier when they postulated that a growth substance (auxin) might be functioning to

Table 9.1
Summary of the Major Plant Hormones and Growth Regulators*

Plant hormone(s) and other growth regulators	Region(s) produced or found	Functions
Auxins	Seed endosperm and, later, embryo; meristematic tissues of apical buds and young leaves.	Affect cell division in meristems of buds; initiation of cambium during development; root growth and tissue differentiation. Stimulate growth of fruit; promote stem elongation and enlargement; control geotropism and phototropism of stem; inhibit lateral bud growth; delay leaf aging and fruit dropping.
Gibberellins	Plants embryos, meristematic regions of apical buds, young leaves, and roots (from which they are often transported to stem).	Affect cell division in meristems of buds; delay leaf aging; affect root growth and differentiation of tissues; especially cell elongation and enlargement; promote stem elongation and seed germination.
Cytokinins	Synthesized in roots and often transported elsewhere; also found in seeds and phloem sap.	Affect cell division in meristems of buds and root growth and differentiation of tissues; stimulate growth of fruits, seed germination, and flowering; delay leaf aging.
Abscisic Acid	Young fruits.	Promotes fall of fruits (abscission) and leaves; retards growth of embryos, coleoptiles, roots, tissue cultures, seed germination, and (possibly) flowering.
Ethylene	Plant tissues such as ripening fruits.	Promotes fruit ripening; inhibits root growth and lateral bud formation. Promotes bulb development and flowering in some plants; inhibits it in others.

*It should be stressed that plant growth and development is generally the result of a complex interaction between all of these substances, as well as environmental factors.

inhibit the differentiation of xylem and phloem into the dormant lateral buds (at that time cytokinin had not yet been shown to occur naturally in green plants).

Recently, investigations by Thimann and his colleague Tsvi Sachs have helped greatly to clarify the mechanism by which lateral buds are released from apical dominance. In one set of experiments, a solution of cytokinin in alcohol and carbowax was applied directly to the lateral buds of pea seedlings. Within two days, the normally dormant lateral buds started growing, but did not elongate as much as uninhibited control buds did. Thimann and Sachs were able to make these buds elongate normally by applying IAA to their apices. They concluded from these experiments that the cytokinin acts to release the lateral buds from the inhibition of the growing apex, while auxin acts directly on the internodes, promoting their elongation only *after* the initial inhibition has been removed.

This experimental clarification of the roles of auxin and cytokinin makes it now possible to explain the importance of nutritional factors and plant vigor in the expression of apical dominance. For example, the release of normally inhibited lateral buds following high nitrogen fertilization observed by Gregory and Veale may be attributed to an increased synthesis of cytokinins in these buds (cytokinin molecules contain several atoms of nitrogen). Thus nutrition influences dominance indirectly by changing the overall growth rate of the plant.

There is also evidence that auxin and gibberellin may interact in apical bud inhibition. In experiments by William P. Jacobs and his associates at Princeton University, a lanolin paste containing 1 percent indoleacetic acid and 1 percent gibberellic acid was applied to the cut surfaces of pea stems immediately following the excision of the apical bud. The resulting inhibition of lateral bud growth was found to be nearly as complete as that produced by the intact apical buds.

The growth inhibitor abscisic acid may also be involved in apical dominance. The dormancy of potato tubers as well as isolated potato buds is prolonged effectively by treatment with abscisic acid. Various species of deciduous trees have been induced to enter the dormant condition by applications of this growth inhibitor. For example, when abscisic acid was applied to the leaves of actively growing birch seedlings, the apical buds soon developed the form and structure of dormant buds.

It seems clear that the phenomenon of apical dominance involves a highly complex interaction and balance among many factors, including auxin, gibberellin, cytokinin, abscisic acid, and nutrition.

9.7
PHOTOPERIODISM

The developmental botanist is interested in trying to discover the mechanisms by which the plant grows, develops, and differentiates: processes resulting in the production of its own species-specific, three-dimensional form and structure. One of the most interesting examples of the manner in which plant form and structure can be changed and controlled is found in the transition from vegetative growth to flowering. During this process, the activity of the stem apex undergoes a remarkable change. While the plant is growing vegetatively, the shoot apical meristem produces new leaves, axillary buds, and the new tissues of the shoot. As flowering begins, however, the apical meristems undergo an extensive reorganization that culminates in the production of a small number of floral organs of characteristic types (sepals, petals, and the like).

What factors, both of the plant and of its environment, are involved in triggering this dramatic change in the development of a plant?

Facts about Flowering . . .

As summarized in Supplement 9.2, experiments from the 1920s through the 1950s suggested that the following elements are involved in the photoperiodic response of flowering:

1. The immediate environmental factor responsible for triggering flowering appears to be light.

2. Plants respond to the length of a dark period broken by a light period. There appear to be three types of plants in this regard: (a) short-day plants (SDP) require long dark periods for flowering; (b) long-day plants (LDP) require shorter dark periods; and (c) independent plants (IP) seem to be independent of the period of darkness and their flowering is triggered by other factors.

3. The photoperiodic response in plants is mediated by a light-detector molecule in the leaves.

4. This detector is a special photosensitive protein, **phytochrome.** Two forms of phytochrome have been observed in SDPs and LDPs: (a) phytochrome 660 (P_{660}), which absorbs maximally in the red end of the light spectrum (wavelength around 660 mμ), and (b) phytochrome 735 (P_{735}), which absorbs maximally in the far-red end of the spectrum, at a wavelength around 735 mμ. Phytochrome is somehow linked to a time-measuring mechanism in the cells.

. . . and a Working Hypothesis

As a result, the following hypothesis was developed to explain the photoperiodic response. Light, acting on the photoreceptive phytochrome pigment, starts a chemical reaction that goes to completion in the dark. In absorbing the red light, molecules of one form of phytochrome are also converted into a second form.

At the end of a dark period, all the phytochrome in a plant is in the 660 form. Sunlight, which contains red light, converts most of the P_{660} to P_{735}. At the end of the

SUPPLEMENT 9.2
THE CAUSES OF FLOWERING: SOME EARLY HYPOTHESES

It has been known for centuries that some species of plants produce their flowers early in the spring, others in the summer, and some, like *Chrysanthemum*, in the fall. Not until the early 1920s, however, was any significant progress made in understanding the flowering phenomenon. At that time plant physiologists W. W. Garner and H. A. Allard at the U.S. Department of Agriculture in Beltsville, Maryland, noticed that a variety of *Nicotiana tabacum*, known locally as Maryland Mammoth tobacco, grew extraordinarily tall but did not flower during the summer. When they grew this tobacco in the greenhouse during the late fall and winter, instead of tall plants they obtained only short plants that flowered very quickly (see Diagram A). While Garner and Allard were not originally investigating flowering (they were studying a disease of this plant), the unusual flowering characteristic of the Mammoth tobacco caught their attention. Why, they asked, should this plant produce flowers during the winter in the greenhouse but not do so outdoors during the summer? They also noticed another significant occurrence: the plants flowered freely during the winter months. As early spring came, however, the flowering stopped and the plants resumed growth to become tall stalks.

It seemed evident that some factor of the environment must have forced the remarkable winter flowering. Garner and Allard hypothesized that temperature was involved. It was soon clear, however, that temperature coud not be an important factor, since the tobacco continued to produce flowers even when the greenhouse temperature was kept as high as that prevailing outdoors during the summer. Light was another possible factor, since the light in the greenhouse during the winter was considerably different from that of the sunlight during the summer. It was discovered, however, that the date of flowering was not materially affected by the intensity of the light.

This left the exposure to light of the plants over periods of time (which was clearly less in the winter than in the summer). Could the length of the daily exposure be responsible for the flowering? Garner and Allard developed the following hypothesis: If length of exposure to daylight during the winter is the crucial factor in regulating the flowering process, then keeping a plant in the light for periods corresponding to those of winter days should stimulate the production of flowers. In testing this hypothesis, they designed and constructed a dark house in which the plants could be placed for various periods of time during the summer months. By shortening the duration of the daily exposure to light, they succeeded in forcing plants of Maryland Mammoth tobacco to flower during midsummer. The control plants did not flower until October.

Would the plants flower during the winter if artificial light were used to increase the length of daily illumination during

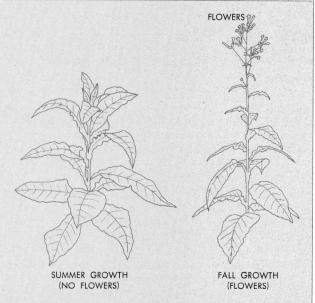

FLOWERS

SUMMER GROWTH
(NO FLOWERS)

FALL GROWTH
(FLOWERS)

Diagram A
(Adapted from W. W. Garner and H. A. Allard, *Yearbook of Agriculture.* Washington, D.C.: United States Department of Agriculture, 1920, p. 399.)

the short days of winter months? In another series of experiments, Garner and Allard found that when they supplemented the natural length of the winter days by 8 hr of artificial light, the plants did not flower, but behaved like typical summer-grown Maryland Mammoth plants. It seemed clear that day length was the most important factor in inducing plants to flower. Garner and Allard suggested that the day length favorable for the flowering of each species be called its photoperiod, and its response to the relative length of day and night photoperiodism.

Experiments with other species of plants led to the grouping of plants into three categories, short-day, long-day, and independent. Short-day plants are those that normally flower either in the late summer or fall or in the very early spring. These include Maryland Mammoth tobacco, cocklebur, dahlia, chrysanthemum, aster, goldenrod, and poinsettia. Long-day plants are just the reverse, generally flowering in midsummer. Some well-known examples of long-day plants are clover, gladiolus, spinach, lettuce, and radish. The flowering of some species of plants, however, is independent of day length; such plants are able to flower whether the days are long or short. Some examples are dandelion, tomato, zinnia, and green beans.

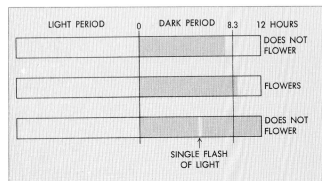

Diagram B
Summary of Hamner and Bonner's experiments showing that flowering in cocklebur (*Xanthium*) is controlled by the length of the dark period rather than the length of the day. Even a single flash of light during the long dark period keeps the plants from flowering.

In the late 1930s, Karl C. Hamner and James Bonner were studying the photoperiodism of cocklebur plants. They wondered whether the length of the day or the length of the night was crucial in the photoperiodic response. They added a dark period to the middle of a day, and a light period to the middle of the night. The former had no effect; the latter did. Hamner and Bonner concluded that the initiation of flowers is controlled not by the day length, *but rather by the duration of the night*, as summarized in Diagram B. Thus the names "long-day plant" and "short-day plant" are actually misleading; it would be more appropriate to speak of "short-night plants" and "long-night plants."

What part of the plant detects the photoperiodic stimulus? When all the leaves are removed from a plant, in most instances it will not flower, even with the proper photoperiod. However, if a single leaf remains on the plant, flowers will be produced following photoperiodic treatment. In fact, soybean plants will flower even if less than 1 cm^2 of a single leaf is exposed to the inducing photoperiod!

For light to be effective in photoperiodic control, it must be absorbed by the plant. In addition, this absorbed light energy must be utilized by the plant in its measurement of the length of the dark period. A long series of experiments begun during the 1940s by F. A. Borthwick, S. B. Hendricks,

and their associates at the U.S. Department of Agriculture, Beltsville, Maryland, has helped resolve these problems. The investigators discovered that cocklebur and soybean plants would not flower if the dark period was interrupted by red light at a wavelength of about 660 millimicrons. Later in their research, they were greatly surprised to discover that the inhibiting effect of the red light could be erased by far-red light at a wavelength of approximately 735 millimicrons. After a brief exposure to far-red light following the red light, the plants would produce flowers as if no light had been turned on them at all (Diagram C).

By 1956 the work of Borthwick and Hendricks enabled them to present evidence supporting a hypothesis that the light was being detected by a pigment that existed in two states, a red-light-absorbing and a far-red-light-absorbing form. Conclusive support for this hypothesis came a few years later with the isolation of a pigment called phytochrome, which proved to be maximally receptive to red light at about 660 millimicrons (mμ) and to far-red light at about 735 millimicrons (mμ). Additional biochemical studies of the isolated pigment showed it to be a protein that occurs in plant tissues at very low levels (about one molecule in 10 million). In this way it became clear over the years that flowering is initiated by the relationship between light and dark periods, and that the photoperiodic response is mediated by a lightsensitive pigment in plant leaves.

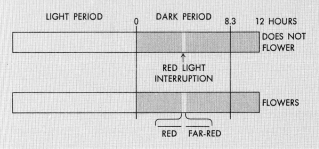

Diagram C
An interpretation of experiments by Borthwick and Hendricks on the effects of red and far-red light on the flowering of cocklebur. Far-red light reverses the effect of red light.

day, therefore, the plant contains mostly P$_{735}$. During the night the P$_{735}$ is converted relatively slowly into P$_{660}$ at a rate that probably varies from one plant to another. The effective dark period for any plant may well be the time required for conversion of all the P$_{735}$ to P$_{660}$. This possible phytochrome mechanism is shown here in schematic form:

RED LIGHT → P$_{660}$ slow nighttime conversion   P$_{735}$ ← FAR-RED LIGHT

A bright period in the middle of a dark period causes a plant to respond as though the dark period were short.

An alternative hypothesis divides the task of time measurement between phytochrome, which would act as an on-off signal, and an autonomous time-keeping process, an endogenous rhythm, that becomes activated by phytochrome. There is strong evidence for the existence of endogenous rhythms in plants, but their link to phytochrome conversion is not understood.

Florists and agriculturists use the technique of interrupting night length in order to force plants to flower out of regular season. For example, poinsettias can be made to bloom at Christmas time and lilies at Easter. Interestingly, although the practical application of the photoperiodic response is known, its theoretical basis is still largely a mystery. There is as yet no evidence showing how the conversion of P_{735} to P_{660} triggers the hormone system that is ultimately responsible for inducing flowering—in itself one of the most remarkable changes of form and structure in the life of a plant. Much research is currently being done in this area (and more is needed) in attempts to understand the exact mechanism involved.

The Problem of Communication

We saw earlier that the plant is able to detect the photoperiodic stimulus with the phytochrome system in its leaf cells. However, it is the shoot apical meristem, located some distance away, that undergoes transformation from the vegetative to the reproductive state. It seems reasonable to hypothesize that the plant may contain some substance or substances that transmit the photoperiodic stimulus from the leaves, down the petiole, and up the stem into the apical meristem. Evidence supporting this hypothesis has been provided by grafting experiments. A portion of the stem of a cocklebur plant that has received a photoperiod favorable for the induction of flowering (plant A) is grafted to the stem of a plant that has not been given the inductive dark period (plant B). Within a week or two, plant B begins to flower. The two plants can then be separated and a third plant (C) grafted to plant A. Soon plant C will flower (Fig. 9.22). This procedure has been continued through eight or ten different grafts over a long period of time, always with the same result. It seems evident that the flowering stimulus is able to pass a graft union from one plant to another.

Further support for the idea of a flowering stimulus transported from the leaves to the apical meristem has been provided by another kind of experiment. Different groups of cocklebur plants were given a favorable photoperiodic stimulus (a single long, dark period). Immediately after this treatment, the leaves of one group were removed. Some hours later, the leaves were removed from a second group. The procedure was continued at regular intervals for several days following the long dark period. Nine days after the dark period, the plants that had their leaves removed immediately after treatment were still vegetative. The plants that had their leaves removed 35 hr after the dark interval, however, were producing flowers almost as well as those that retained their leaves. It seems logical to hypothesize that when the leaves are removed immediately following the inductive dark period, very little (if any) of the flowering stimulus has been transported out of the leaves. However, after about 35 hr, nearly all the flowering

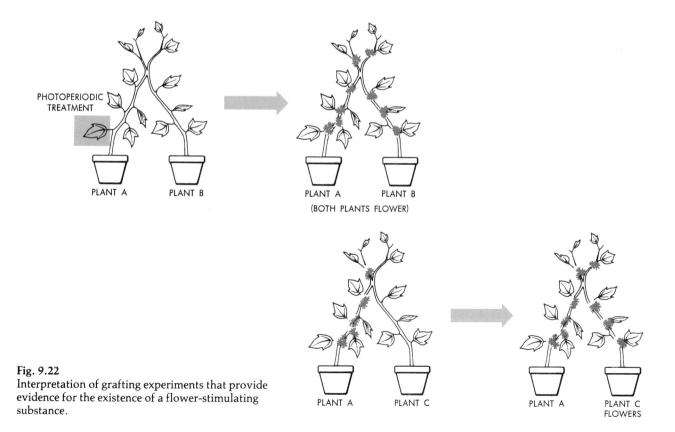

Fig. 9.22
Interpretation of grafting experiments that provide evidence for the existence of a flower-stimulating substance.

The flowering response in plants appears to be triggered by the length of the interval between two exposures to light. Length of exposure to the light itself has little influence on flowering.

stimulus has been translocated from the leaves and is moving through the stem to the apical meristem.

Thus there is evidence that some kind of chemical linkage exists between the reaction to the photoperiod and the production of flowers. Yet despite the efforts of numerous investigators over a period of many years, no such substance has been isolated or extracted. One possible explanation is that such a substance may not actually exist. A flower is a complex structure, and to some it seems impossible for a single substance to induce the many changes necessary to convert a vegetative apex into a floral meristem.

There is good evidence, however, that at least two plant hormones are involved in the flowering process. Applications of gibberellic acid induce flowering in many long-day plants grown under short-day conditions. There is also the probability that a flowering-*inhibiting* substance present in apical meristem cells may keep the apex in the vegetative condition. If this is so, then proper environmental stimulus (for example, a favorable photoperiod) might cause the synthesis of the inhibitor to be inhibited. When the growth inhibitor abscisic acid is injected into the stem cavity of rye grass near the apical meristem, the plants do not flower, even in a favorable photoperiod. The amount of abscisic acid appears to increase in leaves of some long-day plants held under short-day conditions, thus inhibiting the production of flowers. It is possible, therefore, that both abscisic acid and gibberellic acid may be important hormonal intermediaries between the reception of the photoperiodic stimulus and the biochemical events connected with flower production by the shoot apex.

Are environmental factors other than light involved in the induction of flowering? Consider the flowering of wheat. Spring wheat produces flowers in early summer after being planted in spring. Winter wheat, however, does not normally flower until the early summer following planting the previous fall. Wheat farmers during the nineteenth century discovered that winter wheat planted in the spring would also flower, provided the grains had been moistened and exposed to cold temperatures for several weeks prior to planting. This low-temperature treatment has been called vernalization, a word that means "inducing spring-like behavior."

Which portion of the plant perceives the low-temperature effect? By cooling some portions of the plant while keeping the rest at normal temperature, investigators found that the plant would detect the low temperature only if its shoot apical meristem was cooled. Cooling of the leaves or the roots had no effect. By placing flowers of intact plants of winter rye in vacuum bottles with crushed ice for varying lengths of time, researchers discovered that the developing embryos inside the immature seeds could be successfully vernalized. In fact, the young embryos became sensitive to low-temperature treatment within five days after fertilization occurred. When the plants reached maturity weeks or months later, they produced flowers. Thus the cells of the shoot apical meristem of the embryo are changed in some way by the cold, go through normal vegetative growth, and then become reproductive later in the life cycle of the plant.

The mechanism of the plant's response to cold is not yet understood. It has been found, however, that a few plants that ordinarily require low temperature for flowering can often be induced to flower without cold treatment by applications of gibberellic acid over a period of days. Thus gibberellic acid will substitute for the cold requirement, at least in such long-day plants as carrot, cabbage, radish, and spinach. However, it has no effect on short-day plants such as soybean and cocklebur kept under long-day conditions.

9.8
PLANT DEFENSES AGAINST DISEASE

Like animals, plants are subject to diseases. Occasionally an imbalance of power occurs between the attacked and the attacker in the fight against disease; one can still walk through woods in the South and find the decaying trunks of chestnut trees, felled decades ago; blight, a disease caused by fungus, killed them. Similarly, in New England and elsewhere, giant elms, their bark and other protective tissues bored through by beetles, have succumbed by the thousands to Dutch elm disease.

A few years ago, an antiviral factor called **interferon** was discovered in animals. We shall deal with animal interferon in more detail in the following chapters. Of significance here, however, is that in March 1980 Dr. Ilan Seda of Hebrew University in Jerusalem reported the purification of an antiviral factor from tobacco plants that closely resembles animal interferon in both structure and function. For example, the plant substance, like animal interferon, is an acid-resistant, phosphorylated glycoprotein (sugar and protein combination) that is produced in cells only after the plant has been attacked by a virus. Only small amounts are needed—about four to five molecules per cell—to resist infection, and the substance acts against a wide variety of plant viruses. Finally, its mode of action is the same as animal interferon; it induces an enzyme to break apart certain RNA molecules and a small nucleotide polymer is involved in regulating the plant substance's activity. By incorporating a special "N-gene" into a plant's genetic material, botanists have been able to confer resistance to formerly nonresistant plants. This feat would appear to hold promise for the possibility of producing virus-resistant crops.

Since the discovery of animal interferon, wide-ranging hopes for use of the substance in the treatment of various pathological conditions, including cancer, led to a great deal of headline-grabbing news about interferon. The discovery of an interferon-like substance in plants holds great promise, since larger quantities of the substance can be attained from plant tissues than animal. Unfortunately, the plant substance does not appear to be effective against animal viral infections.

If interferon turns out to be the miracle drug its earlier proponents suggested it might be, virologists will be searching for ways to transform the antiviral plant substance so that it acts against animal viruses. The *if* beginning the last sentence is a big one, however; there are currently some signs that animal interferon is simply not living up to its advance billings.

Summary

1. Plants grow in length at the root or stem tip (the apical meristem). This is accomplished by cell division and cell elongation. The process is additive, and thus the position of older cells does not change with respect to the surface of the ground, even though the plant continues to increase in length.

2. Plants grow in width through addition of new cells—principally xylem and phloem—by the cambium (lateral meristem). Cambium continually produces new xylem cells toward the inside of the stem and phloem cells toward the outside.

3. Plant hormones are chemical messengers produced in one part of the plant and distributed to all other parts. Certain target tissues respond to the hormones. Hormones can either stimulate or retard growth of the plant part, depending on the type of hormone and its concentration.

4. Auxin and gibberellin stimulate cell elongation; cytokinin stimulates cell division.

5. The fall of leaves, flowers, or fruits from the main plant stem is regulated by hormones or such hormone-like substances as senescence factor, abscisic acid, and ethylene.

6. Plants are stimulated to grow toward the light (positive phototropism) because of hormones produced in the stem tip. On the light side of a stem tip, auxin production and distribution is inhibited. Auxin is produced on the darker side of the stem, causing the plant stem tip to grow toward the light.

7. A plant hormone like auxin at one concentration will stimulate stem growth but at the same concentration greatly inhibit the growth of roots or lateral buds on a stem. Auxin appears to stimulate certain tissues to produce ethylene, which directly inhibits growth of these tissues. Like auxin, ethylene plays a double role in many plant processes. It stimulates the ripening of fruits and the growth of certain portions of the stem. It inhibits the growth of roots and causes leaves to fall off.

8. The overall shape of a plant is controlled by the apical meristems on one or more branches. This is accomplished by hormonal regulation. A growth-inhibiting auxin, produced by the terminal buds, has varying inhibitory effects on lateral buds. When the terminal bud is removed, the lateral buds usually grow at a much more rapid pace. Cytokinin also appears to be involved and counteracts the inhibition of lateral buds by auxin. Growth patterns of the intact plant are the result of interaction between auxin and cytokinin.

9. Auxin appears to function as an inhibitor of lateral buds by preventing the differentiation of xylem and phloem. Cytokinin serves to unlock the lateral buds so that differentiation can begin. Once differentiation of the lateral bud has begun, auxin accelerates the rate of cell elongation. Thus the role of auxin as growth stimulator and inhibitor is clearly demonstrated.

10. In most plants, light triggers the flowering process. Plants respond not to the light itself, but to the length of the interval between exposures to light (the dark period). In flowering, there appear to be three groups of plants: long-day plants (responding to short dark periods), short-day plants (responding to long dark periods), and independent plants (relatively unaffected by light or dark periods). This response of plants to light is called photoperiodism.

11. Photoperiodism is triggered by the photosensitive pigment phytochrome found in plant leaves. There are two chemical forms of photochrome: P_{660} (which absorbs in the red region of the spectrum, at wavelength 660 mμ), and P_{735} (which absorbs in the far-red region of the spectrum, at wavelength 735 mμ). These two forms are interconvertible. This conversion process triggers the hormone system of the plant and acts as a timing device.

12. Certain plants can also be induced to form flowers by environmental factors such as cold. The sensitive area of the plant in this case appears to be the apical meristem.

Exercises

1. How are annual rings produced in a woody plant?

2. Is the oldest part of a tree toward the center or toward the outside? Explain your answer in terms of the way in which a plant adds new growth to its body.

3. In a cross section of an evergreen tree, a botanist counted fifteen growth rings. How many rings would there be an inch or so back from the very tip of the tree?

4. Tomato plants produce flowers and fruit both outdoors during the summer and in greenhouses during the winter. What does this indicate about their photoperiodic requirements?

5. Refer to Fig. 9.23, a graph that shows the relation between osmotic concentration outside the cell and amount of growth. Explain this graph with reference to how auxin is thought to influence the growth of plant tissues.

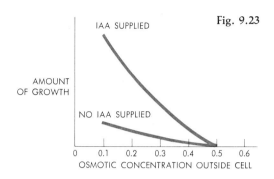

Fig. 9.23

Chapter 10
Homeostasis and the Processing of Food and Waste

INTRODUCTION

In this chapter and the following four, we will discuss the organ systems that the animal tissues, first introduced in Chapter 5, compose, as well as the physiological control mechanisms that coordinate these organ systems. While we will use a few examples involving animals other than the vertebrates (animals with backbones), in general we will present a fuller discussion of the physiology of invertebrates (animals without backbones) in later chapters. Among the vertebrates, primary (though not exclusive) attention will be focused upon a class of animals called Mammalia (the mammals), a group of organisms that includes human beings.

10.2
THE PRINCIPLE OF HOMEOSTATIC CONTROL

As pointed out in Chapter 5, tissues are often united in various combinations to produce organs; the stomach, for example, is a structure composed of muscular, nervous, and various types of epithelial cells, such as glandular cells that secrete acids and enzymes. In complex organisms, organs in turn are united to form organ systems; the stomach, for example, joins with the mouth, pharynx, esophagus, intestines, and so forth, to form the digestive system. The fact that all of these

organ systems—digestive, reproductive, nervous, muscular, and the like—all generally work in a highly coordinated manner suggests strongly that organisms possess distinct physiological control mechanisms that make the coordination of such highly complex processes possible.

One of the most interesting features of such physiological control mechanisms is that they are built into the system they regulate. The controls are self-adjusting; they do not require constant monitoring by an outside agent. Such controls maintain the system in *equilibrium.* A system in equilibrium is one whose overall characteristics are not changing. It is useful to distinguish here between two general types of equilibrium: **static equilibrium** and **dynamic equilibrium.**

In the tank shown in Fig. 10.1(a), the water level remains the same over a long period of time. Because the overall characteristics of this system are not changing, it is said to be in equilibrium. The equilibrium would be upset if we opened the tap at either tube X or tube Y. With both tubes plugged, however, the system

Fig. 10.1
Two mechanical systems, showing (*a*) static equilibrium and (*b*) dynamic equilibrium. The float, connecting rod, and valve mechanism in (*b*) constitute an example of a self-regulating system.

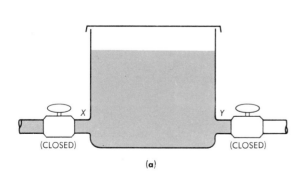

(a)

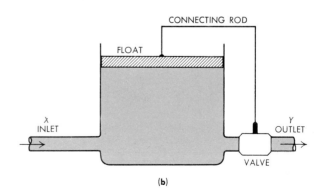

(b)

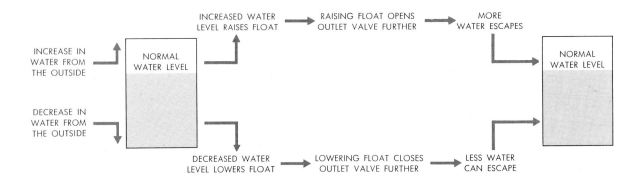

Fig. 10.2
Schematic diagram for the self-controlling system shown in Fig. 10.1(b).

remains in *static* equilibrium; *no water enters or leaves the tank.*

In the tank shown in Fig. 10.1(b), the water level also remains the same for a long period of time, *yet there is a continual flow of water in and out.* The parts of the system are dynamic, or changing—but the overall characteristics remain the same. This is a condition of *dynamic equilibrium.*

Note that the maintenance of this dynamic equilibrium requires a control mechanism, represented by the float, the connecting rod, and the outlet valve. Suppose the inflow of water increases. The level of the water rises, thus raising the float. The connecting rod between float and valve causes the valve to rise also, increasing the size of the valve opening and allowing more water to flow out. As the volume of outflowing water increases, the water level falls, lowering the float and, of course, the valve. This reduces the size of the valve opening and the volume of water leaving the tank. The water level increases and the cycle begins again. The reverse situation creates the same cycle with a minor difference; if the inflow of water decreases, the cycle begins at the point at which the water level falls, lowering the float.

It is possible to have a system in dynamic equilibrium with some outside source of control. For example, if a valve were placed at the outlet pipe (Y), it could be adjusted in such a way as to keep the water level constant. But the important feature of the system in Fig. 10.1(b) is that the amount of water allowed to escape is controlled by a part built into the system itself. In other words, the system contains a built-in self-monitoring control (see Fig. 10.2).

The self-regulating system just examined is analogous in principle to regulatory mechanisms operating in living organisms. Maintenance of a constant water level in the tank is similar to maintenance of a constant level of some substance—carbon dioxide, salt, or glucose, for example—in the body fluids. The float and connecting rod represent body organs stimulated into action by the change in any of these substances in the blood. One function of such systems is to help preserve the constancy of the internal environment despite considerable fluctuations in the external environment. The importance of this function was first pointed out clearly by the great nineteenth-century French physiologist Claude Bernard, who wrote: "It is the fixity of the internal environment which is the condition of free and independent life. . . . All the vital mechanisms, however varied they may be, have only one object, that of preserving constant the conditions of life in the internal environment."

In the early twentieth century, the American physiologist Walter Bradford Cannon coined the term **homeostasis** (from Greek *homoios*, meaning same, and *stasis*, meaning maintenance) to refer to the processes that maintain a constant internal environment in the face of an organism's continually changing external environment. Physiological control mechanisms operate to preserve the status quo. Maintaining a constant internal environment is a continual struggle. This struggle, of course, is one that every physiological system ultimately loses, death being the final failure of

---

All feedback control systems operate on processes that exist in dynamic equilibrium —where an overall balance is preserved although components of the system are constantly in flux.

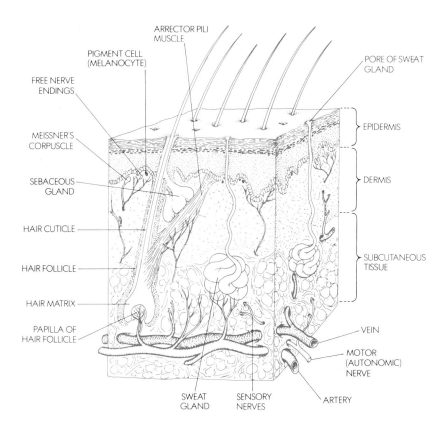

PIGMENT CELL
(MELANOCYTE)

ARRECTOR PILI
MUSCLE

FREE NERVE
ENDINGS

PORE OF SWEAT
GLAND

MEISSNER'S
CORPUSCLE

EPIDERMIS

SEBACEOUS
GLAND

DERMIS

HAIR CUTICLE

HAIR FOLLICLE

SUBCUTANEOUS
TISSUE

HAIR MATRIX

PAPILLA OF
HAIR FOLLICLE

VEIN

MOTOR
(AUTONOMIC)
NERVE

SWEAT
GLAND

SENSORY
NERVES

ARTERY

Fig. 10.3
A sectional view of human skin. The skin is the largest body organ and, together with the hair, nails, and associated glands, makes up the integumentary system. By varying the amount of blood flow into it, as well as the amount of fat deposited within it, an animal's skin plays a major role in maintaining a homeostatic balance of body temperature.

the body systems to maintain constancy in the face of changing external and internal conditions.

An Example of Homeostasis: Thermal Adaptation

Anyone who has lived in a temperate zone accepts, almost without thinking about it, that the presence or absense of certain organisms is related to the season. Probably only an exception to this rule would get our attention; the sighting of a snake or the necessity to slap a mosquito in January are examples. This is because most land-based animals do not possess the full homeostatic ability of temperature regulation possessed by most mammals and the other "warm-blooded" class of animals, Aves (the birds). All animals, of course, produce body heat as the result of the metabolic processes discussed in Chapter 6; indeed, even in the most efficient animal, most of the energy released from energy-rich compounds is lost for useful work. Yet some forms are able to use this waste heat to rise above the restrictions of environmental temperature variation and remain active all year round. Such forms are called **homeotherms** or **endotherms,** the latter term indicating usage of internally produced body heat. The advantages to being homeothermic are considerable; by maintaining a steady body temperature at a level at which the metabolic enzymes operate most efficiently, homeotherms

are free to exploit a wider range of environmental habitats all year round.*

Of course, since temperatures vary widely in temperate zones, homeotherms must be able to increase or decrease body heat retention. Thus arctic mammals such as seals or walruses have fur and/or thick layers of fat for insulation, and blood vessels can be constricted near the surface of their bodies to retain more body heat. Many animals, including humans, undergo involuntary muscle contractions called shivering; the increased activity releases more heat to warm the body. In hot temperatures, the dilation of skin blood vessels allows more heat to be brought to the surface for cooling (see Fig. 10.3). The evaporation of perspiration also has a cooling effect; the panting of a dog not only increases heat dissipation from the inner lung surfaces, but also from the tongue surface. Other animals simply move to cooler areas, such as an underground burrow.

Most animals, however, are "cold-blooded," and are referred to as **poikilotherms** or **exotherms,** the latter

---

*Since heat loss is related to body size (the smaller the animal, the greater the amount of surface area per unit volume; see also Supplement 4.1), some animals, especially small ones, must hibernate during the winter in arctic or temperate zones. During hibernation, the metabolic rate is lowered dramatically, the animal remains dormant, and thus very little energy is consumed. Larger animals do not hibernate, though some species of bears do enter a period of long sleep in which they live off their reserves of body fat.

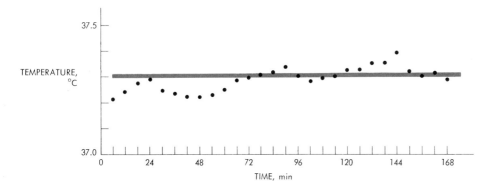

Fig. 10.4
Normal variations in mammalian body temperature with environmental temperature held constant. Body temperature fluctuates but on the average is 37.3°C.

term indicating that these forms must rely on external heat, generally the sun. Fish, amphibians (such as frogs and salamanders), and reptiles (such as lizards, snakes, and turtles) are all poikilothermic; the heat released from their metabolism is not retained and homeostatically balanced within their bodies but dissipates to the outside. Thus, at rest, such animals' body temperatures are the same as the surrounding medium. In the case of fish, for example, this means that their enzymes must be able to operate at much lower temperatures than those of mammals, while land-based reptiles can remain active only when the temperature is high enough to allow them to maintain a high metabolic rate. This is why reptiles are only seen in temperate zones during the warmer months, and why they are often seen "sunning themselves" on heat-retaining surfaces, such as rocks; this is a sort of behavioral homeostasis. In the winter, both reptiles and amphibians must remain dormant.

A good example of homeostasis is the method by which the human body temperature is kept constant.* Human beings display a relatively constant body temperature of 37°C (98.6°F). If the temperature of the external environment rises, a special region of the brain called the hypothalamus stimulates the sweat glands to perspire. The hypothalamus is composed of tissue that is very sensitive to changes in temperature of the blood passing through it. Increased sweating cools the body, because water molecules absorb heat from the body when they evaporate from the skin surface. As the body temperature falls, stimulation of the hypothalamus is decreased. This results in a reduced rate of sweating. As a result of this continual juggling, the body temperature is soon brought back to its normal 37°C from a slight fluctuation above this temperature (see Fig. 10.4).

*The body temperature for any warm-blooded animal never varies more than two or three degrees from an average, unless some major upset occurs in the physiological processes, as in a fever. The temperatures of all warm-blooded animals are not the same, however. While that of humans is normally 37°C, in chickens it is 40–42°C, dogs 39°C, horses 38°C, and sloths 33–34°C.

Sweating still occurs at normal temperatures, but much less than at high temperatures.

If the outside temperature falls below the value of 37°C, cooler blood reaches the hypothalamus. The hypothalamus responds by reducing the amount of sweating. This further reduces the amount of heat the organism loses to the outside environment. By conserving heat within the body, the hypothalamus prevents the temperature of the blood from falling to a critically low level. As the body temperature increases due to this heat conservation, the sweat glands begin to operate again, and again we see the balancing procedure involved in homeostatic systems.

Feedback Mechanisms

Several principles emerge from the example of thermal regulation. The first is that homeostatic mechanisms achieve their effect by *modification of processes that occur continually.* For example, sweating occurs at all times in humans. The amount of perspiration released from the sweat glands depends on the amount of stimulation by the brain's hypothalamus, which in turn is modified by the temperature of the blood that reaches it. Using a continually occurring process as the basis for homeostatic control allows a more immediate response to changes in the environment. The body is continually making small adjustments in amount of sweating in response to local environmental changes. These small immediate adjustments are possible because the regulatory mechanism in the hypothalamus simply increases or decreases the rate of a process already in operation.

A second principle is that *all homeostatic systems have a feedback mechanism.* By means of a feedback mechanism, part of the output of the system is fed back into the system as an instruction, telling it what to do. The effect of sweating is to cool the blood as sweat leaves the outer regions of the body. As this cooler blood flows back to the central regions, some of it passes through the hypothalamus. This conveys a message to the hypothalamus: that the blood temperature has decreased. It responds by reducing its stimulation of the sweat glands. Thus the process of sweating produces an overall cooling of the entire system, and a report that this cooling has occurred is fed back into the control center.

Two types of feedback can be recognized in self-regulating systems. One is called **negative feedback,** the other **positive feedback.** Negative feedback is operating when a change in the system in one direction (for example, increased sweating) is converted into a command to change that system in another direction (for example, decreased sweating). In body temperature control, an increase in sweating lowers the body temperature. A decrease in body temperature is converted into a command (in the hypothalamus) to decrease the amount of water excreted through the sweat glands. This contributes to a rise in body temperature and thus to increased sweating. Such negative feedback helps to keep a system in equilibrium.

Positive feedback, however, may destroy equilibrium. Suppose that the connecting rod from the float (Fig. 10.1b) had been attached to the *inlet* (at X), instead of the outlet. As more water came into the tank and the level rose, the float would have been lifted up, opening the inlet valve further, so that even more water could enter. This, in turn, would have raised the level of the float, opening the inlet tube still further, and so on. If the inlet tube were of a diameter greater than the outlet, the system would overflow. The positive feedback in this case produces a runaway process.

Such positive feedback sometimes occurs when normal control mechanisms in the body break down. If a person's body temperature rises much above 42°C (107.6°F) the negative feedback mechanism breaks down and is replaced by positive feedback. High temperature brings about an increase in the metabolic rate (chemical reactions). This produces more heat, which raises the temperature, and so on. The ultimate result of this vicious cycle is death.

It should be stressed here that not *all* positive feedback is destructive. Indeed, certain kinds of positive-feedback physiological mechanisms may be highly adaptive for organisms. For example, the sighting of a fox by a rabbit produces a dramatic increase of certain substances in the blood causing its heartbeat and blood pressure to increase. These changes, in turn, allow extra blood to be pumped to muscles that may need extra energy in order to escape. Yet, were the positive feedback to continue unchecked, the strain on the rabbit's system would be too much, and the rabbit would die. Once the danger is passed, the system must return to a steady state of dynamic equilibrium.

From what has been said about homeostatic mechanisms and negative feedback, we can guess that the constancy of any system is not perfect. This is true. The body temperature of any warm-blooded animal, for example, does not remain perfectly constant. Instead, it continually fluctuates about an average point that we speak of as "normal" (Fig. 10.4). This is true even if the temperature of the external environment is perfectly constant. Every adjustment in one direction or the other tends to overshoot the mark. This brings the alternative response into play. The result of this overshooting is slight variation of the conditions in the system. Such fluctuations are characteristic of any living system existing in a steady state.

10.3 DIGESTION

Life without energy is a contradiction in terms. No living system can endure long without some kind of fuel—its food—to keep it running. We call the taking in of food **ingestion;** the mechanical and chemical changes brought about in this food for the organism's use we call **digestion.**

The first process, ingestion, may vary widely from one form of organism to another. The one-celled amoeba ingests food at any point in its plasma membrane; most higher animals ingest food through a specialized opening, or mouth. Whatever the means, however, the final result is the same—the taking in of raw materials for digestion and eventual assimilation into the organism's own living material.

Digestion, too, may vary from one animal species to another. Some digestion, such as that of the amoeba or sponge, is *intracellular;* the food passes directly into the cell cytoplasm in food vacuoles and is there attacked chemically by digestive enzymes. In higher animals and many lower ones, however, *extracellular* digestion is the rule. Specialized glandular cells secrete digestive enzymes onto the food, which is usually confined by some means to a specific area of the body during the digestive process. The stomach and intestines perform this confining function in higher animals. After being digested, the food is absorbed and distributed to other body cells.

There are even cases of extracellular digestion occurring outside the organism's body; a starfish, for example, extrudes its entire stomach out of the body and digests the clam it has forced open right in the clam's own shell. Spiders, after paralyzing their prey, inject their digestive juices into the insect's body. After some digestion has taken place, the fluid is drawn back into the body for absorption. The process is repeated until only the dried husk (exoskeleton) of the prey remains.

As has already been implied, digestion is primarily a chemical process, but it is not entirely so. Mechanical digestion, the physical reduction of ingested food to smaller masses, is also important. Such a breaking apart exposes more surface area of the food material to the digestive enzymes. *Teeth* are probably the most familiar instruments of mechanical digestion in animals, but there are many others. Birds, who lack teeth, grind their

SUPPLEMENT 10.1
WHAT ARE HUNGER PANGS?

The problems of hunger are intimately involved with the problems of digestive physiology. What is hunger? What causes hunger pangs?

Experiments on human subjects provide some answers to the last question. By the technique diagrammed for the esophagus in Fig. 10.5 a subject's stomach contractions can also be measured and recorded on a kymograph. The subject, who cannot see the kymograph, presses a button whenever he or she feels hunger pangs. Almost inevitably, the pangs occur when the kymograph is recording strong stomach contractions. These results support a hypothesis attributing the cause of hunger pangs to stomach contractions.

Other observations also support this hypothesis. An empty stomach undergoes periods of quiescence in which no strong contractions occur. During such periods no hunger pangs occur.

If hunger pangs are due to stomach wall contractions, there should be a limit to their intensity, the limit being set by the strength of the stomach wall muscles. Hunger pangs should be no stronger after two weeks of fasting than after a two-day fast. This prediction is verified by a compilation of experimental data, plus records of the experiences of those deprived of food by being marooned or imprisoned.

It seems fairly certain, then, that the hypothesis attributing the cause of hunger pangs to stomach contractions is a correct one. However, hypotheses explaining the *causes* of the stomach contractions are on far less firm ground. The contractions do not depend on nerves, for they still occur after the gastric nerves are cut. The sugar content of the blood may be hypothesized as an influencing factor. In favor of this hypothesis are experimental data showing that when blood sugar levels are reduced by a sizable extent, hunger pangs are initiated. Evidence can be gathered against this hypothesis by showing that the fluctuations occurring in the mammalian eating cycle are very slight, although blood sugar levels may vary considerably. Furthermore, hunger often sets in before the food ingested at the last meal is completely absorbed, an observation that does away with the "reasonableness" of the blood-sugar hypothesis.

Most puzzling is the fact that experimental methods can detect no difference between the stomach contractions that cause hunger pangs and those that occur during the end of gastric digestion, yet the latter contractions are not felt at all. To date, this puzzle remains unsolved. It is virtually certain that brain regions are involved in the hunger and eating patterns of mammals. A cat that has been fed to capacity will continue to eat if a certain region of its brain is stimulated electronically. On the other hand, a cat that has been deprived of food will not eat if a different region of its brain is stimulated.

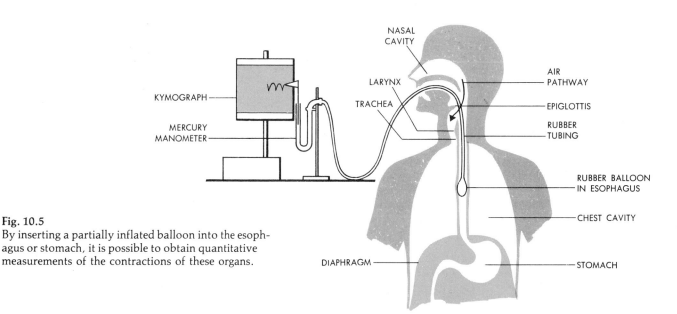

Fig. 10.5
By inserting a partially inflated balloon into the esophagus or stomach, it is possible to obtain quantitative measurements of the contractions of these organs.

food in the crop, a muscular sac often filled with small stones. But for those animals, such as snakes, that swallow their prey whole, the process of digestion must be virtually entirely chemical in nature.

Besides such obvious breaking-apart processes as chewing, from the time food enters the digestive tract until it leaves, it is subjected to a considerable amount of mechanical movement. Some of this movement aids in breaking up a swallowed food mass and exposing more of the surface to the digestive juices. Other movements, termed **peristalsis** (rhythmic contractions of circular muscles in the esophagus and intestines) aid in moving the food mass or **bolus** along the digestive tract. Such movements can be detected either by direct ob-

servation of the organs in an animal whose abdominal cavity has been exposed, or by filling the organs with an opaque substance (such as barium sulfate) and viewing them with X-rays. A third technique, which may be used to study peristalsis in either the esophagus or stomach, is to introduce into these organs a partially inflated rubber balloon attached to a hollow tube (Fig. 10.5). Any peristaltic contractions cause an increase in air pressure within the balloon, which can in turn be recorded on a mercury manometer and/or a kymograph.

A change of major importance in the evolution of animals was the move from a one-opening to a two-opening digestive system. An animal such as *Hydra*, with a one-opening digestive system, must eliminate

SUPPLEMENT 10.2
HOW DIGESTION IS INITIATED

As with most systems, knowledge of the anatomy of the digestive system was acquired through dissection. Digestive physiology, however, is not so easily attacked. It is obviously more difficult to carry out controlled experiments on human beings than on laboratory animals. Information gathered from clinical reports is, of course, quite helpful, and studies on nonhuman subjects can be useful in understanding human physiology as well.

Digestive physiologists noted early that when a hungry animal is exposed to food, its stomach begins to secrete gastric juice. What factors initiate this gastric secretion? It is reasonable to hypothesize that the nervous system of the animal that sights or smells the food causes involuntary impulses to be relayed to the stomach, which then begins to secrete gastric juice. Working with dogs, the Russian physiologist Ivan Pavlov experimentally tested the nerve-impulse hypothesis as an explanation for gastric juice secretion. He reasoned that if gastric juice secretion is due to impulses from the nervous system, then severing the nerves leading to the stomach should cause a cessation of gastric juice secretion. Yet, after the nerves were cut, gastric juice secretion was only reduced to three-fourths of its previous amount. The hypothesis had therefore to be modified; evidently the nervous system exercises only partial control over gastric juice secretion.

What other factors might play a role in initiating the secretion of gastric juices? Pavlov hypothesized that a hormone might be involved. He reasoned that entry of food into the stomach might cause the liberation of a hormone into the bloodstream. This hormone might act with the gastric nerves supplying the stomach to stimulate gastric juice secretion. An ingenious experiment was designed. Again, Pavlov rea-

soned that if a hormone is released into the bloodstream to act with the gastric nerves in stimulating gastric juice secretions, then blood circulating from a well-fed dog into a dog that has not been fed should cause gastric juice secretion in the second animal.

Two dogs, one just fed, were anesthetized and their bloodstream joined. The prediction of the hypothesis was verified; the second dog secreted gastric juice. Since the second dog's nervous system would not be involved in its secretion of gastric juice, it would be predicted that only 75 percent of the normal secretion would be produced. This prediction was also verified.

How do we know that some absorbed product of digestion is not causing gastric juice secretion? If this were the case, the injection of such products into the bloodstream would initiate gastric juice secretion. It doesn't. On the other hand, when extracts of the gastric lining on which protein digestion products have acted are injected into the bloodstream, they cause gastric juice secretion in the stomach. This result is consistent with Pavlov's hypothesis. So, too, is an experiment in which a piece of the stomach wall is transplanted, with blood vessels intact, to the skin of the same organism. The transplant begins to secrete gastric juice soon after proteinaceous substances are placed into the animal's own stomach.

Later work showed that when food enters the stomach, a hormone (gastrin), is produced by glandular cells in the stomach wall. The gastrin is released into the blood, by means of which it travels all over the body. When it returns to the stomach, it stimulates the gastric glands to secrete gastric juice.

SUPPLEMENT 10.3
THE MOVEMENT OF FOOD

Swallowing

Much work in the field of digestive physiology has gone into the study of the causes of movement in the organs of the digestive tract. Swallowing, for example, initiates peristalsis in the esophagus. What controls swallowing? Nerves? Hormones? Or something else? We can begin by hypothesizing the first of these possible causes: If swallowing is controlled by nerves, then severing the nerves innervating the swallowing region should prevent swallowing. The prediction is verified; experimental animals in which the appropriate nerves have been severed are unable to swallow. Clinical observations of human patients deprived of these nerves by accident or disease provide further evidence in support of the nerve hypothesis; such persons are also unable to swallow.

A healthy person, of course, can swallow almost at will. This suggests that swallowing is entirely under conscious control of the nervous system (as opposed to an organ like the heart, which is not). Yet further experimentation casts doubt on this supposition. If the proper nerves are artificially stimulated, swallowing occurs automatically, without conscious action. This indicates that, while swallowing *is* under the conscious control of the nervous system, it is not entirely under the control of the regions of the brain involved in conscious action.

The act of swallowing is far more complex than might be thought. It involves muscles of the tongue, mouth, pharynx, larynx, and esophagus, each of which must relax and contract at the proper time and in the correct sequence. The anatomy of the upper digestive tract seems at first to be designed for asphyxiation: the opening of the trachea is larger than that of the esophagus, and it would seem that swallowed food would more easily pass into it. Yet swallowing "the wrong way" is relatively rare. Anatomical examination of human cadavers and experimental animals reveals the presence of a flap (the epiglottis) that, when folded downward, effectively blocks the opening to the trachea. It is reasonable to hypothesize, then, that during swallowing the epiglottis performs this function. Studies of the swallowing reflex in anesthetized animals support this hypothesis. But since the lowering of the epiglottis is a rather mechanical process, we might suspect that a more reliable safeguard against the possibility of food entering the respiratory tract must have evolved, especially in light of the fact that even a piece of food too small to cause choking could nevertheless initiate a serious bacterial infection in the respiratory system. Another explanation for the rarity of this occurrence must be sought.

It is known that swallowing is a reflex action involving certain nerves; therefore it is reasonable to hypothesize that the same nerves that initiate the swallowing reflex also send to the respiratory center in the brain impulses that inhibit breathing while the food is passing the tracheal opening. Such a hypothesis can readily be tested by experiment. The reasoning is as follows: If the swallowing reflex involves the sending of impulses to the brain that inhibit breathing, then continued stimulation of the nerves that cause the swallowing reflex in an anesthetized animal should continue to prevent its breathing, even when the epiglottis is prevented from closing the opening into the trachea. Experimental results support this hypothesis.

Stomach and Intestine Activity

Since we know the active role played by nerves in the upper digestive tract, it is reasonable to hypothesize that similar roles are played by the nerves supplying the lower digestive tract. Anatomically, the stomach and intestines are supplied with two sets of nerves. Experimental stimulation of one set of nerves in anesthetized animals results in increased activity of the stomach and intestines; peristalsis and other characteristic churning movements increase. Stimulation of the other set of nerves tends to inhibit these movements. These preliminary observations support a hypothesis that the role of the nerves in this region is the same as that of the nerves of the pharyngeal-esophageal region. To test this hypothesis adequately, however, a crucial experiment must be done. If stomach and intestinal movements are entirely controlled by nerves, as are those movements of the swallowing reflex, then severing all the nerves leading to the stomach and intestines should prevent these movements from recurring. Experimental results contradict the hypothesis. Despite cutting of the nerves, digestive movements (slightly modified) still occur. Even isolated strips of intestine will continue to contract when placed in a proper medium. Thus the movements in this area must be controlled in a manner quite different from those of, say, the upper esophagus. In view of the rather unique structure and properties of smooth muscles (see Section 5.19) perhaps this is not too surprising.

Microscopic dissection provides partial answers to the questions posed by these experimental results. The walls of the stomach and intestines contain vast complexes of interconnected nerve cells. Such complexes are also found in the body walls of certain primitive animals. Despite the fact that these animals have no brain whatsoever nor any central nervous system, their body walls are capable of rhythmic contractions. It seems reasonable to hypothesize that a similar kind of control system is involved in the movement of the stomach and intestines of higher animals. Although experiments can theoretically be designed to test this hypothesis, it is not yet possible to separate the nerve network from the smooth muscles of the stomach and intestinal walls without damaging the latter beyond the ability to contract. Here is a case in which the crucial test of a hypothesis must await development of new experimental techniques and technical skills.

Passage from Stomach to Duodenum

The passage of food from the stomach into the duodenum poses another problem for the digestive physiologist. The pyloric valve consists of a ring-shaped muscle (sphincter muscle). When the muscle is contracted, the contents of the stomach are effectively blocked from entering the duodenum. When the stomach is full, the pyloric valve opens intermittently, allowing a spurt of semiliquid food material to enter the duodenum. It is obviously important to the accomplishment of adequate gastric digestion that the pyloric sphincter remain closed most of the time. Also, it is obviously important that the valve open occasionally to let the digestive processes of the lower tract proceed. What factors control the opening and closing of the pyloric sphincter? Preliminary experimental observations reveal the following facts:

1. The physical nature of the stomach contents influences the opening and closing of the pyloric valve. Water passes out of the stomach into the duodenum within ten minutes or less after being swallowed. Solid foods on the other hand, may still be in the stomach five hours after being swallowed.

2. The strength of the stomach's peristaltic contractions seems to be an influencing factor. Weak stomach contractions have little effect, but a powerful wave of peristaltic contractions opens the sphincter just before the wave of contractions reaches the valve.

3. There is a considerable difference in pH between the digestive tract medium on the gastric side of the pyloric valve and that on the intestinal side. As mentioned earlier, the former is quite acid, about pH 2. The latter is slightly basic, between pH 7 and 8. These pH ranges correlate with the optimum range of efficiency for the protein-digesting enzymes. When acid is introduced into the duodenum just beyond the sphincter, the valve closes. When a basic solution is introduced in the same region, the valve opens.

These observations (particularly the last) suggest an attractive hypothesis concerning the controlling factors that open and close the pyloric valve. Strong peristaltic movements of the intestine force open the valve. This allows an acidic fluid to spurt through into the duodenum, closing the valve. Gradually, however, the acidic fluid is neutralized by the basic bile and pancreatic juice. Thus the sphincter muscle relaxes and the pyloric valve opens again. More acidic fluid spurts through, and the cycle is repeated. This is a nice, simple hypothesis, one that seems in accord with what is known about the digestive system.

The philosopher Alfred North Whitehead once wrote, "Seek simplicity and distrust it." This statement is highly pertinent to scientific hypotheses in general and to the one last cited in particular. It is an attractive hypothesis and, probably for this reason, has persisted. Science, however, is a suspicious discipline. More experiments show that while acidic substances do indeed cause the pyloric valve to close, so do strong bases; so, in fact, does almost any stimulation of the duodenal epithelial lining. Furthermore, many apparently healthy persons secrete no acid at all in their stomach juices, yet their pyloric valves appear to operate normally. Further, removal of the pyloric valve and reattachment of the stomach directly to the duodenum does not change the rate of stomach emptying. All of these observations contradict the nice, simple hypothesis mentioned above.

The precise manner by which the opening and closing of the pyloric valve is controlled remains unknown. It seems likely, however, that the first two factors—the consistency of the stomach contents and the strength of stomach peristalsis—are more important than any pH factors that may be involved.

indigestible material through the same opening by which the food entered. Thus, mixing of undigested and indigestible food material inevitably occurs.

A two-opening system avoids this mixing. The human digestive system is such a two-opening system. Food ingested through the mouth passes through the pharynx and esophagus to the stomach. From here it passes into the duodenum, small intestine, and large intestine (colon), with the indigestible materials moving into the rectum for elimination through the anus.

Digestion proceeds in an assembly-line fashion. Such a system is particularly advantageous in the digestion of large and complex food molecules. Proteins, for example, are first digested by hydrolysis into smaller

Control and organization of the digestive system appear to occur through two major processes: neuronal and hormonal communication.

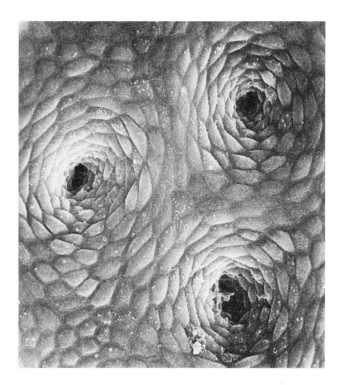

Fig. 10.6
Scanning electron micrograph of the inner surface lining of a monkey stomach (× 1200), showing three of the gastric pits from which the gastric juice emerges. The gastric juice, rich in hydrochloric acid, keeps the pH of the stomach highly acidic (around pH 2). (Micrograph courtesy Dr. Peter Andrews, Georgetown University).

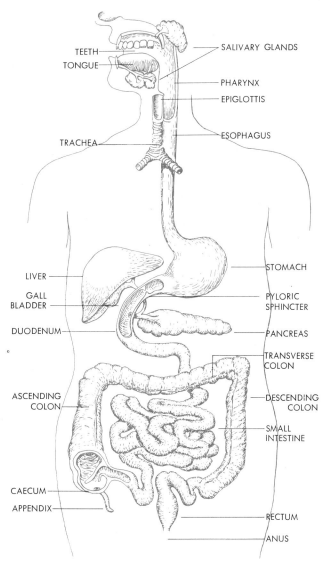

units (proteoses, peptones), which are themselves then hydrolyzed into still smaller units, the peptides. The peptides, in turn, are hydrolyzed into amino acids. In a two-opening digestive system, the appropriate enzymes are secreted at different points along the route. Thus the enzymes that hydrolyze large proteins into small proteoses and peptones are secreted near the beginning of the digestive tract. Those enzymes that hydrolyze the proteoses and peptones into peptides are secreted further along the digestive tract. The hydrolysis of peptides into amino acids occurs still further along the tract. The human digestive system (Fig. 10.7) demonstrates an efficient arrangement of digestive organs and glands.

Glandular Functions

Although ingested food contains a wide variety of minerals and vitamins and a few miscellaneous organic compounds, its main bulk is composed of carbohydrate, fat, and protein. The first of these, being a primary source of energy for most organisms, may compose 80 percent of the diet.

Despite considerable differences in the molecular structure of carbohydrates, fats, and proteins, the digestion of these three classes of nutrients proceeds by the same chemical process, hydrolysis, which, you will recall, involves the chemical addition of a water molecule to key chemical bond linkages. This results in the splitting of the molecule involved into small units. The

Fig. 10.7
The mammalian digestive system, exemplified here by the human digestive apparatus, is a sort of "assembly line" system. The enzymes secreted by the various organs progressively break food substances down into smaller and smaller molecular subunits.

All metabolic processes occur within individual cells. Digestion is the first step in the process of hydrolyzing complex molecules into simple components that can be absorbed into the blood and carried to all cells of the body.

process of hydrolysis can be represented diagrammatically as follows:

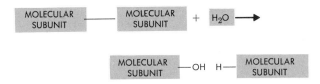

Hydrolysis will proceed spontaneously, releasing a small amount of heat energy as it does so. However, digestive enzymes (a specific one for each reaction) greatly increase the reaction rate by lowering the energy barrier that must be cleared before the reaction will proceed. Many digestive enzymes are highly specific to the substance (substrate) on which they work. Maltase, for example, will catalyze only the reaction converting maltose to glucose; sucrase will catalyze only the reaction converting sucrose to glucose and fructose.

Even with enzymatic catalysis, however, the hydrolysis reactions of digestion are strongly influenced by the environmental medium of the digestive tract. Like most organic reactions, they have an optimum temperature at which they proceed best. Another important factor is pH. The enzyme pepsin, produced in the stomach, works best in a highly acidic environmental medium (about pH 2). The enzyme trypsin, secreted into the duodenum by the pancreas, works best in a slightly basic environmental medium (about pH 8).

The end-products of digestion must meet two criteria. First, they must be sufficiently soluble to be absorbed and carried by the blood. Second, they must have molecular dimensions sufficiently small to pass through the walls of the digestive tract. In the case of carbohydrates, the final digestive products are generally simple sugars, such as glucose and fructose. With fats, the end-products are glycerol and fatty acids. In the case of proteins, the final units are amino acids, while nucleic acids yield simple sugars, nitrogenous bases, and phosphates.

Several glands and other accessory organs are intimately associated with the digestive process. The salivary glands secrete saliva into the mouth. Saliva contains **amylase**, an enzyme that begins the hydrolysis of starch by converting it to the disaccharide maltose. Glands in the lining of the lower region of the stomach (pyloric mucosa) produce the hormone **gastrin** (see

Supplement 10.2), which stimulates the secretion of gastric juice, rich in hydrochloric acid, and the secretion of enzymes by the pancreas. The liver produces **bile** which, after storage in the gallbladder, enters the digestive tract at the duodenum. Bile contains salts that cause ingested fats to break up into tiny droplets (a process called *emulsification*), thereby exposing more surface area for attack by fat-digesting enzyme molecules. Bile salts also increase the water solubility of fatty acids and substances such as vitamin K, which is necessary for proper clotting of the blood.

The pancreas produces three digestive enzymes, which enter the digestive tract via pancreatic ducts opening into the duodenum. One of these is the fat-digesting enzyme, pancreatic **lipase**. The others are **pancreatic amylase,** which continues the hydrolysis of starches begun by salivary amylase in the mouth, and **trypsinogen,** an inactive form of the enzyme *trypsin*, which, when activated in the digestive tract (were it to be produced directly in the active form, the pancreas would digest itself), converts proteins into the intermediate molecular forms proteoses and peptones on the way to producing peptides. The conversion of trypsinogen to trypsin is brought about by another enzyme, *enterokinase*, secreted by the intestinal wall.

Besides the salivary glands, liver, and pancreas, virtually the entire digestive tract performs glandular functions. From esophagus to anus, mucus is produced in specialized glandular epithelial cells to aid in lubrication, thus reducing friction between the food masses and the walls of the digestive tract. The stomach wall secretes hydrochloric acid (which destroys bacteria, dissolves minerals, and establishes the optimum pH for stomach enzyme activity), **pepsinogen** (which, when converted by hydrochloric acid into the active stomach enzyme *pepsin*, begins the digestion of proteins, converting them to the next-smaller molecular units, proteoses and/or peptones), and **rennin** (which curdles milk proteins). To complete the digestive process, the small intestine produces many enzymes besides enterokinase.

The small intestine is also the principal region of food absorption. With some notable exceptions, including alcohol and most other drugs, no foods are absorbed by the stomach. Chemical analysis of material extracted from the first few feet of the small intestine shows considerable amounts of nutrient materials still present; the same analysis run on material taken from the last few feet of the small intestine shows virtually none. The final task of the digestive tract, then, appears not to be concerned with digestion, but rather with the preparation of indigestible materials for elimination. This is primarily the role of the colon, or large intestine. Here most of the water is reabsorbed into the bloodstream. The remaining material consists mostly of bacteria and indigestible materials, such as cellulose. Bacteria are found in large numbers in the colon, where they find a suitably warm, moist, and dark environment in which to live. These bacteria may benefit their host by producing vitamins that can be absorbed and used.

The relationship of bacteria and cellulose in the stomachs of herbivorous ruminant animals, such as cattle, is an important one. The bacteria found there produce **cellulases**, enzymes capable of splitting cellulose into the monosaccharide units of which it is composed. These monosaccharides are consumed by the bacteria which in turn release free fatty acids that are used as a major energy source by the host animal. When some of the bacteria die, their cells are digested by the host, which is thereby provided with those nutrients generally found in such cells, including nucleic and amino acids. Thus the bulk of the plant material on which these animals feed can be assimilated into their body matter, and from there, perhaps, to our own. Without these herbivorous animals—or better, without their bacteria—the vast source of energy available from plant cellulose would go virtually untapped.

In the middle of the nineteenth century, the famous German physiological chemist Justus von Liebig (1803–1873) wrote:

> *Whatever views we may entertain regarding the origin of fatty constituents of the body, this much at least is undeniable, that the herbs and roots consumed by the cow contain no butter; . . . that no hogs' lard can be found in the potato refuse given to swine; and that the food of geese or fowls contains no goose fat. . . . The masses of fat found in the bodies of these animals are formed in their organism.*

Liebig was emphasizing the fact that whatever substances an organism takes in are converted within its body to its own kind of molecules (fats, proteins, steroids, carbohydrate). The function of digestion is thus to break the food substances of the consumed organism (cow, chicken, or carrot) into basic building blocks of the consuming organism such as amino acids, fatty acids, or simple sugars. After all, amino acids are amino acids, wherever they come from. For example, after digestion breaks cow or chicken protein into amino acids in a person's intestines, these molecules are absorbed into the bloodstream and carried to the individual body cells. Here they are reassembled into protein, but human protein rather than cow or chicken protein.

10.4 ABSORPTION

Just as a finger in the hole of a doughnut cannot truly be said to be *inside* the doughnut, so food in the digestive tract of a mammal cannot truly be said to be inside the mammal; the food must first pass through the walls of the digestive tract into the blood by processes known as **absorption**.

Microscopic examination of the intestinal wall reveals it to be composed of tiny, finger-like projections (**villi**), each supplied with vessels for the circulation of blood and lymph. The presence of villi greatly increases the surface area of the small intestine, making it ideally suited for food absorption (Fig. 10.8). There is also considerable evidence that the *microvilli*, which cover the surface of the villi, adsorb enzymes onto their surface, thereby greatly increasing their digestive efficiency.

Food absorbed from the intestine is collected from the capillaries into the **hepatic portal vein**, which runs from the intestines to the liver. This complex is called the hepatic portal system. The hepatic portal vein is unusual: it is one of the few mammalian blood vessels that begin and end in capillaries without its blood passing through the heart. After collecting absorbed food from the capillaries in the intestine wall, the portal vein proceeds to the liver, where it again breaks down into capillaries to exchange materials with liver cells. In the liver, much of the glucose is absorbed out of the blood. To a lesser extent, such other substances as fatty acids and amino acids are also absorbed by the liver. Thus the liver acts to control the levels of various substances, particularly sugar, in the blood. Regardless of the quantity of sugar ingested, the amount of sugar in the hepatic vein leading out of the liver remains quite constant. Food materials passing from the liver are carried to the vena cava, one of the main veins leading to the heart. By this route food materials pass into the general circulation.

10.5 ELIMINATION

In humans, materials pass through the approximately 20 ft of small intestine in from 4 to 6 hr, yet take as much as 24 hr to pass through the 4 ft of the large intestine, or colon. This fact alone indicates that the role of the large intestine may be an important one. Yet, since chemical analyses of materials entering the large intestine from the small intestine reveal very few nutritive materials left to be absorbed, it seems equally evident that the large intestine is less concerned with usable than with unusable materials.

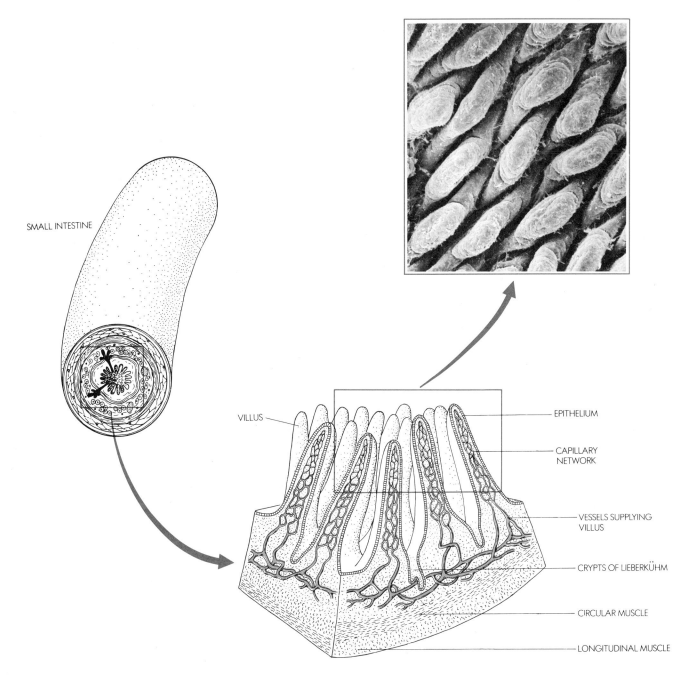

SMALL INTESTINE

VILLUS

EPITHELIUM

CAPILLARY NETWORK

VESSELS SUPPLYING VILLUS

CRYPTS OF LIEBERKÜHM

CIRCULAR MUSCLE

LONGITUDINAL MUSCLE

Material entering the large intestine is quite watery; upon leaving, it has been formed into damp, solid masses, the **feces.** It is reasonable to hypothesize, therefore, that the large intestine returns much of the water used in digestion to the bloodstream. Both analytical and tracer experiments support this hypothesis. Tracer experiments also show that the fecal masses contain many of the same substances that were in the entering food. Such substances seem not to have been digested or absorbed.

On the other hand, chemical analyses of fecal material leaving the large intestine reveal the presence of bacterial cell masses and metallic ions such as calcium.

Fig. 10.8
The absorptive surface area of the small intestine is greatly increased by the presence of fingerlike projections called villi and the valleys (crypts of Lieberkuhm) existing between them. The scanning electron micrograph of villi in the small intestine of a monkey ($\times$ 250) shows the villi as viewed from the top or inner cavity of the intestine. The wrinkles are due to the contractile movement of the villi. Each villus possesses a core of connecive tissue, smooth muscle cells and blood and lymph vessels. (Micrograph courtesy of Dr. Peter Andrews, Georgetown University Department of Anatomy.)

SUPPLEMENT 10.4
THE NATURE OF ABSORPTION

In the small intestine, soluble food materials exist in high concentration. The blood supplying the intestinal lining would show a considerably lower concentration of digested food materials. According to the simple laws of diffusion and osmosis, soluble food molecules would be expected to pass into the bloodstream, and tracer experiments show that they do. It can be hypothesized that diffusion and osmosis are the only factors responsible for the absorption of foods and minerals from the lower digestive tract. An experiment must then be designed to see whether this hypothesis leads to contradictions.

Loops of the small intestine in experimental animals are tied off from the rest of the digestive tract. A 0.5 percent concentration of sodium chloride solution (0.5 g NaCl in 100 cm^3 of distilled water) is injected into some of the loops (series A). A 1.0 percent concentration of sodium chloride is injected into an equal number of other loops (B). A 1.5 percent concentration of sodium chloride is injected into a third set of loops (C). Blood has a sodium chloride concentration of approximately 1.0 percent. Thus the concentration of sodium chloride in the A series is less than that of the blood supplying the same intestinal loops. The concentration in the B series is the same as that of the blood, and the concentration in the C series is greater than that of the blood. A hypothesis proposing that only the physical factors of diffusion and osmosis are involved in the absorption of soluble substances through the intestinal wall must predict that sodium chloride will not pass into the blood in the loop B series, for there is no concentration gradient in either direction. For the loop C series, the hypothesis would predict some passage of sodium chloride into the blood by diffusion, while water would be drawn from the blood by osmosis. The intestinal loops of series A would lose water to the blood by osmosis but gain in salt content by diffusion of sodium chloride molecules from the blood.

Table 10.1 summarizes the results of the experiment. The predictions made by the hypothesis are not verified; there is absorption of sodium chloride into the blood supplying the intestinal loops in all three series. Clearly, diffusion and osmosis alone are not sufficient to explain intestinal absorption. Yet these data do indicate that diffusion and osmosis are at least partly involved, for there is an inverse relationship between the amount of water absorbed from the solutions in the intestinal loops and their sodium chloride concentrations. There is also a direct relationship between the concentration of sodium chloride found in the various solutions and the amount of sodium chloride ultimately appearing in the blood. Both these relationships are predictable on the basis of hypothesized roles of diffusion and osmosis in the absorption process.

Table 10.1

Effect of Concentration of Salt on the Absorption of Salt and Water from the Intestine

| Series | NaCl solution injected into intestine | | | Absorbed | |
	Concentration, %	H$_2$O, ml	NaCl, g	H$_2$O, ml	NaCl, g
A	0.5	100	0.5	75	0.3
B	1.0	100	1.0	50	0.6
C	1.5	100	1.5	25	1.0

If blood serum from an animal is placed in an intestinal loop of the same animal, the serum is soon absorbed. Since this serum would be in perfect concentration with the blood and tissue fluids, a hypothesis proposing that only diffusion and osmosis are involved in digestive-tract absorption would predict that no absorption would occur. Clearly, this hypothesis is no longer tenable.

There must be other forces at work in intestinal absorption of digested food. It is reasonable to hypothesize that active transport may be involved; the intestinal lining cells may be performing work that results in the movement of substances against the concentration gradient. Histological examination reveals the cells of the intestinal lining to be the columnar epithelial type, and cytological investigation shows that such cells are liberally supplied with mitochondria, features often characteristic of metabolically active cells. Further support for the active transport hypothesis comes from the fact that substances pass through the intestinal wall against a concentration gradient only while the intestinal loop is alive. Soon after death an equilibrium is established between the tissue fluids and the medium inside the loop.

From the observation that sodium chloride is quickly absorbed through the intestinal wall while magnesium sulfate (MgSO$_4$) is not, it is tempting to hypothesize that the intestinal wall acts as a semipermeable membrane; only molecules of the proper dimension are able to pass through it (a molecule of MgSO$_4$ is larger than a molecule of NaCl). Such a hypothesis must be quickly rejected, however; amino acids and glucose have molecules far larger than those of magnesium sulfate, yet they are readily absorbed. Thus the problem of intestinal absorption is reminiscent of the one first posed in Chapter 4: What factors are involved in the selective passage of substances across plasma membranes? To date, no complete explanation has evolved.

Closer examination of the large intestine and its contents shows that large colonies of bacteria (mostly *Escherichia coli*) live there, metabolizing those food materials missed by the digestive process further up the tract.* This accounts for the presence of bacterial cell masses in the feces, but it does not account for the presence of the calcium. However, chemical analysis of blood entering the wall of the large intestine shows a higher concentration of calcium ions than is found in the blood leaving that organ. Evidently, then, the large intestine also removes excess calcium ions from the body, releasing them into the materials that pass through it.

Note, however, that there is a distinct difference between the status of material such as cellulose found in fecal material and that of the calcium ions. The fecal matter has never been "inside" the body; it was never a part of the body's metabolism. The calcium ions, however, *were* inside; they are left over fragments of body metabolism. In dealing with cellulose, therefore, the large intestine is acting as an organ of *elimination*. In dealing with calcium ions, however, it acts as one of several organs of *excretion*, the getting rid of the excess or undesirable products of metabolism, a process to which attention will be directed shortly.

10.6
NUTRITION

Today the question of nutrition is a complex one, though it is so only for human beings and the animals and plants they raise for their consumption. In their normal habitat, organisms manage to solve their nutritional needs quite nicely. Human beings, however, often have many cultural biases, taboos, and the like concerning their diet; the Jewish prohibition against eating pork is one example and the Hindu prohibition against eating beef another.

Currently there is widespread interest in western countries in vegetarianism. In the Far East, of course, vegetarianism based primarily on grains such as rice has been common for centuries. Such vegetarianism is usually by necessity; the inevitable loss of energy that is involved in feeding the grain to an animal and then eating the animal is a luxury some nations cannot afford. In the West, and especially the United States, we have until recently enjoyed the luxury of energy to waste. The current dramatic rise in vegetarianism, especially among young people, may signify an awareness that the days of this wealth of energy and our continuing abuse of its abundance may soon be coming to an end. Vegetari-

anism may also exist for other reasons, of course. Some persons simply do not wish to eat the flesh of animals that have been killed for this purpose, and it has been said that the percentage of vegetarians would rise significantly if everyone were forced to visit a slaughterhouse.†

It seems evident from both the fossil record and human digestive anatomy and physiology that human beings evolved either as meat eaters (**carnivores**) or, more likely, as both plant and animal eaters (**omnivores**). The critical question about vegetarianism then is: can one obtain a balanced diet containing all of the essential nutrients needed to remain in good health without including meat? The key words here are "balanced diet" and "essential." As almost everyone knows, a balanced diet is one that contains proper proportions of carbohydrate, fat, protein, vitamins, and minerals; the "proper amount" often varies according to one's school of thought.‡

The term "essential" has a more precise, scientific meaning. When we speak of an essential amino acid in the diet, we mean one that an organism needs for survival and cannot synthesize in its own body. While certain lower organisms are able to synthesize virtually all of their nutritional need from the simple substances they take in, in the course of evolution the higher organisms have lost this ability and thus must obtain certain substances in their diet, generally by consuming other organisms that can synthesize those substances. In terms of amino acids, for example, there are a few we can synthesize for ourselves and others we cannot; those we cannot, therefore, become *essential* amino acids in our diet.

To return then to vegetarianism, in answer to the question concerning the ability of such a diet to provide essential nutrients: it seems to be able to do so *if* the person on the diet is willing to learn what the essential nutrients are and be sure to include foods that contain these essentials.

10.7
HOMEOSTASIS, DIGESTION, AND
FEEDING BEHAVIOR

The homeostatic mechanisms involved in digestion are perhaps not obvious. Yet, on the macroscopic level, such mechanisms are obvious, indeed. If a person is hungry, hunger pangs (see Supplement 10.1) lead to eating. As the stomach becomes increasingly filled with

---

*Recall that in ruminants (for example, cattle), the presence of bacteria in the stomach makes possible the metabolic use of free fatty acids released as one of the byproducts of the bacterial hydrolysis of cellulose. *E. coli* do not have cellulose-splitting enzymes, however, so this material in the human diet only adds bulk to the intestinal contents—a factor that may aid in the proper function of the large intestine.

†In this regard, however, the statement made by vegetarians that they are vegetarian because they "couldn't eat a living thing" is not valid; when one eats a steak, it is thoroughly dead. A reasonably fresh salad, on the other hand, is composed of plants such as lettuce and tomatoes, whose cells are still alive and respiring as they are being consumed.

‡Especially when one is dieting! Some of the so-called crash diets have been proven to be dangerous and even fatal in some individual cases.

food, a negative feedback mechanism causes the hunger slowly to disappear and the eating, too, slows and eventually ceases. Yet, as Supplement 10.2 makes clear, there are wheels within wheels; the presence of protein in the stomach, for example, initiates gastric juice secretion via the hormone gastrin; as the protein becomes digested into its component parts, the production of gastrin and thus gastric juice slows; just the *absence* of protein is sufficient to provide a negative feedback that inhibits gastrin secretion. Digestive enzymes, too, appear in response to the presence of their proper substrate; peptidases for peptides; sucrases for sucrose; lipases for lipids; and so on, until their further release is inhibited by the absence of their substrate and perhaps other physiological changes associated with food satiation. Note that the principle of homeostasis does *not* tell us anything about the precise molecular mechanisms involved in such positive and negative physiological feedback. Rather it provides us with a general mode of operation that clarifies some general guiding principles by which all physiological systems seem to work.

But homeostatic mechanisms are also involved in the regulation of the feeding behavior that results in having food to digest. The previous chapters dealing with cellular physiology and biochemistry make it obvious that living organisms require a wide variety of substances in order to survive. In complex organisms such as human beings, these substances must be obtained in the diet. Indeed, as just pointed out, one of the most obvious homeostatic mechanisms is the one by which our bodies inform us of their need for food via hunger, and we restore nutrient balance by the act of eating. The system is a great deal more intricate than that, of course. As we have seen, living systems need several different substances—carbohydrates, fats, proteins, minerals, and the like—and, equally important, need them in the proper proportions. Thus, as we saw in the previous section, the term "balanced diet" simply refers to the proper mixture of nutrients to keep one's body in the homeostatic equilibrium necessary for its proper functioning.

In a very real sense, it is possible to view food intake as a means of maintaining an energy balance within the body. For, ultimately, the energy acquired through food must equal the amount of energy expended. If more energy is taken in than is immediately needed it can be stored, perhaps as fat; if less energy is taken in than is expended, a loss of weight results as the body burns off its reserves in order to maintain equilibrium as long as it can. In general, however, a healthy individual's weight, like body temperature, hovers around an average or mean without any major fluctuations. This fact alone implies the existence of some sort of homeostatic mechanism at work.

One portion of this homeostatic mechanism appears to lie in the brain. Within the lateral area of the hypothalamus is a region whose stimulation causes rats to eat, even if they have previously eaten to capacity a short time before. In a lower region of the hypothalamus, on the other hand, lies a region whose stimulation leads to a cessation of feeding behavior, even if the animals have not eaten for a long time and are surely hungry. It is easy to picture a system in which hunger stimulates the lateral region of the hypothalamus to cause feeding behavior and the satisfying of that hunger stimulates the lower hypothalamic region to inhibit further feeding for a time. But what physiological factors are involved in the stimulation of these stimulation and inhibition feeding behavior centers in the hypothalamus?

Several hypotheses have been proposed. Some suggest that the neural signals of the hypothalamus are influenced by the level of food stores within the body. It is known, for example, that not only does an animal such as a deer eat when its body food reserves need replenishing but may also selectively seek out specific nonorganic substances temporarily in short supply, such as salt. It has been hypothesized that the levels of certain food substances (such as glucose and/or amino acids) are detected by certain specialized receptor cells in the hypothalamus. These receptor cells, depending upon whether the blood sugar levels are high or low, may inhibit or release the hypothalamic feeding center, respectively. This hypothesis is supported by the observation that hunger and low blood sugar levels may often be in close correlation, and that an increase in blood sugar levels increases the electrical activity of the hypothalamic feeding cessation center and decreases the electrical activity of the feeding center. An increase in the amino acid levels in the blood has a similar, though less marked, effect. Still another hypothesis involves fat in the regulatory process, noting in support of this hypothesis that increases in the amount of stored body fat (adipose tissue) decreases the rate of feeding.

Another hypothesis suggests that body temperature is inversely related to feeding behavior. This hypothesis is based upon the observation that animals exposed to cold temperatures tend to overeat while those exposed to warm temperatures tend to undereat. This hypothesis envisions a sort of physiological thermostat that monitors metabolic rate. The rate of metabolism, of course, is directly related to temperature; the higher the metabolic rate, the higher the body temperature and vice versa.

Still other hypotheses suggest that the mechanical swelling or distention of the stomach and intestinal tract by the food bulk may act indirectly to depress the feeding center in the hypothalamus and slow or stop feeding. Or the body may have some way of measuring the amount of saliva, digestive fluids, chewing, swallowing, and the like and use this information to suppress feeding beyond a certain point.

Of course, as many know all too well, in humans psychological factors play a large role in the amount of food ingested; the hungriest person may lose his or her appetite on receiving bad news, while persons suffering from obesity often overeat when not actually hungry. This may be one of the reasons why diet programs that involve group meetings of overweight persons, where support for each other can be provided, are generally far more successful than individual diets alone.

10.8
EXCRETION

An inevitable result of metabolism is the production of "waste" products. The quotes around the word waste denote that often what is produced as metabolic waste may well be molecules or molecular fragments that are, themselves, useful but are simply present in excess in the body at that particular time. The removal of calcium ions via the large intestine, for example, is homeostatically controlled; when calcium ions are plentiful in the body, more are removed; when calcium ions are in low supply few, if any, may be. Perspiration, too, removes some of the byproducts of metabolism, including salts. In times of dehydration, however, salt retention may be important to prevent further water loss; the waste salt is retained to maintain a proper water balance in the body fluids. Thus what is "waste" at one time may be far from that at another, and this should be kept firmly in mind when dealing with the following material.

As stated at the end of the last section, the process of the removal of metabolic wastes is called excretion. The lungs, therefore, are excretory organs that excrete carbon dioxide and water into the air, and the colon is an excretory organ that excretes excess calcium and iron into the fecal material passing through it. In complex animals, however, the **kidneys** are the principal excretory organs. Within the kidneys the vast majority of the nitrogenous wastes of protein catabolism are excreted. Without kidneys, humans and other complex animals would die from the accumulation of these highly poisonous wastes.

The possession of kidneys is by no means universal in the animal world, however. As with the circulatory and respiratory systems, only animals with cells that are not exposed to a surrounding watery environment have a need for any excretory system at all. A "need" for an excretory system, of course, does not result in its evolution. Natural selection favors any variation allowing a multicellular organism to eliminate its waste products more effectively. The accumulation of such variations over long periods of time would lead to the development of a complex excretory system, such as that found in mammals. The famous renal (kidney) physiologist Homer Smith once wrote a book entitled *From Fish to Philosopher* (see Suggested Readings at the back of the book), in which he discussed the evolutionary history of living organisms based almost entirely upon the evolution of their excretory system. It seems probable that our earliest ancestors evolved in habitats where they were surrounded by water with a composition much the same as their body fluids. Thus they were in balance, osmotically, with their environment. As animals began to move to regions that were not balanced osmotically with their body fluids, they had to develop means of either pumping excess water out of their bodies, like the contractile vacuole of *Paramecium*, or excreting salt out of the body, as do bony fish, using special glands located on the gills where they will be exposed to the surrounding water. (As one might expect, freshwater fish *absorb* salt). Seagulls, which feed almost exclusively on saltwater fish, must also excrete salt, which they do through special tear ducts. In still other cases, the problem is solved by either becoming adapted to having the body fluids be nearly as salty as the surrounding ocean or being able to retain large amounts of urea in the body, thereby remaining isotonic with the environment. Such mechanisms have evolved in the hagfish and the shark, respectively, organisms on which Homer Smith did most of his research. The lungfish is yet another with which Smith worked. This animal survives many weeks in the dried-up mud of Lake Victoria in Africa building up levels of metabolic byproducts in its body that would be fatal to almost any other organism.

In organisms surrounded by air, the problem is compounded by the need to get rid of excretory wastes, yet retain body fluids. Besides that lost in the urine, we lose water in our feces, sweat, and exhaled air. Thus water replacement by drinking it or obtaining it in our foods is absolutely vital; a person can survive for much longer without food than without water.

10.9
THE BIOCHEMISTRY OF EXCRETION

In Chapter 6 we saw that amino acids may enter the main pathways of cellular metabolism. For this to occur the amino acids undergo a series of reactions known as **transamination** (in some specific cases known as **deamination**) in which their amino groups are removed and transferred to an intermediate substance, an α-keto acid such as α-ketoglutarate. Transamination consists of essentially two steps, where R_1 and R_2 stand for two side-groups:

$$
\begin{array}{cccc}
R_1 & R_2 & R_1 & R_2 \\
| & | & | & | \\
H-C-NH_2 + & C=O \rightleftharpoons & C=O + & H-C-NH_2 \\
| & | & | & | \\
COOH & COOH & COOH & COOH \\
\text{AMINO} & \alpha\text{-KETO} & \alpha\text{-KETO} & \text{AMINO} \\
\text{ACID} & \text{ACID} & \text{ACID} & \text{ACID}
\end{array}
$$

Thus the amino group is transferred to an α-keto acid to produce a new amino acid.

Suppose, for example, that the amino acid involved in the above reaction is alanine:

$$
\begin{array}{c}
CH_3 \\
| \\
H-C-NH_2 \\
| \\
COOH
\end{array}
$$

Since the R-group for alanine is a methyl group, CH_3, transamination occurs as follows:

$$
\begin{array}{ccccccc}
\overset{\displaystyle CH_3}{\underset{\displaystyle COOH}{H-\overset{|}{\underset{|}{C}}-NH_2}} & + & \overset{\displaystyle R}{\underset{\displaystyle COOH}{\underset{|}{C}=O}} & \rightleftharpoons & \overset{\displaystyle CH_3}{\underset{\displaystyle COOH}{\underset{|}{C}=O}} & + & \overset{\displaystyle R}{\underset{\displaystyle COOH}{H-\overset{|}{\underset{|}{C}}-NH_2}} \\[2em]
\text{ALANINE} & & \alpha\text{-KETO} & & \text{PYRUVIC} & & \text{AMINO} \\
 & & \text{ACID} & & \text{ACID} & & \text{ACID}
\end{array}
$$

The products of this reaction are another amino acid, depending on the α-keto acid involved, and pyruvic acid. The pyruvic acid may be used as a framework on which to build another amino acid, or it may be oxidized to provide energy for ATP synthesis, just as may the pyruvic acid produced by the splitting of glucose in glycolysis.

However, in producing pyruvic acid from an amino acid rather than from glucose, a different waste product, **ammonia** (NH_3), is formed. (At physiological pH, most of this ammonia becomes converted into ammonium ions, NH_4, which are highly soluble.) Ammonia is also highly toxic, however. Thus if the wastes are to be excreted in the form of ammonia, a large amount of body water must be present to keep it in very dilute solution and the waste-containing fluid or **urine** must pass from the body in an almost continuous flow. It is not surprising, therefore, to find that animals that excrete nitrogenous wastes in the form of ammonia are aquatic in habitat, in no danger of dehydration.

As we have seen, land animals are faced with quite a different situation. Far from having a limitless supply of water surrounding them, they face the problem of conserving water within their bodies. The loss of water necessary to keep the ammonia safely diluted would be fatal. Instead, these animals convert the ammonia to either **uric acid** or **urea**.

Uric acid ($C_5H_4N_4O_3$) is found in the form of solid crystals. Being insoluble, these crystals can be stored in the body for considerable periods of time. Insects, birds, and reptiles excrete ammonia in the form of uric acid. Since these animals lay their eggs on land, the young must have a way of holding nitrogenous wastes within the egg until hatching; uric acid provides a convenient form, since its insolubility prevents reabsorption. But the insolubility of uric acid provides still another benefit; since it passes out of the body with the feces in a dry, crystalline state (as in the white portion of bird droppings), water loss from the body is greatly reduced. In humans high levels of uric acid are associated with the symptoms of gout.

Urea is somewhat less soluble than ammonia but, unlike the latter compound, is not highly toxic unless present in relatively high concentrations. Urea formation is a complex, multistep process that takes place mostly in the liver. The formation of urea occurs in a cyclic metabolic pathway (see Fig. 10.9). In a complex series of reactions involving carbon dioxide and ammonia, a compound called carbamyl phosphate is formed in a process that requires the expenditure of two ATPs. Carbamyl phosphate reacts with the terminal amino group of the amino acid ornithine, which is already present in the cell, to form citrulline. Energized by another ATP, citrulline combines with aspartic acid to yield an intermediate (arginosuccinic acid) that breaks down into the amino acid arginine and fumaric acid (which also occurs in the citric acid cycle). The arginine then interacts with water to form urea and ornithine. Because a molecule of ornithine is regenerated for each one that is used, the pathway is called the **ornithine-citrulline cycle**. The ornithine-citrulline pathway is endergonic, requiring the expenditure of three ATPs in all. The body must expend energy to convert the waste products of protein metabolism into a less toxic state for excretion.

The urea passes out from the liver dissolved in the blood, from which it is filtered and excreted. Carrying out this filtration and excreting the filtrate is the job of a mammalian excretory system. The main filtering organ of this system is the kidney.

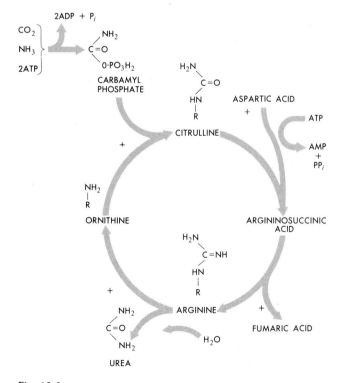

Fig. 10.9
The ornithine-citrulline cycle, by which carbon dioxide and ammonia are combined to form urea. This cycle is the pathway by which ammonia, the waste product of protein metabolism, is made less toxic to cells and prepared for excretion.

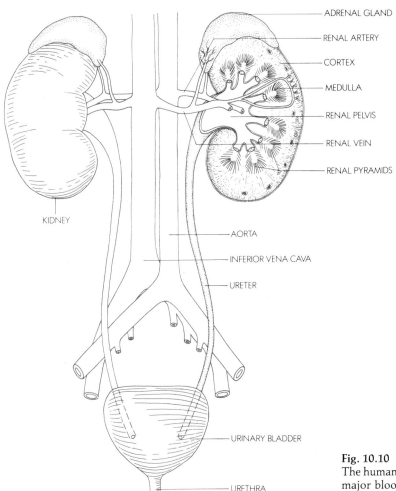

ADRENAL GLAND

RENAL ARTERY

CORTEX

MEDULLA

RENAL PELVIS

RENAL VEIN

RENAL PYRAMIDS

KIDNEY

AORTA

INFERIOR VENA CAVA

URETER

URINARY BLADDER

URETHRA

Fig. 10.10
The human excretory system and its
major blood vessels.

The Kidneys

The kidneys are reddish-brown organs located in the
small of the back, just above the pelvic girdle or hip region. Approximately 11 cm in length, each kidney is
capped by an endocrine gland, the adrenal gland (*ad-*,
toward; *renal*, kidney), and is shaped rather like a lima
bean. The kidneys lie outside of the main body cavity,
the peritoneum, which contains most of the digestive organs. Each kidney is surrounded by three layers of tissue, an inner fibrous connective tissue (renal capsule),
an outer double-layered membranous tissue (renal fas-
cia) and a middle layer of fat (adipose capsule). These
tissues serve to support the kidneys and insulate them
from damage by a heavy blow.

Each kidney is supplied with blood by a renal artery. The renal arteries branch off the main body artery,
the aorta, on either side. Blood leaves the kidney via the
renal vein and passes into the general circulation
through the inferior vena cava. In relation to the size of
the organ they supply, the renal blood vessels are the
largest in the body. More than one-fifth of all the blood
in the human body passes through the kidneys every
few minutes, thereby ensuring that the concentration of
urea does not reach dangerous proportions.

Internally, each kidney in longitudinal section can
be seen to be divided into three general regions, the **cortex, medulla,** and **renal pelvis.** The outermost, the cortex, blends into the next layer, the medulla. Indeed, at
times, the renal columns of the cortex lie beside the renal
pyramids of the medulla (see Fig. 10.10). The tips or
papillae of these pyramids extend into small channels,
or minor calyces (singular, calyx), which merge into
major calyces which, in turn, unite to form the renal
pelvis. The renal pelvis is actually only the extended upper end of the ureter. The general pathway followed by
the metabolic wastes on the way to being excreted in the
urine is: renal artery, cortex and medulla, pelvis, ureter,
bladder and, from there, to the body exterior through
the urethra. Much of the excretory system, including
portions of tubules of the renal cortex, the renal pyra-

SUPPLEMENT 10.5
THE COUNTERCURRENT MULTIPLIER EFFECT

In the interests of saving energy, engineers have long been familiar with the concept of countercurrent exchange. The duct from a furnace, for example, which conveys the hot exhaust gases to the outside will be placed side-by-side with the duct that brings in the cold air from the outside to be heated. By virtue of the two ducts being in contact, some of the heat from the exhaust warms the incoming cold air. Thus heat that would ordinarily be wasted is used to increase the furnace's efficiency; not as much fuel is needed to heat the incoming air as would otherwise be the case.

As usual, however, nature got there first. Precisely the same principle is used to warm appendages such as the fins, tails, and feet of animals living in cold environments; the cold blood returns to the body through veins lying alongside arteries carrying warm blood out, and thus heat is transferred to the incoming cold blood and warms it. The blood vessels in fish gills are also arranged so as to allow countercurrent exchange of oxygen.

The principle of countercurrent exchange also applies to the kidney's loop of Henle, as well as the surrounding blood vessels. In this case, however, the fact that the two systems (the descending and ascending loops) are connected directly

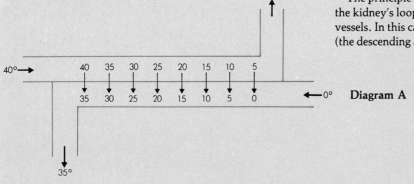

Diagram A

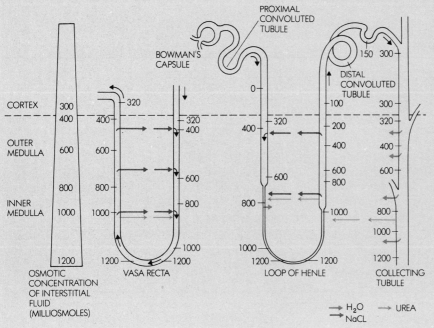

Diagram B
The countercurrent multiplier effect in the kidneys. The numbers indicate relative concentrations of the various interstitial fluids and the arrows the direction of movement. (From A. P. Spence and E. B. Mason, *Human Anatomy and Physiology*. Menlo Park, Calif.: Benjamin/Cummings, 1979.)

to each other allows a *multiplier* effect; the one-way passage of chloride ions, for example, allows them to be used over and over again (much as the "waste" heat is returned to the furnace) and results in a higher concentration of sodium ions at the bend of the loop than would otherwise be possible. The same is true of the blood vessels, which attain a counter-current exchange multiplier effect both with the loop of Henle and the surrounding extracellular fluid of the kidney medulla. The result is to trap solutes in high concentration in the lower loops of the blood vessels, as well.

This concept may be easier to grasp if you imagine one long hallway with a hairpin bend through which people (representing ions) move in only one direction. The two hall-ways are connected to each other on their adjacent sides by one-way doors, through which as many people as possible who have passed the bend are shoved, so that they have to retrace their steps. A moment's reflection will make it clear that the greatest concentration of people must occur at the hairpin bend, and that many of those people will have been there before, perhaps many times.

mids of the medulla, the calyces, pelvis, ureter, bladder, and urethra, is involved with urine collection and con-veyance, with the bladder performing the function of temporary storage. It is clear, therefore, that the criti-cal phases of waste removal from the blood, as well as urine formation, must occur within regions of the renal cortex and medulla that lie in the path between the renal artery and the collecting tubules leaving these regions.

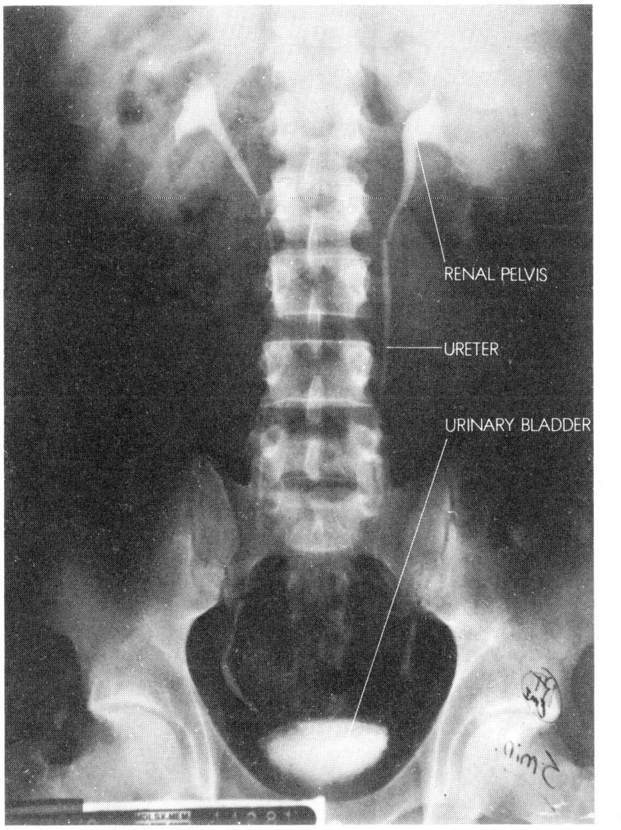

Microscopic examination seems to support this conclusion. The collecting tubules of the medullary re-gion, if traced back into the cortex, are seen to originate in a tiny, cup-like structure, **Bowman's capsule.*** Each capsule surrounds a capillary tuft, or **glomerulus.** (If you close one hand into a fist and grasp it with the other, you will approximate this arrangement.) Tracing the blood flow backwards, it can be seen that the glo-merular capillaries originate from the renal artery. Be-tween the glomerulus and Bowman's capsule, then, much of the vital removal of urea and most of other wastes must take place. Blood leaving the glomerulus, now considerably reduced in the amount of waste it car-ries and carrying its vital plasma proteins, glucose, and the like, eventually passes out of the kidney through the renal vein.

It does not pass directly, however. The arteriole draining blood from the glomerulus first breaks down again into capillaries. These capillaries surround the urinary tubule leading from Bowman's capsule. Shortly after leaving the capsule, this tubule follows a tortuous, twisted path, and at one point, in a section lying in the medullary region, becomes quite thin before continuing as a tubule back into the cortex and then leaving both the cortex and medulla via a collecting tubule. Fig. 10.12 shows the relationship of a glomerulus, Bowman's cap-sule, the various tubule regions, and the associated blood vessels. Together, these parts comprise one se-creting unit of the kidney, or **nephron.**

The functional work of the kidneys, then, is accom-plished in the nephrons (see also Supplements 10.5 and 10.6). In human beings, there are approximately one million nephrons per kidney (a number that seems more

Fig. 10.11
X-ray of the urinary tract. The urine contains a substance, given by injection, that makes it appear on the X-ray film. The bladder is less visible because of the bones of the pelvic area behind it. (X-ray courtesy Herbert Kaufman, M.D.)

RENAL PELVIS

URETER

URINARY BLADDER

---

*Actually, if you follow this pathway in the tubule (see Fig. 10.12), you will note that it loops from the cortex into the medulla again before returning to the cortex. The significance of this will be dis-cussed shortly.

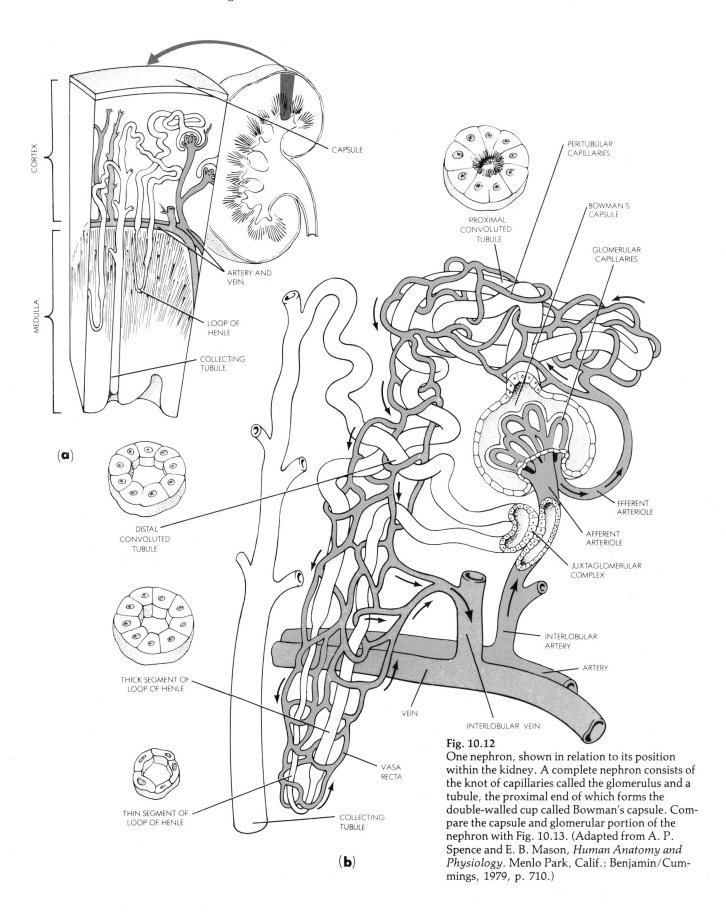

CORTEX

MEDULLA

CAPSULE

ARTERY AND
VEIN

LOOP OF
HENLE

COLLECTING
TUBULE

(a)

DISTAL
CONVOLUTED
TUBULE

THICK SEGMENT OF
LOOP OF HENLE

THIN SEGMENT OF
LOOP OF HENLE

(b)

PROXIMAL
CONVOLUTED
TUBULE

PERITUBULAR
CAPILLARIES

BOWMAN'S
CAPSULE

GLOMERULAR
CAPILLARIES

EFFERENT
ARTERIOLE

AFFERENT
ARTERIOLE

JUXTAGLOMERULAR
COMPLEX

INTERLOBULAR
ARTERY

ARTERY

VEIN

INTERLOBULAR VEIN

VASA
RECTA

COLLECTING
TUBULE

Fig. 10.12
One nephron, shown in relation to its position
within the kidney. A complete nephron consists of
the knot of capillaries called the glomerulus and a
tubule, the proximal end of which forms the
double-walled cup called Bowman's capsule. Compare the capsule and glomerular portion of the
nephron with Fig. 10.13. (Adapted from A. P.
Spence and E. B. Mason, *Human Anatomy and
Physiology.* Menlo Park, Calif.: Benjamin/Cum-
mings, 1979, p. 710.)

SUPPLEMENT 10.6
FILTRATION AND RESORPTION: HOW DO WE KNOW?

Histological investigation reveals that the thin walls of Bowman's capsule closely resemble those of the lung alveoli. Since the alveolar lining cells do little or no work in the movement of materials across them, it is reasonable to hypothesize that osmosis and diffusion might be sufficient to accomplish the passage of waste materials from the glomerulus capillaries to the capsule tubules. Filtration would also be involved, of course, since formed elements suspended in the plasma, such as blood cells and platelets, would not pass through the capillary or tubule wall. Nor, normally, would the proteins of which plasma is partially composed.

The hypothesis can be formulated as follows: If the physical processes of osmosis, diffusion, and filtration are primarily responsible for the movements of materials from the glomerulus capillaries to the capsule, then the urine in Bowman's capsule should be of the same composition as the blood plasma, less its proteins. Here is a case where the testing of an hypothesis had to await the development of technical skills equal to the task of performing the experiment. The cavity within Bowman's capsule is less than 0.1 mm in diameter. The problem of extracting enough fluid for chemical analysis is immense. Nevertheless, such an extraction was eventually made by means of a finely drawn-out quartz tube. As predicted, the capsular fluid has essentially the same composition as the blood plasma, less its proteins. Thus it appears that the membranes of the glomerular capillaries and Bowman's capsule serve simply as selective filters to prevent the passage into the capsular sac of the formed elements of the blood (such as the blood cells and platelets) and plasma proteins.

Yet while the physical factors of osmosis and diffusion are sufficient to account for the movements of molecules across membranes from a region of greater to a region of lesser concentration, they cannot account for the rapid passage of fluid from the blood plasma to the capsule. The lining cells seem to perform no work in accomplishing this movement. Therefore, the mechanical energy for this movement can come from only one other source, the blood pressure. This one remaining hypothesis can be formulated as follows: If blood pressure provides the mechanical energy responsible for the movement of fluid from the glomerular blood plasma to Bowman's capsule, then urine output should show a direct correlation with blood pressure.

Drainage tubes are inserted into the ureter of an anesthetized dog, and the amount of urine collected per unit time is recorded. By removing some blood from the animal, the blood pressure can be lowered; by administering drugs causing constriction of the blood vessels, it can be raised. The results are as would be predicted by the blood pressure–urine output hypothesis; low blood pressure results in a sharply reduced urine output, while high blood pressure results in an increased urine output.

Chemical analysis of urine samples extracted from any point between the tubules leaving the medullary region and the urethra reveals no important differences between them. Evidently, then, the vital urine extraction and preparation processes are completed before this portion of the system is reached. A similar comparative chemical analysis shows the fluid in Bowman's capsule and the fluid that issues from the urethra to be remarkably different indeed. First, with respect to its solutes, the voided urine is far more concentrated than the capsular fluid. To explain this, we can advance the hypothesis that large amounts of water are resorbed from the filtrate as it passes between the capsule and the medullary tubules.

This hypothesis can be tested in several ways. Measurements of the concentrations of various substances within the proximal and distal tubules can be compared to the concentration of those same substances in the capillary network surrounding the tubules. Such comparisons show that it is largely water, along with some small molecules (such as glucose) and ions (Na^+, Ca^{++}), that passes through the membranes. Precise experimental analysis shows that in humans, about 180 liters of fluid pass into the glomerular sac every day. Yet only about 1.5 liters are excreted as urine. These results indicate a resorption quantity of approximately 99 percent.

Anatomical evidence in support of the resorption hypothesis is supplied by the fact that the capillaries leaving the glomerulus do not directly leave the kidney, but rather become closely associated with the urinary tubule along part of its length leading away from the capsule. Also supportive is the thinning of the tubule at one point in its twisted path; such a thinning provides a greater surface area per unit volume for resorption to occur. The rather viscous nature of the blood leaving the glomerulus, in comparison to its more watery nature in the renal vein, is also consistent with the hypothesis. However, precise chemical analyses make it quite clear that water resorption alone is not sufficient to explain the difference in chemical composition between capsular fluid and urine. Certain solutes are more concentrated than others; a hypothesis that water resorption is the *only* factor at work would predict otherwise. For example, bladder urine normally contains no glucose; capsular fluid does. Glucose, therefore, must also be resorbed, *against a concentration gradient.* The cells lining the urinary tubules must perform work and use energy to move glucose back into the blood. Cytological examination of the tubule cells reveals that they are glandular in nature and possess characteristics of work-performing cells, such as containing numerous mitochondria (see also Supplement 10.5).

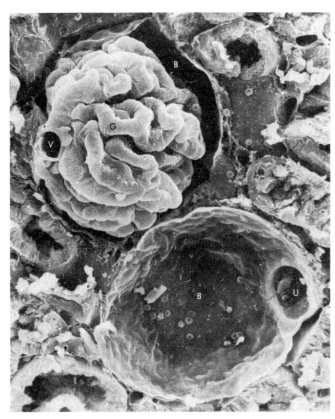

Fig. 10.13
Scanning electron micrograph of two Bowman's capsules (B) in a rhesus monkey kidney (× 500). The one on the left contains its glomerulus (G) while, in the other, the glomerulus was probably pulled out by the blade when the section was cut. The cavities of some of the uriniferous tubules (U) and a blood vessel (B) are also visible. (Micrograph courtesy Dr. Peter Andrews, Georgetown University.)

than sufficient, since an individual can get along quite nicely with just one kidney). Besides the aforementioned primary removal of wastes between the glomerulus and Bowman's capsule, important exchanges involving diffusion, osmosis, active transport, hormonal influences, and the like, all occur before the blood leaves the kidneys.

The precise means of urine formation can be summarized as follows:

1. Blood entering a kidney from the aorta via the renal artery is distributed via small afferent arterioles into the capillary beds we call the glomerulae (see Fig. 10.14). Blood pressure forces much of the fluid from the blood to filter through the permeable membranes of the capillaries and capsules. The fluid remaining in the blood still contains the plasma proteins, cells, and the like, too large to pass through the membranes.

2. Following the fluid collected in one capsule (the filtrate) first , this fluid passes into the tubule of the nephron. The entire length of this tubule is divided into four regions, the *proximal* (next to the capsule), *convoluted* (twisted or bent) *tubule*, the U-shaped loop of Henle (one portion of which is quite thin), the *distal* (away from the capsule) *convoluted tubule*, and the *collecting duct*.

3. The blood that remains, after the filtration of much of its contents into the capsule, leaves via the *efferent arteriole*. The arteriole breaks down almost immediately into *peritubular capillaries* which, as their name suggests, surround the tubule leaving the capsule. In one portion, where they surround the loop of Henle and the collecting ducts, the peritubular capillaries form a capillary complex called the *vasa recta*. From the vasa recta the blood, now changed considerably in content from when it left the capsule, collects into veins that will eventually unite to form the renal vein leading to the inferior vena cava.

4. In the proximal convoluted tubule, chloride ions and glucose are returned to the interstitial ("between tissue") fluid by active transport and pass from this fluid into the blood plasma in the peritubular capillaries; the positively charged sodium ions follow the negatively charged chloride ions by passive transport.* The water is reabsorbed into the interstitial fluid and blood plasma by osmosis.

5. Note from Fig. 10.12 that the loop of Henle and the collecting ducts lie in the medulla of the kidney; all the rest of the nephron parts lie in the cortex. This anatomical perspective is important in understanding what goes on in the loop and collecting ducts: they extend into a region in which the environment has an increasing ion concentration and therefore free-energy gradient. This gradient, as indicated above, results from chloride ions in the filtrate being actively transported outside, with the sodium ions following by passive transport. Since this movement of ions from inside the loop to outside results in a higher concentration of water molecules within the loop, water follows the sodium and chloride ions by osmosis. It is important to emphasize, however, that the active transport of chloride ions and the resulting passive transport of sodium ions *does not occur until the filtrate reaches the ascending portion of the loop leading to the distal convoluted tubule.* The same ascending portion of the loop is fairly permeable to water, and thus the movement of water out of the loop occurs mostly in the *descending* portion of the loop.

6. The division of the loop of Henle into two regions with different characteristics lying side-by-side sets up a

---

*It was once thought that the sodium ions were actively transported, with the chloride ions following passively. Recent studies support a hypothesis proposing that the chloride ions are pumped, and the sodium ions follow passively.

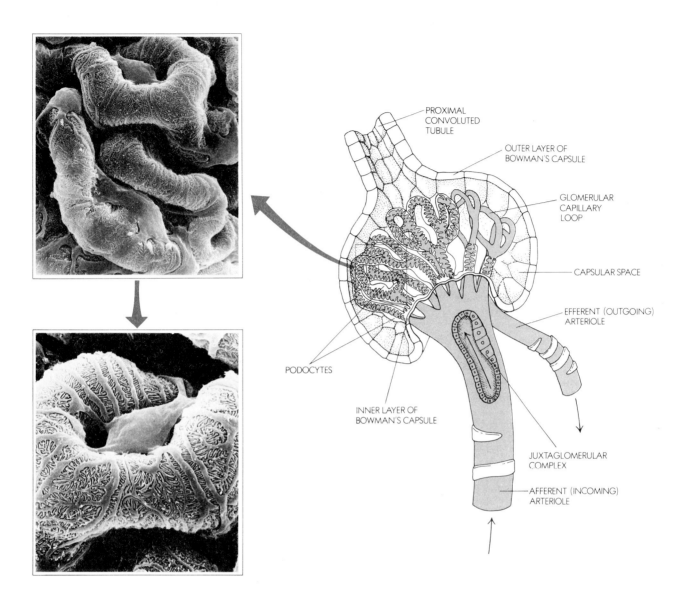

Fig. 10.14
Longitudinal section of a renal corpuscle. The corpuscle is composed of Bowman's capsule
and the knot of coiled capillaries known as the glomerulus. All of the renal corpuscles are
contained in the kidney cortex. The wall of Bowman's capsule is composed of squamous epithelial cells resting on a thin basement membrane. The inner layer of each capsule contains
specialized epithelial cells called podocytes whose processes adhere to the glomerular capillary loops. The space between the capsular inner layer and the glomerular capillaries may
promote the rapid movement of substances from the blood into the capsular space. The scanning electron micrographs show podocyte processes surrounding glomerular capillary loops
in a rat kidney (× 2600 and 4500). (Micrographs courtesy Dr. Peter Andrews, Georgetown
University.)

In the body cells, the metabolism of fats and carbohydrates yields carbon dioxide and water as end-products. The carbon dioxide is eliminated through the lungs. The oxidation of protein, on the other hand, produces ammonia, which is highly toxic. In mammals ammonia is converted into urea, which is removed from the blood by the kidneys and excreted.

countercurrent mechanism (see Supplement 10.5). The chloride ions actively transported out of the ascending loop diffuse back into the filtrate through the walls of the descending loop by this mechanism; thus they are literally recycled. The same is true of sodium ions and, in some animals at least, molecules of urea.

7. The advantages of the proximity of the vasa recta to the loop of Henle now become apparent, since this location means that these capillaries are now surrounded by an interstitial fluid that is highly concentrated in chloride and sodium ions. Thus still more of these ions and some water eventually pass back into the circulatory system and body via the renal vein and inferior vena cava.

8. From the ascending portion of the loop of Henle, the filtrate passes through the distal convoluted tubule in the kidney cortex. By this time the loss of chloride and sodium ions has lowered the filtrate's solute concentration below that of the surrounding interstitial fluid and plasma, and thus some chloride and sodium ions may be resorbed for return to the body. The volume of fluid has now been considerably reduced from the volume that originally filtered into Bowman's capsule.

9. The filtrate next passes from the distal convoluted tubule into the collecting tubule. While the name of this tubule correctly describes its function, it is significant that its walls vary from time to time in their permeability, and thus it plays an important role in maintaining body fluids in proper homeostatic balance. If the water permeability of the tubule is high, water leaves the filtrate, further increasing the solute concentration within the tubule. This concentrated low-volume fluid passes out of the collecting tubule at the tip of a papilla into the renal pelvis, eventually to pass through one of the ureters for storage in the bladder before passing to the body exterior through the urethra.

It may sometimes happen that an abnormal amount of glucose occurs in the blood, as in the case of a person with diabetes mellitus. In such cases, the excess glucose appears in the urine, providing a simple test for the condition. Evidently, if the concentration of glucose in the blood exceeds a certain level, the urinary tubule cells are unable to return all of the glucose to the

blood; the excess remains in the tubular fluid and is thus found in the urine. This observation fits nicely with the concept of facilitated and active transport mechanisms across membranes discussed in Chapter 4; conceivably all the receptor sites of the tubule cells are working to fullest capacity and can handle no more molecules. This same phenomenon is found for other substances normally reabsorbed by the tubule cells, though the *amount* of each substance the cells can handle varies widely from one substance to another—presumably, the receptor sites, too, vary in number and possibly efficiency of action. This fact alone suggests that the reabsorption process is only part of the story, and a complete explanation must await a fuller understanding of cellular work processes in general.

Research with isotopes indicates that the tubule cells may also perform work in moving certain substances *from the blood to the urine*, as well as the reverse. Earlier investigation of the work of the kidney by analysis of the fluids involved at various points in the process appears to have considered only the net difference between opposing flows moving across the urinary tubule walls. Such analytical investigation also indirectly bears on the nature of the reabsorptive forces involved (such as diffusion, active transport, and the like) as well as the factors that trigger the tubule cells to commence their work. By reabsorbing bicarbonate ions, for example, the kidney helps to keep the blood mildly basic, while at the same time secreting hydrogen ions into the filtrate and making the urine slightly acidic. Thus, by working in a delicately balanced coordination with these mechanisms, the kidney plays as large a role in maintaining a proper pH of the blood as any of the respiratory mechanisms.

10.10
HOMEOSTASIS AND THE KIDNEYS

We have seen that the process of urine formation within the nephron is a complex process relying, among other things, upon such physical forces as filtration, simple diffusion, and osmosis as well as active transport in moving ions back and forth across the tubule walls. All of these activities are intricately interrelated. For example, the active transport of chloride ions, with the sodium ions following, influences the direction and extent of water movement by osmosis. In turn, the amount of filtrate the kidneys are called on to process is related to physical and chemical forces within the body; high blood pressure, or hypertension, for example, is a physical force and may be caused by several factors (see Supplement 11.3), one of which involves the kidney itself. Let us see how this occurs by discussing the mechanism of maintaining homeostatic balance of just one substance, the element sodium, which, as we have seen, is found in the blood and interstitial tubular fluid in the ionic (Na^+) state. Keep in mind that similar factors regulate the homeostatic balance of other substances.

An increase in the output of the hormone aldosterone, produced by the cortical region of the adrenal glands located on the kidneys (see Section 14.4), causes the kidneys to retain sodium ions. This retention of sodium, in turn, means that water is also retained. An increase in body water means an increase in blood volume which, of course, raises the blood pressure; thus, persons suffering from hypertension are usually put on low-salt diets.

What causes the increased release of aldosterone by the adrenal cortex? A specialized group of cells in the kidneys produce a substance, renin, which appears to act as an enzyme in converting a blood glycoprotein (sugar-protein complex) produced in the liver, called angiotensinogen into angiotensin I. Angiotensin I is converted in the lungs by angiotensin converting enzyme (ACE) to angiotensin II, which in turn causes the increase in aldosterone secretion by the adrenal cortex, as well as causing blood vessel constriction, further raising the blood pressure.

The concentration of potassium ions in the fluids bathing the adrenal glands also appears directly to influence the release of aldosterone; an increased elevation in potassium ion concentration means more aldosterone is released. Potassium ion concentration, in turn, is linked to sodium ion concentration via the sodium-potassium pump (see Sections 13.3 and 13.4).

But what causes the increase in renin production, which in turn leads to the sequence of events just described? It appears that renin secretion is controlled by the amounts of another adrenal gland hormone, epinephrine, as well as by those branches of the sympathetic nervous system that innervate the kidneys. Note that while the intent here was to deal with the maintenance of homeostatic balance of body sodium, the dis-

cussion automatically had to include many other factors and substances as well. So it is with virtually all physiological homeostatic mechanisms.

If the body is to regulate its sodium content, it must have some mechanism for detecting the amount of sodium ions in the body. In fact, however, there appears to be no mechanism that could detect the total body mass of sodium. Intracellular detectors would also seem to be ruled out; sodium ions for the most part are kept out of cells by the sodium-potassium pump. Yet this fact, alone, suggests a means of detection. Because of the osmotic effects exerted by sodium, the extracellular fluid volume is directly related to the mass of extracellular sodium. Thus, body sodium content may be indirectly measured by sensing extracellular fluid volume. While as a whole this, too, would probably be impossible to sense, the extracellular fluid volume can be divided into interstitial fluid volume and plasma volume. It is possible that pressure receptors called baroreceptors may respond to pressure changes resulting from changes in plasma volume, which in turn is related to sodium ion concentration. The body also possesses osmoreceptors that are sensitive to the osmotic pressure of the extracellular fluid. Regulatory changes brought into play by these receptors can compensate for changes in osmotic concentration of the extracellular fluid, again largely due to sodium, by stimulating or slowing down the drinking of water. This, of course, involves the physiological mechanism of thirst and the resulting drinking behavior it elicits. Thirst itself is thought to be partly under the control of the effects of angiotensin II on the brain.

Besides aldosterone, a second hormone, the peptide vasopressin (often referred to in medicine as the antidiuretic hormone, or ADH), is produced by the hypothalamus of the brain. Vasopressin indirectly affects the concentration of sodium ions by regulating the amount of water in the urine. It does this by increasing the permeability of the nephron collecting ducts, resulting in more water passing back into the blood and a more highly concentrated urine. Since the amount of vasopressin produced is inversely related to both water deprivation and dehydration, the result of which increases the plasma osmotic pressure and lowers the blood pressure, it is evident that vasopressin is also related indirectly to the homeostatic balance of sodium ions in the body.

Summary

1. Individual cells, whether of multicelled or single-celled organisms, are equipped with control mechanisms to keep their internal environment adjusted to changes in the external environment.

2. Homeostasis refers to the processes by which the internal environment of organisms is kept constant in the face of changing external conditions.

3. All homeostatic mechanisms operate by regulating the rates of systems in dynamic equilibrium. (Dynamic equilibrium is contrasted to static equilibrium, in which a system remains constant, but no change at all takes place in its internal elements.) Body temperature, blood sugar level, and plasma pH are themselves constantly changing. Regulation affects their rate of change and minimizes the net change, promoting a "steady state."

4. To be absorbed into the cells of multicellular organisms, food must first be digested. Digestion is a process by which complex macromolecules are broken down into their component building blocks. Digestion takes place in the alimentary canal (gut) of most animals, under the influence of enzymes. These enzymes are produced in the cells of digestive organs and pour into the alimentary canal through ducts.

5. Absorption of digested food takes place in the small intestine. The walls of the intestine contain numerous finger-like projections, villi, which greatly increase the surface area. The villi are equipped with capillaries, where the actual absorption takes place.

6. Undigested solid materials, along with some liquid and ions, are eliminated from the alimentary canal in the form of feces.

7. In both the large and small intestine the process of digestion is aided by the presence of the bacterium *Escherichia coli*. These bacteria also produce some substances, such as vitamins, that are necessary for the existence of the host. Without the presence of certain intestinal bacteria, many animals would die.

8. A balanced diet is one that contains proper proportions of carbohydrate, fat, protein, vitamins, and minerals; precisely what the "proper" amounts are tends to vary from individual to individual. Those amino acids in the diet listed as "essential" are those that an organism cannot synthesize for itself.

9. Excretion is the process by which the organism gets rid of the waste products of cellular metabolism. It is fundamentally different from elimination, in which roughage and other materials that never entered the body's cells are removed. The fundamental metabolic waste in mammals is urea. Most metabolic wastes are the end-products of protein metabolism. In insects, birds, and reptiles, one end-product of protein metabolism is uric acid. Uric acid tends to be excreted in a solid crystalline form. In many aquatic animals, the main end-product of protein metabolism is ammonia.

10. In mammals, urea is produced in the liver by the breakdown of amino acids released from cells all over the body. The urea is then released into the bloodstream where it ultimately passes through the kidneys. The fundamental process of the kidney is one of filtration, carried out by a basic unit called the nephron. Blood enters the glomerulus, where urea, many minerals, ions, and much water are filtered out. Capillaries surround the drainage tube leading from Bowman's capsule and gradually reabsorb the essential substances the body must retain, principally many ions and water. The urea is not reabsorbed; it passes from the distal convoluted tubule into the renal pelvis, where it is collected and passes in the urine via the ureters into the bladder for storage until it is excreted through the urethra.

Exercises

1. What is meant by the term "homeostasis"? To what aspect of living organisms does it apply?

2. What are some examples of homeostatic mechanisms in organisms? Are such mechanisms found in *all* organisms, or just some?

3. Trace through the digestive system the pathway of a morsel of food containing protein, fat, and carbohydrate. Discuss the changes that take place in the types of food as this passage occurs.

4. Observe Fig. 10.15. If we calculate the amount of sugar an animal can absorb from its diet, as well as the rate at which that animal burns glucose during a 12-hr period, the percentage of sugar in the blood should vary as curve A in the graph. Actual measurements, however, show that the amount of blood sugar is considerably less (curve B). Other evidence shows that the amount of blood sugar after a meal is considerably less in the hepatic vein than in the hepatic portal vein. Offer a hypothesis to explain these results. Devise an experiment to test your hypothesis.

5. A pH analysis of urine has revealed that the urine of some animals is highly acid, while that of others is highly alkaline. There is a corresponding difference in the diets of these groups of animals. Which of the following statements offers the best explanation for this difference in urine?

   a) The kidneys keep a balance in the diet by removing certain acids or bases from the environment.

   b) The kidneys maintain a constant pH of the blood by removing excess acids or bases that result from the diet.

   c) The specific pH of the urine must be a particular level regardless of the intake of acids or bases from the environment.

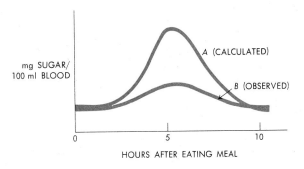

Fig. 10.15

Chapter 11
Circulation, Blood, and the Nature of Respiratory Exchange

11.1
INTRODUCTION

The previous chapter was concerned primarily with the ingestion and absorption of food, as well as the elimination and excretion of the waste products of that food. The situation is analogous to filling the gas tank of a car as well as observing the elimination of unburned residues and exhausts.

Yet putting gasoline into an automobile's tank does little good if there is no fuel pump to move the gas to the cylinders where it can be burned or fuel lines through which it can be moved. Similarly, the digestion and absorption of foods is of little value unless the food products involved are distributed to the body cells that need them. It is this process—that of the distribution of digested and absorbed food products, as well as exhaustion of the metabolic wastes of respiration—with which this chapter is primarily concerned.

11.2
CIRCULATION: THE HEART

The fuel line of a complex organism's body is the **circulatory system,** generally composed of the **cardiovascular system,** which transports the blood, and the **lymphatic system,** which transports a special fluid known as lymph, a watery fluid collected from the tissue fluids. In addition to the heart, the cardiovascular system in humans and other vertebrates consists of **arteries, arterioles, capillaries, venules,** and **veins.** The arteries and veins are relatively large vessels distinguished by the structure of their walls and by the direction of blood flow in them. Arteries carry blood away from the heart, and veins carry blood toward the heart. Arterioles are simply small versions of arteries, and venules are small versions of veins. The capillaries are microscopic, thin-walled vessels located in the tissues, and act as connecting links between arteries and veins.

The heart keeps the blood moving in a continuous circular pathway through the body. Before tracing how food absorbed from the intestine is transported by the circulatory system, let us first examine the structure and function of the heart.

The mammalian heart and that of the bird consist of two quite distinct, but coordinated, pumping systems. The right side of the heart (*right atrium* and *ventricle*) forms one pumping system, whose function it is to move blood to and from the lungs. This is the **pulmonary circulation.** The left side of the heart (*left atrium* and *ventricle*) is a second pumping system, which moves blood throughout the rest of the body. This is the **systemic circulation.** Blood passes between the right and left sides of the heart only indirectly, by way of the lungs. Despite the distinctness of the two pumping systems, the heartbeat is coordinated in such a way that the heart acts as a whole. The two atria contract simultaneously, during which time the ventricles are relaxed. The atria then relax, while the ventricles contract. This regular coordinated rhythm moves blood continually throughout all parts of the body, from well before birth until death. At rest, the human heart pumps about 300 liters (approximately 75 gallons) of blood per hour; given an average level of activity, it pumps an estimated 18 million barrels in a lifetime. Little wonder that from ancient times until very recently the heart has always been regarded as the center of the life force, the most essential organ of the body.

Other Vertebrate Hearts. The hearts of vertebrates such as fish or reptiles reflect the functional role of the two pumping systems in the mammalian heart. Fig. 11.1 compares the cardiac structure of fish, reptiles, and mammals. In the fish, there are no lungs (gills serve the analogous function of gas exchange) and hence no pulmonary circulation. The fish heart consists of a single atrium and a single ventricle (Fig. 11.1a). Blood passing out the aorta in a fish goes to a number of different branches, two of which lead to the gills. Blood is passed from the gills to other parts of the body and eventually back to the heart. In the reptile heart (Fig. 11.1b), there are two atria and a single ventricle; but this ventricle is partially divided down the middle. Leading out of the ventricle are two arteries—a pulmonary artery and an

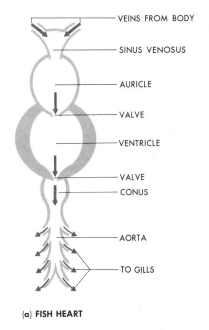

(a) FISH HEART

VEINS FROM BODY
SINUS VENOSUS
AURICLE
VALVE
VENTRICLE
VALVE
CONUS
AORTA
TO GILLS

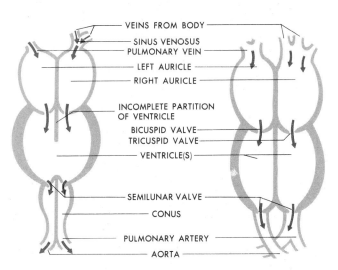

VEINS FROM BODY
SINUS VENOSUS
PULMONARY VEIN
LEFT AURICLE
RIGHT AURICLE
INCOMPLETE PARTITION
OF VENTRICLE
BICUSPID VALVE
TRICUSPID VALVE
VENTRICLE(S)
SEMILUNAR VALVE
CONUS
PULMONARY ARTERY
AORTA

(b) REPTILE HEART   (c) MAMMALIAN HEART

Fig. 11.1
Heart structure in three vertebrates: (*a*) fish; (*b*) reptile; and (*c*) mammal. (Drawing courtesy of Alfred S. Romer, from *The Vertebrate Body*, 4th ed. Philadelphia: W. B. Saunders, 1970. Reprinted by permission of Holt, Rinehart and Winston.)

aorta. The partial septum helps keep the oxygenated and deoxygenated blood somewhat separate, though it is obvious that in the lower part of the ventricle some mixing inevitably takes place. The fact that in the reptile heart (and even more so in amphibians such as the frog) oxygenated and deoxygenated blood mix means that the body tissues of these organisms receive less oxygen per

heartbeat than those of mammals. Compared to mammalian structure (Fig. 11.1c), the heart structure in reptiles is inefficient.

Structure. A cutaway section of a human heart is shown in Fig. 11.2. The heart consists of four chambers, two atria* (singular, atrium) on the upper side, and two ventricles on the lower. The heart is divided into a right and left side, between which there is no direct connection in the adult. Blood flows into the human heart from the body by way of two large veins called the **venae cavae.** The *superior vena cava* brings blood down from the head, arms, and upper body, while the *inferior vena cava* collects blood from that portion of the body below the heart, including the internal organs and legs. Blood entering the heart from the venae cavae is deoxygenated: it has given up most of its oxygen to the body cells and is now heavily laden with carbon dioxide. The venae cavae pour their blood into the right atrium as the latter relaxes its muscular walls.

The atria are very small compared to the ventricles. While both atria and ventricles have about the same internal volume, the ventricular walls are much thicker, since their contraction must drive blood throughout various regions of the body and back again to the heart.

Blood Flow Through the Heart. Venous blood enters the right atrium through three vessels: the superior vena cava, the inferior vena cava, and the coronary sinus, which drains the heart muscle or **myocardium.** From the right atrium, the blood flows into the right ventricle through the **tricuspid valve,** so named because it has three cup-shaped flaps that fold in and out, closing or opening the passageway between atrium and ventricle. These flaps, or cusps, are arranged to fold more or less down into the ventricle. Since they will not fold back into the atrium, they prevent backward flow of blood from the ventricle into the atrium when the ventricle contracts. The presence of valves throughout the heart and in the veins ensures that the blood flows in only one direction.

The relaxed state of the heart is called **diastole.** During diastole the right ventricle fills with the blood it received from the right atrium. The heart soon contracts, entering **systole.** During systole, the right ventricle forces blood through the **pulmonary valve** into the **pulmonary artery.** The pulmonary valve is composed of only two cusps.

The pulmonary artery carries deoxygenated blood from the heart to the lungs. Here the artery become divided into increasingly smaller vessels, arterioles, and finally into **capillaries.** These capillaries are located in the walls of sacs, or **alveoli,** which make up the inner lining

*Because early anatomists thought the heart atria resembled little ears, they were originally called auricles, based on the Latin word for ears. However, this term has fallen into disuse in recent years.

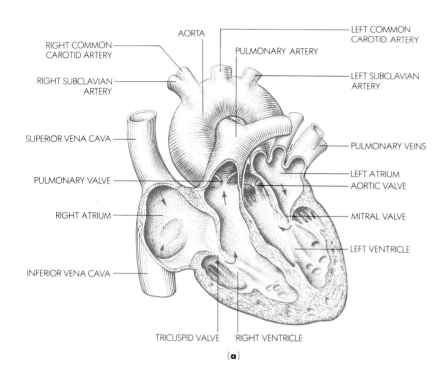

RIGHT COMMON CAROTID ARTERY

RIGHT SUBCLAVIAN ARTERY

SUPERIOR VENA CAVA

PULMONARY VALVE

RIGHT ATRIUM

INFERIOR VENA CAVA

AORTA

PULMONARY ARTERY

LEFT COMMON CAROTID ARTERY

LEFT SUBCLAVIAN ARTERY

PULMONARY VEINS

LEFT ATRIUM

AORTIC VALVE

MITRAL VALVE

LEFT VENTRICLE

TRICUSPID VALVE   RIGHT VENTRICLE

(a)

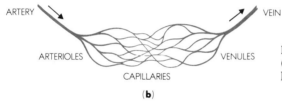

ARTERY

ARTERIOLES

CAPILLARIES

VEIN

VENULES

(b)

Fig. 11.2
(a) The parts of the heart. (b) The connecting links between arteries and veins, the capillaries.

of the lungs. Here the capillaries are in contact with air that has a much higher oxygen and lower carbon dioxide content. Accordingly the blood loses its carbon dioxide and picks up oxygen. Blood then passes from the capillaries into larger vessels, the venules, which eventually empty into the pulmonary vein. The pulmonary vein carries the now-oxygenated blood back to the left atrium from which the blood is pushed into the left ventricle. Backward flow of the blood into the left atrium is prevented by the **mitral valve** between the left atrium and ventricle.

With the contraction of the left ventricle, the blood is forced out of the heart through the aortic valve into the **aorta,** the major artery leaving the heart. The aorta carries the oxygenated blood pumped from the heart to be distributed to all parts of the body. Several branches form immediately off the aortic arch (see Fig. 11.2),

carrying blood to the arms, chest, and head. From the point where the aorta arises from the heart, the coronary arteries emerge and pass down, branching over the heart and feeding the cardiac muscle tissue, the myocardium.

Since blood is forced out into the aorta by the systole contraction of the left ventricle, the blood moving through arteries travels in a kind of pulsating motion. This pulse can be felt in a number of places where an artery comes close enough to the surface for its rhythmic expansion and contraction to be detected. Pulse can be felt on the side of the nose just below the eye socket, on the side of the neck where the carotid arteries pass upward to the head, and on the inner surface of the wrist. The pulse is caused by the muscular contraction of the heart, rather than by any independent contraction of the arteries.

Arteries carry blood away from the heart. Veins carry blood toward the heart.

The human heart is a magnificent pump. At rest, it pumps about 300 liters (approximately 75 gallons) of blood per hour or about 18 million barrels in an average lifetime.

The Heartbeat

No circulation of the blood is possible without a driving force. As we have seen, the origin of the forces involved in circulation are the muscular contractions of the heart. A primary problem of circulatory physiology, therefore, is to discover the cause of the heartbeat.

The fact that the heart is supplied with two main nerve trunks makes it reasonable to hypothesize that nerve impulses are the cause of the heartbeat. Preliminary experiments seem to support this hypothesis, for mild electrical stimulation of these nerve trunks in a frog heart speeds up or slows down the rate of the heartbeat, depending on which trunk is stimulated. The crucial experiment is an obvious one.

We can reason as follows: If the heartbeat is caused by impulses transmitted from the nervous system, then severing of nerves leading to the heart should effectively prevent the heartbeat. The results of the experiment contradict the hypothesis; the heart continues its rhythmic beating after its nerves are severed. The cause of the beat must therefore lie within the heart itself. Further experiments produce another interesting bit of information. Isolated strips of frog heart muscle will continue to contract rhythmically. In tissue culture, microscopically small pieces of heart tissue will continue to beat. One is tempted to agree with Harvey that "the motion of the heart is to be comprehended only by God" (see Supplement 11.1).

Limulus, the horseshoe crab, has a heart composed of muscle and nerve that can be separated easily. We can therefore set up a hypothesis and see whether experimental results support or contradict it: If nerve impulses are necessary to initiate the heartbeat in *Limulus*, then removal of the nerves from the *Limulus* heart should cause cessation of the heartbeat. The hypothesis is supported. Removal of the nerve tissue of the *Limulus* heart results in immediate cessation of the heartbeat.

Can the experimental results obtained with *Limulus* be extrapolated to humans? Unfortunately, no. Ordinarily such extrapolations from lower organisms are justified so long as we remember that a point has been stretched and we may be wrong. In the case of *Limulus*, however, we are dealing with a heart that is substantively different from that of higher forms.

However, quantitative measures of isolated strips of vertebrate heart muscle disclose an interesting fact.

Those strips showing the least activity come from the lower region, or apex, of the ventricles. Those showing the most activity come from the upper region of the atria, particularly in the right atrium near the point at which the superior vena cava enters. Dissection in this region of the heart reveals the presence of a specialized node of tissue, the **sinoatrial node**, sometimes referred to as the "pacemaker." This node, rather than being distinctly nerve tissue as is the case with the fibers found in *Limulus*, is actually composed of modified heart muscle (myocardial) cells. From the sinoatrial node impulses spread into the surrounding atrial muscle. The atrial muscle conducts impulses throughout the atria and to a second node, the **atrioventricular** node. From the atrioventricular node specialized fibers called **Purkinje fibers** conduct impulses throughout the ventricle. Both the atrioventricular node and the Purkinje fibers are modified heart muscle tissue.

When the pattern of branching of these fibers is compared to experimental data on the degree of contraction shown by isolated muscle strips, it becomes evident that the cells surrounded by the smallest number of fibers are capable of the smallest amount of automatic contraction. It would seem, then, that while the nerve tracts leading to the heart are not essential for initiation of the heartbeat, the modified muscle tissue intrinsic in the heart itself probably is. Studies of developing chick embryos reveal that their heart muscle tissue begins to beat well before any nerve cells are present. It would seem, then, that automatic heartbeat is an intrinsic feature of heart muscle cells, though the role of controlling the heartbeat may later reside in the node-fiber network.

It is likely that, in the vertebrate heart, contractility is an intrinsic feature of the cardiac tissue itself. Even granting this assumption, there are several unanswered questions. A heart that has been beating for some time in a calcium-free medium will eventually stop. The addition of more calcium ions results in renewed beating activity. Does this mean that calcium ions are responsible for initiating the heartbeat? Or does it simply mean that calcium ions must be present if the heart is to respond to the true heartbeat initiator? If the latter is the correct interpretation, what *is* this "true" heartbeat initiator? To date, there are no definite answers to these questions, so we can only continue to hypothesize about

SUPPLEMENT 11.1
WILLIAM HARVEY AND THE CIRCULATION OF THE BLOOD

In the seventeenth century the dominant theory of the function of the heart, arteries, and veins was derived from Claudius Galen, who had lived some 1500 years before. Galen's theory held that the blood ebbed and flowed from the arteries and veins and in and out of the heart, in a pulsating manner, like ocean tides. Wastes were expelled from the blood into the heart, and new blood was made by mixing various spirits from the lungs, heart, and liver. As part of this scheme, Galen held that blood and various spirits passed in two directions through the pulmonary veins. He also claimed that blood passed through the wall or septum of the heart separating the right and left ventricles via tiny, invisible pores.

The seventeenth-century physician William Harvey challenged Galen's conception, proposing an alternative hypothesis. Harvey's major contention was that the blood traveled in a circular path around the body, passing from the heart to the arteries, through the veins, and back to the heart again. According to Harvey, the heart acted as a pump that kept the blood moving continually in a single direction. This idea was not entirely original. The concept of a circulatory pathway for the blood had been advanced in the late sixteenth and early seventeenth centuries by several people, including Michael Servetus (1511–1553), Realdo Colombo (1510–1559), Andrea Cesalpino (1519–1603), and Giordano Bruno (1548–1600). All these men, including Harvey, were much impressed by the Copernican notion of the revolution of the planets in a circular path around the sun (indeed, Bruno was burned at the stake by the Church for believing this and other "heretical" ideas); they borrowed the planetary analogy in suggesting that the blood moved in a circular path around the center of its solar system, the heart. However, it was Harvey who, in 1628, first provided experimental evidence to support the idea of a circulatory pathway.

Historical data indicate that Harvey came quite early in his life to the hypothesis that the blood circulates by the action of the heart, as opposed to ebbing and flowing. Once he had reached this conclusion, Harvey proceeded to gather evidence to demonstrate its correctness.

Observational Evidence

The first line of evidence that Harvey advanced was observational. As a young man he had studied medicine at the University of Padua in Italy. His teacher there was Geronimo Fabrizio Fabricius (c. 1533–1619), the surgeon who first pointed out the existence of one-way valves between the atria and ventricles in the heart and within the veins. Fabricius, however, did not understand the significance of these valves. Harvey did. He argued that they would prevent the backward flow of blood from ventricle to atrium or from the heart into the veins. The valves ensured the one-directional motion of blood throughout the body.

Harvey also examined the septum between the right and left ventricles and found that he could observe none of the pores or other openings by which Galen claimed blood could pass from one side to the other. From these observational data, it appeared to Harvey that the septum could not allow blood to pass through. This observation was made, however, before the days of the microscope, so it could be claimed (and was claimed by Galen's supporters) that the pores were too small to be seen with the naked eye. However, Harvey also pointed out that the septum is a relatively thick tissue, supplied with blood vessels of its own. This would hardly be necessary if, in fact, the tissue of the septum were constantly experiencing the passage of blood from the right to the left ventricle. Moreover Harvey noticed that both ventricles contract simultaneously. Mechanically, it would be difficult for liquid to pass from the right to the left ventricle while both were contracted. All these observations suggested to Harvey that blood passed from the right ventricle into the left indirectly—by going through the lungs. This lesser circulatory pathway, called today the pulmonary circulation, was distinct to Harvey from the greater, or systemic circulation. One of Harvey's achievements was in pointing out the difference and the connection between these two circulatory pathways within the body.

Comparative Anatomy

Harvey also used comparative anatomy to support his hypothesis. He noted that in those animals possessing lungs, there is both a right and left ventricle. The right ventricle pumps blood to the lungs; the left ventricle, to the rest of the body. In support of this functional division of labor in the heart, Harvey pointed out that the left ventricle is much larger than the right—a difference that is in accord with the greater size of the systemic circulation. Harvey further showed that in animals without lungs, such as fish, there is only one ventricle, which pumps to the body as a whole. This ventricle, he argued, is analogous to the left ventricle of animals with lungs. Animals intermediate between fish and higher land animals, such as reptiles, have a single ventricle that is partially divided down the middle (see Fig. 11.1). Harvey's comparative evidence argued for a functional and structural relationship between the presence of two ventricles and two circulatory pathways (pulmonary and systemic), and one ventricle and one circulatory pathway (systemic only).

Experimental Evidence

Providing the most important evidence, however, were Harvey's experiments. Beginning with the heart, he demonstrated that passing a probe from the right ventricle through the pulmonary artery to the lungs was very easy. So was passage of a probe from the venae cavae through the right atrium into the right ventricle. On the other hand, passing the probe from the pulmonary artery into the ventricle without damaging the pulmonary valve was impossible. Similarly, passage of the probe from the ventricle into the right atrium was retarded by the tricuspid valve.

Since a mechanical probe is not the same thing as a liquid, Harvey performed another experiment. He attached the venae cavae of a heart removed from a dog to a simple hand pumping device. Water pumped into the right atrium emerged from the pulmonary artery. Conversely, he reasoned, if passage were one-way, then it ought to be impossible to force water from the pulmonary artery back through the heart and out the vena cava. The prediction was confirmed by the experiment. The liquid was blocked by the pulmonary valve. All that happened when he pumped water into the heart through this route was that the pulmonary artery swelled to a size equivalent to the amount of water pumped into the heart.

Another experiment was designed to test Galen's assumption that blood passes through pores in the septum between the right and left ventricles. Harvey reasoned that if liquid could pass through the septum, then pumping liquid into the right ventricle, with the pulmonary artery blocked, should cause seepage into the left ventricle. To test this hypothesis, he used an excised heart in which he tied off the pulmonary artery and pumped water into the right atrium. The more liquid he pumped in, the more the right ventricle became distended, yet no seepage could be observed into the left ventricle. Harvey also made one experiment with a heart-lung preparation, that is, a heart to which the lungs were still attached by the pulmonary artery and veins. By pumping liquid into the right ventricle, Harvey could cause the lungs to become highly distended with the liquid he was pumping into the heart.

To demonstrate that blood was pumped in a circular path through the systemic as well as the pulmonary circulation, Harvey used a number of lines of evidence. When he cut open an artery (such as the descending aorta), the blood poured from it in spurts; when he cut open a vein, the blood flowed from it smoothly, with no spurts. He noted also that distension of the arteries followed immediately upon contraction of the ventricles. This implied that forceful muscular contraction of the heart pumps blood through the arteries. After passing through the smallest arteries (which Harvey couldn't see) and the capillaries (which Harvey never saw but assumed to be present) into the veins, the blood was collected and brought back to the heart. When it passed through the capillaries, the pulse of the heart was lost.

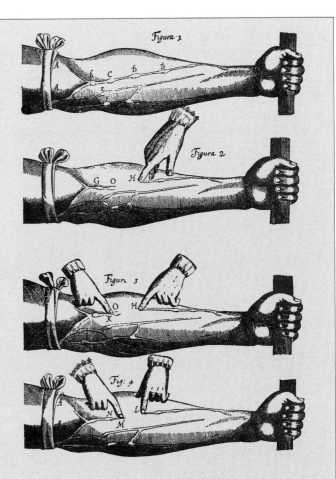

Diagram A

Along the same lines, Harvey performed experiments on a number of living organisms whose hearts he exposed by dissection. When he tied off the vena cava the heart continued to beat, but it became smaller, pumped very little blood through the pulmonary artery or aorta, and in all ways demonstrated that it was deprived of blood. If, on the other hand, he tied off the aorta or the pulmonary artery, the heart became "engorged with blood." To Harvey, the evidence was overwhelming that in all the animals he dissected, blood flowed in a one-way path from the veins, into the heart, and out again through the arteries. Moreover, he emphasized that the heart was really two very separate pumps: one (the right side) for pumping blood through the pulmonary circulation, the other (the left) for pumping blood through the systemic circulation. The two pumps were coordinated, but blood did not pass directly between them.

The question still remained whether Harvey's conclusions about circulation, based on experiments with lower animals, could also be applied to humans. He developed a very simple experiment that would demonstrate the one-way flow of blood through the systemic circulation of a human. Taking a person's arm (see Diagram A), he tied a tourniquet around the upper portion, effectively cutting off the flow of venous blood toward the heart (the veins are closer to the surface of the skin than arteries and hence are affected by the tourniquet, whereas arteries are far less so). The veins consequently fill with blood and become distended. When a finger is run posteriorly along a vein, toward the hand, stopping (and remaining) at position H (Diagram A, Figures 1 and 2), blood flows backward to refill the vein only as far as the first valve it encounters (position O, Figure 2). Posterior to that valve, the vein remains unfilled with blood (that is, between positions O and H, Diagram A, Figure 2), demonstrating that the valves effectively retard the backward flow of blood. Similarly, a second finger can now be applied to the blood in the section of vein anterior to the valve (Diagram A, Figure 3) and used to push the blood toward the valve. The valve is seen to distend, but it blocks the backward movement of blood. If the finger at position L in Diagram A, Figure 4, is then released, the vein immediately fills with blood up to the valve. These demonstrations showed Harvey that the blood clearly moved in a single direction in the human being, just as it did in his experimental animals.

Quantitative Data

But Harvey was not willing to stop with this level of evidence. So far most of his ideas were qualitative. He now sought to apply a quantitative argument. Having estimated that the left ventricle contains approximately 2 or 3 ounces

of blood at its fullest, Harvey went on to make the following calculation:

> . . . *let us suppose as approaching the truth that the fourth, or fifth, or sixth, or even but the eighth part of its charge is thrown into the artery at each contraction; this would give* [approximately] *half an ounce . . . of blood as propelled by the heart at each pulse into the aorta; which quantity, by reason of the valves at the root of the vessel* (aortic valve) *can by no means return into the ventricle. Now, in the course of half an hour, the heart will have made more than one thousand beats, in some as many as two, three, and even four thousand. Multiplying the number of drachms* [a drachm is one-eighth of an ounce] *propelled by the number of pulses, we shall have either one thousand half ounces, or one thousand times three drachms, or a like proportional quantity of blood. . . .* [This represents] *a larger quantity in every case than is contained in the whole body!*

By demonstrating that more blood passes through the heart in an hour than is contained in the whole body, Harvey showed that the blood *must* circulate; otherwise, the body would have to manufacture a much larger quantity of blood than anyone was willing to concede possible.

Harvey's approach to testing his hypothesis is a good example of the way logical reasoning, combined with observation and experiment, can be used to make a case for a particular hypothesis. Harvey began with a definite hypothesis in mind: namely that the blood is pumped in a circular path through the body by the action of the heart. He showed how this hypothesis was as much, if not more, in line with anatomical evidence than its rival (Galen's hypothesis). He also tested his hypothesis by experiments. Finally, he tried to show by a quantitative argument that Galen's hypothesis would require the body to manufacture far more blood in an hour's time than seemed possible.

the nature of the primary causative agent of the heartbeat.

11.3
CIRCULATION: BLOOD VESSELS AND BLOOD

Arteries and Veins.   Cross sections of an artery and a vein are shown in Fig. 11.3. Arterial walls have an outer coat of connective tissue, a middle layer of smooth muscle, and an inner coat of endothelium (epithelial tissue that lines the inside of a duct). The outer coat contains strong fibrous connective tissue that gives the artery the strength not to burst under pressure. At the same time, the connective tissue provides elasticity to the artery so

that it can adjust to changes in pressure. By contracting or relaxing and thereby regulating the size of the arteriole cavity, the middle layer of smooth muscle can determine the amount of blood going to any particular organ, and, of course, affect blood pressure. The walls of veins have the same three layers, but in each case they are thinner than in an artery. Veins are not subject to the same degrees of pressure as arteries, since they do not receive blood under direct pressure from the heart. They are, however, even more elastic.

Capillaries.   Capillaries are the smallest blood vessels in the body. Their walls consist of a single layer of flattened cells, an endothelium continuous with the

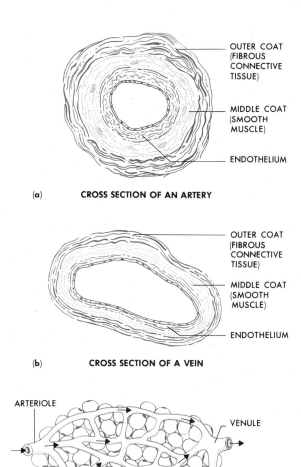

OUTER COAT
(FIBROUS
CONNECTIVE
TISSUE)

MIDDLE COAT
(SMOOTH
MUSCLE)

ENDOTHELIUM

(a)   CROSS SECTION OF AN ARTERY

OUTER COAT
(FIBROUS
CONNECTIVE
TISSUE)

MIDDLE COAT
(SMOOTH
MUSCLE)

ENDOTHELIUM

(b)   CROSS SECTION OF A VEIN

ARTERIOLE

VENULE

CAPILLARIES

BODY CELLS

(c)   DIAGRAM OF A CAPILLARY NETWORK

Fig. 11.3
(*a, b*) Cross section of an artery and vein. (*c*) Diagram of blood flow through the capillary network. In (*c*) the arrows represent the direction of blood flow.

endothelial lining of the arteriole at one end and the venule at the other. Exchange of materials between the blood and the body cells takes place through the walls of the capillaries (see Fig. 11.4). Flow through the capillaries is controlled by the opening or closing of the entrance to the capillary. This is accomplished by small, muscular **precapillary sphincters,** which are rings of muscular tissue located at the arteriole end of a capillary branch. Like the drawstring on a purse, the precapillary sphincters can close the opening into a capillary by contracting and drawing the opening shut. Along with the smooth muscles in the arterial and arteriole walls, these sphincters regulate at least 70 percent of the flow of blood to an organ or region of the body. The collective group of capillaries that permeate the body tissues make up the capillary bed (see Fig. 11.5).

Blood consists of a fluid portion, **plasma,** and a cellular portion consisting of **erythrocytes** (red blood cells), **leukocytes** (white blood cells) of several types, and **platelets.**

Plasma.   Plasma is more than 90 percent water. Its major solute is a complex mixture of proteins, making up about 7 percent of the plasma by weight. Fatty substances are also present in suspension and solution. Other plasma components include salts, glucose, amino acids, vitamins, hormones, and the waste products of cellular metabolism. Most of the plasma proteins are

Fig. 11.4
A capillary, showing red blood cells squeezing through it. Note the thinness of the capillary walls, making it possible for water, ions, and large molecules to pass back and forth between the circulatory system and the fluid surrounding the body cells. (Photo courtesy Thomas Eisner, Cornell University.)

synthesized in special cells in the liver. The major plasma protein is **albumin**, a relatively small molecule whose osmotic effect helps retain water within the vessels. Albumin also serves as a nonspecific carrier protein by binding to certain other small molecules in the plasma. If the albumin content of the blood falls critically low, water moves from the capillaries into the body tissues, producing a swelling known as edema. Other plasma proteins are concerned with the clotting of the blood and the transport of specific substances such as lipids, iron and copper ions, and vitamin B_{12}. Fats and sugars (mostly glucose) are present in the plasma from absorption in the intestine or from having been released from storage somewhere in the body.

Erythrocytes. Erythrocytes, or red blood cells, are highly specialized, carrying oxygen to the cells and removing carbon dioxide from them to the lungs. Red blood cells are generated in the bone marrow. When mature, the mammalian red blood cell lacks a nucleus and hence is incapable of cell division. Fig. 11.6 shows a group of red blood cells; the scanning electron micrograph emphasizes the concave structure of both sides of the cell, a structure that increases their surface area for more oxygen and carbon dioxide takeup. The normal red blood cell count ranges between 4,000,000 and 6,000,000 cells per cubic millimeter. People who live at high altitudes have a higher count, since the air is thin-

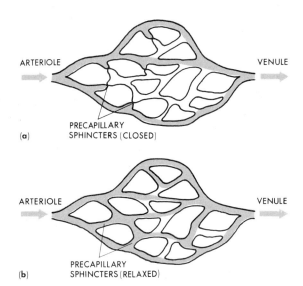

Fig. 11.5
(*a*) With several precapillary sphincters closed, blood flow through the capillary bed in a particular region is restricted. Capillaries without much flowing blood are shown in white. (*b*) With relaxation of precapillary sphincters, blood flow resumes through the bed.

Fig. 11.6
(*a*) Scanning electron micrograph (× 3333) of red blood cells, showing clearly the doubly concave surface of each cell. Red blood cells are essentially sacs containing a single type of protein, hemoglobin. In the adult human they lack a nucleus and most of the common cell organelles, which have degenerated by the time the cell is released from where it is produced in the bone marrow. (Micrograph courtesy Dr. Marion I. Barnhart, Wayne State University School of Medicine. Originally published in R. M. Nalbandian, ed., *Molecular Aspects of Sickle Cell Hemoglobin, Clinical Applications,* 1971. Courtesy of Charles C Thomas, Springfield, Illinois. (*b*) Higher magnification (×13,000) of red blood cell photographed under the scanning electron microscope. (Micrograph courtesy National Physical Laboratory, Royal Postgraduate Medical School, Hammersmith, England.)

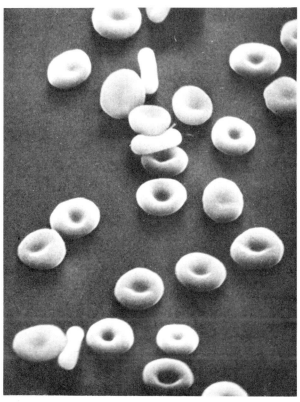

(a)

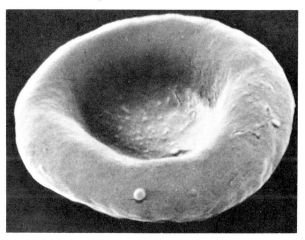

(b)

Blood is a connective tissue containing plasma (water and colloidal proteins) and red blood cells (erythrocytes), white blood cells (leukocytes), and platelets.

ner and there is less oxygen available per unit of air inspired. This adaptive response helps the body continue delivering the same amount of oxygen to the cells.

Leukocytes. Unlike red blood cells, leukocytes, or white blood cells, have a nucleus (see Fig. 11.7). They are independently motile, capable of amoeboid movement within and outside of the vascular system. Most of the leukocytes in the body are actually found *outside* the vascular system proper in the tissue fluid, to which they travel by squeezing out between the cells of the capillary walls. Their functions vary, but in general they help the body fight infections. They are also involved in repair procedures, such as the healing of a wound.

The number of leukocytes in the blood ranges from about 4500 to 11,000 per cubic millimeter. Fluctuations occur even during the course of a single day, the number being higher during periods of activity and lower during periods of rest.

There are three types of leukocytes. **Granulocytes** are larger than red blood cells, have a multilobed nucleus, and contain many granules in their cytoplasm. Their major functions include phagocytosis (see Fig. 11.8), and they play a role in the antibody formation processes involved with the immune response. **Monocytes** are the largest cells of the blood. Their function appears to be largely phagocytic, for they are highly motile. **Lymphocytes** function largely in vital defense mechanisms involving acquired immunity to foreign substances. There are two main types of lymphoctyes, labeled B-cells and T-cells. Arising embryologically from the yolk sac, B-cells develop in the fetal liver and spleen and become a vital part of the body's defense mechanism against infection (which will be discussed shortly), while T-cells, maturing in the thymus gland located in the chest, play a major role in recognizing and rejecting foreign tissue in the body. During later embryonic development, these cells migrate out of the thymus and populate the developing lymph glands. Here they establish themselves, and in the adult they are produced there by descendants of the original embryonic migrants.

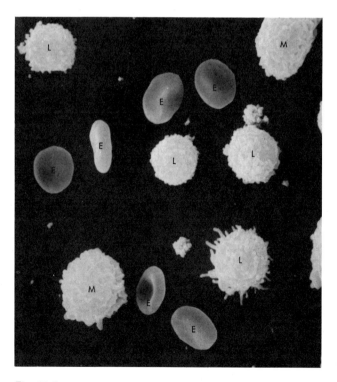

Fig. 11.7
Scanning electron micrograph of two types of human leukocytes, monocytes (M) and lymphocytes (L) (× 3500). Six red blood cells or erythrocytes (E) are also shown. (Micrograph courtesy Dr. Peter M. Andrews, Georgetown University.)

Platelets. Platelets or thrombocytes are cytoplasmic fragments that develop in the bone marrow as pinched-off fragments of large cells called megakaryocytes. From 250,000 to 400,000 platelets, each about 2.5 micrometers in diameter, may be found in a cubic millimeter of blood. Platelets play a major role in the clotting process (see Supplement 11.2). Platelets have a life span of about ten days, as compared with red blood cells which average between 130 and 145 days.

Blood is obviously a complex tissue with many varied functions. It must transport nutrients to, and wastes away from, the body cells. It must serve as one of the body's defenses against infection. It must also be able to prevent its own leakage from the vascular system by clotting in case of injury. Finally, it has to play a principal role in maintenance of water balance in the body.

In recent years, using the information we have been discussing concerning the composition and physiological functioning of blood, researchers have developed artificial blood containing organic compounds in which hydrogen atoms are replaced by fluorine atoms.* Dogs

*These molecules carry oxygen by having the fluorine-bearing side-chains surrounded by an "oxygen cloud," as contrasted with the highly specific interaction between oxygen and hemoglobin.

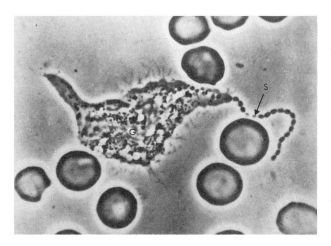

Fig. 11.8
A leukocyte (granulocyte G), surrounded by several erythrocytes, in the process of ingesting infectious streptococci, S (bacteria) by phagocytosis. (Micrograph courtesy Pfizer Inc.)

which had 90 percent of their blood replaced with such fluids remained in perfect health three years later. Rats, subjected under anesthesia to having their blood completely replaced with perfluoro tri-*n*-butylamine, exhibited normal drinking, urinating, and nest-building behavior upon awakening. In ten days, their blood protein levels returned to normal and the animals remained perfectly healthy. The benefits to medicine of eventually being able to develop an effective blood substitute for humans are obvious.

11.4
THE EXCHANGE OF SUBSTANCES ACROSS CAPILLARY WALLS

The exchange of nutrient and other materials between the blood and tissue fluids occurs in the capillaries. Some substances, such as water, glucose, urea, salts, and lipid-soluble substances (including alcohol and anesthetic gases), pass very freely across the walls while others, such as proteins, vary in the degree to which they do so. In the liver, for example, protein moves rather readily across the capillary walls, while it hardly does so at all in the central nervous system.

While most substances pass through the capillary walls via the wall cells themselves, lipid-insoluble materials appear to diffuse *between* the cells, passing through the cementing substance holding the cells together. This cementing substance is kept in a porous

---

SUPPLEMENT 11.2
THE CLOTTING OF BLOOD

When you cut your finger, there is at first profuse bleeding. Gradually, however, the bleeding slows down and stops. Around the wound you can observe a clot of dried blood. This clot prevents further blood loss. Fortunately, however, blood does not normally form clots within the arteries and veins. What prevents blood from clotting within the circulatory system, and what triggers it to clot at the site of a wound?

When you examine a blood clot under the scanning electron microscope, it appears as a network of long threads, or fibers, in which red blood cells are trapped (see Diagrams A

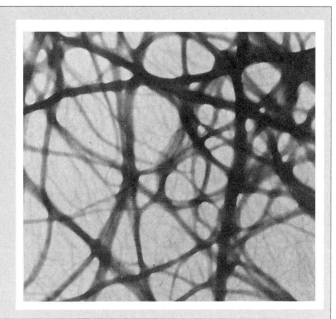

Diagram A
Meshwork of fibrin polymers. (Courtesy Keith Porter.)

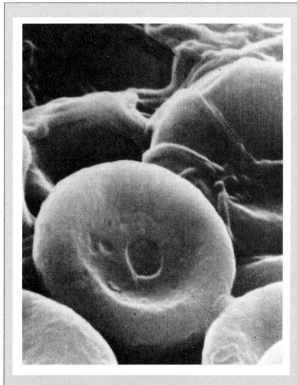

Diagram B
Scanning electron micrograph of red blood cells trapped by
fibrin threads. (Courtesy T. L. Hayes and L. W. Macdonald,
Lawrence Berkeley Laboratory, University of California,
Berkeley.)

and B). This fibrous network is composed of a protein
known as fibrin. Clotting takes place whether the red blood
cells are present or not—it is a property of the plasma alone.
The blood cells simply increase the effectiveness of the clot
by becoming entrapped in it and plugging up spaces between
the fibers.

At the site of a wound, blood platelets begin to break
down. The factors responsible for this breakdown are not
entirely understood. The breakdown of platelets triggers a
series of other reactions, which are outlined in Diagram C.
The breakdown of either the platelets or cells of the damaged
tissues (or both) releases a substance called thromboplastin
at the wound site. The presence of thromboplastin, in conjunction with calcium ions (Ca^{++}), chemically triggers the
conversion of prothrombin into thrombin. Prothrombin is
an inactive form of the clotting enzyme thrombin. Once activated, thrombin catalyzes the conversion of soluble fibrinogen into insoluble fibrin.

The conversion of fibrinogen into fibrin occurs by a process of polymerization. The mechanism for this polymerization has been studied in some detail by Dr. Koloman Laki
and his associates at the United States National Institute

of Arthritis and Metabolic Diseases. Fibrinogen is a medium-
sized protein molecule about 600 Å long, with a molecular
weight of 330,000. Its three-dimensional structure consists of
three areas where the polypeptide chain is folded into a globular shape, connected by two areas of straight alpha helix
(see Diagram D). The middle and end globular regions of
fibrinogen both contain two peptide groups (recall that peptides are small groups of amino acids), for a total of four.
Thrombin catalyzes the hydrolysis of the peptide bonds that
hold these peptides in position. The result is that the four
peptides are cleaved from the fibrinogen (see Diagram E, top
and middle), triggering the unfolding of a fourth globular

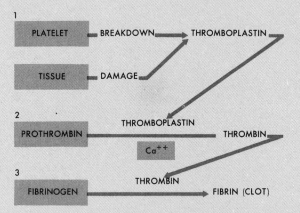

Diagram C
Reactions resulting from platelet breakdown.

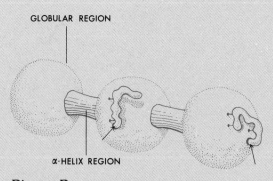

Diagram D
Fibrinogen molecule showing one of the two peptides on the
middle and end globular regions. Arrows point to the area
where thrombin splits off from the peptide.

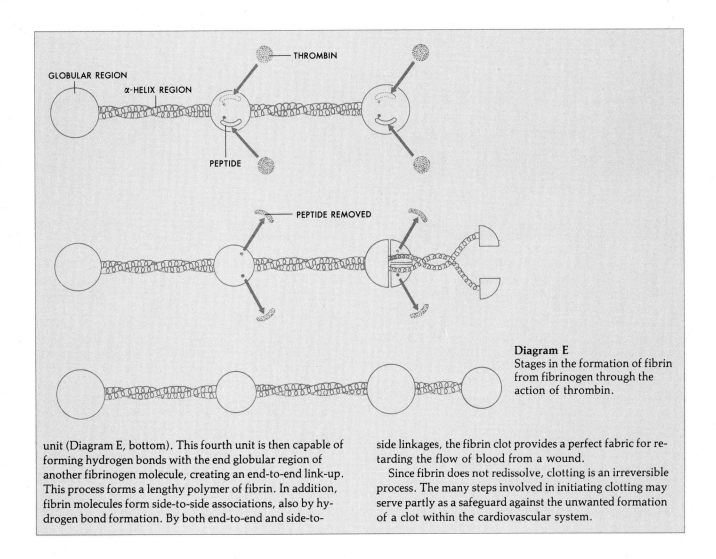

GLOBULAR REGION

α-HELIX REGION

THROMBIN

PEPTIDE

PEPTIDE REMOVED

Diagram E
Stages in the formation of fibrin
from fibrinogen through the
action of thrombin.

unit (Diagram E, bottom). This fourth unit is then capable of forming hydrogen bonds with the end globular region of another fibrinogen molecule, creating an end-to-end link-up. This process forms a lengthy polymer of fibrin. In addition, fibrin molecules form side-to-side associations, also by hydrogen bond formation. By both end-to-end and side-to-side linkages, the fibrin clot provides a perfect fabric for retarding the flow of blood from a wound.

Since fibrin does not redissolve, clotting is an irreversible process. The many steps involved in initiating clotting may serve partly as a safeguard against the unwanted formation of a clot within the cardiovascular system.

state by the endothelial cells. Oxygen and carbon dioxide are able to pass through the capillary wall via either route, the wall cells or the cementing substance, a factor that undoubtedly is important to the efficiency with which the capillaries carry out this important exchange.

As one might expect, a major factor in the movement of substances across the capillary walls is simply diffusion; substances pass out of the cells and tissue fluids being supplied by the capillaries because they are in higher concentration than within the capillaries, while the reverse is true for other substances, such as nutrients. In addition to this diffusion, however, the fluid portion of the blood (with the general exception of proteins) may be forced across the capillary walls by filtration as a result of blood pressure. This pressure, originating from the heart, is called **capillary hydrostatic pressure**. While there is an opposing force, the **tissue fluid hydrostatic pressure**, it is generally far less than the capillary hydrostatic pressure, and thus the

The exchange of materials between the circulatory system and the body cells occurs only through the capillaries. Their walls are but one cell thick.

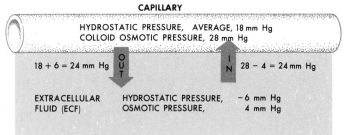

CAPILLARY

HYDROSTATIC PRESSURE, AVERAGE, 18 mm Hg
COLLOID OSMOTIC PRESSURE, 28 mm Hg

18 + 6 = 24 mm Hg   O U T   I N   28 − 4 = 24 mm Hg

EXTRACELLULAR   HYDROSTATIC PRESSURE,   − 6 mm Hg
FLUID (ECF)   OSMOTIC PRESSURE,   4 mm Hg

Fig. 11.9
Diagram of the hydrostatic and osmotic pressures that influence the exchange of materials between capillaries and tissue fluid.

movement of fluid out of the capillaries can occur. Indeed, since the tissue hydrostatic pressure may actually be below that of atmospheric pressure, it may indeed help in this outward movement (see Fig. 11.9).

Besides the hydrostatic pressures exerted by the fluids within the capillaries and tissue spaces, osmotic pressure also plays a role. Due principally to the presence of proteins, the tissue fluid exerts a **colloid osmotic pressure** (a colloid is a solute particle suspended in a medium; such particles do not settle out or pass through cell membranes), due mostly to the proteins it contains in colloidal suspension (see Fig. 11.10). An opposing force is maintained by the colloid osmotic force of the plasma caused by plasma proteins, especially albumin, and it is often referred to as the **oncotic pressure**.

Whether fluid enters or leaves a capillary, therefore, depends on which of the combined forces of diffusion, hydrostatic pressure, and colloid osmotic pressure is the greater. If pressure is higher within the capillaries, substances will leave and enter the tissue fluid for transport into the body cells; if pressure is lower within the capillary, the movement is in the opposite direction. As might be expected, outward moving forces generally predominate at the arterial end of the capillaries, and inward forces at the venous ends. However, not all of the fluid returns. Some, along with any protein it contains,

enters the lymph and returns to the general circulation via the lymphatic system.

11.5
THE LYMPHATIC SYSTEM

Cells of the body exist in the extracellular space outside the capillaries. The extracellular fluid that bathes them is a mixture of water and the various components that have filtered from the plasma inside the capillaries. This material must be recollected and brought back into the general circulation. As we have seen, some of this is accomplished by the capillaries. But some of the water and solutes from the extracellular space are collected by special vessels of the lymphatic system. The system derives its name from the plasma filtrate, or **lymph**. Lymph is collected by capillaries of the lymphatic system (see Fig. 11.11) and drained progressively into larger vessels called **lymphatic ducts**. Ultimately these ducts empty into large veins in the upper part of the thorax, especially the subclavian vein near the heart. Through that route they reenter the cardiovascular system. Because they are structurally more fragile and less distinct than arteries and veins, the lymphatic vessels are difficult to trace. Like veins, the larger lymphatic vessels contain valves.

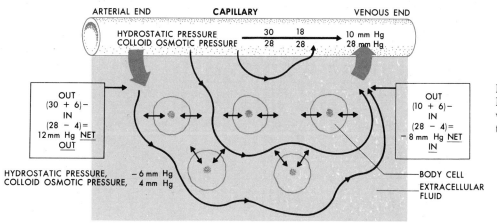

ARTERIAL END   **CAPILLARY**   VENOUS END

HYDROSTATIC PRESSURE   30   18   10 mm Hg
COLLOID OSMOTIC PRESSURE   28   28   28 mm Hg

OUT
(30 + 6) −
IN
(28 − 4) =
12 mm Hg NET
OUT

OUT
(10 + 6) −
IN
(28 − 4) =
− 8 mm Hg NET
IN

HYDROSTATIC PRESSURE,   − 6 mm Hg
COLLOID OSMOTIC PRESSURE,   4 mm Hg

BODY CELL
EXTRACELLULAR FLUID

Fig. 11.10
Factors affecting the flow of water between capillaries and the extracellular fluid.

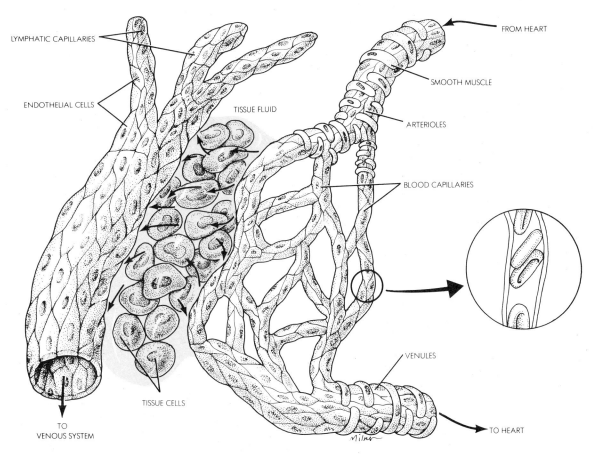

Fig. 11.11
General design of the lymphatic capillaries. Note their relationship to the blood and tissue fluid, and that, unlike the vessels of the blood circulatory system, the lymphatic capillaries begin as dead-ends. The fluids carried in the lymphatic system are eventually returned to the general circulatory system. (Drawing adapted from A. P. Spence and E. B. Mason, *Human Anatomy and Physiology.* Menlo Park, Calif.: Benjamin/Cummings, 1979, p. 557.)

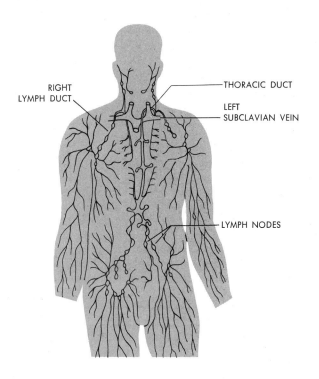

Fig. 11.12
The lymphatic system supplements the more extensive circulatory system. It returns valuable blood plasma proteins to the blood and aids in fighting infection.

A diagram showing the distribution of lymph vessels over the upper region of the body is shown in Fig. 11.12.

Spaced along the larger lymphatic ducts are larger regions called **lymph nodes**. These nodes have three basic functions. They filter out particulate matter (such as bacterial cells) and prevent it from being circulated back into the bloodstream. Second, as pointed out earlier, lymph nodes contain leukocytes that attack and phagocytize foreign cells that enter the body. And third,

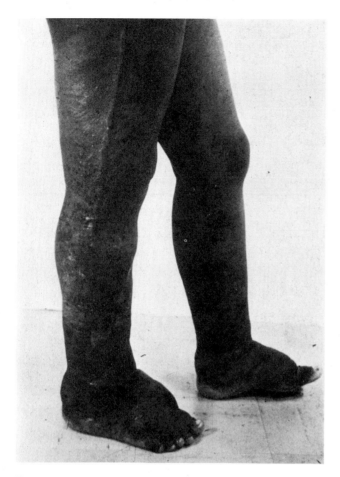

Fig. 11.13
Elephantiasis. The blockage of lymphatic vessels by the roundworm *Wuchereria bancrofti* results in fluid retention and swelling. The name of the condition comes from the immense swelling that may occur in the affected regions, as well as the superficial resemblance of the skin to that of an elephant. (Photograph courtesy the Center for Disease Control, Atlanta, Georgia.)

lymph nodes are the secondary sites for production of **lymphocytes.** Since lymphocytes are the main cells that produce antibodies, lymph nodes are the centers of antibody formation during infection. Lymphocytes are produced in lymph nodes at the rate of about 10 billion per day. When the body is infected, the lymph nodes sometimes become swollen and painful. This is because they are filtering out bacterial or other material from the blood at a great rate and are thereby sites of infection themselves.

Sometimes lymphatic vessels become infected or blocked and are unable to drain fluid away. In one example, blockage is caused by a parasitic roundworm,

Wuchereria bancrofti, producing a condition known as elephantiasis (see Fig. 11.13). The name derives from the fact that infected parts of the body swell to elephant-like proportions.

In addition to the lymph nodes, there are several other organs that are lymphoid in nature, though having no direct connection with the lymphatic system. One of these, the thymus gland, will be discussed in Chapter 14. The **tonsils** are small masses of lymphoid tissue in the throat region that serve as an additional defense against bacterial infection. (One portion of the tonsils, located in the nasopharyngeal region, often becomes enlarged during childhood and is referred to as the adenoids.) At times the tonsils themselves become badly infected with the bacteria they trap (a condition called tonsillitis), which may necessitate their surgical removal (tonsillectomy). Indeed, until the 1940s, it was rare to find a person whose tonsils had not been removed. However, the development and improvement of antibiotics made this operation far less common.

Fig. 11.14
Scanning electron micrograph of a rhesus monkey spleen ($\times$ 1200). Note the porous, spongy nature of the splenic tissue and the many blood cells visible. The spleen is the largest of the lymphoid organs and lies between the stomach and the diaphragm. (Micrograph courtesy Dr. Peter M. Andrews, Georgetown University Department of Anatomy.)

The largest of the lymphoid organs is the **spleen**. This organ lies between the stomach and the diaphragm and is divided into compartments. In each of these compartments, two types of *splenic pulp*, red and white, can be distinguished. The red pulp contains erythrocytes, lymphocytes, and large phagocytic cells, while the white pulp surrounds the splenic arterioles and is thought to be the site of lymphocyte activity (see Fig. 11.14).

During embryological development the spleen plays a major role in the production of erythrocytes, but this capability normally disappears following birth. It also filters the bloodstream of worn-out erythrocytes and serves as a site of immune defense against antigens in the blood (see Section 11.6). Finally, the spleen serves as a sort of reservoir that may contain as much as 350 ml of blood. In times of emergency such as excessive bleeding (hemorrhage), as much as 200 ml of this blood may be squeezed from the spleen into the general circulation.

11.6
DEFENSE MECHANISMS AND THE IMMUNE RESPONSE

The lymphatic system plays a major role in the ability of an organism to respond effectively to the invasion of the body by outside agents such as bacteria or viruses. In the case of bacteria entering an open wound in the skin, for example, both lymphocytes and leukocytes act as phagocytes, literally engulfing the invading bacteria and digesting them. The warming and reddening of the area is due to the damaged cells' release of chemicals such as histamines, which increase the capillaries' permeability and dilate blood vessels, bringing more blood to the scene. Sometimes the phagocytes themselves are killed, and the wound becomes a battlefield with dead and dying bacteria, phagocytes, and tissue fragments littering the area as **pus**. The presence of the bacteria leads to increased production of phagocytic leukocytes, enabling the detection of internal infections by a "blood count," which is actually a determination of the ratio of one blood cell type to another to see if there is a significant variation from the norm.

In 1796 physician Edward Jenner introduced the concept of vaccination to confer immunity to smallpox (see Section 20.12 for details of Jenner's work). While other types of vaccines were developed over the next centuries for several often deadly bacteria-caused diseases such as diphtheria, tetanus, and whooping cough, and virus-caused diseases such as rubella (German measles), chicken pox, measles, mumps, and polio, only recently have we learned how they work. In the case of smallpox, the protein coat of the vaccinia virus is sufficiently similar to the smallpox virus to "fool" the organism into either producing globular proteins called **antibodies** or converting certain B-lymphocytes into

a sensitized form often called a plasma cell. Plasma cells are recognizable by their large size and extensive endoplasmic reticulum with numerous ribosomes, presumably for antibody synthesis. The series of steps leading to antibody production is initiated by the introduction by the foreign agent of molecules called **antigens**. An antigen may be a wide variety of substances, usually protein, but occasionally complex carbohydrates, or lipids as well; indeed, anything that triggers antibody formation on the part of the invaded host organism can be an antigen.

The antigen-antibody reaction is a highly specific one. Indeed, their interaction is often compared to that of an enzyme and its substrate. Once an antigen is recognized as a foreign substance, there are at least three defense strategies that may be employed. One is to render the invader incapable of causing damage to the host organism by modifying its structure in some slight but crucial way. Covering it with some substance that makes it a target for a phagocyte is another. Sometimes, in conjunction with blood components called **complement**, action is taken directly to destroy a foreign cell.

The differentiation of a B-lymphocyte into a plasma cell may occur on contact between the antigen and the antibody specific for that antigen and located on the surface of the B-lymphocyte. More often, recognition of the antigen by the T-lymphocytes is also necessary before the B-lymphocyte-to-plasma-cell differentiation can occur. In either case, the resulting plasma cells produce large amounts of antibody, and the presence of the antigen stimulates them to reproduce, further increasing antibody production. It often becomes a contest between host antibody production and infectious agent antigen to see which can out-produce the other, with the health or even life of the host at stake. In some conditions, such as tetanus, the invading bacteria may take over so rapidly and produce so much toxin that the host organism's defenses cannot produce antibody fast enough to survive. The result can be a painful death in which the muscles of the body contract with great force, bowing the body backwards and clenching the mouth shut (thus the common name for tetanus of "lockjaw"). Today, tetanus vaccination is routine, and booster shots are often given after the suffering of puncture wounds of the sort known to lead to tetanus.

The key to the success of vaccines, such as those for tetanus, smallpox, polio, and the like, rests on the ability of the host organism's system to remember the initial antigen invasion and thus always keep the correct antibody forces in reserve. This is accomplished by specialized cells (called, appropriately, "memory cells") produced during the divisions of the B-lymphocyte to produce plasma cells. The memory cells, unlike the antibodies, remain in the circulation, ready to begin antibody synthesis the moment its specific antigen once again invades the organism.

Vaccination differs from exposure to the actual disease itself in that the infectious agent in the vaccine is carefully weakened (or even killed) by heating before it is introduced into the organism. Or, as in the case of smallpox, the vaccinia virus is used to introduce antigens sufficiently similar to those of the smallpox virus so as to stimulate production of antibodies effective against both.

One of the most exciting developments in the field of vaccine development deals with the disease cholera. The bacterium causing the disease, *Vibrio cholerae*, invades only the intestines, where it produces a two-part toxin. One portion, labeled B, binds to cell receptors of the intestinal wall. The other part, A, then acts as an imposter hormone, causing the endothelial cells to increase greatly their production of fluids. The result is a highly debilitating diarrhea that is good for the bacterium in that it spreads it widely to other humans in areas of poor sanitation, but the resulting dehydration causes an often fatal breakdown of blood circulation in the victim.

Noting that persons who recover from cholera are then immune to further attacks, microbiologists Drs. Takeshi Fondi and Richard Finkelstein of the University of Texas looked for a mutant form of the bacterium that would produce only the B portion of the toxin, reasoning that such a strain might trigger antibody formation yet, lacking the hormone-imitating A portion, not cause the dehydration. By exposing natural cholera bacteria to a mutation-causing agent (mutagen), nitroseguanidine, they examined those who survived the poisoning. By growing these mutant strains they looked for one that had the "B not A" characteristic they desired. Finally the desired form appeared and preliminary tests on the highly susceptible baby rabbit produced no disease.

Human beings, however, are the only natural host for *Vibrio cholerae*. Thus it remains to be seen if the new strain will be effective in our species. While the new strain could revert again to the virulent form, it appears thus far to be quite stable. Its existence provides for some hope that we are on the verge of a breakthrough in preventing isolated cases of cholera leading to huge epidemics.

The T-lymphocytes play a major role in antibody formation. Their name comes from the fact that they must undergo a maturing process in the thymus gland after their production in the bone marrow. Besides their complementary work to that of the B-lymphocytes in antibody formation against foreign antigens, T-lymphocytes also play the major role in recognizing and breaking down tissues foreign to the body. Cancerous tumors are one such foreign tissue and, indeed, the spread of certain types of cancers usually occurs only when the body's immune response system has been suppressed. There is some evidence that carcinogenic (cancer-causing) substances provide such supressors.

It is not yet clear, however, whether their doing so allows cancer already present to develop or whether the carcinogen is the actual cause of the cancer.

As beneficial as the work of T-lymphocytes is, there are times when one wishes they were less efficient. This is when organ transplants (such as kidney, heart, and the like) between those who are not identical twins are necessary. Thus massive radiation of the thymus gland, lymph nodes, and the like, or the giving of certain chemicals may be necessary to supress the immune response until the graft "takes." When this is done the patient must, of course, be kept in a sterile environment, since he or she would be highly vulnerable to infections ordinarily easily resistible.

Human infants are occasionally born with no immune system at all. Such infants must be kept in totally sterile environments; one boy has lived in a special plastic world for several years with the hope that a breakthrough can be found that will enable his system to begin producing an immune response. In other cases, an individual's immune system is not able to recognize "its own" and begins to produce antibodies that attack the healthy tissues of the same individual. A type of anemia called myasthenia gravis results from this autoimmunity, and the damage to the heart often associated with rheumatoid arthritis may also be the result of this phenomenon.

The structure of antibodies in terms of their primary structure (amino acid sequence) as well as the various theories currently put forth concerning the genetic basis of their formation will be considered in Chapter 19.

Interferon

Immune response to bacterial infections is often swift enough to allow recovery; this is usually not the case with viral infections. In 1957 Dr. Alick Isaacs and Jean Lindenmann reported the discovery of a cellular protein they called **interferon** (see also Section 9.9). It has been hypothesized that host cells infected with viruses induce those cells to produce interferon, which then passes to other uninfected cells. Here it is thought to induce the production of another protein that blocks the genetic ability of the virus to infect the cells. While the interferon production comes too late to save the infected cells, immunity is conferred upon those not yet affected; seen poetically, the dying cells expend their final energy warning others of the impending danger.

The logical next step is to produce interferons in large enough amounts to fight off viral infections as they occur; in essence, to make interferon to viral infections what penicillin is to bacterial infections. Two major problems have stood in the way of realizing this dream, however. For one, rather than being specific for the invading virus, interferon is specific to the host. Thus, unlike the blood-clotting protein thrombin, where cattle thrombin can be used to stop bleeding in humans, interferon produced in the cells of another organism will work *only* in that organism, and not in human beings:

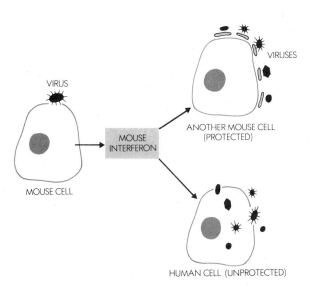

The second problem has to do with amounts. Since interferons are highly active compounds, only a few molecules are necessary to create a powerful antiviral state in a cell. This means, therefore, that they are produced only in minute quantities, difficult to separate from other cellular protein. These minute quantities are measured in "biological units." A biological unit of interferon is the amount that will reduce a virus infection in one million cells in a test tube by 50 percent. Based on estimates of the size of an interferon molecule (determination of interferon amino acid composition is nearly completed, but the molecule is too large to be synthesized by currently available techniques), a biological unit of interferon would have a mass of one picogram, or one-millionth of a millionth of a gram. This has been compared to a truckload of table salt; if the entire load is equivalent to one gram, a picogram of interferon would be one grain of salt. Using improved techniques involving byproducts of blood donations at the Finnish Red Cross Blood Transfusion Center in Helsinki, Dr. Kari Cantell has been able to obtain one-tenth of a gram of pure interferon—still only enough to treat 10,000 patients with minor virus infections or 200 with chronic virus conditions.

Interest in interferons has grown with the discovery that, besides conferring resistance to viral infections, interferons can perform a variety of regulatory roles in the cell. Especially significant is the fact that they can in-

hibit cell growth, and appear to have actually slowed or stopped cancer in laboratory animals.

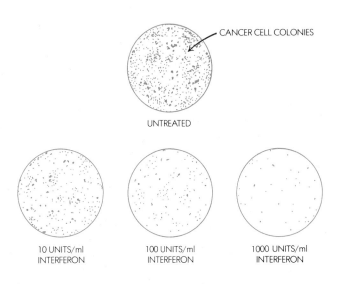

Besides acting directly on the cancer cells themselves, there is reason to believe that interferons may also stimulate the body's own natural defenses against the cancer cells.

Quite recently, however, serious doubts have been raised about interferon's well-publicized potential as a "wonder drug" for cancer. Some studies indicate that interferon has, overall, produced less evidence of tangible improvement in cancer patients than would be expected from conventional treatment. In a form of cancer called multiple myeloma, for example, four out of fourteen patients did experience a remission of the disease after interferon treatment. However, a leader of the study, Dr. Elliot F. Osserman of Columbia University's College of Physicians and Surgeons, pointed out that with chemotherapy the expected number would be eleven out of fourteen. He and others were intrigued, however, that at least something in the interferon preparations had even this limited success in some cases.

The first excitement over the possibility that interferon might be a significant weapon against cancer in humans stemmed from a study at the Karolinska Hospital in Stockholm, Sweden. Dr. Hans Strander and his colleagues reported encouraging results with interferon treatments after surgery in patients with osteogenic sarcoma, a highly dangerous form of cancer of the long bones of the arms and legs. However, a Committee of the National Cancer Institute of the United States reviewed the evidence at the request of the Swedish researchers and judged that the form of cancer in the patients showing improvement was not as severe as was originally thought. When these cases were discounted from the Swedish study, the effect of interferon on the osteogenic sarcoma was not statistically significant. In 1980, Dr. Strander reported that of twelve patients treated with interferon who had been followed for at least four years from the beginning of their interferon treatment, six are free of detectable disease. However, the National Cancer Institute's Dr. Arthur S. Levine, who was on the review committee, points out that some of these successes were in the group that the committee had recommended be eliminated from the study. Thus, though he feels that interferon research is far from valueless, it does not to date appear to be the breakthrough it was once thought to be. The large amount of money recently given to a laboratory specializing in interferon research by the Shell Oil Company may enable the question to be settled once and for all.

Recently it has been announced that the gene responsible for producing human interferon has been successfully "transplanted" into bacteria. Since bacteria are inexpensive to raise, reproduce rapidly, and can be grown in far larger quantities than human cells, as well as being easier to purify, this news is encouraging. However, unlike human cells, bacterial cells are not capable of adding various sugar molecules (glycosylation) to the interferon as would normally be done. These sugars may be vital in protecting the interferon from inactivation or alteration in the body, and thus until large quantities of bacterial-synthesized human interferon are produced, we will not know if it works in human beings. However, according to a report from the First International Congress of Interferon Research held in Washington, D.C., in November 1980, the production of biologically active interferon by inserting human genes into bacteria is moving ahead rapidly. The bacterially produced interferon is not only active in laboratory tests

but seems to protect other species—squirrel monkeys, mice, and hamsters—against a fatal virus, encephalomyocarditis. However, in terms of its effectiveness in treating cancer in humans, interferon still falls far short of being a miracle drug, and produces side effects as well. It remains to be seen if these problems can be overcome by future research.

11.7
THE HOMEOSTATIC REGULATION OF
BLOOD PRESSURE

Because the heart is a pump forcing blood through a closed system, there exists a certain pressure within the cardiovascular system known as **blood pressure.** In a resting adult, the average arterial blood pressure rises to 120 mm mecury (Hg) during ventricular systole and falls to about 80 mm Hg at diastole. However, blood pressure is not constant throughout the arterial system; it is greatest within the arteries, falling progressively lower in the capillaries and in the veins (see Fig. 11.15). In the veins the pressure may drop to only a few mm Hg—a serious problem, since the blood still has a long way to go to get back to the heart. However, the tendency of the blood to pool in the veins is counteracted by the valves that prevent backward flow. Thus, though blood has a tendency to become increasingly sluggish as it moves through the veins, it still continues to move in a single direction.

Maintaining blood pressure is important to several aspects of normal body physiology. It is important in keeping the blood moving through the circulatory system. If the pressure drops too low, certain regions simply do not receive an adequate blood supply and the tissues die. Blood pressure is also critical in the movement

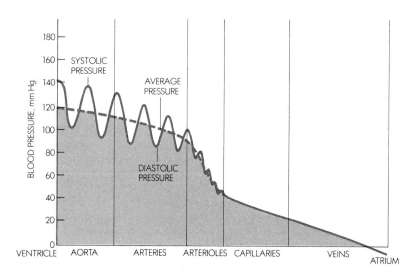

Fig. 11.15
Chart showing blood pressure in various parts of the circulatory system. The systolic and diastolic pressures, as well as the average blood pressure, are shown for the aorta and arteries. Note that the venous pressure drops below zero (that is, below atmospheric pressure) near the right ventricle. This implies that blood is moved from the vena cava into the heart by the pumping action of the right atrium.

of materials into and out of the capillaries (see Section 11.4). Too low a blood pressure can cause the failure of certain substances to pass from the capillaries into the extracellular fluid. It is therefore very important for the body to regulate blood pressure rather precisely in the arteries and veins. Several homeostatic mechanisms are involved in this regulation.

Regulation of Local Blood Flow by Metabolic Products

One kind of blood pressure regulation takes place at the local level, that is, in an area where some "disturbance" occurs. Such disturbance might be in very active tissues, such as a muscle during vigorous contraction, or the application of a tourniquet to an arm or leg to temporarily cut off the blood supply. When a disturbance occurs, the blood vessels in the area are found to **dilate**, or enlarge. This response is advantageous; it allows more blood to pass into an area where metabolic products are accumulating faster than they could normally be removed. Since it is known that the walls of arteries and veins are innervated from the central nervous system, it is reasonable to hypothesize that this dilation of the vessels is controlled by nerves. This hypothesis leads to a prediction: if we cut the nerves leading to an area and then disturb it by tying a tourniquet around the

area, then the blood vessels ought not to dilate. This prediction is contradicted, however. It appears that something in the blood itself triggers vessel dilation.

This second hypothesis can be tested by performing a simple experiment. If vessel dilation is caused by some substance in the blood of disturbed tissues, then injection of blood from a disturbed area of the body into an undisturbed area ought to produce dilation. The experiment confirms the prediction. Further investigation has shown that among the specific substances that elicit vessel dilation are carbonic acid and other acid byproducts of cellular metabolism. Apparently the byproducts of cellular metabolism are themselves the signal devices for dilation of blood vessels. As these products accumulate, they cause the blood vessels to increase in size and thus to increase the amount of blood flow in their location. This normally alleviates the problem: the excess metabolic products are transported away and the resulting decrease in their concentration allows the vessels to contract.

Regulation of Blood Flow by Nervous Controls

There are various pathways by which nerves running from the central nervous system to the cardiovascular system control blood pressure. One of the most impor-

---

SUPPLEMENT 11.3
MEASURING BLOOD PRESSURE

Almost all of us have had our blood pressure taken at one time or another. As we have seen, the body's circulation can be analyzed as essentially a pump pushing a fluid through a system of pipes. Blood pressure is the pressure within the arteries that is developed by the pumping action of the heart. A person's blood pressure is expressed as two numbers, one over the other, that tell how strongly blood is going into and out of the heart. The higher number is the systolic or pumping pressure as the heart contracts. The lower number represents the diastolic or filling pressure which occurs when the heart relaxes.

Blood pressure is measured by a device known as a *sphygmomanometer*, which consists of an inflatable rubber tourniquet (called a cuff) attached to a mercury manometer (see Diagram A).

The blood pressure is expressed in terms of the height of a column of mercury, measured in millimeters, at its greatest (during systole, when the heart is contracted) and least (at diastole, when the heart is relaxed). Thus, for example, one's blood pressure is expressed as two numbers, one over the

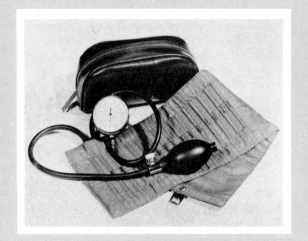

Diagram A
Sphygmomanometer (Courtesy of Turtox, Chicago.)

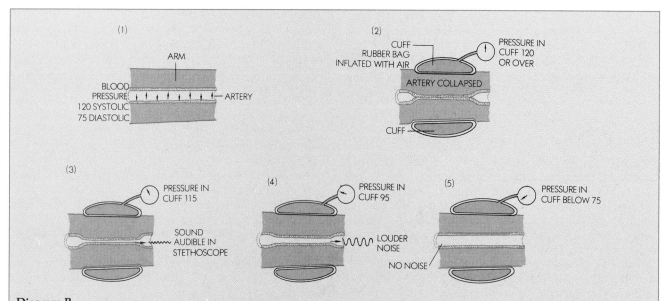

Diagram B
Principle of operation of the sphygmomanometer. (After C. A. Villee, *Biology*, 6th ed. Philadelphia: W. B. Saunders, 1972, p. 357. Reprinted by permission of Holt, Rinehart and Winston.)

other (for instance, 120/80 mm Hg), with the height of the mercury (Hg) column giving the systolic blood pressure at the top and the diastolic blood pressure at the bottom.

The principle on which the sphygmomanometer works is illustrated in Diagram B. Blood normally pushes out on the arterial walls because of the force under which it is being pumped by the heart (part 1). The rubber cuff is wrapped around the arm (usually around the biceps muscle) and it is pumped up, increasing pressure uniformly all around the arm. When the pressure increases beyond a certain point, say 120 mm, the artery is collapsed and no blood can pass through (part 2). The pressure in the cuff is then slowly released until it falls just below 120 mm Hg (or whatever the pressure may be for a particular patient). The artery then remains collapsed except for a short period during systole, when a small amount of blood spurts through (part 3). This

spurt produces a small sound audible in a stethoscope applied to the arm just below where the cuff is attached. When the cuff pressure is reduced to, say, 95 (part 4), more blood passes through and a louder noise can be heard in the stethoscope. When the pressure is reduced to 75, the artery is not collapsed at all and blood flows through continuously, producing no sound. Thus the diastolic pressure is 75.

By listening for the first audible sound as the cuff pressure is lowered, the person taking the blood pressure can determine the point at which the artery is just barely opening up during systole. It is then simply a matter of reading the cuff pressure off the sphygmomanometer (in mm Hg) to determine the patient's systolic blood pressure. The higher the blood pressure, the more cuff pressure must be applied to collapse the artery; the opposite is true for lower blood pressure.

tant involves a simple pathway between the aorta, carotid arteries, and central nervous system. **Stretch receptors** (nerve endings sensitive to stretching) are located in the walls of the aortic branches leading to the brain and in the carotid arteries. When the blood pressure is increased, the walls of the arteries, especially those close to the heart, stretch, stimulating the stretch receptors. A signal travels from the receptors to a reflex center in the lower part of the brain, the medulla. In the medulla, an inhibitory nerve fiber is stimulated, sending impulses to slow the heart down, and to dilate the blood vessels.

Both responses tend to lower the blood pressure. Once this occurs, the aorta and carotid arteries are no longer stretched, reducing or eliminating the impulses sent to the medulla. The medulla, in turn, stops sending out inhibitory messages to the heart, which begins to pick up again. The rate is never perfectly constant from one moment to the next, of course. As we saw in Section 10.2, all homeostatically controlled systems oscillate around a mean. The blood pressure is constantly changing as it is affected by feedback processes operating within the cardiovascular system.

SUPPLEMENT 11.4
HYPERTENSION: A CONTROL SYSTEM GONE AWRY

Our study of animal physiology so far has shown how the various organ systems function individually and in concert to maintain the internal environment of the organism in a steady state, despite constant variations in the external environment. In human terms, this "steady state" might be called health. At times, however, our health is interrupted when our organ systems fail to function normally.

Pathophysiology (from the Greek *pathos*, meaning "suffering") is the study of the abnormal functioning of organ systems, and it holds a central role in the science of medicine. The cause for one disease might be some overwhelming challenge from the environment, such as heatstroke or poisoning, while for another it may be clearly genetic, such as an inherited enzyme deficiency. Usually, however, diseased states, much like the normal steady state, derive from a complex interaction between the internal and external environments of the organism.

Hypertension is the most common life-threatening disease in the United States today. It affects nearly 20 percent of the total population. The name itself refers to the fact that people with this disease have a higher-than-normal blood pressure. In healthy people the circulatory system is adjusted continuously by self-regulatory mechanisms to maintain a level of pressure adequate to deliver oxygenated blood to the various tissues of the body. As a part of the body's overall response to sudden stress or exercise, blood pressure may rise temporarily to higher-than-normal levels. People with hypertension have a blood pressure that is *consistently* above normal, a condition that leads to a number of damaging effects on the organism as a whole.

"Normal" blood pressure is really a statistical concept. In any human population, blood pressures range along a bell-shaped distribution curve similar to the pattern followed by other measured characteristics such as height or weight. Blood pressure is measured by the amount of pressure the blood can exert to hold a column of mercury (Hg) at a particular height, measured in millimeters (see Supplement 11.3). During systole, when the ventricles are contracted, the pressure is greater than at diastole, when they are relaxed. Thus blood pressure is given as two numbers, one over the other. A reading of 140/95 mm Hg, for example (considered to be about a maximum for good health, though it varies according to age), means that systolic pressure supports a mercury column 140 mm high and diastolic 95 mm high. People with pressures above this level have a much higher risk of eventually developing serious complications or dying earlier. For example, a thirty-five-year-old man with a blood pressure only mildly elevated to 150/100 has a life expectancy seventeen years shorter than a man of the same age with a normal pressure. The effects of hypertension can be examined on two levels: (1) in terms of the damage sustained by specific organ systems, and (2) as a statistical impact upon the life expectancy of hypersensitive people.

Damage Due to Hypertension

The pathophysiology of hypertension is seen most strikingly in the heart, brain, and kidneys. As blood pressure rises above normal, the heart must pump against more and more resistance. In particular the left ventricle, the main pumping chamber, must contract more vigorously to push blood through the arterial tubes. As an initial compensatory mechanism, there occurs a progressive thickening of the muscular walls of the left ventricle, enabling it to generate more pressure with contraction. However, under the strain imposed by years of hypertension, the heart weakens and often eventually fails to pump efficiently against the abnormally high arterial resistance. When this occurs, blood backs up from the left ventricle into the lungs, causing pulmonary congestion and difficult breathing, a serious medical condition known as congestive heart failure.

The arteries of the heart are also damaged by prolonged hypertension. The smooth inner lining of the arteries is disrupted and the connective tissue loses its strength and elasticity. The arteries become narrower and less able to deliver blood to the abnormally thickened and strained cardiac muscle (myocardium). Finally the point comes when a coronary artery closes off or is blocked by a blood clot. The part of the myocardium fed by that artery dies for lack of oxygen. The death of a localized area of heart tissue through arterial obstruction is termed a myocardial infarction, or one kind of "heart attack." The frequency and severity of this illness is all too familiar.

The cerebrovascular arteries that supply the brain, as well as the arteries of the kidney, are damaged and narrowed by hypertension. Here, as in the heart, infarction (death) of tissue through obstruction of the arterial blood supply can seriously impair the function of these vital organs. An acute brain infarction is called a stroke. A stroke victim is left with paralysis or loss of sensation in the parts of the body controlled by whatever specific central nerve centers are destroyed when a key cerebrovascular artery becomes blocked. In the kidney, the intricate network of small arteries linked to glomerular filtration is affected by hypertension. This leads to progressive failure in the kidney's ability to filter the blood, excrete wastes, and help maintain homeostasis.

Though hypertension can undermine a person's health in several very important ways, not every person with hypertension necessarily suffers all the possible serious consequences. The disease usually has a long "silent" period of years when the damage to arteries and organs may progress without causing any symptoms, until the damage suddenly surfaces with some catastrophic loss of function such as a stroke. This role as a "silent killer" in part explains why the far-reaching implications of prolonged high blood pressure were not fully appreciated for many years, despite long-standing medical knowledge of its harmful effects on specific organ systems.

Hypertension and Causes of Death

Recently, large-scale studies have established a close connection between hypertension and a number of the leading causes of death and disability. The best known of these is the Framingham study, named for the city in Massachusetts where a carefully selected population of more than 5000 people were followed with regular health examinations for a period of fourteen years. At the start none of the 5000 subjects had any symptoms of either hypertension or organ damage. However, the initial examination showed that a significant percentage had high blood pressure. During the following fourteen years, the part of the population with high blood pressure developed cardiovascular diseases at a much greater rate than the people whose blood pressures were normal at the beginning of the study. Data from this study (see Diagram A) allowed physicians to define the health risks of hypertension precisely for the first time. For example, the figure shows that the risk of stroke is nearly four times greater (184 to 51) in people with hypertension than in people with normal blood pressures. Clearly, then, a person with hypertension can feel well and yet face the probability of serious trouble in the future.

What is the root cause of this terrible health problem, and what can be done about it? Much current research focuses on this question, seeking explanations from genetic, physiological, and environmental lines of investigation. Unfortunately, only in relatively few cases can doctors trace the cause of hypertension to some specific malfunction in the body. In over 95 percent of the cases very little is known about exactly what causes the body's usual control system to go wrong.

One very important fact requiring attention and explanation is that hypertension is much more prevalent in the black population of the United States than in the white, and more prevalent in lower socioeconomic classes of all races than in upper classes. To pursue complete understanding of hypertension, medical researchers must adopt a comprehensive point of view that spans from the internal workings of cell, tissue, and organ to the behavior of the whole human organism in society. Only in this way can they expect to find the mechanisms that may link environmental strains such as racism and poverty to the basic pathophysiology of hypertension.

Despite our ignorance of the causes of hypertension, medical therapy is usually simple and quite effective in lowering

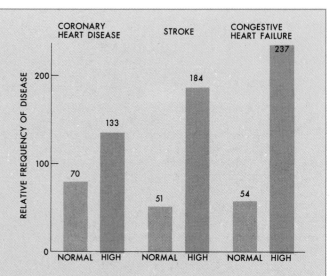

BLOOD PRESSURE STATUS AT INITIAL EXAMINATION

Diagram A
The risk of developing certain cardiovascular diseases over a fourteen-year period according to blood pressure status (normal or high) at initial examination, measured in men aged thirty to sixty-two. The units of the vertical scale and the exact numbers pictured are somewhat arbitrary calculations, but the relative values from one bar to another represent real differences in risk. (After William B. Kannel, et al., "Epidemiologic assessment of the role of blood pressure in stroke: The Framingham Study," *Journal of the American Medical Association* 214, Oct. 12, 1970, pp. 301–310.)

blood pressure and preventing the serious complications and deaths associated with the disease. Since hypertension often produces no warning symptoms for years, only regular medical examinations can alert patients when their blood pressure is elevated. But because many people do not receive regular medical checkups, only about half the people with hypertension know they have the disease. Of these, only half are under adequate treatment. Poor people and those of minority races often have least access to good health care and are thus least likely to receive proper treatment early in the course of the disease. This tragic gap between our capability to detect and treat hypertension and what is actually being done costs tens of thousands of lives each year. The medical profession and society as a whole need to assume with greater determination the task of making the advances of medical research and the benefits of modern medicine available to all people.

11.8
THE RESPIRATORY SYSTEM

In Chapter 6 we saw that respiration—the breakdown of food substances and release of the energy they contain—involves the transfer of electrons from one acceptor molecule to another. The final electron acceptor eventually passes into the surrounding environment, such as the atmosphere or perhaps a liquid culture medium. In anaerobically respiring yeast cells, for example, the final electron acceptor is pyruvic acid. In animals, including humans, the final acceptor is generally oxygen.

Evolution of Respiratory Mechanisms

The primary seat of respiration, of course, is the cell. Therefore, a major evolutionary development was that of respiratory exchange mechanisms capable of supplying enough electron acceptor molecules. For one-celled animals, this problem is solved by simple diffusion, provided that cell size does not exceed certain critical surface-to-volume ratios. Even those forms composed of two layers of cells (such as *Hydra*; see p. 879) can carry out respiratory exchange by diffusion, for the cell surfaces are exposed to a constantly changing environmental medium—in the case of *Hydra*, water.

In animals with more than two cell layers, however, more elaborate respiratory exchange mechanisms have evolved. The earthworm absorbs oxygen and releases carbon dioxide through its skin; the motion of insects aids movement of air to their body tissues through finely divided air tubes; fish exchange oxygen and carbon dioxide through gills; most other vertebrates breathe air into sac-like lungs, bringing about the same exchange.

Despite vast anatomical differences between such external respiratory exchange devices, all contain the

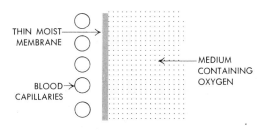

Fig. 11.16
The essential principle of operation for external respiration systems, whether respiration is by tubes, gills, or lungs.

same essential features (Fig. 11.16). In mammals, external respiration involves two main phases. In one, breathing draws oxygen-containing air into close contact with the damp epithelium of the lung, where respiratory exchange occurs. In the other phase, oxygen and carbon dioxide are transported to and from the respiring cells. As we have just seen, this second phase is a primary function of the circulatory system.

A major feature in the evolution of the respiratory system has been an increase in the amount of surface area available for oxygen–carbon dioxide exchange. In fish and other marine animals, the gills may be finely branched to the point of being feathery in appearance. The lungs of progressively complex vertebrates show a corresponding subdivision of inner lung cavities into smaller and smaller chambers. The simple, sac-like lung of the frog, which must be supplemented by cutaneous (or skin) respiratory exchange, is a far cry from the hundreds of thousands of air alveoli found in the mammalian lung. In humans it is estimated that the alveoli provide a respiratory exchange surface fifty times greater than the area of the skin.

SUPPLEMENT 11.5
THE CONTROL OF BREATHING

Anyone who has tried it knows that our rate of breathing can be voluntarily controlled within certain limits. This fact alone is justification for a hypothesis attributing the full responsibility for breathing control to the nervous system.

Neural Control.   We can reason as follows: If breathing is entirely under control of the nervous system, then experimental severing of the nerves supplying the breathing muscles should prevent breathing. This hypothesis is supported by experimental results: the breathing muscles of

an animal in which the appropriate nerves have been severed are immediately paralyzed and unable to function. Obviously, respiratory muscle does not have the cardiac muscle's intrinsic ability to contract.

Several muscles are involved in breathing, of course. This leads us to suspect that somewhere there may be a respiratory control center, either in the brain itself or in the nerves innervating the breathing muscles. Experiments involving the severing of nerve fibers and tracts in various regions pinpoint the medulla of the brain as the location of this control center. Anatomical and physiological experiments indicate that this center sends out periodic impulses along the respiratory nerves to initiate the sequence of events leading to inspiration.

It is reasonable to wonder what causes the cells of the respiratory center to send out these impulses. Preliminary investigation reveals that electrical stimulation of the vagus nerve fibers innervating the lungs inhibits the respiratory center; the breathing movements temporarily cease. Further investigation reveals that the same effect can be achieved by inflating the lungs. This observation leads to a promising hypothesis involving homeostatic feedback. Possibly the inflation of the lungs during normal inspiration causes a stretching or pressure that stimulates the vagus nerve fibers to send inhibitory impulses to the respiratory center. With the cessation of impulses from the respiratory center, muscles controlling inspiration would relax, leading to exhalation. The resulting collapse of the lungs at expiration would cause a cessation of impulses along the vagus nerve to the respiratory center, freeing it to initiate another inspiration-expiration cycle.

This hypothesis is supported. The respiratory center *is* affected by inhibitory impulses coming to the medulla over the vagus nerve. But is this neural control the only one? Why does vigorous activity lead to increased breathing?

An interesting situation throws light on the subject. It is impossible to commit suicide by holding one's breath. Ultimately, the respiratory center takes over involuntary control and breathing resumes. The renewed breathing occurs at an increased rate and depth. It is reasonable to hypothesize that these increases may compensate for the oxygen deprivation that results from holding the breath. This hypothesis is supported when we observe that breathing rate slowly returns to normal, presumably because the oxygen deficiency no longer exists.

Chemical Control.   It is a simple step from such observations to a hypothesis of a chemical basis for the control of breathing, operating in conjunction with nervous control. Either decreased breathing or increased muscular activity (such as in exercise) causes an oxygen deficiency in the blood, as well as an increase in the waste products of muscular activity, primarily carbon dioxide. Either a decrease in oxygen or an increase in carbon dioxide could affect the respiratory center as the blood circulated through it. But which one?

To determine this, we must experimentally increase the concentration of one, while keeping the concentration of the other constant. Oxygen is a promising first choice. Since the result of breathing is bringing oxygen and the body cells into close contact, it is reasonable to hypothesize that the rate of breathing may be related to the degree of oxygen deficiency shown by these cells. Thus as the concentration of oxygen in the blood decreases, the respiratory control center causes the rate of breathing to increase. Conversely, the higher the oxygen content of the blood, the slower the breathing rate. According to this hypothesis, then, there is an inverse relationship between the rate of breathing and the concentration of oxygen in the blood.

If a person breathes into a plastic bag for a few minutes, his or her breathing rate gradually increases. This is consistent with a hypothesis proposing oxygen as the agent whose concentration in the blood affects the breathing rate, for as breathing into the bag continues, the oxygen supply is gradually depleted. This experiment does not adequately test the oxygen hypothesis, however, for the concentration of carbon dioxide in the blood is also increasing; it may be the increase in carbon dioxide, rather than the decrease in oxygen, that affects the breathing rate.

A slight modification of this experiment, however, eliminates the carbon dioxide variable. The air breathed in and out is allowed to pass over calcium hydroxide, which removes the carbon dioxide. Now the oxygen hypothesis can be adequately tested. We can reason that if oxygen concentration in the blood affects breathing rate, we can predict that breathing into a plastic bag containing calcium hydroxide (for removal of CO_2) should cause an increase in breathing rate. The prediction is not verified; no increase in breathing rate occurs.* Evidently the increase in the concentration of carbon dioxide in the blood, rather than any decrease in oxygen concentration, influences the breathing rate, for the rate *is* increased by breathing into the bag when the carbon dioxide is allowed to accumulate. Further supporting evidence comes from experiments in which individuals are provided with above-normal amounts of oxygen, and only a slight excess of carbon dioxide. Since the excess oxygen increases the oxygen concentration of the blood, the hypothesis that there is an inverse relationship between oxygen concentration and breathing rate predicts that the breathing rate will slow down. Instead, it increases. This result is consistent with the alternative hypothesis proposing a *direct*, rather than inverse, relationship between carbon dioxide concentration (or possibly the carbonic acid it forms in the blood) and the rate of breathing.

If an anesthetized animal is given a rapid application of artificial respiration, the carbon dioxide concentration of its blood drops to a point well below normal. The result is a complete cessation of breathing activity. This result extends our original hypothesis considerably. Not only does a high concentration of carbon dioxide in the blood increase breathing; a low concentration effectively blocks it. It is intriguing that the body's physiology places such great importance on a compound which is, in essence, a waste product.

---

*There may be a slight increase. However, it will not occur until the oxygen concentration is lower than it is when no carbon dioxide absorbant is used. It should also be pointed out that any *sudden* drop in oxygen will initiate rapid breathing. We are referring here to the control of breathing rates under normal conditions.

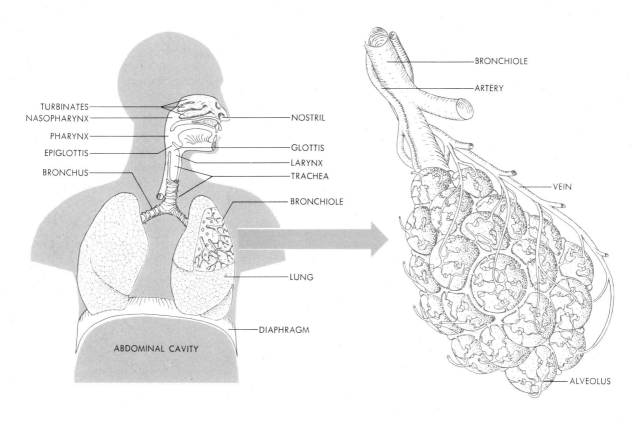

Fig. 11.17
The human respiratory system and its functional units, the alveoli, provide extensive surface area for the exchange of respiratory gases.

The basic anatomy of the human respiratory system is shown in Fig. 11.17. In all mammals, a sheet-like layer of muscle, the **diaphragm,** joins with other tissues to enclose completely the lungs in the chest or thoracic cavity. The movement of the diaphragm downward, combined with upward and outward movements of the ribs, serves to enlarge the thoracic cavity. Since the thoracic cavity is closed, with only one external opening (the trachea), this increase in size results in an inhalation (or **inspiration**) of air from the outside. The lungs, composed of elastic tissue, are stretched by the increased air pressure within them, much like a blown-up balloon. This elasticity results in the elimination of air during exhalation (or **expiration**).

Obviously, not all of the air within the lungs is exhaled at any one expiration. Nor is all the oxygen removed from the air before it leaves the body. The amount of oxygen that *is* removed, however, and the amount of carbon dioxide eliminated, are well within the quantities required by the organism. Should exercise call for additional oxygen, the quantity of air entering the lungs can be increased by a faster breathing rate or deeper inspiration.

Gas Exchange in the Lungs

At first glance, it seems reasonable to equate the absorption of oxygen into the blood in the lungs with the absorption of digested food substances in the small intestine. Here diffusion was only a small part of the story,

with the greater portion of the absorptive process being accomplished by work performed by the intestinal epithelial cells. A histological comparison between the epithelial cells of the intestine and those lining the alveoli throws doubt on this supposition, however. The intestinal epithelial cells are columnar, the epithelial cells of the alveoli, squamous. Furthermore, the intestinal cells are richly supplied with mitochondria; the cells of the alveoli are not. From the general rule in living organisms that structure complements function, we could conclude that the alveoli cells perform little or no work in regard to the movement of oxygen and carbon dioxide across them and hypothesize that only diffusion is involved.

This is a reasonable hypothesis, since the concentration of oxygen in the blood entering the alveoli would be lower than that of the entering air and, conversely, the concentration of carbon dioxide in the same blood would be higher than that in the air. Concerning carbon dioxide removal from the blood to the alveolar air, the alternative active transport hypothesis is further weakened when we note that the concentration of carbon dioxide in the alveolar air is less than in the blood,

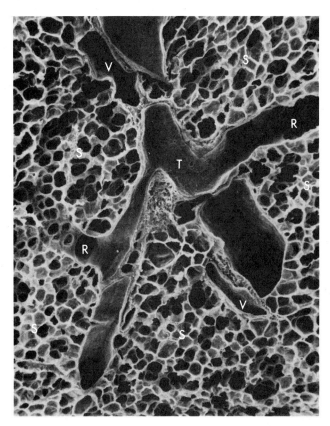

Fig. 11.18
Scanning electron micrograph of rat lung (× 100). Note that as the bronchioles terminate and begin to divide (T), they begin to become respiratory bronchioles (R) as they open into alveolar sacs. Numerous alveolar sacs (S) and a blood vessel (V) are also visible. (Micrograph courtesy Dr. Peter M. Andrews, Georgetown University Department of Anatomy.)

instead of vice versa, as the hypothesis would predict. Indeed, experimentally increasing the concentration of carbon dioxide in the alveolar air actually results in a movement of carbon dioxide molecules *back into the blood*, with a resulting increase in breathing rate. Evidently the alveoli cells can neither concentrate carbon dioxide in the alveoli nor move it against a diffusion gradient. Diffusion must, therefore, be at least primarily responsible for the movement of carbon dioxide molecules from blood to alveoli during normal respiratory activity.

Similar experiments carried out in the capillaries show that diffusion is also responsible for the movement of oxygen from the blood into the body cells and the movement of carbon dioxide in the reverse direction.

Hemoglobin and Respiratory Gas Exchange

Although both oxygen and carbon dioxide will dissolve in water (which constitutes 90 percent of the blood plasma), they will do so only within certain limits. Yet analysis of the oxygen or carbon dioxide content of the blood reveals amounts of these substances far in excess of those limits. Quite obviously, both oxygen and carbon dioxide must be present in the blood in another form in addition to the dissolved state. Chemical analysis of blood provides confirmation of this. The oxygen in blood exists mostly in chemical combination with **hemoglobin**, a protein respiratory pigment, within the red blood cells. This chemical combination is weak enough to allow the hemoglobin to release the bound oxygen to the body cells. Carbon dioxide is found in both the plasma and the red blood cells, mostly in the form of bicarbonate ions (HCO_3^-).

The chemical reaction uniting oxygen and hemoglobin can be expressed as:

$$Hb + 4O_2 \rightleftharpoons Hb(O_2)_4$$
$$\text{HEMOGLOBIN} \quad \text{OXYGEN} \quad \text{OXYHEMOGLOBIN}$$

This reversible reaction can be studied easily outside the body, in a test tube. Hemoglobin is purplish in color; oxyhemoglobin a bright red.

Although it is obviously advantageous to the organism to have hemoglobin combine easily with oxygen in the lungs, while releasing oxygen easily in the capillaries, we might well ask how the reversible reaction behaves so conveniently. Exposing given samples of blood to various concentrations of oxygen reveals that the amount of oxyhemoglobin formed is related, though not directly, to the concentration of oxygen. Further, the amount of oxyhemoglobin present in a blood sample decreases if the sample is placed in air from which most of the oxygen has been removed. We can conclude that the oxygen-hemoglobin reaction, like most biochemical reactions, undergoes equilibrium shifts to either the right or left whenever the concentration of reactants or products is changed. Thus in the lungs, where oxygen concentration is high, the reaction is strongly shifted to the right; that is,

$$Hb + 4O_2 \rightleftharpoons Hb(O_2)_4.$$

In the body tissues, where the oxygen concentration is low, the reaction is strongly shifted to the left; that is,

$$Hb + 4O_2 \rightleftharpoons Hb(O_2)_4$$

and the released oxygen is free to diffuse through the capillary walls into oxygen-poor cells.

Partial Pressure and Tension

The relationships of high oxygen concentration to oxyhemoglobin formation and low oxygen concentration to oxyhemoglobin dissociation may seem quite adequate to explain the transport of oxygen from the lungs to the body tissues. However, a bit of reflection casts doubt on this assumption. The oxygen in the alveolar air is in the form of a gas and diffuses as such. To enter the blood, the oxygen must dissolve and diffuse as a dissolved sub-

Oxygen is transported in the blood almost exclusively by hemoglobin. Carbon dioxide is transported by three mechanisms: by being dissolved in plasma in the gaseous form; by formation of carbonic acid and subsequent dissociation into bicarbonate and hydrogen ions ($HCO_3^{-2} + H^+$); and by attachment to hemoglobin.

stance. In chemical combination with hemoglobin, the oxygen is essentially nondiffusible, for hemoglobin is confined within the red blood cells. Thus it is not sufficient just to consider the percentage concentration of oxygen in cases where the oxygen diffuses from one state to another, such as from a free gas in the alveoli to a dissolved state in the blood plasma.

If the dissolved oxygen content of a body of water is compared with that of the air with which it is in contact, the difference may be as much as 100 to 1, with the air containing the greater amount. Were the physical factors the same in the liquid and gaseous state, equal amounts of oxygen would be expected in both air and water. But the physical factors are *not* the same. There are limits, for example, to the amount of oxygen that can be dissolved in water, limits dependent in turn on such physical factors as temperature and pressure.

To deal with these variable factors effectively, the physiologist calls on the concept of **partial pressures** of a gas in mixture with other gases. The concept of partial pressures can be expressed in the following equation:

$$\begin{matrix} \text{PARTICAL} \\ \text{PRESSURE} \\ \text{GAS } a \end{matrix} = \frac{\begin{matrix}\text{NO. OF PARTICLES} \\ \text{OF GAS } a\end{matrix}}{\begin{matrix}\text{TOTAL NO. OF GAS} \\ \text{PARTICLES IN MIXTURE}\end{matrix}} \times \begin{matrix}\text{TOTAL PRESSURE} \\ \text{OF ALL GASES} \\ \text{IN THE MIXTURE}\end{matrix}$$

Thus the partial pressure of oxygen in the atmosphere at sea level (760 mm Hg) is $1/5 \times 760$ mm Hg (since oxygen composes about one-fifth of the atmosphere), or 152 mm Hg. Partial pressure, then, gives a numerical value for the amount of any given gas in a mixture of gases, such as the amount of oxygen or carbon dioxide in the atmosphere.

If the concept of partial pressure is to be of physiological value, it must be equated with a measure of the same gas that is in solution, such as oxygen dissolved in blood plasma, or in combination in a solution, such as the oxygen in oxyhemoglobin (Fig. 11.19). Such a measure is referred to as the **tension** of a gas. The tension of a given gas in solution is expressed numerically in terms of the partial pressure of the same gas with which it is in equilibrium. Equilibrium in this case refers to the point at which the number of gas molecules changing from the gaseous to the dissolved state is equal to the number changing from the dissolved to the gaseous state. Note that such an equilibrium by no means implies that the concentration of dissolved oxygen is equal to the concentration of oxygen in the gaseous state.

The concepts of partial pressure and tension give us parameters by which the transport of oxygen and carbon dioxide by the blood can be more readily understood. For example, experimental measurements reveal that the actual concentration of oxygen in arterial blood is *greater* than its concentration in the alveolar air. Thus any hypothesis that proposes simple diffusion to explain the entry of oxygen into the blood from the alveoli immediately encounters a contradiction. In this case, oxygen obviously does *not* move from a region of greater to a region of lesser concentration. Indeed, it does just the reverse. The oxygen in the blood *does*, however, decrease in partial pressure, or tension, from alveoli to blood to tissues. Thus the concepts of partial pressures and tension, rather than a concept that considers only concentrations, must be incorporated into any hypothesis that explains the transport of oxygen and carbon dioxide to and from the body tissues. When this is done, the hypothesis leads to experimentally verifiable predictions.

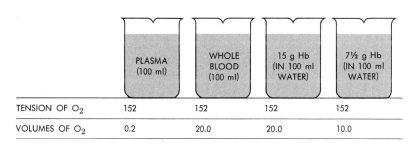

Fig. 11.19
The concepts of partial pressure and tension must be considered in the formulation of any hypothesis dealing with the exchange of oxygen and carbon dioxide in the alveoli. Note the importance of hemoglobin to the amount of oxygen that enters the water.

	PLASMA (100 ml)	WHOLE BLOOD (100 ml)	15 g Hb (IN 100 ml WATER)	7½ g Hb (IN 100 ml WATER)
TENSION OF O_2	152	152	152	152
VOLUMES OF O_2	0.2	20.0	20.0	10.0

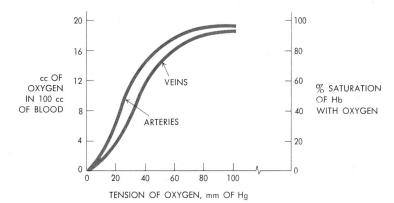

cc OF OXYGEN IN 100 cc OF BLOOD

TENSION OF OXYGEN, mm OF Hg

% SATURATION OF Hb WITH OXYGEN

VEINS

ARTERIES

Fig. 11.20
The oxyhemoglobin dissociation curve. Note that at any given oxygen tension, arterial blood contains a greater volume percentage of oxygen than does venous blood.

Figure 11.20 shows the oxyhemoglobin dissociation curve, which expresses the relation between the volume percentage of oxygen held in loose chemical combination with hemoglobin and the tension of oxygen to which the blood is exposed. Two curves are given, one for arterial blood and one for venous blood. Note that for any given oxygen tension, venous blood combines with oxygen less readily than arterial blood. Experimental investigation reveals that the addition of acid to blood lowers the amount of oxygen it can carry. Thus, yet another factor aids in the dissociation of oxygen from oxyhemoglobin in the capillaries, allowing the oxygen to enter the tissues. Carbon dioxide enters the blood from the tissues in the form of carbonic acid. The increased acidity, in combination with the large difference of oxygen tension that exists between them, provides considerable impetus to the passage of oxygen molecules from blood to tissue.

The transport of carbon dioxide in the blood is different from that of oxygen, but the same factors of partial pressure and tension apply. Some of the carbon dioxide in the blood is simply in a dissolved state, and this is the form in which the gas is released into the alveoli.

However, this is but a small amount of the total carbon dioxide content of the blood; most is in the form of bicarbonate. Considerable quantities of dissolved carbon dioxide are converted by an enzyme (carbonic anhydrase) into carbonic acid. Finally, some of the carbon dioxide travels in combination with plasma proteins or hemoglobin. The reactions bringing about these combinations are reversible. Concentration differences existing in the capillaries or lungs shift these reactions in the appropriate direction, causing them to release the bound carbon dioxide into solution in the blood plasma and, eventually, to the alveolar air for exhalation.

X-ray diffraction and other techniques have now provided good experimental evidence to suggest that the hemoglobin molecule is able to carry out its dual function of carrying oxygen from the lungs to the tissues and aiding in carrying carbon dioxide back to the lungs by switching back and forth between one molecular conformation and another. Recent evidence has also shown that the muscle protein myoglobin, besides acting as an oxygen-storage molecule in muscle cells, helps hemoglobin in transporting oxygen from the capillaries to the muscle cell mitochondria.

Summary

1.   The mammalian heart serves as a pump to keep the blood moving in a circular path throughout the body. Blood passes from the venae cavae into the right atrium, and through the tricuspid valve into the right ventricle. From here it enters the pulmonary arteries and goes to the lungs. The blood returns to the heart from the lungs through the pulmonary veins into the left atrium, and through the mitral valve into the left ventricle. From the left ventricle blood passes through the aortic valve into the aorta, and via various blood vessels to the rest of the body. The right side of the heart causes blood flow to and from the lungs (pulmonary circulation), while the left side causes blood flow to and from the rest of the body (systemic circulation).

2.   The heartbeat is regulated by a combination of the heart's own intrinsic "pacemaker" and its connection to the central nervous system. The pacemaker is localized in two nodes, one on the upper and one on the lower part of the heart.

3.   Besides the heart, the circulatory system is made up of arteries and arterioles carrying blood away from the heart, veins and venules carrying blood to the heart, and capillaries, the very small vessels connecting the arteries and veins. The passage of substances from the blood to the cells and from the cells into the blood occurs through the capillary walls. Blood consists of several components: plasma, erythrocytes, leukocytes, and platelets. The erythrocytes function as hemoglobin containers, and they are the main vehicles for oxygen transport. Leukocytes are involved in the body's defense system. Certain types of leukocytes (lymphocytes) appear to be the production site for antibodies.

Platelets are small cell fragments involved in the formation of blood clots.

4. Exchange of materials between capillaries and the tissues they penetrate is governed by various physiochemical factors. Among these are the osmotic and hydrostatic pressure balances between the fluid within the capillaries and the extracellular fluid outside them.

5. Blood pressure is regulated by several factors: (a) Certain metabolic substances have a direct effect on the muscles of capillary walls, causing dilation. Dilation of the capillary increases circulation, and the metabolic products are carried away. (b) Stretch and pressure receptors in several principal arteries are sensitive to increased pumping of the heart. When stimulated, these receptors cause a center in the medulla to send impulses that slow down the heartbeat. Thus the rate of heartbeat is constantly monitored by various sensing devices that respond to changes in physiological conditions of the blood and extracellular fluid.

6. The lymphatic system has three functions: (a) to collect liquids and various substances from the extracellular fluid and return them to the general circulatory system; (b) to filter out foreign matter, such as bacteria, for phagocytizing by leukocytes, and (c) to serve as the main center for the production of antibodies, special proteins that combine with (and thus remove from circulation) foreign substances in the blood or extracellular fluid.

7. The immune response involves the mobilization of the body defenses against foreign antigens introduced by either bacteria or viruses by producing antibodies. Two kinds of cells, B- and T-lymphocytes, are involved in antibody formation. Immunity to a disease can by attained by either previous infection or vaccination. In the case of virus infections, interferon may play a major role in disease control.

8. The respiratory system is involved in bringing oxygen from the external environment into contact with individual cells. In the air sacs, or alveoli, that make up the spongy tissue of the lung, oxygen is absorbed by hemoglobin-containing red blood cells.

9. Hemoglobin picks up oxygen at a rate dependent upon the partial pressure of oxygen coming in contact with the lung capillaries. At high altitudes, where the partial pressure of oxygen is less, fewer molecules of hemoglobin are saturated with oxygen than at low altitudes. Conversely, dissociation of oxygen from hemoglobin in the tissues is also a function of partial pressure. In the tissues the oxygen pressure is low, since most of it has been used up in cellular respiration; this favors the dissociation of the hemoglobin-oxygen complex. The association of oxygen with hemoglobin is thus a reversible process. In the tissues, hemoglobin that has lost its oxygen picks up carbon dioxide, which is also transported away from the tissues by dissolving in the plasma.

10. The rate of breathing is controlled by nerve centers in the brain (the respiratory center) that are sensitive to changes in pH in the blood. As the body becomes more active, more acidic substances such as lactic acid and carbonic acid are released into the blood. This stimulates the respiratory center, which in turn signals the muscles of the diaphragm more frequently. This process tends to bring CO_2-containing blood to the lungs faster and thus facilitates the intake of oxygen. As the CO_2 concentration in the blood is lowered, the pH returns to normal (about 7.4), and the respiratory center reduces the frequency of its signals to the diaphragm.

Exercises

1. Describe the pathway that blood takes through the mammalian heart.

2. What differences in heart structure are apparent between the frog and a mammal? What is the significance of these differences?

3. All other factors being equal, the degree to which a population of hemoglobin molecules in the blood will become saturated with oxygen depends on the pressure of the gas in the medium to which the hemoglobin is exposed (whether blood in the body or plasma solution in a test tube). The relationship between percent saturation of hemoglobin molecules in a sample and the pressure of oxygen to which the sample is exposed is shown in Fig. 11.21.

Answer the following questions:
   a) Does the curve shown in Fig. 11.21 represent a more adaptive or less adaptive condition as far as the main function of hemoglobin is concerned than a linear relationship would represent? Why?

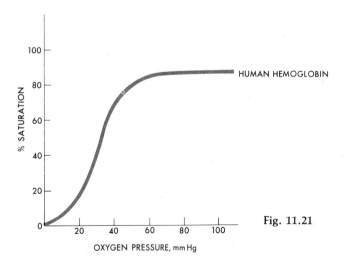

Fig. 11.21

b) Llamas normally live in the Andes at high altitudes, where the oxygen pressure is considerably lower than at sea level. Predict and draw what you think an oxygen-saturation curve for llama hemoglobin would look like. Explain why this difference would help adapt the organism to its specific environment.

Chapter 12
Body Movement:
The Skeleton and the Nature of Muscle Contraction

12.1
INTRODUCTION

The release of energy in an animal is important only insofar as it can be coupled with useful activity. An animal that did nothing more than break down fuel molecules would be like an automobile whose motor is running but whose gears are not engaged. For a living organism, engaging the gears means linking vital energy-releasing processes with the energy-requiring processes associated with life. In this chapter we shall consider the ways in which energy is used in two very important animal activities: movement and coordination.

The ability to move from place to place is a characteristic that has long been associated with the animal kingdom. To move effectively, however, requires a complex system of coordination—a system that has reached its peak of development in the vertebrate body. Neither movement nor coordination could occur if they were not coupled with energy-releasing processes at the molecular level. As we consider the processes of movement and coordination, it will be important to see not only the way in which such coupling is thought to occur, but also the methods by which our understanding of these processes has developed.

12.2
THE SKELETAL SYSTEM AND THE RELATION OF BONE TO MUSCLE

The vertebrate skeleton consists of many bones (in humans, 206), held together by strong fibrous ligaments. The skeleton can be divided into two regions: the **axial skeleton,** consisting of the skull, vertebral column, and thorax (ribs and sternum); and the **appendicular skeleton** consisting of the pectoral girdle (shoulder region), the pelvic girdle (hip region), and the upper and lower limbs (see Fig. 12.1). The skeleton lends support to the soft tissues of the body. It also serves as a protective armor for many of the body's vital organs. The brain, for example, is covered by the skull, the spinal cord by the vertebrae, the heart and lungs by the ribs, and the bladder and reproductive organs by the pelvis. As indicated earlier, the hard portion of the bones

Fig. 12.1
Major bones of the complete human skeleton.

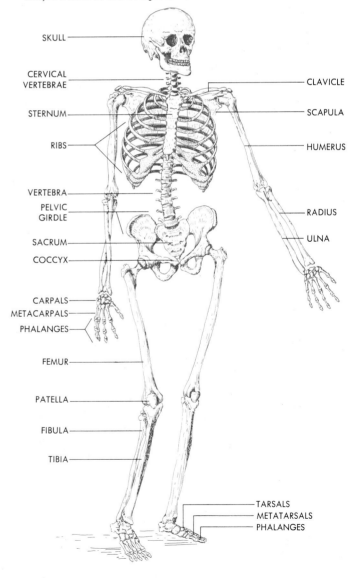

SKULL

CERVICAL VERTEBRAE

STERNUM

RIBS

VERTEBRA

PELVIC GIRDLE

SACRUM

COCCYX

CARPALS

METACARPALS

PHALANGES

FEMUR

PATELLA

FIBULA

TIBIA

CLAVICLE

SCAPULA

HUMERUS

RADIUS

ULNA

TARSALS

METATARSALS

PHALANGES

338

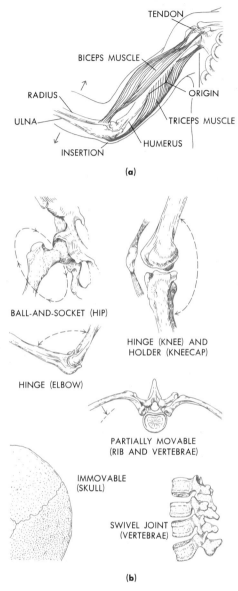

Fig. 12.2
(*a*) The relationship between muscles and bones for two movable joints, the shoulder and elbow. (*b*) Six types of joints and their forms of mobility (or, in one case, immobility).

serves as a mineral reservoir for the body and the marrow for the production of red blood cells in the adult.

The regions where the different bones of the body join, or articulate, with each other are called **joints**. Joints may be classified according to the material that connects them, that is, fibrous connective tissue or cartilage, and the degree and kind of movement allowed by these joints. There are several types of joints in the skeletal system, each specialized for a particular task.

The ribs are attached to the thoracic vertebra by partially movable joints. These joints allow the limited motion necessary for breathing. Ball-and-socket joints, permitting free motion in almost any direction, are found in the hip and upper shoulder. Hinge joints, which permit free swinging movement in one main direction, exist at the elbow and knee. Finally, immovable joints are found in such places as the skull, where rigidity is necessary in order to protect the brain (see Fig. 12.2).

The skeletal system's cooperation with the muscular system to produce coordinated movement is another function. Few muscle contractions, if any, produce such movement directly. Most act indirectly by affecting another body part. A muscle unattached to such a part or parts moves nothing but itself. If useful body movement of the organism is to result, then, the muscle must produce a force upon an object. The jellyfish, for example, contracts muscle fibers in its bell. This action squeezes water out underneath the bell, thus moving the animal in the opposite direction by jet propulsion. An insect moves by means of the force exerted by its muscles against the inner surfaces of its hard outer body covering or exoskeleton.

Jellyfish and insects, however, are invertebrates, animals not possessing bones. In vertebrate organisms, including humans, the skeletal muscles' action on bones, under the control of the nervous system, produces coordinated movement. The muscles are attached to bones by means of tendons, composed largely of white fibrous connective tissue. One end of a tendon is attached to the surface of the bone, the other end to a muscle. The other end of the muscle is attached by a tendon to another bone. In order for an organism to produce movement in one direction, it must somehow exert a force against the surrounding medium (ground, water, air) in the other direction. This can only be accomplished if the muscles are anchored by the tendon to a fairly immovable bone at one end and a movable bone at the other. The anchoring point on the fairly immovable bone is called the muscle's **origin**. The attachment point on the movable bone is called the muscle's **insertion**. Contraction of the biceps muscle, for example, whose origin via its tendon is at two points (the term "biceps" means "two heads") on a shoulder bone, the scapula, results in movement of the lower arm bone, the radius, on which it is inserted. The result is to bend the arm at the elbow. A watery fluid secreted between the joints serves as a lubricant to reduce friction between the bones when they rub against each other. Thus the amount of force that must be exerted by the contracting muscle is considerably lessened.

Muscles can exert a force only by contracting—that is, by pulling. They cannot push. Thus for every muscle whose action pulls a bone in one direction there must be another muscle, attached to the same bone, that works *antagonistically* to the first muscle by pulling the bone in the other direction. When one muscle is contracted, the muscle antagonistic to it must be relaxed. The strength exerted by some muscles, should they accidentally pull full force at the same time, could injure the bone or the muscle. This implies that some system must coordinate muscle activity, causing antagonistic muscles to contract or relax at the appropriate time. The system responsible for this coordination is the nervous system.

12.3
NERVES AND MUSCLES: ELECTROCHEMICAL POTENTIAL

All cell membranes of any sort are to some degree **polarized;** that is, there is a difference in electric potential on one side of the membrane compared to the other. In most cases, the outside of the cell is positively charged in relation to the inside. The difference in potential is due to an unequal distribution of ions—more of certain positively charged ions are on the outside than on the inside. This difference in potential results from the fact that the membrane is differentially permeable to certain ions. For example, the membranes of most cells are slightly permeable to both sodium and potassium ions, but are generally more permeable to the latter. Since there are more potassium ions on the inside, their tendency to leave in greater numbers makes the inside negative. In addition, most cells actively transport ions to the outside, thus setting up a definite difference in ionic concentrations across the cell membrane. This difference in concentration on the two sides of a cell membrane establishes an *electrochemical potential.*

As has been discussed before, whenever a difference in potential exists between two sides of a system, that system is capable of doing work. Doing work involves lowering the difference in potential; as water runs over the dam to turn a generator, or as one kind of ion moves from higher to lower concentration across a membrane, potential energy is lost. In both cases, an expenditure of energy is required to restore the potential. For the dam, water from lower levels must be pumped back up to the other side (or moved back up by the sun's energy through evaporation and precipitation); for the cell, certain ions must be moved out of the cell by active transport.

Thus whenever a cell membrane is altered so as to allow ions that are normally kept out to pass through, the potential is lowered. This means that the concentration of certain ions is more nearly equal on both sides. When the potential drops to zero, the membrane is said to be **depolarized.** The polarized state and the depolarized state of a cell membrane are both shown in Fig. 12.3. Because the potential across cell membranes is associated with the distribution of ions, it is said to be electrochemical in nature. The manner in which nerve and muscle cells put their electrochemical potential to work is their most unique evolutionary adaptation.

12.4
THE STRUCTURE OF MUSCLE TISSUE

Perhaps nowhere in the animal world is the relation between structure and function more apparent than in the study of the molecular basis of muscle contraction. A cross section through a whole muscle (such as the human biceps) shows that it is composed of many individual muscle cells or **muscle fibers** (Fig. 12.4). These fibers are arranged parallel to each other and extend from one end of the muscle to the other. Each fiber possesses a cell membrane, called the *sarcolemma,* from which transverse or *t*-tubules penetrate into the inner regions of the cell or fiber.

A longitudinal section through the same muscle shows the parallel fibers with alternating light and dark

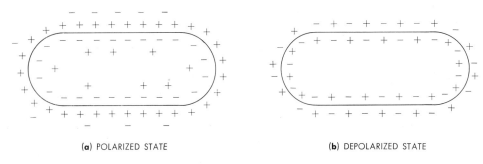

(a) POLARIZED STATE   **(b)** DEPOLARIZED STATE

Fig. 12.3
As shown in (*a*), the outside of a cell is generally positively charged in relation to the inside. This difference can usually be measured and given a value in volts (or millivolts when dealing with small potentials). In (*b*), the membrane potential has fallen to zero; the membrane is thus said to be depolarized. Depolarization occurs when the membrane allows the passage of ions to which it is normally impermeable. In the depolarized state, the distribution of positive and negative ions is the same on each side of the membrane.

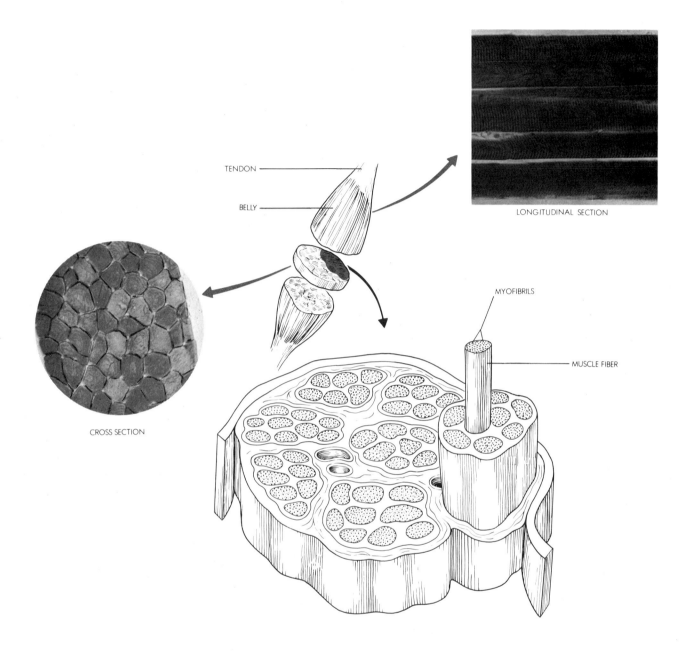

TENDON

BELLY

LONGITUDINAL SECTION

CROSS SECTION

MYOFIBRILS

MUSCLE FIBER

bands. These bands, known as striations, are characteristic of skeletal and cardiac muscle tissue (see Section 5.19).

Skeletal muscles require stimulation from the nervous system in order to contract. Each muscle fiber is supplied by a single nerve cell or neuron called a **motor (movement) neuron.** The nerve cell endings closely approach the muscle cell membranes at several specialized regions, thereby helping to ensure that the entire muscle fiber receives signals from the neuron. These regions of close contact between neurons and muscle fibers are called **neuromuscular junctions.**

The individual muscle fiber is the fundamental unit of structure in the intact muscle. It is not, however, the

Fig. 12.4
The anatomy of skeletal muscle as may be seen by the naked eye and the light microscope. Note the striations clearly visible in the longitudinal section. (Micrograph courtesy Professor H. E. Huxley; drawing adapted from A. F. Spence and E. B. Mason, *Human Anatomy and Physiology.* Menlo Park, Calif.: Benjamin/Cummings, 1979, p. 183.)

fundamental unit of contraction. Each nerve leading to the skeletal muscles branches out so that a number of muscle fibers receive stimuli from the same neuron. This minimum unit of contraction—the group of fibers stimulated by impulses from one neuron—is known as a

Response of virtually all cells to environmental stimuli is the result of a change in membrane polarity.

motor unit. The fewer the fibers stimulated by each neuron, the more precisely movement can be controlled. Thus for the muscles controlling eye movements, over which we have very precise control, a single neuron may lead to only six or ten muscle fibers. On the other hand, for muscles such as the biceps, over which we have less precise control, a single neuron may trigger several hundred muscle fibers.

Observation shows that although the whole muscle is capable of many degrees of contraction, individual muscle fibers show an all-or-none response (Fig. 12.5). Any fiber will respond only to stimuli that exceed a certain **threshold** of intensity. Above this threshold, the fiber responds by full contraction, regardless of how far above the threshold the strength of the stimulus may have been. The fine levels of adjustment displayed by whole muscles result from the different numbers of motor units that may be placed in operation at a given time. The number of units in operation is determined by the number of nerve endings conveying an impulse. In a whole animal, therefore degree of muscular response is directly proportional to the number of motor units activated by the nervous system. It is also directly proportional to the frequency of excitation of these units. Thus two relatively independent means for gradation of the degree of muscular contraction are available to the organism.

Fig. 12.5
Comparison of patterns of contraction (*a*) for a single muscle fiber, or motor unit, and (*b*) for a whole muscle. The former shows the all-or-none response, while the latter shows gradual increase in contraction compared to gradual increase in strength of stimulus.

Each muscle fiber is surrounded by a membrane, across which an electrochemical potential exists. The outside of this membrane is positively charged with respect to the inside. A nerve impulse causes the membrane to become depolarized, and an *action potential* sweeps the length of the entire fiber. Reduction in membrane potential causes the muscle fiber to contract.

The all-or-nothing response of a single muscle cell (we will also encounter a version of this phenomenon in relation to nerve cells) is a good example of the philosophical position that "quantitative changes lead to qualitative changes." This principle states that the accumulation of small quantitative changes in any system can lead ultimately to qualitative changes in the system as a whole. A relaxed muscle is qualitatively different from a contracted one. The two represent very different physical states of muscle tissue. As the stimulus is increased below the threshold, quantitative changes are taking place (increased numbers of motor units are contracting), but qualitative change (overall contraction, a change in state of the muscle) has not occurred. Above the threshold, enough motor units are contracting to produce a change in the muscle's qualitative state: it contracts as a whole. The principle that small changes in quantity can bring about a wholly new state of affairs (a qualitative change) over a period of time is important because it appears to be generally applicable to all biological phenomena.

In the past, the question of how a muscle contracts has been approached from two quite different directions. One approach is biochemical; it depends on the analysis of chemical changes in the muscle as a result of contraction. The other is biophysical; it relies on the use of such techniques as X-ray diffraction and electron

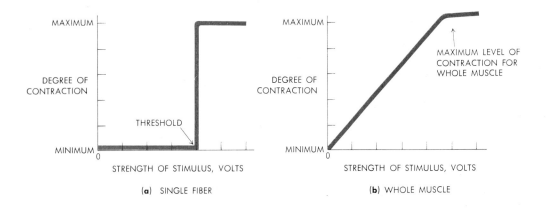

(**a**) SINGLE FIBER   (**b**) WHOLE MUSCLE

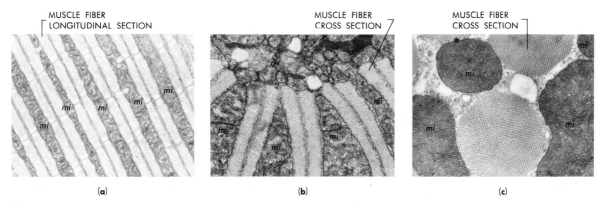

Fig. 12.6
The large number of mitochondria (mi) between these muscle fibers attests to the high degree of activity these particular fibers exhibit. These electron micrographs are (a) longitudinal section through the flight muscle of the damselfly, $\times$ 4700; (b) cross section through damselfly flight muscle, $\times$ 9900; and (c) cross section through the flight muscle of an aphid, $\times$ 11,250. (Photos courtesy D. S. Smith, University of Miami.)

microscopy to study changes in the anatomy of muscle fibers during contraction.

12.5
MUSCLE CONTRACTION:
THE BIOCHEMICAL APPROACH

The first step in unraveling the problem of muscle contraction involves determining what substances disappear during the process and what substances appear. It can be shown that glycogen, a phosphate compound known as **creatine phosphate**, and ATP are consumed by active muscle preparations,* while carbon dioxide, lactic acid, and inorganic phosphate increase in quantity during muscle contraction. On the basis of this informa-

---

*A muscle "preparation" usually consists of a muscle isolated from an organism and placed in a special chamber where it is bathed in a physiological salt solution. Measurements can be made on the substances used up or given off by the muscle's contraction.

tion, a general "equation" can tentatively be established:

These are used up
$$\overbrace{\text{GLYCOGEN, O}_2\text{, CREATINE PHOSPHATE, ATP}}$$

These are produced
$$\overbrace{\text{CO}_2\text{, C}_3\text{H}_6\text{O}_3\text{, P}_i\text{, CREATINE, ADP}}$$

where $C_3H_6O_3$ is lactic acid and P_i inorganic phosphate.

Experimental Observations

In attempting to piece together a picture of the biochemistry of contraction, muscle physiologists had access to the following experimental observations:

1.   It had been observed that a muscle placed in an atmosphere rich in oxygen contracted several hundred times before becoming fatigued. A hypothesis proposing that oxygen is essential for contraction was con-

---

In the contraction of muscle, small increases in strength of stimulus increase quantitatively the number of motor units contracting. At the threshold value, enough quantitative changes have occurred to produce a qualitative change: the contraction of the muscle as a whole.

tradicted, however; a muscle placed in an oxygen-free atmosphere (such as pure nitrogen) continued to contract up to eighty times before fatigue.

2. In the 1940s biochemist Albert Szent-Gyorgyi found that when a solution containing ATP and appropriate ions was added, muscle fibers extracted from whole muscles would contract and the striations would move closer together. In another experiment Szent-Gyorgyi showed the proteins actin and myosin to be responsible for contraction in the whole muscle.

3. Another experiment showed that even when the conversion of glycogen to lactic acid is blocked by a specific inhibitor (iodoacetate), the muscle is still able to contract. Such a "poisoned" muscle, however, is capable of only one-third the number of contractions of a normal muscle.

It is possible to draw some conclusions from this information. ATP, rather than the oxidative processes associated with the breakdown of glucose, seems to be the direct source of energy for muscle contraction. It is reasonable to infer from this that the further oxidation of glucose is responsible for regenerating the ATP consumed during muscle contraction. The consumption of oxygen is thus required to rebuild the exhausted supply of ATP.

This is not the complete picture, however. The compound creatine phosphate is found in plentiful supply in muscle tissue. It seems to be used up during contraction, but what is its exact role?

The following information summarizes the observations concerning probable energy sources of muscle contraction based on studies of extracted muscle preparations where there is no permeability barrier. (There is no way to supply ATP to an intact muscle except by the metabolism of foodstuffs.)

1. A muscle deprived of both ATP and creatine phosphate will not contract.

2. A muscle deprived of creatine phosphate but supplied with ATP will continue to contract until the ATP is used up.

3. A muscle supplied with only creatine phosphate will not contract.

4. A muscle deprived of ATP but supplied with creatine phosphate and ADP will contract. Chemical analysis shows that ATP has been synthesized. When the supply of either ADP or creatine phosphate has been exhausted, the muscle ceases to contract.

5. Calcium ions (Ca^{++}) are essential for contraction to occur. In the absence of calcium ions, a muscle will not contract even if stimulated. Analysis of such muscles shows that in the absence of calcium ions no

ATP is hydrolyzed to ADP and inorganic phosphate (see Supplement 12.1).

Some Conclusions

What conclusions can be drawn from all these data? It seems clear that although ATP provides the direct energy currency for contraction, creatine phosphate acts as a reserve of high-energy phosphate groups. When the ATP supply runs low, high-energy phosphate groups are transferred to ADP, thus regenerating ATP. When the muscle is resting and ATP is being re-formed by fermentation and glycolysis, phosphate groups are transferred back to creatine, synthesizing creatine phosphate. Calcium ions must play a role in the processes leading to splitting of ATP.

The relationship between these processes is summarized in Fig. 12.7. An equilibrium must exist between the high-energy phosphate "pool" of ATP and that of creatine phosphate. The direction of the reaction thus shifts in whichever direction the demand lies.

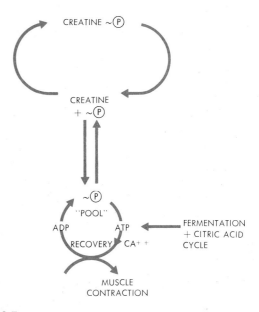

Fig. 12.7
The relation between creatine phosphate and ATP in the energy provision for muscle contraction. Creatine phosphate provides energy-rich phosphate groups to reconvert ADP, produced in muscle contraction, to ATP. High-energy phosphate bonds are ultimately generated by fermentation and the citric acid cycle. When the ATP supply in a cell is high, the equilibrium of reaction is so shifted that the ~℗ are transferred from ATP to creatine. When the ATP supply runs low, the equilibrium shifts back toward the transfer of ~℗ from creatine phosphate to ADP. Calcium ions are necessary for the utilization of ATP in contraction.

SUPPLEMENT 12.1
THE KEY TO CALCIUM—CALMODULIN

Of all the ions that have already been mentioned—or will be—in this book, none are more versatile or wide-ranging in their effects than those of the metal calcium. For years children have been told by parents that they need to drink their milk, rich in calcium, for strong bones and teeth. True as this may be, it is by no means the entire calcium story. Indeed, calcium is there in an individual's life from the very start —the surface of the fertilized egg responds to the sperm by releasing calcium into its interior.

As we have seen in this chapter, a muscle cell responds to a nerve signal by releasing calcium ions, thereby bringing about contraction. But there is still more; calcium is involved in cytoplasmic streaming and cell movement, chromosome movement, neurotransmitter release, endo- and exocytosis, axonal flow in nerve cells, and hormone release in the adrenal glands. Finally, calcium ions play a role in enzyme regulation in the synthesis and degradation in cyclic AMP, a molecule that often acts as the messenger relaying signals from outside the cell to its molecular machinery inside.

In the 1960s, Dr. Wai Yiu Cheung of the University of Tennessee discovered calmodulin, a protein now known to be the molecular link to calcium's versatility, in brain tissue. In turn, the key to calmodulin's function is the way its molecular shape and symmetry is affected when calcium ions bind to it. There are actually four different binding sites on the calmodulin molecule, and the distortion of its shape caused by the calcium ions binding enables it to bind a second molecule, usually an enzyme, which is thereby activated. The enzyme's activation in turn catalyzes the activation of the final target molecule. In the case of the contraction of voluntary (skeletal) muscle, this target molecule is myosin. Calmodulin operates in a similar way in the contraction of involuntary (smooth) muscle cells. As indicated earlier, however, calmodulin's activity is not limited to muscle cells; all kinds of cell movement, from amoebic to the pinching inward (cleavage) of the cell that occurs during division, involve calmodulin. Because of calmodulin, some investigators rank calcium ions, along with hormones and the cyclic nucleotides, as one of the three most important regulators or messengers in mammalian systems. Further, the activities of these three are closely interrelated; the biochemical effects and metabolism of one invariably affects those of the others.

Calmodulin has been identified in the tissues of widely differing animal species and in plants as well. Generally speaking, molecules with important and universal functions and found in cells from widely different sources are strikingly similar, and calmodulin is no exception. This suggests that calmodulin evolved early in the history of eukaryotic cells.

Since very little ATP is actually stored within the muscle itself, a sudden spurt of muscular activity might quickly exhaust the supply. This is where creatine phosphate comes into play; it transfers its high-energy phosphate groups to ADP. This transfer allows the muscle to work at its maximum rate for about 15 seconds.

In very strenuous exercise, however, 15 seconds is not long enough for the breathing and circulatory rates of the body to adapt to the oxygen needs of the tissues. To meet this demand, muscle cells produce ATP by fermentation. This fermentation yields lactic acid, with a net gain of two ATPs per glucose unit in the glycogen molecule. Fermentation represents a poor gain when the total amount of potential energy contained in a glycogen molecule is considered; although fermentation enables the muscle to keep going for a while, the glycogen supply eventually runs low, and lactic acid accumulates in the muscle cells and the extracellular fluid. The accumulation of lactic acid in the muscles produces fatigue.

Muscles are amazingly efficient in the extraction of energy from glycogen. One-fifth of the lactic acid produced is fully oxidized to CO_2 and water, producing enough ATP to reformulate the remaining four-fifths of the partially oxidized lactic acid back into glucose or glycogen.

Fatigue is the result of several interacting factors. One is that the accumulation of lactic acid lowers the pH of the muscle fibers, reducing their ability to contract. A second is that during prolonged and strenuous exercise, the muscle builds up an "oxygen debt." The amount of the debt represents the quantity of oxygen required to oxidize the accumulated lactic acid to carbon dioxide and water. When the muscle cell is very active, it keeps contracting on borrowed energy, that is, on the incomplete breakdown of glycogen. If the cell is to continue functioning it must repay this debt as soon as the body can supply the necessary amount of oxygen. This arrangement has obvious survival value to the organism. It allows the muscles to continue functioning at a moment of stress, even though their oxygen needs cannot be met by the body at that time. It also provides a means by which the reserves can be restored during periods of less strenuous activity.

The lactic acid that has accumulated in the muscles during exercise diffuses into the blood and is carried to the liver. Here, one-fifth of the lactic acid is completely oxidized to water and carbon dioxide, yielding ATP. The ATP is then used to resynthesize glycogen from the remaining lactic acid. Glycogen produced in the liver is transported back to the muscles as glucose, where it is re-formed into glycogen by dehydration synthesis.

Using one-fifth of the lactic acid to resynthesize the remaining four-fifths back into usable fuel is a remarkably efficient process for conserving energy. It would be like using a small amount of gasoline to re-form the ex-haust materials from a car back into usable fuel. The available evidence indicates that muscle contraction has a general efficiency of about 45 percent.

12.6
MUSCLE CONTRACTION: THE BIOPHYSICAL APPROACH

The discussion of Section 12.5 focused on one aspect of the problem of muscle contraction: the nature of the chemical reactions involved. There remains the problem of how the two proteins actin and myosin actually produce a contraction within the muscle fibers. The electron microscope has been of immense value in studying the molecular structure of the muscle and has provided a clue as to how actin and myosin might function to produce contraction.

Microscopic Observations

If a single muscle fiber is viewed in cross section under the electron microscope, it is seen to consist of many **myofibrils** (see Fig. 12.8). Each myofibril consists of alternating thick and thin filaments, identified as myosin and actin, respectively. Figure 12.8(d) shows the arrangement of actin and myosin as seen in a highly magnified cross section.

It was once thought that actin and myosin molecules might produce contraction by coiling or folding.

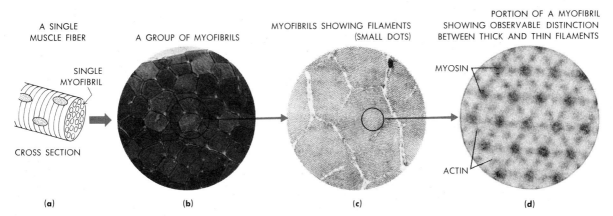

Fig. 12.8
Successively lower levels of organization in vertebrate muscle structure, seen in cross section. (a) Diagrammatic representation of a single muscle fiber, showing myofibrils in cross section. (b) A group of myofibrils. (c) The same view in more detail. (d) The arrangement of thick (myosin) and thin (actin) filaments can be seen in considerable detail in this electron micrograph. [(b) courtesy J. W. Kimball; (c) courtesy Professor H. E. Huxley; (d) from *Revue Canadienne de Biologie* 21 [1962], pp. 279–301, courtesy D. S. Smith, University of Miami.]

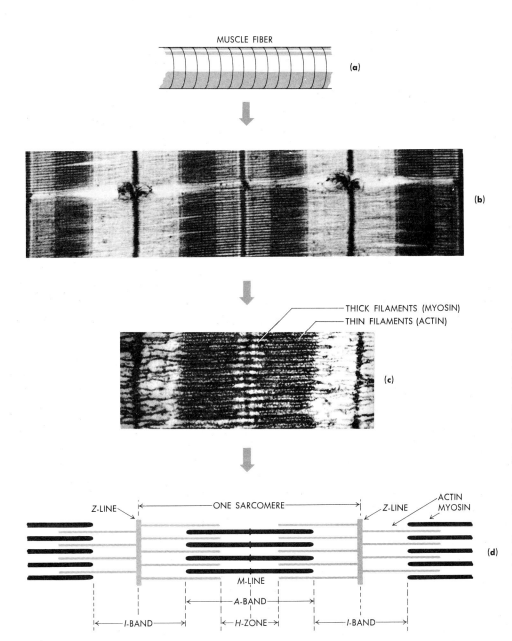

Fig. 12.9
Successively lower levels of muscle organization shown in longitudinal section. (*a*) A single muscle fiber with striations. (*b*) The alternating bands of light and dark that create the striations. (*c*) Electron microscope micrograph showing how the overlapping of thick and thin filaments produces the banding. (*d*) Schematic representation of the arrangement of actin and myosin in one sarcomere (the simplest molecular unit of contraction). Various regions of the sarcomere are labeled (*Z*-line, *A*-band, and so on) for identification. (Micrographs courtesy Professor H. E. Huxley.)

Electron micrographs, however, suggest a more likely theory. Figure 12.9(c) shows a longitudinal section through a muscle fiber. Note that the dark region in the center (a striation) is caused by the overlapping of the thin filaments of actin and the thick filaments of myosin. This arrangement is shown diagrammatically in Fig. 12.9(d).

H. E. Huxley and his co-workers have observed both relaxed and contracted muscle tissue under the electron microscope. Successive stages, from relaxed to contracted, are shown in Fig. 12.10. Several important observations can be made from the photographs. It seems that during contraction:

1. The *H*-zone disappears.

2. A new dense zone appears in the center of the *A*-band where the *M*-line was normally seen.

3. The distance between consecutive *Z*-lines becomes less.

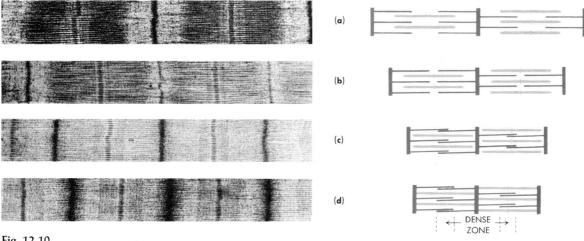

Fig. 12.10
Electron micrographs of contracting muscle and diagrams of
corresponding filament movements. (Micrographs courtesy
Professor H. E. Huxley.)

Conclusions and Questions

All these observations suggest the following scheme for
muscle contraction. When a muscle is stimulated by a
nerve or an electric current, a structure that runs
throughout the muscle fiber, the **sarcoplasmic reticulum**
(see Fig. 12.11), releases calcium ions that flow into the
muscle fibers. In the presence of these ions ATP, which
is bound to myosin, is hydrolyzed to release energy.
Molecules of actin and myosin interact in such a way
that the thin filaments slide over the thick filaments. As
the thin filaments come together the H-zone disappears.
The thin filaments continue sliding until their ends overlap, producing the dense zone in the center of the A-
band. As a result of this sliding, the distance between Z-
bands is greatly reduced. Recovery from contraction
occurs when new molecules of ATP are bound to myosin. The myosin and actin slide apart and the muscle
fiber relaxes. ATP is thus involved in both contraction
and relaxation of muscles.*

Cross sections of relaxed muscle in the region of the
A-band where thick and thin filaments overlap show
that each myosin filament is surrounded by six actin
filaments (Fig. 12.10). The sliding-filament hypothesis

predicts that a cross section through the dense zone of
contracted muscle (Fig. 12.10) should show approximately twice as many actin filaments as were present
before contraction. An electron micrograph of such a
section verifies the prediction (Fig. 12.12). The sliding-
filament hypothesis appears to account well for the
events of contraction.

The question still remains as to the precise nature of
the mechanism by which the thick and thin filaments
slide along each other. Several years ago Huxley and his
co-workers suggested that intermolecular "bridges" exist
between filaments of actin and myosin. They hypothesized a ratchet-like arrangement in which the filaments
were moved across each other much as the teeth of one
set of gears move another. At the time, however, it was
not possible to see such connections. More recently, a
technique known as negative staining has made it possible to get very clear pictures of ultrathin sections of
muscle. Photographs made of muscle treated in this way
show that cross-bridges between actin and myosin filaments do exist in uncontracted muscles and, indeed,
these cross-bridges are an integral part of myosin molecules. Each myosin molecule is known to be composed
of two identical subunits, each shaped rather like a golf
club. The handles or "shafts" of each club are wound
tightly around each other, giving the impression of a
single shaft with two club heads, side-by-side. The H-
zone, then, is composed of an area in which only the
shafts of the many myosin molecules making up a thick
filament are found; the region called the A-band is
where the several doubled club heads of the myosin

---

*This is why muscles show rigor mortis on death of the organism. As
ATP becomes unavailable, the muscles can neither relax nor contract
and the muscle fibers become locked into position, stiffening the
whole body of the organism.

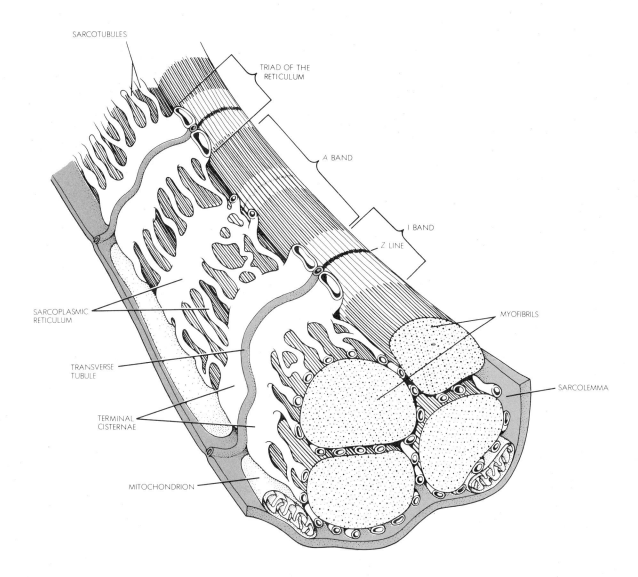

SARCOTUBULES

TRIAD OF THE RETICULUM

A BAND

I BAND

Z LINE

MYOFIBRILS

SARCOPLASMIC RETICULUM

TRANSVERSE TUBULE

TERMINAL CISTERNAE

SARCOLEMMA

MITOCHONDRION

Fig. 12.11
The relationship of the sarcoplasmic reticulum to the myofibrils of muscle cells. From the sarcoplasmic reticulum, calcium ions are released into the cells on stimulation by a nerve or electric current. (From A. P. Spence and E. B. Mason, *Human Anatomy and Physiology*. Menlo Park, Calif.: Benjamin/Cummings, 1979, p. 188.)

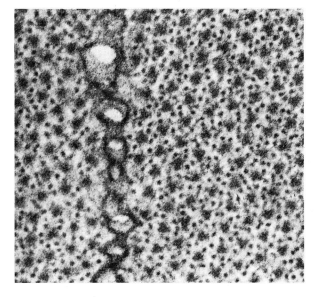

Fig. 12.12
Electron micrograph of contracted muscle cut in cross section through the dense region. As predicted by the sliding-filament hypothesis, the number of thin filaments surrounding each thick filament is approximately twice as great as in relaxed muscle. (Micrograph courtesy Professor H. E. Huxley.)

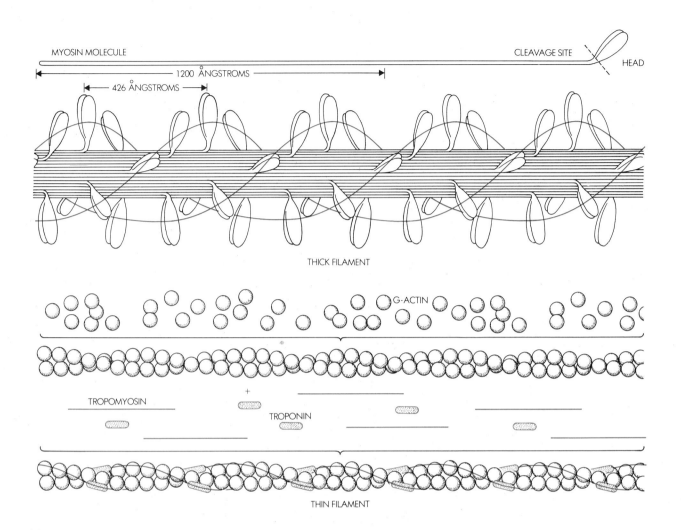

MYOSIN MOLECULE

CLEAVAGE SITE

HEAD

← 1200 ÅNGSTROMS →

← 426 ÅNGSTROMS →

THICK FILAMENT

G-ACTIN

TROPOMYOSIN

+

TROPONIN

THIN FILAMENT

Fig. 12.13
Highly diagrammatic view of myosin and actin molecules. The myosin molecule is rod-shaped and has a head composed of two portions separated by a cleft. In an intact muscle, several hundred myosin molecules are arranged in a sheath to form one thick filament. During contraction, the projecting heads form a myosin-ATP intermediate. The colored line shows how the myosin heads are arranged in a spiral around the thick filaments. The spherical actin molecules (called *G*-actin) form a twisted strand (double helix), with as many as 300 to 400 molecules per single actin myofilament. Each thin filament thus formed is about one micrometer long. Two regulatory proteins, tropomyosin and troponin, are intimately associated with the thin filaments of actin. Tropomyosin is a long (about 400 Å) slender molecule that spirals along the thin filament in contact with about seven *G*-actin molecules. Note that each of the two actin molecules forming the helix has its own tropomyosin filament that partially blocks access to its outer surface, a factor that becomes important during muscle contraction (see Fig. 12.14). The troponin molecules are shorter and thicker than those of tropomyosin and are attached to the latter at their ends.

molecules protrude to form the cross-bridges (see Figs. 12.13, 12.14; legend for Fig. 12.14 appears on p. 352).

Based upon this model and solid experimental evidence that the head regions of the myosin molecules exhibit both actin binding and ATPase activity, the sliding of the filaments across each other that results in muscle contraction is seen as being due to the myosin projections "hooking on" to the actin filaments, thereby temporarily forming an **actomyosin** complex. The result is a conformational change in the myosin molecules, leading to a bending and swiveling of the myosin projections that drag the actin molecules connected to them along. The cross-bridges then break free and quickly reattach to other points on the actin molecules, leading to another bending, swiveling, and dragging-along process. It is this ratchet-like mechanism, repeated many times by each myosin molecule on its actin neighbor, that is seen as causing muscular contraction. The extraction of actomyosin from smooth muscle and the discovery of filaments within it indicate that, despite its lack of

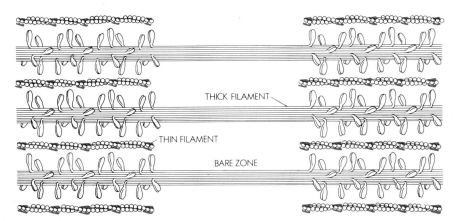

RELAXED MUSCLE

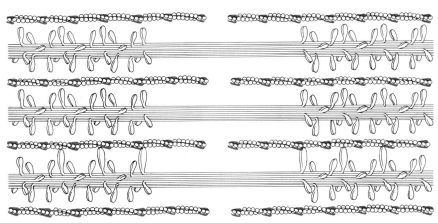

CONTRACTED MUSCLE

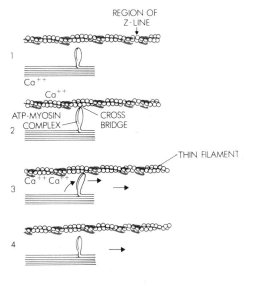

◄ **Fig. 12.14**
Highly diagrammatic representation of muscle contraction at the molecular level. While only a longitudinal sectional view is presented here, remember that each thick filament is surrounded by the several thin filaments with which it makes contact during contraction. Note that on each side of the bare zone (I-band and Z-line), the myosin molecules are arranged in the thick filament with opposite orientation of their doubled heads. The actin molecules are polar, and this polarity is reversed in the strand on either side of the bare zone. Contraction begins when calcium ions released by the sarcoplasmic reticulum in response to a nerve impulse to the sarcolemma of the muscle fiber (see Fig. 12.11) reach the myofilaments. Here the calcium ions bind with the troponin of the thin filaments (see Fig. 12.13). It is thought that this action results in the shifting of the tropomyosin strands from a position blocking an interaction between the myosin of the thick filaments and the actin of the thin filaments. The myosin heads, which have united with ATP to form a charged myosin-ATP intermediate, form temporary cross bridges with the actin-containing thin filaments and then bend and swivel, pulling the thin filaments with them. Since the myosin heads swivel in opposite directions on either side of the bare zone, the overall result is a greater overlap of the thick and thin filaments and a shortening of the muscle. The energy for this process is provided by the hydrolysis of ATP, which changes the myosin-ATP intermediate back to myosin and ADP, as well as bringing about detachment of the myosin heads from the thin filament. This detachment is possible because the calcium ions that initiated the attachment process are removed in a fraction of a second by a calcium pump in the membrane of the sarcoplasmic reticulum, which stores them for immediate or later reuse. If immediate, the cycle begins again, another myosin-ATP intermediate is formed from the pool of ATP available in the cell, and a new myosin-actin cross-bridge is formed farther along the actin fiber before it can slip back to its original position. Thus a molecular ratchet-like mechanism is seen as the actual cause of muscle contraction.

striations and precise filament arrangements characteristic of striated or skeletal muscles, the same mechanism is involved in smooth muscle contraction as well.

Let us now review the complete picture of muscle contraction as established by both biochemical and biophysical studies. It has been shown that ATP exists in muscle fibers attached to filaments of myosin. The passage of the action current down the muscle fiber somehow triggers a chemical reaction in which molecules of myosin combine with molecules of actin. The resulting compound, actomyosin, acts as an enzyme that splits the third phosphate group from the ATP molecule. This produces a complex of ADP and actomyosin, with inorganic phosphate left over. In this way the energy from ATP is specifically released to power the contraction process. The interaction of actin and myosin causes the two filaments to slide together and thus shortens each striated region of the muscle fiber. After contraction, the actomyosin-ADP complex is regenerated to form actin and myosin-bound ATP. After this short "recovery period" the muscle is ready to contract again.

Study of the nature of muscle contraction has shown how biochemistry and molecular anatomy have combined to give a detailed picture of the relation between structure and function. Yet many questions still remain unanswered, or answered incompletely. The model just discussed for the sliding actions of the actin and myosin filaments is only a hypothetical one, and by no means the only one that has been put forth. Further, it is still not understood how actomyosin acts as an enzyme in the contraction process and precisely what happens to it in the process. All that can be said to date is that the sliding filament hypothesis is probably the most comprehensive model yet developed to account for muscular contraction and the one that seems most in accord with experimental observations. Like all useful models or hypotheses, however, it serves mostly to organize known facts and give direction for future research.

12.7
HOMEOSTASIS AND MUSCLE ACTION

We often use our muscular system to remain in a physical homeostatic balance with our environment. If we lose our balance, the proper muscles are immediately brought into action in order to prevent a fall. In this case we are generally aware of the homeostatic interaction between the nervous and muscular systems, and several other systems as well (including endocrine, circulatory, and others).

Usually however, we are not aware of the "fine tuning" interactions that go on all the time to keep our positional movements coordinated ones. Within the skeletal muscles are complex capsules called *muscle spindles*, containing thin *intrafusal fibers* well supplied with sensory neurons. Closely associated with the muscle spindles are structures in the tendons called *Golgi tendon organs*. Acting in unison, these structures monitor muscle tension and prevent excessive stretching. The muscle spindles and Golgi tendon organs are the main **proprioceptors** of the body, enabling us to know the position of our fingers, hands, feet, and so forth without the necessity of looking to see where they are.

Summary

1. Muscles can produce motion by acting on bone. The striated muscles that pull bones are called skeletal muscle. In addition, there is smooth muscle, responsible for much involuntary activity (movement of the intestines, change in size of the iris of the eye, constriction or dilation of blood vessels), and cardiac muscle, found in the heart.

2. All cells display a differential distribution of electric charge (caused by differential distribution of positive and negative ions) across the membrane. The cells of muscle and nerve use this membrane potential in performing their specialized functions.

3. Both muscles and nerves function as a result of temporarily changing potential across their membranes. This is accomplished by the opening of membrane "pores" that allow certain ions to flow in or out. When the flow occurs in such a way as to even out the charge distribution, the membrane becomes depolarized. When the flow occurs in such a way as to accentuate the already-existing charge differential, the membrane becomes hyperpolarized.

4. Muscles function when groups of muscle cells called motor units are activated. A motor unit is a group of muscle cells stimulated by impulses from a single nerve. Motor units function *individually* in an all-or-nothing way. The muscle *as a whole* can show varying degrees of contraction according to how many motor units are brought into play.

5. Individual muscle cells oxidize glucose (stored as glycogen) or produce it from fatty acids to produce the energy necessary for contraction. Oxidation of glucose does not provide the energy directly, however. Muscles use directly the energy from a high-energy phosphate bond of ATP. The ADP produced is regenerated to ATP by interacting with a high-energy phosphate bond from creatine phosphate. Creatine phosphate acts as a reservoir of high-energy phosphate groups for ATP production to power muscle contraction.

6. Muscle cells or fibers are composed of large quantities of protein arranged as myofibrils. Myofibrils are made up of alternating thick protein filaments (myosin) and thin protein filaments (actin). Muscle contraction involves the sliding of the thick and thin filaments over one another, producing a shortening of the fiber as a whole. Electron microscope evidence suggests that filamentous bridges exist between actin and myosin, but *precisely* how such cross-bridges produce sliding is still unknown.

Exercises

1. In terms of motion of the organism, what is the relation of bones to muscles?

2. Describe the anatomy of a representative bone, from the macroscopic to the microscopic level.

3. Describe the anatomy of a representative muscle, from the macroscopic to the microscopic level.

4. How do we know that oxygen is not necessary for the actual contraction of muscles? What *is* the role of oxygen in muscle contraction?

5. Explain the "sliding filament" hypothesis of muscle contraction. What evidence supports this hypothesis?

Chapter 13
The Nervous System

13.1
INTRODUCTION

As indicated in the last chapter, the neuromuscular junction unites the system that brings about body movement with the system that both initiates that movement and helps coordinate it. This system is the **nervous system.** Bringing about the muscular contraction that causes movement is only part of the story, however. The nervous system is also the gateway to the surrounding environment: it detects what is going on in the world around the organism and enables it to respond appropriately. Some philosophers have even suggested that the existence of the world that surrounds us is dependent upon its being perceived. In other words, the nervous system gives the environment reality. While many might question this conclusion, there is little doubt that, to an individual living system without a nervous system, the surrounding environment has no reality at all.

13.2
NERVES, NEURONS, AND COMPLEX INTEGRATION IN THE NERVOUS SYSTEM

An isolated muscle, when stimulated by an electric current, will contract. Most useful muscular activity, however, is performed by intact muscles within a living organism. Here, most of the action of muscle on bone that produces an organized, coordinated movement is triggered by impulses originating within and conveyed by **nerves** of the nervous system. Nerves, in turn, are cable-like structures composed of connective tissue, blood vessels, and the nerve cells or **neurons** that are the functional units of the nervous system (see Fig. 13.1).

Structurally, neurons can be divided into three types, which differ only in the number of processes that originate from the nerve cell body. **Unipolar neurons** have one extending process, though this process becomes divided shortly after leaving the nerve cell body. **Bipolar neurons** have a process extending from each end of the cell body; the occasional fusing together of these two processes during embryological development produces unipolar neurons. Most common, however, are

multipolar neurons, in which one long process arising from the nerve cell body serves as an axon and several other processes serve as dendrites (see Fig. 13.2).

In terms of function also, neurons can be divided into three types. **Sensory** or **afferent neurons** transmit impulses *toward* the central nervous system. These impulses come from such sensory receptors as the eyes, ears, skin, and the like. The sensory neurons enable detection of environmental stimuli such as light, sound, odors, heat, cold, and contact.

Merely sensing an environmental stimulus is only one part of the functioning of a higher organism, of course. An active response, involving muscular contraction, may also be required. Active responses are brought about by impulses conveyed to the muscles by **motor neurons.**

The roles carried out by sensory and motor neurons are quite specialized ones. Sensory neurons are quite incapable of causing motor action, and motor neurons cannot detect environmental stimuli. For a sensory stimulus to be translated into action, it must be connected to motor neurons. The necessary connection is provided in the brain and spinal cord by **association** or **internuncial neurons.** These neurons serve the same function as connector plugs in a telephone switchboard. Incoming stimuli from sensory neurons are hooked up to the appropriate motor neurons in much the same way that a telephone operator uses a plug on the switchboard to connect an incoming call with its appropriate destination.

The entire nervous system of a higher animal can be divided into two regions. The first is the **central nervous system,** consisting of the brain and spinal cord. The second region, the **peripheral nervous system,** consists of sensory and motor neurons that extend to virtually every region of the body.

While some processing of neuronal data does occur in the peripheral nervous system, the most important site for organizing and integrating the myriads of stimuli in all higher organisms is the central nervous system. This function is accomplished by both the spinal cord and the brain, to which considerable attention will be directed later in this chapter.

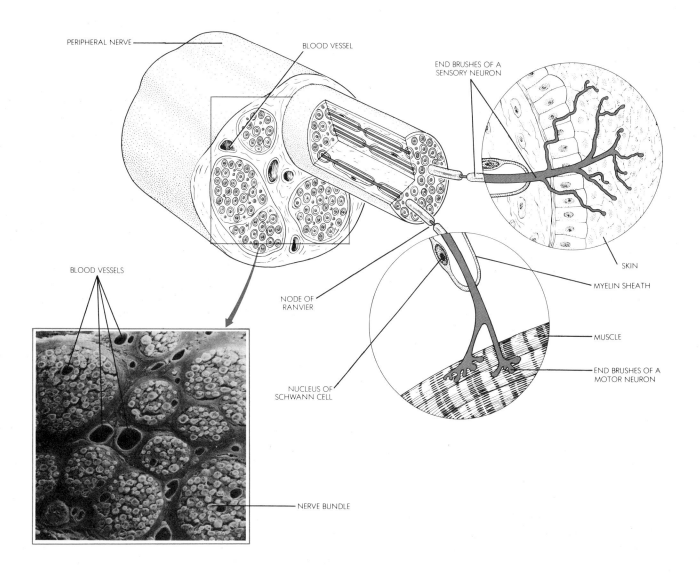

As pointed out in Chapter 5 in our discussion of animal tissues, in many vertebrate neurons the axons are surrounded by a lipid-containing myelin sheath. Such neurons are said to be myelinated. Embryologically, the myelin sheath develops from the Schwann cells. The Schwann cells envelop the axon from both sides, but rather than meeting and fusing, continue to grow over each other (see Fig. 13.3) to form a "jelly roll" arrangement of the sheath. Schwann cells are also intimately involved in the neuromuscular junction.

While, as might be expected, the myelin sheath serves as an insulator, its presence is far more significant than that function alone would indicate. At intervals, the sheath is pinched in to form the **nodes of Ranvier**. In myelinated nerve fibers, the nerve impulse, instead of moving continuously along the axon membrane surface, can jump from node to node, resulting in an almost one-hundredfold increase in velocity.

Fig. 13.1
The relationship of neurons to nerves of the peripheral nervous system (that portion lying outside the brain and spinal cord). The axons of the neurons may terminate in either sensory (such as skin, eye, or ear) or motor (that is, muscle, for movement) regions. The neurons are held together in the nerve by layers of connective tissue supplied by blood vessels. (Micrograph from *Tissues and Organs: A Text-Atlas of Scanning Electron Microscopy* by Richard G. Kessel and Randy H. Kardon. W. H. Freeman and Company. Copyright © 1979; drawing from A. P. Spence and E. B. Mason, *Human Anatomy and Physiology*. Menlo Park, Calif.: Benjamin/Cummings, 1979, p. 270.)

Fig. 13.2
(*a*) Types of neurons; (*b*) some examples of the most common types, bipolar and multipolar. (*a* adapted from A. P. Spence and E. B. Mason, *Human Anatomy and Physiology*. Menlo Park, Calif.: Benjamin/Cummings, 1979, p. 275.)

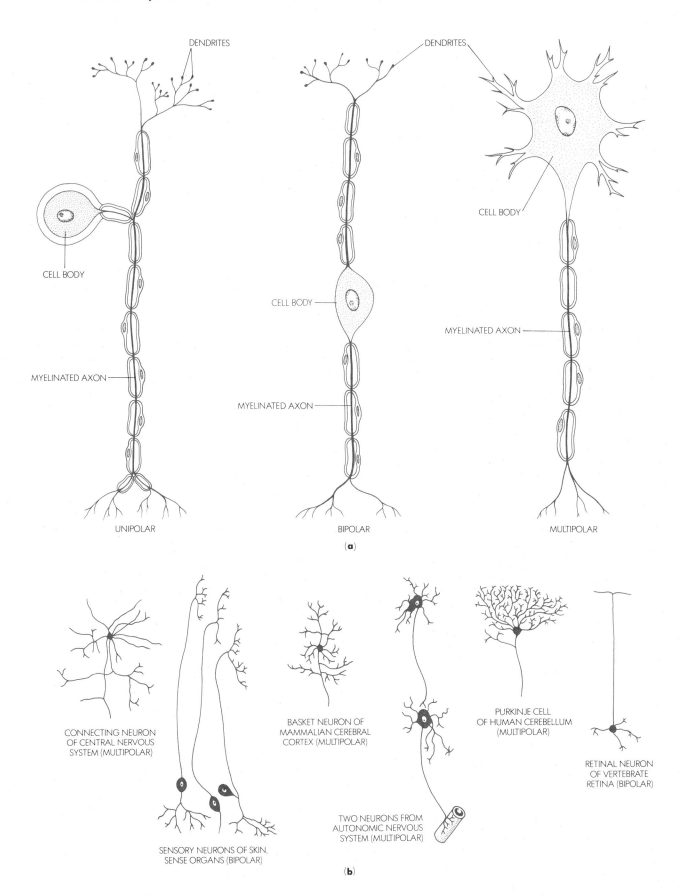

DENDRITES

CELL BODY

MYELINATED AXON

UNIPOLAR

DENDRITES

CELL BODY

MYELINATED AXON

BIPOLAR

DENDRITES

CELL BODY

MYELINATED AXON

MULTIPOLAR

(a)

CONNECTING NEURON
OF CENTRAL NERVOUS
SYSTEM (MULTIPOLAR)

SENSORY NEURONS OF SKIN,
SENSE ORGANS (BIPOLAR)

BASKET NEURON OF
MAMMALIAN CEREBRAL
CORTEX (MULTIPOLAR)

TWO NEURONS FROM
AUTONOMIC NERVOUS
SYSTEM (MULTIPOLAR)

PURKINJE CELL
OF HUMAN CEREBELLUM
(MULTIPOLAR)

RETINAL NEURON
OF VERTEBRATE
RETINA (BIPOLAR)

(b)

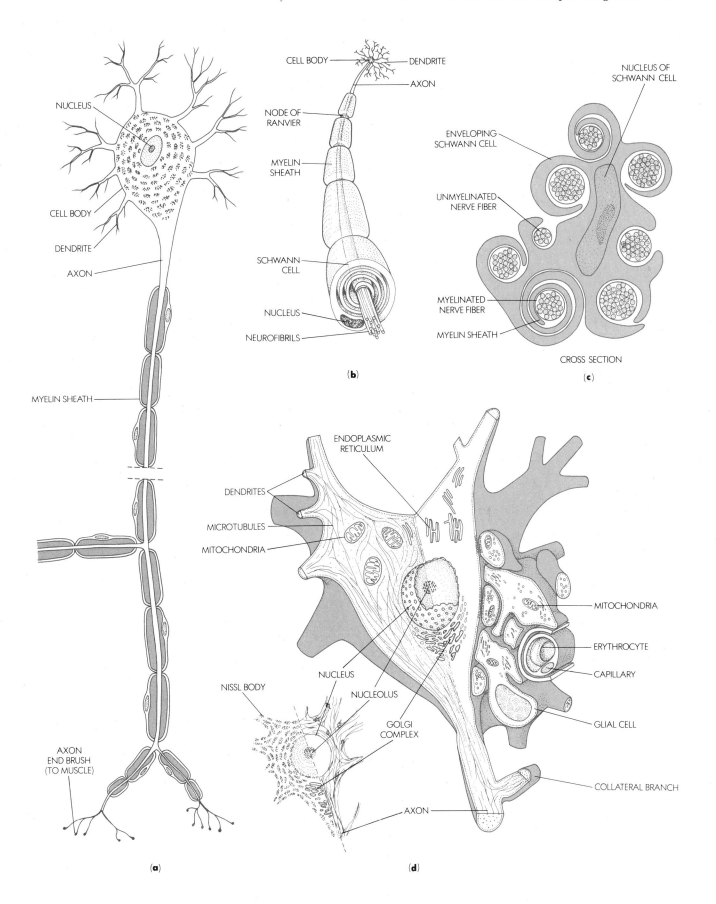

NUCLEUS

CELL BODY

DENDRITE

AXON

MYELIN SHEATH

AXON
END BRUSH
(TO MUSCLE)

(a)

CELL BODY

DENDRITE

AXON

NODE OF
RANVIER

MYELIN
SHEATH

SCHWANN
CELL

NUCLEUS

NEUROFIBRILS

(b)

NUCLEUS OF
SCHWANN CELL

ENVELOPING
SCHWANN CELL

UNMYELINATED
NERVE FIBER

MYELINATED
NERVE FIBER

MYELIN SHEATH

CROSS SECTION

(c)

ENDOPLASMIC
RETICULUM

DENDRITES

MICROTUBULES

MITOCHONDRIA

NISSL BODY

NUCLEUS

NUCLEOLUS

GOLGI
COMPLEX

AXON

MITOCHONDRIA

ERYTHROCYTE

CAPILLARY

GLIAL CELL

COLLATERAL BRANCH

(d)

◄ Fig. 13.3

(*a*) A motor neuron, showing the relationship of the processes, myelin sheath, and end brushes to muscle (see also Figs. 13.1 and 13.11). (*b*) The relationship of the myelin sheath to the axon seen in greater detail. As the nervous system develops, the myelin sheath that surrounds the axon is laid down by specialized cells called Schwann cells. One or a number of Schwann cells may produce the myelin sheath along the length of any one axon, each Schwann cell covering about 1 mm. The exposed portion of the axon between the myelin sheath segments is called the node of Ranvier. In myelinated nerve fibers, the impulse can jump from node to node greatly increasing the speed of conduction. The myelin sheath surrounding the axon is produced by repeated folding of the Schwann cells, one of which (*c*) may surround several axons. In some cases, however, the nerve fiber is not completely surrounded. Such unmyelinated fibers generally show a slower rate of nerve impulse conduction. (*d*) Anatomy of the neuron cell body and its surroundings as determined by electron microscopy. The glial cells have been hypothesized to play a role in providing nutrition for the neurons, especially during development. Certain stains enable the light microscope to reveal structures called Nissl bodies (see small insert, left side). Under the electron microscope, Nissl bodies are found to be flattened cisternae and ribosomes of the endoplasmic reticulum. (In neurons, many of the ribosomes are found floating free between the cisternae rather than attached to them.) The right side of the smaller drawing shows how metallic stains reveal microtubules (neurotubules) and microfilaments (neurofilaments). The structures in (*b*) labeled as neurofibrils are composed of both microtubules and microfilaments. The microtubules are thought to play a role in the transport of materials within the neuron while the latter may provide structural support. (*a* and *c* after A. P. Spence and E. B. Mason, *Human Anatomy and Physiology.* Menlo Park, Calif.: Benjamin/Cummings, 1979, pp. 271, 272.)

13.3
NEURONS AND THE NATURE OF THE NERVE IMPULSE

Like muscle cells, neurons have a high electric potential. The nature of this potential and the properties of the nerve impulse resulting from it have been the subject of considerable investigation.

It was originally thought that the nerve impulse was just a form of electricity. By late in the nineteenth century it had become obvious that the nerve impulse traveled much too slowly to be equated with an electric current passing over a wire (electric current travels at the speed of light, or 186,000 miles per second; the fastest nerve impulse is transmitted at 120 meters per second). What is the nature of the nerve impulse?

One clue is the fact that the rate of nerve conduction is dependent on temperature in precisely the same way chemical reactions are. This suggests that the impulse itself may be *electrochemical* in nature. In other words, propagation of the nerve impulse may be the result of the movement of particles (molecules or ions) bearing positive or negative charges. With this hypothesis in mind, we shall turn to experimental evidence

gathered in recent years—the evidence on which our present concept of nerve conduction is based.

The Voltmeter and the Squid

In the last several decades physiologists have learned a great deal about the nature of the nerve impulse by experiments performed on the giant axon of the squid (see Fig. 13.4). For the nerve physiologist, the squid axon is a remarkable gift of nature. Its very large diameter of 1 mm (the human nerve-cell axon has a diameter of about 0.01 mm) makes possible detailed studies of changes inside and outside the nerve cell during conduction.

The giant axon of the squid has anatomical features characteristic of all nerve cells. The membrane of the axon is composed of layers of protein and lipid whose thickness varies from 50 to 100 Å. The membrane is quite selective for the ions and molecules that pass through it.

Conduction in the nerve cell can be studied with an instrument known as a voltmeter. A voltmeter consists mainly of two recording electrodes connected to a meter that records differences in electric potential between the electrodes. The instrument can be used in two different ways to study nerve impulses. In one case, both electrodes are placed on the outside of the axon. With this method any differences in potential between two parts of the outer axon surface can be measured (Fig. 13.5a). In the other case, one electrode is placed inside the axon while the other is placed on the surface (Fig. 13.5b). This arrangement makes it possible to measure changes

Fig. 13.4

By virtue of its size, the giant nerve fiber of the squid allows experimental testing of the "sodium-potassium pump" hypothesis. It has recently been shown with such nerves that virtually all of the cytoplasm can be removed and replaced with artificial solutions via a small glass tube. Although such a system is apparently quite lifeless, it displays all the essential behavior of a normal living nerve. This photograph shows the cytoplasm already extruded. (Photo courtesy Trevor I. Shaw, from *Discovery*, March 1966.)

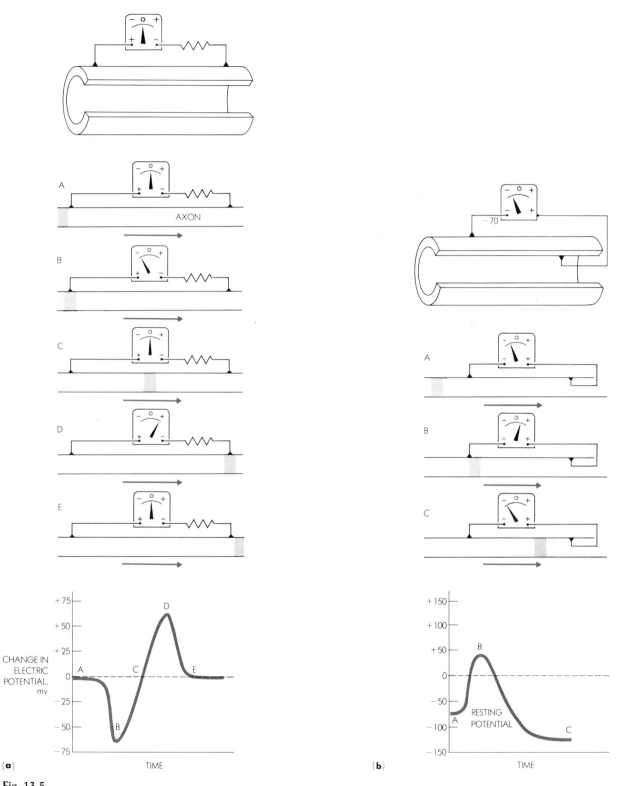

Fig. 13.5
The measurement of difference in electric potential between two points on a neuron. The shading on the axon of the neuron represents an impulse. The capital letters to the left of the voltmeters correspond to those on the graphs below. In (a) the readings are taken with both recording electrodes placed on the outer surface of the axon, as indicated in the drawings at top. In (b) one electrode is placed outside and one inside the axon. The resting potential of most neurons ranges between −60 and −90 millvolts (mv). In the case of this neuron, the difference is −70 mv.

in potential on the inside compared to the outside at any given point on the axon. Thus the voltmeter provides a way of measuring difference in potential between any two points with which the electrodes are in contact.

The data given in Fig. 13.5(b) yield a considerable amount of information about the nature of the nerve impulse. In the lower drawing we see that the nerve impulse itself has not reached the region of the axon where the recording electrodes are located. Hence that part of the axon is in a resting condition and shows, electrically, what is called its **resting potential.** The voltmeter indicates further that in the resting condition the outside of the nerve is positively charged with respect to the inside. The potential is measured as 70 millivolts (mv). On the graph to the right, it is shown as the horizontal part of the graph line. When the nerve impulse reaches the region of the membrane spanned by the two electrodes (middle drawing), some dramatic changes occur. The needle on the voltmeter swings first to zero, then to the positive side of the scale. Thus the outside of the axon becomes not only as negatively charged as the inside, but *even more so.* This is represented on the graph by the sharp upswing of the graph line. The reversal of potential between inside and outside is only temporary, however. The graph line slopes down again, indicating that the original membrane potential is eventually restored (bottom drawing); the outside once again becomes positively charged with respect to the inside.

How can we account for these changes? What is happening in the nerve axon and its surrounding medium to cause the impulse to be generated and propagated down the entire axon? Once started, the nerve impulse travels the length of the axon without further stimulus from the outside. Furthermore, like a muscle cell, a nerve cell has a threshold, and fires in the all-or-nothing manner. Any theory of nerve conduction must account for all these observations.

Ion Flow and Polarity

As has been indicated, in the resting state intact nerve (and muscle) fibers exhibit a steady difference of electrical potential across their membranes, with the exterior being positive in relation to the interior. The reasons for this polarization are revealed by investigations of the chemistry of the fibers themselves as well as the extra-cellular fluids that surround them. These fluids contain several positive ions, principally sodium and potassium, and negative ions, principally chloride. To differing degrees, each of these ions can diffuse across the membrane. Within the nerve cells are negatively charged ions to which the nerve cell membrane is impermeable. As a consequence of the resulting negative-inside, positive-outside electrical gradient, the negative chloride ions are excluded while the positive sodium and potassium ions tend to move through the cell membrane into the cell interior. Despite this latter tendency, however, measurements of ion concentrations show that only the potassium ions are in higher concentrations inside than outside. The sodium ions, on the other hand, are found in higher concentrations outside the cell than within it.

It is evident that since the sodium ions *can* pass through the cell membrane, and since there is present within the cell a negative-charge force that would tend to make them do so, there must be some sort of energy-expending mechanism to keep them out. This mechanism is referred to as the **sodium-potassium pump.** The sodium-potassium pump establishes a strong electrochemical gradient in favor of moving sodium ions into the cell. In turn, the accumulation of positive potassium ions inside the cell (because of the presence of negative ions within that cannot pass out) creates a chemical concentration gradient favoring an outflow of potassium ions. The resulting deficit of positive charges accounts for the negatively charged interior of the nerve cell. In summary, the "resting" nerve cell exhibits an equilibrium of chemical concentration and electrical gradients. The resulting potential difference constitutes the resting potential of the nerve, measured by the voltmeter at −70 mv in Fig. 13.5(b).

With the information just discussed and with the results of voltmeter studies, it is now possible to propose a hypothesis to account for the conduction of impulses along the nerve fiber.

It has already been indicated that the nerve cell membrane differs in its permeability to various kinds of ions. Figure 13.6 shows that immediately after stimulation the nerve cell membrane suddenly becomes highly and specifically permeable to sodium ions, some of which respond by immediately moving into the nerve cell interior. The nerve resting potential has now become an **action potential;** an impulse is now being transmitted. The influx of sodium ions continues to a point

---

The nerve cell membrane, like all cell membranes, normally exists in a polarized state. Polarization is caused by the differential distribution of ions, bearing electric charges, on either side of the membrane. The resting potential of the nerve cell membrane is usually around −60 to −90 millivolts.

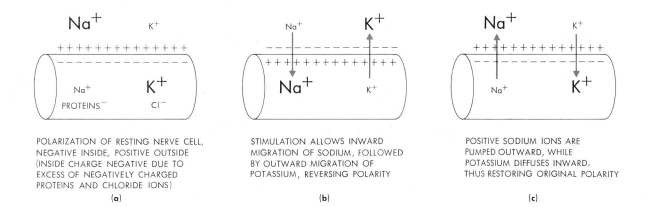

POLARIZATION OF RESTING NERVE CELL, NEGATIVE INSIDE, POSITIVE OUTSIDE (INSIDE CHARGE NEGATIVE DUE TO EXCESS OF NEGATIVELY CHARGED PROTEINS AND CHLORIDE IONS)

(a)

STIMULATION ALLOWS INWARD MIGRATION OF SODIUM, FOLLOWED BY OUTWARD MIGRATION OF POTASSIUM, REVERSING POLARITY

(b)

POSITIVE SODIUM IONS ARE PUMPED OUTWARD, WHILE POTASSIUM DIFFUSES INWARD, THUS RESTORING ORIGINAL POLARITY

(c)

Fig. 13.6
Schematic illustrations of the propagation of a nerve impulse down an axon. After passage of the impulse, the sodium-potassium pump restores the original distribution of sodium and potassium ions.

actually *beyond* the neutral or depolarized state, and the result is the temporary reversal of polarity shown by the voltmeter in Fig. 13.5(a). However, the potential starts to be restored almost immediately, as the gradual downward slope of the graph line in Fig. 13.5(b) indicates.

These observations raise two important questions: How does the nerve impulse, once started, propagate itself down the entire length of the nerve? And, once the membrane has undergone the reversal of polarity that occurs, how is the original potential reestablished?

The hypothesis to answer the first question is based at least partly on the observation that when any portion of a nerve cell axon is stimulated (by a threshold dose of electric current, for example) the membrane seems to lose its impermeability in the region of stimulation. Depolarization results in the polarity reversal of a specific region of the axon. This reversal causes the area next to

it to become permeable, producing reversal at that point, and so on down the axon. It is like a series of chain reactions. Once the process gets going, each stage triggers the next. Thus the nerve impulse can loosely be described as a wave of depolarization followed by polarity reversal sweeping down the axon. If the axon is stimulated in the middle, the wave of polarity reversal spreads in both directions. If the axon is stimulated at either end, the impulse will spread in only one direction the whole length of the nerve cell. Of course, this merely describes what may happen during nerve impulse propagation along the fiber; a precise *explanation* of how this propagation may occur remains for later discussion.

The second question—How is membrane resting potential restored after polarity reversal?—is also difficult to answer with certainty. As Fig. 13.7 indicates, the stimulus-caused change in nerve cell membrane permeability leading to sodium ion influx is followed by an outward movement of potassium ions. That this outward movement of potassium ions first restores the membrane potential after polarity reversal, rather than the sodium-potassium pump, is shown by demonstrating that a nerve cell is still capable of carrying many impulses without any sodium-potassium pump activity. Only after a relatively long period of inactivity does the sodium-potassium pump act to restore the original sodium-potassium ion distribution characteristic of the

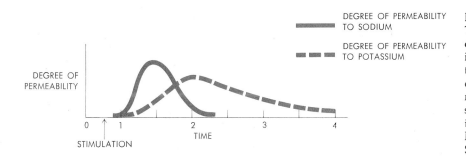

DEGREE OF PERMEABILITY TO SODIUM

DEGREE OF PERMEABILITY TO POTASSIUM

Fig. 13.7
The degree of permeability of the nerve cell membrane to sodium and potassium ions as a function of time after stimulation. These data indicate that, immediately after stimulation, sodium ions rush into the axon; this is followed slightly later by the outflow of potassium ions from the cell. (Adapted from R. D. Keynes, "The Nerve Impulse and the Squid," *Scientific American*, December 1958, p. 86.)

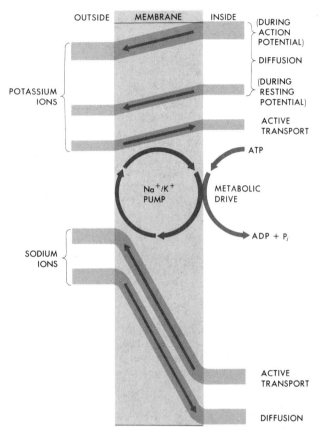

Fig. 13.8
Schematic diagram representing changes in the position of Na$^+$ and K$^+$ ions across a nerve cell membrane in resting state. Both diffusion and active transport of ions are diagrammed here. Arrows indicate the direction of ionic movement. K$^+$ diffuses out of the cell during action potential and even during resting potential (the membrane is said to be "leaky" to K$^+$). Too much leakage during resting potential is retarded by negatively charged proteins and other molecules inside the neuron. The loss of K$^+$ during action potential is restored by active transport. During action potential, Na$^+$ diffuses in very rapidly (shown by the steep downward slope of the diffusion line.) The sodium-potassium pump (active transport) must then move the sodium ions back outside against a concentration gradient (there is still much more sodium outside than inside, even after depolarization). The metabolic drive mechanism, the conversion of ATP to ADP, drives the active transport of both Na$^+$ and K$^+$. (Redrawn from CRM, *Biology Today*, 2d ed. New York: CRM/Random House, 1975, p. 552.)

resting nerve cell. The ionic events involved in membrane depolarization and repolarization are summarized in Fig. 13.8.

There is recent evidence that the sodium ions move into the neurons via ion-specific molecular channels embedded in the nerve membranes. Just how these ion channels are able to discriminate between such similar ions as sodium and potassium is not yet entirely clear. Current hypotheses propose that the channels allowing sodium ions to pass contain energy barriers that are able to discriminate between subtle differences in the chemical structure and other properties of the ions that attempt to enter.

13.4
THE SODIUM-POTASSIUM PUMP

The concept of the sodium-potassium pump is an excellent example of a model constructed to account for a natural phenomenon—in this case, the extrusion and holding of sodium ions from the nerve cell against a concentration gradient. There is no requirement that the pump bear any resemblance to reality; no one visualizes an actual mechanical pump present in the nerve cell. It is only necessary that the model adequately account for the known data concerning sodium ion distribution and allow us to make accurate predictions.

Often a scientific model will also suggest additional experiments to test its accuracy. For example, all mechanical pumps require an expenditure of energy. The sodium-potassium pump hypothetical model predicts, therefore, that, if the energy-yielding processes of the nerve cell are blocked, normal reestablishment of the resting potential will eventually be prevented. The following experiment tests this hypothesis. If we put the giant axon of the squid in a medium containing radioactively labeled sodium, we can measure sodium ion concentration changes inside and outside the cell under various conditions. The results of one set of experiments are shown in Fig. 13.9. After an initial stimulation, the sodium ion concentration inside the axon is noted to rise. Almost immediately, however, the sodium-potassium pump begins to work and the sodium ion level drops. Presumably, energy is being expended in this process. At point *A*, a metabolic inhibitor known to block the production of ATP is injected into the axon. The extrusion of sodium ions soon ceases. A second stimulation causes more sodium ions to rush into the nerve cell. With the inhibitor still present, however, the sodium ions cannot be eliminated. At point *B* the inhibitor is washed away, thus restoring the metabolic processes that produce ATP. The sodium ion level is seen to fall. A third stimulation temporarily increases the sodium ion concentration within the nerve axon, but it is removed by the sodium-potassium pump when the stimulation has ceased. These data support the idea that an active transport mechanism for sodium indeed exists and can be viewed as a form of pump.

Intriguingly, there is evidence that the idea of a molecular pump for ions such as sodium and potassium may not be as farfetched as it sounds. In 1968 Dr. A. G. Lowe of Manchester University in England put forth a hypothesis based on the idea that the ATP-splitting enzyme (ATPase) acts almost like a simple rocking-arm pump. A simplified version follows:

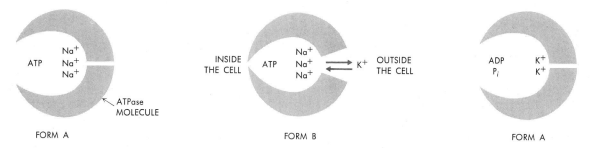

FORM A   FORM B   FORM A

The ATPase molecule is visualized as being embedded in the cell membrane and in contact with both the internal and external environment of the cell. In form A, the ATPase molecule allows sodium ions plus ATP to diffuse into it. As a result of binding with the ATP and sodium, the enzyme's conformation is changed to form B, making the sodium ions available for exchange with the potassium ions in the external environment but isolating the sodium ions from the cell interior. When all the sodium ions have been replaced in the ATPase molecule by potassium ions, the ATP is hydrolyzed and the ATPase returns to form A, thereby releasing the potassium ions and products of ATP hydrolysis back into the cell.

The drugs digitalis and ouabain are known to inhibit both active transport and the ATPase enzyme. Digitalis, obtained from the foxglove plant, *Digitalis purpurea*, is well known for its use in treating certain types of heart disease, and persons with a history of such heart trouble often carry digitalis with them in case of chest pains. The drug regulates the coordination of impulses, thereby affecting the rate and rhythm of the heart muscle's contraction. Lowe has suggested that such drugs may act to inhibit ATPase by "locking" the enzyme into one of its two forms, probably B. The Lowe model is doubly interesting because besides such phenomena as the sodium-potassium pump, it may also explain substrate phosphorylation (see Section 6.2), in which ATP is generated, by picturing the phosphorylation process as a reversal of the steps described for active transport. Such models provide valuable new ideas for future experimentation.

The present picture of the means by which a nerve impulse is conducted along the fiber can be summarized as follows:

1. Nerve stimulation results in a sharply increased permeability to sodium ions in the nerve cell membrane. Sodium ions move in, temporarily reversing the polarity of the nerve cell from negative inside and positive outside to positive inside and negative outside. *This "wave" of depolarization and polarity reversal, sweeping along the nerve fiber, comprises the nerve impulse.*

2. After polarity reversal the membrane is first restored to its original negative-inside, positive-outside polarity by the outward migration of potassium ions.

3. These two processes occur within a few milliseconds. During a relatively long period of inactivity, the sodium-potassium pump actively transports sodium ions from inside to outside the nerve cell. This outward movement of sodium ions causes the cell interior to become highly negative again because of the negative ions within that cannot pass through the cell membrane. The

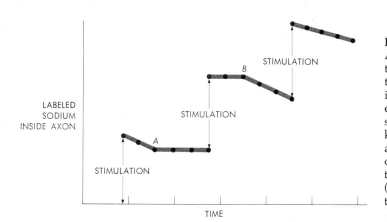

Fig. 13.9
A giant axon is placed in a medium containing radioactive sodium ions as a tracer. When the axon is stimulated the first time, the amount of tracer sodium inside the axon increases markedly. The sodium concentration inside eventually begins to fall as a result of the action of the sodium-potassium pump. At point *A* a metabolic poison known to block the production of ATP is injected into the axon. This stops the active transport of sodium ions out of the cell. At *B* the poison is washed out of the axon, thereby restoring its ability to remove sodium ions. (Adapted from R. D. Keynes, "The Nerve Impulse and the Squid," *Scientific American*, December 1958, p. 90.)

positive potassium ions are thus drawn back into the cell, and the normal resting potential is restored.

The entire picture of nerve impulse conduction has been developed primarily within the last thirty years. Although many details have been worked out, many more problems still await answers from future researchers. More knowledge is needed about the sodium-potassium pump and the nature of the nerve cell membrane. For example, how does the membrane become more permeable to sodium ions when stimulated? How is the original degree of permeability restored? These and other questions will stimulate future hypotheses and experiments.

13.5
NEURON-TO-NEURON CONNECTIONS

The nervous systems of all multicellular animals involve the complex interaction of many neurons. The nerve impulse initiated by one receptor cell (in the eye, for example) must be transmitted to secondary neurons that carry it to the brain or other parts of the central nervous system. Although each individual axon can transmit an impulse in either direction, this is not true of the intact nervous system. Here the nerve impulses always travel on a one-way path. This one-way flow is maintained by the **synapse**, the junction between two neurons. The synapse plays an important role in coordination and function of the complex organism.

The Synapse

Neurons are situated so that the synapse represents the area where the end brush of one neuron lies next to the dendrites of another. The dendrites of any one neuron may form a synapse with the end brushes of a number of other neurons. A number of observations, both anatomical and experimental, have been made concerning the synapse. These provide the basis for formulating a hypothesis to explain how synaptic transmission of the nerve impulse may occur.

Observation 1. Light and electron micrographs of the synapse show that there is no direct connection (that is, no physical contact) between the end brush of one neuron and the dendrites of the next.

Observation 2. The synapse slows down the transmission of the nerve impulse.

Observation 3. Transmission across a synapse is always one-way, from end brush to dendrite, never in the reverse direction.

Observation 4. Successive transmissions across a synapse bring about fatigue of the synapse area itself.

Observation 5. Stimulation of some neurons actually seems to inhibit the neurons that lead away from them at the synapse.

It was once thought that when a nerve impulse reached the end brush of one neuron it created an electric field across the synapse that depolarized the dendrites of the next neuron. However, this hypothesis could not explain adequately some of the observations. Most notably, observations 4 and 5 would not be predicted if synaptic transmission were due to creation of an electric field. There is certainly no reason why an electric field should produce fatigue at the synapse. Nor is there any satisfactory way to explain how the creation of an electric field should in some cases stimulate the postsynaptic neurons and in other cases prevent them from firing.

A more likely hypothesis, which may be called the "chemical transmitter" hypothesis, has been proposed in an attempt to explain more adequately what happens at the synapse. One such chemical transmitter substance, known as **acetylcholine** (abbreviated as **ACh**), is released at the synapse when an impulse reaches the end brush of the stimulated neuron. The structural formula of acetylcholine is shown below:

$$CH_3-\underset{\underset{CH_3}{|}}{\overset{\overset{CH_3}{|}}{N}}-CH_2-CH_2O-\underset{\underset{CH_3}{\diagdown}}{\overset{\overset{O}{\diagup\!\!\!\diagup}}{C}}$$

ACh is capable of diffusing across the synaptic space. When enough molecules of ACh reach the dendrites of a second neuron, they cause that neuron to fire. Because dendrites do not release ACh, transmission of a nerve impulse at the synapse can occur in only one direction. When ACh reaches the dendrite of the postsynaptic neuron, it joins onto the dendrites at certain sensitive regions. Molecules of ACh are thought to attach to the sensitive sites in much the same way as enzymes fit onto substrate molecules (Fig. 13.10). It is not yet known just how the binding of ACh to the postsynaptic neuron changes the permeability of the membrane enough to cause the neuron to fire.

If the chemical transmitter theory is correct, why do postsynaptic fibers not continue firing indefinitely once ACh molecules have attached to the sensitive sites?

The synapse is the connecting link between neurons. All processing of data in the nervous system takes place at the synapses.

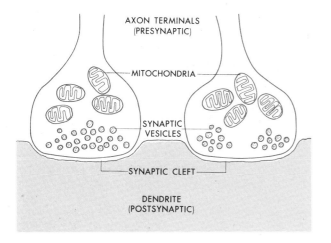

Fig. 13.10
The synapse, showing schematically the relation of presynaptic and postsynaptic neurons. In its end brush, the presynaptic neuron has special vesicles in which molecules of acetylcholine are produced. When the impulse reaches the end brush, these molecules are released. They diffuse across the synaptic space and attach onto sensitive areas of the dendrites of the postsynaptic neuron. If enough molecules of acetylcholine are picked up by the postsynaptic neuron, it will fire. The molecules attached to the postsynaptic fiber are soon destroyed by an enzyme known as cholinesterase.

Neurophysiologists and biochemists have isolated and identified an enzyme known as **cholinesterase,** located in the receiving dendrite. Cholinesterase, as its name implies, acts on acetylcholine, breaking it down into an inactive form. The postsynaptic membrane can thus repair itself and be prepared to fire again if more ACh is released. All this occurs relatively quickly in the intact organism. The distance across the synapse is about one micrometer, and it requires about one or two milliseconds for the ACh released by the presynaptic vesicles to reach the other side. The time required for diffusion explains why impulses that must travel across synapses take slightly longer than those traveling the same distance over single neurons.

Successive stimulation of a series of nerves produces fatigue at the synapse. The chemical transmitter theory accounts for this by assuming that the presynaptic vesicles contain only a limited amount of ACh. Successive stimulation uses up ACh more rapidly than the cell can synthesize it. Thus the synapse becomes fatigued, and a temporary block to further nerve conduction occurs.

In animals such as crayfish, synapses have been discovered which have electrical transmission; there is none of the delay of approximately one-thousandth of a second that characterizes synapses at which the transmission is chemical. Such electrical synapses have also been identified in a few of the lower vertebrates as well.

Not all neurons release the same transmitter substance at the synapse. Besides acetylcholine, many other neurotransmitters have been reported as confirmed in the central nervous system, including norepinephrine, dopamine, serotonin, epinephrine, and glutamic acid (in insects), as well as numerous peptides (including somatostatin, endorphins, and others). Indeed, there have recently been several observations to the effect that more than one known neurotransmitter may occur in a single neuron, thus contradicting the older one-neuron–one-neurotransmitter idea.

The physical strength of nerve cell synapses in the brain is remarkable. If brain tissue is mechanically homogenized in a blender, examination of the resulting homogenate reveals that the synapses survive intact; the two cell surfaces at the synaptic junction remain together, though each membrane has been torn by the homogenization from the cell to which it belonged. Thus it is possible to isolate and study large fractions of synapses as to their structure and composition. For one thing, such studies indicate that the part of the synapse that receives the signals—the postsynaptic membrane—may be built around a framework of a tubulin-like protein. For another, such pinched-off nerve endings show a high degree of a chemical activity associated with the enzyme adenylate cyclase. This is significant because adenylate cyclase activity is responsible for the synthesis of cyclic adenosine monophosphate or cyclic AMP, a fact that, as will be discussed later, closely associates the mode of action of both the nervous and endocrine (hormone-secreting) systems.

EPSP and IPSP

In the living organism, the effects of neurotransmitters at the synapse may be of two types. One is excitatory, and it has the effect of triggering the next neuron to fire. This is the type of synapse we described above, and it gives rise to an **excitatory postsynaptic potential** (**EPSP**). The neurotransmitter for EPSP causes the postsynaptic neuron to depolarize and thus fire. A second kind of synaptic junction, involving other neurotransmitter substances, causes inhibition of the postsynaptic neuron. This is called the **inhibitory postsynaptic potential,** or **IPSP.** At the membrane level, it appears that, during inhibition, the pores for K^+ flow inward are opened selectively, while the Na^+ pores are closed. Consequently, the postsynaptic neuron's membrane becomes hyperpolarized and requires much more stimulation to reach its new threshold. The postsynaptic neuron is "inhibited." Exactly how inhibition works remains a mystery. Some observations indicate that it is a much more complex process than has been realized until recently.

Neurobiologists have been tempted to think that the EPSP and IPSP result from two different neurotransmitters, each affecting a postsynaptic cell in opposite ways. However, it has been observed that the axon terminals (at the opposite end of the neuron from the dendrites) from a single neuron may excite one postsynaptic neuron while inhibiting another, since most neurons synapse with a number of other neurons. Further, while there are increasing examples of more than one neurotransmitter in a single neuron, it is interesting that no neuron has been observed to use more than one kind of transmitter at its synapses. This suggests strongly that whether a postsynaptic cell is excited or inhibited depends upon properties of its cell membrane, not upon the nature of the neurotransmitter.

So far we have been discussing synapses as if most interconnections between neurons were organized like a straight chain, perhaps giving the impression that the firing of one neuron is all that is required to cause the postsynaptic neuron to fire or become inhibited, as the case may be. This is not so. Most neurons synapse with a number of other neurons *simultaneously.* This means that neurons are organized more as networks than as straight chains. (Most neurons in complex organisms receive inputs from between 100 and 10,000 synapses. The principles of organization of such a network will be discussed in the next section.) A single presynaptic stimulus is usually not sufficient to cause the firing of a postsynaptic neuron. Several incoming stimuli from several presynaptic neurons are generally necessary for a postsynaptic neuron to become excited. The postsynaptic neuron adds together (or *summates*) the inputs from a number of incoming stimuli. Depending upon the type of postsynaptic neuron being considered, it is likely that some of its dendrites will show an excitatory postsynaptic potential and others an inhibitory postsynaptic potential. Thus, whether the postsynaptic neuron fires or not depends upon the IPSPs and EPSPs it receives. Through the efficient performance of this highly complex process comes the tremendous integrative capacity of the nervous system in animals.

13.6
NEURON-TO-MUSCLE CONNECTIONS

In the chick embryo, muscle tissue can develop independent of any contact with neural (nerve) tissue. Yet, soon after the cells that will become muscle begin to take on their characteristic appearance, acetylcholine receptors appear, first widely scattered over the muscle tissue surface and, later, in small clusters. Some investigators have interpreted this as a "beckoning" of the muscle to the neurons approaching it during embryological development, seemingly inviting them to form the connections so vital to its normal functioning.

The joining of a nerve end brush with a muscle fiber is called the **neuromuscular junction.** The nerve fiber branches at its end, each branch adhering tightly to the sheath covering a single muscle fiber (see Fig. 13.11). At the tips of the many nerve branches are small pockets thought to contain ACh; these are similar to ACh pockets of the presynaptic end brush. The functioning of the neuromuscular junction and that of the synapse are quite similar. When a nerve impulse reaches the neuromuscular junction, a small amount of transmitter substance is released in the minute region or **end plate** between the end brush and the muscle fiber. Just as at the synapse, the transmitter substance at the neuromuscular junction causes the muscle-cell membrane to become more permeable to sodium and potassium ions. The resulting movement of ions causes depolarization at the motor end plate. This depolarization excites adjacent regions and an impulse comparable to the nerve impulse is generated. An action current thus sweeps down the muscle fiber, stimulating the reaction between actin and myosin discussed in Section 12.6. Soon after the transmitter substance is released at the neuromuscular junction, it is destroyed by an enzyme in the same manner as at the nerve synapse.

In mammals, two types of transmitter substances are known to be produced at different neuromuscular junctions. One type of nerve ending produces ACh; the other produces norepinephrine. The two types of nerves differentiated in this way form the autonomic nervous system, to be discussed in Section 13.9.

A number of drugs are known to act directly on nerves or the neuromuscular junction; nicotine and caffeine are examples. People who have taken in too much caffeine (from coffee or various colas) often feel jittery because their nerves are more sensitive to stimulation and fire more readily. This is why caffeine is called a stimulant. Other drugs serve to inhibit activity at the neuromuscular junction. One powerful inhibitor is the drug curare, long known as a poison by natives of Central and South America. The exact mechanism by which curare acts is not known, but it seems to affect muscle membranes in such a way that the transmitter substance fails to perform its task. An animal poisoned with curare dies by paralysis.

Still another class of substance acts on the neuromuscular junction—but by an entirely different mechanism. An example of such a substance is di-isopropylfluorophosphate (DFP for short). DFP prevents the hydrolysis of ACh by the enzyme cholinesterase, thus allowing the muscle fiber to continue firing once a single stimulus has been applied. A person exposed to DFP has prolonged muscle spasms and convulsions, leading eventually to death.

Other examples are parathion and malathion, two components of certain insecticides. Parathion and malathion are also anticholinesterases, achieving their effects as insecticides by affecting the nervous system of insects. Parathion and malathion can also have the same effects on other animals, including human beings (the symptoms are violent, abnormal muscular contractions or spasms), a factor that has caused many environmentalists to strongly oppose their use.

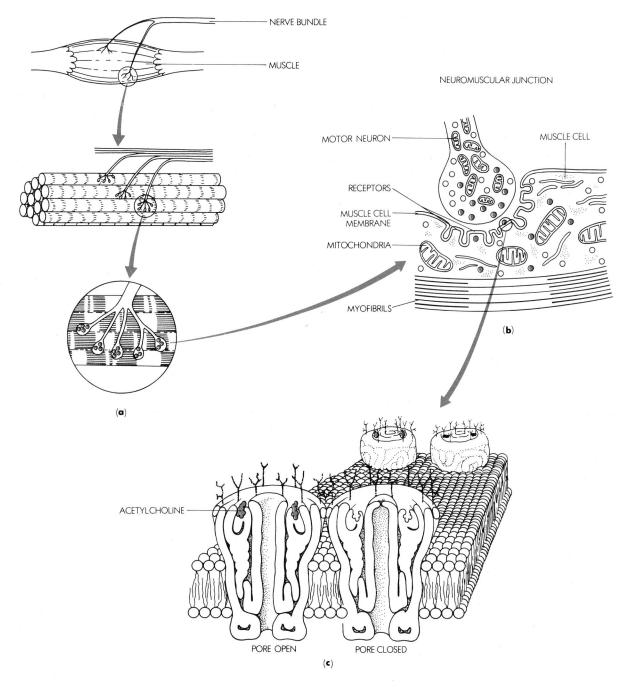

NERVE BUNDLE

MUSCLE

NEUROMUSCULAR JUNCTION

MOTOR NEURON

MUSCLE CELL

RECEPTORS

MUSCLE CELL MEMBRANE

MITOCHONDRIA

MYOFIBRILS

(b)

(a)

ACETYLCHOLINE

PORE OPEN

PORE CLOSED

(c)

Fig. 13.11

(a) A diagrammatic representation of the neuromuscular junction. Acetylcholine released at the neuron ending (b) travels across the 600-Å-wide synaptic gap in the end plate and interacts at specific sites (c) located at the crests of folds of the postsynaptic membrane. The result is to open receptor channels, allowing sodium and potassium ions to flood through the membrane and creating a potential that initiates muscle contraction. It is estimated that an average end plate contains about 50 million receptors and that an average nerve impulse releases about sixty acetylcholine vesicles from the nerve terminal. Each vesicle contains approximately 10,000 neurotransmitter molecules. In the condition known as myasthenia gravis, characterized by weakened and easily tired muscles, the number of receptors appears to be severely reduced.

13.7
THE REFLEX ARC

The nervous system of animals may consist of hundreds of thousands or millions of individual neurons. All these neurons receive and transmit stimuli in the form of action potentials that are similar, regardless of the type of stimulus giving rise to them. If the action potentials generated in a single neuron were measured, it would not be possible to distinguish one stimulus from the other. Yet the nervous system can distinguish between the high-pitch frequency of an operatic soprano and a flute; between the smell of grass and that of a dilute perfume; between a panoramic landscape and an electronics wiring diagram. It is apparent that the nervous system as a whole has characteristics that no individual parts have by themselves. Those characteristics can be summed up in one word: *organization.* The organization of neurons into a nervous system makes it possible for organisms to process information and make the fine distinctions between stimuli.

If a frog is decapitated and its upper body, heart, and other organs removed so that there is little left except the backbone, spinal cord, hind legs, and a few other associated tissues, and acetic acid is applied to the hind-leg skin, the hind legs will make strenuous and well-coordinated scratching movements that seem aimed at removing the irritant (Fig. 13.12). It seems evident, then, that while the brain may play a role in initiating and coordinating muscular activity, its presence is not always essential for such activity to occur. This response can be studied further by other experimentation on freshly prepared frogs, and it can be shown that only the legs, the skin to which the acid is applied, the nerves that supply both, and the region of the spinal cord to which these nerves are attached are necessary for the response to occur.

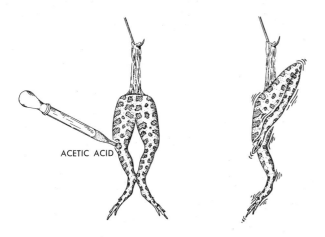

ACETIC ACID

Fig. 13.12
With only the legs and spinal cord intact, a "frog" can still respond appropriately to a stimulus.

If a cat is operated upon and the proper dorsal root of its spinal cord cut (as in Fig. 13.13), the cat will feel no pain in the appendage innervated by the cut nerve but will still be able to move it. If, on the other hand, the ventral nerve fibers are cut, the cat will feel pain in the affected appendage but will be unable to move it. Such studies, combined with clinical case histories of human patients, show that the dorsal root of the spinal cord is composed of incoming sensory neurons, while the ventral root is composed of outgoing motor neurons.

The dissection and microscopic examination of the gray matter of the spinal cord having revealed the presence of association and motor nerve cell bodies, we can offer an explanation for the response shown by the frog and cat. Sensory neurons transmit impulses to the interneurons, which in turn stimulate the appropriate motor neurons—either at the same level of the spinal cord, or via ascending and descending tracts to other levels. This hypothesis can be experimentally tested and supported by showing that an operational blockage of any one of the links in the hypothesized stimulus-response chain effectively prevents the typical response. Operational blockage elsewhere in the nervous system has no effect on the response.

The response shown by the frog is an example of a **reflex action,** in this case a rather complex one. Many reflex actions are known in animals, the most familiar probably being the knee-jerk reflex in humans, which is a routine part of most physical examinations. A tap on the knee just below the kneecap results in a sudden kick by the lower leg. Like the frog's reaction to the acetic acid, the knee-jerk reflex is a stretch reflex that can be shown to be independent of the brain. The tap on the tendon below the kneecap stretches the muscle to which the tendon is attached and excites stretch receptors in the muscle. The impulse is transmitted by sensory nerve fibers to the spinal cord and from there returned to the muscle via a motor nerve, which causes contraction and movement of the lower leg. The impulses that travel to the brain do little more than inform the patient what has happened; the brain plays no role in bringing about the kicking response.

In some reflexes, several interneurons may be involved in conveying the sensory neuron's impulse to the motor neurons. By repetitions of the reflex action, the time that elapses between stimulus and response can be shortened. The fact that the decrease in this time interval occurs in discrete units of time, rather than gradually, supports a hypothesis proposing that shorter pathways, eliminating some of the intermediate association neurons involved, are being followed.

At times, reflex actions may be of considerable advantage to an organism. A hand placed on a hot stove will be jerked away before the brain can possibly become aware of what has happened. The time saved pays off in less damage to the hand. Care must be exercised in

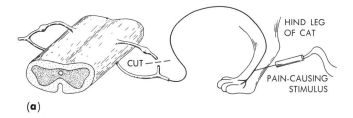

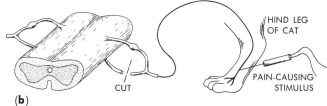

Fig. 13.13
(a) With the dorsal root cut, a cat cannot feel pain, but is able to move its foot. (b) With the ventral root cut, a cat can feel pain from the foot, but is unable to move the appendage in response.

interpreting this fact, however. The reflex action of removing the hand from the stove is brought about by the mechanics of the system involved and *not*, in any conscious sense, for the purpose of preventing more damage. The selective advantage offered by reflex actions is obvious enough to account for their evolution without our having to resort to teleological explanations.

Not all reflex actions are built in, as is the knee-jerk response. Some have to be learned through repetition of a sequence of events, such as those involved in learning to play a musical instrument or to drive a car. The Russian physiologist Ivan Pavlov, for example, fed experimental dogs immediately after ringing a bell. After a time, the dogs would secrete saliva whenever they heard the bell, even though no food was given. Here was a reflex action not ordinarily present among the behavior patterns displayed by dogs. Such experimentally established behavior patterns are known as **conditioned reflexes.**

13.8
THE BRAIN AND CENTRAL NERVOUS SYSTEM

The brain is important in the overall coordination of both voluntary and involuntary processes in the higher animal.

Scientific knowledge of the anatomy of the human brain has come largely through dissection. However, the functions of the parts of the brain are not so easily ascertained. A logical first step in determining the function of a body organ is to remove it and see what effect its removal has on the organism. Needless to say, such experiments cannot be carried out on human beings. Therefore, scientific knowledge concerning brain function is largely based on two sources of information. The first of these is extrapolation from studies made on experimental animals where one part or section of the brain is removed. Changes in the animal's subsequent behavior are a clue as to what part of the animal's activity is governed by that section of the brain. A second source of information is observations made by surgeons

on patients suffering from brain damage (caused by accident, disease, or congenital malformation). A person deprived from birth of the cerebral cortex (the outer region of the cerebrum) is capable of performing only the most basic animal functions, such as excretion, ingestion, and the like, and is in no way capable of any reasoning. It is logical to assume, therefore, that the cerebral cortex is responsible for carrying out that function. Such observations provide much valuable information about the function of all brain parts. They also tend to confirm the validity of extrapolating information gained from the study of other animals' brains to that of the human being.

In order of their size, the four major parts of the mammalian brain are the **cerebrum, cerebellum, pons Varolli (pons)**, and the **medulla oblongata.** As mentioned in Section 13.2, the brain and spinal cord together compose what we refer to as the central nervous system. The **ventricles** within the brain compose a fluid-filled system of cavities that are interconnected with each other as well as with the central canal of the spinal cord. The roof of each ventricle is thin and, instead of neurons, contains a network of capillaries called a *choroid plexus.* In these plexuses, and the cells that cover them, the circulating cerebrospinal fluid that fills the ventricles, spinal cord central canal, and the subarachnoid space surrounding the brain and spinal cord is produced. There are two lateral ventricles, one in each of the cerebral hemispheres, a centrally located third ventricle, and a fourth ventricle located beneath the cerebellum, where it is continuous with the central canal of the spinal cord.

The cerebrum is the largest part of the brain in human beings, representing about 70 percent of the entire nervous system. It is composed of the cell bodies of several billion *association neurons.* It is in the cerebrum that signals for most (though not all) action originate. The cerebrum, then, is an "association center." It receives sensory stimuli and translates them via the motor neurons into an appropriate response. In the cerebrum, too, the intellectual processes and emotions appear to be based.

The cerebrum also stores information gathered through the senses; it is thus the area responsible for memory. If regions of the cerebrum are touched with a needle through which a mild electric shock is applied, a human patient can be made to recall quite minor incidents long since forgotten. This has seemed to some to

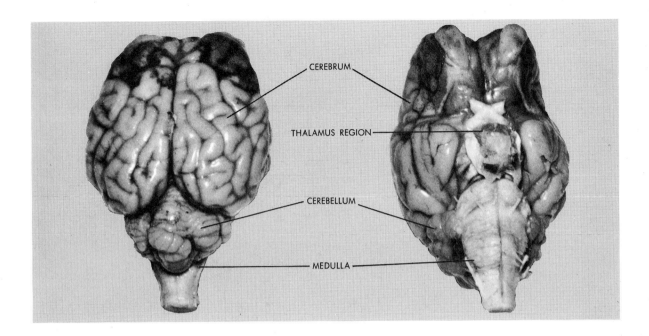

CEREBRUM

THALAMUS REGION

CEREBELLUM

MEDULLA

Fig. 13.14
A sheep's brain: the dorsal, or top, view (left) and the ventral, or bottom, view (right). (Courtesy Schettle Biologicals.)

indicate that all information, no matter how trivial, is stored in the cerebrum. If this is correct, in view of the length of a human lifetime and the many thousands of bits of information gathered each minute by the senses, the storage capacity of the cerebrum must, indeed, be prodigious.

There is as yet no completely satisfactory hypothesis to explain the mechanism of memory storage. Some puzzling facts emerge. For example, surgical removal of one half of the cerebrum may have little or no effect on intelligence or memory—there seems to be little need for intercommunication of the cerebral regions. Moreover, when rats are trained to a maze and then their cerebral hemispheres criss-crossed with several cuts to sever all connections between the regions, such rats remember their previous training as well as control rats do.

Many laboratories are now working to find a molecular basis of memory. While the results of such work have at times been contradictory, there is hope that some clues will be found in the near future. This seems to be one of the most promising areas of research in the field of neurobiology.

The cerebrum is divided into two **cerebral hemispheres** (see Fig. 13.14). These cerebral hemispheres can be further divided into four lobes: *occipital, frontal, parietal,* and *temporal.* The occipital lobe is concerned mainly with sight, including color, size, and distance of

object being viewed, form, and movement. The frontal lobes deal with the highly complex phenomena of reason and emotion. Complex voluntary movements are controlled by a group of large cells in the posterior region of the frontal lobes, the motor cortex. Damage in this region (as well as the cerebellum) results in the uncontrolled muscle spasms we associate with various conditions such as cerebral palsy.

The speech center is also located in the cerebrum. Interestingly, the speech center tends to be predominately located in the right cerebral hemisphere in left-handed people and the left cerebral hemisphere in right-handed people. If the speech center in one half is damaged or diseased, speech is impaired or lost, but can be relearned by the other half.

The frontal lobes are intimately connected with other lobes of the brain as well as the thalamus (to be discussed shortly), and through these connections feelings or emotions are associated with senses such as sight or hearing. Disease or damage to the frontal lobes can lead to personality changes, often of an extreme nature.

The cerebellum lies in the back of the skull and is composed of many foldings arranged into two large masses and a middle portion. Motor impulses from the cerebrum are sent to this region of the brain (see Figs. 13.14 and 13.15). The cerebellum "sorts out" these impulses and sends them to the appropriate muscles at the proper time to effect an orderly muscle response. The cerebellum improves in its performance of these duties as the organism matures and gains practice. The helpless thrashings of a human infant show little sorting of cerebral signals on the part of the cerebellum. As the infant grows, the signals become better coordinated; the infant learns to pick things up and to crawl. Even learning to

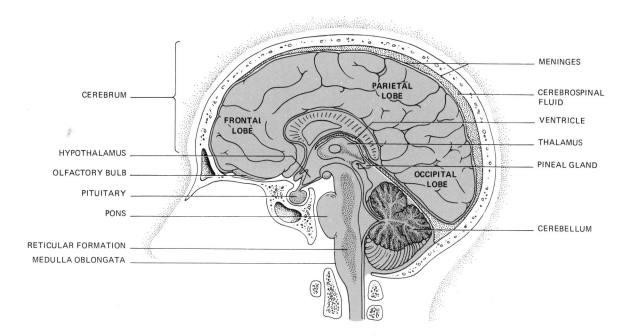

Fig. 13.15
The human brain cut lengthwise between the two cerebral hemispheres. (From John W. Kimball, *Biology*, 2d. ed. Reading, Mass.: Addison-Wesley, 1968.)

ride a bicycle, which calls for muscular coordination in response to the sense of balance, depends primarily on having the proper sequence of nerve impulses sent out from the cerebellum. Accidental damage to the cerebellum may cause a human being to lose coordinated control of muscular activities; such a person's motions may resemble those of an infant.

The pons lies just anterior (above, in the upright human) to the medulla oblongata. It consists of a maze of bundles of nerve fibers that originate in the cerebrum and end in the cerebellum. Via the pons, then, the learning processes of the cerebrum are transmitted to the cerebellum to be translated into the coordination between the input of our senses and the appropriate muscle actions we call responses.

The medulla oblongata is the swollen area of the anterior spinal cord that connects the cord to the brain. The medulla walls are thick and composed mostly of nerve tracts leading up from the spinal cord into the higher regions of the brain. The medulla also contains nerve centers that control many involuntary physiological processes, such as breathing, heart rate, blood vessel contraction or dilation (which, in turn, raises or lowers the blood pressure), and vomiting. Certain viruses, such as those causing polio, may invade the spinal fluid surrounding the spinal cord. If the virus reaches the medulla, death may result through paralysis of important nerve tracts.

Besides these larger main areas of the brain, there are other smaller but equally important brain parts.

The Midbrain. The midbrain lies between the pons and the cerebrum, and serves as a sort of relay station for sensory impulses as well as some involuntary re-

sponses (such as the narrowing of the eye pupil in response to bright light). The midbrain also gives rise to three of the cranial nerves.

The Thalamus. The thalamus is located above the midbrain and appears to play a role in the production of emotions. If the thalamus is electronically stimulated, or diseased, spontaneous laughter or crying may result.

The Hypothalamus. As its name implies, the hypothalamus is located underneath the thalamus. Composed, like the thalamus, of a relatively small group of cells, the hypothalamus nonetheless is a major center of varied and diverse activities. It plays a role in the homeostatic regulation of body temperature, rate of metabolism, and, via its effects on the autonomic nervous system, the heartbeat. The hypothalamus is also concerned with the development of sexual activity, sleep activities, fat metabolism, water balance, and emotions such as fear and rage. Perhaps most significant, as the preceding functions suggest, the hypothalamus is the brain region that, via its interaction with the pituitary gland, perhaps more than any other, ties closely together the functioning of the nervous and endocrine systems.

The Epithalamus. The epithalamus lies above the thalamus, forming a thin roof over the third ventricle of the brain. A small endocrine gland, the epiphysis or pineal gland, extends from the posterior region of the epithalamus. The function of the pineal gland will be discussed later.

One rather surprising feature of the brain is its high metabolic rate, consuming much energy; this is surprising in that, unlike the muscles or kidneys, for example, the brain appears to perform no work. If brain waves are measured by an instrument called an electroencephalograph, however, much activity is recorded even in a sleeping organism. The continuous activity recorded is probably the result of the movement of ions. The brain is the first organ damaged by oxygen deprivation and occasionally, even though a drowning person may be made to breathe again if rescued in time, the individual may show the result of irreversible brain damage. There is recent evidence that an increase in the metabolic rate of the brain enhances learning.

Several years ago it was discovered that the brain contains so-called pleasure centers. Rats with electrodes inserted into a tiny region of the cerebrum would constantly push a lever to cause electrical stimulation of this area, even foregoing food when hungry to enjoy the reward. That the same behavior has been shown in rhe-sus and squirrel monkeys suggests that the same centers exist in the human brain as well. There is some evidence that the regions of the brain sensitive to drugs such as morphine are located in the same regions where the monkey studies suggest the pleasure center should be located. In this case, the pleasure reward to the person taking the drug is offset by the physiological and psychological damage it causes.

Most of the brain, as we have seen, is divided into two halves. From the top, the cerebrum is seen to be divided into right and left hemispheres. Upon entering the brain, nerve tracks from one side of the body cross with those from the other side. Thus the left side of the brain is largely responsible for the right side of the body, and vice versa.

Twelve pairs of cranial nerves extend from the mammalian brain (in the lower vertebrates, only the first ten are found—see Table 13.1). Eleven of these nerves innervate only structures located in the head and neck region, while one, the tenth or vagus nerve, extends to other regions as well. The vagus, for example, innervates the heart (where it serves to slow the heartbeat) and stomach (where it stimulates the release of hydrochloric acid and pepsinogen and causes stomach

Table 13.1
The Twelve Cranial Nerves.*

Nerve No.	Name	Type of function	Innervation site	Causation
I	Olfactory	Sensory	Olfactory epithelium	Sense of smell
II	Optic	Sensory	Eye	Sense of vision
III	Oculomotor	Motor	Four of the six eye muscles	Motion of eyeball
IV	Trochlear	Motor	One of the six eye muscles	Motion of eyeball
V	Trigeminal	Sensory and motor	Three branches: head and face muscles, lower jaw teeth, and jaw muscles	Motion of face muscles causing expression, sensation in teeth and portions of jaw skin, jaw motion
VI	Abducens	Motor	One of six eye muscles	Motion of eyeball
VII	Facial	Sensory and motor	Face	Sensation and movement of facial muscles
VIII	Auditory (Acoustic)	Sensory	Internal ear	Sense of hearing
IX	Glossopharyngeal	Sensory and motor	Tongue and pharynx	Sensation and movement in tongue and pharynx
X	Vagus	Sensory and motor	Viscera (heart, stomach, etc.)	Plays a role in the motion and sensation in heart and visceral organs
XI	Spinal Accessory	Motor	Same as vagus	Same as vagus
XII	Hypoglossal	Motor	Tongue muscles	Movement of tongue

*The hypoglossal is absent in fish, while in the lower vertebrates the spinal accessory is incorporated into the vagus. Students of anatomy have for decades memorized the cranial nerves by the first letters of the following poem (or variations thereof): "On Old Olympus's Towering Top, A Finn And German Viewed Some Hops."

The brain processes most of the "data" (nerve impulses) passing into or out of the central nervous system. It is the clearinghouse and coordinator for most unconscious and all conscious processes.

contractions). Some of the cranial nerves are purely sensory (for example, the eighth or auditory nerve of the ear), some are purely motor (as is the fourth or abducens, which innervates one of the eye muscles), and some are a mixture of both sensory and motor fibers (including the fifth or trigeminal nerve to the head, mouth, jaw, and facial region).

The spinal cord extends from the back of the brain. In vertebrates, the spinal cord runs through the hollow center of the vertebrae and is thus protected from injury. Thirty-one pairs of spinal nerves extend to each side of the body from the spinal cord, which is covered on the outside by three membranes, the **meninges.** Both the spinal cord and brain are bathed by **cerebrospinal fluid,** which contains mineral salts and traces of proteins and sugar. The cerebrospinal fluid helps protect the nervous system from physical shock and probably also plays a role in its nutrition.

The entire nervous system is composed, on one hand, of regions with heavy concentrations of nerve cell bodies and, on the other, of regions entirely composed of nerve cell processes, either axons or dendrites. In regions where nerve cell bodies predominate, the heavy concentration of cell cytoplasm gives the tissue a definite grayish color, hence the term **gray matter.** The cortex of the cerebrum is such a region. Where nerve cell processes predominate, as in the cerebellum, the nerve tissue is called **white matter.** The white appearance is due to the high lipid content of the myelin (see Fig. 13.3) that surrounds the nerve processes (the gray matter is unmyelinated).

The familiar white, thread-like nerves seen in a dissected body are only the axons and dendrites of the centrally located neurons. A cross section of such a nerve looks much like a cross section of a telephone cable carrying several wires. Each individual nerve cell process is insulated from the others by fatty sheaths. The nerve cell bodies are located either within the spinal cord or just outside it in the dorsal root of the ganglion. Some of the axons of the motor neurons and the dendrites of sensory neurons—those, for example, that extend from the big toe to just outside the spinal cord—may be as much as 3 or 4 ft long.

The Spinal Cord

The spinal cord of vertebrates lies within the vertebral column, or backbone. As we saw in Section 13.7, the spinal cord is the site of reflex connections. It is much more than that, however. The spinal cord is also the center for integrating impulses traveling from higher to lower levels in the nervous system. In a word, it is the connecting link not only between sense organ and muscle (as in the simple reflex arc), but also between all parts of the body and the brain. The processing of stimuli occurs at all levels of the spinal cord. Ultimately, however, any processing of data that requires some kind of conscious, rational response takes place in the brain.

Like the brain, the spinal cord consists of gray and white matter. It is bilaterally symmetrical, with a butterfly-shaped region of gray matter in the center that is the focus of synaptic activity. Here, axons from sensory input neurons make contact with motor output neurons. Association neurons are found in this area. Most sensory stimuli do not pass out directly over motor neurons; higher vertebrates do not respond to most stimuli by simple reflex. More often, sensory neurons synapse in the gray matter of the spinal cord with connecting association neurons, which synapse with other association neurons or motor neurons. Association neurons in the gray matter synapse with other neurons running up and down the spinal cord. At the intervening spaces between each vertebra in the backbone, more sensory and motor neurons enter the spinal cord. The job of the association neurons and the other neurons running up and down the spinal cord is to integrate all these various impulses and send them on to their proper destinations. The neuron processes running up and down the length of the spinal cord are located outside the gray matter in the white matter.

How the spinal cord integrates stimuli to help produce appropriate responses is a major unsolved problem in contemporary neurobiology. Only a few clues help us understand how the nervous system attains the fine degree of integration and control it exercises every second of an organism's life.

One of these clues comes from analysis of simple reflex arcs. The knee-jerk reflex actually involves two sets of antagonistic muscles that produce opposite effects on the same structure. (Another pair of antagonistic muscles are the biceps and triceps muscles. The biceps muscle bends the lower arm while the triceps muscle extends it.) The gastrocnemius muscle (which bends the lower leg upward) and the sartorius muscle (which extends it) are two other pairs of antagonistic muscles. Normally,

SUPPLEMENT 13.1
PSYCHOSURGERY

Some behavioral characteristics can be modified by surgery performed on various portions of the vertebrate brain. Such surgery may involve either removing a section or severing neuronal connections, such as cutting the corpus callosum. When such techniques are applied to human beings for the sake of modifying behavior, the process is called *psychosurgery*.

One area of psychosurgery involves electrical stimulation of the brain (ESB), where tiny electrodes are implanted in the brain. The process is painless, since the brain has no pain receptors; indeed, patients undergoing brain surgery are often wide awake and can talk with the surgeon. With the electrodes implanted in certain regions of the brain of a monkey the experimenter, by pushing the appropriate buttons, can make the animal feel hot or cold, hungry (although it has just eaten) or not hungry (although it has not been fed), thirsty, fearless, anxious, peaceful, violent, sexually aroused, sleepy, or alert.

The fact that humans under surgery may experience such feelings without being aware of their cause by an outside electrical stimulus has made some people apprehensive about the prospect of such techniques being used to control people in the future, and a few have even advocated that such research be stopped. Others, however, point out the benefits of such knowledge. A few years ago, for example, an epileptic patient had an electrode implanted in his brain that enabled him to turn off the convulsive attacks with which he had been plagued all his life merely by pushing a button. Proponents of psychosurgery respond to critics by stating that there is nothing in such knowledge that is inherently bad, but rather that the good or evil lies in society's use of it.

Proponents of psychosurgery, such as Dr. O. J. Andy at the University of Mississippi Medical Center and Dr. William H. Sweet of the Massachusetts General Hospital, claim that various techniques now being developed make it possible to operate on precise, localized centers in the brain, thought to be the seat of specific behavior patterns. Psychosurgeons have been especially interested in locating the center for violent behavior, such as the hippocampus or amygdala, located underneath the frontal lobes of the cerebrum. Using these new surgical techniques, neurosurgeons have reported "cures" for hyperkinesis (extreme activity) in children, epilepsy, homosexuality, and various kinds of so-called "violent" behavior. Andy, Sweet, and others maintain that many such behavioral "abnormalities" are the result of pathological brain function. Although psychosurgery is a relatively new field, its advocates maintain that it holds immense promise for curing certain behavioral problems that have heretofore defied treatment by psychiatry and counseling. Considerable faith in this new method

is reflected in the significant increase in public funding of psychosurgery and related fields in recent years. The National Institute of Mental Health (NIMH) awarded grants of over half a million dollars in 1972 for psychosurgery; the Justice Department's Law Enforcement Assistance Administration granted another $108,000 to two Boston psychosurgeons in 1973. In 1974 the California State Department of Health and Welfare, in conjunction with the state Council on Criminal Justice, set aside $1.5 million to establish a psychosurgery research and experimental center, called The Center for the Reduction of Violence, at UCLA.

Critics of psychosurgery oppose it on both technical and moral grounds. On technical grounds, critics such as Dr. Peter Breggin of the Georgetown University School of Medicine claim that new techniques for brain surgery mean nothing when so little is known about the localization of behavioral control centers in the brain. Dr. Breggin and others claim that psychosurgery is just a slightly updated version of frontal lobotomy, an operation in which part of the frontal lobe of the cerebrum is removed and which was used extensively in treating psychiatric cases in the 1930s and 1940s. While lobotomies often did markedly change the behavior and personality of patients, the long-term results have been highly unfavorable. Patients have shown early signs of senility and have often been described as passive nonentities fifteen to twenty years after the operation. With this history in mind, Dr. Breggin argues that today's psychosurgeons should know much more about the long-term effects of their surgery before performing any operations on human patients.

Critics also point out that the clinical reports on cases of patients undergoing psychosurgery are anything but encouraging. They maintain that many psychosurgeons claim success for their patients if the patient simply becomes more controllable. Dr. Breggin has cited a number of cases where surgeons reported a patient as showing great improvement after surgery, even though the original behavior that the operation was performed to correct was not alleviated. He also points to cases of operations on children for so-called behavior disorders. One nine-year-old boy was operated on four times for hyperkinesis. He was reported "much improved," although the surgeon admitted that after a year the boy began to show intellectual deterioration. Homosexuals who underwent psychosurgery were considered improved because they showed no sexual drive at all. Dr. Breggin and others argue that such clinical observations suggest the effects of psychosurgery are massive and often apparent only as long-term and usually destructive changes—and that they cannot be used to predictably change a specific behavior. More often than not, the effects of surgery on the brain appear to be generalized and to have a myriad of unpredictable side effects.

On moral grounds, critics of psychosurgery argue that the changes wrought by tampering with a delicate structure such as the brain affect the patient's total personality. Thus psychosurgeons wield undue influence over a person's entire life. Since brain surgery produces irreversible changes (nerve cells in the brain do not grow back once cut or damaged), these critics maintain that use of this technique should be limited to the most serious cases, where the patient's life is at stake. They further claim that psychosurgery infringes severely on a person's basic right to self-determination and free choice.

Even more alarming to critics are the claims that proponents of psychosurgery make for its use in large-scale social control. For instance, Dr. Sweet has stated that psychosurgery can be used to treat people who commit acts of "senseless violence." "Senseless violence," he writes, "is more likely to be associated with organic brain disease than violence with an ascertainable motive." When he made this statement, Dr. Sweet was referring to ghetto rioters in Watts (1966) or Detroit (1969), whose behavior he regarded as senseless. Dr. O. J. Andy holds similar views: "Criminal aggressive behavior out on the streets? Yes. I think those who are involved in any uprising such as Watts or Detroit could have abnormal brains. Those people should undergo tests with whatever capacity we now have." One of the bluntest critics of this aspect of psychosurgery is Dr. Alvin Poussaint, Associate Professor of Psychiatry at the Harvard Medical School. He claims that Sweet's and Andy's investigations are racist: "The whole concept is vicious. When all these institutions around the country decide to study violence, who do

they go look at? The *black* man. But who's committing all the violence?" Dr. Poussaint goes on to cite what he saw as the violence practiced by white people: that of political leaders waging the Vietnam war, the violence of some white police officers, and the violence of the Ku Klux Klan. He points out that no one has proposed examining the brains of white people involved in such violent activity for "organic brain disease."

What role, if any, psychosurgery is to play in the future of society is perhaps more a social than a biological question. Proponents of psychosurgery see in their work a new technique that could be applied on a large scale to reduce various kinds of "inappropriate" behavior in a population of people. They see psychosurgery as a means by which many people could lead harmonious lives filled with less anxiety and less destruction to themselves and those around them. It thus has the potential, they argue, of great social good. Critics, on the other hand, claim that the definition of what is appropriate or inappropriate behavior is itself very subjective. Who is to say whether the ghetto *rioter* or ghetto *conditions* represent the more abnormal situation? Some would argue that violence against an oppressive environment is very *normal* (witness general approval of the American Revolution)—that abnormality lies in passively accepting such conditions. Furthermore, using any technological device as a means of changing the behavior of large groups of people raises the spectre of a highly regulated society not unlike Aldous Huxley's *Brave New World*. Substitute "psychosurgery" for "soma" (a drug dispensed to all lower-intelligence castes in Huxley's novel to keep them docile), and, in the opinion of certain critics of the practice of psychosurgery, we have arrived all too soon at our own "brave new world."

for a member of an antagonistic pair of muscles to contract, the other member must relax; if the gastrocnemius started to contract and the sartorius did not relax, no lower-leg movement could occur, and the result might be a pulled muscle or bone fracture. Coordination between the two members of an antagonistic pair of muscles is essential for organized movement.

We can see how this integration is possible by examining the two hypothetical schemes illustrated in Fig. 13.16. Both schemes make use of the known distinctions between excitatory and inhibitory neurons. The circuits in both schemes can produce the excitation and inhibition required to effect a coordinated movement such as the knee-jerk reflex. In both cases movement is initiated by tapping the knee and exciting a sensory neuron leading into the spinal cord. In the scheme known as a **recurrent collateral circuit** (Fig. 13.16a), sensory input is fed through association neurons in the spinal cord to motor neurons. Some of these motor neurons, however, branch before leaving the spinal cord and make an ex-

citatory synapse with an association neuron, which then makes an inhibitory contact with the motor neurons going to the antagonistic set of muscles. Thus the same sensory input leads to excitatory impulses to one member of a muscle pair and inhibitory impulses to its antagonist.

A second type of proposed integrating mechanism is called an **afferent collateral circuit** (Fig. 13.16b). Here, the sensory input neuron is visualized as branching before synapsing. One of the branches makes an excitatory synapse with an association neuron, which inhibits motor neurons running to the antagonistic muscle. In this case the motor neuron does not initiate the antagonistic response itself; rather, the incoming sensory neuron simultaneously stimulates both the reflex response and inhibits its opposing movement.

No matter which scheme is correct (and indeed both may be present in different circuits), the result is that simple coordination and integration of responses is possible. A tap on the knee triggers sensory neurons

(a) RECURRENT COLLATERAL CIRCUIT

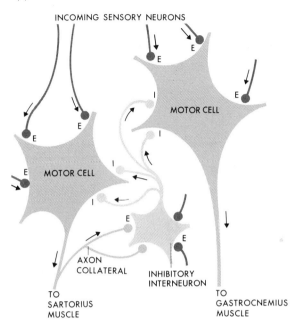

(b) AFFERENT COLLATERAL CIRCUIT

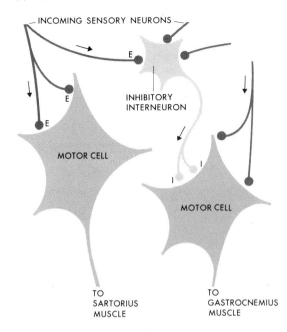

Fig. 13.16
Two schemes for integration of excitatory and inhibitory responses necessary to result in the knee-jerk reflex. (a) Excitation passing from a sensory to a motor neuron, which then branches, sending one branch to the sartorius muscle and another branch, via an inhibitory interneuron, to a second motor neuron that innervates the gastrocnemius muscle. Since the gastrocnemius muscle is antagonistic to the sartorius muscle, its inhibition allows the sartorius muscle to extend the leg. (b) A second situation, wherein the sensory neuron excites both motor neuron and inhibitory interneuron directly. This inhibitory interneuron then inhibits the motor neuron leading to the gastrocnemius muscle. The net result is the same as in (a), but by a different pathway. In the diagrams, E stands for an excitatory stimulus, I for an inhibitory stimulus. Motor neurons are shown in color, inhibitory interneurons in gray.

that produce an excitatory synapse with motor neurons leading to the sartorius and other closely related muscles. The same sensory input produces inhibition of the gastrocnemius, a muscle antagonistic to the sartorius. The result is that the lower leg moves outward. A simple kind of coordination has been effected through the spinal cord.

This model applies to only the simplest kind of reflex activity. Unfortunately, the complex behavior of higher organisms, including voluntary movement and conscious decisions, cannot be understood on such a simple level. While it is always tempting to try to describe all human behavior in terms of hypothetical models of biologically determined neuronal connections, the evidence is so incomplete that such thinking amounts to little more than speculation. As we will see later, the

anatomy, physiology, and evolution of behavior in animals is an exciting but very complex subject.

13.9
THE AUTONOMIC NERVOUS SYSTEM

The portion of the nervous system we have been discussing thus far has, for the most part, been that portion falling under the heading of the somatic nervous system, in which the movements caused are under voluntary control. Many organs of the body, however, function perfectly well without any conscious effort on the part of their owner. For example, the peristaltic movements of the intestinal walls, beating of the heart, and secretion of the pancreas require no conscious action. Nevertheless, all the organs involved *are* under the control of the central nervous system. That portion of central and peripheral nervous system responsible for carrying out such involuntary vital processes is called the **autonomic nervous system.**

The autonomic nervous system consists of two complete sets of nerves, the *sympathetic* and *parasympathetic* systems. Some organs (such as the heart) are innervated by both sympathetic and parasympathetic nerve fibers while others receive only one (such as the adrenal medulla, which receives only sympathetic nerve fibers).

Anatomically, the sympathetic and parasympathetic systems originate at different points in the central nervous system. Those parasympathetic axons whose cell bodies are located in the brain travel with motor and/or sensory tracts of four of the cranial nerves—the oculomotor (III), facial (VII), glossopharyngeal (IX), and vagus (X)—to regions of the head, thorax, and abdomen. In both the sympathetic and parasympathetic systems, motor impulses travel to their effector organs over a relay of two or more neurons, rather than by a single neuron as is the case with nerves of the somatic nervous system. The "relay stations" for the sympathetic system lie within masses of nerve cell bodies or ganglia lying next to the spinal cord, while those ganglia of the parasympathetic system usually lie within or close to the organ being innervated (see Fig. 13.17).

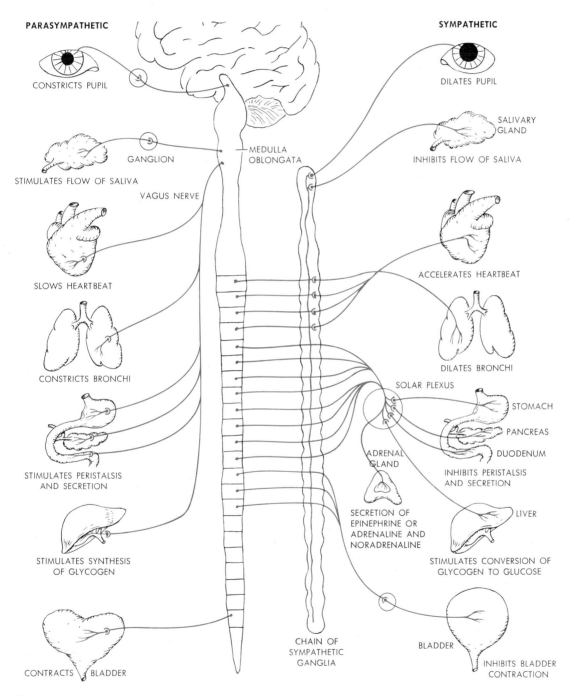

Fig. 13.17
The human autonomic nervous system. The parasympathetic and sympathetic nervous systems work antagonistically to each other. Note that the nerve supplying the adrenal gland does not pass through any synapses after leaving the spinal cord. (From John W. Kimball, *Biology,* 2d ed. Reading, Mass.: Addison-Wesley, 1968.)

The effects of the two autonomic nervous systems are generally opposite. For example, the sympathetic nervous system speeds up the heart and slows down digestion. The parasympathetic system slows down the heart and speeds up digestion. It might be thought that when the heart is beating slowly only the sympathetic nerve is sending impulses. Instruments designed to measure nerve activity, however, contradict this idea; both nerves are sending impulses at the same time. When the heart is beating quickly, more impulses are passing through the sympathetic nerve than through the parasympathetic nerve. Conversely, when the heart is beating slowly, more impulses are passing through the parasympathetic nerve than the sympathetic nerve. Such an arrangement allows an immediate response when the actual need to slow down or speed up arises.

Anatomically, the autonomic nervous system is unique in that in most cases motor impulses travel to their effector organs over a relay of two or more neurons, not by a single neuron. The relay stations for the sympathetic system lie within the lateral portions of the gray matter of the spinal cord; those for the parasympathetic system lie outside the spinal cord, as a series of ganglia, usually within the organs innervated.

Knowing that the heart is a muscle and that its contraction therefore has a chemical basis leads us to wonder just how the parasympathetic and sympathetic nerves bring about such a fine level of regulation. In 1920 the physiologist Otto Loewi performed an experiment to test the hypothesis that the sympathetic and parasympathetic nerves leading to the heart produced different neurotransmitter substances with opposite effects. He removed one frog's heart with its sympathetic and parasympathetic nerves attached, and another heart with no nerves attached, and placed each preparation in a saline solution. After stimulating the parasympathetic nerves of the first preparation for a considerable period of time, Loewi removed the heart and its attached nerves from the saline bath and placed the second heart into this solution. His reasoning was that if parasympathetic fibers release an inhibitory substance that slows down the heart rate, then traces of this substance in the saline solution bathing a heart whose parasympathetic fibers had been stimulated should slow down a fresh heart—one that had been beating after excision in normal saline solution—if it were placed in the solution formerly occupied by the first heart. When this second heart was placed in the saline bath, the rate of heating *was* slowed down, as Loewi had predicted. The opposite effect was found when the experiment was repeated with only the sympathetic fibers stimulated.

The substance released causing these effects is now known to be the neurotransmitters generally characteristic for each system. Like the somatic motor nerves, the parasympathetic nervous system secretes acetylcholine. So, too, do the preganglionic fibers of the sympathetic nervous system. After emerging from the ganglia, how-

ever, the postganglionic fibers of the sympathetic system, with the exception of those that innervate the sweat glands (which produce acetylcholine), secrete the compound norepinephrine and, perhaps, some epinephrine.*

13.10
THE SENSE ORGANS

No living organism is fully aware of its surrounding environment. It can only perceive the part of its world that it is equipped to perceive. Thus for the simple amoeba there is no such thing as noise, for it has no organs of hearing. For most dogs and cats there is no color, for their eyes lack the proper visual components to perceive it.

An environmental stimulus can be viewed as simply an expression of energy transmission. A **sense organ** is a sensory nerve ending specialized for the detection of one form of this energy. A noise, for example, is an expression of sound energy, transmitted as waves of compression and decompression in a medium such as air or water; the ear is a receptor designed to detect such waves. Light is a form of radiant energy, the eye a specialized receptor for its detection. No matter what the receptor, however, its communication with the central nervous system takes place entirely via sensory nerve processes.

Complex organisms are continually in contact with their environment. They need to be able to *sense* conditions in the environment and adjust their behavior accordingly. In all higher animals this process is accomplished with the aid of sense organs: the organs of sight, hearing, taste, touch, and smell. All these sense organs are highly specialized groupings of nerve cells located in anatomical settings that increase their effectiveness in receiving and transmitting stimuli.

The specialization of sense organs and the sensory neurons supplying them is illustrated by the fact that we must often translate one form of energy into another before we are aware of its presence. Television waves are all around us, but without the proper mechanisms we cannot perceive them. In order to receive these waves, we must employ an instrument (a TV set) that translates them into wavelengths of visible light, which our eyes can perceive. Television sets, radios, and most scientific instruments are simply extensions of our biological sensory receptors. The specialization of sensory neurons to the form of energy they detect is further illustrated by the fact that the auditory nerve, which supplies the ear, interprets pain as sound (a ringing in the ears). The optic

---

*Clinically, norepinephrine and epinephrine are generally called by their trade names, noradrenalin and adrenalin, respectively. Both norepinephrine and epinephrine belong to the general group of substances called hormones and will be discussed in greater detail in Chapter 14.

nerve, which supplies the eye, interprets pain in terms of flashes or pinpoints of light ("seeing stars").

In this section we shall examine the structure and function of the sense organs in some detail.

The Sense of Taste

Neurons that detect chemical phenomena are called **chemoreceptors**. Chemoreceptors form the basis of the senses of taste and smell. In all invertebrates and most vertebrates, chemoreceptors are little more than the exposed ends of specialized sensory neurons. Vertebrates have developed specialized chemoreceptors located on the tongue and nasal passages. Both sets of receptors lead to a region of the brain on the lower side of the cerebrum, under the frontal lobes, called the olfactory region, where taste and smell are recorded.

In the vertebrate taste system, chemoreceptors are mostly spaced along the tongue in groupings of cells known as **taste buds,** or **papillae.** Taste buds have hair-like projections exposed at the surface of the tongue. These hair cells (Fig. 13.18) are the actual taste recep-

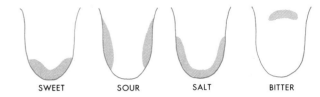

Fig. 13.19
Distribution on the tongue of taste buds sensitive to the four basic tastes: sweet, sour, salt, and bitter. Although there are areas of overlap, the taste buds in each area are particularly homogeneous in the types of chemical substances to which they are sensitive.

tors. In some cases, a single sensory neuron sends its dendrites to the whole group of cells that comprise the taste bud. In others, the taste bud is innervated by more than one neuron. To make the situation still more complex, some neurons may connect with one taste cell, and others with many.

Traditionally, there are four basic tastes: sweet, sour, salt, and bitter. Sensitivity to each of these tastes is greatest on certain areas of the tongue (see Fig. 13.19). Yet it has become clear in recent years that no single taste bud is sensitive to only a single kind of chemical substance. While some are, in fact, specific to sweet, sour, salty, or bitter substances, the majority appear to respond to two or more categories of tastes. The broad range of tastes we associate with foods are really obtained through the sense of smell. This is why taste disappears when a person has a cold.

The Sense of Smell

The organs of the sense of smell are called **olfactory organs.** As indicated in the previous section, much of what we call "taste" is actually due to our sense of smell. In general, the distinction is made between the sense of taste and sense of smell by the medium via which the substance is being transported to the receptor organs. If it travels in solution, it is called taste; in air, smell. It should be stressed, however, that in smell as well as taste the substances being detected probably dissolve in the body fluid before they act on the sensory nerve endings. In some aquatic insects and fish the receptors are enclosed in cavities open to the surrounding water. In this case, they are considered olfactory organs because of their resemblance to such organs in related land-based forms.

The sensitivity of the sense of smell in some organisms is remarkable. A male moth, for example, may detect the sex attractant put out by a sexually receptive female several thousand meters upwind from him by the olfactory receptors on his antennae. It has been estimated that the insect is reacting to no more than a half-dozen molecules of the attractant substance. In verte-

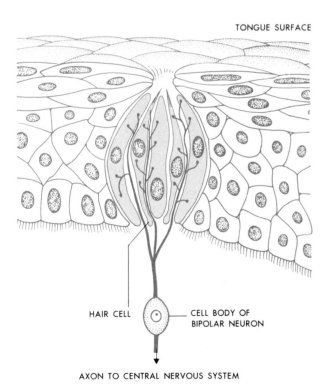

TONGUE SURFACE

HAIR CELL   CELL BODY OF BIPOLAR NEURON

AXON TO CENTRAL NERVOUS SYSTEM

Fig. 13.18
Longitudinal view of a vertebrate taste bud, showing that it is largely a collection of exposed sensory nerve cell endings. Hair cells are the actual taste receptors. Their depolarization is picked up by dendrites of the bipolar neuron that forms synapses with the hair cells. (After E. O. Wilson, *Life on Earth.* Sunderland, Mass.: Sinauer, 1973, p. 499.)

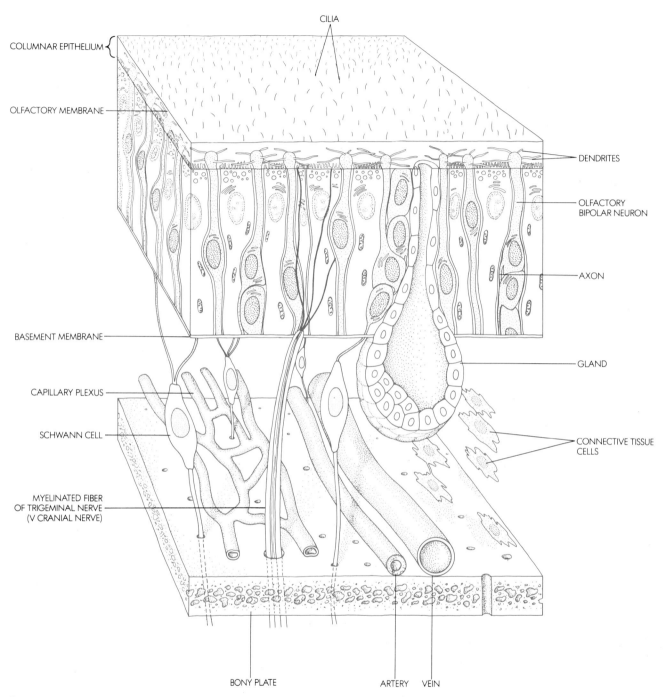

Fig. 13.20
A sectional view of the olfactory epithelium, which consists largely of columnar epithelial cells. The damp surface at top serves as a region where it is believed incoming gases are selectively dissolved and their chemicals trigger nerve impulses that travel via axons to synapses in the olfactory region of the brain to be registered as specific scents. Some of the other tissue present has been removed for greater clarity.

brates, the degree of sensitivity is generally related to the amount of epithelial surface area lining the nasal passages available for olfactory nerve receptors (see Fig. 13.20). Thus the long muzzles of many breeds of dogs provide them with a good sense of smell with which they can track other organisms. Short-muzzled animals (such as cats) generally rely on other senses, especially vision, to locate their prey.

After the substance is detected by the olfactory receptors, the message is conveyed to the brain by the first cranial nerve, the olfactory. In the olfactory lobes of the cerebrum, the message is interpreted in a manner providing the sensation we associate with smell. It is interesting that sensory nerves such as the olfactory interpret other sensations in the "language" of the sensation they generally convey. Thus a sharp blow to the nose carries a distinct scent to the brain, a blow to the eye is interpreted visually by the optic nerve as "seeing stars," and the auditory nerve conveys certain kinds of pain as a "ringing in the ear."

The Sense of Touch

The organs of the body associated with the sense of touch are known as **tactile organs.** The tactile organs are stimulated by pressure and, in humans, are composed of tactile corpuscles consisting of a tiny elliptical bulb enclosed in a sheath of connective tissue and connected to several sensory nerve fibers. Such corpuscles are especially abundant on the fingertips where, in humans, the sense of touch is especially well developed. (It is because of this that the raised dots of the Braille system of reading work so well for the blind.)

In both invertebrates and vertebrates, hairlike projections or bristles serve as sensitive organs of touch via the sensory nerve endings at their bases. The whiskers of a cat are probably the best known example of this in vertebrates, but the sensitivity of human hair to even a light touch (often cited in romantic literature) provides yet another.

There appears to be a separation of the nerve endings in the skin for the detection of pressure and those for the detection of pain. A sharp needle at one point of the skin may elicit the feeling of pressure but no pain; at another, just the reverse. The same sort of distinction appears to be the case for heat and cold receptors.

The Sense of Hearing

To "hear" something means different things to different organisms. To such diverse forms as bacteria or tapeworms it means nothing at all; they simply do not possess structures capable of sensing that which we generally associate with sound. To a bat, to hear something means survival itself, as this animal uses reflection of the ultrasonic sounds it emits to locate barriers in its path and catch the flying insects upon which most species of bats feed. And, quite obviously, the detection of sound may be the difference of life and death for some organisms if it enables them to sense the movements of a predator that might devour them. The most intricate of the sound-detecting devices to evolve is the hearing or **auditory apparatus** we call the ear.

Sound itself is produced and transmitted by the vibratory motion of molecules in a medium such as air.

This motion travels in a wave of alternating compression (increased pressure) and expansion (decreased pressure). What we call hearing, then, is the interpretation of the sensations produced by sound when it stimulates the auditory apparatus. Air is not the only medium through which sound can travel; the sounds of whales, for example, can be heard for many miles underwater. In a sense, the lateral line system of fish, extending down both sides from the head region to the tail, can be considered an auditory apparatus, as it detects alternating changes in water pressure caused by prey or predator and thereby enables attack or escape.

In humans, sound enters the ear opening (auditory meatus) surrounded by the pinna or auricle, and hits the tympanic membrane, often referred to as the "ear drum" (see Fig. 13.21). The tympanic membrane, in turn, is closely associated on its inner surface with the three bones of the middle ear (auditory ossicles); in order, they are the malleus (hammer), incus (anvil), and stapes (stirrup). The stapes transmits the sound vibrations to the oval window, which marks the entry of the sound signals to the fluid-filled inner ear, or cochlea (the name means "snail," referring to its shape). As Fig 13.21 indicates, the cochlea is a rather complex structure; it can be viewed as a closed hydraulic system, like the brake system on a car, where pressure applied to the brake pedal

Fig. 13.21
(a) A sectional view of the human ear, showing the spatial arrangements of the outer (light color), middle (gray), and inner ear (dark color), and the semicircular canals. The Eustachian tubes, usually closed, open during swallowing or yawning and equalize pressure on either side of the tympanic membrane (ear drum). (b) A partially uncoiled view of the cochlea, showing how the ossicles transmit sound-wave-caused vibrations through the middle ear to the oval window. Here the liquid (perilymph) of the cochlea transmits the sound waves to the basilar membrane of the cochlea and its associated organ of Corti. (c) A cross section of the cochlea canals, showing their relationship to the organ of Corti. (d) An enlargement of the organ of Corti showing its relationship to the actual sound-wave-detecting hair cells. Vibrations from the fluid of the vestibular and tympanic canals of the cochlea produce a potential in the hair cells which, in turn, generate nerve impulses in the cochlear nerve which travel to the brain via the auditory nerve (cranial nerve VIII). The basilar membrane is narrower at the base end of the cochlea than it is at the apical end. High-frequency sounds (1500 to 20,000 cycles per second) cause the narrow end to vibrate, and low-frequency sounds (20 to 500 cps) cause the wider portions to vibrate, with the intermediate ranges acting on the membrane in between. Compare with the accompanying scanning electron micrographs of the same region. (Micrograph from *Tissues and Organs: A Text-Atlas of Scanning Electron Microscopy* by Richard G. Kessel and Randy H. Kardon. W. H. Freeman and Company. Copyright © 1979.)

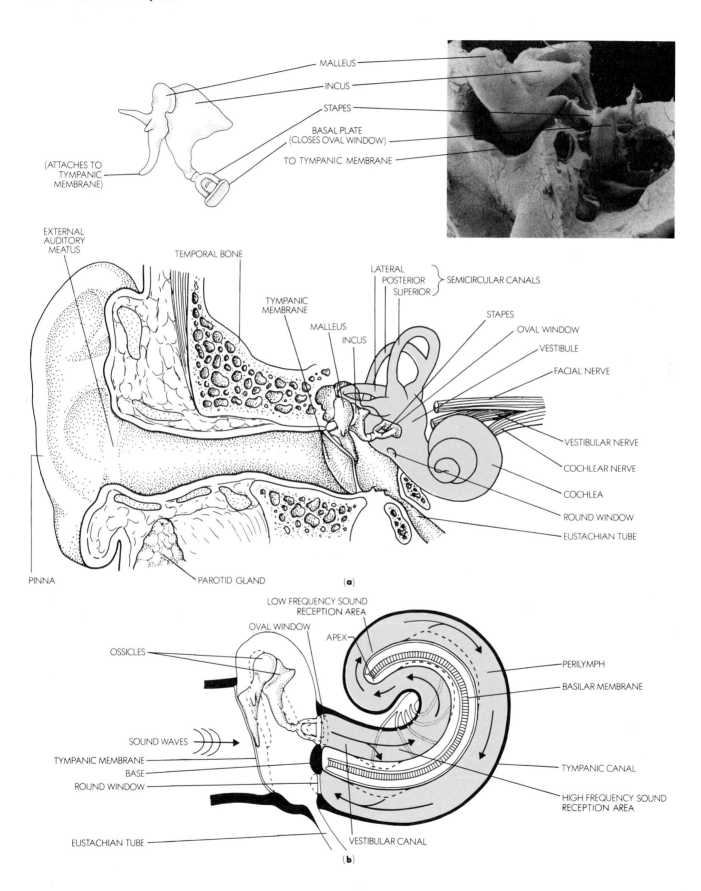

MALLEUS

INCUS

STAPES

BASAL PLATE
(CLOSES OVAL WINDOW)

TO TYMPANIC MEMBRANE

(ATTACHES TO TYMPANIC MEMBRANE)

EXTERNAL AUDITORY MEATUS

TEMPORAL BONE

TYMPANIC MEMBRANE

MALLEUS

INCUS

LATERAL
POSTERIOR
SUPERIOR } SEMICIRCULAR CANALS

STAPES

OVAL WINDOW

VESTIBULE

FACIAL NERVE

VESTIBULAR NERVE

COCHLEAR NERVE

COCHLEA

ROUND WINDOW

EUSTACHIAN TUBE

PINNA

PAROTID GLAND

(a)

LOW FREQUENCY SOUND RECEPTION AREA

OVAL WINDOW

APEX

OSSICLES

PERILYMPH

BASILAR MEMBRANE

SOUND WAVES

TYMPANIC MEMBRANE

BASE

ROUND WINDOW

TYMPANIC CANAL

HIGH FREQUENCY SOUND RECEPTION AREA

EUSTACHIAN TUBE

VESTIBULAR CANAL

(b)

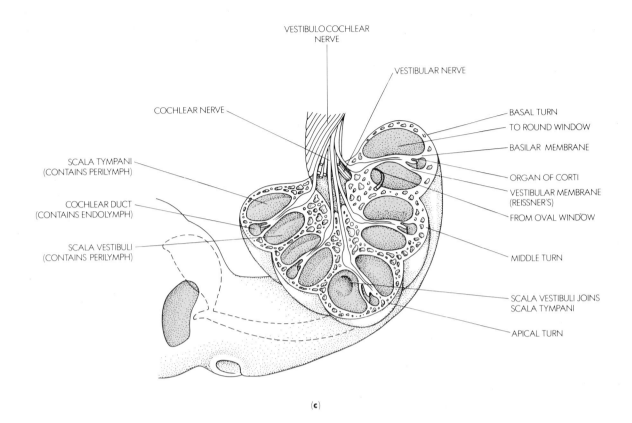

VESTIBULOCOCHLEAR NERVE

VESTIBULAR NERVE

COCHLEAR NERVE

BASAL TURN

TO ROUND WINDOW

BASILAR MEMBRANE

SCALA TYMPANI (CONTAINS PERILYMPH)

ORGAN OF CORTI

VESTIBULAR MEMBRANE (REISSNER'S)

COCHLEAR DUCT (CONTAINS ENDOLYMPH)

FROM OVAL WINDOW

SCALA VESTIBULI (CONTAINS PERILYMPH)

MIDDLE TURN

SCALA VESTIBULI JOINS SCALA TYMPANI

APICAL TURN

(c)

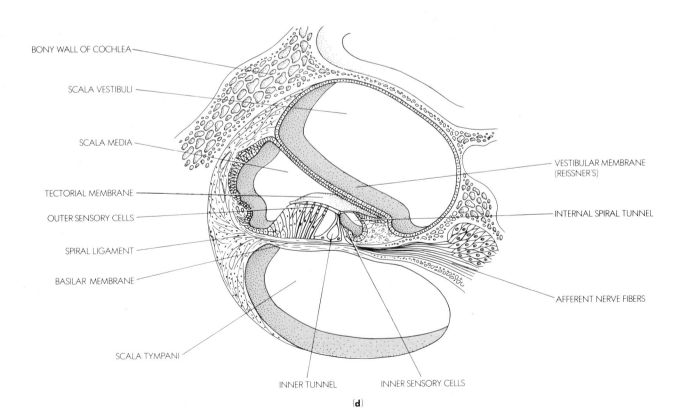

BONY WALL OF COCHLEA

SCALA VESTIBULI

SCALA MEDIA

VESTIBULAR MEMBRANE (REISSNER'S)

TECTORIAL MEMBRANE

OUTER SENSORY CELLS

INTERNAL SPIRAL TUNNEL

SPIRAL LIGAMENT

BASILAR MEMBRANE

AFFERENT NERVE FIBERS

SCALA TYMPANI

INNER TUNNEL

INNER SENSORY CELLS

(d)

results in pressure being applied to the brake drums on the wheels of the car. Similarly, the inward pressure of the stapes caused by sound vibrations applied to the oval window membrane causes a fluid pressure in the cochlea that results in an outward movement into the middle ear of the round window membrane that equalizes the pressure. Thus the sound vibrations travel through the cochlear fluid, from oval to round window. As pointed out, fluid can also convey sound vibrations, and in the inner ear the motion of the fluid in the vestibule (scala vestibuli) of the inner ear causes the basilar membrane to move in a pattern that appears to be related to the frequency and amplitude of the sound. The motion of the basilar membrane, in turn, is transmitted to the organ of Corti, the actual sensory element of the inner ear; from the organ of Corti the nerve action potentials are transmitted to the auditory area of the brain responsible for hearing.

In terms of its sensitivity, few organs can match the ear. Precise measurements have shown that the healthy ear can respond to sound energy only a millionth of a billionth of a watt per square centimeter. Yet the same ear is capable of withstanding the hundred-watt-per-channel or more generated by a rock band—though not, many physicians maintain, without suffering some hearing damage.

The reason for this remarkable sensitivity lies in the transfer of sound waves to the inner ear via the three auditory ossicles. The mechanism has been compared to holding a car door open with one finger while the car is in motion. While the measurement of wind pressure per unit area of the door is quite small, the pressure exerted on the finger is large. In addition, the separate bones provide a lever action. Thus relatively small vibrations of the tympanic membrane arrive considerably amplified at the oval window for transmission to the fluid of the inner ear.

The Eustachian tubes, leading from each middle ear to the upper pharynx, are essential for maintaining equal air pressure inside with atmospheric pressure outside; any great imbalance would tighten the tympanic membrane and make it less sensitive to sounds. Anyone who has descended rapidly in an elevator or airplane is familiar with the often painful results of such a pressure imbalance and finds that relief is often obtained by yawning or chewing gum, both of which act to open the Eustachian tubes. Accompanying the feeling of pressure is the gradual drop in hearing; when the ears "pop," the world seems noisier than usual for a few seconds.

If a nonhandicapped person were given the choice of having to lose either sight or hearing, almost all would choose to save their sight. Yet from birth the choice is not so simple—a blind child is able to function and communicate well and compensate for the handicap of blindness by sharpening the other senses, most especially that of touch, so important in reading Braille. A person deaf from birth, however, loses the ability to learn to speak as well, since we learn this skill by listening. While remarkable progress has been made in educational programs for the deaf, there is little doubt that deafness from birth is a serious handicap.

Deafness itself is usually the result of an injury, aging, or a genetic defect resulting in damage to or defects in the internal parts of the ear. It has been known for more than 200 years that stimulation of the auditory nerve will produce the perception of sound. Thus the development of an artificial ear for the deaf hinges primarily on cracking the code of the electrical impulses by which the ear conveys such information to the brain, in essence tricking the brain into thinking the ear is working even if it is not. A team of scientists headed by Dr. Robert White of Stanford University has developed a device consisting of an external box that converts sound into electrical signals and transmits them across the skin into the mastoid cavity above the ear. There, a surgically implanted receiver and nerve stimulator the size of a quarter sends electrical impulses down a series of electrodes to the auditory nerve. Already such devices have enabled deaf persons at least to be able to tell if the telephone or doorbell is ringing, as well as to distinguish other sounds, such as a fire alarm. The Stanford team is hopeful that they will eventually be able to provide full speech comprehension.

Orientation in Space

Closely associated with the auditory apparatus is the **vestibular apparatus** or **labyrinths**, consisting in part of the *semicircular canals* and the *otolithic organs*, the *utricle* and *saccule*. The semicircular canals are filled with fluid and cause dizziness when one spins quickly on an axis for a minute or so; since the fluid keeps on moving after one stops spinning, the brain interprets the body still to be in motion.

The otolithic organs, on the other hand, detect gravitational effects, centrifugal force, and linear acceleration. The term "otolith" means "ear stone," and refers to the association with the structure of small "stones" (actually particles of calcium carbonate); and their inertia transmits the position of the body to the sensory nerve endings (see Fig. 13.22). In some animals, the stone consists of grains of sand picked up after birth (these include the sand shark or dogfish). If metal filings are substituted for the otoliths of a crayfish, the animal orients itself to the pull of a magnet rather than gravity, even if it means being in a vertical or upside-down position.

The Sense of Vision

Vision is one of the most highly developed of all the senses. The **visual system** of vertebrates, and some invertebrates like the squid, is similar to a camera (Fig. 13.23). Light is collected by a lens and focused on a photosensitive layer (analogous to the film in a camera)

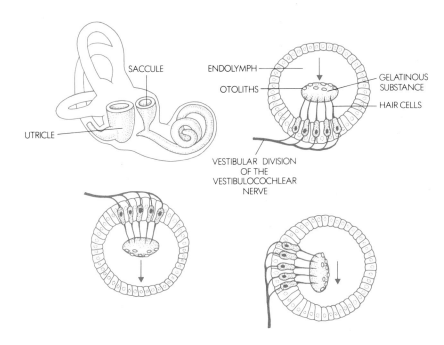

Fig. 13.22
The semicircular canals and their associated sacs, the utricle and saccule. Both the canals and sacs are filled with a fluid (endolymph). Certain regions of both the canals and sacs contain beds of neurosensory hair cells with calcium carbonate crystals called otoliths, embedded in a gelatinous substance, upon them. The otoliths add weight to the gelatinous substance, causing the hair cells to bend depending upon the position of the head and how the force of gravity (arrows) acts upon them. The bending of the hair cells triggers nerve impulses that travel to various motor areas of the cerebellum and medulla to initiate compensatory movements that bring the body back into a balanced position of equilibrium and keep it there. The system is continuous with the cochlear duct, which also contains endolymph (see Fig. 13.21b). (From A. P. Spence and E. B. Mason, *Human Anatomy and Physiology*. Menlo Park, Calif.: Benjamin/Cummings, 1979, p. 423.)

known as the retina. The lens can become shorter and more rounded for close-up focus and longer and thinner for distance focus. This focusing is accomplished by the action of ciliary muscles attached to the lens. The eye can also compensate for the amount of light available by changing the size of the opening, or pupil, through which light entering the eye must pass. The pupil is surrounded by a pigmented muscle tissue, the iris, that regulates the size of the opening. (The iris gives the particular color to a human eye.) The cornea is a protective layer outside the iris and lens. It contains a nutritive fluid, the aqueous humor, which bathes the tissue of the cornea, pupil, and lens. Inside the eyeball is another nutritive fluid, the vitreous humor. These fluids are the only means of supplying necessary nutrient substances to the cornea, iris, and lens, which lack blood vessels.

As people age, the lens becomes less elastic and less able to focus on nearby objects. Older people are often farsighted—that is, they can see well at a distance, but not up close. The opposite condition is called nearsightedness, which means that a person can see nearby ob-

jects quite well but cannot focus adequately on distant objects. People who are nearsighted when they are young often find that their eyesight improves naturally with age, because the normal loss of elasticity by the lens during aging makes it possible to see distant objects more clearly as time goes on.

The analogy between the eye and a camera breaks down when we consider that the photochemical events that occur when light hits a piece of photographic film are very different from those that occur when light hits the retina. The surface of the retina is populated by a large number of highly specialized sensory nerve cells called, according to their shape, rods and cones. Figure 13.24 illustrates the shape of rods and cones and their organization in the retina. In the average human retina, there are about 125,000,000 rods and 6,500,000 cones. In dim light only rods are functional, whereas the cones function in bright light. The rods transmit light in black and white only, while the cones are able to transmit color. The rods are more numerous around the periphery of the retina, cones more in the center, near the fovea.

The view we receive through our sense of sight is not a point-by-point representation of the outside world, as is the case with the image on a photographic film. It is a collection of statements about significant fragments of that world. Through experience, we learn to put that information together into a cohesive and integrated visual pattern.

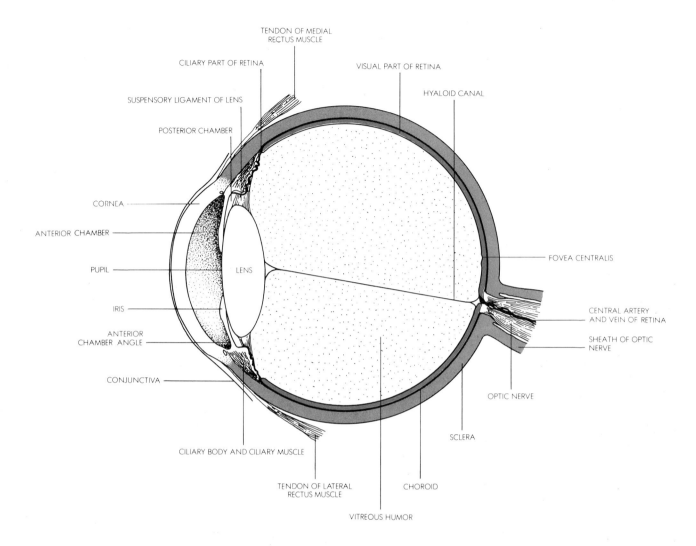

TENDON OF MEDIAL
RECTUS MUSCLE

CILIARY PART OF RETINA

SUSPENSORY LIGAMENT OF LENS

POSTERIOR CHAMBER

VISUAL PART OF RETINA

HYALOID CANAL

CORNEA

ANTERIOR CHAMBER

PUPIL

LENS

IRIS

ANTERIOR
CHAMBER ANGLE

CONJUNCTIVA

FOVEA CENTRALIS

CENTRAL ARTERY
AND VEIN OF RETINA

SHEATH OF OPTIC
NERVE

OPTIC NERVE

SCLERA

CILIARY BODY AND CILIARY MUSCLE

TENDON OF LATERAL
RECTUS MUSCLE

CHOROID

VITREOUS HUMOR

Fig. 13.23
Longitudinal section through the human eye. Light enters through the transparent cornea and bends to pass through the lens, focusing it and causing an upside-down image of what is being viewed to be conveyed to the fovea of the retina. The retina, which can be viewed as merely a cup-like photosensitive extension of the optic nerve (cranial nerve II), contains the rod and cone cells, the latter being responsible for color vision (see Fig. 13.24). The amount of light entering the eye is regulated by an involuntary enlargement or shrinking of the pupil by a pigmented sphincter muscle, the iris, surrounding it. The image on the retina is conveyed to the brain via the optic nerve, where it is perceived as being right side up. Since the optic nerves cross at a region called the optic chiasma, what the right eye perceives is registered by the left side of the brain and vice versa. (Drawing adapted from A. P. Spence and E. B. Mason, *Human Anatomy and Physiology.* Menlo Park, Calif.: Benjamin/Cummings, 1979, p. 399.)

This means, among other things, that a person can perceive color in an object only if its image falls fairly close to the fovea—that is, in the direct line of vision.

To demonstrate this point, take a colored object and slowly move it from the side into your field of vision. You will find that you become aware of the object's size and shape before becoming aware of its color. Most vertebrates do not possess cones, but only rods. This may be because most vertebrates have been nocturnal in habit during their evolutionary history, using their eyes to see under very dim illumination, where color becomes unimportant. Since nocturnal organisms generally sleep during the day, the ability to see clearly in bright light has not been necessary. In the evolution of the human species, however, foraging shifted from nighttime to daytime. Cones, and the color vision they make possible, represent an evolutionary adaptation to bright light. Contrary to some popular opinion, it was not color vision that was advantageous to our early ancestors. Rather the ability to see well in bright light promoted the survival of the species. Color vision is simply a byproduct of the need to see well in bright light.

In addition to rods and cones, the vertebrate retina contains many sensory and connector neurons. These

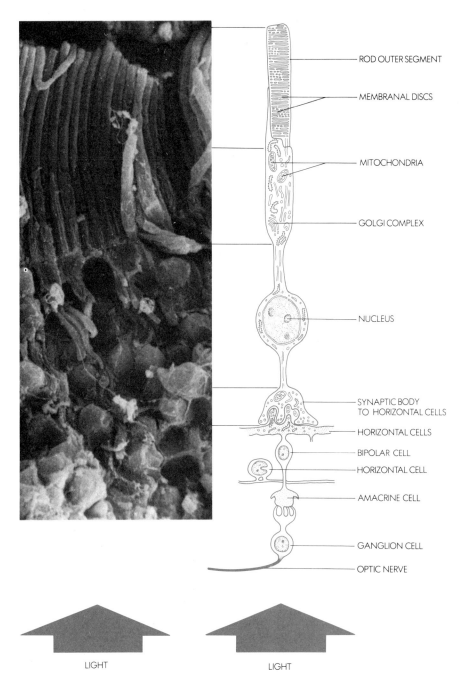

ROD OUTER SEGMENT

MEMBRANAL DISCS

MITOCHONDRIA

GOLGI COMPLEX

NUCLEUS

SYNAPTIC BODY TO HORIZONTAL CELLS

HORIZONTAL CELLS

BIPOLAR CELL

HORIZONTAL CELL

AMACRINE CELL

GANGLION CELL

OPTIC NERVE

LIGHT

LIGHT

Fig. 13.24
A diagrammatic representation of the retina in sectional view. Light energy absorbed by the pigmented layer and the photoreceptor cells (rods and cones) beneath is transmitted via bipolar neurons and ganglion cells to fibers of the optic nerve. Lateral transmission of information also occurs via horizontal cells in the outer layer of synapses and via amacrine cells in the inner layer. Compare the diagram with the actual scanning electron microscope view at the left (× 2305). Note that before the photons of light can be absorbed they must first pass through the layers of optic nerve cells and their processes. (Micrograph from *Tissues and Organs: A Text-Atlas of Scanning Electron Microscopy* by Richard G. Kessel and Randy H. Kardon. W. H. Freeman and Company. Copyright © 1979.)

cells form many synapses with one another. The ganglion cells in the lower levels of the retina ultimately collect all the information obtained by the rods and cones and transmit it, via the optic nerve, to the brain. It is in the brain that the information picked up initially by the photoreceptors is perceived.

Photoreception. What are the events leading to photoreception in the rods and cones? All land and marine vertebrates and some insects have a red pigment, rhodopsin, bound to membranes within the rod cells. Rhodopsin consists of a protein, opsin, combined with a vitamin A derivative, retinaldehyde (or retinal). The retinaldehyde portion of the molecule can exist in a number of configurations, or isomers, in which the shape and thus the chemical properties of the molecule can be changed without changing the number or kinds of atoms that compose it. One configuration is called the 11-*cis* form:

The other, called the all-*trans* form, occurs when there is a rotation on the single chemical bond around the number 11 carbon, which straightens the molecule out (recall that rotation is not possible around a double bond):

It is hypothesized that this configurational, or isomeric, shift is triggered when a photon of light strikes a molecule of rhodopsin when the retinaldehyde portion is in the 11-*cis* form. After the isomeric shift, the retinaldehyde portion of the molecule no longer fits with the opsin portion, and the two units of the rhodopsin molecule separate (see Fig. 13.25). After the retinaldehyde breaks away, the opsin molecule also undergoes a configurational change, uncoiling slightly. A specific enzyme reconverts the all-*trans* retinaldehyde back into the 11-*cis* form that can recombine with opsin. The cycle can now be repeated as soon as another photon is absorbed.

Splitting of the retinaldehyde and opsin molecules causes a change in the membrane potential of the rod cell. Interestingly, in the vertebrate retina such a change in potential does not lead to depolarization of the membrane, but rather to *hyper*polarization. This means that the splitting of opsin and retinaldehyde opens the potassium pores rather than the sodium pores in the membrane. As potassium flows outward, the membrane becomes more, rather than less, polarized. A configurational shift of a single rhodopsin molecule can cause hyperpolarization of the entire rod cell. When this happens, a neurotransmitter substance is released from vesicles at the base of the rod. This triggers nerve pulse action potentials in the postsynaptic bipolar cells and starts the impulse on its way to the brain.

Cones function in very much the same way as rods, but they contain a slightly different visual pigment protein, iodopsin, composed of retinaldehyde bound to a different type of opsin from that found in rods. When

light photons strike the retinal cones, these iodopsin proteins themselves stimulate the formation of unique electrical charges that change 11-*cis* retinaldehyde's sensitivity to different wavelengths of light, or colors. The bond between the 11-*cis* retinaldehyde and the opsins breaks, generating a nerve impulse that carries the color-coded information to the brain.

There are, in fact, three different types of cones, each containing a slightly different version of iodopsin. One kind responds best to blue light, one to red, and one to green. Through the interaction of these three different types of cones, the human eye is capable of extremely fine color discrimination. Our aesthetic abilities, as far as color is concerned, are highly dependent upon the chemical events associated with visual pigments. People who are color-blind for one or another color lack functional iodopsin of one or more types.

The Function of the Retina. How are visual data processed by the retina so as to lead to an organized image in the brain? Although much has been learned about this process in the last decade, we are still far from comprehensive understanding of how stimulation of individual photoreceptor cells leads to formulation of a recognizable pattern in the brain. The events leading to integration occur at all levels in the visual process: in the rods, in the subsurface of the retina, in the ganglion cells, along the optic nerve, and finally in the visual cortex of the cerebrum. Once a photoreceptor (either rod or cone) cell has triggered the release of transmitter substance to the sensory neurons (bipolar cells) that underlie it, the impulse can go in several directions. The bipolar cell can pass the impulse directly to a ganglion cell, or the image can be passed on through horizontal cells (which connect bipolar cells with each other and with rods and cones) to other adjacent bipolar cells. Horizontal cells can connect with several bipolar and several photoreceptor cells. In turn, bipolar cells can synapse directly on one or more ganglion cells, and they can be connected to still more ganglion cells through **amacrine cells** (see Fig. 13.24). As a result of all these hookups, a single ganglion cell usually receives input from several photoreceptor cells at once.

Like all other postsynaptic neurons within the nervous system, the ganglion cells receive both excitatory and inhibitory impulses. To understand how the data-processing system works in the retina, consider the following example. Suppose a very small beam of light is shone on the retina, stimulating one photoreceptor cell at a time. If we then record impulses from the ganglion cells that receive impulses from that photoreceptor, we find that there is no stimulation. If we enlarge the spot of light so that it touches a group of photoreceptors, we find that the ganglion cell will respond once the beam reaches a diameter of about 1 mm. Such a cluster of photoreceptors is called the **receptive field** for that particular ganglion cell. However, it has been found that

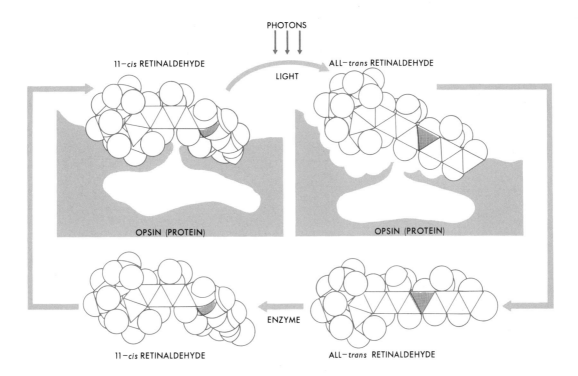

PHOTONS

11−*cis* RETINALDEHYDE

LIGHT

ALL−*trans* RETINALDEHYDE

OPSIN (PROTEIN)

OPSIN (PROTEIN)

ENZYME

11−*cis* RETINALDEHYDE

ALL−*trans* RETINALDEHYDE

Fig. 13.25
The form of 11-*cis* rhodopsin consists of retinaldehyde (a vitamin A derivative) bound to the protein opsin (upper left). When the molecule is struck by a photon of light, retinaldehyde undergoes a configurational shift to the all-*trans* form (upper right). This isomer of retinaldehyde no longer fits into the pocket it once occupied on the opsin portion of the molecule, and the two units fall apart. An enzyme converts the all-*trans* retinaldehyde back into the 11-*cis* form, which rebinds with opsin to begin the cycle over again. (After *Biology Today*, 2d ed. New York: CRM/Random House, 1975, p. 589.)

not all photoreceptors of the receptive field have the same effect. Those photoreceptors in the center (designated "on-center" cells) cause the ganglion cell to fire an impulse. Those on the periphery (designated "off-center" cells) inhibit the ganglion cell. The advantage of this arrangement is that it allows the retina to emphasize contrast between light and dark, yielding a more distinct picture for the brain.

How exactly is the contrast obtained? Consider two ganglion cells next to each other in the lower region of the retina. They synapse with photoreceptor cells that are also close together in the upper part of the retina. Through their synaptic patterns, other ganglion cells synapse with the two photoreceptors by synapsing directly with adjacent bipolar cells. The receptive fields of

the two ganglion cells thus overlap: the center of one's field might be the periphery of the other's, and vice versa. If our small beam of light shines right in the region of overlap, one of the two ganglion cells will be inhibited, the other excited. Thus the contrast between the bright center of the beam of light and the darkness surrounding it will be enhanced by inhibition of the firing of those photoreceptors near the beam's periphery that receive a dimmer light.

Once impulses from ganglia enter the optic nerve, they are transmitted to an area on the underside of the brain, the **lateral geniculate body.** David Hubel and Torsten Wiesel at Harvard Medical School have done much to elucidate the role of the lateral geniculate body. They found, for example, that cells of this structure behave much like ganglion cells: each cell responds to a different small receptive field of the retinal disc. They also have on-center and off-center characteristics. In fact, the inhibitory function of the periphery of the receptive field is even more pronounced in cells of the lateral geniculate body than in the ganglion cells. Thus the tendency to emphasize contrast already present in the message being sent from the retina is further exaggerated in the lateral geniculate body. Among other things, this arrangement means that thin points of light are particularly emphasized in the human visual system, since they can produce excitation without inhibition. At the same time, very small points of light (such as a star) can be completely overlooked if they impinge on a peripheral area and thus produce only inhibitory impulses. To demonstrate this, look up into the sky on a clear night and focus on a particular star. Do not select one of the brightest, but one that is still readily visible. Then move your gaze so that you are looking to the right or left of the star. You may find that the star suddenly becomes invisible. This is because the light is falling on the peripheral area of a receptive field in your retina.

The lateral geniculate body transmits the partially processed data to the visual cortex. Here, incoming neurons bearing messages from widely different regions of the retina synapse at three different levels. Cells on the surface of the cortex are receptive to positional information; cells in the next layer receive inputs from ganglion cells that detect orientation and movement; at the deepest layer of the cortex are the cells that respond to highly complex visual patterns. Response to different visual patterns is therefore a property of the cortical cells in each layer and does not reside in the cells of the lateral geniculate body. Again, a human being's picture of the outside world is not simply a point-by-point transcription of the objects in that world, like a photographic or television image. Rather, it is a collection of signals about lines, points, contours, edges, and other fragments of objects. Nevertheless, through experience and practice, the human learns to interpret these scraps of data as whole and integrated pictures of the outside world.

The Nervous Systems of Other Forms

Not all forms, of course, have a nervous system composed of a brain, spinal cord, and peripheral nerves. Nonetheless, in a multicellular form, the necessity exists for one portion of the organism to "know" what another portion is doing if coordinated activity is to result. In the tentacled cnidarian (or coelenterate) *Hydra*, for example, the trunk-like body has a network or reticulum of nerve fibers extending fairly evenly throughout (though a bit more concentrated around the mouth), while bell-shaped jellyfish, also cnidarians, have a nerve ring around the rim of the bell connected to a nerve network throughout the bell and tentacles. The advent of bilateral symmetry, in which one side of the body is an approximate mirror image of the other, marks the movement toward a concentration of nervous tissue into an anterior mass of neurons called a ganglion, or primitive brain, with nerves running backwards to supply the body. At times, there is one ganglion; at others more than one. We shall have more to say later concerning the nervous systems of other forms.

Summary

1.   The nerve impulse is electrochemical in nature. The impulse is a successive wave of depolarization that travels down the length of the nerve cell axon. Depolarization occurs when the nerve is stimulated. In its resting state, the nerve cell axon has a net positive charge on the outside and a net negative charge on the inside. There are more sodium ions outside, and more chloride and potassium ions on the inside. There are also protein molecules on the inside that bear an overall negative charge. Depolarization involves opening of pores in the cell membrane so that sodium first flows inward, followed by the potassium's flow outward. The result is that for a brief period of time the charge distribution across one small section of the membrane is reversed. Almost immediately active transport, or the "sodium-potassium pump," begins to move sodium ions back outside, restoring the potential. Hyperpolarization occurs when only the pores allowing potassium to leave the axon are opened, increasing the differential charge distribution across the membrane at that point.

2.   Nerve cells (neurons) connect with one another by means of the synapse. At a synapse, the axon end brushes of one neuron come close to the dendrites of another. Vesicles in the axon end brush release neurotransmitter substances when the nerve impulse reaches the end of the axon. The neurotransmitter substance diffuses across the tiny space between the axon end brush and the dendrites, causing depolarization of the postsynaptic neuron. The synapse is the controlling point in the data-processing activity of the nervous system.

3.   Most neurons form synapses with a large number of other neurons. Seldom is a synapse formed between just two neurons.

4.   Nerves innervate muscles through the neuromuscular junction, the point of attachment between the neuron's end brush and the motor units of the muscle. The nerve impulse initiates contraction in the muscle by releasing a neurotransmitter substance at the neuromuscular junction. Acetylcholine is the most common neuromuscular transmitter.

5.   The nervous system as a whole is composed of the peripheral nervous system, consisting of afferent sensory and efferent motor nerves, and the central nervous system, the brain and spinal cord. Integration of information coming into the system as a whole takes place at various levels: at the level of the sense organs themselves (touch, sight, taste, smell, and hearing), within the spinal cord (as in simple reflexes), and in the interaction between the spinal cord and brain.

6.   The simplest reflex arc involves sensory input entering the spinal cord over the dorsal root. The sensory neuron synapses with interneurons, which transmit the impulse to the appropriate outgoing motor neuron. Motor neurons pass out of the spinal cord through the ventral root at the same level at which their associated sensory neurons enter the spinal cord through the dorsal root. More complex reflexes, of which the knee-jerk is a common example, involve both the excitation and the inhibition of certain motor neurons. This is accomplished by the presence of inhibitory as well as

excitatory interneurons in the butterfly-shaped gray matter of the spinal cord. The combination of excitation and inhibition is essential to produce a coordinated response.

7.   Incoming information to the spinal cord is also sent to the brain via nerve tracts passing up and down in the white matter of the cord. Connections between these tracts and signals coming in at any level of the spinal cord are made by interneurons. The brain is thus apprised of all information passing through the central nervous system.

8.   Associated closely with the spinal cord are the nerves that form the autonomic nervous system. This system controls all involuntary processes, such as heartbeat, peristalsis, the secretion of digestive and endocrine glands, and the conversion of glycogen to glucose in the liver. The autonomic system is divided into two parts: the sympathetic and parasympathetic systems. Each internal organ is innervated by branches from the parasympathetic and sympathetic systems. In each case, the effects of the sympathetic and parasympathetic systems are opposite; where one speeds up a process, the other slows it down. Neither system can be characterized as a general accelerator or inhibitor, however.

Two major differences characterize the sympathetic and parasympathetic systems:

a) The sympathetic system is structured as a relay, with two neurons forming a synapse between the central nervous system and the organ innervated. Synapses occur in ganglia outside the central nervous system, close to the target organs.

b) The parasympathetic system also operates as a relay, with the two neurons synapsing *within* the central nervous system. Parasympathetic neurons release acetylcholine at their junctions with the organs they innervate. Basically, the parasympathetic system prepares the body for action by depressing nonessential involuntary functions (such as digestion) that are not essential at the mo-

ment and accelerating necessary functions such as heartbeat.

9.   The brain is the main processing center for incoming and outgoing nerve impulses. It consists of four main parts: the brain stem or medulla, the cerebellum, cerebrum, and pons. Other areas include the thalamus and hypothalamus. The functions of each part are listed briefly below:

a) Cerebrum: The center of rational thought, verbal expression, and the control of most voluntary activity. The sensory centers of the body are located in the cerebrum.

b) Cerebellum: Coordinates impulses sent from the cerebrum to various muscles, thereby helping to coordinate complex motions.

c) Thalamus: A relay center for incoming sensory impulses.

d) Hypothalamus: Containing cells especially sensitive to changes in body temperature, the hypothalamus is the main control center for regulating body temperature. Together, the thalamus and hypothalamus also regulate appetite, water balance, carbohydrate and fat metabolism, blood pressure, certain aspects of the emotional state, and sleep.

e) Medulla: Contains nerve centers controlling many involuntary physiological processes, such as breathing, heart rate, blood vessel contraction or dilation, and vomiting.

f) Pons: A connecting link between the right and left cerebral hemispheres, the pons functions in conjunction with the corpus callosum in providing communicating links between hemispheres.

10.   The sense organs are the gateway between the outside world and the nervous system. All sense organs consist of modified dendrites of sensory nerves. The five senses are those of taste, smell, touch, hearing, and sight.

Exercises

1.   Which of the following observations would be *correct* evidence to show that the nerve impulse is not merely electrical, but electrochemical in nature?

a) The nerve impulse travels more rapidly along the axon than an electric current is transmitted along a suitable conductor.

b) The nerve impulse travels more slowly along the axon.

c) Given a certain minimal electrical excitation, a nerve fiber will always conduct a current of the same magnitude despite the strength of stimulus.

d) The nerve fiber will conduct an impulse with a strength that varies proportionally with the intensity of the stimulus.

2.   Discuss what is meant by the following terms in relation to the passage of a nerve impulse along an axon: (a) threshold, (b) all-or-nothing.

3.   What is the so-called "sodium-potassium pump"? How is it involved in neuron conduction? Do other types of cells show such a phenomenon?

4.   Explain how nerve impulses are thought to be transmitted across a synapse. What evidence supports this hypothesis?

5.   Distinguish between the parasympathetic and sympathetic nervous systems.

Chapter 14
The Endocrine System

14.1
INTRODUCTION

As we saw in the last chapter, sensory and motor impulses pass from neuron to neuron via chemical transmitter substances such as acetylcholine or epinephrine. The latter compound is a **hormone.** Hormones are chemical substances released into the bloodstream by a gland or region that cause an effect elsewhere in the body. Those glands and regions that produce hormones are called, collectively, the **endocrine system** (see Fig. 14.1). As will become clear, the interrelationships between the nervous and endocrine systems are so close and interwoven it is perhaps not too much of an exaggeration to consider one system merely a highly specialized extension of the other.

Perhaps nowhere is this close interrelationship of nervous and endocrine systems so evident as in animal reproductive systems. The successful reproduction of animals depends to a considerable extent upon the proper nerve responses and actions and responses of mating organisms, and to no less an extent upon a proper hormonal environment within the body. Furthermore, particularly in humans, certain psychological factors may enter into reproductive patterns, at times suppressing or completely blocking them.

The gamete-producing organs of both the male and female, the gonads (the testes and ovaries, respectively), are also endocrine glands. For instance, a portion of the mammalian ovary (the corpus luteum) secretes hormones that prepare the female for pregnancy by buildup of the uterine lining for nourishment of the developing embryo. Special cells in the testes (the interstitial cells) produce a hormone (testosterone), which is responsible for the adult development of the reproductive organs, deep voice, growth of beard, and other such characteristic changes associated with male puberty. A group of hormone-like substances called prostaglandins (produced in several different tissues of the body—see Supplement 15.1) and those manufactured in the placenta, a complex organ involved in mammalian embryonic development, also play an important role in reproductive physiology. The precise role of these substances will be postponed until the discussion of reproduction and development in Chapters 15 and 20.

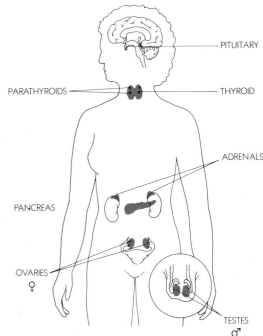

Fig. 14.1
The location of the major endocrine glands in the human female. In the male, the testes replace the ovaries as the endocrine glands associated with the reproductive system. For example, the islets of Langerhans embedded within the pancreas comprise the endocrine portion of this primarily digestive gland. It should be stressed that only those glands that are readily identifiable as distinct anatomical structures are shown here. There are other regions, such as portions of the intestinal wall, that are also endocrine in function; that is, they secrete hormones. Because of its small size, the pineal body, located near the pituitary, is not indicated. The thymus gland is large during embryonic development but shrinks during childhood, and in the adult it may be difficult to find. It is generally located behind the sternum, in the chest region.

Research on the endocrine system generally proceeds from the macroscopic to the microscopic level. If tissue *A* is suspected of having an endocrine gland function, it is surgically removed from an experimental animal to see what deficiency, if any, the animal develops.

A function for gland *A* (assuming that it is a gland) may already be suspected. Thus we can reason that if gland *A* plays a role in physiological process *x*, then surgical removal of gland *A* should result in a cessation of process *x*. The cessation of process *x* is usually detected by certain deficiency symptoms, such as failure to attain sexual maturity or a drop in blood pressure. Should the removal of tissue *A* have the predicted effect, extracts of the tissue may be prepared and injected into the animal's bloodstream. If the deficiency symptoms are relieved, the extract may be analyzed chemically and the appropriate hormone isolated and, perhaps, identified. Studies of its chemical structure may even lead to eventual laboratory synthesis.

14.2
HORMONES OF THE ENDOCRINE SYSTEM: AN OVERVIEW

In terms of their molecular structure, the majority of hormones are peptides (as is insulin) or have molecular structures derived from amino acids (for example, epinephrine). One group, the prostaglandins, are lipids. All other known hormones are steroids (for example, the male and female sex hormones).

Animal hormones can be divided into two types based upon their mode of operation: direct action or by stimulating the production of yet another hormone to cause an action. The latter group are usually referred to as **tropic** hormones. Some exceptions to the general rule that hormones are produced within distinct endocrine glands include cholecystokinin in the intestine, which causes the gallbladder to contract, gastrin in the stomach which, as pointed out in Chapter 10, plays a role in causing the secretion of the enzyme-containing digestive fluids, erythropoietin in the kidney, which controls the rate of red blood cell production, and about a half-dozen more. We shall now discuss the major endocrine glands and the functions of some of the hormones they secrete.

14.3
THE HYPOTHALAMIC-PITUITARY COMPLEX

For years, many textbooks portrayed the entire endocrine system as being under the control of a "master gland," the **pituitary gland** or hypophysis. In truth, however, the pituitary itself is regulated by the hypothalamus of the brain, with which the pituitary is closely associated. Since the hypothalamus is actually a part of the brain, it is dealt with as such in Section 13.8. Here, however, an additional function of the hypothalamus must be stressed, for it contains specialized neural cells called **neurosecretory cells,** which have long axons that pass from their cell bodies in the hypothalamus into the **posterior lobe** of the pituitary (**neurohypophysis** —see Fig. 14.2). The two polypeptide hormones produced by the hypothalamus pass along the neurosecretory cell axons attached to molecules of a carrier protein, neurophysin. Here the hormones are stored for release into the general circulation when needed. In addition to synthesizing and transporting these hormones,

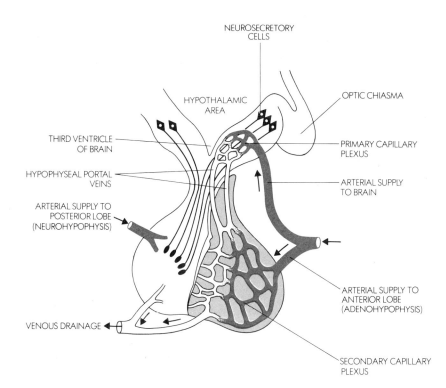

Fig. 14.2
The anatomical relationship of the pituitary gland or hypophysis to the hypothalamus of the brain. As the name neurohypophysis for the posterior lobe of the pituitary indicates, the lobe originates embryologically as part of the nervous system, while the anterior lobe or adenohypophysis originates from a pouch (Rathke's) in the roof of the mouth. The neurosecretory cells produce the antidiuretic hormone (ADH) and oxytocin, which are transported along their axons to the posterior lobe of the pituitary for release into the circulation. The neurosecretory cells also produce releasing substances that travel via the circulation to the anterior lobe of the pituitary.

Hormones are chemical messengers usually produced in specialized glands of the endocrine system. Each hormone has one or more target tissues elsewhere in the body that respond to the hormone in a particular way. Thus the endocrine system is supplementary to the nervous system as a means of communicating between different parts of the body.

neurosecretory cells may also conduct nerve impulses. It is thought that this activity may be intimately involved in the release of these hormones from the pituitary.

The intimate relationship between the hypothalamus, a part of the central nervous system, and the pituitary, a gland in the endocrine system, illustrates a vitally important principle—that the nervous and endocrine systems can be viewed as merely variations on the same theme—that of homeostatic communication and coordination.

14.4
GLANDS OF THE ENDOCRINE SYSTEM

The Pituitary Gland

The **anterior lobe** of the pituitary gland (**adenohypophysis**) secretes tropic hormones, including the *adrenocorticotropic hormone* (ACTH), which stimulates the outer region or cortex of the adrenal glands (adrenal cortex) to produce its hormones (to be discussed shortly); *thyrotropin*, which acts in a similar manner on the thyroid gland; the growth hormone, *somatotropin*; and some hormones associated with sexual functions, to be discussed in Chapter 15. An undersecretion (hyposecretion) or oversecretion (hypersecretion) of somatotropin during development results in the production of midgets, or the reverse condition, giantism, respectively.

Unlike the posterior lobe of the pituitary, which via its neurosecretory cell processes orginating in the hypothalamus has an intimate anatomic association with the brain, the anterior lobe does not. It would *seem*, therefore, that this portion of the pituitary would not be under brain control regarding the synthesis and release of the hormones it produces. In fact, however, a considerable amount of such control exists. Within the brain, specialized neurosecretory cells produce compounds known as hormone releasing or inhibiting substances. (If the substance has been chemically identified, it is called a hormone; if not, it is called a releasing or inhibiting factor). The neurosecretory cells in the brain that produce these substances release them into a capillary network (primary capillary plexus) associated with the third ventricle of the brain. Via connecting veins (hypophyseal portal veins) the released inhibiting substances travel to the anterior lobe of the pituitary.

As already mentioned, the posterior lobe of the pituitary stores and releases two hormones produced by the hypothalamus, the polypeptides **oxytocin** and **vasopressin.** Oxytocin causes the contraction of smooth muscles and plays a special role in the contraction of the uterus during childbirth. It has also been associated with maternal care and secretion of milk in nursing mothers. Oxytocin also has a pronounced effect upon the smooth muscles of the intestine, gallbladder, and urinary tract, though this effect has been hypothesized to be connected with sexual orgasm rather than any physiological function of the affected organs. Vasopressin, often called the *antidiuretic hormone* or *ADH* acts upon the kidneys, where it stimulates the reabsorption of water, constricts the coronary arteries, and raises the blood pressure. Finally, in some animals the mid-portion, or **pars intermedia,** of the pituitary produces a hormone (*melanocyte stimulating hormone, MSH*) which stimulates cells called **melanocytes** to disperse black pigments, bringing about a darkening of the skin.

In 1980 it was reported that some pituitary hormones may be transported directly to the brain as well as being carried there and elsewhere in the general circulation. Working with sheep, a team of researchers under Dr. R. Bergland of the Harvard University Medical School and the Beth Israel Hospital in Boston performed experiments supporting a hypothesis proposing such a direct transport of ACTH, as well as suggesting that this hormone may be produced in and released from the brain as well as the pituitary.

The Thyroid Gland

One of the largest of the endocrine glands, the **thyroid gland** is located just below the larynx, partially encircling the trachea (see Fig. 14.3). In humans, the gland consists of two lobes connected by a narrow portion, the *isthmus;* in most other vertebrates, the lobes are separate. Internally, the thyroid-hormone-producing cells are organized into hollow balls or follicles of cells held together by connective tissues. The central cavity of each follicle contains a protein substance called the colloid, a stored form of the thyroid hormones.

One of the hormones released by the thyroid gland is produced by special cells called C-cells. These cells have a different embryological origin (neuroectodermal)

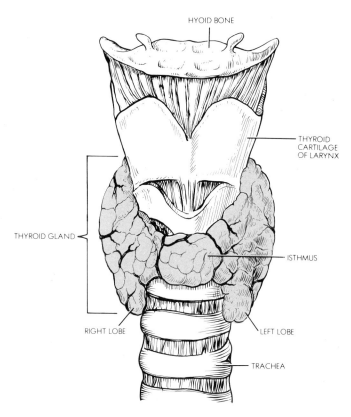

HYOID BONE

THYROID
CARTILAGE
OF LARYNX

THYROID GLAND

ISTHMUS

RIGHT LOBE

LEFT LOBE

TRACHEA

Fig. 14.3
The thyroid gland in relation to the nonendocrinal anatomical parts associated with it. The view is from the front of a human being standing erect; a view from the back is shown in Fig. 14.5. (From A. P. Spence and E. B. Mason, *Human Anatomy and Physiology*. Menlo Park, Calif.: Benjamin/Cummings, 1979, p. 422.)

than the rest of the thyroid gland, which arises from an epithelial thickening in the floor of the pharynx. The hormone produced by the C-cells is called *calcitonin*. Calcitonin, a small polypeptide consisting of thirty-two amino acids, acts to lower blood calcium and phosphate levels by decreasing the resorptive activity of the bone cells (osteoclasts) involved in the development of bone.

The two main thyroid hormones are *tetraiodothyronine*, or *thyroxine*, and *triiodothyronine*. As their

names imply, molecules of these hormones contain four and three atoms of iodine, respectively. Most of the body's iodine is obtained in the diet (see Supplement 14.2) as inorganic iodide (I^-). Within the follicles the iodide is oxidized to elemental iodine and added to the amino acid tyrosine in one step on the way to producing the thyroid hormone.

The effects of the thyroid hormone are several and are generally associated with an increase in metabolic

SUPPLEMENT 14.1
THE ENDORPHINS

In 1971 a team of researchers, headed by Dr. Avram Goldstein of the Stanford University School of Medicine, discovered receptors in nerve cells for opiate drugs such as morphine. The existence of such receptors suggested that the body must produce its own "opiates," and a search for them began. After four years, small proteins or peptides with the chemical properties of morphine were found in animal pituitary glands. Other investigators confirmed the existence and activity of these peptides and called them "endorphins" ("morphine within"). Similar substances, called "enkephalins," had been discovered previously by researchers in Scotland and Sweden.

The discovery of endorphins and enkephalins has generated considerable interest in both the scientific and the medical communities. They seem to provide a possible answer to problems such as pain and stress; they have been found in greatest concentration in the areas of the brain associated with pain and emotion. Further, their occurrence in the pituitary gland suggests that they may either regulate hormone secretion or even act as hormones themselves. This last possibility includes not only the endorphins and enkephalins, but also an entire array of other small protein molecules that have been given the general label of "neuropeptides."

The medical interest in endorphins is also based upon their potential for solutions to age-old medical problems. For example, depression and schizophrenia (split personality) are conditions whose treatments have proven frustratingly difficult for psychiatrists. Intense psychotherapy combined with special drugs to treat depression and other drugs to treat schizophrenia have, at best, only been marginally successful. Now, according to Dr. Nathan S. Kline of the Rockland Research Institute, Orangeburg, New York, the symptoms of both depression and schizophrenia can be alleviated temporarily by the injection of an endorphin called beta-endorphin. In the most dramatic case, one man who had been totally incapacitated for fifteen years is now living on his own after beta-endorphin treatments and now holds a regular job. Unfortunately, the treatments are expensive and must be continued. Research continues now on means to lower the cost of beta-endorphin production.

Beta-endorphin, however, seems to have a split personality of its own. Besides mood control, it has been used successfully (as has an enkephalin) to treat chronic pain. The beta-endorphin is not injected, but its concentration in the blood is increased by electrical stimulation from wires implanted in a patient's brain and controlled by a receiver in the chest. For pain relief, the patient holds a small transmitter over the chest region. The relief from pain occurs within a few minutes and may last for several hours or days. While it may become necessary to use the transmitter with increasing frequency, it has been found that increasing the levels of the amino acid tryptophan in the diet (it is found, for example, in dairy foods and nuts) also somehow acts to reduce the need for transmitter use. Since the increasing use of transmitter "hits" for pain relief is similar to that seen in drug addiction, this last discovery may have significance for its treatment.

There is one other common addiction long known to physiologists—that of overeating. The physiological basis for this phenomenon, often causing obesity, has eluded discovery. Dr. David Margulis of Temple University believes that a division of the autonomic nervous system he terms the "endorphinergic" division, responsible for the release of peptides such as beta-endorphins, is the culprit. He further postulates the existence of yet another autonomic nervous system division, the "endoloxonergic," which signals the body that food is plentiful and conservation not necessary, thereby stimulating activities that use up stored energy and inhibiting the appetite for food and water. (While Dr. Margulis acknowledges that no "endoxolones" have yet been identified in the body, he and his colleagues have identified a peptide in the thyroid and pituitary of obese rats that fits the description of an endoloxonergic hormone.)

The Margulis team finds that the pituitary glands of rats and mice who are genetically obese contain twice the amount of beta-endorphin as the pituitaries of their lean littermates. When given small doses of an opiate antagonist (naloxone), the obese animals ceased their overeating behavior. Thus Dr. Margulis envisions a homeostatic balance between a system that informs an organism when it is time to take in food for energy storage and another system that inhibits this eating behavior and encourages energy usage. Noting further that beta-endorphins also lower the body's normal metabolic rate and temperature as well as causing muscle relaxation and decreased activity, all of which, of course, lower energy expenditure, Dr. Margulis believes that it is this sort of activity that provides the physological base for hibernation. If his hypothesis is correct, obesity in humans (being overweight is the major cause of heart attacks, the number one killer) might be cured by drugs that would restore the homeostatic balance between the endorphinergic and endoloxonergic systems.

Recently Dr. Goldstein has discovered a new endorphin fifty times more potent than beta-endorphin and 200 times more powerful than morphine. Less than one-billionth of an ounce of this peptide can completely halt the movements of a small mass of muscle tissue. This new endorphin has been called "dynorphin," from the Greek word *dynamis*, meaning "power." Dynorphin's powerful nature suggests that it binds with high specificity to its receptor, and thus unlocks a specific function—perhaps, like morphine, pain relief. The problem with the traditional use of drugs such as morphine or codeine to stop severe pain is that they unlock a variety of other side effects, such as dizziness and nausea, to say nothing of being potentially addictive. Goldstein hopes that dynorphin will act at the level of the spinal cord, the first relay station for pain messages from the body, without affecting other higher brain functions.

Finally there is evidence that the key to the age-old mystery of acupuncture may be at least partially provided by endorphins. Known by the Chinese for centuries, but only recently introduced into Western medicine, acupuncture involves the insertion of needles into certain regions of the body in order to block pain. Despite dramatic evidence that acupuncture works, at least with some individuals, not even the Chinese understood why. Now a research team at the National Institute of Mental Health in Bethesda, Maryland, has shown that in rats, acupuncture triggers the release of brain endorphins; following acupuncture, three key regions of the brain, mostly associated with the hypothalamus and thalamus, showed a marked decrease of endorphins while there was a corresponding increase of these compounds in the cerebrospinal fluid surrounding the brain and spinal cord. The study seemed to confirm an earlier one by Swedish researchers that reported evidence for an opiate-like substance in the cerebrospinal fluid of patients after acupuncture and to suggest why acupuncture has been reported successful by Chinese researchers in Hong Kong in blocking withdrawal symptoms in opium addicts.

rate. This increase, in turn, leads to increased oxygen consumption and body heat production by accelerating the oxidation of carbohydrates (and, in general, lipids) and the release of energy. The hormones appear to act indirectly by stimulating the production of RNA in the nucleus. This, in turn, leads to an increase in the production of enzymes, including mitochondrial respiratory enzymes. The increased enzyme concentration means that more substrate can be acted upon, and the **basal metabolic rate (BMR)** is thereby raised.

Interestingly, the effect of the thyroid gland hormone is often dose-related. For example, moderate doses given to rats whose thyroid glands have been removed stimulates protein synthesis, but large amounts inhibit it; small doses of thyroxine in the presence of the pancreatic hormone insulin increase glycogen synthesis while large doses cause a breakdown of glycogen (glycogenolysis).

A number of external factors, such as low environmental temperature or emotional stress may increase or decrease thyroid gland activity respectively. These factors, in turn, are believed to be mediated by communication between the brain and the pituitary and thyroid glands. Since this sort of control is characteristic of many of the endocrine glands and provides an excellent example of homeostatic control, it is worthwhile to expand a bit on the mechanisms of such control.

Recall that homeostatic mechanisms employ both negative and positive feedback (see Section 10.2). In the case of the endocrine system, the production and release into the bloodstream of a hormone by gland A may cause the production of another hormone by gland B. The increased concentration of gland B hormone in the blood may, in turn, influence the production of still more hormone from gland A. This, of course, is positive feedback, and can work only until a certain maximum concentration of gland B hormone is reached (with no such maximum, the system would run wild). On the other hand, the increase in gland B hormone concentration in the blood may decrease production of gland A hormone, and thus the feedback is a negative one.

The control of the thyroid gland provides an example of negative feedback. The production of thyrotropin-releasing hormone (TRH) by the neurosecretory cells of the hypothalamus of the brain causes the release of the thyroid-stimulating hormone (TSH) from the pituitary gland. The TSH travels through the blood to the thyroid gland, where it causes the production and release of the thyroid hormones. Besides the aforementioned effects of these hormones on the basal metabolic rate, their increased concentration in the bloodstream provides negative feedback to either the hypothalamus (to inhibit TRH release) or the pituitary (to inhibit TSH release). The result in either case is the same: less thyroid hormone production and release from the thyroid gland.

Such a feedback system running from target gland to pituitary or hypothalamus is called a **long feedback loop** (see Fig. 14.4). Note, however, that a **short feed-**

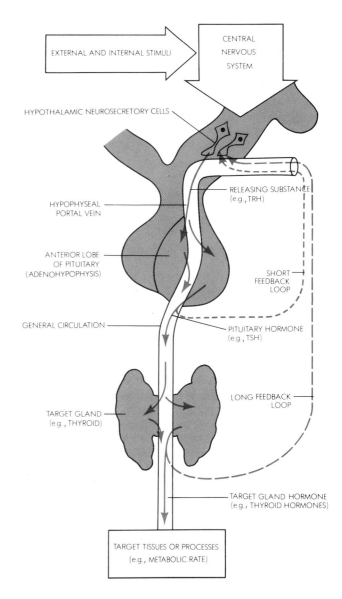

Fig. 14.4
An example of homeostatic feedback in the endocrine system. The production of thyrotropin-releasing hormone (TRH) by the neurosecretory cells of the brain hypothalamus causes the release of thyroid-stimulating hormone (TSH) from the anterior lobe of the pituitary (adenohypophysis). When TSH reaches the thyroid gland via the bloodstream, it causes the production and/or release of the thyroid hormones. The increased concentration of these hormones provides a negative feedback effect on either the hypothalamus or the pituitary, and thus TRH or TSH production is slowed or inhibited. This long feedback loop may be supplemented or replaced by a short feedback loop, in which the pituitary gland's TSH itself may have a negative feedback effect on the production of TRH by the hypothalamus. Such a loop is essential when the target organ is not hormone-producing, such as skin or bone. (Adapted from A. P. Spence and E. B. Mason, *Human Anatomy and Physiology*. Menlo Park, Calif.: Benjamin/Cummings, 1979, p. 441.)

back loop may also exist. The pituitary hormones themselves, in this case TSH, may feed back to the hypothalamus to influence the production of TRH. In cases where the target organ does not produce hormones (for example, bone), a short feedback loop becomes essential.

Both long and short feedback loops allow the endocrine system to play a large role in its own homeostatic regulation. And, as we have seen, the principles of negative feedback, in which the overall response of a system is inhibited, or positive feedback, in which the response is increased, apply to the regulation of many other physiological processes which may or may not involve the endocrine system.

Several physiological conditions result from thyroid gland undersecretion (hypothyroidism) or oversecretion (hyperthyroidism). If undersecretion occurs at an early developmental age, bone and connective tissue growth is retarded, hair growth is lessened, and the skin is dry. The reproductive and nervous tissues are also slowed in their development, leading to a physical and mental retardation called cretinism. The effects of early hypothyroidism can be lessened if thyroid hormones are administered early; otherwise, the condition is generally permanent. Hypothyroidism in the adult causes a condition called myxedema. More common in women than men, myxedema symptoms are somewhat similar to those of cretinism but include puffiness of the facial tissues and swelling of the tongue and larynx. The basal metabolic rate and body temperature is low. Administration of thyroid hormones alleviates the condition.

As might be expected, hyperthyroidism causes a high basal metabolic rate, body temperature, and a quickened pulse. Secondary symptoms include perspiring heavily, nervousness, emotional instability, and insomnia. Treatments include surgery to reduce thyroid tissue mass, the use of radioactive iodine to destroy some of the hormone-producing cells, or drugs that block thyroid hormone function.

The Parathyroid Glands

For years it was noticed that surgical removal of the thyroid gland often led to muscle and nerve irritation leading to cramps and fatal convulsions. It is now known that these resulted from the accidental removal of the parathyroid glands along with the thyroid. There are four parathyroid glands in humans, located in back of (and occasionally within) the thyroid gland (see Fig. 14.5). Thus, they were overlooked by earlier anatomists.

The *parathyroid hormone PTH* (known clinically as parathormone) is believed to exist in two or three forms in the circulation. Its major function is to serve as a controller of calcium and phosphorus metabolism. Since the parathyroid hormone acts to increase calcium concentration in the blood plasma, it works to counter the effects of the thyroid gland hormone, calcitonin; the cooperative homeostatic balance of these two hormones enable bone to be built up and broken down as needed when an organism grows. As might be expected, the levels of calcium excretion in the urine are decreased (and

SUPPLEMENT 14.2
SIMPLE GOITER, IODINE, AND THE THYROID GLAND

For years it was noticed that in certain regions of the world (including Switzerland and the Great Lakes region of the United States), a swelling of the neck in the region of the thyroid gland called simple goiter was prevalent. In 1883 it was noticed that the surgical removal of the thyroid gland caused the condition to occur (as well as other symptoms associated with thyroid hormone deficiency). This led to the practice of giving patients with simple goiter extracts of thyroid gland or sheep thyroid glands in their diets in order to correct the condition.

In 1896 it was discovered that the thyroid gland contained iodine, an element that had not been detected previously in living tissue. In 1905, Dr. David Marine of Western Reserve University in Cleveland, Ohio (located in a region noted for a high occurrence of simple goiter in both humans and ani-

mals) associated the condition with low iodine concentrations in the food and water consumed in the area. In an experiment involving schoolchildren in Akron, Ohio, in which half were given iodized salt and half ordinary salt, Marine showed the iodine in the diet alone was sufficient to control simple goiter. Today, virtually all table salt is iodized (it contains about 0.01 percent potassium iodide). A few years later the discovery that the thyroid hormones are composed of the amino acid tyrosine connected to three or four atoms of iodine provided a molecular basis for Dr. Marine's hypothesis.

A more complex form of goiter, called Graves' disease or exophthalmic goiter, is characterized by glandular enlargement and a protrusion of the eyeballs from their sockets.

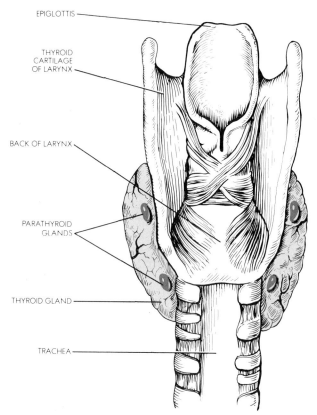

EPIGLOTTIS

THYROID CARTILAGE OF LARYNX

BACK OF LARYNX

PARATHYROID GLANDS

THYROID GLAND

TRACHEA

Fig. 14.5
A view of the parathyroid glands, showing their intimate positional association with the thyroid gland. Since the parathyroid glands are often embedded within the thyroid, or at least not as distinguishable as indicated here, they often used to be removed unknowingly during surgery to remove the thyroid gland (thyroidectomy). (Drawing adapted from A. P. Spence and E. B. Mason, *Human Anatomy and Physiology*. Menlo Park, Calif.: Benjamin/Cummings, 1979, p. 445.)

that of phosphate increased) by the parathyroid hormone. In addition, the parathyroid hormone plays a role in the metabolic transformation of a derivative of vitamin D and, via this derivative (vitamin D_3), stimulates calcium absorption in the intestines.

As might be predicted, hyperparathyroidism results in a weakening of bone structure because of the removal of too much calcium from the bone to the blood plasma (decalcification). As the plasma content rises, calcium salt deposits build up in such soft tissues as the kidneys (calcification). (The results of hypoparathyroidism have already been described.)

In 1980 Drs. M. F. Crass III and Peter K. T. Pang of the Texas Technical School of Medicine showed that a synthetic preparation of bovine (cattle) parathyroid hormone, when injected into dogs, caused a marked dilation of the coronary artery, the main blood vessel sup-

plying the heart. The fact that the parathyroid hormone is vital to the regulation of blood calcium levels, and that calcium is important in the control of vascular smooth muscle contractility, suggests that this finding may provide a useful tool in discovering more about the regulation of smooth muscle tone. It is also interesting to note that only the terminal peptide (containing the first thirty-four amino acids) of the synthetic hormone was present, demonstrating that the parathyroid hormone, like many others, seems to use only certain specific regions of the molecule in triggering its effects.

The Adrenal Glands

In humans, the **adrenal glands** are located on the top of (or *superior to*) the kidneys. Like the kidneys themselves, each adrenal gland consists of two parts, an outer cortex and an inner medulla. Both in embryological origin and function the two regions are quite distinct from each other.

The adrenal medulla secretes two closely related hormones (called *catecholamines*), *epinephrine* (adrenaline) and *norepinephrine* (noradrenaline). Both hormones are synthesized from the amino acid phenylalanine and differ only in that epinephrine has an extra methyl (CH_3) group:

$$HO-\bigcirc-CH(OH)-CH_2-NH-CH_3$$
EPINEPHRINE

$$HO-\bigcirc-CH(OH)-CH_2-NH_2$$
NOREPINEPHRINE

Epinephrine increases both the rate and force of the heartbeat and increases the systolic blood pressure. Epinephrine also lessens the amount of blood going to the skin, mucous membranes, and kidneys by constricting the blood vessels to those regions while at the same time dilating the blood vessels supplying the heart, skeletal muscles, and lungs. Thus epinephrine can be seen as sort of an emergency or stress hormone, when quick physical action may be called for.

Such quick action cannot occur without fuel to burn, of course. Epinephrine also acts to mobilize liver carbohydrate stores via the breakdown of glycogen to lactic acid. The lactic acid can be used by the liver to produce glucose. Thus the indirect effect of epinephrine secretion is a rise in blood sugar levels, thereby providing the needed fuel for energy. Indeed, it is in their action on carbohydrate metabolism that most of us are familiar with the action of our own adrenal medulla hormones; the increase in blood glucose and its use for quick energy, as well as the rapid beating of the heart and the like, are associated with such emotions as fear and anger. In some organisms, the quick response epinephrine action makes possible undoubtedly has selective advantage in escaping from predators.

Epinephrine also stimulates the release of ACTH from the pituitary gland. ACTH, in turn, causes the release of hormones from the adrenal cortex called *glucocorticoids*, which stimulate the production of carbohydrates, especially from proteins.

Norepinephrine acts to constrict blood vessels, increasing their resistance to peripheral flow and thereby also increasing both the systolic and diastolic blood pressure.

It is interesting to note that the adrenal medulla originates during embryological development from neural tissue. This is not by any means unique in the endocrine system; indeed, the posterior lobe of the pituitary develops from an outpocketing of the third ventricle of the brain (it is often referred to as the lobus nervosus), and this is the reason for its intimate anatomical association with the hypothalamus. The adrenal medulla, however, can be considered as a modified part of the sympathetic division of the autonomic nervous system; its cells function in a manner similar to postganglionic sympathetic neurons. Indeed, epinephrine produces effects similar to those caused by stimulation of the entire sympathetic nervous system; the heartbeat and blood pressure are increased, sugar is released by the liver into the bloodstream, and blood is diverted to the heart and muscles by the dilation of the arterioles supplying them and contraction of those supplying areas such as the stomach and intestines. In a sense, then, the adrenal medulla is a specialized ending of the sympathetic nerve fibers that innervate it. This again suggests a very close relationship between the nervous and endocrine systems.

The adrenal cortex makes up the largest amount of the adrenal gland (approximately 80 percent) and is roughly organized into three layers. The outer layer (zona glomerulosa) produces a group of hormones called *mineralocorticoids*, which are involved in the regulation of alkali metal ions (such as sodium, potassium, and lithium). One of these mineralocorticoid hormones, *aldosterone*, has already been discussed in terms of its effects upon the reabsorption of sodium and the excretion of potassium in the kidneys (see Section 10.9).

Another group of adrenal cortex hormones is thought to be produced in the middle layer (zona fasciculata) of cells. These hormones, the aforementioned glucocorticoids, are (as their name suggests) associated with the metabolism of carbohydrates, acting to conserve them by shifting the use of glucose for metabolic energy to fatty acids. As mentioned earlier, they also promote the production of carbohydrates from proteins within the liver. The only major glucocorticoids found in the general circulation are *cortisol* and *corticosterone*; the predominance of one or the other depends upon the species.

Glucocorticoid secretion may be caused by a variety of stressful situations. The stress results in the release of a **corticotropin releasing factor (CRF)** from the brain, which in turn causes the release of ACTH from the pituitary gland. The ACTH triggers the release of the glucocorticoids (such as cortisol) from the adrenal cortex. These may then exhibit an inhibitory influence over ACTH release from the pituitary via negative feedback. Interestingly, the CRF-ACTH-glucocorticoid system is related to the time of day (that is, it is diurnal), with lower levels in the evening than in the morning. Cortisol is bound on release to a sugar-protein complex molecule (glycoprotein) called **transcortin** for transport in the circulation.

All of the adrenal cortex hormones are steroids (see Section 3.4) synthesized from cholesterol, and the term *corticosteroid hormones* is generally applied to both the mineralocorticoids and glucocorticoids. The adrenal cortex also produces androgenic and estrogenic (and progestagenic) substances, bringing about effects similar if not identical to the male and female hormones respectively. (The term corticosteroid hormones is not applied to these substances, which are released only under certain conditions.)

Corticosterone

17α-Hydroxy-11 dehydrocorticosterone
(Cortisone)

17α-Hydroxycorticosterone
(Cortisol)

18-Oxocorticosterone
(Aldosterone)

Hyposecretion of the adrenal cortex hormones leads to a condition called Addison's disease, the symptoms of which vary according to whether a mineralocorticoid or a glucocorticoid hormone is deficient. If a mineralocorticoid such as aldosterone is deficient or lacking, there is a decreased extracellular fluid volume, low blood pressure, weight loss, general weakness, and so forth. If a glucocorticoid hormone is deficient or lacking, a sharp loss of appetite results, leading to excessive thinness, weakness, apathy, and a lowered resistance to stress. (Indeed, the adrenal glands, especially the adrenal cortices, have been closely associated with the body's ability to withstand stress.) The condition can be treated by appropriate adrenal cortex hormone administration.

Hypersecretion of adrenal cortex hormones may also have pathological effects. Too much aldosterone produces an excess of extracellular fluid and is marked by swelling and hypertension, while an excess of cortisol results in Cushing's syndrome, also characterized by weight gain and hypertension as well as by symptoms of diabetes. Tumors on the adrenal cortex may result in an oversecretion of androgenic or extrogenic substances, leading to masculinization or feminization of the body.

The Pancreas

Located behind the stomach between the spleen and the duodenum of the small intestine, the **pancreas** is both a digestive gland producing digestive enzymes and an endocrine gland producing hormones. The endocrine portion is actually composed of aggregations of cells clustered into groups called the **islets of Langerhans.** Each islet is composed of different cells types (see Fig. 14.6). The pancreas is innervated by fibers from both the sympathetic and parasympathetic nervous systems,

which play an important role in regulating the synthesis and release of islet hormones.

Interestingly, while the digestive portion of the pancreas clearly originates embryologically from outgrowths of the duodenum, there is some suggestion that the islet tissues may originate from neuroectodermal tissue that migrated into the deeper intestinal area from the outer regions of the embryo at an early stage of development. Since the entire nervous system originates from such tissue, once again a close relationship between it and the endocrine system is suggested. Indeed, it has even been suggested that the entire endocrine system may have such neuroectodermal origins.

There are four pancreatic peptide hormones, *glucagon, insulin, somatostatin* (growth inhibiting), and *pancreatic polypeptide.* Most significant are the first two, which act in an antagonistic homeostatic balance to each other. Glucagon acts to increase blood glucose levels by increasing the rate of glycogen hydrolysis in the

Fig. 14.6
The pancreas, showing an enlargement of its endocrine portion, the islets of Langerhans. The alpha cells of the islets secrete the hormone glucagon and the beta cells secrete insulin. (Adapted from A. P. Spence and E. B. Mason, *Human Anatomy and Physiology.* Menlo Park, Calif.: Benjamin/Cummings, 1979, p. 451.)

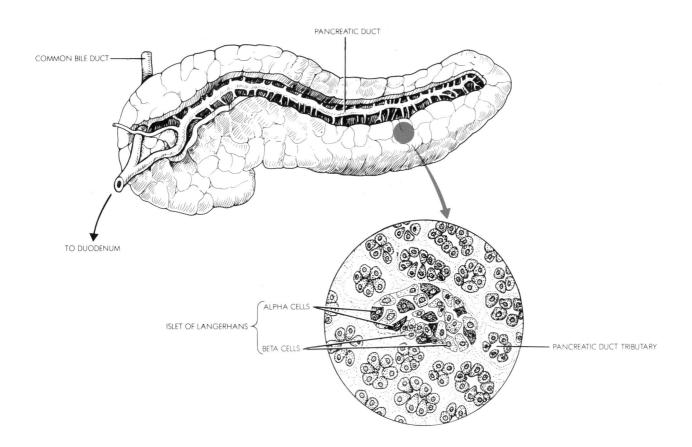

COMMON BILE DUCT

PANCREATIC DUCT

TO DUODENUM

ISLET OF LANGERHANS

ALPHA CELLS

BETA CELLS

PANCREATIC DUCT TRIBUTARY

liver. It also inhibits the synthesis of fatty acids, stimulates the formation of keto compounds from lipids, and increases the rate of breakdown of phosphorus- and nitrogen-containing compounds.

Insulin is probably the most familiar pancreatic hormone, primarily because its deficiency causes diabetes mellitus, a genetically based condition characterized by the presence of high levels of glucose in the urine. Diabetics must compensate for this deficiency by a careful diet and occasional insulin injections. Insulin acts in opposition to glucagon (as well as epinephrine, the glucocorticoids, and the growth hormone) by affecting the permeability of cell membranes to glucose. This leads to increased glucose phosphorylation (see Supple-

SUPPLEMENT 14.3
THE DISCOVERY OF INSULIN

A case study in endocrine research is the discovery of insulin, a hormone known to be produced in the pancreas and to regulate the blood sugar level. In 1886 two physiologists, von Mering and Minkowski, were studying the role of the pancreas in digestion. They removed the pancreas from dogs, expecting to record upsets in the digestion of fats (since the pancreas is the only gland that secretes a fat-splitting enzyme). Instead, the dogs showed a great increase in urine output and died within ten to thirty days. Observant animal caretakers noted that ants were attracted to the urine of such experimental animals but not to the urine excreted by the control animals. Further investigation revealed the cause—the urine of each dog whose pancreas had been removed contained sugar. The resemblance of these symptoms to those of human beings afflicted with diabetes mellitus was evident. Thus an unexpected result and a chance observation led to a rare opportunity for research on a disease fatal to humans, and many investigators turned to the problem.

Histological examination of the pancreas revealed that it is actually composed of two kinds of tissue. One tissue secretes digestive enzymes. The other is found as islet tissue (called the islets of Langerhans) embedded within the digestive tissue. But which of these tissues is the one whose absence causes the diabetes symptoms? It is not possible surgically to remove the islet tissue from the digestive tissue. If the pancreatic ducts are tied off, however, the digestive tissue degenerates, leaving only the islet cells intact. Any hypothesis proposing that the absence of the digestive tissue causes diabetes symptoms would predict that symptoms of the disease would occur progressively as the digestive tissue degenerates. Alternatively, a hypothesis proposing the islet tissue to be the crucial region must predict that no diabetes symptoms will appear. Experimental results have contradicted the first hypothesis and supported the second. No diabetes symptoms were seen in the experimental animals.

But how does the absence of islet tissue cause symptoms of diabetes? It is reasonable to hypothesize that islet tissues secrete a hormone that prevents the effects of diabetes from occurring. Accordingly, experimental animals whose pancreas had been removed were fed pancreas tissue by mouth to see whether the hypothesized hormone could be absorbed in large enough quantities to prevent diabetes symptoms. The results were negative. The hormone, if it existed, was apparently digested and thereby rendered ineffective. Pancreatic extracts dissolved in an appropriate medium were then injected into an organism, again without effect.

About the time of this seeming impasse, a young Canadian physician named F. Banting became interested in the problem. It occurred to him that perhaps the digestive enzymes produced by one type of pancreatic cell might be digesting the hormone produced by the other. To get around this problem, Banting needed to find some way of isolating islet tissue without also triggering the detrimental action of the digestive enzymes of the pancreas.

As was pointed out previously, ligation (tying off) of the pancreatic duct leads to degeneration of the digestive tissue but leaves the islet tissue intact. This enables isolation of pure islet tissue. (Later, the knowledge that in embryonic development the islet tissue begins functioning well before the digestive tissue begins to produce enzymes provided still another means of separating the islet from the digestive tissue.) Banting extracted embryonic pancreas tissue and injected islet tissue extracts into experimental dogs whose pancreases had been removed. Diabetes symptoms subsided dramatically. Attention was then turned to isolation of the hormone produced by the islet cells. The hormone insulin was eventually isolated in pure enough form to be used on human patients. Since the insulin of one mammal is much like that of another, animal insulin can be used to treat human patients.

In the 1950s, Frederick Sanger and his co-workers were able to determine the molecular structure of insulin, which turned out to be composed of two polypeptide chains made up of twenty-one and thirty amino acids and joined together side-by-side by two disulfide bonds (see Supplement 3.2). Laboratory synthesis on a commercially practical scale, however, had to wait for the discovery of proinsulin, a precursor compound in the synthesis of insulin (and a now-known preproinsulin) which made possible the formation of the correct disulfide bonds.

SUPPLEMENT 14.4
CONTROL OF BLOOD SUGAR LEVEL

The way blood sugar level is controlled in the mammalian body is a good example of a hormonally influenced homeostatic control mechanism. It is important that the sugar glucose be maintained at a constant level in the blood if all cells are to function properly. This is particularly true where brain cells are concerned. A number of organs are involved in maintaining glucose at a relatively constant level in the blood. These include the liver, the pancreas, the medullary portion of the adrenal gland, and the hypothalamus. The hypothalamus is especially sensitive to changes in the glucose level of the blood, and it serves as a master regulating center.

When we eat, large amounts of glucose enter the blood through the portal vein, which (as pointed out earlier) runs from the small intestine to the liver. Cells of the liver metabolize glucose to glycogen, which is then stored. To keep the level of blood sugar constant, the liver releases small amounts of glucose into the hepatic vein, which runs from the liver to the heart via the vena cava. What tells the liver how much glucose to send out into the blood? What control mechanisms are involved?

Molecules of glucose in the blood have an effect on a three-part control system. One organ is the pancreas, specifically the cells of the islets of Langerhans, located in the pancreas. The presence of glucose in the blood flowing through these cells stimulates them to secrete insulin. This hormone, released into the blood, is carried to all parts of the body. The major effect of insulin is to lower the amount of glucose in the blood.

Glucose molecules in the blood also act on the hypothalamus itself. An increase of glucose in the blood stimulates the hypothalamus to send inhibitory messages to the liver. As a result, the liver lowers the amount of glucose it releases into the hepatic vein. On the other hand, if the blood sugar level begins to fall below normal, the liver receives a signal to release more glucose. Thus there are two systems in operation, each involving different control centers—the hypothalamus and the islets of Langerhans. Yet both centers are stimulated by the same feedback mechanism—the amount of glucose in the blood.

A third mechanism helps regulate the level of glucose in the blood during periods of stress or heavy exercise. Suppose we are doing very strenuous work. Our body cells are using up large amounts of glucose. The level of glucose in the blood begins to decrease; the liver cannot release glucose as fast as it is being oxidized in the tissues. Now the hypothalamus becomes active. It stimulates the medulla or central portion of the adrenal gland to release the hormone epinephrine. The epinephrine enters the bloodstream, where it has a variety of physiological effects. One of those effects is to activate certain enzymes in the liver cells, the result being an increase in the rate at which glycogen is converted to glucose. Thus more glucose passes out of the liver into the bloodstream.

It is apparent that there are considerable physiological adjustments involved in keeping the blood sugar level constant. Nearly all cells of the body can withstand low glucose levels in the blood for a short period of time; muscle can tolerate low glucose levels for several hours. But when the glucose level drops below a critical value, brain cells cannot remain alive more than a few minutes. The various control mechanisms centered in the hypothalamus and the pancreas help prevent that level from ever being reached.

How do hormones like insulin work? Most studies suggest that insulin is somehow involved in regulating the metabolism of sugars in the body. Diabetics not only have a high concentration of sugar in their urine but also have difficulty storing glucose as glycogen in the liver. It was once hypothesized that insulin acted on some of the enzymes involved in carbohydrate metabolism, in some way ensuring their normal function. Other investigations suggest that insulin acts directly on an enzyme (glycogen synthetase) involved in the synthesis of glycogen from glucose. Insulin also acts on cell membranes, making them permeable to glucose. When the hormone is absent, glucose cannot get into the cell as readily and thus cannot be used as an energy source or stored as glycogen reserves. Instead, it builds up in the blood and subsequently appears in the urine.

ment 14.4) and causes a reduction of blood sugar (hypoglycemia). It also plays a role in increasing the rate of glucose-to-glycogen conversion in the liver, thereby further lowering blood sugar levels.

Besides its effects upon carbohydrates, insulin increases the entry of amino acids into muscle cells, thereby increasing the rate of protein synthesis. It also inhibits the breakdown of lipids (lipolysis) and stimulates free fatty acid uptake from the blood, thereby influencing fat levels in adipose tissue.

The Thymus Gland

The **thymus gland** is a double-lobed structure located in the chest region, behind the sternum. In the human embryo the gland is huge in proportion to the rest of the body, but as the individual matures after birth the thymus beomes progressively smaller. Indeed, in an adult it is often difficult to locate.

As indicated in the discussion of the immune response in Chapter 11, the thymus gland is responsible for the maturation of T-lymphocytes, which play a major role in antibody formation. If the thymus gland is removed from newborn mice, the capacity to produce antibodies is lost. Such experimental mice will also accept grafts of foreign tissue that control animals reject. Interestingly, removal of the thymus only a few weeks after birth has far less effect or none at all; evidently the thymus acts early in life to establish the immune response mechanism. It seems that during development immature T-lymphocytes migrate to the thymus gland from the forming bone marow. After maturing in the thymus, the T-lymphocytes then move to the lymph nodes and spleen. From here they and their descendants will respond to antigens entering the body. That the thymus gland plays a similar time-dependent role in humans seems likely if only from observation of the aforementioned reduction in the gland's size during childhood.

If the thymus gland is removed from a newborn mouse, wrapped in a covering impermeable to cells, and reimplanted, the experimental animals still develop at least some normal immunological capabilities. Such experiments led to the identification in 1966 of the thymus gland hormone *thymosin*—presumably the hormone molecules could pass through the covering, though the T-lymphocytes could not. Thymosin is believed to cause the maturation, proliferation, and immunological competence of lymphocytes. To date, about four other proposed hormones from the thymus gland have been chemically identified and at least four or five more proposed on experimental evidence that have yet to be chemically identified.

The thymus gland itself is controlled by a complex homeostatic relationship with other glands. Hormones from the adrenal cortex and the sex hormones inhibit the thymus—possibly this is why at puberty the gland becomes smaller—while those of the thyroid gland stimulate it. The actions of these thymus-influencing glands, in turn, are controlled by the pituitary gland. An imbalance in this system may cause enlargement of the thymus gland. Such an imbalance seems to occur in cases of the rare muscle disease myasthenia gravis, which, as pointed out in Chapter 12, is characterized by extreme weakness. While the precise cause of this condition is unknown, there does appear to be interference with the transmission of nerve impulses at the neuromuscular junction; not enough acetylcholine receptors seem to be present to cause contraction. For reasons not entirely clear, in many cases of myasthenia gravis the thymus gland is found to be enlarged, and its removal often has a beneficial effect on the condition.

The Pineal Gland

The **pineal gland** or **epiphysis** is attached to the posterior part of the brain epithalamus, close to the pituitary gland. In humans, it is shaped like a pine cone (thus the term "pineal") and is about the size of a pea. The one hormone known definitely to be associated with the pineal gland (often called the pineal body) is *melatonin*. In some organisms, melatonin acts in opposition to the melanocyte-stimulating hormone (MSH) produced by the pars intermedia (middle region) of the pituitary; for example, in frogs, whereas MSH causes a darkening of the skin, melatonin causes it to lighten. Melatonin also influences pigmentation patterns in some mammals. In mice, for example, MSH causes pigmentation to be deposited in the hair and melatonin causes its absence. In weasels, the change of coat color from brown in summer to white in winter is caused by a switchover from the influence of MSH to melatonin.

Possibly acting through some as-yet-undetermined peptides it produces, the pineal gland's influence evidently extends to the rest of the endocrine system as well, since in boys about two to three years of age (but, interestingly, not girls) a tumor of the pineal gland may result in premature sexual development, with growth of pubic hair and the genitals, as well as the other characteristics associated with sexual maturity. It is thought that the pineal gland may help regulate the pituitary, a gland intimately involved in puberty, by acting upon certain hypothalamic releasing centers.

A 1949 edition of the *American College Dictionary* (Random House) defines the pineal gland as "a body of unknown function present in the brain of all vertebrates having a cranium, believed to be a sense organ." The latter phrase points to the fact that the pineal gland has two components, a glandular part and a light-sensitive (photoreceptive) part. In some vertebrates, both are present, while in others (mammals) the photoreceptive portion has been lost. Because of the photoreceptive part, early hypotheses concerning the pineal gland pictured it as a highly modified remnant of a "third eye." This third eye is a photoreceptive structure located in the top of the skulls of some vertebrates (including certain lizards and lampreys). In 1958, Drs. R. C. Stebbins and R. M. Eakins of the University of California at Berkeley suggested that, in reptiles, this third eye measures the intensity of solar radiation. Such measurements would be a matter of great importance to poikilotherms (cold-blooded organisms), as they rely on the sun for their body heat and must move back and forth from shade to sun to keep their temperature constant. Stebbins and Eakins point out that the evolution of homeothermy (warm-bloodedness) in birds and mammals made this function obsolete. It is interesting to note in this regard that the fossil skulls of the earliest amphibians and reptiles often show rather large openings in the top of their skulls where the third eye is located in some modern forms.

Most intriguing are suggestions that the pineal gland may be connected in some animals with **circadian rhythms,** an internal 24-hour cycle of alternating phases of high and low physiological activity in animals and plants. The principal external regulating factor in these rhythms is light. Circadian rhythms represent an example of a biological clock, an internal timing device that keeps on working, seemingly uninfluenced by such factors as body temperature. While the pineal gland is not located in the top of the skull as was its predecessor, the third eye, it may still react to impulses caused by light exposure transmitted to it by eyes or other photoreceptors. In birds, who possess very thin, light, and porous skulls, light may even filter through to the pineal gland. Thus the pineal may be involved directly in measuring day length and informing birds when to begin courtship and build nests in the spring, migrate in the winter, and so on.

In December 1980, a team of scientists headed by Dr. Alfred Lewy at the National Institute of Mental Health in Maryland reported that light mediated through the human eye does appear to affect human endocrine function in much the same way as in other animals. Bright artificial light, similar in intensity to that of daylight, suppressed blood levels of melatonin in experimental subjects. Currently the NIMH team is investigating the possibility that manic depressive patients are more sensitive to light than normal subjects, a relationship suggested by the association of biological rhythm abnormalities with manic depression. (The connection of light with endocrine function in humans has long been suspected on the basis of such observations that, in northern Finland, the majority of pregnancies occur during the summer, when the daylight lasts up to twenty hours.) The NIMH results appear to provide the first clear-cut demonstration of a light-induced hormone secretion in humans.

14.5
THE PROPOSED MECHANISMS OF HORMONE ACTION

Much of the basic research in endocrinology is directed toward determining the precise mechanisms by which hormones work at the molecular level to affect cellular activity. Thus far, as a result of this research, at least two types of hormonal action mechanisms have been established.

One of these types is called the **second messenger hypothesis,** which deals especially with the protein and catacholamine hormones and proposes that the hormones act on cellular function via an intermediate compound, a cyclic nucleotide called **3′,5′ cyclic adenosine monophosphate,** or **cyclic AMP** (abbreviated **cAMP**).

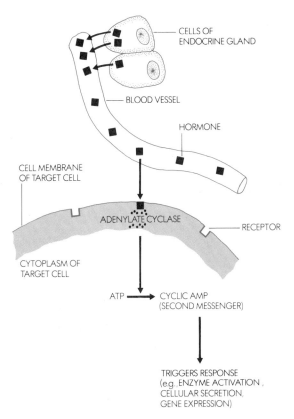

Fig. 14.7
The mechanism of hormone action as postulated by the second messenger hypothesis. The second messenger, cyclic AMP, is formed from ATP in a reaction catalyzed by the enzyme adenylate cyclase. The cyclic AMP, then, acts directly to bring about the hormone's effect, its precise effect depending upon the type of cell being stimulated.

Cyclic AMP is formed from adenosine triphosphate (ATP) in a reaction catalyzed by the enzyme adenylate cyclase (or adenyl cyclase). According to the second messenger hypothesis, the hormone molecules bind to specific receptors on cell membrane surfaces. The result is the triggering of a biochemical mechanism (usually acting through the production of another cyclic nucleotide, cyclic guanosine monophosphate, or cyclic GMP) that increases or decreases the formation of cyclic AMP (see Fig. 14.7).

The second messenger hypothesis, then, views the hormone as a primary messenger and cyclic AMP as a secondary one that acts directly to affect cell function. The precise effect of cyclic AMP depends on the cell type stimulated. In the case of the thyroid gland, for example, the binding of the thyroid-stimulating hormone (TSH) to the thyroid gland cells causes an increase in cyclic AMP levels, which in turn causes the release of the thyroid hormones; in the case of the hormone glucagon binding to liver cells, cyclic AMP concentrations are increased. The result is a higher level of glycogen breakdown and an increase in blood sugar levels. In other cases, some hormones may act to lower the activity of the enzyme adenylate cyclase and thus lower cyclic AMP production.

The way in which cyclic AMP works is complex, yet it provides an excellent example of dual regulation by a single enzyme, protein kinase. This enzyme is activated directly by cyclic AMP. Protein kinase is composed of two different types of subunits, a regulatory (R) subunit and a catalytic (C) subunit. The R subunit can bind cyclic AMP reversibly. When it does so, it becomes dissociated from the C subunit. The C subunit, which is active in the dissociated state, proceeds to catalyze the phosphorylation (via ATP) of the enzyme phosphorylase kinase, thereby activating it. Phosphorylase kinase, in turn, phosphorylates (and thereby activates) the enzyme phosphorylase. Phosphorylase, in turn, catalyzes the breakdown of glycogen, which leads via intermediate forms of glucose to either glycolysis and the Krebs cycle or the release of glucose into the plasma. It is interesting to note that the cyclic AMP-stimulated protein kinase can also *de*activate (by phosphorylation) the enzyme glycogen synthase, which, as its name suggests, catalyzes the synthesis of glycogen from glucose. In other words, protein kinase not only stimulates glycogen breakdown but also inhibits the reverse reaction, glycogen synthesis; thus the term "dual regulation."

Cyclic AMP is known to stimulate a number of different enzyme systems in a wide variety of tissue types (such as bone, muscle, kidney, ovary, and the like). Thus its use as a secondary messenger enables different hormones to produce many different physiological responses (from calcium absorption in the intestine to increased contractility in muscles) by acting through the same molecular mechanism. Moreover, since the molecule involved (ATP) is readily available in the cell, the process is highly efficient; the cell does not have to synthesize a whole set of special internal messengers, each specific to only one kind of external messenger to which the cell is sensitive. Thus the cell achieves a high degree of regulatory response with a minimum of excess molecular machinery.

While it is not clear that all hormones activate only the cyclic AMP system, evidence is mounting that many types of animal cells respond to hormone treatment by increasing or decreasing their internal levels of cyclic AMP. This provides yet another good example of a homeostatic control mechanism, in which hormones or neurotransmitters such as norepinephrine or acetylcholine may act antagonistically to each other in cyclic AMP production. Indeed, the antagonistic action of several different releasing factors on the release of tropic hormones from the pituitary gland is believed related to the way they increase or decrease cyclic AMP levels.

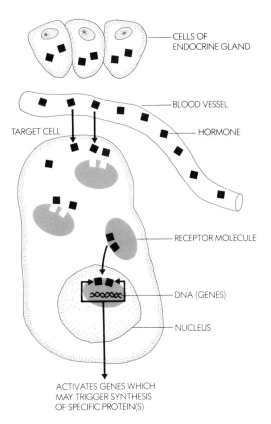

Fig. 14.8
Postulated mechanism of some hormones, especially the steroids. In this case, unlike the mechanism postulated by the second messenger hypothesis, the hormone actually enters the cell (rather than merely binding to its membrane), and then binds to specialized receptor molecules within the cytoplasm or nucleus. The hormone-receptor complex is then transported to the genetic material, deoxyribonucleic acid (DNA), which, via the synthesis of messenger RNA, may induce the synthesis of new proteins to bring about the hormone's effect.

It is significant to note that the mediation in hormone action by the secondary messenger cyclic AMP makes the process directly analogous to the action of the nervous system. In both the endocrine and nervous systems chemical messengers (hormones and neurotransmitters respectively) travel through the extracellular medium to bind to secondary receptors on the surface of a second cell and affect its activity in some way.

In certain cases, peptide hormones have been found within cells and, indeed, it appears that some hormones, especially the steroid hormones, work via a mechanism whereby the hormone, rather than binding to a membrane receptor site, actually enters the cell and binds to either cytoplasmic or nuclear receptor proteins. The binding is thought to cause a structural change in the receptor proteins and the new hormone-receptor complex is transported to the genetic material where it may activate the synthesis of messenger RNA and thereby induce new protein synthesis (see Fig. 14.8).

Finally it has been proposed that some hormones may act by changing the permeability of the cell membrane and thereby change the degree of availability of certain substrate molecules to the cell. Thus, for example, target gland cell A's production or release of hormone X might be increased by the entry into it of larger amounts of the raw materials from which X is synthesized. Or, it might be decreased by a lower amount of raw material entering. A hormone acting to make a cell membrane more permeable to raw materials, or less so, would be expected to produce these effects.

Summary

1. The endocrine system consists of several glands or tissues that produce specialized substances called hormones. The hormones travel via the bloodstream to various organs or tissues where they act directly or indirectly to bring about their specific effects.

2. The major endocrine glands of the body, in terms of their being anatomically distinct structures, include the pituitary, thyroid, parathyroids, adrenals, pancreas (islets of Langerhans), and the gonads (the female ovaries and male testes). Certain other regions, however, also have endocrine functions (these include the wall of the duodenum, which secretes the hormone secretin).

3. The anatomical and physiological relationship between the brain hypothalamus and the pituitary gland via neurosecretory cells emphasizes the close interrelationship between the nervous and endocrine systems. Indeed, at least some portions of the endocrine system (such as the posterior lobe or neurohypophysis of the pituitary and the medulla of the adrenals) develop embryologically from neural tissue.

4. Hormones themselves are either peptides, amino acid derivatives, lipids, or steroids.

5. Some hormones (tropic hormones) act by stimulating the production of other hormones, which then act more directly.

6. Hormones are hypothesized to act either by binding to specific sites on the cell membranes and utilizing cyclic AMP as a "second messenger" within the cell to bring about hormone-specific effects or by binding with receptor molecules within the cell and stimulating the genetic material (DNA) to synthesize specific proteins.

Exercises

1. In terms of coordination, what are the relative advantages to the organism of the nervous and endocrine systems?

2. Physiologists have been concerned with the question of what effect insulin has on rate of fat synthesis (conversion of glucose into fats) in mammals. In the series of experiments shown in Fig. 14.9 (on p. 408) radioactive carbon (C^{14}) was used as a tracer.* Study the figure, then answer the following questions.

    a) Do these experimental results support or refute the hypothesis that insulin *does* speed up synthesis of fats in mammalian tissue? Give reasons for your answer.

    b) What is the function of the experiment series in row 1? What does it show in relation to the experiment as a whole?

    c) It has been shown that insulin does not increase the rate of synthesis of fats from glucose under *in vitro* (test-tube) conditions—that is, if the cells of fat tissue have undergone degradation, yet all the molecules from the cell are retained. Can you suggest a mechanism by which insulin could affect the rate at which glucose is converted into fats in living cells within an organism?

3. The two graphs in Fig. 14.10 (on p. 408) show the effect on blood glucose concentration of administering 50 g of glucose to two different individuals. What differences do you note in the response of the two individuals? Which of the two individuals appears to be normal? What may be wrong with the other? Give reasons for your answer.

---

*Adapted from Vincent Dole, "Body Fat," *Scientific American,* December 1959, p. 74.

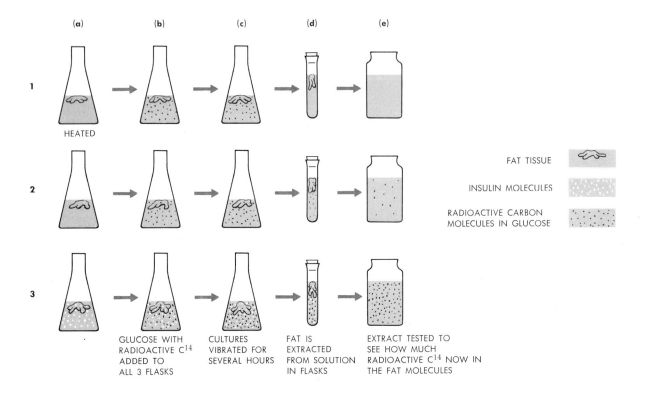

(a)   (b)   (c)   (d)   (e)

1   HEATED

2

3   GLUCOSE WITH RADIOACTIVE C^{14} ADDED TO ALL 3 FLASKS   CULTURES VIBRATED FOR SEVERAL HOURS   FAT IS EXTRACTED FROM SOLUTION IN FLASKS   EXTRACT TESTED TO SEE HOW MUCH RADIOACTIVE C^{14} NOW IN THE FAT MOLECULES

FAT TISSUE

INSULIN MOLECULES

RADIOACTIVE CARBON MOLECULES IN GLUCOSE

Fig. 14.9

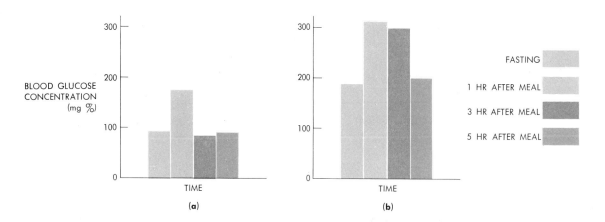

BLOOD GLUCOSE CONCENTRATION (mg %)

TIME
(a)

TIME
(b)

FASTING

1 HR AFTER MEAL

3 HR AFTER MEAL

5 HR AFTER MEAL

Fig. 14.10

Chapter 15
Human Reproduction and Sexuality

15.1
INTRODUCTION

Despite the beauty and fascination that the processes of human reproduction, development, and sexuality have always held, few subjects have been so widely misunderstood or fraught with so much superstition, myth, and ignorance. Indeed, only recently have human sexuality and reproduction come to be considered proper spheres for biological research. It is no exaggeration to state that more has been learned in the past thirty years on these subjects than in the previous 3000!

In this chapter we shall discuss only human sexual anatomy, physiology, and behavior; we shall deal in later chapters with these same topics as they occur in other animals.

15.2
DEVELOPMENT OF SEX ORGANS

Sexual differentiation of human male and female embryos is related to their different chromosome complements, the XY for males, and the XX for females.* However, development of the male or female genitalia and internal sex organs proceeds along remarkably similar lines in both sexes during the first trimester. In fact, the male and female genitalia and gonads derive from many of the same structures in the developing embryo. Thus, despite their genetic differences, in the early stages embryos of both sexes have the potential for differentiating into either a male or a female. The factors governing the actual direction of differentiation are largely hormonal.

The reproductive organs begin to develop in conjunction with the embryonic kidneys. A pair of tubules, the **Wolffian ducts,** grow from the midregion of the young embryo toward the posterior. These are joined by a second pair of tubules, forming a common duct that ultimately develops into the ureter. Just lateral to the Wolffian ducts are a second pair of tubules, the **Mullerian ducts,** which are precursors of the Fallopian tubes, uterus, and part of the vagina. The Mullerian

ducts appear in both male and female embryos. The primordial gonads develop as two thick **germinal ridges** in the body cavity (coelom) but lie near the Mullerian and Wolffian ducts. These ridges contain several types of cells. One type is larger than the others and will become the primordial germ cells. Primordial germ cells reach the germinal ridge by migration from nearby regions. The germinal ridge develops into the indifferent reproductive organ. At this stage, approximately by the beginning of the third month, the primordial germ cells are similar in both ovaries and testes.

In male embryos the primordial germ cells migrate from one region of the germinal ridge to another and become organized into the seminiferous tubules (see Fig. 15.1). In female embryos, the primordial germ cells do not migrate but remain localized in their original position within the germinal ridge. Certain parts of the germinal ridge are then reabsorbed, and the central part of the developing gonad becomes filled with loose tissue and blood vessels. The primordial germ cells are surrounded by masses of cells and swell as they develop into oögonia. These undergo rapid mitotic divisions and after 20 weeks have yielded some 7,000,000 oögonia. Mitosis then ceases and no further ova are produced. The total number of oögonia is actually reduced from this point on until puberty—there are about 1,000,000 oögonia present at birth and about 400,000 by adolescence. Of these, only about 400 to 450 will be released at ovulation as mature ova during a woman's reproductive period. The remainder degenerate.†

From the end of the first trimester, the development of the various ducts and tubules associated with the male and female reproductive systems is quite different for each sex. In male embryos, the Wolffian duct becomes the vas deferens. This development proceeds under the stimulation of testosterone secreted by the already differentiating interstitial cells of the developing testes. The testes appear to secrete a second peptide hormone that inhibits the development of the Mullerian ducts. In the female embryo, the absence of the male hormones and the presence of the female hormones cause the Wolffian ducts to degenerate and the

---

*See Section 18.6, beginning on p. 517.

†See Section 20.7, beginning on p. 601.

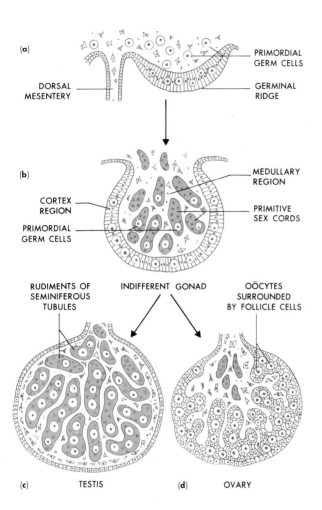

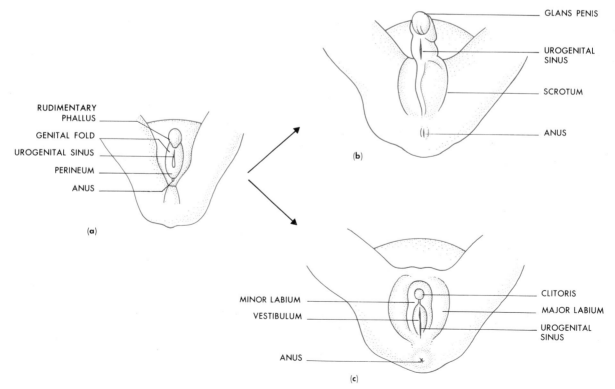

Fig. 15.1
Diagram of development of gonadal tissue (testes and ovaries) in higher vertebrates. (*a*) The undifferentiated primordial germ cells in the germinal ridge; some of the cells making up the future gonad come from the ridge itself and others from the adjacent, less well-defined region. (*b*) Sexually undifferentiated sex organ, similar in both young male and female organisms. (*c*) Differentiation into the male sex organ, the testis; the germ cells are located in the future seminiferous tubules. (*d*) Differentiation into the female sex organ, the ovary.

Fig. 15.2
Development of the male and female external genitalia from a neutral condition. (*a*) The sexually neutral stage in genital development at the beginning of the third month. (*b*) Male embryo at sixteen weeks, showing enlargement of the phallus into the developing penis. (*c*) Female embryo at sixteen weeks, with the phallus becoming incorporated into the clitoris.

Mullerian ducts form the Fallopian tube or oviducts at one end and fuse to form the uterus, cervix, and vagina at the other.

Development of the external genitalia follows much the same course as the two sets of internal reproductive organs (see Fig. 15.2). In both male and female embryos the external opening of the urogenital sinus is flanked by elongated thickenings, the genital folds. The folds meet anteriorly and form an outgrowth, the **rudimentary phallus.** Under the stimulation of various sex hormones, this sexually indifferent stage developed by the begin-ning of the third month undergoes differentiation toward either male or female. In males, the rudimentary phallus develops to become the penis. In females, the phallus grows only slightly and becomes the clitoris. In females, the genital folds do not fuse as in the male, but remain as the labia minora. Thus internal and external sex organs of the human reproductive system develop from the same rudimentary structures in both sexes. The major stimulus for differentiation in one direction or the other is provided by estrogens or testosterone—them-selves produced under genetic influences.

In some cases, hormonal imbalances during em-bryonic development produce individuals in whom both male and female systems develop more or less equally (**hermaphroditism**) or in whom both are pres-ent, though one may be more developed than the other. Such congenital problems of sexual differentiation are the result of genetic and/or chromosomal aberrations. Usually neither gonad is functional. Such conditions can sometimes be corrected by surgery or hormone therapy administered at precise times during the child's life.

15.3
THE MALE REPRODUCTIVE SYSTEM

The human male reproductive system consists of several organs. An internal diagram of these organs is shown in the sectional view in Fig. 15.3. The paired **testes**, contained within a sac, the **scrotum**, serve two very important functions. First, they produce sperm and thus are the organs necessary for male fertility. They are

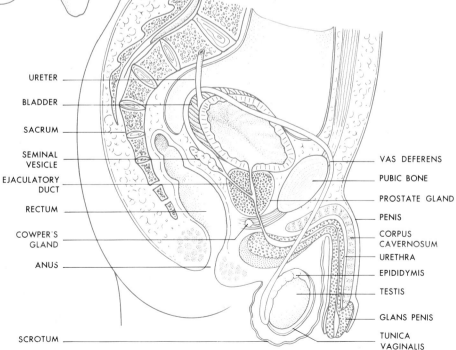

Fig. 15.3
The human male reproductive sys-tem, showing its location within the pelvic girdle.

PELVIC GIRDLE

VAS DEFERENS (SEMINAL DUCT)

DORSAL ARTERY

DORSAL VEIN

CORPUS CAVERNOSUM

URETHRA

CORPUS SPONGIOSUM

RAPHÉ

URETER

BLADDER

SACRUM

SEMINAL VESICLE

EJACULATORY DUCT

RECTUM

COWPER'S GLAND

ANUS

SCROTUM

VAS DEFERENS

PUBIC BONE

PROSTATE GLAND

PENIS

CORPUS CAVERNOSUM

URETHRA

EPIDIDYMIS

TESTIS

GLANS PENIS

TUNICA VAGINALIS

Human development includes all aspects of the sequence of biological and behavioral changes that characterize the human life cycle.

also endocrine glands, producing the steroid male sex hormone, *testosterone.* The sperm-producing part of the testis consists of a series of highly coiled tubes, the **seminiferous tubules.** The seminiferous tubules converge into about a dozen ducts leading into the **epididymis,** a much-coiled collecting tube that lies on the upper part of each testis. Along the lower surface of the testis the epididymis connects to the **vas deferens,** or seminal duct, which leads upward, curves over and behind the front arch of the pelvis, and eventually joins the urethra coming from the bladder. From this point of juncture the two tubes join and follow a common course to the exterior through the penis.

The **penis** is a cylindrical organ composed of three bodies of spongy tissue. Normally there is only a moderate flow of blood through this tissue. During sexual excitation, however, the capillaries open up, allowing greatly increased volumes of blood to enter the spongy tissue. This extends the length and diameter of the penis until it becomes firm, a state known as **erection.** As Fig. 15.3 shows, the three layers of spongy tissue making up most of the penis are not identical. The two side layers, the *corpora cavernosa* (singular, *corpus cavernosum)* are larger; the lower layer, *the corpus spongiosum,* is smaller and surrounds the urethra. The corpus spongiosum expands at the upper end of the penis to form the cap-like *glans penis.* The glans is covered by a fold of skin called the *foreskin,* which is often removed surgically in the process of circumcision shortly after birth. The foreskin serves as a protection for the glans penis. (Originally done for religious purposes, removal of the foreskin is still practiced by many physicians on the grounds that unless the skin beneath the fold is kept clean, it may become infected.)

The mammalian penis, including that of the nonhuman primates, usually contains a bone, the *os penis,* which serves to give the organ part of its rigidity. The human male lacks this bone; the aforementioned flow of blood into spongy tissue alone makes the penis rigid. The rigidity during sexual excitation allows the penis to be inserted into the female.

The Seminal Fluid and Sperm Production

At several points along the vas deferens and urethra, ducts leading from various glands pour their secretions into the sperm suspension. These glands include the paired **seminal vesicles,** the **prostate gland** (in human males the two prostate glands are fused to form a single gland; in other mammals there are two prostates), and

the paired **Cowper's glands.** The seminal vesicles and Cowper's glands contribute a mucous alkaline secretion, while the prostate adds a thin milky fluid. This entire mixture is known as the *seminal fluid.* The seminal fluid contains glucose and fructose that provide energy sources for the sperm, acid-base buffers, and mucous materials that lubricate the passages through which the sperm travel. Sperm formation in the testes (*spermatogenesis*) involves the division and transformation of specialized cells lining the seminiferous tubules (see Fig. 15.4). The precise nature of spermatogenesis will be discussed in Chapter 20. Basically, the transformation phase involves a reduction in the volume of cytoplasm, with the nucleus becoming the head of the sperm cell, and the remaining cytoplasm the flagellum or tail.

In the male of most mammalian species, spermatogenesis continues throughout the animal's lifetime, although there is some decrease in old age. This situation is in marked contrast to the female who, as pointed out earlier, normally has at birth all the eggs she will ever possess as an adult (approximately 400,000). For most human females, only 400 of these will be ovulated and thus be available for fertilization. The male, on the other hand, releases an average of about 480,000,000 sperm in one ejaculation.

The sperm-producing cells in the testes and the spermatids cannot tolerate the higher temperature in the abdomen. Since it is more exposed to circulating air, the scrotum has a lower average temperature than the abdomen. However, testicular tissue is also sensitive to cold. Thus, to help in maintaining a constant temperature for the testes, the scrotum is furnished with a lengthwise set of involuntary smooth muscles. When the testes become cold, as for example when swimming, the muscles contract, drawing them up close to the abdominal cavity. When the testes become too warm, the muscles of the scrotum relax, lowering the testes away from the abdominal cavity.

Male **sterility** is the result of failure of the testes to produce functional sperm. Sterility may arise from a number of different causes. In some cases it may be due to the failure of the testes to develop normally during embryonic and later childhood or adolescent growth. In other cases it may be due to chromosomal disorders, which may impair normal meiotic processes and lead to lowered or nonexistent sperm production. Sterility should be distinguished from **impotence,** which is defined as the inability to copulate. In men, this is most commonly caused by failure to achieve erection, either for physical or, far more commonly, for psychological reasons.

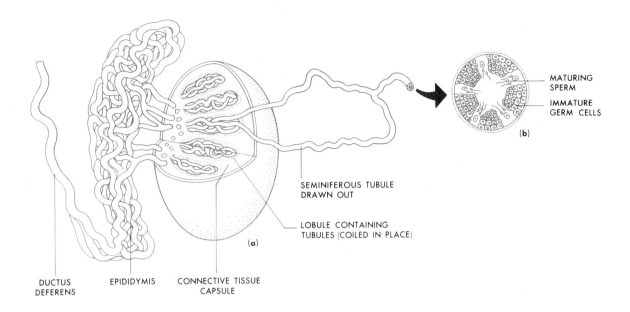

Fig. 15.4
The human testis, showing a cross section of a seminiferous tubule and a detailed view of sperm production. (*a*) One testis with the epididymis viewed from the top, showing the seminiferous tubules. (*b*) Cross section of one seminiferous tubule, showing the cells that give rise to the sperm. The cells are increasingly mature closer to the cavity (lumen) of the tubule; mature sperm remain attached with their tails projecting into the cavity until they break free and move toward the epididymis. (After *Biology Today*, CRM/Random House, 1972.)

The Testes as Endocrine Glands

In addition to their sperm-producing function, the testes also serve as one of the body's important endocrine glands. As might be suspected, the testes produce the male sex hormone, consisting of several different hormones called, collectively, **androgens.** The most well-known of the androgens is **testosterone,** a steroid compound with the following characteristic chemical structure:

The male sex hormones control the development of both the primary and secondary sex characteristics. Primary sex characteristics are those parts of the reproductive system which play a role in the reproductive process (such as the testes). Secondary sex characteristics are those phenotypic traits that differentiate males from females in any species but are not directly related to sex. In deer, for example, antler development is controlled by androgens, as are the plumage patterns and large comb development in roosters. In humans, patterns of body growth, deepening of the voice, hair distribution, and general musculature are all controlled by androgens. The development of the actual reproductive tissue—for instance, the sperm-producing cells of the testes—is not controlled by the androgens directly but by another group of hormones produced in the pituitary gland.

The androgens are synthesized from cholesterol, which is derived from acetate (acetic acid). Thus the production of sex hormones is tied directly to the citric acid cycle and connected to the general pathways of intermediary metabolism. In the testes, androgens are produced in the **interstitial cells of Leydig,** which are different from those producing the sperm. As an organ, then, the testes are differentiated into two quite different cell types: one producing functional sperm, the other producing several androgens. In addition to the interstitial cells, androgens are produced by another endocrine gland, the adrenal. The adrenal gland is located

---

Male sex hormones are collectively called androgens; female sex hormones are collectively called estrogens.

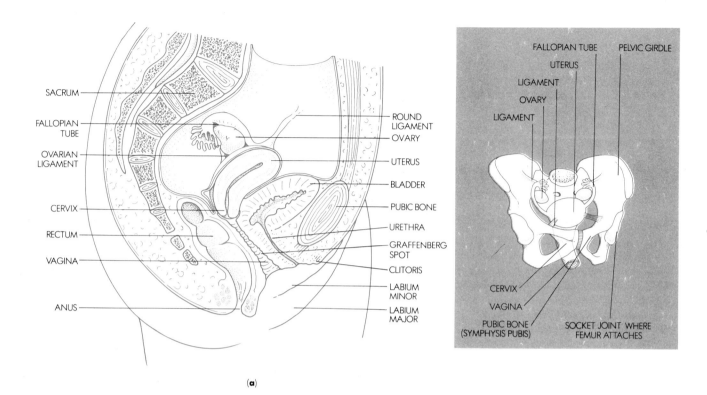

SACRUM

FALLOPIAN TUBE

OVARIAN LIGAMENT

CERVIX

RECTUM

VAGINA

ANUS

ROUND LIGAMENT

OVARY

UTERUS

BLADDER

PUBIC BONE

URETHRA

GRAFFENBERG SPOT

CLITORIS

LABIUM MINOR

LABIUM MAJOR

FALLOPIAN TUBE

UTERUS

LIGAMENT

OVARY

LIGAMENT

PELVIC GIRDLE

CERVIX

VAGINA

PUBIC BONE (SYMPHYSIS PUBIS)

SOCKET JOINT WHERE FEMUR ATTACHES

(a)

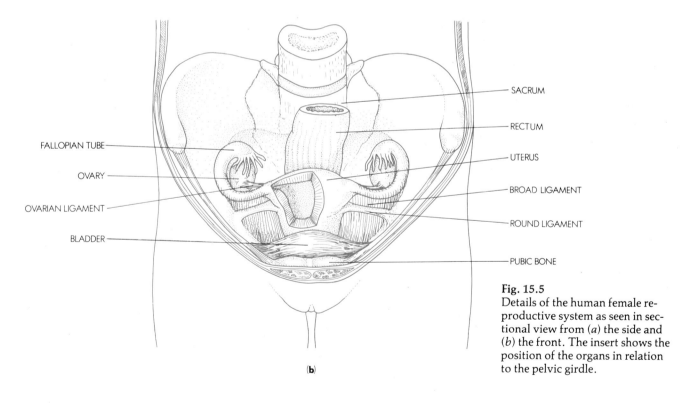

FALLOPIAN TUBE

OVARY

OVARIAN LIGAMENT

BLADDER

SACRUM

RECTUM

UTERUS

BROAD LIGAMENT

ROUND LIGAMENT

PUBIC BONE

(b)

Fig. 15.5
Details of the human female reproductive system as seen in sectional view from (a) the side and (b) the front. The insert shows the position of the organs in relation to the pelvic girdle.

While the human female generally ovulates one egg at a time, a male may release as many as 480,000,000 sperm in one ejaculation.

on top of the kidney in both sexes and in both secretes androgens. In most females, however, the masculinizing effect of the androgens is not enough to override the more powerful influence of the female sex hormones.

Castration in the male (orchidectomy) involves removal of the testes. If castration occurs before **puberty**—the time when males undergo rapid differentiation of secondary sexual characteristics—the secondary sexual characteristics do not develop. The male grows to normal adult size, but the voice does not deepen, and he never develops the male pattern of hair growth and distribution. Though generally showing some sex drive, he is, of course, sexually infertile. If castration occurs after puberty, the already-developed secondary sexual characteristics do not disappear (the voice does not suddenly become high-pitched, for instance). But the individual may tend to put on more weight and show a lessening of sex drive. Castration is routinely used on many domesticated farm animals (a "steer" is a young bull castrated before puberty) to produce leaner meat. It also makes the animals more docile and easier to handle.

15.4
THE FEMALE REPRODUCTIVE SYSTEM

Since it serves a greater variety of functions, the female reproductive system is more complex than that of the male. The following discussion of the anatomy of the female reproductive system is based on Figs. 15.5 and 15.6.

The **ovaries** are paired, usually oval, glands about 3 cm long. Like the testes, they serve a double function, producing sex hormones as well as gametes. The ovaries are held in place within the body cavity by two pairs of ligaments. The mature egg is released from the ovary during ovulation and passes into the abdominal cavity, where it is picked up by the **ostium** (plural ostia), the upper end of the **Fallopian tubes.** The egg is guided into the ostium by the beating of cilia in the epithelial lining of the Fallopian tubes, which serve to conduct the egg down toward the uterus. The egg is usually fertilized in the upper third of the Fallopian tubes.

It is an interesting anatomical fact that, while the ostia tend to envelop the ovaries, thereby decreasing the chances of the ova missing the openings of the Fallopian tubes, it *is* possible for the egg to escape into the abdominal cavity. In rare instances, an ovum fertilized just after leaving the ovary may miss the Fallopian tube opening and begin embryological development within the abdominal cavity. Development cannot proceed beyond a certain point, however, unless the embryo implants in a tissue richly supplied with blood, such as the intestinal wall, which is able to nourish it in a manner approximating that of the uterus. A few instances have been reported where development of a full-scale embryo did occur within the abdominal cavity, and a full-term baby was delivered surgically. Most of

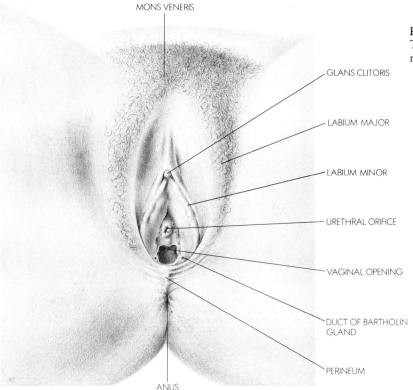

Fig. 15.6
The external anatomy of the human female reproductive system.

MONS VENERIS

GLANS CLITORIS

LABIUM MAJOR

LABIUM MINOR

URETHRAL ORIFICE

VAGINAL OPENING

DUCT OF BARTHOLIN GLAND

PERINEUM

ANUS

In the human being, as in all higher vertebrates, the testes and ovaries serve two functions: they produce germ cells (egg and sperm) and they produce the male and female sex hormones. Two very different groups of cells are responsible for these functions.

these ectopic pregnancies, however, must be removed or abort themselves. Ectopic pregnancies also occur in the Fallopian tubes. This presents a highly dangerous situation for the mother, as the developing embryo may burst the tube and cause fatal internal bleeding.

The Fallopian tubes end in small openings into the wall of the **uterus** or womb, a pear-shaped organ located just behind the urinary bladder. The uterus has thick walls of smooth muscles with a thick mucus-rich inner lining, the *endometrium*, well supplied with blood vessels. The function of the uterus is to contain and provide nourishment for the developing embryo. Normally, the cavity in the center of the uterus is nearly closed. During pregnancy, however, it enlarges enormously as the embryo grows (in later stages the embryo is called a fetus). The uterus terminates in a muscular neck, the **cervix,** which projects a short distance into the **vagina,** a short muscular tube extending from the uterus to the exterior. The vagina serves a double function: it is a receptacle for the penis during sexual intercourse or **coitus**** (and thus for sperm at ejaculation), and it serves as the birth canal.

The external genitalia of the human female are known collectively as the **vulva** (Fig. 15.6). The vulva consists of the *labia majora,* two folds of fatty tissue covered by skin; the *labia minora,* thin folds of tissue lying within the folds of the labia majora and concealed by them, and the **clitoris,** a sensitive structure composed of erectile tissue, which is homologous to the penis in the male. Like the penis, the clitoris becomes engorged with blood during sexual excitement. A large number of nerve endings are found in the spongy tissue of the clitoris, whose function appears to be to provide one of the areas responsive to sexual stimulation in the female.

Just behind the clitoris is the opening of the urethra, which in women serves only for the passage of urine. Behind the urethra is the opening to the vagina, which may be partly or, rarely, entirely occluded by the **hymen,** a thin membrane composed of connective tissue. The hymen is often opened by the first sexual intercourse, but various forms of exercise or such simple activities as riding a bicycle can also bring this about. Opening into the vagina on either side are certain glands (for example, Bartholin glands) that produce a mucoid fluid whose basic function is uncertain.

15.5
THE FEMALE REPRODUCTIVE CYCLE

In addition to its role in the sexual act and gamete production, the female reproductive system must also serve another very important function for which there is no counterpart in the male. Should fertilization of the egg occur, the uterus must be able to receive and nourish a developing embryo.

While most fishes, amphibians, and reptiles lay eggs that develop outside of the body, a few forms retain the developing embryos within the body in the lower end of the oviduct. The energy powering this development is usually supplied by the egg yolk, though in a few forms some nourishment may be obtained from the mother.

In the mammals, however, the developmental process within the mother's body is most refined. Because this development occurs within a specific organ, the uterus, it is referred to as **intrauterine development.** In humans, as in most mammals, the young embryo becomes implanted in the uterine wall. Approximately two weeks after fertilization, a special organ, the **placenta** is formed, serving as the basic region of exchange for nutrients and waste between the embryo and the mother's blood. The details of embryonic development will be discussed in Chapter 20.†

Estrous and Menstrual Cycles

Most mammals exhibit a cyclic recurrence of reproductive physiology and behavior. Periods when the female is ovulating and will copulate with males alternate with periods, often of long duration, when the female is not ovulating and will not copulate with the male. This cycle is called the **estrous cycle** and for most mammalian species is seasonal, with periods of ovulation and reproduction occurring once or twice a year.

---

*In nonhuman animals, coitus is generally called **copulation.**

†One group of mammals, the marsupials or pouched mammals (such as the kangaroo and opossum), shows a primitive form of intrauterine development. The embryo implants in the uterine wall, but does not form a placenta. After a period of time the very immature embryo is "born," crawling out of the uterus and finishing its development in a pouch on the mother's belly. Within the pouch, the young are attached to mammary gland teats that provide the nourishment. In the evolutionary history of the mammalian group, the marsupials have been an offshoot. The major line of development has been toward increasing the effectiveness of the uterine wall in receiving and maintaining the developing embryo.

During this period the female is said to be in estrus or "heat." In human beings and a few higher primates, there is no seasonal estrous cycle, nor any particularly specific relationship between ovulation and sexual drive. The human female has a **menstrual cycle,** occurring approximately on a monthly basis. The name "menstrual" derives from the Latin word *mens* (month) and refers to the discharge of blood and tissue from the uterus (*menstruation*) that occurs every twenty-eight days in adult women. Because it is neither seasonal nor related to specific behavioral patterns associated with intercourse, the menstrual cycle is not viewed simply as the human version of estrus. Obviously, however, the menstrual cycle evolved from the estrous cycle shared by our mammalian ancestors.

The menstrual cycle (and, indeed, the estrous cycle of other mammals) is largely under control of the **hypothalamo-pituitary complex.** The hypothalamus contains two neurosecretory centers, both of which control pituitary gland function via the secretion of neurohormones. One center, termed the **tonic center,** produces a *gonadotropic-releasing hormone (GnRH)* that causes the release of two gonadotropic hormones, *follicle-stimulating hormone (FSH)* and *luteinizing hormone (LH)* from the anterior lobe of the pituitary. These gonadotropic hormones stimulate the female gonad (the ovary) to secrete more female sex hormones (estrogens). There are three estrogens in the female bloodstream: estradiol, estrone, and estriol, of which estradiol appears to be the most active. The combined action of these hormones aids in preparing physiologically for reproduction.

As its name suggests, FSH stimulates the growth of an ovarian follicle. It also determines the sensitivity of specialized cells within the follicle, called **thecal cells,** to LH. Interestingly, while in most mammals the influence of LH appears to first cause the secretion of *male* hormones (androgens), which are then converted to estrogen by follicle cells called **granulosa cells,** in humans the thecal cells are apparently able to produce estrogens directly. The increased production of estrogens by the developing follicle and their circulation in the bloodstream produce a negative feedback effect on the tonic center that inhibits the release of FSH from the pituitary. (There is some evidence that the follicle may secrete a peptide that specifically causes this inhibition within the pituitary.)

The second hypothalamic neurosecretory center is called the **surge center.** When the circulating estrogens produced in the developing follicle reach a critical threshold level, the surge center is stimulated to release a large amount of *gonadotropic hormone-releasing hormone* (GTH-RH). The GTH-RH, in turn, causes a surge of LH release from the pituitary and a lesser release of FSH. The LH surge causes the mature follicle, now called a **Graafian follicle,** to rupture and release the ovum or egg (**ovulation**). The LH surge also induces the

formation of the **corpus luteum,** a temporary endocrine gland that develops within the ruptured Graafian follicle. The corpus luteum in the human female produces estrogens and **progesterone,** a steroid hormone that causes the uterine lining, the **endothelium,** to prepare for the possibility of pregnancy. The progesterone also has a negative feedback effect on both the tonic and surge centers. If pregnancy does not occur, the corpus luteum degenerates; if pregnancy does occur, the corpus luteum continues to function until the placenta takes over the production of progesterone and, by converting a weak androgen produced by the fetal adrenal gland, estrogens as well.*

It is worthwhile pointing out that, while it is true that there is no obvious counterpart to the female menstrual cycle in the male, there *is* a 24-hour rhythm in the secretion of pituitary and gonadal (testicular) hormones (see Fig. 15.7). However, only the tonic neurosecretory center of the hypothalamus appears active; the surge center is apparently suppressed by the presence of high levels of the male hormones. FSH stimulates the early stages of spermatogenesis and causes special cells called **Sertoli cells** to produce androgens that play a role in the maturation of sperm to the point of being able to fertilize the egg (sperm taken directly from the testes are unable to do so). FSH also stimulates the production of a peptide that suppresses FSH release from the pituitary. LH controls androgen synthesis by the interstitial cells of Leydig, and this androgen controls the primary and secondary sex characteristics as well as certain aspects of behavior. One of the androgens, testosterone, inhibits LH release, acting via the hypothalamus in a typical negative feedback pattern.

Phases of the Menstrual Cycle

The menstrual cycle of the human female is actually the result of the close association of two cycles. Technically, the term menstrual cycle refers to the monthly developmental changes that occur in the endometrium of the uterus. This cycle, however, is normally controlled by the **ovarian cycle.** The ovarian cycle involves the development and release of the ovum from the ovarian follicle and the various events just described going on in the ovary and the endocrine system that regulate the cycle. Keeping this close association in mind, it is possible to divide the menstrual cycle into three phases, the *menstrual phase,* the *preovulatory* or *follicular phase,* and the *postovulatory* or *luteal phase.*

---

*In the dog, the corpus luteum continues to function throughout pregnancy. There are many differences between mammals in the hormonal relations during estrus, and it is difficult to generalize about the hormonal relationships during the estrous cycle without doing so to the point of error. The above discussion refers primarily to reproductive physiology in the human female.

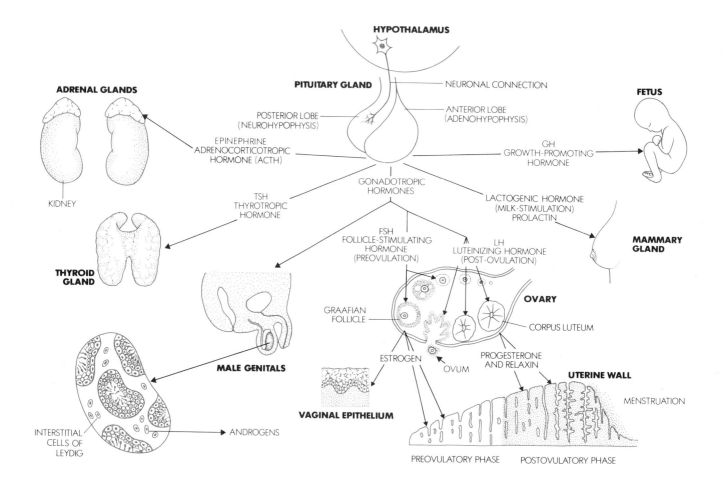

Fig. 15.7
Hormones produced in the pituitary control general body metabolism (by controlling production of epinephrine, and the thyroid hormone), sexual development and maintenance (through production of FSH and LH), and various aspects of fetal growth and development (lactogenic hormone and growth hormone, GH). As the diagram illustrates, the pituitary is particularly important in regulating the female reproductive cycle and for controlling aspects of pregnancy. Note that action of the gonadotropic hormones on the testes results in their production of the male hormones.

As with any cycle, where one begins in describing it is almost arbitrary. In the case of the menstrual cycle, it is customary to refer to the day on which the menstrual phase begins, the first day of menstruation, and we shall follow this convention here. In terms of the physiology of the menstrual cycle, however, the beginning of menstruation actually marks the end of the cycle. For, what occurs during the menstrual phase is the breakdown and loss from the body of the rich endometrial lining of the uterus, consisting of blood, disintegrated endometrial tissue, mucus, and various other uterine secretions.

As mentioned earlier, the menstrual cycle is, *on the average*, about twenty-eight days in length. It should be stressed, however, that only about 30 percent of women actually have a twenty-eight-day cycle; in some it is shorter, in others longer. The same is true of the menstrual phase of the cycle; while the *average* is four days, it generally ranges from three to six days in length.

During the menstrual phase, there is a low level of gonadal hormones in the blood. Since, in the female, the estrogens exert a negative feedback influence on the hypothalamus (and perhaps directly on the pituitary gland), the hypothalamus is freed to release its gonadotropin-releasing hormone, GnRH, which causes the anterior lobe of the pituitary to release FSH and a lesser amount of LH. The cooperative relationship between these two hormones, FSH and LH, results in the initiation of the preovulatory or follicular phase.

In the follicular phase, under the influence of FSH, a primary follicle within one of the two ovaries begins to undergo further development. By the end of the first week of the menstrual cycle, the follicle is large enough (secondary follicle) to produce enough estrogens to induce repair and regrowth of the endometrium. Within the endometrium blood vessels again appear and tubular glands begin to form. This occurrence marks the actual beginning of the preovulatory or follicular phase, which is approximately seven to nine days in length. The secondary follicle, now enlarged to the point of being called a Graafian follicle, continues the increase in estrogen secretion, and as a result the endometrium continues to thicken (at this time the smooth muscles of the uterus may show a rhythmic contraction).

Toward the end of the follicular phase, the amount of LH produced by the pituitary begins to rise, and the Graafian follicle starts to produce progesterone. The combined action of the estrogens and progesterone further enriches the endometrium, while the LH causes the Graafian follicle to rupture and release the egg (ovulation). Ovulation generally occurs around the fourteenth day of the twenty-eight-day menstrual cycle, and signals entry into the postovulatory or luteal phase of the cycle.

The luteal phase lasts approximately thirteen days and, if pregnancy does not occur, ends with the onset of menstruation. As its name implies, the phase is domi-

nated by the corpus luteum, which continues the secretion of estrogens but also greatly increases the production of progesterone. (Since one of the actions of progesterone is to cause the uterine glands to begin the secretion of a fluid rich in glycogen, the luteal phase is also known as the secretory phase.) Progesterone also causes the breasts to prepare for milk secretion should pregnancy occur.

During the last week of the menstrual cycle, if the egg is not fertilized, the corpus luteum begins to disintegrate, causing a sudden drop in the levels of estrogens and progesterone in the blood. As a result, spasms occur in the blood vessels supplying the endometrium, reducing blood flow into it. The lack of adequate amounts of blood causes the endometrial tissues to degenerate. Relaxation of the spastic blood vessels causes the renewed blood flow to carry the degenerated endometrial tissue and its contents out of the uterus through the vagina, to be released as the menstrual flow.

When the level of estrogens and progesterone in the blood is low enough, their negative feedback effect on the hypothalamo-pituitary system is correspondingly

SUPPLEMENT 15.1
THE PROSTAGLANDINS

In 1930, it was discovered that human semen could cause isolated strips of human uterine muscle to contract or relax. In 1932, a lipid substance causing these effects was isolated from semen. The next year, based upon the mistaken assumption that this substance was produced in the prostate gland, it was named "prostaglandin."

Interest in prostaglandin waned until 1962, when it was discovered that there was more than one kind of prostaglandin, and that they were all twenty-carbon fatty acids. To date, fourteen are known, thirteen of which appear in the human seminal plasma. It was also discovered that while the seminal vesicles are the main source of prostaglandins, they could also be isolated from the intestines, liver, pancreas, heart, lung, and brain, as well as both female and male reproductive tissues. Since they are produced in such diverse areas, the usual technique of discovering the function of an endocrine gland by removing it and seeing the effects on the organism is useless. Instead, studies with specific inhibitors for either the production, transport, release, or action of the prostaglandins must be employed.

Prostaglandins occur in extremely small amounts in the body; probably only a few tenths of a milligram are synthesized each day. Yet their effects are wide-ranging and often antagonistic to each other. Besides uterine contractions, among the various and diverse activities linked to prostaglandins are the inhibition of the secretion of progesterone

by the corpus luteum of the ovary, the inhibition of gastric secretion, the relaxation of smooth muscles in the air passages of the lungs, an increase in urine flow and sodium ion excretion in the kidneys, an increase in blood pressure by one prostaglandin, and a decrease in blood pressure by another.

Their role in reproductive physiology, however, makes the prostaglandins of interest here. Their effect upon the uterus, causing it to contract, suggests that prostaglandins play an important role in both birth and spontaneous abortion, and, indeed, high concentrations of two prostaglandins are found in the blood during the latter. The entire physiological state must be taken into consideration when discussing prostaglandin action, however. For example, one prostaglandin inhibits spontaneous activity of human uterine muscle strips when taken from a nonpregnant woman, while another prostaglandin stimulates it. If the strips are taken from a pregnant woman, however, *both* prostaglandins cause stimulation. The action of these prostaglandins has led investigators to study their possible use in inducing abortions.

As previously indicated, some prostaglandins inhibit the secretion of progesterone by the corpus luteum and set into action the events leading to its regression. It would seem, therefore, that at least some prostaglandins must necessarily be involved in any answer to the as yet unanswered ques-

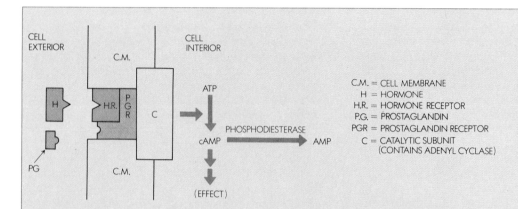

Diagram A
A model proposing that there exists in the cell membrane a specific receptor site for prostaglandins and that this region mediates the effects of the prostaglandins on the production of cAMP. The amount of cAMP in a tissue is normally determined by both its formation through the activity of the enzyme that plays a role in its synthesis (adenyl cyclase) or degradation (phosphodiesterase). Thus, the prostaglandins may act to affect cAMP levels either by stimulating adenyl cyclase activity or inhibiting phosphodiesterase activity. They might also somehow affect the availability of ATP.

tion: What factors control the termination of one estrous cycle and the onset of another? Regardless of how they affect the corpus luteum, the fact that this effect can either prevent or terminate pregnancy provides the possibility for a new concept in fertility control. Much yet remains to be done in discovering the role prostaglandins play in the male reproductive system, as well.

An interesting discovery is that, unlike most hormones, prostaglandins seem to act near the tissue from which they were released. For example, the prostaglandin causing increased urine flow and sodium excretion is produced in the kidney. In one experiment, the transplanting of a sheep's ovary to its neck or even the abdomen resulted in no corpus luteum regression, but when both ovary and uterus were transplanted to the neck, normal regression occurred. Thus it appears that the uterus produces a prostaglandin that acts locally on the ovary. It has been hypothesized that the prostaglandin might reach the ovary via a countercurrent mechanism (see Supplement 10.5) between the uterine vein and ovarian artery, which lie in close proximity.

As is the case with many hormones, ideas concerning the precise mechanism by which prostaglandins carry out their work are still in the highly speculative, model-building stage. The observation that the addition of prostaglandins and gonadotropins to incubated mouse ovaries stimulates the formation of cyclic adenosine monophosphate (cAMP) in a manner characteristic for each of these substances (rather than being the same) has led to a hypothesis that there is a specific prostaglandin receptor in the cell membrane. (Interestingly, an important prostaglandin precursor, arachidonic acid, is found in cell membranes.) One model (see Diagram A) depicts the gonadotropic hormone and prostaglandin approaching the cell membrane together, each having a specific receptor site to which it will bind. It is known that, while prostaglandins increase cAMP formation in most tissues, they actually decrease it in others. A hypothesis proposing the existence of a specific prostaglandin receptor site predicts that an increase or decrease in cAMP production should be mediated at the cell membrane level. Evidence that this is, indeed, the case was reported in the early 1970s.

lessened and the production and release of its releasing hormones initiate a new ovarian and menstrual cycle (see Fig. 15.8).

15.6
THE SCIENTIFIC STUDY OF HUMAN SEXUALITY

Strong interest has always existed in sexuality, and the subject has generated considerable mythology throughout history. Only in the past two or three decades, however, has the study of human sexuality been af-

forded the scientific research it deserves. As long as inquiry into the nature of sexuality was restricted to anecdotal information often obtained from highly biased or unreliable sampling procedures, it served more to perpetuate old myths than to elucidate new truths. Falsehoods such as the claims that masturbation caused insanity, and that homosexuality is an inherited illness have persisted down to the present time. Only as sex research has become an area for legitimate scientific study have the old ideas begun to be dispelled and a more rational view of human sexuality put in their place.

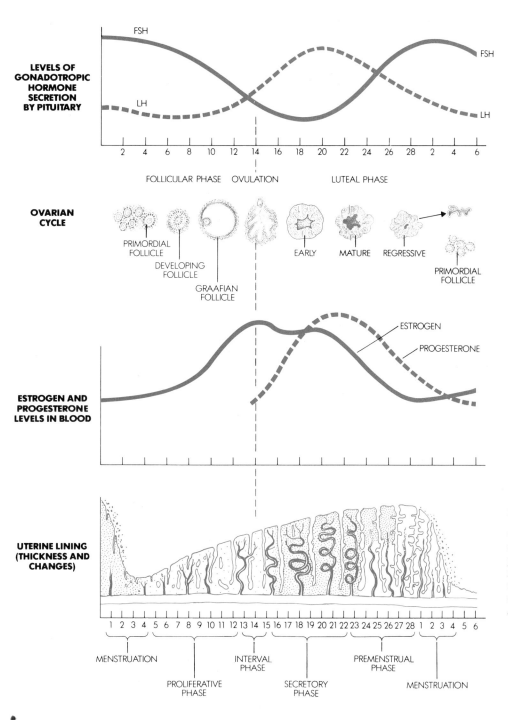

LEVELS OF GONADOTROPIC HORMONE SECRETION BY PITUITARY

FSH

LH

FSH

LH

2 4 6 8 10 12 14 16 18 20 22 24 26 28 2 4 6

FOLLICULAR PHASE   OVULATION   LUTEAL PHASE

OVARIAN CYCLE

PRIMORDIAL FOLLICLE

DEVELOPING FOLLICLE

GRAAFIAN FOLLICLE

EARLY   MATURE   REGRESSIVE

PRIMORDIAL FOLLICLE

ESTROGEN

PROGESTERONE

ESTROGEN AND PROGESTERONE LEVELS IN BLOOD

UTERINE LINING (THICKNESS AND CHANGES)

1 2 3 4 5 6 7 8 9 10 11 12 13 14 15 16 17 18 19 20 21 22 23 24 25 26 27 28 1 2 3 4 5 6

MENSTRUATION

PROLIFERATIVE PHASE

INTERVAL PHASE

SECRETORY PHASE

PREMENSTRUAL PHASE

MENSTRUATION

Fig. 15.8
Diagram of the various events of the menstrual cycle, showing simultaneous changes in pituitary secretions, ovarian follicles, estrogen and progesterone levels in the blood, and thickness of the uterine wall, or endometrium.

Today the scientific study of sexuality (**sexology**) involves data from anthropology, psychology, and biology. From anthropology have come valuable data of comparative approaches to sexuality in different cultures. From psychology have come important data on sexual patterns and beliefs in human subjects. From biology have come comparative studies of sexual behavior in nonhuman species, especially primates, as well as direct physiological studies of sexual functioning in human beings.

Pioneering work in this field was launched in the late 1930s by Indiana University biologist Alfred Kinsey. Applying sampling techniques and methods of statistical analysis learned from his long career as a field zoologist, Kinsey and his co-workers showed for the first time that the field of sex research could be ap-

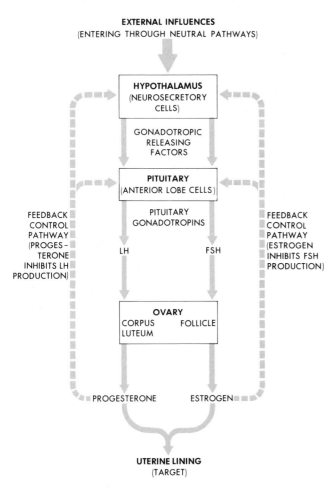

EXTERNAL INFLUENCES
(ENTERING THROUGH NEUTRAL PATHWAYS)

HYPOTHALAMUS
(NEUROSECRETORY CELLS)

GONADOTROPIC RELEASING FACTORS

PITUITARY
(ANTERIOR LOBE CELLS)

PITUITARY GONADOTROPINS

LH   FSH

FEEDBACK CONTROL PATHWAY (PROGES-TERONE INHIBITS LH PRODUCTION)

FEEDBACK CONTROL PATHWAY (ESTROGEN INHIBITS FSH PRODUCTION)

OVARY
CORPUS LUTEUM   FOLLICLE

PROGESTERONE   ESTROGEN

UTERINE LINING
(TARGET)

Fig. 15.9
Summary diagram of control of the estrous (menstrual) cycle in humans and other mammals. Solid lines represent production of hormone or neurosecretory factor; dashed lines represent negative feedback influences by which the cyclic changes of estrus are regulated. In nonhuman mammals the role of the hypothalamus is greater in initiating pituitary action at the onset of estrus. Thus other mammals are much more subject to external influences on their estrous cycles than are humans.

proached in a scientific manner. In two large volumes, *Sexual Behavior in the Human Male* (1948) and *Sexual Behavior in the Human Female* (1953), Kinsey and his associates Wardell Pomeroy, Clyde Martin, and Paul Gebhard summarized their findings and presented scientists and the public with a vast amount of data about sexual practices in the United States. Kinsey's methods involved interviews with American adults, seeking information about a variety of individual sex attitudes and behaviors. The large number of interviews taken and analyzed represented Kinsey's efforts to

avoid the drawback of inadequate sample sizes that characterized the more anecdotal approaches to the study of sex in earlier generations.

To make his surveys as thorough and as free from error and bias as possible, Kinsey took a number of precautions in actually preparing for the interviews. The interviewers were given a thorough training course in interview techniques, so that all subjects were given as nearly as possible a standard interview. Elaborate precautions were used to ensure confidentiality, in order to encourage honest and open responses. Such careful practice probably made Kinsey's findings the most objective scientific data on sex attitudes and practices that had ever been obtained.

Kinsey and his co-workers came up with new and often astounding data. For example, premarital (before marriage) and extramarital (outside of marriage) sex were found to be far more prevalent among United States adults than had previously been thought. Even more startling in many ways were Kinsey's data showing that nearly 40 percent of United States males had engaged in some overt homosexual activity during adolescence or early adulthood. These findings forced sociologists, psychologists, anthropologists, and biologists to reconsider the standard "norms" of sexual behavior in American and other societies and to reorient their thinking about what is adaptive and nonadaptive sexual behavior.

Despite its pioneering achievement, Kinsey's work was not free from errors. Critics (and Kinsey himself) rightly noted that the sampling technique used was likely to produce biased results. For example, some samples of the population might more often than others refuse to submit to interviews about their sex lives, while others might not return the questionnaires. Thus, those who eventually provided information might well have been people who were most open and free in discussing their sex life. The sexual practices of such individuals might well not be the same as those of people who refused to take part in the study. Such valid criticisms undermined some of the specific results Kinsey and his group obtained. They did not, however, detract from one of Kinsey's major achievements: his demonstration that sex research can be approached objectively and scientifically—and it is interesting to note that later studies have, for the most part, confirmed Kinsey's findings.

In more recent studies along the same lines, William Masters and Virginia Johnson at Washington University School of Medicine, St. Louis, have extended the methodology of Kinsey and his group into other areas of sex research. Their book, *Human Sexual Response* (1966), presents large amounts of data about the behavioral and physiological aspects of sex, including the nature of presexual arousal, bodily changes during intercourse and orgasm, the nature of the refractory period, and the physiology and psychology of impotence. By recording changes during intercourse or masturbation in physiological parameters such as heart-

SUPPLEMENT 15.2
A SCIENCE OF LOVE?

The word "love" is probably the best known and least clearly defined term in the human vocabulary. In some cultures there is no word that is an exact counterpart to the way it is used in Western cultures, and the idea of waiting to see whether a man and woman are "in love" with each other before they form a family is never given a thought. Nor was it in Western societies for centuries; royal marriages were arranged for purely political reasons by both Church and State (as was the marriage of England's King Henry VIII with Spain's Catherine of Aragon) while the peasantry was more concerned with the amount of dowry, or material wealth—money, livestock, and the like—the prospective bridegroom would receive from the bride's family than with any romantic feelings of the heart.

In the biological sciences we are used to seeking possible underlying physiological causes for behavioral effects; if a person is hyperactive, for example, one possible cause might be oversecretion of the thyroid or adrenal glands. This oversecretion, in turn, is perhaps caused by some pathological condition, such as a tumor. There is nothing generally pathological associated with the feelings we associate with love, of course, but nonetheless there is a group or syndrome of symptoms many of us can associate with the feeling of being in love—if this were not so, stories, sonnets, plays, and the like dealing with love would be of interest only to their authors. According to Drs. Donald F. Klein and Michael R. Liebowitz of the New York State Psychiatric Institute, "Love brings on a giddy response similar to an amphetamine high and the crash that follows breakup is much like an amphetamine withdrawal." They feel this is because the brain of a person in love pours out its own chemical counterpart to amphetamine—phenylethylamine. Upon being rejected, the same brain stops this production immediately, and thus the "withdrawal" symptoms. They note that rejected "love junkies" often go on binges of eating chocolates, which are rich in phenylethylamine. Dr. Liebowitz has been able to treat depression caused by love failures by a combination of psychotherapy and administering drugs that inhibit the breakdown of phenylethylamine. Unfortunately, these drugs require a special diet (no fermented food such as wine or cheese) and may interfere with orgasm. It is hoped that newer drugs without these side effects may soon be available.

Dr. Liebowitz compares the extremes of elation and depression in his patients (termed hysteroid dysphoria) to being on a roller coaster. He describes such patients as being attractive, competent, and likable, but whose ability to work is related to whether their love situation is on a "high" or "low." In such persons, Dr. Liebowitz states that the levels of phenylethylamine vary widely because of a defective or unstable homeostatic control mechanism that may be either inherited and/or due to an acquired defect. He notes that such patients often pick inappropriate persons with whom to fall in love—the love object is either married, aloof, or not the sort to form a lasting relationship. Further, the patient's constant and intrusive need for attention and praise often drives the love object away.

On the anatomical level, Dr. John Money of Johns Hopkins University School of Medicine has noted that the removal of tumors from the pituitary gland during or prior to puberty often make it difficult for the person to fall in love. Dr. Money hypothesizes that the surgery may occasionally interrupt two sets of pathways essential to "eroto-sexual" behavior, one that signals the pituitary to release the appropriate hormones and the other that signals the time for appropriate mating behavior.

On the psychological level, Dr. Dorothy Tennov of the University of Bridgeport has coined the word "limerence" to go along with "love" and "lust." She maintains that limerence describes a state the other two terms do not, in which the limerent object's qualities are exaggerated, and he or she becomes an obsession, causing an acute longing in the person experiencing limerence. Dr. Tennov sees limerence as being unlike love or lust in that one can love more than one person, have sex with more than one, and be either heterosexual, homosexual, or bisexual, but that no one is limerent about more than one person at a time.

Finally, it might be pointed out that many persons become quite upset at the idea of even attempting to find a physiological basis for the symptoms we associate with falling in love, perhaps along the lines of an "ignorance is bliss" philosophy. Wisconsin's Senator William Proxmire, for example, states that he doesn't want such research funded "because I don't want an answer."

---

beat, breathing rate, and galvanic skin response, they were able to bring to light many aspects of human sexual activity that had been either unknown or badly misunderstood. Among their most important findings, for instance, was the fact that women are capable of multiple orgasms, a fact not scientifically recognized previously.

15.7
THE PHYSIOLOGY OF SEXUAL INTERCOURSE

Sexual intercourse, or coitus, involves a complex set of behavioral and physiological activities in the human female and male. During the preintercourse period of arousal, as well as during intercourse itself, the female

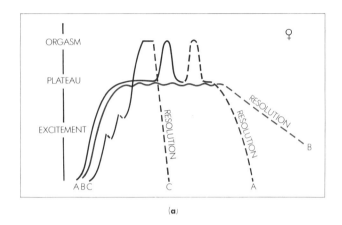

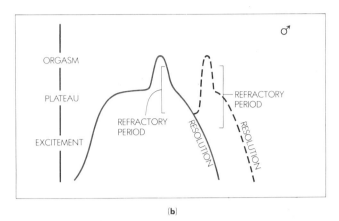

(a)   (b)

Fig. 15.10
The female (a) and male (b) sexual response cycles as reported by William H. Masters and Virginia Johnson in 1966. Note that three distinct patterns are recorded here for the female, but other variations have since been reported. While Masters and Johnson identified only one basic male response pattern, others have been reported. The basic differences usually appear in the shorter length of the plateau phase before orgasm in the male and the longer period of time before a second orgasm occurs.

and male appear to experience approximately the same basic responses. These responses fall into four phases, named in order of their occurrence: *excitement, plateau, orgasm,* and *resolution* (see Fig. 15.10).

The Male Response

Sexual excitement can be initiated by a large number of factors, including visual cues, smell, touch, or fantasy. The excitement phase is accompanied by a variety of extragenital responses such as increased breathing and heart rates and a general level of muscle tension (in both smooth and striated muscles). Genital response in the male begins with lengthening of the penis. As the penis becomes more engorged with blood, it becomes increasingly firm and erect. At the same time, nerve endings at the tip become increasingly sensitive to touch. The process of erection is achieved by blood filling the three areas of spongy tissue (vasocongestion). The scrotum also experiences some vasocongestion; through involuntary contraction of muscles surrounding the vas deferens, the testes are actually lifted.

During the plateau phase and after insertion into the vagina, the penis becomes slightly larger in circumference, especially around the tip, or **glans penis.** The testes are not only lifted up, but may also increase in size by as much as 50 percent. Indeed, if the plateau phase is prolonged, testicular size may increase by up to 100 percent. The plateau phase is essentially that period during which full arousal is achieved and maintained as intercourse progresses.

The sex act culminates in the orgasmic phase. Prior to orgasm, lubricating fluid from Cowper's glands may exude from the tip of the penis. This fluid also serves to neutralize the acidic pH of the urethra caused by urine, thereby making a more hospitable environment for the sperm to pass through. It frequently contains active sperm, though not in high concentration. The approach of orgasm is generally signaled by a vasocongestion reaction in the chest, neck, face, shoulders, and other parts of the body. Breathing rate also increases, as do heart rate and blood pressure. When orgasm is achieved, semen is forcefully pumped out of the epididymis and vas deferens and through the urethra, in a process known as **ejaculation.** During the first part of ejaculation, semen is pushed up into the base of the urethra by contraction of muscles all along the spermatic cord. The semen first collects at the base of the urethra in the *urethral bulb,* which can expand to two or three times its usual size. The sphincter muscles surrounding the bladder close, to prevent sperm being forced back up into the bladder. During the second part of ejaculation, seminal fluid from the prostate gland and seminal vesicles flows into the urethral bulb. Due to muscle contraction initiated by the continual rubbing of the penis, the sperm and various seminal fluids are forced out of the penis. Most of the sperm is ejaculated in the first portion of the semen. The remainder of the semen consists largely of various lubricating fluids.

After ejaculation, the male experiences the resolution phase. The penis begins to shrink in size, losing its rigidity. There is a general decrease in muscle tension, heart and breathing rates, and blood pressure. Resolution occurs relatively rapidly after ejaculation —generally in about 5 min. In some individuals, the penis may sometimes remain rigid for a while after ejaculation, though a period of time is usually required to achieve a second orgasm.

There is no fixed number of orgasms that a male can experience in a given period of time. The number varies from individual to individual, and in any one individual from one period of time to another. Such factors as general state of health, age, fatigue, degree of

Human males usually experience only one orgasm during each act of intercourse, whereas females may experience several orgasms in rapid succession.

relaxation, length of time between the present and the previous sexual act, and degree of attraction to the sexual partner all may influence the intensity of a male's activity on any occasion. In general, however, the resolution phase in males is longer than that for females, a significant number of whom may be multiorgasmic.

The Female Response

The female response to sexual arousal also generally occurs in four phases. The excitement phase is initiated by visual, tactile, or fantasy sensations. Heart rate, breathing rate, and blood pressure are accelerated. As a result of vasocongestion, the breasts may show nipple erection and increase in size. The labia minora and the clitoris show vasocongestion. Muscle tension increases, the particular muscles that show tension depending in part on the woman's position during coitus. During the excitement phase, genital changes include erection of the clitoris and vaginal lubrication by mucoid secretion.* During excitement the vagina itself expands somewhat due to vasocongestion, and the uterus becomes partially elevated.

The plateau phase in females, as in males, is essentially a continuation of processes that began in the excitement phase. As intercourse proceeds, rubbing or movement of the clitoris maintains or increases the female's arousal. Areas of the labia minora and the outer region of the vagina become more engorged with blood and, other than the breasts and clitoris, become the most sensitive areas for sexual stimulation. It is in these areas of the genitalia that repeated stimuli produce orgasm.

At orgasm the woman feels a series of short contractions in the labia minora and lower end of the vagina. The uterus then may follow with irregular uterine contractions. As pointed out earlier, females can often experience several orgasms within a few minutes of one another. During orgasm, muscle tension increases, heart and breathing rates increase, blood pressure goes up, and many areas of the skin show a flush. The general overall feeling, as in the male, is that of explosive release.

The resolution phase in women is much the same as in men. It involves a relatively quick return to normal heartbeat, breathing, muscle tension, and blood pres-

sure. The resolution phase is much longer if the woman has reached plateau phase but did not experience orgasm.

It is often found that orgasm is achieved more easily by men than women. Sometimes a man's orgasm occurs within a short period of time after reaching plateau phase. It usually takes somewhat longer for the rubbing of the female genitals to produce the orgasmic phase. Much of the success of sexual intercourse between any two partners depends upon the degree to which both learn how to relate their own phase responses to those of the other. Both partners need not always achieve orgasm, however. Not only is this an unrealistic goal for all intercourse, but it implies that orgasm is the chief and only pleasurable aspect of intercourse. This attitude belittles the other aspects of sexual interactions, besides their role in the reproductive process, that may in themselves be highly enjoyable and rewarding.

15.8
DEVELOPMENT OF SEX ROLES AND BEHAVIOR

No area of human biology has been the object of greater interest and general curiosity than the study of sexuality. The term "human sexuality" includes a diversity of topics related to such things as the biological, sociological, psychological, and ethical aspects of sex. Subjects ranging from the chromosomal and hormonal aspects of sexual differentiation to the development of sex roles, sexual identity, preference, behavior, and physiology are all aspects of the general area of human sexuality. Because sexuality is manifest in behavioral *and* physiological terms, it represents an important interface between biology, sociology, anthropology and psychology. Further, because human sexuality has both genetic and developmental aspects, it is perhaps fitting to include a brief discussion of the subject at this point.

Sexual behavior, as well as sexual anatomy and physiology, undergoes considerable development during a human being's lifetime. Indeed, we seem to be sexual beings from the very start; baby boys are often born with erections, and similar erections are reported to occur within the uterus. Likewise, female infants experience vaginal lubrication. In the very young child, from birth through three or four years of age, such sexual behavior as fondling of the genitals and interest in both parents' sex organs is evident. But such behavior is largely androgynous—neither consciously male nor female. It is also more restricted in scope and intensity than the behavior that develops later on. As puberty

---

*There have been recent reports of cases of female "ejaculation," in which a large amount of fluid is released during sexual excitement. The fluid differs in composition from the vaginal lubricant and the fact that it contains acid phosphatase has led some to hypothesize the presence of a vestigial female prostate gland (male prostate gland fluid is rich in acid phosphatase).

SUPPLEMENT 15.3
BIRTH CONTROL

Attempts by human beings to prevent conception are as old as recorded history; Egyptian tombs dating back thousands of years before Christ contain writings and artwork attesting to the antiquity of the practice.

The attempts continue today. Basically, these attempts can be divided into two types: those that attempt to prevent conception (thus the term, "contraception") and those that prevent the development of the fertilized egg. In the former category are condoms, diaphragms, "the pill," and surgical techniques. Condoms are simply sheaths, often made of rubber, which fit over the penis and prevent the semen from entering the vagina. The diaphragm, individually fitted to the female wearer and combined with a spermicidal (sperm-killing) jelly, fits over the cervix, thereby denying the sperm access to the uterus. While there are several types of birth control pills, the general effect is to interfere with the normal feedback inhibition process by which ovulation is triggered. By introducing higher-than-normal levels of estrogen and progesterone into the blood, the pill suppresses FSH and LH production by the pituitary gland. Thus the normal cycle of follicular growth and ovulation does not occur. Because each type of pill has its own side effects, some of which may be harmful to health (or even fatal), the pill should only be taken after consultation with a physician. Surgical techniques involve tying or cutting the Fallopian tubes of the woman or cutting the main sperm ducts leading out of each testis in the male (vasectomy). In some cases, the contraceptive effect of tying off the Fallopian tubes and vasectomies can be reversed, though the success in doing so cannot be guaranteed.

Among the methods in the category of those birth control devices that are operative after conception, the intrauterine device (IUD) is probably the most widely used. The principle of the IUD is often attributed to Arabian camel drivers, who inserted pebbles into the uterus of the female camel to prevent pregnancy during long caravan journeys. In humans, IUDs vary widely in size and shape (see Diagram A). They are inserted by a physician or paramedic in an applicator that holds them in a compressed conformation until they pass through the narrow cervical canal. When released in the uterine cavity, they snap into their normal configuration. Strings attached to the IUD are allowed to project through the cervical opening into the vagina to allow for easier removal if necessary.

Just how IUDs work is not clear. Dr. Robert Cleary of the Indiana University Medical Center has hypothesized that the metal ions such as zinc or copper released by the IUDs prevent progesterone from binding to its receptor sites on the endometrial lining, thus making it unreceptive to the prospective embryo.

Like the pill, IUDs entail considerable risk. The uterine wall may become perforated leading to bleeding and infection. As with the pill, the need for close medical supervision makes the IUD unsatisfactory for use in countries with poor

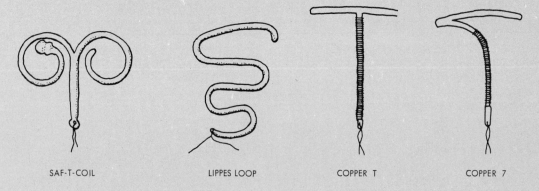

SAF-T-COIL   LIPPES LOOP   COPPER T   COPPER 7

Diagram A
Some types of intrauterine devices (IUDs). IUDs work by preventing implantation after the egg is fertilized, though the precise means by which they do so is not yet completely understood. They are inserted into the uterus by a device that keeps them elongated until the neck of the cervix is passed, whereupon they are released and snap into the conformations shown above. Many early types of IUDs caused internal bleeding and uterine infections. A woman considering their use should obtain competent medical advice before doing so.

medical facilities. Even with medical care available, both the pill and IUDs involve some risk. On the average, however, the risks are less than those involved in becoming pregnant.

Religious opposition to birth control is most strongly (though not entirely) centered in the Roman Catholic Church, which, via three Papal encyclicals (1588, 1930, 1968) views birth control as frustrating the primary purpose of sex, as intended by God, to have children. (It is for the same reason that such acts cannot achieve "finality," i.e., pregnancy, that the Church also labels masturbation and homosexual acts as sinful.) While, especially since 1968, there has been much dissension within the Church among theologians, priests, nuns, and lay persons on this controversial issue (approximately four out of five Catholics in the United States ignore the teaching), it still remains official teaching. In recent years, the Church has approved "rhythm" (and other variations thereof) as a means of birth control. Rhythm involves keeping records of slight variations of the female body temperature (since these variations correlate roughly with certain phases of the menstrual period) and restricting intercourse to the so-called "safe" periods. Because of its unreliability, as well as objections to its "scheduling" aspects, rhythm has not been widely adopted, even among Catholics.

Behavioral Methods of Preventing Fertilization

Obviously the most effective method of preventing union of sperm and egg is to avoid sexual intercourse altogether. In some religions abstinence from sex, or celibacy, is a custom practiced by the clergy and even some lay persons. Effective as celibacy is as a birth control method, it has simply not been a practical method for use on any large scale. Most people are not willing to avoid sex, even if they do not want children.

Another behavioral method for reducing the chances of fertilization is known as *coitus interruptus,* or withdrawal. This method means that the male withdraws his penis from the female shortly before he reaches orgasm. Withdrawal has several problems that make it a less-than-successful method of birth control. One is that sperm can be found in small quantities in some of the secretions from the penis prior to actual orgasm (in the lubricants released from Cowper's and other glands), and one of these sperm may fertilize an egg. Another problem is that withdrawal is sometimes difficult to achieve before orgasm. Orgasm often arrives quickly, making quick withdrawal necessary if the method is to be effective. This aspect of the problem is serious because about 70 percent of the sperm ejaculated are found in the first burst of semen from the penis. Thus, if the male has begun ejaculation before withdrawal, the chances of fertilization are still very high. Coitus interruptus has a high degree of uncertainty.

Other methods of sexual interaction which avoid or minimize fertilization include anal and oral sex, as well as mutual masturbation. In anal sex the penis is inserted into the anus rather than the vagina. Oral sex involves use of the mouth to stimulate the genitalia; it can be practiced by both partners. Mutual masturbation involves manipulation and thus stimulation of each partner's genitalia. All three of these methods are used in both heterosexual and homosexual love-making. Their birth control value in heterosexual relations is obvious. Because such acts cannot achieve "finality" or conception, they are still considered "unnatural" and sinful by many religious organizations, however.

approaches, however, biological and psychological changes occur rapidly and sexual functioning matures. At the same time, sexual identity and sex roles, to whatever extent they exist in the culture, usually crystallize.

The ability to bear children undergoes a rapid decline in women between the early and late forties (a period know as **menopause**), during which ovulation and menstruation cease and hormonal changes occur. In men, production of functional sperm continues well into old age. At advanced ages members of both sexes *may* experience a decrease in sexual urge, though it is seldom totally absent at any age. (There is evidence indicating that the degree of sexual drive and activity is directly related to the amount of this activity when younger.) These patterns of change in sexual drive and functioning from prepuberty through old age parallel certain hormonal changes in the male and female body. They follow a developmental pattern remarkably consistent for most individuals within a given culture.

The study of human sexual behavior and customs has been embroiled for ages in the "nature-nurture" or "genes versus environment" controversy. "Nature" proponents argue that certain patterns of sexual behavior are biologically determined (through genetic control of the nervous system) and are thus basic components of human nature around which social customs and institutions must be molded. "Nurture" proponents argue that nearly all human sexual behavior is environmentally or culturally determined (learned by the child as he or she grows up in a particular society). Biological and anthropological studies within the past several decades show that human sexual behavior is enormously plastic and that most behavior is learned from the social atmosphere in which a child develops.

On what sort of evidence does such a conclusion rest? The degree of variation in sexual behavior among different human societies is so enormous that it is virtually impossible to speak of any behavior as being "normal" or "typical" for the species as a whole. With the possible exception of sexual relations between members of the same family (incest), virtually all forms of sexual behavior are acceptable in one society or another, even though some may be strongly condemned in others. Even within a single society, different subgroups often have remarkably different sexual customs. The existence of such a large range of variations in sexual practices does not, in itself, establish conclusively that such differences are determined by cultural factors. However, if sexual behavior patterns were biologically rather than culturally determined, then we should expect to find the same consistent pattern running through all cultures—such as we find, for example, among the varieties of human hemoglobin. That such consistency cannot be observed contradicts the "nature" hypothesis but is consistent with the "nurture" hypothesis.

Another line of evidence is slightly more experimental in nature. If sexual habits are biologically determined, then we might expect that children raised in cultures different from that into which they were born would exhibit sexual behavior characteristic of their parents' culture. However, studies of the development of children adopted across cultural lines contradict the hypothesis. The children display the sexual customs and mores of their adopted parents. Such evidence lends support to the nurturist view (without demonstrating conclusively, of course, that it is correct).

Still another line of evidence comes from physiological studies on individuals whose endocrine glands, particularly the ovaries or testes, are abnormal or missing. Individuals with hyperactive testes or ovaries secrete larger than usual amounts of testosterone or estrogens, respectively. Other individuals have relatively inactive testes or ovaries, and in some adults these glands are removed surgically for various reasons. In all these cases, hormonal level has only slight effects on sexual behavior. In general, hormonal levels influence extent or intensity of sexual drive. They do not appear to greatly affect the specific kinds of sexual behavior an individual displays. Adults who are castrated (that is, whose testes or ovaries are surgically removed) continue to show all the sexual behavior patterns they displayed prior to castration, though often with decreasing frequency and/or intensity. Of course, such evidence only suggests that amount and kind of sex hormone is not a major biological determinant for sexual behavior. It does not exclude other possible determinants, such as neurophysiological makeup. However, since the male and female sex hormones are crucial in the specific direction of anatomical and physiological determination of sex both during embryogenesis and at puberty, they are likely candidates for determination of sexual behavior as well. All that can be said at present is that hormonal patterns do not appear to be necessary to maintain a developed behavior pattern with regard to sex.

The nature-nurture controversy begs the most important question relating to our understanding of human sexual behavior and attitudes. All human behavior is developmental; that is, it develops in stages during the growth and maturation of the individual. This development can be viewed as arising from two components: biological and environmental, with all behavior developing within a basic biological framework. Biologically, human beings have certain anatomical and physical traits that inevitably influence behavior: number of appendages, metabolic rate, brain structure, bipedalism, and the like. These are very generalized traits, yet they have some effects on the type of behavior the species can display. Furthermore, every individual human being shows slight variations in these broad physiological and anatomical traits. No two individuals (except perhaps identical twins) have exactly the same biological constitution with which to interact with their environment. Yet, it is precisely in this interaction of biological makeup and environmental factors that behavioral patterns develop in human beings. The overriding influences on this developmental process, except in cases of gross physical or mental abnormality, are environmental. With regard to the development of sexual behavior in particular, this environment includes parental influence, influence of peers (other children), and general social attitudes. The general plasticity of our behavior has made these environmental influences the most important factors in determining the specific course of any individual's sexual development.

Behavior patterns develop slowly, over long periods of time, by the constant input the environment offers. Only by repeated exposure to social behaviors and attitudes displayed by others does the young child begin

---

The male and female gonads and genital structures originate from much of the same early embryonic tissue. The young embryo of either sex contains all the same structures among the primordial reproductive tissue. Hormonal differences triggered by the XX or XY chromosome makeup determine the direction in which sexual differentiation will occur.

SUPPLEMENT 15.4
ABORTION

Abortion is the removal of the embryo or fetus from the uterus in order to terminate the pregnancy.

Methods of inducing abortion vary with the stage of pregnancy. The main problem is to remove the fetus and its membranes and placenta without damaging the uterine wall or causing infection in the mother. Two methods are used during the first three months of pregnancy. Both methods involve dilating the cervix to gain entrance into the uterus and differ only in their means of extracting the embryo. The first method involves simply removing the embryo from the uterus with forceps. The placenta is then removed by gently scraping the uterine lining. In the second method, known as aspiration, the embryo is removed by suction through a hollow tube, a cannula. While both of these methods are relatively simple, they require anesthetics and completely sterile procedure, since the major risk is infection of the woman's uterus or vagina. More recently, prostaglandins (see Supplement 15.1), which cause the smooth muscles of the uterus to contract and induce abortion, are being used in a number of places around the world.

After the third month of the pregnancy, these simpler methods will not work, since the embryo has grown to such an extent that it partially blocks entrance into the uterus through the cervix. The alternative is to remove the fetus surgically by cutting through the abdominal wall. This is, of course, a much more serious type of abortion, prone to a number of difficulties. The most common is infection. A somewhat less traumatic process can be carried out in the later stages of pregnancy without surgery. It involves injecting hypertonic saline solutions into the amniotic sac surrounding the fetus. This process induces labor within 8 to 72 hr. After the fetus is expelled, dilation and aspiration are usually performed to ensure that all membranes are removed. In all stages of pregnancy, abortion by any method involves risk of serious infection in the mother's uterus. For this reason it is always preferable for abortions to be carried out in a hospital or clinic under careful medical supervision.

Based primarily (though, it is maintained, not entirely) on the religious belief among some faiths that ensoulment occurs "at the moment of conception," abortion is strongly opposed by the leaders of some religious organizations, who equate it with murder. Some groups, calling themselves "Pro-life" or "Right to Life," are lobbying for a Constitutional amendment to prohibit abortions. Others, however, generally referring to themselves as "pro-choice," feel that such an amendment would violate the principle of separation of Church and State as well as compel a woman either to remain pregnant against her will or to seek a dangerous back-alley abortion, many of which prove fatal or impair the woman's ability to have children at a later date. The historical, theological, legal, and philosophical roots of the controversy are both ancient and complex, and lie well beyond the scope of a biology textbook. For those interested in this issue, literature supporting the Right-to-Life position can be obtained by writing to the Respect Life Committee, National Conference of Catholic Bishops, 1312 Massachusetts Avenue, N.W., Washington, D.C. 20005, and that of the pro-choice from the National Abortion Rights Action League (NARAL), 825 15th Street, N.W., Washington, D.C. 20005.

to internalize certain concepts of his or her sexuality and its "appropriate" expression. Most modern Western societies have been enormously repressive with regard to sexual attitudes. This has had a profound influence on the development of sexual behavior in those societies. As they grow up and attempt to make sense of the world around them, young people in western countries have been presented with mixed messages with regard to sex. For example, sexual attractiveness has been held up as a highly desirable trait (largely, it appears, for commercial purposes), but sexual activity outside of narrowly confined limits has been strongly condemned.

Many anthropologists feel that there is no absolute positive or negative social value inherent in any particular sex act, but rather that the total social context in which the act takes place determines its value. In many non-Western societies, sexual activities normally forbidden are allowed and even encouraged on ceremonial occasions. Such activities have a specific meaning within the total social framework. Many anthropologists feel that some Western societies have created enormous psychological and social problems for themselves by virtually eliminating such "ceremonial" situations. (Remnants of such ceremonial license exist in our own society. On special occasions such as weddings or New Year's Eve parties, it is permissible to kiss another person's spouse or perhaps hug a person of the same sex. Under other circumstances these practices are discouraged.) It is the view of many anthropologists today that such ceremonial license serves not only to relieve ten-

SUPPLEMENT 15.5
BIOLOGICAL DETERMINISM VERSUS ENVIRONMENTALISM:
A CONTROVERSY AND THE ISSUES

The publication in the mid-1970s of Harvard University biologist Edward O. Wilson's book *Sociobiology: The New Synthesis* has led to sharp controversy between the author and a group of biologists representing a political action group called "Science for the People" (SFP), and the "Committee Against Racism" (CAR). The controversy has spilled over into the pages of several scientific journals (*Science*, March 19, 1976, Vol. 191, No. 4232; and *BioScience*, March 1976, Vol. 26, No. 3), as well as the popular press. Interestingly, the leading protagonists are Wilson and biologist Richard Lewontin, who work in the same building, Harvard's Museum of Comparative Zoology. Both Wilson and Lewontin are universally recognized as top-ranking experts in their respective fields.* The issues on which the disagreement is based are both scientific and political, and the entire controversy illustrates the difficulty (if not impossibility) of separating science from its political ramifications.

By coincidence, the authors of this textbook also find themselves on different sides in the controversy, primarily (though not entirely) on political grounds. One of us (Allen) takes the view of SFP and CAR. The other (Baker), though not unsympathetic to some of the important issues being raised by SFP and CAR, takes an opposing side.

The essential position of SFP and CAR is that Wilson's book is simply another (though perhaps more subtle) example of biological determinism, which holds that human beings' social behavior is partly or wholly determined by their genes. Such a position, they argue, tends to supply justification for the existing political or social system by implying that it is genetically determined—the "natural" way of things. But Wilson maintains that his book constantly stresses his belief that culture, not genetics, is the dominant force shaping human behavior, and he has elsewhere written against what he calls the "naturalistic fallacy of ethics which uncritically concludes that what is, should be."† In emphasizing this, he points out that "Even when we can identify genetically determined behavior, it cannot be used to justify a continuing practice in present and future societies." To do so, he says, "would invite disaster."

Wilson does state his belief that there are "a number of unique human characteristics, so distinct that they can be safely classified as genetically based." He lists as examples the drive to develop language, the avoidance of incest by taboo, and the sexual division of labor. But SFP and CAR contend that there is no solid scientific evidence whatsoever to justify attributing *any* human social behavior to genetic bases, and they are concerned that the tendency of the popular press to pick up and publicize hypotheses supporting biological determinism will encourage some people to use such ideas to support repressive political legislation.

As is suggested elsewhere in this book, such concern is not unjustifiable. Biological determinism arguments—such as attributing to genetic causes lower scores on I.Q. tests for blacks, the appropriateness of domestic rather than intellectual occupations for women, homosexuality, and the like—have often been used to justify oppressive laws, cutbacks in essential services, or discrimination against the groups involved.

A study group of SFP has written extensive analyses of Wilson's book, analyses in which Wilson and his defenders maintain that he is quoted out of context. For example, in the *New York Review of Books* the SFP group states, "He [Wilson] asserts that 'humans are absurdly easy to indoctrinate' and therefore 'conformer genes' must exist." They do not mention, however, that Wilson clearly states "if we assume for argument" before postulating the existence of "conformer genes." The SFP defense (Allen) is that merely to postulate the existence of such genes gives the impression, especially to lay persons (to whom the book has been widely advertised), that there is scientific evidence for such postulation.

Such an impression *will* possibly, even probably, be given (Baker). To state, however, that the postulation of certain ideas should be forbidden on the grounds that they may be misused politically or contradict religious or political dogma infringes directly and dangerously on what many claim is central to the very nature of science, in which freedom of inquiry is a vital and integral part. As Wilson puts it, the *perversion* of the idea should be discouraged, not the idea itself. He strikes a responsive chord among many scientists when he refers to the position of SFP as "academic vigilantism." He points out that the position of SFP has led other groups to refer to his book, and thus indirectly to him, as "racist" and that some of his lectures have been interrupted by hostile questioning. Thus as a scientist he finds himself constantly having to defend *himself* rather than being able to deal with the intellectual substance of his ideas.

---

*Wilson is a zoologist with a distinguished research career dealing mostly with the social insects. Lewontin, Alexander Agassiz Professor of Zoology at Harvard, is a population geneticist engaged in both theoretical and experimental research in evolutionary genetics.

†These and the following Wilson quotes are from "Sociobiology: A New Approach to Understanding the Basis of Human Nature," *New Scientist*, May 13, 1976.

It is further maintained (Baker) that the very cases raised by SFP and CAR writers argue, historically, in favor of sociobiological studies being encouraged rather than repressed. SFP and CAR are quite correct that the biologically deterministic concepts that led to the eugenics movement of the late nineteenth and early twentieth centuries did have repressive political effects; indeed, as Allen himself has pointed out elsewhere, leading geneticists testified in favor of restrictive immigration legislation that discriminated heavily against the immigration of Jews and southern Europeans. Had SFP and CAR been around then, and had the course of action they recommend been followed, genetic research applicable to eugenics would have been stopped and we would have been denied the knowledge we now possess, especially in the field of behavioral genetics, that reveals the unscientific base of eugenic hypotheses. It is not learning, but rather *"little* learning" that is the "dangerous thing."

SFP, however, maintains that research into or speculation about the genetic control of human behavior should be discouraged because such investigations (1) cannot by their very nature yield rigorous answers, and will thus inevitably produce pseudoscience; and (2) will inevitably be misused in the present social system, which they see as rampant with injustice and prejudice. The questions cannot be answered scientifically, SFP and CAR argue, because the only way to separate genetic from environmental components of behavior is to carry out controlled human breeding and rearing experiments, which most human beings are unwilling to do. Lewontin sees the question of what contribution genetics makes to human behavior as not a biologically useful one, and SFP and CAR see even the *suggesting* of hypotheses proposing genetic bases for human social behavior as being a political act, regardless of the results obtained. For this and other reasons, SFP has launched attacks against all research falling under the aegis of biological determinism, most notably and successfully in the case of the *XYY* controversy (see Supplement 17.3).

The issues involved are extremely important ones and cannot be done full justice here. Certainly a great deal will be heard about this controversy in the years to come.

sions but also to build strong social bonds and thus greater cohesiveness among members of a group or community.

Masturbation and Sleep-Associated Orgasm

Masturbation is manipulation of the genitals to produce sexual stimulation and arousal, sometimes culminating in orgasm. In males, masturbation involves rubbing the penis with the hands or against some objects including clothing; in females, it involves rubbing the vulva, especially the clitoris, and/or inserting the fingers or an object into the vagina. Masturbation also occurs among many species of mammals, especially in the primates.

In humans, possibly for cultural reasons, masturbation appears to be more common among males than among females. According to data from the United States and Canada, about nine-tenths of all males report that they engage in masturbation on some regular basis, whereas only about two-thirds of women report the same. Masturbation appears to be extremely common among adolescents of both sexes, at a time when sexual drive is increasing but sexual contact with other persons is still limited. As other sexual activity increases, masturbation tends to decrease. However, it rarely ceases entirely, even among married people.

Myths have abounded in the past about masturbation causing insanity, stuttering, pimples, and a whole host of physiological and psychological problems. The term masturbation itself shows a cultural bias, since it means "self-pollution." In order to remove this stigma against the practice, researchers and practitioners (such as sex educators, counselors, and therapists) in the field of sexology generally use the term "self-pleasuring."

In some religions, masturbation is still viewed as being a sinful act. It is now generally accepted, however, by many people that masturbation is something pleasurable in its own right, which produces no ill effects on adolescents or adults and provides a healthy way of dealing with sexual drive. Some psychologists see masturbation as a problem, however, only if it becomes the *sole* means of expressing sexual feelings—that is, when it may be symptomatic of a person totally unable, psychologically, to make sexual contact with another individual.

Occasionally human beings may experience orgasms in sleep. In males this is sometimes referred to as a "wet dream," since it usually involves the ejaculation of semen. Data for the United States suggest that virtually all males and about two-fifths of females experience orgasm in sleep. In males, "wet dreams" occur most frequently during adolescence and early adulthood, falling off by middle to late twenties. In women, however, orgasm in sleep is relatively uncommon during adolescence and increases to a peak during the thirties and forties. Orgasm in sleep is sometimes but not always accompanied by erotic dreams. The frequency of such orgasms is not necessarily related to lack of other sexual activity.

15.9
HOMOSEXUALITY

One of the most misunderstod aspects of sexual behavior is **homosexuality**, or sexual activity between members of the same sex. While the term "homosexual" ap-

SUPPLEMENT 15.6
THE ORIGINS OF VIOLENCE: A CASE OF CULTURAL DETERMINISM?

In the late 1950s and early 1960s Drs. Harry and Margaret Kuenne Harlow, working at the primate laboratory at the University of Wisconsin, performed their now-classic studies of the effects of sensory deprivation on rhesus monkey infants. Removed from their mothers at birth, the infants were deprived of touch and play with peers, and fed by a bottle. Uniformly, such monkeys grew into socially detached, often hostile animals, who would exhibit autistic rocking behavior, self-mutilation, and aggression toward others. Sexually mature isolated females, though sexually receptive (physiologically speaking), would attack prospective mates. If placed in a restraining rack which made copulation possible, these animals became indifferent to their young, often brutalizing or even killing them.

The Harlows followed up their studies by showing that, if given a choice of a wire or soft terry cloth surrogate mother, the infants spent most of their time clinging to the soft mother, even when it was the wire mother that contained a milk-providing bottle. The Harlows interpreted this finding as representing the extreme importance of comforting sensory input to the development of normal social interaction between infant and mother.

Equally significant was the finding that rhesus monkey infants preferred cloth surrogates that moved in a rocking or turning motion to those that were still and that infants raised with motionless surrogates showed autistic self-rocking movements, as if to compensate for their deprivation by providing the motion stimulation themselves.

Such findings have been of great interest to psychologists and psychiatrists associated with institutions for the mentally ill, for precisely the same behavior syndromes are seen in institutionalized humans—autistic rocking movements, emotional and/or social detachment, self-mutilation, and a tendency toward violent behavior (the latter often the reason for institutionalization). The same behaviors are also reported by social workers among some of their noninstitutionalized clients. In many cases, the family histories showed the affected individuals to be unwanted children who received little, if any, attention and what little *was* received was often of a violent, abusive nature. Such observations seemed to have obvious connections to the well-documented observations that child abusers often turn out to have been abused children themselves.

How are such observations to be accounted for on the anatomical-physiological level? Body movements, even passive ones (such as being rocked), direct a stream of impulses toward the brain cerebellum. The cerebellum is unique among the brain regions in that cell development within it continues long after birth. Thus, in infants, such motions probably play a major role in cerebellar development; institutionalized children, often unable to receive such motions and other types of sensory stimulations associated with parental contact, are known to lag behind their noninstitutionalized counterparts in motor coordination and development.

The Harlows' original work with sensory deprivation in rhesus monkeys has been followed up by many scientists. One of these is Dr. James W. Prescott, a developmental neuropsychologist most recently at the National Institute of Child Health and Human Development in Bethesda, Maryland. It is his hypothesis that "the rocking behavior of isolation-reared monkeys and institutionalized children may result from insufficient body contact and movement. Consequently, both touch and movement receptors don't receive sufficient sensory stimulation for normal development and function." Besides coming to this conclusion from his own research data and those of others, Prescott notes the common observation that young children like to be picked up, rocked, swung, whirled through the air, and to go on merry-go-rounds, roller coasters, and the like. Both children and adults generally like the motions enjoyed in warm baths or massages, and the sensory inputs such activities cause. Physical exertions too, such as jogging, often produce a "natural high."

Prescott's hypothesis is unorthodox on at least two counts. For one, as opposed to a hypothesis based upon biological determinism (see Supplement 15.5) in which a particular behavior is hypothesized to be due to genetic influences, the Prescott hypothesis suggests a sort of "cultural determinism" (Prescott himself refers to "ecobiology" as opposed to sociobiology) in which environmentally or culturally imposed behavior patterns of the parents have distinct anatomical and physiological results on their offspring that can influence their behavior. Thus, for example, children raised in a strict puritanical religious environment in which affection, physical touching, and sensory pleasure in general are forbidden may show actual "incomplete or damaged development of the neuronal systems that control affection (for instance, a loss of the nerve cell branches called dendrites)." Prescott believes that there is neurophysiological evidence linking not only the limbic areas of the brain responsible for emotion (called the hippocampus, amygdala, and septal areas) with each other but also these limbic areas with the cerebellum. It is this last connection that is the second unorthodox part of the Prescott hypothesis; traditionally, the cerebellum, except for its role in coordinating muscular movement, has not been linked at all to behavior. Dr. Prescott believes that research done by Dr. Robert G. Heath of Tulane University demonstrates the existence of connections between these emotional centers and the cerebellum. In addition, it is maintained, two-way communication can be shown between the cerebellum and areas of the brain associated with pleasure and displeasure.

Dr. Prescott has done cross-cultural studies (based on studies carried out by anthropologists) that he says show a significant correlation between the physical affection shown human infants and the amount of adult violence. Among the Comanche and Ashanti, for example, he finds the level of violence high, while where physical affection is high, as among the Maori of New Zealand or the Balinese, violence is low. Dr. Prescott writes that "culture is the handmaiden of our neurobiology, and without a proper environment for physical affection, a peaceful, harmonious society may not be possible."

It seems obvious that any relationship between environmental stimuli, brain development, dendrite formation, and the nature of social behavior, whether socially acceptable or not, is going to be a complex one. Most certainly, considering the escalation of violence in our society, the answer to the questions such a relationship poses is an important one.

plies to both men and women, women homosexuals are more often referred to as *lesbians*. The view that has been dominant until very recently claims that adult homosexuality is an abnormality signifying an emotional or psychological disturbance. Such a view sought to explain homosexuality by reference to "abnormalities" in a person's past history. The "abnormality" invoked might be environmental (a disturbed home life) or biological (hormonal imbalance or chromosomal aberration). The fundamental assumption underlying either explanation, however, is that homosexuality is an illness.

Although there have been various attempts in the past to lay the cause of homosexuality to genes, chromosomes, or hormones, there has been no consistent evidence to support such views. Far more prevalent has been an "environmental" theory of homosexuality. The logic employed by proponents of this theory correlates certain patterns of early family life, judged to be abnormal, with the occurrence of homosexuality. One problem with this approach is that what is considered "abnormal" is highly subjective. In some cultures, homosexuality is considered quite normal. Furthermore, the theory has been based on evidence from case studies of individuals who came to a psychiatrist or hospital for counseling and/or therapy about their homosexuality. By definition, such individuals are automatically classified as emotionally disturbed. Data on homosexuals who did not need psychiatric counseling never appeared in the sample, so it was highly biased. More telling studies by Kinsey and others suggest that extremely high percentages of the human population have engaged in same-sex sexual activity at some point in their lives. Since most such people are not emotionally disturbed, they are not identifiable for research purposes.

Today, the sickness theory of homosexuality, whether genetically or environmentally based, is beginning to be discarded. A broader view of homosexuality, embodied in the early studies of Kinsey, has been amplified by other workers more recently. The new theory sees homosexual behavior as simply part of a broad spectrum of normal human sexual response. As Kinsey first put it, there is a continuum of human sexual behavior, ranging from pure homosexuality on the one side to exclusive heterosexuality on the other. The idea of a norm of behavior, according to Kinsey, is fallacious:

> *Human beings do not represent two discrete populations, heterosexual and homosexual. The world is not to be divided into sheep and goats. . . . The living world is a continuum in each and every one of its aspects. The sooner we learn this concerning human sexual behavior the sooner we shall reach a sound understanding of the realities of sex.*

Keeping the continuum of sexual behavior in mind, modern researchers argue that all human beings develop and retain the potential for expression of both homosexual and heterosexual behavior (what is frequently referred to as bisexual or ambisexual behavior) throughout their life. Through social conditioning, expression of homosexual feelings is discouraged and heterosexual responses encouraged in most western cultures. The result is that the capacity for homosexual response is suppressed in all but a few people, and for those who do not or cannot suppress it, culturally induced feelings of guilt, shame, and lack of self-esteem may be the result. For very good reasons, such persons may seek psychiatric counseling. Thus, it is not necessarily their sexual preference *per se*, but rather society's intolerant and oppressive attitude, that is the problem.

Kinsey's data, as shown in Fig. 15.11, indicate that between the ages of ten and fifteen distinct homosexual behavior involving sexual arousal and orgasm occurs in about 25 to 30 percent of United States males. In the late adolescent and early adult periods, the frequency of homosexual activity among United States males appears to decline considerably. This appears to coincide with an increase in heterosexual activity. Such figures suggest that, at least among a substantial proportion of the male population in the United States, overt homosexual be-

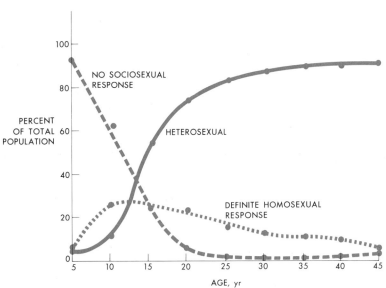

Fig. 15.11
Graph showing the development of sociosexual response in United States males, with specific emphasis on heterosexuality and homosexuality. Data are from Alfred Kinsey's studies of 1948. Although over thirty years old, Kinsey's data are largely confirmed in recent studies. The graph shows that between the ages of five and twenty years, the number of males who show no sociosexual response to any other individual declines rapidly. Between the ages of five and ten, homosexual response increases rapidly, while heterosexual response rises more slowly. Homosexual response shows a slow decline after age ten or fifteen, while heterosexual response shows a dramatic increase after age ten. These data suggest that both homosexual and heterosexual responses are part of the normal developmental processes of the preadolescent and adolescent male. Heterosexual response clearly dominates the sexual behavior of most males by middle to late adolescence.

havior during adolescence is a "normal" response. A specific behavior that occurs in 25 to 30 percent of the population cannot be easily dismissed as "abnormal."

Drawing on Kinsey's findings, some researchers have presented a revised version of the sickness theory. They argue that homosexuality may be a natural stage through which adolescents pass; it is one possible route on the road to heterosexual adult behavior. Proponents of such a viewpoint regard homosexuality as natural among adolescents but unnatural among adults. Commonly, they refer to homosexual behavior among adults as an example of "arrested development."

Others, however, see such an interpretation as culturally biased. It is possible, they argue, that both homosexual and heterosexual behavior among adults is natural in the biological and psychological sense; that is, both are expressions of naturally occurring sexual and emotional feelings of which all adults are theoretically capable. The failure of homosexual behavior to remain at its adolescent level or to increase with maturation could be viewed as the result of strong social pressures, rather than as a biologically determined process of development. There might be nothing "unnatural" about homosexual behavior among adults. It has even been argued that homosexual behavior is biologically adaptive for any population, serving as an avenue of sexual expression without resulting in an increase in population. Evidence for this view comes from studies of sexual behavior in animal populations. Often, as population density increases, the frequency of homosexual behavior is noted to increase, and there is a corresponding decrease in frequency of heterosexual encounters.

Two recent books (see Suggested Readings, at the back of this text) by Bell and Weinstein and Masters and Johnson describe the authors' massive studies of homosexuality and come to the basic conclusion that homosexuals have as wide a range of sexual behavior stereotypes as heterosexuals; have the same kinds of sexual problems; and, when the effects of repressive social attitudes and legislation are taken into account, are as well adjusted psychologically as heterosexuals. As is often the case when scientific studies fly in the face of cultural myths, the works have been widely attacked in some parts of the popular press and, on certain procedural aspects (though not generally on their basic conclusions), in the scientific literature as well. In general, however, there is an increasing tendency in the field of sexology to modify research attempts to find the "causes" of homosexuality and concentrate rather on trying to learn more about the development of sexual behavior of *all* types—after all, we do not know what "causes" heterosexuality either. Most certainly there is no scientific justification whatsoever for denying homosexuals their civil rights and it is appalling to realize that even today in some states it is still perfectly legal to do so. Hopefully, as sex education becomes more widespread, the ignorance that leads to such legislation will disappear.

15.10
COMPARATIVE SEXUAL BEHAVIOR AMONG DIFFERENT SPECIES

The study of comparative sexual behavior among animals is a part of the general study of animal behavior, or ethology. As noted earlier, the use of comparative data from nonhuman species must be approached with caution. Nevertheless, the range of findings about animal sexuality cannot help but be useful in the study of human sexuality, if in no other way than in providing a picture of the diversity of sexual behaviors among different species. Some findings from studies in comparative animal behavior can be briefly summarized as follows:

1. Among all mammalian species there are two general patterns of behavior that appear to be sex-related. One is for "masculine" mounting behavior and the other is for "feminine" being-mounted behavior. The animal that mounts the other is more active and exhibits rhythmic thrusting motions of the pelvis. The mounted animal tends to be less mobile. However, both patterns of behavior are exhibited by both sexes in all species, indicating that the behavior is not completely neurologically differentiated between male and female.

2. In both sexes suitable stimuli (visual, tactile, behavioral) increase sexual excitement and produce various physiological responses, such as increased heart rate and breathing rate, muscular tension, and genital arousal.

3. Patterns of sexual behavior appear in the "play" behavior of young mammals before they mature sexually. Playful mounting behavior is exhibited by young males and, to a lesser extent, by young females. There is increasing evidence that children deprived for various reasons (poverty, being "unwanted," religious prohibition, and the like) of affection, touch, play behavior with parents or peers, and so on, are often unable to relate sexually in adulthood, and may even be prone to violence (see Supplement 15.6).

4. Homosexual activity is known in nearly all mammalian (and some other) species.

5. Considerable oral stimulation precedes or occurs during sexual intercourse in many mammalian species. Particularly common are licking, gentle biting, and nuzzling. Oral stimulation of the genitalia by both sexes has been observed in many species, including rhesus monkeys, chimpanzees, and gorillas. Masturbation is also known among a large variety of mammalian species (for example, horses).

6. A cycle of sexual activity is characteristic of females in all nonhuman mammals; it is also characteristic of the males in some species.

7. Sexual behavior in most mammalian species is typified by a wide range of individual variations. In proceeding up the mammalian evolutionary tree from simpler to more advanced species like the primates, the trend is toward less rigidly programmed behavior of all kinds, including sexual behavior. In addition, the range of individual variation within a species increases as we proceed toward the more advanced nonhuman species.

8. With regard to pair bonding and mate selection, there is no uniform pattern observable among nonhuman species. A few species form lifelong pair bonds (monogamy), most form pair bonds only for a given reproductive season, and others form no pair bonds at all, with care of the young performed by the female or male alone.

Studies of comparative mammalian sexual behavior suggest that specific behavior patterns are programmed into individual organisms. Such findings have led to immediate speculation that human sexual patterns such as sex role patterns, sexual preferences, and even sexual customs are similarly programmed. In general, programmed behavior is that which is biologically (genetically) determined and which can thus be modified only by working against the innate tendencies of the organism. As valuable as such comparative studies have been, they often lead to a methodological fallacy. Analogies between nonhuman and human species at best suggest that similar patterns of behavior *might* arise from the same biological cause (common genetic elements). However, cultural influences have an enormous impact on the development of human behavior. Thus an alternative hypothesis (namely, that human males learn sexual procedures from cultural experience) leads to the same observed behavior. Because it is not possible to experiment directly with human beings under controlled conditions, there is no way of distinguishing clearly between the two hypotheses. While analogy may suggest the *possibility* of similar causes for similar behaviors, it cannot establish such similarities conclusively.

Two other factors make it difficult to draw conclusions about human sexuality from animal analogies. It usually happens that any particular behavior pattern one observes in human beings is, first, not characteristic of *all* human beings, who often show a wide range of variability among different cultures, and, second, can be found in some nonhuman species but not others. To argue, for example, that monogamy is the most natural form of pair-bonding pattern for human beings, one could gain support by referring to certain vertebrate species such as Canada geese and turtle doves. On the other hand, one could just as effectively support the opposite notion, that polygamy is the most natural pair-bond pattern, by choosing such examples as chimpanzees. Virtually no human sexual behavior is not present in some nonhuman species and absent in others. In regard to reasoning by analogy, it has been said that the animal kingdom is like the Bible or Shakespeare; one can find individual quotations that prove any point one wishes to make. Ultimately, the only satisfactory method for determining the causes for human sexual (or other type of) behavior is direct study of human beings themselves. As has already been pointed out, such studies have recently made considerable advances by combining the scientific approach of such workers as

Kinsey, Masters, and Johnson with the anthropological methods of cross-cultural analyses.

Male-Female Differences

Nowhere has the nature-nurture argument been more visible than in the question of whether male and female sex roles and patterns of behavior are innately or culturally determined. In the past, a large variety of behavioral and personality traits have been said to be innate in the male and female psyches. Men were supposed to be strong physically and emotionally, seldom to cry or show their inner feelings, and to be aggressive, competitive, and analytical. Women, on the other hand, were supposed to be physically and emotionally weaker than men, irrational, passive, and nonanalytical.

Following from the arguments presented earlier, it would appear that such behavior is no more innate than any other types of behavior or personality patterns. Yet the idea has persisted for generations. In the late nineteenth century, anthropologists and biologists argued from measurements of skull size and shape that women were intellectually inferior to men. Some claimed that skull size and proportions made women behave more like children or the "savages" of Africa than like white European males. Such arguments were used to buttress opposition to the movements for women's suffrage in Europe in the 1870s and 1880s. The size and shape of the skull, a biologically determined trait, was supposed to determine (limit) women's intellectual development.

It goes without saying that few people today accept any relationship between skull size or shape and any kinds of intellectual or behavioral capacities. Some biologists, however, feel that such arguments are reappearing today in a new area of research termed "sociobiology" (see Supplement 15.5), since a few sociobiologists have suggested that women's position in society has a genetic basis and that the different social roles seen today in our society evolved because they were selectively advantageous. However, besides the problem of ever being able to test such a hypothesis, their historical "batting average" is poor. In the nineteenth century, for example, such characteristics as emotionality, irrationality, passivity, and the like were associated with women, and a sex-associated genetic basis hypothesized. Today we recognize that not only are these characteristics found in both men and women but that they are clearly the result of social conditioning. Women today, especially those young enough to have developed in a society increasingly influenced by the women's movement, are significantly different from their nineteenth-century counterparts. The personality traits assigned to women in the past were largely stereotypes invented by men within the context of a patriarchal, male-dominated society; the women's movement has shown that women can, if given the freedom to do so, defy all such stereotypes. Indeed, if such behavior were genetically controlled, it should be difficult if not impossible to modify it easily. The fact is that over the past decade increasing numbers of women have begun to take their place in important positions in society, and the old stereotypes of "typical" female behavior are fading. In the face of such changes in actual behavior—and accepting that some biologically determined factors *do* influence aspects of male and female sexual behavior—it becomes increasingly difficult to maintain that specific sex-role patterns are innately ingrained in either men or women.

Summary

1.   The same primordial embryonic tissues give rise to the ovaries and the testes. Thus sexual differentiation in male and female embryos begins with the same structures, but development occurs along increasingly divergent lines as the embryo grows. Accessory parts of the male and female reproductive systems develop from different embryonic structures (the male sperm ducts from the Wolffian ducts and the female Fallopian tubes from the Mullerian ducts) that are both present in the young embryo. In males, the Mullerian ducts degenerate under the influence of testosterone; in females, the Wolffian ducts degenerate under the influence of estrogens.

2.   The human male reproductive system consists of the penis, testes, seminal vesicles, ejaculatory ducts, and various fluid-producing glands. The testes serve two functions: production of sperm and secretion of the male sex hormone testosterone by special cells, the interstitial cells.

Testosterone is important in stimulating the development of the male sex organs during embryonic growth and the secondary sex characteristics at puberty. The general name for all male sex hormones is androgens.

3.   The female reproductive system consists of the ovaries, Fallopian tubes, uterus, vagina, and clitoris. The clitoris is anatomically homologous to the male penis and functions as an organ of stimulation. The vagina is a receptacle for the penis during intercourse. The uterus contains the embryo during development and is the site of menstrual build-up every month. The Fallopian tubes connect the uterus with the ovaries which produce, between them, one egg per month. Like the testes, the ovaries are also endocrine

glands, producing the female sex hormones collectively called estrogens.

4. Eggs are produced in the ovaries within structures called Graafian follicles. It takes a follicle about two weeks to complete the final stages of growth prior to release of the egg. After the egg is released, the old follicle becomes the site of production of the hormone progesterone. As an endocrine gland, the follicle is known as the corpus luteum.

5. The egg matures as it is passing down the Fallopian tube toward the uterus. Fertilization, if it occurs, generally takes place in the Fallopian tube.

6. The menstrual cycle is under hormonal and neural control and shows all the characteristics of a negative feedback (feedback inhibition) process. Follicle-stimulating hormone (FSH) stimulates growth of the follicle, which produces estrogens; a high level of estrogen in the blood causes the hypothalamus to inhibit secretion of FSH by the pituitary. Luteinizing hormone (LH) stimulates production of progesterone by the corpus luteum; a high level of progesterone in the blood causes the hypothalamus to inhibit secretion of LH by the pituitary.

The menstrual cycle shows a long-term (twenty-eight-day average) oscillation compared to the more usual short-term oscillation of a few minutes to a few hours for most other negative feedback systems. The long-term response of this system is a result of the fact that it does not begin to release neurotransmitter to the pituitary until well after the levels of estrogen or progesterone have begun to fall.

The estrous cycle in mammals consists of alternating periods of ovulation (during which the female will accept males sexually) and nonovulation (during which the female will not accept a male). In some higher primates and human beings, there is no strict estrous cycle, since the female will usually accept a male for intercourse at any time in her reproductive cycle. Human and anthropoid ape females, on the average, have a twenty-eight- to thirty-day menstrual cycle. During three or four days of this cycle, blood and tissue are given off as the menstrual flow.

During the first two weeks of the menstrual cycle, the egg is developing within the Graafian follicle, and the lining or endometrium of the uterus is building up a very rich capillary supply. If the egg is not fertilized, the endometrium breaks down, producing a loss of blood for approximately the last four days of the cycle. If the egg is fertilized, the uterine lining stays intact and becomes the place where the young embryo implants.

7. Sexual intercourse generally involves sexual arousal by both partners followed by coitus, or insertion of the penis into the vagina. For both male and female, four stages can be identified: excitement, or the beginnings of arousal; plateau, or full sexual arousal and increased sensitivity of the genitalia to touch; orgasm, which in the male includes the ejaculation of sperm and in male and female an explosive feeling of release; and resolution, or decrease in sexual arousal and return to the pre-excitement level. Usually, males have only one orgasm during intercourse, whereas some females may experience several orgasms in relatively rapid succession.

8. While certain structures and functions of the male and female reproductive systems are biologically quite distinct, sexual behavior and attitudes appear to be determined by social rather than biological causes. Such aspects of sexual behavior as sex roles, sex identity, homosexuality, and the like also appear to be the result of social rather than biological development.

Exercises

1. How do the structures of the male and female reproductive systems in humans work to ensure maximum chances of fertilization of the egg?

2. In many mammalian species that regularly have two or more offspring at a time, each fetus normally has its own placenta. Occasionally, however, two fetuses share the same placenta, which means that the circulatory systems of the fetuses are directly interconnected. If the twins are of opposite sex, the female always turns out to be a sterile form called a freemartin, while the male is sexually normal. Given the fact that in male embryos the production of testosterone by embryonic interstitial cells occurs earlier than production of estrogens in female embryos, how can you explain the freemartin case?

3. Many groups have lobbied aggressively against legalization of abortion on the grounds that it is murder. Other groups have worked for legalizing abortion, claiming that it is the right of a woman to decide whether she wants a child or not and that unwanted children create many problems for families. Opponents of abortion claim that the unborn fetus has rights too, and that no one should be allowed to decide whether it is to be born or not.

a) Describe the ways in which the antiabortion argument displays elements of mechanistic philosophy (see Section 1.9).

b) What is the central issue about which the two sides disagree? How might differences on that issue be resolved?

c) Why is it difficult to derive definitive, all-encompassing answers to issues such as abortion?

4. A textbook on human reproduction for college students states that "Homosexuality is biologically absurd and anatomically ridiculous." What does this statement reveal about the author's assumptions and/or value system?

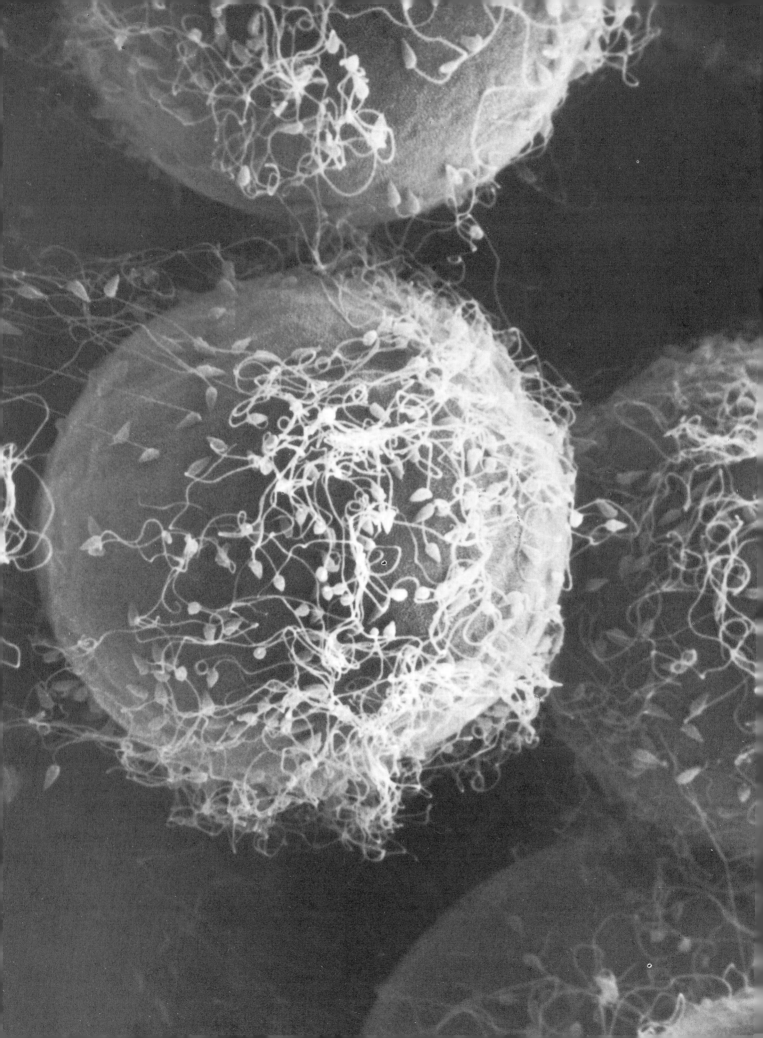

PART IV
HEREDITY
AND
DEVELOPMENT

All organisms, from one- to many-celled, are capable of reproducing their kind. Reproduction involves transmission to the offspring of traits possessed by the parents. There are many likenesses and also some differences between offspring and parents. People do give birth to children who grow up to resemble other members of the human species. Yet offspring differ in many small details from their parents. Two brown-haired parents can have a red- or blonde-haired child, for example. Nevertheless, it is apparent that all organisms inherit physical traits from their parents.

The study of heredity is called **genetics,** that branch of biology concerned with transmission, or how traits are passed on from one generation to the next. Genetics is also concerned with **gene expression,** or how the transmitted hereditary information develops into recognizable adult traits.

In exploring how genes produce their effects on the cellular and molecular levels, the science of genetics merges with the study of development. Developmental biology is concerned with the processes of orderly change that characterize the life cycle of every species. Each organism starts as a fertilized egg; the egg develops as an embryo, growing not only larger but more complex as its cells differentiate and specialize; the adult maintains itself, repairs injured parts, and produces reproductive cells of its own to produce a new generation; the organism eventually grows old, deteriorates, and dies. To a large extent all these changes are programmed into the hereditary makeup of the organism. Developmental biologists seek to understand how these changes are brought about at the organismic, cellular, and molecular levels.

In this unit we will focus attention first on how genes are transmitted from generation to generation, and then on how genes produce their effects at the molecular and biochemical levels. The unit will end with a thorough look at the relationship between modern genetics and developmental biology, two fields occupying a central area within contemporary biological research.

Chapter 16
Genetics I: From
Math to
Mendel

16.1
INTRODUCTION

Modern genetics is the study of the structure and function of genes, the basic units of heredity passed on from parent to offspring in the reproductive process. Genetics has become one of the most important aspects of all of current biology. Indeed, genetics relates directly to virtually every other branch of biology today. Because genes guide the production of proteins, especially enzymes, within the cell, genetics has become an important aspect of biochemistry and cell biology. Because genes are responsible for the progressive differentiation of tissues and organs during development, genetics has become a central focus of modern developmental biology. And because genes are the basic units of selection within populations, genetics has become the foundation for contemporary evolutionary theory. Knowledge of genetics has virtually revolutionized every field of biology today.

16.2
OVERVIEW

The many levels of organization on which we can approach the study of genetics are shown in Fig. 16.1. Proceeding from right to left, the diagram shows successively higher levels of organization. On the far right is the molecular level, or what is called *molecular genetics*. Molecular genetics is the study of the molecular structure of the genetic material, deoxyribonucleic acid (DNA) and ribonucleic acid (RNA), and how they function to determine the characteristics of organisms. For most organisms, DNA is the basic hereditary material,

Fig. 16.1
Levels of organization of modern genetics. Concerned with the study of the gene, modern genetics carries out investigations at a number of levels of organization. See text for details.

LEVELS OF ORGANIZATION IN MODERN GENETICS

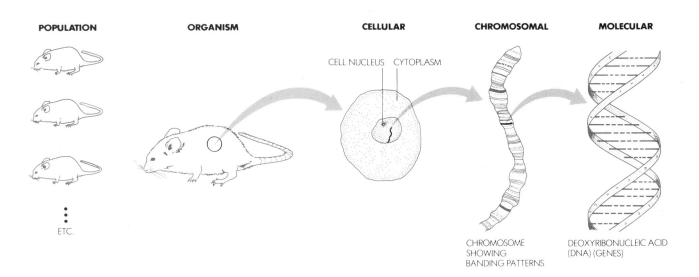

POPULATION   ORGANISM   CELLULAR   CHROMOSOMAL   MOLECULAR

CELL NUCLEUS   CYTOPLASM

ETC.

CHROMOSOME
SHOWING
BANDING PATTERNS

DEOXYRIBONUCLEIC ACID
(DNA) (GENES)

though for a few viruses RNA has this function. DNA determines specific traits in organisms by guiding the production of specific polypeptide chains, one or more of which interact to form a protein molecule. A unit of DNA that specifies the structure of a given polypeptide is called a **gene**; genes are considered the basic structural and functional units of heredity. The units of DNA, the genes, are passed on from generation to generation in the reproductive process. Molecular genetics is concerned with how genes are structured, how they replicate themselves, how they are passed on from one generation to the next, how they vary (that is, undergo mutation), and how they control the synthesis of specific proteins. In short, molecular genetics focuses on the relationship between the molecular structure of nucleic acid and the biochemical or metabolic processes within cells.

Proceeding to the left in Fig. 16.1, we come next to the chromosomal and cellular levels of organization. The DNA of both prokaryotes and eukaryotes is organized into thread-like or rod-shaped structures known as **chromosomes**. In eukaryotic cells chromosomes are composed of nucleic acid and protein and are contained within the cell nucleus. In prokaryotic cells, chromosomes are composed only of nucleic acid and are not enclosed within a nucleus, though often they are localized in certain areas of the cell (sometimes referred to as a "nucleoid"). In both prokaryotes and eukaryotes there are also smaller collections of genes that exist apart from the major chromosomal or nuclear organization. In eukaryotes some genes are found in cell organelles such as mitochondria and chloroplasts; in prokaryotes some genes are found on small, extrachromosomal elements known as plasmids. The inheritance of these nonchromosomal or nonnuclear genes often follows a different pattern from the inheritance of chromosomal and nuclear genes—a difference we will discuss in greater detail later in this and in the following chapter. The entire complement of genes in an organism, regardless of their physical arrangement within the cell, is known as the organism's **genome**.

Following Fig. 16.1 farther to the left, we encounter the organismic level of organization. The organismic level is concerned principally with the ways in which genes are passed on, or transmitted, from one generation to another. It is concerned with breeding patterns of organisms, the number and kind of offspring that can be produced from various crosses, and with how genes affect the overall development of adult traits. Whereas molecular genetics is concerned with the way in which DNA guides the production of specific proteins, organismic genetics is concerned with how genes guide the production of gross anatomical and physiological traits—for instance, comb shape in chickens, height in pea plants, or coat color in guinea pigs. Organismic genetics encompasses classical Mendelian genetics, such as the inheritance of dominant and recessive traits. Organ-

ismic genetics is focused primarily on the reproduction of individual organisms as breeding (or family) lines.

The highest level of organization on which we can study genetics is that of the population, as shown to the far left in Fig. 16.1. Instead of viewing genes merely as parts of individual organisms or cells, population geneticists view genes *collectively,* as groupings of hereditary units that are to some extent separate from the organisms that house them at any moment in time. To the population geneticist, a population of organisms is not merely a grouping of animals or plants, but a collection of genes that have a life of their own and are passed on from generation to generation, mixing and remixing with each new breeding cycle. The importance of population genetics for modern biology lies in the information and perspective it provides about the natural history, ecology, and evolution of species. For example, evolution can be regarded from the population genetics point of view as merely a change in the number and kind of genes found in a population of interbreeding organisms over successive generations. By analyzing such changes, population geneticists can determine the way evolution may be taking place within a given population under a given set of conditions. Because it is amenable to quantification and thus to mathematical treatment, population genetics has contributed enormously to an understanding of the many complex interactions involved in such processes as evolutionary change, or the breeding dynamics within animal and plant communities.

Although we can view genetics on many different levels of organization, biologists today often attempt to integrate two or more levels when studying any particular problem. For example, much exciting current research is based on trying to discover how the structural arrangement of genes on the chromosomes of eukaryotes affects how and when the gene is "turned on" or "turned off" with respect to guiding protein synthesis. Sixty years ago, a major area of genetic research focused on relating breeding patterns studied on the organismic level to visible changes occurring on the chromosomal level. Indeed, today there is much concern with showing how specific mutations in DNA (a gene) affect protein structure and function, and thus the evolutionary fate of individuals within a population. Although at times we may study one or another aspect separately, all levels of organization in modern genetics are interrelated. At various points in this and the next unit we will discuss some of those interrelationships, especially with regard to human heredity (Chapters 18 and 19) and evolution (Chapters 22 and 23).

Some Terms and Definitions

Growth and Reproduction. It is necessary to make a distinction between growth and reproduction. The increase in size which all organisms show during their lifetime is known as **growth.** Growth is characteristic of unicellular as well as multicellular organisms, though it is more dramatic in the latter. Growth is the result of the build-up of structural molecules at a rate more rapid than that at which they are broken down. One-celled organisms grow by increasing the size of their single cell. Growth of a multicellular organism is generally the result of an increase in the size of individual cells, an increase in the number of cells, or both.

Reproduction, on the other hand, is one of the most universally recognized and important characteristics of living things. It is characterized by the formation of a new and separate individual from a previously existing individual or pair of individuals. Reproduction in nearly all animals and plants is either **sexual** or **asexual.** Sexual reproduction involves the union of two special reproductive cells called **gametes,** produced by two distinctly different individual parent organisms. In most cases gametes can be distinguished as belonging to two different sexes, male and female. The male gamete (usually called a **sperm**) fuses with the female gamete (called an **ovum** or **egg**) in a process known as **fertilization.** The result is the formation of a single-celled **zygote.** By repeated cell divisions the zygote eventually develops into a complete organism. Sexual reproduction, though characteristic of multicellular forms, is by no means restricted to them. Most single-celled organisms, both eukaryotic and prokaryotic, are also capable of sexual reproduction at some phase in their lives.

Asexual reproduction occurs when a single organism gives rise to one or more organisms without the process of gamete formation and fertilization. The most common form of asexual reproduction is **cell division** or **fission.** During cell division, a single parent cell splits into two daughter cells, each exactly like the original parent cell. Although asexual reproduction is most common among the one-celled plants and animals, it is by no means limited to simple forms, since the cells of *all* organisms undergo cell division sometime in the course of their existence. Indeed, the process of growth in multicellular forms always involves cell division. Cell division is a reproductive process on the organismic level only in the case of unicellular organisms.

Other forms of asexual reproduction aside from simple cell division includes budding, vegetative reproduction, and parthenogenesis. **Budding** can be observed on the cellular level in one-celled forms such as yeast. Here, a single parent cell produces a small outgrowth, or bud, at one side of the cell. The bud grows to a certain point and splits off, forming a new individual cell. Budding can also be found as a reproductive process in multicellular forms such as the freshwater *Hydra* or sponges. Buds develop as outgrowths of the parent organism, eventually breaking away and becoming independent organisms on their own.

Vegetative reproduction occurs when adult organisms, especially plants, send out growths (from either existing roots or stems) that establish themselves as independent organisms. The new offshoot can either remain attached, through a stem or root, to the parent plant, or become a separate plant in its own right. Vegetative reproduction also occurs when a part of a parent plant breaks off and develops roots or shoots of its own. Vegetative reproduction occurs regularly in willow trees, ivy, wandering Jew, strawberries, iris, and many aquatic vascular plants.

Parthenogenesis is a process in which adult female organisms produce eggs that undergo embryonic development and produce new, adult, female organisms without fertilization by a sperm. Such parthenogenetic individuals are genetically identical to their mothers. Parthenogenesis occurs naturally in some species, especially in certain seasons of the year. For example, rotifers (small invertebrates with a mouth surrounded by beating cilia that look like a rotating wheel) normally reproduce parthenogenetically during the summer, but in the fall produce a generation of males which fertilize the females. The resulting zygotes form spores that remain dormant over the winter and grow into females in the spring. Parthenogenesis is also characteristic of aphids. It is an evolutionary development from normal sexual reproduction (in which the males have become largely superfluous!).

It should be obvious that one major difference between sexual and asexual reproduction lies in the genetic makeup of the offspring in relation to each other and the parents. In sexual reproduction, the offspring are seldom genetically identical to each other or to their parents. In asexual reproduction, just the converse is true. In all forms of asexual reproduction the offspring are genetically identical to each other and to the single parent that produced them.

Genetics involves the study of two different but interconnected processes: *transmission* and *translation.* Transmission is the process by which genetic elements are passed on from one generation to another. Translation is the process by which genetic elements, once transmitted, guide the production of adult traits. We will begin our study of heredity with the process of transmission. Because transmission involves the shuffling and reshuffling of genes from one generation to the next, patterns of transmission follow the basic laws of chance, or probability. It will therefore be appropriate to begin our investigation of the hereditary process with an examination of the principles of probability and the so called "laws" of chance.

16.3
THE PRODUCT PRINCIPLE OF PROBABILITY

When a coin is tossed into the air, the chances that it will fall heads (or tails) are even, or 50–50 (which can be expressed as $\frac{1}{2}$, or one out of two). Suppose that two coins are tossed simultaneously. What *now* would be

The product principle of probability states that the chance that two or more independent events will occur simultaneously is given by the product of the chances of any of the events occurring individually.

the chances that both of them would fall heads (or both tails). The answer is given by the *product principle of probability*. This principle states that *the chance that two or more independent events will occur simultaneously is given by the product of the chances that each of the events will occur individually*. In the example of the simultaneous flipping of two coins, each has a 50–50 chance of coming down heads (or tails). In other words, the probability that either coin will come down heads (or tails) is $\frac{1}{2}$. Therefore, the probability that *both* of them will come down heads (or tails) is the product of $\frac{1}{2}$ times $\frac{1}{2}$ or $\frac{1}{4}$. The probability that the first coin will fall heads while the second comes down tails is $\frac{1}{4}$, and the probability that the second coin will come down heads while the first comes down tails is also $\frac{1}{4}$. Thus the total probability for a head and tail combination is $\frac{1}{2}$ times $\frac{1}{2}$, or $\frac{1}{4}$.

Tossing coins produces the following chance distribution of results:

Three Coins

Distribution:	HHH	HHT	HTT	TTT
Probability:	$\frac{1}{8}$	$\frac{3}{8}$	$\frac{3}{8}$	$\frac{1}{8}$

Four Coins

Distribution:	HHHH	HHHT	HHTT	HTTT	TTTT
Probability:	$\frac{1}{16}$	$\frac{4}{16}$	$\frac{6}{16}$	$\frac{4}{16}$	$\frac{1}{16}$

It would be possible to go on in a similar manner computing the chances of various head-tail distributions among the simultaneous tossings of as many coins as desired. It is obvious, however, that for a problem involving a large number of coins, the arithmetic would get a little tedious.

Binomial Expansions

There is a very simple and convenient algebraic method of computing the probability of occurrence of any combination of heads and tails in any given number of coins. It consists of expanding a binomial $(a + b)$ to the nth power, where a and b are simply symbols for the two possible mutually exclusive results (in this case, heads and tails), and n represents the number of units participating in the event (in this case, the number of

coins involved). In other words, any coin problem of this type is solvable by expanding $(a + b)^n$.

The reader will certainly have had enough algebra to handle simple binomial expansions. One might proceed as follows:

For $(a + b)^2$:

$$
\begin{array}{r}
a + b \\
\times\ a + b \\
\hline
ab + b^2 \\
a^2 + ab \\
\hline
a^2 + 2ab + b^2
\end{array}
$$

For $(a + b)^3$:

$$
\begin{array}{r}
a + b \\
\times a + b \\
\hline
ab + b^2 \\
a^2 + ab \\
\hline
a^2 + 2ab + b^2 \\
\times a + b \\
\hline
a^2b + 2ab^2 + b^3 \\
a^3 + 2a^2b + ab^2 \\
\hline
a^3 + 3a^2b + 3ab^2 + b^3
\end{array}
$$

This procedure is perfectly adequate and quite accurate. But it is very time-consuming. A great deal of paper would be required for a problem involving the expansion of a binomial with $n = 23$. Fortunately, a shortcut is available. To demonstrate this shortcut, a few binomials will be used, giving n increasing values:

1. $(a + b)^2 = a^2 + 2ab + b^2$

2. $(a + b)^3 = a^3 + 3a^2b + 3ab^2 + b^3$

3. $(a + b)^4 = a^4 + 4a^3b + 6a^2b^2 + 4ab^3 + b^4$

4. $(a + b)^5 = a^5 + 5a^4b + 10a^3b^2 + 10a^2b^3 + 5ab^4 + b^5$

5. $(a + b)^6 = a^6 + 6a^5b + 15b^4b^2 + 20a^3b^3 + 15a^2b^4 + 6ab^5 + b^6$.

A careful examination of these expansions reveals several pertinent facts. First, note that the exponential values of a and b have a distinct ordered relationship to each other: as one decreases, the other increases. Thus, for example, in $(a + b)^4$ the a goes to a^4, a^3, a^2, a, and a^0, or 1, while the b begins at zero and proceeds b, b^2, b^3, and b^4. Note also that the exponents in each term of the expansion always add up to a number that is the value of n. This relationship of the exponents to the n number is *not* coincidental. It is based on the fact that *the exponents represent all possible combinations that can arise in any situation given an n number*. In the flipping of four coins, for example, where $(a + b)^4$ is used, the exponents reveal the obvious; one can get four heads, three heads and one tail, two heads and two tails, one head and three tails, or four tails. With this knowledge applied to all binomial expansions, the coefficient of the expansion can be constructed by running the a from n down to zero and the b from zero up to n. For example, in $(a + b)^7$:

$$a^7 + a^6 + a^5 + a^4 + a^3 + a^2 + a + a^0.$$

Next the ascending b's are added as follows:

$$a^7 + a^6b + a^5b^2 + a^4b^3 + a^3b^4 + a^2b^5 + ab^6 + b^7.$$

As predicted, the exponents of each term add to n, or 7.

It now remains only to find the proper coefficients. There are several methods for doing this. Perhaps the simplest is the procedure outlined by the following rule: *To expand any binomial, multiply the exponent of a term by the coefficient of the term, and divide by the number of the term in the expansion.* This procedure gives the coefficient of the next term in the expansion.

In any binomial expansion, the first coefficient is always 1. Here, in the first term, a^7, the exponent is 7. Thus we multiply $7 \times 1 = 7$ and divide by 1 (since a^7 is the first term in the expansion). This gives the second coefficient:

$$a^7 + \mathbf{7}a^6b + \cdots.$$

Working with this coefficient (7) to find the next, we multiply the exponent 6 by the coefficient 7, obtaining 42. Since $7a^6b$ is the second term in the expansion, divide 42 by 2, obtaining:

$$a^7 + \mathbf{7}a^6b + \mathbf{21}a^5b^2 + \cdots.$$

Again we multiply the exponent by the coefficient, getting $5 \times 21 = 105$. Since $21a^5b^2$ is the third term in the expansion, we divide this by 3, which gives us 35, the next coefficient:

$$a^7 + \mathbf{7}a^6b + \mathbf{21}a^5b^2 + \mathbf{35}a^4b^3 + \cdots.$$

Then, $4 \times 35 = 140$, and $140 \div 4 = 35$, the next coefficient:

$$a^7 + \mathbf{7}a^6b + \mathbf{21}a^5b^2 + \mathbf{35}a^4b^3 + \mathbf{35}a^3b^4 + \cdots.$$

Continuing this process, we arrive shortly at the complete expansion:

$$a^7 + \mathbf{7}a^6b + \mathbf{21}a^5b^2 + \mathbf{35}a^4b^3 + \mathbf{35}a^3b^4 + \mathbf{21}a^2b^5 + \mathbf{7}ab^6 + b^7.$$

Note that the coefficient 1, understood in the term b^7, could have easily been derived by use of the rule, for $1 \times 7 = 7$. Since b^7 is the last and seventh term in the expansion, we would divide 7 by 7, which is, of course, 1. The above method is probably the quickest way of determining the coefficients of a binomial expansion, particularly if n is a large number.

What exactly do the coefficients tell us with regard to the laws of probability? *The coefficients tell us the number of ways in which the combinations given by the sum of the exponents can occur.* Suppose, for example, that four pennies are flipped. One of the terms in the expansion of $(a + b)^4$ is $4a^3b$. Letting a stand for the chance that heads will occur and b for the chance of tails, the exponents show that we may get three a's and one b (or three heads and one tail). But the coefficient 4

tells us that there are four possible ways that this combination of three heads to one tail can be obtained. This can be seen easily if we give each penny a number label. For example, three heads and one tail might appear in any of the arrangements shown here:

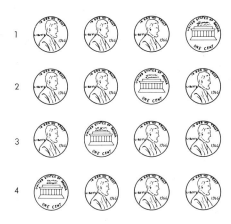

Similarly, the other coefficients indicate that there is only one way to get the pennies to come up all tails or all heads; there are four ways of getting a combination of three tails and one head, and six ways of getting a combination of two heads and two tails.

Suppose we flip five pennies and want to know the chances of getting a combination of three heads and two tails. We will let a represent the chances for heads, which are 50–50, or $\frac{1}{2}$, and b the chances for tails, which are also $\frac{1}{2}$.

Since this problem involves five pennies, $(a + b)^n$ becomes $(a + b)^5$:

$$(a + b)^5 = a^5 + 5a^4b + 10a^3b^2 + 10a^2b^3 + 5ab^4 + b^5.$$

From this expansion we select the term that gives the combination of interest to us—in this case, three heads and two tails. This particular combination is given to us by only one term, $10a^3b^2$. Substituting the known values of $\frac{1}{2}$ for a and b and carrying out the necessary arithmetic, we get:

$$10a^3b^2 = 10(\tfrac{1}{2})^3(\tfrac{1}{2})^2 = 10(\tfrac{1}{8})(\tfrac{1}{4}) = 10(\tfrac{1}{32}) = \tfrac{10}{32} = \tfrac{5}{16}.$$

Thus a distribution of three heads to two tails is predicted to occur $\frac{5}{16}$ of the time. Or to state it another way, out of sixteen tosses, each involving five pennies, five tosses will be expected to show a distribution of three heads to two tails. Many groups of sixteen tosses must be made, of course. The accuracy of a prediction is entirely dependent on the number of trials. Indeed, there is a positive correlation between the accuracy of the prediction and the number of trials performed.

Progeny and Probability

If the binomial expansion works for coins, it should work for anything that involves two chance possibili-

The laws of probability state that given enough events, even the most improbable situation becomes highly probable. For instance, it has been said that given an infinite number of monkeys and an infinite number of typewriters, one of the monkeys would eventually produce Hamlet.

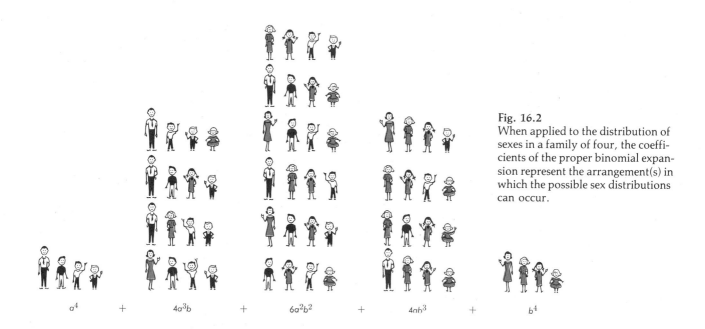

$$a^4 \qquad + \qquad 4a^3b \qquad + \qquad 6a^2b^2 \qquad + \qquad 4ab^3 \qquad + \qquad b^4$$

Fig. 16.2
When applied to the distribution of sexes in a family of four, the coefficients of the proper binomial expansion represent the arrangement(s) in which the possible sex distributions can occur.

ties. Since a and b are merely symbols, they can be maneuvered at will. Consider, for example, the sex distribution of human infants. The chances that a child will be male (or female) are approximately 50–50 or $\frac{1}{2}$. What, then, are the chances that in a family of four children all the children will be girls? We will let a represent the chances for boys, which are 50–50, or $\frac{1}{2}$, and b the chances for girls, which are also $\frac{1}{2}$. Using $(a + b)^n$, with $n = 4$:

$$(a + b)^4 = a^4 + 4a^3b + 6a^2b^2 + 4ab^3 + b^4.$$

Since b represents the chances for girls, we want the term in the expansion that gives us 4 b's. Clearly this is b^4. Substituting, we get

$$b^4 = (\tfrac{1}{2})^4 = \tfrac{1}{16}.$$

In other words, the chances of getting all girls in a family of four children are the same as the chances of each child being a girl multiplied by each other child's chance of being a girl, or $\frac{1}{2} \times \frac{1}{2} \times \frac{1}{2} \times \frac{1}{2} = \frac{1}{16}$ (Fig. 16.2).

We are now ready to see how the mathematics of probability applies to inheritance in sexually reproducing organisms.

16.4 EXPERIMENTAL GENETICS: MENDEL'S WORK

The laws of chance and probability were well known in certain of the natural sciences during the latter part of the nineteenth century. They had been applied to such areas of physics as statistical mechanics and dynamics. However, virtually no application had been made to any areas of biology. It was in just such an application that Gregor Mendel (Fig. 16.3), an Augustinian monk in Brünn, Austria (now Brno, Czechoslovakia), made his outstanding contributions to the study of heredity.

Mendel published his important work on heredity in peas in 1866. Mendel's work made virtually no impression on anyone for the next thirty-five years. Only in 1900 did three investigators more or less independently rediscover Mendel's original paper and publicize it to the biological world.

The Right Person, the Right Time, the Right Place

The study of heredity did not originate with Mendel. From ancient times, questions about how parents influence the appearance of their offspring had attracted considerable attention and speculation. Every sort of hypothesis imaginable has been proposed to account for

Fig. 16.3
Gregor Johann Mendel
(Photo courtesy Burndy
Library, Norwalk, Conn.)

the various patterns or lack of patterns of heredity that people thought they observed in their families, their livestock, and their crops. No scheme had met with any widespread acceptance. With the publication of Darwin's *Origin of Species* in 1859, the issue of how heredity is controlled became even more important. Darwin's theory of natural selection depended upon the fact that certain characteristics of organisms showed slight variations that could be inherited. Darwin assumed that this happened, though he had very little evidence. Many of Darwin's critics as well as his supporters sought to determine whether this assumption was true. Their attempts usually involved experimental breeding. Most of the experiments were carried out with small numbers of organisms, and often there was no systematic method devised for making observations and keeping records of the offspring.

Mendel was familiar with many of the animal and plant breeders, beekeepers, and other agricultural investigators in the rich farming area where he lived. The needs of agricultural breeders appear to have stimulated Mendel to actually undertake some breeding experiments of his own around 1854 or 1855. Later, after 1859, Mendel also read Darwin's *Origin of Species*, and was intrigued by the problems of hereditary transmission it raised.

Mendel was an amateur breeder, but he was not unfamiliar with the natural science of his day. In Brünn

he served as a science teacher in the monastery school. He had studied physics, zoology, and mathematics at the University of Vienna. Although Mendel became conversant with mathematics and natural science, he was never able to pass the examination to qualify for a high school teaching certificate. Particularly significant is the fact that he passed the part of the exam dealing with the more mathematical physical sciences but was unable to pass the natural history (biology) portion.

Most of Mendel's important experiments were done between 1856 and 1863. In 1865 he presented his paper reporting results obtained from thousands of tedious breeding experiments. The report was read aloud at a meeting of the Brünn Society for the Study of Natural Science and was later published in the transactions of that society. At the meeting Mendel received polite attention. However, his application of mathematics to the ratios of plant offspring seemed too much for the audience to take. Attention wandered. It is recorded in the minutes that "there were neither questions nor discussions." The fact that scientific history was made that night remained unnoticed for almost thirty-five years.

Probability and the Pea

Mendel performed breeding experiments on several species of plants and animals, but is best known for his work with the garden pea, a plant that is normally self-pollinating, and hence inbred in the natural state. By close examination of these plants he was able to distinguish several distinct, mutually exclusive inherited characteristics. Among these were the color and shape of the seeds, the positioning of the flowers on the stems (some at the end, others on the side) and the height of the plants (whether they were short or tall). In describing Mendel's experimentation and his conclusions we will, for simplicity, deal with the characteristic of height.

Mendel crossed true-breeding tall plants with true-breeding short plants.* Mendel could conceive of several possible results in the offspring: (1) the plants might be all tall, (2) they might be all short; (3) they might be a mixture of some tall and some short; or (4) they might all be of an intermediate height. Figure 16.4 represents

---

Fig. 16.4
A representation of Mendel's first cross of garden peas, which resulted in an F_1 of all tall plants.

*"True-breeding" simply means that, when crossed, such plants always yield progeny that are exact duplicates of themselves.

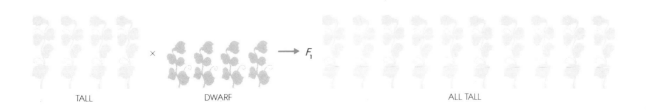

TALL   DWARF   ALL TALL

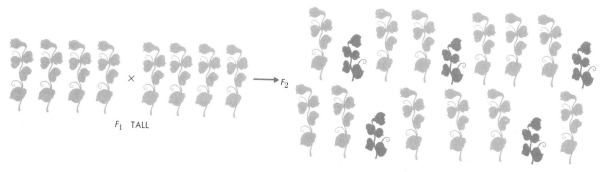

Fig. 16.5
A representation of Mendel's second cross of garden peas, which resulted in an F_2 of 787 tall plants to 277 short plants.

the results that Mendel actually obtained: all of the F_1 (first filial, or offspring, generation) plants were tall.

Mendel found himself faced with many questions. Had the determiner of the shortness characteristic been destroyed, or was it merely hidden? Were the F_1 tall plants genetically similar to their tall parents; that is, would they breed true?

Mendel then crossed two tall plants from the F_1 generation. Figure 16.5 represents the results: out of 1064 F_2 plants, 787 were tall and 277 short.

Mendel explained the results of his first cross by coining two terms still used in present-day genetics. Since the shortness trait was completely hidden by the tallness trait, he postulated tallness to be **dominant** over shortness, which he termed **recessive**.

It was in his interpretation of the second cross, involving the F_1 tall plants, that Mendel's real genius was asserted. Long before he began his work, many people had noticed numbers like those just cited in the above experimental crosses. Few, however, had bothered to look at the significance of these numbers.

Mendel noticed that 787 tall plants to 277 short ones constituted a ratio of 2.84:1. In similar crosses involving other characteristics of the pea plant, he obtained similar results: a dominant characteristic appeared in the F_1 generation, with ratios of 3.15:1, 2.96:1, 2.82:1, 3.14:1, 2.95:1, and 3.01:1 in the F_2 generations.

The suggestion that a ratio of three to one was predictable in the F_2 of such crosses was too strong to be overlooked. To Mendel, this discovery was very exciting. He recognized the ratio as being indicative of the operation of the laws of chance and probability. He therefore hypothesized as follows: Suppose that in true-breeding tall plants there is not just one factor influencing height, but *two*. Suppose further that two other factors are present in short plants. If only *one* of these factors gets into a sperm cell and only *one* into an egg cell, then the union of these cells will produce a plant

with two factors, one from the tall plant and one from the short. Since the factor for tallness is dominant over the factor for shortness, the hybrid in the F_1 will be tall. If we let T stand for the tallness factor and t for the shortness factor, the first cross can be diagrammed as follows:

PARENTS	TALL (TT)	×	SHORT (tt)
	T T		t · t
	(SPERM CELLS)		(EGG CELLS)
FERTILIZATION		Tt	
F_1		TALL (Tt)	

Mendel hypothesized that in the F_1's the two factors for each trait separate in the formation of gametes, so that only one factor for an inherited characteristic gets into each gamete. He also suggested that factors unite during fertilization. In other words, Mendel recognized that the two factors influencing height in pea plants must separate or **segregate** from each other in the production of the germ cells and then be randomly reunited at fertilization. This concept is known as Mendel's first law, the **law of segregation**. For simplicity, Mendel's first law can be stated as follows: the two factors for any trait are separated from one another in the formation of germ cells, and will end up in different gametes and hence in different offspring. Diagramming Mendel's cross between two F_1 plants, as shown below

F_1 CROSS	(TALL) Tt	×	(TALL) Tt
	T t		T t
	(SPERM CELLS)		(EGG CELLS)

we see that each parent plant can produce *two* kinds of gametes instead of only one. Half of the gametes produced will carry the tallness factor T. The other half will carry the shortness factor t. Mendel saw that if the two types of gametes were produced in equal numbers by both parent plants, and if it is assumed that one form of the trait is dominant over the other, his three-to-one ratio could be explained on the basis of chance and probability. Since any type of sperm cell has an equal chance of fertilizing any type of egg, there are four possible fusions that can take place:

1. A sperm cell carrying the T factor may fuse with an egg carrying the T factor, yielding TT.

2. A sperm cell carrying the T factor may fuse with an egg carrying the t factor, yielding Tt.

3. A sperm cell carrying the t factor may fuse with an egg carrying the T factor, yielding Tt.

4. A sperm cell carrying the t factor may fuse with an egg carrying the t factor, yielding tt.

Since the F_1 generation showed that the T factor is dominant over the t factor, the first three fertilization possibilities will result in tall plants. Only the fourth produces a short plant, the result of the combination of two recessive factors for shortness.

16.5
SOME GENETIC TERMINOLOGY

Mendel's "factors" are, of course, what we now call genes, those units of inheritance composed of deoxyribonucleic acid, or DNA.

A brief look at some genetic terminology will facilitate understanding of the material to follow. Alternative or contrasting forms of a gene are called **alleles**. Genes T and t in Mendel's tall and short pea plants are alleles; alleles T and t represent the contrasting inheritance factors for a single trait, height.

The term **homozygous** is used to designate an individual in which both alleles for a trait are identical; if the two alleles are different, the individual is **heterozygous** (heterozygous individuals are sometimes called *hybrids*). For example, both the tall and the short parent pea plants used by Mendel were homozygous, since they both contained like genes for height (TT and tt respectively). However, two-thirds of the F_2 tall plants and all of the F_1 tall plants were heterozygous, since they contained different alleles (Tt). Of course, an individual may be heterozygous for some pairs of genes but homozygous for other pairs.

A distinction must be made between the *appearance* of an organism and the inheritance factors or genes that it will pass along to its descendents, for, as we have already seen, the one may not reveal the other. The appearance registered by means of our senses is called the **phenotype**. The phenotype of the plant TT is tallness.

The phenotype of the plant Tt is also tallness. The plant with tt, however, has a phenotype of shortness.

On the other hand, the description of an organism's genetic makeup (based on data from breeding experiments) is called its **genotype**. While phenotype is a classification according to appearance, genotype is a classification according to genetic composition. Genotype tells you what genes an organism can or cannot pass on to the next generation. For example, the symbols TT represent the genotype for a homozygous tall pea plant; the symbols Tt represent the genotype of a heterozygous tall pea plant. The genotype of a short plant is tt.

It is important to make the distinction between genotype and phenotype in describing ratios given by an experimental genetic cross, since, as we have seen, the two are not always the same. In the cross between the heterozygous F_1 tall pea plants, for example, our phenotypic ratio was three talls to one short, or 3:1. But the genotypic ratio was one homozygous tall to two heterozygotes to one homozygous short, or 1:2:1.

16.6
APPLICATION OF MENDEL'S HYPOTHESIS
TO ANIMALS

As powerful as it is, Mendel's hypothesis would be of considerably less interest if it applied only to peas. Much experimental work in the first three decades of the twentieth century showed that Mendel's basic ideas applied to a wide variety of animals as well as plants.

Mice are often used in genetic studies. They reproduce rapidly, and it is easy to distinguish differences in coat color. Thus, coat color is one genetic trait often studied in mice. Further, differences in coat color are due to simple interacting factors.

A certain well-established strain of mice possesses a solid black coat.* Another strain has a solid brown coat. Since both are true-breeding strains, crosses between black mice always yield black offspring. It is reasonable to wonder, therefore, what would be the color of the offspring resulting from a cross between brown mice and black. Here are a few of the possibilities:

1. An intermediate color between black and brown.

2. Some black mice and some brown.

3. Spotted black and brown mice.

4. A color entirely different from black and brown.

5. All black mice.

6. All brown mice.

---

*Mice that have been mated brother-to-sister for at least twenty generations are considered an established inbred strain.

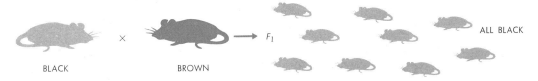

Fig. 16.6
A representation of the first cross of black mice with brown mice, yielding an F_1 of 992 mice, all black.

We can make several *reciprocal matings*, that is, matings where in some cases the male is black and the female brown, while in others the female is black and the male is brown. This eliminates sex as a variable in the experiment, while the large number of offspring produced by many matings most probably allows for accurate generalizations.

Figure 16.6 shows that prediction 5 is the correct one. All 992 mice obtained (the F_1 generation) are solid black and indistinguishable in appearance from their black parents.

As with Mendel, we can ask several questions. What has happened to the factor for brown color? Has it been completely destroyed by the black color, or is it still present but hidden? Does the fact that the F_1 black mice appear identical to their black parent mean that they are also genetically identical, or are they also hybrids, like Mendel's F_1 tall peas, that will not breed true?

To answer these questions, we can breed F_1 black mice among themselves. Figure 16.7 reveals the results of this cross: of the 1278 progeny, 961 are black and 317 brown.

Mendel's hypothesis provides a ready explanation for the results of a cross between two black F_1 mice. Note that 961 black mice to 317 brown mice closely approximates a three-to-one ratio. Therefore, it can be hypothesized that the F_1 male parent produced two kinds of sperm, half carrying a B factor for black, the other half carrying a b factor for brown. The female produced two kinds of eggs, half carrying a B factor, the other half carrying a b factor. Given complete domi-

nance of black over brown, and the fertilization possibilities governed by the laws of chance and probability discussed in Section 16.3, we can readily account for the observed 3:1 ratio.

As has been constantly stressed, a good hypothesis must not only explain the observed, it must also act as a basis for accurate predictions. Thus, we can reason that if in mice, factors for the black and brown coat colors segregate into the gametes and recombine at fertilization according to the laws of chance and probability, and if black is completely dominant over brown, then crosses between an F_1 black mouse described above and a brown mouse should result in offspring of which one-half are black and one-half are brown. This sort of a cross, between an F_1 black mouse and its recessive parent, is called a *testcross* or *backcross;* it is employed frequently to determine the genotype of an unknown organism (or group of organisms). The defining characteristic of a testcross is that the unknown individual is mated with the parent or parent-type that is recessive for the trait in question.

An F_1 black mouse, being the result of crossing a pure black (BB) with a pure brown (bb) mouse, must carry both coat-color factors, B and b. The dominance of black over brown makes the animal's phenotype black. Half the gametes it will produce will carry the B factor, and half will carry the b factor. The brown mouse, on the other hand, can produce only gametes that carry the b factor. Since these gametes have an equal chance of fusing with either of the other gametes, a 50–50 distribution of black to brown mice is the prediction. Of the 833 offspring of such a cross, 412 were

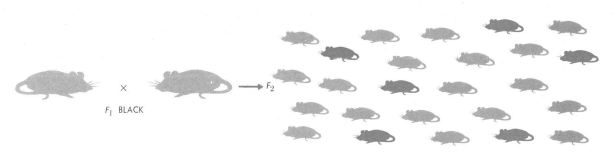

Fig. 16.7
A representation of the second cross of mice, yielding an F_2 of 961 black mice to 317 brown mice.

black and 421 were brown. Since the slight deviation in the predicted ratio is not statistically significant, the data support Mendel's hypothesis. Moreover, Mendel's hypothesis gains stature by this experiment because the original generalization regarding a species of plant has been successfully extrapolated to a species of animal.

16.7
SIMPLE MENDELIAN INHERITANCE IN HUMAN BEINGS

Like other organisms, human beings show a basic Mendelian inheritance pattern for a large number of traits. For example, regardless of their racial origins, most human beings synthesize a certain amount of pigment, known as melanin, which gives color to our skin, hair, and the iris of the eye. Some individuals in every race, however, show a phenotypic variation in which no pigment is produced at all. These are *albinos*, characterized by very light skin (with a pinkish tinge), almost pink eyes, and virtually white hair. Two albino individuals are shown in Fig. 16.8. Albinos are very sensitive to the sun and their skin burns easily. Furthermore, reduced eye pigmentation makes albinos very sensitive to bright light.

How do we study genetic traits in human beings? We cannot mate people at will in the laboratory as Mendel mated peas in his experimental garden. Further, humans produce so few offspring that it would be difficult to get any sort of statistical sampling. Geneticists have developed a tool known as the pedigree analysis chart, or the "family tree," as a means of learning something about the inheritance patterns for certain traits. Pedigree charts trace the occurrence and recurrence of a trait through several generations of a family line. Through such charts it is sometimes possible to learn about the genetics of a particular trait: is it due to a dominant or a recessive gene? Is it the result of the interaction of two or more genes? And so on. A pedigree chart for albinism is shown in Fig. 16.9. A discussion of this figure will serve to illustrate how such charts are constructed, what sorts of information they contain, and how that information can be used for genetic analysis. (The symbols and notations on the chart are explained in the caption.)

The pedigree chart for albinism shown in Figure 16.9 reveals several important pieces of genetic information. First, it is apparent that phenotypically the trait can skip one or more generations: that is, it can disappear from generations II and III, only to reappear in generation IV. These observations suggest that the trait is determined by a recessive gene. The fact that the albino condition also appears relatively infrequently throughout the five generations shown, and that it skips a generation or two, is further evidence for its recessiveness; so, too, is the fact that normally pigmented parents may produce an albino child. A second important piece of genetic information contained in Figure 16.9 is

Fig. 16.8
Two albino individuals, a genetic condition that results in lack of synthesis of the skin pigment melanin. Albinos traditionally have very light (pinkish-white) skin and white hair, and lack pigment in the iris of the eyes. (Photo courtesy A. M. Winchester.)

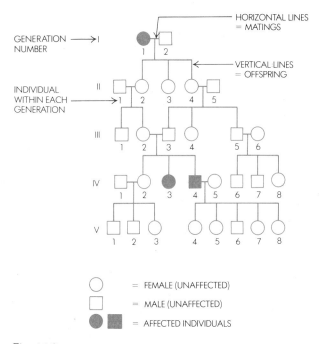

Fig. 16.9
A pedigree chart, or "family tree" showing the distribution of albinism in five generations of a family line. Albinism appeared first in the mother of generation I, and reappeared in two siblings of generation IV. The symbols used in constructing pedigree charts are as follows: squares represent males, circles females; figures for affected individuals (those who show the trait) are solid, while unaffected individuals are represented by open figures. Parents are linked by a horizontal line, with their siblings appearing beneath them, arranged horizontally from left to right in chronological order of birth. Generations are indicated by Roman numerals at the left. Numbers underneath offspring provide a convenient way to identify each child within a given generation.

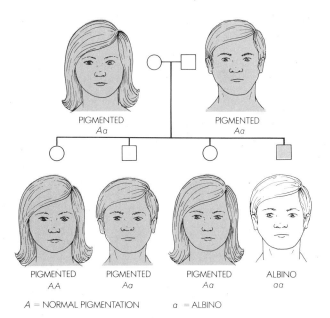

PIGMENTED
Aa

PIGMENTED
Aa

PIGMENTED
AA

PIGMENTED
Aa

PIGMENTED
Aa

ALBINO
aa

A = NORMAL PIGMENTATION   *a* = ALBINO

Fig. 16.10
The inheritance of a form of albinism from a Mendelian point of view. Assuming that the albino condition is recessive to its normal counterpart, pigmentation, albino individuals would be homozygous recessive, or *aa*. Normally pigmented individuals could be either *AA* or *Aa*. Two heterozygous, normally pigmented parents, as shown above, would be expected to produce $\frac{3}{4}$ pigmented, and $\frac{1}{4}$ albino offspring. Of the pigmented individuals, however, $\frac{1}{3}$ would be expected to be homozygous dominant (*AA*) and $\frac{2}{3}$ heterozygous (*Aa*). Such predictions, of course, are statements of probability, not actual outcome. As Figure 16.9 shows, sometimes the results obtained in particular cases are radically different from expectation. The significance of such departures from prediction are meaningful, however, only when statistically large samples are involved.

that when albinism reappears after skipping two generations, it appears as pronounced as in generation I. In other words, the phenotypic expression of the recessive allele for the trait has not been altered by being combined with the gene for normal pigmentation.

If a hypothesis proposing that albinism is recessive and carried by a single pair of genes is correct, then we should be able to predict the inheritance pattern to be expected from a mating such as that between individuals 2 and 3 of generation III. If we let *A* stand for normal pigmentation, and *a* for albinism, then we can predict that the two parents in generation III who produce albino children would both be *Aa*. The normally pigmented offspring of that marriage (individual 2, generation IV) would be either *AA* or *Aa*; the two albinos would be *aa*. But in what proportion should we expect these different types of offspring to appear? The Men-

delian prediction is shown in Fig. 16.10. Note that according to our hypothesis, only one out of every four offspring would be expected to be albino. However, according to the actual pedigree chart in Fig. 16.9, *two out of three offspring of this marriage are albinos.* That is almost the opposite of what we would predict on the Mendelian hypothesis. Does that mean the trait is perhaps inherited in a non-Mendelian way? Or that it is not genetic at all?

The answer to both questions is: *possibly*, but not necessarily. Remember that Mendelian ratios only begin to assume the expected values when large numbers of offspring are involved. Our expectation of one out of every four offspring being an albino could be met only if we were dealing with hundreds of offspring from a single set of parents. With only three offspring, it is quite likely that sampling error alone will produce some unexpected ratios. On the basis of such limited statistical evidence as shown in the above chart, it would not be logical to conclude that albinism is inherited in a non-Mendelian way. In fact, all the evidence gained from study of many different family pedigree charts suggests that the Mendelian hypothesis is most likely the correct one.

The albinism example highlights one very important feature of Mendelian genetics: *its predictions are statistical expectations based only on probabilities.* Mendelian genetics does not allow us to predict with certainty. All we can do is estimate the probability that a trait will or will not appear in a certain number of offspring. There is no way of being absolutely certain what future offspring *as individuals* will be like. We can only estimate, based on past occurrences, what the probability might be of the next offspring, or group of offspring, showing (or not showing) a particular trait, and in what proportion.

The inheritance pattern for dominant Mendelian traits in human beings can be illustrated by brachydactyly, a condition in which the fingers are greatly shortened and have a stubby appearance (Fig. 16.11). A pedigree chart for brachydactyly is shown in Fig. 16.12. Note that the pattern of inheritance is somewhat different from what we found for a recessive trait. In the first generation the female parent is brachydactylous. Of her eight offspring, four show the condition. Offspring number 2 in the second (II) generation displays the trait and of his twelve children, seven (over 50 percent) are brachydactylous. Following the condition on down from the third through the sixth generations, we can see that a large number of individuals are affected.

Note that brachydactyly appears in approximately 50 percent of the offspring in every generation. This is frequently the case with dominant traits. Sometimes, however, the trait may appear in every individual in affected families. An explanation for both situations derives easily from the Mendelian hypothesis. Individuals who show the dominant trait can be genotypically

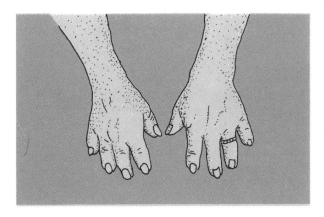

Fig. 16.11
Short, stubby appearance of brachydactylous hands in the heterozygous condition. Homozygous dominant individuals show a somewhat more exaggerated stubbiness, which is accompanied by severe arthritis of the joints.

Fig. 16.12
Pedigree chart for the partially dominant condition, brachydactyly. The symbols and notational scheme are the same as those used for the pedigree of albinism in Fig. 16.9. Note the much greater frequency of appearance of brachydactyly in each generation, than of albinism, and that, unlike albinism, at least one parent of an affected individual is also affected. This is one of the major differences in inheritance patterns between dominant and recessive traits. (Modified from Victor McKusick, *Human Genetics*, 2nd ed. Prentice-Hall, 1969, p. 48. Reprinted by permission of Prentice-Hall, Inc., Englewood Cliffs, New Jersey.)

either heterozygous (*Bb*) or homozygous (*BB*). By contrast, individuals who show the recessive trait must be genotypically homozygous (*bb*). Now, conditions such as brachydactyly are relatively rare in the human population. If we assume that finger shape is not a major factor in mate selection, most brachydactylous persons would probably marry individuals with normal hands. The fact that the vast bulk of the population from whom brachydactylous people choose their mates have normally shaped hands (that is, have the genotype *bb*) means that most matings between affected and unaffected people will be of the type *Bb* × *bb*. Even if the affected parent were homozygous (*BB*), all the offspring would be heterozygous (*Bb*), though showing the brachydactylous phenotype. This means that most affected individuals in any family line would more likely than not be heterozygous.

There are several characteristics of the inheritance pattern of dominant traits which are illustrated by the pedigree in Fig. 16.12. (You may find it useful to compare Fig. 16.12 with the pedigree for recessive traits in Fig. 16.9 as one way of highlighting the differences between dominant and recessive inheritance.)

1. Dominant traits appear far more frequently in a pedigree line than recessive traits. This would be expected since a dominant gene can be expressed in both the homozygous and the heterozygous states, whereas, by contrast, a recessive gene can be expressed only in the homozygous state.

2. Dominant traits do not skip generations, but reappear generation after generation in a family line. This

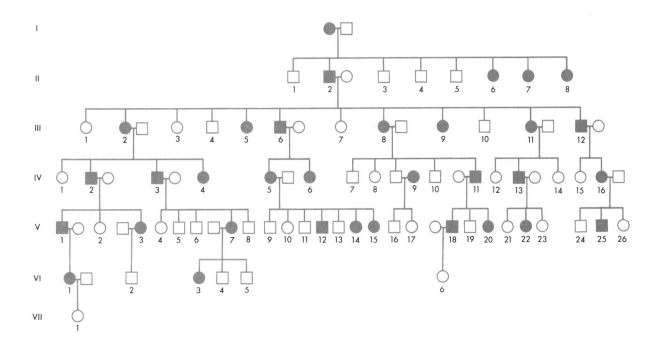

would be expected because of the nature of dominance —that is, you always see a dominant trait when a dominant gene is present (that is, after all, the definition of dominance).

3. In dominant inheritance, every affected individual must have at least one parent who shows the condition. In recessive inheritance, by definition, both parents can appear unaffected, yet still produce affected offspring.

It is an old idea that in a family line, or even a large, interbreeding population, dominant genes will tend to take over, simply because they are "dominant" and are thus expressed more frequently, and will thus increase in percentage over successive generations. Conversely, it is also held that recessive alleles, because they are expressed less frequently, will tend to disappear or be "stamped out" of a population or family line after a period of time. Both of these ideas are incorrect. For example, as we trace the inheritance of brachydactyly down through six generations, note that the relative *proportion* of affected to nonaffected individuals does not change substantially (count up the numbers in each generation and you will see that the ratio of affected to nonaffected individuals remains about 50 percent). Similarly, note that while albinism skips a generation or two, when it reappears, it is neither less frequent (compared to normal pigmentation), nor less intense than in the first generation. The mere fact of a gene being dominant, or recessive, does not tell us anything about how its frequency of occurrence will change during successive generations.

16.8
INCOMPLETE DOMINANCE

Few characteristics are inherited in as simple a manner as the tallness and shortness in peas or the black and brown coat colors in mice. Indeed, simple inheritance in which only one pair of genes is involved is generally very much the exception rather than the rule. Nor is it required for one characteristic to be completely dominant over the other. Quite often the combination of different genes tends to produce varying degrees of partial or *incomplete dominance*, with the latter usually resulting in a blending of the two genes to produce a different phenotype than found in either parent.

An example of incomplete dominance is seen in the breeding of certain types of cattle. If a red animal is crossed with a white one, an intermediate-colored animal, a roan, is produced (Fig. 16.13). Crosses between two roans yield a ratio of one red to two roans to one white. Once again, Mendelian genetics provides a ready explanation for the results (Fig. 16.13).

Note that if red coat had been dominant over white (or vice versa), a 3:1 F_2 ratio would result. However, since the genes interact with each other to produce an

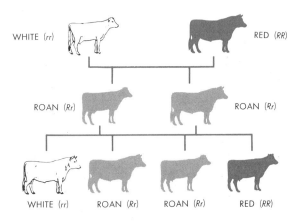

Fig. 16.13
Pattern of inheritance of coat color in cattle. White and red parents produce roan F_1's, two roans interbred produce a 1:2:1 F_2 ratio of white:roan:red. The R and r alleles show incomplete dominance.

intermediate color, a 1:2:1 ratio is obtained. Further, in this case the F_1 animals are easily distinguishable from their parents. Note also that this 1:2:1 ratio is the same obtained in flipping two pennies simultaneously. The number of times we obtain both heads, a head and tail combination, or both tails is 1:2:1, respectively.

Prior to Mendel, it was widely believed that inherited characteristics represented a blend of each individual trait shown by the parents. Skin color of offspring in human interracial marriage lent considerable support to this idea. It might be noted in passing that a hypothesis proposing blending of inherited traits, at least upon a superficial analysis of their appearance, is just as satisfactory as Mendel's particulate hypothesis in explaining the results of the cross between red and white cattle. The F_2 generation, however, yields results that contradict the blending hypothesis and support Mendel's particulate theory. Although the gene for white and the gene for red interacted in the F_1 hybrids to produce the intermediate roan color, each gene was unaffected by this interaction. In the F_2 the red gene produced the red color as intensely as in the original parent; the same is true of the white gene.

16.9
TWO PAIRS OF GENES

In predicting all of the gamete combinations that can occur at fertilization in any given cross, geneticists often use a simple matrix, or what is called a *Punnett square* (named after R. C. Punnett, an early British geneticist).

SUPPLEMENT 16.1
WERE MENDEL'S RESULTS TOO GOOD?

In 1936 the British mathematician Ronald A. Fisher (1890–1962) made a statistical analysis of the data Mendel reported in his 1866 paper. Fisher was trying to answer a basic question about scientific procedure: To what extent did Mendel have his conclusions in mind when he carried out his experiments with peas? Did he know what he expected or hoped to find, or was he approaching the question from a neutral, objective viewpoint?

Looking at Mendel's data for yellow versus green seeds, Fisher found that for a cross between two heterozygotes, Mendel reported 6022 yellows and 2001 greens. Since yellow seed is dominant to green seed, the expected ratio should be 3:1. Mendel's results represent a deviation of 5 from the 3:1 ratio, whereas on statistical grounds a deviation of 26 or more would be expected. Of course, getting a close fit to expectations is not in itself improbable on occasion. For example, in flipping a penny 1000 times, it would be quite possible to get exactly 500 heads and 500 tails. However, if people claimed to get 500 heads and 500 tails *every time*, we might wonder about their honesty. In the case of Mendel, Fisher found the same close fit to expectation throughout all his reported ratios. Fisher calculates that, taking the whole series of experiments, the chance of getting the results Mendel claimed to get would have a probability of .00007—or once out of 14,000 trials. The argument comes close to saying that no such series of experiments as the ones Mendel reported he had done could ever be carried out.

The possibility that Mendel knew what he expected to get is made more apparent by one case where he got an equally close fit to *wrong* expectations! Mendel set about to determine which of the F_2 plants showing the dominant traits were homozygous and which were heterozygous. His expectation was that these should occur in the ratio of 1:2. The method for making this determination was to allow the F_2's to self-fertilize and produce seeds and then to raise the seeds to mature plants and observe their various traits. For these test purposes, Mendel reported that he planted only ten seeds from each of the F_2 dominant plants he wanted to test. If the F_2 dominant was in fact a heterozygote, one-fourth of its offspring would be expected to show up as homozygous recessives; that is, the chance that any one offspring will not be homozygous recessive is .75, or three out of four. Thus the chance that none of the ten will be a homozygous recessive is $(.75)^{10}$, or .0563, which is equal to about 5 or 6 percent. This means that from breeding only ten of their seeds, five or six percent of the actual F_2 heterozygotes could be expected to appear as homozygotes. Mendel did not take this source of error into account and reported a 1.8874 to 1.1126 ratio result that was close to his expected (and uncorrected) 2:1

ratio. Such close agreement would be expected only once in 2000 tries.

Other workers repeated Mendel's experiments and recorded their data. In most cases, the actual ratios came out just as Mendel's did. Indeed one observer, Erich von Tschermak, who in 1900 was among the three discoverers of Mendel's original paper, got results even closer to the expected values than Mendel!

How can we interpret this case study? What does it suggest about the process of scientific investigation? The fact that Mendel's reported results are too much in agreement with expectations should not come as a total surprise, given the subjective elements involved in hypothesis formulation and justification. Mendel spent two years in preliminary experiments before he began the series of experiments reported in his paper of 1866. During this time, it is likely that he came to have certain expectations about what kinds of ratios he could expect from certain crosses. Thus Mendel may have reported only those results that best corresponded to his expectations. In other words, Mendel selectively chose his data, presenting only a small part of the total data he had collected. These data supported quite closely the expectation Mendel had acquired through his previous two years' experience with breeding peas. Of course, it is possible that Mendel fudged the original data or that assistants who knew what his expectations were fudged the counts for him. There is, however, little to support these latter hypotheses.

One hypothesis suggests that Mendel himself may have been unconsciously biased in counting the offspring from his various crosses. Knowing the ratios he expected to get, Mendel may have made subjective decisions concerning the borderline appearances he observed. No two organisms, even of identical heredity, look *exactly* alike. Among tall plants, some are shorter than the average; among wrinkled seeds, some are smoother than others; among green seeds, some are more yellowish-green than others. Classifying such intermediate organisms often involves subjective judgment. There is no way to provide absolute, judgment-free methods of deciding into which category one or another offspring should fall. In making such decisions, the expectations of the observer can often have a profound effect on the final numbers of individuals reported in each category. If you are an observer who expects that the ratio of wrinkled to smooth seeds will be 3:1, you will have a tendency to put all slightly wrinkled seeds into the wrinkled category, just to ensure that the wrinkled dominates the smooth category. It is very difficult to assess the influence of such biases. However, when results come out closer to expectation than normal, it is reasonable to look in this direction for an explanation.

Fisher's analysis of Mendel's work has underscored an important point about scientific investigation. It is virtually impossible to approach a problem in science or any other field with absolutely no expectations. Almost all investigators tend to have some idea of what they think the results of a given experiment will be. Not only is there nothing wrong with this procedure, but it may well be the most effective way in which any problem-solving activity can proceed. The critical quality of scientific thinking lies not in avoiding all prejudgments, but rather in being willing to subject these prejudgments, in the form of hypotheses, to the test of experience. If experience contradicts expectation, then the hypotheses must be revised or discarded. Being "scientific"

involves being able to change one's mind in the face of experience, not being so hesitant as to avoid all prejudgments to begin with.

Mendel's results were very close to expectation—perhaps too close to be taken at face value as raw, unselected data. The genius of discovery, however, often lies in knowing which data to trust and which to ignore. Mendel had enough experience to know what ratios to expect. The exact method by which he got data agreeing so closely with those expectations remains a mystery. Undoubtedly Mendel exercised some subjective interpretation, something researchers do virtually all the time. Science is not, nor can it be, entirely free from subjectivity.

With coat color in mice, for example, the Punnett square would look like this:

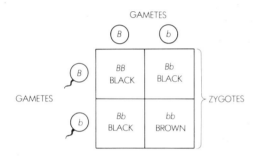

Use of the Punnett square prevents the overlooking of a possible gamete combination. Thus far we have been dealing with the principles of Mendelian inheritance using examples involving only a single pair of genes. Most genetically based traits, however, are produced by *more* than a single pair, and the use of the Punnett square therefore becomes more important. Let us examine some traits whose expression is influenced by two pairs of genes.

In addition to the black and brown coloration in mice, there is another factor that influences coat color. This factor is known as the *agouti* condition. When the agouti condition is combined with black, brown, or

some other coat color, it produces a characteristic muted appearance. If we examine the individual hairs of the coats of agouti mice, the reason for this appearance becomes clear. The hairs of a completely black or brown mouse are solidly colored along their entire length. The hairs of an agouti animal have a distinct band across them, near the end (Fig. 16.14).

As in the case of black or brown coat color, the inheritance of the agouti condition is dependent on one pair of genes and is dominant. Therefore, crosses between homozygous agouti and nonagouti mice yield an F_1 that are all agouti. Crosses of the F_1 agouti animals yield a ratio of three agoutis to one nonagouti.

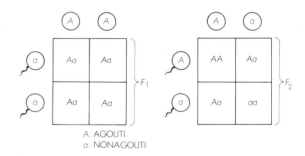

Such F_2 3:1 ratios of black to brown or agouti to nonagouti are obtained when the characteristics are considered *separately*. Suppose, however, that they are considered *together*. In particular, consider a cross involving black agouti mice from a pure inbred strain (genotype *AABB*) with brown nonagouti mice (genotype *aabb*). Black, of course, is dominant over brown, as is agouti over nonagouti. It might be predicted, therefore, that the F_1 mice will show both dominant characteristics—that they will all be black agoutis. This is precisely what occurs. In a series of experiments, many matings of black agoutis with brown nonagoutis produced 1624 F_1 animals, all black agoutis.

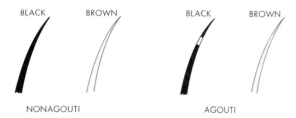

Fig. 16.14

Were these F_1 black agoutis genotypically the same as their dominant parent? If so, then crosses between F_1 mice should produce only black agouti mice. Previous experience, however, indicates that these F_1 individuals will not breed true, despite the fact that their appearance is indistinguishable from that of their black agouti parents. Past experience also leads us to expect the three-to-one ratio. *But three whats to one what?* Three black agoutis to one brown nonagouti? Or three black nonagoutis to one brown agouti? Is there any justification for assuming that the black color always appears linked to the agouti condition and the brown to the nonagouti? If not, then it must be admitted that we can obtain *four* possible kinds of mice instead of just two, namely black agoutis, black nonagoutis, brown agoutis, and brown nonagoutis.

Many such F_1 crosses were made, and a total of 1625 baby mice were obtained. Of these, 909 were black agoutis, 304 were black nonagoutis, 299 were brown agoutis, and 103 were brown nonagoutis, a ratio of 9:3:3:1.

The Law of Independent Assortment

Another examination of Mendel's work provides a hypothesis to explain this 9:3:3:1 ratio. Mendel had noticed that one inherited difference in the garden pea was seed color; some plants had yellow seeds, others green. Crosses of yellow-seed producers with green-seed producers yielded only plants that produced yellow seeds. Mendel hypothesized the dominance of yellow over green, combined with segregation of the factors for these seed colors, and saw that a three-to-one ratio of yellows to greens should result from crosses of the F_1 yellows. Experimental results supported this hypothesis.

However, Mendel also noticed that the pea seeds differed in still another characteristic. Some were round and smooth while others were wrinkled. Mendel found that crosses between plants that produced round seeds and plants that produced wrinkled seeds gave an F_1 of plants that produced round seeds; thus roundness is dominant over wrinkled. Crosses between plants of the F_1 generation produced a 3:1 ratio of round to wrinkled seeds.

Mendel then considered the factors of seed color and seed shape together. He noticed that crosses between two pure varieties of pea plants—one showing the dominant characteristics of round and yellow seeds,

the other the recessive characteristics of wrinkled and green seeds—yielded an F_1 phenotypically identical to the dominant parent; that is, the F_1 plants all produced round yellow seeds. However, crosses of these F_1 plants yielded a ratio of nine plants that produced round yellow seeds, to three that produced round green seeds, to three that produced wrinkled yellow seeds, to one that produced wrinkled green seeds. Notice that the ratios obtained by Mendel for the peas were the same as those obtained with the mice in the experiment mentioned above.

In interpreting his results, Mendel hypothesized about what must be happening in gamete production. Recall that when he was working with just one characteristic, such as tallness/shortness in pea plants, Mendel hypothesized the existence of one pair of "factors" (genes) that segregate at gamete formation. When he began dealing with two characteristics, Mendel hypothesized the existence of *two* pairs of genes, each pair of which was concerned with the inheritance of one of the two contrasting traits (round-wrinkled and yellow-green). The members of each of these pairs also segregate during gametogenesis, with each seed-color gene going to a different gamete. The same is true for the genes for seed shape; these also go into separate gametes. But, Mendel reasoned, if the segregation of the genes for seed color is *entirely independent* of the segregation of the genes for seed shape, and occurs *completely at random*, then once again the mathematics of chance and probability will apply. This is known as Mendel's second law, the *law of independent assortment*.

Mendel's first cross can be diagrammed:

Here, R = round, r = wrinkled, Y = yellow, and y = green. Since each parent can produce only one type of gamete, only one fertilization combination is possible:

Mendel's second law (the law of independent assortment) states that the individual members of pairs of alleles segregate independently of one another during pollen or egg cell formation.

Note that Mendel's F_1 plant (producing round yellow seeds) differs genotypically from the parent that produces round yellow seeds. All the F_1 plants must be doubly heterozygous, or dihybrid. With this in mind, it becomes evident that each F_1 plant can produce four types of gametes:

F_1 MALE PLANT (RrYy) F_1 FEMALE PLANT (RrYy)

GAMETES (RY) (Ry) (rY) (ry)   GAMETES (RY) (Ry) (rY) (ry)

Mendel stressed completely random distribution of the allelic genes and completely independent assortment of the nonallelic genes.

There are now several fertilization possibilities, each determined entirely by chance. With just one pair of genes involved, only two types of gametes could be produced by the F_1 hybrids. There were therefore 2×2, or 4, fertilization possibilities. With complete dominance, a three-to-one ratio is obtained. In this case, however, each F_1 individual produces *four* different types of gametes. The fertilization possibilities are thus 4×4, or 16. The three-to-one ratio in the monohybrid cross yields the total number of fertilization possibilities with one pair of genes, for $3 + 1 = 4$. The same holds true for the ratio obtained when two pairs of genes are involved, for $9 + 3 + 3 + 1 = 16$.

This dihybrid cross can be diagrammed by use of a Punnett square, as shown in Fig. 16.15. A count of the phenotypes reveals the experimentally obtained 9:3:3:1 ratio.

Experimental and Mathematical Evidence

Mendel recognized that his hypothesis had to have predictive value if it was to be accepted by his fellow scientists. Accordingly he crossed plants of known genotypes. Before obtaining his experimental results, he determined theoretically the types of gametes that each would produce and calculated the fertilization possibilities. Mendel then recorded his predictions of the types of plants and the proportion of each that these crosses would be expected to produce. Two such crosses were as follows:

1. *Genotype RrYy with RRYY.* Mendel hypothesized that in terms of these genetic factors, the dihybrid would produce four kinds of gametes, *RY, Ry, rY,* and *ry.* The other would produce only *RY* gametes. Thus no matter which fertilization combination occurred, the seeds of the resulting plants would all be round and yellow. *Results:* 192 plants, all of which produced round yellow seeds.

2. *Genotype RrYy with rryy* (a back cross). Mendel hypothesized that four kinds of gametes would be produced by the dihybrid (*RY, Ry, rY,* and *ry*) and one kind (*ry*) by the other plant. By considering all the possible fertilization combinations, Mendel predicted four different kinds of plants in equal numbers. *Results:* 55 with round yellow seeds, 51 with round green seeds, 49 with wrinkled yellow seeds, and 53 with wrinkled green seeds.

By incorporating the mathematical laws of chance and probability into the science of genetics, Mendel provided a solid foundation upon which all later investigations in genetics could be based. Once it had been deter-

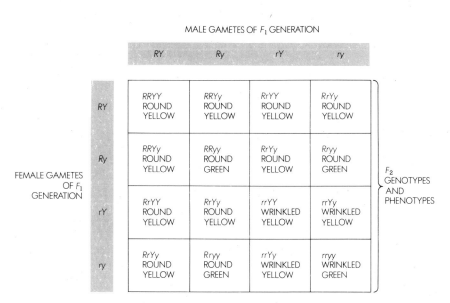

MALE GAMETES OF F_1 GENERATION

	RY	Ry	rY	ry
RY	RRYY ROUND YELLOW	RRYy ROUND YELLOW	RrYY ROUND YELLOW	RrYy ROUND YELLOW
Ry	RRYy ROUND YELLOW	RRyy ROUND GREEN	RrYy ROUND YELLOW	Rryy ROUND GREEN
rY	RrYY ROUND YELLOW	RrYy ROUND YELLOW	rrYY WRINKLED YELLOW	rrYy WRINKLED YELLOW
ry	RrYy ROUND YELLOW	Rryy ROUND GREEN	rrYy WRINKLED YELLOW	rryy WRINKLED GREEN

FEMALE GAMETES OF F_1 GENERATION

F_2 GENOTYPES AND PHENOTYPES

Fig. 16.15
Punnett square for the cross between two pea plants, each heterozygous for two traits: seed color and seed shape. In such a cross, Mendel found that the offspring showed a 9:3:3:1 phenotype ratio.

mined how genes interact with each other (whether they demonstrate dominance, incomplete dominance, or recessiveness), reliable predictions could be made concerning the phenotypic and genotypic ratios of the progeny.

16.10
MULTIPLE ALLELES

Thus far, only genes that exist in two allelic forms have been considered. Many genes, however, have more than one allele. Genes of this sort are called **multiple alleles**. In rabbits, for example, there is a gene C that causes a colored coat; the recessive allele c produces albino. But two other alleles of the same gene have also been discovered. One is symbolized c^h, and yields in homozygous form ($c^h c^h$), the "Himalayan" pattern, in which the body is white but the extremities (the tips of nose, ears, tail, and legs) are colored. Another allele of the C gene is symbolized c^{ch}, and in homozygous form ($c^{ch} c^{ch}$) results in a "Chinchilla" pattern, a light-gray color over the entire body. In this case, the genes can be arranged in a series, C, c^{ch}, c^h, c, with each gene dominant to the genes that follow it and recessive to those that precede it. In other examples of multiple alleles, the genes may be incompletely dominant, so that the heterozygote shows a phenotype intermediate between those of its parents.

Since each allele in a multiple allele system is a variant of the same gene, any individual can have only two of the alleles present at a time (that is, Cc^h, c^h, or $c^{ch} c^{ch}$). In this respect multiple allele systems differ from what is called polygenic inheritance, where groups of different genes (each with its own set of alleles) work together to produce a phenotype different from that observed with any one of the pairs of genes by itself (see Section 16.12).

A good example of a multiple allele system is found in human blood groups, particularly the A-B-O system. In the nineteenth century doctors knew that it was impossible to transfuse blood from certain individuals to other individuals without producing a massive and often fatal reaction in the recipient. In other cases, however, transfusions produced no adverse reaction.

Studies to determine the differences in these cases led to the discovery by Karl Landsteiner in 1900 of blood groups among human populations. As a result of such studies, all human beings are now classified into four basic groups, A, B, AB, or O. It is possible to transfuse blood from an individual of one group into another individual of the same group. But the successful transfusion from individuals of one group to those in another group can occur only in certain specified ways. We now know that differences among the blood groups are due to chemical differences in substances (**antigens**) located on the red blood cell surface, as well as on the surface of other cells in the individual's body. The chemical differ-

Table 16.1

The Six Genotypes of Blood Groups and Their Phenotypic Expression.*

Genotype	Phenotype
I^A/I^A	Type A
I^A/i	
I^B/I^B	Type B
I^B/i	
i/i	Type O
I^A/I^B	Type AB

*Note that genes I^A and I^B are codominant—that is, both genes show their effects (both antigens are present on the red blood cell surface) when combined in the heterozygous condition.

ences result from the presence or absence of certain enzymes in the cells of individuals of each blood type. Thus it is possible to confirm biochemically what was deduced from pedigree studies over seventy years ago: blood groups are determined genetically.

The genetics of the A-B-O blood group system is relatively simple. There are three alleles, symbolized I^A, I^B, and I^O or i. Thus there are six possible genotypes, I^A/I^A, I^A/i, I^B/I^B, I^B/i, i/i, and I^A/I^B. Since i is recessive to both I^A and I^B, only four phenotypes are actually observed: the four blood groups A, B, AB, and O. Using the basic patterns of Mendelian genetics, it is possible to determine from blood group phenotypes of both parents and offspring what their genotypes are. The relationships between blood group phenotype and genotype are shown in Table 16.1. Note that the dominant genes I^A and I^B, when combined, produce a fully heterozygous blood group, AB, in which the effects of *both* genes are equally present. In such situations, where two alleles are both expressed in the heterozygotes, the alleles are said to be **codominant**. Codominance thus differs from both dominance, where one allele masks the effects of the other, and incomplete dominance, where both alleles interact to produce an intermediate phenotype.

Blood group analysis, including many more blood groups than the common A-B-O, is often a means of determining paternity in cases where identification of the father of a child is disputed. Genetic evidence of blood groups is thus frequently introduced in court in paternity suits. However, blood tests can never demonstrate conclusively that a given man *is* a child's father—only whether it is genetically *possible* for him to be. On the other hand, following from the principles of the truth table in Chapter 1, it is possible to demonstrate with greater certainty that a man is probably not a child's father.

Consider the following example. A woman with blood type O has a child with blood type O. She claims

SUPPLEMENT 16.2
BLOOD GROUP DIFFERENCES AND BLOOD TRANSFUSION

Tests to determine which blood group an individual belongs to are quite simple. Required for the test are two drops of the individual's blood and two reagents called anti-A and anti-B serums. These substances contain antibodies that have been generated in the body of an organism, such as a rabbit, into which red blood cells of type A or B have been injected. Type A blood has A substances on its red blood cell surface, while type B blood has B substances on the cell surfaces. The rabbit's immunological system has produced antibodies to the specific substances on the red blood cell surface, and thus the red cell substance acts as an antigen to the rabbit's immunological system. Anti-A reagent contains antibodies that specifically react with A substances on red blood cell surfaces, while anti-B reagent contains antibodies that specifically react with B substances on the red blood cell surface. When an antibody for A substance comes in contact with a blood sample containing type A red blood cells, the antibody binds to the A substance on the red blood cell surface. With appropriate differences for the substances involved, this causes the cells to stick together, a process known as agglutination. The same happens for B substance.

To type blood, one drop of donor blood is mixed with anti-A reagent, the other with anti-B. After a few minutes, one of two reactions will be observed in each drop: (1) the blood will remain whole and unclumped, or (2) the blood will collect in clumps or agglutinate (see Diagram A). The former possibility is called a *negative reaction* (no agglutination), while the latter is called a *positive reaction*. As Diagram A shows, in a test of the blood of any one individual, the reaction can be positive or negative for each reagent; hence there are four possibilities altogether. The four possibilities correspond to the four major groups, into one of which any human's blood will fall. A person with a negative test with both reagents A and B belongs to blood group O; a person with a negative reaction to reagent B, but positive reaction to reagent A, belongs to group A; a person negative

for reagent A but positive for B belongs to the B group; and a person with positive reactions to both reagents A and B belongs to group AB. It is necessary to observe the reaction in both drops of blood sample in order to classify any individual into a blood group.

Blood group substances A and B are present on the cell surface not only of humans but of other organisms as well, including bacteria. Hence it is not surprising to find that from birth, the immune system of human beings contains antibodies against substances A and B. The antibodies are part of the individual's ready-response immune system.

Usually individuals do not contain or produce antibodies against their own specific body's molecules. (When the body does develop immunity to some of its own molecules, a severe condition known as the autoimmune disease develops. The disease is rare, and usually fatal.) With regard to blood group substances, it has been found that an individual's blood plasma or serum contains antibodies against the A or B substances that he or she lacks on the surface of red blood cells at birth. The antigens and antibodies found in the blood of individuals of the different blood groups are listed in Table 1. From this list the possible patterns of blood transfusion can be determined.

One factor that has a considerable influence on the direction of successful transfusions has been neglected. Transfusions involve transferring *both* red blood cells and blood plasma from donor to recipient. Any transfusion thus involves passage of some donor plasma containing the donor's already-existing antibodies of whatever blood type he or she represents. We might expect that this plasma could, in certain cases, cause the recipient's cells to agglutinate. However, in comparison to the recipient's total blood volume, the amount of plasma transferred is very small. Consequently, there will be relatively few antibodies transferred that will agglutinate the recipient's red blood cells. Some small amount of agglutination will occur, but no serious harm will

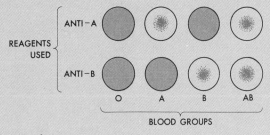

Diagram A

Table 1

Antigens and Antibodies Contained in the Blood of A-B-O Blood Groups

If the red cells contain the antigen	Then the serum contains the antibodies
A	anti-B
B	anti-A
AB	none
O	anti-A and anti-B

be done. So far as the donor is concerned, the important transfusion question is what antigen the donor red blood cells contain; as far as the recipient is concerned, the question is what antibodies the recipient's plasma contains.

Individuals of type A can donate to other individuals of type A or type AB, since these recipients do not contain anti-A antibodies in their serum. Type B individuals can donate to type B or type AB recipients, since neither contains the anti-B antibodies in the blood plasma. Individuals of type AB can donate only to other AB's, since the serum of all other recipients would contain one or the other or both A and B antibodies. Type O individuals, however, can donate to themselves or *any* of the other groups, since type O red blood cells contain neither A nor B antigens. For this reason, type O individuals are called *universal donors.* Their blood can be used for transfusion into anyone else with no risk as far as blood group complications are concerned.

Now let us look at the recipient of blood. Individuals of blood group A can receive blood from individuals of type A

or O, but not from B or AB. Type A blood serum contains anti-B antibodies and, since both red blood cell types contain B substance, would agglutinate the newly transfused red blood cells from B or AB donors. In a similar fashion, type B individuals can receive blood from other B's or from O, but not from A or AB. Type O individuals cannot receive transfusions from any other individuals than O's. People with type O blood have both A and B antibodies in their plasma and would agglutinate the transfused red blood cells from A, B, or AB donors, all of whom have either A or B antigens on their red blood cells. Type AB individuals, on the other hand, can receive transfusions from all blood group types, including their own. AB individuals have no antibodies in their plasma and will not agglutinate donor red blood cells from A, B, or AB types. For this reason, persons with blood type AB are known as *universal recipients.* In practice, however, blood transfusions are carried out between matched donors and recipients.

the father was a man who turns out to have blood type AB. It can be demonstrated that this man is very likely *not* the father. The mother's genotype is obviously *ii*, since that is the only genotype that shows type O blood. The disputed father's genotype is $I^A I^B$. The offspring of such a cross would thus be likely to be AO or BO, as follows:

Parental genotypes	$I^A I^B$			ii	
Gametes	$I^A I^B$			i	
Offspring genotypes		$I^A i$			$I^B i$
Possible phenotypes		Type A	Type B		

No type O offspring could result from the above cross. Thus, barring mutation, the suspected father's paternity can be rejected. If the suspected father had also been type O, however, he *could* be the child's father. Such evidence is only *consistent* with the hypothesis proposing that he is the father; it does not establish that hypothesis with certainty. Another man with type O blood could be the father instead.

Another blood type following the multiple allele pattern is the series of Rh alleles, obtaining its name from the fact that it was first identified in the blood of rhesus monkeys. The Rh factor is an agglutinogen (like the ABO series) and its presence or absence in the blood indicates whether the individual is Rh-positive or Rh-negative. Rh-positive is dominant over Rh-negative. The Rh factor may cause difficulties if an Rh-negative woman bears a child by an Rh-positive man. Since the growing fetus develops its own genetically determined blood type, it will be Rh-positive if the father is homozygous for Rh-positive. (If the father is heterozygous

there will be a 50 percent chance that the fetus will be Rh-positive.) If, as often happens, blood from the fetus contacts the mother's blood (through rupture of capillaries in the placenta), her white blood cells will be stimulated to produce antibodies to the Rh factor. Usually more than nine months are needed for this antibody level to build up high enough to cause trouble for the fetus. In later pregnancies, however, some of these antibodies may pass into the fetus's bloodstream; if this fetus is also Rh-positive—a certainty if the father were homozygous—the anti-Rh-positive antibodies would cause agglutination. The resulting condition, *erythroblastosis fetalis,* is often severe enough to cause death before birth; more often, such babies die after birth.* Increased familiarity with parental blood types and knowledge of their genetic basis, however, have led to the use of massive blood transfusions given at birth to save afflicted babies.

Because blood types are inherited and yet play little or no role in conscious mate selection, they remain important tools in the study of gene frequencies or proportions within populations. It is established that the proportion of blood types in a population remains fairly constant from one generation to the next as long as there is no intermarriage with other groups; a group of Germans who migrated to Hungary around 1700, for example, have kept not only the language and customs

---

*Interestingly enough, the frequency of *erythroblastosis fetalis* is much less than would be predicted from the known frequency of marriages where the male is Rh$^+$ and the female Rh$^-$. There is now some evidence that the A-B-O blood group system is possibly associated with minimizing the effects of the fetus's Rh$^+$ blood on the immunizing system of the mother.

characteristic of the Germans in Germany, but also the same blood-type frequency.

The origin of the natives of the Aleutian Islands was long a subject of debate. Some anthropologists hypothesized that the Aleuts were derived from Eskimo-Indian crosses. Others hypothesized an Asiatic origin. Both Eskimos and Indians have a relatively low frequency of B and AB blood groups, whereas B is common in Asia. Development of a new technique allowing for determination of the blood types of mummies and skeletons paved the way for study of the blood types of thirty Aleutian mummies. They showed a high proportion of B and AB blood groups, thus supporting the hypothesis that the Aleuts were of Asiatic rather than Eskimo-Indian origin.

16.11
PLEIOTROPY

A number of years ago the geneticist Theodosius Dobzhansky compared two groups of the small fruit fly *Drosophila* that differed by a single phenotypic trait, eye color: one group had red eyes, the other white. This eye color difference was known to be the result of a single gene, with white recessive to red. On closer examination, however, Dobzhansky noted that the two groups differed significantly in a number of other visible traits. White-eyed flies had reduced viability, longevity, and fertility and a differently shaped spermatheca (a special organ of the female reproductive tract that functions to store sperm after copulation but prior to fertilization). Although eye color was perhaps the most immediately observable effect of this particular gene, the gene appeared to influence a number of other traits as well. As Dobzhansky remarked, the gene could just as well have been called the spermatheca, viability, or fertility gene. Such a condition, where one gene has a variety of phenotypic effects, is called **pleiotropy**.

An illustration of pleiotropy is provided by the work of Hans Gruneberg, who studied a group of congenital (present at birth)* abnormalities in the laboratory rat, including lack of elasticity in the lungs, enlarged heart, closed nostrils, thickened ribs, and obstruction of the trachea. Through breeding experiments Gruneberg was able to show that all of these phenotypic differences were due to a single gene. Furthermore, physiological and biochemical analyses revealed that the gene affected the formation of a protein essential in the pathway for cartilage synthesis. Since cartilage is one of the most widespread structural components in the body (the ribs, trachea, and nose contain much cartilage), the many diverse effects of mutation in

a single gene responsible for its synthesis are easy to understand.

Geneticists now believe that virtually every gene is to one degree or another pleiotropic. The existence of pleiotropy in cases like the above illustrates that our custom of referring to a gene by a single phenotypic effect ("white eye," "wrinkled seed," "feathered shank," and the like) is in many ways a misleading habit of terminology. From a practical point of view it may be necessary to label genes according to one or a few phenotypic effects. However, it is always important to keep in mind that most genes do, in fact, show pleiotropism when we take the time to study carefully their actual biochemical and physiological effects.

16.12
NONALLELIC GENE INTERACTIONS: EPISTASIS

No sooner had geneticists begun to find examples of the 9:3:3:1 (or dihybrid) ratio in a variety of plants and animals, than they began to find exceptions to that ratio. For example, certain breeds of poultry have feathers on their shanks, while other breeds do not. Crosses between homozygous feathered and unfeathered birds yield all feathered offspring. A hypothesis that one pair of genes influences feathering is untenable, however, for crosses between F_1 feathered birds do not yield the predicted 3:1 ratio of feathered to unfeathered birds. Instead a 15:1 ratio is obtained. The expression of the obtained ratio in sixteenths rather than in fourths makes it likely that two pairs of genes, rather than one, must be involved. We have already dealt with cases of complete dominance involving two pairs of unlinked genes, and in those cases the ratio was 9:3:3:1. A hypothesis to account for the deviant 15:1 ratio must therefore be proposed.

One such hypothesis follows directly from the assumption that Mendel's law of independent assortment holds just as well in this case as in one yielding a 9:3:3:1 ratio, but that a different sort of gene action is in effect. The crosses between the feathered and unfeathered birds can be diagrammed as shown in Fig. 16.16. It will be seen that the 15:1 ratio can be accounted for by assuming that feathering will occur even if just one dominant gene from either pair is present. In these chickens, unfeathered shanks will result only from the complete absence of dominant genes. This hypothesis is supported by crosses between unfeathered birds and F_1 dihybrids. A 3:1 ratio of feathered to unfeathered birds would be predicted, and this is the ratio obtained.

In dogs, a cross between certain homozygous brown animals and homozygous white animals yields all white puppies. Many crosses of F_1 white animals yield an offspring ratio of 12 whites to 3 blacks to 1 brown.

Again a modification of the 9:3:3:1 ratio has occurred and an explanatory hypothesis is needed. The

---

*It is important to keep in mind that the term "congenital" or the phrase "present at birth" does not necessarily imply "genetically determined," though it can. Conditions arising from processes or accidents during development are congenital, but not genetic.

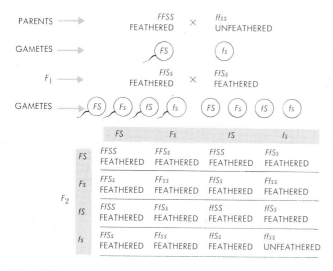

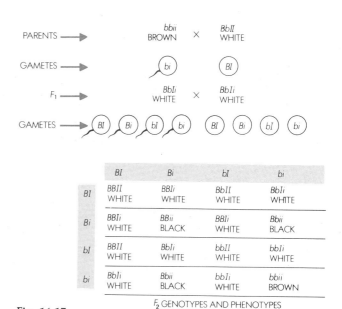

Fig. 16.16
Punnett square for the cross between a feathered and an unfeathered variety of bird. The phenotype of the F_2 offspring is given under each genotype in the Punnett square. Note that the phenotype ratio among the F_2 is not the expected 9:3:3:1, but 15:1.

Fig. 16.17
Punnett square for the cross between a brown and a white dog. The F_1 is all white, but the F_2 shows a modification of the 9:3:3:1 ratio: there are 12 whites to 3 blacks to 1 brown. Hypothesizing that one of the two pairs of alleles involved in this trait is a color-determining gene, I, whose presence in the dominant form inhibits any color formation accounts for the 12:3:1 ratio.

most satisfactory hypothesis is one that makes the following three assumptions:

1. Two pairs of randomly assorted genes are involved; that is, the genes are not linked and Mendel's second law of random assortment is in effect.

2. One of these gene pairs influences coat color to the extent that it causes the coat to be black or brown, with the former being dominant.

3. The second gene pair is one that, when the dominant form is present in either the homozygous or heterozygous state, *inhibits* the production of any pigment at all; that is, the animal is white.

The cross depicting this hypothesis is diagrammed in Fig. 16.17. This hypothesis, proposing a pigment-inhibiting gene, can be tested by crossing brown dogs with F_1 whites. The predicted 1:2:1 ratio of black to white to brown is obtained.

Feathered shanks in poultry and coat color in dogs are but two of many examples of modifications of two-pair ratios. Such examples illustrate two important points. First, what seem to be contradictions of the ratios predicted by Mendelian genetics are often not contradictions at all. Indeed, close analysis reveals that these results provide still further evidence in support of Mendelian principles. Second, and more important, *gene pairs often work together to produce their phenotypic effects, the action of one gene pair often profoundly influencing the action of others.* A situation such as that shown in the two examples described above is known as **epistasis.**

The existence of epistasis, modifiers, and multiple alleles indicates that genes do not function as isolated units, but work together. No gene affects only a single phenotypic character, but may affect all other characters of the organism to one degree or another. Similarly, no phenotypic character is produced by a single gene, but by an interaction of many of the genes in the organism and with factors in the environment.

16.13
NONALLELIC GENE INTERACTIONS: POLYGENIC INHERITANCE

Genetically speaking, the human species is enormously heterogeneous. It is sometimes said that the human species is **polygenic** (many different genes interacting to produce each trait), or **polymorphic** (many phenotypic forms). This simply means that there is a great deal of variation for any given trait, blood group or otherwise. For example, over thirty different blood groups have been identified. While the frequencies of these groups vary from one subpopulation, or group, to another, almost every population shows a large number of the different groups in at least some proportion.

One important aspect of polygenic inheritance is that it does not yield the customarily discrete phenotypic categories of classical Mendelism. Polygenic systems usually show the phenotype of offspring ranging over a series of small, graded differences, or continuous variations. Height in human beings is an example of the sort of continuous variation in phenotype produced by polygenic systems. As Fig. 16.18 shows, height distribution of U.S. males varies over quite a range, from less than 150 to over 190 centimeters (roughly 4 feet, 10 inches to over 6 feet). Note that we can draw the graphs in two ways: as a bar graph emphasizing discrete heights (Fig. 16.18a), or as a line graph, emphasizing the smooth continuity from one height to another (Fig. 16.18b). In either case the point is the same: there is no clear-cut separation of phenotypes into easily distinguishable groups. There are enough intermediates between any particular categories of height to make it impossible to form anything but completely arbitrary groupings.

What is the cause of continuous variation, and how does it differ, genetically, from discrete or discontinuous inheritance? Some of the differences in height are undoubtedly due to environmental factors (nutrition, amount of calcium in the diet as a child, and the like), while others appear to be genetic. The degree of one or the other is usually very difficult to estimate. As a polygenic system, height is controlled by a number of different genes. The effect of the presence of these genes is thus quantitative—that is, the more positive genes present for height, the taller the individual will be; the fewer such genes present, the shorter the individual will be. Such genes are not alleles of one another, but must control different aspects of the total growth process. For example, one gene could control hormone synthesis, another uptake and retention of calcium (both processes necessary for bone growth). Lack of one will retard the amount the other can do, even if the second gene is present in full dose (that is, homozygous). Because the different genes in a polygenic system act in a quantitative way, polygenic inheritance is sometimes called quantitative inheritance. The various phenotypes produced by polygenic systems differ from one another quantitatively, by small gradations, not by discrete and quantitatively different categories.

Fig. 16.18
Distribution of U.S. males for height, shown graphed in two ways, but representing the same data. (a) A bar graph, or histogram, showing five categories of height, chosen arbitrarily. Although this graph appears to show height as a series of discrete, discontinuous categories, in fact all sorts of intermediates fall between each of the groupings shown above. (b) A "bell-shaped" or "normal" curve of height distribution, emphasizing the continuous nature of the differences between samples. Both graphs (a) and (b) picture phenotype variation of a polygenically determined trait, where several pairs of alleles interact to produce the overall phenotype. Polygenic inheritance produces a more or less continuous phenotype distribution curve among offspring, in contrast to simple Mendelian inheritance, which produces real, rather than arbitrary, groupings, with distinct discontinuities between them.

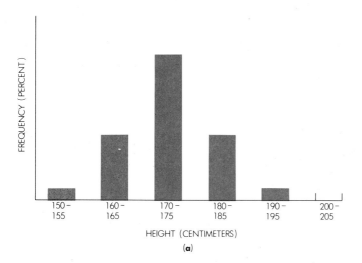

(a)

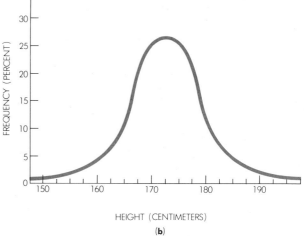

(b)

Table 16.2

Parental genotypes and phenotypes	*AABB* (dark-red)	×	*aabb* (white)		
F_1		*AaBb* (medium-red)			
Cross between two F_1's	*AaBb*	×	*AaBb*		

F_2 phenotypes	(Dark-red)	(Medium-dark red)	(Medium-red)	(Light-red)	(White)
Genotype (and number of each genotype)	1 *AABB*	2 *AABb* 2 *AaBB*	4 *AaBb* 1 *AAbb* 1 *aaBB*	2 *Aabb* 2 *aaBb*	1 *aabb*
Total number of each phenotype expected among progeny	1/16	4/16	6/16	4/16	1/16
Number of color units in each genotype	4	3	2	1	0

The exact genetic components involved in height determination in human beings has not been worked out. However, we can get a better understanding of polygenic systems by considering the case of color inheritance in wheat. In 1909 the Swedish geneticist Hermann Nilsson-Ehle crossed two strains of wheat, one of which had red kernels and one of which had white kernels. The F_1's from this cross were all of a uniform, but intermediate shade. When he crossed these F_1's, the resulting F_2 showed five color classes, ranging from the dark red of one parent to the white of the other. These various phenotypes, their frequencies, and the genotypes responsible, are shown in Table 16.2. Note that if we plotted the frequencies of each color category on a graph (Fig. 16.19), we would get a histogram and a line graph which follow a so-called normal curve of distribution, such as that shown for height, in Fig. 16.18(b). There are measurable gradations between one category and another, but they are not sharply defined differences. Such a distribution pattern is, as we have seen, characteristic of polygenic systems.

What sort of genetic explanation is represented by the list of genotypes given for each phenotype in Table 16.2? Nilsson-Ehle assumed that two pairs of genes were involved in determining kernel color in wheat. For the sake of simplicity, let us call these pairs *A* and *a*, *B* and *b*. The dominant member of each allele pair produces

one unit of color in the wheat kernel; but the effects of *A* and *B* are indistinguishable. This means that a kernel with a unit of color produced by the *A* gene is the same phenotypically as one whose color is produced by the *B* gene. The recessive alleles in each pair produce no color, and hence can be viewed more as lacking the dominant *A* or *B* than as having any effect in their own right. Thus, according to this scheme, the genotypes *AAbb*, *aaBB*, and *AaBb* look alike phenotypically, since each has two units of color. Similarly AABb and AaBB look

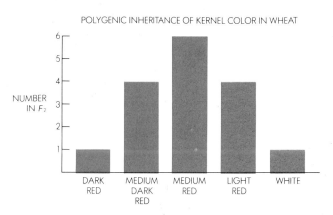

Fig. 16.19

alike phenotypically since each has three units of color. With a Punnett square, all the possible genotypic combinations can be determined, as well as the phenotypic categories, from red to white. The results of such a calculation are summarized in Table 16.2. Note that the frequencies that fall into each color class are those that would be expected on the basis of the interaction of two pairs of alleles. To test his hypothesis further, Nilsson-Ehle made additional predictions and experimental crosses. For example, he predicted that the white and red F_2's would breed true when self-fertilized. When he made the cross, this is exactly what he observed. Now, what predictions would you make if, for example, either of the two light-red phenotypes were self-fertilized? Nilsson-Ehle predicted that these genotypes would produce $\frac{1}{4}$ medium-red, $\frac{1}{2}$ light-red, and $\frac{1}{4}$ white offspring. Again, he observed the expected ratios. The explanation is based on the assumption that the dominant form of each allele contributes one color unit to the wheat kernels. As you can see from the bottom line of Table 16.2, dark-red is produced by the presence of four color units, medium-dark by three, medium-red by two, light-red by one, and white by none. Nilsson-Ehle's hypothesis led to predictions that were borne out by experimental tests. Far from contradicting the basic principles of the Mendelian theory, the concept of polygenic, or quantitative, inheritance proved to be amenable to Mendelian analysis.

All of the examples discussed in this section—epistasis, pleiotropy, multiple alleles, and polygenic inheritance—emphasize one important point: *genes and gene products interact with another.* Genes are not discrete, isolated units acting totally independently within cells or the whole organism. Although we will discuss this subject in greater detail in Chapter 20, the point should be clearly emphasized at the beginning of our study of genetics. We can only begin to understand the way genotype and phenotype are related to one another when we recognize that genes interact with the cellular, organismic, and even the external environment in which they exist.

16.14
BINOMIAL EXPANSIONS AND MENDELIAN THEORIES: GENETIC COUNSELING

Much of the practical usefulness of genetics is based on its ability to predict the phenotypes and genotypes of plant and animal progeny. Certainly commercial breeding experiments would be of little use if it were not possible to predict the types of expected offspring with statistical validity.

Of particular importance to humans is the application of this predictive capability to genetic prognosis (also called "genetic counseling"). Genetic prognosis might be used to predict for a couple what types of children they could expect to have. In some cases the infor-

mation would be of minor interest; a blue-eyed couple probably would not be much concerned to learn that they would not be likely to have any brown-eyed children. In other cases, the questions might be more critical. It is not usually desirable to bring into the world a child with traits that bring about early death or deterioration. Couples who would be likely to have children with such afflictions might want to think twice before doing so. Advice from well-trained genetic counselors can be useful in such situations to guide parents in facing such difficult choices.

Let us take a concrete example. There sometimes occurs in humans a condition known as Tay-Sachs disease, which results in mental deficiency, blindness, paralysis, and early death. The condition is inherited as a simple recessive trait. Thus, unaffected parents whose family histories indicate that they are both heterozygous for the condition may expect one out of four of their children to show the condition. Suppose that such a couple goes to a genetics expert and asks for advice. They would like to have at least three children. However, they recognize the possibility that some or all of their children may be afflicted with this condition, and they want to find out from the geneticist what the chances are that all three of their children will be normal.

Knowing that both these people are heterozygous for Tay-Sachs disease, the geneticist can tell them immediately that three-quarters of their children can be *expected* to be normal. The Punnett square below illustrates this:

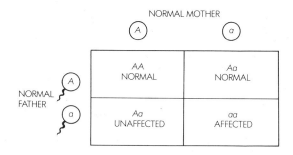

But this tells the couple very little about the matter in which they are most interested—the *chances* that all three of their children *will* be unaffected. Three-fourths of three children is a meaningless figure to most parents, as indeed it is to almost anyone.

The product principle of probability provides the answer, in this case by the binomial expansion $(a + b)^n$ with $n = 3$. Let a = chances for normal children (that is, $\frac{3}{4}$); let b = chances for abnormal children ($\frac{1}{4}$). Since there are three children involved, we expand as follows:

$$(a + b)^3 = a^3 + 3a^2b + 3ab^2 + b^3.$$

That portion of the binomial expansion which is of most interest to us is the one representing three normal chil-

SUPPLEMENT 16.3
SOME HINTS ABOUT SOLVING GENETICS PROBLEMS

The study of genetics has been the nemesis of many biology students because of difficulties they encounter in the solution of problems. On the other hand, some students who may have been doing quite poorly in a biology course suddenly blossom in the study of genetics. It is difficult to see why these differences occur; possibly it is because genetics problems, by their mathematical nature, demand very precise reasoning. Nevertheless, almost all genetics problems encountered in introductory biology can be attacked in a simple and straightforward manner.

The following steps will help in the solution of genetics problems. They are applicable to the problems in Chapters 16, 17, and 18.

1. Determine the type of inheritance dealt with in the problem. Does it show complete dominance, incomplete dominance, or some other inheritance feature? Are there one, two, or more pairs of genes involved? Usually this information is given in the problem. If not, it can be deduced from the phenotypic ratios of the offspring.

2. Determine the genotypes of the individuals involved.

3. Determine the types of gametes each parent can produce. Arrange them into a Punnett square and fill in the possible progeny genotypes.

4. Count the resulting phenotypes and express them as a ratio. Often this is as far as the problem will require you to go.

5. If chance predictions are involved, expand the binomial $(a + b)^n$ to the proper nth power. Substitute the fractions given by the Punnett square in step 4 and solve.

Sample Problem: In humans, the ability to taste the bitter chemical phenylthiocarbamide (PTC) is due to a dominant gene T; inability to taste it is due to the recessive allele t. A man who can taste PTC, but whose father could not, marries a woman who can also taste PTC, but whose mother could not.

a) What proportion of their children will probably have the ability to taste PTC?

b) If they have five children, what are the chances that four will be tasters and one a nontaster?

Step 1. We determine that this is a case of simple dominance and that there is only one pair of genes involved.

Step 2. We determine that the man's genotype must be Tt (since he is a taster, he must have at least one gene T, but since his father was a nontaster (tt), and he is the product of his father's sperm, he must also have a t gene.) Likewise, the woman must also be Tt, since her mother was a nontaster (tt), and she is a product of her mother's egg.

Step 3. We determine that the man and woman can each produce two types of gametes, carrying either a T or t gene. We next put them in a Punnett square, as follows:

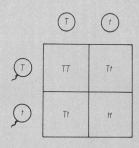

Step 4. We count up the phenotypes. In this case there are two, tasters and nontasters. The phenotypic ratio in this case is 3:1. This gives the answer to part (a); $\frac{3}{4}$ are tasters.

Step 5. We expand $(a + b)^n$ with $n = 5$ as follows:

$$a^5 + 5a^4b + 10a^3b^2 + 10a^2b^3 + 5ab^4 + b^5.$$

Letting a = tasters and b = nontasters, we choose the proper term in the expansion that gives us four tasters and one nontaster: four a's to one b. Clearly this is the term $5a^4b$. Substituting the Punnett square fractions of $\frac{3}{4}$ tasters to $\frac{1}{4}$ nontasters in this term, we obtain $5a^4b = 5(\frac{3}{4})^4(\frac{1}{4}) = 5(\frac{81}{256})(\frac{1}{4}) = 5(\frac{81}{1024}) = \frac{405}{1024}$, or 405 of 1024 chances. This is the answer to part (b).

dren (a^3). Substituting the fraction $\frac{3}{4}$ as given by the Punnett square, we obtain

$$a^3 = (\frac{3}{4})^3 = \frac{27}{64}.$$

Thus in sixty-four three-child marriages of this sort, twenty-seven would be expected to have all three children unaffected. This couple's chances of attaining their desired family size without encountering the unfavorable condition are a little less than 50–50.

As another example, suppose that a biology instructor wishes to demonstrate a 3:1 Mendelian ratio to his or her classes by crossing two black mice, both heterozygous for brown (*Bb*). If a litter of only four mice is obtained, what are the chances that this litter will demonstrate the desired ratio?

Once again we turn to the binomial expansion and proceed as follows. Let a = chances for a black mouse, or $\frac{3}{4}$; let b = chances for a brown mouse, or $\frac{1}{4}$. Since there are four mice involved, we expand $(a + b)^n$, with $n = 4$:

$$(a + b)^4 = a^4 + 4a^3b + 6a^2b^2 + 4ab^3 + b^4.$$

We want to know what chance we have of getting three black to one brown, or, as we have symbolized it, three a's to one b. We therefore look for the term in the expansion that gives us this ratio. Clearly this is $4a^3b$. Removing this term from the expansion, we substitute and solve as follows:

$$4a^3b + 4(\tfrac{3}{4})^3(\tfrac{1}{4}) = 4(\tfrac{27}{64})(\tfrac{1}{4}) = 4(\tfrac{27}{256}) = \tfrac{108}{256} = \tfrac{27}{64}.$$

You can see from this answer that the instructor is a bit of an optimist if he or she expects to get a perfect demonstration with a single cross.*

In the first example discussed here dealing with Tay-Sachs disease, it can be seen that the chances of getting all three children normal can be derived, first, intuitively by showing with the Punnett square that the chance that any one child will be normal is $\frac{3}{4}$, and, second, by showing with the product principle of probability that the chance of all three being normal is $\frac{3}{4} \times \frac{3}{4} \times \frac{3}{4} = (\frac{3}{4})^3 = \frac{27}{64}$. For the second example, dealing with black and brown mice, it can be seen that the chance for the specific 3-black-to-1-brown combination is $(\frac{3}{4})^3 \times \frac{1}{4}$ and that there are four possible ways of getting this combination.

In the above examples it just happens that both predictions come out to $\frac{27}{64}$; this is mere coincidence. We have used simple figures and situations to illustrate how genetic predictions (prognoses) are made using the techniques of Punnett square analysis and binomial expansion. In most genetic counseling the situation is far more complex, however. In many cases the genetics of the diseases are less well-known than Tay-Sachs and the family histories of the parents are less clear. (For example, it is not always easy to determine whether one or both parents are likely to be heterozygous for Tay-Sachs.) Nonetheless, genetic counseling in advance of pregnancy has assumed an increasingly important place in modern medicine as one way of providing prospective parents with more information on the chances of genetic disease in any children they might have.

Summary

1. Genetics, the study of inheritance, can be investigated at five levels of organization:
   (a) *Molecular:* The molecular structure and function of genes;
   (b) *Cellular and Chromosomal:* the cytological organization of genes, where they are located within cells, as parts of what structures, their behavior during cell division and gamete formation;
   (c) *Organismic:* The patterns of inheritance displayed by the breeding of organisms, the determination of dominance, recessiveness, gene interactions, and the like; and
   (d) *Population:* The study of genes within populations, their distribution from generation to generation, viewed independently from the individual organisms that carry them; population genetics is closely allied to the study of ecology and evolution.

2. Breeding experiments in genetics are based on the laws of probability. Even though any one event in itself is unpredictable, given enough events the overall outcome can be predictable.

3. The product principle of probability states that the chance that two or more independent events will occur simultaneously is given by the product of the chances that each of the events will occur individually. In flipping two pennies, for example, the chance of either by itself turning up heads is $\frac{1}{2}$ (1 out of 2, or 50–50). The chance that *both* will turn up heads simultaneously is given as $\frac{1}{2} \times \frac{1}{2}$ or $\frac{1}{4}$.

4. Mendel's original observations on peas suggested that the ratios of different types of offspring from any cross followed the laws of probability. Mendel assumed that every parent organism contains, in its germ cells, two "factors" for every observable trait. One factor derived from that organism's male parent, the other from its female parent. Mendel claimed the factors could be either dominant or recessive with respect to one another. A dominant trait is one that completely masks its recessive counterpart (complete dominance). A recessive trait is one masked by a dominant counterpart. Dominant and recessive are relative terms; that is, one factor is always dominant with respect to some other factor. Incomplete dominance results often in an intermediate phenotype.

5. Mendel's first law, the law of segregation, states that in the process of forming gametes from a parental germ cell, the two factors for any trait always segregate and are distributed to two different gametes. Mendel's second law, the law of independent assortment, states that the segregation of

---

*And even more of an optimist in expecting the mother mouse to cooperate and produce a litter of four or eight that will nicely divide into a perfect 3:1 ratio. If the instructor is blessed with the author's luck, the result will be a litter of 12 mice, all brown.

The fact that both our examples came out to be $\frac{27}{64}$ is coincidental. You should not expect all answers to come out this way.

two or more pairs of factors occurs independently—that is, randomly.

6.   The factors for any given trait can exist in any organism in one of several combinations: homozygous dominant, homozygous recessive, or heterozygous. Homozygous means that both factors are alike, whether both are dominant or both recessive. Heterozygous means that the factors are different. In complete dominance, homozygous dominants and heterozygotes look alike; only the homozygous recessives appear different. The terms homozygous and heterozygous thus refer to the genetic makeup, or genotype, of the individual.

7.   Differences in appearance and genetic makeup have led to geneticists making a distinction between genotype and phenotype. Phenotype is the appearance of the organism for any given trait or combination of traits, while genotype re-

fers to the actual genetic makeup of the individual with respect to any trait or traits. Thus, individuals homozygous and heterozygous for a trait showing complete dominance will have the dominant phenotype. Their genotypes will be different, however, since the heterozygotes can pass on two different alleles for the trait to the next generation, whereas homozygotes can pass on only one allele.

8.   In determining offspring ratios from any given cross, both phenotype and genotype can be used. It is very important to indicate which ratio is meant. Usually ratios are determined by looking at the phenotype only. Mendel's ratio of 3:1 in crosses between two heterozygous tall pea plants is a phenotype ratio. The genotype ratio is 1:2:1 (one homozygous tall; two heterozygous talls; and one homozygous recessive).

Exercises

1.   Suppose you flipped five pennies simultaneously. What are the chances of getting:
   (a) All five heads?
   (b) Four heads and a tail?
   (c) Three heads and two tails?
   (d) Two heads and three tails?
   (e) One head and four tails?
   (f) All five tails?

2.   A spotted rabbit and a solid-colored rabbit were crossed. They produced all spotted offspring. When these F_1 rabbits were crossed, the F_2 consisted of 32 spotted and 10 solid-colored rabbits. Which characteristic is determined by a dominant gene?

3.   A mouse breeder has a pure strain of black mice that normally breed true and produce only black offspring. Occasionally, however, a brown mouse or two has appeared in the litters of some of the mice. What do you hypothesize as the cause of the appearance of brown coat color? If the brown mice are killed as soon as they are born,

how many generations will it take to completely eliminate the brown gene?

4.   In humans, aniridia, a type of blindness, is due to a dominant gene. Migraine, a headache condition, is the result of a different dominant gene. A man with aniridia, whose mother is not blind, marries a woman who suffers with migraine but whose father does not. In what proportion of their children would *both* of these conditions be expected to occur?

5.   In summer squash, colorless fruit is due to a dominant gene W; colored fruit is due to its recessive allele w. Disc-shaped fruit is determined by a dominant gene S, sphere-shaped fruit by its recessive allele s. How many different genotypes may squash plants have in regard to color and shape of fruit? How many categories of phenotypes could be expected from their genotypes? How many different homozygous genotypes are possible? What phenotypic ratio would you expect from a cross between two heterozygous plants?

Chapter 17
Genetics II: Cell
Reproduction

17.1
INTRODUCTION

As we saw in the previous chapter, Mendel's approach to the problem of heredity was to breed organisms, count the number and kind of offspring, and deduce patterns of inheritance from the data. Although his approach was experimental, his conclusions were couched in abstract, or theoretical terms. That is, Mendel had never seen any hereditary particles or "factors" in his pea plants, nor had he seen the "factors" segregate or randomly assort. Mendel *hypothesized* their presence. His hypothesis—formulated as Mendel's first and second laws—was highly predictive in a number of cases. It made sense to a lot of people who, like Mendel, busied themselves with animal and plant breeding. But Mendel's conception displayed a notable lack of concreteness. No one knew what his "factors" might be or of what substances they might be composed. It was equally unclear what sort of actual process could cause the two factors for each trait to segregate from one another during the formation of gametes. Furthermore, as time went on, an increasing number of exceptions to Mendel's basic laws began to appear. (For example, assortment among two or more factors did *not* always appear to be at random.) Ultimately, Mendel's hypothesis raised as many questions as it answered.

While Mendel was experimenting with peas, cytologists were studying the microscopic structure of the cell nucleus and the behavior of the strange, rod-shaped bodies, called chromosomes, within the nucleus. The cytologists did not know of Mendel's work, nor did Mendel know of the cytologists' work. Yet, study of cytology, especially after the rediscovery of Mendel's laws in 1900, provided the elements missing from Mendel's original scheme as it had been proposed in 1865. For it was in the structure and behavior of the chromosomes (viewed as groups of "factors" or "genes") that biologists found a concrete explanation of abstract processes such as independent assortment and segregation. Mendel's ideas and the exceptions to them made more sense when they could be thought of in terms of real, visible structures within living cells.

In the previous chapter we investigated heredity as a property of whole organisms. In the present chapter we will investigate heredity as a property of individual cells. Throughout the living world reproduction of the whole organism, animal or plant, unicellular or multicellular, begins with the reproduction of cells. Thus, to pursue our study of the process of heredity in whole organisms, it is necessary to gain a knowledge of the process of reproduction at the cellular level.

First, however, let us begin with an overview of the general reproductive process, the life cycle, as it occurs in most kinds of organisms.

17.2
GENES AND ORGANISMS: THE LIFE CYCLE

The reproductive process in *all* organisms, unicellular and multicellular alike, encompasses a series of changes that are called its **life cycle**. The details of the life cycle vary from one type of organism to another, or even from one set of conditions to another in the same organism. Figure 17.1 schematically shows life cycles for both multicellular and unicellular organisms. Let us examine the life cycle concept, following the details shown in Fig. 17.1.

We will begin with the life cycle for multicellular forms (Fig. 17.1a) with **gametogenesis** (though by definition a cycle can begin anywhere). Gametogenesis is the process by which the reproductive cells, the gametes, are formed in the adult or parent organism. Each gamete contains genetic information to produce

Fig. 17.1
Schematic diagram of the life cycles of primarily sexually reproducing organisms (*a*), and primarily asexually reproducing organisms (*b*). Diagram (*a*) is also characteristic of multicellular organisms, while (*b*) is characteristic of unicellular organisms. Both types of cycles involve the alteration of a diploid and a haploid phase. In multicellular forms the diploid phase is predominant; in unicellular forms, the haploid phase is dominant. Most unicellular organisms are haploid; the cells of virtually all multicellular forms are diploid.

HAPLOID PHASE

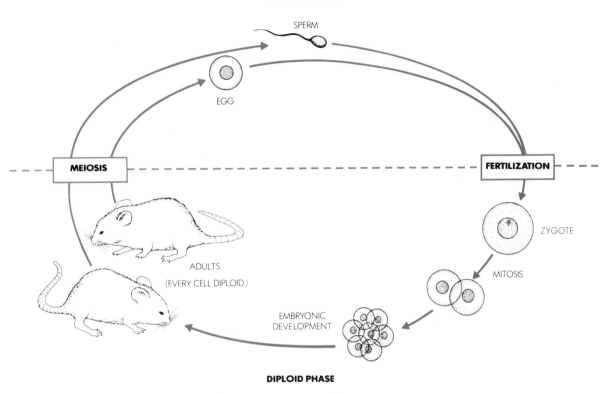

(a)

HAPLOID PHASE

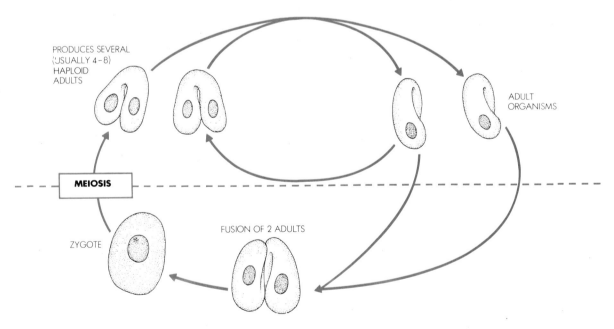

(b)

every trait in the offspring—in other words, one complete blueprint for an organism of the same species. Thus a sperm contains one complete blueprint and an egg another. When fertilization occurs, in animals or plants, the resulting zygote contains two blueprints of the genetic information present in either gamete (that is, it has a double dose). Genetically speaking, zygotes are said to be **diploid,** gametes **haploid.** The diploid zygote undergoes cell division, beginning the process of **development.** Development leads in multicellular forms to differentiation, whereby cells specialize into leaves, roots, and flowers (in plants), or skin, nerves, and muscles (in animals). Eventually development leads to the production of adult organisms, which are, in their own turn, capable of producing gametes and starting the cycle all over again.

Only one thing is missing in this brief outline of the life cycle. How do gametes become haploid again? In the reproductive organs—the *gonads*—of the parents, gametogenesis involves a process called *meiosis*, or reduction division (see Fig. 17.1). In meiosis, the number of genes (and chromosomes) is reduced by one-half, from the diploid to the haploid number. Meiosis, like mitosis, is a highly regularized process. Meiosis occurs only in cells in the diploid phase and, by definition, contain two copies of every chromosome—one from the male and the other from the female parent. In meiosis the two copies of each chromosome are separated from one another (as described in Section 17.5.)

As you can see in Fig. 17.1, the life cycle of unicellular forms differs from that of most multicellular forms in two basic ways. The first is the relative amount of time spent in the diploid compared to the haploid phase. As shown in Fig. 17.1, unicellular forms spend most of their time in the haploid condition, existing as diploids only during the relatively brief period of zygote formation. Soon after the latter process, the zygote undergoes meiosis and produces haploid cells that take up their own independent existence. Thus, the cells of most bacteria, algae, mosses, protozoans, and other simple forms are genetically haploid. By contrast, most multicellular forms are diploid throughout their adult lives, existing in the haploid state only as eggs or sperm. The significance of this difference will be discussed in some detail in Chapter 29.

A second difference is apparent when we compare the life cycles of multicellular and unicellular forms. Most multicellular forms are capable only of sexual reproduction, although asexual processes such as budding or vegetative reproduction can also occur. By contrast, most unicellular forms are capable of both sexual and asexual reproduction, though for the majority the asexual process is the most common. Asexual reproduction obviously has the advantage of producing a large number of offspring in a short period of time. Bacteria have been known to generate daughter cells by the division of a parent cell every twenty minutes.

The life cycle incorporates the two major processes in which genes must engage during an organism's life time: heredity and development. Heredity refers to the process by which genes are transmitted from parent to offspring. In sexual reproduction heredity involves gametogenesis and fertilization. In asexually reproducing forms, it involves mitosis and cell division. In either case, the process of heredity necessarily requires the prior replication of the genes and the chromosomes of which they are a part. The process of gene and chromosome duplication insures that the next generation receives the same number and kind of genes as those of the parental generation.

Development refers to the fact that genes guide the growth and differentiation of the individual organism from zygote to adult—that is, they control the process of embryonic growth and differentiation. Both transmission and development involve the interaction of genes and other parts of cells and the surrounding environment. And both are essential for the existence of life as we know it on this earth.

Let us now turn to the process of reproduction on the cellular level first with a general discussion of the processes of cell division (in both prokaryotes and eukaryotes) followed by a detailed examination of mitosis and meiosis. It will be very important for your study of genetics in Chapters 18 and 19 to gain a thorough understanding of how both prokaryotic and eukaryotic cells divide and how eukaryotic multicellular organisms produce gametes.

17.3
CELL DIVISION: SOME GENERAL PROBLEMS

Chromosomes are not easily visible under the ordinary light microscope. In the nineteenth and twentieth centuries cytologists developed a series of *stains* (see Section 1.3 and Supplement 1.1), which made some of the structural details of chromosomes more apparent. However, to really see the chromosomes clearly, even after staining, the cells had to be fixed and sectioned, which killed them. Thus, in the early days, knowledge of cell division was based on nonliving specimens. This was much like looking at random frames from a movie film, and trying to deduce the sequence of action.

The invention of the phase-contrast microscope helped cytologists see the process of cell division in an ongoing, dynamic way in living cells, with much of the same detail as seen in stained and fixed cells (see Section 4.3). In presenting the details of cell division in this chapter, we are confined to words and still pictures on the printed page. There is the danger of thinking of cell division in a static way; even if we have the sequence of images correct, they still look like frozen movie frames. Try, then, to think of the changes in the chromosomes, the cell nucleus, and the entire cell as a dynamic, ongoing series of events in the reproductive process of living cells.

The Nucleus in Heredity

Throughout most of the latter half of the nineteenth century cytologists debated the role of the nucleus in the cell's life. Today, we know that the nucleus is vital to all the cell's daily activity, as well as to its reproductive capacity. Some simple observations and experiments support this generalization.

1. Cells that normally lack a nucleus at maturity, such as red blood cells, are incapable of reproduction and ultimately degenerate.

2. The cell nucleus can be removed by simple microsurgical techniques. Enucleated ("lacking a nucleus") amoebas are able to live with undiminished activity up to several weeks. Gradually, however, their activity slows down and they cease capturing food, roll up into a sphere, and eventually die.

3. If a new nucleus (from another amoeba of the same species) is inserted into the enucleated cell within a few days after the original nucleus is removed, the cell resumes its normal activity, does not degenerate, and eventually will reproduce. If the nucleus is inserted after about five or six days, however, the cell degenerates.

4. The nucleus from the fertilized egg of one variety of frog can be replaced with the nucleus from a fertilized egg of another variety. When this is done, the resulting adult resembles the variety of frog that donated the nucleus, not the recipient variety.

5. If a nucleus is transplanted from an intestinal cell of the frog (that is, an already highly differentiated cell) into an enucleated egg, the nucleus begins to direct the synthesis of all kinds of enzymes besides those characteristic of intestinal cells. In its original intestinal cell, the nucleus would guide production or produce only those enzymes characteristic of intestinal cells.

These observations point to three important conclusions regarding the role of the nucleus in cell life:

1. The nucleus is responsible for the transmission of hereditary information from one generation to the next. The seat of heredity lies in the cell nucleus of the fertilized egg.

2. A long-term presence of a nucleus is generally necessary for the ongoing, day-to-day activities of a cell. Without the continued presence of the nucleus, a cell cannot maintain itself against the natural process of degeneration.

3. Although the nucleus controls many of the activities of a cell, the cytoplasm also controls or influences the activities of the nucleus. The growth and development as well as the day-to-day metabolic activities of a cell are controlled by the interaction of the nucleus and the cytoplasm. (See Supplement 17.1).

A more detailed analysis of the above generalizations, and one group of experiments on which they are based, can be found in Supplement 17.1.

SUPPLEMENT 17.1 THE NUCLEUS AND THE CONTINUITY OF CELLULAR LIFE

In the early 1950s German biologist Joachim Hämmerling studied the role of the cell nucleus in determining the specific hereditary characteristics of cells. He worked with a single-celled marine alga known as *Acetabularia* (see Diagram A). Each stalk and umbrella-like cap compose just one cell, which may be up to 5 cm in length. The nucleus of the cell is always found in the stalk. This fact, coupled with its large size, makes *Acetabularia* an ideal organism through which to study the role played by the nucleus in the life of the cell.

Hämmerling's first experiment simply involved cutting an *Acetabularia* cell in half, as shown in Diagram A. The half without a nucleus continued to live for a while, but eventually died. The half with a nucleus regenerated a new umbrella, and continued to live and reproduce. Hämmerling performed this same experiment on many *Acetabularia* cells, always with the same results. He therefore concluded that

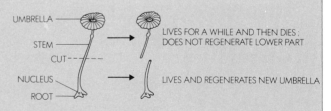

Diagram A

while the presence of the nucleus in *Acetabularia* is *not* essential to the cell's short-term existence, it *is* essential for the continuation of life and regeneration of the lost part.

The fact that the upper part of the *Acetabularia* cell did not regenerate the lower part, while the lower part, with the nucleus, did regenerate the upper part, makes it reasonable.

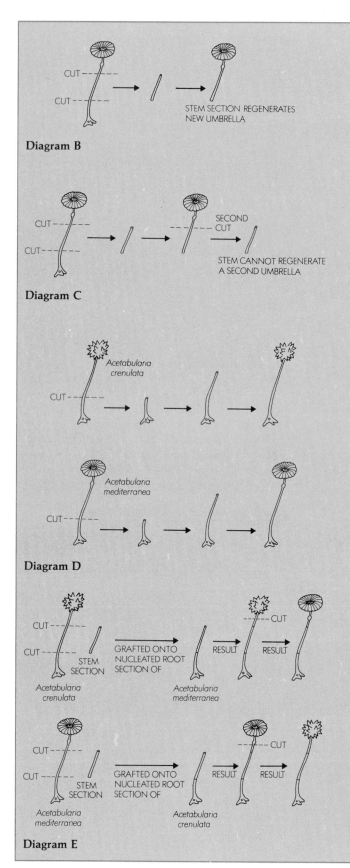

Diagram B

CUT

CUT

STEM SECTION REGENERATES
NEW UMBRELLA

Diagram C

CUT

CUT

SECOND
CUT

STEM CANNOT REGENERATE
A SECOND UMBRELLA

*Acetabularia
crenulata*

CUT

*Acetabularia
mediterranea*

CUT

Diagram D

CUT

CUT

STEM
SECTION

*Acetabularia
crenulata*

GRAFTED ONTO
NUCLEATED ROOT
SECTION OF

*Acetabularia
mediterranea*

CUT

RESULT

RESULT

CUT

CUT

STEM
SECTION

*Acetabularia
mediterranea*

GRAFTED ONTO
NUCLEATED ROOT
SECTION OF

*Acetabularia
crenulata*

RESULT

CUT

RESULT

Diagram E

to hypothesize that the presence of the nucleus is required if regeneration is to occur. To test this hypothesis, Hämmerling performed the experiment shown in Diagram B.

Note here that the stalk, without a nucleus, still managed to regenerate an umbrella. This result contradicted the hypothesis.

However, after a *second* operation on the same stem that regenerated the first umbrella, the stem section could not regenerate a second umbrella (Diagram C). Thus the original hypothesis needed merely to be modified to state that the presence of the nucleus is essential to the *continued* regeneration of the lost umbrella by an *Acetabularia* cell.

It is now possible to suggest a slightly different hypothesis to explain the results of these first experiments. Perhaps the nucleus secreted some sort of "regenerative substance" into the cytoplasm surrounding it. It was the presence of this regenerative substance that enabled the *Acetabularia* stem section to regenerate the first umbrella. The stem section was unable to regenerate a second umbrella because it had used up all the regenerative substance given it by the nucleus and was unable to produce the substance itself.

This hypothesis can be tested by cutting the stem of the cell at varying distances from the nucleus. If the hypothesis is correct, we could predict a correlation between the amount of regenerative power possessed by some section and its closeness to the nucleus; the section closest to the nucleus should have the highest powers of regeneration.

Experiments of this sort, performed on many different types of cells, lead to the conclusion that the nucleus is the control center for virtually all the life processes carried on by the cell. It is true that cells that have had their nuclei removed will continue to live for a while. However, such cells are incapable of reproduction and growth, and sooner or later they die. Certain cytoplasmic elements of the cell are responsible for the synthesis of proteins and the respiratory, energy-supplying processes of the cell. Yet, even this factor does not make the cell independent of the nucleus for its continued life. Even these specialized cytoplasmic structures seem dependent on the nucleus for their continued functioning. Thus the nucleus emerges as a most important single portion of most cells.

There is another feature of the nucleus that can be demonstrated by experiments on two different species of *Acetabularia*. These two species, *Acetabularia mediterranea* and *Acetabularia crenulata*, are chosen for the unique and easily recognized shape of their umbrellas. When the umbrellas are cut off, each species regenerates an umbrella typical of its kind (Diagram D). These results are hardly surprising.

Suppose, however, that a section of the cut stem of one species is grafted onto the cut end of a nucleated piece of the other species. Two such experiments can be performed: the grafting of a stem section from *A. mediterranea* onto a nucleated root section of *A. crenulata*, and the grafting of a stem section from *A. crenulata* onto a nucleated root section of *A. mediterranea*. The object is to see which portion of the cell (that is, cytoplasm or nucleus) determines the characteristics of the regenerated umbrella. The results of these experiments are shown in Diagram E. Note that in both experiments the first umbrella regenerated is identical to that of the species from which the stem section came. However, if this first regenerated umbrella is cut off and a second one allowed to regenerate, the second umbrella will be characteristic of the species from which the nucleated root section came.

How can these results be explained? Consider the hypothesis that the cell nucleus secretes a regenerative substance into the cytoplasm. This hypothesis can be nicely modified and expanded to fit the experimental observations. Perhaps the first regenerated umbrella looks like that of the species from which the stem section came because the grafted stem section still contains some of the regenerative substance of its own species. It uses up this regenerative substance in producing the new umbrella. Then, unable to produce more of the regenerative substance of its own species, it has to rely on the nucleus of the other species for the substance used in building the second umbrella. According to this hypothesis, the second umbrella will resemble the *Acetabularia* species from which the nucleated section came—and it does.

The results of these experiments tell us what might happen if two nucleated root sections from *A. mediterranea* and *A.*

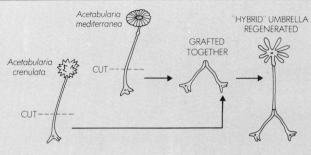

Diagram F

crenulata are fused so that both nuclei contribute to the regeneration of one new umbrella. Diagram F shows the results of such an experiment. A mixed or hybrid type of umbrella is formed with characteristics of the umbrellas of both species. These results fit well with our hypothesis. We can visualize both nuclei specifying different types of regenerative substance. Since both types of these substances are used in umbrella regeneration, a hybrid umbrella would be predicted.

17.4
CELL REPRODUCTION AMONG PROKARYOTES

Cell division, or cell reproduction, is far more simple in prokaryotic than eukaryotic cells. Recall that prokaryotes have no nucleus, but have their genetic material organized into a single, circular chromosome attached to the inner surface of the plasma membrane. Reproduction in prokaryotes involves, first, replication of the circular chromosome, and second, splitting of the cell into two equal parts. This process is known as *binary fission* ("binary" emphasizing the equality of the two division products), and is diagrammed in Fig. 17.2(a) on p. 476.

Stage 1 shows the single, circular chromosome attached to the inner surface of the cell membrane. The chromosome has just begun to replicate. Stage 2 shows the chromosome about 60 percent replicated. By stage 3, the chromosome has been fully replicated, with both daughter chromosomes attached to the cell membrane. The cell wall and membrane have also begun to increase in size by adding a new segment in the midregion. In stage 4 more cell wall and membrane have been added, and both have begun to pinch inward in the middle. Meanwhile, the daughter chromosomes have separated, though each is still attached to the inner cell membrane. Stage 5 shows the two newly formed cells, both equivalent genetically to the original parental cell.

17.5
CELL DIVISION IN EUKARYOTES

The process of cell division in eukaryotes is considerably more complex than in prokaryotes. Two distinct events occur:

1. *Mitosis or karyokinesis*—Division of the cell nucleus and chromosomes in such a way that each daughter nucleus receives the same number and kind of chromosomes as the parent nucleus.

2. *Cytokinesis*—Splitting of the parent cell into two daughter cells, each containing one of the replicated nuclei.

Although these events are obviously interrelated, we will examine them separately.

The Cell Cycle

The entire reproductive life of eukaryotic cells is a four-stage cycle (see Fig. 17.3). A full cycle is that period of time from the end of one mitosis to the beginning of another. Each stage is designated by a given letter; and the space allotted to each stage indicates roughly the percentage of time the cell spends in that phase. We begin with the M stage, which is the process of mitosis

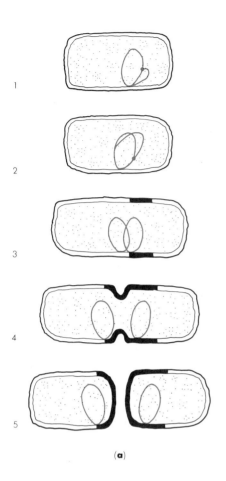

(a)

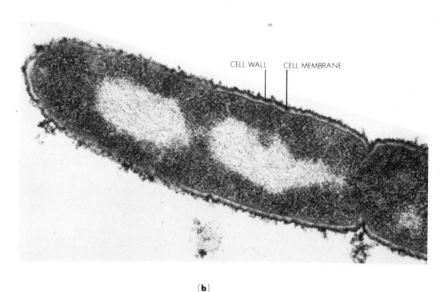

Fig. 17.2
Cell division in a prokaryote such as the rod-shaped bacterium, *Bacillus subtilis.*
(a) Diagram of five stages in binary fission. The colored circle inside the cell represents the replicated daughter chromosome; the black portion of the cell wall represents newly synthesized cell wall and plasma membrane. Both parental and daughter chromosomes remain attached to the inside surface of the cell membrane (see text for details). (b) Electron micrograph of dividing *B. subtilis* (× 120,000). Note the pinching in of the cell wall between the two newly formed daughter cells. The two light areas in the center of the left-hand cell contain the chromatin. (Courtesy Thomas F. Anderson.)

CELL WALL   CELL MEMBRANE

(b)

Fig. 17.3
The cell cycle, consisting of a period of mitosis (M), a gap (G_1), a period of DNA replication and synthesis (S), and another gap period (G_2) before a return to mitosis. The space given over to each period is an average indication of the time duration, as shown by the percentages given on the outside of the circle. The ratios of the four time periods vary considerably from cell type to cell type and species to species.

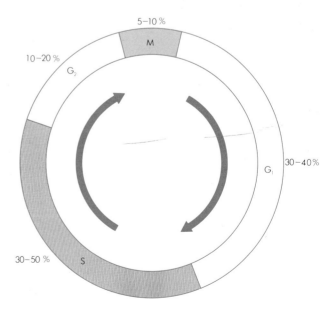

5–10 %

M

10–20 %

G_2

30–40%

G_1

30–50 %

S

Cell division and mitosis are not synonymous terms. Cell division refers to the processes of mitosis and cytokinesis, by which one cell becomes two. Mitosis is one part of that process: the duplication and division of chromosomes in such a way that both daughter cells end up with the same number and kind of chromosomes as the parent cell.

itself. In general, mitosis takes 5 to 10 percent of the cell's total cycle. After mitosis, there is a period designated as G_1, or a gap phase. During this period the cell is metabolically active, and is often growing. In genetic terms, however, it is functioning with the same amount of DNA received from its parent cell. The cell begins to synthesize new DNA during the S, or synthesis phase. The S phase occupies between 30 and 50 percent of the cell cycle. Another gap phase, G_2, follows S. In G_2, the cell prepares for another mitosis, and a new cell cycle. Since the cell cycle is a reproductive one, for every cell that begins the mitotic phase, two cells emerge at the beginning of G_1, each continuing through the remainder of the cycle as an independent entity.

The length of a complete cell cycle varies considerably from one cell type to another, and from one species to another. In most animals the cycle lasts anywhere from 18 to 24 hours while in plants it lasts from 10 to 30 hours. However, in some species it can be as short as 20 minutes and in others as long as several weeks. By contrast, muscle and nerve cells in higher animals neither grow nor divide at all after reaching maturity yet they have nuclei and continue to function throughout the life of the organism.

Mitosis

Mitosis is a nuclear process that insures that daughter cells are genetically identical to the parent cell. Because it is a highly ordered sequence of changes, mitosis accomplishes its major purpose with persistent regularity.

The Mitotic Process. Mitosis consists of four main stages: prophase, metaphase, anaphase, and telophase, with an interphase occurring between. We will follow these stages in sequence through the diagrams shown in Fig. 17.4. Refer back to this diagram at each step to make sure you can picture the chromosomal changes.

Photomicrographs of these stages in the animal cell (whitefish) are shown in Fig. 17.5.

Interphase. The longest stage of mitosis is interphase. Interphase has often been referred to as a "resting stage," because it includes the period a cell spends between one division and the next. However, a cell in interphase is not resting at all but is metabolically quite active.

The cell in Fig. 17.4(a) is in interphase. The nucleus and nuclear membrane are clearly distinguishable. The single centriole in each cell replicates, forming two centrioles. No chromosomes are seen, however, since the chromosomal or chromatin is diffused throughout the nucleus. It is only when the chromatin threads begin to condense into a tight coil that the chromosomes become readily visible.

Prophase. Prophase begins with the coiling of chromatin into rod-shaped chromosomes. (Fig. 17.4b, and c; Fig. 17.5a, b, and c). The centrioles begin slowly to migrate away from one another, toward opposite ends, or "poles" of the cell. The nucleolus becomes less distinct, and often disappears by late prophase. The chromosomes replicate during this period, and are *visibly* doubled by middle or late prophase. If chromosome doubling did not occur before cell division, each daughter cell would end up with only half as many chromosomes as in the parent cell. Obviously, duplication of each chromosome is essential to keep a constant chromosome number from one cell generation to the next.

The two strands of a duplicated chromosome are called **chromatids** and are genetically identical. During prophase each pair of chromatids is held together by a single constricted region, called the **centromere** (see Fig. 17.4c and Fig. 17.6, which is on p. 480.)

As the nuclear membrane disappears during late prophase, the two centrioles reach opposite poles of the cell. In the animal cell a series of microtubules forms

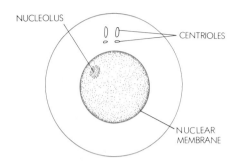

(a) INTERPHASE

CENTRIOLES APPEAR DUPLICATED;
NUCLEOLUS AND NUCLEAR MEMBRANE
VISIBLE;
CHROMOSOMES INDISTINCT.

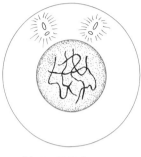

(b) EARLY PROPHASE

CENTRIOLES BEGIN MOVING APART;
NUCLEOLUS BECOMES INDISTINCT;
NUCLEAR MEMBRANE VISIBLE;
CHROMOSOMES APPEAR THREAD-LIKE.

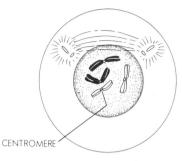

(c) MIDDLE PROPHASE

CENTRIOLES MOVE FARTHER APART;
CHROMOSOMES THICKEN, APPEAR
AS TWO CHROMATIDS JOINED
TOGETHER BY ONE CENTROMERE.

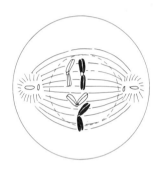

(d) LATE PROPHASE

CENTRIOLES REACH OPPOSITE ENDS
OF NUCLEUS;
SPINDLE FORMS BETWEEN CENTRIOLES;
NUCLEAR MEMBRANE DISAPPEARING;
CHROMOSOMES MOVE TOWARD EQUATOR.

(e) METAPHASE

CENTROMERES OF EACH CHROMATID
PAIR ATTACHED TO SPINDLE;
CHROMATID PAIRS MOVE TO
EQUATOR;
NUCLEAR MEMBRANE HAS DISAPPEARED.

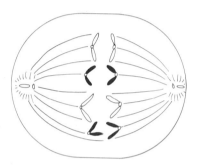

(f) EARLY ANAPHASE

CENTROMERES DUPLICATED AND
BEGIN MOVING APART ON SPINDLE;
SISTER CHROMATIDS PULLED APART
AS SPINDLE MICROTUBULES SHORTEN.

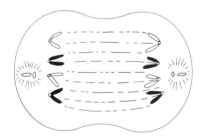

(g) LATE ANAPHASE

THE TWO IDENTICAL SETS OF
CHROMOSOMES MOVE TOWARD
OPPOSITE POLES;
SPINDLE BEGINS TO DISAPPEAR;
PINCHING IN OF CELL MEMBRANE
(CYTOKINESIS) BEGINS.

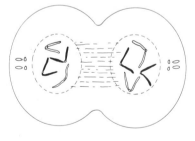

(h) TELOPHASE

NEW NUCLEAR MEMBRANES BEGIN
TO FORM;
NUCLEOLUS REFORMS;
CENTRIOLES DUPLICATE;
CHROMOSOMES BEGIN TO THIN OUT;
CYTOKINESIS ALMOST COMPLETE.

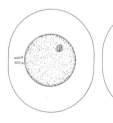

(i) NEW INTERPHASE

CYTOKINESIS COMPLETED;
NUCLEAR AND CELL MEMBRANES
COMPLETELY REFORMED;
CHROMOSOMES INDISTINCT.

Fig. 17.4
Stages of mitosis in the animal cell.

Fig. 17.5
Mitosis in cells of the whitefish embryo. (Photos, Turtox, Chicago.)

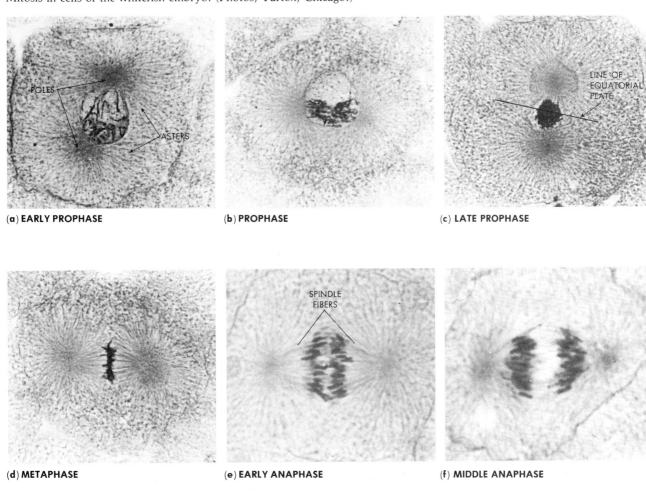

(a) EARLY PROPHASE   (b) PROPHASE   (c) LATE PROPHASE

(d) METAPHASE   (e) EARLY ANAPHASE   (f) MIDDLE ANAPHASE

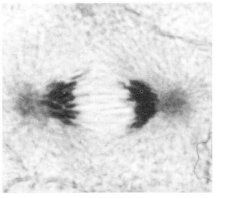

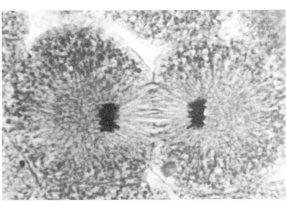

(g) LATE ANAPHASE   (h) TELOPHASE

Fig. 17.6
Photograph of late prophase chromosomes, each consisting of two identical chromatids held together by a single centromere. The drawing at right shows the relationship between chromatids and centromere in greater detail. At this stage the centromere has not replicated, and hence holds the two chromatids together. After replication of the centromere at the end of metaphase, the sister chromatids can separate. Note that the location of the centromere along the length of each chromatid pair varies. In some, it is near the midpoint, in others more toward one end. Since the attachment point of the centromere is constant and a characteristic of each pair of chromosomes of a species, it is one of the identifying features of each chromosome pair.

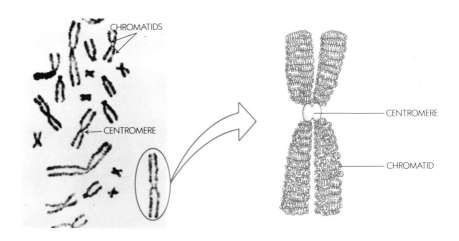

between the centrioles, creating the **spindle** (Fig. 17.7). Each centromere is attached to a spindle microtubule. A second set of microtubules, the **asters**, radiate out in all directions from each centriole. The spindle tubules help pull the chromosomes apart in the anaphase stage of mitosis. The function of the asters is unknown.

Metaphase. During metaphase, one of the shortest portions of the mitotic cycle, the chromosomes line up along the cell's equatorial plate (Fig. 17.4d and e, and 17.5d). If viewed in a two-dimensional section, the chromosomes appear to form a solid line at the cell's midpoint. However, if viewed from one of the poles, the chromosomes would appear as a flat plate, seen face-on.

Fig. 17.7
Electron microscope photograph of rat thymus gland cells in prophase, × 6500. Note that the nuclear membrane is still present and has a duplicated or "stacked" appearance. A centriole can be seen cut in cross section. From the zone around the centriole, spindle fiber microtubules can be seen extending outward. The enlargement at the right (× 11,000) shows more clearly that the microtubules lead to the centromeres of the chromosomes. (Photo courtesy Raymond G. Murray, Assia S. Murray, and Anthony Pizzo, Indiana University. Reproduced by permission of Rockefeller University Press, *Journal of Cell Biology.*)

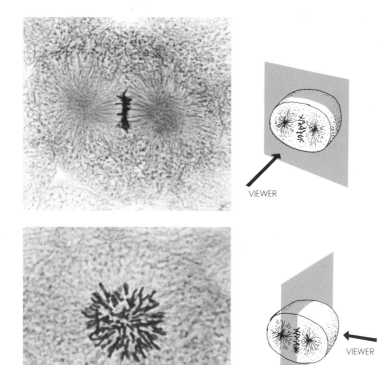

Fig. 17.8
Two views of metaphase. In the top figures the microscope is focused on the plane that runs through the center of both asters (see the sketch at the right). The chromosomes appear as if on a line. In the lower figure the cell has been turned for observation and the microscope focused on the plane of the chromosomes (see sketch). The chromosomes appear to be distributed in a circle. Such microscopic views are only two-dimensional. The reality of three dimensions is best inferred when both views are considered.

The difference in these two views of the metaphase chromosome arrangement is shown in Fig. 17.8.

Toward the end of metaphase the centromeres duplicate, so that each member of a chromatid pair is free to move to opposite ends of the cell. At this stage, each chromatid, with its own centromere, is called a chromosome.

Anaphase. In anaphase the paired chromosomes are pulled to opposite poles of the cell by the spindle microtubules to which the centromeres are attached (Fig. 17.4f and g; Fig. 17.5e, f, and g). The forces that produce the actual movement of the chromosomes toward the poles are not clearly understood. Several current hypotheses are summarized in Supplement 17.2.

SUPPLEMENT 17.2
HOW DO CHROMOSOMES MOVE?

Most cell biologists believe that the spindle and possibly the centrioles are involved in the mitotic movement of chromosomes. The spindle is composed of microtubules, the same structures found in cilia and flagella. Recall that microtubules are composed of a polymerized series of globular subunits called "tubulin" (see Section 5.11). It is tempting to think that the same molecular mechanism may be responsible for movement in all three structures—the mitotic spindle, cilia, and flagella. There is evidence to support this idea.

Edward W. Taylor of the University of Chicago, Gary Borisy of the University of Wisconsin, and Leslie Wilson of

Stanford University have studied the effects of the alkaloid* colchicine on microtubular function. Application of colchicine to cells with beating cilia and flagella, or those undergoing mitosis, stops movement completely. The researchers have hypothesized that colchicine binds to the tubulin sub-

---

*Alkaloids are a class of compounds that are usually basic ("alkaline") and contain one or more nitrogen atoms in a cyclic structure. Many alkaloids are toxic.

units and prevents them from polymerizing. But what about already-formed tubules? Shinya Inoué of the University of Pennsylvania has shown that the spindle in dividing cells will break down if colchicine is added (or if the temperature is lowered to a certain degree). Under such conditions mitosis is arrested. Furthermore, Inoué found that the effects could be reversed by removing colchicine from the cell medium (or by raising the temperature). Recovery of full mitotic activity does *not* require further protein synthesis, since microtubules were able to re-form even when inhibitors of protein synthesis were added to the cell cultures. These observations have led a number of workers to maintain that microtubules form spontaneously when enough tubulin subunits are present—that is, microtubules exist in dynamic equilibrium with a pool of tubulin subunits in the medium. The mitotic spindle could thus be seen as forming when the equilibrium shifted, due perhaps to increased synthesis of tubulin subunits, toward polymerization. This shift could begin to take place at early prophase. What might trigger such a shift is, at this stage, unknown.

How can the theory of microtubular structure account for chromosome movement? Inoué and others have proposed a model for chromosome movements based on the polymerization-depolymerization concept of microtubule formation. According to Inoué, the mitotic movements of chromosomes, and even of the centrioles themselves, are the result of lengthening and shortening of spindle tubules. At the start of mitosis, for example, the two centrioles separate from one another and move toward opposite ends of the cell. Inoué has suggested that this separation is produced by polymerization of tubulin, forming microtubules at "orienting centers" around the centrioles. Growing microtubules push the two centrioles apart. Furthermore, Inoué notes that the centromeres are the regions on the chromosomes where the microtubules connecting the chromosomes to the poles originate. (Some microtubules are chromosome-to-pole, other microtubules are pole-to-pole.) Polymerization, and thus growth in length of these tubules, push the chromosomes to the center of the cell (equator) during late prophase and early metaphase. During anaphase, Inoué proposes, the chromosome-to-pole tubules begin to depolymerize, with one unit of tubulin being removed at a time from the polar end. This produces a shortening of the microtubule, thus pulling the chromosome closer and closer to the pole.

Another intriguing model for how chromosomes are moved toward the poles during anaphase has been proposed by J. Richard McIntosh of the University of Colorado. McIntosh has suggested that the chromosome-to-pole microtubules may slide across the pole-to-pole microtubules by the action of molecular cross-bridges between them. This model is directly derived from knowledge about the mechanism of movement in cilia and flagella. Ian Gibbons of the University of Hawaii found that cross-bridges exist between the microtubules of cilia. The cross-bridges are composed of a protein of high molecular weight, which Gibbon named dynein. These dynein bridges, or "arms," can be seen in electron micrographs as projections from one of each pair of outer tubules in the 9 + 2 structure of flagellar microtubules (see Fig. 5.13). As it turns out, dynein is an enzyme (an ATPase) that cleaves off the third phosphate group from ATP releasing energy. McIntosh, Gibbons, and others have hypothesized that the tubules in both flagella and the mitotic spindle slide across each other in a manner similar to that of the sliding of actin and myosin filaments in muscle contraction. Hydrolysis of ATP, catalyzed by dynein, provides the energy. This model has been strengthened in recent years by the discovery that the protein actin is, in fact, actually present in the mitotic spindle, though its function is unknown.

Another factor that may be involved in chromosome movements during anaphase is the direct repulsion of sister chromatids from one another. This process may be quite independent of, and supplementary to, the action of the spindle tubules. If a dividing cell is poisoned with iodoacetic acid (CH_2ICOOH), which prevents the centromeres from cleaving and thus arrests anaphase, an interesting phenomenon can be observed. The free arms of the sister chromatids sweep out from each other and away from the cell equator. This observation suggests that perhaps the chromosomes themselves provide at least some of the motive force behind their polar migrations at anaphase.

Whatever the exact mechanism that produces chromosome and centriole movements during mitosis, it is clear that some very exciting questions and answers lie ahead in this field of research.

Telophase. By late anaphase or early telophase, the chromosomes have reached their respective poles. Almost as a reversal of prophase, the chromosomes once again become less distinct, the nuclear membrane begins to re-form, and the nucleolus reappears. The centrioles in each new cell replicate, and the cell begins to furrow inward (see Fig. 17.4h and i; Fig. 17.5h). As furrowing becomes complete, two new cells are formed.

Each cell is then prepared to enter the G_1 phase and start the cycle all over again.

The details of mitosis in plant cells (the onion root tip) are shown with photomicrographs in Fig. 17.9, and in drawings in Fig. 17.10. Mitosis in plant cells is largely the same as in animal cells. There are some differences, however. Unlike animal cells, higher plant cells do not have centrioles or asters. The spindle is present and appears to function in the same fashion as in animal cells.

The above has been a general review of "typical" animal and plant cell mitosis. You should keep in mind, however, that some rather noticeable differences occur throughout the various groups of eukaryotes. For example, in some protists the nuclear membrane never disappears, with the spindle forming *inside* the cell nucleus. The nucleus undergoes a "pinching-in" at telophase, which creates two daughter nuclei, each with a full set of chromosomes.

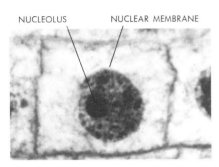

NUCLEOLUS   NUCLEAR MEMBRANE

(a) INTERPHASE

CELL NOT DIVIDING; CHROMOSOMES LONG, THREAD-LIKE; NUCLEOLUS EVIDENT. NEW CHROMATIDS BEING SYNTHESIZED.

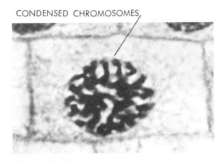

CONDENSED CHROMOSOMES

(b) PROPHASE

CELL NOW DIVIDING; CHROMOSOMES (EACH WITH TWO CHROMATIDS) COILED SHORTER, THICKER; NUCLEAR MEMBRANE AND NUCLEOLUS DISAPPEAR.

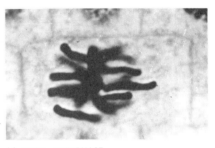

(c) EARLY METAPHASE

CHROMOSOMES BECOMING ALIGNED IN CENTER; SPINDLE FIBERS APPEAR BETWEEN CELL POLES.

(d) METAPHASE

CHROMOSOMES NOW ALIGNED IN CENTER; CHROMATIDS SOON BEGIN TO SEPARATE AT KINETOCHORE.

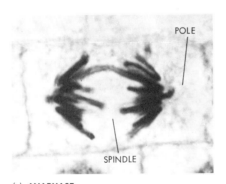

POLE

SPINDLE

(e) ANAPHASE

CHROMATIDS (OF EACH CHROMOSOME) COMPLETE THEIR SEPARATION, MOVE APART ALONG SPINDLE TOWARD OPPOSITE POLES OF CELL; NOTE ABSENCE OF ASTERS.

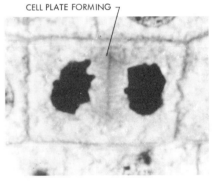

CELL PLATE FORMING

(f) TELOPHASE

CHROMATIDS (EACH NOW A CHROMOSOME) ARRIVE AT POLES, UNCOIL, BECOME LONG, SLENDER; NUCLEAR MEMBRANE AND NUCLEOLUS RE-FORM; CELL WALL BEGINS TO BE PRODUCED ABOUT MIDWAY BETWEEN THE TWO NEW NUCLEI.

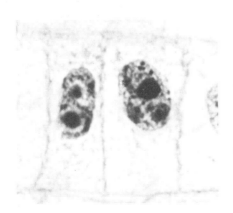

(g) EARLY INTERPHASE

CELL HAS COMPLETED DIVISION CYCLE; TWO NEW DAUGHTER CELLS NOW PRESENT; EACH CELL BEGINS TO ENLARGE AND ELONGATE.

Fig. 17.9
Plant cell division is shown here in the apical meristem of an onion (*Allium*) root. The onion possesses sixteen pairs of chromosomes. Note the lack of aster formations at the poles of the cell. Lack of asters is a result of the fact that the cells of higher plants do not have centrioles, from which the astral formation develops. They do, however, form a perfectly viable spindle, showing that the centrioles are not essential for spindle formation. (Photos courtesy Carolina Biological Supply Company.)

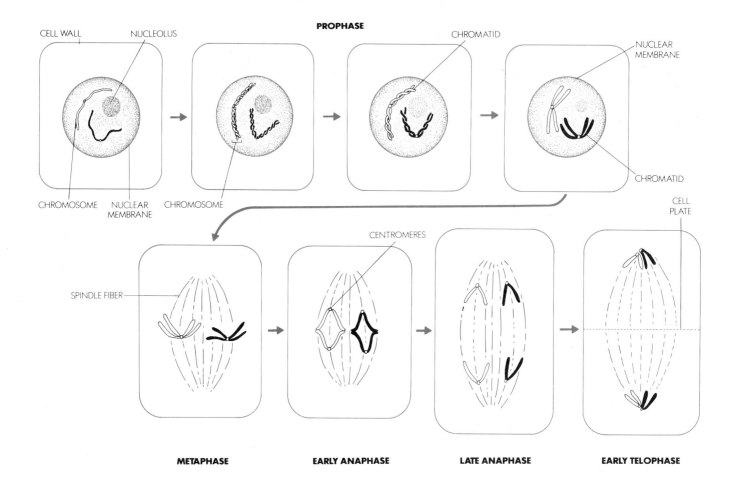

PROPHASE

CELL WALL   NUCLEOLUS   CHROMATID   NUCLEAR MEMBRANE

CHROMOSOME   NUCLEAR MEMBRANE   CHROMOSOME   CHROMATID

CELL PLATE

SPINDLE FIBER   CENTROMERES

METAPHASE   **EARLY ANAPHASE**   **LATE ANAPHASE**   **EARLY TELOPHASE**

Fig. 17.10
Schematic representation of mitosis in plant cells. For simplicity, the cell shown here has a diploid chromosome number only of four (two pairs). Note that each member of the pair is visibly doubled by mid-prophase. The two members of each homologous pair are shown in black and white, indicating that one originated from the organism's female parent, the other from its male parent. Note also that the chromosome of the daughter cells shown in late anaphase–early telophase is identical.

Cytokinesis

In the majority of living organisms mitosis is usually followed by **cytokinesis,** or cell-splitting. Animal cells normally begin cytokinesis in mid- to late telophase. The cell rounds up into a sphere (see Fig. 17.11) and then forms a long groove, the **cleavage furrow,** which grows deeper as telophase progresses (Fig. 17.12). Eventually the furrow cuts through the cell, forming two independent units. Very little is known about cleavage furrow formation. A ring of actin microfilaments, similar to spindle microtubules, has been found in the cell membrane along the line where the cleavage furrow develops and may be involved in the constriction pro-

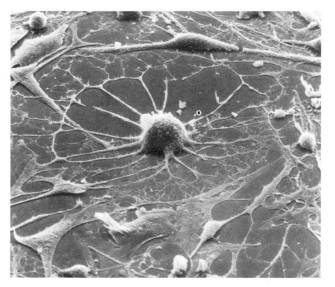

Fig. 17.11
Scanning electron micrograph (× 200) of a mouse cell just before it divides. Note how the cell "rounds up," but still has long processes stretching out to the perimeter of the cell's former boundary. (Courtesy Dr. Paul B. Bell, Jr., University of Oklahoma.)

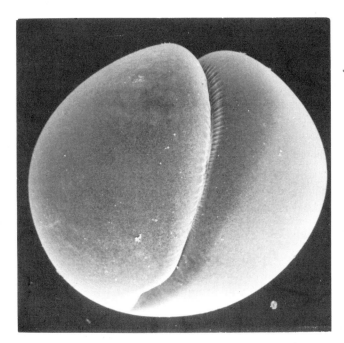

Fig. 17.12
Scanning electron micrograph of a frog egg beginning to undergo cytokinesis. Note the visible, pinched-in cleavage furrow down the center. The stress folds inside the groove may be the result of microfilament constriction, a process which is thought to be responsible for furrow formation (see Supplement 17.2). (Courtesy H. W. Beams and R. G. Kessel, *The American Scientist* 64 [1976]; reprinted by permission of the authors and *The American Scientist*.)

invertebrates. The resulting plant or animal body consists of large, noncellular masses containing multitudes of nuclei. These bodies are said to be *coenocytic*.

17.6
MEIOSIS

With certain exceptions, the number of chromosomes in the body cells of an organism is constant from one cell to the next. Thus, for example, each body cell of a certain gill fungus has 4 chromosomes, of an elm tree 56, and of a stalk of sugar cane 80. In an earthworm, each body cell has 32 chromosomes, in a bull frog 26, in a chicken 18, and in a human 46. Because of the chromosome duplication that occurs during mitosis, each daughter cell is assured of a full complement of chromosomes after each division.

Every multicellular organism is the result of millions of mitotic divisions, starting with the first division of the fertilized egg. For example: the human fertilized egg contains 46 chromosomes, of which 23 were contributed by the egg and 23 by the sperm. Thus half of the 46 chromosomes in a person's body cells are duplicate descendants of the mother's (*maternal*) chromosomes and half of them are duplicates of the father's (*paternal*) chromosomes. The 46 chromosomes are thus more accurately described as 23 pairs of **homologous chromosomes.** Each pair carries many inheritance factors that influence specific traits, such as eye or hair color. One member of the homologous pair carries the maternal factor for the trait, the other the paternal factor. The chromosome numbers given for humans and the other organisms listed earlier are thus double (**diploid** or 2n) numbers. The gametes of these organisms, however, cannot have a diploid number of chromosomes. If they did, their union at fertilization would result in a tetraploid (or 4n) number. With each ensuing generation, the chromosome number would increase geometrically. Obviously, this does not occur. Rather, each gamete contains a **haploid** (or n) number of chromosomes. Thus in dogs, with a diploid number of 56, a sperm and an egg each contain 28 chromosomes; in horses, with a diploid number of 60, a sperm and an egg each contain 30 chromosomes.

The cellular division process by which the daughter cells receive only the haploid number of chromosomes is known as meiosis. In animals this process takes place during the formation of the gametes (gametogenesis), while in most plants meiosis occurs when spores are produced (sporogenesis; refer back to Fig. 17.1a).

cess. This theory is strengthened by the fact that drugs that break down microfilaments (for example, cytochalasin) prevent formation of the cleavage furrow and thus arrest cytokinesis.

Cytokinesis is different in plant cells. Plants have rather rigid cell walls, and hence cannot divide as animal cells do by the simple pinching-in process. In higher plants (see Figs. 17.9f and 17.10g) a membrane known as the **cell plate** begins to form along the equator of the cell from small membranous vesicles that merge together. The vesicles may come from the Golgi apparatus and/or the endoplasmic reticulum. More research is needed before we discover how the cell plate is formed.

It is important to recognize that mitosis and cytokinesis do not always occur together. Mitosis without cytokinesis is quite common in some lower plant groups, such as certain algae and fungi and some lower

---

Meiosis I is a reduction division: it results in the chromosome number being reduced to one-half that of the diploid number. In eukaryotes, each resulting haploid cell contains one member from each pair of chromosomes found in the parent cell.

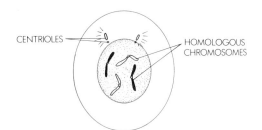

(a) EARLY PROPHASE I

CHROMOSOMES CONDENSE FROM THREADS
INTO RODS;
EACH CHROMOSOME HAS ALREADY
REPLICATED BY THIS TIME.

(b) MIDDLE PROPHASE I

HOMOLOGOUS CHROMOSOMES BECOME
THICKER AND SYNAPSE;
CENTRIOLES MOVE APART.

(c) LATE PROPHASE I

EACH HOMOLOG APPEARS AS DOUBLE-
STRANDED;
SYNAPTIC PAIRS THUS FORM FOUR-STRANDED
STRUCTURE, THE TETRAD;
CENTRIOLES AT OPPOSITE ENDS OF CELL.

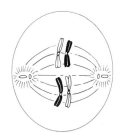

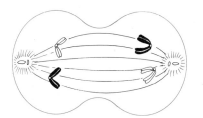

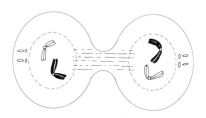

(d) METAPHASE I

EACH TETRAD MOVES TO EQUATOR;
CENTROMERE OF EACH TETRAD ATTACHED TO
SPINDLE MICROTUBULE.

(e) ANAPHASE I

STRANDS OF TETRAD SEPARATE;
TWO MATERNAL STRANDS GO TO ONE
POLE AND TWO PATERNAL TO THE
OTHER POLE;
THIS EVENT HAS EFFECT OF SEPARATING
TWO ORIGINAL HOMOLOGS.

(f) TELOPHASE I

NEW NUCLEAR MEMBRANE BEGINS TO FORM
AT EACH END OF CELL;
CYTOKINESIS BEGINS;
NEW NUCLEI ARE TECHNICALLY CONSIDERED
HAPLOID, SINCE ORIGINAL PARENTAL
HOMOLOGS WERE SEPARATED IN ANAPHASE I.

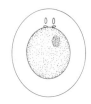

(g) INTERPHASE

NO REPLICATION OF GENETIC MATERIAL OCCURS.

(h) PROPHASE II

PAIRED IDENTICAL STRANDS OF EACH HOMOLOG
BECOME VISIBLE

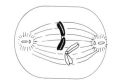

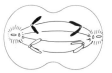

(i) METAPHASE II

(j) ANAPHASE II

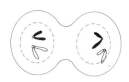
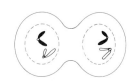

(k) TELOPHASE II

(l) INTERPHASE

Fig. 17.13
The complete process of meiosis for a typical animal cell. Meiosis differs from mitosis in that it has two complete cycles of chromosome segregation but only one cycle of chromosome replication. The end result is cells with one-half the total number of chromosomes as the parent cell. For simplicity's sake, no crossing over is shown.

Meiosis resembles mitosis in that it, too, occurs concurrently with cell division and has the same descriptive phases. As in mitosis, a spindle apparatus is formed, the chromosomes migrate apart at anaphase, and daughter cells result. Yet meiosis differs significantly from mitosis. Generally there are two cell divisions, known as Meiosis I and Meiosis II. Certain of the chromosome behaviors in meiosis are different from those in mitosis. And, of course, the most important difference is that meiosis is a reductional process, yielding a haploid number of chromosomes.

Meiosis is characterized in prophase of the first division (meiosis I) by a process called **synapsis**, in which the homologous chromosomes pair up and lie next to each other (see Fig. 17.13a and b). During the S-phase of interphase in meiosis, DNA replication and histone synthesis have already occurred, and each homologous chromosome has replicated. Thus synapsis results in a structure, the **tetrad**, composed of four chromatids (see Fig. 17.14a). Note, however, that the duplicates remain attached at the region of the centromere.

During tetrad formation sections of the chromatids of different pairs often overlap or wrap around each other, forming patterns called **chiasmata** (singular, **chiasma**). This is shown diagrammatically in Fig. 17.14(b) and (c). Many cytologists now think that chiasma formation provides cytological evidence that a phenomenon called **crossing over** has occurred, in which there is a reciprocal exchange of chromosomal pieces between sister chromatids. Since it allows for further mixing of genetic traits, crossing over has obvious importance to the study of both inheritance and genetic variation, and this additional potential for recombining parental traits is a major advantage of sexual reproduction.*

After tetrad formation, the homologous chromosomes line up at the cell equator in a typical metaphase, with anaphase and telophase following closely (Fig. 17.13d, e, and f). Being joined at the centromere, the chromatid pairs must travel together. Thus for any single pair of homologous chromosomes, the maternal chromatids go to one pole while the paternal chromatids go to the other at metaphase I of meiosis. This does not mean, however, that when two or more pairs of chromosomes are involved, the resulting daughter cells

*In Fig. 17.13(b) and (c), the tetrads are shown as though no chiasmata form. This is merely for the sake of visual clarity and to focus attention on the movements of the chromatids in the tetrad.

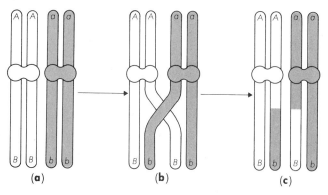

Fig. 17.14
The process of synapsis and chiasma formation during prophase I of meiosis. (a) By mid-prophase I homologous chromosomes (each is already duplicated by this point) pair up, forming a four-stranded association called a tetrad. In the diagram shown here, the maternal member of the homologous pair is shown in white, the paternal in color. The letters on each chromosome stand for specific genetic factors. Thus, A and its allele a are one set of traits; B and its allele b, another. There is no general significance to the fact that in the above diagram the maternal homolog carries both dominant traits, and the paternal both recessives. (b) Two adjacent strands of the tetrad twist around each other, forming a chiasma. (c) The two have broken where they were intertwined, and exchanged parts. Note that, genetically speaking, this exchange, or recombination as it is called, produces two chromosomes with a different combination of traits than was observed previously on any members of the tetrad: Ab and aB. Such recombinations are an important source of variability in populations of organisms. Synapsis and tetrad formation occur regularly in meiosis, while chiasma is a much rarer occurrence.

will carry only maternal or paternal chromosomes. Any combination of maternal and paternal chromosomes is possible, since each pair of chromosomes (tetrad) is independent of the others.

Because they are joined at the centromere, the two chromatids are still considered to be one chromosome. Thus the division that has occurred in anaphase I is a **true reduction division**: the number of chromosomes in each daughter cell has been reduced by one-half, or from diploid (2n) to haploid (n).

The second meiotic division (Fig. 17.13g and h) follows, often with only a very brief interphase period. In this division, however, the centromeres divide at metaphase (Fig. 17.13i), releasing the chromatids to migrate to the opposite poles (anaphase II, Fig. 17.13j) as full-fledged chromosomes. Note that this occurrence merely retains the haploid number of chromosomes, rather than restoring the diploid number. From each original cell at the beginning of the meiotic process, *four* haploid cells result (Fig. 17.13l).

Photomicrographs of meiosis in an animal cell (the grasshopper *Chorthippus parallelus*) are shown in Fig. 17.15. You will find it a good review to follow the stages

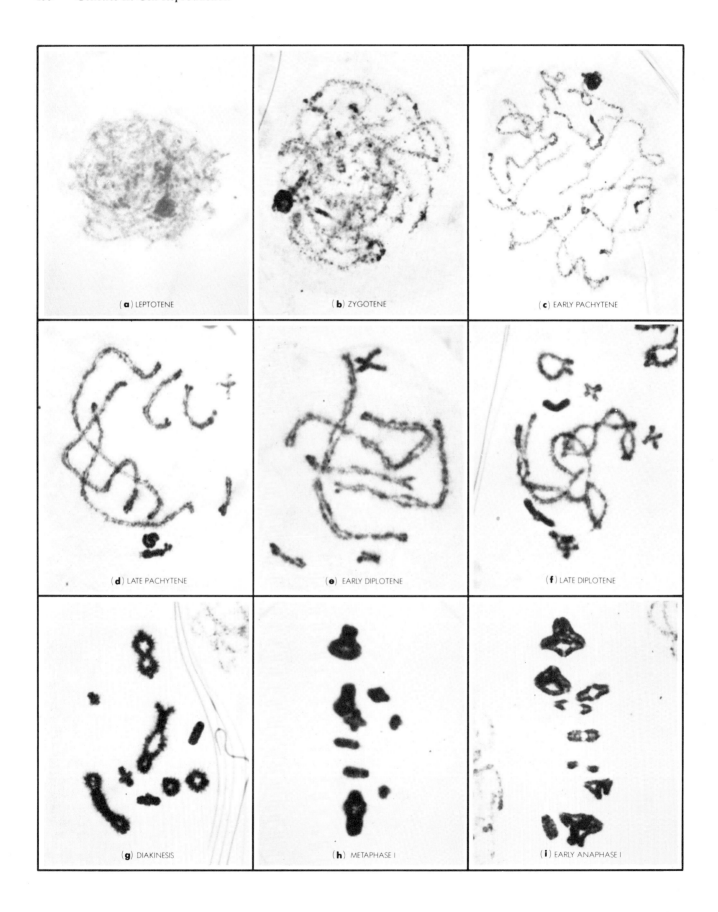

(**a**) LEPTOTENE

(**b**) ZYGOTENE

(**c**) EARLY PACHYTENE

(**d**) LATE PACHYTENE

(**e**) EARLY DIPLOTENE

(**f**) LATE DIPLOTENE

(**g**) DIAKINESIS

(**h**) METAPHASE I

(**i**) EARLY ANAPHASE I

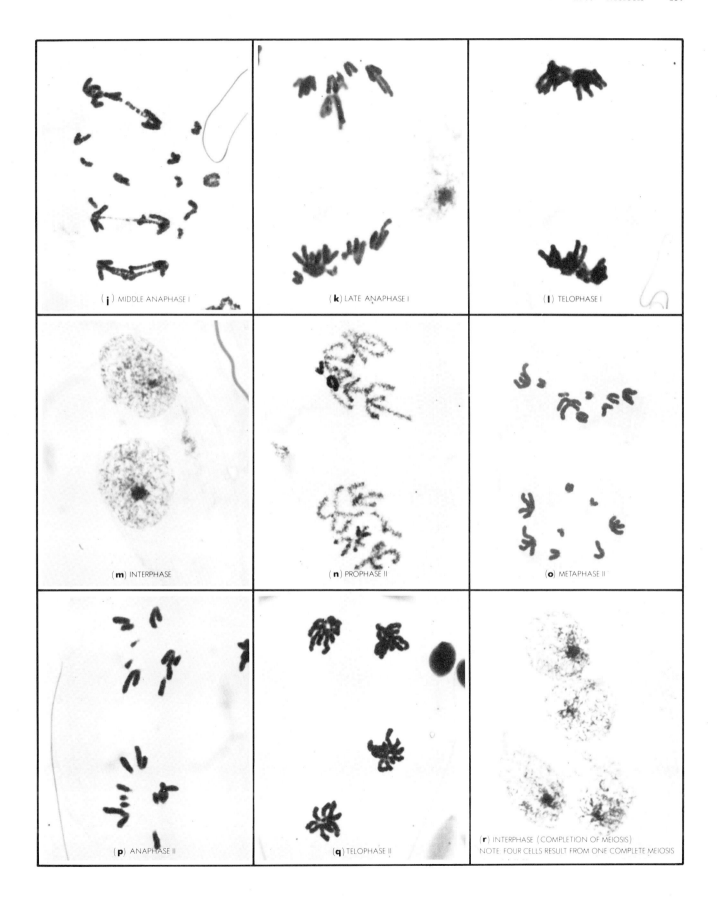

(**j**) MIDDLE ANAPHASE I

(**k**) LATE ANAPHASE I

(**l**) TELOPHASE I

(**m**) INTERPHASE

(**n**) PROPHASE II

(**o**) METAPHASE II

(**p**) ANAPHASE II

(**q**) TELOPHASE II

(**r**) INTERPHASE (COMPLETION OF MEIOSIS)
NOTE: FOUR CELLS RESULT FROM ONE COMPLETE MEIOSIS

◀ **Fig. 17.15**
Light microscope photographs of the various stages of meiosis in the testis of the grasshopper, *Chorthippus parallelus*. The terms for the first five stages shown here—leptotene, zygotene, pachytene, diplotene, and diakenesis—refer to various parts of prophase. As you can see, the whole of prophase involves continual condensation and thickening of the chromosomes. Note that by early diplotene the chromatid strands (in tetrad formation) can be seen clearly twisted around each other (synapsis). Prophase of meiosis is generally much longer in duration than in mitosis; in meiosis prophase may take two to three days, while in mitosis it is usually a few minutes to a few hours for most animals. (Photomicrographs courtesy Professor James L. Walters, University of California, Santa Barbara.)

of meiosis in this series of micrographs. You will then be able to compare the general features of mitosis and meiosis in Table 17.1.

Variations Resulting from Meiosis

Most organisms have many pairs of chromosomes. Given the fact that a meiotically dividing cell is metabolically active, with the chromosomes undergoing considerable shuffling, the fact that mishaps resulting in unequal distribution of chromosomal material occur so rarely is an indication of the highly organized state of cellular processes. Occasionally, however, mishaps do occur. Offspring are sometimes born with deformities that can be traced to a failure of a pair of homologous chromosomes to separate at anaphase. As a result, both members of the pair may migrate to one pole and end up in the same gamete. The other gamete, of course, would receive no member of this chromosome pair. Failure of

homologous chromosomes to segregate at anaphase I of meiosis is known as **nondisjunction**. Nondisjunction of chromosome pair 21 is the cause of Down's syndrome in human children.

Sex chromosomes determine the sex of an organism. In humans, the sex chromosomes consist of an *X* chromosome and a *Y* chromosome. The diploid cells of every woman contain two homologous *X* chromosomes, while those of a man contain one *X* chromosome and one *Y* chromosome. Should nondisjunction occur with the sex chromosomes rather than the other chromosomes (**autosomes**) and the resulting gametes participate in formation of the zygote, the resulting zygote may be triploid with such sex chromosomal combinations as *XXY*, *XYY*, and the like. (For a discussion of some recent questions and controversy about the behavioral effects of nondisjunction in the sex chromosomes, see Supplement 17.3.)

Nondisjunction also occurs in plants. For example, it may occur in each of the twelve pairs of chromosomes in the Jimson weed (*Datura stramonium*). The individual plant is variously affected, depending on which pair of chromosomes is involved. If chromosome pair *A* exists as a triploid, then the plant is small, with abnormally narrow leaves and tiny fruits. If chromosome pair *J* undergoes nondisjunction, the plant has dark puckered leaves. In other plant genera, nondisjunction of chromosomes has resulted in new species with fewer chromosomes than their ancestors. For example, in the false dandelion (*Crepis*), the species *C. fuliginosa* ($n = 3$) has most likely been derived from *C. neglecta* or its ancestor ($n = 4$) as a result of nondisjunction.

Gametogenesis. Gametogenesis refers to the production of gametes in the germ tissue of animals (the similar

Table 17.1
Comparison of Mitosis and Meiosis

Mitosis	Meiosis
1. One division cycle	1. Two division cycles (I and II)
2. Maintains constant chromosome number from parent to daughter cell	2. Reduces chromosome number in half
3. Daughter cells genetically identical	3. Daughter cells of several genetic types; though only four daughter cells are produced at any one time, the number of genetic possibilities is considerable
4. No synapsis or tetrad formation	4. Homologous chromosomes come together and intertwine (synapsis), forming tetrads
5. No crossing over	5. Crossing over occurs (Prophase I)
6. Occurs in growth, regeneration, and embryonic development in sexually reproducing forms; in asexual forms is the basic mode of reproduction	6. Occurs in higher organisms in the production of gametes; in lower forms occurs immediately after formation of the zygote
7. Occurs in less time than meiosis	7. Takes very long time (several days to several weeks)

SUPPLEMENT 17.3
IS THERE A "CRIMINAL CHROMOSOME"?

In 1961 a human male was found showing an XYY chromosome complement, a result of nondisjunction of an X and a Y chromosome (presumably during spermatogenesis in the subject's maternal grandfather). In December 1965, Dr. Patricia Jacobs and her colleagues at Western General Hospital in Edinburgh (Scotland) published a cytogenic study of male inmates in the hospital's security ward. These inmates all had records of what was considered violent criminal behavior. Of the 197 men surveyed, 7 showed the XYY chromosome condition. Later surveys found that the frequency of XYY males in some prisons appeared to be significantly higher than that estimated for the general population.

Despite a report by the original XYY workers published soon after their first report showing that the XYY male prisoners were involved in crimes involving property rather than violence, by the late 1960s the idea of a link between an XYY chromosomal complement and violent behavior had been widely circulated by the media throughout much of Western Europe and the United States. In France and Australia XYY defendants in two murder trials were given light sentences (in one case acquitted) on the grounds that their violent behavior was beyond their control. A report appeared in 1968 that Richard Speck, who killed eight nurses, was XYY. (That report was later shown to be false.) This publicity gradually convinced the public, mistakenly, that biologists, particularly geneticists, had accepted as valid the hypothesis that an extra Y chromosome caused an increased tendency toward violent and criminal behavior. In some circles the extra Y was referred to as the "criminal chromosome."

One of the chief critics of the theory on scientific grounds was Dr. Digamber S. Borgaonkar, of the Johns Hopkins University School of Medicine. Dr. Borgaonkar made an exhaustive study of most of the XYY cases reported, examining both the data and the methods used to obtain the data. His conclusion was that most of the studies were carelessly executed. For instance, he found that the data were often so unreliable as to be virtually meaningless, and that the suggestion of any cause and effect relationship between an extra Y chromosome and criminality was consequently unsupportable.

The criticisms Dr. Borgaonkar directed against the XYY work provide several insights into the problems of data collection and analysis:

1. One of the important assertions of the XYY work is that XYY males are more disposed toward violence than XY males. Dr. Borgaonkar found from analyzing papers reporting on behavior traits of XYY males within penal institutions that at least in these circumstances, XYY males were on the whole more cooperative than their XY counterparts. Thus

the claim that XYY males are more aggressive cannot be considered valid *without specifying the environment involved.*

2. Few physiological or psychological traits have been found that actually distinguish XYY males from other males. Only height appears to be distinctive (XYY males are on the average slightly taller than XY males). Other traits—skeletal structure, electroencephalograms (EEGs), electrocardiograms (EKGs), skin traits, and the like—all seem to be average. Hormonal levels are no different for populations of XYYs and XYs. The I.Q. of XYY males appears to be about the mean for inmates of penal institutions. No significant differences in personality traits distinguish XYYs from XYs. In short, by all significant physical or physiological criteria that might affect behavior, XYY males rate about the same as XY males.

3. Methodologically, the techniques of collecting and analyzing behavorial data about XYY males appeared to Borgaonkar, on the average, to be very unreliable and nonrigorous.

   a) All the studies lacked either a "blind" or "double-blind" procedure. In a blind experiment, an investigator interviewing a subject to determine behavioral and personality traits does not know what is being tested for (that is, that the patient is suspected of displaying violent behavior); the patient may know, however, the purpose of the study. In a double-blind experiment, neither investigator nor subject would know what relationships were being sought. Blind and double-blind procedures help ensure that neither investigators nor subjects will find more of what they are looking for than is actually there.

   b) Virtually none of the studies of XYY males had been conducted with matched control groups against which data on the behavior of the XYY subjects could be compared. That is, in testing the hypothesis that an extra Y chromosome is a significant cause for criminal behavior, it is necessary to eliminate other variables (such as poor socioeconomic status, bad family life, and the like) that may also have profound effects in molding personality. Most of the studies compared the behavior of XYY males to control groups of randomly chosen XY males not matched for social class, family background, or economic status. Thus, several variables are introduced simultaneously. When two or more variables are present it is impossible to say which may be the more important cause for a particular phenomenon.

   c) Researchers placed much reliance for descriptions of violent or criminal behavior on sources such as police records, legal documents, or records from correctional institutions. Not only are these descriptions likely to be highly variable; they are also subjectively biased in very specific ways. For example, there is no standard definition of what constitutes "violent" behavior. Is swearing at a prison authority or police officer evidence of a tendency toward violence? Or, does physical violence have to be involved? Moreover, police and prison administrators are likely to classify as violent any behavior openly disrespectful of or hostile to their own authority. Yet mere resistance to authority is not by itself adequate indication of a propensity to violence and criminality.

d) There was an element of selection involved in the subjects who were investigated in most of the studies. Only a small fraction of violent behavior actually comes to the attention of authorities and is recorded. Much more needs to be known, Dr. Borgaonkar argues, about the kinds of violent behavior displayed by people who do not come to official notice. Is it the same, less, or more than that displayed by those who are actually caught, convicted, and placed in penal institutions to become readily available objects for study? To claim that XYY males are more violent than XY males requires knowing what kind of violence XY males in the general population perpetrate. In fact, Dr. Borgaonkar reports that penal institution records indicate that XYY males have committed no more violent crimes than the XY males in prison with them. In the absence of data about the differences between the kinds of crimes that arouse official attention and those that do not, it could be hypothesized that XYY males are simply more open and honest than XY males and thus get caught more readily. In other words, if there is any genetic basis to the argument at all, it is that perhaps the extra Y chromosome is an "honesty-determining," rather than a "criminality-determining" chromosome!

In concluding his study,* Dr. Borgaonkar points out that some behavioral and developmental conditions (such as Down's syndrome, or "Mongolism") are definitely hereditary in nature. He does not deny the role of heredity in determining, in a general way, some broad patterns of personality development. On the other hand, he emphasizes that behavioral problems once thought to be largely hereditary (certain kinds of epilepsy or abnormalities in EEG) have been strongly linked to social class and family stability. He cautions against assuming a genetic cause for something as specific as behavior, when the obvious environmental influences that can affect such behavior have been largely ignored. He writes:

Inadequate understanding of the phenomena and the premature conclusions about the XYY phenotype, which have

---

*From D. S. Borgaonkar and Saleem A. Shah, "The XYY chromosome male—or syndrome?" *Progress in Medical Genetics X*, 10 (1974), pp. 135–222.

been reported with distressing frequency, have produced remarkably simplistic views of the interactions between XYY genotypes and the almost infinitely varied environments with which they interact. We should always keep in mind that even the demonstration of a genetic contribution to poor impulse control warrants only the conclusion that in certain environments some persons with particular genotypes will respond by developing certain behavioral problems more frequently than others. However, this does not preclude the possibility that in some other environments persons with the very same genotypes (i.e., XYY) may well manifest socially adaptive behaviors.

The XYY case raised other questions about research on human subjects, especially where negative aspects of an individual's makeup are the main focus. The issue came to light dramatically in 1974 and 1975 at the Harvard University Medical School in Boston. A large project screening for XYY babies had been in progress at Boston Hospital for Women (formerly Boston Lying-In Hospital) from 1968 to 1975. The researchers heading the project, psychiatrist Stanley Walzer and geneticist Park Gerald, both from the Harvard Medical School, wanted to identify XYY males born in the hospital and follow their personality and behavioral development through adulthood. Walzer and Gerald explained that the purpose of their study was to identify XYY genotypes early in a child's life, so that psychiatric counseling could be provided to help overcome personality problems if and when they arose. Their research was funded by the Crime and Delinquency Division of the National Institutes of Mental Health.

Dr. Jonathan Beckwith, also of Harvard Medical School, and Dr. Jonathan King of Massachusetts Institute of Technology mounted an extensive campaign, beginning in 1974, to have the XYY project closed down. They argued that parents participating in the project (that is, who had agreed to have their children studied) were not adequately informed of what the project was about. They claimed that parents did not understand the stigma that might be attached to their child, even if he appeared perfectly normal in his behavior, if he were known as an XYY type. (A well-known American geneticist appalled his colleagues at a professional meeting a few years ago by claiming that he "wouldn't invite an XYY home to dinner.")

Given this misinformation the public had received about "criminal chromosomes," Beckwith claimed that Walzer and Gerald did not take adequate steps to inform the parents of how participation in the project might affect their child's future. Moreover, Beckwith and King argued that given what most people now think about the XYY genotype, the hypothesized relationship between an extra Y chromosome and criminal behavior could become a self-fulfilling prophecy. In other words, parents who know their children are XYY will treat them differently, perhaps pushing them almost unconsciously toward violent behavior. However, the medical school's Human Studies Committee found that Walzer and

Gerald's work did comply with their requirements that (1) informed consent be properly obtained, (2) the patient's rights be protected, and (3) the benefits of participating in the study outweigh the risks.

Beckwith and King did not agree. They argued that the committee was composed of established doctors who had a stake in protecting themselves and their colleagues from challenges to the fundamental nature of the research. Beckwith and King aimed their criticism of the Boston project largely at the moral and ethical implications of the *XYY* research. Beckwith and King argued that the patient's rights are considerably more in jeopardy when the conclusions from supposedly scientific work are erroneous, as they maintained was the case with the current *XYY* screening project. To be stigmatized for life is bad under any circumstances, they stated; to be stigmatized erroneously is worse.

As a result of the pressure brought to bear on the project by Beckwith, King, and others, the *XYY* project in Boston was discontinued in the spring of 1975. Some people thought Walzer stopped screening *XYY*s because he finally realized that the risks outweighed the benefits. Walzer denied this. "I hope no one thinks I don't still believe in my research," he declared. "I do. But this whole thing has been a terrible strain. My family has been threatened. I've been made to feel like a dirty person. I was just too emotionally tired to go on." Walzer agrees that talk of a "criminal chromosome" is nonsense, but he still thinks there is enough evidence of certain learning difficulties in *XYY* children to justify an early identification leading to corrective therapy. King and Beckwith, on the other hand, claim that the potential harm to individual people is far greater than the potential good, and that under such conditions stopping the screening project is justifiable.

process in plants is called sporogenesis). In the male, gametogenesis is called **spermatogenesis** and in the female, **oögenesis**. Figure 17.16 shows a general comparison between the cellular and chromosomal events in spermatogenesis and oögenesis.

Spermatogenesis. The starting male cell, the "grandparental" cell of the future sperm, is called a **spermatogonium**. One spermatogonium undergoes mitosis to produce two daughter cells, a primary spermatocyte and a cell that remains a spermatogonium. The primary

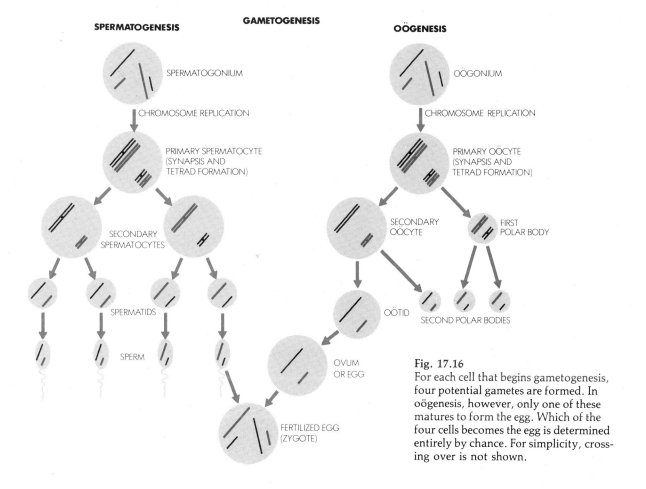

Fig. 17.16
For each cell that begins gametogenesis, four potential gametes are formed. In oögenesis, however, only one of these matures to form the egg. Which of the four cells becomes the egg is determined entirely by chance. For simplicity, crossing over is not shown.

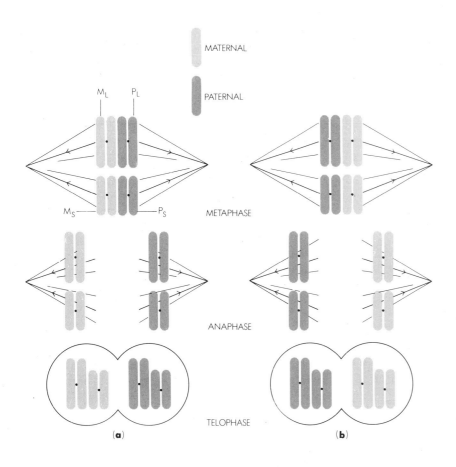

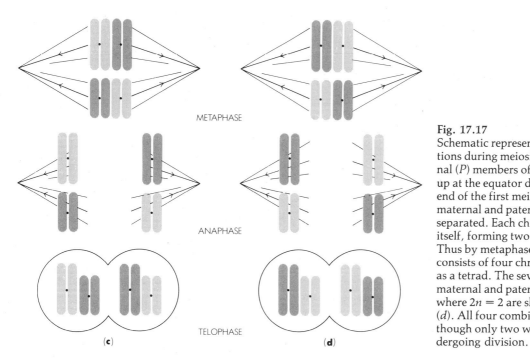

Fig. 17.17
Schematic representation of chromosome positions during meiosis. Maternal (*M*) and paternal (*P*) members of each homologous pair line up at the equator during metaphase I. By the end of the first meiotic division (division I) maternal and paternal homologs have been separated. Each chromosome has duplicated itself, forming two chromatids, by prophase I. Thus by metaphase I, each homologous pair consists of *four* chromatids collectively known as a tetrad. The several combinations of maternal and paternal chromatids for a cell where $2n = 2$ are shown in (*a*), (*b*), (*c*), and (*d*). All four combinations are possible, although only two will occur for any one cell undergoing division.

In the first anaphase of meiosis, the direction of separation of maternal and paternal homologous chromosomes of any one pair is totally independent of the direction of separation of those in any other pair.

spermatocyte undergoes the first meiotic division into two secondary spermatocytes. The secondary spermatocytes undergo the second meiotic division to form four equal-sized spermatids, each with a haploid nucleus. The spermatid then develops without further cell divisions into a mature sperm.

Oögenesis. In all the basic meiotic events oögenesis is identical to spermatogenesis. The chief difference is that only a single egg is produced from a primary oöcyte as compared to four sperm from a single spermatocyte. In Fig. 17.16 we can also see that polar bodies are formed during oögenesis. After the first meiotic division, one of the daughter cells, the first polar body, is smaller. The larger daughter cell, the secondary oöcyte, contains more cytoplasm; both cells, however, contain the haploid number of chromosomes. Each cell, the secondary oöcyte and the first polar body, then undergoes the second meiotic division. Each daughter cell from the division of the first polar body contains very little cytoplasm and hence remains small. Division of the secondary oöcyte produces one large cell, the oötid, containing most of the cytoplasm, and a small cell, the second polar body. The three polar bodies along with their haploid sets of chromosomes are ultimately discarded during oögenesis. A single haploid set is preserved within the oötid, which matures into an ovum, or egg cell.

Polar body formation during oögenesis insures that the remaining egg receives a large amount of cytoplasm. Even in animals such as the human being, where nourishment of the embryo is supplied early in life by the mother, a rich supply of cytoplasm in the egg is necessary until the embryo becomes firmly attached to the uterine wall.

Meiosis and Probability

At metaphase of meiosis I all the chromosome pairs are lined up along the cell equator prior to separation. The segregation of maternal and paternal chromosomes from each pair produces combinations that follow the laws of probability. For simplicity's sake, let's consider a species that has only two pairs of chromosomes, and no crossing over (diagrammed in Fig. 17.17). Each homolog carries only maternal (if it originated from the female parent) or paternal (if it originated from the male) genetic factors. In Fig. 17.17(a), at metaphase, the maternal chromosomes of each pair are arbitrarily placed on the left, and the paternal on the right. However, the paternal chromosomes could also be on the left and the maternal on the right (Fig. 17.17b). In fact, two other combinations are also possible (see Fig. 17.17c and d). This is an important observation: the direction of migration taken by one member of any chromosome pair in no way influences the direction of migration taken by a member of any other chromosome pair. In other words, chromosomes follow Mendel's second law, that of random assortment. The only requirement at anaphase I is that the two members of the original homologous pair (each now doubled) separate from one another and go to opposite poles. Short of aberrations in the distribution process, such as nondisjunction, a gamete can get a set of chromosomes that includes every possibility, from all maternal or all paternal chromosomes to any combination in between.

The number of combinations of maternal and paternal chromosomes in meiosis is a function of the number of chromosome pairs in the species. The number of possible maternal-paternal combinations can be expressed as 2^n, where n refers to the number of chromosome pairs. For example, in organisms where the diploid number is 4, the total number of possible combinations is 2^4, or 16. In organisms where the diploid number is 8, the number of combinations is 2^8, or 512. For human beings, with 23 pairs, the number of possible combinations is 2^{23} (figure it out for yourself!). Thus, meiosis results in enormous variability in the assortment of maternal and paternal chromosomes. Add to this the new combinations of traits that are made possible by crossing over, and it is easy to see why it is extremely unlikely that any two offspring of the same parents in a sexually reproducing organism will ever be alike in all their combinations of traits.

Summary

1. Individual cells reproduce by cell division. Cell division consists of several processes: (a) mitosis, the replication and division of the chromosomes among daughter cells; and (b) cytokinesis, the splitting of the cell into two daughter cells.

2. Although it is convenient to describe it in terms of a number of specific, discrete stages, the entire process of cell division is a dynamic and active one. The stages are:

a) Prophase: The duplicated chromatin strands begin to condense into chromosomes.

b) Metaphase: The chromosomes assemble at the equatorial plate, each chromosome attached by its centromere to a spindle fiber, in turn connected to the centrioles at either end of the cell. (The cells of higher plants lack centrioles, yet mitosis occurs just the same. In this case the spindle fibers appear to be attached to some other structure at the polar region of the cell.)

c) Anaphase: The spindle fibers appear to be the vehicles for pulling the chromosomes away from the equator toward the poles. This separation occurs in such a way that each daughter cell gets a duplicate set (the diploid number) of chromosomes. The daughter cells are now genetically identical to each other and to the parent from which they are derived.

d) Telophase: The chromosomes group around the poles of the cell, and the nuclear membrane begins to form.

3. Following telophase the process of cytokinesis goes to completion. The old cell, now with two identical nuclei arranged at opposite ends, begins to pinch in at the equator (or, in plants, build a cell plate and new membrane), creating two new cells.

4. Among the unanswered questions about mitosis are: (a) What initiates the process of cell division in the first place? (b) What forces are responsible for moving the chromosomes about during early metaphase, and especially at anaphase? (c) What controls the rate or cessation of mitosis? What mechanism is responsible for maintenance of continual mitosis in some tissues (bone marrow, gamete-producing tissues, epithelium, and the like), and for the lack of mitosis in others (nerve cells, muscle cells)?

5. Meiosis is similar in many ways to mitosis. However, it is a reduction not an equational division, producing daughter cells with one-half the number of chromosomes as the original parental cell. Meiosis converts a diploid parental cell into haploid daughter cells.

6. Meiosis occurs in two series of divisions:

a) Prophase I: Duplicated chromatin condenses into rod-shaped chromosomes; each chromosome is duplicated, the two duplicate bodies being held together at the kinetochore. (Each duplicate member of such a pair is called a chromatid. Two duplicated chromatids, bound together, make up a chromosome.) Homologous chromosomes now come together and undergo synapsis. Since four chromatids are involved, the group is called a tetrad. Crossing over (chiasma) can also occur between homologous strands of the tetrad. In chiasma, equivalent sections of homologous strands may interchange.

b) Metaphase I: Tetrads line up at the equatorial plate; centromeres are attached by means of spindle fibers to the polar regions, often to the centriole.

c) Anaphase I: Homologous chromosomes separate from one another, each chromosome still consisting of two chromatids. There is no regularity to the separation of maternal and paternal chromosomes. Each pair of homologous chromosomes separates from the equator without any regard to the direction in which other pairs are separating. Thus the newly formed nucleus might contain all maternal, all paternal, or any possible combination of chromosomes. Normally, however, just as in mitosis, the two members of a homologous pair go to opposite ends of the dividing cell.

d) Telophase I: The new nuclei are surrounded by a nuclear membrane, and cytokinesis proceeds to completion.

e) Prophase, metaphase, anaphase, and telophase II: The duplicated chromatid strands are now separated from one another and go into separate cells.

7. Meiosis consists of one chromosome duplication and two cell divisions. The net result is four haploid cells. In spermatogenesis, four sperm cells are produced; in oögenesis only one haploid product actually becomes a functional egg.

8. Two observations are of primary importance with respect to the role meiosis plays in the process of heredity:

a) During synapsis and tetrad formation, homologous chromosome pairs may break and exchange corresponding segments with one another. This creates the possibility of new genetic combinations.

b) During separation at anaphase, the direction of migration toward a pole taken by a member of one homologous chromosome pair in no way affects the direction taken by a member of another pair. During metaphase and anaphase I, then, the chromosomes assort at random.

9. Meiosis occurs in the process of gametogenesis. It is the principal process by which haploid eggs and sperm are produced from the diploid germ cells, spermatogonia and oögonia.

10. Replication is a property of some cell organelles as well as of whole cells. Mitochondria and chloroplasts both have the ability to replicate; their replication process is governed by DNA contained within the organelles. The details of this process are less clearly understood than is mitosis in whole cells.

Exercises

1. Identify the stages in mitosis in the following photographs.

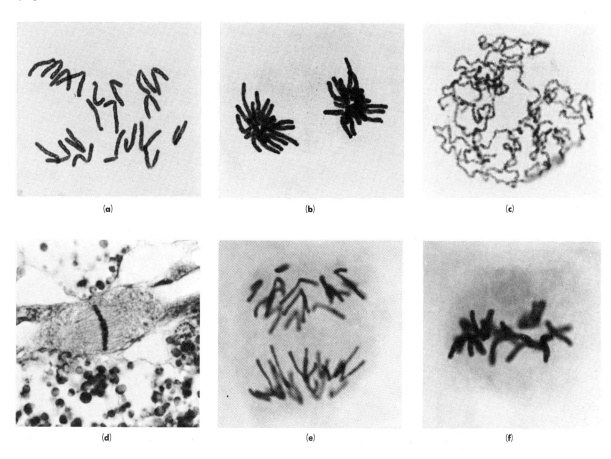

(a)   (b)   (c)

(d)   (e)   (f)

2. Why is duplication of the chromosomes an indispensable step in mitosis?

3. It was once hypothesized that the spindle fibers were not material structures, but merely lines of stress appearing in the cell during mitosis. Suggest an experiment to test this hypothesis.

4. What experimental evidence suggests DNA replication occurs during interphase?

5. Explain why meiosis is a necessary part of the life cycle of any sexually reproducing organism.

Chapter 18
Genetics III: Genes
and
Chromosomes

INTRODUCTION

As we recall from Chapter 16, Mendel proposed a purely abstract model based on segregating and randomly assorting factors to account for the phenotypic ratios observed in his pea plants. And, as we learned in Chapter 17, at about the same time cytologists began investigating the structure and behavior of chromosomes—especially the complex movements associated with mitosis and meiosis. After Mendel's work was rediscovered in 1900, the behavior of chromosomes during meiosis in particular was seen to parallel the behavior of Mendel's hypothesized factors. Correlating Mendel's scheme with the observations of cytology became one of the challenging problems of the early twentieth century. Attempts to make that correlation, however, raised some important questions.

First, it was not clear what role chromosomes played in heredity in higher organisms. Most biologists by 1900 believed that chromosomes influenced hereditary traits but agreement stopped there. For instance, there was considerable debate on the question of whether all the chromosomes in a given species were the same, or different. The work of cytologists such as Theodor Boveri (1862–1915) revealed an answer: each pair of chromosomes in a given cell is qualitatively different from all other pairs in the hereditary traits it affects (see Supplement 18.1).

Second, the close parallel between the behavior of Mendel's factors (genes) and the behavior of the chromosomes during meiosis might lead to a tentative hypothesis that a gene and a chromosome are the same thing. If this is true, then chromosomes should demonstrate segregation during gametogenesis; furthermore, members of separate pairs of homologous chromosomes should show random and independent assortment into the prospective gametes. Both of these predictions are verified.

Other observations, however, seemed inconsistent with this hypothesis. Most obvious is the chromosome number. A red fox, for example, has only 34 chromosomes (17 pairs). It is difficult to imagine that all its inherited characteristics could be controlled by only 17 pairs of genes. Furthermore, its close relative the arctic fox has 52 chromosomes. Such a wide variation in the number of "chromosome-genes" would not be predicted between these closely related forms. Finally, there seems to be no consistent principle underlying the variation in chromosome number between organisms of widely differing evolutionary status; this fact becomes extremely difficult to explain if we are to accept the hypothesis of oneness between chromosome and gene. For example, it is difficult to see why a one-celled radiolarian should have 1600 "chromosome-genes," while a crayfish has 200, and a human only 46.

In this chapter we will explore both the methods used by geneticists to resolve such problems and the ideas that we hold today about the relationships between genes and chromosomes. Most important, we will investigate how biologists now view the relationship between the arrangement of genes on chromosomes and how genes function to determine adult traits.

18.2
MENDEL AND CHROMOSOMES: THE DISCOVERY OF LINKAGE

There is a simple hypothesis to account for the dilemma posed by assuming that every chromosome represents only a single gene. The hypothesis that a chromosome

---

Mendel never observed chromosomes, nor did he know anything about the events of mitosis and meiosis. His hypothesis was based solely on observed phenotypic ratios among the offspring from specific breeding experiments.

SUPPLEMENT 18.1
THE INDIVIDUALITY OF THE CHROMOSOMES

In the middle and latter parts of the nineteenth century, cytologists were convinced that all the chromosomes in the cell nucleus were identical, that is, carried identical hereditary information. Given the state of microscope technology at the time this is not an unreasonable conclusion. In the sea urchin, for example, there are 36 chromosomes (18 pairs); they are small and look physically very much alike under the average microscope available in the latter part of the nineteenth century. The fact that the chromosomes looked alike tended to suggest that they must be alike.

Initially, the German cytologist and embryologist Theodor Boveri (1862–1915) agreed with his contemporaries. But his embryological researches with the early stages of sea urchin development convinced him otherwise. In 1902 he published the results of a highly imaginative set of experiments which suggested that the chromosomes in the nucleus of a cell were qualitatively different from one another in their hereditary makeup.

Investigators before Boveri had observed that if they put a sea urchin egg into a highly concentrated suspension of sperm, frequently two sperm would enter one egg (normally, of course, only one sperm fertilizes an egg). In these doubly fertilized eggs, there are three complete sets of chromosomes, one set from the egg and one set from *each* sperm. This causes complications with the mitotic apparatus as the zygote begins to divide, yielding abnormal distribution of chromosomes to the daughter cells. To understand the nature and the meaning of these abnormalities, we will review briefly the sequence of events during and immediately following normal fertilization.

In normal fertilization, the sperm brings into the egg with it not only a nucleus containing one full set (the haploid number) of chromosomes, but also a centriole. The centriole contributed by the sperm to the egg normally duplicates shortly after fertilization, forming two centrioles, which migrate to opposite poles of the egg cell. The spindle apparatus forms between these two poles. After the maternal and paternal nuclei fuse, the chromosomes duplicate themselves, move onto the spindle, and go through a normal mitotic process. The daughter cells resulting from this division will contain, as in any mitotic division, an equivalent number (normal diploid number) and kind of chromosomes. Embryonic cell divisions are called *cleavages*; thus, what we just described above is the first cleavage in the sea urchin embryo, yielding two cells. The second cleavage would normally yield four cells, the third eight, the fourth sixteen, and so on. The daughter cells from these early cleavages are called *blastomeres.* In normal cleavage, the chromosome number remains constant (the diploid number) in all blastomeres. For the sea urchin this would mean 36 (18 pairs).

When two sperm enter an egg cell, however, the mitotic apparatus is thrown off. Boveri observed that in the case of any one doubly fertilized egg, one of two possible abnormalities can occur:

1. Each sperm's centriole duplicates, producing a total of four centrioles, and four spindle poles; these cells produce *four* blastomeres at first cleavage.

2. Only one sperm's centriole duplicates, yielding three centrioles and three spindle poles; these cells produce three blastomeres at first cleavage.

In both cases, the doubly fertilized egg starts out with a total of 54 chromosomes (18 from the egg, and 18 from each sperm). These 54 chromosomes duplicate themselves during the first mitotic division, producing a total of 108. Because of the existence of either three or four centrioles, and three or four poles, respectively, the distribution of chromosomes becomes irregular. The blastomeres resulting from doubly fertilized eggs usually end up with different, abnormal numbers of chromosomes—that is, numbers and combinations different from the expected full complement of 36 per cell. The possibility of such confusion is clearly shown by the physical arrangement of the chromosomes on the interlocking spindles for both three- and four-blastomere forming embryos (see especially Boveri's original figures, shown in Diagram A, parts c and d).

To understand how this occurs, consider first the case in which both sperm centrioles duplicate, producing a four-celled embryo at first cleavage. With a total of 108 chromosomes, a four-way division would yield, on the average, 27 chromosomes per cell (108/4 = 27), far from the normal complement of 36. What Boveri observed was that each of the four blastomeres resulting from the first cleavage of such a cell possessed quite different numbers of chromosomes, some receiving only a few, others a large number. Now, Boveri found that he could place these embryos into calcium-free seawater and separate the four blastomeres from one another. Each blastomere could then develop on its own, sometimes becoming a full-fledged, or at least partially developed, larva (the sea urchin larva is known as pluteus). Boveri noted that whenever a separated blastomere contained an abnormal number of chromosomes (either too many or especially too few) it developed into an abnormal pluteus. Moreover, abnormal development appeared to be related not merely to abnormal *number* of chromosomes, but to the actual chromosomes present. Critical in this regard was the fact that the most striking abnormalities occurred when a particular blastomere lacked any copy of one particular chromosome. Moreover, when the same

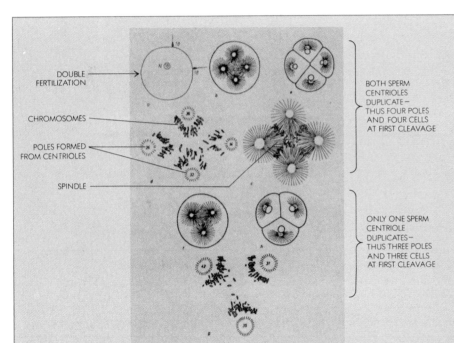

DOUBLE FERTILIZATION

CHROMOSOMES

POLES FORMED FROM CENTRIOLES

SPINDLE

BOTH SPERM CENTRIOLES DUPLICATE— THUS FOUR POLES AND FOUR CELLS AT FIRST CLEAVAGE

ONLY ONE SPERM CENTRIOLE DUPLICATES— THUS THREE POLES AND THREE CELLS AT FIRST CLEAVAGE

Diagram A
From Fritz Baltzer, *Archiv für Zellforschung* 2 (1909), and *Archiv für Entwicklungsmechanik* (1911), schematized.

chromosome was missing from two different blastomeres, the pluteus developed the same sorts of abnormalities.

Boveri's results were remarkably consistent. Out of 1500 embryos raised from four-celled blastomeres, 1499 developed abnormally. Boveri concluded from these results that the abnormal development was directly caused by the lack of certain specific chromosomes in each blastomere, thus indicating that each chromosome contained different hereditary determinants from the others.

But what about the one blastomere out of 1500 that developed normally? Boveri noted that this blastomere contained the normal complement of 36 chromosomes (in our modern terminology it must have contained at least one member of each homologous pair). In terms of probability, 1 out of 1500 normal distributions is about what one would expect in a process which divided 108 chromosomes in four ways.

Boveri's observations of the doubly fertilized eggs that divided only *three ways* (as opposed to four ways) at first cleavage produced similar results. These embryos also showed very abnormal distributions of chromosomes. Now, Boveri made a prediction. In terms of probability, there would be a better chance of getting at least one copy of each type of chromosome in each blastomere when three complete sets of chromosomes are divided three ways than when three sets of chromosomes are divided four ways. Boveri's observational results confirmed this expectation. Out of a total of 719 eggs that divided into three cells at first cleavage, a total of 58 developed normally, and were found to have at least one copy of each of the 36 chromosome types.

Boveri concluded from his experiments that for normal development to occur, every cell in a developing embryo must have the regular complement of 36 chromosomes. Normal development could occur if there were extra chromosomes beyond the 36, but unless they had at least 36, and this meant at least one of each of the 36 types, the embryo would develop abnormally. To Boveri this meant clearly that each chromosome in the regular complement must be qualitatively different in terms of the hereditary determinants it carried. Abnormal development was directly related to the absence of certain chromosomes, and hence the hereditary information they contributed. As Boveri wrote in his 1902 paper:

> . . . [it is] not a definite number but a definite combination of chromosomes [that] is necessary for normal development, and this means nothing other than that the individual chromosomes must possess different qualities.

Although Boveri was correct in an overall sense, there were two problems with his work which were confusing to his contemporaries. First, he did not make a clear distinction between chromosomes and chromosome pairs. That is, he continually discussed abnormality of distribution in terms of the total of 36 chromosomes, not in terms of 18 pairs. Thus, it is not clear whether he thought that the two members of a homologous pair were identical or different. Furthermore, Boveri carried out his double fertilization experiments using three different species; especially in the smaller experimental groups, which formed only three cells at first cleavage, the greater percentage of normal larvae could possibly be ascribed to species rather than to chromosomal differences.

Despite these problems, however, Boveri's work laid the groundwork for our later, more complete understanding of what has been called the doctrine of "the individuality of the chromosomes." It is on this basis that much of genetics in the first three decades of the century is based.

and a gene are one and the same thing can be modified to *a chromosome represents several genes and these genes are located in a linear order along the chromosome's length.*

The first portion of this hypothesis, that a chromosome represents several genes, nicely overcomes the objections just raised to the one-chromosome–one-gene hypothesis. The second portion, dealing with location of genes on the chromosomes, is quite another matter. Such a hypothesis necessarily leads to certain predictions: if there are several genes located on a chromosome, then certain characteristics should tend to be inherited together. The reason for this prediction becomes apparent if the chromosome is visualized as a string of beads, each bead representing one gene. Since these genes are joined (or **linked**) together, this hypothetical model predicts that *wherever one gene on a chromosome goes, so must all the other genes on that chromosome.*

Are there any cases that have been definitely established in which one genetic characteristic always appears with another? It is easy to feel intuitively that there are such cases. In humans, for example, we generally associate the occurrence of freckles with sandy or reddish hair. It might be proposed, therefore, that the genes influencing freckles and those influencing red hair are linked—that they are on the same chromosome.* However, humans have many chromosomes, and cannot be bred experimentally, so such a linkage is difficult to establish.

It is necessary, therefore, to turn to an organism that has fewer chromosomes and adapts more easily to controlled breeding. The tomato plant has 12 chromosome pairs. In tomatoes, tall growth habit is the result of a dominant gene *D*; dwarf growth habit is the result of a recessive allele *d*. Smooth epidermis is due to a dominant gene *P*; pubescent (hairy) epidermis is due to a recessive allele *p*. As would be predicted, crosses between homozygous tall smooth plants (*DDPP*) and dwarf pubescent ones (*ddpp*) yield an F_1 of all tall smooth plants (genotype *DdPp*). Thus far, then, Mendel's first and second laws regarding segregation and random assortment are supported.

Assume now that a test-cross is made between a tall smooth F_1 tomato plant and a dwarf pubescent tomato plant. The F_1, a tall smooth plant with genotype *DdPp*, should produce four types of gametes, *DP*, *Dp*, *dP*, and *dp*, in equal frequencies. The dwarf pubescent plant could produce only one type of gamete, *dp*. The fertilization possibilities are thus:

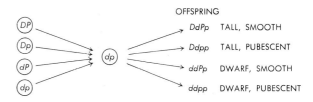

yielding a ratio of one tall smooth, to one tall pubescent, to one dwarf smooth, to one dwarf pubescent.

Such a cross has been made, yielding 112 plants. Of these, 54 were tall smooth plants and 58 were dwarf pubescent plants, a ratio of approximately 1:1. Note that these are parental combinations of the traits.

Clearly Mendel's second law is contradicted by these results; random assortment of the factors for height and skin texture cannot have occurred. Instead, wherever the factor for tallness went, the smooth-skin factor must have followed. Likewise, wherever the factor for dwarfness went, the pubescent factor went also. Such a result, and the 1:1 ratio of tall smooth to dwarf pubescent plants, would be predicted by a hypothesis that the genes for tallness and smooth skin were located on the same chromosome—that they were linked. According to this hypothesis, the genes for dwarfness and pubescent skin must also be linked.

The experimental results just cited are matched by similar results obtained with certain inherited traits in other plants and animals. Such results strongly support the hypothesis of gene linkage and force us to impose a strong qualification on Mendel's second law: *Genes assort at random if and only if they are located on separate, nonhomologous chromosomes.* Thus in many organisms, particularly those with few chromosomes, gene linkage is the rule rather than the exception.

Of course, it may be that the genes for tallness and smooth skin are not only on the same chromosome but are actually two different expressions of the same gene. The same might be true for dwarfness and pubescent skin. As we will see in the next section, however, these hypotheses yield predictions that are sharply contradicted.

---

*It is also possible, of course, that freckles and red hair may have the same underlying genetic cause, in which case they would be an example of pleiomorphism.

When certain traits appear to be inherited together a large part of the time, they are said to be linked. Linkage provides an exception to Mendel's second law (the law of independent assortment).

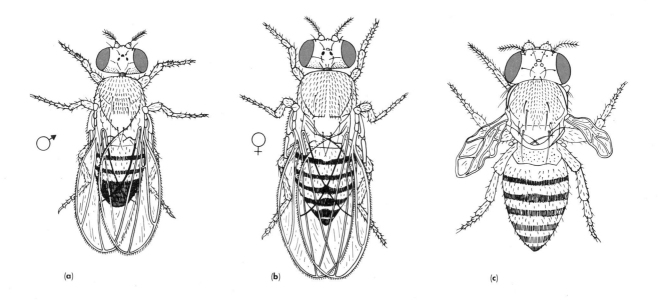

(a)   (b)   (c)

Fig. 18.1
The fruit fly, *Drosophila melanogaster*. Normal (wild-type) male (*a*) and female (*b*) show the wild-type characteristics of straight wings and red eyes. A vestigial-wing mutant is shown in (*c*); the much shorter wings are nonfunctional. The vestigial mutation is recessive to wild-type normal wings and is located on the second chromosome pair.

18.3
THE FRUIT FLY ERA

In the period between 1910 and 1940, geneticists turned to the fruit fly, *Drosophila melanogaster*, for an intensive study of the chromosomal basis of inheritance. This small insect has a short life cycle (approximately two weeks) and is easily raised in the laboratory. *Drosophila* also has a low chromosome number (four pairs), with each pair easily distinguishable from the others. Finally, and most important, *Drosophila* shows hundreds of inherited variant characteristics.

One such variant characteristic is vestigial wings. Flies with this characteristic have only stumps where their wings should be (Fig. 18.1). The condition is caused by the presence of a single pair of recessive genes, for crosses between vestigials yield only vestigials, while crosses of purebred winged flies with vestigials yield all winged. The F_1 intercross yields the familiar 3:1 ratio of dominant to recessive phenotypes.

A major center for fruit fly research was the Columbia University Laboratory of Thomas Hunt Morgan (1866–1945) and his associates. Their work, genetically speaking, made *Drosophila melanogaster* the most thoroughly understood multicellular organism in the world. A part of Morgan's work focused on the inheritance of

eye color in *Drosophila*. The regular, or "wild-type" insect has red eyes. One day a white-eyed male fly appeared in one culture bottle. It was crossed with a red-eyed female and the resulting progeny were all red-eyed. The F_1 intercross yielded a ratio of three red-eyed flies to one white-eyed fly. Morgan noticed, however, that *all the white-eyed flies were male.*

Sex-Linkage

Since the location of the genes on the chromosomes had already been hypothesized, it was reasonable to look for differences in the chromosomes of male and female fruit flies to explain the connection of the white-eyed condition to maleness. Morgan was already familiar with such a difference. In the male fruit fly, one pair of chromosomes is markedly different from the other three pairs. This pair consists of one normal-appearing chromosome, the *X* chromosome, and one short, bent chromosome, the *Y* chromosome. In the female there are two matching *X* chromosomes.*

Morgan saw immediately that the occurrence of white-eyed males in the F_2 generation could be explained by postulating the white-eye gene to be recessive and located on the *X* chromosome. The *Y* chromosome, being shorter, has no homologous position, or what geneticists call a *locus*, for the eye-color gene. Thus the mere occurrence of the white-eye gene would be enough

---

*At the time of Morgan's work on the white-eye condition, the *Y* chromosome had not been detected in *Drosophila*. Morgan therefore assumed the *X* chromosome had no homolog in the male. In terms of the hypothesis he proposed, the presence or absence of a *Y* chromosome makes little difference, so we shall deal with his hypothesis as if he had been familiar with the *Y* chromosome in this insect.

The position or place that a particular gene occupies on a chromosome is called a locus. Genes occupying the same locus on homologous chromosomes govern the same trait.

to cause white eyes in the male, since no dominant gene for red eyes would be present to override its effect:

(The prime ['] symbol on the X indicates the presence of the recessive white-eye gene.) Morgan called this condition **sex-linkage,** or **sex-linked inheritance,** to indicate a condition determined by genes on the X chromosome. Today, it is more precise to reserve the term "sex-linked" for any condition located on either the X or the Y chromosome, and to use the more specific terms "X-linked" or "Y-linked" for specific X- or Y-related conditions, respectively. The reason for this is obvious: the transmission pattern of a Y-linked trait would differ completely from that of an X-linked trait. For reasons to be explained shortly, most cases of sex-linkage are X-linkage rather than Y-linkage.

In the female, with two X chromosomes, white eyes would occur only rarely. Even if the white-eye gene were present on one chromosome, it would usually be masked by the dominant red-eye allele on the other, and this female would still be red-eyed:

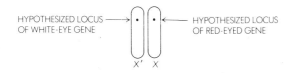

Only occasionally, when a female contained the white-eye mutation on *both* X chromosomes, would it be visible. Morgan thus pictured the first cross of the white-eyed male with the red-eyed female as follows:

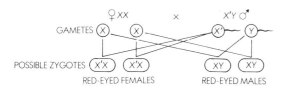

Crossing the F_1 flies resulted in a ratio of three red-eyed flies to one white-eyed fly, with white-eye appearing only in the male flies.

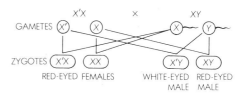

Thus Morgan's hypothesis, placing the locus of the recessive white-eye gene on the X chromosome, nicely accounted for his experimental results. But it also predicted the possibility of obtaining white-eyed females. If the locus of the recessive white-eye gene is on the X chromosome, then crosses between white-eyed males ($X'Y$) and red-eyed females whose fathers were white-eyed ($X'X$) should produce a 1:1 ratio of red-eyed to white-eyed flies. Of the white-eyed flies, half should be females. In terms of the Punnett square, it can be represented:

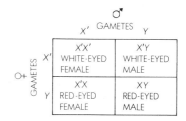

The prediction is verified: one-fourth of the total offspring turned out to be white-eyed females. Thus, the sex-linkage hypothesis is supported.

Sex-Linkage in Human Beings

Sex-linkage has been determined for many traits in human beings, as well as for the fruit fly, most commonly red-green color-blindness and hemophilia. A pedigree chart for hemophilia A, as it appeared in the descendants of Queen Victoria, is shown in Fig. 18.2. Note that the appearance of hemophilia is restricted to males, though a number of females have been identified as carriers. This pattern is typical of all recessive, X-linked traits. The genetic patterns leading to such a pedigree are shown schematically in Fig. 18.3. The pattern is the same for other X-linked traits such as red-green color-blindness. If a woman (such as Queen Victoria) heterozygous for hemophilia A marries a normal man

Fig. 18.2
Pedigree chart showing the recurrence of hemophilia A in European royalty descended from Queen Victoria. Note several important features of this chart: (1) Phenotypic hemophiliacs appear to be limited exclusively to males; (2) Most of the descendants of either sex are *not* hemophiliacs; (3) Hemophilic sons appear to inherit the condition from their mothers; and (4) as the chart shows, it is the mothers, in all cases, who are the direct descendants of Victoria. The mutation for hemophilia, by the way, probably originated with Victoria, since it had not appeared in any of her ancestry. People suffering from hemophilia A cannot synthesize a blood protein known as antihemophilic globulin (AHG), required for the formation of thromboplastin, and thus for normal blood clotting. Numbers within circles or squares indicate more than one offspring of that sex; thus ④ means four daughters. (From John W. Kimball, *Biology*, 4th ed. Reading, Mass.: Addison-Wesley, 1978.)

(for example, Albert), one-half of the sons among their offspring would be expected to show hemophilia, and one-half of the daughters would be phenotypically normal, though carriers of the gene for hemophilia. As in *Drosophila*, sex-linkage in humans follows the pattern in which affected male children always inherit the sex-linked gene from their mothers. Affected female children only show the condition phenotypically when they inherit one sex-linked gene from each parent. For this to happen, the mother must be a carrier and the father must show the sex-linked trait phenotypically. The gene or genes for hemophilia are recessive to the normal allele. Although the inheritance pattern appears to be by-and-large Mendelian, there are many variant phenotypes of each condition. The reason for this will be explained at the beginning of Chapter 19.

Numerous conditions (at least ninety-five as of 1981) are now known to be determined by genes on the X chromosome in human beings. These include a certain form of albinism, production (or lack of production) of the enzyme glucose-6-phosphate dehydrogenase, and ichthyosis, a drying out and scabbing of the skin. A few genes have been tentatively localized on the Y chromosome, but only one, hairy ears, has retained much of a claim for such Y-linkage. It is clear in humans that some

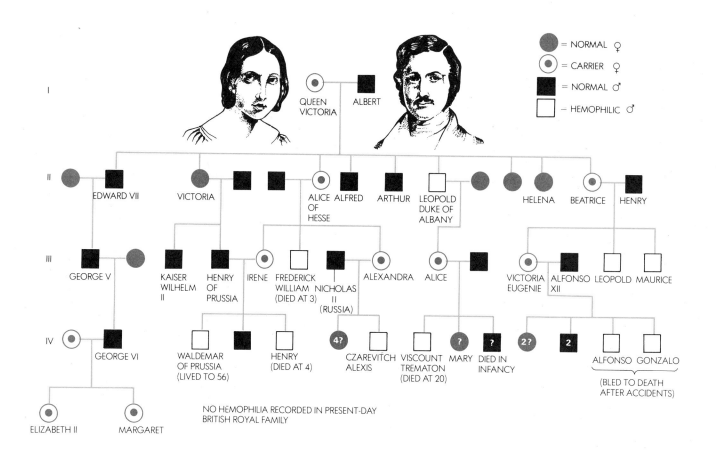

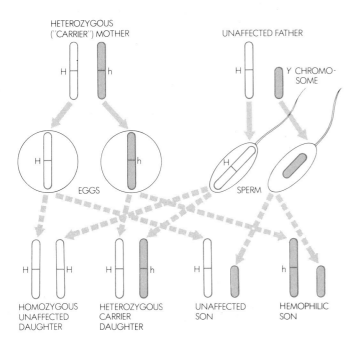

HETEROZYGOUS
("CARRIER") MOTHER

UNAFFECTED FATHER

H   h

H   Y CHROMO-
             SOME

H   h

EGGS

H   SPERM

HOMOZYGOUS
UNAFFECTED
DAUGHTER

HETEROZYGOUS
CARRIER
DAUGHTER

UNAFFECTED
SON

HEMOPHILIC
SON

Fig. 18.3
Diagram of inheritance for a typical *X*-linked characteristic, hemophilia *A*, in human beings. The trait is carried on the *X* chromosome, so that in a marriage between a carrier mother and a normal father, one-half the children will possess the gene for hemophilia. Of these, only half the males will be expected to show the condition. The heterozygous daughters will not show the trait because it is masked by the dominant allele on the other *X* chromosome. Half the sons will be expected to be normal. Only the sons who inherit the single *X* chromosome bearing the recessive gene for hemophilia will show the condition. The same inheritance pattern applies for red-green color-blindness.

genes that affect the development of maleness must be located on the *Y* chromosome, because no one can be a male without a *Y* chromosome. The striking and obvious patterns of inheritance resulting from *Y*-linkage makes it unlikely that the condition would be overlooked.

Thus Morgan's hypothesis, devised solely to explain the inheritance of the white-eyed condition in *Drosophila*, also accounts for the inheritance of red-green color-blindness, hemophilia, and a number of other *X*-linked conditions in humans.

18.4
CHROMOSOME MAPPING IN EUKARYOTES

If the same cross between a tall smooth F_1 tomato plant and a dwarf pubescent plant is carried out many times, some tall pubescent and dwarf smooth plants *do* appear. In a total of 402 progeny, a typical phenotype distribution is as follows: 198 tall smooth, 8 tall pubescent, 6 dwarf smooth, 190 dwarf pubescent. The appearance of these nonparental combinations contradicts the hypothesis proposing that "tall smooth" and "dwarf pubescent" may each be due to one rather than two genes. Thus there seems no way to explain the appearance of these few tall pubescent and dwarf smooth recombinant plants other than to assume that occasionally *linkage can be broken*. By itself, however, this assumption is not enough. We must hypothesize not only that linkage can be broken (that chromosomes break), but that *the broken pieces must actually be exchanged between homologous chromosome pairs*.

During tetrad formation in prophase I of meiosis, homologous chromosomes often form chiasmata. We can visualize what might have happened with the tomato plants as follows:

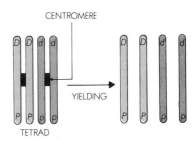

Therefore, only *DP* and *dp* gametes are produced. Occasionally, however, the following occurs:

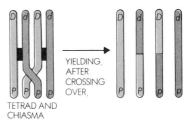

Therefore a few *DP* and *dP* gametes are produced, as well as the expected *DP* and *dp* gametes.

The chiasmata formed by homologous chromosomes within living cells can be observed under the microscope. Once again, a hypothetical model constructed to explain observed genetic ratios has physical reality in actual chromosome behavior.

It is now a well-established idea that chromosomes consist of linear arrays of genes with a specific order characteristic for each chromosome pair. Indirect evidence for this concept comes from chromosome **mapping**. The process of mapping takes advantage of crossing over and reciprocal exchange of parts between homologous chromosomes during meiosis. The mapping procedure gives clear results: gene locations can be determined with great accuracy. However, mapping does not "prove" linear order; it is only logically consistent with it. In fact, without linearity, the procedure for mapping could not work. The mapping process is a logical consequence of linearity.

Direct evidence for the existence of genes on chromosomes, and for a linear order, comes from the field of cytogenetics. Cytogenetics involves the use of cytology to investigate genetic phenomena—for example, attempting to determine the exact location of genes on chromosomes and to correlate changes in chromosome structure with observable phenotypic traits in organisms. Some of the techniques used by cytogeneticists in attempting to locate genes at specific spots, or loci (singular, locus), on a chromosome are explained in Supplement 18.2. To understand cytogenetics methods it is important first to know how genetic maps are constructed (using only crossover data obtained from breeding experiments).

A New System of Symbols

Thus far we have been using the classical Mendelian symbols for genes (capital letter for the dominant form, small letter for the recessive). Twentieth-century geneticists have gradually evolved another, shorthand system. The wild-type form of any characteristic is usually symbolized by +; the mutant form by a letter, such as a, b, c, w, B, and the like. The letter is usually the first letter of the word describing the mutation (such as w for white eye, m for miniature wing, or B for bar eye). If the mutant form is recessive to the wild type, a small letter is used; if the mutant is dominant over the wild type, a capital letter is used.

These symbols are used to indicate either phenotype or genotype. If the symbols are used alone (abc, or mwf, or $+++$), then they refer to phenotype only. If, however, they are on one side of a straight line (for haploid cells or organisms) or both sides of a straight line (for diploid cells or organisms), they indicate genotype; for example:

$$\underline{a \; b \; c}$$

HAPLOID

$$\frac{a \quad b \quad c}{+ \quad + \quad +}$$

DIPLOID

The line represents the chromosome on which the genes are linked. For diploid organisms, the symbols on each side of the line represent the particular combination of genes (for whatever traits are being observed) located on one member of the chromosome pair; the symbols on the other side of the line represent the array of genes on the homologous chromosome. In the case of the above diploid organism, for the three traits symbolized by a, b, and c, one chromosome has all three mutant genes, while the other has all three wild-type genes. Of course, it is not necessary for all the mutant genes to be found on one chromosome and all the wild-type genes on the other. We could have any of the following combinations for diploid organisms where three pairs of linked genes are involved:

$$\frac{a \quad b \quad +}{+ \quad + \quad c}, \frac{a \quad + \quad c}{+ \quad b \quad +}, \frac{+ \quad + \quad c}{a \quad b \quad +}.$$

Each combination represents a different genotype, though all have the same phenotype.

A Simple Two-Point Cross

If genes are located in a consistent linear order along a chromosome, and if crossing-over occurs, then the amount of crossing-over between any two genes should be in direct proportion to the distance between them. We can detect crossing-over when it leads to exchange of chromosome parts and consequently to recombination of traits in the offspring. The more exchange between any two linked genes in the gamete-producing cells of the parents, the greater the proportion of offspring showing the crossover phenotype. This percentage gives a direct measure of the amount of chromosomal crossing-over. This, in turn, gives a direct measure of the distance between the genes involved.

To understand the mapping process, consider a simple case of two genes known to be linked on the X chromosome of *Drosophila*: v (for vermilion eye color), and ct (for cut-wing). We first cross a pure wild-type female

$$\frac{+ \quad +}{+ \quad +}$$

with a recessive male

$$\underline{v \qquad ct}$$

to produce a double heterozygote female:

$$\frac{v \qquad ct}{+ \qquad +}.$$

We can now back-cross these F_1 females to recessive males as shown below:

$$\frac{v \qquad ct}{+ \qquad +} \quad \times \quad \frac{v \qquad ct}{\Longrightarrow}.$$

The results are shown in Table 18.1.

Table 18.1

	Phenotype	Number of individuals	Percentages
Parental categories	Vermilion/cut	425	42.5
	Wild type	445	44.5
Recombinant categories	Vermilion/cut	62	6.2
	Wild type/cut	68	6.8
		1000*	100.0

*Of course, seldom do offspring numbers work out to exactly 1000. However, for simplicity in calculating percents, we use in these examples cases involving exactly 1000 individuals.

Note that the largest two categories, vermilion/cut and wild type, are the same as the two parental categories, and occur in approximately equal numbers (within each category there are also equal numbers of males and females).* The smaller categories are the recombinant groups, vermilion/wild type, and the reciprocal, wild type/cut (also occurring in approximately equal numbers, with equal numbers of males and females in each category). We can assume that the recombinant types are the result of crossover events between the two gene loci. Note that we would expect the recombinant categories to be smaller, because crossover events occur less frequently than non-crossover events. To determine the relative distance between the genes for *v* and *ct* we add the percentages for the two recombinant categories, 6.2 + 6.8, to give a total of 13. We do this because the two recombinant categories are both the result of a single crossover event; each recombinant category is the reciprocal of the other. This means that genes *v* and *ct* (and their wild-type alleles) are 13 units apart on the *Drosophila* X chromosome. A chromosome map for these two genes can be represented as:

The units are called *map units* and they are given as *relative values only*. That is, map units cannot be converted automatically into any actual units of distance, such as microns (μ) or millimicrons (mμ). However, as we will see later, by a series of additional procedures,

*Even though we happen to be looking at X-linked genes in this case, the number of males and females showing any combination of traits will be roughly the same because we are doing a back- or test-cross.

genes can be localized to actual physical positions on chromosomes, thus transforming relative map distances into real physical distances.

Now, suppose we look at a third gene also located on the X chromosome, *f*, for forked bristles (wild-type *Drosophila* have straight bristles). We can map the distance between *ct* and *f* in the same manner described above, using a doubly heterozygous female

$$\frac{ct \quad f}{+ \quad +}$$

and a recessive male

$$\frac{ct \qquad f}{\longrightarrow}.$$

The offspring phenotypes are given in Table 18.2 (again out of 1000 individuals). Again, note that if we add together the two recombinant categories, we get a total percent of 29.5, which translates into 29.5 map units. Thus, we can draw an additional map of the genes for cut-wing and forked bristles as follows:

Since we are dealing with the same chromosome, we would like to place all three genes on the map with respect to one another. This might seem like an easy thing to do, but a moment's reflection will indicate that it isn't. Two different maps will equally well conform to the data we have at hand:

Table 18.2

	Phenotype	Number of individuals	Percentages
Parental categories	Cut/forked	360	36.0
	Wild type	345	34.5
Recombinant categories	Cut/wild type	150	15.0
	Forked/wild type	145	14.5
		1000	100.0

Table 18.3

	Phenotype	Number of individuals	Percentages
Parental categories	Vermilion/forked	370	37.0
	Wild type	395	39.5
Recombinant categories	Vermilion/wild type	120	12.0
	Forked/wild type	115	11.5
		1000	100.0

We can distinguish between these two possibilities and thus determine the order of the three genes on the chromosome by making a third cross, between *f* and *v*. The results from such a cross are shown in Table 18.3. These data show that forked and vermilion are 23.5 map units apart. Since cut and vermilion are only 13 map units apart, and cut and forked are 23.5 map units apart, the only possible sequence of genes is the second alternative above, that is,

<u>ct   v   f</u>,

and the three-gene map should look like this:

If our map for all three genes is correct, then we should expect that the sum of the two shorter distances *ct–v* and *v–f* should equal the longer distance *ct–f*. That is, 13.0 + 23.5 should equal 36.5; however, note from Table 18.2 that the observed distance between *ct* and *f* is 29.5. How can we account for this discrepancy? As early as 1911 A. H. Sturtevant noted that the sum of map distances measured over small intervals would be expected to be more accurate than the sum of values measured over large intervals because of the phenomenon of *double crossing over*. The longer the distance between two genes on a chromosome, the more chance that *two* simultaneous crossover events might occur between them. Now, two crossover events between any two genes will cancel out each other's effects, yielding no visible recombinant classes between the two more distant genes. Thus, it becomes necessary to correct genetic maps for the effects of double crossovers. How do we do this?

It is possible to simplify the whole mapping process and to take double crossovers into account by making what is called a three-point cross—that is, by observing three linked genes simultaneously. For simplicity, we will consider three hypothetical genes, *a*, *b*, and *c*.

Three-Point Cross

If we cross a heterozygous female

<u>a   b   c</u>
+   +   +

with a homozygous recessive male,

<u>a   b   c</u>
⟶

we could observe several categories of crossover phenotypes: *ab*+, ++*c*, *a*+*b*, or +*b*+. Note that the order of the genes is not *known* to be *a*, *b*, *c*; in fact, it is the exact sequence of these three genes and their relative distances apart that we are trying to determine. Table 18.4 shows the number and kind of phenotypes obtained among the offspring. The phenotypes are given here by a single set of symbols for each category. For example, an offspring showing all three wild-type traits is given as + + +, and one showing all three recessives is *abc*. These symbols thus represent phenotypes, not genotypes.

Table 18.4

Phenotype	Number of individuals	Categories
1. + + +	436	Parental types
2. *abc*	741	Parental types
3. *a*+ +	107	Double crossovers
4. +*bc*	122	Double crossovers
5. *ab*+	313	Single crossover I
6. + +*c*	236	Single crossover I
7. *a*+to *c*	390	Single crossover II
8. +*b*+	382	Single crossover II

The first aspect of the data to observe is the double crossover category (lines 3 and 4). We know this category represents the double crossover because *it is noticeably the smallest phenotype category*. This is always the case. Double crossovers occur when an independent crossover event occurs in each of the two intervals separating the three gene loci.

The double crossover class can be used to indicate the sequence of genes on the chromosome. If the order of genes is *a*, *b*, *c*, as suggested above, a double crossover would involve two single crossovers: one between *a* and *b*, and the other between *b* and *c*, as shown in Fig. 18.4. Thus we would expect the double crossover classes to show the phenotype combinations of *a*, +, and *c* on

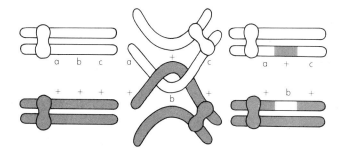

Fig. 18.4
The recombination results of double crossing over.

the one hand, and $+$, b, and $+$ on the other. However, as the data shown in Table 18.4 indicate, this is not the double crossover category, hence the order cannot be *abc*. Only one particular order of any three genes can yield a double crossover combination. By trial and error, we can determine relatively easily that the order *bac* is the only one that could give double crossover categories of $a++$ and $+bc$, as shown below in Fig. 18.5.

Once the order of genes is determined, the relative distances apart can be calculated. Crossovers between *b* and *a* are found in the last two columns of Table 18.4. (Note the reasoning involved here: if *bac* is the correct sequence, then a crossover between *b* and *a* would yield phenotype recombinations of $b++$ and $+ac$; these are indeed the combinations found in the last two columns.) The number of crossovers between *b* and *a* is determined first by adding together the values for each group (390 + 382) to give the total of 772. But this does not represent quite all of the single crossover occurrences between genes *b* and *a*. Every double crossover involves two single crossovers. Thus, to be accurate, the number of double crossovers must also be included when considering the total number of crossovers. To 772, then, we should add 229, since in each of these latter cases a single crossover between *b* and *a* also occurred. The total of 1001 (772 + 229) is then divided by the total number of F_2 counted (for all eight categories). Since this was 2727 (1177 + 229 + 549 + 772), the fraction 1001/2727 = 0.36 or 36 percent recombination between *b* and *a*. Thus *b* and *a* are 36 map units apart on the

chromosome. The crossover percent between *a* and *c* is found in the fifth and sixth columns of Table 18.4. This is calculated in the same way as before: thus, 549 + 229 = 778; 778/2727 = 0.28, 28 percent, or 28 map units. The appropriate genetic map would be as shown in Fig. 18.6.

Validity and Utility in a Concept

The concept of genetic mapping is a highly imaginative one, and it is based on a number of assumptions. One of these is the linearity of gene arrangement along the chromosome. Another is that the chances of breakage are about equal anywhere along the length of the chromosome. If some sections tended to break more easily than others, the fact would tend to alter distance calculations. These assumptions were explicitly recognized by some of the early pioneers in mapping techniques. The model worked—it gave consistent values for map distances. But there was no direct evidence to demonstrate that the model actually corresponded to real events at the chromosomal level. Many critics of the early theories of gene mapping complained that the process was purely hypothetical, since it could not be shown that genetic maps necessarily had anything to do with real chromosomes. In fact, between 1912 and 1930 many biologists felt that a model was useless in science unless it could ultimately be shown to represent reality. However, these critics missed a main point about the use of models. Models can be valuable in suggesting further lines of work even if the material reality of the model itself cannot be demonstrated at the moment. The model of the chromosome as a linear array of genes led to the investigation of many problems: the interaction of genes, the effect of position of genes on their expression, and the concept of sex-linkage, among others. Thus the model served a distinct purpose in the history of our ideas about gene structure and function. The relationship between the model and the material reality of chromosomes was eventually demonstrated directly almost twenty years after the mapping concept was introduced. This demonstration, one of the most exciting in the history of genetics, came through the introduction of a new technique with a more favorable material: the preparation and microscopic study of the extremely large chromosomes of the salivary glands of *Drosophila* larvae (see Supplement 18.2).

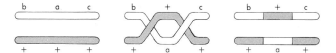

Fig. 18.5
Recombination possibilities from double crossing over if gene order is *bac*.

Fig. 18.6

SUPPLEMENT 18.2
CHROMOSOMES: THE MATERIAL BASIS OF MENDEL'S GENES

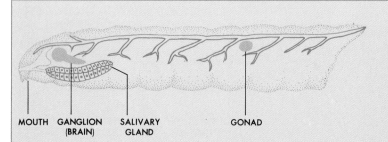

MOUTH GANGLION   SALIVARY   GONAD
       (BRAIN)   GLAND

Diagram A
Lateral (side) view of a *Drosophila* larva, showing position of salivary glands, in whose cells the "giant" chromosomes are located. A photograph of a chromosome from the salivary gland, showing considerable fine structure, appears in Fig. 5.1.

The fact that genes could ultimately be shown to exist as specific *loci* (singular, "locus") or points, at distinct positions on a given chromosome, is due largely to two lines of cytological investigation: (1) the study of chromosomal aberrations and (2) the study of the fine structure of individual chromosomes. Both contribute direct evidence in support of the idea that genes are physically a part of the chromosome structure.

As we have seen, chromosomes are small, though visible, structures within the cell nucleus. Although each pair of chromosomes has its own characteristic size, shape, and banding pattern, the so-called fine structure of the chromosomes remained largely a mystery until the late 1920s.

At that time Theophilus S. Painter (1889–1969), then at the University of Texas, and others found that the salivary gland cells of older *Drosophila melanogaster* (fruit fly) larvae (see Diagram A) contained extremely large chromosomes. Their increased size is due to the fact that although the chromosome replicates numerous times, the replicated strands do not separate. Thus the "giant" salivary gland chromosomes are in reality several hundred duplicate strands bound together. This means that the banding structures are magnified a hundred or more times even before the cytologist turns a microscope on them! This discovery made possible the detailed analysis of structural changes known collectively by the term "chromosomal aberrations."

Chromosomal aberrations include deletions, duplications, translocations, and inversions. These aberrations can occur from time to time during meiosis in the gamete-producing cells of adult organisms. The aberrant chromosome is passed on to the next generation through the egg or sperm. Depending on the nature of the homologous chromosome with which it pairs, the aberrant chromosome may or may not affect the phenotype of the offspring inheriting it. Through analyzing the relationship between phenotypic variations and specific chromosomal aberrations, further support has been found for the idea that chromosomes are the bearers of genes. Let us consider two aberrations, deletions and inversions, to see how such studies support the theory.

Deletions

Occasionally a broken piece of chromosome fails to become attached to another chromosome and is simply lost. Such **deletions** offer a unique chance to identify gene loci. For example, if the absence of a particular chromosome segment is always accompanied by a certain phenotype deficiency, it is reasonable to suppose that the gene normally responsible for preventing this deficiency is located on the missing piece.

A deletion may also produce evidence for the linearity of genes on the chromosomes. Assuming linearity of genes on homologous chromosomes, it follows that genes influencing the same trait will pair off opposite each other during synapsis prior to the first meiotic division (see Diagram B). Suppose that gene *B* is deleted from one of the homologs. In order for similar genes to pair, an unusual synapsis figure must occur, such as shown in Diagram C. Such synaptic figures have been observed in chromosomes after a segment of one chromosome has been deleted, a fact that strongly supports the hypothesis that genes are arranged in a linear fashion on the chromosomes. Moreover, in the above case

```
A      a              A      a
B      b              B      b
C      c   SYNAPSIS   C      c
D      d              D      d
HOMOLOGOUS
CHROMOSOMES
```

Diagram B

```
            A           A              A    a
GENE B  B                               b
DELETED     C   C   C         C  SYNAPSIS   C  c
REJOINING   D   D   D         D          D  d
```

Diagram C

Diagram D

Diagram E

the organism shown in Diagram C will phenotypically display the mutant recessive *b* trait, due to the deletion of the *B* gene from the homologous chromosome. Thus the observation of a phenotypic change in the offspring can be correlated with a deletion in a specific region of a chromosome. This is strong evidence that genes are located on chromosomes.

Inversions

Occasionally an entire midsegment of a chromosome may break and rejoin in a completely reversed position, as shown in Diagram D. Such an occurrence is an **inversion.** Since it is extremely unlikely that the homolog of an inverted chromosome would be/undergo a similar inversion, we can predict that the chromosomes will have to be greatly contorted before proper synapsis can take place. Using line model chromosomes, one of which has the inversion shown in Diagram D, at least one possible synaptic figure may be drawn (Diagram E.) Such postinversion synaptic figures have actually been observed in corn plants.

The existence of inversion supports the idea of a linear arrangement of genes on chromosomes in two ways. First, since chromosomes tend to synapse with corresponding genes across from one another, the unusual synaptic patterns of postinversion chromosomes are in agreement with predic-

tions. If the genes were not parts of chromosomes and if they were not linearly arranged, such synaptic contortions in inverted chromosomes would not occur. Second, inversions seldom show recombination within the inverted region of the chromosome. When inversion strains of the fruit fly are bred, for example, fewer recombinants are found for those very traits contained within the inverted region of the chromosome. Again, support for the idea of genes as physical parts of chromosomes arranged in a linear fashion comes from a correlation between genetic and cytological data.

How has all this been aided by discovery of the salivary gland chromosomes? The availability of large, easily observable chromosomes from an organism about which so much breeding data existed made possible the detailed correlation between phenotypic and chromosomal variation. Without the giant salivary gland chromosomes, it would have been impossible to observe that inversions had actually taken place in the chromosome. Detecting an inversion cytologically requires being able to see the bands in enough detail to actually recognize when they are out of place. Such detailed observations are possible only on such large, visible chromosomes. By making such detailed comparisons between breeding data (crossover frequencies) and cytological study of giant chromosomes, a detailed correspondence between genetic and cytological maps is possible (see Diagram F).

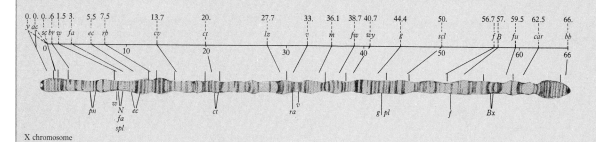

X chromosome

Diagram F
Detailed comparison between the genetic map (above) and a corresponding section of the salivary gland chromosome of *Drosophila* (below). The genetic map is based on analysis of cross-over frequencies from breeding studies. The loci determined in this manner are relative to one another and do not have any distinct physical location. The chromosome map is based on cytological analysis of giant salivary gland chromosomes; rearrangements of banded sections are compared to observed phenotypic changes in particular traits. In this way the corresponding gene locations between the genetic and cytological maps are determined. The letters on the genetic map (above) symbolize specific genes. The letters below the chromosome map show the physical location of the gene on the chromosome.

18.5
CHROMOSOME MAPPING IN BACTERIA AND VIRUSES

In studying the genetic makeup of prokaryotic cells, some different techniques are employed from those used with multicellular organisms such as *Drosophila*. First, let us consider the bacteria.

Bacteria

In 1946 Joshua Lederberg and E. L. Tatum discovered that bacteria such as *E. coli* can mate and exchange genetic material. In the next fifteen years biologists such as François Jacob and Elie Wollman took advantage of the mating process to map the bacterial chromosome.

Mating does not occur between *any* two bacteria, but between members of different strains, or what are known as *mating types*. Two such strains are called the F^+ and F^-, with F standing for the "fertility factor." F^+ bacteria have a fertility factor while F^- bacteria lack it.* The F^+ factor is a piece of DNA that determines, among other things, the ability of the bacterium to form a **conjugation tube,** or bridge through which genetic material is exchanged during mating. When F^+ cells are mixed in a liquid culture medium with F^- cells, mating occurs, and the F^- cells are converted into F^+ cells. The conversion results from the fact that the F factor is transferred from the F^+ to the F^- cells during conjugation. But what is the F factor and how does it relate to the rest of the bacterium's genetic material?

In F^+ *E. coli*, the F factor exists as a separate entity from the rest of the cell's chromosome. Such physically separate segments of DNA are not uncommon in bacteria and are called **plasmids.** In general, mating between bacteria involves replication of all or part of the F^+ cell's DNA (including plasmids) and transfer of some or all of that replicated DNA to the recipient. In matings between the specific strains F^+ and F^-, however, it is most frequently the F^+ factor that is transferred from donor to recipient cell.

There is yet another mating type in *E. coli* known as *Hfr*, or "high frequency of recombination" strain. *Hfr* cells are able to mate with F^-, and thus it is reasonable to assume that, like the F^+ strain, *Hfr* bacteria possess the F factor. But mating between *Hfr* and F^- strains produce quite different results from mating between F^+ and F^- cells. For one thing, many more characteristics are transferred from *Hfr* to F^- cells than from F^+ to F^- cells. For another, F^- cells are only infrequently transformed into F^+ cells by conjugating with *Hfr* cells, whereas most of the time they are transformed by con-

jugation with F^+. How can we interpret these results? *Hfr* bacteria have an F factor that has become physically incorporated into the bacterial chromosome at one specific spot. Plasmids which can exist as either separate entities or as part of the bacterial chromosome are known as **episomes.** (Episomes are thus a subclass of plasmids.)

Whenever *Hfr* and F^- bacteria mate, the *Hfr* chromosome is replicated with one copy passing across the conjugation tube into the F^- cell, the other remaining inside the donor. The entire mating process is depicted in Fig. 18.7. The *E. coli* chromosome is circular and the F factor is integrated into the chromosome at a specific site that is always the same for any given line, or substrain, of cells, but may vary from one substrain of *Hfr* cells to another. In all substrains, regardless of where the F factor is inserted into the bacterial chromosome, the transfer of the newly synthesized chromosome across the conjugation bridge begins just behind the F factor, so that the F factor is always the last portion of the chromosome to be transferred (see Fig. 18.7a to c). If conjugation goes to completion, the entire *Hfr* chromosome, including the F factor, is passed over to the recipient cell (which now becomes diploid for all the traits, as well as becoming an F^+ strain). If conjugation is somehow interrupted before the chromosome is completely transferred, the conjugation tube and chromosome in it are both broken. Consequently, the recipient cell is left with its own complete chromosome and whatever portion of the replicated donor chromosome has been transferred. The recipient will thus be diploid for some traits and haploid for others.

A number of years ago François Jacob and Elie Wollman, then of the Pasteur Institute in Paris, reasoned that by interrupting the process at various points in time they could produce recipient cells with various lengths of donor chromosome (all of course without the F factor, since conjugation was being interrupted before the F factor was transferred). Using substrains of *Hfr* and F^- *E. coli* which have identifiable metabolic mutations, it is possible to stop conjugation at any stage along the way, and determine (by metabolic tests) what portions of the donor chromosome have been transferred per unit of time.

To understand how chromosome transfers are detected, consider the following metabolic test. Suppose that gene A^+ controls ability to synthesize growth substance A from precursors, while B^+ controls ability to synthesize growth substance B from precursors—and on down the line for genes C^+, D^+, and E^+. This means that a bacterium which has gene A^+ does not need to be supplied with substance A, because it can make its own. The mutant forms of each gene (A^-, B^-, C^-, and the rest) are unable to synthesize their respective substances. Thus, bacteria with gene A^- must be supplied with A in the medium. The same would be true for genes B^-, C^-, D^-, and E^-.

---

*Most wild-type *E. coli* are F^- because F^- cells are resistant to certain viruses that completely kill off F^+ cells.

(a)

(b)

Plate I
Warning and protective coloration

(a) Warning coloration of the insect larva of *Sibine stimulans*. The hairs protruding from the larva are sharp and contain a poison that causes skin irritation in humans and other animals. The unusual shape as well as the bright colors of the animal serve to warn predators to keep away. (Photo courtesy of Center for Disease Control, Atlanta/Biological Photo Service.)

(b) Batesian mimicry in the Monarch (top) and Viceroy (bottom) butterflies. Only the Monarch is distasteful to bird predators. However, by resembling the Monarch so closely, the tasty Viceroy escapes a considerable amount of predation. (Photo courtesy of Richard Humbert/Biological Photo Service.)

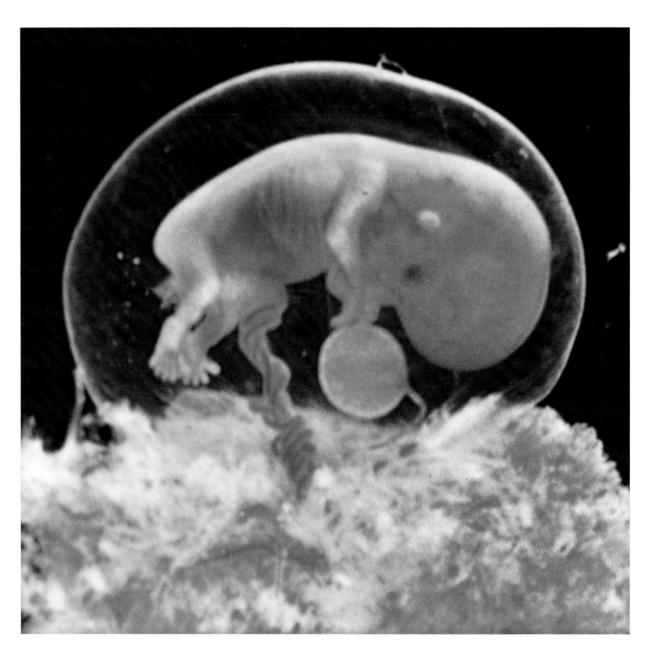

Plate II
Human fetus at 5 weeks in amniotic sac. It is only 8 mm (1/3 in) long from crown to rump and weighs 1/1000 of an ounce. Particularly prominent is the umbilical cord attached to the placenta. The round sphere in the center is the remains of the yolk sac. (From Roberts Rugh and Landrum Shettles, *From Conception to Birth*. New York: Harper & Row, 1971.)

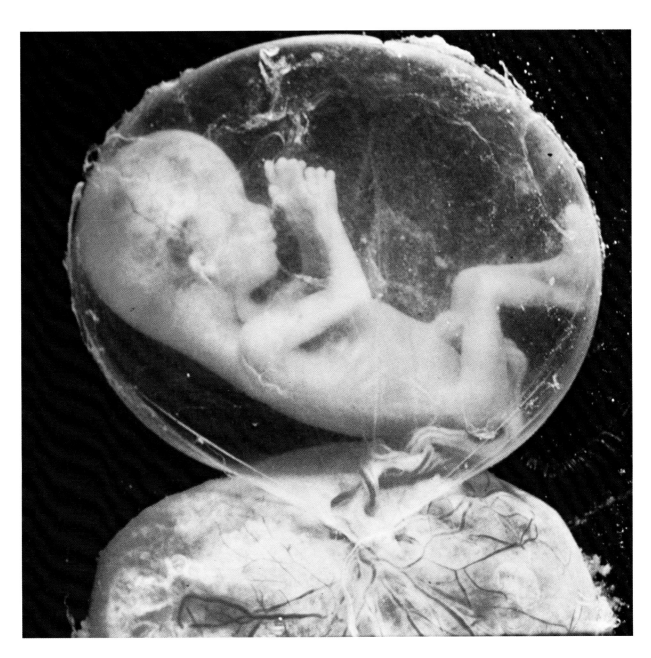

Plate III
Human fetus at 12 weeks in amniotic sac. The fetus
is 150 mm (3 in) long from crown to rump and
weighs 1/2 ounce. The placenta is to the left, with
the umbilical cord passing between the fetus's legs.
(From Roberts Rugh and Landrum Shettles, *From
Conception to Birth.* New York: Harper & Row,
1971.)

(a)

(b)

(c)

(d)

Plate IV
Four of the earth's
major biomes

(a) The taiga biome, characterized by coniferous forests, is extensively spread
over the northern parts of
North America, Europe,
and Asia. This view is from
North America.

(b) Tropical rain forest
biome, showing the highly
dense vegetation with virtually every inch of space
covered. (Courtesy Jack
Diani and the Missouri
Botanical Garden.)

(c) Temperate deciduous
forest biome in autumn
(North Carolina, Great
Smoky Mountains).

(d) Desert biome. View of
the Sonoran desert in the
southwest United States
(Arizona), showing the typical features of an arid biome: cactus and other
plants adapted to conserve
water, relatively sparse
ground cover, and dry, porous soil. Average annual
rainfall is only 9–10 inches.

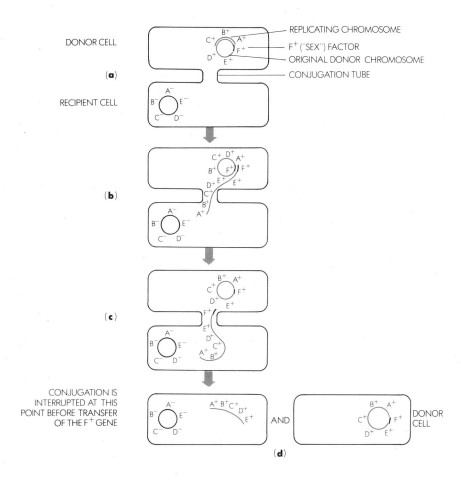

DONOR CELL — REPLICATING CHROMOSOME
— F^+ ("SEX") FACTOR
— ORIGINAL DONOR CHROMOSOME
(a)
— CONJUGATION TUBE

RECIPIENT CELL

(b)

(c)

CONJUGATION IS
INTERRUPTED AT THIS
POINT BEFORE TRANSFER
OF THE F^+ GENE

AND   DONOR
CELL

(d)

Fig. 18.7
Bacterial conjugation and chromosome transfer in *Escherichia coli*. The donor cells are of the strain *Hfr*, which stands for "high frequency recombination." The donor chromosome, which has integrated into it the *F*, or "fertility" factor, is replicated and passed across the conjugation tube or bridge. Since the rate of transfer is constant, time can be used as a measure of how much of a chromosome has been transferred. Interrupting the conjugation at different time periods thus allows different lengths of donor chromosome to be transferred. By various metabolic tests (described briefly in text) it is possible to determine which genes have been transferred and, by comparing these against the time elapsed, to map the linear arrangement of genes on the chromosome.

Following the diagram in Figure 18.7, suppose that interruption occurs after genes A^+ and B^+ have passed over from donor to recipient cell. The recipient will now contain two copies of genes A and B (represented A^+A^- and B^+B^-). The recipient cell will be able to grow on a medium lacking substances A and B, but it must be supplied with substances C, D, and E (because it has only genes C^-, D^-, and E^-). The time period required to get a recipient strain that could grow without A and B is a measure of the amount of chromosome that must have passed through the conjugation tube. Time can thus be translated into distance of genes apart along the bacterial chromosome. Allowing the transfer process to proceed a little longer produces a recipient strain that can be grown without substance C. This process is carried out hundreds of times, in each case allowing the conjugation to proceed a little longer than the time before. In this way, the sequence of genes along the bacterial chromosome can be mapped.

Once inside the bacterial cell, the donor chromosome can also pair with its homologous segment on the recipient chromosome, and an exchange, or *recombina-*

tion, can take place. Recombination does not always occur, but when it does the donor chromosome segment replaces the homologous segment in the recipient chromosome; the replaced segment becomes detached and is eventually degraded.* The new recombinant chromosome is replicated faithfully in future generations. Its progeny are thus available for testing and provide geneticists with the information needed about which genes are transferred per unit time.

From such studies a chromosome map of bacteria such as *E. coli* (see Fig. 18.8) can be constructed. For *E. coli*, the most thoroughly studied bacterium, the chromosome exists most of the time (except during conjugation, for example) as a circle. In principle, however, it preserves the linear arrangement of genes characteristic of rod-shaped chromosomes.

---

*Such recombination is called nonreciprocal, since there is not an even exchange of material between two equivalent chromosomes.

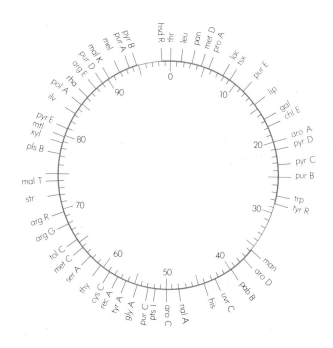

Fig. 18.8
Recently revised circular linkage map of the bacterium *Escherichia coli*, strain K_{12}. Numbers inside the circle represent map positions on the circular chromosome in minutes, determined by interruption studies of the kind explained in the text. Only 52 percent of the total 615 precisely mapped genes of this bacterium are shown here. (Redrawn from B. J. Bachmann, K. B. Low, and A. L. Taylor, "Recalibrated linkage map of *Escherichia coli* K_{12}," *Bacteriological Review* 40 (1976): 116–167. Courtesy of the authors and the publisher.)

Viruses

Viruses are particles of very small size (0.1 to 0.01μ) consisting of a nucleic acid core surrounded by a protein coat (see also Section 26.2). In most viruses the nucleic acid core consists of DNA, but in a few cases it is RNA. The nucleic acid usually consists of one long DNA molecule, the viral chromosome. Since viruses do not conjugate (or undergo any other form of sexual reproduction), at first glance it might appear that they offer no opportunity for chromosome mapping. Such is not

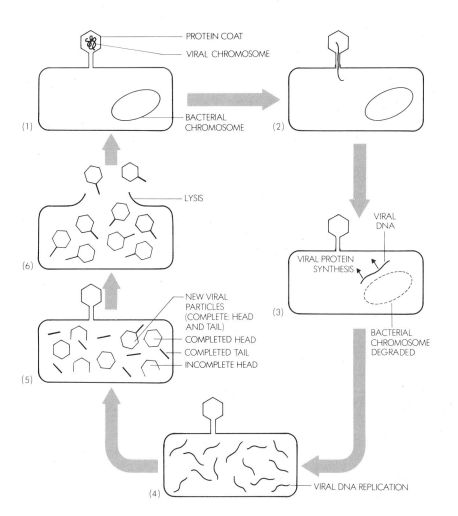

Fig. 18.9
Simplified life cycle of a virus, bacteriophage T_4. (1) The bacteriophage ("phage" for short) adsorbs onto the bacterium (in this case *E. coli*) and punctures the cell wall. (2) The phage injects its chromosome (a DNA molecule) into the bacterial cytoplasm. (3) The phage and bacterial chromosomes exist side-by-side for a short time. The phage DNA first guides the synthesis of a few proteins, one of which is an enzyme which degrades the bacterial chromosome. At the same time, (4) phage DNA begins to replicate itself, forming many viral chromosomes. (5) Phage genes commandeer the cell's machinery to produce phage proteins including those of the coat. Self-assembly of coat and phage DNA yields hundreds of phage particles. The cell is eventually ruptured (6) and progeny virus particles are released, to begin the infection cycle all over again.

the case, however. To understand how geneticists can map viral chromosomes, it will be necessary to digress for a moment to examine the process of viral replication.

The basic life cycle of a virus is diagrammed in Fig. 18.9. The virus shown here is a bacteriophage (or "phage," for short) so named because it parasitizes bacteria. Beginning at stage 1 (upper left-hand corner), the virus adsorbs to the bacterial cell wall and punctures it. The viral nucleic acid (in this case DNA) is injected into the bacterial cell (2). Once inside the bacterial cytoplasm, the viral DNA begins guiding the production first of viral protein, then shortly of more viral DNA (stages 3 to 5). Eventually whole new viruses are produced. The cell ruptures (a process known as **lysis**), releasing the progeny into the environment (see Fig. 18.10).

To map the viral chromosome geneticists take advantage of two well-known facts.

1. Different genetic strains exist for each type of virus, such as T_4. These strains may differ in only one trait: for

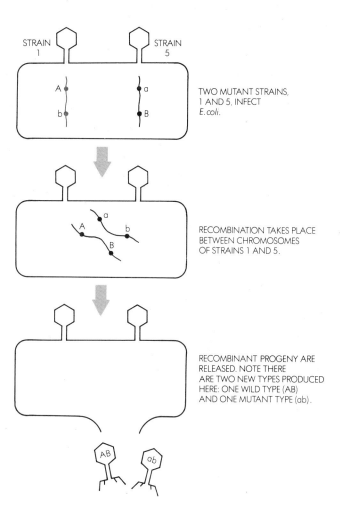

TWO MUTANT STRAINS, 1 AND 5, INFECT *E. coli.*

RECOMBINATION TAKES PLACE BETWEEN CHROMOSOMES OF STRAINS 1 AND 5.

RECOMBINANT PROGENY ARE RELEASED. NOTE THERE ARE TWO NEW TYPES PRODUCED HERE: ONE WILD TYPE (AB) AND ONE MUTANT TYPE (ab).

Fig. 18.11
Recombination in phage T_4, infecting *E. coli*, strain K_{12}. Note that in recombination, two types of progeny virus different from either parent are produced: one that is a wild type (*AB*, both wild-type genes), and one that is a mutant (both mutant genes, *ab*).

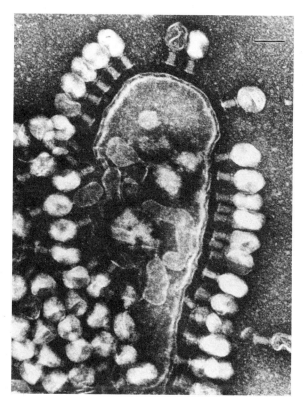

Fig. 18.10
Electron micrograph of T_4 phage lysing an *E. coli*. In this final stage of infection the bacterial cell ruptures and progeny phage are released. Note the phage still attached to the outside. (Micrograph courtesy of Dr. Lee D. Simon, Waksman Institute for Microbiology, Rutgers University.)

example, the ability to make the protein coat for its own tail but not its own head region, or vice versa.

2. More than one virus particle usually attacks each host cell simultaneously. Thus, if the researcher adjusts the concentration of virus particles and bacteria precisely enough, it is possible for most bacterial cells in a culture to be infected by two or more phage.

If equal quantities of two different strains of phage are allowed to infect the bacterial culture at the same time, most bacteria will very likely be infected by at least one representative of each strain. This means that at least two different chromosomes will exist at the same time inside of a single bacterial cell.

The process of recombination and mapping is outlined in Fig. 18.11. Virus strain #1 has the genotype

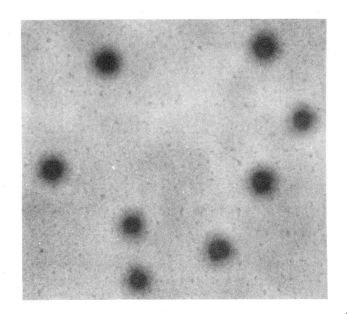

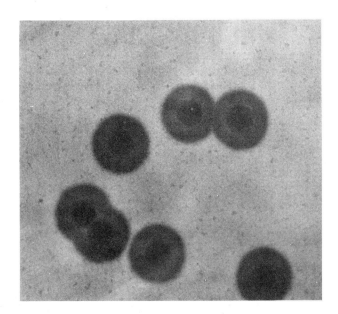

(a)

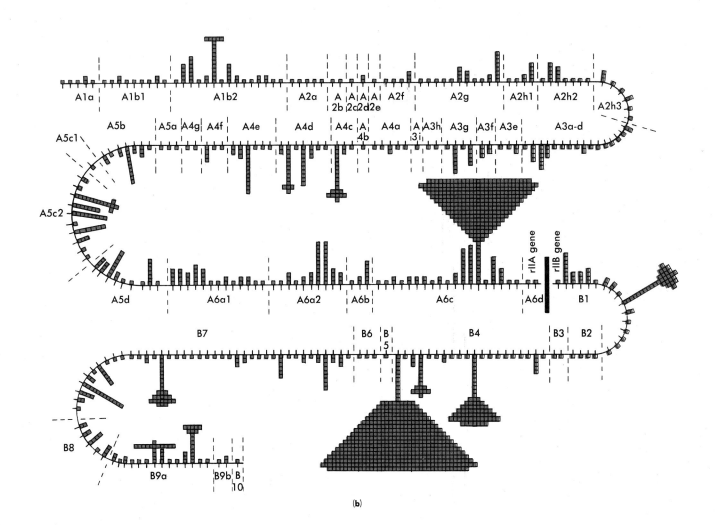

(b)

Ab (*A* is the functional gene for, let us say, making head protein, whereas *b* is the nonfunctional gene for making tail protein). Strain #5 is just the converse. It is genetically *bB*. When a pure culture of strain #1 is placed into a bacterial culture, no new viral progeny are produced, because they have an incomplete protein coat. Similarly, when virus strain 5 is placed by itself in a bacterial culture it fails to produce functional progeny. However, when strains 1 and 5 are placed together into a bacterial culture, both viruses can infect the same bacterium. Once inside the bacterial cell, the two viral chromosomes can undergo chromatid crossing over and genetic recombination, analogous to the recombination that occurs in eukaryotic cells during meiosis. The result of such recombination is that some completely functional (wild-type, or *AB*) virus progeny are produced. The frequency of recombination is proportional to the distance of the genes apart on the viral chromosome.

By other techniques geneticists have been able to analyze the structure of viral genes in great detail. The work of Seymour Benzer (formerly of Purdue University, now at the California Institute of Technology) and others has shown that each gene in the bacteriophage T$_4$ consists of many mutable sites—that is, specific regions, or points, where a mutation affecting some phenotypic trait can occur. For example, in a gene known as RII, which affects rate of viral replication, over 2000 different mutable sites have been identified! These sites fall into two main subregions of the gene (the *A* and *B* regions). Their positions relative to one another can be determined with great accuracy, as shown in Fig. 18.12.

The results of mapping experiments in many species (from viruses and bacteria to eukaryotic plants and animals) show that genes, the basic units of heredity, are arranged in a linear fashion along the chromosomes for most organisms. Although the gross structures of prokaryotic and eukaryotic chromosomes differ considerably (recall that eukaryotic chromosomes consist not only of DNA but also a form of protein called histone), in terms of the arrangement of the units of heredity they are very much the same.

18.6
THE CHROMOSOMAL BASIS OF SEX DETERMINATION

Sex is one of the most important characteristics that an organism inherits. In most species of eukaryotic organisms (including both plants and animals), sex is determined in some way by the presence or absence of the sex or accessory chromosomes (*X* and *Y*). The pattern of sex determination varies from one species to another, however. To illustrate two important patterns of sex determination, consider the cases of *Drosophila* and human beings.

Drosophila

As discussed in Section 18.3, male *Drosophila* are usually *XY* and female *XX*. But, which of the two sex chromosomes actually *determines* sex? At first it seemed logical to assume that the *X* determined femaleness and the *Y* maleness. However, occasionally *Drosophila* embryos are born with only an *X* chromosome and no *Y* (because of nondisjunction in the parental gonads during gamete formation). These embryos developed into male flies! Bridges and others concluded that in the fruit fly the *X* chromosome was the sex determiner. One *X* produced a male and two *X*'s a female. The *Y* chromosome did not appear to affect the determination of sex. Though usually sterile, flies with a single *X* chromosome were phenotypic males. While the *X* chromosome in *Drosophila* contains many non-sex-determining genes such as several for eye color, it was also thought to contain some genes influencing the direction in which sexual development occurs.

As it turned out, however, sex determination in *Drosophila* is not a matter of specific genes on any one chromosome. The factor that seems to trigger a given embryo to develop in the direction of maleness or femaleness is the *ratio* of *X* chromosomes to the number of members in each autosome set. This notion was first fully developed by Calvin Bridges in the 1920s, and became known as the *balance* theory of sex determination. Bridges' scheme is summarized in Fig. 18.13. Over the years Bridges was able to build up stocks of *Drosophila* with varying numbers of autosomes and sex chromosomes (to do this he took advantage of random nondisjunctions in both autosomes and sex chromosomes). That is, he could obtain some flies which had only one *X* chromosome, but two complete sets of autosomes (that is, both members of each homologous pair or group present). Similarly, he could obtain flies with two *X* chromosomes, and three complete sets of

Fig. 18.12
(a) Plaques (growths) of bacteriophage T$_4$ on a "lawn" of bacteria. Each dark spot represents an area where phages are reproducing by lysing bacteria. The rate of growth and shape of the plaque are genetically controlled. Note the differences between the two groups of plaques shown. The left is wild-type T$_4$, while the right is the mutant known as *r*II. *r*II mutants have been used extensively in studying viral genetics because their phenotype (plaque shape and size) is so easy to recognize. (b) Using *r*II mutants, geneticists such as Seymour Benzer have mapped very small regions within the viral chromosome. This map shows the entire *r*II gene, which controls plaque size and shape. This map shows the location of a large number of mutable sites *within* the single *r*II gene. Each square on a given site represents a given frequency of mutation at that site. The greater the number of squares, the more "mutable" the site. As the diagram shows, some sites appear to be highly mutable; these are sometimes called "hot spots." (Photographs courtesy Millard Susman, University of Wisconsin.)

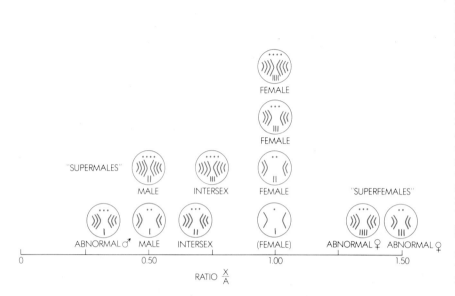

Fig. 18.13
Schematic representation of Bridges' balance theory of sex determination. The horizontal scale indicates ratios of X chromosomes to complete sets of autosomes (or to number of homologs present in each autosomal group). *Drosophila* has three autosomal groups, and one sex group of chromosomes. According to Bridges' theory, normal female sex determination resulted from any combination of X and autosomal chromosomes whose ratio (X/A) equalled 1.0. Any normal male determination would result from a combination of X and autosomal chromosomes whose ratio was 0.5. Flies whose ratios fell between 0.5 and 1.0 were called "intersexes," since they were intermediate in phenotype between male and female. Those flies which fell below 0.5 were exaggerated males, or "supermales"; those which fell about 1.0 were exaggerated females, or "superfemales."

autosomes. Almost any combination, short of fully haploid individuals, could be obtained; many of these types could breed, and when they did they bred true. Bridges observed the phenotypes of these various nondisjunction flies. He found that the degree of "maleness" or of "femaleness" of the flies' phenotypes depended not on the absolute number of X or Y chromosomes, but on the *ratio* of X chromosomes to the autosomes. He represented this as the X/A ratio. Keep in mind that the ratio is expressed as number of X chromosomes to the number of *complete sets* of autosomes, or the number of homologs present in each homologous pair of autosomes. Thus, for example, a fly which had three complete sets of autosomes (three members of each autosomal group, or triploid for the autosomes), but only two X chromosomes (diploid for the sex chromosomes) would have a ratio of 2/3 or 0.66. This numerical value would hold whether the fly also happened to have a Y chromosome or not. Phenotypically, flies with a ratio of 0.66 were *intersexes*—that is, intermediate in appearance (for example in the structure of the sex organs) between males and females. On the other hand, flies with a ratio of 1/1 or 1.0, were phenotypic females, regardless of the actual number of X chromosomes and autosomes present. Thus, as shown in Fig. 18.13, phenotypically normal females resulted from combinations of $1X/1A$, $2X/2A$, $3X/3A$, and the like. (Although the lowest circle in the middle grouping shows a wholly haploid individual, no such fly was ever obtained. Haploid individuals of a normally diploid species are usually nonviable.) On the other hand, diploid females with patches of haploid tissue have been observed. In those cases the phenotype of the haploid tissue is always female—a strong corroboration of Bridges' idea.

Flies with an X/A ratio greater than 1.0 Bridges called "superfemales"—indicating that they have somewhat exaggerated (usually larger-sized or more prominent) female structures. At the other end of the scale, flies with a ratio of 1/2 (0.5) were males, whether they had a Y chromosome or not, though those which lacked a Y were sterile. Flies with an X/A ratio lower than 0.5, such as 1/3 (0.33) were called "supermales"; like their superfemale counterparts, they also had exaggerated sex-related traits.

Biologists do not know how the ratio of X chromosomes to autosomes produces sex differentiation in *Drosophila*. The ratio of X's to autosomal sets must trigger the expression of genes governing the development of male or female traits. Evidence now suggests that male-determining genes are distributed among the autosomes, while the majority of female-determiners reside on the X chromosomes. How these specific genes are triggered by the presence of one or two X chromosomes is still a mystery.

Human Beings

While sex is determined in *Drosophila* largely by the ratio of X chromosomes to autosomes, the situation is different in human beings. In our own species, the Y chromosome appears to be the major sex determiner. In humans, an individual with a Y chromosome is male, even if he also has one, two, or even three X chromosomes. Conversely, an individual lacking a Y chromosome is a female, whether she has one, two, or three X's.

As in *Drosophila*, human beings show a range of variation in sex-chromosome number (See Table 18.4) All of these result from some form of nondisjunction.

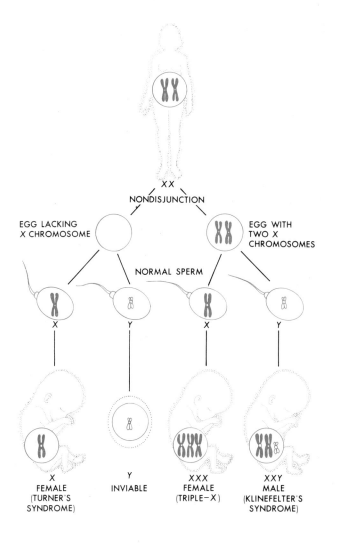

Other than the normal phenotype, only two of the types shown in the table appear to be phenotypically normal though they possess an unusual chromosome complement: XYY males (see Supplement 17.3) and the triple-X females. Both have normal sexual development and both are fertile. Most of the others show some variation in sex development and are infertile, or are at least of doubtful fertility.

Women with Turner's syndrome (XO) have rudimentary ovaries, or in some cases none at all; they have undeveloped breasts, often show "webbing" or folding of skin around the neck, and are unusually short. Men with Klinefelter's syndrome (XXY) are outwardly male (they have male genitalia), but their testes are undeveloped and their breasts enlarged; they have longer limbs than average and sparse body hair. Almost all reported cases have been sterile. Both Klinefelter's and Turner's syndromes are produced by the nondisjunction mechanism shown in Fig. 18.14. Although individuals with Klinefelter's syndrome have two X's, the presence of the Y chromosome determines their generally masculine appearance. Although individuals with Turner's syndrome have only one X, the lack of a Y chromosome determines a generally female appearance.

As Table 18.5 shows, still other abnormalities of sex-chromosome constitution are found in human beings. A few triple-X females have been observed. They

Fig. 18.14
Nondisjunction in the X chromosomes of a human female can result in an egg cell with two X chromosomes and an egg cell with no X chromosome (designated O). Even when fertilized by normal sperm, such eggs always give rise to defective offspring.

Table 18.5
Chromosomal Complements in Sex Determination in Human Beings

Descriptive name of phenotype	Sex phenotype	Fertility	Sex-chromosome complements
Normal male	Male	+	XY
Normal female	Female	+	XX
Turner's syndrome	Female	−	XO
Klinefelter's syndrome	Male	−	XXY
XYY syndrome	Male	+	XYY
Triple X syndrome	Female	±	XXX
Triple X-Y syndrome	Male	−	$XXXY$
Tetra X syndrome	Female	?	XXXX
Tetra X-Y syndrome	Male	−	$XXXXY$
Penta X syndrome	Female	?	XXXXX

Adapted from V. McKusick, *Human Genetics.* Englewood Cliffs, N.J.: Prentice-Hall, 1969, p. 19.

SUPPLEMENT 18.3
LOCALIZING GENES ON HUMAN CHROMOSOMES

Although a great deal has been learned about the location of genes on the chromosomes of many species, including bacteria and viruses, only in recent years have geneticists made much headway in mapping human chromosomes. The reasons for this should be obvious. The various techniques used in traditional mapping—specifically planned matings and the analysis of recombination frequencies—are inapplicable to human beings. Over the past half-century, human geneticists have been able to localize a number of traits to one or another chromosome. Even then, the results have been tentative and often imprecise.

Today, however, several new techniques have been introduced that make possible the mapping of human chromosomes with considerable accuracy. These techniques include new staining procedures that resolve some of the ultrastructure of human chromosomes, and the ability to form hybrids between human and other mammalian cells in tissue culture. With these techniques, over 100 genes have been assigned to specific chromosomes, and often to specific regions on those chromosomes.

In the past, cells for karyotyping have been harvested from their culture medium, fixed (which kills them), stained, and observed under the microscope (see Supplement 1.1). The customary stains cytologists have used (acetocarmine and orcein, for example) show only gross structures such as large breaks, extra chromosome segments, and the like. Fine detail of individual chromosomes (such as banding patterns) are usually not visible because the stains cover the entire chromosome evenly. In recent years, fluorescent dyes have been introduced to show specific banding patterns of each chromosome (Diagram A). Trypsin (a proteinase) or heat can be used to denature the chromosome, "opening up" the otherwise tightly knit structure to fuller view. The fluorescent dyes preferentially stain certain regions, revealing the bands that characteristically appear in the chromosomes of the other, more cytologically favorable species (such as *Drosophila*; see Supplement 18.1, Diagram F). Being able to see the fine structure of the chromosomes has made it possible for cytogeneticists to observe small changes like translocations, deletions (in the middle, as well as at the ends),

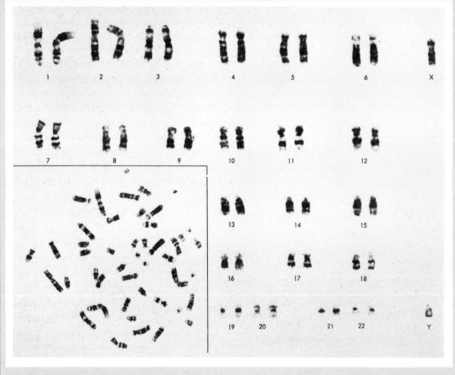

Diagram A
Human chromosome group spread out (bottom left) and arranged into a karyotype (remainder of field) showing the detail revealed by fluorescent staining technique. The individual shown here is a normal male. (Courtesy James L. German III, New York Blood Center.)

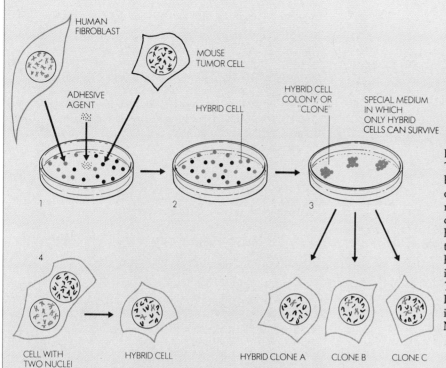

Diagram B
The basic procedure of somatic cell hybridization. Color represents human cells or chromosomes; black represents mouse cells and chromosomes; gray cells or clones represent hybrids. Each hybrid clone developed by the end of this process is the descendant of a single hybrid cell, and thus each cell in a clone is genetically identical. (Redrawn from *The New Human Genetics* by Maya Pines, National Institute of General Medical Sciences, DHEW Publication No. NIH 76-662: p. 34.)

and inversions. Geneticists cannot map chromosomes unless they have visible details to follow as "markers"—that is, for correlating specific structural changes in the chromosome with specific phenotypic changes. Fluorescent dyes have made such observations possible for human chromosomes.

There is a major problem, however, with trying to correlate chromosomal aberrations in humans with distinct, visible phenotypic changes. In most animals, but especially in humans, the effects of large-scale chromosomal changes, in either number or structure, tend to upset the entire developmental process. The result is a whole host of *general*, rather than specific, phenotypic abnormalities. By general is meant a trait, such as mental retardation, that may be brought about by any number of alterations in development. Furthermore, a general condition such as mental retardation may be the result of an imbalance of genes or chromosome parts occasioned by a deletion, translocation, or other chromosomal aberrations. Some more precise technique is needed to relate specific chromosomal changes to specific phenotypic traits (for example, with a specific protein product of the gene).

A useful technique for studying the effects of specific chromosome deletions or abnormalities is known as *somatic cell hybridization*. Although it may sound like something out of science fiction, somatic cell hybridization is a very real and powerful new tool of modern biology. Diagram B illustrates how the technique works. Human cells, usually fibro-

blasts (embryonic connective tissue) are brought together in a culture medium with mouse tumor cells (for example, mouse hepatoma, a liver cancer). By application of agents such as polyethylene glycol (a dehydrating agent found in antifreeze) or inactivated virus (such as Sendai virus), the cells are caused to stick together. As a result, after a period of time some mouse and human cells have fused, producing a doubly nucleated "hybrid" cell (called a **heterokaryon,** meaning cells with two or more *different nuclei*). In a few days, most of the hybrid cells show only a single nucleus, which includes both human and mouse chromosomes. Now, the strain of mice used for these experiments contains 40 chromosomes, while the human has 46. But the hybrid nuclei always have less than the expected total of 86—the usual range being between 41 and 55. Interestingly enough, most of the chromosomes that are lost are human, not mouse, and the loss is strictly *random*.

Dr. Frank H. Ruddle of the Yale University Medical School has seen how this technique can help in an ambitious plan to map human chromosomes. Using the randomness of the somatic fusion process, Ruddle has developed a series of hybrid cell lines, each of which contains only one, or at most a few, human chromosomes. Thus, for example, one line will contain the normal complement of mouse chromosomes plus human chromosomes 1 and 7. Another line will contain all the mouse chromosomes plus human chromosomes 5 and 6. Each gene can be recognized in cell culture by the protein,

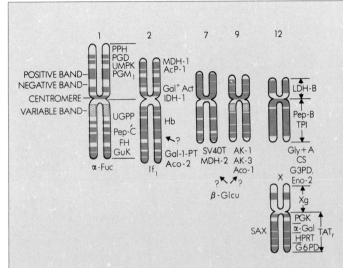

Diagram C
Map locations for genes on the 22 autosomes of human beings. Dark and light bands on each chromosome are made visible by fluorescent staining. These bands are constant and characteristic for each of the 22 pairs and for the X chromosome. Numbers above each chromosome refer to the pair to which the chromosome belongs (there is a standardized numbering system). Symbols represent particular proteins produced by the gene at the designated locus or position. Question marks, such a those found between chromosomes 7 and 9 (for β-Glcu) indicate that the gene may be on either chromosome; evidence to date is insufficient for more precise determination. (Modified from National Institutes of General Medical Sciences Publication No. NIH 76-662, *The New Human Genetics*, by Maya Pines, pp. 40–41. Used with permission.)

or polypeptide, it produces. Using electrophoresis, it is possible to separate many of these protein products—even mouse-specific from human-specific proteins. Thus, if two proteins are consistently found together in a line of hybrid cells where only human chromosome 1 appears, it is clear that the genes for these two proteins must be located on that particular chromosome.

Still, such information tells us nothing about where the genes would be located on the chromosome. Here is where the study of chromosomal aberrations, particularly with the new technique of fluorescent staining, comes in. It is possible to develop hybrid lines that contain only one or a few human chromosomes, each showing deletions, inversions, or other visible markers. (Cells with these aberrations are taken from patients whose chromosomal abnormalities are already known.) For example, suppose a hybrid cell line containing human chromosome 1 is always found to contain protein *A*. Now, if we find a second cell line containing chromosome 1 with a deleted end, and observe that in that cell line no protein *A* is found, we infer that the gene producing *A* is localized on that tip of the chromosome. More precise localization depends on finding successively small deletions—tips of chromosome arms or midregions; of course, such deletions are more rare and thus more difficult to detect.

Using hybridization, fluorescent staining, and electrophoresis, Ruddle and others have begun to map the human genome. Maps of the 6 human autosomes are shown in Diagram C. The banding patterns that characterize each chromosome type are shown in alternating light and dark regions. Genes are designated by the initials of the enzymes or other proteins which they produce (for example, on chromosome 2, short arm, MDH-1 stands for malate dehydrogenase, type 1). The location of some genes has been established precisely—that is, within one part of an arm of a chromosome. The location of other genes has been specified only to a particular arm, but not to a specific locus within that arm. Other genes have been localized to a particular chromosome but not to one or another arm; while still others can be localized only within a group of chromosomes.

are produced by the same type of nondisjunction processes that give rise to Turner's or Klinefelter's syndrome. Although phenotypically female, many show various degrees of mental retardation and are infertile. However, some triple-*X* females show no signs of retardation and *are* fertile. More surprisingly, children born to fertile triple-*X* females have been either normal males or females. This is unexpected, since we would predict that a triple-*X* mother could produce two kinds of eggs: an *XX*-bearing egg, and an *X*-bearing egg. Accordingly, half of her daughters would be expected to have the triple-*X* condition, and half her sons would be *XXY* (Klinefelter's syndrome). Only a few such *XXY* individuals have turned up, and geneticists are uncertain why there have not been more. It could be chance, but it may also be that *XX*-bearing eggs are inviable. The small number of triple-*X* females observed appears to result from a high spontaneous abortion rate among triple-*X* fetuses.

Other sex-chromosome constitutions also occur, including tetra-*X*, tetra-*XY*, and penta-*X*. These are relatively rare and result from several generations of compounded nondisjunctions of the *X* chromosome. Fertility in these individuals is extremely low.

Sex determination in human beings and other mammals thus differs considerably from the process in

Drosophila. Here is another example of the fact that it is not always possible to generalize a mechanism in one species to all species, even when the same elements (such as *X* and *Y* chromosomes) are involved. Indeed, determination of sex in human beings parallels far more closely the process in the plant *Melandrium* than that in the fruit fly.

Through evolutionary history, several different methods of achieving two separate sexes (*sexual dimorphism*) have evolved. The existence of such variations in the chromosomal basis of sex determination leaves modern biologists still very much in the dark about what actually determines the development of an individual into a male or a female. More will be said about the developmental aspects of this process in Chapter 20.

The Lyon Hypothesis

Abnormalities in the number of *X* chromosomes appear to have less severe effects on overall phenotype than similar abnormalities of the autosomes. Triple-*X* females are sometimes retarded, as are individuals with Klinefelter's or Turner's syndromes, but they are not generally so retarded as individuals with Down's syndrome. They are often fertile. Other physical abnormalities associated with different doses of sex chromosomes are also less severe than Down's syndrome.

Why should this difference exist? One hypothesis suggests that since it is normal in a population of any mammalian species for the *X* chromosome to exist in two different doses (a double dose for normal females, and a single dose for normal males), the sex chromosomes have evolved over the generations a "dosage-compensation mechanism."

The first clue as to what this mechanism might be came in 1949. At that time M. L. Barr made the seemingly minor observation that most nondividing nerve cells of female cats contain a dark-staining, small body in the nucleus. This small body is never present in the nerve cells of male cats. Called the Barr body, this small object was found to exist in the cells of females of many other mammalian species, including human beings. The fact that the Barr body was lacking in women with Turner's syndrome, who are *XO*, and present in men with Klinefelter's syndrome, who are *XXY*, suggested that it might be sex chromatin. But where did this sex chromatin come from, and why did it appear as an extra body in the cell nucleus of females only?

A second clue came from observations of the variegated coat color patterns typical of many mammals. It was known, for example, that the condition known as tortoise shell in cats is almost always found only in females. A similar condition is known in mice. These observations suggested to observers that cells in different parts of the animal's body were genetically different, like a genetic mosaic.

In 1961 British investigator Mary Lyon proposed a hypothesis to account for the following observations:

1. The presence of the Barr body in the cells of female, but not male, mammals.

2. The apparent compensatory mechanism that mammalian cells have for dealing with either one or two *X* chromosomes.

3. The appearance of coat color variegation only in female mammals.

Lyon proposed that relatively early in the development of females, one *X* chromosome in each cell is inactivated. This means that DNA on the inactive *X* chromosome is somehow permanently prevented from transcribing mRNA and producing protein. The inactivation is passed on to all replicates of that chromosome. Either the maternally or paternally derived *X* chromosome can be inactivated, probably at random. This means that in about half the body cells of a female, the maternally derived *X* chromosome is transcribing mRNA, while in the other half it is the paternally derived *X* chromosome that is transcribing. Such random inactivation would account for the variegated, patchwork patterns in the heterozygous female. It also suggests that the Barr body is the inactive chromotin (that is, DNA) from the one *X* chromosome. No one is certain how inactivation is brought about.

The Lyon hypothesis can be tested in a simple way. A sex-linked mutation on the *X* chromosome has been discovered that results in a deficiency (lack) of the enzyme glucose-6-phosphate dehydrogenase, the enzyme responsible for one step in the biochemical pathway of glucose oxidation. Women heterozygous for the mutation would be expected to have a normal allele for the enzyme on one *X* chromosome and the mutant form on the other. If the Lyon hypothesis is correct, then heterozygous women should display two populations of cells: one with a normal amount of enzyme activity, and one with very little or no activity. If the Lyon hypothesis is incorrect, then heterozygous women should have only one population of cells, all with the same relatively high level of enzyme activity. Studies of sample tissues from heterozygous females indicate that there are, indeed, two populations of cells: one with high and the other with extremely low glucose-6-phosphate dehydrogenase activity. The evidence supports the Lyon hypothesis.

The Lyon hypothesis provides a way of understanding the gross phenotypic abnormalities that result from extra autosomes, compared to the relatively small abnormalities that result from extra sex chromosomes. Since the presence of one or two *X* chromosomes is a regular occurrence, a compensatory mechanism for double dosage has obviously been selected for and has evolved. But the presence of extra autosomes is a very unusual occurrence, and a similar compensatory mechanism has not evolved. Thus, extra autosomes yield complexities of cell metabolism that alter or completely prevent normal development.

18.7
HUMAN CHROMOSOMAL ABNORMALITIES
AND GENETIC DISEASE

In the previous section we saw that some human phenotypic disorders are associated with unusual numbers of sex chromosomes (such as Turner's syndrome and Klinefelter's syndrome). There are, in addition, a host of other phenotypic problems associated with abnormalities in the autosomes. These autosomal abnormalities fall into two categories: (1) abnormalities in the number of chromosomes; and (2) abnormalities in the structure of individual chromosomes, for example, loss of a piece of a chromosome. While some of these disorders have only minor pheno-

typic effects, the majority produce rather widespread developmental problems that profoundly affect the life of the individual. We will examine just a few such disorders from each category. The discovery and analysis of human chromosomal disorders in recent years has begun to throw light on a number of human genetic defects; in some cases knowledge of the chromosomal basis of such defects has suggested medical remedies.

Disorders of Chromosome Number

Down's Syndrome (Trisomy-21). One of the most commonly encountered chromosomal abnormalities is known as Down's syndrome (also called trisomy-21). The syndrome was first described in 1866 by John Langdon Down (1820–1896), a London physician. Down observed a whole series of traits which, to one degree or another, were always associated with a form of mental retardation. Down at first called the condition "Mongolism" or "Mongolian idiocy," since the eyes of affected individuals superficially resembled those of Orientals (see Fig 18.15). The latter term is, however, no longer used today.

The physical traits Down associated with the syndrome included a prominent forehead, flattened nasal bridge, habitually open mouth, projecting lower lip, skin fold at the inner corner of the eyes, short and broad neck, rough and dry skin, abnormally shaped and aligned teeth, wide gap between the first and second toes, and a broad crease across the palm (called the "simian line"). Children with Down's syndrome are mentally retarded. Affected children are slow learners but often enjoy music and crafts and with proper care and education are often able to learn to read and write.

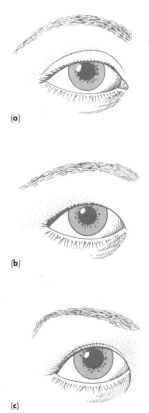

Fig. 18.15
Differences in the appearance of (a) the Caucasian eye, (b) the Oriental eye, and (c) the eye of an individual afflicted with Down's syndrome (trisomy-21). The Down's syndrome eye is quite different from the Oriental eye, and Down's syndrome is not related to Oriental ancestry. Hence "Mongolian," the name originally given to the condition, is really a misnomer. (After E. Peter Volpe, *Human Heredity and Birth Defects.* New York: Pegasus, a Division of Bobbs-Merrill, 1971, p. 72.)

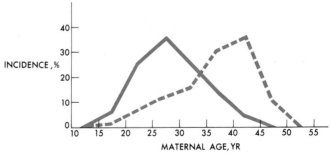

Fig. 18.16
Age distribution of mothers of Down's syndrome patients compared with that of all mothers. Vertical axis of graph shows the relative percent of mothers in each age group. (Based on data from Lionel S. Penrose; from Victor McKusick, *Human Genetics,* 2d ed. Englewood Cliffs, N.J.: Prentice-Hall, 1969, p. 23.)

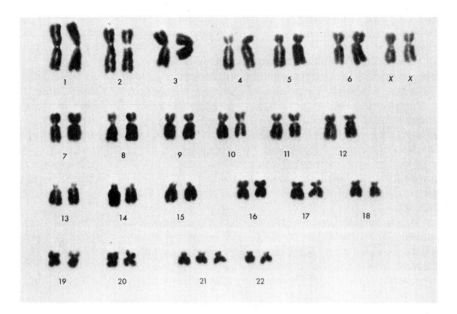

Fig. 18.17
Chromosome complements of a male with Down's Syndrome, with a total of 47 chromosomes, pair 21 having an extra member. The syndrome is sometimes called trisomy-21 in recognition of the role which an extra chromosome of this pair must play in the development of the condition. For comparison, see Supplement 18.3, Diagram A. (Courtesy James L. German III, New York Blood Center.)

However, most Down's syndrome patients are unable to care for themselves, and the trend has been to institutionalize them. Recent evidence suggests, however, that home care for the child produces greater development of mental and physical abilities than the kind of institutional care to which most afflicted children have been exposed.

Because many children with Down's syndrome also have congenital defects of the heart or other organs, their chances of survival are severely reduced. A number of years ago the geneticist Lionel Penrose estimated that the life expectancy of children with Down's syndrome was about 12 years. Sixty percent of all children afflicted with Down's syndrome do not survive beyond 10 years of age. Those who do survive into adulthood generally do not reproduce. Males with Down's syndrome are sterile. Females are not usually sterile; as of 1970, thirteen cases were known where females with Down's syndrome had children. About half of the recorded offspring have shown Down's syndrome, and the other half have been normal.

What is the genetic cause for Down's syndrome? The earliest clinicians describing the syndrome noted that it occurred predominantly (though not always) in children of older mothers. Statistical data suggested that above age 35, a woman's chances of bearing a child with Down's syndrome increase eight-fold: The incidence is one per 1500 for women under 30, one per 750 for women between 30 and 34, one per 600 for women 35 to 39, and one per 300 in women aged 40 to 45 (see Fig. 18.16).

In 1959 three French scientists at the University of Paris provided cytological evidence that Down's syndrome was associated with a chromosomal abnormality in which chromosome pair 21 occurred in triplicate. (See Fig. 18.17.) Studies in the past two decades have

revealed that nondisjunction is responsible for the failure of the two homologous chromosome 21's to separate at meiosis during either oögenesis, as shown in Fig 18.18, or sometimes during spermatogenesis. Cyto-

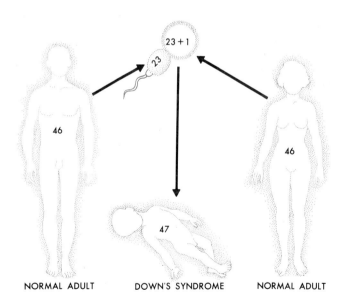

NORMAL ADULT   DOWN'S SYNDROME   NORMAL ADULT

Fig. 18.18
Chromosomal formation of individual afflicted with Down's syndrome born to two normal parents. Both parents have the normal chromosome complement of 46. Nondisjunction of the two members of pair 21 during oögenesis leads to production of an egg with 23 + 1 chromosomes; the extra chromosome is a member of pair number 21, which is thus represented by both homologs from the mother. (After E. Peter Volpe, *Human Heredity and Birth Defects.* New York: Pegasus, a Division of Bobbs-Merrill, 1971, p. 72.)

logical examination of many Down's syndrome patients has shown this pattern to exist in the vast majority of cases.

Occasionally variations in chromosome number other than trisomies occur, though these appear to be quite rare. Variation in chromosome number, regardless of the actual number of chromosomes, whether more than or less than the standard number for the species, is known as *aneuploidy*. Trisomies are just one kind of aneuploidy.

So far we have been discussing only those abnormalities in chromosome number that occur in the germ cells—that is, in the formation of egg and sperm. It is also possible for aneuploids (again, principally trisomices) to occur by nondisjunction in somatic cells during embryonic development. If such somatic nondisjunctions take place early in embryonic life, a large number of cells will be affected, and the embryo may be spontaneously aborted. If, however, aneuploidy occurs later in development, only a small patch of cell (descendants of the cell in which the nondisjunction first occurred) will be affected. Most late-occurring somatic nondisjunctions therefore appear to be relatively unimportant—they affect too few a number of cells, and at too late a stage in the cells' developmental process, to seriously affect function. There is one exception, however. The cells of many malignant tumors (cancers) are often aneuploid. It is possible—though we must emphasize by no means certain—that one of the many possible causes of cancer may be such variations in number of somatic cell chromosomes. Many tumor cells appear to have an extra chromosome 16. Some genes on this chromosome may well be responsible for the control of cell division. Thus, when the balance of these genes is upset due to the presence of an additional chromosome, mitotic controls are lost, and a fast-growing, cancerous tumor results.

Biologists do not know how the presence of an extra member of a chromosome pair (as in trisomies) alters the normal pathway of development. Certainly, the presence of extra chromosomes produces less abnormality in development than the absence of a chromosome (we can conclude that this is true because no cases of autosomal monosomy—having only one member of a pair—have been recorded). But still, why the rather severe and dramatic effects of an *extra* chromosome? The only conclusion biologists can draw at present is that the genome of any cell functions as a well-integrated and well-balanced whole, which appears to be dependent on autosomal *disomy*. Genes occur in certain doses (normally, one for each member of a homologous pair of chromosomes). Disturbing these doses must upset regulatory processes within the cell.

Disorders of Chromosome Structure (Deletions)

Cri-du-Chat (Cat-Cry) Syndrome. Called by the French name *"cri du chat"* (which means, literally, "cry

of the cat"), this syndrome was first described by French physicians. Babies affected by the condition have a cry which sounds like a cat in distress. The sound is the result of the failure of the vocal cords to come together completely. Other symptoms include a rounded face, small cranial vault, and mental retardation. The karyotype of individuals afflicted with this condition show the loss of about one-half of the short arm of chromosome 5. The condition can be transmitted from normal parents to their children if the parent with the deletion still has the deleted chromosome segment present, but attached to a different chromosome (a condition known as *balanced translocation*).

Chromosome Deletions and Cancer.   Several different types of cancer have been linked to deletions of certain chromosomes. Chronic granulocytic leukemia is a malignant condition in which individuals produce too many of the type of leukocytes called granulocytes (see Chapter 11). Various drugs can slow down the proliferation of these granulocytes, thus keeping the malignancy under some control. The occurrence of the condition is frequently, but not always, associated with a deletion from chromosome 22.* It is interesting to note that the shortened chromosome 22's found in this type of cancer patient occur only in cells of the bone marrow, where the granulocytes are produced. To date, no such deletions have been found in gametes of affected persons, and hence chronic granulocytic leukemia appears not to be transmissible from parent to offspring.

Another form of cancer thought to be the possible result of chromosomal deletion is called *retinoblastoma.* A tumor of the retina found mostly in children, retinoblastoma is often associated with loss of a segment of chromosome 13.

Still another form of cancer, known as Wilm's tumor, has led researchers to recognize the value of genetic analysis in prevention of the onset of possibly fatal malignancy. Like retinoblastoma, Wilm's tumor is found mostly in children, where it has often been associated with a condition known as *aniridia*, or lack of irises in the eye (Fig. 18.19), developmental abnormalities of the urogenital tract, and a mild form of mental retardation.

In 1975 cytogeneticists at the University of Minnesota and at Yale Medical School found that some patients with Wilm's tumor, aniridia, and the other associated clinical abnormalities lacked an end-piece of chromosome 11. Normally occurring in the kidney during early childhood, Wilm's tumors can be removed surgically, or treated by chemotherapy, if detected soon enough. Thus, if a child is born without irises, doctors use this signal to search for a chromosome 11 deletion. If the deletion is also present, the chances are signifi-

---

*The deleted chromosome 22 segment is not lost, but is now known to be translocated to one of the larger chromosomes, usually 9.

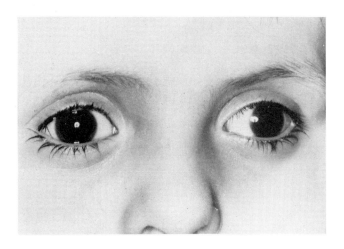

Fig. 18.19
One of a pair of identical twins showing the condition known as aniridia, or lack of irises in the eyes. A number of such individuals (but *not* all) also show a deletion in chromosome 11, and have a high susceptibility to Wilms tumor, a cancer of the kidneys which occurs in young children. (Photograph courtesy of Robert W. Miller, National Cancer Institute. From G. B. Kolata, "Genes and Cancer: The Story of Wilms Tumor," *Science* 207 [Feb. 1980], pp. 970–971. Copyright 1980 by the American Association for the Advancement of Science.)

cantly increased that the child will develop Wilm's tumor. Thus, early detection of the potential for this form of cancer can help to prevent the spread of tumors that could claim the patient's life.

There is a methodological problem involved in attempting to draw a cause-effect relationship between particular chromosome deletions and the occurrence of cancer. In all the types of cancer listed above, chromosomal deletions have been observed in only a small percentage of the actual cancer patients. For example, out of 1200 retinoblastoma patients examined, only twenty-four had chromosome 13 deletions: that is, about one out of every fifty. Wilm's tumor has been directly observed in only eight patients who also have aniridia and the chromosome 11 deletions. These are very small numbers. The statistical errors inherent in small samples make it impossible to draw any firm conclusions about chromosome deletions causing cancer. What is necessary, of course, is to know the number of cases in which the deletion had been observed without a resulting cancer. These figures are difficult to come by because chromosomal screening tests are not routinely carried out in hospitals, or during health examinations. Thus, while it is tempting to think that at least some forms of cancer may be the result of chromosome deletions (and therefore, possibly, detectable early in life) there are good reasons to remain cautious about such inferences at the present time.

SUPPLEMENT 18.4
HUMAN BEHAVIOR AND GENETICS: EUGENICS PAST AND PRESENT

Few people would agree that it is desirable for a child to be born with Down's syndrome, Tay-Sachs Disease, or any of the many other severely crippling genetic conditions known to exist in human beings. Despite the difficult moral questions such disorders present, they have one advantage over other genetic diseases: chromosome abnormalities can be visibly detected early in embryonic life. With this knowledge at hand, it is possible to make decisions whether to bring the child to term or abort the fetus at an early stage.

There has been much discussion over the years about how to eliminate genetic disorders. Where the conditions are clinically definable and universally recognized as genetic in origin, such as hemophilia, albinism, Down's syndrome, and the like, the issue is still a moral and ethical one. Should fetuses with such diagnosed conditions be aborted? Should parents with the potential to have children with genetic defects be discouraged from having children? Who has the right to determine whether particular parents can have children, or a particular fetus is to be born? Many people have argued that social and behavioral phenotypes are all genetic, or largely genetic, and that undesirable behavioral traits should be eliminated from the human population in the same way as clinically definable genetic defects. Behavioral phenotypes which have been claimed to be genetic include personality, mental ability (I.Q.), alcoholism, mathematical ability, criminality, pauperism, laziness, industriousness, and rebelliousness. Not surprisingly, many people have also questioned the claim that any of these behaviors have a hereditary basis.

In the past, whether certain traits usually associated with human behavior and personality are inherited or determined by different environmental conditions has been referred to as the nature-nurture debate. Those who feel that such traits are determined biologically are on the nature side and are often referred to as biological determinists, since they hold that many aspects of human personality are determined by some biological cause. Those who feel that such traits are determined environmentally are on the nurture side and are often referred to as environmentalists.

The importance of the nature-nurture argument over the years has not been so much the intellectual quality of the issues it raises, but rather the social, political, and economic conclusions to which one view or the other tends to lead. For example, some persons on the nature side of the argument often take the statement "The poor shall always be with us" as axiomatic—as a "fact." They *explain* that fact by reference to hereditary factors: the poor achieve a lower socioeconomic status because they are genetically inferior and "don't have what it takes" from the very beginning to be successful. Naturists, or biological determinists, tend to believe that the various socioeconomic strata associated with modern societies are largely biological in origin. Nurturists, or environmentalists, on the other hand, argue that human personality traits, behavior, and position within society are determined largely if not exclusively by environmental factors. They claim that poverty and low socioeconomic status are determined by the social and economic environment in which people grow up, not by their genes. Nurturists have generally approached the solution to social problems by environmental reform; naturists by some form of what has commonly been called eugenics.

The Eugenics Movement in the Early Twentieth Century

Eugenics refers to any attempt to improve the human genetic stock. In general, eugenics programs have consisted of two components: **positive eugenics,** the encouragement of those with "desirable" traits to have more offspring, and **negative eugenics,** the discouragement or even active prevention of reproduction among those thought to have "undesirable" traits. The latter have ranged from educational programs designed to discourage certain classes or groups of people from having children to actual programs for compulsory sterilization.

In the early decades of the twentieth century the "Eugenics Movement," as it was called, became extremely prominent. The movement gained considerable popularity by the mid-1920s and had significant social and political impact. It was responsible for the passage of laws allowing for compulsory sterilization (of criminals, people judged to be "feeble-minded," and sex offenders) in thirty states. It successfully lobbied for antimiscegenation laws (prohibiting nonwhites from marrying whites) in some thirty states. Along with some labor unions and big business, eugenics lobbyists were instrumental in 1924 in the passage of the Johnson Act, or the so-called Immigration Restriction Act, by the United States Congress. The Johnson Act limited annual immigra-

tion from southern and central European countries to one percent of the number from those countries who were living in the United States in 1890. Restriction on immigration was based on the argument that people from these specific countries (Italians, Spaniards, Slavs, Poles) were genetically inferior to the Anglo-Saxon stock that made up the bulk of those considered to be successful in the United States.

Although some supporters of the Eugenics Movement were undoubtedly motivated by humanitarian principles, the main force of the movement was to justify rather than alleviate racism and social oppression. Sweeping statements were made about racially inferior ethnic groups. Marriages between different ethnic groups (eugenicists often made no distinction between race and ethnic group) were considered detrimental because the offspring would be "hybrid inferiors." One prominent eugenicist wrote that:

> *Whether we like to admit it or not, the result of the mixture of the two races, in the long run, gives us a race reverting to the more ancient, generalized and lower type. . . . The cross between a white man and a negro is a negro, and the cross between any one of the three European races and a Jew is a Jew.*

Such feelings were mirrored in the titles of two of the most popular books on eugenics at the time: Madison Grant's *The Passing of the Great Race*, published in 1916, and Lothrop Stoddard's *The Rising of Colour Against White Supremacy*, published in 1920. Both lamented the increasing number of immigrants in the United States and the genetic threat this "inferior horde" posed to American society. The biological inferiority of all non-Anglo-Saxon groups was said to be "proven" by studies in genetics and evolution. Virulent racial and ethnic prejudice was promoted under the guise of scientific authority.

As a result of widespread belief in such ideas, thousands of people were sterilized for alleged "feeblemindedness." Many potential immigrants were unable to join their families in the United States because of the new quotas established after 1924. In 1940, because of the Immigration Restriction Act, a ship carrying Jewish refugees from Nazi Germany was denied entrance to the United States; those refugees were returned to Germany, many later to die in concentration camps. More subtle, but in many ways equally harmful, were the indirect effects of the Eugenics Movement on attitudes about racial and ethnic differences. Over a thirty-year period its propaganda instilled ideas of racial distrust and even hatred. Many of these same ideas still remain with us today.

The Eugenics Movement gained much of its prestige and following by claiming to have a "scientific" basis. Through the Eugenics Record Office, administratively part of the Laboratory for the Study of Experimental Evolution at Cold Spring Harbor (New York), Charles B. Davenport and Harry H. Laughlin directed research into human heredity, especially the inheritance of personality and behavioral traits. Laughlin and Davenport concluded that alcoholism, imbecility, seafaringness (the urge to go to sea), laziness, insubordination, and a host of other traits were determined by simple Mendelian inheritance. They provided mountains of evidence, thousands of pedigrees, and other kinds of "facts" to support such claims. Davenport tried to remain scientifically respectable by refraining from making overtly derogatory racial or ethnic remarks. But Laughlin testified at great length before the House Committee on Immigration and Naturalization between 1922 and 1924. His testimony was instrumental in convincing many congressmen that immigration from certain geographic areas was jeopardizing the quality of America's future genetic stock.

Many geneticists initially favored some of the aims of the Eugenics movement. When the movement became more overtly racist and anti-Semitic, some withdrew their support. They objected not only to the movement's racial overtones, but also to the fact that subjective judgments about superiority and inferiority of one group over another were being made without any sound biological evidence. By 1920 most geneticists realized that the Eugenics movement was based on backward, incorrect biology. Unfortunately, very few geneticists spoke out publicly against the eugenicists or their movement, but withdrew their support from the movement quietly. Some historians have argued that the failure of geneticists to publicly repudiate the Eugenics movement allowed the movement to gain authority in the public eye; it appeared that Davenport and Laughlin spoke for *all* geneticists.

Eugenic Ideas Today

Although there is no longer a formal eugenics movement in the United States, ideas similar to those advanced by the early eugenicists have reappeared in the past fifteen years. Among them are claims that intelligence is hereditary and that blacks are inferior to whites in intelligence; that *XYY* males are predisposed to criminal and violent behavior; that homosexuality is genetically determined; and that male-female sex-role differences are due to heredity. Because these theories place the cause for certain human behavioral traits at the genetic level, they would all be examples of theories founded in biological determinism.

Increasing concern on the part of some people about the resurgence of such ideas falls into two broad areas: scientific and social.

On the scientific side, there has been considerable debate within the biological community about whether available data establish any of the above hypotheses. For example, in 1969 Berkeley psychologist Arthur Jensen advanced the hypothesis that the average difference in I.Q. scores between black and white populations in the United States was due to genetic factors. To substantiate his point, Jensen analyzed I.Q. scores for a number of monozygotic (identical) twins

who had been reared apart. Since monozygotic twins have identical heredity and, if reared apart, are assumed to have substantial differences in environment, analysis of their I.Q. scores might yield some information on the relative influence of genetics or environment on determination of intelligence. Similarity of I.Q. scores would indicate a greater role for heredity; dissimilarity a greater role for environment. Since the publication of Jensen's paper, numerous investigators have criticized his methods and conclusions. Among the objections have been:

1. I.Q. tests are culture-bound, reflecting experience as much as or more than innate brain capacity; hence, such tests are an inadequate tool for measuring the phenotypic trait called intelligence.

2. The I.Q. data on many of Jensen's monozygotic twin pairs were obtained in an unreliable manner: different kinds of tests were used, administered by many different people in many different ways.

3. Many of the twins used in the study did not actually grow up in very separate environments; that is, many twins were raised by different members of the same family, such as aunts and uncles, or grandparents, and often lived very close to one another and visited each other's house frequently.

4. Jensen's statistical analysis of the data contained some errors that made comparison of black and white I.Q. scores invalid.

5. Recently the validity of even the original I.Q. measurements has been questioned. Several investigators have found that Burt, the investigator whose data Jensen quoted, never made most of the interviews with monozygotic twins, nor did he compute their scores as his papers claimed. Indeed, much of the data was taken from a study made in London between 1892 and 1903 (though not referenced by Burt), and a paper from 1926 by two separate authors. Burt simply picked and chose those data which would agree with his predictions—and prejudices! To make matters worse, several co-authors of Burt's papers have been found to be nonexistent.

As long as the data base for Jensen's conclusions is subject to such questions, it is obvious that the social significance of his conclusions cannot be accepted as meaningful.

Historically, most theories of biological determinism have been based on similarly slim, if not actually incorrect or oversimplified, data (see also Chapter 1, Supplement 1.3). Many investigators feel this is because the questions are not really answerable in a rigorous scientific way. For example, where behavioral or mental traits are concerned, any given phenotype is obviously a result of some interaction between genetic and environmental factors. To adequately separate the influence of each, it would be necessary to do the kinds of controlled breeding and rearing experiments that ethical considerations forbid with human beings. Without being able to carry out such controlled experiments, no rigorous answer is possible. Yet in the past, as at present, theories of biological determinism have been brought forward, often with considerable force, despite such limitations of methodology.

On the social side, many investigators are concerned about the mass publicity that theories of biological determinism often receive. Although every major theory of biological determinism in the past has evoked a considerable debate within the scientific community, it is the biological determinist's side that has usually been broadcast to the public at large. Many biologists feel that the public is treated to only one side of what is a highly controversial scientific debate. Thus, it often appears that the scientific community agrees with the biological determinists, thereby adding weight to the latters' arguments. Once a hypothesis has been suggested, its use in the social sphere is beyond the scientist's control. The social cost of being wrong in claims about the attributes or capabilities of individuals or groups is so high that many feel there is no point in continuing to ask the sort of questions to which no rigorous answers are possible.

It should be emphasized that it is certainly *possible* that many human behavioral and mental traits—even fairly specific ones, like the ability to achieve a certain score on I.Q. tests—might be genetically controlled. But as we saw in Chapter 1, science is an attempt to determine what is true, not just what might be possible, in the natural world. All sorts of possibilities exist, but that something is possible does not mean it is true.

The Ethics of Eugenics
From the earliest days of the eugenics movement onward, questions of ethics and morality have continually been raised about the practice of controlled breeding. Thus, even if eugenic ideas were true, there would still be the very real question of whether they should be made the basis of social planning. How far can society go in determining who should marry whom, and who should have children? How far should society go in encouraging, let alone forcing, people to be sterilized? Who is to decide which traits are advantageous, and which harmful? These are difficult moral questions that cannot be decided in an offhand way without producing untold human harm.

Moreover, biologists have pointed out that even the most rigorous programs of negative selection (selecting *against* a particular trait) are ineffective in removing the genes for that trait from the population. Therefore, many people point out that even if there were no moral or ethical problems with eugenics, as a practice it has severe limitations in accomplishing its own ends.

For scientific, practical, and ethical reasons, many people view theories of biological determinism of any kind with some skepticism. Others feel there is considerable danger in not adopting some eugenic programs soon. They feel that the human species is rapidly accumulating an overload of defective and inferior genes that may eventually lead to biological and social deterioration. Some people of this persuasion strongly advocate programs to control the breeding of those who are known or judged to carry undesirable traits. Obviously this question involves complex moral and legal issues that cannot be solved in isolation from a more general social context.

18.8
EXTRA-CHROMOSOMAL INHERITANCE

In 1909 the German botanist Carl Correns (one of the three rediscoverers of Mendel's paper in 1900) noted that certain traits in plants did not follow expected Mendelian inheritance patterns. He was studying a plant known as the four-o'clock (*Mirabilis jalapa*), some of whose leaves show a patchy, or variegated coloration. Correns observed, among other things, that the distribution of green pigment varied enormously from leaf to leaf: some were entirely green over the whole surface, while others had large splotches of white mixed in with patches of green, and some were all white (that is, had no green at all). Microscopic examination of cells in the white areas revealed that no chloroplasts were present.

Correns was curious about the inheritance of the variegated trait, and carried out several breeding experiments. If he took ovules from flowers on a part of the plant (branch) with fully green leaves, the resulting offspring had fully green (nonvariegated) leaves, even if the source of pollen had been a flower from a part of the plant with variegated or totally white leaves. The converse was also true: if the ovule came from an all-white portion of the plant, the offspring would be all white (and nonphotosynthetic) even if the source of pollen had been a flower from a wholly green portion of the plant. When ovules came from a flower on a variegated portion of the plant, however, the offspring turned out to be of three types—wholly green, wholly white, and variegated—again, regardless of the source of pollen. Correns's results are summarized in Table 18.6.

It is clear from the above results that the offspring inherit their leaf coloration *strictly from the female parent*, a phenomenon called *maternal inheritance*. Since virtually all the cytoplasm of any zygote is contributed by the female parent, it has been suggested that maternally inherited traits are passed from generation to generation through the cytoplasm of the egg. In recent years geneticists have shown that cytoplasmic organelles such as chloroplasts and mitochondria have

Table 18.6
Inheritance of Leaf Variegation Pattern in *Mirabilis jalapa* (Four-O'Clock)

Leaf color on portion of plant from which pollen was taken	Leaf color on portion of plant from which ovule was taken	Leaf color exhibited by progeny
Full green	Full green White Variegated	Full green White Green, white, and variegated
White	Full green White Variegated	Full green White Green, white, and variegated
Variegated	Full green White Variegated	Full green White Green, white, and variegated

Modified from Ursula Goodenough, *Genetics*, 2d ed. New York: Holt, Rinehart and Winston, 1978, p. 563.

DNA of their own and are self-replicating. A plant cell will thus pass on to its offspring only the types of chloroplasts that it carries in its own cytoplasm. This generalization can be demonstrated in the following way. If all the chloroplasts are removed from a single-celled organism such as *Euglena*, neither this cell nor its descendants develop new chloroplasts. (*Euglena* can survive heterotrophically without its chloroplasts.) If new chloroplasts of a recognizably different sort are surgically reintroduced into a *Euglena* that lacks chloroplasts, the cell's descendants will contain only the new type of chloroplasts.

Studies of both chloroplasts and mitochondria over the past decade have revealed that both of these organelles have their own DNA and ribosomes. The DNA in these organelles is small, coding for only ten to twenty polypeptides at most (chloroplast DNA tends to be slightly larger in molecular weight than mitochondrial). Given the complex respiratory enzymes found in mitochondria, and the photosystems I and II found in chloroplasts, it is evident that the small DNA segments found in these organelles cannot possibly code for all the necessary components. Where do the rest of the mitochondrial and chloroplast proteins come from?

While some of the molecular constituents of mitochondria and chloroplasts are coded by genes in the organelles themselves, others are coded by genes in the nucleus (See Table 18.7). In mitochondria, for example, the organelle's genome codes for several polypeptide subunits found in mitochondrial enzymes, and for one found in mitochondrial ribosomes. These polypeptides are also synthesized right inside the mitochondrion. Most of the other polypeptides found in mitochondrial protein are coded in the cell nucleus and synthesized in

Table 18.7
Localization of Genetic Information for Major Mitochondrial Enzyme Systems in Yeast

Enzyme complex from mitochondria	Number of polypeptide subunits		
	Total	Coded in nuclear genes	Coded in mitochondrial genes
Cytochrome oxidase	7	4	3
Cytochrome *b, c, a*, and *a₃* complex	7	6	1
Adenosine triphosphatase	9	5	4
Polypeptides in large ribosomal subunits	30	30	0
Polypeptides in small ribosomal subunits	22	21	1

Data drawn from P. Borst, in *International Cell Biology*, B. R. Brinkley and K. R. Porter, eds. New York: Rockefeller University Press, 1977; taken from Goodenough, 1978, p. 555.

the cell cytoplasm. A similar pattern is observed with proteins in chloroplasts. It is obvious that, while possessing the ability to reproduce themselves, the organelles are not self-sufficient. They cannot exist or reproduce themselves outside of living cells.

What is particularly intriguing about the relationship between the nuclear and extra-nuclear genomes in organelles such as mitochondria is the fact that both genomes usually contribute to the coding of each particular enzyme. For example, as shown in Table 18.7, of the total of seven polypeptide chains found as subunits in the various cytochrome oxidase molecules of yeast, four are coded by the nuclear DNA and three by the mitochondrial. In short, the mitochondrial and nuclear DNA in any cell share in determining the chemical and physical characteristics of the organelle. Mitochondria and chloroplasts can replicate themselves fully in the appropriate medium—the cell cytoplasm—which includes, of course, the components of the organelle coded by the nuclear DNA. However, even though the nuclear DNA directs the synthesis of all the mitochondrial or chloroplast components, if the organelle

and its DNA are not present to begin with, not even a recognizable part of the organelle ever develops.

The sharing of genes in eukaryotic cells between the cell nucleus and cytoplasmic organelles has led to many intriguing evolutionary speculations. Among these are the suggestion that mitochondria and chloroplasts evolved from originally free-living bacteria and algae, respectively, that were incorporated into the eukaryotic cell as a sort of internal parasite. (This hypothesis is discussed in more detail in Chapter 27, Supplement 27.1.) If such a hypothesis about the origin of mitochondria and chloroplasts is true, it is interesting to imagine what sort of advantage might be gained by having some organelle proteins coded in the organelle DNA and others in the nuclear DNA. The relationship between chromosomal and extra-chromosomal DNA, though still a largely unexplored area, highlights something geneticists have recognized for a long time: genes and chromosomes are not isolated structures, but function in an integrated way within the whole cell. The division between nucleus and cytoplasm is much less distinct than we have been accustomed to viewing it in the past.

Summary

1. The chromosome theory of inheritance, as developed for eukaryotic organisms, claims that individual genes exist as discrete units on the chromosomes in the cell nucleus. Genes are linked together, existing side-by-side on the same chromosome. In all eukaryotic organisms, the number of traits inherited together (linkage groups) corresponds closely to the haploid number of chromosome groups.

2. Linkage of genes on the same chromosome is sometimes disrupted by crossing over between homologous chromosomes, a process that occurs during synapsis of meiosis I in gamete-producing cells. Breakage is usually followed by recombination, which produces a new combination of alleles.

3. The existence of linkage, crossing over, and recombination has been used to make genetic maps. Genetic map distances are given in arbitrary units (map units) that are purely relative. The genetic map itself has no necessary correspondence to the physical structure of the chromosome. Devising a method to show that genetic maps correspond directly to physical regions of a given chromosome pair required the development of cytological techniques, especially those associated with the large salivary gland chromosomes of *Drosophila.*

4. Morgan and others showed that sex-linkage occurred in *Drosophila.* Sex-linkage is a phenomenon by which certain traits appear most frequently (but usually not exclusively) in members of one sex. For organisms like *Drosophila* and human beings (where males are *XY* and females *XX*), recessive sex-linked traits are phenotypically expressed most

frequently in males. Sex-linked traits are found as genes on the *X* chromosome. Because most mutants are recessive, they will show up only in males, where the *Y* chromosome has no corresponding allele. In females, the recessive sex-linked traits will not show up very frequently, since most of the times the homologous *X* chromosome will contain a dominant allele. Only when the daughter receives two mutant alleles, one from her father, the other from her mother, will she show the *X*-linked recessive trait. White-eye is a recessive, *X*-linked trait in *Drosophila;* red-green color-blindness and hemophilia are two *X*-linked traits in human beings. In many kinds of organisms a few traits are determined by genes on the *Y* chromosome. These would be sex-linked traits also, and they would appear exclusively in males.

5. Plasmids are extra-chromosomal DNA in bacteria. One of the common plasmids is the *F* (or sex) factor, which can exist separately in the cytoplasm (where it is called an episome or episomal plasmid), or integrated into the bacterial chromosome (where it is called a nonepisomal plasmid).

6. While most genes in eukaryotic organisms are contained within the nucleus as part of the chromosomes, some traits are determined by extra-chromosomal genes. Such genes do not follow Mendelian inheritance patterns; extra-chromosomal (or cytoplasmic) genes are contributed to an offspring solely by the mother, through the egg cytoplasm, a process called "maternal inheritance." Chloroplasts and mitochondria, with their own extra-nuclear DNA, show maternal inheritance.

Exercises

1. In rats, pigmentation is determined by a dominant gene *P*, albinism by its recessive allele *p*. Black coat color is due to a dominant gene *B*, whitish color to its recessive allele *b*. For the black gene to express itself, however, the gene *P* must be present. If a rat with the characteristics *PPBB* is crossed with another whose characteristics are *ppbb*, what will be the phenotypes of the first generation?

2. Among the children of a man and woman who are both heterozygous for albinism, the distribution of their offspring will be expected to be three with normal pigmentation to one albino. However, studies of a number of families in which albinism is known reveal an interesting situation: Albinos make up considerably more than 25 percent of the offspring of these families. Furthermore, the fewer the children in each family, the greater the proportion of albinos among them. Geneticists, however, maintain that these results still agree perfectly with Mendelian principles. Offer an explanation for this discrepancy.

3. In mice, yellow coat color is known. Crosses between black-coated and yellow-coated mice yield a 1:1 ratio of black- to yellow-coated mice. These same blacks, mated to each other, produce only black-coated young. These results are consistent with a hypothesis proposing a genotype of *BB* for the black mice and *Bb* for the yellow mice. However, this hypothesis also predicts that crosses between yellow mice should yield a ratio of one black to two yellow to one of whatever color results from genotype *bb*. This ratio, however, is not obtained. Instead, the result is a 1:2 ratio of black- to yellow-coated mice. Further investigation salvages the hypothesis, however. Dissection of the uteruses of the yellow females yielding such 1:2 ratios shows that one-fourth of the young are arrested at an early stage of their embryological development and die before birth. Evidently, then, the *bb* genotype represents a lethal combination, and the organism cannot survive. Suggest how, in terms of the modern concept of gene action, a lethal gene combination might act to produce its effects.

4. In cats, short hair is dominant over long hair. A long-haired male is mated to a short-haired female whose father had long hair. They produce a litter of ten kittens.

   a) What proportion of their kittens would you predict to have long hair?

   b) Give *two* hypotheses that might account for all ten kittens being long-haired.

   c) Devise an experiment that would distinguish between the two hypotheses given in (b).

Chapter 19
Genetics IV: The Molecular Biology of the Gene

19.1
INTRODUCTION

The preceding two chapters have focused largely on the physical nature of heredity: how genetic factors are transmitted from one generation to another and how genes are arranged on chromosomes. The other side of the story of heredity is the question of how genes *function*. This has commonly been referred to as the problem of gene **translation.** It involves studying the molecular nature of genes and their relationship to protein structure. It also involves studying the relationship between genes and biochemical pathways within the cell. In the last analysis, genes can be thought of as molecules containing coded information that is translated into specific protein structure. As enzymes, some of these proteins control biochemical pathways and ultimately play an important part in determining the development of the organism's phenotype.

19.2
GENE STRUCTURE

It is now known that genes are composed of nucleic acid (see Supplement 19.1). For most organisms the nucleic acid is deoxyribonucleic acid (DNA); in some (mostly plant viruses), the gene is composed of ribonucleic acid (RNA). As we saw in the previous chapter, in viruses and prokaryotes such as bacteria, genes are arranged in a linear fashion along a single molecule of nucleic acid (the bacterial or viral chromosome). In eukaryotes, on the other hand, the nucleic acid is organized in a more complex way, as part of chromosomes that also contain protein. In both prokaryotes and eukaryotes, however, the molecular structure of nucleic acid—particularly the DNA—and its basic function is the same. Indeed, the function of DNA is intimately related to its structure. Since DNA is the molecule of heredity in the vast majority of organisms, we will focus our discussion on the molecular structure of the DNA, rather than RNA, gene.

Determining the structure of DNA necessarily entails two steps:

1. The molecular subunits that compose DNA must be identified.

2. The way these parts are fitted together to form the entire DNA molecule must be determined hypothetically.

Step 2 is subject to two limitations. First, if genes are composed of DNA, then DNA itself must be a self-replicating molecule. The hypothesized model of DNA structure must therefore explain how genes are able to form copies of themselves. Second, the hypothesized model of DNA structure must also account for the way that genes are able to carry out their functions leading to the expression of phenotypes they control.

Components of DNA

Step 1 involves a chemical analysis of pure DNA. Such an analysis reveals DNA to be composed of just three different chemical substances.

1. A five-carbon (pentose) sugar, **deoxyribose.**

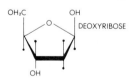

2. Phosphates. Each phosphate group is composed of an atom of phosphorus surrounded by four atoms of oxygen and two of hydrogen (the latter are not shown).

$$-O-\overset{\overset{\displaystyle O}{\|}}{\underset{\underset{\displaystyle O}{|}}{P}}-O-$$

PHOSPHATE

SUPPLEMENT 19.1
DNA, THE MOLECULE OF HEREDITY: HOW DO WE KNOW?

Among the first investigators to be interested in the chemical composition of the cell nucleus was German biochemist Friedrich Miescher (1844–1895). From the white blood cells he found in pus, Miescher isolated a substance that had not been identified in cells before. Acid in nature and rich in phosphorus, the new material was named *nuclein* by Miescher in 1871. (It was subsequently called nucleic acid, or deoxyribonucleic acid.) Later studies with other types of cells, including salmon sperm, indicated that nuclein was present in most, if not all, nuclei.

Recall that at about this time (in the last twenty-five years of the nineteenth century), cytologists were becoming convinced that the chromosomes were the bearers of heredity in the cell. When nuclein was found to be localized on the chromosomes, it would seem logical to conclude that this substance was, perhaps, the molecule of heredity. That was not a foregone conclusion, however. Chromosomes of eukaryotes consist not only of nuclein, but also of a protein called histone and various lipids. One or more of these substances could also be the primary genetic material.

Griffith's Seminal Observations

In 1927 an experiment bearing indirectly on this question was performed by Frederick Griffith. Griffith worked with two strains of pneumonia bacteria (see Diagram A). Strain S forms smooth colonies on a bacterial agar plate and each cell is encapsulated (that is, surrounded by a mucopolysaccharide coat). Strain R forms rough colonies, and R cells are unencapsulated (that is, not surrounded by a mucopolysaccharide coat).

More important to Griffith's experiment, however, was the fact that when injected into mice, strain S causes the mice to die. Progeny bacteria harvested from the dead mouse and injected into the other mice also caused those mice to die (Diagram A, part 2). The injection of strain R cells into mice, on the other hand, does not cause death (Diagram A, part 1). The lack of virulence is also an inherited feature of these strain R bacteria.

Griffith found that when strain S bacteria were heated to 60°C they were killed. Injection of such dead strain S bacteria was no longer fatal to mice (Diagram A, part 3). However, if the dead strain S bacteria were injected along with some living strain R bacteria, some of the mice receiving such injections died (Diagram A, part 4). The blood of these mice was always infected with living virulent bacteria, that is, *bearing capsules*.

The interpretation of these experimental results seems clear: something must have passed from the dead strain S bacteria into the living strain R to transform them from nonvirulent, unencapsulated cells to virulent, encapsulated cells. Further, the fact that these transformed bacterial cells passed

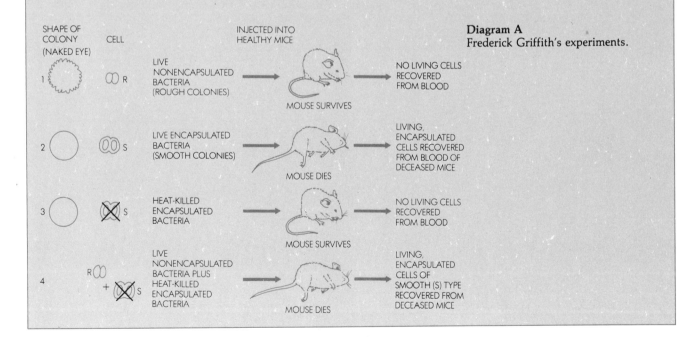

Diagram A
Frederick Griffith's experiments.

on their new characteristics to their descendants indicated it was the genetic material of the dead strain S cells with which they were mixed before injection that must have entered the strain R cells and transformed them.

Isolating the "Active Fraction"

The next step was a logical one—isolate the genetic substance responsible for Griffith's bacterial transformation and chemically identify it. This task proved not to be an easy one. From the fact that the main bulk of chromosomal material is protein, it was generally thought that the genes, too, must be proteins. Yet none of these isolated proteins caused bacterial transformation. Finally, in 1944, O. T. Avery, C. M. MacLeod, and M. McCarty, who were working with pneumococcus bacteria, isolated a "biologically active fraction" from encapsulated strain S bacteria. Under appropriate culture conditions, it was shown that this active fraction could genetically transform the R strain to the S strain. By various chemical and physical techniques, this active fraction was found to be composed primarily of DNA. Avery and his co-workers concluded their historic paper with the words:

> The evidence presented here supports the belief that a nucleic acid of the de[s]oxyribose type is the fundamental unit of the transforming principle of Pneumococcus Type III.

It seemed, then, that DNA was the primary genetic material. By some still-undetermined process, one strain of bacteria had absorbed the DNA of another strain and had thereby acquired characteristics of the donor strain. Further,

and most important, this transformation of the recipient strain of cells was passed on to their descendants.

There were, however, some objections to acceptance of DNA as the primary genetic material on the basis of these experiments. These objections were based on the fact that the transforming DNA fraction prepared by Avery, MacLeod, and McCarty was only about 95 percent pure. It was reasoned that the 5 percent impurities (mostly protein) could be the genetically active fraction, rather than the DNA. This argument was later somewhat weakened when it was shown that highly purified DNA, containing negligible amounts of other compounds and less than 0.02 percent protein, was still capable of causing transformation.

The critical experiment, however, utilized the enzyme deoxyribonuclease. This enzyme hydrolyzes DNA, destroying its function, but leaves other compounds, such as RNA and proteins, unaffected. It could be reasoned that if DNA is the primary genetic material causing bacterial transformation, then addition of the enzyme deoxyribonuclease to the transforming mixture before it is exposed to the recipient bacteria should destroy its ability to cause transformation. On the other hand, if the genetic activity of the transforming mixture is *not* due to the DNA, but rather to the protein or other impurities it contains, then exposure of the enzyme deoxyribonuclease to the transforming mixture should have no effect on its ability to cause transformation.* The results of

---

*This hypothesis is valid only if the assumption is made that *either* DNA *or* protein, but not both interacting with one another, determines hereditary traits. In the present case the assumption turned out to be a safe one.

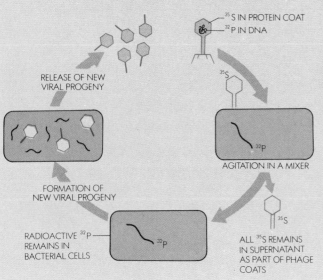

RELEASE OF NEW VIRAL PROGENY

FORMATION OF NEW VIRAL PROGENY

RADIOACTIVE ^{32}P REMAINS IN BACTERIAL CELLS

^{32}P

^{35}S IN PROTEIN COAT
^{32}P IN DNA

^{35}S

^{32}P

AGITATION IN A MIXER

^{35}S

ALL ^{35}S REMAINS IN SUPERNATANT AS PART OF PHAGE COATS

Diagram B

Hershey-Chase experiment with phage T$_2$. The experiment shown here is a later modification of the actual experiment by Hershey and Chase, as published in 1952. Hershey and Chase did not perform the double-labeling shown in the diagram, but labeled only the DNA with radioactive phosphorus.

this experiment clearly supported the first hypothesis. No bacterial transformation was caused by mixtures exposed to deoxyribonuclease.

Further Evidence

The bacteriophage viruses discussed in Chapter 18 provide still further insight into the possibility that DNA is the primary genetic material. Recall that phage become attached to the surfaces of bacteria. After about 40 min the bacteria burst, each cell releasing about 100 complete new T_2 viruses. It is clear that some substance (or substances) must pass from the infecting virus into the bacteria cell and cause the formation of new viruses. This substance, therefore, must contain the genes of the virus.

Chemical analysis of the T_2 virus reveals it to be composed only of DNA and protein:

PROTEIN COAT

DNA

DNA contains phosphorus; protein does not. Conversely, the viral protein contains sulfur; DNA does not. To investigators A. D. Hershey and Martha Chase this fact suggested a critical experiment as outlined in Diagram B. By tagging or labeling the viral DNA with radioactive phosphorus they could tell which part of the virus entered the bacterial cell and participated in the formation of new viruses. The results (see Diagram B) showed that almost all of the DNA entered the bacterial cell on infection. In contrast, only 3 percent of the protein did so. The conclusion to these experiments on bacterial transformation and viruses seems inescapable: genes, at least those of the pneumococci bacteria and T_2 viruses, are made of DNA.

Note, however, that we have not yet generalized beyond the organisms used in these experiments. The crucial question is: Are *all* genes made of DNA? The answer is no, for there are some viruses that do not contain DNA. In such viruses, RNA has been shown to be the primary genetic material. In terms of studies done to date, it seems clear that DNA is the molecule of heredity for virtually all species of plants and animals. DNA appears to be the basis of most hereditary processes in the living world.

3. Usually four nitrogenous (nitrogen-containing) bases—adenine, guanine, cytosine, and thymine. Adenine and guanine are **purines.** Purine molecules are double-ring structures. Cytosine and thymine are **pyrimidines.** Pyrimidine molecules consist of a single ring of atoms.

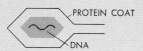

GENERALIZED PURINE

GENERALIZED PYRIMIDINE

Note that the pyrimidine bases are smaller than the purines. The significance of this size difference will be seen shortly.

Structure of DNA

With these substances identified, many biologists turned their attention to Step 2—devising a satisfactory hypothetical model for the structure of DNA. Among these biologists were J. D. Watson and F. H. C. Crick, then at the Cavendish Laboratory in Cambridge, England, and Maurice Wilkins and Rosalind Franklin at Kings College, London. Using data obtained from many different experiments, Watson and Crick proposed a model of DNA structure that has proved highly successful both in its ability to account for gene replication and function and in the accuracy of the predictions that can be derived from it.

When extracted and precipitated in a cold solution of ethanol sodium chloride, DNA rather resembles strands of a spider web or spun glass. Electron microscope photographs taken since the Watson and Crick model was proposed confirm the thread-like nature of

DNA is the molecule of heredity in nearly all organisms, from bacteria and viruses to human beings. Only in certain types of virus, such as tobacco mosaic virus or polio virus, is the hereditary material different. There it is a form of RNA.

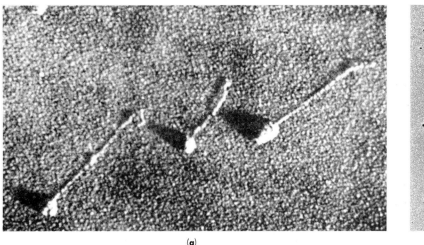
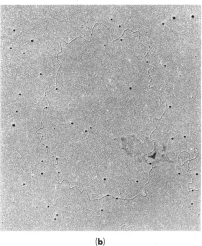

(a)   (b)

Fig. 19.1
Under the electron microscope, DNA appears long and thread-like. (a) DNA is seen extruding from influenza viruses. (b) The circular DNA strand of the lambda viruses that infect *E. coli*. Each such DNA strand is calculated to contain approximately 50,000 nucleotide pairs. (Lambda virus micrograph courtesy Lorne A. McHattie and Vernon C. Bode, Harvard University Medical School.)

DNA (Fig. 19.1). Such photographs show each DNA molecule to be quite long but only about 20 Å wide. This latter figure is an important one, for it reveals that the DNA molecule can only be about ten or twelve atoms across.

A second important consideration in determining DNA structure is that one must know which parts are capable of being chemically united; one does not attempt to do a jigsaw puzzle by forcing together pieces that obviously do not fit. It can be shown that the molecular subunits of DNA are joined together into larger subunits called **nucleotides.** Each DNA nucleotide consists of one of the nitrogenous bases, one molecule of deoxyribose, and a phosphate group. Since four nitrogenous bases are generally found in DNA, there are four different nucleotides (Fig. 19.2).

It can be shown that these nucleotides will join together to form long, polynucleotide strands (Fig. 19.3). Nucleotides can undergo dehydration synthesis between

Fig. 19.2
The four nucleotide bases that form the building blocks of deoxyribonucleic acid, DNA. A nucleotide consists of a purine or pyrimidine base bonded to a sugar and a phosphate group. The sugar-phosphate groups are identical for all nucleotides in DNA. Only the bases differ from one nucleotide to another.

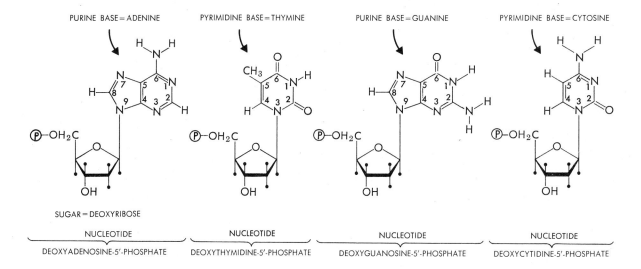

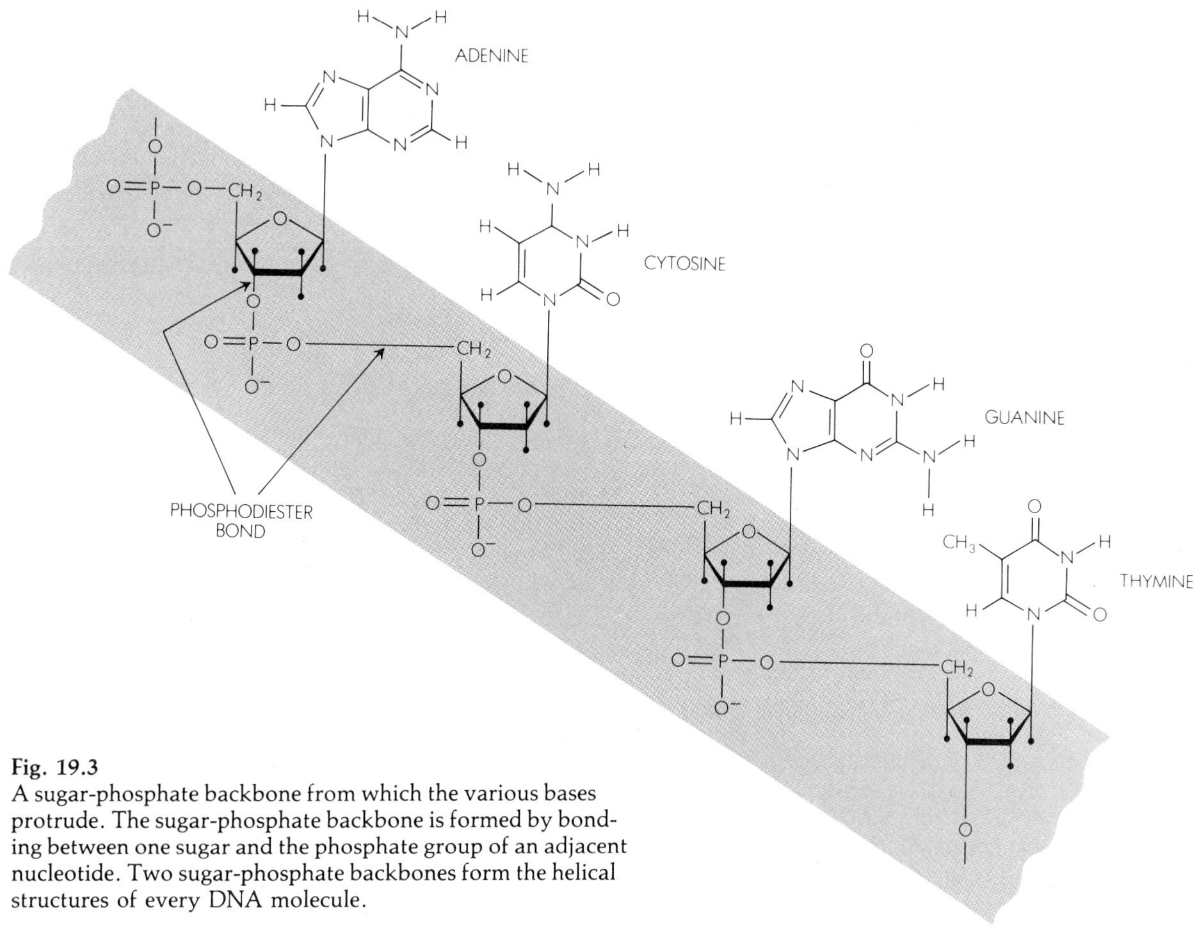

Fig. 19.3
A sugar-phosphate backbone from which the various bases protrude. The sugar-phosphate backbone is formed by bonding between one sugar and the phosphate group of an adjacent nucleotide. Two sugar-phosphate backbones form the helical structures of every DNA molecule.

the phosphate residue on the 5' carbon of the sugar and the 3' hydroxyl group on an adjacent sugar, forming what is called a phosphodiester bond. This fact suggests that the same arrangement might be found in the thread-like DNA molecule. But the distance across one polynucleotide strand is only about 10 Å, and DNA, as already mentioned, is generally about 20 Å wide. This suggests that DNA might be composed of paired polynucleotides lying side by side in chemical union.

Determination of the way in which these two strands might be oriented is based largely on informa-

tion supplied by the physical chemist. First, the molecular structure and configuration of the nitrogenous bases show them to be capable of forming weak hydrogen bonds with each other. Second, the nitrogenous bases are repelled by water (they are *hydrophobic*), while the sugar-phosphate portions of each nucleotide readily form bonds to water molecules (they are *hydrophilic*). These facts favor an arrangement in which the nitrogenous bases face the interior of the DNA molecule (from which the water of the surrounding cellular environment would be essentially excluded) with the

All the processes of life—such as development of the egg into a fully differentiated adult, the chemical functioning of a cell, or the transformation of a normal cell into a cancer cell—can be viewed in terms of the functioning of DNA molecules.

SUPPLEMENT 19.2
THE DOUBLE HELIX: SCIENCE AS A SOCIAL PROCESS

In 1968, fifteen years after J. D. Watson and F. H. C. Crick published their now-famous paper on the structure of DNA in the pages of *Nature* magazine, Watson published a book on his own titled *The Double Helix*. To the casual browser in a bookstore the title might have suggested at best a semipop-ularized discussion of molecular genetics. Such was far from the case, however. *The Double Helix* is Watson's personal account of how he entered genetics, came to collaborate with Crick, and ultimately got involved in a fast and at times underhanded race to solve the molecular configuration of DNA. *The Double Helix* blows to the wind the public image, or stereotype, of scientists as cool and logical individuals, working alone in an ivory tower for the pure love of truth. Watson describes his various colleagues in candid and many times biased terms. Yet his account does portray the intense-ly human side of the scientific enterprise.

Storms of protest from the scientific community greeted the publication of *The Double Helix*. Some of the criticisms were justified. Watson *was* highly biased about some of those he encountered. For example, his portrayal of the one woman who played a major role in the discovery, the late Rosalind Franklin, has been attacked by men and women alike as highly unfair and grossly sexist. He also failed to give the biochemical side of the development a significant place in his narrative. But much of the outcry came from scientists who felt that Watson had undermined the scientific enterprise itself. Many felt he had betrayed the image. But what *is* the scientific enterprise? What *is* the image?

Since the publication of *The Double Helix* a variety of other writers, nearly all somewhat more objective than Wat-son, tried to reconstruct the developments that led to discovery of the structure of DNA. These accounts include Robert Olby's *The Path to the Double Helix* (University of Washing-ton Press, 1974), Franklin Portugal and Jack Cohen's *A Cen-tury of DNA* (MIT Press, 1977), Horace Fredland Judson's *The Eighth Day of Creation* (Simon and Schuster, 1979), and Anne Sayre's *Rosalind Franklin and DNA* (Norton, 1975). While the accounts vary considerably, they all tell something of the same story: of the great importance of human interac-tions and institutional affiliations in scientific work.

The most persistent, and ultimately fruitful interaction was, of course, that between Watson and Crick. Their col-laboration came about quite by accident. Watson had com-pleted his doctoral degree in Indiana University where he had worked on bacteriophages with Salvador Luria, one of the early phage investigators. From his teacher, Watson had gained a curiosity about the nature of the gene—what it was and how it functioned. He was aware of the work of Beadle and Tatum and others who were emphasizing the idea that

genes controlled biochemical pathways. Watson knew that he would have to learn some biochemistry if he were ever to study gene function, and he knew he hated chemistry of all sorts. Obtaining a postdoctoral fellowship, he therefore went to Europe on Luria's advice, to study biochemistry. By his own admission, this seemed absolutely worthless, since biochemists at that time (1951) were by and large totally un-concerned with genetics. Eventually, through a chance trip to Naples, Watson met Maurice Wilkins, an X-ray crystal-lographer from Kings College, London, who had been work-ing for several years on the analysis of crystalline DNA. Watson decided then and there that he would try and go to England to learn about X-ray crystallography and the work going on in London and Cambridge on the structure of nucleic acids and proteins. This, he thought, might provide some insight on the nature of the gene. At first his request to change the locality of his fellowship was denied. But eventu-ally he managed to gain enough funds to move to Cam-bridge. It was here, at the famed Cavendish Laboratory, that he first met Crick.

Ten years older than Watson, Crick was a physicist who returned to Cambridge after World War II to complete his doctoral degree. There he had become interested in biophys-ics, especially X-ray crystallography. Watson described Crick in colorful, if unambiguous terms:

I have never seen Francis Crick in a modest mood. . . . Although some of his closest colleagues realized the value of his quick, penetrating mind, and frequently sought his advice, he was often not appreciated, and most people thought he talked too much. [He] talked louder and faster than anyone else and, when he laughed, his location with-in the Cavendish was obvious.

[*Double Helix*, pp. 7, 9]

Crick was always looking for new ideas and talking excitedly (and, one gathers, loudly) about all sorts of theories. Crick was especially fascinated by proteins. He was working in the Cavendish, whose director, Sir Lawrence Bragg, was one of the foremost, though older, developers of the technique of X-ray crystallography.

From Crick, Watson learned the importance, if not the practice, of the X-ray crystallography technique. Admit-tedly, Watson knew nothing about it when he arrived, but Crick excited him about the prospects. So, too, did a number of other distinguished workers at the Cavendish Laboratory, such as Max Perutz and John C. Kendrew, both studying the three-dimensional structure of hemoglobin and myoglobin. The concept of molecular structure, and its relation to function, was very much in the air in 1951 when Watson arrived at the Cavendish from Europe.

As quick, free-ranging, and imaginative as were the theoretical discussions between Watson and Crick, there were some problems that hindered their work on solving the molecular structure of the gene. Neither knew any biochemistry; Crick knew some X-ray crystallography, but it was not really his field; and Watson knew none. And neither did much in the way of experiments. How could they hope to make any progress on such a difficult multidisciplinary project?

Fortunately for Watson and Crick, they did not have to be dependent on their own X-ray crystallography skills. At Kings College in London, Maurice Wilkins and another investigator in his laboratory, Rosalind Franklin, had been working for several years on getting good X-ray diffraction pictures of various crystalline forms of DNA. It was Wilkins's and Franklin's hope to get accurate enough data of the distances of every atom and group of atoms apart so that the three-dimensional structure of the molecule would become evident. But it was painstaking work; it required the perfection of many techniques and the laborious calculation of huge quantities of data. Wilkins was a mild-mannered man who enjoyed theory but preferred to stay close to data and measurements. Rosalind Franklin was a hard-working biophysicist who forged a career in an area of science even more dominated by men than most others. In *The Double Helix* Watson portrays Franklin as secretive and vindictive. In fact, however, she appears to have been merely cautious. She had collected large amounts of data on different DNA fibers, but was unwilling (even in 1952) to draw conclusions about what the shape of the molecule might be. A year earlier (1951) Linus Pauling had published his theory of the α-helix of proteins, and as a result many people had suggested that DNA might have some sort of helical shape. Rosalind Franklin was antihelical, because her measurements didn't suggest how the helix could be held together by any known chemical forces. So, Franklin and Wilkins sat on much of their data for longer periods of time than was comfortable for impetuous theorizers like Crick and Watson.

During their morning discussions at the Cavendish Lab, Watson and Crick built molecular models of nucleic acid subunits. They had been inspired in this technique by Pauling, who had used actual molecular models to work out his theory of protein structure. Lacking any professionally made molecular models (like those which abound among organic chemistry students today), Watson and Crick cut out shapes from cardboard to represent the actual dimensions of the various nucleotides. Then they tried to fit these together in some sort of orderly pattern. The problem was there were many configurations, all of which would satisfy what data were then available: the molecule could be a single, double,

or triple strand with the bases pointing either inward or outward. To know which of these models was correct it was essential that Watson and Crick have access to an accurate diffraction pattern of the DNA molecule. What were the actual dimensions (In Ångstrom units) of the nucleotides? What was the diameter of the DNA molecule in its crystalline form? With Wilkins and Franklin remaining cautious, the answers seemed far away.

Several other factors influenced the direction of research at this time. One was the rumor which spread that Linus Pauling was now hard at work on the structure of DNA. With his reputation for brilliant insight, Pauling's entry into the field sent shivers down some people's spines. Not Watson and Crick, however. It was exciting, so Watson tells us, to suddenly find themselves in a race, especially against such a formidable opponent as Pauling. In this race Watson and Crick were helped along by the presence of Pauling's son, Peter, at Cambridge, who unwittingly kept them informed about his father's latest ideas, or the ideas he had just scrapped. Pauling did not know his ideas were being passed on by his son, but this information was of great advantage to Watson and Crick because it provided them with shortcuts. They could easily forget about ideas Pauling had seriously considered but rejected. In 1952 Pauling, it turned out, was most intrigued with the notion of a triple-stranded structure for DNA. But from his letters to Peter, it was also clear that he was finding such a structure unsatisfactory. This knowledge spurred Watson and Crick to look to other models, perhaps a two-stranded molecule.

A second factor that greatly enhanced Watson and Crick's work was the arrival in Cambridge of the enigmatic and outspoken American biochemist Erwin Chargaff. Chargaff had studied the base composition of DNA in a number of organisms and had made the remarkable discovery that the ratio of adenine to thymine, and of guanine to cytosine was always roughly 1:1. He did not know the exact significance of these data for the structure of the molecule, but he knew it was important information. When he presented a paper on the subject, and discussed his results, Watson and Crick realized they did not even know about Chargaff's work. They hurriedly looked up his original data, and recognized that perhaps the structure of DNA somehow involved the matching up of adenine to thymine and cytosine to guanine. The regularity of the ratio was too persistent not to play some important role in the molecule's structure. But they could not propose any serious relationship between A-T and C-G because they lacked sound X-ray crystallographic data.

At this point, Watson put to work his near-photographic memory, and paid a visit to Rosalind Franklin's laboratory in London. Purporting simply to want to keep abreast of her work, Watson used every opportunity he could find to look at data tables, and commit them to memory. On the train back to Cambridge he wrote down furiously as much as he could remember. Wilkins's recent data were also as reliable as Franklin's, but they, too, were not published. Then a happy accident occurred (happy, at least, for Watson and Crick). One of the major investigators at Cavendish, Max Perutz, received a mimeographed report of the work going on at the Kings College laboratory; Perutz was on a government commission whose job it was to oversee scientific research work and, presumably, pass judgment on future funding. In the report were the precise measurements that Wilkins and Franklin had not yet published in a journal. Perutz recognized the importance of these data, and he subsequently passed them on to Watson and Crick.

Then, another unexpected clue came to Watson and Crick: a chance remark by a young chemist sharing a desk in Crick's lab suggested that under most conditions the bases of DNA would be in a different chemical form (that is, the keto rather than enol form) than Watson had been picturing them. Changing the chemical configuration suggested how the bases might hook up with each other—that is by hydrogen bonding, which could occur between purines and pyrimidines in the keto (but not enol) forms. This idea would not only explain how the molecule was held together, but also why Chargaff's data were so universally applicable. The most accurate model thus showed the sugar phosphate backbone on the outside of the molecule with the bases facing inward, so that A joined to T and C to G by hydrogen bonds across the center of the helix. This structure had biological implications that went beyond the most optimistic hopes Watson and Crick had initially held. It could explain how DNA replicated as well as how it could contain specific genetic information in a linear sequence of bases.

We can draw several conclusions from this brief run-through of the discovery of DNA. The first is that science is an interactive process: people work together, exchange ideas, and build on each other's data. The free exchange of information is essential to the scientific process. At the same time, this case illustrates the negative effects of secrecy, rivalry, and competition. Not only were a number of people embittered by the lack of recognition that they received for their contributions to development of the double-helix concept, but Watson and Crick themselves suffered the enmity of many of their colleagues who felt used or unfairly dealt with during the "race" for the solution. Such personal rivalries and animosities are unfortunately a product of a competitive approach to science. Such competition is fostered by the system of grants, prizes, and better-paying jobs for those who become famous.

A second generalization that comes from the DNA story is the important interaction between theory and experimentation, between ideas and data. Both theory and data are essential to the pursuit of productive science. There is no way to separate the two. Often, within the scientific community, attempts are made to do so. The Ph.D. does the thinking, the graduate students and technicians do the experimenting. While it is always likely that some people will be more prone

to generalize than others, it is probably unproductive if the division between "thinkers" and "doers" proceeds too far. There was no reason that Watson and Crick could not have learned more of the X-ray diffraction techniques that Wilkins and Franklin had developed with such skill. Had Watson and Crick shown more humility and been willing to learn from Wilkins and Franklin, instead of persisting in bombarding the Kings College group with model building, the whole investigation might have proceeded more smoothly and rapidly. The different orientations of the two groups were not the problem, but their lack of cooperative interaction was. Such persistent division of labor runs counter to the social reality of science as an interactive process.

A third point to come out of the DNA story is the destructive role of sexism in scientific work. Without the careful work of Rosalind Franklin and Maurice Wilkins, Watson and Crick would never have been able to get beyond the speculative stages. They had access to important data that were lacking, say, to Linus Pauling, and that may well have made the difference in their more rapid solution of the double-helix problem. But Watson's attitude toward Rosalind Franklin as displayed in *The Double Helix* was anything but considerate. More important was the constant prejudice that Franklin faced in the stuffy English scientific and academic community of that day. In her biography of Franklin, Anne Sayre describes how after a luncheon meeting at one of the Cambridge undergraduate colleges, the men scientists retired to the Commons Room (where all the college Masters, Tutors, and Dons meet for socializing), where Franklin was not welcome *because she was a woman*. Furthermore, she never had a permanent academic position, and was always working under the direction of one or another established male investigator. Who knows how such instability in her professional life may have slowed down Franklin's X-ray crystallography work, or caused her to be exceptionally cautious about the data she published, or the theories she would accept. And, of course, who knows the amount of personal frustration such entrenched attitudes create. Obviously such attitudes are antithetical not only to the full pursuit of human capabilities, but even to the process of science itself.

Fourth, the story of DNA illustrates the fruitful developments that come from the interaction of various scientific disciplines. Watson was a biologist with an interest in heredity. Crick was a physicist with a theoretical curiosity about macromolecules. Wilkins and Franklin were also physicists, but with a more empirical orientation. The interaction between these different fields and the different orientations to the problem of DNA are one of the characteristics of new and revolutionary developments in science. Historians of science frequently point out that great advances in science come from the interaction of previously separate and distinct fields, or points of view. The DNA story is one of the best modern examples of this generalization.

The DNA case is not unique in the annals of science. It illustrates dramatically the many human aspects—both positive and negative—of science in its social context. We need not conclude that the only or best way to do science is to be competitive, suspicious, or secretive. If anything, we can learn from such case histories how to organize and pursue science in a more realistic and humane atmosphere. Part of that new organization will undoubtedly reflect the growing recognition of science as a fundamentally cooperative venture.

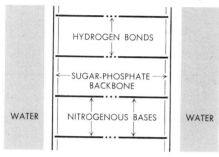

Fig. 19.4
Two-dimensional schematic representation of the double helical DNA molecule; a highly stylized diagram of the sugar-phosphate backbone and the nitrogenous bases joined across the center by hydrogen bonds.

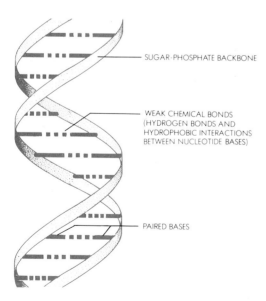

Fig. 19.5
A more three-dimensional representation of DNA, showing the coil of the helix and the position of the bases with respect to the sugar-phosphate backbone.

sugar-phosphate units on the outside. This arrangement can be tentatively represented as shown in Fig. 19.4.

The next clue to DNA structure was provided by X-ray diffraction.* Data obtained from the use of this technique on crystallized DNA showed the model to have a helical rather than a planar structure, with one complete twist of the helix occurring every 34 Å, or about every ten base pairs. The previous representation of DNA structure in Fig. 19.4 must therefore be modified to convey its helical nature as shown in Fig. 19.5.

Obviously, a most desirable way of confirming the hypothesized helical structure of DNA would be to take a "picture" of it. This feat was accomplished in 1970 (Fig. 19.6), and the double-helical nature of DNA given still more concrete reality.

The Base Pairs

Another problem remains, however. Granted that the helical strands of DNA polynucleotides are held together by the hydrogen bonds that form between them, it must still be determined which bases pair with which. Do they, for example, pair purine to purine, pyrimidine to pyrimidine, or purine to pyrimidine?

The answer is partly given by our knowledge of the physical nature of hydrogen bonds and the conditions necessary for their formation and the constraints imposed by a uniform diameter of 20 Å for the helix. Besides calling for the presence of a covalently bonded, positively charged hydrogen atom and a covalently bonded, negatively charged acceptor atom, hydrogen bonds can only form over certain critical interatomic distances. Recall that the purine bases are double-ring structures, larger than the single-ring pyrimidines. If two purine bases formed hydrogen bonds between them, any two pyrimidines in the double strand of DNA would be held too far apart for the formation of their own hydrogen bonds. Conversely, if the pyrimidine

---

X-ray diffraction is a technique that involves passing X-rays through crystallized DNA and determining how they are deflected by the structure through which they pass. X-ray diffraction is *somewhat* analogous to determining the three-dimensional shape of an object by the shadow it casts, but the process is considerably more complex than this analogy indicates.

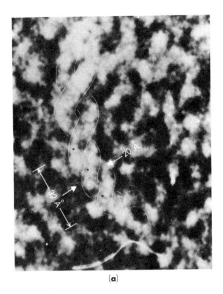
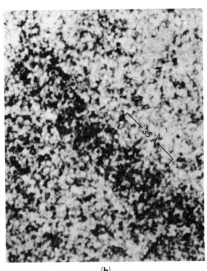

(a)   (b)

Fig. 19.6
These two micrographs dramatically confirm the helical structure of DNA. (a) A short length of one DNA molecule from chromosomes of the pea. The double-stranded helix is plainly discernible. For additional clarity, the artist has added white hairlines around the edge of the helix. (b) A portion of a filament of calf thymus DNA. The double helical substructure has the predicted diameter of 20 Å, with a period of approximately 35 Å for one complete turn. [(a) courtesy Jack Griffith, California Institute of Technology; (b) courtesy F. P. Ottensmeyer, University of Toronto.]

bases were at the proper distance for hydrogen-bond formation, the purine bases would overlap.

An alternative hypothesis, which eliminates these difficulties, is clear. Purine-to-pyrimidine bonding must occur. However, there are two purines involved, adenine and guanine, and two pyrimidines, cytosine and thymine. A new question therefore arises: granting a purine-to-pyrimidine base pairing, which of the two purine bases pairs with which pyrimidine? Again, the answer is provided by the factors leading to the formation of hydrogen bonds between the nitrogenous bases. The only base-pairing combinations that allow the formation of the hydrogen bonds found in DNA are those of adenine-thymine and guanine-cytosine.

It is interesting to note that the limitation of purine-pyrimidine base pairing to adenine-thymine and gua-

nine-cytosine combinations could have been deduced from other data. Careful chemical analyses of DNA had been made earlier by Columbia University biochemist Erwin Chargaff to determine just how much of each kind of base is present in the DNA of a particular species. The results of some of these analyses are presented in Table 19.1. Note that in any one species, the amount of adenine equals or closely approximates that of thymine, while the amount of guanine closely approximates or equals that of cytosine, the deviations in each ratio lying well within the range to be expected from errors in experimental measurements. These results would be predicted by the Watson and Crick model, which specifies an adenine-thymine and guanine-cytosine base pairing within the DNA molecule.

Given these details of structure, an accurate model of the DNA molecule can now be drawn (see Fig. 19.7). This diagram clearly shows the double-helical nature, the purine-to-pyrimidine bonding, and the role of the sugar-phosphate groups in establishing the helical "backbone."

19.3
DNA REPLICATION

Background

The Watson-Crick model of DNA must account for the ability of the molecule both to direct cell activity (that is, gene function) and to replicate itself (reproduce). The beauty of the model is that it suggests ways in which both processes could occur. Subsequent experimentation has provided considerable understanding of both processes. Let us consider here the second question: how the DNA molecule makes a perfect copy of itself. Work on this problem began almost immediately after the

Table 19.1
Adenine-to-Thymine and Guanine-to-Cytosine Ratio in DNA (in molar ratios)

Source of DNA template	Adenine	Thymine	Guanine	Cytosine
Bovine sperm	28.7	27.2	22.2	20.7
Rate bone marrow	28.6	28.4	21.4	20.4
Herring testes	27.9	28.2	19.5	21.5
Wheat	27.3	27.1	22.7	22.8
Yeast	31.3	32.9	18.7	17.1
Escherichia coli	26.0	23.9	24.9	25.2
Phage T2	32.0	32.0	18.0	18.0

From Erwin Chargaff (1955)

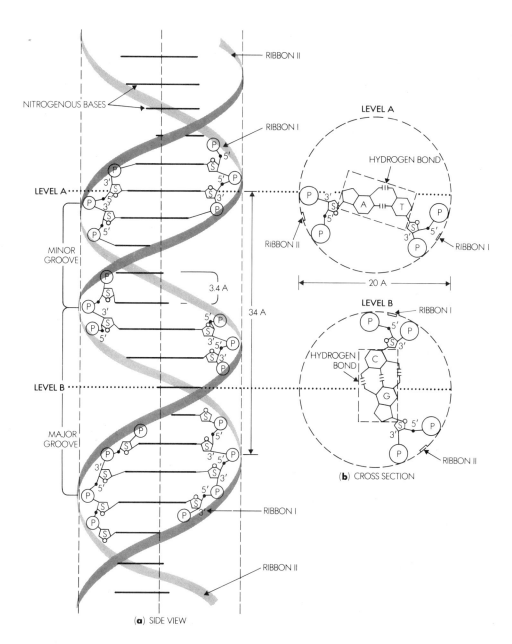

RIBBON II

NITROGENOUS BASES

RIBBON I

LEVEL A

MINOR GROOVE

3.4 A

34 A

MAJOR GROOVE

LEVEL B

RIBBON I

RIBBON II

(a) SIDE VIEW

LEVEL A

HYDROGEN BOND

RIBBON II

RIBBON I

20 A

LEVEL B

RIBBON I

HYDROGEN BOND

RIBBON II

(b) CROSS SECTION

Fig. 19.7
Schematic diagram of the three-dimensional structure of the double-helix DNA molecule. The two intertwined sugar-phosphate backbones are referred to as Ribbons I and II. (a) A side view of the molecule. (b) Two cross sections, one at level A, the other at level B, as indicated on the side view. This diagram has been prepared to represent actual molecular dimensions and thus to give an indication of how the molecule fits together in space. (Reprinted, with permission, from an article by Dr. William Etkin in the November 1973 *Bioscience* published by the American Institute of Biological Sciences.)

publication of Watson and Crick's original paper in 1953.

It is quite easy to imagine that the process of replication involves an unwinding and separation of the two DNA polynucleotide strands, with "unzipping" occurring through the breakage of the hydrogen bonds of each base pair. This mental picture is supported by the fact that the amount of energy needed to separate the DNA strands is equivalent to the amount of energy needed to break hydrogen bonds. Once separated from their partners, the unpaired nucleotides on each strand would attract their complementary nucleotides from the surrounding medium. Thus each unpaired polynucleotide strand would specify the nucleotide sequence of a strand complementary in structure to itself. The result would be two DNA molecules, each an exact replica of the original (Fig. 19.8).

One immediate prediction of the Watson-Crick hypothesis suggested by the diagram is that DNA in the process of replication should have the form of the letter Y. Dr. John Cairns, now at the Harvard School of Public Health, has verified this prediction.

An elegant experimental test of the means of DNA replication suggested by Watson and Crick was per-

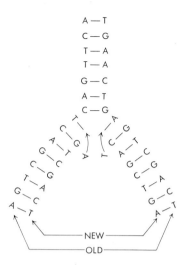

Fig. 19.8
A highly diagrammatic representation of base pairing and DNA replication as hypothesized by Watson and Crick. The four bases—adenine, thymine, guanine, and cytosine—are represented by the letters A, T, G, and C. The total length of the segment of molecule represented here would be approximately 50 Å (0.005 μ).

formed in 1958 by Matthew Meselson and F. W. Stahl, then at the California Institute of Technology. *Escherichia coli* bacteria were grown in a medium containing glucose, mineral salts without nitrogen, and ammonium chloride (NH_4Cl) in which almost all of the nitrogen atoms were the heavy isotope, N^{15}. The bacteria were allowed to grow in this medium until fourteen generations of bacteria had arisen from those used to inoculate the culture. Hence virtually all of the nitrogen in the DNA of these bacteria was heavy nitrogen.

The key to the Meselson-Stahl experiment lay in the experimental technique. When a cesium chloride solution is spun in an ultracentrifuge up to about 60,000 times the force of gravity, the molecules of cesium chloride begin to sediment. However, due to their relatively low molecular weight, they never sediment completely. Instead, the result is a gradation of low-to-high solution density, from the top of the centrifuge tube to the bottom. Any foreign molecule centrifuged in such a cesium chloride gradient will come to rest at a level at which its density equals that of the surrounding solution. Thus DNA extracted from *E. coli* bacteria grown in a medium containing the regular light isotope of nitrogen, N^{14}, will form a band at a higher point in the centrifuge tube than DNA from *E. coli* grown in a medium containing N^{15}.

Meselson and Stahl removed the experimental bacteria from the heavy nitrogen medium and allowed them to undergo just one more generation, one more DNA replication. Note that as a result of this step,

heavy DNA molecules containing N^{15} replicated in a medium in which only light nucleotides containing N^{14} were available. According to the hypothesis of Watson and Crick, the result of such replication should be "hybrid" DNA molecules, each molecule containing one heavy strand and one light strand. After extraction and centrifugation, these hybrid DNA molecules should appear as a new band in the cesium chloride density gradient. This new band should lie in the region between the completely light DNA band formed from bacteria grown in a medium containing N^{14}, and the heavy DNA band formed from bacteria grown in a medium containing N^{15}. This prediction is verified by the experimental results.

The test can be carried further by allowing the N^{15}-labeled bacteria to replicate for *two* rather than one generation in a medium containing N^{14}. Here one would predict two DNA bands in the cesium chloride density gradient tube. Half of the DNA should be found in the area occupied by light DNA. The other half should be in the area occupied by the "hybrid" DNA. Again, the results obtained by Meselson and Stahl verified these predictions and supported the hypothesis of Watson and Crick (see Fig. 19.9). According to these results we would say that the method of DNA replication is *semiconservative*. This means that in each replication, one parental strand is *conserved* and appears intact in one of each of the daughter molecules. The process would be conservative if the two parental strains replicated themselves and then came back together, leaving the daughter molecule to be composed of the two newly synthesized strands.

E. coli, of course, is a bacterium, a fairly primitive organism. What about higher organisms in which the DNA is located in distinct chromosomes? We have only working hypotheses to explain just how DNA is arranged within the chromosome. However, despite this uncertainty, an experiment performed by J. H. Taylor of Columbia University sheds some light on the problem of extrapolation from the replication of DNA in *E. coli* to the replication of DNA in higher organisms.

Taylor worked with plant root tips, in which the cells constantly undergo mitosis. He immersed these root tips in a solution containing the nucleoside thymidine, which had been labeled with radioactive hydrogen (tritium, H^3). The root tips were left in this solution long enough for many of their cells to double their DNA content, but not long enough for it to be doubled again. Any DNA molecules formed during this time would incorporate the radioactive thymidine into their structure and thus themselves become radioactive (see Fig. 19.10).

As soon as the pair of daughter chromosomes became visible, they were tested by autoradiography. The hypothesis of Watson and Crick predicts that the members of each pair of chromosomes should contain radioactive DNA and that they should contain it in equal

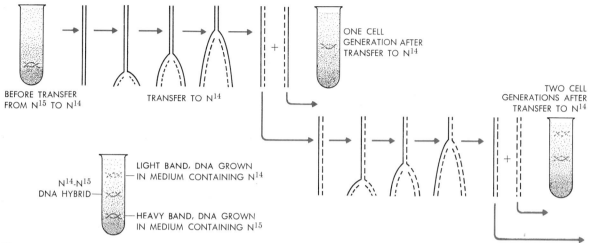

Fig. 19.9
The Meselson-Stahl experiment is based on a density gradient caused by ultracentrifugation for 48 hr in a cesium chloride solution. When DNA molecules are spun in such a solution, they separate into regions of the centrifuge tube at which their density equals that of the CsCl solution.

amounts. This prediction was verified by a count of the black dots found on the photographic film where radiation had fogged it.

If one allows cells containing radioactive daughter chromosomes to undergo another cycle of duplication in a solution containing no radioactivity, the result is a completely different prediction, which provides yet another test of the Watson-Crick hypothesis. Here, autoradiography should reveal one member of each pair of daughter chromosomes to be radioactive and the other nonradioactive. This prediction is verified. Because of the many complexities involved in experimen-

Fig. 19.10
Diagrammatic representation of Taylor's experiment with chromosome duplication in root tip cells of a pea plant. Note that the results are consistent with the Watson-Crick hypothesis concerning DNA replication. (After John W. Kimball, *Biology*, 2d ed. Reading, Mass.: Addison-Wesley, 1978.)

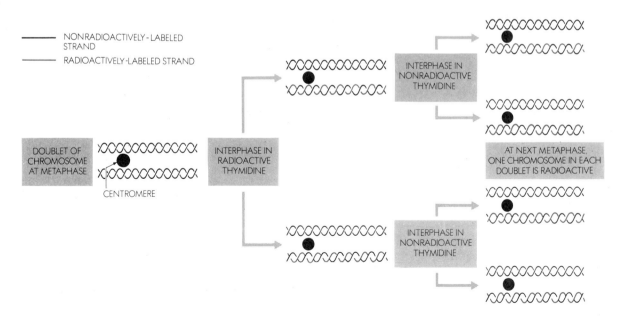

tation of this sort, which introduce the possibility of un-controllable variables, Taylor's observations have been questioned by other investigators. However, the re-ported results are consistent with the Watson-Crick hy-pothesis, and similar experiments on human chromo-somes in tissue culture yield the same results.

From the Meselson-Stahl and Taylor experiments, it is apparent that: (1) DNA does replicate itself, by one strand acting as a template on which a second whole strand is synthesized, and (2) this process appears to be general, from the DNA of bacteria to that of eukaryotic organisms such as plants or human beings. Many of the details of the process are still unclear; its broad outline, however, is now well established.

The Biochemistry of DNA Replication

The Watson-Crick model of DNA makes several addi-tional predictions about how self-replication takes place:

1. Each strand of the double helix is complementary in base structure to the other; hence, if the molecule un-winds, each strand can serve as a template for the exact replication of its partner. The replication of each strand produces two whole, new molecules, each consisting of one strand of old, and one of newly synthesized DNA.

2. The two strands of the DNA molecule lie "head to tail" in the double helix. In chemical terms this means that the end called 3' of one strand lies opposite the end called 5' of the other, as shown in the diagram to the right above. (3' and 5' refer to the 3-carbon and the 5-carbon of the ribose—the two carbons that bond to phosphate and thus link together nucleotides along the sugar-phosphate backbone; for a more detailed picture, refer back to Fig. 19.7a.)

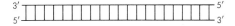

In the intervening years abundant evidence has borne out these two major predictions. A more detailed description of how these predictions were tested can be found in both the remainder of this section and in Supplement 19.3.

In 1956 a team of scientists headed by Arthur Kornberg, then at Washington University in St. Louis, succeeded in isolating an enzyme capable of synthe-sizing DNA in the test tube. Kornberg named this enzyme DNA polymerase. In subsequent years three different DNA polymerase molecules have been identi-fied in *E. coli*; these are called DNA polymerase I, II, and III.* DNA polymerase III appears to have the major

---

*Sometimes abbreviated as Pol I, Pol II, and Pol III.

role in synthesizing new DNA; it is also called replicase. Polymerase I functions largely to fill in gaps between adjacent small polynucleotide fragments in a longer chain. No specific function has yet been identified for polymerase II. Using a mixture that turned out to con-tain DNA polymerase I and III, in 1967 Kornberg (then at Stanford University) and associates were able to syn-thesize a perfect copy of bacteriophage DNA *in vitro*.

In DNA polymerase III, only one site on the enzyme can interact with free nucleotide triphosphates. It is hypothesized that once the DNA polymerase en-counters a given base in the template strand, it can cata-lyze formation of the ester bond (in the sugar-phosphate backbone) in the daughter strand *only* when the correct, complementary base (nucleotide) is present. Presum-ably only the complementary base will have the correct geometry to align properly with the elongating daughter strand and thus be in the correct position to form an ester bond. For example, if a pyrimidine is called for, a purine might stick out too far to align properly. Once an ester bond is formed, the enzyme is presumably free to move on to the next base in the template. Polymerase III adds to a growing DNA chain rapidly, at the rate of more than 1000 bases per second!

However, the bacteriophage DNA synthesized by Kornberg forms a closed circle (similar to that shown in Fig. 19.11) and, more important, consists of a single

Fig. 19.11
Photomicrograph of autoradio-graphed molecule of circular DNA of a bacteriophage, λ, simi-lar to that found in the phage φX-174 used by Kornberg. λ DNA is double-stranded, while φX-174 DNA is single-stranded. The length of the circular strand is about 16 μ.

SUPPLEMENT 19.3
GENES PRODUCE PROTEINS: HOW DO WE KNOW?

Between 1905 and 1908 English physician Archibald Garrod encountered several examples of metabolic diseases, the most prominent of which was alkaptonuria.

Garrod suggested that persons with alkaptonuria were defective in enzymes involved in amino acid metabolism. It was not until 1958 that it was shown conclusively that patients with alkaptonuria had defective homogentistic acid oxidase, which, as shown in Diagram A (right-hand side) normally converts homogentistic acid into fumaric and acetoacetic acids. The defective enzyme means homogentistic acid cannot be oxidized, and hence it accumulates in the tissues. The defective enzyme serves as a "roadblock" in the step-by-step biochemical pathways. As Diagram A shows,

albinism also results from a defective enzyme in this same series of pathways. Although not a metabolic disease in the sense of alkaptonuria, albinism nonetheless results from the same process: a defective enzyme that blocks the pathway by which one substance (in this case tyrosine) is converted into another (melanin, or pigment).

With the help of geneticist William Bateson, Garrod analyzed the appearance of alkaptonuria, albinism, and several other metabolic defects in a number of families. He concluded that the metabolic defects were inherited as recessive autosomal traits, following strictly Mendelian lines.

With his conviction that various metabolic abnormalities were due to single Mendelian genes, and his belief that each abnormality was the result of a defective enzyme, Garrod went on to develop an imaginative hypothesis. Genes have their effects, he claimed, by producing enzymes. Normal genes produce normal enzymes, which carry out normal metabolic reactions. Mutant genes produce abnormal, or defective, enzymes, which block normal metabolic reactions, and lead to build-up of intermediate substances such as homogentistic or phenylpyruvic acids. Garrod's hypothesis did not attract much attention among his contemporaries, however. Despite the fact that his ideas were well known (he delivered the distinguished Croonian Lectures to the Royal Society of England on the subject in 1908) and were published as a book, *Inborn Errors of Metabolism*, in 1909, Garrod's hypothesis had to wait another thirty years for further development.

In 1941 George Beadle and Edward L. Tatum, both then at Stanford University, published an important paper outlining their work on biochemistry and heredity in the red bread mold *Neurospora*. It was in this paper that Beadle and Tatum first enunciated their "one gene, one enzyme" hypothesis: the notion that genes control the production of specific enzymes, and through this route control cell metabolism and, ultimately, an organism's phenotype.

Diagram A
Outline of the pathways involved in tyrosine metabolism in human beings. The heavy black lines indicate a block in the metabolic pathway due to a defective enzyme. Substances synthesized prior to the block in the pathway accumulate. A block at (*a*) produces albinism, since tyrosine cannot be converted into the pigment melanin. A block at (*b*) produces alkaptonuria, a genetic disease characterized by accumulation of homogentistic acid in the body.

Neurospora has proved to be a very convenient organism for the study of biochemical genetics for several reasons:

1. It can be easily grown in the laboratory on a basic (called "minimal") agar medium in test tubes or Petri dishes.

2. It has a short life cycle producing a new generation in only ten days.

3. It is normally haploid, which means that all genes in any individual organism are expressed.

4. *Neurospora* usually reproduces asexually, so that particular strains or new mutants can be perpetuated without the genetic mixing that would occur during sexual reproduction.

5. On occasion, when two different mating strains of *Neurospora* are brought together, they can undergo a sexual fusion, producing genetic recombination.

To understand the work of Beadle and Tatum it will be necessary to look briefly at the reproductive process in *Neurospora*. In its sexual phase, *Neurospora* reproduces by fusion of adjacent thread-like filaments (the mold that appears as fuzz on a moldy cake or piece of bread), called hyphae, which have originated from separate plants. After hyphal fusion, haploid nuclei from each filament fuse, producing a zygote. Enclosed in a spore case formed from the hyphae walls, the zygote undergoes one complete meiosis and mitosis to produce a total of eight haploid spores (*asco-*

spores) within a long tubular spore case, the ascus (see Diagram B). At maturity the ascus ruptures and the spores are expelled. Each spore is able to germinate into a new set of hyphae and thus a new plant.

Beadle and Tatum grew spores from wild-type *Neurospora* on what is called a *minimal medium*, a combination of minerals, salts, nutrients, and the other substances the fungus cannot make for itself. The medium is called "minimal" because it contains the minimum number of substances that wild type *Neurospora* needs to grow. If any one substance is removed from a minimal medium, the wild type plant cannot grow.

Beadle and Tatum hypothesized that the ability of wild type *Neurospora* to convert the basic growth substances in the medium—such as vitamins or amino acids—into other usable metabolic products was genetically determined; that is, genes controlled each step in each specific metabolic pathway. They hypothesized further that a mutation in a gene controlling any one step in a biochemical pathway would produce a defect, or blockage at that step, so that the organism would no longer be able to synthesize the needed end-product. In order for such a mutant *Neurospora* to grow, therefore, it would be necessary to determine what needed substance it was no longer able to make for itself, and supply that substance in the medium.

After raising wild-type cultures, Beadle and Tatum irradiated the asexually produced spores to induce mutations (see Diagram C, step 1). They then transferred these spores to

Diagram B
Segregation (meiosis) and ascospore formation in the red bread mold *Neurospora crassa*. The fact that the meiotic products are physically aligned in the ascus, according to the way segregation took place at meiosis I, has proved to be a valuable asset in the genetic analysis of this species.

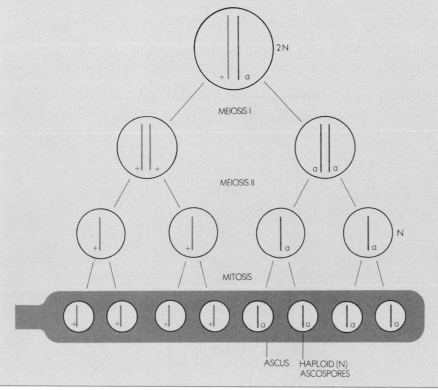

MEIOSIS I

MEIOSIS II

MITOSIS

ASCUS   HAPLOID (N) ASCOSPORES

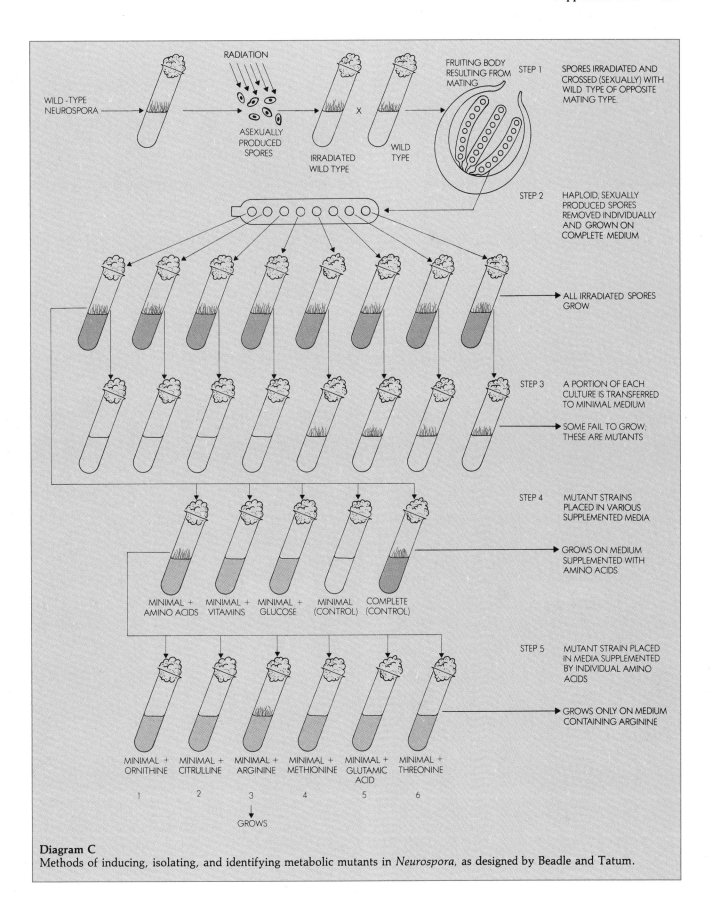

RADIATION

WILD-TYPE NEUROSPORA

ASEXUALLY PRODUCED SPORES

IRRADIATED WILD TYPE

WILD TYPE

FRUITING BODY RESULTING FROM MATING

STEP 1   SPORES IRRADIATED AND CROSSED (SEXUALLY) WITH WILD TYPE OF OPPOSITE MATING TYPE.

STEP 2   HAPLOID, SEXUALLY PRODUCED SPORES REMOVED INDIVIDUALLY AND GROWN ON COMPLETE MEDIUM

ALL IRRADIATED SPORES GROW

STEP 3   A PORTION OF EACH CULTURE IS TRANSFERRED TO MINIMAL MEDIUM

SOME FAIL TO GROW; THESE ARE MUTANTS

STEP 4   MUTANT STRAINS PLACED IN VARIOUS SUPPLEMENTED MEDIA

GROWS ON MEDIUM SUPPLEMENTED WITH AMINO ACIDS

MINIMAL + AMINO ACIDS

MINIMAL + VITAMINS

MINIMAL + GLUCOSE

MINIMAL (CONTROL)

COMPLETE (CONTROL)

STEP 5   MUTANT STRAIN PLACED IN MEDIA SUPPLEMENTED BY INDIVIDUAL AMINO ACIDS

GROWS ONLY ON MEDIUM CONTAINING ARGININE

MINIMAL + ORNITHINE

MINIMAL + CITRULLINE

MINIMAL + ARGININE

MINIMAL + METHIONINE

MINIMAL + GLUTAMIC ACID

MINIMAL + THREONINE

1   2   3   4   5   6

GROWS

Diagram C
Methods of inducing, isolating, and identifying metabolic mutants in *Neurospora*, as designed by Beadle and Tatum.

tubes containing a *complete* medium, that is, one that contained the minimal substances plus a number of other growth factors such as vitamins, amino acids, or other substances that wild-type *Neurospora* normally make for themselves. The complete medium supported the growth of all spores, both wild-type and mutant. To identify what mutations were present, Beadle and Tatum carried out the following procedure: the spores were dissected out of the spore case (ascus) one-by-one, the position of the spore within the case was listed, and each spore was grown on a complete medium (Diagram C, step 2). The purpose of this step was simply to allow every spore to germinate, and thus yield a number of homogeneous cultures for use in the next step of the genetic analysis.

In step 3 the investigators transferred portions of each mutant colony to tubes containing minimal medium. Those that did not grow must contain a mutation. Portions of those mutant colonies were then taken from the Step 2 stock cultures and grown on a number of special media, each containing the minimal requirements plus a single supplemental substance (Diagram C, step 4). This step narrows down the sort of mutation each strain carries. If, for example, a particular mutant grows on minimal medium plus amino acids, as shown in Diagram C, step 5, spores from that culture can be transferred to tubes containing minimal medium plus one or another specific amino acid (glutamic acid, tyrosine, and the like). As Diagram C shows (bottom line), the only culture tube to show growth is number 4, containing minimal medium plus methionine. This indicates that the particular strain isolated in step 4 lacks the ability to synthesize its own methionine from precursors. By this sort of trial-and-error process, it is possible to identify the exact metabolic mutation which each strain processes. Irradiation must have destroyed the mold's ability to synthesize the particular amino acid methionine that it normally needs to grow.

By crossing different mutant strains with each other and with wild-type plants, Beadle and Tatum were able to show that most mutations were inherited as single genes. Their analysis was simple. Returning to Diagram B, note that in the diploid zygote produced by a sexual fusion between a wild-type (+) and mutant (α) strain, each gene is present on its respective chromosome. During meiosis the chromosomes segregate in such a way that the physical arrangement of spores within the ascus would be expected to be (from left to right as shown in Diagram B), +, +, +, +, a, a, a, a. When the workers isolated each spore, grew it into a culture, and tested it for a metabolic mutation, whenever mutations were present the physical arrangement of the spores was in accordance with expectations. That is, the spores are arranged in a 4-4 pattern typical of the segregation products of single genes.*

---

*Crossing over sometimes occurs producing various modifications of the 4-4 arrangement, depending on which strands of the meiotic tetrads are involved. Although not explained here, crossing over events were very important for the genetic side of Beadle and Tatum's argument.

It was now possible for Beadle and Tatum to determine exactly how each mutation affected the metabolic characteristics of *Neurospora*. Special biochemical analyses were able to reveal what particular enzymes were lacking in each of the many different mutant strains. For example, in tube number 3 at the bottom of Diagram C, it is apparent that the particular strain isolated from step 4 lacks the ability to produce ornithine. Biochemical analysis indicated that this strain was unable to synthesize functional argininosuccinase, which catalyzes the transformation of argininosuccinic acid into arginine, as shown in the following pathway:

$$\text{Citrulline} \xrightarrow{\hspace{2cm}} \underset{\text{acid}}{\text{Argininosuccinic}} \xrightarrow{\overset{\text{Arginino-}}{\underset{\text{succinase}}{}}} \text{Arginine}$$

When the mutant is supplied with citrulline or argininosuccinic acid it cannot grow because it cannot convert argininosuccinic acid into arginine, a necessary amino acid. If the mutant is supplied with arginine, however, it can grow by using the substance supplied to it from the outside. Similar analyses showed that for each of the mutants, one particular enzyme was defective. The enzyme was always one involved in the pathway which synthesized the needed substance.

From both the biochemical analyses and the patterns of spore segregation as described above, Beadle and Tatum enunciated their "one gene, one enzyme" hypothesis. They argued that genes specify the structure and activity of enzymes, which in turn catalyze metabolic reactions. Mutations in a gene result in production of altered and usually nonfunctional enzymes, so that the metabolic pathway is blocked at the point where the affected enzyme normally functions.

Beadle and Tatum shared the 1958 Nobel Prize for their work with *Neurospora*. In a speech delivered at the award ceremony, Beadle said:

> In this long, roundabout way, we had rediscovered what Garrod had seen so clearly many years before. By now we know of his work and were aware that we had added little if anything new in principle. We were working with a more favorable organism and were able to produce, almost at will, inborn errors of metabolism for almost any chemical reaction whose product we could supply through the medium. Thus, we were able to demonstrate that what Garrod had shown for a few genes and few chemical reactions in Neurospora.

Where does the "one gene, one enzyme" hypothesis stand today? Several new developments have caused biologists to revise and restate the generalization. First, genes appear to direct synthesis of protein; while all enzymes are proteins, not all proteins are enzymes. Collagen, for example, accounts for approximately one-third of all the protein in the human body. It is a structural protein and not an enzyme; yet it is produced by the action of genes.

Second, and more important, genes do not guide the synthesis of whole proteins, but of smaller components, usually single polypeptide chains. Thus today we speak of the "one gene, one polypeptide" hypothesis. Such a rephrasing corrects both of the oversimplifications contained in the original hypothesis as stated by Beadle and Tatum, while preserving the overall framework of their important idea—that is, showing the relationship between genes and proteins.

rather than a double strand. Most DNA is double-stranded. Like most enzymes, DNA polymerase is quite specific in its action, not only as to substrate but also as to how it acts on this substrate. In particular, Kornberg found that *DNA polymerase could synthesize a strand of DNA only in a manner causing the new polynucleotide chain to grow in the 5'-to-3' direction.*

Kornberg's discovery presented geneticists with an immediate dilemma. Since a DNA molecule unwinds from one end, one strand will be oriented in the 3'-to-5' direction, while the other will be oriented in the 5'-to-3' direction. Since molecules of DNA polymerase can synthesize new strands only in the 5'-to-3' direction, we might well ask how both strands get copied at the same time. There is no problem understanding this process for the strand which is running from the 5' to the 3' end. But the other strand cannot be synthesized from its free end on down the open part of the molecule (the way its complementary partner strand is synthesized). The solution to this problem came from the work of Reiji Okasaki and his colleagues at Nahoyo University in Japan. Exposing cells very briefly to radioactive nucleotides so that only the most recently synthesized DNA was labeled, Okasaki found that the newest DNA existed in extremely short fragments (of only a few nucleotides), held to the parental strand by hydrogen bonds. Soon thereafter, Okasaki found, these short fragments are joined together by a second enzyme, which is now called *ligase,* or the "joining enzyme." Thus, DNA polymerases join nucleotides together to produce short fragments, with ligase catalyzing the joining together of these fragments to make a complete strand of DNA. Thus, it appears that while the two parental strands of any single DNA molecule grow in the same direction in the overall sense (see Fig. 19.12), they grow in opposite directions for very brief periods of time.

The directedness of DNA polymerase activity (always from the 5' to the 3' end) is combined with an unusual and complementary property: 3'-to-5' exonuclease activity. All three forms of DNA polymerase have the ability to degrade (excise nucleotides) a DNA strand in the 3'-to-5' direction. Thus, DNA polymerases can break down, as well as extend, the length of a polynucleotide chain. The energetics strongly favor DNA synthesis, but the fact remains that the same enzymes can catalyze the reaction in both directions. A question immediately arises as to why one and the same enzyme has the ability to carry out both reactions and, indeed,

appears to do so. While it is always true that in theory any enzyme can catalyze a reaction in both forward and reverse directions, in most biochemical examples we are familiar with two different enzymes are usually involved. In studying the exonuclease activity of the various DNA polymerases, investigators have found that the enzymes work preferentially on incorrectly paired bases; for example, on an adenine paired to a cytosine, guanine, or another adenine.* If by chance the wrong base gets added to a growing DNA strand, it has a much greater chance of being removed before the next base is added on, if the same polymerizing enzyme is right there to immediately cut out the mistake. The exonuclease property of the DNA polymerases has given them the name of "proofreading" enzymes. Thus, with one set of remarkable enzymes, the cell can synthesize, degrade, and proofread its DNA molecules.

19.4
PROTEIN SYNTHESIS

Differences in the DNA of different organisms can be traced to differences in the order of base pairs along the DNA molecule. Similarly, differences between proteins can be traced primarily to differences in the kinds and sequence of amino acids along the polypeptide chains. The conclusion seems clear and inescapable: *the sequence of base pairs along DNA molecules must somehow control the kind and order of amino acids found in the proteins of an organism.* DNA, then, carries a **genetic code**—the blueprint establishing the kinds of proteins synthesized by the cellular machinery that make an individual organism unique.

---

*This preference is established by both the directionality of synthesis (5'-to-3') and by energy considerations. In 5'-to-3' synthesis, the growing DNA chain ends up with a terminal —OH group, on which the triphosphate group of the next free nucleotide can be joined. The energy from the triphosphate bond drives the synthesis. If synthesis proceeds from 3' to 5', the triphosphate group is joined onto the end of the growing chain; if a terminal nucleotide were removed by a proofreading enzyme, this would create a 5' monophosphate-ended chain with no high-energy bonds to catalyze the next addition. Thus, if synthesis were in the 3'-to-5' direction, every proofreading event would end up stopping further synthesis of the DNA strand. It is probably for this reason that no DNA polymerase enzymes have evolved that carry out synthesis in the 3'-to-5' direction.

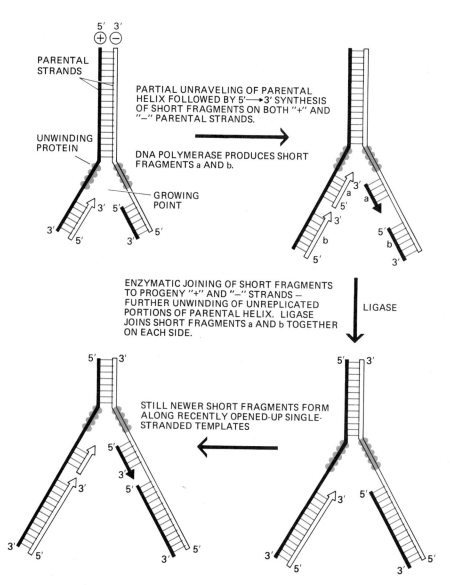

PARENTAL STRANDS

UNWINDING PROTEIN

PARTIAL UNRAVELING OF PARENTAL HELIX FOLLOWED BY 5'⟶3' SYNTHESIS OF SHORT FRAGMENTS ON BOTH "+" AND "−" PARENTAL STRANDS.

DNA POLYMERASE PRODUCES SHORT FRAGMENTS a AND b.

GROWING POINT

ENZYMATIC JOINING OF SHORT FRAGMENTS TO PROGENY "+" AND "−" STRANDS — FURTHER UNWINDING OF UNREPLICATED PORTIONS OF PARENTAL HELIX. LIGASE JOINS SHORT FRAGMENTS a AND b TOGETHER ON EACH SIDE.

LIGASE

STILL NEWER SHORT FRAGMENTS FORM ALONG RECENTLY OPENED-UP SINGLE-STRANDED TEMPLATES

Fig. 19.12
Okasaki's modification of Arthur Kornberg's original hypothesis concerning DNA replication. As Kornberg originally hypothesized, a single enzyme, DNA polymerase, is involved in synthesizing new DNA by moving from the 5' to the 3' end of an open, or single, strand of parental DNA. The dilemma for Kornberg was how to account for two-directionality of synthesis (since both strands of the double helix are copies, yet both run in opposite directions) with the same enzyme. Okasaki has recently demonstrated that Kornberg's enzyme does guide synthesis on both strands, in opposite directions, but for only very short segments. Another enzyme, known as ligase, joins these short strands together. In the diagram to the left, the two strands of the parental DNA are labeled (+) or (−) and shown in black and white, respectively. (Adapted from J. D. Watson, *Molecular Biology of the Gene*, 2d ed. Menlo Park, Calif.: W. A. Benjamin, 1970, p. 286.)

On the basis of the results of many different kinds of experiments performed in laboratories all over the world, a model system has been hypothesized to explain the sequence of events comprising protein synthesis, from gene to final product. The model was first worked out with respect to bacterial and viral DNA. In the absence of much detailed information about protein synthesis in eukaryotic cells, molecular geneticists tended to generalize from prokaryotes to eukaryotes. However, in recent years the entire picture of gene structure and protein synthesis in eukaryotic organisms has been revised. While the general principles of gene structure, transcription, and translation are the same for prokaryotes and eukaryotes, the details are significantly different. In terms of their molecular genetics eukaryotes are not simply more complex versions of prokaryotes.

In molecular terms, the genotype of an organism is the information coded in the base sequence of DNA; the phenotype is the collection of different kinds of protein molecules that DNA produces in the cell.

In order to simplify and clarify our discussion of protein synthesis, we will first focus on the process as it occurs in prokaryotic cells (bacteria). We will then turn to the modifications which appear in the eukaryotes.

To understand fully the model system for protein synthesis, let us first identify a major group of the characters, the various types of RNA, which enact this amazing drama. The types of RNA involved and their basic interaction with ribosomes, DNA, and amino acids are basically the same for both prokaryotes and eukaryotes. Hence, we need make no fundamental distinction between the structure or the function of eukaryotic and prokaryotic RNA.

Chemically, RNA molecules are very similar to those of DNA; they consist of sugar and phosphate units connected to four different kinds of nitrogenous bases. In RNA, however, thymine does not occur, its place being taken by **uracil**. Like DNA, RNA can store genetic information in its base sequence. Like the circular DNA of Kornberg's bacteriophage, RNA molecules are single-stranded. Finally, the sugar in RNA is ribose rather than the deoxyribose of DNA.

Ribonucleic Acid (RNA)

Present ideas concerning the molecular configuration of RNA molecules are much less clear than the Watson-Crick model of DNA. There are good reasons for this. RNA is far more difficult to obtain in pure crystalline form than DNA. Hence RNA is less easily studied by X-ray diffraction techniques. As a result, the type of information this technique can provide has been lacking for RNA, at least until fairly recently. In addition, RNA occurs in at least three forms. Each of these forms has a different structure and function. The three forms of RNA recognized today are *transfer* RNA, *ribosomal* RNA, and *messenger* RNA. Structural differences between these forms are not due primarily to differences in the nucleotides involved, but result from differences in molecular weight and configuration.

Messenger RNA.   Messenger RNA (*m*RNA) carries the genetic code from DNA to the ribosomes, where protein synthesis occurs. As its name implies, messenger RNA carries a message. This message is a complementary transcript of the DNA. Messenger RNA transmits the genetic message in the sequence of its own bases, forming a pattern complementary to that of the DNA that formed it. In the ribosome, the coded message that messenger RNA carries is translated into a specific amino acid sequence. Messenger RNA thus acts as an intermediary between DNA and protein.

Transfer RNA.   Transfer RNA (*t*RNA), often called soluble RNA, is the smallest type of RNA. Each transfer RNA molecule contains only seventy to eighty nucleotides.

Transfer RNA picks up individual amino acids in the cell and carries them to the sites of protein synthesis. Since each RNA molecule will pick up only one type of amino acid, there are many molecular variations of transfer RNA, one for each type of amino acid. Each has a slightly different sequence of bases. This enables each transfer RNA molecule to bind to one and only one specific type of amino acid.

In December 1964, the precise sequence of the seventy-seven nucleotides of the transfer RNA coding for the amino acid alanine was worked out by a team of Cornell University scientists headed by Dr. Robert Holley. Holley's work, published in 1965, gave precise data only on the sequence of nucleotides in one transfer RNA molecule. At that time there was little evidence about the three-dimensional structure and thus about how the molecule could actually work. In recent years, however, X-ray diffraction studies and even electron microscopy have revealed that the *t*RNA molecule has a generally constant shape, although each of the twenty different types (one specifically for each amino acid) have recognizable differences. The generalized shape is shown in the three-dimensional sketch in Fig. 19.13. The molecule tends to assume this shape as its thermodynamically most stable configuration, with hydrogen bonds forming across the strands between complementary bases. The loop containing the anticodon is where the *t*RNA molecule interacts with the messenger RNA; the **anticodon** is a sequence of three bases (triplet) complementary to a specific triplet on the *m*RNA. The $3'(-OH)$ end, or amino acid arm, of the *t*RNA molecule is where the specific amino acid attaches. The distance from the anticodon loop to the amino acid arm appears to be constant for all *t*RNA molecules, ensuring that all amino acids will be brought adjacent to each other when *t*RNA attaches to the messenger. Schematic representations of a few types of *t*RNA molecules are shown in Fig. 19.14. There are many more types of *t*RNA in the cell as there are amino acid types to be incorporated into protein.

Ribosomal RNA.   Most ribosomal RNA (*r*RNA) have a relatively high molecular weight (though some can be as low as 120 nucleotides). As pointed out in Chapter 5, it is contained in ribosomes, located in the cytoplasm of the cell. Ribosomes are composed of two thirds RNA and one third protein. The ribosomes function as centers of protein synthesis. It is to the ribosomes that the transfer RNA molecules carry their amino acids. During this process, some interaction occurs between transfer and ribosomal RNA.

No satisfactory picture has yet been developed for the molecular configuration of ribosomal RNA. For at least part of the length of the molecule, the structure appears to be helical. The rest of the molecule has an unknown shape. Working with this form of RNA presents special problems of technique. Attempts to crystallize

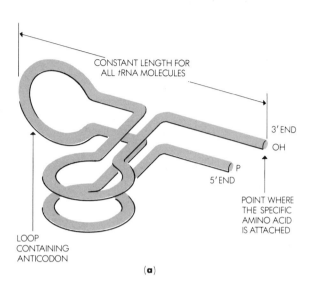

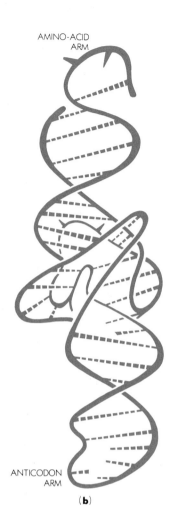

Fig. 19.13
(a) One of the possible three-dimensional models for the structure of transfer RNA. This generalized picture applies to all *t*RNA molecules; though there are many differences between the twenty-odd types, they may all possess at least as much in common as the pattern of folded loops and the constant distance from the anticodon to the amino acid attachment site. (Adapted from Loewy and Siekevitz, *Cell Structure and Function*, 2d ed. New York: Holt, Rinehart, and Winston, 1969, p. 168). (b) The actual physical structure of the *t*RNA molecule is probably not so neat looking as that shown in (a). A representation of one type of *t*RNA molecule, based on X-ray diffraction data, shows that it contains a number of twists and loops, folding back and base-pairing on itself in many places. For purposes of illustrating specific kinds of *t*RNA, however, the more schematic model shown in (a) will be used (for example, see Fig. 19.14).

whole ribosomes have met with only partial success. However, today neutron-scattering techniques have been used with considerable success to characterize with greater accuracy than previously the molecular configuration of ribosomal RNA.

Protein Synthesis in Prokaryotes

The sequence of events involved in producing a specific peptide chain from a genetic code on DNA can be represented as follows:

Transcription: The production of a specific molecule of messenger RNA (*m*RNA) from a given sequence of DNA.

Translation: The production of a specific amino acid sequence (polypeptide) by interaction of *m*RNA, specific transfer RNAs (*t*RNA) to which given amino acids are attached, and ribosomes (including ribosomal RNA).

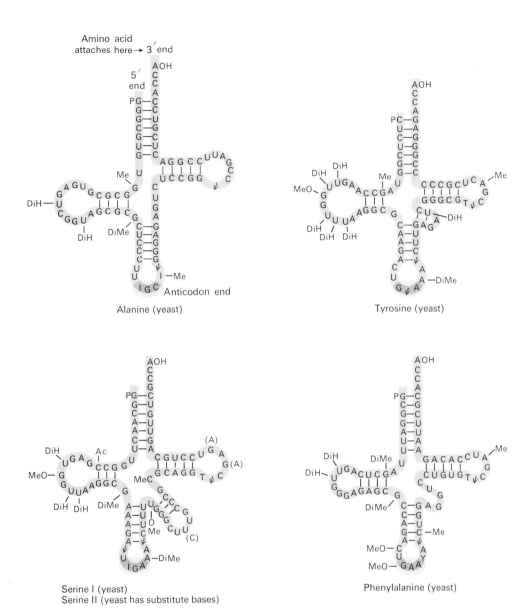

Fig. 19.14
Several representations of the "cloverleaf" pattern of *t*RNA, showing possible variations in the structure of specific types. Each type of *t*RNA attaches to one type of amino acid at the OH arm shown at the top of each molecule. Some slightly modified bases are incorporated into *t*RNA. These are indicated as DiH-U (dihydroxy-uridine), Me-G (methyl-guanidine), MeO-G (methoxyguanidine), I (inosine), and ψ (a form similar to uridine). P is a phosphate group attached to the 5' end of the molecule (after Holley). (Struther Arnott, "The Structure of transfer RNA," *Prog. in Biophys. and Mol. Biol.*, 22 [1971]: 181–213; the various *t*RNAs are diagrammed on pp. 183–185.)

The specific polypeptide produced in this manner may be an enzyme that acts to control one step in a metabolic pathway within the cell. Since all enzymes are thought to be produced in this way, the genetic code of DNA ultimately controls the entire metabolic activity of the cell. The experimentally established details of transcription and translation are described in the following two subsections, and diagrammed in Figs. 19.15 and 19.16.

I. Transcription. By base pairing, DNA directs the synthesis of a molecule of *m*RNA. The enzyme that catalyzes transcription of *m*RNA from DNA is known as

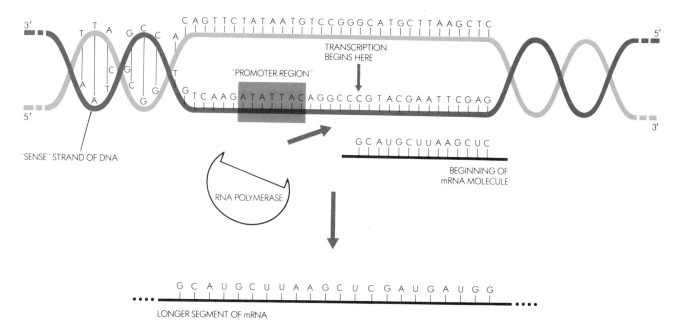

Fig. 19.15

Steps in prokaryotic protein synthesis I: *Transcription.* Messenger RNA is synthesized from a single strand of DNA. The DNA unwinds in an enzymatically controlled reaction and one strand, the "sense" strand of the gene, serves as the template from which *m*RNA is synthesized. Synthesis begins by the enzyme RNA polymerase recognizing the "promoter region" (color), a specific sequence of nucleotides that says "start here." Actual transcription begins a few bases down the strand from the promoter (black arrow). Transcription always proceeds from the 5′ to the 3′ direction along a DNA strand. Thus, in the above diagram, messenger is synthesized from left to right starting at the arrow. If the upper strand were the sense strand, it would be transcribed from right to left starting at its promoter (not shown here because there is no promoter region in the antisense strand). When the completed *m*RNA is synthesized, it falls away from the DNA and is ready to participate in step II, *Translation.*

DNA-dependent RNA polymerase or, more commonly, RNA polymerase. Transcription is initiated at a particular region of the DNA molecule called the promoter region, or promoter sequence. Promoter regions consist of a number of nucleotides, but always have some variant of the sequence 3′ . . . ATATTAC . . . 5′ (remember that 3′ and 5′ indicate the two different ends of a single strand of DNA). The promoter region signals the RNA polymerase to "start here," transcribing the particular strand that contains the ATATTAC sequence, or its variant. Transcription begins about six bases away from the actual promoter region (see Fig. 19.15), so that the promoter region itself is not actually

transcribed. In most cases only one of the two strands of any double-helix DNA molecule contains "sense" information. In some genes it is one strand, and in some genes the other (the two strands can be differentiated, recall, by the directions in which they run—one strand runs from 3′ to 5′, while its complement runs from 5′ to 3′). The important point here is that in one gene the sense strand that is transcribed into messenger RNA may be the 5′-to-3′ strand, while in another gene it is the 3′-to-5′ strand.

The nucleotide sequence of this newly formed messenger RNA molecule will be complementary to that of the DNA from which it was transcribed. In other

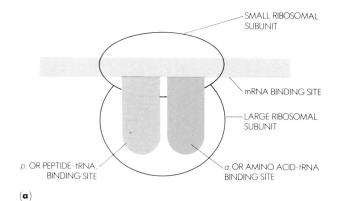

(a)

(b)

Fig. 19.16
Physical arrangement of binding sites on the two subunits of the ribosome: the *p*, or peptide-*t*RNA binding site; the *a*, or amino acid-*t*RNA binding site; and the *m*RNA binding site. The sites are shown unoccupied in (*a*). In (*b*) the sites are occupied, indicating how the spatial arrangement brings into proximity two transfer RNA molecules with their attached amino acids. The two *t*RNA-amino acid complexes that occupy the *p* and *a* sites at any one time carry amino acids that will be adjacent in the completed polypeptide chain. (Modified from David L. Kirk, *Biology Today*. New York: Random House, 1980, p. 521).

words, if the sequence of bases on the active ("sense") strand of the DNA is 3′ . . . ATCCGTGGGA . . . 5′, then the complementary sequence of bases in the new messenger RNA would be 5′ . . . UAGGCACCCU . . . 3′ (keep in mind that RNA molecules substitute the base uracil for thymine). Once transcribed, the messenger RNA is ready to participate in the second major phase of protein synthesis: translation.

II. Translation: The Process of Protein Synthesis. In the cytoplasm of the prokaryotic cell, amino acids are activated by complexing with an energy-rich compound such as ATP to form a highly reactive complex; this complex is then transferred to the adenine base at the end of a *t*RNA molecule specific for that amino acid. Both of these reactions are catalyzed by the same enzyme, amino acyl-*t*RNA synthetase. The two steps are summarized below:

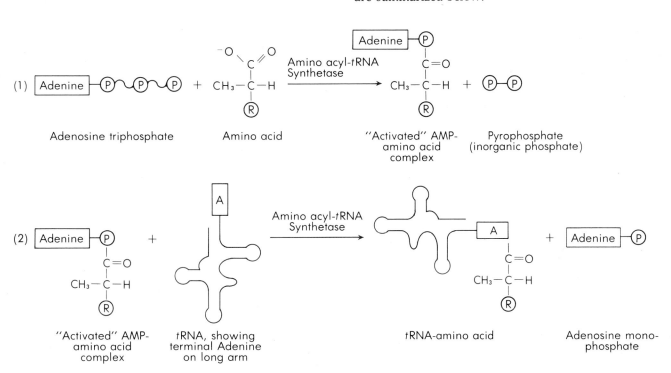

The adenine monophosphate (AMP) that is left over in reaction (2) is restored to its energy-rich form as ATP by two phosphorylation reactions. The energy from the hydrolysis of the two phosphate groups from ATP in reaction (1) is still stored in the amino acid-tRNA complex; the complex is thus "activated," meaning it contains potential energy that will be used later to forge the peptide bond between the amino acid on the tRNA and the growing polypeptide chain.

These two-step reactions are highly specific; each of the twenty amino acids must be joined onto the corresponding tRNA specific for that amino acid. Since there are at least twenty amino acids, and twenty kinds of tRNA, we might predict that there are at least twenty kinds of amino acyl-tRNA synthetase molecules. That prediction turns out to be correct: each enzyme ensures that only the amino acid specific for a given type of tRNA will be joined to it. Once the correct amino acid is bound to its specific type of tRNA, the entire "activated," or energy-rich, complex is ready for the next stage of protein synthesis, interaction with mRNA at the ribosome.

Ribosomes have three binding sites that insure that mRNA and the tRNA-amino acid complexes will be brought together. Two sites are for tRNA and one for mRNA. Figure 19.16 shows the structure of a ribosome, which consists of two subunits: one large subunit known as the 50S subunit, and a smaller one known as the 30S subunit. The two binding sites for tRNA are located mostly on the large subunits, but extend across to the small subunit. The two tRNA binding sites are called the p, or peptide-tRNA binding site (this site is usually occupied by a tRNA molecule to which a growing polypeptide chain is attached), and the a-site, or amino acid-tRNA binding site (usually occupied by a tRNA molecule to which a single, specific amino acid is attached). The binding site for mRNA is located exclusively on the smaller subunit of the ribosome. The binding sites are spatially arranged in such a way that when all three are occupied the specific sequence of bases making up the code for a given amino acid on the mRNA is brought into direct contact with the anticodon arm of the tRNA. In this way the two tRNA molecules carrying two adjacent amino acids in the future polypeptide chain are brought next to one another in a precise spatial arrangement that allows for peptide bond formation. The positioning of the tRNAs and the mRNA on the ribosome binding sites is shown schematically in Fig. 19.16(b).

It is important to emphasize that the sequence of amino acids in a polypeptide chain is determined not by the amino acids themselves (once they are attached to their respective tRNAs), but by the sequence of bases on the mRNA molecule fed to a ribosome. The ribosome likewise does not determine the specific sequence in any polypeptide chain. Ribosomes are like tape recorders. They will play out any sequence of sounds (that is, polypeptide chain) depending upon the coded tape (that is, the mRNA) fed into them. Thus, the sequence of amino acids is determined by the interaction of the anticodon end of the tRNA with its specific complementary sequence of bases on the mRNA. Once attached to the tRNA, the nature of the amino acid has no bearing on its incorporation into a growing polypeptide chain. Experiments have been performed in which an amino acid is chemically converted from one type to another after being attached to its specific tRNA (for example, cysteine is first chemically linked to its appropriate tRNA; the cysteine is then converted into alanine). The modified amino acid is incorporated into the protein as if it were cysteine, indicating that it is the tRNA and mRNA, and not the amino acid itself, that determines where the amino acid is placed in a growing polypeptide chain.

Protein synthesis (translation) occurs in three major steps: initiation, elongation, and termination. Let us examine each of these steps in sequence.

Initiation. This is the most complex of the three steps. The basic problem in initiation is to find the correct starting place on the mRNA molecule. All start signals on mRNA yet investigated (both prokaryotes and eukaryotes) are one of two forms: AUG or GUG. With this regularity of an initiation signal it might seem that there would be no misreading of the message. However, these two triplet sequences could occur elsewhere along the mRNA; for example, every time the triplet CAU preceded another triplet starting with a G. How does the ribosome differentiate between the true initiation sequence AUG, and a random sequence of AUG that occurs somewhere else along the mRNA chain? As it happens, a short piece of RNA, called a "leader," precedes the initiation codon on every mRNA molecule. The leader is transcribed from the DNA, but it is not translated into protein. It is, in short, similar to the leader tape that starts out at the beginning of a reel of recording tape. The leader portion of the mRNA molecule consists of a specific sequence of four bases (usually, but not always, AGGA) called the **finder sequence**, which is separated by seven nucleotides from the start signal (that is, from AUG or GUG). This system is precisely engineered to the structure of the ribosome. There is in the small ribosomal subunit a specially positioned portion of rRNA that projects onto the mRNA bind site (refer again to Fig. 19.16a). This molecule of rRNA contains a nucleotide sequence that is complementary to the finder sequence (it is sometimes called an **antifinder sequence**). When the finder sequence on mRNA binds to the antifinder sequence of the rRNA, the initiation signal on the mRNA is positioned precisely on the p binding site of the ribosome.

At this point a single kind of tRNA, bearing the modified amino acid formyl methionine, can bind to the initiation signal (AUG, or GUG) on mRNA. The formyl methionine-tRNA complex is then bound by weak chemical bonds to the p site on the ribosome. The entire complex of *m*RNA, formyl methionine-*t*RNA and ribosomal RNA, bound together, is known as the **initiation complex.** Protein synthesis can now begin. The initiation process does not have to begin again until another polypeptide is ready to start being translated.

Elongation. Elongation is the process by which individual amino acids are added sequentially to a growing polypeptide chain. The amino acids are added one at a time, moving from the *a* site to the *p* site on the ribosome. The order of amino acids added to the polypeptide is specified precisely by the sequence of bases on the *m*RNA. The events of elongation are shown diagrammatically in Fig. 19.17. With the *m*RNA held into position on the ribosome by the positioning of the formyl methionine-*t*RNA complex (bound onto the *p* site), an exposed messenger codon, a triplet of exactly three bases, occupies the *a* site. An enzyme now selects the appropriate amino acid-*t*RNA complex that corresponds to the exposed triplet at the *a* site. The match is made specific because the *t*RNA has an exposed anticodon region that is complementary to the one triplet on the messenger that is exposed at the *a* site (Fig. 19.17a). For example, if the triplet of bases on *m*RNA that occupied the *a* site were UCC, the only *t*RNA that would bind there (by hydrogen bonding) would be one with an anticodon of AGG (see Fig. 19.17b). It so happens that this is the *t*RNA which specifically binds to the amino acid serine. Thus, with the triplet codon of UCC occupying the *a* site, serine and only serine can be brought into position and inserted into the polypeptide chain. In the process a high-energy phosphate bond is split and transferred to the amino acid-*t*RNA complex, activating it. This activated complex now has its own energy source for forming the peptide bond. (It is an endothermic reaction.)

We now come to the most basic reaction in the process of protein synthesis: formation of the peptide bond. The peptide bond is formed between the amino acid occupying the *p* site and the amino acid occupying the *a* site. The bond is formed while both amino acids are still attached to their respective *t*RNA molecules (see Fig. 19.17c). The energy comes from the high-energy phosphate group (that is, from guanosine triphosphate, GTP) attached to the amino acid occupying the *a* site. This reaction is catalyzed by an enzyme built into the surface of the ribosome. Once the peptide bond has been formed, the amino acid in the *a* site is now a part of the growing polypeptide chain. As the peptide bond is formed between the amino acids in the *p* and *a* sites, the amino acid in the *p* site is removed from its *t*RNA. This leaves an "uncharged" (that is, carrying no amino acid) *t*RNA in the *p* site. Because the *p* site has a specific chemical affinity for a *t*RNA only when it is carrying an amino acid, the "naked" *t*RNA falls away from the *p* site.

Once the *p* site has become empty, the *m*RNA with the amino acid-*t*RNA complex bound to it at the *a* site moves down three bases along the ribosome. This brings the amino acid-*t*RNA complex onto the *p* site, and opens up the *a* site for binding of the next amino acid-*t*RNA. Exactly how the whole messenger-*t*RNA complex moves down precisely three bases along the ribosome is an unsolved problem of molecular genetics. This process of shifting positions is called **translocation,** and it requires the expenditure of another high-energy phosphate bond (again from GTP). Elongation continues as long as there are meaningful codons on the *m*RNA.

Termination. After elongation continues for a while, the end of the messenger comes up and *termination* occurs (see Fig. 19.17d and e). Termination is not a random process, but is highly controlled by special "full-stop" or termination codons. Recent studies have shown that certain triplets or codons (called "nonsense codons") in the *m*RNA chain automatically bring about termination of the peptide chain at that point (see Fig. 19.17d). The nonsense codons are UAA, UAG, and UGA. When the ribosome reaches a nonsense codon, the bond between the final amino acid and the *t*RNA molecule to which it is attached is hydrolyzed. This reaction is mediated by a protein "release factor," which may act as an enzyme or in some other capacity not clearly understood at present. Thus the final amino acid is released from its *t*RNA molecule without forming a peptide bond with another amino acid, as is usually the case. The exact chemistry of termination is still being actively investigated. It appears to be built into the genetic message of the DNA as accurately as the position of each amino acid.

Alexander Rich and his co-workers have found that the long molecule of *m*RNA may have more than one ribosome associated with it at any one time. Electron microscopy can reveal several ribosomes spaced at intervals along the length of *m*RNA. Such a cluster of ribosomes held together by *m*RNA is called a polyribosome, or polysome for short (see Fig. 19.18). Each ribosome of a polysome is involved in protein synthesis. The quantity of growing protein associated with any given ribosome will depend on how far that ribosome has traveled from the starting end of the *m*RNA. Thus it is clear that each molecule of *m*RNA may serve to generate a number of identical proteins from its coded message.

Dramatic evidence in support of the hypothesized roles of DNA, *m*RNA, and ribosomes in protein synthesis has been obtained through electron microscopy (see Fig. 20.15).

The events of DNA replication and protein and RNA synthesis all depend upon the formation of specific, though weak, chemical interactions among the

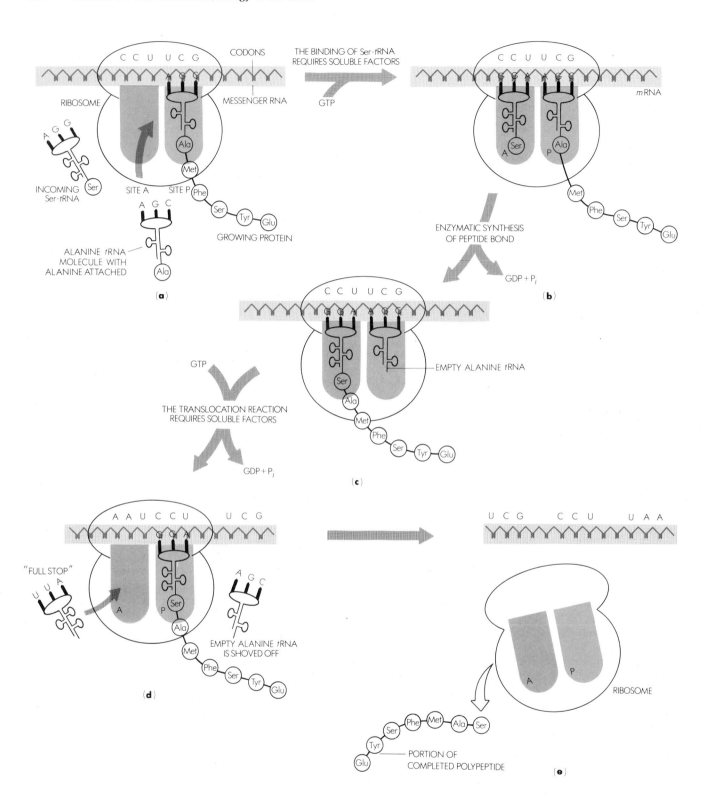

THE BINDING OF Ser-*t*RNA REQUIRES SOLUBLE FACTORS

ENZYMATIC SYNTHESIS OF PEPTIDE BOND

THE TRANSLOCATION REACTION REQUIRES SOLUBLE FACTORS

(a)

CODONS

RIBOSOME

MESSENGER RNA

INCOMING Ser-*t*RNA

SITE A   SITE P

GROWING PROTEIN

ALANINE *t*RNA MOLECULE WITH ALANINE ATTACHED

(b)

*m*RNA

GTP

GDP + P$_i$

(c)

GTP

EMPTY ALANINE *t*RNA

GDP + P$_i$

(d)

"FULL STOP"

EMPTY ALANINE *t*RNA IS SHOVED OFF

(e)

RIBOSOME

PORTION OF COMPLETED POLYPEPTIDE

Fig. 19.17
The sequence of steps involved in peptide formation at the ribosome. For the sake of convenience, only a single ribosome is shown here, though several ribosomes are usually attached to any one messenger molecule. The sequence illustrated here (from *a* to *e*) shows the steps involved in the addition of one amino acid, and the termination reaction by which the peptide chain is hydrolyzed away from the ribosome-*t*RNA complex.

nucleic acids involved. Chief among these weak interactions is hydrogen bonding between specific pairs of complementary bases. These are highly specific, ensuring the accuracy of the processes. Their weak character ensures that the chemical associations formed in determining a sequence will be temporary. In the case of protein synthesis, the weak bonds ensure that the newly created polypeptide can be easily detached from the messenger RNA and the ribosome.

The end-product of the above sequence of reactions is a completed protein molecule. In many cases, however, a protein can acquire all of its functional properties (including enzymatic) only when it is closely bound to other proteins. The fully functional protein is then said to be made up of *protein subunits*. Such a protein may be an enzyme, or it may be a structural or transport protein such as collagen or hemoglobin. The primary structure of all proteins is determined by DNA through the mechanism outlined above. The coiling of a single peptide chain into the alpha helix is determined by intramolecular forces associated with the elements of the peptide linkage. The combination of various coiled proteins into a fully functional protein molecule is determined by the number and location of each amino acid and the interactions that are possible between their side-chains (secondary, tertiary, and quaternary structure). Because these intricate conformations of proteins are de-

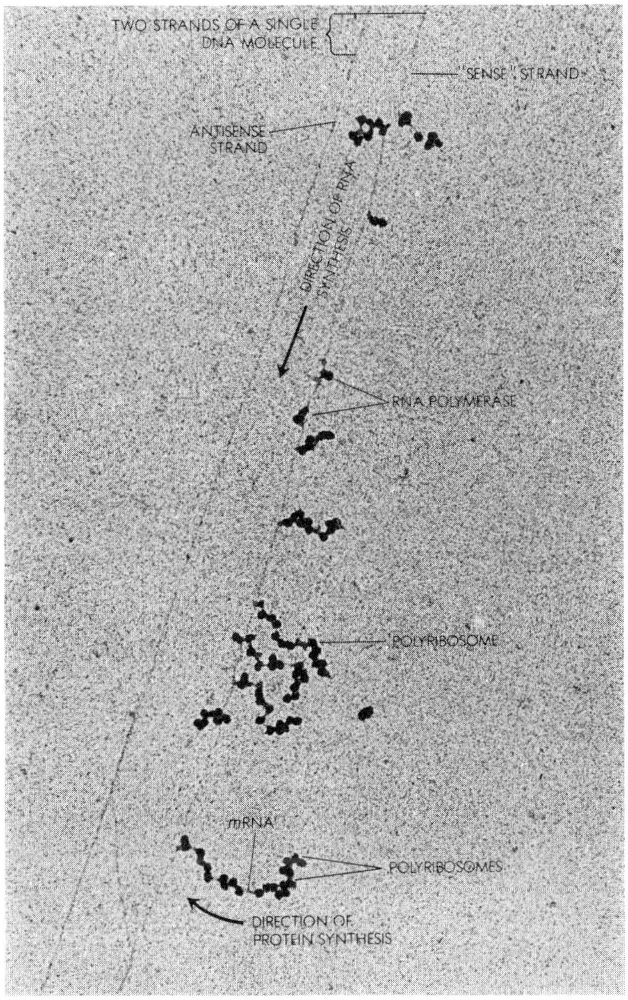

Fig. 19.18
This electron micrograph, magnified 36,000 times, gives dramatic support to the hypothesized roles of DNA, RNA, and the ribosomes in bacterial protein synthesis. Shown is a portion of the circular chromosome of the bacterium *Escherichia coli.* Most of the chromosome is inactive, while the active portions show *m*RNA transcribed from the chromosomal DNA by RNA polymerase. This type of close contact between DNA, RNA, and ribosomes had earlier been predicted in prokaryotic cells, in which there is no organized nucleus.
(Adapted from O. L. Miller, Jr., Barbara A. Hamkalo, and C. A. Thomas, Jr., *Science* 169 [1970]: 392.)

Since the code on a segment of mRNA is complementary to a corresponding segment of DNA, the mRNA code is called an anticodon.

pendent upon the specification of the amino acid sequence, we can say that all information about the cell's structure and function resides in the sequence of amino acids in the proteins, and thus in the bases of DNA.

In prokaryotes, genes and the proteins for which they code are known to be **colinear.** Colinearity means that the sequence of amino acids in the completed polypeptide follows the same order as the sequence of triplets (codons) in the DNA molecule coding for that polypeptide. Thus a single base-pair change at one end of a DNA molecule can result in the change of a single amino acid at one end of the corresponding protein; similarly, a single mutation at the other end of the DNA molecule results in the change of a single amino acid at the other end of the protein. Furthermore, the order of mutations all along the DNA corresponds to the order of the amino acids along the polypeptide. In other words, the genetic map within the limits of a single "gene" (so far as it has been determined) corresponds point-by-point with the amino acid sequence of the polypeptide (see Fig. 19.19).

Protein Synthesis in Eukaryotes

Protein synthesis in eukaryotes is somewhat different from that in prokaryotes, largely as a result of differences in the organization of eukaryotic DNA. When investigators began applying the model of protein synthesis developed with bacteria to higher animals and plants, they encountered unexpected observations that did not fit the classical picture. Some of the anomalies are:

1. It was first discovered in animal viruses, but later in humans and other eukaryotic organisms, that there is far more DNA per genome than was necessary to make up the genes. For example, in humans a single complete gene contains on the average of between 1000 and 3000 nucleotide pairs. The maximum number of imaginable human genes has been placed at 100,000. Thus, we might expect to find that the DNA content of every human cell contains approximately 300 million nucleotides. However, conservative estimates suggest that the human genome (per cell) contains about 5 billion (!) base pairs. This means that over 80 percent of human DNA may not be part of a structural gene (that is, a gene coding for a polypeptide)!

2. A similar overabundance of RNA has also been found in large numbers of eukaryotic cells. James Darnell, working at Rockefeller University, found large strands of RNA in the cell nucleus. He called this excess RNA *heterogeneous nuclear RNA* (symbolized *hn*RNA). The *hn*RNA was found to be the precursor of the *m*RNA that later was found in the cell cytoplasm. However, Darnell noted that only 20 percent of the *hn*RNA produced in the nucleus appeared in the cytoplasm.

3. Both *hn*RNA and the cytoplasmic *m*RNA derived from it contain regularized, constant end-sequences, or markers. For example, Darnell found that a constant poly-adenylate (Poly-A) sequence of 200 nucleotides was attached to the 3' or terminating end of *hn*RNA. A smaller sequence, called a "cap," was located at the 5' end. Darnell and others observed two very interesting and unexpected features of these caps:

(a) The same caps appear in the large *hn*RNA and in its much smaller derivative, cytoplasmic *m*RNA. This would suggest that if the RNA is reduced in size between the time of its formation in the nucleus and its involvement in protein synthesis in the cyto-

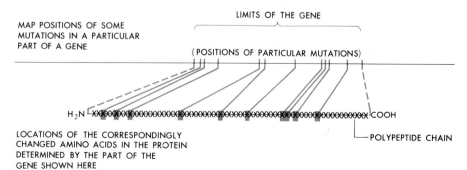

MAP POSITIONS OF SOME MUTATIONS IN A PARTICULAR PART OF A GENE

LIMITS OF THE GENE

(POSITIONS OF PARTICULAR MUTATIONS)

H₂N COOH

POLYPEPTIDE CHAIN

LOCATIONS OF THE CORRESPONDINGLY CHANGED AMINO ACIDS IN THE PROTEIN DETERMINED BY THE PART OF THE GENE SHOWN HERE

Fig. 19.19
Diagrammatic representation of the collinearity of a prokaryotic gene and the polypeptide chain for which it codes. Specific regions of the genetic map are known to code for a certain segment of the polypeptide chain; the arrangement of these identified areas along the DNA corresponds to the sequence of amino acids being coded for in the corresponding sections of the protein.

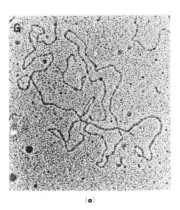
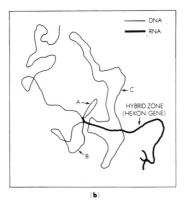

(a)   (b)

Fig. 19.20
(*a*) Electron micrograph of *m*RNA for a specific gene, known as the "hexon" gene, of the animal virus adenovirus 2; the *m*RNA shown here is ready for interaction with *t*RNA and ribosomes to produce a polypeptide. The *m*RNA is shown paired with a fragment of the viral DNA from which it was transcribed. (*b*) Schematic interpretation of the loop structure shown in the micrograph. Because large sections of intervening sequence have been cut out of the *m*RNA, when it pairs with its own original DNA large loops of DNA are formed. These loops presumably represent the intervening sequences on DNA, which find no complementary regions in the "edited" *m*RNA. (Photograph courtesy Susan Berget, Claire Moore, and Phillip Sharp, Massachusetts Institute of Technology; diagram redrawn from *New Scientist*, January 5, 1978, p. 18, with permission).

plasm, midsections, rather than end-pieces, must be "cut out" of the molecule.

(b) The caps at each end of the RNA are remarkably constant, regardless of the particular gene from which they are transcribed. Caps are not specific for particular gene sequences.

4. When the *m*RNA for a protein from adenovirus 2 (a virus that can cause cancer in animals, but not humans) is placed into a solution with the virus DNA and allowed to "hybridize," unusual configurations can be seen in the resulting electron micrographs (see Fig. 19.20a). The *m*RNA derived from a particular piece of DNA contains complementary bases and hence in the hybridization process should pair up wherever complementary sequences exist. In such hybridization the electron micrographs show large loops of DNA. It is thought that these loops represent base sequences on the DNA that have no counterpart on the RNA, and thus cannot be "matched up" or hybridized (Fig. 19.20b).

The data suggest that eukaryotic and animal virus genes occur in pieces rather than as one continuous segment of DNA coding for one continuous polypeptide. That is, the DNA sequence coding for a particular polypeptide is split into two or more segments interrupted by a nonsense group of bases called an **intervening sequence**. The polypeptide-coding, "sense" regions of the DNA are called **encoding sequences**. The existence of encoding sequences interrupted by intervening sequences seems to be very much the rule among eukaryotes and animal viruses. To date, over a dozen animal genes have been carefully analyzed for their DNA sequences. These include human hemoglobin, mammalian insulin, chicken ovalbumin, immunoglobulins (antibodies), and lysozyme. The only eukaryotic genes so far found not to be broken up by intervening sequences are those for human interferon, a protein produced by cells in response to viral infection, and for histone, the protein found in abundance in chromosomes. The number of intervening sequences varies from one type of eukaryotic gene to another. The record is held by one form of chicken albumin, with sixteen. Most others contain between two and four intervening sequences. It is important to keep in mind that we are using the term *gene* here in the sense derived earlier from the work of Benzer and others (see Supplement 19.2): that is, a DNA sequence coding for a complete, functional polypeptide chain. Thus, intervening sequences are interruptions *within* the DNA coding for a single polypeptide. This means that what turns up as a single polypeptide chain in the cytoplasm is composed of two or more shorter

---

The genes of most eukaryotic cells exist in pieces or fragments, called "encoding sequences," which are separated from one another by nonsense messages called "intervening sequences." Each encoding sequence determines a part of a polypeptide chain.

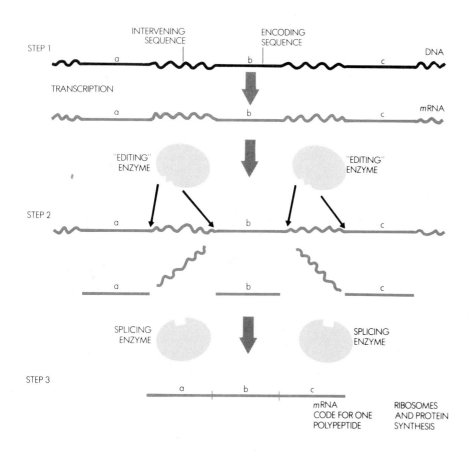

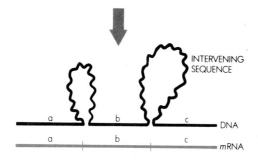

Fig. 19.21
Transcription in eukaryotic and animal viruses which have "split" genes. a, b and c represent subsections of one gene (they could be called "mini-genes") coding for a specific polypeptide.

segments each synthesized from different encoding sequences on the DNA.

From these observations, molecular biologists have constructed a general picture of gene transcription and translation in animal viruses and eukaryotic cells. The following summary is based on the model outined in Fig. 19.21.

During the first step of protein synthesis (Fig. 19.21, step 1), an RNA copy is made of the entire length of the DNA containing gene fragments. That is, transcription includes both encoding and intervening sequences. There is so much extra DNA that the *hn*RNA is often five to ten times longer than the sum of the individual encoding sequences.

Second, while still in the nucleus, the intervening sequences which have been transcribed into the *hn*RNA are snipped out, and the ends of the encoding sequences joined together (Fig. 19.21, steps 2 and 3). This part of the process is very much like editing a movie film. If *hn*RNA is thought of as the negative, then we can think of unwanted sequences snipped out by an editor and the meaningful parts spliced back together before being developed (translated) into the positive film (polypeptide). So far, molecular geneticists have just begun to identify the "editing" enzymes. In 1980, John Abelson, at the University of California-San Diego, identified two polypeptides that appear to be editing enzymes for transfer RNA molecules in yeast. The polypeptides carry out both editing and resplicing processes. It is not known, however, whether the polypeptides are subunits of a single enzyme, or two separate enzymes. Nor is it yet understood how the editing enzyme(s) recognize the intervening sequences to snip them out.

Third, the shortened piece of RNA, now a functional *mRNA* molecule, passes from the nucleus into the cytoplasm. Here it interacts with transfer RNA and ribosomes to guide the synthesis of a specific polypeptide.

The major differences, then, between protein synthesis in pro- and eukaryotes appear to lie in the arrangement of the DNA in the nucleus, and, as a consequence, in the process of transcribing RNA and preparing it for translation in the cytoplasm. Once *mRNA* is in the cytoplasm the process of protein synthesis appears to be the same for all cells.

19.5
THE GENETIC CODE

If the sequence of *t*RNA molecules that attach to any given messenger is to be determined precisely, the messenger RNA molecule must contain a very specific set of information. That information, coming originally from DNA, is in the form of a code, or what is called the "genetic code." The code lies in the specific sequence of nitrogenous bases on DNA (and subsequently on RNA). How many bases are involved in coding for each specific type of *t*RNA molecule and, hence, for each type of amino acid? In other words, what is the nature of the genetic code?

This question was first attacked by simple arithmetic. It was obvious that more than one base was involved. With only one base playing a role, only four amino acids could be selected. After all, there are only four kinds of bases in RNA.

Could there be two bases involved? Again, with only two bases, the possible arrangements are 4^2, or 16. This number is not large enough to allow for the selection of the twenty amino acids then known to be used in protein synthesis.

The minimum number of bases that could be involved in amino acid selection seemed to be three. This number gave a possibility for the selection of 4^3, or 64, amino acids, which is more than enough. Higher numbers than three were possible, of course, but seemed unnecessary and thus, it was assumed, less likely. The relationship between the number of bases involved in the code and the number of words that can be coded for is given in Table 19.2.

Table 19.2

Possible Genetic-Code Letter Combinations as a Function of the Length of the Code Word (A = adenine, G = guanine, U = uracil, C = cytosine)

Singlet code (4 words)	Double code (16 words)				Triplet code (64 words)			
A	AA	AG	AC	AU	AAA	AAG	AAC	AAU
G	GA	GG	GC	GU	AGA	AGG	AGC	AGU
C	CA	CG	CC	CU	ACA	ACG	ACC	ACU
U	UA	UG	UC	UU	AUA	AUG	AUC	AUU
					GAA	GAG	GAC	GAU
					GGA	GGG	GGC	GGU
					GCA	GCG	GCC	GCU
					GUA	GUG	GUC	GUU
					CAA	CAG	CAC	CAU
					CGA	CGG	CGC	CGU
					CCA	CCG	CCC	CCU
					CUA	CUG	CUC	CUU
					UAA	UAG	UAC	UAU
					UGA	UGG	UGC	UGU
					UCA	UCG	UCC	UCU
					UUA	UUG	UUC	UUU

After M. Nirenberg (1963)

The definition of a gene must include not only the DNA sequences coding for protein but also the control sequences—the "stop" and "start" signals—at each end of the gene. Stop and start signals are particular triplets that signal RNA polymerase to stop or start transcribing.

Evidence for the Triplet Code

Work on the genetic code proceeded, therefore, on the assumption that the code was a triplet involving only three bases. But the hypothesis was based largely on what appeared to be logical. What actual evidence exists for the triplet code hypothesis? One of the most direct tests rests on a refined technique that allows the geneticist to insert one or more bases into a synthetic *m*RNA chain. The *m*RNA is then "fed" activated amino acids attached to *t*RNA, and the activating enzymes specific for each type of amino acid. The system can synthesize proteins in a test tube. The resulting protein then reflects the significance (in terms of amino acid sequence) of changes in one or more bases in the DNA code. To understand this experiment, keep in mind the fact that *m*RNA carries the code without any punctuation marks or spaces—the code is read by the ribosome as one sequence of triplets after another. The ribosome thus moves along the *m*RNA molecule (or vice versa) three bases at a time (on the hypothesis that the code is triplet). Thus a mistake in the code caused by adding or deleting one base would be multiplied throughout the remainder of the sequence. It would be read as "nonsense" and would produce an incomplete protein.

An analogy helps illustrate how the production of nonsense codes can be used to indicate whether the code is a triplet or not. Consider the following sentence made up of only three-letter words (a triplet code):

THE MAN SAW TOM BUY THE TOY

This sentence is written with punctuation (in this case spaces) between the triplet codes. On *m*RNA, however, the sentence would be written:

THEMANSAWTOMBUYTHETOY

Even though this looks like nonsense at first glance, it is apparent that the "message" can be read by moving along the line three letters at a time:

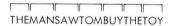
THEMANSAWTOMBUYTHETOY

However, if we add another letter in the line somewhere near the beginning, it is obvious that this will throw off the sense of the sentence from that point on—that is, as long as we continue to read three letters at a time:

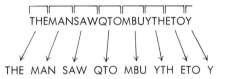
THEMANSAWQTOMBUYTHETOY

THE MAN SAW QTO MBU YTH ETO Y

After the third triplet, the message becomes nonsense. Such experiments, in which one base was added to an already-existing messenger RNA molecule, have produced nonsense messages. F. H. C. Crick, Sidney Brenner, L. Barnett and R. Watts-Tobin used an acridine dye known as proflavin, which produces mutations by causing the addition or deletion of a nucleotide in DNA. They developed a series of "addition" mutants of the rII region of bacteriophage T4. With one or two bases added to the code, the DNA segment from that point on is read as "nonsense." But Crick and his co-workers reasoned that the insertion of three bases should restore the sense of the message on the DNA beyond the mutant region. They set about to test their hypothesis. They combined two additional phage mutants, and observed that no wild-type phage plaques appeared in a "lawn of"

Messenger RNA molecules are like a sequence of letters in a sentence with no punctuation marks. The only signals that approximate punctuation in the mRNA are start-stop indicators. Any given mRNA molecule may have several start and stop indicators. The region between the start and stop signals codes for a continuous polypeptide or protein chain. The region of DNA corresponding to the mRNA segment between start-stop signals is called a gene.

Table 19.3
The Genetic Code

First letter	Second letter				Third letter
	U	C	A	G	
U	Phenylalanine	Serine	Tyrosine	Cysteine	U
	Phenylalanine	Serine	Tyrosine	Cysteine	C
	Leucine	Serine	(End chain) (ochre)*	(End chain)	A
	Leucine	Serine	(End chain) (amber)*	Tryptophan	G
C	Leucine	Proline	Histidine	Arginine	U
	Leucine	Proline	Histidine	Arginine	C
	Leucine	Proline	Glutamine	Arginine	A
	Leucine	Proline	Glutamine	Arginine	G
A	Isoleucine	Threonine	Asparagine	Serine	U
	Isoleucine	Threonine	Asparagine	Serine	C
	Isoleucine	Threonine	Lysine	Arginine	A
	Methionine	Threonine	Lysine	Arginine	G
G	Valine	Alanine	Aspartic acid	Glycine	U
	Valine	Alanine	Aspartic acid	Glycine	C
	Valine	Alanine	Glutamic acid	Glycine	A
	Valine	Alanine	Glutamic acid	Glycine	G

*Ochre and amber are two specific "nonsense" sequences that signal chain termination. The terms "ochre" and "amber" are older usages that still occur extensively in the literature.

E. coli. When they combined *three* such mutants, however, they obtained some wild-type plaques. These results dramatically confirmed the triplet nature of the genetic code.

This experiment is a genetic one: it makes changes in the genetic code at the DNA level and then looks for some phenotypic expression of that change. It is also possible to test the triplet code hypothesis biochemically. Marshall Nirenberg and some of his co-workers have taken advantage of the fact that in cell-free extracts of *E. coli* (extracts in which components of cells have been released from being bound within the cell membrane), *m*RNA and *t*RNA bind to ribosomes. The researchers found that the amount and kind of *t*RNA (carrying specific amino acids) bound to ribosomes depended on the length and sequence of the *m*RNA added to the system. If *m*RNA consisting of only two nucleotides was added, no *t*RNA was bound. If *m*RNA three nucleotides long was added, however, *t*RNA and ribosomes would bind to it. Furthermore, if different types of *m*RNA three nucleotides long were added, *t*RNAs carrying *particular* amino acids were bound (for example, the sequence UUU caused the *t*RNA carrying phenylalanine to bind).

This approach allowed Nirenberg and others to determine systematically which sequences of three nucleotides corresponded to each specific amino acid. These sequences are now called triplets or **codons.**

All the possible codons have now been associated with specific amino acids or a role in polypeptide synthesis. Table 19.3 gives the triplet of base pairs for each of the twenty most common amino acids. To select the proper triplets or codons for any amino acid, simply read in order the letters appearing to the left, above, and to the right of it. Thus the codons for glycine (gly) are GGU, GGC, GGA, and GGG, while those for lysine (lys) are AAA and AAG. Some of the codons assigned here are less certain than others, and those suspected of being connected with the beginning of polypeptide chain synthesis are not included. The two codons labeled nonsense, ochre and amber (UAA and UAG), are related to no amino acid and are hypothesized to represent "punctuation" in the chain. The codon UGA is hypothesized to terminate polypeptide chain synthesis.

From the data in Table 19.3 geneticists say that the genetic code must be *degenerate;* that is, more than one codon can select for a given amino acid. More signifi-

SUPPLEMENT 19.4
RESTRICTION ENDONUCLEASES AND THE SEQUENCING OF DNA

In the early to mid-1970s molecular geneticists discovered a group of bacterial enzymes that cleave DNA molecules at very specific points in the nucleotide sequence. Known as *restriction endonucleases*, these enzymes help bacteria in the natural state ward off infection by bacteriophage. The enzymes cleave viral DNA as soon as it is inserted into the bacterial cell. Every bacterial cell has several kinds of restriction endonucleases, each specific for a certain nucleotide sequence. When a restriction endonuclease encounters the nucleotide sequence to which it is specific, it catalyzes hydrolysis of the phosphodiester bond connecting the sugars and phosphates between specific neighboring bases along the backbone of the DNA molecule. This cleavage begins to reduce the large DNA molecule into a series of smaller fragments. Each restriction endonuclease will cleave a given DNA molecule only in those places where its own specific nucleotide sequence is found. For example, the endonuclease known as Eco RI (which stands for *E. coli*, restriction enzyme 1, that is, the first to be discovered in *E. coli*), is specific for the nucleotide sequence:

$$5' \ldots G \overset{\downarrow}{} A A T T C \ldots 3'$$

(where 5' and 3' indicate the two different ends of a single strand of DNA). The enzyme cleaves the strand between the first G and A, as indicated by the arrow. Several other restriction endonucleases, the DNA sequences they attack, and the organisms from which they are derived, are shown in Table A.

If a restriction endonuclease recognizes only one specific sequence of nucleotides, how can it cleave *both* strands of the double helix at the same level of the molecule? To cut one strand will obviously not break up the whole molecule, since the complementary strand remains intact. A close inspection of each of the nucleotide sequences in Table A, however, will reveal a very interesting pattern. If you determine what the

sections of DNA complementary to those shown in Table A would be, you will note that in each case the complementary strand presents the same order of bases, but *in reverse!* That is, whether read forward or backward the base sequence is the same. Words or phrases that read the same forward or backward are called palindromes, which literally means a "running backward." The fact that restriction endonucleases have evolved to be specific only for base sequences that are palindromes represents an amazing molecular adaptation. It allows the same endonuclease molecule to cut a double helix apart at the same place along the length of the DNA molecule.

Restriction endonucleases have found several important roles in recent research on the structure of genes. For one thing, they have been crucial to the new technology of recombinant DNA (see Supplement 19.5). For another, they have greatly aided in determining for the first time the sequence of nucleotides in the DNA of various organisms. To date, the complete nucleotide sequence of two viruses, ϕX174 and SV40, have been determined using various restriction endonucleases. Without these enzymes, the task of sequencing the lengthy nucleotide chains (over 5000 nucleotides for each) would have been prohibitively time-consuming.

To understand the procedure by which restriction endonucleases are useful in DNA sequencing, consider the following example. The virus SV40 DNA (5220 nucleotide pairs) is first treated with the restriction endonuclease Hind III (from the bacterium *Haemophilus aegyptius*) which cleaves the entire DNA molecule into eleven fragments. These fragments are separated from one another by gel electrophoresis and designated for reference by letters of the alphabet. Investigators then take one fragment, let us say the F fragment, and treat it with a second endonuclease, such as Eco RI. If there is no reaction (no cleavage), then it is clear that the F fragment

Table A
Representative Restriction Endonucleases

Name of enzyme	DNA sequence it attacks (arrow indicates site of cleavage)	Organism from which derived
Eco RI	$5' \ldots G \overset{\downarrow}{} A A T T C \ldots 3'$	*E. coli*
Eco RII	$\overset{\downarrow}{} C C T G G$	*E. coli*
HAE III	$G G \overset{\downarrow}{} C C$	*Haemophilus aegyptius*
Hind III	$A \overset{\downarrow}{} A G C T T$	*Haemophilus aegyptius*
Aval	$C G Pu \overset{\downarrow}{} Py C G^*$	

*Pu = purine; Py = pyrimidine

contains no Eco RI sequence, that is, no GAATTC. If, on the other hand, the F fragment is cleaved into two pieces, investigators know there is one sequence of GAATTC within the fragment. If there are three fragments, then it is clear there are *two* Eco RI sites (GAATTC) within the F fragment, and so on. As it turns out, treatment with Eco RI cleaves the F fragment into two segments.

The cleavage products of the F fragment are then separated by gel electrophoresis and collected into two homogeneous samples. If the fragments are small enough (usually less than 100 nucleotides), they can be analyzed by a variety of chemical methods, as described below. If they are much longer than 100 nucleotides, they can be treated with additional restriction enzymes, and further cleaved into specific short segments.

Small cleavage products of the F fragment can be labeled on one end with radioactive phosphorus (P_{32}). The labeled fragments can then be treated with a reagent that breaks the nucleotide chain wherever a particular base, for example thymine, occurs. The treatment is mild enough so that the chains are cleaved at only a few of the many possible sites where thymine occurs. This step produces a number of much shorter segments, called polynucleotides, that can be separated from each other by electrophoresis. Now, among the various polynucleotide segments, some will retain their P_{32} label, indicating that they are one end of the chain. If, among the labeled fragments, we find some that are 8, 15, 25, and 40 nucleotides long, we know that thymine must be present at positions 8, 15, 25, and 40 from the labeled end of the fragment. The procedure can then be repeated with reagents or enzymes that cleave the DNA fragment next to adenine, guanine, or cytosine, allowing geneticists to determine bit by bit the entire nucleotide sequence of each segment.

The results obtained when the analysis is begun with Hind III can then be compared to those obtained when the whole process is begun with a different restriction enzyme. If the results are comparable, it is clear that the sequence is correct. In this way, the entire genome can be sequenced with great accuracy.

Although restriction endonucleases provide molecular biologists with a powerful new tool for determining gene sequences, so far they have only been used on the smallest genomes known. The 5375 nucleotides that comprise the entire genetic message of the virus ϕX174 make up a small genome, especially when compared to the genome of bacteria, which contains millions of nucleotide pairs, and that of mammalian cells, which contains billions. As important as restriction endonucleases have been in these first stages of unraveling gene sequences, it is likely that other techniques will have to be developed before the method can be applied to organisms above the viral level.

cant, however, is the fact that all the experimental evidence to date points toward the universality of the genetic code throughout the living world. Except for the DNA in mitochondria, which recent evidence suggests have some codon differences from nuclear DNA, the triplets that select particular amino acids in mammals are identical to those that select the same amino acids in *E. coli*. Through genetics, we gain even more insight into the way all living systems share processes in common.

19.6
MUTATIONS

Another result of working out the genetic code is that biologists now have a much clearer idea of the molecular basis of mutation. Recall that we have discussed mutations in the last two chapters largely in terms of phenotypic changes. Not only that, but the phenotypic changes have usually been gross, visible changes—from red to white eyes, from normal to vestigial wing, and the like. To the extent that we have discussed the genotypic basis of mutations, we have simply stated that genes undergo a change, whose existence we infer through observations on the phenotype. But what are mutations? What do they mean on the genetic level?

And, what do they mean on the biochemical level—that is, on the molecular phenotype within individual cells?

Geneticists now define **mutations** as changes in the number or sequence of nucleotide base pairs on DNA. Mutations can have one of three effects:

1. They can make the code so nonsensical that no protein is synthesized at all.

2. They can cause the coding of messenger that contains a more limited amount of nonsense so that protein is produced but is nonfunctional.

3. They can cause the coding of protein that has an altered function; that is, the protein (for example an enzyme) can function either less or more effectively in its particular biochemical role than the nonmutated (normal) form. The degree to which the functioning of the polypeptide is altered depends, of course, on the location of the mutation—whether it is in a "sensitive" or an "insensitive" region of the protein.

The latter are known as structural gene mutations. Such a mutation occurs in a gene that codes for a polypeptide that enters directly into a metabolic reaction related to the phenotype. There is a second kind of gene, known as a regulatory gene, whose function is to code for a protein that regulates the expression of other genes in the genome. We will discuss the nature and role of regulatory genes in more detail in Section 19.8. For the moment, it is sufficient to say that regulatory genes can also show mutations that can alter the phenotype of the organism in a variety of ways.

Mutations are like typographical errors in the genetic code. These errors get translated into further typographical errors in the protein itself. What causes mutations? How do base-pair substitutions take place? What determines the exact nature of any particular kind of base-pair substitution? The answer to these questions is not completely understood by geneticists today, but some answers are available. To understand these answers, it will be necessary to digress momentarily to discuss the kinds of processes known to induce mutations artificially. For it is through these methods that mutations are most frequently studied.

Mutations can be induced by exposing organisms to various physical and chemical processes: to mutagenic chemicals such as nitrous acid or hydroxylamine or to physical factors such as high-energy radiation. Nitrous acid deaminates nucleotides, removing amino groups and substituting in their place a keto group ($=CO$), as follows:

CYTOSINE
(original base)

HNO₂

URACIL
(modified base)

Without going into great detail, suffice it to say that, as in the example above, cytosine can be converted into a modified form that is capable of binding not with guanine, but with adenine. The result of exposure of organisms to mutagenic chemicals, then, is to alter base structure in such a way that new pairing partners get substituted into a DNA molecule. A given triplet sequence is changed, and consequently the genetic message is modified.

Ultraviolet light, X-rays, or other sources of high-energy radiation can cause other kinds of modification

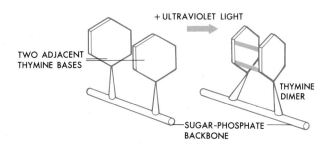

Fig. 19.22
Formation of thymine dimer by configurational shift of two adjacent thymine bases. The shift results from electron rearrangements in the orbitals of the bases; the rearrangement is due to high-frequency, ultraviolet light.

of existing bases in DNA. Ultraviolet (UV) light is known to cause adjacent thymines to undergo configurational shifts (a so-called *tautomeric shift*) so that they form a thymine dimer (see Fig. 19.22). This dimer sometimes causes the daughter strand to have a gap across from the dimer. The cell's DNA-repair enzymes often make mistakes in trying to fill in this gap. That is, the repair enzyme cannot adequately resolve the fact that pairing to the dimer is impossible. Hence the enzyme may cause a different kind of base to be substituted in the daughter strand. This, of course, changes the code for the daughter strand.

One very important feature of mutations is that they are random. That is, mutations occur (mistakes are made) more or less randomly along a DNA strand. The occurrence of a given mutation, such as substitution of bases in one part of the DNA rather than in another, is not made in direct response to any environmental factor. This may sound contradictory to the above discussion of mutagenic chemicals and radiation, which are, after all, environmental factors that cause mutations. There is no contradiction, however. Mutagenic factors such as chemicals or radiation increase the general level of all mutations. Usually they do not selectively favor any one mutation over another. They increase the propensity for mistakes to be made, but they do not selectively determine which mistakes are made. As far back as 1927, geneticist H. J. Muller demonstrated that X-radiation directly causes an increase in the random mu-

Mutations are randomly occurring events. Environmental factors such as high-energy radiation or mutagenic chemicals cause an increase in the overall mutation rate. Mutations do not occur, however, in response to specific environmental conditions such as temperature, food, or specific "needs" of the organism.

Not only does DNA guide the synthesis of RNA, but in some cases the reverse also occurs: RNA can guide the synthesis of DNA. This reaction, which occurs in cases where RNA viruses invade animal and plant cells, is mediated by an enzyme known as reverse transcriptase, which seems to exist ready-made for the convenience of the virus.

tation rate. Through carefully designed experiments he showed that the rate of general mutation in *Drosophila* is proportional to the amount of X-radiation to which the insects were exposed. More evidence that gene mutations are indeed spontaneous, rather than specifically elicited by one or another environmental agent, came from the 1943 experiments of Salvador Luria and Max Delbrück.

19.7
WHAT IS A GENE? THE GENE AND OPERON CONCEPT

What, then, is a gene? We have come a long way from Mendel's genetic "factors." Any meaningful definitions of a gene must deal with the intact working entity represented by the gene concept.

Such a definition is now possible. The relationship of DNA to the protein whose structure it specifies can be represented as follows (the arrow encircling DNA shows that it is a template for its own replication):

$$\underset{\text{(replication)}}{\overset{\curvearrowright}{\text{DNA}}} \xrightarrow{\text{(transcription)}} \text{RNA} \xrightarrow{\text{(translation)}} \underset{\text{(POLYPEPTIDE)}}{\text{PROTEIN}}$$

This representation, often simplified verbally to "DNA, RNA, protein," has been referred to as biology's "central dogma."

A Significant Discovery

In 1965 some interesting results were reported by biochemist Dr. Howard M. Temin, now at the University of Wisconsin. As pointed out earlier, some viruses (such as the tobacco mosaic virus) contain RNA but no DNA. Dr. Temin noted that when such an RNA-only virus invades a cell, strands of DNA complementary to the viral RNA can be found. The implications of this discovery were rather profound: *it suggested that these DNA fragments had been synthesized with the viral RNA as the template*, instead of vice versa. At the time, however, not much attention was paid to Temin's results.

In 1970 Temin and his colleague Dr. Satoshi Mizutani reported that they had found RNA-dependent DNA polymerase that would synthesize DNA using the viral RNA found inside cells infected with an RNA virus. Within days, similar results were reported by David Baltimore of the Cold Spring Harbor laboratories. Still more convincing evidence followed from Dr. Sol Spiegelman's laboratory at the Columbia University Institute for Cancer Research. The four nitrogenous bases of DNA—adenine, guanine, cytosine, and thymine—were labeled with tritium and mixed with viral RNA. The labeled bases soon showed up in intact DNA. Thus the evidence seemed conclusive: viral RNA could synthesize DNA.

Not satisfied, Spiegelman went a step further, reasoning that if the viral RNA had served as a template for DNA synthesis, then the RNA should be complementary to one strand of the DNA and would form a double-stranded hybrid with it. Spiegelman mixed viral RNA and its hypothesized DNA product and spun the mixture in an ultracentrifuge for three days. Since RNA and DNA have different molecular weights, they will form separate layers or fractions in the centrifuge tubes. Upon inspection, both these layers were found—but so too was a third, intermediate layer. Undoubtedly this third layer, lying between the other two, was the RNA-DNA hybrid.

Spiegelman has tested twelve RNA-only viruses for their ability to synthesize DNA. Eight of them can do so; four cannot. Of considerable interest is the fact that the eight that can synthesize DNA from RNA cause tumors in animals, while the four that cannot synthesize DNA from RNA do not cause tumors. With the growing evidence for the role of viruses in at least some forms of cancer, the possibility exists that identification of the enzyme responsible for RNA-directed DNA synthesis might enable the process, and thus perhaps the cancer, to be arrested. Since the transfer of genetic information from DNA to RNA can be blocked by an antibiotic that knocks out the crucial enzyme, this reverse blockage possibility does not seem too remote. In the past five years the enzyme responsible for this process, reverse transcriptase, has been identified. Blocking the enzyme has not, however, proved successful in treating animal tumors.

Usually DNA codes for RNA, which in turn codes for the synthesis of protein. In RNA-containing viruses, however, viral RNA sometimes codes for DNA, which in turn codes for mRNA, which codes for protein. The direction of coding is not universally from DNA to RNA.

It thus seems that the representation of biology's "central dogma" must be modified to include the possibility of reverse transcription of DNA by RNA:

$$\overset{\text{(replication)}}{\overset{\curvearrowright}{\text{DNA}}} \xrightleftharpoons[]{\text{(transcription)}} \text{RNA} \xrightarrow{\text{(translation)}} \underset{\text{(POLYPEPTIDE)}}{\text{PROTEIN}}$$

It can also be modified to include those cases where the starting genetic material is RNA, rather than DNA (such as the RNA viruses, for example TMV, the tobacco mosaic virus):

$$\overset{\text{(replication)}}{\overset{\curvearrowright}{\text{RNA}}} \xrightarrow{\text{(transcription)}} \text{mRNA} \xrightarrow{\text{(translation)}} \text{POLYPEPTIDE}$$

Knowing that proteins are an end-product of gene action enables us to work backward to pinpoint the theoretical size of individual genes. Consider a protein containing 500 amino acids. For the selection of each of these, a triplet of three bases is required. Thus, for this protein, *the gene is a portion of the DNA molecule containing 1500 base pairs.* On the basis of average molecular weights for amino acids, it can be further predicted that this gene will have a molecular weight of approximately 10^6. For other proteins, depending on their size, the gene is correspondingly larger or smaller. It is both interesting and intellectually satisfying to note that this estimate of gene size agrees well with calculations made on the basis of more macroscopic investigation, such as data obtained from genetic mapping experiments.

A more direct means of reaching the same conclusion concerning gene size is simply to divide the number of nucleotides in a chromosome by the number of genes located along it. (For example, there are over seventy known genes on the chromosome of the bacterial virus T4.) The resulting figure, of course, represents an average number of nucleotides per gene (though there is no reason to assume that all the genes are necessarily the same size). Knowing the average size of the nucleotides allows calculations concerning the size of a gene.

But size is not a very useful way of describing a gene. We want to know what a gene does, not just how big or small it is. As a result of work on the relations between DNA, RNA, and proteins, most molecular geneticists now consider a gene to be a unit of DNA, or sometimes RNA, that codes for the production of a biologically significant entity.* That entity can be a polypeptide, or the structure of tRNA and rRNA. The older gene concept was a much more structural concept than the modern-day gene concept. The modern view includes both structural *and* functional characteristics. We continue to use the term gene today, but with its newer rather than its older meaning.

An Enlarged View

With this new concept of the gene, we can take a fresh and meaningful look at some phenomena associated with it. It is possible, for example, to hypothesize that crossing over involves the breakage and rejoining of intact DNA molecules, and there is solid experimental evidence to support this hypothesis. Mutations can be viewed as either a change in the order of base pairs along the DNA molecule or a change in the kinds of bases that occur there. Either change will result in a changed sequence of amino acids in the protein synthesized or a substitution of one amino acid for another. Sickle-cell anemia results from the substitution of valine for glutamic acid in but 1 of 574 amino acids in hemoglobin. A very slight change can have far-reaching effects on the organism.

The new gene concept also enables us to picture ways in which chemical agents, such as nitrous acid, can cause gene mutations. Nitrous acid is capable of converting one nitrogenous base into another. This results in the selection of an amino acid different from the one that would normally have been selected.

Even Mendel's concept of a recessive "factor" takes on new meaning in light of the present gene concept. The recessiveness of a gene can often be viewed as its failure to produce any functional protein at all. The dominant gene, however, produces enough functional protein to hide the recessive gene's failure. There are probably enough good enzyme molecules to catalyze the metabolic reaction, even though the total number may be reduced. Or possibly there is a control mechanism that increases the output of mRNA by the domi-

---

*The term cistron is sometimes used to refer to the structural and functional unit we have been calling a gene. The term was invented to try to avoid the purely structural concept associated with the classical "gene" theory of the Morgan school.

nant gene to compensate for the inactivity of its recessive allele. Without this compensatory increase, an intermediate phenotype may result.

Gene Isolation and Synthesis

Biologists have long wondered if it would ever be possible to isolate an individual gene and, more important, to synthesize one in the laboratory. The answer is obviously important, for the accomplishment of these two feats might open the door to the possibility of "genetic engineering" by which humans could rid themselves of certain undesirable genetic traits (such as hemophilia) and even develop new traits that might have some positive value.

Those who maintained that the feats of gene isolation and synthesis were impossible based their belief on two firm, unassailable convictions. The first was that genes are so complex that their chemical synthesis and isolation are all but impossible. Genes, in other words, cannot be made to order. The second argument, designed to counter suggestions that existing genes might be modified, revolved around the difficulty, if not the impossibility, of identifying and altering specific nucleotides among the hundreds of millions that make up human genes. A chink in the armor of this second argument, however, developed with the demonstration by Dr. Stanfield Rogers of the Oak Ridge National Laboratory that the far simpler DNA of viruses can be chemically modified and then the genes of changed virus transferred to plants. If it can be done with plants, the possibility always exists that it might be done with animals, including humans.

More important, the first "impossibility"—that of gene isolation—has now been accomplished. Successful isolation of gene material from cells of the toad *Xenopus laevis* was accomplished in 1968 by M. Birnsteil and his colleagues at the University of Edinburgh, and isolation of the DNA sequences in *E. coli* homologous to ribosomal RNA was accomplished by D. E. Kohne of the Carnegie Institute of Washington. In 1969 a team of Harvard University biologists (J. Shapiro, L. Machatti, L. Eron, G. Ihler, and K. Ippin) headed by Dr. Jonathan Beckwith announced the isolation of the gene coding for the enzyme β-galactosidase, which catalyzes the breakdown of lactose, or milk sugar. The average length of the purified *lac* gene (see Fig. 19.23) turned out to be 1.4 μm. In purifying the *lac* gene Beckwith and others recognized that the gene consists of more than merely a single DNA sequence for the enzyme β-galactosidase. There are several DNA sequences involved, some of which serve to control the transcription of the structural gene sequence. Thus the functional unit of heredity is not merely a structural gene, but is also the control elements associated with it. More will be said on this very important subject in Section 19.8.

The gene-isolation success was followed only a few months later by announcement of the artificial synthesis

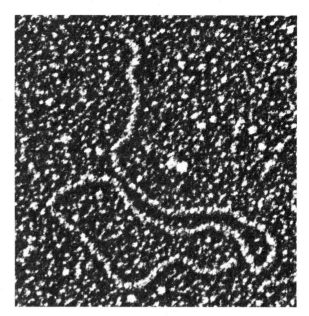

Fig. 19.23
Electron micrograph of an isolated gene. The gene, called the *lac* operon, is responsible for the metabolism of lactose in *E. coli*. Its calculated length is 1.4 μm. (Courtesy Lorne A. McHattie, Harvard University Medical School.)

of a gene by a team of University of Wisconsin scientists headed by Dr. Har Gobind Khorana. Dr. Khorana, who had already received a 1968 Nobel prize for his work in deciphering the genetic code, developed a method of tacking together nucelosides into short lengths of DNA. The synthesized gene was built using the gene for alanine transfer RNA from yeast as a model. This transfer RNA was the first to have its structure determined (by a team of scientists headed by Dr. Robert W. Holley at Cornell University). The Wisconsin workers have shown that their artificial gene has the same parts in the same sequence and spatial arrangement as the naturally produced gene.

It still remained to be shown, however, that the artificial gene is structurally and functionally equivalent to the naturally produced gene. Dr. Khorana, now at Massachusetts Institute of Technology, has devised two approaches to this problem. The first uses his "gene" to direct the synthesis in a test-tube system of a piece of alanine transfer RNA. The second shows that this synthesized alanine transfer RNA will function properly in making proteins.

Early in 1979 Khorana and his associates reported that they had successfully synthesized the tyrosine *t*RNA gene for *E. coli* and shown its complete biological activity, that is, the ability to direct *in vitro* transcription of the synthetic gene to produce tyrosine *t*RNA molecules.

SUPPLEMENT 19.5
RECOMBINANT DNA AND GENETIC ENGINEERING

In recent years the public has been treated to one of the most inflammatory controversies in modern biology: recombinant DNA and the concept of genetic engineering. Genetic engineering has been defined as "the deliberate redesigning of organisms at the gene level to achieve preconceived human goals." Its goal is the same as that of the older plant and animal breeders: to produce new strains of organisms with new traits, or new combinations of traits, not observed before in nature. But its techniques are different. The older procedures relied on selecting from a particular mating those offspring most closely bearing some desired trait. Today's genetic engineering employs the new technology of transferring already-existing genes from one organism to another using special vectors such as bacteria and viruses.

The most important technique employed in modern genetic engineering is recombinant DNA. As the name implies, recombinant DNA puts together DNA from two different sources (for example, a human and a bacterial cell). Here is how it works. The particular gene that one wants to take out of a host organism is identified (let us say, the mammalian gene for producing insulin). The DNA of donor cell cultures is extracted and subjected to a restriction endonuclease (see

Supplement 19.4) as shown in Diagram A. Using Eco RI restriction endonuclease, for example, it is possible to extract DNA segments about 4000 base pairs long on the average (that is because an Eco RI site is found about once in 4000 base pairs). Within these segments, as extracted from the DNA of different organisms, are likely to be one or more complete genes, for example the gene for insulin. These genes, contained on various fragments, provide the basis for recombination with the genome of other organisms.

Use of restriction enzymes has another advantage besides actually cutting out specific genes from the total DNA of a cell. The ends of each restriction fragment have a protruding, single-stranded section produced by the manner in which the restriction enzyme cuts the double helix (see Diagram A, top and middle). When the same restriction enzyme (for example Eco RI) is used on two different DNA sources, note that the two cut fragments are complementary. The two DNA strands can easily rejoin by complementary base-pairing (Diagram A, bottom). The ends of complementary strands are thus said to be "sticky" because they will pair up with other strands and join the ends of two different molecules together. Restriction enzymes cut the DNA molecule

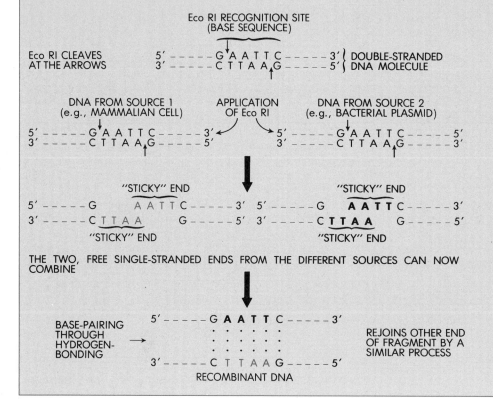

Diagram A

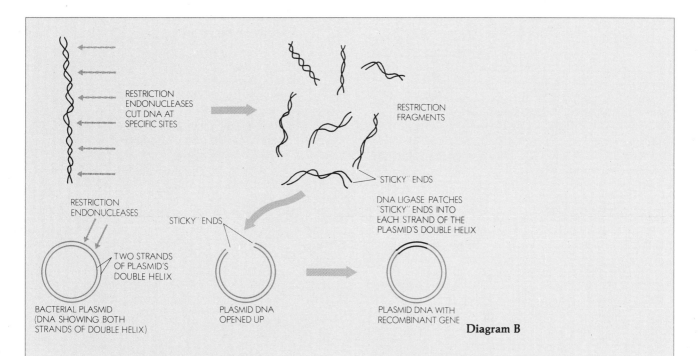

RESTRICTION ENDONUCLEASES CUT DNA AT SPECIFIC SITES

RESTRICTION FRAGMENTS

"STICKY" ENDS

RESTRICTION ENDONUCLEASES

STICKY" ENDS

DNA LIGASE PATCHES "STICKY" ENDS INTO EACH STRAND OF THE PLASMID'S DOUBLE HELIX

TWO STRANDS OF PLASMID'S DOUBLE HELIX

BACTERIAL PLASMID (DNA SHOWING BOTH STRANDS OF DOUBLE HELIX)

PLASMID DNA OPENED UP

PLASMID DNA WITH RECOMBINANT GENE

Diagram B

in half and make the cut in such a way that the new pieces can "glue" themselves back together spontaneously.

Once DNA from a desired source has been fragmented, the next step is to isolate the particular gene to be transferred to another organism. This can be done by using *mRNA* as what is called a *probe*. With special techniques, the *mRNA* for the gene in question is isolated and added to a solution of the random DNA segments produced by exposure to one or another restriction enzyme. The DNA fragment containing the gene in question will base-pair with the *mRNA* and can thus be "pulled" out of solution by centrifugation. The purified "gene" is now ready for the next step —incorporation into a bacterial plasmid that will serve as its "vehicle" or transfer agent. Bacterial cells are lysed and their DNA (plasmids plus chromosome) extracted. The plasmids are then separated from the bacterial chromosome by differential centrifugation, creating a "purified" plasmid solution. Next, the same restriction endonuclease used to produce the gene fragment is also applied to the plasmids, so that the latter are also cut (see Diagram B) and have the same sticky ends. The transferred gene and the plasmid can now form hydrogen bonds between the bases on the sticky ends of their respective strands. DNA ligase then catalyzes the covalent bonding between the backbone of the transferred and plasmid DNA. The final product is a plasmid with a transferred gene "sewn" into one part of the circle (Diagram B, lower right). The plasmid plus its new "gene" insertion makes up recombinant DNA.

The recombinant DNA plasmids can now be added to a culture of bacterial cells that have been treated with calcium to make their membranes more permeable. The cells take up the recombinant plasmids (much as Griffith's pneumococci

bacteria took up DNA, as described in Supplement 19.1) and thus become "transformed." Since the plasmids replicate and transcribe their own genetic material independently within the bacterial cytoplasm, it is now possible to obtain many millions of copies of the recombinant genes.

What are the advantages and uses of the recombinant DNA techniques? First of all, purely in a research sense, transferring eukaryotic genes into the simpler chemical environment of prokaryotic cells makes it possible to study gene activity with greater precision. Second, the possibility of isolating and quickly replicating millions of copies of particular genes means that bacteria can be "designed," or "engineered," as factories for synthesizing large quantities of particular gene products. Such a possibility has many implications for medical research. If, for example, we combine a gene for human insulin with a bacterial plasmid to produce a strain of bacteria that could synthesize human insulin, a plentiful and inexpensive source of this much-needed drug could be obtained. Such recombinant bacterial strains have already been designed to synthesize not only insulin, but also the protein interferon, produced by body cells to fight off viral infection. The availability of large sources of interferon, for example, is seen by some experts as leading to a possible treatment of viral-induced cancers.

Third, recombinant DNA has the potential to revolutionize the medical procedures of immunization (such as vaccination). To prepare a vaccine under standard techniques, it is necessary to grow the disease organism in large quantities, a process which has some inherent dangers—that is, those of growing large quantities of pathogenic bacteria. Furthermore, it is necessary to make the vaccine harmless—the vaccine must be able to elicit the immune response from the body without going so far as to establish an infection. Recombinant DNA offers the possibility of engineering bacteria that can produce only the protein against which the antibody is normally made—the protein that is usually found on the surface of the bacterium or virus that does the infecting. Such a process would eliminate the need to work with the living, pathogenic organism. As an example of this approach, recombinant plasmids containing the gene for the protein coat of the virus causing human serum hepatitis have been inserted into bacteria, where they manufacture large quantities of the coat protein. This protein is "harvested" and injected into patients to confer immunity to the virus.

Both the theoretical and practical implications of recombinant DNA research have been so important that in 1980 the three Nobel Prizes for chemistry were awarded to investigators in this field—Paul Berg of Stanford University, Walter Gilbert of Harvard, and Frederick Sanger, of Cambridge University, England. Berg was the first actually to form a recombinant DNA molecule; Gilbert was the first to get bacteria to synthesize useful, nonbacterial proteins (both insulin and interferon); and Sanger (a Laureate once before, in 1958, for his methods of determining the amino acid sequence of proteins) devised a new and easy method of sequencing the bases along DNA.

Genetic Engineering and the Recombinant DNA Controversy

The development of genetic engineering, especially through recombinant DNA techniques, has been greeted over the past decade with both excitement and concern. The benefits, as described above, are clear. The concerns, however, are not so clear. They have come from two directions. One is the sort of "brave new world" image engendered by the very term "genetic engineering," which has created a fantasy of geneticists designing new types of human beings, or of altering the genetic makeup of people deemed to be "undesirable" (a sort of modern-day eugenics). In this sense, genetic engineering is considered to be the ultimate in the technological control of human social life.

The other area of concern has been the issue of safety in recombinant DNA laboratories. Some biologists themselves, as early as 1974, feared that transferring genes to bacteria and viruses might create pathogenic strains which, if they escaped from the laboratory, could cause widespread infections throughout the population. Between 1974 and 1977 a number of meetings were held by biologists involved in recombinant DNA research to develop guidelines for potentially dangerous research. Some investigators in the field,

such as Paul Berg and Roy Curtis, voluntarily placed a moratorium on their own research for several years until proper guidelines could be adopted. In other instances, such as in Cambridge, Massachusetts, concerned citizens called upon the city council to impose a citizens' moratorium on the very large amount of recombinant DNA work going on at Harvard and Massachusetts Institute of Technology (both located in Cambridge). In recent years the concern over safety seems to have diminished.

There are several reasons for the lessening fears of recombinant DNA. One is that in 1975, nearly 150 molecular biologists met in Asilomar, California, to hammer out a set of guidelines about how recombinant DNA research was to be carried out. These guidelines allowed the research to continue, but under specific sets of instructions about sterility, safety precautions, and the like. With only slight modification, these guidelines were adopted by the National Institutes of Health, which fund much of the recombinant DNA research. The presentation of guidelines helped in some ways to alleviate popular fears that scientists were proceeding with their research mindless of any social responsibility.

A second factor contributing significantly to lessening of the concern about the danger of recombinant DNA is the fact that the biological danger now appears far less real than was originally thought. Many researchers have discovered that the strains of bacteria commonly used for recombinant DNA research are by now so adapted to laboratory conditions that they cannot survive in a natural environment. This point was demonstrated dramatically in the laboratory of recombinant DNA researcher Roy Curtis, who was working with the common strain of *Escherichia coli* known as K-12. Since *E. coli* is a normal and necessary part of human intestinal flora, the fear was that K-12 cells containing recombinant DNA for some toxic substance (being engineered, for example, as one step in producing an antitoxin) might enter the human intestine and outcompete the normal bacteria there. Curtis had volunteers drink a solution of live *E. coli* K-12 cells (without, of course, any recombinant DNA in them), subsequently measuring the content of their intestinal flora for presence of the new strain. All the tests proved negative. *E. coli* K-12 were either killed (in the stomach) before reaching the intestine, or failed to compete successfully with the already-established bacterial flora there. Within three days after ingestion, those K-12 cells that did get into the intestine had been totally eliminated. These and other similar results on both human and animal subjects have convinced most biologists that recombinant DNA research, as currently understood, probably has far less danger to it than originally imagined.

19.8
GENE REGULATION

In multicellular organisms every cell is thought to contain all of the genes for every trait that the organism displays. Muscle cells, for example, contain genes not only for the production of muscle protein, but also the genes for nerve cells, skin, eye, and pancreas proteins. Yet, in muscle cells, only the genes producing muscle-type proteins are active. The others appear to be "turned off." Similarly, in bacteria, certain enzymes are actively synthesized when the medium contains the enzymes' particular substrate. When the substrate is removed from the medium, the bacterium stops synthesizing the enzyme. How is such genetic regulation brought about? How are genes turned on and off selectively in a cell? For over a century these questions, or ones like them, have intrigued biologists the world over.

Today biologists can frame questions about gene regulation in molecular terms—that is, as various interactions between DNA, RNA, and the protein-synthesizing pathways. Two models have been developed to account for how organisms can regulate gene expression: (1) **inducible systems,** where the gene is normally turned off but can be turned on under the proper stimulus; and (2) **repressible systems,** where the gene is normally turned on, but can be turned off under the proper stimulus. In both cases regulation involves an interaction between genes and substances in the cell cytoplasm. The mechanisms by which inducible and repressible systems are thought to function have much in common. However, for the sake of clarity we will discuss each separately.

Inducible Gene Systems: β-galactosidase in *E. coli*

In the late 1950s and early 1960s several clues about what later came to be called "enzyme induction" were provided by Jacques Monod and Francois Jacob at the Pasteur Institute in Paris. Jacob and Monod found that the bacterium *Escherichia coli* does not usually synthesize the enzyme β-galactosidase in significant quantities unless its substrate, lactose, is present in the cell. If lactose is placed in the medium, however, the enzyme begins to be synthesized from amino acids already present in the bacterium's metabolic "pool." It seemed clear to Jacob and Monod that somehow the presence of lactose triggered the gene coding for β-galactosidase to begin transcription.* After many years of work—including studying the β-galactosidase system from both a biochemical and a genetic point of view—a model to explain how lactose *induces* the gene for β-galactosidase to transcribe *m*RNA has been developed.

---

*Historically it was not quite this simple. At the time the β-galactosidase system was being elucidated, several alternative hypotheses as to the mechanism involved in induction were put forward. One suggested that control mechanism occurred on the level of the enzyme itself: that β-galactosidase existed in an inactive form that was activated by lactose. Another suggested that a "masked" messenger RNA molecule for β-galactosidase existed already in the cell cytoplasm, but was simply "unmasked" by lactose. A third, and the one which finally appeared to be most accurate, is the gene induction model described in the text.

Mapping studies have shown that several genes are involved in the induction of β-galactosidase in *E. coli*. These genes are of several different types, and it is well to distinguish clearly between them at the outset:

1. *Structural genes,* which code for the amino acid sequence of several specific proteins (enzymes), some of which, such as β-galactosidase, are involved in the actual breakdown of lactose, and some of which are involved in additional processes connected with lactose utilization, such as transport of the sugar in or out of the cell.

2. An *operator gene,* which is located near the structural genes on the bacterial chromosome, and controls the rate of transcription of the structural genes.

3. *Promotor gene,* a DNA sequence very close to the operator gene on the chromosome, which provides the signal for the beginning of *m*RNA synthesis.

4. A *regulator gene,* which codes for a specific repressor substance controlling the action of the operator gene.

The first three types of genes are collectively referred to as an operon (in this case specifically the *lac* operon). An operon is a coordinated group of genes which act as a unit, usually with an inbuilt regulatory capacity. Let's see how these various types of genes interact with other substances in the cell cytoplasm to regulate the rate of β-galactosidase synthesis.

The basic elements of induction in the lactose operon are outlined in Fig. 19.24. We will begin at (*a*) where the operon is "turned off," that is, when lactose is not present in the cell cytoplasm. The regulator gene transcribes mRNA coding for a specific protein, the repressor substance. The repressor substance is an allosteric protein that can shift back and forth between an active and an inactive form. When lactose is absent, the repressor molecule is in the active form, that is, in a geometrical shape in which it specifically binds to the operator gene. In so doing, the repressor molecule also blocks a binding site on the promoter gene where RNA polymerase attaches. With this binding site blocked, no transcription is possible for any of the structural genes in the operon. As a consequence, β-galactosidase and the other enzymes involved in lactose breakdown and utilization are not synthesized.

When lactose is present, however, molecules of the sugar can bind directly to the repressor protein, causing it to change its shape so that it no longer fits the operator gene site (see Fig. 19.24b). The repressor is then said to be "inactivated." Now, because the repressor is inactive and cannot fit onto the operator site, the binding site for RNA polymerase is no longer blocked. Note in Fig. 19.24(a) that two other molecules, cyclic 3′,5′ AMP (*c*AMP) and *c*AMP receptor protein (CRP), are

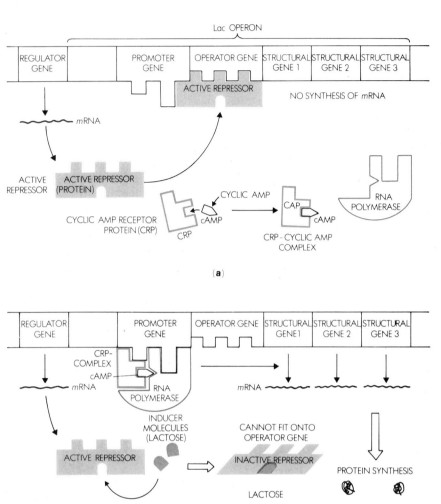

Fig. 19.24
The Jacob-Monod model of gene regula-tion, showing the negative feedback proc-ess in an inducible system (a). The three structural genes to the right, plus an "op-erator" and a "promoter" gene, make up what is called the *lac* (for lactose) operon, which codes for three enzymes involved in lactose metabolism in *E. coli*. Trans-cription of the structural genes is under the control of the operator gene. When the operator is functional, the structural genes are transcribed; when the operator is blocked, the structural genes are not transcribed. Transcription is initiated by RNA polymerase, which binds to the pro-motor gene and begins transcribing a few nucleotides to the right of it, as shown in (b). For the RNA polymerase to recognize and bind to the promoter gene, a molecule of cyclic AMP receptor protein (CRP), combined with a molecule of cyclic AMP (cAMP) must be bound to the left side of the promotor region as shown in (b). Ac-tive repressor coded for by the DNA of the regulator gene, as shown in (a), can bind to the operator and thus block at-tachment of RNA polymerase. The repres-sor can be inactivated by combining with molecules of lactose, or a close derivative of lactose; lactose is the substrate whose metabolism is effected by the enzymes synthesized from the three structural genes.

involved in the normal functioning of RNA polymerase. The CRP-cAMP complex first binds to the promoter site and is necessary for the RNA polymerase to "recognize" the promoter. When this three-molecule complex binds to the promoter gene, mRNA synthesis can begin. The various structural genes are transcribed, their respective enzymes are synthesized, and lactose metabolism can take place. As lactose molecules are broken down, their concentration necessarily decreases and so there are fewer and fewer of them to bind to the repressor protein and convert it to the inactive form. Consequently, more and more repressor molecules become activated and bind to the operator gene, and transcription of the struc-tural genes stops.

The lactose system is said to be *inducible* because, as we have seen, the enzymes for lactose utilization are not synthesized unless lactose is present. Induction is involved in those gene systems whose protein products are used only on occasion, or sporadically in cell metab-olism. For example, *E. coli* normally do not encounter lactose; it would thus be a waste of the cell's energy to synthesize a continual supply of β-galactosidase that was seldom used. Inducible gene systems appear also to be involved most frequently in controlling production of these enzymes involved in catabolic, as opposed to anabolic pathways.

There is another type of genetic control process, in which the structural genes are transcribing most of the time, but can be slowed down or stopped completely on occasion. This sort of control process is called *gene repression*, and in many ways is just the opposite of gene induction.

Repressible Gene Systems: The Histidine Operon in *E. Coli*

The amino acid histidine is used continually in the day-by-day protein synthesis of most organisms, including *E. coli.* Nine different enzymes are involved in the pathway of histidine synthesis, all coded by structural genes within one operon, called the histidine operon (or His operon, for short). As the supply of histidine builds up within the cell, transcription of the structural genes for the various histidine enzymes can be slowed down, or completely stopped. At the same time, as the supply of histidine dwindles, transcription of the histidine operon resumes. The histidine operon is an example of a repressible gene system because the gene functions constantly unless it is signaled to turn off—that is, unless it is repressed. The functioning of the histidine operon is shown schematically in Fig. 19.25.

The repressible gene system for histidine synthesis has all the same elements as the inducible system for lactose: regulator, promoter, operator, and structural genes; the CRP-cAMP-RNA polymerase complex; and an allosteric protein repressor molecule. The chief difference between the repressible and inducible systems lies in the nature and function of the repressor molecule. In repressible systems the repressor protein is normally in the inactive allosteric form—that is, a geometric shape that will not bind to the specific histidine operator gene (Fig. 19.25a). As a result, the operator gene is normally not blocked, which also means that the promoter gene is not blocked. With the promoter gene free, the CRP-cAMP-RNA polymerase complex can fit onto the

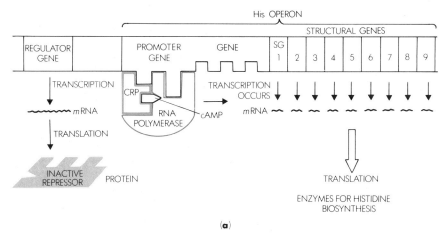

(a)

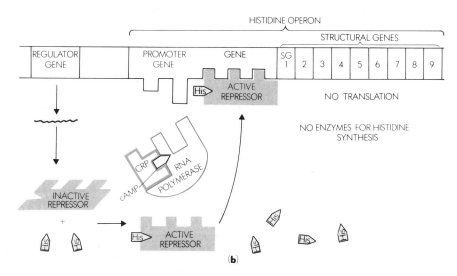

(b)

Fig. 19.25
A repressible gene system, the histidine operon in *E. coli.* (a) When there is no excess of histidine (abbreviated His) in the cytoplasm, the regulator gene produces (through transcription and translation) the repressor protein, which is normally in its inactive allosteric form. In the inactive form the repressor protein cannot bind to the operator gene site, and hence allows room for the CRP-cAMP-RNA polymerase complex to bind to the promoter site. This means that *m*RNA is transcribed for the structural genes, which code for enzymes in histidine synthesis. (b) When an excess of histidine is present, these molecules bind with the repressor protein, shifting it allosterically to the active form, which can bind to the operator and thus prevent transcription of the structural genes.

promoter site and initiate transcription of the structural genes for histidine synthesis. Transcription of the structural genes is the usual state of things in the repressible gene system.

When histidine builds up, however, repression begins to occur at the genetic level. Histidine molecules can bind at a specific binding site on the inactive repressor molecule, shifting it allosterically to the active form. The repressor can then fit onto the histidine operator gene, and thus prevent the CRP-cAMP-RNA polymerase complex from binding to the promoter. The result is that transcription of the structural genes for the enzymes involved in histidine synthesis cannot occur. As the currently existing enzymes become degraded and are not replaced, histidine synthesis slows down, and the gene system coding for histidine production is said to be *repressed.** The system has a built-in regulatory capacity, however. As histidine molecules are used up there are fewer of them to bind to the repressor, and hence the system again becomes turned on. Repressible gene systems appear to be the major controls in anabolic, or synthetic, as opposed to catabolic, or breakdown biochemical pathways.

There is yet a third type of genetic control system, which involves elements of both repression and induction within the same operon. This type of system is referred to as positive control or derepression. The *lac* operon is again a useful example.

Positive Control or Derepression: Catabolite Repression by Glucose on the *lac* Operon

As we have seen, the *lac* operon can be turned off (repressed) by the absence of lactose. However, if glucose is present along with lactose, the *lac* operon still remains turned off. The presence of glucose can override the inductive effect of lactose. In such a situation, we say that *E. coli* preferentially metabolizes glucose over lactose when both sugars are present. How is this preference controlled at the genetic level? The model for such a control mechanism, known as catabolite repression, derepression, or positive control (all three terms are sometimes used) is outlined in Fig. 19.26.

Glucose molecules inhibit the production of cyclic 3',5' AMP (cAMP) in the cell. When glucose is present the level of cAMP in the cells falls sharply. As we saw in Fig. 19.24, RNA polymerase cannot bind to the promoter site and initiate transcription of the structural genes unless cyclic AMP (cAMP) and cyclic AMP receptor protein (CRP) are present and attached to the promoter

site. The presence of cAMP is necessary for the formation of this complex: no cAMP, no CRP-cAMP complex, and no binding of CRP-cAMP to the promoter. When glucose is present, the cAMP level in the cell is low;† thus there is no formation of the active CRP-cAMP-RNA polymerase complex necessary to begin transcription of the *lac* operon. Glucose is said to function in this situation as a catabolite repressor; that is, it represses or turns off a gene system for utilization of another catabolite (a substance that enters a catabolic pathway) than itself. The presence of glucose favors its own catabolism over that of lactose, by repressing genes for lactose breakdown.

In summary, then, we have seen three mechanisms by which the control of gene transcription can be effected in *E. coli.*

1. *Gene induction:* where the regulator gene normally produces an active repressor molecule that binds to the operator gene, at the same time preventing the CRP-cAMP-RNA polymerase complex from binding to the promoter site. This keeps transcription of the structural genes from taking place. Transcription can be initiated by the presence of lactose, which binds specifically to the repressor molecule and inactivates it. Gene induction appears to be most frequently employed in control of catabolic or breakdown pathways.

2. *Gene repression:* where the regulator gene normally produces inactive repressor protein; that is, repressor that cannot bind to the operator gene. By binding with molecules of the end-product of the pathway controlled by the operon's structural genes, however, the repressor shifts to an active allosteric form. When it does so it can bind to the operator gene and repress the transcription of the subsequent structural genes. As the end-product is used up, there are fewer molecules of it to bind to repressors, the latter molecules shift back to their inactive form, and transcription of the structural genes can take place. Repression is most generally found in operons coding for enzymes involved with anabolic, or synthetic, pathways.

3. *Catabolite repression (derepression, or positive control):* where the presence of one substance (usually a catabolite) diminishes the amount of cyclic 3',5' AMP in the cell, and hence prevents the formation of the active CRP-cAMP-RNA polymerase complex. Since the complex is not available to bind to the promoter, initiation of transcription of the structural genes cannot begin. This will be true even if the repressor molecule for the operon in question is fully activated, for example, in the presence of lactose.

---

*Sometimes the term co-repression is used to refer to repressible systems, since a second or co-repressor substance (such as the end-product histidine), in addition to the repressor molecule, is involved in turning the system off.

---

†Glucose inhibits the enzymes involved in cAMP synthesis.

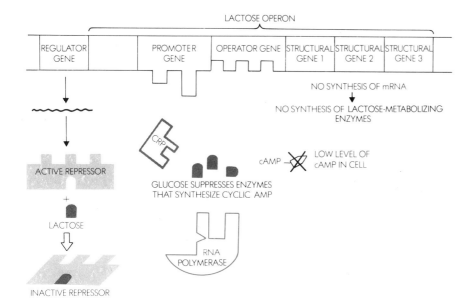

LACTOSE OPERON

| REGULATOR GENE | PROMOTER GENE | OPERATOR GENE | STRUCTURAL GENE 1 | STRUCTURAL GENE 2 | STRUCTURAL GENE 3 |

NO SYNTHESIS OF mRNA

NO SYNTHESIS OF LACTOSE-METABOLIZING ENZYMES

CRP

GLUCOSE SUPPRESSES ENZYMES THAT SYNTHESIZE CYCLIC AMP

cAMP → LOW LEVEL OF cAMP IN CELL

ACTIVE REPRESSOR

+

LACTOSE

INACTIVE REPRESSOR

RNA POLYMERASE

Fig. 19.26
Schematic diagram illustrating the functioning of a positive, derepression, or catabolite repression control system (all three names are used). The lactose operon can be turned off by presence of glucose in the cytoplasm even if lactose is also present. Relatively low levels of glucose in the bacterial cell suppress the enzymes involved in synthesis of 3',5' cyclic AMP. Lack of cAMP in the cell means that the protein CRP cannot bind with it. The CRP-cAMP complex normally binds with RNA polymerase and aids in the polymerase molecule "recognizing" the promoter site. If RNA polymerase cannot recognize, and therefore bind to, the promoter site, transcription of the operon cannot be initiated, and the enzymes for lactose breakdown do not get synthesized. Through a different mechanism the presence of glucose has the same effect as the absence of lactose: repression of the *lac* operon.

What about gene regulation in organisms more complex than bacteria? During the 1950s, Barbara McClintock of the Carnegie Institution of Washington found that different varieties of corn (*Zea mays*) have kernels with many spots of various sizes and frequencies on their surfaces. Through ingenious genetic analyses, McClintock obtained evidence that these spots are due to the action of controlling elements that affect the genes for color. One of these elements directly controls the action of a structural gene for pigment formation, and thus is functionally similar to the operator gene in bacteria. Another controlling element functions in a manner comparable to the regulator gene in bacteria.

Conclusive evidence is still lacking for the existence in multicellular organisms of a gene control system as precise and orderly as the Jacob-Monod model. It is wise to be cautious in extrapolating from unicellular organisms of prokaryotic cell organization to multicellular eukaryotes. Yet it would seem reasonable to expect complex organisms to possess gene-regulating devices at least as sophisticated as those of bacteria—though, of course, they may be of a different kind. Still the bacterial system is valuable as a working hypothesis for use in studies of the growth and development of multicellular organisms.

Summary

1. In nearly all organisms, deoxyribonucleic acid (DNA) is the molecular substance of the gene. Only certain viruses, such as the tobacco mosaic virus and the polio virus, have ribonucleic acid (RNA) as their genetic material.

2. The DNA molecule consists of two spiral polynucleotide chains wound around each other as a double helix. The backbone of the polynucleotide is composed of sugar-phosphate linkages. Joined to each sugar-phosphate group is one of four nitrogenous bases: adenine and guanine (purines), and cytosine and thymine (pyrimidines). The double strands are held together by hydrogen bonding between bases. The bonding pattern is always adenine-thymine (A-T) and gua-

nine-cytosine (G-C). The two complementary strands also lie head-to-tail with regard to the chemical groups. The 3' phosphate is exposed on one end of one strand; lying across from it is the 5' phosphate group of the complementary strand. The reverse is true at the other end of the molecule.

a) DNA replicates by separation of the two helical strands and the enzymatic catalysis of new strand synthesis from each separated strand. Each old strand acts as a template for the formation of the new strand. Each new molecule is thus composed of one old strand against which a new, complementary strand has been built.

b) Several enzymes have been identified as contributing to the replication process (Kornberg enzyme, DNA polymerase I and III). It is now thought that DNA polymerase I serves primarily a repair and gap-filling function, whereas DNA polymerase III may be the most active enzyme in synthesis of new DNA from old.

3. DNA not only replicates itself but also guides the synthesis of proteins. It does this through the synthesis of messenger RNA (mRNA) directly from one strand (in a process called transcription). RNA contains the purine uracil in place of thymine; thus wherever an adenine appears in the DNA template, a uracil appears in the mRNA. mRNA carries genetic information from the DNA of the chromosome to the ribosome. The "message" on mRNA is translated into a specific sequence of amino acids in a newly synthesized polypeptide. A ribosome attaches to the mRNA molecule and moves slowly along the "messenger." This process involves two other types of RNA: transfer RNA (tRNA) and ribosomal RNA (rRNA). There are as many different types of tRNA molecules as there are different types of amino acids (roughly twenty). Each type of tRNA attaches to its particular type of amino acid, a process that requires the expenditure of energy from ATP and involves an "activating enzyme." The activated amino acid and tRNA complex diffuses to a ribosome, where it is incorporated into a protein. This incorporation is determined by the genetic code contained in the mRNA molecule (copied in turn from the DNA).

   a) The genetic code is a triplet code. Information is stored in sequences of three bases each. The code is contained in the linear sequence of bases along the molecule's length. Each three bases on DNA are called a codon. Each codon determines a complementary triplet sequence on mRNA.

   b) The final stage of protein synthesis are as follows. Ribosome moves along the mRNA molecule three bases (one anticodon) at a time. As it stops at each triplet, the appropriate tRNA molecule containing its activated amino acid binds to the triplet by base-pairing. This holds the amino acid in place, and it is joined to the amino acid next to it in the growing polypeptide chain. Joining involves formation of a peptide bond through hydrolysis of the activated phosphate group. Enough energy is provided to drive the uphill reaction of forming the peptide bond. The polypeptide chain grows in this way, one amino acid at a time.

4. Genes and proteins in prokaryotes are colinear. The sequence of amino acids in a protein chain corresponds to the sequence of codons for those specific amino acids in the DNA molecule.

5. In eukaryotes, genes and proteins are not collinear. Some genes are split into encoding sequences and intervening sequences. Encoding sequences contain DNA for amino acid sequences; intervening sequences contain nonsense DNA. Some genes may have as many as sixteen or seventeen intervening sequences. Protein synthesis in eukaryotes follows a somewhat different course as a result of the existence of "split" genes:

   a) All the DNA—encoding and intervening sequences alike—is transcribed.

   b) Intervening sequences are clipped out of the mRNA by one or more "editing enzymes" and the meaningful transcripts joined back together.

   c) The "edited" mRNA passes out of the nucleus into the cytoplasm, where translation takes place.

6. Mutations are typographical errors in the linear sequence of bases in a DNA molecule. If one base is substituted for another, the sense of the genetic message can be altered, at least for one portion of the resulting protein chain. If a base is added or deleted, the entire message beyond that point in the DNA molecule will become "nonsense," and thus result in a nonfunctional or barely functional protein.

   a) Mutations can be induced by agents such as ultraviolet light or certain chemicals. These chemicals or high-energy radiation can cause substitution of one base for another, binding of two neighboring bases, or sometimes a deletion.

   b) Mutations occur at random. Environmental agents such as high-energy radiation or certain chemicals increase the overall mutation rate, but they do not increase the likelihood of one mutation over another.

7. The classical term "gene" now refers to one or more portions of DNA whose products form a complete protein molecule (such as an enzyme).

8. The Jacob-Monod model of enzyme induction accounts for the fact that the enzymes involved in lactose metabolism in E. coli are not synthesized unless lactose is present in the medium. Lactose "turns on" the genes for enzyme synthesis. Jacob and Monod hypothesized that there are several kinds of genes: regulator genes that produce a repressor substance (protein), operator genes to which the repressor can reversibly bind, structural genes that code for specific enzymes active in lactose metabolism, and promoter genes where RNA polymerase attaches to the DNA to initiate transcription. The promoter, operator, and structural genes associated with production of a particular protein are called an operon. The regulator gene controls the whole operon. When repressor binds to the operator gene, transcription of the structural genes is blocked. Appearance of a "signal," in this case lactose molecules in the medium, serves to inactivate the repressor protein. With no repressor able to bind to the operator, the operator becomes free and transcription of the structural genes can take place. This system is an example of *negative* regulation, or derepression, since the function of lactose is only to inactivate a repressor.

9. The Jacob-Monod model also accounts for another type of control, known as gene repression. In contrast to gene induction, in repressible systems the regulator gene

produces an inactive repressor protein—that is, one that cannot bind to the operator gene. By binding to some end-product molecule of the metabolic pathway controlled by structural genes in the operon under consideration, the repressor molecule can assume an active form (allosteric shift). It can thus bind to the operator and prevent further transcription of structural genes in the operon.

10. Positive control also exists. Cyclic 3',5' AMP (cAMP) and its associated cAMP receptor protein (CRP) are necessary components for transcription to take place, even if lactose is plentiful. CRP and cAMP activate RNA polymerase, making it possible for the latter to bind to the promoter gene and start transcription. The amount of cAMP is a function of the amount of glucose present in the cell (glucose metabolism produces energy for converting AMP into ATP; hence the more glucose, the less cAMP). As long as glucose is present in abundance, the bacterial cell does not produce enzymes for breaking down lactose, even when the latter is available. CRP and cAMP are positive regulators, since their presence is necessary for one component, RNA polymerase, to function.

Exercises

1. Describe the major conclusions drawn from the experimental work on DNA performed by:

a) F. Griffith;

b) Avery, McCarty, and MacLeod;

c) Hershey and Chase;

d) Watson and Crick.

2. Two identical cultures of HeLa cells (a human cancer cell line that has been cultured in the laboratory for many years) were infected with equal numbers of polio virus. Radioactive uridine was added to both cultures, while actinomycin D was added to culture Number 2. Actinomycin D inhibits DNA-dependent RNA synthesis. Uridine is a form of uracil and is incorporated into RNA. After a 2-hr incubation at 37°C, two measurements were made on each culture: (1) the amount of radioactivity incorporated into the RNA, and (2) the number of new virus particles that had been produced. The following data were obtained:

Culture number	Actinomycin added	Amount of radioactivity in RNA, counts/min	Number of new virus particles, millions
1	no	530	111
2	yes	23	102

a) Assuming that all the cells were infected with virus, how would you account for these results? Explain your reasoning specifically in relation to the data.

b) How does your answer relate to the "central dogma" concept of molecular biology (DNA→RNA→protein)?

3. How do we know that the genetic code is a triplet?

4. Analyze the following experiment and follow the instructions given.

Four different strains of the yeast *Neurospora* were found by the geneticists Beadle and Tatum to be mutants as the result of X-radiation. In the following table, the symbol + indicates that the strain could grow on the listed medium. The symbol 0 indicates lack of growth.

Medium	Strain	#1	#2	#3	#4
Minimal		0	0	0	0
Minimal plus thiamine (B$_1$)		+	+	+	+
Minimal plus thiazole		+	0	+	0
Minimal plus pyrimidine		0	+	0	0

Assume that in each strain X-rays have caused only one gene to mutate, and that the above is the result of these mutations. Two strains of *Neurospora* can be "fused" so that the nuclei of both are present in the same cytoplasm.

a) Predict whether the sexually produced offspring of the following strains of *Neurospora* would grow on the minimal medium: (i) strain 1 × strain 2; (ii) strain 1 × strain 3; (iii) strain 1 × strain 4; (iv) strain 2 × strain 4; (v) strain 2 × strain 3.

b) Assuming that *at least* one of the above crosses *will* grow on the minimal medium, offer a simple genetic hypothesis to explain the inheritance of this ability.

Chapter 20
Developmental Biology

20.1
INTRODUCTION

Developmental biology is concerned with every aspect of regular change that characterizes a particular species of animals or plants. It is concerned with seed germination, with repair of wounds and tissue replacement, with regeneration, with immunology, with the menstrual cycle, with aging and death. Nothing concerning the regular changes that organisms experience falls outside the sphere of developmental biology. Though it is often equated with embryology (the study of the growth and differentiation of embryos), in actuality developmental biology is a much broader field.

20.2
EMBRYONIC DEVELOPMENT: AN OVERVIEW

The development of embryos consists of two different but interrelated processes: growth and differentiation. **Growth** involves increase in the size of the organism, generally by increase in the number of cells (and, to a limited degree, increase in the size of individual cells).

The biological issues of interest in the study of growth concern the *regulation* of growth activities. What initiates growth in one or several parts of the organism? And what controls the *rate* of growth? How is growth stopped at certain points? Obviously, the control of growth processes is programmed into the genome of most organisms. All members of a given species have certain growth limits, an indication that control processes are operating.

Differentiation is the process by which like cells become specialized. All cells of an adult multicelled organism start from a single fertilized egg, the zygote. The zygote undergoes numerous cell divisions by mitosis, leading to a group of cells, arranged in a ball-like form (see Fig. 20.1). At this stage all the cells in the "ball" look quite similar, though some may be slightly larger or smaller than others. As cell division continues, however, certain cells begin to take on different sizes, shapes, and biochemical properties. Some become embryonic nerve cells, others visual cells, others blood cells, and still others skin or muscle cells. This process of differentiation into specialized cell types is of fundamental interest to developmental biologists. Indeed, the

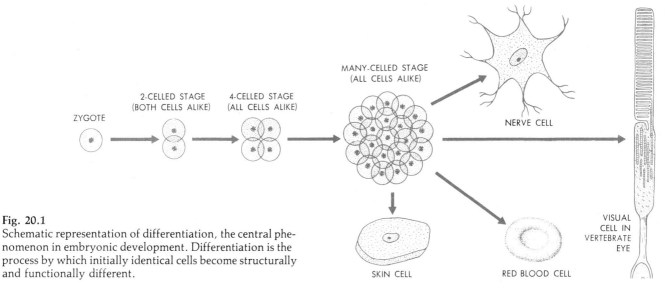

Fig. 20.1
Schematic representation of differentiation, the central phenomenon in embryonic development. Differentiation is the process by which initially identical cells become structurally and functionally different.

problem of differentiation has been one of the key issues in developmental biology since Aristotle (384–322 B.C.).

But *how* does differentiation occur? It is thought that all cells of a multicelled organism are genetically alike; that is, they contain all the same genetic information. Yet chemical analysis reveals that nerve cells contain different proteins from muscle or skin cells. How does this come about? What process is involved in triggering some cells to become nerve and others muscle or skin? What is happening on the level of molecules and genes? Conversely, what is it that causes some cells to lose their special properties, as in the conversion of certain cells into cancer cells? Cancer cells are essentially descendants of once-specialized cells that lost their special functions, and hence *de-differentiated.* Cancer cells resemble cells of embryonic stages, losing, among other things, the controls on their rate of growth. Thus the problem of cancer is fundamentally one of development (see Supplement 20.4).

20.3
GAMETES: THE SPERM AND THE EGG

In both animals and plants the male gamete is called the sperm, and the female gamete the egg, or ovum. In animals, sperm are produced in organs called testes, and eggs in organs called ovaries. In plants, sperm and egg are produced in a variety of structures, depending on the evolutionary stage of each group.

Before considering the processes of fertilization and embryonic development, it will be helpful to examine the structure of the starting elements: the gametes, sperm and egg, which in themselves are among the most highly specialized cells. We will begin with the sperm.

Structure of the Sperm

The sperm of most animal species consists of three morphologically distinct regions: head, middle piece, and tail (or flagellum). A detailed drawing of the overall structure and internal detail of a typical sperm (mammalian) is shown in Fig. 20.2. The sperm head contains the large, prominent nucleus, surrounded by an outer

Fig. 20.2
Drawing of a human sperm (*a*) for comparison wih a scanning electron micrograh (*b*). The drawing shown in (*c*) gives some internal anatomical detail. (Micrograph from *Tissues and Organs: A Text-Atlas of Scanning Electron Microscopy* by Richard G. Kessel and Randy H. Kardon. W. H. Freeman and Company. Copyright © 1979.)

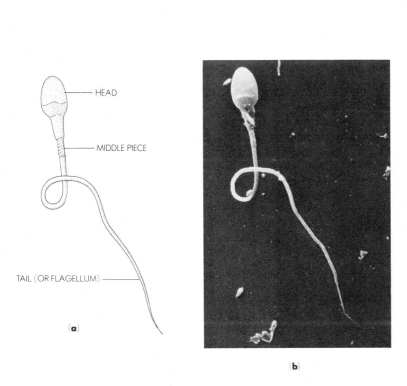

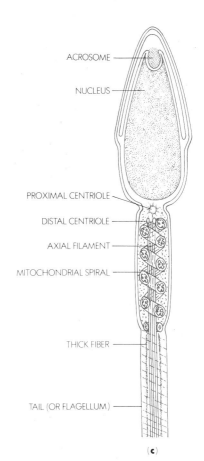

membrane and an inner cone-shaped tip near the front, the **acrosome.** The acrosome is formed from several Golgi bodies that congregate near the tip. At the time of fertilization, it appears to release enzymes that help the sperm head penetrate the outer regions of the egg. Note that the sperm has very little cytoplasm in the head. Just behind the head, in the middle region, are two centrioles, which are part of the motile apparatus, the flagellum. The bulk of the midregion is filled with mitochondria that provide the much-needed energy (ATP) for sperm movement. As shown in Fig. 20.2, the mitochondria are arranged in the middle piece in a spiral. This is typical of the human sperm, but not of all mammalian sperm or the sperm of other animal types. In most species, however, the middle piece is densely packed with mitochondria. The tail consists of microtubular filaments, arranged in the 9 + 2 pattern typical of virtually all flagella in the animal kingdom. The microtubular filaments are derived from the centrioles.

Structure of the Egg

Eggs differ considerably in structure from sperm. Eggs have much more cytoplasm, which contains, among other things, a nutritive substance called **yolk** (see Fig. 20.3), which provides the first food the developing organism must have to start its embryonic development. Yolk is a mixture of fatty compounds and proteins. The fatty compounds provide a rich source of energy, and the proteins supply building materials for growth.

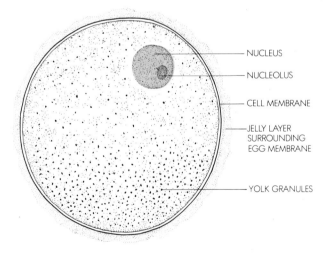

Fig. 20.3
Diagram of a typical ovum (egg) showing distribution of yolk granules within the cytoplasm. Yolk granules provide the nutrient for early stages of embryonic development. Here they are distributed slightly unevenly toward one pole of the cell.

Almost everyone is familiar with the yolk of a chicken's egg. As in all birds' eggs, the amount of yolk is quite large. The eggs of reptiles such as snakes, alligators, turtles, and lizards also contain a great deal of yolk. Eggs of such animals as frogs and insects contain yolk, too, though far less than the eggs of reptiles and birds. The eggs of mammals, such as the cat, dog, rabbit, or human, contain very little yolk.

The amount of yolk stored by eggs of different species depends on how isolated from other sources of food the embryo is. Embryos of birds and reptiles must complete their entire development from fertilized egg to young animal within the confines of their original egg shell. Their yolk supply is correspondingly very large.

Frog and insect eggs with an intermediate amount of yolk have just enough stored energy and building material to carry development part of the way. They produce an intermediate, free-living stage, the **larva,** which is capable of feeding and thus acquiring the material needed for its complete development. Larvae such as the frog tadpole, moth or butterfly caterpillars, beetle grubs, and fly maggots are notoriously voracious. The transformation of larvae into adults is called **metamorphosis** and is typical of organisms that produce eggs with intermediate amounts of yolk.

Mammals have no larval stage. The amount of yolk present, through very small, is sufficient to get the fertilized egg (zygote) through its early cell divisions. These divisions occur as it passes down the oviduct. When it enters the uterus, the embryo becomes implanted in the uterine wall. Until birth, energy and raw materials come from the mother's bloodstream. In a very real sense, the developing embryo is a parasite.

Animals whose developing young are separate from the mother, and who derive their nourishment entirely from the egg yolk, are said to be **oviparous.** Fishes, amphibians, reptiles, and birds are nearly all oviparous. In some cases the young derive their nourishment from the yolk, but still develop within the mother's body. Such animals are said to be **ovoviviparous.** Two examples are the garter snake and the dogfish, or sand shark. In cases in which the embryo derives almost all its food from the mother and develops within her body, the term **viviparous** is used. With a few exceptions, mammals are viviparous.

20.4
FERTILIZATION

Embryonic development for all multicelled animals and plants generally begins with fertilization. There are actually two processes involved in fertilization: activating the egg and uniting of egg and sperm nuclei. As we learned in Chapter 16, which sperm actually fertilizes any given egg is largely a matter of chance. Generally speaking, large numbers of sperm are attracted to each individual egg, thousands of sperm sometimes congre-

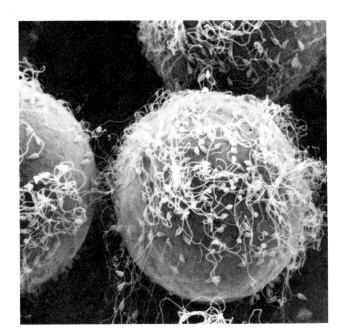

Fig. 20.4
Scanning electron micrograph of the surface of a sea urchin egg, covered with sperm. Many sperm swarm over each individual egg, but only one sperm usually penetrates the membrane. After penetration, the egg is activated, a process that produces changes in the cell surface (see Supplement 20.1) preventing other sperm from entering. (Courtesy Dr. Mia J. Tegner, Scripps Institution of Oceanography.)

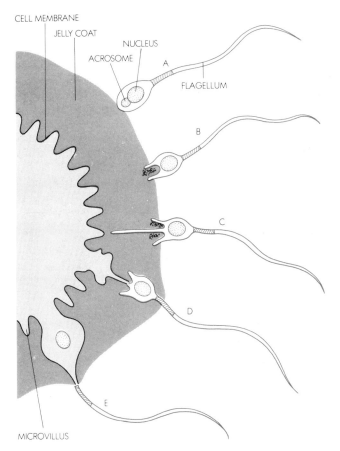

CELL MEMBRANE
JELLY COAT
NUCLEUS
ACROSOME
A
FLAGELLUM
B
C
D
E
MICROVILLUS

gating on or near the surface of a single egg (see Fig. 20.4). As we shall see in later chapters, different species of animals and plants have different ways to ensure that sperm find eggs. The process is least efficient where sperm or eggs are simply released into the surrounding medium and must find each other by chance. It is most efficient when the male inserts the sperm within the female body close to the point of origin of the egg.

That fertilization is not absolutely essential to initiate development is shown by the fact that some animals (aphids, drone bees) develop naturally by **parthenogenesis**, the process whereby an egg develops without first being fertilized. Parthenogenesis can also be induced artificially in the laboratory by pricking an egg with a needle or changing the chemical composition of the water surrounding it. Development of frogs to the tadpole and even adult stages has been obtained parthenogenetically in the laboratory. None of these latter organisms is fertile.

Once contact between the gametes occurs, a chain of chemical and physical events begins. These events are outlined in Fig. 20.5. First the sperm become sticky and clump together (agglutinate) on the egg surface. Actually it is necessary for many sperm to stick to the egg surface in order to break down the layers of material that in many species form external envelopes to the egg cell. Eventually, one sperm comes close enough to the egg cell membrane to make direct contact. By a recently discovered mechanism, the penetration of a single sperm causes immediate changes in the egg's cell membrane, preventing the attachment of additional sperm. When they come close to an egg, sperm show an "acrosome reaction" (Fig. 20.5b), releasing the acrosome contents;

Fig. 20.5
Five steps in fertilization: (a) one sperm, out of many, begins to break through the jelly coat surrounding the egg membrane; (b) as the membranes of the acrosome and that of the sperm itself (the plasma membrane) fuse, the acrosome releases its contents, including enzymes which break down the jelly layer and plasma membrane of the egg; (c) a membrane-like tube protrudes from the sperm head into the jelly-like layer, where (d) it fuses with a microvillar extension of the egg cell membrane; (e) this produces a clear, unimpeded passageway from the sperm head into the egg cytoplasm, through which the sperm nucleus can now pass freely.

the acrosome filament extrudes from the head end of the sperm (c); this filament makes the initial contact between sperm and egg and fuses with the egg cell membrane (d). The sperm nucleus now passes freely into the egg cytoplasm (e). Enzymes from the acrosome then digest the egg cell membrane so that the sperm may enter. It is interesting that this enzyme (hyalurinodase) is the same enzyme secreted by many bacteria when they infect healthy cells.

The egg membrane also responds physically to contact. It rises up to meet the sperm head, forming what is known as a **fertilization cone** (see Fig. 20.6). Formation of the fertilization cone is followed by a general lifting off of the outer egg membrane all around the cell, until a layer (the **hyaline layer**) is formed, separating the inner membrane from an outer membrane (the **fertilization membrane**). The inner membrane (the **vitelline membrane**) forms the new boundary layer of the egg cytoplasm.

The sperm cell maintains contact with the vitelline membrane so that once the sperm cell membrane is broken, the sperm nucleus is released within the egg. The sperm centriole divides to provide the spindle for the first cleavage division. Precisely what happens to the egg cell's centriole is unknown. There is evidence that it is present, because if the egg is caused to divide parthenogenetically, it forms a spindle of its own.

It should be emphasized that the egg contributes considerably to the initiation of development, just as does the sperm. There is, of course, a genetic as well as a structural or physiological contribution from the egg. The egg's genome governs all the early stages of development, for the maternal gene's mRNA is present in the egg cytoplasm at the time of activation. The influence of the maternal genome persists until approximately the time of gastrulation, when the male genome first begins to come into play. Evidence for this notion is derived from analysis of early development in hybrid embryos between two strains of the same species. In these cases, chemical products during blastula formation are almost exclusively of the sort characteristic of the maternal species. More light is thrown on the process by studies of early development in eggs that lack a nucleus and yet are activated by exposure to dilute acid. A series of unfertilized eggs were submitted to certain physical conditions that caused numbers of cells to be cleaved in half. One half of each egg contained a nucleus, the other half lacked a nucleus. The resulting half-cells were then separated into two batches, the enucleated and the nucleated. Both types of cells were activated, the nucleated ones by sperm, the enucleated ones by exposure to dilute acid. The response of the eggs was then determined by measuring the change in the rate of protein synthesis. This criterion was chosen because it is known that during normal fertilization protein synthesis increases rapidly shortly after the activation of eggs.

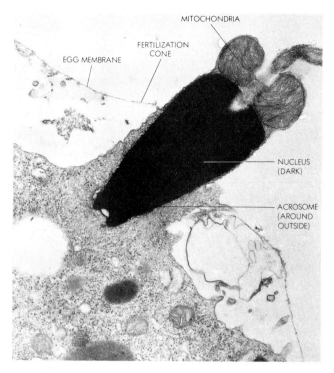

Fig. 20.6
Electron micrograph ($\times$ 17,600) of sperm in the process of penetrating the egg cell membrane. The acrosome enables the sperm to penetrate the egg membrane. Once penetration has occurred, the egg cell membrane lifts up to engulf the entering sperm. (Courtesy Dr. Everett Anderson.)

The results are shown in Fig. 20.7. Data for a control (unaltered) egg are included on the graph for comparison. The graph shows that there is very little difference in the initial stages of protein synthesis between eggs that possess and those that lack a nucleus. Realization of this fact has led to the notion that mRNA for the initial protein synthesis following activation exists in the egg cytoplasm in a masked, or inactive, form. It appears to be transcribed from maternal cell DNA back in early stages of oögenesis. Of course, the half-eggs without a nucleus cannot exist for long—soon they become inactive and eventually die. But the fact that early protein synthesis occurs in the absence of any nucleus shows that this process has already been programmed by the maternal genes, a coding that must take place long before the egg is ready for actual fertilization.

To summarize, fertilization is followed immediately by three major changes in the egg-sperm system.

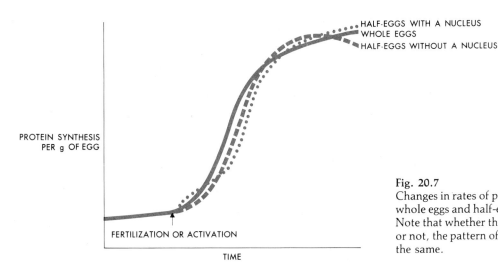

HALF-EGGS WITH A NUCLEUS
WHOLE EGGS
HALF-EGGS WITHOUT A NUCLEUS

PROTEIN SYNTHESIS
PER g OF EGG

FERTILIZATION OR ACTIVATION

TIME

Fig. 20.7
Changes in rates of protein synthesis after activation of whole eggs and half-eggs with or without a nucleus. Note that whether the activated egg contains a nucleus or not, the pattern of protein synthesis remains virtually the same.

First, the egg membrane changes its permeability at contact so that a different sort of two-way movement in and out can take place. Second, there are drastic cytoplasmic changes: there is a considerable rearrangement of material, protein synthesis is initiated (prior to fertilization, protein synthesis activity is either low or entirely absent), and the axes of the future embryo are established. Third, for most eggs meiotic division is finally completed. The sperm centriole divides to form the spindle; there is a fusion of the two nuclei or, as they are termed, male and female *pronuclei*. At this point the genetic material is united and the diploid or $2n$ status restored.

20.5
EARLY EMBRYONIC DEVELOPMENT

The early development of an embryo is almost entirely *epigenetic*; that is, as a process of differentiation and specialization. This fact presents a very complicated situation to the embryologist. If the embryo were completely *preformed*—that is, a mere miniature of the adult form, as many sixteenth-, seventeenth-, and eighteenth-century investigators believed—its development would be little more than simple growth in size. However, in epigenetic development, cellular differentiation and specialization must take place at precisely the right time and place if the individual is to be normal. Work on embryology has therefore been directed toward the problem of just *how* complex and specialized regions of an organism can arise from unspecialized cellular regions—for this is precisely what occurs.

In most higher organisms there are three such cellular regions, called the **primary germ layers.** The primary germ layers are the first distinguishable areas within the early animal embryo. They will give rise to the tissues and organs of the adult animal. The three primary germ layers are the **ectoderm** ("outer skin"), **mesoderm** ("middle skin"), and **endoderm** ("inner skin"). Table 20.1 shows some of the parts of the adult body that arise from each germ layer.

Table 20.1
The Three Primary Germ Layers and Some Body Parts That Arise from Them

Ectoderm	Mesoderm	Endoderm
Skin epidermis	All muscles	Lining of digestive tract, trachea, bronchi, and lungs
Hair and nails	Dermis of skin	
Sweat glands	All connective tissue, bone, and cartilage	Liver
Entire nervous system, including brain, spinal cord, ganglia, and nerves	Dentine of teeth	Pancreas
Nerve receptors of sense organs	Blood and blood vessels	Lining of gallbladder
Lens and cornea of eye	Mesenteries	Thyroid, parathyroid, and thymus glands
Lining of nose, mouth, and anus	Kidneys	Urinary bladder
Teeth enamel	Reproductive organs	Urethra lining

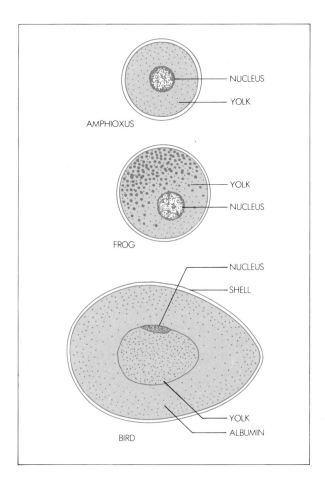

Fig. 20.8
Three egg types showing differences in amount and distribution of egg yolk. Differences in yolk distribution affect the patterns of cleavage in the egg once it is fertilized. Yolk slows down cell division, so that those cells containing much yolk divide more slowly (and thus at any point in time appear larger) than cells containing less yolk. The yolk in *Amphioxus* and frog eggs can be apportioned into individual cells by cleavage; that in the bird's egg cannot, because of the massive amount of yolk in relation to the egg cell cytoplasm.

Early Cleavage

As we have just seen, embryological development generally begins with fertilization. The fertilized egg, or zygote, immediately becomes a beehive of biochemical activity. It soon undergoes its first mitotic division (**cleavage**). The resulting two daughter cells divide again, producing four. Each of these divides, giving eight cells. The eight soon become sixteen. As can be seen, early cleavages result in a geometric increase in cell number. Later, the increase in numbers is less precise. Eventually, however, a hollow ball of many hundreds of cells is formed.

Not all organisms show the same cleavage patterns. The amount of yolk present in an embryonic cell often determines the way in which it divides. Yolk is composed of very dense material. Thus yolk-laden cells divide more slowly than those containing less yolk. In the bird's egg, the yolk does not become divided into the individual cells at all. Three different egg cell types, based on amount and location of yolk, are shown in Fig. 20.8.

Blastula Formation

The hollow ball of cells formed by cleavage is called a **blastula,** the cavity inside the blastula is called the **blastocoel,** and the blastula's cells are called **blastomeres** (see Fig. 20.9). It is interesting to note that the blastula, composed of many blastomeres, is often no larger than the original zygote from which it developed. There exists a critical relationship between cell size and the amount of cell membrane available for respiratory exchange (see Supplement 4.1). The unfertilized egg has a relatively low rate of metabolism; respiratory exchange is low. Once it is fertilized, however, things change. The zygote immediately divides several times, and these cleavages bring surface-volume proportions to a point that will sustain the greatly increased metabolic activity accompanying development. Blastula formation also gives the embryo an adequate number of cells (essentially alike except for yolk content) for the first building blocks of the new organism.

Quite obviously the embryo cannot feed, so each cell must have its own source of energy and raw materials. Cleavage requires a great deal of energy for the mechanics of mitosis and the synthesis of nucleic acid (which must occur during the chromosome replication accompanying each division).

Gastrulation

The formation of a blastula is followed by a process called **gastrulation,** which produces another hollow ball of cells, the **gastrula.** The gastrula often resembles the blastula in appearance. Unlike the blastula, however,

Fig. 20.9
Early embryology (zygote to blastula) of three representative organisms: *Amphioxus*, frog, and bird. The three organisms represent different yolk distribution within the egg: *Amphioxus* has evenly distributed yolk throughout the egg cytoplasm; frog has yolk more toward the vegetal pole of the egg; and birds have the yolk all concentrated at one end of the egg—so much so, in fact, that the embryo forms first as a disc of cells atop the large yolk mass. Presence of yolk in the frog egg slows down cell division in the lower half of the embryo, resulting in larger cells.

EARLY EMBRYOLOGY: EGG TO BLASTULA

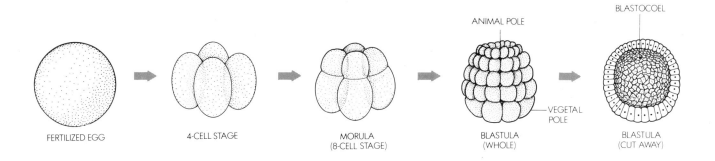

FERTILIZED EGG   4-CELL STAGE   MORULA (8-CELL STAGE)   BLASTULA (WHOLE)   BLASTULA (CUT AWAY)

ANIMAL POLE

VEGETAL POLE

BLASTOCOEL

(a) AMPHIOXUS

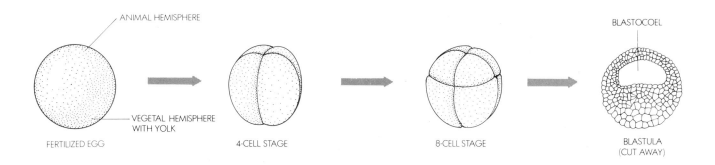

ANIMAL HEMISPHERE

VEGETAL HEMISPHERE WITH YOLK

FERTILIZED EGG   4-CELL STAGE   8-CELL STAGE   BLASTULA (CUT AWAY)

BLASTOCOEL

(b) FROG

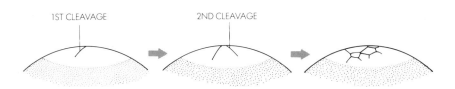

1ST CLEAVAGE   2ND CLEAVAGE

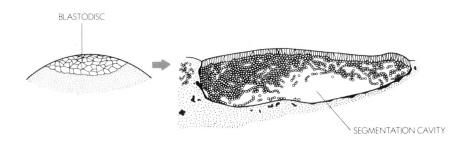

BLASTODISC

SEGMENTATION CAVITY

(c) BIRD

MIDDLE EMBRYOLOGY: GASTRULATION

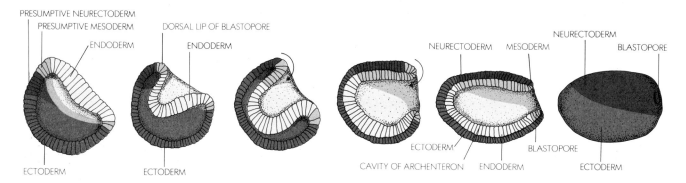

PRESUMPTIVE NEURECTODERM
PRESUMPTIVE MESODERM
ENDODERM
DORSAL LIP OF BLASTOPORE
ENDODERM
NEURECTODERM   MESODERM
NEURECTODERM
BLASTOPORE
ECTODERM
ECTODERM
ECTODERM   BLASTOPORE
CAVITY OF ARCHENTERON   ENDODERM
ECTODERM

(a) **AMPHIOXUS**

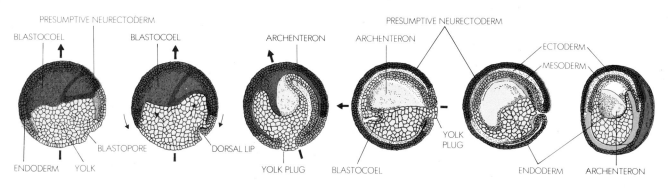

PRESUMPTIVE NEURECTODERM
BLASTOCOEL
BLASTOCOEL
ARCHENTERON
ARCHENTERON
PRESUMPTIVE NEURECTODERM
ECTODERM
MESODERM
BLASTOPORE   DORSAL LIP
YOLK PLUG
YOLK PLUG
ENDODERM   YOLK
YOLK PLUG
BLASTOCOEL
ENDODERM   ARCHENTERON

(b) **FROG**

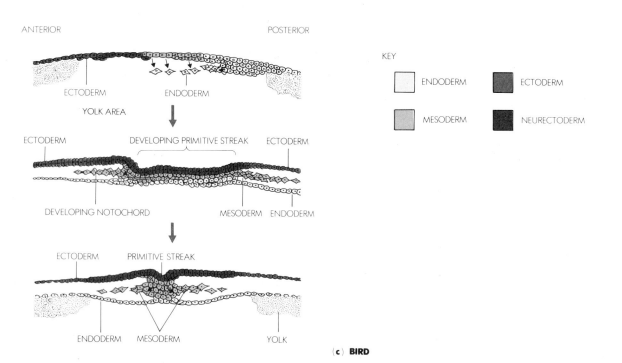

ANTERIOR   POSTERIOR

ECTODERM
ENDODERM
YOLK AREA

KEY

☐ ENDODERM   ■ ECTODERM

▨ MESODERM   ■ NEURECTODERM

ECTODERM   DEVELOPING PRIMITIVE STREAK   ECTODERM

DEVELOPING NOTOCHORD   MESODERM   ENDODERM

ECTODERM   PRIMITIVE STREAK

ENDODERM   MESODERM   YOLK

(c) **BIRD**

Blastula formation occurs without any significant increase in overall size. The cyto-plasm of the original egg is merely portioned out into increasingly smaller compart-ments.

the early gastrula has at least *two* layers of cells, instead of only one. Gastrulation, like blastula formation, differs considerably from one animal to another (see Fig. 20.10). Once again, yolk is an important determining factor. If there is very little yolk, as in *Amphioxus*, the side of the blastula simply pushes in, and the blastocoel is slowly obliterated. A new cavity, the **gastrocoel**, or **archenteron** (primitive gut), is formed. If there is a little more yolk, as in the frog blastula, invagination occurs more toward the top, or **animal pole,** of the blastula. The dense yolk-laden cells are concentrated at the bottom, or **vegetal pole** region. Finally, in yolk-laden eggs (birds, reptiles), invagination is greatly modified. It involves only those cells resting on top of the yolk mass.

An analogy may show how the yolk influences gastrulation. Imagine squeezing a soft rubber ball. If the ball is empty, one side can be pushed against the other. This might correspond to an *Amphioxus* blastula with little dense yolk. If the ball is half-filled with sand, only the top portion of the ball can be pushed in. This is similar to the intermediate yolk content of the frog's blastula. However, the ball almost completely filled with sand has only a small pinch of rubber at the top that can be moved. This ball is similar to the yolk-laden eggs of birds. Invagination must be considerably modified from that which occurs in blastulas containing less yolk. However, no matter how gastrulation takes place, the net result is roughly the same; *three primary germ layers are formed.*

Let us continue the analogy of the empty rubber ball. If you squeeze it so that one wall touches the other, you eliminate the inner cavity. This corresponds to the obliteration of the blastocoel. In doing so, however, your fingers form another cavity, corresponding to the archenteron. Note that this new cavity differs from the old one in two important respects. First, it opens to the exterior. This opening corresponds to the **blastopore.** Second, the new cavity is enclosed within a double- rather than a single-layered wall. Similarly, in the embryo the archenteron becomes surrounded by a double layer of cells, of which the outer layer is the ectoderm. With the formation of the mesoderm (by segregation from the endoderm in amphibians), the three primary germ layers are established. In *Amphioxus*, the mesoderm gradually spreads downward to lie between the endoderm and ectoderm (Fig. 20.10). The result is an embryo with three primary germ layers from which all the later organ systems can develop. In fact, by the time of late blastula or early gastrula, many regions of the young embryo are beginning to be determined for a specific fate in the adult.

Fig. 20.10
Gastrulation in the three representative chordates whose early embryology was shown in Fig. 20.9. (*a*) *Amphioxus;* future head region is to the left. Cells begin to migrate inward pushing the presumptive (future) mesoderm into the center of the former blastocoel. Presumptive tissues are coded as shown in the key. Arrows indicate directions of cell movements (morphogenetic movements) during invagination. (*b*) Frog; presence of the larger amount of yolk (than in *Amphioxus*) restricts invagination. Future mesodermal cells moving into the blastocoel thus have to work their way around the yolk, which is pushed forward and partially out through the blastopore, forming the yolk plug. (*c*) Bird (chick); because of the extreme amount of yolk, gastrulation involves the flattened plate of cells folding over itself on top of the yolk. The layers and cavities of the chick egg are less distinguishable than in either *Amphioxus* or the frog.

Gastrulation produces three primary germ layers, ectoderm, mesoderm, and endo-derm, from which all the later tissues and organ systems of the adult organism are derived.

Morphogenetic Movement

The key to the process of gastrulation, and to many of the patterns that follow in later development, is the phenomenon of **morphogenetic movement**. During invagination of the blastula, for example, certain groups of cells near the region of the dorsal lip begin to move inward, literally migrating down into the indentation that forms the blastopore itself. Later, as the various germ layers organize themselves and begin differentiation into specialized tissues, additional cell migrations occur. These migrations are collectively called morphogenetic movements.

The observation of morphogenetic movements has led to several important questions. How do cells move? What determines the final destiny of particular cells? A variety of studies have thrown light on these questions. Early in the century H. V. Wilson showed that cells of an adult sponge, when separated from one another by being passed through a cheesecloth filter, soon reaggregate in a dish of seawater. More important, the reaggregation is specific: cells of like types clumped together. Even more remarkable, cells rearranged themselves from a random heap into an organization characteristic of normal, adult sponges.

Wilson's findings sparked considerable interest and much subsequent research. For example, Dr. A. A. Moscona, at the University of Chicago, has mixed suspensions of kidney-tubule-forming cells from a chick embryo with cartilage-forming cells from a mouse embryo. When such mixing is done, the chick cells clump together to form kidney tubules. The mouse cells clump together to form cartilage.

The results of this experiment pose an interesting question. Did the chick cells clump with chick cells and the mouse cells with mouse cells because they were from the same species? Or was it because they have similar functional duties? Moscona performed a crucial experiment. He mixed embryonic chick and mouse cartilage-forming tissue and observed the patterns of reaggregation. The results showed the formation of a chick-mouse mosaic of embryonic cartilage tissue. The cells involved seem to be governed more by the role they are to play in the developing embryo (that is, cooperating with their neighboring cells to form cartilage) than by the inherent factors (that is, the fact that they are of different species) that might cause them to segregate.

The experiments of Wilson and Moscona have revealed several important generalizations regarding morphogenetic movements in the developing embryo. First of all, it has become clear that the ability to move is characteristic of almost all types of cells. It occurs constantly during embryonic development. Second, cells appear to move by sending out pseudopodia, much in the manner in which amoebae move. Third, cells display selective adhesiveness; that is, cells of a particular type (either already differentiated or still presumptive) tend to stick together when they come into contact with one another, and there is a trait, characteristic of nearly all embryonic cell types, of being able to recognize their own cell types. Fourth, embryonic cells cannot move through a purely liquid medium, but require a surface over which to move. Thus in gastrulation, the cells that move toward the blastopore and begin to push it inward slide over other cells to reach their destination. In morphogenetic movement, all cell movements occur over other groups of cells. Fifth, not only individual cells, but whole sheets of cells can move from one part of the embryo to another.

What factor or factors guide this movement in a given direction? How do migrating cells "know" where to go along the surface of other cells? There are no clear answers to this question. More and more evidence is suggesting that the surface of embryonic cells plays a key role in structuring their positions and forms in the developing embryo.

While cells of any given tissue tend to recognize and adhere to one another, it has been shown in recent years that cells do not adhere equally well at all points on their surfaces. Furthermore, while cells of like tissue can adhere closely at one point in embryonic life, their patterns of adhesion often change, as in the formation of a cavity where two layers of cells may separate from one another. Patterns of changing adhesion of cells appear to be responsible for the ultimate pattern represented in adult organs. Thus the key to understanding how development of form occurs resides in the properties—and the changing properties—of cell adhesions. The changing cell shapes clearly play a role in cell movements. Cell shape, in turn, relates to the role of microtubules and microfilaments (see Chapter 4) in intracellular development. Electron microscopists have revealed whole series of microtubules and microfilaments throughout the cytoplasm in embryonic cells. The arrangement of these tubules forms a kind of cellular skeleton, suggesting that the arrangement of microtubules and filaments determines the shape the cell assumes. Moreover, the shape of the cell determines a great deal about the shape of cell aggregates, which in turn determines, to some extent, the shape of the resulting organ. By using reagents known to selectively inhibit microtubules, microfilaments, or both, developmental biologists have been able to determine something of the roles these structures play in tissue and organ formation.

20.6

LATER EMBRYONIC DEVELOPMENT: TISSUES INTO ORGANS

In the normally developing embryo, the period following gastrulation includes formation of specific tissues, organs, and systems. At the gastrula stage the basic outline of the adult organism is already laid down. The blastopore will become the future anus, and the archenteron the future gastrointestinal tract. The organization of the gastrula into an embryo involves a variety of processes: cell growth, differentiation, and migration.

Neurulation: Formation of the Nervous System in Chordates

In the following section we will compare the formation of the nervous system in the three representative organisms whose early embryology we have discussed previously. All three organisms are representatives of the major group, or phylum, Chordata. These are organisms which at some time in their lives have a notochord running from front to back along their dorsal (corresponding to our back) region. Although there are differences among the various chordates, the basic principles of neurulation are the same. The three forms—*Amphioxus*, frog, and chick—are shown in Fig. 20.11. We will begin with the late gastrula stage and the formation of the neural tube.

The Neural Tube.   The **neural tube** develops out of ectoderm in a region lying on the dorsal side of the blastopore, or, as in the case of the chick, along the primitive streak. The ectoderm in this region is in contact with the underlying mesodermal tissue. This contact is enough to stimulate the development of the neural plate from the single layer of ectoderm. As this process begins to take place, two intracellular changes can be observed (see Fig. 20.12). The ectodermal cells begin to elongate to form the **neural plate.** The cells of the neural plate begin to take on a wedge-shaped appearance. Since the narrow end of the wedge shape in each cell points upward, the plate as a whole curves downward. As more and more cells of the neural plate become wedge-shaped, the plate's two ends finally meet and fuse to form a closed tube.

While it is clear that the internal changes in cells of the presumptive neural plate are triggered by contact with the underlying mesoderm, it is still not known what specific effect such contact has. Recent evidence, however, indicates that the cause for folding comes from within the cells of the neural plate. For example, if a piece of neural plate is cut out of a developing frog embryo and reinserted upside down, it curves in just the opposite direction (outward, rather than inward) from that in the normal embryo. Thus whatever causes the individual cells of the neural plate to take on the wedge shape is internal to the cells. An interesting experiment

suggests that this internal mechanism is, in fact, the work of microtubules. If an embryo undergoing neural plate formation is treated with colchicine (a plant alkaloid known to cause the disassembly, or prevent the assembly, of microtubules, but does not alter the function of microfilaments), cells that have already elongated collapse and cells that are about to elongate do not. The neural plate does not form. Using the electron microscope, workers have determined that normally elongating cells show a number of microtubules running along the long axis of the cell. Comparison studies of colchicine-treated cells reveal that these microtubules have failed to form.

The antibiotic cytochalasin causes the inactivation or disassembly of microfilaments, but it does not affect microtubules. When embryos developing a neural plate are treated with cytochalasin, cells of the plate elongate normally but do not assume the wedge shape. Consequently, the neural plate never folds. Electron microscopy reveals that normal wedge-shaped neural plate cells have a band of microfilaments along the upper surface (see Fig. 20.12b). Cytochalasin-treated neural plate cells never develop this band of microfilaments. Contraction of the microfilaments in untreated cells causes the neural plate to curve inward, as shown in Fig. 20.12.

A recent, surprising discovery is that microfilaments from many cell types are either chemically identical or very similar to the muscle protein actin and can interact with myosin from muscle cells. Moreover, myosin similar to smooth muscle myosin has been isolated from cells such as those of the neural plate. While it is not absolutely certain that the change in cell shape associated with formation of the neural tube is caused by an actin-myosin reaction, it is quite possible that a similar mechanism is involved.

The Central Nervous System.   Once the neural tube differentiates from the ectoderm, it forms the basis for the development of the central nervous system. The neural tubes shown in Fig. 20.12 are, of course, in cross section. The tube extends along the dorsal region of the embryo and is soon covered over by a layer of ectodermal tissue. The anterior end of the neural tube, farthest from the blastopore, eventually differentiates into the brain. The posterior end, nearest the blastopore, usually differentiates into the lower end of the spinal cord. Experiments have shown that at the time of formation of the neural tube, cells anywhere along the tube are capable of differentiating into brain or spinal cord cells. Which specific differentiation occurs depends on location of the cells within the neural tube; this, in turn, depends on the specific region of archenteron roof underlying a particular region of the tube. The neural tube

LATE EMBRYOLOGY: NEURULATION

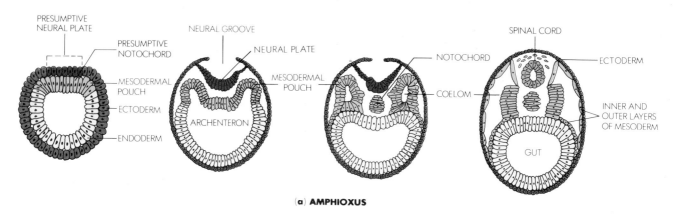

PRESUMPTIVE NEURAL PLATE

PRESUMPTIVE NOTOCHORD

MESODERMAL POUCH

ECTODERM

ENDODERM

NEURAL GROOVE

NEURAL PLATE

MESODERMAL POUCH

ARCHENTERON

NOTOCHORD

COELOM

SPINAL CORD

ECTODERM

INNER AND OUTER LAYERS OF MESODERM

GUT

(a) AMPHIOXUS

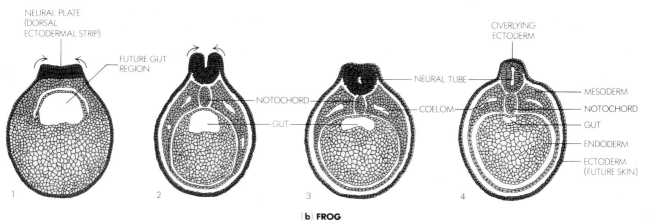

NEURAL PLATE (DORSAL ECTODERMAL STRIP)

FUTURE GUT REGION

NOTOCHORD

GUT

NEURAL TUBE

COELOM

OVERLYING ECTODERM

MESODERM

NOTOCHORD

GUT

ENDODERM

ECTODERM (FUTURE SKIN)

1   2   3   4

(b) FROG

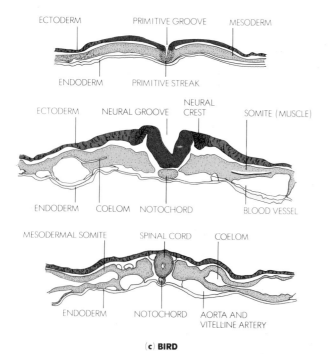

ECTODERM   PRIMITIVE GROOVE   MESODERM

ENDODERM   PRIMITIVE STREAK

ECTODERM   NEURAL GROOVE   NEURAL CREST   SOMITE (MUSCLE)

ENDODERM   COELOM   NOTOCHORD   BLOOD VESSEL

MESODERMAL SOMITE   SPINAL CORD   COELOM

ENDODERM   NOTOCHORD   AORTA AND VITELLINE ARTERY

(c) BIRD

Fig. 20.11
Neurulation (formation of the neural tube) in the three representative chordates: (a) *Amphioxus*, (b) Frog, and (c) Bird (chick). Note that in all three cases the process of neurulation begins with a folding in of the neural crest region of ectoderm, forming the neural groove. This is soon covered by a layer of ectodermal tissue, so that the prospective neural tube is pushed inward. The dorsal (back) side of the organism is up, as shown in these diagrams, and the ventral, or belly side is down. The neural tube is formed from ectoderm, while the notochord (prospective backbone for the frog and chick, prospective notochord for *Amphioxus*) is formed from mesodermal tissue. All three forms show the development of the gut (surrounded by endoderm) and the coelom, a body cavity surrounded by mesoderm. After neurulation the formation of the major organ systems (including the central nervous system from the neural tube) begins.

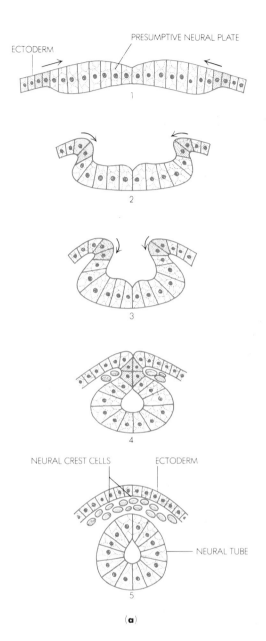

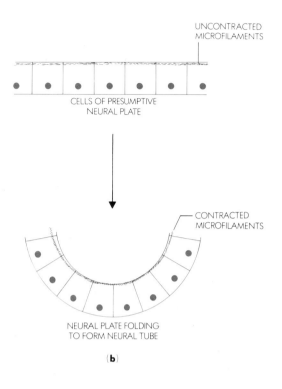

Fig. 20.12
Formation of the neural tube from the neural plate in the frog embryo. (a) Detail of folding process. The neural plate (top) is surrounded on either side by cells (dark color) that form the ectodermal ridges (steps 2 and 3). These ridges eventually fuse to form the enclosed neural tube (steps 4 and 5). (b) Schematic diagram of the folding of the neural plate to form the neural tube. Contraction of microfilaments along the top edges of the cells causes each cell of the plate to become wedge-shaped. This change in shape of the individual cells causes the entire sheet to curve downward.

differentiates in intimate association with the archenteron roof.

When initially formed, the neural tube is a single layer of cells. Cell division begins to occur very soon thereafter. Cells tend to divide more rapidly near the central canal, reducing the size of the original lumen to a slit. All the while, repeated cell divisions cause the tube

to thicken. As these gross changes are taking place, the cells in the walls of the tube undergo a number of transformations. One readily observable transformation is that certain cells begin migrating from the center toward the periphery of the tube. These cells begin a second kind of transformation: they start differentiating into functional neurons. Those that come to lie at the periph-

Fig. 20.13
Overall development of the frog, from first cleavage to neurula stage. These scanning electron micrographs provide a complete and visually dramatic picture of the gross changes in structure as the frog embryo develops to the stage where the central nervous system forms. This sequence of micrographs thus records the earliest stages of development through the appearance of the first major organ system. (*a*) Cleavage furrow in the first division of the egg cell. The first cleavage begins at the animal pole (in the region near where the sperm entered) and extends to the vegetal pole, thus dividing the egg into two hemispheres. Since the yolk is located most prominently in the vegetal pole region, the first cleavage produces two cells with equal amounts of yolk (though concentrated in their lower, vegetal, halves). (*b*) The second cleavage is perpendicular to the first, but still in a longitudinal plane. This produces four cells, with each still containing equivalent amounts of yolk. (*c*) The third cleavage occurs horizontally, dividing the embryo into eight cells, four animal hemisphere and four vegetal hemisphere cells. As the micrograph shows, the vegetal hemisphere cells are larger, because their larger quantity of yolk slows down the division process. (*d*) Blastula stage, a hollow ball of cells surrounding an internal cavity, the blastocoel (not shown). Note that as cleavage continues, the cells become smaller and smaller as the original egg cytoplasm is divided up. There is no overall change in total volume of the embryo from the original egg through the formation of the neurula. (*e*) Through cell migrations from the surface inward, the blastula begins to form the gastrula. The point where the cells begin to migrate inward is known as the blastopore, shown here as a crescent-shaped cleft. The upper surface, or edge, of the blastopore is called the dorsal lip. This area has a primary function in stimulating differentiation in later stages of amphibian development. (*f*) Full gastrula, showing the yolk plug, a group of cells containing yolk from the center of the old blastula. As cells continue to migrate from the animal pole toward, and in through the blastopore, the yolk plug becomes smaller as the cells are pushed back inside the embryo. (*g*) Beginning of the formation of the neurula stage. Two folds (the neural folds) begin to form along the dorsal surface of the embryo from the dorsal lip to the future anterior (mouth) end. The center of the folds is the neural plate, which will eventually become covered as the folds meet (*h*). The neural plate thus becomes housed underneath the surface ectoderm represented by the folds and is ultimately transformed into the neural tube, the future central nerve cord. The entire development from first cleavage through formation of the neurula (*h*) requires about 55 to 58 hours at 18°C. (Micrographs courtesy Professor Richard G. Kessel, University of Iowa, Iowa City; from R. G. Kessel, and C. Y. Shih, *Scanning Electron Microscopy in Biology.* New York: Springer-Verlag, 1974; pp. 338–341.)

ery will become the cells of the white matter of the spinal cord and eventually grow into motor neurons. Between these outside cells and the central canal of the spinal cord are the cells that will become the gray matter, the interneurons, and various types of connecting neurons.

The formation of motor and sensory nerves is one of the most complex aspects of development in the central nervous system. Motor neurons, developing from the ventral roots of the gray matter, produce long fiber-like extensions, which eventually form the axons of the fully differentiated motor neuron. Motor neurons thus grow outwardly from the central nervous system; they do not grow from the peripheral system toward the spinal cord. The properties of cell surfaces in various organs developing in peripheral regions appear to determine where the motor neuron fibers stop growing and form motor end plates. The motor neuron appears to "know" its target cell or tissue; just how it does so is still a matter of great interest and much conjecture.

Sensory neurons form somewhat differently from motor neurons. When the neural plate is forming, a group of cells splits off from the ectoderm and comes to lie just between the neural tube and the overlying ectoderm (see Fig. 20.12). These cells are known as **neural** crest cells. Soon after the neural tube closes and its cells begin dividing, neural crest cells start to migrate. Some of them take up residence in small groups, segmentally arranged, along either side of the developing spinal cord. These groups of cells become the basis for the developing spinal ganglia. The majority of cells in the spinal ganglia differentiate into sensory nerve cells. A long process grows out from these cells toward the peripheral organs. A shorter process grows in the other direction, connecting the ganglion cell to the spinal cord via the dorsal root. Thus sensory neurons have a different mode of origin from motor neurons, though all are derived from ectodermal cells.

Proper development of the nervous system—that is, correct hookups between motor neurons, sensory neurons, and the organs they innervate—is highly dependent on the parallel development of the various body organs. Experiments have shown that if a developing peripheral region is damaged, the nerves growing to that region are much smaller in size, and fewer nerve cells are observed. As in all aspects of development, every part of the embryo differentiates in conjunction with other parts; that is, the embryo develops as a complete interacting system. While adjustments are possible at certain stages, the normal development of any one

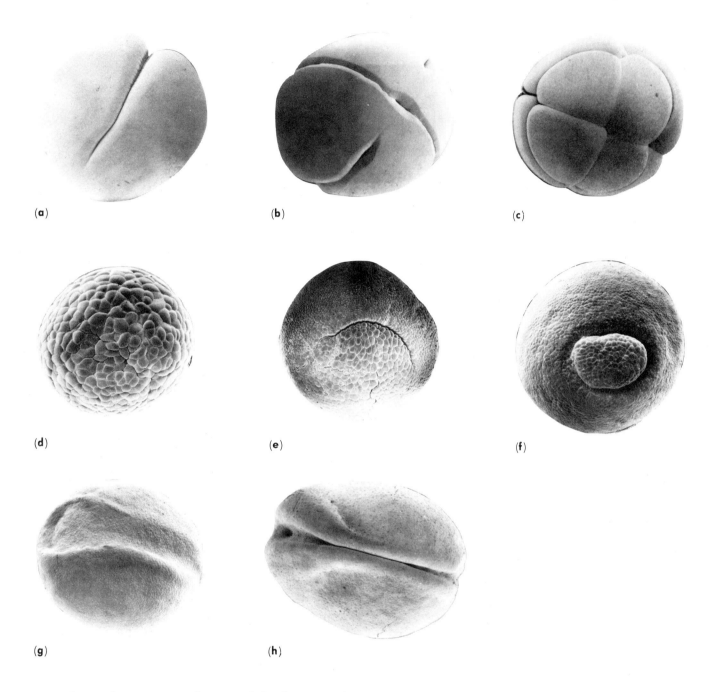

(a) (b) (c)

(d) (e) (f)

(g) (h)

part of an embryo requires the normal development of all other parts. Of course, the more distant the parts physically or functionally, the less direct effect a change in the development of one will have on the development of the other. Nonetheless, it is important to emphasize this interdependence of parts in the normal development of structure and function in embryos.

Fig. 20.13 provides an overview of the entire development of the frog embryo from egg to neurula. Try to relate the overall structures shown here to the detailed cross sections shown earlier in Figs. 20.9, 20.10, and 20.11.

20.7
HUMAN DEVELOPMENT

In its broadest outlines, the development of the human embryo follows the general sequence of events described above for three other members of the phylum Chordata. Human beings, however, as representatives of the class Mammalia, have one major difference from the forms we have examined so far: the embryo develops for a long period of time within the mother's body (intra-uterine development). In contrast, the eggs of *Amphioxus* and the frog develop floating free in the water, and

Human development includes all aspects of the sequence of biological and behavioral changes that characterize the human life cycle. Development of the individual begins with the formation of sperm and egg in the parents' reproductive organs and ends with death.

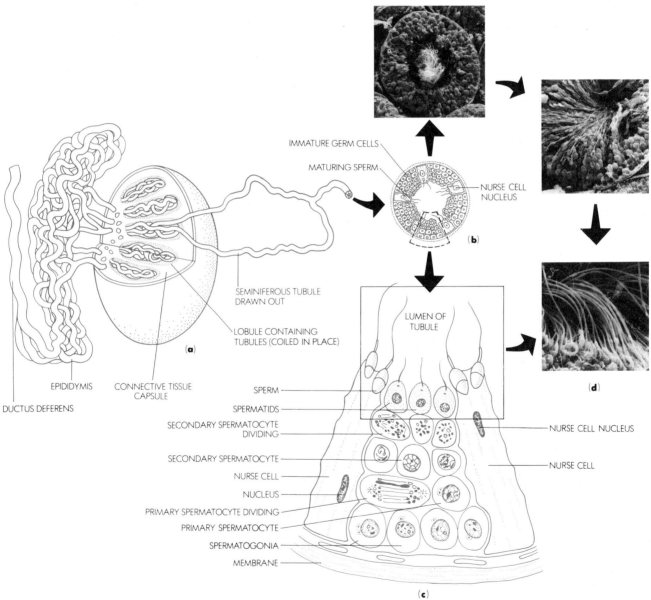

IMMATURE GERM CELLS

MATURING SPERM

NURSE CELL NUCLEUS

(b)

SEMINIFEROUS TUBULE DRAWN OUT

LOBULE CONTAINING TUBULES (COILED IN PLACE)

(a)

EPIDIDYMIS

CONNECTIVE TISSUE CAPSULE

DUCTUS DEFERENS

LUMEN OF TUBULE

(d)

SPERM

SPERMATIDS

SECONDARY SPERMATOCYTE DIVIDING

NURSE CELL NUCLEUS

SECONDARY SPERMATOCYTE

NURSE CELL

NURSE CELL

NUCLEUS

PRIMARY SPERMATOCYTE DIVIDING

PRIMARY SPERMATOCYTE

SPERMATOGONIA

MEMBRANE

(c)

Fig. 20.14
The human testis showing a cross section of a seminiferous tubule and a detailed view of sperm production. (*a*) One testis with the epididymis viewed from the top, showing the seminiferous tubules. (*b*) Cross section of one seminiferous tubule, showing the cells that give rise to the sperm. The cells are increasingly mature as they come closer to the cavity (lumen) of the tubule; mature sperm remain attached with their tails projecting into the cavity until they break free and move toward the epididymis. (*c*) A detailed diagrammatic view of a cross section of a seminiferous tubule showing the meiotic divisions as the cells divide to produce sperm. The nurse or Sertoli cells provide nourishment for the developing cells. (*d*) Scanning electron micrographs of cross section of seminiferous tubules at increasing magnification ($\times$ 330, $\times$ 580, $\times$ 1650). (Micrograph from *Tissues and Organs: A Text-Atlas of Scanning Electron Microscopy* by Richard G. Kessel and Randy H. Kardon. W. H. Freeman and Company. Copyright © 1979.)

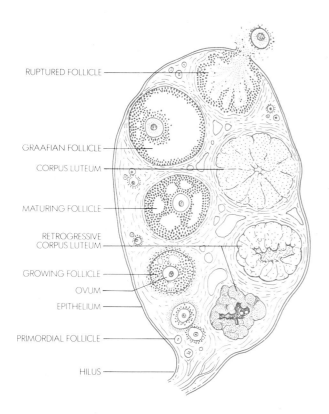

RUPTURED FOLLICLE

GRAAFIAN FOLLICLE

CORPUS LUTEUM

MATURING FOLLICLE

RETROGRESSIVE CORPUS LUTEUM

GROWING FOLLICLE

OVUM

EPITHELIUM

PRIMORDIAL FOLLICLE

HILUS

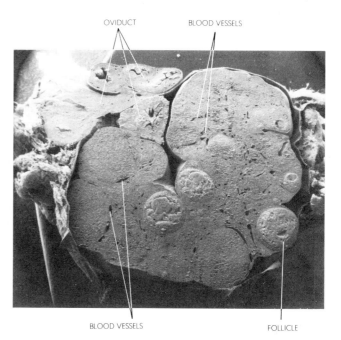

OVIDUCT   BLOOD VESSELS

BLOOD VESSELS   FOLLICLE

Fig. 20.15
Longitudinal section of ovary, showing various stages in the progressive development of ova and Graafian follicles. The drawing is a composite in that the immature eggs and follicles are at the lower left, with progressive development arranged chronologically in a clockwise pattern. In any ovary, follicles in different stages of development bear no spatial relationship to one another. At right is shown a scanning electron micrograph of a rat ovary and portions of the oviduct in section view. Several blood vessels (bv) and developing follicles (df) are shown. (Micrograph from *Tissues and Organs: A Text-Atlas of Scanning Electron Microscopy* by Richard G. Kessel and Randy H. Kardon. W. H. Freeman and Company. Copyright © 1979.)

the eggs of birds develop within an enclosed shell outside of the parent.

Intrauterine development has proved eminently successful in the evolution of mammals. However, it does pose some special problems for the actual growth of the embryo. Chief among these are the problems of nutrition during early development and the switch from a totally dependent to a more independent style of life at birth. We will examine human development as an example of how embryonic growth and differentiation have been modified in response to the intrauterine environment, beginning with the formation of the mammalian sperm and egg: spermatogenesis and oögenesis.

Spermatogenesis in the Human Male

Figure 20.14 shows a sectional drawing of the testis and a more highly magnified picture of a cross section through a seminiferous tubule. Within these tubules gametogenesis takes place. Spermatogenesis is the process by which, through mitosis and meiosis, functional sperm are produced. The actual process of sperm formation is diagrammed in detail in Fig. 20.14(c), and is also compared to oögenesis in Fig. 20.15. The starting cell, the "grand-parental" cell of the future sperm, is a **spermatogonium.** One spermatogonium undergoes mitosis to produce two daughter cells, one a primary spermatocyte and while the other remains a spermatogonium. The primary spermatocyte begins the first

meiotic division and divides into two secondary spermatocytes. The secondary spermatocytes undergo the second meiotic division to form four equal-sized spermatids, cells with a considerable amount of cytoplasm and a haploid nucleus. Each spermatid then undergoes a complicated process of growth and development (but no further cell division) to form a mature sperm.

Oögenesis in the Human Female

Eggs are produced in the ovary in structures known as **Graafian follicles** (see Fig. 20.15). In a cross section of an ovary a number of follicles can be observed in various stages of development. As the follicle is maturing, the egg undergoes meiotic divisions. The follicle consists of a series of cells that line the wall and surround the egg cell. When the egg is ready the follicle bursts, a process known as **ovulation.** Usually the human egg completes its final meiotic stage, including extrusion of the polar body, after ovulation.

Oögenesis has been discussed in considerable detail in Chapter 17. In all the basic meiotic events it is identical to spermatogenesis. The chief difference between the two lies in the fact that starting from a primary oöcyte, only a single egg is produced, as compared to four sperm from a single spermatocyte. In Fig. 18.15 the basic patterns of spermatogenesis and oögenesis are compared. These patterns apply to human beings as well as to other animals.

It is in the formation of the polar bodies that oögenesis differs most from spermatogenesis. In the first meiotic division, one of the daughter cells, the first polar body, is smaller than the other. The larger cell, called the secondary oöcyte, contains more cytoplasm from the parent cell, but both cells contain the chromosomes divided up during the first meiotic division. Each cell, the secondary oöcyte and the first polar body, then undergoes the second meiotic division. Both daughter cells from the first polar body contain little cytoplasm and hence are also small. Division of the secondary oöcyte produces one large cell, the oötid, containing most of the cytoplasm, and a small cell, one of the second polar bodies. Thus three haploid sets of chromosomes are ultimately discarded during oögenesis. A single haploid set is preserved within the oötid, which matures into an ovum, or egg cell.

The function of polar body formation during female gametogenesis is to produce an egg that has a large amount of cytoplasm. Even in animals such as the human being, where nourishment of the embryo is supplied early in life by the mother, a rich supply of cytoplasm in the egg is necessary for those first few days before the embryo becomes firmly attached to the uterine wall.

From Fertilization to Birth

Although both sperm, egg, and uterine wall undergo developmental processes prior to fertilization, development of the embryo can technically be said to begin with formation of the zygote. Fertilization usually takes place in the Fallopian tubes. The precise mechanism by which sperm swimming along the linings of the vagina, cervix, uterus, and Fallopian tubes eventually find the egg is still not clearly understood. Although it would seem logical that some form of chemical stimulus leads the sperm to the egg, no solid evidence yet exists to support this hypothesis.

The human **gestation period** (period of pregnancy) on the average lasts for 266 days, or approximately nine months. It can be divided into three parts, or *trimesters,* each three months in duration. During the first trimester the fetus develops from zygote into organized embryo with recognizable human features and reaches a length of 0.32 cm (about 0.13 in.). It also increases its weight by over 500 times. The first trimester is essentially the period of cell and organ differentiation. The second and

Fig. 20.16
Photographs of early stages in human development. (*a*) A 12-hr human zygote surrounded by the cup-shaped vitelline membrane. Two discarded polar bodies can be seen in the perivitelline space. (*b*) Two-cell human embryo; photomicrograph is of a thin section through the two cells, fixed and stained to show nuclei. (*c*) Sixteen-cell human embryo. (*d*) Human blastocyst at five days. The cluster of cells at the top are embryonic; those at the bottom will form the extraembryonic membranes. (*a* courtesy of Dr. L. B. Shettles, from *Ovum Humanum.* New York: Hafner, 1960. *b, c,* and *d* courtesy Carnegie Institution of Washington, Department of Embryology, Davis Division, and Drs. Hertig, Rock, Adams, and Mulligan.)

third trimesters are characterized more by growth of the differentiated parts into the full-sized fetus with an average weight of about 6 to 8 pounds.

The First Trimester.   Some thirty hours after fertilization, the zygote divides into two cells, or blastomeres (see Fig. 20.16). About ten hours later it divides again, producing four cells, and then eight. By seventy hours there are sixteen cells, each more or less identical in size and appearance. With each successive cleavage the cells become smaller, so that the sixteen-cell embryo is about the same size as the fertilized zygote. The cells at this stage are enclosed in a membrane, the **zona pellucida,** that holds them together. Cleavage takes place as the embryo moves slowly down the Fallopian tubes toward the uterus. Movement through the oviducts and uterus is accomplished by the rhythmic beating of cilia along the walls of these organs; muscular contractions of the walls themselves also aid movement.

After reaching the uterus, the embryo (usually consisting of thirty-two to sixty-four cells) lies free on the uterine surface for two or three days before implantation. By this time the embryo is a hollow sphere, the **blastocyst,** one cell thick on one side and several cells thick on the other (see Fig. 20.16d). The cells on the thin side, known as the **trophoblast,** will give rise to the membranes that surround the embryo, while those on the thick side will develop into the embryo proper. Implantation is brought about with the aid of an enzyme secreted by the trophoblast. The enzyme helps the trophoblast digest away the tissues of the uterine wall in order to become firmly implanted. Implantation is generally complete around twelve days after fertilization. Nourishment of the embryo is made possible because in digesting tissue of the uterine wall the trophoblast also digests some blood vessels, causing the young embryo to lie in a small pool of blood. Soon after implantation is complete, the uterine lining closes over the embryo.

Of course, continued nourishment of the embryo after implantation cannot depend on leakage of minute

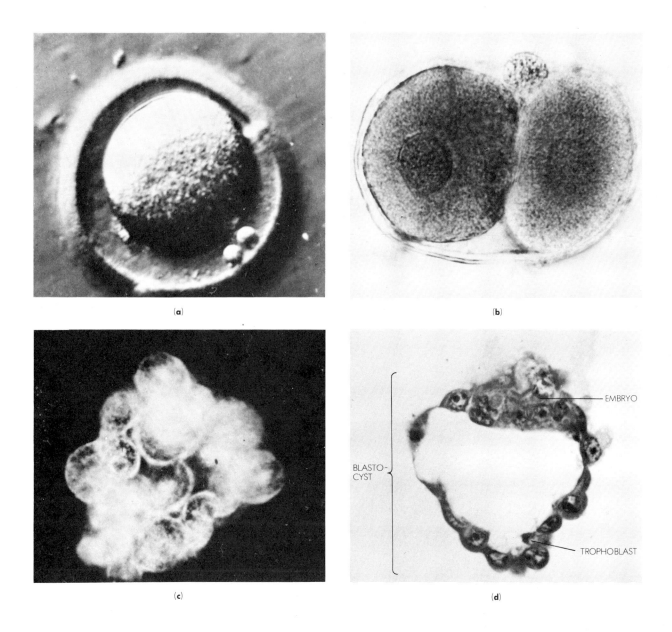

(a)

(b)

(c)

(d)

BLASTO-
CYST

EMBRYO

TROPHOBLAST

drops of blood from the uterine wall. The trophoblast soon begins to develop outgrowths, the **chorionic villi,** that penetrate the tissue of the uterine wall and begin forming the placenta, which thus develops from both the embryo and the uterine tissue (Fig. 20.17). The placenta provides a structural and physiological basis for exchange of nutrient and waste materials between the expectant mother and the embryo. In its fully developed state, it consists of a network of capillaries derived from the embryonic circulatory system. The capillaries from the fetus penetrate the uterine lining tissue and come to lie close to the capillaries of the mother's circulatory system (Fig. 20.18). It is important to note, however, that normally *the blood of the two circulatory systems does not actually mix.* The exchange of materi-

als is accomplished exclusively by diffusion and active transport across capillary membranes. As a result, only small molecules normally pass between mother and child.

By the end of the first month, the embryo is surrounded by membranes formed from the trophoblast; these membranes protect it physically by enclosing it in a bubble of fluid (see Fig. 20.17). During the first month, the embryo shows the development of its longitudinal axis, the so-called **primitive streak.** The tissue of the future reproductive organs is the first to be set aside for differentiation. The nervous system is the first body system to actually develop, appearing by day 24 as the neural plate and neural tube. These structures will become the brain and spinal cord, respectively, and many of the cranial and spinal nerves. On day 26 the four primary brain vesicles begin to form; these vesicles will become specific parts of the mature brain. The sense organs also begin to develop.

During the first month the embryo develops 32 pairs of muscle segments, or **somites,** which appear on each side of the primitive streak. The somites will give rise to most of the voluntary muscles of the body, a large part of the skeleton, and the connective tissue of the skin. The mouth appears by day 27 or 28, and gill slits, or visceral arches, can be found on each side of the throat. The gill slits are thought by some to be vestiges of our ancient vertebrate ancestry, harking back to a time when we were wholly aquatic. The gill slits do not develop into functional gills, of course, but eventually give rise to the Eustachian tube, which connects the throat to the middle ear. The lungs appear by day 30 but only in a rudimentary stage. The tube-shaped heart,

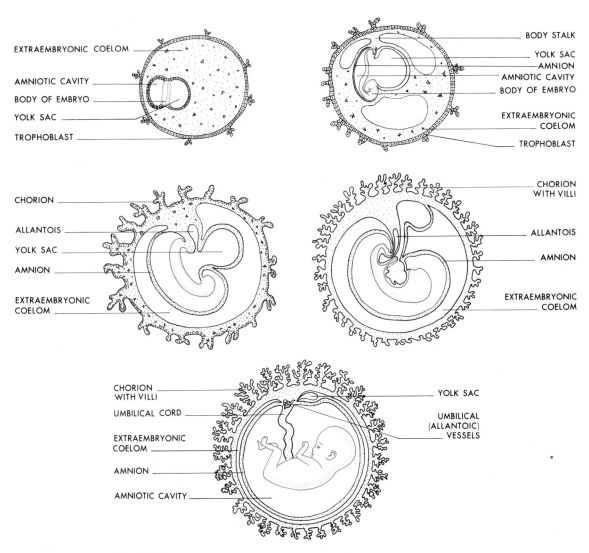

Fig. 20.17

Development of the embryonic membranes. Surrounded by the amnion, the amniotic cavity gradually expands outward, filling up the cavity known as the extraembryonic coelom. The chorion forms outgrowths, the chorionic villi, that penetrate the uterine wall and will eventually form the placenta. As the embryo grows and obtains more and more of its nourishment from the mother, the yolk sac decreases in size. The yolk sac does not serve as a food reservoir as in many other vertebrates, but rather appears to be largely a vestigial organ that develops only after the embryo starts growing. (Modified from B. M. Patten, *Human Embryology.* McGraw-Hill, 1968, p. 104.)

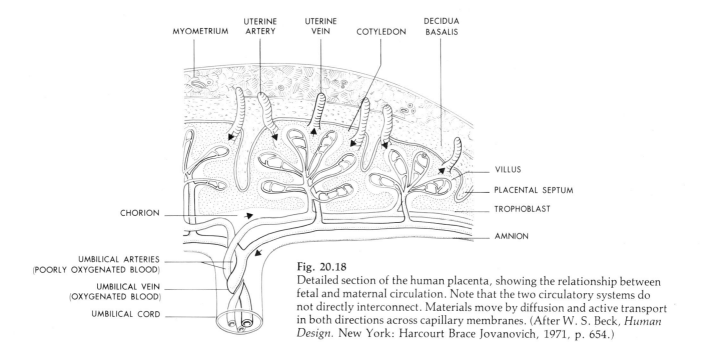

MYOMETRIUM   UTERINE ARTERY   UTERINE VEIN   COTYLEDON   DECIDUA BASALIS

VILLUS

PLACENTAL SEPTUM

TROPHOBLAST

AMNION

CHORION

UMBILICAL ARTERIES (POORLY OXYGENATED BLOOD)

UMBILICAL VEIN (OXYGENATED BLOOD)

UMBILICAL CORD

Fig. 20.18
Detailed section of the human placenta, showing the relationship between fetal and maternal circulation. Note that the two circulatory systems do not directly interconnect. Materials move by diffusion and active transport in both directions across capillary membranes. (After W. S. Beck, *Human Design.* New York: Harcourt Brace Jovanovich, 1971, p. 654.)

which is visible by day 21, shows increasingly rapid pulsations by day 30. Likewise, the primitive kidneys have begun to form by day 30. Thus, by the end of the first month many of the body's organs and systems have begun to differentiate. The embryo is still very small, however, about half the size of a pea, and looks much like the embryo of almost any other vertebrate.

During the second month the germ cells, which have already started to differentiate, migrate toward the region of the primitive kidneys. When they reach their destination, they become incorporated into the primitive urogenital system. Outlines of the skeleton are also visible (see Color Plate II). By the end of the month, the embryo is about 3 cm (1¼ in.) long and is entirely enclosed in the fluid-filled **amniotic sac.** So thorough is the perfusion of the placenta with the mother's blood that the amniotic fluid is completely replaced about every 3 hr. The embryo can swallow and inhale the amniotic fluid without harm and does so until birth.

By the end of the third month the embryo is a little over 7.5 cm (3 in.) long—half of this length being contributed by the head, which shows more rapid growth during the first trimester than any other part of the body. During the remaining two trimesters, however, the rest of the organs and body parts begin to catch up. The term "embryo" is usually used to refer to the developing human organism during the first trimester. From the fourth month onward, it is generally referred to as a **fetus** (see Color Plate III). Although many of its bodily organs are functional by the end of the first trimester, the fetus is totally unable to survive outside the uterus at this time.

The Second and Third Trimesters. During the second trimester, the fetal brain begins to form convolutions and thereby increases its surface area. The eyes are able to perceive light. The fetus is growing in size all the time, and by the end of the fifth month it is approximately 23 to 28 cm (9 to 11 in.) long and weighs about 225 g (one-half pound). It can float freely within the amniotic sac, since its only attachment is via the umbilical cord that connects it to the placenta. The placenta itself is still growing, as it must provide increasing surface area for exchange of nutrients and waste between

The first trimester of human embryonic development completes all the major elements of differentiation. The second and third trimesters are given over largely to growth and preparation for birth.

fetus and mother. By the end of the second trimester, the fetus can usually survive if delivered prematurely at this stage.

The third trimester is characterized largely by further growth of the fetus and refinement of the nervous system and brain. During this period the mother's system must function increasingly for two complete individuals. The added weight and volume of the fetus pushes her stomach and diaphragm upward and often makes breathing particularly difficult. Her blood volume is about 30 percent greater than normal at the seventh month of pregnancy. About 16 percent of an expectant mother's total blood supply at any moment is located in her uterus and placenta. Furthermore, many specific nutrients that she consumes are largely used by the fetus. Only recently has the importance of maternal diet for proper fetal nutrition been recognized. For example, about 85 percent of the calcium and iron in the pregnant mother's diet goes into the fetus for building the skeleton and synthesizing hemoglobin, respectively. Nitrogen is consumed in great quantities in building fetal proteins, especially those going into the nervous system. The claim that protein deficiencies in the mother's diet may retard fetal brain development, however, has only been supported in the case of massive protein deficiencies, where the mother is near starvation.

During the final month or two of pregnancy, the fetus "drops" into position for birth: it moves forward, with its head usually pointing downward in the uterus. This relieves some of the pressure on the mother's diaphragm and abdominal organs. It also prepares the child for delivery (see Fig. 20.19b). During the ninth month the mother's "water" may "break"—with a resulting discharge of liquid through the vagina. This discharge results from the breaking of the amniotic sac and the release of the amniotic fluid within it. Unless it occurs significantly long before birth, this does not harm the fetus. Usually, the break occurs after the onset of labor.

Birth. It is not completely clear what factors initiate the process of birth, or **parturition.** One factor appears to be changes within the placenta, which slows its growth and physiological activity during the seventh month. Just prior to birth, the placenta exhibits regions of degeneration and breakdown of its capillary bed. Other factors include mechanical and hormonal changes that cause contraction of the smooth muscles of the uterine wall. It is well known that stretching smooth muscles increases their tendency to contraction. Hence the growth of the fetus as it stretches the muscles of the uterine wall may be one of the factors triggering the birth process. Complex hormonal changes also occur in the mother's body from the seventh month onward. During the last two months of pregnancy, progesterone secretion begins to slow down while estrogen secretion increases. Immediately before birth, relatively large quantities of estrogens appear to be secreted. Estrogens promote contractility and irritability of the smooth muscles of the uterus. In addition, by the onset of labor the hormone oxytocin is secreted by the posterior lobe of the pituitary in greatly increased quantities. Oxytocin is known to cause contraction of the uterine muscles. In turn, stretching of the uterine muscles is known to trigger an increase in oxytocin secretion. Oxytocin can also be administered to women intravenously to induce, or stimulate, labor artificially. Sometimes this is done if the fetus is large, to induce birth before the head and shoulders reach such proportions that delivery would be difficult.

The process of birth begins with a long series of involuntary contractions of the uterus, known as "labor pains." The period of labor can be said to start with the first uterine contraction and end with birth. It can be conveniently divided into stages. The first, called dilation, lasts about 10 to 12 hr, during which contractions of the uterus force the fetus toward the cervix, causing the latter to dilate. The progress of labor is usually measured by two signposts: the frequency and strength of uterine contractions and the degree of dilation of the cervix. The former can be measured qualitatively by noting the expectant mother's response (uterine contractions are often uncomfortable, even sharply painful in later stages), or quantitatively by placing a pressure-sensitive balloon within the uterus. The latter is measured by optical examination by the attending obstetrician. During the dilation phase, contractions of the uterus come about every 15 to 20 min, and last 25 to 30 sec. At the end of the first period of labor, the amnion ruptures (if it has not already done so prior to the onset of labor), releasing the amniotic fluid. The first period of labor ends with full dilation of the cervix, preparatory to emergence of the baby. This event is dramatically "announced" by *crowning*—the appearance of the baby's head through the cervix.

Birth is initiated by a variety of factors, including stretching of the uterus by the growing fetus, a physiological slowing down of exchange across the placenta, and complex hormonal changes within the expectant mother's body.

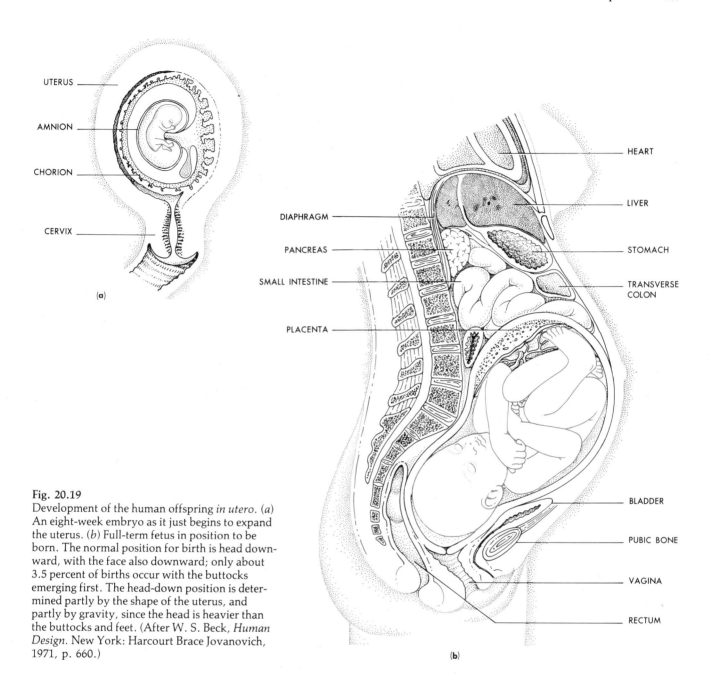

UTERUS

AMNION

CHORION

CERVIX

(a)

DIAPHRAGM

PANCREAS

SMALL INTESTINE

PLACENTA

HEART

LIVER

STOMACH

TRANSVERSE COLON

BLADDER

PUBIC BONE

VAGINA

RECTUM

(b)

Fig. 20.19
Development of the human offspring *in utero.* (*a*)
An eight-week embryo as it just begins to expand
the uterus. (*b*) Full-term fetus in position to be
born. The normal position for birth is head downward, with the face also downward; only about
3.5 percent of births occur with the buttocks
emerging first. The head-down position is determined partly by the shape of the uterus, and
partly by gravity, since the head is heavier than
the buttocks and feet. (After W. S. Beck, *Human
Design.* New York: Harcourt Brace Jovanovich,
1971, p. 660.)

The second stage of labor is called the expulsive
stage and lasts from several minutes to an hour. During
this period, uterine contractions occur more frequently
(every 1 to 2 min), last longer (50 to 90 sec), and are
much stronger than those during the first stage. During
delivery, the mother may push with her abdominal
muscles to aid the uterine contractions in moving the
baby out of the birth canal. Almost immediately after
the head emerges, the shoulders follow. Soon the whole
baby is delivered. The umbilical cord is still attached to
the placenta within the mother. The cord is usually
clamped and cut, stopping the supply of oxygenated

blood to the child and causing a subsequent build-up of
carbon dioxide. The resulting CO_2 levels in the blood
stimulate the respiratory center of the child's brain, and
contractions of the diaphragm begin. The child draws
its first breath only seconds after being born—sometimes before it is fully out of the birth canal. The traditional slap given to a newly born child is another way to
stimulate the respiratory process, though this practice is
declining with the increasing popularity of "gentle" or
"natural" childbirth. Once the first breath is taken, a
whole series of physiological reactions occur, the most
important of which are related to the switch from fetal

SUPPLEMENT 20.1
THE SWITCH FROM FETAL TO ADULT CIRCULATION

Once the umbilical cord is cut, the baby must rapidly adjust to a wholly new way of life. It can no longer rely upon the mother for all nutrients and exchange of wastes. Its first breath produces an expansion of the lungs, which up until this time have been collapsed. For the first time it must oxygenate its own blood, instead of receiving oxygen through the placenta.

With expansion of the lungs, the fetal circulatory system must suddenly adapt to a new direction of blood flow. The

Diagram A
Adult and fetal circulatory pathways. The inset shows two of the crucial changes from fetal to adult form. With expansion of the lungs, blood is diverted to them in greatly increased volume. This causes the ductus arteriosus to squeeze shut and the foramen ovale to be forced shut. (Modified from Jeffrey J. W. Baker, *In the Beginning.* Columbus, Ohio: American Education Publications.)

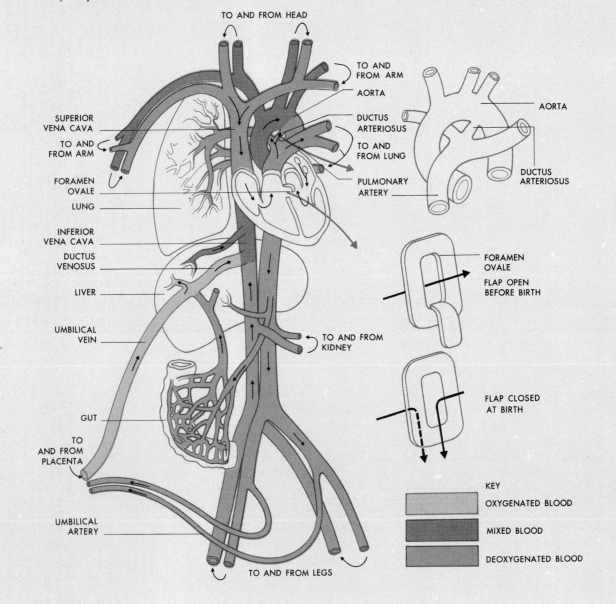

pathway of fetal circulation is very different from that of adult circulation (see Diagram A). The umbilical vein brings oxygenated blood from the placenta to the inferior vena cava. Here it mixes with venous (deoxygenated) blood returning from the fetal tissues and is carried to the right atrium. Since the fetal lungs are not functional as oxygenating organs, most of the blood from the right atrium passes through an opening, the foramen ovale, into the left atrium. From here it passes to the left ventricle and out through the aorta to the fetus's body. Some of the blood in the right atrium traverses the normal route through the right ventricle and out through the pulmonary arteries. However, since the lungs are not expanded, they can receive only a small volume of blood. Some of the blood passing through the pulmonary arteries is shunted directly into the aorta through another direct connection: the ductus arteriosus, a small channel between the pulmonary arteries and the aortic arch. Blood is returned to the placenta for further oxygenation by the umbilical arteries. Thus oxygenated and deoxygenated blood pass through the fetal heart.

At birth the expanding lungs suddenly draw blood from the right ventricle through the pulmonary arteries, establishing for the first time the adult pattern of flow from the heart to the lungs. The pressure of blood coming back for the first time into the left atrium from the pulmonary veins shuts off the foramen ovale by pressing a flap of skin over the opening, as shown in Diagram A. The foramen is thus effectively sealed off by equal pressure of blood coming from both sides—from right and left atria. The flaps of skin eventually grow together as part of the tissue of the atrial walls and seal off the foramen ovale completely. Occasionally, however, the foramen ovale fails to close, leaving an opening between the two atria. Since deoxygenated blood entering the right atrium can thus pass into the left atrium (bypassing the lungs), impure blood is delivered to the general circulation. This condition produces a bluish aspect to the skin and nails, and has given rise to the name "blue baby" for individuals affected by failure of the foramen to close. In the past, "blue babies" often lived until early adolescence, but few survived to adulthood. A simple operation has now been developed that makes it possible to seal the defective foramen surgically.

One other change takes place with the baby's first breath. The increased flow of blood through the pulmonary arteries causes them to expand, squeezing the left pulmonary artery against the aorta and other surrounding tissues. This effectively closes off the ductus arteriosus, ensuring that all blood passing out of the right side of the heart will be diverted to the lungs. The separation of blood flow between the right and left sides of the heart is now complete, having made the dramatic change from fetal to adult pattern in a few short seconds.

to adult circulation (see Supplement 20.1). With this switch, the infant becomes biologically independent of the mother.

The final period of labor is known as the placental phase. It involves contractions of the uterus, with expulsion of fluid and eventually the placenta itself. The placenta and attached portion of the umbilical cord are known collectively as the **afterbirth.** The mother is usually able to hold, and even nurse, the child almost immediately after expulsion of the afterbirth.

Although labor is often a painful and exhausting experience, many woman also describe it as exhilarating. The recent trend in many hospitals to encourage the father to be present and even actively involved during labor and delivery makes it possible for him to also experience part of the thrill of the first moments in the child's life. Far from being the end of development, birth is only one important step toward continued growth and change that will last a lifetime.

Miscarriage

Failure of the embryo or fetus to reach full term, with resultant expulsion from the uterus, is termed a **miscar-** **riage.** Miscarriages can occur anywhere from the zygote-to-trophoblast stage (in which case the mother usually does not know she has ever been pregnant) up through the sixth or seventh month. Most miscarriages, however, occur prior to the fifth month. Some result from physical injuries sustained during accidents to the mother, such as a fall, while others result from the development of genetic defects in the embryo. The mother's uterus appears to be able to "recognize" certain abnormal developments in the fetus and to reject it before it reaches full term. Still other miscarriages, especially in later stages of pregnancy, occur because of a weak cervix—an inability of the cervix to retain the fetus within the uterus. It is possible surgically to correct a weak cervix, since it can be detected long in advance of pregnancy.

Statistics reveal that in the United States there is approximately one miscarriage for each four live births. Among countries or sectors of the United States where medical care is inadequate, the frequency of such miscarriages is even greater. During pregnancy, regular visits to an obstetrician or gynecologist (physicians dealing with female genitals, the reproductive system, and pregnancy) are important in diagnosing problems which, if untended, might lead to miscarriage.

20.8
PLANT DEVELOPMENT

The developmental patterns of plants are considerably different from those of animals. In the following section we will discuss in broad outline the development of a higher plant.

Gametogenesis

Pollen Development. Pollen grains are binucleate cells that act in plants as producers of male gametes, the sperm. Pollen develops in the anther of flowers from diploid **microspore mother cells,** as shown in Fig. 20.20.

After undergoing meiosis, a tetrad of haploid microspores is formed from each microspore mother cell. Each of these microspores develops into a binucleate pollen grain, or what is called **male gametophyte.** A gametophyte is a structure in plants that consists (usually) of one to many cells (usually haploid) that produce gametes. The pollen grain is one of the smallest gametophytes in the entire plant kingdom, consisting really of only one cell with two nuclei when mature. The pollen grain is released from the anther and carried to the female part of the flower (pistil) by wind, insects, or other animals—and this initiates a process called **pollination.**

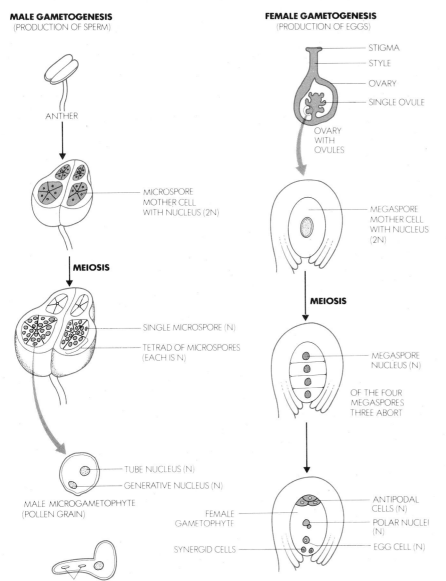

MALE GAMETOGENESIS
(PRODUCTION OF SPERM)

ANTHER

MICROSPORE MOTHER CELL WITH NUCLEUS (2N)

MEIOSIS

SINGLE MICROSPORE (N)

TETRAD OF MICROSPORES (EACH IS N)

TUBE NUCLEUS (N)

GENERATIVE NUCLEUS (N)

MALE MICROGAMETOPHYTE (POLLEN GRAIN)

SPERM

FEMALE GAMETOGENESIS
(PRODUCTION OF EGGS)

STIGMA
STYLE
OVARY
SINGLE OVULE

OVARY WITH OVULES

MEGASPORE MOTHER CELL WITH NUCLEUS (2N)

MEIOSIS

MEGASPORE NUCLEUS (N)

OF THE FOUR MEGASPORES THREE ABORT

ANTIPODAL CELLS (N)

FEMALE GAMETOPHYTE

POLAR NUCLEI (N)

SYNERGID CELLS

EGG CELL (N)

Fig. 20.20
Development of the male (left) and female (right) gametes (sperm and egg, respectively) in a flowering plant. Each begins with a diploid starting cell, the spore mother cell (microspore mother cell for pollen, and megaspore mother cell for egg). In male gametogenesis the microspore mother cell undergoes meiosis, generating a tetrad of microspores, each of which is haploid. The haploid nucleus undergoes mitosis, producing a tube nucleus that will govern the formation of the pollen tube after the pollen lands on the stigma. The mature pollen grain is called the male microgametophyte. In female gametogenesis, the megaspore mother cell within each ovule undergoes meiosis, producing four megaspore nuclei, each of which is haploid. Three of these nuclei degenerate. The one remaining nucleus undergoes a series of mitotic divisions to produce the haploid egg cell, three antipodal cells, two cells called synergids, and one large cell containing two haploid polar nuclei. The process of fertilization of the egg cell is shown in detail in Fig. 20.21.

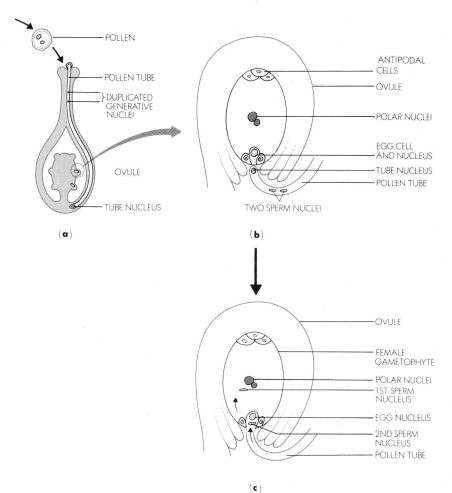

Fig. 20.21
Fertilization in a flowering plant. (a) The pollen cell, containing two nuclei (a tube nucleus and a generative nucleus) lands on the top of the pistil (stigma) and begins to grow a pollen tube through the tissue of the style. At the front of the tube is the tube nucleus; behind follow the two generative nuclei (the original generative nucleus divides). (b) The pollen tube, with its three nuclei, reaches the ovule. One pollen tube is necessary to fertilize each ovule (this means one pollen grain for each ovule). (c) The pollen tube enters the ovule and ruptures. The tube nucleus degenerates, and the two generative nuclei enter the female gametophyte. One nucleus fertilizes the egg, producing a 2n zygote. The other fuses with the two polar nuclei, producing a 3n (triploid) nucleus. The triploid nucleus and the cell of which it is a part (the large cell of the female gametophyte occupying all the space except that occupied by the antipodal, egg, and synergid cells) gives rise to the endosperm, or stored food tissue, that will eventually be included within the seed.

Egg Development. Female gametogenesis in most flowering plants is somewhat more complex than in the male. The female gamete, the egg, is produced in **ovules** that lie within the ovary (see Fig. 20.20, top right). Each ovule consists initially of a single large, diploid **megaspore mother cell.** The megaspore mother cell undergoes meiosis to generate four haploid megaspores, three of which are ultimately cast off. The remaining megaspore enlarges greatly and passes through three mitotic divisions, producing a total of seven cells with eight nuclei. The largest cell is binucleate (remember, each nucleus is still haploid) and occupies the center of the former ovule space. The two nuclei of this cell are called the *polar nuclei.* In addition there are three antipodal cells at one end of the ovule space, and two synergid cells plus the egg cell (ovum) at the other. Together, these seven cells (eight nuclei) comprise the **female gametophyte.**

Fertilization in Plants

Once a pollen grain has been discharged from the anther and lands on the **stigma** (the top part of the pistil, the female reproductive organ in flowering plants), it begins to germinate. The events are depicted in Fig. 20.21. A pollen tube grows out of the pollen cell and continues this growth down through the tissue of the **style** into the ovary. After pollen germination the generative or sperm nucleus duplicates. The two resulting generative, or sperm, nuclei (both haploid) follow behind the tube nucleus down the pollen tube. When the pollen tube reaches an ovule within the ovary, it grows into the female gametophyte and discharges its contents. The tube nucleus disintegrates, while one of the generative nuclei fuses with the egg and the other with the two polar nuclei. The first fusion produces the diploid zygote, while the second produces the triploid **endosperm** nucleus. This nucleus, through repeated mitosis and cytokinesis, gives rise to tissue known as the endosperm, the stored food of the future seed. The process of double-fertilization that occurs in the flowering plant means that the stored food necessary for the early germination of the seed is contained in tissue developed from a cell line separate from that giving rise to the embryo proper. It also means that the endosperm tissue is composed of triploid (3*n*) while embryonic tissue is composed of diploid (2*n*) cells.

Growth and Development of the Plant Embryo

The embryo and endosperm tissue are retained within the ovary of the maternal plant, eventually becoming surrounded by a tough, protective **seed coat.** The resulting structure—the endosperm, embryo, and seed coat —is known as the **seed.** The development of the embryo within the seed begins shortly after fertilization. The major stages in the development of a plant embryo are diagrammed in Fig. 20.22. As the embryo develops (Fig. 20.22a–c) several forms of presumptive or future tissue can be identified: protoderm (future epidermis), group tissue (future cortex), and provascular tissue (future xylem and phloem). By the fourth stage (Fig. 20.22d),

the **cotyledons,** or food-leaves have begun to appear. The cotyledons develop from the endosperm tissue and are thus composed of triploid cells, whereas the remainder of the developing embryo is composed of diploid cells. The entire embryo is held to the wall of the former ovule, or embryo sac, by a chain of cells called the **suspensor.**

Further elongation and development of the cotyledons and of the **hypocotyl** (the portion of the embryo that will become the future stem) causes the embryo to bend over, the position it will ultimately take inside the fully developed seed (Fig. 20.22e). By the time the embryo has begun to arch, the beginnings of vascular tissue differentiation and of root and shoot meristems formation are visible (Fig. 20.22f and g). The suspensor has become much reduced. The seed coat now forms around the embryo and endosperm; when the seed is

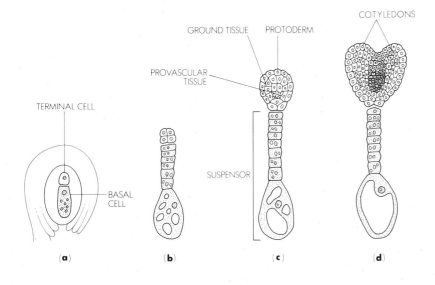

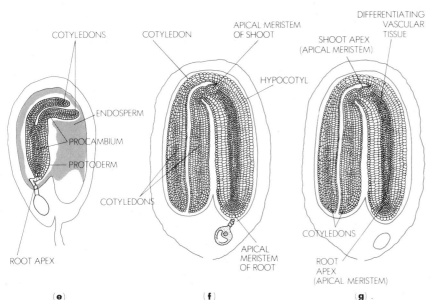

Fig. 20.22
Embryonic development of a flowering plant (the "shepherd's purse," *Capsella bursapastoris*). (*a*) The zygote undergoes division to produce a two-celled structure, consisting of a smaller terminal cell, and a larger basal cell; (*b*) The terminal cell divides, producing a quadrant of four cells at the top and a group of flattened cells underneath; the basal cell grows in size but does not divide; (*c*) The top quadrant of cells becomes the globe-shaped embryo, while the flat cells underneath and the enlarged basal cell become the suspensor, a chain of cells connecting the embryo to the wall of the embryo sac; (*d*) The cotyledons, or "food leaves" begin to develop, giving the spherical embryo a heart-shaped appearance; (*e*) The cotyledons have developed further, and some of the cells have started to differentiate into recognizable cambium; (*f*) The cotyledons elongate, as does the hypocotyl (the portion of the embryo that eventually becomes the stem and the root), causing the embryo to fold over in a kind of inverted horseshoe shape. The cotyledons contain stored food. As the embryo grows further, the cotyledons enlarge and become filled with carbohydrate (*g*), while the plant embryo itself develops more differentiated tissue: for example, foliage leaves, prospective stem, and root.

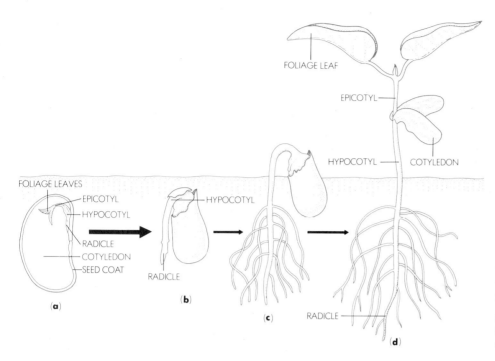

Fig. 20.23
Germination and early development of a flowering plant, the bean (*Phaesolus*). The seed (*a*) begins to develop. The seed absorbs water (imbibition), and the coat bursts. The radicle (young root) emerges first, growing downward into the soil (*b*). The hypocotyl begins to push upward, carrying the old seed coat and cotyledons with it (*c*). After exposure to light, the hypocotyl and foliage leaves begin to synthesize chlorophyll. The developing stem straightens and the young seedling finally becomes self-sufficient (*d*).

ripe it is released from the ovary (which has now become the **fruit**) and is ready for **germination.**

At this point the seed usually enters a dormant state. Several factors may induce dormancy: (1) lack of endosperm, most of whose contents have been absorbed by tissues of the cotyledons; (2) reduction of light penetration into the embryo, brought on primarily by thickening of the seed coat (light appears to be a stimulant to germination); (3) lowered concentration of the plant hormone gibberellin; and, finally, (4) secretion of growth-inhibiting substances by various extra-embryonic tissues of the seed. Dormancy periods can last for weeks, months, years, or even centuries. Seeds taken from 3000-year-old Egyptian tombs have germinated when provided with proper temperature, light, and moisture conditions. The record for dormancy, so far as is known, is held by a collection of seeds of the lupine, found deeply buried below the permafrost line in the Arctic tundra. The seeds were estimated to be at least 10,000 years old. Yet when planted, many germinated after only 48 hours.

Germination and Early Development

The germination of a typical seed (the common bean, *Phaesolus*) is shown in Fig. 20.23. The **radicle** (young root) grows down from the seed, and the hypocotyl (young stem) grows upward, forming an arch (Fig. 20.23a, b, and c). Eventually the young foliage leaves push out of the seed coat and emerge into the light. The empty cotyledons remain attached to the stems, a reminder of the young seedlings' former dependent status.

By the time the hypocotyl has emerged from the ground all the major plant tissues within the young seedling have begun to differentiate. Up until this point the young seedling has been growing on food stored originally in the cotyledons. On exposure of the hypocotyl and foliage leaves to light (Fig. 20.23c), chlorophyll synthesis begins, and the seedling gains the ability to manufacture its own food. Hence, it becomes a full-fledged (though still not fully mature) plant in its own right.

Comparison of Animal and Plant Development

The differences between animal and plant development are not as great as it might seem. Both go through broadly similar processes of gametogenesis, fertilization, cleavage, and differentiation. There are, however, some significant differences that deserve mention:

1. In plants, cleavage and cell growth occur simultaneously, while in animals cleavage occurs first with growth coming much later.

2. In plants, a few cells are set aside early in development to become the future growth regions—the meristems, including cambium. These regions remain unspecialized and embryonic.

3. Plant cells show little morphogenetic movements during embryogenesis, while such movements are a major part of many stages of animal embryogeny.

4. And finally, the fully developed animal ceases to grow and add new organs, whereas the plant always grows and produces new organs (roots, stem, leaves, flowers) throughout its lifetime. It can truly be said that a perennial plant (one that continues to live from one growing season to another) never ceases to grow and develop.

20.9
CAUSES OF DIFFERENTIATION: EMBRYONIC INDUCTION

The primary germ layers (the ectoderm, mesoderm, and endoderm) give rise to the very specialized tissues and organs of the body. Yet we are still left with the question of how this occurs. What factors, for example, cause the descendants of certain endodermal cells to become parts of the intestinal lining, while others end up in the thyroid gland? In Section 20.6, details of the formation of the neural plate and tube from ectoderm were described, including the process that, on a cellular level, was responsible for the folding inward of the neural plate to form the neural tube. The discussion concluded with the statement that internal changes in cells of the presumptive neural plate are triggered by contact with the underlying mesoderm.

Yet, the question remains: what specific effect does such contact have? How would experimental embryologists approach such a question? The neural plate does not begin its development until the gastrula has fully formed. At that time, the infolding of the blastopore is complete and the inner cell layers of endoderm and developing mesoderm come to lie close to the ectoderm layer (see Fig. 20.10). Would the neural tube form from ectoderm that has been isolated from an embryo considerably before it begins neural-tube formation?

To find out, we could remove ectodermal regions destined to form neural tubes from some frog embryos and culture them in a separate medium. This experiment asks an important question: whether the development of a neural tube is caused by intrinsic factors within the ectodermal region involved, or by extrinsic factors contingent on the location of the ectodermal region within the embryo. The results of the experiment seem to provide a definite answer. In no case are neural tubes formed by the isolated pieces of ectoderm. The embryos from which the ectodermal pieces were surgically removed do not develop a neural tube either, presumably because they were deprived of cells that would ordinarily have given rise to this structure.

It seems evident, then, that some stimulus from the surrounding environment of the embryo's body mass is responsible for the formation of a neural tube by the ectodermal region. The next question to be posed is: from what region of the embryo does this stimulus originate? A logical choice is the mesoderm, since the neural-tube-forming ectoderm rests on top of it. An experiment can be performed to test a hypothesis proposing the mesoderm as the source of the stimulus. We can deduce that if the mesoderm contributes something to the overlying ectoderm to cause it to form a neural tube, then mesodermal cells, wrapped in a sheet of ectoderm and isolated from an embryo at the same stage as in the previous experiment, will cause the formation of neural tissue by this ectoderm. Such an experiment was performed by Johannes Holtfreter in the 1940s. The results were as predicted; the sheet of ectoderm enclosing the mesoderm differentiated into neural tissues.

The Primary Organizer

One of the great experimental embryologists of this century was the German Hans Spemann (1869–1941). Working with amphibian embryos (salamander and frog), Spemann set out to answer some of the questions posed by differentiation of the primary germ layer.

Spemann called the action of the mesoderm on the ectoderm one of **embryonic induction**. He hypothesized that the mesoderm *induces* the ectoderm to differentiate into the neural tube. He spoke of such inducing tissues

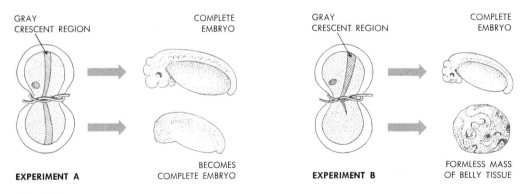

GRAY CRESCENT REGION   COMPLETE EMBRYO

EXPERIMENT A   BECOMES COMPLETE EMBRYO

GRAY CRESCENT REGION   COMPLETE EMBRYO

EXPERIMENT B   FORMLESS MASS OF BELLY TISSUE

Fig. 20.24
The results of these experiments convinced Spemann that the primary organizer would be found near the region of the gray crescent. Although in experiment A only the half containing the nucleus developed at first, two complete embryos were eventually formed. (Adapted from John W. Kimball, *Biology*, 2d ed. Reading, Mass.: Addison-Wesley, 1968.)

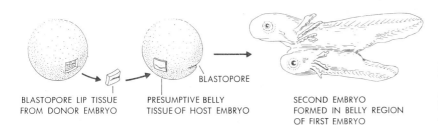

BLASTOPORE LIP TISSUE
FROM DONOR EMBRYO

PRESUMPTIVE BELLY
TISSUE OF HOST EMBRYO

BLASTOPORE

SECOND EMBRYO
FORMED IN BELLY REGION
OF FIRST EMBRYO

Fig. 20.25
This experiment, for which Spemann received a Nobel prize, showed the dorsal lip of the blastopore to have primary organizer capabilities.

as "organizers," since they seemed able to organize other tissues into definite regions of the embryo. For example, Spemann knew that the eye lens in the frog, *Rana fusca*, is formed by a process of induction. Like the neural tube, the eye lens forms from ectoderm, this time ectoderm in the eye region of the embryo. However, Spemann also saw that there was not just one inductive process at work here. In eye formation, the head mesoderm induces formation of the brain, and the middle region of the brain induces the formation of the optic nerve and vesicle; the optic vesicle, in turn, induces the formation of the eye lens from the ectoderm. Working backwards, Spemann could visualize the embryo as being built by progressively less specific inducing tissues, or organizers. Thus the organizer that induces lens formation from the ectoderm is a specific organizer with a limited range of responsibilities. The organizer that induces brain formation has a wide range of responsibilities, including in its chain of command the eye lens, as well as many other head parts. Somewhere in this chain of induction there had to be a tissue responsible for organizing the entire organism.

In the newly fertilized frog and salamander egg, a region called the **gray crescent** appears. Spemann had experimented with the gray crescent before. He knew that if he tied a fertilized salamander egg into two halves so that a portion of the gray crescent extended into each half, two normal embryos eventually resulted. If, on the other hand, he tied the egg so that only one half had the gray crescent, only this half developed normally. The other half produced an unorganized ball of belly tissue (Fig. 20.24).

With these results in mind, Spemann turned his attention to cells associated with the gray crescent. Those cells are located on the dorsal lip of the blastopore, which is formed just below the area of the gray crescent. He considered that these cells might possibly be the primary organizers he was looking for.

They were. Taking blastopore lip cells from one salamander embryo, he transplanted them into presumptive belly tissue of another. The host cells would ordinarily have formed the ventral areas of the salamander. However, when dorsal blastopore tissue was put among them, their future was changed. The result was not just the formation of a brain or an eye lens. Instead, a *complete new embryo* was induced, joined to the other like a Siamese twin (Fig. 20.25).

The importance of this experiment cannot be overemphasized. It, and Spemann's subsequent work, led to a Nobel Prize in 1935. Spemann published the results of his experiment in 1924, in co-authorship with his graduate student Hilde Mangold. In this paper, he gave reality to the concept of epigenesis and supported his hypothesis of inductive organizers. From this point on, embryologists began to picture development of the embryo as a series of inductive processes, with each member of the series being essential for the development of the members that follow it. Thus, for example, the eye lens would not form if the optic vesicle were removed; the optic vesicle would not form if the midbrain were removed; the midbrain would not form if the head mesoderm were removed; and finally, *nothing would form if the primary organizer, the dorsal lip of the blastopore, were removed.*

Having found the primary organizer, Spemann was at a loss to explain its nature. His experiments had not revealed the precise manner by which induction was carried out. Obviously, much more research was needed.

Chemical Induction

In the development of the neural tube, the mesoderm acts as an organizer. It provides the extrinsic factors that induce the ectoderm region to build the portion of the embryo (the neural tube) for which it is responsible.

Embryonic induction involves two-way reciprocal interaction between inducing and induced tissue. The original inducer influences the induced tissue to differentiate. The differentiated tissue can now act as a secondary inducer on other, nondifferentiated tissues, or it can reciprocally induce further differentiation in its own original inducer.

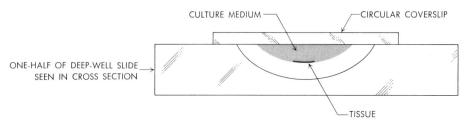

CULTURE MEDIUM ——

—— CIRCULAR COVERSLIP

ONE-HALF OF DEEP-WELL SLIDE →
SEEN IN CROSS SECTION

—— TISSUE

Fig. 20.26
The hanging-drop technique, developed by embryologist Ross G. Harrison.

What is the nature of these extrinsic factors? How does the mesoderm send its instructions to the ectoderm, causing it to differentiate into neural tissue? It could not be a nervous transmission, for nerve fibers have not yet been formed. Might it be some sort of chemical transmission? Could a substance be diffusing from the mesoderm to the ectoderm, inducing it to differentiate into nerve tissue? This hypothesis deserves some testing.

If an underlying organizer tissue is separated by a thin piece of impermeable material from the tissue it is supposed to induce to differentiate, no differentiation takes place. Note, however, that although this experiment may support the hypothesis that a chemical is involved in induction, the results are far from conclusive. It may well be that physical contact between cells, rather than any chemical substance, is necessary for differentiation to occur.

If a synthetic membrane with pores so small that only individual molecules can pass through it is placed between inducing tissue and the tissue it normally induces, differentiation *does* occur. This supports the hypothesis that an organizer substance diffuses from the inducing tissue. Yet this experiment still does not satisfactorily rule out the possibility that physical contact may also play a part.

Two embryologists, V. C. Twitty and M. C. Niu, tried a different approach. They used the hanging-drop technique of tissue culture designed by the embryologist Ross G. Harrison. The hanging-drop technique has been of great value to experimental embryology (Fig. 20.26). Into half of their hanging-drop cultures, Twitty and Niu placed organizing mesoderm. The other cultures, serving as controls, received no mesoderm. After a few hours, the mesoderm was removed from the experimental cultures, and ectoderm was placed into both experimental and control cultures. In the experimental cultures that had held the mesoderm, the ectoderm differentiated into neural tissues. In the control cultures, no differentiation took place. It seemed obvious that the mesoderm had released a substance into the culture medium that induced differentiation in the ectoderm. Chemical analysis of the culture medium that had held mesoderm was a logical next step. The analyses showed traces of nucleic acid. However, subsequent studies have not confirmed the tempting hypothesis that the "organizer" might be some form of nucleic acid.

Another interesting experiment throws light on the relationship between the inducing and the induced tissue. Oscar E. Schotté transplanted a cell region that would ordinarily form frog flank tissue to the mouth region of an older salamander embryo (Fig. 20.27). This experiment asked an important question: Would the piece of frog tissue destined to become flank tissue still do so in the region in which a mouth is needed? In other words, do certain inherited factors within the frog tissue determine its fate, or does the environment (the surrounding salamander tissue) do so?

The experimental results showed that the presumptive flank skin tissue of the frog formed a mouth. Since the inducer from the underlying salamander mesoderm would transmit its own genetic information, we would expect it to instruct the frog tissue to form a mouth. The transplanted cells seem to "listen" to what organizer substances from the salamander tissues tell them to do, rather than following the pattern of their own normal future.

However, close examination of the *kind* of mouth formed showed that it was distinctly one of a frog larva, and *not* one of a salamander! It is obvious that the frog tissue was responding to an inducer in the only way it could: After all, it had genes only for producing a frog mouth. However, it responded to the general message of the inducer to which it was exposed: to form a mouth. The organizer, then, seems to give only general instructions as to what kind of structure is to be formed. The genes of the tissue being induced determine what style of structure will be built.

Other experiments of more recent origin have emphasized a further fact. Induction is a two-way process, involving not only the action of the inducing tissue

Fig. 20.27
Oscar E. Schotté's experiment. The results indicate that while the organizer determines in general what organs are to be formed, the genes of the differentiating tissue control the details of those organs.

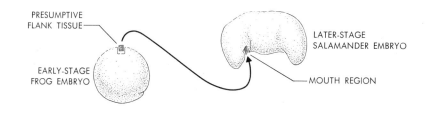

PRESUMPTIVE
FLANK TISSUE ——

EARLY-STAGE
FROG EMBRYO

LATER-STAGE
SALAMANDER EMBRYO

—— MOUTH REGION

on the induced, but also a reciprocal action of the induced tissue on its inducer. The two tissues interact with one another. Spemann's view of induction was much like a military hierarchy, with a series of primary, secondary, and tertiary inducers passing a "command" along the chain. In fact, it is now known that after an induced tissue has begun to differentiate, it can act as a further inducer for the tissue that originally induced it. Thus inducer and induced tissues form an interacting system, developing together along the path of their respective differentiations. The idea of hierarchy of orders passing in only one direction from bosses (inducer) to workers (induced tissue) proves to be erroneous when applied to the induction process in embryonic development.

20.10
TOTIPOTENCY AND THE REVERSIBILITY OF DEVELOPMENT

Embryologists and geneticists believe that every cell of the developing, as well as the adult, organism (animal or plant) contains all the genetic information present in the zygote. In other words, every cell has a copy of every gene originally contributed by the sperm and egg. Yet obviously, through tissue interactions and inductive processes, only certain genes are active in the various differentiated cells as development proceeds. How reversible or irreversible is this process? Can cells which have become differentiated revert to a less specialized state and develop in other directions? In other words, do cells remain *totipotent* as differentiation proceeds? **Totipotency** refers to the total potential of the cell for expressing all of its multitude of genes. The fertilized egg is fully totipotent—it can (and does) differentiate in all possible directions. But what about cells once they have begun to specialize?

In the 1950s, Robert Briggs and Thomas J. King devised the delicate experimental technique shown in Fig. 20.28 to answer this question. Nuclei were transplanted from the cells of older frog embryos into frog

eggs from which the nuclei had been removed (enucleated eggs). Briggs and King first transplanted nuclei from blastula cells into enucleated eggs. Many normal embryos resulted. Evidently, then, the transplanted nucleus was still capable of acting as a "general practitioner" of development. It could still direct (assuming it does so) the differentiation of a complete embryo, rather than merely that part of an embryo it would have acted on if it had remained in the blastula. The experiment was repeated with cell nuclei transplanted from early gastrulas into enucleated eggs. Again, some normal development occurred. However, when nuclei from cells of *late* gastrula stages were transplanted, abnormal development resulted in all cases.

Do these experiments support the hypothesis that irreversible genetic changes—that is, a loss of totipotency—occur in cells as they differentiate? It would be tempting to think so, but we must not jump too quickly. An additional set of experiments by British biologist John B. Gurdon and his associates in 1962 provided some interesting and unexpected results. They transplanted to enucleated host egg cells the nuclei from endodermal cells of the South African clawed toad (*Xenopus laevis*) ranging in age from late gastrula to the free-swimming tadpole stage. The results of Gurdon's experiments are shown in Fig. 20.29. Note that before differentiation ceased late gastrula donor nuclei were able to generate a whole range of developmental stages, including a complete tadpole. Indeed, those that developed to tadpole stage could, for the most part, develop into functional adult frogs. The fact that *any* tadpoles or adult frogs were obtained, however, strongly suggests that the totipotency of all late gastrula donor nuclei is not restricted. Some, at least, can reverse the developmental process up to that point and start over as fully totipotent as the original zygote.

Results similar to those obtained by Gurdon on animals came from the work on plants of A. C. Hildebrandt and Vilma Vasil at the University of Wisconsin. In experiments reported in 1965, they placed small pieces of pith tissue from stems of hybrid tobacco plants

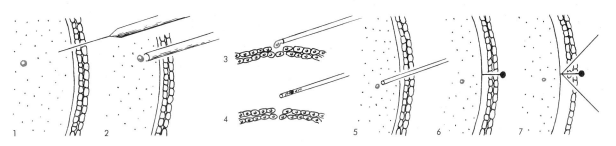

Fig. 20.28
The Briggs and King operational technique. (1) The egg is pricked, causing the nucleus to move to the surface. (2) The nucleus is removed with a micropipette. (3, 4) An ectodermal cell from a blastula is drawn into the micropipette, whose diameter is so small that the cell membrane is ruptured. (5) The cell nucleus, with most of the cytoplasm removed from it, is injected into an enucleated egg. (6, 7) Snipping the protrusion prevents egg cytoplasm from escaping.

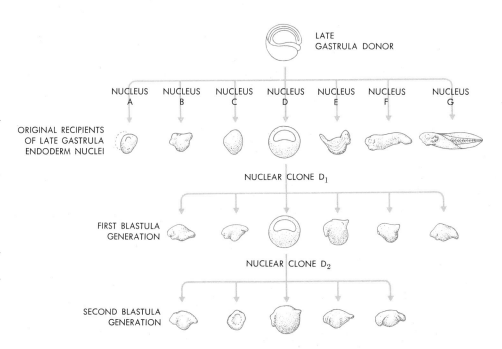

Fig. 20.29
Nuclei removed from various late gastrula endoderm cells vary widely in their ability to control embryonic development when injected into enucleated eggs. Nucleus A, for example, seems to have lost its ability to control normal development, while nucleus G still retains the ability and causes the development of a complete embryo. However, when descendants (clones) of any one particular nucleus are injected into enucleated eggs, they yield embryos that are uniform in the stage of development they attain. This indicates that the genes controlling development are "turned off" in cell nuclei as development proceeds.

on a solid agar culture medium. After approximately a month the tissue, now considerably increased in size, was transferred to a liquid culture medium. Following several days of growth in containers placed on a machine designed to shake them gently, the tissues dissociated into single cells and groups of cells. Using micropipettes and a dissecting microscope, the investigators picked out single cells from the medium and grew each one in a microculture, isolated from all other cells. Out of 150 single cells isolated, thirty-two underwent mitotic divisions, producing in about two weeks irregular, unorganized masses of fifty to seventy-five cells. Each cell was then transferred to a culture bottle containing a solid agar medium (Fig. 20.30). By manipulating the composition of the medium, the investigators obtained hundreds of plantlets with shoots and roots. To show that these plantlets, each the product of a single cell, were completely normal, they were transferred to sterilized soil and grown to mature, flowering plants.

How can we reconcile the apparent differences between the experiments of Briggs and King on the one hand, and Gurdon, Hildebrandt, and Vasil on the other? Their results seem contradictory. In fact, they are not. Nuclear transplantation is a difficult and tricky procedure. Relatively few viable transplants result out of hundreds that are attempted. The older the nuclei transplanted, the more difficult it is to get the cell to live at all. Briggs and King's work came earlier than Gurdon's, and thus encountered more practical difficulties in obtaining viable transplants. As both sets of investigators carried out further studies, their results came more and more to coincide. The general conclusion of

developmental biologists today is that most cells retain the *potential* for totipotency throughout life. The debate is by no means entirely closed, however.

Regeneration and Totipotency

Regeneration is the process by which organisms replace lost parts. Regeneration occurs throughout the animal and plant kingdoms. In animals it can be quite striking. If a sponge is chopped into a hundred parts, a hundred new sponges result. A starfish cut in half gives two starfish. Tear a claw from a lobster, and a new claw grows. A salamander can lose an arm and grow a new one. Most higher animals regenerate their skin constantly. If a human being receives a cut, the wound heals over by regeneration of damaged tissue.

Regeneration is as much a process of development as morphogenesis (the development of form) in the growing embryo. In regeneration, cells differentiate into specialized forms to replace the tissue that was lost or damaged. Moreover, because those cells start out prior to the regeneration process as specialized cells of the adult, and then de-differentiate and respecialize into the needed damaged types of cells, those cells must be totipotent from the outset.

We will briefly summarize what is known about regeneration today:

1. Primitive organisms are better at it: The less highly specialized the organism, the greater its power of regeneration tends to be. Organisms such as *Hydra*, planaria, starfish, or earthworms have considerably greater pow-

Fig. 20.30
Growth of tobacco plants from an undifferentiated mass of cells developed from a single isolated cell. (*a*) Undifferentiated mass of cells. (*b*) Development of several leafy shoots from a mass of cells. (*c*) Plants growing in aseptic culture (two-thirds natural size). (*d*) Formation of roots by the leafy shoots. (Courtesy Vimla Vasil and A. C. Hildebrandt, *Science* 150, pp. 889–892. Copyright 1965 by the American Association for the Advancement of Science.)

ers of regeneration than vertebrates, especially mammals.

2.   Within any given species of animal, the power of regeneration is greater in younger organisms than in older. Experiments conducted on the South African clawed toad, *Xenopus laevis*, showed that the capacity of the hind limb for regeneration declines as the age of the animal increases.

A number of observations and experiments over the years suggest how regeneration occurs; they also explain to some extent the two observations listed above. Consider the regeneration process in the limb of *Xenopus laevis* larvae. A few hours after amputation, migrating epidermal cells form a protective cap over the wound. Damaged and dead cells are demolished and removed. Finally, and most important, a small group of unspecialized regeneration cells appear. These cells are the beginning of a cone-shaped mass of unspecialized cells, collectively called the **blastema**. *The appearance of a blastema is typical and universal in limb regeneration. It is from the cells of the blastema that the new limb originates.*

Since this much is known, histological (tissue) and cytological (cellular) studies of limb stumps give a clue as to why younger animals are capable of more regeneration than older ones. First, as the animals get older, the number of regeneration cells that contribute to blastema formation becomes smaller (relative to the mass of the stump). Correspondingly, the blastema is smaller; thus there are fewer cells available to produce the new limb. Second, the regeneration cells themselves are less "embryonic" in appearance; in other words, *they are less undifferentiated.* Research seems to indicate that *an animal's capacity for natural regeneration decreases as the degree of tissue differentiation increases.* From this, it might follow that a highly specialized cell, such as a neuron, would be poor at regenerating—and neurons generally are. On the other hand, unspecialized mesenchyme cells, such as are found in sponges, have considerable regenerative ability. For a specialized cell to regain regenerative abilities, it must first "de-differentiate," that is, revert to an unspecialized state. In examining the stump cells from *Xenopus laevis* larvae whose limbs have been amputated, such a de-differentiation is observed. For example, muscle cells in a stump lose their striations and seem less specialized in character.

It had been shown as early as 1823 that even the salamander could not regenerate a limb if the nerves supplying it were first cut. Ths demonstrated importance of nerve tissue to regeneration led Case Western Reserve University's Dr. Marcus Singer to hypothesize that the low ratio of nerve tissue to other tissues in the frog's limb (as compared to its relatively high ratio in the salamander's thin limb) might be responsible for the

Regeneration involves covering over of the wound area by epidermal cells followed by accumulation of blastema cells in the area under the wound cover. Blastema cells are relatively nonspecialized cells that differentiate into the specialized tissue of the regenerating organ.

difference in the regenerative ability of the two animals. His reasoning was that if the differences between the regenerative ability of frogs and salamanders are attributable to differences in mass ratio between neural and other types of tissues in their limbs, then experimentally increasing the amount of neural tissue supplying a frog's limb might be sufficient to induce regeneration after amputation.*

In 1954 Dr. Singer diverted the iliac nerve supplying the hind leg of a frog into its arm. He then amputated this doubly innervated appendage. Regeneration occurred. Later, other investigators showed that the same regeneration would take place if adrenal glands were implanted into the stump. Thus neural tissue was not the only factor that could induce regeneration. Even before Singer, L. W. Polezhayev of the U.S.S.R. managed to achieve the regeneration of the forelimb of an adult frog by repeatedly injuring the wound, and S. Meryl Rose of the University of Illinois achieved the same result by immersing the stump many times in a strong salt solution. In 1968 Tulane University's Dr. Merle Mizell showed that the hind legs of an immature opossum can be induced to show regenerative responses if developing brain tissues from the cerebral cortex are implanted in the region to be amputated two to four days before amputation.

Many students of regeneration hold that nerves must secrete some "regeneration substance." This hypothesis has been supported by the observation, on invertebrates, that nerve cells in areas of regenerating tissue have a larger-than-usual number of secretory granules. Other investigators suggest that regeneration is stimulated by increased frequency of very weak electrical currents. Such currents have been observed to occur in tissue that has been, or is continuing to be, physically traumatized (by repeated injury, for example). It has been demonstrated that application of weak electrical currents will actually induce de-differentiation and increased mitotic activity in adult vertebrate cells and thus lead to greater regeneration. Since the amount of weak electric current generated at any wound site is largely a function of the amount of nerve tissue inner-

vating that site, advocates of the electrical hypothesis claim that electrical currents put forth by the nerves trigger regeneration, not a chemical substance. It is not possible to distinguish between these two alternative hypotheses at the present time.

In either case, the existence of regenerative ability is a strong indication that fully differentiated adult cells in higher animals can de-differentiate and re-differentiate into new forms. These cells must be totipotent. We turn next to a consideration of how we might explain, on the genetic level, both totipotency and the ability of cells to differentiate.

20.11
EVIDENCE FOR GENE REGULATION
IN GROWTH AND DEVELOPMENT

Is there experimental evidence that during development some genes are active while others are not?

At certain times during the development of insects, "puffs" or swellings along the chromosomes can be associated with gene activity. The puff areas are thought to be regions of intense *m*RNA transcription. The nonpuff areas are thought to be inactive genes. The existence of puffs in particular regions of particular chromosomes appears to be both *tissue-specific* and *time-specific*. Tissue-specific means that normally, in adult tissue of a particular type, the same chromosomal regions appear puffed, suggesting that these regions are permanently active in that tissue. Generally, only a very small percentage of genes appear to be active in any one type of tissue. Specific puffs appear at particular times in given chromosomal regions during embryonic development. These puffs come and go as insect development (metamorphosis) proceeds. Figure 20.31 shows the changing patterns of chromosome puffs during stages of metamorphosis in the midge, *Rhyncosciara angelae.* Puffing in insect chromosomes can be induced by hormones *in vitro.* The fact that insect metamorphosis is guided by changing patterns of hormone production suggests strongly that changes in gene activity may be either triggered or suppressed by the appearance and disappearance, respectively, of certain hormones.

Similar findings have been made in the developing oöcytes of amphibia. A few years ago workers discovered that the so-called lampbrush chromosomes found

---

*Salamanders and frogs are both members of the class amphibian, yet the salamander *can* regenerate a severed limb whereas the frog normally cannot.

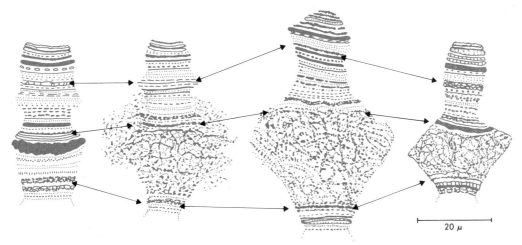

Fig. 20.31
A portion of one giant chromosome from the midge larva *Rhyncosciara angelae* during different stages of larval development. The arrows and connecting lines indicate comparable bands. Changes in puffing (expansion) of particular regions indicate increase or decrease in activity of genes in those areas. Different genes appear to be active at different stages in the insect's development. (From M. F. Breuer and C. Pavin, "Behavior of polytene chromosomes of *Rhyncosciara angelae* in different stages of development." *Chromosoma* 7, pp. 275–280.)

in cells of insects, a number of other organisms, and especially in amphibian oöcytes, were surrounded by brushy fibers (Fig. 20.32). Extensive chemical studies have revealed that these brushy fibers are DNA, each long central fiber being one double-stranded DNA molecule. Extending from each fiber are loops, now known to be *m*RNA—presumably in the process of being transcribed from DNA. Since the oöcyte is known to

Fig. 20.32
Electron micrograph of nucleolar genes coding for *r*RNA isolated from an amphibian oöcyte. The particular genes shown here are in the nucleolus and code for ribosomal RNA (*r*RNA). The long horizontal axis is DNA. The brushy fibers extending from it are *r*RNA molecules being transcribed. Where no brushy fibers extend from the DNA, no transcription is assumed to be occurring. (Micrograph courtesy O. L. Miller, Jr., and Barbara R. Beatty, Biology Division, Oak Ridge National Laboratory; from "Visualization of nucleolar genes," *Science* 164 [May 23, 1969], pp. 955–957).

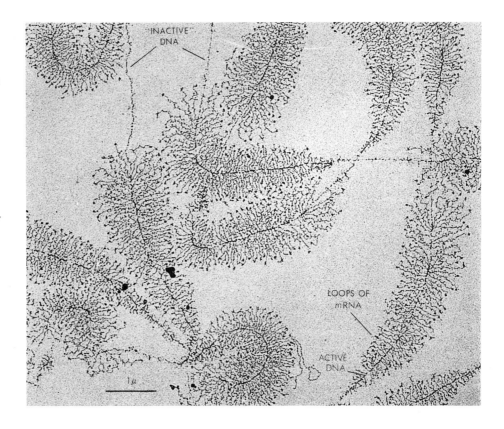

SUPPLEMENT 20.2
THE CELLULAR BASIS OF AGING

One of the phenomena on which developmental biology has begun to throw light is the process of aging, presently being studied at both the organismic and cellular levels. Conventional wisdom in the recent past held that aging was largely a property of whole organisms and not of individual cells. This view was based on the supposition that cells could grow and divide *ad infinitum* outside a multicellular organism's body —that is, in tissue culture. Recent experiments have raised doubts about this supposition, however. Leonard Hayflick at Stanford University has shown that cultured human fibroblasts (embryonic cells that give rise to connective tissue) divide only a limited number of times before they deteriorate, lose their capacity to divide further, and finally die. Hayflick and others have also found that the longer the life span of the species, the greater the number of times its cells will divide in culture. This work raises a number of questions. It suggests at the outset that cells age and die on their own, quite apart from any effect residing within the whole organism may produce. Aging is thus a cellular as well as an organismic phenomenon.

Since one of the general observations all investigators have made is that aging cells lose their ability to divide, cancer cells may provide a clue to the aging process, since cancer cells appear to be able to divide indefinitely in tissue culture. A famous line of human cancer cells named HeLa (after Helen Lane, the woman from whom they were derived after her death more than thirty years ago) is still being cultured for use in standardized cancer cell studies. The study of cancer has suggested that cells are not limited *out of necessity* in the number of times they can divide. What is more important, aging in noncancerous cells follows a regular pattern. Cells of each particular type have definite generational spans. Senescence and death of cells does not appear to be left up to random, or chance, processes. They appear to be programmed in some way into the cell.

What factors might detemine the regular aging of cells? Thought on this matter is divided into two schools. One claims that senescence is genetically programmed into the cell, though its rate can be altered by environmental circumstances. The other claims that senescence is largely a result of the accumulation of accidental changes which occur to cells over a period of time. The apparent regularity of the aging process is thus viewed as a result of statistics. Given enough cells and enough time, a regular pattern emerges from what are largely random, individual accidents to cells.

Let us consider the random accident idea first.

Aging as Accumulated Cellular Accidents

Random mutations may produce aging by causing damage to DNA molecules. Although the cell has DNA-repair mechanisms, mutations may occur too rapidly for them all to be repaired. Thus, as mutations accumulate in the body cells, the cells lose many of their functions, including eventually the function of division. The longer a particular cell lives, the more mutations it would be expected to accumulate, so that for every size and shape of organism there are average rates of deterioration of functional cells.

An interesting set of experiments has been performed by Ronald Hart of Ohio State University and Richard Setlow of the Oak Ridge National Laboratory in Tennessee. They measured the extent of DNA repair in cultured fibroblasts from seven species. The cells were first exposed to ultraviolet light to induce mutations. Of the species examined, human, elephant, and cow fibroblasts were almost five times more active in DNA repair than corresponding cells from mice, rats, and shrews. These results appeared to show an interesting correlation between average life span and efficacy of DNA repair mechanisms. On the other hand, the repair system of hamsters was intermediate between those of human beings and rats, although the hamster's average life span is very close to that of the rat. Setlow's experiments raise some interesting questions, but they do not seem to establish a strict correlation between DNA-repair mechanisms and aging.

Aging as Part of a Genetic Program

Bernard Strehler of the University of Southern California in Los Angeles has suggested that accumulated *loss* of genetic material could cause aging. It is known that for most cells there is considerable redundancy of genetic material: that is, there appear to be repetitions of the same DNA for most known genes. This means the cell does not have to rely upon a single copy of its genetic blueprint for any one trait. As cells age, however, Strehler noted a loss of considerable amounts of this redundancy. He observed that DNA extracted from the brains of 10-year-old beagles bound 30 percent less labeled *m*RNA than did DNA from young animals. This suggested to Strehler that DNA from older animals contained fewer sequences complementary to the RNA than the DNA from younger animals. How the loss occurs, however, is still a matter of much speculation.

Research on aging is just beginning to develop testable hypotheses. Part of this is a result of seeking to answer the question on a cellular as well as an organismic level. Understanding the underlying molecular processes that distinguish old from young cells may raise a host of questions that can be attacked by experimental methods. To date, the most profitable approaches to this long-standing and mysterious process have involved chemical studies at the level of the aging cell.

be very active in synthesizing proteins necessary for the early stages of development, it seems clear that the extended fibers and loops represent selected genes being transcribed. Studies of these chromosomal loops and chromosome puffs suggest strongly that genes in higher organisms are selectively turned on and off during development. The mechanism for this switching process is still a matter of conjecture.

Much of the evidence for gene activation and deactivation during growth and development has come from experiments using plants and animals in which the antibiotic actinomycin D was used to suppress the synthesis of *m*RNA. Without a continuing supply of *m*RNA, the production of enzymes and other protein molecules ceases. By this method, *m*RNA has been demonstrated to be involved in many aspects of growth and development. It has been shown, for example, that the cotyledons of peanut seeds begin synthesizing *m*RNA at the onset of germination, and within a week the content of *m*RNA per cotyledon doubles. Treatment with actinomycin D, however, severely impairs the synthesis of *m*RNA by the cotyledons. (Actinomycin D unites with the guanine of DNA and prevents the RNA polymerase molecule from using the DNA chain in the synthesis of *m*RNA.) Actinomycin D has also been shown to inhibit growth of soybean hypocotyls through suppression of *m*RNA synthesis. This antibiotic has been found to inhibit the synthesis of an enzyme (amylase) by germinating seeds of barley. Sensitivity to actinomycin D is often restricted to specific developmental stages. This suggests that the genes for RNA synthesis are being activated and inactivated in some manner.

Globulin

Further support for the hypothesis that growth and development in plants is based upon changing patterns of gene activation and deactivation has been provided by James Bonner and his colleagues at the California Institute of Technology. Different organs and tissues from pea plants were removed and each one placed separately into a solution containing C^{14}-labeled leucine (an amino acid). In a short time, the radioactive leucine was used by the organ or tissue in synthesizing protein. Each part of the pea plants was then ground up and the proteins were extracted. Using the radioactive leucine as a marker, the quantity of a protein known as globulin was determined for each preparation (Table 20.2). Very little globulin was found to be present in the cells of the roots, stems, and leaves. The protein content of the seed cotyledons, however, had as much as 9.3 percent globulin. Bonner and his co-workers concluded that pea cotyledon globulin was synthesized by the cells in the developing cotyledons but not in any significant amount by cells elsewhere in the plant.

Following this experiment, Bonner and his group sought to discover the mechanism by which the genes for making globulin are able to function in cotyledon

Table 20.2

Synthesis of Pea Seed Globulin in Different Organs of Pea Plants*

Organ	Pea seed globulin as percent of total protein
Cotyledon	4.7–9.3
Roots	0.18
Apical bud	0.15
Leaves	0.11

*After J. Bonner, R. C. Huang, and R. Gilden, *Proceedings of the National Academy of Science* 50, 1963, p. 893.

cells, but are apparently suppressed in all other cells. To analyze the manner in which the genes of the pea plant exert control over globulin synthesis, Bonner and his colleagues isolated chromosomes from young, growing cotyledons and from vegetative buds of pea plants. They ground the tissues in a blender to disrupt the cell walls and nuclear membranes. By repeated centrifugation, chloroplasts and other nonchromosomal organelles were removed, leaving the chromosomes virtually alone. The biochemical composition of the isolated chromatin was determined to be about 36 percent DNA, 37 percent histone (a type of protein), 10 percent RNA, 10 percent nonhistone protein, and small amounts of the enzyme RNA polymerase.

From their earlier experiments, Bonner and his colleagues knew that the cells of pea cotyledons synthesize globulin protein, whereas the cells of vegetative buds do not. They supplied additional RNA polymerase and the four nucleotide building blocks of *m*RNA to chromosome preparations of buds and cotyledons, thus producing two *m*RNA-generating systems (Fig. 20.33). Ribosomes were then added to each preparation to establish the necessary condition for protein synthesis by the isolated chromosomes. After 30 minutes, the chromosomes and ribosomes were removed by centrifugation and the proportion of globulin protein in each preparation determined. The newly synthesized proteins in the pea cotyledon preparation consisted of about 7 percent globulin protein (just a bit less than was present in the intact plant). As before, very little globulin protein was present among the abundant proteins synthesized by the pea buds.

It is clear from these experimental results that the procedures used to isolate chromosomes do not harm the gene or genes controlling the production of *m*RNA for globulin synthesis. In addition, the results demonstrate that while chromosomes from both tissues support the synthesis of *m*RNA, only those of pea cotyledons support the synthesis of the *m*RNA responsible for synthesizing globulin protein.

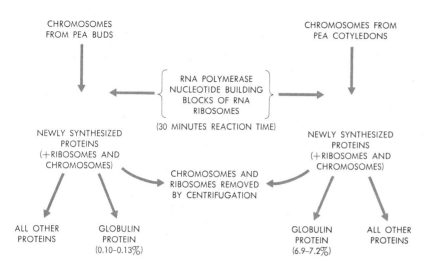

CHROMOSOMES
FROM PEA BUDS

CHROMOSOMES FROM
PEA COTYLEDONS

RNA POLYMERASE
NUCLEOTIDE BUILDING
BLOCKS OF RNA
RIBOSOMES

(30 MINUTES REACTION TIME)

NEWLY SYNTHESIZED
PROTEINS
(+RIBOSOMES AND
CHROMOSOMES)

CHROMOSOMES AND
RIBOSOMES REMOVED
BY CENTRIFUGATION

NEWLY SYNTHESIZED
PROTEINS
(+RIBOSOMES AND
CHROMOSOMES)

ALL OTHER
PROTEINS

GLOBULIN
PROTEIN
(0.10–0.13%)

GLOBULIN
PROTEIN
(6.9–7.2%)

ALL OTHER
PROTEINS

Fig. 20.33
Summary of Bonner's experiments on synthesis of globulin protein by chromosomes isolated from pea buds and pea cotyledons. (Based on J. Bonner, R. C. Huang, and R. Gilden, *Proceedings of the United States National Academy of Science* 50, 1963, p. 893.)

Histone

Thus we can hypothesize that the gene for globulin-making in the pea bud chromosomes must be suppressed in some way. But what is the suppressing agent? Since histone is present in the chromosomes in essentially the same quantity as DNA, this protein is a logical choice for the suppressor. If the histone is functioning in the suppression of the globulin-synthesizing gene system in the pea bud chromosomes, then chromosomes from which histone has been removed should be able to synthesize globulin. Bonner and his co-workers designed an experiment to verify this prediction. They dissociated the histone and other proteins from the chromatin preparation and separated out the heavier DNA by centrifugation. The pure DNA remaining was then used to support the synthesis of *m*RNA and ribosomal proteins in the test tube. In the absence of histone, the bud DNA now supported as much pea globulin synthesis as did the DNA from the cotyledon cells. These results strongly support the histone-gene suppression hypothesis. Thus histone could be the long-sought suppressor for the synthesis of globulin protein.

While the histone hypothesis of gene suppression is very attractive, in actuality the problem has simply been moved at least one step backward and we must now ask: What controls the histones? There is some evidence supporting a hypothesis that the histones might be under the control of some small molecule coming from outside the cells, just as the genes of the bacterium *E. coli* are influenced by the presence of lactose in their environment (see Section 19.8). Bonner and his group discovered that the buds on freshly harvested tubers of the potato plant do not grow, even if supplied with favorable conditions. Such structures are said to be dormant. When chromosomes are isolated from the cells of these dormant buds, they synthesize very little RNA, even if additional RNA polymerase is supplied. However, after treatment with gibberellic acid, a naturally occurring growth-promoting substance (see Section 9.5), isolated chromosomes supplied with the proper enzymes and building blocks synthesize RNA rapidly. It can be hypothesized that the gibberellic acid molecules may chemically bind to the histones and somehow act to prevent the histones from suppressing the activity of the genes for *m*RNA synthesis. Thus, the dormancy of the freshly harvested potatoes might be broken, enabling them to commence growth. The relation of histones to the regulation of gene action is currently one of the most challenging and controversial topics in developmental biology.

It cannot be emphasized too strongly that modern developmental biologists no longer seek to understand (or explain) processes such as embryonic differentiation, regeneration, or aging in terms of either genetic or environmental causes alone. It is clear that to varying extents, both are involved. The phenotype of any organism is the result of a constant interaction between genome and environment during adult life as well as embryonic development. No organism, no developing embryonic cell, is without a genome; nor is any organism or cell without an environment. Development is an integrated and interactive process between the two.

20.12
THE IMMUNE RESPONSE AS
A DEVELOPMENTAL SYSTEM

Throughout most of this chapter we have examined the ways in which embryological development is thought to occur, focusing mostly on the anatomical or morphological features of developing organisms. However, other less-evident attributes of living organisms are also subject to developmental processes. For example, biologists today recognize the behavior of an organism to be as much a product of developmental processes as an anatomical structure; we shall have more to say about this in Chapter 30. In the present section, however, we

SUPPLEMENT 20.3
CANCER: EMBRYONIC DEVELOPMENT IN REVERSE

Cancer is a major cause of death in the United States today. It affects all ages and all socioeconomic groups, though statistically people between the ages of forty and sixty are the hardest hit. What is cancer, and what have we learned in recent years about its cause and its cure?

Cancer is a disease in which cells lose the ability to control the basic metabolic processes, including their rate of growth. Cancer cells grow rapidly, producing clusters of cells called tumors. Cancer cells also break away from the tissue of which they are a part and drift off throughout the body producing secondary tumors in a process known as *metastasis*. Metastasis is a particularly difficult problem because it provides a means by which cancer spreads to all areas of the body. Once a tumor has begun to metastasize, it becomes increasingly difficult to treat.

What are the characteristics of cancer cells, and how do those characteristics help us in trying to find a cure for this disease?

1. *Metabolism.* Most cancer cells have a very high metabolic rate, consuming large amounts of glucose and producing large quantities of waste products, including lactic acid. Cancer cells also spend a considerable amount of metabolic energy in random protein synthesis, especially the synthesis of proteolytic enzymes. In general, the metabolic processes associated with cancer cells are those characteristic of any relatively nonspecialized tissue.

2. *Cell Growth and Reproduction.* Cancer cells lack the usual controls on mitosis and cytokinesis characteristic of most adult tissue. Given sufficient nourishment, they divide almost continuously. Such unmitigated reproduction has several negative effects on the whole organism: it uses large quantities of the body's energy reserves, making these unavailable for other vital functions; it reduces the ability of the affected tissue or organ to carry out its main function; and, in many cases, it produces mechanical problems, such as blockage of the gastrointestinal tract, or generates pressure (as in brain tumors), either of which can lead to additional problems of their own.

3. *Cell Surface.* Both microscopic and chemical analysis reveal that cancer cells have quite different surface (cell membrane) properties than normal cells. Scanning electron micrographs reveal the physical differences dramatically. Cancer cells have many blebs, or projections, on their surfaces, whereas normal cells are smooth when cultured outside the body. Cancer cells tend to grow in layers, whereas normal cells spread out in a single sheet across the culture plate. And, finally, cancer cells in culture tend to be more rounded in overall shape, whereas normal cells are flatter. All of these differences appear to be due to altered cell membrane properties in cancer cells. Uncontrolled protein synthesis means that many unusual proteins and protein derivatives are present in the cell and can become embedded in the membrane. Cancer cell membranes are rich in mucopolysaccharides (repeating double glucose units with an amine group) and glycoproteins (a combined carbohydrate and protein). At the same time, many of the cell's normal membrane receptor molecules are crowded out, or pushed down to the inner surface of the membrane.

4. *Cell-Cell Interactions.* Added to these changes is a modification in all behavior of cultured cancer cells called contact inhibition. When normal cells in culture medium come into physical contact with one another, their movement slows down, they tend to adhere to each other, and their mitotic rate decreases. Experimental evidence demonstrates that all these changes are mediated through cell surface contact. Cancer cells are not inhibited in any of these activities by surface contact. Their lack of contact inhibition leads to the formation of multiple layers of cells, the assumption of peculiar cell shapes and membrane configurations (blebs), and continous cell division. In all these ways cancer cells are very similar to embryonic cells of late blastula and early gastrula stages. Recall that in these stages cell movement (morphogenetic movements) is an important process in generating the three main germ layers (endoderm, mesoderm, and ectoderm). Embryonic cells in the late blastula and early gastrula, like cancer cells, also lack the contact inhibition characteristic of more mature, differentiated cells.

From the description above it should be clear that cancer cells have many characteristics in common with embryonic cells. These shared traits have led many biologists to view cancer cells as a product of cell de-differentiation, that is, a reversal of the embryonic process. What causes this reversion of cancer cells to the embryonic state is not known. Oncologists—medical researchers who study cancer—are convinced there is probably no single cause of the disease. Certain individuals and family lines may show a propensity to cancer, but it is also clear that many environmental factors may determine the degree to which that propensity is realized.

In recent years viruses have been implicated as causative agents in many common forms of cancer. Many mammalian cancer cell lines in culture have been shown to have a portion of a virus genome inserted into the cell's own DNA, as a provirus (see chapter 19). In addition to virus insertion into the genome, other alterations of the genetic machinery of cells by environmental factors such as chemical mutagens and radiation have been shown to result in cancer. In addition a weakened immunological response may also result in proliferation of cells normally kept in check.

It is clear that finding methods of treating cancer will become one of the most significant medical breakthroughs of the present century. To do so will require, and involve, the solution to a number of basic biological problems: the genetic control of development (including rate of mitosis), and the nature of the immune response, to name only a few. Indeed, an understanding of the nature of cancer lies at the very heart of modern biology.

shall look in some detail at the development of the immune system (first introduced in Section 10.6) that protects animals against disease. The immune response is currently becoming one of the most useful systems for understanding the cellular and molecular basis of development.

Most organisms have some mechanism for warding off infection by agents from the outside. Bacteria use restriction enzymes to cut up the DNA of virus invaders. Insects secrete a generalized antibacterial protein to ward off bacterial infection. Vertebrates, and particularly mammals, respond to foreign invasion by synthesizing a special group of proteins called immunoglobulins which bind to the invader and help to remove it from the body. This process is called the immune response, and is a function of the body's immune system. The mammalian immune system is the most complex and elaborate defense mechanism that has evolved in the animal kingdom. The study of the immune system—its cellular, biochemical, and genetic aspects—comprises the science of **immunology**.

Although biologists have been studying various aspects of immunology for almost 200 years, new and exciting developments have occurred particularly rapidly in the past decade. While the study of immunology has been vital to understanding medical problems (such as allergies and organ transplantation), it has also begun to throw light on a large number of problems in the area of developmental biology. For example, the entrance of foreign material into the body triggers special cells in the immune system to the synthesis of proteins, not just any proteins, but proteins whose molecular shape fits precisely that of molecules on the surface of the invader. This process has a counterpart in embryonic development and in enzyme induction in bacteria; in all three cases genes in particular cells are "turned on" or "turned off" by interacting with some molecules outside the cells in question. How genes are triggered so specifically is one of the central questions in both immunology and developmental biology.

The Immune Response: Overview

During the latter half of the eighteenth century, smallpox ravaged much of Europe. Those who were not killed by the disease were left with "pocks" or pits on their bodies. In the 1780s a young London physician, Edward Jenner, noted that those who survived the disease never contracted it again. Jenner and others had concluded that exposure to smallpox induced the body to become resistant, or *immune* to the disease. Jenner had also made another interesting observation. There was one group of people who seemed immune to smallpox though they themselves had never had it. These were milkmaids, young women who milked cows. Cows suffer from a disease similar to but much milder than smallpox, called (appropriately) cowpox. Jenner noted that milkmaids often got pus from pox sores on the cows' udders into cuts on their hands. He reasoned that exposure to cowpox had rendered the young women immune to smallpox without actually giving them a serious case of the disease (we now know that smallpox and cowpox are caused by two different viruses, variola and vaccinia, respectively).

In 1796 Jenner performed a bold experiment. He took pus from a milkmaid who had contracted cowpox, and using a needle scratched it into the skin of an eight-year-old boy. In two months he injected the boy with an otherwise lethal dose of smallpox. (One wonders why a child was used, and who the child was.) After two months no disease had developed. Jenner went on to develop a general procedure for producing artificial immunity, a process which became known as "vaccination," named for the cowpox disease.* The process of "vaccination" became widespread throughout England, and curbed future outbreaks of the disease.

Consider a second case. In the summer of 1979 a medical student was admitted to a hospital in Philadelphia complaining of a rash on her skin. The doctors could find nothing wrong, and assumed she had simply picked up a mild allergy. The medical student went home but returned to the hospital the next morning with the rash now covering her entire body. She was feeling somewhat nauseous, and was admitted to the hospital for further examination. No one had a clue to the cause of the rash, but the patient's condition worsened as the day progressed. She died less than twenty-four hours

---

*The word "vaccination" itself is derived from the French *vache*, which means cow. After Jenner's discovery, vaccinations against smallpox were quite common in England. The French, looking down on most things English, coined the term *vaccine*, which means "cowlike." The reference is, of course, to the fact that material from cows was injected into humans, thus "cow-izing" them.

later. The final diagnosis was that the young woman had developed an immune reaction to her own skin. Her body was busy making immunoglobulins against the cells of her skin, causing the latter to be rejected as if it had been a transplant from another individual. Such a process is known as **autoimmunity,** meaning that the immune system begins to reject the self (*auto* = self). Fortunately, autoimmunity is relatively rare, though there are a number of known autoimmune diseases that affect human beings.

The brief case histories described above highlight two sides of the coin of the immune process. On the one hand, the immune system is able to recognize foreign material in the body and synthesize specific immunoglobulins against it. In this sense one of the system's major functions is to recognize "nonself." On the other hand, the system must also be able to recognize itself, and *not* build antibody against its own tissues. In other words, the immune system must be able to distinguish between self and nonself.

As an aspect of developmental biology, the study of immunology raises a number of important questions. For example, in what part (or parts) of the body does the immune response take place? How is the response triggered? What is the chemical and molecular nature of the immune response? How is the immune response genetically determined? By what chemical and/or molecular mechanism does the immune system distinguish between self and nonself? How does the immune system develop during the individual's life? And, finally, what is the relationship between the immune response and other developmental/physiological problems people encounter, such as allergy, autoimmunity, rejection of grafts and organ transplants, and cancer? These questions will form the focus of our discussion of the immune system and its functioning.

Definitions. Let's begin by reviewing definitions. When a foreign substance invades the body, it can stimulate cells of the immune system to produce a specific class of proteins called **immunoglobulins.** Immunoglobulins specifically synthesized against a foreign substance are known as **antibodies.** Antibodies produced to any foreign substance are highly specific in their molecular configuration and are able to bind to the foreign substance with a kind of lock-and-key fit.

Foreign substances which interact specifically with an antibody are called **antigens.** Antigens can be anything from a single molecule to molecular surface markers on the membrane of a bacterial cell or the protein coat of a virus. In short, any substance that can combine with an antibody is called an antigen.

In most cases, antigens are also responsible for triggering the immune system to begin synthesizing specific antibodies against themselves. However, there are some substances that can bind to antibodies yet cannot stimulate the immune system to produce those antibodies.

Morphine is one such example. If injected in a pure form into the blood, morphine will not trigger the immune system to make antibodies against morphine. However, if the morphine is first combined with a larger molecule such as a protein and then injected, the immune system will make antibodies specific to both the protein and to the morphine portion of the morphine-protein complex. By combining with antibody, morphine acts as an antigen. But because it cannot elicit antibody production on its own, it is called a **hapten,** rather than an antigen.

The first time a particular antigen invades the body, it elicits a **primary immune response,** which is usually slow to develop (several days to several weeks). During the primary response, antibody specific to the antigen is being slowly synthesized and built up with the organism. If the same antigen enters the body at some later date, in what is called a secondary invasion, it evokes a **secondary immune response.** The secondary response is more rapid and intense than the primary, producing antibodies in a matter of one to a few days. All future invasions by the same antigen produce responses similar to the secondary response.

Cellular Basis of the Immune Response

The cells that synthesize specific antibodies in response to presence of an antigen are the lymphocytes, those blood cells found in a variety of places in the mammalian body: the lymphoid tissues of the lymph nodes, spleen, bone marrow, thymus, and liver. Lymphocytes are derived from precursors called hemopoietic stem cells, located in the bone marrow of adults. As shown in Fig. 20.34, hemopoietic stem cells differentiate into a number of types of blood cells: erythrocytes (red blood cells), lymphocytes, macrophages, and three types of white blood cells. All these cells are circulating throughout the blood and lymph systems. White blood cells, and especially the type of white blood cell known as macrophages, engulf foreign matter such as bacterial cells, or antibody-bound antigens, and thus remove them from the system. Indeed, the name "macrophage" comes from two Greek terms meaning "large" (*macro*) and "eater" (*phage*), describing appropriately the voracious appetite these cells have for debris or foreign matter in the blood.

There are two types of lymphocytes: the so-called B-lymphocytes and T-lymphocytes. B-lymphocytes, which are produced from lymphoid stem cells in the bone marrow, reside in lymph nodes and pass throughout the general circulation. T-lymphocytes, which also arise in bone marrow, mature in the tissue of the thymus gland, and reside in the thymus; but, like the B-lymphocytes, T-lymphocytes move constantly through the blood and lymphatic systems. B-lymphocytes are responsible for what is called the **humoral immune response,** while T-lymphocytes are responsible for the

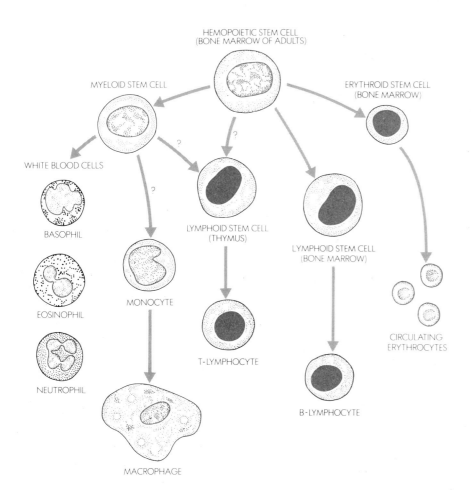

HEMOPOIETIC STEM CELL
(BONE MARROW OF ADULTS)

MYELOID STEM CELL

ERYTHROID STEM CELL
(BONE MARROW)

WHITE BLOOD CELLS

BASOPHIL

EOSINOPHIL

NEUTROPHIL

MONOCYTE

LYMPHOID STEM CELL
(THYMUS)

LYMPHOID STEM CELL
(BONE MARROW)

T-LYMPHOCYTE

B-LYMPHOCYTE

CIRCULATING
ERYTHROCYTES

MACROPHAGE

Fig. 20.34
The origin of the various blood cells from hemopoietic stem cells. The hemopoietic stem cells differentiate into at least four types of specialized cells: erythroid stem cells, lymphoid stem cells of the bone marrow and of the thymus gland, and myeloid stem cells. These, in turn, give rise, respectively, to erythrocytes (red blood cells), B-lymphocytes, T-lymphocytes, and the various white blood cells including macrophages, neutrophils, eosinophils, and basophils. All the cells involved in the immune response thus share a common origin.

cellular immune response. What are the functions of these two different immune responses?

B-lymphocytes are stimulated by freely circulating antigen to undergo a series of changes that eventually lead to the synthesis of specific antibodies (the details of this response will be discussed in the next section). The antibodies are released into the blood and pass throughout the circulatory system. When they encounter the specific antigen that stimulated their production, the antibodies bind to the antigen forming a kind of precipitate or antigen-antibody complex.

T-lymphocytes are stimulated to produce specific antibodies by cell-bound antigens that invade the circulatory system. Unlike the antibodies produced by B-lymphocytes, those produced by T-cells remain bound to the lymphocyte cell surface. T-cells thus bind antigens to their own cell surfaces, forming larger complexes of cells (the details of this process will be discussed in a later section).

Several experiments provide insight into the different roles of B- and T-lymphocytes in the immune response. Because of their rapid division rate, all lymphocytes are highly sensitive to X-radiation. If we remove the thymus gland from an animal (thus removing its source of T-cells), and then expose the animal to X-radiation (killing off any already circulating T-cells, plus all the B-lymphocytes), we would predict that the animal should display no immune response in the future—to either molecular or cellular antigens. This prediction is borne out: such animals show neither humoral nor cellular immune response. If we then re-inject T-cells into the animal, the cellular response is restored, but the humoral response is not. These data strongly suggest that T-cells control cellular immunity.

The above experiment also gives support to the idea that B-cells control humoral immunity. We can test this hypothesis by removing the thymus from an animal that is then irradiated and injected with B-lymphocytes. Here, we would predict that humoral immunity would be restored, but cellular would not. Our prediction is not completely borne out, however; while there is no restoration of cellular immunity (as predicted), there is only partial restoration of humoral immunity (an unexpected result). "Partial restoration" means in this case that the animal can make free-floating antibodies to some antigens, but not to all.

These data have been interpreted as follows: while B-lymphocytes largely control the humoral immune response and T-lymphocytes the cellular response, the two systems must interact in a very particular way. Full B-cell activity appears to require the presence of T-cells (the converse is not true, however—full T-cell activity seems to take place whether B-cells are present or not). In some way the presence of T-cells appears to facilitate the function of B-cells. Recent research has shown that there is a category of T-lymphocytes called "helper T-cells," that help B-lymphocytes respond to particular antigens. There is also a category of "suppressor T-cells" that can, under certain conditions, suppress the response of B-lymphocytes to antigens. It is thus obvious that the humoral and cellular immune systems interact with one another and that neither has exclusive control over one or the other aspect of the immune response.

The basic principles of the immune response apply, with minor modifications, to both the humoral and cellular systems. For simplicity, we will discuss these principles using the humoral system as an example. Later, we will compare the systems in some detail.

How Antigens Interact with Lymphocytes to Stimulate Antibody Production

As already mentioned, there are literally millions of different types of substances that can act as antigens. Regardless of the wide variety of molecular shapes that these various antibodies display, most can serve to trigger the production of antibodies. This triggering appears to be the result of the interaction of specific parts of the antigen with receptor sites on the surface of the lymphocyte. Antigen molecules have specifically recognizable regions of the molecule (whether the molecule is a protein, which most of them are, or a different type of molecule such as a carbohydrate) called **antigenic determinants.** In protein antigens, these determinant regions are usually about ten amino acid units in length. There may be two or more determinant sites, each of a different structure, on each antigen; and the determinant sites on different antigens may be very different from one another. It is important to remember that the part of an antigen—that is, the determinant region—recognized by the body's lymphocytes is usually small. In other words, lymphocytes do not necessarily recognize antigens as totalities, but rather as specific determinant regions.

When an antigen invades the body, how does it interact with lymphocytes? As mentioned before, we will focus our attention for the moment only on the humoral immune response. As shown in Fig. 20.35, B-lymphocytes contain specific receptor sites on their surface—apparently embedded in the cell membrane. These receptors are very specific antibodies that the lymphocyte is genetically programmed to produce. It is thought by most immunologists today that each B-lymphocyte in the lymphoid tissue is able to produce one, and only one, type of antibody. The lymphocyte can be thought of as wearing a sample of its antibody on the

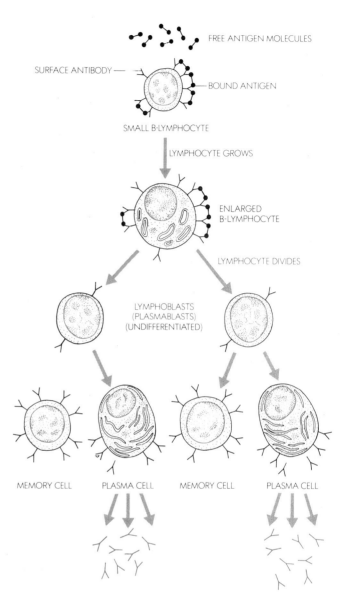

Fig. 20.35
Diagram of a B-lymphocyte being stimulated by presence of an antigen to undergo growth and differentiation into two types of specialized cells, memory cells and plasma cells. Plasma cells synthesize antibody of the type specific for that B-lymphocyte cell; memory cells are poised for action should a secondary invasion by the same antigen occur.

surface, as a sort of "advertisement" that traps the unsuspecting antigen "customer." The cells are lying in wait, so to speak, for an antigen whose determinants will fit the surface antibody marker. As the blood and tissue fluid carries antigens past these cells, the lymphocyte that displays an antibody specific for a particular antigen will bind the antigen. This binding is due to

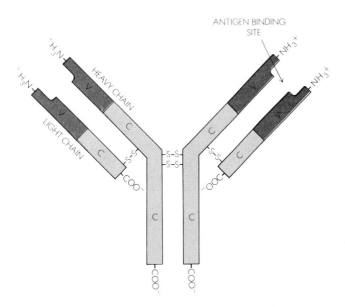

Fig. 20.36
Schematic representation of an antibody molecule. All antibodies are tetramers of four polypeptide chains: two longer, heavy chains, and two shorter, light chains. The chains are held together by four disulfide (S—S) bridges. Each heavy and each light chain is divided into two functionally different regions: the constant region (symbolized C) and the variable region (symbolized V). The constant region is similar from one type of antibody molecule to another, whether the antibodies are to different antigens or belong to different classes. The variable region differs from one antibody molecule to another. The variable region gives each antibody type its specific uniqueness for a given antigen. At the amino end of the two Y-arms of the antibody is the specific antigen-combining site.

chemical bonds formed between the antibody and the determinant region on the antigen.

Once the antigen has bound to the lymphocyte surface, the cell undergoes growth and then division (see Fig. 20.35). How the binding of the antigen to surface antibodies triggers this cell response is not yet understood. As Fig. 20.35 shows, the enlarged lymphocyte divides first into two undifferentiated cells, the lymphoblasts or plasmablasts. Each of these cells then undergoes a division into two different types of cells: a **plasma cell** and a **memory cell**. Both types of cells are capable of synthesizing and secreting the specific antibody that the original B-lymphocyte contained on its surface. The plasma cells are the ones that proliferate and secrete large quantities of antibody into the bloodstream. The memory cells do not secrete antibody, but remain "on call." If the same antigen appears in the body again at a later date, they can begin proliferating and secreting antibodies rapidly. The presence of the memory cells ac-

counts for long-term immunity. On a second invasion by an antigen, the memory cells are primed. They do not have to undergo the same kind of prolonged growth and differentiation characteristic of B-lymphocytes in a primary invasion. Thus, the response to a secondary invasion can be more rapid and effective than the response to a primary invasion.

How specific is the antibody response by B-lymphocytes to particular antigens? Consider the case of a very simple antigen, polyglycine. This molecule is nothing more than a group of glycine molecules joined together to produce a small polypeptide chain (composed, as the name polyglycine implies, of many glycines). When polyglycine is injected into an animal, several different antibodies are produced in response. Each antibody appears to be synthesized by a different plasma cell. But even a very simple antigen, which can have only a single determinant region (since it is made up of the same amino acid), can nonetheless elicit the synthesis of a heterogeneous group of antibodies. The simplest explanation for this phenomenon is that even a simple antigen can be recognized by several different B-lymphocytes, each of which proliferates and secretes slightly different antibodies. As you might imagine, the more complex the antigen, the more heterogeneous the population of antibodies produced. Seldom, then, is it possible to inject one antigen, however simple, and expect to produce only one antibody.

The examples above do not contradict the idea that antibodies are specific to the antigens that stimulate their synthesis. Each of the antibodies produced in response to injection of polyglycine binds specifically with polyglycine. The main point is that several different antibody types may bind specifically (though some more readily than others) with a single antigen. This is especially true for complex antigens composed of several molecules bound together. In such cases one or more different antibodies are made for each component of the molecular complex. In other words, one type of antibody is made for each antigenic determinant site.

Antigen-Antibody Interactions

The Structure of Antibody Molecules. Regardless of what antigen stimulates their secretion, all antibodies have a similar molecular architecture. This is as true of the antibodies produced by T-lymphocytes as of those produced by B-lymphocytes. Fig. 20.36 shows a schematic view of a typical antibody molecule, as elucidated in the early 1970s by Rodney Porter and his co-workers at Oxford University in England. Each molecule consists of four polypeptide chains, two lengthier "heavy" chains, and two shorter "light" chains. The chains are held together by disulfide bridges. The amino ($-NH_2$) end of each chain is known as the "variable" region, while the carboxyl end is known as the "constant" region. The constant regions of all antibody molecules are more nearly the same in amino acid sequence, whereas the variable regions are, as the name implies, quite different from each other in each type of antibody. The variable region of each antibody molecule is specific for the kind of antigen against which the antibody acts. The constant region of the antibody chains plays a role in the molecule's overall functioning, too. The constant region interacts with other molecules in the blood and helps to make the antigen-antibody complex more easily recognizable to macrophages. The constant region also facilitates the transport of antibodies across cell membranes.

The variable region of the antibody chain is much like the active site of an enzyme molecule. It is specific for a particular antigen. This means that the antigen can not only fit into, but can bind with, the antibody molecule. Here the analogy to an enzyme-substrate interaction ends, however. The antibody binds the antigen molecule over 1000 times more strongly than an enzyme binds a substrate, and the interaction is largely irreversible. The slight differences in structure, then, between an antibody produced in response to antigen A and one produced in response to antigen B, resides in the slightly different structures of the active, or combining site on the two Y-arms of the molecule. How these combining sites are synthesized specifically to fit particular antigens will be discussed farther on.

When an antibody encounters the antigen for which it is specific, the antigen binds to the antibody's combining site. This leaves part of the antigen molecule protruding from the combining site, as shown in Fig. 20.37(a). One free Y-arm of another antibody molecule of the same type can now combine with the free region of the antigen, thus binding the antigen from both sides, as shown in Fig. 20.37(b). Interaction with a third antibody molecule can thus bind three antigens, producing a triangular complex such as that shown in Fig. 20.37(c). That these binding figures actually occur is confirmed by the unusual electron micrograph shown in Fig. 20.37(d). The formation of these lattice-works produces large antigen-antibody complexes that become more easily recognized by circulating macrophages. The formation of antigen-antibody complexes initiates the clean-up responses by macrophages and other white blood cells.

The clumping of antigens by formation of antigen-antibody complexes is called **agglutination**. Recall that a similar process is observed when red blood cells come in contact with an incompatible serum. The clumping of red blood cells is due to the antibodies in the plasma binding to the antigens on the red blood cell surfaces. Because each antibody in the serum can bind to two or more antigens, often on different cells, the cells can be held together in a tightly bound complex. The agglutination or clumping of red blood cells is observed in blood-typing tests.

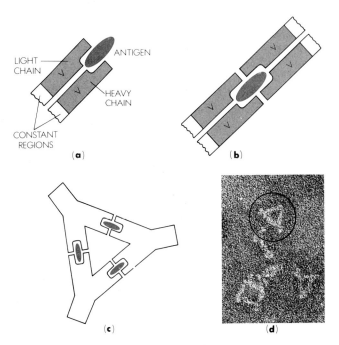

Fig. 20.37
The formation of an antigen-antibody complex. (a) The variable regions of a light and heavy chain on one of the Y-arms of an antibody molecule are shown bound to an antigen (color). The antigen fits specifically into the antigen-combining site, much like a substrate into the active site of an enzyme. The binding between antigen and antibody is much stronger than that between enzyme and substrate. (b) Binding of antigen by two adjacent antibody molecules (one arm of each Y-shaped molecule is shown here). (c) The binding of antigen molecules by adjacent antibody molecules can lead to the formation of complexes such as the triangular forms shown here. Such complexes form precipitates, or agglutinations, that help signal passing macrophages to swallow up the complex. (d) Electron micrograph showing the actual formation of the triangular antigen-antibody complexes diagrammed in (c). (Micrograph courtesy N. M. Green of the National Institute for Medical Research, London, and the late Robin Valentine.)

Classes of Immunoglobulins. Immunologists distinguish between five different classes of immunoglobulin molecules, based on differences in the constant regions of their heavy chains. These five classes are designated by the symbols Ig (for "immunoglobulin"), followed by a letter indicating the class—for example, IgA, IgG, IgM, IgD, and IgE (see Table 20.3). There is no necessary relationship between the kind of antigen against which an antibody is made and the class of immunoglobulins to which it belongs. Thus, antibodies to a very large range of antigens can all belong to the same immunoglobulin class. Conversely, antibodies belonging to all five classes can be generated in response to a single antigen. The different classes of immunoglobulins

Table 20.3

Immunoglobulin class	Where synthesized and/or localized	Major characteristics and functions
IgM	B-lymphocytes and T-lymphocytes in lymph nodes and other lymphoid tissues	The predominant class of immunoglobulin found in blood during the primary immune response. IgM is a pentamer consisting of five Y-shaped units bound together; IgM is relatively unstable and cannot pass from blood into body tissues. Its major function is to provide early protection against invaders that, at least initially, are localized in the bloodstream.
IgG	B-lymphocytes and T-lymphocytes in lymph nodes and other lymphoid tissues	IgG usually replaces IgM in the bloodstream during late part of a first infection, and throughout a secondary infection by the same antigen. IgG is a monomer, produced in larger quantities than IgM, is more stable, and accumulates in the bloodstream to a greater extent than IgM.
IgA	B-lymphocytes in the lacrymal (tear), mucous, and salivary glands	Since it is found in the mucous film lining the respiratory, urogenital, and digestive tracts, IgA is the "first line" of defense against reinvasion by a previously encountered antigen. It is also present in mothers' milk and thus provides infants with certain immunities before their own immune system starts to function. IgA is a dimers (two Y-shaped molecules bound together).
IgD	Synthesized by cells in lymphoid tissue lining intestine; found on B-lymphocytes throughout body	Serves as principal receptors on surface of all B-lymphocytes in body. Also may function to ward off re-invasion by an antigen through the intestine (this latter role is only speculative). IgD is a monomer. Less is known about IgD than the other classes of immunoglobulins.
IgE	Synthesized by B-lymphocytes; found principally on surface of mast cells	Principal immunoglobulins involved in allergy reactions. IgE molecules attach permanently to mast cell surfaces; when triggered by contact with their appropriate antigen, they cause mast cells to release their contents, including histamine. The latter causes capillaries to dilate and become more permeable. This produces the "runny" nose and eyes associated with hay fever and other allergies. IgE molecules are monomers.

differ primarily in three characteristics: (1) the time sequence in which they appear in the immune response; (2) the location in the body where they are produced and where they function; and (3) the consequences that occur after they bind to antigens. The five classes of immunoglobulins and their respective functions are summarized in Table 20.3.

The Specificity of Antibodies to Antigens

Immunologists have long wondered how lymphocytes are able to synthesize antibodies specific to whatever particular antigen happens to enter the body at any given time. The basic problem with which researchers have had to contend is a theoretical one: since there are literally millions, if not billions, of foreign substances that could possibly enter the body, how can the immune system produce as many specific types of antibodies as there are possible antigens in the world? In response to this dilemma two hypotheses have emerged in the past

several decades: the **instructional theory**, and the **clonal selection theory**. Let us examine these two theories and their predictions in some detail.

The Instructional Theory. The instructional theory begins with the notion that lymphocytes are genetically programmed to synthesize only rather generalized antibody molecules—that is, the amino acid sequence of the antibody molecule does not predispose it to fit any particular antigen. The "fit" is determined by the presence of the antigen itself. According to the instructional theory, when an antigen comes in contact with a lymphocyte, it serves as a sort of mold or template, causing a shift in the conformation of the antibody molecule to fit the structure of the antigen. In other words, the antibody protein is "instructed" by the antigen to fold into a highly specific tertiary structure, one that is specific to the particular antigen in question. The chief attraction of the instructional theory is that it avoids the necessity of the body having millions of genetic messages programmed for every conceivable antigen. There is a kind of genetic economy to the instructional theory.

If the instructional theory is correct, then we might expect antibody molecules to have an enormous amount of flexibility in the secondary and tertiary structure of their polypeptides. That is, we would expect their final three-dimensional shape to be determined in each case specifically by the shape of whatever antigen is present when the antibody molecule takes its shape. We can perform a simple experiment to verify this prediction. If we take a sample of antibody molecules all synthesized in response to the same antigen and gently heat-denature them (so that the proteins unfold from their tertiary and secondary configurations), and then allow them to refold in the absence of antigen, we find that they all take the same shape—the shape specific for the antigen that elicited their production in the first place. Furthermore, the antibody molecules take the same specific shape even when allowed to refold in the presence of a different antigen. These results are in agreement with the notion, discussed in Chapter 2, that the three-dimensional structure of proteins is determined largely by the sequence of amino acids, and is therefore preset in the primary structure of the molecule.

The Clonal Selection Theory.   According to the clonal selection theory, B-lymphocytes are each genetically programmed to synthesize a single, specific antibody, different from that synthesized by other lymphocytes. Each B-lymphocyte wears on its surface a "sample" of the antibody it can synthesize; these samples act as the

receptor sites for antigen molecules. Antigens interact with the one (or the few) B-lymphocytes that bear the antibodies specific to them. Each stimulated lymphocyte grows and divides, producing a clone of cells that begin synthesizing an antibody. An outline of the clonal selection theory is shown in Fig. 20.38. Note that the theory is called clonal *selection* for a good reason. The circulating antigen *selects* the appropriate lymphocyte out of the billions it must encounter by binding specifically to the antibody displayed on the cell's surface. Since each lymphocyte wears its own unique antibody on its surface—the very antibody it has the genetic program to synthesize—the antigen has little trouble selecting out the appropriate cell to which it can bind. The result of the selection process is a clone of cells that now pro-

Fig. 20.38
Diagrammatic representation of the clonal selection hypothesis. Out of a very large number of small B-lymphocytes, each with its own unique, specific antibody receptor on the surface, one is stimulated by an antigen that fits its antibody. The one selected lymphoctye (F, in this case) grows and divides, forming plasma (P) and memory (M) cells. Plasma cells produce F antibody in large quantities. Memory cells remain in the circulatory system (especially concentrated in lymph nodes), poised to begin rapid synthesis of F antibody in the event that F antigen enters the system again. All the plasma cells producing antibody F are derived from the same original lymphocyte, and thus form a clone.

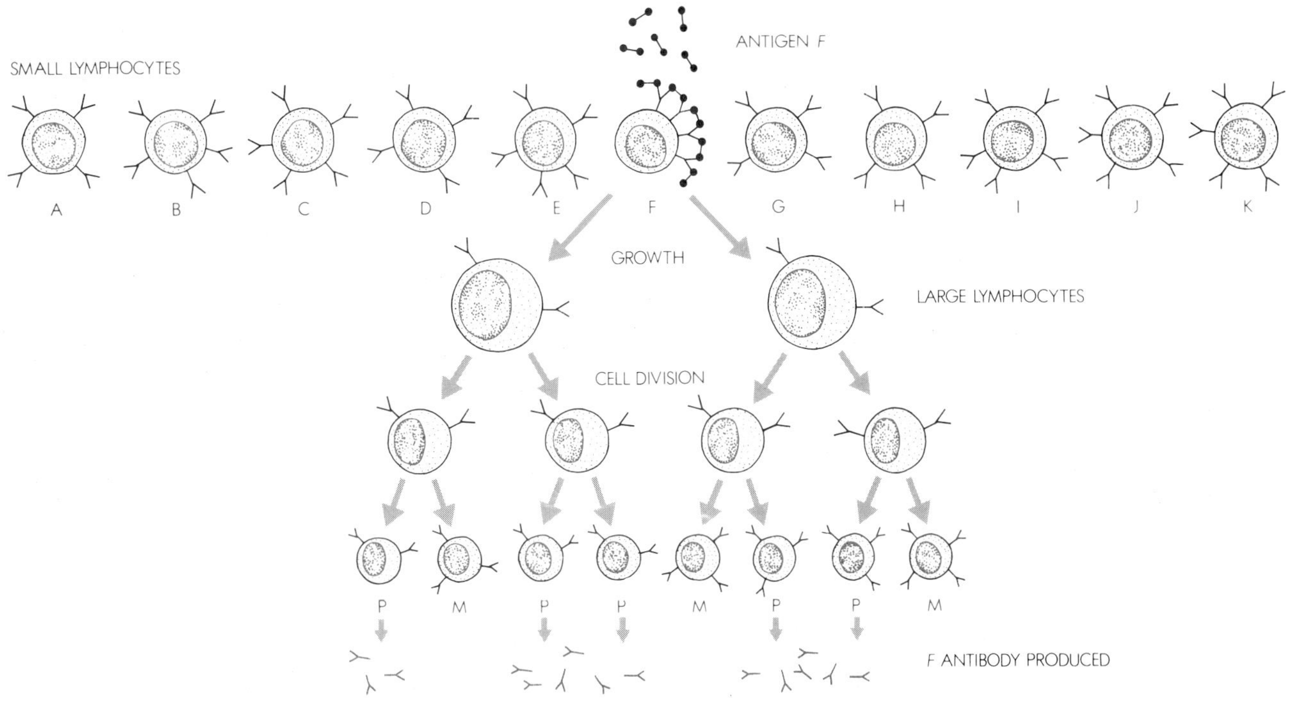

duces large quantities of antibody against the specific antigen. The formation of a clone merely amplifies the antibody-producing capability of the B-lymphocytes.

Several lines of evidence support the clonal selection hypothesis. Suppose we inject an animal with enough radioactively labeled antigen to kill the lymphocyte cells that take it up. If the clonal selection theory is correct, we would predict that after the injection such an animal would lose the ability to produce antibodies to that particular antigen. This prediction is verified. Interestingly, the animal does *not* lose the ability to produce antibodies against other antigens. These results suggest that each type of lymphocyte is preprogrammed to generate antibodies against only one antigen.

Another experiment verifies this conclusion. If we pass a heterogeneous sample of lymphocytes down a column of glass beads to which a specific antigen has been attached, we would predict that the cells that finally pass out at the bottom should be unable to react with the antigen present in the column. That is, the beads should bind to, and thus trap, the lymphocytes that carry on their surface antibodies specific for that antigen. Those lymphocytes should not come out at the bottom of the column, whereas all others should. The prediction is borne out: lymphocytes that pass through the column will not agglutinate or form the typical clumping response when exposed to the antigen present in the column.

To summarize, then, the clonal selection theory states that lymphocytes are genetically programmed to produce one type of antibody, whose primary, secondary, and tertiary structures are highly fixed, and whose combining sites are specific to one particular type of antigen molecule. Copies of these specific antibodies are carried on the lymphocyte surface and serve to bind antigen, thus triggering the processes that lead to lymphocyte growth, proliferation, and increased antibody production.

Although the clonal selection theory appears to be more in line with available evidence, it has one major drawback—namely, how the genetics of such a system must work. How can the genome of any mammalian species possibly contain individual genes for all the possible antibodies in the world? New evidence regarding the structure of the eukaryotic chromosome has suggested a possible answer to this intriguing problem.

The Genetics of Immunoglobulin Diversity

Two hypotheses have been advanced in recent years to account for how genetically based diversity in the immune system originates. One is called the **somatic mutation hypothesis,** the other the **germ-line hypothesis.** These two hypotheses are based on quite different assumptions about the origin of genetic diversity, and lead to very different predictions.

The Somatic Mutation Hypothesis. This hypothesis is based on the assumption that each lymphocyte contains genes for only a few varieties of antibody molecules, but that these genes are highly susceptible to mutation. The term "somatic" indicates that the mutations take place in somatic, not germ cells (that is, in the hemopoietic stem cells that give rise to mature lymphocytes). According to the somatic mutation hypothesis, lymphocytes mutate frequently as the immune system is developing during early childhood. The mutations are random, but occur frequently enough so that by the time the individual is mature the immune system contains lymphocytes for virtually every antigen conceivable.

At first glance this hypothesis sounds highly unlikely. How could genes be thought of as mutating at a greater rate in some tissues than others? Is this not a totally arbitrary hypothesis, invoking a speculative mechanism for which there is no evidence? Actually, not completely. Geneticists know that some genes can be mutational "hot spots"—that is, can mutate far more frequently than others. Mutation rate is not the same from gene to gene within the same organism, or from organism to organism. Presupposing a very high rate of mutation in lymphocyte-producing cells may thus not be so far-fetched after all. Still, there is not much evidence that lymphocyte-producing cells mutate as extensively as the somatic mutation hypothesis would require.

The Germ-Line Hypothesis. The germ-line hypothesis, in the form it was originally stated a decade or so ago, assumed that every lymphocyte contains genes for all possible antigens. As the immune system matures, all but one of those genes in each cell is turned off. Which genes are turned off (or, rather, which gene is allowed to stay turned on) is a matter of chance. In this way, any organism will be likely to have lymphocytes that, collectively, can cover the entire range of antigen possibilities.

Admittedly, it seems preposterous that every lymphocyte (indeed, every cell in the body) could contain *all* the genes for every conceivable antigen in the world. A little calculation, however, suggests that the number of genes required is not as great as we might imagine. For the sake of argument, let us suppose that there are 10 billion (10^{10}) possible antigens in the world. If the specificity of each antibody is determined by a combination of the heavy and light chains, then we would need only 200,000 (10^5) different genes for each type of chain, to get the total of 10 billion ($10^5 \times 10^5 = 10^{10}$) different combinations. Now, 400,000 genes represent less than 20 percent of the cell's total DNA. While this is a significant percentage, given the evolutionary importance of the immune system it is not an inconceivably large portion of the genome to be devoted to warding off infection. Such calculations, of course, do not demonstrate that so many antibody-producing genes *are* actually present in cells. It does suggest, however, that far fewer genes than we might at first imagine would still be able to produce a very large variety of antibody molecules. Thus we cannot rule out the germ-line hypothesis simply on the grounds of improbability.

The Present View. If the germ-line hypothesis is correct, we would expect to find a relatively large number of antibody-determining genes—as we saw, somewhere in the vicinity of 200,000 genes for each (light and heavy) chain. However, genetic evidence suggests that the constant regions of all the basic immunoglobulin classes are determined by one or a few genes (perhaps as few as one gene per class of antibodies: IgA, IgC, IgM, and the like). The variable regions appear to be coded by a similarly small number of genes. How can we explain this difference between the expected and the observed? Today, immunologists have made a major revision in the germ-line hypothesis.

Consider that each immunoglobulin polypeptide chain (light or heavy) is coded by two genes: one for the variable region (symbolized V) and one for the constant region (symbolized C). Consider further that each germ cell, or each hemopoietic stem cell, contains one C gene, and many V genes for each of the two chains (heavy and light) making up antibody molecules. Recall from Chapter 19 that the DNA for most eukaryotic genes appears to exist in interrupted sequences—that is, regions coding for specific amino acid sequences interspersed with nonsense sequences. We can visualize that, during lymphocyte development, various gene regions are cut and respliced so that each C gene becomes physically linked to one of the many V genes. This cutting and resplicing process occurs randomly, so that many different combinations of V with C are possible. Once such a combination is effected, the V gene determines the antibody structure to be synthesized by that cell and its progeny. The antibody differences from one lymphocyte to another are determined by which V gene is spliced to which C gene. The different V gene–C gene combination can also account for the very wide range of genetically determined antibodies that a single organism can generate.

Since new and simpler methods of sequencing the DNA of genes have been used (beginning in 1975), it has been found that a third gene, in addition to the C and V genes, is also involved. Known as the J, or "joining" gene, this third component resides at some physical distance from the C and V genes on the DNA of hemopoietic cells. As differentiation into lymphocytes occurs, DNA is cut and respliced in such a way that C, V, and J genes are brought close together. The three genes thus form a functional genetic unit, a supragene if you will, determining one of the two chains (heavy or light) of the immunoglobulin molecule. This model for generating antibody genes is diagrammed in Fig. 20.39.

Geneticists concerned with the immune response feel that in each hemopoietic stem cell there are one or two C genes, a larger number of J genes, and a still

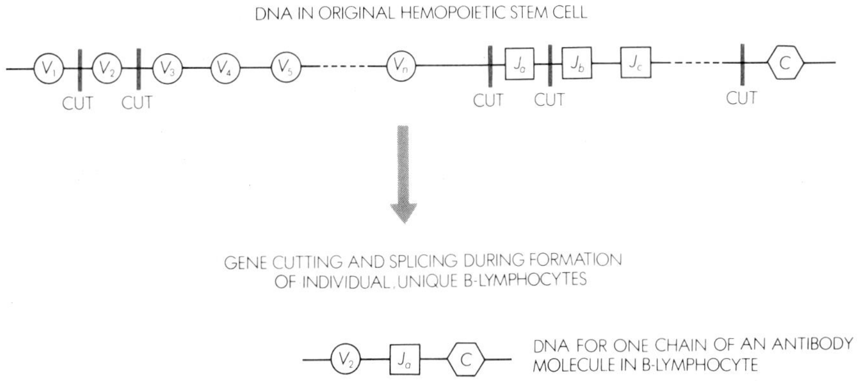

Fig. 20.39

Schematic representation of gene splicing and rearrangement in the maturation of B-lymphocytes from hemopoietic stem cells. In the DNA of the original hemopoietic stem cells (and, indeed, in all the cells of the body), are a number of genes for the variable region of the antibody molecule (symbolized V_1 through V_n). In addition, there are a smaller number of "joining genes" (symbolized J_a, J_b, and the like), and a single gene for the constant region of the antibody chain (there are different C genes for the heavy and the light chains). Through cutting and resplicing of this lengthy DNA chain, one or another V gene is joined with one or another J gene, both ultimately being spliced to the C gene. The final result is a single DNA sequence, or gene, for one heavy or light chain. Through random cutting and splicing of the V and J genes, a large number of variants of the same basic antibody molecule can be obtained. Indeed, with only 200,000 V genes for the light chain, and 200,000 V genes for the heavy chain (forgetting about the J genes for the moment), over 10 billion (10^{10}) different antibody molecules, each with a unique combining site for just one sort of antigen, could be produced.

DNA IN ORIGINAL HEMOPOIETIC STEM CELL

GENE CUTTING AND SPLICING DURING FORMATION OF INDIVIDUAL, UNIQUE B-LYMPHOCYTES

DNA FOR ONE CHAIN OF AN ANTIBODY MOLECULE IN B-LYMPHOCYTE

larger number of V genes. Exactly how many of each is still unknown. It appears, then, that mammals are able to generate a wide variety of antibody-producing lymphocytes by cutting and resplicing a relatively small number of genes within the cell's genome during growth and differentiation of hemopoietic stem cells. Recombinant DNA, therefore, does not appear to be an invention only of today's molecular geneticists. It has been around for at least 200 million years, since the evolution of the mammalian immune system.

The Cell-Mediated Immune Response

In its molecular details, the cell-mediated immune response is similar to the humoral response: the basic structure of antibody molecules, the interaction of antigen and antibody, and the mechanism for generating antibody diversity are all virtually identical. The major differences lie in (1) the general class of antigens against which the cell-mediated system operates; (2) the fact that T-lymphocytes do not release their antibodies into the bloodstream, but retain them in their cell surfaces; and (3) the way T-lymphocytes interact with other cells, such as macrophages, to attack and dispose of antigens.

The cell-mediated immune response is a major line of defense against large-scale antigenic invaders: parasites, other multicellular agents, tumors, and even the body's own cells that have been infested with viruses. In each of these cases surface molecules on the invaders serve as the antigenic markers, signaling the system that a foreign substance is present. The same is true of the body's own cells, that is, those infected by viruses or display tumorous growth. We find cell surface markers on these to be different from those on cells of normal tissue. These different antigenic markers signal to the immune system that something has gone wrong within the cells. It is these different markers that are recognized by T-cells.

There appear to be two ways in which T-lymphocytes become aware that foreign or foreign-appearing cells are present in the body. One is through the accidental release of some of the protein embedded in the cell membrane. This protein passes from the cell surface into the bloodstream, and is thus brought into contact with T-cells wherever they may be located in the body. The other method is through the constant wanderings or surveillance activity of T-lymphocytes throughout the body. As these cells move endlessly through the capillaries into the body tissues and back again, they occasionally encounter a group of cells with markers that are foreign to the body (that is, are antigenic). The T-lymphocytes bind to those cells (through a surface-to-surface antigen-antibody reaction) and begin to form cell complexes. These complexes signal other components of the body's defense system to rush to the scene and attack the foreign material. These "other com-ponents" of the immune system include primarily the macrophages.

The Graft Rejection Response. To understand just how the cell-mediated immune process works, consider the case of skin grafting between two strains of mice. Suppose we graft patches of skin from mouse strains A and B onto mice of strain A. These strains are highly inbred, and so are genetically almost identical. After two weeks we find that the patch of skin from strain A remains intact, has become vascularized, and is in fact growing. The patch of skin from strain B, on the other hand, has not vascularized and is eventually rejected. If we follow up this experiment by another, in which we now transplant skin from mice of strains B and C onto the *same* A mouse we used for the first experiment, we get some interesting, but not unexpected, results. This time the mouse rejects the graft of skin from strain B within *two days*, while it still takes two weeks to reject the graft from strain C. We can conclude that the cell-mediated immune response, like the humoral response, is able to distinguish self from nonself, and also possesses memory.

Now, let us see how the graft-rejection response works. T-lymphocytes, like B-lymphocytes, wear on their cell surfaces copies of the antibodies they are able to synthesize. When skin from mouse strain B is grafted onto an individual of strain A, T-lymphocytes that bear antibodies specific for the foreign cell antigens become "sensitized," that is, they develop more antibodies on their surfaces and increase their metabolic processes. These sensitized and activated cells are called rather ominously "killer lymphocytes," or "natural killer" (NK) cells. This epithet merely refers to the higher level of activity that these cells display. The killer lymphocytes move to the area where the graft is located, and attach to the foreign cells (perhaps by an antigen-antibody reaction, though the process of attachment is not completely understood). After attaching to the graft cells, the killer lymphocytes proceed to break down the cell membranes of the graft tissue. In addition, killer lymphocytes cause several other defense mechanisms to come into play. They release a substance that immobilizes passing macrophages, thus detaining them in the area of the graft. The killer lymphocytes also stimulate macrophage "appetite" (in a way not yet understood), causing these cells to begin devouring the graft cells with particular gusto. And finally, killer lymphocytes release another signal substance that calls other forms of white blood cells to the graft area. The result of this unified onslaught is that not only are the graft cells killed directly, but the blood supply to the graft tissue is cut off, and the foreign tissue atrophies.

The graft rejection response is part of the body's defense mechanism against large-scale invaders such as parasites. Yet, in our age of medical technology, it has posed a serious barrier to the process of organ and tissue transplantation. The problem with transplantation is, of course, that the host organism recognizes the tissue as foreign, and unleashes the cell-mediated immune response against it. The only exception to this response is when the donor and the host are identical twins. In that case, successful transplantation is possible. To circumvent the immune response in transplantation, doctors have tried first suppressing the recipient's own immune system. This technique has two major drawbacks, however. Suppressing the immune system opens the individual to innumerable infections. Quite often people who survive the organ transplantation process succumb to one or another disease. Another problem is that such suppression cannot be maintained indefinitely. Once the immune system is allowed to come back into play the transplanted tissue is recognized as foreign, and the rejection process begins.

In 1979 immunologists announced a promising new development in organ transplantation. In experiments with laboratory animals, researchers found that if a donor organ was freed of all macrophages and B-lymphocytes contained within its tissues, it could be transplanted and maintained indefinitely in a host. Techniques have now been developed to free donor organs of lymphocyte cells. This exciting development has not yet been tried with human beings. Its success to date with laboratory animals, however, suggests that successful organ transplantation and maintenance may well be a reality in the not-too-distant future.

Histocompatibility Antigens and Their Genetic Basis. In recent years immunologists have begun to recognize not only the cellular but also the genetic basis of the graft-rejection response. From research carried out first with mice and later with human beings, biologists now know that the cells of any particular individual are distinguishable from the cells of any other individual by a series of unique surface protein markers. These markers, called collectively histocompatibility antigens (the human leukocyte antigen, or HLA), are determined by a series of genetic loci, each of which has a number of alleles. Although any two individuals may share a number of histocompatibility antigens in common, it is extremely unlikely that they would have all their antigens the same. To understand why this is so, we must consider the genetics of the HLA system.

Prior to 1975 two HLA loci were known in human beings; in 1975 two more loci were discovered (there appear to be others, as well). One of these loci has twenty known alleles, while another has fifteen. With just these two loci, meiosis can generate 300 different combinations (20×15), and sexual reproduction over 25,000 different genotypes. If we consider the two other loci and their alleles, the total number of possible combinations is well over 300 million. This means that it would be *very* unlikely for any two people to have the same exact combination of histocompatibility antigens.

Of course, such possibilities do exist, but they are exceedingly unlikely. For all practical purposes, and especially for organ transplantation, we are all antigenetically unique.

So important are the implications of the work on the HLA system and its genetics that the 1980 Nobel Prize in Physiology or Medicine was granted to three workers in this field: Jean Dausset of the University of Paris, and St. Louis Hospitals; Baruj Benacerraf of Harvard Medical School; and George Snell of the Jackson Laboratories in Bar Harbor, Maine. Determination of the exact genetic basis of HLA diversity will be of great help in the future as researchers tackle the problem of grafting and organ transplantation.

The graft-rejection response raises another side of the coin regarding the function of the immune system: Not only does the system recognize foreign material and mobilize against it, it also recognizes the body's own tissues and does *not* normally mobilize against them. This is the basis of saying that the system has the basic ability to distinguish self from nonself. How does the immune system make this distinction?

The Distinction Between Self and Nonself

In 1945 R. D. Owen at the California Institute of Technology made an interesting observation. He noted that two fraternal twin calves had on their red blood cells not only their own, but their twin's antigens (recall that fraternal twins come from two different eggs and are no more genetically alike than any two siblings). The calves were healthy, and while being able to make antibodies to all sorts of other antigens, did *not* make antibodies to their twin's cell surface protein. After investigating a number of such cases, Owen demonstrated that at the time the twin calves were being carried *in utero* their circulatory systems had been interconnected and they had exchanged blood. What was surprising to Owen was that, apparently as a result of such exchange, the two calves' immune systems recognized each other's antigens, and tolerated them as "self."

Owen's remarkable discovery suggested at first that the ability of the immune system to distinguish between self and nonself was a product of what substances it came into contact with early in life. Molecules encountered in fetal life or early life might be tolerated as self by the immune system simply because they were there from the outset. It was argued by some theorists that presence of a molecule (that is, a potential antigen) in early life prevented development of the lymphocyte-precursor cells capable of producing antibody against those molecules. In this way all antibody-producing cells capable of responding to self molecules would be weeded out early on. By definition, then, the lymphocyte-precursor cells that survived would respond only to foreign substances.

However attractive such a notion was, other data suggested the matter was not quite so simple. For instance, some proteins that are synthesized for the first time only later in life are nonetheless recognized as self. No antibodies are made against them even though the immune system has never encountered them. More strikingly, E. L. Triplett at the University of California showed that tolerance to self can be only temporary. He removed the pituitary glands from tree frogs and some weeks later re-implanted them *in the same individuals.* The frogs now rejected their own tissues as if they were foreign transplants. Obviously the frogs' immune system had not lost the ability to recognize and produce antibodies against itself. Under normal circumstances, however, the lymphocytes capable of synthesizing antibodies against self must be suppressed. That suppression disappears when exposure to self is interrupted.

Autoimmunity. Under certain conditions, even without interruption of exposure to self, the immune system can lose its ability to distinguish between self and nonself. Such a breakdown is known as an autoimmune disease. It is not always clear what initiates an autoimmune response. Once initiated, however, the body's lymphocytes begin producing antibodies against one or another of the organism's own proteins (or other molecules). This leads to an immune reaction against the body's tissues that can lead to disastrous consequences. Several autoimmune diseases are rather well known. Rheumatoid arthritis results from the immune system developing antibodies against collagen molecules produced by connective tissues in the joints. Rheumatic heart disease is the result of building antibodies against the material making up the valves of the heart. Ulcerative colitis results from antibodies specific against the individual's own intestinal epithelium.*

Immunologists hope that the study of autoimmunity will lead to a better understanding not only of how to cure these sometimes tragic diseases, but also of the mechanism by which the entire immune system develops its ability to distinguish self from nonself.

The Fetus as a Foreign Tissue. One of the intriguing questions that the study of immunology has raised is: what prevents a mother's immune system from developing antibodies against her own fetus? We know that cell surface molecules are constantly passed across the placenta from fetus to mother. Since the fetus has half its genetic information from the father, many of its proteins and other molecules would be foreign to the mother's system. Yet, mothers do not produce antibodies against their own fetuses. How is this possible?

Several observations provide clues to this problem:

1. Skin grafted to the wall of the uterus in an experimental animal was rejected about as rapidly as when it was grafted to any other part of the body. The uterus is thus not a "privileged" site that the immune system cannot reach.

2. Certain cells from the fetus often fall off the placenta, pass into the mother's bloodstream, and are carried to parts of the body (usually the lungs) where they lodge and even grow into small, benign tumors. They remain there throughout pregnancy, but are rejected after birth. The mother's immune system does not produce antibodies to these cells during the period of pregnancy.

3. If lymphocytes from a woman who has had several children by the same husband are mixed with his blood cells, the lymphocytes quickly produce antibodies against the man's antigens. The mother's immune system does develop antibodies (that is, becomes "sensitized") against those antigens the fetus shares with the father, yet does not reject the fetus.

A major breakthrough in understanding this seeming paradox came from the following experiment. If the procedure described for (3) above is repeated in the presence of the mother's own serum, her lymphocytes do not produce effective antibodies against her husband's cells. Her blood serum must contain some factor that protects his cells from being recognized. Further investigation shows that the factor is IgG—in fact, the IgG antibodies are the very ones the mother has generated against her husband's antigens.

What keeps these antibodies from attacking the husband's cells? It turns out that the IgG synthesized in this case belongs to a subclass of IgG molecules that is incapable of triggering the final destructive phase of the immune response. Thus the antigens on the husband's cells or those on the fetus's cells remain masked by antibody, and do not become the focus of further attack; the destructive process is halted. The IgG antibodies produced by the mother in these cases are known as "blocking antibodies"; they block completion of the entire immune response. Other factors, including placental hormones that block the action of T-lymphocytes in the area of the uterus, also aid in preventing the mother's immune system from rejecting the fetus as if it were a foreign graft. This example, better perhaps than any others, illustrates the enormous subtlety of the immune system.

The Immune Process as a Developmental System

We have introduced a discussion of the immune system in this chapter because it represents, among other

---

*At least in the case of rheumatic heart disease, the triggering mechanism appears to be infection of the body by the strep throat bacteria *Streptococcus pyogenes.* The strep infection itself is not what produces ultimate damage to the heart. Rather, one of the antigenic determinants on the bacterium is very similar to a molecule found in the heart valves. After bacterial invasion, the immune system produces antibodies to *S. pyogenes* that also can bind to the material of the heart valves, damaging them irreversibly.

things, a model for understanding some basic aspects of development. It is time now to ask exactly what the immune process reveals about problems of animal development.

As we have stressed throughout the chapter, the central problem of development has always been, and continues to be, differentiation: how cells become different through the turning on and off of genes. In Section 19.7 we examined one model for regulation of gene expression: the Jacob-Monod theory of suppression and activation. Another model may well be the one by which lymphocytes, in their development, have come to specify the amino acid sequence for a single antibody: that is, gene splicing and rearrangement. Using the

immune system as a model, we might speculate that in the development of the embryo, genes are cut and spliced in such ways as to determine that one DNA sequence over another will guide the production of protein in that cell. Whatever the exact mechanism, developmental biologists recognize that cells of the immune system *are* able to alter their genetic messages as the cells differentiate and mature. If differentiation is defined on the most fundamental level as differential protein synthesis from one time period to another, then cells of the immune system may well represent at least one way in which development and differentiation occur. Thus the immune system may well be a useful experimental model for all cellular development.

Summary

1.   Developmental biology comprises the study of all regular changes within the life history of the organism from fertilization through embryonic development to growth, aging, and death. In addition it studies such processes as the immune response, aging, and cancer. Embryology is simply one aspect of developmental biology.

2.   The central problem of developmental biology is that of cell differentiation: how cells that are alike at a very early stage become progressively different, ending up as highly specialized cells such as muscle, nerve, or red blood cells.

3.   Embryonic development begins with the formation of gametes: sperm from primary spermatocytes and eggs from primary oöcytes in the germ tissues of the parents. Fertilization of the egg by the sperm, however, actually initiates the process of embryonic development. Many sperm are required to ensure that one actually penetrates the egg cell membrane. Once this has happened, the egg cell surface undergoes certain physical changes, usually ensuring that a second sperm does not get through. The egg membrane lifts up, forming an outer hyaline layer around the cell. The outermost boundary of this layer is the fertilization membrane; the innermost layer closest to the egg cytoplasm is called the vitelline membrane.

4.   Activation of the egg follows fertilization. A certain number of the egg's genes have been transcribed into *m*RNA and translated into protein prior to fertilization; after fertilization this material is activated, which leads to intense metabolic activity. Thus maternal genes determine the characteristics of the initial stages of development.

5.   Early embryonic development passes through the following stages: single-celled zygote, two-cell (each cell is called a blastomere), four-cell, eight-cell, and so on to the formation of the blastula, a hollow ball of unspecialized cells surrounding a cavity, the blastocoel. Invagination of the blastula (at the blastopore) eventually produces the three-layered gastrula composed of ectoderm (outer layer),

mesoderm (middle layer), and endoderm (inner layer). These are called the primary germ layers.

6.   Each primary germ layer gives rise to certain groups of tissues in the adult organism: ecotoderm to skin, nervous system, lens and cornea of eye, lining of nose, mouth, and anus, hair, nails, and sweat glands; mesoderm to muscles, connective tissue, blood and blood vessels, kidneys, and the reproductive organs; endoderm to lining of the digestive tract, liver, pancreas, various endocrine glands, the bladder, and the lining of the urethra.

7.   Early divisions of the egg cell are called cleavages. Patterns of cleavage are determined largely by the amount and distribution of yolk in the egg cell. The presence of a great deal of yolk slows down cleavage. In eggs such as the frog's, with only a moderate amount of yolk concentrated at one end, a blastula forms with larger blastomeres at the end where the yolk was concentrated. In the bird's egg, where there is a massive amount of yolk, cleavage is restricted to the cytoplasmic area that resides as a disc at the top of the egg.

8.   Gastrulation occurs by morphogenetic movements of cells from the outside of the blastula to the inside. The movements themselves and the process by which migrating cells find their proper destinations are determined by properties of the cell surfaces.

9.   Differentiation of cells into tissues and organs occurs primarily by a process known as embryonic induction. In the formation of the nervous system ectodermal cells lying just to the dorsal side of the blastopore come into contact with mesodermal cells of the underlying archenteron roof. Contact induces ectodermal cells to begin elongation to form the neural plate. Soon cells of the neural plate begin to take on a wedge-shaped appearance due to contraction of microfilaments at the upper end of the cells, causing the plate to buckle or fold inward. Eventually buckling occurs to such an extent that the two ends of the plate meet, forming the neural tube. The primary induction responsible for this se-

quence is contact between archenteron roof (the inducer) with cells of the overlying ectoderm (the induced).

10. Induction usually starts a chain of events known as cell determination. The initial inducer is called the primary inducer; once induced, a differentiating tissue may itself become a secondary inducer, and so on. Embryologist Hans Spemann found that the tissue of the dorsal lip of the blastopore was the primary inducer for differentiation of the whole embryo, and he called it the "organizer."

11. The causes of induction are still largely unclear, especially at the molecular level. It is known that induction can only begin after contact between inducer and induced tissue. However, if ectodermal tissue is placed in a medium in which underlying mesoderm has been allowed to incubate but has been removed, the ectoderm begins to form neural plate. Inducer tissues appear to produce a substance or substances that diffuse into induced tissues and initiate differentiation. To date, no one has identified what such a substance or substances might be.

12. Induction appears to selectively stimulate the genes present in the induced tissue. The inducer determines the general direction of differentiation. The induced tissue responds within the limits or capabilities of its own genome. Frog mesoderm can induce differentiation in salamander ectoderm, but the induced tissue responds genetically like salamander, not frog.

13. Embryonic induction is a reciprocal process. Inducer causes changes in the induced tissue. The induced tissue can then "induce" changes in the inducer. Inducer and induced tissue thus undergo differentiation together by a constant interaction.

14. In human beings eggs are produced in the ovaries within structures called Graafian follicles. It takes a follicle about two weeks to complete the final stages of growth prior to release of the egg. After the egg is released the old follicle becomes the site of production of the hormone progesterone. As an endocrine gland, the follicle is known as the corpus luteum.

15. The human egg completes its meiotic divisions as it is passing down the Fallopian tube toward the uterus. As it makes this passage, the egg extrudes the third polar body. Fertilization, if it occurs, usually takes place in the Fallopian tube.

16. When fertilization occurs, implantation of the embryo in the uterine wall serves to keep the uterine lining from breaking down. Soon the embryo develops its organ of nourishment, the placenta, through which nutrients and waste are exchanged with the mother's blood. The placenta is also an endocrine organ, secreting estrogen and progesterone. These hormones negatively inhibit FSH and LH oscillation and serve to maintain the uterine lining throughout pregnancy (approximately 266 days, or 9 months).

17. The period of human gestation, or pregnancy, is divided into three periods, or trimesters, of approximately three months each.

a) *First trimester:* The embryo becomes enclosed in its various membranes, becoming the trophoblast, by the end of the first month. The primitive streak appears, marking the anterior-posterior axis. The nervous system appears, as do the muscle segments and the rudimentary sense organs and the gill slits. The germ cells begin differentiation by the end of the first month, and by the middle of the second they migrate to form the future testes or ovaries. The bones become visible in outline during the second month, and the head shows rapid growth. During the first trimester, the future individual is referred to as an embryo.

b) *Second trimester:* The brain begins to form by the fourth month, and by the end of the second trimester the future individual looks like a miniature human being, with all the major body parts differentiated. From the fourth month on it is called a fetus.

c) *Third trimester:* The final stage is characterized largely by growth of the already-differentiated parts. As it reaches the end of term, the fetus "drops" into position within the uterus; the head is usually pointed downward and is normally the first part to emerge.

18. Birth, or parturition is initiated by changes within the placenta, whose physiological activities begin slowing down from the seventh month onward. Hormonal and mechanical changes (such as crowding within the uterus) also appear to influence the onset of birth. Labor pains are contractions of smooth muscles in the uterine wall prior to birth. During labor, contractions of the uterine muscles become increasingly stronger and more frequent. The cervix also dilates to allow passage of the fetus.

19. Failure of an embryo or fetus to reach full term is called a miscarriage. Miscarriages can occur at any time during pregnancy but are most common during the first trimester.

20. Regeneration is the process by which an organism replaces a lost, injured, or worn-out part. Human beings constantly regenerate new skin, but we cannot regenerate fingers, arms, or legs. Many lower animals, such as salamanders, can regenerate whole appendages.

Some observations have been made regarding ability to regenerate:

a) The more highly differentiated the tissue, the less likely it can regenerate.

b) The older the organism, the less ability its various tissues and organs have for regeneration.

c) In vertebrates, regenerative ability of a tissue appears to be related in some way to amount of innervation of that tissue.

21. As regeneration of a limb bud begins, blastema cells appear at the wound site. Blastema are unspecialized cells

that congregate at an area where regeneration is occurring; they ultimately differentiate into the specialized tissues of the regenerating organ. Differentiation of blastema cells has been useful as a model system for studying many aspects of the general differentiation process.

22. Experiments with frog eggs, tobacco plants, and other organisms indicate that highly specialized cells from embryos or adults still retain the ability to form a whole new organism. These cells are genetically totipotent; they still carry all the genetic information for forming a complete adult. Differentiation appears to be a process of selectively turning off (or on, as the case may be) particular genes.

23. The study of development has left more unanswered than answered questions. The basic problem of cell differentiation, at the tissue as well as the cellular level, is still largely unsolved.

Exercises

1. What evidence exists for the theory of preformation and the theory of epigenesis? From the type of observation that can be made with only a magnifying glass, which theory seems more reasonable?

2. How can the amount of yolk available in a fertilized egg determine the course of embryonic development?

3. Describe the changes that occur in an egg immediately after fertilization.

4. If a blastula nucleus from one species of frog is transplanted into an enucleated egg of another species, the development of the resulting embryo stops at an early stage. If the nuclei are taken from embryos of this kind before they die and transplanted back to enucleated eggs of the original donor species, the resulting transplant embryos develop abnormally. What do these results indicate about the relationship between the nucleus and the cytoplasm?

5. Spinal cord and notochord cells induce cartilage formation in amphibian embryos. They will also induce cells to form cartilage when cultured outside the embryo. However, they will only induce cartilage formation in cells that would ordinarily form cartilage. Does this fact support or refute the conclusions drawn from Schotté's experiment?

PART V
EVOLUTIONARY AND POPULATION BIOLOGY

It is interesting to contemplate an entangled bank, clothed with many plants of many kinds, with birds singing in the bushes, with various insects flitting about, and with worms crawling through the damp earth, and to reflect that these elaborately constructed forms, so different from each other, and dependent on each other in so complex a manner, have all been produced by laws acting around us. . . . There is a grandeur in this view of life, with its several powers, having been originally breathed into a few forms or into one; and that, whilst this planet has gone cycling on according to the fixed law of gravity, from so simple a beginning endless forms most beautiful and most wonderful have been, and are being evolved.

—*Charles Darwin, "On the Origin of Species," First Edition, 1859, pp. 489-490.*

The abundant and luxurious animal and plant life of the tropics has always been a source of inspiration and insight into the natural history of life on the entire planet. In the nineteenth century it was of great significance to naturalists as they tried to understand the origin and history of species. To naturalists such as Charles Darwin and Alfred Russell Wallace from the temperate zones of Europe, the tropics offered an unprecedented display of exotic adaptations, of habitats filled by every sort of flora and fauna imaginable. Young Darwin wrote that the tropical forest was a revelation: "Delight itself, however, is a weak term to express the feelings of a naturalist who, for the first time, has been wandering by himself in a Brazilian forest." Wallace reported on "the wonderful variety and exquisite beauty of the butterflies and birds . . . ever new and beautiful, strange and even mysterious." The tropics were profoundly influential for both naturalists in the formulation of the theory of evolution by natural selection.

(Illustration from J. B. von Spix and C. F. P. Martius, Atlas zur Reise in den Jahren 1817–1820 gemacht (1823–1831). Courtesy David Dwyer and the Missouri Botanical Garden.)

All living things function as parts of populations, local groups of organisms of the same species that share similar living habits and resource requirements, can interbreed with each other, and interact with other species around them. Populations of all organisms undergo changes over time. Some of these changes are short-term and local—for example, migration or expansion or contraction of population size. Others are long-term and affect the history of the entire population and/or species—for example, the formation of a new species from an old species, or extinction. It is at the population level that the major interactions between organisms take place in nature. Investigating biological interactions at this level of organization comprises what we call today *population biology.*

Population biology consists of several very different, but closely related subjects. These include taxonomy (the process of categorizing living organisms into groups with similar traits and past histories), the transmutation of species, or what we call evolution (how the many forms of life we know today came to exist), ecology (the interrelations among different populations and their natural environments), and biogeography (how animal and plant groups are distributed across the face of the earth). None of these topics can be approached by the study of single organisms. The modern biologist interested in the diversity and spread of life on our planet must study organisms in terms of interacting groups: from small ant colonies to an entire ecosystem such as the lush tropical rain forest shown on the opening page of this unit.

Central to the modern study of population biology is the study of evolution and ecology. Evolution is a general term for the process by which one population gives rise to one or more populations different from itself—that is, to diversity—over time. The concept of evolution is one of the most important organizing principles in biology. In many ways all other biological concepts—from the molecular to the cellular and organismic— can best be understood in terms of the evolutionary process. Evolution not only accounts for the apparent diversity of all living forms on earth but also relates that diversity to the history of life on our planet.

Ecology is the study of the interrelationships among many diverse groups of organisms with each other, and with their physical environment. Ecological systems are the result of evolution, and thus the organisms composing them display many interrelationships, many adaptations to each other and to the system as a whole. Being a product of evolution, ecological systems are also continuing to evolve in the present. The dramatic and sometimes sudden change that human beings wreak within the delicate balance of an ecosystem has become of major concern in recent years as we continue to pollute and ravage our environment.

The process of evolution itself was thoroughly elucidated in the theory of natural selection proposed by Charles Darwin in 1859. Darwin's theory was the first comprehensive attempt to account for the mechanism of evolution in terms of processes known to work around us all the time. Darwin's theory grew out of his own observations as a naturalist and his reading of contemporary literature in natural history, geology, and political economy. From these diverse sources, Darwin inferred a mechanism for evolution—natural selection—which, though modified and considerably expanded since his day, forms the basis for our current understanding of how life has developed on earth. This theory, probably biology's most important unifying concept, offers an explanation for how diversity and change can occur.

Once organisms appeared and started evolving, their diversification took many paths. Much of biology has been concerned with tracing the actual course of evolution of various animal and plant groups. Using the evidence of comparative anatomy, behavior, biochemistry, and paleontology, this study has provided a reasonably detailed picture of the evolutionary history of animals and plants on earth. Yet, figuring out which organisms may have evolved from which ancestors requires much speculation, deduction, and circumstantial reasoning. The process of tracing evolutionary relationships provides insight into some of the most challenging processes of logic in the field of biology.

To understand the evolutionary process, we must first understand how organisms are grouped together into meaningful categories, such as species, families, orders, or phyla. The process of ordering groups of organisms into categories is called taxonomy, and the general field that encompasses taxonomy is known as systematics. Systematics and taxonomy are more than housekeeping operations. Behind them lie fundamental philosophical and biological problems: What *is* a species? Are species *real* units in nature or arbitrary groupings invented by biologists? Are higher taxonomic categories, such as genera, classes, families, or phyla, real or arbitrary? How is it possible to devise a taxomonic system that reflects past historical reality and at the same time encompasses groups of animals and plants which are constantly changing and evolving at the present time? In trying to resolve some of these problems, biologists have come to view taxonomic groups, particularly the most fundamental group, the species, on a population level. Species are not viewed, as they used to be not very long ago, as idealized categories represented by one or a few type specimen. Rather, taxonomists today look at species as collections—populations—of organisms showing a range of characters and traits. Taxonomy is a populational concern. It is therefore appropriate that we begin this unit with a look at the concepts and methods of modern taxonomy.

Chapter 21
Taxonomy and the
Classification of
Living Organisms

21.1
INTRODUCTION

Most sciences attempt to find orderly patterns in nature. Chemists, for example, try to find relationships between the elements and compounds with which they work. They speak of metals and nonmetals, acids and bases, and carbohydrates and proteins. The periodic table is one result of attempts to find meaningful order among the known chemical elements. Indeed, much human progress through history can be seen as the result of the search for order and patterns.

We classify things because it makes them more orderly and easy to understand, and because it simplifies our experience. However, the biologist is faced with a staggering task when it comes to sorting out the world of living organisms. In no other group of objects does so much variation exist from individual to individual and from group to group.

21.2
SOME DEFINITIONS

What is taxonomy? Quite often in textbooks (including earlier editions of this one) three terms—classification, systematics, and taxonomy—have been used almost interchangeably. However, they are not synonymous; a closer look at their definitions will help elucidate the many aspects of modern taxonomic work.

Classification

Classification, the simplest of these terms to define, is the "ordering" of organisms (or any other objects) into groupings, or categories. The organisms in each category are bound together by some common features —some systematic relationship—between them. There are two basic methods or procedures involved in any classification process. The first procedure is defining the categories themselves: groupings such as "fruits," "vegetables," "cars," and "bicycles" are all classes—assemblages of individual entities with at least one common

feature (by which the group is defined in the first place). As much as possible, it is desirable to have unambiguous criteria for establishing each category so that any object being classified will fall into one or the other already established group, but not into two (or more).

The second procedure in drawing up classes of objects is to group categories together in such a way that some are subordinate to others, or some are included within others. For example, we have a general class of objects known as "automobiles." Within that class we have a subclass, called makes or brands (Ford, Chevrolet, Volkswagen, and the like); within each make we have yet another subset, called "models" (Pinto, Mustang, Falcon, all subsets of Ford); and within each model we have colors, year, and accessories. The example above represents a *hierarchical ordering* of categories—that is, categories or groupings arranged from top to bottom according to degree of inclusiveness. In a hierarchical system, objects at any one level partake in all the characteristics of the levels above it, but not necessarily in the levels below it. In a hierarchical system, if you know into which lower category an object falls, then you immediately know all the higher categories in which it is included. Conversely, if you want to determine into which lower category an object should be placed, you can start with the most general groupings (for example, automobiles or Fords) and work downward to more specific ones.

Systematics

Systematics is "the comparative study of any group of organisms and of any and all relationships among them by means of the techniques of one or more branches of biology."[*] Systematics is really the process of examining organisms from many points of view, in order to determine their range of similarities and differences. Systematic studies generally result in classification, but not

---

*Otto Solbrig and Dorothy Solbrig, *Introduction to Population Biology and Evolution* (Reading, Mass.: Addison-Wesley, 1979), p. 444.

647

always. Figure 21.1 shows the various procedures involved in systematic studies in biology. Note that systematics cannot be defined in terms of any one method of examination, but only by a large number. For example, organisms collected in the field are grown or raised in the laboratory and crossed with one another to determine something about their reproductive patterns and their genetics; they are studied anatomically,

physiologically, and biochemically; they are compared to forms collected previously (either as living or preserved specimen); and they are compared to descriptions of similar organisms published by other biologists. Evolutionary relationships are also a concern of systematic studies.

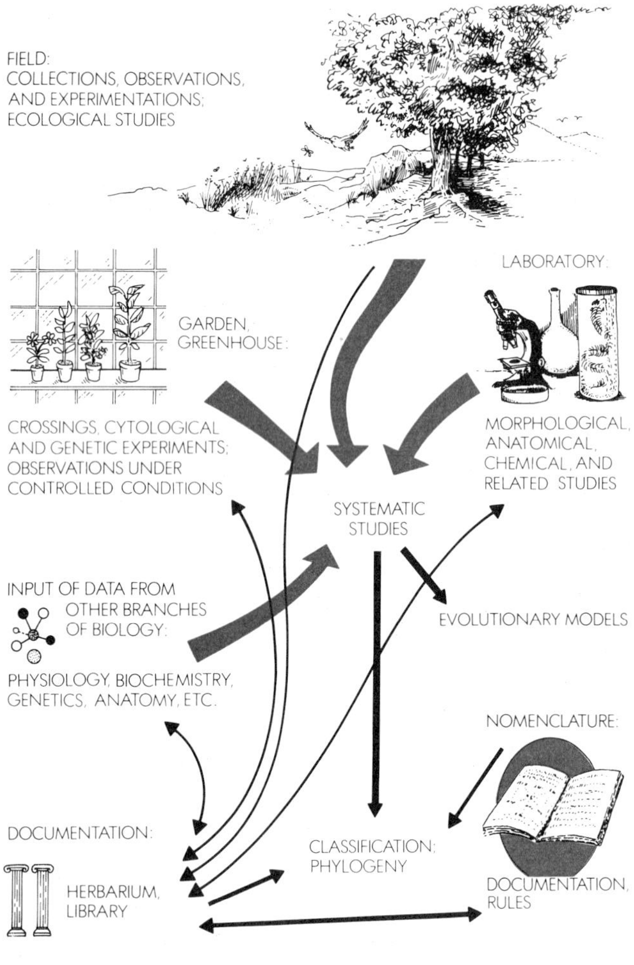

Taxonomy

Taxonomy is "the theoretical study of classification, including its bases, principles, procedures and rules."* Taxonomy is really the theory of classification, the procedures and processes by which classification is devised and carried out. Taxonomy is to systematics what theoretical physics is to all of physics. Taxonomy includes both classification and systematics, but derives many of its concepts from systematics. Thus, for example, when comparative biochemical data began to be used in systematics, the theory of taxonomy had to be somewhat altered to account for this new type of information—that is, how does the information gained from biochemical studies of different organisms compare to that gained from anatomical or immunological studies?

21.3
DEVELOPMENT OF TAXONOMIC SCHEMES

Because people in different parts of the world and at different times have had varying purposes for devising classification systems, there was no uniformity for a long time in the criteria by which organisms were grouped. For example, Theophrastus (370–285 B.C.) tried to classify all the known plants on the basis of form, life span, and habitat. Criteria used by other investigators included habitat (land or water) and geographical distribution (Asian, European, African). In some cases the classes were organized along practical lines, such as medicinal effects (those which cure fevers, induce vomiting, or act as antidotes to specific poisons).

Criteria for Classification

Of course, any of these criteria are perfectly adequate bases for developing a classification system. Which system a given naturalist chooses depends on his or her *purpose* for making the system in the first place. In a system of classifying plants based on medicinal functions, the plants in any given group might have nothing in common as far as size, shape, and color are concerned. But they would have closely related medicinal functions. Many classification schemes were developed according to such criteria in the period before the eighteenth century.

Fig. 21.1
Systematics occupies a central position in all aspects of recognizing, describing, and comparing organisms in nature. Systematics uses data from the field; the greenhouse and/or breeding laboratory; anatomical, physiological, and biochemical studies; and evolutionary studies (data on the change in organisms over time). While systematic studies usually result in classification of organisms (and are usually undertaken for that purpose), classification is not a necessary outcome of systematics. Arrows represent the directional flow of information. (Modified from O. T. Solbrig, *Evolution and Systematics.* New York: Macmillan, 1966.)

---

*George Gaylord Simpson, *Principles of Animal Taxonomy* (New York: Columbia University Press, 1961), p. 11.

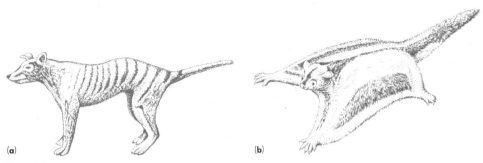

(a)   (b)

Fig. 21.2
Taxonomic problems are often caused by superficial anatomical resemblances between two or more fundamentally distinct groups of organisms. This is usually the result of a similarity between the ecological positions the organisms occupy, rather than common lines of descent. Thus organism (a), the Tasmanian wolf (Thylacinus), looks very much like a member of the dog family, whereas (b), the flying phalanger (Petaurus), looks like it belongs to the squirrel family (most notably the flying squirrels). Both the dog and squirrel families are members of the class Mammalia. However, the Tasmanian wolf and flying phalanger are marsupials (not true mammals) and are thus more closely related to each other than to their mammalian counterparts.

Today we recognize that the many varieties of organisms on earth have developed from common ancestors by modification, and biologists have therefore come to agree that the most useful overall classification scheme should reflect phylogenetic relationships. In general, anatomical structures seem to provide the basis for a classification system of this type. Taxonomists have found, however, that it is never valid to use a single criterion to establish the position of any group of animals in the classification scheme. For example, organisms that resemble each other in outward appearance are not *necessarily* closely related in evolutionary development.

Figure 21.2 shows two organisms, one of which looks like a member of the dog family and the other a member of the order Rodentia (which includes squirrels, chipmunks, and rats). On the basis of outward characteristics alone, it would seem valid to so classify them. However, the reproductive physiology of these animals reveals that they are not typical mammals (like the dog and squirrel), but marsupials. Unlike most mammals, marsupials give birth to young in a very immature form and carry them in pouches during the time needed for the offspring to complete their development. In lines of descent, then, the two organisms shown in Fig. 21.2 are more closely related to each other than to either the dog or the flying squirrel. But evolutionary adaptation has brought about a chance resemblance between them and certain other, quite distinct groups of animals—organism (a), the so-called Tasmanian "wolf," occupied the same ecological niche in its native territory (Australia) as a wolf or dog in other geographic regions; organism (b) occupies a niche similar to that of flying squirrels in North America.

Similar examples can be found in the plant kingdom. Figure 21.3 shows three plants that look as though

Fig. 21.3
Three plants similar in vegetative structure but not very closely related. Left: *Euphorbia milii* Des Moulins (family Euphorbiaceae, order Euphorbiales), a native of South Africa commonly known as "Crown of Thorns" or, in the commercial trade, "Euphorbia splendens." Center: *Alluaudia procera* Drake (family Didiereaceae), of southern Madagascar. Right: *Fouquieria splendens* Engelmann (family Fouquieriaceae), a plant of the southwestern United States commonly known as ocotillo. (Photo courtesy Ladislaus Cutak and Claude Johnston, Missouri Botanical Garden, St. Louis.)

they might be classified within the same family. On the basis of external characteristics of the vegetative body alone, it would seem valid to so classify them. However, study of the flowers, fruits, and other aspects of anatomy reveals differences of such magnitude as to justify separation of these plants into three separate families, each in a different order. The most likely explanation for the close similarity in vegetative features is evolutionary adaptation to similar arid environments.

The need for agreed-on criteria in any classification scheme is obvious. In addition, there must be some plan of nomenclature and grouping, one that will accommodate not only all animals and plants known at present, but also those that will be discovered in the future. This need was met in 1735 by Carolus Linnaeus, when he published the first of ten editions of his book *Systema Naturae*. Linnaeus did not originate the system he presented; it was the work of several men before him. By publishing *Systema Naturae*, however, he organized existing knowledge about classification and presented it in a clear and comprehensive way.

The Linnaean System

The Linnaean system as it is used today consists of seven basic hierarchical classifications or groupings. They are listed below from largest to smallest:

Kingdom

  Phylum (plural, phyla) or division

   Class

    Order

     Family

      Genus (plural, genera)

       Species (plural, species)

Note that the groups listed in this scheme form a *hierarchy*, as described above in Section 21.2. Let us see how the hierarchical organization of biological categories works in the process of classification.

You are probably familiar with the game of "twenty questions," in which one person tries to discover the identity of an object that another has in mind by asking twenty questions about the object. The opponent must answer these questions honestly; the fewer questions it takes to discover the identity of the object, the better the questioner's score. Now, suppose that the questioner started out by asking the opponent such questions as, "Is it a cat? . . . a dog? . . . a doorknob? . . . a nail? . . . a chessboard? . . . a bicycle?" Obviously, at this rate the twenty questions would be used up very quickly. There would be little chance of quessing the right answer, because each question would eliminate only one

possibility instead of several. On the other hand, if the first question were one such as, "Is it animal, vegetable, or mineral?" (a division, incidentally, into which Linnaeus divided all natural objects), the reply would eliminate two-thirds of the objects in the universe. Given the answer to this first question (let us say that the object is an animal), the questioner could then proceed along the same line of inquiry, trying to determine what *kind* of animal it is (bird, reptile, mammal). Later questions would become progressively more definitive until the correct answer was obtained. In this way, it is usually possible to get the answer well before the twentieth question.

The biologist who tries to identify a new animal or plant is simply playing the game of twenty questions. He or she begins by observing the specimen and determining the phylum (or kingdom, in some ambiguous cases) to which it belongs, In many cases, it may be obvious where the organism belongs from the level of phylum down to the level of order or even genus. In such cases, the biologist's past experience helps narrow down the possibilities very quickly. She or he still must try to place the organism in a species, however. This involves asking specifically more and more restricted questions, until a category is finally reached below which one cannot go. Usually—but not always—this is the species.

Genera, orders, families, classes, and phyla exist only as classificatory groupings designated by the biologist. The species is the only taxonomic category that has any reality in nature—especially the animal world—because members of one species are usually reproductively isolated from members of another species. This is a real, testable criterion, whereas the higher taxa are formulated according to nontestable criteria—that is, similarity of origin, or similarity of structure and function. All other categories above the species level are more or less arbitrary groupings, designed by taxonomists for the purposes of arranging and ordering species with respect to one another. All categories above the species level are referred to as *higher taxa* (singular, *taxon*).

Binomial Nomenclature

In addition to advocating the hierarchical grouping system, Linnaeus supported the adoption of other rules for taxonomy. Organisms were to be identified by their genus and species names, with the former capitalized and the latter uncapitalized. Thus human beings are classified as *Homo sapiens* and the lion as *Felis leo*. This system of naming in taxonomy is known as *binomial nomenclature*, which means "naming with two names."

Linnaeus also insisted that all taxonomic names be latinized. He even did this to his own name (which was Karl von Linné in its Swedish form). There were two main reasons for the choice of Latin. First of all, it was the language used by all scholars in Linnaeus's day. Second, since it was a "dead" language, the meanings of its words were less likely to change. The use of Latin in taxonomy does not mean that names for all organisms must come from Latin words. Many names are chosen from Greek, French, English, and other languages. But whatever their origin, the words are given Latin endings.

Certain rules of writing taxonomic names have become generally accepted. The genus and species names are always written in italics, or underlined, as in *Fasciola hepatica*, the sheep-liver fluke. A trinomial nomenclature is occasionally used. For example, Linnaeus classified the Caucasian human group as *Homo sapiens sapiens*. The third term in these cases is a subspecies designation.

Such rules as these are written and enforced by international commissions. In zoology, the first commission met in 1842 and included among its members Charles Darwin. The rules adopted by this commission must be observed by every biologist who wishes recognition for work. Some of the rules are quite legalistic and complex; in fact, most professional biologists are not entirely familiar with them. No such board exists for plant taxonomy, though a set of written rules of nomenclature exists for botanists.

Due to the influence of the International Commission on Taxonomy, the discovery and attainment of international recognition of a new animal species follows a fairly standard pattern. Suppose an entomologist captures an insect and believes it to represent a new species. The procedure to have the find recognized might well be as follows:

Recognition and Description of the Organism. This includes studies of the anatomy, physiology, morphology, and other pertinent information that will enable the entomologist to figure out what relationship the species bears to other known groups.

Classification of the Organism. Assigning the specimen where it belongs in the heirarchical system of Lin-

naeus depends on the evidence obtained in the first step. It is accomplished by "keying out" the organism—comparing its description with printed descriptions of this general type of organism down to the point at which the specimen either does or does not fit the description of a species already entered in the taxonomic key. If it does, of course, it is not a new species. The use of a key in biology is just a formalized game of twenty questions.

A complication in classification is encountered in species having members that exist in many different anatomical forms. Such **polymorphism** (*poly* = many; *morph* = form) is shown by many colonial insects—for example, ants, bees, and termites (see Fig. 21.4). The worker bee, the queen, and the male (drone) are all quite different; yet they are all the same species. The taxonomist must make sure that newly discovered types are not classified as new species when in reality they are simply different forms of a known species.

If it is revealed that the organism represents not a new species, but an old one already described, the entomologist simply assigns the specimen its proper place within the grouping. If the organism represents a new species, however, it will not fit into the key. The entomologist must now describe it very carefully, point out how it fails to fit into the established keys, present illustrations of the organism, and explain why he or she thinks it represents a new, undescribed species.

It is important to point out that classifying an organism, or "keying it out" does not *necessarily* say anything about its phylogenetic relationship to other organisms or groups with which it is placed. Some keys and some classification schemes are based on presumed phylogenetic relationships and some on pure similarity of form and/or function. Most do, however, bear some relationship to evolutionary history in the broadest sense (for example, most keys are for broadly related groups such as flowering plants, mammals, or insects).

Choice of nomenclature. The naming of the new species is the simplest step of all. The entomologist chooses a name, latinizes it, and publishes the findings in a short

(a)

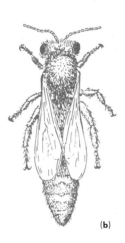

(b)

(c)

Fig. 21.4
A well-known example of polymorphism is shown by the three forms of the honeybee, *Apis mellifera*. (*a*) The sterile female worker bee. (*b*) The fertile queen (female). (*c*) The drone (male). Although quite different in appearance, all three represent the same species and are genetically members of the same breeding population.

paper devoted to a description of the newly discovered species. Special emphasis is put on those respects in which the species differs from the most nearly similar type. Illustrations are given with attention to *precision*.

The biologist has long been the subject of many jokes—some good and some bad—about choices of names for living organisms. They seem, to the uninitiated, needlessly complex. What the average person calls a rat, the zoologist calls *Rattus rattus rattus*.

Yet, there are excellent reasons for the biologist's choice of the longer road. First, the use of common names for organisms leads to a great deal of confusion. Often the common name varies from locality to locality, within even a few square miles. For example, members of the insect order Odonata are commonly called dragonflies, but they are also known as horse stingers, mosquito hawks, snake feeders, snake doctors, witch doctors, and devil's darning needles. By contrast, only one name for this order of insects is used by biologists all over the world.

Second, a single common name is often used indiscriminately for a wide range of species. In fact, in common usage the name "rat" can be given to many different animals (for example, muskrat). Similarly, the term "wolf" is applied not only to the familiar wild relative of the dog, but also to the Tasmanian "wolf," which is in fact a marsupial, as we saw in Fig. 21.2. On the other hand, the biologist's term *Rattus rattus rattus* pinpoints precisely the genus, species, and subspecies being considered.

Biology students often become firmly convinced that taxonomic names are chosen on the basis of being difficult to learn. However, if the words seem difficult, they are not deliberately made so. On the contrary, most biological naming is done with an eye to its being both simple and descriptive. For example, the lily named *Tigrinum giganteum* is the giant tiger lily. The one called *Tigrinum speciosum rubrum* is a red variety.

The Problem of Species Definition

The definition of a species was far simpler in the time of Linnaeus than it is today. In the eighteenth century, it was believed that species were fixed and unchanging.

Therefore, it was thought that detailed descriptions of plants and animals would apply equally well at any point in time, past, present, or future.

However, as ideas of evolution became widely accepted and paleontological work brought to light fossil forms with no counterpart among living organisms, it became obvious that the old idea of unchanging species was incorrect. Today, we think of living species as merely the top branches of a constantly growing tree. If we were to cut across the branches at any level, we would see the species distribution as it existed *at that time*.

The procedure followed in early taxonomy was to describe a given species in terms of a detailed anatomical study of a few "representative" specimens of that species. During the eighteenth and nineteenth centuries, examples of new and undescribed species discovered in remote areas of the world were sent back to professional taxonomists whose job it was to describe and classify the organisms. After examining a new organism, the taxonomist would decide whether it was merely a subdivision of an existing species or was different enough to be rated as a new species. The number of specimens the taxonomists studied, however, was usually limited to one or two of each species, and the study did not take them outside the laboratory or workroom. Later taxonomists took to the field to study specimens and thus were able to study species not only on a larger scale, but also on the basis of a different criterion—the presence or absence of cross-breeding.

As the following example illustrates, breeding patterns reveal species distinctions that would be overlooked in a simple anatomical examination. About forty different species of the small fruit fly *Drosophila* live in the vicinity of Austin, Texas. Many of these species are quite similar anatomically, so much so that it would take a real expert to note any distinguishing features. The differences are far less extreme, for example, than between a bulldog and a Great Dane, which are members of the *same* species. Yet, study of these flies in their natural habitat shows that none of these species crossbreed with each other. Their differences are great enough so that cross-breeding, or hybridization, does not produce viable offspring. If the older species concept had been applied to this problem, and a few flies

---

Biologists have long attempted to devise a single criterion, such as ability to interbreed, for distinguishing between species. But no single criterion has ever been successful in defining species across the animal, plant, and microbial kingdoms.

studied anatomically under the microscope, taxonomists would have come to different conclusions concerning the number of species and the distinctions between them. By observing the organisms in relation to their environment, the investigator can obtain a more biologically meaningful grouping of the flies into species.

Cross-breeding is used more frequently than almost any other criterion for defining species. Yet, this criterion also poses problems that make it difficult to apply to all organisms. Consider the following examples:

1. Cross-breeding has always been more useful to zoologists than botanists or microbiologists as a species criterion. Even with animals, however, the matter is by no means clear-cut. Lions and tigers have produced viable offspring in captivity (the cubs are called "ligers"). Yet no case of cross-breeding between the two species has ever been reported in nature. It would seem from this observation that *Felis leo* (lion) and *Felis tigris* (tiger) are biologically distinct species, despite the physiological capability of interbreeding.

2. Two plant populations can sometimes hybridize yet, by all anatomical, physiological, or ecological criteria, appear to be completely separate and distinct species.

3. Many microorganisms seldom, if ever, show a sexual phase in their life cycles. For these forms the cross-breeding criterion is meaningless. Yet, species of microorganisms do exist as clearly and unmistakably as species among multicellular forms.

These are just a few of the problems encountered in attempting to arrive at a satisfactory species definition. We will now look at some of the ways modern taxonomists have attempted to develop a species concept applicable to all types of organisms.

Biologists think that closely related species have diverged from a common ancestor. Divergence can be seen in various stages throughout nature today. What may be only varieties of a single species at one point in time may become distinct species at another. For example, all housecats (Siamese, Persian, Angora) are today considered members of one species, *Felis domestica*. In the distant future, however, if there is no gene flow between existing varieties, they may diverge enough to be considered members of separate species. The species concept developed by modern biologists tries to account for this dynamic quality of species.

The most workable species concept considers the species as an active population of organisms with many anatomical, physiological, and behavioral characteristics in common. In many cases, physiological or behavioral differences can distinguish species where anatomical ones cannot. This way of looking at species may be called the *multidimensional species concept*. It is based on the idea that *no single criterion* is sufficient to define a species. In accordance with the multidimensional concept, the taxonomist tries to take all an organism's aspects—morphology, behavior, physiology, and reproductive patterns—into account in asking whether two groups of organisms represent the same or different species.

21.4 SOME MODERN TAXONOMIC APPROACHES

Modern taxonomic techniques differ considerably from those of Linnaeus's day. Part of this difference is the result of vast improvements in instruments. Computers, for example, have greatly facilitated the handling of taxonomic data. Most of the change, however, results from recognition of new criteria for establishing taxonomic relationships between living organisms.

Behavioral Criteria

Differences in behavior patterns have become one such important criterion in modern animal taxonomy. For example, differences in the mating behavior of the three- and nine-spined stickleback fish are a species-distinguishing characteristic. There is nothing else to prevent them from mating, for when their eggs are artificially inseminated, they produce healthy hybrid offspring. But the three-spined male has one pattern of courtship behavior (a form of swimming "dance") that lures only three-spined females. The nine-spined stickleback has a slightly different dance, one that attracts only nine-spined females. It is interesting to note that the dances of both males are similar at the start. Therefore, a three-spined male can hold the attention of a nine-spined female until the moment his behavior deviates from the pattern of her species. As soon as it does, she "loses interest." In this manner, hybridization is prevented from occurring in nature.

Another example of the use of behavior patterns in taxonomy can be seen in Fig. 21.5. Here, analysis of sound records of cricket calls provides the basis for separating the members of a single genus into different species. Cricket calls are emitted by males to attract females. A female will respond only to a call that closely resembles the pattern for her species. Mating does not occur between species whose patterns differ greatly from each other.

Biochemical Criteria

Biochemical differences are an important means by which species may be distinguished. For example, some types of protein, such as pigments or the blood protein hemoglobin, vary in their composition from species to

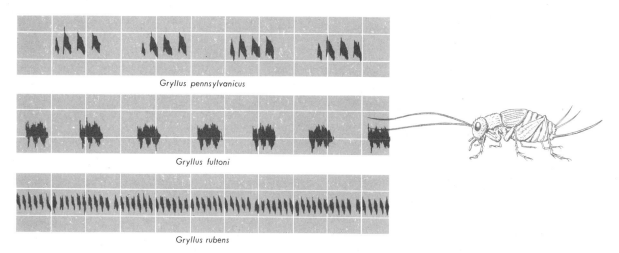

Fig. 21.5
Sound patterns recorded for three species of common field cricket. Such data reveal that six different species of field cricket exist in the eastern United States alone. The use of traditional taxonomic procedures had earlier led taxonomists to believe that all the crickets in the western hemisphere were of a single species.

species. The composition of proteins can be determined by the process of electrophoresis, and the electrophoretic patterns for the same proteins in different species can then be compared (see Fig. 21.6).

Biochemical and molecular studies are an important new aid to taxonomists working with all groups of

animals or plants. But they are especially useful to those working with classification of microorganisms. Two bacterial populations may appear completely identical in terms of morphological traits. However, the sum total of their biochemical differences may be great enough to consider them a separate and distinct species.

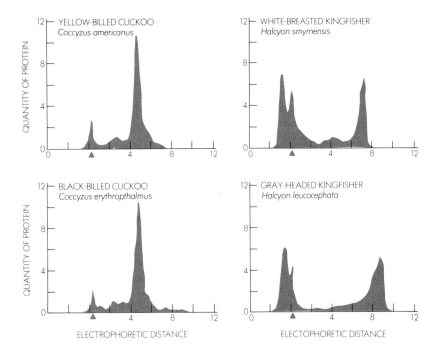

Fig. 21.6
Electrophoretic patterns of egg white proteins. Note particularly the similarities between the species of a single genus, in this case the cuckoo (*Coccyzus*) and the kingfisher (*Halcyon*), and the obvious differences between the two species. (Based on photos in C. G. Sibley, "The Electrophoretic Patterns of Avian Egg-White Proteins as Taxonomic Characters," *The Ibis* 102, pp. 215–284.)

SUPPLEMENT 21.1
NUMERICAL TAXONOMY

Numerical taxonomy is a new approach to taxonomy that has flourished with the development of the digital computer that allows the manipulation of large data banks. Taxonomists have always used measuring and counting as a principal means to identify characters that distinguish individual organisms. But classical taxonomists could handle only a few quantitative characters at a time in the absence of the computer. Furthermore, with the computer it is also possible to use sophisticated statistical methods to analyze the data.

In its operational form, numerical taxonomy involves several procedures, or steps, shown in summary form in Diagram A. First, a sample of the organisms to be classified is collected in the field. Second, traits to be used in classifying the organisms into categories are selected and measured quantitatively. Third, the quantitative data are entered onto data cards or recorded on computer tape. Fourth, the data

are fed into a computer that is programmed to compare the individuals or groups for degrees of similarity or difference on the basis of the measured traits. And finally, the computer output is used to set up taxonomic categories, or place the specimen into already existing categories.

In general, taxonomists prefer to use a number of different traits of varying sorts for constructing classification schemes. For example, a classification based on external characteristics of a group of organisms might be very different from one based on internal characteristics, such as arrangement of the nervous system, structure of the heart, and the like. Similarly, anatomical characteristics might yield a very different classification from physiological characteristics. Thus, the more traits used the greater the possibility of achieving maximum consistency and constancy of groupings. However, until the advent of the computer the

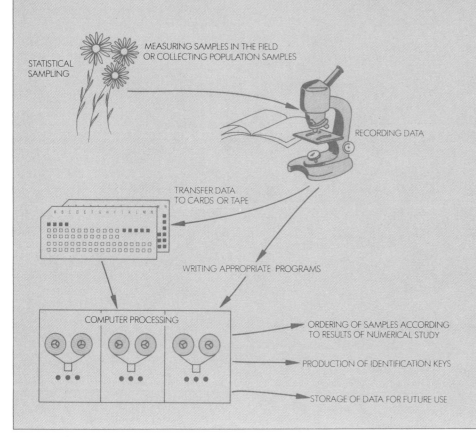

STATISTICAL SAMPLING

MEASURING SAMPLES IN THE FIELD OR COLLECTING POPULATION SAMPLES

RECORDING DATA

TRANSFER DATA TO CARDS OR TAPE

WRITING APPROPRIATE PROGRAMS

COMPUTER PROCESSING

ORDERING OF SAMPLES ACCORDING TO RESULTS OF NUMERICAL STUDY

PRODUCTION OF IDENTIFICATION KEYS

STORAGE OF DATA FOR FUTURE USE

Diagram A
(From O. Solbrig and D. Solbrig, *Introduction to Population Biology and Evolution.* Reading, Mass.: Addison-Wesley, 1979.)

use of many traits as the basis of a taxonomic system was simply not practical. The human mind is not capable of comprehending a vast array of different traits simultaneously. Thus numerical taxonomy would not be possible without modern computers.

Obtaining sufficient quantitative measurements of phenotypic traits is a major problem that numerical taxonomists encounter. In the case of easily quantifiable traits (such as number of petals on flowers, or bristles on an insect's legs) numerical data can be collected straightforwardly. In cases of less easily quantified traits (such as color), a special technique can be employed. Varying states of a phenotypic trait can be given numerical values and entered into the computer, being treated as actual measures. For example, leaves of many plants are "hairy"—that is, covered with innumerable small hairs like those found on the leaves of African violets. It would be difficult, or impossible, to count all the hairs on all the leaves, yet for many related species or subspecies variation in hairiness is an obvious trait. In this case the numerical taxonomist makes a ranking to which numbers are assigned—for example, 0 = hairless leaf, 1 = sparsely haired, 2 = regularly haired, and 3 = densely haired. (The same procedure could be applied to even more qualitative characteristics such as color.) In this way numerical taxonomists can enter values for any trait, however difficult to actually measure.

The function of any taxonomic system is to group together in the same category organisms that are more similar to each other than they are to any other group of organisms. To do

this each organism must be compared to every other and some measure of their similarity established. For example, we may simply count the number of characters that they have in common, and decide that the more characters the two organisms share, the more similar they are. But since some organisms have more characters than others (a lizard has legs but a snake does not), we might want to use a proportion of all characters that are shared rather than the absolute number. Other statistical techniques have been developed that take into account not only identity but also degree of similarity, in determining relationships between organisms. Using the computer, numerical taxonomists plot the degree of similarity and/or difference.

To understand the principle of clustering, consider Diagram B. This graph represents three different traits—symbolized as A, B, and C—being considered simultaneously. The value for each trait is plotted on the appropriate axis. Where the three axes meet in space we can plot a point. This point represents a single organism, compared for traits A, B, and C. The procedure is repeated for other individual organisms from the same or similar populations. This produces clusters of points in a three-dimensional space. These clusters represent natural groupings of organisms based on a set of compared similarities. As shown in Diagram B, there are two clusters, which might be considered two species or subspecies based on other criteria such as reproductive isolation or geographical distribution.

The geometric representation shown in Diagram B, however, can accommodate only three traits at a time. With the

Diagram B

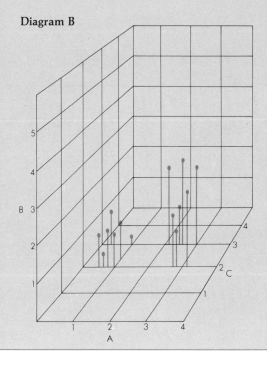

computer, where representation is not limited to visual space, there is no restriction on the number of traits that can be compared at one time; thus, clusters based on many characters can be produced.

Compared to classical taxonomy, numerical taxonomy allows far more characters to be used in arriving at taxonomic conclusions. It also presents data in a more quantitative way. However, it does present several basic problems. The first is that of giving quantitative values to qualitatively varying characters (colors, for example). This introduces a kind of artifact. The characters may seem to have more quantitative differences than are real. (How, for example, is it possible to quantify colors?) A more important problem is that there is no agreement regarding the best criterion of similarity. Decisions are still subjective in the last analysis, a problem that no taxonomic system or technique can fully avoid. What numerical taxonomy does provide is a way of making a number of objective evaluations of traits that *can* be measured, and on whose importance (as characters for taxonomic purposes) there is general agreement.

An Example of the Species Problem: Leopard Frogs in North America

An example of the complex problem of using a single criterion for taxonomic purposes (even one as important as the ability or lack of ability to interbreed), is illustrated by the case of the common grass or leopard frog. As shown in Fig. 21.7(a), there is a continuous distribution of this frog from Vermont to Florida, and westward through Louisiana, Texas, and the midwest (Kansas, Missouri, and Illinois). The original taxonomic studies of these frogs began in the late 1930s and early 1940s with the work of John A. Moore, then at Columbia University. An embryologist, Moore was interested in the effects of the sperm nucleus on rate of development in frogs. Quite by accident, he found that when he used sperm and eggs from geographically diverse populations (such as from Florida and Vermont), he obtained abnormal, and usually nonviable embryos. Moreover, there appeared to be a direct relationship between geographic distance apart of the parental populations and degree and frequency of abnormal hybrid embryos. The farther the two parental populations were apart geographically, the less successful the hybridization was. On the other hand, crosses between neighboring populations of frogs—for example between Vermont and upstate New York populations—produced relatively few abnormal embryos. Each population seemed to be able to interbreed with its nearest neighbor with no problems, even though the more extreme regions of the range of distribution seemed incompatible. Moore concluded that the different populations, especially from north to south, contained different variations and that the more geographically distant the two populations, the more genetic variability existed between them. These diffferences produced genetic incompatibilities which emerged as abnormal embryonic development.

On the basis of these observations, Moore argued that the entire population of leopard frog, from Vermont to Florida, Louisiana, and the midwest was a single species subdivided into local geographic populations or subspecies. He argued, but did not demonstrate with any data, that since genes could flow from Vermont to Florida by passing through each neighboring population, gene flow could persist from Vermont to Florida, Louisiana, Texas, and Missouri. The single species name *Rana pipiens* was given to all these forms. Moore's taxonomic proposals became standard throughout the literature, and were repeated as an example of the population genetics approach to taxonomy in many textbooks (including previous editions of this one).

In the mid- to late 1970s, however, Moore's scheme was challenged. It was pointed out that in analyzing his different geographic variants, Moore had considered all the frogs from any single state (with the exception of New York, which he divided into northern and southern) as one population, summarizing all their variations as numerical averages. This procedure masked much of the important variation in frog populations found within states. It also smoothed over abrupt changes *between* adjacent populations when these did not happen to coincide with state boundaries. The resulting picture was one of a series of gradual population differences, running both north-south and east-west. Subsequent field studies have shown that there are often sharp discontinuities in traits from one population to another.

Modern studies of the east coast and midwestern leopard frogs have revealed some important new data bearing on the taxonomic issue. Hybridization between neighboring populations of frogs does occur, but far less frequently than Moore and others thought. Zones of hybridization exist where two population boundaries meet, but the density of individuals in those zones is often low. As a result the number of chance interactions is greatly reduced. Gene flow through the leopard frog populations from Vermont all the way to Florida or the midwest by way of local hybridizations is probably a rare event. In addition, data from other sources—for example, studies of mating calls and enzyme differ-

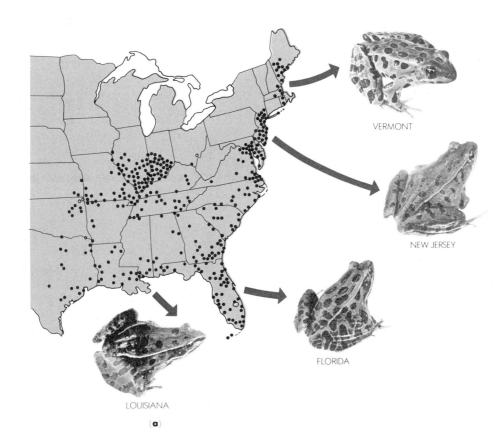

VERMONT

NEW JERSEY

LOUISIANA

FLORIDA

(a)

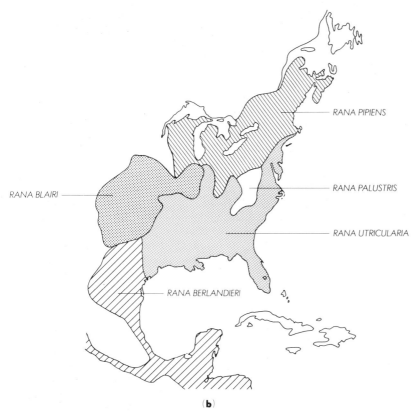

RANA PIPIENS

RANA PALUSTRIS

RANA BLAIRI

RANA UTRICULARIA

RANA BERLANDIERI

(b)

Fig. 21.7
Maps of the eastern, southern, and midwestern United States showing distribution of various populations of the leopard or grass frog. The question has been for some years: Are these all the same species, or members of different species? (*a*) Distribution map showing local populations (dots), treated as a single species, *Rana pipiens*. Photographic insets show basic external structure and spotting patterns for representatives of four geographic populations. This map represents the scheme proposed by John A. Moore and others in the 1940s. (*b*) Revised distribution map showing ranges for population treated as four separate species (there are more; only four are shown here). Different hatching distinguishes geographic ranges of each species. At the boundaries between populations hybridization has been observed occasionally. The entire group of species is now known as the *Rana pipiens* complex. Even though some hybridization has been observed at various boundary points, other data (mating calls, enzyme differences) suggest that the various groups deserve to be ranked as separate species. With limited hybridization, gene flow from widely separate geographic regions is likely to be insignificant. (Photos courtesy John A. Moore.)

ences—indicate that local populations are quite a bit more distinct, in many cases, than morphological traits have suggested. As a result, taxonomists now believe that *Rana pipiens* in the eastern and midwestern regions of North America is not a single species, but a number of separate species, as shown in the second map in Fig. 21.7(b).

The example of the taxonomy of *Rana pipiens* illustrates the importance of a multidimensional approach to defining species. To most biologists today a combination of data from many fields—morphology, behavior, biochemistry, physiology, and geographic distribution—must be used to understand the concept of species. Such data must also be used in classifying any particular organisms being studied in the laboratory or museum. The concept of species is still a useful one in biology. But if the concept is used in a rigid, static way, it can obscure our understanding of the dynamics of life. If species are nothing else, they are constantly evolving, changing groupings. Our species concept must reflect that fact.

21.5
TAXONOMY AND PHYLOGENY: DIFFERING SCHOOLS OF THOUGHT

The central problem of modern taxonomy is how to produce a classification system that will group species together by similarity of appearance and ecological positions, while also reflecting their true evolutionary

history. All taxonomists are in agreement that this is the ideal aim of any classification scheme.

However, achieving this goal is not always easy. There are two general reasons for this. The main one is that taxonomists usually lack direct evidence about evolutionary history—especially recent history—of most species. Fossil evidence is not always available, or if it is, does not always tell a very complete story. In general, biologists *infer* evolutionary history from observing the characters of modern-day forms. Such inferences involve a good deal of guess-work. Inferences are, by definition, not direct observations, and thus constitute *secondary*, rather than *primary* information. Thus a dilemma exists for the taxonomist: to base a classification scheme directly on inferred evolutionary relationships would be bypassing first-order information and basing a whole scheme on second-order information. In other words, it would be like grouping people into categories according to inferences about their intentions or motivations (which must usually be inferred from their behavior) rather than from direct observation of the behavior itself. Many taxonomists feel uncomfortable about basing a taxonomic scheme primarily on second-order information.

It is important to emphasize that when good fossil or other direct historical evidence is available (for example, past records of frequency and distribution of forms, such as game management or insect control data from the past), and the evolutionary history of any group is well established, all taxonomists would agree that this information should form the basis for classification. The problem is that the direct evidence for doing so is often lacking.

There is another problem as well. As two or more species evolve from closely related ancestors, they often diverge considerably in their phenotypic (including biochemical) characters. But they may evolve at different rates, so that of three forms that are thought to have diverged from a common ancestor, two may resemble each other much more closely than either resembles the third. Yet a solid evolutionary *inference* suggests (but cannot demonstrate) common ancestry. How is the taxonomist to judge the most appropriate classification scheme when direct observation of traits suggests one grouping, but indirect evolutionary inference suggests another? In a converse situation, sometimes two or more species from quite different ancestry may independently evolve to look very much alike (a phenomenon known as evolutionary *convergence*; see Chapter 24). Unless the evolutionary history is known with any reliability, it is quite possible that historically unrelated forms could end up being placed in the same taxonomic category (species, genus, order, or the like), based solely on phenotypic similarity. Modern taxonomists are in a dilemma over just this issue, as the following example should make clear.

Phenetics and Cladistics

At the present, there are two major schools of thought about how best to construct classification schemes. Each school emphasizes a different set of criteria and ultimately reflects a different sense of the purpose of taxonomy as a science.

The Phenetic School.   Adherents to this school (called *pheneticists*) argue that in most cases it is impossible to determine evolutionary relationships with any accuracy, and hence taxonomists should not attempt to construct classification schemes to reflect phylogeny. The major thrust of phenetic classification schemes is to group together, using objective, repeatable procedures, organisms that are similar to each other in phenotypic appearance, behavior, physiological processes, or biochemical composition. Pheneticists look at the totality of such phenotypic traits—that is, they are not contented with examining just one or two traits. But their main interest is a logical, convenient, and useful classification scheme based on more objective, primary information (that is, direct observation and measurement of the phenotypic traits). Pheneticists have been particularly attracted by the approach and techniques of numerical taxonomy (see Supplement 21.1) because of its more quantitative and objective character. Emphasizing these latter criteria, pheneticists place less weight on inferences from the fossil record or from comparative anatomy and embryology. Field manuals for identifying species are good examples of the pheneticists' approach to taxonomy and classification.

The Cladistic School.   Adherents to this school (called cladists) emphasize that classification schemes should reflect as much as possible historical (phylogenetic) relationships. Cladists feel that any classification scheme that ignores or treats secondarily evolutionary relationships is biologically meaningless. To cladists, the major criterion for determining which group or groups to place together in higher taxonomic categories is whether the groups seem to share a common ancestry. Cladists think that the history of related species can be indicated by branching points in a cladogram, a chart combining some elements of a classification chart and some elements of an evolutionary diagram (see Fig. 21.8). From a variety of data, but including inferences about evolutionary history, branch-point relationships are determined and a cladogram prepared. Once the cladogram for a particular group has been worked out, an appropriate classification (grouping) scheme can be developed from it.

To understand the cladistic approach, consider the hypothetical example diagrammed in Fig. 21.8. A, B, C, and D represent four individual species. The taxonomic problem is to decide how to group them into higher categories, such as genera or families. From available data, as the cladogram shows, it is inferred that A and B

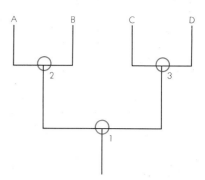

Fig. 21.8
Hypothetical cladistic diagram (cladogram) for the taxonomy and ancestral relationships among a group of organisms (A, B, C, and D). The groups could be individual species or higher taxa such as genera or families. Cladograms emphasize taxonomic position as a function of both phenotypic characteristics and evolutionary relatedness. Relationships are shown schematically as branch-points. The lines are drawn as square angles to emphasize that at each branch-point the divergence is equivalent taxonomic; that is, the divergence does not produce a main line of development on the one hand and a minor offshoot on the other (of course, in terms of modern descendants, one group may go on to develop many descendants, and the other become only minor or die out altogether). Cladograms do not try to reflect such developments. The cladogram shows only degree of relatedness. The more closely two groups are related, the fewer branch-points between them on the cladogram. Thus A and B would be more closely related to one another than either would be to C or D. From such a cladogram taxonomists would tend to group A and B into one category (genus, species, and the like) and C and D into another.

share a common ancestor (a common branch-point, labeled 2) as do C and D (a common branch-point labeled 3). From the ancestral relationships shown in the cladogram, cladists would group A and B together, and C and D together, each (perhaps) as a separate genus. The two genera might then be grouped into the same family (consisting of *four* species), because they all share a more remote common ancestor two branch-points away (branch-point 1). The degree of relationship between any two existing groups is determined from a cladogram by counting the number of branch-points back to the common ancestor. The more branch-points back, the more distant the relationship will be.

It should be emphasized that cladists' branching diagrams are not equivalent to evolutionary trees, such as those we are used to seeing in discussions of animal and plant evolution (for example, see Supplement 24.1). Evolutionary trees attempt to represent some sense of actual time scale for various divergences; cladistic diagrams do not. Although branch-points at the top of the diagrams are thought to have occurred more recently than those at the bottom, cladograms are *not* meant to represent historical time charts. Keep in mind that cladists are first and foremost concerned with classification, with grouping organisms. What cladists try to do is use phylogenetic inferences to make their classification schemes compatible with evolutionary history. For this reason the emphasis on the branching pattern in cladists' diagrams is a guide to assigning taxonomic position for various groups.

Cladists recognize that their approach involves more complex and subjective evaluaton of data than the pheneticists' approach. Pheneticists have, in fact, set themselves a simple task. On the one hand, because evolutionary inferences are subjective, and more often subject to great differences of opinion, pheneticists do not base their schemes on such guesswork. By downplaying second-order information (inferences), they are able to obtain more quantitative, objective, and repeatable groupings (that is, groupings on which more people could agree).

However, the simplicity of the pheneticists' approach has also turned out to be its greatest limitation. Most biologists feel that any classification scheme is incomplete if it does not reflect, as nearly as evidence will allow, phylogenetic relationships. There is the strong conviction that, like it or not from a taxonomic point of view, all species on earth have originated from common ancestors, and that any taxonomic scheme will be unsatisfactory unless it reflects those relationships. As a result of this long-standing evolutionary perspective, the cladistic approach is gaining more ground among today's younger population biologists.

Cladistics and Phenetics in Practice: Classification of the Higher Vertebrates

The cladogram in Fig. 21.9 shows the hypothesized ancestral relationships among some of the most prominent modern vertebrates (amphibians, reptiles, birds, and mammals). We can use this chart as an example of how taxonomists might approach the problem of assigning groups of organisms to higher taxonomic categories—in this case to that higher category called a "Class."

Consider first the case of egg-laying (for example, duck-billed platypus), marsupial (pouched), and placental mammals. Phenotypically, all three groups share several very important traits: they are warm-blooded, have hair covering their bodies, and nurse their young with milk produced in mammary glands. On a phenotypic level, the marsupials and placental mammals share many more traits in common (for example, giving birth to live, as compared to egg-encased, young) than either shares with the egg-laying mammals. Fossil evidence confirms this basic view, so that on the cladogram (which, by definition, includes evolutionary information, or inference) the placental and marsupial mammals are shown derived from a common branch-point. Together, they share yet another branch-point one step

further back in time and taxonomic distance with the egg-laying mammals.

Note that although the placental mammals are a more widespread and important group than either the marsupials or the egg-laying mammals today, the lines on the cladogram are drawn equidistant and with equivalent degrees of branching. (The more traditional evolutionary charts tend to show the placental mammals as a main line, with the egg-laying and marsupial mammals as off-shoots). The form of the cladogram emphasizes the taxonomic equivalence of each group, irrespective of its evolutionary importance or its current prominence.

On the basis of both phenotypic similarity and commonality of branch-points, the three groups could logically be placed, and with a minimum of controversy, into the same higher taxon, the class Mammalia. The grouping is easy—a point on which cladists and pheneticists in this case would easily agree—because phenotypic and evolutionary evidence are in close agreement.

The matter is not so easy in the case of the reptiles and birds, however. For many years biologists have

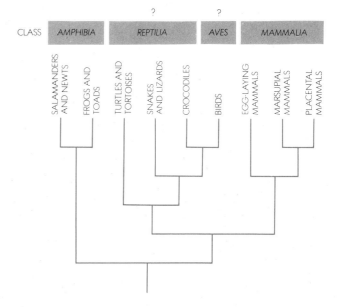

Fig. 21.9
Cladistic diagram (cladogram) showing the branch-point relationships between the major groups of vertebrates (terrestrial). Traditional grouping of these categories into higher taxa such as Class is shown in the colored boxes above the top of the chart. The ones with question marks are those where some modern controversy exists between the cladistic and phenetic approach. (Modified from Colin Patterson, *Evolution*. Ithaca, N.Y.: Cornell University Press, 1978, p. 126. Copyright Trustees of the British Museum (Natural History), 1978.)

noted that birds share a number of phenotypic characters with reptiles, such as presence of scales on the legs and modified scales (feathers) on the body, a four-chambered heart, and reproduction with very similar egg types. Evolutionarily, birds have long been considered derived from ancestral reptiles, most notably a group of small dinosaurs, the theocodonts (see Section 29.13). However, among living reptiles, evidence from fossils and from comparative anatomy suggests that birds share their most immediate common ancestry with the crocodiles (as shown in the cladogram in Fig. 21.9).

From a taxonomic point of view, grouping the birds, crocodiles, and other modern reptiles into higher categories (class) presents a problem. Birds have evolved a number of specializations for their mode of life (flying) that phenotypically separate them from any modern reptiles. For one thing, they are warm-blooded, whereas all modern reptiles are cold-blooded.* For another, feathers are really a rather specialized modification of scales—the similarity is certainly there, but it is much obscured. And, of course, flying is a very divergent form of locomotion from the walking, crawling, or swimming that the reptiles use. For these reasons, birds have traditionally been grouped on a sort of phenetic basis into a separate class, Aves, from the reptiles (Reptilia).

From the cladogram in Fig. 21.9, however, a good case can be made for grouping the birds with the class Reptilia, rather than creating it as a separate category. Indeed, this is what some cladists today are advocating. Such a grouping would more clearly stress the evolutionary relation of birds to reptiles, and particularly that of birds to their nearest living reptilian relatives. On the grounds of the many phenotypic differences between birds and reptiles, however, pheneticists might argue somewhat more strongly for keeping the traditional distinction between Aves and Reptilia. Since the evolutionary relationships between birds, dinosaurs, and modern-day crocodiles involve much inference from an incomplete fossil record, pheneticists would argue that the important evidence on which to base a classification should be phenotypic traits. Pheneticists' arguments are particularly strong in cases like this where phenotypic differences are great anyway. Note that pheneticists do not deny the evolutionary data. In fact, most pheneticists would agree strongly with the hypothesis that birds are derived from the dinosaurs. What they would argue is that such hypotheses should be given less weight (or in some cases no weight at

all, if the evolutionary data are sparse) in assigning taxonomic positions than the evidence of objectively observed and measured phenotypic traits.

21.6
CONCLUSION: TAXONOMIC SCHEMES AND EVOLUTION

Modern taxonomy is based on the attempt to construct a system of classification of modern and extinct species that will reflect both existing differences among organisms and their phylogenetic or evolutionary history. A phylogenetic chart of all the major phyla of the living world is shown in Fig. 21.10. Such charts are different from the cladogram shown in Fig. 21.9. The difference is that phylogenetic charts usually contain some element of time span, whereas cladograms, as pointed out on p. 660, do not. Phylogenetic charts try to suggest actual paths of divergence and relatedness in a more historical way than cladograms. Cladograms express relationships of affinity; phylogenetic charts show more of the actual historical development: when certain groups diverged in time, how closely two lines remain to one another after they diverge, and the like. While there are elements in common between cladograms and phylogenetic charts, it is important to keep in mind that they are not equivalent.

Both cladograms and phylogenetic charts are at best scientific hypotheses, or models, showing a certain presumed relationship among a group of species. As such, they are subject to constant revision and modification as new data—from biochemical analyses, comparative anatomy and physiology, or biochemistry—become available. The major purpose of taxonomy is not only to provide an orderly method of accounting for, and understanding relationships among, the many forms of life on earth; it is also to highlight evolutionary relationships, and therefore provide an insight into the history of life on earth.

The phylogenetic system we have chosen to use throughout the remainder of this book, in dealing with the various major groups of organisms, is called the "five-kingdom system" (see Fig. 21.1). The five-kingdom system breaks the living world down into five major groupings, called kingdoms: the Monera (all prokaryotic organisms, including viruses and other very small, cell-like forms), the Protista (unicellular eukaryotic organisms, both those containing and those lacking chlorophyll), the Plantae (unicellular and multicellular eukaryotes with both chlorophylls a and b), the Fungi (nongreen organisms that reproduce by means of spores), and the Animalia (multicellular eukaryotes lacking chlorophyll). The five-kingdom system has almost universally replaced the older two-kingdom or three-kingdom systems much in use ten or twenty years ago.

---

*Recent evidence suggesting that at least some of the dinosaurs were warm-blooded, or exhibited some degree of temperature control, strengthens the proposed relationship between birds and this group of dinosaurs. However, from a pheneticist's point of view such information is too hypothetical to be used for taxonomic purposes.

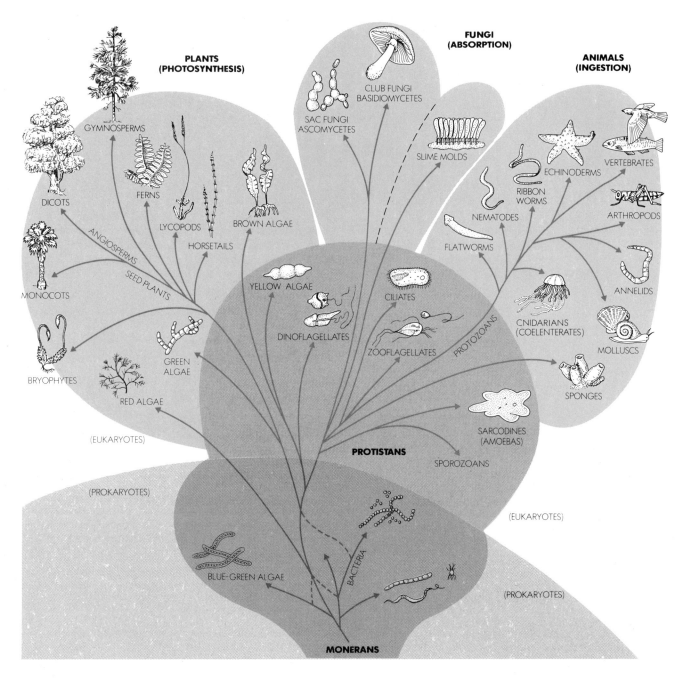

Fig. 21.10
Diagrammatic representation of the
five-kingdom system of classification.
The five basic kingdoms are designated
by the large oval groupings: the Monerans, Protistans, Plants, Fungi, and
Animals. Only the Monerans are prokaryotic in their cell structure; members
of the other four kingdoms are composed of eukaryotic cells. (Redrawn
from CRM, *Biology Today*, 2nd ed.,
1975, pp. 36–37).

Summary

1. Taxonomy, systematics, and classification are distinct terms referring to different aspects of the process of describing and ordering the relationship among groups of organisms.

    a) Systematics is the comparative study of any group or groups of organisms and of any and all relationships among them by a number of different methods of observation. These methods can include field sampling, laboratory study, breeding experiments, biochemical analyses, and the like.

    b) Classification is the process of ordering organisms into categories, from species and variety to higher taxa. It is the process of actually assigning organisms to particular groups in a hierarchical system.

    c) Taxonomy is the theoretical study of classification, including its basic foundation, its principles, procedures, and rules.

2. Taxonomic schemes have several purposes:

    a) They provide a method of keeping track of the many known groups of animals and plants.

    b) They provide a method of determining relationships between newly discovered groups and groups already known.

    c) They show a variety of similarities between existing groups.

    d) They provide some insight into possible evolutionary relationships between existing and extinct forms.

3. The species is the fundamental unit among populations of organisms in nature. Therefore the species grouping is the fundamental unit in taxonomy. Contrary to opinion in the past, biologists are now agreed that species represent *real* groupings in nature. They are not just arbitrary categories designed by human beings for their own convenience.

4. Biologists used to look for a single criterion for defining the differences between species at all levels of the animal and plant kingdoms. It is now recognized that no single criterion, or rule, can be applied throughout the organismic world in deciding whether two groups of organisms belong to the same species or not. Yet, one criterion continues to be most important: whether the groups can or do interbreed in nature. If interbreeding does occur, it is likely that they are the same species; if interbreeding cannot occur or does not usually occur, it is likely that the groups belong to different species.

5. The modern species concept is called multidimensional because it emphasizes the use of many kinds of evidence in deciding whether two populations belong to the same species. In addition to the criterion of interbreeding, taxonomists also use morphological traits (anatomy), biochemical similarities and differences, behavioral traits (especially applicable in animals), ecological traits (like what the different populations consume for food, what organisms they serve as food, and how they interact with their natural environment), and geographic distribution (like how far the two groups are living apart and how often they might intermix). It is often necessary to use all these criteria to decide whether two groups belong or do not belong to the same species.

6. Species are grouped together into higher taxonomic categories such as genus, family, order, class, phylum, and kingdom. All these higher categories are essentially arbitrary creations. They do not represent groupings with biological reality. Thus more subjective criteria are generally used to group species into genera, families, orders, and the like.

7. Organisms are given taxonomic names on the basis of a system of binomial nomenclature. The first name is the generic and the second the species name. Thus the scientific name for the common housecat, *Felis domestica*, gives the species as *domestica* and the genus as *Felis*. The genus *Felis* contains other cats of different species, such as *Felis tigris* (the tiger) and *Felis leo* (the lion). Both names are necessary to identify the species.

8. Behavioral and biochemical criteria for taxonomy have been employed in recent years to reveal similarities or differences that cannot be detected by examination of morphology alone. Two species of cricket may look very much alike. But because their chirps are quite different, they do not interbreed in nature and thus exist, even side-by-side, as two separate species. Biochemical differences are very useful in establishing possible evolutionary relationships. Evolutionary divergence among populations manifests itself first at the DNA level; hence differences in DNA (or their product, protein) among populations may reveal otherwise hidden similarities or differences.

9. Two schools of taxonomy exist today with regard to the criteria on which a classification system should be based:

    a) The phenetic school maintains that classification schemes should be based solely on phenotypic similarity and/or differences; classification should not attempt to represent phylogenetic history.

    b) The cladistic school maintains that evolutionary relationships alone should guide the development of classification schemes. Evolutionary relationships are judged by the nearness of branch-points in a cladistic diagram: the nearer the branch-point to the present, the lower the position the population will probably be assigned in the classification hierarchy (that is, species, genus, family). Conversely, the further away the branch-points from the present, the higher will be the position in the hierarchy.

Exercises

1. What is the purpose of classification as a process in any area of human activity? Can classification schemes be arbitrary? What essential problem do biologists face that makes their efforts at classification difficult?

2. This chapter discussed at length how biologists view the concept of "species" and how they distinguish one species from another. Why are biologists so interested in the question of what a species is?

3. Describe the differences between the pheneticists and cladists as two schools of thought in modern taxonomy.

4. The hypothetical insects shown below belong to four different species of the same genus. For convenience these species are labeled A, B, C, and D. On the basis of the data given in Table 21.1 and the "additional information from the fossil record," determine where each species belongs on the blank phylogenetic chart. Explain your reasons for placing each species in the position you placed it.

INSECT SPECIES

A

B

C

D   "HORNS"

Additional information from the fossil record:
   a) All early fossils of members of this genus are completely white and hornless and possess normally developed wings;
   b) More recent fossil specimens of this genus have two spots on the abdomen (like species B);
   c) No fossil specimens with four spots on the adbomen have yet been found.

Table 21.1*

Species	Body entirely white	Horns present on head and thorax	Wings very short	Exactly two spots on abdomen
Hypotheticus albus	+	−	−	−
Hypotheticus cornutus	−	+	−	−
Hypotheticus brachypterus	−	−	+	−
Hypotheticus bimaculatus	−	−	−	+

*Example and data modified from E. O. Wilson, Thomas Eisner, Winslow Briggs, Richard E. Dickerson, Robert Metzenberg, Richard O'Brien, Millard Susman, and William E. Boggs, *Life on Earth*, 2d ed. Sunderland, Mass.: Sinauer, 1978, p. 527.

Now fill in the following cladistic diagram showing the presumed taxonomic and evolutionary relationships:

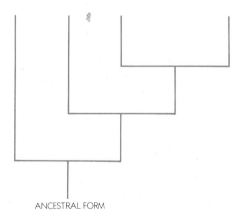

ANCESTRAL FORM

Chapter 22
Diversity and
Its Explanation:
The Origin of *The Origin*

22.1
INTRODUCTION

As we look at the world around us, one of its most striking features is the wide diversity of species which inhabit our planet. There are literally millions of kinds of living organisms of many different sizes, shapes, and modes of life. There is hardly any area of the earth that does not harbor some living organisms, adapted to the peculiar conditions of the local environment. Not only do we observe diversity between species of organisms, we also observe diversity between the individual members of any one species. Diversity seems to be a canon of the biological world.

Even more significant than our awareness of this diversity has been our attempt to explain its origin. Numerous theories have been advanced over the centuries to account for the origin of species. These fall into three broad categories: theories of *special creation*, theories of *spontaneous generation*, and theories of *transmutation of species*, or *evolution*. The three types of theories represent very different views about the nature of explanation in science, and very different ways of looking at living organisms.

Special creation theorists hold that species arose in the past as the result of a supernatural act (see Supplement 22.1). At the time of creation, each species possessed exactly the same characteristics as it does today. The theory of spontaneous generation, on the other hand, holds that many or all species have arisen more or less full-blown from nonliving matter, such as tadpoles from mud, or primitive cells from proteinaceous globules (see Supplement 26.1). Spontaneous generation assumes that once spontaneously generated, a species remains unchanged or fixed in its traits.

Explanations of diversity based on spontaneous generation and special creation hold that each species was created at a point in time and remained unchanged, or at least changed only within certain limits (but never into a completely new species). Adherents of both special creation and spontaneous generation tend to see the

world as static and fixed rather than dynamic and changing.*

Despite these similarities, spontaneous generation and special creation are different in two important aspects. First, special creation requires the action of a supernatural force, or creator; spontaneous generation, on the other hand, can be thought of as occurring by the action of chemical and physical processes that are part of our everyday world. In this sense, spontaneous generation is based on philosophical materialism, while special creation is based on philosophical idealism (see Section 1.7). Second, spontaneous generation, as a theory, can be tested, as described in Supplement 26.1, whereas special creation cannot. In science, that is a very important difference.

Adherents of theories of transmutation argue that species as we know them today have arisen by modification from previously existing species. However the first species originated, the constant process of transmutation has produced all the array of species that now exist. Modern species are literally descended from species that existed in the past. Not only has transmutation of species been going on in the past, but also it continues at present. It is, indeed, a dynamic concept.

From prehistoric times through the middle part of the nineteenth century the idea of special creation was the most widely accepted explanation for the origin of species. As science and philosophical materialism spread throughout Europe and America in the nineteenth century, the idealistic notion of a literal creation came increasingly into question. Naturalists were beginning to accept the idea that species arose over the course of time by modification of previously existing species. To many naturalists, such a belief was not necessarily incompatible with traditional religion; they could still

---

*Some versions of the spontaneous generation view in the past have held that old species die out and new ones are spontaneously generated, but still, from this point of view, change within a species is largely nonexistent.

SUPPLEMENT 22.1
EVOLUTION AS A HYPOTHESIS

On April 30, 1973, Senate Bill 394 became law in the state of Tennessee. The bill requires that all textbooks used for teaching biology within the state of Tennessee must give equal time and emphasis to theories of the origin of species that are alternatives to the concept of evolution. The law refers not only to secular theories such as spontaneous generation, but also to religious accounts such as that found in Genesis. In 1980 a similar bill was passed in Arkansas. Urged on largely by religious groups, equal-time bills are currently being considered by the legislatures in a number of other states. Tennessee's former anti-evolution law, under which school teacher John Scopes was tried in 1925, was repealed in 1967. More recently, a new movement called "scientific creationism" has arisen, continuing to challenge the teaching of evolution as the only explanation for the origin of diversity. For example, in February 1981 a case was launched in California by two parents in behalf of their children, requiring the state to teach that special creation is a scientifically valid concept, equivalent to Darwin's theory of natural selection. The grounds for the case were that by teaching Darwinism to the exclusion of special creation, the state was making evolution into a religion. If this premise were accepted, the state of California could be charged with failure to uphold the constitutional separation of church and state. By and large biologists have felt that such a demand is both unnecessary and unreasonable.

What can this controversy indicate about the nature of science and the process of accepting or rejecting hypotheses in any area of rational thought? Let us keep in mind first that the bill refers to the teaching of the theory, or hypothesis, of *evolution*, rather than to Darwin's (or anyone else's) hypothesis for how evolution is brought about. Evolution is one hypothesis to account for the universally accepted observation that there are many different species of animals and plants inhabiting the earth. Other hypotheses attempt to account for these same observations, such as theories of "special creation" or of spontaneous generation (see Section 22.1). There are two ways to approach the issue of whether special creation and evolution (transmutation of species) are both legitimate scientific hypotheses and should thus be included together in textbooks. One way is to examine the logic of the two hypotheses; the other is to ask in what way or ways each corresponds to people's daily experience.

Logic

With regard to logic, anti-evolutionists argue that most modern biology textbooks present the hypothesis of evolution as an accepted *fact* and thus do not emphasize its hypothetical nature. This criticism has some justification, in that too often hypotheses are presented in science as if they had no controversial elements. In biologists' views, however, evolution is still the best *hypothesis* developed to date to account for the origin of widespread diversity in the living world.

On what basis can such a claim rest? Biologists prefer the hypothesis of evolution to belief in special creation because it is testable. The hypothesis of evolution can be tested against the available fossil record. By definition, special creation puts the issue out of the range of testability. Creation occurred once in the time before human beings existed; it occurred by supernatural processes of which, by definition, we can never have knowledge. Thus, to accept the notion of special creation requires an act of faith. While acceptance of the notion of evolution requires faith, too, it is faith of a sort that the human mind can rationally evaluate by examination of sensory data.

Another reason most biologists prefer the hypothesis of evolution to that of special creation is that evolution accounts for more of the observed facts about living organisms than special creation. For example, Huxley's remark that "God must have been inordinately fond of beetles" points up the difficulty of imagining a creator creating separately the several hundred thousand species of beetles on earth today. Or consider the observation that many species of organisms show a large number of anatomical and physiological characteristics in common. By the theory of special creation, one must assume that the creator was either lazy or simply liked certain anatomical and physiological schemes and thus used them repeatedly. Such assumptions require a further stretch of faith. Or consider the observation that fossil remains often show some linear progression over time, from more ancient species to their modern counterparts. An inseparable part of that series of observations is the fact that all the more ancient forms have become extinct—they have ceased to exist. By the theory of special creation, it would have to be assumed that the creator came to dislike certain species in the past and allowed them to become extinct. Such an assumption implies imperfection or caprice on the part of the creator. One way out of such a dilemma was to assume, as did numerous eighteenth-century naturalists and religious zealots, that the creator placed fossils in the strata as decoys, to lead arrogant human beings astray if they attempted to unravel the mysteries of creation!

The hypothesis of evolution can account for each of the above observations in a far less complex and capricious way. Similarity of body plans for a group of organisms can be explained as a result of descent with modification from a common ancestor. Extinction and continuity among certain fossil lines can be explained as a natural outcome, in each circumstance, of the modifiability of species and their environments. Special creation requires that new assumptions be invented specifically for each new set of observations.

Evolution can account for each set of observations by logical extensions of its most basic premises, that species are modifiable and pass some of their modifications on to their offspring.

Experience

More important, in each case where the hypothesis of evolution requires further subhypotheses or extensions, those extensions are in agreement with people's everyday experience or knowledge from other fields. It does not require a new and particular set of assumptions with no basis in experience. For example, we all know that no two organisms of the same species are alike, but that each has its own individual differences. We also know that weather and environmental conditions change in areas over time; few people would doubt the geological evidence of an ice age that capped much of the northern hemisphere with glaciers. We also know from agricultural experiments that selective breeding for certain characteristics (such as high milk production in cows) produces considerable change over a number of generations. All these observations from people's everyday experience lend support to the idea of species modification with descent.

Given the fact that most biologists today have a logical reason for preferring the hypothesis of evolution to that of special creation, is there any justification for opposing the inclusion in textbooks of the theory of special creation as an equally viable hypothesis? Anti-evolutionists argue that this attitude on the part of biologists is dogmatic and does not give students a chance to come to grips with all sides of the issue for themselves. Anti-evolutionists also claim that biologists indoctrinate students with their side of the story, systematically excluding the other side.

Biologists have responded to such criticism in several ways. Some argue simply that special creation is a religious belief and thus has no place in a scientific textbook. Even were special creation true, since it cannot be tested, it is beyond the realm of science. These arguments are the most practical ways to approach the controversy from the scientific side, but they avoid one of the most intriguing issues that the controversy raises.

The long-standing debate between biologists and religious advocates over evolution is a reflection of two very different world views. It boils down to a debate between rational and nonrational philosophy. It is not a debate that can be settled in a few words, nor can either viewpoint be conclusively demonstrated to the satisfaction of its opponents. The rational view of the world is based on philosophical materialism, as described in Chapter 1. It maintains that people's ideas are developed from their concrete daily experience rather than being "revealed" to them from some unknown and unknowable source. A fundamental premise of the rational, materialist view of the world is that all phenomena are eventually understandable. While not claiming that everything is already known or that there will ever be a time when everything is known, the rational view argues against the existence of unknowable forces, or beings, or spirits at work in the universe. Nothing is by nature unknowable.

In contrast, the nonrational view of the world, as reflected in most traditional religions and in secular movements such as astrology, is essentially idealist. It maintains that some aspects of human experience are by nature unknowable and that all events in the world do not conform to the same basic laws. The nonrationalist view is unwilling to see all aspects of the world explained in rational terms. The origin of species has historically been one realm in which nonrationalists have urged most vehemently against the introduction of rationalist methods. No doubt this is because findings in this area are related to the origins of human beings themselves.

The rationalist and nonrationalist views of the world are in fundamental conflict. The very methods of the one are considered inapplicable by advocates of the other. The two can be reconciled only by partitioning off human experience into two categories: an area in which rational thought prevails and an area in which nonrational thought prevails. The fact that in the past many scientists have held strong religious beliefs indicates that some people are able to partition off their experience in this way. Yet the history of the past 300 years has shown gradual reduction in application of nonrational methods and gradual increase in the application of rational methods to all areas of human experience. The reason is that the rational methods of observation, logic, and testing of ideas against experience have proven to be effective for most people, not just scientists, in dealing with all aspects of the world around them. As earlier parts of this book have tried to suggest, this method is not peculiarly "scientific." It is a method applicable in all areas of human experience, from psychology to history to art.

Because belief in special creation is based on nonrational methods, most biologists do not think it should be included on equal footing with the idea of evolution in textbooks or courses. They argue that to include it as an equally valid way of explaining the origin of species would be philosophically untenable and revert to a way of thinking that is outmoded in the present world.

postulate a role for God, who created the first life forms from which everything else evolved.

The question at stake for many naturalists was not "Does transmutation occur?" but, "By what *mechanism* does it occur?" The view we hold today, in its most general form, was first put forward during this period—the theory of evolution by natural selection. The theory of natural selection is undoubtedly one of the most important generalizations in modern biology. It was introduced to the world in comprehensive form in 1858 by the English naturalists Charles Darwin (1809–1882) and Alfred Russel Wallace (1823–1913). Darwin's book *On the Origin of Species by Means of Natural Selection* (generally referred to as *The Origin of Species*), published in 1859, had significant effects on the intellectual world of the mid-nineteenth century. *The Origin of Species* contained a massive amount of data supporting the concept of evolution in general and the theory of natural selection in particular. No matter how much religious leaders and biologists might have disagreed with Darwin, they could not ignore his work.

At this point we should distinguish between several terms: evolution, speciation, and natural selection. **Evolution** (which we have been using synonymously with transmutation) is the process by which a population of organisms undergoes some sequential changes in genotype over successive generations. Evolution may or may not give rise to new species. Its defining feature is change in genotype over time. **Speciation** is the process by which one species gives rise to (or is transformed into) two or more new species. **Natural selection** is the particular *mechanism* that Darwin proposed for how evolutionary change (including speciation) takes place. It is important to keep these distinctions clear. It is possible, for example, to accept the theory of evolution without accepting either speciation or natural selection. (The converse, however, is not true: that is, it is not possible to accept either speciation or natural selection without accepting evolution.)

In this chapter we will view the development of Darwin's theory of natural selection historically. We will begin by examining the evidence for transmutation of species that had begun to accumulate by 1831, the year Darwin graduated from Cambridge University. Next we will survey several kinds of explanations of transmutation that had been advanced prior to Darwin's work. Then we will look at the particular experiences in Darwin's own background that brought him to the idea of natural selection. And, finally, we will examine the structure of Darwin's theory as an example of an explanation in biology. With this approach, we hope not only to present some idea of "how we know what we know"—a continuing theme of this book—but also a clearer understanding of the process of natural selection itself.

22.2
EVIDENCE FOR TRANSMUTATION OF SPECIES BEFORE 1830

Evidence from Taxonomy

The period between about 1450 and 1750 has been referred to by European historians as the "Age of Expansion." During this 300-year period, armies, navies, missionaries, and entrepreneurs from the Western mercantile nations, seeking easier trade routes to various parts of the new world, explored the major continents of Asia, Africa, and North and South America. During the course of these explorations, naturalists, who either went on the voyages themselves, or inherited the specimens collected by those who did, became aware of a vast array of previously unknown organisms. This unprecedented expansion of knowledge made it necessary to find some way of systematizing the types of animals and plants on earth. As we saw in the previous chapter, organisms have been classified in one way or another since the time of the ancient Western civilizations— Persia, Egypt, Greece, and Rome—but these systems had been based largely on local populations and were largely designed for utilitarian purposes (such as classifying plants for medicinal purposes). The old systems of classification could not accommodate the many new forms. More importantly, by the eighteenth century naturalists recognized that the older systems of classification did not always reflect the obvious similarities observed between certain forms. As we saw in Chapter 21, by the 1740s Linnaeus began to devise a system of classification, or taxonomy, that was based on more true, biological relationships. From the work of Linnaeus and subsequent taxonomists, two generalizations became clear that had considerable bearing on the question of the origin of the species:

1. Not all individuals of a species are alike; there are many variations within any single species, sometimes complicating the task of assigning organisms to a specific species category.

2. Between any two distinct species there often appeared a graded series of intermediate forms—for example, the degree of ornateness in snail shells among species in the Caribbean Islands. Although each group represented a separate and distinct species, the changes from one to another gave the appearance of being small variations on a common theme. The more examples people looked at, the less distinct the boundaries between established species often became.

Taxonomy itself in the eighteenth or early nineteenth century did not automatically lead to evolutionary thinking. But it did bring to light many problems that only an evolutionary hypothesis could ultimately explain.

Comparative anatomy demonstrates the many similar patterns of form that exist in the animal and plant kingdoms. One way of explaining the existence of similarities of pattern is through descent with modification from a common ancestor.

Evidence from Comparative Anatomy

A second kind of evidence came from comparative anatomy. Comparative anatomy is the study of similarities and differences between the anatomical structures of two or more species.

Comparative anatomy emphasizes the difference between homologous and analogous structures. **Homologous** structures have basic anatomical features in common and are derived from the same part of the embryo, but do not necessarily perform the same function. For example, the wing of a bat, the leg of a horse or dog, and the hand of a human being show homologous bone structures (see Fig. 22.1). **Analogous** structures, on the other hand, are used for similar purposes but are of different origin and are not necessarily built on the same anatomical plan. The wing of a bird and the wing of a butterfly are analogous but not homologous, since they do not have any basic structural or embryologically derived parts (bones, muscles) in common.

The idea of evolution from a common ancestor is strongly supported by the study of homologous structures. Divergence produces basic modification of similar structures, so that over long periods of time the differences become pronounced. However, the same basic plan (as in the forelimbs and hindlimbs of vertebrates shown in Fig. 22.1) remains apparent. Comparative anatomy, and the study of homologous structures, became an important aspect of natural history in the period between 1775 and 1820.

The stimulus for increased interest in comparative anatomy was two-fold. (1) As with taxonomy, the expansion of European commercial interests to far-distant continents brought to light new forms which, on careful analysis, appeared to be modifications (often quite exotic, to be sure) of already-known structural plans. Naturalists studied the new forms brought back from long voyages and compared their anatomical structures with those of familiar animals and plants. Such comparisons led to ideas about the similarity of body plan in species recognized to belong to groupings, such as the mammals, that held a number of basic traits in common. (2) Several European countries, most notably France and Germany, experienced an immense growth in building and mining during the eighteenth and early nineteenth centuries. These activities unearthed many fossils, which often had striking similarities in structure to modern-day forms (see Fig. 22.2). Comparative anatomy grew into a full-fledged science under the leadership of such men as Georges Cuvier (1769–1832) and Etienne Geoffroy Saint-Hilaire (1772–1844). Cuvier in particular studied fossil as well as living forms, and emphasized that throughout the animal kingdom, many modifications exist of a few basic body plans. Cuvier believed that all the parts of an animal were so highly interrelated that one part could not be modified without *all* the other interrelated parts being modified to some degree as well. Curiously, Cuvier used this belief to argue against the idea of transmutation, claiming that major variations in any structure would be so disharmonious to the body's integrated anatomical plan that the organism could not function.

Cuvier steadfastly refused to see any of the evolutionary potentials in his own work. Although he was virtually the founder of the modern science of comparative anatomy, Cuvier remained a staunch believer in special creation throughout his life.

One of the important discoveries of comparative anatomy was the presence of **vestigial organs.** Vestigial structures are those that do not appear to have any useful role at present, and yet seem to have had a function

Homologous structures are derived from the same basic embryological structures in different species; the adult structures may or may not perform the same function (such as the arm of a person and the front leg of a horse). Analogous structures do not have any embryological history in common but serve the same function in the adult (such as the wing of a butterfly and the wing of a bird).

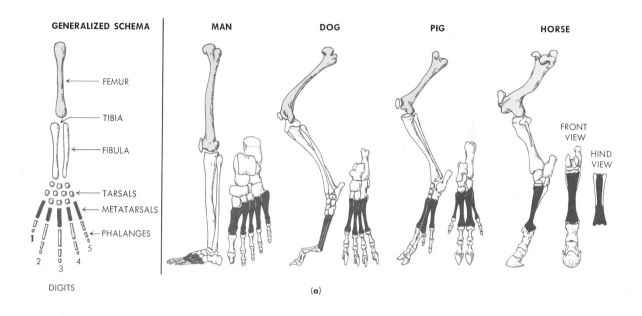

GENERALIZED SCHEMA — MAN — DOG — PIG — HORSE

FEMUR

TIBIA

FIBULA

TARSALS

METATARSALS

PHALANGES

DIGITS

FRONT VIEW

HIND VIEW

(a)

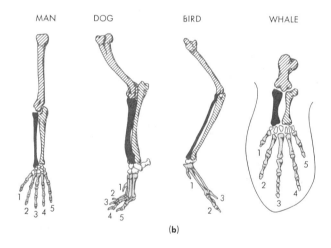

MAN — DOG — BIRD — WHALE

(b)

Fig. 22.1
Homologous bones in the forelimb and hind-limb of several vertebrates. (*a*) Hindlimb homology. To the left is a generalized scheme for the vertebrate limb. (*b*) Forelimb homology. Note that though the forelimbs of these four vertebrates are all used for quite different functions, they nevertheless show a similar basic pattern. (Redrawn by permission from G. G. Simpson, C. S. Pittendrigh, and L. H. Tiffany, *Life: An Introduction to Biology.* New York: Harcourt, 1957, p. 465.)

Fig. 22.2
Fossil *Megatherium* (the great ground sloth), said to have been one of the most popular fossils of the late eighteenth and early nineteenth centuries. The first fossil skeleton was found in South America in 1789. These fossils were astonishing in their size: they were fourteen feet long and over eight feet tall at the shoulders! The ground sloths are related to modern-day bears, with many similarities in anatomical structures (such as structure of the fore-paws). But prior to the development of a comprehensive theory of evolution, it was difficult to understand exactly what sort of relationships fossil sloths and modern bears might have to one another. (Original from *Illustrierte Zeitung,* Vol. 14 [Leipzig 1850], p. 256; taken from L. Pearce Williams, *Album of Science: The Nineteenth Century.* New York: Charles Scribner's Sons, 1978, Fig. 278, p. 189. Reproduction courtesy Mr. James Maurer, Charles Scribner's Sons.)

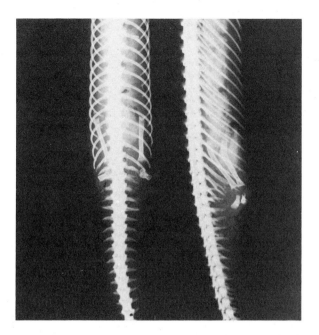

Fig. 22.3
Vestigial leg bones in snakes (in this case, the python). These rudimentary bones are literally "leftovers" from the evolutionary process. The ancestors of modern snakes had appendages for movement. Today, snakes move solely by oscillations of the rib cage. In the embryonic development of each python, as well as other species of snakes, rear limb bones display a partial growth, as shown in this X-ray photograph. (Photo from Gavin de Beer, *Atlas of Evolution.* London: Thomas Nelson and Sons, Ltd., 1964, Fig. 113, p. 43. Used with permission of Elsevier Nederland B. V., publishers.)

in the past. Examples of vestigial organs are hip and leg bones in snakes (see Fig. 22.3), rudimentary wings in the ostrich, and the partial development of an egg-tooth (a special projection on the beak that egg-laying vertebrates, such as reptiles, have for breaking the egg at hatching) in early embryonic development of marsupials (pouched mammals such as the kangaroo). The existence of vestigial organs was difficult to explain with the theory of special creation, especially since in many modern-day forms, vestigial organs appear only in early embryonic development, disappearing as the embryo matures. According to the theory of transmutation, however, vestigial organs could be viewed as the remains of organs no longer needed, the leftovers of evolutionary development.

Evidence from the Fossil Record

A third kind of evidence comes from the fossil record. As mentioned above, excavation and mining in the eighteenth and nineteenth centuries had a stimulating effect on the growth of geology, and particularly **paleontology**, that aspect of geology concerned with the study of fossils. Studying the sequence of fossils from ancient to more recent rock layers disclosed several important items of information.

1. The sequence showed, in general, an increase in diversity as well as complexity of fossil forms. Slight modifications in fossil forms from one stratum to another seemed indicative of the slow modification of forms over time. While this evidence does not exclude the ideas of special creation or spontaneous generation, it certainly supports the idea of transmutation.

2. Some fossil forms appeared to have no living relatives; these forms are said to be **extinct**. The phenomenon of extinction showed naturalists that some species have been unable to perpetuate themselves. By extension, if old species die out, cannot new species originate from previous ones?

3. The fossil record sometimes showed forms intermediate between two presently living types. For example, as we saw in Section 21.5, it is now well established that birds and reptiles diverged from a common ancestor. One of the most convincing items of information leading to this conclusion was the discovery of *Archaeopteryx* (Fig. 22.4), a fossil form intermediate between these two groups. *Archaeopteryx* had teeth and a long tail (characteristic of reptiles), and it also had feathers (characteristic of birds). Such genuine "missing links," though found rarely, have come to be strong evidence for the transmutation concept.

Naturalists in the eighteenth and nineteenth centuries noted that the fossil record contained distinct gaps (Fig. 22.5). Many times the fossils in one layer were totally different from those in the layer just above or beneath it. These gaps were thought to be the result of great cataclysms that swept the earth, destroying all or nearly all the life forms. It was supposed that after each successive cataclysm, the earth was repopulated with new forms by acts of special creation.* This view of the geological record was called catastrophism.

More recent geological evidence has shown why such gaps occur. Changes in the level of the earth's surface cause regions to be raised or lowered from one period to another. This elevation of fossil-laden sediments exposes them to such eroding forces as wind and water. Thus whole layers, along with the fossils they contain, may be lost forever. Since such layers may represent vast periods of time, considerable evolutionary history

---

*An alternative version, with more direct evolutionary implications of the catastrophist theory, runs as follows: after the catastrophe a few living forms manage to survive, though most are wiped out. These forms proliferate and become different through transmutation, giving rise to a large number of descendant forms.

The fossil record shows that many species of the past are now extinct. It also shows that many of those species have modern counterparts with intermediate forms progressing up the geological strata.

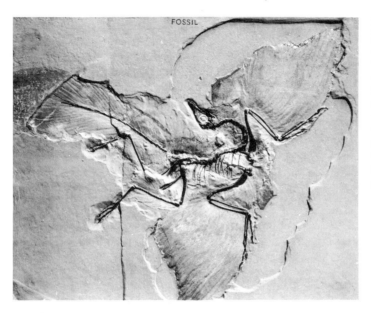

Fig. 22.4
Fossil of the primitive bird *Archaeopteryx*, clearly representing the evolutionary development of modern birds from primitive reptiles. Note the long tail and the claws on the wing, which are reptilian characteristics, contrasted with the feathers and the general structure of the appendages, which are avian features. (Courtesy American Museum of Natural History.)

will be missing. Furthermore, a layer of sediment deposited when an area is under water will be quite different in appearance and composition from a layer deposited when the area is dry. The fossil forms will obviously be different as well. There are thus very sharp discontinuities in the fossil record that mask the more continuous, gradual geological processes that most geologists today believe are responsible for producing all changes in the earth's surface.

Evidence from Biogeography

As a result of the expansion of shipping and voyages of discovery in the eighteenth and nineteenth centuries, naturalists were able to obtain much wider evidence about the geographic distribution of animals and plants than had ever been available before. As naturalists arranged the new specimens in the great museums of Europe, they noted geographic patterns in where organisms of certain types were found. For example, they observed that there were many similarities among three species of rhinoceros inhabiting close geographic regions: India, Sumatra, and Java. These three forms were noticeably different from the rhinoceros of Africa. An even more striking example came from observing the flora and fauna on the small island of Tristan da Cunha,

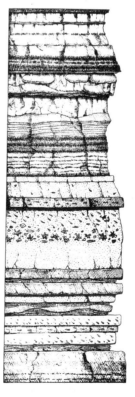

Fig. 22.5
An illustration from a study in paleontology and geology published early in the nineteenth century, showing the distinct breaks in strata, with similar changes in their fossil content, that led many naturalists to the idea of catastrophism. Distinct divisions between layers were taken to indicate great upheavals by fire or flood.

which lies in the South Pacific roughly halfway between Africa and South America. Naturalists found that the flora of Tristan da Cunha had many similarities to that of both major continents. Of twenty-eight species of flowering plants on the island, nineteen were found to exist in both South America and Africa as well; seven species were shared by the island with South America, and two by the island with Africa.

It was difficult to explain these sorts of observations on the basis of the theory of special creation. Why would the creator go to such pains to spread variant types along what amounted to geographical gradients? On the theory of transmutation, however, it was much easier to explain the facts of geographic distribution. As ancestral forms migrated, they moved out from an original home to a more distant area. As their descendants invaded new territories, change was brought about through adaptation to new conditions. The farther away the migrants went, the more differences they would show from the ancestral forms, since they would, on the whole, have been separated from the original group for a much longer time.

Evidence from Animal and Plant Breeding

Animal and plant breeding has been practiced for thousands of years, with only limited success. By the beginning of the nineteenth century, however, much information and practical wisdom had been accumulated on the methods of improving domesticated plants and animals, from dogs and cats to pigeons, wheat, cows, and chickens. All breeders knew that the basic method for producing a new strain of animal or plant was to *select* as future parents the few individuals from each generation who displayed most prominently the particular characteristics desired (such as high milk or egg production, largest tail feathers). These individuals would be used to breed *all* of the next generation. Further selection would enhance the trait even more in the third generation, and so on.

By 1800 the work of practical breeders had emphasized that:

1. All organisms show slight, individual differences from their parents and siblings (recall that taxonomists like Linnaeus were coming to the same conclusion).

2. Some of these variations are passed on to the next generation—that is, they appear to be hereditary in nature.

3. Selection of particular variations can intensify traits in a particular direction (for example, toward greater size or darker color), thus demonstrating the possibility of progressive, linear change.

It should be emphasized that domesticated breeds are not different species from one another. For example, all

breeds of dog from chihuahua to Great Dane are members of the same species. They are capable of interbreeding, though when brought together they do not always do so.

Practical breeders demonstrated two important principles of significance for evolutionary theory in the early nineteenth century: (1) major, inherited, or stable changes could be produced in a population of organisms after only a few generations; and (2) such changes were achieved by selectively breeding only certain individuals from each generation. These principles clearly emphasized that some transmutation was possible, at least within the boundaries of a species. Species appeared *not* to be so immutable and fixed in their characteristics as had previously been thought.

22.3
SPECIES CHANGE AND ADAPTATION
BEFORE DARWIN

Probably the most comprehensive and best-known view of how species became so specially adapted to the conditions of life, prior to Darwin's, was that made prominent by the French zoologist and naturalist Jean Baptiste de Lamarck (1744–1829). Lamarck held that change had been progressive in the history of life on earth, beginning with the simplest forms and leading eventually to what he called the "most perfect" plants and animals. At the top of this progression were, of course, human beings. To explain this particular sequence of changes, Lamarck put forward three general premises, or assumptions (and a fourth, the existence of spontaneous generation, which we will mention only briefly later):

1. The ability of organisms to become adapted to their environments over the course of their own lifetimes.

2. The ability to pass on those acquired adaptations, such as well-developed muscles, to their offspring.

3. The existence within organisms of a built-in drive toward perfection of adaptations.

Lamarck drew from these premises a conclusion about how evolution or transmutation could occur. We should point out that Lamarck was mostly interested in what we would call "vertical" evolution—the process by which one form is slowly modified over many generations into a new and different form. This kind of evolutionary change is "vertical" in the sense that there is a steady progression from one form to another in a *linear* fashion, much like proceeding up the rungs of a ladder. Such a process accounts for change, a transformation, of one species into another.

To understand how Lamarck's scheme worked, consider the example for which he became so famous: how giraffes evolved from a short-necked to a long-necked species (see Fig. 22.6). Lamarck supposed that, like most other mammals, the ancestral giraffes had

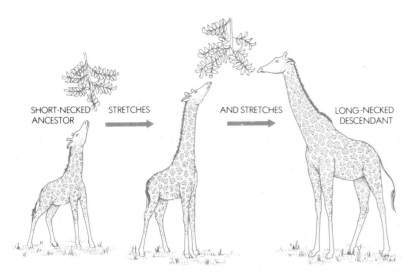

SHORT-NECKED ANCESTOR   STRETCHES   AND STRETCHES   LONG-NECKED DESCENDANT

Fig. 22.6
Lamarck's theory of evolution by the inheritance of acquired characteristics. According to Lamarck, the ancestral giraffes stretched their necks in order to obtain leaves higher up on the trees. As the giraffes' will to perfection came into operation, the appropriate neck muscles began to respond to this stretching by increasing in length. This increase was passed on to the next generation, so that individuals were born with slightly longer necks from the outset. In this way, the evolution of long necks was achieved in direct response to an adaptive need.

short necks. As the population grew and ate up the vegetation close to the ground, including the lowest branches of trees, food became scarce. According to Lamarck, as giraffes began stretching their necks to reach leaves higher on the trees, their neck muscles and bones responded, through a "perfecting principle," and grew in length. When these giraffes reproduced, they passed on the tendency for slightly longer necks to their offspring. The inheritance of acquired characteristics was fundamental to Lamarck's theory of evolution, for without it the variations—say, for long neck—would die with the individuals who acquired them. Thus, according to Lamarck, organisms responded to their environment by developing certain adaptations; these adaptations were passed on to the next generation, which thus began with slightly more adaptive traits than their parents. In this view, evolution was a truly progressive, "perfecting" process. Lamarck's theory was one of gradual change, though he envisioned it occurring much more rapidly than we would predict today, or than Darwin recognized a half-century later.

Lamarck's scheme was compelling in several aspects. First, it embraced the idea of transmutation, which was becoming more and more accepted by naturalists in the early years of the nineteenth century. Lamarck demonstrated how, by gradual changes, species could be altered in significant ways over the course of several generations.

Second, Lamarck's theory had a strong element of rationalism in it, despite the notion of organisms "willing" adaptive changes into existence, or invoking a "perfecting principle." At a time when naturalists were beginning to seek more rational explanations for the origin of species, Lamarck's theory appeared to be a considerable improvement over ideas of special creation.

Third, Lamarck's concept of the inheritance of acquired characteristics was one that was generally accepted, in one form or another by most naturalists at the time. For example, it had long been a common belief that cutting off the tails of dogs would cause their puppies to be born with shorter tails (this is not, of course, the case). Thus, Lamarck provided a believable mechanism for how transmutation could actually take place.

Fourth, Lamarck accounted nicely for the progressive appearance of life on earth, as revealed by the fossil record. With his notion of perfection, Lamarck's theory would have predicted that as one went up the geological scale, organisms would show greater and greater complexity and diversity, as they adapted more fully to their modes of life. This is just what the geological record appeared to show.

There were two main problems with Lamarck's theory, however, which brought it into disfavor even before Darwin's work came on the scene. One was the notion of a "drive to perfection," which many naturalists came to see as mystical and unnecessary.

Another problem was that Lamarck's theory really did not account for diversity, except by invoking the very questionable idea of spontaneous generation. According to Lamarck each species was more or less spontaneously generated, and evolved new adaptations over time. There was no true speciation, however, no actual transmutation of one species into another. That is, Lamarck's theory represented an attempt to explain how giraffes (and other organisms) developed specific adaptations, like long necks, not how giraffes might have evolved from some other species such as horses or zebras. Lamarck's theory was of the origin of adaptation, not the origin of new species. It was not a theory of transmutation *per se*, and hence did not answer the basic question of how diversity originates.

Thus, by the 1830s or 1840s, Lamarck's evolutionary ideas had fallen into disrepute. Lamarck died a pauper, despite the fact that his work in nonevolutionary subjects (principally invertebrate zoology) was recognized as the best of its kind in early nineteenth century Europe.

22.4
DARWIN'S EARLY BACKGROUND AND THE VOYAGE OF THE *BEAGLE*

Charles Darwin was born in 1809 of an upper-class, wealthy English family. Darwin was exposed to the best education available. He attended a private secondary school, spent two years (1825–1827) studying medicine at the University of Edinburgh, and, after abandoning this profession (because the sight of blood nauseated him), entered Cambridge University where he received his undergraduate degree in 1831. Darwin did not do particularly well in any of his studies. However, outside of the classroom, he was an avid field naturalist with an extremely keen sense of observation. He collected all sorts of animals and plants, and became acquainted with several noted geologists, botanists, and zoologists. On graduation, however, Darwin was not certain what profession he would undertake.

Because his family was well connected, and because Darwin had come to know several prominent naturalists at Cambridge, in the fall of 1831 he was offered a position as naturalist on the H.M.S. *Beagle*. The *Beagle*

was preparing for a lengthy, round-the-world voyage to collect hydrographic, geographic, and meteorological data that would be of future use to expanding British shipping and Navy operations. The twenty-two-year-old naturalist set sail on the *Beagle* on the bleak morning of December 27, 1831. He would not see England again for five years.

When Darwin left on the *Beagle* he believed in special creation. But he took with him as reading material the first volume of Charles Lyell's just-published (1830) *Principles of Geology*. This work was a landmark in modern geology because it advanced a new viewpoint for understanding how changes occurred in the earth's surface. As a viable alternative to catastrophism, Lyell proposed what became known as the principle of uniformitarianism. He argued that all the major changes

Fig. 22.7
Map tracing the voyage of the *Beagle*, from December 27, 1831, through October 2, 1836. Darwin learned much natural history during the voyage but did not find traveling by sea much to his liking. His cabin was so small that he had to remove a drawer from his locker to accommodate his feet when he lay down on the bunk. He wrote back home after less than a month at sea, "The misery I endured from seasickness is far beyond what I ever guessed at. . . . The real misery only begins when you are so exhausted that a little exertion makes a feeling of faintness come on—I found nothing but lying in my hammock did any good." (Adapted from D. D. Ritchie and R. Carola, *Biology*. Reading, Mass.: Addison-Wesley, 1979, p. 509.)

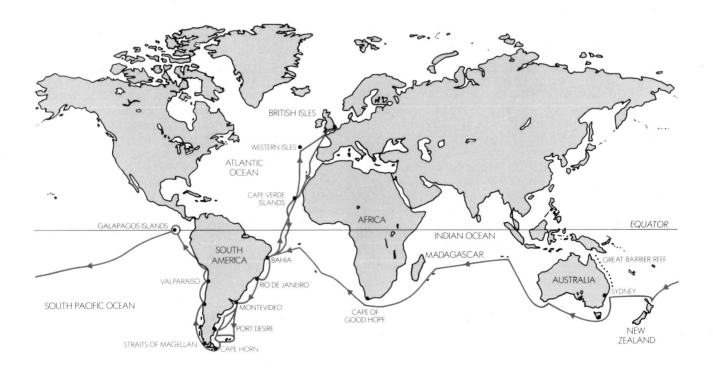

Fig. 22.8
The *Beagle* at the foot of Mt. Sarmiento, at the southern tip of South America. The stormy voyage and the desolate conditions of Tierra del Fuego impressed Darwin with the fact that living organisms, including human beings, seemed "infinitely adaptable" to extremes of environmental conditions. (Engraving courtesy of the National Maritime Museum, Greenwich, England.)

observed in the earth's geological history were the result of the continuing action of small-scale, persistent forces of the kind we observe every day, such as wind and water erosion.

The Principles of Geology made an immense impression on Darwin because it explained large changes by knowable small-scale and persistent forces. Lyell's ideas suggested that geological change is an ongoing process, not one that occurs periodically or only once in the past. Lyell's book started Darwin thinking about other kinds of natural changes and how they might be brought about.

As he visited many different geographic regions (see Figs. 22.7 and 22.8) with their very different forms of life, Darwin made several important observations and began to ask several equally important questions:

1. In South America, fossils of the extinct giant sloth, *Megatherium*, were extremely similar to present-day living forms. Darwin asked: "Have the modern forms developed by slow modification from the now-extinct ones?"

2. Two species of South American large, flightless birds, the rheas (one, *Rhea americana*, living on the central and southern east coast, the other, *Pterocnemia pennata*, living on the southern west coast), shared many more similarities than either shared with the ostriches of Africa or Australia (see Fig. 22.9). Could the greater similarity of geographically adjacent forms suggest possible descent from a common ancestor?

3. On various oceanic islands, such as the Galápagos (off the west coast of South America), the Society Islands and New Zealand (in the South Pacific), and the Cape Verde and Canary Islands (in the North Atlantic), there was always considerable similarity between the island species and their mainland counterparts. African-like species were found in abundance on the Cape Verde and Canary Islands, South-American–like species on the Galápagos. What could account for this geographic pattern around the world?

4. Darwin's own geological observations of the chalk cliffs and pebble beaches along the west coast of South America, and the Great Barrier Reef of Australia suggested that these structures were formed by slow, gradual changes, in agreement with Lyell's principle of uniformitarianism (see above). Could, in fact, great mountain ranges and deep ravines be produced by con-

"Animals, our fellow brethren in pain, disease, death, suffering, and famine—our slaves in the most laborious works, our companions in our amusements—they may partake of our origin in one common ancestor—we may be all netted together."
—Charles Darwin, 1836.

Fig. 22.9
Three living representatives of the walking, or flightless birds (superorder Palaeognathae): (*a*) common rhea (*Rhea americana*) and (*b*) lesser rhea (*Pterocnemia pennata*), both from South America; and (*c*) the common ostrich (*Struthio camelus*) from Australia. Although the two rheas belong to one order, and the ostrich to another, their similarities are striking. Two observations about the rheas of South America struck Darwin on his visit there during the *Beagle* voyage. One was that the two species occupied distinct and separate geographic ranges: *Rhea americana* on the central and southern coast, and *Pterocnemia pennata* on the central west coast and throughout the whole southern tip of the continent. In both geographic localities, however, rheas occupied the same ecological position. A second observation was that the two species of South America resembled each other much more closely than either resembled the ostriches of Africa. To Darwin this suggested that geographic proximity might indicate a close common ancestry.

stant, almost imperceptible geological forces acting over millions of years?

5. Darwin and other naturalists knew that all organisms of any given species vary slightly among themselves. No two animals or plants are identical. Many of these "slight, individual differences," as Darwin called them, appeared to be inherited, and thus could be passed on to the next generation. Could some of these variations, Darwin asked, be better adapted to particular environments than others?

The answer to the questions that Darwin's observations had raised finally began to become clear. Species were not immutable, but changed into other species with time. Because they had a common ancestor living in South America, modern sloths resembled fossil sloths; and for the same reason the two species of South American rhea were more similar to each other than either was to the African ostrich. What struck Darwin also was that the modifications species appeared to have undergone were adaptive—that is, they fit the organism to the particular mode of life suitable to its locality.

The *Beagle* returned to England on October 2, 1836. Soon thereafter, Darwin began organizing his observations, sorting his specimens, and reading a wide variety of books, from natural history to political economy. It was during the first year or so after his return that Darwin hit upon the idea of natural selection.

22.5
THE THEORY OF NATURAL SELECTION

Although he had become an evolutionist during his travels around the world, Darwin was still confronted with a crucial question: by what mechanism does evolution occur?

In considering this question, Darwin drew on two kinds of information: the experience of animal and plant breeders, and the observations of naturalists (including himself) about the extent of individual variation within and between species.

The main practice throughout human history for producing new breeds of agricultural plants and animals was *selection*. Farmers deliberately chose as parents for the next generation those individuals with the particular traits—phenotypes in our present-day language—that they wanted in the offspring. Breeding was not random but selective. Selection in this case was *directed* by human beings making conscious choices. Could something like selection occur in nature without a "chooser" directing the course of events?

By 1837 all the ingredients were there for Darwin to state clearly the principle of natural selection, explaining how selection of the sort human beings practiced in breeding domesticated animals and plants could occur in nature. Then in late September 1838 he happened to be reading, as he said "for amusement," a book titled *An Essay on the Principle of Population*, by the clergyman and political economist Thomas Robert Malthus (1766–1834). Malthus's book, written primarily in opposition to the egalitarianism of the French Revolution, attempted to demonstrate that poverty was caused by the poor themselves. Malthus argued that poverty was a result of high birthrate among the urban poor. The reason for this was simple according to Malthus. The human population grows at a geometric rate—that is, increase is on the order of 2, 4, 8, 16, 32, 64, 128, and so on—while its food supply grows only at an arithmetic rate—2, 4, 6, 8, 10, and the like. This meant that populations *always* tended to outgrow their food supplies, and thus exerted a relentless pressure on each member for survival. If the poor could recognize this basic "law" and, like the wealthy, cut back their birthrate, according to Malthus they could escape from endless poverty (see also Supplement 25.1).

In a flash of insight Darwin saw the implications of Malthus's argument for the process of evolution. If every species, like human beings, tended to produce more offspring than could be fed, the result would be constant *competition* between organisms of the same kind, for limited food. The concept of competition supplied the previously missing element to Darwin's theory. In the competitive struggle for life, some organisms would be able to survive, and others would not. What is it, Darwin asked, that determines which organisms will survive and which will not? The answer was inherited variations, the slight individual differences that he knew existed among members of every species on earth. Some variations would allow the organism to get more food faster, or in some way survive better—that is, compete more favorably; these would be called *favorable variations*. Other variations would cause the organism to obtain less food or other resources—that is, to compete less favorably; these would be called *unfavorable*, or *deleterious variations*. The selection of favorable variations from unfavorable ones was not accomplished by a conscious human being or a deity, but by the conditions of the environment—that is, the availability of food, water, and other resources, the presence of predators, climatic changes, and the like.

The outcome of the selection process in nature was, to Darwin, **adaptation.** But how do we determine what is better adapted and less well adapted? Darwin's criterion for adaptation was not simply survival or longevity, but *differential fertility*. Those individuals that are best adapted are those that leave the most surviving, reproductively active offspring in succeeding generations. The individuals that are least well adapted are those that leave the least number of reproductively active offspring in the next generation. The prize in natural selection goes to those who leave the most copies of themselves—including their favorable variations—for the

Darwin's theory of natural selection is based not on the idea of death for the unfit, and longevity for the fit, but on the notion of differential fertility. Differential fertility is the claim that the more fit individuals in every generation will leave a few more off-spring than the unfit. Whether an organism lives a long time or not is immaterial; what counts is the number of offspring it leaves—and hence the number of copies of a favorable variation—for the next generation.

next generation. They are the best adapted, the *most fit* individuals. It is important to note that we do not have to conceive of the less fit individuals as dying off, being killed by ferocious competitors, or leaving *no* offspring at all. They simply leave *fewer* offspring, and thus *fewer* copies of themselves than their counterparts with the "more favorable" variations.

We can summarize the basic structure of Darwin's theory of natural selection as a series of four observations and two conclusions:

Observation 1: All organisms have a high reproductive capacity. A population of organisms with an unlimited food supply and not subject to predation could quickly fill the entire earth with its own kind. Under optimal conditions, all organisms have an enormous reproductive potential, or capacity for population growth.

Observation 2: The food supply for any population or organisms is *not* unlimited. The growth rate of the population tends to outrun the growth rate of the food supply.

Conclusion 1: The result must be continual competition among organisms of the same kind (organisms that have the same food or other resource requirements).

Observation 3: All organisms show heritable variations. No two individuals in a species are exactly alike.

Observation 4: Some variations are more favorable to existence in a given environment than others.

Conclusion 2: Those organisms possessing favorable variations for a given environment will be better able to survive than those that possess unfavorable variations; thus, organisms with favorable variations will be better able to leave more offspring in the next generation. As the number and kind of phenotypes change from one generation to the next, evolution occurs.

The observations given here are not simple types such as we described in Chapter 1. They are really inductive generalizations based on a large number of individual observations by Darwin and other naturalists.

Nonetheless they are the concrete evidence on which the theory of natural selection is built. They become the premises from which the theory of natural selection is deduced.

One of the most important aspects of Darwin's theory was his emphasis that selection acted on small, individual differences between members of the same species. He did not believe that very large-scale variations, such as the sudden appearance of an extra finger or toe, a dramatic color variation such as an albino, or other large changes, were the raw material of evolution. For one thing such variations occurred rarely in nature; for another they were usually so conspicuous or disruptive to the total functioning of the organism that they were almost invariably selected against. Rather, small, "almost imperceptible" (in Darwin's words) differences among individuals—a slight variation in the number of abdominal bristles in the fruit fly, a slightly greater ability to retain water in desert mice, for example—provided most of the raw material for natural selection. Lacking any reliable mechanism of heredity, Darwin *assumed* that many of the slight individual differences that he observed were, in fact, inherited. But he did not know for sure—a point on which many of his critics pounced. It was only with the rediscovery of Mendelian theory after 1900 (and not really until 1930) that Darwin's theory of natural selection was fully and successfully provided with a workable theory of heredity.

Another important point in Darwin's view of variations was that they arose randomly—that is, by chance. According to Darwin, variations did not arise because organisms felt the need for them, but arose spontaneously without regard to need. According to Darwin, if a particular variation happens to occur in an environment where it confers an advantage, it will be selected for. If it occurs in an environment where it confers a disadvantage, it will be selected against. The notion of highly adaptive structures, such as the vertebrate eye, evolving by the accumulation of *chance* variations was extremely difficult for many of Darwin's contemporaries to understand. In fact, several of his strongest critics used the seeming impossibility of evolution by chance to argue that Darwin's theory had to be rejected. Obviously,

SUPPLEMENT 22.2
DARWIN AND THE ECONOMISTS

Charles Darwin's work in natural history has often been held up as one of the classic examples of a scientist working on a problem for its own sake, of science uninfluenced by practical ends or by the world around it. Darwin, it is said, was a "pure" scientist, ahead of his times, envisioning what others had not been able to envision. Implicit in this view is the value judgment that science somehow can, and ought, to be free from outside—meaning cultural, political, or economic—influences.

Such a view of science itself has been called into question by numerous historians, philosophers, and sociologists of science over the past two decades. Historians Robert Young and Bernard Norton in England have made a convincing case for the idea that Darwin's theory of natural selection was directly a product of early to mid-nineteenth century social and economic thought. According to Young and Norton, Darwin's theory of natural selection was largely a projection of the social ethics of mid-nineteenth century industrial capitalism onto the world of nature.

By all historical accounts, the nineteenth century was a period of ruthless, even brutal economic expansion in England. The early industrial period was characterized by fierce competition between rival manufacturers and mercantilists, low wages, poor working conditions, the exploitation of child and female labor, and the purposeful lack of economic regulation of everything from wages and prices to imports and exports. Companies and industries rose and fell according to how well they competed for cheap labor, raw materials, and markets. What is the possible connection between these sorts of economics and the development of Darwin's theory of natural selection? The connections that Young and Norton make between Darwin and economic thought illustrates some of the subtle, but perhaps important ways in which the very nature of scientific theories can be determined by social influences.

Cultural historians have long recognized that ideas of economic competition were pervasive throughout the nineteenth century. In the early 1850s, according to historian Richard Hofstadter, Herbert Spencer (1820–1903), a contemporary of Darwin's and founder of the school of thought later known as "social Darwinism," coined the phrase "survival of the fittest." Spencer's phrase referred to competition both in human economics and in animal life. The notion that economic and social progress were the result of struggle and competition—with some individuals winning and others losing—was a generally accepted idea among educated middle- and upper-middle-class English society. Darwin matured and was educated during this very period (1820s and

1830s) and was undoubtedly aware of the ideas of struggle and competition as a fundamental economic process.

More explicitly, Darwin was himself quite interested in political economy, the study of the relationships between economic and political processes. He read the works of such nineteenth century political economists as Adam Smith (1723–1790), David Riccardo (1772–1823), and Thomas Robert Malthus (1776–1834). All of these writers attempted to develop a rational explanation for how the capitalist system (mercantile and industrial) functioned. They pinpointed three major characteristics of capitalism as it existed in the early years of the nineteenth century:

1. First and foremost capitalism was based, they pointed out, on competition, on the right of one entrepreneur to get business away from another. Competition and the struggle it produced among individuals were said to yield the best outcome for all people. The public, the buyers, selected the best products. Those manufacturers who made the best products thrived; their less successful competitors fell by the wayside.

2. According to Smith, Riccardo, and Malthus, the price and availability of products were based on the "law of supply and demand"—that is, the more something is in demand, the less of it there will be to go around, hence the higher the price. Conversely, the less something is in demand, the greater supply there is of it, and hence the lower the price. The implication of this relationship was that economic systems regulate themselves without outside management or overseeing. This is, in brief, the philosophy of *laissez-faire* (from the French meaning "to allow to act"), which formed the foundation of nineteenth century capitalism. The concept of *laissez-faire* embodies the notion that governments should not impose tariffs, control imports and exports, levy taxes, or in any other way regulate the economic system. According to early nineteenth century capitalist theorists, if left alone the economic system will regulate itself—albeit with harsh effects on some individuals. In contrast, government regulation supports individuals (as in welfare), businesses (as in tax incentives), or other groups that cannot make it on their own in the competitive struggle. Such support goes against the natural order of things and against the welfare of society as a whole.

3. A third characteristic of early nineteenth century economic writers was the general sense of *progress* that they saw emerging from the *laissez-faire* system. Not all political economists at the time had the same degree of faith in inevitable progress. For example, Adam Smith saw factory

owners as oftentimes trying to deceive their workers and the public; and David Riccardo emphasized the harshness and brutality of industrial capitalism. But both men were united in their belief that through competition and *laissez-faire*, the best would always win out in the long run. New ideas, inventions, or innovations would all be tried out in the economic marketplace. Those that were beneficial would win out; those that were useless or detrimental would fail. Progress toward better living standards and quality of life was thought to be the inevitable and ultimate outcome of the free workings of the capitalist system.

In *The Origin of Species* Darwin used a number of terms or phrases borrowed from classical political economy: terms such as "the economy of nature," the "war of nature," "worker" and "slave" insects, the "struggle for existence," "competition," the "war of all against all," and "division of labour." Where Darwin explicitly learned these terms is difficult to say. He may have picked them up from general usage, since the language of political economy was prominent in the writings of naturalists and nonnaturalists alike from the late eighteenth century onward. Karl Marx said of Darwin, "It is remarkable how he recognizes among beasts and plants his English society with its division of labour, competition, opening up of new markets, inventions and the Malthusian struggle for existence." Historian Donald Worster has per-

haps best summed up the social influences on Darwin's thinking:

> *The emphasis Darwin gave to competitive scrambling for place simply could not have been so credible to people living at another place and time. It is absolutely impossible to conceive such a view of nature coming from say, a Hopi in the American southwest. . . . Even in the limited realm of nineteenth century western science, it is striking how much of Darwin's work and the social response to his ideas were the products of the Victorian frame of mind.*

It is a truism to say that people speak the language with which they grow up, and further, that the kind of language people use both reflects and determines what they think. Yet, this is precisely the point we wish to emphasize here. Darwin *was* a product of his own times; his language and his theory of natural selection both reflect the economic and social world, as he experienced it, in the mid- to late nineteenth century.

Cultural biases influence the generation as well as public reaction to scientific ideas. Sometimes those influences provide a special insight, as they undoubtedly did for Darwin. Sometimes they restrict insight and vision. In either case, scientific ideas are never generated in a cultural vacuum, but are, like all other ideas, the product of a particular time and place in history.

those critics did not understand the laws of probability and chance or the process of statistical thinking (see Section 16.3).

To see how Darwin's theory works, consider again Lamarck's own case of the evolution of long necks in giraffes (see Fig. 22.10). In the ancestral population of giraffes, some individuals were born with longer necks than others. (The differences in length were probably

not very great, on the order of a few centimeters or less.) These variations in neck length occurred by chance; an equal number of variations for shorter neck probably also occurred within the population. The variations are to some degree inherited. Now, as long as vegetation near the ground is plentiful, giraffes with longer necks are at no greater advantage than those with shorter necks, and hence, at least with regard to neck

DARWIN'S GIRAFFE

ANCESTORS OF
VARYING NECK LENGTH

NATURAL SELECTION

NECK LENGTHS
STILL VARY, BUT
ARE LONGER

Fig. 22.10
Evolution of the giraffes' long neck according to Darwin's theory of natural selection (for comparison, see Lamarck's version, Fig. 22.6). Ancestral giraffes were born, by chance, with varying neck lengths, some longer, some shorter. As leaves on the lower limbs of trees became scarce, a selection pressure was generated, favoring those organisms with slightly longer necks. Those organisms were able to eat more and thus leave more offspring to the next generation than their shorter-necked counterparts. Thus the frequency of longer-necked individuals increased in the next generation; it is this change in numbers of certain traits in a population that we call evolution.

The evolutionary process produces change in species through two processes: the occurrence of inherited variations and the action of natural selection.

length, will leave approximately the same number of offspring in the next generation. However, if giraffes increase in number faster then vegetation can grow, the ground-level food supply will become scarce. A *selection pressure* is thus generated, favoring those individuals with longer necks, who will have access to a greater supply of food—that is, leaves higher up in the trees. As a result, those individuals with longer necks will be able to eat more, live longer, be more robust, and leave a greater number of offspring to the next generation. Thus, assuming that some part of the variation in neck length is genetically determined, the frequency of long

necks would be expected to be greater in the next generation. If the same selection pressures continued to exist generation after generation, the giraffe population would show an evolutionary development toward longer and longer necks.

In contrast to Lamarck's explanation, Darwin's does not invoke any will or "desire" on the part of giraffes to lengthen their necks. Nor does it postulate the inheritance of acquired characteristics. The outcome of both Darwin's and Lamarck's scheme is adaptation of the giraffe population to changing environmental conditions (such as food supply), but the mechanism is very

SUPPLEMENT 22.3
LAMARCK REVISITED—OR, ANOTHER PERSPECTIVE ON THE INHERITANCE OF ACQUIRED CHARACTERISTICS

It is commonplace to point out today that the idea of the inheritance of acquired characteristics is invalid. Since the experiments of August Weismann in the 1890s, increasing numbers of biologists have rejected the "Lamarckian" view that changes in the body cells or tissues of an adult (or even embryo) can cause a change in the organism's germ cells and be transmitted to the offspring. In a series of celebrated experiments, Weismann cut off the tails of many generations of mice to see what effect such a "somatic change" could have on tail length in succeeding generations. He found no effect: future generations of mice were born with just as long tails as those with which their ancestors had begun. Weismann and other "neo-Darwinians" claimed that the Lamarckian theory of inheritance of acquired characters was a dead issue.

In the 1950s English biologist C. H. Waddington introduced the concept of *genetic assimilation* as a more sophisticated alternative to both the neo-Darwinian and neo-Lamarckian viewpoints. According to Waddington, for his day, Lamarck may not have been so completely off the track as today's commentators often portray him.

Waddington agreed fully with the idea that acquired characteristics are not directly inherited. A giraffe that grows a long neck in response to stretching will not as a direct result have offspring with longer necks at birth. However, giraffes might differ genotypically in how effective stretching will be

in altering their own phenotype (that is, neck length). In other words, some genotypes might be more sensitive to the effects of stretching than others; stretching by animals with "sensitive genotypes" would produce phenotypically longer necks than stretching by giraffes with "less sensitive" genotypes. In a competitive environment (where ability to stretch for higher leaves on the trees would be advantageous), the more sensitive genotypes will yield phenotypically longer necks, and thus be selectively favored. Note, however, that the act of stretching the neck does not in itself cause a change in the genes for neck length. The genes allowing for greater lengthening of the neck if and when the giraffe stretched are already present in at least some (if not all) giraffes in the population.

It is important to emphasize that, according to Waddington's idea, the more sensitive genotypes (for stretching) would not be selectively favored unless the giraffe had cause to stretch its neck and permit those genes to express themselves. If giraffes lived in an environment where food was abundant near the ground, the sensitive genotypes for response to stretching would not show up as an altered phenotype. Thus, the act of stretching in such cases could be an essential part of the evolutionary process—a feature that Lamarck appreciated, clearly, though with a different hereditary mechanism in mind.

Table 22.1

Lamarck	Darwin
Variations occur in a given direction because of an environmentally induced *need*.	Variations occur randomly, regardless of environmental conditions.
Variations acquired by the organism during its own lifetime are passed on to the next generation	Acquired variations cannot be passed on to the next generation.
Changes in characteristics in a species over time occur as a response to the organism's felt need to adapt.	Changes in characteristics in a species over time occur by selection weeding out the less fit and favoring the most fit.
Evolution occurs relatively rapidly since organisms respond directly to the environment by varying in the appropriate direction.	Evolution occurs relatively slowly, since selection acts almost exclusively on the randomly occurring individual differences among members of a population.

different. The differences are summarized and compared in Table 22.1.

Darwin wrote down his gradually developing ideas about natural selection in notebooks, diaries, and several unpublished manuscripts over a period of twenty years. In 1856 he received a letter from a thirty-three-year-old naturalist in Malaysia, Alfred Russel Wallace (1823–1913), outlining what was essentially Darwin's own theory of natural selection. The two men, quite independently of one another, had hit on the same basic idea. Needless to say, Darwin was somewhat disturbed. After all, he had been working on the theory of natural selection, collecting data, and organizing his thoughts for twenty years. Now, it appeared he might be scooped by a very competent, but far less experienced investigator than himself. At the same time he recognized that Wallace *had* come upon the same idea, and did deserve credit. A compromise was quickly found. Darwin and Wallace would each write papers describing their theo-

ries. Their papers would then be read together at a forthcoming meeting of the Linnaean Society of London. Simultaneously, Darwin agreed to begin work on his major treatise outlining the theory of natural selection and presenting all his accumulated data. Wallace fully agreed that Darwin's experience and data were so vast, and his insights so unique, that he, Darwin, should write the major work presenting the new theory to the scientific world. This Darwin did between 1857 and 1858. *The Origin of Species* was published in November 1859; all copies were sold out by the end of the first day—quite a record for a hefty treatise on natural history.

From that time down to the present day, the Darwin-Wallace theory of natural selection has become a major organizing principle for all of biology. In the next two chapters we will examine the modern-day extensions and ramifications of Darwinian theory.

Summary

1.   Three major types of theories have been proposed in the past to account for the presently observed diversity of species in the natural world: special creation (species were each specially created by a supernatural power unknowable to human beings); spontaneous generation (species have been in the past, and possibly still are, formed from nonliving organic and mineral material); transmutation, or evolution (species today have arisen from previously existing species by a process of descent with modification). The first of these ideas is, by definition, beyond human investigation; the second has never been demonstrated in any experimental situation; the third, evolution, is the most widely accepted today.

2.   Many observations accumulated from 1650 to 1800 about the natural history of plants and animals around the world. This was stimulated in large part by increased geographic exploration associated with mercantile and colonial expansion of the great European powers (England, France, Spain, Portugal) around the world. Many of these observations began to undermine the theory of special creation and supported, directly or indirectly, the idea of transmutation. These include:

   a) The work of Linnaeus and other taxonomists, which showed that
      (i) Not all individuals of a species are alike; there are many individual variations within a given species.
      (ii) Between any two distinct species there often appear a graded series of intermediate forms. There are often no sharp large-scale dividing lines between related species.
   b) Evidence from comparative anatomy: In the early 1800s comparative anatomists, such as Georges Cuvier in France, had begun to show that many animals were built on the same basic body plan. Cuvier distinguished be-

tween *homologous structures* (where the same basic structure or tissue has been modified for different functions) and *analogous structures* (where very different structures or tissues have been modified to serve similar functions). Particularly striking was the similarity often seen between modern and fossil forms. Comparative anatomists also discovered a number of vestigial organs, apparently degenerate structures that had lost most of their original function.

c) Evidence from the fossil record: Paleontology, the study of fossils, showed that many graded series of forms could be found between fossil and modern-day forms. Generally, the most primitive forms were found in the older, lowest strata, and the more advanced forms in the younger, less deep strata. Also, breaks or gaps in the geological record were often accompanied by drastic changes in the fossil content. The geological record was thus somewhat ambiguous: in some cases it suggested gradual change and modification in forms through many successive strata; in other cases it showed sharp breaks, or discontinuities. The sharp breaks were particularly noticeable to some geologists, who developed a theory of catastrophism, the idea that the earth's history was punctuated from time to time with gigantic upheavals that quickly, and thoroughly, changed the entire landscape and with it the existing fauna and flora.

d) Evidence from biogeography: Because of the expansion of European trade many naturalists in the eighteenth and early nineteenth centuries had the opportunity to explore the continents of Asia, Africa, and South America. Evidence about the geographic distribution of plants and animals—what is called biogeography—showed that often the closer two related populations or species were to each other in space, the more similarities they often displayed.

e) Evidence from animal and plant breeding: The practical cultivation of agricultural organisms had demonstrated the principle of modification through selection. From each generation, plant and animal breeders selected only one or two individuals showing the desired traits. These individuals acted as parents for the next generation. Such selective procedures showed that over many generations organisms could be greatly modified; they were not so fixed and immutable as special creationists claimed.

3. A major explanation for the transmutation of species prior to Darwin's theory of natural selection was that advanced by the French naturalist Jean Baptiste de Lamarck in 1809. Lamarck assumed that organisms had a built-in drive toward perfection, toward adapting themselves to the environment; he also assumed that modifications acquired by an organism during its own lifetime could be passed on to its offspring (the inheritance of acquired characteristics). Through the interaction to these two basic principles, Lamarck envisioned species being modified (in adaptive ways) over successive generations. However, Lamarck's theory was of how adaptations originate, not how one species becomes another; Lamarck believed that each species type was originally formed through spontaneous generation.

4. Charles Darwin developed an alternative theory to Lamarck's, one that did not rest on unproved assumptions about perfection and yet could account for true transmutation. Darwin explained the theory of natural selection in his book, *On the Origin of Species by Natural Selection*, published in 1859.

5. Darwin's theory of natural selection can be summarized as follows:

a) More organisms are always born than can survive (because of limited food supplies or some other resource).

b) Competition exists between all members of the same species, since these individuals share all the same requirements for life.

c) All individuals of a species show slight, individual variations among themselves. These variations occur spontaneously, and by chance, they do not occur in response to a direct environmentally induced need. Some of these variations are hereditary.

d) Any variations that allow an individual to compete more successfully for limited resources will be favored; that is, the organism will be able to leave more offspring in the next generation than individuals who lack the favorable variation. Successful variations increase an organism's differential fertility, and hence fitness.

e) As a result of differential fertility, the proportion of the favorable to other variations in the species will be increased from generation to generation. That is evolution.

f) The end result of evolution is adaptation of the organism to its environment.

6. Darwin's theory of how evolution occurred was eventually more acceptable than Lamarck's because:

a) It brought together more data, of many different kinds: from geology, biogeography, animal and plant breeding, paleontology, comparative anatomy, and embryology. It was a truly *synthetic* theory.

b) It did not depend upon a mystical explanation such as "will power" of the organism, or the "will to perfection."

c) It did not require that acquired characteristics be inherited, as did Lamarck's theory. (Though Darwin did develop a modified version of this idea in later editions of *The Origin.*)

d) It explained not only adaptation, but also how one species could be transformed into another over time.

7. Although Darwin's theory was not completely accepted in the mid-nineteenth century, even by biologists, it has become the foundation for all modern evolutionary thinking. Biologists today have greatly modified many of Darwin's specific ideas. But the basic core of his theory—the existence of natural selection acting on slight individual variations among members of a species—still remains as valid today as 100 years ago.

Exercises

1. In what sense is the idea of special creation an example of philosophical idealism as discussed in Chapter 1?

2. Naturalists have observed that the fish inhabiting underground rivers, such as Mammoth Cave, Kentucky, are blind. They have eye sockets and outward appearance of vestigial eyes, but these are no longer functional as organs of sight. Explain how (a) a Lamarckian and (b) a Darwinian would account for the origin of sightlessness in cave animals (such as fish).

3. What specific contribution to Darwin's thinking on evolution can be attributed to each of the following?

    a) His experiences on the voyage of the *Beagle*

    b) Lyell's *Principles of Geology*

    c) The experience of animal and plant breeders

    d) Malthus's *Essay on Population*

4. In what way is Darwin's theory "synthetic" rather than strictly inductive or deductive?

5. Distinguish between the terms "evolution," "natural selection," and "speciation." Which terms apply most directly to Darwin's theory?

Chapter 23
Microevolution: Variation and Selection at the Population Level

23.1
INTRODUCTION

Since the publication of Darwin's *Origin of the Species* in 1859, many exciting problems have come to light within the field of evolutionary theory. Among the most important are those that involve the application of concepts of heredity—classical Mendelian and, more recently, molecular genetics—to the Darwinian theory of natural selection. Genetics has also provided the basis for viewing evolution at the level of populations, a vantage point that was not so readily available to Darwin and his contemporaries. Molecular genetics has made it possible to understand the chemical basis of gene replication and mutation, the evolution of macromolecules such as proteins, and even the evolution of entire biochemical systems such as the cytochromes, or the membrane transport proteins.

In recent years the terms macroevolution and microevolution have been used to refer to two aspects of the evolutionary process. **Microevolution** consists of processes that lead to genetic changes within a population of organisms over two or more generations. These processes include the origin of variation, the nature of selection as it acts on different kinds of variations, and the dynamics of population—including size, growth rate, and structure—that affect the selection process. Microevolution is not particularly concerned with the origin of new species *per se*, but rather with the small-scale (hence the prefix "micro") processes that ultimately, but not necessarily, can lead to the formation of new species. By contrast, **macroevolution** refers to large-scale processes that ultimately do give rise to new groups. The study of macroevolution is concerned with such questions as how one species diverges into two or more, with the factors leading to rapid versus slow evolutionary change, with interactions among organisms and between organisms and their environment, and with those large-scale changes over long periods of geological time that produce higher taxa—that is, genera, classes, and phyla.

Chapters 23 and 24 will be concerned with the processes of, respectively, microevolution and macroevolution. In Chapter 23 we will focus on the evolu-

tionary process at the population level. We will examine the nature of populations—their structure and their biological characteristics. Next, we will explore the nature of variation within populations, both its extent and the various ways it is produced. Then, we will discuss the genetics of populations viewed as a whole (as opposed to the genetics of individual organisms)—that is, the methods of studying genetics at the population level, the factors that contribute both to genetic equilibrium and to genetic change from one generation to another. The chapter will close with a detailed examination of two cases of microevolution: melanic moths in England over the past 100 years, and sickle-cell anemia in human beings in Africa and the United States.

23.2
SPECIES AND POPULATIONS: AN OVERVIEW

In Chapter 21 we examined the biological species concept, which defined species in terms of a number of criteria (ability to interbreed, morphological similarities, geographic distribution in nature, behavioral similarities, and a common evolutionary history). What is the distinction between a species and a population? Every species consists of one or more **populations,** a group of potentially interbreeding organisms which interact *as a group* (they may or may not inhabit the same region; migrating organisms, for example, remain part of the same population as long as they migrate together).

We can understand the distinction between species and populations easily by considering our own case as human beings. Human beings are spread over the whole earth. We are capable of interbreeding, and by this and other criteria are considered to be one large species. Yet we live in numerous, more localized populations, such as those clustered in particular cities or towns, islands, states, even countries, and continents. Populations are most frequently described in terms of breeding (or potentially breeding) groups—the collection of individuals within which mating and reproduction is most likely to occur. While such a group is most frequently associated with a specific geographic locality, it is important to recognize that this is not necessarily so (as in migrating forms), and is not the defining criterion of a population.

Populations may consist of two or more subpopulations, called **demes**. A deme is a local breeding population that can exchange genes with other demes of the same population and/or species in the same vicinity. Demes often differ among themselves only slightly, quite often more physiologically than anatomically. Nonetheless, each deme in a larger breeding population will have its own characteristic distribution of genes.

The boundaries between populations, whether geographical or reproductive, are not always easy to draw. Sometimes two populations may overlap to the extent that it is difficult to determine where one ends and the other begins. In human beings, for example, the boundary between urban and suburban populations is fuzzy and cannot be drawn rigidly. We can define populations as real and meaningful biological groups without always having to draw hard-and-fast lines between them.

An important point to keep in mind about populations is that while the individuals who make them up come and go, the populations persist as real groupings from generation to generation. The population has a life, a continuity of its own, that transcends the finite life of its individual members. This continuity is established through passing on of genes—of DNA, to be more precise—from generation to generation. It is on this basis that the individuals who compose any population at any moment in time can be said to share a common ancestry. The fact that populations have a genetic integrity has made it possible to develop a special field of evolutionary study known as **population genetics**. Population geneticists investigate the genetic structure of whole populations, both at one point in time, and over successive generations. They seek to determine the ways in which the population as a whole changes or remains the same over periods of time. This can be done by looking at the frequencies, or percentages, of alleles in the population from generation to generation. A change in allele frequencies over several generations indicates that evolution is occurring.

Evolution is a result of the dynamic interaction between two opposing tendencies in the reproductive process: faithful replication and variation. Both of these processes are absolutely necessary for evolution as we know it to occur. If every organism reproduced itself faithfully, there could be no evolution. If there were no faithful replication, there could be no continuity. Indeed, given the constant interaction between these two opposing processes, evolution is bound to occur. In the next section we will examine the process of genetic variation on which evolution depends.

23.3
VARIATION: ITS ORIGIN, EXTENT,
AND SIGNIFICANCE

The Origin of Variations

Genetic variants in a population originate generally in one of three ways. The first is **mutation**, or **point mu-** **tation**, which involves a change in the fundamental genetic information contained within a single gene. The second is **chromosomal mutations**, or **chromosomal aberration**, which involve rearrangements of chromosome structure in such a way as to affect either the expression of certain genes or the recombinations that can occur during gametogenesis. The third is **polyploidy**, which involves doubling, tripling, quadrupling, and so on of the entire set of an organism's chromosomes, a process that often leads to variation in phenotype.

Point Mutation.   In Chapter 19 we saw that point mutations result from the substitution of one or more nucleotide pairs in the base sequence of DNA. Substitutions alter the genetic message on the DNA, in turn altering the complementary sequence on the corresponding *m*RNA. Ultimately the effect of mutation is to produce an altered protein molecule, which may or may not affect the adult organism's phenotype.

Point mutations produce very specific effects: a change in a specific type of protein molecule. The ramifications of these changes on the organism's phenotype, however, are quite varied. Some point mutations, such as the alteration of a single base pair in the gene for normal hemoglobin, producing sickle-cell hemoglobin, have vast and drastic effects. In other cases, change in a single base pair produces effects that are noticeable but of no serious consequences, such as the change from straight to forked bristle in *Drosophila*. In still other cases, point mutations produce no observable effect in gross phenotype and can often be detected only by methods, such as chromatography or electrophoresis, that distinguish directly between differently structured protein molecules.

Most point mutations are deleterious—that is, they produce a phenotype that is nonfunctional, or at least less functional than the original, given the organism's usual environment. The genome of any organism has evolved over millions of years into a precise assemblage of genetic blueprints. "Mistakes" in copying that blueprint would, for the most part, be disruptive rather than beneficial. On occasion, however, some mutations lead to change that is favorable, at least in certain specified environments. Though favorable point mutations are rare, they occur frequently enough to provide some, but not all, the raw material for evolution.

Chromosomal Mutations.   Chromosomal mutations are changes in the order of genes on a chromosome, resulting from a gross change in the structure of the chromosome itself. The various types of chromosomal mutations are discussed in some detail in Supplement 18.2. To summarize here, these include: deletions (loss of a chromosomal segment), duplications (addition of a segment already present in the same chromosome to produce two tandem segments), inversions (removal of a segment from one part of a chromosome, and its reinsertion in the reverse direction at the same position in

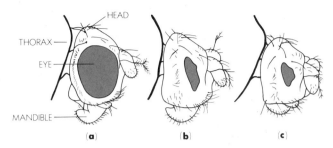

Fig. 23.1
The "bar" eye condition in *Drosophila*, illustrating the phenomenon of position effect—the alteration in expression of a phenotype by the position or location of the gene(s) on a chromosome. Bar eye is due to a duplication in one portion of the X chromosome. (*a*) Normal, wild-type eye shape in *Drosophila*; the genotype of this individual would be +/+ if a female, and +/y if a male. (*b*) "Bar" eye phenotype for a female that has two "bar" duplications, one on each X chromosome; her genotype is thus written as *B/B* (bar eye is a dominant mutation over the wild-type alleles). (*c*) A female with "ultrabar," a phenotype produced by having two bar duplications, but located on the same X chromosome rather than on homologous chromosomes; the genotype of ultra-bar is thus written as *BB/+*. In both (*b*) and (*c*) the same number of mutant chromosome regions exist, but the phenotypes are different because of the spatial location of the bar regions. "Position effect" is the result of the fact that genes and their products interact to produce the phenotype; this interaction is apparently affected by where the genes are physically located within the genome.

the chromosome), frame shifts (removal of a segment and its reinsertion at a different place in the same chromosome), and translocation (removal of a segment and its reinsertion in a different chromosome). All of these large-scale changes in chromosome structure can greatly affect adult phenotype. The exact way in which some chromosome mutations—for example, inversion or duplication—affect phenotype is not clear. We do not know why a slight change in the sequence of genes in a chromosome should alter the way the genetic message is expressed, nor why having a second copy of a chromosome segment should sometimes produce a gross phenotypic change. Yet such chromosomal changes do have their effects, and as such provide variations on which natural selection acts (for one example of the phenotypic effects of a chromosomal mutation—duplication—see Fig. 23.1).

Polyploidy. Polyploidy refers to a condition where the chromosome number is some multiple of the normal diploid number (3*n*, 4*n*, 5*n*, and so on). Polyploidy results from chromosomal duplication without subsequent cell division. Under natural conditions it occurs more frequently in plants than animals. Polyploidy can be artificially induced by various chemicals, such as colchicine. Polyploid individuals are often phenotypically larger in size than their diploid counterparts, though it is

not always possible to predict the sorts of changes in phenotype that polyploidy will induce. Many varieties of decorative and agricultural plants are the results of natural or induced polyploidy. As we saw in Chapter 18, additional chromosomes seem to present major developmental problems in animals, and as such polyploidy is probably a minor source of favorable variations in animals. In plants, however, the formation of new subspecies sometimes begins with the occurrence of polyploidy in one or a few individuals.

The Distribution of Variation in a Population

Most traits in any species of organism are variable. Even strains of laboratory rats that have been inbred for thirty or forty generations are not genetically identical. Variation is the rule, not the exception, in biological populations. The nature and extent of variation for any given trait is a defining characteristic of any breeding population.

There are two types of variation in nature, and it is important to distinguish between them. One is genotypic variation that results from actual changes in the genetic code on DNA, or from changes in chromosomal structure that alter the way genetic messages are expressed. This is obviously the most fundamental sort of variation and is the type that is passed on to the next generation. Another is phenotypic variation—that is, even with the same genetic information, variations in the development process produce changes in adult phenotype. These alterations are caused by environmental factors such as temperature, amount of food, light, water, and minerals available, and as such cannot be passed on to the next generation. It is genotypic variation that is the raw material of evolution. How do we measure the extent of variation in any population, and how, in particular, do we distinguish between genotypic and phenotypic variation?

To study variation within a population it is usually necessary to begin by measuring a given phenotypic trait in a representative sample of the population. We measure phenotypic traits because they are the easiest to detect. It is often difficult, or impossible, to measure genotypic variability directly, because we cannot observe it except when expressed as phenotype. The measurements of individual phenotype characters, such as the tail length in a population of deer mice, yield a statistical description of the character. This consists of several statistical parameters, such as the **mean** or average value of the character (obtained by summing the value of all the observations and dividing by the number of observations), the **mode** (the most common value, not necessarily equal to the mean), and the **range** (obtained by subtracting the smallest value from the largest). These concepts are explained in more detail in Supplement 23.1.

A useful way to obtain a graphic appreciation of the variation of a phenotypic character in a population

SUPPLEMENT 23.1
THE ANALYSIS OF DISTRIBUTIONS WITHIN POPULATIONS

Many types of measurements in biology involve sampling small amounts of data from the vast collections potentially available. For instance, it would be impossible from a practical point of view to measure the height of all the human beings in a large city in order to determine the average height of the city's population. Not only would such a procedure be time-consuming and laborious, it would also be unnecessary, since by choosing sample individuals from among the population at large, we can obtain acceptably accurate results with a minimum of effort. This means that if the sampling is unbiased (that is if all the individuals are not taken from one neighborhood, age group, or the like), the average height calculated from a fraction of the total population should be nearly the same as that of the entire population. Thus average height of 1000 adult males between the ages of 25 and 35 out of a population of 500,000 should be equivalent to the average height of all adult males between 25 and 35 in the population. One of the chief problems in any sampling technique, however, is the possibility that the chosen few measurements will be biased—that they will not be representative of the population as a whole. There are several ways in which the degree of bias affecting the validity of sampling data can be determined. Some of the ways in which such data can be treated will be discussed in this section. First, however, let us consider how sampling is done and how the raw data collected from field or laboratory measurements are converted into a useful form.

Several years ago a group of biologists set out to measure a specific characteristic, tail length, in two closely related species of deer mice (genus *Peromyscus*). They suspected from previous reports that the two populations were different in this characteristic and wanted to see whether this difference was statistically significant. The first collecting trip in the field yielded only fifteen specimens from population *A*. Although this was a small sample, the organisms were brought back to the laboratory and their tail lengths measured. The values that were recorded are shown in Table 1.

Note that for the sake of accuracy three different observers measured each organism. The slight discrepancies in the values recorded for each organism reflect a type of error that inevitably results when measurements are made. All measurements involve some estimation, and thus some differences will always arise. Despite the fact that measurements are objective and quantitative, they still involve human judgment and hence an element of subjectivity.

There are a number of ways in which such data can be analyzed. One important step in the analysis is to calculate the **mean** or **average** value of tail length for the sample at hand. The mean (symbolized $\bar{X}$) is determined by adding up

Table 1

Organism no.	Tail length, mm		
	Observer 1	Observer 2	Observer 3
1	60.5	60.2	60.3
2	61.0	59.9	61.1
3	62.2	62.0	63.0
4	68.1	68.0	67.9
5	60.7	60.6	60.2
6	58.3	58.4	58.5
7	66.6	66.0	66.3
8	56.7	56.6	56.6
9	62.5	62.6	62.5
10	60.8	50.9	60.7
11	58.0	58.2	58.1
12	52.4	54.5	54.5
13	56.7	56.2	56.1
14	58.9	58.8	58.7
15	60.2	60.3	60.2

(symbolized by Σ) all the individual values (1 through 15) and dividing by the total number of values (15) in the sample. This can be expressed mathematically as

$$\bar{X} = \frac{\Sigma X}{N},$$

where ΣX represents the sum of all the individual measurements, and N the total number in the population. Using the values from observer 1, we find that $\bar{X} = 60.3$. The mean value is useful in comparing one sample with another.

The data collected in Table 1 represent a survey of fifteen organisms from a large natural population. The average of these fifteen individuals, 60.3, may not be typical of the whole population. The smaller the sample, the greater the chance of sampling error—results not typical of the whole. After all, finding the mean of a sample of data is nothing more than making a generalization and it is dangerous to generalize from a small sample of data. It is therefore important to have as large a sample of measurements as possible. For instance, if organisms 14 and 15, as recorded in Table 1, had values of 40.0 (and, let us say, represented mutations), the mean would be lowered from 60.3 to 57.6. This is a large difference if one is talking about the average for the entire population. The question would then become: Are two out of every fifteen mice mutants for tail length, or did two mutants just happen to turn up in the present sample?

Sampling error is a significant problem in the collection of data. A large sample helps to reduce the misleading effects of such error.

In their present form, the data in Table 1 yield only a small amount of information. Calculation of the mean is one step toward analysis of the results. A second step is organization of the data into a graphical form that will tell us not only what the average value is, but how each measurement relates to that average; that is, the *distribution* of the data within the sample.

One of the most common ways to graph data such as those just given is to construct a histogram. A histogram is a form of bar graph. The height of each bar measures the number of individuals. The placement of the bar along a horizontal line indicates the specific range of value for each measurement. For example, the extremes of tail length given in Tables 1 are 52.4 and 68.1. On this basis nine categories, each representing a range of 2 mm, can be set up along the horizontal axis of the bar graph. Along the vertical axis the number of organisms can be measured off, 1 to 23. The histogram is constructed by making the bar for each category correspond in height to the number of organisms in that category. The completed histogram, taken from a large number of measurements, is shown in Diagram A.

The histogram has the advantage of showing immediately on inspection the category containing the largest number of organisms. This is called the *mode*. The histogram also has the advantage of showing immediately the way in which other categories are grouped around the mode. In Diagram A, the other categories are grouped roughly evenly on both sides of the mode; thus they represent a *normal distribution*.

In a perfectly normal distribution, the mean and mode are identical; however, the mean is not necessarily the same as the mode, and the two should not be confused. In principle they represent different quantities. It is true that in many of the sampling procedures biologists deal with, the two are identical or very close together. To determine the mean accurately, however, it is still necessary to make the calculations discussed in the previous section.

If a line is drawn through the bars on the histogram, the data can be shown as a line graph (see Diagram B). This graph shows a normal distribution curve. It is more or less symmetrical, with the value having the greatest frequency in the center and with values decreasing equally on both sides.

We have already discussed two statistics relating to graphs; the mean and the mode. On a graph, the mean is a measure of the degree to which the values fall in the middle of the range of observed values. The mode, as in the histogram, represents the category containing the largest number of individual items of data.

One other statistic is of importance—the *median*. The median defines the recorded value that is the midpoint of a number of observations. Thus in a class of 101 students, the

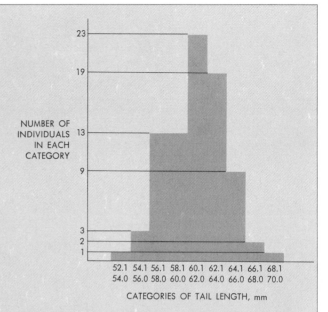

Diagram A
Histogram of data compiled from a large sample of measurements of tail length in deer mice. Histograms show clearly the number of individuals in each category. The largest single category (in this case, 60.1–62.0 mm) is called the mode.

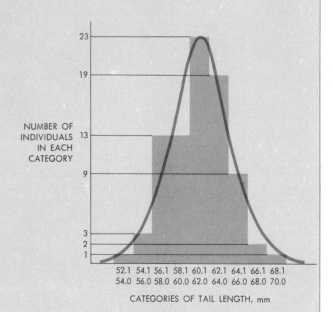

Diagram B
Line graph of data presented in Diagram A superimposed on the histogram of these data.

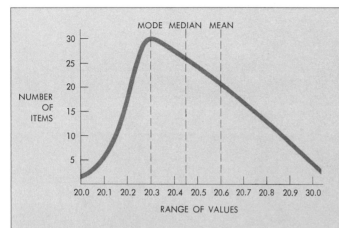

Diagram C
Asymmetrical curve showing mode, median, and mean, each having a separate value.

median grade is the one that 50 students scored higher than, and 50 lower. If a distribution curve is symmetrical (that is, if it is a normal curve) the median and the mean are the same. If the curve is asymmetrical, the two are different. Diagram C shows an asymmetrical curve with the mean and the median marked. The median occurs between the values of 20.4 and 20.5. The mean, however, is found at 20.59, that is, nearly at the value 20.6. The mean, median, and mode

thus represent different ways of describing the degree of symmetry of a curve.

Despite the methods developed thus far to describe the distribution of data, we still have no adequate way to estimate the *dispersion* of the observations. For example, all three curves shown in Diagram D are symmetrical and have identical means, modes, and medians. Yet it is obvious that in each case the data are dispersed quite differently around the mean.

One statistic that provides an estimate of dispersion is called the **variance**, symbolized by s^2. The greater the value for the variance, the more widely dispersed are the data around the mean. Of the three curves shown in Diagram D, curve C obviously has the greatest variance. To relate this to our earlier example, we should expect that the variance for a distribution of tail length would be much greater in a natural population of mice than in an inbred laboratory strain. Both types of mice will show approximately normal distribution curves, but the laboratory mice, because they are genetically more similar to each other and are raised under more uniform conditions, will generally show less variance in the dispersion of the measurements. The dispersion could also be measured by the *range*, that is, the difference between the minimum and maximum measurements in the data. But then we would know only the lower and upper limits of measurement, irrespective of how great the frequency of the various measurements might be. Variance has the advantage of taking frequency of distribution into account.

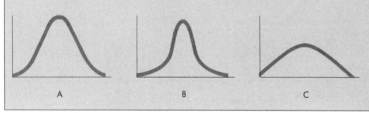

Diagram D
Three curves of frequency distribution with the same mean, but different dispersion of data around the mean. The variance for curve C would be the greatest, whereas that for B would be the least.

is to draw a frequency distribution curve. Such a curve presents a graphic representation of the mode (the category under the highest peak) and the range. The shape of the curve provides information about the distribution of the trait within the population. The curve in Fig. 23.2 shows a so-called bell-shaped or normal distribution curve. Additional discussion of such curves and their significance is found in Supplement 23.1

From the evolutionary point of view, we might ask how much of the total variation, as shown in Fig. 23.2, is due to heredity and how much to environment? There are several ways to answer this question, but all involve some form of genetic experimentation.

1. Breeding experiments. Individuals from a natural field population can be brought into the laboratory and

bred for several generations under as nearly identical conditions as possible.

2. Selection experiments. If the variation is genetic, or at least largely so, selection toward one or another extreme of the range of distribution for the trait will move the mean of the next generation in the direction of selection (see Fig. 23.3). If, however, the variation present in a population is largely a result of environmental causes (and hence is not genetic), selection in a given direction will not alter the mean value for the trait in future generations.

From the evolutionary point of view, the important variations in a population are those which have a genetic base. It is on these variations that natural selection

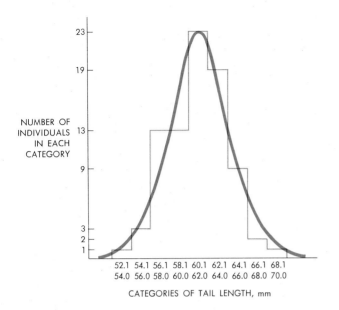

Fig. 23.2
Bell curve of normal distribution for a phenotypic trait in a population. The trait being measured here is tail length in a sample of eighty-five deer mice (genus *Peromyscus*). The curve is shown superimposed on a bar graph in which tail length is plotted by categories, one vertical column representing each category (that is, tail length of 52.1 through 54.0 mm). For further details see Supplement 23.1.

Fig. 23.3
The effect of selection on variation within a population under three different sets of conditions: (*a*) when the variation is caused exclusively by environmental factors; (*b*) when variation is caused by a combination of environmental and hereditary factors; and (*c*) when variation is caused only by hereditary factors. (Modified from Simpson and Beck, *Life*, 2d ed. New York: Harcourt Brace Jovanovich, 1965, p. 472.)

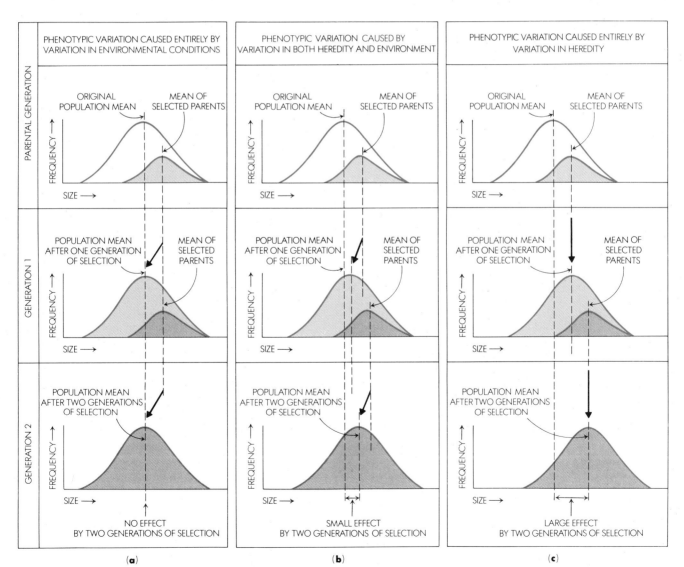

can ultimately act. However, for the biologist it is important to remember that the expressed phenotypic variation observed in any population is always a result of some combination of genetic and environmental factors. There is no formula for distinguishing one from the other in any individual situation. Only by careful field measurements and laboratory genetic investigation (including biochemical analysis of protein structure) is it possible to gain an idea as to how much of the observed variability in any natural population is genetic, and how much environmental.

The Persistence of Variations in a Population: Polymorphism

If certain distinct variations in a population persist from generation to generation, the population is said to be **polymorphic** (or, to express **polymorphism**). Polymorphism is the occurrence of two or more distinct forms within a single population or within a species. Examples of polymorphism include right-handed and left-handed coiling is snail shells or blood groups in the human population (see Fig. 23.4). Most polymorphisms involve two, three, or four different forms, though in one rather extreme case, 120 different forms of body pattern have been recorded for a single species of platyfish. Polymorphic traits, then, are nothing more than several distinctive forms of one trait that persist within a population from one generation to another. Most polymorphisms are a result of genetic differences—the presence of two or more alleles for a trait within a population (multiple alleles).

The existence of polymorphism illustrates an important point about the process of evolution. According to the tenets of the theory of natural selection, we would expect one form of a particular phenotypic trait to be functionally more advantageous, more adaptive than others; we would therefore expect all members of the population to show that trait in time. Although some variation within a population is to be expected, how can we explain the existence of discrete polymorphism?

Biologists identify two types of polymorphism: transitional and balanced. **Transitional polymorphism** refers to the situation where the polymorphism is transient, because the frequency of one form is on the decline and that of the other is on the rise (though the former may never completely disappear). The best known case of a transitional polymorphism is that of the peppered moth (and other so-called melanic moths) in England, a case that we will analyze in detail in Section 23.7. A century ago most of the moths were variegated, that is, basically light-colored with black speckles; only a few showed an overall dark, "melanic" coloration. By the 1950s, however, in certain regions of England the light form was almost nonexistent, and the dark form had almost completely taken over. The existence of polymorphism—presence of both a dark and a light form—had almost disappeared in a few localities by the middle of the present century. Transitional polymorphism is usually the result of selection acting on one of the forms more than the other, gradually shifting allele frequencies so that one phenotype becomes predominant while the other is gradually eliminated.

Balanced polymorphism is characterized by the relatively steady-state persistence of two (or more) forms in a population over successive generations. Consider the following examples. In the land snail *Cepaea nemoralis* (see Fig. 23.5) three different alleles affect the

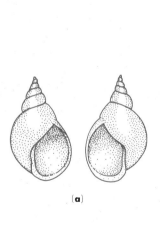

(a)

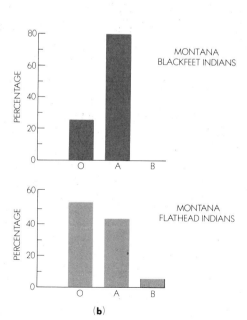

(b)

Fig. 23.4
Examples of polymorphism. (*a*) Left-handed and right-handed coiling of snail shells. Both phenotypes persist in the population generation after generation. (*b*) Polymorphism in blood-group frequencies in two populations of western American Indians. Note that the Blackfeet are dimorphic, whereas the Flatheads are trimorphic. The adaptive significance of the different blood groups is not clear. It may be that people of any one blood group are slightly advantaged over people from other groups in terms of disease resistance, though this idea is not firmly established.

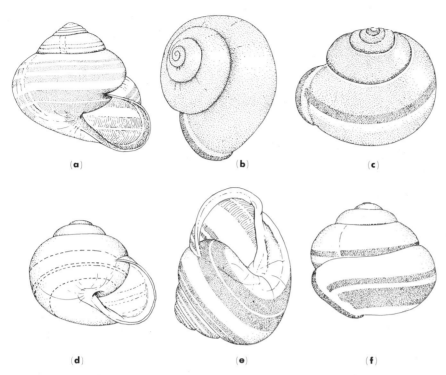

(a)   (b)   (c)

(d)   (e)   (f)

Fig. 23.5
Polymorphic forms of the land snail *Cepaea nemoralis* from England. Polymorphism exists for basic shell coloration: (a) yellow, (b) pink, and (c) brown; and for banding patterns: (a) five bands and a dark lip around the opening, (b) no bands and dark lip, (c) one band, (d) outlines of at least one band, but no pigmentation, (e) five bands but light-colored lip, and (f) three bands. Banding pattern and shell coloration are due to two independently assorting genetic loci, so that all combinations of color and band number are possible. The persistence of the various forms from one generation to another within the same population is an example of balanced polymorphism. The variety of forms reflects differences in background coloration throughout the population's range. Selection has favored the preservation of different alleles to provide maximum degrees of camouflage. (From P. M. Sheppard, *Natural Selection and Heredity*. New York: Harper Torchbooks, 1960, p. 85. Used with permission of Hutchinson Publishing Group Ltd.)

ground color of the shell: it can be brown, pink or yellow. In addition, the shell may display any one of six banding patterns—from zero to five bands. In a given population snails are often found with different shell colors and banding patterns.

In England, where the genetics and evolution of *Cepaea nemoralis* have been studied extensively, naturalists demonstrated that the color and banding of the shell camouflage the snail and hide it from thrushes, which are its natural predator. So, for example, unbanded brown and pink shells blend best with a uniform brown background of dead leaves, while yellow banded forms blend best with the varied color patterns of field vegetation. But the background of field and forest changes with the seasons and the years. Furthermore, snails move back and forth between forest-like and field-like habitats, for example along hedgerows. Given such changes in season and habitat, no one color form provides a superior camouflage *for all situations*. Thus, selection works to preserve several different forms (that is, polymorphs) within the larger breeding population. The polymorphism remains in balance so long as there are two or three different habitats over which the snail population as a whole ranges. It is advantageous for the population as a whole to preserve this variability. The presence of several alleles for shell color has one disadvantage, however. Since the various subpopulations of snail (forest and field) interbreed, some individuals will inherit genotypes and hence phenotypes that are inappropriate to their habitats. For example, some forest

forms will be born with yellow color, and some field forms with brown. These forms will be more subject to predation, and hence those individuals will be selected against. However, because the environment is so varied, selection will favor preservation of the alleles for different color and banding patterns within the population as a whole.

Once a variation of any sort, whatever its origin, has been introduced into a population, what is its fate? How is it selected for or against? Are all variations either positive or negative? Or, are there neutral variations which are neither favored, nor opposed, by natural selection? To answer these and other questions, we need first to understand the basic principles of population genetics.

23.4
THE EFFECT OF SELECTION ON VARIATION WITHIN A POPULATION

As pointed out earlier, population genetics is the field of biology that studies the genetic composition of whole animal or plant populations. The population itself, rather than individual cells or organisms, becomes the basic unit of biological study.

How is it possible to describe the genetic makeup of an entire population? This is not quite as difficult as it may sound. In the first place, the genetic characteristics of the population can be treated as the sum of the genotypes of all the individual members. Thus we may speak of the **gene pool** of a population. A gene pool is defined as the total genetic information possessed by the members of a population of sexually reproducing forms. The number and kind of each allele found in the population determine the unique characteristics of the gene pool.

In the second place, population genetics uses statistical methods. Individual variations in organisms do not enter the picture unless a given variation becomes statistically significant. Only if enough organisms turn up with the same variation in a given generation will this variation become apparent as a definite and nonrandom change in the composition of the gene pool. Such methods of analysis are very important in studying the genetic changes involved in the evolutionary process.

By operating on individual organisms, natural selection ultimately brings about a change in the composition of the gene pool to which these individual organisms contribute. The gene pool of a population is divided up each generation and parceled out to new individual members (the offspring). Inevitably new mutations or combinations of genotypes occur that express themselves as specific individual phenotypes. As a result of natural selection, certain genotypes will leave more offspring than others. Certain alleles will be passed on to the next generation in greater numbers, other in less. The composition of the gene pool will be changed accordingly.

Genetic Equilibrium

Two opposing factors are at work on the gene pool of a population. One is natural selection, which tends to alter the composition of the gene pool from one generation to another. The other factor is expressed in the concept of **genetic equilibrium**, which holds that under very specific conditions the ratio between various alleles in a population tends to remain constant from one generation to the next. This concept applies regardless of the proportions of the genes in the initial population. The idea of genetic equilibrium was introduced into biology in 1908 by G. H. Hardy, an English mathematician, and Wilhelm Weinberg, a German physician; hence it is usually referred to as the Hardy-Weinberg law. Although it is sometimes thought by nonbiologists that recessive genes must eventually be wiped out of a population, according to the concept of genetic equilibrium this is not necessarily the case. For example, despite the fact that blue eyes are recessive to brown, the number of blue-eyed people in the human population remains relatively constant from one generation to another.

We mentioned earlier that genetic equilibrium would be maintained in a population *only* under very specific conditions. These conditions are as follows:

1. *Mating in the population must be random.* That is, each phenotype must have an equal chance of reproducing. If this were not the case—if one male phenotype were more acceptable to the females than another—then the favored males would leave more offspring. For example, if female birds of a given species always choose the males with red plumes, the genes producing this phenotype will increase in the population.

2. *All matings must yield, on the average, the same number of offspring.* If mating between two particular phenotypes consistently yielded fewer offspring—then the alleles for those phenotypes would decline in number with each generation.

3. *The population must be sufficiently large that chance variations in a small number of organisms do not affect the statistical average.* In small populations the effects of random mutations or immigration or emigration of a few organisms can produce statistically significant changes in gene ratios. By analogy, a failing grade in a class of three affects the class average far more than in a class of thirty.

4. *The mutation rate must have reached its own equilibrium.* This means that the number of A alleles mutating to a must be equaled by the number of a alleles mutating to A. As a result, the overall ratio of A to a will remain the same from one generation to the next.

5. *There is no migration into or out of the population.* This means simply that no new alleles are brought into the population by immigration, nor are existing alleles leaving the population by emigration.

These five factors represent abstract idealized conditions seldom met in natural populations. Why, then, do we bother to discuss them at all? First, to the extent that any of these factors operate in a population, they tend to check or oppose the dynamic processes of change. Second, to understand the dynamics of change in a population it is necessary to understand also the factors opposing that change. The course of any and every evolutionary change is the result of a dynamic interaction between the factors contributing to stability and those contributing to change.

The Hardy-Weinberg law is a cornerstone of the modern population approach to evolution. It can be expressed in quantitative terms that emphasize the exact nature of genetic equilibrium. In its simplest form, the law speaks to the case of a single gene locus with two alleles, A and a. There are three possible genotypes represented by a two-allele system: AA (homozygous dominant), Aa (heterozygous), and aa (homozygous recessive). The Hardy-Weinberg law states that the frequencies of these genotypes will remain constant from generation to generation, as long as the conditions outlined in the last paragraph prevail. In other words, the Hardy-Weinberg law states that:

In a large (theoretically infinite) population where mating is random, the frequency of AA, Aa, and aa will remain constant from generation to generation in the absence of selection or migration.

SUPPLEMENT 23.2
DETECTING VARIATION IN A POPULATION

One of the persistent problems evolutionary biologists have faced in recent years is that of finding ways to determine accurately how much genetic variability exists in any population of organisms.

This is not an easy task. Dominant genes mask the presence of recessives, thus reducing our estimate of the number of mutant loci present. In addition, many variations have subtle and pleiotropic effects making their presence difficult to detect. Two classical approaches to this problem have made use of artificial selection experiments and Mendelian breeding experiments.

That selection *can* produce very profound effects on a given phenotypic character, if variability is present, is illustrated by experiments in which egg-laying capacity in hens was selected for over many generations. In one classic case, between 1933 and 1965, the egg-laying capacity of a commercial flock of chickens was increased from 125.6 eggs to 249.6 eggs per hen per year—an increase of almost 100 percent in thirty-two years! This amount of change could not have occurred if the population of chickens did not have an extensive reservoir of genetic variability (for egg-laying) to begin with. The very fact that artificial selection experiments work most of the time is a strong indication that there is much genetic variation in populations for virtually every phenotypic trait.

Mendelian breeding experiments, under controlled conditions, permit the detection in the laboratory of mutations that generally remain hidden in natural populations (so-called cryptic variation). For example, most mutations are recessive to their wild-type alleles; they are also more often than not deleterious—that is, they lead to organisms that are not well-adapted overall. Both of these factors contribute to the relative "invisibility" of many mutations. Recessive mutations only show up phenotypically in the homozygous state, though they may be present in a significant number of heterozygotes. Futhermore, since most of these mutations are deleterious, selection, under natural conditions, would keep their frequency very low. As a result, the chances of homozygous recessive combinations occurring in the first place would be very low. In the laboratory, however, less harsh living conditions would increase the chances of homozygous recessive individuals surviving, even if they might be less hardy in nature. The visibility of such mutants in a laboratory (as opposed to field) environment provides a rough estimate of gene frequency for recessive variations in the population at large.

Mendelian analyses, however, have one major drawback. By definition, Mendelian breeding experiments involve only those traits that show at least one variant allele. Mendelian analyses are powerless to study variations in alleles that do not produce visible phenotypic effects. Yet, most geneticists and evolutionists feel sure that there are many variations at the DNA level whose phenotypic effects are so slight, or so unrecognizable, that they simply go undetected. This same low-level variation may be extremely important, however, in providing the innovations for the evolutionary process.

A way out of this seeming impasse has been provided by recent developments in molecular biology. As we know, genes code for proteins—the specific sequence of base pairs on DNA determining the specific sequence of amino acids in the polypeptide product. Proteins are, in one sense, the initial and most direct phenotypic expression of particular genotypes. If a certain protein is invariant in the population it is probably true that the gene coding for that protein is also invariant in the population. Similarly, if the protein shows variation, then it is likely that the gene is represented by more than one allele in the population. Using the technique of gel electrophoresis (see Chapter 5), it is possible to estimate the number of different forms of proteins present in a sample drawn from different individuals in a population. By focusing on several specific proteins representing an unbiased sample, it is possible to estimate the number of variant alleles in a population for any trait or group of traits.

Diagram A is a photograph of an electrophoretic gel showing several variants for a single enzyme, malate dehydrogenase, or MDH for short (it is involved in the oxidation of car-

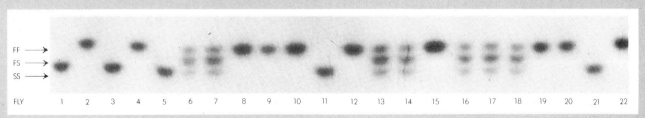

Diagram A
Courtesy Professor Ayala.

bohydrates in the Krebs cycle). A single molecule of MDH consists of two polypeptide chains synthesized from the two alleles present on a pair of homologous chromosomes. In other words, the loci on the maternal and the paternal chromosome both synthesize polypeptides, any two of which, regardless of their chromosomal origin, can join together to form an MDH molecule. If any variations in these alleles occur within the population, they will show up as variant MDH molecules on an electrophoretic gel. For example, the gel shown in Diagram A was prepared by extracting samples of MDH from twenty-two individual *Drosophila*. Enzyme extract from each individual was placed at a spot along the bottom of the gel (indicated by numbers 1 through 22, each number corresponding to one fly). The proteins were then allowed to migrate in an electric field, that is, "upward" along the gel. The spots higher up represent the "fast"-moving protein molecules (symbolized F); the spots lower down represent the "slow"-moving molecules (symbolized S). Recall that the rate of movement of proteins in an electrical field is a consequence of and proportional to their electric charge.

If we read each numbered column upward, we will get some idea of the genetic constitution of each individual fly in the sample of twenty-two. For example, numbers 1 through 4 have only one type of MDH each: 1 and 3 have S molecules, while 2 and 4 have F. Flies number 6 and 7, on the other hand, have three types of molecules: one F, one S, and an "intermediate." We can account for these types in simple genetic terms. Since each molecule of MDH consists of *two* polypeptides, flies with only the slow-moving or fast-moving forms (for example, 1 or 2) would be genetically homozygous. That is, fly 1 would be SS, meaning that its molecules of MDH consisted of two polypeptides of the slow-moving form, and 2 would be FF, or homozygous for two polypeptides of the fast-moving form. Flies 6 and 7 must

be heterozygous: they have alleles for both F and S polypeptide chains, and consequently can produce both an F and an S polypeptide. This means they can have three types of protein molecule, depending on the combination that occurs: FF, FS, and SS (we assume all three forms occur in equal numbers). Each of these three forms separates out on the gel at a different spot, with the heterozygous molecule falling in between the two homozygous, as might be expected.

Whether these three forms of MDH have any selective value over each other is not known. Electrophoretically, the molecules differ from each other, and in most cases electrophoretic differences mean some differences in biochemical function. But such differences in function in the biochemists' test tube may not necessarily translate into a selective value, positive or negative, in nature. What is important about the electrophoretic technique in this situation is that it provides a way of detecting the amount of genetic variability in a population that is otherwise invisible. Electrophoresis helps to reveal cryptic variability in a subtle and precise way. However, it does not tell us whether such variations are of positive, negative, or neutral adaptive value as far as selection is concerned.

Using the gel electrophoresis process to survey populations, biologists have found that species differ widely in their range of variability. Birds show very little, whereas *Drosophila* shows three to four times as much. Invertebrates as a general group show more variability (about 13 percent on the average) than vertebrates (about 7 to 8 percent). All in all, it appears that there is much more genetic variation present in most animal and plant populations than biologists have previously suspected.

Since variation is the raw material on which natural selection acts, this discovery has important consequences for our concepts of how fast, and in what directions, evolution can take place.

By definition, the three genotypes possible will always represent 100 percent of the combinations (for that particular allele pair) in the population. Whatever the original frequency of each genotype, the proportions of each genotype will remain the same from one generation to the next. Thus, if 60 percent of the original population consisted of genotypes *aa*, 30 percent of genotypes *Aa*, and 10 percent of *AA*, we would expect to find the same proportions in the next generation as well given the conditions specified by Hardy-Weinberg.

A corollary of the concept of genetic equilibrium is that the frequencies of the individual alleles (either *A* or *a*) will tend to remain the same from one generation to another. In the population-genetics approach to evolution, biologists prefer to speak about the frequencies of individual alleles rather than the frequency of individual genotypes. Thus, we can say that in any given popula-

tion where *A* and *a* represent the only two alleles for a given trait,

frequency of A + frequency of a = 100%, or 1.0.

It is customary to express frequency as some fraction of one, rather than as a percentage. So, if 100 percent is equivalent to a frequency of 1.0, then 70 percent is equivalent to a frequency of 0.7, and 15 percent to a frequency of 0.15. Thus, according to the basic premises of Hardy-Weinberg, if the frequency of *A* in the first generation is 0.8, and that of *a* is 0.2, the frequency of each allele in the second (and subseqeunt) generation(s) would also be expected to be 0.8, and 0.2, respectively.

Now, what is the relationship between the frequencies of certain alleles in a population's gene pool, and the frequencies of certain genotypes? The two are directly related. Consider the following example. Suppose we begin with a population of mice with the two alleles A and a, showing the genotypic frequency of 49 percent AA, 42 percent Aa, and 9 percent aa (see Fig. 23.6). We can ask two questions: (1) What are the frequencies of the two alleles A and a in the initial population? and (2) What will be the genotype frequencies expected in the next generation? As Fig. 23.6 shows, each organism in the parental population contributes gametes to the gene pool for the next generation, but in proportion to the frequency of each parental genotype in the original population. Let us assume (for convenience) that the population consists of 100 organisms, and that each contributes 10 gametes to the gene pool. Then, 49 out of every 100 mice will have the genotype AA and will con-

tribute to the gene pool 490 gametes containing the dominant allele A; 42 out of every 100 mice will have the genotype Aa and will contribute 210 gametes with allele A and 210 gametes with allele a (that is, $210 + 210 = 420$) to the gene pool; and 9 out of every 100 mice will have the genotype aa and will contribute 90 gametes with allele a to the gamete gene pool. The distribution of each allele would thus be as follows:

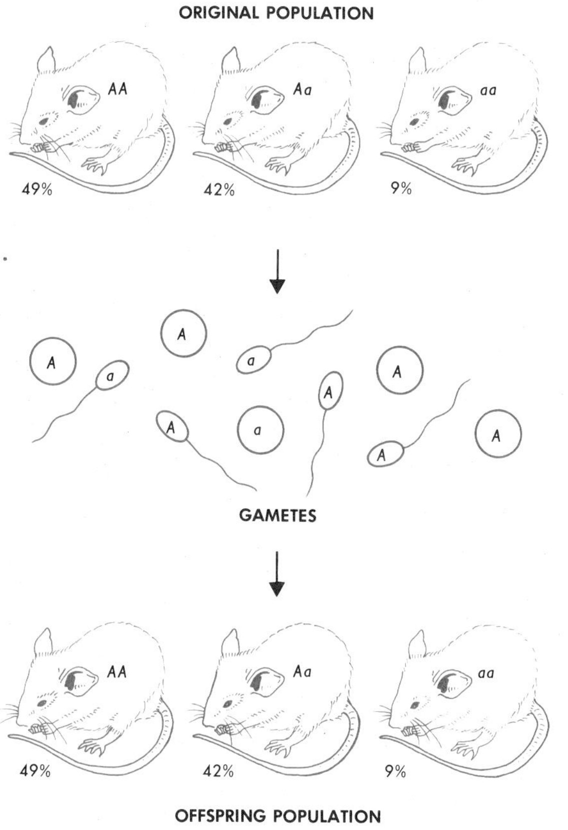

	Gametes		
	A	a	Total
49 are AA and produce	490	—	490
42 are Aa and produce	210	210	420
9 are aa and produce	—	90	90
	700	300	1000

We see that there are a total of 700 A alleles and 300 a alleles, or a ratio of 7:3 in the population's gametic gene pool at the time of mating. The frequency of the A allele is thus 0.7, and that of the a allele is 0.3 (that is, $700/1000 = 70$ percent, or 0.7; and $300/1000 = 30$ percent or 0.3).

The expected frequency of each genotype in the next generation can be calculated once the frequencies of each allele in the parental population are known. This can be done with the aid of the Punnett square, as described in Section 16.9. Since mating is random, the probability of any two alleles combining is given as the product of the frequency of each allele in the gamete gene pool (refer back to the product principle of probability, Section 16.3), as shown in the following Punnett square:

ORIGINAL POPULATION

GAMETES

OFFSPRING POPULATION

Fig. 23.6
If 49 percent of the original parental population has a genotype of AA, 42 percent a genotype of Aa, and 9 percent of a genotype of aa, the Hardy-Weinberg law states that the same percentages will be found in the next generation. With 49 percent of the original population AA, 42 percent Aa, and 9 percent aa, the frequency of the A allele is 0.7, and that of the a allele 0.3. Thus, of the total gametes produced by the parental population (ten gametes are shown here), seven are A while three are a. Random combinations of these alleles will yield pairings in the same proportions as existed in the original population.

Thus, the probability of getting the genotype AA is given as 0.7×0.7. The probability of getting the genotype Aa is given as $2 \times (0.7 \times 0.3)$; and the probability of getting genotype aa is given as 0.3×0.3. Thus the probability of getting genotype AA is 0.49; that of getting genotype Aa is 0.42; and that of getting genotype aa is 0.09. It should be apparent that the values given for each genotype are identical to the original percentages of each genotype in the parental population.

The Hardy-Weinberg law states that gene frequencies in a population will remain constant from one generation to the next as long as certain conditions are met: the members of the population mate randomly, the offspring of all combinations are equally successful, the population is large enough to avoid sampling error, and the mutation rate has reached equilibrium.

The Hardy-Weinberg law can be stated as a simple mathematical equation from which predictions can be made. If we let the symbol p stand for the frequency of allele A, and the symbol q stand for the frequency of allele a, then the frequency of genotype AA would be equivalent to $q \times q$, or q^2 (refer back to the Punnett square above), the frequency of genotype Aa would be equivalent to $p \times q$ (however, as the Punnett square shows, there are two combinations of Aa, hence we would show this by writing $2pq$); and finally, the frequency of genotype aa would be written as q^2. Since the frequencies of all these genotypes equal 1.0, we can write the Hardy-Weinberg law as:

$$p^2 + 2pq + q^2 = 1.0.$$

Using the above example of the A and the a alleles in mice, we can see that $p^2 = 0.7 \times 0.7$, or 0.49; $2pq = 2 \times (0.7 \times 0.3)$, or 0.42; and $q^2 = 0.3 \times 0.3$, or 0.09. Expressed in the equation form, we have

$$\underset{(0.7 \times 0.7 = 0.49)}{p^2} + \underset{(2 \times 0.7 \times 0.3 = 0.42)}{2pq} + \underset{(0.3 \times 0.3 = 0.09)}{q^2} = 1.0.$$

It should be apparent that this formula is simply an expansion of the binomial equation, as discussed in Section 16.3.

The Hardy-Weinberg law and its mathematical expression are an extremely useful approach to population genetics and its relationship to evolutionary problems. Using the formula, it is possible to determine the frequency of alleles in a population by sampling a limited number of organisms. If, for example, the gene A is completely dominant over a in the foregoing example, then the frequency of homozygous recessives (the only ones whose phenotype automatically indicates the genotype) can be used to calculate the frequency of each allele. If the frequency of aa is found to be 9 out of 100 (frequency = 0.09), then we can calculate that the frequency of a is 0.3 (since $0.3 \times 0.3 = 0.09$). This also establishes the frequency of A is 0.7, since the frequencies of $p + q$ must equal 1.0. If A is incompletely dominant over a, the task is even simpler, since the phenotypes of homozygous dominant, homozygous recessive,

and heterozygote will be different. Each can then be counted for sampling purposes.

If, in sampling gene frequencies in one generation, we note a shift in the frequency for the next generation, we can look for conditions in the population that might be working against the maintenance of genetic equilibrium. These conditions might be nonrandom mating or natural selection acting upon one or another genotype in the offspring. In some cases both conditions may be at work simultaneously. Whatever the case, the failure to maintain a Hardy-Weinberg equilibrium indicates that some evolutionary process is at work. Fundamentally, evolution is nothing more than a shift in gene frequencies (either random or nonrandom) in successive generations. In Section 23.6 we will consider a specific example of how evolution produces a shift in gene frequencies. First, however, it will be useful to review the various agents that contribute to a shift in gene frequencies between generations.

23.5
AGENTS OF EVOLUTION

A number of factors can produce a systematic change in gene frequencies within a natural population. These include natural selection, mutation pressure, genetic drift, and gene flow. Although all of these factors undoubtedly interact in almost every natural population, it will be helpful to examine them separately.

Natural Selection*

Natural selection acts directly on certain phenotypes in a population to produce a shift in the frequency of phenotypes (and underlying genotypes) in the next generation. Selection acts on the sum total of the variability within a population, allowing some phenotypes to reproduce more readily than others. Evolutionists have identified three different ways in which selection can act on the variability within a population: stabilizing, directional, and disruptive selection. These forms of selec-

---

*The discussion here focuses on natural selection; however, the same principles would apply to artificial selection as well.

tion, and their consequences, are summarized in Fig. 23.7. Let us consider each.

Stabilizing Selection. Stabilizing selection refers to a situation where those organisms whose phenotypes fall somewhere in the midrange, or around the mean and modal classes, are favored. As Fig. 23.7(a) indicates, when selection favors the mean and modal groups, the extremes are selected against. Gradually, after several generations, the range of phenotypes for the particular trait in question is narrowed, as a result of stabilizing selection. Stabilizing selection works to increase the number of individuals within the population showing the mean or modal phenotype. In the process, of course, it also increases the frequency of the genes that determine the mean, or modal condition.

Directional Selection. Directional selection (Fig. 23.7b) occurs when one phenotype, for example large bean size, is favored over another phenotype, such as small bean size. Directional selection shifts the mean of the subsequent generations in the direction of the favored phenotype. It is possible for the mean of future generations to continue moving in the direction of selection as long as some genetic variability for the trait in question persists. As we saw in the example of Wilhelm Johannsen's experiments with beans (Supplement 23.2) selection can only continue altering the frequency of phenotypic expression as long as genotypic differences for a trait exist within the population. As soon as a population becomes genotypically homogeneous, selection ceases to alter the mean of the population for that trait.

Disruptive Selection. Disruptive selection (Fig. 23.7c) occurs when both extremes of a phenotypic distribution are favored, and the mean or modal groups are not. For example, consider a population of deer mice where both long (LL) and short (ll) tails are particularly advantageous, but medium-length (Ll) tails are not, and where L is incompletely dominant over l. If the population starts out with a normal distribution for tail length, over several generations disruptive selection would produce a bimodal distribution of both long- and short-tailed mice. Disruptive selection thus results in polymorphism, where the two homozygotes (LL and ll) would show greater superiority than the heterozygote (Ll). Disruptive selection is less common than either stabilizing or directional selection.

Mutation Pressure

As we have seen, mutations—both point and chromosomal—produce much of the variability on which natural selection acts. Since both kinds of mutations tend to occur at certain rates, some biologists have suggested that there is a kind of mutation pressure which can direct evolutionary developments along its own lines. For example, we could have a series of point mutations:

$$A_1 \rightleftarrows A_2 \rightleftarrows A_3 \rightleftarrows A_4 \rightleftarrows A_5.$$

If the forward mutation rates were greater than the back, a directionality would be established that could possibly account for some kinds of evolutionary development. However, the majority of biologists do not believe that mutation pressure alone, without selection, can produce any significant evolutionary developments.

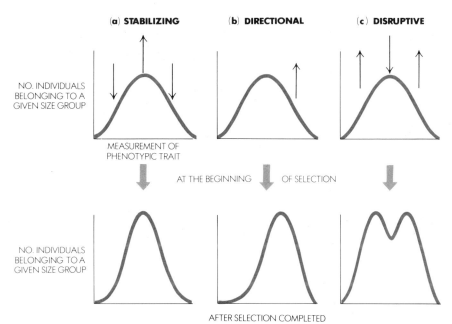

Fig. 23.7
The effects of three different types of selection—stabilizing, directional, and disruptive—on distribution of phenotypic variations in a population.

The main reason for this conclusion is that, as we have seen, mutation rates are very low. Their effects would almost always be overridden by even the mildest forms of selection. Only in the case of a totally neutral trait (that is, one that had absolutely no selective advantage or disadvantage whatsoever) would mutation pressure be able to show any effect on gene frequencies in successive generations. If electrophoretic studies on protein variability in a population are valid (see Supplement 23.2), and if the many variant forms of most proteins are selectively neutral, then their current frequencies are entirely the result of mutation, and specifically of the equilibrium established between the forward and reverse mutation rates. There is much debate today, however, on the degree to which any concrete examples of evolutionary change are the result of mutation pressure *per se*.

Genetic Drift

The subject of **genetic drift** has been widely (and sometimes heatedly) discussed among evolutionary biologists for nearly half a century. The term genetic drift refers to the process by which gene frequencies are altered from one generation to the next by chance events alone—that is, where no selection is involved. The cases in which genetic drift is most frequently thought to occur are usually small populations, where only a few organisms (sometimes fewer than 100) are involved. Under such circumstances, it is possible that by chance some genotypes might leave more offspring than others in the next generation. For example, suppose we started with a population of ten mice of three genotypes, *AA*, *Aa* and *aa*. If only six individuals (three pairs) were actually successful in breeding, it might well be that all six were of one genotype (for example *AA* or *aa*). This could happen only if the phenotypes determined by the genotypes *AA* or *aa* were selectively neutral—that is, had little to do with reproductive potential. With such a small population, one genotype could become completely fixed, and another completely eliminated in a population in one or a few generations. Such "sampling error" (on the part of the organisms doing the breeding) produces what is called genetic drift. The word "drift" emphasizes the fact that the process is the result of random, nondirected events.

The **founder effect** is a special case of genetic drift. Consider the following example. After a river floods its banks and then recedes, fish are often trapped in smaller pools or ponds that are now isolated from the river itself. The fish in such isolated pools have happened there by chance; they represent an isolated founder population that may or may not represent all the genotypes present in the larger fish population inhabiting the river. Genotypically and phenotypically it could thus show a marked departure from the parental population from which it is an offshoot. The frequency of genes in the founder population's gene pool (no pun intended here) would very likely be different from the original population. If the founder population maintained itself over many generations (assuming that the pond does not dry up), it could end up with the distinctive set of traits differentiating it from the larger population in the river.

Founder populations can also become established through migration. For various reasons (overcrowding, scarcity of food) groups of organisms may leave an original population and found a new population of their own. Often the migrants are few in number, and may not represent the characteristic distribution of genotypes found in the population from which they come. In a new environment, with different conditions, after a few generations natural selection can end up producing a very different distribution of genes in the founder, as opposed to the original, parental population.

Gene Flow

Gene flow is defined as the movement of genes from one part of a population to another or from one population to another, through interbreeding. Within populations, gene flow simply means that genes located in one part of the breeding group will be able to pass or "flow" to other parts of the same group, or to another, within a larger breeding population. With respect to the larger population, gene flow between subpopulations does not alter the frequency of genes within the population as a whole. It merely redistributes them.

Gene frequencies in a population *can* be altered, however, by gene flow between two quite distinct breeding populations. Usually these populations represent the same species, though in plants gene flow can sometimes occur between neighboring populations of different species. Where the territories of nearby but distinct populations overlap, some interbreeding usually does occur. Such interbreeding can introduce new alleles into each population. In populations that are more geographically distant from one another, interbreeding can occur through migration of organisms from one population to another. On the whole, the greater the geographic distance between two breeding populations, the less frequent interbreeding and the exchange of genes will be. Sometimes gene flow can occur between rather distant breeding populations by passing through neighboring, geographically intermediate, populations.

Let us review briefly the main points we have established about evolution by natural selection.

1. We have emphasized the population concept of evolution, stating that populations may be considered as interbreeding groups with a common gene pool. Individual organisms are carriers of alleles; although the organisms die, the gene pool persists from generation to generation.

2. We have listed the various mechanisms by which genetic variations are introduced into a population's

gene pool: point mutation, chromosomal mutation, and polyploidy.

3. We have discussed the distribution of variations within a population, including normal distributions; the methods of distinguishing genetically caused from environmentally caused variations; and the persistence of different variants of a given trait, or polymorphism, within populations from one generation to another.

4. We have discussed the Hardy-Weinberg law of genetic equilibrium. We have seen how the Hardy-Weinberg concept, mathematically formulated, allows us to estimate the frequency of certain alleles in a population. Evolution can thus be expressed quantitatively as the shift in allele frequency from one generation to the next. It is the *lack* of genetic equilibrium.

5. We have listed, and briefly discussed, some of the major agents of evolution: natural selection (stabilizing, directional, and disruptive), mutation pressure, genetic drift, and gene flow. Although selection is probably the most important in the long run, all four agents no doubt act to some extent to alter gene frequencies within populations.

With this general background, we are now in a position to examine in detail two case histories of how evolution by natural selection actually works: industrial melanism in moths and sickle-cell anemia in human beings.

23.6
EVOLUTION IN ACTION: THE CASE OF THE PEPPERED MOTHS

Two forms of the peppered moth, *Biston betularia*, have been known to exist in various parts of England. One form, *Biston betularia typica*, is light-colored with small dark spots irregularly scattered over its wings and body. The other form, *Biston betularia carbonaria*, is much darker, due to much greater quantities of the pigment melanin (see Fig. 23.8). The latter is often called the "melanic" form.

In the past, samples collected in the field showed that the light form was far more prevalent than the dark. This was explained along Darwinian lines by claiming that the light form was *cryptically colored,* or camouflaged. Organisms that are cryptically colored inherit certain patterns of pigmentation that allow them to blend with their backgrounds. On tree trunks covered with lichens* light varieties of the peppered moth are perfectly camouflaged, but the darker melanic form stands out prominently (see Fig. 23.9).

---

*Lichens are greenish-gray growth found covering many rocks and trees. A lichen is actually a combination of an alga and a fungus (for more details and photograph, see Fig. 28.8). The peppered moths often rest on the lichen-covered tree trunks during the day.

Fig. 23.8
Two phenotypic forms of the British peppered moth, *Biston betularia.* The light form, *Biston betularia typica,* is shown above; its mottled appearance has led to the name "peppered moth." Two hundred years ago it was the most prevalent form. The dark form, *Biston betularia carbonaria,* has, until recently, been relatively rare. The two forms of the moth are members of the same species, differing only in basic coloration. As a result of increased pollution, with a progressive darkening of the background, the light form has diminished in numbers, and the dark form increased. Bird predation is thought to be the major selective agent determining survival rates of each type.

Fig. 23.9
On a lichen-encrusted tree trunk in nonindustrial regions of England, melanic forms of the peppered moth stand out conspicuously. The light form of the moth, however, is barely discernible (center). Under such conditions, the melanic form is subject to heavy predation. (Photo courtesy Dr. H. B. D. Kettlewell.)

Fig. 23.10
Distribution map of forms of the peppered moth in England. Each circle represents a population sample taken in a given area. The colored portion of each circle indicates the percentage of melanic forms found in the area, while the white portion indicates the percentage of light-colored forms. (Adapted from H. B. Kettlewell, *The Evolution of Melanism.* Oxford: Clarendon Press, 1973, p. 135.)

In the past hundred years, however, the number of melanic moths has increased quite drastically. In some areas the dark form has almost totally replaced the light. A map (Fig. 23.10) comparing distribution of melanics and light-colored moths show that the darker form predominates in the industrial centers and the regions to the southeast.

How might we explain the change in frequency of dark, compared to light, forms? As industrialization polluted the natural environment of many areas of the British Isles with smoke and soot, selection pressures changed. Light color became increasingly disadvanta-

geous. Light-colored forms were now more easily visible (Fig. 23.11) to bird predators, and hence were able, on the whole, to reproduce less. Conversely, the dark forms became increasingly adapted—their dark color shifted from being a liability to being a benefit. Selection pressure had changed, and with it the selective advantage of certain color patterns in the moth population. What had been adaptive in the past became non-adaptive at the present, and vice versa. Data showed that in some areas the melanic form had virtually replaced the light-colored form in the previous sixty years.

Darwin's theory accounts nicely for these results. But how do we *know* that the concept of natural selection is really adequate to explain real changes in the natural world? Even if we accept Darwin's basic notion of natural selection, how can we determine whether the decline in light-colored moths is due to bird predation, to pollution itself, or to some other, as yet unobserved factor? How do we know that the color patterns in the peppered moth are genetic? Do birds find it as difficult to distinguish camouflaged moths as human beings do? British naturalist H. B. D. Kettlewell set out a number of years ago to try to answer some of these questions. In doing so, he developed one of the clearest case histories available of the evolutionary process in action.

First, a review of available genetic studies indicated that the color patterns—dark and light forms—were determined by a single Mendelian gene pair. Dark forms were either Cc or CC, while light forms were cc. The melanic condition was thus one of simple dominance.

Two more very important questions centered around predation on the moths by birds. Does the moth's coloration actually serve as a camouflage to protect it against predation? In other words: do birds perceive moths against different backgrounds the same way that we do? At first glance this might seem like an unnecessary question, but in fact it is not. Different species of animals have very different visual perceptions. For example, cats can see very well at levels of illumination too low for humans, while bumblebees can see the sun through a cloud layer that completely obscures it for us. There is no *necessary* basis for believing that camouflage patterns would appear the same to another species—such as birds—as they do to us. Since our explanation of the change in frequency of the melanic form of moth is based on the adaptive value of protective coloration, it is absolutely essential to know whether, in fact, birds are fooled by the camouflage.

To answer the above questions, Kettlewell first conducted a systematic field survey, examining the stomach contents of birds thought to prey upon peppered moths. He found, indeed, that the stomachs of hedge sparrows, robins, and great tits contained numerous specimen of *Biston betularia.* Moreover, in the more polluted regions, he found a disproportionate number of light forms (compared to dark forms) of the

Fig. 23.11
On a tree trunk darkened by soot from industrial areas, the
light-colored form of the peppered moth is much more visible
than the melanic form. In this situation, the melanic form has
greater survival value than the lighter variety. (Photo courtesy
Dr. H. B. D. Kettlewell.)

moth in birds' stomachs; and, in unpolluted areas, just
the reverse. This evidence suggested that the color patterns of the moths are adaptive in making them less visible—the dark forms being selectively disadvantaged in
the unpolluted, the light forms selectively disadvantaged in the polluted areas.

Kettlewell went on, however, to get even more direct evidence that birds are selective predators of the
moths. He set up a movie camera in a polluted and in a
nonpolluted forest. In each area, he placed specimens of
both the dark and the light forms of the moth on tree

trunks, and (with a great deal of patience) waited for
birds to pick off the moths. The results were quite clear-
cut. In the polluted areas the film showed birds selectively taking light-colored forms from the trees, and
leaving dark-colored ones alone, even though they
might be only a few inches away. The pattern was just
the opposite in nonpolluted areas. Of course, birds did
take some dark-colored forms from dark-colored backgrounds and some light-colored forms from light-
colored backgrounds. But the percentage of conspicuous forms eaten was significantly higher than that of the
protectively colored forms.

Kettlewell then set about to determine if selective
predation could account for the change in gene frequencies of light and dark alleles on a population level
and under natural (field) conditions. He made the following prediction:

If: natural selection favors darker moths in industrial
regions and lighter moths in nonindustrial regions; and

If: equal numbers of light and dark moths are released
in both regions, and recaptured after a period of time,

Then: in the industrial regions far more dark forms
should be recovered than light forms, and in the nonindustrial areas far more light forms should be recovered
than dark forms.

In the early 1950s Kettlewell designed an experiment to test these predictions. Dark and light forms of
the moth were raised in the laboratory and both kinds
were released in the two types of environments. Before
release, the moths were carefully marked with cellulose
paint, a different color being used for each day of the
week. In this way it was possible to determine when a
given moth had been released and how long it had survived.

After the moths had been released for a period of
time, large lanterns were set up in the woods at night
and surviving moths recaptured. Table 23.1 contains
the results. The first column on the left gives the percentage of each form present in the respective environments before the experiment began. This is only a reference figure, indicating that indeed in the unpolluted area
the light form prevailed, and in the polluted area the
dark form prevailed. The second column from the left
gives the number of marked moths Kettlewell and his
associates released. The third and fourth columns show,

As Darwin emphasized in The Origin of Species, *the only criterion for success in an
evolutionary sense is differential fertility: the ability to leave behind more offspring
than other members of one's own species.*

Table 23.1

Area	Unmarked moths present in area before experiment (%)		Number of marked moths released		Number of marked moths recaptured		Percent of marked moths recaptured	
	Light	Dark	Light	Dark	Light	Dark	Light	Dark
Dorset Woods (unpolluted)	95%	5%	496	488	62	34	12.5%	6.4%
Birmingham (polluted)	10%	90%	137*	447	18	123	13.0%	27.5%

*The value of 137 here is curious. Note that for the Dorset Woods part of the experiment, approximately equal numbers of light- and dark-colored moths were released. But in the Birmingham experiment, over three times as many dark forms were released as light forms. In his research reports Kettlewell does not explain the reason for this difference. It could simply be that, for reasons of experimental difficulty, the laboratory breeding process had produced only 137 light-colored forms by the time the experiment was ready to begin. Rather than postpone the study, it may have been considered best to go ahead with uneven numbers of moths. Of course, the uneven number would not necessarily invalidate the results, since the important figure, percent recapture of marked moths, is calculated as percent of the moths released, not of the total number of moths in the environment.

respectively, the actual number of marked moths recaptured, and percent recapture of marked moths. As the fourth column shows clearly, the predictions of Kettlewell's hypothesis are borne out. In the unpolluted area the rate of recovery of light forms was about twice as high as that of dark forms; in the polluted area the rate of recovery of the dark form was about twice as great as that of light forms. In a field situation, under as natural a set of conditions as possible, melanic forms are favored where dark-colored backgrounds predominate, and nonmelanic forms are favored where light-colored (lichen-covered) backgrounds predominate. Because of continued selection over the years, in unpolluted areas the c allele has become most common; in polluted areas the C allele has become most common. As the background has changed, so has the gene frequency for color patterns in *Biston betularia*.

The case of industrial melanism illustrates how a previously less-fit organism (the dark form) has been rendered *more* fit by a change in the environment. This change likewise caused a well-adapted form (the light-colored moths) to become *less* fit. The mutant gene for melanism did not occur, of course, in response to the environmental change. As we have seen, mutations (for example, from c to C) are not direct responses to a specific environmental change. The mutation from c to C occurred anyway; in fact, it has probably been occurring as long as the species of peppered moth has been

around. But, prior to the environmental change it was kept at a low frequency by being selected against. Spread of the gene was possible only when changed conditions favored those organisms that carried it.

23.7
HUMAN EVOLUTION IN ACTION:
SICKLE-CELL ANEMIA

Although many factors serve to upset the genetic equilibrium of a population, perhaps none is so important as the combined effect of mutation and natural selection. How this may occur in natural populations is illustrated by the human condition known as sickle-cell anemia. This condition occurs primarily, but not exclusively, among blacks and affects the oxygen-carrying capacity of red blood cells. The biochemistry and genetics of sickle-cell anemia have been carefully studied in recent years. It is apparent that mutation of the gene for normal hemoglobin molecule produces the sickle-cell allele. The mutation may be represented as:

$$Hb^A \rightarrow Hb^S$$

where Hb^A stands for normal hemoglobin and Hb^S for sickle-cell hemoglobin. When oxygen tension in the blood gets low, Hb^S molecules fold up, or collapse. As a result, the red blood cells, composed of almost 100 per-

Fundamentally, evolution is nothing more than a nonrandom shift in gene frequencies in successive generations.

SUPPLEMENT 23.3
SEXUAL SELECTION

In later editions of *The Origin of Species*, Darwin gave a great deal of emphasis to the idea of sexual selection. According to this concept, the female or male (depending on the species) selects a mate on the basis of certain visible criteria, such as color markings, elaborate plumage, horns (all called secondary sexual characteristics), and even behavioral displays (for example, the strutting of male birds before females). Darwin held that the male or female of a species would consistently select a mate that possessed certain specific characteristics; in our terms today we would say that under such conditions mating was nonrandom. Darwin distinguished sexual selection from natural selection. Today, however, we see sexual selection as just a special case of natural selection.

Darwin argued that the evolution of sexual dimorphism was the result of two different selective forces: (1) competition among males of a species for females; and (2) female choice of some males over others. According to Darwin, when males compete with one another any one individual's fitness, or adaptiveness, depends on his ability to win conflicts with other males (of the same species), and thus gain access to females. Having access to females, however, does not insure that mating will occur; the male's fitness also depends on his ability to encourage discriminating females to choose him over others.

An interesting and illustrative example of sexual selection (both male competition and female choice) comes from the recent work (1976–1980) of Professor Randy Thornhill, formerly of the University of Michigan, now of the University of New Mexico. While a graduate student at the University of Michigan, Dr. Thornhill began to study mating behavior in the black-tipped hangingfly (*Hylobittacus apicalis*), a relatively primitive insect in which males and females are morphologically very much alike, with a slim brown body, long thin legs, and four narrow, dragonfly-like wings. Both males and females fly about foraging. On the average, males fly twice as far per unit time as females, partly because the females are usually burdened down with a heavy "freight" of eggs, and also partly because, through evolution and sexual selection, the males have become the chief food-getters. Hangingflies feed mostly on other insects and arthropods. Foraging is a necessary activity, but it is not without risks, the chief one of which is spider webs (spiders are one of the hangingfly's predators). The more foraging a male hangingfly does, the more likely that he will encounter a spider web and become entangled. Taken by itself the extensive foraging activity of the males (compared to females) seems nonadaptive. How could this extensive and risky behavior have evolved? The answer lies in seeing foraging as a part of a

Diagram A
Male and female hangingflies with "nuptial gift." (Photo courtesy Randy Thornhill, University of New Mexico.)

larger, ritualized courtship behavior that ultimately maximizes the male's chance of mating and leaving progeny for the next generation.

The mating sequence of the hangingflies begins with male foraging. When a male captures an insect it flies to a bush, holding the prey with its rear legs, and grabs onto a twig with its front legs. The prey is killed by injection of enzymes through the male's proboscis (a projection from the mouth region) into the prey's body. The enzyme paralyzes the prey, and then dissolves its tissues. After feeding for a while the male, with the prey still clasped between his rear appendages, will begin flying through the low-lying bushes in search of females (this journey is called the male's "nuptial flight"). At the end of the nuptial flight the male again attaches to a twig by his front appendages and protrudes from his abdomen a pair of glandular sacs that emit a chemical messenger, called a pheromone. The pheromone serves to attract the attention of a female, which flies over and comes to rest opposite the male on the same twig (see Diagram A). No contact between genitalia, located in the tip of both the male and female abdomens, occurs at first. The male presents the female with the prey. She usually takes it in her two front appendages and begins feeding. During this process the male begins to mate with her, making contact between her abdomen and his (Diagram A). If the female finds the prey satisfactory, she remains for the duration of the mating period (which can last between twenty and twenty-five minutes) feeding all the while. After intercourse the male and female struggle for possession of the prey, a battle that the male wins much more often than the female. The male will then fly off with the prey in his front appendages and, if it is not completely devoured, offer it to another female; if it has no more nourishment left in it, the male will drop the carcass to the ground and begin seeking another prey. Within four hours after mating the female lays her eggs (on the average of three from each mating); after several more hours she is ready to copulate with another male.

Thornhill noted that after making initial contact with males not all females would remain long enough to copulate, even though they had already begun to feed on the "nuptial gift." The critical factor appeared to be the *size* of the prey which the male offered the female. If the prey was small (under approximately 16 square millimeters, a figure determined by multiplying the prey's length by its width), females would fly away; but if the prey were larger than 16 mm^2 the females would remain. This clearly suggests that the female is making a choice, that is, a discrimination between sample offerings. Why should the female make such a choice? We can hypothesize that the female gains two advantages from this process. First, she gets food brought to her, and thus minimizes her own risk in getting caught in spider webs during foraging flights. Second, since females expend much more metabolic energy in producing eggs than males do in producing sperm, it is to her advantage to select larger prey; she gains more nourishment per mating from large prey than from small ones. Moreover, in his investigation, Thornhill showed that females that regularly select those males bringing larger prey lay more eggs per unit of time than those that regularly mate with the first male they encounter (Thornhill called the latter females "nondiscriminating").

We may now reasonably ask: What advantage is there for the male to play the female's adaptive game of capturing larger prey? This is an especially important question since the male undergoes more risk in his prey-capturing flights than the female who sits and waits for nourishment to be brought to her. Thornhill first established clearly that males do in fact select out the largest insects from the total group of insects they encountered. He measured the sizes of the prey that various male hangingflies had captured, showing that males consistently selected the larger prey, often ignoring the smaller ones that came along. By contrast, females, when they did forage for themselves, chose prey randomly—that is, their samples were very close to the distribution of sizes of insects available in the surrounding environment. For the male, selecting the larger prey confers advantage since it will

significantly increase his chances of a successful mating. Especially if a male can use the same prey as inducement for a second female, he may greatly increase the number of offspring he sires in the next generation. Measurements of the sperm content of sperm-storage organs in female hangingflies showed that those males who brought undersized prey left few if any sperm, even if copulation began. By contrast, males with large prey left considerably more sperm.

Thornhill also observed that competition between male hangingflies has taken an interesting twist. Rather than going out through the brush to find their own prey, many males attempt to steal prey from one another. There are two ways that "robber" hangingflies attempt to do this: by direct assaults on the other male, and by mimicking female mating behavior. The female mimic behavior is particularly fascinating. When a potential robber male spots a male with a nuptial gift, he will fly toward the other male and lower his wings in exact imitation of a female who is attracted by the male pheromone. The robber male will attach to a twig opposite the first male, just as females do, but keeps his abdomen from touching that of the first male (presumably keeping the first male from recognizing that he is an impostor!). The robber hangingfly begins to feed on the prey. This lasts for a minute or two, after which the robber attempts to wrest the prey away from its holder, an effort that is successful 22 percent of the time. (By contrast, of all the attempts to steal prey by direct assault, only about 14 percent are successful.) The advantage of robbery is obvious for the male hangingfly. By stealing another fly's prey, the robber avoids the dangers of foraging journeys, and hence increases his reproductive capacity while minimizing his risk. According to Thornhill's evidence, robber males are more successful in leaving offspring than nonrobber males.

The case of the hangingflies emphasizes an important aspect of sexual selection: what is called "male resource control." In many species male behaviors are geared to controlling the availability of some resource, usually food or space (for example, nesting sites), necessary to successful rearing of young. Male resource control is an important component in attracting females, a necessary first step to producing offspring. There is a real biological advantage to presenting females with abundant resources, since in one way or another such resources are necessary for successful production of young. The female's metabolic needs, and the male's response to those needs, represent an interacting set of behaviors that have evolved together through sexual selection. They have resulted in sexual dimorphisms, especially in behavior, which ensure that members of both sexes will maximize their genetic contribution to subsequent generations.

cent hemoglobin, also collapse. The change in shape of the cells is a diagnostic trait of the disease (Fig. 23.12). When this folding-up occurs, it is very difficult for the affected cells to pick up oxygen in the lungs and trans-

port it to the tissues. In addition the collapsed red blood cells block capillaries and in other ways interfere with circulation. In many cases, reduced vigor and ultimately death are the results of sickle-cell anemia.

The case of sickle-cell anemia shows that adaptiveness of traits can only be understood in the context of all aspects of the population's environment. What is adaptive in one environment is nonadaptive or maladaptive in another.

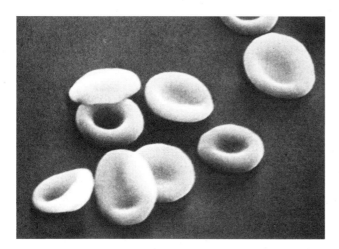

(a)

(b)

Fig. 23.12
(*a*) Normal red blood cells showing basic concave circular shape (magnification × 6750). (*b*) Red blood cells that have sickled due to low oxygen tension. The deformed shape is a result of collapse of hemoglobin molecules; such cells have a greatly reduced ability to transport oxygen. (Scanning electron micrographs courtesy Dr. Marion I. Barnhart, Wayne State University School of Medicine. Originally published in R. M. Nalbandian, ed., *Molecular Aspects of Sickle Cell Hemoglobin, Clinical Applications.* Springfield, Ill.: Charles C Thomas, 1971.)

Allowing Hb^S to represent the sickle-cell gene and Hb^A the normal allele, let us see how natural selection acts on this gene in a population. With these two alleles, three combinations are possible: $Hb^S Hb^S$ (individuals with sickle-cell anemia), $Hb^A Hb^S$ (individuals with the sickle-cell trait, a mild form of the disease), and $Hb^A Hb^A$ (individuals with normal hemoglobin). The sickle-cell alleles show incomplete dominance, so each of the genotypes also shows a characteristic phenotype.* Persons homozygous for the sickle-cell gene generally die very young and leave few offspring, if any. Persons heterozygous for the sickle-cell trait usually live to sexual maturity, since they are affected by this condition only in cases of extreme exercise or high altitudes.

The frequency of these three genotypes among American blacks has been studied by J. V. Neel and others. In the United States the frequency of the Hb^S allele is 0.05, while that of the Hb^A is 0.95.† According to the Hardy-Weinberg equilibrium law, we should expect these frequencies to remain constant from generation to generation. However, natural selection is highly unfavorable to individuals of the $Hb^S Hb^S$ genotype. It has been estimated that a child born with sickle-cell anemia has one-fifth as much chance as other children of surviving to sexual maturity. This indicates that roughly 80 percent of the $Hb^S Hb^S$ genotypes fail to survive beyond infancy or childhood.

In the United States, natural selection thus limits the number of mutant Hb^S genes passed on to the next generation. If the frequency of Hb^S is 0.05 in one generation, then calculations show that, based on the present rate of selection, the frequency will be reduced by about 16 percent in the next generation; the frequency will thus decline from 0.05 to 0.042 in one generation. (The figure of 16 percent is taken from calculations based on medical records.) The theoretical "cutting down" effect of natural selection in this instance is rep-

---

*Individuals with the $Hb^S Hb^S$ genotype have all of their hemoglobin in every RBC as the sickle type. Individuals with the $Hb^A Hb^S$ genotype have half sickle hemoglobin in every cell; the other half is normal. Individuals with the $Hb^A Hb^A$ genotype have all normal hemoglobin.

†In other words, the Hb^S gene makes up 5 percent of the gene pool for this allele, and Hb^A about 95 percent.

resented diagrammatically in Fig. 23.13. Note that the *rate* at which the mutant Hb^S allele is eliminated from the population becomes slightly lower in each generation. The fewer Hb^S alleles in the population, the less chance they have of coming together to form the lethal Hb^{S}Hb^S homozygote. Thus selection against a particular gene can reduce the frequency with which it occurs in the population. However, since the rate of decrease becomes less each generation, a very long time is generally required to effectively eliminate an unfavorable recessive gene from the population.

The case of sickle-cell anemia illustrates very well how gene frequencies vary due to differences in environmental conditions. The frequency of the Hb^S allele is much greater in certain parts of Africa than in other parts of the world. The areas in which the sickle-cell allele is most frequent are also the areas of high malaria incidence. It has been discovered that heterozygous carrier (Hb^{A}Hb^S) individuals possess selectively greater resistance to malaria than homozygous nonsickling individuals.* Thus in areas in which malaria exists, the sickle-cell trait has a high adaptive value to the individuals who possess it. The disadvantage of the sickle-cell trait is more than compensated for by the greater resistance to malaria that the heterozygote confers.

Sickle-cell trait is selectively disadvantageous under one set of conditions and advantageous under another. Therefore, natural selection tends to preserve both the Hb^S and Hb^A alleles in the population. Either homozygote is at a decided disadvantage in malaria-infested regions. Superiority of the heterozygote therefore keeps the frequency of either allele from declining to a very low level.

On the genetic level, the existence of sickle-cell hemoglobin is a result of a single mutation in the genes responsible for synthesizing the β-chain of hemoglobin. The α-chain of every hemoglobin molecule is normal. The mutant β-chain has the substitution of valine for

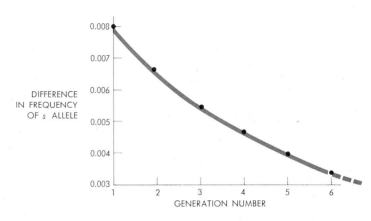

Fig. 23.13
Graph showing the difference in frequency of Hb^S allele in successive generations.

glutamic acid in one position along the lengthy polypeptide chain. Since the DNA triplet codon for glutamic acid is cytosine-thymine-thymine (CTT) and that for valine is cytosine-adenine-thymine (CAT), it is reasonable to conclude that the mutation from normal to sickle-cell hemoglobin comes about by the substitution of a single base pair, adenine for thymine, in DNA. This small substitution produces an enormous impact on the individual's chances for survival, and on the course of evolution in a population.

In this chapter we have seen how natural selection, acting on individual variations within populations, can produce a shift in gene frequencies, and a corresponding shift in phenotypes, over several generations. The question we are now left with is: How does this process account for the origin of new species? And how is the origin of species related to adaptation? These two questions will form the focus of the next chapter.

Summary

1. Species are defined as groups of anatomically and physiologically similar organisms that share a common evolutionary history and that are capable of interbreeding (share a common gene pool). Species are composed of one or more local populations, called demes.

2. Variation within populations provides the "raw material" of evolution. There are two types of variation—genotypic and phenotypic. Genotypic variations are actual changes in the genetic code of DNA or changes in chromosomal structure that are transmitted to following generations. Phenotypic variations result from a combination of genetic and environmental factors.

3. Polymorphism is the persistence of two or more forms of a trait from one generation to another within a population. Transitional polymorphism refers to situations where two or more forms exist for a while, but with gradually changing frequencies over successive generations. Eventually one phenotype predominates and the other(s) gradually disappear. In balanced polymorphism, the two or more forms persist at steady frequencies from generation to generation.

*The way in which the sickle-cell trait appears to confer resistance to malaria seems related to the fact that the infectious phase of the malarial protozoan enters the red blood cells as a parasite. The metabolic activity of the protozoan uses up oxygen, thus reducing the oxygen tension within the cell (that is, the amount of oxygen). Molecules of sickle-cell hemoglobin collapse when the oxygen tension gets low, thereby causing the entire red blood cell to fold inward, assuming the sickle shape. Cells that have thus collapsed are, along with the parasitic protozoans within them, more readily destroyed by phagocytes. In this way the malarial parasite is prevented from spreading throughout the bloodstream.

4. Genetic variations in a population occur in one of two ways. The first is by point mutation, resulting from the substitution of at least one nucleotide pair for another in the base sequence of DNA. The effect of a mutation is to produce an altered protein molecule that may affect the organism's phenotype. Point mutations occur spontaneously and at random. Any particular point occurs at a predictable rate, called the mutation rate and calculated as the number of mutations per gene replication, per cell division (or per gamete). Mutation rates are low, on the order of one per 10,000 to one per 100 million gametes.

5. Chromosomal mutations are a second means of creating variation in a population. Such mutations involve changes in the physical arrangements of genes on a chromosome, such as deletions, duplications, and inversions of chromosomal segments. Since gene interaction is an important factor in translating genotype into phenotype, chromosomal mutations that alter the position of one gene with respect to others on the same chromosome will also alter the expression of that gene.

6. Population genetics is concerned with the fate of variation within populations. The collective genes in a population are known as the population's gene pool, the figurative arena into which each organism casts its gametes during sexual reproduction. As the "pool" concept implies, it is theoretically possible for a gene coming from any one individual in the population to become combined with its counterpart from any other individual, as long as random mating occurs. Natural selection, by favoring certain genes over others, can ultimately change the composition of the gene pool. If a variation is favorable, genotypes containing that variation will contribute more offspring (hence more copies of that variation) than others to the next generation. If the variation is unfavorable, the individuals possessing it will contribute fewer offspring (and less copies of the variation) to the next generation. In either case, the composition of the gene pool will change accordingly.

7. The Hardy-Weinberg law of genetic equilibrium states that the frequencies of any alleles in a large population will tend to stay the same from one generation to the next, whatever the original frequencies might be, provided that no selection, mutation, or gene flow occurs, and provided that mating is random. If gene frequencies are found to be shifting from one generation to another, one or more of the above conditions are not being met; the Hardy-Weinberg law does not apply.

8. Evolution occurs through the action of several factors, including natural selection, mutation pressure, genetic drift, and gene flow. Natural selection acts on the phenotypic variation that exists in a population by allowing some phenotypes to reproduce more readily than others. Stabilizing selection favors the reproduction of organisms having the most common phenotypes and selects against the extremes. Directional selection, which is probably the most common form of selection, favors one of the extreme phenotypes over all the others. The mean of future generations shifts toward the favored phenotype because individuals exhibiting that trait reproduce more frequently than other individuals. Disruptive selection favors both extremes of a distribution, and acts against the mean value for the character under consideration. If disruptive selection continues over several generations, the population will eventually show a bimodal distribution.

9. Mutations, both point and chromosomal, produce the variability on which natural selection acts. Biologists have suggested that a mutation pressure exists that can direct evolution along its own lines. However, because mutation rates are very low, mutation pressure may be of only minimal importance in producing evolutionary change.

10. Genetic drift refers to the process by which gene frequencies are altered from one generation to the next by random change alone. Genetic drift usually plays a role in directing evolution only in small populations, where the effects of sampling error are most observable. Or, some organisms may migrate or form a new colony that does not fully represent the variability present in the larger population.

11. Gene flow refers to the movement of genes either within a population or from one population to another through the migration of organisms. In the case of some plants, genes can be carried from one population to another by agents such as wind, water, or animal vectors. Gene flow can generally occur only between different populations of the same species. Gene flow is another mechanism, in addition to mutation, for introducing variation into a population.

12. The case of the melanic moth, *Biston betularia*, illustrates several important principles of evolution by natural selection:

   a) The adaptive value of genotypes (and hence phenotypes) is relative to the environment (background color in this case) in which the organism lives. Light-colored moths are adaptive in a lighter (nonpolluted) environment; dark-colored moths are adaptive in a darker (polluted) environment.

   b) Predation is one of many kinds of selective agent acting differentially on various genotypes and phenotypes within a population.

   c) As environmental conditions change, allele frequencies can change due to changing selection pressures (differential predation of moths by birds).

   d) If a selection pressure is rigorous enough, one allele can virtually replace another in a population over a number of generations.

13. The case of sickle-cell anemia in human beings illustrates several additional principles of evolution by natural selection:

   a) Genotypes can have very different selective values in different environments. Sickle-cell trait (the heterozygote) is something of a disadvantage in nonmalarial environments. In malarial environments, however, it confers an overall greater advantage and thus is found in greater frequency.

b) In a nonmalarial environment the frequency of the Hb^S allele can be reduced slowly by selection, but it can never be totally eliminated. Eventually the gene frequency will become so low that the chances of the strongly unfavorable homozygous recessive condition ($Hb^S Hb^S$) occurring are so slight that a low level of the mutant allele will always remain in the population. A dominant mutant allele, however, can be effectively eliminated, at least to the level where it is present only as a result of continual mutation.

Exercises

1. Occasionally individual fruit flies are born with shortened, stubby wings, the so-called vestigial wing. The condition is inherited. These organisms cannot fly and do not survive in nature. Design an experimental environment (in a laboratory) that would selectively favor these flies over winged ones.

2. One criticism of Darwinian natural selection has centered on the evolution of such adaptive features as cryptic (camouflage or protective) coloration. An American biologist, W. L. MacAtee, undertook a thorough study of camouflage to see if, in fact, insects that appeared camouflaged to the human eye were camouflaged to their predators (birds). He analyzed the contents of birds' stomachs for the number (and kind) of so-called camouflaged, or protectively colored, species of insects. The results are given in the table below:

	Highly camouflaged to human eye	Moderately camouflaged to human eye	Noncamouflaged to human eye
Total sample of 1000 insects from stomachs of 10 birds	107	390	503

a) Does it appear that birds see the camouflage in the same way as humans? How do you know?

On the basis of the above results MacAtee concluded that the theory of natural selection, as applied to camouflage, was insufficient. He reasoned that, since a number of highly camouflaged insects were in fact eaten by birds, camouflage must not be adaptive.

b) Was MacAtee's reasoning valid in light of his results?

c) A number of Darwinians strongly disagreed with MacAtee's conclusions. On what grounds might a convinced selectionist argue that MacAtee's own results support, rather than contradict, the theory of natural selection?

d) Another argument against the evolution of protective coloration states that because many animals survive well without it, protective coloration is not essential to survival. Hence such coloration is not adaptive. What is the fallacy in this argument?

3. Sickle-cell anemia, a genetic disease affecting hemoglobin molecules, exists in two forms: (a) sickle-cell disease is a homozygous condition, symbolized as $Hb^S Hb^S$, and normally causes early death of its possessor before the age of 20; (b) sickle-cell trait is a heterozygous condition, symbolized as $Hb^A Hb^S$, which is less severe, and may cause reduced vitality and some shortening of the life span, with possessors usually living to the age of 40 or beyond. The normal individual, symbolized $Hb^A Hb^A$, has normal longevity. Those individuals with the $Hb^S Hb^S$ genotype have two abnormal hemoglobin chains per molecule, while those with $Hb^A Hb^S$ have half their hemoglobin with two abnormal chains and half with two normal chains. The sickle-cell condition can be studied on a number of different levels, and brings together many areas of study (for example, biochemical, genetic, ecological, evolutionary). Discuss the sickle-cell condition in terms of the specific problems or situations described below relating, where called for, one level of study to another (that is, biochemical to physiological to evolutionary and so on).

a) Describe the molecular differences between the normal and abnormal hemoglobin; that is, between Hb^S and Hb^A individuals' blood. What is the genetic basis of the change?

b) The Hb^S gene is found far more frequently in the black population in the United States than in the white. The gene is also found in black populations of East Africa far more frequently than in the black population of the United States, as shown in the table below:

Gene	Frequency	
	Among U.S. blacks	Among East African blacks
Hb^S	0.05	0.2
Hb^A	0.95	0.8

Account for these differences. What is the relationship, historically and genetically, of the two populations? What is happening to the Hb^S gene among blacks in the United States? Will the gene ever be eliminated? Why or why not?

Chapter 24
Macroevolution: Divergent Speciation and Adaptation

24.1
INTRODUCTION

Two of the evolutionary problems of greatest concern to Charles Darwin in *The Origin of Species* were how diversity of species arose and how each new species became adapted to its particular environment and mode of life. Ironically, given the title of his book, Darwin never adequately explained the first of these problems. The theory of natural selection accounts well for how a population of organisms becomes modified, genotypically and phenotypically, over numerous generations; but the theory Darwin proposed in 1859 simply does not offer any explanation for divergence—that is, the splitting of one species into two or more over time.

The problem of adaptation was somewhat different. Darwin felt adaptations were the critical test of his theory. He observed that some adaptations in nature were so marvelous and intricate that it seemed they must have arisen by an act of special and purposeful creation. With regard to the vertebrate eye, for example, he wrote:

> To suppose that the eye, with all its inimitable contrivances for adjusting the focus to different distances, for admitting different amounts of light, for the correction of spherical and chromatic aberration, could have been formed by natural selection seems, I freely confess, absurd in the highest degree.

If the theory of natural selection could explain such adaptations, it might well be accepted as the best account yet for how evolution occurs.

These two subjects—divergent speciation and adaption—form the focus of the present chapter. This chapter is thus concerned with macroevolution, with large-scale evolutionary processes: how two or more species arise from a single pre-existing species, how evolutionary convergence and parallelism take place, and how selection acts to maximize an organism's adaptation at every step in the course of its evolutionary history.

It should be pointed out that all macroevolution depends upon and is a direct product of microevolution. So far as we know, natural selection, acting on variations within populations, is responsible for the large-scale events we call macroevolution as much as for the small-scale events we call microevolution. Nonetheless, there are additional circumstances and factors involved in macroevolution that lead to different overall outcomes from those involved in microevolution—for example, the divergence of one species into two, or the origin of higher taxa such as genera, classes, and phyla. For this reason, macroevolution is treated as a separate topic although it builds directly on the principles of microevolution discussed in Chapter 23.

24.2
SPECIATION: PRINCIPLES AND PATTERNS

Speciation is the process by which one or more new species arise from a previously existing species. The great diversity of life on earth as we know it today has arisen through this process, which is a fundamental aspect of the evolution of all populations. Most speciation results in the formation of two or more new species, or what is called **divergent speciation**. The basic pattern of divergent speciation is shown below, where one species, called the original or ancestral species (species 1) gives rise to two (or sometimes more) descendant species (species 2 and 3).

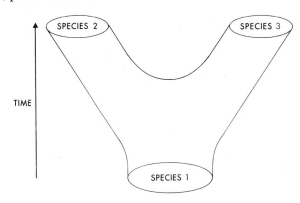

The process of divergent speciation (or divergence) is usually slow, and may occur by one of several hypothesized mechanisms (discussed in detail in this section): **allopatric** speciation (which involves some form of geographic isolation between parts of the ancestral population), **parapatric** and **sympatric** speciation (without direct geographic isolation of parts of the ancestral population), and **abrupt speciation** (by polyploidy, large-scale chromosome rearrangements, and the like). The first three forms of speciation are gradual; the latter, as the name implies, occurs through sharp discontinuities, sometimes as quickly as between one generation and the next. We will briefly examine each of these modes of speciation.

Allopatric Speciation

Allopatric speciation is probably the most common form of speciation. It involves several different steps, or processes. The first and most fundamental step is geographic isolation, where a group of interbreeding subpopulations, or a single breeding population, becomes separated into two or more isolated groups. The chief way in which such separation can occur is through some change in the physical environment, such as the diversion of a river or creek through a field.

The second step is independent evolution of the gene pools of the separate isolated populations. Since it would be highly improbable that each separate population would have the same number and kinds of variations occurring at the same time, and since each subpopulation inhabits slightly different geographic regions and is subject to slightly different selection pressures, the changes in gene frequencies over time would be expected to become different. The differences would gradually accumulate to such an extent that the separated goups would be reproductively isolated from one another. That is, representatives of two such populations could not produce viable offspring.

A third step (which often occurs, but not always) involves a merger between the formerly divergent populations. Merger becomes possible either because of expansion of the separate populations so that their ranges overlap, or through environmental changes, such as the drying up of a river, which allow for migration. Merger of the populations can lead to one of several outcomes, depending on the particular species involved (especially depending on whether the species are animals or plants) and the conditions of the environment. In both animal and plant species mergers generally result in increased competition, since the two divergent populations, having descended from a common ancestor, are still likely to share a number of common traits and thus require many of the same resources. Competition can lead to extinction of one of the species or to further and more rapid divergence as one way in which the two species reduce direct competition between themselves. In the case

KAIBAB SQUIRREL

ABERT SQUIRREL

Fig. 24.1
The Abert and Kaibab squirrels represent two distinct natural populations, separated by a significant geographic barrier, the Grand Canyon. The Abert squirrel inhabits the southern rim of the Grand Canyon, while the Kaibab squirrel inhabits the northern rim. In nature the two do not interbreed. It is thought that both modern populations are descendants of a single ancestral population that was divided by deepening and widening of the Colorado River over the ages. Phenotypically, the two squirrel populations have diverged. The Kaibab squirrels have darker underbodies and whiter tails; the Abert squirrels have light underbodies, darker tails, and longer, darker ears. (After G. G. Simpson, C. S. Pittindrigh, and L. H. Tiffany, *Life: An Introduction to Biology.* New York: Harcourt, Brace and World, 1957, p. 481; redrawn with permission.)

of plants, merger can sometimes result in hybridization between the two populations if they are genetically compatible. In any event, the coming together of two species, whether they have diverged from a commmon ancestor or not, almost always accelerates the evolutionary process.

The essential feature of allopatric speciation is that it begins with some form of physical separation between two or more subpopulations of the same species. Isolation is a necessary but not sufficient condition for allopatric speciation to occur.

The evolution of two populations of tuft-eared squirrels inhabiting the north and south rims of the Grand Canyon (Fig. 24.1) provides a simple example of allopatric speciation. The Abert squirrel inhabiting the southern rim and the Kaibab squirrel inhabiting the northern rim are similar, yet have visible physical differences. The two populations are usually considered separate species, since little or no gene flow exists between them in nature.

Biologists hypothesize that both the Abert and Kaibab squirrels arose in the past from the same ancestral population. This was apparently a freely breeding population that inhabited the entire region now represented by the north and south rims of the Canyon. Widening of the bed of the Colorado River and the subsequent formation of a deep canyon created a geographic barrier that effectively prevented gene flow between the separate groups. (Of course, not all geographic barriers are as spectacular as the Grand Canyon; a marshy area between two field habitats can also be an effective barrier for some species.)

Like all populations in nature, the original squirrel population contained a range of variations—for example, of tail size, ear length, and skull shape. These already-existing variations, plus newly occurring variations in the two separate populations, provided the raw material on which selection acted. Even though the environmental differences between the north and south rims of the Canyon are slight, selection pressure would be expected to be somewhat different for the two populations. Divergence over time becomes inevitable. The two populations have not yet merged in any significant manner, so we do not know what the outcome would be if they were to inhabit the same environment. Given that neither is highly specialized, competition might be severe, with either elimination of one population or rapid divergence of the two populations being most likely.

Parapatric and Sympatric Speciation

These two forms of speciation have in common the fact that neither requires actual geographic isolation as a precondition for divergence.

Parapatric Speciation. Parapatric speciation occurs when (1) genetically unique organisms arise in a population and are able to exploit some unoccupied ecological position within the population's normal range; and (2) when those individuals are from the start reproductively isolated, or at least significantly so, from the rest of the population. Exactly how these differences might arise

without any apparent geographical isolation is not well understood. One example of new species which appear to have arisen by parapatric speciation is the Old World mole rat, *Spalax ehrenbergi,* which lives in Israel. Moles are a particularly likely form in which parapatric speciation could occur, because of their burrowing, and thus relatively immobile, lifestyle. Adult moles live during the day in burrows. They emerge only at night to forage, and even then, because of their awkward methods of moving about, seldom stray far from their homes. Consequently, an adult mole may spend the vast portion, if not all, of its life in an extremely limited area. Migration and other forms of contact between moles in neighboring populations are often quite minimal.

Four species of moles exist in an area of Israel extending from the cooler, more humid region of Mt. Hermon and the Golan Heights in the north to the hotter, more arid region of Jerusalem and the Sinai Desert in the south. The population ranges of several of the species directly overlap, so that some contact is possible. The populations differ in chromosome number (two northern species have 52 and 54 chromosomes; a species from central Israel and the Lower Galilee Mountains has 58; and the southern species has 60). The four species also differ in the metabolic rates, which are higher in the humid northern species and lower in the arid southern species. On the other hand, the four species are remarkably similar in their proteins: 96 percent of those tested were virtually identical. Hybrids with reduced viability can be produced in the laboratory, but these are not known to occur in nature. Furthermore, fossil evidence suggests that the present four species were derived from a single ancestral species (*Spalax mimtus*) that lived throughout the same area a half-million years ago.

The best explanation for all these observations is to suppose that the four species of mole rat evolved parapatrically from the ancestral form. The evolutionary scenario might go like this: as different subpopulations of the ancestral population moved into different climatic areas (cool and humid, or dry and arid), random chromosomal changes took place that could have produced in one or a few generations both reproductive isolation and metabolic differences. Mode of life contributed to limited contact between subpopulations, furthering the chances of accumulation of many small differences and thus increasing the divergence between the groups. As time went on, variations continued to accumulate in the separate populations until they became as different as we see them today.

The defining characteristics of parapatric speciation, then, is that divergence of traits occurs at the same time reproductive isolation is being established. By contrast, in allopatric speciation, divergence of traits (because of isolation) occurs long before reproductive isolation is achieved.

Sympatric Speciation. Sympatric speciation is said to occur when two (or more) species become reproductively isolated before any other anatomical or physiological divergence has occurred. The most common example used to illustrate the possible mechanism of divergence by sympatric speciation is found among parasites. Because parasitic animals tend to mate and remain on their individual host organism, often for generations, a relatively self-contained, individual colony can come to exist on each host. That colony would be adapted to the particular individual host. On moving to a new host, however, where the chemical (or other) factors might be somewhat different, the parasite will have to adapt—a process that could select for rapid morphological and physiological changes over just a few generations. For instance, the new host organism (especially if it is an animal) might be a nocturnal individual (active at night) whereas the majority of its species is diurnal (active in the day). Although living in the exact same locality, this difference in habit of the host organism could effectively make the parasite reproductively isolated from other members of its species. After just a few generations, the parasites might be unable to reproduce, or show lower offspring viability when mating with parasites from other hosts. This divergence in reproductive viability can be accomplished without any noticeable divergence of other characters. According to Guy Bush at the University of Texas, the evolution of many species of the North American hawthorne fly (*Rhagoletis pomonella*), which parasitizes plants such as the hawthorne and apple, appears to be an example of divergence by sympatric speciation. New species have been reported to appear rather suddenly in an apple orchard; the new species do not have different chromosome numbers or other noticeable differences in anatomy or physiology. They may have arisen by short-term localization on, and adaptation to, the specific conditions (biochemical, physiological) existing on a particular apple tree within the orchard.

It should be clear that in any specific case it may be difficult to draw a line between parapatric and sympatric speciation. Such a distinction revolves about the biologist's ability to know whether reproductive isolation actually precedes or occurs simultaneously with differentiation of other morphological traits. Most biologists agree that both parapatric and sympatric speciation are relatively rare phenomena, occurring much less frequently than allopatric speciation. However, evolutionists used to think that no divergence could ever occur unless two populations were geographically or spatially isolated from one another for long periods of time. It is now recognized that such total geographic separation of two or more populations is not the only way that the populations can diverge. What all three forms of speciation—allopatric, parapatric, and sympatric—have in common is that however divergence originates, it cannot be perpetuated if there is a significant amount of continual gene flow between the diverging populations. For divergent speciation to occur, the one absolutely essential element is complete or at least greatly increased reproductive isolation between the populations. As long as gene flow is frequent between any two or more populations, those populations will never diverge significantly enough to achieve the status of new species.

Abrupt Speciation: Polyploidy

There are several mechanisms of abrupt speciation—that is, speciation that occurs in one or two generations by a sharp break or discontinuity in physiology or morphology. We will focus our attention here on what is perhaps the most frequently occurring mechanism, polyploidy (see also Section 23.3). Recall that polyploidy is the condition created when a cell multiplies its chromosome complement (usually by undergoing mitosis without cytokinesis). A polyploid organism is generally incapable of forming fertile offspring with members of the ancestral diploid organisms (parents) from which it was derived. Such infertility is a result of an odd number of chromosomes produced when a polyploid fertilizes a diploid. An uneven chromosome number greatly complicates the process of spindle-formation, and thus of cell division in the embryonic cells. Consequently, development is usually arrested. Polyploids can often self-fertilize, however, or fertilize other polyploids with the same chromosome number. In this way, a new polyploid organism may be virtually isolated reproductively from its parental population in one generation. If the polyploid is capable of reproducing and of finding an ecological position it can exploit, biologists usually say it has formed a new

---

For divergent speciation to occur, the one absolutely essential element is complete, or at least greatly increased, reproductive isolation between the populations. As long as gene flow is frequent between any two or more populations, they will never diverge significantly enough to achieve the status of new species.

---

species. As mentioned in Section 23.3 polyploidy is far more common as a method of forming new species in plants than in animals. In fact, it has been estimated that over 40 percent of all species of flowering plants have originated by polyploidy.

In summary, then, there are several modes of speciation—that is, divergence between populations leading to the formation of new species. These modes or mechanisms include allopatric, parapatric, and sympatric speciation, which occur to varying degrees relatively gradually; and polyploidy, which provides new species quite abruptly, often in a single generation. It is important to emphasize that for true speciation of any sort to occur gene flow between subgroups or subpopulations within

a larger population must be highly restricted. If gene flow occurs readily between two populations the populations will never diverge enough to become separate species.

24.3
MECHANISMS OF REPRODUCTIVE ISOLATION

For divergence between two or more populations to proceed to the level of new species formation, the populations must become and remain reproductively isolated. There are many forms of reproductive isolation, but all yield the same effect: failure to produce viable

---

SUPPLEMENT 24.1
PHYLOGENETIC TREES AND THEIR MEANING

Divergence of species over periods of time can be represented diagrammatically by a phylogenetic "tree" (a reconstruction of the evolutionary history of a group or groups of organisms). Phylogenetic charts are sometimes referred to as "family trees," but that term is misleading—the related groups are not restricted to taxonomic families. Phylogenetic trees give some indication of the actual path of evolution that biologists think particular groups followed. For example, two phylogenetic trees are shown in Diagram A. One of these charts, part a, representing the pedigree of "man," comes from the nineteenth century and shows some of the fallacies often involved in constructing and interpreting phylogenetic trees: for example, seeing evolution as proceeding along one main trunk with most species as "side branches." The other, part b, is a more modern example of the phylogeny of the dinosaurs. There are several important points of comparison between the two diagrams.

Ernst Haeckel (1834–1919), who prepared the chart for the "Pedigree of Man," was a strong believer in the progressive nature of evolution. That is, he saw evolution as always progressing toward greater perfectibility of forms. He also thought about this progress teleologically, that is, as goal-oriented or directed toward some final end. These beliefs had some distinct bearing on the way Haeckel drew his phylogenetic tree of Homo sapiens. First, note that the tree has a large main trunk, going straight upward (like most conifers, and distinctly unlike elms or maples), with a single pinnacle, or tip. At this tip resides Homo sapiens, the prince of creation, the goal toward which all evolution (that is, growth of the tree) has been directed. Thus, in Haeckel's view human beings represented not only the most advanced and perfected form in all the animal world, but also the "goal" of evolutionary development. Second, note that all other animal

groups are represented distinctly as secondary offshoots of the main evolutionary development toward human beings. Haeckel's tree pictures evolution as a kind of branching ladder.

By contrast, the modern representation of phylogenetic trees is more clearly branching than ladder-like. Divergence at certain points in time produces two or more lines that are equivalent to each other, that is, develop simultaneously but in different directions. In part b, for example, it is impossible to claim that of the two main lines branching off from Coelophysis in the Triassic, neither the Tyrannosaurus or the Anatosaurus lines were more important. The branching was simply divergent, leading to two quite different forms. There is also no hint of teleology in the modern diagram. Divergence from Coelophysis was no more "directed" toward Tyrannosaurus than toward Anatosaurus or Triceratops. Phylogenetic trees represent a review of a growth pattern after the fact. At no point along the way is the end-point of the growth (or even the present position along the route) predictable.

An important feature that both phylogenetic trees in Diagram A share is the concept of evolution as divergence from common ancestors. For example, in Haeckel's diagram the Amphibia (located a little more than halfway up the main trunk) are the common ancestor of the main portion of the stem leading to primitive mammals (and ultimately human beings) and the side-branch leading to reptiles and birds. In the modern "tree" Coelophysis is the common ancestor of Anatosaurus and Tyrannosaurus, as well as of other forms such as Brontosaurus, Stegosaurus, Ankylosaurus, and Triceratops. The concept of common ancestor is quite different from that of steps along a ladder. In the ladder-view of evolution, modern-day forms are thought of as placed at various

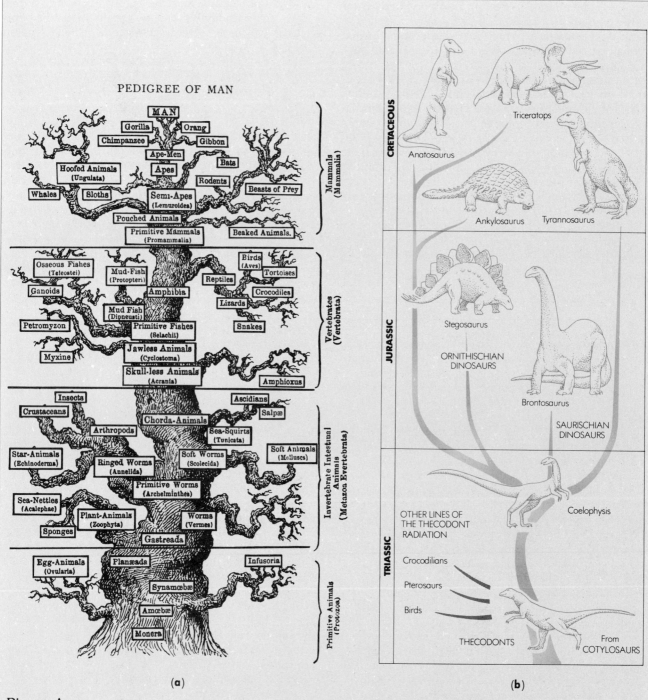

Diagram A

rungs according to their degree of complexity. In the branching view of evolution, common ancestors generally have no modern counterparts. For example, by the time *Brontosaurus* or *Anatosaurus* was roaming the earth, their smaller common ancestor, *Coelophysis*, had disappeared. Phylogenetic trees attempt to show that two or more modern (or co-existing) forms have diverged not from each other, nor from some simpler co-existing form, but from a common ancestor that was not quite like any modern species. Even if the common ancestor did not become extinct, it also evolved during the intervening time.

offspring, or greatly reduced viability of offspring that are produced. What are the various mechanisms by which reproductive isolation is achieved? These mechanisms fall into two broad categories: **premating isolating mechanisms**, and **postmating isolating mechanisms**.

Premating Isolating Mechanisms

Temporal Isolation. Although two populations of organisms may inhabit the same geographic region, their reproductive periods may occur at different seasons of the year or times of day. As a result, members of one population do not mate with members of the other, even though physical contact between the two is possible. This type of isolation is particularly common in plants, but it is also found in certain types of insects, snails, and frogs.

Ecological Isolation. If two populations are capable of interbreeding, yet do not do so because they live in different habitats, they are said to be ecologically isolated. For example, two populations of the deer mouse *Peromyscus* are capable of interbreeding in the laboratory. One population comes from the forest and the other from a nearby field habitat. In nature, each population remains almost completely within its own habitat. The forest population rarely if ever enters the field, and vice versa. In this situation, divergence has reached the point where the two populations are close to being completely separate species.

Physiological or Gametic Isolation. Among the first differences to appear in isolated populations are physiological and/or biochemical ones. Since genes produce enzymes, slight changes in genetic code show up first as changes in enzyme molecules. These changes need not be great enough to produce noticeable outward (phenotypic) effects, and they will not usually prevent interbreeding should the populations come back together. However, if the changes are in some way connected with the chemistry of the gametes themselves or of the seminal fluids, fertilization may be hindered. In sea urchins, for example, biochemical differences allow the sperm of one species to have greater success with eggs of its own species. Such physiological differences prevent or greatly inhibit successful hybridization.

Behavioral Isolation. In many animals, specific behavioral traits are known to be inherited. Particularly important to evolution are those behavioral patterns connected with mating. In certain birds, fish, and insects, the male and female of a given species perform highly specific and elaborate courtship rituals. Male birds may strut about in a precise pattern, displaying their plumage in hopes of attracting a particular female. Once the female becomes interested, the male begins a new behavioral ritual that ultimately leads to copulation. If either animal fails to perform the expected behavior at the right time, the other may not respond further, and mating is not successfully completed.

That behavioral patterns can serve as isolating mechanisms between species is shown by the courtship of several species of fish. The male three-spined stickleback (genus *Spinachia*) builds a nest attached to water plants. Then, by performing a series of zigzag swimming motions, he induces the female of the species to lay eggs in the nest. These swimming motions are highly specific. If the male does not perform the ritual precisely, the female immediately loses interest and swims away.

In many of the rivers in which three-spined sticklebacks live, there also exists a second species, the nine-spined sticklebacks. When males of this species court females, the behavioral pattern is somewhat different. Instead of performing the zigzag dance, male nine-spined sticklebacks bounce along vertically in the water, bobbing up and down in front of the female. The number of bounces and the position of the male are extremely important in bringing about the egg-laying response in the female. In both species the behavior of the male sets up a reflex action in the female. In some way not yet understood, this reflex pathway in both male and female is determined by the genetic makeup of the individual. If a male three-spined stickleback courts a female nine-spined stickleback, the female fails to show the egg-laying response. The female seems to lose interest as soon as the male begins the unfamiliar zigzag dance. Such divergence in behavioral patterns, which probably first arose from geographical isolation, serves as an effective isolating mechanism even though geographical isolation is no longer in effect. Divergence from some ancestral stickleback allowed two courtship patterns to develop; speciation has taken place, and interbreeding has become difficult or impossible.

Mechanical Isolation. Mechanical isolation occurs when organisms are unable to mate successfully because of anatomical differences in the structure of their sex organs that prevent successful union of male and female gametes. That is, it is quite possible that the sperm and egg, if allowed to come together, would be able to fertilize and undergo normal development. However, if differences in size or structure of the genitalia are involved, the gametes will never have a chance to unite. For example, among dogs, Great Danes and chihuahuas are members of the same species (*Canis familiaris*), yet their great differences in body size and hence reproductive organs tend to isolate them from one another. If Great Danes and chihuahuas were the only dogs in existence, they would more than likely evolve to represent two separate species.* Mechanical isolation is also an impor-

---

*Because there are many breeds of dog intermediary in size between Great Danes and chihuahuas, however, gene flow between the two breeds probably occurs to some extent today. Without these intermediaries, however, such gene flow would be eliminated or greatly reduced by mechanical factors.

SUPPLEMENT 24.2
SPECIATION AT THE MOLECULAR LEVEL:
HEMOGLOBIN AND MOLECULAR EVOLUTION

Hemoglobin has been the subject of much research as an example of molecular evolution. While molecular evolution could be studied in any protein, hemoglobin is a particularly good example. It is widespread within the animal (and even plant) world, easily obtainable in pure form, and small enough to make complete amino acid sequencing a relatively easy task. Mammalian hemoglobins have a molecular weight of approximately 68,000; the range of weights from different phyla, however, extends from 17,000 to 3,380,000. In almost all the organisms in which it is found, hemoglobin serves as an oxygen carrier. Structurally and functionally it is similar to myoglobin, an oxygen storage molecule in the muscles. While a careful study of the hemoglobins from different phyla is currently yielding new evidence on general animal phylogeny, hemoglobin speaks most directly to the evolution of the chordates in general and vertebrates in particular.

How can we use data on hemoglobin types to determine vertebrate phylogenetic relationships? Even more specifically, how can we use this information to affix a time scale to vertebrate evolution?

In Supplement 3.2 we discussed the three-dimensional structure of the single polypeptide chain of myoglobin. Vertebrate hemoglobins consist of four polypeptide chains bound together by weak interactions establishing a quaternary structure. A number of important observations can be made in using hemoglobin as an evolutionary marker. The molecule consists of two pairs of identical polypeptides: called the alpha (α) and beta (β) chains. The two α-chains have 141 amino acids and the two β-chains 146, numbers that are constant for all vertebrates except a few species of fish. The α- and β-chains are folded into a tertiary structure very much like that of myoglobin, with its 153 amino acids. The α- and β-chains in human beings differ from one another by 84 amino acids. Furthermore, many of the higher vertebrates (especially the mammals) have fetal hemoglobin synthesized by the fetus. Fetal hemoglobin consists of two α-chains and two chains of a very differnt sort, the gamma (γ) chains; fetal hemoglobin has no β-polypeptides. Thus all mammals can manufacture three types of polypeptide chains: α, β, and γ. In addition, the primates can synthesize a fourth hemoglobin chain, the delta (δ) chain, which replaces the β-chains in a small percentage of the hemoglobin molecules.

Differences between the hemoglobins of the human being and the horse, and between these two and the myoglobin of the sperm whale, are summarized in Table 1. Over twenty types of hemoglobins have been studied in detail, but for simplicity we will consider only these three. As Table 1

Table 1
Differences between Human and Horse Hemoglobins and Myoglobin from the Sperm Whale.

	Horse α	Human α	Horse β	Human β	Human δ	Human γ	Whale Mb
Horse α	0	18	84	86	87	87	118
Human α		0	87	84	85	89	115
Horse β			0	25	26	39	119
Human β				0	10	39	117
Human δ					0	41	118
Human γ						0	121
Whale Mb							0

Note: Reading across, the numbers at any intersection indicate the differences in amino acids in the chains being compared. For example, between horse α and horse α there are 0 differences; between horse α and human α there are 18; between horse α and horse β there are 84; between horse α and human β there are 86, and so on.

shows, the most similar chains are the human β and δ, with only ten differences. The next closest degree of similarity exists between the α-chains of the human and the horse (eighteen differences). Human and horse β-chains are more similar to one another than human β and γ (fetal) chains. And the α-chains of these two species differ considerably from any of the other chains found in either species. Finally, whale myoglobin differs markedly from all the other chains—a finding that would be predicted on the basis of the whale's great divergence in other traits from most mammals.

Using these data, we can construct a simple phylogenetic chart (Diagram A). The differences between myoglobin and hemoglobin are significantly great enough in all respects (amino acid sequence, quaternary structure) to suggest that these two diverged quite long ago in the history of vertebrates. It can be hypothesized that a very early ancestor of modern animals was an oxygen user who had a single-chain molecule serving both functions for which hemoglobin and myoglobin are now individually specialized: carrying oxygen to the cells and storing it in the cells. About 900 million years ago differentiation occurred, producing an ancestral hemoglobin (oxygen-carrying) molecule and an ancestral myoglobin (oxygen-storing) molecule. This hypothesis is supported by the fact that primitive chordates as well as many marine worms and insects living today have a one-chain hemoglobin molecule.

During the evolution of the fish, a new variation occurred that produced two types of hemoglobin chains: the ancestors of the modern α and β. Biochemically, these two chains were able to associate into a four-chain quaternary structure. The great number of amino acid differences between α- and β-chains can be accounted for by hypothesizing an early divergence of these chains from the ancestral hemoglobin. The fewer differences between β- and γ-chains, as well as the fact that γ-chains are found only in mammals, suggests that γ diverged from β more recently, during the evolution of the mammals. In even more recent times, the β-line split again during the evolution of the primates, yielding the δ-chain. The function of the δ-chain is not known.

How is it possible for one protein to "give rise" to another? The answer, of course, is that proteins do not produce other proteins. Variations must occur within the DNA that codes for the proteins. The scenario for the evolution of four hemoglobin chains can be described as follows. Initially, one DNA sequence (a gene) coded for the single oxygen-binding molecule—say up to about 900 million years ago. By accident, that whole sequence of DNA could have been duplicated during cell division, so that one offspring would receive two identical DNA sequences (two copies of the gene) for the primitive hemoglobin. Such accidental duplication of whole genes is not uncommon. Geneticists now know of many cases wherein genes or even larger segments of chromosomes are duplicated and passed on in double doses. As time passes, each DNA sequence will undergo its own independent rate of mutation. Here is where the importance of the accidental mutations comes in; mutations that might be lethal if they occurred where only one copy of the gene exists are harmless if they occur in a duplicate gene, since the other gene can still make functional protein. Eventually, however, a mutated gene might produce a protein with a different, but perhaps related, function. In the present case, myoglobin could have originated by mutation from a primitive hemoglobin. Like hemoglobin, it interacted with oxygen; but myoglobin became specialized within certain cells to store, rather than transport, oxygen. Further duplications could occur in the gene producing hemoglobin, so that at least two types of chains were produced simultaneously (the α- and β-chains or the α- and γ-chains, for instance).

How can biochemists determine the rate at which proteins like hemoglobin evolved? When they compare the amino acid sequences for a number of different species whose dates of phylogenetic divergence are known from geological or morphological data, they can make some quantitative estimates. For example, it is possible to plot the number of changes per 100 amino acid residues in any two species-specific proteins against the time that has elapsed since the two

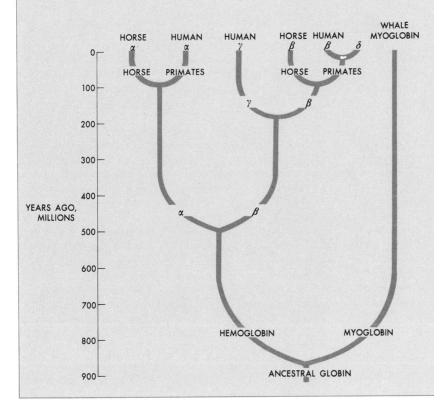

Diagram A

Evolutionary chart of myoglobin and hemoglobin. The data from Table I serve as the basis for this chart. Basically, the greater the difference between the two hemoglobin chains in terms of amino acid sequence, the longer ago they are presumed to have diverged from a common ancestor. Determining the *actual* dates at which divergence may have taken place, as shown on the vertical scale, requires independent information, especially the fossil record.

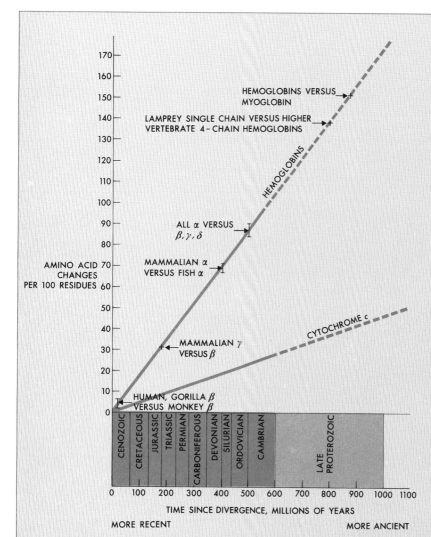

Diagram B
Graph showing ratios of amino acid differences per 100 residues for a series of comparisons between various hemoglobins, or hemoglobin and myoglobin. Each point on the graph represents a ratio of the numbers of amino acid changes per 100 residues plotted against the point in time when divergence between the two chains being compared is thought to have taken place. As in Diagram A, points of divergence in time are determined from the fossil record and other independent evidence. The dotted lines represent extrapolations.

species split (see Diagram B). Each data point on the graph represents a comparison of two hemoglobin chains (γ versus β or mammalian α versus fish α). Ratios obtained in this way for a whole series of hemoglobins, including myoglobin, fall along a straight line on the graph. This line shows a direct and quite constant relationship between the number of amino acid changes in the protein and the distance back in time that any two chains must have diverged. The slope of the line indicates *rate* of evolutionary change.

Note from the graph that the rate for hemoglobin evolution appears to be more than three times that for another protein such as cytochrome *c*. How might we explain this difference? One suggestion is that hemoglobin, unlike cytochrome *c* or other enzymes, reacts with a small molecule (oxygen), whereas most enzymes react with larger molecules with highly intricate surface configurations. The interaction between oxygen and hemoglobin depends less on the fitting together of surfaces than on interaction between the electron

clouds of the oxygen and the iron atom of the porphyrin ring. Thus, the three-dimensional structure of the hemoglobin polypeptide chain has less constraints put upon it, and variations in amino acid sequence would have less effect on the function of hemoglobin than on a molecule like cytochrome that must react in a more specialized way. Although mutations probably appeared at the same rates in the genes for hemoglobin and cytochrome *c*, the rate of survival of mutant forms was perhaps much greater for hemoglobin.

It is important to emphasize that a graph like that shown in Diagram B can be constructed only if the dates of divergence between any two proteins can be established by independent information, such as reference to the fossil record. For example, the divergence between the α- and β-chains is thought to have occurred during the evolution of higher fish; thus from fossil evidence we can date this split in the late Ordovician period, or about 450 million years ago. Sometimes, though, it is difficult to date exactly when a particular type of hemoglobin chain appeared—the γ, or fetal type, for example. The degree of difference between the γ- and β-chains suggests theoretically that these two types might have diverged about 200 to 225 million years ago, during the mid-Triassic period. This calculation makes sense: mammals first appeared during the Triassic and γ-hemoglobin is found only in mammals.

Recent studies by Allan Wilson of the University of California, Berkeley, have attempted to refine the picture of molecular evolution even more precisely. Wilson and his co-workers have studied variability in mitochondrial DNA, which shows far more mutations per unit time than nuclear DNA. The reason for this appears to be two-fold. First, mitochondrial DNA is replicated every time a mitochondrion replicates, which is more often than the cell in which it is found. Second, mitochondrial DNA is not subject to the action of repair enzymes, which exist in abundance in the nucleus. As a result, mutations that are ordinarily "repaired" in nuclear DNA remain visible in mitochondrial DNA. What Wilson and his associates have found is that mitochondrial DNA mutates, and therefore evolves, about ten times faster than nuclear DNA. This means that it is posssible to study evolutionary rates, at least within mitochondrial DNA types, over a far shorter time scale than is possible for nuclear DNA. Whereas nuclear DNA changes (as measured, for example, through protein changes) on the scale of millions of years, mitochondrial DNA may change on the scale of thousands of years. This means that, using mitochondrial DNA changes among related species, it is possible to study divergences that may have taken place less than 10,000 years ago. As the technique of mitochondrial DNA analysis becomes refined, it may well prove to be the "electron microscope" of evolution—allowing biologists to resolve minute levels of phylogenetic divergence that were invisible by the more standard molecular method of comparing related protein structures.

tant factor in many insect-pollinated flowers. Plants of one species may have flowers whose structure (size and shape) allow pollination by only one species of insect; members of other insect species simply cannot enter the flower, and therefore cannot deliver pollen from other species of flowers. Thus, two closely related plant species may be mechanically isolated from one another because their flowers do not allow the entrance of diverse species of insect pollinators who might be carrying pollen from other species or varieties.

Postmating Isolating Mechanisms

Mortality of the Hybrid. Even if fertilization is accomplished, hybrid embryos between two different species sometimes display reduced viability because of incompatibility of the genomes that have been combined. Some embryos begin development normally, but die before reaching maturity. In this case the two populations of organisms would be isolated because their offspring cannot reach sexual maturity.

Sterility of the Hybrid. In certain cases fertilization and embryonic development of a hybrid are normal, but the hybrid is itself reproductively sterile. This is generally due to differences in chromosome number between the parental species, and thus inability of the chromosomes to pair correctly at meiosis. As a result, the hybrid cannot form functional gametes. One of the best known examples of hybrid sterility is the mule, which is a hybrid between the donkey and the horse. Mules are physiologically normal, but generally cannot produce viable gametes, and hence are sterile. Genet-

ically speaking, sterile hybrids, like hybrids that die off in the embryonic stage, are genetic dead-ends. The two species producing such hybrids are said to be reproductively isolated.

Reduced Fitness of the Hybrid. In some cases hybrids are fertile, but produce offspring that are less fit than the original parents. This means that the F_1 hybrids leave fewer offspring (F_2) of their own, and hence fewer copies of their genes for the next generation. By virtue of their reduced fertility, such hybrid forms are said to be less fit than their nonhybrid counterparts. In a competitive environment, such hybrids, even if they occur, would be selected against. In time, they would either tend to be eliminated from the population, or be kept at a relatively low frequency.*

In general, once two populations have begun to diverge and show some reduction in viability of hybrid offspring between them, natural selection favors development of premating over postmating isolating mechanisms. A great deal of metabolic energy is used in producing gametes, in effecting fertilization, and in the development of embryos. If the embryo that results from a hybrid between two species is not going to pro-

---

*It should be pointed out here that many interspecific hybrids in plants are perfectly fit, leave many offspring, and sometimes form quite distinct hybrid populations on their own. Some plant hybrids can even achieve the status of a separate species. Reduced fitness of the hybrid includes by definition only those hybrids that are less fit (leave fewer offspring) than other hybrids or than nonhybrids. Reduced fitness of hybrids is far more common in animals than in plants.

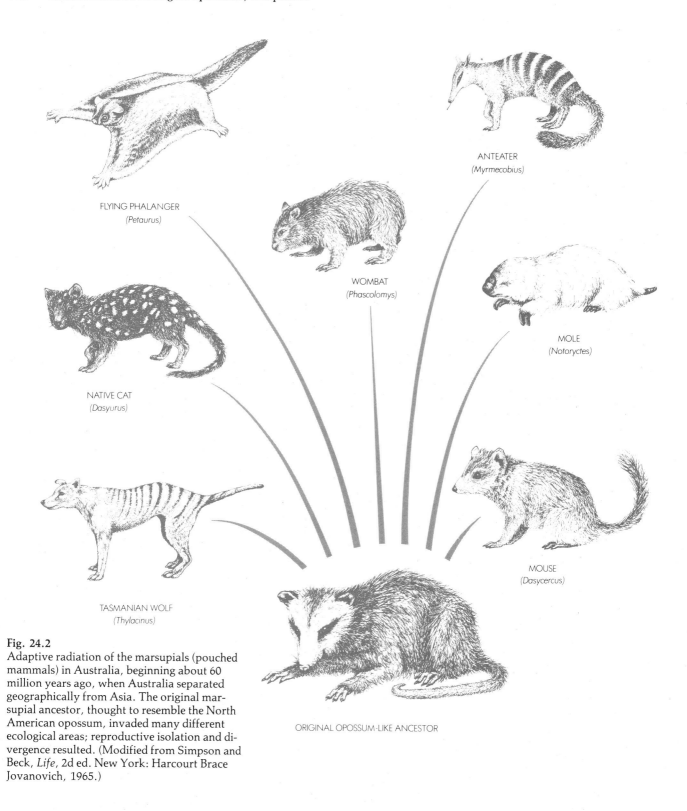

FLYING PHALANGER
(Petaurus)

ANTEATER
(Myrmecobius)

WOMBAT
(Phascolomys)

MOLE
(Notoryctes)

NATIVE CAT
(Dasyurus)

MOUSE
(Dasycercus)

TASMANIAN WOLF
(Thylacinus)

ORIGINAL OPOSSUM-LIKE ANCESTOR

Fig. 24.2
Adaptive radiation of the marsupials (pouched mammals) in Australia, beginning about 60 million years ago, when Australia separated geographically from Asia. The original marsupial ancestor, thought to resemble the North American opossum, invaded many different ecological areas; reproductive isolation and divergence resulted. (Modified from Simpson and Beck, *Life*, 2d ed. New York: Harcourt Brace Jovanovich, 1965.)

duce viable offspring of its own, the energy consumed in gamete production, fertilization, and embryonic growth is wasted. Thus, in any two species where nonproductive hybridization occurs, those individuals with variations that discourage or prevent reproduction at the premating stage will have a distinct advantage over those where isolation occurs at the postmating stage. As we might expect, most of the mechanisms of reproductive isolation found in nature turn out to operate at the premating level.

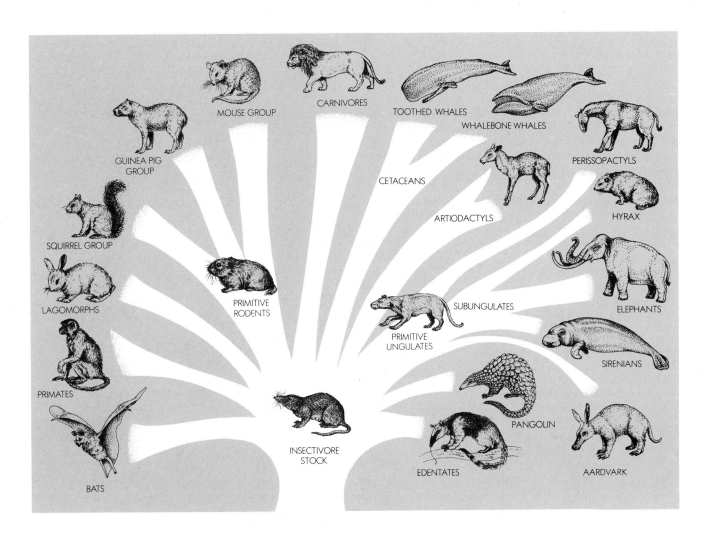

Figure labels (clockwise): MOUSE GROUP · CARNIVORES · TOOTHED WHALES · WHALEBONE WHALES · PERISSODACTYLS · CETACEANS · HYRAX · ARTIODACTYLS · ELEPHANTS · SUBUNGULATES · SIRENIANS · PRIMITIVE UNGULATES · PANGOLIN · EDENTATES · AARDVARK · INSECTIVORE STOCK · BATS · PRIMATES · LAGOMORPHS · SQUIRREL GROUP · GUINEA PIG GROUP · PRIMITIVE RODENTS

24.4
DIVERGENCE AND ADAPTIVE RADIATION

Somewhere around 60 million years ago, ancestral marsupials (pouched mammals; see Section 29.13) invaded the continent of Australia by an already rapidly disappearing land connection to the Asian subcontinent. The original ancestor was similar to the opossum. In Australia it found an enormously varied environment free of large animal competitors. Although placental mammals were in abundance throughout the European and Asian continents, none had managed to reach Australia by this period.* With no competitors, the ancestral marsupial population grew and spread over the continent, until they occupied a number of different environments. In the different environments very different selection pressures existed. As a result, different popula-

---

*Only later did some placental mammals make it across the relatively wide stretch of ocean that came to separate Australia from the Asian subcontinent. These mammals included a rat (making the migration by island-hopping), bats (by wind dispersal), dogs (brought by the aboriginal people who migrated to Australia), and the aborigines themselves (who came by boat).

Fig. 24.3
Adaptive radiation of the mammals, which began at about the same time or slightly before the marsupials invaded Australia (that is, about 60 to 70 million years ago). (Redrawn from A. S. Romer, *The Vertebrate Body*. Philadelphia: W. B. Saunders, 1960, p. 72. Reprinted by permission of Holt.)

tions of the original marsupial invaders rapidly became adapted to the many ecological conditions they encountered. The marsupial line diverged in a number of quite different directions (Fig. 24.2): the kangaroo (a plains-dwelling herbivore), the Tasmanian wolf (a forest- and field-dwelling carnivore), the flying phalanger (a tree-dwelling herbivore), the wombat (a ground-burrowing herbivore), and the marsupial mole (a ground-burrowing insectivore, that is, an insect-feeder), and a rodent-like mouse (a terrestrial and burrowing herbivore).

Divergence from a single ancestral form into a variety of new species adapted to very different modes of life is known as **adaptive radiation.** Adaptive radiation is usually characterized by rapid divergence made possible by exploitation of numerous unused or underused environmental resources in a given area. It is an exceedingly important evolutionary process and has occurred many times throughout the history of life. One major adaptive radiation followed the first conquest of land by vertebrates (350 to 400 million years ago). A new unexploited territory became available; there were no other large animals present. The fish-like ancestors that made the invasion gave rise to a whole host of new species that diverged sharply from each other as they occupied different parts of the land. The reptilian radiation was another great period of evolutionary expansion among vertebrates, taking place 250 to 300 million

years ago, producing, among other groups, the great dinosaurs. Still a third adaptive radiation among vertebrates was that of the mammals, which began about 60 million years ago producing such terrestrial forms as the bats, carnivores, rodents, hoofed animals, and primates, and such aquatic forms as the whales and dolphins (see Fig. 24.3).

A Case Study in the Mechanism of Adaptive Radiation: Darwin's Finches

When Darwin visited the Galápagos Islands in 1832, he collected some specimens of ground finch belonging to

Fig. 24.4
Natural selection favors those variations that allow each organism to adapt successfully to a specific habitat. All the organisms shown occupy similar ecological niches, but each species has evolved still further to minimize competition for resources. (From David Lack, "Darwin's Finches." Reprinted with permission. Copyright © 1953 by Scientific American, Inc. All rights reserved.)

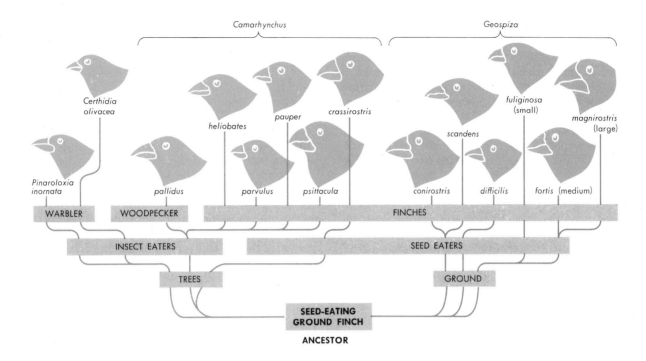

Fig. 24.5
Evolutionary history of the finches on the Galápagos. This diagram illustrates the basic radiation that occurred at each point. The first radiation was between the two major habitats, ground and tree. Within each of these areas, subradiations have taken place.

the subfamily Geospizinae. Although Darwin himself did not draw any special conclusions from study of these birds, later investigators have found in them excellent sources of data on adaptive radiation and other evolutionary principles.

Geological evidence shows the the Galápagos are of rather recent volcanic origin (having been formed about 2,000,000 years ago) and have never been connected to the South American mainland. As soon as vegetation developed on the islands, immigrant animals began to move in and colonize.

It is hypothesized that the present-day Galápagos finches (the male and the female of each species are shown in Fig. 24.4) are descended from a single type of finch that immigrated to the islands from the mainland sometime in the distant past. All attempts to find any such a mainland ancestor have, however, been unsuccessful. Perhaps this is because the ancestral finch has become extinct, or perhaps it is because the Galápagos finches have diverged so far from the ancestral traits that their relation to the mainland form is unrecognizable.

An evolutionary diagram of the adaptive radiation of the finches throws light on the probable way in which divergence has occurred (Fig. 24.5). The fourteen species of finch living today can be roughly grouped into two categories: the ground finches (genus *Geospiza*) and the tree finches (genus *Camarhynchus*). In addition, there is a single "warbler" species and one isolated species peculiar to the most distant island of the group.

Figure 24.4 indicates that the ancestral finch population split up into ground-dwellers and tree-dwellers soon after arriving on the Galápagos. Variation and selection produced adaptations not only in foot structure,

but also in the shape and size of beaks. Seeds form the major diet of ground finches, while the tree finches feed mostly on insects. Competition for food in each major category (insects, seeds) produced further adaptive radiation within each area; all six species of ground-dwelling finches have slightly different diet preferences, even though all may be classified as seed-eaters. For example, three of the six species have differences in beak size and shape that make them specialized for specific sizes of seed (*G. magnirostris*, large seeds; *G. fortis*, medium seeds; and *G. fuliginosa*, small seeds). It is interesting that these three forms are found together on nearly all the Galápagos Islands. Because these species have adapted to specific sizes of seed, competition for food between them is reduced. This subradiation occurs because organisms are able to exploit numerous unused resources. The other three species of ground-dwellers, on the other hand, combine seed diets with flowers or cactus pulp.

Among the tree finches, one type feeds on seeds while the rest feed on insects. Like the ground finches, the tree finches have shown further radiation, especially in size and shape of beak. An especially interesting adaptation has occurred in one species of tree finch, *Camarhynchus pallidus*. This bird exploits essentially the same niche inhabited by woodpeckers in other geographic regions. Although lacking the hard beak and long tongue of true woodpeckers, this finch uses a cactus spine or small stick to probe insects from cracks in trees or from under the bark. In this case, an adaptation in behavior has allowed the organism to exploit an unused resource.

On the South American mainland, no group of finches has been able to undergo the great adaptive radiation seen on the Galápagos. It seems clear that on the mainland most available ecological positions are already filled by other types of birds. The mainland forms lacked the "ecological opportunity" that makes adaptive radiation possible. The founder population that inhabited the Galápagos, however, had little competition from other birds. As a result, its members could reproduce rapidly and spread out to exploit numerous environments. Adaptive radiation is, then, partly dependent on the availability of unused resources that organisms can exploit.

What role may the various modes of speciation discussed in Section 24.2 have played in the adaptive radiation of the Geospizinae? The most traditional explanation is that divergence occurred by allopatric speciation. Because there are almost exactly the same number of species of finch (fourteen) on the Galápagos as there are islands in the group (thirteen), it seems likely that the first phase of adaptive radiation carried representatives of the ancestral ground finch to all the islands. On each island, in relative geographic isolation from each other, the various populations, through natural selection, became adapted to the local environments. For example, on an island where most of the available seeds were large, selection would favor large-billed finches. On an island where most seeds were small, small-billed finches would be favored. At the same time, of course, other variations were accumulating, so that, over time, the two populations came to differ not only in beak size, but in many other traits as well. After a while, the two populations had become reproductively isolated in one way or another.

Eventually, however, population expansion on each island or accidental migrations during storms brought finches from one island to the others. Here, the immigrant finches encountered the already-established populations. In most cases the two groups were reproductively isolated, so there was very little if any interbreeding. But there may have been competition, especially if the immigrant happened to share the same food and other environmental needs as the already-established species. When the newcomer was able to exploit some new or slightly different food source or living space on the new island, it could establish itself with, perhaps, a minimum of competition from the original inhabitants. This must have happened frequently enough that several different species of finch were eventually able to live on each island. According to the most

commonly held view, the various species of finch on each island did *not* originate there by parapatric or sympatric speciation, but as a result of allopatric speciation.

This is a likely hypothesis, with the advantage of explaining why there are almost exactly as many species of finches as there are islands of the Galápagos group. In 1979, however, B. R. Grant and P. R. Grant of the University of Michigan questioned whether many of the finch species co-existing on particular islands might not have arisen there by some form of parapatric or sympatric speciation similar to that described for the mole rat in Israel (see Section 24.2). The Grants argue that subgroups within some finch species on a single island show polymorphism for songs and other behavioral traits with no evidence of geographic isolation. Such divergences within a single population, they suggest, could provide the basis on which reproductive isolation developed. Little in the way of concrete data exists to support or refute this interpretation, however. For instance, it would be quite useful to know the chromosome counts of each species and subspecies of finch. As with the mole rat, examination of the finches' chromosome numbers might indicate something about the possible mechanism by which such parapatric or sympatric divergence could have occurred.

24.5
CONVERGENT EVOLUTION, PARALLEL EVOLUTION, AND CO-EVOLUTION

One of the striking features of divergence of genotype between two species is that it does not always produce divergence of phenotypes. It is not uncommon for two groups of organisms with quite different ancestry to end up looking very much alike. Such evolutionary developments are called *convergence*, or *convergent evolution*. The similar phenotypic traits of body shape and tail structure in fish and whales is one example of convergence. Another is the overall similarities between quite different species of desert mammals from North and South America (see Fig. 24.6).

Of all the examples of convergence known, however, one of the most extensive is that between the marsupials of Australia and the placental mammals of the rest of the world. Figure 24.7 shows the close resemblance between a number of marsupials and their placental counterparts. The Tasmanian wolf (or "Tasmanian devil" as it is often called) and the placental mammals are both carnivores and live in partly forested regions. The native cat, a marsupial, is very similar in appearance to members of the true cat family; like them, it is also a carnivore and feeds off small rodents or rodent-like forms. A particularly striking series of convergences (striking because of their highly specialized nature) can be found between the flying phalanger (marsupial) and the flying squirrel (placental), the marsupial and placental anteaters, and the marsupial and placental

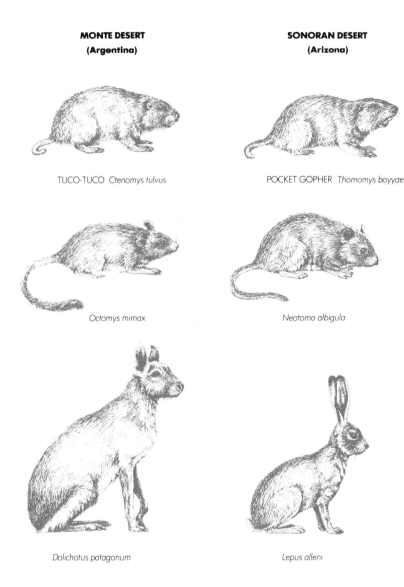

MONTE DESERT
(Argentina)

SONORAN DESERT
(Arizona)

TUCO-TUCO *Ctenomys tulvus*

POCKET GOPHER *Thomomys boyyae*

Octomys mimax

Neotoma albigula

Dolichotus patagonum

Lepus alleni

Fig. 24.6
Convergence of three species of desert mammals from two quite different localities: the Monte desert in Argentina, and the Sonoran desert near Tuscon, Arizona. Top row: the gopher-like tuco-tuco of the Monte is very similar to the Arizona pocket gopher. Middle row: two species of desert rat showing highly convergent body shape and long tail. Bottom row: convergence of rabbit-like form of the Monte with the true rabbit of the Sonora. All three pairs of species began from virtually unrelated ancestry. (Redrawn from M. A. Mares et al., "The strategies and community patterns of desert animals," p. 161.)

moles. In each of these cases the corresponding forms have many of the same food and living requirements, though they have arisen from quite different ancestry.

Given that in the previous chapter we emphasized the random processes involved in variation and natural selection, it might seem that convergent evolution would be an extremely unlikely phenomenon. Yet, a moment's reflection will suggest why it is not so unusual after all. Plants or animals of different origin inhabiting similar habitats, or living under similar ecological conditions, will undergo selection for similar sorts of adaptations. In desert mammals, for example, there are only so many ways in which water conservation can be accomplished (for example, by storage mechanisms in specialized tissues or by reabsorption in the kidneys). That many desert mammals of different ancestry have evolved the same or nearly the same water conservation mechanisms should not be so surprising. Among the

marsupials, convergence with the placental mammals along so many lines is a result of adaptive radiation over a virtually uninhabited (by other large animals) continent. Many of the habitats and ecological areas found in Australia are similar to habitats found elsewhere. In adapting to an arboreal existence, for example, the flying phalanger came to have many characteristics of the placental flying squirrel (the stretched skin between the appendages and the body, for instance). In adapting to feeding on ants, the marsupial anteater underwent many of the same selection pressures for elongated snout (to suck ants out of narrow anthills) that the placental anteaters faced in their evolution in South America. Although the kangaroo has no phenotypic counterpart on other continents, in Australia it occupies the same ecological position (a grazing, plains-dwelling runner) that the hoofed placentals (horse, zebra) occupy elsewhere. Convergence is the result of similar selection

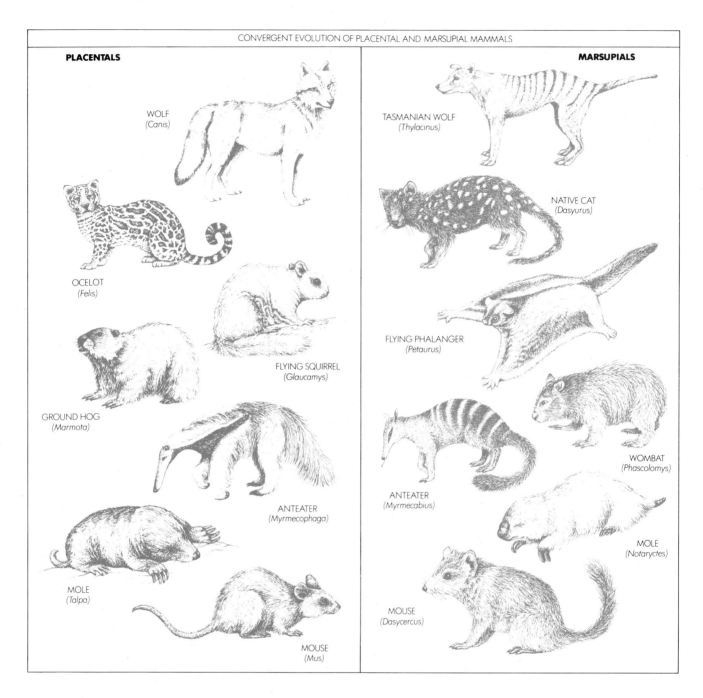

CONVERGENT EVOLUTION OF PLACENTAL AND MARSUPIAL MAMMALS

PLACENTALS

WOLF
(Canis)

OCELOT
(Felis)

GROUND HOG
(Marmota)

FLYING SQUIRREL
(Glaucamys)

ANTEATER
(Myrmecophaga)

MOLE
(Talpa)

MOUSE
(Mus)

MARSUPIALS

TASMANIAN WOLF
(Thylacinus)

NATIVE CAT
(Dasyurus)

FLYING PHALANGER
(Petaurus)

WOMBAT
(Phascolomys)

ANTEATER
(Myrmecabius)

MOLE
(Notaryctes)

MOUSE
(Dasycercus)

Fig. 24.7
Convergent evolution between a number of marsupial mammals and their placental counterparts. Note the similarity of phenotypes between comparable species. (Redrawn from G. G. Simpson, C. S. Pittindrigh, and L. S. Tiffany, *Life. An Introduction to Biology.* New York: Harcourt-Brace, 1965, p. 500.)

pressures acting on random variations within two genetically separate and ancestrally unrelated (or only distantly related) populations.

Biologists sometimes speak of **evolutionary parallelism,** as well as evolutionary convergence. Parallelism occurs when two species that are closely related but evolve in separate places (and hence are reproductively isolated) diverge but follow something of the same general evolutionary course. For example, it used to be thought that the porcupines of North America and those of the Old World had evolved from a common, spine-bearing ancestor, their divergence (and phenotypic differences) being merely a result of geographic isolation. However, evidence now suggests that both were derived from a common nonspiny ancestor, and though evolving in quite different geographical localities, both developed quills as a protective device. Their evolution was parallel. In parallel evolution two species start out the same (common ancestry and thus common gene pool) and evolve independently along similar lines. In convergent evolution, on the other hand, two species start out very different (from different ancestry) and independently come to resemble each other more and more (converge means, literally, to "come together"). The common denominator in both convergence and parallelism is that the species involved are showing adaptation to very similar ecological conditions. Natural selection thus favors variations, however different in genetic origin, that solve similar problems.

Co-Evolution

Co-evolution is one of the most remarkable aspects of the evolution of adaptations. Co-evolution is defined as the process in which two (rarely more) species evolve adaptations to one another, such that evolutionary change in one species produces an evolutionary change in the other. Co-evolution is mutually beneficial to each species. There are many examples of co-evolution (mimicry, described in Supplement 24.3, is an example), but two will suffice here.

Most flowering plants depend upon external agents for pollination—that is, to transport pollen from one flower to another. Some flowers use water or wind, whereas others use one or another animal pollinator, such as insects or birds. In cases of animal pollination there is often a striking correlation between pollinators and the plants they pollinate. For example, bees are attracted by blue or yellow colors (they cannot detect red at all) and by sweet odors; they forage only during the day, and land on the flower to feed (as opposed to hummingbirds, for example, which suck nectar while hovering in the air in front of the flower). Flowers that are pollinated by bees usually have brightly colored yellow or blue petals (seldom red, because bees cannot distinguish red very easily); secrete a sweet, fragrant odor; are open during the day but closed at night; and have a lower petal that protrudes forward as a kind of "landing strip."

By contrast, flowers pollinated by hummingbirds are quite often red or yellow, but seldom blue (hummingbirds cannot see blue light as well), and frequently have little or no fragrance, and no landing strip. An example of a flower that has co-evolved with insects for pollination is the orchid. In co-evolution of pollinators and the flower species they pollinate, natural selection has acted on the gene pool of both populations to favor any variations that would favor successful interaction of insect and flower. Success for the flowering plant means getting as much pollen dusted onto the insect as possible, and hence ensuring a maximum dispersal of its

pollen. Success for the insect means obtaining a maximum supply of nectar (sugar for making honey).

Another example of co-evolution is that between rabbits and the myxoma virus in Australia. There were no rabbits in Australia until 1859, when a wealthy Englishman imported a dozen animals from Europe to decorate the grounds of his estate. Six years later he had more rabbits than he knew what to do with: 30,000 on his estate alone! By 1887 over 20 million rabbits had been killed in the state of New South Wales, and by 1950 Australia was on the verge of losing most of its pasture land to rabbits. Then, as a form of biological control, rabbits from South America infected with myxoma virus were released into Australia. The virus causes only mild disease in its normal South American host, but is fatal to the European rabbit. The effect was dramatic—the rabbit population began to decline almost immediately. After a period of time, however, it was noticed that some few European rabbits were surviving. What is more, their offspring appeared to be resistant to myxoma. What had happened?

At first it appeared that, through mutation or some other genetic variation, a few Australian rabbits had become resistant to the virus. A closer look, however, showed that this was only half the story. A double selection had taken place. The original virus was so rapidly fatal to the European rabbit that often the rabbits died before they could pass the virus on to another host (the viruses are normally transmitted by a mosquito vector). As a result, the virus in that host would never have a chance to reproduce and spread its genes throughout the population. Originally, then, both host and parasite were ill-adapted to each other. However, as the European rabbit underwent selection for resistance to the virus, the virus also underwent selection for reduction in virulence (disease-causing ability). Selection favored those virus strains that could infect and reproduce within the European rabbit without killing it. (A similar co-evolution between the myxoma virus and its South American rabbit host must have occurred already in the past.) Through co-evolution, then, the Australian rabbit and the myxoma virus adapted to one another such that continued reproduction of both was favored.

24.6 ADAPTATION

One of the outcomes of evolution by natural selection is that species become adapted to their surroundings. The organisms making up any population show similar adaptations to the environment in which they are living.

Adaptations are defined as the modification of an organism in structure or function that allow it effectively to utilize resources in the environment. In the evolutionary sense, adaptations have been produced over many generations by selection acting on variations at the genetic level. Thus, specific evolutionary adapta-

SUPPLEMENT 24.3
NATURE'S COSTUME BALL—OR, HOW DO I RECOGNIZE YOU AT THE PARTY?

Of the many wonderful types of adaptations in the living world, none are so marvelous, or arouse so much amazement, as those that in one way or another protect animals from their predators by either camouflage or dramatic advertisement. Many organisms in nature come with a "costume" that helps them conceal themselves or advertise that they have something unpleasant to offer if attacked. Although we now recognize that protective costuming has developed in various species through natural selection, the presence of such amazing adaptations was for many years held up as inexplicable by traditional Darwinian theory.

Consider first adaptations for concealment. The most common form of concealment found in the animal world is through color variations that blend the organism into its background; this is called *cryptic coloration*, or *camouflage*. The light-colored melanic moths of England (see Fig. 23.8, top) are a very good example of cryptic coloration. Another and more unusual form of camouflage combines cryptic coloration with a particular morphology so that the organism (usually in the resting position) resembles something its predators normally do not eat. One of the best examples of this sort of concealment is found in insects that so resemble thorns that they are completely missed by most bird predators (see Diagram A).

The opposite sort of costume from those that conceal is those that advertise. In this category are some of the most unlikely sorts of color adaptations. Some insects, for example, are brightly colored and stand out conspicuously against their backgrounds. These insects also turn out to be harmful, or at least distasteful to predators: the insects sting, are poisonous, or taste or smell bad (see Color Plate I). The irritating hairs of *Sibine stimulae* cause a skin eruption and fairly severe itching (in humans). The bright coloration makes it easier for predators to learn to avoid these particular species. The animal's bright colors, therefore, serve as protection because they warn the predators to stay away. Appropriately, this type of adaptation is called *warning coloration*.

In 1862 the British naturalist H. W. Bates proposed the notion of *mimicry* (or, as it is called today, *Batesian mimicry*) as an extension of the principle of warning coloration. Bates discovered that sometimes a harmless, or perfectly tasty insect mimics in color or shape the appearance of a harmful species that displays warning coloration. The mimic species gains protection because it looks like the distasteful species (called the *model*); wary predators avoid the mimic as well as they avoid the model.

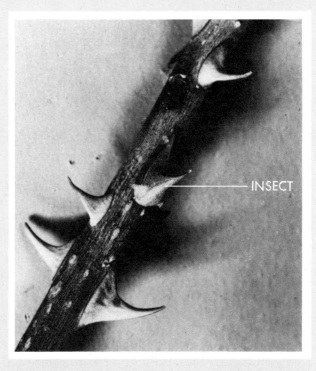

Diagram A
Insect (can you tell which one?) camouflaged as a thorn. (Photo from Gavin De Beer, *Atlas of Evolution*. Used by permission of Elsevier Nederland, Amsterdam.)

An example of Batesian mimicry is the viceroy butterfly which closely resembles the highly distasteful monarch butterfly (Color Plate I). Jane van Z. Brower demonstrated the effectiveness of Batesian mimicry in an experiment with viceroy and monarch butterflies. She first fed viceroy butterflies to blue jays, which ate them readily. Next, Brower fed monarch butterflies to the jays. Since monarchs are distasteful, the jays refused to eat them. When viceroys were again offered to the jays, the jays refused the viceroys, too. Brower then repeated this experiment with artificial models and mimics. She painted one set of mealworms (hard-shelled beetle larvae) with a green stripe and another set with an orange stripe. She dipped some of the worms with a green stripe into a quinine solution, giving them a bitter taste. The rest of the green-striped and the orange-striped mealworms were dipped in distilled water, which didn't change their taste. The green-striped worms were chosen as the distasteful species because in nature green would usually be a cryptic rather than a warning color, and the predators would, therefore, not have already learned to avoid green worms. After feeding the worms to caged starlings, Brower found that the birds avoided all green-striped worms. The green-striped worms dipped in distilled water (mimics) were protected by their resemblance to bitter-tasting worms dipped in quinine solution (models).

Originally, it was thought that Batesian mimicry could be successful only if the number of models exceeded the number of mimics, ensuring that a predator would have a greater chance of eating a distasteful model than an agreeable mimic and thereby learn to avoid the model's color pattern. Brower's experiment with the mealworms, however, disproved this hypothesis. She showed that jays avoided most of the mimics even when mimics made up 60 percent of the green-striped worm population. In other words, even when apparently tasty mimics were present in larger numbers than the distasteful models, the jays avoided them. It is now thought that the success of Batesian mimicry depends on (a) how distasteful the model is, (b) how tasty the mimic is, and (c) how many alternative food sources are available.

A second type of mimicry—Müllerian mimicry—was described by Fritz Müller in the late nineteenth century. Müllerian mimicry differs from Batesian mimicry in one significant way—namely, all the species (two or more) that look alike are distasteful to their predators. What, one may then ask, is the survival value of their resemblance? Since they already have one protection, why do they need another? The answer is very simple. For predators to *learn* that a particular warning coloration means a bad taste, or a sting, the predator must eat at least a few individuals. If every distasteful species had its own individual warning coloration, predators would have to sample individuals from each species before learning their lessons. However, if several species in a given area (thus exposed to the same predators) not only are distasteful, but look sufficiently alike, the predators will learn to avoid all the species after sampling only a few individuals. In short, Müllerian mimicry is advantageous because it reduces the loss of individual prey while the predators are becoming educated.

Nature's costume ball has many variations in both coloration, pattern, and form. It provides a striking example of the diversity to which selection and adaptation can lead.

tions are built into the population's gene pool, and into the genotypes of some, if not all, members of the population. It is important not to confuse the concept of evolutionary adaptation with a more colloquial and frequently encountered use of the term adaptation. "Adaptation" sometimes refers to short-term, somatic responses that organisms make (physiologically, anatomically, and behaviorally) to new conditions they encounter (for example, when mammals grow thicker hair in winter or humans produce more red blood cells in response to living at high altitudes). Biologists prefer to use the term "acclimate" or "acclimatization" to refer to such short-term modification.* Throughout the remainder of this book we will use the term adaptation in its strict evolutionary sense.

Adaptations take every conceivable form—some are anatomical, others physiological, and still others biochemical and behavioral. All types of adaptations interact to contribute to the total welfare of the organism in its natural habitat.

Anatomical adaptations are usually the most obvious and well-documented. They include the peculiar foot structure of birds for perching, grasping, wading, or swimming, the modification of anterior appendages in insects into mouthparts for feeding, and bright and/or distinctive color pattern on certain insects as warning signals to predators. **Physiological adaptations** include presence of certain enzymes for digesting particular foods (as in a wood-digesting enzyme in the ship worm *Teredo*), or complex osmoregulatory mechanism in aquatic, and especially marine, organisms. **Biochemical adaptations** include the many interactions among biochemical pathways (such as the interconnections between carbohydrate and protein synthesis pathways) that make it possible to shift from one to the other depending on the available food source; they also include the precise structure of enzyme molecules allowing them precisely to fit the shape of specific substrate molecules. **Behavioral adaptations** are found only among animals and include the inherited behavioral responses of males and females of a species to each other's specific courtship patterns, the responses of the females of many species to threats against their young (so-called maternal instinct), and the ability of migratory birds to navigate by the stars. In all cases, these adaptations contribute to the overall survival of the species by increasing the differential fertility of the organisms that display them.

In thinking about adaptations as the outcome of evolution it is important to keep in mind that *all* adaptations are relative to one another, that is, to adaptations of the same *kind*. Thus, the degree of effectiveness of a mechanism for conserving water in desert mammals or plants is relative only to another such mechanism. There is no absolute scale of effectiveness of adaptations. This also means that no perfect adaptation is ever possible. No animal can run infinitely fast from its

---

*Indeed, the ability to acclimate is itself an adaptation in the evolutionary sense of the word.

predators, nor can any desert plant become 100 percent efficient in conserving water (that is, lose no water whatsoever by evaporation). There is thus always room for perfecting any given adaptation, however effective it may appear at the moment. Evolution of any particular adaptation can have no end unless, of course, the species bearing it becomes extinct.

There is another important point along this line. All adaptations in any organism are interrelated. That is, a given anatomical structure or physiological process is adapted not only to specific external processes (locomotion, water conservation, and the like) but also to the other structures and functions the organism possesses. Thus, a desert rodent's behavioral trait of nocturnal life aids its physiological adaptations to water conservation (desert animals reabsorb most of the liquid that other mammals excrete in the urine; at the same time, by moving outside their burrows only at night, desert rodents minimize water loss due to heat evaporation). Any particular adaptation always evolves in conjunc-

tion with a number of other adaptations in the same organism. We often speak of natural selection as if it acts on one trait at a time. For example, it is common to hear biologists speak of individual traits such as tooth size or shape, or leg length, as adaptations produced by natural selection. To some extent such language is necessary; we cannot conceive of a large number of interacting traits simultaneously. However, it is important to keep in mind that natural selection does not act to increase the adaptiveness of isolated character traits, but rather to increase overall adaptiveness of the organism as a whole. Organisms and their adaptations are wholes, acted upon in an integrated way by natural selection. Differential fertility is ultimately the main judge of evolutionary success. No matter how effective a particular individual adaptation may be, it will have little selective value if it does not contribute to the overall reproductive success of the organism that bears it. (For a further, intriguing aspect of the problem of interrelated adaptations, see Supplement 24.4.)

SUPPLEMENT 24.4
EVOLUTION AND ARCHITECTURE: A COMMON THEME

The great cathedral of St. Mark's in Venice consists of a large central dome mounted on four archways that intersect each other at right angles (see Diagram A). The tapering triangles formed by the 90° intersection of the arches are called *spandrels*, and they are necessary, in fact unavoidable, byproducts of mounting a dome on arches. In St. Mark's, each spandrel is beautifully decorated with a mosaic depicting a major scene, or symbolic representation, from Christian scripture. For example, in the design shown below in Diagram A, the figure at the top is that of an evangelist sitting at his writing table recording the scriptures; below is a man, representing one of the four great Biblical rivers (the Tigris, Euphrates, Indus, and Nile) pouring water from a pitcher into the tapering space at the bottom of the spandrel. The design is so elaborate, and so admirably suited to its space, that we might well be tempted to view it as the starting point for the entire architectural surroundings. It is almost as if the cathedral were built in such a way as to accommodate the mosaics within each spandrel.

Of course, in an architectural sense, it is easy to recognize that spandrels were not designed to meet the needs of the mosaics, but the other way around. Spandrels are a byproduct of the physical constraints of building archways that intersect one another at right angles. It is true that the mosaics are admirably suited—that is, *adapted*—to the space created

by the architecture. But in this case we know which came first. However, if we transfer the same kind of thinking to the evolutionary process, particularly to the evolution of adaptations, the situation is not quite so clear.

Evolutionary theorists Stephen Jay Gould and Richard Lewontin have used the example of the spandrels of St. Mark's to highlight a very important feature of the modern theory of adaptation. Criticizing what they call the "adaptationist program," which they argue has dominated evolutionary thinking for decades, Gould and Lewontin point out that many phenotypes in the animal or plant world may exist and persist not because they are particularly adaptive in their own right, but because they are necessary byproducts of something else that *is* more directly adaptive. Such traits would be "secondarily adaptive," or perhaps, not really adaptive at all. The "adaptationist program" that Gould and Lewontin criticize represents an attempt to find a primary adaptive purpose for every structure, physiological process, or biochemical characteristic of a species. Gould and Lewontin feel that this way of thinking greatly distorts our view of how evolution actually works and of the way organisms are put together.

The followers of the adaptationist program assume that nothing exists in nature without having some general, primary purpose, *for which it was selected*. Thus, for example,

Diagram A
(Photograph courtesy of Stephen Jay Gould.)

if horns and antlers, once thought to be primarily weapons for combat, are shown to be used seldom for fighting, adaptationists suddenly propose that their major function is heat transfer; if the facial structure of Eskimos, once claimed to be adapted to withstand cold, is now found to lose heat at the same rate as other facial structures, adaptationists claim the Eskimo's face is adapted to support the musculature for chewing raw meat; if butterfly color patterns are not camouflage, they are warning patterns against predators; and so on. The problem with this mode of thinking, according to Gould and Lewontin, is that it leads to endless speculation, as well as the failure to consider alternative, nonadaptive explanations. For example, any of the above adaptive interpretations might be true, but how do we tell? Modern-day evolutionists are not concerned with what might possibly be true, but what is true—with what forces have shaped the structure and function of organisms as we know them today. The adherent to the adaptationist program may be in danger

of becoming like Dr. Pangloss in Voltaire's satire *Candide*, who says:

> *Things cannot be other than they are. Everything is made for the best purpose. Our noses were made to carry spectacles, so we have spectacles. Legs were clearly intended for breeches, and we wear them.*

Translated into our architectural analogy, spandrels were constructed to house paintings of evangelists, and so have come into being.

Gould and Lewontin attack the adaptationist program on another ground. Not only is it speculative, it fails to see the evolutionary process working on organisms as wholes. Rather, it breaks the organism down into a mosaic, a group of isolated traits, each being acted upon by selection more or less independently. Now we know that organisms function, and develop embryologically, as wholes, not as a collage of separate traits. There is no reason to think the evolutionary process is any different. We know that genes are both pleiotropic (affect many different traits simultaneously) and epistatic (interact to affect a particular phenotype). Thus, selection for a particular phenotypic effect of a gene (or group of genes) could, in fact probably does, produce a lot of other side effects on associated traits in the organism. Selection may operate primarily on one trait, for example physiological processes for water conservation in desert mammals, but bring along with it a lot of modifications in other traits— what we might think of as "excess baggage"—at the same time. There is so much association and interaction among parts of a whole organism, that selection for any one part is very likely to have multiple effects on many, if not literally all, the others.

We may well admit that Gould and Lewontin are correct, and that we should stop trying to find *the* major reason why every trait that exists in any group of organisms is adaptive. But, we might ask, why is this such an important issue to begin with? Isn't it merely a refinement in thinking? No, not really; one of the major problems with the adaptationist program is that it tends to preclude other possible explanations for the origin or maintenance of traits in a population. For example, genetic drift (see Section 23.5) has usually been relegated to a minor position in evolutionary theory. How many textbooks (including earlier editions of this one!) have allowed that traits might become fixed in a population by genetic drift only if the population is so small that it is likely to become extinct before any major changes of future importance can occur anyway. However, as Gould and Lewontin point out, genetic drift may be a far more significant aspect of evolution by natural selection than heretofore recognized, especially if certain genes drift along with (perhaps "are pulled along with" would be a better phrase) other genes in the course of selecting particular gene complexes. Like thinking of the spandrels of St. Mark's as designed for the mosaics that fill them, evolutionists who adopt too adaptationist a program may end up inverting the natural order of cause-and-effect.

24.7
A CASE STUDY IN EVOLUTION AND ADAPTATION: THE EVOLUTION OF THE HORSE

The origin and evolution of the modern horse provides an excellent example of divergent speciation and adaptation over geological time. Abundant fossil evidence exists to document the 60-million-year evolutionary history of this eminently successful group (family Equidae; genus *Equus*). During its history, the horse family has passed through stages equivalent to about thirty species. What selection pressures helped direct the evolutionary pathway of the horse? What major adaptations did this group of mammals undergo? How did these adaptations inteact with one another in fashioning the sort of animal the modern horse has become?

Adaptations of Modern Horses

Horses are hoofed mammals (a group called the ungulates), with a foot and leg structure adapted for rapid running on the hard, fast plains that form its natural habitat. Members of the horse family living today include not only the common horse but also the donkey (ass) and zebra. The leg structure, as shown in Fig. 24.8(a) (top), consists of one central digit (indicated as digit III, homologous to the middle finger of our own hands), with the other digits greatly reduced. The fingernail of digit III has become the hoof, and provides a hard, durable support for the entire leg. The muscles of each leg are attached near the proximal end (near the body) thus moving the leg something like a semi-free-swinging pendulum. This arrangement is admirably suited to light, rapid motions, and thus allows for great speed and agility.

Horses obtain their food by grazing, that is, by feeding on such vegetation as grass or other plants covering the surface of the ground. To graze, however, horses have to get their heads down to the ground, a problem compounded by their adaptation for running, which favors long legs. Now, one way to solve the problem of getting the head down to the ground is through lengthening of the neck—and to some extent horses have moderately long necks when compared to most other mammals. However, at the same time horses have relatively large skulls, and particularly large jaw bones, an adaptation to chewing found in most grazers. A large skull places a constraint on the length the neck can have, since a large head becomes unwieldy at the end of a long neck. To some extent horses have solved the problem of reaching the ground by also lengthening the anterior part of the skull itself—that is, by elongating the snout into a muzzle. One interesting consequence of this lengthening of the muzzle is that it has literally produced a gap in the teeth between the incisors and the premolars (see Fig. 24.8c, 5). Humans have utilized this gap for inserting a "bit" to control a horse's movements.

The horse's tooth structure is admirably suited to its grazing lifestyle. The premolars and molars are massive, acting as very effective grinders. Grasses contain a good deal of silica, and thus are quite abrasive, wearing down the horse's teeth continually. The growth pattern of horse's teeth circumvents this problem remarkably well. The surface of the molars and the premolars is ridged by a complex set of hard enamel lines (see Fig. 24.8b), set in a matrix of softer dentine and cement, which wear away more rapidly than the enamel. The differential wearing produces a roughened, lined surface not unlike the ridged surface of an old-fashioned millstone. The enamel ridges protrude above the softer material, and act as little chisels in grinding the tough grasses.

If space permitted we could elaborate further adaptations of the horse to its life on the plains, but our purpose will be served best by concentrating on three areas:

1. Formation of the thin leg and single hoof by elongation of digit III, with corresponding loss of the other digits.

2. Elongation of the frontal region of the skull (anterior to the eyes) into a muzzle.

3. Development of the premolars and molars into high-crowned, continuously growing grinders.

The Earliest Horse, *Eohippus*

Figure 24.9 shows a schematic phylogenetic tree of the horse family. The earliest mammal to show true signs of being a horse appeared in the Eocene period, about 60

Fig. 24.8
Evolutionary developments in three adaptive traits of the horse family: foot structure, shape of molar teeth, and structure of the skull. Skulls, teeth, and feet are identified by numbers: 1, *Eohippus*; 2, *Miohippus*; 3, *Merychippus*; 4, *Pliohippus*; 5, *Equus* (modern horse). Note the gradual, but recognizable changes in these three traits as the horse evolved from overall adaptation to a forest ecology (*Eohippus*) to overall adapation to a plains ecology (from *Merychippus* onward). This change in habitat (due to climate-induced changes in the environment) led to different selection pressures, ultimately selecting for very different adaptations. (From G. De Beer, *Atlas of Evolution*. London: Thomas Nelson, 1964, p. 50; from Th. Dobzhansky, *Evolution Genetics and Man*. New York: Wiley, 1961, p. 300; and from Scott, *A History of Land Mammals in the Western Hemisphere*. New York: Macmillan, 1937, by permission of American Philosophical Society.) *Note:* Drawings not to scale.

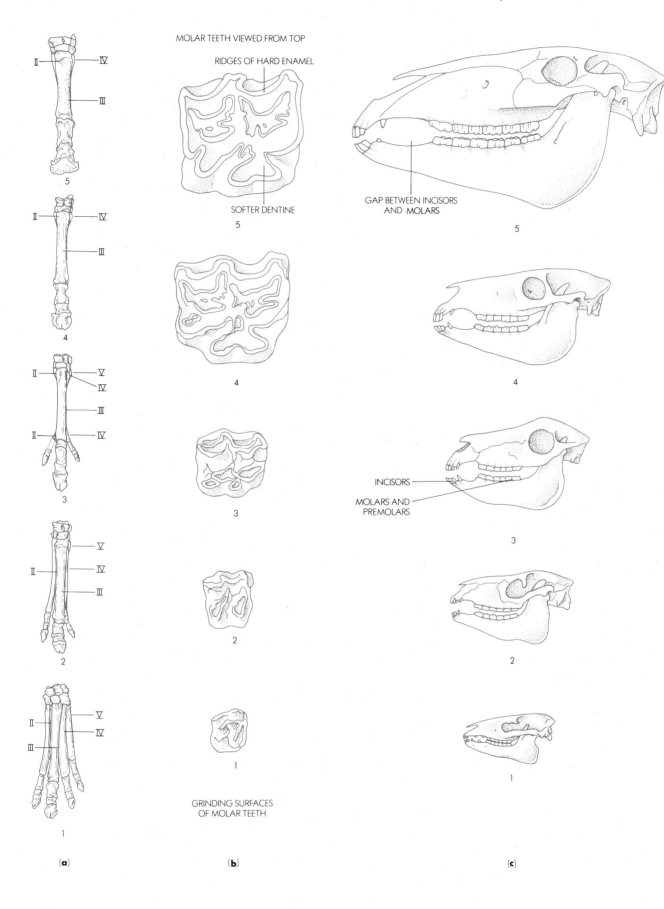

MOLAR TEETH VIEWED FROM TOP

RIDGES OF HARD ENAMEL

SOFTER DENTINE

GAP BETWEEN INCISORS
AND MOLARS

INCISORS

MOLARS AND
PREMOLARS

GRINDING SURFACES
OF MOLAR TEETH

(a)   (b)   (c)

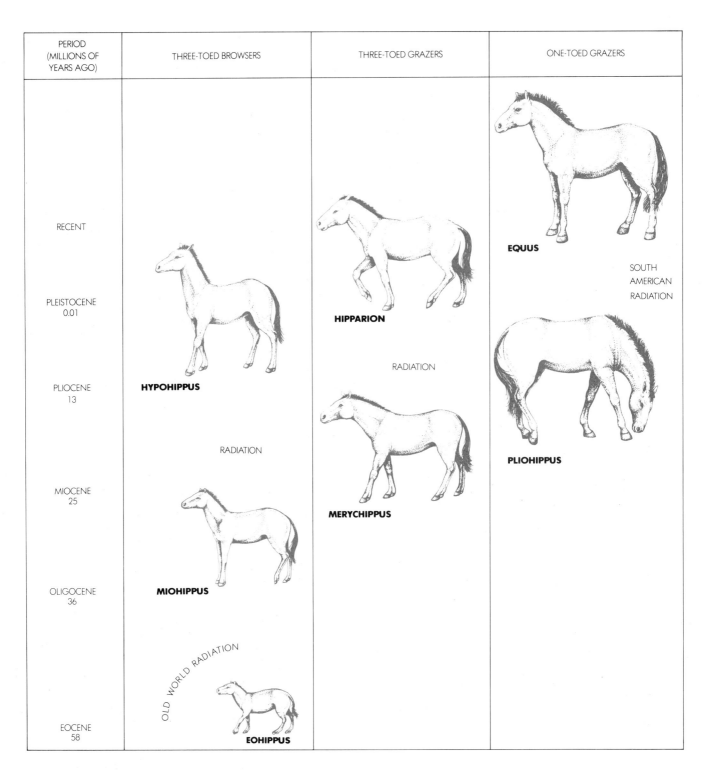

PERIOD (MILLIONS OF YEARS AGO)	THREE-TOED BROWSERS	THREE-TOED GRAZERS	ONE-TOED GRAZERS

Fig. 24.9
Simplified evolutionary scheme (phylogeny) of the horse family (Equidae). The first "horse," *Eohippus*, appeared in the Eocene period, about 60 million years ago. It was actually much more closely related to a group of mammals, the condylarths and creodonts, which became extinct at the beginning of the Eocene, than to modern horses. Note the continued development of new lines from radiations that occurred periodically in the history of this family. The path from *Eohippus* to modern *Equus* is not straight and directional, but quite varied, taking many twists and turns—only a few of which are shown here. Each radiation involved exploitation of a new environment. (After Simpson and Beck, *Life*, 2d ed. New York: Harcourt Brace Jovanovich, 1965.)

million years ago. This small animal, *Eohippus*,* was not much bigger than a fox terrier, with four toes on the front feet and three on the back. Each toe ended in a small modified nail, or hoof. The snout was elongated, but did not have the extended muzzle characteristic of modern-day horses. *Eohippus* had simple teeth, the molars were much like our own, with low crowns, developed, pronged roots, and surfaces covered with cusps rather than ridges. *Eohippus* lived in forests, browsing on soft vegetation. Its multi-toed feet, by spreading the weight out over a broader area, supported the animal on the soft forest floor. The foot structure of *Eohippus*, however, was distinctly horse-like, with digit III showing a clear predominance over the others on both front and rear legs. However, in its largely forest habitat *Eohippus* probably escaped its enemies more by hiding than by running away.

From *Eohippus* to the Modern Horse

The much-simplified phylogeny of the horse shown in Fig. 24.9 highlights several intermediate forms as well as evolutionary dead-ends between *Eohippus* and the modern horse. Two intermediates, *Miohippus* and *Merychippus*, represent the first major divergences in the evolution of the horse family. Both originated from *Eohippus*. *Miohippus* remained in the forests and continued life as a browser, while *Merychippus* became adapted for life on the plains. This shift was not a mere matter of chance. In the Eocene and Oligocene eras, most of the earth enjoyed a warm and humid climate. Tropical evergreen and temperate forests were widespread, providing ample food for browsers. But as time went on, the earth began to cool off and become drier. Abundantly grassed prairies and plains began to appear where lush forests had once stood. *Merychippus* was able to adapt (in the evolutionary sense) to this change —a truly momentous development in the history of the horse family. The plains had few occupants of a size to compete with *Merychippus*. Here, then, was a relatively new and unexploited environment. Anatomically, *Merychippus* showed distinct adaptations to life on the plains. Although still possessing three toes, *Merychippus*'s digit III predominated more clearly than in either *Eohippus* or *Miohippus*. In addition, *Merychippus* was distinctly a grazer. Its molars had distinct ridges and its muzzle was elongated and thickened.

Once established on the plains, *Merychippus* underwent adaptive radiation, giving rise to a number of new species, particularly *Hipparion* and *Pliohippus*. Much of this radiation occurred rapidly in the Miocene (about 25 million years ago) and Pliocene (13 million

years ago) periods. Because they were highly mobile, Miocene and Pliocene horses migrated over considerable portions of the earth's surface. While *Merychippus* remained "at home" in North America, its descendent *Hipparion* migrated from North America to the Old World, colonizing not only in Europe and Asia, but Africa as well. Such wide-ranging migrations were made possible, of course, by the much closer association of the continents at that time (see Section 24.6). *Hipparion* was about the size of a small pony and was still three-toed. Meanwhile, in North America, *Pliohippus*, directly descended from *Merychippus*, was the first horse with a single-toed foot. Evolutionists think that *Pliohippus* is the ancestor of the entire genus *Equus*, to which belong the modern horses (*Equus caballus*, *Equus przewalskii*, and *Equus burchelli*), three species of zebras (*Equus zebra*, *E. grevyi*, and *E. quagga*), and two species of donkeys, or asses (*E. asinus*, the African donkey, and *E. hemionus*, the Asiatic donkey).

The speciation that took place from *Pliohippus* onward may have occurred according to the following scenario. Small populations of *Pliohippus* migrated to South America, Europe, Asia, and Africa. A radiation took place in South America giving rise to several subgroups of modern horses that roamed the flat pampas regions in the area now occupied by Argentina. These groups never developed as separate genera. In Eurasia a general radiation took place; one line appears to have formed a separate group, the donkeys. Another part radiated into Africa and, isolated from its Eurasian relatives, eventually formed still another group, the zebras. The radiation developing from *Merychippus* and later *Pliohippus* was the most extensive in the history of the horse family.

Despite invasion and adaptive radiation throughout Europe by *Pliohippus* in the Pliocene, and by various of its predecessors during earlier epochs, the true or modern horse (*Equus caballus*) evolved only in North America. With the exception of the lines giving rise to donkeys and zebras, most ancestors of the horse that invaded the Old World left no modern descendents. Yet in North America, new radiations and innovations in the horse family continually occurred from the Eocene through the Pleistocene. The major developments of the one-toed hoof and formation of the high-crowned ridged teeth and elongated muzzle all took place in North America. Migrations during the various epochs carried these innovations to Eurasia and Africa. Once there, however, no major adaptive innovations occurred.

About 10,000 or 20,000 years ago the population of horses that had formerly been so widespread throughout North and South America died out. Meanwhile, representatives of the true horses had invaded Europe during the upper Pleistocene. *Equus caballus* ranged over the Eurasian continent and was eventually domesticated. From this European stock, all modern horses

---

Eohippus is the older name more frequently encountered in the popular literature. A taxonomic revision has assigned the generic name *Hyracotherium* to the same group.

have descended. *Equus caballus,* in its domesticated form, was reintroduced into North and South America in the fifteenth and sixteenth centuries by the Spanish conquistadores. The modern groups of so-called wild horses of the United States (referred to as "mustangs") are actually descendents of the original domesticated horses that escaped from captivity and reverted to the wild state some three or four hundred years ago.

It is possible to reconstruct such detailed phylogenetic histories as that of the horse family only when abundant fossil evidence for the many intermediate stages exists. Thus, for example, we can trace the various migrations of *Eohippus, Miohippus, Merychippus,* and their relatives because their fossils appear in large quantities in the Eocene, Miocene, and Pliocene strata of Eurasia, Africa, and the Americas. The widespread appearance of the fossils suggests that at any point in time gene flow must have been possible among populations that lived very far apart (the highly mobile nature of horses may have aided this flow). At the same time, some of the populations must have become isolated, since a number of different species of the genus *Equus* diverged from *Pliohippus* (including one species of modern horse, two of donkeys, and three of zebras). How such isolation occurred, or exactly when it occurred, is not known.

The evidence from geographic distribution of fossil horses raises some interesting and as-yet-unanswered questions about evolutionary history of this important family of mammals. One is the remarkable observation that all the major innovations in horse evolution seem to have occurred in North America, although horses and their ancestors certainly existed elsewhere in the world during most of the last 60 million years. A second interesting question is why horses suddenly disappeared in the Americas some 10,000 years ago. It has been hypothesized that they were killed off by the human population that, at about that time, invaded North America from the Bering Strait. Even though so many details of the evolution of the horse family have been uncovered in past years, there are still major questions yet to be answered.

Horses and the Process of Evolution

The evolution of the horse demonstrates clearly some of the more subtle but important principles of the macroevolutionary process.

First, it is clear from the diagram in Fig. 24.9 that evolution is not a linear process. The path from *Eohippus* to modern *Equus caballus* was not a straight line, like a ladder, with intermediates perched as rungs along the way. The phylogeny of horses consists of a series of branching lines, representing various periods of adaptive radiation. There were many branch-points along the way in which one form diverged in several direc-

tions, most of which ultimately became extinct. Evolutionary developments consist of a more or less haphazard series of trials and errors. Only a very few of all the trials are ultimately successful today. The evolutionary process is not gifted with foresight; it is not teleological, having a goal in mind, a purpose or directionality to it. Looking back on any given phylogenetic history, we can trace the stages through which it passed. But looking ahead, we cannot tell where the next developmental stage will lead. In this fact lies both the mystery and the excitement of evolutionary studies.

Second, the evolutionary history of any species follows a path of ecological adaptation. The modern-day horse evolved from its little *Eohippus* ancestor because of changing ecological (climatic) conditions (disappearance of the humid forests of the Oligocene). Adaptation to changing conditions was necessary; the alternative was extinction. As evolution of the horse family occurred, selection favored the variations that were adaptive to the new conditions of life on the plains. At first, the adaptations were the most basic—*Merychippus,* the first form to live completely out of the forest, showed the most radical changes from its ancestors: a clearly predominant middle digit, bridged teeth, and distinctly elongated anterior skull. Later developments were extensions of these basic initial innovations.

Third, evolution proceeds at varying rates of speed, or tempos. For example, *Eohippus* appears to have been around for a long period of time (perhaps as long as 20 million years) with only slight modifications in its basic phenotype. As *Miohippus* and *Merychippus* moved onto the plains, however, the rate of change appears to have quickened. Within a bare 9 or 10 million years some major trends had developed: reduction of all but the central digit (III), formation of the ridged molars, and distinct elongation of the muzzle. Conversely, the development from *Merychippus* to the now-extinct *Hipparion,* which occupied about 25 million years, was slow and produced relatively little change in physical structure. Paleontologist George Gaylord Simpson has said that evolution occurs in fits and starts, with periods of rapid and often dramatic change alternating with periods of slower change—periods of "revolution" alternating with periods of "evolution." Periods of revolution are almost always associated with major ecological changes—either the climate changes or the organisms migrate to a new habitat (or both). Periods of evolution are associated with more stable ecological relationships, where new innovations in structure and function are perfected, refined, and elaborated. It is clear, however, that both of these processes associated with different tempos of change are integral and inevitable parts of the evolutionary process as we know it.

Fourth, the case of the horse shows clearly how important geographic migrations often are in determining the course of evolutionary history. The continued migration and radiation of horses from North America to

SUPPLEMENT 24.5
THE IRISH ELK: AN EVOLUTIONARY CASE STUDY

Preserved in many museums and private trophy rooms throughout Europe, the British Isles, and to a lesser extent the United States are specimens of a magnificent extinct mammal known as the Irish elk (see Diagram A). The animal was the largest deer the world has ever known. Its shoulder height was about 6 ft and its antlers, with a spread of up to 12 ft, weighed 90 lb.

It has been a matter of much conjecture during the nineteenth and twentieth centuries why this commanding and impressive beast should have become extinct. Over the years a number of hypotheses have been suggested to explain this phenomenon. One of the first was that the Irish elk's extinction was caused by the appearance of *Homo sapiens*. Human beings, as efficient hunters, were thought to have killed the large animals for food. This hypothesis was tested by a review of the paleontological record. If human beings could have caused the deer's extinction, then one would expect to find human bones in the same strata as elk bones, indicating

Diagram A
Romanticized portrait of the Irish elk painted in 1904 by J. G. Millais. The painting highlights the impressive antlers, which often had a full spread of 12 ft. Facing forward, the animal displayed the full palm of the antlers. The painting inaccurately depicts early humans hunting the deer (lower right). Evidence shows that the Irish elk became extinct before human beings invaded its northern European habitats. (Photograph courtesy of Stephen Jay Gould.)

contemporaneous existence. In fact, the geological record shows that the Irish elk had become extinct in certain areas, for example, northern Ireland, before human beings inhabited that region. Humans therefore could not have been responsible for the animal's disappearance.

A second hypothesis has been somewhat more difficult to test. In the late nineteenth century, many naturalists observed that certain extinct species appeared to have what observers saw as "bizarre" structures: exaggerated traits that seemed by all human judgment to be nonadaptive or even harmful. The extremely long teeth of the sabre-toothed tiger and the immense antlers of the Irish elk have been the traits most commonly cited in the past as bizarre structures. One hypothesis for how such bizarre structures developed is called the theory of *orthogenesis*. Orthogenesis holds that, in the early evolutionary history of certain traits, there is a selective advantage in some particular direction—for example, increase in size. After several generations of adaptive increase, however, the structures are able to take on an evolutionary momentum of their own. They continue to increase in size in succeeding generations even though such increase becomes less and less adaptive. According to the theory of orthogenesis, the Irish elk became extinct because in due time its antlers became so unwieldy that the animal's body was less able to support them. Mired down in bogs or clumsily stumbling about, the deer was more easily caught and devoured by its predators.

At the time it was introduced in the 1880s, the theory of orthogenesis was opposed to the Darwinian theory of natural selection on several counts. Strong adherents of orthogenesis maintained that an increase in antler size continued to occur even beyond the point where such an increase was adaptive. Darwin's theory maintains that any trend observed in evolutionary history occurs because that trend is adaptive under a given set of conditions. According to Darwin, the moment any particular structure reached a nonadaptive or harmful state it would be selected against. Darwin and his followers argued that nonadaptive structures would never develop in a species. Darwinians also discounted the orthogeneticists' idea of a structure developing "evolutionary momentum." No one had any idea what such a momentum was; to Darwinians it sounded like a vague name for some unmeasurable force.

Against the theory of orthogenesis, Darwinians advanced the idea of *allometry*. Allometry is the view that as one part of an animal's body changes in size, either during growth of the individual or during evolutionary development through successive generations, other parts change proportionally. The allometric theory is consistent with the Darwinian idea of natural selection in that it holds that all changes that show a steady line of development occur because they are adaptive. To be adaptive, such changes require correlated changes in other parts of the body; accordingly, a series of structures are selected together.

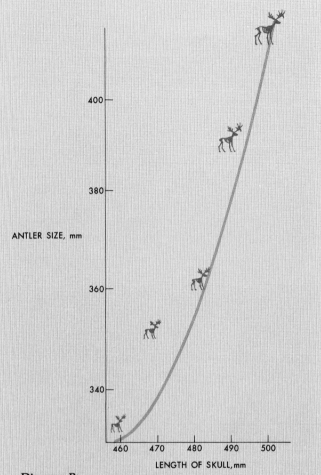

Diagram B
Graph showing relationship between antler size and overall body size as measured by skull length. The graph line indicates that antler size and body size increase together in a regular way. Antler size increases about 2½ times as fast as body size over a series of 5 specimen groups. Antler size is a composite measure of antler length, antler width, and the length of certain tines, or points. (After Stephen J. Gould, "The misnamed, mistreated, and misunderstood Irish elk," *Natural History* 82, 1973, pp. 10–19.)

It is possible to determine whether the allometric or the orthogenetic interpretation of the Irish elk case is more likely. If the allometric theory is correct, then we should expect to find that as the Irish elk's antlers increased in size over evolutionary time, its general body size should show a proportional increase in size. Conversely, if the orthogenetic theory is true, then increase in antler size should not have any necessary relationship to body size. Stephen Gould of Harvard University obtained data that distinguished between the two hypotheses. He noted first that the older versions of the allometric theory were based on a general, qualitative assumption among animal biologists that the size of the Irish elk's antlers increased proportionally to increase in the size of the body. But there was no quantitative data to support the notion. So Gould measured antler size and skull length (skull length is the most reliable measurement Gould could obtain for body size) in 79 skeletons. The results are shown in Diagram B. As the graph indicates, there is a constant, and thus allometric, relationship shown between increase in antler size and increase in general body size. That relationship is not, however, on a 1:1 proportion; rather, antler size increases approximately 2½ times faster than body size. To determine how unusual this proportion might be, Gould checked data for allometric relationships among a number of species of living deer. The same general relationship held: if species of deer are arranged in order of increasing size, antler size increases about 2½ times faster than body size. Since even the largest deer known today have no difficulty holding up their heads, and are in fact very agile and graceful animals, Gould concluded that the Irish elk's body was quite adequate to support even a 12-ft, 90-lb pair of antlers.

If the allometric theory is correct, however, it raises an interesting question for the Darwinian concept of evolution. Of what selective advantage could such a large antler spread be? It is too much to believe that such a massive structure, regrown every year, would have no function. The first and most prevalent hypothesis in the nineteenth century was that the antlers served as weapons in the eternal struggle for existence. This was a natural assumption during a century that glorified the concepts of competition and the "struggle for existence." However, Gould and others have made an observation that throws doubt on this interpretation. The tines, or points, on the antlers are pointed *backwards*—hardly a very useful weapon! In fact, Gould has emphasized that for the tines to be pointed forward to pierce an enemy, the deer would have to hold its head down between its legs.

If the antlers of the Irish elk are not weapons, what function might they serve? Darwin hinted at the possibility that antlers in general were display devices for attracting females. This hypothesis is quite plausible, but recent field studies among modern deer species suggest another, more interesting idea. According to Valerius Geist of the University of British Columbia, antlers and horns may serve as "visual dominance-rank symbols" among males of a species. Large antlers confer high status on their bearers; they intimidate other males and thus provide more ready access to the female (whether or not the larger antlers in themselves are specific attractants to the female is still a matter of conjecture). In situations of potential conflict between two males over a female, antlers and horns may serve to prevent fighting and possible injury, rather than being used as actual weapons.

If Geist's interpretation is correct, the Irish elk's antlers are admirably constructed to serve the purpose of symbolic display. The smaller deer of today must rotate their heads from side to side in order to show the full antler palm. Such a motion would have created problems for the Irish elk. The torque (rotational motion) produced by swinging 90 lb spread out over a 12-ft spread would have been enormous. But the Irish elk had only to look straight ahead to display its full palm (see Diagram A).

What, then, may have caused the Irish elk to become extinct? Gould's hypothesis is based upon paleontological data. It appears that the Irish elk became extinct at about the time of the last glacial period, some 11,000 or 12,000 years ago. The large deer normally inhabited the open tundra and fields of northern Europe and the British Isles. When it was driven south by the advancing ice cover, the deer encountered dense forests, where its large antlers were a decided disadvantage. What had been adaptive as a ritual display mechanism in the old environment became a distinct disadvantage in the new. Thus the Irish elk—to Gould— became extinct because of changed environmental conditions under which one of its specialized traits, giant size, was no longer adaptive.

other continents greatly affected the rate and direction of evolution. In fact, had forerunners of modern *Equus* not migrated to the Old World some 10,000 or more years ago, horses as a group might be extinct now. For, as we have seen, although horses flourished for 60 million years in North America, they appear to have died out on this continent in the recent past. It is quite frequently the case that widespread geographic distribution of a species not only produces a variety of evolutionary lines, but also is inadvertently a safety factor against extinction. Species that live in a single, restricted geographic area have little protection against rapid environmental changes to which they cannot adapt.

Fifth, as we have emphasized, evolution acts on the totality of the phenotype, and thus on the totality of the genotype. The evolution of the hoof, the tooth structure, and the muzzle did not occur as three isolated developments, each evolving separately. Selection, as it favored one organism's phenotype over another, was really acting on the total set of genotypic processes within that individual that had contributed to the particular phenotype. In turn, by acting on the totality of an individual's genotype, selection acts on the totality of the population's genotype. Genes that are incompatible with one another are theoretically subject to negative selection as much as genes that are incompatible with the environment. In the evolution of horses, selection pressures favored those combinations of traits that better adapted their bearer to a changed ecological habitat. The evolution of teeth cannot be separated from the evolution of leg, neck, and jaw. After all, the most effective grinding teeth in the world are of little adaptive value if food cannot be brought into contact with them effectively, or if the organism cannot escape its predators.

Sixth, and last, organisms can be viewed as evolving between two opposing forces: specialization and flexibility. Natural selection ultimately weeds out those who tend too much in either direction. The fact that in the history of life so many species have become extinct emphasizes how difficult it is for organisms to maintain a balance between these two pressures.

Summary

1. Speciation is the process by which one or more species arise from previously existing species. Divergent speciation involves the splitting up of a species into two or more subpopulations which are, or become, reproductively isolated from one another. After a long period of separation, each subpopulation will undergo changes and evolve into a new species.

2. There are several modes of divergent speciation: (a) allopatric speciation, where two (or more) subpopulations become geographically isolated, and diverge along separate lines; (b) parapatric speciation, where two or more populations develop a reproductive barrier, at the same time they are also diverging in other respects, for example ecologically, but without geographic barriers; (c) sympatric speciation, where two or more populations become reproductively isolated prior to any other major divergence, but while still living in the same geographic area; and (d) abrupt divergence such as that resulting from polyploidy.

3. Once reproductive isolation has been achieved it is maintained in several ways. Premating isolating mechanisms include: (a) temporal isolation—two species reproduced at different times of the year or times of the day; (b) ecological isolation—two species capable of interbreeding live in very different niches; (c) physiological isolation—copulation is possible but the sperm and egg of the two species are incompatible; (d) behavioral isolation—specific behavioral cues (such as courtship rituals) are required to bring about copulation; and (e) mechanical isolation—copulation is physically impossible because of anatomical differences.

4. Postmating isolating mechanisms include: (a) hybrid mortality—the hybrid embryo does not develop to full term; (b) hybrid sterility—the hybrid is viable, but cannot produce offspring; and (c) reduced hybrid fitness—the hybrid can reproduce, but its offspring are nonfertile or significantly less fertile than nonhybrid offspring.

5. Adaptive radiation is the process by which an ancestral species undergoes considerable divergent speciation in an environment where there are many resources to exploit. The number of species that evolve depends on the number of resources available. A common example of adaptive radiation is found in Darwin's finches on the Galápagos Islands.

6. A new approach to animal (and plant) phylogeny consists of investigations at the molecular level. Based on the concept that changes in genes represent changes in DNA, and that chances in DNA are observed as changes in the amino acid sequence of the proteins coded for, molecular evolutionists study the differences in amino acid structure of related protein molecules from different species in an attempt to discover how closely the species are related. For example, differences in the amino acid sequence of the respiratory protein cytochrome c, as observed in a variety of vertebrate species, can be used to determine the evolutionary divergence of these species. The greater the difference in amino acid sequence, the longer ago the species diverged.

7. Natural selection favors survival of the fittest organisms—those best adapted to their conditions of life. The characteristics that serve to adapt an organism to its environment are referred to as adaptations. Some adaptations are anatomical, others physiological, and still others biochemical and behavioral. All types of adaptations interact with one another to contribute to the fitness of the organism in its natural habitat. In addition, both physiological and biochemical adaptations have a structural or anatomical basis.

8. Evolution is a continual process; no group of organisms is ever perfectly adapted to its environment, hence there is always the tendency and potential for greater adaptation. In addition, environments are continually changing; therefore traits that are adaptive today may not be adaptive tomorrow. Clearly, no organism can ever be said to have reached the end of its evolutionary development, until it has become extinct.

9. The origin and development of the horse provides an example of the evolution of adaptation over geological time. Modern horses are admirably adapted to the flat, treeless plains and fields that form their natural habitat. Included among the adaptations that suit their lifestyle so well are: (1) formation of a long, thin leg with muscles attached to the proximal end and formation of a hoof, by elongation of digit III and loss of the other digits, to support the leg, permitting rapid running on hard ground; (2) elongation of the anterior portion of the skull into a muzzle enabling the horse to reach down to the grasses on which it grazes; and

(3) development of premolars and molars into massive, high-crowned, ridged, and continuously growing grinders, which are well-suited to a grazing lifestyle.

10. The evolution of the horse illustrates a number of important principles of the evolutionary process.

   a) Evolution is not linear or goal-directed. The evolution of the horse consisted of a haphazard series of trials and errors that attempted to adapt the horse to conditions of a changed environment.

   b) The evolution of any species follows a path of ecological adaptation. Changing ecological conditions (the disappearance of lush forests) made adaptation necessary for *Eohippus*.

   c) Evolution occurs at varying rates of speed. *Eohippus* existed in the forest for a long period of time with only slight modifications in its phenotype. As *Miohippus* and *Merychippus* were forced onto the plains, however, the rate of change quickened substantially.

   d) Geographic migrations are important in determining the course of evolutionary history. If the forerunners of the modern horse had not migrated from America to the Old World, they may well have become extinct.

   e) Evolution acts on the totality of the phenotype and genotype of both the individual organism and of the population. Selection pressures in the evolution of the horse favored not one or another single trait (toes, teeth surfaces), but combinations of traits that better adapted their bearer overall to a change in the environment. The evolution of teeth suited to grazing cannot be separated from the evolution of an elongated muzzle for reaching the ground.

Exercises

1. List the major types of premating and postmating isolating mechanisms. Why is isolation in some form absolutely essential for divergent speciation to occur?

2. Define adaptive radiation.

3. Explain how the Galápagos finches seem to have developed from a single common ancestor. Discuss the stages by which this could have occurred. In your discussion include the role(s) played by the following: (a) adaptive radiation; (b) competition; (c) isolation (of various types); (d) availability of various food and habitat patterns on the islands.

4. Describe the process of determining phylogenetic relationships by comparative studies of protein structure. Where such studies have been done to date, are the results similar to or different from those obtained by more conventional methods of comparative anatomy and paleontology?

5. In what way, or ways, does the phylogenetic history of the horse illustrate the principles of evolution of adaptations?

Chapter 25
Ecological Relationships:
The Interaction of Organisms
and Their Environments

25.1
INTRODUCTION

All organisms live in nature in a close relationship with a great many other organisms. The living things in any given area (pond, lake, river, ocean, or forest) form part of a biotic community. In a biotic community the existence of each population, as well as that of each individual, is governed to some extent by the presence of all the others. The biotic community forms a web the structure of which depends on every single strand. If just one strand of this web is changed, the structure might assume an entirely different character.

Introduction of nonnative organisms often allows for rapid growth of the new species while some native species decline in abundance. Without natural enemies, the new species reproduces rapidly and can do extreme damage to the native biotic community. The natural history of many parts of Australia, for example, has been altered temporarily, or perhaps even permanently, by the rapid spread of jackrabbits after their introduction in the last century (see Section 24.5).

The study of the various relationships between organisms and their environment is known as **ecology.** The environment of any organism includes two major aspects, the abiotic (nonliving or physiochemical components) environment and the biotic (living components, other organisms) environment. The abiotic environment includes the amount of light, moisture, wind, water current, pressure, temperature, and acidity, and the presence or absence of various minerals. The biotic environment, on the other hand, includes all the living organisms with which an animal or plant comes into contact.

Before proceeding further with our discussion of the interactions among organisms in ecological systems, it will be helpful to consider several important definitions.

25.2
DEFINITIONS
Populations, Communities, and Ecosystems

The biotic environment is not merely a random array of species and populations, but a highly organized set of interacting groups of organisms. It is useful to examine the biotic community at three levels of organization: populations, communities, and ecosystems (see Fig. 25.1).

Populations, which we have already defined in the previous chapter as interbreeding groups of individuals of the same species, represent the lowest level of organization in the ecological hierarchy (other, of course, than individuals). **Communities,** the next higher level in the hierarchy, consist of all the populations living in a given geographic area (it could be as small as a tidepool or rotten log, or as large as a lake or open prairie). The term community is restricted just to the groups of organisms in a given area. **Ecosystems,** the highest level of organization, are composed of all the communities in an area and their physical environments, taken as a whole. Ecosystems tend to be defined in such a way as to include all the components necessary for self-sufficiency. Ecosystems can be stable and self-maintaining, or they can be evolving and changing. Like the terms "population" and "community," the term "ecosystem" can be applied to small or large spatial or geographic units. On one end of the size scale, a closed, balanced aquarium is a small but complete ecosystem. It is self-sufficient to a large degree, and recycles its material contents (minerals, organic material) and receives only radiant energy (light) from the outside (that is, outside the system itself, the aquarium). On the other end of the size scale, a lake, large forest, or the south Pacific ocean can also be considered an ecosystem, with many more interacting parts.

Ecology is the branch of biology that studies the interactions between organisms and their environments.

Fig. 25.1
The three levels of organization in an ecological system: the population, composed of a group of interbreeding organisms with all the same basic living requirements (food, habitat, moisture, light), shown to the left; the community, consisting of all the populations living in a given area (here, a field, shown at left and center); and finally the ecosystem, consisting of all the communities in an area and their physical environments (here, showing both field and forest communities).

Populations, communities, and ecosystems form a hierarchy in the natural world—that is, with populations as the least inclusive and ecosystems the most inclusive units. Ecosystems consist of interacting communities (plus the sum total of abiotic factors), and communities consist of interacting populations (of one or more species). We can visualize the hierarchical relationship as follows:

Ecosystem
  Community
    Population

Ecological Habitat and Niche

The concepts of habitat and ecological niche are very useful in describing the various relationships between organisms.

The **habitat** is the place in which an organism lives in the biotic community. The term may refer to an area as large as the ocean or desert, or as small as the underside of a lily pad or the intestine of a termite. Ecologists sometimes speak of habitat as the "address" of an organism within the community. It is a place that contains suitable supplies of such required resources as food, water, and nesting sites.

The **ecological niche** occupied by an organism is harder to pinpoint than its habitat. The term "niche" refers to the role an organism plays within the biotic community. To what organisms does it serve as food; upon what organisms does it feed? What are the upper and lower limits to its toleration of changing temperatures? What minerals does it require from the environment? What minerals does it return to the environment? Answers to such questions help establish the exact niche an organism occupies. Just as an organism's habitat is spoken of as its "address" within the biotic community, its ecological niche is called its "profession." Unlike its habitat, an organism's niche is defined by all the physical, chemical, and biotic factors that influence an organism's maintenance and reproduction.

Because of the different ways they use the habitat, organisms may live together in the same general habitat yet have quite different ecological niches. Tidal pools near the ocean are a good example. They contain a wide variety of organisms: animals such as starfish and sea anemones and plants such as "seaweed" and smaller filamentous algae, all of which have roughly the same habitat. Yet within a single tidal pool the algae serve as producers, since they can manufacture carbohydrates by using energy from the sun, and the animals serve as consumers. The larger ones, such as starfish, feed on a variety of smaller animals that ultimately feed on the plants.

Consider another example. In the shallow water along the shore of a lake it is possible to observe a large variety of water insects, all of which have the same habitat. Some of these, such as the "backswimmer" (genus *Notonecta*), serve as predators and feed on other small animals. Other backswimmers, of the genus *Corixa*, serve as decomposers and feed on dead or decaying material. Both organisms have the same habitat. Their ecological niches within that habitat, however, are quite different. It is a general rule that each species in a given habitat occupies a different ecological niche from all others.

25.3
THE NATURE AND METHODS OF STUDY OF POPULATIONS

We can begin the study of ecological principles with an examination of populations in nature and the methods of studying them. In the previous two chapters we have been concerned with populations largely in a genetic sense; that is, as groups of interbreeding organisms characterized by certain gene frequencies. In this chapter we will be concerned also with populations, but in a much larger context. We will focus our attention on how organisms interact with other individuals within their own population, with members of other populations, and with the abiotic environment. In particular,

we will approach the study of populations in terms of the factors that regulate their growth and size.

Population Distributions

Two of the most important characteristics of populations and their study are population size (number of individuals) and the distribution of individuals within the populations' geographic area.

Population size (that is, total number of individuals), has an important bearing on both the evolutionary and ecological characteristics of a population. Small populations generally harbor less genotypic and phenotypic variation, and hence have less of a range of adaptations or potential adaptations to a wide variety of environments than larger populations. As a result small populations are more restricted in the community and ecosystem interactions in which they can take part. Because larger populations generally harbor more genetic variability, contain a greater range of actual or potential adaptations, they can take advantage of a more varied set of environmental conditions. Population size can thus be a crucial factor in the ecological history of a population.

There are two important aspects of distribution within a population that we will consider separately: **density** and **spacing**. In many ways these two are related, but they are not synonymous terms.

Density. Density refers to the number of organisms of the same kind per unit of space (expressed as square areas, such as mm^2 or km^2). It is an objective and quantifiable measurement. But its meaning and significance in an ecological sense depend upon other factors, most particularly the kind of organism and the kind of environment involved. For example, it is obvious that the same density of deer and of mice per square kilometer would have vastly different consequences for the vegetation in the area. Similarly, high density of a given kind of organism has one ecological consequence in an environment with plentiful food, quite another in an environment with limited resources.

Spacing. Spacing refers to where the organisms within a given area are actually located. There may be 1000 organisms within a square kilometer, but they are not likely to be evenly spaced throughout. Spacing may be related to localization of resources (water sources, living spaces, food supply), to changing environmental conditions (wind velocity on different parts of a mountain slope), or to social interactions among the organisms (as in herding among many mammals). Knowledge of density without knowledge of spacing, and vice versa, can lead to incomplete or erroneous interpretations of how a population of organisms is interacting with its environment. Knowledge of both spacing and density thus provides a starting place for understanding the structure of

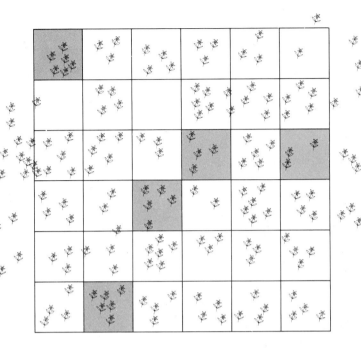

Fig. 25.2
Quadrature method of sampling the size and density of a population within a given area. The entire area to be included in the sample is contained within the large square. This area is subdivided into a number of equal squares, or quadrats (thirty-six are shown here). Several of the quadrats from different parts of the larger area are chosen for study (color). The number of organisms (of whatever type are being studied) within each quadrat are counted and the totals averaged for all the sampled quadrats. The total number of organisms is averaged among the six quadrats (there are twenty-four organisms in the six colored quadrats, or an average of four per quadrat); if there are six quadrats with an average of four organisms each, and a total of thirty-six quadrats, the estimated population size is 4 × 36 = 144; the density is expressed as 144 organisms per 36 square meters, kilometers, or whatever.

populations and the interactions that occur both within and between them.

Methods of Determining Population Density and Spacing

In most cases, population density and spacing are determined by estimates made from population samples. It is seldom feasible to actually count the number of organisms in a natural population (it is, of course, easier to do this with a laboratory population), or to chart precisely the spatial location of individuals within an area. Most frequently, population biologists will take samples from a given population and from these data form an estimate of the size, density, and distribution of the entire population. Consider briefly how this can be accomplished.

Estimating Population Density and Spacing. There are two basic methods for estimating population density: sampling and mark-and-recapture methods. Sampling involves staking out an area for survey or study and dividing it into a gridwork of equal parts (called squares or quadrats as shown in Fig. 25.2). Several of the quadrats are then examined, the number of organisms in each counted, and an estimate of the whole made from the sample (for further details see caption to Fig. 25.2).

Mark and recapture techniques involve first capturing a sample of the organism whose numbers are to be estimated (for example, fifty individuals). The fifty individuals are then marked in some way (a tag or spot of paint) and released back into the environment. Later (a day or two, or a week later) a second sample of the same organism is collected in the same place, and the number of marked *versus* unmarked individuals is recorded. Suppose that, of a second sample of fifty individuals, ten marked forms were recaptured (ten is 20 percent of fifty). Since we know we released fifty marked individuals, we could estimate the total population in the area to be approximately 1000. (Fifty marked individuals were released; 20 percent of the recapture sample was marked; fifty is 20 percent of 1000.)

The resulting data from such sampling procedures give an estimate of the number of individuals. To convert that number into density we simply have to specify the area over which the sample was taken. Thus, density is always expressed as n/area, where n = number of organisms.

The spacing of organisms throughout an area can be determined by the same sampling methods as described above for density. In this case, however, the most meaningful data could be actual numbers within each quadrat, rather than averaged out over a group of quadrats or over the area as a whole. A spacing pattern might look exactly like the diagram shown in Fig. 25.2, with the samples from each quadrat indicated on a grid map of the area.

Distribution studies have shown that the spacing of organisms throughout an area is seldom either uniform or entirely random. The most common form of distribution is called **clumping,** in which organisms (of the same kind) tend to be found clustered together in some parts of an area, and absent from or in a lower density in other areas. There are several reasons why clumping is so prevalent in most populations: (1) Reproductive patterns in both animals and plants favor clumping. In animals, males and females must come together to breed, and often stay together in large groups (herds, flocks, or colonies) throughout much of the year. In plants, vegetative reproduction (nonsexual, such as formation of runners along the ground, from which a new plant develops) or patterns of seed distribution (in which seeds often tend to fall close to the parent plant) also lead to large-scale clumping. (2) Environmental conditions, especially distribution of resources, are seldom uniform throughout a particular area. Plants grow where soil, water, and light conditions are favorable; animals live where there are places to build nests, or to find water and food. Even a relatively small field or woodland area contains a number of different habitats or micro-environments that can greatly affect distribution of organisms within the area as a whole. (3) Animals often show degrees of social organization that produce clumping: for example, they group together for protection, foraging, reproduction, and rearing of young.

25.4
POPULATION GROWTH AND LIMITATION

Under optimal conditions, a single pair of fruit flies, producing an average of fifty to sixty offspring per generation, could yield 1.4×10^{55} offspring in the course of a year (approximately twenty-six generations/year). One hundred starfish could produce 10^{79} offspring in just fifteen generations. (That's more than the number of electrons in the visible universe.) A single bacterium, multiplying by fission, could produce 250,000 descendants in six hours! And, Charles Darwin estimated that a single pair of elephants, the slowest breeders of all animals known,[*] would be capable of reproducing 19 million (19×10^6) offspring in 750 years!

Obviously populations do not grow to such extremes. Starfish, bacteria, and elephants do not overrun the earth. Population size is regulated by the constant interaction between two opposing factors: the biotic potential of the species, and environmental resistance of the ecosystem. Environmental resistance is, in turn, broken down into two different forms: (1) density-dependent factors, that is, factors that derive from the increasing density of the population as it grows; and (2) density-independent factors, environmental changes such as floods or climate shifts, which happen totally independent of the density of the population, but obviously can affect its growth. We will examine each of these topics: biotic potential first, then both forms of environmental resistance.

---

[*]Darwin assumed each female reproduced between the ages of 40 and 90, giving birth to six offspring during that time.

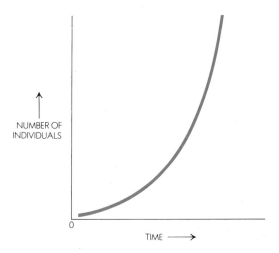

Fig. 25.3
Exponential growth curve for a population under ideal conditions (that is, unlimited food and other growth requirements, unlimited space, and with maximum offspring survival). The curve is said to be "exponential" because the population continues to grow at an increasing rate until the rate of growth becomes infinite. Reaching this "infinite" rate of increase is, of course, only a theoretical consideration.

Biotic Potential, or Intrinsic Rate of Increase

Under optimal conditions all organisms have the potential for extremely rapid population growth. The growth curve for organisms under optimal conditions is shown in Fig. 25.3 and is said to represent *exponential growth*. Note that as time goes on the curve becomes increasingly steep—indicating that the *rate of increase* in numbers is becoming faster and faster. The rate can be expressed as an exponential series—that is, with numbers increasing by the exponent rather than base value of a number. For example, in the expression 2^3, 2 is the base number and 3 the exponent, so $2^3 = 8$. An exponential series with 2 as the base number and the exponents increasing as 1, 2, 3, 4 . . . would thus yield 2^1, 2^2, 2^3, 2^4 . . . and the value of each expression would be, respectively, 2, 4, 8, 16. . . . The value for each number in the series goes up **exponentially,** instead of merely arithmetically (that is, 1, 2, 3, 4, 5 . . .). Note that the *rate of increase* (not merely the total numbers) becomes greater and greater with succeeding generations. Eventually the rate of increase approaches infinity—the graph line becomes almost vertical.

The rate at which a population of organisms reproduces itself under optimal conditions is an expression of the species' **biotic potential.** Some organisms such as jackrabbits have a high biotic potential; others, such as whooping cranes, have a low biotic potential. In the abstract, neither a high nor a low biotic potential (except at either extreme, near infinity or zero) is necessarily more adaptive than the other. How adaptive a certain biotic potential is depends upon many factors of the environment, including the presence of other species as food sources or as predators, and temperature and moisture cycles. In some environments it might be more adaptive for a species to produce very large numbers of offspring (most of which die before reaching reproductive age) while in other environments it might be more adaptive to produce fewer offspring, many or most of which survive to reproduce.* Despite the tendency to think that a high biotic potential is always more adaptive, recall that the chief criterion of adaptations in Darwinian terms is always differential fertility: the passing on of genes not only to the next, but to succeeding generations. A number of strategies under different environmental conditions can all be seen as adaptive in this general, evolutionary sense.

The biotic potential of a species determines the rate of increase that a population shows from generation to generation. The rate of increase, I, for any growing population can be expressed mathematically as

$$I = rN \qquad (25.1)$$

where r is the average birth rate minus the average death rate (or, $b - d$)†, and N is the number of individuals in the population at any moment in time. The equation states simply that rate of increase is equal to the average birth rate minus average death rate multiplied by the number of organisms in the population at that moment. If empirically derived measurements for any population are fed into the above equation for successive time periods (for example, successive generations) the value for I can be determined for successive time periods.

It is obvious from the above equation that any population will show an increase in numbers only if average birth rate exceeds average death rate (that is, if $r > 0$). A population will be stable, or maintain its numbers at a constant value, if average birth rate equals average death rate (that is, if $r = 0$). And of course, a population will actually decline if average birth rate is less than average death rate (that is, if $r < 0$). The value for r, then, tells us whether a population will grow, remain stable, or decline. With numerical values which go into calculating r in any stance (that is, the values for $b - d$), we can learn something about how that growth,

stability, or decline is brought about. For example, a perfectly stable population can be achieved by a high death rate matched by a high birth rate or a low birth rate and a low death rate. Similarly a fast-growing population can be achieved by an extremely high birth rate matched by a moderate death rate; or by a moderate birth rate matched by a very low death rate. Thus the term r is referred to as the **intrinsic rate of increase** of a population and is synonymous with biotic potential. The term r is *intrinsic* to the population (species) under consideration because it is a feature of the species' biological makeup.

Note that the equation $I = rN$ tells us why the exponential growth curve shown in Fig. 25.3 turns upward with increasing sharpness. The value for I is dependent not only on r, but also on N. While r can remain more or less the same throughout a population's growth, N is constantly changing. As more individuals are reproducing in a population, N continues to increase, and hence the value of N increases. The curve turns upward faster and faster.

Environmental Resistance, or Carrying Capacity

Opposing the effects of intrinsic rate of increase (r) is what biologists call **environmental resistance**. Environmental resistance comprises all those factors in the environment that tend to prevent a population of organisms from multiplying at an unlimited rate: limitation of food supply, competition with other organisms, predation, or the effects of climate. All these limiting factors act together to restrict population growth.

As a convenient measure of environmental resistance, every environment has what is known as a **carrying capacity**. Carrying capacity refers to the maximum number of individual organisms of any certain type that the environment can support over a long period (without degrading the environment itself). Carrying capacity includes all those factors in any particular environment (such as predation, competition with similar types of organisms, available food supply, or light, minerals, and water) that tend to restrict birth rate. Thus, environmental resistance is an intrinsic property of neither the environment nor the species, but of the *interaction* between them. If either changes, the carrying capacity may change.

Most populations show a growth curve similar to that in Fig. 25.4. This type of curve is known as a **logistic growth curve**, and is characterized by the generalized S-shape, a result of the effects of environmental resistance on intrinsic rate of increase. Note that the curve starts out very much like an exponential curve, turning upward only slowly at first (acceleration phase), but with an increasing rate as time goes on. Eventually it reaches a maximum, and momentarily stable rate of increase (inflection point), after which there is a gradual decline (deceleration phase), and eventual leveling off (steady state). As various forms of en-

---

*Ecologists and evolutionists are fond of designating two very different adaptive strategies in the living world with regard to biotic potential: what is called r and K strategies (or r and K selection). High biotic potential characterizes r strategy; these are organisms that produce many offspring quickly, and are successful in "boom and bust" environments, like deserts where rainfall is infrequent but copious when it comes, or flood plains of rivers which experience periodic disturbances (such as flooding). K strategy is more adaptive in stable environments, and is characterized by a lower biotic potential, but with greater survival rate of the offspring that are born.

†Birth and death rates are always given in per capita figures, for example, birth rate per 1000, or 10,000. This allows for birth and death rates to be compared between populations regardless of differences in population size.

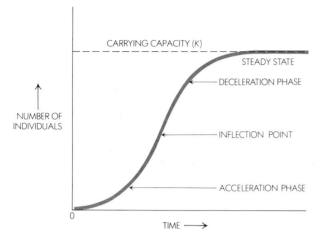

Fig. 25.4
The logistic growth curve, showing the early acceleration phase, where birth rate exceeds death rate; the inflection point where rate of growth of the population is constant; and the deceleration phase where death rate approaches the value of birth rate. At equilibrium or steady state, birth rate and death rate are equal. The carrying capacity (dotted line), K, indicates the upper limit in population (of a specific type of organism) that a given environment imposes. K occurs at various positions on the vertical axis depending upon the type of organism and the particular environment involved.

vironmental resistance come into play, the birth rate begins to decline, so reducing the value of r in our previous equation. As the value for r comes closer and closer to one, the curve levels off and we reach the equilibrium point. The equilibrium point indicates that birth rate and death rate are approximately equal. The occurrence of the equilibrium point at or near the carrying capacity is a result of organisms using the resources of the environment close to the limit.

We can express the *dynamics* of the logistic growth curve also by an equation:

$$I - r\left(\frac{K-N}{K}\right)N \tag{25.2}$$

where

I = rate of increase in the number of individuals
r = intrinsic rate of increase or biotic potential
N = number of individuals in the population at any moment
K = environmental resistance, or carrying capacity of the environment

Equation (25.2) is nothing more than Eq. (25.1) with the insertion of a new *limiting term*, $(K-N)/K$, which is equivalent to environmental resistance, or carrying capacity. It is limiting in the following way: when N is very small relative to K (when the population density is low relative to the carrying capacity), the value of $(K-N)/K$ is close to one. Therefore the population can

grow at almost its maximum rate (that is, rN). But as N approaches K in value (as population size approaches the carrying capacity of the environment), the value of $(K-N)/K$ falls; it becomes a fractional number, reducing the value of the two terms, r and N. This means, of course, the value for I decreases—the population growth rate slows down. A population that has reached its equilibrium point would have $r = 0$ and is said to show *zero population growth*.

By inserting the new term $(K-N)/K$ into Eq. (25.1), we have shown how a logistic growth rate (or curve) is a product of the factors generating an exponential curve (intrinsic rate of increase) plus environmental resistance. These two opposing factors constantly interact in a dynamic way throughout all phases of any population growth. In the earlier phases of population growth, biotic potential (r) is the dominant factor, with environmental resistance, $(K-N)/K$, being secondary. The two factors balance out at the inflection point, while environmental resistance comes more and more into play as the population size nears the carrying capacity, K.

The value of $(K-N)/K$ is said to be **density-dependent**. This means that environmental resistance comes increasingly into play as the density (a function of population size in a finite area) of a population increases. Population growth rate decelerates and eventually reaches a plateau (zero population growth) because density reaches a point where environmental capacity (resources) becomes limiting, that is, the environment can support no greater number of organisms (of that particular kind). The effects of environmental resistance on the value of r is thus density-dependent.

The dynamic interaction between biotic potential and environmental resistance can be seen in the actual growth curve for a population of sheep in Australia for a ninety-five-year period (1840–1935), as shown in Fig. 25.5. The data points give the actual number at each count; the black line the general trend. The colored line shows the generalized direction of the population growth—as it clearly follows a logistic pattern. Note, however, that the actual numbers tend to oscillate back and forth around the plateau, which may represent the carrying capacity, or somewhere below the carrying capacity, of the environment (the plateau value here is equal to a population density of 6500 sheep). Consider, for example, the data for the years 1885 onward. In that year the population rose to about 7500; in the following fifteen years the number steadily declined to about 5000 before beginning to rise again. It reached the plateau level by about 1905, only to fall again to less than 4000 by 1915. By 1925, the population was back above carrying capacity, down again by 1930, and back up by 1935. As the population goes over the carrying capacity, environmental resources (food supply, among other factors) become limiting, competition becomes severe, some or all organisms in the population reproduce less well, and the population size declines. Decrease in

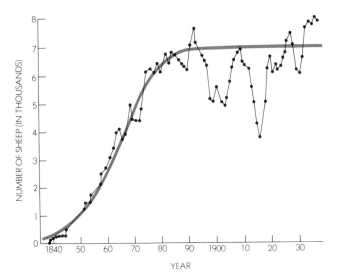

Fig. 25.5
Actual changes in population density of the sheep population of South Australia from 1840–1935. Note that the population growth follows the general pattern of a logistic curve. Note also that after the population begins to level off (presumably near the carrying capacity of the environment), the actual density oscillates around the plateau value (about 7000 sheep). This oscillation is a result of the interactions between biotic potential and environmental resistance (expressed here as carrying capacity). See text for a detailed analysis. (Data originally from J. Davidson, in *Transactions of the Royal Society of South Australia*, 62 (1938); modified from William T. Keeton, *Biological Science*, 3d ed. New York: W. W. Norton 1980, p. 827.)

could easily reach a point where recovery would be difficult. One of the most important jobs of those who work in game and fish management is to keep tabs on population growth rates and sizes. Limits on cropping—the process of harvesting excess animals from a population—can be increased or decreased according to how much the estimated density exceeds the point of maximum sustained yield. Thus, a knowledge of population dynamics is essential to sound management of game and commercial animal resources.

Idealized Curves and Real Populations: A Cautionary Note

We should add a cautionary note at this point with regard to generalizing from the logistic curve to real populations in nature. Few populations follow an idealized growth curve such as that shown in Fig. 25.4. One of the reasons is that an idealized curve shows population growth for only a single species. In reality, of course, populations of many kinds of organisms interact in a natural ecosystem; hence the growth for any one is usually influenced by that of at least some, if not all, of the others. Growth curves of real populations are shaped by more than a single species' biotic potential and a given environment's carrying capacity.

Another reason why few population growth curves fit the ideal logistic pattern is that such a pattern assumes a more or less uniform, or constant, age distribution within the population. For example, two populations of the same size will follow very different growth patterns if one has a large number of young organisms and another a large number of old organisms. The human population in different countries at present is a good case in point. In the United States in 1970, for example, approximately 30 percent of the population was under 20 years of age, and about 55 percent between 20 and 60. By contrast, in 1970, 55 percent of India's population was under age 20, and 40 percent was between 20 and 60. The growth rates of the two populations subsequently reflected this difference in age structure. Because it had more people in the prime reproductive age group in 1970, India's population growth rate continued to be much greater between 1970 and 1980 than that of the United States. Other factors, such as N, or total numbers (which in India are also much greater), as well as socioeconomic factors, have also contributed to India's continually high population growth rate.

With regard to projections about future human population growth rates it is especially important to avoid making wholesale generalizations from either idealized exponential and logistic curves, or existing population growth data (see Supplement 25.1). Even when the more quantitative data such as age structure and present growth rates are known, other factors, primarily cultural (social, political, and economic) are important determiners of population growth rates in human beings. For example, the shift from urban to

numbers allows more resources per animal, and hence biotic potential once more manifests itself. The remaining organisms produce more offspring, and the population density begins to increase once again. By this kind of seesaw, or oscillating pattern, the population density can often oscillate around some plateau density at or below the carrying capacity of the environment.

At the inflection point in a logistic curve, the graph line shows its steepest incline, indicating that the *rate* of population increase is greatest (for that particular population and environment). The inflection point is also called the point of **maximum sustained yield,** a term that indicates that the population is growing fastest at this point in time. Knowing when a population is at its point of maximum yield is important in the management of game and fish populations. Experience has shown that for a population continually to yield a harvestable crop, it is important to keep the population numbers from falling below the inflection point. If the total number falls much below the point of maximum sustained yield and especially if it is kept there (for example, by continual overkill by hunting or fishing), the population

SUPPLEMENT 25.1
POVERTY, HUNGER, AND OVERPOPULATION: A DEBATE

In recent years much publicity has been accorded to what many people consider a major problem confronting the human race: too many people. Newspapers, magazines, and a number of national and international organizations have argued that much of the world's poverty and hunger is related to overpopulation, and they view the exponential growth of the human population with alarm (Diagram A). A leading exponent of this view is Stanford University biologist Paul Ehrlich, who claimed in *The Population Bomb* (1968) that, unless steps were taken immediately to reduce the birth rate, starvation on a mass scale would result. And, indeed, starvation does exist in many parts of the world today.

The issue has been subject to considerable debate and controversy. Some persons in the wealthy nations, as well as large number of representatives of the poorer countries of the world, maintain that poverty is not necessarily a result of overpopulation, but rather derives from specific social and economic conditions.

What are the two sides of this debate? First consider the observations, or "facts," upon which both sides agree.

1. The growth rate of the human population as a whole is on the upswing and has been for several generations (see Diagram A).

2. The birth rate in "Third World" countries of Asia, Africa, and Latin America is often considerably higher than in the Western, industrialized countries (Diagram B).

3. All these Third World countries have primarily agricultural economies.

4. Many of these countries experience chronic poverty and hunger.

The problem to be explained is: What is the relationship, if any, between population growth rates and the chronic hunger and poverty that affect so many of the Third World countries? Several hypotheses have been advanced over the past few years to account for this chronic poverty and hunger. For purposes of comparison only, these hypotheses will be dealt with as if they fitted neatly and exclusively into one of two categories that will be termed "neo-Malthusian" and "socioecological." Keep in mind, however, that in reality most contemporary hypotheses recognize the complexity of the phenomena they are attempting to explain and thus are themselves complex, generally containing elements of *both* neo-Malthusian and socioecological viewpoints.

The Neo-Malthusian View

The neo-Malthusian view is historically the older, deriving from the writings of Thomas Robert Malthus (1766–1834), a clergyman and political economist in England in the late eighteenth and early nineteenth centuries. Malthus's *Essay on Population* (1798) not only influenced Charles Darwin, but served as the basic textbook on human population growth for the next century. In his *Essay* Malthus stated two laws: (1) that population size tends to grow faster than food supply, and (2) that population size in humans is controlled by disasters such as famine, pestilence, or war.

Malthus also argued that the more food that became available to a population, the more the birth rate would be expected to go up. Thus the neo-Malthusian view suggests that an alternative to mass starvation as a way of controlling the human population is to institute large-scale birth control programs, the reasoning being that if poverty and hunger are

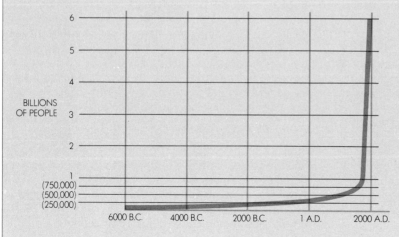

Diagram A
Typical graph showing exponential growth rate projected for the human population to the year 2000. The rapid upswing in the number of people is a source of alarm for neo-Malthusians.

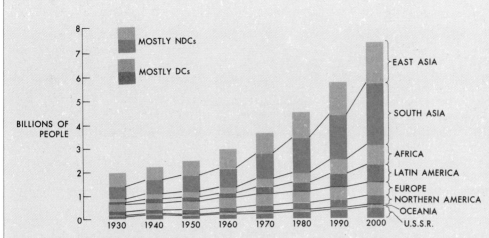

Diagram B
Graph showing growth of the human population by geographic region. Developed countries (DCs) are shown in gray; nondeveloped countries (NDCs) are shown in color. (Modified from *Population Bulletin* 21, No. 4.)

a result of overpopulation, then reducing the population growth rate should reduce the prevalence of poverty and hunger. Persons subscribing to the neo-Malthusian view base their programs for action on this hypothesis and its predictions. For example, biologist Garret Hardin has argued that the wealthier countries should not provide aid to any country that does not institute birth control measures. Whatever program of action they advocate, all neo-Malthusians agree that reducing the present growth rate of the world's population is an absolute necessity for the future survival of the species.

The Socioecological View

The socioecological hypothesis is historically more recent. Although not original with him, it was first publicized and developed in an 1844 essay by Friedrich Engels, a colleague of Karl Marx. Modern proponents of the socioecological view argue that population growth rate is affected by a complex interaction of biological, social, and economic factors that cannot be explained by Malthus's simplistic laws. They maintain that the neo-Malthusian approach is mechanistic in that it tries to explain the social phenomena of hunger and poverty by reference to a single biological cause—the tendency of the reproductive rate to outrun the food supply. Hunger and poverty, the socioecologists believe, are caused more by inequities in the economic system and by the maldistribution of world resources than by people having too many children. For example, some socioecologists argue that it is incorrect to claim that poverty in a country such as Guatemala is due to overpopulation when fully 50 percent of its usable land is owned by a single American company that grows bananas and other fruits for consumption in the United States. Under such economic conditions, the socioecological advocates argue, reducing the population of Guatemala would not in itself necessarily eliminate hunger and poverty.

Socioecologists point out further that in the agricultural economies of less developed countries, it is usually an economic *necessity* to have a large number of children. Farming in a nonmechanized economy requires a large number of hands; one or two children are not enough to provide the help necessary to survive. Even though it means more mouths to feed, it is ultimately more efficient, in terms of total per capita production of food, to have a large family. Since no Social Security or other programs exist to care for people in their old age, children also become a necessary form of insurance for the elderly. In a poor country with high infant mortality, one or two children might not provide enough survivors to care adequately for elderly and infirm parents. People often see an advantage in getting one child in the family out of the small village for further education, hopefully at the college level. He (it is almost always a male) can then be expected to get a good job and help the rest of the family escape the poverty of rural life. However, to give one child the opportunity for further education requires a number of other children to work at home and produce a little extra to augment the family income. All these factors, the socioecologists argue, are responsible for holding birth rate at a high level in many poor countries; the high birth rate is seen as a simple logical response to certain economic and social realities.

The socioecological hypothesis that social and economic inequality cause poverty and hunger predicts that readjusting social and economic conditions will be necessary to eliminate poverty and hunger. Socioecologists tend to favor a large-scale reordering of economic policies and systems throughout the world, even though they recognize that this is a very difficult goal.

Testing the Neo-Malthusian Hypothesis

One might hypothesize from the neo-Malthusian argument that if overpopulation produces poverty, then we might expect to find some correlation in various countries between population density and amount of poverty and hunger. Those who support the socioecological hypothesis point out that, according to the data, no such relationship exists. Among the most densely populated countries known are Bangladesh (425 people/square kilometer), England (324 people/square kilometer) and Belgium (317 people/square kilometer); among the least densely populated countries are Pakistan, India, and Indonesia (67, 164, and 81 people/ square kilometer, respectively). Among the densely populated countries, Bangladesh has severe hunger and poverty, while England does not. Lower-density Pakistan, India, and Indonesia have a considerable amount of hunger and poverty. Neo-Malthusians counter this argument by pointing out that density is not a very meaningful measure of "overpopulation," since it says nothing about land productivity or import-export relationship. One square mile of rich farmland, for example, can support more people adequately than a square mile of rocky or arid desert.

A second assumption of the neo-Malthusian position is that there is simply not enough food in various poor countries to support the population that currently lives there. The assumption is supported by reference to the chronic hunger and malnutrition of certain countries in Africa or Latin America, where food productivity appears to be low. Socioecologists admit that not all regions of the world are equally productive agriculturally and that different areas will probably always continue to have differential agricultural outputs. However, socioecologists point out that neo-Malthusians often ignore the economic relationships that control the production and use of food throughout the world. For example, the Humboldt Current off the west coast of South America (Peru and Chile), and the Benguela Current off the west coast of Africa (Nigeria, Gambia, the Camerouns, and Angola) are a rich source of fish, a form of food high in protein. Ironically, there is considerable malnutrition and starvation in Peru, Chile, and especially in the African countries of Angola, Nigeria, and the Cameroons. Socioecologists point out that most of the fish are caught by foreign fishing vessels, including those of the United States and Great Britain. Especially in the United States, most of the fish is then converted into fishmeal and fed to cattle in feedlots, fattening them just prior to sale. Socioecologists point out that not only does the fish not feed the people who live where it originates, but it is used in an ecologically inefficient way by being fed to cattle, thereby pushing its consumption by humans further up on the food chain (see Section 23.6). In defense of this fishing, one biologist stated that the meal made from these fish "smells bad and tastes worse!" implying that it is unfit for human consumption. However, what does and does not taste good is culturally determined: what is considered a delicacy in one country may be looked upon with disgust in another. Furthermore, the argument is ecologically unsound, because these so-called "junk" fish may be a part of a food chain leading to fish that *are* highly ap-

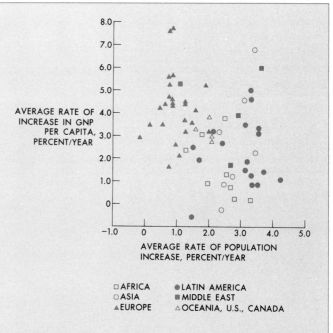

Diagram C

Graph showing the lack of any relationship between overall economic growth of a country (measured as gross national product) and average rate of population increase. Each symbol represents one country, and the graph includes data collected over the period 1958–1966 for population and 1960–1965 for GNP. These data suggest that rapid population growth does not necessarily mean reduced economic development. (Source: *U. N. Demographic Yearbook, 1966;* and the Agency for International Development Statistics and Reports Division, Office of Program and Policy Coordination, 1970.)

propriate for human consumption. Finally, why might the fishmeal not be used to fatten cattle for consumption by residents of the nations whose natural resources are being exploited by other countries? According to socioecologists, hunger in these countries on the west coast of Africa and Latin America is due not to overpopulation, but to the fact that these countries do not control their own natural resources economically or politically.

Another basic neo-Malthusian assumption relates to the natural factors that control population growth. Recall that the neo-Malthusian hypothesis predicts that factors such as famine and pestilence are inevitable unless population growth is controlled on a large scale. Proponents of the socioecological view point out that the population growth rate has begun to slow down in the United States, Europe, and a number of the more highly industrialized countries, even without massive birth control programs, suggesting that more complex and subtle factors affect the birth rate. These data appear to support the socioecological view that as social and economic conditions change, the birth rate also changes. To counter this argument, neo-Malthusians point out that such changes in birth rate take place long after economic changes have come about, and that such factors cannot, therefore, have any immediate usefulness in curbing large-scale increases in population. While waiting for the subtle factors to begin operating, the neo-Malthusians argue, the population rapidly grows to a critical level, resulting in the deaths of many thousands. Subtle factors, even if they do have an effect, cannot avert present-day results of overpopulation.

A final test to which the neo-Malthusian argument might be put is formulated as follows: if increase in population causes hunger and poverty, then a distinct relationship between economic growth rate and population growth rate of a country might be expected—the greater the population growth rate of a country, the lower its economic growth rate. Data collected by the United Nations and by the United States Agency for International Development (AID) show no correlation whatsoever—positive or negative—between percent increase in population and percent increase in economic growth as measured by "gross national product," or GNP. A graph correlating these two factors shows data points scattered all over the grid (Diagram C). Of course, the very general measurement entailed in determining the GNP often hides many crucial aspects of economic growth—for example, it says nothing about actual distribution of income, only the overall amount. It does, however, give a very rough measure of whether an economy is growing or not. Such data suggest to some people that a high rate of population increase does not *necessarily* retard overall economic growth.

Testing the Socioeconomic Hypothesis

It is more difficult to put the socioecological view to a direct test. To do so would require the experimental design of large-scale economic and social changes in specific countries and subsequent observation of the effects on birth rate. Such "experiments" are not really possible in the scientific sense. However, the social and economic revolutions that have taken place in such countries as China and Cuba have had results similar to the carrying out of such experiments. Whatever else they may feel about the social revolutions in those countries, all observers agree that the revolutions redistributed a considerable amount of wealth and provided far more effective health care, jobs, social security, and education for more people than before these revolutions. If we observe patterns of birth rate in these countries prior to and following their revolutions (prior to 1949 for China, and prior to 1960 for Cuba), the data show that birth rates in both

countries went up for a few years and then began to drop. Both China and Cuba now show declining rates of population growth. The causes behind this change are obviously complex, involving an interaction of governmental, psychological, sociological, and economic factors. Cuba, for example, lost many people through emigration after 1960. In both countries, however, lower birth rates have been officially encouraged, largely through the practice of delayed marriage and, more recently, through economic penalties imposed on families with more than two children.

Many neo-Malthusians agree that such social and political changes *can* bring about eventual stabilization of population size. They disagree with socioecologists on two points however: (1) the optimum size of the world population at stability, and (2) the kinds of social and political changes (social revolution, for example) that might be involved. Many neo-Malthusians feel that the successes achieved by countries such as China in areas of health care, education, employment, or decline in birth rates were attained at a political cost of certain kinds of liberties for certain people—for example, the right of private ownership. Some socioecologists have argued that this liberty usually results in the control of very important commodities, such as land or food, by relatively few individuals. Since such control is exercised largely for private profit, it is in part responsible for the poverty and hunger that the neo-Malthusians claim is caused by overpopulation. Both sides agree that all societies necessarily impose certain restrictions on individual liberty. The nature of these restrictions forms the crux of the points of disagreement between them.

It is obviously impossible to test scientifically either the neo-Malthusian or the socioecological hypotheses rigorously. Quite obviously, the problem of poverty and hunger is not entirely a scientific one, solvable by scientific means. Thus while biologists may agree with the scientific principles embodied in the field of population biology and ecology and recognize their important relevance to the problems of poverty and hunger, as individual citizens they may well disagree on which social policy or policies will best alleviate the problem. Indeed, as was indicated in the preface of this book, the co-authors find themselves in different camps. There is still much to be learned about the factors affecting human population growth.

One point should be stressed, however. Neither the neo-Malthusians nor the socioecologists claim that the number of human beings on earth in the future is an unimportant issue. Nor do the socioecologists claim that the more people the better. Certainly, there is some satisfactory number where the human population is not pushing too hard against the carrying capacity of the environment. Where the two schools of thought most disagree is on the question of the role population growth at present plays in *causing* social problems, as opposed to the role social problems may play in *causing* a population growth rate that is out of kilter with the environment.

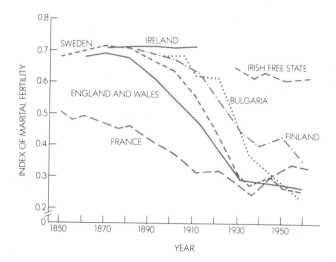

Fig. 25.6
The demographic transition in several European countries, expressed as changes in the index of fertility of married women. The basis for computing the index does not concern us here, except to point out that the highest value—that is, highest fertility—on the scale could not exceed 1.0. The graph clearly shows a decline in fertility rate among married women of child-bearing age over a 100-year period. This was the period during which most European countries became industrialized and urbanized. (From A. Coale, "Decline of fertility in Europe," in Behrman et al., *Fertility and Family Planning*, University of Michigan Press, 1969.)

rural life has had a marked effect on fertility rates throughout the world, and hence population growth (it affects the value of *r*). Similarly, economic status, sense of social well-being, religious beliefs (regarding sex, age of marriage, procreation, and the like) and a host of other factors may render invalid projections of human population growth based on idealized logistic curves.

An example of the effects of such complex factors shows up in a phenomenon demographers (those who study population changes such as size and growth rates) call the **demographic transition**. The demographic transition is defined as the decline in birth rates that has characteristically followed industrialization (including urbanization) in all the major western countries (see Fig. 25.6). Although no demographers are sure what factors most influence such a regular decline in birth rates, the decline is clearly related in part to the fact that children are less able to earn their keep at an early age in an industrial, compared to an agricultural economy. (Children can feed farm animals and help with a large variety of agricultural chores at the age of five or six, whereas the dreadful experiences of child labor in industry in the nineteenth and early twentieth centuries in Europe and the United States have attested to the impossibility of incorporating children into the industrial workforce at comparable ages.) Yet, the demographic transition is

real and is being found even in the newly industrialized economies of countries such as Taiwan and Japan. In most Western industrialized countries, the demographic transition has proceeded so far that population growth rates have declined to the zero population growth level or even below (for example, as in France, in Fig. 25.6).

25.5
THE REGULATION OF POPULATION GROWTH

In nature a number of factors in the biotic environment interact to influence and regulate the growth of population. Chief among these are competition and predation. We will examine each of these processes in some detail.

Competition

Competition occurs when organisms use the same resources, such as food or space, under circumstances where the resource is limited, or where use of the resource by one species or individual limits access to that same resource by another. Competition involves negative interaction between individuals or populations such that the presence of one reduces the fitness of the other. There are two basic types of competition in the natural world: intraspecific and interspecific competition.

Intraspecific competition occurs between members of the same species. For example, many tadpoles in a freshwater pond hatch and begin to feed. Only a few, however, will manage to survive to reproductive maturity. Those that survive are the ones that manage to find enough food, escape predation, and live long enough to undergo metamorphosis into frogs. In the early stages, each tadpole requires a small amount of resources, but as they get larger the tadpoles interfere with each other, the larger ones crowding out the small ones from feeding areas and depriving them of needed nutrients. Some individuals are able to grow faster and become more sturdy (this may be due to genetic differences, or there may be elements of chance involved) at an earlier stage. This differential growth gives some individuals in the population a competitive edge over others where resources (plant material for food) are finite. Through intraspecific competition, the general character of the species may change over periods of time as the more successful competitors leave behind a greater number of offspring than the less successful.

Interspecific competition results when individuals of two or more different species living in the same general area compete for the same limited resource in such a way that changes in the population density of one affects the population density of the other. A classic example of interspecific competition was originally investigated by G. F. Gause in the 1930s. Gause studied interspecific competition among several species and genera of protists, among them two species of *Paramecium: P. aurelia and P. caudatum*. Gause set up several cultures; in some he placed *P. aurelia* by itself, and in some he placed *P. caudatum* by itself. In still others he placed the two species together. For food, Gause provided the *Paramecia* with bacteria. He then kept daily records on the population density of each species under each set of conditions. The results are shown in Fig. 25.7. Note that when each species is grown separately, it exhibits the basic logistic (*S*-shaped) growth curve; that is, each species appears able to grow up to the environment's carrying capacity (approximate density value of 200), at which point the growth curve levels off.

When the two species are grown together their nearly identical habits and requirements produce considerable competition. For *P. aurelia*, competition increases the time required for the growth curve to reach its plateau. But the *P. aurelia* population does reach a plateau and eventually becomes stable.

The story is quite otherwise for *P. caudatum*. In competition with *P. aurelia*, the population size of *P. caudatum* begins to increase (days 0 through 6) but then suddenly declines. *P. caudatum* must in some ways be less successful in getting bacteria or other resources when *P. aurelia* is present. In this case interspecific competition results in the virtual elimination of one of the two competing species. Such a drastic result is not always the outcome of interspecific competition. Sometimes one or both species co-exist, but each at a lower density. Sometimes the two species restrict each other's ranges and end up inhabiting different physical microenvironments within the larger environment. An interesting example of the latter case is shown in the interspecific competition between two genera of flour beetles *Oryzaephilus* and *Tribolium*. The details of this example are outlined in Fig. 25.8.

From his experiments with *Paramecium* and other species, Gause advanced what he called the **competitive exclusion principle.** This idea can be stated simply: two species with extremely similar ecological requirements cannot live together for more than a short time without one species eliminating the other, or the two greatly restricting their individual habitats. In other words, no two species can occupy the same ecological niche at the same place and at the same time. No two species are identical in their ability to use environmental resources. So, in a competitive situation where two species occupy the same niche (that is, use virtually the same basic resources), one is almost invariably bound to have a slight edge over the other from the outset. This edge, or difference, can become increasingly magnified as competition proceeds.

Switching microenvironments is not the only way two similar species can avoid the worst effects—that is, extinction—of interspecific competition. One or both species can also switch foods, especially in a heterogeneous environment where there are numerous sources

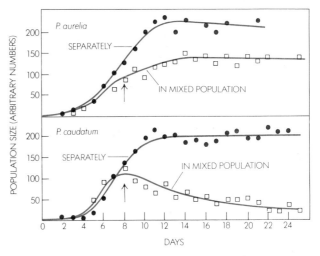

Fig. 25.7

Gause's competition experiments with *Paramecium aurelia* and *Paramecium caudatum*. (*a*) *P. aurelia* grown separately and with *P. caudatum*. (*b*) *P. caudatum* grown separately and with *P. aurelia*. When both species are grown alone they show fairly typical logistic growth curves. When grown together, *P. aurelia* outcompetes *P. caudatum*, whose population density takes a sudden dip after about six days. Both species compete (in this experiment) for the same food supply: bacteria. (From G. F. Gause, *The Struggle for Existence.* Baltimore: Williams and Wilkins, 1934.)

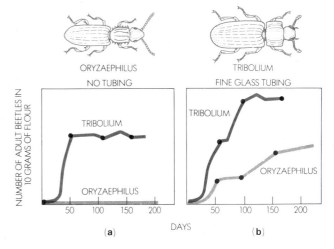

Fig. 25.8

Interspecific competition among two flour beetles, *Oryzaephilus* and *Tribolium*. In the graph to the left (*a*) the two species were placed in the same environment with a plentiful supply of flour. *Tribolium* is the better competitor, among other things directly attacking the smaller *Oryzaephilus*. Through such direct competition, *Tribolium* eliminates *Oryzaephilus*. If *Oryzaephilus* is provided with a group of small glass tubes (large enough for it to fit into, but small enough to keep *Tribolium* out), *Oryzaephilus* can escape its competitor, and still have enough food to survive (*b*). With this slight change in the structure of the environment, the effects of competition can be drastically altered. (Adapted from Wilson et al., *Life on Earth,* 2nd ed. Sunderland, Mass.: Sinauer, 1978.)

The regulation of population size, like the regulation of physiological events, occurs through modulation of ongoing processes that tend to establish a dynamic equilibrium. Consequently, the size of any given population is constantly oscillating around a mean and never remains absolutely fixed.

available. In switching behavior, when any food source becomes scarce (through interspecific competition or for some other reasons) one organism or species can switch to another sometimes similar and sometimes quite different food source. Such behavior is frequently observed in birds, whose insect prey often die off rapidly after a temperature or climatic change. John L. Brooks of Yale University has demonstrated recently that trout and yellow perch can switch from small, free-floating aquatic organisms (plankton) such as the copepod *Daphnia* to larger organisms such as insect larvae and even adult insects when *Daphnia* becomes scarce. Switching provides a way of insuring flexibility in the feeding resources of a species. Not all species, however, have this flexibility built in. Some are obliged to remain on one basic food source, even if that source becomes scarce.

In summary then, competition—both interspecific and intraspecific—is a major factor in determining and regulating population density. It can cause the elimination of individuals within a species, or of a whole species from an area. Interspecific and intraspecific competition are major factors in keeping most populations at a maximum density well below the carrying capacity of the environment. Both intraspecific and interspecific competition are important in affecting the

future course of a population's development. Both forms of competition can, and usually do, lead the population in new evolutionary directions.

Predation as a Regulator of Population Growth

Predation is another factor that can limit population size. Just as the amount of any food-producing population determines the size of the population that feeds on it, the reverse is also true. In other words, the size of a food population itself is determined by the size of the population that preys on it. In a stable ecosystem, the relative ratios of prey and predator will not change much over the course of time. The graph in Fig. 25.9 illustrates dramatically how the sizes of prey and predator populations are completely interdependent. This graph shows estimated cyclic changes in population of lynx and snowshoe hares in Canada for a period of ninety years. Three things are important. First, note that on any one occasion the peak of the curve for the snowshoe hare (the prey) is always a good deal higher than that for the lynx (the predator). Second, note that the peak for the lynx population always occurs a little later than that for the hare population. As the number of hares increases, so does the number of lynx. This relationship results because more lynx can survive to have

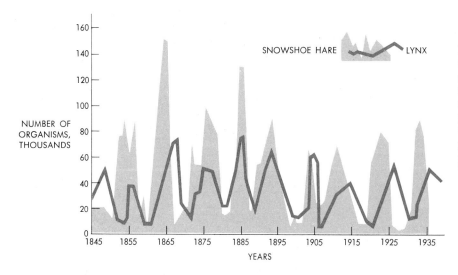

Fig. 25.9
Cyclic changes in populations of snowshoe hares and lynx in Canada from 1845 to 1935. This graph is constructed from records of the number of pelts received per year by the Hudson's Bay Company. (After C. A. Villee, *Biology.* Philadelphia: W. B. Saunders, 1957, p. 577. By permission of Holt, Rinehart and Winston.)

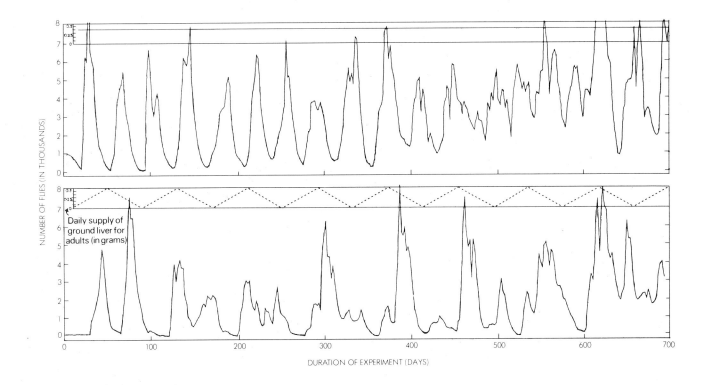

kittens when the supply of snowshoe hares is abundant. Third, note that when the number of hares decreases, so does the number of lynx.

It is worth pointing out that in predation, the very young, old, sick, or weak individuals in the prey population most often fall victim to the predators. Not only does this eliminate the less functional members of the prey population, but by culling out what may be defective genotypes, predators become an important agent of natural selection.

As simple and obvious as it might seem, there is more complexity to the interpretation of the lynx-hare graph than meets the eye. For example, ecologists are not certain that predation by the lynx *causes* the hare population to oscillate. There is good evidence that the hare population has its own internal cycling pattern with the lynx population following passively behind. For example, on Anticosti Island and in the southern plains region of Canada, lynx are absent but the snowshoe hare populations still show the same approximate ten-year cycle. At least in this case it appears that the existence of predators is not the primary *cause* of population oscillation among hares. That such intrinsic cycles may be characteristic of most natural populations is indicated by data for population growth in the sheep blowfly shown in Fig. 25.10. The top graph shows the effect of feeding the organisms a regular amount of food each day; the bottom graph shows the effect of cycling the food supply. With constant food supply an ex-

Fig. 25.10
Cyclic population changes in the sheep blowfly, *Lucilla cuprina*. The top graph represents the population raised in the laboratory with a constant supply of food each day (0.4 g of ground liver). The population density still cycled rather markedly. The bottom graph shows what happened when the food supply was cycled regularly (see the dotted line at the top of the graph). Although cycling of the food supply affected the periodicity and extent of the blowfly population cycles, it is clear that cycling of food (prey) does not necessarily *cause* cycling of predator populations. (From A. J. Nicholson, "The self-adjustment of populations to change," in *Cold Spring Harbor Symposium in Quantitative Biology* 22 [1958]: 153–173.)

tremely regular cyclic pattern is still visible. When food supply is varied, it affects to some degree the periodicity and amplitude of the response, but not the basic cyclic pattern itself.

Predation is real, and does indeed affect population growth rates. But much more needs to be learned about the dynamics of population growth itself, in the absence of predation, before it can be claimed that oscillations in population size, such as those shown in Figs. 25.9 and 25.10, are directly a result of predation itself.

Knowledge (or lack of knowledge) about prey-predator relationships can be crucial in determining policies for protecting certain species in the wild. An in-

SUPPLEMENT 25.2
THE KAIBAB DEER CASE: MYTH OR REALITY?

For many years ecologists have repeated an account of how predators control their prey populations, using as an example the history of the deer on the Kaibab Plateau, a wilderness area near the Grand Canyon. Shortly after 1907, when a national park was created in this area, conservationists attempted to protect the wild deer by prohibiting hunting and simultaneously eliminating natural predators.

In the period from 1920 to 1930 it was reported that the deer herd had undergone a population "explosion" (Diagram A, part a). No detailed census data were obtained, but various visitors and game wardens estimated that the population tripled or perhaps quadrupled over a period of twenty years. Although it seemed obvious at the time that the population had increased, it was far from clear what mechanisms produced the increase.

It was first assumed that the increase resulted from hunters shooting most of the predators and thereby allowing the prey, the deer herd, to increase. Other variables may have been equally important, however. For example, before the land became a national park, parts of it had been used by ranchers for grazing cattle. Furthermore, natural and accidental fires had often caused much burning of the grasslands on the plateau. Prior to 1907, fire and grazing kept the vegetation available to the deer at a minimum. The explosion in the deer population by 1930 could as plausibly have been explained by patterns of land use as by a release from predation. The important point to emphasize is that it was simply *assumed* that a lack of predators *caused* the rise in deer population. The hypothesis was never carefully tested. As more ecologists and textbooks (including early editions of this one) repeated the hypothesis, it became accepted as a fact.

Even more unfortunate is the way in which the original estimates were continually reinterpreted to produce the appearance of stronger "evidence," as indicated by Diagram A, graphs b and c. This process of smoothing data to provide more convenient diagrams occurs all too frequently in both biological and social sciences. You will recall the earlier dis-

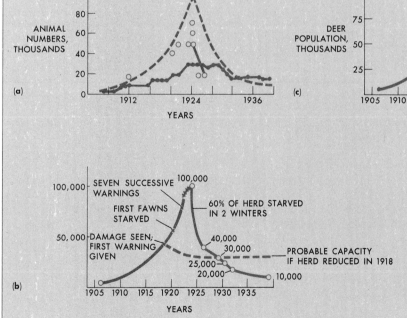

Diagram A
Population changes in the Kaibab deer herd. (*a*) Population estimate of the Kaibab deer herd, from Rasmussen (1941). Linked solid circles are the forest supervisor's estimates, circles give other people's estimates, and the dashed line is Rasmussen's own estimate of a trend. (*b*) Interpretation of the trend by the ecologist Leopold in 1943. (*c*) The trend represented by Davis and Golley (1963). (After Caughley, 1970.)

cussion of how the number of chromosomes in human cells had long been incorrectly observed because people "knew" what to expect. It is only through renewed study by other investigators who provided *critical* reexamination of previous reports that these myths are identified. The 1970 reexamination of the Kaibab deer case by G. Caughley is an example of the importance of studying original data, rather than accepting them at face value from textbooks. Textbooks (the present one not excepted) are often notorious for perpetuating accepted dogma in a field without reexamining the original data.

In ecology, because of the long periods of time over which natural events occur, it is extremely difficult to repeat some studies. Thus it becomes much more crucial that field experiments be extremely well designed from the beginning. Field ecologists cannot usually assume great potential reproducibility of their experiments by other investigators if their studies require decades to complete and if the natural communities are undergoing large-scale responses to changes in climate, successional development, or human interference. Likewise, students must be especially critical of ecological reports that overinterpret limited sets of data.

teresting example of such effects, with all the problems involved in collecting reliable data and drawing valid conclusions, is found in the Kaibab deer case (earlier in this century) described in Supplement 25.2.

Other Factors Regulating Population Growth

In addition to the two major population regulators —competition and predation—several other factors are also involved. One is **emigration**, the departure of individuals and groups from a dense population. The studies of John C. Calhoun and others (see Section 30.9) have shown that crowding induces a variety of behavioral and physiological changes in laboratory animals. These changes can result in mass emigrations from the crowded population areas. There are many examples of such emigrations caused by overcrowding. One of the most frequently encountered is that of the small rodents known as lemmings, which live chiefly in Scandinavia. When a lemming population becomes too large, great hordes of lemmings begin to migrate from the overpopulated region. This migration is impulsive. Lemmings on the march stop neither to rest nor to eat. At some point in their migration, lemmings usually encounter a lake, river, or fjord. Ignoring the barrier, they sometimes plunge right into the water and drown. This famous spectacle has given rise to a great number of romantic and exaggerated stories of the lemmings' "march to the sea." The march to the sea is of course an artifact of Scandinavian topography. Because there are so many lakes and fjords in this area of the world, a lemming emigration does not have to proceed very far before encountering a body of water. There is, however, no suicidal urge in lemmings induced by the stress of overcrowding.

Another equally dramatic example is the migratory phase of locusts, so classically described in the Old Testament (Exodus) as a locust plague. Many species of locusts have two developmental forms—a solitary and a migratory form. In low population density most of the individuals exist in the solitary form. As population density increases, a greater and greater proportion of young locusts develops into the migratory form. The sight and smell of other locusts seem to have a triggering effect on the developmental process. These locusts are a darker color, are more gregarious, and have longer wings than the solitary forms. When there are a certain critical number of migratory forms in a population, a mass exodus takes place, and swarms of individuals begin emigrating. They cover the land, eat all the vegetation they can find, and eventually die of both stress and starvation. In the process of reducing their numbers, however, the locusts wreak untold destruction on the environments through which they pass.

Another mechanism for controlling population growth is **physiological control.** As we have seen, crowding produces distinct responses among individuals in a population. Ecologists have found that in overcrowded rabbit populations in the wild, at certain times many individuals appear to be in a state of physiological shock: they have low blood pressure and low blood sugar, and usually die. The shock is thought to be the result of hormonal imbalances induced by the stress of overcrowding. The role of hormones in behavioral

Habitat is an organism's home, the place where it lives. Niche is an organism's occupation, what it "does for a living" within the ecosystem.

changes associated with high density populations is not clear, however. Such changes have not been observed as frequently in nature as in laboratory experiments, and their significance remains doubtful as natural regulators of population growth. However, high density in animals is known to cause reduction in sexual activity, including delayed or repressed spermatogenesis and oögenesis.

It has also been found that disease becomes rampant in dense populations. This is partly due, of course, to the easier transmission of germs in crowded situations; but it is also due to an increased physiological susceptibility to disease induced by high density. In dense populations, for example, animals have been noted to show a significant reduction in antibody formation and other of the body's front-line defense mechanisms, thus making them more vulnerable to infection by outside agents.

25.6
COMMUNITY AND ECOSYSTEM STRUCTURE
AND ENERGY FLOW

In an earlier section (25.2) we briefly defined the terms community and ecosystem. We will expand on those definitions here and examine the flow of energy (and later materials) through these systems. We will focus particularly on the ecosystem level of organization.

Recall that in an ecological sense "community" refers to the populations of various species inhabiting a given geographic area. Communities interact, especially in their food relationships, since every community (except perhaps green plants) depends upon some other community for a supply of energy.

The term ecosystem has been used in several previous sections to refer to many interacting biotic communities and their associated physicochemical environments. Any community or group of communities can be treated as an ecosystem. For example, the boundaries of a pond ecosystem can be defined in terms of the entire watershed, or only in terms of the shoreline vegetation, depending on the limits one wishes to draw. If a study deals with the nutrient dynamics of the entire pond, then the whole watershed of the surrounding terrestrial communities must be included. If the study is limited to the role of certain species of aquatic snails as grazers on particular aquatic plants, the boundaries of the ecosystem need consist only of the shallow portions of the pond where those snails and plants live. Generally, ecosystems will be defined so that specific inputs and outputs can be identified, and the boundaries will be determined in a manner that allows for measurement of these flows. Thus all ecosystems are abstractions; they can be as large and complex as our abilities to measure them or as small as required by a given set of interactions. All ecosystems are characterized, however, by self-sufficiency in terms of the flow of materials (which are recycled) and the presence of photosynthesizing green plants or their equivalents (as sources for energy capture into the system). In this sense an ecosystem is different from a collection of biotic communities.

Members of a community and an ecosystem can be classified according to their mode of obtaining energy. They may be producers, primary consumers, secondary consumers, tertiary consumers, or decomposers. These are convenient designations for major groups of organisms that differ greatly in their sources of energy. The *producers* are primarily green plants that use solar

SUPPLEMENT 25.3
CONCENTRATION OF TOXINS IN FOOD WEBS

An appreciation of energy flow is necessary to understand how certain synthetic compounds move through natural food webs. Initially many people assumed that the continued use of artificial pesticides would not pose problems for wildlife because the total amount introduced was relatively small and would be rapidly diluted. Yet as more data became available, a startling pattern emerged. Small doses accumulated and became concentrated in the bodies of consumer species especially at the uppermost trophic levels. (Remember that there are fewer consumers at high trophic levels and these consume many more organisms at the next lower level, and so on.) The poisons are not excreted but accumulate,

with deadly effect, in the tissues of higher predator organisms. In some areas excessive amounts of pesticides were applied and entered aquatic ecosystems, either through indirect runoff and rainfall or through direct application.

One of the surprising outcomes of the introduction of these synthetic pesticides was the long distances they were transported from the application sites. Pesticides were found in birds, fish, and seals in Antarctica, for example, though the nearest place these chemicals were used was thousands of miles away. Oceanic and atmospheric currents had carried pesticides far beyond the localized sites of their application. Evaporating water vapor and dust from croplands can con-

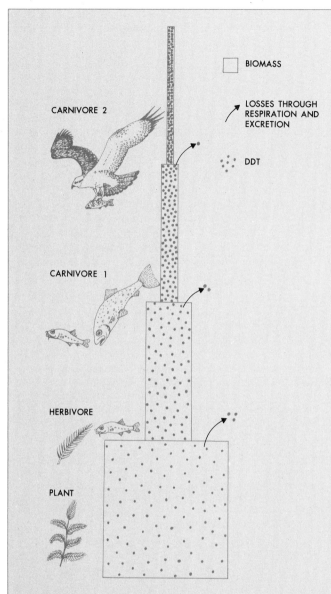

BIOMASS

LOSSES THROUGH RESPIRATION AND EXCRETION

DDT

CARNIVORE 2

CARNIVORE 1

HERBIVORE

PLANT

Diagram A
The concentration of DDT at different parts of a food chain. Note that as it proceeds along the chain, the concentration of DDT increases. This is partly because much of the biomass at any one level is oxidized to yield energy, leaving the non-oxidizable part behind; it is also partly because DDT is not readily excreted by living systems and tends to build up and be passed along from prey to predator.

tain relatively high doses of pesticides which are then redeposited when rain falls in other locations. It is estimated that 40 tons or more of dichlorodiphenyltrichloroethane (DDT) reach England through rainfall each year. Because many

pesticides are not broken down by oxidation or microbial metabolism, they can persist in the environment for many years. These compounds can enter the cycles of the global ecosystem and reside in certain tissues of consumer species. Concentrations of these synthetic toxins can increase over a long period of continued ingestion if the compounds are not excreted. One study of DDT in Lake Michigan showed that even though concentrations were relatively low in the lake sediments (0.0085 parts per million), there was a forty-eight times higher concentration in the aquatic invertebrates (0.41 parts per million). The fishes feeding on these invertebrates showed a further twenty-fold increase in DDT concentration (ranging from 3.0 to 8.0 parts per million) in their bodies. Predatory birds that fed on these fish accumulated as much as 3177 parts per million of DDT in their fatty tissue. These high concentrations in the uppermost consumers were found to have very toxic effects. A number of different predatory birds showed seriously impaired ability to reproduce because DDT and other synthetic chemicals interfered with the formation of eggshell. There were high mortality rates among the young birds because very thin eggshells frequently broke before the birds were ready to hatch.

One of the first ecologists to report these effects was Rachel Carson. Her book *The Silent Spring* set off a major effort to document further the indirect effects of pesticides and eventually led to some global regulations on the use of synthetic chemicals. A new impetus was given to exploring ways of controlling pests by biological means, such as maintenance of natural predators that consume noxious insects and rodents and use of natural hormones to change reproductive behavior. Ecologists have found that many species use complex chemical signals to attract their mates or to defend themselves from competitors and predators. Currently, a great deal of research focuses on how these natural chemicals are synthesized and broken down in food webs so that their effectiveness can be maintained in balancing predator-prey interactions without leading to the disastrous results of the introduction of synthetic compounds that may destroy a food web. Diagram A illustrates how persistent compounds such as DDT were concentrated through continued recycling in natural food webs until the ultimate consumers were severely poisoned.

The effect of increasing concentration of toxins at higher trophic levels of a food web is an example of "biological amplification." This term stresses that through concentration, the effects of a toxin as well as the toxin itself are magnified. The toxin becomes increasingly effective—that is, amplified—through concentration.

The large scale use of pesticides continues, despite our increasing knowledge of its deleterious effects. The July 1981 spraying of large areas of northern California to control the Mediterranean fruit fly provides yet another example of powerful, short-term economic interests (in this case, the fruit growers) taking precedence over the long-term interests of the larger society, including farm workers.

energy to convert inorganic substances into living tissue. While some bacteria are chemosynthetic others are photosynthetic (see Section 6.14). These green plants and specialized bacteria may differ greatly in their modes of reproduction, longevity, and size, but they share the same attribute of converting sunlight and chemical sources of energy into living substances which, in turn, provide energy for the consumers and decomposers. (Some plants also produce toxic substances that inhibit growth by their competitors or repel consuming species. These intricately evolved chemical interactions can be disrupted when human-made compounds are introduced into natural ecosystems, see Supplement 25.3.)

Generally **primary consumers** are herbivores, organisms that eat only plants. These "vegetarians" may be microscopic in size and consumers of small algae, or they may be very large mammals that forage on grasses and shrubs. For example, the small "water flea" *Daphnia major*, is a small herbivorous organism. These "zooplankton" swim and filter floating algae in suspension in the upper, well-lighted water of a lake. The **secondary consumers** are carnivores that consume the herbivores. Similarly, the **tertiary consumers** are other carnivores that consume the secondary consumers. These distinct groups interact to form a **food chain.** Several interconnected groups produce a **food web** (Fig. 25.11). The movement of matter through food webs is *cyclic*, because the **decomposers** (fungi and some bac-

teria) convert dead organic tissue of producers and consumers back into mineralized matter that can once again be recycled through the food web.

Energy Flow

The movement of energy is *not* cyclic but unidirectional; it flows from the producers to the consumers and decomposers. Energy must continually be added to the food web in order for the recycling of matter to continue. In some communities, where there is no direct input of solar energy, the source of energy is the organic debris and breakdown products from other communities that are dependent upon solar energy. The sunless environment of cave communities and the very deep waters in the ocean receive no direct solar energy, but they do obtain energy indirectly by using the debris or byproducts coming down to them from other communi-

Fig. 25.11
Generalized food web of Gatun Lake in the Panama Canal. Heavy arrows indicate connections between major consumers and producers, while thin arrows denote connections of minor food flows. Arrows point from item eaten to the eater. (After Zaret and Paine, "Species introduction in a tropical lake," *Science* 182, 1973, pp. 449–455.)

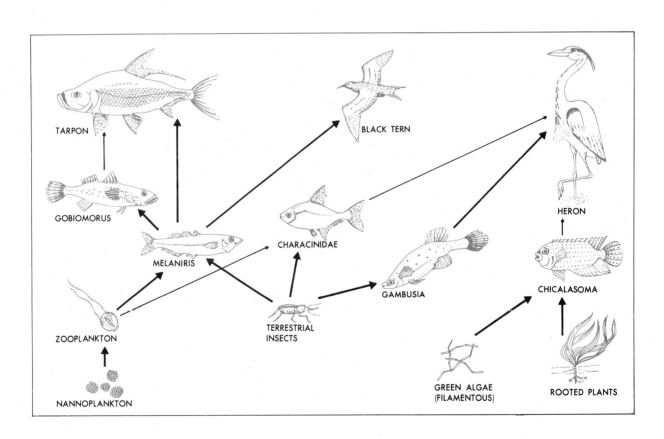

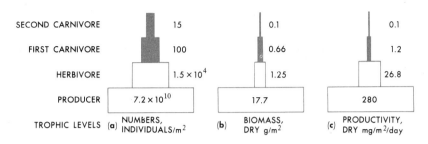

ties exposed to the sun. In these specialized "dark-adapted" communities, the producers may be greatly reduced in number or completely absent.

The structure of a community can be depicted graphically by counting the number of individuals or by determining the **biomass** of all members of each species. Biomass is the total mass of organic material of a species per unit of area or volume. From an energetic point of view, the various groups of organisms that successively feed on one another are grouped into **trophic levels.** The producers are the **autotrophs,** or the first trophic level, while the various consumers are the **heterotrophs.** Each group constitutes a distinct level. The diagram in Fig. 25.12 is called a **pyramid of numbers** (Fig. 25.12a) if number of individuals is used, or a **pyramid of mass** (Fig. 25.12b) if biomass is used. If the mass is converted to units of energy, then the diagram is termed an **energy** or **productivity pyramid** (Fig. 25.12c). In many communities, the pyramid is steeply truncated; the greatest mass and energy content occurs in the base among the producers. However, it is possible to have an inverted pyramid of biomass if the flow of energy is very rapid and the size and life cycle of the producer organisms is reduced relative to those of the consumers. For example, the biomass of microscopic algae in open ocean water is often very small relative to the grazing herbivores. The only way this low abundance of algae can supply adequate energy to the herbivores is if there is a rapid "turnover" of algae: the algae reproduce quickly and are eaten quickly by the consumers. In such cases there is an inverted pyramid of biomass (Fig. 25.13).

In no community can a flow of energy greater than that available in the first trophic level persistently occur in the upper trophic levels. In fact, in all communities decreasing amounts of energy are available to the organisms in the upper trophic levels. This decrease occurs because all organisms are somewhat inefficient in their metabolic use of energy and all lose energy through the generation of heat. In addition, no organism can digest or use all parts of its food (for example, bones, hair, or cellulose). At each step up the pyramid, the organisms are able to use only 10 to 20 percent of the total energy available from the preceding level (although a few species may be more highly adapted and much more efficient than others, none have been observed to use more than 60 percent).

Why does each transfer of energy extract only a small fraction of the total potential energy from the previous level? The decrease in available energy along a food chain is in agreement with the first and second laws of thermodynamics. The first law holds that energy cannot be created or destroyed, but only changed in form, in ordinary chemical reactions. None of the energy in the universe is lost. The second law states that the total amount of *usable* energy in any system tends to decrease with time. This is because no transformation of energy is 100 percent efficient. In a given transformation, some energy is always converted into heat. (For a more detailed discussion of this material see Baker and Allen, *Matter, Energy, and Life,* 4th ed., Sections 1.4 and 5.6. Reading, Mass.: Addison-Wesley, 1980.)

The recent discovery of chromosynthetic bacteria as primary producers at the great ocean depths of the

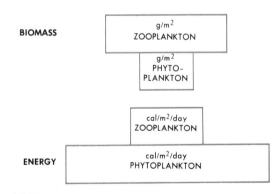

Fig. 25.13
Inverted pyramids of mass in the sea. The values are in grams per square meter. A small crop of phytoplankton (plants) supports a larger crop of zooplankton (animals) and bottom animals that presumably all derive their nutrients from the small mass of plants. The low concentration of microscopic organisms occurs because they "turn over" quickly, so that the rate of of energy flowing through the phytoplanktonic producer level is fast enough to be capable of sustaining the large biomass of animal trophic levels that feed on them. Obviously, the inverted pyramid is an artifact of different generation times between producers and consumers. In one year a generation of consumers feeds on many generations of phytoplankton. (After Odum 1959, from data of Riley 1956 and Harvey 1950.)

Galápagos rift off the coast of South America, however, has shown that at least some food webs can begin with other than solar energy. Great fissures along the rift serve as vents for the release of geothermal energy. Certain bacteria use this energy to synthesize carbohydrates and thus form the beginning of a food web. A whole fauna, including giant tubeworms, albino crabs, and the like, are consumers ultimately dependent on the chemosynthetic bacteria. These organisms and their deep ocean habitat, where sunlight never penetrates, form their own ecosystem that is totally independent of energy from sunlight, which powers (directly or indirectly) virtually all of the earth's other ecosystems.

Three important principles emerge from this discussion of trophic dynamics. First, to be complete and self-containing, any food chain must always have photosynthesis at the beginning and decay at the end. In generalized form, then, a food chain or web may be represented as:

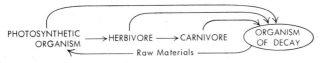

Energy must be constantly supplied from the outside to keep the food chain operating. The ultimate source for most photosynthetic systems in most cases is the sun.

Second, in general, the shorter a food chain the more efficient it is. Conversely, the more steps in such a chain, the greater the loss of usable energy that results.

Third, the size of any population is ultimately determined by the number of steps in the food chain, which, in turn, determines the amount of energy available for that population. With the decrease in useful energy at each step along the chain, very little energy is available for a population of quaternary consumers. The size of a population of quaternary consumers is typically less than that of tertiary consumers; a population of tertiary consumers is less than one of secondary consumers, and so on.

Movement of Materials

When an animal or plant dies, the materials within its body are acted on by organisms of decay—bacteria and various other decomposers. These organisms break down the complex proteins, fats, and nucleic acids of the dead organism and release many of the inorganic components back into the environment. At the same time, the decomposers gain their own nourishment. Decomposer organism thus fill an extremely important niche within an ecosystem. Through the metabolic activity of decomposers, vital inorganic materials are prevented from remaining locked up in the bodies of dead organisms.

Consider the freshwater lake ecosystem shown in Fig. 25.11. Solar radiation is the driving force that keeps

any such system going. This energy is harnessed by the producers: rooted water plants and free-floating algae. Small organisms feed on the algae and are eaten in turn by larger forms. If these larger forms have no natural enemy, they ultimately die and are then acted on by organisms of decay, which break down the large organic molecules and absorb the substances they need. The rest is returned to the water. As a result, most of the materials that plants withdraw from the water in the course of their normal metabolism are returned.

Productivity as a Measure of Energy Flow

One way in which very different ecosystems can be compared is by measuring the rate of energy flow through the biological community on a per-unit-area basis. For example, when solar energy falls on a freshwater lake, a certain amount will be reflected from the water surface and the remainder will penetrate into the water. The amount of energy actually taken up by plants in the spring will be dependent on the transparency of the water and the abundance of the plants. In clear springwater with high densities of green plants, much of the sunlight will be intercepted by the plants, and a portion of that amount will be used for photosynthesis. Conversely, in a turbid pond with few plants growing in the water, the flow of energy into this ecosystem directly from the sun will be relatively less on a per-unit-area basis.

The amount of energy taken up by the producers in any ecosystem is called **primary productivity**, expressed in calories per unit area per unit time. The total amount of energy taken up by the producers is the **gross primary productivity.** When the energy expended by the producers during respiration is measured and subtracted from the total, the result is **net primary productivity.** Only this latter value is available to the consumers, and only a small fraction of this energy will reach the upper trophic levels. The productivity of different ecosystems varies greatly, depending on the latitude and amount of solar energy available throughout the year, as well as on other factors such as temperature, nutrient availability, and rainfall. The ecological measure of energy flow does *not* refer to the quality or type of material being produced, especially with regard to human consumption. In other words, the ecologist measuring "productivity" of a cornfield might include all of the corn plant (not just the ears of edible corn) as well as the associated weeds. This complete compilation would then stress the relative productivity of the cornfield as an ecosystem so that it could be compared with, say, a forest ecosystem. Measurement of productivity is also useful for comparing the stages of a given ecosystem as it develops over time (see Section 25.7). The former tries to convert the total productivity of the land into useful productivity by eliminating from the crop undesirable competitors (weeds) and consumers (pests and diseases).

The movement of materials through an ecosystem is cyclic. The movement of energy is unidirectional.

25.7
THE CYCLIC USE OF MATERIALS

To operate for any length of time, an ecosystem requires a constant input of energy. As we have seen, that energy is captured in the food-making processes of green plants. It is released again in the metabolism of both plants and animals. The matter (atoms and molecules) involved in an ecosystem, however, does *not* have to be continually replenished from the outside. The chemical elements composing living organisms may be recirculated within a given ecosystem. All materials are of course present on the planet, despite the continual loss of small amounts of hydrogen and oxygen into outer space. Several such cycles can be traced.

The Carbon Cycle

Nearly every compound involved in the metabolic activity of living things contains the element carbon. The availability of this element in the environment is therefore a crucial factor in the maintenance of animals and plants. The continued existence of any ecosystem requires that the carbon "locked up" within organisms

ultimately be returned to the environment, unless there is periodic input from new materials. Thus atoms of carbon in the global ecosystem are passed around the carbon cycle.

A carbon cycle is pictured in Fig. 25.14. Let us begin with free carbon dioxide in the atmosphere. The level of atmospheric carbon dioxide is maintained by

Fig. 25.14
Diagrammatic representation of the carbon cycle. Carbon from the atmosphere, in the form of carbon dioxide (CO_2), is incorporated into carbohydrate in the process of photosynthesis. Carbohydrate is oxidized for energy by plants, animals, and microorganisms. This process returns CO_2 to the atmosphere. Some carbon is taken out of circulation for varying periods of time in what can be called a "carbon sink" (lower right) as coal, petroleum, gas, or limestone. In time, much of the carbon locked in the carbon sink will be returned to the atmosphere as CO_2 through one or another form of oxidation. The numbers given after each source refer to the total number of grams of carbon circulated to or from the source per square meter of the earth's surface per year.

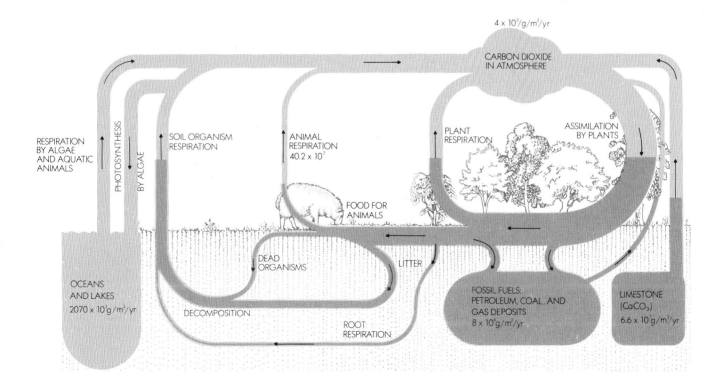

animals and plants, both of which release this gas as an end-product of respiration, the process of releasing energy. Besides releasing carbon dioxide, plants have the ability to use carbon dioxide through photosynthesis in the manufacture of carbohydrates.

The carbohydrates in plant material can follow one of several courses. If the plant is eaten by an animal, the carbohydrate is burned in the animal tissues to yield energy and release carbon dioxide back into the atmosphere. Some carbon may pass out in the waste products of animal metabolism, such as urine or feces. The carbon in these waste products, as well as that which is part of the animal material at death, is acted on by organisms of decay. One end-product of these decay processes is CO_2, which is released back into the atmosphere.

On the other hand, the plant may simply die before being devoured by an animal and in this case its organic substances (including carbohydrates, fats, proteins, nucleic acids, and vitamins) are acted on directly by the bacteria and fungi of decay. Sometimes, however, the plant materials may undergo a quite different process at death. By being deposited at the bottom of lakes in tightly packed layers, covered over with mud or other organic debris, and subjected to great pressure, plant parts may be turned into coal. This process requires very long periods of time and normally occurs less frequently than decomposition by bacteria. The carbon locked in coal is taken out of the global ecosystem for a long period of time. In due course, however, it is returned to the atmosphere as carbon dioxide through burning or weathering.

In addition to coal, there is a second very large "reservoir" for carbon that recycles very slowly. Calcium carbonate (limestone) is precipitated chemically in tropical oceans and in hardwater lakes and streams, and a chalky material is deposited as sediment. In addition, $CaCO_3$ is also produced biologically by microorganisms such as algae and by large reef-building invertebrates, as well as many vertebrates. These shells, bones, and microscopic bodies accumulate over long periods and are gradually transformed into sedimentary rock. Massive outcrops of limestone may contain billions of fossils. A close examination of a single piece of blackboard chalk reveals many different species preserved in these ancient carbonates that were deposited in former shallow seas. This carbon "reservoir" contains the largest amount of carbon of any of the various potential carbon sources on earth. Carbonic acid is a major link in the carbon cycle as it forms from carbon dioxide in the atmosphere being dissolved in rain water. Through the slow weathering of limestone by carbonic acid, this soft rock releases calcium and bicarbonate ions into rivers, lakes, and oceans.

The accumulation of carbon dioxide in the atmosphere produces what is called the "greenhouse" effect. CO_2 reflects back to earth heat that is radiated upward

from the surface of the earth, much as the windows of a greenhouse trap the radiant energy that has entered as light, been absorbed by the soil inside, and radiated back as heat. CO_2 thus maintains the earth's temperature somewhat higher than it would otherwise be without this gas in the atmosphere.

In all these ways, carbon taken from the atmosphere by plants in photosynthesis is ultimately returned. In the course of time, a single atom of carbon in an ecosystem may have existed in a variety of compounds in many different organisms.

The Nitrogen Cycle

The element nitrogen is no less essential to life than carbon. In living organisms, nitrogen is found chiefly in amino acids and proteins. Since these molecules are constantly being built up and broken down in normal metabolic activity, it is essential that new sources of nitrogen always be available to an organism. Nitrogen is cycled from environment to organism back to environment by one of several paths (Fig. 25.15).

Atmospheric nitrogen is the principal "reservoir" of nitrogen, but it is not as easily available, since plants and animals cannot incorporate gaseous nitrogen into their metabolic processes (for example, in protein synthesis). Only some kinds of bacteria and blue-green algae can convert atmospheric nitrogen into a usable form (that is, NH_3 or NO_4).

Four types of bacteria are involved as key parts of the nitrogen cycle. Before considering the details of the entire cycle, it will be helpful to examine the specific nature of each of these bacteria.

Nitrogen-Fixing Bacteria. Nitrogen fixing bacteria live in the soil and on the roots of plants, especially legumes such as alfalfa (legumes are plants that bear their seeds in pods, such as beans or peas), in little swellings known as **nodules.** They also occur in several other types of plants, either in root nodules or in dense colonies in leaf tissues. Nitrogen-fixing bacteria have the ability to take free nitrogen gas from the atmosphere and convert it into soluble nitrates (compounds containing NO_3 such as potassium nitrate, KNO_3). Because they are soluble, nitrates can be taken into the roots of higher plants. Only nitrogen fixing bacteria and some types of blue-green algae can make use of atmospheric nitrogen.

Putrefying Bacteria. Putrefying bacteria are found chiefly in the soil and in the mud at the bottom of lakes, rivers, or oceans, where they break down animal and plant proteins, converting them into ammonium compounds such as ammonium phosphate $(NH_4)_3PO_4$. These compounds are released into the soil or water where they can be acted on by other types of bacteria.

Nitrosofying Bacteria. Nitrosofying bacteria (genus *Nitrosomonas*) act on ammonium compounds such as

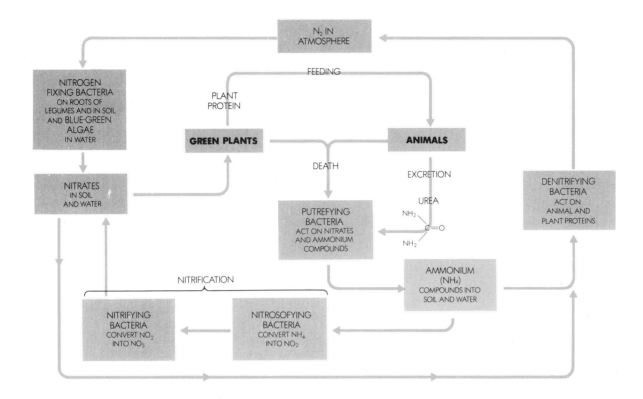

Fig. 25.15

The nitrogen cycle. Nitrogen from the atmosphere is converted into a biologically usable form (that is, rendered soluble as nitrates) by nitrogen-fixing bacteria found on the roots of leguminous plants. From there, the nitrates can follow a number of paths and pass through a series of conversions. Denitrifying bacteria convert nitrates and ammonium back into atmospheric nitrogen, to complete the large cycle. Nitrates are used in the synthesis of amino acids by bacteria, green plants, and fungi. Most animals get their nitrogen by eating plant proteins, but corals have blue-green algae which live in conjunction with them and provide a direct source of usable nitrates into the water.

those produced by the process of putrefaction. By various chemical processes, these compounds are converted into **nitrites,** molecules containing NO_2. Like nitrates, nitrites are soluble.

Nitrifying Bacteria. Nitrifying bacteria (genus *Nitrobacter*) are able to convert nitrites, produced by nitrosofying bacteria, into nitrates.

Denitrifying Bacteria. Dentrifying bacteria convert either nitrates or ammonium compounds into molecular nitrogen (N_2). Thus the denitrifying bacteria serve as a means of returning molecular nitrogen to the atmosphere.

Consider now how all these components fit together to form the complete nitrogen cycle. It is con-

venient to begin again with the atmosphere. Molecular nitrogen composes about 78 percent (by volume) of the earth's atmosphere, but neither animals nor green plants use nitrogen in this form. It must first be converted into soluble nitrate compounds. This is accomplished by free-living blue-green algae and bacteria and by the nitrogen-fixing bacteria on the roots of leguminous plants. The nitrates produced there pass into the soil and are absorbed by the roots of plants. Once inside the plant, nitrates can be converted into amino acids that are the building blocks of proteins. In this way, atmospheric nitrogen becomes incorporated into protein. Or it may become converted into bacterial protein without passing into green plants.

The nitrogen in plant protein may now take two different routes. If the plant is eaten by an animal, the protein is broken down and reconstituted as animal protein. In higher animals, the breakdown of proteins yields the nitrogen-containing compound urea. Lower animal forms excrete excess nitrogen as other compounds, such as ammonia (NH_3), or uric acid. In any case, such nitrogenous wastes are acted on by either putrefying or nitrosofying bacteria. The ultimate result is that the nitrogenous wastes are returned to the cycle as either nitrates (if acted on by nitrifying bacteria) or molecular nitrogen (if acted on by denitrifying bacteria).

If the plant dies, on the other hand, the proteins may be acted on by putrefying bacteria (the bacteria of decay). This changes the protein into various ammo-

SUPPLEMENT 25.4
ACID RAIN

Acid rain is produced by the combination of pollutants such as sulfur oxides (SO_2), nitrogen oxides (NO and NO_2), nitrates (NO_3), and carbon oxides (CO and CO_2), with water vapor in the atmosphere. Equations for some of these reactions are given below:

$$SO_2 + H_2O \rightarrow H_2SO_4 \text{ (sulfuric acid)}$$
$$NO_2 + H_2O \rightarrow HNO_3 \text{ (nitric acid)}$$
$$CO_2 + H_2O \rightarrow H_2CO_3 \text{ (carbonic acid)}$$

These acids accumulate in clouds in the upper atmosphere. They may either be rained down immediately or carried aloft by the prevailing winds to be precipitated in some faraway place when the conditions are right for rain. Acid rain is very destructive to the natural and human-made environments. On the east coast of the United States, for example, the average pH of rainfall has been measured at between 4.0 and 4.2 (remember neutral is 7.0 and that the pH scale is logarithmic). A survey in 1975 showed that more than 100 lakes in upstate New York had a pH of less than 5, with some as low as 3! More than 80 of these lakes had totally lost their fish populations due to increasing acidity of the waters. In 1980 acid rain was reported in other areas of the country, such as Colorado. Acid rain also destroys building surfaces (especially those made of soft materials such as limestone) and the paint coverings on houses and automobiles. Acid rain has become a serious problem in recent years in areas, such as the northeast, that have a heavy concentration of industry. Only strict attention to pollution control by both automobile owners and big industry can lower the amount of acid rain generated in the atmospheres above our larger cities.

nium compounds, as well as some nonnitrogenous waste products. The ammonium compounds thus produced can be acted on by either of two types of bacteria. They may be converted into atmospheric nitrogen by the action of denitrifying bacteria, or they may be converted into nitrite compounds by nitrosofying bacteria. Either of these processes occurs in soil or water, as bacteria act on ammonium compounds released by the bacteria of decay. The nitrogen taken from the green plants by herbivorous animals may be returned to the environment by the death of the animal. In this case, the putrefying bacteria are again the agents by which proteins are converted into ammonium.

The nitrogen cycle illustrates the fact that animals are unnecessary for the operation of an ecosystem. Only green plants and bacteria are essential. Bacteria are needed because they can make use of the nitrogen in the atmosphere. Plants are needed because they can use sunlight to synthesize the organic compounds that provide the bacteria with a source of energy when they decay. Animals only enter the picture by feeding on plants. The nitrogen cycle (as well as the carbon cycle) would operate perfectly well without them. Blue-green algae are both nitrogen fixers and photosynthesizers. An ecosystem of only blue-green algae is therefore conceivable, and in fact existed for millions of years before more advanced forms evolved.

These cycles are by no means the only ones that may be traced in nature. Other important cycles are the oxygen, the calcium, and the sulfur cycles. This last one has been strongly affected by industrial wastes and threatens many natural equilibria through so-called acid rain (see Supplement 25.4).

25.8
SUCCESSION: CHANGES IN ECOSYSTEMS THROUGH TIME

Anyone who has planted a garden knows what happens if it is not cultivated: the desired plants are soon hidden by various kinds of weeds. When a farmer allows a cultivated field to lie unused, a crop of annual weeds generally grows on it during the first year and perennial herbs may appear the second. Gradually, the perennials are replaced by shrubs and eventually by trees.

Such a sequence of changes in plant communities occupying a single region is called **succession**. Primary succession occurs when a previously unoccupied area, such as a sand dune or a lake, is gradually occupied by vegetation. Similarly, primary succession occurs when bare rocky surfaces are slowly covered by lichens and mosses that build up a soil layer that eventually can support small seedlings. Secondary succession occurs in an area where some plant communities previously existed, but were destroyed: for example in an abandoned farm field or a burned-out forest.

Primary Succession

Much of the modern thinking concerning succession was developed though the pioneering efforts of several ecologists, including Henry C. Cowles of the University of Chicago. Cowles described in great detail the successional changes in the vegetation of the sand dunes at the southern end of Lake Michigan. Many centuries ago, the shore of the lake extended much farther south than at present. Over the years the lakeshore slowly retreated

BEACH GRASSES   COTTONWOODS BEACH GRASSES   JACK-PINE FOREST   OAK FOREST   BEECH MAPLE FOREST   HUMUS SOIL   SAND

northward, leaving behind a series of progressively older sand dunes and beaches. Walking from the present-day lakeshore, Cowles observed a series of different plant communities (Fig. 25.16). Near the edge of the water there were no plants because of the destructive action of the waves in this large lake. Higher up on the beach, where the sand is dry in summer and frequently buffeted by the waves of winter storms, a few species of succulent annual plants managed to survive. Behind the beach, the sand dunes began. The sand dunes are rigorous environments, very hot in the day and cooler at night. Cowles noted that beach grasses survived on the dunes and actually helped to secure them from the action of wind with their extensive underground stem (rhizome) systems. Various species of insects were the principal animals among the dunes.

Once the dunes had been stabilized by the grasses, Cowles found that various species of shrubs, including cottonwoods, became established. The matted roots of these plants added to the stability of the dunes. On the slightly older dunes behind the cottonwood community, shrubs of other genera, including junipers and jack pine, flourished. Farther back from the pine woods was a forest dominated by oak trees. Finally, several miles from the present lakeshore, Cowles observed forests of sugar maple and American beech growing in deep, rich soil.

Cowles interpreted his observation by hypothesizing that the series of communities represent different

Fig. 25.16
A highly diagrammatic portrayal of plant succession on the dunes at the southern end of Lake Michigan. (After Eugene P. Odum, *Fundamentals of Ecology.* Philadelphia: W. B. Saunders, 1959, p. 261. By permission of Holt, Rinehart and Winston.)

stages in ecological succession beginning with bare beach and culminating in a well-established forest of sugar maple and beech trees. The oldest and presumably stable community in such succession became known as the **climax community.** The farther a given area was from the shore, the older a stage of development it represented. Space could be transformed, for the ecologist, into time.

Cowles's original outline of the subsequent steps these communities followed was essentially correct, but further study demonstrated that the dune successional sequence was more complex than Cowles had appreciated. Other investigators found there was no single linear sequence leading from pine to black oak to beech and maple. Instead a network of different assemblages could occur, depending upon specific characteristics of the soil, drainage, and biotic interactions.

These early ideas about succession, as expressed by Cowles and F. E. Clements, among others, tended to identify successional stages with one or a few types of species. They often *typed* stages of succession rigidly,

SUPPLEMENT 25.5
INTIMATE RELATIONSHIPS—COMMENSALISM, MUTUALISM, AND PARASITISM

In the various interrelationships among organisms in an ecosystem, some of the most fascinating are those under the general heading of "symbiosis." The term symbiosis has been used by biologists in quite different ways over the years. We shall use it here to refer to all the various sorts of interactions involved in "living together" (*sym* = same, together; *biosis* = living) that are beneficial to at least one, or all, of the species involved. There are several kinds of symbiotic relationships that exist in nature categorized on the basis of how

each species is benefited, harmed, or unaffected by the association. These forms of symbiosis are: parasitism, mutualism, and commensalism.

Parasitism

Parasitism is a form of symbiosis in which one member of the association benefits, while the other is harmed. Tapeworms in mammals, mistletoe on trees, or ticks on vertebrates are all examples of parasitic relationships. The tapeworm gets all its nourishment at the expense of the animal in whose intestine it lives; mistletoe takes sap from the tree on which it grows; and ticks feed on vertebrate blood. In parasitic relationships the feeder is called a parasite, and the organism on which it feeds is called the host. The hosts not only get nothing positive from such a relationship, but are actually harmed. Sometimes the hosts are killed; other times they are simply weakened. Parasites can live either inside their host (internal parasites), or outside (external parasites). Examples of internal parasites are liver flukes, tapeworms, the malarial protozoan (which lives in the blood of mammals and in the stomach of mosquitos), and pathogenic bacteria (bacteria that are harmful, such as pneumonia bacteria). Examples of external parasites are skin fungi (animals) or leaf fungi (plants), leeches, ticks, and fleas, or wasps that lay their eggs inside other insects, such as caterpillars. Some parasites actually live inside a host's cells (for example, some protozoans, and all plant and animal viruses).

How does one draw the line between parasitism and predation? It would seem that the definition for a parasite could apply equally well to a predator. The answer is that there is no sharp distinction. The two modes of existence grade into one another, and there may be cases where it is actually difficult to decide which term best characterizes particular relationships. In general, however, we do make some useful distinction, centering around the length of the association. For example, a lynx that preys upon showshoe hares attacks its prey, eats it quickly, and goes on its way. However, the tapeworms, which also attack the hare, live inside the host for a long period, slowly debilitating the host, maybe even eventually killing it just as the lynx did. But the process and the time required in the two cases are quite different. Therein lies an essential difference, in concept at least, between parasitism and predation.

Commensalism

Commensalism is a form of symbiosis in which one member benefits, and the other is neither harmed nor benefited. What the commensal organism gains from its host can be quite varied from case to case: food, protection, shelter, or some combination of these. Examples of commensalism are the small crustaceans that live in the mouths of fishes such as menhaden, where they pick up bits of the fish's leftover food as it goes by. The crustacean gets free food, but the fish is neither harmed nor helped by this situation. Orchids that live on tree limbs in the tropics are another example of commensalism. The orchid gains support from the tree, which raises it off the ground so that its roots can dangle and absorb moisture from the air. Orchids do not gain any nourishment from the trees on which they rest, and therefore are true commensals. One dramatic example involves anemone fish, which live around and in between the tentacles of sea anemones. Normally the anemones sting fish, paralyzing

them and digesting them (if they are small). The anemone fish, however, appears immune to the sting, and on occasion even pokes its head down into the digestive cavity of the anemone to steal small portions of food. An example of behavioral commensalism involves the small staphylinid beetle, which lives in a termite colony as a tolerated scavenger. Sometimes it even rides on the head of a worker termite, intercepting food as it is passed from one worker to another. The beetle gains food and protection from the relationship, but the termite appears to be neither harmed nor helped.

Mutualism

Mutualism refers to associations that are beneficial to both species involved. Such relationships are common among both plants and animals. Lichens represent a mutually beneficial relationship between an alga and a fungus. The alga (being photosynthetic) provides food and the fungus provides support and protection, and helps to hold water. So complete and equal is the mutualistic relationship in this case that for years many people thought lichens were single organisms. Another example of mutualism is the cleaning activity of wrasse fish. These fish eat food particles from the teeth of larger fish such as the squirrelfish, to the mutual benefit of both. The squirrelfish gets rid of food that could decay and cause a health problem; the wrasse does not have to go foraging for its food. An example of behavioral mutualism is found in the culturing of aphids by ants. Ants bring aphid larvae into their colonies and raise the young. By stroking the aphids, the ants cause the aphids to release a sweet secretion, which the ants eat. The ants thus gain some nourishment from the relationship while the aphids get protection within the colony (in addition to having all their food brought to them).

The relationships described above—parasitism, commensalism, and mutualism—are intimate associations that have evolved together over millions of years. Symbiotic relationships are a good example of co-evolution. Most such relationships demonstrate a high degree of specificity—for example, most parasites can live on only one or a few closely related species, and many predators have only one or a few species of prey. Such precision is an indication of the fine-tuning that adaptations can have.

The relationships among the various forms of symbiosis are summarized in Table A.

Table A

Relationship	Species A	Species B
Parasitism	Benefits	Harmed
Commensalism	Benefits	Neutral*
Mutualism	Benefits	Benefits

*Neutral means that the species is neither helped nor harmed.

and did not pay enough attention to the processes that caused succession. Debates over whether succession resulted in a single type of climax (monoclimax theory) for each region, or several possible climaxes (polyclimax theory), raged from the 1920s through the 1950s. Only with the introduction of populational thinking into ecological issues such as succession did a more dynamic picture of community changes come about. The newer version of successional theory is the ecological counterpart of population genetics in post-Darwinian evolutionary theory. This more recent view in the study of succession is called **population pattern theory**. It sees each stage of succession in terms of varying physical and biological conditions that can give rise to a number of different plant communities. Most important, whereas the older theories saw each succession as eventually ending in a stable, climax community—a community that perpetuated itself and did not undergo further change—population pattern theory sees succession as an ongoing process (where even the climax stage evolves, though much more slowly than some of the other stages). Like evolution, succession has no stopping place.

Secondary Succession

The vegetation of an area may be destroyed by fire, grazing, cultivation, or road building. If the soil is not eroded away by rainwater or wind, revegetation of the area will take place relatively rapidly. The process of secondary succession on such artificially modified habitats varies considerably, depending on the slope and climate. One well-studied example occurs in abandoned farmland in Georgia and the Carolinas (Fig. 25.17). As soon as cultivation stops, the fields are colonized during late summer and autumn by several species of herbaceous plants, including horseweed and crabgrass. The horseweed lives through the winter as a dwarf rosette. The following spring growth resumes, producing a tall, many-branched plant that flowers during the summer. While the horseweed is growing to maturity, other herbs, including white aster, invade the field and develop flowers during the second summer. By the third

summer a tall-growing type of grass called broomsedge appears, replacing the aster and horseweed as the dominant plants.

By the time broomsedge appears, pine seedlings are noticeable, and within five to ten years they start to form a forest where once the farmer grew corn and crops. Once the pines grow tall enough to cast a shade over the soil beneath their crowns, however, their own seedlings can no longer grow (young pine plants require almost full sunlight). Not only is there competition between the older pines and the seedlings for light, but there is also intense competition between these two growth stages for the available water in the soil. Thus there is little or no continued reproduction of the pines in this same area.

The seedlings of several other tree species *can* compete successfully in the shaded environment beneath the pines, however. Various species of oaks and hickory and broadleafed trees such as sweet gum are able to use the dim light for photosynthesis and to obtain water from deep in the soil (their roots extend several meters below the roots of the pines, which are confined mostly to the upper 10 to 20 m). Thus sweetgum, oak, and hickory seedlings come up under the pines, and within 100 years after a field has been abandoned they develop into the dominant and overstory trees of the forest. During this interval of time, an understory of dogwood and red maple is also developing. Eventually the old pines begin to die and are replaced by oaks and hickories, so that after about 150 to 200 years the forest consists mostly of broadleaf deciduous trees.

To observe the various states of ecological succession is one problem; to *explain* why such successional changes occur is another. It seems clear that major changes in the vegetation of a region can only follow changes in the environment. Plants and other organisms

Fig. 25.17
Plant succession following abandonment of crop land in the Piedmont region of southeastern United States. (After Eugene P. Odum, *Fundamentals of Ecology.* Philadelphia: W. B. Saunders, 1959, p. 263. By permission of Holt.)

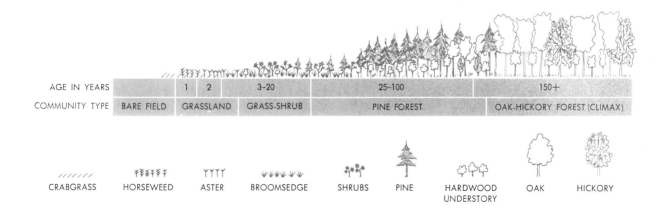

AGE IN YEARS		1	2	3–20	25–100	150+
COMMUNITY TYPE	BARE FIELD	GRASSLAND		GRASS-SHRUB	PINE FOREST	OAK-HICKORY FOREST (CLIMAX)

CRABGRASS   HORSEWEED   ASTER   BROOMSEDGE   SHRUBS   PINE   HARDWOOD UNDERSTORY   OAK   HICKORY

living in a community modify their environment. For example, trees shade the ground beneath them, affecting soil temperature and humidity. Leaves fall to the soil surface and undergo decay. The resulting material affects the runoff of rainwater, soil temperature, and the formation of humus in the soil. These factors in turn affect soil development and change the quantity and type of available nutrients, soil pH, and aeration. These modifications of the environment by the organisms usually make it less favorable for themselves and more favorable for species that could only have survived there earlier with great difficulty. As organisms change the environment, they make it possible for other organisms to compete successfully with the established species. Such behavior seems unlikely to be selected for during evolution; it occurs because it is unavoidable. Permanence and lack of change do not exist in the natural world.

The environment may also be changed by forces other than living organisms. An overflowing stream, for example, may deposit fertile silt on bottomland. The bottom of a pond or lake may be raised because silt has been washed in, as often happens in bodies of water impounded by dams. The chemical content of the soil may change because of leaching. These and similar modifications of the environment are usually followed by changes in the vegetation.

Ecological succession eventually results in the formation of vegetation existing in a steady-state equilibrium with the soil, climate, and herbivorous animals. Although no natural community is static, such a relatively stable community has less tendency than the earlier successional stages to modify its environment in a way that is injurious to itself. The plants of such a "climax community" are able to perpetuate themselves because their seedlings can survive in competition with older plants. If environmental factors do not change appreciably, the climax vegetation will continue for centuries without being replaced by another stage. A climax community may change somewhat in the kind of plants that comprise it. Until about fifty years ago, for example, the climax forests of eastern North America contained abundant chestnut trees. Chestnut trees have now been virtually eliminated by a fungal disease.

In fact, the original climax vegetation that once existed over most of North America (and the entire earth) has been mostly destroyed by humans. The normal stages in succession have been set back, modified, or stopped by such human activities as lumbering, grazing domestic animals, cultivation, urbanization, industrialization, and even high-energy radiation. In settled regions, many of the plant communities that now exist do so not because of the natural process of ecological succession but because of deliberate human interference. People maintain such types of vegetation as crops, pastures, golf courses, and lawns, as well as wildlife preserves and managed forests, because of economic inter-

ests. These communities require some effort to maintain, however. Managed forests, for example, must be periodically thinned by cutting some of the trees.

Succession and the Abundance of Species

As succession proceeds, what happens to the number and kind of species represented? Does the number of species remain the same, with the major change being mere replacement of one species by another? Or does the actual number of species change, either increasing or decreasing the diversity of the community? These and many other questions have been studied quantitatively by ecologists. A typical pattern of secondary succession that bears directly on the rate of diversity has been provided in a study by Fakri Bazzaz of the University of Illinois. Bazzaz collected data on the number of plant species inhabiting an abandoned farm in Illinois over a forty-year period. The results are shown in Fig. 25.18, with each bar representing one species, and each color one of the three vegetation types. For the purposes of the study, he grouped plants into three categories: herbs (nonwoody, especially grasses), shrubs and trees (woody). Note that one year after abandonment the field contained only herbs, mostly grasses but also forms like goldenrod and Queen Anne's lace. After only four years some shrubs appeared, and by twenty-five and forty years later there were trees present in increasing abundance. By the fortieth year each type of vegetation is present in more nearly equal quantities than at the beginning.

There are two additional pieces of information to glean from this graph. The first is that the numbers of species *in all three categories*, but especially of shrubs and trees, increased as succession proceeded. The data from the graph are shown numerically in Table 25.1. Note that the number of herb species does not change markedly throughout the entire period, even though the number of shrubs and trees is increasing. The addition of shrubs and trees does not drive out the herbs in this successional process. From their first appearance, the number of shrubs and trees tended to increase rather steadily (shrubs from three, four, seven, nineteen; trees from fourteen to twenty-three). The fact that so many

Table 25.1

Type of vegetation	Years (from beginning) in which samples taken				
	1	4	15	25	40
Herbs	31	27	26	30	34
Shrubs	0	3	4	7	19
Trees	0	0	0	14	23
Total species	31	30	30	51	76

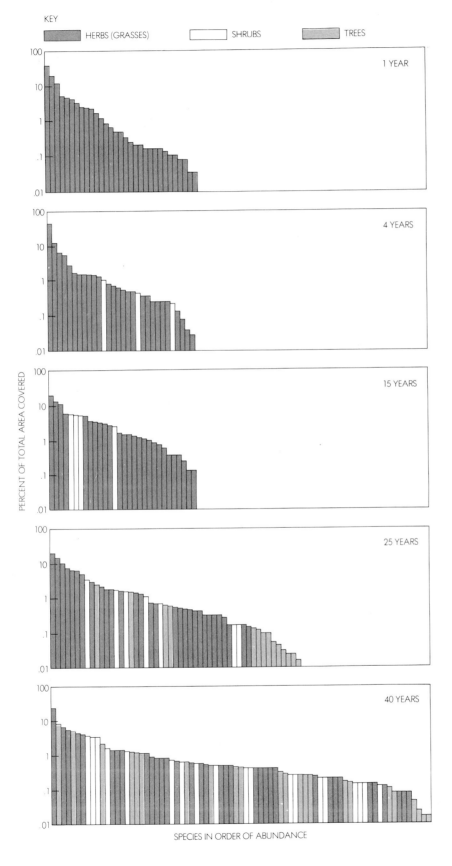

KEY

HERBS (GRASSES)   SHRUBS   TREES

1 YEAR

4 YEARS

15 YEARS

25 YEARS

40 YEARS

PERCENT OF TOTAL AREA COVERED

SPECIES IN ORDER OF ABUNDANCE

Fig. 25.18
The relative abundance of three classes of vegetation over a forty-year period in an abandoned field in southern Illinois. The relative abundance of each type is indicated by the height of the bars. Each bar represents a different species, and each color a different overall kind of vegetation: grasses and herbs (grey), shrubs (white), and trees (color). (Redrawn from Robert M. May, "The evolution of ecological systems," *Scientific American,* September 1978, p. 170.)

herbs appeared rapidly at first (within a year thirty-one species—all herbs—became established) is a result of the fact that unoccupied soil is extremely susceptible to inhabitation. Grasses and herbs are very well adapted to getting a foothold when no other plants are there. Therefore the first stages of succession are the most rapid, with the expansion of new species into a basically unoccupied group of niches. The establishment of the grasses in particular has two effects on the future course of succession. The first is short-term. Because of their rapid expansion into an area and relatively rapid growth, the grasses retard the invasion of other plants. Second, grasses and other herbs form root systems that hold the soil and make it possible for other, more slowly growing plants (such as shrubs and trees) to gain a foothold. Thus, the rate of invasion of new species slows down after a while, for predictable reasons.

A second important generalization we can make from the data in Fig. 25.18 is that succession proceeds generally from a few species with very dense populations (high percentage of area covered in the graphs in Fig. 25.18), to a large number of middle-range species —that is, species that all have about the same population sizes—that are less dense than in the earlier stages. Note that as the spread on the graph becomes larger (especially in the forty-year sample), the heights of the bars become more nearly the same, indicating that the abundance of each species becomes more nearly equal as the diversity of species increases. In the study of succession, this is what ecologists describe as the trend "from dominance to diversity."

Why should this trend exist? In the open, abandoned field (considered an ecosystem), there are initially some unfilled niches: on top of the soil, just below the surface, gullies. Seeds of many plant species are distributed by wind—and it would be highly unlikely for only one species of seed to reach the field. As we have seen, grasses and herbs are fast-growing and become established quickly. This begins to change the ecosystem as a whole, primarily by holding the soil, preventing the immediate invasion of other species, and by providing sources of food and habitats for other species, primarily animals. As new species enter the system, they too alter the conditions. Shrubs provide new habitats and refuges for larger animals, such as rabbits or pheasant; and trees provide habitats and a food source for a whole host of species, from fungi to insects, squirrels, and birds. Unless the ecosystem is accidentally disturbed by human beings or by natural events such as flooding or fire, once succession begins, diversity is almost an inevitable outcome.

Succession: Summary of Major Developments

As modern studies have increasingly shown, succession is not a uniform process in various ecosystems. Many differences exist between geographic regions, individual localities, soil and topographical areas, and the like.

Nonetheless, a number of general trends can be identified; with some modifications they apply to most examples of succession.

1. Species composition changes constantly during succession; the rate of change is usually more rapid at the outset, becoming more slow as time goes on.

2. The number and kind of species increases at the beginning of succession, and becomes more or less stable as the ecosystem matures. Succession tends to progress from large populations of a few species, to smaller populations of many species. This is especially true of heterotrophic organisms, which exist in much greater diversity in most older, compared to most younger, ecosystems.

3. During the earlier stages of succession primary productivity (by autotrophs) increases. By the time the ecosystem has begun to stabilize, productivity has reached a high, and usually stable, level.

4. Food webs become more complex during succession; at the same time the relationships among the various communities (species) involved in a food web tend to become more specialized as time goes on. This increases the efficency of resource utilization within the ecosystem as a whole.

5. As succession progresses, the interactions among various species alter the environment, setting the stage for later developments within the system. Organisms in an ecosystem are constantly evolving, and thus changing the biotic and physical environment with which each species has to contend. Although ecologists speak of succession more or less ending with a climax community, in reality there is no final stage in the process. Rate of change may slow down at the climax stage, but change, and evolution, within the system will always continue.

25.9
BIOMES

In addition to their interactions with one another, organisms also interact with the physical and chemical environment where they live. Various physicochemical factors—sunlight, water, minerals, and pH, among others—have a strong effect on the kind of life that inhabits any particular area. This point can best be illustrated by considering several types of environments found throughout the world.

Various ecological systems on land have been separated into broad categories known as biomes. A **biome** is a very large community of species that inhabit a particular region; each biome has a distinctive and easily observable physical appearance.

Color Plate IV shows several of the more common biomes on the earth: temperate deciduous forest, tropical rain forest, desert, and taiga (coniferous for-

SUPPLEMENT 25.6
THE CLEANING OF LAKE WASHINGTON

The importance of Lake Washington relates not only to how the people of Seattle solved a major water pollution problem but also to the general way in which the removal of a major nutrient, phosphorus, was found to be a key factor in the excessive growth of blue-green algae. The results of studies on Lake Washington have played a major role in formulating policies of controlling water pollution in many lakes around the world, including Lakes Erie and Ontario. Lake Washington is relatively large with a surface area of 50 square miles (128 km^2), a maximum depth of 194 ft (59 m), and an average depth of 59 ft (18 m). Originally the lake had clean water and low productivity of algae; transparency of the water was relatively high.

In 1865 only 300 inhabitants lived in Seattle. By 1922 there were an estimated 50,000 people, with thirty pipes draining raw sewage into the lake. The first attempt to stop pollution of the lake was construction of a diversion system so that sewage outflows were transferred to more open, marine waters off Puget Sound, an arm of the Pacific Ocean. By 1963, however, eleven sewage plants processing wastes from an estimated 100,000 people had been constructed, and effluent from these plants was flowing into the lake. Although the treatment facilities did improve the quality of water coming into the lake, about 300 kg of phosphorus entered the lake each day. Dissolved phosphorus is usually present in only very small concentrations in waters from natural drainage areas. In the 1950s Dr. W. T. Edmondson, a biologist at the University of Washington, began to arouse city officials and citizens. He noted that high population densities of *Oscillatoria rubescens* (a species of blue-green algae known to be indicative of polluted water) and other algae had caused the

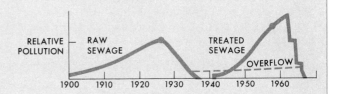

Diagram A
Sewage history of Lake Washington. Graph shows relative amounts of sewage entering the lake at different periods of history.

transparency of the water in 1957 to be about half what it was in 1950.

The citizens adopted a plan to restore the lake, and a new diversion system was begun in 1961. All nutrient-rich effluents were discharged to the sea by 1968. There were sharp indications that algae growths were being reduced while the flow of nutrients into the lake was declining. The yearly values of dissolved phosphorus showed a steady decline and the water clarity increased. By 1970 growth of algae had reverted to 1950 levels.

The long-term trends of these dramatic changes in pollution (Diagram A) are reflected in changes in the organic content preserved in the sediments of Lake Washington (Diagram B). Cores of sediments were obtained from the bottom of the lake using a specially designed tube for extracting continuous deposits of soft mud. Samples were burned in a very hot furnace (500°C). Greater loss of weight on ignition indi-

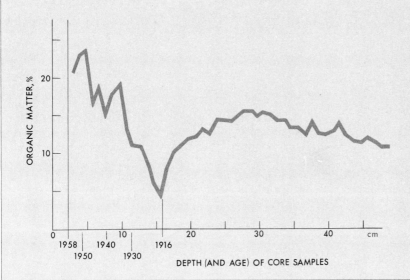

Diagram B
Content of organic matter at different depths from a region of Lake Washington. Amount of organic matter at any depth is determined by loss of mass on ignition at 500°C. The more organic matter present, the greater the amount of algal and other life forms present in a particular year, and hence the more polluted the lake could be said to be.

cated more organic material present in the particular sample. For instance, a core sample from a depth of 16 cm showed a low value (4 percent). This is correlated with the period beginning in 1916 when sand was washed into the lake due to construction of a canal. After 1916 there was a marked increase (to 25 percent) in organic matter until the 1950s. At the same time there were marked changes in the species composition of algae and microcrustaceans in Lake Washington. Remains of diatoms (distinctive silicious cell walls) and zooplankton (chitinous carapaces) were extracted from the sediment cores, and their distributions showed a shift in abundance with dominance of different species. The types of organisms indicative of polluted water were most numerous when the organic content of the sediment was highest.

In lakes around the world ecologists have been able to trace historical changes in organic production so that the impact of even ancient settlements can be documented. These studies suggest that many lakes are capable of rapid recovery following relatively long periods of nutrient flow into their basins. It appears that many nutrients can be washed out of the lakes, especially if flow in and out is relatively rapid as it is in areas with high rainfall. Much of the inorganic phosphorus is bound to the lake sediments and is eventually buried beneath the active surface of nutrient exchange with the open water of the lake. These findings have led some ecologists to be optimistic regarding plans to restore many of our once-polluted lakes. However, even though Lake Washington was effectively "saved" from excessive growths of algae, the solution of transferring dissolved phosphorus to the sea assumes there will be adequate mixing and dilution of this nutrient so as not to cause disruption of marine ecology. Monitoring the possible consequences of a "solution" for one problem and determining the effects on related ecosystems is intrinsic to ecological studies of complex natural systems.

est). In addition, four of the biomes are shown with data on their temperatures and precipitation levels in Fig. 25.19. From the diagram in Fig. 25.19 we can see that the differences between the kinds of organisms that comprise each biome type are obvious—particularly among the plants. For instance, the hot desert biome consists of an assemblage of plants that are all adapted to the same general physical conditions of high temperature and low moisture. The species of plants and animals inhabiting the Sahara desert are different from those inhabiting the Mojave desert; yet both regions would be spoken of as desert biomes.

The biome concept is useful in emphasizing one of the most important ideas in modern ecology: that the physicochemical environment determines the total variety of life that can exist in any broad geographic area. The physicochemical environment operates primarily by "dictating" the kinds of plants that can exist; the kinds of plants operate to determine the kinds of animals that inhabit an area. If the major physical and chemical factors prevalent in a given region are known, a trained ecologist can predict fairly accurately the kind of biome that will be found there. Among the most important of the physicochemical factors are temperature (range as well as average) and amount of water. Each of the four biomes pictured in Fig. 25.19 is differentiated from the others primarily in terms of these two factors. As the graphs show, average temperature in a tropical rain forest stays about 80°F all year, but the amount of rainfall may vary considerably from early March (1 inch/month) to early December (16 inches/month). On the other hand, the average temperature in a grassland biome in the United States varies considerably (from 25 to 70°F) from midwinter to midsummer, while the amount of precipitation is low all year (between 1 and 2 inches/month). Differences in these two factors produce markedly different types of biomes.

Of course, other physical and chemical factors are also important, such as presence of certain minerals, pH, type of soil (amount of clay, sand, and the like), and amount of wind and sunlight. Limitations in each of these other factors may modify the kind of biome that might be predicted for a given area on the basis of temperature and water conditions alone. Thus an area where temperature and rainfall patterns would otherwise produce grasslands might produce a semidesert biome if the soil were sandy and allowed rapid runoff of water. It cannot be emphasized too strongly that the physical factors in the environment are the ultimate determiners of the type of biotic community that can exist in any particular place.

A dramatic example of the effects of physicochemical environment on biome type can be seen by simply climbing up a high mountain (see Fig. 25.20). There is a strong parallel between the biomes inhabiting certain altitudes on the mountain and those inhabiting certain latitudes on the surface of the earth as one proceeds from the equator northward. Ascent of the mountain presents, in telescoped form, a biological journey across a large segment of the earth's surface. At the bottom of the mountain (suppose for the sake of completeness that the mountain is in a fairly tropical climate to begin with) is the very dense tropical rain forest, with its high humidity, tall trees, and large number of species. A little way up the mountain the tropical rain forest gives way to a forest of deciduous trees (which lose their leaves in winter), corresponding to the dominant forests of the north-temperate zone of North America

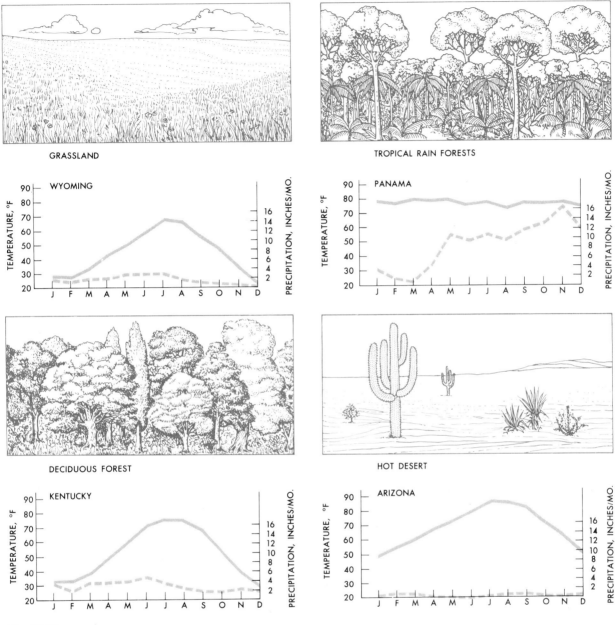

Fig. 25.19
Four types of land biomes found throughout the major continents (except the arctic and antarctic regions) of the earth. Each biome type is largely determined by temperature and amount of rainfall per year. Graphs showing temperature and rainfall for each biome type accompany the drawings. On the graphs the solid line traces temperature, the broken line precipitation.

(hence called the temperate deciduous forest). Deciduous forests require less rainfall and humidity and can withstand lower temperatures than tropical forests. They also have a less extensive assemblage of species.

Proceeding up the mountain (or northward along the earth's surface) we encounter the coniferous forest, consisting of various spruces, pines, and firs (known as the taiga). Conifers require still less rainfall and can

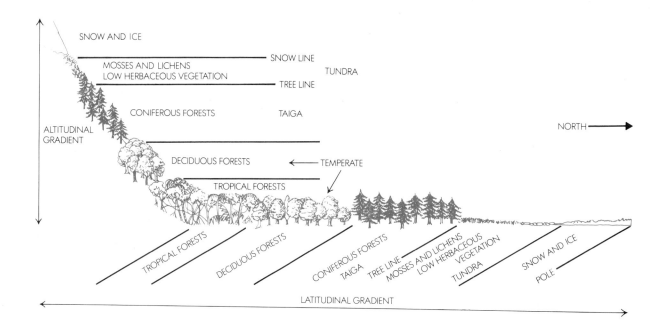

Fig. 25.20
Biome types proceeding up a mountain correspond to biome types proceeding northward from the equator toward the pole. In North America, average temperature falls as one proceeds up a mountain or toward more northern latitudes. The different biome types found in each area of latitude or altitude are a consequence of temperature, moisture, type of soil, wind, and sunlight. These physical factors determine the total assemblage of life that can inhabit each particular area.

withstand more severe cold than deciduous trees. They can also live more readily in rocky and sandy soil, since they have taproots that penetrate deeply into the earth. Proceeding on upward, or northward, the coniferous forest begins to disappear. First the trees that remain are scrubby, showing the effects of high wind and cold temperatures. Finally trees disappear completely. On the mountainside we say we have reached the "tree line." The biome to occupy the next highest altitude, or northerly latitude, is often called the **tundra**. It is cold and windy, with vegetation limited to low-lying shrubs, mosses, and lichens. Tundra usually has a permanent stratum of frost, the **permafrost**, the exact depth of which varies according to latitude (or altitude). Because of the permafrost, when the tundra thaws in warmer weather, water cannot soak into the ground and run off. Consequently, tundras are marked during the summer with bogs and small lakes or pools. Lying still beyond the tundra, at the top of the mountain or around the arctic regions of the earth, is permanent ice and snow. Such areas are not necessarily lifeless, but contain algae, bacteria, fish, and a few larger animals (for example, penguins and polar bears).

The distribution of biomes on the surface of the earth is a consequence of the various physicochemical (including climatic) factors characterizing each geographic area. As we will point out in a later section, the presence of living systems in a given area can, after some period of time, alter physicochemical conditions. There is thus a constant interplay between the physicochemical (abiotic) factors that determine the kind of life that can exist, and the life forms (biotic factors) that do exist at any point in time.

25.10
THE GEOGRAPHICAL DISTRIBUTION OF LIFE: BIOGEOGRAPHY

The principles that govern the establishment of biomes can be applied more broadly to an understanding of the large-scale distribution of life across the face of the earth. Like the diversity within biomes, the diversity of life throughout the many geographical regions of the earth has both a pattern and a history. Organisms are no more spread randomly through the land and seas than they evolved randomly without regard to climate and other environmental factors. The study of the patterns of geographic distribution of life forms is called **biogeography**. Over the past century, the data from biogeography have become increasingly important in our understanding of both ecology and evolution, and of the relationship between them.

Modern biogeographic studies have two different complementary aspects: the present and the past distribution of organisms. We might think of the history of life in terms of a large, branching tree, as shown in Fig. 25.21. If we looked straight down on the tree from

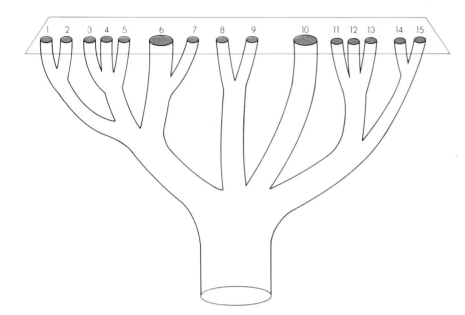

Fig. 25.21
Branching pattern or "phylogenetic tree," showing the relationship between horizontal distribution of the tips of the tree branches (numbered circles) and the branching pattern established by their growth. The present-day position of the tips is a direct consequence of the growth patterns of each branch. Yet, by looking straight down on the tips (imagine an opaque plane through which all the tips protrude, but below which you can see nothing) you would not be able to determine accurately which tips represented closely related branches. The development of phylogenetic charts of species is similar to the growth pattern of the tree shown above. The modern geographic distribution of forms is similar to looking down on the tips of the tree branches. In both cases we are seeing only the present-day outcome of a long process of historical development. If we look at the tree from the side, or look at fossil organisms and their past distribution, we can get a clearer picture of the actual branching, or developmental, pattern—the history of life.

above, we would see only the tips—the ends—of the branches. The number of tips would tell us something about the number of divergences that had occurred from the main stem—though it would not tell us exactly at what point the divergences had taken place, or their pattern. We must look at the tree both from above and from the side for the complete view of not only where the tips of the branches are today, but also how they came to be that way. The same is true of biogeography. Looking at where organisms are found on the earth today is comparable to looking at only the tips of the tree branches of the tree; but looking at fossil evidence of animal and plant distribution is comparable to seeing the whole history of the tree's growth and development. Biogeography is thus an aspect of biology that is very much concerned with history as well as the present. In this way it unites ecology and evolution in a direct way.

A number of observations by biogeographers today and in the past have raised some interesting questions about the evolutionary process. One is why the geographic range of animals or plants—that is, where they are found—is as it is at the present; and conversely, why it has changed over past geological history. Another is how the geographical distribution of species is related to the various environments where organisms are found—that is, adaptations of organisms and communities to various geographic environments. Still another is why there are often discontinuities in distribution, or range, of a particular species. A fourth question is: How do the numbers of species relate to the size

of the area, and how does elimination or addition of species to an area affect the other species that are there?

Biogeographic Regions

Figure 25.22 shows a biogeographic map of the world at present, divided into six major regions (Nearctic, Palaearctic, Neotropical, Ethiopian, Oriental, and Australian). Each region has a distinct climate, which in turn affects the kinds of plants and animals that can live there. For example, the Oriental region is centered on the equator, and gets high average amounts of rainfall (between 80 and 100 inches per year, though some subareas receive much less). These conditions support rapid and continuous growth of vegetation, thus favoring the types of plant life that we recognize as characteristic of tropical forests. The conditions also favor plants that live in close relationship with certain kinds of fungi that recycle nutrients immediately, thus keeping nutrients in the standing crop of vegetation rather than in the soil. This leads to very slow build-up of soil, since all organic matter is broken down to its elemental parts quickly and recycled. Thus, tropical forests are characterized by shallow soils, which, in turn, affect the kinds of new plant species that can gain a foothold.

In turn, the nature of tropical vegetation greatly affects the kinds of animals that can live there. If temperature conditions are favorable for life the year round, tropical animals will not usually show the annual yearly cycles of activity, such as hibernation, distinct breeding periods, or migration, found in animal

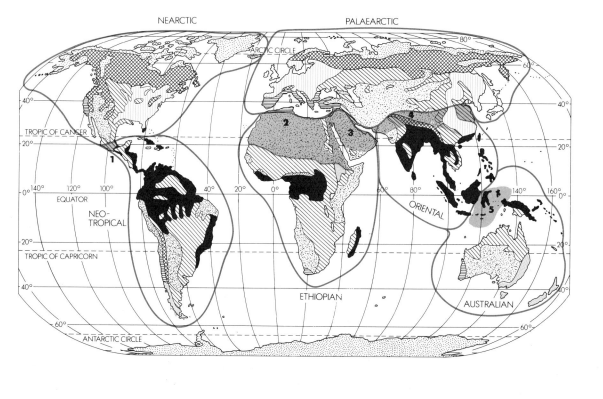

Fig. 25.22
Biogeographic map of the world. The major biogeographic regions are outlined in black (Nearctic, Palaearctic, Neotropical, Ethiopian, Oriental, and Australian). Vegetation within each region is shown by various degrees of shading and hatching, according to the key shown to the left. Geographic barriers, such as mountain ranges, deserts, or large ocean expanses separating the major regions are shown in color and numbered. Designation of the barriers is as follows: (1) Length of the dry season in Guatemala separates the Mexican plateau to the north from the tropical lowlands to the south; (2 and 3) The Great Saharan Desert of Northern Africa, and the Arabian Desert of the Middle East, which together separate the northern Palaearctic region from the southern Ethiopian region; the Himalayan Mountains (and their eastward extension), separating the Palaearctic and Oriental regions; and (5) the marine waterway barrier of Indonesia separating the Oriental and the Australian regions. Note the general correspondence between geographic location and predominant types of vegetation. An obvious causal link exists between geography, climate, vegetation types, and animal life in any area. As this relationship evolves for any area, shifts occur in biogeographic patterns. (After Simpson and Beck, *Life*, 2d ed., Harcourt Brace, 1965, pp. 708–709 and 728; Wilson et al., *Life on Earth*, Sinauer, 2d ed., 1978, p. 702).

species in the temperate (Nearctic) regions. In tropical forests, the diversity of vegetation allows for a much greater specialization in animals' feeding habits. For example, species of tropical birds can specialize in seeds, fruits, or insects that live on the ground, in the understory, or in the crown of trees, respectively, producing a vertical distribution instead of the more common horizontal distribution of niches and therefore of species. In tropical jungles some mammals are terrestrial, but many more are arboreal (some, such as the tree sloths, seldom descending to the ground). In tropical savannas, on the other hand, such as the famous Serengetti plain in

Kenya and Tanzania where grass is abundant the year round and there are few trees, many more animals are terrestrial, and many, such as wildebeasts, antelopes, and zebras, have evolved leg and hoof structures well adapted to running along the ground (see Section 24.7). In desert and semidesert areas, such as the southwestern United States and Mexico, where vegetation is sparse and shows a pronounced cyclical growth pattern dependent on rainfall, animals are mostly nocturnal; if diurnal, they have special coloration to decrease the heat load, and show special adaptations for water retention. Good examples of adaptation to desert conditions are the kangaroo rats of the Sonoran desert (northern Mexico and Arizona). They are mainly active at night and spend the day in cooler underground burrows. Because they can extract more water than other animals from relatively dry plant tissues kangaroo rats can subsist on a diet of seeds with little or no water supplementation. In addition, they are anatomically adapted to hop on their hind legs, decreasing body contact with the hot desert surface. They also tend to be lean, exposing more body surface per unit of body mass; this increases the radiation of body heat and prevents temperature build-up.

The six major biogeographic regions shown in Fig. 25.22 are separated from each other in an east-west direction by large oceans: the Atlantic separates the Nearctic and Neotropical from the Palaearctic and Ethiopian; the Indian Ocean separates the Ethiopian and Oriental, and the Pacific separates the Palaearctic, Oriental, and Australian from the Nearctic and Neotropical. The regions are also separated from one another in a north-south direction by several kinds of natural barriers, as shown in color on the map in Fig. 25.22. In both north-south and east-west directions these major barriers have restricted the migration of animals and the dispersal of plant forms in recent times. As a result of these barriers and of the specific climatic and other environmental conditions prevailing in each major biogeographic region, each region has developed its own characteristic fauna and flora.

Despite the existence of actual barriers separating the six regions, the boundaries indicated in Fig. 25.22 are not as rigid and fixed as the map lines. Everywhere climates grade slowly from one to the other. Animal and plant populations intergrade in much the same way. Obviously, in rather specific cases (for example where land meets water) there are sudden and sharp discontinuities both in topography and in species distributions. In most cases, however, the dividing lines are far less abrupt. Even proceeding from a plain up the side of a mountain range (see Fig. 25.20) there are slow intergradations of environment and animal and plant communities, making it impossible to draw sharp lines. A map such as that shown in Fig. 25.22 helps us identify the large-scale relationships between organisms and different geographic regions, but should not be taken as consisting of hard, fixed boundary lines.

What are the characteristics of the major biogeographic regions? The Nearctic and Palaearctic (sometimes called, collectively, Holarctic) are the two regions which are most physically alike. Both contain the northernmost vegetation type, the tundra, characterized by low-growing bushes and ground cover with no trees; the taiga, or coniferous forest; grasslands and temperate deciduous forests (hardwoods); and some desert (the deserts of southwestern United States and the deserts of central Asia north of the Himalayas). The fauna of the Nearctic and Palaearctic regions includes such animals as wolves, hares, moose (called "elk" in Europe), deer, caribou, mountain goat, and bison. The Oriental region is the habitat of the tiger, Indian elephant (a different genus from its African counterpart), gibbons, water buffalos, antelopes, and many other mammals found almost nowhere else. The Ethiopian region is particularly prominent because of its distinctive forms: the giraffe, zebra, African elephant, and antelopes (some related to the Oriental forms and some wholly different).

The flora and fauna of the Neotropical and Australian regions is the most unique. The Neotropical is characterized by heavy rainfalls (up to 80 inches per year in some places, and more in others) and dense vegetation taking the form of the tropical "rain forest." Tropical rain forests are spread over parts of Central America, the northern part of South America, central Africa, southern Asia and parts of India, the East and West Indies, the south Pacific islands, and even some parts of northern Australia. For the person coming from the temperate north, nowhere does life appear more abundant and exotic than in tropical rain forests. By comparison, as Darwin found, the temperate zones appear barren. Whereas the latter will often consist of one or two species of trees (rarely more than 12), the rain forest may have over 100—and in at least one location over 500 species of trees alone have been catalogued! In terms of mammals, the distinctive forms of the Neotropical region are the New World monkeys (as distinct from the Old World monkeys of the Oriental and Ethiopian region), the sloths, true anteaters, guinea pig, and other related rodents and armadillos.

The fauna, especially mammalian, of the Australian region is even more unique than that of the Neotropical. A good deal of its climate is virtually desert-like. The vast central regions get very little rainfall (5 to 20 inches per year), with only the northern and eastern coastal areas having anything like the rainfall we know in the temperate zones of North America (40 to 60 inches). The mammalian fauna of Australia consists of more marsupials than found in any other region. These include the kangaroo, wombat, marsupial squirrel, and Tasmanian wolf (a marsupial dog). In addition, the peculiar group known as Monotremes (such as the duckbilled platypus), which lay eggs but nurse their young with milk, is also found only in Australia. As we saw in Chapter 24, only two groups of marsupials are found outside Australia. One of these is the familiar opossum of North America which is, incidently, absent from Australia. Thus the Australian biogeographic region, which is physically more isolated than perhaps any of the other major regions, has also the most specialized fauna.

Questions and Problems

The descriptive data of animal and plant biogeography raise some interesting questions and problems for the evolutionist. One is the peculiar patterns of continuity and discontinuity that exist among particular groups of animals and plants as we pass from one region to another. For example, most of the mammalian fauna of North America resembles that of northern Asia much more than it does that of South America. An exception is the porcupine of North America, which has its closest relatives in South America. Given the land bridge that currently exists between North and South America it is difficult to explain why most North American fauna does not follow the porcupine pattern. How did such distributions come to be the way they are?

A second problem that biogeographic patterns reveal is that of *disjunctive distribution.* Frequently, two closely related groups of organisms (two species, or even two populations of the same species) occur in widely separated regions with no closely related forms in between. Common examples are the tapirs, which live in Central and South America on the one hand, and in southeast Asia on the other; and the camels, which live as wild animals only in South America (the llamas and their relatives) and Asia (the familiar desert camel). Fossil and other evidence strongly suggests that such disjunctive groups must have had a common ancestry at some point in the past. But how did their distribution become so disjointed?

Biogeographers and evolutionists turn to evidence from paleontology and geology, particularly studies in the changes in the location and structure of continents, to determine how the exact patterns of animal and plant distribution could have come about. While our knowledge of this problem is still far from complete, some exciting new progress has been made within the past decade.

Continental Drift

In the history of science it frequently happens that new developments in one area of work are stimulated by developments in a totally different area. Such is the case with modern biogeography. For decades most interpretations of biogeography have been based on the idea that the continents have always been pretty much in the same positions, relative to one another, that we see them in today.

Geologists now recognize that the land areas of the earth have gradually been shifting their positions over the whole of the earth's history. This shift results from the position of the land masses of the earth on top of flattened plates that gradually shift over one another in lateral directions. The theory of **plate tectonics,** as it is called, has replaced the older geological picture of the earth's crust as a more or less continuous, solid sheet of rock that changes only by rising and falling, folding and buckling. Using plate tectonic theory, geologists have reconstructed the migration of the present continents from a single land mass, called Pangaea, in the Paleozoic era (some 200 million years ago), into the six continents that exist today (see Fig. 25.23). This migration has largely involved a break-up of Pangaea and the dispersal of its split-off sections. It has long been recognized that parts of today's continents appear to fit quite closely together, like the slightly modified pieces of a jigsaw puzzle (for example the east coast of Africa and the west coast of South America). Without the mechanism of plate tectonics to explain how the continents could actually move apart from one another, the suggestion that perhaps South America and Africa once existed side-by-side as part of a larger land mass always seemed highly speculative. Now, however, most geologists agree that plate tectonic theory explains how such changes are physically possible, and could help explain both the shape of the modern continents and their present positions with respect to one another.

The concept of continental drift can help explain some of the perplexing problems of biogeography posed earlier. Consider, for example, the wide dispersal of the dinosaurs and other reptiles common to the Mesozoic era, about 180 to 130 million years ago. During this period, dinosaur fossils were distributed throughout what is now Europe, Asia, and North and South America. As Fig. 25.23 indicates, at about this time the supercontinents of Pangaea had split into two major land masses, Laurasia and Gondwanaland. Given the theory of continental drift, it is easy to see how dinosaurs could have spread freely over these areas, and achieved a kind of widespread dispersal that is otherwise difficult to explain. Continental drift suggests that dispersal of groups of land animals was much more global in earlier eras than it is today. Even 60 or 70 million years ago, when the continents had begun to assume more nearly their present positions, land connections between Eurasia and North America made possible the exchange of animal forms between these areas to a degree much greater than could occur today (see Fig. 25.23c). (The theory of continental drift has also been useful in understanding and explaining the evolution of the horse in the Americas and Eurasia, as referred to in Section 24.7.)

Land Areas and the Distribution of Species

What determines the number and kind of species that can inhabit a given geographical area, be it an island or an entire continent? Is there a fixed number of species,

(**a**) 200 MILLION YEARS AGO

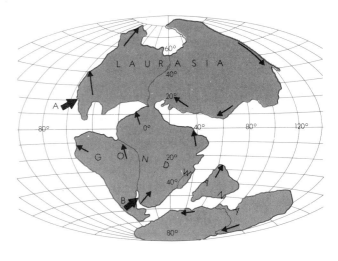

(**b**) 180 MILLION YEARS AGO

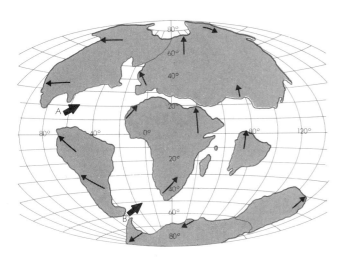

(**c**) 70 MILLION YEARS AGO

Fig. 25.23
General scheme for continental drift. (*a*) At the beginning of the Palaeozoic era, some 200 million years ago, most of the present continental land mass existed as a single large supercontinent, Pangaea. Thin lines within the supercontinent show stress regions where land masses are breaking apart. The geometric figures A and B serve as geographic reference points for all three maps. Ancient oceans, panthalassa (ancestral to the Pacific) and Tethys (ancestral to the Mediterranean) surrounded Pangaea on all sides. (*b*) As plates within the earth's crust shifted, the land mass of Pangaea began to break up, forming two separate supercontinents, Laurasia and Gondwanaland. (*c*) About 70 or 80 million years ago Laurasia began to split up into what has become North and South America on the one hand, and Eurasia on the other; Gondwanaland split into the present continents of South America, Africa and Australia. Continental drift was finally accepted by geologists when the mechanism of plate tectonics explained how large land masses could "move" around the earth's surface. (Adapted from W. T. Keeton, *Biological Science*, 3d ed. New York: W. W. Norton, 1980.)

that any given geographic area can support? Biogeographers, led by E. O. Wilson and the late Robert McArthur, have come to the conclusion that a distinct relationship exists between the size of a land area and the number of species that land area can support. Moreover, this relationship appears to be a relatively fixed, or stable quantity over long periods of time—barring, of course, major climatic or geological changes. As long

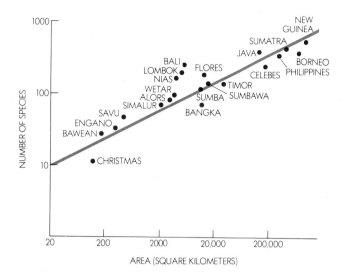

Fig. 25.24
Area-species curve for freshwater and land birds of the East Indies. Note that both axes are plotted on a logarithmic scale. The straight-line curve suggests that the numbers of species on each individual island are near their saturation point—that each island contains about as many different species as it can ecologically support. The area-species curve suggests a major generalization in biology: for any given region there is a maximum amount of species diversity that can co-exist at any point in time. The number of species thus remains remarkably stable unless climatic or other environmental changes are occurring in a region. (From Wilson et al., *Life on Earth*, 2d ed., Sinauer, 1978, p. 703.)

as the physiographic and climatic features of a region remain the same, the diversity and the quantity of life that region can support is relatively fixed.

The conclusion that a relationship exists between land area and number of species has been arrived at by studying the question empirically—that is, by counting the number of species inhabiting given land areas of different sizes. Figure 25.24 shows data for the distribution of species of freshwater and land birds among the islands of the East Indies (including New Guinea, Borneo, the Philippines, and Java). The islands range in size from New Guinea (over 200,000 square kilometers) to Christmas Island (slightly under 200 square kilometers), with all sizes in between. Island size (in square kilom-

eters) is plotted on the horizontal x-axis, and number of species per island on the vertical y-axis. Both axes are plotted on a logarithmic scale, where each equal linear increment represents a ten-fold numerical increase. This graph represents what is called an **area-species curve**. The straight, upward slope of the line indicates a regular relationship between land area and species number. If you examine the graph carefully, however, you will note that the relationship is not arithmetical, that is, the increase is not one-to-one. As land area increases tenfold, the number of species present approximately doubles. Thus, for example, an island with an area of 2000 square kilometers (such as Simalur) supports about twice as many species as one with an area of 200 square kilometers (such as Christmas).

Islands are particularly useful for studies of area-species relationships because they represent discrete localities with easily definable boundaries. Although much of the field work on land area and species distribution has been done on islands, it is thought that the general principles emerging from such work apply to nonisland areas as well. In fact, some biogeographers argue that land areas such as a desert or a mountain rising out of a plain can be treated as if they were islands, at least for the purpose of studying their species distribution. This argument assumes that such land units are biologically very much like an island: they have more or less defined limits, a moderately uniform environment, and their own particular collection of species that do not migrate excessively in and out of the area.

The area-species curve is not a theoretical formulation, but an empirically derived, inductive generalization. It expresses a fundamental and stable relationship between organisms and their environment. What is the basis of this relationship and especially its stability? Although Wilson and McArthur did not attempt to assign causes to their equilibrium concept, others have.

As we have seen earlier, evolution tends to sort out the relationships between organisms so as to ultimately *reduce* competition. Thus, if a new species migrates to an island that is already filled to capacity with species, the invader will necessarily face stiff competition with whatever form has the lifestyle most similar to its own. As we have seen earlier (Sections 24.3 and 25.5), this competition will ultimately result in either the extinction of the least fit (comparatively speaking) of the species or

---

Biogeographers have come to the conclusion, from many empirical studies, that a regular relationship exists between the size of a land area and the number of species that land area can support. As land area increases ten-fold, the number of species approximately doubles.

their divergence in ways that minimize direct competition. However, species-equilibrium theory should not be construed as saying that extinction of species in an area is always, or even largely, caused by immigration. The overall average extinction rate for species can be calculated in general; this rate seems to apply to species whether immigration occurs or not. Thus, as established species in an area become extinct, immigrant species can find a niche into which they can move. Sometimes immigrants cause an established species to become extinct; more frequently, however, established species become extinct for other reasons, leaving a niche into which immigrants can move.

In addition to the area-species relationship, biogeographers have developed three other empirically verified generalizations:

1. The smaller the island (or any defined land area) the higher the extinction rate of species observed on it. It is easy to understand why this should be so. If a dozen species of animals colonize each of two islands, one with an area of 2000 square kilometers and the other with an area of 20 square kilometers, each species on the larger island will have much more room to expand than its counterpart on the smaller island. On the smaller island, even if each species survives the initial migration and reproduces successfully, it will quickly reach its maximum population size, one that will be considerably smaller than on the larger island. As a consequence of small population size, sharp fluctuations in external conditions (such as climate, weather, or introduction of predators, disease or parasitism) and internal fluctuations (such as genetic drift) can much more easily wipe out the whole population. Since each species on the smaller island has fewer members, it is more vulnerable to the vagaries of external conditions than is each species on the larger island.

2. The second generalization is known as the **distance effect**. It states that a more remote island (or a more geographically isolated land area) is able to be reached by fewer invaders; hence, a smaller number of new species will colonize it in a given period of time. This is an obvious effect of distance. Thus, isolated or remote islands may have fewer species than they can in reality support simply because not enough colonizers have reached these areas in recent times. Ireland, for example, lacks several basic life forms that exist elsewhere in Europe. The well-known lack of snakes in Ireland is not a result of the powers of St. Patrick, as myth has it. Rather, it appears that snakes *did* exist in Ireland before the recent ice age, but were wiped out by advancing glaciers. When the glaciers receded snakes never managed to recolonize the island from England or mainland Europe, where snakes do still exist. In time, perhaps, snakes will make it back to Ireland.

3. A third generalization is known as the **feeder effect**, and states that larger land areas tend to be the source of new colonizers that invade smaller, more remote land areas. For example, colonization of the smaller Pacific islands off the east coast of Australia and New Guinea (the Solomon Islands, New Hebrides, New Caledonia, Fiji Islands, Samoa, Society Islands) appears by and large to start principally from New Guinea (though some also spread out from Australia). New Guinea is the source, or "feeder" region for species that disperse eastward, hopping, stepping-stone fashion, from nearer to more distant islands in the group. The feeder effect relates directly to the distance effect. It is obvious that if migrations start at one end of a chain of islands, the more remote islands in the group will receive colonizers later and probably less extensively than the closer islands.

The three generalizations listed above form the basis of what biogeographers call **species-equilibrium theory**. It can be represented graphically as in Fig. 25.25. This theory states that through colonization,

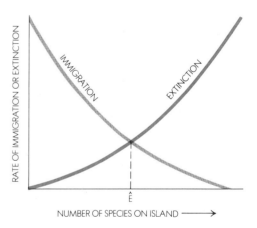

Fig. 25.25
Species-equilibrium curve, showing the relationship between extinction and immigration rates and the number of species on an island. As the number of species on an island increases, the rate of successful immigration falls, while the rate of extinction increases. The size of the island area, and therefore the number of available niches, appear to be the limiting factor(s). The number of species on an island is, in turn, a function of the dynamic interaction between immigration and extinction rates. Under any circumstances species have a tendency to become extinct. Similarly, species are always migrating and therefore ready to become immigrants into a specific area. At point *E*, where the two curves cross, the population is at equilibrium, balanced between the two opposing tendencies of immigration and extinction. It is important to keep in mind that the species-equilibrium concept does not imply that immigration *causes* extinction, or vice versa. Both are constant tendencies in natural populations. For different sizes of islands the two curves will have different slopes, so that every island has its own *E* value.

migration, and evolutionary adaptation, species tend to invade and occupy all the available ecological places open to them in a given area, be it an island or an entire continent. The total number of species that exist in an area remains relatively stable, or at equilibrium. The number of species an area can harbor depends upon size, climate, physiography, and other factors; but for any given area (barring major upheavals) this number remains relatively stable over time. Species-equilibrium theory relates species number in an area to overall land area and distance of an area from a source of immigrant species.

An interesting test of species-equilibrium theory has been carried out by Dan Simberloff of Harvard University. Just off the coast of southern Florida are thousands of small islands, characterized by one type of flora, red mangrove trees (some of the islands are so small they can hold only a single tree). A survey of the fauna of these islands showed that a relatively constant number of species of arthropods (insects, spiders, mites) existed for each island size. For instance, where an island was about 40 feet in diameter, the number of species tended to be between twenty-five and forty-five. Such constancy did not, by itself, constitute confirmation of species-equilibrium theory. However, to test the theory experimentally, Simberloff fumigated six different islands with methylbromide, killing off all the arthropods, but leaving the red mangroves unharmed. The process of recolonization was then observed. Within ten months after fumigation, each of the experimental islands was occupied by approximately the same number of species that had lived there before. The islands most distant from the mainland were slower to become recolonized than those that were closer.

What was interesting to note, however, was that the kinds of species present after recolonization were quite different from those present before fumigation. Different species of insects and spiders colonized the islands once they were empty. In all ways these experimental results were in complete accord with predictions based on species-equilibrium theory. It is not often in field biology and evolutionary theory that it is possible to pose hypotheses and test them so directly.

25.11
ECOLOGY, THE ENVIRONMENT, AND ECONOMICS

Most people are familiar with the environmental movement that began to flourish between the late 1960s and the early 1970s. By 1968 it had become increasingly apparent to many people in the United States that, if not checked, pollution of lakes, rivers, the oceans, and the atmosphere of certain regions of the earth (usually highly industrialized areas) could make the whole planet uninhabitable. Central to the growth of the environmental movement was the study of ecology. As we have seen in this chapter, ecology is the general study of the interrelationships among organisms and the physico-chemical environment in which they exist. A new breed of ecologists, the environmentalists, studied ecology in order to predict how certain human activities would affect large-scale ecosystems. The environmental movement emerged as an attempt to apply ecological principles and understandings to the specific kinds of problems human beings were creating in their daily activities.

However, the environmental movement was something more than the academic study of ecology. It included not only an attempt to understand the sources of environmental problems, but also the attempt to do something about them. Environmental activists made surveys and traced pollutants to their sources. Angry environmental groups picketed and demonstrated at industrial sites, power plants, politicians' offices—anywhere that they could spot a problem—and sometimes achieved a positive response. In the United States, recognition of increasing pollution problems eventually led to federal action, including establishment of the Environmental Protective Agency and passage of the Clean Air Standards Acts of 1970 and 1971. State legislatures also passed a variety of bills designed to clean up air and water and preserve land from the worst forms of degradation.

However, pollution itself turns out to be only part of the total spectrum of environmental problems modern industrial society faces. As many environmental groups pointed out, the demands of an ever-expanding production system were causing excessive wastes on the one hand and excessive depletion of future resources on the other. Moreover, it was noted that the problems of pollution and resource depletion were not evenly spread over the earth's surface. Pollution proved to be a problem for large industrial and relatively wealthy countries of the western hemisphere. Depletion of resources occurred mostly in the poorer countries of Asia, Africa, and Latin America, where financial interests of corporations in the wealthy countries controlled the rate at which many natural resources were used. There was an irony, environmentalists pointed out, in the fact that resources such as copper or petroleum were being taken from the ground in relatively poor countries (such as Chile or Venezuela, respectively) and used at such an enormous rate in rich countries (such as the United States and Britain) that the latter were becoming polluted. Approaching the situation in a truly ecological fashion, environmentalists followed the flow of materials and energy over the whole planetary ecosystem, from where they were obtained to where they were used. The problems of pollution in one locality could not be separated from the problems of resource use in another.

One result of this concept is the recognition that environmental issues are intimately related to economics: human "rules" about how resources are to be controlled and distributed within a society. The pollution problems of the industrialized countries are often a direct consequence of the uneven distribution and use of the world's resources. For example, the United States has 6 percent of the world's population but consumes between 40 percent and 50 percent (higher by some estimates) of the world's natural resources (such as metals and petroleum). If the entire population of the earth used resources at a similar rate, the world's ecosystem could sustain neither the accumulated waste products nor the depletion of natural resources for very long. Recognizing this fact has caused increasing numbers of environmentalists to insist that a prerequisite to solving our ecological problems is reordering our economic priorities in a more rational way. This suggestion has meant not only readjusting rates of production and consumption in the wealthy countries to a level that the world ecosystem can sustain, but also redistributing wealth so as to put the consumption levels of all people on a more equal footing.

The relationship between ecology and economics suggests that environmental legislation in countries such as the United States treats only part of the problem. The effect of such laws as the Clean Air Act is to require that industries clean up their waste-disposal processes—put filters on smoke stacks or pollution control devices on exhaust systems of automobiles. While this is important, it is relatively inefficient to make a mess and then spend time and energy to clean it up. In fact, as some critics have emphasized, the pollution control industry that has sprung up in the last six or seven years itself creates an enormous amount of pollution! A much more efficient procedure would be to reduce consumption of raw materials. This would not only reduce the amount of pollution, but would also alleviate the burden on the world's natural resources.

Private versus Public Transportation

To comprehend this point more clearly, consider the following example. Meeting a society's transportation needs by privately owned automobiles is considerably less efficient than systems of mass transit (buses, subways, or monorails) such as those that exist in Paris, Mexico, Moscow, and Toronto. The widespread use of private automobiles in the United States consumes far greater quantities of raw materials than any form of mass transit currently available. For example, automobile production in the United States accounts for 66 percent of all domestic rubber consumption, 60 percent of all domestic iron consumption, and 10 percent of all copper, aluminum, and manganese consumption. Even more dramatic is the fact that 25 percent of the entire world's annual petroleum output is used to operate privately owned passenger cars in the United States!

Comparing automobile transportation to other forms of passenger travel yields some interesting figures. In terms of fuel (that is, energy) use, automobiles consume an average of about 5400 Btu's (British thermal units, a standard measure of energy) per passenger mile traveled, while trains consume 2620 and buses 1700. In terms of safety per passenger mile traveled, automobiles cause several times as many deaths as trains.

In addition to its use of raw materials, the privately owned automobile is the major cause of urban air pollution. Data for 1971 indicate that about 60 percent of all urban air pollution comes from tailpipe exhaust of automobiles. If the pollution caused by all the industries related to automobile production (at its 1971 rate) were also included, the figure would be much higher. While the use of pollution control devices has reduced tailpipe emissions considerably since 1971, it has not reduced the rate at which the automobile consumes natural resources. In fact, the use of pollution control devices has actually increased fuel consumption by lowering efficiency.

Thus the introduction of mass transit systems would be a far sounder practice ecologically, both in terms of resource use and pollution, than the efforts currently underway to install pollution control devices on a still-increasing number of individual vehicles. Why has this not happened in pollution-ridden countries such as the United States? In fact, instead of putting large amounts of money into development for mass transit systems, over the past decade the United States has invested billions of dollars in its elaborate interstate highway system, encouraging the use (and thus production) of even more automobiles.

Why has this ecologically less efficient path been followed? Certainly one factor has been the enormous lobbying power, and thus political influence, of the automobile-related industries, including petroleum, steel, and rubber. Since the early 1930s the automobile manufacturers have been relentless in their opposition to mass transit systems. Between 1932 and 1951, holding companies created by large automotive manufacturers (General Motors, Chrysler) bought up trolley systems in major United States cities and then dismantled them! This practice indirectly forced the public to buy, and ultimately come to rely on, the privately owned vehicle for most or all their transportation needs. Such a practice has wasted vast quantities of the earth's resources that might otherwise have been used for more productive and socially useful items, such as trucks and tractors to increase agricultural output in less industrialized countries. It has also produced high levels of pollution in the industrialized countries themselves.

Here is but one example of where environmental considerations meet economic and political issues. The production of automobiles and related materials occupies a central position in the economy of such industrialized countries as the United States. Both industry and government have argued that by providing jobs and stimulating the economy, the manufacture and operation of privately owned vehicles yields far more social good than the environmental problems it causes. Many people argue that converting the automobile industry to the production of mass transit systems would greatly increase the level of unemployment, because, as a more efficient system of transportation, mass transit would not require as many workers either to produce or maintain it.

The Social Costs

There are two ways to look at this issue. Both involve asking: What is the social cost of doing something a particular way? Social cost is the total cost society will ultimately have to pay for a certain set of activities. There are essentially two forms of social cost: short-term costs and long-term costs. Changing over the automobile industry from producing individual cars to producing mass transit systems would involve short-term costs: for example, it would cost the companies a great deal of money to retool their factories and retrain their workers. Long-term costs, on the other hand, are the result of activities whose consequences are not usually seen until long after the fact. Ecological degradation through pollution or resource depletion is an example of a long-term social cost; its effects are often not felt until many years or even generations later. Designing a form of transportation that minimizes environmental degradation would be socially more economical in the long run than continuing to use a very expensive form just because the short-term cost of a changeover seems large.

Quite often protecting the environment has been juxtaposed with increasing unemployment. The argument has been advanced, largely by those who speak for industry, that to use up fewer resources and produce less pollution it would be necessary to lower total industrial or agricultural output. Inevitably, this would mean that people would lose their jobs. For example, mass transit vehicles do not have the resale value and thus the capacity for continually increasing profit that the industry now enjoys with passenger cars. If the automobile industry were to switch to the production of mass transit systems, the industry would employ fewer workers in the long run. Industry representatives argue that the long-term social costs of unemployment benefits might be greater than the long-term environmental costs.

It is obvious that posing the choice in this way makes any decisions extremely difficult. Few people would be willing to accept environmental strictures if it meant taking jobs away from others. However, many environmentalists do not see the choice as between sound environmental policies on the one hand and increased unemployment on the other. They point out that there are many socially useful types of jobs—such as those in education or health care—for which people could be retrained. Such jobs could be made available, environmentalists point out, if certain political and economic decisions were made. The reason for unemployment is not a lack of work to be done; it is rather a lack of money flowing in the proper place. Most environmentalists do not see protection of the environment as a necessary threat to full employment.

Environmentalists also make the point that if the use of natural resources is not managed in a rational way, massive depletion and massive pollution will produce widespread unemployment anyway. Ecologically speaking, our entire productive enterprise is based upon the availability of raw materials. To fail to recognize this point is to lead irreversibly toward that day when production is no longer possible because the environment simply cannot supply the resources.

No study of the human environment and its management can ignore the importance of economics and politics. In the rules of economics reside the patterns of resource use and distribution within any society. Of course, economic rules are designed by human beings and can be changed. If, as some environmentalists argue, the private enterprise system threatens ecological destruction by its appetite for expanding profit and production, it may be possible to modify that system in order to develop sound environmental policies.

Summary

1.   All organisms live in a closely interconnected web of relationships with each other and the physicochemical environment. Alteration of any single strand of this web may have profound effects on all other strands.

2.   Ecology is the study of the various relationships between organisms and their environment.

3.   All organisms compete to one degree or another for the limited supply of resources available. Competition is a result of the finite energy supply and other resources in any ecological system.

4.   Interaction among organisms takes several forms:
   a) Direct competition, where one organism gets something that another does not.

   b) Predation, where one organism consumes another, using the prey's body material (biomass) as a source of building materials and energy.

   c) Parasitism, where one organism consumes part of the biomass of another without directly killing it (at least immediately).

   d) Mutualism, where two organisms exploit one another in such a way as to benefit both without killing or impairing either.

   e) Commensalism, where one organism gains something while the other is unharmed; commensalism, mutualism, and parasitism are forms of symbiosis (living together).

5. Biotic potential is the ability of a species to increase in numbers in a given habitat, in the absence of population control by predation, competition, or interference from other species, and in the absence of crowding from members of its own species. Environmental resistance is the combination of those aspects of the environment that prevent organisms from achieving maximum biotic potential (examples are limited food supply and disease). The interaction of biotic potential with environmental resistance determines the size of any population in nature.

6. All environments have a carrying capacity: the ability to support a certain population size of a certain organism (species). Carrying capacity is represented by the symbol K, and is different for each species-environment system.

7. The growth of populations of organisms generally follows some variation of the logistic growth curve. This curve is characterized by having an S-shape overall. The early period of population growth is geometric; after that the rate of growth slows down, producing a leveling-off at the K value for the particular environment. The population size tends to oscillate around K.

8. Regulation of population size at equilibrium is the result of several kinds of interactions:

a) Competition: as more organisms of a particular kind reproduce, competition for limited food increases, killing off increasing numbers of organisms.

b) Predation: as the prey population increases, the predator can more easily find and kill prey. This counterbalances the increased reproduction rate of the prey and keeps its population down. As a result of the availability of more prey, the predator population can increase in size. But as prey become scarcer, the predators must compete more sharply, limiting their population size.

c) Migration: as the density of organisms increases in a population, certain responses can be triggered in individuals that cause them to migrate or refrain from reproductive behavior. These responses are physiological and developmental, often triggered by sight or smell stimuli.

9. Ecological interrelationships such as competition, predation, symbiosis, and parasitism have all evolved by natural selection.

10. Interspecific competition is competition among populations of two or more species. Its net effect is often to make the two species less and less alike over time in terms of their ecological requirements. The effect of such competition is to further reduce the amount of competition between the species in any particular locality, leading to character displacement.

11. Intraspecific competition is competition among members of the same species.

12. Within ecosystems, organisms occupy a habitat and an ecological niche.

a) An organism's habitat is the place where it lives, such as in fresh or salt water, on a desert, under rotting logs, or high in the trees.

b) Ecological niche refers to the *role* the organism plays within the community. For what organisms does it serve as food? What organisms does it eat? What minerals does it require from, and contribute to, the environment? Ecological niche is sometimes compared to a human being's profession in the economic community.

13. Organisms within an ecosystem obtain energy as producers, as primary, secondary, or tertiary consumers, or as decomposers.

a) Producers are those organisms that can convert the sun's energy directly into carbohydrate through the photosynthesis of green plants and certain photosynthetic bacteria.

b) Consumers (primary, secondary, tertiary, and so on) are those organisms that must obtain their energy by eating other organisms. Primary consumers are herbivores that eat plants directly. Secondary and tertiary consumers are various forms of carnivores that eat herbivores or other carnivores, respectively.

14. The movement of *materials* through an ecosystem is largely cyclic. That is, most material used by organisms is eventually returned to the ecosystem through natural processes such as respiration, excretion, and decomposition. The carbon, nitrogen, water, oxygen, and other cycles are familiar illustrations of this point.

15. The flow of *energy* through an ecosystem is not cyclic, but unidirectional. Energy flows from producers to consumers and decomposers. Energy keeps moving through an ecosystem because there is a continual supply of energy entering the system from the sun.

16. Organisms in an ecosystem exist at various trophic levels. The primary producers, or autotrophs, represent the first level. The various consumers, or heterotrophs, represent the various other levels. In terms of trophic level, the number of organisms decreases in passing from the first to successively higher levels. The same is true most of the time for actual mass of organic material, or biomass, and for energy. Thus trophic levels can be depicted in terms of "pyramids" of mass and energy. This means that there are fewer organisms, or less energy available, the further one goes up the trophic levels of an ecosystem. This observation is in accordance with the first and second laws of thermodynamics.

17. The various trophic levels can also be described as a food web for any particular ecosystem.

18. The carbon cycle can be thought of as starting and ending with atmospheric carbon dioxide (CO_2). This CO_2 enters the living world through photosynthesis, during which it becomes fixed into carbohydrate. The oxidation of carbohydrate during cellular respiration returns most of the carbon to the atmosphere as CO_2. Some CO_2 is given off from decomposition of dead organic matter, a process carried out by microorganisms. A small amount of carbon becomes locked into a "carbon sink" where it is trapped as coal, petroleum, natural gas, or limestone. This carbon is

withdrawn from the ecosystem for some period of time, but it eventually gets oxidized into CO_2 (when people burn coal or gasoline, for instance).

19.   The nitrogen cycle can be thought of as starting with atmospheric nitrogen. Nitrogen-fixing bacteria living in nodules on the roots of legumes (such as pea or alfalfa) can convert this atmospheric nitrogen into soluble nitrates. Nitrates enter the plant and are incorporated into protein. Protein from dead organisms can be broken down by the action of putrefying bacteria, which convert the protein into compounds. Ammonium compounds are acted on by nitrosofying bacteria that convert them into nitrites (NO_2 compounds), which are soluble. Nitrites can be converted into nitrates by nitrifying bacteria. Finally, nitrates and ammonium compounds both can be broken down into molecular nitrogen, which is released back into the atmosphere. The whole function of the nitrogen cycle is to convert relatively inactive molecular nitrogen into a form (nitrates) that is soluble and can easily enter the plant or animal cell. Nitrates, of course, are necessary for the continued production of amino acids and proteins.

20.   Ecosystems can exist perfectly well without animals. All the cycles and processes can be carried out by plants and unicellular autotrophs. The same is not true for animals, however.

21.   Ecosystems are not static relationships; they are constantly changing through long periods of time. The regular, even highly predictable, stages through which ecosystems grow and develop are called succession. Primary succession occurs when aquatic ecosystems become filled in and eventually turn into land. Secondary succession occurs when a land ecosystem undergoes developmental changes into another kind of land ecosystem (a sand dune to a grassland to a pine forest). Successional changes occur because each stage changes the environment and paves the way for a new grouping of organisms. The general pattern of secondary succession begins with an open area (sand or meadow) inhabited with grasses. The grasses are replaced by pines or other evergreens, whose seeds need a lot of sunlight to grow. As a pine forest matures it shades the ground: neither the grasses nor future pine seedlings can grow well. Seeds of hardwood trees such as oak or hickory can grow in shady

areas, so these plants move in and eventually come to dominate the pines. Since the oaks and hickories can replenish themselves, they represent a relatively stable or "climax" community in the successful process.

22.   The term biome refers to approximately six major climax communities that are found around the world. These are: tundra (where the dominant plants include mosses, lichens, and small shrubs; there is relatively less species diversity than other biomes; subsoil is permanently frozen; and there are large plains and bogs); taiga (dominated by coniferous forests, with lakes and bogs, moderate precipitation, and very cold winters); deciduous forests (in the temperate zones, with varying rainfalls and more species diversity than in tundra or taiga); tropical rain forests (dense jungle with abundant rainfall and much diversity); grasslands (not necessarily farther south than the tropical rain forests; relatively low annual rainfall, or seasonal rainfall, therefore few trees but much luxuriant grass); deserts (low annual rainfall, with extreme temperature and moisture fluctuations).

23.   Biogeography is the study of the present (and past) distribution of life on the surface of the earth. The earth is divided into five major biogeographic, primarily faunal, regions: Nearctic, Palaearctic Neotropical, Ethiopian, and Australian. The flora and fauna of each region are products of a variety of nonbiotic, but especially biotic (particularly plants) factors. Historical biogeography, combined with evidence from present-day distributions, helps evolutionists understand how various species and higher taxa have evolved. Considerable evidence from the geological study of continental drift has aided biogeographers in piecing together and explaining the discontinuities and other data regarding present-day geographical distribution.

24.   The environmental movement is an attempt to apply ecological principles to understand how human activities (such as factory or automobile pollution) are altering the natural ecosystem in which we live. The environmental movement seeks to change the improper use of natural resources and the unnecessary piling up of waste materials. Environmentalists recognize that in order to accomplish this task they must understand the relationships between ecological and economic processes.

Exercises

1.   What is a food chain?

2.   Why is there such a great decrease in the amount of available matter or energy along each step of a food chain? How does this relate to the second law of thermodynamics?

3.   On the basis of the following assumptions, calculate the overall efficiency of the energy chain leading from the sun to you.

a)   Of the sun's energy, 99% goes unused.

b)   Of the 1% used by plants, only about 0.6% ends up in glucose (assumed here to be the *only* form of usable energy from photosynthesis).

c)   Half of this glucose is used up by the plant for its own life processes.

d)   Of the glucose reaching your cells, 60% is formed into ATP.

e) Your usage of this energy is 55% efficient.
Can you detect any "energy leakages" other than those listed here?

4. Distinguish between interspecific and intraspecific competition.

5. Nature's most unusual organisms are generally found in extreme environments (such as ocean bottoms, deserts, or Arctic wastes). The grotesqueness and uniqueness of these organisms are generally anatomical, but they may also be physiological. What reasons can you give for this phenomenon?

PART VI
ORIGIN AND DIVERSITY OF LIFE

The great dinosaurs of the Jurassic Period (180 million years ago) are among the most exotic forms of life to have inhabited the earth. In their heyday, they assumed myriads of forms and occupied a large number of ecological niches. Some were aquatic, others lived in swamps, some inhabited drier deserts and plains, and still others flew through the air. Their diversity reflects the wider diversity of life as it has existed on our planet throughout geological time. The fact that the dinosaurs became extinct also reflects the broader fact that extinction is the rule rather than the exception in biological evolution. While geological evidence shows that the total number of families of organisms has increased throughout time, it also shows that the diversity of life on our planet over its whole history is infinitely greater than the number of species existing at any one point in time. In fact, the number of species living today has been calculated to be between 1/10 of 1 percent and 1/1000 of 1 percent of all the species that have ever lived! This amount of diversity is truly staggering.

(*Painting by Robert Kane, courtesy American Museum of Natural History.*)

As we look around us in the living world we are immediately impressed with its enormous diversity. Within a tiny drop of pond water or the immense canopy of a tropical rain forest, we can discover an almost unimaginable variety of living beings. Organisms inhabiting the world's biotic regions range from tiny microorganisms, so small that some of them can be seen with only the most high-powered microscopes, to elephants, redwood trees, and the huge blue whales that dwarf human beings. Organisms live in every conceivable environment: from the near-zero waters under the polar ice caps to hot springs whose temperatures approach the boiling point of water; from depths of nearly ten kilometers below the ocean surface to heights three miles above sea level.

Diversity of organisms is closely related to an important theme that we can discern from a study of life on this planet. Diversity is a result of adaptation. As organisms evolve adaptations to particular environments, they necessarily diversify. Seldom are two environments identical in the selection pressures to which they expose living organisms. With variation and natural selection, diversification is a necesssary byproduct of adaptation. Not only are organisms different (that's what diversity is all about—differences), but as we know they also share varying degrees of similarity, or phylogenetic relatedness. Many adaptations are merely variations of a theme that the organisms share with many other groups to which they are historically related.

A recent biology textbook, in discussing diversity, suggested that after the origin of autotrophs on the primitive earth, the evolution of life might easily have stopped. The authors argue that gradually the surface of the seas could have become covered by primitive green algae, existing in an equilibrium such that eventually the number of new cells produced (by mitosis) would just equal the number of old cells that died. Life's development on earth might just have stopped at the steady-state equilibrium. However, there is an important point missing from this interesting suggestion. The appearance of diversity was not a matter of accident; it was not simply one of several roads that the history of life on earth could have followed. Because of competition, natural selection, and the presence in populations of genetic variability, diversity is an invevitable and unavoidable outcome of the evolutionary process. Diversity is part of life itself.

To understand the history and diversity of forms that now exist, however, it is first necessary to understand how life came to exist on earth. Evolution builds on a chemical base that comprises what biologists call the "prebiotic" era of the earth's history. What was happening during that time that made it possible for associations of molecules to form something that had at least a few characteristics of what we call "life"? How did events in the prebiotic era influence the subsequent history of life on our planet? We will begin this part with an examination of the current state of knowledge about chemical evolution and the origin of life by chemosynthesis (Chapter 26). We next turn to a survey of the major kingdoms of the living world: the Monera, Protista, Fungi, plants, and animals (Chapters 27, 28, and 29). A final chapter deals with animal behavior, its adaptive nature and its evolution. Behavior has played such an important role in animal evolution that its inclusion as the final chapter of the book seems highly appropriate.

In these chapters we will follow two interrelated lines of thought. One will be descriptive, a comparison of the major groups of organisms that inhabit the earth today, a sort of walk through the portrait gallery of the animal and plant phyla. The other will be theoretical, inferring something of the phylogenetic relationships thought to exist among the various groups we survey. We will describe the various problems that modern biologists face and the methods they use to deduce possible evolutionary developments among major groups of organisms. In each of these chapters, then, we will raise evolutionary questions regarding how transitions from one form and mode of life to another (for example, from prokaryotes to eukaryotes; from unicellularity to multicellularity; from water to land) might have occurred, the adaptations involved, and the diversity that resulted.

Chapter 26
The Origin of Life

INTRODUCTION

The origin of life almost certainly was a gradual step-by-step process, an evolution occurring slowly over thousands or perhaps millions of years. Yet, no one can say precisely when life began to evolve. Our concept of the evolutionary process now presupposes the presence of complex components: a highly structured genome (DNA, RNA), intricate cellular and metabolic pathways, and highly differentiated cell structures such as membranes and organelles. As the simplest prebiotic or protobiotic globules were formed, however, they could not have contained these complex elements; indeed, they may have lacked all of them. Yet life did develop, expand, and diversify on the earth, whatever its simplest beginnings may have been. In this chapter we will look at the evidence suggesting that life on earth could have originated according to the basic principles of physics and chemistry. In addition to the evidence and clues, we shall also look at some of the speculative ideas put forward by scientists seeking to answer the question: How did life originate on earth?

26.2
THE ORIGIN OF LIFE BY CHEMOSYNTHESIS

The Russian biochemist A. I. Oparin, in 1924, and the English biologist J. B. S. Haldane, in 1929, independently outlined how life might have originated on earth by chemosynthesis. Both Oparin and Haldane maintained that life originated from organic chemicals in the ancient oceans of the prebiotic (prior to life) earth. The earth's atmosphere, they said, contained relatively little free oxygen and much hydrogen. It was a "reducing" atmosphere, because the plentiful amount of hydrogen chemically reduced (that is, donated electrons to) other atoms or molecules. Lightning, heat from the earth's crust, ultraviolet radiation, natural radioactivity and volcanic activity (including hot geysers) provided energy to form simple organic compounds. Basic chemical building blocks such as amino acids, carbohydrates, purines, and others accumulated in the oceans and even-

tually coalesced into droplets that may have been surrounded by simple hydrocarbon chain "membranes." When the droplets reached a certain size they became physically fragile and "divided" into two. Some molecules within these droplets became catalytic, and self-replicating molecules began to guide the daily chemical reactions, as well as the "reproductive" process of the simple cells. Life as we know it had "arrived."

But what evidence exists for such a fanciful, speculative scheme? Remember, science deals not merely with what is possible, but with what, to the best of our knowledge, actually *is*. Most of the steps postulated above can actually be observed to take place in the laboratory under simulated environmental conditions of the prebiotic earth. But what do we know about the earth's surface then? And how could life have begun to form under these prebiotic conditions?

The earth and its atmosphere are thought to have been formed about 4.5 billion years ago. Geochemists believe that at that time components of the earth's atmosphere included hydrogen, water vapor, methane, ammonia, and some hydrogen-sulfur compounds (such as hydrogen sulfide gas, H_2S). The universe as a whole was, and is, mostly hydrogen (about 92.8 percent). As the earth formed in the universe of swirling gases, the heavier elements became concentrated at the core, leaving the lighter elements, such as hydrogen, oxygen, and nitrogen on the outside to form the crust. Because hydrogen is highly reactive under certain conditions (high temperature, electric discharges), it combined readily with elements such as oxygen (to form water), carbon (to form methane), and nitrogen (to form ammonia).

The early atmosphere of the earth contained relatively little oxygen, much of it combined into mineral silicates such as olivine ($FeMgSiO_4$), concentrated in the earth's crust. The light weight of the uncombined oxygen allowed it to escape earth's gravity. Gases such as ammonia, methane, and hydrogen sulfide were produced when carbon and sulfur fumes were spewed into the atmosphere from the earth's depth either spontaneously or during volcanic activity.

Thus by about four billion years ago, the earth had a warm mantle, large oceans, active volcanos, and an

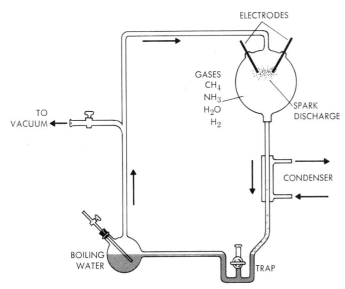

Fig. 26.1
Apparatus designed by Stanley Miller for the circulation of methane, ammonia, water vapor, and hydrogen. Water is boiled in the flask at lower left. The products of chemical reaction are collected in the trap at lower right. Energy for chemical reaction comes from an electric discharge in the flask at upper right.

atmosphere of water vapor, hydrogen, methane, ammonia, and perhaps hydrogen sulfide. The scene was set for the first of what we can consider as six phases in the origin of life.

The First Phase: Formation of Building Blocks

In 1952 Harold Urey and Stanley Miller, then at the University of Chicago, began a series of laboratory investigations of the origin of life. Miller, a graduate student at the time, set up an apparatus (Fig. 26.1) into which he introduced the gases ammonia, methane, and hydrogen, along with water kept boiling in a flask. Boiling and condensation kept the substances circulating through the system. Two electrodes provided a periodic electric discharge into a large bulbous chamber at the top of the apparatus. After a week, Miller collected and analyzed the residue. It included several amino acids, a carbohydrate (succinic acid, an intermediate compound associated with the Krebs cycle), and several other organic compounds. Following Miller's procedure, other workers have produced simple components of nucleic acids and even ATP, by varying the type and amounts of the reactants used. An example of a known reaction leading to the production of an amino acid (serine), a purine (adenine) and a pentose sugar (ribose), are shown in Fig. 26.2. Miller's experiments and current knowledge of the steps involved in such chemical syntheses support the hypothesis that in a reducing atmosphere of the primitive earth, the basic building blocks of life could have been formed spontaneously.

However, although the Miller experiment yields many of the amino acids found in the proteins of living organisms today, it also yields related organic compounds that are *not* present. Thus, the Miller-type synthesis produces three isomers of the amino acid alanine (all with the formula $C_3H_7NO_2$), yet of the three, only one, alanine, has become part of living systems. In one case, the Miller experiment yielded seven amino acid isomers (all with the formula $C_4H_9NO_2$), none of which occur in modern organisms! Apparently more building blocks were available than used. We are challenged by the question: why were only twenty amino acids built into proteins when many more than that were available? There is a second problem in the Miller experiment. Equal amounts of the two optical isomers (a pair with mirror-image molecular structures) of every amino acid type were formed, for example the two optical isomers of the amino acid alanine (*D*-alanine and *L*-alanine). Yet with some exceptions, all living organisms today use only *L*-amino acids. If both *D*-amino acids and *L*-amino acids are synthesized, why have only *L*-amino acids been generally incorporated in living systems?

A third problem is that high-energy radiation such as ultraviolet rays or electric discharges not only build up but also break down complex molecules such as

Fig. 26.2
Two biochemical pathways by which some of the known building blocks of life can be synthesized without the presence of living cells or already-existing enzymes. (*a*) Synthesis of the amino acid serine from condensation of two formaldehyde molecules; (*b*) Formation of adenine, a principal component of both ATP and nucleic acids, by several steps: it is thought that four hydrogen cyanide molecules combine to form a tetramer, which can rearrange itself to form a five-member ring; a fifth molecule of hydrogen cyanide is added, closing the second ring.

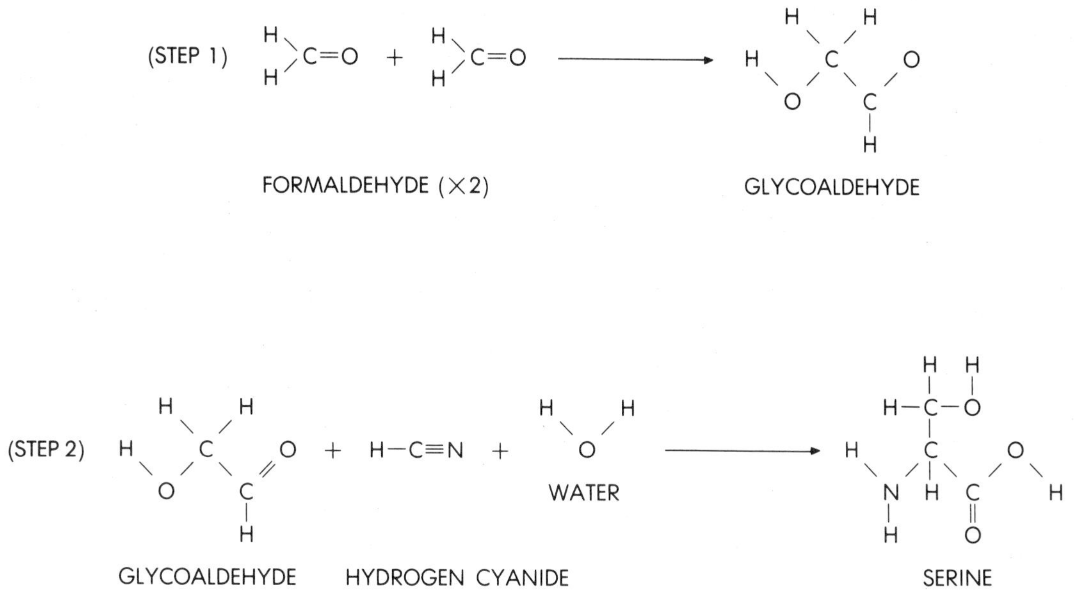

(STEP 1)
FORMALDEHYDE (×2)
GLYCOALDEHYDE
(STEP 2)
GLYCOALDEHYDE
HYDROGEN CYANIDE
WATER
SERINE

(**a**)

HYDROGEN CYANIDE
Light energy
DIAMINOMALEONITRILE (HCN)4
(INTERMEDIATE 1)
Light energy
(INTERMEDIATE 2)
Light energy
ADENINE

(**b**)

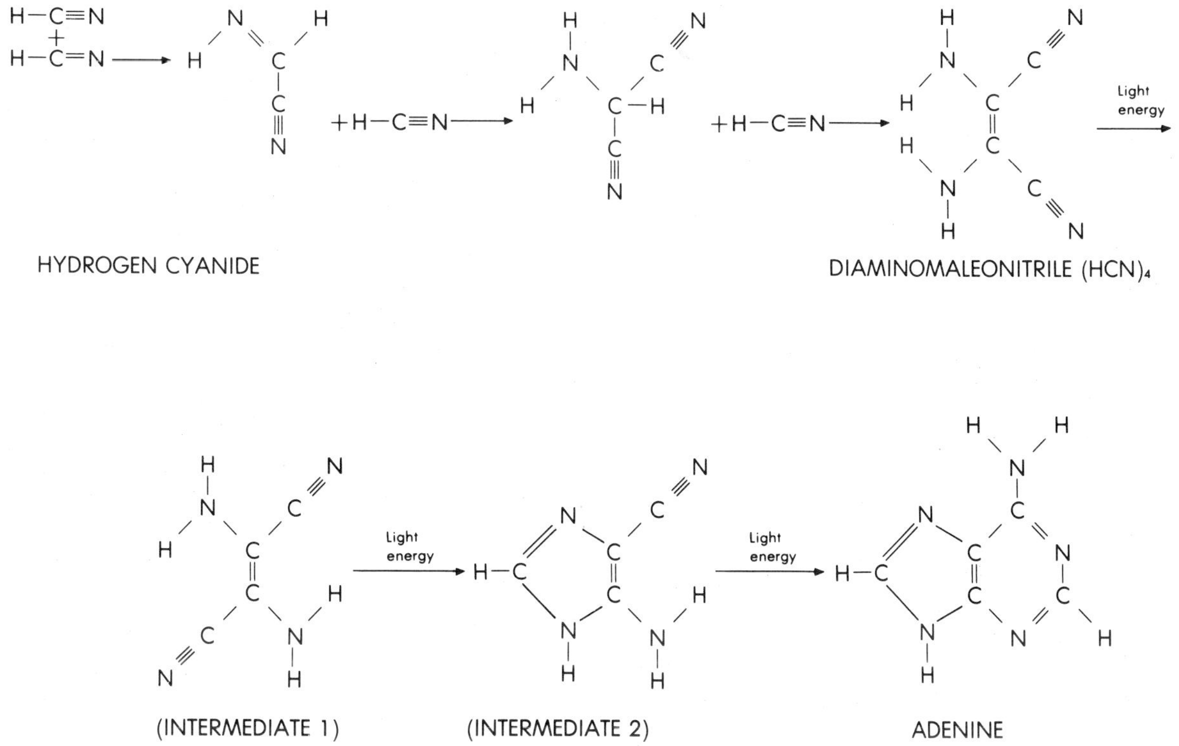

amino acids, purines, and simple carbohydrates. How could a pool of building block molecules accumulate if the molecules were subject to continual breakdown, perhaps as rapidly as they were formed?

The Second Phase: Formation of Polymers

As soon as amino acids, simple carbohydrates, and fatty acids or hydrocarbon chains were present, they probably began to join together to form larger macromolecules. Amino acids formed peptides, while glucose and other smaller carbohydrate units formed larger sugar and starch molecules. And hydrocarbon chains, uniting with three-carbon sugars, could have yielded primitive fat molecules. What were the conditions that allowed these larger molecules to form? And were these molecules similar to modern proteins, carbohydrates, and fats?

These questions prompted Professor Sidney W. Fox of Florida State University to conduct a series of experiments. Fox hypothesized that the warm mantle of earth's environment at that very early period might have provided the thermal energy necessary to drive the subunits together. Fox heated mixtures of different amino acids for varying periods of time. Dipeptides and even long-chain peptides were produced, supporting Fox's hypothesis. An example of how polymerization

might have occurred on the prebiotic earth is shown in Fig. 26.3. Such polymerizations, involving dehydration, must have occurred originally without catalysts (enzymes), though "coupling agents"—molecules or ions that attach to one reactant, thereby facilitating the combination of that reactant with another—may have been involved.

But two really important questions remain: First, how similar are thermally produced proteins to those we find in living organisms today? When Fox tested the thermally produced proteins he found they gave positive reactions for protein. They could be broken down by several proteolytic enzymes, such as pepsin, and they could be used as nutrients by bacteria. (This implies that bacterial enzymes break the proteins down into amino acids.) Indeed, thermally produced proteins, if formed on the primitive earth, might have closely resembled some of the proteins produced today by living organisms.

The second question is: How could polymerization between subunits take place under prebiotic conditions if opposing hydrolytic reaction is favored in the oceans? There are at least three possible answers. One speculation is that the basic building blocks, the monomers (amino acids, purines or pyrimidines, glucose units) had negatively charged groups, such as phosphate, attached to them that allowed them to compete successfully with

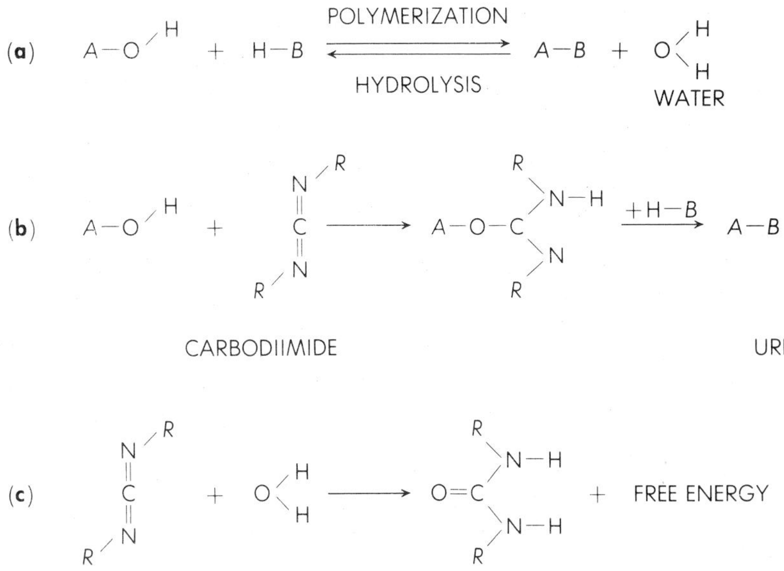

Fig. 26.3

Possible chemical model for formation of a polymer chain by dehydration. The overall reaction is shown at the top (a), where subunits A and B are joined together (dehydration), to the right, and split apart (hydrolysis), to the left. In the presence of water, hydrolysis is thermodynamically favored—that is, it proceeds spontaneously because it releases more energy than it requires. Because hydrolysis is favored, polymerization can occur only if some "coupling agent" is present to which one of the subunits can bind, and thus increase the likelihood that it will encounter the other in an energy-favorable way. In the example shown here, cyanogen serves as the coupling agent. On the prebiotic earth carbodiimide could have served as the coupling agent, producing the polymer, A—B and a urea derivative (b). The uptake of water by the carbodiimide produces the energy necessary for the reaction (c).

water for the coupling agents. (It is interesting to note that all the major monomers involved in biosynthetic pathways today usually become activated by having a high-energy phosphate group transferred to them as the reaction begins. Could this mechanism be a holdover from the earliest stages of the history of life?) Another speculation is: The polymerization reactions took place in small, evaporating pools around the shore or in the moist sand of some Archean beach where, by their high concentration, they counteracted the tendency of water to hydrolyze the products once formed. Still a third speculation is that the reactions took place on the surface of clays and silica sheets, where water is less prevalent, and where there is an enormous surface area.

Although thermal energy may well have been responsible for forming the first macromolecules in the organic soup, other forms of energy have been shown to cause the formation of polymers. In 1974 Dr. Cyril Ponnamperuma at the University of Maryland's Laboratory of Chemical Evolution showed that sound energy (from volcanic eruption), the formation and collapse of bubbles in mud, and high-energy radiation (X-rays, ultraviolet rays) could all generate the formation of complex polymers.

The Third Phase: Coacervates and Droplets; Prebiotic Cells

Once formed, the various organic macromolecules had two possible fates: degradation by energy sources such as ultraviolet radiation, or combination with other organic molecules. Although degradation did occur, the formation of molecular aggregates was favored, probably because of the polarity of the macromolecule (see Section 2.4). How did these molecular aggregates set the stage for the origin of living organisms?

In 1936, A. I. Oparin proposed what he called the coacervate theory. Oparin theorized that protein-like substances in the early broth formed aggregates that tended to develop a simple membrane around them, due to surface tension. Oparin showed that a mixture of gelatin and gum arabic produced small spherules or coacervates. A surface layer separated each coacervate from the external liquid in which it was suspended. Fatty acids may have collected at the surface (recall that one end of a fatty acid chain is hydrophilic—that is, strongly attracted to water) and thus formed a primitive lipid membrane around the coacervate.

Oparin carried out several imaginative and revealing experiments with his coacervates. For example, he added the modern-day enzyme phosphorylase to a coacervate solution. Phosphorylase carries out dehydration synthesis of glucose to produce starch. The phosphorylase diffused through the coacervate's exterior surface into the interior of the droplet. Oparin then added glucose-1-phosphate to the medium. It, too, diffused into the coacervate. Once inside the coacervate, the glucose-1-phosphate was polymerized into starch by the phosphorylase! Moreover, the larger starch molecules could not diffuse out as easily and accumulated within the coacervate. Oparin had produced a small starch factory inside each coacervate. He then added the enzyme amylase, which, you will recall, hydrolyzes starch molecules. Again, Oparin found that the amylase diffused into the coacervates. Once inside it began to "digest" the starch molecules. These coacervates were behaving like simplified metabolic units. Oparin found that other molecules could diffuse into the coacervates, including chlorophyll. When various reducing and oxidizing dyes diffused into them, the coacervates behaved like primitive electron transport systems.

Sidney Fox has done further work based on Oparin's coacervate idea. He has shown that thermally produced proteins will form spherules with diameters about the same size as bacterial cells (see Fig. 26.4). These spherules, like Oparin's coacervates, are separated from the external medium by a surface layer. Adding such salts as sodium chloride to the medium causes the spherules to shrink in size. Fox attributes this to the

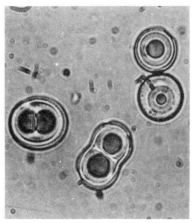

(a)

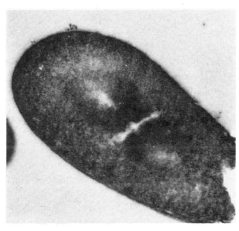

(b)

Fig. 26.4
(a) Photograph showing both single and aggregate microspheres produced in Fox's laboratory at Florida State University. Note the regular appearance of the spherules. In general they resemble spherical bacteria. It is thought that such aggregates (mostly of protein) could represent the beginnings of organization that eventually led to life. (b) Proteinoid microspheres produced by Dr. Fox. Note the close resemblance to the cellular structure shown by some bacteria. (Photos courtesy Professor Sidney W. Fox.)

movement of water out of the spherule. Like living cells, the spherules show osmotic properties. In addition, two spherules can combine with each other, or one spherule can split apart upon reaching a certain critical size. These observations suggest strongly that something like Oparin's coacervates or Fox's spherules could have been the precursors of the first living organisms.

The formation of a surface layer, later followed by an actual membrane, was an important development. Not only did a membrane provide some protection for the macromolecules within it; it also kept the molecules it contained in close contact with each other, thus increasing the chances that chemical reactions would take place. The molecular complexes became more effective chemical combinations. One possible result would have been that chemical combinations such as amino acids forming into proteins could use the potential energy from the breakdown of other molecules (such as phosphates or carbohydrates) as a driving force.

Thus, in the third phase, macromolecules such as protein or carbohydrate became organized into bodies with definite shape and unity, properties resembling those of living things.

The Fourth Phase

Up to this point, molecular aggregates lacked a well-developed internal organization and a reproductive capacity. Although an aggregate could reproduce simply by splitting in two, there was no way of determining the characteristics that the resulting parts might possess.

A real breakthrough in the origin of life occurred when nucleic acids became the major molecular "organizers" within aggregates. Since genetic information is transmitted in the same way (by the triplet code) in all known organisms, nucleic acids must have appeared at an early stage. Nucleic acids provided reproductive continuity and directed the immediate activities of the coacervate complex. Energy-capturing systems could become more efficient as they also became a hereditary part of the molecular aggregates. By the end of phase 4, these molecular aggregates had become true "living" organisms.

The Fifth Phase

The fifth phase is characterized by the beginning of evolutionary development. Once continuity through genetic control was introduced, natural selection as we know it could begin to operate. One of the first results of evolution was increased efficiency in capturing energy from carbohydrate breakdown. Those organisms that could use energy most efficiently could reproduce their kind most rapidly; hence there was a simple selection pressure. It was probably at this time that the processes of electron transfer and storage of energy in phosphate bonds first appeared.

An increase in the efficiency with which organisms could use energy sources in the environment led to increased reproductive capacity. Thus the number of organisms in the early oceans began to rise, and competition became more intense.

It must be pointed out that to date there is very little experimental data to suggest how nucleic acids —even if formed by thermal processes—could have come to control biochemical processes such as protein synthesis in the primitive cell. The evolution of nucleic acids has been one area of the study of the origin of life in which, as of the present time, biologists have made the least breakthroughs. Perhaps during this decade substantial experimental results will be obtained in this frontier of biological research.

The Sixth Phase

Up through the fifth phase organisms had subsisted on carbohydrates and other energy sources already present in the environment. These organisms were heterotrophs, as are all animals today.

As time went on, the carbohydrate supply diminished. Competition became more and more pronounced. Under such conditions, any variation in an organism in the direction of manufacturing its own carbohydrate supply would be greatly favored. Variations must have occurred that allowed organisms to use light energy to produce carbohydrates. Thus a new form of metabolism, the process of photosynthesis, evolved.

The appearance of self-sufficient autotrophs marked another major innovation in the early history of life. The earliest autotrophs may have resembled certain present-day bacteria. Bacteria have very simple cell structures, but some of them can use light or heat energy to produce carbohydrates. The appearance of a photosynthetic organism provided a balance that has been characteristic of the biotic community ever since.

Once autotrophs began to evolve, the history of life on earth took a dramatic turn. Photosynthesis produced oxygen, which then made possible the pathway of aerobic respiration. As the ozone layer formed, life on land and the surface of the seas became more possible, since ozone in the atmosphere (today concentrated in a layer in the stratosphere approximately 25–35 miles above the earth's surface) absorbs ultraviolet radiation, which is so destructive to living systems. Thus, in a sense, oxygen set in motion a "revolution" in the development of life (this revolution is described in more detail in Supplement 26.1).

Oparin holds that the origin of life was neither a lucky accident nor a miracle, but the result of perfectly natural and ordinary scientific laws. He believes that life developed chemically from simple hydrocarbons, as suggested by the preceding scenario. Once introduced, hydrocarbons evolved with increasing complexity within evolving molecular aggregates. There is some geological evidence to support this claim: the older the rocks, the simpler the hydrocarbons.

SUPPLEMENT 26.1
OXYGEN AND EVOLUTION

An examination of the fossil record indicates that at various periods in the history of life, there has been a sudden and rapid increase in the number and kinds of organisms on earth. In the early history of life, two such rapid "bursts" occurred: the first at the beginning of the Cambrian and the second at the beginning of the Silurian (600 million and 425 million years ago, respectively). The earlier period saw an enormous proliferation of eukaryotic organisms and the appearance of the first multicellular forms. In the later period, there was an enormous spread of life from the oceans onto the land. What might account for these two periods of change?

One theory holds that both developments were made possible by increases in the oxygen concentration of the

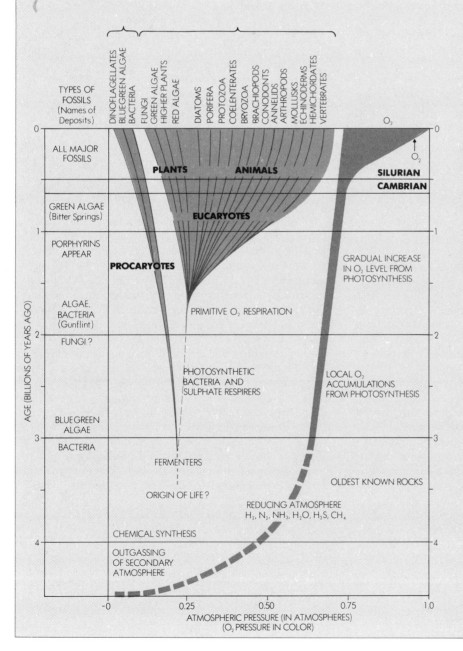

Diagram A
Chart showing the evolutionary history of major life forms, combined with an indication of the changes in the oxygen concentration in the earth's atmosphere (color) over the past 4½ billion years. The column to the left lists the age of the various precambrian fossils (i.e., those more than a half-billion years old). Concentration of oxygen in the atmosphere (measured as oxygen pressure) increased dramatically about a half-billion years ago at the beginning of the Cambrian. This change was accompanied by a rapid expansion of life. (Modified from E. O. Wilson, T. Eisner et al, *Life on Earth*, 2d ed., Sinauer, 1978, p. 520.)

earth's atmosphere. This theory, known after its founders as the Berkner-Marshall theory, is based upon geochemical evidence suggesting that the quantity of oxygen in the earth's atmosphere reached 1 percent of present value by the early Cambrian and 10 percent of present value by the beginning of the Silurian (see Diagram A). These increases were both due to the development of photosynthetic activity in primitive plants.

Increase in oxygen concentration in the atmosphere by the early Cambrian made possible the evolution of aerobic respiration (citric acid cycle). As oxygen became more available, variations leading to incorporation of oxygen into the respiratory process would have had increased selective advantage. That advantage, of course, derives from the greater efficiency, in terms of total energy yield, of using oxygen rather than an intermediate like pyruvic acid as a final electron acceptor. With the development of aerobic processes, the stage was set for the evolution of multicellular forms. These forms, with their greater energy requirements, are a product of the increased availability of oxygen half a billion years ago.

During the late Ordovician and early Silurian, the invasion of land began. An increase in oxygen concentration to up to 10 percent of present value made existence on land possible for several reasons. It provided enough oxygen for organisms to survive out of the water. More important, however, oxygen becomes converted to ozone (a molecular form with the formula O_3), a layer high up in the stratosphere which effectively filters out ultraviolet and other forms of high-energy radiation from the sun and outer space. The ozone screen today protects all forms of life on the land from the effects of damaging high-energy radiation. (The supersonic transport, or SST, and Freon spray cans are opposed strongly by environmentalists today because both are known to have some damaging effects on the ozone layer.) In the very earliest period of the development of life on earth, water filtered out high-energy radiation. Only with the appearance of the ozone layer, however, did land become habitable. Once land dwelling became possible, life spread very quickly (that is, showed an adaptive radiation) into the new environment.

Like any hypothesis attempting to explain how one specific set of events took place in the past, the Berkner-Marshall theory cannot be tested rigorously. It is in agreement with geochemical data. It is certainly an attractive means of accounting for the two rapid bursts of evolution during the early eras of life on earth.

In Oparin's opinion, the problem of the "primal soup"—the sea containing dissolved organic substances, including simple polymers of amino acids and nucleic acids—has now been largely solved by the innumerable experiments of the sort performed by Stanley Miller. To Oparin, a more difficult question is how enzymes first came into existence, since an enzyme's structure is adapted in a very specific way to its particular catalytic task. Oparin and others, however, have developed several intriguing scenarios of how enzymes may have evolved from simple inorganic catalysts.

As seen earlier, to Oparin the best model to simulate the evolution of enzymes is the coacervate drop, an isolated part of the primeval soup separated from the solution by surface boundaries but able to interact with the environment. A sort of prebiological natural selection allowed the more "successful" drops (those with a greater catalytic ability to incorporate their surroundings) to grow at the expense of the less successful.

Catalysts, which could only have come from the environment, thus become the telling requirement. The first of the catalysts must have been rather inefficient. However, through modification by the addition of organic molecules, their catalytic activity could have been improved a thousand-fold. An enormous number of molecular alternatives must have arisen, of which only the most efficient remained. These may possibly be today's coenzymes.

Meanwhile, the random joining of amino acids could have led to primitive proteins with some form of active center. These would soon disappear, however, unless a further controlling mechanism evolved that could maintain their structure. Even at these early stages the polynucleotides, later to become DNA and RNA, might have had some organizing ability and, if they produced favorable catalysts, would have been selected.

Oparin maintains that all these processes are subject to experimental testing and his laboratory is still

The chemosynthetic theory of the origin of life raises as many questions as it answers. Like any scientific theory, its value lies in the degree to which it gives rise to testable hypotheses.

actively engaged in research even though such experiments can never *exactly* duplicate the way things actually must have happened when life first appeared on earth.

Opposing Views

Hypotheses about the origin of life are at best tentative and subject to continual and frequent revision. Indeed, a fundamental aspect of the explanation outlined above has been called into question, for new geological evidence suggests that the chemical composition of the earth's early atmosphere contained very little ammonia or methane, and hence was not as "reducing" an atmosphere as the classical model of Oparin, Haldame, and Miller supposed. Since ammonia is known to dissolve very rapidly in water, the presence of free ammonia in the atmosphere would be very unlikely. A new hypothesis proposes that a number of elements necessary for life (carbon, hydrogen, and nitrogen) were spewed into the atmosphere by volcanos. Other studies suggest that bombarding these elements with sunlight of a specific wavelength (2536 Å) can produce molecules such as amino acids. At the present time there is no crucial test to decide which of these two hypotheses better explains how life actually did arise on the earth. New evidence will have to come from studies on the chemical composition of the earth's oldest rocks, a field that is currently developing rapidly.

26.3
THE GEOLOGICAL TIME SCALE AND PRIMITIVE FOSSILS

Establishing a date for such events as the origin of life is philosophically and technically impossible. It is *philosophically* impossible because life did not originate at any one time in history. The entire process was evolutionary, beginning with the evolution of the elements themselves and proceeding to the formation of simple and then complex molecules and finally their aggregation into proto-cells. The task of dating the origin of life is *technically* impossible, at least with any accuracy, because the earliest true organisms left little or no fossil records. The oldest traces of life are the remains of primitive plants in strata at least three billion years old. Life forms must have been developing long before these primitive plants developed. Once fossils exist, geological dating procedures make it possible to establish their age with a fair degree of certainty. One prominent method is based on the rate of **radioactive decay** of the isotopes of certain elements—for example, U^{238}, an isotope of uranium. Radioactive elements undergo transmutation to more stable forms by emitting subatomic particles, such as β-particles (that is, electrons) and/or neutrons. They may also emit radiant energy in the form of gamma rays. This "decay" procedure continues until all the atoms of a given sample have been transmuted to the stable form. For example, atoms of U^{238} undergo radioactive emission and decay ultimately to the stable form, Pb^{206} (lead).

Radioactive decay occurs at a steady, measurable rate. The rate differs for each isotope of each element. Yet the process is so regular that, for example, we can say it takes 4.5 billion years for half of the atoms in a sample of U^{238} to become Pb^{206}, or that it takes 15 days for half the atoms in a sample of P^{32} to decay to P^{30}. This period is termed the **half-life** of the isotope.

Many rock samples can be accurately dated by comparing the ratio of U^{238} to Pb^{206} contained in them. U^{238} is a relatively common isotope and is found in many rocks and fossils. The basic assumption of this method is that the older the rock, the less U^{238} it will contain and the more Pb^{206}. The oldest rocks known are estimated by this method to be more than 3.2 billion years old. The age of the earth is estimated to be between four and five billion years. A general scheme of geological time, with indications of the history of life in each period, is shown in Table 26.1.

For any accurate idea about the appearance of true cells or living systems on earth, it is necessary to have fossil remains. Unfortunately, cells have no hard parts (like shells or bones) that can be preserved easily. The problem is even more acute for flimsy associations of

Table 26.1
Kinds of Life in Various Geological Eras*

Eras†	Periods†	Epochs	Aquatic life	Terrestrial life
Cenozoic 63±2	Quaternary 0.5±3	Recent Pleistocene	Periodic glaciation	Humans in the new world First humans
	Tertiary 63±2	Pliocene Miocene Oligocene Eocene Paleocene	All modern groups present	Hominids, monkeys, and *pongids* Adaptive radiation of birds Modern mammals and herbaceous angiosperms

(Continued)

Table 26.1 (Continued)

Eras†	Periods†	Epochs	Aquatic life	Terrestrial life
Mesozoic 230±10	*Cretaceous* 135±5		*Mountain building (e.g., Rockies, Andes) at end of period* Modern bony fishes Extinction of ammonites, plesiosaurs, ichthyosaurs	Extinction of dinosaurs, pterosaurs Rise of woody angiosperms, snakes
	Jurassic 180±5		*Inland seas* Plesiosaurs, ichthyosaurs abundant Ammonites again abundant Skates, rays, and bony fishes abundant	Dinosaurs dominant First lizards: *Archeopteryx* Insects abundant First mammals, first angiosperms
	Triassic 230±10		*Warm climate, many deserts* First plesiosaurs, ichthyosaurs Ammonites abundant at first Rise of bony fishes	Adaptive radiation of reptiles (theo-codonts, therapsids, turtles, croc-odiles, first dinosaurs, rhyncho-cephalians)
Paleozoic 600±50	*Permian* 280±10		*Appalachian Mountains formed, periodic glaciation and arid climate* Extinction of trilobites, placoderms	Reptiles abundant (cotylosaurs, pelycosaurs); cycads and conifers; gingkoes
	Pennsylvanian 310±10	*Carboniferous*	*Warm humid climate* Ammonites, bony fishes	Insects abundant, centipedes First reptiles; coal swamps
	Mississippian 345±10	*Carboniferous*	*Warm humid climate* Adaptive radiation of sharks	Forests of lycopods, club mosses, and seed ferns Amphibians abundant, land snails, first insects
	Devonian 405±10		*Periodic aridity* Placoderms, cartilaginous and bony fishes Ammonites, nautiloids	Forests of lycopods and club mosses, ferns, first gymnosperms Millipedes, spiders, first amphibians
	Silurian 425±10		*Extensive inland seas* Adaptive radiation of ostracoderms Eurypterids	First land plants Arachnids (scorpions)
	Ordovician 500±10		*Mild climate, inland seas* First vertebrates (ostracoderms) Nautiloids, Pilina, other mollusks Trilobites abundant	None
	Cambrian 600±50		*Mild climate, inland seas* Trilobites dominant First eurypterids, crustaceans Mollusks, echinoderms Sponges, cnidarians, annelids Tunicates	None
Precambrian 4600			*Periodic glaciation* Fossils rare but many protistan and invertebrate phyla present First bacteria and blue-green algae	None

*After John W. Kimball, *Biology*, 2d ed. Reading, Mass.: Addison-Wesley, 1968.

†With approximate starting dates in millions of years ago.

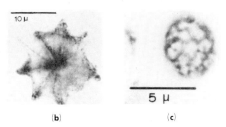

(a)   (b)   (c)

Fig. 26.5
Three examples of "fossil" microorganisms from the Gunflint formation, Ontario. These are thin rock sections, photographed in transmitted light, of specimens estimated to be about 1.9 billion years old. (a) *Gunflintia grandis*, a filamentous alga ($\times 2135$); note cell walls dividing the filament into units. (b) *Kakabekia umbellata*, a microorganism of uncertain taxonomic relationships. This organism appears in a number of different but related structural forms in the Gunflint formation ($\times 1763$). (c) *Huroniospora macroreticulata*, a spheroidal, sporelike body showing a thick wall and sculptured "trimmings" ($\times 3132$). (Photos courtesy E. S. Barghoorn and S. A. Tyler, from *Science* 147, 1965, pp. 563–575. Copyright © 1965 by the American Association for the Advancement of Science.)

macromolecules like coacervates. In the past two decades, however, a field known as micropaleontology has shown remarkable progress. By very special techniques, light and electron microscopy have been applied to the study of thin sections from sedimentary rock. These sections show many very primitive cells whose impression has been left in mud now turned to rock. Several remarkably clear micrographs of "fossil" cells are shown in Fig. 26.5. They come from the so-called Gunflint deposits in Ontario and are approximately 1.9 billion years old. In older rocks, primitive bacteria-like structures have been dated at about 3.1 billion years.

Radioactive dating indicates that the rocks in which the earliest fossil cells have been found are at least 3 billion years old. If cells existed that long ago, the events that formed the necessary foundation (roughly phases 1 through 4) must have been under way long before that. Even a conservative estimate could well set the formation of molecular aggregates at 4 billion years ago. Life has been developing on earth for a very long time.

A measurement scale of time like billions or millions of years is difficult to imagine. We can more easily compare the whole history of life on earth to a 24-hr scale, the appearance of various forms being designated in terms of time of day. If we set the origin of life at one minute after midnight, the first fossils do not begin to appear until 6 P.M. Even more striking, the age of mammals begins at 11 P.M., and humans appear at 11:59. The short history of our species on earth is some indication of the vast time over which the origin of life and evolution to its present stage are thought to have occurred.

In conclusion, we can ask: if the above hypothesis for the chemical origin of life is valid, should we not expect living forms to have originated elsewhere in the universe than on our planet? Is it not likely life has arisen many times in the past? The answer is obviously yes. But when and where, or what forms it has taken, still remain a mystery. The recent visit of Viking I (Fig. 26.6) to the surface of Mars failed to reveal any signs of life on the other planet in our own solar system most likely to support life. Further space explorations may indeed reveal life forms elsewhere—only the future will tell.

Fig. 26.6
A 100° angle view of the surface of Mars, as revealed by a photograph taken from Viking I in 1977. The martian surface is shown to be rocky and barren, with small mounds and hills (or dunes) 6 to 8 feet high. The hopes that Viking I would find life on Mars were not realized. As the photograph shows, the surface appears arid and lifeless. Soil analyses confirmed these visual impressions. (Photograph courtesy National Aeronautics and Space Administration.)

Summary

1. Life is thought to have originated by a process of chemosynthesis: The idea that all components of earliest living systems were synthesized by ordinary chemical and physical processes in the very early history of the earth.

2. The most widely accepted theory of chemosynthesis maintains that the early atmosphere of the earth contained water vapor, methane, ammonia, and free hydrogen. These gases contain all the elements essential for life as we know it: carbon, hydrogen, oxygen, and nitrogen.

3. One version of the chemosynthetic theory sees life originating in six phases:

   a) First Phase: Random synthesis of organic molecules such as amino acids, simple sugars, hydrocarbon chains, and purines. The work of Stanley Miller showed that these could be produced at random from water vapor, methane, ammonia, and hydrogen if energized by sources such as an electric current or discharge (lightning) or high-energy radiation.

   b) Second Phase: Formation of macromolecules (polymers) from free amino acids, sugars, nucleotides, and the like. Sidney Fox has shown that heat and other relatively low-energy sources could cause such combinations. Fox and others found that proteins randomly synthesized in this manner are selective in the frequency with which they incorporate specific amino acids. The resulting proteins were not composed of the same proportions of each type of amino acid as the surrounding "broth."

   c) Third Phase: Association of macromolecules into semiorganized droplets or other units. Factors such as polarity and the existence of hydrophilic and hydrophobic regions of the macromolecules could have been involved in forming such associations. Oparin's coacervate theory is one way of accounting for the tendency to form higher-level associations among macromolecules. Coacervate droplets are a suspension of one liquid within another as a kind of colloidal suspension. Fox has shown that thermally produced proteins will form spherules about the same size as bacterial cells. Spherules can combine with one another or split, suggesting very simple cell-like units.

   d) Fourth Phase: Appearance of molecular "organizers" within spherules or coacervate-like droplets. These organizers, nucleic acids or their precursors, controlled the internal organization of the spherules; in some way they also came to control the division of internal material when a droplet split in two. In this way division of the droplet changed from a random to a more organized process.

   e) Fifth Phase: The beginnings of evolutionary development. This became possible only after the introduction of an organizer molecule into living systems allowing for heredity. Once continuity could be guaranteed between generations, variation and natural selection could operate. Among the first processes selectively favored must have been those involved with more efficient methods of harnessing energy. The similarity of energy-harnessing systems throughout the living world (glycolysis and the citric acid cycle) suggests a very early origin for this process in the history of life.

   f) Sixth Phase: Introduction of photosynthetic mechanisms. Up until this phase all "organisms" were heterotrophic. With the evolution of photosynthesis, autotrophism became possible. The appearance of autotrophs meant that life could become self-sufficient, depending only on solar energy to keep the system going.

4. One of the important developments in the early history of life was the introduction of catalysis. The first catalysts were probably no more than inorganic ions (such as those of iron) incorporated into coacervate droplets or proteinoid spherules. Association of inorganic ions with small peptides enhanced catalytic function—hence the origin of enzymes.

5. Once aggregates of proteins and nucleic acid developed, they may have followed one of two very different evolutionary courses. Some may have given rise to more cell-like structures similar to modern-day bacteria. Others may have given rise to virus-like particles that parasitized the primitive cells.

6. The time scale for the origin of life on earth covers well over 3 billion years (possibly 4 billion). The first true fossil remains of primitive cells are at least 1.6 billion years old, whereas some fossil stromatolites (mats of microbial colonies or cell groupings) are dated at 2.6–3.1 billion years. Obviously the more primitive organisms and molecular aggregates must have preceded the appearance of cells by many millions or billions of years.

Exercises

1. Explain briefly how living, cellular forms could have originated from simple organic molecules such as amino acids, carbohydrates, lipids, and nucleic acids.

2. What is the significance of viruses and experiments that can be done with them to modern hypotheses concerning the origin of life?

3. Dr. Sidney Fox performed a number of experiments that were a continuation of the work of Stanley Miller. He heated a dry mixture of a number of amino acids to about 90°C. Analysis of the results showed that a number of polypeptides had been formed. He also found that if he heated the amino acids in the presence of phosphoric acid, polypeptides were formed at a temperature as low as 71°C. The

polypeptides produced in this manner could be broken down by specific enzymes from animals.

In addition, bacteria could use these polypeptides for food.

a) What does this experiment show about the idea that organic substances could have been formed on the primitive earth by random action? Are living organisms essential for the synthesis of organic molecules?

b) From a thermodynamic point of view, what is the significance of the fact that polypeptides form at a lower temperature in the presence of phosphoric acid?

c) What evidence from Fox's experiments suggests that the proteins formed by experimental methods are similar to native protein found in living organisms?

d) Would it be logical to conclude that proteins formed on the primitive earth (before any life appeared) were similar to the proteins that later became incorporated into the living matter of organisms? Why or why not?

4. Identify the selection pressures that may have been operating at each of the points of development listed below:

a) Formation of a membrane around nucleic acid–molecular aggregate complexes.

b) Development of a definite respiratory process.

c) Development of a photosynthetic mechanism.

5. Describe the process of radioactive dating. On what principles is it based?

Chapter 27
The Diversity of Life I: The Monera and Protista

27.1
INTRODUCTION

The kingdoms Monera and Protista are collectively called "microorganisms," and are all unicellular forms. Some members of the Monera, especially some of the blue-green algae, exist as semicolonial groupings of cells, often arranged in long filaments; but each cell of such a grouping is still capable of independent existence. Microorganisms are classified as Monera or Protista on the basis of cell structure—particularly the presence or absence of a true nucleus. The kingdom Monera includes all cells lacking a true nucleus, the prokaryotes. The Protista, by contrast, all have truly nucleated cells; that is, they are eukaryotes.

At the boundary between the Monera and the non-living world of organic molecules lie several less easily defined groups: the **rickettsias**, a heterogeneous group of cells smaller than bacteria, all of which are obligate intracellular parasites; the **mycoplasmas**, which are the smallest free-living cell-like units known, and are also obligate intracellular parasites (in humans they cause a kind of pneumonia); and the viruses, which are sometimes not classified as living organisms at all, since they consist only of two types of molecules: a nucleic acid core surrounded by a protein coat; and finally the **viroids**, naked molecules of RNA that cause at least one known disease, scrapie in sheep. For convenience, we will consider all of these "twilight" organisms in this chapter; technically there is no agreement yet among biologists as to the best way to classify them.

An important consideration in this chapter will be the question of how the Monera, Protista, and the "twilight" organisms evolved—that is, their possible phylogenetic relationships. It seems pretty clear that the Monera, with their simple cell structure, must have evolved first. The Protista, with their more complex cell structure, came later. Despite their simplicity the "twilight" organisms may actually have evolved most recently of all these groups since as intracellular parasites they depend on the pre-existence of cells for their basic life processes.

The Kingdom Monera

The kingdom Monera consists of three major subgroups, or phyla: the Schizophyta, or bacteria; the Cyanophyta, or blue-green algae, and a newly discovered group, the Prochlorophyta. A summary of the major groups composing the kingdom Monera is given in Table 27.1.

The most general characteristic of the Monera is the fact that they are all prokaryotes (details of the cell structure of prokaryotes are given in Chapter 4). To review briefly, prokaryotic cells lack true nuclei; that is, they have no nuclear membrane (so that hereditary material such as DNA is not localized within the cell), plastids, or mitochondria, though they do possess ribosomes. Some of the Monera have flagella, but these organelles lack the typical 9 + 2 microtubular structure characteristic of flagella in eukaryotic cells. The most easily observed trait within the cytoplasm of Monera is the presence of chromatic material; in all cases studied to date this has been found to exist as a single continuous or "circular" molecule of DNA. Some Monera, such as the blue-green algae and some bacteria, are autotrophs (can make their own food) but most are heterotrophic and depend on some outside source of nutrient molecules for energy.

All members of the kingdom Monera are prokaryotes—that is, they have no nuclear membrane separating their hereditary material from the cell cytoplasm. They also lack mitochondria and plastids, though they do possess ribosomes.

Table 27.1
Characteristics of the Kingdom Monera

Phylum/class (with examples)	Morphology	Method of movement	Method of reproduction	Method of nutrition	Ecological role	Other major characteristics
Schizophyta Class: Eubacteria (*Escherichia coli, Streptococcus, Staphylococcus, Myobacterium tuberculosis*)	Rod-shaped, spirillum, and spherical	Flagella; many nonmotile	Binary fission	Chemautotrophs, bacterial photosynthesis; some heterotrophs	Decomposers; some pathogenic (lockjaw, tuberculosis)	Have rigid cell walls; form internal spores (see Fig. 27.3)
Class: Myxobacteria Gliding bacteria (*Thiothrix, Simonsiella; Myxococcus fulvus*)	Filaments, some rods (large and small)	Slow gliding along surface; method of gliding unknown	Binary fission, forming long chains, or filaments	Chemautotrophic, but some heterotrophic	Decomposers, especially of polysaccharides	Some reproduce by forming fruiting bodies similar to slime molds
Class: Actinomycetes Bacteria that do not form internal spores (endospores) (*Actinomyces, Mycobacterium, Streptomyces, Staphylococcus, Lactobacillus*)	Rods and spheres; some unicellular, others linked as long mycelia (long filamentous strands)	Mostly nonmotile	Spores (external), binary fission, and fragmentation of mycelium	Heterotrophic	Decomposers; some pathogenic; (tuberculosis, leprosy)	Some form special reproductive structures for spore formation (e.g., conidiospores)
Class: Spirochetes (*Spirocheta, Treponema, Leptospira*)	Elongate, flexible, coiled, or spiral	Twisting about by use of axial filaments inside cell wall	Binary fission	Heterotrophic	Decomposers; live symbiotically in Molluscs; pathogenic (syphilis)	Many forms are obligate anaerobes
Class: Mycoplasmas (also called PPLO)	Smallest cells that can be free-living	Nonmotile	Binary fission	Heterotrophic	Pathogenic (mycoplasmal pathogens)	Many are intracellular parasites
Class: Microtatobiotes Rickettsiae	Very small (1/10 size of average bacterium); elongated cocci	Nonmotile	Binary fission	Heterotrophic	Pathogenic (typhus; Rocky Mountain spotted fever)	All are obligate intracellular parasites; very heterogeneous group
Viruses	Very small (filterable)	Nonmotile	Molecular replication	Heterotrophic	Many are pathogenic (common flu, tobacco mosaic disease)	Noncellular; nucleic acid "core" surrounded by a protein coat
Viroids	Very small molecules of naked RNA 300–400 nucleotides long	Nonmotile	Molecular replication	Heterotrophic	Pathogenic (scrapie in sheep)	
Cyanophyta Blue-Green Algae (*Oscillatoria, Nostoc*)	Unicellular, filamentous, and colonial	Gliding; some nonmotile	Binary fission	Photosynthetic autotrophs	Carbon and nitrogen fixation	
Prochlorophyta (*Prochloron*)	Unicellular	Nonmotile	Binary fission	Autotrophic	Photosynthesizers	Have photosynthetic pigments of chlorophyta, but are prokaryotic

The biological success of the Monera undoubtedly stems from their rapid rate of reproduction and their diverse methods of metabolism. Under the best of growing conditions, a population of Monerans can double its size about every half hour. Metabolically, their diversity is astonishing. They can survive in the near-freezing waters of Antarctica, and hot springs at temperatures up to 92°C; their spores can withstand both high and low temperatures as well as drying for extended periods of time. Some bacteria are able to survive without the presence of any free oxygen, obtaining their energy completely from anaerobic glycolysis (see Chapter 5).

The Monera, especially the bacteria, are important ecologically as decomposers, since they break down organic material into a form that can be absorbed, and thus used, by plants. They are also involved in the process of nitrogen fixation, in which the abundant nitrogen gas of the air is converted by bacteria in the soil into soluble ammonia (NH_3) or ammonium (NH_4^+) (for details of nitrogen fixation, see Chapter 25). Despite their structural and to some extent biochemical simplicity, Monerans have played, and continue to play, a profoundly important role in the history and perpetuation of life.

27.2
PHYLUM SCHIZOPHYTA

Following the classification scheme outlined in Table 27.1, there are six major groups or classes of the phylum Schizophyta. In this section we will examine the major groups of the Schizophyta: the bacteria, the mycoplasmas, the rickettsiae, and the viruses.

The Bacteria: Classes Eubacteria, Myxobacteria, Actinomycetes, and Spirochetes

The classes of bacterial organisms are distinguished on the basis of their cell shapes and modes of movement. These groups can be differentiated into three basic morphological types: the spherical or **cocci** bacteria, the rod-shaped or **bacillus** bacteria, and the elongated, flagellated **spirochetes**. These three basic morphological types are shown in Fig. 27.1.

One of the most distinctive features of bacteria is the presence of a cell wall surrounding the plasma membrane. The cell wall gives the different forms of bacteria their characteristic shapes. Many bacteria have firm, rigid cell walls; some have quite flexible cell walls, and none (except the mycoplasmas) lack a cell wall completely. The cell wall is composed of many kinds of molecules, but principally an association of complex polymers called peptidoglycans. In some bacteria, molecules of another polymer, lipopolysaccharide, are laid down over the peptidoglycan layer. Bacterial cells that lack the lipopolysaccharide layer combine with such dyes as gentian violet, and thus stain readily. These bacteria are referred to as **gram-positive bacteria**. Bacterial cells with the lipopolysaccharide layer do not combine with gentian violet and thus do not stain. Such bacteria are called **gram-negative**. (The designation "gram" comes from the Danish bacteriologist Hans Christian Gram, who first discovered this basic distinction between bacterial types.) The use of dyes is thus one important method in identifying bacteria. It also turns out that gram-positive bacteria are more susceptible than gram-negative bacteria to most antibodies, and to lysozyme (see Chapter 5), an enzyme found in nasal secretions and saliva (it can digest cell walls).

Fig. 27.1
Light and electron micrographs of the three morphological types of bacteria. (a) Spherical or coccus form (top, light micrograph of *Micrococcus* × 1000; bottom, electron micrograph of *Staphylococcus aureus* × 55,000). (b) Rod-shaped or bacillus form (top, light micrograph of *Azobacter* × 1000; bottom, electron micrograph of *Bacillus subtilis* × 80,000); (c) Spiral or spirillum form (top, light micrograph of *Spirillus volutans* × 1232; bottom, electron micrograph of *Aquaspirillum bengal* × 7700. (Light micrographs courtesy Turtox, Chicago. Electron micrographs: (a) courtesy Eli Lilly and Co.; (b) courtesy Thomas F. Anderson; (c) courtesy Noel R. Krieg and the American Society for Microbiology.)

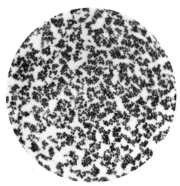

(a)

The function of bacterial cell walls is not completely clear. It has been hypothesized that the wall helps keep the cell from rupturing. Most bacteria are hypertonic to the environment in which they live, a feature that would tend to cause water to flow from the medium *into* the cell. Thus, the cell wall might function much like the hard rubber of an automobile tire, which prevents the softer inner tube inside from bursting when it is pumped with very high pressures of air.

Bacteria have evolved several diverse solutions to the problem of obtaining energy. Some are autotrophs, synthesizing their own carbon compounds by fixing carbon dioxide. Of evolutionary interest in this regard are the different sources of energy that bacteria have exploited in carrying out carbon fixation. Some, the photosynthetic bacteria, use sunlight captured by their own form of chlorophyll, which differs in certain slight ways from that found in green plants and blue-green algae. Still others, the chemosynthetic bacteria, obtain energy by degrading compounds such as ammonia and hydrogen sulfide; they then use this energy to reduce carbon dioxide to carbohydrates. The biochemical

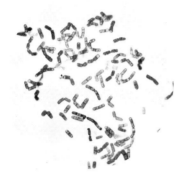

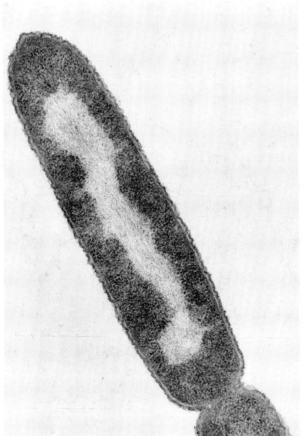

(b)

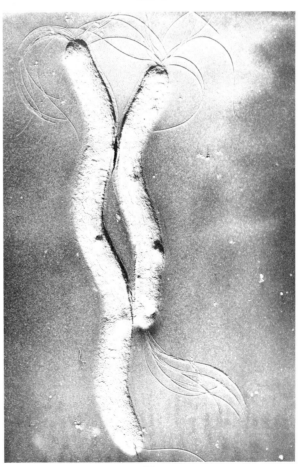

(c)

Bacteria have evolved several diverse solutions to the problem of obtaining energy. Most are heterotrophs, deriving their energy from already-synthesized carbon compounds. Many are autotrophs, using a variety of energy sources to fix carbon dioxide into carbohydrates: Photosynthetic bacteria use sunlight as the energy source, whereas chemosynthetic bacteria degrade compounds such as ammonia and hydrogen sulfide to provide energy.

processes by which the various autotrophic bacteria synthesize carbohydrates are discussed in some detail in Section 5.15.

As Table 27.1 shows, most bacteria are capable of some form of motion—either by gliding, by the use of flagella, or by the use of axial filaments that stretch along the inside of the cell wall attached at either end of the cell. By contracting, the axial filaments cause the cell to twist and turn and thus move through the medium (albeit somewhat randomly). The latter form of motion, characteristic of the spirochetes in particular, might really be considered a form of flagellar motion.

Bacterial flagella function in much the same way as eukaryotic flagella, but are structured quite differently. Recall that eukaryotic flagella consist of two central contractile fibrils surrounded by nine pairs of microtubules running the length of the flagellum. The bacterial flagellum consists of a single chain of contractile protein, composed of monomers of the protein flagellin, (see Section 5.12), which are assembled in chains and wound around a central core. Recent evidence from electron microscopy shows that bacterial flagella are attached to the cell wall and cell membrane by a complex series of structures. The flagellin filament ends at the cell wall in a hook, made of a different protein. The hook fits into a series of rod and ring-like structures, the basal body, allowing the whole structure to rotate and providing forward motion. Unlike the basal body in eukaryotic flagella, that in bacterial flagella does not seem to connect to a known or observable organelle (such as the Golgi complex) within the cytoplasm.

An additional extension found on bacterial cells are **pili** (singular, *pilus*). These are rigid, cylindrical rods that extend out from the cell wall; they are, however, thinner and usually shorter than flagella. Like flagella, they are also composed of a series of protein monomers (called pilin) arranged into polymer chains. Pili serve to fasten the bacterial cell to other bacteria (as, for example, during conjugation), or to a food source.

Bacterial reproduction consists of two basic forms: binary fission (cell division) and budding (with subsequent fragmenting of a budding chain). Barring mutation, bacteria, like all asexually reproducing forms, produce identical daughter cells, or clones, of themselves. In addition, many bacteria engage in genetic recombination by processes such as conjugation, transformation, and transduction (see Chapter 18).

When a single bacterial cell undergoes fission, it produces two daughter cells. These two cells, in turn, each produce two daughter cells, yielding a total of four. As this process continues (proceeding at least for a while at a geometric rate), a bacterial **colony** develops. All members of a colony are descended from the original progenitor cell, and are thus (barring mutation) genetically identical. Because bacterial cells are not highly motile, the cells of a colony tend to stay clumped together. When growing on an artificial medium in a petri dish (Fig. 27.2), colonies are easily distinguishable as discrete dots. The shape of the colony is a characteristic of each species of bacterium. It is, in fact, one of the ways in which bacteriologists identify particular species, or even strains of bacteria.

Because bacteria are agents of many diseases, we tend to think of them as mostly harmful organisms. Actually, the reverse is true. Human and most animal life would be unable to survive without bacteria. E. coli in our intestine provides us with needed vitamins. Nitrogen-fixing bacteria provide nitrates for plant nutrition. Finally, bacteria are major organisms of decay.

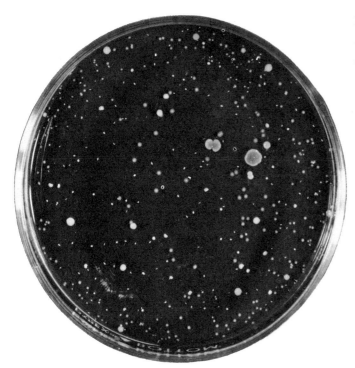

Fig. 27.2
Bacterial culture plate (petri dish) containing several colonies. Each colony is descended from a single bacterium, so all the cells in any one colony are genetically identical.

Bacteria carry out the process of **sporulation,** or spore formation. Sporulation is not a reproductive process, but one that helps to sustain the cell during unfavorable conditions (see Fig. 27.3). When conditions become unfavorable, the bacterial cell manufactures a very strong, virtually impermeable membrane—the spore coat—from materials in its cytoplasm. This coat surrounds the bacterial DNA and a small part of the cytoplasm. The remaining parts of the bacterial cell disintegrate, leaving the well-protected spore containing the organism's DNA. Spores can withstand temperatures up to 100°C (boiling) for several hours; nonsporulated bacterial cells of the same species are killed quickly by such temperatures.

(a)

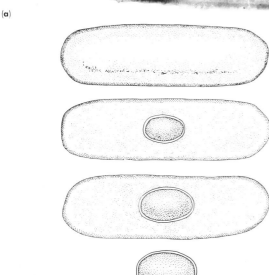

(b)

Fig. 27.3
The process of sporulation in *Bacillus cereus.* The electron micrograph (a) shows the large spore being formed toward one end of the cytoplasm of the bacterial cell. The spore has a very tough protective coat enabling the cell to survive through unfavorable conditions (× 46,000). (b) Drawing of sporulation process. The spore is formed from elements in the cytoplasm and encased in its coat while still within the original cell. (Micrograph courtesy Dr. George B. Chapman and *The Journal of Bacteriology.*)

Mycoplasmas (PPLO Organisms)

A very special group of the schizophyta are known as Mycoplasmas. They are the smallest cells known, and they range in diameter from 0.1 to 0.25 μ (about one-tenth the size of an average bacterium). One kind of Mycoplasma is known as PPLO, or "pleuropneumonia-like organism," because it produces a pneumonia-like infection in cattle. About thirty different species of Mycoplasmas have been recognized to date.

The taxonomic position of Mycoplasmas has been in some doubt for years. In many ways they physically resemble bacteria, but their small size and the fact that they can be isolated only with the use of filters small enough to trap viruses suggested they should be classified with the latter group. In recent years, however, investigations regarding the cell wall chemistry of Mycoplasmas, including discovery of a peptidoglycan component, has caused microbiologists to reassign the group. Mycoplasmas are now generally included firmly within the Schizophyta, as a small, but *bona fide* bacterial group, the class Mycoplasmata.

The existence of Mycoplasmas was first suspected in the nineteenth century in connection with a contagious cattle pneumonia. The causative agent could not be isolated at the time, however: it could neither be seen with a light microscope, nor could it be trapped by the available porcelain filters used to remove bacteria from blood. In 1898 researchers showed that Mycoplasmas would grow on a culture medium, and did not require living host cells. Finally, in 1931 a special filter capable of trapping these tiny organisms was developed. It was only then that Mycoplasmas could be observed, their actual size adequately estimated, and something of their structure determined (see Fig. 27.4).

PPLO cells lack most of the organelles that characterize eukaryotic cells, but they contain most components found in prokaryotic cells. They have what appears to be a typical lipid-protein membrane surrounding the cell. They have DNA loosely organized within the cytoplasm, and they have ribosomes for protein synthesis. However, their internal organization appears even less highly structured than that of bacteria. PPLO, for example, do not appear to have their nuclear material localized in the way that is common for most bacteria.

PPLO reproduction involves a complex life cycle that takes two days to complete. The species of PPLO that has been studied most intensively forms small "elementary bodies" that grow in size for several days until they are the size of a bacterium (about 1 μ in diameter). The large cell may divide into two daughter cells, or it may form elementary bodies within itself that are eventually released. PPLO may not represent the very smallest existing cells—there may still be smaller ones yet undetected. However, it seems unlikely that there will be discovered structures qualifying as true cells that are very much smaller than PPLO. There simply would not be enough room for the minimum number of organelles needed to sustain life.

Rickettsiae

Rickettsiae and viruses are sometimes grouped together into a single taxonomic unit (the order Microtatobiotes) within the phylum Schizophyta.* Rickettsiae and viruses have two features in common: both are obligate intracellular parasites (they cannot live outside other living cells), and both have at least one stage in their life cycle that is so small it can pass through the porcelain filters used to trap bacteria. Structurally, Rickettsiae and viruses are very different. Rickettsiae are small, cell-like organisms, while viruses are even simpler, consisting merely of nucleic acid surrounded by a protein coat. It is unclear whether either of these forms should technically be called living.

Despite their small size and biological simplicity, rickettsiae and viruses have had, and continue to have, a significant and profound influence on human life, mostly as agents of disease.

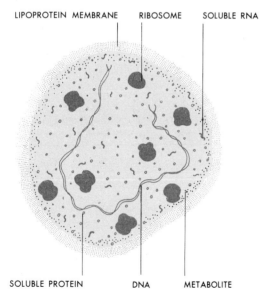

LIPOPROTEIN MEMBRANE   RIBOSOME   SOLUBLE RNA

SOLUBLE PROTEIN   DNA   METABOLITE

Fig. 27.4
Diagram of a typical mycoplasma, the PPLO (pleuropneumonia-like organism). These extremely small cells contain all the types of molecules found in most living things (carbohydrate, lipids, protein, and nucleic acid) as well as ribosomes. Note the much larger relative size of ribosomes shown here; this indicates something of the tiny size of PPLO. Some mycoplasmas like PPLO are highly pathogenic in humans and other mammals. (After *Biology Today.* New York: CRM/Random House, 1972, p. 157.)

---

*Some biologists place the rickettsiae and viruses in a completely separate category apart from any basic taxonomic scheme, on the grounds that they are not living entities but large molecular aggregates ("twilight organisms," as described earlier).

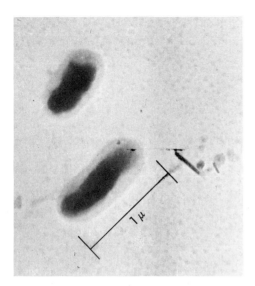

Fig. 27.5
Electron micrograph (× 20,000) of two rickettsiae, in this case the type that causes typhus. Note that even under the electron microscope very little detail of these cells is visible. (Courtesy Thomas F. Anderson.)

Rickettsiae are obligate intracellular parasites living in such arthropods as fleas, mites, ticks, and lice, that serve as **vectors** (carriers) of disease. When the arthropod bites an animal, such as a mouse or a human, rickettsiae are transferred to the host. They can cause severe infection; such deadly diseases as Rocky Mountain spotted fever and typhus are due to infection by rickettsial cells, transferred to human hosts by ticks and lice, respectively. In general, there is a specificity between each unique rickettsial type and the vector that transmits it. Fortunately, today antibiotic therapy is very effective against rickettsial diseases.

Rickettsiae are similar in shape and appearance to spherical (cocci) bacteria, but are somewhat more elongated (see Fig. 27.5). They have a cell wall, but both the wall and underlying membrane are "leaky"; that is, they allow the passage back and forth of much larger molecules than most cell walls and membranes allow. Rickettsiae contain all the essential metabolic machinery of life: ribosomes, DNA, RNA, and enzymes. They depend on the host cell to supply them with coenzymes and ATP. The group was named for Howard Taylor Ricketts, who pioneered in the study of rickettsial diseases. Rickettsiae are often considered intermediate

in complexity (and possibly phylogeny) between the true bacteria and the viruses.

Viruses

General Features. Viruses are associations of two macromolecular components: an inner "core" of nucleic acid (either DNA or RNA, but not both, depending on the type of virus) and an outer "coat" of protein. The nucleic acid carries the genetic program for producing more viruses; the protein is a protective structure that also is instrumental in the virus attacking its host. Figure 27.6 shows several types of viruses. Viruses function by invading living cells. Inside the host cell, the virus reproduces itself by using the cell's chemical machinery. Often, this destroys the host cell (though not always, depending on the virus type and infection pattern.)

As previously mentioned, viruses cannot reproduce outside living cells. They can be chemically crystallized and stored for years; yet, when resuspended in a liquid medium and brought in contact with their specific host cell, they can infect the cell and reproduce more viruses. But while they are crystals, they do not carry out any of the chemical reactions associated with living cells. In view of these characteristics, many biologists have found it impossible to classify viruses as either living or nonliving. The terms simply have no meaning when applied to them. If the ability to reproduce constitutes "living," then viruses are alive. On the other hand, if the ability to carry out respiration and synthesize one's own proteins is considered "living," then viruses are not alive. Functioning only with the aid of intact living cells, they are like a set of blueprints with no factory of their own where their plans can be put into action. Yet, once inside a living cell, viruses can direct the synthesis of proteins, reproduce virus parts, and assemble those parts into new, intact viruses.

Since all viruses live within already-existing cells, they are for the most part harmful to the host cell. Viruses are major causes of disease in both animals and plants. A mottled, deteriorating condition affecting tobacco plants is caused by the tobacco mosaic virus, TMV. Smallpox and polio are caused by viruses, as are many varieties of the common "cold." Viruses are particularly difficult to combat because common antibiotics do not usually affect them. The most effective medical approach to viral diseases is prevention of the disease through immunization (for example, vaccination). Exposure of the human (or animal) body to a small dosage of attenuated virus allows the body's defenses against that specific virus to build up without the

---

Viruses cannot reproduce outside of living cells. They can be chemically crystallized and stored on a shelf for years. Yet, when resuspended in liquid and brought into contact with their specific host cell, they can infect the cells and reproduce themselves.

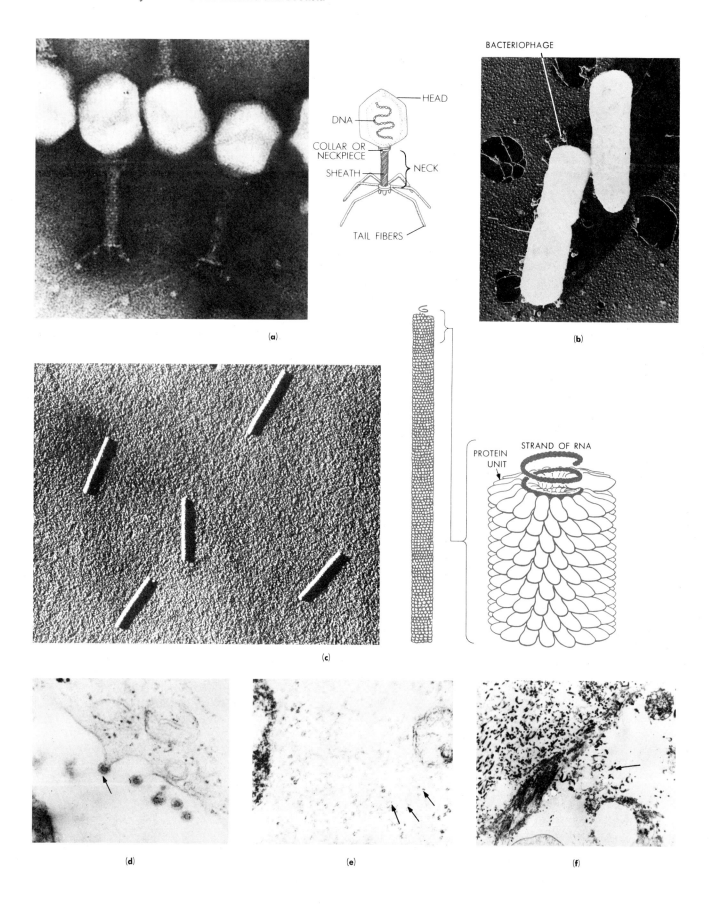

(a)

(b)

(c)

(d)

(e)

(f)

Fig. 27.6
Electron micrograph of a group of T₂ bacteriophages. Each phage consists of a head, neck, and tail region. The head outer coat and the neck and tail consist of protein. Inside the head is DNA carrying the virus's genetic information. Phages without the thickened neck-piece are "triggered," meaning they have released their DNA. DNA is shot out through the neck and tail into a bacterium to which the phage has attached itself. (Micrograph courtesy Dr. Sidney Brenner and Dr. R. W. Horne.) (*b*) The bacterial cell at the left shows several bacteriophages attached to it. (*c*) Electron micrograph (× 67,000) of rod-shaped tobacco mosaic virus. The photograph was taken after the sample had been "shadowed" with chromium to increase contrast. (Micrograph courtesy U.S. Public Health Service.) To the right is a diagrammatic representation of the structure of the same virus. Its single coiled strand of RNA is seen emerging from the stack of spirally arranged protein units in which it is embedded. All the protein units in this virus are identical. Each unit is composed of a sequence of 158 amino acids. The complete TMV has 2200 protein units. (Adapted from W. M. Stanley and E. G. Valens, *Viruses and the Nature of Life.* New York: E. P. Dutton, 1961.) (*d,e,f*) Three viruses causing pathological conditions in human beings. The rubella virus (*d*) causes German measles and may cause malformations or death in developing fetuses if their mothers are infected during the early stages of pregnancy. (Micrographs courtesy of the Center for Disease Control, Atlanta, Georgia 30333).

organism's succumbing to the disease. Then when the virus is encountered in the environment, it can be eliminated from the body early because the body "recognizes" it as foreign. Once the virus has invaded the body's cells in an infection, it is much more difficult to eliminate.

The reproductive cycle of a typical virus is shown in Fig. 27.7. The diagram pictures a bacteriophage

Fig. 27.7
Stages in the life cycle of the virus, bacteriophage T₄, infecting a bacterium such as its normal host, *Escherichia coli.* T₄ undergoes two different but interrelated phases in its life cycle, the lytic (left) and lysogenic (right). In the lytic phase, phage particles inject their nucleic acid (in this case DNA) into the host cell and produce new viral particles. Progeny viruses are released when the cell breaks open, or "lyses." In the lysogenic phase, phage particles inject their nucleic acid into the host cells. The viral nucleic acid then becomes incorporated as a "prophage" into the host cell's chromosome. The viral nucleic acid replicates at the same rate as the host cell's, giving rise to many progeny each with a prophage inserted into its chromosome. Occasionally the prophage is released and begins to make virus progeny. The host cells are eventually lysed as the phage moves from the lysogenic to the lytic phase. (Redrawn from J. D. Watson, *The Molecular Biology of the Gene,* 2d ed. W. A. Benjamin, 1970, p. 205.)

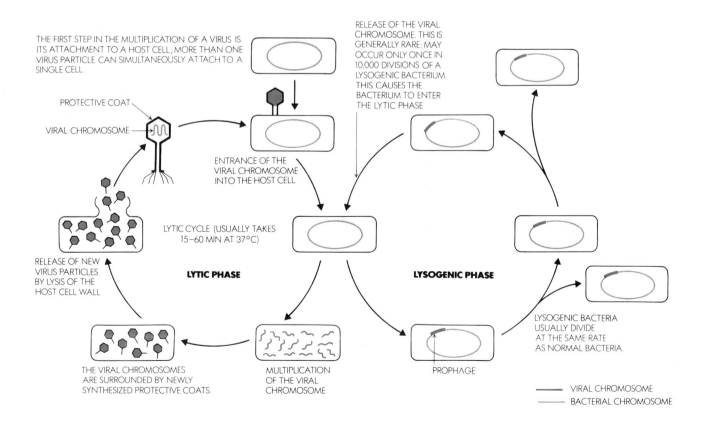

THE FIRST STEP IN THE MULTIPLICATION OF A VIRUS IS ITS ATTACHMENT TO A HOST CELL; MORE THAN ONE VIRUS PARTICLE CAN SIMULTANEOUSLY ATTACH TO A SINGLE CELL.

PROTECTIVE COAT

VIRAL CHROMOSOME

ENTRANCE OF THE VIRAL CHROMOSOME INTO THE HOST CELL

LYTIC CYCLE (USUALLY TAKES 15–60 MIN AT 37°C)

LYTIC PHASE

RELEASE OF NEW VIRUS PARTICLES BY LYSIS OF THE HOST CELL WALL

THE VIRAL CHROMOSOMES ARE SURROUNDED BY NEWLY SYNTHESIZED PROTECTIVE COATS.

MULTIPLICATION OF THE VIRAL CHROMOSOME

RELEASE OF THE VIRAL CHROMOSOME. THIS IS GENERALLY RARE; MAY OCCUR ONLY ONCE IN 10,000 DIVISIONS OF A LYSOGENIC BACTERIUM. THIS CAUSES THE BACTERIUM TO ENTER THE LYTIC PHASE

LYSOGENIC PHASE

LYSOGENIC BACTERIA USUALLY DIVIDE AT THE SAME RATE AS NORMAL BACTERIA.

PROPHAGE

—— VIRAL CHROMOSOME
—— BACTERIAL CHROMOSOME

(called "phage" for short) invading its normal host cell, a bacterium (*E. coli*). The details of the life cycle are explained in the figure. Note, however, that many viruses, such as the one shown here, can reproduce in either a **lytic** or a **lysogenic** phase. In the lytic phase, the viral DNA is injected into the host cell, where it reorganizes the cell's biochemical machinery to produce new virus particles. This cycle can be completed in as short a time as one half hour. At the end of the lytic phase the cell is ruptured and viral progeny are released. In the lysogenic phase, however, the viral DNA becomes incorporated into the host cell's chromosome, replicating only as fast as the host cell DNA replicates. This process allows the slow spread of virus through many daughter bacterial cells. On occasion, as the diagram shows, lysogenic viruses can suddenly become lytic. The virus DNA dissociates from the bacterial chromosome and begins guiding the production of viral DNA and protein. New virus progeny arise and rupture the cell, completing the lytic phase.

A familiar example of the transformation of the lysogenic to lytic phase in the viral life cycle is found in the human disease known as "shingles." Shingles involves irritation at nerve endings in the skin caused by viral lysis at the cell periphery. The virus responsible for shingles is known as *Herpes zoster,* and is the same virus that produces chickenpox. One theory proposes that after a childhood infection of chickenpox, *Herpes* virus remains dormant in neural ganglia or peripheral nerves. Possibly, the virus could remain in nerve cells as an episome (a separate circular piece of DNA not attached to the host cell's chromosome) or as a provirus (that is, integrated into the host cell chromosomes), though the latter is unlikely. In either case, at some later point in time, the virus particles become active again, and enter a lytic phase. This produces the eruptions of the peripheral nerve endings associated with shingles. Some biologists believe that, in a similar way, the shift from a lysogenic to a lytic phase by certain viruses may be one of the causes of cancer. Some cancers are known to be associated with viruses, but how the presence of the virus triggers uncontrolled cell division is not known (see Supplement 20.3).

Some interesting experiments that have been performed with tobacco mosaic virus (TMV) throw light on questions of the origin of life. By delicate chemical procedures it is possible to separate the RNA from the

protein coat of TMV. It is also possible to break the protein coat down into its component units. Simple adjustments in the acidity of the medium are enough to bring the protein units back together to form long unit aggregates. These aggregates are rods that look exactly like the intact virus. Such "empty" protein shells, however, are unable to infect plant tissues.

If the protein shells are placed in a solution of viral RNA, they form particles that behave like normal, intact TMV. The fact that a solution of this material is able to infect tobacco leaves is a good indication that the protein shells and nucleic acid organized themselves to form complete TMV particles. It is known that in nature neither nucleic acid nor protein alone is capable of producing infection. Both parts must be present for the virus to reproduce.

A second type of experiment provides even more interesting results. Heinz Fraenkel-Conrat and his associates at the University of California, Berkeley separated the protein coats and the RNA cores from TMV viruses and from another strain of rod-shaped virus (called HR). In each case, the protein coats were carefully decomposed into their individual units in such a way that the molecules of protein were not denatured. The researchers then placed the protein coats from the TMV strain with the RNA cores of the HR strain (see Fig. 27.8). The acidity was adjusted so as to allow the units of protein to aggregate. The result of this mixture was the self-organization of a virus with a TMV coat and an HR core of nucleic acid (RNA). Here was a new type, or hybrid virus. When administered to plant tissues, the newly formed virus caused an infection characteristic of the HR strain in every case. In other words, the infection was characteristic of the type of RNA involved. This strongly supports the hypothesis that the nucleic acid of viruses is the major infective agent. The protein may play a passive role, perhaps as a protective cover for the nucleic acid.

As the hybrid virus was allowed to duplicate, it was noted that the new viruses developed coats of the type of protection characteristic of the HR strain. This experiment showed that in viruses, RNA contains the full genetic complement—it is coded with all the information needed for building complete viruses.

Recent experiments by P. Jonathan G. Butler and Aaron Klug at the Medical Research Council Laboratory of Molecular Biology in Cambridge, England, have

---

Viruses illustrate an important principle: much biological organization is the result of spontaneous formation of components of a structure, given the right environmental (physical and chemical) conditions. Viruses taken apart down to their protein subunits will spontaneously reaggregate.

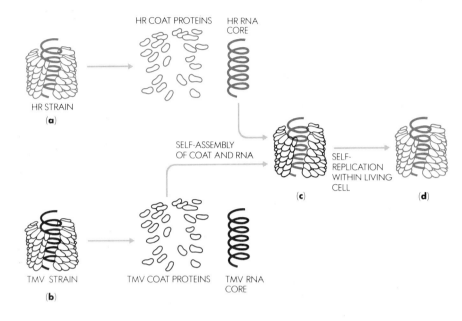

Fig. 27.8
Summary diagram of Fraenkel-Conrat's experiments, showing the hybridization of viruses. The progeny at (*d*) have become slightly modified from their parents at (*a*) and (*b*). (Adapted from H. Fraenkel-Conrat, "Rebuilding a Virus." *Scientific American*, June 1956, p. 47.)

provided a more detailed picture of the process of reassembly in TMV. They have shown, for example, that the form of aggregation of the discs produced by combinations of protein subunits varies with pH. Furthermore, they have provided a detailed picture of the actual tertiary structure of the polypeptide chains composing the subunits. Among other things, this detailed picture has shown that the polypeptides have "binding sites" for the RNA that help give the RNA stability within its central core.

The above experiments lead to two important generalizations. The first is that many complex biological structures are capable of self-assembly given the right chemical and physical (temperature) environment. Achieving highly organized molecules or even organelles does not neccessarily demand the pre-existence of complex biochemical assembly systems (recall that the subunits of ribosomes can dissociate or reassemble with changes in pH). The second is that viral components are fully functional on reassembly. Being taken apart does not alter the ability of either the viral nucleic acid or the protein coat to carry out its normal function when put back together.

The Body's Defenses against Viruses: Interferon. While the incidence of some viral diseases (including poliomyelitis and smallpox) has been dramatically reduced by vaccines, most viruses are not susceptible to the drugs and antibiotics that are so successful against bacteria. The immune response does occur (if you have had the mumps or measles once, you are generally immune for life), but is usually too slow to effect a cure when the disease first occurs. However, interferon is also produced in response to virus invasion, and is thought

to play a role in recovery (see also Chapter 20). One hypothesis proposes that viral infection of a cell induces interferon production. The interferon is then transported to other as yet uninfected cells, conferring on them a resistance to infection by the same or even other viruses. Interferon acts possibly by inducing the cell to synthesize another protein that blocks the transcription of *m*RNA from viral DNA. Interferon also stops cell multiplication, a fact that makes it of great interest in cancer research. Just how it does so is not certain. However, a team of researchers at Rockefeller University, headed by Dr. Lawrence Pfeffer, have evidence that interferon may act by "aging" cells, since young cells exposed to the antiviral protein act similarly to those that have gone through fifty generations *in vitro*. Aging cells concentrate their energy on growing larger rather than on reproducing.

As mentioned previously, when interferon was first discovered in the mid- to late 1950s, hopes were high that it could become a powerful agent against viral disease. This potential has never been realized, however. One of the main problems has been the fact that in living tissues interferon is produced in extremely small amounts. The job of extracting it from animal tissues is extremely costly and laborious. In the past few years the possibility of using interferon as not only an antiviral but also an anticancer agent has redoubled efforts to find more effective ways of producing the protein easily. New strides in the field of genetic engineering (see Supplement 19.5) have demonstrated the possibility of isolating the mammalian gene for interferon and transferring it to bacteria. By using techniques of recombinant DNA researchers have produced strains of bacteria that are able to synthesize large quantities of interferon. It is just now becoming possible to experiment in a more practical and systematic way with interferon as both an anticancer and an antiviral agent. The early high hopes for this protein may not be so unrealistic after all.

27.3
PHYLUM CYANOPHYTA: THE BLUE-GREEN ALGAE

Blue-green algae share many characteristics with bacteria. As in the bacterial cell, the DNA of blue-green algae cells is not contained within a true nucleus, but is localized in certain regions (see Fig. 27.9a). Like bacteria, blue-green algae have ribosomes and lack most of the other organelles found in eukaryotic cells. And, like bacteria, blue-green algae cells are surrounded by a protective carbohydrate cell wall. Unlike most bacteria, however, all the 1500 or so species of blue-green algae are able to carry on photosynthesis. This process is accomplished without the chloroplasts typically found in cells of higher plants. Blue-green algae have both photosystems I and II. They also have chlorophyll *a* (but none of the other forms of chlorophyll found typically in higher plant cells) along with several other pigments

(such as phycocyanin, a bluish pigment and phycoerythrin, a reddish pigment) involved in the photosynthetic process. Lacking chloroplasts, blue-green algae cells have chlorophyll bound to infoldings of the cell membrane (Fig. 27.9b). These folded membranes give the internal structure of the blue-green algae the appearance of considerably greater complexity and organization than the bacterial cell. In reality, the dif-

Fig. 27.9
Blue-green algae are similar in many organizational features to bacteria. Nuclear material is not localized in an enclosed nucleus. There are few organelles, with the exception of ribosomes and a cytoplasmic membrane system for photosynthesis. The cell is surrounded by a thick cell wall of carbohydrate; outside this there is sometimes another, more amorphous layer called a "sheath." The sheath is usually a semi-fluidlike material, similar to thick jelly. It prevents water loss from the algal cell, as well as affording increased protection. The electron micrograph shown in (*b*) gives a good indication of the cytoplasmic membrane system in one alga, *Anabaena* (× 6800). (Micrograph courtesy William T. Hall, Electron-Nucleonics Laboratories, Inc.)

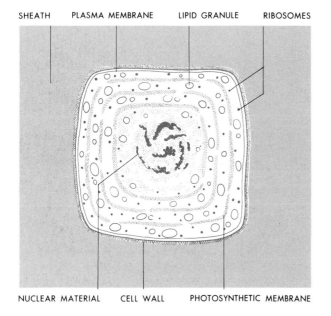

SHEATH   PLASMA MEMBRANE   LIPID GRANULE   RIBOSOMES

NUCLEAR MATERIAL   CELL WALL   PHOTOSYNTHETIC MEMBRANE

(a)

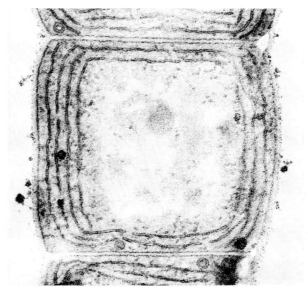

(b)

Blue-green algae share many characteristics with bacteria: they lack a nuclear membrane and all cell organelles except ribosomes. Both groups possess a rigid cellulose cell wall. Unlike bacteria, however, blue-green algae all have photosystems I and II.

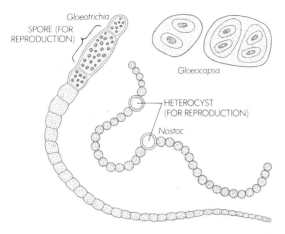

Fig. 27.10
Three genera of the phylum Cyanophyta, the blue-green algae. All these organisms possess chlorophyll *a*, plus several other pigments, such as phycocyanin, which is responsible for the blue-green color characteristic of the cyanophytes. Reproduction is largely by fission; on occasion special spore-like cells are formed that can germinate to produce a new filament.

ference is not great. With the exception of the infolded cytoplasmic membrane system and the presence of photosynthetic pigments, the blue-green algae are organized in a manner very much like bacteria. Many biologists think the two groups share a close evolutionary rela-

tionship, and indeed, opinion among many biologists now holds that blue-green algae should be classified with the bacteria.

Many, though not all, species of blue-green algae exist as groups of cells arranged in long filaments (see Figs. 27.10 and 27.11). Some of these live in the moist earth, others in water. A few species can endure high temperatures—up to 70°C—such as those found in hot springs. The variety of colors observed in the hot springs of Yellowstone National Park is due to the accessory pigments of many species of blue-green algae that live in the springs. Like the bacteria, the blue-green algae are very widely distributed throughout the world.

Because they are autotrophs, the blue-green algae are able to live on bare areas of rock, sand, and soil. After the massive volcanic eruption of 1883 on the Indonesian island of Krakatoa, the first living organisms to appear on the barren volcanic ash and pumice were filamentous blue-green algae. Within a few years they had gained a strong foothold and had covered the rocks with a gelatinous film. This film layer eventually became so thick that the seedlings of higher plants could become rooted in it. It is tempting to hypothesize that, in a similar way, the blue-green algae may have prepared the way for the invasion of land during the early evolutionary history of the earth.

Many species of blue-green algae are capable of nitrogen fixation. They are able, like the nitrogen-fixing bacteria, to convert atmospheric nitrogen (N_2) into soluble nitrates. In blue-green algae such as *Nostoc* (see Fig. 27.11) this process takes place in special thick-

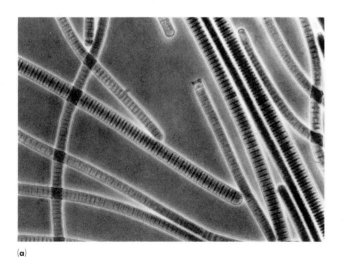

(a)

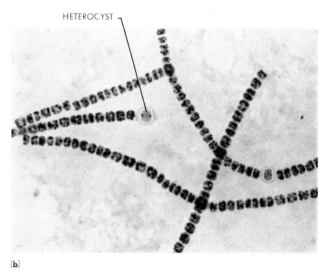

(b)

Fig. 27.11
Two filamentous species of blue-green algae. (*a*) *Oscillatoria,* showing the regular arrangement of cells end-to-end to form long chains, or filaments. The name derives from the fact that under the microscope the filaments themselves can be observed gliding slowly back and forth ("oscillating") along the substratum (× 1000). (*b*) *Nostoc,* showing the beadlike cells arranged in short filaments punctuated with larger, more transparent heterocysts; the latter are centers for nitrogen fixation. [(*a*) from John W. Kimball, *Biology,* 2nd ed. Reading, Mass.: Addison-Wesley, 1968; (*b*) courtesy Carolina Biological Supply Co.]

walled cells called heterocysts. Nitrogenase, a special enzyme within the heterocyst, is responsible for carrying out a major part of the reaction. When blue-green algae are grown in media containing a rich source of nitrates, they do not produce heterocysts, a good indication that the heterocyst is the seat of nitrogen fixation. Study of the thick wall of heterocysts may provide a clue to the possible evolutionary origin of blue-green algae. It is now known that the thick walls help exclude oxygen from the heterocyst interior. Nitrogenase is highly sensitive to molecular oxygen—so much so that it is irreversibly poisoned by prolonged contact with the gas. These observations suggest that blue-green algae evolved very early in the earth's history, perhaps in pre-Cambrian times (3.5 to 4.0 billion years ago), when the oxygen concentration of the earth's atmosphere was extremely low (see Section 26.2).

27.4
PHYLUM PROCHLOROPHYTA

In 1976 a scientist at the Scripps Institution of Oceanography in California, Dr. Roger A. Lewin, reported the discovery of a representative of what appeared to be a whole new phylum of prokaryotic cells, the Prochlorophyta. These are unicellular forms containing pigment systems very similar to those in green algae and the higher plants. Prochlorophyta have both chlorophylls *a* and *b* (recall that blue-green algae have only chlorophyll *a*), xanthophylls and carotenoids (the same pigments found in higher plants), and lack the blue and red pigments, phycocyanin and phycoerythrin, found in blue-green algae. Yet, prochlorophyta lack any trace of a structured nucleus. They seem to be exactly intermediate between the blue-green algae on the one side and the green algae on the other.

The discovery of the Prochlorophyta is more than a biological curiosity. It has important implications for our understanding of the origin of eukaryotic cells. A number of years ago Lynn Margulis and others postulated that chloroplasts of higher plant cells (including green algae) originated by incorporation of the cells of prokaryotic blue-green algae into their cytoplasm (the endosymbiont theory, see Supplement 27.1). A problem with this theory has always been that the pigment systems in blue-green algae, as we have seen, are very different from those in the chloroplasts of higher plant cells. Discovery of the Prochlorophyta has eliminated this difficulty. Possibly the Prochlorophyta, rather than the blue-green algae, are the progenitors of chloroplasts in eukaryotic plant cells. Such a suggestion is, of course, pure speculation at this point; too little is yet known about the Prochlorophyta to jump to a conclusion about their role in evolution. Nonetheless, their discovery has opened up a new problem area associated with understanding the transition from the prokaryotic to the eukaryotic cell.

The Kingdom Protista

The kingdom Protista is characterized by eukaryotic (with a true nucleus) mostly unicellular organisms, which reproduce both asexually (by mitosis) and sexually (by fusion), with the original chromosome number being restored at some subsequent time by meiosis. Apart from these similarities, however, the Protista are an extremely diverse group of organisms. Their modes of nutrition vary considerably—some are free-living heterotrophs, others are parasitic heterotrophs, and still others are autotrophs, capable of photosynthesis. The kingdom includes organisms well known to everyone

SUPPLEMENT 27.1
THE ORIGIN OF EUKARYOTES

It is reasonable to hypothesize that the eukaryotes rose from the prokaryotes, and thus share, at a very fundamental level, a common evolutionary history. After all, both groups use DNA (for the most part) as the genetic material; employ a triplet genetic code; transcribe DNA into *m*RNA, and produce proteins with ribosomes (including similar types of ribosomal and transfer RNAs). Such biochemical similarities suggest a common, if long-distant, ancestry. How might this evolutionary development have taken place? By what possible steps could prokaryotes give rise to the far more complex, organelle-rich eukaryotic cells?

A number of similarities have been noted between the mitochondrial, and chloroplast structure and function of bacteria. These similarities have suggested the intriguing hypothesis that both these cell organelles may have arisen from various types of prokaryotic cells that became incorporated into the structure of primitive eukaryotic cells. Thus, mitochondria could have once been aerobic bacteria, and chloroplasts once prokaryotic algae, such as the blue-greens. What is the evidence on which this suggestion is based?

Mitochondria and chloroplasts are roughly the same size as bacteria and have a similar membrane structure. The inner membrane of chloroplasts resembles the cell membrane of blue-green algae in that it is the site of photophosphorylation (blue-green algae do not have chloroplasts, only infoldings of the cell membrane to which chlorophyll is attached). Similarly, the inner membrane of the mitochondrion resembles the cell membrane of bacteria in that both are sites of oxidative phosphorylation.

A number of antibiotics exist that inhibit the growth of prokaryotic cells by interfering with their protein-synthetic pathways. These same antibiotics interfere with protein synthesis within the mitochondria and chloroplasts. They do not inhibit protein synthesis in the cytoplasmic ribosomes (those bound to the endoplasmic reticulum) of eukaryotes. Conversely, drugs that inhibit protein synthesis by cytoplasmic ribosomes do not inhibit mitochondrial or chloroplast ribosomes.

Both chloroplasts and mitochondria contain their own DNA and ribosomes; the DNA is very similar in its general structure to that found in bacteria. The ribosomes are not bound to membranes and are a slightly smaller size and density than cytoplasmic ribosomes of eukaryotes.

The permeability characteristics of bacterial, mitochondrial, and chloroplast membranes are very similar. They also resemble the eukaryotic cell membrane very closely.

Putting all these observations together, it is not difficult to speculate on how prokaryotic cells could have become incorporated into eukaryotic cells. If the original prokaryotes were devoured by the primitive eukaryotes in a process similar to endocytosis (engulfing), then the similarities of the outer membranes of eukaryotes, prokaryotes, mitochondria, and chloroplasts would be conveniently explained. If the prokaryotes were not digested away, they could have taken up an independent existence within the cytoplasm of the larger host cell. Such occurrences are by no means rare in the biological world. Bacteria live in the digestive tract of all mammals and are essential to mammalian physiology. A number of small invertebrate animals are known to incorporate algal cells into their body (so that the animals actually appear green); and at least one invertebrate, a slug, has been discovered recently that incorporates chloroplasts directly into some of its cells! There is ample evidence to suggest that foreign bodies can be taken into cells and remain intact after incorporation.

Over time, according to this hypothesis, as primitive prokaryotes continued their existence within eukaryotic cells, they lost a number of their general functions and became more and more specialized. In support of this idea it should be noted that the DNA in mitochondria no longer contains genetic information for the electron transport system (such as the cytochromes or NAD); this information is coded in the eukaryote's nuclear DNA. Mitochondrial DNA appears to code only for the structural proteins of mitochondrial membranes and other components. Today mitochondria and chloroplasts are dependent on eukaryotic cells for their existence. They have lost their ability to live independently.

As noted above, the molecular structure of DNA within chloroplasts and mitochondria also resembles that in prokaryotic cells more closely than the nuclear DNA in eukaryotes. For example, in chloroplasts and mitochondria, the DNA is present as a single molecule, as it is in blue-green algae and bacteria. It is also not complexed with histones, as in nuclear DNA. Furthermore, the relative proportions of guanine plus cytosine to adenine plus thymine in mitochondrial DNA is like that of bacterial, rather than nuclear, DNA. Likewise, chloroplast DNA (from *Euglena*) shows extensive sequence similarities with the genes for ribosomal RNA in blue-green algae.

An interesting recent suggestion has expanded the discussion of the origin of eukaryotic organelles beyond chloroplasts and mitochondria. It has been proposed that symbiotic spirochetes could account for the origin of flagella, cilia, and the basal bodies from which the spindle for mitosis could have been constructed. Evidence for this view comes from an interesting analogy. The flagellate *Myxotricha*, which lives in the gut of termites (and, like its relative *Trichonympha*, can digest cellulose), appears to be covered with many flagella, but actually only four of them are its own. The rest are spirochetes that attach themselves to the cell surface. The lashing about of the spirochetes helps *Myxotricha* move; its own flagella serve merely for steering!

Recently, however, this whole view of the evolution of the eukaryotes has been thrown into some doubt by discoveries in the biochemistry of messenger RNA formation. It turns out that very different mechanisms operate in eukaryotes and prokaryotes (see Chapter 19). James E. Darnell from the Rockefeller University and others have found that eukaryotes have what are called "divided genes." Divided genes are noncontiguous sections of DNA coding for an individual protein; that is, the DNA coding for a single, long messenger RNA molecule (which in turn codes for a single polypeptide chain) is in several pieces on the eukaryotic chromosome. The pieces are separated from each other by what are called "intervening sequences" of DNA. The great difference in molecular mechanisms involved in transcribing prokaryotic and eukaryotic DNA has thus led some biologists to suggest that the two groups of organisms must have had separate and independent origins. This suggestion clearly goes against the idea that eukaryotes evolved from modified prokaryotes. However, it would not necessarily argue against the subtheory that the organelles of eukaryotes (such as mitochondria and chloroplasts) are derived from endosymbiotic prokaryotes. It *would* say, however, that eukaryotes may have had their own line of development from coacervates or other molecular associations in the early oceans and were not derived at all from a prokaryotic ancestor. At the moment there is no way to determine whether this hypothesis is correct, but it is a radical idea that is forcing biologists now to rethink their more traditional, and seemingly logical, theory of the origin of eukaryotes from prokaryotes.

Table 27.2
Characteristics of the Protista (Eukaryotes)

Phylum (examples)	Class (examples)	Cell morphology	Motility	Method of nutrition	Other defining characteristics
Euglenophyta (*Euglena*)		Cells flagellated; no cellulose cell walls, but with outer pellicle; chloroplast present; unusual in having contractile vacuoles in a photosynthetic cell	Flagellum	Photosynthetic autotrophs; some heterotrophs	Contain chlorophylls *a* and *b*; mostly freshwater
Pyrrophyta "Fire algae" or Dinoflagellates (*Gonyaulax*)		Cells with two furrows: transverse and horizontal, each with a flagellum; cellulose cell walls; chloroplasts present	Flagellum	Photosynthetic autotrophs; some heterotrophs	Chlorophylls *a* and *c* and carotenoids; both marine and freshwater; reproduction mostly asexual
Chrysophyta Golden algae and diatoms	Xanthophyceae Yellow-green algae (*Characium, Vaucheria*)	Mostly unicellular, though some filamentous; cell walls contain pectose and silicon	Adults nonmotile but free-swimming zoospores have flagella	Photosynthetic autotrophs	Chlorophylls *a* and *c* and carotenoids; store food as oils rather than starch
	Chrysophyceae Yellow-brown algae (*Dinobryon*)	Flagellated cells common; most motile forms have contractile vacuole; chloroplasts present	Flagella	Photosynthetic autotrophs	In addition to chlorophylls *a* and *c*, have xanthophylls; store food as oils
	Bacillariophyceae diatoms (*Cytotella, Navicula*)	Protoplast enclosed in fine double shell of two parts; shells highly symmetrical and etched; chloroplasts; cell wall contains silicon	Flagella	Photosynthetic autotrophs	Same pigments as yellow-greens; shells pile up in ocean as "diatomaceous earth"; also store food as oils
Protozoa	Mastigophora or Flagellata (*Trypanosoma* and *Trichonympha*)	Protozoa with flagella, which have the 9 + 2 microtubular structure; no cell wall or chloroplasts; some species can form pseudopods	Flagella	Heterotrophic	Most primitive of protozoa, possibly evolving from *Euglena*-type ancestor by loss of chloroplast; some, such as *Trypanosoma* pathogenic (African sleeping sickness)
	Sarcodina (*Amoeba*)	No cell wall or covering; move by pseudopods; large nucleus; one group, the foramanifera, have brightly colored shells.	Pseudopodia	Heterotrophic; feed by engulfing prey or food particles	Complex cells, which can even "react" to outside stimuli, as in sensing presence of food particle and pursuing it
	Ciliata (*Paramecium*)	Well-defined cells, covered with cilia (9 + 2 microtubules) arranged in rows; two nuclei, one large, one small, present; cytoplasm highly differentiated; has "oral groove" into which food is swept	Cilia, which are highly coordinated in beating	Heterotrophic; sweep in food and also engulf	Most complex of all protozoans; can carry out highly coordinated movements; can "learn"; both sexual and asexual reproduction
	Suctoria (*Podophyra*)	Adult cells have tentacles; young cells have cilia	Young cells move by cilia; adults nonmotile and attached to substrate by stalk	Heterotrophic	
	Sporozoa (*Plasmodium vivax*)	Cells undergo complex morphological changes during the life cycles; all forms parasitic and live in several hosts and vectors during a cycle	Certain stages in life cycle have cilia for movement; adults nonmotile	Parasitic	All forms are parasites; most well known is *Plasmodium vivax*, which causes malaria in humans

The kingdom Protista consists of mostly unicellular eukaryotic organisms, which reproduce both asexually (by mitosis) and sexually (by fusion).

who has any familiarity with biology: *Euglena, Amoeba, Paramecium,* golden algae, yellow-green algae, diatoms, and slime molds, to name only a few. A summary chart of the phyla comprising the kingdom Protista is shown in Table 27.2.

Protista represent the simplest of all the eukaryotes. The dividing line between the prokaryotes and eukaryotes is one of the most significant in the biological world. As a group, prokaryotes employ an enormous variety of chemical pathways in gaining energy, yet each one is relatively simple in structure. In contrast, eukaryotes largely employ the same biochemical pathways, lacking the versatility of prokaryotes (for example, no eukaryote can metabolize petroleum or hydrogen sulfide the way certain bacteria can), but their structures are highly diverse and complex. In the history of life, eukaryotic cells opened up an evolutionary panorama that has never ceased to yield new forms, new differentiations, and new adaptations.

One of the most important characteristics of the eukaryotic cell is its compartmentalization. In prokaryotes and eukaryotes alike, most important metabolic reactions occur on cell membranes, or in organelles (such as ribosomes) attached to cell membranes. In prokaryotes, all such reactions occur on the same membrane: the inner surface or infoldings of the plasma membrane. Compartmentalization in prokaryotes is minimal.

In eukaryotes, by contrast, intracellular membrane systems (such as endoplasmic reticulum, or more highly organized organelles such as mitochondria) not only serve as surfaces on which reactions can occur but also segregate one set of reactions from another. From an evolutionary point of view this segregation, or compartmentalization, provides several advantages. One is that the components of each reaction system (metabolic pathways) can be kept closer together, and thus increase the efficiency of chemical interaction. A second is that when the elements of a particular pathway (including enzymes, substrates, intermediates, and end-products) are together in the same place, control of the rate of reaction is more precise. If a pathway is subject to end-product inhibition (as many are), the end-product will

have a greater chance of interacting with (and inhibiting) the appropriate enzyme if the components of the entire pathway are localized in a small area. Thus, it may be that the evolution of compartmentalization in the eukaryotic cell has been so successful because it opened up the possibilities of much greater efficiency and regulation. As we have seen, the latter in particular has become a critical aspect of the life of all higher organisms (from cell differentiation during embryogenesis to enzyme induction and the production of immunoglobulins).

We turn now to a brief discussion of each of the major phyla within the kingdom Protista.

27.5
THE ALGAE

Phylum Euglenophyta

Characteristics of the phylum Euglenophyta can be illustrated by its most well known member, *Euglena*, a single-celled photosynthetic organism that lives in fresh water. The cell of *Euglena* is highly complex, as shown in Fig. 27.12. *Euglena* possesses a well-defined nucleus and contains chloroplasts with chlorophylls *a* and *b*. *Euglena* is thus photosynthetic; the product of its CO_2 reduction is the starch paramylum, an unusual molecule found almost exclusively in this group. The cells of *Euglena* lack a cell wall but have a series of flexible protein strips, making up what is called the **pellicle** inside the cell membrane. *Euglena* reproduces asexually, dividing down the middle to form two new cells that are thus mirror images of each other.

One of the most important characteristics of *Euglena* is its large, prominent flagellum (there are actually two flagella, one of which is small and inactive). The flagella are attached at the base of an egg-shaped opening called the reservoir, at the anterior part of the cell. The circular, spinning motion of the flagellum "pulls" the *Euglena* through the water. Some mud-dwelling forms of *Euglena* take advantage of their lack of a cell wall and move by wriggling. A contractile vacuole empties its contents of excess water collected from all parts

The dividing line between the prokaryotes and eukaryotes is one of the most significant in the biological world. The appearance of eukaryotic cells opened up an evolutionary panorama that has never ceased to produce new forms and new adaptations.

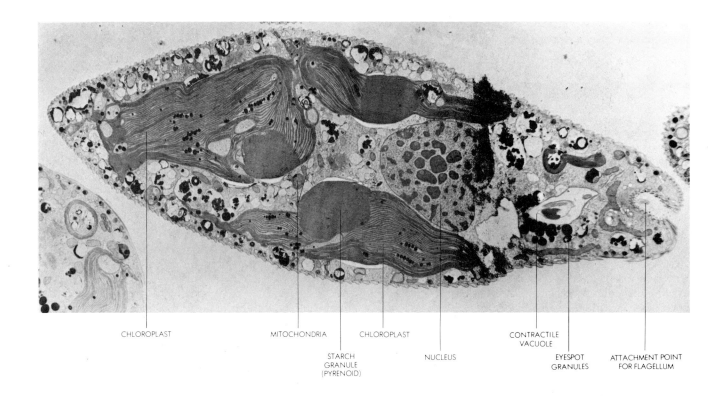

CHLOROPLAST   MITOCHONDRIA   CHLOROPLAST   NUCLEUS   CONTRACTILE VACUOLE   EYESPOT GRANULES   ATTACHMENT POINT FOR FLAGELLUM

STARCH GRANULE (PYRENOID)

Fig. 27.12
Euglena, a representative of the phylum Euglenophyta, of the kingdom Protista. The electron micrograph shows and identifies the major organelles. Compare the internal complexity of the eukaryotic protozoa to that of the prokaryotic Moneran, the bacterium *E. coli*, or the blue-green alga *Anabaena* (Fig. 27.9). (Micrograph courtesy P. L. Walne and J. H. Arnott, from *Planta* 77: 325–354, 1967.)

of the cell into the reservoir. Only a few of the photosynthetic organisms have contractile vacuoles, which are otherwise common among the Protozoa.

If *Euglena* is grown in the dark it loses its photosynthetic pigments and feeds heterotrophically on dead organic material floating in the water around it. If the organism is grown in a very rich medium the cells will undergo fission more rapidly than the chloroplasts, so that some daughter cells end up without the photosynthetic organelles. The same effect can be obtained by treating *Euglena* with streptomycin or ultraviolet light. In either case, the nonphotosynthetic *Euglenas* are quite functional as heterotrophs. In fact, a nonphotosynthetic *Euglena* is virtually indistinguishable from the protozoan *Astasia*, in the class Mastigophora (see Fig. 27.15). It is tempting to speculate that at least some, perhaps, of the present-day protozoans may have evolved from an autotrophic *Euglena*-like ancestor by loss of chloroplasts and the taking up of a heterotrophic existence.

Euglena reproduces by longitudinal, binary fission, usually starting at the anterior end. During mitosis, the chromosomes within the nucleus replicate, forming pairs that split longitudinally. Since the original *Euglena* is haploid, the cell becomes diploid for a while. Cytokinesis (longitudinal fission) begins to occur. The old flagellum is retained by one of the two halves of the dividing cell. The other half will generate its own flagellum later. As in all binary fission, the two daughter cells are genetically identical.

Occasionally *Euglena*, like most protists, grows forms that are called cysts. The cell rounds up and becomes spherical, developing a thick gelatinous covering around itself. Cysts may serve to tide the organism over a period of drought, or to provide a kind of "resting stage" before resumption of further cell divisions. If *Euglenas* are brought into the laboratory in a dish of water, some cysts can be found on the drying sides of the dish. As the cyst stage ends, the cells usually reproduce, giving rise to two (or occasionally more) new cells, which break out of the cyst. As many as thirty small *Euglenas* have been observed to emerge from a single cyst.

Phylum Pyrrophyta (the "Fire Algae" or Dinoflagellates)

The phylum Pyrrophyta (*Pyro* is Greek for fire) takes its name from the fact that many of its species are red (from the abundance of xanthophylls and carotenoids.)

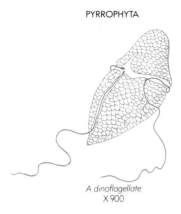

PYRROPHYTA

A dinoflagellate
X 900

Fig. 27.13
A dinoflagellate, *Gelnodinium cinctum*, of the phylum Pyrrophyta. These single-celled protists have two flagella and stiff cellulose cell walls. Dinoflagellates are thought to be intermediate in many ways between prokaryotes and eukaryotes. Like prokaryotes they lack histones in their DNA; like eukaryotes they have the traditional 9 + 2 microtubular structure in their flagella. Several species of dinoflagellates produce a toxin that can kill fish by the millions. Heavy concentrations of these dinoflagellates in the oceans off Florida, southern California, and New England form the infamous and highly poisonous "red tides."

The dinoflagellates are one-celled organisms, about 1000 different species of which are known today. The cells of many species are covered by a hard but porous cell wall with markings that have caused one author to compare them to the helmets of ancient warriors (see Fig. 27.13). The dinoflagellates usually have two flagella that lie partially within grooves in the cell wall. One groove runs horizontally around the equator of the cell; the other vertically. The beating of the flagella causes the cell to spin like a top as it moves through the water (hence the name "dino," or "spinning" flagellate).

Many dinoflagellates are photosynthetic, containing chlorophylls *a* and *b* along with the carotenoids and xanthophylls. The nonphotosynthetic dinoflagellates live on small bits of organic matter suspended in the water, or as parasites on a variety of marine invertebrates. Taken as a whole, the dinoflagellates are second only to another algal protist, the diatoms (phylum Chrysophyta; see discussion below), as primary photosynthesizers in the marine environment. Dinoflagellates also live symbiotically with many invertebrates (for example, with coral animals), providing them with much of their needed carbon and oxygen. If the corals lacked their dinoflagellate symbionts, it has been estimated they would be able to grow only about one-tenth as fast.

If you have ever been to the Pacific Ocean along the coast of southern California, or in the Caribbean, you undoubtedly noticed the amazing phosphorescence of the waters at night. This is due mostly to bioluminiscent dinoflagellates, which grow in such abundance that their glowing makes the ocean surface shimmer.

Dinoflagellates also play a somewhat less romantic and aesthetically pleasing role in the marine ecosystem. When the species *Gonyaulax catanella* grows rapidly and produces a large population (what is called an algal "bloom"), some of its metabolic byproducts are poisonous to fish and to invertebrates such as the mussel. Known as the "red tides," these blooms of dinoflagellates have killed fish in massive quantities over the years. Red tides are common periodically in the Gulf of Mexico, and off the coasts of southern California and New England. *Gonyaulax* produces a metabolic poison that is a nerve toxin. One gram of this toxin has been estimated to be sufficient to kill 5 million (!) mice in fifteen minutes. Mussels can concentrate *Gonyaulax* toxin without harm to themselves, but if humans eat the mussels they can become seriously, even fatally, ill. Another species of dinoflagellate produces a toxin in oysters, especially in warmer weather, hence the old warning: "Never eat oysters in a month without an *r*."

Phylum Chrysophyta (Golden Brown Algae and Diatoms)

Organisms in the phylum Chrysophyta derive their common name from the fact that their cells contain a large quantity of carotenoids (especially the carotenoid fucoxanthin) and hence impart to the organisms a yellowish-brown color. The Chrysophyta are all photosynthetic and contain chlorophylls *a* and *c*, but lack chlorophyll *b*. There are over 10,000 species of Chrysophyta. Although the majority are marine, there are also a number of freshwater species. Chrysophyta make up the vast majority of the plankton in marine food chains.

The Chrysophyta store most of their energy reserves as oils, rather than starch. Because of this ability, it has been suggested that the golden brown algae might be cultivated as oil producers for human consumption.

Taken as a whole, the dinoflagellates (phylum Pyrrophyta, or "fire algae") are second only to the diatoms (phylum Chrysophyta) as primary photosynthesizers in the marine ecosystem.

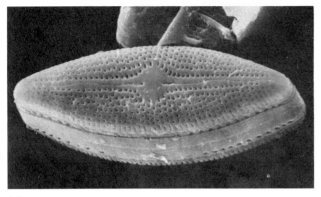

(a)

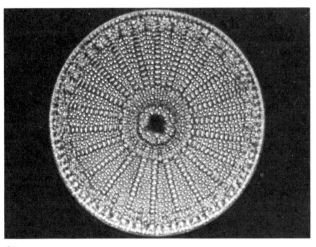

(b)

Fig. 27.14
Two diatoms, representatives of the phylum Chrysophyta. The silicious shells of these protists are highly sculptured with many ridges and tracings, which provide pores between the protoplast and external environment. (*a*) Scanning electron micrograph of *Navicula* (× 3080); (*b*) Light micrograph of *Stephanopyxis*. (*a* Courtesy Professor Richard B. Kessel; *b* Courtesy Professor Harold Levin.)

The cells of chrysophytes are unusual among the plant-like protists in being diploid. Only their gametes, during the sexual reproducing phase, are haploid. Most Chrysophyta reproduce asexually by cell division, though a sexual phase occurs occasionally.

The majority of species composing the phylum Chrysophyta are the diatoms. The unusual and beautiful organisms have their cells surrounded by a fine two-part shell made of calcium. The shells are highly sculptured and appear as intricate designs under the microscope (see Fig. 27.14). Electron microscopy has revealed that the shells consist of minute pores that connect the protoplast of the cell with the exterior environment. The classification of diatoms is based almost en-

tirely on the shape and appearance of the shells, which are remarkably uniform for any one species.

When cells of the diatoms die, they sink to the bottom of the ocean, often accumulating in thick layers over the years. Such deposits are called diatomaceous earth. Uplifted sediments of diatomaceous earth form the spectacular White Cliffs of Dover on the southeast coast of England. Diatomaceous earth has proved to be of enormous commercial value. It is used in preparing detergents, polishes, paint removers, and fertilizers. Its most important general use is as an abrasive, being a major component of silver polish and some toothpastes.

27.6
PHYLUM PROTOZOA (THE ANIMAL-LIKE PROTISTS)

Protozoans are all one-celled heterotrophs lacking photosynthetic apparatus. Most protozoa reproduce asexually by cell division, though many have a sexual cycle as well. The nuclei of most protozoans are haploid. In the sexual cycle, the gametes are also haploid. Fusion of gametes produces a diploid zygote, called a **zygospore**, which forms a tough wall around itself and can survive cold or dry conditions.

The five classes of Protozoa are distinguished largely by their modes of locomotion: by flagella (the Mastigophora), by pseudopodia (the Sarcodina), by cilia (the Ciliata), and by a combination of cilia and tentacles (the Suctoria, which we will not discuss). A fifth class, the Sporozoa, are all internal parasites and lack any motile stage. Their more complex life cycle will be discussed separately.

Class Mastigophora

The flagellates, or Mastigophora, are considered to be the most primitive of the Protozoa. Almost all have one or two flagella, like the genus *Astasia* shown in Fig. 29.15(a); while some, such as *Trichonympha*, shown in Fig. 27.15(b), have a large number. Most Mastigophora either are free-living or live as symbionts in the bodies of higher animals or plants. Among the most interesting symbionts is *Trichonympha*, a genus containing several species that live in the gut of termites. There, the flagellates produce an enzyme that digests the cellulose of wood that the termites eat. Without the *Trichonympha* in its gut, the termite could gain no energy from the wood it ingests.

Among the parasitic zooflagellates is *Trypanosoma gambiense* (see Fig. 27.15c), which causes African sleeping sickness (not to be confused with what is called ordinary sleeping sickness, encephalitis, which is caused by a virus). Trypanosomes live in their host's blood where they multiply, consume energy, and release poisonous waste products. In humans or less-resistant domesticated animals, the trypanosomes can eventually invade the nervous system. They produce first lethargy in the

The flagellates may be crucial organisms in the history of higher forms of plant and animal life on earth. Phytoflagellates such as Euglena may have given rise to all the higher (multicellular) plants; zooflagellates such as Astasia (which is virtually identical to Euglena without the chloroplasts) may have given rise to all the higher, multicellular animals.

MASTIGOPHORA

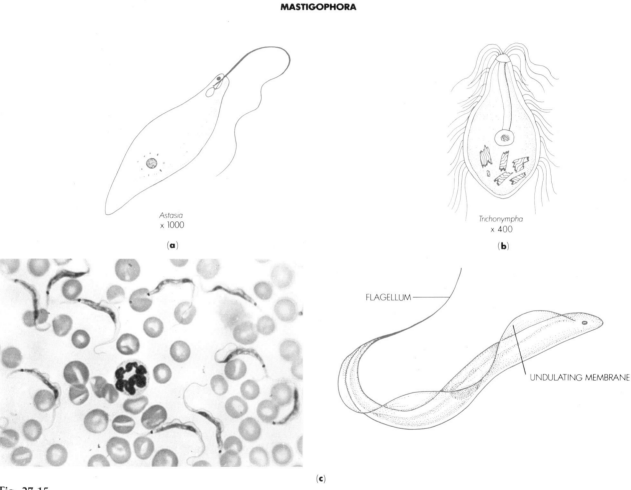

Astasia
x 1000

(a)

Trichonympha
x 400

(b)

FLAGELLUM

UNDULATING MEMBRANE

(c)

Fig. 27.15
Three representatives of the class Mastigophora (phylum Protozoa). The Mastigophora are a diverse group, including *Astasia* (a), which is, for all intents, *Euglena* without its chloroplasts; the symbiont *Trichonympha* (b), which lives in the gut of termites and produces the enzyme necessary for digesting the cellulose of wood; and the parasite *Trypanosoma* (c), which infects humans and is the agent causing African sleeping sickness. Trypanosomes have evolved an interesting evolutionary adaptation to living in mammalian blood, where they are subject to attack by the body's immune system. The body makes antibodies to antigens on the *Trypanosoma's* cell surface. Just when the antibody response is about to become effective (when antibody levels in the blood begin to rise) the *Trypanosoma* changes the antigenic determinants on its surface and thus escapes attack. This process can be repeated any number of times, thus allowing an infection, once established, to persist. (Photo, Eric Grave.)

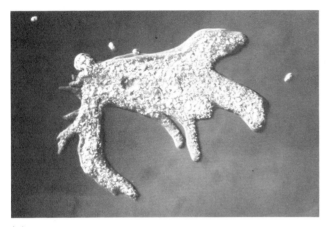

(a)

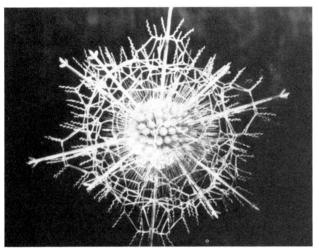

(b)

Fig. 27.16
Representatives of the class Sarcodina, the Amoebas, of the phylum Protozoa. (*a*) The genus *Amoeba* (species *proteus*), showing the formation of several large pseudopods at the upper and lower right. Pseudopods serve for both locomotion and feeding. (*b*) A Radiolarian, a form of partially shelled Sarcodinas. In the glass and wire model shown here, the long spoke-like projections outward represent thin pseudopods. The Radiolarian shell is composed of silicates. (*a* Courtesy Professor Robert D. Allen; *b* Courtesy Professor Harold Levin.)

host (hence the descriptive name "sleeping sickness") and eventually death. The trypanosomes are carried from host to host by blood-sucking insects known as tsetse flies. They pick up the trypanosome from the blood of an infected individual and the flagellate passes into the fly's intestine where it undergoes several stages of its life cycle before passing to the fly's salivary gland. Here, after further development, the adult form is produced and can pass into a new host at the fly's next bite. Trypanosomes can live in the blood of a large number of vertebrates, although the wild mammals of Africa appear to be largely resistant. These animals, however, can harbor the adult parasites, thus providing a continual reservoir for transmission by the tsetse fly's bite. Public health programs in Africa have attempted to eradicate sleeping sickness by preventing the tsetse fly from reproducing. So far, unfortunately, such efforts have made little progress.

The flagellates are thought to occupy an extremely important position in the evolution of multicellular life on earth. As we shall see at the end of this section, both zoologists and botanists hold that the flagellates may have given rise to *all* the other eukaryotic organisms.

Class Sarcodina

The class Sarcodina includes the amoeboid Protozoa. Sarcodina lack cell walls or coats, although some species have shells. Sarcodina feed by the formation of pseudopodia, flowing extensions of the cell membrane which form arms that can surround ("engulf") a small organism or bit of organic matter. One of the most common examples in all biology, the amoeba (Fig. 27.16a), is a member of the class Sarcodina. The nonshelled amoebas can become quite large for single-celled organisms; *Chaos chaos* can grow to 1 mm in diameter, and is clearly visible to the naked eye. Among the shelled Sarcodina are the Foraminifera and Radiolaria. Their shells are composed of calcium carbonate, or silicates, and have a highly intricate, delicate appearance (Fig. 27.16b). Like the other Sarcodina, the Radiolarians and Foraminiferans produce pseudopods, but these serve only a feeding, not a locomotor, function. The pseudopods of the shell forms are restricted in their motion and are not able to engulf their prey. They capture prey by secreting a sticky mucous compound on the surface of the cell membrane. Once entrapped, the prey is digested by proteolytic enzymes and taken into the protoplast by an infolding of the membrane (producing a food vacuole).

The Sarcodina are thought to have originated directly from the Mastigophora. Many flagellates have an amoeboid stage in their life cycle, and many amoebas have a flagellated stage in their life cycle. The Sarcodina have been on earth an extremely long time. Fossil Foraminiferans have been found in Cambrian rocks, while those of Radiolarians have been uncovered in pre-Cambrian deposits over one billion years old.

Class Ciliata

Organisms of the class Ciliata are the largest and most advanced of all the Protista. The ciliates have the greatest number and variety of cell organelles of any microorganisms known. The cilia are usually arranged in some sort of rows over the cell surface; on examination under the electron microscope the cilia are found to show the typical 9 + 2 microtubular pattern. About

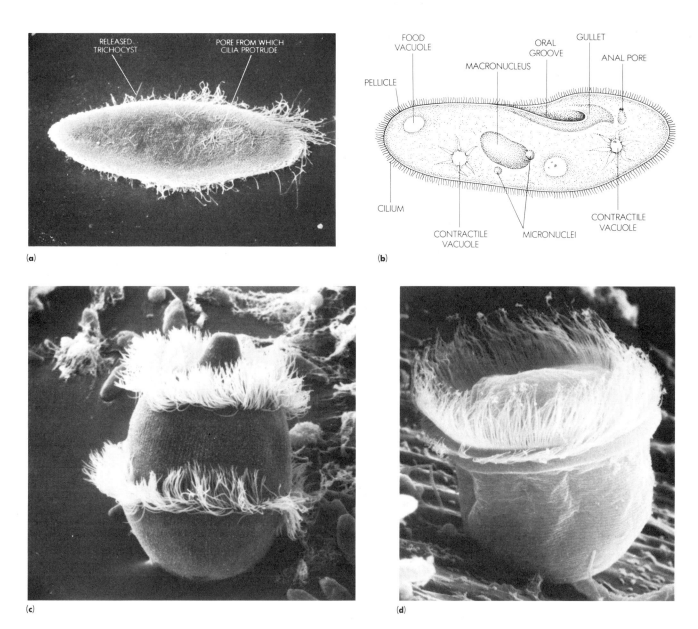

Fig. 27.17

Representatives of the class Ciliata, which includes those Protozoa having cilia on their surfaces. Cilia have the 9 + 2 microtubular structure and are used both for movement and for creating a current of water that brings food toward the oral region. (*a*) and (*b*) *Paramecium*, a classic representative of the ciliates. The scanning electron micrograph in (*a*) shows the regular rows of cilia that make up the surface texture of the organism's covering, or pellicle. The long extended fibrils are discharged trichocysts. Internal detail of cell structure is shown in (*b*). (*c*) The barrel-shaped *Didinium*, which feeds on other ciliates such as *Paramecium* (× 1130). The cilia are shown in two rings, one around the mouth (to the left), the other around the equator. (*d*) *Vorticella*, showing a single row of cilia around the buccal oral cavity. The stalk is seen to the lower right. (Micrographs courtesy Professor Richard G. Kessel.)

6000 species of ciliates have been identified. Almost all are free-living, though there are a few parasitic forms. Ciliates live in both fresh and salt water. Several representatives of the class Ciliata are shown in Fig. 27.17.

The slipper-shaped *Paramecium* is one of the best representatives of the ciliates (Fig. 27.17a and b). *Para-*

mecium has a tough outer "skin" called a **pellicle**, underneath which lies the cell membrane. Some species have stiff plate-like structures embedded in the skin. The cilia are arranged in rows along the pellicle, each cilium protruding through an opening, or pore, in the surface (Fig. 27.17a). The cilia are connected at the base by a series of

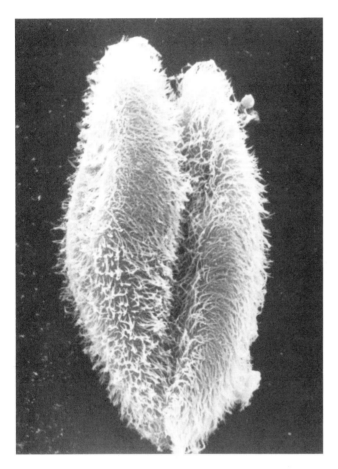

Fig. 27.18
Scanning electron micrograph of two *Paramecia* conjugating. The oral grooves are toward the center. The thread-like structures on the surface are exploded trichocysts. (Photo courtesy Professor Richard G. Kessel.)

A unique feature of the ciliates in general is the fact that they all have two kinds of nuclei, a large **macronucleus** and one or more smaller **micronuclei**. The macronucleus is polyploid, containing many copies of the chromosomes; in one species of *Paramecium* it is estimated that the macronucleus contains over 800 copies of the organism's chromosome complement. The macronucleus appears to control the day-to-day metabolic functions of the organism. The micronucleus is concerned solely with reproduction and with generating the macronucleus. A *Paramecium* can live without its micronuclei for indefinite periods of time as long as some portion of the macronucleus is still present. It cannot reproduce, however. Conversely, a *Paramecium* cannot maintain itself for very long without a macronucleus even if it has one or more functional micronuclei.

During asexual reproduction the micronuclei divide by a regular form of mitosis. The macronucleus divides in a very different way by a mechanism that is not clear. DNA replication precedes each division of the macronucleus, and genetic analysis has revealed that each daughter *Paramecium* receives a full complement of genetic material. These observations suggest that some regular divisional process occurs within the macronucleus. Asexual reproduction is completed, following mitosis, by longitudinal fission.

Paramecium, like many ciliates, can undergo a form of sexual reproduction known as **conjugation**. Two *Paramecia* come together with their oral grooves facing (Fig. 27.18). The micronuclei (which are diploid) divide meiotically, producing a number of haploid micronuclei (four from each micronucleus). All but two of these in each cell disintegrate. The macronucleus also disintegrates. In each cell, one micronucleus remains stationary; the second micronucleus moves across the oral groove into the other cell where it fuses with the stationary nucleus there. A simple fertilization has taken place, and the diploid number restored. Each cell has contributed one of its own micronuclei to and received a micronucleus from its partner. After fertilization, the new recombinant nucleus in each cell undergoes several mitotic divisions. Some of the daughter nuclei remain micronuclei, but at least one develops into a macronucleus. Several cell divisions occur before the appropriate number of macro- and micronuclei (characteristic of the species) is restored to each cell. The process of conjugation, like all sexual reproduction, ensures the exchange of genetic material, and hence increases variability within the population.

An unusual feature of *Paramecium*, and indeed of many ciliates, is its possession of small, thread-like harpoons, or spears, called **trichocysts** (see Fig. 27.19). Trichocysts are coiled in small vesicles below the pellicle. When the organism is stimulated, the trichocysts explode and release their coiled-up barbs. These remain protruding from the surface. Some species' trichocysts

conductile fibrils, through which coordinated beating of the cilia is effected. There may also be a group of contractile fibers at the base of the cilia that effect the beating movement. Watching a *Paramecium* held in place on a microscope stage in a drop of water, you can see the rows of cilia beating in a rhythmic, undulating fashion. Ciliary motion is highly regulated and coordinated in *Paramecium*, but how this coordination is brought about is not entirely understood.

Paramecium feeds on small bits of organic matter, smaller protozoa, and algae that are swept into its oral groove (see Fig. 27.17b) by a current of water generated by the beating cilia. The particles are captured in an infolded pocket called the **gullet**. When food particles reach the bottom of the gullet, the end of the gullet is pinched off as a kind of food vacuole, which moves into the cytoplasm where digestion takes place.

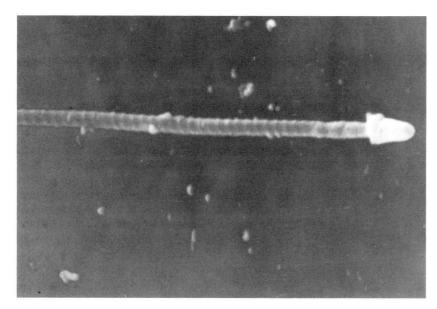

Fig. 27.19
Exploded trichocyst of *Paramecium*, showing the barbed head and the banded thread below. The exploded threads may be more than 20 μm in length. Once trichocysts have been discharged, new trichocysts must be resynthesized in their place. The exact function of trichocysts is not known, since *Paramecia* can apparently get along very well without them. (Photo courtesy Professor Richard G. Kessel.)

contain a toxin that can paralyze prey. Trichocysts may also serve to hold the *Paramecium* to a surface while it feeds.

Class Sporozoa

The final group of the Protozoa that we will discuss is the class Sporozoa. These forms are all internal parasites with complex life cycles involving at least two hosts. For many of the Sporozoans one of the hosts is a higher animal (a vertebrate, usually a mammal). As the name of the class implies, these Protozoa have a spore-like stage at some point in their life cycles. They also lack cilia, flagella, tentacles, or other locomotor organelles. The best known and most notorious of the Sporozoans is *Plasmodium vivax*, the infective agent of malaria. The life cycle of *Plasmodium* is shown in detail in Fig. 27.20. *Plasmodium* holds the world's record for having killed more human beings than any other single infectious organism in the course of human history. In World War II alone, more lives were lost by malaria than through all the armed conflict combined.

27.7
EVOLUTIONARY RELATIONSHIPS OF THE EUKARYOTES: A LOOK AHEAD

However exactly eukaryotic cells arose (and there is much debate about this), once they came into existence they could exploit many new ecological niches. The most important evolutionary advances of the eukaryotes are: (1) enclosed organelles that compartmentalize biochemical pathways; (2) appearance of a true nucleus (also a compartmentalization); (3) development of dip-

loidy, so that deleterious mutations do not always show their effects; (4) appearance of various forms of cell motility (such as cilia, flagella, and pseudopods); (5) great expansion of autotrophism, especially with the introduction of chloroplast-based photosynthesis. Once these features had become established, an enormous radiation of life took place in the pre-Cambrian period. This radiation led to the establishment of all the modern lines of organisms on earth today.

Lines developing during the vast pre-Cambrian radiation are diagrammed in Fig. 27.21. The ancestral organisms that gave rise to this radiation are thought to have been photosynthetic flagellates. Several lines of evidence suggest this possible development. First, virtually all modern forms of eukaryotic organisms have some sort of 9 + 2 flagellated cells appearing somewhere in the life cycle; in higher animals and plants today such cells appear in the form of male gametes. Second, virtually identical flagellated forms appear in two major groups of the Protista: *Euglena* (Euglenophyta), and *Astasia* (Mastigophora). Indeed, as pointed out previously, microscopic study of *Astasia* (see Fig. 27.15a) reveals that it is little more than *Euglena* without chloroplasts. It is not difficult to imagine an evolution in which an ancestral photosynthetic flagellate similar to *Euglena*, by losing its chloroplasts, gave rise to a line of cells that became the Mastigophora. The Mastigophora, in turn, gave rise to the other Protista (Sarcodina, Ciliata, and Sporozoa) and to the multicellular animals. The Euglenophyta may have given rise to the algae and higher plants. The Protista thus set the stage for the two great developments of life to follow on earth in the course of the next billion years: the evolution of plants and animals. We will follow this course of development in the next two chapters.

SEXUAL REPRODUCTION PHASE IN MOSQUITO

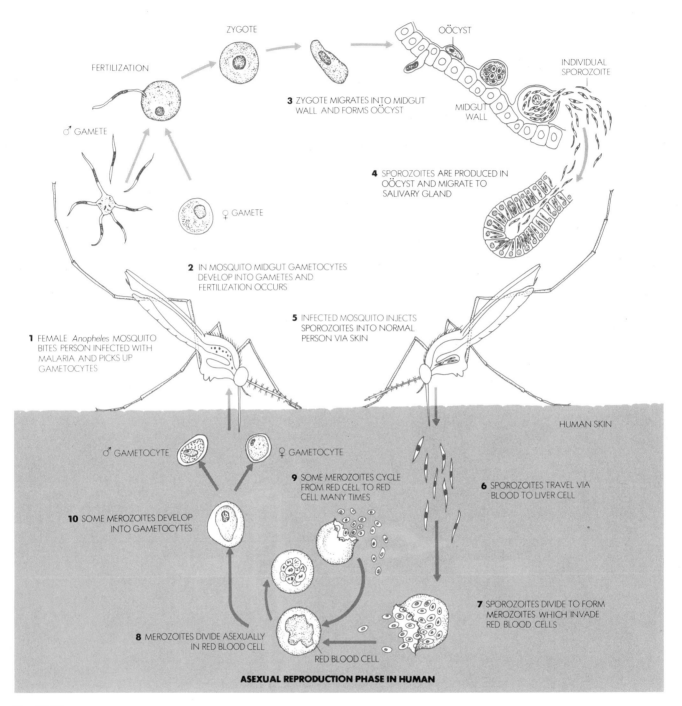

3 ZYGOTE MIGRATES INTO MIDGUT WALL AND FORMS OÖCYST

4 SPOROZOITES ARE PRODUCED IN OÖCYST AND MIGRATE TO SALIVARY GLAND

2 IN MOSQUITO MIDGUT GAMETOCYTES DEVELOP INTO GAMETES AND FERTILIZATION OCCURS

5 INFECTED MOSQUITO INJECTS SPOROZOITES INTO NORMAL PERSON VIA SKIN

1 FEMALE *Anopheles* MOSQUITO BITES PERSON INFECTED WITH MALARIA AND PICKS UP GAMETOCYTES

9 SOME MEROZOITES CYCLE FROM RED CELL TO RED CELL MANY TIMES

10 SOME MEROZOITES DEVELOP INTO GAMETOCYTES

6 SPOROZOITES TRAVEL VIA BLOOD TO LIVER CELL

7 SPOROZOITES DIVIDE TO FORM MEROZOITES WHICH INVADE RED BLOOD CELLS

8 MEROZOITES DIVIDE ASEXUALLY IN RED BLOOD CELL

ASEXUAL REPRODUCTION PHASE IN HUMAN

Fig. 27.20
Life cycle of the malarial parasite, the Sporozoan *Plasmodium vivax.* The cycle involves two hosts, the mosquito (usually of a single genus, *Anopheles*), where sexual reproduction occurs, and the human, where asexual reproduction occurs. Male and female gametocytes produced during the sexual phase in the human are picked up when a mosquito bites an infected person (step 1). Fertilization occurs in the mosquito's gut, producing a zygote (steps 2 and 3). The zygote forms a cyst in the gut wall, eventually giving rise to sporozoites, which migrate to the mosquito's salivary glands (step 4). An infected mosquito bites another human, and the sporozoites pass out of the salivary glands into the person's blood stream (step 5). Sporozoites travel to human liver cells and undergo several divisions, producing merozoites (steps 6 and 7). Merozoites can reproduce asexually in red blood cells, a new generation being produced about every twenty-four hours. Release of the merozoites in this process produces the cyclic chills malaria patients experience. Some merozoites develop into gametocytes in the human blood, and the infection cycle begins again.

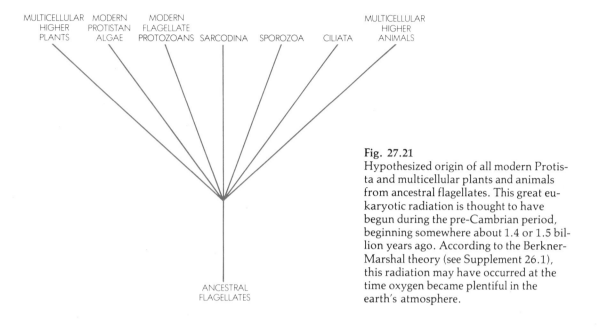

MULTICELLULAR HIGHER PLANTS   MODERN PROTISTAN ALGAE   MODERN FLAGELLATE PROTOZOANS   SARCODINA   SPOROZOA   CILIATA   MULTICELLULAR HIGHER ANIMALS

ANCESTRAL FLAGELLATES

Fig. 27.21
Hypothesized origin of all modern Protista and multicellular plants and animals from ancestral flagellates. This great eukaryotic radiation is thought to have begun during the pre-Cambrian period, beginning somewhere about 1.4 or 1.5 billion years ago. According to the Berkner-Marshal theory (see Supplement 26.1), this radiation may have occurred at the time oxygen became plentiful in the earth's atmosphere.

Summary

Because the major characteristics of each group discussed in this chapter are summarized in Tables 27.1 and 27.2, we will not provide additional summaries here. The summary points listed below will focus on general characteristics of the Monera and Protista, or larger subgroups (such as the Schizophyta or the viruses).

1. Classification of unicellular organisms is based on the presence or absence of a nucleus. Organisms lacking a true nucleus are called prokaryotes, and are classified in the kingdom Monera. Organisms which have a true nucleus are called eukaryotes and are classified in the Protista kingdom. The hereditary material of the Monera exists as a single continuous or circular molecule of DNA or, in some viruses, RNA.

2. The Monera lack many of the common organelles found in eukaryotic cells, such as plastids and mitochondria. They do, however, have ribosomes, and some have flagella, though not containing the 9 + 2 microtubule structure characteristic of eukaryotes.

3. Most of the Monera are heterotrophs, though some are autotrophic, carrying out photosynthesis or chemosynthesis. Heterotrophic bacteria can be either aerobes or anaerobes. The Monera are important ecologically as decomposers. They also play an important part in nitrogen fixation. Though they are structurally very simple, the Monera have been very successful due to their rapid potential rate of reproduction and their diverse metabolic pathways.

4. Some bacteria move by gliding (Myxobacteria), some by the use of flagella (Eubacteria), and others by the use of contractile filaments along the inside of the cell wall (Spirochetes).

5. The bacterial flagellum consists of a single chain of contractile protein (flagellin) wound around a central core and attached to the cell wall and cell membrane. Other extensions found on bacterial cells are pili, which are rigid, cylindrical rods composed of protein monomer (pilin). They are thinner and shorter than flagella and serve to fasten a bacterial cell to a food source, the surface of water, or another bacterial cell.

6. Bacteria reproduce asexually by means of binary fission (mitotic cell division) or budding. Their daughter cells are therefore genetically identical to the parent cell. Since they are nonmotile, bacterial cells clump together forming colonies. The shape and color of the colonies are characteristic of each species of bacterium. Bacteria are capable of a simple form of sexual reproduction and genetic exchange known as fusion.

7. A special class of the phylum Schizophyta is known as Mycoplasmas. Ranging in diameter from 0.1 to 0.25μ, they are the smallest cells known. For many years taxonomists did not know whether to classify them as bacterial cells or as viruses. Their placement among the bacteria is based on the recent discovery of a peptidoglycan component in their cell walls that is very similar to that found in bacterial cell walls.

8. Viruses and Rickettsia are grouped together in the class Microtatobiates within the phylum Schizophyta (bacteria). Both are obligate intracellular parasites and are powerful agents of disease.
    a) Rickettsia are small, cell-like organisms that live in arthropods (fleas, mites, ticks, and lice) that serve as vectors or carriers. When a vector bites a human or an ani-

mal, the rickettsial cells are transferred to the host and can cause such severe diseases as typhus.

b) Viruses are composed of an inner "core" of RNA or DNA surrounded by an outer coat of protein. The nucleic acid core carries a genetic program for producing more viruses. Viruses cause disease by invading a living host cell and using that cell's chemical machinery to reproduce themselves. This action can kill or badly damage the host cell. Experimental evidence shows that both a protein coat and the nucleic acid core are necessary to invade a host. The protein coat serves mainly to protect the viruses. The nucleic acid, however, is the major infecting agent. The viruses, because they require a host cell's machinery to reproduce, raise an interesting question as to whether they are living or nonliving. Antibiotic treatment is not effective against viruses but immunization therapy is. Some human viral diseases include smallpox and polio.

9.   The second phylum in the kingdom Monera is the Cyanophyta or blue-green algae, which shares several characteristics with bacteria. Their DNA, as in bacteria, is not contained within a nucleus, but, like bacteria, is localized in the cell; they also have a protective cell wall. Unlike bacteria, however, blue-green algae have no flagella or other motile organs. Also, all species of blue-green algae are aerobic, while only some bacteria are. All species of blue-green algae are photosynthetic. The blue-green algae have both photosystems I and II and chlorophyll *a*. The chlorophyll, however, is not compartmentalized into chloroplasts, which the blue-green algae lack, but is bound to infoldings of the cell wall. Photosynthetic bacteria differ from blue-green algae in that they do not possess chlorophyll *a* and they do not release O_2 as a byproduct. Many species of blue-green algae aggregate into long filaments which are found in moist earth or water, and on the leaves of tropical (wet) forest trees.

10.   The kingdom Protista is composed of an extremely diverse group of organisms that represent the simplest eu-

karyotes. Most can reproduce both sexually and asexually. The different Protists vary considerably in their modes of nutrition. Some species are heterotrophs, either free-living or parasitic, and others are autotrophs. Their cell structure, when compared to prokaryotes, is complex; but unlike prokaryotes they all use largely the same metabolic pathways. These metabolic reactions are compartmentalized into cell organelles: mitochondria, chloroplasts, and endoplasmic reticulum. The evolutionary advantage of such compartmentalization lies in the greater efficiency and better control of reactions that is possible when all steps in a reaction sequence are together in a small place.

11.   The kingdom Protista consists of four phyla: the Euglenophyta (flagellated cells with chloroplasts), the Pyrrophyta (two flagella at right angles to each other; the dinoflagellates), the Chrysophyta (some flagellated, all photosynthetic; the Golden Algae and the Diatoms), and the Protozoa (all nonphotosynthetic).

12.   The Protozoa are divided into five classes, based largely on method of motility: the Mastigophora (by flagella), the Sarcodina (by pseudopods), the Ciliata (by cilia), the Suctoria (by tentacles and cilia), and the Sporozoa (nonmotile; also parasitic). Some species in each class are parasitic. The worst human parasite of all is the Sporozoan *Plasmodium vivax*, which causes malaria.

13.   Many biologists now feel that all modern eukaryotic organisms, including the Protists as well as the multicellular plants and animals, arose from a flagellated ancestral Protist. Evidence for this view comes from the fact that most eukaryotic organisms today have a flagellated stage in their life cycles (the flagellated sperm of most species of either animal or plant); the fact that all flagella show the typical $9 + 2$ microtubular structure, implying a very early, common origin; and the fact that one common protozoan, *Astasia*, is virtually identical to *Euglena*, but lacks chloroplasts.

Exercises

1.   What evidence supports the hypothesis that the bacteria and blue-green algae may have evolved from a common ancestor?

2.   On what grounds might the argument be made that the distinction between "animal" and "plant" is meaningless on the microorganismic level?

3.   What are the arguments for and against the idea that eukaryotes and prokaryotes share a common evolutionary

origin (that is, that eukaryotes developed from ancestral prokaryotes)?

4.   What is the evolutionary advantage to eukaryotes of having metabolic pathways compartmentalized in separate organelles?

5.   Why would it be incorrect to say that viruses as we know them may represent the most primitive form of life in evolutionary history, the "missing link" in the historical development from the nonliving to the living world?

Chapter 28
The Diversity of Life II:
Fungi and Plants

28.1
INTRODUCTION

While the Monera and Protista diversified to exploit the evolutionary opportunities in the water, the fungi and plants took advantage of the opportunities on land. Simply by being able to raise themselves above the surface of the ground, they found a new environment, a new ecological niche. As the first colonizers of the land, they deserve a special place in our consideration of diversity and its origin.

The Kingdom Fungi

To many people, the word "fungus" has a rather unpleasant connotation. It implies something slimy and decaying—in short, a low-life sort of organism. However, fungi are not always slimy, nor are they low on the scale of life. Many of them are elaborately constructed, and their life cycles are among the most complex in the entire living world. The group is very diverse, ranging from the simple one-celled yeasts to the large and complex mushrooms and bracket or gill fungi which grow on dead trees and logs. The Fungi are all heterotrophic, gaining their nourishment by breaking down organic matter in the environment. Most common fungi live off decaying plant material, such as leaf mold on the forest floor. However, many live on the skins of animals, in the roots or leaves of plants, and even on moist paper. Fungi are literally everywhere.

Table 28.1 summarizes the major characteristics of the fungi. Except for the one-celled yeast, almost all fungi consist of a mass of filaments, called **hyphae**, which grow into a cottony mass, the **mycelium**. The mycelium is **coenocytic**; that is, it is not divided into separate cells, but is a long tube filled with cytoplasm and nuclei. The walls of the hyphae contain a variety of polysaccharides, but especially prominent is the polymer chitin, composed of molecules of N-acetylglucosamine held together by β-glycosidic bonds. (N-acetylglucosamine is found in bacterial cell walls, but not as the polymer chitin, and also in the exoskeletons of insects and other arthropods.) In fungi it provides a firm protective covering for the hyphal wall. By breaking down dead organic tissue, fungi contribute to the general process we call decay. The fungus accomplishes this breakdown by secreting digestive enzymes through the hyphal walls. These enzymes digest the organic matter, which is then absorbed into the hyphae. Despite their soft delicate nature, the hyphae of a fungus can penetrate hard wood without being damaged: they literally digest their way through a food source.

Evolutionarily the fungi present a problem. While it is tempting to think they may have arisen from a single ancestral group, or form, evidence suggests they probably did not. The various fungi we know today may well be the result of several quite different, but converging evolutionary lines.

The kingdom Fungi is divided into five phyla, three of which we will discuss in detail: the Myxomycota (slime molds), the Eumycota (true molds), and the Mycophycophyta (the lichens).

28.2
PHYLUM MYXOMYCOTA (THE SLIME MOLDS)

As Table 28.1 shows, there are two major classes of slime molds: the Myxomycetes, the plasmodial slime molds; and the Acrasiomycetes, the cellular slime molds (two other classes exist, but we will not discuss them here). The two major classes demonstrate the fundamental features of this curious and often colorful group.

Slime molds take their name from the fact that at one stage in their life cycle they consist of a spreading slimy mass. Normally, slime molds live in the woods on decaying logs, dead leaves, or other organic matter in cool, wet places. They come in a large variety of colors. Biologists have hypothesized that the pigments may be photoreceptive, since only pigmented slime molds require light for spore production.

A plasmodial or "true" slime mold, in its nonreproductive phase, is a large diploid, multinucleate, amoeboid mass called a plasmodium (Fig. 28.1). It feeds by moving in amoeboid fashion slowly across a surface, engulfing and digesting small particles of decaying organic matter, bacteria, and fungal spores along the way.

Table 28.1
Characteristics of Major Groups of the Kingdom Fungi

Phylum	Class (examples)	Form cell/hyphal structure	Distinctive characteristics	Reproduction	Ecological importance
Myxomycota (the slime molds)	Myxomycetes (Plasmodial slime molds: *Physarum*)	Noncellular; body consists of multinucleate plasmodia that differentiate into sporangia for reproduction	Moves by flowing across substrate; many species have highly pigmented protoplasts; in these species light is necessary for spore formation	Heterotrophic; feed on bacteria, decaying organic matter, yeast, etc.	Organic decomposers
	Acrasiomycetes (Cellular slime molds: *Dictyostelium*)	Body similar to Myxomycetes; sporangia formation very different; spores born into tall stalk of partially differentiated cells	Same as Myxomycetes	Individual amoeboid masses swarm together as multicellular mass prior to forming sporangia	
Eumycota (true molds)	Oömycetes (water molds, white rusts, downy mildew)	Body ranges from a single cell to a much branched filamentous mass of hyphae (mycelium).	Produce motile reproductive cells, each with one flagellum directed forward and one whip-like flagellum directed backward		Saprophytes or parasites (many cause diseases of plants)
	Zygomycetes (black bread mold)	Body mycelial in form	Produce a sexual spore; no flagellated cells; some trap and digest small animals such as amoebae and nematodes in the soil	Some disperse spores forcefully	Saprophytes or parasites
	Ascomycetes (*Neurospora*, yeasts, morels, truffles)	Sac fungi; body mycelial in form (one-celled in some); mushroom-like and large	Hyphae divided by perforated septa; dikaryons; no flagellated cells	Formation of asexual spores (in conidia) and sexual spores (ascospores) in ascus that are haploid	Powdery mildews of fruits, chestnut, blight, Dutch elm disease, ergot; food
	Basidiomycetes (toadstools, mushrooms)	Club fungi, body mycelial in form (one-celled in some) numerous hyphae closely coherent	Fruiting body of many species is large (up to three feet or more in diameter); hyphae divided by perforated septa; dikaryons; no flagellated cells	Sexual spores produced in basidia	Food (especially mushrooms); rusts, smuts (cause many plant diseases, such as wheat rust)
	Deuteromycetes or Fungi Imperfecti (*Penicillium*)	Have no flagellated cells at any point in their life cycle	Fungi with no sexual reproduction		Making of Camembert and Roquefort cheeses; ringworm; useful in antibiotic production
Mycophycophyta	Lichens (*Lecidea atrata*)	Symbiotic combination of an alga and a fungus; fungus is typically hyphal; alga is unicellular	Symbiotic existence	Fungus produces spores; algae divide asexually	Provide first stages of colonization of bare rock or other solid non-oil areas; provide food for some Arctic animals

As a plasmodial slime mold feeds, its diploid nuclei divide mitotically. Because cytokinesis, or cell division, does not occur, the plasmodium of a true slime mold is multinucleate. Plasmodial slime molds usually grow to be 2 to 3 inches long, but can sometimes reach a length of 10 inches and can weigh 50 grams or more.

When plasmodial slime molds' food supply dwindles, they enter a reproductive phase, becoming sta-

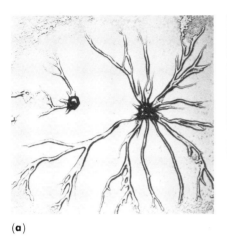

(a)   **(b)**

Fig. 28.1
(a) Multinucleate plasmodium of "true" or plasmodial slime mold (courtesy K. B. Raper); (b) Sporangia of plasmodial slime mold during reproductive phase (Photo courtesy Charles W. Mims.)

tionary and separating into many little mounds of living matter which develop into **sporangia,** structures atop a stalk in which spores develop (see Fig. 28.2). Meiosis occurs in the sporangia, while cell walls form around the resulting haploid nuclei to form individual spores. When conditions are again favorable for the plasmodial slime molds, the spores will germinate. Depending on the species, each spore will produce one to four haploid flagellated cells, or gametes. Some of these gametes will fuse, forming diploid zygotes which lose their flagella and develop into plasmodial slime molds.

The life cycle of a cellular slime mold is shown in Fig. 28.2. The spores of cellular slime molds germinate not into flagellated gametes (as in the plasmodial slime molds) but into individual, free-living amoeboid cells (Fig. 28.3a). During the feeding stage the nuclei of each amoeboid cell divides mitotically, shortly thereafter undergoing cytokinesis. The two independent daughter cells that are produced by this division become separate amoeboid cells. When the food supply of a cellular slime mold decreases, the amoeboid cells become attached to one or two centrally located cells that release cyclic AMP (cAMP). As the cells become coated with cAMP, they become sticky, merge, and attract other cells (Fig. 28.3b). This aggregate, often called a "slug," moves along a substrate, leaving behind it a trail of slime (Fig.

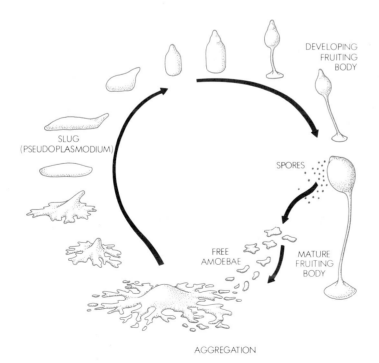

DEVELOPING
FRUITING
BODY

SLUG
(PSEUDOPLASMODIUM)

SPORES

FREE
AMOEBAE

MATURE
FRUITING
BODY

AGGREGATION

Fig. 28.2
Life cycle of a cellular slime mold, *Dictyostelium.* Details are in the text. (After J. T. Bonner.)

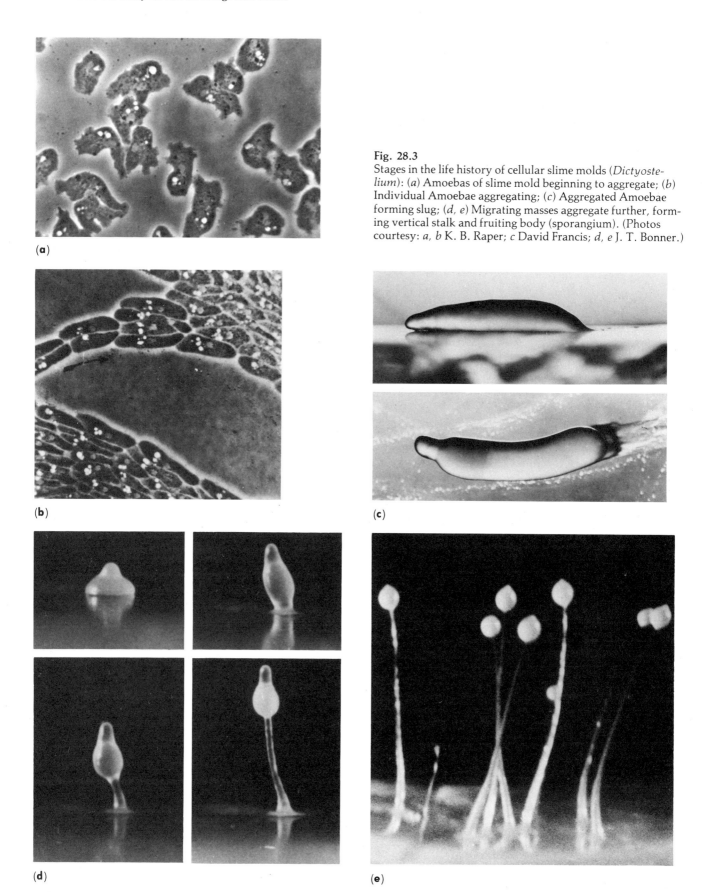

(a)

(b)

(c)

(d)

(e)

Fig. 28.3
Stages in the life history of cellular slime molds (*Dictyostelium*): (*a*) Amoebas of slime mold beginning to aggregate; (*b*) Individual Amoebae aggregating; (*c*) Aggregated Amoebae forming slug; (*d, e*) Migrating masses aggregate further, forming vertical stalk and fruiting body (sporangium). (Photos courtesy: *a, b* K. B. Raper; *c* David Francis; *d, e* J. T. Bonner.)

28.3c). The cells of the slug eventually differentiate and form a stalked sporangium (see Fig. 28.3d and e).

Since cellular slime molds have no flagellated stage and no sexual process—that is, no fusing of gametes—they are thought to have originated directly from free-living, soil-dwelling amoebae, many species of which share similar reproductive characteristics.

28.3
PHYLUM EUMYCOTA: TRUE MOLDS

Table 28.1 lists five of the major classes of the phylum Eumycota. We will focus on two representative groups: the Ascomycetes, the "sac fungi," including yeasts; and the Basidiomycetes, the club fungi, including toadstools and mushrooms. This will provide a detailed glimpse into the nature of the fungi as a whole.

The Ascomycetes and Basidiomycetes differ from most of the other fungi in that their hyphae are usually subdivided into cellular segments by **septa.** Since the septa are perforated, however, each segment may contain one to many nuclei. (The hyphal filaments of most other fungi are essentially long tubes with very few or no transverse walls.)

Although similar in possessing a basically filamentous body organization and chitinous cell walls, the Ascomycetes and Basidiomycetes differ from each other in certain features of their reproductive systems (Fig. 28.4). In the former group, the meiospores (spores produced by meiosis) are formed inside special elongated, terminal hyphal cells known as **asci** (singular, *ascus;* from the Greek word for "sac"). Each of the four meiospores divides once by mitosis, producing eight ascospores in each ascus. At maturity the ascospores are forcibly ejected through an opening that forms in the top of the ascus and are blown about by the wind. The Basidiomycetes, however, produce their meiospores on

a minute stalk called a **basidium** (plural, *basidia;* from the Greek word for "base of pedestal") on the outside of a special terminal hyphal cell. Usually only four meiospores (basidiospores) are developed from each basidium. When mature, the basidiospores are forcibly detached from the stalks and dispersed by air currents.

The Ascomycetes

The reproductive cycles of an Ascomycete (the bread mold *Neurospora*) and a Basidiomycete (mushroom) are shown in Figs. 28.5 and 28.7 for comparison. Let us consider the Ascomycete first. Spores released from the ascus fall onto the ground and germinate. They are usually one of two different "mating types," designated + or −. Each spore grows into a mass of hyphae (mycelium). It is impossible to distinguish a + hyphal mass from a − hyphal mass by observation. However, + hyphae will grow toward − hyphae (and vice versa); and where the two hyphae touch, they can fuse—that is, the cell walls can be broken down by enzymes and nuclei exchanged. The invading or donor nuclei begin to divide inside the filament of the receiving or host hypha, forming a different type of filament, or hyphal growth, called a *heterokaryon* (literally, "different nuclei"). Note that the two nuclear types (+ and −) have not fused at this point. For that reason, the hypha containing + and − nuclei is not said to be diploid, a term reserved to describe nuclei only after fusion (the heterokaryon mycelium is said to be $N + N$ rather than $2N$). The mycelium growing from the heterokaryon eventually forms a fruiting body from which asci develop. Pairs of nuclei—one of each mating type that formed the original fusion—enter each ascus. Eventually, the paired nuclei fuse and form a zygote nucleus (which is now truly diploid or $2N$). The zygote nuclei undergo meiosis to form four haploid ascospores, each of which undergoes a further mitosis, producing a total of eight spores per ascus. Of these eight, four will be of the + and four of the − mating type.

The Basidiomycetes

The Basidiomycetes have evolved special structures not only for bearing the spores (that is, basidia), but also for aiding in their dispersal. Consider the umbrella type of fruiting body of Basidiomycetes such as the mushroom (see Fig. 28.6). The undersurface of the umbrella is modified into flat vertical plates of tissue (gills), tubes, or spines covered with basidia; these are adaptations for increasing the spore-producing area. Since they are located on the underside of the umbrella fruiting body, the spore-producing structures are protected from the rain. In many species, the umbrella is elevated above the substratum by a slender stalk (the mushroom's stem). This slight elevation enables the spores to be picked up by air currents immediately after they are ejected from the basidia. In certain species (including those in the

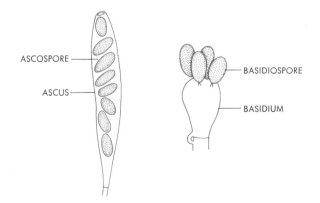

Fig. 28.4
Reproductive structures of two representative fungi. Left, the ascus of an Ascomycete such as the bread mold *Neurospora crassa.* Right, the basidium of a Basidiomycete such as a mushroom or toadstool.

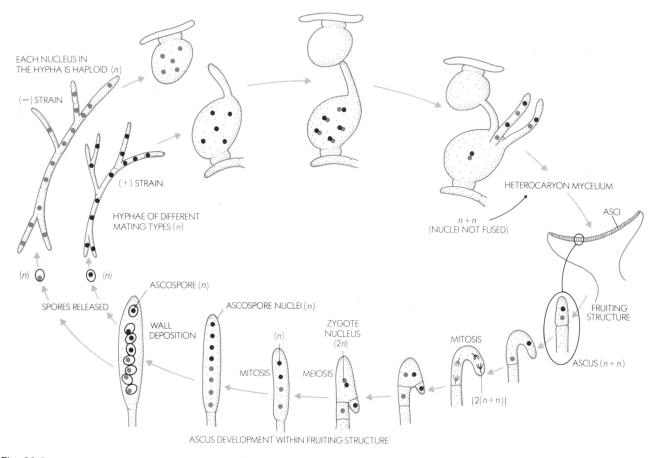

Fig. 28.5
Reproductive cycle of an ascomycete such as the red bread mold *Neurospora*. See text for details.

Fig. 28.6
Mushroom (Basidiomycete) showing stem, cap, and gills lined with basidia. (Photo, Harold Hungerford.)

family that includes the bracket fungi) the stalk has been lost. Since these fungi usually grow on the trunks of standing trees or fallen logs, the fruiting bodies are already elevated enough for efficient spore dispersal.

The reproductive cycle of a basidiomycete (mushroom) is shown in Fig. 28.7. Basidiospores germinate to produce mycelia; as with the ascomycetes, two different mating strains (+ and −) exist. Hyphae from the two different types can fuse to form a secondary mycelium (middle, right), the heterokaryon, containing both + and − nuclei (unfused). The hyphae grow into a heterokaryon mycelium $(N + N)$ which eventually sends up a fruiting body, the basidiocarp, or familiar mushroom umbrella. The hyphal mass under the soil is quite extensive. The familiar mushroom is only a small part of the entire plant mass. Gills on the underside of the cap contain basidia, each of which, like the ascus, consists of a pair of opposite-type nuclei. These fuse, forming a diploid zygote nucleus. Meiosis follows almost immediately, resulting in four haploid nuclei, two of each mating type. These haploid spores are called

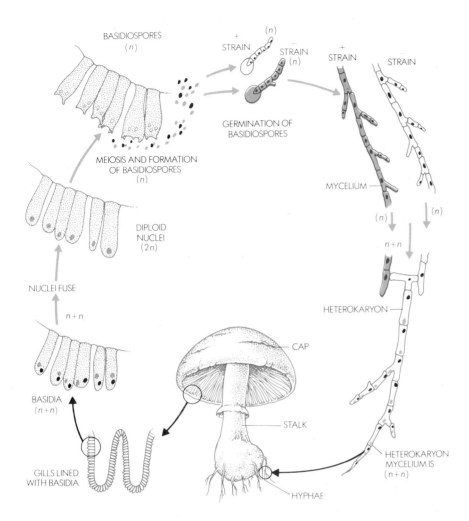

Fig. 28.7
Life cycle of a basidiomycete, such as a mushroom (see text for details).

basidiospores; on release from the basidia they will each germinate, and the cycle begins again.

28.4
THE PHYLUM MYCOPHYCOPHYTA (LICHENS)

No discussion of the fungi would be complete without mention of that special and curious group known as lichens. Lichens are actually dual or composite organisms, composed of a fungus and an alga living in a symbiotic association (see Fig. 28.8a).

At least twenty-six genera of algae are known to be involved in lichen associations, including eight blue-green algae, seventeen green algae, and one yellow-green alga. The most common blue-green algae found in lichens are *Nostoc* and *Syctonema*. The unicellular alga *Trebouxia* is the most common green alga of lichens. More than 80 percent of all lichens involve genera of green algae. Fungi classified in the Ascomycetes and a few species of Basidiomycetes produce lichens.

The lichen body is highly diverse in form and structure. Many species simply produce a powdery layer over rocks or tree bark. Others develop a thin crust that adheres to the substratum so tightly that the lichen appears to be painted on. Still other species of lichens grow into leafy forms much less tightly attached to the bark or rock surface. The most complex lichen bodies resemble tiny shrubs with flattened or cylindrical branches. Some species of these lichens stand stiffly erect, while others hang in long streamers from the branches of trees. A typical lichen is shown in cross section in Fig. 28.8(b).

Internally, the lichen body may exhibit considerable differentiation. The algal cells may be scattered throughout or they may be restricted mainly to a specific layer. The fungal hyphae may produce tissue-like layers, including an epidermis-like configuration on the surface, a cortex-like layer just beneath, and a central core or medulla. Bundles of hyphae often form root-like absorbing and anchoring structures on the lower surface of the lichen body. The fungal hyphae form a close net-

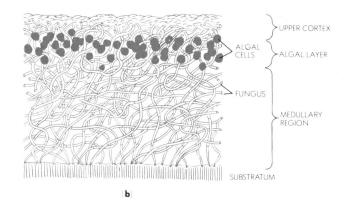

Fig. 28.8
Lichens, a combination of an alga and a fungus. (*a*) Lichens grow on various kinds of substrata, including tree trunks (as shown here), rocks, or any other solid material to which they can attach. (*b*) The organization of a lichen shows that the algal cells live completely surrounded by the fungal hyphae. The algae provide nutrient organic materials through photosynthesis; the fungus provides anchorage and protection for the algae against environmental conditions. Lichens are able to survive in some of the harshest environments on earth. They are particularly able to withstand the cold and wind of the arctic tundra or at high altitudes. (Photo courtesy Jack Diani; drawing adapted from John W. Kimball, *Biology,* 4th ed. Reading, Mass.: Addison-Wesley, 1978.)

work enclosing the algal cells, often embedding the alga in a tissue-like mass. In most lichens, projections of the fungal hyphae penetrate the algal cells.

Investigations of lichens by numerous botanists, including intensive physiological studies by Vernon Ahmadjian at Clark University, have demonstrated that a lichen is not a simple mixture of an alga and a fungus. Not only does the lichen represent a physiological interaction between alga and fungus, but the association produces a distinctive morphological entity with features considerably different from those of the individual fungus or alga. Thus the composite plant is different from either of its two components.

The exact nature of the interaction between the fungus and the alga comprising a lichen is only now beginning to be understood. One widely accepted hypothesis proposes that the alga, being photosynthetic, provides the fungus with organic compounds. The fungus, in turn, provides the alga with water and minerals, as well as protection from desiccation and high light intensities. Some support for this hypothesis has come from autoradiography. Pieces of a lichen were provided with C^{14}-labeled sodium bicarbonate as a source of carbon dioxide for photosynthesis. The C^{14} rapidly appeared in various organic compounds, first in the algal layer of the lichen and, fifteen minutes later, in the hyphae of the fungus. The movement of materials from the alga to the

fungus probably occurred by diffusion from the alga and subsequent absorption by the fungus. It appears that both the alga and the fungus benefit from the association—that the relationship is a mutualistic one. Further, the algal cells within the lichen appear to be healthy, and in many areas lichens are known to grow where neither the alga nor the fungus could survive alone.

Lichens are important land flora. In the forests of Scandinavia and in subarctic Canada, they form extensive growth covers, providing pasture for grazing reindeer and caribou. Lichens also grow on well-lighted areas such as rock surfaces and tree trunks throughout most of the world and through breakdown of materials in the substratum help in the formation of soil.

The Kingdom Plantae

The kingdom Plantae consists of multicellular eukaryotic organisms that contain photosynthetic pigments in plastids. The major characteristics of the kingdom Plantae are summarized in Table 28.2. Plants were the first successful colonizers of the land, beginning in the Ordovician or Silurian periods about 500 to 450 million years ago. In the five-kingdom classification system, the kingdom Plantae includes red algae (Rhodophyta), brown algae (Phaeophyta), and green algae (Chlorophyta), among aquatic forms; the mosses (Bryophyta) and ferns, conifers, and flowering plants (Trachaeophyta), among primarily terrestrial forms.

28.5
THE ALGAL GROUPS

The three algal groups included with the kingdom Plantae—the red, brown, and green algae—show more complexity and cell differentiations than the algae groups included among the Protists. For the most part, Protists are unicellular, with virtually no cell specialization. Because some of the red, brown, and green algae show some degree of cell differentiation akin to that of higher plants—for example, the possession of special "anchor" or holdfast organs, primitive vascular elements, or specialized reproductive cells—botanists have found it

Table 28.2
General Overview of the Kingdom Plantae

Phylum	Examples	Major characteristics	Vascular tissue present (and type)	Seed present
Rhodophyta Red algae		Plants chiefly reddish due to red pigment phycoerythrin; bodies vary from microscopic single cells through simple filaments to large plants with some tissue differentiation; very complex sexual reproductive systems (but no flagellated cells of any sort); attached to rocks along ocean shores from high intertidal regions to the depth of over 100 meters in clear water	No	No
Phaeophyta Brown algae (*Ecotocarpus, Fucus, Sargassum*)		Plants brownish due to the xanthophyll pigment fucoxanthin; bodies vary from simple unbranched filaments only 1 mm long to massive plants 200 ft long with well-developed tissues; grow attached to rocks along ocean shores, often forming dense subtidal forests	Sievelike elements similar to those of vascular plants; serve for food conducting	No
Chlorophyta Green algae (*Chlamydomonas, Volvox, Spirogyra*)		Plants bright green due to chlorophyll in plastids; body diversity is considerable, ranging from one-celled motile or nonmotile species to motile and nonmotile colonies, and from simple filaments to massive plants over 3 m long (with tissues) with diversity of life cycles and modes of sexual reproduction; abundant in fresh and marine waters, on tree trunks, moist rocks, leaf surfaces, and soil	No	No
Bryophyta	Liverworts, hornworts, and mosses	Multicellular terrestrial plants without vascular tissues or true roots; life cycle includes a conspicuous gametophyte with smaller sporophyte permanently attached; sex organs (archegonia and antheridia) exist; multicellular reproduction by gametes and spores	No	No
Tracheophyta	Lycopodium, horsetail ferns, conifers, and flowering plants	Vascular terrestrial plants with differentiation of organs into leaves, roots, and stem; only male gametes of some species are motile; vascular plants have well-developed conducting tissues for transport of organic materials and minerals; progressive reduction in the gametophyte in passing from the lower (ferns) to higher (flowering plants) groups	Yes: primitive vessels, tracheids, sieve cells	In some forms (conifers and flowering plants)

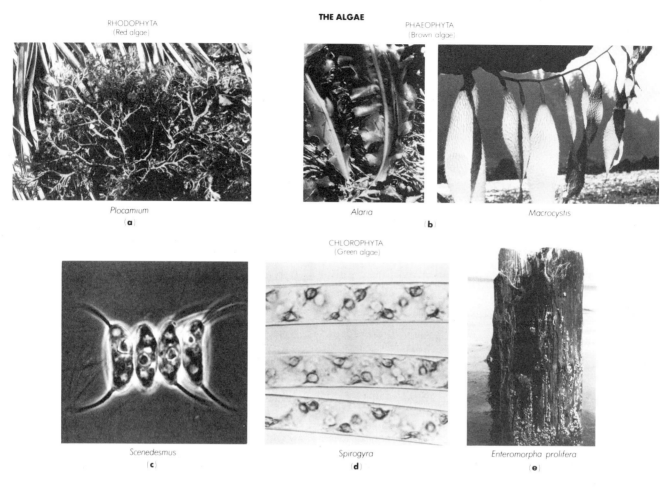

THE ALGAE

RHODOPHYTA
(Red algae)

Plocamium

(a)

PHAEOPHYTA
(Brown algae)

Alaria   *Macrocystis*

(b)

CHLOROPHYTA
(Green algae)

Scenedesmus   *Spirogyra*   *Enteromorpha prolifera*

(c)   **(d)**   **(e)**

more appropriate to include them with the other "true" plants.

Rhodophyta (Red Algae)

With few exceptions, the Rhodophyta are a marine group. Most are multicellular, growing attached to rocks, pilings, or wherever they can hold on in ocean waters to a depth of 100 to 175 meters (see Fig. 28.9a). Growing attached to a substrate is highly adaptive for red algae, since the plant body needs to be in constant contact with circulating water on all sides for proper gas exchange. Floating masses of red algae usually die from improper aeration. Red algae use chlorophyll *a* to carry on photosynthesis; they lack chlorophyll *b*, though some have a third, or different kind of chlorophyll (called chlorophyll *d*). In addition to chlorophyll, red algae have two other pigments, phycoerythrin and phycocyanin (recall that these same two pigments occur in blue-green algae as well), which impart their reddish color to some of them. The red pigments are accessories in photosynthesis, absorbing wavelengths of light that penetrate the deeper waters where the algae sometimes live (shorter wavelengths in the blue end of the spectrum penetrate farther through water). The red algae lack flagella at any stage of their life cycle. Their photosyn-

Fig. 28.9

Three algal phyla of the kingdom Plantae. (*a*) Rhodophyta (red algae), plants that can vary in size from microscopic single cells to multicellular forms over 3 meters in length, with some cellular differentiation apparent. The form shown here, *Plocamium,* usually attains a size of 30 to 60 cm. (*b*) The Phaeophyta (brown algae), showing two species: left, *Alaria,* a kelp, showing the long, flowing body which in some species can attain a length of 25 meters. *Alaria* and other brown algae attach to a substrate by a specialized holdfast organ (bottom). Considerable tissue differentiation is sometimes apparent in the Phaeophyta, including proto-vascular elements similar in some respects to vascular tissues in higher plants. Right, the genus *Macrocystis,* a marine form from the west coast of North America. The large bulbous structures are the blades growing from a stalk. The brownish-yellow color is due to the pigment fucoxanthin. (*c–e*) Representatives of the Chlorophyta (green algae). (*c*) The microscopic, four-celled form, *Scenedesmus,* showing cell nuclei; (*d*) Another microscopic, filamentous form, the freshwater *Spirogyra,* so named because of its spiral chloroplasts; (*e*) The marine form, *Enteromorpha,* shown growing in a mat on wharf pilings. (Photos *a–c* and *e*: J. Robert Waaland, Biological Photo Service; *d*, Carolina Biological Supply Company.)

thetic pigments are attached to membranes that form primitive kinds of chloroplasts, resembling the infolded plasma membrane of blue-green algae. Because they resemble blue-greens in many ways (though they are eukaryotic), the red algae are thought to be a very old group.

Phaeophyta (Brown Algae)

The Phaeophyta are the principal seaweeds of temperate and polar regions (see Fig. 28.9b). Like the red algae, the browns are almost exclusively marine. They dominate the rocky shores throughout the cooler regions of the world. Some, like the kelps, also form extensive beds offshore. *Sargassum nitans,* one species of brown alga, is the principal component of the dense floating masses of seaweeds that literally cover the surface of the Sargasso Sea, a vast, 4- to 6-million-square-kilometer area of ocean stretching from the Bahamas to the Azores. Brown algae contain chlorophylls *a* and *c,* along with fucoxanthin, which masks the green of their chlorophylls, and thus accounts for the brownish color (and their name).* Unlike the higher green plants, the brown algae store their photosynthetic products as an unusual polysaccharide called laminarin instead of as starch; occasionally they may store food products as oils. Like other higher green plants, brown algae contain cellulose in their cell walls.

The brown algae can be very large. Some of the giant kelps along the Pacific coast reach lengths up to 60 meters. Because these are annual plants—that is, they live for only one growing season—they must attain this giant size in a matter of months. The kelps show clear evidence of specialization of parts and cell differentiation. The long body of a kelp, as shown in Fig. 28.9(b), has a holdfast organ (holdfast organs are not found in all brown algae), a leaf-like blade (making up the bulk of the wavy part of the plant), and specialized "conducting tissues" down the center of a primitive stalk, called the *stipe.* Carbohydrates produced in the blade, which rises toward the surface of the water, are transported through these vascular tissues into the stipe and the holdfast.

Chlorophyta (Green Algae)

The Chlorophyta are the most diverse of all the algae, comprising at least 7000 species. Most green algae are aquatic; the vast majority live in fresh water, although a few species are marine. Some green algae live in terrestrial habitats, including the melting surface of snow, as green patches on tree trunks, and symbiotically with fungi as lichens. In size, the green algae are considerably smaller than most of the red or brown algae. Many are microscopic, though filamentous forms such as *Spirogyra* (Fig. 28.9c) can produce a visible mass when they are

clumped together in great numbers. The largest green alga is a marine form *Codium,* which lives in the Gulf of Mexico; specimens have been known to attain a breadth of 25 cm (9 or 10 in.) and a length of over 8 meters (25 ft).

The green algae have both chlorophylls *a* and *b,* and rigid, cellulose-containing cell walls. Some of the green algae are flagellated (including *Chlamydomonas,* discussed below), and the others have flagellated gametes or zoospores. Although most green algae are unicellular, some show varying degrees of colonial or multicellular existence. *Spirogyra* grows as a filament, each cell of which is independent of the others. *Volvox,* on the other hand, shows the extreme of colonial existence within the green algae (see Fig. 28.9). In *Volvox* some cells specialize to become vegetative (photosynthesizing) cells, which *cannot* reproduce themselves. Indeed, because of this specialization, a case can be made for considering *Volvox* a truly multicellular form.

The cell structure of the Chlorophyta is highly organized and as complex as that found in any higher plant cell. A detailed diagram of the unicellular green alga *Chlamydomonas* is shown in Fig. 28.10. The cell of an individual *Chlamydomonas* is usually less than 25 micrometers (μ) long, with two flagella that protrude through the cell wall at the anterior end. Inside each cell is a single, large chloroplast, usually cup-shaped, with a pigmented area, the "eye spot," near the anterior rim. Note that the chloroplast of *Chlamydomonas* is very similar in general structure and shape to the higher plant chloroplast. *Euglena* has a large, prominent nucleus with a nucleolus. The flagella terminate in basal bodies within the cytoplasm, and have the typical 9 + 2 structure. All the organelles we associate with typical eukaryotic cells—endoplasmic reticulum, contractile vacuoles, mitochondria, ribosomes, and Golgi complex—are found in *Chlamydomonas* and other green algae.

Reproduction in the Chlorophyta. Members of the phylum Chlorophyta reproduce by both sexual and asexual means. The life cycles of three green algae, *Chlamydomonas, Spirogyra,* and *Oedogonium* are shown in Figs. 28.11, 28.12, and 28.13, respectively. An examination of these life cycles will not only illustrate two somewhat different modes of reproduction within the Chlorophyta but also indicate an important direction that occurred early in plant evolution: differentiation of sexual types.

We will consider *Chlamydomonas* first. In the asexual phase, shown at the bottom, middle, a single haploid adult cell undergoes mitosis to produce two daughter cells. Since the adult cell of *Chlamydomonas,* like that of virtually all the green algae, is haploid, the mitotic process produces haploid daughter cells. The daughter cells take up an independent existence of their own. In the sexual phase, shown on the left and upper part of Fig. 28.11, two adult cells come together with their flagellar ends (anteriors) facing. There are two

---

*It is worth pointing out that not all brown or all red algae look brown or red to the human eye. Many bear some color resemblance to their name, but not all.

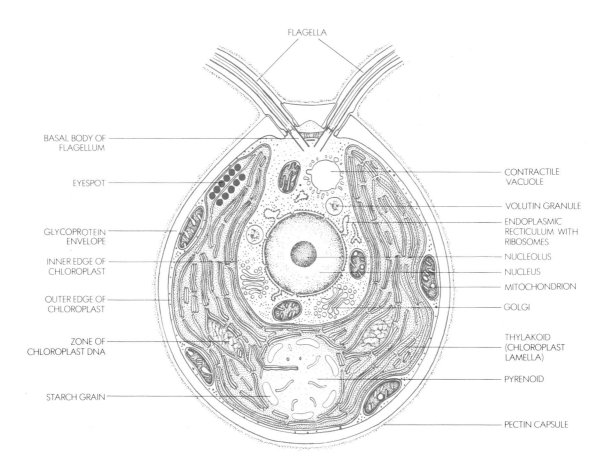

FLAGELLA

BASAL BODY OF
FLAGELLUM

EYESPOT

GLYCOPROTEIN
ENVELOPE

INNER EDGE OF
CHLOROPLAST

OUTER EDGE OF
CHLOROPLAST

ZONE OF
CHLOROPLAST DNA

STARCH GRAIN

CONTRACTILE
VACUOLE

VOLUTIN GRANULE

ENDOPLASMIC
RECTICULUM WITH
RIBOSOMES

NUCLEOLUS

NUCLEUS

MITOCHONDRION

GOLGI

THYLAKOID
(CHLOROPLAST
LAMELLA)

PYRENOID

PECTIN CAPSULE

Fig. 28.10
Diagram showing structure of the unicellular green algae
Chlamydomonas as revealed by the electron microscope.
Only one of the two contractile vacuoles is shown. Volutin
granules are probably stored phosphates. (From Keith Vicker-
man and Francis E. G. Cox, *The Protozoa.* Boston: Houghton
Mifflin, 1967, p. 20. By permission of John Murray, Ltd.)

mating strains, a + strain and a − strain. When a +
strain fuses with a − strain cell, the result is a diploid
zygote, which usually forms a protective spore, the
zygospore, which can tide the organism over unfavorable conditions (drought, cold). When the zygospore is
stimulated to continue the cycle, meiosis occurs, and
anywhere from two to thirty-two haploid cells emerge
and develop into mature individuals. The fusion of two
individual cells, acting as gametes, is known as **isogamy.** In isogamy the two gametes are morphologically
alike and cannot be differentiated in any visible way.
Moreover, their behavior shows no differentiation—
both participate in the fusion process in an equal and
identical way. Isogamy is characteristic of many Chlorophyta.

In *Spirogyra,* which is a filamentous alga, the reproductive process is slightly different. Each cell of a
filament is haploid. Asexual reproduction (not shown in
Fig. 28.12) is by simple mitosis and cell division, producing long filaments of similar haploid cells. Sexual
reproduction involves conjugation, which occurs between two filaments that come to lie side-by-side. A
conjugation tube develops between each pair of opposite cells (see Fig. 28.12(a), stages 2–3) across the filaments. Then the cell body (excluding the cell wall) of
each cell in the two filaments rounds up into a spherical
mass. Each cell mass in one of the two filaments begins
to pass over its respective conjugation tube into the cell
of the other filament. Thus, one filament acts as a donor
and the other as a *recipient.* Within the recipient cell,
the two cell masses form a diploid zygote (also called a
zygospore), which also serves as a protection during unfavorable conditions. When the zygospore begins to develop it undergoes meiosis, forming four haploid nuclei.
Three of these nuclei degenerate, while the fourth grows
into a mature *Spirogyra* filament. It should be apparent
that the mating process in *Spirogyra* is not exactly isogamy. While morphologically there is no visible difference between the two filaments or the rounded-up
protoplasts within them prior to conjugation, there is a
behavioral difference. One of the filaments acts as a

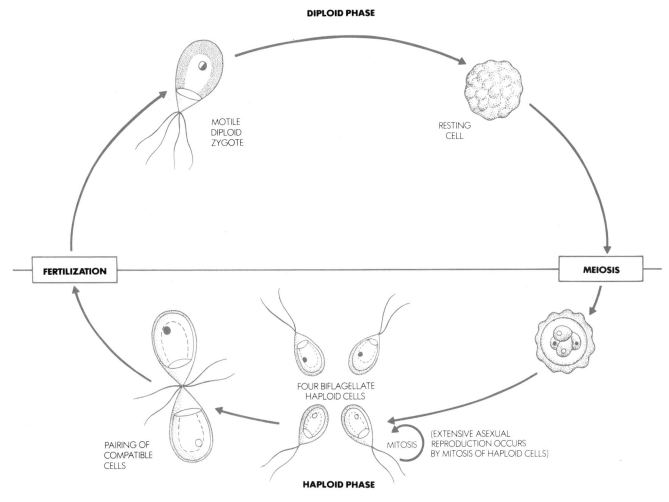

DIPLOID PHASE

MOTILE
DIPLOID
ZYGOTE

RESTING
CELL

FERTILIZATION

MEIOSIS

FOUR BIFLAGELLATE
HAPLOID CELLS

MITOSIS

(EXTENSIVE ASEXUAL
REPRODUCTION OCCURS
BY MITOSIS OF HAPLOID CELLS)

PAIRING OF
COMPATIBLE
CELLS

HAPLOID PHASE

Fig. 28.11
Life cycle of the green alga *Chlamydomonas*, usually found in pools, lakes, and damp soil. Asexual reproduction is shown on the left and sexual reproduction on the right. At center bottom is an enlarged view of one individual.

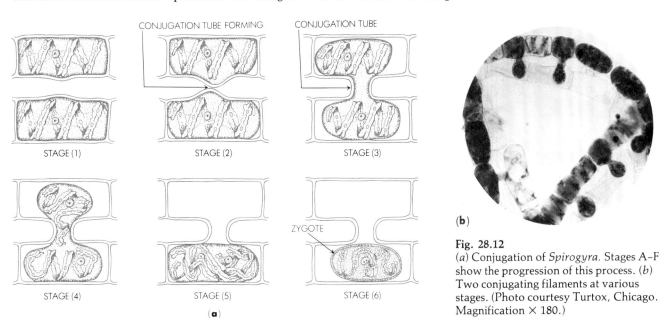

CONJUGATION TUBE FORMING

CONJUGATION TUBE

STAGE (1)   STAGE (2)   STAGE (3)

STAGE (4)   STAGE (5)   STAGE (6)

ZYGOTE

(a)

(b)

Fig. 28.12
(*a*) Conjugation of *Spirogyra*. Stages A–F show the progression of this process. (*b*) Two conjugating filaments at various stages. (Photo courtesy Turtox, Chicago. Magnification × 180.)

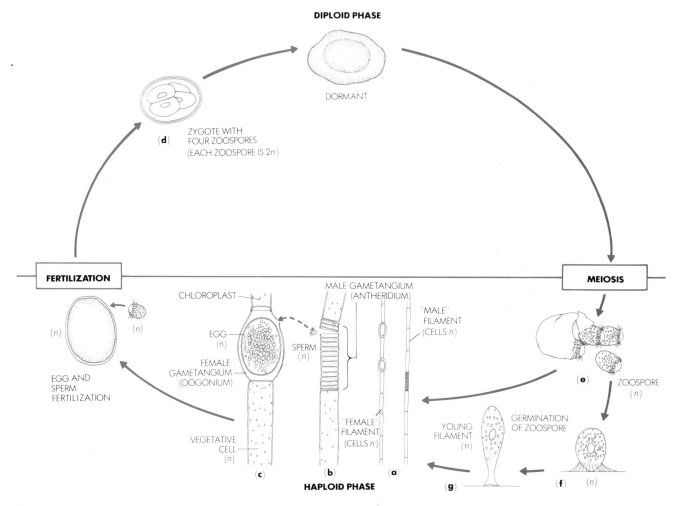

Fig. 28.13
True heterogamy in the filamentous green alga *Oedogonium*.
See text in details.

donor, the other as a recipient. Such a differentiation between gametes, however slight it may be in this case, is called **heterogamy.**

Heterogamy can be seen carried to a further and more recognizable degree by another filamentous green alga, *Oedogonium* (see Fig. 28.13). All cells of the filament are haploid. In asexual reproduction cells divide by mitosis, producing a lengthening of the filament. After reaching a certain length the filaments tend to break, the free-floating portions eventually reattaching

and starting an independent growth of their own. In sexual reproduction, two separate filaments produce gametes from specialized cells called **gametangia,** which are morphologically different (Fig. 28.13b and c). The male gametangium is called an **antheridium,** and the female gametangium an **oögonium.** The antheridium produces many haploid sperm by mitosis. Each oögonium produces one haploid egg cell, also by mitosis. The antheridia break open when ripe, and the sperm swim out. When the sperm find an oögonium, fertilization

---

Sexual reproduction has one immensely important evolutionary advantage: it allows the mixing and recombination of genetic elements that make for greater diversity.

can take place, producing a diploid zygote. The zygote forms into a zygospore (Fig. 28.13d) that can remain dormant for a year or more before beginning to grow. When it starts to grow the zygospore undergoes meiosis to produce four haploid zoospores (Fig. 28.13e), each of which can germinate into a young haploid filament (Fig. 28.13f and g).

The three green algae discussed above show a distinct trend, within a single phylum, from complete isogamy to complete heterogamy. It seems likely that the evolution of sexual differentiation as we know it today must have begun with the ancient aquatic algae by some simple, gradual process with far more intermediate stages than represented in the three examples chosen here. The adaptive function of sexual differentiation is that it allows one of the gametes, the egg cell, to conserve its energy supplies and thus give the zygote a better chance for survival until it can become self-sufficient. The three species used in this illustration are not, of course, stages in an actual evolutionary sequence; we do *not* mean to say that the modern species *Oedogonium* evolved from *Spirogyra* which in turn evolved from *Chlamydomonas*. This series only illustrates that the possible intermediary stages from isogamy to heterogamy could have existed in the past, and thus could have provided a basis for gradual evolution by natural selection.

28.6
THE EVOLUTION OF PLANT LIFE ON LAND

Up until now, we have been discussing plant groups that live virtually all of their life in an aquatic environment. Yet as we know from looking around us, plants have spread over the face of the land until they have become, visibly at least, the dominant life forms. Several hypotheses have been advanced over the years to suggest how plant life evolved adaptations to terrestrial existence.

Compared to an aquatic environment, the environment on land is harsh and uninviting. The major problems of life for plants on land all derive from the scarcity of water. These problems can be summarized as:

1. Supplying a liquid environment for the cells of the plant body. This problem has two aspects:
   a) Obtaining a constant enough supply of water in the first place for the plant's total needs.

   b) Transporting the available liquid throughout a multicellular body, so that all cells receive enough for their metabolic function.

2. Preventing desiccation (drying out).

3. Providing physical support for the plant body to have greater access to sunlight (water bouys up the plant body and allows it to float toward or on the surface where sunlight is plentiful).

4. Withstanding the more extensive changes in the environmental conditions that occur on land (water provides less extreme temperature changes, and humidity is always at 100 percent). Air is a less effective insulator, fluctuating radically in north temperate zones where, for example, land plants must be able to withstand 50° to 60°C changes in temperature from summer to winter. Similarly, humidity of the air changes a great deal on land, whereas it is constant in the water.

5. Reproducing without the medium of water for transporting sperm and egg.

Terrestrial plants have evolved various adaptations to solve each of these problems. Among the plants we have studied so far, it is safe to say that few of the algae (Monera or Protista) have evolved a truly terrestrial existence. Those that live on land usually do so by living in very moist places or existing within a film of water in the soil or on the surface of leaves, logs, or rocks. The fungi are truly terrestrial, but their lifestyle is adapted to a very narrow range of conditions: They can live only in relatively moist places and tend to have short (and very rapid) growth periods to take advantage of available water (some mushrooms will grow to full size within a matter of hours after a rainstorm). Only the bryophytes (mosses and liverworts) and the tracheophytes (vascular plants) have truly developed adaptations to meet all of the major problems of terrestrial life. Evolutionarily, the tracheophytes appear to have made the first major adaptations to terrestrial life, with the bryophytes coming somewhat later. However, following the basic classification system of the kingdom Plantae, we will consider the bryophytes first.

As we survey the bryophytes and tracheophytes, keep in mind the various ways in which their morphological and physiological adaptations solve some of the problems of terrestrial existence identified above. At the

---

The evolution of multicellularity was favored because it allowed for cell specialization. Specialized cells can work cooperatively to extract energy more efficiently from the environment.

SUPPLEMENT 28.1
THE ORIGIN AND ADAPTIVE SIGNIFICANCE OF MULTICELLULARITY

Multicellularity has one major advantage that has opened up innumerable evolutionary doors: the possibility of cell specialization resulting in an increase in the ability of the organism to exploit its environment and increase its own chances for survival.

Concerning when and how multicellularity evolved, biologists can only speculate. Complex multicellular animals were already firmly established by the beginning of the Cambrian period, some 600 million years ago. Algae with highly differentiated multicellular bodies had evolved into at least two major groups (the green and the red algae) by the Ordovician period, some 500 million years ago. Thus the fossil record is of little help in dating the origin of multicellularity. We must be satisfied with examining organisms living today to draw reasonable and fruitful inferences about the nature of the first multicellular organisms.

It seems reasonable to hypothesize that the multicellular state originated through the aggregation of separate cells into a colony. Several surviving species of green algae demonstrate likely stages in this evolution. *Chlamydomonas*, for example, generally exists in a unicellular motile state (Diagram A, part a). When conditions become unfavorable for activity, however, each cell loses its flagella and becomes aggregated with other individuals into temporary associations. On the return of favorable conditions, the individual cells regain their flagella and the aggregation breaks up.

However, a species known as *Gonium* represents a permanent association of several *Chlamydomonas*-like cells (Diagram A, part b). For *Gonium* the multicellular colony is the normal way of life. Each colony consists of four, eight, sixteen or thirty-two cells held together loosely in a flattened mass of jelly. All the cells in the colony are identical in both structure and function and appear to have no interactions with one another. Each cell is able to reproduce itself asexually, by mitotic cell division, or sexually by becoming freeswimming and mating with another cell.

In the genus *Pandorina* (Diagram A, part c) the colonies are larger and more tightly knit than in *Gonium*. There are eight, sixteen, or thirty-two cells closely bound in a sphere of jelly. All cells in *Pandorina* retain full reproductive capacity (either asexual or sexual). They are structurally and functionally similar to each other, and to *Chlamydomonas*. There is a slight differentiation in size of the gametes, one being smaller than the other. However, both are flagellated and free-swimming. Essentially *Pandorina* shows isogamy.

The genus *Eudorina* (Diagram A, part d) shows the first stages of cell differentiation, and a consequent division of labor. The cells are actually interconnected by cytoplasmic strands. The colonies are somewhat larger than those of *Gonium* or *Pandorina*, consisting of sixteen, thirty-two, or

Diagram A
Sequence of stages of colonial existence among present-day green algae (making up the so-called Volvocine series). (*a*) The simplest unicellular form, the beginning of the series, *Chlamydomonas* (drawing). Subsequent stages in the series are seen as progressively larger colonial associations of *Chlamydomonas*-like cells. (*b*) The genus *Gonium pectorale* typically consists of from four to thirty-two cells (here sixteen cells), arranged in a highly regular, close-packed manner, forming a kind of curved sheet (not a sphere) with the two flagella of the *Chlamydomomas*-like cells protruding from the convex surface of the sheet. (*c*) *Pandorina morum* resembles *Gonium* in cell number, but differs in that the individual cells are always arranged in a sphere, with flagella protruding outward from all sides. The colony shown here contains sixteen cells. (*d*) *Eudorina elegans* form somewhat larger colonies than *Gonium* or *Pandorina*, with cells less tightly packed. This species normally consists of either thirty-two or sixty-four cells, but some species have as many as 128. (*e*) *Pleodorina californica*, a larger colony still than the preceding examples, consists of anywhere from 32 to 128 and (rarely) 256 biflagellated cells. The cells are arranged around the surface of a sphere, and are held together (as in the other examples as well) by a sheath of mucilage-like material. The specimen shown here contains sixty-four cells, but the colony has a distinct morphological and functional division of labor. The upper, left-hand part of the colony is the somatic or vegetative end, containing cells that carry out all functions except reproduction. One or a few cells in the other hemisphere, on the lower left-hand surface, will actually divide to form new colonies. The reproductive cells carry out all other functions characteristic of the vegetative cells (such as swimming and photosynthesis). (*f*) *Volvox*. Of the series represented in this diagram, only *Volvox* shows true division of labor among its cells; hence, it is the only one to be truly a multicellular "organism." *Volvox* colonies consist of two completely interdependent cell types. The approximately 1000 to 2000 small "somatic" cells are *Chlamydomonas*-like, and carry out all vegetative functions; they never reproduce once they have been formed. The few large reproductive cells, all located in one hemisphere, are not *Chlamydomonas*-like, lack flagella, and are dependent on the somatic cells to keep them in a favorable (including nutritive) environment, for both photosynthesis and growth. (Photos courtesy of Regina Birchem; captions with the aid of David Kirk.)

sixty-four cells, all of which look more or less alike. In some species, cells in only one half of the colony can reproduce; the others are incapable of cell division once the colony has reached maturity. These cells, which might be called "vegetative" or *somatic* cells (as distinct from *reproductive* cells),

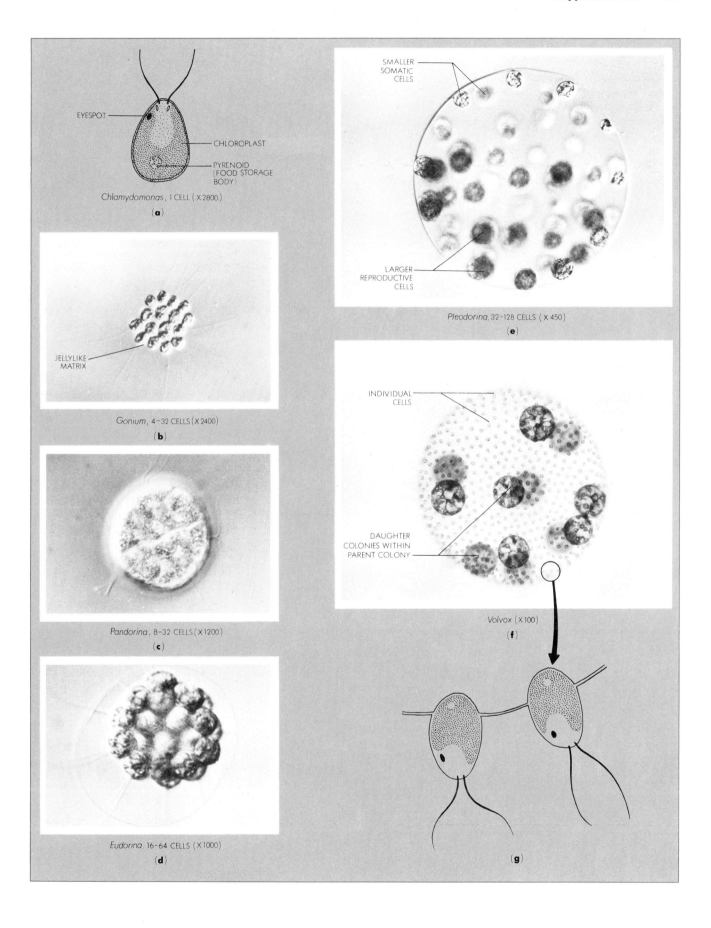

EYESPOT

CHLOROPLAST

PYRENOID
(FOOD STORAGE
BODY)

Chlamydomonas, 1 CELL (×2800)

(a)

JELLYLIKE
MATRIX

Gonium, 4–32 CELLS (×2400)

(b)

Pandorina, 8–32 CELLS (×1200)

(c)

Eudorina, 16–64 CELLS (×1000)

(d)

SMALLER
SOMATIC
CELLS

LARGER
REPRODUCTIVE
CELLS

Pleodorina, 32–128 CELLS (×450)

(e)

INDIVIDUAL
CELLS

DAUGHTER
COLONIES WITHIN
PARENT COLONY

Volvox (×100)

(f)

(g)

carry out photosynthesis and all the other major processes of life. The reproductive cells differentiate at a certain point in time to form sex cells of two types: either small, free-swimming sperm, or large, round eggs that do not swim but remain embedded in the jelly matrix of the colony. *Eudorina* thus shows a simple form of heterogamy.

In *Pleodorina* (Diagram A, part e) the division of labor is more pronounced. Each colony contains 32, 64, or 128 cells of two observably distinct sizes: the smaller cells are entirely somatic and only the larger ones can reproduce. Furthermore, the more or less spherical colony is regionally specialized. The small somatic cells lie in one hemisphere, while the larger reproductive cells lie in the other. Reproduction is either sexual or asexual. Asexual reproduction involves mitotic cell division by reproductive cells (not all the reproductive cells divide, but those that do, do so in synchrony), each cell giving rise to a daughter colony. The daughter colonies break away from the parent and take up an independent existence. Sexual reproduction involves differentiation of reproductive cells into either eggs or sperm. Usually, a colony will produce all male or all female gametes, though occasionally a single colony will give rise to both at the same time. Thus, we can even see the beginnings of sexual dimorphism (two types) among colonies: some that are "male" (sperm-producing), and others that are "female" (egg-producing).

The final stage in this hypothetical evolutionary development is found in *Volvox*. A *Volvox* colony is a large, well-integrated, and highly differentiated "organism" consisting of interconnected *Chlamydomonas*-like cells embedded in a layer of jelly resembling a hollow sphere (Diagram A, part g). The term organism is placed in quotation marks here because *Volvox* is not usually regarded as an organism in the strictest sense of the word. Its levels of cell differentiation are not as great as that found even in the simplest multicellular

animals, the sponges. Nonetheless *Volvox* does display some very important steps toward true organismic characteristics—even more so than *Pleodorina*. For example, *Volvox* colonies are considerably larger, on the average, than *Pleodorina*. In addition, *Volvox* reproductive cells are much larger than the somatic cells; like *Pleodorina*, the reproductive cells of *Volvox* are located only in one hemisphere. The *Volvox* colony is able to move about as a single group with the flagella of the individual cells. The reproduction of *Volvox* is similar to that of *Pleodorina*. Reproductive cells undergo fission to produce daughter colonies within the parent colony. When the young colonies have completed their division they break out of the old colony, which then dies. Occasionally reproductive cells enter a sexual phase; instead of producing daughter colonies, they will produce eggs or sperm. After fertilization a zygote divides to produce a new colony.

This sequence of stages in the evolution of multicellularity is of course hypothetical. It is not clear exactly how the various stages themselves may actually have arisen. While it is likely that the six organisms shown in Diagram A are in fact evolutionarily related, we have no direct evidence for this. By observing the above sequence, it is not difficult to imagine the cells of *Chlamydomonas* sticking together after cell division, thus producing an aggregate of four (*Gonium*), eight or sixteen (*Pandorina*), thirty-two or sixty-four (*Eudorina*), and so on. The similarity of the individual cells making up all these colonial forms, both to each other and to *Chlamydomonas*, is striking. Whatever the exact evolutionary mechanism involved, it seems clear that multicellularity must have evolved by the adherence of unicellular forms into an association that conferred some selective advantage on the colony. Once associated together, groups of cells had opened to them the possibility of division of labor and hence specialization.

---

same time, consider how their lifestyle is limited by the realities of terrestrial conditions.

28.7
PHYLUM BRYOPHYTA (MOSSES AND LIVERWORTS)

Bryophytes are simple terrestrial plants that form a dominant part of the vegetation in some parts of the world—for example, on forest floors and in bogs in temperate regions, and at higher tropical altitudes. The major characteristics of the phylum Bryophyta and its three classes are summarized in Table 28.3. The most common examples of bryophytes are the mosses (see Fig. 28.14a) and the liverworts (see Fig. 28.14b). Most bryophytes inhabit moist, shaded areas and have few adaptations for resisting desiccation. Some species live on tree trunks, and others inhabit shallow rock crevices

where they are not exposed to extreme temperatures and drying winds. Bryophytes usually grow only during periods of moist weather and manage to survive long periods of drought by becoming dormant. Thus two of the bryophytes' most important characteristics indicate their need for a moist environment, and hence their lack of full adaptation to the land:

1. The lack of vascular tissue (except in a few species where phloem-like and very simple tracheid-like cells are present), and hence dependence on moisture in the air or on the leaf surface (from dew or rain) to provide individual cells with water.

2. The sperm must swim across the surface of one part of the plant to reach the egg; this can occur only if a thin film of moisture is present on the surface. Thus bryophytes cannot reproduce in a moderately dry environment.

Table 28.3
Characteristics of the Phylum Bryophyta

Class	Examples	Major characteristics	Vascular tissue (and type)	Seed
Hepaticae	Liverworts (*Ricciocarpus*)	Alternation of generations with conspicuous gametophyte plant; sporophyte without stomata; short-lived, dying after spores mature; unicellular rhizoids	No	No
Anthoceros	Hornworts (*Anthoceros*)	Gametophyte similar to liverwort; sporophyte with stomata; continues growth for a long time by the addition of new cells from a basal meristem	No	No
Musci	Mosses (*Minum, Polytrichium*)	Gametophyte filamentous at first, later developing an erect stem with leaves; sporophyte growing out of gametophyte; long-lived; multicellular rhizoids	No	No

BRYOPHYTA

(a)

(b)

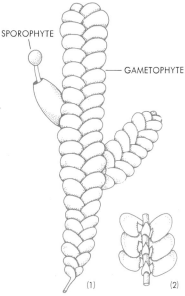

SPOROPHYTE

GAMETOPHYTE

(1)   (2)

Fig. 28.14
Moss. The leafy portion below is the haploid gametophyte; the slender stalks with an apical cap are the diploid sporophytes. Inside the caps of the sporophytes haploid spores are produced. Each spore grows into a haploid gametophyte (see life cycle in Fig. 28.15). (Photo, Jack Diani.) (*b*) Liverworts. Left: gametophytes of *Ricciocarpus natans*, growing floating in pond water. Right: a leafy liverwort, *Porella bolanderi*, with dorsal view (1) of a large leafy gametophyte with small sporophyte. In the ventral view (2), the small leaves on the underside of the gametophyte are visible. (Photo courtesy William C. Steere, New York Botanical Garden; drawings from Arthur W. Haupt, *Introduction to Botany*, New York: McGraw-Hill, 1956, p. 321.)

The evolution of the reproductive organs in plants has resulted in an increase in the likelihood of cross-fertilization as opposed to self-fertilization.

Bryophytes reproduce both by asexual (growth and fragmentation) and by sexual means. In their sexual reproduction, bryophytes show two distinct phases in their life cycle: a **sporophyte** and a **gametophyte** generation, which follow one another in a regular, alternating sequence. To one degree or another all higher plants (and some algae, too), from the bryophytes to the most complex tracheophytes, show an *alternation of generations* (that is, an alternation of the sporophyte with the gametophyte stage) in the completion of a single life cycle. The gametophyte stage in all plants is always haploid, the sporophyte always diploid. As the names imply, the sporophyte generation gives rise to spores; the

gametophyte generation gives rise to gametes. Let us follow briefly the life cycle of a moss, as shown in Fig. 28.15.

A haploid spore (stage a) germinates into a filamentous structure called a **protonema** (stage b). The protonema grows along the surface of the soil (or other surfaces such as tree trunks) sending down small root-like structures (but not true roots), rhizoids (stage c), which hold the protonema in place and absorb moisture. The developing protonema also sends up a leafy portion, the gametophyte, which is capable of photosynthesis. The leafy gametophyte forms the typical green carpet of moss we see on the forest floor. All the cells of the pro-

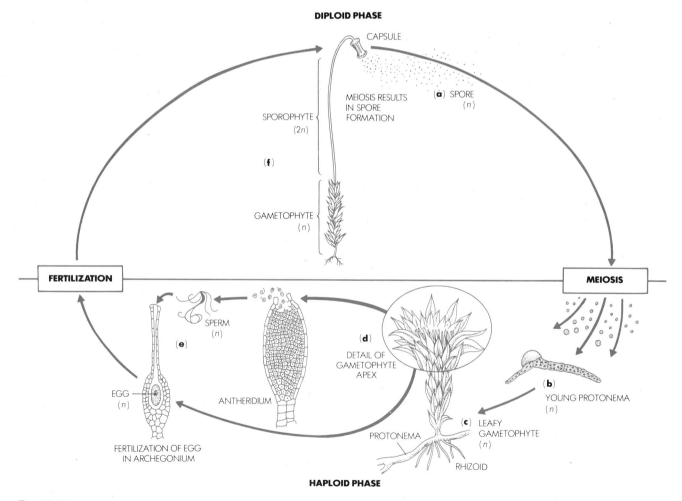

Fig. 28.15
Schematic representation of the life cycle of a moss.

tonema as well as those of the leafy gametophyte are haploid. If we look carefully at the apex of a mature gametophyte we see two specialized structures, the antheridia and archegonia. Antheridia produce the male gamete, or sperm; archegonia produce the female gamete, or egg (stages d and e). Sperm from the antheridia must swim across parts of the leaves at the tip of the gametophyte to find the archegonia. Swimming down the neck of the archegonia the sperm fertilizes the egg, producing a diploid zygote. This is the sporophyte; its cells are all diploid (stage f). Inside the capsule, spore mother cells (2N) undergo meiosis to produce haploid spores, and the life cycle begins again.

28.8
PHYLUM TRACHEOPHYTA (VASCULAR PLANTS)

Of all groups of plants the tracheophytes have evolved the most sophisticated adaptations to terrestrial life. The defining characteristics of the phylum is the presence of vascular tissue, specialized cells that conduct water with dissolved materials and wastes to all the plant cells. These adaptations have allowed the tracheophytes to exploit the land environment to the fullest, invading virtually every part of the globe. They are a truly versatile group, including species as different from one another as the tiny duckweed that floats on the surface of ponds, to the ferns, club-mosses, and the giant sequoia. The groups composing the phylum Tracheophyta and their major characteristics are summarized in Table 28.4.

The most important traits of the vascular plants are:

1. The presence of vacular tissue, at least in the form of xylem—especially tracheids.

2. A layer of somatic tissue around the reproductive organs, providing protection and preventing desiccation.

3. Retention of the multicellular embryo within the female reproductive organ (archegonium).

4. Presence of a waxy layer, the cuticle, on the surface of plant parts exposed to the air (principally leaves and stems).

In the following section we will survey the major groups of tracheophytes. We will discuss each group briefly and then focus on a detailed discussion of only three of the most important representatives: the ferns, the conifers, and the flowering plants.

Subphylum Psilopsida

The Psilopsida are a very simple group of tracheophytes. They lack true roots or leaves, but have stems

Table 28.4
Characteristics of Members of the Phylum Trachaeophyta

Subphylum	Class (with examples)	Major characteristics	Type of vascular tissue present	Seed
Psilopsida	*Psilotum*	Dichotomously branching plants with no true roots, or leaves; stems contain vascular tissue and carry out photosynthesis; no cambium, hence no secondary growth; spores produced at tip of some branches	Tracheids	Not present in modern forms
Sphenopsida	*Equisetum* (horsetails)	Vascular plants with jointed stems marked by conspicuous nodes and elevated silicious ribs; sporangia in a strobilus at the tip of the stem; leaves are scale-like; sperm are motile	Tracheids, in primitive forms; cells with thin areas (pits) in their lateral walls	Not present in modern forms
Lycopsida	*Lycopodium* (club mosses)	Vascular plants with microphylls (little leaf-like structures) in spirals; extremely diverse in appearance; all have motile sperm	Tracheids, in primitive forms; cells with thin areas (pits) in their lateral walls	Not present in modern forms
Pteridopsida	Ferns	Leaves organized as large fronds; underground stem (rhizome); gametophyte more or less free-living and usually photosynthetic; multicellular gametangia and free-swimming sperm present; leaves called megaphylls	More highly specialized, especially vascular tissue	Not present in modern forms

(Continued)

Table 28.4 (Continued)

Subphylum	Class (with examples)	Major characteristics	Type of vascular tissue present	Seed
Spermopsida	Pteridospermae (seed ferns)	Extinct ferns that bore true seeds and had true leaves, stems, and roots	Tracheids, vessels; well-developed phloem	Yes
	Cycadinae (cycads)	Seed plants with slow cambium growth and pinnately compound, palm-like, or fertile leaves; ovules not enclosed; sperm are flagellated and motile, but are carried to the vicinity of the ovule in a pollen tube	With sluggish cambium growth	Present; naked seed
	Ginkgoinae (Ginkgo tree)	Seed plants with active cambium growth and fan-shaped leaves with dichotomous venation; sperm carried to the vicinity of the ovule in a pollen tube but are flagellated and motile	Active cambium growth	Present; naked seed
	Coniferinae (conifers: pine, fir, spruce, hemlock)	Megaphylls often reduced to needles or scales; seed plants with cambium and simple leaves, in which the ovules are not enclosed and the sperm are not flagellated	More highly specialized, well-developed conducting tissue for the transport of water and organic materials and minerals	Present; naked seed
	Angiospermae (flowering plants)	Multicellular plants with vascular tissues; gametophyte small and dependent upon sporophyte, which is large, becoming herbaceous, shrubby, or woody; leaves usually broad; sperm without flagella, transferred to ovule by a pollen tube; seed enclosed within an ovary that develops into a fruit	Highly specialized, well-developed conducting tissue for the transport of water and organic material and minerals	Present; enclosed within ovary (fruit)
	Subclass 1: Dicotyledonae (dicots) (carrots, oak, maple)	Embryo inside seed with two cotyledons; flower parts mostly in fours or fives; vascular tissue in distinct strands or bundles arranged in a cylinder or circle; leaves net-veined	Highly specialized, well-developed conducting tissue for the transport of water and organic materials and minerals	Present; enclosed within ovary (fruit)
	Subclass 2: Monocotyledonae (monocots) (lily, iris)	Embryo with one cotyledon; flower parts mostly in threes or multiples thereof; vascular tissues usually in scattered bundles; leaves with parallel veins	Highly specialized conducting tissue	Present; enclosed within ovary (fruit)

with dichotomous branching (dividing into twos), and true vascular tissue in the form of primitive tracheids. An example of the psilopsids, *Psilotum*, is shown in Fig. 28.16. *Psilotum* has an underground, horizontal stem with small, unicellular rhizoids that bear some resemblance to the root hairs of higher plants. Photosynthesis is carried out by the vertical, above-ground stem. The simple scales on the stems of some species are not true leaves and play only a minor part in photosynthesis. Lacking any cambium, *Psilotum* cannot produce secondary growth.

Reproduction is by spores, produced in sporangia at the tips or on the sides of some of the branches. Through meiosis, haploid spores are produced that germinate on the ground (or just under the surface) into small but independent gametophytes. Each gametophyte produces both antheridia and archegonia. Fertilization takes place within the archegonium, yielding a diploid zygote that develops into the sporophyte (Fig. 28.16 shows the sporophyte). The sporophyte generation is thus far more visible and predominant in the life cycle of *Psilotum* than it was in the bryophytes. Reduc-

Fig. 28.16
The genus *Psilotum,* modern-day representative of the ancient group Psilopsida. Note the dichotomous branching (two-pronged) of the stems, the lack of leaves, and the bulbous sporangia that occur along the main branches or on the tips of side branches. (Photo, Runk/Schoenberger for Grant Heilman Photography.)

tion in size of the gametophyte is a trend that appears with increasing prominence among the higher groups of plants.

The Psilopsida are descended from an ancient group of vascular plants. In fact, they are the oldest group of fossil vascular plants known, dating from the late Silurian period, 450 to 425 million years ago. These ancestors of the Psilopsida that remain became quite widespread during the Devonian period, but have since become extinct, leaving no fossil remains in the intervening 400 million years.*

Subphyla Lycopsida (Lycopods) and Sphenopsida (Horsetails)

These two groups contain several modern representatives, among them the common lycopods (such as the so-called Christmas fern *Lycopodium*) and the horsetails. A representative of each group is shown in Fig. 28.17.

The Lycopods. The lycopods have true roots and leaves, the latter possibly originating from scale-like structures found on the stems of ancestral forms such as the psilopsids. The lycopods have distinct vascular tissue and bear their spores on the underside of specially modified leaves known as **sporophylls.** Sporophylls are closely arranged at the top of specialized stems, forming a cone-shaped structure, the *strobilus* (Fig. 28.17a). The club-shaped appearance of the strobilus has given the

---

*This and other evidence has led David W. Bierhorst of the University of Massachusetts and others to argue that the organisms we call Psilopsids today are descended not from the original Psilopsida, but from a line that gave rise to modern-day ferns. The matter is still one of much debate, however.

(a)   (b)

Fig. 28.17
Members of the subphyla Lycopsida (*a*) and Sphenopsida (*b*). (*a*) *Lycopodium,* also known as the club moss, running pine, ground pine, or Christmas pine, possesses true roots, stems, and leaves. The cone-shaped strobili at the top of certain stalks produce spores that germinate into small gametophytes. (*b*) Horsetail, *Equisetum,* also has true roots, stems, and leaves. The leaves are arranged on the stem in whorls. Strobili are produced at the top of leaf-bearing stalks in some species, or at the top of special reproductive stalks in others. (Photos, *a* Runk/Schoenberger for Grant Heilman Photography; *b* Grant Heilman Photography.)

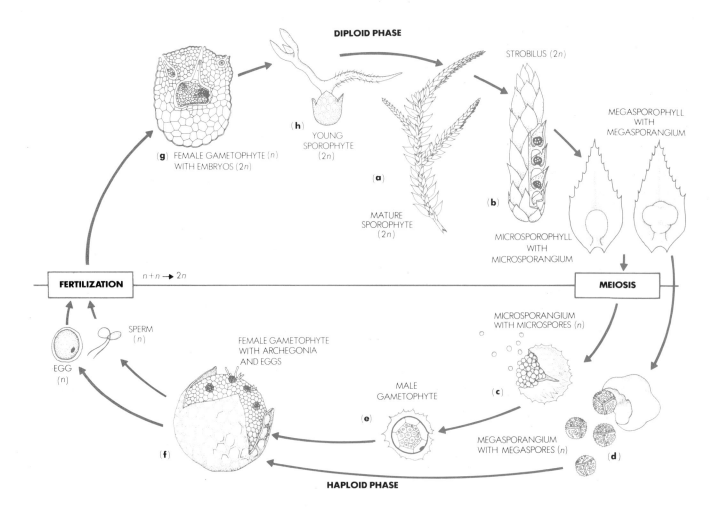

Fig. 28.18
Reproduction in a contemporary heterosporous vascular plant, the spike-moss *Selaginella*. (After E. W. Sinnott and K. S. Wilson, *Botany: Principles and Problems*. New York: McGraw-Hill, 1963, p. 417.)

lycopods their common name, "club mosses." Sporophylls contain sporangia. Spores produced within the sporangia of *Lycopodium* give rise to gametophytes that produce both antheridia and archegonia. Since there is no sexual differentiation between spores or the gametophytes to which they give rise, *Lycopodium* is said to be **homosporous** (*homo*, same; *sporous*, spore).

Another lycopod, *Selaginella*, is **heterosporous** (*hetero*, different; *sporous*, spores). The strobilus contains two types of sporophylls: megasporophylls and microsporophylls (see Fig. 28.18 stages a and b). Megasporophylls contain megasporangia, which produce large megaspores, while microsporophylls contain microsporangia and produce smaller microspores (stages c and d). Each type of spore germinates into a separate (and morphologically distinct) type of gametophyte: the megaspores to female gametophytes, which produce eggs; the microspores to male gametophytes, which pro-

duce sperm (stages e and f). The sperm must swim (or be splashed by raindrops) from the male to the female gametophyte, where fertilization occurs. The embryo (stages g and h) grows into a mature plant (a diploid sporophyte). During evolution, heterospory has come to predominate in the life cycles of the higher plants.

The Sphenopsida (Horsetails). Horsetails are found in relatively moist localities such as swamps or river flood plains. There is only one genus living today, *Equisetum*. The horsetails consist of a straight stalk or stem with true leaves arranged in whorls (see Fig. 28.17b) around the "stem." The horsetails also have true roots. The stems are hollow and jointed but contain chlorophyll and, with leaves, carry out photosynthesis. *Equisetum* lacks cambium and cannot produce secondary growth. The stalks of *Equisetum* contain silicon, and hence have an abrasive quality. This feature made the plants par-

Fig. 28.19
Reconstruction of a carboniferous forest during the Mississippian and Pennsylvanian periods (350 to 320 million years ago). The large tree-like forms on the left are giant lycopsids, now-extinct vascular plants related to today's club mosses (*Lycopodium* and *Selaginella*). Giant ferns are also visible. The trees to the lower right are sphenopsids, primitive relatives of today's "horsetails" (*Equisetum*). One of the geological periods takes its name from the discovery of large fossil deposits of giant tracheophytes in the coal beds of Pennsylvania. Much of our modern coal was deposited during this period of lush, heavy vegetative growth. (Courtesy Field Museum of Natural History, Chicago).

ticularly useful in pioneer days for scouring (in fact, the horsetails are sometimes called "scouring rushes"). Although limited in number and distribution today, the sphenopsids were once a dominant form of life on earth. During the coal-forming Mississippian and Pennsylvanian periods (300 to 200 million years ago), sphenopsids were among the most prominent plant groups (see Fig. 28.19). Most species of sphenopsids appear to be homo-

Fig. 28.20
A fern, showing leafy frond and stem. (Photo courtesy Jack Diani.)

sporous, although in at least one species half the spores develop into male-producing gametophytes whereas the other half yield female-producing gametophytes. In this species, however, if fertilization does not take place the female gametophyte also begins producing male gametes along with eggs!

Subphylum Pteridopsida (Ferns)

The ferns are an advanced group of tracheophytes, containing well-developed vascular tissue, along with true roots, stems, and leaves (see Fig. 28.20). The stems of most ferns are horizontal, lying either on or under the ground forming a **rhizome**. In some species in the tropics, however, the stem stands upright and forms a trunk. "Tree ferns" were common during the Mississippian and Pennsylvanian periods, as shown in Fig. 28.19. The leaves of ferns, where photosynthesis takes place, are called **fronds**. Ferns, like all higher plants, show alternation of generations. The typical fern we see, especially the leafy frond, represents the sporophyte. Spores are produced on the undersurface or around the edges of fronds, which thus act as both photosynthetic organs and as sporophylls (in some species all the fronds are sporophylls; in other species, only certain fronds are). Some species of fern are homosporous, while others are heterosporous; when the spores are released, they germinate into gametophytes. The life cycle of a homosporous fern is diagrammed in Fig. 28.21.

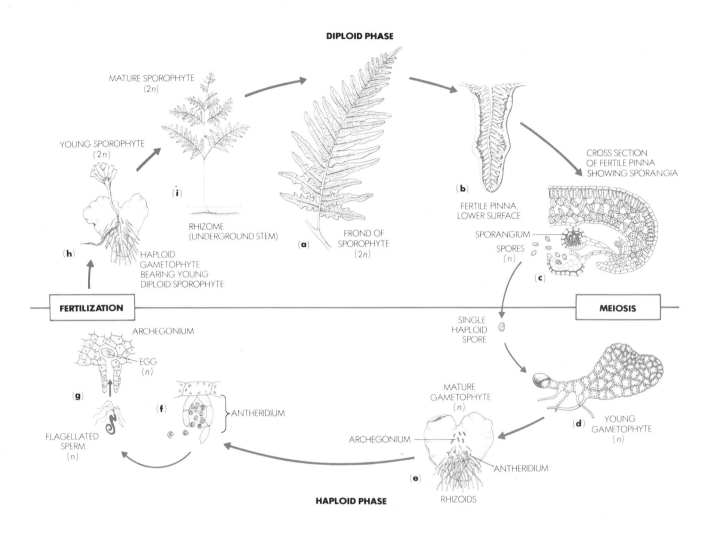

DIPLOID PHASE

MATURE SPOROPHYTE
(2n)

YOUNG SPOROPHYTE
(2n)

(i)

RHIZOME
(UNDERGROUND STEM)

FROND OF
SPOROPHYTE
(2n)

(a)

(b)

FERTILE PINNA,
LOWER SURFACE

CROSS SECTION
OF FERTILE PINNA
SHOWING SPORANGIA

SPORANGIUM

SPORES
(n)

(c)

(h)

HAPLOID
GAMETOPHYTE
BEARING YOUNG
DIPLOID SPOROPHYTE

FERTILIZATION

MEIOSIS

SINGLE
HAPLOID
SPORE

ARCHEGONIUM

EGG
(n)

(g)

(f)

ANTHERIDIUM

FLAGELLATED
SPERM
(n)

MATURE
GAMETOPHYTE
(n)

ARCHEGONIUM

(e)

ANTHERIDIUM

RHIZOIDS

(d)

YOUNG
GAMETOPHYTE
(n)

HAPLOID PHASE

ANTHERIDIUM

RHIZOIDS

Fig. 28.21
Life cycle of a fern. The mature frond is really a sporophyte, since it bears sporangia on the undersurface of the leafy portion. The spores are released and germinate into a haploid gametophyte. The mature gametophyte produces male and/or female gametes. Fertilization produces a young diploid sporophyte that grows right out of the gametophyte. Spore formation in the sporangia involves a meiotic division, producing haploid spores.

Fig. 28.22
Scanning electron micrograph of a gametophyte of the Boston fern (*Ceratopteris thalictroides*). The vegetative cells of the gametophyte are shown at the upper part and the rhizoids at the lower right. Spherical structures on the lower part of the gametophyte are antheridia, which produce sperm. The Boston fern is heterosporous, though the majority of ferns are homosporous. (Courtesy Professor Richard G. Kessel.)

The fern gametophyte is an interesting structure in its own right, as shown in the scanning electron micrograph in Fig. 28.22. The large, heart-shaped upper region is the vegetative portion. The gametophyte is photosynthetic, so that it lives as an organism independent of the sporophyte. The gametophyte also forms rhizoids as shown in the lower right. These anchor the gametophyte into the soil and also absorb nutrients. The spherical structures on the lower part of the gametophyte are the antheridia (the species shown in Fig. 28.22 is heterosporous). After fertilization, the stem of the sporophyte grows directly out of the gametophyte, forming its own roots for support and absorption. Whereas some species of ferns, such as the one shown in Fig. 28.22, are heterosporous, others are homosporous.

The fern sporophyte is better adapted to terrestrial life than the moss sporophyte. Their well-developed vascular tissue and root system allow ferns to withstand drier conditions without injury. However, in its fertilization process—the swimming of the sperm to the egg—the fern is as dependent on a moist environment as the moss. For this reason ferns have never really been able to move away from humid environments. Even in their heyday during the Mississippian and Pennsylvanian periods (300 to 200 million years ago), the ferns flourished only in very humid swamps.

Subphylum Spermopsida (the Seed Plants)

The Spermopsida include all plants that produce a distinct embryo enclosed within a seed: the cycads, ginkgos, conifers, and flowering plants (see Fig. 28.23). The

Fig. 28.23
Some representatives of the subphylum Spermopsida of the phylum Tracheophyta: (*a*) a living cycad; (*b*) a conifer, the bristlecone pine, that produces its seeds in a cone; and (*c*) the flowering plant from the family Scrophulariaceae. In flowering plants (Angiospermae) the seed is contained within the female reproductive organ, which matures to surround the seed within a fleshy fruit. (Photos: *a, c* Jack Diani).

(b)

(a)

(c)

Spermopsida have been by far the most successful group of tracheophytes in exploiting the land environment. The subphylum consists of five classes, as listed in Table 28.4. Four of these, however, have been traditionally grouped together as the Gymnospermae (*Gymno*, naked; *spermae*, seed). The seeds are said to be naked because they are not protected by a developed ovary wall, as in the case of angiosperms, but are borne directly on the scales of the cones, or seed-bearing structures. However, many botanists today feel that the differences between these five classes are as great or greater than their similarities, and hence prefer to drop the general term "gymnosperm" and deal with each group individually. In this section we will summarize the characteristics of the various groups briefly, and then examine an example, a conifer (pine tree), in detail.

Several classes of the subphylum Spermopsida consist of groups represented by only a few, or in one case, no living species. For example, the class Pterido-

spermae, the "seed ferns," prominent during the Carboniferous period, is wholly extinct (refer again to Fig. 28.19). Their leaves resemble those of ferns, but fossil evidence indicates they produced true seeds, and hence should be classed separately. Two other ancient groups are the Cycadinae and the Ginkgoinae. The Cycadinae include palm-like plants known as cycads (Fig. 28.23a), which were prominent during the Jurassic period, 200 to 180 million years ago, one of the three periods comprising the "age of dinosaurs." Cycads are sometimes called palms, but they are not a true palm, which is an angiosperm. The Cycadinae include about 100 species living today. The Ginkgoinae include only one living species, the Ginkgo tree, which has gained great prominence as an ornamental tree around the world. Ginkgos, however, are very scarce in the wild, and the distribution of this once widespread group is restricted almost completely to landscape plantings by human beings.

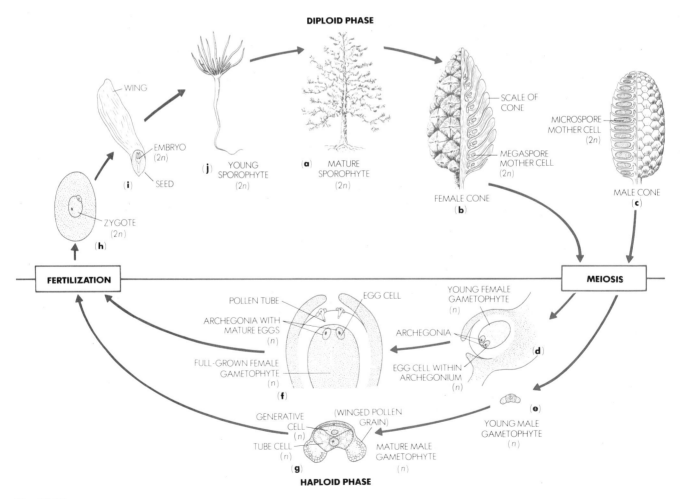

Fig. 28.24
Life cycle of a pine tree (a conifer).

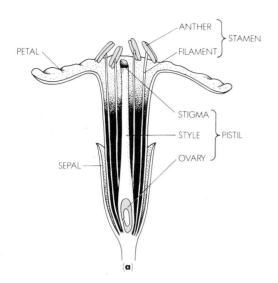

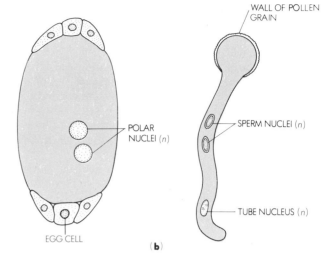

Class Coniferinae (the Conifers)

The most widespread and familiar group of the gymnosperms is the conifers. The term "conifer" means cone-bearer; the conifers characteristically produce their seeds on special structures called **cones.** Woody, seed-bearing plants, containing large amounts of xylem and some phloem, the conifers began to appear in the Devonian period (about 380 million years ago). The woody tissue of conifers is composed only of tracheids (the vessels found in angiosperm stems being a later evolutionary development), but there is considerably more xylem in conifers than in the ferns or primitive gymnosperms. The increased amount of xylem provided greater conduction possibilities and increased the sturdiness of the entire plant. The sporophyte generation of the conifers is dominant (the pine tree represents the sporophyte), the gametophytes being reduced to a small number of cells retained on the sporophyte itself.

Conifers have flattened or rounded leaves called **needles,** which, like all leaves, are equipped with vascular tissue and carry out photosynthesis. The needle is an anatomical adaptation to the cold and/or dry climates in which conifers often live. The cylindrical shape of the needle reduces the amount of surface area per unit of volume exposed to the air. As a result, the needle loses less water and freezes less easily than a flat leaf. As a group, conifers are hardy, and among the most widely distributed and successful plants on earth. Conifers can live at altitudes of 10,000 feet, where winds and the cold would kill virtually any other form of higher plant. They also live in arid regions, in sandy soil, and under other conditions not acceptable to many plant species because of rapid freezing and/or desiccation.

The life cycle of a typical conifer (pine tree) is shown in Fig. 28.24. Like the bryophytes and other tracheophytes, the conifers show a distinct alternation of generations, but in this case the gametophyte is even

Fig. 28.25
Reproduction in flowering plants (Angiospermae). (a) Structure of the flower; the female gametophyte is contained within the ovule, which in turn is part of the ovary. (b) The female (left) and male (right) gametophytes in angiosperms. The female gametophyte consists of seven cells; the large central cell contains two haploid nuclei (the other cells each contain only one haploid nucleus each). One of those cells is the egg cell. The male gametophyte, or pollen grain, consists of a single cell with two nuclei (a tube nucleus and one sperm nucleus) surrounded by a protective coat. As the pollen grain begins to germinate on the stigma, the sperm nucleus divides to produce two sperm nuclei (both haploid), which follow the tube nucleus down the pollen tube.

more reduced and inconspicuous than in ferns. The mature pine tree is the diploid sporophyte, with gametophytes developing in the female cone. All conifers are heterosporous, producing male spores (microspores) in male cones and female spores (megaspores) in female cones (Fig. 28.24, stages a to c). The male cones are relatively small (about 1.5 cm long) and are located on the lower branches of the tree. The female cones are much larger (they average 8 to 12 cm long, though some, as of the Norway spruce, can be up to 20 to 25 cm) and are located on the upper branches of the trees (these are the familiar pine cones). Although conifers are capable of self-fertilization, the positioning of male and female cones on the tree makes it less likely that this will occur, since the pollen grains would have to travel virtually straight upward to reach the female cones.

The undersurface of the scales on the female cones bears diploid megaspore mother cells (stages b and d). Through meiotic division these cells produce a multicellular haploid structure, the female gametophyte, containing from two to five archegonia, in each of which a small egg cell develops (stage f). The undersurface of the male cone contains microspore mother cells (stages c and e) which undergo meiosis to produce young, haploid male gametophytes (stage e). When the male gametophyte matures, it develops small wings for aerial distribution and becomes a *pollen grain* (stage g). Pollen grains are released from the male cones and carried by the wind to nearby trees. When a pollen grain lands on the scale of a female cone, near the female gametophyte, a pollen tube grows from it toward the archegonia (stage f).

The pollen tube is guided by a tube nucleus; as the tube grows the generative cell of the male gametophyte begins to follow. While traveling down the tube the generative cell undergoes a mitotic division, producing two haploid daughter cells. One of these undergoes another mitotic division, producing two haploid sperm cells. One of these latter cells will accomplish the actual fertilization. As you can see, the developing male gametophyte is a very small structure, consisting at most of two cells and six nuclei.

The pollen tube grows down through the tissue surrounding the female gametophyte until it reaches one of the archegonia. The tube nucleus disintegrates, and the two sperm cells are discharged into the archegonium. Both sperm nuclei enter an egg cell, though only one fertilizes the egg nucleus. The resulting diploid zygote (stage h) develops into a young embryo within the female gametophyte. At the same time, the female gametophyte begins to grow and accumulate a quantity of stored food, which will help the young sporophyte during the early stages of germination. The embryo with its stored food, surrounded by a scaly protective coat, is called the seed (stage i). The seeds of most conifers are winged to aid in wind dispersal. On germination, the seed grows into a young sporophyte (stage j), and the cycle is ready to begin anew.

The reproductive pattern of the gymnosperms has several distinct advantages over that of the ferns:

1. Fertilization is independent of outside conditions of moisture.

2. The sperm nucleus is carried to the egg by the formation of a pollen tube. The pollen tube not only guides the sperm nucleus toward the egg but also, by forming a passage through the female gametophyte tissue, provides a more protected environment for the crucial period leading up to fertilization. (In the ferns, sperm must swim on the exposed surface of the gametophyte to reach the archegonia, greatly reducing the chances of successful fertilization.)

3. Because the female gametophyte is retained on the parent sporophyte, the fertilized egg is given a greater amount of protection in its early stages of development. Protection of the female gametophyte and of the developing zygote is of great selective advantage. From

the early gymnosperms onward, this adaptation is a universal characteristic of all higher plants.

4. The embryo is protected by the formation of a seed. This, indeed, represents a major evolutionary advance. The tough seed coat not only protects the embryo from mechanical damage; it also prevents drying out. Thus the embryo sporophyte may be able to survive until it reaches a favorable spot for germination.

5. The seed also allows for the storage of food. Recall that young fern sporophytes can call on very few food reserves. They must be autotrophic almost from the very start. Providing the germinating sporophyte with a small "push"—as all seed plants do—greatly increases the survival capacity of offspring.

Class Angiospermae (Flowering Plants)

The term "angiosperm" means "seed borne in a vessel," that is, enclosed by distinct layers of specialized tissue derived from the ovary wall. This is one way in which angiosperms differ from gymnosperms. Another way is that all angiosperms produce flowers of some kind. Although the first angiosperms appear in the fossil record in the early Cretaceous period, about 135 million years ago, they did not become dominant plant forms until the late Mesozoic and early Cenozoic eras (about 65 million years ago). During that time a great adaptive radiation took place. It appears from the fossil record that angiosperms quickly displaced gymnosperms as the most numerous and widespread land plants.

Through variation and natural selection, angiosperms have been able to develop greater adaptation to land conditions than any other plant group. Their roots and stems, with well-developed xylem and phloem, provide anchorage, support, and efficient conduction (see Chapter 8). The efficiency of conduction is further improved with the appearance of fibers and vessels as additional types of xylem cells. Flowers provide a very effective means of pollen dispersal by attracting pollinating agents such as bees and moths. Conifers, relying only on the wind, are much less efficient in this regard. Finally, the angiosperms produce seeds enclosed or embedded in a fruit. The fruit serves primarily to aid in seed dispersal, again many times with the help of various animal vectors. All these factors make the angiosperms eminently successful in the struggle for life.

The reproductive process in the angiosperms is somewhat more elaborate than that in the conifers. The mature plant represents the sporophyte generation. The flower contains pistils and stamens, which produce the female and the male spores, respectively (see Fig. 28.25a). The male spore (microspore) is produced in the anther, the top part of the stamen, and matures to become the pollen grain. The **pistil** is composed of three parts: the **stigma** (top), **style** (thin, neck-like region), and **ovary**. Within the ovary are *ovules*, where the female spore, the megaspore, is located. Surrounding the

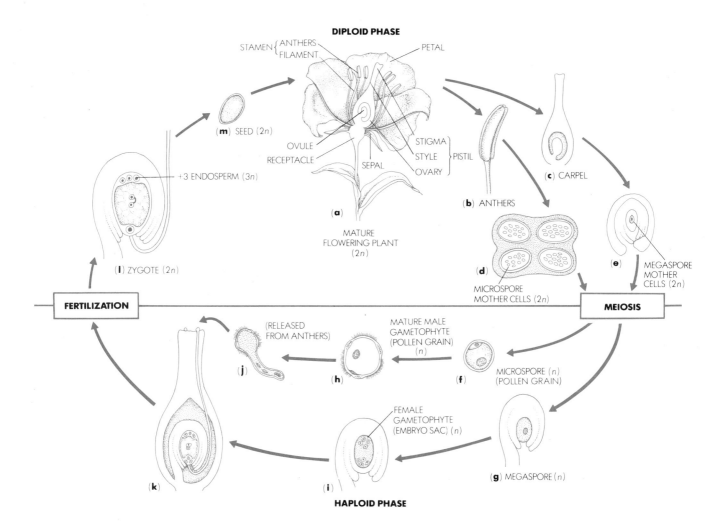

Fig. 28.26
The complete life cycle of an angiosperm (flowering plant). Note the much-reduced size of the male and female gametophytes, each of which consists of only a few cells completely dependent on the sporophyte (adult plant) for nourishment and protection (details are described in the text in conjunction with Figs. 28.25 and 28.27).

pistil are the more familiar flower parts, the **petals**, outside of which are the **sepals**. Collectively, the sepals are known as the **calyx**, and the petals as the **corolla**. Both sepals and petals are leaves that have become modifed over evolutionary history to form protective coverings for the male and female sporangia (anthers and ovary), respectively.

Within the ovary, each ovule contains one megaspore mother cell. Through one meiotic division, the megaspore mother cell forms four haploid megaspores; three of these disintegrate, the one remaining undergoing several mitotic divisions to produce the seven-celled female gametophyte (Fig. 28.25b). One of the cells of the female gametophyte is much larger than the others, and contains two nuclei (both haploid), the polar nuclei. One of the smaller cells is the egg cell. Note that the female gametophyte is completely dependent on the sporophyte plant for nourishment and protection; it has no independent life of its own.

Within the anther, haploid microspores develop by meiosis from diploid microspore mother cells. As the microspores mature into pollen grains, their cell walls thicken, and the nucleus within each cell undergoes mitosis, producing a tube nucleus and a generative nucleus (Fig. 28.5b).* This binucleate structure is the male gametophyte. As we saw in the life cycle of the conifer, the generative nucleus will divide mitotically on germination, producing two generative nuclei. In the flowering plant *both* of these nuclei are involved in fertilization. Note that in angiosperms the male gametophyte is even smaller than in conifers (two as opposed to six nuclei).

---

*The generative nucleus is surrounded by a plasma membrane, and is thus technically a cell, though it has virtually no cytoplasm. We will continue to refer to it as a nucleus.

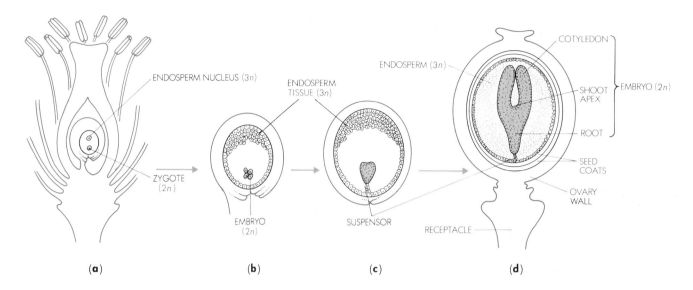

Fig. 28.27
Development of the fruit in angiosperms. The endosperm is tissue that develops from the fertilization of the two polar nuclei by one of the two sperm (generative) nuclei. Cells of endosperm tissue are thus triploid. The embryo itself develops from the fertilized egg cell, and is therefore diploid. It is attached to the wall of the female gametophyte. The embryo consists of a small root, a shoot apex (protomeristematic tissue), and food leaves (one or two, depending on whether the plant belongs to the subclass Monocotyledonae or Dicotyledonae, respectively). The food leaves are for food storage; they also produce enzymes that digest the food reserves in the endosperm. The seed coat develops from the female gametophyte and is surrounded by the thickened ovary wall. Together, the seed and thickened ovary wall produce the fruit.

Let us now follow the life cycle of an angiosperm as outlined in Fig. 28.26. Pollen is carried by the wind, insects, or other means to the stigma, where it is held by a sticky secretion (stages a–j). A pollen tube begins to form, growing down the style into the ovary (stage k). A tube nucleus leads the way, followed by two sperm nuclei. On reaching the ovule, the contents of the pollen tube are discharged into the female gametophyte. One sperm nucleus unites with the egg to produce the zygote, while the other unites with the two polar nuclei to form the beginning of the endosperm tissue (stage l). Endosperm tissue is triploid, as the result of the fusion of *three* nuclei: two polar nuclei and one sperm or generative nucleus. The endosperm tissue will reproduce and provide the stored food for the young embryo.

After fertilization, the development of the zygote and of the female gametophyte contributes to formation of the seed (see Fig. 28.27). The ovary wall begins to grow and, as its cells increase in number, becomes filled with sugary and starchy materials. The seed plus the ripened ovary wall form the fruit. An apple's flesh develops from the floral tube. For comparison to the other tracheophytes we have examined, the complete life cycle of an angiosperm is shown in Fig. 28.27.

The fleshy part of a fruit does not serve as food for the young seed. But the formation of a fruit has two evolutionary advantages over the naked seed of the conifers. The first is that the fruit offers a food enticement to animals. This aids in dispersal of the seeds over a wider geographic area. Second, the fruit offers some protection to the seed until it is mature. The flesh of such fruits as the apple or pear, for example, does not ripen until the seed is mature. Animals are thus discouraged from eating such fruit until the seeds are mature enough to germinate. In both these ways, the angiosperms ensure a greater number of surviving offspring.

The angiosperms thus seem to represent the most successful terrestrial plants. They show adaptations leading to greater efficiency in water retention, sturdiness, conduction, and exposure to sunlight, as well as remarkable adaptations in distribution of pollen and seeds. The intricate adaptations of flowers to insects and vice versa has long excited the wonder of naturalists. One such remarkable adaptation is shown by the flower *Salvia* (a member of the mint family) to honeybees (see Fig. 28.28).

28.9
EVOLUTION OF PLANTS

In this chapter we have surveyed the major groups of the kingdom Plantae. Along the way we have made brief mention of some possible evolutionary relationships. Let us examine, in conclusion, some of the broad generalizations that can be drawn about the origins and possible evolutionary relationships among the various groups of plants on earth today.

Once eukaryotic cells arose, they diversified into the Protists and the algal plants. The first step in that

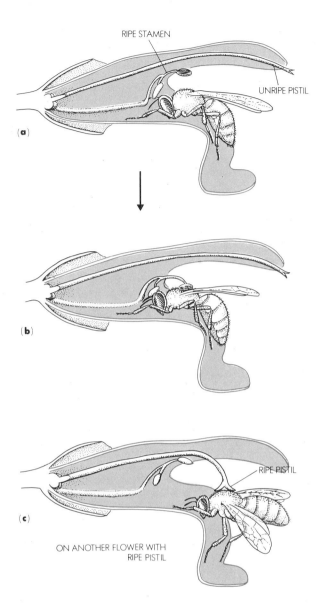

RIPE STAMEN

UNRIPE PISTIL

(a)

(b)

RIPE PISTIL

(c)

ON ANOTHER FLOWER WITH
RIPE PISTIL

Fig. 28.28
Pollination in *Salvia*. Wide distribution of pollen and cross-fertilization are ensured by: (*a*) maturing of the stamen before the pistil; (*b*) "dusting" of pollen onto the bee by a trigger mechanism; and (*c*) subsequent growth of the pistil so that it brushes the backs of the bees who enter the flower a few days later. Pollination is also carried out.

diversification was the adaptive radiation of the algae. From the early eukaryotes evolved the lines leading to modern-day Chlorophyta, Pyrrophyta, Euglenophyta, Chrysophyta, and Phaeophyta. Of these five major groups only one, the Chlorophyta, seems likely to have given rise to the other plant groups. Evidence for this comes from studies of cell structure (especially chloroplast structure), types of photosynthetic pigments, and the biochemistry of food storage. The filamentous green algae may have been the first forms to make any inva-

sion of the land, living in tidal zones where they grew along the moist soil and were periodically inundated by water.

The fossil record clearly shows that the first plants to develop a true land existence were the tracheophytes. Their success depended on two general factors: the development of multicellularity and, as a consequence, the opportunity for cell specialization. Both of these factors aided plants in adapting to the harsh effects of a terrestrial habitat.

Several trends can be noted in surveying the spread of plant life over the face of the earth, all deriving from the ability of plant cells to specialize. One is the development of structures and processes for obtaining, retaining, and transporting water. Protective cell walls and the cuticle that covered the outer layers of the plant body helped reduce excess evaporation. Specialized organs such as roots brought increased amounts of water into the plant and also anchored the plant. And finally vascular tissue allowed the rapid movement of water and dissolved materials throughout the plant body. Indeed, one important trend we have observed in the Tracheophyta, from the Pteridopsida through the angiosperms, is the increase in amount of xylem and phloem tissue. Along with that trend came an increase in vascular tissue specialization: from the simple tracheids present in the older groups of tracheophytes (such as the psilopsids), to the more complex tracheids and vessels found in the angiosperms. All of these anatomical developments helped the plant transport and retain a high concentration of water around its individual cells.

At the same time, the development of vascular tissue helped provide the plant body with support, so that it was not confined, as its algal ancestors had been, to lying flat on the ground. As the early tracheophytes began to stand more erect, they competed more successfully for sunlight. Selective pressure undoubtedly gave an advantage to the forms that stood a little higher or remained standing a little longer than their neighbors. A firm and sturdy plant body became one important adaptation to life on land.

Perhaps the most significant feature of plant evolution on land is the trend among the tracheophytes toward a reduction in the size and duration of the haploid (gametophyte) phase and a corresponding increase in the size and duration of the diploid (sporophyte) phase of the life cycle. This trend is observable only within the tracheophytes; it is not true of groups that continued their evolution within the water (that is, the algae), or those that developed only a partial adaptation to terrestrial life (such as the bryophytes). Since the gametophyte (haploid) phase of the tracheophyte life cycle is concerned directly with fertilization, its reduction in size and retention as part of the sporophyte plant body helps effect union of egg and sperm cells. Moreover, retention of the female gametophyte, especially, on the mature sporophyte, reduces the dependency of the fertilization process on water (through which sperm must otherwise swim).

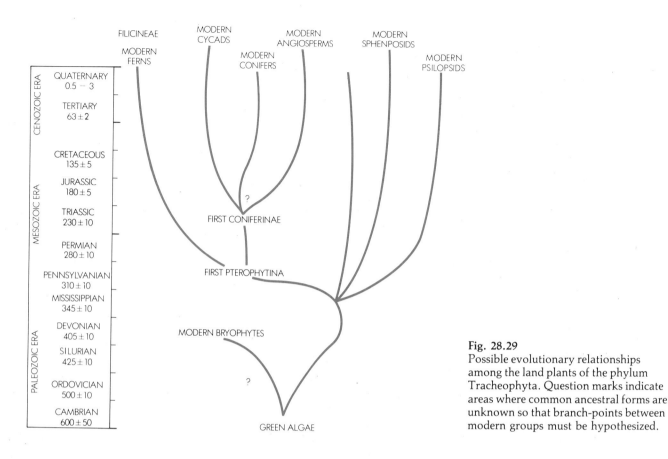

Fig. 28.29
Possible evolutionary relationships among the land plants of the phylum Tracheophyta. Question marks indicate areas where common ancestral forms are unknown so that branch-points between modern groups must be hypothesized.

Once primitive tracheophytes became established on land, they diverged in several directions, as shown by the simplified phylogenetic chart in Fig. 28.29. The modern ferns and seed plants may have shared a common ancestor somewhere around the beginning of the Triassic period, about 230 million years ago. Later, the seed plants diverged into several lines, leading on the one hand to the cycads and on the other to the conifers and flowering plants. The latter two diverged from one another sometime during the Cretaceous period, about 135 million years ago. Among the flowering plants, evidence suggests that the dicots evolved first, with the monocots developing from them.

Summary

As in the preceding chapter, since the major characteristics of each group of organisms are summarized in the tables, a listed summary would be repetitive here. The numbered items below will emphasize only major generalizations about the relationships among living groups today, or their evolutionary history.

1.   Eukaryotic organisms, such as the one-celled green alga *Chlamydomonas*, possibly originated from prokaryotic forms. Chlamydomonas is photosynthetic, having true chloroplasts and mitochondria, and yet moves about rapidly by means of flagella. A very primitive eukaryote, *Micromonas*, is much smaller than *Chlamydomonas* and may represent an intermediate form between prokaryotic and eukaryotic cells.

2.   Multicellular plants were undoubtedly favored by natural selection over purely unicellular forms, especially as plants moved onto the land. As in animals, multicellularity allows for the development of specialized tissues and organs. Multicellularity may have originated by the sticking together of daughter cells after division. Such a process would by itself produce nothing more than a string or clump of similar cells, such as is found in the filamentous green or blue-green algae of today. Even in such simple systems, some specialization of cells can be observed. Certain cells in a filament of green algae can differentiate and become the reproductive cells. As long as cell groups adhere together, specialization is possible.

3.   Differentiation of filaments or even single cells into two mating types (primitive "sexes") can be observed in many forms of algae. In *Chlamydomonas* two seemingly identical cells (of different mating types) fuse to form a zygote. This is isogamy. In the green alga *Spirogyra* two filaments mate (conjugate) by lying side-by-side and transfer-

ring the contents of cells across a bridge into the cells in the opposite strand. In such cases one strand is the donor, the other the recipient. This is the beginning of heterogamy. In the green alga *Oedogonium,* male and female filaments produce distinct sperm and egg cells, thus showing complete heterogamy.

4. The evolutionary advantage of differentiation into mating types, or sexes, is that it provides a mechanism for exchange of genetic material and for genetic recombination. Mixing of characters provides new combinations of traits that can be adaptive, and thus favored by natural selection.

5. Plants first invaded the land approximately 400 million years ago, probably beginning with a clustering of forms around the shores or in tidal regions. The land offered many advantages: less crowding, a rich supply of oxygen, more direct sunlight for photosynthesis. New hazards also placed survival value on adaptations such as: structures to prevent drying out (desiccation) and supporting tissues to raise the plant body above the ground, and motility.

6. Green algae were probably the first to invade land and thus gave rise to all future land-dwelling plants. All land plants use the same basic forms of chlorophyll as green algae (chlorophylls *a* and *b*) and store reserve food as starch.

7. The earliest multicellular land plants were probably very primitive forms resembling modern-day horsetails or club mosses. There was probably little differentiation of these forms into true leaves, stems, or roots. They did appear to have some specialized branching structures that attached them into the ground and specialized fruiting bodies containing gamete-producing cells. Presence of thick cuticle may have prevented excessive water loss. Fossil evidence suggests that such plants had simple vascular (conducting) tissues (tracheids) in the central part of the plant body. The origin of vascular tissues probably came very early in the evolution of land plants.

8. The evolution of specialized reproductive organs demonstrates the adaptation to land by plants. In water, male gametes can swim to female gametes for fertilization. On land, fertilization is not so easily accomplished. Modern-day mosses suggest how early land plants may have begun to solve this problem.

9. The mosses show alternation of generations: a gametophyte generation that produces gametes alternates with a sporophyte generation producing spores. The leafy portion of mosses is the gametophyte generation. Gametophytes produce male gametes (sperm) in specialized organs, antheridia, and female gametes (eggs) in archegonia. The sperm swims in a film of water along the surface of the gametophyte into the archegonium. The zygote develops into a diploid sporophyte. Within the cap, meiosis occurs, producing haploid spores. When ripe the spores are distributed by the wind or water. Each spore grows into a gametophyte, completing the life cycle. Alternation of generations means, among other things, the alternation of a diploid with a haploid generation to complete one life cycle.

10. Alternation of generations can be seen in the life cycles of all major land plants, including modern flowering plants. In general, the evolutionary trend has involved increasing dependence of the female gametophyte on the sporophyte, thus reversing the situation found in mosses, where the dominant stage was the haploid gametophyte. In modern-day pine trees, for example, the tree itself is the diploid sporophyte generation; the gametophyte generation is reduced to a small group of cells within the male or female cones.

11. In the evolution of plants, the trend toward increasing prominence of the sporophyte and decreasing prominence of the gametophyte generation has the advantage of greater protection for the gamete-producing organs.

Exercises

1. In this chapter it was noted that plant evolution showed a trend from isogamy to heterogamy. What adaptive value might heterogamy have over isogamy?

2. Compare the life cycle of an alga such as *Chlamydomonas* with that of a fern. What evolutionary developments are present? Compare both of these with the life cycle of a gymnosperm. What trends are established in the sequence from alga to conifer.

3. What is the possible advantage to a plant of having the diploid generation as the dominant portion in its life cycle?

4. What evolutionary advantages to angiosperms are offered by the formation of (a) flowers and (b) fruits?

5. Darwin was one of the first to try to explain the intricate adaptations of some insects and flowers for the purpose of pollination (co-evolution). Explain how natural selection could give rise to such well-adapted and ingenious modifications on the part of both the flower and the insect (as in the case of *Salvia*).

Chapter 29
The Diversity of Life III:
Animals

INTRODUCTION

The kingdom Animalia includes forms that are multicellular heterotrophs, lack chlorophyll, are capable of motion, and show some degree of cell specialization. Most animals reproduce mainly by sexual means, though some reproduce by the asexual means as well. Most digest their food in an internal digestive tract or cavity, and store food as both glycogen and fat. The difference in nutritional modes between the plant and animal kingdoms has led to enormous differences in overall lifestyles. Plants have evolved a nonmotile existence. They anchor themselves in one place and spend their lives photosynthesizing. They have no need to move. Animals, on the other hand, do not have such an abundant source of usable energy (the sun) immediately at hand. They must seek out their energy sources (food), not in the form of energy itself, but as organic matter—usually plants or other animals. This means they must move about (sometimes quite quickly, if their potential food is itself a fast mover), and possess special mechanical and chemical adaptations (such as a digestive system) for making the material they consume available to their own cells. The heterotrophic lifestyle has demanded a far greater number of special adaptations than the autotrophic style. For this reason the animals, and particularly the higher animals, are the most complex and highly specialized of all organisms.

The animal kingdom is generally divided into approximately twenty-eight to thirty phlya, falling into two major groupings:

1. Parazoa—Animals without organized tissues and no digestive cavity (sponges)

2. Metazoa—Animals with organized tissues and a digestive cavity
   a) Radiata: Animals with radial symmetry (the jellyfish; and *Hydra*)
   b) Protostomia: Animals in which the mouth, and sometimes the anus, develops from the blastopore of the embryo (annelids, molluscs, and arthropods)

c) Deuterostomia: Animals in which only the anus develops from the blastopore, while the mouth originates from an opposite sphere of the embryo (echinoderms and chordates)

In this chapter we will survey the major phyla composing these two large groupings of animals. Our main emphasis will be on structural, functional, and ecological characteristics of the various phyla, but, as with the plants, a major concern also will be possible phylogenetic relationships.

29.2
SOME MAJOR FEATURES OF
ANIMAL ORGANIZATION

There are several important features of the animal world—particularly body plan and its embryonic development—which will be important to keep in mind as we survey (compare and contrast) various phyla of the animal kingdom.

The Presence of Germ Layers

Recall from Chapter 20 that the bodies of most animals are composed of at least two (and in most cases, three) germ layers. Germ layers are tissues in the animal body that are laid down early in embryonic development and out of which major parts of the adult body are formed. In all animals more complex than the flatworms (Platyhelminthes) the body consists of three germ layers: the ectoderm (outer layer), mesoderm (middle layer) and the endoderm (inner layer). (For a fuller discussion of how these germ layers are formed in embryogenesis, and the organ systems to which they give rise, see Section 20.5.) It is important to recognize that the bodies of animals in almost all phyla of the animal kingdom are formed from these same three germ layers. The same basic development plan remains embedded in the embryological processes of the rest of the animal kingdom, the Metazoa; this is a suggestive feature of our common evolutionary origins.

Body Symmetry

Another important feature of the various animal phyla is the kind of symmetry their body plans show. Symmetry refers to the relative size, shape, and distribution of parts that are on opposite sides of a line that can divide any object into halves. The two most common forms of symmetry exhibited by animals are shown in Fig. 29.1: **bilateral** and **radial symmetry**. Bilateral symmetry occurs when an object (in this case an animal body) can be divided in only one way—by a single plane of division—into two equal halves that are mirror-images of one another. Radial symmetry occurs when an object can be divided into two equal halves (that are also mirror-images) by any number of planes of division along the major axis. All animals except the sponges have one or the other form of symmetry to some degree (not always easily recognizable, however). For the most part, the higher one goes in the animal kingdom, the more bilateral symmetry prevails.

Coelom

A third important feature of animal form is the presence or absence of a **coelom**. A coelom is technically defined as a body cavity lined on all sides by epithelial tissue derived from the mesoderm (middle germ layer, as shown in Fig. 29.2). In most animals with a true coelom, the cavity houses the visceral organs. The animal kingdom

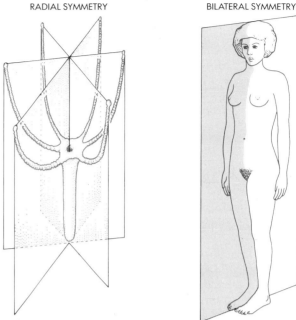

RADIAL SYMMETRY   BILATERAL SYMMETRY

Fig. 29.1
Radial and bilateral symmetry. Bilateral symmetry is characteristic of animals that move about actively: such forms have a right and left side, a front and a back. Animals with radial symmetry are more likely than not to be sessile, spending their life anchored to one place; at best, they move about very slowly.

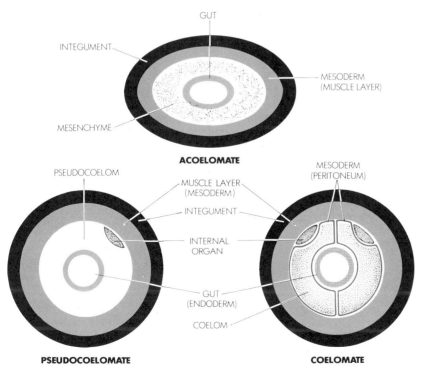

Fig. 29.2
The three major forms of body cavity in the animal kingdom (shown in cross section). Acoelomates have no true coelom—for example, sponges, cnidarians, and flatworms. Pseudocoelomates such as Rotifers and nematodes have a body cavity, called a pseudocoelom, lying between a layer of mesoderm and the gut—but the gut itself is not surrounded by mesodermal tissue. Cnidarians—all higher animals from the molluscs upward—have a true coelom. Mesodermal tissue is shown in color.

can be divided into the "acoelomate" phyla, that is, those lacking a true coelom (for example, the Porifera or sponges, the Cnidaria, and the Platyhelminthes or flatworms); the "pseudocoelomate" phyla, which have a body cavity (pseudocoelom) surrounded on only one side by mesoderm (such as the Rotifera and Nematoda or roundworms); and a very large group of true "coelomate" phyla, which have a true coelom. The coelom provides a large, well-protected cavity in the animal body in which organs and organ systems can be suspended.

Segmentation

A fourth feature of animal organization is the presence or absence of **segmentation**. Segments are definable anatomical units that repeat one another for some portion (or all) of an animal's body length. An earthworm is a good example of a completely segmented animal. From the anatomical and evolutionary point of view segmentation is important because each segment is a duplicate of others, having the same basic structural plan, and sometimes even containing the same internal structures. The term **metamerism** is used to refer to segments that are linear repetitions of one another, with no significant modification or specialization. Segmentation is also important in the evolution of animals. Many organisms appear to have evolved by adding new segments to their bodies. Once a basic external and internal anatomical plan has been established, organisms can evolve greater size and/or degrees of specialization by repeating those body units in a linear fashion. Segmentation is found primarily in the annelid, arthropod, and chordate phyla.

Embryonic Cleavage Patterns

Several features of animal organization are embryonic in nature. Two refer to patterns of early cleavage, determinate or indeterminate, and spiral or radial; a third re-

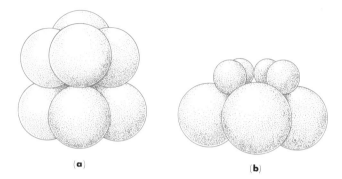

Fig. 29.3
Radial (*a*) and spiral (*b*) cleavage patterns. In radial cleavage the four cells on the top layer (after second cleavage) rest directly on top of the four cells below. In spiral cleavage, the four cells on the top layer rest in the angles between the four cells below. (After William Keeton, *Biology*, 3d ed. W. W. Norton, 1980, p. 1016.)

fers to the way in which mesoderm is formed in the early gastrula.

All animals show either **determinate** or **indeterminate** cleavage. In embryos where the fate of each cleavage product (blastomeres) can be predicted, the cleavage is said to be determinate. Determinate cleavage is found in all the Radiata and Protostomia (Cnidarians, Platyhelminthes, Aschelminthes, Molluska, Annelida, and Arthropoda). In embryos where the fate of individual cells is not fixed from early cleavage onward, the cleavage pattern is said to be indeterminate. Indeterminate cleavage is found in the Deuterostomia (Echinoderms, Hemichordata, and Chordata).

All animals also show one of two geometric arrangements of the cell products of the first two embryonic cleavages. As shown in Fig. 29.3(a), one arrangement is when the cells of the top layer (following second cleavage) rests directly on top of the cells of the bottom layer. This is known as **radial** cleavage. A second type of cleavage is when the top layer of cells rests in the

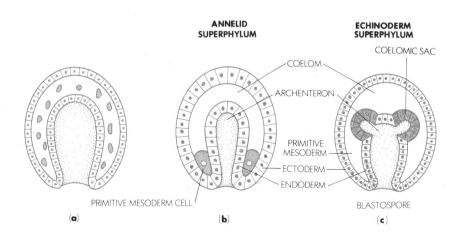

Fig. 29.4
Patterns of mesoderm formation in the animal kingdom. (*a*) Radiata: mesoderm forms from inward migration of ectodermal cells; (*b*) Protostomia: mesoderm originates from several cells near the juncture of the endoderm and ectoderm in the blastopore region; (*c*) Deuterostomia: mesoderm originates from pockets of cells from the archenteron. (After Ralph W. Buchsbaum, *Animals without Backbones.* Chicago: University of Chicago Press, 1948, p. 341.)

angles between the cells of the lower layer. This is known as **spiral** cleavage (Fig. 29.3b). Radial cleavage is characteristic of the deuterostomes, while spiral cleavage is characteristic of the protostomes.

Within the animal kingdom three basic patterns of mesoderm formation can be found. These are shown in Fig. 29.4. In the Radiata, mesoderm is formed in the gastrula by the inward migration of ectodermal cells (Fig. 29.4a). By contrast, in the protostomes and deuterostomes, mesoderm forms from endoderm, not ectoderm. In most protostomes, mesoderm originates from several endodermal cells located near the blastopore, where the infolded endoderm joins the ectoderm (Fig. 29.4b); in others, however, the mesoderm originates from pouches of ectoderm located at the far end of the archenteron from the blastopore (Fig. 29.4c).

Why is this information important? Biologists assume that similar patterns of development, especially in the early stages of embryogenesis, suggest certain broad phylogenetic relationships. On this assumption, embryological evidence is used to construct phylogenetic and taxonomic arrangements of the animal phyla.

Parazoa and Radiata

29.3
THE PHYLA PORIFERA (SPONGES) AND CNIDARIA

The Phylum Porifera

Porifera, also called sponges, are the simplest multicellular animals known (Fig. 29.5a). They spend their lives anchored to rock or other solid surface beneath the water. Some are freshwater, but most are marine. The sponge body consists of numerous small pores through which water is drawn into a central cavity called the **spongocoel** (see Fig. 29.5c). The sponge body maintains its shape by the presence of a "skeleton" of typically silicon- or calcium-containing (calcareous) structures known as **spicules.** These are branched or unbranched structures secreted by special cells in the middle jelly layer of the sponge wall. Some sponges do not produce calcareous spicules, but are supported by a network of tough fibers. These sponges, which come from shallow tropical seas, are harvested for commercial purposes (and the general household word "sponge" is now derived from them).

The sponge body is divided into two cell layers, separated by the mesoglea. The outer layer is made up of what are called **dermal amoebocytes,** relatively unspecialized cells that form a covering surface. The inner layer is composed of special collar cells, **choanocytes,** which have long flagella. The motion of these flagella creates a current that pulls water into the spongocoel through the pores, and forces it out through the **osculum** at the top. Organic material, small microorganisms, and algae are brought into the spongeocoel by the water currents and are engulfed by choanocytes.

Sponges reproduce both sexually and asexually. In asexual reproduction, mature sponges produce a small outgrowth from one side of their body, near the base. This outgrowth, called a **bud,** eventually breaks off and takes up a separate existence. In sexual reproduction, unspecialized cells, the **amoebocytes** in the mesoglea, produce eggs and sperm within a single individual. Sperm are released by one sponge and captured by another. A unique pattern of fertilization and early development lead to the production of a ciliated, free-swimming larva that escapes from the parent sponge, attaches to a new surface, and eventually grows into a new sponge colony. Freshwater sponges reproduce asexually by forming **gemmules,** aggregations of amoeba-like cells that are covered by a hard, outer coating. Gemmules can winter over, an obvious adaptation to the harsher conditions of freshwater (where water temperatures change drastically from summer to winter).

Sponges may have evolved from collared flagellates (Protists), since collar cells are found nowhere in the animal kingdom except in the Porifera. It is possible also that sponges evolved entirely separately from the rest of the animal kingdom. Fossil evidence indicates that sponges were among the earliest forms of animal life on the planet. Most biologists favor the hypothesis that the sponges originated very early from a protistan ancestor, but an ancestor different from the one (or ones) that gave rise to the rest of the animal kingdom.

The Phylum Cnidaria (Formerly Coelenterata)

Cnidarians are all radially symmetrical animals, with a body wall consisting of two layers, an outer **epidermis** and an inner **gastrodermis,** separated by a **mesoglea** similar to that found in sponges (see Fig. 29.6). Jellyfish, Portuguese man-of-war, and the freshwater *Hydra* are all representatives of the phylum Cnidaria. All members of this phylum have specialized stinging cells, **cnidoblasts,** containing a coiled-up, poison-filled barbed thread, the **nematocyst.** When the trigger of the cnidoblast is touched, the nematocyst is shot forth, much like a harpoon. It is used for paralyzing and then trapping prey, as well as for defense. As anyone who has ever encountered a Portuguese man-of-war can attest, the toxin produced by a multitude of nematocysts can be very painful, even to a large animal like a human being.

Cnidaria have two general body forms, the **polyp** and **medusa** (see Fig. 29.7a), which can occur during their life cycle. The polyp, characterized by *Hydra,* is an elongated cylinder like a soft drink bottle with tentacles around the top. The medusa has basically the same form, but is more flattened out and often has a larger amount of mesoglea. Also, polyps attach onto substrates with their tentacles up, while medusae float freely in the water with their tentacles down. The jelly

PORIFERA

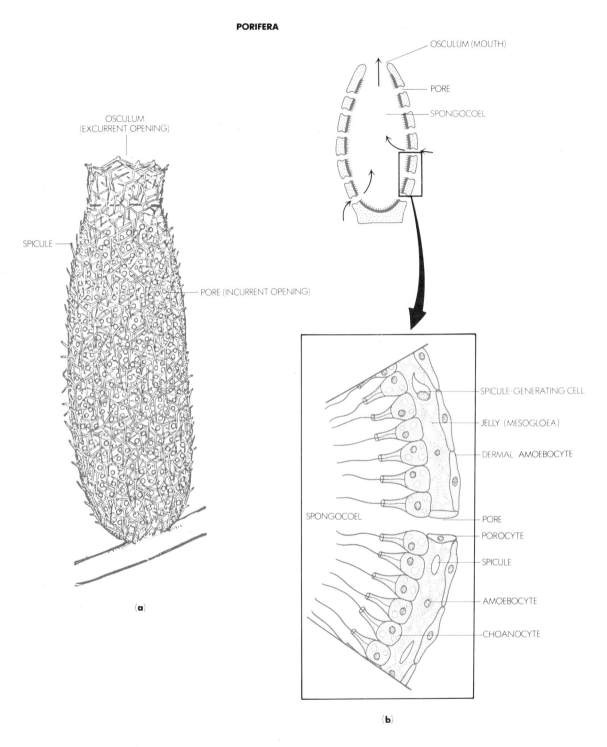

Fig. 29.5
The phylum Porifera (sponges). (*a*) A sponge (simple sac type) showing water circulation pores, and hard, calcareous spicules. Water flows *in* through the pores and *out* through the osculum. (*b*, top) Schematic view of the ascon type sponge, showing pattern of water flow. Not all sponges have exactly this body plan but all have some variation of it. (*b*, bottom) Detail of wall of a simple sponge, showing the various types of cells. The inner body layer is made up of choanocytes, or collar cells, whose flagella pull water into the spongocoel and force it out the osculum.

CNIDARIA

(a)

(b)

(c)

(d)

(e)

(f)

◄ Fig. 29.6

Representatives of the phylum Cnidaria. (*a*) Freshwater *Hydra* (class: Hydrozoa) showing basic polyp form, with tentacles (prey object, the small crustacean *Daphnia*, at lower right). The larger parent *Hydra* is shown with a young polyp on the left. (Photo, Carolina Biological Supply Company.) (*b*) Small portion of a colony of the marine colonial hydrozoan, *Campanularia flexuosa*, showing several developmental stages of the polyps. A full colony forms a kind of "bush" structure. (Photo, Charles R. Wyttenbach, University of Kansas, Biological Photo Service.) (*c*) Portuguese man-of-war, a colonial hydrozoan composed of both polyps and medusae, with an air-filled sac on top for flotation. Contraction of fibrils in the float expels air and the colony sinks; excretion of air into the sac causes the colony to rise. Long nematocyst-containing tentacles (up to 15 meters) can paralyze fish, which are then drawn up toward the feeding polyps (a captured fish is shown here). (Photo, Runk/Schoenberger, Grant Heilman Photography.) (*d*) True jellyfish, *Chrysora Quinquecirrha* (class Scyphozoa), showing the basic medusa form (bell and tentacles) that predominates in members of this class. The long, lacy structure in the center is the manubrium, a tube with extensive arms and a mouth at the tip, used in feeding. (Photo, Runk/Schoenberger, Grant Heilman Photography.) (*e*) Corals (colonial Anthozoans) showing (left) branching calcareous skeleton of the fire coral *Millepora alciornis* with small, delicate polyps protruding, and (right) close-up of polyps of the coral *Montastrea cavernosa* with polyps expanded to capture small drifting microscopic animals (zoöplankton). Polyps often remain tucked in cup-shaped depressions on the calcareous skeleton during the day to avoid predation, but are fully extended at night for feeding. (Photo, James W. Porter, University of Georgia, Biological Photo Service.) (*f*) Sea anemones are polyp-shaped forms that generally have no medusa stage in their life cycles. Unlike the simple *Hydra* polyp, sea anemones have a complex set of internal compartments and a tubular pharynx that leads into a compartmentalized gastrovascular cavity. (Photo, Carolina Biological Supply Company.)

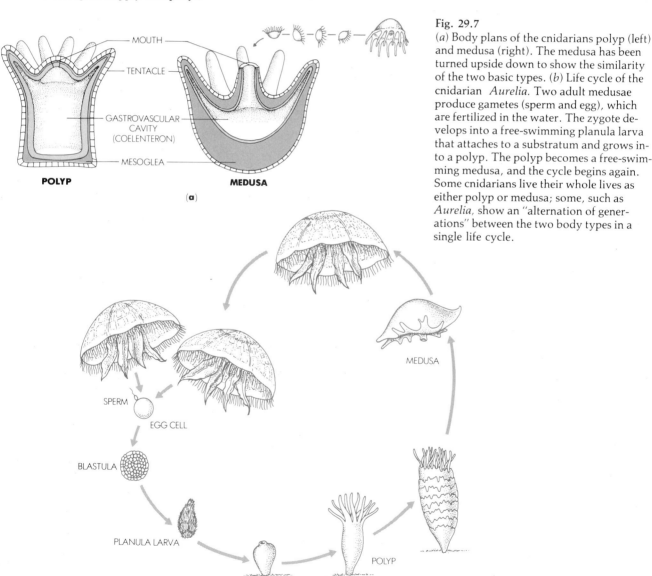

Fig. 29.7

(*a*) Body plans of the cnidarians polyp (left) and medusa (right). The medusa has been turned upside down to show the similarity of the two basic types. (*b*) Life cycle of the cnidarian *Aurelia*. Two adult medusae produce gametes (sperm and egg), which are fertilized in the water. The zygote develops into a free-swimming planula larva that attaches to a substratum and grows into a polyp. The polyp becomes a free-swimming medusa, and the cycle begins again. Some cnidarians live their whole lives as either polyp or medusa; some, such as *Aurelia*, show an "alternation of generations" between the two body types in a single life cycle.

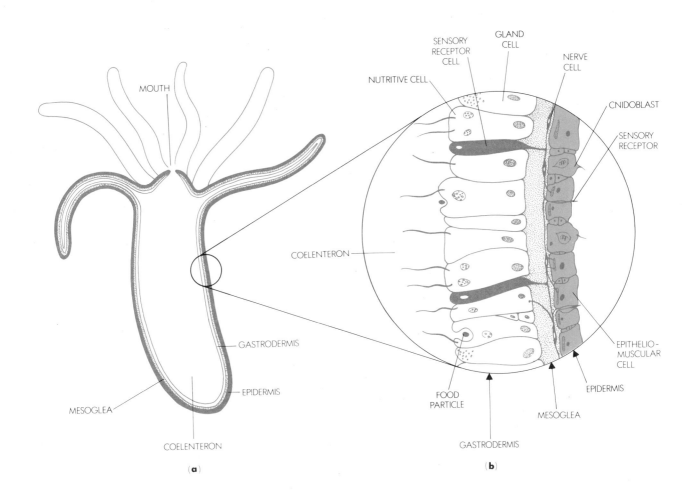

(a)

(b)

Fig. 29.8
Cnidaria such as the freshwater *Hydra* (a), showing basic body structure, tentacles, and internal digestive cavity, the coelenteron. (b) Detail of the body wall of *Hydra*, showing several cell types. The inner layer, or endoderm, is composed of sensory receptors, gland cells, and nutritive cells; the outer layer or ectoderm consists of sensory receptors, epitheliomuscular cells and cnidoblasts (sting cells). Between the two layers is a layer of jelly, the mesoglea.

fish is an example of the medusa form of cniderians —the "jelly" being mesoglea.

The body structure of both the medusa and polyp stages of Cnidaria consists of a hollow cylinder with a single opening at one end, the mouth (Fig. 29.8). Unlike the sponge colony, this one opening in the cnidarians serves both for intake and expulsion of water and food material. All cniderians have tentacles of some form surrounding the mouth. Cnidarians are predators and use their tentacles, heavily equipped with cnidoblasts, to capture prey. When a prey is paralyzed, the tentacles close in around it and force it into the mouth. From there the food passes into the central cavity, the **gastrovascular cavity** or **coelenteron**. Here it is digested extracellularly by enzymes poured into the cavity from special **gland cells.** When digestion is completed, the molecular breakdown products can be used directly or transported to other cells in the animal body.

Cnidarians show considerable cell specialization. Figure 29.8(b) shows the variety of cells that can be found in the body wall of a typical Cnidaria, such as *Hydra*. At least six different types of cells are visible. Sensory (receptor) cells allow the organisms to receive stimuli, and nerve cells make coordinated behavior possible. Movement is accomplished by epitheliomuscu-

lar cells whose contraction constricts the organism around its diameter. Nutritive cells have flagella to keep the water moving within the gastrovascular cavity and also engulf food by forming pseudopods. As mentioned above, gland cells produce digestive enzymes that are poured into the gastrovascular cavity.

Reproduction in the Cnidaria is both asexual and sexual. In asexual reproduction, germ cells in certain regions of the body begin mitotic division and produce a small **bud** on the side. The bud eventually forms into a small organism whose gastrovascular cavity is continuous with that of the parent. Eventually the bud breaks off and becomes an independent organism.

Sexual reproduction in Cnidaria is more complex, often involving a multistage life cycle with alternation of a polyp with a medusa generation. The life cycle of one species of Cnidaria, the jellyfish *Aurelia*, is shown in Fig. 29.8(b). The medusa stage is the sexual stage, with the sexes separate in this species (some species are hermaphroditic, however, with one medusa producing both eggs and sperm). Sperm are released directly into the water; eggs remain attached to the female medusa, where fertilization takes place. After development has proceeded to the gastrula stage, the embryo separates from the parent and develops into a free-swimming ciliated larva the **planula.** The planula eventually settles down, attaches to some object, and develops into a polyp. The polyp produces one or more medusae from its top, beginning the life cycle again. This life cycle may be described as a phenotypic alternation of generations (polyp form alternating with medusa), but is not a genetic alternation of generations (haploid with diploid) as found in plants. Cells of both the polyp and medusa stages of the cnidarian life cycle are diploid.

The phylum Cnidaria is divided into three classes: the Hydrozoa, the Scyphozoa, and the Anthozoa (see Fig. 29.6). The Hydrozoa include the freshwater *Hydra*, a solitary organism with only a polyp stage in its life cycle (Fig. 29.6a) and the colonial form represented by the Portuguese-Man-of-War (Fig. 29.6c). The Scyphozoa are a group of colonial forms in which several kinds of cells live in close, to some extent dependent, association with one another (such as the genus *Obelia*, which has both a polyp and medusa stage; Fig. 29.6b). The Scyphozoa include the true jellyfish (such as *Aurelia*; Fig. 29.6d). In the Scyphozoans the medusa stage of the life cycle tends to predominate, though a polyp stage, much reduced in size, occurs in all members of the class. The Anthozoa include strictly polyp forms such as the sea anemones and corals (Fig. 29.6e and f).

Some Anthozoans are solitary (sea anemone) while others are colonial (corals). The corals are among the most beautiful and intriguing members of the phylum Cnidaria. Coral polyps secrete a carbonate skeleton within which they live. As these skeletons accumulate, they build up **coral reefs,** huge masses of intricately branching calcareous skeletons. The Great Barrier Reef along the eastern coast of Australia is a coral reef 2000 kilometers long. The Marshall Islands in the Pacific are also coral reefs. Charles Darwin wrote an entire book on coral reefs and proposed a theory for how their various shapes are determined. Corals are ecologically important to the marine ecosystem. They generally live in symbiotic relationship with green algae, for which they provide protection and a substrate on which they can grow, and from which they obtain oxygen and nutrients. Such algal-coral associations are often net photosynthetic producers; that is, despite what the animals consume of the carbohydrate or oxygen produced by the algae, the system as a whole generates more than is used. In the marine environment this symbiosis contributes much to productivity and thus to the ocean's food web.

The Cnidaria are significant in the history of animal diversity because they represent an early stage in cell specialization and tissue formation. They also show the beginnings of organismic coordination—indeed, of behavior. Their ability to capture prey in an active, as opposed to a passive (such as filtering), way shows that this special form of heterotrophic nutrition was introduced early into animal evolution.

The Protostomia

29.4
THE PHYLUM PLATYHELMINTHES

The Platyhelminthes are a group of flat-bodied worms. In terms of overall body plan, they are the simplest animals to show bilateral symmetry, and a dorsal (top) and ventral (bottom) surface. Like most bilaterally symmetrical organisms, flatworms have a distinct "head" and "tail" region—that is, anterior and posterior ends. Both bilateral symmetry and the presence of the anterior-posterior axis are characteristic of organisms that move about. Typical flatworms are the free-living *Planaria* (Fig. 29.9a and b), and the parasitic flukes and tapeworms (Fig. 29.9c and d). There are three classes of Platyhelminthes: Turbellaria, the free-living form, which includes the planarians; Trematoda, parasitic flukes such as *Schistosoma* and *Fasciola;* and Cestoda, the parasitic tapeworms. Thus, of three classes, two are composed almost exclusively of internally parasitic forms (endoparasites).

Flatworms have the three distinct tissues layers—ectoderm, mesoderm, and endoderm—characteristic of all animals above the cnidarian level. Not only are these tissues specialized for various functions (ectoderm for protective skin, mesoderm for muscles and reproductive organs, and endoderm for lining and absorption), but two or more types of tissue cells can combine to form complex organs—for example, the muscular pharynx that projects from the ventral surface of the planarian *Dugesia* (Fig. 29.10a) is composed of ectoderm, mesoderm, and endoderm. Thus, we can see a distinct trend: Porifera are composed only of aggregations of cells in a kind of colonial association; the Cnidaria show only tissue levels of specialization; but Platyhelminthes show, for the first time in the animal kingdom, development of the organ level of complexity.

Class Turbellaria

The Turbellaria are freshwater flatworms, many of which are large enough to be visible with the naked eye. The planarian *Dugesia* is a common example of this

PLATYHELMINTHES

(a)

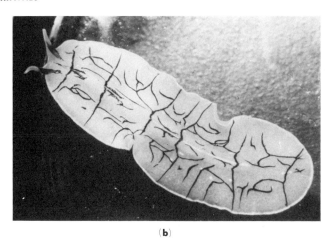

(b)

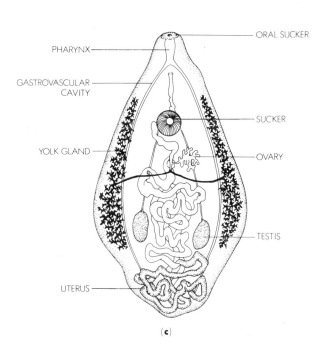

PHARYNX

GASTROVASCULAR CAVITY

YOLK GLAND

UTERUS

ORAL SUCKER

SUCKER

OVARY

TESTIS

(c)

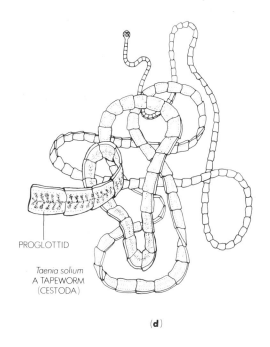

PROGLOTTID

Taenia solium
A TAPEWORM
(CESTODA)

(d)

Fig. 29.9
Representatives of the phylum Platyhelminthes. (*a* and *b*)
Class Turbellaria: (*a*) Freshwater planarian; (Photo, Carolina
Biological Supply Company.); (*b*) A marine polyclad,
Prosthecereus. (Photo, Biological Photo Service.) (*c*) Class
Trematoda, the parasitic fluke *Prosthogonimus* lives in the
oviducts of hens. (*d*) Class Cestoda, the tapeworms. Mature
tapeworm from human intestine (*Taenia solium*). This species
has the pig as its intermediary host. *See also* Figs. 29.11 and
29.12.

class. As can be seen in Fig. 29.10(b), the internal anatomy of *Dugesia* is considerably more advanced than
that of the cnidarians. Planarians have a prominent
nerve cord running down each side. There is a heavy

concentration of nerve cells at the anterior end
(**ganglion**), providing a simple coordination center.
Two **ocelli** or eye-spots are present; with their cross-
eyed appearance, the "eyes" have become one of the

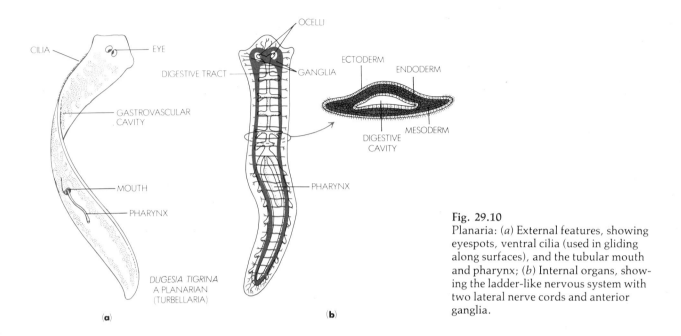

Fig. 29.10
Planaria: (*a*) External features, showing eyespots, ventral cilia (used in gliding along surfaces), and the tubular mouth and pharynx; (*b*) Internal organs, showing the ladder-like nervous system with two lateral nerve cords and anterior ganglia.

Planaria's most prominent and endearing traits. The eye-spots are light-sensitive but do not form images. The two nerve cords are connected by cross-bridges down the entire length, providing a kind of coordination that makes movement for the planarian easy and smooth. Coordinated beating of cilia along the ventral surface of the animal causes *Planaria* to glide along any substratum.

Free-living flatworms (such as the planarians) are carnivorous and ingest food through the muscular pharynx. A planarian will attach itself to its prey, or a dead piece of meat, by means of suction through the pharynx. By rapid muscular movements of the pharynx, combined with suction, the food source is torn into little shreds that are sucked into the internal digestive tract. Here the food particles are subject to phagocytosis by endodermal cells lining the digestive tract. Like the Cnidaria, the flatworms have only one opening (the mouth) to the gastrovascular cavity. Waste materials from metabolism and excess water are eliminated through a well-differentiated excretory system, consisting of a network of fine tubules that run the length of the animal's body and terminate in small openings, or pores, on the surface. Wastes are collected in these tubules and moved along toward the outside by the constant beating of special ciliated cells (flame cells)* located at side branches of the tubules.

Reproduction in free-living flatworms like planaria is both asexual and sexual. *Dugesia*, for example, has re-

markable powers of regeneration. A piece of tail can regenerate a whole new head region; and sometimes the animal divides longitudinally into two. The most common form of reproduction, however, is by sexual means. Planarians are hermaphroditic, both eggs and sperm being produced in the same individual. At mating, two planarians come together ventrally. Each inserts a muscular organ, the penis, into a hollow cavity, the copulatory sac, of the other. A vas deferens leads into each penis, and an oviduct into each copulatory sac. Sperm from the testes pass down the vas deferens, through the penis, and into the copulatory sac. Fertilization occurs internally. The embryo develops directly into a young flatworm, with no larval stage (as is typical of most freshwater species).

Class Trematoda (the Flukes)

The flukes are a group of largely endoparasitic (internal parasites) flatworms that are a source of human and animal disease. All the endoparasitic flatworms have an epidermal layer covered by a thick **cuticle**. Resistant to enzyme action, the cuticle is an adaptation to living within the body of a host (especially in the digestive tract). As shown in Fig. 29.9(c), the body of a trematode is largely composed of reproductive organs: ovaries, testes, uterus, and yolk glands. The reproductive life cycle of the endoparasitic flukes is often quite complex, involving at least two and sometimes three hosts. For example, the so-called Chinese liver fluke, *Clonorchis sinensis*, lives as an adult in human liver tissue (it resembles closely *Prosthogonimus*; see Fig. 29.9c). The eggs are laid in the liver and pass out through the bile duct

---

*They are called "flame cells" because the beating of the cilia within the small capsule formed by indentation of one side of the cell looks very much like a flickering flame.

into the intestine. From there they go into the environment with the feces. Once on the outside, the eggs must be eaten by a particular species of snail if they are to hatch. If the right snail finds them, the eggs develop into small larvae that bore from the intestine into the lymph tissue of the snail. Here, they reproduce asexually. These new forms develop into a second set of offspring that passes out of the snail's body into the water (these are all freshwater forms). If they find a fish, the larvae bore into its tissues, where they form a cyst. If a human eats the fish when it is raw or insufficiently cooked, the cysts will still be alive. The human's digestive enzymes break down the cyst walls and young flukes emerge, finding their way to the liver to start the cycle anew. Humans afflicted with infection by flukes are often weak and debilitated. Trematode infection is quite common in the tropical countries of Asia.

Class Cestoda (the Tapeworms)

Adult tapeworms live exclusively as endoparasites in the intestines of vertebrates (especially mammals). Like the flukes, they have a resistant cuticle surrounding their epidermis. Sharp hooks and oral suckers on the head end (see Fig. 29.11a and b) allow the tapeworms to hold onto the intestinal wall of their host. The cestodes lack both a mouth and digestive tract, an adaptation to life in an environment surrounded by digested

Fig. 29.11
Class Cestoda, the tapeworms. (a) Anterior end of the tapeworm *Hymenolepis diminuta* (× 182). (Photo courtesy Professor Richard G. Kessel.) (b) Head (scolex) of the dog tapeworm, *Taenia pisiformis*, showing hooks and oral suckers (× 100). (Photo courtesy Richard G. Kessel.) (c) Mature segments or proglottids of *Taenia pisiformis*, containing reproductive organs. Sperm and eggs are both produced in each proglottid. Sperm are released from one proglottid by way of the genital pore into the host's intestine, and find their way into the vagina of the same or another proglottid; fertilization then occurs. Ripe proglottids with fertilized eggs pass out of the host animal's intestines with the feces (for details of life cycle, see Fig. 29.12). (Photo, Carolina Biological Supply Company.)

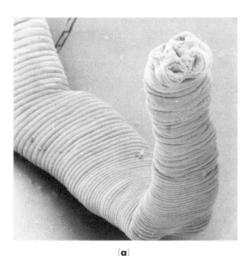

(a)

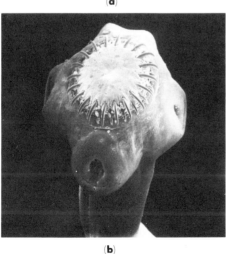

(b)

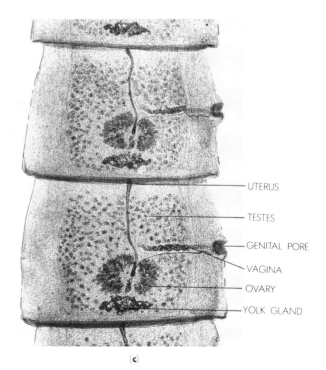

UTERUS

TESTES

GENITAL PORE

VAGINA

OVARY

YOLK GLAND

(c)

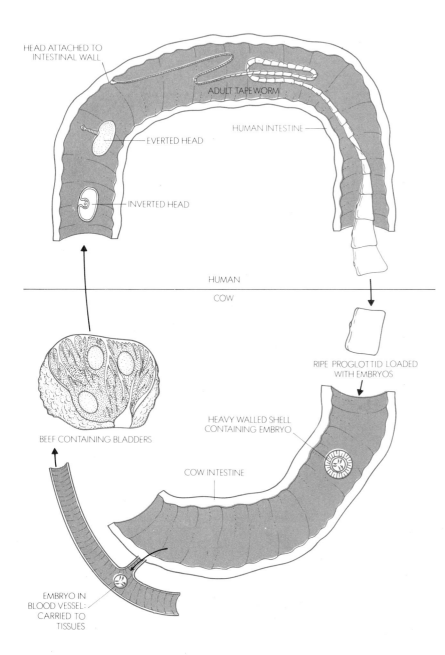

HEAD ATTACHED TO
INTESTINAL WALL

ADULT TAPEWORM

HUMAN INTESTINE

EVERTED HEAD

INVERTED HEAD

HUMAN

COW

RIPE PROGLOTTID LOADED
WITH EMBRYOS

BEEF CONTAINING BLADDERS

HEAVY WALLED SHELL
CONTAINING EMBRYO

COW INTESTINE

EMBRYO IN
BLOOD VESSEL:
CARRIED TO
TISSUES

Fig. 29.12
Life cycle of tapeworm, *Taenia saginata.*
This parasitic flatworm invades two hosts
during its life cycle. The adult lives in the
human intestine, absorbing food from the
host. Ripe proglottids, filled with fertilized eggs, pass out of the intestine in the
feces. If these proglottids are eaten by a
cow, the egg cases break open in the cow's
intestine, releasing a heavy-walled shell
containing the embryo. The embryo bores
through the intestine wall into the cow's
blood stream, where it is carried to the
muscles. If a human eats poorly-cooked
beef, cysts from the muscles enter the human intestine. The inverted head of the
cyst everts, and it hooks onto the human
intestinal cell wall, beginning the cycle
again. (After Buchsbaum, *Animals without Backbones.* University of Chicago
Press, 1948, p. 145.)

food. Absorption of food occurs directly across the
tapeworm's body wall. The vast bulk of the tapeworm's
body is composed of segments called **proglottids,** which
are almost exclusively sacs of reproductive organs (see
Fig. 29.11c). The accumulation of many proglottids
makes some tapeworms quite lengthy (over thirty feet in
Diphyllobothrium latum). Tapeworms not only consume nutrients, but also produce wastes and form a mechanical obstruction in the intestine. Once an animal

contracts a tapeworm, it is extremely difficult to remove
from the digestive tract. Sometimes powerful drugs
have been successful in killing the head, but in many
cases even this treatment has proved only partially successful.

Like the trematodes, most of the cestodes require
several hosts to complete their life cycles. The life·cycle
of the beef tapeworm *Taenia saginata,* is shown and described in some detail in Fig. 29.12.

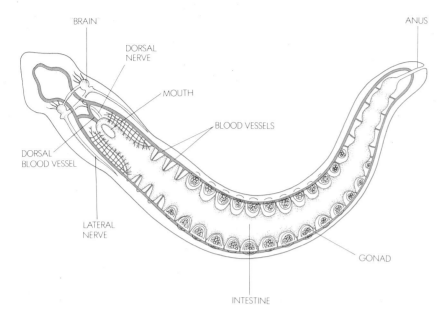

Fig. 29.13
Internal structure of a Nemertea, such as *Tetrastemma.* Note the presence of a one-way digestive tract (mouth, left; anus, right), a blood-vascular system with lateral vessels, and two lateral nerve cords. The Nemertea are free-living marine forms that live buried or partially buried in the sand in shallow water off the Pacific and Atlantic coasts of North America.

29.5
PHYLUM NEMERTEA (RHYNCHOCOELA; RIBBON WORMS OR NEMERTINE WORMS)

The nemertine worms have sometimes been classified with the flatworms, and sometimes as a separate phylum of their own. They are long, free-living marine flatworms, much like a ribbon in shape (hence their common name, "ribbon worms,"), with a proboscis enclosed in a tubular cavity at the anterior end (Fig. 29.13). In some species the proboscis is armed with a barb (and toxin) used for capturing prey and for defense. When inverted, the proboscis is often several times longer than the entire animal. Nemertines commonly live in shallow water under rocks or other protective structures, burrowing into the sand.

The nemertines share with the flatworms the following characteristics: a general flattened body structure; the presence of mesoderm (that is, all three germ layers); and the presence of two lateral nerve cords and a flame-cell excretory system. At the same time, nemertines differ from flatworms in two respects of particular interest to the biologist. (1) They possess a blood vascular system, consisting of one dorsal and two lateral blood vessels. This is the first group of animals in which any distinct circulatory vessels have been encountered. (2) They also have a one-way digestive system, with both mouth and anus. The advantage of a one-way system for digestion and food processing is that it allows the arrangement, in sequential order, of a series of food-digesting organs, and thus of a kind of "assembly line" process for food breakdown. The one-way digestive system predominates throughout all the higher phyla of the animal kingdom. Primarily because of these two important features, the Nemertea are more generally classified as a separate phylum on their own. However, the nemertines are thought to have originated from a flatworm (Turbellarian) ancestor.

29.6
THE PHYLUM NEMATODA (ROUNDWORMS)

The number of species of Nematoda (roundworms) has been estimated at anywhere between 10,000 (conservative estimate) and 400,000 or 500,000 (liberal estimate). One teaspoonful of coastal mud is estimated to contain over 1000 individuals representing 36 different species! Whatever the exact number, it is clear that nematodes are considerably more diverse than any group we have examined thus far. Most are free-living, microscopic forms (see Fig. 29.14), with a distinctly rounded body (as opposed to the flat body of the platyhelminthes or the nemertines).

Nematodes are important to humans in many ways, but principally as parasites. The species that parasitize humans (estimated at about fifty) can be dreadful pests; they include the hookworms, pinworms, and *Trichinella* (which produce trichinosis, an infection humans can get from eating partially cooked pork).

Adult parasitic nematodes live in the small intestine of many mammals, including pigs and humans. The pattern of infection is the same regardless of the host species. Female *Trichinella* are fertilized in the host's intestine, then bore through the intestine wall and deposit their young (already hatched inside the female's uterus) in the lymphatic vessels. The young larvae pass through the lymphatic system into the blood, and are carried to all parts of the body. The larvae infect virtually every organ, but only those entering skeletal muscle survive.

Fig. 29.14
Scanning electron micrograph of the nematode *Foleyella*
(× 130). Note the rounded shape and basically unsegmented
body. *Foleyella* lives parasitically as an adult in the mesentary
and body cavity of the frog. (Photo courtesy Professor
Richard G. Kessel.)

During this process of boring and migration, the parasite does its worst damage to the host, who feels the pain, cramps, nausea, and fatigue from the infection (the adult worms in the intestine are relatively harmless by comparison). The larvae encyst in the skeletal muscle; this is the form in which they are most easily transmitted to another animal. If the cysts are in pig muscle (pork) that is only partially cooked and eaten by a human, the cysts begin developing in the human intestine, repeating the cycle.

Trichinosis can be fatal to humans. Roundworm infections have been, and still are, far more common than most people realize. It has been estimated that over 4,000,000 people throughout the United States today have been heavily infected by *Trichinella*.* Hookworm,

---

*The United States is one of the few countries in the world rich enough (or wasteful enough?) to routinely feed leftovers, including pork scraps, to pigs. Scraps containing cysts can reinfect pigs, whose partially cooked meat can then reinfect more humans, and so on.

another nematode disease, comes from infectious stages that bore through the foot and get into the bloodstream. Both hookworm and trichinosis have been reduced in the United States by the wearing of shoes and by improving sanitary conditions.

Like platyhelminthes, the body of nematodes is constructed from the three basic germ layers. And, like the nemertines, the digestive tract of nematodes is one-way, starting with a mouth and ending with an anus. Nematodes have a pseudocoel—a distinct body cavity surrounded (lined) on one, but not all sides with mesodermal tissue. Nematodes are not segmented, and are covered by a thick cuticle; when they grow they must shed this cuticle, a process called **molting**. One unique feature of the anatomy of roundworms is that they have muscle fibers running only along the long axis of their bodies; they have no muscle fibers running around the body. Consequently, they move by lashing their bodies back and forth, like a whip.

In addition to infecting humans and other animals, nematodes also infect most species of plants including orange trees, tobacco, and strawberry plants. The nematode infection does not necessarily kill the plant, but it weakens the host to such an extent that the host often succumbs to other parasites or infections.

29.7
THE PHYLUM MOLLUSCA

With over 50,000 living species, the Mollusca is one of the largest phyla, in numbers of species, within the whole animal kingdom. Members of this phylum have soft bodies usually contained within a hard, calcified shell. In some notable exceptions, such as slugs, squid, and octopuses, the shell has been lost or internalized over the course of evolution (though it is thought that ancestral forms of these organisms once possessed shells).

There are six living classes within the phylum Mollusca; one of these classes (Monoplacophora), discovered about twenty years ago, is a living representative of a group of primitive molluscs known previously only through the fossil record. Of the six classes, three are particularly prominent and are shown in Fig. 29.15: (1) the Gastropoda, with distinct heads, eyes, and sometimes tentacles, and a one-piece (usually coiled) univalve shell (snails, slugs, whelks, and conchs); (2) the Pelecypoda, lacking distinct head, with a two-part (bivalve) shell (oysters, clams, and mussels); and (3) the Cephalopoda, marine forms with grasping tentacles and no external shell (squid, octopus). Of the other groups, the Scaphopoda has only a few representatives; the Amphineura (represented by chitons) and the Monoplacophore (represented by the *Neopilina*) are not common but are of particular interest in regard to the possible evolutionary origin of the molluscs.

(a)

(b)

(c)

The basic mollusc body plan, with several variations, is shown diagrammatically in Fig. 29.16. The body is bilaterally symmetrical, with three distinct body zones: a **head-foot**, which contains both sensory and motor organs; a central **visceral area** (or mass) containing the organs of reproduction, digestion, and excretion; and finally, a **mantle**, a fold of the body wall, which covers the visceral mass and enfolds it, and also secretes the shell. Posterior to the visceral mass and ventral to (beneath) the mantle is the **mantle cavity**, where the gills are housed. Gills are the organs used for respiration, to extract oxygen from the water and return CO_2 to it. In respiration, water is swept into the mantle cavity by cilia, where it aerates the gills and then flows past the anus, nephridia, and reproductive organs (gonads), and on to the outside. The excretory and reproductive systems discharge their contents into the water current downstream from the gills thus avoiding pollution of the respiratory apparatus.

The digestive tract of molluscs is far more coiled than that in annelids, and so provides additional surface area for absorption. In molluscs, food is mechanically shredded by a special molluscan organ, the **radula**,

Fig. 29.15
Phylum Mollusca (soft-bodied animals). The three classes represented here are also the most diverse. (*a*) Gastropods (snails and slugs; sometimes called univalves). (Photo, Runk/Schoenberger, Grant Heilman Photography.) (*b*) Pelecypods (oysters and clams; sometimes called bivalves). (Photo, Runk/Schoenberger, Grant Heilman Photography.) (*c*) Cephalopods (octopus, squid, and chambered nautilus). Squid and octopuses have some internal cartilaginous support, including, in some species of squid, a cartilaginous brain case. (Photo, Carolina Biological Supply Company.)

which works like a file, rasping particles off the substrate. Shredded material then passes into the mouth. As it moves through the digestive tract (moved by means of cilia in most cases), it is chemically broken down. Food is absorbed by cells lining the stomach and the front part of the intestine, and is then passed directly into the blood.

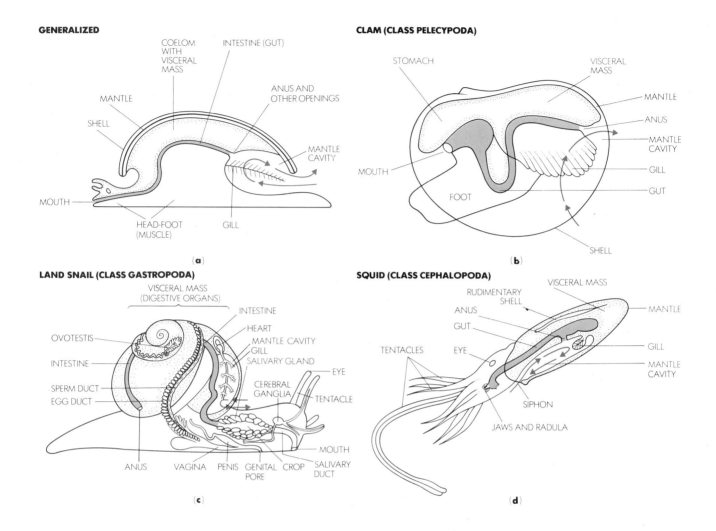

Fig 29.16
Basic body plan of the phylum Mollusca and its modification in three major classes: the Pelecypods (clams and other bivalves), the Gastropod (snails and slugs), and the Cephalopoda (squid, octopus). In the generalized mollusc (a) the basic features are the muscular foot (for locomotion), the visceral mass (containing the major body organs), the intestine, the mantle (and mantle cavity), and the gills. The arrows in (a), (b), and (d) indicate the direction of water flow in and out of the mantle cavity. (b) General structure of the clam. (c) A sea snail, showing highly coiled intestine. In land snails gills are lost and replaced by the blood-vessel-enriched mantle as a respiratory organ. In aquatic snails, gills serve the normal respiratory function. (d) In the Cephalopoda (squid) water enters the mantle cavity through a siphon, which is also found in some gastropods.

Molluscs have a three-chambered heart: two of the chambers collect blood from the gills, and the third pumps it to the body tissues. All groups of molluscs except the cephalopods have an **open circulatory system.**

The blood is collected by the heart and then pumped directly into the body cavity, so that the blood bathes the tissues directly. It is returned from the tissue spaces to the heart, largely by flow patterns created by the mechanical pumping of the heart itself. This system of circulation is far less efficient in terms of delivering oxygen to the tissues than the closed circulatory system characteristic of higher animals such as humans; but for slow-moving molluscs such as the clam or snail it is apparently sufficient. In the cephalopods, however, where active movement is necessary, the circulatory system is closed. They have accessory hearts, as well, that propel blood directly to the gills.

Primitive Molluscs

The Mollusca are an extremely successful group of animals. Interestingly, the vast majority of modern-day species are contained in only two classes: the Gastropoda (snails) and Pelecypoda (bivalves). In the past, however, other molluscan forms were far more promi-

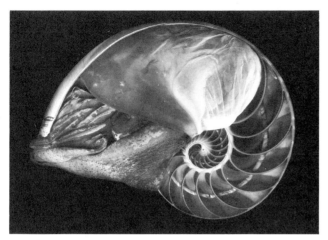

Fig. 29.17
The chambered nautilus, a cephalopod, with cutaway of shell showing internal sections. As the organism grows, it continues to add new sections on at the open (to the left) end. This graceful mollusc has inspired poets and artists alike. (Photo, American Museum of Natural History.)

nent. During the Ordovician and Silurian periods (500 to 400 million years ago) the nautiloids, similar to the chambered nautilus (Fig. 29.17), were extremely common, with shells up to three meters in diameter! In the Mesozoic era, some 100 to 80 million years ago, the ammonites, a group of now-extinct, giant cephalopods, were the dominant invertebrates. By comparison, most of the modern cephalopods are greatly reduced in size. However, one modern representative has actually exceeded its extinct ancestors in size: the giant squid attains a length of 17 or 18 meters (50 to 60 ft), and has often been regarded in the past as a "sea monster" (see Fig. 29.18).

29.8
THE PHYLUM ANNELIDA

The phylum Annelida consists of nearly 9000 species of marine, freshwater, and terrestrial (soil-dwelling) worms. Representatives of the four classes of the phylum Annelida are shown in Fig. 29.19. The fourth class, the Arachiannelida, are primitive forms with a limited distribution. We will discuss each of these classes briefly after describing some general traits of the entire phylum.

All the annelids are segmented, meaning that their bodies are formed from a series of repeating units (segments) with the same basic structural plan. All annelids show bilateral symmetry.

Annelids have a three-layered body, a tubular digestive tract (gut) running most of the body length; and an efficient, closed circulatory system with blood vessels and tubular hearts that pump the blood. A closed

Fig. 29.18
Giant squid, measuring up to 50 or 60 feet, were periodically encountered by whaling and other vessels in the eighteenth and nineteenth centuries. Many times they were regarded as dangerous sea monsters. In 1860, a case was reported by the French battleship *Alecton* in the Atlantic. This individual was 50 feet long exclusive of the tentacles, and was 20 feet in circumference at the largest part. Its weight was estimated at 2 tons! It thus classifies as one of the largest invertebrates known. A description of the attempt of the ship's crew to capture the animal (which was probably sick) gives an indication of its powers of resistance: "The commandant, wishing in the interests of science to secure the monster, actually engaged it in battle. Numerous shots were aimed at it, but the balls traversed its flaccid and glutinous mass without causing it any vital injury. . . . They succeeded at last in getting a harpoon to bite, and in passing a bowling hitch round the posterior part of the animal. But when they attempted to hoist it out of the water the rope penetrated deeply into the flesh, and separated it into two parts, the head with the arms and tentacles dropping into the sea and making off, while the fins and posterior parts were brought on board: they weighed about forty pounds." (From Louis Figuier, *The Ocean World*, Plate XXIV).

(a)

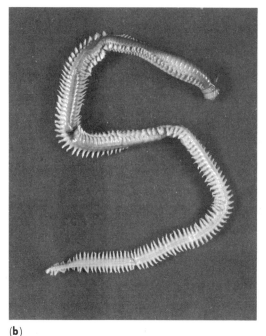

(b)

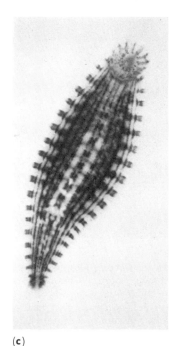

(c)

Fig. 29.19
Representatives of three classes of the phylum Annelida: (*a*) Oligochaeta, the earthworm (*Lumbricus terrestris*), which lacks a distinct head, eyes, or antennae. (Photo courtesy Lynwood M. Chace.) (*b*) Class Polychaeta, the genus *Nereis*, a marine worm with well-developed eyes and antennae. (*c*) Class Hirudinea, the leech, which specializes in sucking blood from vertebrates. (Photo, Carolina Biological Supply Company.)

circulatory system is contained within a continuous set of tubes, not emptying into a body cavity. Oxygen diffuses through the skin into capillaries, which form part of the circulatory system. Similarly, food is absorbed from the gut into the blood by capillaries. The excretory system is composed of **nephridia,** a pair of which is located in most segments; wastes are excreted to the outside by excretory pores. The annelids show clear advancement over the roundworms and flatworms in their highly developed nervous system. Annelids possess a number of special sensory cells, such as taste receptors, light-sensitive cells, and touch cells, whereas some also have well-developed eyes and antennae.

Class Polychaeta

The polychaetes are all marine segmented worms with well-defined heads, eyes, and antennae. As shown in Fig. 29.19(b), the segments of the polychaete *Nereis* are more or less similar in structure. Each bears a pair of locomotor appendages called **parapodia** that also function in gas exchange. The class takes its name from the many bristles, or **chaetae,** that are located on the parapodia. Some species of polychaetes swim, others crawl about on the bottom of the ocean, and still others spend

Fig. 29.20
A polychaete worm, *Chaetopterus*, in its burrow, a U-shaped tube that is cut away to show the position and appearance of the worm inside. Note the many segments and parapodia, whose beating forces a current of water through the tube. Outside the polychaete tube is another worm, a sipunculid, a representative of a small phylum thought to be related to the annelids. (Photo, American Museum of Natural History.)

most of their time in tubes buried in the mud or sand (see Fig. 29.20). In the tube-dwelling forms, water currents bearing organic material for food are moved through the tubes by the beating of the parapodia.

In the majority of polychaetes the sexes are separate—some worms are male and others female. In some species, gametes are produced in virtually all the segments, while in others they are produced in a few specialized segments. Sexual maturity generally occurs at some definite season of the year, varying from species to species and locality to locality. When the females are ready to reproduce, they leave their burrows or the bottom of the ocean and swim to the surface. The males are attracted to the females, and as the females shed their eggs into the water, the males discharge their sperm. Fertilization is external. In *Nereis* and many other species of polychaetes, the fertilized egg develops into a ciliated larva called a **trochophore** (Fig. 29.21). In a few species the trochophore is free-living, but in many (such as *Nereis*) its development occurs wholly within the egg membrane. By the time the larva emerges from the egg, it has already passed the trochophore stage and has elongated into three segments with bristles.

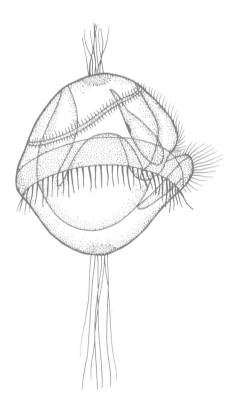

Fig. 29.21
Trochophore larva of the annelid *Nereis*. Note the ciliated bands around the equatorial regions. The trochophore, whose evolutionary significance we will discuss in a later part of this chapter, is found in various annelids and molluscs. (After Ralph Buchsbaum, *Animals without Backbones*. University of Chicago Press, 1948, p. 216.)

Class Oligochaetae

The class Oligochaetae is best represented by the common earthworm (*Lumbricus terrestris*). In order to provide some insight into the internal anatomical organization of an annelid, we will discuss some general features of earthworm anatomy. First, however, consider several important ecological features of earthworms that differentiate them from other annelids.

Earthworms are terrestrial annelids, and hence have many adaptations for terrestrial life. They gain their nutrition from the soil, which they ingest as they burrow. Soil passes into the earthworm's mouth and through the gut, where the organic nutrient is extracted from it. The indigestible material passes out the anus.

Earthworm Structure and Function. A cutaway diagram of an earthworm is shown in Fig. 29.22. The earthworm possesses a true coelom, a large, fluid-filled cavity, surrounded by mesodermal tissue (see Fig. 29.22b). The special adaptive significance of the coelom is more apparent in the annelids than in the molluscs. The coelom makes greater space available for the more complex body organs. Because the coelom is filled with fluid, the body organs (such as gut and heart) are greatly protected from the blows of life and can move about over themselves easily and with little damage due to friction. Without the development of the coelom, the evolution of complex organ systems (which need room while at the same time providing room for body muscle) would not have been possible.

Because the earthworm is segmented (like all annelids), the body is compartmentalized, with each segment containing a number of the same structures: external bristles or **setae** (usually four pairs), two nephridia (excretory organs), and four nerve trunks branching off from the main, ventral nerve cord. Among other components, the body wall consists of two sets of muscles, one running longitudinally and the other circularly around the inside of each segment. An intestine runs throughout the length of the body, from mouth to anus. The mouth leads into a strong, muscular pharynx, which can suck organic material and soil into the gut. From the pharynx food passes into a **crop** and a **gizzard**, where material is (respectively) stored, and then ground up. After grinding in the gizzard, the digestive material moves into the intestine. Along with indigestable soil particles, organically rich waste materials pass out the anus into the soil. Thus, earthworms not only aerate the ground by burrowing through it, but also enrich the soil as it passes through their bodies and mixes with organic wastes.

The earthworm has a well-developed circulatory and nervous system. The circulatory system consists of a dorsal and several ventral blood vessels, connected to five tubular "hearts." The ventral vessels form smaller vessels and capillaries leading from the intestine, and thus serve to absorb and transport nutrients. Material is kept moving through the system by the irregular beating

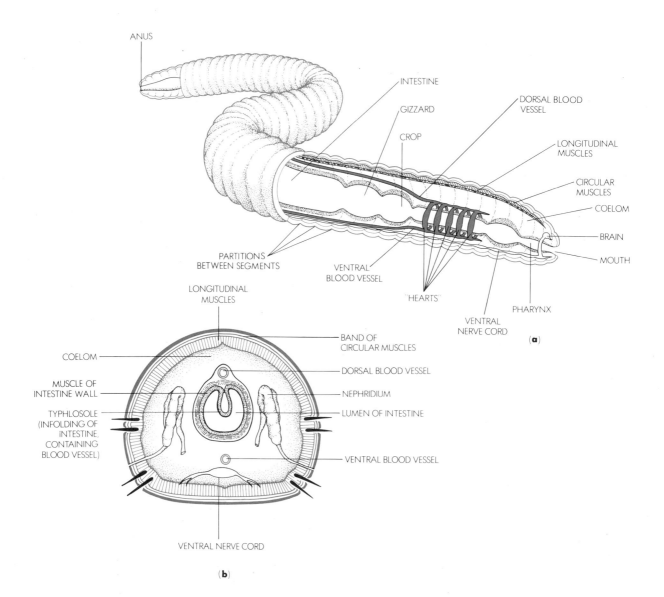

Fig. 29.22
Internal anatomy of the earthworm, showing (a) longitudinal arrangement of parts through various segments, and (b) cross section of one segment. The earthworm (*Lumbricus terrestris*) has a distinct brain, digestive system, circulatory system including a series of five tubular hearts, an excretory system consisting of a pair of nephridia in each segment, and a ventral nerve cord. The true coelom (color) permits a space in which internal organs are protected and given room to move over one another without friction.

of the hearts. Both the heart tubes and the major vessels have valves that keep fluids moving in one direction. The blood of many annelids, including the earthworm, contains hemoglobin, the same basic oxygen-binding

molecule found in our own bodies.* The hemoglobin is dissolved directly in the blood (unlike our bodies, where it is contained in red blood cells). Like all annelids the earthworm has a **closed circulatory system,** which provides a more efficient means of delivering oxygen and nutrients to the body cells than the open circulatory system of molluscs or arthropods. For one thing, a closed system with pumps (that is, hearts) can generate more pressure and hence move blood around the body faster.

The nervous system of the earthworm consists of a mass of ganglionic cells (a brain) at the anterior end,

*Hemoglobin appears to have been independently evolved several different times during the history of animal life. It is found in a number of annelids, insects, and, of course, in virtually all the vertebrates. In all cases the basic molecular structure is surprisingly similar, but the more distinct the groups, on anatomical and other grounds, the more different are the details of their hemoglobin structures (see Supplement 24.2).

with a ventral nerve cord running the length of the entire animal. In each segment the ventral nerve cord enlarges into a ganglion that gives off several pairs of nerves, both afferent and efferent. The earthworm thus has a basic division between a central and a peripheral nervous system. As a result of its higher degree of structural organization, the nervous system of the earthworm is capable of coordinating more complex behavior than we have seen in forms such as the molluscs and annelids. Indeed, experiments have shown that earthworms are capable of learning, and have distinct, and sometimes subtle stimulus discrimination (for example, between two closely related chemicals).

The excretory system of the earthworm in each segment consists of pairs of **nephridia**—ciliated, coiled tubes with one end opening into the coelom and the other end opening to the outside. Coelomic fluid is pulled into the nephridium by the beating of the cilia. As the fluid moves through the coiled tube, water, salts, sugars, and other nutrients are absorbed by cells lining the wall and returned to the coelom. Waste materials are retained inside the nephridium and passed to the outside through a pore. The nephridia in the earthworm (and in all annelids) are responsible not only for elimination of wastes, but also for maintaining water and salt balance.

The earthworm has no special organs for respiration. It exchanges gases directly with the air around it through its skin. Gases from the air dissolve in the liquid layer on the surface of the earthworm's body. Movement of oxygen inward and carbon dioxide outward appears to be largely by diffusion.

Reproduction in the earthworm is sexual; it also has remarkable powers of regeneration. *Lumbricus* is hermaphroditic, that is, each individual has both testes and ovaries, and produces both sperm and egg. Cross-fertilization is the rule, however. In mating, two individuals come together, and are held next to each other by mucous secretions from the **clitellum** (a special collection of glandular cells usually visible on the exterior of the earthworm as a slightly enlarged, lighter-colored group of about six segments). The individuals exchange sperm and separate. The sperm go into each partner's **seminal receptacles**. Fertilization occurs several days later, as the sperm and eggs are brought into contact by the sliding of the clitellum forward. After fertilization, the clitellum, with young fertilized eggs, passes over the anterior end of the worm and is deposited in the ground. Its ends become enclosed, forming a sort of cocoon within which the fertilized eggs develop. There is no larva stage of the earthworm; miniature, perfectly formed worms hatch from the cocoon.

Class Hirudinea (Leeches)

The Hirudinea are probably the most specialized of the annelids. The body is tapered on both ends, with suckers on the first and last segments. Leeches are segmented externally, but not internally, the individual partitions between segments seen in the earthworm being completely absent. The most prominent characteristic of the leeches is their parasitic mode of existence as bloodsuckers. However, it should be clearly stated that not all leeches are parasitic. Many live by capturing small organisms such as worms and insect larvae, and ingesting them whole.

When a bloodsucking leech attacks a host, it attaches first by the posterior sucker and then by the anterior. Some leeches open the skin by making a small cut with sharp jaws on their anterior (oral) end; others use enzymes that digest away the layers of skin down to the capillaries. All leeches secrete an anticoagulant into the blood to prevent clotting at the site of the wound. They then absorb a large quantity of blood, engorging themselves to many times their original volume. After a good meal, leeches can go for long periods of time before feeding again; in some cases for a year or more. In the nineteenth and early twentieth centuries, leeches were routinely sold at drug stores for medicinal purposes—to cure ailments by "bloodletting," supposedly purging the body of toxic substances. These leeches were kept in apothecary jars on the shelf for extended periods of time without feeding and with no apparent harm to the animals.

29.9
THE PHYLUM ONYCOPHORA

Onycophorans are unprepossessing little organisms looking somewhat like caterpillars, with a segmented body each segment of which contains a pair of unjointed appendages (Fig. 29.23). They feed on vegetable matter

Fig. 29.23
Peripatus (phylum Onycophora), showing features of both the annelids and arthropods. Like an annelid, *Peripatus* is long, slender, and segmented; like an arthropod it has antennae and appendages (unjointed) with insect-like claws at the tips (one pair of appendages per segment). Once a widespread group geographically, *Peripatus* is now found in only a few isolated tropical localities. (Photo courtesy of Ward's Natural Science Establishment, Inc.)

and decaying organic material on the forest floor under rocks and logs. The onycophorans are found in tropical regions of the world, particularly the southern hemisphere, but in widely separate places: Australia, New Zealand, South Africa, and South America (primarily in the Andes). This disconnected geographic pattern suggests that the group is quite old and was at one time very widely distributed. The phylum has probably undergone a gradual shrinkage so that it is found only in isolated pockets, or areas where conditions are still satisfactory for its survival.

Onycophorans such as *Peripatus* occupy an unusual place in the animal world. They contain a mixture of arthropod and annelid characteristics. For example, like the annelids, the Onycophora have a pair of nephridia in each segment and a thin, flexible, soft cuticle permeable to water, much more like the cuticle of an earthworm than the exoskeleton of a typical arthropod. At the same time, like the arthropods, the Onycophora have an open circulatory system and appendages with claws at the tip. Because of the mixture of traits, the Onycophora are thought to have evolved from a line of arthropods that was an offshoot of a common ancestral form between annelids and arthropods. In other words, the Onycophora are not themselves, nor are they directly descended from, a common ancestor between annelids and arthropods. They appear, however, to have split off very early from the line leading from the common ancestor of the arthropods. The Onycophora are a small phylum, containing approximately sixty-five species.

29.10
THE PHYLUM ARTHROPODA

For both diversity of species and total numbers of individuals on earth, the Arthropoda win all contests hands down. There are over 765,000 identified species in this phylum, and probably as many again that have not yet been recorded. It has been estimated that there are probably at least one million species of insects alone.

The phylum Arthropoda consists of three subphyla and seven classes, which are listed and their characteristics summarized in Table 29.1.

General Characteristics

All members of the arthropod phylum have a segmented body enclosed in a tough **exoskeleton** (external skeleton) composed primarily of the protein chitin. Arthropods are bilaterally symmetrical, and all show distinct cephalization. The appendages are jointed, one pair per body segment. In fact, the name arthropod means "jointed" (*arth*) "foot" (*pod*). In the anterior region, where the segments have been reduced in many cases to very small size, the appendages associated with each segment have been modified into mouth parts. The

appendages all show considerable adaptation to diverse functions—some for motion, some for support, some for food-getting, some for sensation, and some for offense or defense. The exoskeleton in arthropods serves not only for protection, but also for body support, and it is to the inside of this tough "armor" that the muscles are all attached. The exoskeleton of arthropods has a distinct advantage over the shells of molluscs: the exoskeleton moves at its joint, allowing the appendages to move independently, thus providing greater mobility to the whole organism.

There is one major drawback to the exoskeleton, however. Being hard and tough, it cannot increase in size as the animal grows inside it. Periodically, therefore, arthropods must "molt"—that is, shed their exoskeleton completely and produce a new, larger model. During the molting period, before the new exoskeleton has hardened, the organism is soft and delicate, thus subject to predation on the one hand and to water loss on the other. Molting is also a metabolically costly process, though some insects and freshwater crustaceans recoup at least some of their loss by eating the old exoskeleton.

An important feature of the exoskeleton is that it is waterproof. This is due to its thickness, to the fact that chitin is particularly nonporous, and to the presence of a waxy lipoprotein outer layer (cuticle) that covers the exoskeleton of all arthropods except the Crustacea. The waterproof exoskeleton has been a major feature allowing arthropods to invade the land. Arthropods are, in fact, the only invertebrate group to diversify extensively in nonmoist terrestrial habitats. While many invertebrates live in the soil, and are thus technically terrestrial (earthworms and roundworms, among others), they are shielded from the drying effects of sunlight and direct exposure to the air. Many arthropods, however, have been able to inhabit the surface of the soil and the air because their chitinous exoskeleton is highly effective in preventing water loss.

The prevalence of segmentation within the arthropods suggests a possible common origin from an annelid-like ancestor. The arthropods and the annelids are the only groups where segmentation is universal and readily apparent. Even in some arthropods, such as the insects, where the segments have become fused, a close examination still reveals the underlying segmentation. Of course, segmentation is readily apparent in insect larvae, such as the caterpillar (larva of the moth and butterfly).

Respiration in the arthropods varies considerably, the major differences being between those that are terrestrial and those that are aquatic. In terrestrial forms, the respiratory system consists of a series of cuticle-lined air ducts, the **tracheae**, that bring air directly into the inner parts of the body. The tracheae ramify into very small passageways that are not cuticle-lined and penetrate the body tissues. In aquatic arthropods, such as the

Table 29.1
Major Subdivisions of the Phylum Arthropoda

Subphylum	Class (with examples)	Habitat	Body divisions	Head appendages	Number and location of legs	Wings	Respiration	Excretion
Trilobitomorpha	Trilobita (Trilobites)	Aquatic (marine)	Three (head, thorax, and abdomen)	Four pairs of legs (unspecialized); one pair antennae	Numerous (one pair per segment; all unspecialized)	None	—	—
Chelicerata	Merostomata (*Limulus*, the horseshoe crab)	Aquatic (marine)	Two (cephalothorax and abdomen)	Chelicerae, pedipalps (for chewing and grinding food)	Numerous; five on cephalothorax	None	Gills (attached to abdominal appendages)	No Malpighian tubules
	Arachnida (spiders, scorpions)	Terrestrial (some secondarily aquatic)	Two (cephalothorax and abdomen)	Chelicerae, pedipalps (for chewing and grinding food)	Four pairs (on cephalothorax)	None	Book lungs, tracheae, or both	Malpighian tubules
Mandibulata	Crustacea (crabs, crayfish, lobsters, shrimp, barnacles)	Aquatic (marine and freshwater), some terrestrial	Two (cephalothorax and abdomen)	Antennae (two pairs), mandibles, and maxillae	Numerous (on both thorax and abdomen)	None	Gills	Green glands (head region)
	Chilopoda (Centipedes)	Terrestrial	No real division	One pair antennae and assorted mouthparts	Numerous (one pair per segment)	None	Tracheae	Malpighian tubules
	Diplopoda (Millipedes)	Terrestrial	No real division	One pair antennae and assorted mouthparts	Numerous (two pairs per segment)	None	Tracheae	Malpighian tubules
	Insecta (Insects)	Terrestrial and aquatic (secondarily)	Three (head, thorax, and abdomen)	Antennae, one pair mandibles, and two pair maxillae	Three pairs (on thorax only)	One or two pairs	Tracheae	Malpighian tubules

crayfish or lobsters, gills are located in the cephalothorax (the anterior part of the body, consisting of the fused head and thorax regions) and protrude into ventral spaces where they come into contact with the water, providing a surface for exchange of gases.

The circulatory system of arthropods, like that of molluscs, is open—that is, the blood circulates openly throughout the body cavity rather than being contained in a closed system of tubes. In all arthropods an elongated tubular structure, the heart, is located on the dorsal surface of the body, just inside the exoskeleton. It receives the blood from the open body cavity, and, by its pumping action, pushes the blood forward into arteries and then back, eventually into the body cavity, the **hemocoel,** where it bathes the tissues. Despite the relative inefficiency of an open, compared to a closed, circulatory system, arthropods can be extremely active (for example, a housefly or moth). Arthropods have kept a high metabolic efficiency at the expense of size. Most arthropods, especially terrestrial ones, are quite small, allowing both an open circulatory system and a tracheid system for respiration. Neither blood nor air has to be moved over great distances.

Excretion in arthropods is different in the aquatic and terrestrial species. In the aquatic species nitrogenous wastes, usually in the form of ammonia, are secreted through the gills. A somewhat less important organ of excretion, known as the **green gland,** is located just anterior to the mouth of aquatic forms such as the lobster. It releases its waste materials to the outside through a pore above the eye. In terrestrial arthropods, nitrogenous wastes are eliminated through special excretory structures, the **Malpighian tubules.** These are groups of small, dead-end sacs arranged around the gut so that their blind ends protrude into the open areas of the animal's body (the hemocoel). Fluid is absorbed through the walls of the sacs, or tubules, and becomes concentrated as insoluble uric acid. Water and other salts are reabsorbed from the tubules back into the blood. The concentrated uric acid (still suspended in liquid) passes from the lower end of the tubules directly into the gut, from which it passes into the rectum. Here, water is further reabsorbed and the nitrogenous material, now as dry uric acid, passes out with the feces. The Malpighian tubules are special physiological adaptations to living on land; their role in water conservation for terrestrial arthropods has been one of the fea-

tures (in addition to the exoskeleton) that has placed arthropods among the few truly successful invertebrate colonizers of land.

The arthropod nervous system is structured in a ladder-like fashion, with two parallel nerve cords running along the ventral surface from the brain and repeated cross-connections (like the rungs of a ladder) innervating each segment or region. The brain in arthropods is located in the head, but it is apparent that ganglia spaced periodically along the ventral nerve cords within the segments coordinate much of the animal's general behavior. For example, the grasshopper can jump, walk, or even fly, with its brain completely removed. It has been said that the brain in arthropods appears not so much as an initiator as an inhibitor of impulses.

Arthropods are equipped with elaborate sense organs, particularly the **compound eye,** composed of a number of six-sided units called **facets.** Facets form the top of a long visual tube known as an **ommatidium** (see Fig. 29.24). At the bottom of the ommatidium is the **rhabdom,** the photosensitive part of the actual sensory

Fig. 29.24
(*a*) Compound eye of the arthropods, as shown in the housefly (class Insecta). (*b*) Two ommatidia (visual units, or cells). The facets, or lens surfaces, at the top are hexagonal in shape (six-sided), but appear five-sided in the drawing because they are shown cut away in longitudinal section. (Photo courtesy Professor Richard G. Kessel.)

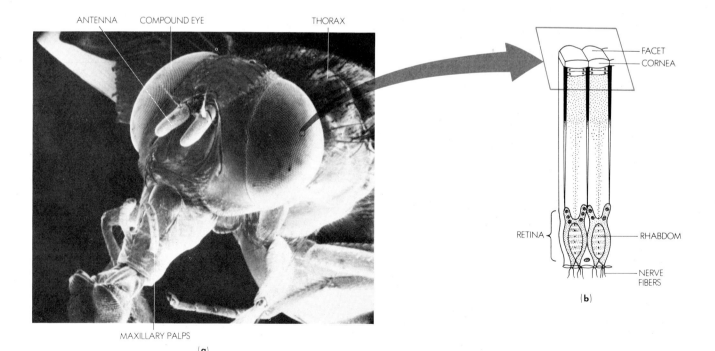

neuron that hooks up to fibers leading to the brain. Each ommatidium and facet, as a single visual unit, is capable of forming an image. Of course, we cannot be sure what the arthropod actually sees in the way of an image with this multifaceted eye, but we do know that the arthropod eye *is* especially sensitive to small changes in light intensity, or "flickers," so that it is highly effective in detecting motion. A blowfly, for example, can detect the flickering of a light at a frequency of 250 flashes per second. By comparison, in humans, flickering at a frequency greater than fifty flashes per second appears continuous. This special physiological adaptation is obviously of great value in helping arthropods (especially insects) escape capture by fast-moving predators such as birds, or the swat of a human hand.

Subphylum Trilobitomorpha, Class Trilobita

The class Trilobita consists of extinct forms that enjoyed great prominence in the early Paleozoic era, about 600 to 200 million years ago. Fossils from that period exist in great abundance, and show the many forms that trilobites took and the widespread distribution they had. A fossilized trilobite is shown in Fig. 29.25. Trilobites were flat with oval-shaped bodies and three distinct body regions: a head, thorax, and abdomen. Like all arthropods, trilobites were segmented; the thorax consisted of a variable number of segments, the abdomen of several fused segments. Each segment contained a pair of jointed appendages that show very little specialization from head to abdomen. The name "trilobite" means

Fig. 29.25
Fossil trilobites. Note the presence of two furrows running down the length of the organism on both sides, dividing the body into three "lobes": one median and two lateral. Fossils of these organisms are prevalent throughout strata from the Cambrian (600 million years ago) to the Permian (280 million years ago) periods. They were once the dominant arthropod on earth. Each segment of the body bore a pair of nondifferentiated appendages. (Photo courtesy Dr. Harold Levin.)

"three-lobed," referring to the division of the body into a central and two lateral lobes by furrows running longitudinally along the dorsal surface (see Fig. 29.25).

Trilobites were the most primitive arthropods of which we have any record, living or fossil. Among their most primitive traits is the lack of specialization among the appendages. There were no appendages specialized for food-getting, grinding (mouth parts), or locomotion. While the front appendages may well have been used more frequently for food-getting or tearing, and the posterior appendages for locomotion, structurally they all appear to be very much the same.

Subphylum Chelicerata

The subphylum Chelicerata contains four classes, of which we will discuss only two: the Merostomata (horseshoe crab) and the Arachnida (spiders, scorpions, mites, and ticks). All members of this subphylum are characterized by special appendages modified into mouth parts, the **chelicerae** (from which the subphlyum's name is derived). Chelicerae are located just in front of the mouth and serve in some cases (as in mites) for grinding and breaking up food, in other cases for picking up and holding food; in some groups, such as spiders, they are modified as fangs.

All the chelicerates, except for the Merostomata, ingest their food in a purely liquid form. After the food particles are shredded to a small size by the chelicerae, enzymes are poured from the midgut region of the digestive tract over the food particles, producing a kind of soup. The soup is pumped into the animal's stomach by muscles in the pharynx.

The body of chelicerates is divided into two main regions: the cephalothorax (fused head and thorax) and abdomen. The chelicerates lack antennae. In some species the cephalothorax contans five pairs of appendages, which usually serve as walking legs. All arachnids, however, use only the posterior four pairs for walking. The front pair is specialized into **pedipalps**. These are also feeding structures, and are much longer than the chelicerae. The abdomen contains no legs whatsoever.

Class Merostomata. The class Merostomata (sometimes called Xiphosura) contains an organism familiar to anyone who has frequented an Atlantic coast beach or the coasts of Japan, Korea, and Malaysia: the horseshoe crab (*Limulus polyphemus*). *Limulus* has existed on earth for millions of years virtually unchanged (see Fig. 29.26). Despite its name, the horseshoe crab is not really a crab at all, but a cousin of the arachnids whose other close relatives have all become extinct. *Limulus* is perhaps best known to biologists today because it has a very large optic nerve with a very prominent and easily accessible eye. It has thus become a major organism for research on visual stimulation, and the relationship between patterns of stimuli and processing of information by the optic nerve and brain (see Section 13.10).

Fig. 29.26
The horseshoe crab, phylum Arthropoda, subphylum Chelicerata, class Merostomata. These ancient forms are still plentiful today. Fossil forms show that the horseshoe crab has remained virtually the same for millions of years.

Class Arachnida. The arachnids, a prominent member of the subphylum Chelicerata, contain such groups as the spiders, scorpions, ticks, mites, and daddy longlegs (see Fig. 29.27). Although the various groups of arachnids differ among themselves in some rather important ways, they all have several features in common: simple (as opposed to compound) eyes; no antennae; and six pairs of appendages on the cephalothorax, four for walking and two for food grinding and/or holding. Spiders and scorpions carry out respiration by means of **book lungs,** a series of parallel, leaf-like sheets in a chitin-lined chamber. Other arachnids use tracheae. Many members of this class (including some species of scorpions and some spiders) are predators who use poison to immobilize their prey. Ticks suck animal blood and are also vectors for a serious rickettsial disease, Rocky Mountain spotted fever (see Section 27.2).

Subphylum Mandibulata

As the name implies, Mandibulata possess **mandibles** instead of chelicerae as their first pair of mouth parts. Mandibles are modified base segments of appendages, serving in the more primitive forms as chewing and biting structures. In most members of this subphylum there is an additional set of mouth parts, the **maxillae,** which also serve for grinding. All members of this subphylum have antennae.

Class Crustaceae. Crustaceans first appeared in the Cambrian period (600 million years ago), and may have shared a common ancestor with the trilobites. As shown in Fig. 29.28, the most prominent modern-day members of this class include crayfish, lobsters, crabs, shrimp, and copepods; but there are also a large number of crustaceans that hardly seem to resemble the common forms at all: barnacles, *Daphnia* (water fleas), and isopods ("sow bugs"), among others.

Many crustaceans, like the chelicerates, have their head and thorax fused into a cephalothorax that bears two pairs of antennae, a pair of mandibles, and two pairs of maxillae. The shape, size, and function of the remaining appendages of both the cephalothorax and abdomen vary greatly from species to species. Some have a high degree of differentiation among the appendages (for example, the lobster), while others show considerably less (for example, the isopods, or "pill bug"). Almost all crustaceans are aquatic, though the pill bug is terrestrial, living under old logs where it is moist. Except for the terrestrial isopods, which have trachea-like structures, all members of this class breathe by means of gills. Crustaceans are important ecologically, serving not only as food for human beings (shrimp, lobster, crayfish), but also for the blue whale and many other large aquatic vertebrates, which feed on tiny swimming crustaceans in the surface plankton.

Class Chilopoda and Diplopoda. The centipedes (Chilopoda) and millipedes (Diplopoda) are terrestrial arthropods, elongated in appearance with numerous segments (Fig. 29.29). Appendages are attached to each segment. In centipedes each segment bears one pair of legs, while in millipedes each segment bears two pairs.

CLASS ARACHNIDA

Fig. 29.27
Representatives of the class Arachnida (phylum Arthropoda): (*a*) Black widow spider; (*b*) Black widow spider emerging from egg case; (*c*) Tick before feeding (left) and engorged with blood after feeding (right); (*d*) Daddy longlegs; (*e*) Scorpion. Arachnids have four pairs of legs and no antennae. While some species of spiders and scorpions are poisonous, the extent and danger of such poisons has been greatly exaggerated. (Photos *a* and *e* Carolina Biological Supply Company; *b* courtesy Professor J. Norman Grim, Northern Arizona State University; *c* and *d* Ron West.)

(a)

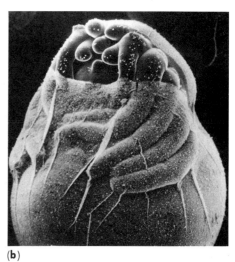

(b)

(c)

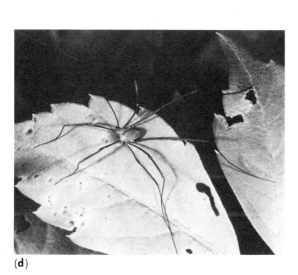

(d)

(e)

CLASS CRUSTACEA

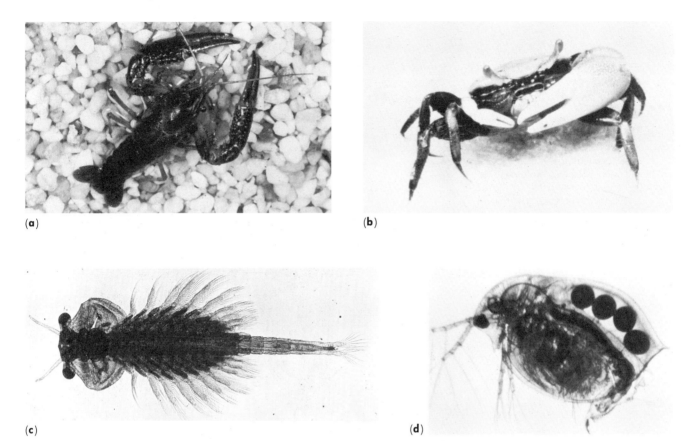

(a)

(b)

(c)

(d)

Fig. 29.28
Representatives of the phylum Arthropoda, class Crustacea: (*a*) photograph of a crayfish, dorsal view; (*b*) Fiddler crab, showing much-enlarged claw on one side; (*c*) Freshwater shrimp, with much less specialization of appendages than the crayfish; (*d*) Freshwater copepod, *Daphnia*, photographed in polarized light. (Photos, Carolina Biological Supply Company.)

CLASSES CHILOPODA AND DIPLOPODA

Fig. 29.29
Examples of the phylum Arthropoda, classes Chilopoda (centipedes) and Diplopoda (millipedes). Centipedes have one pair of appendages per segment, while millipedes have two. These segmented forms show relatively little serial specialization of appendages from front to back. Although modern centipedes and millipedes are not ancestral to the Arthropods, it is thought that they might be descendants of an early, nonspecialized arthropod ancestor.

Neither centipedes nor millipedes usually have as many appendages as their names imply: centipedes have between 15 and 173 segments, while millipedes range between 25 and 200 segments. Centipedes are carnivorous, with one anterior pair of appendages modified as poison claws. Most millipedes, on the other hand, are not carnivorous (one order is, however) and have no poison claws. Both centipedes and millipedes have antennae,

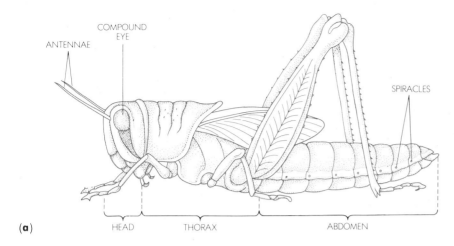

(a)

ANTENNAE

COMPOUND EYE

SPIRACLES

HEAD THORAX ABDOMEN

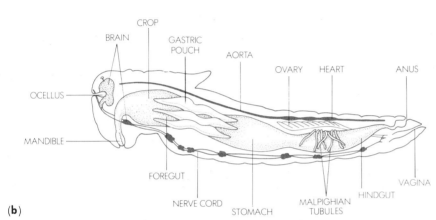

(b)

BRAIN

CROP

GASTRIC POUCH

AORTA

OVARY HEART ANUS

OCELLUS

MANDIBLE

FOREGUT

NERVE CORD

STOMACH

MALPIGHIAN TUBULES

HINDGUT

VAGINA

Fig. 29.30
Internal and external body organization of an insect. The grasshopper is a good representative of the insects because it is less specialized than many other species. (*a*) External anatomy, showing the division of the body into head, thorax, and abdomen, and the three pairs of legs. (*b*) Internal anatomy, showing the dorsal blood vessel (aorta), ventral nerve cord, heart, and excretory system (Malpighian tubules). There is a well-developed brain and several ventral ganglia, providing for well-coordinated motor control. Respiration is through tracheae, which exit to the outside through the spiracles, shown in (*a*) above.

and both respire by tracheae. Excretion is by Malpighian tubules. In the millipedes each double-legged segment has resulted from the fusion of two single-legged segments during the course of evolution.

Class Insecta. The insects are the largest group of organisms on earth. There are more species of insects than of all other animal groups combined. The total number is estimated at about 10^{18}, about a billion billion, or a billion insects for every human being alive. It is clear that in terms of numbers alone, we do not live in the age of humans, or even of mammals, but the age of insects. Their diversity is as great as their numbers; they inhabit virtually every ecological niche and every habitat imaginable. Among the arthropods, insects have diversified into terrestrial niches as successfully as the crustaceans have diversified in aquatic environments.

Insects became prominent during the Mississippian and Pennsylvanian periods. Insect evolution is associated with two great periods of adaptive radiation: one about 350 million years ago, the other about 135 million years ago in the Cretaceous period, along with the appearance and spread of flowering plants. As pollinators of flowering plants, some insects (ancestors of modern butterflies, bees, and wasps) found a wholly

new opportunity to exploit, which provided an important service for plants. In this new wave of evolution, insects and plants co-evolved in ways that are both intricate and, from today's perspective, marvelously adaptive for both species.

The insect body is divided into three distinct parts: the **head, thorax,** and **abdomen** (see Fig. 29.30a). The head consists of a series of completely fused segments, bearing a number of sensory receptors: a pair of antennae, compound eyes, and three pairs of mouth parts (mandible, palp, and labium) derived from ancestral appendages (refer back to Fig. 29.23). The thorax is composed of three segments, each with a pair of locomotor appendages (walking legs). In many insects (but not all) the second and third thoracic segments each bear a pair of wings. Insects are the first organisms on earth to employ flight as a means of locomotion. The abdomen is composed of various numbers of segments (generally twelve or less), none of which bear true appendages. The remnants of appendages may be found in some of the egg-laying and mating structures on the posterior portion of the abdomen.

The internal anatomy of the grasshopper is shown in Fig. 29.30(b). Insects breathe through tracheae, air tubes running throughout the body and opening to the

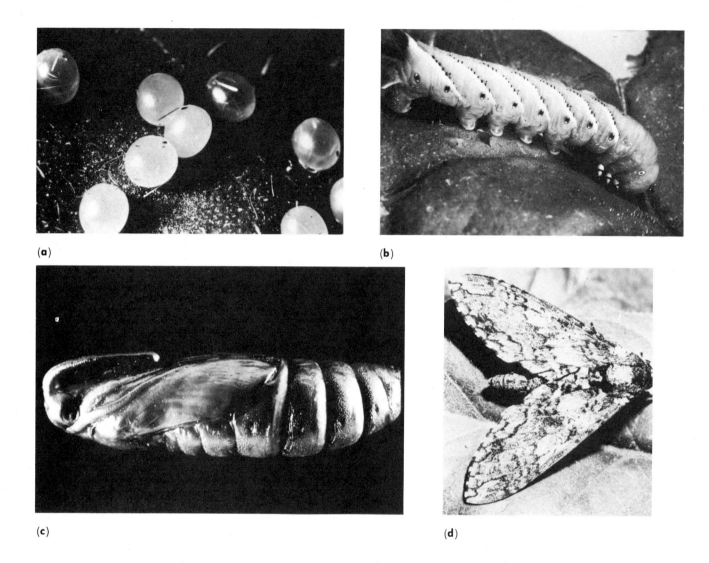

(a)

(b)

(c)

(d)

outside through small openings, the **spiracles** (see Fig. 29.30a). Like the annelids and other arthropods, the insects have a dorsal blood vessel, the aorta, and a ventral nerve cord. They have a true coelom, and an open circulatory system. Blood is moved about through the dorsal aorta and the coelom by the contraction of enlarged portions of the dorsal vessel, referred to as the "heart." Because of the organization of their nervous system and the interaction of the nervous system with hormonal controls, insects are capable of a variety of complex and sometimes subtle behaviors—especially those associated with mating (see Chapter 30, especially Sections 30.2, 30.5, and 30.8 and Supplement 30.1).

Most insects reproduce sexually, though some (such as aphids) have recurrent parthenogenetic generations—that is, will reproduce several generations of females without fertilization; however, generations of males are usually produced at least once a season, so that sexual reproduction also occurs regularly in these groups. The insect life cycle is particularly interesting, because it represents clearly the process of metamorpho-

Fig. 29.31
Metamorphosis in the life cycle of the tobacco hornworm. (a) Eggs ready to hatch. (b) Mature larva, similar to a caterpillar (largely a feeding stage). (c) After feeding for a period of time the larva becomes a pupa (for many insects this pupal case is called a cocoon); inside the pupal case, guided by hormones, the basic larval form is transformed into the adult form. This stage is particularly dramatic, and is sometimes erroneously called *the* metamorphosis stage. (d) Adult hornworm after it has emerged from the pupal case. (Photos, Carolina Biological Supply Company.)

sis. The stages of a typical metamorphosis are shown in Fig. 29.31. In species that show a complete metamorphosis (or **homometabolous development**), the egg hatches into a segmented larva, such as a caterpillar (Fig. 29.31b). The caterpillar is a feeding stage, eating voraciously (that is why they are so destructive to vegetation) for a period of time while it stores up food reserves for the next stage, pupation. After a period of time the larva settles down, and either attaches onto

Table 29.2
The Insects Are the Largest Group of Animals. There Are about 26 Different Orders, Consisting of over 1,000,000 Species, Including the following Common Orders.

Order Orthoptera.   Grasshoppers and cockroaches

Order Isoptera.   Termites

Order Odonata.   Dragonflies and damsel flies

Order Anopleura.   Lice

Order Hemiptera.   Water boatmen, bedbugs, backswimmers

Order Homoptera.   Cicadas, aphids, and scale insects

Order Coleoptera.   Beetles, weevils

Order Lepidoptera.   Butterflies and moths

Order Diptera.   Flies, mosquitos, gnats

Order Hymenoptera.   Ants, wasps, bees, and gallflies

some object or becomes partially buried in the soil and forms a protective covering about itself. The larva then develops into the pupa (Fig. 29.31c). Although the pupa seems inactive, inside the tissues are undergoing considerable reorganization to form a wholly new body structure, the adult (or, sometimes referred to as the **imago**). When the imago is mature the pupal case splits open and the adult emerges. It is still soft and crumpled. But blood is soon pumped into its wings and within hours it has dried out and hardened, and is able to fly.

Insects such as the grasshopper do not undergo metamorphosis. Rather, what hatches from the egg is a young form, the nymph, which resembles the adult in all its details (except for proportion, with a somewhat larger head and pair of back appendages). This type of development is known as **incomplete metamorphosis,** or **hemimetabolous development.**

There are over twenty-six orders of insects, some of which are listed in Table 29.2.

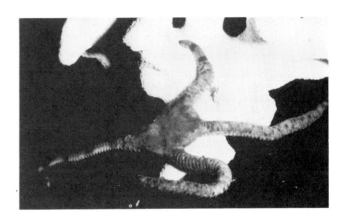

The Deuterostomia

29.11
THE PHYLUM ECHINODERMATA

The echinoderms (spiny-skinned animals) are bottom-dwelling marine animals, found from very shallow intertidal zones to depths of over 1000 ft. The phylum includes a number of easily recognized forms: the starfish, sea urchins, and brittle stars, among others (see Fig. 29.32). The adults are generally radially symmetrical, with calcareous, spine-bearing plates on the outer surface. The larvae of echinoderms are bilaterally symmetrical, a feature which gives an indication of this group's affinity to the chordates. Radial symmetry is thought to be an adaptation to a largely sessile, or at least slow-moving way of life. Fossil evidence suggests that the ancestors of the echinoderms were sessile; their modern

Fig. 29.32
Representatives of the phylum Echinodermata. Echinoderms are mostly radially symmetrical as adults, but bilaterally symmetrical as larvae, and are exclusively marine. Some members of this phylum share with the primitive chordates a common larval type. (Photos, Carolina Biological Supply Company.)

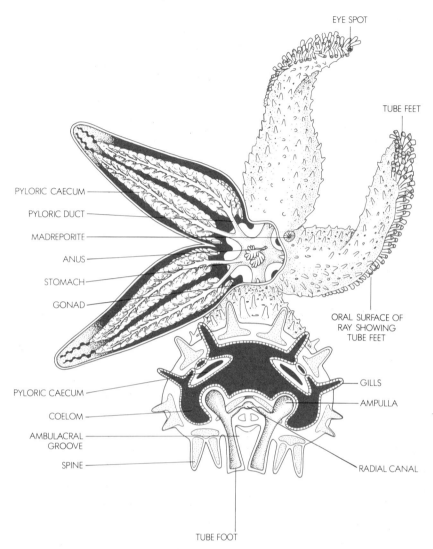

EYE SPOT

TUBE FEET

PYLORIC CAECUM

PYLORIC DUCT

MADREPORITE

ANUS

STOMACH

GONAD

ORAL SURFACE OF
RAY SHOWING
TUBE FEET

PYLORIC CAECUM

COELOM

AMBULACRAL
GROOVE

SPINE

GILLS

AMPULLA

RADIAL CANAL

TUBE FOOT

Fig. 29.33
Cutaway and cross section of a typical echinoderm, the starfish, showing all the major internal and external organs. Note that the cross section of the arm (lower center, facing directly out) shows a distinct coelom, and the downward projecting tube feet. Each arm also has two lateral digestive canals, the pyloric caeca, which carry nutrient to all parts of the animal; digested food material diffuses from the pyloric caecum into the coelom, where it is circulated by the beating of cilia. Echinoderms have no well-developed or definite circulatory system. (After Hegner and Stiles, p. 378).

descendants *do* move about, but so slowly that one might easily think they were stationary.

The structure of the body of echinoderms consists of an internal skeleton of many calcareous (calcium-containing) plates integrated into the animal's body wall. Inside the body wall is a well-developed coelom where the various body organs are located. A diagrammatic representation of the internal structure of one major echinoderm type, the starfish, is shown in Fig. 29.33. The digestive system is short and highly specialized. The mouth, in the center of the underside of the organism, opens into a short esophagus which in turn leads into a thin-walled sac, the **stomach.** Above the stomach is a very slender intestine, which opens to the outside through the anus. The digestive system ramifies from the stomach into each of the arms. Starfish are carnivorous, feeding on larger marine forms such as clams, oysters, or small fish, and on smaller forms such as crus-

taceans, worms, and snails (including sea urchins and sand dollars). In feeding, echinoderms such as the starfish force clams (or other bivalves) open and extrude a portion of their stomach *over* the prey. Enzymes produced by the stomach walls digest the unfortunate bivalve within its own shell; such echinoderms need huge digestive glands (see Fig. 29.33). Other echinoderms, such as the crinoids, are filter-feeders, and have much smaller digestive organs.

Echinoderms have no special excretory system, and their circulatory sytem is not extensively developed—an adaptation to both an aquatic environment, where wastes can be eliminated directly into the water, and to a very slow-moving existence, which does not require an extensive, fast-delivery vascular system. They have no heart; rather water and body fluid within the coelom are kept in motion by the beating of cilia. The nervous system of most echinoderms is simple. A ring of nervous

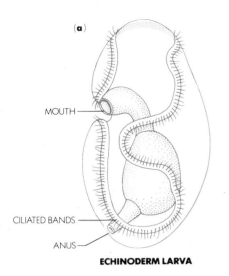

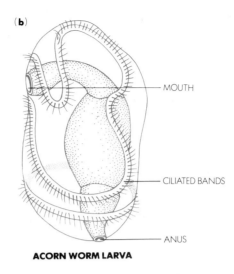

MOUTH

CILIATED BANDS

ANUS

ECHINODERM LARVA

MOUTH

CILIATED BANDS

ANUS

ACORN WORM LARVA

Fig. 29.34
Comparison of the bipinnaria larva of an echinoderm (*a*) with the similar tornaria larva of the acorn worm *Balanoglossus* (*b*). Both larvae are of the so-called Dipleurula type.

tissues encircles the mouth, with five radial nerves extending outward and, in starfish, into the arms. The sense organs are poorly developed.

A special feature of echinoderms is the presence of a **water-vascular system,** and its associated **tube feet.** As shown in Fig. 29.33 numerous tube feet are arranged on either side of a central groove (the ambulacral groove). Tube feet are hollow with a sucker on the outer end. Echinoderms use tube feet for locomotion and for feeding. By attaching its tube feet to both sides of a bivalve shell, for example, and exerting a strong and continous pull, starfish can eventually force a clam shell open (simply by exhausting the clam's strong adductor muscles, which normally hold the shell shut). Studies have shown that a clam can resist a sudden pull against its shell of up to 4000 grams (as everyone knows who has tried to open clam shells). However, the clam finally yields to a pull of 900 grams applied over a continual period of time. An average-size starfish can exert a pull of 1350 grams for several hours—quite enough to overcome its prey.

Locomotion in echinoderms operates by an unusual mechanism involving the tube feet and the water-vascular system, which functions on a kind of hydrostatic pressure principle. The inner end of every tube foot connects to a radial canal that projects out from a central ring canal surrounding the esophagus (refer again to Fig. 29.33). In starfish a single radial canal passes down the center of each arm just underneath the ambulacral groove. Water enters this more or less closed system through a prominent sieve-like plate near the center on the upper surface (called the **sieve plate** or **madreporite**). The circular and radial canals and the tube feet connected to them are thus filled with water under a slightly hydrostatic pressure. In movement, the tube feet and water-vascular system function as follows: At the proximal end (base) of each tube foot (inside the

organism) is a muscular chamber known as the **ampulla** that projects above but is connected to the radial canal. When the ampulla contracts (by muscles in its wall) the liquid within it is forced into the tube foot (but is prevented by a valve from flowing back into the radial canal). This causes the foot to extend; it can then attach to the surface through its suction mechanism. Muscles arranged longitudinally in the walls of the tube foot then contract, shortening the structure and pulling the organism forward. As the tube foot shortens, water is forced back into the ampulla, and a step has been taken.

Echinoderms have separate sexes, but sperm and eggs are shed into the water, where external fertilization occurs. (In fact, this process can be easily induced in a beaker of sea water in the laboratory, and the early stages of embryonic development can be observed under the microscope.) Cleavage results in a ciliated, free-swimming larva, the **bipinnaria** (sometimes called *dipleurula*; see Fig. 29.34a), which has its own complete digestive tract.

Like arthropods, the echinoderms have been a dominant form at times in the past. In origin they go back at least to the Cambrian, 600 million years ago. Fossil echinoderms were sessile, but otherwise show all the major features of their modern descendants. Because of their hard calcareous external coverings, echinoderms have left an abundant fossil record.

29.12
THE PHYLUM HEMICHORDATA

The hemichordates are marine organisms, the most prominent group of which is known as the acorn worms. A typical adult acorn worm, *Dolchiglossus* (shown in Fig. 29.35), lives in U-shaped burrows in the mud along shore lines. It can become relatively large, up to 45 cm in length. The name of the phylum is itself de-

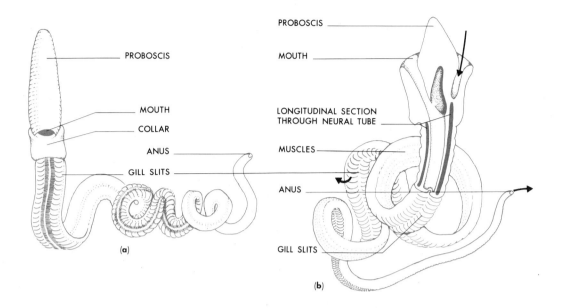

Fig. 29.35
Representative of the Hemichordata, the acorn worm *Dolchi-glosus.* (a) External view showing the major anatomic features: proboscis, collar, and long, worm-like body. (b) Internal view, showing gill slits, neural tube, and the pathway of water (arrows) in through the mouth and out through the gill slits. Once classified as Chordates, the acorn worms (and relatives in the Hemichordata) are now given a phylum of their own.

scriptive of the group's evolutionary position: they are truly "half-chordates," and bear resemblances to both the chordates and the nonchordates (in this case echinoderms).

The body of *Dolchiglossus* consists of an anterior proboscis that vaguely resembles an acorn (hence the common name), a collar region, and a long worm-like body. The mouth is located on the ventral surface at the junction between the proboscis and the collar. Like the chordates, the acorn worms have gill slits in the pharynx, a trait shared only by the hemichordates and the chordates. These slits serve as the exit for water drawn in through the mouth. Like echinoderms, the acorn worms have a ciliated, free-swimming bipinnaria

larva, the **tornaria** (see Fig. 29.34b). At the same time, they have a dorsal *and* a ventral nerve cord—the former being a chordate characteristic, the latter a nonchordate characteristic. There is a cartilaginous structure in adult acorn worms, just posterior to the mouth, that was once thought to represent a primitive notochord—that is, a structural element composed of connective tissue running along the dorsal axis; however, this primitive notochord is no longer thought to be homologous to chordate notochords at all. It is not extensive in an adult, and its embryological origins appear to be different from those of a true notochord. Thus the hemichordata lack one of the major defining characteristics of the chordate group, the presence at some stage in the life cycle of a dorsal notochord.

29.13
THE PHYLUM CHORDATA

The major groups of chordates and their characteristics are summarized in Table 29.3. All chordate embryos have a notochord, a relatively rigid supporting structure made of cartilage, and a dorsal, hollow nerve chord. The notochord lies under the dorsal nerve cord and

Table 29.3
Characteristics of the Phylum Chordata

Subphylum and class	Notochord and/or vertebral column	Respiration	Segmentation	Circulation	Skeletal support	Other traits
Urochordata (Tunicates)	Dorsal notochord in tail region of larva	Gill slits	None	Simple one-chambered heart; open circulation (hemocoel)	None	Filter feeders; gill slits function to take O_2 from water; free-swimming larvae

(Continued)

Table 29.3 (Continued)

Subphylum and class	Notochord and/or vertebral column	Respiration	Segmentation	Circulation	Skeletal support	Other traits
Cephalochordata (Lancelets, e.g., *Amphioxus*)	Full dorsal notochord throughout adult life	Gill slits	Yes (muscles of body wall)	Open circulatory system, blood flowing from vessels into lacunae feeds tissues; no single heart; several parts of vascular system provide irregular and slow (once every 2 min) beats	None	Filter feeders; though capable of swimming, spend much time with tail burrowed in sand; dorsal nerve cord
Vertebrata Agnatha (Jawless fish, e.g., lamprey)	Dorsal, notochord only in embryo; cartilaginous vertebrae in adult	Gill slits	Yes; vertebral column (muscles of body wall and nerves from neural tube)	Closed circulatory system; well-developed S-shaped heart, with three consecutive chambers; some contractile "hearts" in the venous system	Cartilage	Adults feed by attaching onto larger fish with an oral sucker and sucking blood; young hatch from eggs; has free-swimming larva that resembles *Amphioxus*; very important freshwater group during Ordovician to Silurian; probably evolved in freshwater
Placodermi	Dorsal, notochord only in embryo; bony (?) vertebral column in adult	Gills	Yes		Bone (?)	All extinct; armored fish with hinged jaws; prominent during Devonian; ancestral to both the Chondrichthyes and Osteichthyes
Chondrichthyes (Sharks, skates, rays)	Dorsal, notochord only in embryo; cartilaginous vertebral column in adult	Gills	Yes	Closed circulatory system; S-shaped heart of four chambers arranged linearly	Cartilage	Young hatch from eggs; cartilaginous skeleton evolved from a previously bony one in placoderms; carry out osmoregulation by retaining high concentration of urea in body fluids
Osteichyes (Bony fish)	Dorsal notochord only in embryo; bony vertebral column in adult	Gills	Yes	Closed circulatory system; heart of four chambers, but two distinctly differentiated into atrium and ventricle	Bone	Young hatch from eggs; probably evolved in freshwater; contains true bony fish and lobe-fin fishes (which are now restricted to six species); thought to be ancestral to higher vertebrates
Amphibia (Frogs, toads, salamanders)	Dorsal notochord only in embryo; bony vertebral column in adult	Gills as larvae; lungs as adults	Yes	Closed circulatory system; three-chambered heart (two atria, one ventricle)	Bone	Young hatch from eggs; undergo extensive metamorphosis from larva to adult; limbs without claws; poikilothermic
Reptilia (Snakes, lizards, crocodiles)		Lungs	Yes	Closed circulatory system; three-chambered heart with partial septum in ventricle	Bone	Young hatch from eggs (some hatch inside mother's body); poikilothermic
Aves (Birds)		Lungs	Yes	Closed circulatory system; four-chambered heart (two atria, two ventricles)	Bone	Young hatch from eggs; bones hollow; homeothermic
Mammalia		Lungs	Yes	Closed circulatory system; four-chambered heart	Bone	Except for platypus (which lays eggs) young born direct (not encased in egg shell); nourished with milk from mammary glands; homeothermic.

above the gut. Although in most groups the notochord disappears and is replaced by a backbone before the organism reaches adulthood, all chordates possess a notochord as embryos. Another characteristic that all chordates share is the presence of gill slits at some time during their life. The gill slits are a series of openings in the pharynx (throat). In the invertebrate chordates and the lower vertebrates such as fish, the gill slits remain into adulthood and permit the flow of water from the mouth through the pharynx and gills to the outside. In the higher vertebrates the gill slits appear only in the young embryo. As the embryo develops, one of the gill slits is transformed into the Eustachian tube, a canal that connects the middle ear with the pharynx.

The Nonvertebrate Chordates: Subphyla Cephalochordata and Urochordata

The so-called nonvertebrate chordates are represented today by a relatively small group of marine organisms including the lancelet *Amphioxus* (subphylum Cephalochordata) and the tunicates (subphylum Urochordata). They are called nonvertebrate chordates because they have many characteristics (especially as adults) that are reminiscent of various lower groups; yet, as larvae they have some distinguishing vertebrate traits, especially the presence of a distinct dorsal notochord.

Cephalochordata. The subphylum Cephalochordata includes a group of small, marine, fish-like organisms, the lancelets represented by the genus *Amphioxus* (sometimes known by the subgeneric name *Branchiostoma*) as shown in Fig. 29.36. *Amphioxus* bears many chordate characteristics, and in many ways can be viewed as a prototype of the chordate group. Throughout its life *Amphioxus* possesses a notochord, a dorsal nerve cord, and **gill slits.** It also shows distinctly segmented muscular arrangement all along its midregion.

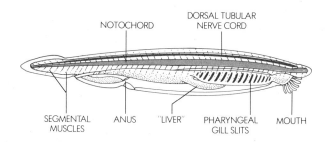

Fig. 29.36
The lancelet *Amphioxus*, showing the basic structure of the chordate phylum. The defining traits of chordates are the presence of a dorsal notochord and a dorsal nerve cord, along with gill slits appearing at some point in the life cycle.

Amphioxus is capable of swimming, but spends a good portion of its adult life partially buried in the sand. It is a filter feeder, drawing water in through the mouth and straining it in the pharynx. The water then passes out through the pharyngeal gill slits. In this passage, oxygen is removed by the gills.

Urochordata. The subphylum Urochordata contains organisms known as tunicates or "sea squirts." The tunicates are all marine, and derive their popular name from the fact that, if disturbed, they contract their bodies

Fig. 29.37
Representative of the nonvertebrate chordates, the Urochordate *Molgula*. (*a*) Internal structure of the adult; the arrows indicate the direction of water flow through the organism, which is mostly a filter-feeder. (*b*) Internal structures of *Molgula* larva, showing the dorsal nerve cord and notochord. Arrows indicate direction of water flow in both adult and larvae.

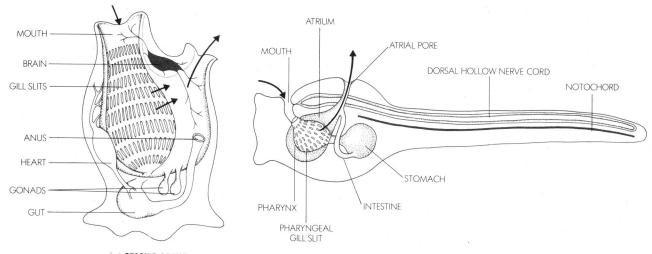

(a) SESSILE ADULT   **(b) FREE-SWIMMING LARVAE**

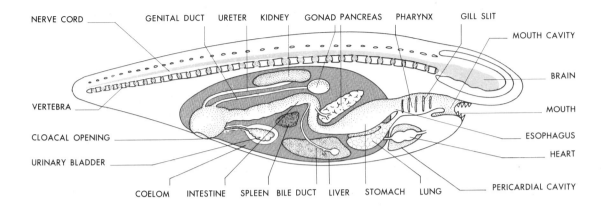

NERVE CORD   GENITAL DUCT   URETER   KIDNEY   GONAD PANCREAS   PHARYNX   GILL SLIT

MOUTH CAVITY

BRAIN

VERTEBRA

MOUTH

CLOACAL OPENING

ESOPHAGUS

URINARY BLADDER

HEART

PERICARDIAL CAVITY

COELOM   INTESTINE   SPLEEN   BILE DUCT   LIVER   STOMACH   LUNG

Fig. 29.38
Schematic diagram of a generalized vertebrate, showing the major characteristics of brain, dorsal nerve cord, vertebral column, heart, and remnant of notochord (which often appears as a full-fledged notochord only temporarily during embryonic development).

forcefully, ejecting streams or squirts of water. One example of the tunicates, *Molgula*, is shown in Fig. 29.37. As adults, these organisms are sessile, remaining attached to the ocean bottom or some other substrate; they are filter feeders, entrapping food particles on a sheet of mucus in a ciliated groove along the pharynx known as the **endasty**. Aside from the presence of gill slits, it seems difficult at first to see how adult tunicates qualify as chordates. However, the possible affinity between the chordates and the invertebrates can be found in the tunicate larva (shown in Fig. 29.37b), which is very much in the chordate mold—in many ways resembling *Amphioxus*. The larvae are free-swimming and bilaterally symmetrical, possessing a well-developed dorsal nerve cord and a notochord beneath it in the posterior half of the body.

The Vertebrates: General Characteristics

The vertebrates are by far the most common representatives of the chordate phylum. The generalized body plan of all vertebrates is shown in Fig. 29.38. Vertebrates possess a notochord at some time during embryonic development. In the lower vertebrates it persists relatively unchanged throughout adult life. In the higher vertebrates, from the bony fish to the mammals, the notochord is replaced progressively during embryonic development by the vertebrae, which develop around it and provide greater strength. In most adult vertebrates the notochord is present only as the gelatinous material of the intervertebral disc. The dorsal nerve cord, which in lower vertebrates and other chordates lies above the notochord, is completely surrounded by the vertebral column in the higher vertebrates. The nerve cord runs through each individual vertebra in the hollow central canal.

The vertebrates comprise the subphylum Vertebrata of the phylum Chordata. There are many kinds of vertebrates, generally falling into nine classes: the Agnatha (jawless fish), Placodermae (primitive jawed fish), Chondrichthyes (fish with cartilage skeletons), Osteichthyes (fish with bony skeletons), Amphibia (which live their early life in the water and their adult life largely on land), Reptilia (mostly land dwellers), Aves (birds), and Mammalia (mammals).

Class Agnatha (Jawless Fish).   Once a large and widely distributed group in the Ordovician period (500 million years ago), the Agnatha are reduced today to two groups, the hagfish and the lampreys. As the name of this class implies, these fish lack hinged jaws. As shown in Fig. 29.39, the lamprey has a rounded mouth adapted for sucking. It feeds by attaching itself to other fish and, with teeth and a rasping tongue, chewing away at the host's tissues. By virtue of their parasitic habits, lampreys have become a major problem for the fishing industry today. For example, after the construction of the Welland Canal around Niagara Falls as part of the St. Lawrence Seaway, lamprey were able to migrate from the sea (their normal habitat) into freshwater (the Great Lakes). Here, in a very few years, they adapted to the nonsaline environment and became a major parasite, virtually eliminating the trout-fishing industry in the Great Lakes. There is now hope that a chemical that is toxic to young lampreys, but not to other fish, will prove effective in eliminating this major aquatic pest.

The skeleton and vertebrae of the Agnatha are composed of cartilage, having lost all trace of bone (although they evolved from bony ancestors). Lampreys have prominent gill slits and lack paired fins (see Fig. 29.39). The lamprey larva, which is a filter feeder, bears a striking resemblance to *Amphioxus*.

Today's lampreys and hagfish are quite different from their Ordovician ancestors. The latter were covered with thick plates of bone-like material. These now-extinct forms, called ostracoderms, often had bizarre shapes and ornamentation. They were small fish, being mostly less than 30 cm in length, and lacked fins. The

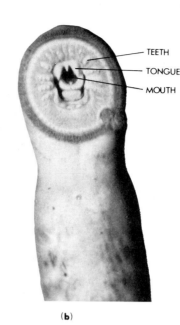

TEETH

TONGUE

MOUTH

Fig. 29.39
(*a*) A sea lamprey. Note the prominent gill slits and the lack of paired fins. (*b*) Close-up of lamprey mouth showing rasping teeth (around outside) and tongue in center. (*c*) Longitudinal section of lamprey showing notochord, dorsal nerve cord, gill slits, and sucking mouth. (Photos, Carolina Biological Supply Company)

(a)

(b)

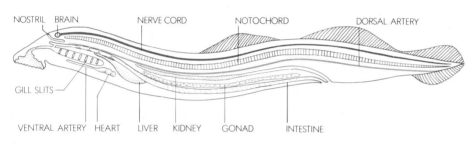

NOSTRIL   BRAIN   NERVE CORD   NOTOCHORD   DORSAL ARTERY

GILL SLITS

VENTRAL ARTERY   HEART   LIVER   KIDNEY   GONAD   INTESTINE

(c)

ostracoderms had a bony skeleton and vertebral column. They appear to have given rise on the one hand to the line that led to today's lampreys and hagfish, and on the other to a second group of armored, now-extinct fish, the placoderms (see below).

Class Placodermi. The placoderms are extinct armored fish with a bony skeleton. They first appeared in the Silurian, rising to prominence as the ostracoderms declined. During the Devonian period, the placodermi became an important group, but died out by the end of the Permian period (280 million years ago). They differed from their ostracoderm ancestors in two basic ways, both relating in part to a change in diet: they had paired fins and hinged jaws (see Fig. 29.40). Paired fins aided mobility by giving the animals a stability in the water that nonfinned fish lack. The earlier agnaths had probably been bottom or filter feeders, spending little time actively swimming through the water. With the stability that fins conferred, swimming and maneuvering became more feasible, making possible a more active, predatory existence. Predation was made

Fig. 29.40
An extinct placoderm from the Ordovician. This reconstruction shows the hinged jaws and the covering of armored plates. (Photo courtesy Field Museum of Natural History, Chicago.)

more effective by the innovation of hinged jaws; fish could then capture their prey instead of merely sucking up organic debris from the ocean or river bottoms. Because fins and jaws appear to have evolved together in the early placoderms, the entire ecology of the fish, and ultimately the vertebrate line, was revolutionized. If we think about the ramifications of the hinged jaw today, it should be apparent that it has made possible not only a wider range of diets for all vertebrates, but also, for humans, a highly precise order of communication through spoken language.

By the Devonian period the placoderms had made an extensive adaptive radiation throughout freshwater rivers and streams. However, by the end of this period, (about 350 million years ago), significant climatic changes had already set in on earth. Many lakes, streams, and rivers dried up or shrank in size, and the temperature became much warmer. Rigorous selection forced the ancestral placoderms to evolve new adaptations. Some groups migrated to the oceans and became adapted to salt water, ultimately giving rise to the Chondrichthyes (cartilaginous fish: sharks, skates, and rays). Another group remained in fresh water and gave rise there to the Osteichthyes (bony fish: perch, trout). This group eventually split, also in response to drying conditions. One group again made its way to the oceans. Another evolved an air-storage organ, or lung, as a means of utilizing atmospheric oxygen when oxygen dissolved in the stagnant water became exhausted (see discussion of Chondrichthyes below).

Class Chondrichthyes (Cartilaginous Fish).   The Chondrichthyes (sharks, rays, skates; see Fig. 29.41) are characterized by a completely cartilaginous skeleton; bone does not appear at any time in the development of these organisms. Since the Chondrichthyes descended from the placoderms, which had a bony skeleton, a cartilaginous skeleton is not necessarily a more primitive condition than a bony skeleton. The skin of members of the class Chondrichthyes is composed of small pointed "teeth" called **denticles,** which give the body a rough abrasive texture. Like the placoderms, the Chondrichthyes have hinged jaws. Most species of the Chondrichthyes are predators, but a few are plank-

ton feeders. Fertilization of the egg occurs within the female.

When the cartilaginous fish returned to the sea they faced a critical problem of osmoregulation: how to keep their body fluids from moving outward into the hypertonic seawater. The Chondrichthyes show one major solution to this problem. Sharks, skates, and rays retain high concentrations of urea in their body fluids, making their blood plasma more isotonic to the sea water. Today's Chondrichthyes have a urea concentration of 2.5 percent compared to approximately 0.02 percent found in other vertebrates.

The water-retention problem brought about two other adaptations in the Chondrichthyes. Even more than the adult body, the eggs of sea animals are subject to the hypertonicity of the water. In the cartilaginous fish the eggs are fertilized *internally,* thus providing a protective environment in the earliest stages of development. Furthermore, by the time the eggs are laid (with the embryos developing inside them), they are covered with a tough, leathery skin that greatly retards the outflow of water.

Class Osteichthyes (Bony Fish).   The Osteichthyes are the second class of fish to arise from the placoderms and include most of the fish with which we are familiar today. They are a large class (more than 17,000 species are known, and it is estimated that the total in existence may exceed 40,000). They all have paired fins and hinged jaws, with bodies covered by scales (see Fig. 29.42). They range in size from one or two centimeters to six meters, and some can swim as fast as 80 km per hour. As the name (*osteo*) implies all members of the class have a skeleton composed of bone.

During the drought that occurred toward the end of the Devonian period, the freshwater bony fish (derived from placoderm ancestors) developed a pair of pouched outgrowths of the pharynx that served as air-storage

Fig. 29.41
A representative of the class Chondrichthyes (sharks, skates, rays): the sand shark (note the prominent gill slits). (Photo, Runk/Schoenberger, Grant Heilman Photography.)

Fig. 29.42
A member of the class Osteichthyes, true, or bony fish. (Photo courtesy Northeast Fisheries Center, Woods Hole, Mass.)

SUPPLEMENT 29.1
THE TRANSITION TO LAND—OR, VERTEBRATES' DAY AT THE BEACH

Most evolutionists now believe that the land-dwelling vertebrates evolved from freshwater crossopterygians, lobe-fin fish that arose from the Osteichthyes during the Devonian. How could such an unlikely-seeming event have occurred? What environmental factors may have favored such a transition? Several factors may have been responsible for the primary invasion of land by vertebrates. One was the fact that the land was covered with vegetation and was already inhabited by certain invertebrate forms, all of which could serve as new food supplies. Another was the fact that the aquatic habitats may have been getting crowded, producing a pressure on organisms to migrate into less-dense surroundings. A third was the fact that many fish lived in small lakes or ponds that often shrank or dried up altogether during certain seasons of the year. There was an advantage to be gained if a fish could migrate across narrow strips of land from smaller to larger bodies of water. All three of these factors may have combined to produce selective pressures toward terrestrial invasion by ancestral fish, all or most of which appear to be freshwater forms.

Whatever the exact pressure(s) involved, the ancestral vertebrates that began the invasion of land faced enormous problems of transition. The most prominent was certainly respiration—keeping the tissues of the body supplied with oxygen during periods out of the water. Other problems were locomotion and avoiding desiccation. Water buoys up the body, making it much easier to move. On land, fish would be virtually immobile. Drying out is no problem for animals in freshwater but on land it can quickly become a major environmental hazard.

How could ancestral fish ever give rise, through the small variations suggested by Darwin's theory, to creatures that inhabit such a varied and hostile environment as the land? Fossil evidence and examination of at least one living species suggest how this development could have taken place. Paleontologists have found fossil forms of primitive fish known as crossopterygians, or lobe-fin fish, that possess lungs as well as gills (see Diagram A). These early fish may have used lungs to keep a ready supply of air available in case they encountered shortages in the water where they lived. The later fish to descend from the crossopterygians converted the lungs into a swim bladder (see text). The crossopterygians also possessed lobe fins, meaning the fins were larger than those of modern-day fish and had bones within them (Diagram A). It is hypothesized that these primitive lobe-fin fish were the ancestors of all terrestrial vertebrates. The lobe-fin fish could have used their bony fins to walk across small tracts of land from pond to pond; they could have stored oxygen for this journey in their "lungs." Natural selection would favor those individuals that could stay longest out of water, since they had more opportunity to find favorable ponds by ranging farther afield in their exploration of land.

Dramatic verification of the anatomy of crossopterygians as deduced from the fossil record came in 1938 when fishermen caught a *living* crossopterygian off the coast of Madagascar. This fish, called a coelocanth, was literally a "living fossil"; its kind had been thought to have been extinct since the Devonian period, or for about 400 million years (Diagram B). Since 1938 a number of other coelocanths have

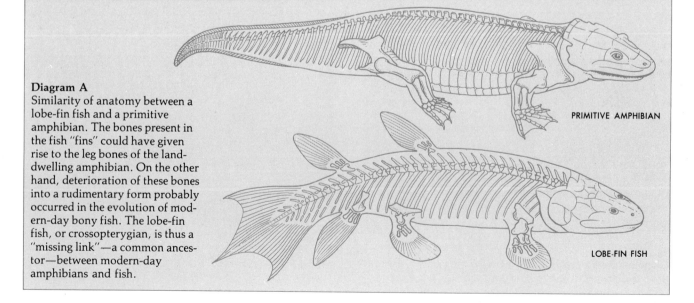

Diagram A
Similarity of anatomy between a lobe-fin fish and a primitive amphibian. The bones present in the fish "fins" could have given rise to the leg bones of the land-dwelling amphibian. On the other hand, deterioration of these bones into a rudimentary form probably occurred in the evolution of modern-day bony fish. The lobe-fin fish, or crossopterygian, is thus a "missing link"—a common ancestor—between modern-day amphibians and fish.

PRIMITIVE AMPHIBIAN

LOBE-FIN FISH

been caught and studied. The studies have confirmed the anatomically intermediate position of coelocanths between modern bony fish and modern amphibians. The coelocanth is truly a "missing link." It appears likely that from the crossopterygians originated the amphibians, which in turn gave rise to all other terrestrial vertebrates, including human beings.

Diagram B
A coelocanth, discovered in 1938, measured at a little over one meter in length. Coelocanths are distinctly fish-like in external appearance; however, their fins are lobed, and unlike modern bony fish, coelocanths have vestigial lungs. Coelocanths are not the particular lobe-fin fish that are thought to have given rise to terrestrial vertebrates, but they do resemble these forms in many ways. (Photo, Popperphoto Ltd.)

and absorption organs (that is, primitive lungs). The lungs were located ventrally, and provided the fish with a source of oxygen when ponds dried up and/or became stagnant. Some of these fish eventually found their way to the ocean where, in a more stable environment with plenty of oxygen, the lungs became unnecessary. Gradually over the course of evolution, these ventral lungs became transformed into a dorsal swim bladder, a gas-filled organ that helps fish keep their balance, and, by inflating or deflating, allows them to change the depth at which they swim. The swim bladder of modern fish thus probably evolved during the Devonian period from primitive lungs of bony fish.

After the Osteichthyes arose from the placoderms they split into three groups: (1) the paleoniscoids, (2) the lungfish, and (3) the crossopterygians (lobe-finned fish). The paleoniscoids are characterized by the transformation of the lung into the swim bladder and by the presence of a fin that contains neither muscle nor bone. The paleoniscoids are the ancestors of most of the common (especially commercial) fish today, both fresh- and saltwater forms, including the salmon, trout, bass, tuna, and mackerel. The lungfish, of which only a few species survive in Africa, Australia, and South America, retain the use of their lungs for air storage. In addition, they show another significant innovation: a connecting passageway between the pharynx and the nostrils. In the paleoniscoids (including today's representatives) the nostrils are used only for smelling (see the discussion of homing in salmon in Chapter 1), but the passageway from the lungfish's nostrils to its pharynx makes it possible for it to breath air with its mouth closed like modern terrestrial vertebrates.

The term "Age of Fishes" has rightly been applied to the Devonian period. An immense adaptive radiation took place during that time among both salt- and freshwater species. Although many of those fish are now ex-

tinct, all the major groups alive today had ancestors who were thriving 400 million years ago.

Classes Amphibia and Reptilia.   The amphibians and reptiles were among the first true vertebrate colonizers of land (see Fig. 29.43). The amphibians (today's frogs, toads, and salamanders) came first and, as their anatomy and life cycle show, they are true transitional forms between the fish and land dwellers. The bodies of amphibia, especially in their larval stages (for example a tadpole, larval stage of a frog) are very fish-like. The amphibian life cycle is tied completely to the water, with eggs hatching and larvae living a completely aquatic existence (including gills) until metamorphosis into the adult, land-dwelling form. Many reptiles, such as the lizards, also have the streamlined, fish-like body structure, though reptiles do not pass through an aquatic phase in their life cycle. The amphibians became prominent during the Carboniferous era, and indeed dominated the land at that time. They declined noticeably during the Permian as they were replaced by the reptiles. This replacement coincided with a period of great geological and climatic change, of mountain-building and gradual warming. The only groups of amphibians that survived gave rise to the modern frogs, toads, and salamanders.

Amphibians.   The amphibians show numerous adaptations to their life on land: fully developed lungs, a three-chambered heart which to some extent separates oxygenated from deoxygenated blood, and a highly developed nervous system. Frogs and salamanders are excellent predators, with lightning-quick reflexes (a frog can catch a flying insect with a flick of its tongue). Their shift from water to land in the course of metamorphosis is accompanied by a change in water balance and excretory mechanisms. For example, as tadpoles, frogs ex-

(a)   **(b)**   **(c)**

(d)   **(e)**

Fig. 29.43
Representatives of the classes Amphibia (*a*) and Reptilia (*b*–*e*): (*a*) Frog (*Rana pipiens*); (*b*) Hog-nosed snake; (*c*) Desert turtle, Mojave, California; (*d*) Alligator; and (*e*) Lizard (genus *Cnemidophorus*). Note reptile scales. (Photo *a* Courtesy John A. Moore.)

crete nitrogenous wastes as ammonia; as adults, they secrete the same wastes as urea. While frogs have lungs as adults, some salamanders do not. Lungless salamanders breathe directly through their skin and through the mucous membranes lining their throats, thus depending completely on a moist environment for gas exchange. If their skins dry out beyond a certain point, they can no longer breathe. Frogs also carry out some gas exchange through their skin. Such processes indicate clearly the dependency of the amphibia on a moist, water-laden environment—a relic of their direct aquatic ancestry.

Reptiles.   By the end of the Carboniferous period, the first reptiles had evolved from the amphibians. During the early Mesozoic era (somewhere between 280 and 250 million years ago) a great adaptive radiation of the reptiles had begun to take place, driving the once-dominant amphibians into a distinctly subordinate position. One possible reason for this change was that the earth entered a second period of great dryness toward the end of the Paleozoic era (280 million years ago; the first dry period came at the end of the Devonian, about 350 million years ago), favoring the evolution of forms that

could withstand an arid climate. A similar evolution was observed among plant life, as conifers began to replace the more water-dependent mosses and ferns. The group that came to dominate the land during the Triassic period (the beginning of the Mesozoic era, about 230 to 200 million years ago) were mammal-like reptiles. This group gave rise on the one hand to the reptiles, including the great dinosaurs, and on the other to the mammals. The reptile group was the first to evolve into a position of dominance, giving to the Mesozoic era (from 230 to 130 million years ago) its designation as the "Age of Reptiles."

Reptiles have evolved many characteristics that show both their relationship to amphibians and their complete transition to a land existence.

1.   The extremely long body was mounted on small legs that projected from the side of the torso, rather than downward. Thus the ventral surface of the animal often dragged along the ground, and walking was frequently slow and awkward.

2.   The reptiles have an almost four-chambered heart, with a partition partway down the center of the ven-

tricle (recall that amphibians have a single ventricle with no partition). The partition is more effective in maintaining a separation of oxygenated and deoxygenated blood after both enter the ventricle. In fish and amphibians the two are mixed. (One reptile, the crocodile, does have a virtually complete four-chambered heart.)

3. All reptiles have fully developed lungs throughout their lives, along with a strong rib and diaphragm musculature that enhances ventilation.

4. Perhaps the most significant adaptation in the reptiles is the fact that they lay an amniotic egg.

What is an "amniotic egg" and why is it important? All embryonic development in the animal world occurs within a watery environment. The embryos literally float in a pool of liquid. For all animals up through the amphibians, development occurs in an external source of water: a pond, tidal pool, or river. With the reptiles, however, we find for the first time an egg that carries along its own water supply within a shell. The shell is porous to gases but not to most liquids, hence is highly adapted for developing in a terrestrial environment. Indeed, the reptilian egg, which is much like the chicken egg in overall structure, is a little pond in its own right. The reptilian egg contains an air space, a large yolk, and several membranes which serve (1) to enclose the embryo (amnion), and (2) for gas exchange and excretion of wastes (**chorioallantoic** membrane). The development of the shell emancipated the reptiles from dependence on large bodies of water, and thus opened to them a much greater range of habitats and geographic locations.

Among the most important reptiles to predominate in the Mesozoic era were the Archosauria, that is, the dinosaurs or ancient (*arch*) lizards (*sauria*). These organisms radiated spectacularly during the Triassic and Jurassic periods (225 to 135 million years ago). Occurring in a wide variety of forms (see Fig. 29.44), the dinosaurs were the most impressive of the animals that had inhabited the earth up to that time. Although most people are familiar only with the largest and most bizarre examples (*Tyrranosaurus rex*, *Brontosaurus*, and *Stegosaurus*), the dinosaurs were generally small and unimpressive.

Recent evidence has suggested that many dinosaurs were homeothermic (warm-blooded). This conclusion is based largely on the observation that the degree of vas-

Fig. 29.44
Reconstructions of three Jurassic-Cretaceous dinosaurs, from paintings by Charles R. Knight in the Field Museum of Natural History, Chicago. (*a*) *Brontosaurus*, a giant dinosaur that lived both on land and in the water. Adults are estimated as having weighed between 25 and 35 tons. (*b*) *Stegosaurus* was a herbivore with large flat plates on its back. Some paleontologists hypothesize that these plates may have aided in thermal regulation. (*c*) *Tyrannosaurus*, a giant carnivorous dinosaur that could be up to 50 feet long and 20 feet high. (Drawings courtesy Field Museum of Natural History.)

(a)

(b)

(c)

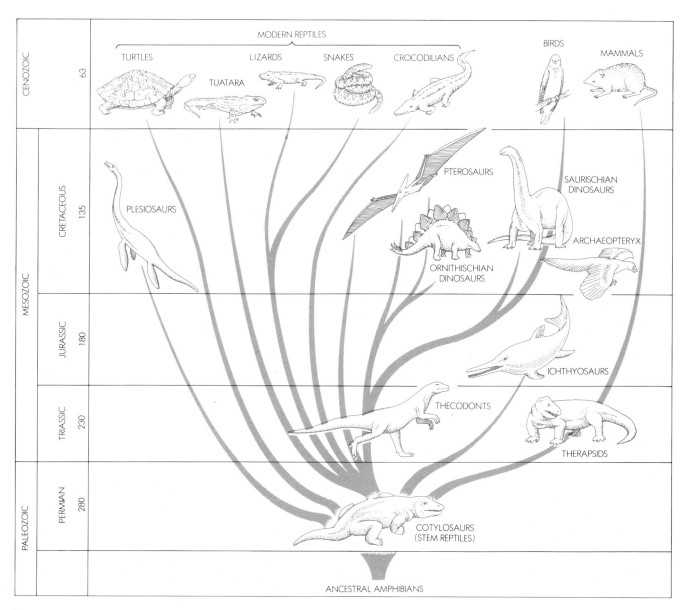

Fig. 29.45
Phylogenetic chart of the major groups of extinct and modern reptiles, including their relationships to the birds and mammals. Note especially the two large groups of dinosaurs, the Saurischians (bipedal carnivores such as *Tyrannosaurus*) and the Ornithischians (mostly bipedal but some quadripedal herbivores such as *Anatosaurus* and *Stegosaurus*), which diverged from the ancestral theocodonts. The birds originated from the saurischian line. Note also that flight evolved in two independent lines in the evolution of the reptiles: in the pterosaurs (flying reptiles such as Pterodactyl) and in the birds.

cularization of dinosaur bones approximates much more closely that of today's mammals and birds (which are homeothermic), than it does today's reptiles (which are poikilothermic). Not all paleontologists agree with this conclusion, however, some claiming that homeothermy arose in the descendants of the dinosaurs (mammals and birds) but not in the dinosaurs themselves. Homeothermy was an important adaptation. Poikilothermic animals become active only when they can absorb heat from the outside. When the ambient temperature is cold, they become sluggish, and must go into burrows or hide until warmer weather. Homeothermic animals, by contrast, can move at the same pace (more or less) regardless of ambient temperature. Such a dependence on outside sources of heat for general physiological activity greatly restricts an animal's ecological adaptability.

While dominating the Jurassic and the early part of the Cretaceous periods, the dinosaurs disappeared by the end of the Mesozoic era (about 65 million years ago). Exactly why this enormously successful group became extinct is unclear. Some theories suggest that perhaps some great geological or climatic cataclysm occurred toward the end of the Cretaceous. In 1980, geological findings suggesting that a giant meteorite hit the earth in this period sparked renewed interest in the cataclysmic theory. According to this idea, the meteorite stirred up huge amounts of dust, reducing the intensity of sunlight, producing cooler weather and thus bringing about a rapid shrinking of the tropical habitats in which dinosaurs lived. A cataclysmic theory is further supported by the fact that a number of other animal species also disappeared around this time—for example the Ammonites. However, such extinctions are not observed in plants and many other animals. On the other side of the coin, the geological record shows that most dinosaur groups became extinct rather slowly, over a 20- or 30-million-year period, rather than very rapidly as the meteorite theory would predict. Whatever the exact cause for their extinction, the dinosaurs disappeared except for one group that had already evolved an important new feature: feathers. This group gave rise to the modern-day birds. The evolution of the reptiles is diagrammed in Fig. 29.45.

Class Aves. Aves, the birds, are highly adapted animals characterized by the presence of feathers, scales on their legs, toothless beaks, a fully developed four-chambered heart, and the ability to maintain a constant body temperature (homeothermy) (see Fig. 29.45). Although some birds, such as the ostrich, penguin, and kiwi, are flightless, the vast majority are able flyers. The birds show many adaptations for flight. One of the chief adaptations is a very light body weight, achieved by several kinds of structural modifications, including having hollow bones. For example, in the seagoing frigate bird, with a wing-span of 2 meters, the entire skeleton weighs only a little over 100 grams (around 4 ounces)! In addition, birds have numerous air sacs attached to their lungs as a way of lightening their overall body weight. In most birds the female reproductive tract has become streamlined, being reduced to a single ovary that diminishes to a very small size except during the reproductive season.

The four-chambered heart is a critical adaptation for flight. Complete separation of the oxygenated and deoxygenated blood (by having separate pulmonary and systemic circulation) leads to maximum efficiency in delivery of oxygen to the tissues. Such efficiency is crucial since flying is a metabolically demanding ac-

tivity that requires the continuous and rapid expenditure of large amounts of energy.

Like their reptilian ancestors, birds reproduce by internal fertilization, with the embryo contained inside a shelled egg. The egg is incubated externally. The young are born less able to take care of themselves than their reptilian counterparts, however. Indeed, most young birds are blind and totally helpless. Consequently, we see in the birds for the first time among vertebrates the appearance of elaborate parental care of the young (see also Section 30.2 and Fig. 30.5).

As shown in Fig. 29.45, the birds evolved from dinosaurs somewhere in the late Triassic or early Jurassic periods (about 230 to 200 million years ago). At that time two different groups of reptiles had already evolved the power of flight. The pterosaurs were a group of reptiles with a membrane-like wing of skin stretched between a greatly elongated arm, the fourth digit, and the body. Pterosaurs were enormous. *Pteranodon*, for example, which lived during the Cretaceous period (about 135 million years ago), had an average wing-span of 7 to 8 meters (some going as high as 12), making it one of the largest animals ever to fly. The pterosaurs, however, became extinct by the end of the Cretaceous period (about 64 million years ago). The other line of reptiles to develop flight had a completely different wing structure: elongated arm bones, less exaggerated digits, and wing and body surface covered with a form of modified scales called feathers. This line gave rise to modern birds, which are thus the closest living relatives of the dinosaurs.

How might flight in birds have evolved? A long-standing theory maintains that some of the ancestral feathered reptiles were tree-dwellers that gradually evolved gliding and then flying mechanisms to aid in moving swiftly from limb to limb or tree to tree. However, another view, proposed recently by Professor John Ostrom of Yale University, has gained considerable support in recent years. Ostrom studied fossils of the five known species of *Archaeopteryx*, a primitive bird that lived about 150 million years ago (see Fig. 22.4). *Archaeopteryx* had teeth and a long, jointed tail—both reptilian characteristics. It also had distinct feathers and hollow bones, both bird characteristics. Ostrom found that the breastbone and keel (the keel is the extended median ridge of the breastbone), so prominent in all modern birds, is lacking in *Archaeopteryx*. Since the breastbone and keel are the points of attachment of the wing muscles, Ostrom's observation suggests that *Archaeopteryx* had very little flight capacity. Furthermore, Ostrom noted that the feathers in *Archaeopteryx* were not attached to the wing bones, as in modern birds, but to the skin. This observation suggests that originally feathers may not have been adaptations for flight at all, but rather were insulators against heat loss. Ostrom believes that *Archaeopteryx* was a ground-

Fig. 29.46
Vulture (*Psyudogyps africanus*), a predatory bird. Note the
position of the wings, extended almost as if to envelope the
prey. One hypothesis on the origin of flight suggests that
feathered, ground-dwelling dinosaurs used their extended
arms to gain speed by hopping, and to enfold the prey in a sort
of net. (Photo, R. S. Virdee, Grant Heilman Photography.)

dweller that at best had only a very limited capacity for
flight.

According to Ostrom's scenario, the ancestors of
Archaeopteryx were feathered, warm-blooded dinosaurs that chased their prey along the ground. Feathered
arms may have served two functions in addition to heat
conservation. As *Archaeopteryx* ran it may have
flapped its extended, feathered arms, helping the primitive bird make small leaps into the air and thus overtake
its prey. Furthermore, the long wing-feathers may have
served as a kind of natural trap, or net, into which the
prey could be enfolded. Indeed many modern predatory
birds, such as the African vulture shown in Fig. 29.46,
use their wings in just such a manner. According to
Ostrom, flight could have evolved as a secondary adaptation to warm-bloodedness and a predatory life style.

Class Mammalia. The mammals, like the birds,
evolved from the reptiles, though from a completely different line and starting at a much earlier period. As
shown in Fig. 29.45, the mammalian line began diverging from the stem reptiles (cotylosaurs) in the late
Permian, about 225 to 230 million years ago. Ther-
apsids, the now-extinct reptiles that eventually gave rise
to the mammals, were plentiful during the Permian and
Triassic. The therapsid reptiles probably were homeothermic, like the reptiles that gave rise to the birds, and
some evidence suggests they had hair. Exactly when the
therapsid line branched and gave rise to mammals is difficult to say. Certainly, by the Cretaceous period
mammals of various sorts were beginning adaptive
radiation and began to supplant some of the dominant
reptiles.

The major characteristics of mammals are:

1. They are homeothermic.

2. Their young are born alive (not in an enclosed egg).

3. The young are nursed with milk produced in glands
called **mammae,** or **mammary glands.**

4. They have hair on their bodies (to some degree).

5. They have a four-chambered heart with complete
separation between right and left ventricles.

6. They have a more complete separation (the palate)
between the alimentary tract and the digestive tract,
making it possible to chew and breathe at the same time.

7. The lower jaw is composed of a single bone, and
the upper jaw is no longer movable with respect to the
skull.

Mammals have radiated throughout a number of
different ecological niches. There are tree-dwelling and
cave-dwelling mammals; aquatic and terrestrial mammals; vegetarian, carnivorous, and omnivorous mammals; mammals that live together all or part of the year
in social groupings, and mammals that are predominantly solitary. Because their mode of reproduction is
not dependent on either a body of water (as in amphibians and fish), or relatively defenseless externally
hatched eggs, mammals have been highly mobile while
still providing a considerable degree of protection for
the young. Indeed, long-term parental care is a hallmark
of mammalian development.

There are about 4500 species of mammals, divided
into three subclasses: the monotremes (Prototheria), the
marsupials (Metatheria), and the placentals (Eutheria).

---

*"The speculative nature of any attempt to reconstruct an animal phylogeny is ob-
vious. Yet such attempts should not be considered pointless. They have as much
value as attempts to fit together any unknown biological pattern."—Robert D. Barnes
in* Invertebrate Zoology *(1968)*

Fig. 29.47
The duckbill platypus, *Ornithorhyncus anatinus*, which lays eggs like a reptile, yet nurses its young after they hatch. In its physiological characteristics, too, the platypus combines reptilian and mammalian features. (Photo courtesy American Museum of Natural History.)

The monotremes are represented by the duckbill platypus (see Fig. 29.47) and the spiny anteater. While monotremes characteristically nurse their young from mammary glands, the young are born and hatch from an egg. They represent a special combination of mammalian and reptilian traits. The duckbill platypus is highly adapted for an aquatic existence though, like all mammals, it is an air-breather. The monotremes are found largely in Australia, although the spiny anteaters are also found in New Guinea.

The marsupials are sometimes referred to as the "pouched mammals," though not all of them have true pouches. The group includes the kangaroo (see Fig. 29.48) and opossum. Marsupials are born in a very immature state. For example, young kangaroos emerge from the womb when they are only a few centimeters in length, crawling up the mother's abdomen until they reach the pouch. Inside the pouch where the young are kept warm and protected, they nurse from the mammae until they can venture more and more frequently from the pouch. Marsupials are geographically distributed throughout the world, but nowhere have enjoyed greater diversity than in Australia. In addition to the well-known kangaroo, there are the marsupial dog (Tasmanian wolf), mole, bat, mouse, and anteater (see also Section 24.5). Outside of Australia the most common marsupial is the opossum.

The placental mammals, as the name implies, bear the embryos attached to the **placenta** (See Section 20.7), an organ for the exchange of nourishment and wastes between mother and offspring. Intrauterine development provides the embryo with maximum protection during the earliest and most vulnerable stages of development. When the young are born, the placenta (along with the umbilical cord, which attaches the pla-

Fig. 29.48
Mother kangaroo and young (called a "joey") in pouch. Marsupials are born in an extremely immature state and continue development in the pouch, where they nurse from the mother's teats. (Photo courtesy Australian Information Service.)

centa to the fetus) is shed as the **afterbirth.** The placental mammals are geographically very widespread and show a number of unusual and highly specialized adaptations.

The main line of mammals split during the Jurassic period, one branch leading to the marsupials and the other to the placentals. It was once thought that the marsupials represented an older, more primitive stage of development of the placental mammals, but evidence now suggests the two lines diverged from a common ancestor early in mammalian history. The earliest placental mammals seem to have been relatively shy creatures that lived in trees—perhaps to escape predation from carnivorous dinosaurs. The modern-day shrews are considered by many to be similar to these earliest mammals.

The subclass of placental mammals is divided into approximately sixteen orders (depending on the classification system used). Eleven of these orders, along with their defining characteristics and examples, are listed in Table 29.4.

From our point of view as humans, the most important group of placental mammals is the primates, which includes along with ourselves the apes, monkeys (New and Old World), gibbons, orangutans, lemurs, tarsiers, and tree shrews (not directly related to the small shrews referred to above). We will defer momentarily our discussion of the primates and their evolution, to review the broad phylogenetic history of the animal groups we have discussed so far.

29.14
EVOLUTIONARY RELATIONSHIPS

Figure 29.49 shows the broad evolutionary relationships that some biologists see between the major groups of animals. This scheme suggests at least two major lines of evolutionary development:

1. The origin of the Metazoa from a flagellated ancestor.

Table 29.4
Some Major Orders of the Placental Mammals

Order	Examples	Major characteristics
Insectivora	Moles, shrews	Long, tapered snout; feet five-toed with claws; sharp-pointed teeth
Chiroptera	Bats	Flying mammals; forelimbs and second to fifth digits greatly elongated supporting a thin membrane or wing; sharp teeth; mostly nocturnal
Primates	Lemurs, monkeys, apes, human beings	Elongated limbs and enlarged hands with five digits and flattened or cupped nails; innermost toe or thumb (or both) opposable; eyes directed forward and capable of full binocular (three-dimensional) vision
Edentata	Sloths, armadillos, anteaters	Teeth not sharp, but reduced to molars in anterior part of mouth, and without enamel
Lagomorpha	Hares, rabbits	Toes with claws; stubby tails; chisel-like incisors that grow continually; no canine teeth; jaw motion only lateral, producing a circular chewing motion
Rodentia	Mice, rats, squirrels, gophers, beavers, porcupines	Chisel-like incisors that grow continually; no canines; jaw capable of lateral and vertical motion
Cetacea	Whales, dolphins, porpoises	Medium to very large size; body usually spindle-shaped, with long, pointed head; no neck region; forelimbs (flippers) are broad, paddle-like; nostrils on top of head; no claws; no hind limbs; teeth present in some, absent in others, but always alike with no enamel; all marine
Carnivora	Dogs, cats, bears, racoons, weasels, skunks, minks, hyenas, seals, walruses, otters, badgers, civets	Usually five-toed (sometimes four), all clawed; limbs highly mobile, with complete tibia, fibia, radius, and ulna (as separated bones); incisors small, with canines as slender "fangs"; teeth with enamel and rooted
Proboscidea	Elephants	Massive bodies; large, flat ears; nose and upper lip modified into slender, long proboscis with nostrils at tip; two upper incisors elongated as tusks; three, four, or five toes
Perissodactyla	Horses, zebras, tapir, rhinoceros	Hoofed mammals with odd number of toes; large body; long legs
Artiodactyla	Pigs, hippopotamuses, camels, deer, giraffes, antelopes, cattle, pigs, sheep, goats, bison	Hoofed animals with even-numbered toes; usually two functional toes (rarely four) on each foot, covered with hoofed nail; many with antlers or horns on the head; all but pigs with reduced dentition; a four-compartment stomach for rumination

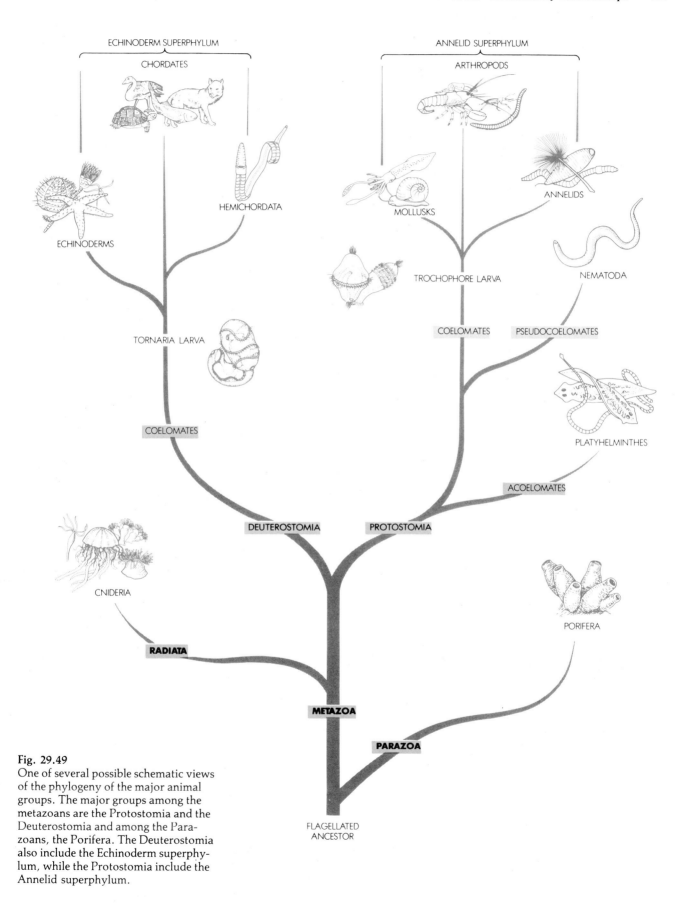

ECHINODERM SUPERPHYLUM

CHORDATES

HEMICHORDATA

ECHINODERMS

TORNARIA LARVA

COELOMATES

DEUTEROSTOMIA

ANNELID SUPERPHYLUM

ARTHROPODS

MOLLUSKS

ANNELIDS

TROCHOPHORE LARVA

NEMATODA

COELOMATES   PSEUDOCOELOMATES

PLATYHELMINTHES

ACOELOMATES

PROTOSTOMIA

CNIDERIA

PORIFERA

RADIATA

METAZOA

PARAZOA

FLAGELLATED
ANCESTOR

Fig. 29.49
One of several possible schematic views
of the phylogeny of the major animal
groups. The major groups among the
metazoans are the Protostomia and the
Deuterostomia and among the Para-
zoans, the Porifera. The Deuterostomia
also include the Echinoderm superphylum, while the Protostomia include the
Annelid superphylum.

2. The divergence between the Deuterostomia and Protostomia (including what have been called the Annelid and Echinoderm Superphyla).

Let us examine briefly the evidence on which the phylogenetic chart shown in Fig. 29.49 is constructed.

The Origin of the Metazoa

Most biologists today believe that the Metazoa (multicellular organisms) evolved from a flagellated ancestor. As we saw in Chapter 28, by adhering together after cell division, primitive flagellates could become colonies that eventually, through varying degrees of cell specialization (see Supplement 29.1), became truly multicellular. The crucial question is how could a simple, spherical colony such as *Volvox* (as explained in Supplement 29.1) ever lead to the complex, multilayered metazoans of today?

One hypothesis suggests that a primitive spherical colony was something like the blastula stage of higher organisms—that is, that through some transformation analogous to gastrulation in modern embryos, the hollow sphere of cells making up the blastula could become invaginated into a two-layered gastrula. This idea was originally suggested in the late nineteenth century by German naturalist Ernst Haeckel, who claimed that Cnidarians such as *Hydra* represented nothing more than organisms whose development never went beyond gastrulation. However, as we saw earlier in this chapter, gastrulation in the Cnidarians does not take place by invagination at all; rather, cells from the ectoderm wander into the center of the blastocoel and form the gastroderm from within. Thus, Haeckel's original idea of multicellular organisms forming from a primitive,

free-swimming gastrula-like organism is highly unlikely.

Another hypothesis, advanced in recent years by K. G. Grell of the University of Tübingen, in Germany, suggests that a multicellular, two-layered form was generated from the blastula as shown in Fig. 29.50. The blastula existed first as a flattened, two-layered organism that crept about the ocean floor, its ventral surface composed of cells that regularly ingested bits of organic matter. The top layer of cells served a protective function. Now, if this organism, called a plakula, occasionally reared up, the cavity formed by the buckling of the ventral surface would become a sort of gastrocoel, or archenteron. This hypothesis sounds believable enough, but what evidence is there that it could in any remote sense be the way the metazoa originated? In 1969 Grell observed a small plakula-like organism, *Trichoplax*, in a sample of water from the Red Sea. *Trichoplax* showed cell specialization in its ventral and dorsal layers, and would occasionally be seen to buckle up into a gastrula-like shape. Of course, observations of *Trichoplax* do not mean that two-layered metazoans actually originated in such a way. But they do lend considerable credence to Grell's hypothesis.

The Split Between the Protostomia and Deuterostomia

A major split can be observed in Fig. 29.49 between the Protostomia and Deuterostomia. What is the evidence for such a large division among animal phyla? Recall from Section 29.1 that one of the chief distinctions between the Protostomia and Deuterostomia lies in the fate of the blastopore. In the Protostomia the blastopore becomes the future mouth, whereas in the Deuterostomia it becomes the anus. Furthermore, the cleavage

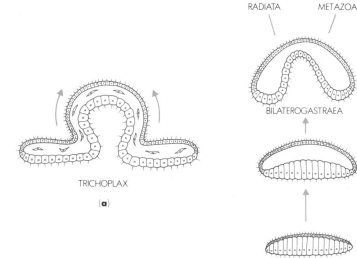

Fig. 29.50
A current hypothesis, advanced by K. G. Grell, to account for the origin of the bilayered gastrula of the Metazoa. (*a*) Observing the small multicellular form *Trichoplax*, Grell noted that the ventral surface was specialized for protection. Occasionally, Grell noticed, *Trichoplax* would buckle up, forming a cavity within the folded ventral surface. This could be the forerunner of a gastrovascular cavity. (*b*) The plakula, the hypothetical form that may have been similar to *Trichoplax*, is shown folding up into a gastrula-like form, a possible starting point for the origin of all the Radiata and Metazoa. (Drawing from William Keeton (illus. Paula Di Santo Bensadoun), *Biological Science*. By permission of W. W. Norton. Copyright © 1980 by W. W. Norton & Company, Inc.; modified from K. G. Grell, 1974.)

patterns of protostomes is determinate and spiral, that of deuterostomes indeterminate and radial. The mesoderm in protostomes is formed from the endoderm of the gastrula by growth of cells inward from the area of the blastopore, whereas in deuterostomes the mesoderm forms from the endoderm by pocketing of cells of the archenteron lining (see Fig. 29.4). There are also significant differences between the two groups in the method of coelom formation, which suggest that a split occurred between these two groups a long time ago. Biologists generally assume that the earliest embryological stages of any species are the oldest in terms of their evolutionary history, and the least likely to be changed by subsequent evolution. Although this assumption may not be a safe one in every case (since embryonic development is an evolving process in its own right), it is likely to be true more often than not. There is thus some justification for considering embryonic similarities and/or differences as significant features for determining evolutionary relationships.

Despite these lines of suggestive evidence there are problems with making fundamental divisions such as between the Protostomia and Deuterostomia, listed above. As most zoologists will quickly point out, all the characteristic distinctions listed above have exceptions. Moreover, the scheme outlined in Fig. 29.50 suggests that the presence of a coelom must have originated twice in animal evolution (since both protostomes and deuterostomes have a coelom)—a somewhat unlikely, though not impossible, event. Our picture of animal evolution is at best tentative and is always subject to major revisions in the future.

Another feature of embryonic development that supports the fundamental distinction between protostomes and deuterostomes involves the structure of larval forms. Among the deuterostomes, the echinoderms and the hemichordates both possess the ciliated bipinnaria larva, as shown in Fig. 29.34, thus indicating an affinity between them. The fact that some hemichordate larvae have a dorsal notochord (like adult chordates) is used to link the three phyla together into a single superphylum.

On the other hand, some members of the Protostomia are linked through the presence of the trochophore larva (see Fig. 29.21), which appears in the development of some annelids and molluscs. Furthermore, the existence of one type of mollusc with a segmented shell (chiton) and another with internal segmentation (*Neopilina*) suggests a relationship between the molluscs

and the annelids. In addition, the arthropods and annelids show several traits in common. For one thing, both show clear segmentation. The possible existence of an intermediate form, such as the Onycophora (for example, *Peripatus*; see Fig. 29.23), suggests that the two groups may have shared a common ancestor at some point in the distant past. We must remember that in determining the actual evolutionary course of any group of organisms, the final arbiter of all theories is the fossil record. Anatomical, molecular, or developmental homologies allow us only to draw general inferences about how evolution *might have* proceeded. Only fossils can help us ultimately determine the actual pattern by which one group may have given rise to another.

29.15
PRIMATE AND HUMAN EVOLUTION

To many of his contemporaries the implication of Darwin's theory that human beings, like all other living organisms, have evolved from simpler forms of life was unsettling, posing religious, philosophical, and social questions of enormous magnitude. In essence the questions focus on the degree to which human behavior, social structure, and customs are a part of our "biological nature." Many of the questions that Darwin's work originally raised are still with us today. By studying human evolution, modern biologists are trying to answer some of these questions.

Characteristics of the Primates

All human beings belong to a single species with the generic and specific designation *Homo sapiens* (*Homo*, human being; *sapiens*, wise, intelligent). That species belongs to the subphylum Vertebrata, the class Mammalia, and the order Primata. The primates are a special group of mammals with a number of living representatives, including tree shrews, tarsiers, Old World (African-Asian) and New World (South American) monkeys, and apes (see Fig. 29.51).

The primates are a varied group of animals, appearing at first glance to have little in common. However, they do share a number of similar features. They have eyes on the front of the face, rather than at the side of the skull. The large overlap this causes between the two fields of vision produces binocular, three-dimensional sight. Primates have functional digits (fingers and toes), and many have an opposable thumb—one digit

---

Fossils are the ultimate form of evidence on which secure evolutionary relationships are based. Without recourse to fossils and the geological strata in which they are found, all phylogenetic schemes remain largely speculative.

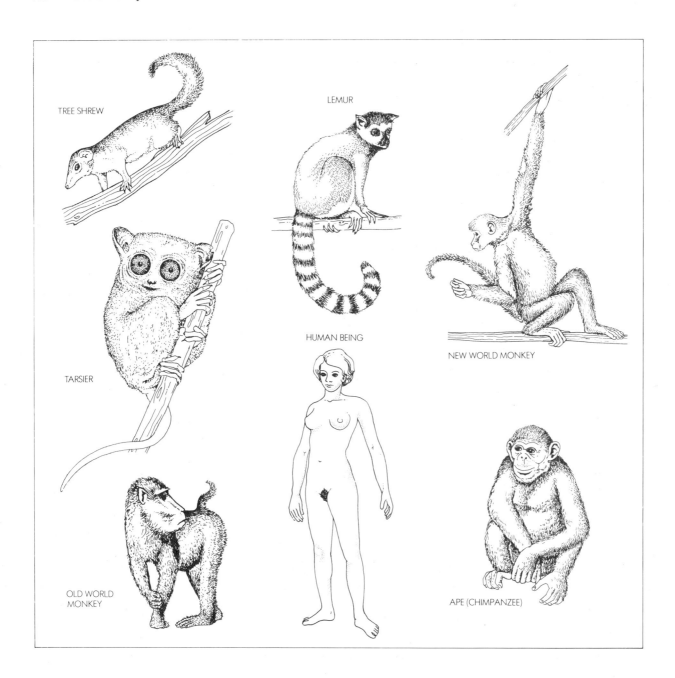

Fig. 29.51
Representatives of the major groups of primates, an order of the class Mammalia. Characteristic of the primates are frontal eyes and binocular vision, opposable thumb, collarbones, functional digits on the hands, flattened nails, and three basic types of teeth.

range. In addition, primates have flattened finger- and toenails.

The Primate Phylogenetic Chart

From anatomically and biochemically analyzing modern representatives of the primate group, as well as from fossil evidence, paleontologists have constructed a phylogenetic chart for the primates (Fig. 29.52). This chart suggests that the most primitive ancestors of the primate group resembled the modern-day prosimians such as the tree shrew or lemur. By analogy with today's tree shrews and other living primates, biologists hypo-

set more or less opposite to the other four on the hands and/or feet—that makes it possible to grasp objects. Primates have collarbones and highly flexible shoulder joints that allow the arms to be rotated over a wide

thesize that about 75 million years ago, a small group of mammals was driven from a solely terrestrial to an arboreal or tree-dwelling environment. In the trees these future primates would have clung carefully to tree branches to avoid falling; their major diet was probably insects, leaves, and fruits or berries.

Among these primitive ancestors of the primates, selection would have favored those who could hold onto tree limbs best and move along by jumping from limb to limb. Traits such as the opposable thumb, making effective grabbing possible, and forward-facing eyes for accurate three-dimensional sight became highly favored in the new environment. In addition, a high degree of muscular coordination and reflex capability must have made for more effective locomotion. Undoubtedly this factor played a major role in the evolution of the primate brain.

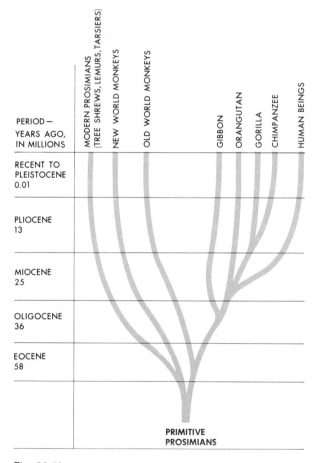

Fig. 29.52
Phylogenetic chart of the major lines of primates. All are thought to have evolved from a primitive prosimian ancestor, similar perhaps to the tree shrews of today. The primates originated from a branch of the mammals by moving into an aboreal (tree) environment.

Data for Studies of Human Evolution

The Fossil Record. The fossil record for the human species is in many ways quite scanty. Not only have human populations always been small by comparison to those of many other species, but their habitat has ranged over so vast an area that fossil remains are widely scattered. Thus the study of human evolution has often been plagued with woefully inadequate fossil data.

Where are human fossils found? Like fossils of many other species, those of humans are found in the rock strata, in caves, various burial sites, and gullies or gorges. However, because of their relatively recent origin, human fossils often lie closer to the top layers of the ground than those of geologically older forms. This has meant that human fossil remains are more subject to being dislodged, damaged, or carried away than fossils of more ancient forms.

Despite these obstacles, human fossils have yielded some extremely important evidence. Anthropologists can determine from certain ridges and lines on a bone how and where a muscle was attached. They can deduce something about posture and pattern of locomotion from examining bones of the limbs or position of the skull on the vertebral column. From the study of size, arrangement, and wear patterns of the teeth, anthropologists can deduce much about lifestyles and habits of ancestral hominids.

One way to determine the fossil's age is from the geologic stratum in which it is found. This works as long as the fossils are firmly embedded in the strata and not, as is so often the case, loosely scattered about among rubble. A more certain method involves mineral accumulation and radioactive dating. For the relatively recent past, C^{14} data have proved particularly helpful for dating fossil remains. Recently, a process known as the potassium-argon method has been employed with success. In this method a fossil is dated by reference to the volcanic material deposited either above or below the strata in which the fossil was found. Other methods for dating fossils also exist, based on sophisticated analyses of changes in the earth's magnetic field at various points in geological time and on the use of U^{238} fission tracks that can be detected in volcanic or other igneous rock deposits. All these methods give reasonably accurate and, most important, relatively *consistent* estimates of fossil age.

Study of Living Primates. Living primates yield two very interesting kinds of data for the study of human evolution. One is data from comparative anatomy—useful in trying to understand how one organ or structure was transformed into another. The comparative anatomist has one very important advantage over the paleontologist; he or she can study a functioning, intact organism. Here the relationship between bones, muscles, visceral organs, and nerves can be studied directly. It is not necessary to speculate or try to deduce such relationships from a few fragmentary parts. Since the higher primates, especially the anthropoid apes, appear to represent our closest living relatives, comparative anatomy of these forms yields much valuable information about how the hominid line could have evolved from an ape-like ancestor.

Study of the living primates also yields behavioral evidence. In the long run nothing differentiates *Homo sapiens* from other creatures on earth so much as the wide range of behaviors of which human beings are capable. Such behaviors include our physical capacity to manipulate tools as well as our intellectual and creative ability to invent symbols and myths. While many of the primates carry out such behaviors to a limited extent, the ability of human beings to do so is much greater. The study of primate behavior suggests how primitive hominids may have behaved and suggests the kinds of behavioral capacities that went along with such characteristics as hip placement or brain size.

Of course, there is considerable danger in using analogies from animal behavior to trace the possible evolutionary development of certain human behaviors. As we will see in more depth in Chapter 30, it is often tempting to try to "explain" human behavior patterns such as aggression by reference to what appears to be innate aggressive behavior in other species. By and large, however, biologists are cautious about making such speculations. We know too little about the genetics of behavior in the human species (as well as other animals) to hypothesize whether any specific behavior patterns could in fact have been selected for by means of evolution. The study of primate behavior is valuable for students of human evolution only to the degree that it may indicate how our more primitive ancestors *might* have behaved. It does not necessarily tell us anything about how they actually *did* behave, or about the origin of any of the behavior traits observed among modern human beings.

Data from Molecular Biology. Study of molecular and biochemical similarities between species can yield useful evidence for establishing, verifying, or revising phylogenetic relationships. For example, studies of blood protein (albumin) and cytochrome *c* structure in modern primates carried out by Drs. Vincent Sarich and Allan Wilson of the University of California at Berkeley establish a closer relationship in time between human beings, gorilla, and chimpanzee than had previously been suspected.

The most controversial aspect of their findings, however, is the belief that the divergence of human from ape leading to the hominid line occurred only around five million years ago. Dr. David Pilbeam of Cambridge University in England states that this belief flies directly in the face of the available fossil evidence. Pilbeam maintained that the fossil called *Ramapithecus* and another discovered in 1962, called *Kenyapithecus*

whickevi, actually belong to the same genus and push back the evolutionary divergence of ape and human to at least 14 million years ago. Furthermore, Pilbeam suggested that *Ramapithecus* itself may have evolved from another form, *Dryopithecus,* which is distinctly hominid and dates back some 30 to 35 million years. This conflict between the biochemical evidence and the fossil evidence is far too great to be overlooked.

Some workers have suggested that Sarich and Wilson's data are correct but that their time determination is off, perhaps because of different rates of evolution in different types of molecules. Indeed, this has proved to be the case: some molecules evolve faster or slower than others. No standard *rate* of molecular evolution can be applied to all types of molecules at all times. Thus, it now appears that the divergence of humans and apes is probably somewhat longer ago than postulated by Sarich and Wilson, that is, more than about 5 million years ago.

Further Evidence on the Human-Ape Evolutionary Divergence

Early in 1978 two new discoveries came to light, both supporting the divergence of the human and ape lines somewhere between 5 and 7 million years ago. Don Johanson, of the Cleveland Museum of Natural History, and his colleague Tim White worked in the Afar region of Ethiopia from 1972 to 1977, uncovering an outstanding series of hominid remains dated at approximately 2.9 to 3.3 million years of age. Of chief importance among them was the skeleton of a single Australopithecine (affectionately named "Lucy"). Lucy is the most complete Australopithecine skeleton (more than 2.5 million years old) that has so far been discovered: she is 40 percent complete (see Fig. 29.53). Given the fact that most early hominid skeletons consist, at most, of a few fragments of jaw and scraps of skull, Lucy represents a major find. Because Lucy's skeleton included an almost complete half pelvis and attached femur, along with a top portion of the tibia, Johanson was able to have an orthopedic specialist examine the entire walking apparatus. The findings suggested that Lucy walked as fully upright as modern-day humans. Yet, at the same time, the palate, also preserved, appeared to be distinctly ape-like.

Further indication of the more recent emergence of the human-ape divergence comes from the amazing discovery, also in 1978, of a set of footprints made by a protohominid about 3.5 million years ago. Found by British anthropologist Andrew Hill in Nairobi, the prints appeared in strata with teeth identified as about 3.5 million years old by Mary Leakey, eminent anthropologist and widow of the late Louis Leakey, the man who opened up many of the East African fossil sites. The hominid appears to have walked across the soft mud or sand near a water hole (the whole track consists

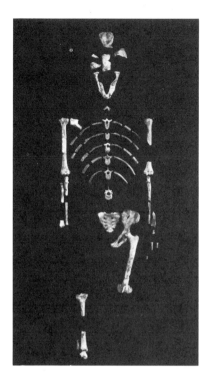

Fig. 29.53
Reassembled skeleton of Lucy, an approximately 2.9- to 3.3-million-year-old hominid skeleton uncovered in Ethiopia in the mid-1970s by Don Johanson and Tim White. Note the particularly complete bones of the appendages, useful in determining Lucy's posture and walking habits. (Courtesy Don Johanson and the Cleveland Museum of Natural History.)

of six footsteps). Although the sun baked the prints dry, they were soon filled up by ash from a nearby erupting volcano. Gradually the prints, now preserved as ash-prints, were covered by sediment, to be exposed again just recently in the eroding Nairobi strata. An analysis of the footprints shows that the hominid walked very much as we do, with big toe pointed straight ahead (as compared to apes whose big toes are pointed outward). Along with Lucy, these finds suggest that fully upright posture was a far earlier development in human evolution than previously imagined, existing at least 4 million years ago. It appears to have preceded significant development of the brain by millions of years. Accustomed as we are to thinking of brain development as the major feature of human evolution (as in many ways it is), it may come as something of a surprise to learn that it was not the earliest of our defining traits to emerge.

Various evolutionists, most particularly Stephen Jay Gould of Harvard, have argued not only that the development of bipedalism occurred earlier in our history than the development of increased brain size, but also that bipedalism was the more significant innovation in that it was the prelude to all later developments.

Gould presents a picture in which early hominids developed upright posture as they moved from the forests onto the plains. Their brains were still small, but their bipedalism gave them a totally new ecological niche to explore, one in which they could be (and were, as we now know) successful. That niche had at its basis not complex intellectual demands, but demands for ways of spotting and transporting food.

Gould points out that as an innovation, the change from being a quadruped to a biped involved a series of evolutionary changes of no small magnitude; indeed, a fundamental rearrangement of anatomy, particularly of the foot and pelvis. Bipedalism also involved innovations in the suspension and organization of the internal organs, particularly those of the abdomen. Upright posture would force all the abdominal organs down (due to gravity), whereas they normally rested against the ventral abdominal wall. By contrast to all these associated changes, development of the brain involved a mere increase in size. While ultimately increase in brain size produced qualitative changes of untold importance in human evolution, Gould argues it was probably accomplished more easily than becoming bipedal. As Gould remarks, "our greatest evolutionary step" was not development of our brains, but the transition from a four-footed to a two-footed existence.

The Origin of the Hominid Line

Evidence from comparative anatomy of living primates, as well as from biochemistry and from the fossil record, suggests very strongly something that human intuition grasped centuries ago: namely, that human beings are very closely related to the apes. The seventeenth-century playwright William Congreve wrote: "I confess free to you. I could never look long upon a monkey without very mortifying reflections." So clear did the relationship appear that it became commonplace in the latter part of the nineteenth century to claim that humans descended directly from modern-day apes.

However tempting it might be to think that modern *Homo sapiens* evolved from modern chimpanzees or apes, what Darwin actually stated was that human beings and modern apes evolved from the same common ancestor. Just as modern human beings have evolved, so have the apes. It may be a matter of judgment whether *Homo sapiens* or the apes have changed more from the common ancestor, even in purely physical terms.

Very little is known about the common ancestor between the hominid (human) and pongid (ape) lines. It is clear from examining the many different adaptations of modern apes and human beings that divergence between them occurred somewhere between 5 to 15 million years ago. A possible candidate for this common ancestor is *Dryopithecus*, an ape with tooth characteristics found only in the gibbons, apes, and humans, and

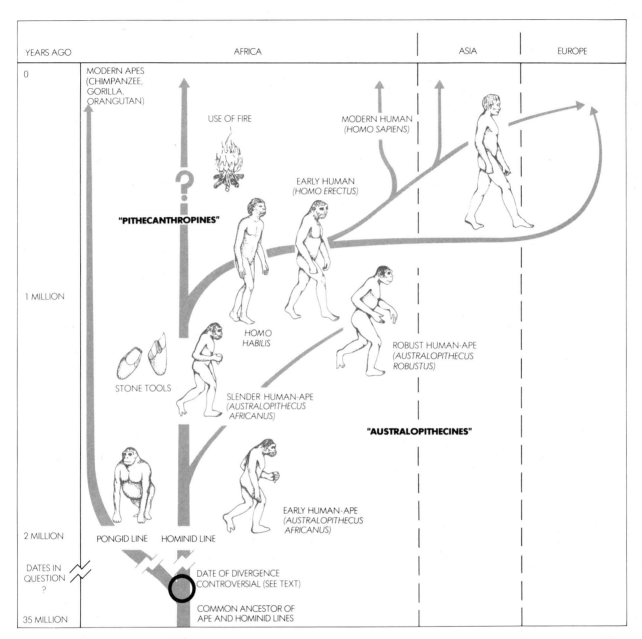

Fig. 29.54
Phylogenetic chart showing the divergence of the pongid (ape) and hominid (human) lines from a common ancestor that existed some 35 million years ago. (The actual date of the split is a matter of considerable controversy. Some argue from biochemical data that the pongid-hominid line diverged only 5 to 7 million years ago. Others, arguing from fossil data, feel the split occurred much further back in time, 14 to 25 or 30 million years ago.) Within the hominid line, modern human beings have derived from a line including the fossil forms *Australopithecus* and *Homo erectus*. Horizontal positions on the chart represent different geographic regions, indicating the possible migratory patterns of hominid populations. The actual course of migration and divergence between the African, Asian, and European subpopulations is still a matter of much speculation. It is certain that divergence was fairly recent and that full geographic isolation never occurred; gene flow was always maintained among all three subgroups.

other characteristics, such as wrist structure, found only in the monkeys. *Dryopithecus* may thus be a transitional form between the lines that gave rise to the monkeys on the one hand and the ape-human line on the other.

Whatever the exact form of the common ancestor, it was probably much less specialized for tree-living than modern apes. Likewise, it was probably not an exclusive ground-dweller as are modern human beings.

It may have spent some of its time in trees and some on the ground to take advantage of both habitats, foraging on the ground but retreating to trees for safety. Early in the Miocene era (about 60 million years ago) plate motion brought Africa and Eurasia together, allowing some ancestral ape populations to spread from Africa into the forests of Europe and Asia. As the climate became cooler and drier, the forests were replaced by grasslands, opening a new environment for the pre-hominid ancestors to explore. Whatever the exact course evolution took, it seems clear that the hominid line evolved on the plains and the pongid line in the forests.

A scheme for the evolutionary development of the hominid line is shown in Fig. 29.54. This chart shows the points of divergence of the pongid and hominid lines and further divergences among various hominid groups. It also pictures something of the migratory patterns among early hominid populations. Much about migration is highly speculative and the subject of intriguing guesswork. The oldest fossil remains—a few teeth and a jawbone—were first discovered in Africa (Kenya). Since the first remains were described, other examples of what are thought to be the same protohominid have been found distributed from Africa to India. Known as *Ramapithecus*, this early hominid possessed many obviously apelike characteristics. Its tooth structure is more human than ape in appearance, but because no

arm or leg bones have been found, it is difficult to determine whether *Ramapithecus* had an upright posture. There appears to be some evidence that it was at least *capable* of upright posture, whether or not it often assumed that position. The jaw fragments that exist suggest that *Ramapithecus* was a transitional form between more ape-like ancestors and what some consider to be the first true hominid, *Australopithecus*.

Skulls and Teeth

While physical anthropologists and paleontologists seek to find as complete skeletal remains of hominid forms as possible, such specimens are rare. In addition, certain bones of the body yield much more information than others about possible evolutionary relationships, lifestyle, and habits. Among the most useful remains are skull, jawbones, and teeth. Luckily these bones are made of harder, less fragile bone material and preserve quite well.

Figure 29.55(a) and (b) show a comparison between the skull of a modern ape and that of a modern human being. The major points of comparison are indicated by arrows. The most important data come from measurements of facial angle and cranial capacity, shape and structure of the jawbone, and form of the connection between skull and vertebral column. In the gorilla the facial angle is acute, while in the human being it is more

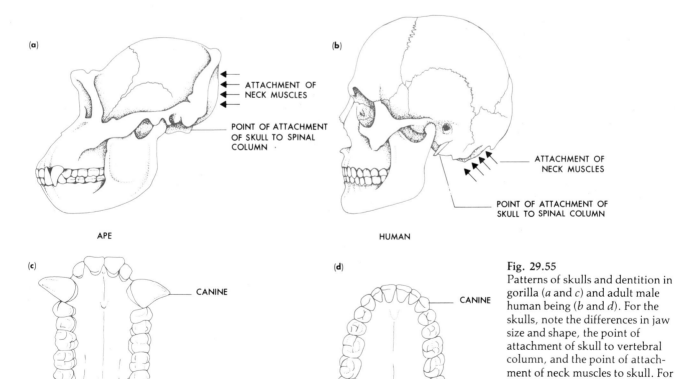

(a)

ATTACHMENT OF NECK MUSCLES

POINT OF ATTACHMENT OF SKULL TO SPINAL COLUMN ·

APE

(b)

ATTACHMENT OF NECK MUSCLES

POINT OF ATTACHMENT OF SKULL TO SPINAL COLUMN

HUMAN

(c)

CANINE

(d)

CANINE

Fig. 29.55
Patterns of skulls and dentition in gorilla (*a* and *c*) and adult male human being (*b* and *d*). For the skulls, note the differences in jaw size and shape, the point of attachment of skull to vertebral column, and the point of attachment of neck muscles to skull. For the teeth plates (upper), note the differences in overall shape and in canine teeth.

open (vertical). The ape's cranial capacity is between 600 and 650 cm³, while that of modern *Homo sapiens* ranges between 1300 and 1600 cm³. The jawbone of the ape is wider and protrudes much more; its more massive appearance is associated with a large amount of grinding activity. The point of attachment to the vertebral column is much farther back in the ape than in the human. This indicates that the ape head sits more forward on the backbone, while the human head rests more or less directly on top of the vertebral column. Also, there are differences in point of attachment of the neck muscles in the two specimens. In the ape the area of attachment is much larger and farther back on the skull, while in the human the area is smaller and lies more underneath the cranium.

These differences in skull structure are directly related to other anatomical differences. The ape walks more frequently on all fours, so its skull is thrust forward. Such an arrangement keeps the head up, looking ahead rather than to the ground. In the human the skull is attached to the vertebral column in a more vertical position, an adaptation for walking upright (bipedalism). For the same reason, the neck muscles of the ape are attached farther back on the skull than in the human. Differences in jaw shape and size relate to differences in general diet. Apes consume uncooked vegetable matter, so they must do more grinding, an activity that requires a larger jaw and stronger jaw muscles. Humans eat more varieties of food, some of which are softened by cooking (apparently an early development in hominid evolution), so a less massive chewing apparatus has been selected for.

Patterns of dentition also reveal much information about diet and habits. As Fig. 29.55(c) and (d) show, there are considerable differences in overall size between the upper teeth plates of the gorilla and human being. In part, this reflects simple differences in overall size between the two organisms. However, the gorilla teeth are arranged in a U-shape, while those of the human form more of an arc. In particular, there is a decrease in size of the canines between the gorilla and human. Being vegetarians, the apes have evolved specialized canines that enable tearing and ripping tough plant materials such as bark or rind. Being omnivores, humans have more generalized canines. Such comparisons show the correlations that can be made between specific anatomical features and more general modes of life in primate forms.

The Australopithecines

In 1924 a physical anthropologist named Raymond Dart found in South Africa a fossil of what is now recognized as one of the earliest undisputed hominids. Although initially referred to as *Australopithecus*, this form and the 200 to 300 similar remains that have been discovered in the nearly sixty intervening years probably do not constitute a single genus. It should be emphasized that

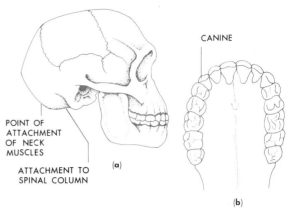

Fig. 29.56
Skull shape and dentition of *Australopithecus*. (*a*) The skull shows the typical pongid facial angle and massive jaw, but has the human point of attachment between skull and vertebral column and between skull and neck muscles (compare to ape and human skulls in Fig. 29.55a and b). (*b*) The upper dental plate shows a basically human arrangement of the teeth around the plate as well as the human shape and size of the teeth, particularly the molars and canines (see Fig. 29.55c and d for comparison).

physical anthropologists and biologists know very little about the actual phylogenetic or taxonomic relationships existing between the many individuals classified as *Australopithecus*. There is much variation among the many skeletons that have been found in East Africa. Not just one or two, but a number of different forms may have coexisted on the African or Asian continents.

Among the most prominent Australopithecine remains are those found in East Africa in the Olduvai Gorge, which runs through northern Tanzania into the area of Lake Rudolf. The various fossils representing *Australopithecus* show a small brain size (about 440 cm³), not much larger than that of a chimpanzee. The skull sits more upright on the vertebral column (see Fig. 29.56a) than is found in the apes, suggesting that the Australopithecines were bipedal. The teeth of Australopithecines are larger than those of modern humans but very similar in shape and arrangement (see Fig. 29.56b). Comparing the skull and dentition patterns of the Australopithecines shown in Fig. 29.56 with those of the modern human and ape shown in Fig. 29.55 shows that these early forms possessed decidedly human traits.

Among the forms of Australopithecines that existed, two seem to have differentiated enough to be considered different species. For lack of a better system of nomenclature we will call these *Australopithecus africanus* and *Australopithecus robustus* (see Fig. 29.54). Both appeared to coexist in the same region of Africa. *A. africanus* was smaller and less specialized than *A. robustus*. The dentition of the two species was sufficiently different to suggest that they had different diets. *A. africanus* appears to have eaten a more general diet, while *A. robustus* seems to be specialized as a seed-eater.

In terms of their culture, very little is known about the Australopithecines. They appear to have used only simple tools; we have no way of knowing whether they had a language. Although crude stone tools have been found with Australopithecine remains, it is not safe to conclude that the Australopithecines themselves made the tools. They were vegetarians and thus probably did little hunting—an activity in which communication and tool use are obviously advantageous.

A. africanus may have led to the later evolution of the genus *Homo*, while *A. robustus* became an evolutionary dead end. Most workers in the field concur that *A. robustus* did not give rise to any obvious descendants. The picture of early hominid evolution is changing daily. A fossil recently uncovered at Olduvai has a large cranial capacity (600 to 700 cm^3) but otherwise has all the characteristics of *Australopithecus*. And, in 1972 Richard Leakey found an even older fossil skull in East Africa with a cranial capacity of more than 800 cm^3. Deposited under a lava flow, it was dated at 2.6 million years. While agreement on these data is uncertain at the moment, if Leakey's find is in fact that old, the phylogenetic charts would have to be considerably revised. In Fig. 29.54, for example, the line leading to the modern genus *Homo* would have to branch off much earlier than currently shown. This would make the modern line contemporaneous with some of the earliest Australopithecines, and not evolved directly from them at all. This hypothesis is consistent with the data suggesting that the ape and hominid line diverged much more recently than previously thought.

The Pithecanthropines

Two of the earliest representatives of the genus *Homo* were *Homo habilis* and *Homo erectus* (an older name is *Pithecanthropus erectus*). Because of the greater abundance of evidence, we will focus our discussion here on *Homo erectus* only. The term "Pithecanthropines" refers to the groups generally collected under the taxonomic name *Homo erectus*.

The Pithecanthropines appeared about one million years ago and were distinctly different from the Australopithecines. They were larger, with brain sizes ranging from 700 to 1200 cm^3, thus approaching the brain size of modern human beings. The teeth of *Homo erectus* were smaller than those of the Australopithecines, suggesting an omnivorous diet. The Pithecanthropines ranged from Africa to Asia, Java, China, northern Africa (Algeria), central Europe, and Germany. They were geographically widespread, indicating a high degree of adaptation to differing climates and conditions.

Culturally speaking, the Pithecanthropines used tools, appear to have developed fire, and probably cooked food. The fact that many remains are found in caves suggest that they were cave dwellers. That some populations lived so far north also suggests that the Pithecanthropines developed some form of clothing, most likely animal skins.

The Pithecanthropines appear to have been hunters, and their enormous migrations may have been connected with the pursuit of animals they preyed upon for food and clothing. Although originating in the hot belt of the tropics, through cultural adaptation the Pithecanthropines were able to adjust to and survive in the harsher climates of the north temperate zone. With its brain capacity and use of tools, *Homo erectus* introduced into hominid evolution a feature that has come uniquely to characterize human beings: the ability to change the environment to enhance the chances of survival. By seeking out caves or building shelters, cooking foods, preparing clothes, and making tools, the Pithecanthropines insulated themselves from the worst features of a harsh and variable environment.

Cultural adaptation also increased the Pithecanthropines' chances of survival and gave them a competitive edge over other hominid forms such as the Australopithecines. Cultural adaptation has several advantages over biological adaptation. The latter occurs by natural selection over many generations; hence it is a slow process. It is also rigid. Once biological adaptation to certain conditions has developed, change is difficult and slow. Cultural adaptation, on the other hand, can occur rapidly—literally overnight, as when a group moves into a cave for shelter. It is also less rigidly prescribed and can be modified according to changed conditions. Much of the evolution of the hominid line can be best understood as the gradual replacement of purely biological adaptation with cultural adaptation. This means the gradual replacement of programmed, biologically determined behaviors with learned, culturally developed activities.

When and where modern human beings, the genus and species *Homo sapiens*, arose is still shrouded in mystery. In the early 1960s anthropologist Carlton Coon of the University of Pennsylvania proposed that in various parts of the world the different populations of *Homo erectus* underwent a parallel evolution through various intermediate forms, including the Neanderthals, to *Homo sapiens*. Coon argued that such parallel development occurred at different rates and without constant gene flow between the various populations. The chance of five or six different, isolated populations evolving at different rates but ending up at the same place (that is, all becoming the same new species, *Homo sapiens*) is highly unlikely, and most evolutionists do not accept Coon's thesis. A more likely hypothesis is that from the Australopithecines onward, a single species of hominids existed, composed of many geographic subspecies, but with constant gene flow between them. From *Australopithecus* onward, hominid evolution involved phyletic, not divergent, speciation.

Probably the oldest representatives of *Homo sapiens* are the Neanderthals, who appear to have flourished about 250,000 years ago. Neanderthals have been the focus of much popular literature and are often depicted as brutish and slow-witted. This popularized view appears to do the Neanderthals considerable disservice. While their cranial capacity (1100 to 1300 cm³) was somewhat less than that of modern human beings, they seem to have been good makers and users of tools. Geographically, Neanderthals ranged over much of Europe and co-existed with a number of other populations of the same species, each having its own combination of traits. For example, Neanderthals existed contemporaneously with another population, the Cro-Magnons (named for the site in France where the first remains were found), in northern Europe. Both were *Homo sapiens*, and they were in continual contact until the ice sheets that pushed southward about 75,000 to 50,000 years ago partially separated them. Contact was maintained by a narrow migratory path along the Mediterranean.

Throughout its evolution in the past million years or so, *Homo sapiens* has always been a *polytypic* species, (consisting of "many types"). The great diversity of human phenotypes that exists today is largely a result of the widespread geographic distribution of the species throughout its history. Genetic differences between populations reflect accumulation of biological adaptations to different local environmental conditions.

Factors Influencing Human Evolution

What selective pressures could have produced the kind of evolutionary history found in the hominid line? As with so many other questions about human evolution, only deductions based on scanty evidence can be offered. Yet some hypotheses appear to be reasonably sound.

The evolution of the human skull (see Fig. 29.57) suggests that changes in form of locomotion, jaw shape, and dentition all occurred simultaneously and in a uniform direction. As cranial capacity increased, for example, jaw and tooth size decreased. During this same time the teeth were becoming less highly specialized.

How can we account for these simultaneous developments? The Australopithecines may have moved to the plains under duress as a result of unsuccessful competition in the overcrowded or gradually shrinking forests. On the plains they were forced to become much less specialized in their diet. Among other things, hunting—at first small animals and later large ones—became an important alternative source of food. Since it was almost exclusively a cooperative effort when large game was involved, hunting required communication. This change in diet favored variations leading to a shortening of the jaw and reduction in size and specialization of the teeth.

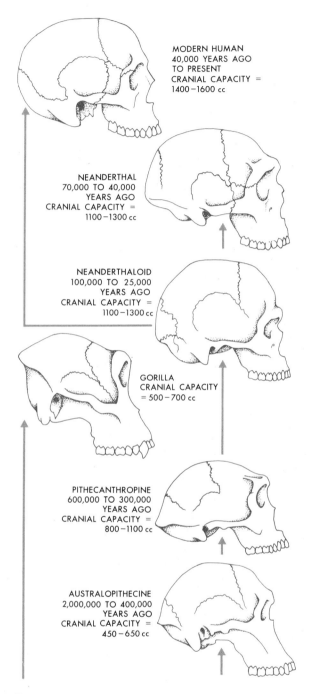

MODERN HUMAN
40,000 YEARS AGO
TO PRESENT
CRANIAL CAPACITY =
1400-1600 cc

NEANDERTHAL
70,000 TO 40,000
YEARS AGO
CRANIAL CAPACITY =
1100-1300 cc

NEANDERTHALOID
100,000 TO 25,000
YEARS AGO
CRANIAL CAPACITY =
1100-1300 cc

GORILLA
CRANIAL CAPACITY
= 500-700 cc

PITHECANTHROPINE
600,000 TO 300,000
YEARS AGO
CRANIAL CAPACITY =
800-1100 cc

AUSTRALOPITHECINE
2,000,000 TO 400,000
YEARS AGO
CRANIAL CAPACITY =
450-650 cc

Fig. 29.57
Evolution of the hominid skull, showing for comparison the line leading to the modern gorilla. Ages and average cranial capacities are given for each.

Change in the material conditions of life for the early Australopithecines (from forest to plains in habitat, and from fruits and berries to both plants and meat in diet) brought additional selection pressures to bear on hominid anatomy. Perhaps the most important changes were in the form of locomotion (and thus posture) and brain size.

Bipedalism. All primates are capable of standing on their hind legs for extended periods of time, although for locomotion all except the human being move most effectively on all fours. Several hypotheses have been advanced to explain why the hominid line may have developed bipedalism to such an advanced degree. One suggests that bipedalism and the erect posture that goes with it were advantages to life on the plains where savannah grasses were very tall (4 to 5 ft). Standing up on hind legs allowed the Australopithecines to see farther and spot prey as well as predators. This hypothesis, however, would account only for temporary bipedalism, such as that demonstrated by most other modern primates. Occasional standing up and looking around would be an advantage that would not in itself account for the development of a specialized ability to walk on two feet for great distances.

A second hypothesis maintains that bipedalism was of adaptive value in carrying freshly killed game to a campsite. Especially among the more primitive hominids, such as the Australopithecines, hunting was probably carried out by a lone individual, or at the most by small groups. Animals were probably caught and killed more by accident than by elaborate planning. One or two individuals would not have been able to consume all the meat from a carcass on the spot. At the same time, to abandon a carcass only partially consumed would waste food.* Bipedalism would have made it possible for even a single individual to carry a carcass, slung over its shoulders, for several kilometers. While such a hypothesis is quite reasonable, there are few ways to test it directly. As in so much of evolutionary theory, there are several possible hypotheses to explain the same phenomenon, and it is impossible to distinguish among them directly.

Still a third hypothesis maintains that erect posture was particularly adaptive because it freed the front appendages (the arms) for tool use. This hypothesis is attractive but leads to a falsifiable prediction. If tool use were the major adaptation that bipedalism conferred, then in the evolutionary history of the hominids we would expect to find bipedalism emerging simultaneously with the development of tools. However, tools of any sort (other than random rocks or sticks) did not appear in the fossil record of early hominids until well after the emergence of bipedalism. This strongly indicates that freeing the hands for tool use was probably a secondary, although in the long run an extremely important, byproduct of erect posture. Comparing the

three hypotheses above, then, it is *most likely* that a major force favoring bipedalism was the ability it gave primates to walk long distances carrying food.

Brain Size. The selective pressures favoring increased brain size provide even more interesting arenas for speculation. Two major hypotheses can be distinguished among the many that have been advanced to explain why the hominid line developed such a large brain and such a high level of intelligence. Both hypotheses take note of the fact that as the brain underwent evolution from primitive vertebrates through lower and higher mammals and the primates, the most dramatic changes occurred in the cerebrum—the thinking part of the brain (see Fig. 29.58). Within the cere-

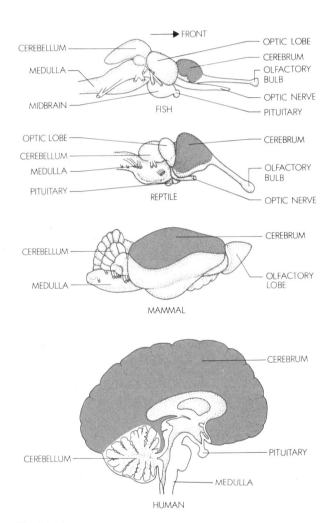

Fig. 29.58
Change in size of the brain (especially the cerebrum) in various vertebrates. The cerebrum (dark areas) contains the centers for rational thought. The other portions of the brain, especially the cerebellum, medulla, and brain stem, are concerned with integration and control of the more reflex activities.

---

*To recognize the advantages of transporting a carcass back to the larger group, and to employ cooperative behavior to accomplish the task implies a high level of social organization. By standards of modern primates, however, this is not an unreasonable task.

brum, the greatest amount of development occurred in the frontal lobes containing, among other things, the speech centers.

One hypothesis suggests that the ability to communicate, especially over distance, was of great selective advantage for hunting. In the more developed hominids, such as the Pithecanthropines, hunting appears to have been regularly carried out by small groups who may have operated in concert and with a preconceived plan. Ability to communicate from one outpost to another would obviously have been of great value in informing other members of the hunting group where the prey was located or in what direction it might be moving. Obviously, language had other advantages once it was developed. According to this hypothesis, however, the primary selective pressure favoring language development was for communication in hunting.

A second hypothesis maintains that language, and thus the brain structure that makes it possible, arose as a cultural adaptation—as an adaptation to living in social groups. According to this theory, the selective advantage would arise from ability to communicate in a variety of situations, of which hunting might be only one, and a relatively unimportant one at that.

It is not possible at the present time to distinguish adequately between these two hypotheses. Little is known about the social habits of any of the early hominid groups. It is not even known how important hunting was to the diets of early hominids. If cross-cultural studies on modern societies offer any evidence, the work of anthropologists such as Allen Johnson of Columbia University suggests that in hunting-gathering societies of today about 90 percent of the nutrition comes from "gathered" materials—plants, fruits, nuts—while only 10 percent comes from "hunted" material—animal meat. How much time the early hominids spent in hunting is totally unknown. It could have been an extremely important part of the population's time budget. On the other hand, it could have been a peripheral activity relegated to infrequent occasions. Without knowing the actual role hunting played, it is not possible to test the hypothesis that brain development occurred primarily as a response to a hunting style of life.

Sex Roles and Biological Determinism

In discussing the social organization of early hominid groups, a current biology textbook makes the following statement:

> *Hunting groups were primarily or exclusively male. An elaborate language was developed to coordinate the actions of the hunters and the complicated social relations necessitated by division of labor between the sexes at home.*

There are three methodological problems implicit in such a statement. They apply not only to questions of human evolution, but to all aspects of hypothesis formulation and testing in biology and other sciences.

The first problem arises from the form and style of the statement. As written, it is dogmatic. It makes a flat statement and gives no indication that other hypotheses also exist. It implicitly draws a connection between hunting and the development of language as if there were no disagreement between anthropologists about this subject. As we have seen, there is more than one hypothesis about the origin of language and about the importance of hunting as a major activity among early hominids.

A second problem is that neither the statement itself nor the surrounding paragraphs in the book give any indication of the evidence on which the statement rests. As anthropologists are quick to point out, it is extremely difficult to determine even the sex of fossil remains, much less more subtle characteristics, such as the social role played by one sex compared to the other. Virtually nothing is known about whether males, females, or both may have made up the hunting groups among Pithecanthropines or primitive *Homo sapiens*. Kill sites containing the bones of slaughtered animals exist, but no evidence survives about whether men or women slaughtered them, cleaned them, or cooked them. No concrete evidence about division of labor exists. The statement might be true, but given the lack of available evidence, its presentation in such a dogmatic form is misleading.

A third problem lies in the implications such a statement has for society today. The statement implies that, at least with regard to certain social activities, a division of labor quite similar to that observed in modern societies arose very early in hominid evolution. The classic picture is of the male assuming the aggressive role, ranging away from the family and fighting for food, with the female staying at home preparing food and minding the children. The further implication is that such a division of labor has been selectively advantageous and has thus become a part of the biological heritage of *Homo sapiens*. It suggests that such male-female stereotypes are part of "human nature."*

It is quite *possible* that such a division of labor *did* occur in primitive hominids. Women were probably

---

*The term "human nature" is an extremely vague and much misused one, referring at different times to everything from love and hate to aggressiveness and altruism. Inherent in the concept most of the time is some idea of innate, nonlearned behaviors or tendencies. Biologically the term "human nature" has virtually no meaning, since none of the so-called traits often attributed to it (for example, selfishness, competitiveness) can be shown to have any *specific* biological basis. Usually "human nature" is brought in to explain that which cannot be readily explained in any other way.

often pregnant and/or nursing a child or two. Such a biological necessity may well have restricted the role women could play in those activities that required moving, sometimes rapidly, over long distances. But what about women who did not have children? Or women with children old enough to take care of themselves for short periods of time? There is simply no evidence. While it is perfectly justifiable to hypothesize that men did the hunting while women spent more time "at home with the kids," it is also important to present the ideas as merely a hypothesis, not as stated fact.

Yet there is another point which is often overlooked in discussions of these areas of overlap between biology, sociology, and anthropology. Even if our surveys of other cultures were to show that they all perpetuated the same sex-role stereotypes, and even if we could show that there was a biological tendency for males to be more "aggressive" (substitute here whatever stereotypic trait you want) than females, such findings would provide no basis for deciding how to deal with the persistence of such stereotypes in our own society. One thing experience teaches us very clearly is the enormous flexibility of human behavior. We can learn vastly different kinds of behaviors, depending on what we are taught and the concrete circumstances we face in our daily lives. For example, in many societies, including England and the United States in the nineteenth century, women were excluded from higher education, voting, and jobs of any responsibility. Biological arguments were brought forward claiming that the structure of women's brains (skull shape, cranial capacity, and the like; see Supplement 1.3) precluded their being granted equal social and political status with men. Sex-role stereotypes were far more overt and rigid 100 years ago than they are today. Our behavior as a social organism has undergone a change that reflects social, rather than biological evolution. With our almost infinite capacity to learn new behaviors, any presumed or even actual biological tendencies and constraints on our behavior seem minuscule by comparison. More important, even if such biologically based constraints on our behavior exist, they should not be used as a smokescreen, excusing us from the effort to continue building toward more equal and humane societies.

Summary

Characteristics of the various animal phyla are summarized in Tables 29.1, 29.3, and 29.4. Consult those charts for a review of the characteristics of the major groups. The summary below focuses mostly on major principles of animal organization and broad evolutionary pathways, or methods of deducing phylogenies among members of the animal kingdom.

1. During the course of evolution, extinction is the rule rather than the exception. As animals have evolved, the total number of species that have become extinct is far greater than the total number living today.

2. Establishing phylogenetic relationships among different animal groups involves considerable guesswork. Important information suggesting that two species might be related comes from comparative anatomy of living groups, such as the birds and reptiles. Another kind of evidence is the discovery of fossil forms representing characteristics of both groups.

3. Establishing broad phylogenetic relationships among the invertebrates (animals without backbones) has involved a variety of kinds of evidence:
    a) Fossils, where gradual changes in particular form, such as are found in the ammonites, can be traced from earlier through later strata in an almost unbroken series. Such a graded series of fossils provides ideal evidence that a certain evolutionary development has taken place.
    b) Early embryonic development, where relationships between forms that appear very different as adults can be deduced by observing similar patterns in embryological development. For example, patterns of cleavage of the fertilized egg (radial versus spiral), of mesoderm formation, and of the larval stages can be used to trace invertebrate phylogenetic relationships.
    c) Adult anatomy, where relationships can be determined by reference to structural and functional homologies.

4. The chordates are hypothesized to share a common ancestor with the echinoderms.

5. The chordates are characterized by a notochord (a dorsally located, firm, supporting structure) and the presence of gill slits at some stage during their life. The simpler chordates such as the tunicate are called "invertebrate chordates." The vertebrates have a vertebral column replacing the cartilaginous notochord.

6. Life on land presented many new problems, especially delivering oxygen to the tissues, maintaining body moisture, and locomotion. The transitional vertebrates solved the first problem with the conversion of fish-like lungs or swim bladders into gills and the second by a series of adaptations restricting water loss. The third was solved by the modification of fins into stronger appendages that could support the entire body. The first land dwellers were probably amphibians living the major portion of their life in direct contact with the water.

7. A new approach to animal (and plant) phylogeny consists of investigations in molecular evolution. Based on the

concept that changes in genes represent changes in DNA and that changes in DNA show up initially as changes in amino acid sequence of a protein, biologists study the differences in amino acid structure of related proteins as a way of measuring evolutionary divergence. Differences in amino acid sequence of the respiratory protein cytochrome c, as observed among a variety of vertebrate species, can be used to determine phylogenetic divergence. Greater difference in amino acid sequence indicates that two species may have diverged longer ago.

8. The animal kingdom as a whole is divided into several large groupings:

    a) Parazoa: Animals without organized tissues and no digestive cavities (for example, sponges).

    b) Metazoa: Animals with organized tissues and a digestive cavity:

        i) Radiata: Animals with radial symmetry (for example, Cnidaria);

        ii) Protostomia: Animals in which the mouth and sometimes the anus develop from the blastopore of the embryo (for example, Annelida, Mollusca, and Arthropoda, the Annelid Superphylum).

        iii) Deuterostomia: Animals in which the anus develops from the blastopore, while the mouth originates from a new opening on an opposite side of the embryo (for example, Echinodermata and Chordata, the Echinoderm Superphylum).

9. The Metazoa are hypothesized to have originated from a flagellated ancestor that took up a colonial existence as a kind of flagellated blastula-like structure. A bilayered gastrula-like form is thought to have been formed by the buckling of a plakula organism, a flattened form that crept about on the ocean floor. The ventral surface of the plakula was specialized for feeding; if the organism buckled upward, forming a cavity between its ventral surface and the substrate, it would look something like a primitive gastrula. The cavity would be analogous (or perhaps homologous) to the archenteron. The buckled plakula-like organism could have been the common ancestor to the Radiata on the one hand and all the other metazoans on the other.

10. Within the Deuterostomia, the echinoderms, invertebrate chordates, and chordates are linked in several ways. Echinoderms and invertebrate chordates are linked through a common larval type appearing in some species of both phyla: the bipinnaria (dipleurula) larva. The invertebrate chordates and the chordates are linked through the presence of a notochord in the early embryonic stages of species in each phylum.

11. Phylogenetic trees postulating evolutionary divergences are often quite hypothetical and are constantly being revised. The only solid evidence on which ancestral lines can be reconstructed comes from the fossil record. Other forms of evidence—comparative anatomy and embryology, biochemical and molecular affinities—are only circumstantial. They suggest a likely evolutionary history, but they can never indicate with any certainty the actual course of that history.

12. The human species (*Homo sapiens*) belongs to the order Primata. Primates are characterized by having an opposable thumb, binocular vision, flattened nails on functional digits, a collarbone, and highly flexible shoulder joints. The primates, an order of the class Mammalia, include human beings, tree shrews, lemurs, Old World (Africa and Asia) and New World (South America) monkeys, and apes.

13. It has been hypothesized that the primates evolved from a group of mammals that took to the trees some 60 or 65 million years ago. The hypothesis views the basic primate characteristics of opposable thumb for grasping and binocular vision for precise judging of distance as adaptations to life in the trees.

14. Fossil remains such as skulls, even fragments, can yield considerable information. Cranial capacity can indicate something about the intellectual capabilities of early hominids. Placement of the skull on the vertebral column indicates posture (walking upright or on all fours). Shape of the back of the skull tells about attachment of muscles of the shoulder, also an indicator of posture. Structure of the teeth yields information about diet and general mode of life.

15. Comparative anatomy of living primates indicates degrees of divergence between modern human beings and the common ancestor of the hominid and pongid lines. Comparative behavior studies indicate something of what behavior patterns go hand-in-hand with certain anatomical characteristics.

16. The first group to emerge as early modern humans is known collectively as Pithecanthropines (a grouping that includes the genus *Homo* and thus encompasses the modern human). The Pithecanthropines ranged from Africa to Asia, Java, China, and Europe. They had larger cranial capacities than the Australopithecines (700 to 1200 cm^3) and very similar tooth structure; they appear to have used tools. They also seem to have developed the use of fire and, at least in some localities, lived in caves. They probably hunted and cooked their food. The older forms of the Pithecanthropines roamed the earth about a million years ago. They are sometimes designated *Homo erectus*, and as far as can be determined, they were probably a different species from modern *Homo sapiens*.

17. The oldest known representatives of *Homo sapiens* are known as Neanderthals, who lived about 300,000 years ago. Neanderthals existed contemporaneously with other subpopulations of *Homo sapiens*, such as the Cro-Magnons. Both had cranial capacities almost equal to that of modern human beings (about 1500 to 1600 cm^3) and appear to have had relatively highly developed cultures.

18. Two hypotheses (out of a number of others) have received particular attention in recent years as explanations for the selective pressures influencing human evolution. One

claims that hunting was a major impetus, especially in its demand for language and communication. The other claims that the advantages of culture—being able to communicate all sorts of ideas—was a major selective pressure. Both recognize the great increase in brain size, especially in the size and functional capacity of the cerebrum, as one of the most important developments in human evolution. The cerebrum contains the centers for rational thought and speech. It is not possible to test these hypotheses rigorously, since very little is known about the amount of hunting primitive hominids did or about other aspects of their culture. Anthropologists are agreed that speech must have played an important role in the evolution of hominids. In what exact context speech was selected for is still not known.

Exercises

1. What are the difficulties inherent in the construction of a wholly satisfactory phylogenetic chart?

2. What reasons can you suggest for the adaptive radiation undergone by the mammals at the end of the age of reptiles?

3. In terms of the exchange of materials with their environment, what advantages do protozoans have over multicellular organisms?

4. What evidence suggests that the birds evolved from the reptiles?

5. Which came first in evolutionary history, freshwater fish or marine fish? On what evidence is your answer based?

Chapter 30
Animal Behavior

30.1
INTRODUCTION

As we saw in Chapter 12, the movement of an appendage can be explained in terms of the contraction of certain key muscles; the contraction of these muscles, in turn, is brought about by the contraction of striated tissue of the muscle. These tissues contract because the muscle cells of which they are composed contract; the muscle cells contract because their myofibrils contract. The contraction of the myofibrils is explained by the hypothesis proposing that actin and myosin filaments slide over each other to cause contraction. What causes this sliding? Investigations are still proceeding to attempt explanations on the molecular, atomic, and subatomic levels.

Yet even given 100 percent success in investigating muscle contraction at the most microscopic level, we still intuitively feel there is something more. We know, for example, that the muscle did not just contract on its own in moving the appendage; a nerve impulse triggered the action. What triggered the nerve impulse? The brain, most likely. But what triggered the brain? Probably it was an outside stimulus perceived by the organism—a noise perhaps, or the sighting of a predator. In other words, the entire system of levels of investigation has now been incorporated into one topic: **animal behavior.***

Early in this century the field of animal behavior became divided into two prominent schools of thought. One school was given great impetus around 1924 with the publication of J. B. Watson's *Behaviorism.* The leading figures in this group, termed "behaviorists," were mostly American experimental psychologists. The second school, mostly European, began around 1909 with

the work of Jacob von Uexküll and O. Heinroth. Later, in the 1930s, important refinements and additions were made by Austria's Konrad Lorenz and Holland's Niko Tinbergen, who shared with Karl von Frisch the 1973 Nobel Prize in Physiology and Medicine for their work in this field. Many scientists, mostly zoologists, became associated with this school of thought and termed themselves "ethologists." **Ethology** may be defined simply as the scientific study of behavior and its adaptive and evolutionary significance.

It is important to recognize the existence of both the "behaviorist" and the "ethologist" orientations, for the conclusions each drew from experimental work were often widely divergent and even contradictory. The behaviorists were trained in psychology and were mostly (if not exlusively) interested in learning theory. Furthermore, this work was done almost entirely in the laboratory. The maze to be learned, the positive reinforcement of food or negative one of electric shock, the "Skinner box" in which an animal learns to press a lever or peck a button for food or drink—these were the classic tools of the behaviorists. For behaviorism, ultimate success was seen in terms of accurately describing an organism's behavior patterns after training; in turn, these descriptions would enable the experimenter to increase learning efficiency and the degree of predictability. Naturally the behaviorists preferred organisms that lent themselves well to such experimental studies: the white rat became the behaviorist's *Drosophila.*

The ethologist, trained for the most part in biology rather than psychology, approached and interpreted things quite differently. Instead of working with just a few species, often highly inbred for laboratory study, the ethologist studied large numbers of species under both laboratory *and* natural environmental conditions. The differences between the behavior of the wild rat and that of the white rat are considerable but hardly surprising. The former is genetically heterogeneous and ex-

---

*Plants, lacking nervous systems, do not exhibit what one would generally call "behavior."

Ethology is the scientific study of animal behavior and its evolutionary significance.

posed constantly to highly nonuniform sequences of environmental stimuli, while the latter may be genetically homogeneous and exposed to as uniformly controlled sequences of stimuli as the human experimenter can devise. With such very different approaches to such widely differing subjects, differing conclusions were inevitable.

In the early twentieth century, another variation of learning theory received considerable attention. The Russian physiologist Ivan P. Pavlov (1849–1936) performed a now-classic experiment. Every time he offered food to a group of experimental dogs, Pavlov also rang a bell. The dogs soon learned to associate the sound of the bell with food. After a sufficient number of repetitions, Pavlov showed that the dogs would secrete saliva in response to the bell alone; the sight, taste, or smell of food was no longer necessary. Such dogs were said to be "conditioned." Pavlov's work attracted much interest. Many variations of conditioning experiments were attempted and were for the most part very successful. Furthermore, the fact that training programs based on the principles of conditioning enable animals to learn quite complicated behavioral routines quickly suggested that the phenomenon might also account for the complex patterns of behavior found in nature. Indeed, Pavlovian conditioning became an extremely influential school of thought in the study of animal behavior.

One of the important differences between behaviorists and ethologists was the degree of emphasis each placed on the role of **instinct** or **innate behavior*** in guiding animal behavior. Ethologists are most interested in behaviors that appear to be innate, that is, are programmed into the organism from the beginning and are therefore at some level genetically determined. Behaviorists are more interested in learned, or "conditioned" behavior. Let us consider the development of instinct theory.

30.2
THE INNATENESS OF CERTAIN BEHAVIORS: SOME EXAMPLES

Web-Building in Spiders

Though interesting, the concepts of learning or conditioning nevertheless provided no satisfactory explanation for many types of observed behavior. The young orb-web spinning spider *Araneus diadematus* Cl., for example, spins a perfect web, in a pattern characteristic

---

*In recent years ethologists have tended to replace the word "instinct" with the term "innate behavior." The reason for this is that the term "instinct" has acquired a number of different meanings over the years. In everyday language, for example, people use "instinct" to refer to learned as well as truly innate reactions, as for example, in the statement "when I saw the dog cross the road I *instinctively* stepped on the brakes." Obviously braking a car is a learned, not an innate response. Because of such variations in use, the term innate is being used more regularly to refer to nonlearned, genetically programmed behavior.

of the species, on her first try—despite the fact that she has never seen her mother perform this remarkably intricate procedure (see Fig. 30.1). *A. diadematus* raised to maturity in glass tubes so small that movement is entirely restricted, on being released but still kept in isolation, can spin perfect webs. The web-spinning feat of *A. diadematus* is performed every morning by the female, who devours the old web before she spins a new one. The orb web is characteristic of several species of spiders. It is an economical one to produce; the spider covers the greatest area with as little material as possible.

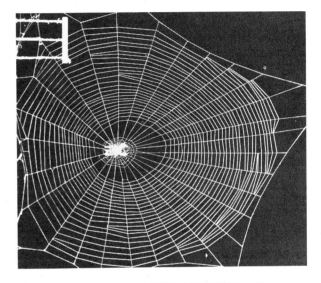

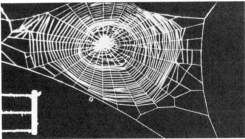

Fig. 30.1
Top photo shows the web of a nine-month-old female cross spider. *Araneus diadematus* Cl. (body mass 115.5 mg), raised in the laboratory and allowed to build a web every day. Its littermate was prevented from building any webs by being isolated in a glass tube. Upon release (body mass 77.8 mg), the experimental animal built the bottom web. Note that only the size, not the pattern, is the distinguishing feature. By the time this animal had spun a fourth web, the product of its spinning was the same as that of the control's. Further investigation revealed that the silk glands of isolated spiders slow to a low level of productivity as a result of lack of use, and this factor seems to account for the smaller web size. When the glands regain their normal rate of silk synthesis, the webs spun become the normal size. (Photos courtesy Peter N. Witt, North Carolina Department of Mental Health, Raleigh, N.C.)

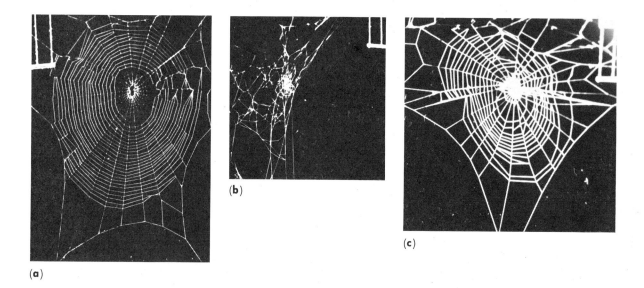

(a)

(b)

(c)

Fig. 30.2
Spider web-spinning behavior is so characteristically precise that it provides a means to study the effects of drugs on behavior. Shown are three webs of an adult female *Araneus diadematus* Cl. (body mass 157.9 mg) built on different days. (*a*) The control was built in approximately 20 min in the early morning. At 4 P.M., the spider was given 0.1 ml of sugar water containing 1 mg of dextroamphetamine, also known as "speed." (*b*) This web was built approximately 12 hr later by the drugged animal; note that the web consists of only some remnants of a hub, a few irregular and frequently interrupted radii, and some erratic strands of sticky spiral. (*c*) This web, built 24 hr later, shows some signs of recovery. However, several more days were required to restore the web to normal. (Photos courtesy Peter N. Witt.)

Fig. 30.3
The effect of other drugs on web-building. Each drug produces a characteristic, predictable change in spider motor behavior, and thus has been useful as a bioassay for drug identification. (Photos courtesy Peter N. Witt.)

All other things being equal, the larger the area of the web, the greater the probability of capturing flying prey.

The web-spinning behavior of *A. diadematus* provides an excellent subject for behavioral study. For one thing, the path taken by the spider in spinning the web can be observed and faithfully recorded. For another, the webs can be removed and sprayed with glossy white paint, spread over a dark box, illuminated from the side, and photographed (Figs. 30.1 and 30.2 are examples). Projection of the resulting films allows precise measurements of the web's proportions. The size, number, and regularity of the web parts reveal a great deal about the motor behavior of the spider that spins the web (see Figs. 30.2 and 30.3).

One such revelation is the fact that web construction occurs in distinct phases. Each is executed in a different pattern of movements. The radii and frame threads are spun first, in a way that is distinctly different from the way the spirals are built. When the radii-

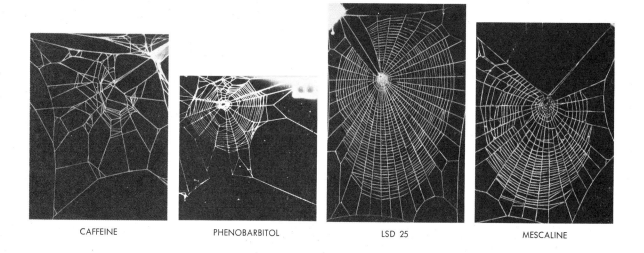

CAFFEINE   PHENOBARBITOL   LSD 25   MESCALINE

Few ethologists use the concept of "instinct" today. For behavior that is programmed into the organism the term "innate behavior" is more commonly used.

building phase has been completed, the spider goes into the next phase, working on the interconnections between the radii that form the spirals. This, too, is performed in a precise sequence of individual steps.

While the glass-tube isolation experiments *seem* to rule out the role of any learning or experience in web construction, the intricacies of the process make it tempting to believe that some sort of reasoning might be involved. Teleological explanations (explanations assuming goal-directedness) are also tempting; the spider spins the web *in order to* trap insects most efficiently. Yet this hypothesis predicts that if some of the web strands are destroyed, the animal will stop to repair the damage—otherwise there would be no purpose to the web. This prediction is not verified. The spider does not stop to repair the damage, even if it is great enough to make the web completely useless for the capture of prey.

Faced with observations of this sort, ethologists have turned to the concept of innate behavior as an explanation. Innate behavior has been defined as a natural involuntary, inherited tendency to perform a specific action. The inherited, instinctive response was seen as grounded in the central nervous system. *Differences* in the various kinds of central nervous systems (and thus the innate behavior coordinated therein) are, of course, to a large extent genetically determined. To most ethologists, therefore, innate behavior was "inherited" behavior, genetically programmed into the organisms, and thus the opposite of learned behavior. The spider web case was seen as a good example of innate behavior.

Further experiments showed that even highly innate, instinctive behavior can be modified. As Figs. 30.2 and 30.3 show, various environmental agents, such as drugs, can greatly modify the details of innate behaviors. However, such external agents usually affect the details, not the fundamental pattern.

"Releaser Mechanisms" as Innate Behavior

Consider, for example, the behavior of the female digger wasp, *Ammophila*. Within a few weeks of summer, she mates with a male, digs a nest hole, constructs cells within it to hold the future developing pupae, hunts and paralyzes with her sting such prey as caterpillars and spiders, puts the prey into the nest (where their state of suspended animation ensures that the young larvae will have fresh meat on which to feed before pupation), lays her eggs, and seals up the hole—a series of very complex behaviors carried out in a precise sequence. Since her parents died long before she was born and there is no

time to "practice," it would *appear* self-evident that her behavior could not be learned.

In 1935, Konrad Lorenz introduced a hypothetical model to account for such highly stereotyped behavior. Lorenz pictured an organism as possessing a very specific receiving center in the central nervous system to which he applied the term **innate releasing mechanism (IRM)**. Lorenz envisioned the innate releasing mechanism as being triggered by some specific stimuli, which he termed "releasers," from the environment. With this hypothetical model, the wasp's behavior can be seen as a sequence of highly coordinated responses (mating, nest building, prey capture, and the like) put forth in response to releasers. For example, the releaser for the example cited might be the appearance and approach of the male. The mating process, once completed, now becomes the stimulus, or releaser, for the next stage in the behavior, digging a nest hole. This, in turn, becomes the releaser for the next behavior, constructing cells within the hole, and so on. Thus, an initial innate releaser mechanism can trigger a whole chain of responses, each link in the chain serving as the IRM for the next.

Note that this sort of thinking allows for a regular scientific sequence of hypothesizing and testing of prediction; for example, the female wasp can be deprived of exposure to the male to see whether the rest of the sequence will follow (it doesn't). Obviously many different interpretations can be placed on this experimental observation, but the point is that at least some progress has been made. The main question at this point is whether we can establish that such innate behavior has a genetic basis. It is thus important at this point in our discussion of ethology to turn to the subject of behavioral genetics.

30.3
BEHAVIORAL GENETICS

In the honeybee, *Apis mellifera* L., certain strains are referred to as "hygienic." The name comes from the fact that if a larva dies within its enclosing cell, the workers uncap the cell and remove the corpse. Other strains are "unhygienic"; when a larva dies, its corpse is left to decay in what has now become its tomb. The descendants of hygienic bees all exhibit the same hygienic behavior pattern; those of unhygienic strains remain, like their progenitors, unhygienic.

Certainly these are sharply contrasting phenotypes, as distinct as Mendel's tall and short peas. We can, in fact, study the genetics of behavior using these two types of bees. As the following scheme shows, we can make a cross between two "pure-bred," or homozygous, bees.

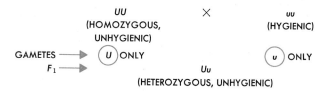

The resulting F_1 is all unhygienic; the dead larvae are left untouched, their cells capped. Clearly, the unhygienic behavior pattern (or better, perhaps, the absence of the hygienic behavior pattern) is the dominant character here. On the basis of the results thus far, tentative genotypes can be assigned as shown (let the letter U stand for "unhygienic" and the letter u stand for "hygienic").

If a backcross is now made between the F_1 unhygienic hybrid, genotype Uu, and the recessive hygienic strain, genotype uu, Mendelian genetics predicts a 50–50 distribution of unhygienic and hygienic bee colonies.

However, in 1964 the animal behaviorist W. C. Rothenbuhler showed that in twenty-nine colonies of bees resulting from such a cross, the following distribution was obtained:

8 colonies: workers left cells capped and did not remove dead larvae (were unhygienic)

6 colonies: workers uncapped cells and removed dead larvae (were hygienic)

9 colonies: workers uncapped cells, but left corpses of larvae untouched

6 colonies: did not uncap cells, but would remove larvae if the caps were removed by the experimenter

It is reasonable to propose that two pairs of genes are operating here, not one, and that the data represent a 1:1:1:1 ratio. Each pair of genes controls one of the two steps in the hygienic behavioral sequence: (1) uncapping the cells and (2) removing the larvae. Thus the original cross of hygienic with unhygienic must be written as shown below:

In terms of Mendelian genetics, this situation works out most happily. Not surprisingly, however, such neat results are rare in behavioral genetics because behaviors are often complex, consisting of many components, and not accurately described as single entities such as eye color or other morphological traits. The example does show that *differences* in behavior can have a genetic basis and, further, that behavioral patterns can be broken down into their component parts, as the hygienic behavior pattern can be broken into (1) uncapping and (2) corpse removal. The hygienic behavior, as we have seen, shows four phenotypes, but it appears to be controlled by gene segregation at only two loci. Yet a seemingly simple behavior such as the reaction of the fruit fly *Drosophila* to gravity (geotaxis) shows complex differences and appears to be influenced by many genes scattered across all the chromosomes. Geotaxis is thus said to be a **polygenic character.**

Lovebirds are members of the parrot family. Within the genus *Agapornis* distinct evolutionary stages of nest building can be shown. In particular, two types of nest-building behavior are known. In one, strips are torn from leaves for nest-building material. The strips are then transported to the building site by being tucked underneath the rump feathers. In the other nest-building behavior pattern, however, the leaf strips are carried back to the nest one at a time in the beak or bill, not tucked into the feathers.

Ethologist William C. Dilger at Cornell University wondered what would happen if he crossed "feather-carrier" lovebirds with "bill-carriers" and kept the young birds isolated so that there was no opportunity for learning. Assuming complete dominance is involved, one would predict an F_1 of either all feather-carriers or all bill-carriers. With incomplete dominance, some sort of behavioral hybrid would be expected. The latter proves to be the case, and the result is a group of hopelessly confused young birds. All are completely incapable of building a nest, because they attempt a compromise between bill-carrying and feather-carrying—which means no carrying at all! A bird might begin to tuck a strip between its rump feathers, but then the bill-carrying "urge" would take over, and the strip would not be released until it was pulled back out of the feathers and dropped on the floor. Then the whole pro-

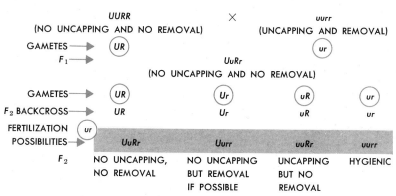

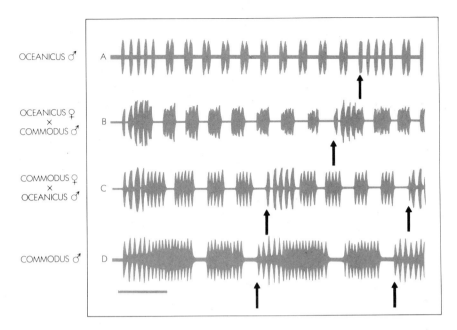

OCEANICUS ♂   A

OCEANICUS ♀
×
COMMODUS ♂   B

COMMODUS ♀
×
OCEANICUS ♂   C

COMMODUS ♂   D

Fig. 30.4
Audiograms for two species of cricket, *Teleogryllus oceanicus* (line A), and *Telegryllus commodus* (line D), and their F_1 male offspring hybrids (lines B and C). Audiograms are produced by recording the cricket's call into a microphone attached to a recording, or acoustical oscilloscope. The frequency and intensity of the "bursts" of sound are clearly visible. Note that the oscillogram pattern for the F_1 hybrids is clearly intermediary between that for the two parental species. (From Lee Ehrman and Peter Parsons, *The Genetics of Behavior*. Sunderland, Mass.: Sinauer, 1976: p. 234. Used with permission.)

cess would begin again. While the hybridization results could be explained by the assumption of incomplete dominance, the hypothesis of polygenic inheritance has also been suggested. To date there is no evidence by which to distinguish between these two hypotheses.

The hypothesis proposing that nest-building behavior is entirely innate is contradicted by the fact that occasional successes were recorded when a hybrid managed to keep a strip in its beak after failing to tuck it into the feathers. After months of practice, success was achieved in more than a third of the trials. Two years later, nearly complete success was attained by the hybrid birds. What seemed to be strictly innate behavior revealed, upon further examination, considerable susceptibility to learning. The "genetic factors" were not entirely suppressed, however; even hybrid birds attaining complete success in nest building still made a preliminary movement toward tucking the strip underneath their feathers before flying off with it held in their bills.

There are numerous other examples of genetically controlled behavior in animals. Consider the case of calling song in male crickets, *Teleogryllus*. Reproductive pairing is brought about by long-range acoustical signals produced by sexually mature adult males when they rub their wings together. Each movement of the wings is precisely timed by the concerted contraction of two sets of wing muscles that function antagonistically (see Chapter 12). Contraction is initiated by neural discharge in the motor neurons leading to each set of antagonistic muscles. The behavior is thus a direct response to the firing pattern of the neurons involved. Evidence suggests that this firing pattern, and consequently the behavior it produces, are innate and species-specific. Not only does each species of cricket produce its own characteristic pattern in terms of frequency and pattern, but also members of each species produce their species' characteristic calling song even when reared in isolation.

The genetics of calling song can be studied in two species, *Teleogryllus commodus* and *Teleogryllus oceanicus*. The basic patterns of both species' calling song are visible in the audiogram shown in Fig. 30.4, lines A and D. Hybrids can be formed in the laboratory by mating members of the two species reciprocally, so that in one cross the male parent is *T. commodus* and the female *T. oceanicus*; and in the other cross the male parent is *T. oceanicus* and the female parent *T. commodus*. The results of each of these crosses, in terms of calling-song patterns of the F_1 male hybrids, are visible in Fig. 30.4, lines D and C. Even a quick visual inspection will indicate that the audiogram patterns for the F_1's are truly hybrids. In both cases, the calling song of the hybrid male is truly intermediate between the song of *both* parental species.

In mammals, many behavioral patterns appear to have a distinct genetic basis. For example, in mice, "waltzing," a wobbling, staggering motion when walking, is known to be due to a single-gene mutation that produces neurological disorders. Brown coat color mutation in mice is associated with increased grooming activity. Strains of mice that are fast learners, and others that are slow learners have been produced by selection over ten to fifteen generations. Mice that have distinct temperature preferences have also been produced by selection, and at least one strain of mice with a distinct preference for alcohol-flavored water (instead of plain water) has been developed. All of these clearly behavioral traits have a genetic basis, and in the case of the

SUPPLEMENT 30.1
MORE ABOUT CRICKET SONGS

In the experiments with cricket calling songs described in the text, we were concerned only with the nature of the calling song of the hybrid F_1 males. What sort of response do the female F_1 hybrids have to male calling songs? Do they respond more to the call of one parent or another, or to the call of their hybrid brothers? In 1973 R. R. Hoy and R. L. Paul set out to investigate this question.

The first problem Hoy and Paul encountered was to design an experimental setup that would allow them to measure precisely (as quantitatively as possible) the hybrid female's response to one call or the other. The ingenious device they came up with is shown in Diagram A. Hoy and Paul mounted a female cricket (called a "tethered" female) to a short metal rod which in turn was mounted in a holder. Out of styrofoam they designed a Y-maze with two sets of branch-points at 120° angles. When the maze was placed in contact with the tethered female's feet she latched onto it and held the styrofoam in mid-air (as shown in Diagram A). When the tethered female walks, the styrofoam maze is propelled backward beneath her. Periodically the female comes to a branch-point. Here she turns either left or right. Two loud-speakers are mounted on each side of the female. Calls of male crickets are played out of one speaker at a time. Thus, when the cricket hears a sound coming from the right, and she reaches a Y-point, she must decide whether to go toward the sound or away from it. Hoy and Paul regarded these choices as "positive" (toward the sound) or "negative" (away from the sound).

Hoy and Paul first tested *T. oceanicus* females with calling songs of *T. oceanicus* males, and *T. commodus* females with calling songs of *T. commodus* males. Each female was presented with twenty signals from the right and twenty from the left. In all, Hoy and Paul tested twenty-two individual females of each species. They found that no females made the correct turn 100 percent of the time, even when presented with the call of their own species. It was thus necessary to establish a *criterion* for what constituted a strong response. In other words, what might be considered an average, or expected, frequency of response of a female to a male of her own species. After repeated trials, the authors found that most females will respond to male calls of their own species about fifteen out of twenty times. This value was then adopted as a base level of "attractiveness" of a male song to a female of the same species. If a female reached this level of responsiveness she was scored as having a positive response, or as having "met criterion." On the average, of the twenty-two females tested with male calls of their own species, fourteen individuals met the criterion.

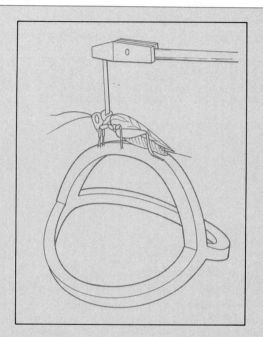

Diagram A

Table A
Responses of 147 Female Crickets to Male Calling Songs

Calling song	Females at criterion/ total tested	Females at criterion (%)
	T. oceanicus on maze	
T. oceanicus	14/22	63.6
T. commodus	4/22	18.0
	T. commodus on maze	
T. oceanicus	3/15	20.0
T. commodus	21/28	75.0
	Hybrid (T-1) on maze	
T. oceanicus	3/11	27.3
T. commodus	8/21	38.0
Hybrid (T-1)	21/28	75.0

Criterion was 15/20 correct choices in direction both for sound played from the right and for sound played from the left.

From Hoy and Paul (1973). Copyright 1973 by the American Association for the Advancement of Science.

Now, Hoy and Paul tested females with the male call of the opposite species. With *T. oceanicus* females on the maze, and *T. commodus* male calls being played over the loudspeaker, only four out of twenty-two females met the criterion; and with the converse, only three out of fifteen females met criterion. These are significantly lower rates of performance than found with females responding to males of their own species. Specificity of calling songs and the response to them is a real phenomenon.

The most interesting and perhaps least predictable response came, however, in testing the hybrid females from *T. oceanicus* × *T. commodus* crosses. Hybrid females were first tested against the calling song of each parental species. When the song was that of *T. oceanicus*, only three out of eleven hybrid females met criterion. When the song was that of *T. commodus*, only eight out of twenty-one met criterion. However, when the song was that of the hybrid males, twenty-one out of twenty-eight females met criterion! The data from these experiments are summarized in Table A. It appears that not only is the behavior of the males in emitting a specific calling song genetically programmed, but so is the response to the song. Hybrid females prefer the calling song of their hybrid brothers over that of their mothers or their fathers.

waltzing and grooming phenotypes, the chromosome on which the gene is located (and even where on the chromosome) is known.

We could give numerous additional examples of genetically controlled behavior, but the above are enough to establish clearly the idea that species-specific, genetically controlled behavior *does* exist in various species throughout the animal kingdom. Thus, it appears that ethologists are justified in talking about the existence of innate behavior patterns that are more or less determined in a given organism from the time of fertilization onward. But the fact that innate behavior exists in one species does not mean it can or does exist in another species, even a closely related one. Here an important warning is in order. Most of the examples of clear-cut genetically determined behavior come from the lower animals—arthropods, in particular. As we go farther up the phylogenetic chart, we can observe less and less behavior that can be shown to be genetically determined. Higher animals, particularly the mammals, incorporate increasing amounts of *learning* into their behavior. In human beings and other primates, the effect of learning appears to be so great as to mask any genetic effects on behavior that might be present. Thus it is very difficult to make statements about the inheritance of human behavioral traits unless we are willing to breed people selectively and raise the offspring under highly controlled conditions. As much has been said in recent years about the genetic basis of human behavior (especially in conjunction with the much-advertised field of sociobiology), it is wise to recognize how difficult it is to collect evidence for such claims in our own species. It is, of course, not impossible that human behaviors *could* be genetically determined to a fairly high degree—the discussion above of the genetics of animal behavior shows that such determination is possible. However, the existence of genetically determined behavioral traits must be determined for each behavior and each species in question. Reasoning by analogy from one species to another can at best be only suggestive.

There is a reason why higher animals have a greater degree of learning ability, with a corresponding decrease in the amount of genetically programmed behavior. Programmed behavior is relatively inflexible. Once a bird starts a courtship dance, it tends to finish it unless active signals cease coming from its potential mate. Learned behavior, on the other hand, is far more flexible—it can be modified for any number of different occasions. With their more complex nervous systems, higher animals are capable of extensive learning, thus reducing the need for programmed behavior to a few simple reflexes (such as the sucking reflex) and patterns in the early stages of life.

30.4
ANIMAL BEHAVIOR AS A COMBINATION OF
LEARNED AND INNATE RESPONSES

Consider a well-known experiment performed by Tinbergen and Perdeck on the feeding behavior sequence occurring between herring gulls and their chicks. The parent arrives at the nest, lowers its head, and points its beak downward in front of the chick. The chick then pecks at the bill, occasionally grasping it and pulling it downward. After a few repetitions of this pecking and pulling, the parent regurgitates the partially digested food. The chick then pecks at the food, breaking it apart and eating it (see Fig. 30.5).

Now it is possible to explain this parent-chick feeding sequence on the basis of releasers and innate releasing mechanisms (such as was done with digger wasp behavior); the behavior of the parent is interpreted as a releaser bringing about the proper response on the part of the chick. This is precisely what Tinbergen and Perdeck did. By building cardboard models, which varied both in color of beak and in the position of a character-

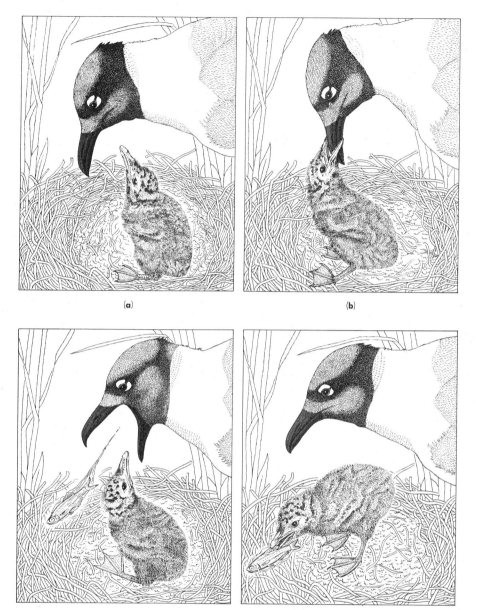

(a)

(b)

(c)

(d)

Fig. 30.5
Normal feeding behavior of the
three-day-old laughing gull chick.
Components of the behavior include
(a) "begging" peck at parent's beak
when lowered, (b) grasping and
stroking of beak, (c) regurgitation of
partially digested food, and (d) the
feeding peck. (From Jack P. Hailman,
"How an Instinct Is Learned." *Scientific American*, December 1969.
Copyright © 1969 by Scientific
American, Inc. All rights reserved.)

istic red patch on the lower jaw, they showed that certain features (such as the shape of the head) were unimportant, whereas the shape of the bill, its motion, and the position and color of the red patch on the lower beak yielded significant differences in the experimenter's ability to elicit the chick's feeding response. All these observations are completely consistent with a hypothesis proposing that an interacting series of releasers and innate releasing mechanisms bring about the proper responses, and it again appears quite unnecessary to suggest the involvement of any learning component in the bird's behavior.

However, care must be exercised in accepting such an interpretation without reservation. In careful studies of the laughing gull, J. P. Hailman and his associates have shown that gull chicks raised in darkness (so that no visual stimuli and thus no pecking practice was possible) differed significantly in pecking accuracy from control chicks exposed to model heads from hatching. Furthermore, a significant increase in pecking accuracy was demonstrated by chicks reared in the wild. It was found that, while a certain improvement in pecking accuracy is achieved without practice, visual experience and practice are necessary if the animal is to attain full pecking accuracy. Finally, with continuing practice and maturation in the nest, the chicks gradually responded more to accurate models of the parent gull's head and less to inaccurate ones. The results strongly suggest that in this case, and quite possibly others, what at first appears to be instinctive or innate behavior may actually be behavior in which learning is one of the causal factors producing the behavior. It seems likely that other sequences of behavior heretofore assigned entirely to the category of innate behavior can and should be reexamined.

The foregoing example suggests that we must use extreme caution in separating behavior into learned and innate components; perhaps we should even question whether such divisions between these two components are likely to be helpful. It will be necessary to return to this problem.

Insight and Imprinting

Animals may learn in a variety of ways. A turtle will avoid an object moved toward it by withdrawing into its shell. Continued exposure to the moving stimulus results in less and less withdrawal until the response is entirely extinguished. Such learning is called **habituation.** There are clearly advantages for organisms in being able to distinguish insignificant stimuli in their environment from those that are highly significant, such as the appearance of prey or a predator.

Trial-and-error learning is also well known in animals. Experimentally, animals learn to move through a maze by trial-and-error; in nature, paramecia utilize the same behavior in swimming around an obstacle in their path. In each case, the technique is the same; the animals moves until progress is blocked, then backs off and tries again in a different direction until the obstacle is passed.

Insight Learning.   Far more complex is learning that seems to involve "reasoning." For a variety of reasons, some animal behaviorists have used the term **insight learning** to refer to learning on the part of an animal when something more than simple conditioning is obviously involved. This does not always mean simply learning a difficult task. Chimpanzees can be trained to ride bicycles and play tunes on musical instruments, and the learning feats of the porpoise are well documented. But the vast majority of such learning can be accomplished by means of simple conditioning arranged in complex sequential patterns.

Insight learning occurs when an animal is able to solve a problem on the first encounter by using experience from a prior, but different situation. A classic example of the presence and absence of insight learning is shown in Fig. 30.6(a) and (b). In Fig. 30.6(a), the raccoon is unable to figure out that by going back around the distant post it would unwind the rope and be able to

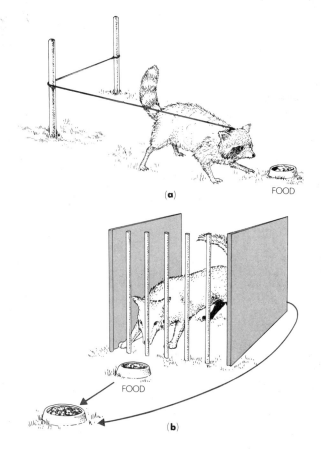

Fig. 30.6
Examples of insight learning and lack of insight learning in two animals. In (a) the raccoon cannot figure out how to go around the second stake in order to lengthen the leash and reach the food. In (b) the dog is able to go outside the cage and reach the food when the food is moved back far enough (arrow) from the bars. When the food is very close the dog tries to press through the bars and does *not* go around and outside. The first behavior is an example of insight learning. (Adapted from E. O. Wilson et al., *Life on Earth.* Sunderland, Mass.: Sinauer, 1978.)

reach the food. In this same situation, a chimpanzee or a human being would be able to make the correct assessment and walk around the distant stake, even though never having confronted that same situation before. In this case the chimpanzee or human would be said to display insight learning. Another example of insight learning comes from Fig. 30.6(b). The dog tries at first to get to the food by pushing against the bars of the cage. If the food is moved farther back, however, as indicated by the arrow, the dog suddenly gains an "insight" and goes around the back of the cage and directly to the food. In the first situation the dog displayed no insight learning; in the second situation it did.

It is important to keep in mind that insight learning is quite different from an inductive generalization. That is, insight learning is more than the application of accumulated past experience of one kind to a new example of the same kind of situation. Insight learning involves putting together several different and separate past experiences to solve a new problem.

Imprinting.   A startling contrast to "insight learning" is a type of learning first described in 1890 by ethologist Douglas Spalding. Spalding noted that a baby chick that had not heard the call of its mother until eight or ten days of age would then refuse to recognize her at all, despite her coaxings. This observation was followed in 1910 by zookeeper O. Heinroth's report that young ducklings follow the first relatively large object they see moving. In 1935, Konrad Lorenz stressed the uniqueness of this form of learning, which he termed, in translation, "**imprinting.**" Lorenz noted that, in geese, imprinting occurs within a short time early in life, lasts a lifetime, and unlike other forms of learning does not seem to involve any "reward."

Imprinting is obviously of selective value, since the first moving and sound-emitting object a young bird is likely to see is its mother. Interestingly, however, young birds also become imprinted to objects such as toys, boxes with ticking clocks inside, and even the experimenter. Such objects, of course, bear little resemblance to the natural mother, and it is interesting to note that no matter what the object to which the animal becomes attached, the attachment is a long-lasting one. Certain male birds imprinted to the human hand may prefer to attempt copulation with the hand than with a receptive female.

Careful studies of maternal imprinting in goats have been carried out by P. H. Klopfer and M. S. Klopfer of Duke University. It was found that contact with the young (kid) *directly after birth* is essential for the mother (the doe) to show normal maternal care patterns of behavior. If this contact is denied, the doe will reject her kid, even if it is returned to her after only an hour's separation. However, if the doe is allowed five minutes with her kid *first*, and then separated for an hour or more, she reaccepts her kid and its littermates (if any) but will not accept a kid that is not her own.

The converse is not the case, however. Denied her own young but allowed five minutes with an alien kid directly after birth, a doe will not only accept that alien kid as her own but will also accept her own young. However, only the alien kid to which she has been exposed is accepted; other alien kids are not. Neither the natural nor the alien kid is able to evoke the response if denied contact with the doe immediately after birth.

What is the biological basis of the differences between natural and alien kids in their ability to evoke certain maternal responses? A hypothesis proposing differences in scent of the young predicts that, if the olfactory pathways to the brain are blocked, the doe should no longer be able to discriminate between her own and alien young. This prediction is verified; application of cocaine to the olfactory membranes results in loss of the discriminatory ability. But what is the precise nature of this olfactory cue, and how is each kid individually "coded"? Also, how does this scent come to be known by the doe?

If a doe is allowed contact with her kid directly after birth and then the kid is taken from her, the doe shows obvious distress. If postbirth contact is denied, however, the doe continues as if she had never mated and given birth. It seems likely that while the stimuli provided by the kid are obviously important, the changing internal physiological environment of the doe may be equally so. For example, although the experiment involving anesthesia of the olfactory sense of the doe shows that a scent (or scents) emanating from the kid does enhance the relative attractiveness of her own kid over aliens, elicitation of maternal behavior itself is clearly not dependent on any special olfactory cues produced by her own kid. Thus it is reasonable to look for physiological differences in the doe during the crucial five minutes after birth that might account for the fact that this period is so important to the eliciting of maternal behavior. Noting that there is a large increase in the blood level of a hormone (oxytocin) as the head of the fetus squeezes through the cervix (neck of the uterus) and passes down the vagina, and noting also that this high level of oxytocin in the blood falls to a normal level within a few minutes after birth, the Klopfers speculated that oxytocin, which apparently plays a role in bringing on the final uterine spasms that deliver the kid, may also activate maternal "centers" in the hypothalamus (or elsewhere). Or, the oxytocin may change the thresholds of peripheral receptors so as to temporarily sensitize the doe to certain elements of the world about her (such as a kid). In the Klopfers' own words, "Even if oxytocin is not the keystone of mother love, its action provides us with a most useful and exciting model."

Certainly the differences between insight learning and imprinting are considerable. It is perhaps not surprising, therefore, that attempts to come up with a general theory of learning covering all such situations have so far been notably unsuccessful.

30.5
COMMUNICATION

Ethologist Martin W. Schein of West Virginia University points out that there is a tendency to look so closely at the causes of a behavioral action and/or the behavioral action itself that the *consequences* of that behavioral action are ignored.

One form of behavior specifically geared to producing consequences is **communication.** Communication between animals may, of course, occur between different species (*interspecific*) or between members of the same species (*intraspecific*). Alarm signals given by African gazelles are usually detected by baboon troops foraging for food in the same area; the alarm is communicated throughout the troop and responded to accordingly. The alarm call of a crow communicates one thing to a hawk and another to other members of the flock to which the crow belongs.

At times, communication between animals of the same species appears limited to only one sense, while at others two or more senses may be involved. A mother turkey will peck at silent models of her young, yet will accept a stuffed mammalian predator if a loudspeaker within it utters the baby turkey's cries. Clearly sound, not sight, is the communication medium here. On the other hand, the complex mating behavior sequences seen in many bird species prior to copulation seem to involve both sound and visual communication. In many mammals, olfactory and visual communication is used in mating; the female first produces a chemical with a distinct odor that communicates her sexual responsiveness to the male and then performs the proper behavior patterns necessary to ensure copulation.

On some occasions, the message communicated through the channels of one sensory system is so "strong" that equally strong contradictory messages conveyed through another system are ignored. Such is the case in ants. When a member of an ant colony dies, certain chemicals produced by its decay are detected by worker ants, who remove the dead ant from the nest. Living ants coated with the same decay products are likewise carried out and, despite their struggles, discarded. Upon each return to the nest by these "dead" ants, they are carried out again. No matter how strongly the visual and tactile communications flash "I am alive," the olfactory message "Remove me" is the predominant one.

Chemicals synthesized within a living organism's body that serve to release a response on the part of another organism to which the chemicals are directed are called **pheromones**. In a sense, pheromones are "ectohormones" in that they cause their effects outside instead of inside an organism's body. The parallel with hormones breaks down, however, in that some pheromones appear to "release" (rather than motivate) behavior, and thus act like other types of taste and smell stimuli. While common in such animals as insects, pheromones are known to play an important communication role in mammals, fish, crustaceans, and spiders, among others.

Honeybee Dance Language

One of the most remarkable examples of communication postulated for any of the "lower" animals is that of the "dance language" of honeybees. This language hy-

pothesis was a logical interpretation derived from the results of a series of ingenious experiments conducted by Karl von Frisch in the 1930s and 1940s. Von Frisch noted that when a food source left near the hive was discovered by one worker bee, there would shortly be hundreds at the site transporting the food back to the hive. Since the location of the food by so many bees in such a short time could obviously not be ascribed to random searching, von Frisch hypothesized that the forager bee that first found the food somehow communicated its discovery to the hive. By using glass-walled hives and marking the forager bees with colored paint, von Frisch was able to follow their movements upon their return to the hive. He noted that each animal ran in circles, first to the right and then to the left, making a kind of C-shaped figure (see Fig. 30.7a). This pattern was repeated many times, with much excited activity on the part of both the dancer and the surrounding workers. The latter would press close to the dancer and touch it with their antennae. Following this, they would leave the hive and fly directly to the food source.

Further observations obtained by placing the food at varying distances from the hive convinced von Frisch that the dance of the bees also communicated distance. When the food is placed some distance (more than 100 meters) from the hive, the dance varies. The bee traverses a figure 8 rather than a C (see Fig. 30.7b). When traversing the central straight line of the figure 8, the bee wags its abdomen rapidly from side to side. Von Frisch compared the number of turns per unit time to the distance of the food source from the hive and found

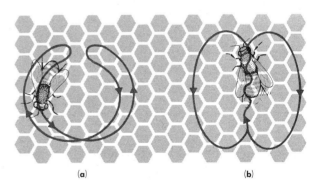

(a)   (b)

Fig. 30.7
Two forms of the "dance" of the honeybee used to communicate the whereabouts of food in relation to the hive: (a) The "round dance," used when food is located close to the hive; (b) The "waggle dance," used when food is more than 100 meters from the hive. Karl von Frisch and others observed that the speed of the dance (the number of times the bee completes the entire sequence per minute) indicates just how far away the food is. Also, the orientation of the central waggle portion of the dance on the surface of the hive indicates direction of the food. (Adapted from John W. Kimball, *Biology*, 4th ed. Reading, Mass.: Addison-Wesley, 1978.)

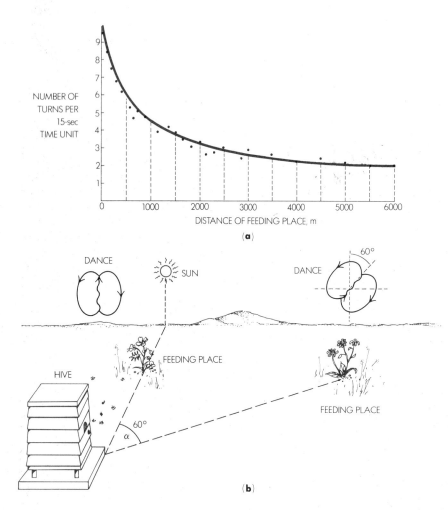

NUMBER OF TURNS PER 15-sec TIME UNIT

DISTANCE OF FEEDING PLACE, m

(a)

DANCE

SUN

DANCE

60°

HIVE

FEEDING PLACE

FEEDING PLACE

60°

α

(b)

Fig. 30.8
Information communicated by the "waggle dance" of honeybees. Bees are able to indicate both distance and direction of a food source. (*a*) Graph showing relationship between the number of turns per unit of time and the distance from hive to food source: the farther away the food, the slower the pace of the dance. (*b*) Diagram indicating how bees communicate direction of a food source. The angular distance of the food from a straight line connecting the sun and the hive (either to the right or left), shown as angle α, is indicated by the orientation of the straight or "waggle" portion of the bee's dance from the vertical, when danced on the upright portion of the comb within the hive.

a distinct inverse correlation (see Fig. 30.8a). He also noted that the straight portion of the dance gave information concerning the position of the food in relation to the sun. For example, if the food were directly in line between the hive and the sun, the waggle part of the dance would be straight up and down (perpendicular to the ground) on the surface of the hive. (Bees build their hives with perpendicular comb surfaces; it is on the comb surface that the dance takes place.) On the other hand, if the food source were 60° to the right of a line connecting the hive and the sun, the straight ("waggle") line part of the dance would be 60° to the left of the perpendicular (Fig. 30.8b). If von Frisch's hypothesis is true, we might wonder if bees are able to communicate directions for food sources on cloudy days. The answer is yes. It turns out that bees do not "see" the same as humans do. The eyes of insects are able to see polarized light, which means that, among other things, they can see the sixty-cycle flickering of ordinary light bulbs, which our eyes cannot see; they can also see the position of the sun (because some of its light is polarized) through a cloud layer.

Von Frisch's "language hypothesis" was challenged by A. M. Wenner of the University of California at Santa Barbara; P. H. Wells of Occidental College, Los Angeles; and D. O. Johnson of the Air Force Academy, Colorado Springs, Colorado. These authors claimed that several observations contradicted the "dance" hypothesis:

1. Worker bees often take much longer to find a food source located by a forager, even after being exposed to the latter's dance, than would be expected if the dance had conveyed precise information about distance and location (some workers took up to 10 minutes to find food only a 20-second flight away).

2. When a food source that bees had already visited was heavily scented for three days by the addition of oil of clove, the worker bees "remembered" the location and found it directly on the fourth day—with no dance information necessary.

3. If a scented food source is set up in an area well outside the range of already-known sources, worker bees find it readily, even if no forager has reached the spot previously. The authors claim that the language hypothesis predicts bees should not visit the new source —unless perhaps one or two find it completely by accident.

Von Frisch, however, challenged these objections in his 1973 Nobel Prize speech. He first pointed to several additional experiments that support the notion that the bee dance conveys precise information about food location. When the bee hive is placed on its side so that the dancing surface is now disoriented in space, worker bees have much more difficulty finding an already-located food source. Moreover, if the sky in the vicinity of a hive is shielded from the sun (by placing a large tent-like covering over it), the workers also fail to pick up the necessary information. Neither of these experimental modifications could be expected to have much effect if bees did not use the sun as a means of orientation, and dancing as a means of communication.

Most important to von Frisch was a logical flaw in his critics' argument. The olfactory hypothesis is not incompatible with the dance hypothesis. Von Frisch never claimed that bees used dance information exclusively in locating food. His hypothesis was perfectly compatible with the use of other clues such as olfaction or sight in locating or relocating food sources. Also, the dance hypothesis speaks to a slightly different aspect of the whole food location phenomenon than the olfaction hypothesis. Dancing was the method by which forager bees *communicated* information to workers about where food sources were found. The hypothesis did not say how foragers found food in the first place, nor did it claim that workers who had gained information from a forager's dance could not be distracted by olfactory or visual clues *en route* from the hive to the communicated source. Finally, it is not unusual that some bees took longer to find the food source than might be expected if the dance conveyed accurate information. A process does not have to work 100 percent of the time to be valid. Not every bee has to find the food source by the shortest possible route in order for us to accept the dance hypothesis. There is always a statistical margin of error in any process. Forager bees may not give accurate information all the time, nor may workers pick up the information correctly. No one would argue that, because you can still get lost after someone has given you directions, giving directions is useless.

One significance of the bee dance hypothesis is that it appears to be a largely innate (genetically controlled) behavior that is species-specific. Different species or subspecies of bees have variations in their dance patterns that resemble variations—dialects, if you will—in human language. For most species the basic pattern is the same—the circular "round" dance for close distances, and the figure-8-shaped "waggle" dance for farther distances. Variations occur in the ways in which the speed of the round or waggle dance indicates

actual distance, or in the methods of indicating direction. For example, the Austrian honeybee (*Apis mellifera carnica*) does the round dance for distances up to 200 feet from the hive, whereas the Italian honeybee (*Apis mellifera ligustica*) does the round dance for distances of 50 feet or less, switching to the waggle for food sources farther away. Moreover, at a distance of 1000 feet, the Austrian bee makes 7.4 runs per 15-second period, whereas the Italian does only 6.4. When Italian workers are put in contact with Austrian foragers, or vice versa, the communication system begins to break down. An even more interesting experiment occurs when Italian and Austrian honeybees are crossed with one another. The F_1 hybrids have a mixed or intermediate dance pattern, which communicates accurately to neither parental species. These experiments suggest strongly that in bees the dance behavior, like all other genetically determined traits, shows variations or dialects between partially isolated or separated populations.

30.6
GENE-ENVIRONMENT INTERACTIONS

There is now a strong tendency in ethology to regard the old controversy over innate versus learned behavior as no longer a fruitful one. Only with complete recognition that the answer is probably age-specific is it considered a worthwhile goal to establish "how much" of a particular behavioral sequence is innate and "how much" is learned. In other words ethologists now recognize that the effects of learning and genetically determined differences in structural factors differ not only from component to component of the behavioral pattern, *but also from developmental stage to developmental stage*. Attention is therefore more apt to be directed toward analysis of the characteristics of each developmental stage and of the transition from one stage to the next. As ethologist D. S. Lehrman says: "The interaction out of which the organism develops is *not* one, as is so often said, between heredity and environment. It is between *organism* and environment! And the organism is different at each different stage of its development."

In systematically challenging the concept of innateness as viewed historically in their own field, ethologists such as Peter Klopfer of Duke University are also challenging the concept of the gene as the repository of data or "blueprint" from which the organism is constructed. Their work suggests that the gene is rather an information-generating device that exploits the predictable and ordered nature of its environment. This view fits well with the model of gene action advanced by Jacob and Monod (see Section 19.8). In their study of insect development, H. A. Schneiderman and L. I. Gilbert (1964) have shown that hormones whose synthesis can be traced ultimately to the action of particular segments of the DNA helix activate genetic transcription at other portions of the helix. As was seen in Chapter 19, the transcription products, in turn, may further exert a feedback and regulatory effect on development.

There are parallel situations at the cultural level. In discussing human development, Erik Erikson (1968) writes:

> *The human infant is born preadapted to an average expectable environment. Man's ecology demands constant and natural historical and technological readjustment which makes it at once obvious that only a perpetual, if only ever so imperceptible, restructuring of tradition can safeguard for each new generation of infants anything approaching an average expectability of environment.*

Thus human behavior, as is the case with development of the embryo, is seen as the outcome of epigenetic rather than preformational processes. Epigenetic development, in turn, assures greater developmental stability, since an epigenetic system is buffered and self-correcting at several points. Like aggression, behavior is, in Klopfer's terms, a *process* not a noun, and the hope of finding the blueprint for a specific innate behavior on a chromosome is illusory. As Lehrman cogently points out,

> *To say a behavior pattern is inherited throws no light on its* development *except for the purely negative implications that certain types of learning are not directly involved. Dwarfism in the mouse, nest-building in the rat, pecking in the chick, and the "zig-zag dance" of the stickleback's courtship are all "inherited" in the sense and by the criteria used by Lorenz. But they are not by any means phenomena of a common type, nor do they arise through the same kinds of developmental processes. To lump them together under the rubric of "inherited" or "innate" characteristics serves to block the investigation of their origin just at the point where it should leap forward in meaningfulness.*

30.7
AGONISTIC BEHAVIOR

This Land Is My Land

As has been stressed more than once in this book, the acceptance of scientific hypotheses is quite frequently related to particular historical periods.* One hypothesis wholly harmonious with the culture of its day was put forth to explain the fighting often observed among males of the same species during the breeding season. The reasons for this fighting seemed perfectly clear; the

---

*It can be correctly pointed out, of course, that the concept of ideas having "their time" is based on whether or not they are accepted; that is, if they are accepted, the time was ripe for this acceptance, and if not, the time was not ripe. The reasoning is circular.

males were competing for a female, with the strongest claiming the prize. The observation of such fighting was one fact among many that led Charles Darwin to his concept of "sexual selection," whereby nature allowed only the strongest to contribute to the species' future gene pool. This concept fit in well with the stereotyped views of manhood and womanhood fashionable in the Victorian era, as it also did with the capitalist ethic of competition as the most effective mechanism for ensuring progress.

In the late nineteenth century Henry Eliot Howard, an English businessman, began a detailed study of British warblers and moorhens. An avid bird-watcher, Howard spent almost thirty years carefully studying warbler behavior and making copious notes. The result, in 1920, was a book entitled *Territory in Bird Life*. Its message was simple: it is *territory*, not females, for which the male birds compete.

Slowly at first, and then with increasing rapidity, Howard's idea gained acceptance. The aggressive behavior of other birds was observed in the fresh new light of **territoriality**, rather than sexuality, as the stimulus that provoked intraspecific fighting. According to the new territorial concept, the males fight for space—a tree, a meadow, a certain portion of a forest—with the winner taking a distinct and definable portion. The established territory, rather than any direct attribute of the male, became the attractant that won the female.

Of course, it should be pointed out that Howard's concept of territoriality may be no less culturally bound than was that of sexuality. He advanced his idea at the time the British empire had spread its territory across much of the globe. It was perhaps not merely by chance that Howard emphasized territoriality as the major factor in animal competition. After all, Britain's territorial expansion was fought against and opposed throughout the nineteenth century, involving Britain in a series of wars and conflicts for over a century (from the American Revolution in 1776 to the Boer War in 1899–1902, and including World War I, 1914–1918).

The basic concept of territoriality proves to have extensions beyond the class Aves (birds). On the Rio Piedras campus of the University of Puerto Rico, lizards, not squirrels, scamper through the trees and along the sidewalks. The animals are strongly territorial; the experimental introduction of one male lizard into another's territory leads to characteristic warning gestures on the part of the owner that, if unheeded, lead to combat. When more than one male lizard is kept in captivity, the cage must be large enough to allow each male to have his own territory; if the cage is too small, the loser of the fight cannot retreat out of the disputed territory to safety, and his death is the result.

Invertebrates also show territoriality, though less commonly. Male crickets, for example, fight vigorously to defend their areas against trespassers; indeed, cricket-fight matches are a popular sport in parts of the Orient. Most interesting, perhaps, is the contention of some re-

The spectrum of behaviors from overt attack to overt flight comprise what is called agonistic behavior. In general agonistic behavior involves ritualized interactions that avoid, rather than lead to, bloodshed.

searchers that territoriality can be demonstrated among some of our closest primate relatives (for example, the howler monkeys of Central America). The trespassing of one troop of howler monkeys into an area claimed by another leads to a nonviolent but deafening vocal contest that continues until the trespassers retreat. The claim that howler monkeys are territorial animals is somewhat weakened, however, by the observation that howling contests may occur even when the troops are some distance from each other. It has also been shown that the territories of howler monkeys are not exclusive and that considerable overlapping occurs.

What is the adaptive significance of territoriality? Most recent authorities subscribe to a sort of ecological hypothesis. Once a male bird has staked out his territory (a tree, for robins, or a rocky nesting site, for gulls) he does not have to defend it all the time. A male robin singing in the morning reestablishes his territorial control of a given tree (or portion thereof), at least for the better part of the day. Despite popular stories to the contrary, a relatively small portion of an animal's time and energy budget goes into territorial defense.

Research on territoriality has established one basic fact: territory is different things to different species. A hawk or eagle defends a rather wide area in which it does all its feeding. Gulls, on the other hand, show no territoriality in feeding, and it is possible that what has been interpreted as territoriality in some gulls may be little more than nest defense, especially in those species in which the nesting "territory" is only two feet or so in diameter.

Agonistic Behavior and Aggression

Ethologist Aubrey Manning of the University of Edinburgh in Scotland points out that "the first essential for a territorial animal is that it be aggressive towards others of its kind." The defense of a territory can be accomplished only if the defender shows **aggression** —initiates or gives signs of initiating an attack— against any individual entering territory claimed by the defender. Note here an almost complete reversal in the use of the term aggression in ethology compared to its use in international conflict. In the latter case, the nation violating the territory of another is referred to as the aggressor. On the other hand, when the invaded nation attacks the trespassing nation's forces, the action is called defensive, not aggressive. This differing use

of the same term is but one of several difficulties encountered when we move from considering aggression in nonhuman animals to aggression in humans.

Clearly the study of aggressive behavior must be approached with Professor Schein's warning (see Section 30.5) in mind, that we too often focus attention on the *causes* of behavior and devote two little attention to the *consequences* of behavior. Merely to focus on the behavior patterns of an aggressive organism is not sufficient; due consideration must be given the effects of this behavior on both the aggressor and the animal that is the target of the aggression.

Actually, there is a spectrum of behavior ranging in intensity from overt attack to overt fleeing. This spectrum of behavior is termed **agonistic behavior.** All displays of agonistic behavior (for example, the raised and compressed threat stance of a territorial lizard) are combinations of motivations to attack and to flee, in various absolute and relative strengths (see Fig. 30.9).

Several years of ethological research on agonistic behavior have established certain "facts" concerning it. For example, the claim has been made that the production and release into the bloodstream of certain hor-

Fig. 30.9
Agonistic behavior in mice. As a result of many encounters, the mouse at the left has assumed a dominant status over the one at the right, which adopts a characteristic defense posture of holding out the forefeet and moving only when attacked.

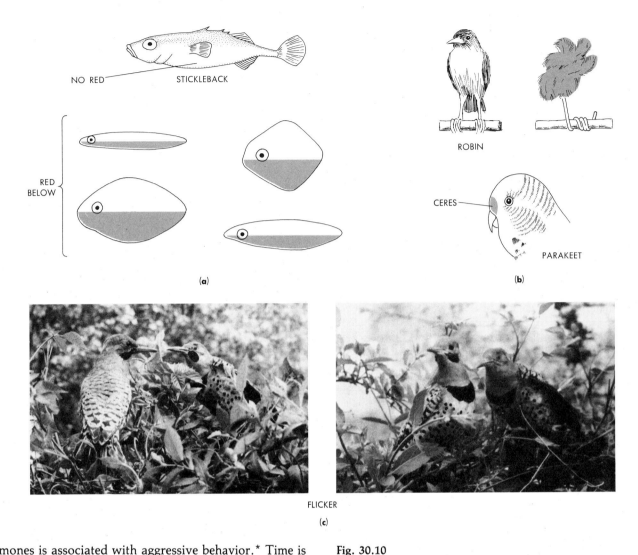

NO RED   STICKLEBACK

RED
BELOW

(a)

ROBIN

CERES

PARAKEET

(b)

FLICKER

(c)

mones is associated with aggressive behavior.* Time is also often a factor; some animal species show agonistic behavior only at certain periods, such as the mating season. Finally, it has been established that aggression may be triggered by certain external stimuli. The male three-spined stickleback, for example, reacts violently to the visual stimulus of red on the underside of a rival male. If presented with a crude model of a rival male with a red underside, the male attacks viciously. One case has even been reported in which males attempted to attack British mail trucks (which are red) driving by the window in which aquariums containing the fish were kept. Yet a more lifelike model, without the red underside, is not attacked. In parakeets and flickers, certain colorations or markings distinguish males from females.

Fig. 30.10
Four examples of "releaser" stimuli. (a) The crude models of stickleback fish painted red on the underside are threatened by a male stickleback, while the true male at top with no red is ignored. (b), Top: the robin will ignore an immature male without the characteristic red breast, but may attack a tuft of red feathers. (b), Bottom: if the characteristic color of the female parakeet's ceres is changed, her mate attacks her as if she were a rival male. (c), If the black spot or "moustache" of the male flicker (*Coleoptus aurates*) is painted on his mate (left), he attacks her as he would a rival male; when it is removed, he accepts her as his mate again (right). (Stickleback models after N. Tinbergen, *The Study of Instinct.* London and New York: Oxford University Press, 1951.)

If the markings are changed experimentally, the males will attack their own mates (see Fig. 30.10).

"Aggression Centers" in the Brain

Also postulated, but far less certain, is the existence of certain "aggression centers" in the brain which, when

---

*While there is little question that the release of certain hormones is associated with agonistic behavior, it *is* still possible to challenge the many cause-and-effect hypotheses stemming from the consequences of increased hormone output, and some ethologists have done so.

Although it is possible to produce an aggressive response by electrically stimulating certain centers in the mammalian brain, there is no evidence that the usual range of human aggressive behavior is produced in the same way. Human aggression—from war and crime to "temper tantrums"—appears to be attributable to environmentally induced rather than built-in biological factors.

stimulated, initiate and control aggressive behavior. Supporting this "aggression center" hypothesis is the observation that electrodes implanted into certain regions of the brain release violently aggressive behavior in normally nonaggressive animals when a mild electric current is applied. As a result of such experiments on cats, Dr. John P. Flynn of Yale University's School of Medicine believes that internal conditions, by themselves, are sufficient to produce aggressive behavior. But psychologist R. Plotnick and neurophysiologist José M. R. Delgado, of the same institution, have noted in their work with rhesus monkeys (Fig. 30.11) that aggression elicited by electrode implantation and stimulation occurs *after* the stimulation, not during it. A hypothesis proposing that brain stimulation is the direct cause of aggression would predict no such delay. The contradictory results seem to suggest that the monkeys may perhaps have been hostile because they connected the

stimulus with a prior disagreeable experience associated with reward-punishment training. It should also be pointed out that aggressive behavior can be halted as well as initiated by electrode implantation and stimulation (see Fig. 30.12).

In 1970, Princeton University psychologist B. G. Hoebel showed similar aggression initiation and suppression using chemicals applied to a specific region of the rat brain's lateral hypothalamus. Rats that normally kill mice ceased to do so when methyl atropine was applied. Conversely, rats that were normally peaceful became killers when carbachol was applied. Carbachol is known to mimic the action of the neurohumor acetylcholine, but it is not broken down by the enzyme cholinesterase. Thus high concentrations of carbachol lead to the killing response. Atropine, by blocking the action of acetylcholine, seems to stop the killing response. Thus, in this case, the difference between aggression and nonaggression is seen in neurophysiological terms—the relative amounts of acetylcholine and cholinesterase within the brain of the rat. The social significance of such interesting findings is self-evident, and Dr. Hoebel states that "if a similar kill center can be identified in humans, it might be possible to cure pathological aggression in man."

Fig. 30.11
Normally placid monkey (left) shows aggressive behavior (right) when stimulated through electrodes implanted in the brain (note thick gloves on handler's hand). (Photos courtesy Dr. José M. R. Delgado.)

Fig. 30.12
José M. R. Delgado of Yale University controls a normally aggressive bull by radio transmission of a stimulus to electrodes in the animal's brain. (Photos courtesy Dr. José M. R. Delgado.)

The Function of Agonistic Behavior

Since the appearance of agonistic behavior in animals is fairly widespread, there must be some selective value to such behavior. In the case of species exhibiting territoriality, agonistic behavior may be necessary for defense. It may also lead to a spacing out of animals so as to prevent overexploitation of the environment.

However, while agonistic behavior may at times be closely linked to territoriality, it is by no means inseparably so. Many animal species are not territorial but exhibit agonistic behavior. In certain birds such as chickens, so-called **peck orders** are found. The members of a flock compete to establish a sort of class system, in which those higher in the social register get first choice of food, space, mates, and the like. Once established, however, peck orders actually serve to reduce conflict, and attempts on the part of an animal to improve its standing in the order are relatively rare. Peck orders are known in primates, also, where they are usually referred to as **dominance hierarchies.** The dominant male of a baboon or rhesus monkey troop is easy to spot, not only by his own behavior but by the differing degrees of "respect" shown him by animals lower down in the social order. If he so desires, the dominant male is the one who drinks first at the water hole, who samples new food first, and who indicates when it is time for the troop to move to the safety of its sleeping quarters for the night. (See, however, Supplement 30.2)

Aggression in Humans

Even if aggressive behavior seems to have a genetic basis in many lower forms of animal life. What about in human beings? *Do* humans fight and kill their own kind because of an innate tendency to do so, or is our aggressiveness purely the result of past or present environmental influences? Or, is some other explanation possible? The thought once prevalent in ethology and such related fields as psychology and anthropology can perhaps best be traced in the following two statements:

There can not be any doubt, in the opinion of any biologically minded scientist, that . . . aggression is in man just as much a spontaneous instinctive drive as most other higher drives.

Konrad Lorenz

. . . in the course of human evolution the power of instinctual drive is gradually withered away, till man has virtually lost all his instincts. If there remains a residue of instinct in man, they are, possibly, the automatic reaction to a loud noise, and in the remaining instances to a sudden withdrawal of support; for the rest, man has no instincts. . . . Evil is not inherent in human nature, it is learned . . . aggressiveness is taught, as are the forms of violence which human beings exhibit.

M. F. Ashley Montagu

Certainly these two quotations illustrate a very wide gap in thinking concerning the origin of aggression in *Homo sapiens.* It is a matter of much interest and concern which, if either, of these two schools of thought is correct. War seems always to have been part and parcel of the human heritage, and we stand now at a point in history where our next great war will very likely be our last.

If human aggressiveness is a highly unacceptable form of behavior in today's society, then clearly, like a disease, we should want to prevent it. But disease control is most effective when the causes for the disease are understood. One who believes human aggressiveness toward fellow humans is innate may well recommend a quite different treatment for the condition from one who believes aggression to be entirely the result of environmental influences. So, indeed, do the two men just quoted differ in their approach to the problem. Lorenz, believing aggression to be an innate drive that must be expressed, recommends athletic events as a specific and harmless outlet for this aggressive drive. Montagu, on the other hand, would concentrate more on finding the proper physical and cultural environment for humankind—an environment in which aggressive behavior would not be rewarded or reinforced.

One Side of a Crucial Issue

British anthropologist Geoffrey Gorer points out that the Latin proverb, *homo homini lupus*—man is a wolf to man—has been taken over by virtually every society that derives its customs, laws, or language from ancient Rome. Actually, this proverb is grossly unfair to wolves. More appropriate ethologically would be *homo homini rattus*—man is a rat to man. Rats do occasionally kill their own kind.* According to Lorenz, rats are quite merciless killers of members of alien packs. Certainly there are close similarities here to humans. Killing others within our own society is generally considered immoral and is subject to punishment, while killing members of other societies or groups (as in war and lynchings) may be not only condoned but even glorified. But here the differences between the human species and other species become much greater.

How can we explain this behavior pattern in our species? Niko Tinbergen, now Professor of Animal Behavior at Oxford University in England, believes that humans still carry the animal heritage of group territoriality. He sees the human as a social ape who found it selectively advantageous to turn carnivore. In a 1968 speech, Tinbergen stated that "Ethologists tend to believe that we still carry with us a number of behavioral characteristics of our animal ancestors which cannot be eliminated by different ways of upbringing, and that group territorialism is one of these ancestral characters." Clearly, in the "nature versus nurture" controversy, Tinbergen seems here to have come down squarely on the side of nature.

Tinbergen also makes plain his feeling that the majority of ethologists agree with Lorenz. A rough counting of ethologists' noses casts some doubt on this supposition, however. Indeed, despite the close parallels he sees between rat packs and human packs, Gorer states unequivocally that "Man has no 'killer instinct'—he merely lacks inhibition." Ethologist J. P. Scott of Bowling Green State University states bluntly that "the physiological evidence is against Lorenz's notion of the spontaneity of aggression and, indeed, it is hard to imagine how natural selection would lead to the de-

velopment of a mechanism of continuous internal accumulation of energy which could unnecessarily put an animal into danger." Scott maintains that the weight of evidence indicates that aggression is produced by an internal mechanism stimulated by external influences. While he admits that the hypothalamus may magnify and maintain the internal aggression mechanism, Scott denies that it initiates the mechanism's action.

Scott goes on to make another important point. He writes that "If Lorenz is right, then man can never lead a happy, peaceful existence, but must continually be sublimating the spontaneous 'drive' which accumulates within him. If the physiologists are correct, then it is theoretically possible for man to lead a happy and peaceful existence provided he is not continually stimulated to violence." Supporting Scott's pessimistic interpretation of Lorenz is Tinbergen's statement that "Lorenz, in my opinion, is right in claiming that elimination, through education, of the internal urge to fight will turn out to be very difficult, if not impossible." In other words, not only is nature powerful, but *nurture may be powerless to counteract it!*

The Other Side

Others are not so sure. Indeed, at a 1968 conference held in Milan, Italy, on aggressive behavior in animals, the only universal agreement to be found among the one hundred or so biologists attending (representing fourteen nations) was that the concept of "aggressive behavior" is poorly defined and that what passes for aggression differs markedly from species to species. For example, S. A. Barnett of the University of Glasgow in Scotland pointed out that the term aggressive behavior tends to lump together legal and political aspects of war, crimes against human beings, and the social behavior of animals. Thus aggressive behavior has been approached from the standpoint of human ethical evaluations rather than objective analyses. Barnett pointed out that when two male wild Norwegian rats meet, a pattern of behavior characterized by an arched back and straight legs, with the flank turned toward the "opponent," occurs. Barnett finds it quite difficult to decide whether this behavior reflects aggression or amicability!

The Milan conference on aggression in animals also revealed a general consensus of opinion that fairly drastic measures are often necessary to induce aggression in animals. For example, Dr. Pierre Karli of the Institute for Medical Biology in Strasbourg, France, reported failure to induce a killing response in nonkiller rats, even by stimulation of those areas of the brain believed to be associated with aggressive "instincts." As was pointed out earlier, in natural killer rats, stimulation of the lateral hypothalamus of the brain does inhibit the killing behavior, a fact that led to the proposal of the existence of aggression centers. But the ablation or destruction of the lateral hypothalamus of nonkillers did

---

*More animal species kill their conspecifics than is popularly believed. In dogs, for example, aggressive encounters may go too far and result in death for one or both of the protagonists. Crabs sometimes kill other crabs' young for food, and gulls often kill the young of others, though seemingly not for food. Cases of cannibalism among conspecifics of the same age have been reported in owls, and killing occurs among bees when they are invaded by members of other hives. Finally, stress due to overcrowding leads lemmings in the wild to kill conspecifics.

not provoke a killer reflex. Nor did destruction of the frontal lobes, long thought to act as inhibitors of the killer rat's murderous impulses. Only cutting the nerves ascending from the nose to the olfactory region of the brain induced a killing response in natural nonkiller rats. Even then, the killing behavior pattern was random biting that eventually led to the death of the victim rather than the swift, directed attack of the natural killer. Starving nonkiller rats would ravenously devour mice killed for them and placed in the cage. The same animals, however, would starve to death with a living mouse in the cage with them, although they might easily have killed and eaten it.

The general consensus of the biologists' meeting at Milan appeared to be that if aggression was indeed innate, it was very difficult to arouse. It was further agreed that aggressive behavior varies markedly among different organisms, and extrapolations from one species to another may well be invalid. As previously mentioned, ethologist P. F. Klopfer has suggested that the dilemma concerning aggression comes from our conceiving of it as a noun, and thus as a distinct entity in itself with an equally distinct causation. But such is obviously not the case; behavior that would be labeled as aggressive can stem from several different emotional states, such as fear, rivalry, hunger, and so on.

Many ethologists believe that the concept of human aggression as innate is scientifically untenable and potentially dangerous. Like the Judeo-Christian concept of original sin, which it closely resembles, the concept of innate aggressiveness provides a simple excuse for our species' bad behavior. This belief in turn might hinder efforts to design an environment to promote the best and suppress the worst aspects of human behavior. On the other hand, if, as J. P. Scott and others maintain, human conflict results from defensive behavior and not from predation, lack of aggressive outlets will not, as Lorenz maintains, cause emotional damage. As Scott puts it: "This leads us to believe that we can work toward an essentially nonviolent world. We can no longer excuse human violence on the basis of our ancestry."

Altruistic Behavior

In contrast to competition and aggression, evolutionists have long recognized a class of behavior known as altruism. As the name implies, altruism is defined as behavior which, while it may reduce the personal reproductive success of the individual performing the behavior, increases the fitness or differential fertility of others. Altruistic behavior is observed in a number of animals, including parent organisms defending their young against predators, childless adults among primates or birds helping in the rearing of other adults' offspring, and wild canids bringing food to other adults who remain behind caring for young pups. One of the most difficult tasks of Darwinian theory has been to explain how such behavior could be favored by natural

selection. If genes for altruistic behavior even exist, they would appear to be selected against, since by definition such behavior benefits the receiver, not the originator. According to classical evolutionary theory, genes for selfish behavior, rather than those for altruism, should be favored.

It is necessary to point out that terms like "altruism" and "selfishness" are highly anthropomorphic (that is, derived from human experience) and thus fraught with social bias. The term is derived from descriptions of human behavior where connotations of consciousness and willing intent are implicit. Of course, when we describe animal behavior as "altruistic" no such consciousness or intent is implied. With animals, only the outcome, not the origin of the behavior, is described as altruistic. As a consequence of the inevitable anthropomorphism that comes from using terms such as altruism, we may be prejudiced against seeing a sound biological basis for such behavior. What we call altruistic behavior—in terms of outcomes, at least—does exist. The question becomes one of how its existence can be explained in terms of the theory of natural selection.

Kin selection begins with the assumption that altruistic behavior is genetically determined. Any genetically determined behavior leading to the perpetuation of genes for the same behavior in the breeding population will be favored by natural selection. In any sexually reproducing species an individual has genes in common not only with its parents and offspring, but also with its genetic relatives, such as siblings, cousins, nephews, and nieces. Thus, altruistic behavior that benefits a genetic relative can lead to perpetuation of genes for altruism even though the individual doing the act may not be producing offspring of its own. For example, when a breeding pair of Florida jays is assisted in rearing its brood by helpers that have not mated and have no territories of their own, those helpers may actually be contributing to the increased frequency of their own genes in the next generation. This would be a valid interpretation if, and only if, the helpers were genetically related to those they were helping (that is, share genes in common, especially genes for altruistic behavior).

In discussing kin selection it is important to distinguish between an organism's narrowly defined **individual fitness,** and its more broadly defined **inclusive fitness.** Individual fitness is customarily defined as the ability of an individual to pass on its own genes directly to the next generation by mating and producing offspring. Inclusive fitness is the sum of the individual's own reproductive success and the success of all its relatives, proportional to their genetic relatedness. A strictly populational concept, inclusive fitness is a measure of the degree to which a gene is represented in the gene pool of future generations, regardless of the mechanism by which it actually gets passed on.

The notion of kin selection can be more clearly stated if we consider how we can express the genetic relatedness of individuals. Any sexually reproducing organism shares 50 percent of its genes with its own offspring, 25 percent with its grandchildren, and 12.5 percent with its great-grandchildren. An individual also shares 25 percent of its genes with its nieces and nephews, and 12.5 percent with its cousins. These percentages are a measure of relatedness or what is called **genetic distance.** Looked at in terms of genetic distance, an individual with no offspring of its own but four nieces and nephews (four nonlineal descendants) can have as many or more of its genes represented in the next generation than an individual with two direct offspring of its own. If, therefore, genes for altruistic behavior lead genetically related individuals to show altruistic behavior toward one another in such a way that the total number of offspring from one genetically related line, taken collectively, is greater than that from another genetically related line, then genes for such behavior will increase in frequency in the one line compared to the other. In this way, in the long run, altruistic behavior could be favored over selfish (maximizing only the number of one's own offspring) behavior.

To date, the theory of kin selection is highly speculative; for example, no one has yet demonstrated that altruism for one's own kin, or anyone else, is a genetic trait. To the degree that the speculation is true, the rest of the theory is likely to be true. Similarly, to the degree that the speculation is false, the rest of the theory becomes false. We are likely never to know whether genes for altruistic behavior actually exist. In that context the theory of kin selection remains merely interesting and intriguing.

One problem with the theory is that it rests on the idea that so-called altruistic acts are directed by an organism toward a genetic relative. There is little evidence that this is always, or even frequently, the case. This objection is even more true when the theory of kin selection is applied to human beings than when it is applied to other animals. In many instances it is difficult to determine the genetic relatedness between organisms involved in what appears to be altruistic behavior.

In addition there are often other explanations for altruistic acts than appear at first glance. For example, an act which seems to be altruistic may actually turn out to be selfish. The helper jays discussed above may actually be waiting on the sideline to take the place of the male who has fathered a brood. This individual could in fact appear to be maximizing its inclusive fitness but actually be maximizing its individual fitness by mating with the female bird.

30.8
THE EVOLUTION OF BEHAVIOR

If some behavioral traits are genetically controlled, it is logical to view them as evolving traits, like any others. To the degree that any behavior is genetically determined it will be under some selective pressure, and its genetic base can be modified by selection. Consider the following example of a behavior and its possible adaptive significance, and the way this behavior could have evolved by natural selection.

Herring gulls remove the eggshells from the nest after their chicks have hatched. What is the selective value of such behavior? The speckled coloring of the gull eggs before hatching gives them excellent protective coloration against the pebble-strewn background on which they are laid. The chicks, too, are protectively colored. The *inside* of the eggshell, however, is white. When the shell is broken at hatching, some fragments may lie with their inner surface exposed. This observation led Niko Tinbergen to hypothesize that the white fragments might enable predators to locate the nest by sight (or even by the odor of the decaying portions of the remaining egg parts adhering to the shell fragments). This hypothesis leads to certain predictions: if removal of eggshells from the nest after hatching aids in protecting the young from predators, then nests with eggshells should be preyed upon more frequently than those without eggshells; and nests with eggshells left closer to them should suffer a higher degree of successful predation than nests with eggshells farther away. Tinbergen tested his hypothesis with experiments conducted in the field (outside the laboratory). Both predictions were verified. Thus the seemingly energy-wasting behavior of eggshell removal on the part of parent gulls was shown to have considerable selective value.

Indeed, it seems as though *behavioral changes may be of major importance in the process of evolution by natural selection.* While stressing the primary importance of geographical isolation in the formation of new species, Ernst Mayr states in his book *Animal Species and Evolution:* "A shift into a new niche or adaptive zone is, almost without exception, initiated by a change in behavior. The other adaptations to the new niche, particularly the structural ones, are acquired secondarily."

How much evidence is there that behaviors actually evolve toward adaptive ends by the gradual modification of programmed responses? An interesting and dramatic example of what may have been the evolutionary pathway for a specific behavior pattern is found among several genera of the "balloon" or "dancing flies" (family Empedidae). In 1875 the German naturalist Baron von Osten-Sacken observed that in two species of balloon fly belonging to the genus *Hilara*, the male approaches the female during courtship carrying a large, empty ball of silk that he has spun. Osten-Sacken was completely puzzled by this behavior, and could offer no explanation for its adaptive significance. Today, however, a whole sequence of courtship behaviors within the family Empedidae suggests not only how the male *Hilara*'s gift-giving is adaptive, but also how the complex pattern might have evolved among the balloon flies by gradual modification of a simple behavior displayed by a primitive ancestral species.

SUPPLEMENT 30.2
SEX AND THE DOMINANT MALE

In all discussions about social hierarchy and dominance among male animals, it is assumed that dominance confers a reproductive advantage on its bearers. Following Darwin's idea of sexual selection, it is assumed that a more sexually aggressive male will mate more frequently and hence he will leave more offspring behind. Making the further assumption that aggression is somehow controlled by genetic factors, it can be concluded that more genes for aggressive behavior (at least with regard to sexual behavior) will be passed on to the next generation than genes for nonaggressive behavior.

Recently this assumption has been criticized on several grounds. Irwin Bernstein of the Yerkes Primate Research Center at Emory University in Atlanta, Georgia, claims that the whole concept of dominance hierarchies, at least among primate males, is vague and ill-defined. Bernstein notes that various measures of dominance, as determined by different observers, do not agree with one another. If captive monkeys are ranked three times according to three different criteria—mounting, grooming, and agonistic behavior—different males come out on top each time. Thelma Rowell of the University of California, Berkeley, agrees with Bernstein. She claims that dominance hierarchies may exist only in the minds of the observers, who make primates compete in the laboratory or field for rewards such as food or water. Such procedures are unnatural and highly anthropomorphic. Primates seldom have to scramble for a limited supply of food in the wild. To create a human-like situation for animals, and then describe their reactions to it in such human terms as "dominant," "aggressive," "meek," "passive," and the like, is not the same as observing natural behavior patterns of the animals involved. It has not escaped the attention of some critics that the concept of dominance hierarchies and the laboratory tests designed to support their existence are highly reminiscent of the social hierarchies that characterize the very societies in which the biological concept of hierarchy was invented.

Not all ethologists agree with Bernstein and Rowell on this matter, however. Sandy Richards, of Cambridge University in England, reports that when captive rhesus monkeys are ranked according to Bernstein's criteria or other criteria, the same males come out as most dominant every time. Critics of Richards's work are quick to point out that her monkeys also lived in the very kind of unnatural circumstances that all observers agree lead to predictable competition for food and water. Further, Richards's groups consisted of a male and several females (rather than the more common multimale group), which could greatly influence the results.

An even more fundamental criticism of the older concept of dominance hierarchies has been made by Bernstein, Thomas Gordon, and Susan Duvall. They tested the assumption that the so-called aggressive or dominant males in a population have a selective advantage over less dominant males in terms of reproduction. Bernstein and others noted that in the past, primate ethologists were satisfied with counting the number of times a male monkey mounted a female as an indication of reproductive success. The greater the frequency of mounts, the more reproductively successful the male was judged to be. Several observations throw this standard of "success" into question. Lee Drickamer of Williams College has pointed out that observers often judge certain male rhesus monkeys to be high-ranking simply because they display the more conspicuous behavior. This was found to be true in judging reproductive success. The number of times a male mounts a female is not an accurate measure of reproductive success.

Bernstein and his co-workers set out to determine whether dominance hierarchies, however they are determined, really bear any relationship to reproductive success. They performed biochemical tests of paternity in a group of rhesus monkeys at the Yerkes Primate Center. They found that no direct relationship existed between dominance of a male, judged by a variety of different criteria, and the number of offspring the male produced. So-called nonaggressive males often left behind as many or more offspring than so-called aggressive males.

Bernstein and others further expressed the opinion that sexist biases may have crept into most past studies on this subject. Investigators had been assuming that the female was a totally passive participant in the reproductive process, willingly accepting whatever male made the most aggressive response toward her. Bernstein and his associates noted that females played an important part in the process itself, often rejecting the most aggressive "dominant" males.

Similar results have been obtained observing baboons in the wild. Glenn Hausfater of the University of Virginia studied agonistic behavior among baboons in Kenya in order to test a hypothesis developed in the 1940s stating that dominant males have the first access to females who enter estrus. According to this hypothesis, if only one female in a group is in estrus at a given time, the top-ranking male will copulate with her; if two females are in estrus, the top two-ranking males would be expected to copulate with them, and so on. During the first 400 days that Hausfater observed baboons, only one or no females were in estrus at any one time. If the older hypothesis were correct, he should have observed the vast majority of matings to be with the top-ranking male. The prediction was not fulfilled, however. The top-ranking male did not mate with three successive females who were in estrus, even though no other estrus females were present. At the same time, males who were not top-ranking *did* mate with the estrus females. It may be, Hausfater concluded, that the females chose the males, rather than vice versa.

The observations of these various workers do not necessarily mean that dominance hierarchies never exist in primate social groups. They *do* indicate that the descriptive definitions of such hierarchies (1) may not be based on the same criteria that ethologists choose as important, (2) may not have much if any relationship to reproductive success, and (3) may be very fluid, changing from day to day under natural environmental conditions. It is obvious from these studies that our current picture of primate and other nonhuman animal social systems may be a primitive one at best, fraught with problems introduced by our own subjective biases and the difficulty of making solid, statistically significant observations under natural conditions. Such findings make it even more imperative that we avoid facile comparisons between nonhuman and human societies.

The courtship of balloon flies follows a well-defined pattern: the male approaches the female and goes through a series of dance motions before beginning to copulate. However, because balloon flies are predatory, the female often devours the male either just before or during copulation. Indiscriminate predation by the female might be adaptive as a feeding mechanism, but it is obviously nonadaptive as a reproductive behavior. The peculiar gift-giving behavior of the male *Hilara* can be seen as an adaptation to avoid predation. The silk ball which the male offers to the female serves as a distraction—as she is preoccupied with it the male can mate with her and avoid being eaten.

But how could such a behavior evolve? In the most primitive genera of Empididae (as judged by other criteria than behavior), such as *Empis* and *Empimorpha*, males first catch a prey—some kind of small fly—and present it directly to the female at the beginning of courtship. While the female feeds on the victim the male mates with her in safety. In several other species, the males add a few strands of silk to the prey, making the gift more visible to the female. In a third group of species of *Empis*, the entire prey is covered by a sheet of silk, producing a kind of "balloon" (hence the name "balloon flies"). In some species of *Empis* and *Empimorpha*, the size of the prey is greatly reduced (because the male has fed on it first), thus making the gift mostly balloon. In *Hilara*, the male does not bother to capture a prey at all, but merely presents the female with an empty balloon!

We can envision an evolutionary sequence through time not unlike the graded series of behaviors described above in the modern species. The earliest stage of the behavior could have originated by chance. Suppose that a male *Empis*, on his way to court a female, encountered another kind of fly and by chance responded reflexively to capture it and carry it with him. The presence of the captured fly could occupy the female's attention, allowing the male to mate. If we assume that this male's reflex response (that is, capturing the prey) was a genetically based neurological variation in reflex behavior, then that reflex behavior could be passed on to the male's offspring. The behavior variation obviously has selective value, in the Darwinian sense, since it allows the male to mate, perhaps more than once, as opposed to being eaten. Males with this variation in neurological function would thus leave more offspring and the behavioral trait would gradually increase in frequency within the population.

Natural selection is undoubtedly affecting female, as well as male, behavior during this evolutionary development. Both are co-evolving an integrated and reciprocal mating behavior that ensures more effective reproduction. Selection favors, first and foremost, completion of the reproductive act; but in the female it also favors a voracious appetite, since she has many eggs to produce. If the female's appetite can be maintained without decreasing the chances of successful reproduction, there will be an overall increase in fitness of the species. So, it is likely that females who respond effectively to the distraction of a gift at mating time will be favored—they will pass on this trait (assuming of course it is genetic) to their offspring. We might assume that the females of successive generations are evolving a higher receptivity to a releaser stimulus—that they are evolving an innate releaser mechanism (IRM) at the same time the male is evolving a behavior that serves as the stimulus itself. As we know with most IRMs, as the stimulus becomes more exaggerated the innate response is triggered more readily. Thus, the increasing size of the balloons manufactured by males of *Hilara* may be selectively favored by the existence of an IRM system that has co-evolved in the female. Males that by chance produce bigger balloons will have more chance of mating and even less chance of being devoured.

The above example of a possible evolutionary pathway for a specific behavior is highly speculative. The existence of a sequence of behavior patterns, however logical, among a group of living species does not mean that the behavior evolved along such a progressive line. Yet, the existence of the intermediate series of behavior patterns, from the simplest presentation of a naked prey to a large and empty silk balloon, suggests that the complex behavior of *Hilara* might stand at the end of a long and gradual evolutionary sequence.

30.9
SOCIOBIOLOGY: EXTRAPOLATING FROM ANIMAL TO HUMAN BEHAVIOR

The peck order shown in certain animal communities is based on the "rights" of certain individuals to be aggressive toward others. The dominance hierarchy of a baboon troop, for example, usually has a dominant male and three or so subordinates who comprise a sort of central governing body within the troop. Outside this central governing body, the dominance hierarchy continues further, with young males approaching prime usually standing higher in the order than females and aging males. The females, in turn, stand higher than the young juveniles and infants.

It would seem that a social structure built upon the absolute right of aggressive dominance over subordinates could hardly be a peaceful one. Yet for animal societies based on peck orders, peace generally reigns. Furthermore, it would seem obvious that the overthrow of a superior by a subordinate could occur only by violent means. Yet this, too, proves not to be the case. Rarely, for example, is a dominant male deposed by a major battle; instead, he falls from power gradually as he ages and fails to meet routine challenges and tests of his ability to carry out his responsibilities. Studies of baboon troops over a period of several years reveal a gradually changing structure, with dominant males being replaced slowly by younger animals who are later replaced by those younger still. The result is a strong and continuing social order allowing high efficiency in ensuring adequate food for all and maximum protection for the weaker.

However, one might observe a baboon troop for a long time without detecting the peck-order framework on which it is constructed. This is because in nature the animals forage widely for food and there is often enough for all. Thus the competitive situations necessary to demonstrate dominance and subordination do not occur often. Rarely, perhaps, only one or two females may be in estrus (receptive for mating), in which case the males may compete for her. Or an ethologist, wishing to study the peck order, may feed the troop by scattering grain in a small area or, perhaps, by throwing it to only one animal. Under these artificially imposed conditions the troop must condense into a smaller area, and chances for contact between individual animals are greatly enhanced. Though mere threats are usually enough to prevent large-scale violence, fights under such conditions can and do occur.

Thus population density, the number of individual organisms per unit of space, definitely affects behavior. Precisely *how* it does so has recently been the object of much research involving many different species of organisms.

One of the earliest studies of this type was carried out by John B. Calhoun of Rockefeller University in New York. Calhoun first worked outdoors with the wild Norway rat. He confined a population in a quarter-acre enclosure, where the animals had no escape from the behavioral consequences of population density increase. The animals, however, were spared the conditions of starvation and disease that normally accompany overpopulation, for they were given an abundance of food and places to nest, in addition to being protected from predation. Under these "ideal" conditions, the observed reproductive rate predicted an adult population of 5000 individuals. Actually, it numbered 150! Investigation revealed an extremely high infant mortality rate; even with only 150 adults in the enclosure, the stress of enforced social interaction led to disruption of maternal behavior. The result was death for most of the young.

Calhoun later moved his study inside and performed more sophisticated investigations dealing with the effects of population density on behavior. Again infant mortality climbed, with levels of from 80 to 96 percent not uncommon. Females stopped building nests for their young; eating habits were changed; normally peaceful reproductive behavior patterns were disrupted and fights often resulted. Homosexual interaction increased, possibly as a means of allowing sexual outlets without increasing the population. Cannibalism, total social disorientation, and high adult mortality rates also became common. A startling number of these deaths occurred in females as a result of either pregnancy or giving birth.

Results such as these are troublesome, for they seem uncomfortably close to the human situation today. In heavily overpopulated Latin America, for example, despite the fact that most such deaths undoubtedly go unreported, pregnancy and childbirth deaths are high. Chile, for example, has 271.9 deaths per 100,000 births, as opposed to Sweden's 11.3. Calhoun concludes a 1962 *Scientific American* report of his work by stating:

> It is obvious that the behavioral repertory with which the Norway rat has emerged from the trials of evolution and domestication must break down under the social pressures generated by population density. In time, refinement of experimental procedures and of the interpretation of these studies may advance our understanding to the making of value judgments about analogous problems confronting the human species.

Just how valid *are* extrapolations of such results to the human situation? It is difficult to judge. The results of Calhoun's studies concerning the effects of high population density on behavior are admirably supported by studies of other species both higher and lower on the evolutionary scale than the Norway rat. Most important, they seem to hold true for primate societies. When primate species normally peaceful in the wild are kept under high-population-density conditions in a zoo, violence leading to maiming and death becomes quite common. Under such conditions, animals of all ranks in the

SUPPLEMENT 30.3
MICKEY MOUSE, NEOTENY, AND THE EVOLUTION OF JUVENILE TRAITS

For over fifty years Mickey Mouse has ranked as one of the great folk heroes of American culture. From the moment they first view Walt Disney's lovable mouse, children and adults alike respond with affection and sympathy to the sometimes mischievous and irascible comic-strip character. But why this instantaneous reaction? Biologist Steven Jay Gould has explained our response to Mickey's appearance in terms of one of the most basic principles of animal behavior: Konrad Lorenz's theory of innate releasing mechanisms (IRMs).

Gould starts by analyzing Mickey's facial and bodily proportions as he has appeared in Disney cartoons and comic strips more or less unchanged since 1953. One of the first things Gould noted was that Mickey possesses a number of what Lorenz calls "juvenile traits." As Lorenz defines them, juvenile traits include "a relatively large head, predominance of the brain capsule, large and low-lying eyes, bulging cheek region, short and thick extremities . . . and clumsy movements." All but the last of these traits are easily visible in the comparison of juvenile and adult traits in four species of animals, as shown in Diagram A. Note that in each case, as Lorenz has argued, we not only recognize the left-hand form immediately as juvenile; we also respond to it with a sympathetic, almost protective attitude. Lorenz views this response as part of our human programmed behavior, adaptive as a means of nurturing and caring for the young. While Gould does not accept the "innateness" part of Lorenz's theory, he argues that Lorenz is correct in claiming that juvenile traits act as a releasing mechanism to adults, eliciting from them a protective response.

Now, an analysis of Mickey Mouse's appearance shows that he possesses all of the basic juvenile traits which Lorenz

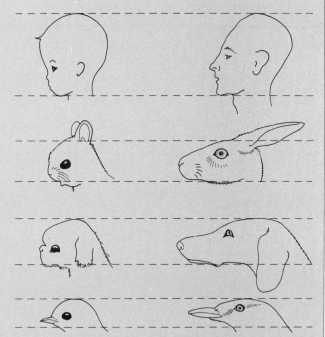

Diagram A
(Konrad Lorenz, *Studies in Animal and Human Behavior,* vol. 2. Cambridge, Mass.: Harvard University Press, 1971.)

has identified. For example, Mickey's head is large for the rest of his body (his head is 48 percent of his body length); his nose (snout) is small (only 49 percent of head length, compared to his comic-book compatriot Mortimer, a real

STAGE 1   STAGE 2   STAGE 3

Diagram B
(© Walt Disney Productions. Used by permission.)

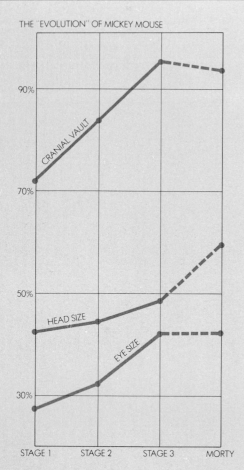

THE "EVOLUTION" OF MICKEY MOUSE

CRANIAL VAULT

90%

70%

HEAD SIZE

50%

EYE SIZE

30%

STAGE 1   STAGE 2   STAGE 3   MORTY

Diagram C
(From Steven J. Gould, *Natural History*, May, 1979, p. 34.)

scoundrel, whose snout is 80 percent of his head length); Mickey's eyes are large (42 percent of head length); and his cranial vault is quite expansive (cranial vault is measured as that portion of the cranium lying above a line connecting the nose to the ear; the higher the forehead and the lower and farther back the ears are pushed, the greater the cranial vault). These juvenile traits, Gould argues, give Mickey Mouse his instant appeal.

What is interesting, however, is that Mickey has not always been so juvenile looking during his sixty-plus years of life! Instead of aging, as most of us do, Mickey has grown younger. As Diagram B shows, Mickey has undergone a considerable evolution in appearance since he was first introduced to the public in 1928. This evolution in appearance has paralleled an evolution in personality and behavior. In his earliest days (Diagram B, stage 1) Mickey was not quite the lovable, well-behaved mouse whom we know today. He was mischievous and, according to one essayist, "even displayed a streak of cruelty." When Disney found he had a profitable character in Mickey, however, the mouse's personality and appearance changed, becoming more conventional and

charming. He was less mischievous and cruel and his amorous relationship with Minnie, while never resolved, at least became less openly feisty and combative.

At the same time Mickey's physical appearance became progressively more youthful (see Diagram B, stages 1, 2, and 3, showing Mickey as he appeared in 1928, 1940, and 1953). Gould measured Mickey's evolving physiognomy for three basic characteristics: size of cranial vault, head size (as percentage of body length), and eye size (as percentage of head size). The data are summarized in the graph in Diagram C. Note for all three characteristics, the changes "evolved" were toward youthfulness: larger cranial vault, larger head in proportion to body, and larger eyes in proportion to head. Whether consciously or not, the Disney artists were responding to what Lorenz called the IRM for youthfulness. What Hollywood's market researchers came up with in studying human responses was the same as Lorenz and other ethologists had come up with in studying animals.

There is another point of biological interest to emerge from Gould's study of Mickey Mouse. The "evolution" of Mickey's appearance over the past fifty-odd years follows the same course as the evolution of the human species: greater retention of juvenile traits. The retention of juvenile traits into adulthood is known as *neoteny*. Many species have evolved neotenous patterns. Some of the more conspicuous examples are salamanders who become adults—that is, reach sexual maturity—without ever leaving the water (normally salamanders start off life as gill-bearing aquatic forms, metamorphosing into terrestrial, lung-bearing adults). These neotenous salamanders retain all their juvenile traits, including gills, throughout their lives. In the same way, Gould sees neoteny as an important factor in human evolution. Indeed, we have been retaining many youthful traits until late in adult life. Like Mickey Mouse in 1928, our earliest forebearers (Australopithecines) had low cranial vaults, small eyes, and elongated snouts and jaws. Also the evolution of the processes of human growth and development has been toward prolongation rather than shortening of periods of infant dependence, thus providing greater opportunities for learning. Accordingly, it could be argued that the simultaneous evolution of stronger parental responses to juvenile traits would help to extend the parents' innate protective behavior further into the child's life. As adults, then, our innate response to youthfulness may be an adaptive trait accompanying our evolution of greater ability to learn during early life.

Of course, much of this is speculation. The evolution of neoteny in human beings may have been the result of selection for a wholly different set of traits—perhaps greater ease in getting the fetus through the birth canal—rather than simply to provide greater parental protective behavior. Nonetheless, it is interesting to wonder how much of our response as adults to youthfulness in our own or other species could be a result of selection for a positive response to an IRM in the young of our own or, by extension, other species as well.

peck order are thrown into close proximity, and there is no escape. Subordinates direct their aggression toward animals of the next lower peck-order status, and a chain reaction of violence results. But it is erroneous to ascribe the *cause* of such behavior to density. Confinement in artificial surroundings would be an equally, or more important, variable. The problem is that there has been relatively little direct and systematic observation of animals under conditions of great density in the wild. Thus problems exist at the simplest level of observation in attempting to determine the true effects of density on animal behavior.

Again, the sticky question: can we extrapolate to humans? *Is* there something significant about the similarity between the abuse in captivity of his subordinates by a male baboon and the treatment of an economically poor person by his or her slightly less poor neighbor, who in turn feels put upon by others still higher on the economic ladder? *Is* there a relationship between the abuse of infant monkeys by their mothers when kept in crowded conditions and the increase in the "battered baby syndrome" (child beating and torture) reported in areas of high human population density?

It can be truthfully said that we simply do not know. After all, Calhoun's rats, though overcrowded, were relatively "wealthy," being kept well supplied with food and water. In humans, overcrowding tends to be associated with poverty and poor education. If nothing else, the example points out how fallible simple-appearing cause-effect relationships may often be. Thus some scientists, such as molecular geneticist and Nobel laureate Joshua Lederberg, urge more caution in extrapolation from crowding experiments in lower animals to humans. Others feel that limiting density by limiting population growth is absolutely essential to prevent war, poverty, and the social pathology that seems to be associated with overcrowding.

In recent years attempts to make such extrapolations the basis of a new science have given rise to sociobiology. According to its chief advocates, such as Harvard's E. O. Wilson, sociobiology is a grand synthetic theory aimed at bringing all of human culture and history under the explanatory umbrella of Darwinian evolution. While sociobiology is concerned with the scientific and evolutionary study of all animal social interactions, it differs from ethology especially in its focus on human behavioral evolution. A more detailed discussion of sociobiology and its methods is found in Section 29.15. It is sufficient to say here that sociobiologists are faced with all the problems of people in any field who try to argue about innate genetic determinants of human behavior by extrapolation from lower animals, or assuming the existence of behavioral genes that have never been demonstrated.

Summary

1. Biology has come to be organized into various levels of investigation, from the molecular to the population level. As a *concept*, however, the idea of levels of organization in living organisms can be misleading. In animal behavior, for example, the organism must be seen as a member of a population and as an individual organism, and of course the component parts of which it is composed affect its behavior.

2. The field of biology dealing with animal behavior is called ethology. The early development of the field was characterized by "heredity versus environment" or "nature versus nurture" debates.

3. Anthropomorphism—the ascribing of human characteristics to nonhuman organisms—and teleological explanations, which ascribe conscious purpose to an organism's actions, were early impediments to ethology. Their elimination put studies in ethology on a far more solid scientific footing.

4. The term "instinctive" behavior was a widely used and misused term in ethology and was later replaced with the term "innate." Both terms refer to behavior that is genetically based.

5. Lorenz's innate releasing mechanism (IRM) concept pictures a genetically based mechanism controlling a specific behavior activated only by exposure to the proper "release" stimulus. Thus, for example, the sight of a female turkey "releases" strutting behavior in the male; in turn, his strutting behavior "releases" her crouching behavior, and so on. While the IRM concept was extremely important as a scientific model that stimulated much important research, more recent studies have raised doubts about the "innateness" of some of the behavior elicited by the releasing stimuli.

6. Behavioral changes are now seen as being as important as anatomical changes in the process of evolution by natural selection.

7. Learning in nonhuman organisms spans a very wide range, from the duplication of "human" behaviors (such as abstract reasoning and symbolic communication) in the insight learning of the chimpanzee to the imprinting of animals such as birds. In the latter, early exposure to a stimulus object often (though not always) causes a lifetime attachment to the object, even when it may bear no resemblance to the natural object to which the young organism would ordinarily be exposed.

8. Communication in animals may be of various types, utilizing sound, movement, sight, odor, or taste. Chemicals that operate in communication are called pheromones.

9. Territoriality refers to the tendency of some organisms to defend a certain space (a tree, bush, plot of ground, or whatever) from intrusion by others of their own species. Not all organisms are territorial, and even in those that

are, the territorial boundaries are not always well defined. The precise reasons for the evolution of territoriality are unclear, though various hypotheses, ranging from spacing in accordance with food supply to protection from predation or disturbance during mating behavior, have been formulated.

10.   A territorial animal must be aggressive toward another animal of its species that enters its territory if it is to hold the territory successfully. The aggression of one animal toward another often triggers a characteristic behavioral reaction in the organism being attacked (defensive posture, retreat, and the like). Aggression and the range of behaviors it elicits in the organisms fighting are collectively called agonistic behavior.

11.   Historically, the hypotheses put forward for the causes of aggression have ranged from aggression's being innate to its being entirely the result of environmental factors. More recent work has emphasized the difficulty of satisfactorily delineating aggressive behavior in different species and casts doubt on the value of hypotheses based upon the concept of innateness for such a complex phenomenon as agonistic behavior.

12.   Peck orders or dominance hierarchies are found in certain organisms (chickens, cows, baboons). In such systems, the population is seen as having a highly structured class system, where those organisms higher on the ranking scale appear to get first access to food, space, mates, and the like. More recent research, however, calls into doubt some of the

interpretations of earlier investigations. For example, blood tests of newborn baboons indicate that as many or more are fathered by adult males or even juveniles, supposedly far down on the dominance hierarchy scale, than are fathered by the supposedly dominant male.

13.   The concept of aggression as innate in humans has resulted from often unjustified or highly speculative extrapolations from aggressive behavior in lower animals. Ethologists such as Konrad Lorenz have often gone beyond their truly scientific data to express speculations with little or no scientific bases. Such speculations have then been picked up and extended by writers who are not scientists, such as playwright Robert Ardrey (see Suggested Readings).

14.   Modern ethology looks upon behavior much as a developmental biologist looks upon the developing limb bud of a chick embryo—that is, as a system whose general features are genetically controlled but whose actual expression may be greatly affected by its internal or external environment. Just as exposure to a particular chemical at an initial stage of development may produce a deformed wing in the adult bird, so may aversive environmental stimuli or deprivation at critical times cause behavioral aberrations of a pathological nature. It is generally to the environment rather than the genome that we must look for means to ensure the potential for healthy behavior for all. The old "nature versus nurture" controversy is not a meaningful debate in terms of promoting progress in the field of ethology.

Exercises

1.   Young Mediterranean cuttlefish (*Sepia*) raised in isolation will only attack and feed upon crustacea of the genus *Mysis* (probably their normal prey in this area of the Mediterranean). Propose hypotheses in the following three categories to answer the question: Why do young *Sepia* attack mysids?

   a) An anthropomorphic hypothesis.

   b) A teleological hypothesis.

   c) An ethological hypothesis (propose a test for this hypothesis).

2.   A mother turkey normally accepts and raises her own young with no problems. A hen turkey deafened several weeks before hatching of the eggs invariably kills all her young when they hatch. A hen turkey deafened 24 hours after hatching will raise her chicks normally. Propose a hypothesis to explain these observations and a test for this hypothesis.

3.   A male and a female dove are put into adjoining cages separated by a glass pane. At a certain period of the year, the male "displays." Following completion of the display, the female builds a nest, lays infertile eggs, and sits on them. A control female, not exposed to a male, does not exhibit this behavior.

Interpret or explain these observations. What do you think this experiment demonstrates?

4.   Read the information given in the following paragraphs, suggest a hypothesis (or hypotheses) to account for this information, design a *controlled* experiment to test each hypothesis, and state which prediction(s) is (are) being checked.

   a) In the early morning, animal *A* reacts to its detection of animal *B* by giving behavior pattern *XYZ*. Animal *B* reacts to its detection of *A* by giving reaction pattern *DEF*. Animal *A* reacts to its detection of animal *C* by giving behavior pattern *XYZ* and *C* reacts to *A* in a *DEF* manner. Animals *B* and *C* do not show any discernible reaction pattern when they detect each other in the absence of *A*.

   b) Later in the day, animal *A* reacts to its detection of animal *B* by giving reaction *XYZ*, but *B* reacts to its detection of *A* by giving either no reaction at all or only *D*. Animal *C*, however, continues to react to *A*'s *XYZ* reaction with *DEF*. Animals *B* and *C* continue to ignore each other in the absence of *A*.

   c) Still later, *A* and *B* detect each other. Animal *A* gives the *XYZ* behavior, *B* gives *DE* behavior, but not *DEF*. Animal *C* now appears, gives no reaction to *A*, but gives reaction series *MNO* to *B*; *B* responds initially with *MNO* but soon switches to *PQR*.

Propose a hypothesis to explain these behavior interactions and the types of behavior you feel may be symbolized here. Suggest a test for your hypothesis in terms of field or laboratory experimentation.

5. Discuss briefly the concept of "innateness" of behavior in comparison to environmental effects on behavior ("nature versus nurture") as it applies to:

a) the formation of the neural tube in amphibian embryos from a strip of the dorsal ectoderm (see Section 20.9);

b) early views on the nature of the ability (or lack of ability) to regenerate shown by organisms at "higher" or "lower" places on the evolutionary ladder or phylogenetic tree (see Section 20.9) as contrasted with contemporary views of the same phenomenon.

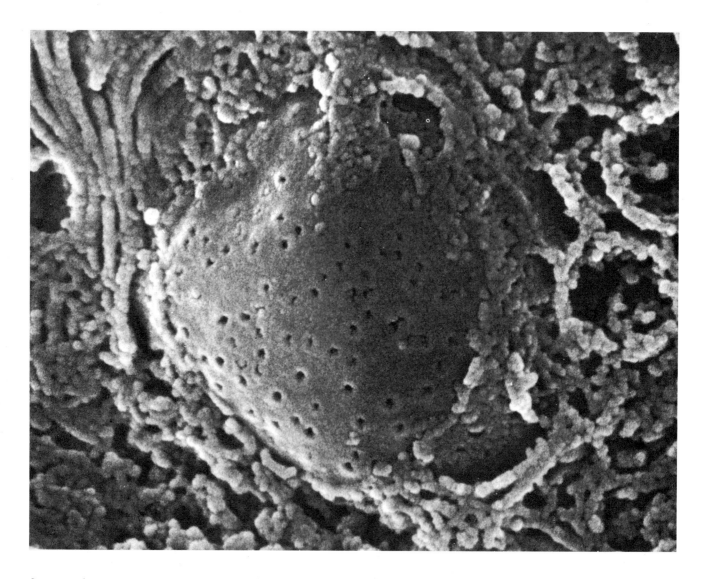

Scanning electron micrograph of a portion of the cell interior,
showing the nucleus and portions of the endoplasmic retic-
ulum. The pores in the nuclear membrane are clearly visible.
(Micrograph courtesy of Dr. Keiichi Tanaka, Professor of
Anatomy, Tottori University School of Medicine, Yonago,
Japan.)

Suggested Readings

CHAPTER 1

Allen, Garland E., "Dialectical materialism in modern biology," *Science and Nature* 3 (Fall, 1980): 43–57. A discussion, with many examples, of nonmechanistic (and also non-vitalistic) thinking in biology. An amplification of the ideas discussed in Section 1.9.

Baker, Jeffrey J. W., and Garland E. Allen, *Hypothesis, Prediction and Implication in Biology.* Reading, Mass.: Addison-Wesley, 1968. Part II contains further analyses of the type applied in this chapter to the work on the precise causes of death in mice exposed to radiation. One of these analyses involves a controversy between researchers. Part III deals with a "Letters-to-the-Editor" debate on the need (or lack of need) for a scientific investigation of racial differences and intelligence.

Bishop, O. N., *Statistics for Biology.* Boston: Houghton Mifflin, 1966. A good treatment, presented in an elementary way (paperback).

Campbell, R. C., *Statistics for Biologists.* New York: Cambridge University Press, 1967. Similar to the Bishop book, but somewhat more advanced. Descriptions are sometimes abbreviated, though more topics are covered (paperback).

Gardner, M., *Fads and Fallacies in the Name of Science.* New York: Dover Publications, 1957. A well-written, very interesting account of a number of pseudoscientific theories, how they gained adherents, and how most of them can be shown to be invalid. Discusses Bridey Murphy, Atlantis, flying saucers, Lysenkoism, Creationism, and extrasensory perception (ESP).

Goldstein, Martin, and Inge Goldstein, *How Do We Know?* New York: Plenum Press, 1978. A very well-written and organized book, which takes the same general approach to the nature of science found in the present textbook. After discussing many aspects of theory formation, observation, and logic, the book contains several case histories (from biology, chemistry, physics, and psychiatry) illustrating the principles of scientific process.

Huff, Darrell, *How to Lie with Statistics.* New York: Norton, 1954. Amusing, illustrated introduction to statistics. Focusing less on actual statistical procedures than most of the other books listed here, Huff's work provides a witty insight into the advantages and limitations of statistical thinking (paperback).

Hull, David, *Philosophy of Biological Science.* Englewood Cliffs, N.J.: Prentice-Hall, 1974. A concise account of some of the many philosophical problems encountered in theory development and testing in the biological sciences. Not easy reading, but a good introduction for the serious student.

Kottler, Malcolm, "From 48 to 46: cytological technique, preconception, and the counting of human chromosomes," *Bull. Hist. Med.* 48 (1974): 465. This excellent article gives a fuller account of the case history described in Section 1.3.

Munson, Ronald, *Man and Nature: Philosophical Issues in Biology.* New York: Dell, 1971. This is a collection of essays by many different authors. One of the best is Ernst Mayr's "Cause and effect in biology," which includes our example of the migrating warblers.

Schefler, William C., *Statistics for the Biological Sciences.* Reading, Mass.: Addison-Wesley, 1979. A thorough introduction to statistical methods.

Van Norman, Richard W., *Experimental Biology,* 2d ed. Englewood Cliffs, N.J.: Prentice-Hall, 1971. A very useful compendium of information about the use of biological literature, setting up experiments, special techniques, measurement, statistics, and the writing of scientific papers.

Weill, Andrew T., Norman E. Zinberg, and Judith M. Nelson, "Clinical and psychological effects of marihuana in man." In *The Process of Biology: Primary Sources,* ed. Jeffrey J. W. Baker and Garland E. Allen. Reading, Mass.: Addison-Wesley, 1970. An excellent example of a paper showing the strengths as well as the limitations of scientific investigation.

CHAPTERS 2 AND 3

The following books contain more detailed treatment of atomic structure, chemical bonding, and formation of molecules:

Baker, J. J. W., and G. E. Allen, *Matter, Energy, and Life,* 4th ed. Reading, Mass.: Addison-Wesley, 1981.

Cheldelin, Veron H., and R. W. Newburgh, *The Chemistry of Some Life Processes.* New York: Reinhold, 1964. An introduction to organic chemistry with special reference to biochemical processes. Assumes more background in chemistry and atomic structure (bonding and the like) than *Matter, Energy, and Life,* but covers some biochemical processes in great detail.

The following books contain some valuable information on enzymes and the principles of biochemistry as applied to living systems:

Koshland, D. E., "Protein shape and biological control." *Scientific American,* October 1973, p. 52. A good discussion of the way enzymatic control is obtained through shifts in conformational structure of the protein.

Lehninger, A. L., *Biochemistry,* 2d ed. New York: Worth, 1975. This is probably the single best biochemistry textbook available at the present time. Very thorough and detailed; excellent for reference.

Loewy, Ariel, and Philip Siekovitz, *Cell Structure and Function,* 2d ed. New York: Holt, Rinehart and Winston, 1969. Not too detailed, a good general textbook for review.

Moss, D. W., *Enzymes.* London: Oliver and Boyd, 1968. This small paperback is an excellent simple introduction to enzymes, their variety, structure, and function.

CHAPTERS 4 AND 5

A number of useful paperback and hardcover books now exist that summarize recent findings in cell biology. Only a few are listed here as guides for more detailed information on many aspects of cell structure and function.

Dyson, Robert D., *Cell Biology: A Molecular Approach,* 2d ed. Boston: Allyn and Bacon, 1978. A well-illustrated authoritative account of many aspects of cell structure and function. It is quite complete and assumes some background on the part of the reader.

Fawcett, D. W., *The Cell, Its Organelles and Inclusions: An Atlas of Fine Structure.* Philadelphia: W. B. Saunders, 1966. An outstanding collection of electron micrographs, arranged by organelle type, with thorough descriptions of each organelle's functions.

Kessel, R. G., and C. Y. Shih, *Scanning Electron Microscopy in Biology.* New York: Springer-Verlag, 1974; and Kessel, R. G., and C. Y. Shih, *Tissues and Organs: A Text-Atlas of Scanning Electron Microscopy.* San Francisco: W. H. Freeman, 1979. Two superb collections of micrographs.

Kimball, John W., *Cell Biology,* 2d ed. Reading, Mass.: Addison-Wesley, 1978. Written for the beginning student, this book contains clear and simple descriptions of many cell structures and functions.

Novikoff, A. B., and E. Holtzman, *Cells and Organelles,* 2d ed. New York: Holt, Rinehart and Winston, 1976. Simple and well-presented descriptions of cell structures and functions. Contains a large section on nerve cells as membrane models.

Wolfe, Stephen, *Modern Cell Biology.* Belmont, Calif.: Wadsworth, 1972. Handsomely illustrated.

Articles on various cell components are listed below for further reference:

Albrecht-Buehler, Guentes, "The tracks of moving cells," *Scientific American,* April 1978. This article contains a fascinating discussion of the tracks left by migrating cells. Particularly significant is its discussion of the role of microfilaments in cell organization as revealed by indirect immunofluorescence. The possible role of the centrioles in cell migration is also discussed.

Bretscher, Mark, "Membrane structure: some general principles," *Science* 181 (August 1973), p. 622. A good review article, somewhat detailed and technical.

Bryan, Joseph, "Microtubules," *Bioscience* 24, no. 12 (Dec. 1974), pp. 701–711. A review of the literature on microtubules. Especially good on the topic of assembly and work with mitotic inhibitors. Also contains a brief discussion of microfilaments.

Capaldi, Roderick A., "A dynamic model of cell membranes," *Scientific American,* March 1974, p. 26. An interesting treatment of membrane structure.

De Duve, Christian, "The lysosome," *Scientific American,* May 1963. An older but well-written article by a pioneer in the study of lysosomal structure and function.

Mahler, Henry S., "Mitochondria: molecular biology, genetics, and development." No. 1 in the Addison-Wesley Module in Biology (Reading, Mass.: Addison-Wesley, 1973). A summary of the information available about mitochondria, including structure and biochemical function.

Morowitz, Harold J., and Mark E. Tourtellotte, "The smallest living cells," *Scientific American,* March 1962, p. 117. A discussion of PPLO, a mycoplasma.

Satir, Peter, "How cilia move," *Scientific American,* October 1974, p. 44. An interesting article on the role of microtubules in ciliary movement. An essay in the interrelationship between structure and function.

Stein, Wilfred, and William Lieb, "How molecules pass through membranes," *New Scientist,* January 10, 1974, p. 77. A simplified review article dealing largely with facilitated and active transport.

CHAPTER 6

A great deal of material has been published recently concerning cellular metabolism, ATP, electron transport, and the like. Below are listed some of those articles and books that might be most useful in providing additional background and recent information in the field.

Baker, J. J. W., and G. E. Allen, *Matter, Energy, and Life,* 4th ed. Reading, Mass.: Addison-Wesley, 1981. A more detailed account of the background material discussed in this chapter. Various chapters of the book deal with such topics as the structure of matter, formation of molecules, the course and mechanism of chemical reactions, acids and bases, the chemical composition of living matter, carbohydrates and lipids, proteins, enzymes, and nucleic acids. The book assumes no formal knowledge of chemistry or physics.

Becker, Wayne. *Energy and the Living Cell.* Philadelphia: Lippincott, 1977.

Lehninger, A. O., *Bioenergetics,* 2d ed. Menlo Park, Calif.: W. A. Benjamin, 1973. An excellent book, though difficult reading. Presupposes some college-level background in chemistry and biology, though it is written clearly and for the most part is comprehensible to the diligent beginning student.

Rosenberg, Eugene, *Cell and Molecular Biology.* New York: Holt, Rinehart and Winston, 1971. Chapter 3 deals with metabolic pathways. This book is well written and aimed at the beginning student, but not "watered down." Provides easier access to additional detail on metabolism than Lehninger.

CHAPTER 7

Bassham, J. A., "The path of carbon in photosynthesis," *Scientific American,* June 1962, p. 88. Although an older article, this one remains an excellent account of the basic methods involved in unraveling the steps of the dark reactions.

Björkman, Olle, and Joseph Berry, "High-efficiency photosynthesis," *Scientific American,* October 1973, p. 80. A detailed discussion of the discovery and mechanism of the C_4 pathway in desert plants such as *Atriplex*.

Govindjee, and Rajni Govindjee, "The absorption of light in photosynthesis," *Scientific American*, December 1974, p. 68. A summary of then-current knowledge of the light reactions. Well written, but technical and highly detailed in places.

Lehninger, A. L., *Bioenergetics*, 2d ed. Menlo Park, Calif.: W. A. Benjamin, 1973. Chapter 6 deals with "Photosynthesis and the Chloroplast." It is an authoritatively but simply written account of the energetics of the light reactions. One of the best summaries available.

Levine, R. P., "The genetic dissection of photosynthesis," *Science* 162 (Nov. 15, 1968), p. 768. A very good (though technical) article dealing with the question of how photosynthesis is controlled genetically.

Levine, R. P., "The mechanism of photosynthesis," *Scientific American*, December 1969, p. 58. An account of both the light and dark reactions, but with emphasis on the light reactions. Older, but more simply presented than the Govindjee article.

Mooney, H. A., O. Björkman, A. E. Hall, and P. B. Tomlinson. "The study of the physiological ecology of tropical plants—current status and needs," *BioScience*, 30, no. 1, January 1980, pp. 22–26. A discussion of the role of tropical rain forests in the ecosystem, which clearly delineates what is and especially what is *not* known about the environmental-limiting factors of photosynthesis in rain forests.

Woodell, George M. "The carbon dioxide question," *Scientific American*, January 1978, pp. 34–43. An excellent analysis of the carbon cycle in terms of the worldwide fixation of carbon by green plants and the effect of deforestation on the concentration of atmospheric carbon dioxide.

CHAPTER 8

General Plant Structure and Function

Esau, Katherine, *Anatomy of Seed Plants*. New York: Wiley, 1960. A comprehensive account of plant structure. Somewhat less technical than her more recently published *Plant Anatomy*, second edition.

Naggle, G. R., and J. F. Fritz, *Introductory Plant Physiology*. Prentice-Hall, 1976. An excellent overall textbook dealing with all aspects of plant physiology.

Laetsch, Watson M., *Plants: Basic Concepts in Botany*. Little, Brown, 1979. A fine introductory textbook in botany, somewhat unique for its emphasis on plant ecology and its environmental and social significance without in any way distracting from its coverage of the more basic biology of plants.

Steward, F. C., *Plants at Work*. Reading, Mass.: Addison-Wesley, 1964. An excellent short treatment of plant anatomy and physiology.

Torrey, John G., *Development in Flowering Plants*. New York: Macmillan, 1967. Treats the development and structure of flowering plants from the fertilized egg to the formation of flowers and fruits. Some background in basic plant biology is necessary for adequate understanding.

Translocation and Transpiration

Gates, David M., "Heat transfer in plants," *Scientific American*, December 1965, p. 77. A general account of the ways in which plants regulate their temperature.

Ray, P. M., *The Living Plant*, 2d ed. New York: Holt, Rinehart and Winston, 1972. This excellent book contains a discussion of long-distance transport in plants.

Zimmermann, M. H., "How sap moves in trees," *Scientific American*, March 1963, p. 133. A well-illustrated account of the use of aphids in studying transport through the phloem. A follow-up of more recent work using the same techniques can be found in: C. H. Bornman and C. E. J. Botha, "The role of aphids in phloem research," *Endeavour* 32, no. 117, September 1973, p. 129.

CHAPTER 9

Boysen-Jensen, P., "Transmission of the phototropic stimulus in the coleoptile of the oat seedling." Reprinted in *Great Experiments in Biology*, ed. M. L. Gabriel and S. Fogel. Englewood Cliffs, N.J.: Prentice-Hall, 1955, pp. 146–148. Boysen-Jensen demonstrates that a layer of gelatin separating the tip of coleoptile from its base does not interfere with the phototropic response.

Darwin, Charles, and Francis Darwin, "Sensitiveness of plants to light: its transmitted effects." Reprinted in *Great Experiments in Biology*, ed. M. L. Gabriel and S. Fogel. Englewood Cliffs, N.J.: Prentice-Hall, 1955, pp. 142–146. The Darwins demonstrate that the phototropic response originates in the apex of the coleoptile.

Galston, Arthur W., and Peter J. Davies, *Control Mechanisms in Plant Development*. Englewood Cliffs, N.J.: Prentice-Hall, 1970. A readable discussion of the photoperiodic response, plant hormones, and general growth phenomena in plants.

Galston, Arthur W., and Peter J. Davies, "Hormonal regulation in higher plants," *Science* 163 (1969): 1288. A review of the then-current status of research on the regulation of plant growth and development by interactions between promotive and inhibitory hormones.

Letham, David S., "Cytokinins and their relation to other photohormones," *BioScience* 19 (1969): 309. An excellent summary; well illustrated.

Steeves, T. A., and I. M. Sussex, *Patterns in Plant Development*. Englewood Cliffs, N.J.: Prentice-Hall, 1972. A very thorough-going review of many aspects of plant growth and development.

Steward, F. C., *About Plants: Topics in Plant Biology*. Reading, Mass.: Addison-Wesley, 1966. An excellent short treatment of plant structure and function.

Van Overbeek, J., "The control of plant growth," *Scientific American*, July 1968, p. 75. A well-written discussion of the manner in which the growth of plants is studied in the laboratory through treatments with the proper combinations of promotive and inhibitory hormones.

CHAPTER 10

Listed below are a few general physiology textbooks suitable for reading by the student interested in probing deeper into the field of animal physiology, and which apply to the following Chapters on this topic as well. The chapters that follow will list only readings pertinent to the topics each includes.

Beaumont, William, "Experiments and observations on the gastric juice and the physiology of digestion." In *The Process of Biology: Primary Sources*, ed. Jeffrey J. W. Baker and Garland E. Allen. Reading, Mass.: Addison-Wesley, 1970. An account of the fascinating observations made in the 1820s by physician Beaumont on Alexis St. Martin after St. Martin sustained a gunshot wound in the stomach. Also included is "Gastric juice of the buzzard" by René Antoine Reaumur.

Cannon, W. B., *The Wisdom of the Body*. New York: Norton, 1960. This is a new edition of Cannon's 1939 work, in which he coined the term "homeostasis" and discussed the types of regulatory mechanisms found in the body. Cannon was one of the important people to pick up Bernard's doctrine of constancy of the internal environment and apply it to physiological studies.

Davenport, H. W., "Why the stomach does not digest itself," *Scientific American*, January, 1972 (offprint no. 1240). A superb article on gastric physiology.

Eckert, Roger, and David Randall, *Animal Physiology*. San Francisco: W. H. Freeman and Company, 1978. An excellent general textbook that covers other animal forms as well as human beings.

Langley, L. L., *Homeostasis*. London: Chapman and Hall, 1965. An excellent short introduction to homeostatic mechanisms. Clearly written and illustrated.

Morton, John, *Guts; The Form and Function of the Digestive System*. New York: St. Martin's Press, 1967. A short (58-page) booklet describing the comparative anatomy and physiology of the intestinal systems in a number of animals.

Ruch, Theodore, and Harry C. Patton, *Medical Physiology and Biophysics*, 20th ed. Philadelphia: W. B. Saunders, 1973. Undoubtedly the best single complete text on modern human physiology, this work should serve primarily as a reference.

Selkurt, Ewald E., ed., *Physiology*, 4th ed. Boston: Little, Brown, 1976. A collection of technical but readable essays on human physiology by several different authors.

Smith, Homer, *From Fish to Philosopher*. Summit, N.J.: CIBA Foundation, 1959. An excellent discussion of the evolution of renal structure and physiology.

Spence, Alexander P., and Elliot B. Mason, *Human Anatomy and Physiology*. Menlo Park, Calif.: Benjamin/Cummings, 1979. An excellent, highly readable general textbook on the subject.

Vander, Arthur J., James H. Sherman, and Dorothy S. Luciano, *Human Physiology—The Mechanisms of Body Function*, 2d ed. New York: McGraw-Hill, 1975. A textbook with a heavy emphasis on physiological mechanisms.

CHAPTER 11

Adolph, E. F., "The heart's pacemaker," *Scientific American*, March, 1967 (offprint no. 1067). A description of research on the regulation of the heartbeat by its nodal tissue.

Boyle, E., "Biological patterns in hypertension by race, sex, body weight, and skin color," *Journal of the American Medical Association* 213 (1970): 1637.

Kannel, William B., et al., "Epidemiologic assessment of the role of blood pressure in stroke; The Framingham Study," *Journal of the American Medical Association* 214 (1970): 301.

The Boyle and Kannel et al. articles in a leading medical journal describe how two key population studies were carried out and the data analyzed. They are somewhat technical.

Mayerson, H. S., "The lymphatic system," *Scientific American*, June 1963, p. 80. A general discussion of the lymphatic system and its role in maintaining the steady state of body tissue fluids.

Perutz, M. F., "The hemoglobin molecule," *Scientific American*, December, 1978 (offprint no. 1413). An up-to-date review of the molecular structure of hemoglobin and its role in respiratory exchange.

Schmidt-Nielsen, K., *How Animals Work*. New York: Cambridge University Press, 1972. A small book, highly readable, dealing with gas exchange in a wide variety of animal species. Available in paperback.

Zweifach, Benjamin, "The microcirculation of the blood," *Scientific American*, January 1959, p. 54. This article discusses the capillary bed and how the opening and closing of capillaries is achieved.

CHAPTER 12

Cohen, C., "The protein switch of muscle contraction," *Scientific American*, November 1975 (offprint no. 1329). A detailed account of muscle contraction at the molecular level.

Gordon, M. S., G. A. Bartholomew, A. D. Grinnell, C. B. Jorgenson, and F. N. White, *Animal Function: Principles and Adaptations*, 2d ed. New York: Macmillan, 1971. The chapter on muscle is very thorough and well written. It covers the anatomical, biochemical, and physiological aspects of muscle structure and function.

Merton, P. A., "How we control the contraction of our muscles," *Scientific American*, May 1972, p. 30. A fine discussion of the feedback control mechanisms involved in regulating muscle contraction.

Murray, J. M., and A. Weber, "The cooperative action of muscle proteins," *Scientific American*, February 1974 (offprint no. 1290).

Ross, Russell, and J. A. Glomset, "Atherosclerosis and the arterial smooth muscle cell," *Science* 180 (June 1973): 1332. A good discussion of the muscular changes involved in hardening of the arteries.

"The brain," *Scientific American*, September 1979. This special issue contains eleven up-to-date reviews by several leading brain researchers.

CHAPTER 13

Bylinsky, Gene, "New clues to the causes of violence," *Fortune*, January 1973, p. 135. An example of the popularized literature suggesting that violent behavior is rooted in specific regions of the brain. Makes a case for altering such behavior by psychosurgical methods.

Galvani, Luigi, "Effects of electricity on muscular motion." Translated by Margaret G. Foley, in *The Process of Biology: Primary Sources*, ed. Jeffrey J. W. Baker, and Garland E. Allen. Reading, Mass.: Addison-Wesley, 1970. This is the most modern and readable translation of Galvani's fascinating experiments. It is well worth the time of any student interested in these early researches.

Hubbard, John, *The Biological Basis of Mental Activity*. Reading, Mass.: Addison-Wesley, 1975. Explores such topics as the sleep-waking cycle, mind-altering drugs, hemispherical dominance, and consciousness.

Keynes, R. D., "The nerve impulse and the squid," *Scientific American*, December, 1958 (offprint no. 58). Of historical interest, this article details the role of the giant axon of the squid in the development of our modern understanding of the conduction of nerve impulses.

Mason, B. J., "Brain surgery to control behavior," *Ebony* 28 (February 1973): 62. A popularized article opposing the use of psychosurgery and the concept that violent behavior is usually due to a pathological condition of the brain.

Schmitt, F. O., *The Neurosciences: Third Study Program.* Cambridge, Mass.: MIT Press, 1974. A hefty volume containing review papers by contemporary neurobiologists. Requires some technical background, but the individual papers are well written and illustrated. This is the single best source for study of the nervous system and neurophysiology.

CHAPTER 14

Eakins, Richard M., *The Third Eye.* Berkeley, Calif.: University of California Press, 1973. A fascinating account of the history of the pineal gland and the research that associates it with the third eye of amphibians, reptiles, and other forms.

Guillimin, R., and R. Burgus, "The hormones of the hypothalamus," *Scientific American*, November 1972 (offprint no. 1260). A discussion of the discovery of the releasing factors of the hypothalamus and their effects upon the pituitary gland.

Le Baron, Ruthann, *Hormones: A Delicate Balance.* New York: Pegasus, 1972. A very readable book that introduces the reader to hormones via a historical approach. Requires no special technical background.

O'Malley, B. W., and W. T. Shrader, "The receptors of steroid hormones," *Scientific American*, February, 1976 (offprint no. 1334). A discussion of the mechanisms by which steroid hormones are hypothesized to act on their target cells.

Sutherland, Earl W., "Studies on the mechanism of hormone action," *Science* 177 (August 1972): 401. The author's Nobel laureate address on the role of hormones in the biochemistry of cellular processes.

Truman, James W., "How moths 'turn on': A study of the action of hormones in the nervous system," *American Scientist* 61 (November–December 1973): 700. A readable and well-illustrated account of how hormones affect the development of the nervous system during metamorphosis in moths.

CHAPTER 15

As is the case with many relatively new areas of research, the literature on human reproduction, development, and sexuality is plentiful and highly diverse in terms of both content and quality. For the interested student, extensive bibliographies and information about other sources are available from the Sex Information and Education Council of the United States (SIECUS), 122 East 42d St., New York, N.Y. 10017.

Addiego, F., E. Belzer, J. Comalli, W. Moger, J. O. Perry, and B. Whipple, "Female ejaculation: a case study," *The Journal of Sex Research*, 17, no. 1 (Feburary 1981). The authors discuss the ejaculation of fluid reported by some women and hypothesize that a previously described area, the "Graffenburg spot,"

may be a rudimentary prostate gland. This issue also contains other articles on this topic.

Bayer, Ronald, *Homosexuality and American Psychiatry: The Politics of Diagnosis.* New York: Basic Books, 1981. A highly readable history of scientific and political issues involved in declassification of homosexuality as a mental illness. A classic example of how "objective" science can be used to reinforce cultural and/or religious myths.

Bell, Alan P., and Martin S. Weinberg, *Homosexualities.* New York: Simon and Schuster, 1978. The authors describe their extensive study of homosexual behavior and conclude that, like heterosexual behavior, it has a wide range of expression.

Boston Women's Health Book Collective. *Our Bodies, Ourselves,* rev. 2d ed. New York: Simon and Schuster, 1976.

Brecher, Edward M., *The Sex Researchers.* Boston: Little, Brown, 1969; revised and reissued by Specific Press, San Francisco, 1979. A highly readable account of the history of sex research.

Brown, Rita Mae, *Rubyfruit Jungle.* Plainfield, Vt.: Printer's Ink Press, 1975. Much work has been reported on male homosexuality, but little has been done on lesbianism. This is an amusing account of a woman's discovery that she is a lesbian.

Churchill, Wainwright, *Homosexual Behavior Among Males.* Englewood Cliffs, N.J.: Prentice-Hall, 1971. A cross-cultural and cross-species account of the prevalence of homosexuality and various attitudes toward it.

Golanty, Eric, *Human Reproduction.* New York: Holt, Rinehart and Winston, 1975. An easy-to-understand book of human sexual and reproductive biology.

Goldberg, Steven, *The Inevitability of Patriarchy.* New York: William Morrow, 1973. A perfect example of an argument based on biological determinism, this well-written book maintains that the arguments of women's liberation run counter to biological reality. The author, a sociologist, is to the "innateness" theory of male domination what Robert Ardrey is to the "innateness" theory of human aggression.

Haeberle, Erwin J., *The Sex Atlas: A New Illustrated Guide.* New York: The Seabury Press, 1978. This book provides an excellent one-volume reference for the field of sexology. The author has a strong historical bent, and the work has excellent synopses of the evolution of our cultural and religious views of sex and sexuality through the centuries.

Katchadourian, H. A., and D. T. Lunde, *Fundamentals of Human Sexuality.* New York: Holt, Rinehart and Winston, 1972. An excellent, clearly written book discussing many aspects of human reproduction from physiology to psychology.

Masters, W. H., and V. E. Johnson, *Human Sexual Response.* Boston: Little, Brown, 1966. Aside from Kinsey's reports of 1948 and 1953, the work of Masters and Johnson is considered by many to be the most thorough-going and scientific account of human sexual behavior in the past thirty-five years.

Page, E. W., C. A. Villee, and D. B. Villee, *Human Reproduction.* Philadelphia: W. B. Saunders, 1972. A comprehensive volume that straightforwardly discusses all aspects of reproduction and sexual development.

Pengelley, Eric T., *Sex and Human Life,* 2d ed. Reading, Mass.: Addison-Wesley, 1978. A simply written and concise general textbook of human reproduction, development, and sexuality.

Shapiro, Howard I., *The Birth Control Book*. New York: Avon Books, 1978. Written in question-and-answer form, this book is a comprehensive and detailed up-to-date review of modern birth control methods.

Weinberg, George, *Society and the Healthy Homosexual*. New York: St. Martin's Press, 1972. Also published in paperback by Anchor-Doubleday, 1973.

Weinberg, Martin S., and Colin J. Williams, *Male Homosexuals*. New York: Penguin Books, 1975. A detailed study of male homosexuality in three different countries (the United States, Denmark, and the Netherlands), with an analysis of data on numerous homosexuals sampled from the homosexual community rather than from psychiatric records.

CHAPTER 16

A number of general textbooks of genetics have been published in recent years, providing a rich source of more extensive discussions of some of the material included in this (and the following) chapter. Several of the most recent and most clearly written are:

Burns, George W., *The Science of Genetics*, 4th ed. New York: Macmillan, 1980. This is a very complete and up-to-date text, well but simply illustrated.

Farnsworth, M. W., *Genetics*. New York: Harper & Row, 1978. A particularly well-organized text; each chapter deals with a limited topic. The organization of topics in this book follows a parallel course with topics in the three genetics chapters of *The Study of Biology*.

Goodenough, Ursula, *Genetics*, 2d ed. New York: Holt, Rinehart and Winston, 1978. A thorough introductory text that can be used as a reference for individual topics raised in this chapter. Quite up-to-date and clearly written.

Levine, Louis, *Biology of the Gene*, 3d ed. St. Louis: C. V. Mosby, 1980. Available in paperback, this text is complete and very precisely illustrated.

Winchester, A. M., *Genetics: A Survey of the Principles of Heredity*, 4th ed. Boston: Houghton Mifflin, 1972. More simply written than any of the others listed above, this book has easy-to-follow explanations and good practice problems.

Several textbooks of human genetics provide more information on this important subject:

Cavalli-Sforza, L. L., *Elements of Human Genetics*. Reading, Mass.: Addison-Wesley; An Addison-Wesley Module in Biology, No. 2, 1973. A short booklet approaching human genetics from the chromosomal, molecular, and population points of view.

Rothwell, Norman V., *Human Genetics*. Englewood Cliffs, N.J.: Prentice-Hall, 1977. A complete, yet not overly detailed survey of topics in human genetics.

Winchester, A. M., *Human Genetics*, 2d ed. Columbus, Ohio: Charles E. Merrill, 1975. A simply written paperback with good illustrations and clear explanations.

Special Topics:

Clarke, C. A., "The prevention of 'rhesus' babies," *Scientific American*, November 1968. A discussion of recent advances in immunological techniques that allow doctors to prevent the debilitating effects of Rh disease.

CHAPTER 17

Gibor, A., "*Acetabularia*: a useful giant cell," *Scientific American*, November 1966 (offprint no. 1057). A description of Hämmerling's and others' studies of the giant algal cell.

Goodenough, Ursula, *Genetics*, 2d ed. New York: Holt, Rinehart and Winston, 1978. Chapters 2 and 3 contain a detailed and well-illustrated discussion of mitosis and meiosis.

Mazia, Daniel, "The cell cycle," *Scientific American*, January 1974 (offprint no. 1288). In an experimentally oriented paper, the author describes studies to determine what factors control the onset of mitosis, the duration of its various stages, and the like.

Swanson, Carl P., and P. L. Webster, *The Cell*. Englewood Cliffs, N.J.: Prentice-Hall, 1977. A good elementary text surveying most aspects of cell life, including reproduction. Chapters 7 and 8 cover mitosis and meiosis.

CHAPTER 18

General Genetics

Crow, James F., "Genes that violate Mendel's rules," *Scientific American*, February 1979. An interesting article that discusses how genes interact to avoid being subject to strict natural selection. A good review of the chromosome theory with some interesting, unexpected sidelights.

Moore, John A., *Heredity and Development*. New York: Oxford University Press, 1963. A semihistorical approach to understanding the problems of classical chromosome theory. Approaches the experiments of Morgan, Sturtevant, Bridges, and others in terms of the questions the investigators were asking and the data they obtained.

Watson, J. D., *The Molecular Biology of the Gene*, 3d ed. Menlo Park, Calif.: W. A. Benjamin Co., 1976. Chapter 1 is an excellent summary, with modern perspective, on Mendelian genetics and the development of the chromosome theory.

Somatic Cell Hybridization

Davidson, Richard L. "Somatic cell hybridization: studies on genetics and development." Reading, Mass.: An Addison-Wesley Module in Biology, No. 3, 1973. A good self-contained discussion of this important technique, and its many ramifications.

Lewin, Roger, "Unnatural union of plants and animals," *New Scientist*, August 5, 1976, p. 270. A takeoff from the somatic cell hybridization experiments discussed in the text, this article shows how hybrids have been made even between animal and plant cells.

Extra-Nuclear Inheritance

Beale, Geoffrey, and Jonathan Knowles, *Extranuclear Genetics*. Baltimore: University Park Press, 1978. A concise study that examines a wide range of examples of extranuclear heredity in higher organisms and bacteria. Well written and reasonably advanced.

Human Chromosomal Inheritance

Bodmer, W. F., and L. L. Cavalli-Sforza, *Genetics, Evolution and Man*. San Francisco: W. H. Freeman, 1976. An excellent introduction to human genetics from an evolutionary perspective.

Stern, Curt, *Principles of Human Genetics*, 3d ed. San Francisco: W. H. Freeman, 1973. The most comprehensive textbook now available on human genetics. Readable, but requiring a solid genetics background.

Winchester, A. M., *Human Genetics*, 2d ed. Columbus, Ohio: Charles E. Merrill, 1975. A simply written, well-illustrated, and straightforward introduction to human genetics, with special emphasis on chromosomal and biochemical disorders.

CHAPTER 19

General Molecular Genetics

Goodenough, Ursula, *Genetics*, 2d ed. New York: Holt, Rinehart and Winston, 1978. Referred to in previous chapters as an excellent general genetics text; very strong in molecular genetics.

De Robertis, E. D. P., and E. M. F. De Robertis Jr., *Cell and Molecular Biology*. Philadelphia: W. B. Saunders, 1980. A thorough discussion, detailed but clear, of all aspects of modern cell biology. Chapters 21 through 25 deal explicitly with topics of molecular genetics.

Watson, J. D., *The Molecular Biology of the Gene*, 3d ed. Menlo Park, Calif.: W. A. Benjamin, 1976. Although somewhat out-of-date, this text still ranks as one of the best summary discussions of molecular genetics in print.

Specific Topics

Brown, Donald. "The isolation of genes," *Scientific American*, August 1973, p. 20. A description of RNA use as a "probe" to locate and retrieve the DNA segment that coded for it.

Cohen, Stanley N., and James A. Shapiro, "Transposable genetic elements," *Scientific American*, February 1980. A discussion of naturally occurring recombinant DNA events, which account for much genetic variability and perhaps for rapid evolution.

Darnell, James E. "Implications of RNA-RNA splicing in evolution of eukaryotic cells," *Science* 202 (December 22, 1978): 1257. A look at gene structure and protein synthesis in pro- and eukaryotic cells; the author suggests that the differences in these processes between the two cell types are so profound as to imply early evolutionary divergence between the two cell types.

Gilbert, Walter, and Lydia Villa-Komaroff, "Useful proteins from recombinant bacteria," *Scientific American*, May 1980, p. 74. A summary of the recombinant DNA technique and discussion of the potential benefits the technique may reveal.

Rich, Alexander, and Sung Hou Kim, "The three-dimensional structure of transfer RNA," *Scientific American*, January 1978, p. 52. A detailed discussion of how the structure of *t*RNA was worked out, and how a knowledge of its structure has clarified its role in protein synthesis.

CHAPTER 20

General Descriptive and Experimental Embryology

Balinsky, B. I., *An Introduction to Embryology*, 3d ed. Philadelphia: W. B. Saunders, 1970. A thorough, well-illustrated, classic introductory text in embryology.

Davenport, Richard, *An Outline of Animal Development*. Reading, Mass.: Addison-Wesley, 1979. Imaginative and well-written, this book tries to approach the problems of development from a philosophical and large-scale point of view. Its subject matter, as the title implies, is limited to animals.

De Robertis, E. M., and J. B. Gurdon, "Gene transplantation and the analysis of development," *Scientific American*, December 1979, p. 74. Discussion of new techniques for introducing purified (isolated) genes into amphibian oöcytes and the ways the genetic message is "read." Focuses on using these methods to study gene regulation in development.

Ebert, James D., and I. M. Sussex, *Interacting Systems in Development*, 2d ed. New York: Holt, Rinehart and Winston, 1970. An excellent modern discussion of the main outlines of development. Focuses especially well on tissue and cell interactions at the molecular level. Available in paperback.

Epel, D. "The program of fertilization," *Scientific American*, November 1977, p. 128. Recent findings on interaction between egg and sperm.

Steward, F. C., and A. D. Krikorian, *Plants, Chemicals and Growth*. New York: Academic Press, 1971. A thorough review of plant development including molecular, cellular, and organism levels.

Trinkaus, J. P., *Cells into Organs*. Englewood Cliffs, N.J.: Prentice-Hall, 1969. A well-written summary review of theories of how cells form into organs in embryonic development.

Wessels, Norman K., *Tissue Interactions and Development*. Menlo Park, Calif.: Benjamin/Cummings, 1977. A very lively book, available in paperback, approaching animal development from a problems point of view.

Wolpert, Lewis, "Pattern formation in biological development," *Scientific American*, October 1978, p. 154. Using development of the chicken wing as a model, the author discusses the problem of how cells recognize their position in the embryo as a whole, and determine the direction of future morphogenetic movements.

Human Development

Beaconfield, Peter, George Birdwood, and Rebecca Beaconfield, "The placenta," *Scientific American*, August 1980, p. 95. An anatomical and physiological discussion of the placenta, and its role (after birth) as an "experimental organism."

Fuchs, Fritz, "Genetic amniocentesis," *Scientific American*, June 1980, p. 47. A general discussion of the technique of amniocentesis and its function in genetic analysis and counseling.

Hayflick, Leonard, "The cell biology of human aging," *Scientific American*, January 1980. An updating on earlier work about the inbuilt mechanisms of aging in human laboratory cell lines.

Rugh, Roberts, and Landrum B. Shettles, *From Conception to Birth*. New York: Harper & Row, 1971. This well-written, simple, and beautifully illustrated work is a most valuable source of information on human embryology and birth. There is virtually no other book available that assumes no previous knowledge of biology yet is so authoritative and interesting.

Immunology

Golub, Edward S., *The Cellular Basis of the Immune Response*. Sunderland, Mass.: Sinauer Associates, 1977. A comprehensive, yet not overly complex treatment of the cellular and molecular (including genetic) basis of the immune response.

Koffler, David, "Systemic Lupus Erythematosus," *Scientific American*, July 1980, p. 52. Discussion of a rheumatic disease caused by autoimmunity to such self substances as DNA.

Jerne, Niels Kaj, "The immune system," *Scientific American*, July 1973, p. 52. One of the best introductions to the general nature of the immune response.

Milstein, Cesar, "Monoclonal antibodies," *Scientific American*, October 1980, p. 66. A discussion of the use of cell-hybridization techniques to produce cells capable of synthesizing large quantities of a highly specific antibody.

Wilson, David, *Body and Antibody: A Report on the New Immunology*. New York: Alfred A. Knopf, 1972. A well-written popularized account of the history and recent developments in immunology. An enjoyable introduction to immunology for the layperson.

CHAPTER 21

Mayr, Ernst, "Difficulties and importance of the biological species concept." In *The Process of Biology: Primary Sources*, ed. Jeffrey J. W. Baker and Garland E. Allen. Reading, Mass.: Addison-Wesley, 1970. A detailed and thorough coverage of the species concept as dealt with by a specialist. The problem of "What is a species?" is discussed in relation to animals, plants, and microorganisms. This work is suggested only for those who are especially interested in this topic.

Sibley, Charles G., "The comparative morphology of protein molecules as data for classification." In *Systematic Zoology* (part of a symposium on "The Data of Classification"). A relatively nontechnical review article on biochemical taxonomy.

Sokal, Robert R., "Numerical taxonomy," *Scientific American*, December 1966, p. 106. A good introduction to the modern quantitative aspects of taxonomy, especially the role computers can play in the taxonomic process.

Solbrig, Otto, and Dorothy Solbrig, *Introduction to Population Biology and Evolution*. Reading, Mass.: Addison-Wesley, 1979. Contains a simple and well-written chapter (Chapter 18) on the relationship between taxonomy and evolution.

Steussy, Tod F., and George F. Estabrook, "Introduction" to Vol. 3 (part 2) of *Systematic Botany* (1978): pp. 145 *ff.* This introduction and the series of papers that follow make up the bulk of this issue of the journal, devoted to cladistics. Although some of the papers are technical, together they present probably the most modern and up-to-date discussion of this new field of taxonomy.

CHAPTER 22

Volumes have been written on the history of evolutionary theory, especially the work of Darwin. Several of the most readable and informative sources are:

Darwin, Charles, *The Illustrated Origin of Species*. New York: Hill and Wang, 1979. An abridged version of Darwin's famous book, with an introduction by Richard E. Leakey. About half of Darwin's original text is reprinted here, with beautiful modern, as well as nineteenth century, illustrations.

De Beer, Gavin, *Charles Darwin*. New York: Thomas Nelson, 1964. This book combines the biology of Darwin's theory with an accurate and readable historical account of Darwin's life and the "evolution" of his ideas on the origin of species. It is well illustrated and written clearly for the nonspecialist.

Eisley, Loren, *Darwin's Century*. New York: Doubleday-Anchor, 1961. A very well-written account of the predecessors and followers of Darwin, as well as of Darwin's own work. Gives a good, comprehensive picture of the entire ambiance of nineteenth century evolutionary thought.

Moorehead, Alan, *Darwin and the Beagle*. New York: Harper & Row, 1969. An interesting and magnificently illustrated chronicle of Darwin's five years aboard the H.M.S. *Beagle*. Moorehead has captured the flavor of Darwin's voyage and portrayed the events that influenced him through a fine collection of contemporary paintings, engravings, and woodcuts.

For those interested in a more detailed and technical account of Darwin's work and the evolution of his method, the following, relatively recent, studies are highly interesting and useful:

Ghiselin, Michael T., *The Triumph of the Darwinian Method*. Berkeley: University of California Press, 1969. A detailed study of Darwin's approach to the problem of species and their origin, including the issues of transmutation, sexual reproduction, selection, and Darwin's philosophy.

Russett, Cynthia Eagle, *Darwin in America: The Intellectual Response*. San Francisco: W. H. Freeman Co., 1976. A very well-written study of the reception of Darwin's work by intellectuals of all sorts—religious leaders, sociologists, biologists, historians, economists, and literati. A very good, relatively short book, relating Darwinian theory to intellectual history.

Vorzimmer, Peter J., *Charles Darwin: the Years of Controversy: The Origin of Species and Its Critics, 1859–1882*. Philadelphia: Temple University Press, 1970. Takes a number of the points on which Darwin's work was criticized and discusses the nature of the opposition, and the response of Darwin (or his defenders, such as Huxley and Wallace). Detailed, but clearly written.

Young, Robert, "Malthus and the evolutionists: the common context of biological and social history," *Past and Present, a Journal of Historical Studies* no. 43 (May, 1969): 109–145. A detailed discussion of the influences of late eighteenth and early nineteenth century political economy on the development of evolutionary thinking—principally in the work of Darwin.

CHAPTER 23

Particularly up-to-date and useful is the *Scientific American* booklet *Evolution* (San Francisco: W. H. Freeman Co., 1978), a collection of articles about many aspects of the evolutionary process. An introductory essay by Ernst Mayr (pp. 2–13) provides a general overview. These articles were originally issued as the September 1978 issue of *Scientific American*. Specific articles of interest are referred to below, and at the end of the following chapter.

Ayala, Francisco J., "The mechanisms of evolution," *Scientific American*, September 1978, p. 57 (also, the Scientific American booklet, *Evolution*, op. cit., p. 14). An up-to-date survey of modern thinking on evolution; especially addresses the topics of variation—its origin and detection—and selection.

Clarke, Bryan, "The causes of biological diversity," *Scientific American*, August 1975, p. 50. A very well-illustrated article, its main point is that diversity within a species is maintained by natural selection.

Futuyma, Douglas, *Evolutionary Biology*. Sunderland, Mass.: Sinauer Associates, 1978. A very well-written, thorough textbook of modern evolutionary theory; would be particularly useful as a reference. Its special advantage is its textual clarity and its sensible down-to-earth approach to complex problems, especially in areas of human evolution (sexuality, race and I.Q., and the like).

Kettlewell, H. B. D., "Darwin's missing evidence," *Scientific American*, March 1959, p. 48. A description of Kettlewell's early experiments, including the mark-release experiments in Dorset Woods and near Birmingham. A more updated and detailed description of over twenty years of work on this subject can be found in Kettlewell's *The Evolution of Melanism*, Oxford, Clarendon Press, 1973. The latter work is over 300 pages and highly detailed; but it is remarkable for the thoroughness of the work; the author leaves no stones unturned.

Solbrig, Otto, and Dorothy Solbrig, *Introduction to Population Biology and Evolution*. Reading, Mass.: Addison-Wesley, 1979. A thorough, well-written treatment of evolution, with a special emphasis on the population nature of evolutionary processes. A somewhat more concise and selective monograph than Futuyma's, listed above.

Thornhill, Randy, "Sexual selection in the Blacktipped Hangingfly," *Scientific American*, June 1980, p. 162. A detailed discussion of the methods and results in Thornhill's study of sexual selection, described in Supplement 23.4.

CHAPTER 24

In addition to those general references given for the preceding chapter, the following will be of interest.

Cavalli-Sforza, L. L., "'Genetic drift' in an Italian population," *Scientific American*, August 1969. Reprinted in *Ecology, Evolution and Population Biology*, San Francisco: W. H. Freeman, 1974, p. 26. An example of the more expanded view of genetic drift, applied to human populations.

Lewontin, Richard, "Adaptation," in *Evolution*. W. H. Freeman, 1978, pp. 114–128. Another well-presented article in the *Scientific American* paperback on evolution. A highly instructive article.

Two *Scientific American* articles by Richard E. Dickerson on the evolution of cytochrome *c* are complementary and useful:

Dickerson, R. E., "The structure and history of an ancient protein," *Scientific American*, April 1972; and "Cytochrome *c* and the evolution of energy metabolism," *Scientific American*, March 1980, p. 137. The first article is more directed toward the evolution of molecular structure, the latter toward the evolution of molecular function, *vis-à-vis* the evolution of energy metabolism. The former article is reprinted in the *Scientific American* book cited above, *Ecology, Evolution and Population Biology*, p. 34.

In the same *Scientific American* book, *Ecology, Evolution . . .*, are two articles on continental drift and evolution:

Hallam, A., "Continental drift and the fossil record," *Scientific American*, November 1972, op. cit., p. 81.

Kurten, Björn, "Continental drift and evolution," *Scientific American*, March 1969, op. cit. p. 91.

On the Evolution of the Horse:

Simpson, George Gaylord, *Horses*. New York: Oxford University Press, 1951. This work is still the classic study on the evolution of horses. It discusses every major aspect, including adaptation, fossil evidence, geographic distribution, and evolutionary tempo.

CHAPTER 25

General

Brill, Winston J., "Biological nitrogen fixation," *Scientific American*, March 1977, p. 68. A general discussion of the role that bacteria and blue-green algae have in nitrogen fixation. Excellent scanning electron micrographs of the intimate relationship between the cells of the root nodules of legumes and the bacterial cells inhabiting them.

Dovring, Folke, "Soybeans," *Scientific American*, February 1974, p. 14. A discussion of the general ecology, agricultural development, and biology of soybeans, including their role in nitrogen fixation.

Gosz, James R., Richard T. Holmes, Gene E. Likens, and F. Herbert Bormann, "The flow of energy in a forest ecosystem," *Scientific American*, March 1978, p. 93. An interesting article on the quantitative energetics of a forest. The authors show how energy is utilized by different segments of a forest ecosystem.

Krebs, Charles J., *Ecology. The Experimental Analysis of Distribution and Abundance*, 2d ed. New York: Harper & Row, 1978. A good general introduction with an emphasis on data from field and laboratory studies. Not easy reading but an excellent resource book.

Luckinbill, Leo S., "*r* and *K* selection in experimental populations of *Escherichia coli*," *Science* 202 (December 15, 1978): 1201. A very good example of the hypothesis of *r* and *K* selection being put to the experimental test. Not easy reading but a short, very worthwhile article.

Pianka, Eric R., *Evolutionary Ecology*, 2d ed. New York: Harper & Row, 1978. A simple but well-written book dealing with many aspects of ecology in evolutionary terms. An especially useful chapter (Chapter 4) deals with physiological ecology.

Ricklefs, Robert E., *Ecology*. Newton, Mass.: Chiron Press, 1973. A very thorough treatment relating ecology to general principles of genetics and evolution. Heavy reading (more difficult than Krebs) with a great deal of detail. Good reference book.

Turk, J. J., T. Wittes, R. Wittes, and A. Turk, *Ecosystems, Energy, Population*. Philadelphia: W. B. Saunders, 1975. A paperback for the general student. Introduces the concepts of ecosystems, energy transfer, and population growth. On the latter subject, the authors take a neo-Malthusian approach.

Environmental Issues

Commoner, Barry, *The Closing Circle*. New York: Bantam Books, 1972. A popular book by one of America's most vociferous environmentalists. Commoner lays blame for the majority of environmental problems in the United States on the economic system, particularly rampant industrialization. A challenging book that is easy reading.

Ehrlich, Paul R., and Anne H. Ehrlich, *Population, Resources, Environment*, 2d ed. San Francisco: W. H. Freeman, 1972. Focuses on a number of human environmental problems including population growth; food production; air, water, and pesticide pollution; and social changes involved with population stabilization.

Lockeretz, William, "The lessons of the dust bowl," *American Scientist* 66 (September–October 1978): 560. The role of environmental degradation through agricultural mismanagement in the 1920s and 1930s in creating the hazard of dust storms.

Murdoch, W. W., *Environment: Resources, Pollution and Society.* Sunderland, Mass.: Sinauer, 1975. A hefty collection of chapters, each written by an expert in the field. There is a wealth of material in this book, particularly on the human environmental problem.

Vallentyne, J. R., *The Algal Bowl—Lakes and Man.* Miscellaneous Special Publication no. 22, pp. 1–185. Ottawa: Canadian Department of the Environment. A well-written, illustrated paperback book about the problem of algal growth on aquatic ecosystems. Although dealing with a specialized topic, this book touches on many important problems in ecology and the human environment.

Watt, K. E. F., *Principles of Environmental Science.* New York: McGraw-Hill, 1973. A basic textbook that takes an ecological approach to environmental problems. Simple, scholarly, and well written. Deals with a range of problems including resource use, self-regulation in ecosystems, fossil fuels and energy, pollution of different ecosystems, agricultural ecosystems, infectious disease, and urban and regional planning.

Ecology and Biogeography

MacArthur, R. H., *Geographical Ecology. Patterns in the Distribution of Species.* New York: Harper & Row, 1972. A specialized book that discusses a subject partly covered in Chapter 24. The subject, an extremely important one in ecology, is the origin and distribution of different species on earth in terms of the ecosystems of which they are a part. It treats such subjects as the relationship between the stability of ecosystems and the number of species comprising them.

Human Population

Barclay, William, "Population control in the Third World," *NACLA (North American Congress on Latin America) Newsletter* 4, no. 8 (December 1970), pp. 1–18. A brief analysis of population control from the socioecologist's point of view. Contains an analysis of what people and groups in the United States strongly support in terms of population control programs in other (especially Latin American) countries.

Chase, Allan, *The Legacy of Malthus.* New York: Knopf, 1977. A very interesting and challenging book detailing the history of Malthusian ideas in the nineteenth and twentieth centuries. A controversial, but highly stimulating treatment.

Ehrlich, Paul R., *The Population Bomb.* New York: Ballantine, 1968. Though an older work, this book is a classic that launched the new wave of neo-Malthusian arguments about the future catastrophe of overpopulation.

Mass, Bonnie, *Political Economy of Population Control in Latin America.* Montreal: Editions Latin America, 1972. Although not always well organized, this short booklet presents some criticisms of the major neo-Malthusian arguments as they apply to Latin America. Written from a socioecological perspective.

Westoff, Charles F., "Marriage and fertility in the developed countries," *Scientific American,* December 1978, p. 51. A detailed look at the decline in birthrate in the industrialized countries, with some speculations as to its cause and the nature of the so-called demographic transition.

CHAPTER 26

Bernal, J. D., *The Origin of Life.* Cleveland: World Publishing Co., 1967. An extremely well-written account of current hypotheses about the origin of life. Assumes no special background and discusses philosophical, biological, and biochemical aspects.

Dickerson, Richard E. "Chemical evolution and the origin of life," *Scientific American,* September 1978. p. 70. Reprinted in *Evolution.* San Francisco: W. H. Freeman, 1978, p. 30. An excellent summary of prebiotic conditions.

Dickerson, Richard E., "Cytochrome *c* and the evolution of energy metabolism," *Scientific American,* March 1980, p. 137. A very interesting and clearly presented discussion of recent evidence regarding the evolution of proteins.

Folsome, Clair Edwin, *Life: Origin and Evolution.* San Francisco: W. H. Freeman, 1979. This is a collection of articles about the origin and evolution of life, reprinted from issues of *Scientific American* over the past twenty-five years. It includes articles on the formation and early evolution of the earth, prebiotic chemistry, protocells, fossils, and the early evolution of life, and concludes with a section on extraterrestrial life. Some of the articles are excellent; others a bit dated.

Fox, Sidney W., and Klaus Dose, *Molecular Evolution and the Origin of Life.* New York: Marcel Dekker, 1977. A very thorough compendium of recent experimental and speculative work. Technical, but clearly written.

Keosian, John, *The Origin of Life.* New York: Reinhold, 1964. A popular account with accurate and well-presented scientific details. Less stimulating than Bernal's treatment because it deals with few of the broader philosophical questions.

Oparin, A. I., *Genesis and Evolutionary Development of Life.* New York: Academic Press, 1968. A useful summary of some of the more recent thinking of one of the pioneers in the field of chemosynthesis. This book replaces Oparin's older classic, *The Origin of Life on Earth,* first published in 1938.

Orgel, L. E., *The Origins of Life.* New York: Wiley, 1973. As the preface states, "This book is not written for professional biologists or chemists, but rather for college or advanced high school students and general readers who have a limited background in chemistry or biology." The author emphasizes the role of natural selection even during the earliest phases of the origin of life.

Schopf, J. William, "The evolution of the earliest cells," *Scientific American,* September 1978, p. 110. Reprinted in *Evolution* op. cit., p. 48.

Wald, George, "The Origins of Life," in *The Scientific Endeavor.* New York: Rockefeller University Press, 1964. A short and very clear exposition of aspects of the chemosynthetic theory. The same paper is also available in *Proceedings of the National Academy of Sciences (U.S.)* 52 (1964): 595.

CHAPTER 27

The Monera

The works listed below by Frobisher et al., Stanier et al., and Wilkinson deal with bacteria, viruses, blue-green algae, and the eukaryotic protists, including protozoa. The focus of each, however, is on the Monera.

Butler, P. Jonathan G., and Aaron Klug, "The assembly of a virus," *Scientific American,* November 1978, p. 62. An up-to-date discussion of work initiated by Fraenkel-Conrat two decades ago with the tobacco mosaic virus. Describes the latest work on the structure of the protein monomers making up the coat of TMV, as well as the chemical association between the monomers and the RNA of the viral "core."

Costerton, J. W., G. G. Geesey, and K. J. Cheng, "How bacteria stick," *Scientific American*, January 1978, p. 86. A discussion of the glycocalyx fiber covering of bacteria, which helps the cells stick to other cells and thus provides the crucial first step to infection.

Echlin, Patrick, "The blue-green algae," *Scientific American*, June 1966, p. 74. A well-illustrated and informative discussion of the blue-green algae.

Frobisher, Martin, Ronald D. Hinsdill, Koby T. Crabtree, and Clyde R. Goodheart, *Fundamentals of Microbiology*. Philadelphia: W. B. Saunders, 1974. A more comprehensive text than the one listed above, including sections on microbes and the environment; techniques of sterilization; the nature and use of antibiotics; microbes and host defense mechanism; soil, water, and air microbiology; microbiology of foods; and industrial microbiology. A very complete and diverse book.

Stanier, Roger Y., Michael Doudoroff, and Edward A. Adelberg, *The Microbial World*, 4th ed. Englewood Cliffs, N.J.: Prentice-Hall, 1976. This classic text is an eminently readable, well-illustrated, and complete but not encyclopedic account of all the major groups of Monera and Protista. An outstanding reference.

Wilkinson, J. F., *Introduction to Microbiology*, 2d ed. New York: Wiley, 1975. A short, well-written introduction to the major groups of Monera, including the viruses; very little discussion of any protists, however. A quick, readable introduction.

The Evolution of Prokaryotes and Eukaryotes

Curtis, Helena, *The Marvelous Animals: An Introduction to the Protozoa*. Garden City, N.Y.: The Natural History Press, 1968. An extremely interesting, popularized but nonetheless detailed discussion of the protozoa. Well illustrated.

Darnell, James E., "Implications of RNA-RNA splicing in the evolution of eukaryotic cells," *Science* 202 (December 22, 1978): 1257. A relatively technical but intriguing article about the differences between eukaryotic and prokaryotic cells in terms of the continuity or discontinuity in DNA sequences making up a single "gene." Presents the case for making a sharp distinction between prokaryotes and eukaryotes on DNA discontinuity.

Margulis, Lynn, "Symbiosis and evolution," *Scientific American*, August 1971, p. 48. A well-presented discussion of the hypothesis that cell organelles such as mitochondria and chloroplasts originated by endosymbiosis from once free-living prokaryotes.

Schopf, J. William, "The evolution of the earliest cells," *Scientific American*, September 1978, p. 110. An up-to-date discussion of the evolution of cellular life.

CHAPTER 28

General Works

Bold, H. C., *The Morphology of Plants*, 3d ed. New York: Harper & Row, 1973. A very good survey of the entire plant kingdom in terms of structure and its relation to function.

Raven, P. H., R. F. Evert, and Helena Curtis, *Biology of Plants*, 2d ed. New York: Worth, 1976. An excellent and readable general introduction to plant biology.

On the Green Algae and the Fungi

Large, E. C., *The Advance of the Fungi*. New York: Dover Publications, 1962. Although first published over forty years ago, this book is worthwhile because it deals with all aspects of the natural history of the fungi. It makes very interesting, down-to-earth, even humorous reading.

Pickett-Heaps, Jeremy D., *Green Algae*. Sunderland, Mass.: Sinauer Associates, 1975. An outstanding general survey of all the groups of green algae, with detailed diagrams of life cycles, and excellent up-to-date photography, including many electron micrographs.

On the Vascular Plants

Arditti, Joseph, "Orchids," *Scientific American*, January 1966, p. 70. An excellent short account of orchids and some of their intricate adaptations (especially to insect pollination).

Bierhorst, D. W., *Morphology of Vascular Plants*. New York: Macmillan, 1971. A very detailed but up-to-date coverage of all the groups of vascular plants. A great advantage of this work is that it treats fossil plants as well as the modern groups.

CHAPTER 29

General

Alexander, R. M., *The Chordates*. New York: Cambridge University Press, 1975. A thorough account of the phylum Chordata, particularly the vertebrates. Serves as an excellent reference work.

Barnes, R. D., *Invertebrate Zoology*, 3d ed. Philadelphia: W. B. Saunders, 1974. A really superb general textbook. It includes discussions of all the major groups, from the Protozoa to the invertebrate chordates.

Buchsbaum, R., *Animals Without Backbones*. Chicago: University of Chicago Press, 1948. A highly readable, well-illustrated book, which has undergone many editions since its first appearance in 1938. Deals exclusively with the invertebrates.

Hanson, Earl D., *Animal Diversity*, 3d ed. Englewood Cliffs, N.J.: Prentice-Hall, 1972. A good, tightly written paperback that gives considerable insight into how the biologist approaches the problem of animal diversity.

Langston, Wann, "Pterosaurs," *Scientific American*, February 1981, p. 122. A fascinating article about the aerodynamics of flight of these giant flying reptiles and their evolutionary history.

Romer, A. E., *The Vertebrate Story*, rev. ed. Chicago: University of Chicago Press, 1959. Comprehensive coverage of vertebrate comparative anatomy.

Human Evolution

Dodonov, Andrey E., "Early Man in Soviet Central Asia," *Scientific American*, December 1980, p. 130.

Howells, William, *Evolution of the Genus Homo*. Reading, Mass.: Addison-Wesley, 1973. A very carefully written account of human evolution treated largely from evidence of the fossil record.

Leakey, Richard E., *Origins*. New York: E. P. Dutton, 1977. A superbly written and even more superbly illustrated semipopular account of modern thinking on the origin of *Homo sapiens*. This book is particularly valuable because the author goes into considerable detail about the methods of locating, removing, and preparing fossils, as well as the difficulties in interpreting the various remains of the hominids.

Reed, Evelyn, *Woman's Evolution*. New York: Pathfinder Press, 1975. A sensitive and well-presented analysis of the evolution of the human species in terms of women's roles. Argues against the theory of inevitable male dominance. (See by way of contrast Goldberg, *The Inevitability of Patriarchy*, cited for Chapter 17.)

CHAPTER 30

Alland, Alexander, *The Human Imperative*. New York: Columbia University Press, 1972. A book countering the ideas advanced by Robert Ardrey in the two selections listed below.

Ardrey, Robert, *African Genesis*. New York: Dell, 1961. The author, a journalist with considerable writing skill, makes the case for the innateness of human aggression toward fellow humans. Good reading, if accompanied by a massive grain of salt.

_____, *The Territorial Imperative*. New York: Dell, 1966. Like *African Genesis*, a masterful job of extrapolation from lower animals to human societies and political systems. Fascinating, but not to be taken too seriously.

Aronson, L. R., E. Tobach, D. C. Lehrman, and J. S. Rosenblatt, *Development and Evolution of Behavior: Essays in Memory of T. C. Schneirla*. San Francisco: W. H. Freeman, 1970. An excellent collection of essays dealing with the evolution and development of behavior, behavioral processes, social behavior, and human behavior.

Callan, H., *Ethology and Society*. New York: Oxford University Press, 1970. An excellent introduction to the problems of using animal behavior to understand human social behavior.

Dilger, William C., "The behavior of lovebirds," *Scientific American*, January 1962, p. 88. An interesting case study of how behavior affects divergence in a species and thus is related to its evolutionary position.

Gilliard, E. T., "The evolution of bowerbirds," *Scientific American*, August 1963, p. 38. A case study in the evolution of exceedingly complex and elaborate mating behavior. This article shows how such activities can be adaptive for the species.

Hardin, Garrett (ed.), *Science, Conflict, and Society*. San Francisco: W. H. Freeman, 1969. A collection of readings in the area denoted by the book's title. Included is the Calhoun study of the relation of rat population density to abnormal behavior. The Calhoun paper is also available as a *Scientific American* offprint from W. H. Freeman.

Hinde, R. A., *Animal Behavior*, 2d ed. New York: McGraw-Hill, 1970. A text for the serious student of behavior that provides a synthesis of ethology and comparative psychology.

Jolly, A., *The Evolution of Primate Behavior*. New York: Macmillan, 1972. An introduction to the evolution of primate behavior with some discussion of the implications for human behavior.

Kandel, Eric R., *The Cellular Basis of Behavior*. San Francisco: W. H. Freeman, 1976. An excellent and comprehensive treatment of behavior in terms of known cellular, and especially neural, properties.

Klopfer, Peter, and Jack P. Hailman, *An Introduction to Animal Behavior: Ethology's First Century*. Englewood Cliffs, N.J.: Prentice-Hall, 1967. An excellent study of animal behavior from the biological viewpoint. The book is especially good in its historical treatment of the subject.

Lehrman, Daniel S., "A critique of Konrad Lorenz's theory of instinctive behavior," *The Quarterly Review of Biology* 28, no. 4 (December 1953): 337–363. An excellent presentation of the case against the concept of instinct as used by Lorenz.

Lorenz, Konrad, *On Aggression*. New York: Harcourt, 1966. Lorenz's most controversial book, now available in paperback. Fascinating reading, but the case the author makes for the innateness of aggression seems not to be accepted by most ethologists.

_____, "The evolution of behavior," *Scientific American*, December 1958, p. 67. An excellent article describing the modern concept of the evolution of specific behavior patterns.

_____, *Evolution and Modification of Behavior*. Chicago: University of Chicago Press, 1965. Lorenz's reply to criticism directed against his arguments for the innateness of certain forms of behavior, but especially that of D. S. Lehrman cited here.

_____, *King Solomon's Ring*. New York: Crowell, 1952. A paperback book by a classic investigator in the realm of animal behavior. This delightfully written and excellent book covers many topics in animal behavior.

Manning, Aubrey, *An Introduction to Animal Behavior*, 2d ed. Reading, Mass.: Addison-Wesley, 1972. A good paperback introduction to ethology.

Marler, P., and W. J. Hamilton, *Mechanisms of Animal Behavior*. New York: Wiley, 1966. A textbook for the serious student of behavior. Of particular interest is Chapter 17, dealing with the embryology of behavior.

Montagu, M. F. Ashley (ed.), *Man and Aggression*. London and New York: Oxford University Press, 1968. A collection of writings by fourteen ethologists, anthropologists, and the like, with the stated intent to counter misleading impressions given to the general public by the wide distribution of Konrad Lorenz's *On Aggression* and Robert Ardrey's *African Genesis* and *The Territorial Imperative*.

Scott, J. P., *Animal Behavior*, 2d ed. Chicago: University of Chicago Press, 1972. An inexpensive and highly readable description of the experimental study of animal behavior.

Tinbergen, Niko, "The evolution of behavior in gulls," *Scientific American*, December 1960, p. 118. An excellent case study in the evolution of behavior.

Von Frisch, K., "Dialects in the language of the bees," *Scientific American*, August 1962, p. 78. An excellent article describing a communication pattern in honeybees. Comparison of the system in closely related varieties and species shows the divergence of communication patterns due to geographic isolation.

Wilson, E. O., *Sociobiology: The New Synthesis*. Cambridge, Mass.: Harvard University Press, 1975.

Glossary

Abortion. Spontaneous or induced termination of a pregnancy before full term.

Abscission layer. Special layer of small cells that appears at the base of the leaf petiole, fruit stalk, or flower stem. As this layer develops, it weakens the stalk or stem, allowing the leaf, fruit, or flower to fall. The abscission layer is formed only after the cessation of auxin production.

Absorption. The passage of food through the walls of the digestive tract for transportation to other parts of the body.

Acetylcholine (Ach). Neurotransmitter secreted at the axon ends of neurons, both at the neuromuscular junctions and in the parasympathetic nervous system. Acetylcholine transmits a nerve impulse across a synapse.

Acid group. The carboxyl (COOH) group located on many organic molecules, especially amino acids and fatty acids. The acid group ionizes at physiological pH (pH = 7.4) to yield a proton in solution as a hydronium ion: $-COOH + H_2O \rightarrow -COO^- + H_3O^+$ (hydronium).

Acrosome. A caplike structure covering the head of the sperm cell; it appears to help the sperm penetrate the egg membrane.

Actin. Contractile protein that interacts with myosin to produce muscle contraction.

Activation energy. The amount of energy necessary to initiate an exergonic reaction.

Active site. The portion of an enzyme molecule into which a given substrate fits. When the active site is blocked, the enzyme cannot catalyze a reaction with its substrate.

Active transport. The movement of molecules against a free-energy gradient, requiring the expenditure of energy.

Adaptation. For an organism, any change (usually somatic) in its structure or function that allows the organism to better cope with conditions in the environment. For a species, any change (usually genetic, selected for by natural selection) that allows the species as a whole to better cope with its environment.

Adaptive radiation. The evolution by natural selection of a variety of types from one ancestral species.

Adenosine diphosphate (ADP). An ester of adenosine that is reversibly converted to ATP for the storing of energy by the addition of a high-energy phosphate group.

Adenosine triphosphate (ATP). A molecule consisting of a purine (adenine), a sugar (ribose), and three phosphate groups. A great deal of energy for biological function is stored in the high-energy bonds that link the phosphate groups, and it is liberated when one or two of the phosphates are split off from the ATP molecule. The resulting compounds are called adenosine diphosphate (ADP) and adenosine monophosphate (AMP), respectively.

Adrenal gland. A gland on the anterior surface of the kidney.

Aerobe. An organism that requires oxygen to carry on the process of respiration.

Aerobic respiration. A series of reactions for the breakdown of glucose in which the element oxygen serves as the ultimate electron acceptor.

Agglutination. The clumping of red blood cells when they are exposed to agglutinogens in blood of an incompatible type.

Agglutinogen. Blood substance that causes agglutination when introduced into blood of an incompatible type.

Aggressive behavior. Animal behavior in which attack is either initiated or threatened.

Agonistic behavior. The entire sequence of behavioral events in animals associated with aggressive behavior on the part of both the aggressor and the organism against which the aggression is directed.

Alkaptonuria. A relatively benign hereditary disease caused by an autosomal recessive gene. People afflicted with the disease cannot make the liver enzyme homogentistic acid oxidase, so that the intermediate compound homogentistic acid accumulates and is excreted in the urine. On exposure to air, homogentistic acid is rapidly oxidized to a dark brown color, an identifying feature of the disease.

Allantois. In bird and reptile embryos, an extraembryonic membrane for the storage of solid, nondiffusible nitrogenous wastes.

Alleles. Genes that occupy similar loci on homologous chromosomes but carry contrasting inheritance factors. For example, the gene for blue eyes in human beings is said to be an allele to the gene for brown eyes. Also, alleles are two or more genes capable of mutating into one another.

Allometry. Study of the relationship between the growth rate of a part of an individual and the growth rate of the whole individual.

Alternation of generations. A characteristic of the life cycle of certain plants in which a sexual generation alternates with an asexual generation.

Alveolus. An air sac in the lungs, thin-walled and surrounded by blood vessels. The hundreds of thousands of alveoli in each lung serve as the major vehicles for gas exchange in the mammalian body.

Amacrine cells. Retinal cells that act as an intermediary between bipolar cells and ganglion cells.

Amino acid. The basic structural unit of proteins, having the general formula:

$$
\begin{array}{ccc}
H & R & O \\
\diagdown & | & \diagup\!\!\!\diagup \\
N-C-C & & \\
\diagup \quad | & \diagdown & \\
H \quad H & OH &
\end{array}
$$

The name "amino acid" is derived from the fact that a basic amino ($-NH_2$) group and an acidic carboxylic group ($-COOH$) are attached to the same carbon skeleton. The R group varies from one amino acid to another, giving each amino acid its particular characteristics.

Ammonia. A highly toxic and soluble waste product resulting from the deamination of amino acids. In aquatic animals, ammonia passes from the body almost continuously, so that a harmlessly low concentration is maintained. In terrestrial animals, ammonia is converted to other less toxic materials, such as urea, which can be safely stored in the body until excretion.

Amnion (amniotic sac). Transparent, thin, but tough membrane making up the sac that encloses and protects the embryos of mammals, birds, and reptiles. In humans the amniotic sac is the "bag of waters" containing amniotic fluid in which the fetus develops. It acts as a shock absorber and in other ways protects the fetus. In humans, spontaneous rupture of the amnion is usually an indication that labor is about to begin.

Anabolism.   The build-up of more complex substances from simpler ones within a living organism (or single cell).

Anaerobe.   An organism that can carry on respiration in the absence of oxygen. Two types of anaerobes can be distinguished: facultative and obligate. Facultative anaerobes (such as yeasts) respire aerobically or anaerobically, depending on environmental conditions. Obligate anaerobes can carry on only anaerobic respiration, regardless of whether or not there is oxygen in the environment.

Anaerobic respiration.   A series of reactions involving the breakdown of fuel molecules (glucose) and the generation of ATP in the absence of oxygen. The end products of anaerobic respiration can be lactic acid or alcohol, depending on the type of cell.

Analogous.   Term applied to body parts that are similar in function but not in structure, such as the wing of a bird and the wing of a bee.

Anaphase.   The phase of mitosis characterized by the separation and movement of homologous chromosomes toward opposite poles of the dividing cell.

Androgens.   General name for the various male sex hormones, or any substance that has "masculinizing" effects on an organism. Testosterone is the most common naturally occurring androgen. Androgens are produced in the testes of males and in the adrenal glands of both males and females.

Anemia.   A state of deficiency of either the number of circulating red blood cells or the amount of hemoglobin in the red blood cells.

Angiosperm.   Flowering plant bearing ovules within a closed organ, the ovary. During the plant's development, the ovary becomes the fruit and the ovules become the seeds.

Animal pole.   The surface of an egg close to the nucleus where the yolk density gradient within the egg is smallest.

Antheridium.   The male gamete-producing organ of the gametophyte generation in lower plants (some algae, the mosses, and ferns).

Anthropomorphism.   The assigning of human characteristics to non-human forms.

Antibody.   A globular protein produced within leukocytes and other cells in animals in direct response to the presence of a foreign substance (antigen) within the body. Antibodies are specific to the particular antigen that elicited their production. Once produced, many antibodies remain in low but persistent concentration in the blood for a long period of time, conferring on the organism immunity to that particular antigen.

Anticodon ("Nodoc").   Triplet of bases on transfer RNA complementary to the codon of messenger RNA.

Antifinder sequence.   A sequence of four nucleotides on the rRNA of the small ribosomal subunit. The antifinder sequence enables the messenger RNA to correctly bind to the ribosome during protein synthesis.

Antigens.   The general name for any substance that enters the blood stream of vertebrates and other chordates and becomes recognized as a foreign substance. Antigens elicit the production of antibodies, proteins specific for precipitating the antigen from the blood. Antigens may be any class of substances—proteins, carbohydrates, etc.

Anus.   Posterior opening of the digestive system of most animals.

Aorta.   The largest artery of the vertebrate body. It leaves directly from the left ventricle of the heart.

Apical meristem.   The growth region in a plant, located at the tip of each branch, where new cells are produced by rapid cell division.

Apoenzyme.   A protein that forms an active enzyme system by combining with a coenzyme. The apoenzyme usually has the determining effect on the specificity of the enzyme complex.

Archegonium.   The female gamete-producing organ of the gametophyte in all plants. Particularly prominent in gametophytes of mosses, ferns, and their relatives.

Archenteron (gastrocoel).   In embryology, the hollow interior of the gastrula stage forming a primitive gut.

Area-species curve.   A graph showing the number of species plotted against the area they inhabit (an island, mountain plateau, and so forth).

Arteries.   Tubular branching vessels that carry blood away from the heart to various other organs.

Arterioles.   The smallest arteries in the circulatory system.

Asexual reproduction.   Development of new organisms without the fusion of gametes. This may occur in plants by either spore formation or vegetative reproduction. Some animals may reproduce asexually by fission or budding.

Association neurons.   Neurons within the central nervous system that connect sensory neurons to motor neurons. Also called internuncial neurons.

Asters.   Fiberlike processes in the cytoplasm of dividing cells producing "starlike" radiations from the poles, or centrioles.

ATP.   See Adenosine triphosphate.

Atrioventricular node.   Specialized muscle tissue of the heart that conducts impulses from the sinoatrial node to the ventricle walls.

Atrium.   An upper chamber of the vertebrate heart that receives blood from the body and pumps it to the ventricle. Also called auricle.

Auricle.   See Atrium.

Autoimmunity.   Having an immune response to one's own tissues. The immune system responds to "self" as "foreign" tissue.

Autonomic nervous system.   That portion of the central nervous system responsible for carrying out involuntary vital processes. The autonomic system is composed of two parts. The sympathetic system is responsible for integrating the body's many functions during an emergency. The parasympathetic system counteracts the effects of the sympathetic system. Both systems operate to some extent at all times, controlling such functions as the size of the iris diaphragm, salivary secretion, heart rate, peristalsis, and secretion in the stomach and duodenum.

Autoradiography (radioautography).   A process whereby the location of radioactive materials is determined by use of photographic film. When a radioactive emission (such as a β-particle) hits a photographic film, it produces an exposure. The more particles, the "brighter" the exposure. In biology this process is especially useful in tracing substances throughout an organism. The organism is fed or injected with a substance containing radioactive atoms. Parts of the organism (a tissue section from the liver, or a leaf) are then exposed to film. Bright spots on the film reveal the distribution of the radioactive substance in the organ or tissue under observation. The exposed film is called an autoradiograph (radioautograph).

Autosome.   Any chromosome that is not a sex chromosome.

Autotroph.   An organism that can generate its own food supply from simple organic and inorganic elements and some external energy source such as sunlight. Green plants are autotrophs.

Auxin.   The name given to a whole group of growth-regulating substances in plants. A number of different molecules serve as auxins, but all have in common the presence of one or two carbon rings. Auxins have profound effects on the elongation of plant cells and other growth phenomena.

Avogadro's number.   The number of atoms or molecules in a gram atomic weight or a gram molecular weight: 6.023×10^{23} particles.

Axon.   A process of a nerve cell that carries the nerve impulse away from the nerve cell body.

Bacteriophage (phage).   A virus that attacks bacteria. The infecting phage causes the bacterium to produce a new generation of phages, destroying the bacterium in the process.

Basal body.   Structure at the base of a bacterial or eukaryotic flagellum that anchors it to the cell and enables it to rotate.

Basal metabolic rate (BMR).   The rate at which metabolism (metabolic reactions) occurs when the body is at rest. Measured usually as a function of O_2 consumption.

Basidium.   Stalk upon which meiospores of *Basidiomycetes* (club fungi) are produced.

Bilateral symmetry.   A common type of animal body form in which the body can be divided (only) into two equal halves by a single plane of division.

Binomial nomenclature.   The system of naming in taxonomy introduced by Linnaeus. Names consist of both the genus name and the species name for the organism. The human being is classified *Homo sapiens* (the genus is capitalized, the species is not, and both are italicized).

Biogenetic law. A nineteenth-century theory devised by Fritz Müller and Ernst Haeckel. This theory held that the stages of embryological development of a given organism repeat the evolutionary stages through which the species passed. This "law" is often stated "Ontogeny recapitulates phylogeny." At present most biologists question the validity of the biogenetic law.

Biogeography. The study of the patterns of geographic distribution of organisms.

Biomes. Large, easily distinguished community units arising as a result of complex interactions of physical and biotic factors. For example, grasslands or deciduous forests constitute two distinct biomes.

Biotic community. A varied aggregate of organisms existing in a common environment, less extensive biologically or geographically than a biome. Division of labor or competition for food may be internal characteristics of a biotic community.

Biotic environment. The sum total of living organisms with which a given plant or animal comes in contact.

Biotic potential. The inherent power of a population to increase in numbers under ideal environmental conditions.

Bipinnaria. Ciliated, free-swimming larva of echinoderms.

Bisexuality (ambisexuality). Sexual orientation to or possessing characteristics of both sexes (in certain instances synonymous with hermaphroditic).

Blastema. A group of unspecialized cells that cover the area of wound, such as in limb amputation in amphibians. These cells give rise to a new tissue (in the case of amphibians with a limb amputation, a new limb).

Blastocoel. The cavity inside the hollow blastula stage of the animal embryo.

Blastocyst. An early stage of embryonic development that consists of a hollow ball of cells one to two cell layers thick.

Blastomere. A cell of the blastula in animal embryos.

Blastopore. The opening of archenteron to the exterior in the gastrula stage in animal embryos.

Blastula. The hollow, single-layered, ball-like structure forming the first identifiable phase of embryonic development in animals. The blastula is the same size as the original zygote, but it is the result of multiple cell divisions.

Blood. A fluid connective tissue composed of living cells and a nonliving matrix, the plasma. The blood carries oxygen, food, and waste products through the body.

Bulk transport. The transportation of a large collection of molecules at one time across the plasma membrane. The processes of phagocytosis and pinocytosis are examples of bulk transport.

Calmodulin. A protein that binds with calcium ions to activate certain enzymes responsible for nerve and muscle action, cell motility, blood clotting, cell division, and other calcium-mediated activities.

Calorie. The amount of heat required to raise the temperature of one gram of water from 14.5 degrees to 15.5 degrees Celsius.

Calvin cycle. A cycle comprising the dark reactions of photosynthesis. It is a major part of the biochemical pathway by which green plants reduce carbon dioxide to sugars.

Calyx. Collectively, the sepals of a flowering plant.

Cambium. A layer of meristematic cells located between the xylem and phloem in plant stems. The cambium causes increase in width of stems and roots by production of secondary xylem and phloem.

Capillaries. The smallest blood vessels in the vertebrate body, having walls one cell thick.

Capillarity. An effect of the adhesive and cohesive properties of water, by which a water column may be raised in a very narrow tube.

Carbohydrates. Compounds such as sugars, starches, cellulose, glycogen, etc., containing carbon, hydrogen, and oxygen, generally in a ratio that can be expressed as $(CH_2O)n$. Carbohydrates are primary energy foods.

Cardiac muscle. Heart muscle, composed of long, striated cells which often branch and fuse together. The nuclei in cardiac muscle cells are centrally located.

Cardiovascular system. In vertebrates, the system consisting of the heart, arteries, capillaries, and veins.

Carnivore. An organism whose diet consists of meat. Such organisms usually display structural adaptations for meat-eating, such as sharp claws and/or teeth.

Carotenoids. Various colored pigments found closely associated with the chlorophyll in green plants and believed to be accessory to the photosynthetic process.

Carotid arteries. Principal arteries of the body. There are two carotid arteries that branch off the aortic arch; the common carotid arteries branch off the aortic arch and split into the external carotid and the internal carotid.

Carrying capacity. The maximum number of organisms a given environment can support over a long period of time, without damaging the environment.

Catabolism. The degradation of complex organic compounds to simpler ones within living organisms or cells.

Catalyst. Any substance that lowers the activation energy of a system, allowing a given chemical reaction to proceed more rapidly. In living systems, enzymes are the main catalysts.

Catastrophism. A late eighteenth- and early nineteenth-century idea accounting for the changes in flora and fauna indicated by fossil records. According to this idea, from time to time great catastrophes destroyed all life on earth, and after each cataclysm a new special creation populated the earth with new forms of life.

Cell. A discrete mass of living material surrounded by a membrane. The basic structural and functional unit of life in nearly all types of organisms.

Cell division. The splitting of a parent cell into two daughter cells. This process consists of two separate phenomena: division of the cytoplasm (cytokinesis) and division of the nucleus (mitosis). The events of the nuclear mitosis follow a regular pattern of four phases: prophase, metaphase, anaphase, and telophase. Between succeeding nuclear divisions the nucleus is in interphase.

Cell membrane. The phospholipid protein bilayer forming the outer surface of every cell. The membrane is flexible, almost fluidlike, and it regulates the nutrients entering the cell and the waste products or secretions leaving the cell.

Cell plate. A cytoplasmic figure formed during plant cell mitosis at the site where a new cellulose partition will be synthesized to separate the two daughter cells.

Cell wall. A rigid structure composed of cellulose, surrounding plant cells.

Cells of Leydig. Cells of the testes that produce androgens such as testosterone.

Cellular metabolism. The total processes in which food and structural materials are broken down (catabolism) and built up (anabolism) within the cell.

Cellulase. An enzyme capable of splitting cellulose into its monosaccharide components.

Cellulose. A large, insoluble polysaccharide of repeating β-linked glucose molecules. Cellulose is the major component of plant cell walls.

Central nervous system. That part of the complete nervous system consisting of the brain and the spinal cord.

Centriole. A small, deeply staining cytoplasmic structure with a $9 + 0$ complex of microtubules. It is thought that the centriole performs a function in cell division; however, many higher plants that seem to have no centrioles still manage cell division.

Centromere. See Kinetochore.

Cerebellum. A region of the brain that receives and sorts out all the impulses originating in the cerebrum and sends them to the appropriate muscle at the proper time to effect an orderly muscular response.

Cerebrospinal fluid. A fluid lubricating the brain and the spinal cord. The fluid contains mineral salts and traces of protein and sugar, and it may be involved in the nutrition of the nervous system.

Cerebrum.   A part of the anterior portion of the brain which receives sensory stimuli and translates them into the appropriate motor response. The cerebrum also stores information gathered through the action of the senses.

Cervix.   A "neck" of the uterus, which protrudes and opens into the vagina.

Chaetae.   Bristles on marine segmented worms used for moving through the water.

Chemical bonds.   The forces of attraction that hold two or more atoms together in a molecule. Formation of chemical bonds is thought to be due to rearrangement of electron clouds.

Chemoreceptors.   Sensory neurons that respond to a specific type of molecule (such as CO_2, hydrogen ions, hormones etc.). Chemoreceptors are found in the aortic arch, in the hypothalamus, in the nasal epithelium, and on the tongue.

Chemosynthesis.   Synthesis of organic compounds using energy derived from other chemical reactions. For example, the chemosynthesis of carbohydrates or proteins occurs in living cells using energy from the oxidation of foodstuffs.

Chemotropism.   The growth response of a plant toward a chemical substance.

Chiasma (plural, chiasmata).   Figure formed by the intertwining of chromosomes during prophase I of meiosis. Crossing over can occur during chiasma formation.

Chlorophyll.   A molecule based on the same ring structure (porphyrin) as hemoglobin, but with magnesium replacing the central iron atom. Chlorophyll is found in all green plants and gives them their color. The molecule functions in photosynthesis by absorbing specific wavelengths of sunlight. It is now known that light raises electrons of the chlorophyll molecule to higher energy levels. As the electrons return to their original level through a series of acceptor molecules, ATP is generated to serve as the direct energy source for reducing carbon dioxide to carbohydrate.

Chloroplast.   A small plastid present in the cells of green plants. The chloroplast contains chlorophyll, which is essential for the photosynthetic activities of the plant.

Chorioallantois (chorioallantoic membrane).   A membrane found in amniote eggs that functions in gas exchange and excretion of wastes.

Chorion.   In intrauterine development, the outermost membrane surrounding the embryo.

Chorionic villi.   In mammals, projections of the chorion that extend into the uterine wall. These are outgrowths of the embryonic sac (chorion), and they provide the basis for surface exchange of materials between the mother and the very young embryo.

Chromatid.   A term applied to each of the two parts of a chromosome after replication as long as these parts remain connected at the kinetochore.

Circulatory system.   System for transport of nutrients, wastes, water, and secretions through the body.

Cleavage.   A series of successive mitotic divisions and cytoplasm distributions in a newly fertilized egg. It accomplishes an increase in cell number with no change in embryonic mass.

Cleavage furrow.   Groove formed between two daughter nuclei in animal cells during cytokinesis.

Clitoris.   Sensitive structure composed of erectile tissue in female mammals; homologous to the penis.

Clones.   A group of genetically identical cells descended from an original progenitor cell.

Codominance.   When the two alleles for a trait are both expressed in a heterozygote, producing an intermediate phenotype.

Codons.   Series of three nucleotides on either DNA or mRNA, transcribed from DNA, that together code for one amino acid in a polypeptide.

Coelom.   A body cavity in many animals that is lined with mesoderm-derived epithelial tissue. The coelom usually contains the visceral organs.

Co-evolution.   An evolutionary process in which two or more species evolve mutually advantageous adaptations to each other.

Coitus.   Sexual intercourse.

Colinear.   In prokaryotes, a situation in which the sequence of amino acids in the completed polypeptide follows the same sequence of triplet codons in the DNA that coded for that polypeptide.

Colon.   Posterior part of the large intestine that removes water from undigested material and, with the rectum, serves as a repository for feces.

Commensalism.   Relationship between two different species of organisms in which one derives benefit and the other suffers no harm.

Communities.   All of the populations living and interacting together in a given geographical area.

Companion cells.   Nucleated cells located adjacent to sieve tubes in plant phloem tissue and containing the sieve tube cytoplasm.

Competition.   Struggle between organisms for the necessities of life. There are two types of competition: intraspecific (between members of the same species), and interspecific (between two or more different species).

Competitive exclusion principle.   Also known as Gause's hypothesis, it states that two species with similar ecological requirements cannot successfully coexist.

Competitive inhibitor.   A substrate that reversibly combines with the active site of an enzyme and lowers the capacity of the enzyme to interact with its regular substrate.

Complementation.   In genetics, a test to determine whether two mutations occur within the same gene or within the same cistron. If two strains have the same mutant phenotype, the mutation may occur within the same gene or in a different gene altogether. In the complementation test, two mutant chromosomes are introduced into the same cell. If mutations occur within the same gene, the mutant phenotype will still be expressed. If the mutations are in different, nonallelic genes (or different cistrons), the wild-type phenotype can be expressed, since each chromosome contains (or "complements" for) the defective gene in the other.

Conditioned reflex.   A behavior pattern learned through repetition of a sequence of events.

Congestive heart failure.   A condition due to inability of the heart to pump blood forcefully back and forth to and from the lungs. The result is backup of blood in the lungs, resulting in a fatal blockage of breathing.

Conjugation.   A physical association and exchange of materials, leading to reproduction in certain organisms such as the green alga *Spirogyra* and the protozoan *Paramecium*.

Connective tissue.   Tissue composed of isolated cells embedded within a nonliving matrix. Connective tissues support the organism and hold its several parts together. Bone, cartilage, and ligaments are examples of connective tissue.

Control group.   In a biological experiment, the organisms maintained under "normal" conditions to serve as a basis for comparison with the experimental group of organisms, in which some variant condition has been introduced.

Copulation.   Sexual intercourse.

Cork.   The outer protection layer of woody plants, made up of dead cells.

Cork cambium.   A special thin layer of meristematic tissue that produces the cork region on the stems and roots of woody plants.

Corolla.   Collectively, the petals of a flowering plant.

Corpora cavernosa.   Two distinct tracts of spongy erectile tissue, situated longitudinally along the length of the human penis (and that of many other mammals) surrounding the single, central area of erectile tissue, the corpus spongiosum.

Corpus callosum.   Neural tissue that connects the right and left cerebral hemispheres.

Corpus luteum.   The yellow, glandular structure in mammals that develops from an ovarian follicle after the egg has been discharged (ovulated). The corpus luteum is the site of progesterone production and secretion. If the egg is fertilized, the corpus luteum persists throughout pregnancy; if the egg is not fertilized, the corpus luteum deteriorates beginning about the third week of the menstrual cycle.

Corpus spongiosum.   A central tubular mass of erectile tissue in the human penis surrounding the urethra and situated within the outer two masses of erectile tissue, the corpora cavernosa.

Cortex. In plants, the storage tissue of the root or stem. In animals, the outer area of an organ such as the kidney or brain.

Cotyledon. A seed leaf present in embryonic plants used for food storage.

Cowper's glands. A pair of glands that lie at the base of the erectile tissue of the penis and contribute the final liquid component, a mucous alkaline secretion, to the seminal fluid.

Cristae. Projections into the central matrix of a mitochondrion, produced by the repeated invagination of the inner mitochondrial membrane and serving to increase the membrane surface area within the mitochondrion.

Crossing over. Exchange of chromosome segments between maternal and paternal chromatids during tetrad formation.

Cuticle. This epidermal covering of almost all plants and some animals (such as flatworms, and arthropod exoskeletons) that protects them from drying out or from being digested (in the case of parasites).

Cutin. The waxy secretion from leaf epidermal layers that forms the leaf cuticle.

Cyclic 3′,5′ AMP. Substance produced from ATP that mediates the effects of hormones on enzyme function and cellular change.

Cytochrome. A molecule in the respiratory assembly with the characteristic porphyrin ring structure found also in hemoglobin, myoglobin, and chlorophyll. By contrast with hemoglobin, in cytochrome the central iron atom is easily oxidized and reduced. This allows the cytochrome to pass electrons in the electron transport chain.

Cytokinesis. Cytological changes, usually occurring along with mitosis, through which the cytoplasm of one cell is divided to form two cells.

Cytology. The study of cells.

Cytoplasm. All the liquid colloidal material in the cell that is enclosed within the plasma membrane, excluding that of the nucleus. Cell organelles reside in the cytoplasm of the cell.

Deamination. The removal of the amino group ($-NH_2$) from an amino acid by chemical oxidation.

Decomposers. Fungi and some bacteria that enable recycling of materials within a food web by converting organic remains of dead organisms into usable inorganic matter.

De-differentiation. The reversion of a cell from a condition of specialization to a nonspecialized, embryonic type of state, as is often the case in cancer cells.

Deductive logic. A process whereby a conclusion is reached by proceeding from a generalization to specific instances.

Degradation. The process of breaking down complex molecules to simpler ones, generally accompanied by a liberation of energy.

Dehydration synthesis. The joining together of small units (such as amino acids or glucose) into a single, large molecule by the elimination of water. One of the units contributes the H^+, the other the OH^-.

Deletion. The loss of a segment of a chromosome during crossing over, resulting in a certain phenotypic deficiency in the developing organism. Such aberrations can provide significant clues in the mapping of gene loci.

Deme. A small, local subgroup within a population that actually interbreeds and can exchange genes with other demes of the same population or species.

Demographic transition. A phenomenon occurring in major Western countries in which the human birth rate decreases following industrialization and urbanization.

Dendrite. A process of a nerve cell body that conducts the nerve impulse toward the nerve cell body.

Density. The number of organisms of the same type in a given unit of space.

Density-dependent factors. Factors operating on individuals to limit or reduce a population when it reaches some critical size. Density-dependent factors include physiological or behavioral changes in individuals that cause migration, reduced mating practices, etc.

Deoxyribose. The five-carbon sugar in DNA (deoxyribonucleic acid).

Depolarization. The condition of a membrane when the distribution of charged ions is approximately the same on both sides, so that that potential across the membrane is zero.

Desiccate. To dry out by losing water (desiccation).

Determinate cleavage. Embryonic cleavage of lower animals in which the fate of the daughter cells (blastomeres) can be predicted.

Dialysis. The process whereby compounds or substances in a heterogeneous solution are separated by the difference in their rates of diffusion through a semipermeable membrane. For example, if a solution of sodium chloride and albumin is placed inside a dialysis bag immersed in water, the sodium chloride ions will diffuse outward about twenty times faster than the albumin. Thus, after a short time, the solution inside the dialysis bag will be mostly protein.

Diaphragm. A contraceptive device used by the female for birth control. The diaphragm is a rubber-covered ring designed to fit over the cervix, preventing sperm from entering the uterus.

Diastole. The phase of the heart's pumping cycle in which the ventricles relax.

Diatom. A form of marine alga living within a tiny, silicon-containing shell. Diatoms produce huge amounts of organic materials by carrying on photosynthesis with the brown pigment fucoxanthin.

2,4-D (2,4-dichlorophenoxyacetic acid). One of a class of growth-promoting substances known as auxins, composed of a single carbon ring and a side chain:

The auxin, 2,4-D, which may be synthesized artificially, acts as a weed killer by selectively stimulating rapid growth in broad-leaved plants (most weeds are broad-leaved). Growth is so rapid that the plants are eventually killed.

Dicotyledonae (dicots). One of the two classes of angiosperms. Distinguishing characteristics are two cotyledons (food leaves) in the embryo, netted venation in the leaves, flower parts in twos, fours, or fives, and fibrovascular bundles in an orderly array in the stem.

Dictyosome. A stack of membranous vesicles similar to the Golgi complex and found in certain animal and higher plant cells. The function of dictyosomes is uncertain, but at least in plants it appears to be related to synthesis of polysaccharides.

Differential fertility. A measure of the success of an inherited variation in terms of its effects on the reproductive capacities of an organism. A variation that increases reproductive capacity is considered a successful one.

Differentiation. The structural or functional changes that occur in cells during the embryonic development of an organism.

Digestion. The enzymatic breakdown of food from large molecules into small ones capable of entering the bloodstream, and eventually the cells, by a process of absorption. Important organs of the digestive system include the stomach, pancreas, gall bladder, and small intestine.

Diploid. Term applied to a cell that contains a pair of each type of chromosome. The number of chromosomes usually given for an organism is the diploid number; hence the human being has 46 chromosomes (23 pairs).

Dispersion. In statistics, the spread of values surrounding the mean.

Distance effect. A generalization of population biology which states that the more geographically isolated an area is, the less it is likely to be colonized by new species.

Dominance hierarchies. "Peck orders" in animals, in which social hierarchies serve to maintain order. A baboon troop manifests a dominance hierarchy.

Dominant. In genetics, a term used to refer to a gene that always expresses itself over its recessive allele in the heterozygous condition.

Dorsal. The upper side of an animal.

Duodenum.   That part of the small intestine closest to the pyloric valve of the stomach. Under the influence of hydrochloric acid from the stomach, the cells of the duodenum produce secretin, which stimulates enzyme release in the pancreas. A large part of the enzymatic breakdown of food occurs in the duodenum, which in the human being is approximately ten inches long.

Dynamic equilibrium.   A state in which the concentration of reactants and products in a chemical reaction remains constant, though not necessarily in equal quantities over time.

Ecological niche.   See Niche.

Ecology.   The study of the relationship between plants and animals, and their environment.

Ecosystem.   All the interacting factors, both physical and biological, forming a biotic community.

Ectoderm.   In an embryo, the outer germ layer giving rise to the epidermis, the neural tube, and the epithelial lining of vertebrates.

Electrophoresis.   In biochemistry, a process used to separate different kinds of organic molecules from each other in a mixture. Electrophoresis takes advantage of the differences in overall net electric charge on different kinds of molecules. When a mixture of molecules is placed on a moist surface (such as a gel) through which an electric current passes, molecules with an overall net positive charge move toward the negative pole (cathode) and those with a net negative charge move toward the positive pole (anode). Rates of migration also vary: molecules with more negative charges migrate toward the positive pole more rapidly than those with less negative charges, etc. In this way different types of molecules tend to separate out at different regions along the gel or other surface.

Elongation region.   The region of a root behind the embryonic area, where cells increase in length.

Embryology.   The study of the structural and functional development of an organism during its early life.

Embryonic induction.   The ability of one type of embryonic germ layer to trigger or specifically influence the differentiation of another germ layer (which usually lies in direct contact with it).

Embryonic region.   The growth region in roots, located just behind the root cap, where new cells are produced by rapid cell division.

Emerson enhancement effect.   Principle stating that the rate of photosynthesis in the presence of two wavelengths of light is greater than the sum of the rates when the two wavelengths are applied separately.

Encoding sequence.   Two or more segments of DNA, separated by a nonsense intervening sequence, that together code for a particular polypeptide in eukaryotic and animal virus genes.

Endergonic.   Term applied to those chemical reactions that result in an overall increase in energy among the formed products and hence the storage of energy. Photosynthesis is an endergonic reaction.

Endocrine system.   All the hormone-secreting glands and tissues in the body.

Endoderm.   In an animal embryo, the innermost germ layer, which gives rise to the lining of the gut.

Endodermis.   In plants, a layer of cells separating the cortex of a root from the central cylinder (stele).

Endometrium.   The thick, vascular, mucus-rich inner lining of the uterus.

Endoplasmic reticulum.   In cells, a maze of membranes in the cytoplasm, at places continuous with the nuclear envelope. The endoplasmic reticulum may serve to increase the surface area of the cell and thus aid in exchange of material.

Endorphins.   Substances involved in the body's intrinsic mechanism for inhibiting pain sensation.

Endosperm.   In plants, a triploid cell or cells developing within the embryo sac immediately after fertilization and serving as the nutrient for the developing embryo.

Endothelium.   A layer of smooth, thin, flattened epithelial cells that line the heart and the blood vessels in all mammals.

Energy barrier.   The amount of energy that any non-excited atom or molecule must gain in order to become "excited" and enter into a chemical reaction.

Enthalpy.   The heat content of matter.

Environmental resistance.   Factors in the environment that oppose or limit the increase in numbers of a given population.

Enzyme.   A protein, the synthesis of which is controlled and directed by a specific gene. Enzymes act as catalysts, directing all major chemical reactions in the living organism.

Epidermis.   In animals, the outer covering of the body, usually several cell layers thick. In herbaceous plants, the thin (one cell thick) outer covering, which secretes a noncellular, waxy cuticle.

Epididymis.   A single coiled tube lying on top of the mammalian testes in which sperm is stored.

Epigenesis.   In embryology, the idea that an entire organism develops from an originally undifferentiated mass of living material.

Epiglottis.   In mammals and other vertebrates, a fleshy flap on the ventral wall of the pharynx that blocks the opening to the trachea during swallowing.

Epinephrine (adrenaline).   The primary hormone of the adrenal medulla. Epinephrine functions in cardiac stimulation, dilation of blood vessels leading to skeletal muscles, and constriction of blood vessels leading to visceral organs.

Epistasis.   A situation in which the products of a gene pair at one locus alters the expression of a gene pair at another locus.

Epithelial tissues.   Surface and lining tissues of the animal body, such as the lining of the digestive tract and the lining of the air passage to the lungs.

Equilibrium phase.   The stage where growth rate in a population of cells or organisms has leveled off, so that the appearance of new cells or organisms just equals the disappearance of old ones.

Erythrocyte.   Red blood cell. Erythrocytes contain hemoglobin (hence their red color) and serve as oxygen carriers.

Esophagus.   The structure conveying ingested material from the pharynx directly to the stomach.

Ester bond.   The anhydrous bonds in lipids formed by removing a hydroxide from the carboxyl group of a fatty acid and a hydrogen from an alcohol group of gycerol.

Estrous cycle.   The recurrent, restricted periods of sexual receptivity in the nonhuman mammalian female, marked not only by egg production but also by increased sexual drive.

Ethology.   The study of animal behavior.

Eugenics.   The attempt to improve the human genetic stock by encouraging breeding of those presumed to have "desirable" genes (positive eugenics) and discouraging breeding of those presumed to have "undesirable" genes (negative eugenics).

Eukaryotes (also eucaryotes).   Cells characterized by true nuclei bounded by a nuclear membrane. The cells of all protozoa and higher animals, most algae (except blue-greens) and higher plants are eukaryotes (that is, they are eukaryotic).

Eutrophication.   The successional process leading from an aquatic system with low productivity through increasingly greater productivity to the ultimate develment of a terrestrial system on the same site.

Evolution.   Gradual, sequential, genotypic changes that occur in populations of organisms over successive generations (also known as transmutation).

Excitatory postsynaptic potential (EPSP).   Type of synaptic response in which the presynaptic neuron releases a neurotransmitter that causes a postsynaptic neuron to depolarize.

Excretion.   The removal of the waste products of metabolic activity. In higher organisms, the blood bathes each cell and carries away waste. The waste material is removed from the blood by the kidneys, the sweat glands, and the lungs.

Exergonic.   Term applied to those chemical reactions in which the end products have less energy than the reactants. Exergonic reactions give off energy. Respiration is an exergonic reaction.

Exoskeleton.   Chitinous external skeleton found in arthropods.

Experimental group.   In an experiment, the objects or organisms that are subjected to alteration.

Expiration.   Process of exhaling air from the lungs.

Exponent.   An integer written slightly above and to the right of a number to indicate how many times the number is to be multiplied by itself in the given expression; for example, $x^2 = x \cdot x$, $4^3 = 4 \cdot 4 \cdot 4$.

Extinct. No longer present in the world population of organisms.

Extrapolation. The calculation of a value or prediction of an event beyond a given series of values or events, based on observing the trend up to a certain point (for example, extrapolating tomorrow's weather based on the trend of the weather over the past 5 days).

Facilitated transport. The movement of substances across a cell membrane from an area of higher concentration to one of lower concentration more rapidly than would occur by simple diffusion, but without the expenditure of energy.

Fallopian tubes. In humans and other mammals, the name for the oviduct, a connecting passageway by whch the ova (eggs) from the ovary are carried to the uterus. In human beings, fertilization usually takes place in the Fallopian tubes.

Fats. Lipid compounds composed of glycerol and fatty acids. Fats are energy-rich compounds, often stored in adipose (fat) tissue.

Fatty acid. Organic molecule composed of a long hydrocarbon chain and terminal acid (carboxyl) group.

Feces. Bodily waste that passes through the colon and is discharged through the anus.

Feedback mechanism. A self-regulating mechanism within all homeostatic systems. Part of the output of the system is cycled back into the system itself in order to regulate further function and output, as, for example, in a thermostat–furnace system.

Ferredoxin. An iron-rich protein found in photosynthetic organisms. During the light reactions this compound is capable of accepting free electrons and passing them through a reducing system to generate reduced NADP (that is, NADPH) and ATP.

Ferredoxin-reducing substance (FRS). Hypothesized primary electron acceptor after light absorption by chlorophyll. FRS passes electrons from chlorophyll to ferredoxin.

Fertilization. In sexual reproduction, the union of the male (sperm) and female (ovum) gametes to form a diploid cell (the zygote) capable of developing into a new organism.

Fertilization membrane. A protective membrane that surrounds the egg once it has been fertilized by a single sperm and prevents multiple fertilizations.

Fertilizin. A carbohydrate and protein substance produced by the egg, capable of causing the agglutination of sperm or the binding of sperm to egg.

Fetus. Term applied to a human embryo after eight weeks of development.

Fibrin. A long protein polymer composing the fibrous part of blood clots. Fibrin is produced from the soluble blood protein fibrinogen by the action of the enzyme thrombin.

Fibrous root. A fairly diffuse network of roots well suited for absorption of water near the surface of the ground. Fibrous systems, characteristic of the grasses, do not penetrate deeply into the soil.

Fibrovascular bundle. Units of vascular tissue grouped together. Fibrovascular bundles include xylem and phloem elements, cambium, and usually some supporting tissue.

Finder sequence. A sequence of four nucleotides that precede the initiator codon on messenger RNA. They enable the *m*RNA to correctly bind to the ribosome (*r*RNA) during protein synthesis.

Fission. A rapid and efficient method of reproduction, found in many microorganisms, which involves the splitting by mitosis and cell division of one cell into two, each of which is genetically identical to the other.

Flagellin. Contractile protein found in the flagella of bacteria.

Flagellum (plural, **flagella**). A long, whiplike extension of cytoplasm from unicellular organisms such as *Chlamydomonas*, or most animal sperm. An outer membrane encloses a highly structured matrix which surrounds microtubules arranged in the familiar 9 + 2 pattern found also in cilia (a 9 + 0 arrangement is found in centrioles). The function of flagella is movement of the cell.

Follicle-stimulating hormone (FSH). A gonadotropin, one of the hormones secreted by the anterior pituitary gland which in mammalian females stimulates growth of ovarian follicles into estrogen-secreting glands (that is, corpus luteum). In males FSH stimulates spermatogenesis.

Food chain. The transfer of energy from one organism to another, starting with the primary producers (photosynthetic organisms), or through primary consumers (herbivores), secondary consumers (carnivores), and decomposers.

Food web. A complex set of interactions within an ecosystem in which any given species usually serves as both prey and predator. All living forms in the ecosystem have a variety of functions. Thus there is no single one-way street of food consumption hierarchies.

Foramen ovale. An opening in the fetal heart allowing blood to pass directly from the right to the left atrium, bypassing the pulmonary transit.

Foreskin. Also called the prepuce, the fold of skin covering the head of the penis in human males.

Founder effect. When a subgroup of a population becomes isolated from the rest of the population (through migration or geological change) and begins to diverge from it genotypically and phenotypically.

Frond. The leaves of a fern growing above ground from the rhizome.

FRS. See Ferredoxin-reducing substance.

Fruit. The ripened ovary of a flower, aiding in seed dispersal.

Furrowing. The infolding of the cell membrane by an animal cell during telophase.

Gametangia. Specialized cells of certain primitive plants that are morphologically differentiated to produce male and female gametes.

Gamete. Male or female reproductive cell containing half the total number of chromosomes (that is, the haploid number) for any given species. These germ cells are formed by the process of meiosis (reduction division) from diploid cells.

Gametogenesis. The production of haploid-differentiated gametes from specialized cells of gonads.

Gametophyte. The small, photosynthetic, haploid stage in the life cycle of lower plants such as mosses and ferns. The gametophyte contains male and female gamete-producing organs.

Gastrin. A hormone secreted by cells in the stomach wall in response to the presence of food in the stomach. The hormone stimulates the secretion of gastric juices into the stomach for the breakdown of food.

Gastrocoel. The primitive gut. It is a cavity formed by the infolding of the side of a blastula-stage embryo.

Gastrodermis. The inner cell layer of cnidarians (coelenterates).

Gastrula. The embryonic stage produced by gastrulation. A hollow structure generated by infolding of the blastula and consisting of two embryonic germ layers.

Gastrulation. The process of embryonic development, produced by infolding of the blastula to form the next embryonic stage, the gastrula.

Gause's hypothesis. The principle of competitive exclusion, which states that two species with similar ecological requirements cannot successfully live together for any length of time because of their competition for basic requirements for life.

Gemmule. Reproductive cells of freshwater sponges that consist of amebocytes covered by a hard outer coating that resists cold, drought, and so forth.

Gene. A part of the hereditary material located on a chromosome. The term "gene" was first used by Johannsen to mean something in a gamete which determined some characteristic of an adult. The gene concept has been refined to the idea of a single gene as the source of information for the synthesis of a single polypeptide.

Gene pool. The total genetic makeup of a population, consisting of all the alleles existing in that population at any given time, regardless of their proportions. This concept provides a way of looking at the possible genetic changes that may occur in a freely breeding population from one generation to another. If 100 organisms (composing a hypothetical population) have 10 genes each (with 2 alleles for each gene), then the gene pool of that population consists of 2000 genes.

Genetic code.   The linear sequence of bases along a DNA molecule which in turn determines the sequence of amino acids in a polypeptide chain. The code itself consists of triplet groups, each specific three bases coding one amino acid. It has been found that more than one triplet can code for the same amino acid; thus the term "degenerate" is used in reference to the genetic code.

Genetic distance.   A measure of how closely related two individuals are, based on genotypic similarity.

Genetic drift.   A condition in which one allele becomes fixed in a population. Genetic drift is thought to be a factor in many small or migratory populations, and perhaps under certain conditions in large populations as well.

Genetic equilibrium.   The maintenance of a more or less constant ratio between the different alleles in a gene pool from generation to generation (often stated as the Hardy–Weinberg Law).

Genotype.   The genetic makeup of an organism; what alleles it actually contains and can pass on to its offspring.

Geographic isolation.   The physical division of an original population into geographically separate groups. Such isolation is usually followed by divergence and eventually speciation.

Geotropism.   The growth response of a plant to gravity.

Germ plasm.   The term developed by August Weismann to describe the reproductive cells (of testes and ovaries) which directly produce the gametes for the next generation. Weismann visualized the germ plasm as being immortal, having continuity with the germ plasms of the preceding and succeeding generations through the process of sexual reproduction. Changes in somatoplasm (body cells) do not affect the germ plasm.

Germinal ridges.   Precursors of the primordial gonads in human and most mammalian embryos. Germinal ridges develop as thickenings of the mesodermal epithelium lining the primitive coelom.

Germination.   Growth of a seed under favorable conditions.

Gestation period.   The period required for a mammalian embryo to develop from fertilization until birth.

Gibberellic acid.   The active agent, extracted from the fungus *Gibberella*, which can cause hyperelongation of the stem in certain kinds of plants. Several structural variations of gibberellic acid are known and are called, collectively, the gibberellins.

Gill slits.   Series of openings in the pharynx of all chordate embryos that, in fish, will develop into gills.

Glandular epithelial cells.   Cells of columnar epithelium specialized for secretion, as in the case of mucus- or wax-secreting cells.

Glans penis.   The central mass of erectile tissue at the tip of the penis.

Glomerulus.   A mass of capillaries enclosed within each Bowman's capsule in the kidney. This structure is responsible for the filtration of waste materials from the circulating blood.

Glycerol (glycerine).   A compound with the formula

$$
\begin{array}{c}
CH_2OH \\
| \\
CHOH \\
| \\
CH_2OH
\end{array}
$$

which combines with fatty acids to form fats.

Glycolysis.   The initial stage of respiration, involving the breakdown of glucose to pyruvic acid. In aerobic respiration, glycolysis yields pyruvic acid for the citric acid cycle. In anaerobic respiration the pyruvic acid is converted to lactic acid (as in bacteria and the muscle cells of higher organisms), or into ethyl alcohol (as in yeasts).

Glycosidic bond.   The anhydrous bonds of carbohydrates formed by removing a hydrogen from an alcohol group of one sugar and a hydroxide from an aldehyde group of the other. Glycosidic bonds hold together the subunits (such as glucose) making up long-chain polysaccharides.

Golgi complex (Golgi body).   A cluster of flattened, parallel, smooth-surfaced membranous sacs found within the cytoplasm. The Golgi complex appears to function in isolating, packaging, and transporting molecules out of the cell.

Gonadotropic-releasing hormone.   Neurohormone produced by the hypothalamus that regulates the release of gonadotropic hormones from the anterior lobe of the pituitary gland.

Gonadotropins.   General name for the group of hormones produced and secreted either by the pituitary gland of both sexes or by the placenta during pregnancy in the mammalian female. Gonadotropins act as stimulators of the reproductive organs; their increased production by the pituitary at puberty is responsible, in part, for the growth of the ovaries or testes.

Graafian follicles.   Vesicles in the ovary in which the eggs are produced. One egg is produced per follicle. After ovulation, a follicle becomes transformed into the corpus luteum, producing progesterone which builds up the uterine lining.

Grana (singular, granum).   Dense stacks of membranes that are part of the lamellar system within chloroplasts. Grana were the first structures discovered within the chloroplast by means of an electron microscope. Chlorophyll molecules are attached to the lamellae and the grana.

Granulocytes.   One of the three types of leukocytes in vertebrate blood. Granulocytes have a bilobed nucleus and their major function is phagocytosis.

Gray crescent.   A surface region in the newly fertilized egg of frogs and salamanders from which the orientation of the developing embryo can be ascertained.

Gross primary productivity.   The total amount of radiant energy taken up by the producers in an ecosystem in a given area and time.

Guard cells.   Cells surrounding leaf stomata and serving to regulate the size of the opening.

Gullet.   An infolded pocket in the paramecium used for capturing food particles.

Gymnosperms.   The conifers and related species of plants that have naked seeds (not completely enclosed, usually borne in a cone).

Habitat.   The surroundings in which an organism resides—the organism's "address" in the biological community.

Habituation.   A simple form of learning in which an animal is repeatedly presented with a stimulus. No punishment or reward results, so the animal stops responding to the stimulus entirely.

Half-life.   The amount of time required for half the atoms of a given radioactive sample to decay to a more stable form.

Haploid.   A condition in which an organism (or a single cell) bears only one copy of each gene. Most higher organisms have two copies (alleles) of the gene for any given character and thus are called diploid. Many microorganisms, however, such as bacteria, *Paramecium*, and most algae, are haploid during most of their life span.

Hardy–Weinberg law.   The generalization that the frequencies of both genes and genotypes will remain constant from generation to generation in a large, freely breeding population in which there is no selection, migration, or mutation. The Hardy–Weinberg law is stated mathematically as the expansion of the binomial, where one member of a pair of alleles is designated p, and the other q: $p^2 + 2pq + q^2 = 1.0$. This mathematical expression means that the frequencies of the alleles symbolized as p and q remain constant (at whatever their initial values) from generation to generation as long as the conditions mentioned above prevail. The Hardy–Weinberg law is sometimes referred to as the law of genetic equilibrium.

Heme.   A complex molecular structure (a porphyrin ring) in which a central ion is capable of undergoing repeated oxidation and reduction. The heme structure is the basis of such important biological molecules as hemoglobin, myoglobin, and the cytochromes.

Hemimetabolous development.   Incomplete metamorphosis in insects. The immature form differs conspicuously from the adult as, for example, in certain insects with aquatic larvae that have compound eyes and external wings like the adult but possess other immature traits.

Hemocoel.   Cavity where blood collects in animals with an open circulatory system.

Hemoglobin.   A red, iron-containing protein pigment in erythrocytes that transports oxygen and carbon dioxide.

Hemorrhage. Bleeding; loss of blood from the circulatory system.

Hepatic portal vein. In most vertebrates, the vein that carries blood from the intestines to the liver. Like all portal systems, the hepatic portal system begins in capillaries (in the intestinal walls) and ends in capillaries (in the liver).

Hepatocyte. An epithelial parenchymatous cell of the liver.

Herbaceous. Term applied to plants that do not contain lignin within their cell walls. These plants are less sturdy than woody plants, which do contain lignin.

Herbivore. An organism whose diet consists exclusively of vegetation.

Hermaphroditism. A state characterized by the presence of both male and female sex organs in the same organism. An individual organism with both male and female sex organs is called a hermaphrodite.

Heterogametes. Two gametes, structurally dissimilar, capable of fusion to form a zygote. The sperm and egg are examples of heterogametes.

Heterogamy. The condition in which gametes are differentiated into two distinct forms (generally male and female).

Heterosexuality. Sexual attraction to the opposite sex.

Heterotroph. An organism that depends on its environment for a supply of nutritive material to build up its own organic constituents and also for general energy requirements.

Heterozygous. Term for the condition in which the two members of a pair of genes located on homologous chromosomes and influencing a given characteristic are different.

Histogram. A graphical representation of statistical data showing frequency distribution by means of a series of rectangles. A bar graph is an example of a histogram.

Histology. The study of tissues.

Homeostasis. The dynamic equilibrium processes that maintain a relatively constant internal environment in the face of variations in the external environment.

Homologous. In anatomy, the term applied to body parts that are similar in stucture but not necessarily in function, such as the arm of a human being and the front leg of a horse.

Homologous chromosomes. The pair of structurally similar chromosomes within a diploid cell which carry inheritance factors influencing the same traits.

Homometabolous development. Complete metamorphosis; for example, in insects, egg to larva to pupa to adult.

Homosexuality. Sexual attraction to the same sex. Female homosexuality is referred to as lesbianism.

Homozygous. Term for the condition in which the two members of a pair of genes located on homologous chromosomes and influencing a given characteristic are identical.

Horizonal cells. Cells in the retina which connect a single photoreceptor cell to several bipolar cells.

Hormone. A chemical substance produced in small quantities in one part of an organism and profoundly affecting another part of that organism. Chemically, hormones may be proteins (insulin), steroids (estrogens), or small metabolites (thyroxin).

Hyaline layer. A layer formed between the inner and outer membranes of an egg. It results from the movement of the outer membrane toward the sperm.

Hydrolysis. The chemical breakdown of a larger molecule into smaller units by the addition of water.

Hydrostatic pressure. In the circulatory system of animals, pressure created by the pumping action of the heart. Hydrostatic pressure aids in forcing water out of the capillaries and into interstitial spaces.

Hymen. A thin membrane that often partially encloses the vaginal opening. The hymen has no known function.

Hypertension. A physiological state characterized by blood pressure that is consistently above normal.

Hypocotyle The part of the plant embryo that will become the stem in the adult.

Hypothalamo-pituitary system. Control centers of hypothalamus and pituitary gland that regulates production and secretion of certain hormones.

Hypothalamus. A region of the brain below the thalamus controlling certain vital bodily functions, predominantly those involving the autonomic nervous system, such as regulation of body temperature.

Hypothesis. A tentative explanation suggested to account for observed phenomena.

Immunology. The field of biology dealing with the process whereby organisms develop a chemical resistance within their bodies to various types of foreign substances (bacteria, viruses, pollen, etc.).

Imprinting. The rapid fixing of social preferences; for example, the tendency of a duckling to follow the first moving object it sees after hatching.

Inclusion. Any of a number of nonliving structures occurring within the cytoplasm of a cell.

Indeterminate cleavage. Embryonic cleavage of higher animals in which the fate of the daughter cells (blastomeres) is not fixed.

Indole-3-acetic acid (IAA). A plant hormone. The most common of the auxins, often referred to merely as auxin.

Induction, embryonic. See Embryonic induction.

Inductive logic. A process whereby a conclusion is reached by proceeding from specific cases to a generalization.

Ingestion. The intake of food into an organism.

Inhibitory postsynaptic potential (IPSP). Type of synaptic response in which the presynaptic neuron releases a neurotransmitter that inhibits depolarization of a postsynaptic neuron.

Initiation complex. The complex of messenger RNA—formyl methionine—transfer RNA with ribosomal RNA, correctly bound together to initiate the translation phase of protein synthesis.

Innate behavior. Ethological term that replaces the concept of instinct. In general, innate behavior is that which is inherited.

Innate releasing mechanism (IRM). Hypothesized receiving center in the central nervous system which is triggered by some external stimulus and elicits a specific sequence of behavioral events.

Insertion. The point of attachment of a muscle to a bone whose movement is controlled by the muscle. Usually at the distal end of the muscle.

Inspiration. The intake of air from the outside due to an imbalance of air pressure created by an expansion of the thoracic cavity.

Instinct. Term applied to behavior that is primarily "genetic" in nature and seems less amenable to change through learning. Most behaviorists prefer to use the term "innate behavior" instead of "instinctive behavior."

Internal environment. The conditions generated inside an organism by the functions and interactions of cells, tissues, organs, and systems.

Interphase. The longest individual phase of mitosis, often considered the resting phase between two active cell divisions, during which the genetic material of the cell is being duplicated.

Interpolation. Filling in a predicted value on a graph between two determined values.

Interstitial cells. In the mammalian testis, cells that produce and secrete the male sex hormones.

Intervening sequence. Group of nonsense condons between two or more segments of DNA that code for a particular polypeptide in eukaryotic and animal virus genes.

Intrauterine device (IUD). Birth control device made of soft, flexible plastic or metal such as copper. It is not known precisely how the IUD works, but it seems to interfere with the implantation of the young embryo in the uterine wall.

Intrinsic rate of increase. The biotic potential of a population or species. The rate of increase a population would have under hypothetical conditions of unlimited resources.

Inversion. The process of chromosome breakage and rejoining in such a way that a whole segment breaks from a chromosome and is replaced in reverse order. Thus, if the sequence of genes on a normal chromosome were represented as abcdefghi, and two breaks occurred, one between c and d and the other between g and h, the middle segment defg might turn and rejoin in reverse direction, so that the sequence of genes on the chromosome would read: abcgfedhi.

Invertebrates.   Organisms characterized by the lack of a notochord. Invertebrates compose approximately nineteen animal phyla.

IRM.   See Innate releasing mechanism.

Islets of Langerhans.   Groups of cells in the pancreas that secrete insulin. Lack of function of the islets causes diabetes.

Isogametes.   Two gametes which are morphologically alike and capable of fusion to form a zygote. Isogametes are not visibly differentiated into male or female forms.

Isogamy.   The fusion of two morphologically similar haploid cells that act as gametes. This occurs in some simple plants.

Isolating mechanisms.   Differences introduced by speciation and divergence during geographic isolation of two populations that prevent interbreeding should the populations come back together. Such mechanisms include seasonal isolation, ecological isolation, physiological isolation, and behavioral isolation.

Karyotype.   Characterization of the chromosome complement of an individual organism with regard to the number, size, and shape of chromosomes present. Karyotypes are usually displayed as a series of chromosomes lined up in a specific order, showing each of the chromosome pairs present (including extra or missing chromosomes).

Kilocalorie (Calorie).   The amount of heat required to raise the temperature of one kilogram of water from 14.5 to 15.5 degrees Celsius. The kilocalorie is used to measure the metabolism (energy turnover) of animals.

Kinetic energy.   Energy of motion; energy in the process of doing work.

Kinetochore (centromere).   The region of the chromosome at which the chromatids remain connected following chromosome replication and to which the spindle microtubules appear to be connected during cell division.

Krebs cycle (citric acid cycle).   A series of reactions in the oxidation of pyruvic acid in which large amounts of energy are released. The hydrogens from pyruvic acid supply electrons for the electron transport chain.

Labia.   In human female genitalia, the thin, pink folds of epithelial tissue that lie lateral to the vaginal opening. Homologous to the male scrotum. There are two labial structures: labia minora and labia majora.

Lamarckism.   An evolutionary theory proposed by Jean Baptiste Lamarck in the eighteenth century, consisting of several basic ideas: Because of new physical needs, new structures arise and old structures are modified; use or disuse of parts causes variation in a structure; and acquired characters can be transmitted to succeeding generations.

Lamellae.   Membranous sacs stacked parallel to each other within the chloroplasts of green plant cells. There are two types of lamellae: stroma lamellae and grana lamellae. Together they are the site of the light reactions of photosynthesis.

Larva.   An intermediate free-living form of some organisms. In this stage the organism eats extensively and stores food. It then undergoes metamorphosis into the adult form, using the stored food as an energy source.

Lateral geniculate body.   A structure located in the thalamus which receives numerous axons from the optic nerve and passes them to the visual area of the cortex.

Leaf primordia.   Structure within the terminal bud of a plant from which the leaves develop.

Legumes.   A certain class of vegetables (including peas, beans, and clover) that bear seeds in pods. Legumes act as hosts for nitrogen-fixing bacteria.

Leukocyte (also **leucocyte**).   A white blood cell.

Leukoplast (also **leucoplast**).   A colorless plastid, thought to serve as a cytoplasmic center for the storage of certain materials, such as starch.

Lignin.   A substance deposited along with the cellulose in the cell walls of woody plants, which gives added strength to these walls.

Linkage.   The location of two or more genes on the same chromosome, so that the two characteristics are passed on together from parent to offspring.

Lipids.   A group of organic compounds including the fats and fatlike compounds and the steroids.

Logarithm.   The power to which a given number (the base) must be raised in order to equal a certain number. Ten is the most commonly used base number. The logarithm of 100 to the base 10 is 2. This can be written: $\log_{10} 100 = 2$, or $10^2 = 100$. The base number is given as a subscript following the term "log."

Logarithmic phase.   The phase of most rapid increase in size for a population; that is, the area of exponential increase on a sigmoid curve.

Logistic growth curve.   Graphical representation of the growth pattern of a population. It includes an acceleration phase, inflection point, deceleration phase, and steady state.

Long-day plants.   Plants that flower when exposed to cycles of light and dark in which the dark phase is shorter than 8½ hours.

Luteinzing hormone (LH).   A hormone secreted in vertebrates by the anterior lobe of the pituitary gland of both sexes. In females LH stimulates formation of the corpus luteum; in males it regulates production of testosterone.

Lymph.   A colorless fluid traveling freely throughout most of the body. It aids in the maintenance of proper osmotic balance, functions in the control of disease, and serves as the main medium of circulation for lipids.

Lymph nodes.   Lymphocyte-rich structures scattered over the body which act as filters for the blood, contain white cells that attack and digest foreign cells, and are secondary production sites for lymphocytes.

Lymphatic system.   A system of vessels that returns protein, lipid, and other macromolecules from the tissue fluid into the general circulatory system.

Lymphocytes.   One type of leukocyte, produced in lymphatic tissue such as the lymph glands, thymus, spleen, or bone marrow. Lymphocytes have a rounded nucleus and appear to function in wound repair and in "remembering" the immune response once an organism has developed immunity to a given foreign substance.

Lysins.   A class of substances produced by sperm to dissolve the protective egg membranes and allow the sperm to enter and fertilize the egg.

Lysis.   The chemical breakdown of a cell, usually under the influence of enzymes released by the rupture of a lysosome or by reproduction of viruses within the cell.

Lysosome.   A sac-like structure containing enzymes that catalyze the breakdown of fats, proteins, and nucleic acids. The membranes of lysosomes protect the cell from being digested by its own enzymes (autolysis). Lysosomes also serve as defense mechanisms, ingesting and digesting foreign toxic agents within the cell.

Macromolecule.   A large molecule built up from small repeating units. Cellulose is a macromolecule built of repeating β-glucose units.

Macronucleus.   A large polyploid nucleus of the *Paramecium* that controls the metabolic functions of the organism.

Mammae.   Mammary glands, where milk is produced in female mammals.

Mass spectrometry.   A process that sorts streams of electrically charged particles according to their different masses by using deflecting magnetic fields. The device that accomplishes this, a mass spectrometer, generally consists of a long tube that generates a magnetic field. The particles of varying masses are passed through the tube and the degree of deflection by the magnetic field is recorded on a photographic film at the other end. Given the strength of the field and the amount of deflection, investigators can calculate the relative masses of the particles. Isotopes of various elements can be detected by mass spectrometry.

Masturbation.   Self-manipulation of the genitals to produce sexual stimulation, arousal, and sometimes orgasm.

Maturation region.   The area within a young root (several centimeters from the tip) where the cells reach their full size and become completely specialized.

Maximum sustained yield.   The period of maximum growth rate of a population; represented by the inflection point on a logistic growth curve.

Mechanism.   The philosophical view that life is explicable in terms of physical and chemical laws, and that the whole is equal to nothing more than the sum of its parts. Also called mechanistic materialism.

Medulla.   The central portion of a gland or organ.

Meiosis.   A process of cellular division which results in each of the daughter cells containing half as many chromosomes as the parent cell (that is, they are haploid). Meiosis occurs primarily in the formation of gametes (in sexually reproducing organisms) or in spore formation in organisms such as ferns or mosses.

Melanin.   Any of a group of dark brown or black pigments occurring in the skin and other parts of the body.

Menopause.   The period (usually between 40 and 50 years of age) in the reproductive life of human females when the recurring menstrual cycle including ovulation ceases. Changes in hormonal balance and control within the female body also take place.

Menstrual cycle.   A cyclic event approximately 28 days in duration in the human female. The function of the cycle is to prepare the endometrium (the lining of the uterus) for implantation of a young embryo should fertilization occur. If a fertilized egg is not received, much of the endometrium is sloughed off and the cycle starts again.

Meristem.   An area of rapidly dividing cells. The meristem may be a single cell, as in ferns, or it may include many cells, as in the flowering plants. Most plants have root and stem meristems.

Meristematic tissue.   In plants, any area of rapid cell division, forming a true growth region, such as the tips of branches and roots or the lateral meristem (the cambium).

Mesenchyme.   An embryonic tissue derived from the mesodermal layer of an animal embryo. Mesenchyme appears in the embryo as a mass of scattered angular or pointed cells with long, protruding processes. In vertebrates the mesenchyme forms the connective tissues (bone, cartilage, etc.).

Mesoderm.   In vertebrate embryos the germ layer lying between ectoderm and endoderm and giving rise to connective tissue, muscle, the urogenital system, the vascular system, and the lining of the coelom.

Mesoglea.   The middle cell layer of sponges and cnidarians (coelenterates).

Messenger RNA (mRNA).   A strand of RNA synthesized in the nucleus of a cell with one DNA strand as a template. Thus mRNA has a base sequence directly complementary to the base sequence on the DNA molecule. After its formation, mRNA migrates from the nucleus to the cytoplasm, where it becomes associated with the ribosomes.

Metabolism.   The sum total of chemical and physical processes within the body related to the release of energy by the breakdown of chemical fuel and the use of that energy by the cells for their own work.

Metamorphosis.   That type of development in which the organism exists in an intermediate free-living stage (larva) prior to changing into the adult form. Metamorphosis also refers to the actual change from larva to adult.

Metaphase.   The phase of mitosis characterized by the lining up of the chromosome pairs along the equatorial plane of the cell.

Metaphysics.   A method of thinking that tries to go beyond the present physical reality and postulate the unknown. In metaphysics. phenomena are regarded as final, immutable, and independent of one another. Metaphysical thinking is generally contrasted to scientific thinking.

Microfilament.   A minute protein fiber in the cytoplasm of certain cells; its function appears to be providing support and shape to the cell.

Micronucleus.   Small nucleus in the *Paramecium* that functions only in reproduction and in regenerating the macronucleus.

Microspores.   Haploid cells in the anther that will develop into pollen grains.

Microtubules.   Tiny intracellular tubes composing such diverse structures as the spindle apparatus and the centrioles.

Minimal medium.   A medium containing only those elements absolutely essential for the growth of a particular microorganism, and which the organism cannot synthesize itself. Generally, minimal media contain a carbohydrate source, various inorganic salts, and sometimes a growth factor such as biotin.

Miscarriage.   An accidental or spontaneous abortion. Miscarriages can occur anywhere from the trophoblast stage through the ninth month of pregnancy.

Mitochondria.   Cytoplasmic organelles of a characteristic structure containing the enzymes for the citric acid cycle.

Mitosis.   The series of changes within a cell nucleus by which two genetically identical daughter nuclei are produced.

Mode.   In a set of data, the item or group of items that appears most often. For example, when a set of grades for a whole class is compiled, the mode is the single grade, or group of grades, that the largest number of students obtained.

Mole.   A gram molecular mass (often called gram molecular weight) of a substance; the molecular mass of a substance in grams. One mole contains Avogadro's number of particles. (6.023×10^{23}).

Molting.   Periodic or seasonal shedding of skin or exoskeleton to enable growth, adapt to climatic change, or signal the onset of maturity.

Monera.   The kingdom that includes all those organisms that lack nuclei in their cells; the monera are all prokaryotes.

Monocotyledonae (monocots).   One of the two classes of angiosperms. Monocots are identified by a single seed leaf (cotyledon) in the embryo, parallel veins in the leaves, flower parts generally in threes or sixes, and scattered fibrovascular bundles in the stem.

Monocytes.   One of three types of leukocytes; large, highly mobile cells whose main function is phagocytosis.

Morphogenesis.   The development of form in embryos.

Morphogenetic movement.   Cell movements (migrations) that change the shape of differentiating tissues in an embryo.

Mosaic theory.   An embryological idea, most often ascribed to Wilhelm Roux, which holds that certain regions of the egg are designated to become specific parts of the organism.

Motor neurons.   Those neurons conveying impulses from the central nervous system to muscles, triggering contraction.

Motor unit.   The minimum unit of contraction in a muscle, composed of all the muscle fibers activated by a single nerve.

Muellerian ducts.   Tubules just lateral to the Wolffian ducts that develop in the mammalian embryo and are the precursors of the Fallopian tubes, uterus, and part of the vagina.

Multiple alleles.   Sets of alleles that contain more than two contrasting members for a given locus.

Mutagen.   Any factor (chemical, radiation, and so forth) that can cause mutation.

Mutualism.   The mutually beneficial association between different kinds of species.

Mutation.   An inherited structural or functional variation of an offspring in relation to its parents. Mutations are due to a change in the chemical structure of DNA, the molecule bearing hereditary information. Once a mutation has occurred, it is transmitted to future generations.

Mycoplasmas.   Smallest cell-like organisms known. They are obligate intracellular parasites that can cause pneumonia in humans.

Myocardial infarction.   Medical term for a heart attack. When arteries supplying the heart wall become choked, blood cannot pass. Certain regions of the heart may be starved for oxygen and cease to contract, or contract arhythmically, producing the "attack."

Myocardium.   The muscular wall of the heart.

Myosin.   Contractile protein that interacts with actin to produce muscle contraction.

Natural selection.   Theory for the mechanism of evolutionary change proposed by Darwin. It states that the environment "selects" for the most adaptive of all the variations that naturally exist within a population.

Negative acceleration phase.   The period of decreasing growth rate in a population, following the exponential increase that occurs during the logarithmic phase.

Negative feedback.   A mechanism of self-regulation whereby a change in a system in one direction is converted into a command for a change in the opposite direction. A means of helping maintain a biological system in dynamic equilibrium.

Nematocyst.   Poison-filled barbed thread ejected from stinging cells of cnidarians (coelenterates) in order to paralyze their prey.

Nephridia.   Excretory organs found in each body segment of annelids.

Nephron.   The functional unit in the kidney, composed of a glomerulus, Bowman's capsule, and associated blood vessels. There are approximately one million nephrons in each human kidney.

Nerve.   Bundle of neuronal axons.

Net primary productivity.   The total amount of radiant energy taken up by the producers in an ecosystem in a given area and time minus the energy they expend in respiration.

Neural crest.   During mammalian embryogenesis, cells that become detached from the forming neural tube, move laterally, and then move dorsally between the closed tube and the overlying ectoderm. The spinal and cranial nerve ganglia are developed from neural crest cells.

Neural plate.   A layer of ectoderm in a primitive streak stage embryo that will develop into skin and nerve tissue.

Neural tube.   In vertebrate embryos the tube formed along the dorsal surface, produced by the infolding of a large mass of ectodermal tissue. The neural tube develops into the entire central nervous system and parts of the peripheral system as well.

Neuromuscular junction.   The point at which a nerve end brush comes into contact with a muscle fiber.

Neuron.   One of the individual cells that make up nervous tissue. Each neuron has a cell body, composed of a nucleus and surrounding cytoplasm, and processes varying in length and number that carry the nerve impulse and make contact with other neurons.

Neurosecretory cells.   Cells in the hypothalamus that secrete tropic hormones, whose target is the pituitary gland.

Niche.   The ecological position of an organism—the organism's "occupation" within the biological community.

Nitrogen-fixing bacteria.   Bacteria capable of drawing nitrogen from the atmosphere and converting it into soluble nitrates. These nitrates can then be used by plants.

Nodes of Ranvier.   Constrictions at regular intervals on myelinated nerve fibers. This is where depolarization occurs, causing a faster rate of impulse conduction.

"Nodoc".   See Anticodon.

Nodules.   Swellings on the roots of certain leguminous plants where nitrogen-fixing bacteria reside in a symbiotic relationship with the plant.

Nondisjunction.   The failure of a pair of chromatids to separate at metaphase, creating an abnormality of chromosome number in both daughter cells. Nondisjunction usually leads to deformed offspring.

Normal distribution curve.   A bell-shaped curve, more or less symmetrical, with the greatest value in the center (mean) and with values decreasing equally on both sides. Also called a normal curve.

Notochord.   A rod-shaped body located dorsally and serving as an internal supporting structure in the embryos of all chordates and in the adults of some; replaced by a vertebral column in vertebrates.

Nuclear membrane.   The unit membrane that separates the nuclear material from the surrounding cytoplasm in a cell. The nuclear membrane is not continuous, but rather is broken at different intervals by nuclear pores that provide a physical passage between the nucleus and surrounding cytoplasm.

Nucleic acid.   A polymer composed of ribose sugar rings and phosphate groups, with organic bases of thymine, guanine, cytosine, adenine, and/or uracil. Both DNA and RNA are nucleic acids.

Nucleotide.   The molecule formed from the combination of a purine or pyrimidine, an appropriate sugar (ribose or deoxyribose), and a phosphate residue. Nucleotides are the basic units of nucleic acid structure.

Nucleus (atomic).   The dense area in the central part of an atom where all protons and neutrons are located.

Nucleus (cellular).   A body found in nearly all cells which contains most of the hereditary information of the cell and acts as the control center of cell function.

Oligotrophic (lake).   Lacking in biological productivity; not yielding much biomass.

Oöcyte.   An immature egg cell undergoing meiosis.

Oögenesis.   The process by which haploid female gametes (eggs) are produced.

Operon.   A group of structural and control genes, usually adjacent to each other on a single DNA segment, that act in a coordinated function.

Opsin.   The protein component of rhodopsin, the visual pigment found in vertebrate retinal cells. Rhodopsin is composed of retinene (a derivative of vitamin A) conjugated to the protein opsin.

Organ.   A unit composed of various types of tissues grouped together to perform a necessary function. The liver and a plant leaf are examples of organs.

Organelle.   A small body appearing within the cell cytoplasm, with a characteristic structure and a definite, though perhaps not always clearly defined, function. Cytoplasmic organelles include mitochondria, ribosomes, the Golgi complex, and the endoplasmic reticulum.

Origin.   The more fixed point of attachment of a muscle to the skeletal structure. The origin is usually the more proximal point of attachment.

Ornithine.   An intermediate compound formed during the conversion of ammonia to urea. Ornithine is also an end product of the urea production process.

Ornithine-citrulline cycle.   A stepwise pathway in which urea is formed from ammonia and carbon dioxide.

Orthogenesis.   An erroneous conception, originating in the nineteenth century, that evolution progresses in a given, straight-line direction. The evolution of the antlers of the Irish elk is often cited as an example of orthogenesis.

Osculum.   A large pore at the top of sponges through which water and wastes exit.

Osmosis.   The passage of a solvent from a region of greater concentration to a region of lesser concentration through a semipermeable membrane.

Ostium.   The upper end of the Fallopian tubes where eggs released from the ovaries enter for transport to the uterus.

Ovary.   In plants, the basal portion of the pistil that encloses the ovules. The ovules, in turn, contain the female sex cell. After fertilization, the ovary becomes the fleshy part of the fruit and the ovules become the seeds. In animals, the organ (usually paired) of the female that produces the ovum, or egg.

Oviparous.   Term applied to organisms that lay eggs in which the embryo continues to develop for some period of time, deriving nourishment from the yolk.

Ovoviviparous.   Term applied to organisms whose young develop within the body of the mother but derive most or all their nourishment from the egg yolk.

Ovulation.   The release of an unfertilized egg from the ovary. In human beings ovulation involves the discharge of a mature ovum from a Graafian follicle of the ovary.

Ovum.   The female reproductive cell containing the haploid number of chromosomes, derived by meiosis from a diploid germ cell. The ovum is fertilized by the sperm, producing the zygote.

Oxidation.   A type of chemical reaction involving the loss of electrons. Frequently, but by no means necessarily, the element oxygen is involved.

Oxidative phosphorylation.   The process whereby electrons removed from substrate molecules are passed through the electron transport chain (cytochromes), in such a way that their potential energy is coupled to the formation of ATP (from ADP and inorganic phosphate).

Oxygen debt.   The amount of oxygen required to oxidize the excess lactic acid accumulated in muscle cells during strenuous exercise.

Oxytocin.   A hormone produced in the hypothalamus and secreted by the pituitary gland, regulating uterine contractions.

Paleontology.   The branch of geology that deals with the study of fossils.

Parasite.   Organism that derives its food from another species of organism by living in or on the host organism, usually to the detriment of the host.

Parasitism.   An association between two species in which one is benefitted and the other is harmed. A type of heterotrophic nutrition found among both plants and animals. A parasitic organism lives in or on the body of a living plant or animal (host) and attains nourishment from it.

Parasympathetic nervous system.   That portion of the central nervous system which arises from the brain and lower tip of the spinal cord (in the region of the lumbar vertebrae), generally without passing through ganglia. It acts functionally to counteract the effects of the sympathetic nervous system; that is, it returns the body to normal after emergency. The transmitter substance released at the nerve endings of parasympathetic fibers is acetylcholine. The vagus nerve (the tenth cranial nerve) is the most prominent parasympathetic tract.

Parthenogenesis.   The development of an egg without fertilization into a new individual. Parthenogenesis occurs naturally in some organisms (such as aphids, rotifers, bees, and ants) but can be induced artificially in higher forms (such as frogs) by applying chemical or physical stimuli to the egg.

Partial pressure.   The pressure exerted by a given component gas in a mixture of gases.

$$\frac{\text{Partial pressure}}{\text{of gas A}} = \frac{\text{Volume of gas A}}{\text{Total volume}} \times \frac{\text{Total pressure}}{\text{of entire sample}}$$

Parturition.   The process of childbirth.

Pathogenic.   Capable of causing disease.

"Peck order".   The establishment of dominance hierarchies, first noted in domestic chickens.

Pellicle.   Tough outer coat of paramecia.

Penis.   The male sex organ through which sperm are ejaculated.

Peptide bond.   The bond formed by dehydration synthesis (elimination of water), which links together two amino acids.

Pericycle.   A thin layer of cambium parenchyma or sclerenchyma one or two layers thick, immediately surrounding the area containing the vascular tissue and cambium in plant roots and stems.

Periosteum.   The thin outer layer of connective tissue covering a bone.

Peripheral nervous system.   The sum total of sensory and motor neurons.

Periplastic envelope.   In bacteria, the area between the plasma membrane and the cell wall that contains a variety of enzymes. The envelope space is not usually visible unless the protoplast is shrunken by plasmolysis.

Peristalsis.   Undulations of the digestive tract that aid in the movement of food.

Peroxisome.   A cytoplasmic cell organelle containing enzymes for the production and decomposition of hydrogen peroxide.

Petals.   Modified leaves of flowering plants that function to protect the reproductive parts and attract pollinators.

Petiole.   The leaf stalk that serves to support the leaf as well as to transport water and minerals from stem to blade and products of photosynthesis from blade to stem.

pH.   Symbol for the logarithmic scale running from 0 to 14, representing the concentration of hydrogen ions or protons (actually hydronium ions) per liter of solution. On the pH scale, 7 represents neutrality, the lower numbers acidity (acids), and the higher numbers alkalinity (bases).

Phagocyte.   A cell in the body capable of engulfing particles from the surrounding medium into its own cytoplasm for enzymatic breakdown (phagocytosis). Phagocytes are found in large numbers lining the walls of lymph node sinuses; they destroy bacteria that have entered the body and been picked up by the lymphatic system. Leukocytes and macrophages are examples of phagocytes in human beings.

Phagocytosis.   The engulfing of microorganisms, other cells, or foreign particles by a cell. For example, phagocytosis occurs when an amoeba engulfs its prey or when a white blood cell engulfs a bacterium.

Phenotype.   In genetics, the outward appearance of an organism, as contrasted with its genetic makeup (genotype).

Pheromones ("ectohormones").   Substances secreted outside the body and eliciting a particular response, such as the attraction of male moths by the sex attractant released by the female.

Phloem.   Plant vascular tissue that transports food throughout the plant, especially from the leaves and stems to the storage areas in the roots.

Phlogiston theory.   A widely accepted seventeenth-century theory of combustion, which held that the substance phlogiston was contained in all combustible bodies and was released from these bodies upon their burning (producing "phlogisticated" air). This theory remained popular until the later eighteenth century, when experimental work by Black, Priestley, Lavoisier, and others led to the oxygen theory of combustion.

Phosphoglyceric acid (PGA).   An intermediate product in carbohydrate metabolism, composed of a 3-carbon backbone with a phosphate group attached to one (usually position 3) of the carbons:

$$
\begin{array}{c}
\quad\;\; O \\
\quad\;\; \| \\
C{-}OH \\
| \\
H{-}C{-}OH \\
| \\
H{-}C{-}O{-}PO_3 \\
| \\
H
\end{array}
$$

Phospholipids.   An important structural part of cellular membranes, the phospholipids contain phosphorus, fatty acids, glycerine, and a nitrogenous base.

Phosphorylation.   The addition of a phosphate group (such as H_2PO_3) to a compound, as in oxidative phosphorylation of ADP (producing ATP) during respiration.

Photophosphorylation.   The process whereby phosphorylation is coupled to the transport of electrons that have been moved to higher energy levels by the absorption of light energy (that is, ATP production during the light reaction of photosynthesis).

Photoperiodism.   The germination or flowering response of plants to day length.

Photosynthesis.   Food-making process of green plants in which they convert the sun's radiant energy into potential chemical bond energy of sugars and starches.

Phototropism.   The growth response of a plant to light stimulus.

Phylogenetic chart.   A diagram showing the evolutionary relationships among a group of species, or within a single species. The so-called family tree is an example of a phylogenetic chart.

Phylogeny.   The study of the evolutionary history of species.

Physical environment.   All the elements surrounding an organism, excluding other living organisms.

Phytochromes.   Special plant pigments capable of light absorption at two distinct wavelengths. Phytochromes are thought to mediate the photoperiodic response of plants.

Pinocytosis.   The process by which materials can be taken into the interior of a cell without passing through the plasma membrane.

Pili.   A cylindrical protein extension found on some bacteria used for anchoring, conjugation, or feeding.

Pistil.   The part of flowering plants that includes the stigma, style, and ovary.

Pituitary.   A "master" endocrine gland consisting of two lobes located beneath the floor of the brain. The anterior lobe produces tropic hormones (those that control the activity of other endocrine glands), as well as certain other hormones that act directly upon body function. The posterior pituitary is the source on an antidiuretic hormone (vasopressin) and oxytocin.

Placenta.   A structure created by the fusion of the chorion from the young embryo with the wall of the uterus. Respiratory, excretory, and nutritional functions of the fetus are carried on by exchanges across this structure. The placenta also secretes hormones regulating certain aspects of fetal development. Presence of a placenta is characteristic of all mammals except the marsupials.

Plasma.   A protein-containing fluid, the liquid portion of blood.

Plasma membrane.   A lipoprotein of a definite structure, which surrounds and contains the living matter within a cell. The membrane has three layers: two outer protein surfaces surrounding an inner core of lipid. The polar part of each lipid molecule is associated with the protein on the surface, while the nonpolar portion points into the middle of the "sandwich." The average plasma membrane diameter is 75 Å.

Plasmagene.   A term often applied to a gene of cytoplasmic rather than nuclear origin.

Plastids.   Small bodies occurring in the cytoplasmic portion of plant cells. Plastids are classified according to color.

Platelets.   Irregularly shaped blood cells of vertebrates, involved in the formation of blood clots. It is thought that breakdown of platelets initiates the series of reactions in clotting.

Pleiotropy.   A condition in which one gene has a variety of phenotypic effects, probably because it codes for a protein involved in several metabolic pathways.

Polar body.   The small daughter cells produced during meiotic divisions of the oöcyte. From the first meiotic division one polar body is produced; from the second meiotic division three polar bodies result. A single primary oöcyte thus gives rise to one large, mature ovum and three small polar bodies. The polar bodies contain the extra sets of chromosomes produced by meiosis. They contain very little cytoplasm, however, since most of this is reserved for inclusion in the single ovum.

Polarization.   Term applied to an unequal distribution (separation) of charged ions, producing an electric potential. Cell membranes are said to be in a state of polarization when they have a greater concentration of positive ions on the outside than inside, or vice versa.

Pollen tube.   An extension of the pollen grain that grows through the stigma and style to the ovary during reproduction in flowering plants. The male pronucleus travels down this tube and is discharged into the ripened egg.

Pollination.   The transport of pollen by wind, water, or animals from an anther to a pistil.

Polygenic character.   A quantitatively variable phenotypic trait determined by interaction of numerous genes.

Polymorphism.   The existence within a single species of members showing many different, but distinct and recurring forms; for example, the drone, queen, and workers occurring in the honeybee.

Polyribosome (polysome).   A cluster of connected ribosomes, usually arranged along a strand of *mRNA*.

Polysaccharides.   Complex carbohydrate molecules built up from simpler sugar units (such as glucose) into long-chain polymers. Polysaccharides are the major constituents of the cell walls and capsules of various microorganisms.

Polysome.   See Polyribosome.

Polyspermy.   The fertilization of a single egg by more than one sperm at a time.

Pons.   A thick bundle of nerve fibers that connect the hemispheres of the cerebellum.

Population.   A group of potentially interbreeding organisms (of the same species) that interact as a group.

Population genetics.   The application of genetic principles to a large number of breeding organisms, that is, to a population.

Positive acceleration phase.   The first section of a sigmoid growth curve for a population, where the system described is just beginning to increase.

Positive feedback.   Biologically speaking, an abnormal state in which a change in a system in one direction serves as a command for continued change in that same direction. This can create a severe physiological imbalance leading to the death of an organism.

Potential energy.   Energy capable of doing work.

Precapillary sphincters.   Sphincter ("drawstring-like") muscles at the beginning of capillaries. Contraction of these muscles (like pulling a drawstring) closes off the capillary, while relaxation opens up the capillary. Concentration or relaxation of the precapillary sphincters provides a means of regulating blood flow into given tissue areas.

Precipitate.   An insoluble product of a chemical reaction in a solution.

Predation.   The process by which one species uses another species for food.

Preformation.   The idea that an already formed, miniature individual exists within the egg and merely increases in size during embryological development.

Primary consumers.   Herbivores; organisms that eat the producers (green plants) in a biotic community.

Primary germ layers.   The first distinguishable areas within the developing embryo, namely the ectoderm, mesoderm, and endoderm. These areas give rise to the tissues and organs of the mature organism.

Primary productivity.   The amount of radiant energy taken up by the producers in an ecosystem, expressed in calories per unit area per unit of time.

Primary spermatocytes.   The cells in the testes derived from spermatogonia and ultimately undergo meiosis to produce haploid sperm.

Primitive streak.   A longitudinal groove that develops on the surface of the embryo of fishes, reptiles, birds, and mammals. Formation of the streak is a consequence of the movement of cells and the formation of mesoderm. The primitive streak marks the future longitudinal axis of the embryo.

Producers.   The organisms in a biotic community that use solar energy to convert inorganic substances into living tissue. Producers are usually green plants.

Progesterone.   A hormone produced and secreted by the corpus luteum of the ovary. The corpus luteum promotes continuous development of the uterine lining and development of mammary glands if pregnancy occurs. If pregnancy occurs the corpus luteum also prevents maturation of new follicles and eggs.

Prokaryotes (also **procaryotes**).   Cells that have no true nucleus and no nuclear membrane. Prokaryotes lack membrane-bound subcellular organelles such as mitochondria and chloroplasts. Bacteria and blue-green algae are examples of prokaryotes.

Prophase.   The first visible phase of mitosis, marked by the condensation of the chromosomal material into chromosomes and the disappearance of the nuclear membrane.

Proprioceptor.   A sensory nerve ending found in tendons or muscles that is sensitive to changes in tension.

Prosecretin.   The inactive form of the hormone secretin, which can be converted into the active form by the action of dilute HCl.

Prostate gland.   A glandular and muscular organ surrounding the urethra just below the bladder in the male mammal. The prostate secretes a major portion of the seminal fluid.

Protective coloration.   Patterns of surface pigmentation that blend with the environment, allowing an organism to remain unobserved and therefore camouflaged from predators.

Protein.   A complex organic molecule composed of amino acids joined in specific sequence by peptide bonds. Proteins serve both structural and enzymatic functions.

Protista.   One of the five kingdoms in the system outlined by Whittaker. Protista includes unicellular and some colonial organisms.

Pseudopod.   An extension of the streaming cytoplasm of amoebae or other one-celled organisms which gives them their irregular shape. In the amoeba, pseudopods function as a means of locomotion and as the tool for the intake of food from the environment.

Pulmonary artery.   Artery conveying venous blood from the right ventricle of the heart to the lungs.

Purine.   One of a class of nitrogen-containing compounds including the bases, adenine and guanine, important components of the nucleic acids DNA and RNA. The two-ringed purine structure is:

Pyrimidine.   One of a class of nitrogen-containing compounds, including the bases cytosine and thymine (and uracil, found in RNA). The single-ring pyrimidine structure is:

$$
\begin{array}{c}
N \\
HC \quad\quad CH \\
HC \quad\quad CH \\
C \\
H
\end{array}
$$

Purkinje fibers.   Structures that receive impulses from the atrioventricular node and conduct them throughout the ventricles, causing the ventricles to contract.

Pus.   The accumulation of white blood cells, bacterial cells, and damaged tissue cells at the site of a bacterial infection.

Pyruvic acid.   The final product of glycolysis, with the formula:

$$
\begin{array}{c}
COOH \\
| \\
C = O \\
| \\
CH_3
\end{array}
$$

Quantasome.   One of the individual membranous structures arranged in columns within a granum. Molecules of chlorophyll are aligned on the quantasomes.

Quantum.   A tiny energy packet in which light travels.

Quantum theory.   The model which holds that light is composed of tiny energy packets (quanta, or photons), which are given off by any light emitter and travel intact through space.

Radial cleavage.   The type of animal cell cleavage in which the four-cell embryo consists of the top layer of cells resting directly on the bottom layer.

Radial symmetry.   A type of animal body form in which the body can be divided into two equal halves by any plane of division that goes through the center of the animal.

Radicle.   The part of the plant embryo that will become the root in the adult.

Radioactive decay.   The decrease in mass of certain unstable elements by emission of elementary particles, continuing until a stable isotopic form has been reached.

Range.   The distribution of a group of numerical values describing the spread between the lowest and highest of the included values.

Receptive field.   A cluster of photoreceptors in the vertebrate retina, all of which must be stimulated in order for a particular ganglion cell (to which all the particular photoreceptors are connected) to be stimulated.

Recessive.   In genetics, a term referring to the relative lack of phenotypic effect of a gene in the presence of its dominant allele. Thus the gene for blue eyes is said to be recessive to that for brown in the human population, because when one allele for blue eyes and one for brown eyes are present in the individual, the blue condition is masked and the individual has brown eyes.

Recombination.   In genetics, the formation of new genotypes (a combination of genes not present in either of the parents) in offspring due to independent assortment of genes and chromosomes during gamete formation.

Reduction.   A chemical reaction involving a gain of electrons.

Reduction division.   A cell division during which the number of chromosomes in each daughter cell is reduced to one-half that found in the parent cell. This is accomplished by nuclear division without previous chromosome duplication.

Reflex action.   A nerve response that occurs without the nerve impulses passing through the brain of the organism (that is, involuntary response).

Regeneration.   The developmental process by which a lost part of an organism (such as an organ, part of an organ or tissue, the skin, etc.) is replaced after damage. Repair of wounds and regrowth of an amputated salamander's arm are examples of regeneration.

"Releasers".   Environmental stimuli that trigger an innate releasing mechanism in animal behavior.

Respiration.   The process whereby the energy of glucose and other fuel molecules is captured by the cells in the form of ATP.

Respiratory assembly.   A series of complex molecules (including cytochromes), found on the inner membranes of mitochondria, and capable of oxidation and reduction. Such assemblies accept electrons from reduced acceptors during the citric acid cycle and pass them along to the final acceptor, oxygen. The enzymes for oxidative phosphorylation are also components of the respiratory assembly.

Response.   A change in behavior by an organism (or tissue) as a result of some chemical or physical change in the environment.

Resting potential.   The normal electric potential difference created by ion distribution across a cell membrane. The term is particularly applied to nerve cells, where it refers to the potential existing across the membrane of a nonfiring cell.

Reversible reaction.   A reaction system where reactants and products are interconvertible. If left to themselves, reversible chemical reactions reach an equilibrium point where just as much reactant is being converted into product as product into reactant at any time. Nearly all chemical reactions are to some degree reversible.

Rh factor.   An antigen found on the red blood cells of certain human beings (designated Rh^+). The Rh factor acts as an antigen, so that individuals with this factor cannot donate blood to individuals who normally lack it (Rh^- individuals). The factor derives its name from having been first discovered in the rhesus monkey.

Rhizoid.   Small rootlike structure extending from a rhizome and functioning in uptake of minerals and water. Found especially in ferns and mosses.

Rhizome.   An underground stem, such as is found in iris.

Rhythm method.   A type of birth control relying on abstinence from coitus during the time of the suspected fertile period of the female.

Ribonucleic acid (RNA).   A complex, single-stranded molecule consisting of repeating nucleotide bases: adenine, guanine, cytosine, and uracil. At least three types of RNA are known, all of which are involved in transcribing the genetic code into protein. Messenger RNA (*m*RNA) carries the genetic code for amino acid sequence from the DNA to the ribosomes; soluble or transfer RNA (*t*RNA), of which there is a specific type for each amino acid, carries each amino acid to the ribosome where it is incorporated into protein in a specific place; ribosomal RNA (*r*RNA) is found only in the ribosomes.

Ribosomal RNA (*r*RNA).   A type of ribonucleic acid found as part of ribosomes. The function of *r*RNA is not completely understood, though it appears to play a role in positioning *m*RNA on the ribosome in a precise manner.

Ribosomes.   Small particles found either free in the cytoplasm or attached to the outer surface of the endoplasmic reticulum in the cells of all eukaryotes and many prokaryotes. Ribosomes contain high concentrations of ribonucleic acid (RNA) and are centers of protein synthesis.

Rickettsiae.   A group of organisms smaller than bacteria that are obligate intracellular parasites and can cause serious human diseases.

Roentgen.   A standardized unit measuring the amount of energy delivered in X-ray (or gamma ray) beams. One roentgen is the amount of radiation that under ideal conditions ($0°C$ and 760 Hg) liberates 2.083×10^9 ion pairs per cubic centimeter of air.

Root cap.   The protective area at the very end of the root, composed of several layers of loosely arranged cells.

Root hair.   A structure projecting from the surface layer of cells in the maturation region. Such structures increase the absorption capacity of the root.

Root pressure.   Osmotic pressure of water diffusing into the root hairs from the soil.

Rudimentary phallus.   Genital fold in both male and female embryos. In males, the rudimentary phallus develops into the penis; in females, it grows only slightly and develops into the clitoris.

Salivary glands.   Three pairs of glands (parotid, sublingual, and submaxillary) that are located in the mouth and secrete saliva. Saliva contains the digestive enzyme salivary amylase, which begins the hydrolysis of carbohydrates.

Sarcolemma.   The cell membrane of a muscle fiber.

Sarcoplasmic reticulum.   The endoplasmic reticulum of muscle fibers. It releases calcium ions in response to depolarization, initiating muscle fiber contraction.

Scrotum.   A pouch at the base of the penis containing the testes and parts of their spermatic cords.

Secondary consumers.   Carnivorous organisms in a biotic community that eat the primary consumers (herbivores).

Secondary spermatocyte.   The daughter cell(s) produced by the first meiotic division of the primary spermatocytes during spermatogenesis.

Secretin.   A hormone secreted by the cells of the duodenum under the stimulus of hydrochloric acid from the stomach. Secretin in turn causes the pancreas to secrete certain digestive enzymes into the duodenum.

Seed ferns.   Extinct organisms of the Devonian period which gave rise to both the conifers and the angiosperms.

Segregation (of alleles).   The separation during gametogenesis of paired factors influencing a single condition.

Semen.   A secretion of the male reproductive organs, composed of the spermatozoa and liquid secretions of various other glands.

Seminal vesicles.   The portion of the male reproductive duct in which sperm are stored prior to copulation.

Seminiferous tubules.   Tubules within the testes where the male sperm are produced. Each testis contains about 1,000 highly coiled seminiferous tubules.

Semipermeable (differentially permeable).   Term applied to a membrane that allows some substances to pass through while prohibiting the passage of others.

Sensory neurons.   Those neurons capable of detecting environmental stimuli and carrying these stimuli to appropriate centers in the central nervous system.

Sepals.   Modified leaves outside the petals of flowering plants that function to protect the reproductive parts.

Sex chromosome.   The chromosomes, commonly referred to as X and Y, whose presence in certain combinations determines the sex of an organism.

Sex-linkage.   Phenotypic characteristics unrelated to sex, caused by genes carried on sex chromosomes.

Shoot-tension (transpiration pull) hypothesis.   The concept that upward movement of water in plants is induced by a vacuum created when water evaporates from the leaves.

Short-day plants.   Plants that flower when exposed to cycles of light and dark where the dark phase is longer than 8½ hours.

Sickle-cell anemia.   A hereditary disease caused by a mutant form of hemoglobin. Under low oxygen tension in the blood, red blood cells containing sickle-cell hemoglobin collapse, assuming a half-moon, or sickle, shape. The disease is mild in the heterozygous form but greatly shortens lifespan in the homozygous dominant form.

Sieve tube.   The most prominent type of phloem, a long tubular structure similar to xylem, but composed of living cells. The cells of the tube are connected by perforations of the cell wall.

Simple sugar.   A molecule composed of a single five- or six-carbon sugar.

Sinoauricular node.   A specialized mass of tissue on the right atrium of the heart near the entry of the superior vena cava. Impulses to the atria originate in the sino-auricular node and spread over the atria to a second node, the auriculoventricular node, located between the atria and the ventricles. This node then stimulates the ventricles.

Skeletal muscle.   A muscle within the body that moves the appendages. Because the cells of this type of muscle show striations running the width of the cell, skeletal muscles are often called striated muscles.

Smooth muscle.   Muscle composed of elongated cells lacking striations and therefore having a smooth appearance under the microscope. The contraction of smooth muscle is under the control of the autonomic nervous system.

Sodium-potassium pump.   The active transport mechanism that functions to concentrate sodium ions on the outside and potassium ions on the inside of a cell membrane.

Soluble RNA.   See Transfer RNA.

Somatoplasm.   The term used by August Weismann for all cells of an organism except reproductive cells ("germ plasm"). In each generation, the somatoplasm is derived from the germ plasm of the preceeding generation, but is distinct from it in that changes in somatoplasm (body cells) will not be passed on to the next generation.

Somites.   Paired, block-like masses of mesoderm arranged in a longitudinal series along the side of the neural tube of the embryo. Each somite will form one vertebra and its associated muscles.

Spacing.   Refers to the distribution of organisms within a given area.

Special creation.   An account of the origin of life and its diverse forms by some act of divine creation.

Specialization (of cells).   The change in cell capability from the performance of a wide range of functions to concentration on one activity or set of activities.

Speciation.   The process by which the accumulated effects of variation within a population make crossbreeding between two given organisms difficult or impossible.

Species.   The smallest unit of taxonomic classification, referring for the most part to a group of individuals capable of breeding among themselves. Species are defined by morphological, ecological, physiological, and biochemical criteria.

Sperm.   The male reproductive cell containing the haploid number of chromosomes, produced from a diploid germ cell by meiosis (reduction division). The sperm fertilizes the egg, producing the diploid zygote.

Spermatids.   After the second meiotic division in spermatogenesis the secondary spermatocytes undergo development into spermatids, which in turn develop directly into functional sperm.

Spermatogonium.   The starting cell, or the "grandparental" cell, of future sperm. A spermatagonium undergoes mitosis a number of times, eventually producing a primary spermatocyte. The latter undergoes meiosis to eventually produce four sperm.

Spermatogenesis.   The process by which haploid male gametes (sperm) are produced from diploid primary spermatocytes.

Sphincter muscle.   A ring-shaped muscle surrounding a tubular organ or a narrow opening and capable of contraction. Examples are the iris, pyloric sphincter, and anal sphincter.

Sphygmomanometer.   Instrument used to measure blood pressure.

Spicules.   Non-cellular skeletal support structures of sponges made of silicon or calcium carbonate.

Spinal cord.   A part of the central nervous system running down the back through the hollow center of the vertebrae, and from which pairs of peripheral nerves emerge.

Spindle fibers.   Fiberlike processes, formed during prophase, extending from the asters. Spindle fibers seem to function in determining the direction followed by the separating chromosomes in anaphase.

Spiral cleavage.   The type of animal cell cleavage in which the four-cell embryo consists of the top layer of cells resting at angles between the cells of the bottom layer.

Spontaneous generation.   A concept according to which living organisms develop from nonliving matter.

Sporangium.   A structure in plants that produce spores. Sporangia are most commonly found in the ferns, mosses, and their relatives.

Spore.   An often thick-walled asexual reproductive cell capable of surviving adverse environmental conditions. Found particularly in bacteria, algae, mosses, and ferns. Limited in animals to one protozoan group, the Sporozoans (for example, the malarial parasite).

Sporophyte.   The diploid phase in the life cycles of ferns, mosses, and flowering plants. The sporophyte grows from the fertilized egg, and produces haploid spores that develop into gametophytes.

Sporulation.   The process of spore production, in bacteria and other prokaryotes, wherein the parent cell subdivides to form a number of smaller active cells that begin a new life cycle.

Standard deviation.   A statistical calculation defined as the square root of the variance. A method of showing the limits within which all items of a distribution should occur relative to the mean. The greater the standard deviation, the wider the "normal," or bell, curve of distribution for the set of data.

Statistical analysis.   The use of mathematics to determine whether deviations from a pattern as predicted by a hypothesis are significant.

Stele.   Central cylinder of a root, separated from the cortex by the endodermis. The stele contains both xylem and phloem elements.

Stigma.   The top of the pistil, upon which the pollen grain lands during the events leading to fertilization.

Stipule.   A small structure located on both sides of a leaf stalk that protects the young leaf before it unfolds.

Stoma (plural, stomata).   An opening on the leaf surface through which gas exchange and water loss take place. Guard cells control the size of the opening and thereby regulate these exchanges.

Striated muscle.   Contractile tissue in which the cells show striations running across their width. Both skeletal and cardiac muscles show such striations, but the term striated muscle is usually considered synonymous with skeletal muscle only. Striated muscle contraction is under the control of the conscious portion of the brain.

Stroma.   The interlamellar medium inside chloroplasts which contains the enzymes involved in the dark reactions of photosynthesis.

Style.   In flowers, the thin, necklike region of the pistil. The style is topped by the stigma and has the ovary at its base. The pollen tube must grow through the style to fertilize the egg within the ovary.

Substrate.   The molecule upon which an enzyme acts during an enzyme-catalyzed reaction.

Succession.   Ecological development that begins in a habitat or area not previously occupied by the given community. Primary succession refers to the invasion by one community (for example, grasses) of a previously unoccupied area (such as a sand dune along the shore of a lake). Secondary succession refers to the replacement of one community (such as grasses) by another community (such as conifers and scrub vegetation).

Symbiosis.   An association of two species in which each may derive benefits from the other.

Syllogism.   A logical scheme or analysis of a formal argument, consisting of three propositions called respectively the *major premise* (major hypothesis), the *minor premise* (minor hypotheses), and the *conclusion*. For example: every virtue is commendable; generosity is a virtue; thus, generosity is commendable.

Symmetry.   The body form of an animal.

Sympathetic nervous system.   That portion of the central nervous system which arises from the thoracic region of the spinal cord and whose fibers pass through one synapse before reaching their sites of action. Functionally, the sympathetic nerves prepare the body for emergencies. They innervate the iris diaphragm of the eye (dilate the pupil); the salivary glands (inhibit secretion); the heart (accelerate beat); the bronchi (dilate tubes); the stomach, pancreas, and duodenum (inhibit peristalsis and secretion); the adrenal glands (stimulate secretion of epinephrine and norepinephrine); the liver (stimulate conversion of glycogen to glucose); and the bladder (inhibit bladder contraction). At their sites of action, sympathetic fibers generally release epinephrine or norepinephrine as transmitter substances.

Synapse.   A small gap separating two neurons where the nerve impulse is transferred from the axon of the first neuron to the dendrite of the second neuron. Synaptic function is thought to be carried on by chemical means.

Synapsis.   The process of the pairing of homologous chromosomes during meiosis.

Synthesis.   The process by which larger molecules can be built up from smaller molecules or atoms.

System.   An association of independent organs throughout the body for the performance of a necessary body function. Some systems in higher animals are the circulatory, digestive, muscular, skeletal, and excretory systems.

Systematics.   The branch of biology that tries to group all of the organisms on earth today by their evolutionary relationships. Also called taxonomy.

Systole.   The contraction phase of the heart pumping cycle, caused by the contraction of the two ventricles.

Taproot.   A root system composed primarily of a single (primary) root that grows downward, with smaller (secondary) roots growing from it.

Taxis (plural, taxes).   Term applied to the simple orientation movements of animals in response to external stimuli; for example, the avoidance reactions of planaria to light.

Taxonomy.   The science of classification; in biology, this refers to the classification of organisms into kingdom, phylum, class, order, family, genus, and species.

Teleology.   Assigning purpose to an action, such as saying the cell takes in calcium ions "in order to . . ."

Telophase.   The final phase of mitosis, in which the cytoplasm of the dividing cell is cleaved and two daughter cells are formed.

Territoriality.   The tendency of some organisms to defend a section of space surrounding them and/or their family.

Tertiary consumers.   Carnivorous organisms in a biotic community that eat the secondary consumers.

Testis.   The male gonad; the glandular organ in which male gametes (sperm cells) and sex hormones are produced.

Testosterone.   An androgen; a hormone produced in the interstitial cells of the testes and responsible for the characteristic changes associated with puberty.

Tetrad.   In meiosis, the four-part structure resulting from the duplication of each pair of homologous chromosomes.

Thalamus.   Area of the brain beneath the cerebrum, serving as a relay center for incoming sensory impulses.

Theory.   A term often used to refer to a hypothesis that has undergone repeated verification.

Thromboplastin.   A lipoprotein released from damaged tissue. When combined with calcium ions and several other proteins in the blood plasma, thromboplastin acts as an enzyme converting prothrombin to thrombin. Thrombin acts to convert fibrinogen to the insoluble fibrin, thus forming a blood clot.

Thermodynamics.   That branch of physical science which deals with heat as a form of energy. It is concerned with such problems as the exchange of energy (measured always in respect to the gain or loss of heat from a system) during chemical or physical processes.

Thoracic cavity.   The chest cavity containing the heart and lungs and enclosed by the ribs and diaphragm.

Threshold.   The lowest level of intensity of stimulation required to elicit a response. Usually applied to muscles and nerves, but also to such phenomena as specific behavioral responses in animals.

Thylakoid discs.   Flattened pancake-like "sacks" with double-layered walls, formed from the inner membrane system of the chloroplast. The lamellae and grana inside chloroplasts are composed of thylakoid discs.

Thymus.   An endocrine gland in the chest, behind the sternum, which produces lymph cells and is thus important in development and maintenance of immunity.

Tissue.   An aggregate of similar cells bound together in an ordered structure and working together to perform a common function.

Tracheae.   Series of ducts found on the abdomen of terrestrial arthropods used for bringing air directly into inner parts of the body.

Tracheids.   A type of xylem tube in which each cell has a tapered end and the ends of many cells overlap to form a continuous tube. Tracheids are nonliving in their mature form.

Transamination.   The stepwise series of reactions in which the amino group from one amino acid type is transferred to an intermediate substance, thus producing another type of amino acid.

Transfer RNA (soluble RNA).   A type of RNA in the cytoplasm of which there are probably more than 20 varieties, one specific for each amino acid. Transfer RNA (tRNA) unites with its specific amino acid and draws it to the ribosome during protein synthesis.

Translation.   An aspect of protein synthesis in which the genetic information coded on messenger RNA is used to specify the order of amino acids in a polypeptide. Translation occurs in the ribosomes, where messenger RNA, ribosomal RNA, and activated amino acids meet and peptide bonds are formed.

Translocation.   The movement of materials from one part of a plant to another.

Transmutation (of species).   An older term for the idea that over a long period of time new species arise through modification of old species.

Transpiration.   The loss of water through the leaves of a plant.

Transpiration pull.   The upward pull that results from the evaporation of water from the leaves of plants.

Tricuspid valve.   Three-flap (cusp) valve between right auricle and right ventricle ensuring that the flow of blood will occur in a single direction through the right side of the heart.

Trochophore.   A ciliated larval form found in some annelids and mollusks.

Trophic level.   Levels of nourishment in a food chain. A food chain is the transfer of energy from its ultimate source in plants through a series of organisms each of which eats the preceding organism.

Trophoblast.   In mammals, the thin-walled side of a blastocyst which gives rise to the placenta and the membranes that surround the embryo.

Tropic hormone.   A hormone that has an endocrine gland as its target tissue.

Tropism.   The growth response of a plant to an outside stimulus such as light, a chemical, or gravity.

Unit membrane.   The model that sees cell membranes (plasma, nuclear, etc.) as consisting of a mosaic of small "sandwiches"; that is, phospholipid molecules between an inner and outer layer of protein.

Uracil.   Nitrogenous base found in RNA that takes the place of thymine in transcription.

Urea.   A highly soluble conversion product of ammonia, formed mainly in the liver. This material is nontoxic if maintained in moderate concentration. It serves as a storage form of ammonia until it is filtered out in the kidneys and is passed from the body in the urine. Urea has the following chemical structure:

$$O=C \begin{array}{c} NH_2 \\ \diagup \\ \diagdown \\ NH_2 \end{array}$$

Uric acid.   A purine found in many organisms as the waste product of ammonia metabolism. Uric acid is highly insoluble, and its nontoxic crystals can be stored in the organism until excretion is possible. Organisms such as birds and reptiles excrete uric acid directly from the body. Mammals convert it into soluble urea.

Urine.   A watery waste containing urea in solution. Molecules of urea are produced from nitrogen-containing compounds (mostly uric acid) in the body by the liver. Urea circulating in the blood is filtered out by the kidneys and stored in the bladder until voided.

Uterus.   The womb; the hollow muscular organ of the female reproductive tract in which the fetus undergoes development.

Vacuolar membrane.   A unit membrane structure that separates the contents of a vacuole from the surrounding cytoplasm.

Vacuole.   A bubble-like structure surrounded by a membrane, occurring in the cytoplasm, and serving as a reservoir to hold food and waste products.

Vagina.   In mammals, a muscular tube that extends from the uterus to the exterior and serves both as a receptacle for the sperm during coitus and as a birth canal when the fetus completes its development.

Variance (s^2).   A numerical calculation describing the extent of dispersion of data around a mean.

Vas deferens.   A long tube-like structure at the top and in back of the testes, in which sperm are stored and through which they pass out through the penis.

Vasopressin.   Hormone produced by the posterior lobe of the pituitary gland that acts to increase blood pressure by causing retention of body fluids. (Also called ADH.)

Vegetal pole.   The surface of an egg opposite the animal pole, on an axis running through the nucleus. The yolk density within the egg is greatest at the vegetal pole.

Veins.   Vessels that carry blood from the various organs to the heart.

Venae cavae.   Two major veins: the superior vena cava, which returns blood from the head, arms, and shoulder; and the inferior vena cava, which returns blood from the lower part of the body.

Ventral.   The lower or belly side of an animal.

Ventricle.   One of the two chambers of the heart that pump blood to the parts of the body.

Venules.   Small blood vessels that carry blood from capillaries to a vein.

Vessel.   A type of xylem formed by cells with thickened walls stacked end-to-end to form cylindrical tubes.

Vestigial organ.   A structure in a degenerate state that remains in an organism but has little or no present function. The appendix in human beings is vestigial.

Villus.   A finger-like projection from the intestinal wall that increases the surface area and thus facilitates absorption of food materials passing down the digestive tract. Villi are generously supplied with both blood and lymph vessels to transport absorbed materials to other parts of the body for use or storage.

Viroids.   Molecules of RNA that behave as viruses in organisms and can cause diseases.

Virus.   A noncellular, submicroscopic particle composed of a protein coat surrounding a nucleic acid core. Viruses can reproduce only inside living cells (eukaryotes or prokaryotes).

Vitalism.   The view that life is an expression of something above and beyond the chemical and physical interactions of a group of molecules.

Vitamins.   Chemical substances, required in only trace amounts, that are thought to aid enzymes in catalyzing specific chemical reactions.

Vitelline membrane.   The inner membrane of an egg that surrounds the egg cytoplasm.

Vitreous humor.   A nutritive fluid filling the main chamber of the eye.

Viviparous.   Term applied to organisms whose embryos develop within the body of the mother and derive their nourishment from the mother.

Vulva.   The external female sex organs (genitalia).

Wavelength.   The distance between a given position on one wave and the same position on the following wave. Wavelength is often symbolized by the Greek letter lambda (λ).

Wave theory of light.   The model that depicts light as demonstrating all the properties of wave motion, analogous to waves on the surface of water.

White matter.   Regions of the brain and spinal cord that are rich in nerve fibers.

Wolffian ducts.   A pair of tubules which, in conjunction with the mesonephric ducts, form the primitive urogenital system of the male mammalian embryo. Both male and female embryos develop Wolffian ducts, though in females these ducts degenerate by the beginning of the third month.

Xylem.   Plant vascular tissues that conduct water and minerals from roots to leaves.

Yolk.   Fatty compounds and proteins stored within the egg that serve as the first food source for the developing embryo.

Yolk sac.   An extraembryonic membrane in many kinds of eggs. It functions to gradually supply food material from the yolk to the developing embryo.

Zygospore.   A diploid zygote of certain types of algae that has formed a protective spore.

Zygote.   A diploid cell, the product of fertilization formed from the union of male (sperm) and female (egg) reproductive cells (gametes).

Index